MACHINE DESIGN
DATA HANDBOOK

Other Handbooks of Interest from McGraw-Hill

Avallone and Baumeister • Marks' Standard Handbook for Mechanical Engineers
Bhushan and Gupta • Handbook of Tribology
Brady and Clauser • Materials Handbook
Bralla • Handbook of Product Design for Manufacturing
Brink • Handbook of Fluid Sealing
Brunner • Handbook of Incineration Systems
Corbitt • Standard Handbook of Environmental Engineering
Ehrich • Handbook of Rotordynamics
Elliot • Standard Handbook of Powerplant Engineering
Freeman • Standard Handbook of Hazardous Waste Treatment and Disposal
Ganíc and Hicks • The McGraw-Hill Handbook of Essential Engineering Information and Data
Gieck • Engineering Formulas
Grimm and Rosaler • Handbook of HVAC Design
Harris • Handbook of Noise Control
Harris and Crede • Shock and Vibration Handbook
Hicks • Standard Handbook of Engineering Calculations
Hodson • Maynard's Industrial Engineering Handbook
Jones • Diesel Plant Operations Handbook
Juran and Gryna • Juran's Quality Control Handbook
Karassik et al. • Pump Handbook
Kurtz • Handbook of Applied Mathematics for Engineers and Scientists
Mason • Switch Engineering Handbook
Nayyar • Piping Handbook
Parmley • Standard Handbook of Fastening and Joining
Rosaler and Rice • Standard Handbook of Plant Engineering
Rothbart • Mechanical Design and Systems Handbook
Schwartz • Composite Materials Handbook
Schwartz • Handbook of Structural Ceramics
Shigley and Mischke • Standard Handbook of Machine Design
Townsend • Dudley's Gear Handbook
Tuma • Engineering Mathematics Handbook
Tuma • Handbook of Numerical Calculations in Engineering
Wadsworth • Handbook of Statistical Methods for Engineers and Scientists
Walsh • McGraw-Hill Machining and Metalworking Handbook
Woodruff and Lammers • Steam-Plant Operation
Young • Roark's Formulas for Stress and Strain

MACHINE DESIGN DATA HANDBOOK

K. Lingaiah, Ph.D.
Bangalore University
Professor Emeritus
Adichunchanagiri Institute of Technology

McGraw-Hill, Inc.

New York   San Francisco   Washington, D.C.   Auckland   Bogotá
Caracas   Lisbon   London   Madrid   Mexico City   Milan
Montreal   New Delhi   San Juan   Singapore
Sydney   Tokyo   Toronto

Library of Congress Cataloging-in-Publication Data

Lingaiah, K.
    Machine design data handbook / K. Lingaiah.
        p.   cm.
    Includes bibliographical references.
    ISBN 0-07-037933-5
    1. Machine design—Handbooks, manuals, etc.   I. Title.
TJ230.L718   1994
621.8'15—dc20
                              93-26349
                                 CIP

   4 5 6 7 8 9 0 DOC/DOC 9 0 9 8 7 6

ISBN 0-07-037933-5

The sponsoring editor for this book was Robert W. Hauserman. This book was set in Times Roman. It was composed by Alden Multimedia.

Printed and bound by R. R. Donnelley & Sons Company.

This book is printed on acid-free paper.

To Professor B. R. Narayana Iyengar

To my parents: Kenchaiah and Madamma

To my wife: M. P. Madamma
and my children: L. Susheela,
L. Nagarathna, L. Vasanth Kumar,
L. Suma, and L. Vivekananda

CONTENTS

PREFACE

In the teaching of machine design in our universities it has been the experience of the author that a considerable amount of data by way of formulae, charts, graphs, and so forth are frequently required in the solution of problems. Several handbooks and textbooks, no doubt give these with particular reference to the topic under consideration. It is not possible to get all the necessary information in one handbook or textbook for solving a project or problem, hence an attempt has been made to compile the various formulae (emperical, semi-emperical, and otherwise), charts, and graphs in the form of a handbook which, it is hoped, will fulfill the above need. It is obvious that it is important to give all the data required for solving design problems. The present compilation is aimed primarily to help in conducting the design classes, both for the under-graduate and post-graduate level to a reasonably satisfactory extent. Incidentally, the handbook may, it is hoped, be useful in setting more practical problems for the closed-book examinations which are at present in vogue in many colleges. If the handbook containing these additonal data is supplied to the student, the problems of a more involved and practical nature could be set to be solved in 3 or 4 hours given for the examination.

In a work of the above nature, it has been necessary to reproduce the data given in various texts and handbooks. The author is grateful to the various publishers and authors who have given their consent to reproduce such materials. Individual acknowledgment has been made in the Acknowledgments.

I am grateful to the Director of Printing and Stationery, Government of Mysore, Bangalore, for undertaking the work of such complication and the Engineering College Co-operative Society, for the assistance in bringing out the book.

I owe the University of Mysore, a debt of gratitude for granting the necessary permission to publish the book.

In a work of this nature there are bound to be errors and the author would be grateful if these are brought to his notice for future guidance. Any suggestion for the improvement will be gratefully received.

I am very grateful to Prof. B. M. Belgaumkar, who has encouraged me in getting up this work.

It is sincerely hoped that this handbook will be useful to students, teachers, and practicing engineers.

K. Lingaiah

ACKNOWLEDGMENTS

With many thanks I acknowledge the permissions granted by the following publishers for reproducing the matter shown against each.

By Permission from McGraw-Hill, Inc.

1. "Variation of Journal Eccentricity Co-efficient with KC_L," from *Design of Machine Members* by Vallance and Doughtie, 3rd ed., 1951, p. 215.
2. "Factors for Horsepower rating of V-belts," from *Machine Design* by Black, 2nd ed., 1955, p. 202.
3. "Cures for Stress-Concentration Factors," from *Machine Design* by Black, 2nd ed., 1955, Appendix X, pp. 393–407.
4. "Comparison of Symmetrical Curves," *Cams* by Harold A. Rothbart, John Wiley and Sons, Inc., 1956, p. 184.

Machine Design by Maleev and Hartman, International Book Company, 3rd Edition, 1957.

5. "Influence of Size on Elastic Limits," p. 118.
6. "Endurance Diagram for Three Types of Stresses," p. 131.
7. "Reciprocals of Stress-Concentration Factors Caused by Surface Conditions," p. 132.
8. "SAE Standard Parallel-Side Splines," p. 282.
9. "Diagram of Disk Constants," p. 311.
10. "Stress-Factors for Helical Springs," p. 313.
11. "Viscosity Temperature Chart," p. 448.
12. "Life Curves of Ball Bearings," p. 482.
13. "Load Capacity of Case Hardened Gleason Straight Bevel Gears Operating at 90° Shaft Angle," p. 602.
14. "Maximum Permissible Errors in Gear-Tooth Profiles," p. 573.

Design of Machine Elements by Virgil Moring Faires, 3rd Edition, Macmillan Company, New York.

Gears

15. "Alignment Chart for Solving Dynamic Load Equation," p. 399.
16. "Number of Teeth for Straight and Zerol Bevel Gears," p. 437.
17. "Number of Teeth for Spiral Bevel Gears," p. 437.
18. "Geometry Factor, J, for Gleason Straight Bevel Gears, 20° Pressure Angle, 90° Shaft Angle," p. 439.
19. "Transmittance, Gear Cases," p. 453.

ABOUT THE AUTHOR

K. Lingaiah was a former Principal, a former Head
of the Department of Mechanical Engineering and
Professor of Mechanical Engineering at the
University Visvesvaraya College of Engineering,
Bangalore University, Bangalore. At present he is a
Professor Emeritus of Mechanical Engineering at
Adichunchanagiri Institute of Technology,
Chikmagular, Karnataka State. He is a consultant to
various industries and has delivered seminars and
lectures at NASA, University of Toronto, Canada,
National University of Singapore, and numerous
universities and engineering colleges in India.

CHAPTER
1

PROPERTIES OF ENGINEERING MATERIALS

SYMBOLS

a	area of cross section, m^2 (in^2)*
	original area of cross section of test specimen, mm^2 (in^2)
A_j	area of smallest cross section of test specimen under load F_j, m^2 (in^2)
A_f	minimum area of cross section of test specimen at fracture, m^2 (in^2)
A_o	original area of cross section of test specimen, m^2 (in^2)
A_r	percent reduction in area that occurs in standard test specimen
Bhn	Brinell hardness number
d	diameter of indentation, mm
	diameter of test specimen at necking, m (in)
D	diameter of steel ball, mm
E	modulus of elasticity or Young's modulus, GPa [Mpsi(Mlb/in^2)]
f_ε	strain fringe (fri) value, μm/fri (μin/fri)
f_σ	stress fringe value, kN/m fri (lbf/in fri)
F	load (also with subscripts), kN (lbf)
G	modulus of rigidity or torsional or shear modulus, GPa (Mpsi)
H_B	Brinell hardness number
l_f	final length of test specimen at fracture, mm (in)
l_j	gauge length of test specimen corresponding to load F_j, mm (in)
l_o	original gauge length of test specimen, mm (in)
Q	figure of merit, fri/m (fri/in)
R_B	Rockwell B hardness number
R_C	Rockwell C hardness number
υ	Poisson's ratio
σ	normal stress, MPa (psi)

---

*The units in parentheses are **US Customary units** [i.e., fps/(foot-pound-second)].

σ_b	transverse bending stress, MPa (psi)
σ_c	compressive stress, MPa (psi)
σ_S	strength, MPa (psi)
σ_t	tensile stress, MPa (psi)
σ_{Sf}	endurance limit, MPa (psi)
σ'_{Sf}	endurance limit of rotating beam specimen or R R Moore endurance limit, MPa (psi)
σ'_{Sfa}	endurance limit for reversed axial loading, MPa (psi)
σ'_{Sfb}	endurance limit for reversed bending, MPa (psi)
σ_{Sc}	compressive strength, MPa (psi)
σ_{St}	tensile strength, MPa (psi)
σ_u	ultimate stress, MPa (psi)
σ_{uc}	ultimate compressive stress, MPa (psi)
σ_{ut}	ultimate tensile stress, MPa (psi)
σ_{Su}	ultimate strength, MPA (psi)
σ_{Suc}	ultimate compressive strength, MPa (psi)
σ_{Sut}	ultimate tensile strength, MPa (psi)
σ_y	yield stress, MPa (psi)
σ_{yc}	yield compressive stress, MPa (psi)
σ_{yt}	yield tensile stress, MPa (psi)
σ_{Syc}	yield compressive strength, MPa (psi)
σ_{Syt}	yield tensile strength, MPa (psi)
τ	torsional (shear) stress, MPa (psi)
τ_S	shear strength, MPa (psi)
τ_u	ultimate shear stress, MPa (psi)
τ_{Su}	ultimate shear strength, MPa (psi)
τ_y	yield shear stress, MPa (psi)
τ_{Sy}	yield shear strength, MPa (psi)
τ'_{Sf}	torsional endurance limit, MPa (psi)

SUFFIXES

a	axial
b	bending
c	compressive
f	endurance
s	strength properties of material
t	tensile
u	ultimate
y	yield

ABBREVIATIONS

AISI	American Iron and Steel Institute
ASA	American Standards Association
AMS	Aerospace Materials Specifications
ASM	American Society for Metals
ASME	American Society of Mechanical Engineers
ASTM	American Society for Testing Materials
BIS	Bureau of Indian Standards
BSS	British Standard Specifications
DIN	Deutsches Institut für Normung
ISO	International Standards Organization

SAE   Society of Automotive Engineers
UNS   Unified Numbering System
Note: σ and τ with subscript S designates strength properties of material used
in the design which will be used and observed throughout this *Machine Design
Data Handbook*.
   Other factors in performance or in special aspects are included from time to
time in this chapter and, being applicable only in their immediate context, are
not given at this stage.

Particular	Formula	
The unit strain	$$\varepsilon = \frac{l_j - l_o}{l_o} = \frac{A_o - A_j}{A_j}$$	(1–1)
Percent elongation in a standard tension test specimen	$$\varepsilon_{100} = \frac{l_f - l_o}{l_o}(100)$$	(1–2)
Percent reduction in area that occurs in standard tension test specimen in case of ductile materials	$$A_r = \frac{A_o - A_f}{A_o}(100)$$	(1–3)
Bridgeman's equation for actual stress (σ_{act}) during r radius necking of a tensile test specimen	$$\sigma_{act} = \frac{\sigma_{cal}}{\left(1 + \dfrac{4r}{d}\right)\left[\ln\left(1 + \dfrac{d}{4r}\right)\right]}$$	(1–4)
The standard gauge length	$$l_o = 5.65\sqrt{a}$$	(1–5)
The modulus of toughness	$$T_m = \int_o^{\varepsilon_f} \sigma \, d\varepsilon$$	(1–6a)
	$$\approx \frac{\sigma_{Sy} + \sigma_{Su}}{2}\varepsilon_f$$	(1–6b)
The approximate relationship between stress and strain in the plastic region of the stress-strain curve	$$\sigma = \sigma_o \varepsilon^n$$	(1–7)
The Brinell hardness number	$$H_B = \frac{2F}{\pi D(D - \sqrt{D^2 - d^2})}$$	(1–8)
The relationship between the Brinell hardness number H_B and Rockwell C number R_C	$$R_C = 88H_B^{0.162} - 192$$	(1–9)
The relationship between the Brinell hardness number H_B and Rockwell B number R_B	$$R_B = \frac{H_B - 47}{0.0074 \, H_B + 0.154}$$	(1–10)
The approximate relationship between ultimate tensile strength and Brinell hardness number of carbon and alloy steels which can be applied to steels with a Brinell hardness number between 200 H_B and 350 H_B only (1,2)	$\sigma_{Sut} = 3.45 \, H_B$ MPa **SI** $= 500 \, H_B$ psi **US Customary System units**	(1–11a) (1–11b)

Particular	Formula
The relationship between the minimum ultimate strength and the Brinell hardness number for steels as per ASTM	$\sigma_{Sut} = 3.10\,H_B$ MPa **SI** (1–11c) $= 450\,H_B$ **US Customary System units** (1–11d)
The relationship between the minimum ultimate strength and the Brinell hardness number for cast iron as per ASTM	$\sigma_{Sut} = 1.58\,H_B - 86.2$ MPa **SI** (1–11e) $= 230\,H_B - 12500$ **US Customary System units** (1–11f)
The relationship between the minimum ultimate strength and the Brinell hardness number as per SAE minimum strength	$\sigma_{Sut} = 2.60\,H_B - 110$ MPa **SI** (1–11g) $= 237.5\,H_B - 16{,}000\,\text{psi}$ **US Customary System units** (1–11h)
In case of stochastic results the relation between H_B and σ_{Sut} for steel based on Eqs. (1–11a) and (1–11b)	$\sigma_{Sut} = (3.45, 0.152)\,H_B$ MPa **SI** (1–11i) $= (500, 22)\,H_B$ psi **US Customary System units** (1–11j)
In case of stochastic results the relation between H_B and σ_{Sut} for cast iron based on Eqs. (1–11e) and (1–11f)	$\sigma_{Sut} = 1.58\,H_B - 62 + (0,\ 10.3)$ MPa **SI** (1–11k) $= 230 H_B - 9{,}000 + (0,\ 1500)$ **US Customary System units** (1–11l)
Relationships between hardness number and tensile strength of steel in SI and US Customary units [7]	Refer to Fig. 1–1.
The approximate relationship between ultimate shear stress and ultimate tensile strength for various materials	$\tau_{Su} = 0.82\,\sigma_{Sut}$ for wrought steel (1–12a) $\tau_{Su} = 0.90\,\sigma_{Sut}$ for malleable iron (1–12b) $\tau_{Su} = 1.30\,\sigma_{Sut}$ for cast iron (1–12c) $\tau_{Su} = 0.90\,\sigma_{Sut}$ for copper and copper alloys (1–12d) $\tau_{Su} = 0.65\,\sigma_{Sut}$ for aluminum and aluminum alloys (1–12e)
The tensile yield strength of stress-relieved (not cold-worked) steels according to Datsko (1,2)	$\sigma_{Sy} = (0.072\,\sigma_{Sut} - 205)$ MPa **SI** (1–13a) $= 1.05\,\sigma_{Sut} - 30\ \text{kpsi}$ **US Customary System units** (1–13b)
The equation for tensile yield strength of stress-relieved (not cold-worked) steels in terms of Brinell hardness number H_b according to Datsko (2)	$\sigma_{Sy} = (3.62\,H_B - 205)$ MPa **SI** (1–14a) $= 525\,H_B - 30\ \text{kpsi}$ **US Customary System units** (1–14b)
The approximate relationship between shear yield strength (τ_{Sy}) and yield strength (tensile) (σ_{Sy})	$\tau_{Sy} = 0.55\,\sigma_{Sy}$ for aluminum and aluminum alloys (1–15a) $\tau_{Sy} = 0.58\,\sigma_{Sy}$ for wrought steel (1–15b)

The approximate relationship between endurance limit (also called *fatigue limit*) for reversed bending polished specimen based on 50 percent survival rate and ultimate strength for nonferrous and ferrous materials	**For students' use**
	
	$\sigma'_{Sfb} = 0.50\,\sigma_{Sut}$ for wrought steel having
	    $\sigma_{Sut} < 1380$ MPa (200 kpsi) (1–16)
	
	$\sigma'_{Sfb} = 690$ MPa for wrought steel having
	    $\sigma_{Sut} > 1380$ MPa **SI** (1–17a)

Particular	Formula

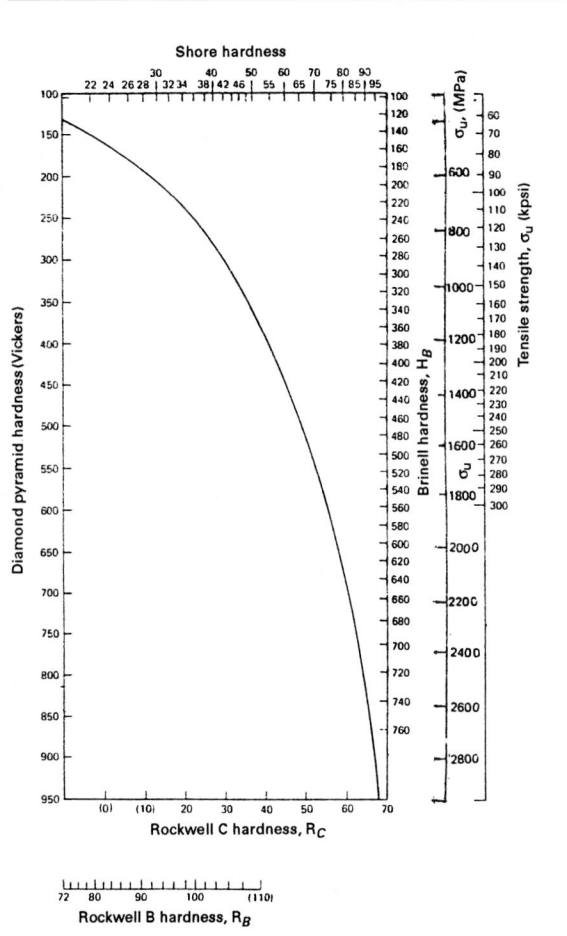

FIGURE 1–1 Conversion or hardness number to ultimate tensile strength of steel σ_{Sut}, MPa (kpsi). (*Technical Editor Speaks, courtesy of International Nickel Co., Inc., 1943.*)

$= 100$ kpsi   for wrought steel having
$\qquad \sigma_{Sut} > 200$ kpsi   **US Customary System units**
$\qquad\qquad\qquad\qquad\qquad$ (1–17b)

For practicing engineer's use

$\sigma'_{Sfb} = 0.35\,\sigma_{Sut}$ for wrought steel having
$\qquad \sigma_{Sut} < 1380$ MPa (200 kpsi) (1–18)

$\sigma'_{Sfb} = 550$ MPa   for wrought steel having
$\qquad \sigma_{Sut} > 1380$ MPa   **SI**   (1–19a)
$\quad = 80$ kpsi   for wrought steel having
$\qquad \sigma_{Sut} > 200$ kpsi   **US Customary System units** (1–19b)

$\sigma'_{Sfb} = 0.45\,\sigma_{Sut}$ for cast iron and cast steel when
$\qquad \sigma_{Sut} \leq 600$ MPa (88 kpsi)   (1–20)

$\sigma'_{Sfb} = 275$ MPa   for cast iron and cast steel when
$\qquad \sigma_{Sut} > 600$ MPa   **SI**   (1–21a)

$\quad = 40$ kpsi   for cast iron and cast steel when
$\qquad \sigma_{Sut} > 88$ kpsi   **US Customary System units**
$\qquad\qquad\qquad\qquad\qquad$ (1–21b)

$\sigma'_{Sfb} = 0.45\,\sigma_{Sut}$ for copper-based alloys
$\qquad$ and nickel-based alloys   (1–22)

$\sigma'_{Sfb} = 0.38\,\sigma_{Sut}$ for wrought aluminum alloys
up to tensile strength of
275 MPa(40 kpsi) based on
5×10^8 cycle life   (1–23)

$\sigma'_{Sfb} = 0.16\,\sigma_{Sut}$ for cast aluminum alloys
up to tensile strength of
300 MPa (50 kpsi) based
on 5×10^8 cycle life   (1–24)

$\sigma'_{Sfb} = 0.38\,\sigma_{Sut}$ for magnesium casting alloys
and magnesium wrought
alloys based on 10^6
cyclic life   (1–25)

Particular	Formula	
The relationship between the endurance limit for reversed axial loading of a polished, unnotched specimen and the reversed bending for steel specimens	$\sigma'_{Sfa} = 0.85\,\sigma'_{Sfb} = 0.43\,\sigma_{Sut}$	(11–26)
The relationship between the torsional endurance limit and the reversed bending for reversed torsional tested polished unnotched specimens for various materials	$\tau'_{Sf} = 0.58\,\sigma'_{Sfb} = 0.29\,\sigma_{Sut}$ for steel $\tau'_{Sf} \approx 0.8\,\sigma'_{Sfb} \approx 0.32\,\sigma_{Sut}$ for cast iron $\tau'_{Sf} \approx 0.48\,\sigma'_{Sfb} \approx 0.22\,\sigma_{Sut}$ for copper	(1–27a) (1–27b) (1–27c)
For additional information or data on properties of engineering materials	Refer to Tables 1–1 to 1–41	

TABLE 1–1
Hardness conversion (approximate)

| Brinell 29.42 kN (3000 kgf) load 10 mm ball | | Vickers or Firth hardness number | Rockwell hardness numbers | | | | Shore scleroscope hardness number | Tensile strength, σ_{Sut} (approximate) | |
Diameter (mm)	Hardness number		A scale 0.588 kN (60 kgf) load	B scale 0.98 kN (100 kgf) load	C scale 1.47 kN (150 kgf) load	15-N scale 0.147 kN (15 kgf) load		MPa	kpsi
2.25	745	840	84		65	92	91	2570	373
2.30	712	783	83		64	92	87	2455	356
2.35	682	737	82		62	91	84	2350	341
2.40	653	697	81		60	90	81	2275	330
2.45	627	667	81		59	90	79	2227	323
2.50	601	640	80		58	89	77	2192	318
2.55	578	615	79		57	88	75	2124	309
2.60	555	591	78		55	88	73	2020	293
2.65	534	569	78		54	87	71	1924	279
2.70	514	547	77		52	87	70	1834	266
2.75	495	528	76		51	86	68	1750	254
2.80	477	508	76		50	85	66	1675	243
2.85	461	491	75		49	85	65	1620	235
2.90	444	472	74		47	84	63	1532	222
2.95	429	455	73		46	83	61	1482	215
3.00	415	440	73		45	83	59	1434	208
3.05	401	425	72		43	82	58	1380	200
3.10	388	410	71		42	81	56	1338	194
3.15	375	396	71		40	81	54	1296	188
3.20	363	383	70		39	80	52	1255	182
3.25	352	372	69	110	38	79	51	1214	176
3.30	341	360	69	109	37	79	50	1172	170
3.35	331	350	68	109	36	78	48	1145	166
3.40	321	339	68	108	34	77	47	1103	160
3.45	311	328	67	108	33	77	46	1069	155
3.50	302	319	66	107	32	76	45	1042	151
3.55	293	309	66	106	31	76	43	1010	146
3.60	285	301	65	106	30	75	42	983	142
3.65	277	292	65	105	29	74	41	955	138
3.70	269	284	64	104	28	74	40	928	134
3.75	262	276	64	103	27	73	39	904	131
3.80	255	269	63	102	25	73	38	875	127
3.85	248	261	63	101	24	72	37	855	124
3.90	241	253	62	100	23	71	36	832	120
3.95	235	247	61	99	22	70	35	810	117
4.00	229	241	61	98	21	70	34	790	114
4.05	223	234		97	19			770	111
4.10	217	228		96	18		33	748	108
4.15	212	222		96	16		32	730	106
4.20	207	218		95	15		31	714	103

TABLE 1–1
Hardness conversion (approximate) (*Cont.*)

Brinell 29.42 kN (3000 kgf) load 10 mm ball			Rockwell hardness numbers						Tensile strength, σ_{Sut} (approximate)	
Diameter (mm)	Hardness number	Vickers or Firth hardness number	A scale 0.588 kN (60 kgf) load	B scale 0.98 kN (100 kgf) load	C scale 1.47 kN (150 kgf) load	15-N scale 0.147 kN (15 kgf) load	Shore scleroscope hardness number		MPa	kpsi
4.25	201	212		94	14				690	100
4.30	197	207		93	13		30		680	98
4.35	192	202		92	12		29		662	96
4.40	187	196		91	10				645	93
4.45	183	192		90	9		28		631	91
4.50	179	188		89	8		27		617	89
4.55	174	182		88	7				600	87
4.60	170	178		87	5		26		585	85
4.65	167	175		86	4				576	83
4.70	163	171		85	3		25		562	81
4.80	156	163		83	1		24		538	78
4.90	149	156		81			23		514	74
5.00	143	150		79			22		493	71
5.10	137	143		76			21		472	68
5.20	131	137		74					451	65
5.30	126	132		72			20		435	63
5.40	121	127		70			19		417	60
5.50	116	122		68			18		400	58
5.60	111	117		65			17		383	55

TABLE 1–2
Poisson's ratio (ν)

Material	ν	Material	ν
Aluminum, cast	0.330	Molybdenum	0.293
Aluminum, drawn	0.348	Monel metal	0.320–0.370
Beryllium copper	0.285	Nickel, soft	0.239
Brass	0.340	Nickel, hard	0.306
Brass, 30Zn	0.350	Rubber	0.450–0.490
Cast steel	0.265	Silver	0.367
Chromium	0.210	Steel, mild	0.303
Copper	0.343	Steel, high carbon	0.295
Douglas fir	0.330	Steel, tool	0.287
Ductile iron	0.340–0.370	Steel, stainless (18–8)	0.305
Glass	0.245	Tin	0.342
Gray cast iron	0.210–0.270	Titanium	0.357
Iron, soft	0.293	Tungsten	0.280
Iron, cast	0.270	Vanadium	0.365
Inconel x	0.410	Wrought iron	0.278
Lead	0.431	Zinc	0.331
Magnesium	0.291		
Malleable cast iron	0.230		

TABLE 1-3
Mechanical properties of typical cast ferrous materials[a]

Material, class, specification	Class or grade	Tension σ_{Sut} MPa	Tension kpsi	Compression σ_{Suc} MPa	Compression kpsi	Shear τ_{Su} MPa	Shear kpsi	Torsional/shear τ_S MPa	Torsional kpsi	Yield σ_{Sy} MPa	Yield kpsi	Endurance σ_{Sfb} MPa	Endurance kpsi	Brinell H_B	Tension E GPa	Tension E Mpsi	Compression E GPa	Compression E Mpsi	Shear G GPa	Shear G Mpsi	Elong. 50 mm (2 in), %	Impact J	Impact ft-lbf	Typical applications
Gray cast iron[b] ASTM class	SAE																							
20	110	152	22	572	83	220	32	179	26			69	10	156	66–97	9.6–14.0			27–39	3.9–5.6		75	55	Soft iron castings
25		179	26	669	97	255	37	220	32			79	11.5	174	79–102	11.5–14.8			32–41	4.6–6.0		75	55	Cylinder blocks and heads, housing
30	111	214	31	752	109	303	44	276	40			97	14	210	90–113	13.0–16.4			36–45	5.2–6.6		81	60	Flywheels, brake drums and clutch plates
35	120	252	36.5	855	124	338	49	334	48.5			110	16	212	100–119	14.5–17.2			40–48	5.8–6.9				Heavy-duty brake drums, clutch plates
40	121	293	42.5	965	140	393	57	393	57			128	18.5	235	110–138	16–20.0			44–54	6.4–7.8		95	70	Cam shafts, cylinder liners
50		362	52.5	1130	164	448	65	503	73			148	21.5	262	130–157	18.8–22.8			50–55	7.2–8.0		108	80	Special high-strength castings
60		431	62.5	1293	187.5	496	72	610	88.5			169	24.5	302	141–162	20.4–23.5			54–59	7.8–8.5		156	115	Special high-strength castings
Malleable cast iron: Ferrite ASTM A47-52, A338	32510	345	50	1434	208	324	47			220	32	193	28	156 max	172	25	172	25			10	22	16.5	General purpose at normal and elevated temperature, good machinability, excellent shock resistance
ANSI G 48-1; FED QQ-I-66e	35018	365	53	1517	220	352	51			241	35	214	31	156 max	172	25	172	25			18	22	16.5	
ASTM A197		276	40							207	30			156 max							5			Pipe flanges, valve parts
Pearlite and martensite: ASTM A220	40010	414	60							276	40			149–197							10			General engineering service at normal and elevated temperatures
ANSI G48-2	45008	448	65							310	45			156–197							8			
MIL-I-11444B	45006	448	65							310	45			156–207							6			
	45010	448	65	1670	242	338	49			310	45	220	32	185	180	26	180	23.2			10	19	14	
	50005	483	70							345	50			179–229							5			
	50007	517	75	1670	242	517	75			345	50	255	37	204	183	26.5	183	23.2			7	19	14	
	60004	552	80							414	60			197–241							4			
	60003	552	80	1670	242	552	80			414	60	270	39	226	186	27	186	23.2			3	19	14	
	70003	586	85							483	70			217–269							3			
	80002	655	95	1670	242	689	100			552	80	276	40	241–285	186	27	186	23.2			2	19	14	
	90001	724	105							621	90			269–321							1			

TABLE 1-3
Mechanical properties of typical cast ferrous materials[a] (Cont.)

Material, class, specification	Grade	UNS No.	Ultimate strength — Tension, σ_{ut} MPa	kpsi	Compression, σ_{uc} MPa	kpsi	Shear, τ_{su} MPa	kpsi	Torsional/shear strength, τ_s MPa	kpsi	Yield strength, σ_y MPa	kpsi	Endurance limit in reversed bending, σ_{sb} MPa	kpsi	Brinell hardness, H_B	Modulus of Elasticity — Tension, E GPa	Mpsi	Compression, E GPa	Mpsi	Shear, G GPa	Mpsi	Elongation in 50 mm (2 in), %	Impact strength (Charpy) J	ft-lbf	Typical applications
Automotive ASTM A602 SAE J158	M3210[c]		345	50							224	32			156 max							10			Steering gear housing, mounting brackets
	M4504[d]		448	65							310	45			163–217							4			Compressor crankshafts and hubs
	M5003[d]		517	75							345	50			187–241							3			Parts requiring selective hardening, as gears
	M5503[e]		517	75							379	55			187–241							3			For machinability and improved induction hardening
	M7002[e]		621	90							483	70			229–269							2			Connecting rods, universal joint yokes
	M8501[e]		724	105							586	85			269–302							1			Gears with high-strength and good wear resistance
Nodular (ductile) cast iron[f]																									
ASTM A395-76 ASME SA 395	60-40-18	F32800	414	60							276	40			143–187							18			Valves and fittings for steam and chemical plant equipments
ASTM A476-70(d) SAE AMS5316	80-60-03	F34100	552	80							414	60			201 min							3			Paper-mill dyer rolls
ASTM A536-72 MIL-I-11466 B(MR)	60-40-18[h]	F32800	461	66.9	359	52.0	472	68.5			329	47.7	241	35	167–178	169	24.5	164	23.8	63–65.5[g]	9.1–9.5[g]	15			Pressure-containing parts such as valve and pump bodies
	65-45-12[h]	F33100	464	67.3	362	52.5	475	68.9			332	48.2			167	168	24.4	163	23.6	64–65[g]	9.3–9.4[g]	15			Machine components subjected to shock and fatigue loads
	80-55-06[h]	F33800	559	81.1	386	56.0	504	73.1			362	52.5	345	50	192	168	24.4	165	23.9	62–64[g]	9.0–9.3[g]	11.2			Crankshafts, gears and rollers
	100-70-03[h]	F34800	758	110	1515	220.0					500	72.5	379	55	257	162	23.5					6–10			High-strength gears and machine components
	120-90-02[h]	F36200	974	141.3	920	133.5	875	126.9			864	125.3	434	63	332	164	23.8	164	23.8	63.4–64[g]	9.2–9.3[g]	1.5			Pinions, gears, rollers and slides
SAE J 434C	D4018		414	60							276	40			170							18			Steering knuckles
	D4512		448	65							310	45			156–217							12			Disk brake calipers
	D5506		552	80							379	55			187–255							6			Crankshafts
	D7003		689	100							483	70			241–302							3			Gears

TABLE 1-3
Mechanical properties of typical cast ferrous materials[a] (*Cont.*)

Material, class, specification	Ultimate strength										Yield strength, σ_{Sy}		Endurance limit in reversed bending, σ_{Sjb}		Brinell hardness, H_B	Modulus of Elasticity						Elongation in 50 mm (2 in), %	Impact strength (Charpy)	
	Tension, σ_{Sut}		Compression, σ_{Suc}		Shear, τ_{Su}		Torsional/shear strength, T_S									Tension, E		Compression, E		Shear, G				
	MPa	kpsi	MPa	kpsi	MPa	kpsi	MPa	kpsi	MPa	kpsi	MPa	kpsi	MPa	kpsi		GPa	Mpsi	GPa	Mpsi	GPa	Mpsi		J	ft-lbf
Alloy cast irons																								
Medium-silicon gray iron	170–310	25–45	620–1040	90–150											170–250								20–31	15–23
High chromium gray iron	210–620	30–90	690	100											250–500								27–47	20–35
High nickel gray iron	170–310	25–45	690–1100	100–160											130–250								80–200	60–150
Ni-Cr-Si gray iron	140–310	20–45	480–690	70–100											110–210								110–200	80–150
High-aluminum gray iron	235–620	34–90													180–350									
Medium-silicon ductile iron	415–690	60–100													140–300								7–155	5–115
High-nickel ductile iron (20Ni)	380–415	55–60	1240–1380	180–200											140–200								16	12
High-nickel ductile iron (23Ni)	400–450	58–65													130–170								38	28
Duriron	110	16													520	158	23						4	3
Mechanite	241–380	35–55							193–241	28–35					190	83	12					10		

Source: Compiled from *ASM Metals Handbook*, American Society for Metals, Metals Park, Ohio, 1988.

[b] Minimum values of σ_u in MPa (kpsi) are given by class number.

[c] Annealed.

[d] Air-quenched and tempered.

[e] Liquid-quenched and tempered.

[f] Heat-treated and average mechanical properties.

[g] Calculated from tensile modulus and Poisson's ratio in tension.

1.11

TABLE 1-4
Typical mechanical properties of gray cast iron

Grade	Tensile strength, σ_{St} MPa	kpsi	Compressive strength, σ_{Sc} MPa	kpsi	Shear strength, τ_S MPa	kpsi	Fatigue limit, σ_{Sf} MPa	kpsi	E Tension GPa	Mpsi	E Compression GPa	Mpsi	Modulus of rigidity, G GPa	Mpsi	Notched tensile strength, σ_{Snt} MPa	kpsi	Elastic strain at failure, percent	Total elastic strain at fracture, percent
FG 150	150	21.8	600	87.0	173	25.1	68[e]	9.9	100	14.5	100	14.5	40	5.8	120[a]	17.4	0.15	0.6–0.75[g]
	42[c]	6.0	84	12.2			68[f]	9.9							150[b]	21.8		
	98[d]	14.2	105	15.2														
FG 200	200	29.0	720	104.4	230	33.4	90[e]	13.1	114	16.5	114	16.5	46	6.7	160[a]	23.2	0.17	0.48–0.67[g]
	56[c]	8.1	112	16.2			87[f]	12.6							200[b]	29.0		
	130[d]	18.8	260	37.7														
FG 220	220	32.0	768	111.4	253	36.7	99[e]	14.4	120	17.4	120	17.4	48	7.0	176[a]	25.5	0.18	0.39–0.63[g]
	62[c]	9.0	123	17.8			94[f]	13.6							220[b]	32.0		
	143[d]	20.7	286	41.5														
FG 260	260	37.7	864	125.3	299	43.4	117[e]	17.0	128	18.6	128	18.6	51	7.4	208[a]	30.2	0.20	0.57[g]
	73[c]	10.6	146	21.2			108[f]	15.7							260[b]	37.7		
	169[d]	24.5	338	49.0														
FG 300	300	43.5	960	139.2	345	50.0	135[e]	19.6	135	19.6	135	19.6	54	7.8	240[a]	34.8	0.22	0.50[g]
	84[c]	12.2	168	24.4			127[f]	18.4							300[b]	43.5		
	195[d]	28.3	390	56.6														
FG 350	350	50.8	1080	156.6	403	58.5	149[e]	21.6	140	20.3	140	20.3	56	8.1	280[a]	40.6	0.25	0.50[g]
	98[c]	14.2	196	28.4			129[f]	18.7							350[b]	50.8		
	228[d]	33.1	455	66.0														
FG 400	400	58.0	1200	174.1	460	66.7	152[e]	22.0	145	21.0	145	21.0	58	8.4	320[a]	46.4	0.28	0.50[g]
	112[c]	16.2	224	32.5			127[f]	18.4							400[b]	58.0		
	260[d]	37.7	520	75.4														

[a] Circumferential 45° notch-root radius 0.25 mm (0.04 in), notch depth 2.5 mm (0.4 in), root diameter 20 mm (0.8 in). Notch depth 3.3 mm (0.132 in), notch diameter 7.6 mm (0.30 in).
[b] Circumferential notch radius 9.5 mm (0.38 in), notch depth 2.5 mm (0.4 in), notch diameter 20 mm (0.8 in).
[c] 0.01 percent proof stress.
[d] 0.1 percent proof stress.
[e] Unnotched—8.4 mm (0.336 in) diameter.
[f] V-notched [circumferential 45° V-notch with 0.25 mm (0.04 in) root radius, diameter at notch 8.4 mm (0.336 in), depth of notch 3.4 mm (0.136 in)].
[g] Value depends on the composition of iron.
[h] Poisson's ratio $\nu = 0.26$.

Note: The typical properties given in this table are the properties in a 30 mm (1.2 in) diameter separately cast test bar or in a casting section correctly represented by this size of test bar, where the tensile strength does not correspond to that given. Other properties may differ slightly from those given.

Source: IS (Indian Standards) 210, 1978.

TABLE 1–5
Mechanical properties of spheroidal or nodular graphite cast iron

Grade	Typical casting thickness		Tensile strength, σ_{St} min		0.2% proof stress, σ_{Sy} min		Elongation,[a] percent min	Brinell hardness, H_B	Impact values, min, (23±5°C)		Predominant structural constituent
	mm	in	MPa	kpsi	MPa	kpsi			J	ft-lbf	
			Measured on Test Pieces from Separately Cast Test Samples								
SG 900/2			900	130.5	600	87.0	2	280–360			Pearlite
SG 800/2			800	116.0	480	69.6	2	245–335			Pearlite
SG 700/2			700	101.5	420	61.0	2	225–305			Pearlite
SG 600/2			600	87.0	370	53.7	2	190–270			Ferrite and pearlite
SG 500/7			500	72.5	320	46.4	7	160–240			Ferrite and pearlite
SG 450/10			450	65.3	310	45.0	10	160–210	9.0[b] (4.3)[c]	6.6(3.2)	
SG 400/15			400	58.0	250	36.3	15	130–180	17.0[b] (15.0)[c]	12.5(11.0)	Ferrite
SG 400/18			400	58.0	250	36.3	18	130–180	14.0[b] (11.0)[c]	10.3(8.1)	
SG 350/22			350	50.8	220	32.0	22	≤ 150	17.0[b] (14.0)[c]	12.5(10.3)	Ferrite
			Measured on Test Pieces from Cast-on Test Samples								
SG 700/2A	30–60	1.2–2.4	700	101.5	400	58.0	2	220–320			
	61–200	2.44–8.0	650	94.3	380	55.1	1				
SG 600/3A	30–60	1.2–2.4	600	87.0	360	52.2	2	180–270			
	61–200	2.44–8.0	550	79.8	340	49.3	1				
SG 500/7A	30–60	1.22–2.4	450	65.3	300	43.5	7	170–240			
	61–200	2.44–8.0	420	61.0	290	42.0	5				
SG 400/15A	30–60	1.2–2.4	390	56.6	250	36.3	15				
	61–200	2.44–8.0	370	53.7	240	34.8	12	130–180			
SG 400/18A	30–60	1.2–2.4	390	56.6	250	36.4	15	130–180	14[b] (11)[c]	10.3(8.1)	
	61–200	2.44–8.0	370	53.7	240	34.8	12		12[b] (9)[c]	8.8 (6.6)	
SG 350/22A	30–60	1.2–2.4	330	47.9	220	31.9	18	≤ 150	17[b] (14)[c]	12.5 (10.3)	
	61–200	2.44–8.0	320	46.4	210	30.5	15		15[b] (12)[c]	11.1 (8.8)	

[a] Elongation is measured on an initial gauge length $L = 5d$ where d is the diameter of the gauge length of the test pieces.
[b] Mean value from 3 tests on V-notch test pieces at ambient temperature.
[c] Individual value.
Source: IS 1865, 1991.

TABLE 1-6
Carbon steels with specified chemical composition and related mechanical properties

| Designation | | % C | % Mn | Tensile strength, σ_{St} | | Elongation, % (gauge length 5.56 $\sqrt{a^*}$ round test piece) | Izod impact value, min (if specified) | |
New	Old			MPa	kpsi		J	ft-lbf
7 C 4	(C 07)	0.12 max	0.50 max	320–400	46.5–58.0	27		
10 C 4	(C 10 †)	0.15 max	0.30–0.60	340–420	49.4–70.0	26	55	40.6
14 C 6	(C 14 †)	0.10–0.18	0.40–0.70	370–450	53.6–65.0	26	55	40.6
15 C 4	(C 15)	0.20 max	0.30–0.60	370–490	53.6–71.0	25		
15 C 8	(C 15 Mn 75)	0.10–0.20	0.60–0.90	420–500	61.0–72.5	25		
20 C 8	(C 20)	0.15–0.25	0.60–0.90	440–520	63.5–75.4	24		
25 C 4	(C 25)	0.20–0.30	0.30–0.60	440–540	63.5–78.3	23		
25 C 8	(C 25 Mn 75 +)	0.20–C.30	0.60–0.90	470–570	68.2–82.7	22		
30 C 8	(C 30 +)	0.25–0.35	0.60–0.90	500–600	72.5–87.0	21	55	40.6
35 C 4	(C 35)	0.30–0.40	0.30–0.60	520–620	75.4–90.0	20		
35 C 8	(C 35 Mn 75 +)	0.30–0.40	0.60–0.90	550–650	79.8–94.3	20	55	40.6
40 C 8	(C 40 +)	0.35–0.45	0.60–0.90	580–680	84.1–98.7	18	41.35	30.5
45 C 8	(C 45 +)	0.40–0.50	0.60–0.90	630–710	91.4–103.0	15	41.35	30.5
50 C 4	(C 50 +)	0.45–0.55	0.60–0.90	660–780	95.7–113.1	13		
50 C 12	(C 50 Mn 1)	0.45–0.55	1.10–1.40	720 min	104.4 min	11		
55 C 8	(C 55 Mn 75 +)	0.50–0.60	0.60–0.90	720 min	104.4 min	13		
60 C 4	(C 60)	0.55–0.65	0.50–0.80	750 min	108.8 min	11		
65 C 6	(C 65)	0.60–0.70	0.50–0.80	750 min	108.8 min	10		

Notes: a^*, area of cross section; † steel for hardening; + steel for hardening and tempering; Mn $\underline{75}$ = average content of Mn is 0.75%.

Source: IS 1570, 1979.

TABLE 1–7
Carbon and carbon – manganese free – cutting steels with specified chemical composition and related mechanical properties

| Designation | | % C | % Si | % Mn | % S | % P (max) | Tensile strength, σ_{St} | | Minimum elongation, percent (gauge length $5.65\sqrt{a^{**}}$) | Izod impact value, min (if specified) | | Limiting ruling section, mm (in) |
New	Old						MPa	kpsi		J	ft-lbf	
10 C 8 S 10	(10 S 11)	0.15 max	0.05–0.30	0.60–0.90	0.08–0.13	0.060	370–490*	53.7–71.0	24*	55	40.6	30(1.2)
14 C 14 S 14	(14 Mn 1 S 14)	0.10–0.18	0.05–0.30	1.20–1.50	0.10–0.18	0.060	440–540*	63.8–78.3	22*			30(1.2)
25 C 12 S 14	(25 Mn 1 S 14)	0.20–0.30	0.25 max	1.00–1.50	0.10–0.18	0.060	500–600*	72.5–87.0	20*			
40 C 10 S 18	(40 S 18)	0.35–0.45	0.25 max	0.80–1.20	0.14–0.22	0.060	550–650*	79.8–94.0	17*	41	30.2	60(2.4)
11 C 10 S 25	(11 S 25)	0.08–0.15	0.10 max	0.80–1.20	0.20–0.30	0.060	370–490*	53.7–71.0	22*			
40 C 15 S 12	(40 Mn 2 S 12)	0.35–0.45	0.25 max	1.30–1.70	0.08–0.15	0.060	600–700*	87.0–101.5	15*	48	35.4	100(4.0)

Notes: a^{**}, area of cross section; *steel for case hardening. Minimum values of yield stress may be required in certain specifications, and in such cases a minimum yield stress of 55 percent of minimum tensile strength should be satisfactory.

Source: IS 1570, 1979.

TABLE 1-8
Mechanical properties of selected carbon and alloy steels

AISI[a] no.	UNS no.	Treatment	Austenitizing temperature		Tensile strength, σ_{St}		Yield strength, σ_{Sy}		Elongation in 50 mm (2 in), %	Reduction in area, %	Brinell hardness, H_B	Izod impact strength	
			°C	°F	MPa	kpsi	MPa	kpsi				J	ft-lbf
1010	G10100	Hot-rolled	—	—	320	47	180	26	28	50	95		
		Cold-drawn			370	53	300	44	20	40	105		
1015	G10150	As rolled			420.6	61.0	313.7	45.5	39.0	61.0	126	110.5	81.5
		Normalized	925	1700	424.0	61.5	324.1	47.0	37.0	69.6	121	115.5	85.2
		Annealed	870	1600	386.1	56.0	284.4	41.3	37.0	69.7	111	115.0	84.8
1020	G10200	As-rolled			448.2	65.0	330.9	48.0	36.0	59.0	143	86.8	64.0
		Normalized	870	1600	441.3	64.0	346.5	50.3	35.8	67.9	131	117.7	86.8
		Annealed	870	1600	394.7	57.3	294.8	42.3	36.5	66.0	111	123.4	91.0
1030	G10300	As-rolled			551.6	80.0	344.7	50.0	32.0	57.0	179	74.6	55.0
		Normalized	925	1700	520.0	75.5	344.7	50.0	32.0	60.8	149	93.6	69.0
		Annealed	845	1550	463.7	67.3	341.3	49.5	31.2	57.9	126	69.4	51.2
1040	G10400	As-rolled			620.5	90.0	413.7	60.0	25.0	50.0	201	48.8	36.0
		Normalized	900	1650	589.5	85.5	374.0	54.3	28.0	54.9	170	65.1	48.0
		Annealed	790	1450	518.8	75.3	353.4	51.3	30.2	57.2	149	44.3	32.7
1050	G10500	As-rolled			723.9	105.0	413.7	60.0	20.0	40.0	229	31.2	23.0
		Normalized	900	1650	748.1	108.5	427.5	62.0	20.0	39.4	217	27.1	20.0
		Annealed	790	1450	636.0	92.3	365.4	53.0	23.7	39.9	187	16.9	12.5
1060	G10600	As-rolled			813.7	118.0	482.6	70.0	17.0	34.0	241	17.6	13.0
		Normalized	900	1650	775.7	112.5	420.6	61.0	18.0	37.2	229	13.2	9.7
		Annealed	790	1450	625.7	90.8	372.3	54.0	22.5	38.2	179	11.3	8.3
1095	G10950	As-rolled			965.3	140.0	572.3	83.0	9.0	18.0	293	4.1	3.0
		Normalized	900	1650	1013.5	147.0	499.9	72.5	9.5	13.5	293	5.4	4.0
		Annealed	790	1450	656.7	95.3	379.2	55.0	13.0	20.6	192	2.7	2.0
1117	G11170	As-rolled			486.8	70.6	305.4	44.3	33.0	63.0	143	81.3	60.0
		Normalized	900	1650	467.1	67.8	303.4	44.0	33.5	63.8	137	85.1	62.8
		Annealed	825	1575	429.5	62.3	279.2	40.5	32.8	58.0	121	93.6	69.0
1144	G11440	As-rolled			703.3	102.0	420.6	61.0	21.0	41.0	212	52.9	39.0
		Normalized	900	1650	667.4	96.8	399.9	58.0	21.0	40.4	197	43.4	32.0
		Annealed	790	1450	584.7	84.8	346.8	50.3	24.8	41.3	167	65.1	48.0
1340	G13400	Normalized	870	1600	836.3	121.3	558.5	81.0	22.0	62.9	248	92.5	68.2
		Annealed	800	1475	703.3	102.0	436.4	63.3	25.5	57.3	207	70.5	52.0

TABLE 1-8
Mechanical properties of selected carbon and alloy steels (*Cont.*)

AISI[a] no.	UNS no.	Treatment	Austenitizing temperature		Tensile strength, σ_{Sr}		Yield strength, σ_{Sy}		Elongation in 50 mm (2 in), %	Reduction in area, %	Brinell hardness, H_B	Izod impact strength	
			°C	°F	MPa	kpsi	MPa	kpsi				J	ft-lbf
3140		Normalized	870	1600	891.5	129.3	599.8	87.0	19.7	57.3	262	53.6	39.5
		Annealed	815	1500	689.5	100.0	422.6	61.3	24.5	50.8	197	46.4	34.2
4130	G41300	Normalized	870	1600	668.8	97.0	436.4	63.3	25.2	59.5	197	86.4	63.7
		Annealed	865	1585	560.5	81.3	360.6	52.3	28.2	55.6	156	61.7	45.5
4150	G41500	Normalized	870	1600	1154.9	167.5	734.3	106.5	11.7	30.8	321	11.5	8.5
		Annealed	815	1500	729.5	105.8	379.2	55.0	20.2	40.2	197	24.7	18.2
4320	G43200	Normalized	895	1640	792.9	115.0	464.0	67.3	20.8	50.7	235	72.9	53.8
		Annealed	850	1560	579.2	84.0	609.5	61.6	29.0	58.4	163	109.8	81.0
4340	G43400	Normalized	870	1600	1279.0	185.5	861.8	125.0	12.2	36.3	363	15.9	11.7
		Annealed	810	1490	744.6	108.0	472.3	68.5	22.0	49.9	217	51.1	37.7
4620	G46200	Normalized	900	1650	574.3	83.3	366.1	53.1	29.0	66.7	174	132.9	98.0
		Annealed	855	1575	512.3	74.3	372.3	54.0	31.3	60.3	149	93.6	69.0
4820		Normalized	860	1580	750.0	109.5	484.7	70.3	24.0	59.2	229	109.8	81.0
		Annealed	815	1500	681.2	98.8	464.0	67.3	22.3	58.8	197	92.9	68.5
5150	G51500	Normalized	870	1600	870.8	126.3	529.5	76.8	20.7	58.7	255	31.5	23.2
		Annealed	825	1520	675.7	98.0	357.1	51.8	22.0	43.7	197	25.1	18.5
6150	G61500	Normalized	870	1600	939.8	136.3	615.7	89.3	21.8	61.0	269	35.5	26.2
		Annealed	815	1500	667.4	96.8	412.3	59.8	23.0	48.4	197	27.4	20.2
8630		Normalized	870	1600	650.2	94.3	429.5	62.3	23.5	53.5	187	94.6	69.8
		Annealed	845	1550	564.0	81.8	372.3	54.0	29.0	58.9	156	95.2	70.2
8740	G87400	Normalized	870	1600	929.4	134.8	606.7	88.0	16.0	47.9	269	17.6	13.0
		Annealed	815	1500	695.0	100.8	415.8	60.3	22.2	46.4	201	40.0	29.5
9255	G92550	Normalized	900	1650	932.9	135.3	579.2	84.0	19.7	43.4	269	13.6	10.0
		Annealed	845	1550	774.3	112.3	486.1	70.5	21.7	41.1	229	8.8	6.5
9310	G93100	Normalized	890	1630	906.7	131.5	570.9	82.8	18.8	58.1	269	119.3	88.0
		Annealed	845	1550	820.5	119.0	439.9	63.8	17.3	42.1	241	78.6	58.0

[a] All grades are fine-grained except for those 1100 series, which are coarse-grained. Heat-treated specimens were oil-quenched unless otherwise indicated.
Values tabulated were averaged and obtained from specimen 12.75 mm (0.505 in) in diameter which were machined from 25 mm (1 in); rounded gauge lengths were 50 mm (2 in).
Source: ASM Metals *Handbook,* American Society for Metals, Metals Park, Ohio, 1988

TABLE 1–9
Mechanical properties of standard steels

Designation		Tensile strength, σ_{St}		Yield stress, σ_{Sy}		Elongation, percent, min (gauge length 5.65 $\sqrt{a^*}$)
New	Old	MPa	kpsi	MPa	kpsi	
Fe 290	(St 30)	290	42.0	170	24.7	27
Fe E 220	—	290	42.0	220	32.0	27
Fe 310	(St 32)	310	45.0	180	26.1	26
Fe E 230	—	310	45.0	230	33.4	26
Fe 330	(St 34)	330	47.9	200	29.0	26
Fe E 250	—	330	47.9	250	36.3	26
Fe 360	(St 37)	360	52.2	220	32.0	25
Fe E 270	—	360	52.2	270	39.2	25
Fe 410	(St 42)	410	59.5	250	36.3	23
Fe E 310	—	410	59.5	310	50.0	23
Fe 490	(St 50)	490	71.1	290	42.0	21
Fe E 370	—	490	71.1	370	53.7	21
Fe 540	(St 55)	540	78.3	320	46.4	20
Fe E 400	—	540	78.3	400	58.0	20
Fe 620	(St 63)	620	90.0	380	55.1	15
Fe E 460	—	620	90.0	460	66.7	15
Fe 690	(St 70)	690	100.0	410	59.5	12
Fe E 520	—	690	100.0	520	75.4	12
Fe 770	(St 78)	770	111.7	460	66.7	10
Fe E 580	—	770	111.7	580	84.1	10
Fe 870	(St 88)	870	126.2	520	75.4	8
Fe E 650	—	870	126.2	650	94.3	8

Note: a^*, area of cross-section of test specimen.

Source: IS 1570, 1978.

TABLE 1–10
Chemical composition and mechanical properties of carbon steel castings for surface hardening

Designation	Chemical composition (in ladle analysis) max, percent									
	C	Si	Mn	S	P	Cr	Ni	Mo	Cu	Residual elements
Gr 1	0.4–0.5	0.60	1.0	0.05	0.05	0.25	0.40	0.15	0.30	0.80
Gr 2	0.5–0.6	0.60	1.0	0.05	0.05	0.25	0.40	0.15	0.30	0.80

Designation	Tensile strength, σ_{St}		Yield strength, σ_{Sy}		Elongation, percent, min (gauge length 5.65 $\sqrt{a^*}$)	Brinell hardness, H_B
	MPa	kpsi	MPa	kpsi		
Gr 1	620	90.0	320	46.4	12	460
Gr 2	700	101.5	370	53.7	8	535

Notes: a^*, area of cross section of test specimen. All castings shall be free from distortion and harmful defects. They shall be well-dressed, fettled, and machinable. Unless agreed upon by the purchaser and the manufacturer, castings shall be supplied in the annealed, or normalized and tempered condition.

Source: IS 2707, 1973.

TABLE 1–11
Chemical composition of alloy steel forgings for general industrial use

New	Old	% C	% Si	% Mn	% Ni	% Cr	% Mo	% V	% Al
20 C 15	20 Mn 2	0.06–0.24	0.10–0.35	1.30–1.70					
15 Cr 3	15 Cr 65	0.12–0.18	0.10–0.35	0.40–0.60		0.50–0.80			
16 Mn 5 Cr 4	17 Mn 1 Cr 95	0.14–0.19	0.10–0.35	1.00–1.30		0.80–1.10			
20 Mn 5 Cr 5	20 Mn Cr 1	0.18–0.22	0.10–0.35	1.00–1.40		1.00–1.30			
21 Cr 4 Mo 2	21 Cr 1 Mo 28	0.26 max	0.10–0.35	0.50–0.80		0.90–1.20	0.15–0.30		
07 Cr 4 Mo 6	07 Cr 90 Mo 55	0.12 max	0.15–0.60	0.40–0.70	0.30 max	0.70–1.10	0.45–0.65		
10 Cr 9 Mo 10	10 Cr 2 Mo 1	0.15 max	0.50 max	0.40–0.70	0.30 max	2.00–2.50	0.90–1.10		
13 Ni 13 Cr 3	13 Ni 3 Cr 80	0.10–0.15	0.10–0.35	0.40–0.70	3.00–3.50	0.60–1.00			
15 Ni 16 Cr 5	15 Ni 4 Cr 1	0.12–0.18	0.10–0.35	0.40–0.70	3.80–4.30	1.00–1.40			
15 Ni 5 Cr 4 Mo 1	15 Ni Cr 1 Mo 12	0.12–0.18	0.10–0.35	0.60–1.00	1.00–1.50	0.75–1.25	0.80–0.15		
15 Ni 7 Cr 4 Mo 2	15 Ni Cr 1 Mo 15	0.12–0.18	0.10–0.35	0.60–1.00	1.50–2.00	0.75–1.25	0.10–0.20		
16 Ni 8 Cr 6 Mo 2	16 Ni Cr 2 Mo 20	0.12–0.20	0.10–0.35	0.40–0.70	1.80–2.20	1.40–1.70	0.15–0.25		
36 S 17	37 Si 2 Mn 90	0.33–0.40	1.50–2.00	0.80–1.00					
37 C 15	37 Mn 2	0.32–0.42	0.10–0.35	1.30–1.70					
35 Mn 6 Mo 3	35 Mn 2 Mo 28	0.30–0.40	0.10–0.35	1.30–1.80			0.20–0.35		
40 Cr 4	40 Cr 1	0.35–0.45	0.10–0.35	1.60–0.90		0.90–1.70			
40 Cr 4 Mo 3	40 Cr 1 Mo 28	0.35–0.45	0.10–0.35	0.50–0.80		0.90–1.20	0.20–0.35		
35 Ni 5 Cr 2	35 Ni 1 Cr 60	0.30–0.40	0.10–0.35	0.60–0.90	1.00–1.50	0.45–0.75			
40 Ni 6 Cr 4 Mo 3	40 Ni 2 Cr 1 Mo 28	0.35–0.45	0.10–0.35	0.40–0.70	1.25–1.75	0.90–1.30	0.20–0.35		
40 Ni 10 Cr 3 Mo 6	40 Ni 3 Cr 65 Mo 55	0.36–0.44	0.10–0.35	0.40–0.70	2.25–2.75	0.50–0.80	0.40–0.70		
25 Cr 13 Mo 6	25 Cr 3 Mo 55	0.20–0.30	0.10–0.35	0.40–0.70	0.30 max	2.90–3.40	0.45–0.65		
55 Si 7	55 Si 2 Mn 90	0.50–0.60	1.50–2.00	0.80–1.00					
50 Cr 4 V 2	50 Cr 1 V 23	0.45–0.55	0.10–0.35	0.50–0.80		0.90–1.20		0.15–0.30	
20 Ni Cr Mo 2		0.18–0.23	0.20–0.35	0.70–0.90	0.40–0.70	0.40–0.60	0.15–0.25		
37 Mn 5 Si 5		0.33–0.41	1.10–1.40	1.10–1.40					

Notes: (1) Sulfur and phosphorus can be ordered as per following limits: (i) S and P — 0.30 max; (ii) S — 0.02–0.035 and P — 0.035 max. (2) When the steel is Al killed, total Al contents shall be between 0.02–0.05 percent.

Source: IS 4367, 1991.

TABLE 1-12
Mechanical properties of alloy steel forgings for general industrial use

Designation	Condition	Tensile strength,[b] σ_{Sr} MPa	kpsi	Yield strength,[b] σ_{Sy} MPa	kpsi	Brinell hardness in soft annealed condition, max, H_B	Elongation[b] percent, min (gauge length $5.65\sqrt{a*}$)[a]	Izod impact[b] value J	ft-lbf	Brinell[b] hardness number, H_B	Limiting ruling section mm	in
20 C 15	H and T	600–750	87.0–108.8	400	58.0	200	18	50	36.9	178–221	63	2.52
		700–850	101.5–123.3	460	66.7		16	50	36.9	208–252	30	1.20
30 C 15	H and T	600–750	87.0–108.8	440	63.8	220	18	50	36.9	178–221	150	6.00
		700–850	101.5–123.3	540	78.3		18	48	35.4	208–252	100	4.00
		800–950	116.0–137.8	600	87.0		16	48	35.4	235–280	63	2.52
15 Cr 3	R, Q and S.R	600 min	87.0 min			170	13	48	35.4		30	1.20
16 Mn 5 Cr 4	R, Q and S.R	800 min	116.0 min			207	10	35	25.8		30	1.20
20 Mn 5 Cr 5	R, Q and S.R	1000 min	145.0 min			217	8	38	28.0		30	1.20
21 Cr 4 Mo 2	H and T	650–800	94.3–116.0	420	61.0	210	16	60	44.3	190–235	150	6.00
		700–850	101.5–123.3	460	66.7		15	55	40.6	208–252	100	4.00
		800–950	116.0–137.8	580	84.1		14	50	36.9	235–280	40	1.60
7 Cr 4 Mo 6	N and T	380–550	55.1–79.8	225	32.6	170	19	60	44.3		40	1.60
10 Cr 9 Mo 10	N and T	410–590	59.5–85.6	245	35.5	187	18	55	40.6		50	2.00
		520–680	75.4–98.6	310	45.0		18	50	36.9			
13 Ni 13 Cr 3	R, Q and S.R	850 min	123.3 min			229	12	48	35.4		60	2.4
15 Ni 16 Cr 5	R, Q and S.R	1350 min	195.8 min			241	9	35	25.8		30	1.20
15 Ni 5 Cr 4 Mo 1	R, Q and S.R	1000 min	145.0 min			217	9	41	30.2		30	1.20
15 Ni 7 Cr 4 Mo 2	R, Q and S.R	1100 min	159.5 min			217	9	35	25.8		30	1.20
16 Ni 8 Cr 6 Mo 2	R, Q and S.R	1350 min	195.8 min			229	9	35	25.8		30	1.20
20 Ni Cr Mo 2	R, Q and S.R	900 min	130.5 min			213	11	41	30.2			
36 Si 7	H and T	800–950	116.0–137.8			217						
37 Mn 5 Si 5	H and T	780–930	113.1–134.9	590	85.6	217	14			235–280	100	4.00
55 Si 7	H and T	1300–1500	188.6–217.6			245				380–440	100	4.00
35 Mn 6 Mo 3	H and T	700–850	101.5–123.3	540	78.3	220	18	55	40.6	208–252	150	6.00
		900–1050	130.5–152.3	700	101.5		15	50	36.9	268–311	63	2.52
		1000–1150	145.0–166.8	800	116.0		13	45	33.2	295–341	30	1.20
40 Cr 4	H and T	700–850	101.5–123.3	540	78.3	220	18	55	40.6	208–252	100	4.00
		900–1050	130.5–152.3	700	101.5		15	50	36.9	266–311	30	1.20
40 Cr 4 Mo 3	H and T	700–850	101.5–123.3	540	78.3	220	18	55	40.6	208–252	150	6.00
		900–1050	130.5–152.3	700	101.5		15	50	36.9	266–311	63	2.52
		1000–1150	145.0–166.8	800	116.0		13	45	33.2	295–341	30	1.20
35 Ni 5 Cr 2	H and T	700–850	101.5–123.3	540	78.3	220	18	55	40.6	208–252	150	6.00
		900–1050	130.5–152.3	700	101.5		15	50	36.9	266–311	63	2.52
25 Cr 13 Mo 6		900–1050	130.5–152.3	700	101.5	230	15	55	40.6	266–311	150	6.00
		1100–1250	159.5–181.3	880	127.6		12	41	30.2	325–370	100	4.00

TABLE 1–12
Mechanical properties of alloy steel forgings for general industrial use (*Cont.*)

Designation	Condition	Tensile strength,[b] σ_{St}		Yield strength,[b] σ_{Sy}		Brinell hardness in soft annealed condition, max, H_B	Elongation[b] percent, min (gauge length $5.65\sqrt{a*}$)[a]	Izod impact[b] value		Brinell[b] hardness number, H_B	Limiting ruling section	
		MPa	kpsi	MPa	kpsi			J	ft-lbf		mm	in
40 Ni 6 Cr 4 Mo 3	H and T	900–1050	130.5–152.3	700	101.5	230	55	55	40.6	266–311	150	6.00
		1100–1250	159.5–181.3	880	127.6		11	41	30.2	325–370	63	2.52
		1200–1350	174.0–195.8	1000	145.0		10	30	22.1	355–399	30	1.20
40 Ni 10 Cr 3 Mo 6	H and T	1000–1150	145.0–166.8	800	116.0	250	12	48	35.4	295–341	150	6.00
		1200–1350	174.0–195.8	1000	145.0		10	35	25.8	355–399	150	6.00
		1550 min	224.8 min	1300	188.5		8	15	11.1	450 min	100	4.00
50 Cr 4 V 2	H and T	900–1100	130.5–159.5	700	101.5	240	12	45	33.2	266–325	100	4.00
		1000–1200	145.0–174.0	800	116.0		10	45	33.2	295–355	40	1.60

[a] $a*$, area of cross section.
[b] Mechanical properties in heat-treated conditions.

Notes: H and T = hardened and tempered; N and T = normalized and tempered; R,Q and S.R = refined quenched and stress-relieved. All properties for guidance only. Other values may be mutually agreed on between the consumers and suppliers.

Source: IS 4367, 1991.

1.21

TABLE 1–13
Chemical composition and mechanical properties of alloy steels**

Designation	C	Si	Mn	Ni	Cr	Mo	V/Al	Tensile strength, σ_{St} MPa	kpsi	0.2 percent proof stress, min, σ_{Sy} MPa	kpsi	Minimum elongation (gauge length = $5.65\sqrt{a*}$), %	Minimum Izod impact value J	ft-lbf	Brinell hardness number, H_B	Limiting ruling section, mm (in)+
20 C 15 (20 Mn 2)§	0.16–0.24	0.10–0.35	1.30–1.70					590–740	85.6–107.3	390	56.6	18	48	35.4	170–217	63 (2.5)+
								690–840	100.0–121.8	450	65.3	16	48	35.4	201–248	30 (1.2)
27 C 15 (27 Mn 2)	0.22–0.32	0.10–0.35	1.30–1.70					590–740	85.5–107.3	390	56.6	18	48	35.4	170–217	100 (4.0)
								690–840	100.0–121.8	450	65.3	16	48	35.4	201–248	63 (2.5)
37 C 15 (37 Mn 2)	0.32–0.42	0.10–0.35	1.30–1.70					590–740	85.5–107.3	390	56.6	18	48	35.4	170–217	150 (6.0)
								690–840	100.0–121.8	490	71.1	18	48	35.4	201–248	100 (4.0)
								790–940	114.6–136.3	550	79.8	16	48	35.4	229–277	30 (1.2)
								890–1040	129.0–150.8	650	94.3	15	41	30.2	255–311	15 (0.6)
35 Mn 6 Mo 3 (35 Mn 2 Mo 28)	0.30–0.40	0.10–0.35	1.30–1.80			0.20–0.35		690–840	100.0–121.8	490	71.1	14	55	40.6	201–248	150 (6.0)
								790–940	114.6–136.3	550	79.8	12	50	36.8	229–277	100 (4.0)
								890–1040	129.0–150.8	650	94.3	12	50	36.8	255–311	63 (2.5)
								990–1140	143.6–165.3	750	108.8	10	48	35.4	285–341	30 (1.2)
35 Mn 6 Mo 4 (35 Mn 2 Mo 45)	0.30–0.40	0.10–0.35	1.30–1.80			0.35–0.55		790–940	114.6–136.3	550	79.8	16	55	40.6	229–277	150 (6.0)
								890–1040	129.0–150.8	650	94.3	15	55	40.6	255–311	100 (4.0)
								990–1140	143.6–165.3	750	108.8	13	48	35.4	285–341	63 (2.5)
40 Cr 4 (40 Cr 1)	0.35–0.45	0.10–0.35	0.60–0.90		0.90–1.20			690–840	100.0–121.8	490	71.1	14	55	40.6	201–248	100 (4.0)
								790–940	114.6–136.3	550	79.8	12	50	36.8	229–277	63 (2.5)
								890–1040	129.0–150.8	650	94.3	11	50	36.8	255–311	30 (1.2)
40 Cr 4 Mo 2 (40 Cr 1 Mo 28)	0.35–0.45	0.10–0.35	0.50–0.80		0.90–1.20	0.20–0.35		700–850	101.5–123.3	490	71.1	13	55	40.6	201–248	150 (6.0)
								800–950	116.0–137.8	550	79.8	12	50	36.8	229–277	100 (4.0)
								900–1050	130.5–152.3	650	94.3	11	50	36.8	255–311	63 (2.5)
								1000–1150	145.0–166.8	750	108.8	10	48	35.4	285–341	30 (1.2)
15 Cr 13 Mo 6 (15 Cr 3 Mo 55)	0.10–0.20	0.10–0.35	0.40–0.70	0.30 max	2.90–3.40	0.45–0.65		690–840	100.0–121.8	490	71.1	14	55	40.6	201–248	150 (6.0)
								790–940	114.6–136.3	550	79.8	12	50	36.8	229–277	150 (6.0)
25 Cr 13 Mo 6 (25 Cr 3 Mo 55)	0.20–0.30	0.10–0.35	0.40–0.70	0.30 max	2.90–3.40	0.45–0.65		890–1040	129.0–150.8	650	94.3	11	50	36.8	255–311	150 (6.0)
								990–1140	143.6–165.3	750	108.8	10	48	36.8	285–341	150 (6.0)
								1090–1240	158.1–179.8	830	120.4	9	41	30.2	311–363	100 (4.0)
								1540 min	223.4 min	1240	179.8	8	14	10.3	444 min	63 (2.5)

TABLE 1–13
Chemical composition and mechanical properties of alloy steels (Cont.)**

Designation	C	Si	Mn	Ni	Cr	Mo	V/Al	Tensile strength, σ_{Sr} MPa	kpsi	0.2 percent proof stress, σ_{Sy} min MPa	kpsi	Min. elongation (gauge length = 5.65 $\sqrt{a*}$), %	Izod J	Izod ft-lbf	Brinell# hardness number, H_B	Limiting ruling section, ( ) mm (in)+
40 Cr 13 Mo 10 V 2 (40 Cr 3 Mo 1 V 20)	0.35–0.45	0.10–0.35	0.40–0.70	0.30 max	3.00–3.50	0.90–1.10	V: 0.15–0.25	1340 min	194.4 min	1050	152.2	8	21	15.5	363 min	63 (2.5)
								1540 min	223.4 min	1240	179.8	8	14	10.3	444 min	30 (1.2)
40 Cr 7 Al 10 Mo 2 (40 Cr 2 Al 1 Mo 18)	0.35–0.45	0.10–0.45	0.40–0.70	0.30 max	1.50–1.80	0.10–0.25	Al:0.90–1.30	690–840	100.0–121.8	490	71.1	18	55	40.6	201–248	150 (6.0)
								790–940	114.6–136.3	550	79.8	16	55	40.6	229–277	100 (4.0)
								890–1040	129.0–150.8	650	94.3	15	48	35.4	255–311	63 (2.5)
40 Ni 14 (40 Ni 31)	0.35–0.45	0.10–0.35	0.50–0.80	3.20–3.6	0.30 max			790–940	114.6–136.3	550	79.8	16	55	40.6	229–277	100 (4.0)*
								890–1040	129.0–150.8	650	94.3	15	55	40.6	255–311	63 (2.5)
35 Ni 5 Cr 2 (35 Ni 1 Cr 60)	0.30–0.40	0.10–0.35	0.60–0.90	1.00–1.50	0.45–0.75			690–840	100.0–121.8	490	71.1	14	55	40.6	201–248	150 (6.0)+
								790–940	114.6–136.3	550	79.8	12	50	36.8	229–277	100 (4.0)
								890–1040	129.0–150.8	650	94.3	10	50	36.8	255–311	63 (2.5)
30 Ni 16 Cr 5 (30 Ni 4 Cr 1)	0.26–0.34	0.10–0.35	0.40–0.70	3.90–4.30	1.10–1.40			1540 min	223.4 min	1240	179.8	8	14	10.3	444 min	(air-hardened)
																150 (6.0)
																(air-hardened)
40 Ni 6 Cr 4 Mo 2 (40 Ni Cr 1 Mo 15)	0.35–0.45	0.10–0.35	0.40–0.70	1.20–1.60	0.90–1.30	0.10–0.20		790–940	114.6–136.3	550	79.8	16	55	40.6	229–277	150 (6.0)
								890–1040	129.0–150.8	650	94.3	15	55	40.6	255–311	100 (4.0)
								990–1140	143.6–165.3	750	108.8	13	48	35.4	285–341	63 (2.5)
								1090–1240	158.1–179.8	830	120.4	13	41	30.3	311–363	30 (1.2)
40 Ni 6 Cr 4 Mo 3 (40 Ni 2 Cr 1 Mo 28)	0.35–0.45	0.10–0.35	0.40–0.70	1.25–1.75	0.90–1.30	0.20–0.35		790–940	114.6–136.3	550	79.8	16	55	40.6	229–277	150 (6.0)
								890–1040	129.0–150.8	650	94.3	15	55	40.6	255–311	150 (6.0)
								990–1140	143.6–165.3	750	108.8	11	48	36.8	285–341	100 (4.0)
								1090–1240	158.1–179.8	830	120.4	11	41	30.3	311–363	63 (2.5)
								1190–1340	172.6–194.4	930	134.9	10	30	22.1	341–401	30 (1.2)
								1540 min	223.4 min	1240	179.8	6	11	40.6	444 min	30 (1.2)
31 Ni 10 Cr 3 Mo 6 (31 Ni 3 Cr 65 Mo 55)	0.27–0.35	0.10–0.35	0.40–0.70	2.25–2.75	0.50–0.80	0.40–0.70		890–1040	129.0–150.8	650	94.3	15	55	40.6	255–311	150 (6.0)
								990–1140	143.6–165.3	750	108.8	12	48	35.4	285–341	150 (6.0)
								1090–1240	158.1–179.8	830	120.4	11	41	30.3	311–363	100 (4.0)
								1190–1340	172.6–194.4	930	134.9	10	35	25.8	341–401	63 (2.5)
								1540 min	223.4 min	1240	179.8	8	14	10.3	444 min	63 (2.5)

1.23

TABLE 1–13
Chemical composition and mechanical properties of alloy steels** (*Cont.*)

| Designation | Percent | | | | | | | Tensile strength, σ_{St} | | 0.2 percent proof stress, σ_{Sy} min, | | Minimum elongation (gauge length = $5.65\sqrt{a*}$), | Minimum Izod impact value | | Brinell# hardness number, H_B | Limiting ruling section, ( ) |
	C	Si	Mn	Ni	Cr	Mo	V/Al	MPa	kpsi	MPa	kpsi	%	J	ft-lbf		mm (in)+
40 Ni 10 Cr 3 Mo 6	0.36–	0.10–	0.40–	2.25–	0.50–	0.40–		990–1140	143.6–165.3	750	108.8	12	48	35.4	285–341	150 (6.0)
(40 Ni 3 Cr 65 Mo 55)	0.44	0.35	0.70	2.75	0.80	0.70		1090–1240	158.1–179.8	830	120.4	11	41	30.3	311–363	150 (6.0)
								1190–1240	172.6–194.4	930	134.9	10	35	25.8	341–401	150 (6.0)
								1540 min	223.4 min	1240	179.8	8	14	10.3	444 min	100 (4.0)

Notes: $a*$, area of cross section; ** hardened and tempered condition—oil-hardened unless otherwise stated; # hardness given in this table is for guidance only; § steel designations in parentheses are old designations; + numerals in parentheses are in inches.

Source: IS 1750, 1988.

TABLE 1–14
Mechanical properties of case hardening steels in the refined and quenched condition (core properties)

Steel designation	Tensile strength, σ_{St}		Minimum elongation, % (gauge length = 5.65 $\sqrt{a*}$)a	Izod impact value, min (if specified)		Limiting ruling section, mm (in)	Brinell hardness number, max, H_B
	MPa	kpsi		J	ft-lbf		
10 C 4 (C 10)	490	71.1	17	54	39.8	15 (0.6)	130
14 C 4 (C 14)	490	71.1	17	54	39.8	> 15 (0.6) ≤30 (1.2)	143
10 C 8 S 11 (10 S 11)	490	71.1	17	54	39.8	30 (1.2)	143
14 C 14 S 14 (14 Mn 1 S 14)	588	85.4	17	40	29.7	30 (1.2)	154
11 C 15 (11 Mn 2)	588	85.4	17	54	39.8	30 (1.2)	154
15 Cr 65	588	85.4	13	47	34.7	30 (1.2)	170
17 Mn 1 Cr 95	784	113.8	10	34	25.3	30 (1.2)	207
20 Mn Cr 1	981	142.3	8	37	27.5	30 (1.2)	217
16 Ni 3 Cr 2 (16 Ni 80 Cr 60)	686	99.6	15	40	29.7	90 (3.6)	184
16 Ni 4 Cr 3 (16 Ni 1 Cr 80)	834 784 735	121.0 113.8 106.7	12	40	29.7	30 (1.2) 60 (2.4) 90 (3.6)	217
13 Ni 13 Cr 3 (13 Ni 3 Cr 80)	834 784	121.0 113.8	12	47	34.7	60 (2.4) 100 (4.0)	229
15 Ni 4 Cr 1	1324 1177 1128	192.0 170.7 163.2	9	34	25.3	30 (1.2) 60 (2.4) 90 (3.6)	241
20 Ni 2 Mo 25	834 686	121.0 99.6	12	61	44.8	30 (1.2) 60 (2.4)	207
20 Ni 7 Cr 2 Mo 2 (20 Ni 55 Cr 50 Mo 20)	882 784 735	128.0 113.8 106.7	11	40	29.7	30 (1.2) 60 (2.4) 90 (3.6)	213
15 Ni 13 Cr 4 (15 Ni Cr 1 Mo 12)	981 932	142.3 135.1	9	40	29.7	30 (1.2) 90 (3.6)	217
15 Ni 5 Cr 4 Mo 2 (15 Ni 2 Cr 1 Mo 15)	1079 932 932	156.5 142.3 135.1	9	34	25.3	30 (1.2) 60 (2.4) 90 (3.6)	217
16 Ni 8 Cr 6 Mo 2 (16 Ni Cr 2 Mo 20)	1324 1177 1128	193.0 170.7 163.6	9	34	25.3	30 (1.2) 60 (2.4) 90 (3.6)	229

a $a*$, area of cross section.

Source: IS 4432, 1967.

TABLE 1–15
Typical mechanical properties of some carburizing steels[a]

AISI no.	Ultimate tensile strength, σ_{Sut} MPa	kpsi	Tensile yield strength, σ_{Sy} MPa	kpsi	Elongation in 50 mm (2 in), %	Reduction of area, %	Core Brinell, H_B	Hardness Rockwell, R_C	Case Thickness mm	in	J	Izod impact energy ft-lbf	Machin-ability
					Plain carbon								
C1015	503	73	317	46	32	71	149	62	1.22	0.048	123	91	Poor
C1020	517	75	331	48	31	71	156	62	1.17	0.046	126	93	Poor
C1022	572	83	324	47	27	66	163	62	1.17	0.046	110	81	Good
C1117	669	97	407	59	23	53	192	65	1.14	0.045	45	33	Very good
C1118	779	113	531	77	17	45	229	61	1.65	0.065	22	16	Excellent
					Alloy steels								
4320[b]	100	146	648	94	22	56	293	59	1.91	0.075	65	48	
4620[b]	793	115	531	77	22	62	235	59	1.52	0.060	106	78	
8620[b]	897	130	531	77	22	52	262	61	1.78	0.070	89	66	

[a] Average properties for 25-mm (1-in) round section treated, 12.625-mm (0.505-in) round section tested. Water-quenched and tempered at 177°C (350°F), except where indicated.

[b] Core properties for 14.125-mm (0.565-in) round section treated, 12.625-mm (0.505-in) round section tested. Oil-quenched twice, tempered at 232°C (450°F)

Source: *Modern Steels and Their Properties*, Bethlehem Steel Corp., 4th ed., 1958 and 7th ed., 1972.

TABLE 1–16
Minimum mechanical properties of some stainless steels

UNS no.	AISI no.	Tensile strength, σ_{St}		Yield strength,[a] σ_{Sy}		Brinell hardness, H_B	Elongation, %	Reduction in area, %	Weldability	Machinability	Application
		MPa	kpsi	MPa	kpsi						
				Annealed (Room Temperatures)							
Austenitic											
S30200	302	515	75	205	30	88	40		Good	Poor	General purpose, springs
S30300	303[b]	585[b]	85[b]	240[b]	35[b]		50[b]	55[b]	Poor	Good	Bolts, rivets, and nuts
S30400	304	515	75	205	30	88	40		Good	Poor	Welded structures
S30500	305	480	70	170	25	88	40		Good		General purpose
S30800	308	515	75	205	30	88	40				
S30900	309	515	75	205	30	95	40				
S31000	310	515	75	205	30	95	40		Good	Poor	Heat-exchange parts
S31008	310 S	515	75	205	30	95	40		Good	Poor	Turbine and furnace
S34800	348	515	75	205	30	88	40				Jet engine parts
S38400	384	415–550	60–80								Fasteners and cold-worked parts
				Annealed High-Nitrogen							
Austenitic											
S20200	202	655	95	310	4560		40				
S21600	216	690	100	415	50	100	40				
S30452	304 HN	620	90	345		100	30				
Ferrite											
S40500	405	415	60	170	25	88 max	20		Excellent		
S43000	430	450	65	205	30	88 max	22[c]		Fair	Fair to good	Screw machine parts, muffler
S44600	446	515	75	275	40	95 max	20		Fair	Fair	Machine parts subjected to high-temperature corrosion
Martensite											
S40300	403	485	70	205	30	88 max	25[c]				Bolts, shafts, and machine parts
S41000	410	450	65	205	30	95 max	22[c]				Bolts, springs, cutlery, and machine parts
S41400	414	795	115	620	90		15	45			
S41800[d]	418[d]	1450[b]	210[b]	1210[b]	175[b]		18[b]	52[b]			
S42000[e]	420[e]	1720	250	1480[b]	215[b]	$52R_C$[b]	8[b]	25[b]			
S43100[d]	431[d]	1370[b]	198[b]	1030[b]	149[b]		16[b]	55[b]			High-strength parts used in aircraft and bolts
S44002	440 A	725[b]	105[b]	415[b]	60[b]	95[b]	20[b]				Cutlery, bearing parts, nozzles, and ball
S44003	440 B	740	107[b]	425[b]	62[b]	96[b]	18[b]				
S44004	440 C	760[b]	110[b]	450[b]	65[b]	97[b]	14[b]				
S50200	502[b]	485[b]	70[b]	205[b]	30[b]		30[b]	70[b]			

[a] At 0.2% offset.

[b] Typical values.

[c] 20% elongation for thickness of 1.3 mm (0.050 in) or less.

[d] Tempered at 260°C (500°F).

[e] Tempered at 205°C (400°F).

Source: *ASM Metals Handbook*, American Society for Metals, Metals Park, Ohio, 1988.

TABLE 1–17
Chemical composition and mechanical properties of some stainless, heat resisting and high alloy steels

Designation of steel	C	Si	Mn	Ni	Cr	Mo	Ti	Nb	S max	P Max	Tensile MPa	Tensile kpsi	Proof MPa	Proof kpsi	Brinell H_B	Rockwell R_B	Elongation in 50 mm (2 in), min, %	Reduction of area, min, %
Chromium Steels																		
X 04 Cr 12*	0.08 max	1.0 max	1.0 max		11.5/13.5				0.030	0.040	415 (445#)	60.2 (64.5#)	205 (276#)	29.7 (40.0)	183	88	22(20)#	(45)#
X 12 Cr 12*	0.80/0.15	1.0 max	1.0 max	1.0 max	11.5/13.5				0.030	0.040	450 (483)	65.3 (70.0)	205 (276)	29.7 (40.0)	217	95	20(20)	(45)
X 07 Cr 17	0.12 max	1.0 max	1.0 max	1.25/2.50	15.0/17.0				0.030	0.040	450 (483)	65.3 (70.0)	205 (276)	29.7 (40.0)	183	88	22(20)	(45)
X 40 Cr 13	0.35/0.45	1.0 max	1.0 max	1.0 max	12.0/14.0				0.030	0.040	600–700	87.0–101.5			(225)	—	—	
X 15 Cr 25 N	0.20 max	1.0 max	1.5 max		23.0/27.0				0.030 and N = 0.25 max	0.045	515 (490)	74.7 (71.1)	275 (280)	39.9 (40.6)	217 (212)	—	20(16)	(45)
Chromium–Nickel Steels																		
X 02 Cr 19 Ni 10	0.03 max	1.0 max	2.0 max	8.0/12.0	17.5/20.0				0.030	0.045	485 (483)	70.3 (70.0)	170 (172)	24.7 (25.0)	183	88	40(40)	(50)
X 04 Cr 19 Ni 9	0.08 max	1.0 max	2.0 max	8.0/10.5	17.5/20.0				0.030	0.045	515 (517)	74.7 (75.0)	205 (207)	29.7 (30.0)	183	88	40(40)	(50)
X 07 Cr 18 Ni 9	0.15 max	1.0 max	2.0 max	8.0/10.0	17.0/19.0				0.030	0.045	515 (517)	74.7 (75.0)	205 (207)	29.7 (30.0)	183	88	40	
X 04 Cr 18 Ni 10 Nb	0.08 max	1.0 max	2.0 max	9.0/12.0	17.0/19.0			10XC-1.0	0.030	0.045	515 (517)	74.7 (75.0)	205 (207)	29.7 (30.0)	183	88	40(40)	(50)
X 04 Cr 18 Ni 10 Ti	0.08 max	1.0 max	2.0 max	9.0/12.0	17.0/19.0		5XC-0.80**		0.030	0.045	515 (517)	74.7 (75.0)	205 (207)	29.7 (30.0)	183	88	40(40)	(50)
X 04 Cr 17 Ni 12 Mo 2	0.08 max	1.0 max	2.0 max	10.0/14.0	16.0/18.0	2.0/3.0			0.030	0.045	515 (517)	74.7 (75.0)	205 (207)	29.7 (30.0)	217	95	40(40)	(50)
X 02 Cr 17 Ni 12 Mo 2	0.03 max	1.0 max	2.0 max	10.0/14.0	16.0/18.0	2.0/3.0			0.030	0.045	485 (483)	70.3 (70.0)	170 (172)	24.7 (25.0)	217	95	40(40)	(50)
X 04 Cr 17 Ni Mo 2 Ti 2	0.08 max	1.0 max	2.0 max	10.0/14.0	16.0/18.0	2.0/3.0	5XC-0.80		0.030	0.045	515 (517)	74.7 (75.0)	205 (207)	29.7 (30.0)	217	95	40(40)	(50)
X 04 Cr 19 Ni 13 Mo 3	0.08 max	1.0 max	2.0 max	11.0/15.0	18.0/20.0	3.0/4.0			0.030	0.045	515 (517)	74.7 (75.0)	205 (207)	29.7 (30.0)	217	95	35(40)	(50)
X 20 Cr 25 Ni 20	0.25 max	2.5 max	2.0 max	18.0/21.0	24.0/26.0				0.030	0.045	515 (490)	74.7 (75.0)	210 (210)	30.5 (30.5)	217	95	40(40)	(50)
X 07 Cr 17 Mn 12 Ni 4	0.12 max	1.0 max	10.0/14.0	3.5/5.5	16.0/18.0				0.030	0.045	550	79.8	250	36.3	217	88	45	
X 40 Ni 14 Cr 14 W 3 Si 2 Ni 4	0.35/0.50	2.0 max	1.0 max	12.0/15.0	12.0/15.0			and W 2.0/3.0	0.035	0.045	(785)	(113.9)	(345)	(50.0)	(269)		(35)	(40)

Notes: Annealed quenched or solution-treated condition; * for free-cutting varieties sulfur and selenium content shall be as agreed to between the purchaser and the manufacturer; ** for electrode steel Nb –10C to 1.0 in place of Ti; # the mechanical properties in parentheses are for bars and flats and the properties without parentheses are for plates, sheets, and strips.

Source: compiled from IS 1570 (part 5), 1985.

TABLE 1–18
Mechanical properties of high-strength low-alloy steels

ASTM specification	Type, grade, or condition	UNS designation	Minimum tensile[a] strength,[a] σ_{St} MPa	kpsi	Minimum yield[a] strength,[a] σ_{Sy} MPa	kpsi	Minimum elongation,[a] % In 200 mm (8 in)	In 50 mm (2 in)	Intended Uses
A242	Type 1	K11510	435–480	63–70	290–345	42–50	18	21	Structural members in welded, bolted, or riveted construction
A440		K12810	435–485	63–70	290–345	42–50	18	21	Structural members, primarily in bolted or riveted construction
A441		K12211	415–485	60–70	275–345	42–50	18	21	Welded, bolted, or riveted structures but primarily welded bridges
A572	Grade 42		415	60	290	42	20	24	Welded, bolted, or riveted structures, but used mainly in bolted or riveted bridges and buildings
	Grade 50		450	65	345	50	18	21	
	Grade 60		520	75	415	60	16	18	
	Grade 65		550	80	450	65	15	17	
A606	Hot-rolled		480	70	345	50		22	Structural and miscellaneous purposes where weight saving or added durability is important
	Hot-rolled and annealed or normalized		450	65	310	45		22	
	Cold-rolled		450	65	310	45		22	
A607	Grade 45		410	60	310	45		22–25	Structural and miscellaneous purposes where greater strength or weight saving is important
	Grade 50		450	65	345	50		20–22	
	Grade 60		520	75	415	60		16–18	
	Grade70		590	85	485	70		14	
A618	Grade I	K02601	483	70	345	50	19	22	General structural purposes including welded, bolted, or riveted bridges and buildings
	Grade II	K12609	483	70	345	50	18	22	
	Grade III	K12700	448	65	345	50	18	20	
A656	Grade 1 and 2		655–793	95–115	552	80	12		Truck frames, brackets, crane booms, railcars, and other applications where weight saving is important
A690		K12249	485	70	345	50	18		Dock walls, sea walls, bulkheads, excavations, and similar structures exposed to sea water
A715	Grade 50		415	60	345	50		22–24	Structural and miscellaneous applications where high strength, weight savings, improved formability, and good weldability are important
	Grade 60		485	70	415	60		20–22	
	Grade 70		550	80	485	70		18–20	
	Grade 80		620	90	550	80		16–18	

[a] May vary with product size and mill form.

Source: ASM Metals Handbook, American Society for Metals, Metals Park, Ohio, 1988.

TABLE 1-19

Mechanical properties of some cast alloy, cast stainless, high-strength and iron-based super alloy steels

Materials classification	Grade	Tensile strength, σ_{St} MPa	kpsi	Yield strength, σ_{Sy} MPa	kpsi	Fatigue[e] endurance limit, σ_{Sf} MPa	kpsi	Elongation in 50 mm (2 in), %	Modulus of elasticity, E GPa	Mpsi	Impact Charpy J	ft-lbf	Brinell hardness, temperature, °C (°F), H_B	Rupture strength 100h at 538°C (1000°F) GPa	Mpsi
						Cast Alloy Steels									
ASTM	Grade														
A352-68a	LC1[a]	448	65	241	35	138	20	24			81	60			
A219-6	WC4[a]	483	70	276	40	159	23	20			75	55			
A148-65	80-50[a]	552	80	345	50	172	25	22			65	48			
A148-1	90-60[a]	620	90	414	60	214	31	20			54	40			
A148-65	105-85[b]	724	105	586	85	244	34	17			79	58	217		
A148-65	150-125[b]	1034	150	862	125	303	44	9			41	30	311		
A148-65	120-95[b]	827	120	655	95	255	37	14			61	45	262		
A148-65	175-145[b]	1207	175	1000	145	331	48	6			32	24	352		
						Cast Stainless Steels									
ACI[d]	CB-30[d]	655	95	414	60			15			3	2			
	C-50[d]	483-669	70-97	448	65			18	200	29	61	45			
	CE-30[d]	600-669	87-97	448	65			18	172	25	14	10			
	CF-8[d]	517-586	75-85	241-276	35-40			55	193	28	95	70			
	CH-20[d]	552-607	80-88	345	50			38	193	28	20	15			
						Ultra-High-Strength Steels									
Medium carbon low alloys 4 140 M, 4330V, D 6AC, 4340 Mod. 5 Cr-Mo-V tool steels:		To 2068	To 300	To 1724	To 250			10			23	17			
H-11 (Mod), H-13 (Mod) Maraging steels (high nickel):		To 2144	To 311	To 1703	To 247			6.6-12			20-30	15-22			
18 Ni (350) Almar 302		1758	255	1689	245			8			31	23			
						High-Strength Low-Alloy (HSLA) Steels									
ASTM	SAE												*Composition*		
A607	J410C[g]	414	60	310	45			25					Cb and/or V		
A606 Types 2, 4[h]		448-483	65-70	310-345	45-50			22					(Proprietary) Cu, Cr, Mn, Ni, P, and other additions		
A607	J410C[g]	448	65	345	50			22					Cb and or V		
715 (sheet)[j]		414	60	345	50			24					(Proprietary) Cb, Ti, Zr, Si, N, V, and others		
A656 (plate)															
A607	J410C[g]	483	70	379	55			20					Cb and/or V		
A607	J410C[g]	586	85	483	70			14							

1.30

TABLE 1-19
Mechanical properties of some cast alloy, cast stainless, high-strength and iron-based super alloy steels (*Cont.*)

Iron-Based Superalloys

Materials classification	Tensile strength, σ_{St} MPa	kpsi	Yield strength, σ_{Sy} MPa	kpsi	Fatigue[c] endurance limit, σ_{Sf} MPa	kpsi	Elongation in 50 mm (2 in), %	Modulus of elasticity, E GPa	Mpsi	Impact Charpy J	ft-lbf	Brinell hardness, temperature, H_B — Temperature °C	°F	Rupture strength 100h at 538°C (1000°F) GPa	Mpsi
Martensitic AISI															
601 17-22A	827	120	689	100			30	21	3.08			21	70	338	49
	531	77	372	54			20					538	1000		
604 Chromalloy	682–896	125–138	655–745	95–108			7	22	3.17			21	70	517	75
	758	110	586	85								538	1000		
610 H-11	931–2137	135–310	689–1655	100–240	896	130	3–17	21	3.05	14–43	10–32	21	70	655–793	95–115
	1241	180	965	140			10					538	1000		
616 422	1034–1655	150–240	682–1207	125–175	621–758	90–100	16–19	20	2.90	14–52	10–38	21	70	400	58
	1172	170	869	126			16					538	1000		
Austenitic															
633 AM 350	103–1413	160–205	414–1207	60–175	482–689	70–100	12–38	20.3	2.94	19	14	21	70	710	103
	1130	160	745	108			9					538	1000		
635 Stainless W	1517–1551	220–225	1482–2000	215–290	372–662	54–96	1.5	20.8	3.02	5–144	4–106	21	70	220	32
	517–552	75–80	255–345	37–50			58					538	1000		
650 16-25-G	758–965	110–140	345–689	50–100			20–45	19.5	2.85	20	15	21	70	538	78
	621	90	228	33			58					538	1000		
653 17-24 Cu Mo	593–772	86–112	276–620	40–90			30–45	19.3	2.80	11–35	8–26	21	70	330	48
	448	65	200	29			37					538	1000		
660 A-286	1007–903	146	655	95			25	20	2.88	56–81	41–60	21	70	689	100
		131	607	88			19					538	1000		

[a] Normalized and tempered.
[b] Quenched and tempered.
[c] Polished specimen.
[d] Corrosion resistance.
[e] Heat resistance.
[f] Heat and corrosion resistance.
[g] Semikilled or killed.
[h] Semikilled or killed—improved corrosion resistance.
[i] Inclusion control-improved formability, killed.

Source: Machine Design, 1981 Materials Reference Issue, Penton/IPC, Cleveland, Ohio, Vol. 53, No. 6 (March 19, 1981).

TABLE 1–20
Mechanical properties of high tensile cast steel

Grade	Designation	Tensile strength, min, σ_{St}		Yield strength (or 0.5% proof stress), min, σ_{Sy}		Reduction in area, min, %	Elongation, min, % (gauge length $5.65\sqrt{a*}$)[a]	Brinell hardness, min, H_B	Izod impact strength, min	
		MPa	kpsi	MPa	kpsi				J	ft-lbf
1	CS 640	640	92.8	390	56.7	35	15	190	30	22.1
2	CS 700	700	101.5	560	81.2	30	14	207	30	21.1
3	CS 840	840	121.8	700	101.5	28	12	248	28	20.6
4	CS 1030	1030	149.4	850	123.3	20	8	305	20	14.5
5	CS 1230	1230	178.3	1000	145.1	12	5	355		

[a] $a*$, area of cross section.

Source: IS 2644, 1979.

TABLE 1–21
Chemical composition of tool steels

Steel designation	% C	% Si	% Min	% Cr	% Mo	% V	% W	% Ni	% Co
T 140 W 4 Cr 50	1.30–1.50	0.10–0.35	0.25–0.50	0.30–0.70			3.50–4.20		
T 133	1.25–1.40	0.10–0.30	0.20–0.35						
T 118	1.10–1.25	0.10–0.30	0.20–0.35						
T 70	0.65–0.75	0.10–0.30	0.20–0.35						
T 85	0.80–0.90	0.10–0.35	0.50–0.80						
T 75	0.70–0.80	0.10–0.35	0.50–0.80						
T 65	0.60–0.70	0.10–0.35	0.50–0.80						
T 215 Cr 12	2.00–2.30	0.10–0.35	0.25–0.50	11.0–13.0	0.80 max[a]	0.80 max [a]			
T 160 Cr 12	1.50–1.70	0.10–0.35	0.25–0.50	11.0–13.0	0.80 max[a]	0.80 max[a]			
T 110 W 2 Cr 1	1.00–1.20	0.10–0.35	0.90–1.30	0.90–1.30			1.25–1.75		
T 105 W 2 Cr 60 V 25	0.90–1.20	0.10–0.35	0.25–0.50	0.40–0.80	0.25 max[a]	0.20–0.30	1.25–1.75		
T 90 Mn 2 W 50 Cr 45	0.85–0.95	0.10–0.35	1.25–1.75	0.30–0.60		0.25 max	0.40–0.60		
T 105 Cr 1	0.90–1.20	0.10–0.35	0.20–0.40	1.00–1.60					
T 105 Cr 1 Mn 60	0.90–1.20	0.10–0.35	0.40–0.80	1.00–1.60					
T 55 Cr 70	0.50–0.60	0.10–0.35	0.60–0.80	0.60–0.80					
T 55 Si 2 Mn 90 Mo 33	0.50–0.60	1.50–2.00	0.80–1.00		0.25–0.40	0.12–0.20[a]			
T 50 Cr 2 V 23	0.45–0.55	0.10–0.35	0.50–0.80	0.90–1.20		0.15–0.30			
T 60 Ni 1	0.55–0.65	0.10–0.65	0.50–0.80	0.30 max				1.00–1.50	
T 30 Ni 4 Cr 1	0.26–0.34	0.10–0.35	0.40–0.70	1.10–1.40				3.90–4.30	
T 55 Ni 2 Cr 65 Mo 30	0.50–0.60	0.10–0.35	0.50–0.80	0.50–0.80	0.25–0.35			1.25–1.75	
T 33 W 9 Cr 3 V 38	0.25–0.40	0.10–0.35	0.20–0.40	2.80–3.30		0.25–0.50	8.0–10.0		
T 35 Cr 5 Mo V 1	0.30–0.40	0.80–1.20	0.25–0.50	4.75–5.25	1.20–1.60	1.00–1.20			
T 35 Cr 5 Mo W 1 V 30	0.30–0.40	0.80–1.20	0.25–0.50	4.75–5.25	1.20–1.60	0.20–0.40	1.20–1.60		
T 75 W 18 Co 6 Cr 4 V 1 Mo 75	0.70–0.80	0.10–0.35	0.20–0.40	4.00–4.50	0.50–1.00	1.00–1.50	17.50–19.00		5.00–6.00
T 83 Mo W 6 Cr 4 V 2	0.75–0.90	0.10–0.35	0.20–0.40	3.75–4.50	5.50–6.50	1.75–2.00	5.50–6.50		
T 55 W 14 Cr 3 V 45	0.50–0.60	0.20–0.35	0.20–0.40	2.80–3.30		0.30–0.60	13.00–15.00		
T 16 Ni 85 Cr 60	0.12–0.20	0.10–0.35	0.60–1.00	0.40–0.80				0.60–1.00	
T 10 Cr 5 Mo 75 V 23	0.15 max	0.10–0.35	0.25–0.50	4.75–5.25	0.50–1.00	0.15–0.30			

[a] Optional

Source: IS 1871, 1965.

TABLE 1–22
Mechanical properties of some tool steels

AISI steel designation	Condition[a]	Tensile strength, σ_{St} MPa	kpsi	Yield strength, σ_{Sy} MPa	kpsi	Elongation, %	Hardness	Hardening temperature °C	°F	Quenched media	Impact strength Charpy V-notch J	ft-lbf	Machinability
H-11	Annealed 870°C (1600°F)[b]	690	100	365	53	25	96 R_B	1010	1850	air	14	10	Medium to high
	Tempered 540°C (1000°F)	2034	295	1724	250	9	55 R_C						
L-2	Annealed 775°C (1425°F)	710	103	510	74	25	96 R_B	855	1575	oil	28	21	High
	Tempered 205°C (400°F)	2000	290	1793	260	5	54 R_C						
L-6	Annealed 775°C (1425°F)[a]	655	95	380	55	25	93 R_B	845	1550	oil	12	9	Medium
	Tempered 315°C (600°F)	2000	290	1793	260	4	54 R_C						
P-20	Annealed 775°C (1425°F)	690	100	517	75	17	97 R_B	855	1575	oil	20	15	Medium to high
	Tempered 205°C (400°F)	1860	270	1413	205	10	52 R_C						
S-1	Annealed 800°C (1475°F)	690	100	414	60	24	96 R_B	925	1700	oil	250	184[c]	Medium
	Tempered 205°C (400°F)	2068	300	1896	275	4	57.5 R_C						
S-5	Annealed 790°C (1450°F)	724	105	440	64	25	96 R_B	870	1600	oil	206	152[c]	Medium to high
	Tempered 205°C (400°F)	2344	340	1930	280	5	59 R_C						
S-7	Annealed 830°C (1525°F)	640	93	380	55	25	95 R_B	940	1725	air	244	180	Medium
	Tempered 205°C (400°F)	2170	315	1448	210	7	58 R_C						
A-8	Annealed 845°C (1550°F)[b]	710	103	448	65	25	97 R_B	1010	1850	air	7	5	Medium
	Tempered 565°C (1050°F)	1827	265	1550	225	9	52 R_C						

[a] Single temper, oil-quenched unless otherwise indicated.
[b] Double temper, air-quenched.
[c] Charpy impact unnotched tests made on longitudinal specimens of small cross-sectional bar stock. The heat treatments listed were to develop nominal mechanical properties for hardened and tempered materials for test purposes only and may not be suitable for some applications.

Source: Machine Design, 1981 Materials Reference Issue, Penton/IPC, Cleveland, Ohio, Vol. 53, No. 6 (March 19, 1981).

TABLE 1–23
Properties of representative cobalt-bonded cemented carbides

Nominal composition	Grain size	Brinell hardness, H_B	Density		Transverse strength, σ_{Sb}		Compressive strength, σ_{Sc}		Proportional limit, compression, σ_{Se}		Modulus of elasticity, E		Tensile strength, σ_{St}		Impact strength	
			Mg/m³	lb/in³	MPa	kpsi	MPa	kpsi	MPa	kpsi	GPa	Mpsi	MPa	kpsi	J	in-lbf
94WC–6Co	Fine	92.5–93.1	15.0	0.54	1790	260	5930	860	2550	370	614	89			1.02	9
	Medium	91.7–92.2	15.0	0.54	2000	290	5450	790	1930	280	648	94	1450	210	1.36	12
	Coarse	90.5–91.5	15.0	0.54	2210	320	5170	750	1450	210	641	93	1520	220	1.36	12
90WC–10Co	Fine	90.7–91.3	14.6	0.53	3100	450	5170	750	1590	230	620	90			1.69	15
	Coarse	87.4–88.2	14.5	0.52	2760	400	4000	580	1170	170	552	80	1340	195	2.03	18
84WC–16Co	Fine	89	13.9	0.50	3380	490	4070	590	970	140	524	76			3.05	27
	Coarse	86.0–87.5	13.9	0.50	2900	420	3860	560	700	100	524	76	1860	270	2.83	25
72WC–8TiC–11.5TaC–8.5Co	Medium	90.7–91.5	12.6	0.45	1720	250	5170	750	1720	250	558	81			0.90	8
64TiC–28WC–2TaC–2Cr$_3$C$_2$–4.0Co	Medium	94.5–95.2	6.6	0.24	690	100	4340	630								

Source: ASM Metals Handbook, American Society for Metals, Metals Park, Ohio, 1988.

TABLE 1–24
Typical uses of tool steel

Steel designation	Type	Typical uses
	Cold-Work Water-Hardening Steels	
T 140 W 4 Cr 50	Fast finishing tool steel	Finishing tools with light feeds, marking tools, etc.
T 133	Carbon tool steels	Engraving tools, files, razors, shaping and wood-working
T 118		tools, heading and press tools, drills, punches, chisels, shear
T 70		blades, vice jaws, etc.
	Cold-Work Oil and Air-Hardening Steels	
T 215 Cr 12	High-carbon high-	Press tools, drawing and cutter dies, shear blade thread
T 160 Cr 12	chromium tool steels	rollers, etc.
T 110 W 2 Cr 1	Nondeforming tool steels	Engraving tools, press tools, gauge, tape, dies, drills, hard
T 105 W 2 Cr 60 V 25		reamers, milling cutters, broaches, cold punches, knives, etc.
T 90 Mn 2 W 50 Cr 45		
T 105 Cr 1	Carbon-chromium tool	Lathe centers, knurling tools, press tools
T 105 Cr 1 M 60	steels	
T 85		Die blocks, garden and agricultural tools, etc.
T 75	Carbon tool steels	
T 65		
T 55 Cr 70	Shock-resisting tool steels	Pneumatic chisels, rivet shape, shear blades, heavy-duty
T 55 Si 2 Mn 90 Mo 33		punches, scarfing tools, and other tools under high shock
T 50 Cr 1 V 23		
T 60 Ni 1	Nickel-chrome-	Cold and heavy duty punches, trimming dies, scarfing tools,
T 30 Ni 4 Cr 1	molybdenum tool steels	pneumatic chisels, etc.
T 55 Ni 2 Cr 65 Mo 3		
	Hot-Work and High-Speed Steel	
T 33 W 9 Cr 3 V 38	Hot-work tool steels	Castings dies for light alloys, dies for extrusion, stamping,
T 35 Cr 5 Mo V 1		and forging
T 35 Cr 5 Mo W 1 V 30		
T 75 W 18 Co 6 Cr 4 V 1 Mo 75	High-speed tool steels	Drills, reamers, broaches, form cutters, milling cutters,
T 83 Mo W 6 Cr 4 V 2		deep-hole drills, slitting saws, high-speed and heavy-cut
T 55 W 14 Cr 3 V 45[a]		tools
	Low-Carbon Mold Steel	
T 16 Ni 80 Cr 60	Carburizing steels	After case hardening for molds for plastic materials
T 10 Cr 5 Mo 75 V 23		

[a] May also be used as hot-work steel.
Source: IS 1871, 1965.

TABLE 1–25
Mechanical properties of carbon and alloy steel bars for the production of machine parts

Steel designation	Ultimate tensile strength, σ_{Sut}				Minimum elongation (gauge length = 5.65 $\sqrt{a^*}$), %
	MPa §	kpsi	MPa ‡	kpsi	
14 C 4 (C 14)**	363	52.6	441	64.0	26
20 C 8 (C 20)	432	62.6	510	74.0	24
30 C 8 (C 30)	490	71.1	588	85.3	21
40 C 8 (C 40)	569	82.5	667	96.7	18
45 C 8 (C 45)	618	89.6	696	101.0	15
55 C 8 (C 55 Mn 75)	706	102.4			13
65 C 6 (C 65)	736	106.7			10
14 C 14 S 14 (14 Mn 1 S 14)	432	62.6	530	76.8	22
11 C 10 S 25 (13 S 25)	363	52.6	481	69.7	22

Notes: a* area of cross section; § minimum; ‡ maximum; ** steel designations in parentheses are old designations.
Source: IS 2073, 1970.

TABLE 1–26
Recommended hardening and tempering treatment for carbon and alloy steels

Designation	Hot-working temperature		Normalizing		Hardening		Quenching		Tempering	
	K	°C	K	°C	K	°C	K	°C	K	°C
30 C 8 (C 30)	1473–1123	1200–850	1133–1163	860–890	1133–1163	860–890	Water or oil		823–923	550–650
35 C 8 (C 35 Mn 75)	1473–1123	1200–850	1113–1153	850–880	1113–1153	840–880	Water or oil		803–1033	530–760
40 C 8 (C 40)	1473–1123	1200–850	1103–1133	830–860	1103–1133	830–860	Water or oil		823–933	550–660
50 C 8 (C 50)	1423–1123	1150–850	1083–1113	810–840	1083–1113	810–840	Oil		823–933	550–660
55 C 8 (C 55 Ma 75)	1423–1123	1150–850	1083–1113	810–840	1083–1113	810–840	Oil		823–933	550–660
40 C 10 Si 8 (40 S 18)	1473–1123	1200–850	1103–1133	830–860	1103–1133	830–860	Oil		823–933	550–660
40 C 15 Si 2 (40 Mn 2 S 12)	1473–1123	1200–850	1113–1143	840–870	1113–1143	840–870	Oil		833–933	560–660
220 C 15 (20 Mn 2)	1473–1123	1200–850	1133–1173	860–900	1133–1173	860–900	Water or oil		823–933	550–660
27 C 15 (27 Mn 2)	1473–1123	1200–850	1133–1153	840–880	1133–1153	840–880	Water or oil		823–933	550–660
37 C 15 (37 Mn 2)	1473–1123	1200–850	1123–1143	850–870	1123–1143	850–870	Water or oil		823–933	550–660
40 Cr 4 (40 Cr 1)	1473–1123	1200–850	1123–1153	850–880	1123–1153	850–880	Oil		823–993	550–720
35 Mn 6 Mo 3 (35 Mn 2 Mo 28)	1473–1173	1200–900			1113–1133	840–860	Water or oil		823–873	550–600
35 Mn 6 Mo 4 (35 Mn 2 Mo 45)	1473–1173	1200–900			1113–1133	840–860	Oil		823–933	550–660
40 Cr 4 Mo 3 (40 Cr 1 Mo 28)	1473–1123	1200–850	1123–1153	850–880	1123–1153	850–880	Oil		823–993	550–720
40 Ni 14 (40 Ni 3)	1473–1123	1200–850	1103–1133	830–860	1103–1133	830–860	Oil		823–923	550–650
35 Ni Cr 2 Mo (35 Ni Cr Mo 60)	1473–1123	1200–850			1093–1123	820–850	Water or oil		823–933	550–660
40 Ni 6 Cr 4 Mo 2 (40 Ni Cr Mo 15)	1473–1123	1200–850			1103–1123	830–850	Oil		823–933	550–660
40 Ni 6 Cr 4 Mo 3 (40 Ni 2 Cr 1 Mo 28)	1473–1173	1200–900			1103–1123	830–850	Oil		823–933	550–660
									or	or
									423–473	150–200
									(depending on hardness required)	
15 Ni Cr 1 Mo 12 (31 Ni 3 Cr 65 Mo 55)	1473–1123	1200–850			1103–1123	830–850	Oil		≤ 933	≤ 660
30 Ni 13 Cr 5 (30 Ni 4 Cr 1)	1473–1123	1200–850			1083–1103	810–820	Air or oil		≥ 523	≥ 250
15 Cr 13 Mo 6 (15 Cr 3 Mo 55)	1473–1123	1200–850			1163–1183	890–910	Oil		823–973[a]	550–700[a]
25 Cr 13 Mo 6 (25 Cr 3 Mo 55)	1473–1123	1200–850			1163–1183	890–910	Oil		823–973[a]	550–700[a]
40 Cr 13 Mo 10 V 2 (40 Cr 3 Mo 1 V 20)	1472–1123	1200–850			1173–1213	900–940	Oil		843–923	570–650
40 Cr 7 Al 10 Mo 2 (40 Cr 2 Al 1 Mo 18)	1473–1123	1200–850			1123–1173	850–900	Oil		823–973	550–700
55 Cr 70	1473–1123	1200–850	1073–1123	800–850	1073–1123	800–850	Oil		773–973	500–700
105 Cr 4 (105 Cr 1)	1373–1123	1100–850			1093–1133	820–860	Water or oil		> 423	> 150 in oil
105 Cr 1 Mn 60	1373–1123	1100–850			1073–1113	800–840	Water or oil		403–453	130–180

[a] Stabilization 823 K (550°C).

Source: IS 1871, 1965.

TABLE 1–27
Mechanical properties of some as-cast austenitic manganese steels

Composition, %				Section			Tensile strength, σ_{St}		Yield strength, σ_{Sy} (0.2% offset)		Brinell hardness, H_B	Elongation in 50 mm (2 in), %	Reduction in area, %	Impact strength Charpy[b]	
C	Mn	Si	Other	Form	mm	in	MPa	kpsi	MPa	kpsi				J	ft-lbf
							Plain Manganese Steels								
0.85	11.2	0.57	—	Round	25	1	440	64	—	—	—	14.5	—	—	—
1.11	12.7	0.54		Round	25	1	450	65	360	52		4			
							1 Mo Manganese Steels								
0.83	11.6	0.38	0.96 Mo	Round	25	1	695	101	345	50	163	30	29		
1.16	10.6	0.60	1.10 Mo	Round	25	1	560	81	400	58	185	13	15		
0.93	13.6	0.67	0.96 Mo	Plate	25	1	510	74	365	53	188	11	16	72	53
							2 Mo Manganese Steels								
0.52	14.3	1.47	2.4 Mo	Round	25	1	600	87	370	54	220	15.5	13		
0.75	14.1	0.99	2.0 Mo	Round	25	1	745	108	365	53	183	34.5	27		
							3.5 Ni Manganese Steel								
0.75	13.0	0.95	3.65 Ni	Round	25	1	655	95	295	43	150	36	26		
							6 Mn-1 Mo Alloys								
0.89	6.3	0.6	1.20 Mo	Plate	100	4	330[a]	48[a]				1[a]			

[a] Properties converted from transverse bend tests on 6-by-13 mm ($\frac{1}{4}$-by-$\frac{1}{2}$ in) bars cut from castings and broken by center loading across 25-mm (1-in) span.
[b] Charpy V-notch.

Source: ASM Metals Handbook, American Society for Metals, Metals Park, Ohio, 1988.

TABLE 1-28
Mechanical properties, fabrication characteristics,[a] and typical uses of some aluminum alloys[b]

Alloy no.	Ultimate tensile strength σ_{Sut} MPa	kpsi	Tensile yield strength,[d] σ_{Syt} MPa	kpsi	Compressive yield strength,[d] σ_{Syc} MPa	kpsi	Shear strength, τ_S MPa	kpsi	Endurance limit in reversed bending,[c] σ_{Sfb} MPa	kpsi	Brinell hardness 4.9 kN (500 kgf) load on 10-mm ball, H_B	Modulus of elasticity,[e] E GPa	Mpsi	Elongation in 50 mm (2 in), %	Corrosion resistance	Machin-ability	Welding Gas	Arc	Resis-tance	Uses
Sand Casting Alloys																				
201.0 -T43	414	60	255	37	386	56	290	42			130			17	4	1	2			Aircraft structural components
-T6	448	65	379	55	207	30					90			8	4	1	2			
240.0 -F	235	34	200	29							90			1.0	4	3	4			Crankcases, spring hangers, housing, wheels
295.0 -T4	221	32	110	16	117	17	179	26	48	7	60	69	10.0	8.5	3	2	2			
-T6	250	36	165	24	172	25	217	31	52	7.5	75	69	10.0	5.0	3	3	2			
319.0 -F	186	27	124	18	131	19	152	22	69	10	70	74	10.7	2.0	3	3	2			Air compressor fitting, crankcase, gear housing
-T6	250	36	164	24	172	25	200	29	76	11	80	74	10.7	2.0	3	3	2			Cylinder heads, impellers, timing gears, water jackets, meter parts
C 355.0 -T6	269	39	200	29	172	25					85	72	10.5	5.0	3	3	2			
356.0 -T6	228	33	164	24	172	25	179	26	59	8.5	70	72	10.5	3.5	3	3	2			
A 390.0 -F	279	40	278	40	186	27	234	34			140	82	11.9	< 1.0	2	4	2			Automotive engine blocks, pulleys, brake shoes, and pumps
520.0 -T4	331	48	179	26					90	13	75	65	9.5	16	1	1	5			Aircraft fittings, and components, levers, brackets
A 535.0 -F	250	36	124	18					55	8	65			9.0	1	1	4			
Permanent Mold Casting																				
355.0 -T6	290	42	185	27	185	27	235	34	69	10	90			4.0	3	3	2			Timing gears, impellers, compressor and aircraft and missile components requiring high strength
C 355.0 -T61	303	44	234	34	248	36	221	32	97	14	90			3.0	3	3	2			
A 356.0 -T61	283	41	207	30	221	32	193	28	90	13	90	72	10.5	10.0	2	3	2			Machine-tool parts, aircraft wheels, pump parts, marine hardware, valve bodies
513.0 -F	186	27	110	16	117	17	152	22	69	10	60			7.0	1	1	5			Ornamental hardware and architectural fittings
Wrought Alloys																				
1100 -O	90	13	35	5			60	9	35	5	23			35	A	E	A	A	B	Sheet metal work, spun hollowware, fin stock
-H 14	125	18	115	17			75	11	50	7	32				A	D	A	A	A	
-H 18	165	24	150	22			90	13	60	9	44			5	A	D	A	A	A	
2011 -T3	380	55	295	43			220	32	125	18	95			15	D	A	D	D	D	Screw machine products
-T6	395	57	270	39			235	34	125	18	97			17						
2014 -O	185	27	95	14			125	18	90	13	45			18		D	D	D	B	Truck frames, aircraft structures
-T4, -T451	425	62	290	42			260	38	140	20	105			20	C	B	D	B	B	
-T6, -T651	482	70	415	60			290	42	125	18	135			13	C	B	D	B	B	
2017 -O	180	26	70	10			125	18	90	13	45			22						
-T4, -T451	425	62	275	40			260	38	125	18	105			22						
2024 -O	185	27	75	11			125	18	90	13	47			20	C	D	D	D	D	Truck wheels, screw-machine products, aircraft structures
-T4, T351	470	68	325	47			285	41	140	20	120			20	—	B	C	C	B	
-T3	485	70	345	50			280	40	140	20	120			18	C	B	C	B	B	
-T86	515	75	490	71			310	45	125	18	135			6	D	E	D	C	B	
3003 -O	110	16	40	6			75	11	50	7	28			30	A	E	A	A	B	Pressure vessels, storage tanks, heat-exchanger tubes, chemical equipments, cooking utensils

TABLE 1-28
Mechanical properties, fabrication characteristics,[a] and typical uses of some aluminum alloys[b] (Cont.)

Alloy no.	Ultimate tensile strength σ_{Su} MPa	kpsi	Tensile yield strength[d] σ_{Syt} MPa	kpsi	Compressive yield strength[d] σ_{Syc} MPa	kpsi	Shear strength, τ_S MPa	kpsi	Endurance limit in reversed bending,[c] σ_{Sfb} MPa	kpsi	Brinell hardness 4.9 kN (500 kgf) load on 10-mm ball, H_B	Modulus of elasticity,[e] E GPa	Mpsi	Elongation in 50 mm (2 in), %	Corrosion resistance	Machin-ability	Gas	Arc	Resis-tance	Uses
-H 14	150	22	145	21			95	14	60	9	40			8	A	D	A	A	A	
-H 18	200	29	185	27			110	16	70	10	55			4	A	D	A	A	A	Trailer panel sheet, storage tanks, sheetmetal works
3004 -O	180	26	70	10			110	16	95	14	45			20	A	D	B	A	B	
-H 34	240	35	200	29			125	18	105	15	63			9	A	C	B	A	A	
-H 38	285	41	250	36			145	21	110	16	77			5	A	C	B	A	A	
5052 -O	195	28	90	13			125	18	110	16	47			25	A	D	A	A	B	Hydraulic tube, appliances, bus body sheet, sheetmetal work, welded structures, boat sheet
6061 -H 34	260	38	215	31			145	21	125	18	68			10	A	C	A	A	A	
-H 38	290	42	255	37			165	24	140	20	77			7	A	C	A	A	A	
-O	125	18	55	8			80	12	60	9	30			25	B	D	A	A	B	Heavy-duty structures requiring good corrosion resistance, truck and marine, railroad car, furniture, pipeline applications
-T 6	310	45	275	40			205	30	95	14	95			12	B	C	A	A	A	
6063 -O	90	13	50	7			70	10	55	8	25			—	A		A	A	A	Pipe, railing, furniture, architectural extrusions
-T 6	240	35	215	31			150	22	70	10	73			12	A	C	A	A	A	
7075 -O	230	38	105	15			150	22	115	17	60			17	—	D	D	C	B	Fin stock, cladding alloy
-T 6	570	83	505	73			330	48	160	23	150			11	C	B	D	C	B	Aircraft and other structures

[a] For ratings of characteristics, 1 is the best and 5 is the poorest of the alloys listed. Ratings A through D are relative ratings in decreasing order of merit.

[b] Average of tensile and hardness values determined by tests on standard 12.5-mm (½-in) diameter test specimens.

[c] Endurance limits on 500 million cycles of completely reversed stresses using rotating beam-type machine and specimen.

[d] At 0.2 percent offset.

[e] Average of tension and compression moduli.

Key: Temper designation: F, as cast; O, annealed; Hxx, strain-hardened; T1, cooled from an elevated temperature shaping process and naturally aged; T3, solution heat-treated and cold worked and naturally aged; T4, solution heat-treated and naturally aged; T2, cooled from an elevated temperature shaping process, cold-worked and naturally aged; T5, cooled from an elevated temperature shaping process and artificially aged; T6, solution heat treated and artificially aged; T7, solution heat-treated and stabilized; T8, solution heat-treated, cold-worked, and artificially aged; TX 51, stress-relieved by stretching.

Source: ASM *Metals Handbook*, American Society for Metals, Metals Park, Ohio, 1988.

TABLE 1-29
Chemical composition and mechanical properties of cast aluminum alloy

IS New	IS Old	BS	Cu	Mg	Si	Fe	Mn	Ni	Zn	Ti	Ti+Nb	Pb	Sn	Also specified Al	Condition	σ_{St} MPa	σ_{St} kpsi	Elongation %	Brinell H_B	Test piece
2447	A-1	LM1	6.0–8.0	0.15	2.0–4.0	1.0	0.6	0.5	2.0–4.0	0.2**		0.3	0.20		As cast	124*	18.0			Sand-cast
																154*	22.3			Chill-cast
4520	A-2	LM2	0.7–2.5	0.3	9.0–11.15	1.0	0.5	1.0	1.2		0.2*	0.3	0.20		As cast	124	18.0			Sand-cast
																147*	21.3			Chill-cast
4223	A-4	LM4	2.0–4.0	0.15	4.0–6.0	0.8	0.3–0.7	0.3	0.5	0.2*		0.1	0.05	R	As cast	139*	20.2	2		Sand-cast
																154*	22.3	2		Chill-cast
5230	A-5	LM5	0.1	3.0–6.0	0.3	0.6	0.3–0.7	0.1	0.1	0.2**	0.2*	0.05	0.05	E	As cast	139*	20.2	3		Sand-cast
																170*	24.6	3		Chill-cast
4600	A-6	LM6	0.1	0.1	10.0–13.0	0.6	0.5	0.1	0.1	0.2**		0.1	0.05	M	As cast	162*	23.5	5		Sand-cast
																185*	26.9	5		Chill-cast
4250	A-8	LM8	0.1	0.3–0.8	3.5–6.0	0.6	0.5	0.1	0.1	0.2**	0.2	0.1	0.05	A	As cast	124	18.0*	7		Sand-cast
																162	23.5*	2		Chill-cast
														I		232	33.6*	3		
																147	21.3*	2.5		
														N		185	26.9*	5		
														D		232	33.6*	1		
4635	A-9	LM9	0.1	0.2–0.6	10.0–13.0	0.6	0.3–0.7	0.1	0.1	0.2*	0.2	0.1	0.05	E	Precipitation-treated	278	40.3*	2		Sand-cast
5500	A-10	LM10	0.1	9.5–11.0	0.25	0.35	0.1	0.1	0.1	0.2**	0.2	0.05	0.05	R	Solution-treated	251	36.4	1.5		Chill-cast
2280	A-11	LM11	4.0–5.0	0.1	0.25	0.25	0.1	0.1	0.1	0.3*	0.05–0.3	0.05	0.05		Solution-treated	278	40.3	2		Sand-cast
																309	44.8	8		Chill-cast
																216	31.3	12		Sand-cast
															WP	263	38.1	7		Chill-cast
																278	40.3*	13		
																309	44.8*	4		
2585	A-12	LM12	9.0–10.5	0.15–0.35	2.0	0.5–1.5	0.6	0.5	0.1			0.1	0.10		Fully heat-treated	278	40.3**	9	100	
	BS 1490 (LM12)		9.0–11.5	0.2–0.4	2.5	1.0	0.6	0.5	0.8	0.2**		0.1	0.1		WP	278	40.3**		100	
4685	A-13	LM13	0.5–1.3	0.8–1.5	11.0–13.0	0.8	0.5	2.0–3.0	0.1	0.2**	0.2	0.1	0.10		Fully heat-treated	170	24.6		100	Sand-cast
																247	35.9		100	Chill-cast
	A-13 (special)														WP	170	24.6			Sand-cast
																278	40.3			Chill-cast
															WP	139	20.2		65	Sand-cast
																201	29.2		65	Chill-cast
2285	A-14	LM14	3.5–4.5	1.2–1.7	0.6	0.6	0.6	1.8–2.3	0.1	0.2*		0.05	0.05		Fully heat-treated	216	31.3		100	Sand-cast
															WP	278	40.3		100	Chill-cast
	A-14 (special)															185	26.9		75	Sand-cast
															WP	232	33.6		75	Chill-cast
4225	A-16	LM16	1.0–1.5	0.4–0.6	4.5–5.5	0.6	0.5	0.25	0.1	0.2**		0.05 / 0.1**	0.05		Solution-treated	170	24.6			Sand-cast
															WP	201	29.2			Chill-cast
															WP	232	33.6			Sand-cast
															WP	263	38.1			Sand-cast
															WP	232	33.6	2		Sand-cast
															WP	278	40.3	3		Sand-cast
4300	A-18	LM18	0.1	0.1	4.5–6.0	0.6	0.5	0.1	0.1	0.2**	0.2	0.1	0.05		As Cast	116	16.8	3		Sand-cast
																139	20.2	4		Chill-cast

TABLE 1-29
Chemical composition and mechanical properties of cast aluminum alloy (Cont.)

IS New	IS Old	BS	Cu	Mg	Si	Fe	Mn	Ni	Zn	Ti	Ti+Nb	Pb	Sn	Also specified Al	Condition	Tensile strength, σ_{St} MPa	kpsi	Elongation %	Brinell hardness, H_B	Test piece
4223	A-22	LM22	2.8–3.8	0.05	4.0–6.0	0.7	0.3–0.6	0.15	0.15	0.2**	0.2	0.05	0.05		Solution (W)-treated	247	35.9	8		Chill-cast
4420	A-24	LM24	3.0–4.0	0.1	7.5–9.5	1.3	0.5	0.5	1.0 3.0**	0.2**		0.1** 0.3	0.20		As Cast	177	25.7	1.5		Chill-cast

Notes: IS: Sp-1-1967 Specification of Aluminum Alloy Castings and BS 1490 (from LM 1 to LM 24) are same.
* Refer to both Indian Standards and British Standards; ** refer to British Standards, BS 1490 only.

Source: IS Sp-1, 1967.

1.42

TABLE 1–30
Chemical composition and mechanical properties of wrought aluminum and aluminum alloys for general engineering purposes

Designation	Al	Cu	Mg	Si	Fe	Mn	Zn	Ti or others	Cr	Condition	Size Over mm (in)	Size Up to and including mm (in)	σ_Sy MPa	σ_Sy kpsi	σ_St MPa	σ_St kpsi	Elong. % (Min)
19000	99 min	0.1	0.2	0.5	0.7	0.1	0.1	—	—	M[a]			20	2.9	65	9.4	18
										O					110#	16.0	25
19500	99.5 min	0.05	—	0.3	0.4	0.05	0.1	—	—	M[a]			18	2.6	65	9.4	23
										O					100#	14.5	25
19600										M[a]			17	2.5	65	9.4	23
										M[a]			90	13	150	21.6	12
										O			175	25.4	240#	34.8	12
24345	Remainder	3.8–5.0	0.2–0.8	0.5–1.2	0.7	0.3–1.2	0.2	0.3*	0.3*	W		10 (0.4)	225	32.6	375	54.4	10
											10 (0.4)	75 (3.0)	235	34.1	385	55.8	10
											75 (3.0)	150 (6.0)	235	34.1	385	55.8	8
											150 (6.0)	200 (8.0)	225	32.6	375	54.4	8
										WP		10 (0.4)	375	54.4	430	62.4	6
											10 (0.4)	25 (1.0)	400	58.0	460	66.7	6
											25 (1.0)	75 (3.0)	420	60.9	480	69.6	6
											75 (3.0)	150 (6.0)	405	58.7	460	66.7	6
											150 (6.0)	200 (8.0)	380	55.1	430	62.4	6
24534	Remainder	3.5–4.7	0.4–1.2	0.2–0.7	0.7	0.4–1.2	0.2	0.3	—	M[a]			90	13.0	150	21.0	12
										O			175#	25.0	240	34.8	12
										W		10 (0.4)	220	31.9	375	54.4	10
											10 (0.4)	75 (3.0)	235	34.1	385	55.8	10
											75 (3.0)	150 (6.0)	235	34.1	385	55.8	8
											150 (6.0)	200 (8.0)	225	32.6	375	54.4	8
43000	Remainder	0.1	0.2	4.5–6.00	0.6	0.5	0.2	—	—	M[a]		15 (0.6)			90	13.0	18
										O		15 (0.6)			130#	18.9	18
46000	Remainder	0.1	0.2	10.0–13.0	0.6	0.5	0.2	—	—	M[a]		15 (0.6)			100	14.5	10
										O					150#	21.8	12
52000	Remainder	0.1	1.7–2.6	0.6	0.5	0.5	0.2	0.2	0.25	M[a]		150 (6.0)	70	10.2	160	23.2	14
53000	Remainder	0.1	2.8–4.0	0.6	0.5	0.5	0.2	0.2	0.25	M[a]		150 (6.0)	100	14.5	240#	34.8	18
										M[a]	50 (2.0)	150 (6.0)	100	14.5	215	31.2	14
										M[a]		150 (6.0)			200	29.0	14
54300	Remainder	0.1	4.0–4.9	0.4	0.7	0.5–1.0	0.2	0.2	0.25	O		150 (6.0)			260#	37.7	10
										M[a]		150 (6.0)	130	18.9	275	40.0	11
										O		150 (6.0)	125	18.1	350	50.8	13
63400	Remainder	0.1	0.4–0.9	0.3–0.7	0.6	0.3	0.2	0.2	0.1	M[a]	All sizes		80	11.6	110	16.0	13
										O	150 (6.0)	150 (6.0)	80	11.6	130#	18.8	18
										W		200 (8.0)	140	20.3	140	20.3	14
										P		23 (0.12)	110	16.0	125	18.1	13
												12 (0.5)	150	21.8	170	24.7	7
										WP	150 (6.0)	150 (6.0)	130	18.9	150	21.8	7
												200 (8.0)	125	18.1	185	26.8	7
64423	Remainder	0.5–1.0	0.5–1.3	0.7–1.3	0.8	1.0	—	—	—	M[a]			130	18.9	150	21.8	6
										O					120	17.4	10
										W			125#	18.1	215#	31.2	15
										WP			155	22.5	265	38.4	13
													265	38.4	330	47.9	7

TABLE 1-30
Chemical composition and mechanical properties of wrought aluminum and aluminum alloys for general engineering purposes (*Cont.*)

Designation	Al	Cu	Mg	Si	Fe	Mn	Zn	Ti or others	Cr	Condition	Over mm (in)	Up to and including mm (in)	Proof MPa	Proof kpsi	Tensile MPa	Tensile kpsi (19500)	Elongation, percent (Min)
64430	Remainder	0.1	0.4–1.2	0.6–1.3	0.6	0.4–1.0	1.0	0.2	0.25	M[a]	All sizes	—	80	11.6	110	16.0	12
										O	—	150 (6.0)	—	—	150#	21.8	16
										W	—	200 (8.0)	120	18.1	185	26.8	14
										W	150 (6.0)	200 (8.0)	100	14.5	170	24.7	12
										WP	—	5 (0.2)	255	37.0	295	42.8	7
											5 (0.2)	75 (3.0)	270	39.2	310	45.0	7
											75 (3.0)	150 (6.0)	255	37.0	295	42.8	7
											150 (6.0)	200 (8.0)	240	34.8	280	40.6	6
65032	Remainder	0.15–0.4	0.7–1.2	0.4–0.8	0.7	0.2–0.8	0.2	0.2	0.15–0.35	M[a]	All sizes	15 (0.6)	50	7.3	110	16	12
										O	—	150 (6.0)	115#	16.7	150#	21.8	16
										W	—	200 (8.0)	115	22.5	185	26.8	14
											150 (6.0)	200 (8.0)	100	14.5	170	24.7	12
										WP	150 (6.0)	200 (8.0)	235	34.0	280	40.6	7
74530	Remainder	0.2	1.0–1.5	0.4–0.8	0.7	0.2–0.7	4–5	0.2	0.2	W (Naturally aged for 30 days)	—	6 (0.24)	200	29.0	245	35.5	6
											6 (0.24)	75 (3.0)	220	31.4	255	37.0	9
											75 (3.0)	150 (6.0)	230	33.6	275	40.0	9
													220	31.4	265	38.4	9
										WP	—	6 (0.24)	245	35.5	285	41.3	7
											6 (0.24)	150 (6.0)	260	37.7	310	45.0	7
													245	35.5	290	42.1	7
76528	Remainder									WP	All sizes	—	—	—	290#	42.1	10
											6 (0.24)	75 (3.0)	430	62.4	500	72.5	6
											15 (0.6)	150 (6.0)	455	66.6	530	78.9	6
													430	62.4	500	72.5	6

[a] Properties in M (as-cast) temper are only typical values and are given for information only.

Key: # Maximum; M—as-cast condition; R—stress-relieved only; P—precipitation-treated; W—solution-treated; WP—solution-treated and precipitation-treated; WPS—fully heat-treated plus stabilization.

Source: IS 733, 1983.

TABLE 1–31
Typical mechanical properties and uses of some copper alloys

Alloy name	UNS no.	Composition,[a] %	Ultimate tensile strength, σ_{Sut} MPa	kpsi	Tensile yield[b] strength, σ_{Syt} MPa	kpsi	Elongation in 50 mm (2 in), %	Hardness Brinell, 4.9 kN (500-kgf load) H_B	Hardness Rockwell,[a] R	Machinability rating[c]	Typical uses
							Cast Alloys				
Leaded red brass	C 83600	85 Cu, 5 Sn, 5 Pb, 5 Zn	255	37	117	17	30	60		84	Valves, flanges, pipe fittings, pump castings, water pump impellers and housings, small gears, ornamental fixtures
Leaded yellow brass	C 85400	67 Cu, 1 Sn, 3 Pb, 29 Zn	234	34	83	12	25	50		80	General-purpose yellow casting alloy, furniture hardware, radiator fittings, ship trimmings, clocks, battery clamps, valves, and fittings
Manganese bronze	C 86300	63 Cu, 25 Zn, 3 Fe, 6 Al, 3 Mn	793	115	572	83	15	225[d]		8	Extra-heavy-duty, high-strength alloy, large valve stems, gears, cams, slow heavy-load bearings, screw-down nuts, hydraulic cylinder parts
Silicon bronze	C 87200	89 Cu min, 4 Si	379	55	172	25	30	85		40	Bearings, bells, impellers, pump and valve components, marine fittings, corrosion-resistant castings
Silicon brass	C 87500	82 Cu, 14 Zn, 4 Si	462	67	207	30	21	115 134[d]		50	Bearings, gears, impeller, rocker arms, valve stems, small boat propellers
Tin bronze	C 90500	88 Cu, 10 Sn, 2 Zn	310	45	152	22	25	75		30	Bearings, bushings, piston rings, valve components, steam fittings, gears
Leaded tin bronze	C 92200	88 Cu, 6 Sn, 1.5 Pb, 4.5 Zn	276	40	138	20	30	65		42	Valves, fittings and pressure-containing parts for use up to 288°C (550°F), bolts, nuts, gears, pump piston, expansion joints
Leaded tin nickel bronze	C 92900	84 Cu, 10 Sn, 2.5 Pb, 3.5 Ni	324	47	179	26	20	80		40	Gears, wear plates, cams, guides
High-leaded tin bronze	C 93700	80 Cu, 10 Sn, 10 Pb	241	35	124	18	20	60		80	Bearings for high-speed and heavy-pressure pumps, impellers, pressure-tight castings
Aluminum bronze	C 95500	81 Cu, 4 Ni, 4 Fe, 11 Al	689–827	100–120	303–469	44–68	12–10	192–230[d]		50	Valve guides and seats in aircraft engines, bushings, rolling mill bearings, washers, chemical plant equipment, chains, hooks, marine propellers, gears, worms
Copper-nickel	C 96300	79.3 Cu, 20 Ni, 0.7 Fe	517	75	379	55	10	150		15	Marine fittings, sleaves and seawater corrosion resistance parts
Nickel-silver	C 97800	66 Cu, 5 Sn, 2 Pb, 25 Ni, 2 Zn	379	55	207	30	15	130[d]		60	Valves and valve seats, musical instrument components, sanitary and ornamental hardware
Special alloy	C 99400	90.4 Cu, 2.2 Ni, 2.0 Fe, 1.2 Al, 1.2 Si, 3.0 Zn	455–545	66–79	234–372	34–54	25	125–170[d]		50	Valve stems, marine uses, propeller wheels, mining equipment gears
							Wrought Alloys				
Cadmium copper	C 16200	99.0 Cu, 1.0 Cd	241–689	35–100	48–476	7–69	57–1			20	Trolly wire, spring contacts, railbands, high-strength transmission lines, switch gear components, and ware-guide
Beryllium copper	C 17000	99.5 Cu, 1.7 Be, 0.20 Co	483–1310	70–190	221–1172	32–170	45–3		R_B98	20	Bellows, diaphragms, fuse clips, fasteners, lock washers, springs, valves, welding equipments, bourdon tubing
Leaded beryllium copper	C 17300	99.5 Cu, 1.9 Be, 0.4 Pb	469–1479	68–200	172–1255	25–182	48–3		R_B77	50	Bellows, diaphragms, fuse clips, fasteners, lock washers, springs, valves, welding equipments, switch parts, roll pins

TABLE 1–31
Typical mechanical properties and uses of some copper alloys (*Cont.*)

Alloy name	UNS no.	Composition,[a] %	Ultimate tensile strength, σ_{Sut} MPa	kpsi	Tensile yield[b] strength, σ_{Syt} MPa	kpsi	Elongation in 50 mm (2 in), %	Hardness Brinell, 4.9 kN (500-kgf load) H_B	Rockwell,[a] R	Machinability rating[c]	Typical uses
Guilding brass (95%)	C 21000	95.0 Cu, 5.0 Zn	234–441	34–64	69–400	10–58	45–4		$64R_B$–$46R_F$	20	Coins, medals, bullet jackets, fuse caps, primers, jewelry base for gold plate
Commercial bronze (90%)	C 22000	90.0 Cu, 10.0 Zn	255–496	37–72	69–427	10–62	50–3		$70R_B$–$53R_F$	20	Etching bronze, grillwork, screen cloth, lipstick cases, marine hardware, screws, rivets
Red brass (85%)	C 23000	85.0 Cu, 15.0 Zn	269–724	39–105	69–434	10–63	55–3		$77R_B$–$55R_F$	30	Conduit, sockets, fasteners, fire extinguishers, condenser and heat-exchanger tubing, radiator cores
Cartridge brass (70%)	C 26000	70.0 Cu, 30.0 Zn	303–896	44–130	76–448	11–65	66–3		$82R_B$–$64R_F$	30	Radiator cores and tanks, flashlight shells, lamp fixtures, fasteners, locks, hinges, ammunition components, rivets
Yellow brass	C 26800	65.0 Cu, 35.0 Zn	317–883	46–128	97–427	14–62	65–3		$80R_B$–$64R_F$	30	Radiator cores and tanks, flashlight shells, lamp fixtures fasteners, locks, hinges, rivets
Muntz metal	C 28000	60.0 Cu, 41.0 Zn	372–510	54–74	145–379	21–55	52–10		$85R_F$–$80R_B$	40	Architectural, large nuts and bolts, brazing rods, condenser plates, heat-exchanger and condenser tubing, hot forgings
Medium leaded brass	C 34000	65.0 Cu, 1.0 Pb, 34.0 Zn	324–607	47–88	103–414	15–60	60–7			70	Butts, gears, nuts, rivets, screws, dials, engravings
Free-cutting brass	C 36000	61.5 Cu, 3.0 Pb, 35.5 Zn	338–469	49–68	124–310	18–45	53–18			100	Gears, pinions, automatic high-speed screws, machine parts
Forging brass	C 37700	59.0 Cu, 2.0 Pb, 39.0 Zn	359	52	138	20	45			80	Forgings and pressings of all kinds
Admiralty brass	C 44300 C 44400 C 44500	71.0 Cu, 28.0 Zn, 1.0 Sn	331–379	48–55	124–152	18–22	65–60			30	Ferrules, condenser, evaporator and heat-exchanger tubing, distiller tubing
Naval brass	C 46400 to C46700	60.0 Cu, 39.25 Zn, 0.75 Sn	379–607	55–88	172–455	25–66	50–17		90–$82R_B$	30	Aircraft turn buckle barrels, balls, bolts, nuts, marine hardware, propeller, rivets, shafts, valve stems, welding rods, condenser plate
Phosphor bronze (5% A)	C 51000	95.0 Cu, 5.0 Sn, trace P	324–965	47–140	131–552	19–80	64–2			20	Bellows, bourdon tubing, clutch disks, cotter pins, diaphragms, fasteners, lock washers, chemical hardware, textile machinery
High-silicon bronze—A	C 65500	97.0 Cu, 3.0 Si	386–1000	56–145	145–483	21–70	63–3			30	Hydraulic pressure liners, anchor screws, bolts, cap screws, machine screws, nuts, rivets, U-bolts, electrical conduits, welding rod
Manganese bronze—A	C 67500	58.5 Cu, 1.4 Fe, 39.0 Zn, 1.0 Sn, 0.1 Mn	448–579	65–84	207–414	30–60	33–19			30	Clutch disks, pump rods, shafting, balls, valve stems and bodies
Copper-nickel (30%)	C 71500	70.0 Cu, 30.0 Ni	372–517	54–75	138–483	20–70	45–15			20	Condensers, condenser plates, distiller tubing, evaporator and heat-exchanger tubing, ferrules, salt water piping
Nickel-silver 55-18	C 77000	55.0 Cu, 27.0 Zn, 18.0 Ni	414–1000	60–145	186–621	27–90	40–2			30	Optical goods, springs, and resistance wires

[a] Nominal composition, unless otherwise noted.
[b] All yield strengths are calculated by 0.5 percent offset method.
[c] Machinability rating expressed as a percentage of the machinability of C 36000, free-cutting brass, based on 100 percent for C 36000.
[d] 29.4-kN (3000-kgf) load.
[e] R_A, R_B, R_F are Rockwell numbers in A, B, F scales.

Note: Values tabulated are average values of test specimens.

Source: ASM Metals Handbook, American Society for Metals, Metals Park, Ohio, 1988.

TABLE 1–32
Nominal compositions and typical room-temperature mechanical properties of some magnesium alloys

| Alloy | Composition | | | | | | Tensile strength, σ_{St} | | Yield strength, σ_{Sy} | | | | | | Elongation in 50 mm (2 in), % | Shear strength, τ_S | | Brinell[b] hardness, H_B |
| | Al | Mn(a) | Th | Zn | Zr | Others | MPa | kpsi | Tensile | | Compressive | | Bearing | | | MPa | kpsi | |
									MPa	kpsi	MPa	kpsi	MPa	kpsi				
Sand and Permanent Mold Castings																		
AZ63A-T6	6.0	0.15		3.0			275	40	130	19	130	19	360	52	5	145	21	73
AZ81A-T4	7.6	0.13		0.7			275	40	83	12	83	12	305	44	15	125	18	55
AZ92A-T6	9.0	0.10		2.0			275	40	150	22	150	22	450	65	3	150	22	84
HK31A-T6			3.3		0.7		220	32	105	15	105	15	275	40	8	145	21	55
HZ32A-T5			3.3	2.1	0.7		185	27	90	13	90	13	255	37	4	140	20	57
ZE41A-T5				4.2	0.7	1.2 RE	205	30	140	20	140	20	350	51	3.5	160	23	62
ZH62A-T5			1.8	5.7	0.7		240	35	170	25	170	25	340	49	4	165	24	70
ZK61A-T6				6.0	0.7		310	45	195	28	195	28			10	180	26	70
Die Castings																		
AM60A-F	6.0	0.13					205	30	115	17	115	17			6			
AS41A-F[c]	4.3	0.35				1.0 Si	220	32	150	22	150	22			4			
AZ91A and B-F[c]	9.0	0.13		0.7			230	33	150	22	165	24			3	140	20	63
Extruded Bars and Shapes																		
AZ31 B and C-F[d]	3.0			1.0			260	38	200	29	97	14	230	33	15	130	19	49
AZ80A-T5	8.5			0.5			380	55	275	40	240	35			7	165	24	82
HM31A-F		1.2	3.0				290	42	230	33	185	27	345	50	10	150	22	
ZK60A-T5				5.5	0.45[a]		365	53	305	44	250	36	405	59	11	180	26	88
Sheets and Plates																		
AZ31B-H24	3.0			1.0			290	42	220	32	180	26	325	47	15	160	23	73
HK31A-H24			3.0		0.6		255	33	200	39	160	23	285	41	9	140	20	68
HM21A-T8		0.6	2.0				235	34	170	25	130	19	270	39	11	125	18	

a Minimum.
b 4.9-kN (500-kgf) load, 10-mm ball.
c A and B are identical except that 0.30% max residual Cu is allowable in AZ91B.
d Properties of B and C are identical, but AZ31C contains 0.15 min Mn, 0.1 max Cu, and 0.03 max Ni.

Source: ASM Metals Handbook, American Society for Metals, Metals Park, Ohio, 1988.

TABLE 1-33
Mechanical properties[a] of some nickel alloys

Name of alloy	Condition	Ultimate tensile strength, σ_{Sut}		Tensile yield strength, σ_{Syt} (0.2% offset)		Elongation in 50 mm (2 in)	Hardness number	Impact strength notched Charpy		Typical uses
		MPa	kpsi	MPa	kpsi			J	ft-lbf	
Nickel 200	Bar, cold-drawn	448–758	65–110	276–690	40–100	35–10				Corrosion-resistant parts
	Annealed	379–517	55–75	103–207	15–30	35–40				
Nickel 270	Strip, cold-drawn	655	95	621	90	4	95R_B			
	Annealed	345	50	110	16	50	35R_B			
Durnickel 301	Bar, cold-drawn, annealed	620–825	90–120	205–415	30–60	55–35				High strength and hardness, corrosion resistance
	Age-hardened	1275	185	910	132	28	35R_C			
Monel 400	Bar, annealed, 21°C (70°F)	517–621	75–90	172–345	25–50	60–35				Corrosion-resistant parts
	Wire, annealed	483–655	70–95	205–380	30–50	45–25				
	Spring temper	1000–1240	145–180	862–1172	125–170	5–2				Springs
Monel K-500	Bar, drawn, age-hardened	965–1172	140–170	724–1034	105–150	30–20	24R_C	53	39	Corrosion-resistant parts
Inconel 600	Rod, annealed	624	91	210	30.4	49	75R_B			Jet engines, missiles, etc. where corrosion resistance and high strength are required
	As rolled	672	98	307	45	46	86R_B			
Inconel 825	Bar, annealed, 21°C (70°F)	1276	185	910	132	28				Superalloy, jet engine, turbine, furnace
	871°C (1600°F)	135	19.6	117	17.0	102				
Inconel X-750	Bar, 21°C (70°F)	1120	162	635	92	24				
	760°C (1400°F)	485	70	455	66	9				
Incoloy 800	Bar, annealed	512–690	75–100	207–414	30–60	60–30				
	Hot-finished	552–827	80–120	241–621	35–90	50–25				
	Wire, spring temper	965–1207	140–175	896–1172	130–170	5–2				
Hastelloy W	Bar, solution-treated 425°C (800°F)	725	105	260	38	56.0				
	900°C (1650°F)	352	52	220	32	14.5				
Hastelloy G-3	Sheet, 6.4–19 mm (0.25–0.75 in) thick	740	107	365	53	56	87R_B			
Hastelloy B	Bar, cast	924	134	462	67	52				
Udimet 700	Bar, 21°C (70°F)	1410	204	965	140	17				Jet engines, missiles, turbines where high-temperature strength and corrosion resistance are important
	870°C (1600°F)	690	100	635	92	27				
Unitemp AF2-1DA	Bar, 21°C (70°F)	1290	187	1050	152	10				
	870°C (1600°F)	830	120	715	104	8				
Rene 95	Bar, forging 21°C (70°F)	1620	235	1310	190	15				
	650°C (1200°F)	1460	212	1220	177	14				
Waspaloy	Bar, 21°C (70°F)	1280	185	795	115	25				
	870°C (1600°F)	525	76	515	75	35				

[a] Values shown represent usual ranges for common sections.
[b] Values tabulated are approximate average ones.
Source: ASM Metals Handbook, American Society for Metals, Metals Park, Ohio, 1988.

TABLE 1–34
Mechanical properties of some zinc casting alloys

Grade	Designation of alloy			Ultimate tensile strength, σ_{Su}		Tensile yield strength, σ_{Syt}		Brinell hardness, H_B	Elongation in 50 mm (2 in),	Impact strength Charpy		Fatigue endurance limit, σ_{Sf}, 10^8 cycles	
	ASTM	SAE	UNS	MPa	kpsi	MPa	kpsi	H_B	%	J	ft-lbf	MPa	kpsi
				Die-Casting Alloys									
Alloy 3	AG 40 A	903	Z 33520	283	41			82	10	58	43	47	6.8
Alloy 5	AC 41 A	925[a]	Z 35531	324	47			91	7	65	48	56	8.2
Alloy 7		903		283	47				14	54	40		
				Zinc Foundary Alloys									
ZA–12													
Sand-cast				276–310	40–45	207	30	105–120	1–3				
Permanent Mold				310–345	45–50	214	31	105–125	1–3				
Die-cast				393	57	317	46	110–125	2				
ZA–27													
Sand-cast				400–440	58–64	365	53	110–120	3–6				
Sand-cast				310–324	45–47	255	37	90–100	8–11				
Die-cast				448	65	434	63	110–125	1				

[a] Die-cast.

Note: Values given are average values.

Source: Machine Design, 1981 Materials Reference Issue, Penton/IPC, Cleveland, Ohio, Vol. 53, No. 6 (March 19, 1981); *SAE Handbook,* pp. 11–123, 1981.

TABLE 1–35
Mechanical properties of some wrought titanium alloys

Name of alloy	UNS no.	Designation	Ultimate tensile strength, σ_{Sut} MPa	kpsi	Tensile yield strength,[a] σ_{Syt} MPa	kpsi	Elongation in 50 mm (2 in), %	Strength impact Charpy J	ft-lbf	Hardness	Machinability	Uses
Commercially pure titanium	R 50520	ASTM Grade 1	240	35	170	25	24	35	26[c]		40[b]	Resistance to temperature effect of structures, easy to fabricate, excellent corrosive resistance, cryogenic applications
Commercially pure titanium		ASTM Grade 2	340	50	280	40	22					
Commercially pure titanium		ASTM Grade 3 (Ti-65A)	450	65	380	55	20			$30R_C$		
Alpha alloy	R 54520	Ti-5AL-2.5Sn	790	115	760	110						Gas turbine engine casting and rings, aerospace structural members, excellent weldability, pressure vessels, excellent corrosive resistance, jet engine blades and wheels, large bulkhead forgings
Alpha alloy	R 54521	Ti-5AL-2.5Sn-ELI	690	100	620	90	19				30[b]	
Alpha alloy	R 54790	Ti-2.25AL-11Sn-5Zr-1Mo	1000	145	900	130						
Alpha-beta alloy	R 56400	Ti-6Al-4V[b]	900	130	830	120	10	24.5	18[e]	$34R_C$	22[b]	Most widely used alloy, aircraft gas turbine disks and blades, turbine disks and blades, air frame structural components, gas turbine engines disks and fan blade, components of compressors
	R 56260	Ti-6Al-2Sn-4Zr-6Mo[c]	1170	170	1100	160						
Beta alloy	R 58010	Ti-13V-11Cr-3Al[c]	900	130	830	120	4	13.5	10[e]		40[b]	Missile applications such as solid rocket motor cases, advanced manned and unmanned airborne systems, springs for airframe applications

[a] At 0.2 percent offset.
[b] Mechanical and other properties given for annealed conditions.
[c] Mechanical and other properties given for solution-treated and aged condition.
[d] Based on a rating of 100 for B1112 resulfurized steel.
[e] Approximate values of annealed bars at room temperature.

Source: ASM Metals Handbook, American Society for Metals, Metals Park, Ohio, 1988.

1.50

TABLE 1-36
Mechanical properties of some lead alloys

UNS no.	Ultimate tensile strength, σ_{Sut} MPa	psi	Yield strength, σ_{Sy} MPa	psi	Shear strength, TS MPa	psi	Fatigue strength at 10^7 cycles, σ_{sf} MPa	psi	Hardness number, H_B	Elongation in 50 mm (2 in)	Creep	Uses
50042	12–13	1740–1885	55	7975	12.5	1810	3.2	464	3.2–4.5	30	7.5 MPa for 1000 h at 1000°C (1832°F) 0.15% per year at 2.07 MPa 3% per year at 2.07 MPa	Low-melting-point chemical process application, used as solder for the jobs
51120	16–19	2320–2755	6–8	870–1160			4.3	624	4–6	30–60		Lead alloyed with tin, bismuth cadmium, indium forms alloys with low melting point; some of these are fusible alloys, used in automotive devices, fire extinguishers, sprinkler heads
52901	27.6	4002					10.33	1495	8.1	48		
54520	30	4350								10 30% at 100°C	33.5 MPa for 1000 h 1.1 MPa for 1000 h at 100°C (212°F)	
54820	34	4930			28	4060				18	0.790 MPa for 0.01% per day	Wiping solder for joining lead pipes and cable sheaths; for automobiles radiator cores and heating units
54915	37 6 at 100°C (212°F)	5365 870			32	4640			12 12	25 130	2.1 MPa for 1000 h	
55030	40.7	5900	33	4785	36	5200			14	60	2.9 MPa for 1000 h	For general purpose; most popular of all
55111	52.5 19 at 100°C (212°F)	7610 2756			37	5380			16	30–60 135–100 at 100°C (212°F)	0.45 MPa for 1000 h at 100°C (212°F)	

Note: Values tabulated are average values obtained from standard test specimens.

Source: ASM Metals Handbook, American Society for Metals, Metals Park, Ohio, 1988.

TABLE 1-37
Mechanical properties of bronzes

Property	Mode of casting test pieces	Railway, bronze					Aluminum bronze			Tin bronze	Silicon bronze
		Class I phosphor* bronze**	Class II gun metal ††	Class III leaded	Class IV bronze †	Class V leaded ‡ gun metal	Grade I	Grade II	Grade III		
Ultimate strength, min.	Sand-cast (cast-on)	186 (27.0)#	196 (28.4)	137 (19.9)	157 (22.8)	186 (27.0)				216 (31.3)	
	Sand-cast (separately cast)	206 (29.9)#	216 (31.3)	157 (22.8)	176 (25.5)	206 (29.9)	647 (93.8)	490 (71.0)	446 (64.7)	226 (32.8)	309 (44.8)
σ_{Sut}, MPa (kpsi)	Chill-cast						647 (93.8)	539 (78.2)	196 (28.4)	245 (35.5)	
Elongation percent, min	Sand-cast (cast-on)	3.0	8.0	2.0	2.0	8.0				8.0	
	Sand-cast (separately cast)	5.0	12.0	4.0	4.0	12.0	15.0	20.0	20.0	12.0	20.0
	Chill-cast						12.0	20.0			

Key: * Brinell hardness, H_B for phosphor bronzes: 60 for sand cast (cast-on) test pieces and 65 for sand-cast (separately cast) test pieces; ** used for locomotive side valves, oil-lubricated side rod, pony pivot bushes, steel axle box, oil-lubricated connecting rod; †† used for fusible plugs, relief valves, whistle valve body, stuffing box, nonferrous boxes, oil-lubricated connecting rod, large end bearings; † used for locomotive grease lubricated non-ferrous axle boxes, side rod and motion bushes; ‡ used for castings for carriage and wagon bearings shells; # σ_{Sut} given in parentheses are the units in US Customary Units (kpsi).

TABLE 1-38
Mechanical properties of rubber and rubber-like materials

Material	Specific gravity	Compressive strength, σ_{Sc}		Tensile strength, σ_{St}		Transverse strength, σ_{Sb}		Hardness shore durometer	Maximum temperature			Effect of heat
		MPa	kpsi	MPa	kpsi	MPa	kpsi		K	°C	°F	
Duprene	1.27–3.00			1.4–28	0.2–4.0			15–95	422	149	300	Stiffens slightly
Koroseal (hard)	1.30–1.40			14–62	2.0–9.0			80–100	373	100	212	Softens
Koroseal (soft)	1.20–1.30			3.4–17	0.6–2.6			30–80	361	88	190	Softens
Plioform (plastic)	1.06	88	12.8	28–34	4.0–5.0	48	7.0		344–393	71–120	160–250	
Rubber † (hard)	1.12–2.00	758	110.0	7–69	1.0–10.0	62	9.0	50*/80	328–344	55–71	130–160	Softens
Rubber ‡ (soft)	0.97–1.25	14	2.0	3.5	0.6	62	9.0		339–367	65–94	150–200	Softens
Rubber (linings)	0.98–1.35	103	15.1	103		103	15.1		361	88	190	Softens

Key: * sclerscope; † coefficient of linear expansion from 0 to 333 K (60°C = 140°F) is 35×10^{-6}; ‡ coefficient of linear expansion from 0 to 333 K (60°C = 140°F) is 36×10^{-6}.

TABLE 1–39
Properties of some thermoplastics

Name of plastic	Tensile strength, σ_{ut}		Elongation in 50 mm (2 in), %	Modulus of elasticity, E		Izod impact strength		Hardness, Rockwell	Resistance to		Coefficient of friction, μ		Application
	MPa	kpsi		GPa	Mpsi	J	ft-lbf		Heat	Chemical	With plastic	With steel	
ABS (general purpose)	41	6	5–20	2.3	0.33	8.8	6.5	103	Available	Fair			Light-duty mechanical and decorative eyeglass frames, automobile-steering wheels, knobs, handles, camera cases, battery cases, phone and flashlight cases, helmets, housing for power tools, pumps
Acrylics	37–72	5.4–10.5	5–50	1.5–3.1	0.22–0.45	0.5–1.6	0.4–1.2	92–100 M	Available	Fair			Light-duty mechanical knobs, pipe fittings, automobile-steering wheels, eyeglass frames, tool handles, camera cases, optical and transparent parts for safety glasses, snowmobile windshields, refrigerator shelves
Acetal	55–69	8–10	40–60	2.8–3.6	0.4–0.52			80–94 M	Good	High			Mechanical gears, cams, pistons, rollers, valves, fan blades, washing-machine agitators, bushings, bearings, chute liners, wear strips, and structural components
Cellulosic (cellulose accetate)	15.2–47.5	2.2–6.9		0.5–2.8	0.065–0.40	1.4–9.9	1.0–7.3	122 R					Decorative knobs, handles, camera cases, pipe fittings, eyeglass frames, phone and flashlight cases, helmets, pumps and power tool housings, transparent parts for safety glasses, lens signs, refrigerator shelves and snowmobile windshields, extruded and cast film, and sheet for packaging
Epoxy resin (glass-fiber filler)	69–138	10–20	4.0	21.0	3.04	2.7–41	2–30	100–110 M					Filament wound structures, aircraft pressure bottles, oil storage tanks and high-performance tubing, reinforced glass-fiber composites
Fluoroplastic group	3–48	0.5–7.0	100–300			4.1	3	50–80 D	Excellent	Excellent		0.05	Gears, bearings, tracks, bushings, roller-skate wheels, chute liners
Nylon	55–83	8–12	60–200	1.2–2.9	0.18–0.42	1.4–4.5	1.0–3.3	114–120 R	Poor	Good	0.04–0.13		Structural, mechanical components such as gears, fan blades, washing-machine agitator, valve, pump, impeller, pistons, and cams
Phenolic (general purpose)	45–48	6.5–7.0		7.6–9.0	1.1–1.3	0.4–0.5	0.30–0.35	70–95 E					Wall plates, industrial switch gears, handles for appliances, housing for vacuum cleaners, automatic transparent rings, housing for thermostats, small motors, small tools, communication instruments, components for aircraft and computers, used as synthetic rubber for tires
Phenylene oxide	48–123	7.0–17.3	4–60	2.4–6.4	0.35–0.93	2.7–6.8	2–5	115–119 R 106–108 L	Good	Fair			Small housing for power tools, pumps, small appliances, hollow shapes for telephones, flashlight cases, helmets, TV cabinets, cable protective cover, bush-bar sleeves, scrubber-vane mist eliminators
Polycarborate	55–110	8–16	10–125	2.3–5.9	0.34–0.86	2.7–21.7	2–16	62–91 M	Excellent	Fair	0.52	0.39	Mechanical gears, pistons, rollers, pump impellers, fan blades, rotor housings for pumps, power tools, phone cases, transparent parts for safety glasses, lenses and snowmobile windshields, refrigerator shelves, and flashlight cases

1.54

TABLE 1-39
Properties of some thermoplastics (*Cont.*)

Name of plastic	Tensile strength, σ_{Sut} MPa	kpsi	Elongation in 50 mm (2 in), %	Modulus of elasticity, E GPa	Mpsi	Izod impact strength J	ft-lbf	Hardness, Rockwell	Resistance to Heat	Chemical	Coefficient of friction, μ With plastic	With steel	Application
Polyimide	25–345	3.6–50	<1			0.3–23	0.25–17	88–120 M	Excellent	Excellent			Molded polyimides are used in jet-engine vane bushings, high load bearings for business machines and computer printout terminals, gear pump gaskets, hydraulic valve seals, multilayer printed-circuit boards, tubes for oil-well exploration.
Polyester	55–159	8–23	1–300	1.9–11.7	0.28–1.7	0.7–2.6	0.5–1.9	65–100 M	Excellent	Poor	0.12–0.22	0.12–0.13	Flashlight and phone cases, housings for pumps, power tools and other appliances, gears, bushings, bearings, tracks, roller-skate wheels and chute liners
Polyethylene	4–38	0.6–5.5	20–1000	0.1–1.2	0.014–0.18	0.7–27.1	0.5–20	10–65*R*					Decorative knobs, automobile steering wheels, eyeglass frames, tool handles, camera cases, phone and flashlight cases, housings for pump and power tools
Polypropylene	34–100	5–14.5	10–500	0.7–6.2	0.1–0.9	0.7–3.0	0.5–2.2	50–110*R*					Mechanical cams, pistons, washing machine agitators, fan blades, valves, pump impellers, gears, bushings, chute liners, bearings, tracks, wear strips and other wear-resisting parts
Polysulfone	70	10.2	50–100	25	0.36	1.8	1.3	120*R*	Excellent	Excellent			Pipe fittings, battery cases, knobs, camera and handle cases, trim moldings, eyeglass frames, tool handles, housings for pumps, power tools, phone cases, transparent parts, safety glasses, lenses, snowmobile windshields, signs, refrigerator shelves, and vandal-resistant glazing

Source: Machine Design, 1981 Materials Reference Issue, Penton/IPC, Cleveland, Ohio, Vol. 53, No. 6 (March 19, 1981).

TABLE 1-40
Properties of some thermosets

Name	Tensile strength, σ_{St} MPa	kpsi	Modulus of elasticity, E GPa	Mpsi	Hardness, Rockwell	Elongation in 50 mm (2 in), %	Impact strength Izod J	ft-lbf	Resistance Heat	Chemical	Application
Alkyd	21–66	3–9.5	2–21	0.3–3.0	98E–99 M		0.5–14	0.3–10	Good	Fair	Military switch gear, electrical terminal strips, and relay housings and bases, automotive ignition parts, radio and TV components, switch gear, and small-appliance housings
Allylic	28–69	4–10			103–120 M		0.3–16	0.2–12	Excellent	Excellent	Switch gear and TV components, insulators, circuit boards, and housings, tubing and aircraft parts, copper-clad laminate for high-performance printed-circuit boards
Amino	34–69	5–10	9–16	1.3–2.4	110–120 M	0.3–0.9	0.4–24	0.27–18	Excellent	Excellent	Electrical wiring devices and switch housings, toaster and other appliance bases, push buttons, knobs, piano keys and camera parts, dinnerware, utensile handles, food-service trays, housing for electric shavers and mixers, metal blocks, connector plugs, automotive and aircraft ignition parts, coil forms, used as baking enamel coatings, particle-board binders, paper and textile treatment materials
Epoxy	28–138	4–20	2.5–21	0.35–3.04	80–120 M	1–10	0.3–41	0.2–30	Excellent	Excellent	Filament wound structures, aircraft pressure bottles, oil storage tank, used with various reinforcements, glass fibers, asbestos, cotton, synthetic fibers, and metallic foils, imprinted circuits, graphite and carbon-fiber-reinforced laminates are used for radomes, pressure vessels, and aircraft components requiring high modulus and light weight, potting and encapsulating electrical and electronic components ranging from miniature coils and switches to large motors and generators
Phenolics	34–62	5–9	7–17	1.0–2.5	70–95 E		0.4–1.5	0.26–1.05	Excellent	Good	Handles for appliances, automotive power-brake systems and industrial terminal strips, industrial switch gear, housing for vacuum cleaners, handles for pots and pans, automotive transmission rings, and electrical components, thermostat housings, housing for small motors and heavy-duty electrical components, small power tools, electrical components for aircraft and computers, pump housing, synthetic rubber for tires and other mechanical rubber goods, dry ingredients for brake linings, clutch facings and other friction products
Silicones	3–45	0.4–6.5			80–90 M	15	0.5–14	0.3–10	Excellent	Excellent	Refrigerator equipment, used as a washing, sealant, laminating parts, injection mold silicon rubber

Source: *Machine Design*, 1981 Materials Reference Issue, Penton/IPC, Cleveland Ohio, Vol. 53, No. 6 (March 19, 1981).

TABLE 1-41
Optical and mechanical properties of photoelastic materials

Material	Elastic limit, σ_{Se} MPa	kpsi	Tensile strength, σ_{St} MPa	kpsi	Young's modulus, E GPa	kpsi	Poisson's ratio, ν	Stress fringe value, f_σ kN/m fri	lbf/in fri	Strain fringe value, f_ε μm/fri	μin/fri	Figure of merit, $Q = E/f_\sigma$ fri/m	fri/in	$S = \dfrac{\sigma_e}{f_\sigma}$ fri/m	fri/in	Remarks
Glass	60.0	8.68	69.0	10.0	69.0	10,000	0.20	304.724–423.812	1740–2420	4.83	191	226,000–163,000	5747–4132	1970–1415	5–3.5	Low optical sensitivity; rarely used.
Cataline (61-893)	38.0–62.0	5.55–9.0	88.2–117.2	12.5–17.0	4.2–4.3	615.0–628.0	0.365	15.236	87	4.83	191	27,600–280,000	7069–7218	2500–4070	63.8–100	Used for 2-dimensional (2-D) and 3-dimensional (3-D) models; susceptible for time edge effect
Methyl Methacrylate (unplasticized)			48.2	7.0	2.8	400.0	0.38	154.113	880	91.00	3582	18,200	455			Low optical sensitivity
Polystyrene	27.6	4.0						54.290	310			56,000	1428	643	16.3	Free from time edge effect; used for photoplasticity
Cellulose nitrate			51.7	7.55	2.4	350.0	0.33	42.907	245	25.40	1000	135,000	3423			
Castolite			56.5	8.25	4.3	623.0	0.35	31.873	182	1.00	39	264,000	6750			
Kriston			48.0–41.4	7.0–6.0	3.7	540.0		14.010	80	4.83	191					
CR-39 (Columbia Resin)	20.6	3.0			1.7–2.6	250.0–380.0	0.42	14.623–17.338	83.5–99	11.90–12.50	468–492	116,000–150,000	2994–3838	1408–1188	36.0–30.0	Used for 2-D models; free from time edge effect
Epoxy Resin: Araldite CN-501	28.3 at 299 K	4.12 at 77°F			3.10 at 298 K; 3.10 at 433 K	452.0 at 77°F; 445.0 at 320°F		10.770 at 298 K	61.5 at 77°F	4.57 at 298 K	180 at 77°F	290,000	7350	2628	67	Used for 2-D and 3-D models
Araldite 6020 at 299.7 K (80°F)					3.10	445.0	0.35	10.157	58.0	4.57	180	305,000	7672			Used most commonly for 2-D and 3-D models
Araldite 6020 at 277.4 K (40°F)								10.595	60.5							Used for 2-D and 3-D models
Araldite B			68.9	10.0	3.2 at 294 K	460 at 70°F	0.362	10.332	59.0	4.83	191	310,000	7796			Used most commonly for 2-D and 3-D models
Bakelite ERL 2774 (50 phthalic anhydride)	55.2	8.0	82.7; 1.38	12.0; 0.20	3.30; 0.036 at 433 K	478.5; 5.2 at 320°F	0.38	10.30; 0.435 at 433 K	58.75; 2.5 at 320°F	4.3	170	320,000; 83,000	8145; 2080	5360	136	Used for 2-D and 3-D models
Hysole 4290: at 296.9 K (75°F)	48.3	7.0			3.45	500.0	0.34	10.508	60.0			328,000	8333	4596	16	Used for 2-D and 3-D models
at 405.2 K (270°F)					0.014	2.0	0.50	0.245	1.4			57,000	1428			
Aramstrong C-6	22.4 at 296.3 K	3.25 at 74°F			3.3 at 294 K	475 at 70°F		13.222 at 293.6 K	75.5 at 70°F	5.59 at 293.6 K	220 at 70°F	290,000	6291	1694	43	Used for 2-D and 3-D models
Polycarbonate	34.5	5.0			2.55	370.0	0.38	7.355	42.0	4.00	157	347,000	8810	4690	119	Little optical and mechanical creep
Marblette (annealed 72 h at 356 K (181°F))					3.79–4.13	550.0–600.0	0.41	14.535	83.0	5.33	210	261,000	6626			Good stress-optical relationship; susceptible for time edge effect
Marblette (phenoformaldehyde)	18.9	2.70	31.0	4.50	1.65	240.0	0.40	9.982	57.0	7.87	310	165,000	4210	893	47.4	Good stress-optical relationship; susceptible for time edge effect
Catalin 800	6.9	1.0	46.2	6.70	1.72	250.0	0.38	10.087	57.6	5.84	230	170,000	4340	684	17.4	
Natural rubber Hard					75.8×10^{-6}	11×10^{-3}	0.50	1.752	10.0							
Soft								0.289	1.7	431.80	17000					
Urethane rubber Hysole 8705 at 296.9 K (75°F)	0.17	2.85×10^{-2}			0.003	0.425	0.467	0.084	0.5	40.60	1598	35,700	850			Used for preparation of models in stress wave propagation and models of dam
Hysole 4485					0.003–0.004	0.425–0.625	0.46	0.158	0.9	82.00	3228	19,000–25,000	4722–694	1076	31.6	
Gelatin (15% gelatin, 25% glycerene and 60% water)							0.50	0.025	0.14	483.00	19016	3032	78			Great optical sensitivity; used for model study of body forces

Sources: K. Lingaiah, Machine Design Data Handbook, Vol. I (SI and Customary Metric Units), Suma Publishers, Bangalore, India, 1986 and K. Lingaiah and B. R. Narayana Iyengar, Machine Design Data Handbook, Vol. I (SI and Customary Metric Units), Suma Publishers, Bangalore, India, 1986.

REFERENCES

1. Datsko, J., *Material Properties and Manufacturing Process*, John Wiley and Sons, New York, 1966.
2. Datsko, J. *Material in Design and Manufacturing*, Malloy, Ann Arbor, Michigan, 1977.
3. *ASM Metals Handbook*, American Society for Metals, Metals Park, Ohio, 1988.
4. Machine Design, 1981 Materials Reference Issue, Penton/IPC, Cleveland, Ohio, Vol. 53, No. 6, March 19, 1981.
5. Lingaiah, K., Machine Design Data Handbook, Vol. II (SI and Customary Metric Units), Suma Publishers, Bangalore, India, 1986.
6. Lingaiah, K., and B. R. Narayana Iyengar, Machine Design Data Handbook, Vol. I (SI and Customary Metric Units), Suma Publishers, Bangalore, India, 1986.
7. Technical Editor Speaks, the International Nickel Company, New York, 1943.
8. Shigley, J. E., *Mechanical Engineering Design*, Metric Edition, McGraw-Hill Book Company, New York, 1986.
9. Deutschman, A. D., W. J. Michels, and C. E. Wilson, *Machine Design—Theory and Practice*, Macmillan Publishing Company, New York, 1975.
10. Juvinall, R. C., *Fundaments of Machine Components Design*, John Wiley and Sons, New York, 1983.
11. Lingaiah, K., and B. R. Narayana Iyengar, *Machine Design Data Handbook*, Engineering College Co-operative Society, Bangalore, India, 1962.
12. Lingaiah, K., *Machine Design Data Handbook*, Vol. II (*SI and Customary Metric Units*), Suma Publishers, Bangalore, India, 1986.
13. Lingaiah, K., and B. R. Narayana Iyengar, *Machine Design Data Handbook*, Vol. I (*SI and Customary Metric Units*), Suma Publishers, Bangalore, India, 1986.
14. *SAE Handbook* 1981.
15. Lessels, J. M., *Strength and Resistance of Metals*, John Wiley and Sons, New York, 1954.
16. Siegel, M. J., V. L. Maleev and J. B. Hartman, *Mechanical Design of Machines*, 4th ed., International Textbook Company, Scranton, Pennsylvania, 1965.
17. Black, P. H., and O. Eugene Adams, Jr., *Machine Design*, McGraw-Hill Book Company, New York, 1983.
18. Niemann, G., *Maschinenelemente*, Springer-Verlag, Berlin, Erster Band, 1963.
19. Faires, V. M., *Design of Machine Elements*, 4th ed., Macmillan Company, New York, 1965.
20. Norman, C. A., E. S. Ault, and I. F. Zarobsky, *Fundamentals of Machine Design*, the Macmillan Company, New York, 1951.
21. Spotts, M. F., *Design of Machine Elements*, 5th ed., Prentice-Hall of India Private Ltd., New Delhi, 1978.
22. Vallance, A., and V. L. Doughtie, *Design of Machine Members*, McGraw-Hill Book Company, New York, 1951.
23. Karl-Heinz Decker, *Maschinenelemente, Gestalting and Bereching*, Carl Hanser Verlag, Munich, Germany, 1971.
24. Decker, K. H., and Banrat Karlheinz Kabus, *Maschinenelemente-Aufgaben*, Carl Hanser Verlag, Munich, Germany, 1970.
25. ISO and BIS standards.

BIBLIOGRAPHY

Black, P. H., and O. Eugene Adams, Jr., *Machine Design,* McGraw-Hill Book Company, New York, 1983.
Decker, K. H., *Maschinenelemente, Gestalting and Bereching*, Carl Hanser Verlag, Munich, Germany, 1971.
Decker, K. H., and Banrat Karlheinz Kabus, *Maschinenelemente-Aufgaben*, Carl Hanser Verlag, Munich, Germany, 1970.
Deutschman, A. D., W. J. Michels, and C. E. Wilson, *Machine Design Theory and Practice*, Macmillan Publishing Company, New York, 1975.
Faires, V. M., *Design of Machine Elements*, 4th ed., Macmillan Company, New York, 1965.
Horger, O. S. (ed.), *(ASME) Handbook for Metals Properties*, McGraw-Hill Book Company, 1954.
ISO standards
Juvinall, R. C., *Fundaments of Machine Components Design*, John Wiley and Sons, New York, 1983.
Lessels, J. M., *Strength and Resistance of Metals*, John Wiley and Sons, New York, 1954.

Lingaiah, K., and B. R. Narayana Iyengar, *Machine Design Data Handbook*, Engineering College Co-operative Society, Bangalore, India, 1962.

Mark's Standard Handbook for Mechanical Engineers, 8th ed, McGraw-Hill Book Company, New York, 1978.

Niemann, G., *Maschinenelemente*, Springer-Verlag, Berlin, Erster Band, 1963.

Norman, C. A., E. S. Ault, and I. E. Zarobsky, *Fundamentals of Machine Design*, the Macmillan Company, New York, 1951.

SAE Handbook, 1981.

Shigley, J. E., *Mechanical Engineering Design*, Metric Edition, McGraw-Hill Book Company, New York, 1986.

Siegel, M. J., V. L. Maleev, and J. B. Hartman, *Mechanical Design of Machines*, 4th ed., International Textbook Company, Scranton, Pennsylvania, 1965.

Spotts, M. F., *Design of Machine Elements*, 5th ed., Prentice-Hall of India Private Ltd., New Delhi, 1978.

Vallance, A., and V. L. Doughtie, *Design of Machine Members*, McGraw-Hill Book Company, New York, 1951.

CHAPTER
2

STATIC STRESSES IN MACHINE ELEMENTS

SYMBOLS

A	area of cross section, $m^2 (in^2)$
A_w	area of web, m^2 (in^2)
a	constant in Rankine's formula
b	radius of area of contact, m (in)
	bandwidth of contact, m (in)
	width of beam, m (in)
c	distance from neutral surface to extreme fiber, m (in)
D	diameter of shaft, m (in)
C_1	constant in straight-line formula
F	load, kN (lbf)
F_c	compressive force, kN (lbf)
F_t	tensile force, kN (lbf)
F_τ	shear force, kN (lbf)
F_{cr}	crushing load, kN (lbf)
e	deformation, total, m (in)
	eccentricity, as of force equilibrium, m (in)
	unit volume change or volumetric strain
e_t	thermal expansion, m (in)
E	modulus of elasticity, direct (tension or compression), GPa (Mpsi)
E_c	combined or equivalent modulus of elasticity in case of composite bars, GPa (Mpsi)
G	modulus of rigidity, GPa (Mpsi)
M_b	bending moment, N m (lbf ft)
M_t	torque, torsional moment, N m (lbf ft)
i	number of turns
I	moment of inertia, area , m^4 or cm^4 (in^4)
	mass moment of inertia, $N s^2 m$ (lbf s^2 ft)
I_{xx}, I_{yy}	moment of inertia of cross-sectional area around the respective principal axes, m^4 or cm^4 (in^4)
J	moment of inertia, polar, m^4 or cm^4 (in^4)
k	radius of gyration, m (in)
k_o	polar radius of gyration, m (in)
k_t	torsional spring constant, J/rad or N m/rad (lbf in/rad)

l	length, m (in)
l_o	length of rod, m (in)
L	length, m (in)
n	speed, rpm (revolutions per minute)
	coefficient of end condition
n'	speed, rps (revolutions per second)
l, m, n	direction cosines (also with subscripts)
P	power, kW (hp)
	pitch or threads per meter
T	temperature, °C (°F)
ΔT	temperature difference, °C (°F)
r	radius of the rod subjected to torsion, m (in)
q	shear flow
Q	first moment of the cross-sectional area outside the section at which the shear flow is required
v	velocity, m/s (ft/min or fpm)
V	volume, m³ (in³)
	shear force, kN (lbf)
ΔV	volume change, m³ (in³)
Z	section modulus, m³ (in³)
α	deformation of contact surfaces, m (in)
	coefficient of linear expansion, m/m/K or m/m/°C (in/in/°F)
γ	shearing strain, rad/rad
$\gamma_{xy}, \gamma_{yz}, \gamma_{zx}$	shearing strain components in xyz coordinates, rad/rad
δ	deformation or elongation, m(in)
ϵ	strain, μm/m (μin/in)
ϵ_T	thermal strain, μm/m (μin/in)
$\epsilon_x, \epsilon_y, \epsilon_z$	strains in x, y, and z directions, μm/m (μin/in)
θ	angular distortion, rad
	angle, deg
	angular twist, rad (deg)
	angle made by normal to plane nn with the x axis, deg
κ	bulk modulus of elasticity, GPa (Mpsi)
ν	Poisson's ratio
ρ	radius of curvature, m (in)
σ	stress, direct or normal, tensile or compressive (also with subscripts), MPa (psi)
σ_b	bearing pressure, MPa (psi)
	bending stress, MPa (psi)
σ_c	compressive stress (also with subscripts), MPa (psi)
	hydrostatic pressure, MPa (psi)
σ_{Sc}	compressive strength, MPa (psi)
σ_{cr}	stress at crushing load, MPa (psi)
σ_e	elastic limit, MPa (psi)
σ_s	strength, MPa (psi)
σ_t	tensile stress, MPa (psi)
σ_{St}	tensile strength, MPa (psi)
$\sigma_x, \sigma_y, \sigma_z$	stress in x, y, and z directions, MPa (psi)
$\sigma_1, \sigma_2, \sigma_3$	principal stresses, MPa (psi)
σ_y	yield stress, MPa (psi)
σ_{Sy}	yield strength, MPa (psi)
σ_u	ultimate stress, MPa (psi)
σ_{Su}	ultimate strength, MPa (psi)
σ'	principal direct stress, MPa (psi)

σ''	normal stress which will produce the maximum strain, MPa (psi)
σ_θ	normal stress on the plane nn at any angle θ to x axis, MPa (psi)
τ	shear stress (also with subscripts), MPa (psi)
τ_s	shear strength, MPa (psi)
$\tau_{xy}, \tau_{yz}, \tau_{zx}$	shear stresses in xy, yz, and zx planes, respectively, MPa (psi)
τ_θ	shear stress on the plane at any angle θ with x axis, MPa (psi)
ω	angular speed, rad/s

Other factors in performance or in special aspects are included from time to time in this chapter and, being applicable only in their immediate context, are not given at this stage.

(*Note*: σ and τ with initial subscript S designates strength properties of material used in the design which will be used and observed throughout this *Machine Design Data Handbook*.)

Particular	Formula

SIMPLE STRESS AND STRAIN

The stress in simple tension or compression (Fig. 2–1a, 2–1b)

$$\sigma_t = \frac{F_t}{A}; \; \sigma_c = \frac{F_c}{A} \qquad (2\text{--}1)$$

The total elongation of a member of length l

$$\delta = \frac{Fl}{AE} \qquad (2\text{--}2)$$

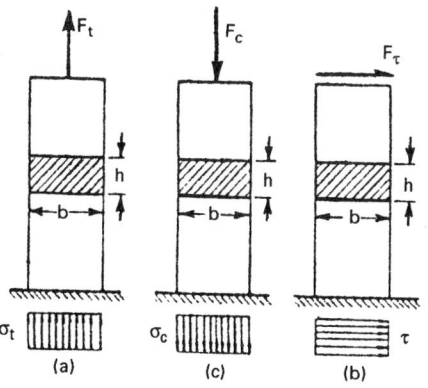

FIGURE 2–1

(Fig. 2–2a)

Strain, deformation per unit length

$$\epsilon = \frac{\delta}{l} = \frac{\sigma}{E} \qquad (2\text{--}3)$$

Particular	Formula

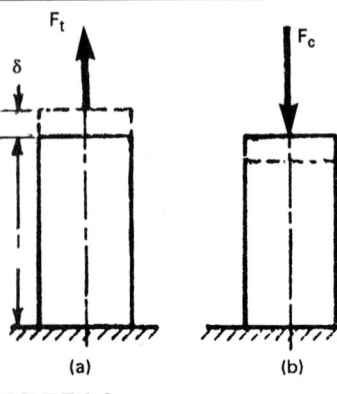 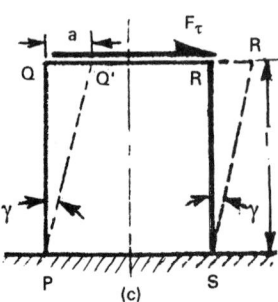

(a)   (b)   (c)

FIGURE 2–2

Particular	Formula	
Young's modulus or modulus of elasticity	$E = \dfrac{\sigma}{\epsilon}$	(2–4)
The shear stress (Fig. 2–1c)	$\tau = \dfrac{F_\tau}{A}$	(2–5)
Shear deformation due to torsion	$\theta = \dfrac{\tau L}{\rho G}$	(2–6)
Shear strain (Fig. 2–2c)	$\gamma = \dfrac{\tau}{G} = \dfrac{a}{l}$	(2–7)
The shear modulus or modulus of rigidity from Eq. (2–7)	$G = \dfrac{\tau}{\gamma}$	(2–8)
Poisson's ratio	$\nu = \text{lateral strain}/\text{axial strain} = \dfrac{\epsilon_t}{\epsilon_a}$	(2–9)
Poisson's ratio may be computed with sufficient accuracy from the relation	$\nu = \dfrac{E}{2G} - 1$	(2–10)
The shear or torsional modulus or modulus of rigidity is also obtained from Eq. (2–10)	$G = \dfrac{E}{2(1 + \nu)}$	(2–11)
The bearing stress (Fig. 2–3c)	$\sigma_b = \dfrac{F}{bd_2}$	(2–12)

STRESSES

Unidirectional stress (Fig. 2–4)

The normal stress on the plane at any angle θ with x axis	$\sigma_\theta = \sigma_x \cos^2 \theta$	(2–13)

Particular	Formula

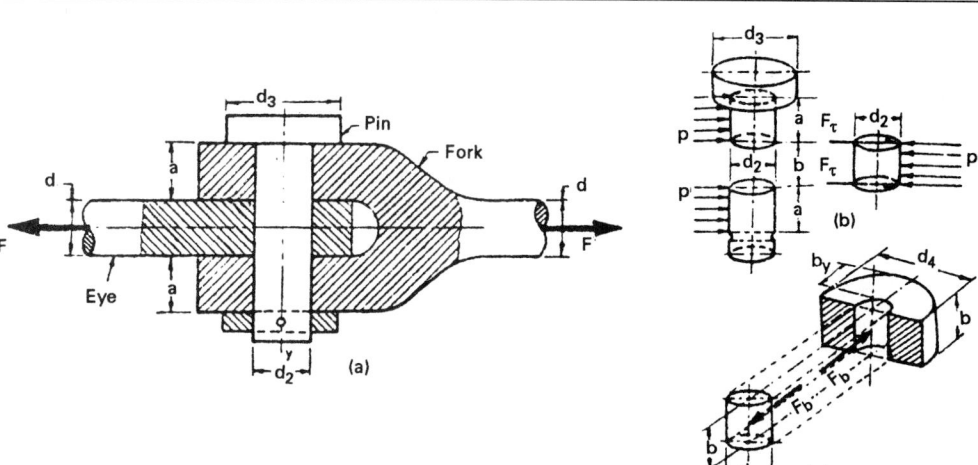

FIGURE 2–3   Knuckle joint for round rods.

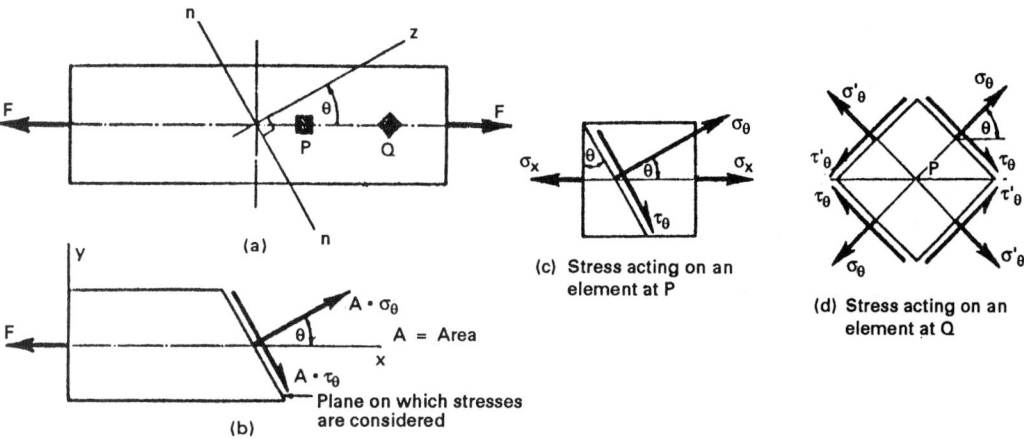

FIGURE 2–4   A bar in uniaxial tension.

The shear stress on the plane at any angle θ with x axis	$\tau_\theta = \dfrac{\sigma_x}{2} \sin 2\theta$	(2–14)
Principal stresses	$\sigma_1 = \sigma_x$ and $\sigma_2 = 0$	(2–15)
Angles at which principal stresses act	$\theta_1 = 0°$ and $\theta_2 = 90°$	(2–16)
Maximum shear stress	$\tau_{max} = \dfrac{\sigma_x}{2}$	(2–17)
Angles at which maximum shear stresses act	$\theta_1 = 45°$ and $\theta_2 = 135°$	(2–18)

Particular	Formula	
The normal stress on the plane at an angle $\theta + (\pi/2)$ (Fig. 2–4d)	$\sigma_\theta' = \sigma_x \cos^2\left(\theta + \dfrac{\pi}{2}\right) = \sigma_x \cos^2\theta$	(2–19)
The shear stress on the plane at an angle $\theta + (\pi/2)$ (Fig. 2–4d)	$\tau_\theta' = \sigma_x \sin\left(\theta + \dfrac{\pi}{2}\right)\cos\left(\theta + \dfrac{\pi}{2}\right) = \tfrac{1}{2}\,\sigma_x \sin 2\theta$	(2–20)
Therefore from Eqs. (2–13) and (2–19), (2–14), and (2–20)	$\sigma_\theta = \sigma_\theta'$ and $\tau_\theta = -\tau_\theta'$	(2–21)

PURE SHEAR (FIG. 2–5)

Particular	Formula	
The normal stress on the plane at any angle θ	$\sigma_\theta = \tau_{xy} \sin 2\theta$	(2–22)
The shear stress on the plane at any angle θ	$\tau_\theta = \tau_{xy} \cos 2\theta$	(2–23)
The principal stress	$\sigma_1 = \tau_{xy}$ and $\sigma_2 = -\tau_{xy}$	(2–24)
Angles at which principal stresses act	$\theta_1 = 45°$ and $\theta_2 = 135°$	(2–25)
Maximum shear stresses	$\tau_{max} = \tau_{xy} = \sigma$	(2–26)
Angles at which maximum shear stress act	$\theta_1 = 0$ and $\theta_2 = 90°$	(2–27)

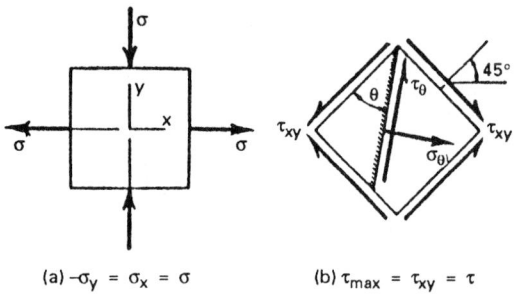

(a) $-\sigma_y = \sigma_x = \sigma$ (b) $\tau_{max} = \tau_{xy} = \tau$

FIGURE 2–5   An element in pure shear.

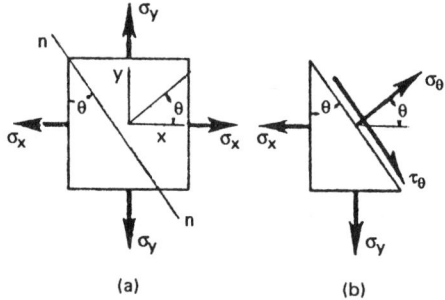

(a)   (b)

FIGURE 2–6   An element in biaxial tension.

BIAXIAL STRESSES (FIG. 2–6)

Particular	Formula	
The normal stress on the plane at any angle θ	$\sigma_\theta = \dfrac{\sigma_x + \sigma_y}{2} + \dfrac{\sigma_x - \sigma_y}{2}\cos 2\theta$	(2–28)
The shear stress on the plane at any angle θ	$\tau_\theta = \dfrac{\sigma_x - \sigma_y}{2}\sin 2\theta$	(2–29)
The shear stress τ_θ at $\theta = 0$	$\tau_\theta = 0$	(2–30)
The shear stress τ_θ at $\theta = 45°$	$\tau_{max} = (\sigma_x - \sigma_y)/2$	(2–31)

Particular	Formula

BIAXIAL STRESSES COMBINED WITH SHEAR (FIG. 2–7)

The normal stress on the plane at any angle θ

$$\sigma_\theta = \frac{\sigma_x + \sigma_y}{2} + \frac{\sigma_x - \sigma_y}{2} \cos 2\theta + \tau_{xy} \sin 2\theta \qquad (2\text{--}32)$$

The shear stress in the plane at any angle θ

$$\tau_\theta = \frac{\sigma_x - \sigma_y}{2} \sin 2\theta + \tau_{xy} \cos 2\theta \qquad (2\text{--}33)$$

The maximum principal stress

$$\sigma_1 = \frac{\sigma_x + \sigma_y}{2} + \left[\left(\frac{\sigma_x - \sigma_y}{2} \right)^2 + \tau_{xy}^2 \right]^{1/2} \qquad (2\text{--}34)$$

The minimum principal stress

$$\sigma_2 = \frac{\sigma_x + \sigma_y}{2} - \left[\left(\frac{\sigma_x - \sigma_y}{2} \right)^2 + \tau_{xy}^2 \right]^{1/2} \qquad (2\text{--}35)$$

Angles at which principal stresses act

$$\theta_{1,2} = \tfrac{1}{2} \arctan \frac{2\tau_{xy}}{\sigma_x - \sigma_y} \qquad (2\text{--}36)$$

where θ_1 and θ_2 are 180° apart

Maximum shear stress

$$\tau_{max} = \left[\left(\frac{\sigma_x - \sigma_y}{2} \right)^2 + \tau_{xy}^2 \right]^{1/2} \qquad (2\text{--}37)$$

Angles at which maximum shear stress acts

$$\theta = \tfrac{1}{2} \arctan \frac{\sigma_x - \sigma_y}{2\tau_{xy}} \qquad (2\text{--}38)$$

The equation for the inclination of the principal planes in terms of the principal stress (Fig. 2–8)

$$\tan \theta = \frac{\sigma_1 - \sigma_x}{\tau_{xy}} \qquad (2\text{--}39)$$

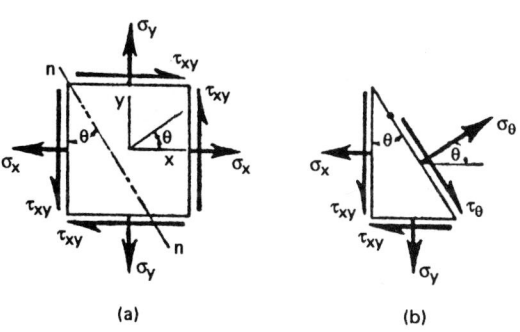

FIGURE 2–7 An element in plane state of stress.

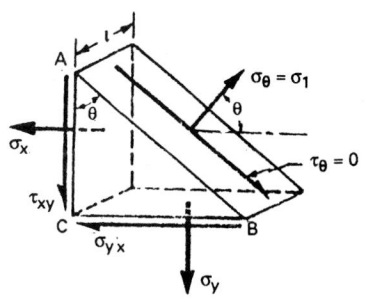

FIGURE 2–8

Particular	Formula

MOHR'S CIRCLE

Biaxial field combined with shear (Fig. 2–9)

Maximum principal stress σ_1	σ_1 is the abscissa of point F
Minimum principal stress σ_2	σ_2 is the abscissa of point G
Maximum shear stress τ_{max}	τ_{max} is the ordinate of point H

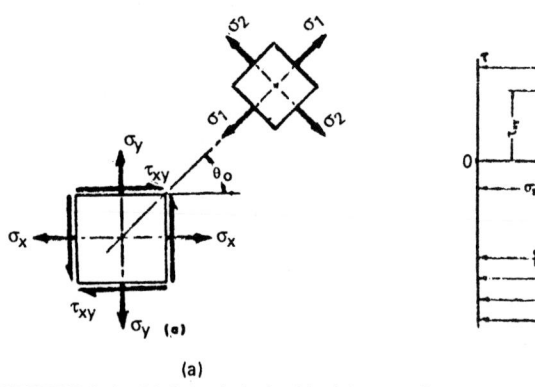

(a) (b)

FIGURE 2–9 Mohr's circle for biaxial state of stress.

TRIAXIAL STRESS (FIGS. 2–10 AND 2–11)

The normal stress on a plane nn, whose direction cosines are l, m, n	$\sigma_\theta = \sigma_x l^2 + \sigma_y m^2 + \sigma_z n^2$	(2-40)
The shear stress on a plane normal nn, whose direction cosines are l, m, n	$\tau_\theta = \sqrt{\sigma_x^2 l^2 + \sigma_y^2 m^2 + \sigma_z^2 n^2}$	(2-41)
The principal stresses	$\sigma_{1,2,3} = \sigma_x, \sigma_y, \sigma_z$	(2-42)
The cubic equation for general state of stress in three dimensions from the theory of elasticity	$\sigma^3 - (\sigma_x + \sigma_y + \sigma_z)\sigma^2 + (\sigma_x\sigma_y + \sigma_y\sigma_z + \sigma_z\sigma_x$ $-\tau_{xy}^2 - \tau_{yz}^2 - \tau_{zx}^2)\sigma$ $-(\sigma_x\sigma_y\sigma_z + 2\tau_{xy}\tau_{yz}\tau_{zx} - \sigma_x\tau_{zy}^2 - \sigma_y\tau_{zx}^2$ $-\sigma_z\tau_{xy}^2) = 0$	(2-43)

The three roots of this cubic equation give the magnitude of the principal stresses σ_1, σ_2, and σ_3.

The maximum shear stresses on planes parallel to x, y, and z which are designated as	$(\tau_{max})_1 = \dfrac{\sigma_2 - \sigma_3}{2}; (\tau_{max})_2 = \dfrac{\sigma_1 - \sigma_3}{2};$ $(\tau_{max})_3 = \dfrac{\sigma_1 - \sigma_2}{2}$	(2-44)

Particular	Formula

MOHR'S CIRCLE

Triaxial field (Figs. 2–10 and 2–11)

Normal stress at point (Figs. 2–11*b*) on one octahedral plane

$$\sigma_t = \tfrac{1}{3}(\sigma_1 + \sigma_2 + \sigma_3) = \tfrac{1}{3}(\sigma_x + \sigma_y + \sigma_z) \qquad (2\text{--}45)$$

or σ_t is the abscissa of point T

Shear stress at point T (Figs. 2–11*b*) on an octahedral plane

$$\tau_t = \tfrac{1}{3}[(\sigma_x - \sigma_y)^2 + (\sigma_y - \sigma_z)^2 + (\sigma_z - \sigma_x)^2 \qquad (2\text{--}46a)$$
$$+ 6(\tau_{xy}^2 + \tau_{yz}^2 + \tau_{zx}^2)]^{1/2}$$

$$= \tfrac{1}{3}\sqrt{[(\sigma_1 - \sigma_2)^2 + (\sigma_2 - \sigma_3)^2 + (\sigma_3 - \sigma_1)^2]}$$

or τ_t is the ordinate of point T

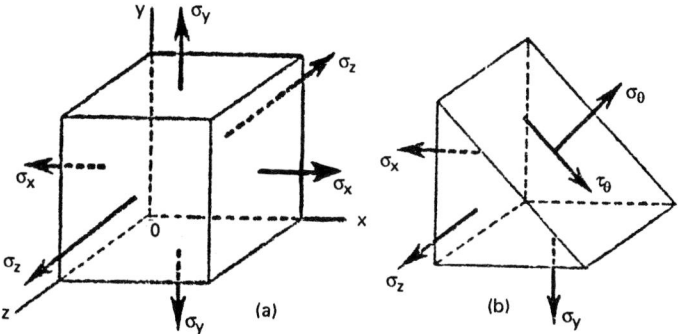

FIGURE 2–10   An element in triaxial state of stress.

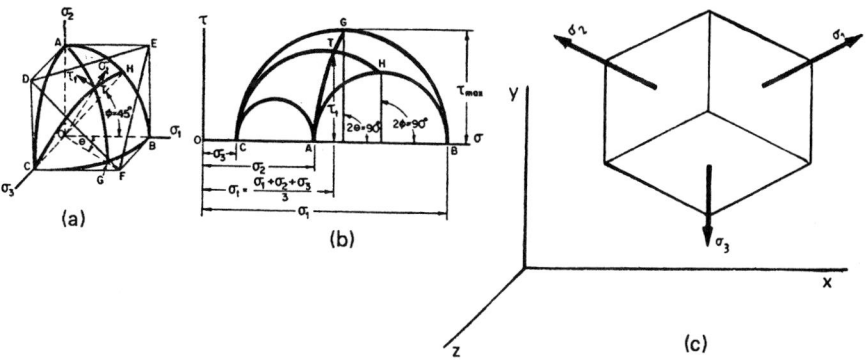

FIGURE 2–11   Mohr's circle for triaxial octahedral stress state.

Particular	Formula

STRESS-STRAIN RELATIONS

Uniaxial field

Strain in principal direction 1

$$\epsilon_1 = \frac{\sigma_1}{E} ; \epsilon_2 = -\nu\frac{\sigma_1}{E} ; \epsilon_3 = -\nu\frac{\sigma_1}{E} \qquad (2\text{-}47)$$

The principal stress

$$\sigma_1 = E\epsilon_1 \qquad (2\text{-}47a)$$

The unit volume change in uniaxial stress

$$\frac{\Delta V}{V} = \frac{(1 - 2\nu)\sigma_1}{E} = \epsilon(1 - 2\nu) \qquad (2\text{-}48)$$

Biaxial field

Strain in principal direction 1

$$\epsilon_1 = \frac{1}{E}(\sigma_1 - \nu\sigma_2) \qquad (2\text{-}49)$$

Strain in principal direction 2

$$\epsilon_2 = \frac{1}{E}(\sigma_2 - \nu\sigma_1) \qquad (2\text{-}50)$$

Strain in principal direction 3

$$\epsilon_3 = -\frac{\nu}{E}(\sigma_1 + \sigma_2) \qquad (2\text{-}51)$$

The principal stresses in terms of principal strains in a biaxial stress field

$$\sigma_1 = \frac{E}{1 - \nu^2}(\epsilon_1 + \nu\epsilon_2) \qquad (2\text{-}52)$$

$$\sigma_2 = \frac{E}{1 - \nu^2}(\epsilon_2 + \nu\epsilon_1) \qquad (2\text{-}53)$$

The unit volume change in biaxial stress

$$\frac{\Delta V}{V} = \frac{(1 - 2\nu)}{E}(\sigma_1 + \sigma_2) \qquad (2\text{-}54)$$

Triaxial field

Strain in principal direction 1

$$\epsilon_1 = \frac{1}{E}[\sigma_1 - \nu(\sigma_2 + \sigma_3)] \qquad (2\text{-}55)$$

Strain in principal direction 2

$$\epsilon_2 = \frac{1}{E}[\sigma_2 - \nu(\sigma_3 + \sigma_1)] \qquad (2\text{-}56)$$

Strain in principal direction 3

$$\epsilon_3 = \frac{1}{E}[\sigma_3 - \nu(\sigma_1 + \sigma_2)] \qquad (2\text{-}57)$$

The principal stresses in terms of principal strains in triaxial stress field

$$\sigma_1 = \frac{E}{(1 - \nu - 2\nu^2)}[(1 - \nu)\epsilon_1 + \nu(\epsilon_2 + \epsilon_3)] \qquad (2\text{-}58)$$

$$\sigma_2 = \frac{E}{(1 - \nu - 2\nu^2)}[(1 - \nu)\epsilon_2 + \nu(\epsilon_3 + \epsilon_1)] \qquad (2\text{-}59)$$

$$\sigma_3 = \frac{E}{(1 - \nu - 2\nu^2)}[(1 - \nu)\epsilon_3 + \nu(\epsilon_1 + \epsilon_2)] \qquad (2\text{-}60)$$

Particular	Formula
The unit volume change or volumetric strain in terms of principal stresses for the general case of triaxial stresss (Fig. 2–12)	$e = \dfrac{dV}{V} = \dfrac{(1-2\nu)}{E}(\sigma_x + \sigma_y + \sigma_z)$ (2–61a)
	$= \dfrac{(1-2\nu)}{E}(\sigma_1 + \sigma_2 + \sigma3)$ (2–61b)

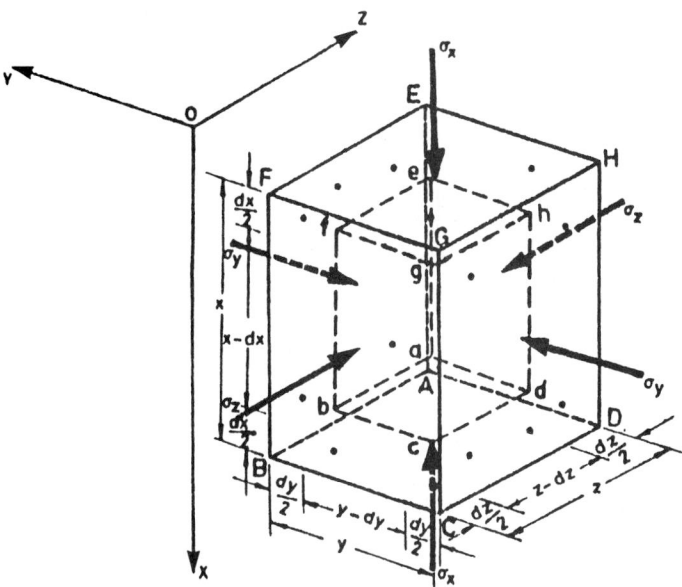

FIGURE 2–12   Uniform hydrostatic pressure.

The volumetric strain due to uniform hydrostatic pressure σ_c acting on an element (Fig. 2–12)	$\dfrac{\Delta V}{V} = \dfrac{-3(1-2\nu)\sigma_c}{E} = -\dfrac{\sigma_c}{\kappa}$ (2–62)
The bulk modulus of elasticity	$\kappa = \dfrac{E}{3(1-2\nu)}$ (2–63)

STATISTICALLY INDETERMINATE MEMBERS (FIG. 2–13)

The reactions at supports of a constant cross-section bar due to load F acting on it as shown in Fig. 2–13	$R_A = \dfrac{FL_b}{L_a + L_b} = \dfrac{FL_b}{L}$ (2–64a)
	$R_B = \dfrac{FL_a}{L_a + L_b} = \dfrac{FL_a}{L}$ (2–64b)
The elongation of left portion L_a of the bar	$\delta_a = \dfrac{R_A L_a}{AE} = \dfrac{FL_a L_b}{LAE}$ (2–65)

Particular	Formula
The shortening the right portion L_b of the bar	$$\delta_b = -\frac{R_A L_a}{AE} = -\frac{FL_a L_b}{LAE} \qquad (2\text{-}66)$$

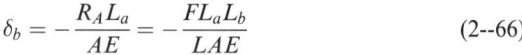

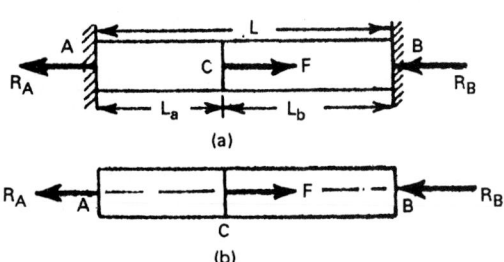

(a)

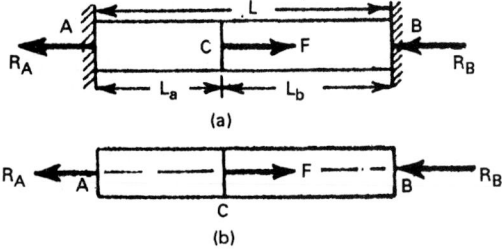

(b)

FIGURE 2–13

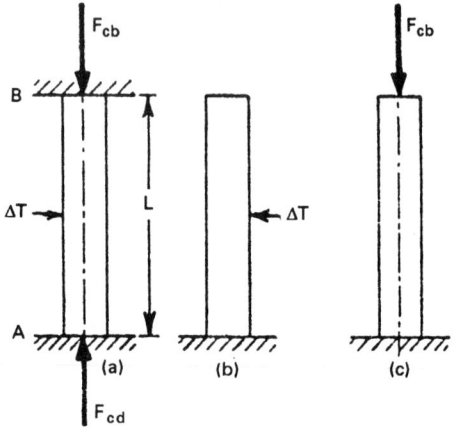

FIGURE 2–14

THERMAL STRESS AND STRAIN

The normal strain due to free expansion of a bar or machine member when it is heated	$$\epsilon_T = \alpha(\Delta T) \qquad (2\text{-}67)$$
The free linear deformation due to temperature change	$$\delta = \alpha L(\Delta T) \qquad (2\text{-}68)$$
The compressive force F_{cb} developed in the bar fixed at both ends due to increase in temperature (Fig. 2–14)	$$F_{cb} = \alpha EA(\Delta T) \qquad (2\text{-}69)$$
The compressive stress induced in the member due to thermal expansion (Fig. 2–14)	$$\sigma_{cT} = \frac{F_{cb}}{A} = -\alpha E(\Delta T) \qquad (2\text{-}70)$$
The relation between the extension of one member to the compression of another member in case of rigidly joined compound bars of the same length L made of different materials subjected to same temperature (Fig. 2–15)	$$\frac{\sigma_s L}{E_S} + \frac{\sigma_c L}{E_c} = (\alpha_c - \alpha_s)L(\Delta T) \qquad (2\text{-}71)$$
The forces acting on each member due to temperature change in the compound bar	$$\sigma_c A_c = \sigma_s A_s \qquad (2\text{-}72)$$
The relation between compression of the tube to the extension of the threaded member due to tightening of the nut on the threaded member (Fig. 2–16)	$$\frac{\sigma_t L}{E_t} + \frac{\sigma_s L}{E_s} = [\text{number of turns }(i) \qquad (2\text{-}73)$$ $$\times \text{ (threads/meter) or pitch }(\boldsymbol{P})] = i\boldsymbol{P}$$
The forces acting on tube and threaded member due to tightening of the nut	$$\sigma_s A_s = \sigma_t A_t \qquad (2\text{-}74)$$

Particular	Formula

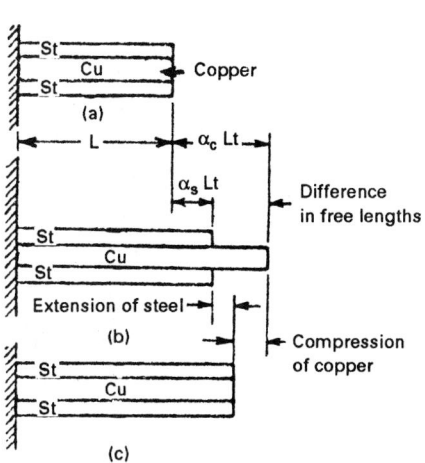

FIGURE 2–15

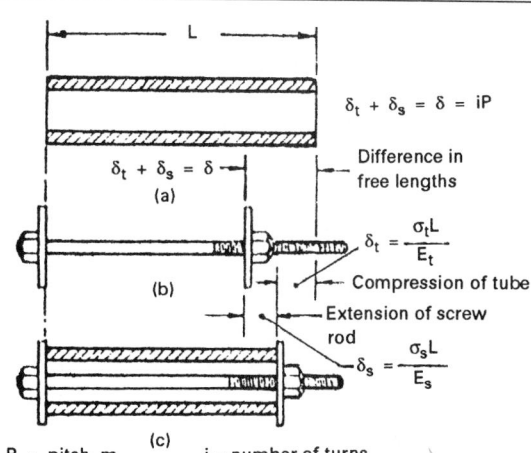

FIGURE 2–16

COMPOUND BARS

The total load in the case of compound bars or columns or wires consisting of i members, each having different length and area of cross section and each made of different material subjected to an external load as shown in Fig. 2–17

$$F = \sum \frac{E_i A_i \delta_i}{L_i} = \delta \sum \frac{E_i A_i}{L_i} \qquad (2\text{–}75)$$

An expression for compression of first bar (Fig. 2–17)

$$\delta_1 = \frac{F}{\Sigma (E_i A_i / L_i)} \qquad (2\text{–}76)$$

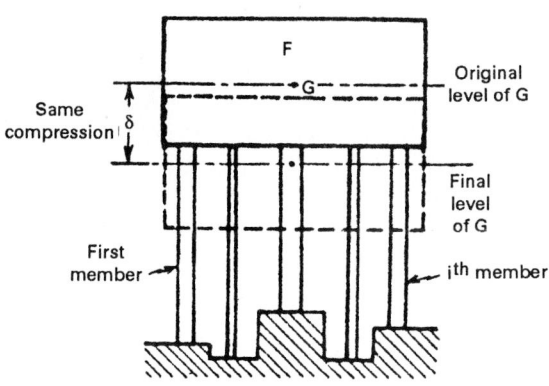

FIGURE 2–17

Particular	Formula

The load on first bar (Fig. 2–17)

$$F_1 = \frac{(E_1 A_1 / L_1)}{\Sigma (EA/L)} F \qquad (2\text{–}77)$$

The load on ith bar (Fig. 2–17)

$$F_i = \frac{E_i A_i \delta_i}{L_i} \qquad (2\text{–}78)$$

EQUIVALENT OR COMBINED MODULUS OF ELASTICITY OF COMPOUND BARS

The equivalent or combined modulus of elasticity of a compound bar consisting of i members, each having a different length and area of cross section and each being made of different material

$$E_c = \frac{E_1 A_1 + E_2 A_2 + E_3 A_3 + ... + E_n A_n}{A_1 + A_2 + A_3 + ... + A_n} \qquad (2\text{–}79a)$$

$$= \frac{\Sigma E_i A_i}{\underset{i=1,2,..n}{\Sigma A_i}} \qquad (2\text{–}79b)$$

The stress in the equivalent bar due to external load F

$$\sigma = \frac{F}{\underset{i=1,2,3...}{\Sigma A_i}} \qquad (2\text{–}80)$$

The strain in the equivalent bar due to external load F

$$\epsilon = \frac{F}{\underset{i=1,2,3...}{E_c \Sigma A_i}} = \frac{\delta}{L} \qquad (2\text{–}81)$$

The common extension or compression due to external load F

$$\delta = \frac{FL}{\underset{i=1,2,3..n}{E_c \Sigma A_i}} = \epsilon L \qquad (2\text{–}82)$$

POWER

The relation between power, torque and speed

$$P = M_t \omega \qquad (2\text{–}83)$$

where M_t in N m (lbf ft), ω in rad/s (rad/min), and P in W (hp)

$$= \frac{M_t n'}{159} \qquad \textbf{SI} \qquad (2\text{–}84a)$$

where M_t in kN m, n' in rps, and P in kW

$$= \frac{M_t n}{9550} \qquad \textbf{SI} \qquad (2\text{–}84b)$$

where M_t in kN m, n in rpm, and P in kW

$$= \frac{M_t n}{63030} \qquad \textbf{US Customary System units}$$
$$(2\text{–}84c)$$

where M_t in lbf in, n rpm, and P in hp

Particular	Formula
Another expression for power in terms of force F acting at velocity ν	$P = \dfrac{F\nu}{1000}$ **SI** (2–85a) where F newtons (N); ν in m/s; and P in kW $= \dfrac{F\nu}{33000}$ **US Customary System units** (2–85b) where F in lbf; ν in fpm (feet per meter), and P in hp (horsepower)

TORSION (FIG. 2–18)

The general equation for torsion (Fig. 2–18)	$\dfrac{M_t}{J} = \dfrac{G\theta}{L} = \dfrac{\tau}{\rho}$ (2–86)
Torque	$M_t = \dfrac{159\,P}{n'}$ **SI** (2–87a) where M_t in kN m, n' in rps, and P in kW $= \dfrac{9550\,P}{n}$ **SI** (2–87b) where M_t in kN m, n in rpm, and P in kW $= \dfrac{63030\,P}{n}$ **US Customary System units** (2–87c) where M_t in lbf in, n in rpm, and P in hp
The maximum shear stress at the maximum radius r of the solid shaft (Fig. 2–18) subjected to torque M_t	$\tau_{max} = \dfrac{16 M_t}{\pi D^3}$ (2–88)
The torsional spring constant	$k_t = \dfrac{M_t}{\theta} = \dfrac{GJ}{L}$ (2–89)

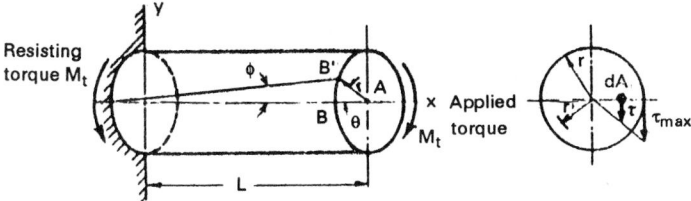

FIGURE 2–18 Cylindrical bar subjected to torque.

Particular	Formula

BENDING (FIG. 2–19)

The general formula for bending (Fig. 2–19)

$$\frac{M_b}{I} = \frac{\sigma_b}{c} = \frac{E}{\rho} \tag{2–90}$$

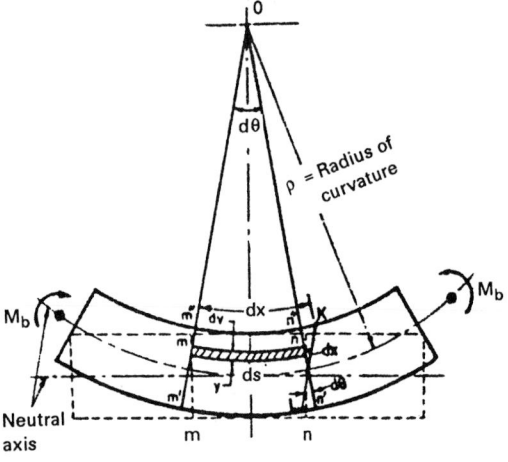

FIGURE 2–19   Bending of beam.

The maximum values of tensile and compressive bending stresses

$$\sigma_b = \frac{M_b c}{I} \tag{2–91}$$

The shear stresses developed in bending of a beam (Fig. 2–20)

$$\tau = \frac{V}{Ib} \int_{y_o}^{c} y\, dA \tag{2–92}$$

The shear flow

$$q = \frac{VQ}{I} \tag{2--93}$$

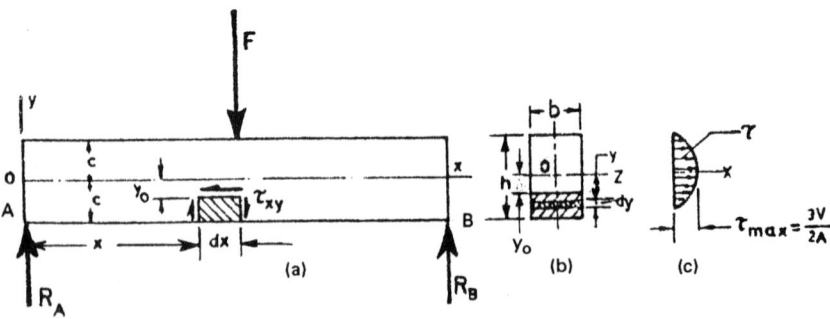

FIGURE 2–20   Beam subjected to shear stress.

Particular	Formula	
The first moment of the cross-sectional area outside the section at which the shear flow is required	$Q = \displaystyle\int_{y_o}^{c} y\, dA$	(2–94)
The maximum shear stress for a rectangular section (Figs. 2–20 and 2–21)	$\tau_{max} = \dfrac{3V}{2A}$	(2–95)

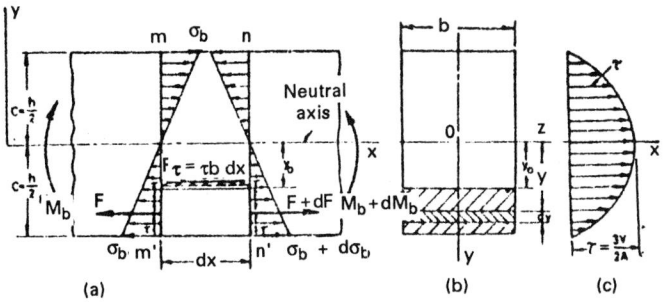

FIGURE 2–21   Element cut out from a beam subjected to shear stress.

For a solid circular section beam, the maximum shear stress	$\tau_{max} = \dfrac{4V}{3A}$	(2–96)
For a hollow circular section beam, the expression for maximum shear stress	$\tau_{max} = \dfrac{2V}{A}$	(2–97)
An appropriate expression for τ_{max} for structural beams, columns and joists used in structural industries	$\tau_{max} = \dfrac{V}{A_w}$ where A_w is the area of the web	(2–98)

ECCENTRIC LOADING

The maximum and minimum stresses which are induced at points of outer fibers on either side of a machine member loaded eccentrically (Figs. 2–22 and 2–23)	$\sigma_{max} = \dfrac{F}{A} + \dfrac{M_b}{Z}$ and $\sigma_{min} = \dfrac{F}{A} - \dfrac{M_b}{Z}$	(2–99)
The resultant stress at any point of the cross section of an eccentrically loaded member (Fig. 2–24)	$\sigma_z = \pm\dfrac{F}{A} \pm \dfrac{M_{bx}e_y}{I_{xx}} \pm \dfrac{M_{by}e_x}{I_{yy}}$	(2–100)

COLUMN FORMULAS (FIG. 2–25)

Euler's formula (Fig. 2–26) for critical load	$F_{cr} = \dfrac{n\pi^2 EA}{(l/k)^2} = \dfrac{n\pi^2 EI}{l^2}$	(2–101)

Particular	Formula

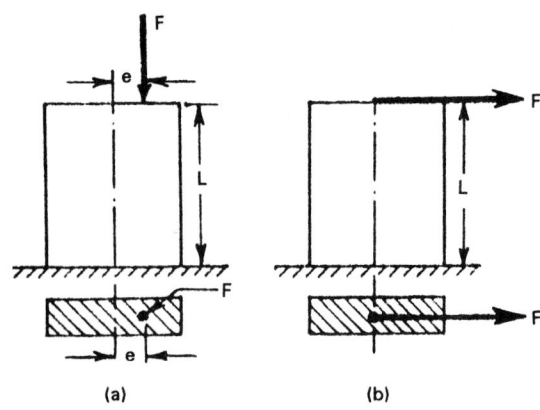

FIGURE 2–22   Eccentric loading.

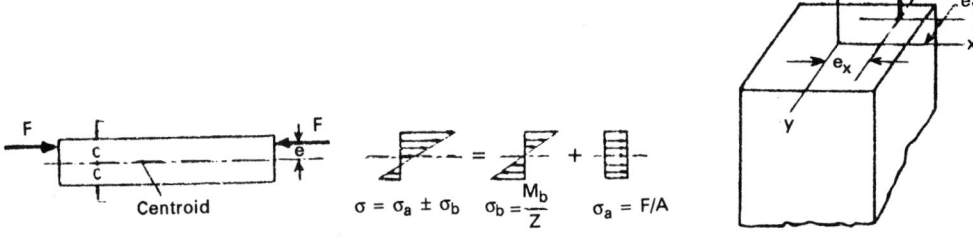

FIGURE 2–23 Eccentrically loaded machine member

$$\sigma = \sigma_a \pm \sigma_b \quad \sigma_b = \frac{M_b}{Z} \quad \sigma_a = F/A$$

FIGURE 2–24

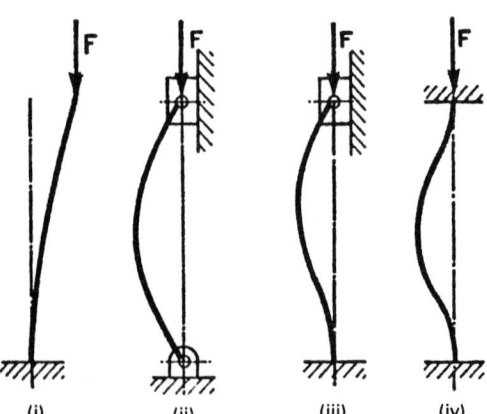

(i) One end is fixed and other is free. (ii) Both ends are rounded and guided or hinged. (iii) One end is fixed and other is rounded and guided or hinged. (iv) Fixed ends.

FIGURE 2–25   Column-end conditions.

Particular	Formula	

Johnson's parabolic formula (Fig. 2–26) for critical load

$$F_{cr} = A\sigma_y \left[1 - \frac{\sigma_y}{4n\pi^2 E} \left(\frac{l}{k} \right)^2 \right]$$

(2–102)

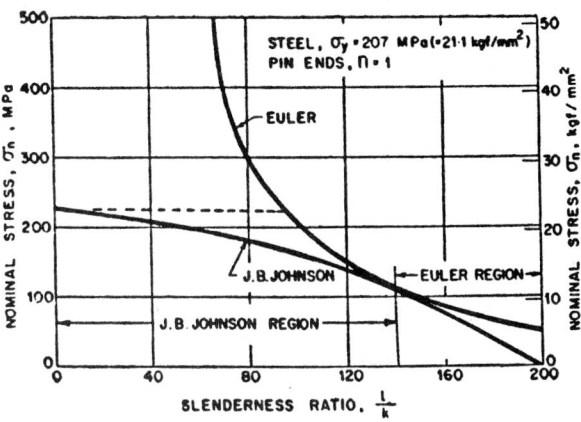

FIGURE 2–26   Variation of critical stress with slenderness ratio.

Straight-line formula for critical load

$$F_{cr} = A \left[\sigma_y - \frac{2\sigma_y}{3\pi\sqrt{(\sigma_y/3nE)}} \left(\frac{l}{k} \right) \right]$$

(2–103)

Straight-line formula for short column of brittle material for critical load

$$F_{cr} = A \left(\sigma - C_1 \frac{l}{k} \right)$$

(2–104)

Ritter's formula for induced stress

$$\sigma_c = \frac{F}{A} \left[1 + \frac{\sigma_e}{n\pi^2 E} \left(\frac{l}{k} \right)^2 \right]$$

(2–105)

Ritter's formula for eccentrically loaded column (Fig. 2–23) for combined induced stress

$$\sigma_c = \frac{F}{A} \left[1 + \frac{\sigma_e}{n\pi^2 E} \left(\frac{l}{k} \right)^2 + \frac{ce}{k^2} \right]$$

(2–106)

Rankine's formula for induced stress

$$\sigma_c = \frac{F_{cr}}{A} \left[1 + a \left(\frac{l}{k} \right)^2 \right]$$

(2–107)

The critical unit load from secant formula for a round-ended column

$$\frac{F_{cr}}{A} = \frac{\sigma_y}{1 + \frac{ec}{k^2} \sec \frac{l}{k} \sqrt{(F_{cr}/4AE)}}$$

(2–108)

Particular	Formula

HERTZ CONTACT STRESS

Contact of spherical surfaces

Sphere on a sphere (Fig. 2–27a)
The radius of circular area of contact

$$a = 0.721 \left[F \frac{\dfrac{1 - v_1^2}{E_1} + \dfrac{1 - v_2^2}{E_2}}{\left(\dfrac{1}{d_1} + \dfrac{1}{d_2} \right)} \right]^{1/3} \qquad (2\text{–}109)$$

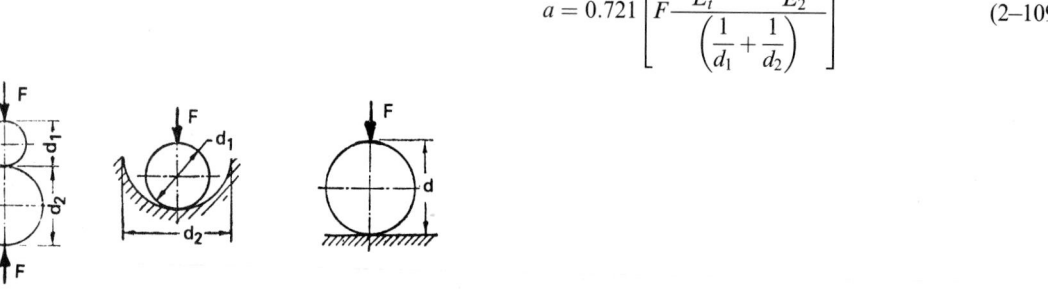

(a)   **(b)**   **(c)**

FIGURE 2–27   Hertz contact stress.

The maximum compressive stress

$$\sigma_{c(\max)} = 0.918 \left[F \frac{\left(\dfrac{1}{d_1} + \dfrac{1}{d_2} \right)^2}{\left(\dfrac{1 - v_1^2}{E_1} + \dfrac{1 - v_2^2}{E_2} \right)^2} \right]^{1/3} \qquad (2\text{–}110)$$

Combined deformation of both bodies in contact
along the axis of load

$$\alpha = 1.04 \left[F^2 \frac{\left(\dfrac{1 - v_1^2}{E_1} + \dfrac{1 - v_2^2}{E_2} \right)^2}{\left(\dfrac{d_1 d_2}{d_1 + d_2} \right)} \right]^{1/3} \qquad (2\text{–}111)$$

Spherical surface in contact with a spherical socket
(Fig. 2–27b)
The radius of circular area of contact

$$a = 0.721 \left[F \frac{\left(\dfrac{1 - v_1^2}{E_1} + \dfrac{1 - v_2^2}{E_2} \right)}{\left(\dfrac{1}{d_1} - \dfrac{1}{d_2} \right)} \right]^{1/3} \qquad (2\text{–}112)$$

The maximum compressive stress

$$\sigma_{c(\max)} = 0.918 \left[F \frac{\left(\dfrac{1}{d_1} - \dfrac{1}{d_2} \right)^2}{\left(\dfrac{1 - v_1^2}{E_1} + \dfrac{1 - v_2^2}{E_2} \right)^2} \right]^{1/3} \qquad (2\text{–}113)$$

Particular	Formula
Combined deformation of both bodies in contact along axis of load	$$\alpha = 1.04 \left[F^2 \frac{\left(\dfrac{1 - v_1^2}{E_1} + \dfrac{1 - v_2^2}{E_2} \right)^2}{\left(\dfrac{d_1 d_2}{d_2 - d_1} \right)} \right]^{1/3} \quad (2\text{–}114)$$
Distribution of pressure over band of width of contact and stresses in contact zone along the line of symmetry of spheres	Refer to Fig. 2–28a.
Sphere on a flat surface (Fig. 2–27c) The radius of circular area of contact	$$a = 0.721 \left[Fd_1 \left(\frac{1 - v_1^2}{E_1} + \frac{1 - v_2^2}{E_2} \right) \right]^{1/3} \quad (2\text{–}115)$$
The maximum compressive stress	$$\sigma_{c(max)} = 0.918 \left[\frac{F}{d_1^2 \left(\dfrac{1 - v_1^2}{E_1} + \dfrac{1 - v_2^2}{E_2} \right)^2} \right]^{1/3} \quad (2\text{–}116)$$ where $d = d_1$ (Fig. 2–27c)

Contact of cylindrical surfaces

Particular	Formula
Cylindrical surface on cylindrical surface, axis parallel (Fig. 2–27a) The width of band of contact	$$2b = 1.6 \left[\frac{F}{L} \frac{\left(\dfrac{1 - v_1^2}{E_1} + \dfrac{1 - v_2^2}{E_2} \right)}{\left(\dfrac{1}{d_1} + \dfrac{1}{d_2} \right)} \right]^{1/2} \quad (2\text{–}117)$$
The maximum compressive stress	$$\sigma_{c(max)} = 0.798 \left[\frac{F}{L} \frac{\left(\dfrac{1}{d_1} + \dfrac{1}{d_2} \right)}{\left(\dfrac{1 - v_1^2}{E_1} + \dfrac{1 - v_2^2}{E_2} \right)} \right]^{1/2} \quad (2\text{–}118)$$
Cylindrical surface in contact with a circular groove (Fig. 2–27b) The width of band of contact	$$2b = 1.6 \left[\frac{F}{L} \frac{\left(\dfrac{1 - v_1^2}{E_1} + \dfrac{1 - v_2^2}{E_2} \right)}{\left(\dfrac{1}{d_1} - \dfrac{1}{d_2} \right)} \right]^{1/2} \quad (2\text{–}119)$$
The maximum compressive stress	$$\sigma_{c(max)} = 0.798 \left[\frac{F}{L} \frac{\left(\dfrac{1}{d_1} - \dfrac{1}{d_2} \right)}{\left(\dfrac{1 - v_1^2}{E_1} + \dfrac{1 - v_2^2}{E_2} \right)} \right]^{1/2} \quad (2\text{–}120)$$
Distribution of pressure over band of width of contact and stresses in contact zone along the line of symmetry of cylinders	Refer to Fig. 2–28b.

Particular	**Formula**
Cylindrical surface in contact with a flat surface (Fig. 2–27c): The width of band of contact The maximum compressive stress	$$2b = 1.6\left[\frac{Fd_1}{L}\left(\frac{1-v_1^2}{E_1}+\frac{1-v_2^2}{E_2}\right)\right]^{1/2} \quad (2\text{–}121)$$ $$\sigma_{c(max)} = 0.798\left[\frac{F}{Ld_1}\frac{1}{\left(\frac{1-v_1^2}{E_1}+\frac{1-v_2^2}{E_2}\right)}\right]^{1/2}$$ $$(2\text{–}122)$$ where $d = d_1$ (Fig. 2–27c)
Deformation of cylinder between two plates	$$\Delta d_1 = \frac{4F}{L}\left(\frac{1-v_1^2}{\pi E}\right)\left(\frac{1}{3}+\log_e\frac{2d_1}{b}\right) \quad (2\text{–}123)$$

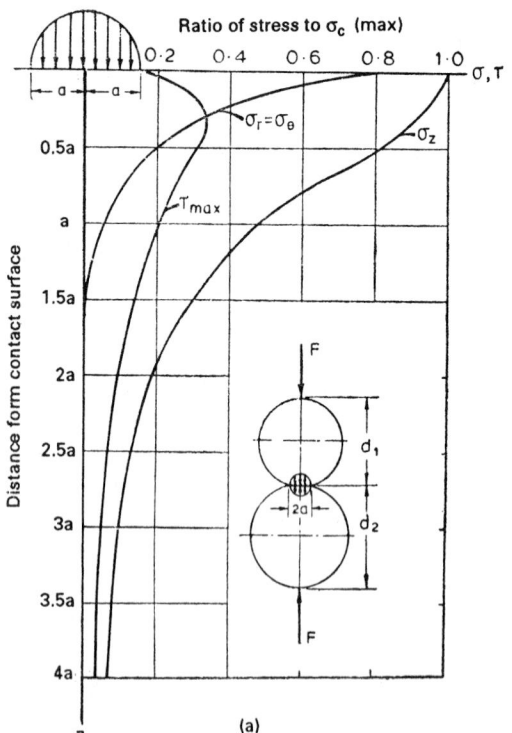

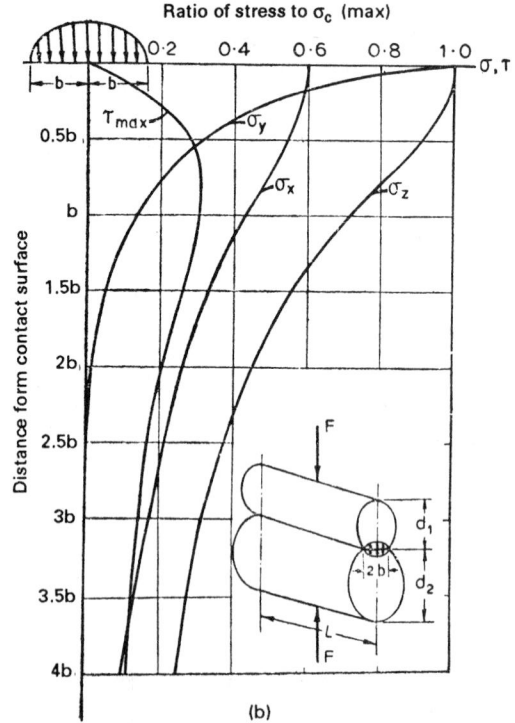

FIGURE 2–28   Distribution of pressure over bandwidth of contact and stresses in contact zone along line of symmetry of spheres and cylinders.

FORMULAS AND DATA FOR VARIOUS CROSS SECTIONS OF MACHINE ELEMENTS

For further data on static stresses, properties and torsion of shafts of various cross sections; shear moments, and deflections of beams, strain rosettes, and singularity functions

Refer to Tables 2–1 to 2–11.

REFERENCES

1. Maleev, V. L., and J. B. Hartman, *Machine Design*, International Textbook Company, Scranton, Pennsylvania, 1954.
2. Shigley, J. E., *Mechanical Engineering Design*, 3d. ed., McGraw-Hill Book Company, New York, 1977.
3. Lingaiah, K., and B. R. Narayana Iyengar, *Machine Design Data Handbook*, Vol. I (*SI and Customary Metric Units*), Suma Publishers, Bangalore, India, 1986.
4. Lingaiah, K., *Machine Design Data Handbook*, Vol. II (*SI and Customary Metric Units*), Suma Publishers, Bangalore, India, 1986.

BIBLIOGRAPHY

Black, P. H., and O. Eugene Adams, Jr., *Machine Design*, McGraw-Hill Book Company, New York, 1965.

Lingaiah, K., and B. R. Narayana Iyengar, *Machine Design Data Handbook*, Engineering College Co-operative Society, Bangalore, India, 1962.

Norman, C. A., E. S. Ault, and I. F. Zarobsky, *Fundamentals of Machine Design*, The Macmillan Company, New York, 1951.

Vallance, A. E., and V. L. Doughtie, *Design of Machine Members*, McGraw-Hill Book Company, New York, 1951.

Timoshenko, S., and J. N. Goodier, *Theory of Elasticity*, McGraw-Hill Book Company, New York, 1951.

Timoshenko, S., and J. M. Gere, *Mechanics of Materials*, Van Nostrand Reinhold Company, New York, 1972.

TABLE 2-1
Torsion of shafts of various cross sections

Cross section	Polar section modulus, $Z_o = \dfrac{J}{c}$	Polar radius of gyration, k_0	Angular deflection, θ	
			In terms of torsional moment, M_t	In terms of maximum stress, τ
	$\dfrac{\pi D^3}{16}$	$\dfrac{D}{\sqrt{8}} = 0.354D$	$\dfrac{32l}{\pi D^4}\dfrac{M_t}{G}$	$\dfrac{2l}{D}\dfrac{\tau}{G}$ τ at circumference
	$\dfrac{\pi(D_1^4 - D_2^4)}{16D_1}$	$\sqrt{\dfrac{D_1^2 + D_2^2}{8}} = 0.354\sqrt{D_1^2 + D_2^2}$	$\dfrac{32l}{\pi(D_1^4 - D_2^4)}\dfrac{M_t}{G}$	$\dfrac{2l}{D_1}\dfrac{\tau}{G}$ τ at outer circumference
	$\dfrac{\pi b^2 h^a}{16}$ $h > b$	$\dfrac{1}{4}\sqrt{b^2 + h^2}$	$\dfrac{16(b^2 + h^2)l\,M_t}{\pi b^3 h^3\,G}$	$\dfrac{(b^2 + h^2)l\,\tau}{bh^2\,G}$ τ at A^b
	$\dfrac{2b^2 h^a}{9}$ $h > b$	$\sqrt{\dfrac{b^2 + h^2}{12}} = 0.289\sqrt{b^2 + h^2}$	$\dfrac{m(b^2 + h^2)l\,M_t}{b^3 h^3\,G}$	$\dfrac{n(b^2 + h^2)l\,\tau}{bh^2\,G}$ τ at A^c
	$\dfrac{b^{3a}}{20}$	$0.289b$	$\dfrac{46.2l\,M_t}{b^4\,G}$	$\dfrac{2.31l\,\tau}{b\,G}$ τ at center of side
	$0.92b^{3a}$	$0.645b$	$\dfrac{0.967l\,M_t}{b^4\,G}$	$\dfrac{0.9l\,\tau}{b\,G}$ τ at center of side

For the rectangular section:

$\dfrac{h}{b} =$	1	2	4	8
$m =$	3.56	3.50	3.35	3.21
$n =$	0.79	0.78	0.74	0.71

[a] This value is not true value of Z_o but is the value of Z_o for a circular section of equal strength and may be used for determining the maximum stress by the formula $\tau = M_t/Z_o$.
[b] At B, shear stress = $10\,M_t/\pi bh^2$.
[c] At B, shear stress = $9M_t/2bh^2$.

Source: V. L. Maleev and J. B. Hartman, Machine Design, International Textbook Company, Scranton, Pennsylvania, 1954.

TABLE 2–2
Shear stress in beams, caused by bending

Section	Shear stress at a distance y from neutral axis, τ, MPa (psi)	Maximum shear stress, τ_{max} MPa (psi)
	$\dfrac{3F}{2bh}\left[1-\left(\dfrac{2y}{h}\right)^2\right]$	$\dfrac{3F}{2bh}=1.5\dfrac{F}{A}$ (for $y=0$)
	$\dfrac{4F}{3\pi r^2}\left[1-\left(\dfrac{y}{r}\right)^2\right]$	$\dfrac{4F}{3\pi r^2}=1.33\dfrac{F}{A}$ (for $y=0$)
	$\dfrac{F\sqrt{2}}{b^2}\left[1+\dfrac{y\sqrt{2}}{b}-4\left(\dfrac{y}{b}\right)^2\right]$	$1.591\dfrac{F}{A}\left(\text{for } y=\dfrac{c}{4}\right)$
		$\dfrac{3F}{4a}\left[\dfrac{bc^2-(b-a)d^2}{bc^3-(b-a)d^3}\right]$ (for $y=0$)

TABLE 2–3
The values of constants a in Eq. (2–107)

Material	Yield stress in compression, σ_{yc}		Values of a for various end-fixity coefficients			
	MPa	kpsi	1	4	2	n
Timber	49	7	$\dfrac{1}{750}$	$\dfrac{1}{3000}$	$\dfrac{1}{1500}$	$\dfrac{1}{n\times750}$
Cast iron	549	80	$\dfrac{1}{1600}$	$\dfrac{1}{6400}$	$\dfrac{1}{3200}$	$\dfrac{1}{n\times1600}$
Mild steel	324	47	$\dfrac{1}{7500}$	$\dfrac{1}{30000}$	$\dfrac{1}{15000}$	$\dfrac{1}{n\times7500}$

TABLE 2–4
End condition coefficients n (Fig. 2–25)

Particular	n
One end fixed and other end free	0.25
Both ends rounded and guided or hinged	1
One end fixed, and the other end rounded and guided or hinged	2
Both ends fixed rigidly	4
Both ends flat	1 to 4

TABLE 2–5
End-fixity coefficients for cast iron column to be used in Eq. (2–104)

End conditions	C_1	Maximum l/k
Round	175	90
Fixed	88	160
One fixed, one round	116	115

TABLE 2–6
Properties of various cross sections

Section	Area A	Moment of inertia I	Distance to farthest point c	Section modulus $Z = \dfrac{I}{c}$	Radius of gyration $k = \sqrt{\dfrac{I}{A}}$
	bh	$\dfrac{bh^3}{12}$	$\dfrac{h}{2}$	$\dfrac{bh^2}{6}$	$0.289\,h$
	$(H-b)b$	$\dfrac{b}{12}(H^3 - h^3)$	$\dfrac{H}{2}$	$\dfrac{b(H^3 - h^3)}{6H}$	$\sqrt{\dfrac{H^3 - h^3}{12(H-h)}}$
	$BH - bh$	$\dfrac{BH^3 - bh^3}{12}$	$\dfrac{H}{2}$	$\dfrac{BH^3 - bh^3}{6H}$	$\sqrt{\dfrac{BH^3 - bh^3}{12(BH-bh)}}$
	$\left(\dfrac{2b + b_0}{2}\right)h$	$\dfrac{(6b^2 + 6bb_0 + b_0^2)h^3}{36(2b + b_0)}$	$\dfrac{(3b + 2b_0)h}{3(2b + b_0)}$	$\dfrac{(6b^2 + 6bb_0 + b_0^2)h^2}{12(3b + b_0)}$	$\sqrt{\dfrac{I}{A}}$
	$\dfrac{\pi D^2}{4}$	$\dfrac{\pi D^4}{64}$	$\dfrac{D}{2}$	$\dfrac{\pi D^3}{32}$	$\dfrac{D}{4}$
	$\dfrac{\pi}{4}(D_1^2 - D_2^2)$	$\dfrac{\pi}{64}(D_1^4 - D_2^4)$ $= \dfrac{\pi}{4}(R_1^4 - R_2^4)$	$\dfrac{D_1}{2} = R_1$	$\dfrac{\pi(D_1^4 - D_2^4)}{32D_1}$	$\dfrac{\sqrt{D_1^2 + D_2^2}}{4}$ $= \dfrac{\sqrt{R_1^2 + R_2^2}}{2}$
	πab	$\dfrac{\pi bh^3}{64}$	$\dfrac{h}{2}$	$\dfrac{\pi bh^2}{32}$	$\dfrac{h}{4}$
	$\dfrac{bh}{2}$	$\dfrac{bh^3}{36}$	$\cdots$	$\dfrac{bh^2}{24}$	$0.236\,h$

TABLE 2-7
Shear, moment, and deflection formulas for beams

Loading, support, and reference number	Reactions R_1 and R_2, vertical shear V	Bending moment M_b, maximum bending moment	Deflection y and maximum deflection
1. Cantilever, end load 	$R_2 = +F$ $V = -F$	$M_b = -Fx$ Max $M_b = Fl$ at B	$y = -\dfrac{1}{6}\dfrac{F}{EI}(x^3 - 3l^2x + 2l^3)$ Max $y = -\dfrac{1}{3}\dfrac{Fl^3}{EI}$ at A
2. Cantilever, intermediate load 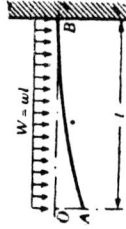	$R_2 = +F$ A to $B: V = 0$ B to $C: V = -F$	A to $B: M_b = 0$ B to $C: M_b = -F(x - b)$ Max $M_b = -Fa$ at C	A to $B: y = -\dfrac{1}{6}\dfrac{F}{EI}(-a^3 + 3a^2l - 3a^2x)$ B to $C: y = -\dfrac{1}{6}\dfrac{F}{EI}[(x - b)^3 - 3a^2(x - b) + 2a^3]$ Max $y = -\dfrac{1}{6}\dfrac{F}{EI}(3a^2l - a^3)$
3. Cantilever, uniform load	$R_2 = +W$ $V = -\dfrac{W}{l}x$	$M_b = -\dfrac{1}{2}\dfrac{W}{l}x^2$ Max $M_b = -\dfrac{1}{2}Wl$ at B	$y = -\dfrac{1}{24}\dfrac{W}{EIl}(x^4 - 4l^3x + 3l^4)$ Max $y = -\dfrac{1}{8}\dfrac{Wl^3}{EI}$
4. End supports, center load	$R_1 = +\tfrac{1}{2}F, R_2 = +\tfrac{1}{2}F$ A to $B: V = +\tfrac{1}{2}F$ B to $C: V = -\tfrac{1}{2}F$	A to $B: M_b = +\tfrac{1}{2}Fx$ B to $C: M_b = +\tfrac{1}{2}F(l - x)$ Max $M_b = +\tfrac{1}{4}Fl$ at B	A to $B: y = -\dfrac{1}{48}\dfrac{F}{EI}(3l^2x - 4x^3)$ Max $y = -\dfrac{1}{48}\dfrac{Fl^3}{EI}$ at B

2.27

TABLE 2–7
Shear, moment, and deflection formulas for beams (*Cont.*)

Loading, support, and reference number	Reactions R_1 and R_2, vertical shear V	Bending moment M_b, maximum bending moment	Deflection y and maximum deflection
5. End supports, intermediate load	$R_1 = +F\frac{b}{l}, R_2 = +F\frac{a}{l}$ A to $B: V = +F\frac{b}{l}$ B to $C: V = -F\frac{a}{l}$	A to $B: M_b = +F\frac{b}{l}x$ B to $C: M_b = +F\frac{a}{l}(l-x)$ Max $M_b = +F\frac{ab}{l}$ at B	A to $B: y = -\frac{Fbx}{6EIl}\left[2l(l-x)-b^2-(l-x)^2\right]$ B to $C: y = -\frac{Fa(l-x)}{6EIl}\left[2lb-b^2-(l-x)^2\right]$ Max $y = -\frac{Fab}{27EIl}(a+2b)\sqrt{3a(a+2b)}$ at $x = \sqrt{\frac{1}{3}a(a+2b)}$ when $a > b$
6. End supports, uniform load $W = wl$	$R_1 = +\frac{1}{2}W, R_2 = +\frac{1}{2}W$ $V = \frac{1}{2}W\left(1-\frac{2x}{l}\right)$	$M_b = \frac{1}{2}W\left(x-\frac{x^2}{l}\right)$ Max $M_b = +\frac{1}{8}Wl$ at $x=\frac{1}{2}l$	$y = -\frac{1}{24}\frac{Wx}{EIl}(l^3 - 2lx^2 + x^3)$ Max $y = -\frac{5}{384}\frac{Wl^3}{EI}$ at $x=\frac{1}{2}l$
7. One end fixed, one end supported, center load	$R_1 = \frac{5}{16}F, R_2 = \frac{11}{16}F$ $M_2 = \frac{3}{16}Fl$ A to $B: V = +\frac{5}{16}F$ B to $C: V = -\frac{11}{16}F$	A to $B: M_b = \frac{5}{16}Fx$ B to $C: M_b = F\left(\frac{1}{2}l - \frac{11}{16}x\right)$ Max $+M_b = \frac{5}{32}Fl$ at B Max $-M_b = -\frac{3}{16}Fl$ at C	A to $B: y = \frac{1}{96}\frac{F}{EI}(5x^3 - 3l^2x)$ B to $C: y = \frac{1}{96}\frac{F}{EI}\left[5x^3 - 16\left(x-\frac{1}{2}\right)^3 - 3l^2x\right]$ Max $y = -0.00932\frac{Fl^3}{EI}$ at $x = 0.4472l$
8. One end fixed, one end supported, uniform load $W = wl$	$R_1 = \frac{3}{8}W, R_2 = \frac{5}{8}W$ $M_2 = \frac{1}{8}Wl$ $V = W\left(\frac{3}{8}-\frac{x}{l}\right)$	$M_b = W\left(\frac{3}{8}x - \frac{1}{2}x^2\right)$ Max $+M_b = \frac{9}{128}Wl$ at $x=\frac{3}{8}l$ Max $-M_b = -\frac{1}{8}Wl$ at B	$y = \frac{1}{48}\frac{W}{EIl}(3lx^3 - 2x^4 - l^3x)$ Max $y = -0.0054\frac{Wl^3}{EI}$ at $x = 0.4215l$

TABLE 2–7
Shear, Moment, and Deflection Formulas for Beams (Cont.)

Loading, support, and reference number	Reactions R_1 and R_2, vertical shear V	Bending moment M_b, maximum bending moment	Deflection y and maximum deflection
9. Both ends fixed, center load	$R_1 = \frac{1}{2}F, R_2 = \frac{1}{2}F$ $M_1 = \frac{1}{8}Fl, M_2 = \frac{1}{8}Fl$ A to $B: V = +\frac{1}{2}F$ B to $C: V = -\frac{1}{2}F$	A to $B: M_b = \frac{1}{8}F(4x - l)$ B to $C: M_b = \frac{1}{8}F(3l - 4x)$ Max $+ M_b = \frac{1}{8}Fl$ at B Max $- M_b = -\frac{1}{8}Fl$ at A and C	A to $B: y = -\frac{1}{48}\frac{F}{EI}(3lx^2 - 4x^3)$ Max $y = -\frac{1}{192}\frac{Fl^3}{EI}$ at B
10. Both ends fixed, intermediate load	$R_1 = \frac{Fb^2}{l^3}(3a + b)$ $R_2 = \frac{Fa^2}{l^3}(3b - a)$ $M_1 = F\frac{ab^2}{l^2}, M_2 = F\frac{a^2 b}{l^2}$ A to $B: V = R_1$ B to $C: V = R_1 - F$	A to $B: M_b = -\frac{ab^2}{l^2} + R_1 x$ B to $C: M_b = -F\frac{ab^2}{l^2} + R_1 x - F(x - a)$ Max $+ M_b = -F\frac{ab^2}{l^2} + R_1 a$ at B Max $- M_b = -M_1$ when $a < b$ Max $- M_b = -M_2$ when $a > b$	A to $B: y = -\frac{1}{6}\frac{Fb^2 x^2}{EIl^3}(3ax + bx - 3al)$ B to $C: y = -\frac{1}{6}\frac{Fa^2(l - x)^2}{EIl^3}[(3b + a)(l - x) - 3bl]$ Max $y = -\frac{2}{3}\frac{F}{EI}\frac{a^3 b^2}{(3a + b)^2}$ at $x = \frac{2al}{(3a + b)}$ if $a > b$ Max $y = -\frac{2}{3}\frac{F}{EI}\frac{a^2 b^3}{(3b + a)^2}$ at $x = l - \frac{2bl}{(3b + a)}$ if $a < b$
11. Both ends fixed, uniform load	$R_1 = \frac{1}{2}W, R_2 = \frac{1}{2}W$ $M_1 = \frac{1}{12}Wl, M_2 = \frac{1}{12}Wl$ $V = \frac{1}{2}W\left(1 - \frac{2x}{l}\right)$	$M_b = \frac{1}{2}W\left(x - \frac{x^2}{l} - \frac{1}{6}l\right)$ Max $+ M_b = \frac{1}{24}Wl$ at $x = \frac{1}{2}l$ Max $- M_b = -\frac{1}{12}Wl$ at A and B	$y = \frac{1}{24}\frac{Wx^2}{EIl}(2lx - l^2 - x^2)$ Max $y = -\frac{1}{364}\frac{Wl^3}{EI}$ at $x = \frac{1}{2}l$

Source: J. E. Shigley, Mechanical Engineering Design, 3d. ed., McGraw-Hill Book Company, New York, 1977.

TABLE 2–8
Some Equations for Use with Method of Castigliano

Type of load	General energy equation	Energy equation	General deflection equation
Axial	$U = \int_0^l \dfrac{F^2}{2EA}\,ds$	$U = \dfrac{F^2 l}{2EA} = (\sigma^2 AL)/2E$	$\delta = \int_0^l \dfrac{F(\partial F/\partial Q)}{EA}\,ds$
Bending	$U = \int_0^l \dfrac{M_b^2 l}{2EI}\,ds$	$U = \dfrac{M_b^2 l}{2EI}$	$\delta = \int_0^l \dfrac{M_b(\partial M_b/\partial Q)}{EI}\,ds$
Combined axial and bending	$U = \int_0^l \dfrac{F^2}{2EA}\,ds + \int_0^l \dfrac{M_b^2 l}{2EI}\,ds$	$U = \dfrac{F^2 l}{2EA} + \dfrac{M_b^2 l}{2EI}$	Sum of axial and bending load
Torsion	$U = \int_0^l \dfrac{M_t^2}{2GJ}\,ds$	$U = \dfrac{M_t^2 l}{2GJ}$	$\delta = \int_0^l \dfrac{M_t(\partial M_t/\partial Q)}{GJ}\,ds$
Transverse shear	$U = \int_0^l \dfrac{v^2 ds}{2GA}$	$U = \dfrac{V^2 L}{2GA} = \dfrac{\tau^2}{2G} AL$	$\delta = \int_0^l \dfrac{V(\partial V/\partial Q)}{GA}\,ds$
Transverse shear (rectangular section)	$U = \int_0^l \dfrac{3V^2}{5GA}\,ds$	$U = \dfrac{3V^2 L}{5GA}$	$\delta = \int_0^l \dfrac{6V(\partial V/\partial Q)}{5GA}\,dS$
Open-coiled spring subjected to axial load F	$U = \int_0^l \dfrac{M_t^2}{2GJ}\,ds + \int_0^l \dfrac{M_b^2 l}{2EI}\,ds$	$U = \int_0^l \dfrac{M_t^2 L}{2GJ} + \dfrac{M_b^2 L}{2EI}$ $= \dfrac{LFR^2}{2}\left[\dfrac{\cos^2\alpha}{GJ} + \dfrac{\sin^2\alpha}{EI}\right]$	$\delta = 2\pi nFR^2 \sec\alpha \left[\dfrac{\cos^2\alpha}{GJ} + \dfrac{\sin^2\alpha}{EI}\right]$

TABLE 2–9
Mechanical and physical constants of some materials

Material	Modulus of elasticity, E		Modulus of rigidity, G		Poisson's ratio, ν	Density, ρ^a	Unit weight, γ^b				
	GPa	Mpsi	GPa	Mpsi		Mg/m³	kgf/m³	kN/m³	lbf/in³	lbf/ft³
Aluminum	69	10.0	26	3.8	0.334	2.69	2,685	26.3	0.097	167
Aluminum cast	70	10.15	30	4.35			2,650	26.0	0.096	166
Aluminum (all alloys)	72	10.4	27	3.9	0.320	2.80	2,713	27.0	0.10	173
Beryllium copper	124	18.0	48	7.0	0.285	8.22	8,221	80.6	0.297	513
Carbon steel	206	30.0	79	11.5	0.292	7.81	7,806	76.6	0.282	487
Cast iron, gray	100	14.5	41	6.0	0.211	7.20	7,197	70.6	0.260	450
Mallable cast iron	170	24.6	90	13.0			7,200			
Inconel	214	31.0	76	11.0	0.290	8.42	8,418	83.3	0.307	530
Magnesium alloy	45	6.5	16	2.4	0.350	1.80	1,799	17.6	0.065	117
Molybdenum	331	48.0	117	17.0	0.307	10.19	10,186	100.0	0.368	636
Monel metal	179	26.0	65	9.5	0.320	8.83	8,830	86.6	0.319	551
Nickel-silver	127	18.5	48	7.0	0.332	8.75	8,747	85.80	0.316	546
Nickel alloy	207	30	79	11.5	0.30	8.3			0.300	518
Nickel steel	207	30.0	79	11.5	0.291	7.75	7,751	76.0	0.280	484
Phosphor bronze	111	16.0	41	6.0	0.349	8.17	8,166	80.1	0.295	510
Steel (18–8), stainless	190	27.5	73	10.6	0.305	7.75	7,750	76.0	0.280	484
Titanium (pure)	103	15.0				4.47	4,470	43.8	0.16	279
Titanium alloy	114	16.5	43	6.2	0.33	6.6				
Brass	106	15.5	40	5.8	0.324	8.55	8,553	83.9	0.309	534
Bronze	96	14.0	38	5.5	0.349	8.30	8,304	81.4		
Bronze cast	80	11.6	35	5.0			8,200			
Copper	121	17.5	46	6.6	0.326	8.90	8,913	87.4	0.322	556
Tungsten	345	50.0	138	20.0		18.82	18,822	184.6		
Douglas fir	11	1.6	4	0.6	0.330	4.43	443	4.3	0.016	28
Glass	46	6.7	19	2.7	0.245	2.60	2,602	25.5	0.094	162
Lead	36	5.3	13	1.9	0.431	11.38	11,377	111.6	0.411	710
Concrete (compression)	14–28	2.0–4.0				2.35	2,353	23.1		147
Wrought iron	190	27.5	70	10.2			7,700			
Zinc alloy	83	12	31	4.5	0.33	6.6			0.24	415

[a] ρ—mass density.
[b] γ (w —weight density; w is also the symbol used for unit weight of materials.

Sources: K. Lingaiah and B. R. Narayana Iyengar, Machine Design Data Handbook, Vol. I (SI and Customary Metric Units), Suma Publishers, Bangalore, India and K. Lingaiah, Machine Design Data Handbook, Vol. II (SI and Customary Metric Units), Suma Publishers, Bangalore, India, 1986.

TABLE 2–10
Relations between strain rosette readings and principal stresses

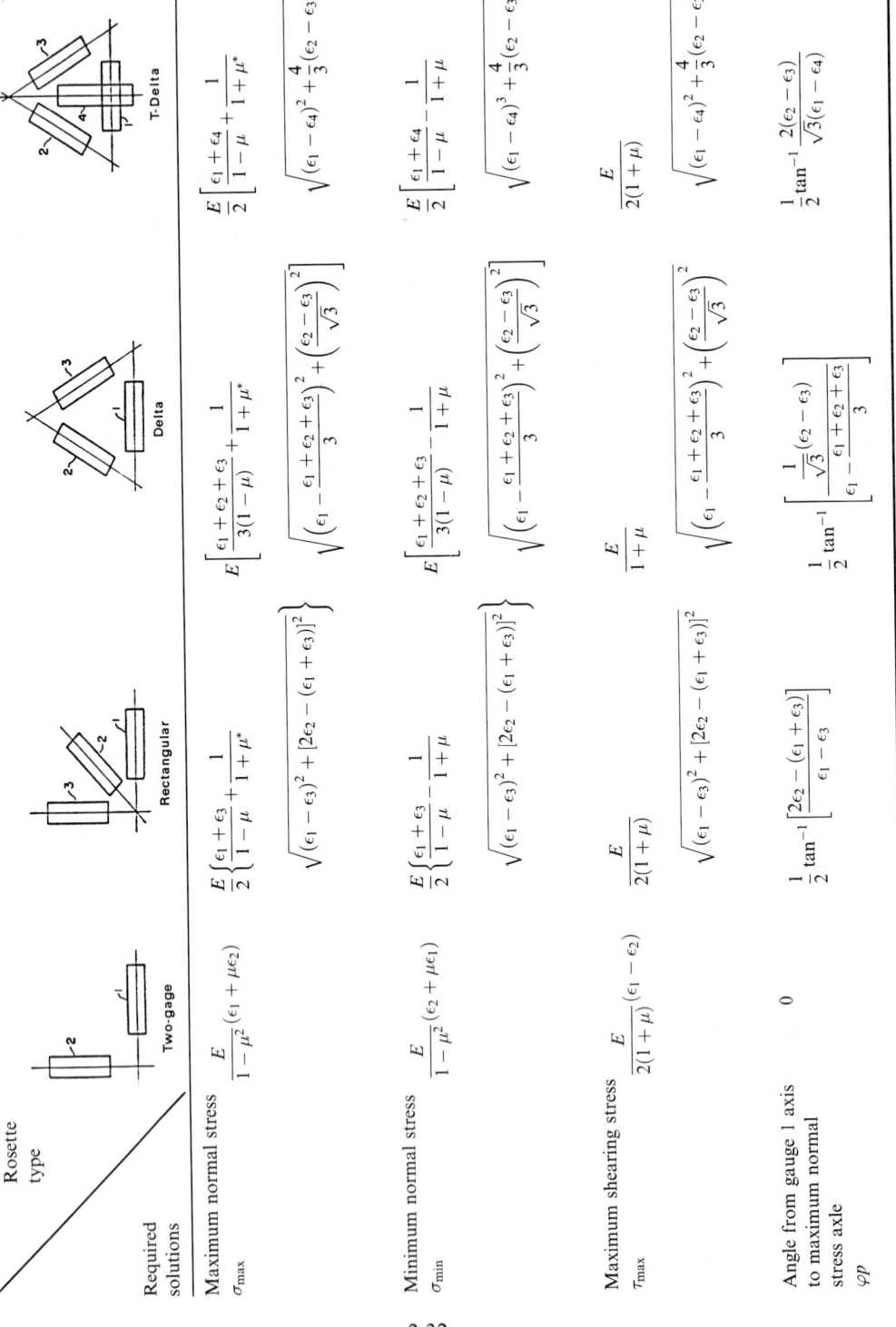

Required solutions	Two-gage	Rectangular	Delta	T-Delta
Maximum normal stress σ_{max}	$\dfrac{E}{1-\mu^2}(\epsilon_1+\mu\epsilon_2)$	$\dfrac{E}{2}\left\{\dfrac{\epsilon_1+\epsilon_3}{1-\mu}+\dfrac{1}{1+\mu^*}\sqrt{(\epsilon_1-\epsilon_3)^2+[2\epsilon_2-(\epsilon_1+\epsilon_3)]^2}\right\}$	$E\left[\dfrac{\epsilon_1+\epsilon_2+\epsilon_3}{3(1-\mu)}+\dfrac{1}{1+\mu^*}\sqrt{\left(\epsilon_1-\dfrac{\epsilon_1+\epsilon_2+\epsilon_3}{3}\right)^2+\left(\dfrac{\epsilon_2-\epsilon_3}{\sqrt{3}}\right)^2}\right]$	$\dfrac{E}{2}\left[\dfrac{\epsilon_1+\epsilon_4}{1-\mu}+\dfrac{1}{1+\mu^*}\sqrt{(\epsilon_1-\epsilon_4)^2+\dfrac{4}{3}(\epsilon_2-\epsilon_3)^2}\right]$
Minimum normal stress σ_{min}	$\dfrac{E}{1-\mu^2}(\epsilon_2+\mu\epsilon_1)$	$\dfrac{E}{2}\left\{\dfrac{\epsilon_1+\epsilon_3}{1-\mu}-\dfrac{1}{1+\mu}\sqrt{(\epsilon_1-\epsilon_3)^2+[2\epsilon_2-(\epsilon_1+\epsilon_3)]^2}\right\}$	$E\left[\dfrac{\epsilon_1+\epsilon_2+\epsilon_3}{3(1-\mu)}-\dfrac{1}{1+\mu}\sqrt{\left(\epsilon_1-\dfrac{\epsilon_1+\epsilon_2+\epsilon_3}{3}\right)^2+\left(\dfrac{\epsilon_2-\epsilon_3}{\sqrt{3}}\right)^2}\right]$	$\dfrac{E}{2}\left[\dfrac{\epsilon_1+\epsilon_4}{1-\mu}-\dfrac{1}{1+\mu}\sqrt{(\epsilon_1-\epsilon_4)^3+\dfrac{4}{3}(\epsilon_2-\epsilon_3)^3}\right]$
Maximum shearing stress τ_{max}	$\dfrac{E}{2(1+\mu)}(\epsilon_1-\epsilon_2)$	$\dfrac{E}{2(1+\mu)}\sqrt{(\epsilon_1-\epsilon_3)^2+[2\epsilon_2-(\epsilon_1+\epsilon_3)]^2}$	$\dfrac{E}{1+\mu}\sqrt{\left(\epsilon_1-\dfrac{\epsilon_1+\epsilon_2+\epsilon_3}{3}\right)^2+\left(\dfrac{\epsilon_2-\epsilon_3}{\sqrt{3}}\right)^2}$	$\dfrac{E}{2(1+\mu)}\sqrt{(\epsilon_1-\epsilon_4)^2+\dfrac{4}{3}(\epsilon_2-\epsilon_3)^2}$
Angle from gauge 1 axis to maximum normal stress axle φp	0	$\dfrac{1}{2}\tan^{-1}\left[\dfrac{2\epsilon_2-(\epsilon_1+\epsilon_3)}{\epsilon_1-\epsilon_3}\right]$	$\dfrac{1}{2}\tan^{-1}\left[\dfrac{\dfrac{1}{\sqrt{3}}(\epsilon_2-\epsilon_3)}{\epsilon_1-\dfrac{\epsilon_1+\epsilon_2+\epsilon_3}{3}}\right]$	$\dfrac{1}{2}\tan^{-1}\dfrac{2(\epsilon_2-\epsilon_3)}{\sqrt{3}(\epsilon_1-\epsilon_4)}$

* Poisson's ratio. The author has used μ as symbol for Poisson's ratio.
Source: Perry, C. C., and H. R. Lissner, *The Strain Gauge Primer*, 2nd ed., McGraw-Hill Publishing Company, New York, p. 147.

2.32

TABLE 2–11
Singularity functions

Function	Graph of $f_n(x)$	Meaning
Concentrated moment	$\langle x - a\rangle^{-2}$	$<x-a>^{-2}=\begin{cases} 1 & x = a \\ 0 & x \neq a \end{cases}$ $\displaystyle\int_{-\infty}^{x} <x-a>^{-2}\,dx = <x-a>^{-1}$
Concentrated force	$\langle x - a\rangle^{-1}$	$<x-a>^{-1}=\begin{cases} 1 & x = a \\ 0 & x \neq a \end{cases}$ $\displaystyle\int_{-\infty}^{x} <x-a>^{-1}\,dx = <x-a>^{0}$
Unit step	$\langle x - a\rangle^{0}$	$<x-a>^{0}=\begin{cases} 0 & x < a \\ 1 & x \geq a \end{cases}$ $\displaystyle\int_{-\infty}^{x} <x-a>^{0}\,dx = <x-a>^{1}$
Ramp	$\langle x - a\rangle^{1}$	$<x-a>^{1}=\begin{cases} 0 & x < a \\ x-a & x \geq a \end{cases}$ $\displaystyle\int_{-\infty}^{x} <x-a>^{1}\,dx = \frac{<x-a>^{2}}{2}$
Parabolic	$\langle x - a\rangle^{2}$	$<x-a>^{2}=\begin{cases} 0 & x < a \\ (x-a)^{2} & x \geq a \end{cases}$ $\displaystyle\int_{-\infty}^{x} <x-a>^{2}\,dx = \frac{<x-a>^{3}}{3}$

Reproduced from J. E. Shigley and L. D. Mitchell, *Mechanical Engineering Design*, McGraw-Hill Book Company, 1983.

CHAPTER
3

DYNAMIC STRESSES IN MACHINE ELEMENTS

SYMBOLS

a, b	coefficient to be used in Eq. (3–23)
c	distance from neutral axis to extreme fiber, m (in)
F_{cv}	centrifugal force per unit volume, kN/m^3 (lbf/in^3)
δ_{st}	static deformation under the action of weight W, m (in)
δ	deformation under impact action, m (in)
E	modulus of elasticity, GPa (Mpsi)
F	force, kN (lbf)
g	acceleration due to gravity, 9.80 66 m/s^2 (32.2 ft/s^2 or 386 in/s^2)
G	modulus of rigidity, GPa (Mpsi)
h	height of fall of weight W, m (in or ft)
k	radius of gyration, m (in)
k_0	radius of gyration, polar, m (in)
K'	kinetic energy, J (ft lbf)
l	length, m (in)
n	speed, rpm
P	power, kW (hp)
r	radius of curvature of the path of motion of mass, m (in)
	the moment arm of the load, m (in or ft)
u	modulus of resilience, J/m^3 ($in\ lbf/in^3$)
U	resilience, J (ft lbf)
	internal elastic energy, J (ft lbf)
U_{max}	maximum internal elastic energy, J (ft lbf)
U_p	potential energy, J (ft lbf)
v	velocity, m/s (fpm)
V	volume, m^3 (in^3)
w	specific weight of material, kN/m^3 (lbf/in^3)
W	weight, kN (lbf)
y	static deflection, m (in)
y'	deflection under impact action, m (in)
σ	static stress (also with subscripts), MPa (kpsi)
σ'	impact stress (also with subscripts), MPa (kpsi)
τ	static shear stress, MPa (kpsi)
τ'	impact shear stress, MPa (kpsi)

α angle of velocity, v, to the horizontal plane, deg
θ static angular deflection, rad
θ' angular deflection under impact action, rad

Particular	Formula

INERTIA FORCE

Power

$$P = \frac{Fv}{1000} \qquad \text{SI} \qquad \text{(3–1a)}$$

where F is in newtons (N), v in m/s, and P in kW

$$= \frac{Fv}{33000} \qquad \textbf{US Customary System units}$$

$$\text{(3–1b)}$$

where F is in lbf, v in ft/min, and P in hp

Velocity

$$v = \frac{2\pi rn}{12} \ (r \text{ in inches}) \qquad \text{(3–2a)}$$

$$= \frac{2\pi rn}{60} \ (r \text{ in meters}) \qquad \text{(3–2b)}$$

Centrifugal force per unit volume

$$F_{cv} = \frac{wv^2}{rg} \qquad \text{(3–3)}$$

IMPACT STRESSES

Impact from direct load

Kinetic energy

$$K = \frac{Wv^2}{2g} \qquad \text{(3–4)}$$

Impact energy of a body falling from a height h

$$K = Wh \qquad \text{(3–5)}$$

The height of fall that would develop the velocity v

$$h = \frac{v^2}{2g} \qquad \text{(3–6)}$$

The equation for energy balance for an impact by a falling body (Fig. 3–1)

$$(U_p + K + U)_a = (U_p + K + U)_b$$
$$= (U_p + K + U)_c \qquad \text{(3–7)}$$

Internal elastic energy of W whose velocity v is horizontal

$$U = \frac{Wv^2}{2g} \qquad \text{(3–9)}$$

Internal elastic energy of W whose velocity has random direction

$$U = \frac{Wv^2}{2g} + W\delta \sin \alpha \qquad \text{(3–10)}$$

Particular	Formula				

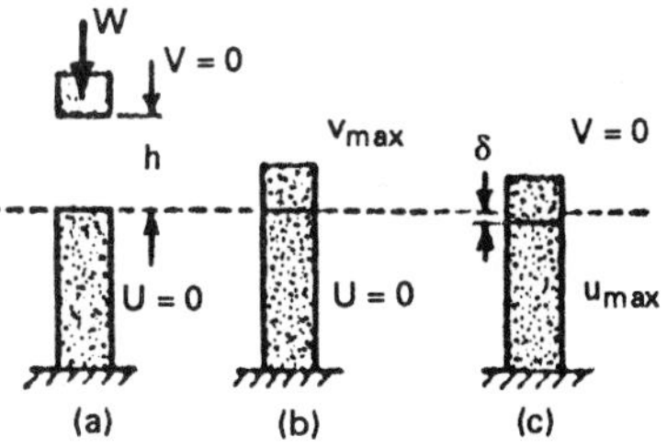

	Fig. Energy	(3–1a)	(3–1b)	(3–1c)	
	U_p	$W(h + \delta)$	$W\delta$	0	(3–8a)
	K	0	$\dfrac{Wv^2}{2g}$	0	(3–8b)
	U	0	0	$W(h + \delta)$	(3–8c)

FIGURE 3–1   Impact by a falling body.

Impact stress due to static load W

$$\sigma' = \sigma\left(1 + \sqrt{1 + \frac{2h}{\delta_{st}}}\right) \qquad 3\text{–}11)$$

Impact stress from a direct load

$$\sigma' = \frac{W}{A}\left(1 + \sqrt{1 + \frac{2hEA}{Wl}}\right) \qquad (3\text{–}12)$$

The general equation for deformation under impact action

$$\delta = \delta_{st}\left(1 + \sqrt{1 + \frac{2h}{\delta_{st}}}\right) \qquad (3\text{–}13)$$

The static deformation or deflection

$$\delta_{st} = \frac{W}{k} \qquad (3\text{–}14)$$

where k is spring constant, kN/m (lbf/in)

Another form of equation for deformation or deflection in terms of velocity at impact v

$$\delta = \delta_{st}\left(1 + \sqrt{1 + \frac{v^2}{g\,\delta_{st}}}\right) \qquad (3\text{–}15)$$

The equivalent static force that would produce the same maximum values of deformation or deflection due to impact δ

$$F_{eq} = W\left(1 + \sqrt{1 + \frac{2h}{\delta_{st}}}\right) \qquad (3\text{–}16a)$$

$$= W\left(1 + \sqrt{1 + \frac{v^2}{g\,\delta_{st}}}\right) \qquad (3\text{–}16b)$$

Sudden load

The stress produced due to sudden load

$$\sigma' = 2\sigma \qquad (3\text{–}17)$$

The deformation due to sudden load

$$\delta = 2\delta_{st} \qquad (3\text{–}18)$$

Particular	Formula

Impact and bending

Impact stress due to bending

$$\sigma_b' = \sigma_b\left(1 + \sqrt{1 + \frac{2h}{y}}\right) \tag{3–19}$$

Deflection of the end of cantilever beam under impact (Fig. 3–2)

$$y' = y\left(1 + \sqrt{1 + \frac{2h}{y}}\right) \tag{3–20}$$

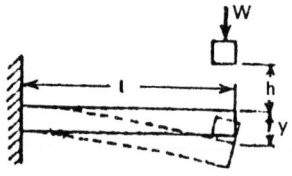

FIGURE 3–2   Impact of falling body on a cantilever beam.

Impact and torsion

Impact shear stress

$$\tau' = \tau\left(1 + \sqrt{1 + \frac{2h}{r\theta}}\right) \tag{3–21}$$

Angular deflection under impact action

$$\theta' = \theta\left(1 + \sqrt{1 + \frac{2h}{r\theta}}\right) \tag{3–22}$$

When a body having weight W strikes another body that has a weight W', according to law of collision of two perfectly inelastic bodies, impact energy Wh is reduced to nWh, the value of n

$$n = \frac{1 + am}{(1 + bm)^2} \tag{3–23}$$

where $m = \dfrac{W'}{W}$

a and b are taken from Table 3–2.

RESILIENCE

Resilience in tension or compression

$$U = \frac{\sigma^2 V}{2E} \tag{3–24}$$

Resilience in bending

$$U_b = \frac{\sigma_b^2 k^2\, Al}{6Ec^2} \tag{3–25}$$

where $\left(\dfrac{k}{c}\right)^2 = \frac{1}{8}$ for rectangular cross section

Particular	Formula
	$= \frac{1}{4}$ for circular section
Resilience in shear	$U_r = \dfrac{\tau_e^2 V}{2G}$ (3–26)
Resilience in torsion	$U_\tau = \dfrac{\tau_e^2 k_0^2 Al}{2Gc^2}$ (3–27)
	where $k_0 = \sqrt{(D_i^2 + D_0^2)/8}$ and $c = \frac{1}{2}D_0$
	for hollow shaft

GENERAL

For further information on resilience and stresses.   Refer to Tables 3–1 to 3–3.

TABLE 3–1
Maximum resilience per unit volume (2, 1)

Type of loading	Modulus of resilience, J (in lbf)
Tension or compression	$\dfrac{\sigma_e^2}{2E}$
Shear, simple transverse	$\dfrac{\tau_e^2}{2G}$
Bending in Beams	
With simply supported ends	
Concentrated center load and rectangular cross section	$\dfrac{\sigma_e^2}{18E}$
Concentrated center load and circular cross section	$\dfrac{\sigma_e^2}{24E}$
Concentrated center load and I-beam section	$\dfrac{3\sigma_e^2}{32E}$
Uniform load and rectangular section	$\dfrac{4\sigma_e^2}{45E}$
Uniform—strength beam, concentrated load, and rectangular section	$\dfrac{\sigma_e^2}{6E}$

TABLE 3–1
Maximum resilience per unit volume (2, 1) (*Cont.*)

Type of loading	Modulus of resilience, J (in lbf)
Fixed at both ends	
Concentrated load and rectangular cross section	$\dfrac{\sigma_e^2}{18E}$
Uniform load and rectangular cross section	$\dfrac{\sigma_e^2}{30E}$
Cantilever beam	
End load and rectangular cross section	$\dfrac{\sigma_e^2}{18E}$
Uniform load and rectangular cross section	$\dfrac{\sigma_e^2}{30E}$
Torsion	
Solid round bar	$\dfrac{\tau_e^2}{4G}$
Hollow round bar with D_0 greater than D_i	$\left[1 + \left(\dfrac{D_i}{D_0}\right)^2\right]\dfrac{\tau^2}{4G}$
Springs	
Laminated with flat leaves of uniform strength	$\dfrac{\sigma_e^2}{6E}$
Flat spiral with rectangular section	$\dfrac{\sigma_e^2}{24E}$
Helical with round section and axial load	$\dfrac{\tau_e^2}{4G}$
Helical with round section and axial twist	$\dfrac{\sigma_e^2}{8E}$
Helical with rectangular section and axial twist	$\dfrac{\sigma_e^2}{6E}$

Sources: K. Lingaiah and B. R. Narayana Iyengar, Machine Design Data Handbook, Vol. I (*SI and Customary Metric Units*), Suma Publishers, Bangalore, India, 1986, K. Lingaiah, Machine Design Data Handbook, Vol. II (*SI and Customary Metric Units*), Suma Publishers, Bangalore, India, 1986, and Maleev, V. L., and J. B. Hartman, *Machine Design*, International Textbook Company, Scranton, Pennsylvania, 1954.

TABLE 3–2
Coefficients in Eq. (3–23) (1)

Type of impact	a	b
Longitudinal impact on bar	$\frac{1}{3}$	$\frac{1}{2}$
Center impact on simple beam	$\frac{17}{35}$	$\frac{5}{8}$
Center impact on beam with fixed ends	$\frac{13}{35}$	$\frac{1}{2}$
End impact on cantilever beam	$\frac{4}{17}$	$\frac{3}{8}$

TABLE 3–3
Resilience in tension

Material	Elastic limit σ		Modulus of elasticity, E		Modulus of resilience, u		Impact strength (Izod no.)
	MPa	kpsi	GPa	Mpsi	J	in lbf	
Cast iron							
Class 20 (ordinary)	42.8[a]	6.2	68.9	10	0.22	1.9	
Class 25	68.9[a]	10.0	89.2	13	0.43	3.8	7.9
Nickel, Grade II	117.2[a]	17.0	24.5	18	0.90	8.0	
Malleable	137.9	20.0	172.6	25	0.90	8.0	2.7
Aluminum alloy, SAE 33	48.3	7.0	66.7	9.7	0.28	2.5	
Brass, SAE 40 or SAE 41	68.9	10.0	82.4	12	0.45	4.0	
Bronze, SAE 43	193.0	28.0	110.8	16	2.77	24.5	
Monel metal							
Hot-rolled	206.9	30.0	176.5	25.5	1.96	17.6	120
Cold-rolled, normalized	482.6	70.0	176.5	25.5	10.79	96	100
Steel							
SAE 1010	206.9	30.0		30.3	1.69	15	
SAE 1030	248.2	36.5	206.9	30	2.45	22	20
SAE 1050, annealed	330.9	48.5	204.8	29.7	4.27	38	
SAE 1095, annealed	413.7	60.0	204.8	29.7	6.77	60	
SAE 1095, tempered	517.1	75.0	204.8	29.7	16.08	94	
SAE 2320, annealed	310.3	45.0	204.8	29.7	3.82	34	52
SAE 2320, tempered	689.5	100.0	204.8	29.7	18.83	167	40
SAE 3250, annealed	551.6	80.0	213.7	31	21.58	193	
SAE 3250, tempered	1379.0	200.0	213.7	31.0	72.57	645	30
SAE 6150, annealed	427.6	62.0	213.7	31	6.96	62	
SAE 6150, tempered	1102.3	160.0	213.7	31	52.47	466	
Rubber	2.1	0.3	1034×10^{-9}	150×10^{-6}	33.89	300	

[a] Cast iron has no well-defined elastic limit, but the values may be safely used anyway for all practical purposes.

Source: Reproduced courtesy of V. L. Maleev and J. B. Hartman, *Machine Design*, International Textbook Co., Scranton, Pennsylvania, 1954.

REFERENCES

1. Maleev, V. L., and J. B. Hartman, *Machine Design*, International Textbook Co., Scranton, Pennsylvania, 1954.
2. Lingaiah, K., and B. R. Narayana Iyengar, *Machine Design Data Handbook*, Vol. I (*SI and Customary Metric Units*), Suma Publishers, Bangalore, India, 1986.
3. Lingaiah, K., *Machine Design Data Handbook*, Vol. II (*SI and Customary Metric Units*), Suma Publishers, Bangalore, India, 1986.

BIBLIOGRAPHY

Lingaiah, K., and B. R. Narayana Iyengar, *Machine Design Data Handbook*, Engineering College Co-operative Society, Bangalore, India, 1962.
Norman, C. A., E. S. Ault, and E. F. Zarobsky, *Fundamentals of Machine Design*, The Macmillan Company, New York, 1951.

STRESS CONCENTRATION AND STRESS INTENSITY IN MACHINE MEMBERS

SYMBOLS

a	diameter of circular hole (cut-out), m (in)*
	semimajor axis of elliptical hole (cut-out), m (in)
	half of the length of the slot, m (in)
$2a$	length of straight crack, m (in) (Figs. 4–32 and 4–33)
A	area of cross section, m (in)
b	semiminor axis of elliptical hole (cut-out), m (in)
	maximum breadth of section of curved bar, m (in)
	width of notch at the edge, m (in)
$b = (w - a)$	effective width of plate across the hole, m (in)
	or net width of plate, m (in)
$2b$	total width of plate with a crack, m (in) (Fig. 4–33)
B	constant in curved bar equation
	outside diameter of reinforced ring in an asymmetrically reinforced circular hole
c	distance from centroidal axis to extreme fiber of beam or inside edge of curved bar, m (in)
C	spring index
d	effective width of plate, m (in)
	width of U-piece at dangerous section, m (in)
	diameter of shaft at reduced section, m (in)
	reduced width of shoulder plate, m (in)
	diameter of hole (cut-out), m (in)
D	total diameter of shaft, m (in)
	total width of plate, m (in)
F	load, kN (lbf)
	force, kN (lbf)
	diametrically opposite concentrated loads on ring or hollow roller, kN (lbf)
h	thickness of plate, m (in)
	thickness of ring or roller, m (in)
	lever arm from critical section of tooth, m (in)
$2h$	length of plate with a crack, m (in) (Fig. 4–33)

h_1 depth of groove, m (in)

depth of shoulder, m (in)

I moment of inertia, area, m^4 (in^4)

J moment of inertia, polar, m^4 (in^4)

K_I opening mode or mode I of stress intensity factor, $MPa\sqrt{m}$ (kpsi $\sqrt{in}$)

K_{II} mode II of stress intensity factor, $MPa\sqrt{m}$ (kpsi $\sqrt{in}$)

K_{IC} opening mode or mode I of critical stress intensity factor or fracture toughness factor, $MPa\sqrt{m}$ (kpsi $\sqrt{in}$)

$K_\sigma = \dfrac{\sigma_{max}}{\sigma_{nom}}$ theoretical stress-concentration factor for normal stress

K'_σ combined stress-concentration factor (K'_σ is a theoretical factor)

K_τ theoretical stress-concentration factor for shear stress (torsion)

$K_{f\sigma}$ fatigue stress-concentration factor for normal stress or fatigue notch factor (normal) (Fig. 14–17) or fatigue strength reduction factor

$K_{f\sigma} = \dfrac{\sigma_f}{\sigma_{nf}} = \dfrac{\text{fatigue limit of unnotched specimen (axial or bending)}}{\text{fatigue limit of notched specimen (axial or bending)}}$

$K_{\sigma n}$ stress-concentration factor (normal) based on net area (nominal) of cross section (i.e., net nominal stress)

$K_{\sigma g}$ stress-concentration factor (normal) based on gross area of cross section (i.e., gross stress)

$K_{f\tau}$ fatigue stress-concentration factor for shear stress (torsion) or fatigue notch factor (torsion)

$= \dfrac{\tau_f}{\tau_{nf}} = \dfrac{\text{fatigue limit of unnotched specimen (torsion)}}{\text{fatigue limit of notched specimen (torsion)}}$

$K_{\sigma u}$ stress-concentration factor for U-grooved plate

$K_{\sigma v}$ stress-concentration factor for a V-grooved plate

$K_{\sigma e}$ effective stress-concentration factor under a static load, equivalent stress-concentration factor

MF magnification factor

M_b bending moment, N m (lbf ft)

M_t torsional moment, N m (lbf ft)

n safety factor

$q = \dfrac{K_{f\sigma} - 1}{K_\sigma - 1}$ index of sensitivity or notch sensitivity factor

r radius of curvature of groove or notch of curved bar, m (in)

minimum radius of curvature of an ellipse, m (in)

polar coordinate

r distance of a point in a plate from the crack tip (Fig. 4–32), m (in)

r_f minimum fillet radius of gear tooth, m (in)

r_t cutter tip radius, m (in)

R_o outside radius of ring or hollow roller, m (in)

R_i inside radius of ring or hollow roller, m (in)

s thickness of the tooth at critical section, m (in)

w depth of U-piece-arm, m (in)

W total width of flat plate, m (in)

$\sigma_x, \sigma_y, \tau_{xy}$	stress components in x, y coordinates, MPa (kpsi)
σ_{max}	maximum normal stress, MPa (kpsi)
σ_{nom}	nominal normal stress computed from F/A or M_bc/I or from an elementary formula which does not take into account the stress concentration, MPa (kpsi)
σ_0	average stress at the root of gear tooth, MPa (kpsi)
τ_{max}	maximum shear stress, MPa (kpsi)
τ_{nom}	nominal shear stress computed from $M_t r/J$, MPa (kpsi)
α	angle of a shallow U-groove, deg
β	angle of V-groove, deg
θ	polar coordinate, deg
	angle made by r the distance of a point from tip of crack with x axis (Fig. 4–32)
	Other factors in performance or in special aspects are included from time to time in this chapter, and being applicable only in their immediate context, are not given at this stage

Particular	Stress concentration factor theoretical/empirical or otherwise	
	Extreme value	**Formula**
(a) Keyway (Fig. 4–1 and Tables 4–1 and 4–2: Profile keyway	$K_{fr} = 1.68$	
Sled-runner keyway	$K_{fr} = 1.44$	

TABLE 4–1
Shear stress-concentration factor for a keyway in a shaft subjected to torsion (by Leven)

r/d	0.0052	0.0104	0.0208	0.0417	0.0833
K_r	3.92	3.16	2.62	2.30	2.06

TABLE 4–2
Stress-Concentration Factor K_{fr} for Keyways

Type of keyway	Annealed		Hardened	
	Bending	Torsion	Bending	Torsion
Profile	1.6	1.3	2.0	1.6
Sled runner	1.3	1.3	1.6	1.6

(b) Curved bar (Fig. 4–2):
For curved bar

$$K_\sigma = 1.00 + B\left(\frac{I}{bc^2}\right)\left(\frac{1}{r-c} + \frac{1}{r}\right) \tag{4–1}$$

where

$B = 1.05$ for circular or elliptical cross section
$ = 0.5$ for other cross section

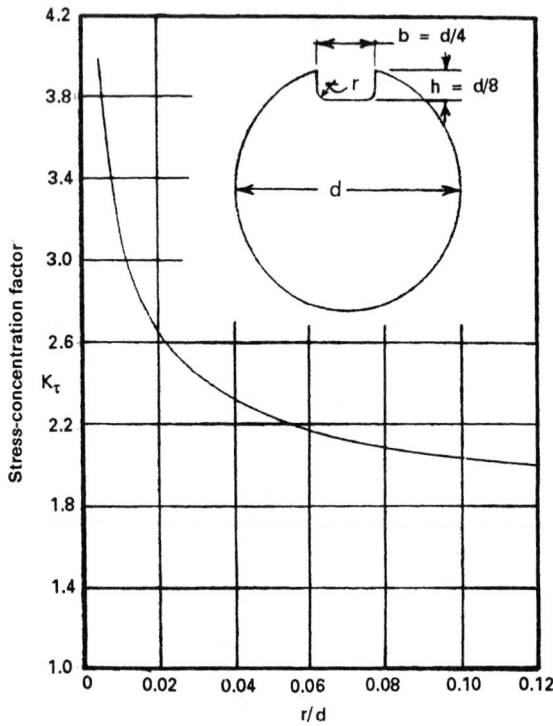

FIGURE 4–1 Stress-concentration factor for the straight portion of a keyway in a shaft in torsion

Particular	Stress concentration factor theoretical/empirical or otherwise	
	Extreme value	**Formula**
(c) Spur gear tooth (Figs. 4–2 and 4–3, Table 4–3): At root fillet of an involute tooth profile of 14.5° presure angle	$K_\tau = 2 \text{ to } 2.5$	$K_\sigma = 0.22 + \dfrac{1}{\left(\dfrac{r_f}{s}\right)^{0.2}\left(\dfrac{h}{s}\right)^{0.4}}$ (4–2)

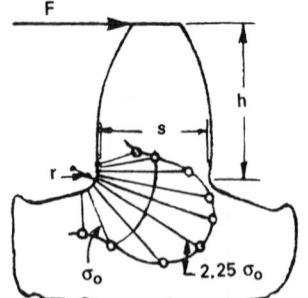

FIGURE 4–2 Stress-concentration factor at root of gear tooth.

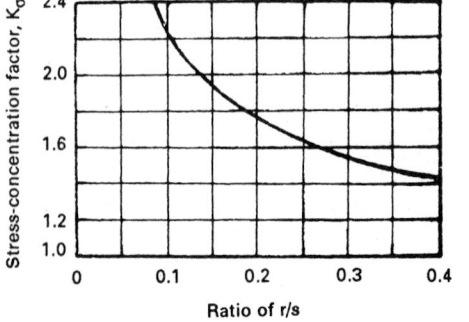

FIGURE 4–3 Effect of fillet radius on stress-concentration at root of gear tooth.

Particular	Stress concentration factor theoretical/empirical or otherwise	
	Extreme value	**Formula**

TABLE 4–3
Stress-concentration factors at root fillet of gear tooth

m, mm	$K_{\sigma t}$	$K_{\sigma c}$
6.24	1.47	1.61
5	1.47	1.61
4.25	1.42	1.57
3.5	1.35	1.50
3	1.345	1.50

At root fillet of a full-depth involute tooth profile of 20° pressure angle

$$K_\sigma = 0.18 + \frac{1}{\left(\frac{r_f}{s}\right)^{0.15} \left(\frac{h}{s}\right)^{0.45}} \quad (4\text{–}3)$$

At root fillet of an involute tooth profile of 25° pressure angle

$$K_\sigma = 0.14 + \left(\frac{s}{r_f}\right)^{0.11} \left(\frac{s}{h}\right)^{0.50} \quad (4\text{–}3a)$$

(*d*) Circular cut-outs (holes) in plates (Fig. 4–4*c*):
For infinite plate in:

  (i) Uniaxial tension (Fig. 4–4*c*) $K_\sigma = 3$ $K_\sigma = \frac{1}{2}\left(2 + \frac{a^2}{r^2} + \frac{3a^4}{r^4}\right)_{r=a} = 3$

$$\hspace{10cm} (4\text{–}4)$$

  (ii) Biaxial tension $K_\sigma = 2$ $K_\sigma = \left(1 + \frac{a^2}{r^2}\right)_{r=a} = 2 \quad (4\text{–}5)$

  (iii) Pure shear $K_\tau = 4$ $K_\tau = \left(1 + \frac{3a^4}{r^4}\right)_{r=a} = 4 \quad (4\text{–}6)$

For stress concentration factor for a semi-infinite plate with a circular hole near the edge under tension. $K_\sigma = 3$

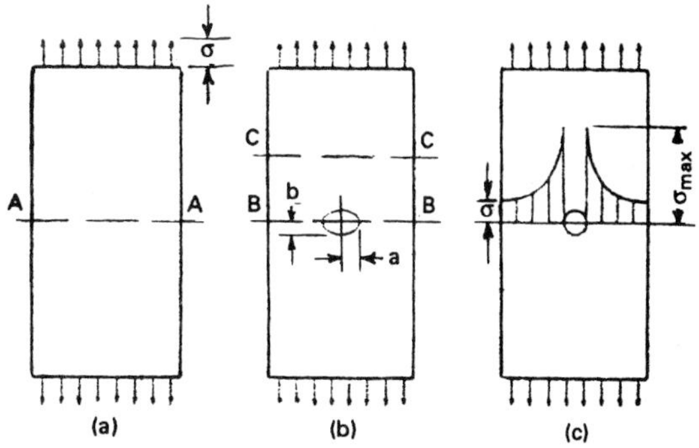

FIGURE 4–4 Stress distribution around holes (cut-outs) in plate in tension.

Particular	Stress concentration factor theoretical/empirical or otherwise	
	Extreme value	Formula
For finite plate in:		
(i) Uniaxial tension (Fig. 4–5)	$K_\sigma = 3$	$K_\sigma = 2 + \left(\dfrac{b}{w}\right)^3$ (4–7)
(ii) Bending (Fig. 4–6)	$K_\sigma = 2$	$K_\sigma = 2\dfrac{b}{w}$ (4–8)
(e) Filled circular hole: For filled circular holes in plate subjected to tension	$K_\sigma = 2.5$	
(f) Reinforced circular holes:		
(i) For stress-concentration factor for a symmetrically reinforced circular hole in a flat plate under uniform uniaxial tension		
(ii) For stress-concentration factor for an asymmetrically reinforced circular hole in a flat plate under uniform uniaxial tension		Refer to Fig. 4.7a, b and c.

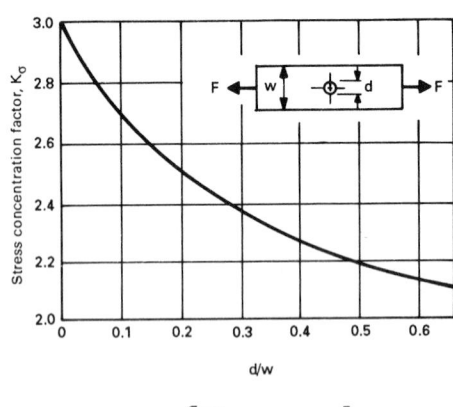

$$K_\sigma = \frac{\sigma_{max}}{\sigma_{nom}} \; ; \; \sigma_{nom} = \frac{F}{(w-d)h}$$

FIGURE 4–5 Stress-concentration factor for a plate of finite width with a circular hole (cut-out) in tension. ("Design Factors for Stress Concentration," *Machine Design*, Vol. 23, Nos. 2 to 7, 1951 and McGraw-Hill, 1977.)

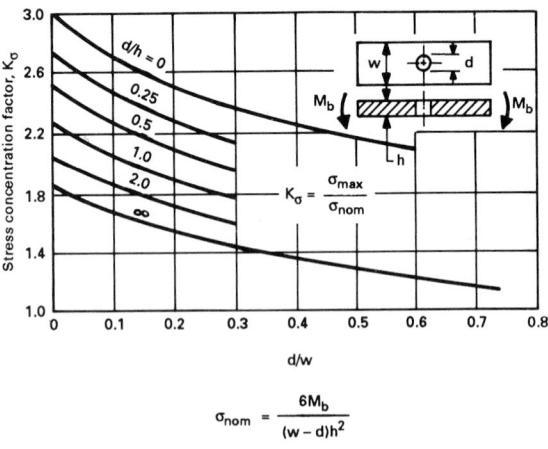

$$\sigma_{nom} = \frac{6M_b}{(w-d)h^2}$$

FIGURE 4–6 Stress-concentration factor for a plate of finite width with transverse circular hole (cut-out) subjected to bending. ("Design Factors for Stress Concentration," *Machine Design*, Vol. 23, Nos. 2 to 7, 1951 and McGraw-Hill, 1977.)

Particular	Stress concentration factor theoretical/empirical or otherwise	
	Extreme value	**Formula**

(g) Noncircular holes (cut-outs) in a plate:

(i) An infinite plate with an elliptical hole (cut-out) in uniaxial tension (load at right angles to major axis) (Fig. 4–4b)

$$K_\sigma = 1 + 2\frac{a}{b} \qquad (4\text{–}9a)$$

$$K_\sigma = 1 + 2\sqrt{\frac{a}{r}} \qquad (4\text{–}9b)$$

(ii) An infinite plate with elliptical hole (cut-out) in uniaxial tension (load parallel to major axis)

$$K_\sigma = 1 + \frac{2b}{a} \qquad (4\text{–}9c)$$

(iii) An infinite plate with elliptical hole (cut-out) in pure shear

$$K_\tau = 2\left(1 + \frac{a}{b}\right) \qquad (4\text{–}10)$$

(iv) An infinite plate with an elliptical cut-out in biaxial tension

$$K_\sigma = \frac{2a}{b} \qquad (4\text{–}11)$$

(v) Transverse bending of a plate containing an elliptical cut-out (or hole)

$$K_\sigma = \frac{\left(1 + v\right)\left(3 - v + \dfrac{2a}{b}\right)}{(3 + v)} \qquad (4\text{–}12)$$

(vi) Slotted plate loaded in tension or bending

$$K_\sigma = 1.064 + 0.788\frac{a}{b} \text{ for } v = 0.3$$
$$(4\text{–}13a)$$

$$K_\sigma = 2 + \left(\frac{b}{w}\right)^3 \text{ for } \frac{a}{r} = 1 \quad (4\text{–}13b)$$

For reduction of endurance strength of steel

Refer to Fig. 4.8.

(h) U-shaped member subjected to bending (Fig. 4–9):

(1) At 0° with horizontal axis

$$K_{\sigma A} = 1 + \frac{d}{4r} \qquad (4\text{–}14)$$

(2) At 70° with horizontal axis

$$K_{\sigma B} = 1 + \frac{w}{5r} \qquad (4\text{–}15)$$

(i) Helical spring:

Stress concentration or Wahl's correction factor for helical spring

$$K_\tau = \frac{4c - 1}{4c - 4} + \frac{0.615}{c} \qquad (4\text{–}16)$$

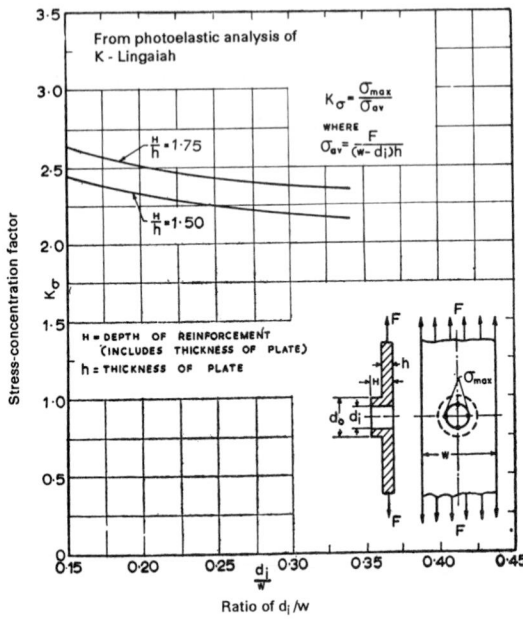

FIGURE 4–7(a) Stress-concentration factor for an asymmetrically reinforced circular hole (cut-out) in a flat plate subjected to tension. (Ph.D work of the author.)

FIGURE 4–7(c)

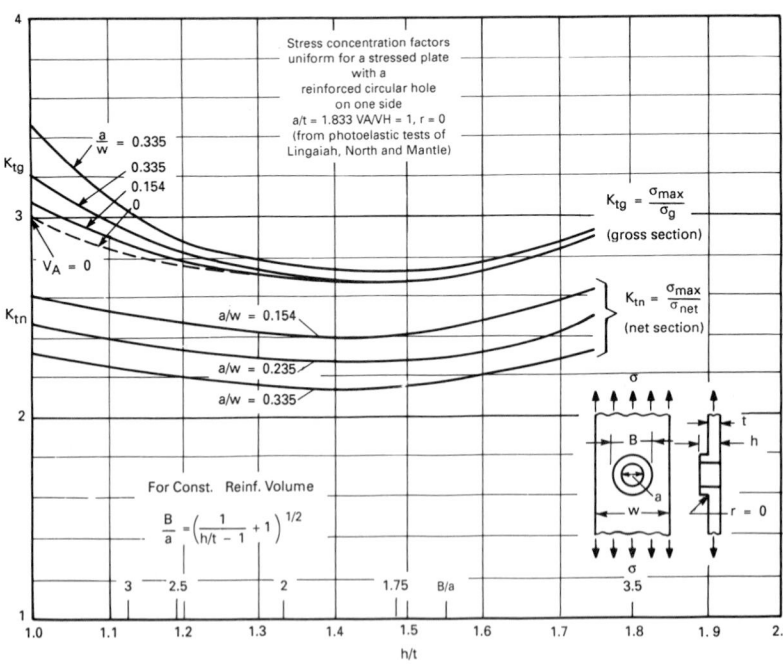

FIGURE 4–7(b) Stress-concentration factor for an asymmetrically reinforced circular hole in a flat plate subjected to uniform unidirectional tensile stress. (Ph.D work of the author, and R. E. Peterson, *Stress Concentration Factors*, John Wiley and Sons, Inc., 1974.)

Particular	Stress concentration factor theoretical/empirical or otherwise	
	Extreme value	**Formula**

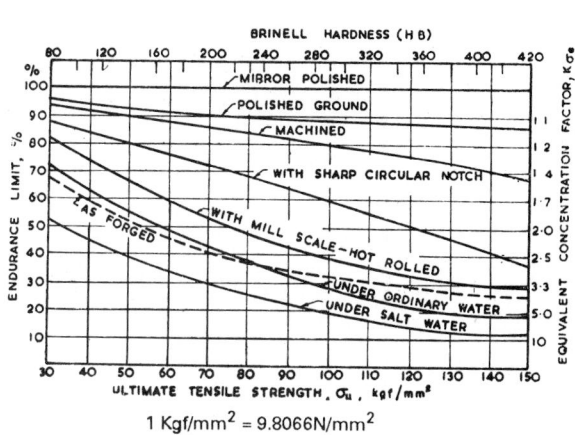

1 Kgf/mm^2 = 9.8066N/mm^2

FIGURE 4–8 Reduction of endurance strength of steel, σ_f.

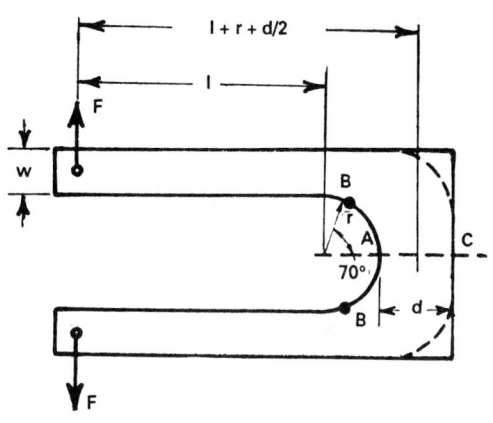

FIGURE 4–9 U-shaped member. (R. E. Peterson, *Stress Concentration Factors*, John Wiley and Sons, 1974.)

(*j*) Ring or hollow roller:

For the ring loaded internally

$$K_\sigma = \frac{\sigma_{max}[2h(R_o - R_i)]}{F\left[1 + 3\dfrac{(R_o + R_i)\left(1 - \dfrac{2}{\pi}\right)}{R_o - R_i}\right]}$$

(4–17)

For the ring loaded externally

$$K_\sigma = \frac{\sigma_{max}[\pi h(R_o - R_i)]}{3F(R_o + R_i)}$$

(4–18)

(k) Shafts with transverse holes (Fig. 4–10):

    (i) Shaft with a circular hole subjected to transverse bending for $a/d \to 0$ $K_\sigma = 3.0$

    (ii) Shaft with a circular hole subjected to torsion for $a/d \to 0$ $K_\tau = 2.0$

(*l*) Shafts with grooves:

Shaft with U and V circumferential groove in:

    (i) Tension or bending (Figs. 4–11 to 4–17)

$$K_\sigma = 1 + 2\sqrt{\frac{h_1}{r}}$$

(4–19)

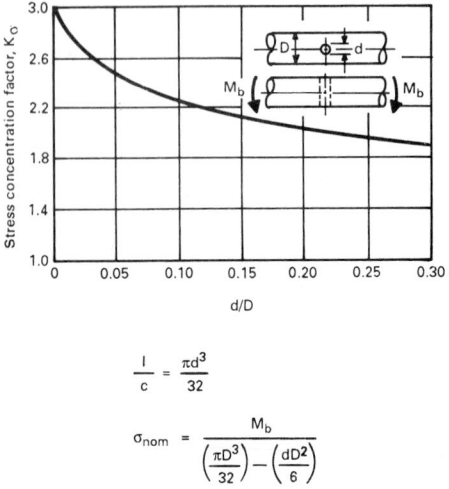

$$\frac{I}{c} = \frac{\pi d^3}{32}$$

$$\sigma_{nom} = \frac{M_b}{\left(\frac{\pi D^3}{32}\right) - \left(\frac{dD^2}{6}\right)}$$

FIGURE 4–10 Stress-concentration factor K_σ for a shaft with transverse circular hole subjected to bending. (R. E. Peterson, "Stress Concentration Factors," *Machine Design*, Vol. 23, Nos. 2–7, 1951 and McGraw-Hill, 1977.)

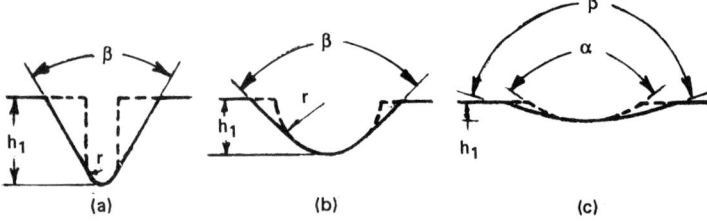

FIGURE 4–11 Types of V-grooves.

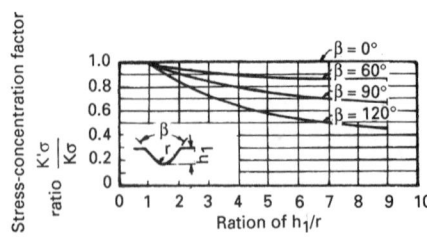

FIGURE 4–12 Stress-concentration factor ratio due to notches of various shapes.

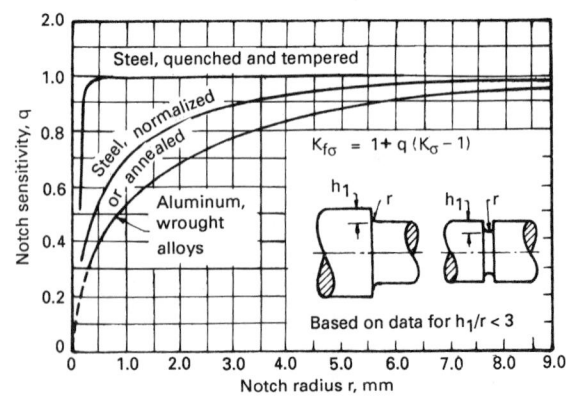

FIGURE 4–13 Average notch sensitivity.

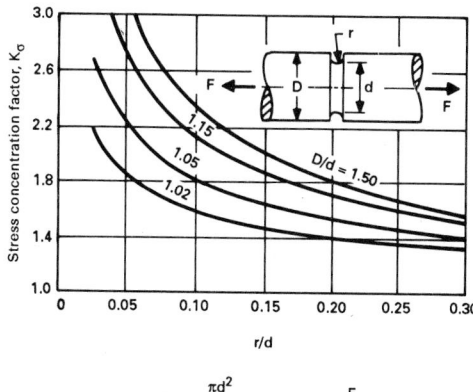

$$A = \frac{\pi d^2}{4} \; ; \quad \sigma_{nom} = \frac{F}{A}$$

FIGURE 4–14 Stress-concentration factor K_σ for a grooved shaft in tension. (R. E. Peterson, "Design Factors for Stress Concentration," *Machine Design*, Vol. 23, Nos. 2–7, 1951 and McGraw-Hill, 1977.)

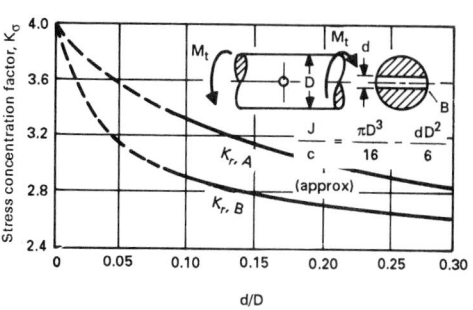

FIGURE 4–15 Round shaft in torsion with transverse hole. (R. E. Peterson, "Design Factors for Stress Concentration," *Machine Design*, Vol. 23, Nos. 2–7, 1951 and McGraw-Hill, 1977.)

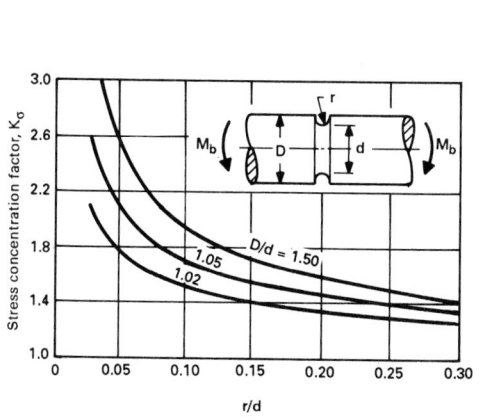

$$\frac{I}{c} = \frac{\pi d^3}{32} \; ; \quad \sigma_{nom} = \frac{M_b c}{I}$$

FIGURE 4–16 Stress-concentration factor K_σ for grooved shaft in bending. (R. E. Peterson, "Design Factors for Stress Concentration", *Machine Design*, Vol. 23, Nos. 2–7, 1951 and McGraw-Hill, 1977.)

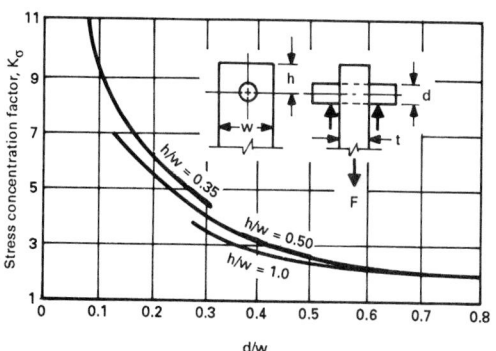

Plate loaded in tension by a pin through a hole, $\sigma_0 = F/A$, where $A = (w - d)t$. When clearance exists, increase K_t 35 to 50 percent. (M. M. Frocht and H. N. Hill, "Stress Concentration Factors around a Central Circular Hole in a Plate Loaded through a Pin in Hole," J. Appl. Mechanics, vol. 7, no. 1, March 1940, p. A-5.)

FIGURE 4–17 McGraw-Hill, 1977.

| Particular | Stress concentration factor theoretical/empirical or otherwise | |
	Extreme value	Formula

(ii) Torsion (Fig. 4–22)

$$K_\tau = 1 + \sqrt{\frac{h_1}{r}} \qquad (4\text{–}20)$$

(iii) Shaft with a small elliptical groove in torsion

$$K_\tau = 1 + \sqrt{\frac{h_1}{r}} \qquad (4\text{–}21a)$$

or

$$K_\tau = 1 + \frac{h_1}{b} \qquad (4\text{–}21b)$$

$$K_\tau = 1 + \frac{b}{r} \qquad (4\text{–}21c)$$

(*m*) Shouldered shaft in torsion (Fig. 4–19):

$$K_\tau = 1 + \left(S\frac{d}{r}\right)^{0.65} \qquad (4\text{–}22a)$$

where S is some function of $\dfrac{D}{d}$

$$K_\tau = 1 + \frac{d}{12r}\left[1 - \frac{\left(1 + 2\frac{r}{d}\right)}{\frac{D}{d}\left(1 + 6\frac{r}{d}\right)}\right] \qquad (4\text{–}22b)$$

For stress-concentration factor and combined factor for stepped-shaft in tension and bending

Refer to Figs. 4–24 to 4–27.

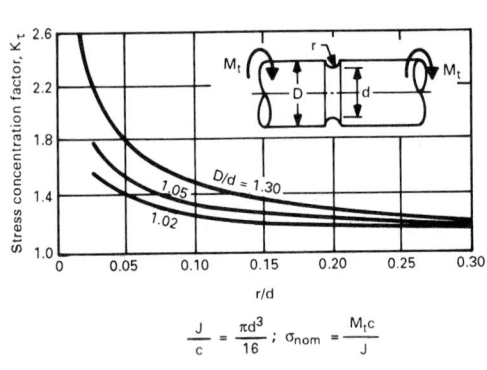

$$\frac{J}{c} = \frac{\pi d^3}{16}; \ \sigma_{nom} = \frac{M_t c}{J}$$

FIGURE 4–18 Stress-concentration factor K_τ for grooved shaft in torsion. (R. E. Peterson, "Design Factors for Stress Concentration", *Machine Design*, Vol. 23, Nos. 2–7, 1951 and McGraw-Hill, 1977.)

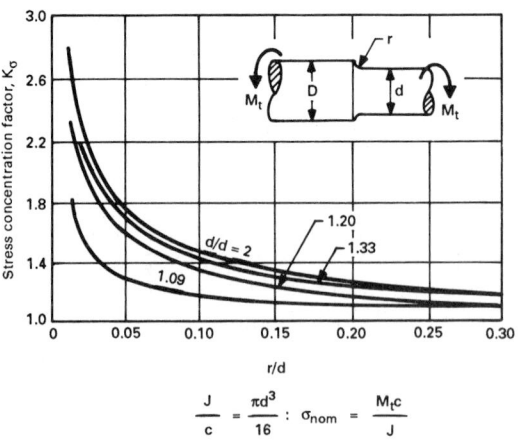

$$\frac{J}{c} = \frac{\pi d^3}{16} : \ \sigma_{nom} = \frac{M_t c}{J}$$

FIGURE 4–19 Stress-concentration factor K_τ for stepped shaft in torsion. (R. E. Peterson, "Design Factors for Stress Concentration," *Machine Design*, Vol. 23, Nos. 2–7, 1951 and McGraw-Hill, 1977.)

Particular	Stress concentration factor theoretical/empirical or otherwise	
	Extreme value	Formula

(*n*) Bar containing grooves:

(i) Bar with U, semicircular or shallow grooves symmetrically placed in tension (Figs. 4–11, 4–12, 4–22)

$$K_\sigma = 1 + \left[\frac{1}{\left(1.55\dfrac{D}{d} - 1.3\right)} \frac{h_1}{r} \right]^n \qquad (4\text{–}23a)$$

or

$$K_\sigma = 1 + \left[\frac{\left(\dfrac{D}{d} - 1\right)}{2\left(1.55\dfrac{D}{d} - 1.3\right)} \frac{d}{r} \right]^n \qquad (4\text{–}23b)$$

$$\text{where } n = \frac{\left(\dfrac{D}{d} - 1\right) + 0.5\sqrt{\dfrac{h_1}{r}}}{\left(\dfrac{D}{d} - 1\right) + \sqrt{\dfrac{h_1}{r}}}$$

(ii) Bar with deep V-groove in tension for

$$\frac{r}{h_1} < 1 \,(\text{Fig. } 4\text{–}11a)$$

$$K_{\sigma v} = 1 + (K_{\sigma v} - 1)\left\{ 1 - \left(\frac{\beta}{180}\right)^{1 + 2.4\sqrt{\frac{r}{h_1}}} \right\} \qquad (4\text{–}24)$$

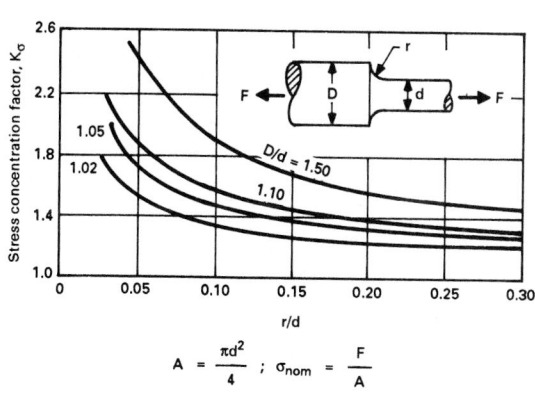

$$A = \frac{\pi d^2}{4} \; ; \; \sigma_{nom} = \frac{F}{A}$$

FIGURE 4–20 Stress-concentration factor K_σ for stepped shaft in tension. (R. E. Peterson, "Design Factors for Stress Concentration," *Machine Design*, Vol. 23, Nos. 2–7, 1951 and McGraw-Hill, 1977.)

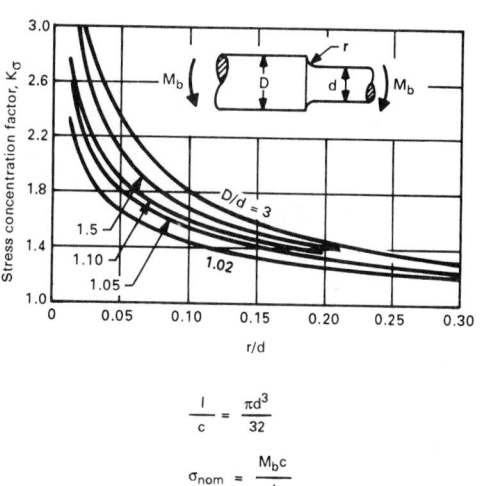

$$\frac{I}{c} = \frac{\pi d^3}{32}$$

$$\sigma_{nom} = \frac{M_b c}{I}$$

FIGURE 4–21 Stress-concentration factor K_σ for stepped shaft in bending. (R. E. Peterson, "Design Factors for Stress Concentration," *Machine Design*, Vol. 23, Nos. 2–7, 1951 and McGraw-Hill, 1977.)

Particular	Stress concentration factor theoretical/empirical or otherwise	
	Extreme value	Formula

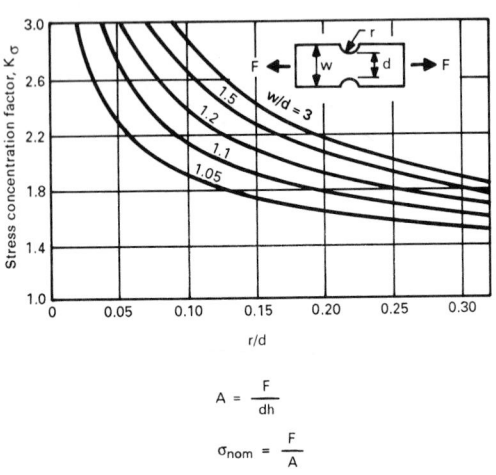

$$A = \frac{F}{dh}$$

$$\sigma_{nom} = \frac{F}{A}$$

h = thickness of flat bar

FIGURE 4–22 Stress-concentration factor K_σ for notched flat bar in tension. (R. E. Peterson, "Design Factors for Stress Concentration," *Machine Design*, Vol. 23, Nos. 2–7, 1951 and McGraw-Hill, 1977.)

(iii) Bar with shallow V-groove in tension for

$$\frac{r}{h_1} > 1$$

$$K_{\sigma v} = 1 + (K_{\sigma v} - 1)\left\{1 - \left(\frac{\beta - \alpha}{180 - \alpha}\right)^{1+2.4\sqrt{\frac{r}{h_1}}}\right\} \tag{4–25}$$

(iv) Elliptical groove at the edge of plate in tension

$$K_\sigma = 1 + \frac{2h_1}{b} \tag{4–26a}$$

$$K_\sigma = 1 + 2\sqrt{\frac{h_1}{r}} \tag{4–26b}$$

(v) Bar with symmetrical U, semicircular shallow grooves in bending (Fig. 4–23).

$$K_\sigma = 1 + \left[\frac{1}{4.27\frac{D}{d} - 4}\frac{h_1}{r}\right]^{0.85} \tag{4–27a}$$

or

$$K_\sigma = 1 + \left[\frac{\left(\frac{D}{d} - 1\right)}{2\left(4.27\frac{D}{d} - 4\right)}\frac{d}{r}\right]^{0.85} \tag{4–27b}$$

	Stress concentration factor theoretical/empirical or otherwise	
Particular	**Extreme value**	**Formula**

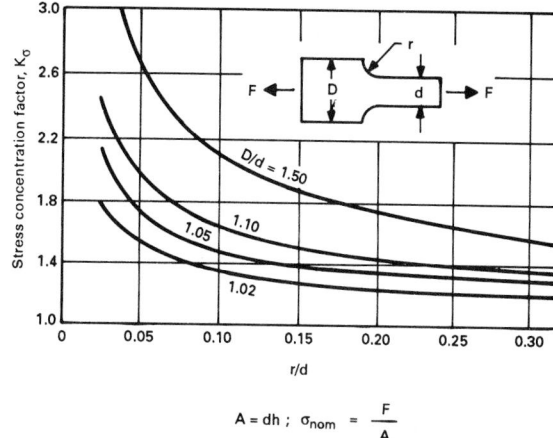

FIGURE 4–23 Stress-concentration factor K_σ for notched flat bar in bending. (R. E. Peterson, "Design Factors for Stress Concentration," *Machine Design*, Vol. 23, Nos. 2–7, 1951 and McGraw-Hill, 1977.)

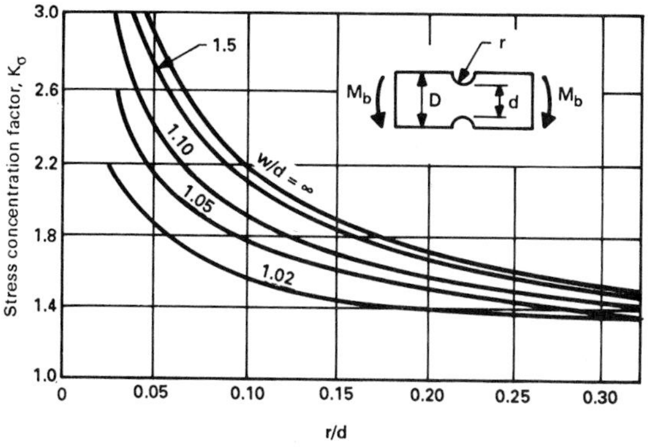

$$\frac{I}{c} = \frac{hd^2}{6} \; ; \; \sigma_{nom} = \frac{M_b c}{I}$$

FIGURE 4–24 Stress-concentration factor K_σ for filleted flat bar in tension. (R. E. Peterson, "Design Factors for Stress Concentration," *Machine Design*, Vol. 23, Nos. 2–7, 1951 and McGraw-Hill, 1977.)

$$A = dh \; ; \; \sigma_{nom} = \frac{F}{A}$$

$$h = \text{thickness of bar}$$

For stress-concentration factors for small grooves in a shaft subjected to torsion.

(*o*) Bar containing shoulders:

   (i) Bar with shoulders in tension (Fig. 4–24)

Refer to Table 4–4.

$$K_\sigma = 1 + \left[\frac{1}{\left(2.8 \dfrac{D}{d} - 2\right)} \frac{h_1}{r} \right]^{0.85} \qquad (4\text{–}28a)$$

or

$$K_\sigma = 1 + \left[\frac{\left(\dfrac{D}{d} - 1\right)}{2\left(2.8 \dfrac{D}{d} - 2\right)} \frac{d}{r} \right]^{0.85} \qquad (4\text{–}28b)$$

TABLE 4–4
Stress-concentration factors for relatively small grooves in a shaft subject to torsion, K_σ

Included angle of V, deg.	$\dfrac{h_1}{r}$				
	0.5	1	3	5	2
0	1.85	2.01	2.66	3.23	4.54
60	1.84	2.00	2.54	3.06	3.99
90	1.81	1.95	2.40	2.40	3.12
120	1.66	1.75	1.95	2.00	2.13

Particular	Stress concentration factor theoretical/empirical or otherwise	
	Extreme value	**Formula**

(ii) Bar with shoulders in bending (Fig. 4–25)

$$K_\sigma = 1 + \left[\frac{1}{\left(5.37\dfrac{D}{d} - 4.8\right)} \frac{h_1}{r} \right]^{0.85} \qquad (4\text{–}29a)$$

or

$$K_\sigma = 1 + \left[\frac{\left(\dfrac{D}{d} - 1\right)}{2\left(5.37\dfrac{D}{d} - 4.8\right)} \frac{d}{r} \right]^{0.85} \qquad (4\text{–}29b)$$

(*p*) Press-fitted or shrink-fitted members (Table 4–5):

   (i) Plain member $K_\sigma = 1.95$

   (ii) Grooved member $K_\sigma = 1.34$

   (iii) Plain member $K_{f\sigma} = 2.00$

   (iv) Grooved member $K_{f\sigma} = 1.70$

(*q*) Bolts and nuts (Tables 4–6 and 4–7)

  Bolt and nut of standard proportions $K_\sigma = 3.85$

  Bolt and nut having lip $K_\sigma = 3.00$

TABLE 4–5

Stress-concentration factors in shrink-fitted members

Particular	K_σ	$K_{f\sigma}$
Plain	1.95	2.00
Grooved	1.34	1.70

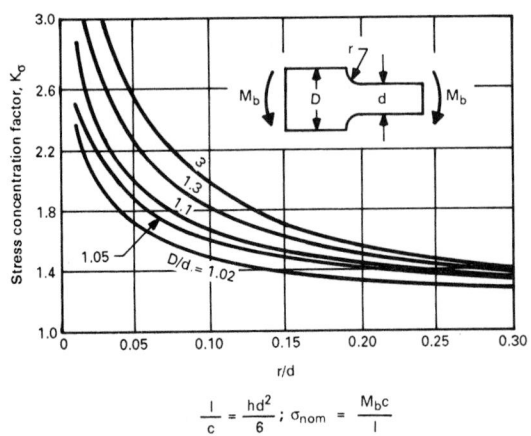

$$\frac{I}{c} = \frac{hd^2}{6} ; \quad \sigma_{nom} = \frac{M_b c}{I}$$

FIGURE 4–25   Stress-concentration factor K_σ for stepped bar in bending. (R. E. Peterson, "Design Factors for Stress Concentration," *Machine Design*, Vol. 23, Nos. 2–7, 1951 and McGraw-Hill, 1977.)

TABLE 4–6
Stress-concentration factors for screw threads

	Analysis			
Types of thread	Seely and Smith	Black (8)	Peterson (1)	Suggested value
Square	2.0			
Sharp V	3.0			
Whitworth	2.0		3.35	5 to 6
US standard Medium Carbon Steel	2.5			
National coarse thread Heat-treated Nickel steel		2.84 3.85		

TABLE 4–7
Stress-concentration factors $K_{f\sigma}$ for screw threads

	Annealed		Hardened	
Type of thread	Rolled	Cut	Rolled	Cut
Sellers, American National, square thread	2.2	2.8	3.0	3.8
Whitworth rounded roots	1.4	1.8	2.6	3.3

TABLE 4–8
Stress-concentration factors for welds

Location	K_σ
End or parallel fillet weld	2.7
Reinforced butt	1.2
Tee of transverse fillet weld	1.5
T-butt weld with sharp corners	2.0

TABLE 4–9
Index of sensitivity for repeated stress

	Average index of sensitivity q		
Material	Annealed or soft	Heat-treated and drawn at 921 K (648°C)	Heat-treated and drawn at 755 K (482°C)
Armco iron, 0.02% C	0.15–0.20		
Carbon steel			
0.10% C	0.05–0.10		
0.20% C (also cast steel)	0.10		
0.30% C	0.18	0.35	0.45
0.50% C	0.26	0.40	0.50
0.85% C		0.45	0.57
Spring steel, 0.56%, 2.3 Si, rolled		0.38	
SAE 3140, 0.73 C; 0.6 Cr; 1.3 Ni	0.25	0.45	
Cr–Ni steel 0.8 Cr; 3.5 Ni		0.25	0.70
Stainless steel, 0.3 C; 8.3 Cr; 19.7 Ni	0.16		
Cast iron	0–0.05		
Copper, electrolitic	0.07		
Duraluminum	0.05–0.13		

	Stress concentration factor theoretical/empirical or otherwise	
Particular	**Extreme value**	**Formula**
(r) Crane hook:		
For crane hook under tensile load	$K_\sigma = 1.56$	
(s) Rotating disk:		
For rotating disk with a hole for		
$\dfrac{R_i}{R_o} \to 0$	$K_\sigma = 2$	
For thin disk (ring)	$K_\sigma = 1$	
(t) Eye bar:		
For eye bar subjected to tensile load	$K_\sigma = 2.8$	
Stress concentration factors for welds	Refer to Table 4–8.	
(u) Notch sensitivity factors (Table 4–9):		
(i) Notch sensitivity factor for normal stress	$q_\sigma = \dfrac{K_{f\sigma} - 1}{K_\sigma - 1}$	(4–30a)
	$q_\sigma = \dfrac{K_{f\sigma} - 1}{K'_\sigma - 1}$	(4–30b)
For index of sensitivity for repeated stresses.	Refer to Table 4–9.	
(ii) Fatigue stress concentration factor for normal stress	$K_{f\sigma} = 1 + q_\sigma(K_\sigma - 1)$	(4–31a)
	$K_{f\sigma} = 1 + q_\sigma(K'_\sigma - 1)$	(4–31b)
(iii) Notch sensitivity factor for shear stress	$q_\tau = \dfrac{K_{f\tau-1}}{K_\tau - 1}$	(4–32)
(iv) Fatigue stress-concentration factor for shear stress	$K_{f\tau} = 1 + q_\tau(K_\tau - 1)$	(4–33)

STRESS INTENSITY FACTOR OR FRACTURE TOUGHNESS FACTOR

The localized stress components at the vicinity of "opening mode" or "mode I" crack tip in a flat plate subjected to uniform applied stress σ at infinity from the theory of fracture mechanics (Fig. 4–26)

$$\sigma_x = \frac{K_I}{\sqrt{2\pi r}} \cos\frac{\theta}{2}\left[1 - \sin\frac{\theta}{2}\sin\frac{3\theta}{2}\right] \tag{4–34a}$$

$$\sigma_y = \frac{K_I}{\sqrt{2\pi r}} \cos\frac{\theta}{2}\left[1 + \sin\frac{\theta}{2}\sin\frac{3\theta}{2}\right] \tag{4–34b}$$

Particular	Stress concentration factor theoretical/empirical or otherwise	
	Extreme value	**Formula**

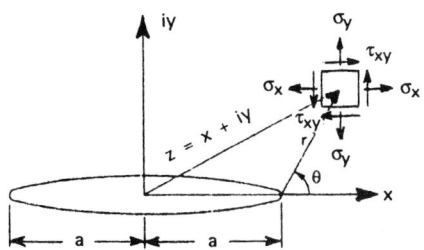

FIGURE 4–26   State of stress in the vicinity of a crack tip.

$$\tau_{xy} = \frac{K_I}{\sqrt{2\pi r}} \sin\frac{\theta}{2} \cos\frac{\theta}{2}\cos\frac{3\theta}{2} \tag{4–34c}$$

The stress intensity factor for a centrally located straight crack in an infinite plate subjected to uniform uniaxial tensile stress, σ perpendicular to the plane of the crack	$K = \sqrt{\pi}\,\sigma\sqrt{a}$	(4–35)
The definition and unit of critical stress intensity factor K_{Ic}	K_{Ic} is the critical stress intensity factor for static loading and plane-strain conditions of maximum constraints and is also referred to as the fracture toughness factor of the material at the onset of rapid fracture and has dimension of (stress $\sqrt{\text{length}}$, i.e., MPa$\sqrt{\text{m}}$ (kpsi $\sqrt{\text{in}}$)	
For values of critical stress-intensity factor (K_{Ic}) for some engineering materials	Refer Table 4–10	

TABLE 4–10
K_{IC} **for some engineering materials**

Material		K_{IC}		Yield strength, σ_{sy}	
Previous designation	**UNS designation**	**MPa $\sqrt{\text{m}}$**	**kpsi $\sqrt{\text{in}}$**	**MPa**	**kpsi**
Aluminum					
2024–T851	A92024–T851	26	24	455	66
7075–T651	A97075–T651	24	22	495	72
7178		33	30	490	71
Titanium					
Ti–6AL–4V	R56401	115	105	910	132
Ti–6AL–4V*	R56401*	55	50	1035	150
Steel					
4340	G43400	99	90	860	125
4340*	G43400*	60	55	1515	220
52100	G52986	14	13	2070	300

Particular	Stress concentration factor theoretical/empirical or otherwise	
	Extreme value	**Formula**
The stress-intensity factor for a centrally located straight crack in an infinite plate subjected to uniform shear stress τ	$K_I - iK_{II} = -i\sqrt{\pi}\,\tau\sqrt{a}$	(4–36)
The stress-intensity magnification factor for a centrally located straight crack of length $2a$ in a flat plate whose length $2h$ and width $2b$ are very large compared with the crack length subjected to uniform uniaxial tensile stress σ.	$MF = \dfrac{K_I}{\sqrt{\pi}\,\sigma\sqrt{a}}$	(4–37)
For stress-intensity magnification factors of plates with straight crack located at various positions in the plate and cylinders subjected to various types of rate of loadings and for various values of $a/b, a/d, a/h, a/(r_o - r_i)$ amd other ratios	Refer to Figs. from 4–27, 4–29 to 4–34	
The factor of safety	$n = \dfrac{K_{Ic}}{K}$	(4–38)

FIGURE 4–27   Stress intensity magnification factor $K_I/\sqrt{\pi}\,\sigma\sqrt{a}$ for various ratios of a/b of a flat plate with a centrally located straight crack under the action of uniform uniaxial tensile stress σ. (J. E. Shigley and L. D. Mitchell, *Mechanical Engineering Design*, McGraw-Hill, 1983.)

FIGURE 4–28   Modes of deformation of crack tip.

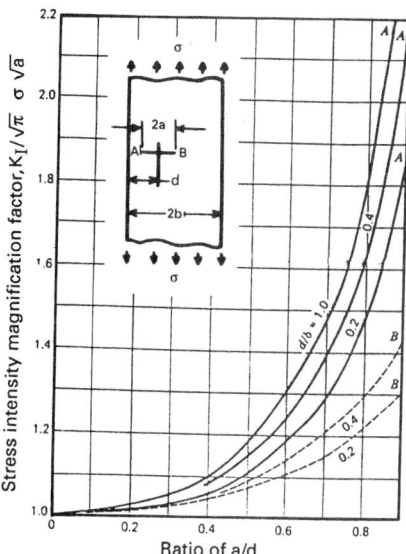

FIGURE 4–29 Stress intensity magnification factor $K_I/\sqrt{\pi}\,\sigma\sqrt{a}$ for an off-center straight crack in a flat plate subjected to uniform unidirectional tensile stress σ; solid curves are for the crack tip at A; dashed curves for tip at B. (J. E. Shigley and L. D. Mitchell, *Mechanical Engineering Design*, McGraw-Hill, 1983.)

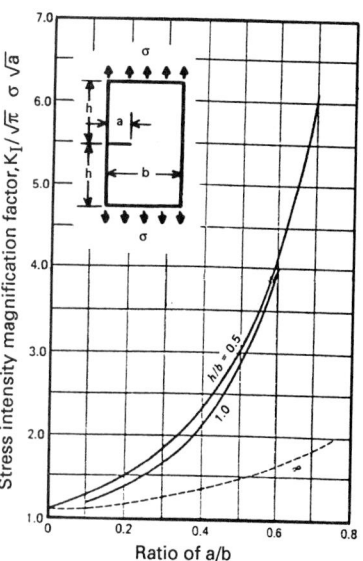

FIGURE 4–30 Stress intensity magnification factor $K_I/\sqrt{\pi}\,\sigma\sqrt{a}$ for an edge straight crack in a flat plate subjected to uniform uniaxial tensile stress σ; for solid curves there are no constraints to bending; the dashed curve was obtained with bending constraints added. (J. E. Shigley and L. D. Mitchell, *Mechanical Engineering Design*, McGraw-Hill, 1983.)

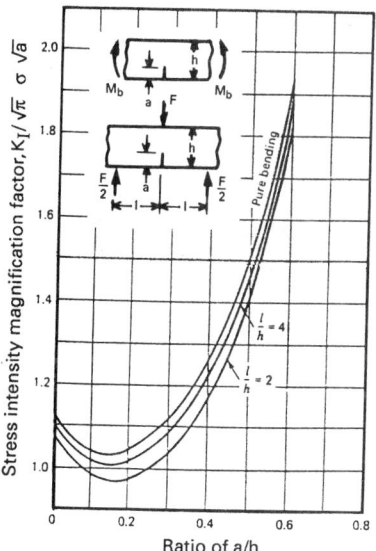

FIGURE 4–31 Stress intensity magnification factor $K_I/\sqrt{\pi}\,\sigma\sqrt{a}$ for a rectangular cross-sectional beam subjected to bending M_b. (J. E. Shigley and L. D. Mitchell, *Mechanical Engineering Design*, McGraw-Hill, 1983.)

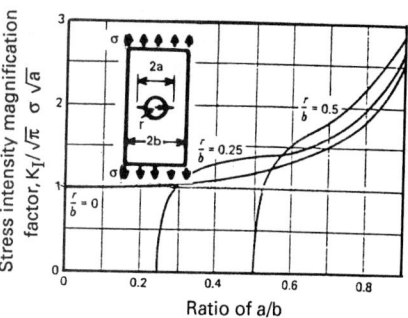

FIGURE 4–32 Stress intensity magnification factor $K_I/\sqrt{\pi}\,\sigma\sqrt{a}$ for a flat plate with a centrally located circular hole with two straight cracks under uniform uniaxial tensile stress σ. (J. E. Shigley and L. D. Mitchell, *Mechanical Engineering Design*, McGraw-Hill, 1983.)

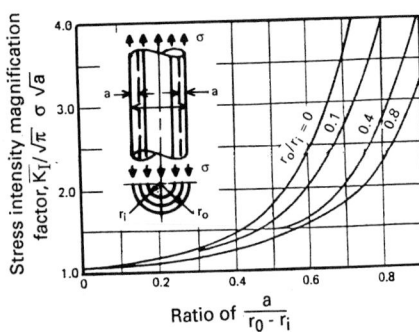

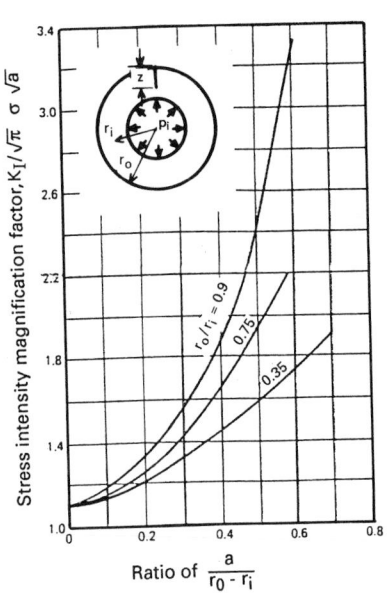

FIGURE 4–33 Stress intensity magnification factor $K_I/\sqrt{\pi}\,\sigma\sqrt{a}$ for axially tensile loaded cylinder with a radial crack of a depth extending completely around the circumference of the cylinder. (J. E. Shigley and L. D. Mitchell, *Mechanical Engineering Design*, McGraw-Hill, 1983.)

FIGURE 4–34 Stress intensity magnification factor $K_I/\sqrt{\pi}\,\sigma\sqrt{a}$ for a cylinder subjected to internal pressure p_i having a radial crack in the longitudinal direction of depth a. Use equation of tangential stress of thick cylinder subjected to internal pressure to calculate the stress σ_θ at $r = r_o$. (J. E. Shigley and L. D. Mitchell, *Mechanical Engineering Design*, McGraw-Hill, 1983.)

REFERENCES

1. Peterson, R. E., *Stress Concentration Factors*, John Wiley and Sons, New York, 1974.
2. Lingaiah, K., "Asymmetrically Reinforced Circular Cutout in a Flat Plate Subjected to Uniform Uniaxial Tension," Ph.D. Thesis, Department of Mechanical Engineering, University of Saskatchewan, Saskatoon, Sask., Canada, 1965.
3. Lingaiah, K., W. P. T. North, and J. B. Mantle, Photoelastic Analysis of an Asymmetrically Reinforced Circular Cutout in a Flat Plate, Subjected to Uniform Unidirectional Stress, *Proc. SESA*, Vol. 23, No. 2, 1966.
4. Lingaiah, K., and B. R. Narayana Iyengar, *Machine Design Data Handbook*, Engineering College Co-operative Society, Bangalore, India, 1962.
5. Lingaiah, K., *Machine Design Data Handbook*, Vol. II (*SI and Customary Metric Units*), Suma Publishers, Bangalore, India, 1986.
6. Shigley, J. E., and L. D. Mitchell, *Mechanical Engineering Design*, McGraw-Hill Book Publishing Company, New York, 1983.
7. Black, P. H., and O. Eugene Adams, *Machine Design*, 2d ed., McGraw-Hill Book Publishing Company, New York, 1955.

DESIGN OF MACHINE ELEMENTS FOR STRENGTH

SYMBOLS

A	area of cross-section, m^2 (in^2)
b	a shape factor ($b > 0$)
B	a constant
e_{sz}	size coefficient
e_{sT}	surface coefficient in case of tension and bending
e'_{sT}	surface coefficient in case of torsion
E	Young's modulus, GPa (Mpsi)
F	normal load (also with suffixes and primes), kN (lbf)
F'_m	static equivalent of cyclic load, kN (lbf)
G	modulus of rigidity, GPa (Mpsi)
h	thickness, m (in)
k_{sz}	size factor
k_{sT}	surface factor
K_σ	theoretical normal stress-concentration factor
K_τ	theoretical shear stress-concentration factor
$K_{f\sigma}$	fatigue normal stress-concentration factor
$K_{f\tau}$	fatigue shear stress-concentration factor
M_b	bending moment (also with suffixes and primes), N m (lbf in)
M'_{bm}	static equivalent of cyclic bending moment, N m (lbf in)
M_t	twisting moment (also with suffixes and primes), N m (lbf in)
M'_{tm}	static equivalent of cyclic twisting moment, N m (lbf in)
n	safety factor
	a constant
n_a	actual safety factor (also with suffixes)
n_d	design safety factor (also with suffixes)
q	index of sensitivity
q_f	index of notch sensitivity for alternating stresses
r	notch radius, mm (in)
t	time, h
x_0	the guaranteed value of x ($x_0 \geq 0$)
y_{max}	maximum deflection
Z_b	flexural section modulus, m^3 or cm^3 (in^3)
Z_t	polar section modulus, m^3 or cm^3 (in^3)
θ	characteristic or scale value ($\theta \geq x_0$)

σ	normal stress (also with suffixes and primes), MPa (psi)
σ_o	initial stress, MPa (psi)
σ_u	ultimate strength, MPa (psi)
σ_e	elastic limit for standard specimen for 12.5 mm ($\frac{1}{2}$ in), MPa (psi)
σ_d	design stress (also with suffixes), MPa (psi)
σ_x	normal stress in x direction, MPa (psi)
σ_y	yield stress, MPa (psi)
	normal stress in y direction, MPa (psi)
σ_{nom}	nominal normal stress, MPa (psi)
σ_{max}	maximum normal stress, MPa (psi)
σ_e'	elastic limit for any thickness h between 12.5 mm ($\frac{1}{2}$ in) and 75 mm (3 in), MPa (psi)
σ_e''	elastic limit for 75 mm (3 in) specimen, MPa (psi)
σ_{fb}	endurance limit in bending, MPa (psi)
τ	shear stress (also with suffixes and primes), MPa (psi)
τ_e	elastic limit in shear, MPa (psi)
τ_{sy}	yield strength in shear, MPa (psi)
τ_{xy}	shear stress in xy plane, MPa (psi)
τ_{nom}	nominal shear stress, MPa (psi)
τ_f	endurance limit in torsion, MPa (psi)
ϵ	engineering or average strain, μm/m (μin/in)
ϵ'	true strain, μm/m (μin/in)
ϵ_t	total creep, after a time t, μm/m (μin/in)
ϵ_o	initial creep, μm/m (μin/in)
$\dot{\epsilon}$	creep rate (μm/m)/h, (μin/in)/h
v_o	a constant

Suffixes   for

s	static strength (σ_u or σ_y)
u	ultimate strength
y	yield strength
e	elastic limit
a	amplitude
b	bending
m	mean
t	tension
max	maximum
min	minimum
f	endurance limit (also used for reversed cycle)
o	endurance limit repeated cycle

Primes   for

$'$ (single)	static equivalent
$''$ (double)	combined stress

Particular	Formula

STATIC LOADS

Influence of size

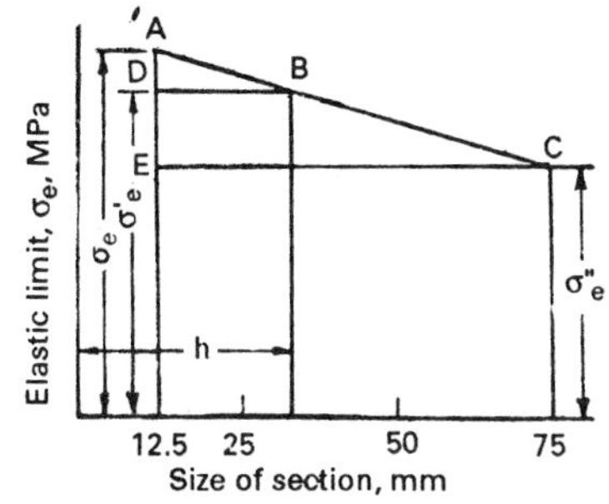

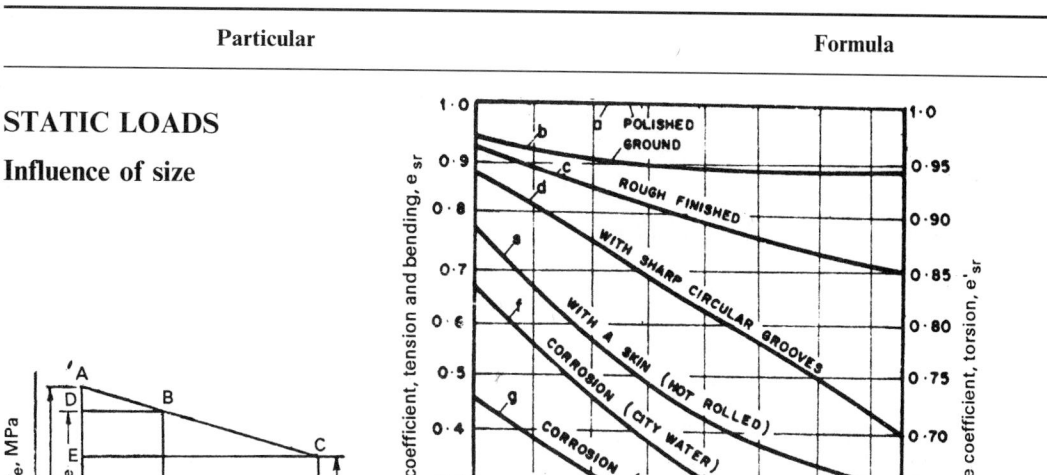

FIGURE 5–1 Change of elastic limit with size of section.

FIGURE 5–2 Influence of size on elastic limits.

The size coefficient (Fig. 5–1, Fig. 5–2, and Table 5–1)

$$e_{sz} = 1 - 0.016\left(1 - \frac{\sigma_e''}{\sigma_e}\right)(h - 12.5) \qquad (5\text{–}1)$$

where σ_e = elastic limit for 12.5 mm (0.5 in)
σ_e'' = elastic limit for 75 mm (3.0 in)

TABLE 5–1
Strength ratios of various materials for use in Eqs. (5–1) and (5–2)

Material	Values of $\frac{\sigma_e''}{\sigma_e}$				
	Natural state	Annealed	Drawn at 650°C	Drawn at 535°C	Drawn at 425°C
Aluminum, strong, wrought	0.93	—	—	—	—
Tobin bronze	0.90	—	—	—	—
Monel metal, forged	0.80	—	—	—	—
Ductile iron	0.80	0.98	—	—	—
Low-carbon steel, $C < 0.20\%$	0.84	—	—	—	—
Medium-carbon steel, 0.30 to 0.50% C	—	0.85	0.72	0.59	0.53
Nickel steel, SAE 2340	—	0.86	0.80	0.74	—
Cr-Ni steel, SAE 3140	—	0.80	0.75	0.70	0.65
Cast iron, Class no. 20	0.55	—	—	—	—
Cast iron, Class no. 25	0.73	—	—	—	—
Cast iron, Class no. 35	0.60	—	—	—	—
Wrought iron	0.55	—	—	—	—

Particular	Formula	
The size factor	$$k_{sz} = \dfrac{250}{300 - 4h + \dfrac{\sigma_e''}{\sigma_e}(4h - 50)}$$	(5–2)
The relation between size coefficient and size factor	$$e_{sz} = \dfrac{1}{k_{sz}}$$	(5–3)
The elastic limit for any thickness h between 12.5 mm and 75 mm can be determined from the relation (Fig. 5–1)	$$\sigma_e'' = \sigma_e - \dfrac{(\sigma_e - \sigma_e'')(h - 12.5)}{(75 - 12.5)}$$	(5–4)

INDEX OF SENSITIVITY

The index of sensitivity	$$q = \dfrac{K_{\sigma a} - 1}{K_\sigma - 1}$$	(5–5)
The actual or real stress-concentration factor	$$K_{\sigma a} = 1 + q(K_\sigma - 1)$$	(5–6)

SURFACE CONDITION (FIG. 5–3)

The surface factor for the case of tension and bending	$$k_{ST} = \dfrac{1}{e_{ST}}$$	(5–7)
The surface coefficient in case of torsion	$$e_{ST}' = 0.425 + 0.575\, e_{ST}$$	(5–8)

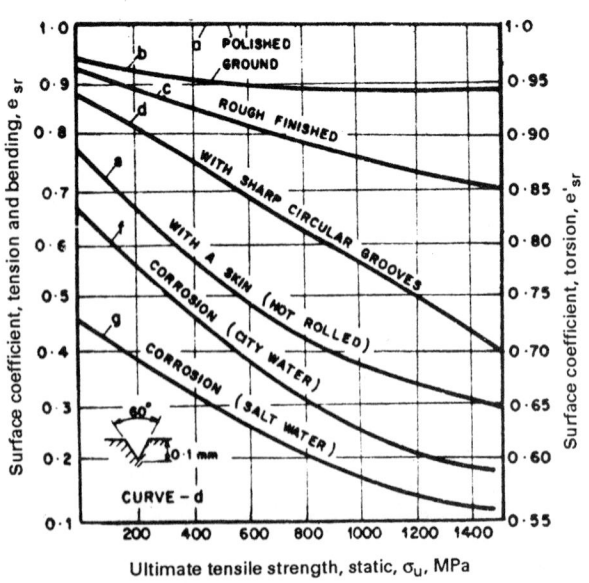

FIGURE 5–3 Reciprocals of stress-concentration factors caused by surface conditions.

Particular	Formula

SAFETY FACTOR

The general equation for design safety factor (Table 5–2)

$$n = k_1 k_2 k_3 k_4 \ldots k_m n_a \tag{5–9}$$

where $k_1 = k_{sz}$ = size factor
$k_2 = k_{s_T}$ = surface factor
$k_3 = k_l$ = load factor
k_4 = material factor

· · · · · · · · · · · · · ·

· · · · · · · · · · · · · ·

· · · · · · · · · · · · · ·

n_a = actual safety factor (Table 5–2)

TABLE 5–2
Reliability factor or actual safety factor[a]

Circumstance	Actual factor of safety or reliability factor, n_a
Strength properties of material well known, load accurately predictable, parts produced with close dimensional control and brought to close tolerance specifications, and low-weight criteria	1
Load accurately predictable and low-weight and low-cost criteria	1.1–1.5
Load accurately predictable and low-cost criteria (low-weight–no criteria)	1.5–2
Overloads expected, materials ordinary but reliability important	2–3
Strength properties not well defined, loading uncertain, human life at stake if failure occurred, high maintenance and shutdown cost	≥3

[a] These values are recommended for use in design, in the absence of specific reliability data.

The design safety factor based on ultimate strength

$$n_{ud} = k_{sz} K_{\sigma a} n_{ua} \tag{5–10}$$

The relationships between allowable stress and specified minimum yield strength using the AISC Code are given here:

Tension

$$0.45\,\sigma_{sy} \leq \sigma_a \leq 0.60\,\sigma_{sy} \tag{5–11}$$

Shear

$$\tau_a = 0.40\,\sigma_{sy} \tag{5–12}$$

Bearing

$$\sigma_a = 0.90\,\sigma_{sy} \tag{5–13}$$

Bending

$$0.60\,\sigma_{sy} \leq \sigma_a \leq 0.75\,\sigma_{sy} \tag{5–14}$$

The expression for forces or loads used to find stresses in machine members or structures as per AISC Code.

$$F = \Sigma\, W_d + \Sigma\, W_l + \Sigma\, K F_l + F_w + \Sigma\, F_{me} \tag{5–15}$$

where

$\Sigma\, W_d$ = sum of dead loads
$\Sigma\, W_l$ = sum of all stationary or static live loads
F_l = impact or dynamic live load
F_w = wind load on the structure
$\Sigma\, F_{me}$ = load which accounts for earthquakes, hurricanes, etc.
K = service factor obtained from Table 5–3

The value of design normal stress

$$\sigma_d \leq \sigma_a \tag{5–16}$$

Particular	Formula

TABLE 5–3
AISC service factor K for use in Eq. (5–15)

Particular	K
For support of elevators	2
For cab-operated traveling-crane support girders and their connections	1.25
For pendant-operated traveling-crane support girders and their connections	1.10
For support of light machinery, shaft- or motor-driven	≥ 1.20
For supports of reciprocating machinery or power-driven units	≥ 1.50
For hangers supporting floors and balconies	1.33

Particular	Formula	
The value of design shear stress	$\tau_d \leq \tau_a$	(5–16a)
The design safety factor	$n_d = \dfrac{\text{strength}}{\text{stress}}$	(5–17)
	$= n_s n_L$	
	where	
	$n_s =$ safety factor to take into account the uncertainty of strength	
	$n_L =$ safety factor to take into account the uncertainty of load	
The equation for design safety factor	$n_d = \dfrac{\text{strength in force units}}{\text{applied force or load}}$	(5–18)
The realized safety factor	$n_r = \dfrac{\sigma_s}{\sigma} \quad \text{or} \quad n_r = \dfrac{\tau_s}{\tau}$	(5–19)
The design safety factor based on elastic limit	$n_{ed} = k_{sz} K_{\sigma a} n_{ea}$	(5–20)
The design safety factor based on yield strength	$n_{yd} = k_{sz} K_{\sigma a} n_{ya}$	(5–21)
The design safety factor based on endurance limit on bending	$n_{fd} = k_{sz} k_{sT} k_l n_{fa}$	(5–22)
	where $k_l =$ load factor	
Design stress based on elastic limit	$\sigma_{ed} = \sigma_e / n_{ed}$	(5–23)
Design stress based on ultimate strength	$\sigma_{ud} = \dfrac{\sigma_{su}}{n_{ud}}$	(5–24)
Design stress based on yield strength	$\sigma_{yd} = \dfrac{\sigma_{sy}}{n_{yd}}$	(5–25)
Design stress based on yield strength in shear	$\tau_{yd} = \dfrac{\tau_{sy}}{n_{yd}}$	(5–26)

Particular	Formula
Static design stress	$\sigma_{sd} = \dfrac{\sigma_{su}}{n_{ud}}$ or $\dfrac{\sigma_{sy}}{n_{yd}}$ as the case may be $\hspace{2em}$ (5–27)
Design stress based on endurance limit	$\sigma_{fd} = \dfrac{\sigma_{sf}}{n_{fd}}$ $\hspace{2em}$ (5–28)

THEORIES OF FAILURE

Particular	Formula
The maximum normal stress theory or Rankine's theory	$\sigma_e = \frac{1}{2}\left[(\sigma_x + \sigma_y) + \sqrt{(\sigma_x - \sigma_y)^2 + 4\tau_{xy}^2}\right]$ $\hspace{2em}$ (5–29)
The maximum shear stress theory or Guest's theory	$\sigma_e = \sqrt{(\sigma_x - \sigma_y)^2 + 4\tau_{xy}^2}$ $\hspace{2em}$ (5–30)
The shear-energy theory or constant energy-of-distortion or Hencky–von Mises theory	$\sigma_e = \sqrt{(\sigma_x - \sigma_y)^2 + 3\tau_{xy}^2}$ $\hspace{2em}$ (5–31)
The maximum strain theory or Saint Venant's theory	$\sigma_e = \frac{1}{2}\Big[(1 - \nu)(\sigma_x + \sigma_y)$ $\hspace{2em}$ $+ (1 + \nu)\sqrt{(\sigma_x - \sigma_y)^2 + 4\tau_{xy}^2}\Big]$ $\hspace{2em}$ (5–32)
The bearing stress which causes failure for no friction at the surface of contact	$\sigma_b = 1.81\,\sigma_e$ $\hspace{2em}$ (5–33)
The bearing stress which causes failure for the friction at the surface of contact	$\sigma_b = 2\,\sigma_e$ $\hspace{2em}$ (5–34)

CYCLIC LOADS (FIGS. 5–4 AND 5–5)

Particular	Formula
The fatigue stress-concentration factor for normal stress	$K_{f\sigma} = q_f(K_\sigma - 1) + 1$ $\hspace{2em}$ (5–35)
The fatigue stress-concentration factor for shear stress	$K_{f\tau} = q_f(K_\tau - 1) + 1$ $\hspace{2em}$ (5–36)
The empirical formula for notch sensitivity for alternating stress of steel	$q_f = 1 - \exp\left[-\dfrac{r\sigma_u^2}{0.904 \times 10^6}\right]$ $\hspace{2em}$ (5–37)
Notch sensitive curves for steel and aluminum alloys	Refer to Fig. 5–6.
The empirical formula for notch sensitivity for alternating stress for high-strength aluminum alloys having $\sigma_u = 415$ to 550 MPa (60 to 80 kpsi)	$q_f = 1 - \exp\left(\dfrac{-r}{0.01}\right)$ $\hspace{2em}$ (5–38)
Endurance strength for finite life	$\sigma_f' = \sigma_f\left(\dfrac{10^6}{N}\right)^{0.09}$ $\hspace{2em}$ (5–39)

Particular	Formula

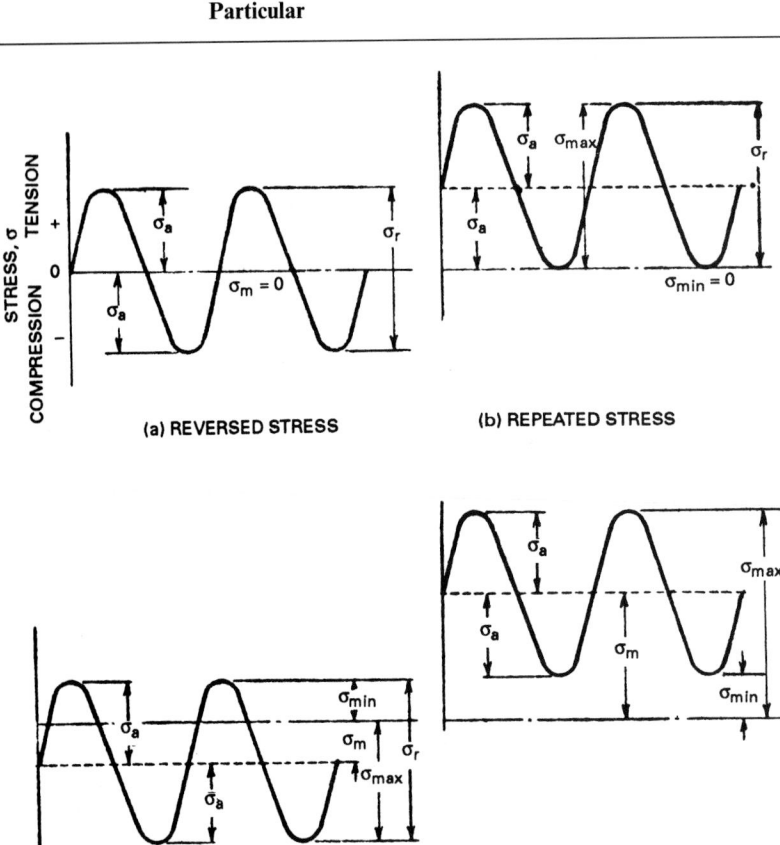

FIGURE 5–4 Types of fatigue stress variations.

The empirical relation between ultimate strength and endurance limits for various materials

where N = required life in cycles

Refer to Tables 5–4 and 5–5.

STRESS-STRESS AND STRESS-LOAD RELATIONS

Axial load

The maximum stress

$$\sigma_{max} = \frac{F_{max}}{A} \qquad (5\text{–}40)$$

The minimum stress

$$\sigma_{min} = \frac{F_{min}}{A} \qquad (5\text{–}41)$$

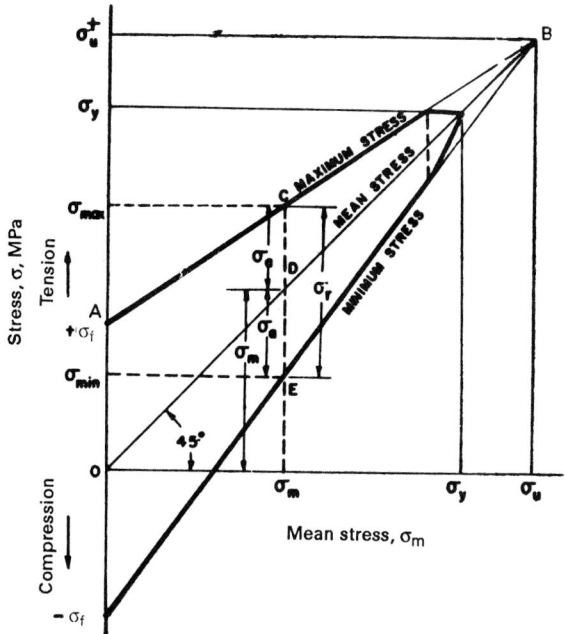

FIGURE 5–5   Modified Goodman diagram.

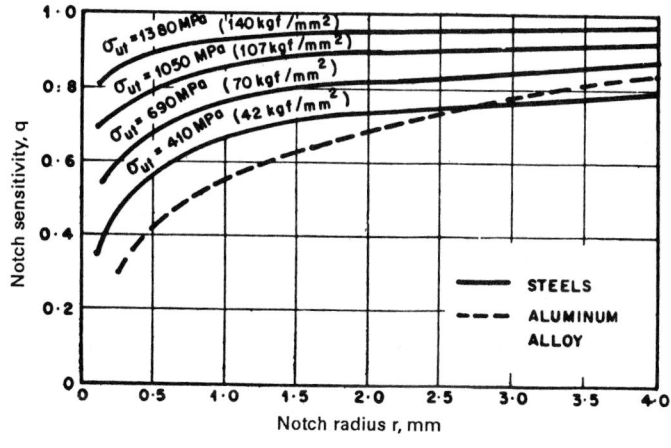

1 kgf/mm^2 = 9.8066 N/mm^2

FIGURE 5–6   Notch-sensitivity curves for steel and aluminum alloys.

TABLE 5–4
Empirical relationship between ultimate strength and endurance limits for various materials (approximate)

Material	Tension, compression, and bending (reversed or repeated cycle)[a]	Torsion (reversed or repeated cycle)[a]
Gray cast iron	$\sigma_{ft} = 0.6\,\sigma_{fb}$ to $0.7\,\sigma_{fb}$ $\sigma_b = 1.2\,\sigma_{fb}$ to $1.5\,\sigma_{fb}$	$\tau = 0.75\,\sigma_{fb}$ to $0.9\,\sigma_{fb}$ $\tau = 1.2\,\tau_f$ to $1.3\,\tau_f$
Carbon steels	$\sigma_{ot} = 1.6\,\sigma_{fb}$ $\sigma_{ob} = 1.5\,\sigma_{fb}$	$\tau_o = 1.8\,\tau_f$ to $2\,\tau_f$
Steels (general)	$\sigma_{ft} = 0.7\,\sigma_{fb}$ to $0.8\,\sigma_{fb}$ $\sigma_{ft} = 0.36\,\sigma_u;\ \sigma_{ot} = 0.5\,\sigma_u$ $\sigma_{fb} = 0.46\,\sigma_u;\ \sigma_{ob} = 0.6\,\sigma_u$	$\tau_f = 0.55\,\sigma_{fb}$ to $0.58\,\sigma_{fb}$ $\tau_f = 0.22\,\sigma_u$ $\tau_o = 0.3\,\sigma_u$
Alloy steels	$\sigma_{ft} = 0.95\,\sigma_{fb}$ $\sigma_{ot} = 1.5\,\sigma_{ft}$ to $1.6\,\sigma_{ft}$ $\sigma_{ob} = 1.6\,\sigma_{fb}$	$\tau_o = 1.8\,\tau_f$ to $2\,\tau_f$
Aluminum alloys	$\sigma_{ot} = 0.7\,\sigma_{fb}$ $\sigma_{ob} = 1.8\,\sigma_{fb}$	$\tau_f = 0.55\,\tau_{fb}$ to $0.58\,\tau_{fb}$ $\tau_o = 1.4\,\tau_f$ to $2\,\tau_f$
Copper alloys		$\tau_f = 0.58\,\sigma_{fb}$ $\tau_o = 1.4\,\tau_f$ to $2\,\tau_f$
Endurance strength for finite life		$\sigma_f' = \sigma_f \left(\dfrac{10^6}{N}\right)^{0.09}$

[a] f — endurance limit (also for reversed cycle); o — endurance for repeated cycle; t — tension; b — bending; u — ultimate; N — number of cycles.

TABLE 5–5
The empirical relation for endurance limit

Material	Endurance limit, σ_f		
	Bending	Axial	Torsion
For steel and other ferrous materials [(for $\sigma_u < 1374$ MPa (199.5 kpsi)]	$1/2$–$5/8\,\sigma_u$	$7/20$–$5/8\,\sigma_u$	$7/80$–$5/32\,\sigma_u$
For nonferrous materials	$1/4$–$1/3\,\sigma_u$	$7/40$–$1/3\,\sigma_u$	$7/160$–$1/12\,\sigma_u$

Particular	Formula	
The load amplitude	$F_a = \dfrac{F_{max} - F_{min}}{2}$	(5–42)
The mean load	$F_m = \dfrac{F_{max} + F_{min}}{2}$	(5–43)
The stress amplitude (Figs. 5–4 and 5–5)	$\sigma_a = \dfrac{F_a}{A}$	(5–44)

Particular	Formula	
The mean stress	$$\sigma_m = \frac{F_m}{A}$$	(5–45)
The ratio of amplitude stress to mean stress	$$\frac{\sigma_a}{\sigma_m} = \frac{F_a}{F_m}$$	(5–46)
The static equivalent of cyclic load $F_m \pm F_a$	$$F'_m = F_m + \frac{\sigma_{sd}}{\sigma_{fd}} F_a$$	(5–47)
The static equivalent of mean stress $\sigma_m \pm \sigma_a$	$$\sigma'_m = \frac{F'_m}{A}$$	(5–48)
The Gerber parabolic relation (Fig. 5–7)	$$\frac{\sigma_a}{\sigma_{fd}} + \left(\frac{\sigma_m}{\sigma_{ud}}\right)^2 = 1$$	(5–49)

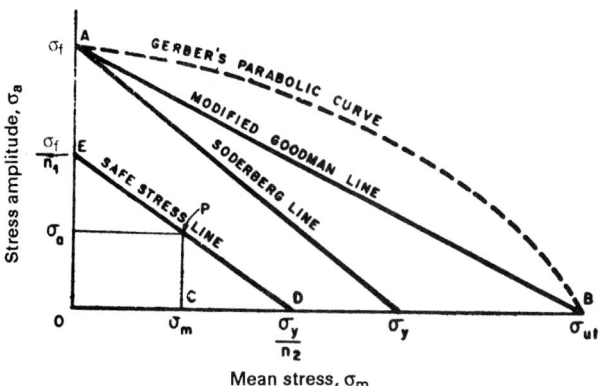

FIGURE 5–7   Graphical representation of steady and variable stresses.

The Goodman relation (Figs. 5–5, 5–7, and 5–9)	$$\frac{\sigma_a}{\sigma_{fd}} + \frac{\sigma_m}{\sigma_{ud}} = 1$$	(5–50)
The Soderberg relation (Figs. 5–7 and 5–8)	$$\frac{\sigma_a}{\sigma_{fd}} + \frac{\sigma_m}{\sigma_{yd}} = 1$$	(5–51)

Bending loads

The maximum stress	$$\sigma_{max} = \frac{M_{b(max)}}{Z_b}$$	(5–52)
The minimum stress	$$\sigma_{min} = \frac{M_{b(min)}}{Z_b}$$	(5–53)

Particular	Formula

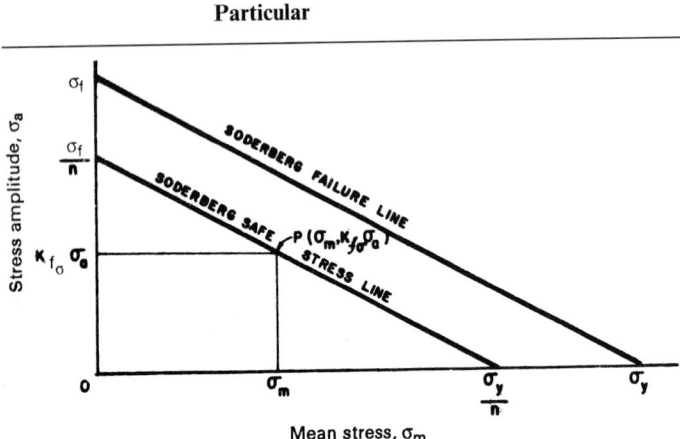

FIGURE 5–8 Representation of safe limit of mean stress and stress amplitude by Soderberg criterion.

The bending moment amplitude	$M_{ba} = \dfrac{M_{b(max)} - M_{b(min)}}{2}$	(5–54)
The mean bending moment	$M_{bm} = \dfrac{M_{b(max)} + M_{b(min)}}{2}$	(5–55)
The bending stress amplitude	$\sigma_{ba} = \dfrac{M_{ba}}{Z_b}$	(5–56)
The mean bending stress	$\sigma_{bm} = \dfrac{M_{bm}}{Z_b}$	(5–57)
The ratio of stress amplitude to mean stress	$\dfrac{\sigma_{ba}}{\sigma_{bm}} = \dfrac{M_{ba}}{M_{bm}}$	(5–58)
The static equivalent of cyclic bending moment $M_{bm} \pm M_{ba}$	$M'_{bm} = M_{bm} + \dfrac{\sigma_{sd}}{\sigma_{fd}} M_{ba}$	(5–59)
The static equivalent of cyclic stress	$\sigma'_{bm} = \dfrac{M'_{bm}}{Z_b}$	(5–60)
The Gerber parabolic relation (Fig. 5–7)	$\dfrac{\sigma_{ba}}{\sigma_{fd}} + \dfrac{\sigma_{bm}^2}{\sigma_{ud}^2} = 1$	(5–61)
The Goodman straight-line relation (Figs. 5–5, 5–7, and 5–9)	$\dfrac{\sigma_{ba}}{\sigma_{fd}} + \dfrac{\sigma_{bm}}{\sigma_{ud}} = 1$	(5–62)
The Soderberg straight-line relation (Figs. 5–7 and 5–8)	$\dfrac{\sigma_{ba}}{\sigma_{fd}} + \dfrac{\sigma_{bm}}{\sigma_{yd}} = 1$	(5–63)

Particular	Formula

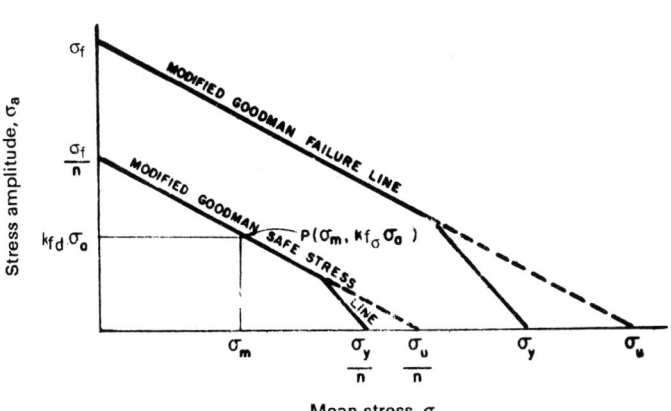

FIGURE 5–9 Representation of safe limit of mean stress and stress amplitude by Goodman criterion.

Torsional moments

The maximum stress	$$\tau_{max} = \frac{M_{t(max)}}{Z_t}$$	(5–64)
The minimum stress	$$\tau_{min} = \frac{M_{t(min)}}{Z_t}$$	(5–65)
The load amplitude	$$M_{ta} = \frac{M_{t(max)} - M_{t(min)}}{2}$$	(5–66)
The mean load	$$M_t = \frac{M_{t(max)} + M_{t(min)}}{2}$$	(5–67)
The stress amplitude	$$\tau_a = \frac{M_{ta}}{Z_t}$$	(5–68)
The mean stress	$$\tau_m = \frac{M_{tm}}{Z_t}$$	(5–69)
The ratio of stress amplitude to mean stress	$$\frac{\tau_a}{\tau_m} = \frac{M_{ta}}{M_{tm}}$$	(5–70)
The static equivalent of cyclic twisting moment $M_{tm} \pm M_{ta}$	$$M'_{tm} = M_{tm} + \frac{\tau_{sd}}{\tau_{fd}} M_{td}$$	(5–71)
The static equivalent of cyclic stress	$$\tau'_m = \frac{M'_{tm}}{Z_t}$$	(5–72)

Particular	Formula

The Gerber parabolic relation (Fig. 5–7)

$$\frac{\tau_a}{\tau_{fd}} + \frac{\tau_m^2}{\tau_{ud}^2} = 1 \tag{5–73}$$

The Goodman straight-line relation (Figs. 5–5, 5–7, and 5–9)

$$\frac{\tau_a}{\tau_{fd}} + \frac{\tau_m}{\tau_{ud}} = 1 \tag{5–74}$$

The Soderberg straight-line relation (Figs. 5–7 and 5–8)

$$\frac{\tau_a}{\tau_{fd}} + \frac{\tau_m}{\tau_{yd}} = 1 \tag{5–75}$$

THE COMBINED STRESSES

Method 1

The static equivalent of $\sigma_m \pm \sigma_a$

$$\sigma_m' = \sigma_m + \frac{\sigma_{sd}}{\sigma_{fd}}\sigma_a \tag{5–76}$$

The static equivalent of $\tau_m \pm \tau_a$

$$\tau_m' = \tau_m + \frac{\tau_{sd}}{\tau_{fd}}\tau_a \tag{5–77}$$

The maximum normal stress theory or Rankine's theory

$$\sigma_e = \tfrac{1}{2}\left[\sigma_m' + \sqrt{\sigma_m'^2 + 4\tau_m'^2}\right] \tag{5–78}$$

The maximum shear theory or Coulomb's or Tresca criteria or Guest's theory

$$\tau_e = \sqrt{\sigma_m'^2 + 4\tau_m'^2} \tag{5–79}$$

The distortion energy theory or Hencky–von Mises theory

$$\sigma_e = \sqrt{\sigma_m'^2 + 3\tau_m'^2} \tag{5–80}$$

The maximum strain theory or Saint Venant's theory

$$\sigma_e = \tfrac{1}{2}\left[(1-\nu)\sigma_m' + (1+\nu)\sqrt{\sigma_m'^2 + 4\tau_m'^2}\right] \tag{5–81}$$

Method 2

The combined maximum normal stress

$$\sigma_{max}'' = \tfrac{1}{2}\left[\sigma_{max} + \sqrt{\sigma_{max}^2 + 4\tau_{max}^2}\right] \tag{5–82}$$

The combined minimum normal stress

$$\sigma_{min}'' = \tfrac{1}{2}\left[\sigma_{min} + \sqrt{\sigma_{min}^2 + 4\tau_{min}^2}\right] \tag{5–83}$$

The combined maximum shear stress

$$\tau_{max}'' = \tfrac{1}{2}\sqrt{\sigma_{max}^2 + 4\tau_{max}^2} \tag{5–84}$$

The combined minimum shear stress

$$\tau_{min}'' = \tfrac{1}{2}\sqrt{\sigma_{min}^2 + 4\sigma_{min}^2} \tag{5–85}$$

The combined maximum normal strain stress

$$\sigma_{max}'' = \tfrac{1}{2}\left[(1-\nu)\sigma_{max} + (1+\nu)\sqrt{\sigma_{max}^2 + 4\tau_{max}}\right] \tag{5–86}$$

The combined minimum normal strain stress

$$\sigma_{min}'' = \tfrac{1}{2}\left[(1-\nu)\sigma_{min} + (1+\nu)\sqrt{\sigma_{min}^2 + 4\tau_{min}^2}\right] \tag{5–87}$$

The combined maximum octahedral shear stress

$$\tau_{max}'' = \tfrac{1}{2}\left[\sqrt{\sigma_{max}^2 + 3\tau_{max}^2}\right] \tag{5–88a}$$

The combined minimum octahedral shear stress

$$\tau_{min}'' = \tfrac{1}{2}\left[\sqrt{\sigma_{min}^2 + 3\tau_{min}^2}\right] \tag{5–88b}$$

Particular	Formula
The combined mean stress	$$\sigma''_m = \frac{\sigma''_{max} + \sigma''_{min}}{2} \qquad (5\text{--}88c)$$
The combined stress amplitude	$$\sigma''_a = \frac{\sigma''_{max} - \sigma''_{min}}{2} \qquad (5\text{--}88d)$$
The Gerber parabolic relation (Fig. 5–7)	$$\frac{\sigma''_a}{\sigma_{fd}} + \left(\frac{\sigma''_m}{\sigma_{ud}}\right)^2 = 1 \qquad (5\text{--}88e)$$
The Goodman straight-line relation (Figs. 5–5, 5–7, and 5–9)	$$\frac{\sigma''_a}{\sigma_{fd}} + \frac{\sigma''_m}{\sigma_{ud}} = 1 \qquad (5\text{--}88f)$$
The Soderberg straight-line relation (Figs. 5–7 and 5–8)	$$\frac{\sigma''_a}{\sigma_{fd}} + \frac{\sigma''_m}{\sigma_{yd}} = 1 \qquad (5\text{--}88g)$$

COMBINED STRESSES IN TERMS OF LOADS

Method 1

Maximum shear stress theory

$$\frac{\sigma_e}{n_{ed}} = \sqrt{\left(\frac{M'_{bm}}{Z_b} + \frac{F'_m}{A}\right)^2 + 4\left(\frac{M'_{tm}}{Z_t}\right)^2} \qquad (5\text{--}89a)$$

The shear energy theory

$$\frac{\sigma_e}{n_{ed}} = \sqrt{\left(\frac{M'_{bm}}{Z_b} + \frac{F'_m}{A}\right)^2 + 3\left(\frac{M'_{tm}}{Z_t}\right)^2} \qquad (5\text{--}89b)$$

where, for solid shafts

$$Z_b = \frac{\pi d^3}{32}; Z_t = \frac{\pi d^3}{16}$$

$$A = \frac{\pi d^2}{4}$$

Method 2

Maximum shear theory

$$\left[\sqrt{\left(\frac{M_{b(max)}}{Z_b} + \frac{F_{max}}{A}\right)^2 + 4\left(\frac{M_{t(max)}}{Z_t}\right)^2}\right]\left[\frac{1}{\tau_{fd}} + \frac{1}{\tau_d}\right]$$

$$+ \left[\sqrt{\left(\frac{M_{b(min)}}{Z_b} + \frac{F_{min}}{A}\right)^2 + 4\left(\frac{M_{t(min)}}{Z_t}\right)^2}\right]$$

$$\times \left[-\frac{1}{\tau_{fd}} + \frac{1}{\tau_d}\right] = 2 \qquad (5\text{--}90a)$$

Particular	Formula

The shear energy theory

$$\left[\sqrt{\left(\frac{M_{b(max)}}{Z_b}+\frac{F_{max}}{A}\right)^2+3\left(\frac{M_{t(max)}}{Z_t}\right)^2}\right]\left[\frac{1}{\tau_{fd}}+\frac{1}{\tau_d}\right]$$

$$+\left[\sqrt{\left(\frac{M_{b(min)}}{Z_b}+\frac{F_{min}}{A}\right)^2+3\left(\frac{M_{t(min)}}{Z_t}\right)^2}\right]\left[-\frac{1}{\tau_{fd}}+\frac{1}{\tau_d}\right]$$

$$=2 \tag{5-90b}$$

CREEP

Creep in tension

When the curve for total creep ϵ_t is approximated as a straight line its equation is

$$\epsilon_t = \epsilon_o + \epsilon t \tag{5-91a}$$

The creep rate $\dot{\epsilon}$ can be approximated by the equation

$$\dot{\epsilon} = B\sigma^n \tag{5-91b}$$

Refer to Table 5–6 for creep constants B and n.

TABLE 5–6
Creep constants for various steels for use in Eqs. (5–91b) to (5–95)

Steel	Temperature °C	B	n
0.39% C	400	14×10^{-36}	8.6
0.30% C	400	44×10^{-30}	6.9
0.45% C	475	—	6.5
2% Ni, 0.8% Cr, 0.4% Mo	450	10×10^{-19}	3.2
2% Ni, 0.3% C, 1.4% Mn	450	21×10^{-22}	4.7
12% Cr, 3% W, 0.4% Mn	550	24×10^{-14}	1.9
Ni-Cr-Mo	500	12×10^{-16}	2.7
Ni-Cr-Mo	500	16×10^{-12}	1.3
12% Cr	455	12×10^{-22}	4.4

Creep rate $\dot{\epsilon}$, when extrapolated into the region of lower stresses, can be determined with greater accuracy by the hyperbolic sine term

$$\dot{\epsilon} = \nu_o \sinh\left(\frac{\sigma}{\sigma_1}\right) \tag{5-91c}$$

True strain

$$\epsilon' = \ln(1+\epsilon) \tag{5-91d}$$

Creep life of aluminum

$$\epsilon_{cr} = \frac{1}{\dot{\epsilon}^n} \tag{5-92}$$

Time for the stress to decrease from an initial value of σ_o to a value of σ

$$t = \frac{1}{EB(n-1)\sigma_o^{n-1}}\left[\left(\frac{\sigma_o}{\sigma}\right)^{n-1}-1\right] \tag{5-93}$$

Particular	Formula

Creep in bending

The maximum stress at the extreme fibers in case of bending of beam is given by the relation

$$\sigma = \left(\frac{C_1}{BD}\right)^{1/n} M_b \qquad (5\text{--}94)$$

The maximum deflection of a cantilever beam loaded at free end by a load F

$$y_{max} = \frac{tF^n l^{n+2}}{D(n+2)} \qquad (5\text{--}95)$$

where $D = \dfrac{1}{B}\dfrac{(2b)^n \left(\dfrac{h}{2}\right)^{2n+1}}{\left(2+\dfrac{1}{n}\right)^n}$

Creep constants B and n are taken from Table 5–6.

RELIABILITY

The probability function or frequency function

$$p = f(x) \qquad (5\text{--}96)$$

The cumulative probability function

$$F(x_j) = \sum_{x_i \le x_j} f(x_i) \qquad (5\text{--}97)$$

where $f(x)$ is the probability density

The sample mean or arithmetic mean of a sample

$$\bar{x} = \frac{x_1 + x_2 + x_3 + x_4 + \ldots + x_n}{n} \qquad (5\text{--}98a)$$

$$= \frac{1}{n}\sum_{i=1}^{n} x_i \qquad (5\text{--}98b)$$

where x_i is the ith value of the quantity
n is the total number of measurements or elements

The population mean of a population consisting of n elements

$$\mu = \frac{x_1 + x_2 + x_3 + x_4 \ldots + x_n}{n} \qquad (5\text{--}99a)$$

$$= \frac{1}{n}\sum_{i=1}^{n} x_i \qquad (5\text{--}99b)$$

The sample variance

$$s_x^2 = \frac{(x_1 - \bar{x})^2 + (x_2 - \bar{x})^2 + \ldots + (x_n - \bar{x})^2}{n - 1} \qquad (5\text{--}100a)$$

$$= \frac{1}{n-1}\sum_{i=1}^{n}(x_i - \bar{x})^2 \qquad (5\text{--}100b)$$

A suitable equation for variance for use in a calculator

$$s_x^2 = \frac{\sum x^2}{n} - \bar{x}^2 \qquad (5\text{--}101)$$

Particular	Formula
The sample standard deviation (the symbol used for true standard deviation is $\hat{\sigma}$)	$$s_x = \left[\frac{1}{n-1}\sum_{i=1}^{n}(x_i - \bar{x})^2\right]^{1/2} \qquad (5\text{–}102)$$
A suitable equation for standard deviation for use in a calculator	$$s_x = \left\{\frac{\sum x^2 - \dfrac{\left(\sum x\right)^2}{n}}{n-1}\right\}^{1/2} \qquad (5\text{–}103)$$
The coefficient of variation	$$cv = (s_x/\bar{x})100 \qquad (5\text{–}104)$$
The normal, or Gaussian, distribution (Fig. 5–10)	$$f(x) = \frac{1}{\hat{\sigma}\sqrt{2\pi}}e^{-(x-\mu)^2/2\hat{\sigma}^2} \qquad -\infty < x < \infty \quad (5\text{–}105)$$
The normal distribution as defined by parameters, the mean μ and standard deviation $\hat{\sigma}$ according to the relation for the relative frequency $f(t)$, which is the ordinate at t	$$f(t) = \frac{1}{\sqrt{2\pi}}e^{-(t^2/2)} \qquad (5\text{–}106)$$ where $t = (x - \mu)/\hat{\sigma}$ Refer to Table 5–7 for ordinate $f(t)$ [i.e. $y = f(t)$] for various values of t. Refer to Table 5–8 for area under the standard normal distribution curve.
The area under normal distribution curve to the right of t (Fig. 5–11)	$$B(t) = 1 - A(t) \qquad (5\text{–}107)$$ where $A(t)$ is the area to the left of t. The area under the entire normal distribution curve is $A(t) + B(t)$ and is equal to unity. The term $B(t)$ can be found from Table 5–8 or by integrating the area under the curve.
Error function or probability integral	$$erf(x) = \frac{2}{\sqrt{\pi}}\int_0^x e^{-t^2}\,dt \qquad (5\text{–}108)$$ Refer to Table 5–9 for $erf(x)$ for various values of x.

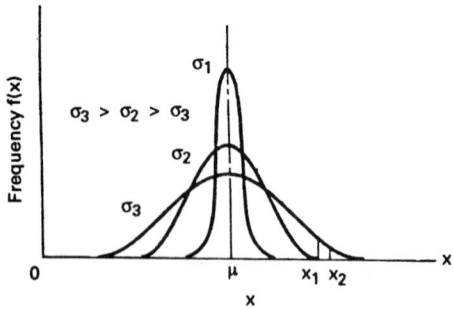

FIGURE 5–10 The shapes of normal distribution curves for various σ and constant μ.

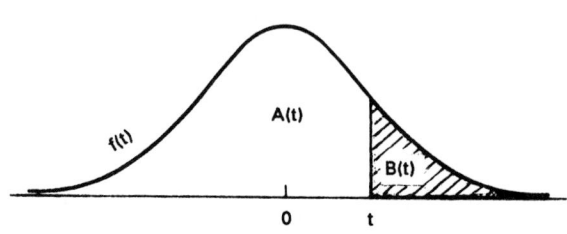

FIGURE 5–11 The Gaussian (normal) distribution curve.

TABLE 5–7
Standard normal curve ordinates

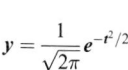

$$y = \frac{1}{\sqrt{2\pi}} e^{-t^2/2}$$

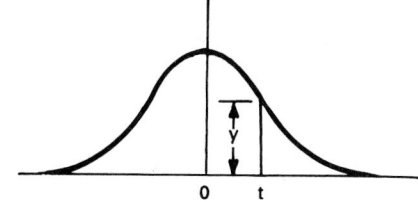

t	0	1	2	3	4	5	6	7	8	9
0.0	.3989	.3989	.3989	.3988	.3986	.3984	.3982	.3980	.3977	.3973
0.1	.3970	.3965	.3961	.3956	.3951	.3945	.3939	.3932	.3925	.3928
0.2	.3910	.3902	.2894	.3885	.3876	.3867	.3857	.3847	.3836	.3815
0.3	.3814	.3802	.3790	.3778	.3765	.3752	.3739	.3725	.3712	.3697
0.4	.3683	.3668	.3653	.3637	.3621	.3605	.3589	.3572	.3555	.3538
0.5	.3521	.3503	.3485	.3467	.3448	.3429	.3410	.3391	.3372	.3352
0.6	.3332	.3312	.3292	.3271	.3251	.3230	.3209	.3187	.3166	.3144
0.7	.3123	.3101	.3079	.3056	.3034	.3011	.2989	.2966	.2943	.2920
0.8	.2897	.2874	.2850	.2827	.2803	.2780	.2756	.2932	.2709	.2685
0.9	.2661	.2637	.2613	.2589	.2565	.2541	.2516	.2492	.2468	.2444
1.0	.2420	.2396	.2371	.2347	.2323	.2299	.2275	.2251	.2227	.2203
1.1	.2179	.2155	.2131	.2107	.2083	.2059	.2036	.2012	.1989	.1965
1.2	.1942	.1919	.1895	.1872	.1849	.1826	.1804	.1781	.1758	.1736
1.3	.1714	.1691	.1669	.1647	.1626	.1604	.1528	.1561	.1539	.1518
1.4	.1497	.1476	.1456	.1435	.1415	.1394	.1374	.1354	.1334	.1315
1.5	.1295	.1276	.1257	.1238	.1219	.1200	.1182	.1163	.1145	.1127
1.6	.1109	.1092	.1074	.1057	.1040	.1023	.1006	.0989	.0973	.0957
1.7	.0940	.0925	.0909	.0893	.0878	.0863	.0848	.0833	.0818	.0804
1.8	.0790	.0775	.0761	.0748	.0734	.0721	.0707	.0694	.0681	.0669
1.9	.0656	.0644	.0632	.0620	.0608	.0596	.0584	.0573	.0562	.0551
2.0	.0540	.0529	.0519	.0508	.0498	.0488	.0487	.0468	.0459	.0449
2.1	.0440	.0431	.0422	.0413	.0404	.0396	.0387	.0379	.0371	.0363
2.2	.0355	.0347	.0339	.0332	.0325	.0317	.0310	.0303	.0297	.0290
2.3	.0283	.0277	.0270	.0264	.0258	.0252	.0246	.0241	.0235	.0229
2.4	.0224	.0219	.0213	.0208	.0203	.0198	.0194	.0189	.0184	.0180
2.5	.0175	.0171	.0167	.0163	.0158	.0154	.0151	.0147	.0143	.0139
2.6	.0136	.0132	.0129	.0126	.0122	.0119	.0116	.0113	.0110	.0107
2.7	.0104	.0101	.0099	.0096	.0093	.0091	.0088	.0086	.0084	.0081
2.8	.0079	.0077	.0075	.0073	.0071	.0069	.0067	.0065	.0063	.0061
2.9	.0060	.0058	.0056	.0055	.0053	.0051	.0050	.0048	.0047	.0046
3.0	.0044	.0043	.0042	.0040	.0039	.0038	.0037	.0036	.0035	.0034
3.1	.0033	.0032	.0031	.0030	.0029	.0028	.0027	.0026	.0025	.0025
3.2	.0024	.0023	.0022	.0022	.0021	.0020	.0020	.0019	.0018	.0018
3.3	.0017	.0017	.0016	.0016	.0015	.0015	.0014	.0014	.0013	.0013
3.4	.0012	.0012	.0012	.0011	.0011	.0010	.0010	.0010	.0009	.0009
3.5	.0009	.0008	.0008	.0008	.0008	.0007	.0007	.0007	.0007	.0006
3.6	.0006	.0006	.0006	.0005	.0005	.0005	.0005	.0005	.0005	.0004
3.7	.0004	.0004	.0004	.0004	.0004	.0004	.0003	.0003	.0003	.0003
3.8	.0003	.0003	.0003	.0003	.0003	.0002	.0002	.0002	.0002	.0002
3.9	.0002	.0002	.0002	.0002	.0002	.0002	.0002	.0002	.0001	.0001

TABLE 5–8
Areas under the standard normal distribution curve

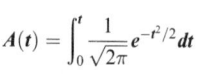

$$A(t) = \int_0^t \frac{1}{\sqrt{2\pi}} e^{-t^2/2} dt$$

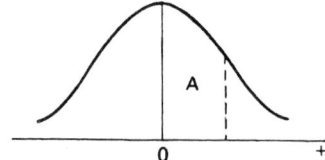

t	0	1	2	3	4	5	6	7	8	9
0.0	.0000	.0040	.0080	.0120	.0160	.0199	.0239	.0279	.0319	.0359
0.1	.0398	.0438	.0478	.0517	.0557	.0596	.0636	.0675	.0714	.0754
0.2	.0793	.0832	.0871	.0910	.0948	.0987	.1026	.1064	.1103	.1141
0.3	.1179	.1217	.1255	.1293	.1331	.1368	.1406	.1443	.1480	.1517
0.4	.1554	.1591	.1628	.1664	.1700	.1736	.1772	.1808	.1844	.1879
0.5	.1915	.1950	.1985	.2019	.2054	.2088	.2123	.2157	.2190	.2224
0.6	.2258	.2291	.2324	.2357	.2389	.2422	.2454	.2486	.2518	.2549
0.7	.2580	.2612	.2642	.2673	.2704	.2734	.2764	.2794	.2823	.2852
0.8	.2881	.2910	.2939	.2967	.2996	.3023	.3051	.3078	.3106	.3133
0.9	.3159	.3186	.3212	.3238	.3264	.3289	.3315	.3340	.3365	.3389
1.0	.3413	.3438	.3461	.3485	.3508	.3531	.3554	.3577	.3599	.3621
1.1	.3643	.3665	.3686	.3708	.3729	.3749	.3770	.3790	.3810	.3830
1.2	.3849	.3869	.3888	.3907	.3925	.3944	.3962	.3980	.3997	.4015
1.3	.4032	.4049	.4066	.4082	.4099	.4115	.4131	.4147	.4162	.4177
1.4	.4192	.4207	.4222	.4236	.4251	.4265	.4279	.4292	.4306	.4319
1.5	.4332	.4345	.4357	.4370	.4382	.4394	.4406	.4418	.4429	.4441
1.6	.4452	.4463	.4474	.4484	.4495	.4506	.4515	.4525	.4535	.4545
1.7	.4554	.4564	.4573	.4582	.4591	.4599	.4608	.4616	.4625	.4633
1.8	.4641	.4649	.4656	.4664	.4671	.4678	.4686	.4693	.4699	.4706
1.9	.4713	.4719	.4726	.4732	.4738	.4744	.4750	.4756	.4761	.4767
2.0	.4772	.4778	.4783	.4788	.4793	.4798	.4803	.4808	.4812	.4817
2.1	.4821	.4826	.4830	.4834	.4838	.4842	.4846	.4850	.4854	.4857
2.2	.4861	.4864	.4868	.4871	.4875	.4878	.4881	.4884	.4887	.4890
2.3	.4893	.4896	.4898	.4901	.4904	.4906	.4909	.4911	.4913	.4916
2.4	.4918	.4920	.4922	.4925	.4927	.4929	.4931	.4932	.4934	.4936
2.5	.4938	.4940	.4941	.4943	.4945	.4946	.4948	.4949	.4951	.4952
2.6	.4953	.4955	.4956	.4957	.4959	.4960	.4961	.4962	.4963	.4964
2.7	.4965	.4966	.4967	.4968	.4969	.4970	.4971	.4972	.4973	.4974
2.8	.4974	.4975	.4976	.4977	.4977	.4978	.4979	.4979	.4980	.4981
2.9	.4981	.4982	.4982	.4983	.4984	.4984	.4985	.4985	.4986	.4986
3.0	.4987	.4987	.4987	.4988	.4988	.4989	.4989	.4989	.4990	.4990
3.1	.4990	.4991	.4991	.4991	.4992	.4992	.4992	.4992	.4993	.4993
3.2	.4993	.4993	.4994	.4994	.4994	.4994	.4994	.4995	.4995	.4995
3.3	.4995	.4995	.4995	.4996	.4996	.4996	.4996	.4996	.4996	.4997
3.4	.4997	.4997	.4997	.4997	.4997	.4997	.4997	.4997	.4997	.4998
3.5	.4998	.4998	.4998	.4998	.4998	.4998	.4998	.4998	.4998	.4998
3.6	.4998	.4998	.4999	.4999	.4999	.4999	.4999	.4999	.4999	.4999
3.7	.4999	.4999	.4999	.4999	.4999	.4999	.4999	.4999	.4999	.4999
3.8	.4999	.4999	.4999	.4999	.4999	.4999	.4999	.4999	.4999	.4999
3.9	.5000	.5000	.5000	.5000	.5000	.5000	.5000	.5000	.5000	.5000

TABLE 5–9
Error function or probability integral

$$erf(x) = \frac{2}{\sqrt{\pi}} \int_0^x e^{-t^2} dt$$

x	0	1	2	3	4	5	6	7	8	9
0.0		.01128	.02256	.03384	.04511	.05637	.06762	.07886	.09008	.10128
0.1	.11246	.12362	.13476	.14587	.15695	.16800	.17901	.18999	.20094	.21184
0.2	.22270	.23352	.24430	.25502	.26570	.27633	.28690	.29742	.30788	.31828
0.3	.32863	.33891	.34913	.35928	.36936	.37938	.38933	.39921	.40901	.41874
0.4	.42839	.43797	.44747	.45689	.46623	.47548	.48466	.49375	.50275	.51167
0.5	.52050	.52924	.53790	.54646	.55494	.56332	.57162	.57982	.58792	.59594
0.6	.60386	.61168	.61941	.62705	.63459	.64203	.64938	.65663	.66378	.67084
0.7	.67780	.68467	.69143	.69810	.70468	.71116	.71754	.72382	.73001	.73610
0.8	.74210	.74800	.75381	.75952	.76514	.77067	.77610	.78144	.78669	.79184
0.9	.79691	.80188	.80677	.81156	.81627	.82089	.82542	.82987	.83243	.83851
1.0	.84270	.84681	.85084	.85478	.85865	.86244	.86614	.86977	.87333	.87680
1.1	.88021	.88353	.88679	.88997	.89308	.89612	.89910	.90200	.90484	.90761
1.2	.91031	.91296	.91553	.91805	.92051	.92290	.92524	.92751	.92973	.93190
1.3	.93401	.93606	.93807	.94002	.94191	.94376	.94556	.94731	.94902	.95067
1.4	.95229	.95385	.95538	.95686	.95830	.95970	.96105	.96237	.96365	.96490
1.5	.96611	.96728	.96841	.96952	.97059	.97162	.97263	.97360	.97455	.97546
1.6	.97635	.97721	.97804	.97884	.97962	.98038	.98110	.98181	.98249	.98315
1.7	.98379	.98441	.98500	.98558	.98613	.98667	.98719	.98769	.98817	.98864
1.8	.98909	.98952	.98994	.99035	.99074	.99111	.99147	.99182	.99216	.99248
1.9	.99279	.99309	.99338	.99366	.99392	.99418	.99443	.99466	.99489	.99511
2.0	.99532	.99552	.99572	.99591	.99609	.99626	.99642	.99658	.99673	.99688
2.1	.99702	.99715	.99728	.99741	.99753	.99764	.99775	.99785	.99795	.99805
2.2	.99814	.99822	.99831	.99839	.99846	.99854	.99861	.99867	.99874	.99880
2.3	.99886	.99891	.99897	.99902	.99906	.99911	.99915	.99920	.99924	.99928
2.4	.99931	.99935	.99938	.99941	.99944	.99947	.99950	.99952	.99955	.99957
2.5	.99959	.99961	.99963	.99965	.99967	.99969	.99971	.99972	.99974	.99975
2.6	.99976	.99978	.99979	.99980	.99981	.99982	.99983	.99984	.99985	.99986
2.7	.99987	.99987	.99988	.99989	.99989	.99990	.99991	.99991	.99992	.99992
2.8	.99992	.99993	.99993	.99994	.99994	.99994	.99995	.99995	.99995	.99996
2.9	.99996	.99996	.99996	.99997	.99997	.99997	.99997	.99997	.99997	.99998
3.0	.99998									

Particular	**Formula**
The resultant mean of adding the means of two populations (Fig. 5–12)	$\mu = \mu_s + \mu_\sigma$ (5–109)

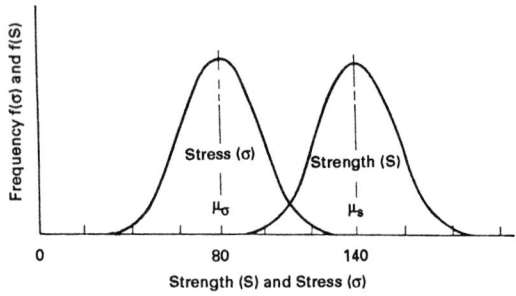

FIGURE 5–12   Distribution curves for two means of populations.

The resultant mean of subtracting the means of two populations	$\mu = \mu_s - \mu_\sigma$ (5–110)
The resultant standard deviation for both subtraction and addition of two standard deviations $\hat{\sigma}_s$ and $\hat{\sigma}_\sigma$	$\hat{\sigma} = \sqrt{\hat{\sigma}_s^2 + \hat{\sigma}_\sigma^2}$ (5–111)
The standard variable t_R (deviation multiplication factor) in order to determine the probability of failure or the reliability	$t_R = \dfrac{\mu_s - \mu_\sigma}{\sqrt{\hat{\sigma}_s^2 + \hat{\sigma}_\sigma^2}} = \dfrac{\mu_s - \mu_\sigma}{\hat{\sigma}}$ (5–112) where subscripts s and σ refer to strength and stress, respectively.
The reliability associated with t_R	$R = 0.5 + A(t_R)$ (5–113) where $A(t_R)$ is the area under a standard normal distribution curve. Refer to Table 5–10 for typical values of R as a function of standardized variable t_R.

TABLE 5–10
Reliability R as a function of t_R

Survival rate (R) %	t_R
50	0
90.00	1.288
95.00	1.645
98.00	2.050
99.00	2.330
99.90	3.080
99.99	3.700

Particular	Formula
	A safety factor of 1 is taken into account in determining the reliability from Eq. (5–113).
The fatigue strength reduction factor based on reliability	$C_R = 1 - 0.08(t_R)$ $\qquad$ (5–114) where t_R is also called the *deviation multiplication factor* (DMF), taken from Table 5–10.
If a factor of safety n' is to be specified together with reliability, then Eq. (5–112) is rewritten to give a new expression for t_R	$t_R = \dfrac{\mu_s - n'\mu_\sigma}{\sqrt{\hat{\sigma}_s^2 + \hat{\sigma}_\sigma^2}} = \dfrac{\mu_s - n'\mu_\sigma}{\hat{\sigma}}$ $\qquad$ (5–115)
The expression for safety factor n' from Eq. (5–115)	$n' = \dfrac{1}{\mu_\sigma}\left[\mu_s - t_R\sqrt{\hat{\sigma}_s^2 + \hat{\sigma}_\sigma^2}\right]$ $\qquad$ (5–116a) $= \dfrac{1}{\mu_\sigma}(\mu_s - t_R\hat{\sigma})$ $\qquad$ (5–116b)
The best-fitting straight line which fits a set of scattered data points as per linear regression	$y = mx + b$ $\qquad$ (5–117) where m is the slope and b is the intercept on the y axis
The equations for regression	$m = \dfrac{\sum xy - \dfrac{\sum x \sum y}{n}}{\sum x^2 - \dfrac{(\sum x)^2}{n}}$ $\qquad$ (5–118a) $b = \dfrac{\sum y - m\sum x}{n}$ $\qquad$ (5–118b)
The correlation coefficient	$r = \dfrac{ms_x}{s_y}$ $\qquad$ (5–119) where r lies between -1 and $+1$ If r is negative, it indicates that the regression line has a negative slope If $r = 1$, there is a perfect correlation, and if $r = 0$, there is no correlation
The equation for frequency or density function according to Weibull	$f(x) = \dfrac{b}{\theta - x_0}\left(\dfrac{x - x_0}{\theta - x_0}\right)^{b-1}\left\{\exp\left[-\left(\dfrac{x - x_0}{\theta - x_0}\right)^b\right]\right\}$ $\qquad$ (5–120)
The cumulative distribution function	$F(x) = \displaystyle\int_{x_0}^x f(x)dx = 1 - \exp\left[-\left(\dfrac{x - x_0}{\theta - x_0}\right)^b\right]$ $\qquad$ (5–121)

Equation (5–121) after simplification

$$F(x) = 1 - \exp\left[-\left(\frac{x}{\theta}\right)^b\right] \qquad (5–122)$$

REFERENCES

1. Maleev, V. L., and J. B. Hartman, *Machine Design*, International Textbook Company, Scranton, Pennsylvania, 1954.
2. Shigley, J. E., and L. D. Mitchell, *Mechanical Engineering Design*, McGraw-Hill Book Company, New York, 1983.
3. Faires, V. M., *Design of Machine Elements*, The Macmillan Company, New York, 1965.
4. Lingaiah, K., and B. R. Narayana Iyengar, *Machine Design Data Handbook*, Engineering College Co-operative Society, Bangalore, India, 1962.
5. Lingaiah, K., and B. R. Narayana Iyengar, *Machine Design Data Handbook*, Vol. I (*SI and Customary Metric Units*), Suma Publishers, Bangalore, India, 1986.
6. Lingaiah, K., *Machine Design Data Handbook*, Vol. II (*SI and Customary Metric Units*), Suma Publishers, Bangalore, India, 1986.
7. Juvinall, R. C., *Fundamentals of Machine Component Design*, John Wiley and Sons, New York, 1983.
8. Deutschman, A. D., W. J. Michels, and C. E. Wilson, *Machine Design—Theory and Practice*, Macmillan Publishing Company, New York, 1975.

CHAPTER 6

CAMS

SYMBOLS

a	radius of circular area of contact, m (in)
A	acceleration of the follower, m/s²
	follower overhang, m in (in/s²)
A_c	arc of pitch circle, m (in)
b	half the band of width of contact, m (in)
B	follower bearing length, m (in)
$a_o = \rho_o + \rho_i$	distance between centers of rotation, m (in)
d	diameter of shaft, m (in)
d_h	hub diameter, m (in)
D_o	minimum diameter of the pitch surface of cam, m (in)
E_1, E_2	moduli of elasticity of the materials which are in contact, GPa (Mpsi)
f	cam factor, dimensionless
$f(\theta)$	the desired motion of follower, as a function of cam angle
F	applied load, kN (1bf)
F_n	force normal to cam profile (Fig. 6–5), kN (1bf)
h	depth to the point of maximum shear, m (in)
K_i, K_o	constants for input and output cams, respectively
L	length of cylinder in contact, m (in)
	total distance through which the follower is to rise, m (in)
n	cam speed, rpm
N_1, N_2	forces normal to follower stem, kN (lbf)
F	total external load on follower (includes weight, spring force, inertia, friction, etc.), kN (1bf)
F_t	side thrust, kN (1bf)
r	radius of follower, m (in)
R_c	radius of the circular arc, m (in)
R_o	minimum radius of the pitch surface of the cam, m (in)
R_p	pitch circle radius, m (in)
R_r	radius of the roller, m (in)
R, S	functions of θ_i and θ_o, in basic spiral contour cams
S	displacement of the follower corresponding to any cam angle θ, m (in)
S_1	initial compression spring force with weight w at zero position, kN (1bf)
v	velocity of the follower, m/s (in/s)
w	equivalent weight at follower ends, kN (1bf)

x, y	cartesian coordinates of any point on the cam surface
y	actual lift at follower end, m (in)
y_c	rise of cam, m (in)
ρ	radius of curvature of the pitch curve, m (in)
ρ_1, ρ_2	radii of curvature of the contact surfaces, m (in)
α	pressure angle, deg
α_m	maximum pressure angle, deg
β	angle through which cam is to rotate to effect the rise L, rad
θ	cam angle corresponding to the follower displacement, S, rad
θ_o	angle rotated by the output-driven member, deg
θ_i	angle rotated by the input driver, deg
ω	angular velocity of cam, rad/s
μ	coefficient of friction between follower stem and its guide bearing
ν_1, ν_2	Poisson's ratios for the materials of contact surfaces
$\sigma_{c,max}$	maximum compressive stress, MPa (kpsi)
τ	shear stress, MPa (kpsi)

Particular	Formula	
Cam factor	$$f = \frac{A_c}{L}$$	(6–1)
The length of arc of the pitch circle	$$A_c = R_p\beta$$	(6–2)
The pitch circle radius	$$R_p = \frac{fL}{\beta}$$	(6–3)

RADIUS OF CURVATURE OF DISK CAM WITH ROLLER FOLLOWER

The displacement of the center of the follower from the center of cam (Fig. 6–1)	$R = R_o + f(\theta)$	(6–4)
For pointed cam, the radius of curvature of the pitch curve to roller follower	$\rho = R_r$	(6–5)
For roller follower, the radius of curvature of the pitch curve must always be greater than the roller radius to prevent points or undercuts	$\rho > R_r$	(6–6)

The radius of curvature for concave pitch curve

$$\rho = -\frac{\{R^2 + [f'(\theta)]^2\}^{3/2}}{R^2 + 2[f'(\theta)]^2 - R[f''(\theta)]} \qquad (6\text{–}7)$$

where

$$R = R_o + f(\theta); \quad \frac{dR}{d\theta} = f'(\theta); \quad \frac{d^2R}{d\theta^2} = f''(\theta) \qquad (6\text{–}7a)$$

The minimum radius of curvature

$$\rho_{min} = \frac{R_o^2}{R_o - f''(\theta)_o} \qquad (6\text{–}8)$$

where $f''(\theta)_o$ is the acceleration at $\theta = 0$

Particular	Formula

The minimum radius of curvature of the cam curve ρ_c $\qquad \rho_{c\,min} = \rho_{min} \pm R_r$ $\qquad\qquad$ (6–9)

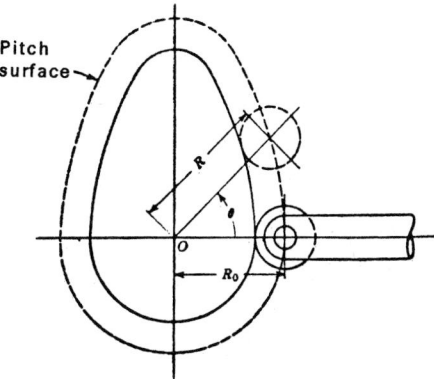

FIGURE 6–1

The radius of curvature for convex pitch curve (Fig. 6–2)

$$\rho = \frac{\{R^2 + [f'(\theta)]^2\}^{3/2}}{R^2 + 2[f'(\theta)]^2 - R[f''(\theta)]} \qquad (6\text{–}10)$$

The minimum radius of a mushroom cam for harmonic motion

$$R_o = \left(\frac{16200}{\beta^2} - 1\right)L \qquad (6\text{–}11)$$

The minimum radius of a mushroom cam for uniformly accelerated and retarded motion

$$R_o = \left(\frac{13131}{\beta^2} - \frac{1}{2}\right)L \qquad (6\text{–}12)$$

For cast-iron cam, the hub diameter

$$d_h = 1\tfrac{3}{4}d + 13.75 \text{ mm } (1.75d + 0.55 \text{ in}) \qquad (6\text{–}13)$$

Plate cam design ÷ radius of curvature:

For eight-power polynomial motion $\qquad$ Refer to Fig. 6–6.

For cycloidal motion $\qquad$ Refer to Fig. 6–7.

For harmonic motion $\qquad$ Refer to Fig. 6–8.

RADIUS OF CURVATURE OF DISK CAM WITH FLAT-FACED FOLLOWER

The displacement of the follower from the origin (Fig. 6–2)

$$R = a + f(\theta) \qquad (6\text{–}14)$$

The parametric equations of the cam contour (Fig. 6–2)

$$x = [a + f(\theta)]\cos\theta - f'(\theta)\sin\theta \qquad (6\text{–}15a)$$

$$y = [a + f(\theta)]\sin\theta + f'(\theta)\cos\theta \qquad (6\text{–}15b)$$

Particular	Formula	
The cam contour given by equations will be free of cusps if	$a + f(\theta) + f''(\theta) > 0$	(6–16)
Half of the minimum length of the flat-faced follower or the minimum length of contact of the follower	$b = f'(\theta)$	(6–17)

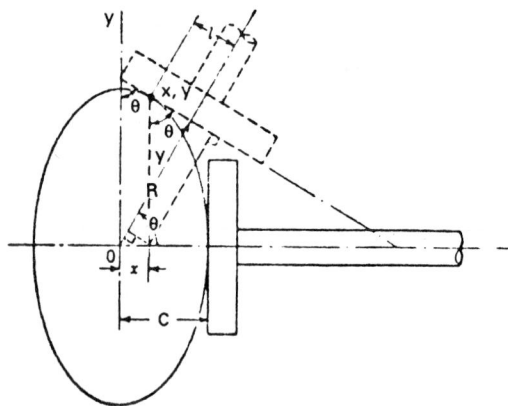

FIGURE 6–2   (Courtesy of H. H. Mabie and F. W. Ocvivk, *Dynamics of Machinery*, John Wiley and Sons, 1957.)

PRESSURE ANGLE

The pressure angle for roller follower	$\alpha = \tan^{-1} \dfrac{1}{R} \dfrac{dR}{d\theta}$	(6–18)
The pressure angle for a plate cam or any cylindrical cam giving uniform velocity to the follower	$\tan \alpha = \dfrac{360L}{2\pi\beta R_o} = \dfrac{360L}{\pi\beta D_o}$	(6–19)
The pressure angle for a plate cam giving uniformly accelerated and retarded motion to the follower	$\tan \alpha = \dfrac{360 \times 2L}{\pi\beta(D_o + L)}$ when $L > D_o$	(6–20a)
	$= \dfrac{180 \times 2}{\pi\beta}\sqrt{\dfrac{L}{D_o}}$ when $L > D_o$	(6–20b)
A precise pressure angle equation for a plate cam giving harmonic motion to the follower or a tangential cam	$\tan \alpha = \dfrac{90L}{\beta\sqrt{R_o^2 + R_o L}}$	(6–21)

Particular	Formula

RADIAL CAM-TRANSLATING ROLLER-FOLLOWER FORCE ANALYSIS (FIG. 6–3)

The forces normal to follower stem (Fig. 6–3)

$$F_R = \frac{l_r}{l_g} F_n \sin \alpha \qquad (6\text{–}22)$$

$$F_L = \frac{l_r + l_g}{l_g} F_n \sin \alpha \qquad (6\text{–}23)$$

The total external load

$$F = F_n \left[\cos \alpha - \mu \left(\frac{2l_r + l_g}{l_g} \right) \sin \alpha \right] \qquad (6\text{–}24)$$

The force normal to the cam profile

$$F_n = \frac{F}{\cos \alpha - \mu \left(\dfrac{2l_r + l_g}{l_g} \right) \sin \alpha} \qquad (6\text{–}25)$$

The maximum pressure angle for locking the follower in its guide

$$\alpha_m = \tan^{-1} \frac{l_g}{\mu (2l_r + l_g)} \qquad (6\text{–}26)$$

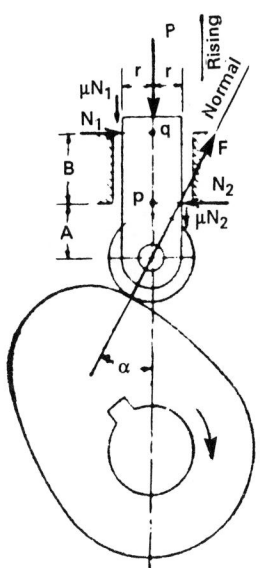

FIGURE 6–3   Radial cam-translating roller-follower force analysis. (Courtesy of H. A. Rothbart, *Cams*, John Wiley and Sons, 1956.)

Particular	Formula

SIDE THRUST (FIG. 6–3)

The side thrust produced on the follower bearing

$$F_t = F \tan \alpha \qquad (6\text{–}27)$$

BASIC SPIRAL CONTOUR CAM

The radius to point of contact at angle θ_o (Fig. 6–4)

$$\rho_o = \frac{a_o}{1 + \dfrac{d\theta_o}{d\theta_i}} \qquad (6\text{–}28)$$

The radius to point of contact at angle θ_i (Fig. 6–4)

$$\rho_i = \frac{a_o \dfrac{d\theta_o}{d\theta_i}}{1 + \dfrac{d\theta_o}{d\theta_i}} \qquad (6\text{–}29)$$

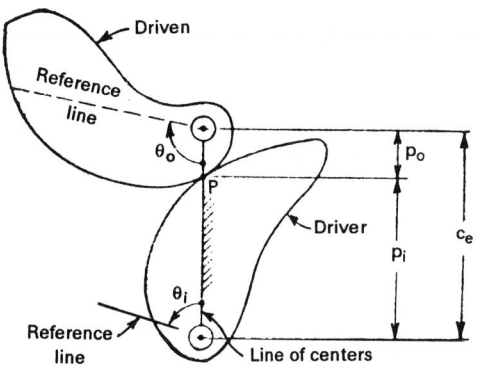

FIGURE 6–4   Basic spiral controur cam.

BASIC SPIRAL CONTOUR CAM CONSTANTS

The radius to point of contact at angle θ_o

$$\rho_o = \frac{a_o}{1 + \dfrac{K_o}{K_i}\left(\dfrac{dS}{dR}\right)} \qquad (6\text{–}30)$$

The radius to point of contact at angle θ_i

$$\rho_i = \frac{a_o \dfrac{K_o}{K_i}\left(\dfrac{dS}{dR}\right)}{1 + \dfrac{K_o}{K_i}\left(\dfrac{dS}{dR}\right)} \qquad (6\text{–}31)$$

where

$$R = \frac{\theta_i}{K_i}; \ S = \frac{\theta_o}{K_o}$$

and $\dfrac{d\theta_i}{dR} = K_i; \ \dfrac{d\theta_o}{dS} = K_o$

Particular	Formula

HERTZ CONTACT STRESSES

Contact of sphere on sphere

The radius of circular area of contact

$$a = 3\sqrt{\dfrac{3F\left[\left(\dfrac{1 - v_1^2}{E_1}\right) + \left(\dfrac{1 - v_2^2}{E_2}\right)\right]}{4\left(\dfrac{1}{\rho_1} + \dfrac{1}{\rho_2}\right)}} \qquad (6\text{–}32)$$

The maximum compressive stress

$$\sigma_{c,max} = \dfrac{3F}{2\pi a^2} \qquad (6\text{–}33)$$

Contact of cylindrical surface on cylindrical surface

Width cf band of contact

$$2b = \sqrt{\dfrac{16F\left[\left(\dfrac{1 - v_1^2}{E_1}\right) + \left(\dfrac{1 - v_1^2}{E_2}\right)\right]}{\pi L\left(\dfrac{1}{\rho_1} + \dfrac{1}{\rho_2}\right)}} \qquad (6\text{–}34)$$

The maximum compressive stress

$$\sigma_{c,max} = \dfrac{2F}{\pi b L} \qquad (6\text{–}35)$$

The maximum compressive stress for $v_1 = v_2 = 0.3$

$$\sigma_{c,max} = \sqrt{\dfrac{0.35F\left(\dfrac{1}{\rho_1} + \dfrac{1}{\rho_2}\right)}{L\left(\dfrac{1}{E_1} + \dfrac{1}{E_2}\right)}} \qquad (6\text{–}36)$$

The maximum shear stress

$$\tau_{max} = 0.295\,\sigma_{c,max} \qquad (6\text{–}37)$$

The depth to the point of maximum shear

$$h = 0.786b \qquad (6\text{–}38)$$

For further data on characteristic equations of basic curves, different motion characteristics, cam factors, materials for cams and followers, and displacement ratios

Refer to Tables 6–1 and 6–2.
For materials of cams refer to Chapter 1 on "Properties of Engineering Materials"

TABLE 6–2
Cam factors for basic curves

Pressure angle α, deg	Types of motion			
	Uniform	Modified uniform	Simple harmonic	Parabolic and cycloidal
10	5.67	5.84	8.91	11.34
15	3.73	3.99	5.85	7.46
20	2.75	3.10	4.32	5.50
25	2.14	2.58	3.36	4.28
30	1.73	2.27	2.72	3.46
35	1.43	2.06	2.24	2.86
40	1.19	1.92	1.87	2.38
45	1.00	1.82	1.57	2.00

REFERENCES

1. Rothbart, H. A., *Cams*, John Wiley and Sons, New York, 1956.
2. Marks, L. S., *Mechanical Engineers' Handbook*, McGraw-Hill Book Company, New York, 1951.
3. Lingaiah, K., and B. R. Narayana Iyengar, *Machine Design Data Handbook*, Engineering College Co-operative Society, Bangalore, India, 1962.
4. Lingaiah, K., *Machine Design Data Handbook*, Vol. II (*SI and Customary Metric Units*), Suma Publishers, Bangalore, India, 1986.
5. Rothbart, H. A., *Mechanical Design and Systems Handbook*, McGraw-Hill Book Company, New York, 1964.
6. Shigley, J. E., *Theory of Machines*, McGraw-Hill Book Company, New York, 1961.
7. Mabie, H. H., and F. W. Ocvirk, *Mechanisms and Dynamics of Machinery*, John Wiley and Sons, New York, 1957.
8. Kent, R. T., *Mechanical Engineers' Handbook—Design and Production*, Vol. II, John Wiley and Sons, New York, 1961.

CHAPTER

7

PIPES, TUBES, AND CYLINDERS

SYMBOLS

d	diameter of cylinder, m (in)
d_c	diameter of contact surface in compound cylinder, m (in)
d_i	inside diameter of cylinder or pipe or tube, m (in)
d_o	outside diameter of cylinder or pipe or tube, m (in)
e	factor for expanded tube ends
E	modulus of elasticity, GPa (Mpsi)
h or t	thickness of cylinder or pipe or tube, m (in)
I	moment of inertia, area, m^4 or cm^4 (in^4)
K	constant
L	maximum distance between supports or stiffening rings, m (in)
p	maximum allowable working pressure, MPa (psi)
p_c	unit pressure between the compound cylinders, MPa (psi)
p_{cr}	collapsing pressure, MPa (psi)
p_i	internal pressure, MPa (psi)
p_o	external pressure, MPa (psi)
r_i	inside radius of tube or pipe, m (in)
σ	permissible working stress, from Table 7–1, MPa (psi)
σ_c	crushing stress, MPa (psi)
σ_r	radial stress (also with primes), MPa (psi)
$\sigma_{r(max)}$	maximum radial stress, MPa (psi)
σ_{sa}	maximum allowable stress value at design condition, MPa (psi)
σ_{su}	ultimate strength, MPa (psi)
σ_θ	tangential stress (also with primes), MPa (psi)
$\sigma_{\theta(max)}$	maximum tangential stress, MPa (psi)
τ_{max}	maximum shear stress, MPa (psi)
ν	Poisson's ratio
η	efficiency, from Table 7–3

Note: The initial subscript S, along with σ, which stands for strength, is used throughout this book.

Particular	Formula

LONG THIN TUBES WITH INTERNAL PRESSURE

The permissible steam pressure in steel and iron pipes (Table 7–1) according to *ASME Power Boiler Code*

$$p = \frac{2\sigma_{sa}}{d_o}(h - 1.625 \times 10^{-3}) - 0.9 \qquad \textbf{SI} \qquad (7\text{–}1a)$$

where h, d_o in m, and p and σ in MPa

$$p = \frac{2\sigma_{sa}}{d_o}(h - 0.065) - 125 \quad \textbf{US Customary System units}$$
$$(7\text{–}1b)$$

where h, d_o in in, and p and σ in psi
For tubes from 6.35 mm (0.25 in) to 127 mm (5 in) nominal diameter

$$p = \frac{2\sigma_{sa}}{d_o}(h - 2.54 \times 10^{-3}) \qquad \textbf{SI} \qquad (7\text{–}2a)$$

where h, d_o in m, and p and σ in MPa

$$p = \frac{2\sigma_{sa}}{d_o}(h - 0.1) \qquad \textbf{US Customary System units}$$
$$(7\text{–}2b)$$

where h, d_o in in, and p and σ in psi
For over 127 mm (5 in) diameter

The minimum required thickness of ferrous tube up to and including 125 mm (5 in) outside diameter subjected to internal pressure as per *ASME Power Boiler Code*

$$h = \frac{pd_o}{2\sigma_{sa} + p} + 0.005d_o + e \qquad (7\text{–}3)$$

where σ_{sa} is the maximum allowable stress value at design condition and e is the thickness factor for expanded tube ends.
Refer to Table 7–1 for σ_{sa}.
Refer to Table 7–2 for e.

The maximum allowable working pressure (MAWP) from Eq. (7–3) as per *ASME Power Boiler Code*

$$p = \sigma_{sa}\left[\frac{2h - 0.01d_o - 2e}{d_o - (h - 0.005d_o - e)}\right] \qquad (7\text{–}4)$$

The minimum required thickness of ferrous pipe under internal pressure as per *ASME Power Boiler Code*

$$h = \frac{pd_o}{2\sigma_{sa}\eta + 2yp} + C = \frac{pr_i}{\sigma_{sa}\eta - (1 - y)p} + C \qquad (7\text{–}5)$$

where η = efficiency (refer to Table 7–4 for η)
y = temperature coefficient (refer to Table 7–3 for y)
C = minimum allowance for the threading and structural stability, mm (in) (refer to Table 7–5 for h values and Table 7–6 for C values)

TABLE 7–1
Maximum allowable stress values in tension of metals for tubes and pipes, σ_{sa}

Specification number	Grade, alloy designation and temper	UNS number	Nominal composition and size	Product form	Specified minimum yield strength σ_{sy} MPa	kpsi	Specified minimum tensile strength σ_{st} MPa	kpsi	38 (100) MPa	kpsi	93 (200) MPa	kpsi	150 (300) MPa	kpsi	205 (400) MPa	kpsi	260 (500) MPa	kpsi
1	2	3	4	5	6	7	8	9	10	11	12	13	14	15	16	17	18	19
(A) Carbon and Low Alloy Steels																		
Carbon Steel:																		
SA-106[c]	A		C-Si	Smls.+·,Pp*	207	30	331	48										
SA-210[c]	C		C-Mn-Si	Smls.Tb**	276	40	483	70										
SA-557[b,f]	C		C-Mn	Smls.Tb	276	40	483	70										
Low Alloy Steel:																		
SA-209[g]	T1		1Cr-½Mo	Smls.Tb	207	30	379	55										
SA-213	T12		1Cr-½Mo	Smls.Tb	207	30	414	60										
SA-369	Fp11		1¼Cr-½Mo-Si	Smls.Pp	207	30	414	60										
SA-250	T1		C-½Mo*	Wld*·Pp&Tb	207	30	379	55										
(B) High Alloy Steels																		
SA-268	TP410	S41000	13Cr.	Smls.Tb,	207	30	414	60	103	15.0	99	14.3	95	13.8	92	13.3	89	12.9
SA-268	TP405	S40500	12Cr-1Al.	Wld.Tb[f]	207	30	414	60	88	12.8	84	12.2	81	11.8	78	11.3	75	10.9
SA-268	TpxM-8	S43035	18Cr-Ti	Wld.Tb[d,f]	207	30	414	60	88	12.8	84	12.2	81	11.8	78	11.3	75	10.9
SA-268	TpxM-8	S43035	18Cr-Ti	Smls.Tb[d,f]	207	30	414	60	103	15.0	98	14.3	95	13.8	92	13.3	89	12.9
SA-268	18Cr-2Mo	S44400	18Cr-2Mo	Wld.Tb[c,d]	276	40	414	60	88	12.8	84	12.2	81	11.8	78	11.3	75	10.9
SA-268	18Cr-2Mo	S44400	18Cr-2Mo	Smls.Tb[c,d]	276	40	414	60	103	15.0	99	14.3	95	13.8	92	13.3	88	12.8
SA-249, SA-312	TP304L	S30403	18Cr-8Ni	Wld.Tb&Pp	172	25	483	70	92	13.3	78	11.4	70	10.2	64	9.3	60	8.7
SA-213, SA-312	TP304H, TP304	S30409, S30400	18Cr-8Ni	Smls.Tb & Pp[g,h]	207	30	517	75	130	18.8	123	17.8	115	16.6	112	16.2	110	15.9
SA-213, SA-312	TP304N	S30451	18Cr-8Ni-N	Smls.Tb & Pp[g,h]	241	35	552	80	138	20.0	138	20.0	131	19.0	126	18.3	123	17.8
SA-249, SA-312	TP304N	S30451	18Cr-8Ni-N	Wld.Tb & Pp[f,g,h]	241	35	552	80	117	17.0	117	17.0	111	16.1	108	15.6	104	15.1
SA-213, SA-312	TP316L	S31603	16Cr-12Ni-2Mo	Smls.Tb & Pp	172	25	483	70	108	15.7	92	13.3	82	11.9	75	10.8	69	10.0
SA-249, SA-312	TP316L	S31603	16Cr-12Ni-2Mo	Wld.P & Tb[f]	172	25	517	75	92	13.3	78	11.3	70	10.1	63	9.2	59	8.5
SA-312, SA-688	TP316H	S31609	16Cr-12Ni-2Mo	Cast.Pp[g]	207	30	517	75	130	18.8	130	18.8	127	18.4	125	18.1	124	18.0
SA-452	XM-15	S31800	18Cr-18Ni-2Si	Wld.Tb[f,g]	207	30	517	75	110	15.9	104	15.1	97	14.1	95	13.7	93	13.5
SA-312	XM-15	S38100	18Cr-18Ni-2Si	Smls.Tb[g]	207	30	517	75	110	15.9	104	15.1	95	14.1	93	13.7	93	13.5
SA-213	TP316N	S31651	16Cr-12Ni-2Mo-N	Smls.Tb[g,h]	241	35	552	80	130	18.8	122	17.7	115	16.6	111	16.1	110	15.9
SA-312	TP316N	S31651	16Cr-12Ni-2Mo-N	Wld.Pp[f,g,h]	241	35	552	80	138	20.0	138	20.0	132	19.2	130	18.8	128	18.6
SA-312, SA-688	XM-29	S24000	16Cr-3Ni-12Mn	Wld.Pp[f] & Tb	379	55	689	100	117	17.0	117	17.0	112	16.3	110	16.0	109	15.8
SA-213, SA-312	TP321	S32100	18Cr-10Ni-Ti	Smls.Tb[g,h] & Pp[f]	207	30	517	75	146	21.2	143	20.8	132	19.2	119	17.3	110	16.0
SA-249, SA-312	TP321H	S32109	18Cr-10Ni-Ti	Wld.Tb & Pp[f]	207	30	517	75	130	18.8	127	18.4	119	17.3	118	17.1	118	17.1
SA-430	FP347H	S34700	18Cr-10Ni-Cb	Smls.Pp[g]	207	30	483	70	110	16.0	93	13.5	83	12.1	76	11.0	70	10.2
SA-213	TP348	S34800	18Cr-10Ni-Cb	Smls.Tb[g,h]	207	30	517	75	130	18.8	123	17.9	113	16.4	107	15.5	103	14.9
SA-249, SA-312	TP348H, TP347H	S34809, S34709	18Cr-10Ni-Cb	Wld.Tb&Pp[f,g]	207	30	517	75	130	18.8	123	17.9	113	16.4	107	15.3	103	14.9
SA-213, SA-312	S30815	S30815	21Cr-11Ni-N	Smls.Tb&Pp[f]	310	45	600	87	149	21.6	145	21.0	141	20.4	135	19.6	127	18.4
SA-789, SA-790	S32550	S32550	25.5Cr-5.5Ni-3.5Mo-Cu	Smls.Tb&Pp[d]	552	80	758	110	190	27.5	189	27.4	177	25.7	170	24.7	170	24.7
SA-789, SA-790	S32550	S32550	25.5Cr-5.5Ni-3.5Mo-Cu	Wld.Tb&Pp[d]	552	80	758	110	161	23.4	161	23.3	151	21.9	145	21.0	145	21.0
SA-789, SA-790, SA-669	S31500	S31500	18Cr-5Ni-3Mo	Smls.Tb[d,f]	441	64	634	92	159	23.0	153	22.2	147	21.3	146	21.2	146	21.2
SA-789, SA-790	S31500	S31500	18Cr-5N-3Mo	Wld.Tb[d,f]	441	64	634	92	135	19.6	130	18.9	125	18.1	124	18.0	124	18.0

7.3

TABLE 7-1
Maximum allowable stress values in tension of metals for tubes and pipes, σ_{sa} (Cont.)

for metal temperature, °C (°F), not exceeding

315 (600)		370 (700)		427 (800)		482 (900)		538 (1000)		593 (1100)		650 (1200)		704 (1300)		760 (1400)		815 (1500)		Specification number
MPa	kpsi	MPa	kpsi	MPa	kpsi	MPa	kpsi	MPa	kpsi	MPa	kpsi	MPa	kpsi	MPa	kpsi	MPa	kpsi	MPa	kpsi	
20	21	22	23	24	25	26	27	28	29	30	31	32	33	34	35	36	37	38	39	40
(A) Carbon and Low Alloy Steels																				
Carbon Steel:																				
83	12.0	81	11.7	64	9.3	45	6.5	17	2.5											SA-106[c]
121	17.5	115	16.6	83	12.0	35	5.0	10	1.5											SA-210[c]
103	15.0	97	14.1	70	10.2	38	5.5	15	2.1											SA-557[b,f]
Low Alloy Steels																				
95	13.8	95	13.8	93	13.5	86	12.7	33	4.8	27	4.0	8	1.2							SA-209[g]
103	15.0	103	15.0	99	14.4	76	11.0	38	5.5	18	2.6	7	1.0							SA-213
		98	14.2	93	13.5	86	12.5	48	6.2											SA-369
88	12.8	88	12.8	84	12.2	76	11.0	98	4.1	20	2.9	7	1.0							SA-250
86	12.4	83	12.1	77	11.1	67	9.7	44	6.4											SA-268
(B) High Alloy Steels																				
73	10.6	71	10.3	65	9.4	57	8.2	23	3.4											SA-268
73	10.6	71	10.3	65	9.4															SA-268
86	12.4																			SA-268
72	10.5																			SA-268
86	12.4																			SA-268
57	8.3	55	8.0	53	7.7															SA-249, SA-312
110	15.9	110	15.9	105	15.2	101	14.7	95	13.8	68	9.8	42	6.1	25	3.7	16	2.3	10	1.4	SA-213, SA-312
120	17.4	118	17.1	115	16.6	110	15.9	103	15.0	67	9.7	41	6.0	28	4.1	16	2.3	9	1.3	SA-213, SA-312
102	14.8	101	14.6	98	14.2	93	13.5	86	12.7	57	8.3	35	5.1							SA-249, SA-312
51	7.4	62	9.0	59	8.6															SA-213, SA-312
55	8.0	52	7.6	50	7.3															SA-312, SA-688
117	17.0	112	16.3	110	15.9	107	15.5	106	15.3	85	12.4	51	7.4	12	1.7	5	0.8	2	0.3	SA-452
93	13.5	93	13.5	89	12.9	85	12.4	95	13.7	85	12.4	51	7.4	19	2.7	11	1.6	7	1.0	SA-312
110	15.9	110	15.9	104	15.1	101	14.6	120	17.4	72	10.5	43	6.3							SA-213
128	18.6	128	18.6	127	18.4	125	18.1	102	14.8	48	6.9	24	3.6							SA-312
109	15.8	109	15.8	108	15.6	106	15.4	95	13.8	52	7.5	32	4.6							SA-312, SA-688
106	15.4	101	14.7	97	14.1															SA-213, SA-312
112	16.4	109	15.8	107	15.5	106	15.3	61	8.9	90	13.0	55	7.9	30	4.4	17	2.5	9	1.3	SA-249, SA-312
67	9.7	64	9.3	63	9.2	62	9.0	99	14.4	63	9.1	30	4.4	15	2.2	8	1.2	5	0.8	SA-430
101	14.7	101	14.7	101	14.7	101	14.7	97	14.0	76	11.1	46	6.7	25	3.7	15	2.1	8	1.1	SA-213
101	14.7	101	14.7	101	14.7	101	14.7	84	12.3	62	9.0	36	5.2	21	3.1	13	1.9	9	1.3	SA-249, SA-312
86	12.5	86	12.5	86	12.5	86	12.5													SA-312, SA-213
122	17.7	119	17.3	116	16.8	112	16.3	103	14.9											SA-789, SA-790
146	21.2	146	21.2																	SA-789, SA-790, SA-669
124	18.0	124	18.0																	SA-789, SA-790

7.4

TABLE 7-1
Maximum allowable stress values in tension of metals for tubes and pipes, σ_{sa} (Cont.)

Specification number	Grade, alloy designation and temper	UNS number	Nominal composition and size mm (in)	Product form	Specified minimum yield strength σ_{sy} MPa	kpsi	Specified minimum tensile strength σ_{st} MPa	kpsi	Maximum allowable stress 38 (100) MPa	kpsi	93 (200) MPa	kpsi	150 (300) MPa	kpsi	205 (400) MPa	kpsi	260 (500) MPa	kpsi
1	2	3	4	5	6	7	8	9	10	11	12	13	14	15	16	17	18	19
(C) Non-ferrous Metals																		
Aluminum and Aluminum Alloys:																		
SB-210	1060-H14[d]		0.250–12.500 (0.010–0.500)	Drawn	69	10	83	12	21	3.0	21	3.0	18	2.6	8	1.2		
SB-210	6061-T6[e]		0.625–12.50 (0.025–0.50)	Smls.Tb	241	35	290	42	72	10.5	72	10.5	58	8.4	31	4.5		
SB-241	3003-H118[e]		Under 25 (under 1)	Smls.Pp	165	24	186	27	47	6.8	46	6.7	37	5.4	17	2.5		
SB-241	5083-H111[d,p]			Smls. extruded Tb	131	19	228	33	57	8.3	57	8.3	38	5.5	21	3.0		
SB-234	1060-H14[e]		Up to 125 (up to 5.00)	Condenser and heat exchanger Tb	69	10	83	12	21	3.0	21	3.0	18	2.6	8	1.2		
SB-234	3003-H25[e]		0.250–12.50 (0.010–0.5000)	Condenser and heat exchanger Tb	131	19	145	21	38	5.5	38	5.5	30	4.3	17	2.4		
SB-234	6061-T6[e]		0.625–6.225 (0.025–0.249)	Condenser and heat exchanger Tb	241	35	290	42	72	10.5	72	10.5	58	8.4	31	4.5		
Copper and Copper Alloys:																		
SB-111	102, 120, 122, 142[i]		Ann	Smls. Copper condenser, Tb.	62	9	207	30	41	6.0	33	4.8	32	4.7	21	3.0		
			LD"		207	30	248	36	62	9.0	62	9.0	60	8.7	57	8.2		
			HD**		276	40	310	45	78	11.3	78	11.3	78	11.3				
SB-111	192 Ann			Smls. Copper, iron alloy condenser Tb.	83	12	262	38	52	7.5	46	6.7	42	6.1	30	4.3		
SB-315	655, Ann[g]			Smls. Cu-Si. Alloy Pp and Tb	103	15	345	50	69	10.0	69	10.0	69	10.0	35	5.0		
SB-466	C71500 Ann[p]			Smls. Cu-Ni 70/30 Pp & Tb.	124	18	345	50	83	12.0	78	11.3	75	10.8	71	10.3	68	9.9
SB-467	C71500 Ann[pp]		(Up to 112.5 inc) (up to 4½ incl)	Wld. Cu-Ni-70/30 Pp	138	20	345	50	87	12.6	61	8.9	61	8.8	61	8.8	61	8.8
SB-543	C700-Ann[p]			Wld.Cu-Ni-90/10Tb	103	15	270	40	59	8.5	56	8.1	52	7.6	50	7.2	43	6.3
	LCW***				241	35	310	45	59	8.5	56	8.1	52	7.6	50	7.2	43	6.3
Nickel and High Nickel Alloys:																		
SB-161	201 Ann	N02201	Ni Low C	Pp & Tb	69	10	345	50	46	6.7	44	6.4	43	6.3	43	6.2	43	6.2
SB-163	800H Ann[k]	N08810	Ni-Fe-Cr	Pp & Tb	172	25	448	65	112	16.2	112	16.2	112	16.2	112	16.2	110	16.0
SB-163	825 Ann[k]	N08825	Ni-Fe-Cr-Mo-Cu	Pp & Tb	241	35	586	85	146	21.2	146	21.2	146	21.2	146	21.2	146	21.2
SB-144	625 Ann[p]	N06625	Ni-Cr-Mo-Cb		414	60	827	120	207	30.0	207	30.0	207	30.0	194	28.2	186	27.0
SB-468	20 cb.Wld. Ann[k,p]	N08020	Cr-Ni-Fe-Mo-Cu-Cb	Pp & Tb	241	35	552	80	117	17.0	117	17.0	115	16.8	110	15.9	107	15.5
SB-619	C-276 Sol. Ann[p]	N10276	Ni-Mo-Cr (All sizes)	Pp & Tb	283	41	689	100	146	21.2	146	21.2	146	21.2	143	20.7	140	20.3
SB-619	G. Sol. Ann[k,p]	N06007	Ni-Cr-Fe-Mo-Cu (All sizes)	Pp & Tb	242	35	620	90	132	19.1	132	19.1	132	19.1	128	18.6	126	18.3

TABLE 7-1
Maximum allowable stress values in tension of metals for tubes and pipes, σ_{sa} (*Cont.*)

for metal temperature, °C (°F), not exceeding

315 (600)		370 (700)		427 (800)		482 (900)		538 (1000)		593 (1100)		650 (1200)		704 (1300)		760 (1400)		815 (1500)		Specification number
MPa	kpsi	MPa	kpsi	MPa	kpsi	MPa	kpsi	MPa	kpsi	MPa	kpsi	MPa	kpsi	MPa	kpsi	MPa	kpsi	MPa	kpsi	
20	21	22	23	24	25	26	27	28	29	30	31	32	33	34	35	36	37	38	39	40

(C) Non-ferrous Metals: Aluminum and Aluminum Alloys

MPa	kpsi	MPa	kpsi	MPa	kpsi	MPa	kpsi	MPa	kpsi	MPa	kpsi	MPa	kpsi	MPa	kpsi	MPa	kpsi	MPa	kpsi	Spec.
																				SB-210
																				SB-210
																				SB-241
																				SB-241
																				SB-234
																				SB-234
																				SB-234

Copper and Copper Alloys

MPa	kpsi	MPa	kpsi	MPa	kpsi	MPa	kpsi	MPa	kpsi	MPa	kpsi	MPa	kpsi	MPa	kpsi	MPa	kpsi	MPa	kpsi	Spec.
																				SB-111
																				SB-111
67	9.6																			SB-315
61	8.8																			SB-466
30	4.3																			SB-467
30	4.3																			SB-543

Nickel and High Nickel Alloy

MPa	kpsi	MPa	kpsi	MPa	kpsi	MPa	kpsi	MPa	kpsi	MPa	kpsi	MPa	kpsi	MPa	kpsi	MPa	kpsi	MPa	kpsi	Spec.
		43	6.2	41	5.9	31	4.5	21	3.0	14	2.0	8	1.2							SB-161
		110	15.7	105	15.3	102	14.8	99	14.4	80	11.6	51	7.4							SB-163
		146	21.0	143	20.8	141	20.5	136	19.7											SB-163
		182	26.0	179	26.0	179	26.0	179	26.0	179	26.0	91	13.2	32	4.7	21	3.0	13	1.9	SB-444
		104	14.7	99	14.3															SB-468
		138	19.6	134	19.4	130	18.9	128	18.5	88	12.7	57	8.3							SB-619
		123	17.8	120	17.4	117	17.0	111	16.1											SB-619

Source: The American Society of Mechanical Engineers, *Boiler and Pressure Vessel Code*, Section VIII, Division I, July 1986.

* Pp = Pipe; ** Tb = Tube; + + Smls. = Seamless; • Wld. = Welded; '' LD = Light drawn; ***LCW = Light cold worked; ** HD = Hard drawn; Ann = Annealed; Soln. Ann = Solution Annealed.

Notes: The raised alphabets a, b, c, d, e......, etc. refer to notes under each category of (A) Carbon and Low Alloy Steels; (B) High Alloy Steel, and (C) Non-ferrous Metals. Refer to notes for detailed information under category of (A) Carbon and Low Alloy Steels; (B) High Alloy Steels, and (C) Non-ferrous Metals in Tables 8–9, 8–10 and 8–11 in Chapter 8.

TABLE 7–2
Thickness factor for expanded tube ends e for use in Eqs. (7–3) and (7–4)

Particular	Value of e
Over a length at least equal to the length of the seat plus 25 mm (1 in) for tubes expanded into tube seats, except	0.04

For tubes expanded into tube seats provided the thickness of the tube ends over a length of the seat plus 25 mm (1 in) is not less than the following:
  2.375 mm (0.095 in) for tubes ≤31.25 mm (1.25 in) OD
  2.625 mm (0.105 in) for tubes >31.25 mm (1.25 in) OD and ≤50 mm (2 in) OD, including
  3.000 mm (0.120 in) for tubes >50 mm (2 in) and ≤75 mm (3 in) OD, including
  3.375 mm (0.135 in) for tubes >75 mm (3 in) OD and ≤100 mm (4 in) OD, including
  3.75 mm (0.150 in) for tubes >100 mm (4 in) and ≤125 mm (5 in) OD, including
	0

For tubes strength-welded to headers and drums	0

Source: *ASME Boiler and Pressure Vessel Code*, Section I, 1983.

TABLE 7–3
Temperature coefficient y

Material	Temperature, °C (°F)[a]					
	≤482 (900)[a]	510 (950)	540 (1000)	565 (1050)	595 (1100)	≥620 (1150)
Ferrite steels	0.4	0.5	0.7	0.7	0.7	0.7
Austenitic steels	0.4	0.4	0.4	0.4	0.5	0.7
For nonferrous materials	0.4	0.4	0.4	0.4	0.4	0.4

[a] Temperatures in parentheses are in Fahrenheit (°F). Values of y between temperatures not listed may be determined by interpolation.
Source: *ASME Boiler and Pressure Vessel Code*, Section I, 1983.

TABLE 7–4
Efficiency of joints, η

Particular	Efficiency, η
Longitudinal welded joints or of ligaments between openings, whichever is lower Seamless cylinders	1.00
For welded joints provided all weld reinforcement on the longitudinal joints is removed substantially flush with the surface of the plate	1.00
For welded joints with the reinforcement on the longitudinal joints left in place	0.90
Riveted joints	Refer to Table 13–4 (Chap. 13)
Ligaments between openings	Refer to Eqs. under Ligament (Chap. 8)
Welded joint efficiency factor	Refer to Table 8–3 (Chap. 8)

Source: *ASME Boiler and Pressure Vessel Code*, Section I, 1983.

TABLE 7–5
The depth of thread h (formula $h = 0.8/i$)

Number of threads per mm (in), i	Depth of thread, h mm (in)
0.32 (8)	2.5 (0.100)
0.46 (11.5)	1.715 (0.0686)

Source: *ASME Boiler and Pressure Vessel Code*, Section I, 1983.

Particular	Formula

The maximum allowable working pressure from Eq. (7–5) as per *ASME Power Boiler Code*

$$p = \frac{2\sigma_{sa}\eta(h - C)}{d_o - 2y(h - C)} \text{ or } \frac{\sigma_{sa}\eta(h - C)}{r_i + (1 - y)(h - C)} \qquad (7\text{–}6)$$

The minimum required thickness of nonferrous seamless tubes and pipes for outside diameters 12.5 mm (0.5 in) to 150 mm (6 in) inclusive and for wall thickness not less than 1.225 mm (0.049 in) as per *ASME Power Boiler Code*

$$h = \frac{pd_o}{2\sigma_{sa}} + C \qquad (7\text{–}7)$$

Refer to Table 7–6 for values of C.

The maximum allowable working pressure as per *ASME Power Boiler Code*

$$p = \frac{2\sigma_{sa}}{d_o}(h - C) \qquad (7\text{–}8)$$

The minimum required thickness of tubes made of steel or wrought iron subjected to internal pressure which are used in watertube and firetube boilers as per *ASME Power Boiler Code*

$$h = 0.0251d_o \qquad (7\text{–}9)$$

The minimum required thickness of tubes made of nonferrous materials such as copper, red brass, admiralty and copper-nickel alloys used in watertube and firetube boilers with a design pressure over 207 kPa (30 psi) but not greater than 414 kPa (60 psi)

$$h = \frac{d_o}{30} + 0.75 \qquad \textbf{SI} \qquad (7\text{–}10\text{a})$$

$$h = \frac{d_o}{30} + 0.03 \qquad \textbf{US Customary System units}$$
$$(7\text{–}10\text{b})$$

The minimum required thickness of tubes made of nonferrous materials such as copper, red brass, admiralty and copper-nickel alloys used in steam boilers of less than 103 kPa (15 psi) and water boilers of less than 207 kPa (30 psi)

$$h = \frac{d_o}{45} + 0.75 \qquad \textbf{SI} \qquad (7\text{–}11\text{a})$$

$$h = \frac{d_o}{45} + 0.03 \qquad \textbf{US Customary System units}$$
$$(7\text{–}11\text{b})$$

The minimum required thickness of tubes when made of nonferrous materials but assembled with fittings, which are based on materials used, and based on whether the pressure is over 207 kPa (30 psi), but not in excess of 1013 kPa (160 psi) or whether the pressure does not exceed 207 kPa (30 psi)

$$h = \frac{d_o}{factor} + 0.75 \text{ except for copper} = 0.027 \quad \textbf{SI}$$
$$(7\text{–}12\text{a})$$

$$h = \frac{d_o}{factor} + 0.03 \quad \textbf{US Customary System units}$$
$$(7\text{–}12\text{b})$$

Particular	Formula
The formula for permissible pressure in wrought-iron and steel tubes for watertube boilers according to *ASME Power Boiler Code*	

$$p = 125 \left(\frac{h - 1 \times 10^{-3}}{d_o} \right) - 0.32 \qquad \textbf{SI} \qquad (7\text{--}13a)$$

where h, d_o in m, and p in MPa

$$p = 18000 \left(\frac{h - 0.039}{d_o} \right) - 250 \quad \textbf{US Customary System units} \qquad (7\text{--}13b)$$

where h, d_o in in, and p in psi

$$p = 96.5 \left(\frac{h - 1 \times 10^{-3}}{d_o} \right) \qquad \textbf{SI} \qquad (7\text{--}14a)$$

where h, d_o in m, and p in MPa

$$p = 14000 \left(\frac{h - 0.039}{d_o} \right) \quad \textbf{US Customary System units} \qquad (7\text{--}14b)$$

where h, d_o in in, and p in psi

$$p = 73 \left(\frac{h - 1 \times 10^{-3}}{d_o} \right) \qquad \textbf{SI} \qquad (7\text{--}15a)$$

where h, d_o in m, and p in MPa

$$p = 10600 \left(\frac{h - 0.039}{d_o} \right) \quad \textbf{US Customary System units} \qquad (7\text{--}15b)$$

where h, d_o in in, and p in psi

Formula (7–13) applies to seamless tubes at all pressures, to welded steel tubes at pressure below 6 MPa (875 psi), and to lap-welded wrought-iron tubes at pressures below 2.5 MPa (358 psi).
Formula (7–14) applies to welded steel tubes at pressures of 6 MPa (875 psi) and above.
Formula (7–15) applies to lap-welded wrought-iron tubes at pressures of 2.5 MPa (358 psi) and above.

ENGINES AND PRESSURE CYLINDERS

The wall thickness of engines and pressure cylinders

$$h = \frac{p d_i}{2 \sigma_{sta}} + 7.5 \times 10^{-3} \qquad \textbf{SI} \qquad (7\text{--}16a)$$

where p, σ_{st} in MPa, d_i and h in m

Particular	Formula
	$h = \dfrac{pd_i}{2\sigma_{sta}} + 0.3$ **US Customary System units**
	(7–16b)

where p, σ_t in psi, and d_i and h in in

$\sigma_{sta} = 9$ MPa (1250 psi) for ordinary grades of cast iron.

OPENINGS IN CYLINDRICAL DRUMS

The largest permissible diameter of opening according to D. S. Jacobus

$$d' = 0.81\sqrt[3]{d_o h(1.0 - K)} \qquad \text{SI} \qquad (7\text{–}17a)$$

where d_o and h in m

$$d' = 2.75\sqrt[3]{d_o h(1.0 - K)} \quad \text{US Customary System units}$$

where d_o and h in in

$$K = \left(\frac{pd_o}{2h}\right)\left(\frac{5}{\sigma_{su}}\right) \qquad (7\text{–}17b)$$

The maximum diameter of the unreinforced hole should be limited to 0.203 m (8 in) and should not exceed $0.6d_o$.

THIN TUBES WITH EXTERNAL PRESSURE

Professor Carman's formulas for the collapsing pressure for seamless steel tubes

$$p_{cr} = 346120\left(\frac{h}{d_o}\right)^3 \qquad \text{SI} \qquad (7\text{–}18a)$$

where h, d_o in m, and p_{cr} in MPa

$$p_{cr} = 50200000\left(\frac{h}{d_o}\right)^3 \quad \text{US Customary System units}$$

$$(7\text{–}18b)$$

where h, d_o in m, and p_{cr} in psi

when $\dfrac{h}{d_o} < 0.025$

$$p_{cr} = 658.5\left(\frac{h}{d_o}\right) - 1.50 \qquad \text{SI} \qquad (7\text{–}19a)$$

where h, d_o in m, and p_{cr} in MPa

Particular	Formula

$$p_{cr} = 95520\left(\frac{h}{d_o}\right) - 2090 \text{ US Customary System units}$$

(7–19b)

where h, d_o in in, and p_{cr} in psi

when $\dfrac{h}{d_o} > 0.03$

Professor Carman's formula for the collapsing pressure for lap-welded steel tubes

$$p_{cr} = 574\left(\frac{h}{d_o}\right) - 0.72 \qquad \textbf{SI} \qquad (7\text{–}20a)$$

where d_o, h in m, and p_{cr} in MPa

$$p_{cr} = 83290\left(\frac{h}{d_o}\right) - 1025 \text{ US Customary System units}$$

(7–20b)

where h, d_o in in, and p_{cr} in psi

when $\dfrac{h}{d_o} > 0.03$

Professor Carman's formula for the collapsing pressure for lap-welded brass tubes

$$p_{cr} = 173385\left(\frac{h}{d_o}\right)^3 \qquad \textbf{SI} \qquad (7\text{–}21a)$$

where h, d_o in m, and p_{cr} in MPa

$$p_{cr} = 25150000\left(\frac{h}{d_o}\right)^3 \quad \textbf{US Customary System units}$$

(7–21b)

where h, d_o in in, and p_{cr} in psi

when $\dfrac{h}{d_o} < 0.025$

$$p_{cr} = 644\left(\frac{h}{d_o}\right) - 1.75 \qquad \textbf{SI} \qquad (7\text{–}22a)$$

where h, d_o in m, and p_{cr} in MPa

$$p_{cr} = 93365\left(\frac{h}{d_o}\right) - 2474 \qquad \textbf{US Customary System units} \quad (7\text{–}22b)$$

where h, d_o in in, and p_{cr} in psi

when $\dfrac{h}{d_o} > 0.03$

Particular	Formula

SHORT TUBES WITH EXTERNAL PRESSURES

Sir William Fairbairn's formula for collapsing pressure for length less than six diameters

$$p_{cr} = 66580 \left(\frac{h^{2.19}}{Ld_o} \right) \qquad \textbf{SI} \qquad (7\text{--}23a)$$

where h, L, d_o in m, and p_{cr} in MPa

$$p_{cr} = 9657600 \left(\frac{h^{2.19}}{Ld_o} \right) \quad \textbf{US Customary System units}$$

$$(7\text{--}23b)$$

where $h, L,$ and d_o in m, and p_{cr} in psi

Thickness of tubes, and pipes when used as tubes under external pressure as per Indian Standards

Refer to Fig. 7–1 to determine the thickness of tubes and pipes; see also Table 7–7.

FIGURE 7–1   Thickness of tubes and pipes under external pressure.

TABLE 7–6
Values of C for use in Eqs. (7–5) to (7–8)

Type of pipe	Value of C^b, mm (in)
Threaded steel, wrought iron, or nonferrous pipe[a]	
19 mm (0.75 in), nominal and smaller	1.625 (0.065)
25 mm (1 in), nominal and larger	Depth of thread h^c
Plain-end[d] steel, wrought iron, or nonferrous pipe	
87.5 mm (3.5 in), nominal and smaller	1.625 (0.065)
100 mm (4 in), nominal and larger	0

[a] Steel, wrought iron, or nonferrous pipe lighter than schedule 40 of the American National Standard for wrought iron and steel pipe, ANSI B36. 10-1970, shall not be threaded.
[b] The values of C stipulated above are such that the actual stress due to internal pressure in the wall of the pipe is no greater than the value of S (i.e. σ_{sa}) given in Table PG 23.1 of ASME Power Boiler Code as applicable in the formulas.
[c] The depth of thread h in inches may be determined from the formula $h = 0.8/i$, where i is the number of threads per inch or from the Table 7–5.
[d] Plain-end pipe includes pipe joined by flared compression coupling, lap (Van Stone) joints, and by welding, i.e., by any method which does not reduce the wall thickness of pipe at the joint.
Source: *ASME Boiler and Pressure Vessel Code*, Section I, 1983.

Particular	Formula

LAME'S EQUATIONS FOR THICK CYLINDERS

General equations

The tangential stress in the cylinder wall at radius r when subjected to internal and external pressures

$$\sigma_\theta = \frac{p_i d_i^2 - p_o d_o^2}{d_o^2 - d_i^2} + \frac{d_i^2 d_o^2 (p_i - p_o)}{4r^2(d_o^2 - d_i^2)} \qquad (7\text{–}24a)$$

$$= a + \frac{b}{r^2} \qquad (7\text{–}24b)$$

The radial stress in the cylinder at radius r when subjected to internal and external pressures

$$\sigma_r = \frac{p_i d_i^2 - p_o d_o^2}{d_o^2 - d_i^2} - \frac{d_i^2 d_o^2 (p_i - p_o)}{4r^2(d_o^2 - d_i^2)} \qquad (7\text{–}25a)$$

$$= a - \frac{b}{r^2} \qquad (7\text{–}25b)$$

where

$$a = \frac{p_i d_i^2 - p_o d_o^2}{d_o^2 - d_i^2} \qquad (7\text{–}25c)$$

$$b = \frac{d_i^2 d_o^2 (p_i - p_o)}{4(d_o^2 - d_i^2)} \qquad (7\text{–}25d)$$

Particular	Formula

Cylinder under internal pressure only

The tangential stress in the cylinder wall at radius r

$$\sigma_\theta = \frac{p_i d_i^2}{d_o^2 - d_i^2}\left(1 + \frac{d_o^2}{4r^2}\right)$$

(7–26)

The radial stress in the cylinder wall at radius r

$$\sigma_r = \frac{p_i d_i^2}{d_o^2 - d_i^2}\left(1 - \frac{d_o^2}{4r^2}\right)$$

(7–27)

The maximum tangential stress at the inner surface of the cylinder at $r = d_i/2$

$$\sigma_{\theta(max)} = \frac{p_i(d_i^2 + d_o^2)}{d_o^2 - d_i^2}$$

(7–28)

The maximum radial stress

$$\sigma_{r(max)} = -p_i$$

(7–29)

The maximum shear stress at the inner surface of the cylinder under internal pressure

$$\tau_{max} = \frac{p_i d_o^2}{d_o^2 - d_i^2}$$

(7–30)

Cylinder under external pressure only

The tangential stress in the cylinder wall at radius r

$$\sigma_\theta = -\frac{p_o d_o^2}{d_o^2 - d_i^2}\left(1 + \frac{d_i^2}{4r^2}\right)$$

(7–31)

The radial stress in the cylinder wall at radius r

$$\sigma_\tau = -\frac{p_o d_o^2}{d_o^2 - d_i^2}\left(1 - \frac{d_i^2}{4r^2}\right)$$

(7–32)

DEFORMATION OF A THICK CYLINDER

The radial displacement of a point at radius r in the wall of the cylinder subjected to internal and external pressures

$$u = \left(\frac{1 - \nu}{E}\right)\frac{p_i d_i^2 - p_o d_o^2}{d_o^2 - d_i^2}r + \left(\frac{1 + \nu}{E}\right)\frac{d_i^2 d_o^2(p_i - p_o)}{4r(d_o^2 - d_i^2)}$$

(7–33)

Cylinder under internal pressure only

The radial displacement at $r = d_i/2$ of the inner surface of the cylinder

$$u_i = \frac{p_i d_i}{2E}\left(\frac{d_i^2 + d_o^2}{d_o^2 - d_i^2} + \nu\right)$$

(7–34)

The radial displacement at $r = d_o/2$ of the inner surface of the cylinder

$$u_o = \frac{p_i d_i^2 d_o}{E(d_o^2 - d_i^2)}$$

(7–35)

Particular	Formula

Cylinder under external pressure only

The radial displacement at $r = d_i/2$ of the inner surface of the cylinder

$$u_i = -\frac{p_o d_i d_o^2}{E(d_o^2 - d_i^2)} \qquad (7\text{--}36)$$

The radial displacement at $r = d_o/2$ of the outer surface of the cylinder

$$u_o = -\frac{p_o d_o}{2} E\left(\frac{d_i^2 + d_o^2}{d_o^2 - d_i^2} - \nu\right) \qquad (7\text{--}37)$$

COMPOUND CYLINDERS

Birnie's equation for tangential stress at any radius r for a cylinder open at ends subjected to internal pressure

$$\sigma_\theta = (1 - \nu)\frac{p_i d_i^2}{d_o^2 - d_i^2} + (1 + \nu)\frac{d_i^2 d_o^2 p_i}{4r^2(d_o^2 - d_i^2)} \qquad (7\text{--}38)$$

The tangential stress at the inner surface of the inner cylinder in the case of a compound cylinder (Figs. 11–1 and 11–2)

$$\sigma_{\theta-i} = \frac{-2p_c d_c^2}{d_c^2 - d_i^2} \qquad (7\text{--}39)$$

The tangential stress at the outer surface of the inner cylinder

$$\sigma_{\theta-ic} = -p_c\left(\frac{d_c^2 + d_i^2}{d_c^2 - d_i^2} - \nu\right) \qquad (7\text{--}40)$$

The tangential stress at the inner surface of the outer cylinder

$$\sigma_{\theta-oc} = p_c\left(\frac{d_o^2 + d_c^2}{d_o^2 - d_c^2} + \nu\right) \qquad (7\text{--}41)$$

The tangential stress at the outer surface of the outer cylinder

$$\sigma_{\theta-o} = \frac{2p_c d_c^2}{d_o^2 - d_c^2} \qquad (7\text{--}42)$$

THERMAL STRESSES IN LONG HOLLOW CYLINDERS

The general expressions for the radial σ_r; tangential σ_θ, and longitudinal σ_z stresses in the cylinder wall at radius r when the temperature distribution is symmetrical with respect to the axis and constant along its length, respectively

$$\sigma_r = \frac{\alpha E}{(1 - \nu)r^2}\left[\frac{4r^2 - d_i^2}{d_o^2 - d_i^2}\int_{r_i}^{r_o} Tr\,dr - \int_{r_i}^{r} Tr\,dr\right] \qquad (7\text{--}43)$$

$$\sigma_\theta = \frac{\alpha E}{(1 - \nu)r^2}\left[\frac{4r^2 + d_i^2}{d_o^2 - d_i^2}\int_{r_i}^{r_o} Tr\,dr + \int_{r_i}^{r} Tr\,dr - Tr^2\right] \qquad (7\text{--}44)$$

$$\sigma_z = \frac{\alpha E}{1 - \nu}\left[\frac{8}{d_o^2 - d_i^2}\int_{r_i}^{r_o} Tr\,dr - T\right] \qquad (7\text{--}45)$$

where $d_o = 2r_o$ and $d_i = 2r_i$

Particular	Formula
The expressions for radial (σ_r), tangential (σ_θ), and longitudinal (σ_z) stresses in the cylinder at radius r when the cylinder is subjected to steady-state temperature distribution, i.e., logarithmic temperature distribution throughout the wall thickness of the cylinder by using equation $T = T_i[\ln R_o/\ln R]$	$\sigma_r = \dfrac{\alpha E T_i}{2(1-v)\ln(R)}$ $\times \left[-\ln(R_o) - \dfrac{I}{R^2-1}(1-R_o^2)\ln(R)\right]$ (7–46) $\sigma_\theta = \dfrac{\alpha E T_i}{2(1-v)\ln(R)}$ $\times \left[1-\ln(R_o) - \dfrac{1}{R^2-1}(1+R^2)\ln R\right]$ (7–47) $\sigma_z = \dfrac{\alpha E T_i}{2(1-v)\ln R}\left[1-2\ln(R_o) - \dfrac{2}{R^2-1}\ln R\right]$ (7–48) where $R = \dfrac{d_o}{d_i} = \dfrac{r_o}{r_i}$; $R_o = \dfrac{r_o}{r} = \dfrac{d_o}{2r}$; $R_i = \dfrac{r_i}{r} = \dfrac{d_i}{2r}$ $T_i =$ temperature at inner surface of cylinder, °C (°F)
The expressions for maximum values of tangential (hoop) and longitudinal stresses at inner and outer surfaces of the cylinder under logarithmic temperature distribution, respectively	$\sigma_{\theta i} = \sigma_{zi} = \dfrac{\alpha E T_i}{2(1-v)\ln R}\left[1-\dfrac{2R^2}{R^2-1}\ln R\right]$ (7–49) $\sigma_{\theta o} = \sigma_{zo} = \dfrac{\alpha E T_i}{2(i-v)\ln R}\left[1-\dfrac{2}{R^2-1}\ln R\right]$ (7–50)
The simplified expressions for maximum values of tangential and longitudinal stresses at inner and outer surfaces of the cylinder under logarithmic temperature distribution when the thickness of cylinder is small in comparison with the inner radius of the cylinder, respectively	$\sigma_{\theta i} = \sigma_{zi} = -\dfrac{\alpha E T_i}{2(1-v)}\left(1+\dfrac{n}{3}\right)$ (7–51) $\sigma_{\theta o} = \sigma_{zo} = \dfrac{\alpha E T_i}{2(1-v)}\left[1-\dfrac{n}{3}\right]$ (7–52) where $d_o/d_i = 1+n$ and $\ln(d_o/d_i) = \ln(1+n)$
The simplified expressions for maximum tangential and longitudinal stresses for thin cylinders under the logarithmic temperature distribution, respectively	$\sigma_{\theta i} = \sigma_{zi} = -\dfrac{\alpha E T_i}{2(1-v)}$ (7–53) $\sigma_{\theta o} = \sigma_{zo} = \dfrac{\alpha E T_i}{2(1-v)}$ (7–54)
The expressions for radial (σ_r), tangential (σ_θ) (hoop), and longitudinal stresses (σ_z) in a cylinder at radius r subject to linear thermal temperature distribution throughout the wall thickness of the cylinder by using	$\sigma_r = \dfrac{\alpha E T_i}{(1-v)r^2}\left[\dfrac{(r^2-r_i^2)(r_o+2r_i)}{6(r_o+r_i)}\right.$ $\left.+\dfrac{2(r^3-r_i^3)-3r_o(r^2-r_i^2)}{6(r_o-r_i)}\right]$ (7–55)

Particular	Formula
equation $T = T_i\left(\dfrac{r_o - r}{r_o - r_i}\right)$ when the thickness of the cylinder wall is small in composition with the outside radius	$\sigma_\theta = \dfrac{\alpha E T_i}{(1-\nu)r^2}\left[\dfrac{(r^2 + r_i^2)(r_o + 2r_i)}{6(r_o + r_i)}\right.$ $\left. -\dfrac{2(r^3 - r_i^3) - 3r_o(r^2 - r_i^2)}{6(r_o - r_i)} - \dfrac{(r_o - r)r^2}{r_o - r_i}\right]$ (7–56)
	$\sigma_z = \dfrac{\alpha E T_i}{1-\nu}\left[\dfrac{r_o + 2r_i}{2(r_o + r_i)} - \dfrac{r_o - r}{r_o - r_i}\right]$ (7–57)
The expressions for maximum tangential (hoop, σ_θ) and longitudinal (σ_z) stresses at inner and outer surfaces of the cylinder under the linear thermal gradient as per equation $T = T_i(r_o - r)/(r_o - r_i)$	$\sigma_{\theta i} = \sigma_{zi} = -\dfrac{\alpha E T_i}{1-\nu}\left[\dfrac{2r_o + r_i}{3(r_o + r_i)}\right]$ (7–58)
	$\sigma_{\theta o} = \sigma_{zo} = \dfrac{\alpha E T_i}{1-\nu}\left[\dfrac{r_o + 2r_i}{3(r_o + r_i)}\right]$ (7–59)
The expressions for maximum tangential and longitudinal stresses at inner and outer wall surfaces of thin cylinder (i.e., $r_o \approx r_i$) under the linear thermal gradient as per equation $T = T_i(r_o - r)/(r_o - r_i)$	$\sigma_{\theta i} = \sigma_{zi} = -\dfrac{\alpha E T_i}{2(1-\nu)}$ (7–60)
	$\sigma_{\theta o} = \sigma_{zo} = \dfrac{\alpha E T_i}{2(1-\nu)}$ (7–61)
	Eqs. (7–60) and (7–61) for the linear thermal gradient are the same as Eqs. (7–53) and (7–54) for a logarithmic thermal gradient.
The wall thickness of a cylinder made of brittle materials	$h = \dfrac{d_i}{2}\left\{\left(\dfrac{\sigma_\theta + p_i}{\sigma_\theta - p_i}\right)^{\frac{1}{2}} - 1\right\}$ (7–62)
The wall thickness of a cylinder made of ductile materials	$h = \dfrac{d_i}{2}\left\{\left(\dfrac{\sigma_\theta}{\sigma_\theta - 2p_i}\right)^{\frac{1}{2}} - 1\right\}$ (7–63)
	where σ_θ = permissible working stress in tension, MPa (psi)

CLAVARINO'S EQUATION FOR CLOSED CYLINDERS

(Based on the maximum strain energy)

The general equation for equivalent tangential stress at any radius r	$\sigma'_\theta = (1 - 2\nu)a + \dfrac{(1+\nu)b}{r^2}$ (7–64)

Particular	Formula
The general equation for equivalent radial stress at any radius r	$$\sigma_r' = (1 - 2\nu)a - \frac{(1+\nu)b}{r^2} \qquad (7\text{--}65)$$ where a and b have the same meaning as in Eqs. (7–25c) and (7–25d)
The wall thickness for cylinders with closed ends	$$h = \frac{d_i}{2}\left[\left\{\frac{\sigma_\theta' + (1 - 2\nu)p_i}{\sigma_\theta' - (1+\nu)p_i}\right\}^{\frac{1}{2}} - 1\right] \qquad (7\text{--}66)$$ where σ_θ' = permissible working stress in tension, MPa (psi)

BIRNIE'S EQUATIONS FOR OPEN CYLINDERS

The equation for equivalent tangential stress at any radius r	$$\sigma_\theta' = (1 - \nu)a + (1 + \nu)\frac{b}{r^2} \qquad (7\text{--}67)$$
The equation for equivalent radial stress at any radius r	$$\sigma_r' = (1 - \nu)a - (1 + \nu)\frac{b}{r^2} \qquad (7\text{--}68)$$ where a and b have the same meaning as in Eqs. (7–25c) and (7–25d)
The wall thickness of cylinders with open ends	$$h = \frac{d_i}{2}\left[\left\{\frac{\sigma_\theta' + (1 - \nu)p_i}{\sigma_\theta' - (1+\nu)p_i}\right\}^{\frac{1}{2}} - 1\right] \qquad (7\text{--}69)$$

BARLOW'S EQUATION

The tangential stress in the wall thickness of cylinder	$$\sigma_\theta = \frac{p_i d_o}{2h} \qquad (7\text{--}70)$$ For σ_θ refer to Table 7–1.

TABLE 7–7
Standard thickness of tubes

Diameter, mm (in)	Minimum thickness, mm (in)
25 (1) and over but less than 62.5 (2.5)	2.37 (0.095)
62.5 (2.5) and over but less than 87.5 (3.25)	2.625 (0.105)
87.5 (3.25) and over but less than 100 (4)	3.000 (0.120)
100 (4) and over but less than 125 (5)	3.375 (0.135)
125 (5) and over but less than 150 (6)	3.750 (0.150)
150 (6) and over	$h = 0.0251d_o$

Source: *ASME Boiler and Pressure Vessel Code*, Section I, 1983.

REFERENCES

1. "Rules for Construction of Power Boilers," Section I, *ASME Boiler and Pressure Vessel Code*, American Society of Mechanical Engineers, New York, 1983.
2. "Rules for Construction of Pressure Vessels," Section VIII, Division 1, *ASME Boiler and Pressure Vessel Code*, American Society of Mechanical Engineers, New York, July 1, 1986.
3. "Rules for Construction of Pressure Vessels," Section VIII, Division 2—Alternative Rules, *ASME Boiler and Pressure Vessel Code*, American Society of Mechanical Engineers, New York, July 1, 1986.
4. Nicholas, R. W., *Pressure Vessel Codes and Standards*, Elsevier Applied Science Publications, Crown House, Linton Road, Barking, Essex, England.
5. Lingaiah, K., and B. R. Narayana Iyengar, *Machine Design Data Handbook*, Engineering College Co-operative Society, Bangalore, India, 1962.
6. Lingaiah, K. *Machine Design Data Handbook*, Vol. II (*SI and Customary Metric Units*), Suma Publishers, Bangalore, India, 1986.

CHAPTER
8

DESIGN OF PRESSURE VESSELS, PLATES, AND SHELLS

SYMBOLS

a	long span of noncircular head, m (in)
	length of the long side of a rectangular plate, m (in)
	pitch or distance between stays, m (in)
	major axis of elliptical plate, m (in)
	long span of noncircular heads or covers measured at perpendicular distance to short span, m (in)
A	factor determined from Fig. 8–3
A	total cross-sectional area of reinforcement required in the plane under consideration, m² (in²) (see Fig. 8–17) (includes consideration of nozzle area through shell for $\sigma_{Sna}/\sigma_{Sva} < 1.0$)
A	outside diameter of flange or, where slotted holes extend to the outside of the flange, the diameter to the bottom of the slots, m (in)
A_1	area in excess thickness in the vessel wall available for reinforcement, m² (in²) (see Fig. 8–17) (includes consideration of nozzle area through shell if $\sigma_{Sna}/\sigma_{Sva} < 1.0$)
A_2	area in excess thickness in the nozzle wall available for reinforcement, m² (in²) (see Fig. 8–17)
A_3	area available for reinforcement when the nozzle extends inside the vessel wall, m² (in²) (see Fig. 8–17)
A_{41}, A_{42}, A_{43}	cross-sectional area of various welds available for reinforcement (see Fig. 8–17), m² (in²)
A_5	cross-sectional area of material added as reinforcement (see Fig. 8–17), m² (in²)
A_b	cross-sectional area of the bolts using the root diameter of the thread or least diameter of unthreaded position, if less, m (in)
A_m	total required cross-sectional area of bolts taken as the greater of A_{m1} and A_{m2}, m² (in²)
$A_{m1} = W_{m1}/\sigma_{sb}$	total cross-sectional area of bolts at root of thread or section of least diameter under stress, required for the operating condition, m² (in²)

$A_{m2} = W_{m2}/\sigma_{sa}$	total cross-sectional area of bolts at root of thread or section of least diameter under stress, required for gasket seating, m² (in²)
b	length of short side or breadth of a rectangular plate, m (in) short span of noncircular head, m (in)
b	effective gasket or joint-contact-surface seating width, m (in)
b_o	basic gasket seating width, m (in) (see Table 8–21 and Fig. 8–13)
B	factor determined from the application material–temperature chart for maximum temperature, psi
B	inside diameter of flange, m (in)
c	corrosion allowance, m (in)
c	basic dimension used for the minimum sizes of welds, mm (in), equal to t_n or t_x, whichever is less
c_1	empirical coefficient taking into account the stress in the knuckle [Eq. (8–68)]
c_2	empirical coefficient depending on the method of attachment to shell [Eqs. (8–82) and (8–85)]
c_4, c_5	empirical coefficients depending on the mode of support [(Eqs. (8–92) to (8–94)]
C	bolt-circle diameter, mm (in)
d	finished diameter of circular opening or finished dimension (chord length at midsurface of thickness excluding excess thickness available for reinforcement) of nonradial opening in the plane under consideration in its corroded condition, m (in) (see Fig. 8–17)
d	diameter or short span, m (in) diameter of the largest circle which may be inscribed between the supporting points of the plate (Fig. 8–11), m (in) diameter as shown in Fig. 8–9, m (in)
d	factor, m³ (in³)
$d = \dfrac{U}{V} h_o g_o{}^2$	for integral-type flanges
$d = \dfrac{U}{V_L} h_o g_o{}^2$	for loose-type flanges
d'	diameter through the center of gravity of the section of an externally located stiffening ring, m (in); inner diameter of the shell in the case of an internally located stiffening ring, m (in) [Eq. (8–55)]
d_e	outside diameter of conical section or end (Fig. 8–8d), m (in)
d_i, D_i	inside diameter of shell, m (in)
d_o, D_o	outside diameter of shell, m (in)
d_k	inside diameter of conical section or end at the position under consideration (Fig. 8–8d), m (in)
D	inside shell diameter before corrosion allowance is added, m (in)
D_p	outside diameter of reinforcing element, m (in) (actual size of reinforcing element may exceed the limits of available reinforcement)
e	factor, m⁻¹ (in⁻¹)

$e = \dfrac{F}{h_o}$ for integral-type flanges

$e = \dfrac{F_L}{h_o}$ for loose-type flanges

E modulus of elasticity at the operating temperature, GPa (Mpsi)

E_{am} modulus of elasticity at the ambient temperature, GPa (Mpsi)

f hub stress correction factor for integral flanges from Fig. 8–25 (When greater than one, this is the ratio of the stress in the small end of the hub to the stress in the large end. For values below limit of figure, use $f = 1$.)

f_r strength reduction factor, not greater than 1.0

f_{r1} $\sigma_{Sna}/\sigma_{Sva}$

f_{r2} (lesser of σ_{Sna} or σ_{Spa})/σ_{Sva}

f_{r3} $\sigma_{Spa}/\sigma_{Sva}$

F total load supported, kN (lbf)

total bolt load, kN (lbf)

F correction factor which compensates for the variation in pressure stresses on different planes with respect to the axis of a vessel (A value of 1.00 shall be used for all configurations, except for integrally reinforced openings in cylindrical shells and cones.)

F factor for integral-type flanges (from Fig. 8–21)

F_L factor for loose-type flanges (from Fig. 8–23)

g_a thickness of hub at small end, m (in)

g_1 thickness of hub at back of flange, m (in)

G diameter, m (in), at location of gasket load reaction; except as noted in Fig. 8–13, G is defined as follows (see Table 8–22):
When $b_o \leq 6.3$ mm (1/4 in), G = mean diameter of gasket contact face, m (in).
When $b_o > 6.3$ mm (1/4 in), G = outside diameter of gasket contact face less $2b$, m (in).

h distance nozzle projects beyond the inner or outer surface of the vessel wall, before corrosion allowance is added, m (in) (Extension of the nozzle beyond the inside or outside surface of the vessel wall is not limited; however, for reinforcement calculations the dimension shall not exceed the smaller of $2.5\,t$ or $2.5\,t_n$ without a reinforcing element and the smaller of $2.5\,t$ or $2.5\,t_n + t_e$ with a reinforcing element or integral compensation.)

h hub length, m (in)

h, t minimum required thickness of cylindrical or spherical shell or tube or pipe, m (in)

thickness of plate, m (in)

thickness of dished head or flat head, m (in)

h_a actual thickness of shell at the time of test including corrosion allowance, m (in)

h_c thickness for corrosion allowance, m (in)

h_D radial distance from the bolt circle, to the circle on which H_D acts, m (in)

$h_G = (C - G)/2$ radial distance from gasket load reaction to the bolt circle, m (in)

$h_o = \sqrt{Bg_o}$ factor, m (in)

h_T	radial distance from the bolt circle to the circle on which H_T acts as prescribed, m (in)
$H = \pi G^2 P/4$	total hydrostatic end force, kN (lbf)
$H_D = \pi B^2 P/4$	hydrostatic end force on area inside of flange, kN (lbf)
$H_G = W - H$	gasket load (difference between flange design bolt load and total hydrostatic end force), kN (lbf)
$H_P =$	
$2b \times \pi GmP$	total joint-contact-surface compression load, kN (lbf)
$H_T = H - H_D$	difference between total hydrostatic end force and the hydrostatic end force on area inside of flange, kN (lbf)
I_s	required moment of inertia of the stiffening ring cross section around an axis extending through the center of gravity and parallel to the axis of the shell, m^4 or cm^4 (in^4)
I'_s	required moment of inertia of the combined ring-shell cross section about its neutral axis parallel to the axis of the shell, m^4 (in^4)
I	available moment of inertia of the stiffening ring cross section about its neutral axis parallel to the axis of the shell, m^4 (in^4)
I'	available moment of inertia of combined ring shell cross section about its neutral axis parallel to the axis of the shell, m^4 or cm^4 (in^4)
k_1, k_2, k_3, k_4, k_5	coefficients
k_6	factor for noncircular heads depending on the ratio of short span to long span b/a (Fig. 8–10)
$K = A/B$	ratio of outside diameter of flange to inside diameter of flange
K	ratio of the elastic modulus E of the material at the design material temperature to the room temperature elastic modulus, E_{am}
K_1	spherical radius factor (Table 8–18)
l	length of flange of flanged head, m (in)
L	effective length, m (in)
	distance from knuckle or junction within which meridional stresses determine the required thickness, m (in)
	perimeter of noncircular bolted heads measured along the centers of the bolt holes, m (in)
	distance between centers of any two adjacent openings, m (in)
	length between the centers of two adjacent stiffening rings, m (in) (Fig. 8–1)

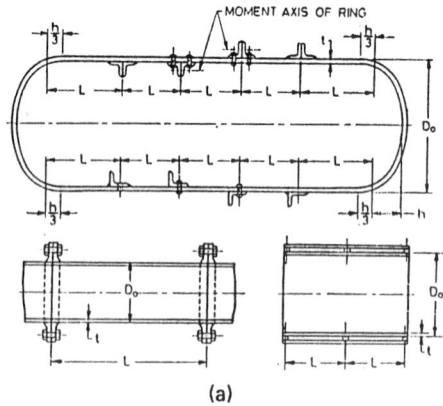

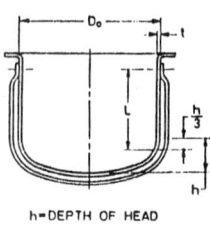

FIGURE 8–1 Cylindrical pressure vessels under external pressure.

$L = \dfrac{te + 1}{T} + \dfrac{t^3}{d}$ factor

m	gasket factor, obtained from Table 8–20
$m = \frac{1}{v}$	reciprocal of Poisson's ratio
M_b	longitudinal bending moment, N m (lbf in)
M_t	torque about the vessel axis, N m (lbf in)
$M_D = H_D h_D$	component of moment due to H_D, m N (in - lbf)
$M_G = H_G h_G$	component of moment due to H_G, m N (in - lbf)
M_o	total moment acting on the flange, for the operating conditions or gasket seating as may apply, m N (in - lbf)
$M_T = H_T h$	component of moment due to H_T, m N (in - lbf)
N	width, m (in), used to determine the basic gasket seating with b_o, based on the possible contact width of the gasket (see Table 8–21)
p_i	internal design pressure, MPa (psi)
p	maximum allowable working pressure or design pressure, MPa (psi)
p_o	load per unit area, MPa (psi)
	external design pressure, MPa (psi)
P	total pressure on an area bounded by the outside diameter of gasket, kN (lbf)
	design pressure (or maximum allowable working pressure for existing vessels), MPa (psi)
P_a	calculated value of allowable external working pressure for assumed value of t or h, MPa (psi)
r	radius of circle over which the load is distributed, m (in)
r_i	inner radius of a circular plate, m (in)
	inside radius of transition knuckle which shall be taken as $0.01 d_k$ in the case of conical sections without knuckle transition, m (in)
R	inner radius of curvature of dished head, m (in)
R_i	inner radius of shell or pipe, m (in)
r_o, R_o	outer radius of a circular plate, m (in)
	outer radius of shell, m (in)
$R = [(C - B)/2]$ $\quad -g_1$	radial distance from bolt circle to point of intersection of hub and back of flange, m (in) (for integral and hub flanges)
R	inside radius of the shell course under consideration, before corrosion allowance is added, m (in)
R_n	inside radius of the nozzle under consideration, before corrosion allowance is added, m (in)
t or h	minimum required thickness of spherical or cylindrical shell, or pipe or tube, m (in)
t	flange thickness, m (in)
t	nominal thickness of the vessel wall, less corrosion allowance, m (in)
t_e	weld dimensions
t_c	thickness or height of reinforcing element, m (in)
t_n	nominal thickness of shell or nozzle wall to which flange or lap is attached, less corrosion allowance, m (in)
t_r	required thickness of a seamless shell based on the circumferential stress, or of a formed head, computed by the rules of this chapter for the designated pressure, m (in)
t_{rn}	required thickness of a seamless nozzle wall, m (in)

t_s	nominal thickness of cylindrical shell or tube exclusive of corrosion allowance, m (in)
t_w	weld dimensions
t_x	two times the thickness g_o, when the design is calculated as an integral flange, m (in), or two times the thickness, m (in), of shell nozzle wall required for internal pressure, when the design is calculated as a loose flange, but not less than 6.3 mm (1/4 in)
T	factor involving K (from Fig. 8–20)
U	factor involving K (from Fig. 8–20)
V	factor for integral-type flanges (from Fig. 8–22)
V_L	factor for loose-type flanges (from Fig. 8–24)
w	width, m (in), used to determine the basic gasket seating width b_o, based on the contact width between the flange facing and the gasket (see Table 8–21)
W	weight, kN (lbf)
W	total load to be carried by attachment welds, kN (lbf)
W	flange design bolt load, for operating conditions or gasket seating, as may apply, kN (lbf)
W_{m1}	minimum required bolt load for the operating conditions, kN (lbf) (For flange pairs used to contain a tubesheet for a floating head for a U-tube type of heat exchanger, or for any other similar design, W_{m1} shall be the larger of the values as individually calculated for each flange, and that value shall be used for both flanges.)
W_{m2}	minimum required bolt load for gasket seating, kN (lbf)
y	gasket or joint-contact-surface unit seating load, MPa (psi)
y	deflection of the plate, m (in)
y_{max}	maximum deflection of the plate, m (in)
Y	factor involving K (from Fig. 8–20)
Z	factor involving K (from Fig. 8–20)
α, α_1, α_2	angles of conical section to the vessel axis, deg (Fig. 8–8d)
ψ	difference between angle of slope of two adjoining conical sections, deg (Fig. 8–8d)
σ	normal or direct stress, MPa (psi)
σ_{sy}	0.2 percent proof stress, MPa (psi)
σ_{sa}	maximum allowable stress value, MPa (psi)
σ_e	equivalent stress (based on shear strain energy), MPa (psi)
σ_{sam}	allowable stress at ambient temperature, MPa (psi)
σ_{sd}	design stress value, MPa (psi)
σ_{Sa}	allowable stress value as given in Tables 8–9 to 8–12, MPa (psi)
σ_{Sna}	allowable stress in nozzle, MPa (psi)
σ_{Sva}	allowable stress in vessel, MPa (psi)
σ_{Spa}	allowable stress in reinforcing element (plate), MPa (psi)
σ_{Sbat}	allowable bolt stress at atmospheric temperature, MPa (psi)
σ_{Sbd}	allowable bolt stress at design temperature, MPa (psi)
σ_{Sfd}	allowable design stress for material of flange at design temperature (operating condition) or atmospheric temperature (gasket seating), as may apply, MPa (psi)
σ_{Snd}	allowable design stress for material of nozzle neck, vessel or pipe wall, at design temperature (operating condition) or atmospheric temperature (gasket seating), as may apply, MPa (psi)
σ_H	calculated longitudinal stress in hub, MPa (psi)

σ_R	calculated radial stress in flange, MPa (psi)
σ_θ	calculated tangential stress in flange, MPa (psi)
σ_0	hoop stress, MPa (psi)
σ_r	radial stress, MPa (psi)
σ_s	strength, MPa (psi)
σ_{su}	ultimate strength, MPa (psi)
σ_z or σ_l	longitudinal stress, MPa (psi)
σ_{zl}	tensile longitudinal stress, MPa (psi)
σ_{zc}	compressive longitudinal stress, MPa (psi)
τ	shear stress (also with subscripts), MPa (psi)
ν	Poisson's ratio
η	joint factor (Table 8–3) or efficiency
$\eta = 1$	(see definitions for t_r and t_{rn})
$\eta_1 = 1$	when an opening is in the solid plate or joint efficiency obtained from Table 8–3 when any part of the opening passes through any other welded joint

Note: σ and τ with initial subscript S designates strength properties of material used in the design which will be used and observed throughout this *Machine Design Data Handbook*.

Other factors in performance or in special aspect are included from time to time in this chapter and, being applicable only in their immediate context, are not given at this stage.

Particular	Formula

PLATES

For maximum stresses and deflections in flat plates   Refer to Table 8–1.

Plates loaded uniformly

The thickness of a plate with a diameter d supported at the circumference and subjected to a pressure p distributed uniformly over the total area

$$h = k_1 d \left[\frac{p}{\sigma_{sd}} \right]^{1/2} \tag{8–1}$$

Refer to Table 8–2 for values of k_1.

The maximum deflection

$$y = k_2 d^4 \frac{p}{Eh^3} \tag{8–2}$$

Refer to Table 8–2 for values of k_2.

Plates loaded centrally

The thickness of a flat cast-iron plate supported freely at the circumference with diameter d and subjected to a load F distributed uniformly over an area $(\pi d_o^2/4)$

$$h = 1.2 \left[\left(1 - \frac{0.67 d_o}{d} \right) \frac{F}{\sigma_{sd}} \right]^{1/2} \tag{8–3}$$

TABLE 8-1
Maximum stresses and deflections in flat plates

Form of plate	Type of loading	Type of support	Eq.	Total load, F	Maximum stress, σ_{max}	Location of σ_{max}	Maximum deflection, y_{max}
	Distributed over the entire surface	Edge supported	8-129	$\pi r_o^2 p$	$\sigma_r = \sigma_\theta = \dfrac{-3F(3m+1)}{8\pi m h^2}$	Center	$\dfrac{3F(m-1)(5m+1)r_o^2}{16\pi E m^2 h^3}$
		Edge fixed	8-130	$\pi r_o^2 p$	$\sigma_r = \dfrac{3F}{4\pi h^2}$	Edge	$\dfrac{3F(m^2-1)r_o^2}{16\pi E m^2 h^3}$
	Distributed over a concentric circular area of radius r	Edge supported	8-131	$\pi r^2 p$	$\sigma_r = \sigma_\theta = \dfrac{-3F}{2\pi m h^2}\left[(m+1)\log_e\dfrac{r_o}{r}\right.$ $\left.-(m-1)\dfrac{r^2}{4r_o^2}+m\right]$	Center	$\dfrac{3F(m^2-1)}{16\pi E m^2 h^3}\left[\dfrac{(12m+4)r_o^2}{m+1}\right.$ $\left.-4r^2\log_e\dfrac{r_o}{r}-\dfrac{(7m+3)}{m+1}r^2\right]$
					$\sigma_r = \dfrac{3F}{2\pi h^2}\left(1-\dfrac{r^2}{2r_o^2}\right)$	Edge	$\dfrac{3F(m^2-1)}{16\pi E m^2 h^3}\left[4r_o^2-4r^2\log_e\dfrac{r_o}{r}-3r^2\right]$
		Edge fixed	8-132	$\pi r^2 p$	$\sigma_r = \sigma_\theta = \dfrac{-3F}{2\pi m h^2}\left[(m+1)\log_e\dfrac{r}{r_o}\right.$ $\left.+(m+1)\dfrac{r^2}{4r_o^2}\right]$	Center	$\dfrac{3F(m^2-1)r_o^2}{4\pi E m^2 h^3}$ when r is very small (concentrated load)
	Distributed on circumference of a concentric circle of radius r	Edge supported	8-133	$2\pi r p$	$\sigma_r = \sigma_\theta = \dfrac{-3F}{2\pi m h^2}\left[\dfrac{m-1}{2}\right.$ $\left.+(m+1)\log_e\dfrac{r_o}{r}-(m-1)\dfrac{r^2}{r_o^2}\right]$	All points inside the circle of radius r	$\dfrac{3F(m^2-1)}{2\pi E m^2 h^3}\left[\dfrac{(3m+1)(r_o^2-r^2)}{2(m+1)}\right.$ $\left.-r^2\log_e\dfrac{r_o}{r}\right]$
		Edge fixed	8-134	$2\pi r p$	$\sigma_r = \sigma_\theta = \dfrac{-3F}{4m\pi h^2}\left[(m+1)\left(2\log_e\dfrac{r_o}{r}\right.\right.$ $\left.\left.+\dfrac{r^2}{r_o^2}-1\right)\right]$	Center when $r<0.31r_o$ Edge when $r>0.31r_o$	$\dfrac{3F(m^2-1)}{2\pi E m^2 h^3}\left[\tfrac{1}{2}(r_o^2-r^2)-r^2\log_e\dfrac{r_o}{r}\right]$
					$\sigma_r = \dfrac{3F}{2\pi h^2}\left(1-\dfrac{r^2}{r_o^2}\right)$		

TABLE 8-1
Maximum stresses and deflections in flat plates (Cont.)

Form of plate	Type of loading	Type of support	Total load, F	Eq.	Maximum stress, σ_{max}	Location of σ_{max}	Maximum deflection, y_{max}
	Distributed over a concentric circular area of radius r	Uniform pressure over entire lower surface	$\pi r^2 p$	8–135	$\sigma_r = \sigma_\theta = \frac{-3F}{2\pi m h^2}\left[(m+1)\log_e\frac{r_o}{r}\right.$ $\left.+\frac{m-1}{4}\left(1-\frac{r^2}{r_o^2}\right)\right]$	Center	$\frac{3F(m^2-1)}{16\pi Em^2h^3}\left[4r^2\log_e\frac{r_o}{r}+2r^2\left(\frac{3m+1}{m+1}\right)\right.$ $\left.-r_o^2\left(\frac{7m+3}{m+1}\right)+\frac{(r_o^2-r^2)^4}{r^2r_o^2}+\frac{r^4}{r_o^2}\right]$ where r is very small (concentrated load) $3F(m-1)(7m+3)r_o^2/16\pi Em^2h^3$
	Distributed over the entire surface	Outer edge supported	$F=\pi(r_o^2-r_i^2)p$	8–136	$\sigma_\theta = \frac{-3p}{4mh^2(r_o^2-r_i^2)}\left[r_o^4(3m+1)\right.$ $+r_i^4(m-1)-4mr_o^2r_i^2$ $\left.-4(m+1)r_o^2r_i^2\log_e\frac{r_o}{r}\right]$	Inner edge	$\frac{3F(m^2-1)}{2Em^2h^3}\left[\frac{r_o^4(5m+1)}{8(m+1)}\right.$ $+\frac{r_i^4(m+3)}{8(m+1)}-\frac{r_o^2r_i^2(3m+1)}{2(m+1)}$ $+\frac{r_o^2r_i^2(3m+1)}{2(m-1)}\log_e\frac{r_o}{r_i}$ $\left.-\frac{2r_o^2r_i^4(m+1)}{(r_o^2-r_i^2)(m-1)}\left(\log_e\frac{r_o}{r}\right)^2\right]$
	Distributed over the entire surface	Outer edge fixed and supported	$F=\pi(r_o^2-r_i^2)p$	8–137	$\sigma_\theta = \frac{-3p(m^2-1)}{4mh^2}$ $\left[\frac{r_o^2-r_i^4-\frac{1}{2}r_o^2r_i^2\log_e\frac{r_o}{r_i}}{r_o^2(m-1)+r_i^2(m+1)}\right]$	Inner edge	.....
	Distributed over the entire surface	Outer edge fixed and supported, inner edge fixed	$F=\pi(r_o^2-r_i^2)p$	8–138	$\sigma_r = \frac{3p}{4h^2}\left[(r_o^2+r^2)-\right.$ $\left.\frac{4r_o^2r^2}{r_o^2-r^2}\left(\log_e\frac{r_o}{r}\right)^2\right]$	Inner edge	$\frac{3p(m^2-1)}{16Em^2h^3}\left[r_o^4+3r_i^4-4r_o^2r_i^2\right.$ $\left.-4r_o^2r_i^2\log_e\frac{r_o}{r_i}+\frac{16r_o^2r_i^4}{r_o^2-r_i^2}\left(\log_e\frac{r_o}{r_i}\right)^2\right]$

TABLE 8–1
Maximum stresses and deflections in flat plates (*Cont.*)

Form of plate	Type of loading	Type of support	Eq.	Total load, F	Maximum stress, σ_{max}	Location of σ_{max}	Maximum deflection, y_{max}
	Distributed over the entire surface	Inner edge fixed and supported	8–139	$F = \pi(r_o{}^2 - r_i{}^2)p$	$\sigma_r = \dfrac{3p}{4h^2}\left[4r_o{}^4(m+1)\log_e\dfrac{r_o}{r} - \dfrac{r_o{}^4(m+3)+r_o{}^4(m-1)+4r_o{}^2 r_i{}^2}{r_o{}^2(m+1)+r_i{}^2(m-1)}\right]$	Inner edge	.....
	Uniform over entire surface	All edges supported	8–140	$F = abp$	$\sigma_b = \dfrac{-0.75b^2 p}{h^2\left(1+1.61\dfrac{b^3}{a^3}\right)}$	Center	$\dfrac{0.1422b^4 p}{Eh^3\left(1+2.21\dfrac{b^3}{a^3}\right)}$
	Uniform over entire surface	All edges fixed	8–141	$F = abp$	$\sigma_b = \dfrac{0.5b^2 p}{h^2\left(1+0.623\dfrac{b^6}{a^6}\right)}$	Center of long edge	$\dfrac{0.0284b^4 p}{Eh^3\left(1+1.056\dfrac{b^5}{a^5}\right)}$
	Uniform over entire surface	Short edges fixed, long edges supported	8–142	$F = abp$	$\sigma_b = \dfrac{0.75b^2 p}{h^2\left(1+0.8\dfrac{b^4}{a^4}\right)}$	Center of short edge	.....
	Uniform over entire surface	Short edges supported, long edges fixed	8–143	$F = abp$	$\sigma_b = \dfrac{-b^2 p}{2h^2\left(1+0.2\dfrac{a^4}{b^4}\right)}$	Center of long edge	.....

Note: Positive sign for σ indicates tension at upper surface and equal compression at lower surface; negative sign indicates reverse condition.

TABLE 8–2
Coefficients in formulas for cover plates

Material of cover plate	Methods of holding edges	Circular plate		Rectangular plate		Elliptical plate
		k_1	k_2	k_3	k_4	k_5
Cast iron	Supported, free	0.54	0.038	0.75	1.73	1.5
	Fixed	0.44	0.010	0.62	1.4; 1.6[a]	1.2
Mild steel	Supported, free	0.42	...	0.60	1.38	1.2
	Fixed	0.35	...	0.49	1.12; 1.28	0.9

[a] With gasket.

Particular	Formula
The deflection	$$y = \frac{0.12d^2F}{Eh^3} \qquad (8\text{–}4)$$
Grashof's formula for the thickness of a plate rigidly fixed around the circumference with the above given type of loading	$$h = 0.65\left(\frac{F}{\sigma_{sd}} \ln \frac{d}{d_o}\right)^{1/2} \qquad (8\text{–}5)$$
The deflection	$$y = \frac{0.055d^2F}{Eh^3} \qquad (8\text{–}6)$$

Rectangular plates

UNIFORM LOAD The thickness of a rectangular plate according to Grashof and Bach

$$h = ab\,k_3\left[\frac{p}{\sigma_{sd}(a^2 + b^2)}\right]^{1/2} \qquad (8\text{–}7)$$

where k_3 = coefficient, taken from Table 8–2

CONCENTRATED LOAD The thickness of a rectangular plate on which a concentrated load F acts at the intersection of diagonals

$$h = k_4\left[\frac{ab\,F}{\sigma_{sd}(a^2 + b^2)}\right]^{1/2} \qquad (8\text{–}8)$$

where k_4 = coefficient, taken from Table 8–2

Elliptical plate

The thickness of uniformly loaded elliptical plate

$$h = ab\,k_5\left[\frac{p}{\sigma_{sd}(a^2 + b^2)}\right]^{1/2} \qquad (8\text{–}9)$$

where k_5 = coefficient, taken from Table 8–2

Particular	Formula

SHELLS (UNFIRED PRESSURE VESSEL)

Shell under internal pressure — cylindrical shell

CIRCUMFERENCE JOINT The minimum thickness of shell exclusive of corrosion allowance as per Bureau of Indian Standards

$$h = \frac{pd_i}{2\sigma_{sa}\eta - p} = \frac{pd_o}{2\sigma_{sa}\eta + p} \tag{8–10}$$

Refer to Tables 8–3 and 8–8 for values of η and σ_{sa}, respectively.

Note: A minimum thickness of 1.5 mm is to be provided as corrosive allowance unless a protective lining is employed.

The design pressure or maximum allowable working pressure

$$p = \frac{2\sigma_{sa}\eta h}{d_i + h} = \frac{2\sigma_{sa}\eta h}{d_o - h} \tag{8–11}$$

TABLE 8–3
Joint efficiency factor (η)

Requirement	Class 1	Class 2	Class 3		
Weld joint efficiency factor (η)	1.00	0.85	0.70	0.60	0.50
Shell or end plate thickness	No limitation on thickness	Maximum thickness 38 mm after adding corrosion allowance	Maximum thickness 16 mm before corrosion allowance is added	Maximum thickness 16 mm before corrosion allowance is added	Maximum thickness 16 mm before corrosion allowance is added
Type of joints	Double-welded butt joints with full penetration excluding butt joints with metal backing strips which remain in place	Double-welded butt joints with full penetration excluding butt joints with metal backing strips which remain in place	Double-welded butt joints with full penetration excluding butt joints with metal backing strips which remain in place	Single-welded butt joints with backing strip not over 16 mm thickness or over 600 mm outside diameter	Single full fillet lap joints for circumferential seams only
	Single-welded butt joints with backing strip $\eta = 0.9$	Single-welded butt joints with backing strip $\eta = 0.80$	Single-welded butt joints with backing strip $\eta = 0.65$	Single-welded butt joints without backing strip $\eta = 0.55$	

Source: K. Lingaiah and B. R. Narayana Iyengar, Machine Design Data Handbook, Engineering College Cooperative Society, Bangalore, India, 1962; K. Lingaiah and B. R. Narayana Iyengar, Machine Design Data Handbook, Vol. I (SI and Customary Metric Units), Suma Publishers, Bangalore, India, 1983; K. Lingaiah, Machine Design Data Handbook, Vol. II (SI and Customary Metric Units), Suma Publishers, Bangalore, India, 1986; and IS: 2825-1969.

Particular	Formula
The minimum thickness of shell exclusive of corrosion allowance as per *ASME Boiler and Pressure Vessel Code**	$t = \dfrac{p R_i}{2\sigma_{sa}\eta + 0.4p}$ (8–12) when the thickness of shell does not exceed one-half the inside radius (R_i)
The maximum allowable working pressure as per *ASME Boiler and Pressure Vessel Code** [from Eq. (8–12)] [1,2]	$p = \dfrac{2\sigma_{sa}\eta t}{R_i - 0.4t}$ (8–13) when the pressure p does not exceed $1.25\ \sigma_{sa}\eta$. σ_{sa} is taken from Tables 8–9, 8–11, and 8–12.
LONGITUDINAL POINT The minimum thickness of shell exclusive of corrosive allowance as per *ASME Boiler and Pressure Vessel Code** [1-10]	$t = \dfrac{pR_i}{\sigma_{sa}\eta - 0.6p} = \dfrac{pR_o}{\sigma_{sa}\eta + 0.4p}$ (8–14) when the thickness of shell does not exceed one-half the inside radius R_i
The maximum allowable working pressure as per *ASME Boiler and Pressure Vessel Code* [from Eq. 8–14)]	$p = \dfrac{\sigma_{sa}\eta t}{R_i + 0.6t} = \dfrac{\sigma_{sa}\eta t}{R_o - 0.4t}$ (8–15) when the pressure p does not exceed $0.385\ \sigma_{sa}\eta$
The design stress for the case of welded cylindrical shell assuming a Poisson ratio of 0.3	$\sigma_d = 0.87\dfrac{p_i r_o}{h}$ (8–16)
The allowable stress for plastic material taking into consideration the combined effect of longitudinal and tangential stress (*Note*: The design stress for plastic material is 13.0 percent less compared with the maximum value of the main stress.)	$\sigma_a = \dfrac{p_i d_o}{2.3\,h}$ (8–17)
The thickness of shell from Eq. (8–17) without taking into account the joint efficiency and corrosion allowance	$h = \dfrac{pd_o}{2.3\,\sigma_{sa}}$ (8–18)
The design thickness of shell taking into consideration the joint efficiency η and allowance for corrosion, negative tolerance, and erosion of the shell (h_c)	$h_d = \dfrac{pd_o}{2.3\,\sigma_{sa}\eta} + h_c$ (8–19)
The design formula for the thickness of shell according to Azbel and Cheremisineff [11]	$h_d = \dfrac{pd_i}{2.3\,\eta\sigma_{sa} - p} + h_c$ (8–20)
The factor of safety as per pressure vessel code, which is based on yield stress of material used for shell	$n = \dfrac{\sigma_{sy}}{\sigma_a}$ (8–21) The factor of safety n should not be less than 4, which is based on yield strength σ_{sy} of material.

*Rules for construction of pressure vessel, section VIII, Division 1, *ASME Boiler and Pressure Vessel Code*, July 1, 1986.

Particular	Formula

Shell under internal pressure — spherical shell

The minimum thickness of shell exclusive of corrosion allowance as per Bureau of Indian Standards

$$h = \frac{pd_i}{4\sigma_{sa}\eta - p} = \frac{pd_o}{4\sigma_{sa}\eta + p} \tag{8-22}$$

The design pressure as per Bureau of Indian Standards

$$p = \frac{4\sigma_{sa}\eta h}{d_i + h} = \frac{4\sigma_{sa}\eta h}{d_o - h} \tag{8-23}$$

The minimum thickness of shell exclusive of corrosion allowance as per *ASME Boiler and Pressure Vessel Code*

$$t = \frac{pR_i}{2\sigma_{sa}\eta - 0.2p} \tag{8-24}$$

when thickness of the shell of a wholly spherical vessel does not exceed $0.356R_i$

The design pressure (or maximum allowable working pressure) as per *ASME Boiler and Pressure Vessel Code*

$$p = \frac{2\sigma_{sa}\eta t}{R_i + 0.2t} \tag{8-25}$$

when the maximum allowable working pressure p does not exceed $0.665\sigma_{sa}\eta$

Shells under external pressure — cylindrical shell (Fig. 8–1)

(a) The minimum thickness of cylindrical shell exclusive of corrosion allowance as per Bureau of Indian Standards

$$h = d_o\left[\frac{1.15p}{\sigma} + 1.1570 \times 10^{-4}\left(\frac{K\sigma L}{d_o}\right)^{2/3}\right]$$

$$\text{SI } (8-26a)$$

where h, d_o, and L in m; σ and p in MPa

$$= d_o\left[\frac{1.15p}{\sigma} + 4.19 \times 10^{-6}\left(\frac{K\sigma L}{d_o}\right)^{2/3}\right]$$

$$\textbf{US Customary units } (8-26b)$$

where h, d_o, and L in in; σ and p in MPa

The design pressure as per Bureau of Indian Standards

$$p = \frac{\sigma}{1.15}\left[\frac{h}{d_o} - 1.157 \times 10^{-4}\left(\frac{K\sigma L}{d_o}\right)^{2/3}\right] \quad \text{SI } (8-27a)$$

where p and σ in MPa; h, d_o, and L in m

$$= \frac{\sigma}{1.15}\left[\frac{h}{d_o} - 4.19 \times 10^{-6}\left(\frac{K\sigma L}{d_o}\right)^{2/3}\right]$$

$$\textbf{US Customary units } (8-27b)$$

Particular	Formula
	where p and σ in psi; h, d_o and L in in

$$\text{for } \frac{L}{d_o} < \frac{5.7\,(10p/\sigma)^{5/2}}{pK} \quad \text{or} \quad < \frac{372.65 \times 10^3\,(h/d_o)^{3/2}}{K\sigma}$$

$$\text{SI} \qquad (8\text{--}27\text{c})$$

where σ and p in MPa; d_o, h, and L in m

$$\text{for } \frac{L}{d_o} < \frac{5.7\,(10p/\sigma)^{5/2}}{pK} \quad \text{or} \quad < \frac{5.41 \times 10^7\,(h/d_o)^{3/2}}{K\sigma}$$

$$\text{US Customary units}\,(8\text{--}27\text{d})$$

where σ and p in psi; d_o and h in in

(b) The minimum thickness of cylindrical shell exclusive of corrosion allowance according to Bureau of Indian Standards [12]

$$h = 2.234 \times 10^{-4} d_o\,(pK)^{1/3} \text{ but not less than}$$

$$(3.5/2)\,(pd_o/\sigma) \qquad \text{SI} \qquad (8\text{--}28\text{a})$$

where d_o and h in m and p in MPa

$$= 4.25 \times 10^{-3} d_o\,(pK)^{1/3} \text{ but not less than}$$

$$(3.5/2)\,(pd_o/\sigma) \qquad \text{US Customary units}\,(8\text{--}28\text{b})$$

where d_o and h in in and p in psi

or

The design pressure as per Bureau of Indian Standards from Eq. (8–28)

$$p = \frac{8.97 \times 10^{10}}{K}\left(\frac{h}{d_o}\right)^3 \text{ but not greater than } \frac{2h\sigma}{3.5 d_o}$$

$$\text{SI}\,(8\text{--}29\text{a})$$

where p in MPa and h and d_o in m

$$= \frac{13 \times 10^6}{K}\left(\frac{h}{d_o}\right)^3 \text{ but not greater than } \frac{2}{3.5}\frac{h\sigma}{d_o}$$

$$(8\text{--}29\text{b})$$

$$\text{US Customary units}$$

where p in psi and h and d_o in in

$$\text{for } \frac{L}{d_o} > \frac{97.78}{(pK)^{1/6}} \quad \text{or} \quad > \frac{14.6}{(100h/d_o)^{1/2}} \qquad \text{SI}$$

$$\text{for } \frac{L}{d_o} > \frac{22.4}{(pK)^{1/6}} \quad \text{or} \quad > \frac{1.46}{(h/d_o)^{1/2}}$$

$$\text{US Customary units}$$

$$\text{or} \quad 5.7\frac{(10p/\sigma)^{5/2}}{pK} > \frac{97.78}{(pK)^{1/6}} \qquad \text{SI}$$

Particular	Formula
	$0.58\dfrac{(10p/\sigma)^{5/2}}{pK} > \dfrac{22.4}{(pK)^{1/6}}$ **US Customary units**
	or $372.65 \times 10^{3}\dfrac{(h/d_o)^{3/2}}{K\sigma} > \dfrac{1.46}{(h/d_o)^{1/2}}$ **SI**
	$54.1 \times 10^{6}\dfrac{(h/d_o)^{3/2}}{K\sigma} > \dfrac{1.46}{(h/d_o)^{1/2}}$ **US Customary units**

(c) In other cases, the minimum thickness of the shell exclusive of corrosion allowance as per Bureau of Indian Standards

$$h = 3.576 \times 10^{-5}\, d_o \left(p\frac{L}{d_o}K\right)^{2/5} \qquad \text{SI} \quad (8\text{–}30a)$$

where h, d_o MPa and L in m; p in

$$= 1.227 \times 10^{-3}\, d_o \left(p\frac{L}{d_o}K\right)^{2/5} \text{US Customary units}$$

$$(8\text{–}30b)$$

where h, L, and d_o in in; p in psi

or

The design pressure as per Bureau of Indian Standards

$$p = \frac{3.162 \times 10^{12}\,(h/d_o)^{5/2}}{LK/d_o} \qquad \text{SI} \quad (8\text{–}31a)$$

where h, L, and d_o in m; p in MPa

$$= \frac{189.58 \times 10^{6}(h/d_o)^{5/2}}{LK/d_o} \qquad \text{US Customary units}$$

$$(8\text{–}31b)$$

where h, d_o, and L in in; p in psi

Reference Chart for *ASME Boiler and Pressure Vessel Code*, Section VIII, Division 1 [13]

Refer to Fig. 8–2.

(d) Maximum allowable stress values

(1) The maximum allowable stress values in tension for ferrous and nonferrous materials (σ_{sa})

Refer to Tables 8–8 to 8–13 for σ_{sa}.

The maximum allowable stress values (σ_{sa}) for bolt, tube, and pipe materials

Refer to Tables 7–1, 8–8, 8–12, and 8–17.

(2) The maximum allowable longitudinal compressive stress (σ_{ac}) to be used in the design of cylindrical shells or tubes, either seamless or butt-welded subjected to loadings that produce longitudinal compression in shell or tube, shall be as given in either Eq. (a) or (b).

$\sigma_{ac} < \sigma_{sa}$ from Tables 7–1,8–9 to 8–13 (a)

$\sigma_{ac} < B$ (b)

where $B =$ a factor determined from the applicable material/temperature chart for maximum design temperature, psi Figs. 8–4, 8–5.

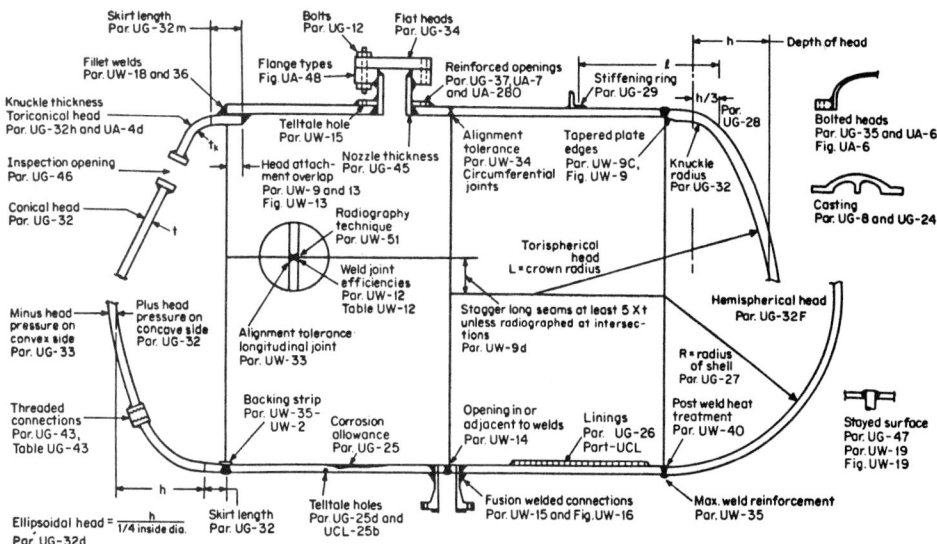

FIGURE 8–2 Reference chart for *ASME Boiler and Pressure Vessel Code*, Section VIII, Division 1. (*By permission, Robert Chuse*, Pressure Vessels—The ASME Code Simplified, *5th ed., McGraw-Hill, 1977.*)

Particular	Formula
	[*Note*: **US Customary units** (i.e., fps system of units) were used in drawing Figs. 8–3 to 8–5 of *ASME Pressure Vessel and Boiler Code*, which is now used to find the thickness of walls of cylindrical and spherical shells and tubes, unless it is otherwise mentioned to use both **SI** and **US Customary units**. Figures 8–3 to 8–5 are in US Customary units. The values from these figures and others can be used in the appropriate equation to find the values or results in SI units, if these values and equations are converted into SI units beforehand.]
(3) The procedure for determining the value of the factor *B*	Select the thickness t (*or h*) and outside diameter D_o or outside radius R_o of a cylindrical shell or tube in the corroded condition. Then calculate the value of A from Eq. (8–32)
The value of factor *A*	$$A = \frac{0.125}{R_o/t} \qquad (8\text{–}32)$$
	Using this value of A enter the applicable material/ temperature chart for the material (Figs. 8–4 and 8–5) under consideration to find B. In case the value of A falls to the right of the end of the material/ temperature line (Figs. 8–4 and 8–5), assume an intersection with the horizontal projection of the upper end of the material/temperature line. From the

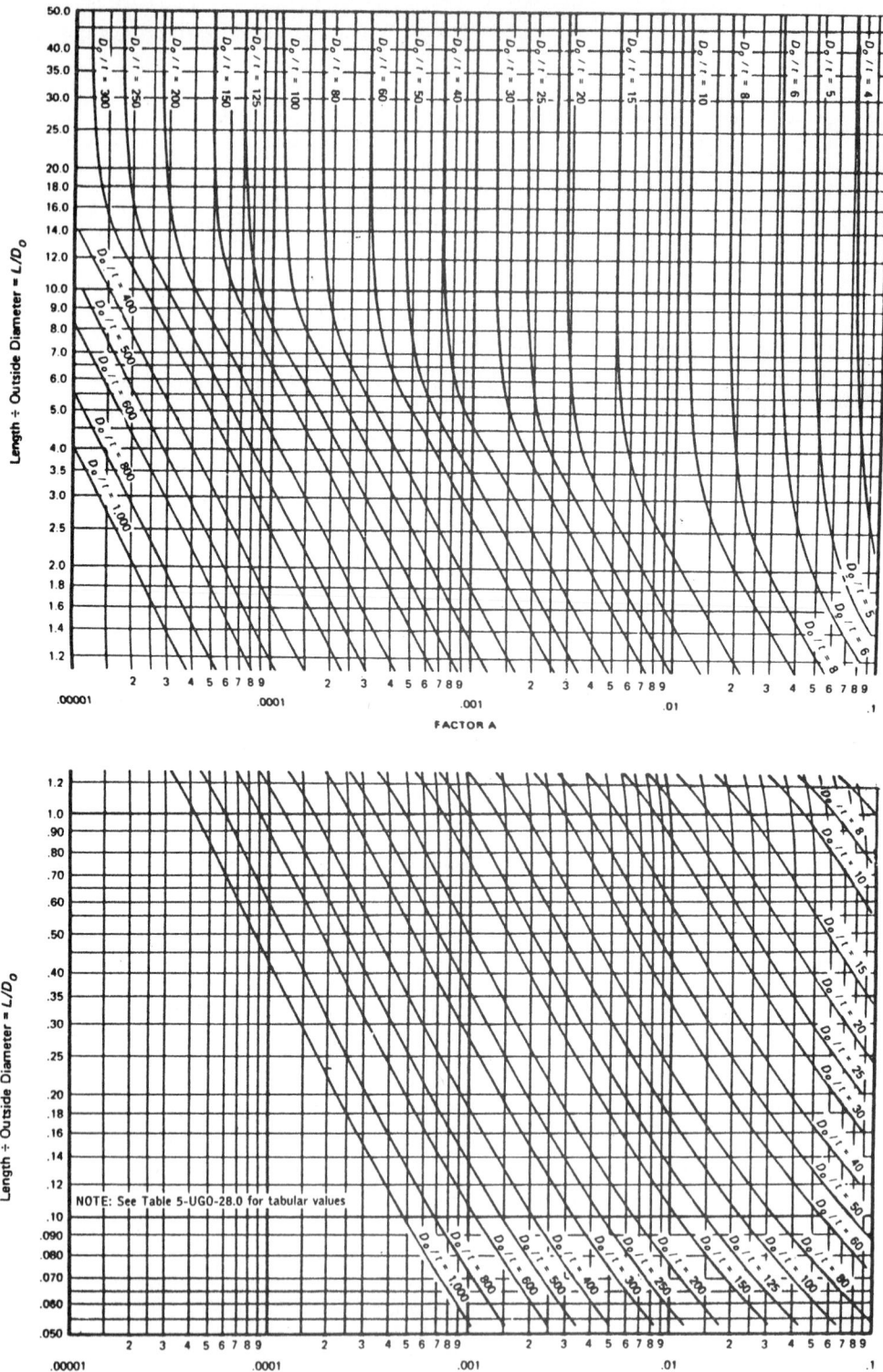

FIGURE 8–3 Geometric chart for cylindrical vessels under external or compressive loadings (for all materials). (*American Society of Mechanical Engineers*, ASME Boiler and Pressure Vessel Code, *Section VIII, Division 1, July 1, 1986.*)

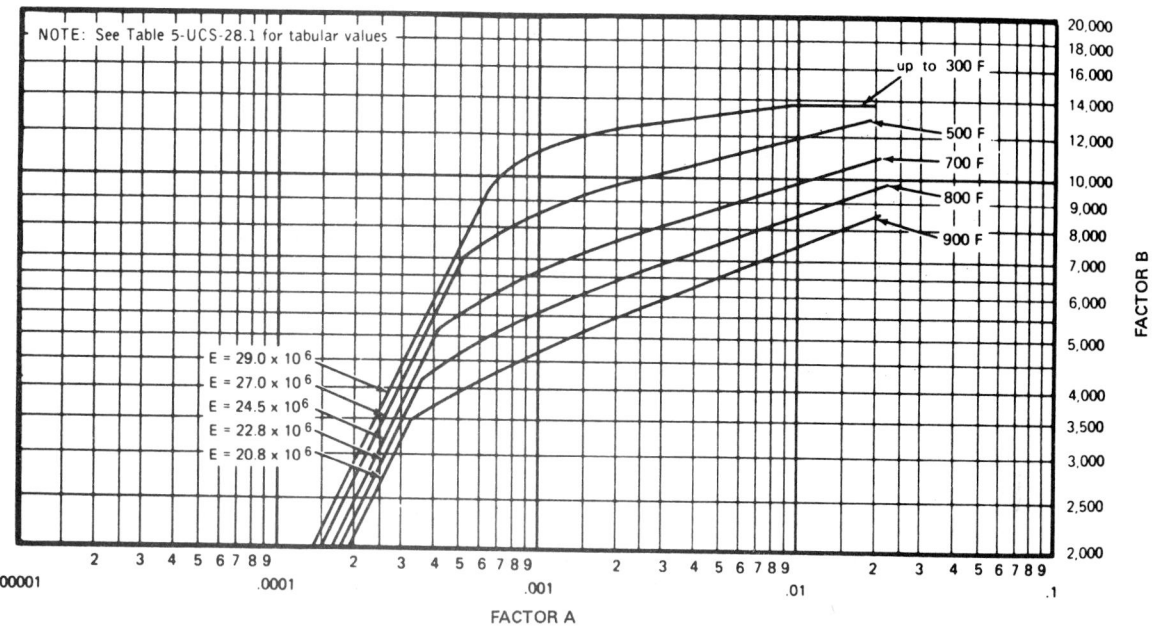

FIGURE 8–4 Chart for determining shell thickness of cylindrical and spherical vessels under external pressure when constructed of carbon or low-alloy steels [specified minimum yield strength 24,000 psi to, but not including, 30,000 psi]; [1 kpsi = 6.894757 MPa].

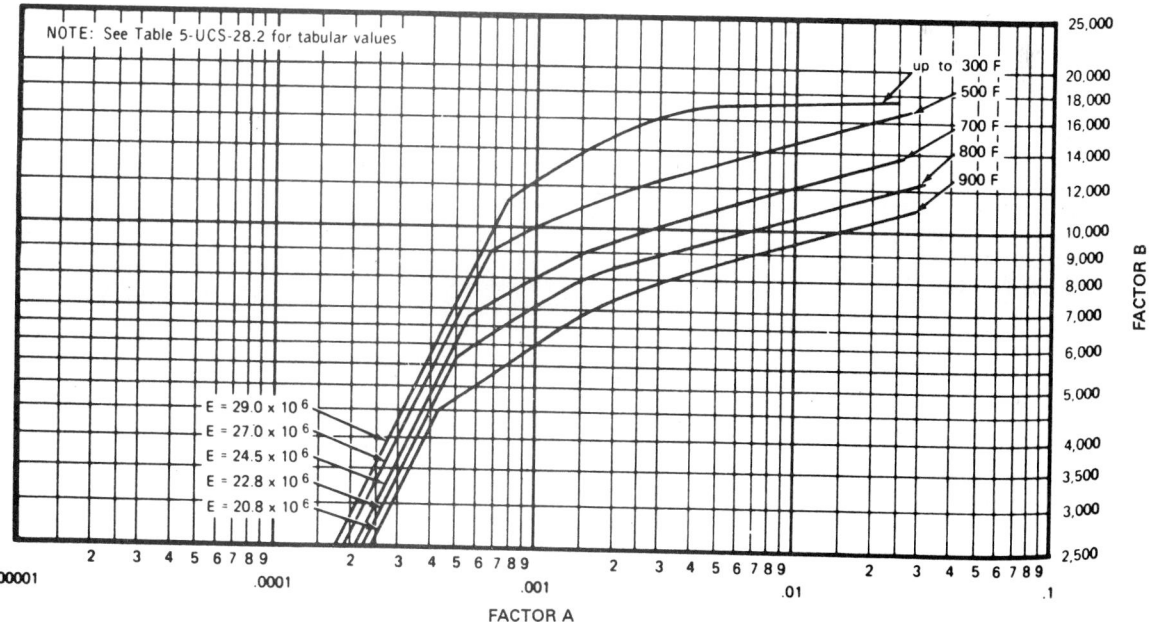

FIGURE 8–5 Chart for determining shell thickness of cylindrical and spherical vessels under external pressure when constructed of carbon or low-alloy steels (specified minimum yield strength 30,000 psi and over except for materials within this range where other specific charts are referenced) and type 405 and type 410 stainless steels (1 kpsi = 6.894757 MPa). (*Source:* American Society of Mechanical Engineers Boiler and Pressure Vessel Code, *Section VIII, Division 1, July 1, 1986.*)

Particular	Formula
	intersection move horizontally to the right and find the value of B. This is the maximum allowable compressive stress for the value of t and R_o assumed.
	If the value of A falls to the left of the applicable material/temperature line, the value of B, psi, shall be calculated from Eq. (8–33).
The expression for value of factor B	$$B = \frac{AE}{2} \qquad (8\text{–}33)$$
	where $E =$ modulus of elasticity of material at design temperature, psi
	Compare the value of B determined from Eq. (8–33) or from the procedure outlined above with the computed longitudinal compressive stress in the cylindrical shell or tube using the selected values of t and R_o. If the value of B is smaller than the computed, compressive stress, a greater value of t must be selected and the procedure outlined above is repeated until a value of B is obtained, which is greater than the compressive stress computed for the loading on the cylindrical shell or tube.
(e) Cylindrical shells and tubes. The required thickness of cylindrical shell or tube exclusive of corrosion allowance under external pressure either seamless or with longitudinal butt-welded joint as per *ASME Boiler and Pressure Vessel Code* can be determined by the following procedure:	
(1) Cylinders having (D_o/t) values ≥ 10. Assume the thickness t of shell or tube. Determine D_o/t and L/D_o. Use Fig. 8–3 to find A. Find the value of A from Fig. 8–3 by following the procedure explained in paragraph (d) (3)	In cases where the value of A falls to the right of the end of the material/temperature line, assume an intersection with the horizontal projection of the upper end of the material/temperature line. Using this value of A enter the applicable material/temperature chart for material (Figs. 8–4 and 8–5) under consideration and find the value of B. This value of B is the maximum allowable compressive stress for the value of t and R_o assumed, Pa (psi).
The equation for maximum allowable external pressure (P_a) by using this value of B	$$P_a = \frac{4B}{3(D_o/t)} \qquad (8\text{–}34)$$
The equation for maximum allowable external pressure P_a for values of A falling to the left of the applicable material/temperature line.	$$P_a = \frac{2AE}{3(D_o/t)} \qquad (8\text{–}35)$$
	where P_a obtained from Eq. (8–35) is equal to or greater than P. P is the external design pressure, psi. This external allowable pressure is 15 psi (103.4 kPa) or less. The maximum external pressure is 15 psi (103.4 kPa) or 25% more than the maximum possible external pressure, whichever is smaller.

Particular	Formula
(2) Cylinders having (D_o/t) values < 10. Using the procedure as outlined in *section* d(3), obtain the value of B. For values of (D_o/t) less than 4, the value of A can be calculated using Eq. (8–36)	$A = \dfrac{1.1}{(D_o/t)^2}$ (8–36) For values of A greater than 0.10, use a value of 0.10
The formula to calculate the value of P_{a1}	$P_{a1} = \left[\dfrac{2.167}{D_o/t} - 0.0833\right] B$ (8–37)
The formula to calculate the value of P_{a2}	$P_{a2} = \dfrac{2\sigma_s}{D_o/t}\left[1 - \dfrac{1}{D_o/t}\right]$ (8–38) where σ_s is the lesser of two times the maximum allowable stress value at design metal temperature, from the applicable Tables 8–9 to 8–12 or 0.9 times the yield strength of the material at design temperature. The yield strength values are twice the B value obtained from the applicable material/temperature chart. The smaller of the values of P_{a1} or P_{a2} shall be used for the maximum allowable external pressure P_a. Thus P_a obtained is equal to or greater than the design pressure P.

Shell under external pressure — spherical shell

The thickness of a spherical shell as per Bureau of Indian Standards	$h = \dfrac{p d_o}{0.80\sigma_{sa}}$ (8–39)
The design pressure as per Indian Standards	$p = \dfrac{0.80\sigma_{sa} h}{d_o}$ (8–40)
The minimum required thickness of a spherical shell exclusive of corrosion allowance under external pressure, either seamless or of built-up construction with butt joints, shall be determined by the following procedure as per *ASME Boiler and Pressure Vessel Code*.	
Select a value for t. Determine D_o/L and D_o/t. Find the value of A by using Fig. 8–3.	
The value of the factor A is also calculated from Eq. (8–41). Using this value of A, find the value of B from the applicable material/temperature chart as done in case of the cylindrical shell	$A = \dfrac{0.125}{R_o/t}$ (8–41)

Particular	Formula
	where R_o is the outside radius of spherical shell in the corroded condition, in
The maximum allowable external pressure P_a for values of A falling to the right of the applicable material/temperature line	$$P_a = \frac{B}{R_o/t} \qquad (8\text{--}42)$$ where P_a is the calculated value of allowable external working pressure for the assumed value of t, Pa (psi), and P is the external design pressure, Pa (psi)
The maximum allowable external pressure P_a for values of A falling to the left of the applicable material/temperature line	$$P_a = \frac{0.0625E}{(R_o/t)^2} \qquad (8\text{--}43)$$ The smaller of value of P_a from Eq. (8–42) or (8–43) shall be used for the maximum allowable external pressure P_a. P_a obtained is equal to or greater than the design pressure P.
For finding the thickness of a shell in the design of a longitudinal lap joint in a cylindrical or any lap joint in a spherical shell under external pressure	The thickness of the shell shall be determined by the rules already narrated for the longitudinal butt joint of the cylindrical and spherical shell, except that $2P$ shall be used instead of P in the calculations for the required thickness.

Cylindrical shell under combined loading as per Indian Standards

The longitudinal stress	$$\sigma_z = \frac{\frac{\pi}{4}pd_i^2 + W \pm 4\dfrac{M_b}{d_i}}{\pi h(d_i + h)} \qquad (8\text{--}44)$$
The hoop stress	$$\sigma_\theta = \frac{p\,(d_i + h)}{2h} \qquad (8\text{--}45)$$
The shear stress	$$\tau = \frac{2M_t}{\pi h\, d_i\,(d_i + h)} \qquad (8\text{--}46)$$
The Huber-Hencky equation for equivalent stress based on the shear strain energy criterion	$$\sigma_e = \sqrt{\sigma_\theta^2 - \sigma_\theta\sigma_z + \sigma_z^2 + 3\,\tau^2} \qquad (8\text{--}47)$$
The basic design stress based on distortion energy theory	$$\sigma_d = [\sigma_\theta^2 + \sigma_z^2 + \sigma_r^2 - 2(\sigma_\theta\sigma_z + \sigma_z\sigma_r + \sigma_r\sigma_\theta)]^{1/2} \qquad (8\text{--}48)$$
Requirements are	
(a) At design conditions	$\sigma_e \le \sigma_{sa} \qquad (8\text{--}49)$ $\sigma_{zt} \le \sigma_{sa} \qquad (8\text{--}50)$ $\sigma_{zc} \le 0.125\,E\,(h/d_o) \qquad (8\text{--}51)$

Particular	Formula
	Refer to Table 8–14 for values of E.
(b) At test conditions	$\sigma_e \leq 1.3\sigma_{sam}$ (8–52) $\sigma_{zt} \leq 1.3\sigma_{sam}$ (8–53) $\sigma_{zc} \leq 0.125\, E_{sam}(h_a/d_o)$ (8–54)

Stiffening rings for cylindrical shells under external pressure

Particular	Formula
The moment of inertia of the stiffening rings as per Indian Standards	$I_s = 0.714 \times 10^{-6} p\, L_s\, d'^3 K$ **SI** (8–55a) where I_s in m⁴, p in Pa, L_s and d' in m $= 4.92 \times 10^{-3} p\, L_s\, d'^3 K$ **US Customary units** (8–55b) where I_s in in⁴, p in psi, L_s and d' in in
The required moment of inertia of a circumferential stiffening ring shall be not less than that determined by one of the formulas given in Eqs. (8–56) and (8–57) as per *ASME Boiler and Pressure Vessel Code*	$I_s = \dfrac{D^2{}_o L_s[t + (A_s/L_s)A]}{14}$ **US Customary units** (8–56) $I'_s = \dfrac{D^2{}_o L_s[t + (A_s/L_s)A]}{10.9}$ **US Customary units** (8–57)
Select a member to be used for stiffening a ring after knowing D_o, L_s, and t of a shell designed already. Then calculate factor B using Eq. (8–58)	
The expression for factor B	$B = \dfrac{3}{4}\left(\dfrac{PD_o}{t + A_s/L_s}\right)$ (8–58)
For calculating factor A	Use the applicable material/temperature chart to find A
For values of B falling below the left end of the material/temperature chart line for the design temperature the value of A can be determined from Eq. (8–59)	$A = \dfrac{2B}{E}$ (8–59)

Particular	Formula

FORMED HEADS UNDER PRESSURE ON CONCAVE SIDE

For domed ends of hemispherical, semiellipsoidal, or dished shape

Refer to Figs. 8–6 for domed end.

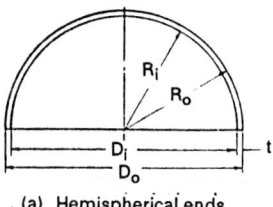
, (a) Hemispherical ends

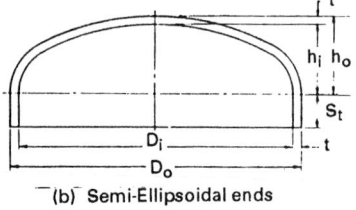
(b) Semi-Ellipsoidal ends

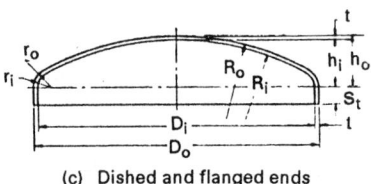
(c) Dished and flanged ends

FIGURE 8–6 Domed ends.

The required thickness at the thinnest point after forming of ellipsoidal, torispherical, hemispherical, conical, and toriconical heads under pressure on the concave side of the shell shall be computed by the appropriate formulas

The thickness of the ends and/or heads under pressure on concave side (plus heads) as per Indian Standards

$$h = \frac{p \, d_o C}{2\eta\sigma_{sa}} \tag{8–60}$$

where C is a shape factor taken from Fig. 8–7

The allowable pressure as per Indian Standards

$$p = \frac{2\eta h \sigma_{sa}}{d_o C} \tag{8–61}$$

Ellipsoidal heads

The required thickness of a dished head of semiellipsoidal form, in which half the minor axis (inside depth of the head minus the skirt) equals one-fourth of the inside diameter of the head skirt, shall be determined by Eq. (8–62) as per *ASME Boiler and Pressure Vessel Code*

$$t = \frac{PD}{2\sigma_{sa}\eta - 0.2P} \tag{8–62}$$

The maximum allowable working pressure or design pressure as per *ASME Boiler and Pressure Vessel Code*

$$p = \frac{2\sigma_{sa}\eta t}{D + 0.2t} \tag{8–63}$$

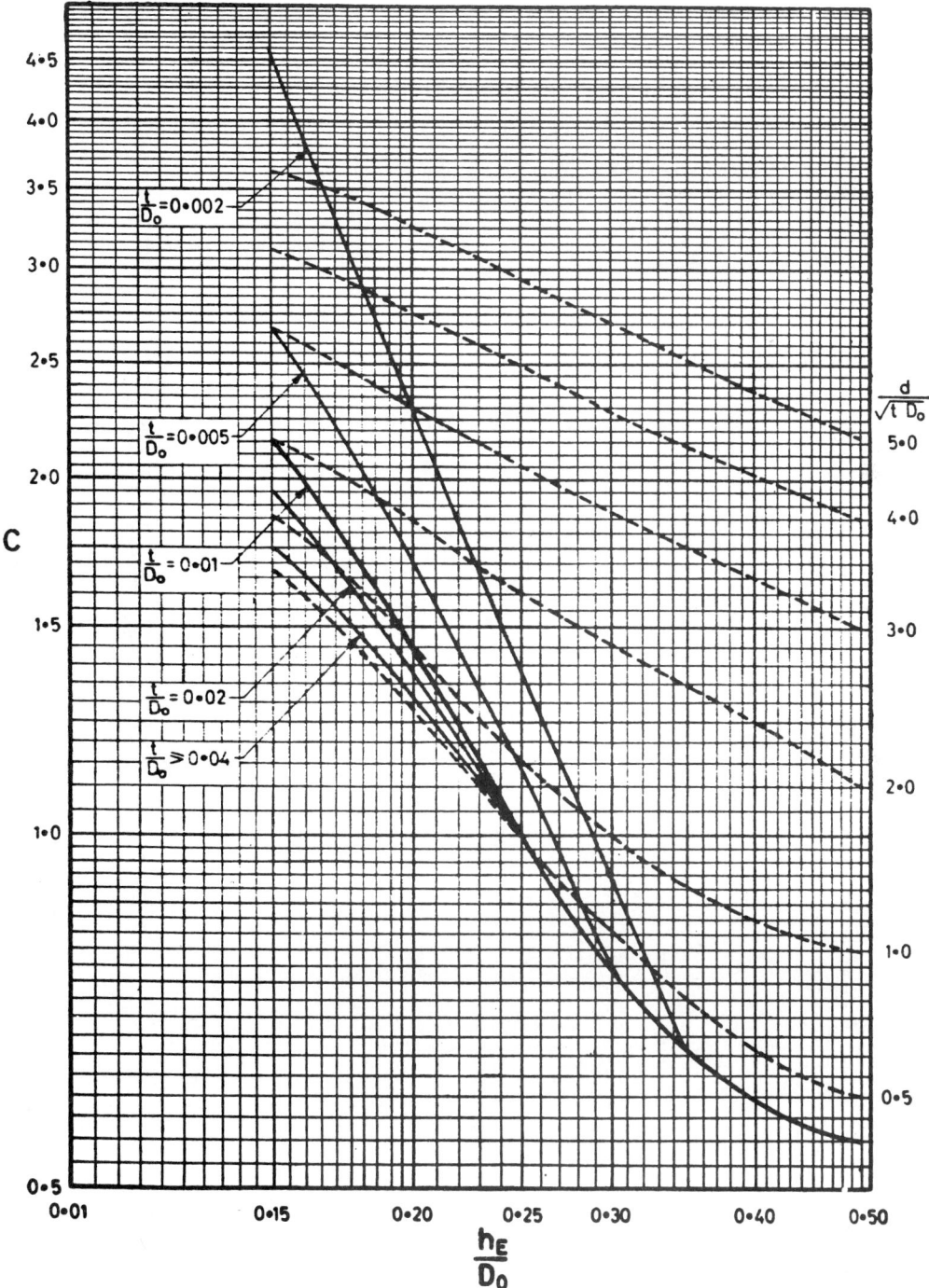

FIGURE 8–7   Shape factor C for domed ends.

Particular	Formula	

Torispherical heads

The required thickness of a torispherical head for the case in which the knuckle radius is 6 percent of the inside crown radius and the inisde crown radius equals the outside diameter of the skirt, shall be determined by Eq. (8–64) as per *ASME Boiler and Pressure Vessel Code*

$$t = \frac{0.885PL}{\sigma_{sa}\,\eta - 0.1P}$$

(8–64)

The maximum allowable working pressure as per *ASME Boiler and Pressure Vessel Code*

$$P = \frac{\sigma_{sa}\,\eta\,t}{0.885L + 0.1t}$$

(8–65)

Hemispherical heads

The required thickness of a hemispherical head when its thickness does not exceed $0.36L$ or P does not exceed $0.665\,\sigma_{sa}\,\eta$, shall be determined by Eq. (8–66) as per *ASME Boiler and Pressure Vessel Code*

$$t = \frac{PL}{2\sigma_{sa}\eta - 0.2P}$$

(8–66)

The design pressure

$$P = \frac{2\sigma_{sa}\eta t}{L + 0.2t}$$

(8–67)

Conical ends subject to internal pressure (Fig. 8–8*d*) as per Indian Standards

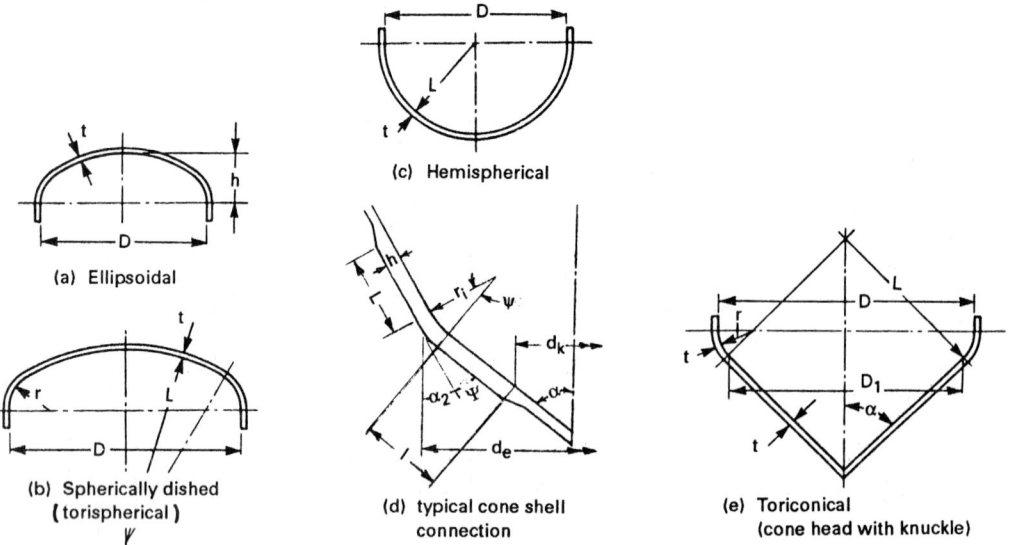

(a) Ellipsoidal

(b) Spherically dished (torispherical)

(c) Hemispherical

(d) typical cone shell connection

(e) Toriconical (cone head with knuckle)

FIGURE 8–8(a) (*a*) Ellipsoidal; (*b*) spherically dished (torispherical); (*c*) hemispherical; (*d*) typical conical shell connection; and (*e*) toriconical (cone head with knuckle).

Particular	Formula

KNUCKLE OR CONICAL SECTION The thickness of cylinder and conical section (frustrum) within the distance L from the junction $(L = 0.5\sqrt{d_e h}/\cos\alpha)$

$$h = \frac{p\, d_e\, c_1}{2\sigma_{sa}\,\eta} \tag{8–68}$$

Refer to Table 8–4 for values of c_1.

The thickness of those parts of conical sections not less than a distance L away from the junction with the cylinder or other conical section

$$h = \frac{p\, d_k}{2\sigma_{sa}\,\eta - p}\left(\frac{1}{\cos\alpha}\right) \tag{8–69}$$

SHALLOW CONICAL SECTIONS The thickness of conical sections having an angle of inclination to the vessel axis more than 70°

$$h = 0.5\,(d_e - r_i)\,\frac{\alpha}{90}\,\sqrt{p/\sigma_{sa}} \tag{8–70}$$

The lower of values given by Eqs. (8–69) and (8–70) shall be used.

TABLE 8–4
Values of c_1 for use in Eq. (18–68) (as function of Ψ and $\frac{r_i}{d_e}$)

$\frac{r_i/d_e}{\Psi}$	0.01	0.02	0.03	0.04	0.06	0.08	0.10	0.15	0.20	0.30	0.40	0.50
10°	0.70	0.65	0.60	0.60	0.55	0.55	0.55	0.55	0.55	0.55	0.55	0.55
20°	1.00	0.90	0.85	0.80	0.70	0.65	0.60	0.55	0.55	0.55	0.55	0.55
30°	1.35	1.20	1.10	1.00	0.90	0.85	0.80	0.70	0.55	0.55	0.55	0.55
45°	2.05	1.85	1.65	1.50	1.30	1.20	1.10	0.95	0.90	0.70	0.55	0.55
60°	3.20	2.85	2.55	2.35	2.00	1.75	1.60	1.40	1.25	1.00	0.70	0.55
75°	6.80	5.85	5.35	4.75	3.85	3.50	3.15	2.70	2.40	1.55	1.00	0.55

Source: IS 2825, 1969.

Conical heads (without transition knuckle) as per *ASME Boiler and Pressure Vessel Code*

The required thickness of conical heads or conical shell sections that have a half-apex angle α not greater than 30° shall be determined by Eq. (8–71)

$$t = \frac{PD}{2\cos\alpha\,(\sigma_{sa}\,\eta - 0.6P)} \tag{8–71}$$

where D = inside diameter
η = minimum joint efficiency, percent

The design pressure

$$P = \frac{2\sigma_{sa}\,\eta t\,\cos\alpha}{D + 1.2t\,\cos\alpha} \tag{8–72}$$

Particular	Formula

Toriconical heads

The required thickness of the conical portion of a toriconical head, in which the knuckle radius is neither less than 6 percent of the outside diameter of the head skirt nor less than three times the knuckle thickness and pressure shall be determined by Eqs. (8–73) and (8–74)

$$t_c = \frac{PD_i}{2 \cos \alpha(\sigma_{sa} \eta - 0.6P)} \qquad (8\text{–}73)$$

$$P = \frac{2\sigma_{sa} \eta t \cos \alpha}{D_i + 1.2 t \cos \alpha} \qquad (8\text{–}74)$$

where D_i = inside cone diameter at point of tangency to knuckle

The required thickness of the knuckle and pressure shall be determined by Eqs. (8–75) and (8–76)

$$t_k = \frac{PLM}{2\sigma_{sa} \eta - 0.2P} \qquad (8\text{–}75)$$

or refer to Eqs. (8–66) and (8–67)
where
M = factor depending on head proportion, L/r
$L = D_i/2 \cos \alpha$

Toriconical heads may be used when the angle $\alpha \leq 30°$

The design pressure

$$P = \frac{2\sigma_{sa} \eta t_k}{ML + 0.2 t_k} \qquad (8\text{–}76)$$

FORMED HEADS UNDER PRESSURE ON CONVEX SIDE

The thickness of heads and ends under pressure on convex side (minus heads) as per Indian Standards

(a) Spherically dished ends and heads

The thickness of the spherically dished heads and ends shall be the greater of the following thicknesses:

(1) The thickness of an equivalent sphere, having a radius r_o or R_o equal to the outside crown radius of the end as determined from Eq. (8–39)

(2) The thickness of the end under an internal pressure equal to 1.2 times the external pressure

(b) Ellipsoidal heads

The thickness of ends of a semiellipsoidal shape shall be the greater of the following:

(1) The thickness of an equivalent sphere, having a radius r_o or R_o calculated from the values of r_o/d_o or R_o/D_o in Table 8–5, determined as per Eq. (8–39)

TABLE 8–5
Values of spherical radius factor $K_o = R_o/D_o$ for ellipsoidal head with pressure on convex side as a function of h_o/D_o for use in Eq. (8–41)

$\frac{h_o}{D_o}$	0.167	0.178	0.192	0.208	0.227	0.25	0.276	0.313	0.357	0.417	0.5
$K_o = \frac{R_o}{D_o}$	1.36	1.27	1.182	1.08	0.99	0.90	0.81	0.73	0.65	0.57	0.5

Particular	Formula

(2) The thickness of the end under an internal pressure equal to 1.2 times the external pressure

(c) Conical heads under external pressure: For a conical end or conical section under external pressure, whether the end is of seamless or butt-welded construction as per Indian Standards.

Use Eqs. (8–68), (8–69), and (8–70).
Equations (8–68) to (8–70) are applicable, except that the shell thickness shall be no less than as prescribed below:

(1) The thickness of a conical end or conical section under external pressure, when the angle of inclination of the conical section to the vessel axis is not more than 70°, shall be made equal to the required thickness of cylindrical shell, in which the diameter is $(d_e/\cos\alpha)$ and the effective length is equal to the slant height of the cone or conical section, or slant height between the effective stiffening rings, whichever is less.

(2) The thickness of conical ends having an angle of inclination to the vessel axis of more than 70° shall be determined as a flat cover.

The thickness of formed heads under pressure on convex side (minus heads) as per *ASME Boiler and Pressure Vessel Code*

The required thickness at the thinnest point after forming of ellipsoidal or torispherical heads under pressure on the convex side (minus heads) shall be the greater of the thicknesses given here

(1) The thickness as computed by the procedure given for heads with the pressure on the concave side of the previous section using a design pressure 1.67 times the design pressure of the convex side, assuming the joint efficiency $\eta = 1.00$ for all cases   *or*

(2) The thickness as determined by the appropriate procedure given in *Ellipsoidal heads* or *Torispherical heads* as per *ASME Boiler and Pressure Vessel Code*

HEMISPHERICAL HEADS The required thickness of a hemispherical head having pressure on the convex side shall be determined by Eqs. (8–41) to (8–43)

$$A = \frac{0.125}{R_o/t} \qquad (8–41)$$

Particular	Formula

$$P_a = \frac{B}{R_o/t} \qquad (8\text{-}42)$$

$$P_a = \frac{0.0625\eta}{(R_o/t)^2} \qquad (8\text{-}43)$$

where R_o = outside radius in corroded condition

The procedure outlined in this section for finding the thickness of a spherical shell can be used to find the thickness of a hemispherical head by using Eqs. (8–41) to (8–43).

ELLIPSOIDAL HEADS The minimum required thickness of ellipsoidal head having pressure on the convex side either seamless or of built-up construction with butt joints shall not be less than that determined by the procedure given here

The factor A is given by Eq. (8–41)

Assume a value of t and calculate the value of factor A using the following equation:

$$A = \frac{0.125}{R_o/t} \qquad (8\text{-}41)$$

Using the value of A calculated from Eq. (8–41) follow the procedure as that given for spherical shells to find the thickness of ellipsoidal heads.

where R_o = the equivalent outside spherical radius taken as $K_o D_o$ in the corroded condition
K_o = factor depending on the ellipsoidal head proportion R_o/D_o (Table 8–5)

TORISPHERICAL HEADS The required minimum thickness of a torispherical head having pressure on the convex side, either seamless or of built-up construction with butt joint

The required thickness shall not be less than that determined by the same design procedure as is used for ellipsoidal heads given in the *Ellipsoidal heads* section, using appropriate value for R_o. For torispherical head, the outside radius of the crown portion of the head (R_o) in corroded condition is taken for design purposes.

CONICAL HEADS AND SECTIONS The required minimum thickness of a conical head or section under pressure on the convex side, either seamless or the built-up construction with butt joints

The symbols involved in design calculations are

This thickness can be determined following the procedure outlined under cylindrical shells in the *Torispherical heads* section to find factors A and B by assuming a value for t_e.

t_e = effective thickness of conical section
L_e = equivalent length of conical section = $(L/2)(1 + D_s/L)$
L = axial length of cone or conical section (Fig. 8–8b)
D_s = outside diameter at small end of conical section under consideration

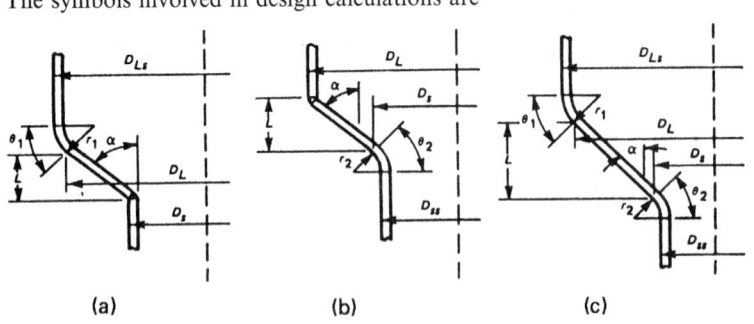

(a)   (b)   (c)

FIGURE 8-8(b) Typical conical section for external pressure.

Particular	Formula
(1) When $\alpha \leq 60°$ and for cones having D_L/t_e values ≥ 10: Assume a value of t_e and determine ratios L_e/D_L and D_L/t_e. The equation for calculating the maximum allowable external pressure P_a for the case of values of factor A falling to the right of the end of the material/temperature line.	$$P_a = \frac{4B}{3(D_L/t_e)} \qquad (8\text{--}76a)$$
Equation for calculating the maximum allowable external pressure P_a for the case of values of factor A falling to the left of the applicable material/temperature line	$$P_a = \frac{2A\,E}{3(D_L/t_e)} \qquad (8\text{--}77)$$ where $D_L =$ outside diameter at large end of conical section under consideration (Fig. 8–8b) $\alpha \;=$ one-half the apex angle in conical heads and section, deg (Compare the value of P_a with design pressure P. If $P_a < P$, then select a new value for t_e and repeat the design procedure.)
(2) When $\alpha \leq 60°$ and for cones having D_L/t_e values < 10: For values of D_L/t_e less than 4, the value of factor A can be calculated by using Eq. (8–78).	$$A = \frac{1.1}{(D_L/t_e)^2} \qquad (8\text{--}78)$$
For values of factor A greater than 0.10, use a value of 0.10	
The equation for calculating P_{a1} using the value of factor B obtained from material/temperature chart	$$P_{a1} = \left[\frac{2.167}{D_L/t_e} - 0.0833\right] B \qquad (8\text{--}79)$$
The equation for calculating P_{a2}	$$P_{a2} = \frac{2\sigma_s}{D_L/t_e}\left[1 - \frac{1}{D_L/t_e}\right] \qquad (8\text{--}80)$$ where σ_s is less than two times the maximum allowable stress value at design metal temperature, from the applicable table or 0.9 times the yield strength of the material at design temperature. The yield strength is twice the value of B determined from the applicable material/temperature chart. The smaller of the values of P_{a1} or P_{2a} shall be used for the maximum allowable external pressure P_a. P_a is equal to or greater than P (Design pressure P is obtained from appropriate table for material.)
(3) When $\alpha > 60°$	The thickness of the cone shall be the same as the required thickness for a flat head under external pressure, the diameter of which equals the largest diameter of the cone.

Particular	Formula

Toriconical head having the pressure on the convex side

The required thickness of a toriconical head having pressure on the convex side, either seamless or of built-up construction with butt joints within the head

The thickness shall not be less than that determined using the procedure followed in the case of a cone having D_L/t_e values ≤ 10 for conical heads and sections with exception that L_e shall be determined using Eqs. (8–81):

The length L_e (Fig. 8–8B, panel a)

$$L_e = r_1 \sin \theta_1 + \frac{L}{2} \left(\frac{D_L + D_s}{D_{Ls}} \right) \tag{8–81a}$$

The length L_e (Fig. 8–8B, panel b)

$$L_e = r_2 \frac{D_{ss}}{D_L} \sin \theta_2 + \frac{L}{2} \left(\frac{D_s + D_L}{D_L} \right) \tag{8–81b}$$

The length L_e (Fig. 8–8B, panel c)

$$L_e = r_1 \sin \theta_1 + r_2 \frac{D_{ss}}{D_{Ls}} \sin \theta_2 + \frac{L}{2} \left(\frac{D_L + D_s}{D_{Ls}} \right) \tag{8–81c}$$

To find the thickness when lap joints are used in formed head construction or for longitudinal joints in a conical header under external pressure

The rules in this section, except the design pressure $2P$ shall be used instead of P in the calculations for the design of required thickness.

UNSTAYED FLAT HEADS AND COVERS (FIG. 8–9, TABLE 8–6)

The thickness h of that unstayed circular heads, covers, and blind flanges as per Indian Standards

$$h = c_2 d \sqrt{(p/\sigma_{sa})} \tag{8–82}$$

Refer to Table 8–6 for values of c_2.

The minimum required thickness t of unstayed circular heads, covers and blind flanges as per *ASME Boiler and Pressure Vessel Code*

$$t = d \sqrt{(CP/\sigma_{sa} \eta)} \tag{8–83}$$

Refer to Table 8–6 for values of C.

The minimum required thickness t of flat unstayed circular heads, covers, and blind flanges which are attached by bolts, causing an edge moment as per *ASME Boiler and Pressure Vessel Code*

$$t = d \left[\frac{CP}{\sigma_{sa} \eta} + 1.9 \frac{W h_G}{\sigma_{sa} \eta d^3} \right]^{1/2} \tag{8–84}$$

where C is taken from Table 8–6

W = flange design bolt load, lbf

$W = W_{m1}$ = the minimum bolt load for operating condition, lbf

$$= 0.785 D_G^2 P + (2b \times 3.14 D_G \, mp) \tag{8–84a}$$

$= W_{m2}$ = the minimum required bolt load for gasket seating, lbf

$$= 3.14b \, D_G y \tag{8–84b}$$

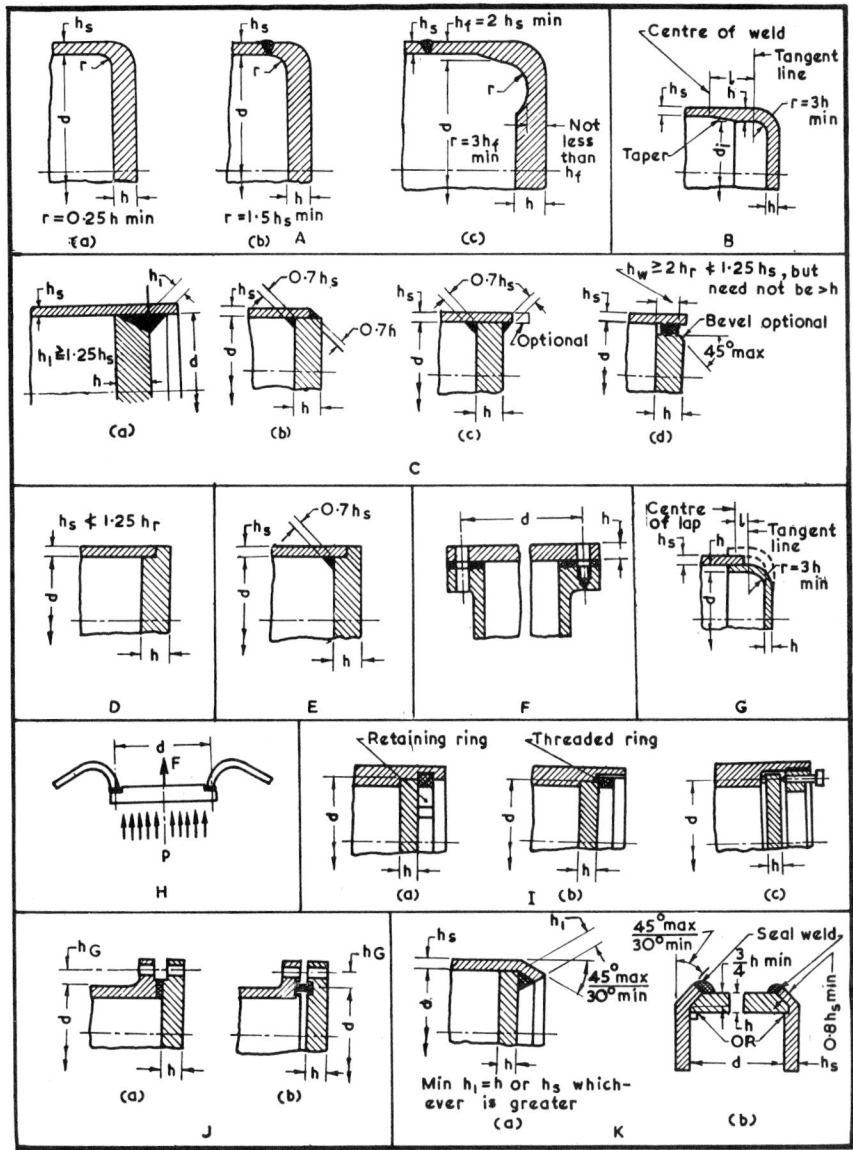

FIGURE 8–9 Types of unstayed flat heads. (*K. Lingaiah and B. R. Narayana Iyengar*, Machine Design Data Handbook, *Engineering College Cooperative Society, Bangalore, India, 1962; K. Lingaiah and B. R. Narayana Iyengar*, Machine Design Data Handbook, *Vol. I*, SI and Customary Metric Units, *Suma Publishers, Bangalore, India, 1983; K. Lingaiah*, Machine Design Data Handbook, *Vol. II*, SI and Customary Metric Units, *Suma Publishers, Bangalore, India, 1986.*)

TABLE 8–6
Coefficients c_2 and C for determining head thickness for typical unstayed flat heads (Fig. 8–9)

Type of head		Coefficient, c_2 and C		
IS (Fig. 8–9)[a]	ASME	c_2, IS (Fig. 8–9)	C, ASME	Remarks
$A(a)$	(b-2)	0.50	0.33 m but $\not< 0.20$	Forged circular and noncircular heads integral with or butt-welded to the vessel
$A(b)$		0.50		
$A(c)$	(b-1)	0.45	0.17	
B	(a)		0.17	Flanged circular and noncircular heads forged integral with or butt-welded to the vessel with an inside corner radius not less than three times the required head thickness
		0.35	0.10	For circular heads when the flange length: 1. $l \geq (1.1 - 0.8\, h_s^2/h^2)\,\sqrt{d_i h}; r \geq 2h,\ d = d_i - r$ and taper is 1:4 (Fig. 8–9) 2. $l = [1.1 - 0.8\,(t_s/t_h)^2]\,\sqrt{t_h d}$ and taper is 1:3 (1)
		0.45		When $r \geq 2h$, $d = d_i - r$ and taper is 1:4 (Fig. 8–9)
		0.50		When $d = d_i$ and $0.25h \leq r < 2h$.
			0.1	For circular heads, when the flange length $l : l < [1.1 - 0.8\,(t_s/t_h)^2\,\sqrt{t_h d}]$ but the shell thickness : $t_s \not< 1.12\,t_h\,\sqrt{1.1 - 1/\sqrt{t_h d}}$; taper is at least 1:3 (2)
$C(a)$ to $C(d)$	(e), (f) and (g)	$0.7\,\sqrt{h_r/h_s}$ but $\not< 0.55$	0.33 m but $\not< 0.20$	Circular plates welded to inside of the pressure vessel
			0.33	Noncircular plates, welded to the inside of a vessel and otherwise meeting the requirements for the respective types of welded vessels
D	(h)	0.7	0.33	For circular plates welded to the end of the shell when t_s is at least $1.25 t_r$
E	(i)	$0.7\,\sqrt{h_r/h_s}$ but $\not< 0.55$	0.33 m but $\not< 0.20$	For circular plates, if an inside fillet weld with minimum throat thickness of $0.7\, t_s$ is used
F	(p)	0.42	0.25	For circular and noncircular covers bolted with a full-face gasket, to shells, flanges or side plates
G	(c)	0.45	0.13	Circular heads lap welded or brazed to the shell with corner radius not less than the $3h$ or $3t$ and l not less than required by formula (2)
		0.55 in other cases	0.20	Circular and noncircular lap welded or brazed construction as above, but with no special requirement with regard to l
			0.30	Circular flanged plates screwed over the end of the vessel, with inside corner radius not less than $3t$ or $3h$ in which the design of threaded joint is based on a safety factor of 4
H		$\sqrt{0.31 + 95(F/Pd^2)}$ ⁻		Autoclave manhole covers $d \not> 610$ mm (24 in)
$I(a)$	(m)	0.55	0.30	Circular plates inserted in the end of a pressure
$I(b)$	(n)	0.55	0.30	vessel and held in place by a positive mechanical
$I(c)$	(o)	0.55	0.30	locking arrangement, and when all possible means of failure are resisted with a safety factor of at least 4; seal welding may be used, if desired

TABLE 8–6
Coefficients c_2 and C for determining head thickness for typical unstayed flat heads (Fig. 8–9) (*Cont.*)

Type of head		Coefficient, c_2 and C		
IS (Fig. 8–9)[a]	ASME	c_2, IS (Fig. 8–9)	C, ASME	Remarks
$J(a)$	(j)	$\sqrt{0.31 + 190(Fbh_G/Pd^3)}$	0.3	Circular and noncircular heads and covers
$J(b)$	(k)		0.3	bolted to the vessel as shown in Fig. 8–9
$K(a)$	(r)	0.7	0.33	Circular plates having a dimension d not
$K(b)$	(s)	0.7	0.33	exceeding 450 mm (18 in) inserted into the vessel as shown in Fig. 8–9 and the end of the vessel shall be crimped over at least 30°, but not more than 45°; the crimping may be done cold only when this operation will not injure the metal in case of (r); in case of (s) the crimping shall be done when the entire circumference of the cylinder is uniformly heated to the proper forging temperature for the material used

[a] Symbols in this column refer to Fig. 8–9.
[b] Where F (or W) is load on bolt.
Sources: K. Lingaiah and B. R. Narayana Iyengar, *Machine Design Data Handbook*, Engineering College Cooperative Society, Bangalore, India, 1962; K. Lingaiah and B. R. Narayana Iyengar, *Machine Design Data Handbook*, Vol. I (SI and Customary Metric Units), Suma Publishers, Bangalore, India, 1983; and K. Lingaiah, *Machine Design Data Handbook*, Vol. II (SI and Customary Metric Units), Suma Publishers, Bangalore, India, 1986.

Particular	Formula
The minimum thickness of noncircular heads and covers as per Indian Standards	$h = c_2 k_6 b \sqrt{(p_i/\sigma_{sa})}$ (8–85) Refer to Fig. 8–10 for values of k_6 and Table 8–6 for values of c_2.
The minimum heads, covers, or blind flanges of square, rectangular, oblong, segmental, or otherwise noncircular shape as per *ASME Boiler and Pressure Vessel Code*	$t = d\sqrt{(ZCP)/\sigma_{sa}\,\eta}$ (8–86a) where Z = a factor for noncircular heads depending on the ratio of short span to long span a/b $Z = 3.4 - \dfrac{2.4d}{D}$ (8–86b) Refer to Fig. 8–10 for values of Z ($Z = k_6$). Z need not be greater than 2.5. d = diameter or short span as indicated in Fig. 8–9. t, d in in and p and σ_{sa} in psi

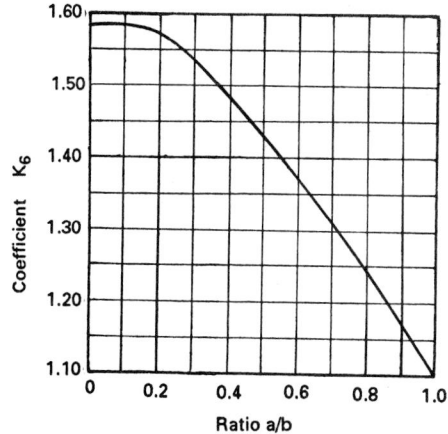

FIGURE 8–10 Coefficient k_6 for noncircular flat heads.

Particular	Formula

The minimum thickness of flat unstayed noncircular heads, covers, or blind flanges attached by bolts causing a bolt load edge moment as per *ASME Boiler and Pressure Vessel Code*

(*Note*: A stayed flat plate and types of stays are shown in Figs. 8–11 and 8–12.)

$$t = d\left[\frac{ZCP}{\sigma_{sa}\eta} + \frac{6Wh_G}{\sigma_{sa}\eta Ld^2}\right]^{1/2} \tag{8–87}$$

where h_G = gasket moment arm, equal to the radial distance from the center line of the bolts to the line of the gasket reaction as shown in Fig. 8–13

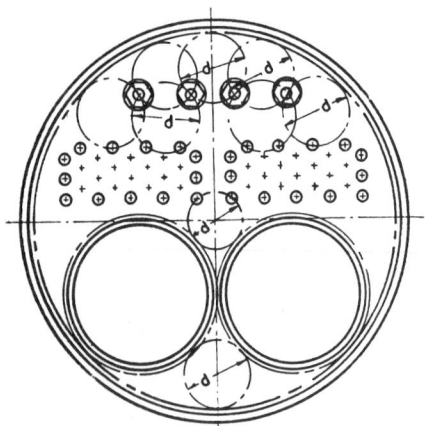

FIGURE 8–11   Stayed flat plate.

The net cover plate thickness under the groove or between the groove and outer edge (t_g) of the cover plate

$$t_g \not< d\sqrt{1.9Wh_G/\sigma_{sa}d^3}$$

for circular heads and covers   (8–88)

$$t_g \not< d\sqrt{6Wh_G/\sigma_{sa}Ld^2}$$

for noncircular heads and covers   (8–89)

The thickness of spherically dished ends and heads secured to the shell through a flange connection by means of bolts as per Indian Standards

$$h = \frac{3pd_i}{2\sigma_{sa}\eta} \text{ for } R \not> 1.3d_i \text{ and } \frac{100h}{R} \not> 10 \tag{8–90}$$

The thickness of a dished head that is riveted or welded to a cylindrical shell according to *ASME Boiler and Pressure Vessel Code*

$$h = \frac{8.33PR}{2\sigma_{Su}} \tag{8–91}$$

STAYED FLAT AND BRACED PLATES OR SURFACES (Figs. 8–11 and 8–12)

The thickness of stayed and braced plate as per Indian Standards

$$h = c_4 d\sqrt{(p/\sigma_{Sa})} \tag{8–92}$$

Refer to Table 8–7 for values of c_4 and Tables 8–8, 8–9, and 8–11 for allowable stress values σ_{sa}.

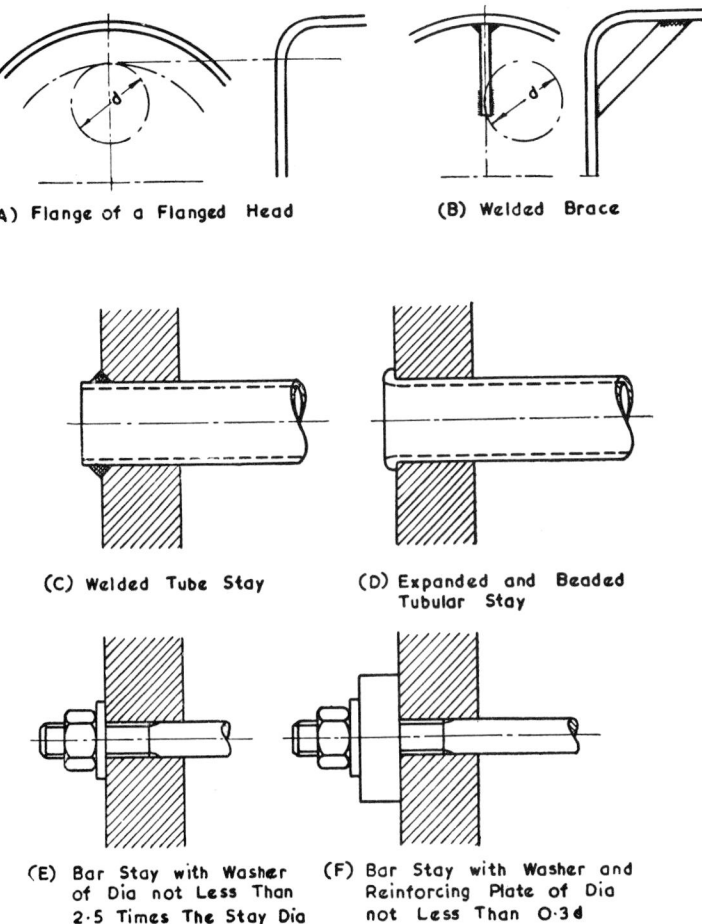

(A) Flange of a Flanged Head

(B) Welded Brace

(C) Welded Tube Stay

(D) Expanded and Beaded Tubular Stay

(E) Bar Stay with Washer of Dia not Less Than 2·5 Times The Stay Dia

(F) Bar Stay with Washer and Reinforcing Plate of Dia not Less Than 0·3 d

FIGURE 8–12 Types of stays.

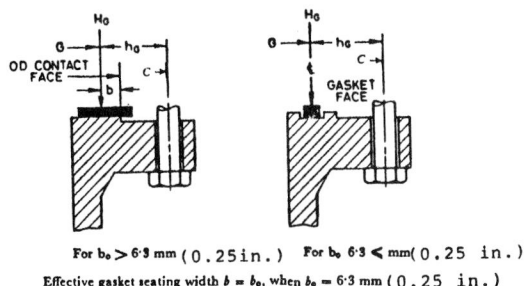

For $b_o > 6\cdot3$ mm (0.25 in.) For $6\cdot3 \leqslant$ mm(0.25 in.)

Effective gasket seating width $b = b_o$, when $b_o = 6\cdot3$ mm (0.25 in.)
and $= 2\cdot5\sqrt{b_o}$, when $b_o > 6\cdot3$ mm(0.25 in.)

NOTE — The gasket factors listed only apply to flanged joints in which the gasket is contained entirely within the inner edges of the bolt holes.

FIGURE 8–13 Location of bolt load reaction.

TABLE 8–7
Coefficients c_4 for determining head thickness for stayed and braced plates

Type of stay	c_4	Remarks
A	0.45	Flange of a flanged head
B	0.45	Welded brace
C	0.55	Welded tube stay
D	0.55	Expanded and beaded tubular stay
E	0.45	Bar stay with washer of diameter not less than 2.5 times the stay diameter
F	0.40	Bar stay with washer and reinforcing plate of diameter not less than 0.3d

Particular	Formula
The minimum thickness for braced and stayed flat plates with braces or stay bolts of uniform diameter symmetrically spaced as per *ASME Boiler and Pressure Vessel Code*	$t = p_t \sqrt{(P/\sigma_{sa} c_5)}$ (8–93)

where

t = minimum thickness of plate, exclusive of corrosion allowance, in

σ_{sa} = maximum allowable stress, MPa (psi), taken from Tables 8–9 to 8–13 for shell plates and Table 8–23 for bolts

The maximum allowable working pressure for braced and stayed flat plates as per *ASME Boiler and Pressure Vessel Code*

$$P = \frac{t^2 \sigma_{sa} c_5}{p_t^2} \qquad (8\text{–}94)$$

where c_5 = a factor depending on the plate thickness and type of stay taken from Table 8–15

Stayed flat plates with uniformly distributed load

Grashof's formula for maximum stress

$$\sigma = 0.2275 \frac{p_t^2 p}{h^2} \qquad (8\text{–}95)$$

The deflection

$$y = 0.0284 \frac{p_t^4 p}{E h^3} \qquad (8\text{–}96)$$

OPENINGS AND REINFORCEMENT

For flanged-in and unreinforced openings in cylindrical or conical shell or spherical shell or heads and ends

Refer to Figs. 8–14 and 8–15. Holes cut in domed ends shall be circular, elliptical, or oblong. The radius r of flanged-in openings (Fig. 8–14) shall not be less than 25 mm. Flanged-in and other openings shall be arranged so that the distance from the edge of the end is not less than that shown in Fig. 8–14. In all cases the projected width of the ligament between any

Particular	Formula

two adjacent openings shall be at least equal to the diameter of the smaller openings as shown in Fig. 8–15.

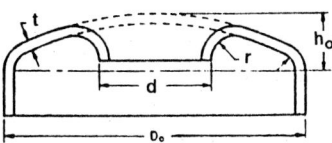

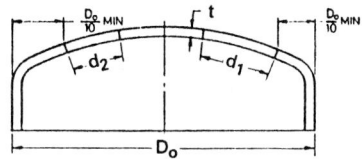

FIGURE 8–14 Flanged-in unreinforced opening.

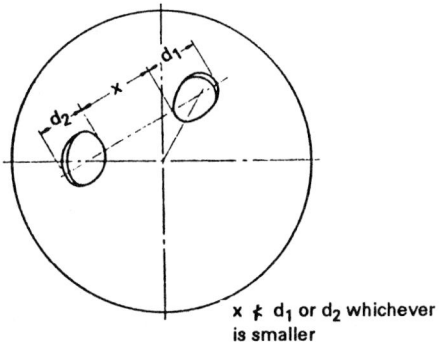

x ≮ d₁ or d₂ whichever is smaller

FIGURE 8–15 Unreinforced opening.

The distance between openings spaced apart, L_o, in a cylindrical or conical or spherical shell as per Indian Standards

$$L_o \nleq L = \frac{d(h_a/h_r)}{(h_a/h_r - 0.95)} \qquad (8\text{–}97)$$

Refer to Tables 8–16 and 8–17 for flange bolting.

For size of openings in cylinders or conical shells or spherical shells subject a maximum of 200 mm (8 in) which do not require reinforcement

Refer to Fig. 8–16.

where

h = distance between centers of any two adjacent openings, mm (in)

d = diameter of largest opening

  = mean value of the major and minor axes in case of elliptical or obround openings, mm (in)

h_a = actual thickness of vessel before corrosion allowance is provided, mm (in)

h_r = required thickness of vessel putting $\eta = 1.0$ before corrosion allowance is added, mm (in)

TABLE 8-8
Allowable stresses σ_{sa} for various ferrous and nonferrous materials

Values are given as MPa (kpsi). The first three data columns are Mechanical properties; the remaining columns are Allowable stress, σ_{sa}, at design temperature, K (°C).

Materials with grade or designation and product	Tensile strength R_m min, σ_{ar} min MPa (kpsi)	A	Yield stress σ_{sy} min MPa (kpsi)	323 K (50 °C)	373 K (100 °C)	423 K (150 °C)	473 K (200 °C)	≤523 K (250 °C)	≤573 K (300 °C)	≤623 K (350 °C)	≤648 K (375 °C)	≤673 K (400 °C)	≤698 K (425 °C)	≤723 K (450 °C)	≤748 K (475 °C)	≤773 K (500 °C)	≤798 K (525 °C)	≤823 K (550 °C)	≤848 K (575 °C)
Plates																			
Carbon and Low-Alloy Steel in Tension																			
1	363 (52.6)	26	0.55 R_{20}[a]	121 (17.5)	121 (17.5)	113 (16.4)	102 (14.8)	93 (13.5)	85 (12.3)	77 (11.2)	74 (10.7)	71 (10.3)	58 (8.4)	42 (6.1)	35 (5.1)				
2A	412 (59.8)	25	0.50 R_{20}	137 (19.8)	126 (18.3)	117 (17.0)	106 (15.4)	96 (13.9)	86 (12.5)	79 (11.5)	76 (11.0)	73 (10.6)	58 (8.4)	42 (6.1)	35 (5.1)				
2B	510 (74.0)	20	0.50 R_{20}	170 (24.7)	156 (22.6)	144 (20.8)	130 (18.8)	119 (17.3)	109 (15.8)	98 (14.2)	93 (13.5)	81 (11.8)	58 (8.4)	42 (6.1)	35 (5.1)				
20 Mo 6	471 (68.3)	20	275 (39.9)	157 (22.8)	157 (22.8)	157 (22.8)	151 (21.9)	140 (20.3)	129 (18.7)	121 (17.6)	108 (15.7)	113 (16.4)	110 (16.0)	106 (15.4)	95 (13.8)	55 (8.0)	36 (5.2)		
20 C 15	510 (74.0)	20	294 (42.6)	170 (24.7)	170 (24.7)	167 (24.2)	151 (21.9)	137 (19.9)	126 (18.3)	114 (16.5)	108 (15.7)	81 (11.8)							
15 Cr 4 Mo 6	490 (71.1)	20	294 (42.6)	163 (23.6)	163 (23.6)	163 (23.6)	163 (23.6)	157 (22.8)	149 (21.6)	141 (20.5)	135 (19.6)	131 (19.0)	128 (18.6)	124 (18.0)	114 (16.5)	84 (12.2)	57 (8.3)	34 (5.0)	
15 C 8	412 (59.8)	25	226 (32.8)	137 (19.9)	137 (19.9)	127 (18.4)	116 (16.8)	105 (15.2)	96 (13.9)	87 (12.6)	82 (11.9)	79 (11.5)							
Forgings																			
20 Mo 6	471 (68.3)	20	275 (39.9)	157 (22.8)	157 (22.8)	157 (22.8)	150 (21.8)	140 (20.3)	130 (18.9)	121 (17.6)	107 (15.5)	113 (16.4)	110 (16.0)	106 (15.4)	76 (11.0)	55 (8.0)	36 (5.2)		
15 Cr 4 Mo 6	490 (71.1)	20	294 (42.6)	163 (23.6)	163 (23.6)	163 (23.6)	163 (23.6)	157 (22.8)	149 (21.6)	141 (20.5)	135 (19.6)	131 (19.0)	128 (18.6)	124 (18.0)	114 (16.5)	84 (12.2)	57 (8.3)	34 (5.0)	
10 Cr 9 Mo 10	490 (71.1)	20	314 (45.5)	163 (23.6)	163 (23.6)	163 (23.6)	163 (23.6)	176 (25.5)	170 (24.6)	161 (23.4)	158 (22.9)	155 (22.5)	150 (21.8)	146 (21.2)	125 (18.1)	94 (13.6)	69 (10.0)	48 (7.0)	31 (4.5)
Tubes and pipes																			
1% Cr ½% Mo[+]	432 (62.7)	20	235 (34.1)	143 (20.7)	143 (20.7)	139 (20.2)	133 (19.3)	127 (18.4)	119 (17.3)	113 (16.4)	109 (15.8)	105 (15.2)	102 (14.8)	98 (14.2)	95 (13.8)	84 (12.2)	57 (8.3)	34 (5.0)	
20 Mo 6	451 (65.4)		245 (35.5)	150 (21.8)	150 (21.8)	143 (20.7)	133 (19.3)	127 (18.4)	115 (16.7)	108 (15.7)	104 (15.1)	101 (14.7)	98 (14.2)	94 (13.6)	76 (11.0)	55 (8.0)	36 (5.2)		
Fe 170	310 (45.0)		173 (25.1)	103 (15.0)	103 (15.0)	97 (14.0)	88 (12.8)	80 (11.6)	74 (10.7)	66 (9.6)	63 (9.1)	61 (8.8)	58 (8.4)	42 (6.1)	35 (5.1)				
Fe 240	414 (60.0)		241 (35.0)	137 (19.9)	137 (19.9)	136 (19.7)	124 (18.0)	113 (16.4)	103 (15.0)	93 (13.5)	88 (12.8)	81 (11.8)	58 (8.4)	42 (6.1)	35 (5.1)				
Fe 290	414 (60.0)		290 (42.1)	137 (19.9)	137 (19.9)	137 (19.9)	137 (19.9)	135 (19.6)	124 (18.0)	113 (16.4)	106 (15.4)	81 (11.8)	58 (8.4)	42 (6.1)	35 (5.1)				
Castings																			
Grade 1	539 (78.2)	17	343 (49.8)	134 (19.4)	134 (19.4)	139 (20.2)	132 (19.1)	120 (17.4)	110 (16.0)	99 (14.4)	94 (13.6)	61 (8.8)	43 (6.2)	31 (4.5)	26 (3.8)	41 (5.9)	27 (3.9)		
Grade 2	461 (66.9)	17	245 (35.5)	115 (16.7)	114 (16.5)	108 (15.7)	100 (14.5)	86 (12.5)	80 (11.6)	79 (11.5)	79 (11.5)	76 (11.0)	74 (10.7)	71 (10.3)	57 (8.3)	41 (5.9)	27 (3.9)		
Grade 3	510 (74.0)	15	304 (44.1)	127 (18.4)	127 (18.4)	127 (18.4)	125 (18.1)	117 (17.0)	100 (14.5)	98 (14.5)	98 (14.5)	91 (13.2)	89 (12.9)	82 (11.9)	57 (8.3)				
Grade 4	481 (69.8)	17	275 (39.9)	120 (17.4)	120 (17.4)	120 (17.4)	117 (17.0)	110 (17.3)	104 (15.1)	99 (14.4)	95 (13.8)	91 (13.2)	89 (12.9)	86 (12.5)	83 (12.0)	64 (9.3)	43 (6.2)	25 (3.6)	
Grade 5	510 (74.0)	17	304 (44.1)	127 (18.4)	127 (18.4)	127 (18.4)	128 (18.6)	127 (18.4)	117 (17.0)	117 (17.0)	115 (16.7)	106 (16.9)	109 (15.8)	106 (15.4)	94 (13.6)	71 (10.3)	52 (7.5)	36 (5.2)	33 (4.8)
Grade 6	618 (89.6)	15	422 (61.2)	154 (22.3)	154 (22.3)	154 (22.3)	154 (22.3)	169 (24.5)	160 (23.2)	152 (22.0)	146 (21.2)	141 (20.5)	137 (19.9)	132 (19.1)	106 (15.4)?	48 (7.0)	34 (4.9)	25 (3.6)	16 (2.3)
Sections, plates, and bars																			
Grade 1	363 (52.6)	26	0.55 R_{20}	121 (17.5)	111 (16.1)	102 (14.8)	93 (13.5)	84 (12.2)	76 (11.0)	70 (10.2)	67 (9.7)	64 (9.3)	58 (8.4)	42 (6.1)	35 (5.0)				
Grade 2	412 (59.8)	25	0.55 R_{20}	137 (19.9)	126 (18.3)	117 (17.0)	106 (15.4)	96 (13.9)	88 (12.8)	79 (11.5)	76 (11.0)	73 (10.6)	58 (8.4)	42 (6.1)	35 (5.0)				
Grade 3	432 (62.7)	22	0.55 R_{20}	143 (20.7)	131 (19.0)	122 (17.7)	111 (16.1)	100 (14.5)	91 (13.2)	83 (12.0)	79 (11.5)	76 (11.0)	58 (8.4)	42 (6.1)	35 (5.0)				
Grade 4	461 (66.9)	22	0.55 R_{20}	153 (22.2)	141 (20.5)	130 (18.9)	119 (17.3)	108 (15.7)	94 (13.6)	89 (12.9)	89 (12.9)	81 (11.8)	58 (8.4)	42 (6.1)	35 (5.0)				
Grade 5	491 (71.2)	21	0.55 R_{20}	163 (23.6)	150 (21.8)	138 (20.0)	126 (18.3)	119 (17.3)	109 (15.8)	98 (14.2)	93 (13.5)	81 (11.8)	58 (8.4)	42 (6.1)	35 (5.0)				
Grade A-N	432 (62.7)	23	235 (34.0)	143 (20.7)	143 (20.7)	133 (19.2)	121 (17.5)	96 (13.9)	88 (12.8)	79 (11.5)									
Grade B-N	490 (71.1)	20	280 (40.6)	163 (13.6)	163 (23.6)	158 (23.0)	143 (20.7)	115 (16.7)	105 (15.2)	94 (13.6)									
Plates, bars, forgings, seamless tubes																			
High-Alloy Steels in Tension																			
X04 Cr 19 Ni 9	540 (78)	28	235 (34)	157 (22.8)		139 (20.2)		122 (17.7)	104 (15.0)	104 (15.0)	97 (14)	92 (13.3)	85 (12.3)	79 (11.5)	35 (5.0)				
X04 Cr 19 Ni 9 Ti 20	540 (78)	28	235 (34)	157 (22.8)		140 (20.3)		124 (18.0)	106 (15.4)	106 (15.4)	104 (15)	104 (15)	104 (15.1)	101 (14.6)	35 (5.0)				
X04 Cr 19 Ni 9 No 40	540 (78)	28	235 (34)	157 (22.8)		140 (20.3)		124 (18.0)	106 (15.4)	106 (15.4)	104 (15)	104 (15)	104 (15.1)	104 (15.1)	35 (5.0)				
X05 Cr 18 Ni 11 Mo 3	540 (78)	28	235 (34)	157 (22.8)		142 (20.6)		127 (18.4)	113 (16.4)	113 (16.4)	110 (16)	110 (16)	110 (16.0)	109 (15.8)	35 (5.0)				
X05 Cr 19 Ni 9 Mo 3	540 (78)	28	235 (34)	157 (22.8)		142 (20.6)		127 (18.4)	113 (16.4)	113 (16.4)	110 (16)	110 (16)	110 (16.0)	109 (15.8)	35 (5.0)				
Ti 20				137 (19.9)		127 (18.4)		117 (18.4)		94 (13.6)									
Castings																			
Grades 7, 8	461 (66.9)	21	205 (30)	137 (19.9)	127 (18.4)	127 (18.4)	117 (18.4)		106 (15.4)	106 (15.4)	104 (15.1)	104 (15.1)	104 (15.1)	104 (15.1)					
Plates																			
Aluminum and Aluminum Alloys in Tensions																			
P1B-M	64 (9.3)	30		12 (1.7)	13 (1.9)	11 (1.6)	10 (1.5)	9 (1.3)	8 (1.2)	7 (1.1)									
NP4-M	186 (27.0)	12		43 (6.2)	42 (6.1)	42 (6.1)	41 (5.9)	37 (5.4)	32 (4.6)	24 (3.5)									
Sheet, strip																			
S1B-½H	98 (14.2)	8		21 (3.0)	20 (2.9)	19 (2.8)	18 (2.6)	16 (2.3)	14 (2.0)	11 (1.6)									
NS4-½H	196 (28.4)	8		54 (7.8)	53 (7.7)	52 (7.5)	49 (7.1)	44 (6.4)	37 (5.4)	24 (3.5)									

TABLE 8-8
Allowable stresses σ_{sa} for various ferrous and nonferrous materials (*Cont.*)

Materials with grade or designation and product	Tensile strength, σ_u min, R_{20}, MPa (kpsi)	Yield stress, σ_{sy} min, MPa (kpsi)	E_{20}	323 K (50°C) MPa (kpsi)	373 K (100°C) MPa (kpsi)	423 K (150°C) MPa (kpsi)	473 K (200°C) MPa (kpsi)	≤523 K (250°C) MPa (kpsi)	≤573 K (300°C) MPa (kpsi)	≤623 K (350°C) MPa (kpsi)	≤648 K (375°C) MPa (kpsi)	≤673 K (400°C) MPa (kpsi)	≤698 K (425°C) MPa (kpsi)
Bars, rods, and sections													
NE5–M	216 (31.3)		88+	54 (7.8)									
NE6–M	265 (38.4)		18+	66 (9.6)									
NE8-0	265 (38.4)		16	69 (10.0)									
HE30—W	186 (27.0)	108 (15.7)	18	51 (7.4)	49 (7.1)	47 (6.8)	46 (6.7)	44 (6.4)	39 (6.7)	28 (4.1)			
HE30—WP	294 (42.6)	245 (35.5)	10	71 (10.3)	70 (10.2)	67 (9.7)	65 (9.4)	54 (7.8)	43 (6.2)	30 (4.4)			
Drawn tubes													
HT30—W	216 (31.3)	108 (15.7)	16	51 (7.4)	50 (7.3)	48 (7.0)	46 (6.7)	44 (6.38)	39 (5.7)	28 (4.1)			
HT30—WP	309 (44.8)	245 (35.5)	7	72 (10.4)	70 (10.2)	67 (9.7)	65 (9.4)	55 (8.0)	43 (6.2)	30 (4.4)			
				Copper and Copper Alloys									
Plate sheet and strips													
Cu Zn 30	275 (40.0)		45	69 (10.0)	69 (10.0)	69 (10.0)	69 (10.0)	68 (9.9)	56 (8.1)	38 (5.5)			
Cu Zn 40	275 (40.0)		30	86 (12.5)	85 (12.3)	81 (11.7)	77 (11.2)	71 (10.3)	53 (7.7)	19 (2.8)			
Bars and rods													
	392 (56.9)		22	69 (10.0)	69 (10.0)	69 (10.0)	69 (10.0)	68 (9.9)	56 (8.1)	38 (5.5)			
Tubes													
Alloy 1	284 (41.2)			69 (10.0)	69 (10.0)	69 (10.0)	69 (10.0)	68 (9.9)	56 (8.1)	38 (5.5)			
Alloy 2	284 (41.2)			86 (12.5)	85 (12.3)	81 (11.7)	77 (11.2)	71 (10.3)	53 (7.7)	19 (2.8)			

Mechanical Properties — *Allowable stress, σ_{sa} at design temperature, K(°C)*

[a] These values have been used on a quality factor of 0.75.

[b] $0.55R_{20} = 0.55 \times 363 = 199.7$ MPa (29.0 kpsi).

Notes: + The elongation values are based on 50.8-mm test piece; a^* area of cross-section; † tube normalized and tempered.

Sources: K. Lingaiah and B. R. Narayana Iyengar, *Machine Design Data Handbook.* Engineering College Cooperative Society, Bangalore, India, 1962; K. Lingaiah and B. R. Narayana Iyengar, *Machine Design Data Handbook,* Vol. I (SI and Customary Metric Units), Suma Publishers, Bangalore, India, 1983; and K. Lingaiah, *Machine Design Data Handbook,* Vol. II (SI and Customary Metric Units), Suma Publishers, Bangalore, India, 1986.

TABLE 8-9
Maximum allowable stress values, σ_{sa}, in tension for carbon and low-alloy steel

Specification no. Grade	Nominal composition	Specified minimum yield strength, σ_{sy} MPa	kpsi	Specified minimum tensile strength, σ_{st} MPa	kpsi	−19 to 345 (−30 to 650) MPa	kpsi	370 (700) MPa	kpsi	400 (750) MPa	kpsi	427 (800) MPa	kpsi	455 (850) MPa	kpsi	482 (900) MPa	kpsi	510 (950) MPa	kpsi	538 (1000) MPa	kpsi	566 (1050) MPa	kpsi	593 (1100) MPa	kpsi	620 (1150) MPa	kpsi	650 (1200) MPa	kpsi	Specification no.
Carbon Steel																														
Plates and sheets																														
SA 285[b,c,d] A	C	165	24	310	45	78	11.3	76	11.0	71	10.3	62	9.0	54	7.8	45	6.5	(Fig. 8-5)												SA 285[b,c,d]
SA 299[c]	C-Mn-Si	276/290	40/42	517	75	130	18.8	122	17.7	108	15.7	87	12.6	66	9.6	45	6.5	31	4.5	17.0	2.5									SA 299[c]
SA 414[b,c] F	C-Mn	290	42	483	70	121	17.5	114	16.6	101	14.7	83	12.0	63	9.2	45	6.5													SA 414[b,c]
SA 515[c] 60	C-Si	220	32	413	60	103	15.0	99	14.4	90	13.0	75	10.8	60	8.7	45	6.5	31	4.5	17.0	2.5									SA 515[c]
70	C-Si	262	38	483	70	121	17.5	115	16.6	102	14.8	83	12.0	64	9.3	45	6.5	31	4.5	17.0	2.5									
SA 516[c] 65	C-Mn-Si	241	35	448	65	115	16.3	107	15.5	96	13.9	79	11.4	62	9.0	45	6.5	31	4.5	17.0	2.5	(Fig. 8-5)								SA 516[c]
70	C-Mn-Si	262	38	483	70	121	17.5	114	16.6	102	14.8	83	12.0	64	9.3	45	6.5	31	4.5	17.0	2.5	(Fig. 8-5)								
SA 537[c] Cl-1 up to 62.5 mm (2.5 in.) incl.	C-Mn-Si	345	50	483	70	121	17.5	115	16.6	102	14.8	83	12.0	64	9.3	45	6.5	31	4.5	17.0	2.5	(Fig. 8-5)								SA 537[c]
SA 620 1 and 2	C-Mn	138	20	276	40	69	10.0	(Fig. 8-5)																					SA 620
SA 812 80	C-Mn-Si-Cb-V	552	80	689	100	147	21.3	(Fig. 8-5)																					SA 812
Carbon steel forgings, castings, and bars																														
SA 36[b,c] bars and shapes	C-Mn-Si	248	36	400	58	100	14.5	(Fig. 8-5)				72	10.5	57	8.5	45	6.5	31	4.5	17.0	2.5									SA 36[b,c] bars and shapes
SA 216[c] cast WCA	C-Si	207	30	413	60	103	15.0	99	14.4	90	13.0	75	10.8	60	8.7	45	6.5	31	4.5	17.0	2.5									SA 216[c] cast
WCC	C-Mn-Si	276	40	483	70	121	17.5	115	16.6	102	14.8	83	12.0	64	9.3	45	6.5	31	4.5	17.0	2.5									
SA 350[f] forge LF1	C-Mn-Si	207	30	413	60	103	15.0	99	14.4	90	13.0	75	10.8	60	8.7	35	5.0	21	3.0	10.0	1.5									SA 350[f] forge
LF2	C-Mn-Si	248	36	483	70	121	17.5	115	16.6	102	14.8	83	12.0	64	9.3	35	5.0	21	3.0	10.0	1.5									
SA 675[b,c] bar 45	C	155	22.5	310	45	78	11.3	76	11.0	71	10.3	62	9.0	54	7.8	45	6.5	31	4.5	17.0	2.5									SA 675[b,c] bar
60	C	190	27.5	379	55	95	13.8	92	13.3	83	12.1	70	10.2	60	8.4	45	6.5	31	4.5	17.0	2.5									
70	C	207	30.0	483	70	121	17.5	115	16.6	102	14.8	83	12.0	64	9.3	45	6.5	31	4.5	17.0	2.5									
Low-Alloy Steel																														
Plate																														
SA 202 A	0.5 Cr-1.25 Mn-Si	310	45	517	75	130	18.8	122	17.7	108	15.7	83	12.0	54	7.8	35	5.0	21	3.0	10.0	1.5									SA 202
B	0.5 Cr-1.25 Mn-Si	324	47	586	85	147	21.3	136	19.8	122	17.7	83	12.0	54	7.8	35	5.0	21	3.0	10.0	1.5									
SA 203 F	3.5 Ni, ≤ 50mm (2 in.)	379	55	552	80	138	20.0																						SA 203
SA 204[g] A	0.5 Mo	255	37	448	65	112	16.3	112	16.3	112	16.2	109	15.8	105	15.3	94	13.7	56	8.2	33.0	4.8									SA 204[g]
C	0.5 Mo	296	43	517	75	130	18.8	130	18.8	121	17.5	130	18.8	118	17.1	94	13.7	56	8.2	33.0	4.8									
SA 225[h] C	Mn-0.5 Ni-V	483	70	724	105	181	26.3	181	26.3	130	18.8	130	18.8	118	17.1	94	13.7	56	8.2	33.0	4.8									SA 225[h]
SA 302 C	Mn-0.5 Mo-0.5 Ni, 0.5 Ni	345	50	552	80	138	20.0	138	20.0	135	19.6	130	18.8	123	17.9	94	13.7	56	8.2	33.0	4.8									SA 302
SA 387 11 Cl2	1.25 Cr-0.5 Mn-Si	310	45	517	75	130	18.8	130	18.8	130	18.8	130	18.8	126	18.3	110	15.9	76	11.0	48	6.9	32	4.6	19	2.8	15	2.1	8	1.2	SA 387
5 Cl1	5 Cr-0.5 Mo	207	30	413	60			95	13.7	91	13.2	88	12.8	83	12.1	75	10.9	55	8.0	40.0	5.8	29	4.2	20	2.9	14	2.0	9	1.3	
Forgings, castings, and bars																														
SA 182 forge F12	1 Cr-0.5 Mo	276	40	483	70	121	17.5	121	17.5	121	17.5	121	17.5	118	17.1	110	15.9	76	11.0	45	6.6	30	4.3	18	2.6	10	1.4	7	1.0	SA 182 forge
F11b	1.25 Cr-0.5 Mo-Si	276	40	483	70	121	17.5	121	17.5	121	17.5	121	17.5	118	17.1	110	15.9	76	11.0	48	6.9	32	4.6	19	2.8	15	2.1	8	1.2	
SA 217 cast WC1	C-0.5 Mo	241	35	448	65	112	16.3	112	16.3	112	16.2	109	15.8	106	15.3	90	13.0	56	8.2	33.0	4.8	35	5.0	23	3.3	15	2.2	10	1.5	SA 217 cast
WC4	1 Ni-0.5 Cr-0.5 Mo	276	40	483	70	121	17.5	121	17.5	121	17.5	121	17.5	118	17.1	103	15.0	63	9.2		5.9	32	4.6	19	2.8					
SA 336 forge C12	9 Cr-1 Mo	413	60	620	90			141	20.5	136	19.8	132	19.1	125	18.2	114	16.5	76	11.0	51	7.4	35	5.0	23	3.3	15	2.2	10	1.5	SA 336 forge
F11	1.25 Cr-0.5 Mo-Si	276	40	483	70	121	17.5	121	17.5	121	17.5	121	17.5	118	17.1	110	15.9	76	11.0	48	6.9	32	4.6	19	2.8					
SA 487 cast 4N	0.5 Ni-0.5 Cr-0.25 Mo-V	413	60	620	90																									SA 487 cast
SA 541 forge 3	0.5 Ni-0.5 Mo V	345	50	552	80	138	20.0	138	20.0	138	20.0	132	19.1																	SA 541 forge
SA 739 bar B11	1.25 Cr-0.5 Mo	310	45	483	70	121	17.5	121	17.5	121	17.5	121	17.5	118	17.1	110	15.9	76	11.0	48	6.9	32	4.6	19	2.8	15	2.1	8	1.2	SA 739 bar
B22	2.25 Cr-1 Mo	310	45	517	75	121	17.5	121	17.5	119	17.2	116	16.9	113	16.4	109	15.8	76	11.0	52	7.6	40	5.8	30	4.4	17	2.5	9	1.3	

8.42

Notes:

[a]These stress values are one-fourth the specified minimum strength multiplied by a quality factor of 0.92, except for SA 283, grade D and SA-36.

[b]For service temperature above 455°C (850°F), it is recommended that killed steels containing not less than 10% residual silicon be used.

[c]Upon prolonged exposure to temperature above 426°C (800°F) the carbide phase of carbon steel may be converted to graphite.

[d]The material shall not be used in thickness above 50 mm (2 in).

[e]The material shall not be used in thickness above 62 mm (2.5 in).

[f]Only killed steel shall be used above 455°C (850°F).

[g]Upon prolonged exposure to temperature above 468°C (875°F), the carbide phase of carbon molybdenum steel may be converted to graphite.

[h]The maximum nominal plate thickness shall not exceed 14.75 mm (0.58 in).

[i]These stress values apply to normalized and drawn materials only.

[j]For other conditions and specifications, the reader is referred to the general notes given for Table UCS-23 of *ASME Boiler and Pressure Vessel Code*, Section VIII, Division 1, July 1, 1986.

Source: The American Society of Mechanical Engineers, *ASME Boiler and Pressure Vessel Code*, Section VIII, Division 1, July 1, 1986.

TABLE 8–10
Maximum allowable stress values, σ_{sa}, in tension for nonferrous metals

Specification no.	Alloy designation	Temper condition	Nominal composition	UNS no.	Size or thickness mm (in)	Specified minimum tensile strength, σ_{st} MPa	kpsi	Specified minimum yield strength, σ_{sy} MPa	kpsi	38 (100) MPa	kpsi	65 (150) MPa	kpsi	93 (200) MPa	kpsi
					Aluminum and Aluminum Alloys										
Sheet and plate															
SB 209	1100[d]	-H 12			1.275–50.0 (0.051–2.000)	96	14	76	11	24	3.5	24	3.5	24	3.5
		-H 14			0.225–25.0 (0.009–1.000)	110	16	96	14	26	4.0	26	4.0	26	4.0
SB 209	3003[d]	-H 14			0.15–25.00 (0.006–1.000)	138	20	117	17	35	5.0	35	5.0	35	5.0
		-H 112			6.25–12.475 (0.250–0.499)	117	17	69	10	30	4.3	30	4.3	30	4.3
SB 209	3004[d]	-H 32			1.275–50.00 (0.051–2.000)	193	28	145	21	48	7.0	48	7.0	48	7.0
SB 209	5052[d]	-H 34			1.275–25.00 (0.051–1.000)	234	34	179	26	58	8.5	58	8.5	58	8.5
SB 209	5454[d]	-O			1.275–75.00 (0.051–3.000)	214	31	83	12	54	7.8	54	7.8	54	7.8
		-H 32			1.275–50.00 (0.051–2.000)	248	36	179	26	62	9.0	62	9.0	62	9.0
SB 209	6061[e,f]	T 4			1.275–6.225 (0.051–0.249)	207	30	110	16	52	7.5	52	7.5	52	7.5
		T 6 Wld			1.275–6.225 (0.051–0.249)	165	24			41	6.0	41	6.0	41	6.0
Rods, bars, and shapes															
SB 221	2024[e]	-T 4			3.125–12.475 (0.125–0.499)	427	62	290	42	107	15.5	107	15.5	107	15.5
					162.54–200.00 (6.501–8.000)	400	58	262	38	100	14.5	100	14.5	100	14.5
SB 221	5086[e]	-H 112			≤125.00 (5.000)	241	35	96	14	61	8.8	61	8.8		
SB 221	3003[d]	-H 112			All	96	14	35	5	23	3.4	23	3.4	23	3.4
SB 221	5456[d]	-O			≤125.00 (5.00)	282	41	131	19	71	10.3	71	10.3		
		-H 111			≤125.00 (5.00)	290	42	179	26	72	10.5	72	10.5		
SB 308	6061[e]	-T 6				262	38	241	35	65	9.5	63	9.2	62	9.0
		-T 6 Wld				165	24			41	6.0	40	5.9	39	5.7
Die and hand forgings															
SB 247	2014 Die[e]	-T 4			≤100.0 (4.000)	379	55.0	207	30	95	13.8	95	13.8	92	13.3
		-T 6			≤50.0 (2.00)	441	64.0	379	55	110	16.0	110	16.0	109	15.9
					50.0–100.00 (2.001–4.000)	434	63.0	372	54			15.8		15.8	
SB 247	6061 Die[e]	-T 6			≤100.0 (4.00)	262	38.0	241	35	65	9.5	65	9.5	65	9.5
SB 247	6061 Hand[e]	-T 6			≤100.0 (4.00)	255	37.0	228	33	64	9.3	64	9.3	64	9.3
					100.025–200.0 (4.001–8.000)	241	35.0	220	32	61	8.8	61	8.8	61	8.8
Castings															
SB 26	SG 70 A(356)[e]	-T 6				207	30.0	138	20	52	7.5	52	7.5	52	7.5
		-T 71				172	25.0	124	18	43	6.3	43	6.3	43	6.3
SB 108	204.0	-T 4			≤50.0 (2.000)	331	48.0	200	29	65	9.5	52	7.5		
					Copper and Copper Alloys										
Sheet and plates															
SB 96[g]	655	Annealed	Cu-Si alloy		≤50 mm (2 in)	345	50.0	124	18	83	12.0	83	12.0	82	11.9
SB 169	610	Annealed	Al-bronze		≤50 mm (2 in)	345	50.0	138	20	86	12.5	86	12.5		
	614	Annealed	Al-bronze		≤12.5 mm ($^1/_2$ in)	496	72.0	220	32	124	18.0	124	18.0	124	18.0
SB 171	C 36500, C 36600 Annealed		Lead-Muntz metal		≤50 mm (2 in)	345	50.0	138	20	86	12.5	86	12.5	86	12.5
	C 36700, C 36800 Annealed				> 50 mm (2 in) –87.5 mm (3.5 in)	310	45.0	103	15	69	10.0	69	10.0	69	10.0
SB 171	443,444,445	Annealed	Admiralty		≤100 mm (4 in)	310	45.0	103	15	69	10.0	69	10.0	69	10.0
SB 171	C 46400, C 46500 Annealed		Naval brass		> 75(3)–125(5)	345	50.0	138	20	86	12.5	86	12.5	86	12.5
	C 46600, C 46700 Annealed					345	50.0	124	18	83	12.0	83	12.0	83	12.0
SB 171	715	Annealed	Cu-Ni 70/30		≤62.5 (2.5)	345	50.0	138	20	86	12.5	78	11.3	72	10.5
		Annealed			≥62.5 (2.5) –125(5) incl	310	45.0	124	18	78	11.3	70	10.1	65	9.4
SB 402	706	Annealed	Cu-Ni 90/10		≤62.5 (2.5)	276	40.0	103	15	70	10.1	67	9.7	66	9.5
Die forgings (hot pressed)															
SB 283[h]	C 37700[h]	As forged	Forging brass		≤37.5 (1.5)	345	50.0	124	18	83	12.0	78	11.3	75	10.9
					> 37.5 (1.5)	317	46.0	103	15	69	10.0	66	9.5	63	9.1
	C 64200	As forged	Forgings, Al-Si bronze		≤37.5 (1.5)	482	70.0	172	25	115	16.7	100	14.5	97	14.0
					> 37.5 (1.5)	469	68.0	159	23	105	15.3	93	13.5	90	13.0
Rods and bars															
SB 98[g]	655,661[g]	Soft[h]	Cu-Si			358	52	103	15	69	10.0	69	10.0	69	10.0
		Half hard[i]				482	70	262	38	121	17.5	121	17.5	121	17.5
SB 98	651[j]	Soft	Cu-Si			276	40	83	12	55	8.0	55	8.0	55	8.0
		Half hard				379	55	138	20	92	13.3	92	13.3	92	13.3

TABLE 8–10
Maximum allowable stress values, σ_{sa}, in tension for nonferrous metals (*Cont.*)

Maximum allowable stress, σ_{sa}, for metal temperature, °C (°F), not exceeding																				
120 (250)		150 (300)		176 (350)		205 (400)		232 (450)		260 (500)		288 (550)		315 (600)		343 (650)		370 (700)		Spec. Number
MPa	kpsi	MPa	kpsi	MPa	kpsi	MPa	kpsi	MPa	kpsi	MPa	kpsi	MPa	kpsi	MPa	kpsi	MPa	kpsi	MPa	kpsi	
22	3.2	19	2.8	14	2.0	8	1.2													SB 209
25	3.7	19	2.8	14	2.0	8	1.2													SB 209
34	4.9	30	4.3	21	3.0	16	2.4	(Fig. 8–9)												SB 209
28	4.0	25	3.6	21	3.0	16	2.4	(Fig. 8–8)												SB 209
48	7.0	40	5.8	26	3.8	16	2.4													SB 209
58	8.5	43	6.2	32	4.1	16	2.4													SB 209
51	7.4	38	5.5	28	4.1	21	3.0													SB 209
52	7.5	38	5.5	28	4.1	21	3.0													
51	7.4	48	6.9	43	6.3	31	4.5													SB 209
41	5.9	38	5.5	32	4.6	24	3.5													
95	13.7	72	10.4	49	6.5	31	4.5													SB 221
88	12.8	69	9.7	42	6.1	29	4.2													
21	3.0	16	2.4	12	1.8	10	1.4													SB 221
																				SB 221
																				SB 221
59	8.5	50	7.2	39	5.6	28	4.0													SB 308
37	5.4	35	5.0	29	4.2	22	3.2													
86	12.5	79	11.5	47	6.8	27	3.9													SB 247
102	14.8	79	11.5	47	6.8	27	3.9													
102	14.8	79	11.5	47	6.8	27	3.9													
63	9.1	54	7.9	43	6.3	31	4.5													SB 247
61	8.8	53	7.7	43	6.3	31	4.5													SB 247
58	8.4	51	7.4	42	6.1	31	4.5													
43	6.3																			SB 26
42	6.1	37	5.4	28	4.1	16	2.4													SB 108
81	11.7	69	10.0	38	5.0															SB 96[g]
																				SB 169
124	18.0	124	18.0	124	18.0	121	17.5	117	17.0	114	16.5									SB 171
86	12.5	85	12.3	75	10.8	36	5.3													
69	10.0	69	10.0	69	10.0	36	5.3													SB 171
69	10.0	69	10.0	68	9.8	24	3.5	14	2.0											SB 171
86	12.5	86	12.5	43	6.3	17	2.5													
83	12.0	83	12.0	43	6.3	17	2.5													
72	10.4	72	10.4	72	10.4	72	10.4	72	10.4	72	10.4	72	10.4	72	10.4	72	10.4	72	10.4	SB 171
64	9.3	64	9.3	64	9.3	64	9.3	64	9.3	64	9.3	64	9.3	64	9.3	64	9.3	64	9.3	SB 402
64	9.3	62	9.0	60	8.7	59	8.5	57	8.2	55	8.0	48	7.0	41	6.0					SB 283[h]
93	13.5	93	13.5	90	13.0	76	11.0	52	7.5	36	5.2									
86	12.5	86	12.5	83	12.0	76	11.0	52	7.5	36	5.2									
69	10.0	69	10.0	35	5.0															SB 98[g]
121	17.5	121	17.5	69	10.0															
55	8.0	48	7.0	35	5.0															SB 98
88	12.8	69	10.0	55	8.0															

TABLE 8–10
Maximum allowable stress values, σ_{sa}, in tension for nonferrous metals (*Cont.*)

Specification no.	Alloy designation	Temper condition	Nominal composition	UNS no.	Size or thickness mm (in)	Specified minimum tensile strength, σ_{st} MPa	kpsi	Specified minimum yield strength, σ_{sy} MPa	kpsi	38 (100) MPa	kpsi	65 (150) MPa	kpsi	93 (200) MPa	kpsi
Castings															
SB 61	922	As Cast				234	34	110	16	59	8.5	59	8.5	59	8.5
SB 148	954	As Cast				517	75	207	30	13	18.8	130	18.8	129	18.7
SB 271	952	As Cast				448	65	172	25	108	15.7	108	15.7	103	14.9
SB 584	976	As Cast				276	40	117	17	52	7.5	52	7.2	48	7.0
Titanium and Titanium Alloys															
Sheet, strip, plate, bar, billet, and casting															
SB 265	Grade 1 (F1)		Sheets, strips, plate			241	35	172	25	61	8.8	59	8.1	50	7.3
SB 381	2 (F2)	Annealed	Forging			345	50	276	40	86	12.5	82	12.0	75	10.9
SB 348	3 (F3)	Annealed	(F stands for forging)			448	65	379	55	112	16.3	107	15.6	99	14.3
	12 (F12)		Bar, billet			482	70	345	50	121	17.5	121	17.5	113	16.4
SB 367[h]	Grade C-2		Casting[b]			345	50	276	40	86	12.5	81	11.7	74	10.7
Zirconium															
Flat-rolled products and bars															
SB 551	Grade R 60702		Hot-rolled products			358	52	207	30	90	13.0			76	11.0
SB 550	R 60705		Bars			552	80	379	55	138	20.0			114	16.6
Nickel and Nickel Alloys															
Plate, sheet, and strip										38°C (100°F)		93°C (200°F)		150°C (300°F)	
SB 127[j]	400	Annealed[j]	Ni-Cu	N04400		482	70	193	28	128	18.6	113	16.4	106	15.4
		Hot-rolled				517	75	276	40	129	18.7	129	18.7	129	18.7
SB 168	600	Annealed	Ni-Cr-Fe	N06600		552	80	241	35	138	20.0	138	20.0	138	20.0
SB 168[j]	600[j]	Hot-rolled[j]	Ni-Cr-Fe	N06600		586	85	241	35	146	21.2	146	21.2	146	21.2
SB 333[k]	B2	Sol. ann.[k]	Ni-Mo	N10665	All	758	110	352	51	190	27.5	190	27.5	190	27.5
SB 424[k]	825	Annealed	Ni-Fe-Cr-Mo-Cu	N08825		586	85	241	35	148	21.5	148	21.5	141	20.4
SB 435[k]	X	Annealed[k]	Ni-Cr-Mo-Fe	N06002	0.063 (1/16)–0.188 (3/16)	689	100	276	40	161	23.3	144	20.9	132	19.2
SB 435	X	Annealed	Ni-Cr-Mo-Fe	N06002	0.063 (1/16)–0.188 (3/16)[k]	689	100	276	40	161	23.3	161	23.3	116	23.3
SB 435[k]	X	Annealed	Ni-Cr-Mo-Fe	N06002	> 0.188 (3/16)	655	95	241	35	161	23.3	144	20.9	132	19.2
SB 443	625	Annealed	Ni-Cr-Mo-Cb	N06625	> 100 (4)	758	110	379	55	190	27.5	190	27.5	190	27.5
SB 463	20Cb	Annealed	Cr-Ni-Fe-Mo-Cu-Cb	N08020		552	80	241	35	138	20.0	138	20.0	136	19.8
SB 575[k]	C22	Sol. ann.[k]	Ni-Mo-Cr	N06022		689	100	310	45	172	25.0	172	25.0	171	24.8
SB 582	G	Sol. ann.	Ni-Cr-Fe-Mo-Cu	N06007	≤ 19.3 (3/4)	620	90	241	35	155	22.5	144	22.9	134	19.5
SB 582[k]	G	Sol. ann.[k]	Ni-Cr-Fe-Mo-Cu	N06007	> 19.3 (3/4)[k]	586	85	207	30	138	20.0	138	20.0	138	20.0
SB 709	28	Annealed	Ni-Fe-Cr-Mo-Cu Low C	N08028		503	73	213	31	125	18.2	125	18.2	117	17.0
Bars, rods, shapes, and forgings															
SB 164	400	Annealed	Ni-Cu	N04400	All sizes	482	70	172	25	114	16.6	101	14.6	94	13.6
SB 166[k]	600[k]	Annealed[k]	Ni-Cr-Fe	N06600		552	80	241	35	138	20.0	138	20.0	138	20.0
SB 166	600	Hot fin	Ni-Cr-Fe	N06600		586	85	241	35	146	21.2	146	21.2	146	21.2
SB 425[k]	825	Annealed[k]	Ni-Fe-Cr-Mo-Cu	N08825		586	85	241	35	146	21.2	146	21.2	141	20.4
SB 462[k]	20Cb	Annealed[k]	Cr-Ni-Fe-Mo-Cu-Cb	N08020		552	80	241	35	138	20.0	138	20.0	136	19.8
SB 511[l]	330		Ni-Fe-Cr-Si	N08330[l]		482	70	207	30	121	17.5	121	17.5	112	16.3
SB 564	625	Annealed	Ni-Cr-Mo-Cb	N06625		758	110	345	50	190	27.5	190	27.5	190	27.5
SB 574[k]	C-4	Sol. ann.	Ni-Mo-Cr	N06455[k]	All sizes	689	100	276	40	172	25.0	172	25.0	172	25.0
Castings															
SA 494[h]	B	Annealed[h]	Ni-Mo	N-12 WV		524	76	276	40	131	19.0	123	17.8	123	17.8
SA 494	C	Annealed	Ni-Mo-Cr	CW-12MW		496	72	276	40	124	18.0	118	17.1	112	16.2

[a]The stress values in this table may be interpolated to determine values for intermediate temperatures.

[b]Stress values in restricted shear shall be 0.8 times the values in this table.

[c]Stress values in bearing shall be 1.60 times the values in this table.

[d]For weld construction, stress values for o material shall be used.

[e]The stress values given for this material are not applicable when either welding or thermal cutting is employed.

[f]Allowable stress values shown are 90 percent those for the corresponding core material.

[g]Copper-silicon alloys are not always suitable when exposed to certain media and high temperature, particularly steam above 100°C (212°F).

[h]No welding is permitted.

[i]If welded, the allowable stress values for annealed condition shall be used.

[j]For plates only.

TABLE 8–10
Maximum allowable stress values, σ_{sa}, in tension for nonferrous metals (*Cont.*)

Maximum allowable stress, σ_{sa}, for metal temperature, °C (°F), not exceeding

120 (250)		150 (300)		176 (350)		205 (400)		232 (450)		260 (500)		288 (550)		315 (600)		343 (650)		370 (700)		Spec. Number
MPa	kpsi	MPa	kpsi	MPa	kpsi	MPa	kpsi	MPa	kpsi	MPa	kpsi	MPa	kpsi	MPa	kpsi	MPa	kpsi	MPa	kpsi	
59	8.5	59	8.5	59	8.5	57	8.3	53	7.7	50	7.2	34	5.0							SB 61
129	18.7	129	18.7	125	18.1	120	17.4	110	16.0	96	13.9	76	11.0	59	8.5					SB 148
100	14.5	98	14.2	98	14.2	98	14.2	98	14.2	98	14.2	81	11.7	51	7.4					SB 271
48	6.9	46	6.7																	SB 584
45	6.5	40	5.8	36	5.2	33	4.8	31	4.5	28	4.1	25	3.6	21	3.1					SB 265
68	9.9	62	9.0	58	8.4	53	7.7	50	7.2	46	6.6	43	6.2	39	5.7					SB 381
90	13.0	81	11.7	72	10.4	64	9.3	57	8.3	52	7.5	46	6.7	41	6.0					SB 348
105	15.2	98	14.2	92	13.3	86	12.5	82	11.9	79	11.4	77	11.1	75	10.8					SB 367[h]
68	9.8	61	8.9	55	8.0	50	7.2													
		64	9.3			48	7.0			42	6.1			41	6.0			33	4.8	SB 551
		98	14.2			86	12.5			78	11.3			72	10.4			68	4.9	SB 550

205°C (400°F)		260°C (500°F)		315°C (600°F)		370°C (700°F)		426°C (800°F)		482°C (900°F)		538°C (1000°F)		593°C (1100°F)		648°C (1200°F)		704°C (1300°F)		760°C (1400°F)		Spec. Number
MPa	kpsi	MPa	kpsi	MPa	kpsi	MPa	kpsi	MPa	kpsi	MPa	kpsi	MPa	kpsi	MPa	kpsi	MPa	kpsi	MPa	kpsi	MPa	kpsi	
102	14.8	101	14.7	101	14.7	101	14.7	98	14.2	55	8.0											SB 127[j]
129	18.7	129	18.7	129	18.7	124	18.0	98	14.2	28	4.0											
138	20.0	138	20.0	138	20.0	135	19.6	132	19.1	110	16.0	48	7.0	21	3.0	14	2.0					SB 168
146	21.2	146	21.2	146	21.2	145	21.1	141	20.4	135	19.6	100	14.5	50	7.2	38	5.5					SB 168[j]
190	27.5	190	27.5	189	27.2	187	27.1	137	19.8													SB 333[k]
132	19.2	126	18.3	123	17.8	119	17.3	118	17.1	116	16.8	115	16.6									SB 424[k]
123	17.8	114	16.5	108	15.6	103	15.6	101	14.7	100	14.5	99	14.3	98	14.2	78	11.3	53	7.7	33	4.8	
158	22.9	154	22.3	146	21.1	140	20.3	136	19.7	135	19.6	131	19.3	121	17.5	78	11.3	53	7.7	33	4.8	
123	17.8	114	16.5	108	15.6	103	15.0	101	14.7	100	14.5	99	14.3	98	14.2	78	11.3	53	7.7	33	4.8	
185	26.8	180	26.1	175	25.4	172	25.0	170	24.6	165	24.0	163	23.7	166	23.4	91	13.2					SB 443
129	18.7	125	18.2	121	17.5	119	17.3	116	16.8													SB 463
165	23.9	160	23.2	157	22.7	154	22.4	153	22.2													SB 575[k]
125	18.2	120	17.4	116	16.8	113	16.4	111	16.1	110	16.0	109	15.8									SB 582
138	20.0	138	20.0	134	19.4	131	19.0	128	18.6	127	18.4	126	18.3									SB 582[k]
109	15.8	100	14.5	92	13.3																	SB 709
91	13.2	98	13.1	98	13.1	98	13.1	88	12.7	55	8.0											SB 164
138	20.0	138	20.0	138	20.0	138	20.0	138	20.0	110	16.0	48	7.0	21	3.0	14	2.0					SB 166[k]
146	21.2	146	21.2	146	21.2	146	21.1	141	20.4	134	19.5	100	14.5	50	7.2	38	5.5					SB 166
132	19.2	126	18.3	123	17.8	119	17.3	118	17.1	116	16.8	114	16.6									SB 425[k]
129	18.7	125	18.2	122	17.7	119	17.3	116	16.8													SB 462[k]
105	15.3	101	14.6	94	13.7	92	13.4	89	12.9	85	12.3	82	11.9	54	7.8	32	4.7	21	3.1	12	1.8	12
185	26.8	180	26.1	175	25.4	172	25.0	170	24.6	165	24.0	163	23.7	166	23.4	91	13.2					SB 564
172	25.0	170	24.7	168	24.4	165	24.0	158	23.0													SB 574[k]
123	17.8	123	17.8	123	17.8	122	17.7	119	17.3	114	16.6	108	15.7									SA 494[h]
112	16.2	112	16.2	112	16.2	111	16.1	105	15.2	99	14.4	95	13.8									SA 494

[k]Nickel alloys have low yield strength. The stress values of these alloys used are slightly on the high side. These higher stress values exceed 2/3 but do not exceed 90 percent of the yield strength at temperature. These stress values are not recommended for the flanges of gasket joints where a slight amount of distortion can cause leakage. Sol. ann. = Solution annealed.

[l]At temperature above 538°C (1000°F), these stress values may be used only if the material is annealed at a minimum temperature of 1038°C (1900°F) and has a carbon content of 0.04% or higher.

[m]These stress values multiplied by a joint efficiency factor of 0.85.

[n]A joint efficiency factor of 0.85 has been applied in arriving at the maximum allowable stress values in tension for this material.

[o]Alloy NO6225 in the annealed condition is subject to severe loss of impact strength at room temperature after exposure in the range of 538° to 760°C (1000° to 1400°F).

[p]For other conditions and specifications, it is suggested to refer to the General Notes given for Table UNF-23.1 of *ASME Boiler and Pressure Vessel Code*, Section VIII, Division 1, July 1, 1986.

Source: The American Society of Mechanical Engineers, *ASME Boiler and Pressure Vessel Code*, Section VIII, Division 1, July 1, 1986.

TABLE 8–11
Maximum allowable stress values (σ_{Sa}) in tension for high-alloy steel

Spec. no.	Grade	UNS no.	Nominal composition	Product form	Specified minimum yield strength, σ_{Sy} MPa	kpsi	Specified minimum tensile strength, σ_{St} MPa	kpsi	−30 to 38 (−20 to 100) MPa	kpsi	93 (200) MPa	kpsi	150 (300) MPa	kpsi	205 (400) MPa	kpsi
SA-240, SA-479	405	S 40500	12 Cr-1 Al[d]	Plate, bar	172	25	414	60	103	15.0	99	14.3	95	13.8	92	13.3
SA-240	410S	S 41008	13 Cr	Plate	207	30	414	60	103	15.0	99	14.3	95	13.8	92	13.3
SA-240	TP 409	S 40900	11 Cr-Ti	Plate	207	30	379	55	95	13.8	90	13.1	87	12.7	84	12.2
SA-240	18 Cr-2 Mo	S 44400	18 Cr-2 Mo[d]	Plate	276	40	414	60	103	15.0	99	14.3	95	13.8	92	13.3
SA-240	430	S 43000	17 Cr[d]	Plate	207	30	448	65	112	16.3	107	15.4	103	15.0	99	14.4
SA-479	410	S 41000	13 Cr	Bar, forge	276	40	483	70	111	16.2	106	15.4	103	14.9	99	14.4
SA-182	F6 ACl.1	S 41000	13 Cr	Bar, forge	276	40	483	70	111	16.2	106	15.4	103	14.9	99	14.4
SA-217	CA 15	J 91150	13 Cr[d]	Cast	448	65	620	90	155	22.5	148	21.5	143	20.7	138	20.0
SA-479	430, XM8	S 43000, S 43035	17 Cr[d,e];18 Cr-Ti[d,e]	Bar[e,g]	276	40	483	70	121	17.5	114	16.6	111	16.1	107	15.5
SA-412	201	S 20100	17 Cr-4 Ni-6 Mn	Plate	310	45	655	95	158	23.0	143	20.8	132	19.1		
SA-182	F 304 L	S 30403	18 Cr-8 Ni	Forge[g]	172	25	448	65	108	15.6	106	15.4	98	14.2	94	13.6
SA-240, SA-479	304 L	S 30403	18 Cr-8 Ni	Plate[g], bar[e,g]	172	25	483	70	108	15.7	108	15.7	105	15.3	101	14.7
SA-351	CF 3	J 92500	18 Cr-8 Ni	Cast[g]	207	30	483	70	121	17.5	114	16.6	105	15.3	104	15.1
SA-351	CF 8	J 92600	18 Cr-8 Ni	Cast[g,h]	207	30	483	70	121	17.5	114	16.6	104	15.1	103	15.0
SA-351	CF 8 M	J 92900	18 Cr-9 Ni-2 Mo	Cast[g,h]	207	30	483	70	121	17.5	121	17.5	118	17.1	116	16.8
SA-336	Cl-F 304 H	S 30409	18 Cr-8 Ni	Forge[g]	207	30	483	70	121	17.5	114	16.6	107	15.5	104	15.1
SA-240, SA-479	302	S 30200	18 Cr-8 Ni	Plate, bar[e,g]	207	30	517	75	130	18.8	123	17.8	114	16.6	112	16.2
SA-182	F 304	S 30400	18 Cr-8 Ni	Forge[g,h]	207	30	517	75	130	18.8	123	17.8	114	16.6	112	16.2
SA-479	304 H	S 30400	18 Cr-8 Ni	Bar[g,e]	207	30	517	75	130	18.8	123	17.8	114	16.6	112	16.2
SA-240	304	S 30400	18 Cr-8 Ni	Plate	207	30	517	75	130	18.8	123	17.8	114	16.6	112	16.2
SA-351	CF 3A	J 92500	18 Cr-8 Ni	Cast[g]	241	35	534	77.5	134	19.4	125	18.2			116	16.9
SA-240	304 N	S 30451	18 Cr-8 Ni-N	Plate[g,h]	241	35	552	80	138	20.0	138	20.0	131	19.0	126	18.3
SA-336	F 304 N	S 30451	18 Cr-8 Ni-N	Forge	241	35	552	80	138	20.0	138	20.0	131	19.0	126	18.3
SA-240	316 L	S 31603	16 Cr-12 Ni-2 Mo	Plate[g]	172	25	483	70	108	15.7	108	15.7	108	15.7	107	15.5
SA-182	F 316 L	S 31603	16 Cr-12 Ni-2 Mo	Forge[g]	172	25	448	65	108	15.7	108	15.7	108	15.7	107	15.5
SA-479	316 L	S 31603	16 Cr-12 Ni-2 Mo	Bar[g,f]	172	25	483	70	108	15.7	108	15.7	108	15.7	107	15.5
SA-351	CF 8 M	J 92900	16 Cr-12 Ni-2 Mo	Cast	207	30	483	70	121	17.5	121	17.5	118	17.1	116	16.8
SA-182	F 316	S 31600	16 Cr-12 Ni-2 Mo	Forge[g,h,i]	207	30	483	70	121	17.5	121	17.5	118	17.1	116	16.8
SA-336	Cl-F 316 H	S 31609	16 Cr-12 Ni-2 Mo	Forge	207	30	483	70	121	17.5	111	16.2	100	14.6	92	13.4
SA-240	316 Ti	S 31635	16 Cr-12 Ni-2 Mo	Plate[i,h,g]	207	30	517	75	130	18.8	130	18.8	127	18.4	125	18.1
SA-182	F 316 H	S 31609	16 Cr-12 Ni-2 Mo	Forge[g]	207	30	517	75	130	18.8	130	18.8	127	18.4	125	18.1
SA-479	316	S 31600	16 Cr-12 Ni-2 Mo	Bar[g,h,e]	207	30	517	75	130	18.8	130	18.8	127	18.4	125	18.1
SA-240	317 L	S 31703	18 Cr-13 Ni-3 Mo	Plate[g]	207	30	517	75	130	18.8	112	16.2	98	14.2	92	13.4
SA-240	XM-15	S 38100	18 Cr-18 Ni-2 Si	Plate[g]	207	30	517	75	130	18.8	122	17.7	114	16.6	111	16.1
SA-240	316 M	S 31651	16 Cr-12 Ni-2 Mo-N	Plate[g,h]	241	35	552	80	138	20.0	138	20.0	132	19.2	130	18.8
SA-479, SA-240	XM-29	S 24000	18 Cr-3 Ni-12 Mn	Plate, bar[g,f]	379	55	689	100	172	25.0	169	24.5	156	22.6	149	21.6
SA-182, SA-336	F 321 H	S 32100	18 Cr-10 Ni-Ti	Forge[g,i]	207	30	483	70	121	17.5	118	17.1	111	16.1	110	16.0
SA-240, SA-479	321	S 32100	18 Cr-10 Ni-Ti	Plate[g,h], bar[g,h,e]	207	30	517	75	130	18.8	127	18.4	119	17.3	118	17.1
SA-182, SA-336	F 347	S 34700	18 Cr-10 Ni-Cb	Forge[g,h,i]	207	30	483	70	121	17.5	115	16.7	105	15.3	99	14.4
SA-351	CFBC	J 92710	18 Cr-10 Ni-Cb	Cast[g,h]	207	30	483	70	121	17.5	114	16.6	105	15.3	96	13.9
SA-240, SA-182	347, 348	S 34700	18 Cr-10 Ni-Cb	Plate[g,h], forge[g,h]	207	30	517	75	130	18.8	123	17.9	113	16.4	107	15.5
SA-479	F 347, F 348	S 34800	18 Cr-10 Ni-Cb	Bar[g,h,e]	207	30	517	75	130	18.8	123	17.9	113	16.4	107	15.5
SA-351	CG 8 M		19 Cr-11 Ni-Mo	Cast[g]	241	35	517	75	121	17.5	121	17.5	118	17.1	116	16.8
SA-182, SA-240	F 44	S 31254	20 Cr-18 Ni-6 Mo	Forge, plate	303	44	648	94	162	23.5	162	23.5	147	21.4	137	19.9
SA-182, SA-240,	F 45	S 30815	21 Cr-11 Ni-N	Forge, plate, bar	310	45	600	87	150	21.8	149	21.6	141	20.4	135	19.6
SA-479		S 30815	21 Cr-11 Ni-N	Forge, plate, bar	310	45	600	87	150	21.8	149	21.6	141	20.4	135	19.6
SA-240, SA-479		S 32550	25.5 Cr-5.5 Ni-3.5 Mo	Plate, bar	552	80	758	110	190	27.5	189	27.4	177	25.7	170	24.7
SA-351	CH 8	J 93400	25 Cr-12 Ni	Cast[g,h]	193	28	448	65	112	16.3	103	14.9	98	14.2	95	13.8
SA-351	CH 20	J 93402	25 Cr-12 Ni	Cast[h]	207	30	483	70	121	17.5	111	16.1	105	15.3	102	14.8
SA-240	309 S, 309 Cb	S 30908, S 30940	23 Cr-12 Ni	Plate[h,j]	207	30	517	75	130	18.8	118	17.2	113	16.4	110	15.9
SA-240, SA-182	310 Cb; Cl-F310	S 31040, S 31000	25 Cr-20 Ni	Plate[g,k,h,j], forge[g,h]	207	30	517	75	130	18.8	118	17.2	113	16.4	110	15.9
SA-479	310 S	S 31008	25 Cr-20 Ni	Bar[g,h,e]	207	30	517	75	130	18.8	118	17.2	113	16.4	109	15.8
SA-240	TP 329	S 32900	26 Cr-4 Ni-Mo	Plate[d]	483	70	620	90	155	22.5	151	21.9	141	20.5	136	19.8
SA-182, SA-336	FXM-27 Cb	S 44625	27 Cr-Mo	Forge[d]	241	35	414	60	103	15.0	103	15.0	101	14.6	98	14.2
SA-240, SA-479	XM-27	S 44627	27 Cr-Mo	Plate[d], bar, shape[d,e]	276	40	448	65	112	16.2	112	16.2	110	15.9	110	15.9
SA-240	XM-33	S 44626	27 Cr-Mo-Ti	Plate[d]	310	45	469	68	117	17.0	117	17.0	116	16.8	114	16.6
SA-240, SA-479	844800		29 Cr-4 Mo-2 Ni	Plate[d], bar[d,e]	414	60	552	80	138	20.0	134	19.4	126	18.3	125	18.1
SA-564	630 H 1100	S 17400	17 Cr-4 Ni-4 Cu	Bar[d,l]	793	115	965	140	241	35.0	241	35.0	241	35.0	235	34.1
SA-182, SA-336, SA-41Z	FXM-11, XM-11	S-21904	20 Cr-6 Ni-9 Mn	Forge, plate	345	50	620	90	155	22.5	154	22.4	148	21.4	136	19.7

TABLE 8–11
Maximum allowable stress values (σ_{Sa}) in tension for high-alloy steel (*Cont.*)

For metal temperature, °C (°F), not exceeding

260 (500)		315 (600)		370 (700)		427 (800)		482 (900)		538 (1000)		593 (1100)		650 (1200)		704 (1300)		760 (1400)		815 (1500)		Spec. Number
MPa	kpsi	MPa	kpsi	MPa	kpsi	MPa	kpsi	MPa	kpsi	MPa	kpsi	MPa	kpsi	MPa	kpsi	MPa	kpsi	MPa	kpsi	MPa	kpsi	
89	12.9	85	12.4	83	12.1	76	11.1	69	9.7	27	4.0	(Fig. 8–5)										SA-240,
89	12.9	85	12.4	83	12.1	76	11.1	69	9.7	44	6.4	20	2.9	7	1.0	(Fig. 8–5)						SA-479
81	11.8	79	11.4	76	11.1	70	10.2	(Fig. 8–5)														SA-240
88	12.8	85	12.4																			SA-240
96	13.9	93	13.5	90	13.1	82	12.0	72	10.5	45	6.5	22	3.2	12	1.8							SA-240
96	13.9	92	13.4	90	13.1	82	12.0	72	10.4	44	6.4	(Fig. 8–5)										SA-240
96	13.9	92	13.4	90	13.1	82	12.0	72	10.4	44	6.4	(Fig. 8–5)										SA-479
133	19.3	129	18.7	125	18.1	115	16.7	76	11.0	34	5.0	15	2.2	7.0	1.0	(Fig. 8–5)						SA-182
103	15.0	100	14.5	97	14.1	89	12.9	76	11.0	45	6.5											SA-217, SA-479
92	13.4	92	13.3	90	13.1	89	12.9															SA-412
99	14.4	96	14.0	93	13.5	90	13.0															SA-182, SA-240, SA-479
102	14.8	102	14.8	102	14.8	100	14.6															SA-351
102	14.8	102	14.8	102	14.8	100	14.6	92	13.4	83	12.0	52	7.5	33	4.8	23	3.3	16	2.3	12	1.7	SA-351
116	16.8	116	16.8	112	16.3	109	15.8	107	15.5	103	14.9	61	8.9	37	5.4	23	3.4	16	2.3	11	1.6	SA-351
102	14.8	102	14.8	102	14.8	100	14.6	98	14.2	92	13.4	68	9.8	42	6.1	25	3.7	16	2.3	10	1.4	SA-336, SA-240, SA-479
110	15.9	110	15.9	110	15.9	105	15.2	101	14.7	95	13.8	68	9.8	42	6.1	25	3.7	16	2.3	10	1.4	SA-182
110	15.9	110	15.9	110	15.9	105	15.2	101	14.7	95	13.8	68	9.8	42	6.1	25	3.7	16	2.3	10	1.4	SA-479
114	16.5	112	16.3	112	16.3	100	14.6															SA-240
123	17.8	120	17.4	118	17.1	114	16.6	110	15.9	103	15.0	67	9.7	41	6.0							SA-351
123	17.8	120	17.4	118	17.1	114	16.6	110	15.9	103	15.0	67	9.7	41	6.0							SA-240
99	14.4	93	13.5	89	12.9	85	12.4	83	12.1													SA-336
99	14.4	93	13.5	89	12.9	85	12.4	83	12.1													SA-240
99	14.4	93	13.5	89	12.9	85	12.4	83	12.1													SA-182
116	16.8	116	16.8	112	16.3	109	15.8	107	15.5	103	14.9	65	9.4	41	6.0	27	4.0	16	2.4	10	1.5	SA-479
116	16.8	116	16.8	112	16.3	110	15.9	107	15.6	103	15.0	85	12.4	51	7.4	28	4.1	17	2.5	8	1.2	SA-351
86	12.5	81	11.8	78	11.3	76	11.0	74	10.8	73	10.6	71	10.3	51	7.4	28	4.1	16	2.3	9	1.3	SA-182
124	18.0	117	17.0	112	16.3	110	15.9	103	15.5	105	15.3	85	12.4	51	7.4	28	4.1	16	2.3	9	1.3	SA-336
124	18.0	117	17.0	112	16.3	110	15.9	103	15.5	105	15.3	85	12.4	51	7.4	28	4.1	16	2.3	9	1.3	SA-240
124	18.0	117	17.0	112	16.3	110	15.9	103	15.5	105	15.3	85	12.4	51	7.4	28	4.1	16	2.3	9	1.3	SA-182
86	12.5	81	11.8	78	11.3	76	11.0															SA-479
110	15.9	110	15.9	110	15.9	104	15.1	101	14.6	94	13.7											SA-240
128	18.6	128	18.6	128	18.6	127	18.4	125	18.1	120	17.4	85	12.4	51	7.4							SA-240
148	21.4	144	20.9	138	20.0	131	19.0															SA-240
110	16.0	110	16.0	109	15.8	107	15.5	105	15.3	96	14.0	62	9.0	37	5.4	22	3.2	13	1.9	8	1.1	SA-479, SA-240, SA-182, SA-336
118	17.1	113	16.4	109	15.8	107	15.5	105	15.3	95	13.8	48	6.9	25	3.6	12	1.7	5	0.8	2	0.3	SA-240, SA-479
96	13.9	94	13.7	94	13.7	94	13.7	94	13.7	91	13.2	63	9.1	30	4.4	15	2.2	8	1.2	5	0.8	SA-182, SA-336
94	13.7	94	13.7	94	13.7	94	13.7	94	13.7	91	13.2	72	10.5	34	5.0	19	2.7	11	1.6	7	1.0	SA-351
103	14.9	101	14.7	101	14.7	101	14.7	101	14.7	96	14.0	63	9.1	30	4.4	15	2.2	8	1.2	5	0.8	SA-240, SA-182
103	14.9	101	14.7	101	14.7	101	14.7	101	14.7	96	14.0	63	9.1	30	4.4	15	2.2	8	1.2	5	0.8	SA-479
116	16.8	123	17.9	121	17.5																	SA-351, SA-182, SA-240
127	18.4	122	17.7	86	12.5	116	16.8	112	16.3	103	14.9	62	9.0	36	5.2	21	3.1	13	1.9	9	1.3	SA-240, SA-182
127	18.4	122	17.7	86	12.5	116	16.8	112	16.3	103	14.9	62	9.0	36	5.2	21	3.1	13	1.9	9	1.3	SA-240, SA-479
170	24.7																					SA-240, SA-479
93	13.5	92	13.3	90	13.0	90	13.0	86	12.5	72	10.5	45	6.5	26	3.8	16	2.3	9	1.3	5	0.8	SA-351
97	14.1	92	13.4	88	12.7	84	12.2	81	11.7	70	10.6	45	6.5	26	3.8	16	2.3	9	1.3	5	0.8	SA-351
107	15.5	105	15.3	104	15.1	103	14.9	96	13.9	72	10.5	45	6.5	26	3.8	16	2.3	9	1.3	5	0.8	SA-240
107	15.5	105	15.3	104	15.1	103	14.9	96	13.9	76	11	59	8.5	41	6.0	24	3.5	11	1.6	5	0.8	SA-240, SA-182
107	15.5	105	15.3	104	15.1	103	14.9	95	13.8	76	11											SA-479
136	19.8																					SA-240
98	14.2	98	14.2																			SA-182, SA-336
110	15.9	110	15.9																			SA-240, SA-479
113	16.4	111	16.1																			SA-479
125	18.1	125	18.1																			SA-240
230	33.3	226	32.8																			SA-240, SA-479, SA-564
123	17.9	117	17.0																			SA-182, SA-336

TABLE 8–11
Maximum allowable stress values (σ_{Sa}) in tension for high-alloy steel (*Cont.*)

Spec. no.	Grade	UNS no.	Nominal composition	Product form	Specified minimum yield strength, σ_{Sy}		Specified minimum tensile strength, σ_{St}		Maximum allowable stress, σ_{Sa}							
									−30 to 38 (−20 to 100)		93 (200)		150 (300)		205 (400)	
					MPa	kpsi	MPa	kpsi	MPa	kpsi	MPa	kpsi	MPa	kpsi	MPa	kpsi
SA-351	CG 6 MMN	J 93790	22 Cr-13 Ni-5 Mn	Cast	241	35	517	75	130	18.8	116	16.9	103	14.9	94	13.6
SA-240, SA-412, SA-479, SA-182	XM-19	S 20910	22 Cr-13 Ni-5 Mn	Plate, bar, forge[f]	379	55	689	100	172	25.0	172	24.9	163	23.6	156	22.7

[a]The stress value in this table may be interpolated to determine values for intermediate temperatures.

[b]Stress values in restricted shear shall be 0.8 times the values in this table.

[c]Stress values in bearing shall be 1.60 times the values in this table.

[d]This steel may be expected to develop embrittlement after service at moderately elevated temperature.

[e]Use of external pressure charts for material in the form of barstock is permitted for stiffening rings only.

[f]These stress values are the basic values multiplied by a joint efficiency factor of 0.85.

TABLE 8–11
Maximum allowable stress values (σ_{Sa}) in tension for high-alloy steel (*Cont.*)

260 (500)		315 (600)		370 (700)		427 (800)		482 (900)		538 (1000)		593 (1100)		650 (1200)		704 (1300)		760 (1400)		815 (1500)		Spec. Number
MPa	kpsi	MPa	kpsi	MPa	kpsi	MPa	kpsi	MPa	kpsi	MPa	kpsi	MPa	kpsi	MPa	kpsi	MPa	kpsi	MPa	kpsi	MPa	kpsi	
90	13.0	87	12.6	85	12.3	83	12.0	81	11.8	79	11.4											SA-351
154	22.3	151	21.9	149	21.6	146	21.2	142	20.6	137	19.9	131	19.0	57	8.3							SA-240, SA-412, SA-479, SA-182

[g]Alloy steels have low yield strength. The stress values of these alloy steels used are slightly on the high side. These higher stress values exceed 2/3 but do not exceed 90 percent of the yield strength at temperature. These stress values are not recommended for the flanges of gasket joints where a slight amount of distortion can cause leakage.

[h]At temperature above 540°C (1000°F), these stress values apply only when carbon is 0.04% or higher on heat analysis.

[i]These stress values shall be applicable to forging over 125 mm (5 in) in thickness.

[j]For temperature above 540°C (1000°F), these stress values may be used only if the material is heat-treated by heating it to a minimum temperature of 1040°C (1900°F) and quenching in water or rapidly cooling by other means.

[k]These stress values at 565°C (1050°F) and above shall be used only when the grain size is ASTM 6 or coarser.

[l]These stress values are established from a consideration of strength only and shall be satisfactory for average service.

Source: The American Society of Mechanical Engineers, *ASME Boiler and Pressure Vessel Code*, Section VIII, Division 1, July 1, 1986.

TABLE 8-12
Maximum allowable stress values, σ_{Sa}, in tension for ferrite steels with properties enhanced by heat treatment

Specification no.	Grade and size	Specified minimum yield strength, σ_{sy} MPa	kpsi	Specified minimum tensile strength, σ_{su} MPa	kpsi	≤66 (150) MPa	kpsi	93 (200) MPa	kpsi	(250) MPa	kpsi	150 (300) MPa	kpsi	205 (400) MPa	kpsi	260 (500) MPa	kpsi	315 (600) MPa	kpsi	345 (650) MPa	kpsi	370 (700) MPa	kpsi	400 (750) MPa	kpsi	427 (800) MPa	kpsi
															Plates												
SA-353[a,b]		517	75	690	100	172	25.0	161	23.4	157	22.7																
SA-517	A, B, D, J 31; 25 mm (1.25 in); 62.5 mm (2.5 in)	690	100	792	115	198	28.8	198	28.8	198	28.8	198	28.8	198	28.8	198	28.8	198	28.8	197	28.7	197	28.7				
SA-517	E ≤ 62.5 (2½ in); >150 mm (6 in); >100 mm (4 in)	690	100	792	115	198	28.8	198	28.8	198	28.8	198	28.8	198	28.8	198	28.8	198	28.8	197	28.7	197	28.7				
SA-533	A, B, C, D, Cl 2	482	70	620	90	155	22.5	155	22.5	155	22.5	155	22.5	155	22.5	155	22.5	155	22.5	155	22.5	155	22.5	152	22.2	146	21.2
	B, D, Cl3; >62.5 mm (2½ in)	572	83	690	100	172	25.0	172	25.0	172	25.0	172	25.0	172	25.0	172	25.0	172	25.0	172	25.0	172	25.0	169	24.5		
SA-553[a]	I, II[a,b,d]	586	85	690	100	172	25.0	161	23.4	156	22.7																
SA-645[a]		448	65	655	95	164	23.8	163	23.7	161	23.3																
SA-724	B	517	75	655	95	164	23.8	164	23.8	164	23.8	164	23.8	163	23.5	163	23.5	163	23.5	161	23.4	159	23.1				
	A, C	482	70	620	90	155	22.5	155	22.5	155	22.5	155	22.5	154	22.3	154	22.3	154	22.3	153	22.2	151	21.9				
												Castings															
SA-487	Cl. 4 Q[e]	586	85	724	105	181	26.3	181	26.3	181	26.3	181	26.3	181	26.3	181	26.3	181	26.3	181	26.3	181	26.3				
SA-487	Cl. 4 QA[e]	655	95	792	115	198	28.8	198	28.8	198	28.8	198	28.8	198	28.8	198	28.8	198	28.8	198	28.8	198	28.8				
SA-487	Cl.C4 6 NM[e]	551	80	758	110	190	27.5	190	27.5	187	27.2	185	26.9	181	26.3	178	25.8	174	25.3	172	24.9	168	24.4	165	23.9	161	23.3
										Pipes and tubes																	
SA-333	g[a,b]	517	75	690	100	172	25.0	161	23.4	156	22.7																
SA-334	g[e,f]	517	75	690	100	146	21.3	137	19.9	133	19.3																
												Forgings															
SA-508	Cl4	586	85	724	105	181	26.2	181	26.2	181	26.2	181	26.2	179	26.0	178	25.8	175	25.4	173	25.1						
SA-522	I[a]	517	75	690	100	163	23.7	153	22.2	148	21.5																
SA-592	A ≤ 37.6 mm (1½ in); E, F ≤ 62.5 mm (2½ in)	690	100	792	115	198	28.8	198	28.8	198	28.8	198	28.8	198	28.8	198	28.8	198	28.8	197	28.7						
SA-592	E, F > 62.5[h] mm (2½ in)	620	90	724	105	181	26.3	181	26.3	181	26.3	181	26.3	181	26.3	181	26.3	181	26.3	180	26.2						

[a] Minimum thickness after forming any section subject to pressure shall be 4.6875 mm (3/16 in).
[b] Not welded or welded if the tensile strength of the Section IX reduced section tension test is not less than 600 MPa (100 kpsi).
[c] Welded with the tensile strength of the Section IX reduced section tension test less than 690 MPa (100 kpsi) but not less than 655 MPa (95 kpsi).
[d] Grade II of SA-533 shall not be used for minimum allowable temperature below −170°C (275°F).
[e] To these stress values a quality factor as specified in UG-24 shall be applied for castings.
[f] These stress values are the basic values multiplied by a joint efficiency factor of 0.85.
[g] The maximum section thickness shall not exceed 75 mm (3 in) for double normalized and tempered forgings, or 125 mm (5 in) for quenched and tempered forgings.
[h] The maximum thickness of non-heat-treated forgings shall not exceed 93.75 (3¾ in). The maximum thickness as heat treated may be 100 mm (4 in).
Source: The American Society of Mechanical Engineers, ASME Boiler and Pressure Vessel Code, Section VIII, Division 1. July 1, 1986.

TABLE 8-13
Maximum allowable stress values in tension for cast iron

Specification no.	Class	Specified minimum tensile strength		Maximum allowable stress for metal temperature, °C (°F), not exceeding			
				Subzero to 232 (450)		345 (650)	
		MPa	kpsi	MPa	kpsi	MPa	kpsi
SA-667	—	138	20	13.8	2.0		
SA-278	20	138	20	13.8	2.0		
SA-278	25	172	25	17.2	2.5		
SA-278	30	207	30	20.7	3.0		
SA-278	35	241	35	24.1	3.5		
SA-278	40	276	40	27.6	4.0	27.6	4.0
SA-278	45	310	45	31.0	4.5	31.0	4.5
SA-278	50	345	50	34.5	5.0	34.5	5.0
SA-47	(Grade 3-2510)	345	50	34.5	5.0	34.5	5.0
SA-278	55	379	55	37.9	5.5	37.9	5.5
SA-278	60	414	60	41.4	6.0	41.4	6.0
SA-476	—	552	80	55.2	8.0		
SA-748	20	138	20	13.8	2.0		—
SA-748	25	172	25	17.2	2.5		—
SA-748	30	207	30	20.7	3.0		
SA-748	35	241	35	24.1	3.5		

Source: ASME Boiler and Pressure Vessel Code, Section VIII, Division 1, July 1, 1986.

TABLE 8-14
Modulus of elasticity for various materials

Material	73 K (−200°C) GPa	Mpsi	173 K (−100°C) GPa	Mpsi	273 K (0°C) GPa	Mpsi	293 K (20°C) GPa	Mpsi	323 K (50°C) GPa	Mpsi	348 K (75°C) GPa	Mpsi	373 K (100°C) GPa	Mpsi	398 K (125°C) GPa	Mpsi	423 K (150°C) GPa	Mpsi	473 K (200°C) GPa	Mpsi	573 K (300°C) GPa	Mpsi	673 K (400°C) GPa	Mpsi	773 K (500°C) GPa	Mpsi	873 K (600°C) GPa	Mpsi	973 K (700°C) GPa	Mpsi	1023 K (750°C) GPa	Mpsi
Ferrous Materials																																
Low-carbon steel C ≤ 0.03%					192	27.8	192	27.8					191	27.7					186	27.0	179	26	169	24.5								
High-carbon steel C > 0.3%					206	29.9	206	29.9					203	29.4					195	28.3	186	27.0	170	24.7								
Carbon molybdenum and chrome molybdenum steel up to 3% Cr					206	29.9	206	29.9					203	29.4					197	28.6	190	27.6	181	26.3	17	2.5						
Aluminum and Aluminum Alloys																																
1B, N3, N4	77	11.2	73	10.6	70	10.2			69	10.0	69	10.0	68	9.9	67	9.7	66	9.6	65	9.4												
H9	73	10.6	70	10.2	67	9.7			65	9.4	65	9.4	64	9.3	64	9.3	63	9.1	59	8.6												
H15	81	11.7	78	11.3	74	10.7			73	10.6	73	10.6	72	10.4	71	10.3	70	10.2	67	9.7												
A6	87	12.6	84	12.2	80	11.6			79	11.5	79	11.5	78	11.3	77	11.2	76	11.0	75	10.9												
Nickel and Nickel Alloy																																
Nickel							207	30.0													200	29.0	184	26.7	162	23.5	137	19.9	115	16.7	107	15.5
Nickel-copper alloy—Ni 70%, Cu 30%							184	26.3													176	25.5	173	25.0	166	24.0	159	23.0	152	22.0	147	21.3
Nickel-chromium ferrous alloy—Ni 75%, Cr 14%, Fe 10%							214	31.0													203	29.4	197	28.6	172	25.0	157	22.8	128	18.6	118	17.0
Copper and Its Alloys																																
Copper—Cu 99.98%							110	16.0	109	15.8			108	15.7			106	15.4	104	15.0	99	14.4										
Commercial brass—Cu 66%, Zn 34%							96	13.9	95	13.8			94	13.6			93	13.5	89	12.9	87	12.6										
Leaded tin bronze —Cu 88%, Sn 6%, Pb-1.5%, Zn-4.5%							89	12.9	88	12.8			87	12.6			85	12.3	82	11.9	85	12.3										
Phosphor bronze—Cu 85.5%, Sn 12.5%, Zn 10%							103	14.9	101	14.6			100	14.5			96	13.9	93	13.5	83	12.0										
Muntz—Cu 59%, Zn 39%							105	15.2	100	14.5			96	13.9			89	12.9	81	11.7												
Cupronickel—Cu 80%, Ni 20%							130	18.8	128	18.6			127	18.4			124	18.0	122	17.7	116	16.8										

TABLE 8–15
Values of coefficient c_5

	Coefficient c_5	Types of stays
1	112	Stays screwed through plates ≤ 1.1 cm thick, with the ends riveted over
2	120	Stays screwed through plates > 1.1 cm thick, with the ends riveted over
3	135	Stays screwed through plates and provided with single nuts outside the plate or with inside and outside nuts, but no washers
4	150	With heads $\nless 1.3$ times the stay diameter, screwed through the plates, or made with a taper fit and having heads formed before installing and not riveted over; these heads have a true bearing on the plate
5	175	Stays with inside and outside nuts and outside washers, when the washer diameter is $\geq 0.4a$, and the thickness $\geq n$

TABLE 8–16
Design stresses for bolted flanged heads, σ_d

Maximum temperature		Minimum of specified range of tensile strength of flange material at room temperature										Alloy bolt steel	
		3170		3520		3870		4220		4920			
K	°C	MPa	kpsi	MPa	kpsi	MPa	kpsi	MPa	kpsi	MPa	kpsi	MPa	kpsi
643	370	74.0	10.5	81.9	12.0	90.7	13.2	97.6	14.2	115.2	16.5	97.6	14.2
672	399	63.3	8.2	73.1	10.6	77.0	11.2	87.3	12.7	102.0	14.8	87.3	12.6
696	423	55.8	8.0	62.8	9.0	68.6	10.0	75.5	11.0	87.3	12.0	75.5	11.0
727	454	47.3	6.9	52.5	7.6	57.4	8.3	62.8	9.0	73.5	10.6	62.8	9.0
755	482	38.3	5.5	41.4	6.0	45.5	6.5	50.5	7.3	58.8	8.5	50.5	7.3
783	510	27.9	4.0	31.1	4.5	31.2	4.5	38.3	5.5	44.2	6.4	38.3	5.5

Particular	Formula
The total cross-sectional area of reinforcement, A_r, as per Indian Standards	$A_r \nless A = d.h_r \qquad (8\text{–}98)$ where d = nominal internal diameter of the branch plus twice the corrosion allowance, mm (in) h_r = thickness an unpierced shell or end calculated from Eq. (8–10)
The expression for K factor	$K = pD_o/1.82\sigma_a h_a \qquad (8\text{–}99)$ Refer to Fig. 8–16a and b, where K has a value of unity or greater, the maximum size of an unreinforced opening shall be 50 mm (2 in)

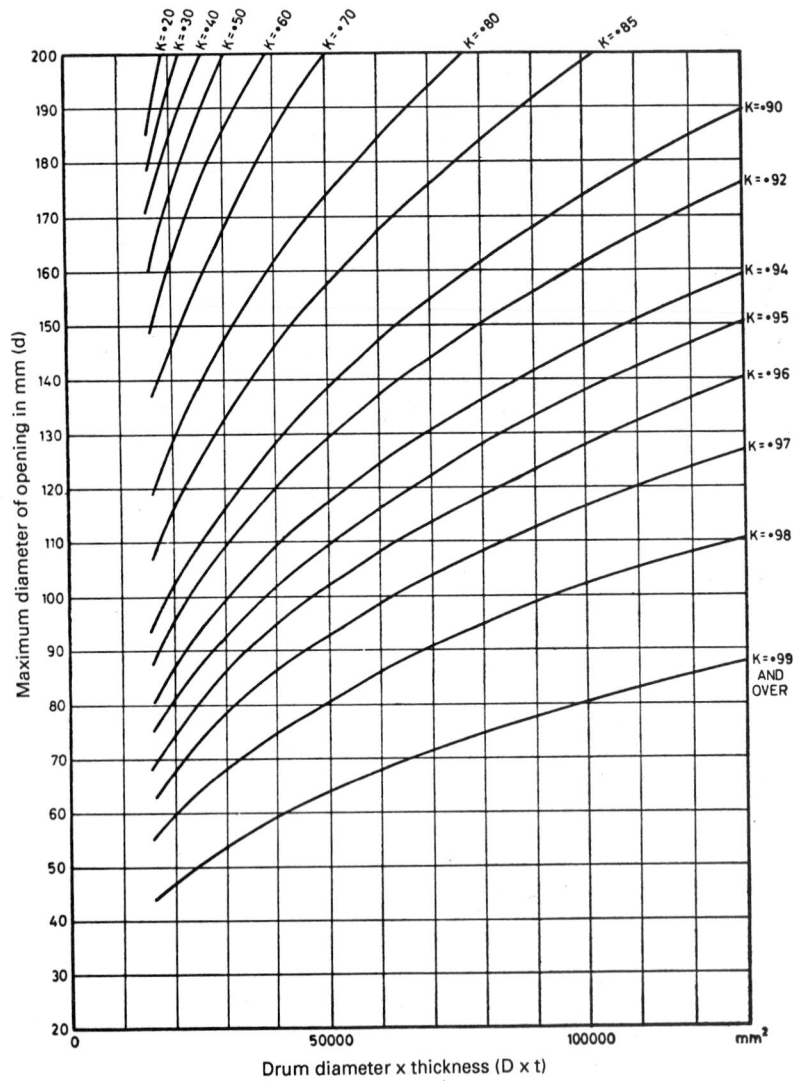

FIGURE 8–16(a**)** Maximum diameter of nonreinforced openings. (*Source: IS 2825, 1969.*)

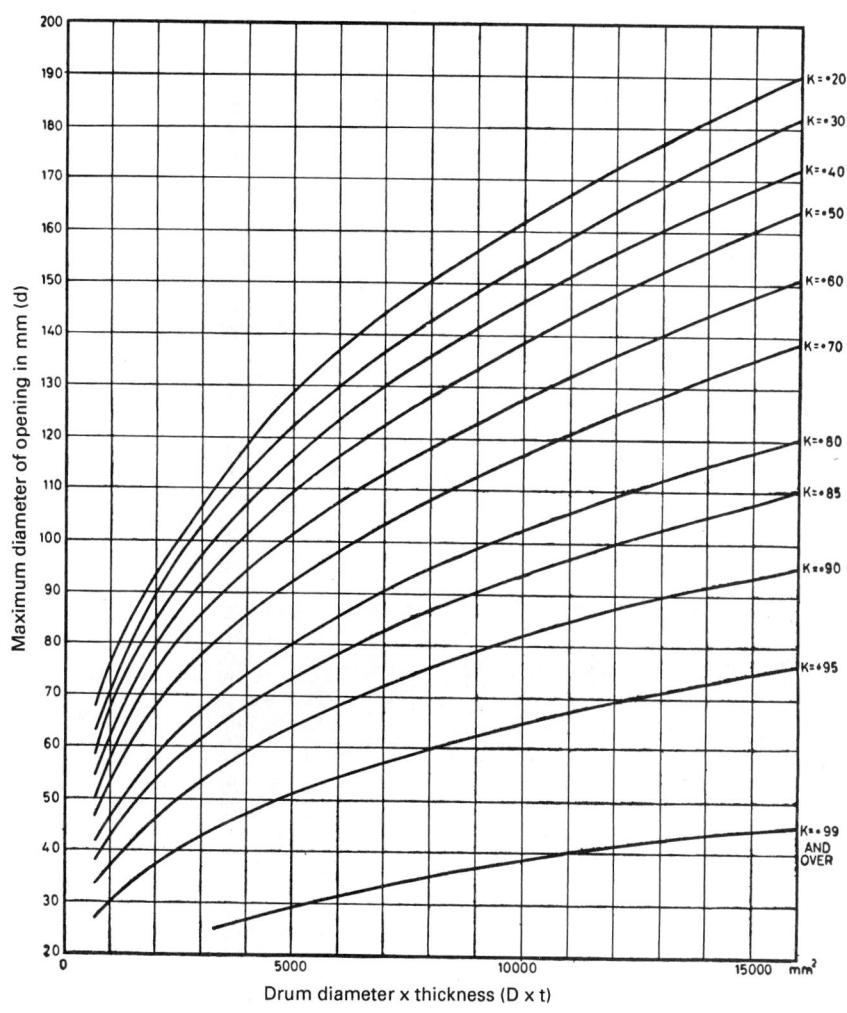

FIGURE 8–16(*b*) Maximum diameter of nonreinforced openings. (*Source: IS 2825, 1969.*)

TABLE 8-17
Allowable stresses (σ_{Sa}) for flange bolting material

Material	Diameter, mm (in)**	Specified tensile strength, σ_{St} MPa (kpsi)	Allowable stress, σ_{Sa}, for design metal temperature not exceeding (°C)						
			50°C MPa (kpsi)	100°C MPa (kpsi)	200°C MPa (kpsi)	250°C MPa (kpsi)	300°C MPa (kpsi)	350°C MPa (kpsi)	400°C MPa (kpsi)
Hot-rolled carbon steel	≤ 150 (6)**	431–510 (62.5–74.0)*	57 (8.3)*	55 (8.0)*	53 (7.7)*	48 (6.9)*			
1% Cr Mo steel	≤ 63.5 (2.5)	843 min (122.3)	193 (28.0)	181 (26.3)	168 (24.3)	159 (23.0)	154 (22.4)	148 (21.5)	140 (20.0)
	> 63.5 (2.5) to 102 (4)	775 min (112.4)	174 (25.2)	163 (23.6)	152 (22.0)	145 (21.0)	141 (20.5)	134 (19.4)	127 (18.4)
5% Cr Mo steel	≤ 63.5 (2.5)	896 min (130.0)	138 (20.0)	138 (20.0)	138 (20.0)	138 (20.0)	138 (20.0)	138 (20.0)	138 (20.0)
	> 63.5 (2.5) to 102 (4)	647 min (93.8)							
1% Cr V steel	≤ 63.5 (2.5)	843 min (122.3)	193 (28.0)	187 (27.1)	181 (26.3)	176 (25.5)	170 (24.7)	165 (23.9)	157 (22.8)
	> 63.5 (2.5) to 102 (4)	804 min (116.6)	174 (25.2)	169 (24.5)	163 (23.6)	159 (23.1)	152 (22.0)	150 (21.8)	143 (20.7)
13% Cr Ni steel	≤102 (4)	696 min (101.0 min)	176 (25.5)	161 (23.4)	141 (20.5)	134 (19.4)	126 (18.3)	119 (11.3)	104 (15.1)
18/8 Cr Ni steel	All (1) (2)	539 min (78.2 min)	129 (18.7)	109 (15.7)	85 (12.3)	78 (11.3)	76 (11.0)	73 (10.6)	72 (10.4)
18/8 Cr Ni Ti stabilized steel	All (1) (2)	In softened condition or ≤ 863 min (125.2) if cold-drawn	129 (18.7)	113 (16.4)	100 (14.5)	93 (13.5)	90 (13.0)	86 (12.5)	84 (12.2)
18/9 Cr Ni Nb stabilized steel	All (1) (2)		129 (18.7)	113 (16.4)	100 (14.5)	93 (13.5)	90 (13.0)	86 (12.5)	84 (12.2)
17/10/2½ Cr Ni Mo steel	All (1) (2)		129 (18.7)	110 (16.0)	94 (13.6)	87 (12.6)	83 (12.0)	79 (11.5)	78 (11.3)
18/Cr 2 Ni steel	≤ 102 (4)	843 min (122.3 min)	212 (30.8)	195 (28.3)	169 (24.5)	160 (23.2)	152 (22.0)	144 (20.9)	127 (18.4)

1. Austenitic steel bolts for use in pressure joints shall not be less than 10 mm in diameter.
2. For bolts of up to 38 mm diameter use torque spanners.
3. High strength is obtainable in bolting materials by heat treatment of the ferritic and martensitic steels and by cold working of austenitic steels.
* Values in parentheses are in US Customary units (i.e., fps system of units).
** Sizes in parentheses are in inches and outside parentheses are in millimeters.
Source: IS 2825, 1969.

8.58

Particular	Formula

Design for internal pressure

The total cross-sectional area of reinforcement A required in any given plane through the opening for a shell or formed head under internal pressure shall not be less than as per *ASME Boiler and Pressure Vessel Code*

The total cross-sectional area of reinforcement in flat heads that have an opening with a diameter that does not exceed one-half of the head diameter or shortest span, shall not be less than that given by Eq. (8–100) as per *ASME Boiler and Pressure Vessel Code*

$$A = 0.5\,td \qquad (8\text{--}100)$$

where t = minimum required thickness of flat head or cover, exclusive of corrosion allowance, m (in)

For nomenclature and formulas for reinforced openings as per *ASME Boiler and Pressure Vessel Code*

Refer to Fig. 8–17.

For values of spherical radius factor K_1

Refer to Table 8–18.

The length of tapped hole l_s to engage a stud

$$l_s = 0.75\,d_s\ \frac{\text{maximum allowable stress value of stud material at design temperature}}{\text{maximum allowable stress value of tapped material at design temperature}} \qquad (8\text{--}101)$$

and also $l_s \nless d_s$, where d_s = nominal diameter of the stud, except that the thread engagement need not exceed $1.5d_s$

TABLE 8–18
Values of spherical radius factor K_1 equivalent to spherical radius = $K_1 D$, $D/2h$ = axis ratio

$\frac{D}{2h}$	3.0	2.8	2.6	2.4	2.2	2	1.8	1.6	1.4	1.2	1.0
K_1	1.36	1.27	1.18	1.08	0.99	0.90	0.81	0.73	0.65	0.57	0.50

LIGAMENTS

The efficiency η of the ligament between the tube holes, when the pitch of the tube holes on every row is equal

$$\eta = \frac{p - d}{p} \qquad (8\text{--}102)$$

where p = longitudinal pitch of tube holes, m (in)

d = diameter of tube holes, m (in)

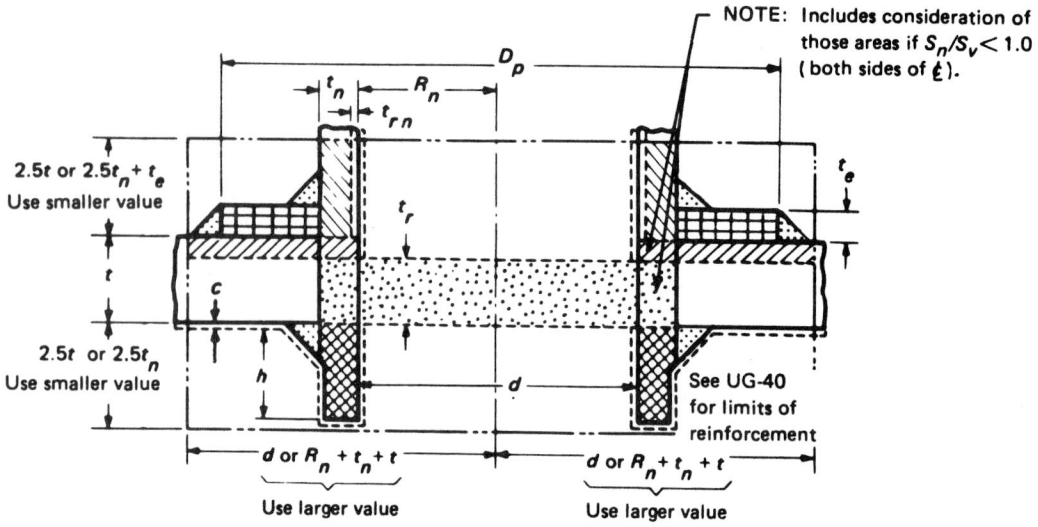

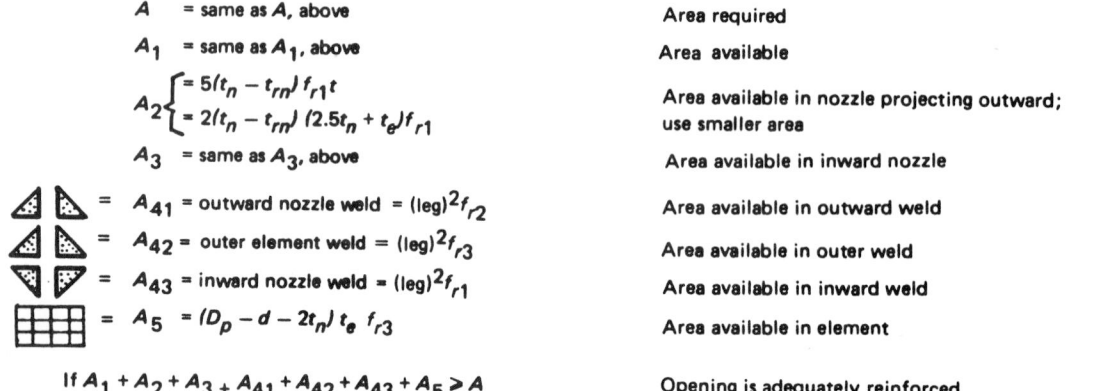

FIGURE 8–17 Nomenclature and formulas for reinforced openings. (This figure illustrates a common-nozzles configuration and is not intended to prohibit other configurations permitted by the code.) (*American Society of Mechanical Engineers*, ASME Boiler and Pressure Vessel Code, *Section VIII, Division 1, July 1, 1986.*)

Particular	Formula
The efficiency η of the ligament between the tube holes, when the pitch of tube holes on any one row is unequal (Fig. 8–18)	$$\eta = \frac{p_1 - nd}{p_1} \qquad (8\text{–}103)$$ where p_1 = unit length of ligament, m (in) n = number of tube holes in length, p_1

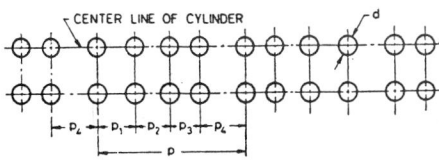

FIGURE 8–18 Irregular drilling.

The efficiency η of the ligament, when bending stress due to weight is negligible and the tube holes are arranged along a diagonal line with respect to the longitudinal axis or to a regular sawtooth pattern as shown in Fig. 8–19a to d	$$\eta = \frac{2}{A + B + \sqrt{(A-B)^2 + 4C^2}} \qquad (8\text{–}104)$$ where $A = \dfrac{\cos^2 \alpha + 1}{2[1 - (d\cos\alpha)/2a]}$ $B = \dfrac{1}{2}\left(1 - \dfrac{d\cos\alpha}{a}\right)(\sin^2\alpha + 1)$ $C = \dfrac{\sin\alpha\cos\alpha}{2\left(1 - \dfrac{d\cos\alpha}{a}\right)}$ $\cos\alpha = \dfrac{1}{\sqrt{1 + (b^2/a^2)}}; \sin\alpha = \dfrac{1}{\sqrt{1 + a^2/b^2}}$

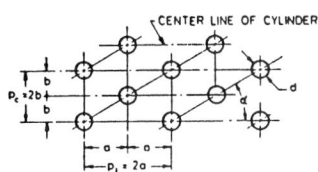

FIGURE 8–19(a) A regular staggering of holes.

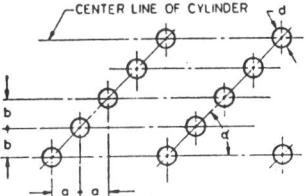

FIGURE 8–19(b) Spacing of holes on a diagonal line.

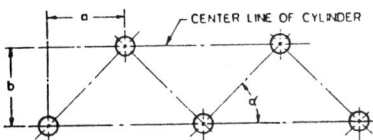

FIGURE 8–19(c) Regular sawtooth pattern of holes.

FIGURE 8–19(d)

Particular	Formula
The smallest value of efficiency η of all the ligaments (longitudinal, circumferential, and diagonal) in the case of regular staggered spacing of tube holes	$\eta = \dfrac{p_c}{p_L} = \dfrac{P_L - d}{P_L}$ or $\dfrac{d}{a}$ (8–105) The symbols are as shown in Fig. 8–19d.
For minimum number of pipe threads for connections as per *ASME Boiler and Pressure Vessel Code*	Refer to Table 8–19.

TABLE 8–19
Minimum number of threads for connections

Size of pipe connection, mm (in)	12.5 and 18.75 ($\frac{1}{2}$ and $\frac{3}{4}$)	25.0, 31.25, and 37.5 (1, 1$\frac{1}{4}$, and 1$\frac{1}{2}$)	50.0 (2)	62.5 and 75 (2$\frac{1}{2}$ and 3)	100–150 (4–6)	200 (8)	250 (10)	300 (12)
Threads engaged	6	7	8	8	10	12	13	14
Minimum plate thickness required, mm (in)	10.75 (0.43)	15.25 (0.62)	17.50 (0.70)	25.0 (1.0)	31.25 (1.25)	37.50 (1.5)	40.5 (1.62)	43.75 (1.75)

BOLTED FLANGE CONNECTIONS

Bolt loads

The required bolt load under operating conditions sufficient to contain the hydrostatic end force and simultaneously to maintain adequate compression on the gasket to ensure seating	$W_{m1} = H + H_P = \dfrac{\pi}{4}G^2 P + 2b\pi\, GmP$ (8–106)
For additional gasket criteria	Refer to Tables 8–20 to 8–21.
The required initial bolt load to seat the gasket joint-contact surface properly at atmospheric temperature condition without internal pressure	$W_{m2} = \pi bGy$ (8–107) Refer to Table 8–20 for y.
Total required cross-sectional area of bolts at the root of thread	$A_m > A_{m1}$ or A_{m2} (8–108)
Total cross-sectional area of bolt at root of thread or section of least diameter under stress required for the operating condition	$A_{m1} = \dfrac{W_{m1}}{\sigma_{Sbd}}$ (8–109) Refer to Tables 8–17 and 8–23 for σ_{Sbd}.
Total cross-sectional area of bolt at root of thread or section of least diameter under stress required for gasket seating	$A_{m2} = \dfrac{W_{m2}}{\sigma_{Sbat}}$ (8–110)

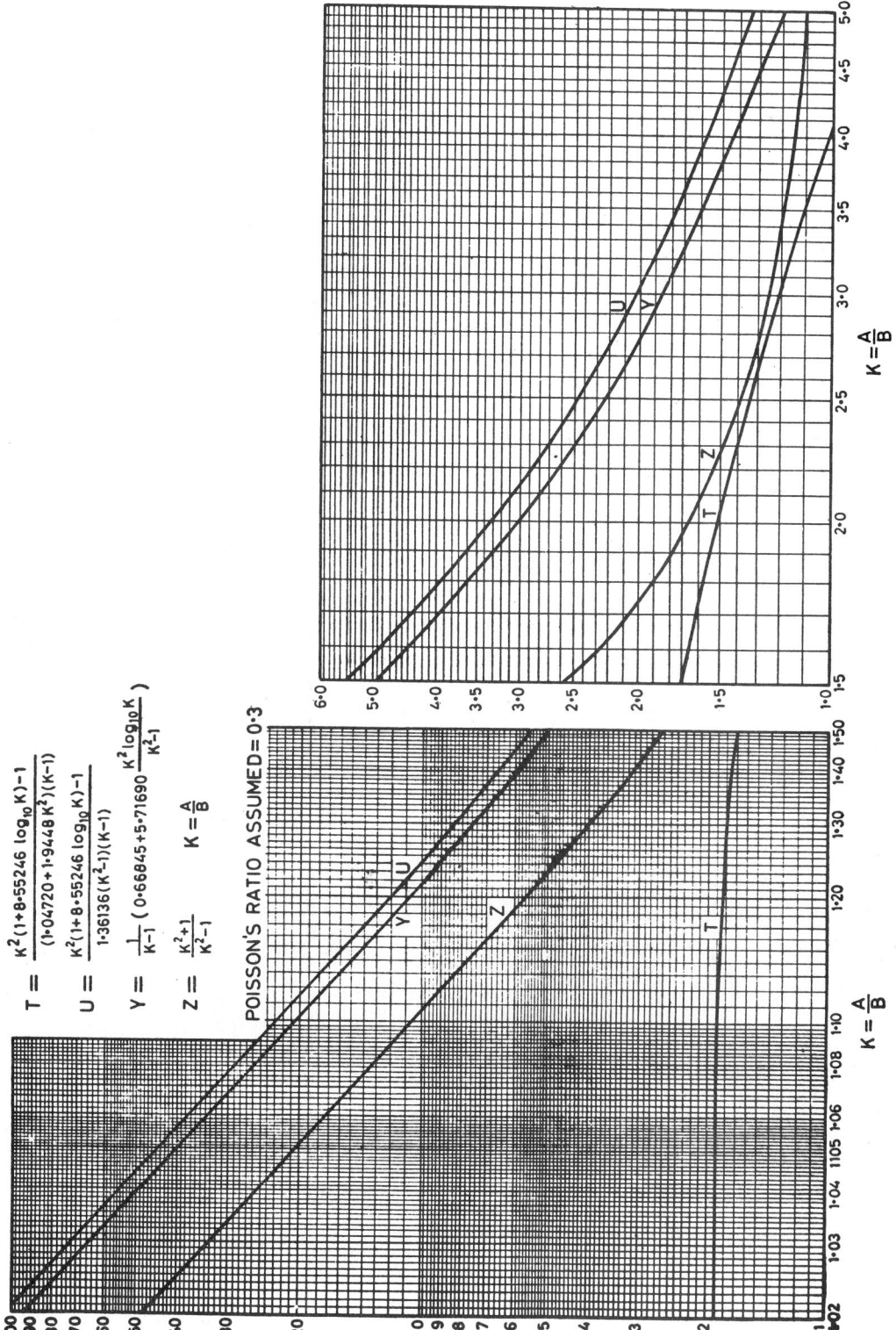

$$T = \frac{K^2(1+8 \cdot 55246 \log_{10} K) - 1}{(1 \cdot 04720 + 1 \cdot 9448 \, K^2)(K-1)}$$

$$U = \frac{K^2(1+8 \cdot 55246 \log_{10} K) - 1}{1 \cdot 36136(K^2-1)(K-1)}$$

$$Y = \frac{1}{K-1}\left(0 \cdot 66845 + 5 \cdot 71690 \, \frac{K^2 \log_{10} K}{K^2-1}\right)$$

$$Z = \frac{K^2+1}{K^2-1} \qquad K = \frac{A}{B}$$

POISSON'S RATIO ASSUMED = 0·3

FIGURE 8-20   Values of T, U, Y, and Z for $K = (A/B) > 1.5$. (*Source: IS 2825, 1969*)

TABLE 8–20
Gasket materials and contact facings[a]

Dimension N mm (in) (min)	Gasket material			Gasket factor, m	Minimum design seating stress, y MPa (kpsi)	Sketches and notes	Refer to Table 8–21 Use facing sketch	Use column
10	Rubber without fabric or a high percentage of asbestos fiber:							
		<70 IRHD* (75A Shore Durometer)		0.50	0		1 (a, b, c, d), 4, 5	II
		70 IRHD (75A)* or higher		1.00	1.37 (0.2)			
	Asbestos with a suitable binder for the operating conditions	3.2 mm (0.125 in)		2.00	11.0 (1.6)			
		1.6 mm (0.062 in)		2.75	25.5 (3.7)			
		0.8 mm (0.031 in) thickness		3.50	44.8 (6.5)			
	Rubber and elastomers with cotton fabric insertion			1.25	2.75 (0.40)			
	Rubber and elastomers with asbestos fabric insertion, with or without wire reinforcement	3-ply		2.25	15.2 (2.2)		1 (a, b, c, d), 4, 5	II
		2-ply		2.50	20.0 (2.9)			
		1-ply		2.75	25.5 (3.7)			
10	Vegetable fiber			1.75	7.55 (1.1)			
	Spiral-wound metal, asbestos-filled	Carbon steel, stainless steel or monel metal		2.50	68.9 (10.0)		1 (a, b)	II
				3.00	68.9 (10.0)			
	Corrugated metal, asbestos inserted or	Soft aluminum		2.50	20.0 (2.9)			
		Soft copper or brass		2.75	25.5 (3.7)			
		Iron or soft steel		3.00	31.0 (4.5)			
	Corrugated metal, jacketed asbestos filled	Monel metal or 4-6% chrome steel		3.25	38.0 (5.5)			
		Stainless steels		3.50	44.8 (6.5)			
	Corrugated metal	Soft aluminum		2.75	25.5 (3.7)		1 (a, b, c, d)	II
		Soft copper or brass		3.00	31.0 (4.5)			
		Iron or soft steel		3.25	38.0 (5.5)			
		Monel metal or 4-6% chrome steel		3.50	44.0 (6.5)			
		Stainless steel		3.75	52.4 (7.6)			
	Flat-metal-jacketed, asbestos-filled	Soft aluminum		3.25	38.0 (5.5)		1a, 1b, 1c*, 1d*, 2*	II
		Soft copper or brass		3.50	44.0 (6.5)			
		Iron or soft steel		3.75	52.4 (7.6)			
		Monel metal or 4-6% chrome steel		3.50	55.1 (8.0)			
		Stainless steels		3.75	62.1 (9.0)			

TABLE 8-20
Gasket materials and contact facings (*Cont.*)

Dimension N, mm (in) (min)	Gasket material		Gasket factor, m	Minimum design seating stress, y MPa (kpsi)	Sketches and notes	Refer to Table 8–21 Use facing sketch	Use column
10	Grooved metal	Soft aluminum	3.25	3.80 (5.5)		1 (*a*, *b*, *c*, *d*), 2, 3	II
		Soft copper or brass	3.50	44.8 (6.5)			
		Iron or soft steel	3.75	52.4 (7.6)			
		Monel metal or 4–6% chrome steel	3.75	62.1 (9.0)			
		Stainless steels	4.25	69.6 (10.1)			
6	Solid flat metal	Soft aluminum	4.00	60.7 (8.8)		1 (*a*, *b*, *c*, *d*), 2, 3, 4, 5	I
		Soft copper or brass	4.75	89.6 (13.0)			
		Iron or soft steel	5.50	124.2 (18.0)			
		Monel metal or 4–6% chrome steel	6.00	150.3 (21.8)			
		Stainless steels	6.50	179.3 (26.0)			
	Ring joint	Iron or soft steel	5.50	124.2 (18.0)		6	
		Monel metal or 4–6% chrome steel	6.00	150.3 (21.8)			
		Stainless steels	6.50	179.3 (26.0)			
	Rubber O-rings:						
	<75 IRHD (75A Shore Dur)		3^c	0.69 (0.10)		7 only	
	75 (75A) to 85 IRHD (85A)		6^c	1.47 (0.2)			
	Rubber square section rings:					8 only	II
	<75 IRHD (75A Shore Dur)		4^c	0.98 (0.14)			
	75 (75A) to 85 IRHD (85A)		9^c	2.75 (0.40)			
	Rubber T-section rings:					9 only	
	Below 75 IRHD (75A Shore Dur)		4^c	0.98 (0.14)			
	Between 75 (75A) and 85 IRHD (85A)		9^c	2.75 (0.40)			

[a] Gasket factors (m) for operating conditions and minimum design seating stress (y).

[b] or * The surface of a gasket having a lap should not be against the nubbin.

[c] These values have been calculated.

Note: This table gives a list of many commonly used gasket materials and contact facings with suggested design values of m and y that have generally proved satisfactory in actual service when using effective gasket seating with b given in Table 8–21 and Fig. 8–13. The design values and other details given in this table are suggested only and are not mandatory.

Source: IS 2825, 1969.

TABLE 8–21
Effective gasket width

Facing sketch (exaggerated)	Basic gasket seating width, b_o	
	Column I	Column II
1a	$\dfrac{N}{2}$	$\dfrac{N}{2}$
1b[a]		
1c	$\dfrac{w+25T}{2}; \left(\dfrac{w+N}{4}\,\text{max}\right)$	$\dfrac{w+25T}{2}; \left(\dfrac{w+N}{4}\,\text{max}\right)$
1d[a]		
2	$\dfrac{w+N}{4}$	$\dfrac{w+3N}{8}$
3	$\dfrac{w}{2}; \left(\dfrac{N}{4}\,\text{min}\right)$	$\dfrac{w+N}{4}; \left(\dfrac{3N}{8}\,\text{min}\right)$
4[a]	$\dfrac{3N}{8}$	$\dfrac{7N}{16}$
5*	$\dfrac{N}{4}$	$\dfrac{3N}{8}$
6	$\dfrac{w}{8}$	—
7	—	$\dfrac{N}{2}$
8	—	$\dfrac{N}{2}$
9	—	$\dfrac{N}{2}$

[a]Where serrations do not exceed 0.4 mm depth and 0.8 mm width spacing, sketches 1b and 1d shall be used.

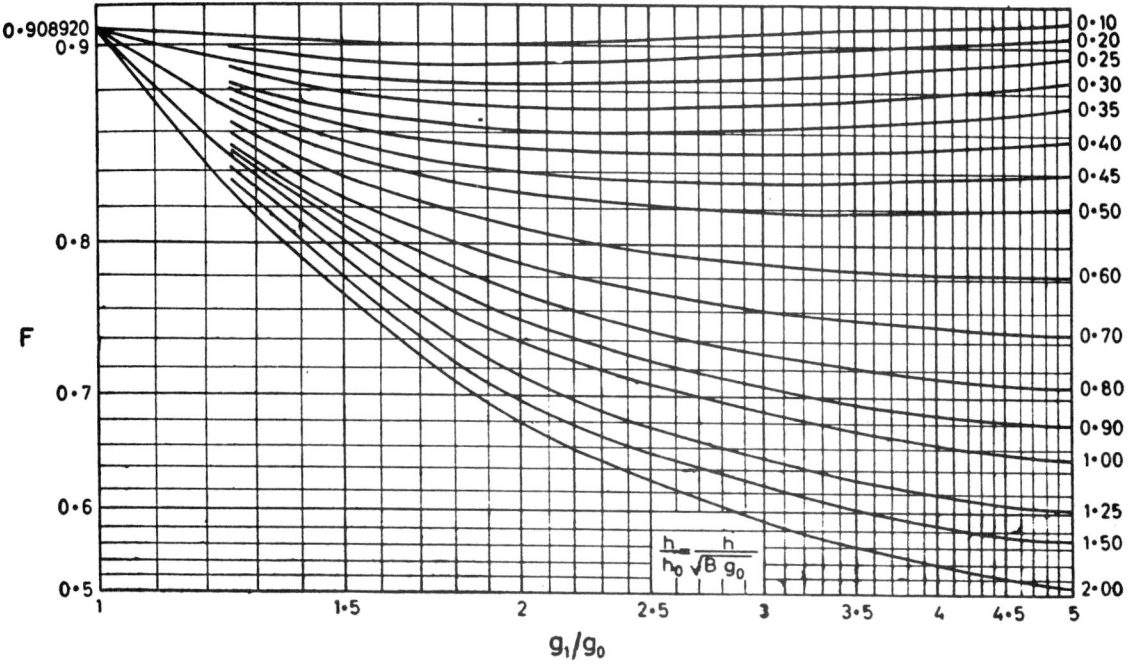

FIGURE 8–21 Values of F (integral flange factors). (*Source: IS 2825, 1969.*)

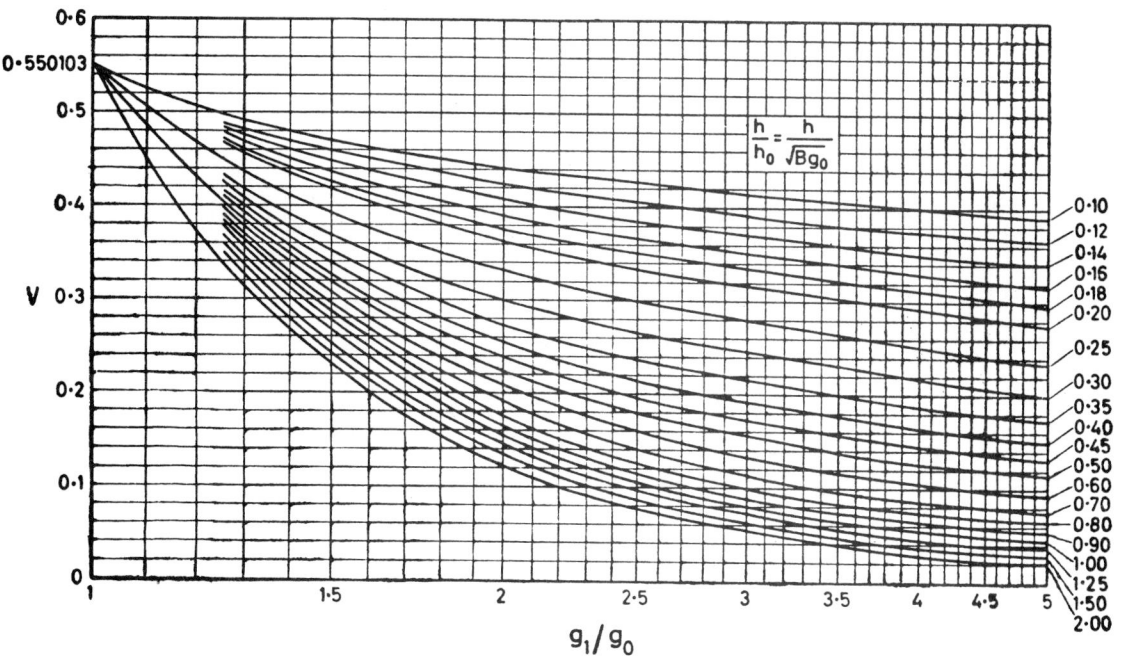

FIGURE 8–22 Values of V (integral flange factors). (*Source: IS 2825, 1969.*)

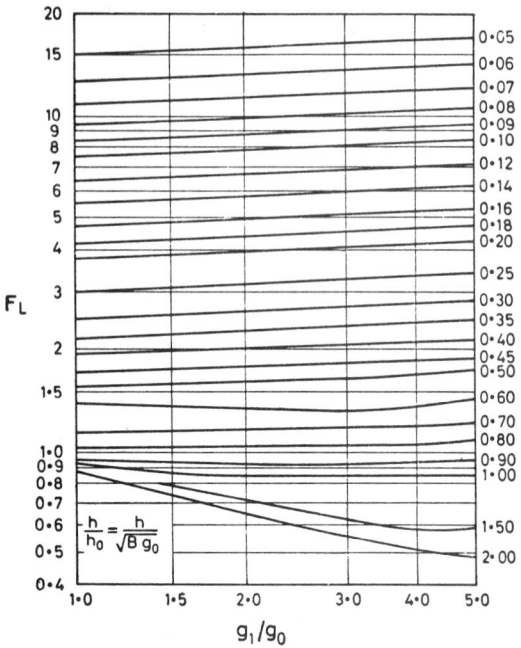

FIGURE 8–23 Values of F_L (loose hub flange factors). (*Source: IS 2825, 1969.*)

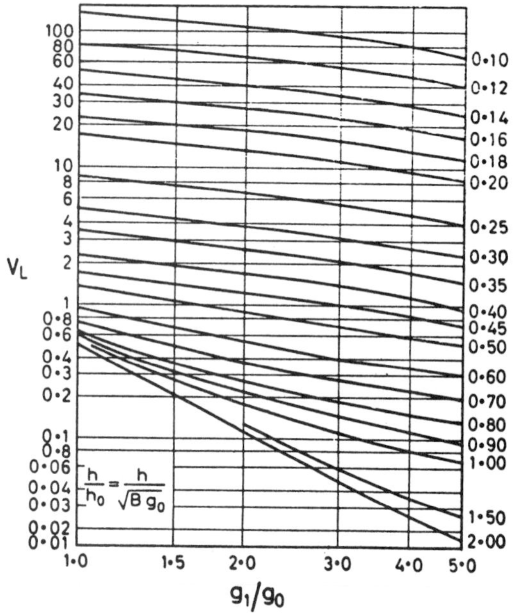

FIGURE 8–24 Values of V_L (loose hub flange factors). (*Source: IS 2825, 1969.*)

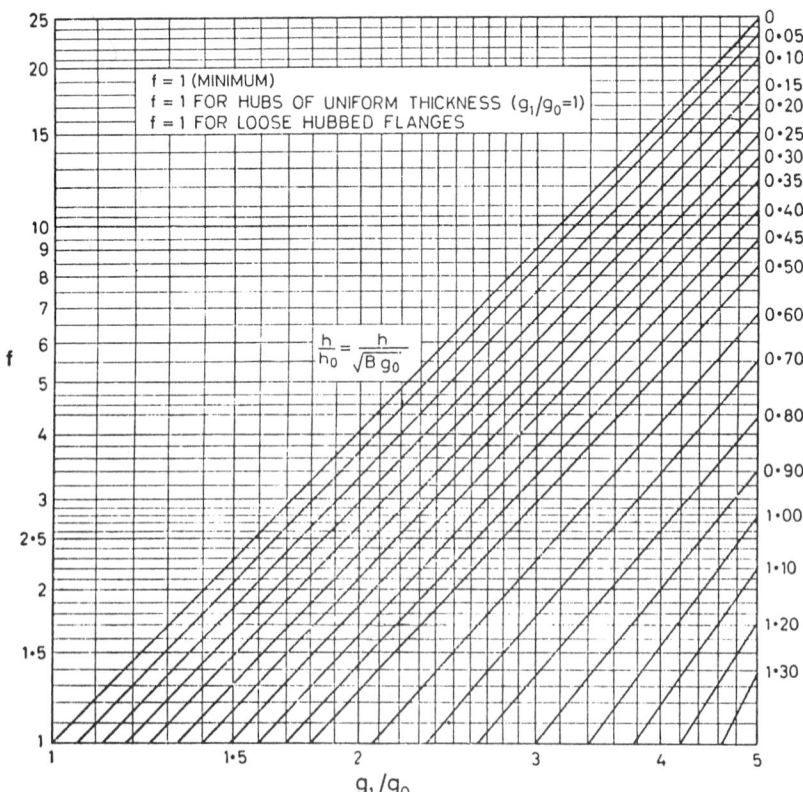

FIGURE 8–25 Values of f (hub stress correction factor). (*Source: IS 2825, 1969.*)

Particular	Formula
The actual cross-sectional area of bolts using the root diameter of thread or least diameter of unthreaded portion (if less), to prevent damage to the gasket during bolting up	$A_b = \dfrac{2\pi y\,GN}{\sigma_{Sbat}} \not< A_m$ (8–111)

Flange design bolt load W

The bolt load in the design of flange for operating condition	$W = W_{m1}$ (8–112)
The bolt load in the design of flange for gasket seating	$W = \left(\dfrac{A_m + A_b}{2}\right)\sigma_{Sbat}$ (8–113)
The relation between bolt load per bolt (W_b), diameter of bolt D and torque M_t	$W_b = 0.17 D\,M_t$ for lubricated bolts (8–114a)

US Customary units

where W_b in lbf, D in in, M_t in lbf in

$= 263.5 D\,M_t$ **SI**   (8–114b)

where W_b in N, D in m, M_t in N m

$= 0.2 D\,M_t$ for unlubricated bolts   (8–114c)

US Customary units

where W_b in lbf, D in in, M_t in lbf in

$= 310 D\,M_t$ **SI**   (8–114d)

where W_b in N, D in m, M_t in N m

Flange moments

The total moment acting on the flange M_o for operating condition	$M_o = M_D + M_t + M_G$ (8–115a)
	$= H_D h_D + H_T h_T + H_G h_G$ (8–115b)

This is based on the flange design load of Eq. (8–112) with moment arms as given in Table 8–22.

The total flange moment M_o for gasket seating, which is based on the flange design bolt load of Eq. (8–113)	$M_o = W h_G = \left(\dfrac{A_m + A_b}{2}\right)\left(\dfrac{C - G}{2}\right)\sigma_{Sbat}$ (8–116)

TABLE 8–22
Moment arms for flange loads under operating conditions

Type of flange	h_D	h_T	h_G
Integral-type flanges	$R + 0.5g_1$	$\dfrac{R + g_1 + h_G}{2}$	$\dfrac{C - G}{2}$
Loose-type except lap joint flanges and optional-type flanges	$\dfrac{C - B}{2}$	$\dfrac{h_D + h_G}{2}$	$\dfrac{C - G}{2}$
Lap joint flanges	$\dfrac{C - B}{2}$	$\dfrac{C - G}{2}$	$\dfrac{C - G}{2}$

TABLE 8–23
Maximum allowable stresses in stays and stay bolts, σ_{Sa}

	Stress			
	For lengths between supports not exceeding 120 × diameter		For lengths between supports exceeding 120 × diameter	
Type of stay	MPa	kpsi	MPa	kpsi
(a) Unwelded or flexible stays less than 20 × diameter long, screwed through plates with ends riveted over	51	7.5		
(b) Hollow steel stays less than 20 × diameter long, screwed through plates with ends riveted over	55	8.0		
(c) Unwelded stays and unwelded portions of welded stays, except as specified in (a) and (b)	66	9.5	58	8.5
(d) Steel through stays exceeding 38 mm diameter	71	10.4	62	9.0
(e) Welded portions of stays	41	6.0	51	6.0

Source: ASME Boiler and Pressure Vessel Code.

Particular	Formula

Flange stresses

The stress in the flange shall be determined for both the gasket seating condition and the operating condition.

INTEGRAL-TYPE FLANGES AND LOOSE-TYPE FLANGES WITH A HUB.
There are three types of stress:

The larger of these two controls with the following formulas:

Longitudinal hub stress

$$\sigma_H = \frac{fM_o}{Lg_1^2 B} \tag{8–117}$$

Radial flange stress

$$\sigma_R = \frac{(1.33te + 1)\,M_o}{Lt^2 B} \tag{8–118}$$

Particular	Formula
Tangential stress	$$\sigma_\theta = \left(\frac{YM_o}{t^2 B}\right) - Z\sigma_R \qquad (8\text{--}119)$$
For flange factors values	Refer to Figs. 8–20 to 8–25.

LOOSE-TYPE FLANGES WITHOUT HUB AND LOOSE-TYPE FLANGES WITH HUB WHICH THE DESIGNER CHOOSES TO CALCULATE

(a) Stresses without considering the hub

   (1) Tangential stress

$$\sigma_\theta = \frac{YM_o}{t^2 B} \qquad (8\text{--}120)$$

   (2) The radial and longitudinal stress

$$\sigma_H = \sigma_R = 0 \qquad (8\text{--}121)$$

(b) Allowable flange design stresses: The flange stresses calculated by Eqs. (8–117) to (8–121) shall not exceed the values of stresses given by Eqs. (8–122) to (8–126).

   (1) The longitudinal hub stress

$$\sigma_H \not< \sigma_{Sfd} \qquad \text{for cast iron} \qquad (8\text{--}122a)$$
$$\sigma_H \not> 1.5\sigma_{Sfd} \qquad \text{for other materials} \qquad (8\text{--}122b)$$

     (i) The longitudinal hub stress for optional-type flanges designed as integral and also integral type where the neck material constitutes the hub of the flange

$$(a)\ \sigma_H \not> 1.5\sigma_{Sfd} \quad \text{or} \quad 1.5\sigma_{Snd} \qquad (8\text{--}123a)$$

The smaller of σ_{Sfd} and σ_{Snd} is to be selected.

     (ii) The longitudinal hub stress for integral-type flanges with hub welded to the neck, pipe, or vessel wall

$$(b)\ \sigma_H \not> 1.5\sigma_{Sfd} \quad \text{or} \quad 2.5\sigma_{Snd} \qquad (8\text{--}123b)$$

The smaller of σ_{Sfd} and σ_{Snd} is to be selected.

   (2) The radial stress

$$\sigma_R \not> \sigma_{Sfd} \qquad (8\text{--}124)$$

   (3) The tangential stress

$$\sigma_\theta \not> \sigma_{Sfd} \qquad (8\text{--}125)$$

   (4) The average of σ_H and σ_R, and σ_H and σ_θ

$$(\sigma_H + \sigma_R)/2 \not> \sigma_{Sfd} \qquad (8\text{--}126a)$$
$$(\sigma_H + \sigma_\theta)/2 \not> \sigma_{Sfd} \qquad (8\text{--}126b)$$

Flanges under external pressure

The design of flanges for external pressure only shall be based on the formulas given for internal pressure except that for operating conditions.

$$M_o = H_D (h_D - h_G) + H_T(h_T - h_G) \qquad (8\text{--}127a)$$

$$\text{for operating conditions}$$

$$M_o = Wh_G \qquad \text{for gasket seating} \qquad (8\text{--}127b)$$

$$\text{where } W = \sigma_{Sbat} (A_{m2} + A_b)/2 \qquad (8\text{--}128)$$

REFERENCES

1. "Rules for Construction of Pressure Vessels," Section VIII, Division 1, *ASME Boiler and Pressure Vessel Code*, The American Society of Mechanical Engineers (ASME), New York, 1986 ed., July 1, 1986.
2. "Rules for Construction of Pressure Vessels," Section VIII, Division 2, Alternative Rules, *ASME Boiler and Pressure Vessel Code*, ASME, New York, 1986 ed., July 1, 1986.
3. "Rules for Construction of Power Boiler," Section 1, *ASME Boiler and Pressure Vessel Code*, ASME, New York, 1983 ed., July 1, 1971.
4. "Recommended Rules for Care of Power Boilers," Section VII, *ASME Boiler and Pressure Vessel Code*, ASME, New York, 1983.
5. "Rules for in Service Inspection of Nuclear Power Plant Components," Section XI, *ASME Boiler and Pressure Vessel Code*, 1971.
6. "Heating Boilers," Section IV, *ASME Boiler and Pressure Vessel Code*, ASME, New York, 1983.
7. "Recommended Rules for Care and Operation of Heating Boilers," Section VI, *ASME Boiler and Pressure Vessel Code*, ASME, New York, 1983.
8. "Part A: Ferrous Materials," Section II, *ASME Boiler and Pressure Vessel Code*, ASME, New York, 1983.
9. "Part B: Non-ferrous Materials," Section II, *ASME Boiler and Pressure Vessel Code*, ASME, New York, 1983.
10. Azbel, D. S., and N. P. Cheremisinoff, *Chemical and Process Equipment Design — Vessel Design and Selection*, Ann Arbor Science Publishers, Ann Arbor, Michigan, 1982.
11. Bureau of Indian Standards, ZS 2825-1969 (under revision).
12. Chuse, R., *Pressure Vessels — The ASME Code Simplified*, 5th ed., McGraw-Hill Book Company, New York, 1977.
13. Lingaiah, K., and B. R. Narayana Iyengar, *Machine Design Data Handbook*, Engineering College Cooperative Society, Bangalore, India, 1962.
14. Lingaiah, K., and B. R. Narayana Iyengar, *Machine Design Data Handbook*, Vol. I (*SI and Customary Metric Units*), Suma Publishers, Bangalore, India, 1983.
15. Lingaiah, K., *Machine Design Data Handbook*, Vol. II (*SI and Customary Metric Units*), Suma Publishers, Bangalore, India, 1986.

DESIGN OF POWER BOILERS

SYMBOLS

C	smoke area consisting of the total internal transverse area of the tube, m^2 (ft^2)
d	diameter of cylinder or shell, m (in)
	diameter or short span, measured as shown in Fig. 8–9 (Chap. 8)
d_o	maximum allowable diameter of opening, m (in)
	outside diameter of cylinder or shell or tube or pipe, m (in)
D_o	outside diameter of furnace or flue, m (in)
$D.S.$	disengaging surface or area of water surface through which steam bubbles must be discharged, the water being considered at the middle-gauge cock, m^2 (ft^2)
E	modulus of elasticity, GPa (Mpsi)
G	area of the grate as finally adopted, m^2 (ft^2)
h or t	thickness of tube or shell wall, m (in)
H	total heating surface in contact with the fire, m^2 (ft^2)
l	length of the flue sections, m (in)
n	factor of safety to be taken as 5 for usual cases
L	radius to which the head is formed, measured on the concave side of the head, m (in)
P	rated power of boiler
p or P	maximum allowable working pressure, Pa or MPa (psi)
R_i	inside radius of cylindrical shell, m (in)
S	volume of steam included between the shell and a horizontal line through the position of the central gauge as finally determined, m^2 (ft^2)
t (or h)	thickness of tube or pipe or cylinder or shell or plate, m (in)
$S.H.S.$	total area of superheating surface based on the actual area in contact with the fire, m^2 (ft^2)
W	net water volume in the boiler below the line of the central gauge cock, m^2 (ft^2)
$W.H.S$	total area of water heating surface based on the actual area in contact with the fire, m^2 (ft^2)
σ_{Sa}	maximum allowable stress value, MPa (kpsi) from Tables 7–1, 8–9 to 8–11, and 8–17 (Chap. 8)
η	efficiency of joint

Other factors in performance or in special aspects are included from time to time in this chapter and, being applicable only in their immediate context, are not given at this stage.

Note: σ and τ with initial subscript S designates strength properties of material used in the design, which will be used and observed throughout this *Machine Design Data Handbook*.

Particular	Formula

BOILER TUBES AND PIPES

For calculation of the minimum required thickness (t) and maximum allowable working pressure (p or P) of ferrous and nonferrous tubes and pipes from 12.5 mm ($\frac{1}{2}$ in) to 150 mm (6 in) outside diameter used in power boilers as per *ASME Boiler and Pressure Vessel Code* [2,3]

Refer to Eqs. (7–1) to (7–15) (Chap. 7).

For efficiency of joints (η), temperature coefficient (y), minimum allowance for threading, and structural stability (C) as per *ASME Boiler and Pressure Vessel Code*

Refer to Tables from 7–2 to 7–6 (Chap. 7).

For maximum allowable stress value (σ_{Sa}) for the materials of tubes and pipes as per *ASME Boiler and Pressure Vessel Code* [3]

Refer to Table 7–1

The maximum allowable working pressure for steel tubes or flues of fire tube boilers for different diameters and gauges of tubes as per *ASME Power Boiler Code* [2]

$$p = \frac{96.5}{d_o}(h - 1.625 \times 10^{-3}) \quad \textbf{SI} \qquad (9\text{–}1a)$$

where p in MPa; h and d_o in m

$$= \frac{14000}{d_o}(h - 0.065) \qquad \begin{array}{c}\textbf{US Customary System}\\\textbf{Units}\end{array} \quad (9\text{–}1b)$$

where p in psi, h and d_o in in

For maximum allowable working pressure and thickness of steel tubes

Refer to Tables 7–7, 9–1, 9–2 and 9–4

The maximum allowable working pressure for copper tubes for firetube boilers subjected to internal or external pressure as per *ASME Power Boiler Code* [2]

$$p = \frac{83}{d_o}(h - 1 \times 10^{-3}) - 1.7 \quad \textbf{SI} \qquad (9\text{–}2a)$$

where p in MPa; d_o and h in m

$$= \frac{12000}{d_o}(h - 0.039) - 250 \;\begin{array}{c}\textbf{US Customary System}\\\textbf{units}\end{array} \quad (9\text{–}2b)$$

where p in psi, d_o and h in in

Refer to Table 9–3

For maximum allowable working pressure and thickness of copper tubes

Refer to Tables 9–3 and 9–5.

TABLE 9-1
Maximum allowable working pressures for seamless steel and electric resistance welded steel tubes or nipples for watertube boilers [from Eq. (7-4)]

Wall Thickness			Tube outside diameter, mm (in)																										
			12.5 (0.5)		19.0 (0.75)		25.0 (1.0)		31.25 (1.25)		37.5 (1.5)		43.75 (1.75)		50.0 (2.0)		62.5 (2.5)		75.0 (3.0)		87.5 (3.5)		100.1 (4.0)		112.5 (4.5)		125.0 (5.0)	
mm	in	Nearest Bwg no.	MPa	psi	MPa	psi	MPa	psi	MPa	psi	MPa	psi	MPa	psi	MPa	psi	MPa	psi	MPa	psi	MPa	psi	MPa	psi	MPa	psi	MPa	psi
1.375	0.055	17−	3.38	590	2.41	350	3.24	470	2.42	350	3.0	430	3.38	490	2.83	410	2.76	400	2.34	390	2.90	420						
1.625	0.065	16	7.52	1090	4.62	670	5.00	720	3.80	550	4.06	590	4.34	630	3.65	530	3.45	500										
1.875	0.075	15+	11.03	1600	6.90	1000	6.62	960	5.10	740	5.24	760																
2.125	0.085	14+			9.24	1340																						
2.375	0.095	13							12.13	1760																		
2.625	0.105	12−							13.65	1980	11.03	1600	9.24	1340	7.93	1150												
3.000	0.120	11									12.90	1870	10.82	1570	9.24	1340	7.17	1040	5.80	840								
3.375	0.135	10+											12.34	1790	10.62	1540	8.20	1190	6.62	960	5.52	800	4.68	680				
3.750	0.150	9+											13.92	2020	12.00	1740	9.24	1340	7.52	1090	6.27	910	5.38	780	4.62	670	4.07	590
4.125	0.165	8													13.38	1940	10.34	1500	8.34	1210	7.03	1020	6.00	870	5.24	760	4.62	670
4.500	0.180	7															11.45	1660	9.24	1340	7.72	1120	6.62	960	5.80	840	5.10	740
5.000	0.200	6−															12.90	1870	10.48	1520	8.76	1270	7.52	1090	6.55	950	5.80	840
5.500	0.220	5																	11.65	1690	9.80	1420	8.34	1210	7.31	1060	6.48	940
6.000	0.240	4+																	12.90	1870	10.68	1550	9.24	1340	8.07	1170	7.17	1040
6.500	0.260	3+																			11.86	1720	10.14	1470	8.90	1290	7.86	1140
7.000	0.280	2−																			12.90	1870	11.03	1600	9.65	1400	8.55	1240
7.500	0.300																			13.92	2020	12.00	1740	10.48	1520	9.24	1340
8.000	0.320																					12.90	1870	11.24	1630	10.00	1450
8.500	0.340																					13.78	2000	12.06	1750	10.68	1550
9.000	0.360																							12.90	1870	11.45	1660
9.500	0.380																							13.72	1990	12.13	1760
10.000	0.400																									12.90	1870
10.500	0.420																									13.65	1980

* Bwg = Birmingham wire gauge

Source: *ASME Power Boiler Code*, Section I, 1983.

TABLE 9-2
Maximum allowable working pressures for steel tubes or flues for firetube boilers [from Eq. (9-1)]

Wall thickness mm	in	Nearest Bwg no.	25.00 (1)		37.50 (1.50)		50.00 (2)		62.50 (2.50)		75.00 (3)		87.50 (3.50)		100.00 (4)		112.50 (4.50)		125.00 (5)		137.50 (5.50)		150.0 (6)	
mm	in	Bwg no.	MPa	psi	MPa	psi	MPa	psi	MPa	psi	MPa	psi	MPa	psi	MPa	psi	MPa	psi	MPa	psi	MPa	psi	MPa	psi
2.375	0.095	13	2.90	420	1.93	280	1.45	210	1.17	170														
2.625	0.105	12	3.86	560	2.62	380	1.93	280	1.59	230	1.31	190	1.10	160										
3.000	0.120	11	5.31	770	3.58	520	2.69	390	2.14	310	1.80	260	1.52	220	1.38	200	1.24	180						
3.375	0.135	10+	6.76	980	4.55	660	3.38	490	2.76	400	2.28	330	1.93	280	1.72	250	1.52	220	1.38	200				
3.750	0.150	9+			5.52	800	4.14	600	3.30	480	2.76	400	2.34	340	2.06	300	1.86	270	1.65	240	1.52	220		
4.125	0.165	8			6.48	940	4.83	700	3.86	560	3.24	470	2.76	400	2.41	350	2.21	320	1.93	280	1.80	260	1.65	240
4.500	0.180	7					5.58	810	4.48	650	3.72	540	3.17	460	2.83	410	2.48	360	2.28	330	2.07	300	1.86	270
5.000	0.200	6−					6.55	950	5.24	760	4.34	630	3.72	540	3.31	480	2.90	420	2.62	380	2.41	350	2.21	320
5.500	0.220	5					7.52	1090	6.00	870	5.03	730	4.27	620	3.79	550	3.38	490	3.03	440	2.76	400	2.55	370
6.000	0.240	4+					8.40	1230	6.83	990	5.65	820	4.83	700	4.28	620	3.79	550	3.38	490	3.10	450	2.83	410

Source: ASME Power Boiler Code, Section I, 1983.

Particular	Formula
The external working pressure, for plain lap-welded or seamless tubes up to and including 150 mm (6 in) external diameter, and if the thickness is greater than the standard one	$p = \dfrac{1}{n}\left[\dfrac{596h}{d_o} - 9.6\right]$ **SI** (9–3a)

where p in Pa, h and d in m

$$= \frac{1}{n}\left(\frac{86670h}{d_o} - 1386\right)$$ **US Customary System units** (9–3b)

where p in psi, h and d in in

For proportion of standard boiler tubes   Refer to Table 9–6.

TABLE 9–3
Maximum allowable working pressure for copper tubes for firetube boilers[a] [from Eq. (9–2)]

Outside diameter of tube		Gauge, Bwg																	
		12		11		10		9		8		7		6		5		4	
mm	in	MPa	psi	MPa	psi	MPa	psi	MPa	psi	MPa	psi	MPa	psi	MPa	psi	MPa	psi	MPa	psi
50.00	2	1.17	170	1.65	240	1.72	250	1.72	250	1.72	250	1.72	250	1.72	250	1.72	250	1.72	250
81.25	3.25					0.76	110	1.03	150	1.52	220	1.72	250	1.72	250	1.72	250	1.72	250
100.00	4									0.90	130	1.10	160	1.72	250	1.72	250	1.72	250
125.00	5													1.03	150	1.31	190	1.59	230

[a] For use at pressure not to exceed 1.7 MPa (250 psi) or temperature not to exceed 208° C (406°F).
Source: ASME Power Boiler Code, Section I, 1983.

TABLE 9–4
Maximum boiler pressures for use of ANSI B16.5 standard steel pipe flanges and flanged valves and fittings

Primary service pressure rating		Maximum allowable boiler pressure			
		Steam service at saturation temperature		Boiler feed and blow-off line service	
Mpa	psi	MPa	psi	MPa	psi
1.14	164.7	1.41	204.7	1.20	174.7
2.17	314.7	4.44	644.7	3.65	529.7
2.86	414.7	5.75	834.7	4.68	679.7
4.23	614.7	8.10	1174.7	6.79	984.7
6.30	914.7	11.40	1654.7	10.10	1464.7
10.44	1514.7	17.23	2514.7	16.13	2339.7
17.33	2514.7	22.10	3206.0	22.20	3220.7

Notes: Adjusted pressure ratings for steam service at saturated temperature corresponding to the pressure, derived from Table 2 to 8 ANSI B16.5–1968. Pressures shown include the factor for boiler feed and blow-off line service required by ASME corrected for saturation temperature corresponding to this pressure.
Source: ASME Power Boiler Code, Section I, 1983.

TABLE 9–5
Maximum external working pressures for use with lap-welded and seamless boiler tubes[a]

Nominal diameter, external diameter, mm (in)	Standard thickness, mm	Maximum allowable pressure		Nominal diameter, external diameter, mm (in)	Standard thickness, mm	Maximum allowable pressure	
		MPa	psi			MPa	psi
51 (2)	2.4	2.84	427	89 (3.5)	3.1	2.16	308
58 (2.25)	2.4	2.55	380	96 (3.75)	3.1	1.96	282
64 (2.5)	2.8	2.65	392	102 (4)	3.4	2.06	303
70 (2.75)	2.8	2.45	356	115 (4.5)	3.4	1.67	238
76 (3)	2.8	2.26	327	127 (5)	3.8	1.67	235
83 (3.25)	3.1	2.26	327	153 (6)	4.2	1.37	199

[a] External diameter 50 to 150 mm (2 to 6 in).

TABLE 9–6
Proportions of standard boiler tubes

Nominal diameter, actual external diameter mm (in)	Actual internal diameter, mm	Thickness, mm	External circumference, mm	Internal circumference, mm	External transverse area, mm^2	Internal transverse area, mm^2	Length of tube m^{-2} of internal heating surface, m	Weight per meter	
								N	lbf
45 (1.76)	38	2.4	140	125	1600	1200	7.58	24.5	1.679
51 (2)	46	2.4	160	144	2000	1700	6.58	28.2	1.932
58 (2.25)	50	2.4	181	165	2000	2100	5.78	32.0	2.186
64 (2.5)	56	2.8	200	183	3200	2600	5.24	40.7	2.783
70 (2.75)	64	2.8	220	200	3800	3200	4.74	44.9	3.074
76 (3)	71	2.8	240	221	4500	3900	4.38	49.1	3.365
83 (3.25)	76	3.0	260	241	5400	4500	3.98	58.5	4.011
89 (3.5)	81	3.0	280	260	6200	5400	3.71	63.0	4.331
96 (3.75)	89	3.0	300	280	7000	6200	3.45	68.0	4.652
102 (4)	94	3.3	320	290	8000	6900	3.25	80.8	5.532
115 (4.5)	107	3.3	360	340	10000	9000	2.86	91.2	6.248
127 (5)	120	3.8	400	370	12800	11100	2.58	112.3	7.669
153 (6)	142	4.2	480	450	18300	16300	2.15	150.0	10.282

Particular	Formula
The external pressure, for plain lap-welded, or seamless tubes or flues over 50 mm (2 in) and not exceeding 150 mm (6 in) external diameter	Refer to Table 9–5.
The minimum required thickness of component when it is of riveted construction or does require staying as per *ASME Power Boiler Code* [2]	$$h = \frac{pR_i}{0.8\sigma_{Sa}\eta - 0.6p} \qquad (9\text{–}4)$$
The maximum allowable working pressure as per *ASME Power Boiler Code*	$$p = \frac{0.8\sigma_{Sa}\eta}{R_i + 0.6h} \qquad (9\text{–}5)$$

DISHED HEADS

The thickness of a blank unstayed dished head with the pressure on the concave side, when it is a segment of a sphere as per *ASME Power Boiler Code*	$$h = \frac{5pL}{4.8\sigma_{Sa}\eta} \qquad (9\text{–}6)$$

where

L = radius to which the head is dished, measured on the concave side of the head, m (in)

η = efficiency of weakest joint used in forming the head (Refer to Table 8–3 for η.)

The minimum distance between the centers of any two openings, rivet holes excepted, shall be determined by Eq. (9–7)	$$L = \frac{A + B}{2(1 - K)} \qquad (9\text{–}7)$$

where

L = distance between the centers of the two openings measured on the surface of the head, m (in)

A, B = diameters of two openings, m (in)

K = same as defined in Eqs. (9–8a) and (9–8b)

The expression for K	$$K = \frac{pd_o}{1.6\sigma_{Sa}h} \qquad (9\text{–}8a)$$
	$$K = \frac{pd_o}{1.82\sigma_{Sa}h} \qquad (9\text{–}8b)$$

Equation (9–8a) shall be used with shells and headers designed by using Eqs. (9–4) and (9–5).

Equation (9–8b) shall be used with shells and headers designed by using Eqs. (9–9) and (9–10):

The minimum required thickness of ferrous drums and headers based on strength of weakest course as per *ASME Power Boiler Code*	$$h = \frac{pd_o}{2\sigma_{Sa}\eta + 2yp} + C \text{ or } \frac{pR_i}{\sigma_{Sa}\eta - (1 - y)p} + C \quad (9\text{–}9)$$
The maximum allowable working pressure as per *ASME Power Boiler Code*	$$p = \frac{2\sigma_{Sa}\eta(h - C)}{d_o - 2y(h - C)} \text{ or } \frac{\sigma_{Sa}\eta(h - C)}{R_i + (1 - y)(h - C)} \quad (9\text{–}10)$$

Particular	Formula
	For values y, C, and σ_{Sa} refer to Tables 7–1, 7–3, and 7–6
The thickness of a blank unstayed full-hemispherical head with the pressure on the concave side	$$h = \frac{pL}{1.6\sigma_{Sa}\eta} \qquad (9\text{–}11a)$$ $$h = \frac{pL}{(2\sigma_{Sa}\eta - 0.2p)} \qquad (9\text{–}11b)$$ Equation (9–11b) may be used for heads exceeding 12.5 mm (0.5 in) in thickness that are to be used with shells or headers designed under Eqs. (9–9) and (9–10) and that are integrally formed on seamless drums or are attached by fusion welding and do not require staying.
The formula for the minimum thickness of head when the required thickness of the head given by Eqs. (9–9) and (9–10) exceeds 35 percent of the inside radius	$$h = L(y^{1/3} - 1) \qquad (9\text{–}12)$$ where $y = \dfrac{2(\sigma_{Sa}\eta + p)}{2\sigma_{Sa}\eta - p} \qquad (9\text{–}12a)$

UNSTAYED FLAT HEADS AND COVERS

The minimum required thickness of flat unstayed circular heads, covers and blind flanges as per *ASME Power Boiler Code*	$$h = d\sqrt{Cp/\sigma_{Sa}} \qquad (9\text{–}13)$$ where $C =$ a factor depending on the method of attachment of head on the shell, pipe or header (refer to Table 8–6 for C) $d =$ diameter or short span, measured as shown in Fig. 8–9
The minimum required thickness of flat unstayed circular heads, covers or blind flange which is attached by bolts causing edge moment Fig. 8–9(*j*) as per *ASME Power Boiler Code*	$$h = d[Cp/\sigma_{Sa} + 1.78\, Wh_G/_{Sa}d^3]^{1/2} \qquad (9\text{–}14)$$ where $W =$ total bolt load, kN (1bf) $h_G =$ gasket moment arm, Fig. 8–13 and Table 8–22. Refer to Tables 8–20 to 8–22 and Fig. 8–13
For details of bolt load H_G, bolt moments, gasket materials, and effect of gasket width on it	
The minimum required thickness of unstayed heads, covers, or blind flanges of square, rectangular, ellipitcal, oblong segmental, or otherwise noncircular as per *ASME Power Boiler Code*	$$h = d\sqrt{ZCp/\sigma_{Sa}} \qquad (9\text{–}15)$$

Particular	Formula
	where
	$Z = 3.4 - 2.4d/D$ $\qquad$ (9–15a)
	$D =$ long span of noncircular heads or covers measured perpendicular to short span, m (in)
	Z need not be greater than two and one-half (2.5)
	Equation (9–15) does not apply to noncircular heads, covers, or blind flanges attached by bolts causing bolt edge moment
The minimum required thickness of unstayed noncircular heads, covers, or blind flanges which are attached by bolts causing edge moment Fig. 8–9 as per *ASME Power Boiler Code*	$h = d\,[ZCp/\sigma_{Sa} + 6Wh_G/\sigma_{Sa}Ld^2]^{1/2}$ $\qquad$ (9–16)
The required thickness of stayed flat plates (Figs. 8–10 and 8–11) as per *ASME Power Boiler Code*	$h = p_t\sqrt{[p/\sigma_{Sa}c_5]}$ $\qquad$ (9–17)
	where
	$p_t =$ maximum pitch, m (in), measured between straight lines passing through the centers of the stay bolts in the different rows
	(Refer to Table 9–7 for pitches of stay bolts.)
	$c_5 =$ a factor depending on the plate thickness and type of stay (Refer to Table 8–15 for values of c_5.)
	For σ_{Sa} refer to Tables 8–8, 8–23, and 8–11
The maximum allowable working pressure for stayed flat plates as per *ASME Power Boiler Code*	$p = \dfrac{h^2\sigma_{Sa}c_5}{p_t^2}$ $\qquad$ (9–18)
For all allowable stresses in stay and stay bolts	Refer to Chapter 8

Also for detail design of different types of heads, covers, openings and reinforcements, ligaments, and bolted flanged connection

COMBUSTION CHAMBER AND FURNACES

Combustion chamber tube sheet

The maximum allowable working pressure on tube sheet of a combustion chamber where the crown sheet is suspended from the shell of the boiler as per *ASME Power Boiler Code*	$P = 27000\,\dfrac{h(D - d_i)}{wD}$ $\quad$ **US Customary System units** $\quad$ (9–19a)
	where
	$h =$ thickness of tube, in
	$w =$ distance from the tube sheet to opposite combustion chamber sheet, in

TABLE 9–7
Maximum allowable pitch for screwed staybolts, ends riveted over

| Pressure | | Thickness of plate, mm (in) | | | | | | | | | | | | | |
MPa	psi	7.8125	(0.3125)	9.375	(0.375)	10.9375	(0.4375)	12.500	(0.50)	14.0625	(0.5625)	15.6250	(0.625)	17.1875	(0.6875)
						Maximum pitch of staybolts, mm (in)									
0.67	100	131.25	(5.25)	159.375	(6.375)	184.375	(7.375)								
0.76	110	125.000	(5.000)	150.000	(6.000)	175.000	(7.000)	209.375	(8.375)						
0.83	120	118.750	(4.75)	143.750	(5.75)	168.750	(6.75)	200.000	(8.000)						
0.86	125	118.750	(4.75)	140.625	(5.625)	165.625	(6.625)	193.750	(7.75)						
0.90	130	115.625	(4.625)	137.500	(5.50)	162.500	(6.50)	190.625	(7.625)						
0.96	140	112.500	(4.50)	134.375	(5.375)	156.250	(6.25)	184.375	(7.375)	209.375	(8.375)				
1.03	150	106.250	(4.25)	128.125	(5.125)	150.000	(6.000)	178.125	(7.125)	200.00	(8.000)				
1.10	160	103.125	(4.125)	125.000	(5.000)	146.875	(5.875)	171.875	(6.875)	193.750	(7.75)				
1.17	170	100.000	(4.000)	121.875	(4.875)	140.625	(5.625)	168.150	(6.75)	187.500	(7.500)	209.375	(8.375)		
1.24	180			118.750	(4.75)	137.500	(5.50)	162.500	(6.50)	184.375	(7.375)	203.125	(8.125)		
1.31	190			115.625	(4.625)	134.375	(5.375)	159.375	(6.375)	178.125	(7.125)	196.875	(7.875)		
1.38	200			112.500	(4.50)	131.25	(5.25)	153.125	(6.125)	175.000	(7.000)	193.750	(7.750)	212.500	(8.50)
1.55	225			106.25	(4.25)	121.875	(4.875)	146.875	(5.875)	162.500	(6.50)	181.250	(7.25)	200.000	(8.00)
1.72	250			100.000	(4.000)	115.625	(4.625)	137.50	(5.50)	156.250	(6.25)	171.875	(6.875)	175.625	(7.625)
2.07	300					106.250	(4.25)	125.000	(5.000)	140.625	(5.625)	156.250	(6.25)	175.00	(7.000)

Source: *ASME Power Boiler Code*, Section I, 1983.

Particular	Formula
	D = least horizontal distance between tube centers on a horizontal row, in
	d_i = inside diameter of tube, in
	P = maximum allowable working pressure, psi
	$$= 186 \frac{h(D - d_i)}{wD} \quad \textbf{SI} \qquad (9\text{--}19b)$$ where p in MPa; h, D, d_i, and w in m
The vertical distance between the center lines of tubes in adjacent rows where tubes are staggered	$$D_{va} = (2d_i D + d_i^2)^{1/2} \qquad (9\text{--}20)$$ where d_i and D have the same meaning as given under Eq. (9–19)
For minimum thickness of shell plates, dome plates, and tube plates and tube sheet for firetube boiler	Refer to Table 9–8.
For mechanical properties of steel plates of boiler	Refer to Table 9–9.

TABLE 9–8
Minimum thickness of shell plates, dome plates, and tube sheet for firetube boiler

Diameter of				Minithickness			
Shell and dome plates		Tube Sheet		Shell and dome plates		Tube sheet	
m	in	m	in	mm	in	mm	in
≥ 0.9	≥ 36	1.05	42	6.25	0.25	9.375	0.375
> 0.9–1.35	> 36–54	> 1.05–1.35	> 42–54	7.81	0.3125	10.94	0.4375
> 1.35–1.8	> 54–72	> 1.35–1.8	> 54–72	9.375	0.375	12.5	0.500
> 1.8	> 72	> 1.8	> 72	12.5	0.50	14.06	0.5625

Source: *ASME Power Boiler Code*, Section I, 1983.

TABLE 9–9
Mechanical properties of steel plates for boilers

Grade	Tensile strength		Yield stress, percent min of tensile strength	Elongation percent gauge length, $5.65 \sqrt{a*}$ [a]
	MPa	kpsi		
1	333.4–411.9	48.4–59.7	55	26
2A	362.8–480.5	52.6–69.7	50	25
2B	509.9–608.0	74.0–88.2	50	20

[a] $a*$ area of cross section.
Source: IS 2002–1, 1962.

Particular	Formula

Plain circular furnaces

FURNACES 300 MM (12 IN) TO 450 MM (18 IN) IN OUTSIDE DIAMETER, INCLUSIVE. Maximum allowable working pressure for furnaces not more than $4\frac{1}{2}$ diameters in length or height where the length does not exceed 120 times the thickness of the plate

$$p = \frac{0.36(18.75T - 1.03L)}{D} \quad \text{SI} \qquad (9-21a)$$

where p in MPa, T, D, and L in m

$$p = \frac{51.5(18.75T - 1.03L)}{D} \quad \begin{array}{l}\textbf{US Customary System} \\ \textbf{units} \quad (9-21b)\end{array}$$

where p in psi

D = outside diameter of furnace, in
L = total length of furnace between centers of head rivet seams, in
T = thickness of furnace walls, sixteenths of an inch

The maximum allowable working pressure for furnaces not more than $4\frac{1}{2}$ diameter in length of height where the length exceeds 120 times the thickness of the plate

$$p = \frac{29.3T^2}{LD} \quad \text{SI} \qquad (9-22a)$$

where p in MPa; T, L, and D in m

$$= \frac{4250T^2}{LD} \quad \textbf{US Customary System units} \qquad (9-22b)$$

where p in psi; T, L, and D in in

Circular flues

The maximum allowable external pressure for riveted flues over 150 mm (6 in) and not exceeding 450 mm (18 in) external diameter, constructed of iron or steel plate not less than 6 mm (0.25 in) thick and put together in sections not less than 600 mm (24 in) in length

$$p = \frac{56h}{d} \quad \text{SI} \qquad (9-23a)$$

where p in Pa; h and d in m

$$= \frac{8100h}{d} \quad \textbf{US Customary System units} \qquad (9-23b)$$

where p in psi; h and d in in

d = external diameter of flue, in

The formula for maximum allowable external pressure for riveted, seamless, or lap-welded flues over 450 mm (18 in) and not exceeding 700 mm (28 in) external diameter, riveted together in sections not less than 600 mm (24 in) nor more than $3\frac{1}{2}$ times the flue diameter in length, and subjected to external pressure only

$$p = \frac{6.7h - 0.4l}{d} \quad \text{SI} \qquad (9-24a)$$

where p in Pa; h, l, and d in m

$$= \frac{(966h - 53l)}{d} \quad \textbf{US Customary System units}$$

$$(9-24b)$$

Particular	Formula
	where p in psi and d in in
	h = thickness of wall in 1.5 mm (0.06 in)
	l > 600 mm (24 in) and < $3\frac{1}{2}d$
The maximum allowable working pressure for seamless or welded flues over 125 mm (5 in) in diameter and including 450 mm (18 in)	
(a) Where the thickness of the wall is not greater than 0.023 times the diameter as per *ASME Power Boiler Code*	$$p = \frac{68948\,h^3}{D^3} \qquad \textbf{SI} \qquad (9\text{–}25a)$$
	where p in MPa; h and D in m
	p = maximum allowable working pressure
	D = outside diameter of flue
	h = thickness of wall of flue
	$$= \frac{10^7 h^3}{D^3} \qquad \textbf{US Customary System units}$$
	$(9\text{–}25b)$
	where p in psi; h and D in in
(b) Where the thickness of the wall is greater than 0.023 times the diameter.	$$p = \frac{119\,h}{D} - 1.9 \qquad \textbf{SI} \qquad (9\text{–}26a)$$
	where p in MPa; h and D in m
	$$= \frac{17300\,h}{D} - 275 \quad \textbf{US Customary System units}$$
	$(9\text{–}26b)$
	where p in psi; h and D in in
Equations (9–24) and (9–25) may applied to riveted flues of the size specified provided the section are not over 0.91 m (3 ft) in length and the efficiency (η) of the joint	$$\eta \nleftarrow \frac{pD}{138\,h} \qquad \textbf{SI} \qquad (9\text{–}27a)$$
	where p in MPa; D and h in m
	$$\nleftarrow \frac{pD}{20000\,h} \qquad \textbf{US Customary System units}$$
	$(9\text{–}27b)$
	where p in psi; D and h in in

Particular	Formula

THE MAXIMUM ALLOWABLE PRESSURE FOR SPECIAL FURNACES HAVING WALLS REINFORCED BY RIBS, RINGS, AND CORRUGATIONS

(a) Furnaces reinforced by Adamson rings

$$p = \frac{6.7\,h - 0.4\,l}{d} \qquad \text{SI} \qquad (9\text{--}28a)$$

where p in Pa; h and d in m

$$= \frac{1080\,h - 59\,l}{d} \qquad \text{US Customary System units}$$

$$(9\text{--}28b)$$

where p in psi

h = thickness of wall, 1.5 mm (0.06 in) not to be less than 8 mm ($\frac{5}{16}$ in)

l = length of flue section, not to be less than 450 mm (18 in)

(b) Another expression for the maximum allowable working pressure when plain horizontal flues are made in sections not less than 450 mm (18 in) in length and not less than 8 mm ($\frac{5}{16}$ in) in thickness (Adamson-type rings)

$$p = \frac{0.4(300\,h - 1.03\,L)}{D} \qquad \text{SI} \qquad (9\text{--}29a)$$

where p in MPa; h, L, and D in m

$$= \frac{57.6(300\,h - 1.03\,L)}{D} \qquad (9\text{--}29b)$$

where p in psi; h, L, and D in in

(c) Corrugated rings

$$p = 68.5\,\frac{h}{d} \qquad \text{SI} \qquad (9\text{--}30a)$$

where p in Pa; h and d in m

$$= 10000\,\frac{h}{d} \qquad \text{US Customary System units}$$

$$(9\text{--}30b)$$

where p in psi; h and d in in

h = thickness of tube wall, mm (in) not to be less than 11 mm (0.44 in)

(d) Plain circular flues riveted together in sections

$$p = \frac{6.7\,d - 0.4\,l}{d} \qquad \text{SI} \qquad (9\text{--}31a)$$

where p in Pa; d and l in m

$$= \frac{966\,h - 53\,l}{D} \qquad \text{US Customary System units}$$

$$(9\text{--}31b)$$

where p in psi; l and d in in

Particular	Formula

Ring-reinforced type

The required wall thickness of a ring-reinforced furnace of flue shall not be less than that determined by the procedure given here

Assume a value for h (or t) and L. Determine the ratios L/D_o and D_o/t.

Following the procedure explained in Chap. 8, determine B by using Fig. 9–1. Compute the allowable working pressure P_a by the help of Eq. (9–32)

The allowable working pressure (P_a)

$$P_a = \frac{B}{(D_o/t)} \qquad (9\text{–}32)$$

where D_o = outside diameter of furnace or flue, in

Compare P_a with P. If P_a is less than P select greater value of t (or h) or smaller value of L so that P_a is equal to or greater than P, psi

The required moment of inertia (I_s) of circumferential stiffening ring

$$I_s = \frac{LD_o^2\left(t + \dfrac{A_s}{L}\right)A}{14} \qquad (9\text{–}33)$$

where

I_s = required moment of inertia of stiffening ring about its neutral axis parallel to the axis of the furnace, in^4

A_s = area of cross section of the stiffening ring, in^2

A = factor obtained from Fig. 9–1

The required moment of inertia of a stiffening ring shall be determined by the procedure given here

Assume the values of D_o, L, and t (or h) of furnace. Select a rectangular member to be used for stiffening ring and find its area A_s and its moment of inertia I. Then find the value of B from Eq. (9–34)

The expression for B

$$B = \frac{PD_o}{t + A_s/L} \qquad (9\text{–}34)$$

where P, D_o, t, A_s, and L are as defined under Eq. (9–33)

The value of factor A

After computing B from Eq. (9–34), determine the value of factor A by the help of Fig. 9–1 and B. If the required I_s is greater than the moment of inertia I, for the section selected above, select a new section with a larger moment of inertia and determine a new value of I_s. If the required I_s is smaller than the moment of inertia I selected as above, then that section should be satisfactory.

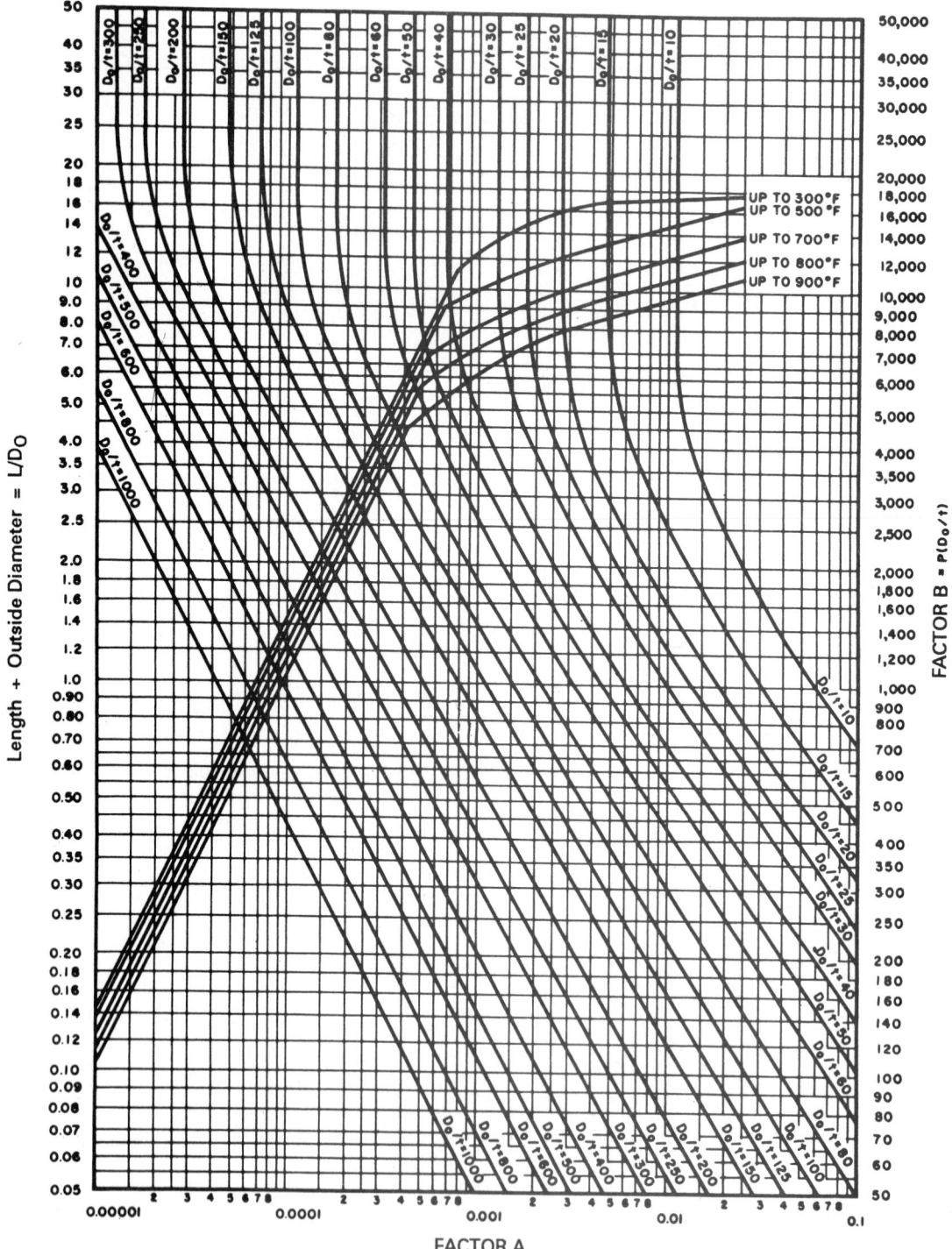

FIGURE 9–1 Chart for determining wall thickness of ring reinforced furnaces when constructed of carbon steel (specified yield strength, 210 to 262 MPa (30 to 38 kpsi) (1 kpsi = 6.894757 MPa). (*Source*: "Rules for Construction of Power Boilers,"ASME Boiler and Pressure Vessel Code, *Section I, 1983* and "Rules for Construction of Pressure Vessels," *Section VIII, Division I*, ASME Boiler and Pressure Vessel Code, July 1, 1986.)

Particular	Formula

Corrugated furnaces

The maximum allowable working pressure (P) on corrugated furnace having plain portion at the ends not exceeding 225 mm (9 in) in length

$$P = \frac{tC_6}{D} \qquad (9\text{–}35)$$

where

t = thickness, in, not less than 7.8 mm ($\frac{5}{16}$ in) for Leeds, Morison, Fox and Brown, and not less than 11 mm ($\frac{7}{16}$ in) for Purves and other furnaces corrugated by sections not over 450 mm (18 in) long.

D = mean diameter, in

Values of C_6 are taken from Table 9–10

TABLE 9–10
Values of C_6 for use in Eq. (9–35)

	C_6
1. For Leeds furnaces, when corrugations are not more than 200 mm (8 in) from center and not less than 56.25 mm (2.25 in) deep	17,300
2. For Morison furnaces, when corrugations are not more than 200 mm (8 in) from center to center and the radius of the outer corrugation is not more than one-half of the suspension curve	15,600
3. For Fox furnaces, when corrugations are not more than 200 mm (8 in) from center to center and not less than 37.5 mm (1.5 in) deep	14,000
4. For Curves furnaces, when rib projections are not more than 225 mm (9 in) from the center to center and not less than 34.375 mm (1.375 in) deep	14,000
5. For Brown furnaces, when corrugations are not more than 225 mm (9 in) from center to center and not less than 40.625 mm (1.625 in) deep	14,000
6. For furnaces corrugated by sections not more than 450 mm (18 in) from center to center and not less than 37.5 mm (1.5 in) deep, measured from the least inside greatest outside diameter of the corrugations and having the ends fitted into the other and substantially riveted together, provided the plain parts at the ends do not exceed 300 mm (12 in) in length	

Source: *ASME Power Boiler Code*, Section I, 1983

Stayed surfaces

The maximum allowable working pressure (P) for a stayed wrapper sheet of a locomotive-type boiler

$$P = \frac{11000 \, t\eta}{R - \Sigma s \sin \alpha} \qquad \textbf{US Customary System units}$$
$$(9\text{–}36a)$$

where

t = thickness of wrapper sheet, in
R = radius of wrapper sheet, in
η = minimum efficiency of wrapper sheet through joints or stay holes

Particular	Formula
	$\Sigma\, s \sin \alpha$ = summated value of transverse spacing ($s \sin \alpha$) for all crown stays considered in one transverse plane and on one side of the vertical axis of the boiler
	s = transverse spacing of crown stays in the crown sheet, in
	α = angle any crown stay makes with the vertical axis of boiler
	$$P = \frac{76\, t\eta}{R - \Sigma s \sin \alpha} \qquad \textbf{SI} \qquad (9\text{--}36b)$$
	where P in MPa; s, t, and R in m
The longitudinal pitch between stay bolts or between the nearest row of stay bolts and the row of rivets at the joints between the furnace sheet and the tube sheet or the furnace sheet and the mud ring	$$L = \left(\frac{56320\, t^2}{PR}\right)^2 \qquad \textbf{US Customary System units}$$ $$(9\text{--}37a)$$ where
	t = thickness of furnace sheet, in
	R = outside radius of furnace, in
	P = maximum allowable working pressure, psi
	$$= \left[\frac{2.535 \times 10^9\, t^2}{PR}\right]^2 \qquad \textbf{SI} \qquad (9\text{--}37b)$$
	where P in Pa; t, L, and R in m
Cross-sectional area of diagonal stay (A)	$$A = \frac{aL}{l} \qquad (9\text{--}38)$$ where
	a = sectional area of direct stay, m (in)
	L = length of diagonal stay, m (in)
	l = length of line drawn at right angles to boiler head or a projection of L on a horizontal surface parallel to boiler drum, m (in)
The total cross-sectional area of stay tubes which support the tube plates in multitubular boilers	$$A_t = \frac{(A - a)P}{\sigma_{Sa}} \qquad (9\text{--}39)$$ where
	A = area of that portion of tube plate containing the tubes, m (in)
	a = aggregate area of holes in the tube plate, m² (in²)
	P = maximum allowable working pressure, Pa (psi)
	σ_{Sa} = maximum allowable stress value in the tubes, MPa (psi) ⊅ 48 MPa (7 kpsi)
	σ_{Sa} is also taken from Table 8–23

Particular	Formula
	The pitch of stay tubes shall conform to Eqs. (9–17) and (9–18) and using the values of C_7 as given in Table 9–11.
The pitch from the stay bolt next to the corner to the point of tangency to the corner curve for stays at the upper corners of fire boxes shall be as given in Eq. (9–40)	$$p_t = \frac{90[C_7(T^2/P)]^{1/2}}{\text{angularity of tangent lines } (\beta)} \qquad (9\text{–}40a)$$ where **US Customary System units**
	T = thickness of plate in sixteenths of an inch P = maximum allowable working pressure, psi C_7 = factor for the thickness of plate and type of stay used $$= 7592 \, \frac{\sqrt{C_7(T^2/p)}}{\text{angularity of tangent lines } (\beta)} \quad \textbf{SI} \quad (9\text{–}40b)$$ where p_t and T in m, and p in p_a
For various values of C_7	Refer to Table 9–11.

TABLE 9–11
Values of C_7 for determining pitch of stay tubes

Pitch of stay tubes in the bounding rows	When tubes have nuts not outside of plates	When tubes are fitted with nuts outside of plates
Where there are two plain tubes between two stay tubes	120	130
Where there is one plain tube between two stay tubes	140	150
Where every tube in the bounding rows is a stay tube and each alternate tube has a nut.	—	170

Source: *ASME Power Boiler Code*, Section I, 1983.

FINAL RATIOS

Design of a *horizontal return tubular boiler*

$\dfrac{H}{G}$ ranges from 35 to 45 in firetube boilers; 37 is a good working value (9–41a)

$\dfrac{S}{W}$ lies between $\frac{1}{2}$ and $\frac{1}{3}$ for most types of cylindrical boilers (9–41b)

$\dfrac{C}{G}$ varies from $\frac{1}{6}$ to $\frac{1}{8}$ (9–41c)

$\dfrac{S}{P} = 16.7 \times 10^{-3}(0.6)^*$ to $19.5 \times 10^{-3}(0.7)^*$ (9–41d)

$\dfrac{H}{P} = 0.92$ to 1.12 m^2 (10 to 12 ft^2) for externally fired boiler per hp

$= 0.74$ m^2 (8 ft^2) for Scotch boiler per hp (9–41e)

---

* The units in parentheses are in US Customary System units.

Particular	Formula	
	$\dfrac{P}{G} = 53 \, (5.22)^*$	(9–41f)
	$\dfrac{DS}{N} = 64 \times 10^{-3}$ to 73×10^{-3}	(9–41g)
Design of a vertical *straight shell multitubular* boiler	$\dfrac{H}{G} = 60$	(9–42a)
	$\dfrac{WHS}{G} = 45$	(9–42b)
	$\dfrac{SHS}{WHS} = \dfrac{1}{3}$	(9–42c)
	$\dfrac{S}{W} = \dfrac{1}{3}$	(9–42d)
	$\dfrac{S}{P} = 22.3 \times 10^{-3} \, (0.80)^*$	(9–42e)

A = Total area of steam segment
D = Diameter of shell or drum
h = Height of the segment to be occupied by steam

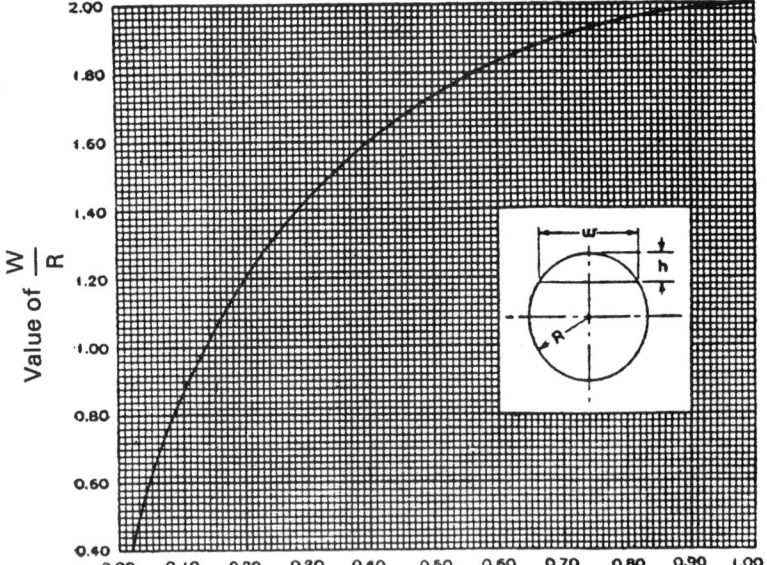

FIGURE 9–2 Disengaging surface in horizontal cylindrical shell. (*Source: Reproduced from G. B. Haven and G. W. Swett*, The Design of Steam Boilers and Pressure Vessels, *John Wiley and Sons, Inc., 1923.*)

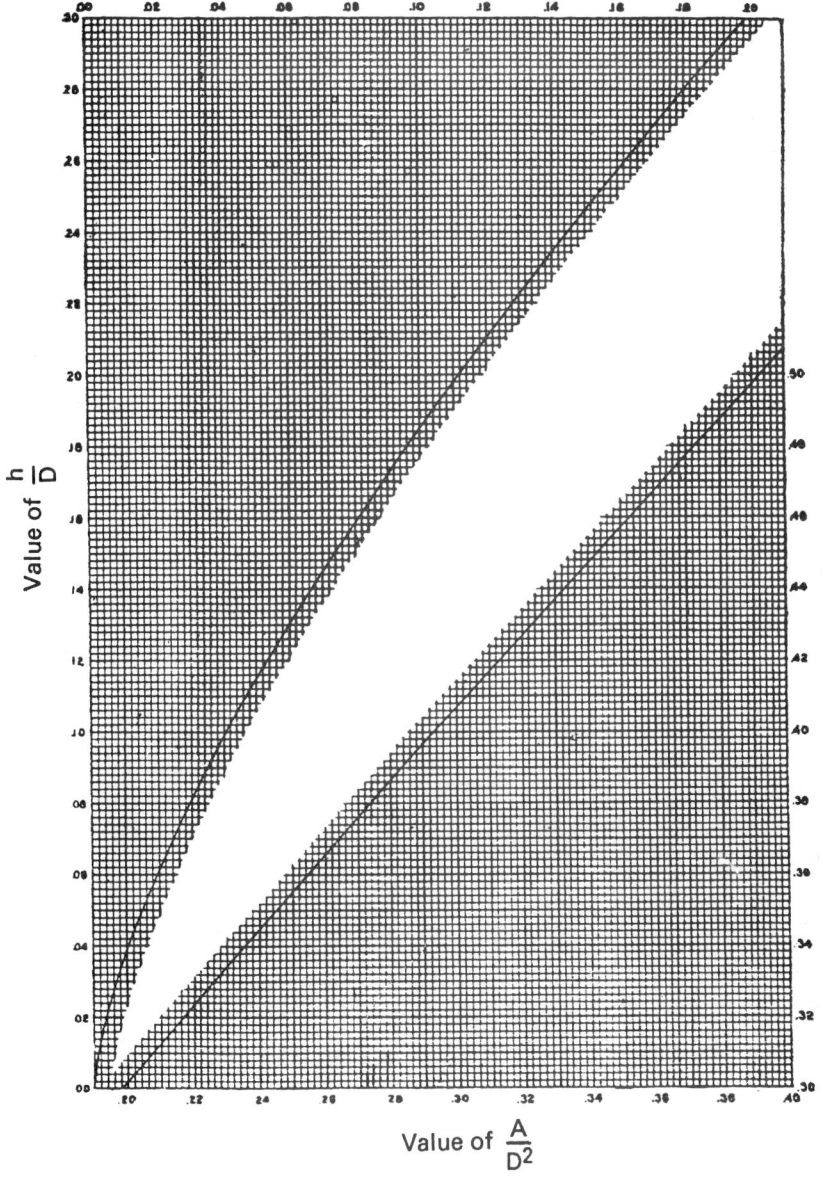

FIGURE 9–3 Areas of circular segments. (*Reproduced from G. B. Haven and G. W. Swett*, The Design of Steam Boilers and Pressure Vessels, *John Wiley and Sons, Inc., 1923.*)

Particular	Formula	
	$\dfrac{C}{G} = \dfrac{1}{5.5}$	(9–42f)
	$\dfrac{H}{P} = 1.12 \ (12)^*$	(9–42g)
	$\dfrac{G}{P} = 18.3 \times 10^{-3} \ (20)^* \ \text{or} \ \dfrac{P}{G} = 51 \ (5)$	(9–42h)
	$\dfrac{P}{DS} = 51 \ (5)^*$	(9–42i)
Watertube boiler design	$\dfrac{H}{G} = 50$	(9–43a)
	$\dfrac{S}{P} = 11.2 \times 10^{-3} \ (0.424)^*$	(9–43b)
	$\dfrac{H}{P} = 0.92 \ (10)^*$	(9–43c)
	$\dfrac{P}{G} = 51 \ (4.67)^*$	(9–43d)
	$\dfrac{DS}{P} = 27.5 \times 10^{-3} \ (0.308)^*$	(9–43e)
For mechanical properties of carbon and carbon manganese steel plates, sections and angles for marine boilers pressure vessels and welded machinery and mechanical properties of steel plates for boilers	Refer to Table 9–12.	

TABLE 9–12
Mechanical properties of carbon and carbon manganese steel plates, sections, and angles steel for marine boilers, pressure vessels, and welded machinery

Grade	Tensile strength		Elongation percentage min on gauge length		Bond test diameter of former
	MPa	kpsi	$5.65\sqrt{a^*}$ [a]	200 mm	
1	362.8–441.3	52.6–64.0	26	25	$2t$
2	411.9–490.3	59.7–71.1	25	23	$2t$
3	431.5–529.6	62.6–76.8	23	21	$3t$
4	460.9–559.0	66.8–81.0	22	20	$3t$
5	490.3–588.4	71.1–85.3	21	19	$3\frac{1}{2}t$

[a] Area of cross section.
Source: IS 3503, 1966.

For properties of boilers	Refer to Table 9–13.
For evaporation of water, average rate of combustion of fuels, and minimum rate of steam produced	Refer to Tables 9–14 to 9–16

TABLE 9–13
Properties of boilers
Horizontal return tubular boilers

Diameter of shell, mm	Diameter of tubes, mm			Length of tubes, m				
910	64	76		2.44	3.66			
1070	64	76		3.05	3.66	4.28		
1220	64	76		3.66	4.28	4.58	4.88	
1370	76	89		3.97	4.28	4.58	4.88	5.5
1520	76	89		4.28	4.88	5.50		
1680	76	89		4.88	5.18	5.50	6.10	
1830	76	89		4.88	5.50	6.10		
1980	76	89	102	4.88	5.18	5.50	6.10	
2130	76	89	102	4.88	5.50	6.10		
2290	76	89	102	4.88	5.18	5.50	6.10	
2440	76	89	102	4.88	5.18	5.50	6.10	

Dry-back scotch boilers

Diameter of shell, m	Diameter of tubes, mm	Length of tubes, m	Inside diameter of furnace, mm	Length of grate, m
		Short Types		
1.19	76	2.90	920	1.22
1.99	89	3.50	920	1.53
2.14	89	3.80	970	1.93
2.29	89	3.80	1150	2.03
2.44	89	3.97	1270	2.21
2.90	89	3.80	970	1.93
3.20	89	3.80	1150	2.03
3.50	89	3.89	1270	2.21
		Long Types		
2.06	102	4.88	970	1.93
2.21	102	4.88	1040	2.24
2.36	102	4.88	1140	2.44
2.84	102	4.88	970	1.93
3.05	102	4.88	1040	2.24
3.28	102	4.88	1140	2.44

Locomotive-type boilers without dome				**Vertical firetube boiler for power plant use**	
Diameter of waist, mm	Length of 75 mm tubes, m	Dimensions of grate		Diameter of tubes, mm	Length of tubes, m
		Width, mm	Length, m		
925	2.14	0.76	1.22	50	3.96
1070	2.44	0.92	1.27	62	4.27
1220	3.20	1.07	1.38		4.57
1370	3.36	1.22	1.53		4.88
1520	3.97	1.38	1.53		
1680	4.58	1.53	1.68		
1830	4.58	1.68	1.83		

Particular	Formula
For permissible strain rates of steam plant equipments	Refer to Table 9–17.
For water level requirements of boilers	Refer to Table 9–18.
For minimum allowable thickness of plates for boilers	Refer to Table 9–19.
For disengaging surface per horsepower	Refer to Table 9–20.
For heating boiler efficiency	Refer to Table 9–21.

TABLE 9–14
Evaporation kg (lb) of water per kg (lb) of fuel reduced to standard condition [from and at 373 K (100°C)]

Type of fuel	Approximate		Evaporation per kg (lb) of fuel, kg (lb)
	kJ/kg	Btu/lb	
Anthracite	29,038.3–27842.2	12,500–12000	9.5–9
Coke	30,228.7	13,000	9.5
Semibituminous	33,703.7	14,500	10
Bituminous	29,098.3	12,500	9
Lignite	22,106.3	9,500	6
Fuel oil	41,868.0	19,000	14.5

TABLE 9–15
Average rates of combustion (kg/m^2 (lb/ft^2) of grate surface per hour) draft 12.55 mm ($\frac{1}{2}$ in) water column

Fuel used	Stationary grate
Anthracite	44–68.5 (9.14)
Semibituminous	98 (20)
Bituminous	68.5 (14)
Lignite	58.5 (12)

TABLE 9–16
Minimum kilograms (pounds) of steam per h per ft^2 of surface

Particulars	Firetube boilers		Watertube boilers	
	kg	lb	kg	lb
Boiler heating surface				
Hand-fired	11.0	5	13.2	6
Stoker-fired	15.4	7	17.6	8
Oil-, gas-, or powder-fired	17.6	8	22.1	10
Water wall heating surface				
Hand-fired	17.6	8	17.6	8
Stoker-fired	22.1	10	26.5	12
Oil-, gas-, or powder-fired	30.9	14	35.3	16

Source: *ASME Power Boiler Code*, Section I, 1983.

TABLE 9–17
Permissible strain rates for steam plant equipment

Machine part	Strain rate per hour
Turbine disk (pressed on shaft)	10^{-9}
Bolted flanges, turbine cylinders	10^{-8}
Steam piping, welded joints, and boiler tubes	10^{-7}
Superheated tubes	10^{-6}

TABLE 9–18
Water level requirements[a]

Horizontal return tubular boilers		Vertical firetube boilers	
Boiler diameters, mm	Distance between gauge cocks, mm	Boiler diameters, mm	Distance between gauge cocks, mm
910, 1070, 1220	75	910–1220	100
1370, 1520	100	1250–1680	125
1680, 1830,		1700–2410	150
1980, 2130	125	2460–3100	175

Dry-back Scotch boilers	Locomotive-type boilers
Low water level 89 mm above surface of tubes for all diameters; distance between gauge cocks may be reduced to a minimum of 75 mm	Low water level must be 75–125 mm above the water surface of the crown sheet; distance between gauge cocks is usually 75 mm for all diameters

[a] Low water level 890 mm above surface of tubes.

TABLE 9–19
Minimum allowable thickness of plates for boilers (all dimensions in mm)

	Power boilers		Heating boilers	
Minimum thickness	Shell and dome plate diameter	Tube sheet diameter	Shell or other plate diameter	Tube sheet or head diameter
6.5	≤ 910		≤ 1065	
8.0	> 910–1370		> 1065–1530	≤ 1065
9.5	> 1370–1830	≤ 1065	> 1530–1980	> 1065–1530
11.0		> 1065–1370	> 1980	> 1530–1980
12.5	> 1830	> 1370–1830		> 1980
14.5		> 1830		

TABLE 9–20
Disengaging surface per horsepower mean water level

| Type of boiler | Disengaging surface | |
	m²/kW	m²/hp
Horizontal return tubular	0.087–0.10	0.065–0.0745
Dry-back Scotch	0.075–0.087	0.056–0.0650
Vertical straight shell	0.020–0.025	0.0149–0.0186
Vertical (Manning)	0.011–0.013	0.0084–0.0093
Locomotive type	0.100–0.125	0.0745–0.093
Sectional water tube	0.037–0.0500	0.0279–0.0372

TABLE 9–21
Heating boiler efficiency

Firing method	Efficiency, %
Hand-Fired Coal	
Lignite	49
Subbituminous	44–63
Bituminous	50–65
Low-volatile bituminous	44–61
Anthracite	60–75
Coke	75–76
Stoker Conversion	
Bituminous	55–69
Anthracite	63
Burner Conversion	
Natural gas	69–76
Oil	51; 65; 70
Designed for Burner	
Stoker	60–75
≤ 45 kg	65
> 45 kg	70
Gas	70–80
Oil	70–80
Cast-iron boilers	68
Steel boilers	70
Package units	75

REFERENCES

1. Haven, G. B., and G. W. Swett, *The Design of Steam Boilers and Pressure Vessels*, John Wiley and Sons, Inc., New York, 1923.
2. "Rules for Construction of Power Boilers," *ASME Boiler and Pressure Vessel Code*, Section I, 1983.
3. "Rules for Construction of Pressure Vessels," *ASME Boiler and Pressure Vessel Code*, Section VIII, Division I, July 1, 1986.
4. *Code of Unfired Pressure Vessels, Bureau of Indian Standards*, IS 2825, 1969, New Delhi, India.
5. Nichols, R. W., *Pressure Vessel Codes and Standards*, Elsevier Applied Science Publishing Ltd., Barking, Essex, England, 1987.
6. Lingaiah, K., and B. R. Narayana Iyengar, *Machine Design Data Handbook*, Engineering College Cooperative Society, Bangalore, India, 1962.
7. Lingaiah, K., *Machine Design Data Handbook*, Vol. II (*SI and Customary Metric Units*), Suma Publishers, Bangalore, India, 1986.

CHAPTER
10

ROTATING DISKS AND CYLINDERS

SYMBOLS

g	acceleration due to gravity, m/s^2
r	any radius, m (in)
r_i	inside radius, m (in)
r_o	outside radius, m (in)
h	thickness of disk at radius r from the center of rotation, m (in)
h_2	thickness of disk at radius r_2 from the center of rotation, m (in)
σ	uniform tensile stress in case of a disk of uniform strength, MPa (psi)
σ_θ	tangential stress, MPa (psi)
σ_r	radial stress, MPa (psi)
σ_z	axial stress or longitudinal stress, MPa (psi)
ρ	density of material of the disk
ω	angular speed of disk, rad/s
ν	Poisson's ratio

Particular	Formula

DISK OF UNIFORM STRENGTH ROTATING AT ω RAD/S (FIG. 10–1)

The thickness of a disk of uniform strength at radius r from center of rotation

$$h = h_2 \exp\left[\frac{\rho\omega^2}{2\sigma}(r_2^2 - r^2)\right] \qquad (10–1)$$

SOLID DISK ROTATING AT ω RAD/S

The general expression for the radial stress of a rotating disk of uniform thickness

$$\sigma_r = \frac{3 + \nu}{8}\rho\omega^2(r_o^2 - r^2) \qquad (10–2)$$

Particular	Formula

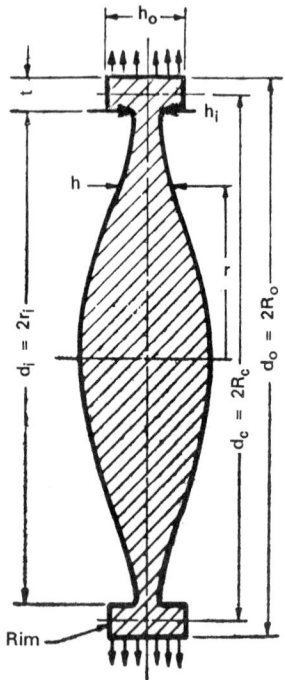

FIGURE 10–1   High-speed rotating disk of uniform strength.

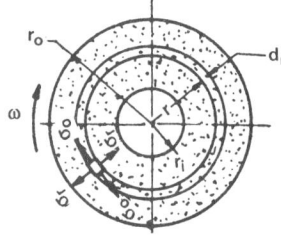

FIGURE 10–2   Rotating disk of uniform thickness.

The general expression for the tangential stress of a rotating disk of uniform thickness

$$\sigma_\theta = \frac{3+\nu}{8}\rho\omega^2\left(r_o^2 - \frac{1+3\nu}{3+\nu}r^2\right) \qquad (10\text{–}3)$$

The maximum values are at the center, where $r = 0$, and are equal to each other

$$\sigma_{r(max)} = \sigma_{\theta(max)} = \frac{3+\nu}{8}\rho\omega^2 r_o^2 \qquad (10\text{–}4)$$

HOLLOW DISK ROTATING AT ω RAD/S (FIG. 10–2)

The general expression for the radial stress of a rotating disk of uniform thickness

$$\sigma_r = \frac{3+\nu}{8}\rho\omega^2\left(r_i^2 + r_o^2 - \frac{r_o^2 r_i^2}{r^2} - r^2\right) \qquad (10\text{–}5)$$

The general expression for the tangential stress of a rotating disk of uniform thickness

$$\sigma_\theta = \frac{3+\nu}{8}\rho\omega^2\left(r_i^2 + r_o^2 + \frac{r_i^2 r_o^2}{r^2} - \frac{1+3\nu}{3+\nu}r^2\right) \quad (10\text{–}6)$$

The maximum radial stress occurs at $r^2 = r_o r_i$

$$\sigma_{r(max)} = \frac{3+\nu}{8}\rho\omega^2(r_o - r_i)^2 \qquad (10\text{–}7)$$

Particular	Formula
The maximum tangential stress occurs at inner boundary where $r = r_i$	$$\sigma_{\theta(max)} = \frac{3+\nu}{4} \rho\omega^2 \left(r_o^2 + \frac{1-\nu}{3+\nu} r_i^2 \right) \qquad (10\text{--}8)$$

SOLID CYLINDER ROTATING AT ω RAD/S

Particular	Formula
The tangential stress	$$\sigma_\theta = \frac{\rho\omega^2}{8(1-\nu)} [(3-2\nu)r_o^2 - (1+2\nu)r^2] \qquad (10\text{--}9)$$
The radial stress	$$\sigma_r = \frac{\rho\omega^2}{8} \left(\frac{3-2\nu}{1-y} \right) (r_o^2 - r^2) \qquad (10\text{--}10)$$
The maximum stress occurs at the center	$$\sigma_{r(max)} = \sigma_{\theta(max)} = \frac{\rho\omega^2}{8} \left(\frac{3-2\nu}{1-\nu} \right) r_o^2 \qquad (10\text{--}10a)$$
The axial strain in the z direction (ends free)	$$\epsilon_z = \frac{-\nu}{2} \frac{\rho\omega^2 r_o^2}{E} \qquad (10\text{--}11)$$
The axial stress under plane strain condition (ends free)	$$\sigma_z = \frac{\rho\omega^2}{4} \left(\frac{\nu}{1-\nu} \right) (r_o^2 - 2r^2) \qquad (10\text{--}12a)$$
The axial stress under plane strain condition (ends constrained)	$$\sigma_z = \frac{\rho\omega^2 \nu}{4(1-\nu)} \left[\frac{1}{2}(3-2\nu)r_o^2 - 2r^2 \right] \qquad (10\text{--}12b)$$

HOLLOW CYLINDER ROTATING AT ω RAD/S

Particular	Formula
The tangential stress at any radius r	$$\sigma_\theta = \frac{\rho\omega^2}{8} \left(\frac{3-2\nu}{1-\nu} \right) \left[r_i^2 + r_o^2 + \frac{r_i^2 r_o^2}{r^2} - \left(\frac{1+2\nu}{3-2\nu} \right) r^2 \right] \qquad (10\text{--}13)$$
The radial stress at any radius r	$$\sigma_r = \frac{\rho\omega^2}{8} \left(\frac{3-2\nu}{1-\nu} \right) \left[r_i^2 + r_o^2 - \frac{r_i^2 r_o^2}{r^2} - r^2 \right] \qquad (10\text{--}14)$$
The axial stress (ends free) at any radius r	$$\sigma_z = \frac{\rho\omega^2}{4} \left(\frac{\nu}{1-\nu} \right) [r_i^2 + r_o^2 - 2r^2] \qquad (10\text{--}15)$$
The axial stress under plane strain conditions (ends constrained) at any radius r	$$\sigma_z = \frac{\nu\rho\omega^2}{4} \left(\frac{3-2\nu}{1-\nu} \right) \left[r_i^2 + r_o^2 - \frac{2r^2}{(3-2\nu)} \right] \qquad (10\text{--}16)$$
The maximum stress occurs at the inner surface where $r = r_i$	$$\sigma_{\theta(max)} = \frac{\rho\omega^2}{4} \left(\frac{3-2\nu}{1-\nu} \right) \left[r_o^2 + \left(\frac{1-2\nu}{3-2\nu} \right) r_i^2 \right] \qquad (10\text{--}17)$$

Particular	Formula

The axial strain in the z direction (ends free)

$$\epsilon_z = -\frac{\nu\rho\omega^2}{2E}[r_i^2 + r_o^2] \tag{10–18}$$

The displacement u at any radius r of a thin hollow rotating disk

$$u = \left[\frac{\rho\omega^2 r}{E}\frac{(3 + \nu)(1 - \nu)}{8}\right.$$

$$\left.\left(r_o^2 + r_i^2 + \frac{1 + \nu}{1 - \nu}\frac{r_o^2 r_i^2}{r^2} - \frac{1 + \nu}{3 + \nu}r^2\right)\right] \tag{10–19}$$

SOLID THIN UNIFORM DISK ROTATING AT ω RAD/S UNDER EXTERNAL PRESSURE p_o (FIG. 10–3)

The radial stress at any radius r

$$\sigma_r = -p_o + \rho\omega^2\left(\frac{3 + \nu}{8}\right)(r_o^2 - r^2) \tag{10–20}$$

The tangential stress at any radius r

$$\sigma_\theta = -p_o + \rho\omega^2\left(\frac{3 + \nu}{8}\right)\left(r_o^2 - \frac{1 + 3\nu}{3 + \nu}r^2\right) \tag{10–21}$$

The maximum radial stress at $r = 0$

$$\sigma_{r(max)} = -p_o + \rho\omega^2\left(\frac{3 + \nu}{8}\right)r_o^2 \tag{10–22}$$

The maximum radial stress at $r = r_o$

$$\sigma_r = -p_o \tag{10–23}$$

The maximum tangential stress at $r = 0$

$$\sigma_{\theta(max)} = \sigma_{r(max)} \tag{10–24}$$

The displacement u at any radius r

$$u = \frac{r}{E}(1 - \nu)\left\{-p_o + \frac{\rho\omega^2}{8}\left[(3 + \nu)r_o^2 - (1 + \nu)r^2\right]\right\} \tag{10–25}$$

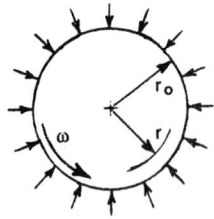

FIGURE 10–3 Rotating disk of uniform thickness under external pressure.

CHAPTER
11

METAL FITS, TOLERANCES, AND SURFACE TEXTURE

SYMBOLS

A	area of cross section, m^2 (in^2)
d	diameter of shaft, m (in)
	diameter of cylinder, m (in)
E	modulus of elasticity, GPa (Mpsi)
E_c	modulus of elasticity of cast iron, GPa (Mpsi)
E_s	modulus of elasticity of steel, GPa (Mpsi)
F	force, kN [lbf or tonf (pound force or tonne force)]
l	length, m (in)
	length of hub, m (in)
	effective length of anchor, m (in)
L	original length of slot, m (in)
M_t	torque or twisting moment, N m (lbf in)
p	pressure, MPa (psi)
p_c	contact pressure MPa (psi)
t	temperature, °C (°F)
α	coefficient of linear expansion, $(m/m)/°C$ $[(in/in)/°F]$
δ	total change in diameter (interference), m (in)
Δd	change in diameter, m (in)
ν	Poisson's ratio
σ	stress, MPa (psi)
μ	coefficient of friction
n	factor of safety

SUFFIXES

a	axial
b	bearing surface
c	contact surface, compressive
d	design
f	final
h	hub
i	internal inner
o	original external, outer

r	radial, rim
s	shaft
θ	tangential or hoop
1	initial
2	final

Particular	Formula

PRESS AND SHRINK FITS

Change in cylinder diameter due to contact pressure

The change in diameter

$$\Delta d = d\epsilon_\theta \tag{11-1}$$

The change in diameter of the inner member when subjected to contact pressure p_c (Fig. 11–1)

$$\Delta d_i = -\frac{p_c d_c}{E}\left(\frac{d_c^2 + d_i^2}{d_c^2 - d_i^2} - \nu\right) \tag{11-2}$$

The change in diameter of the outer member when subjected to contact presure p_c (Fig. 11–1)

$$\Delta d_o = \frac{p_c d_c}{E}\left(\frac{d_o^2 + d_c^2}{d_o^2 - d_c^2} + \nu\right) \tag{11-3}$$

The original difference in diameters of the two cylinders when the material of the members is the same

$$\delta = \Delta d_o + \Delta d_i$$

$$= \frac{p_c d_c}{E}\left(\frac{d_o^2 + d_c^2}{d_o^2 - d_c^2} + \nu\right) + \frac{p_c d_c}{E}\left(\frac{d_c^2 + d_i^2}{d_c^2 - d_i^2} - \nu\right) \tag{11-4}$$

The total change in the diameters of hub and hollow shaft due to contact pressure at their contact surface when the material of the members is the same

$$\delta = \Delta d_s + \Delta d_h = d_s - d_h$$

$$= \frac{p_c d_s}{E_s}\left[\frac{d_s^2 + d_i^2}{d_s^2 - d_i^2} - \nu_s\right] + \frac{p_c d_h}{E_h}\left[\frac{d_o^2 + d_h^2}{d_o^2 - d_s^2} + \nu_h\right]\text{exactly} \tag{11-5a}$$

$$\delta = p_c d_c\left[\frac{d_c^2 + d_i^2}{E_s(d_c^2 - d_i^2)} + \frac{d_o^2 + d_c^2}{E_h(d_o^2 - d_c^2)} - \frac{\nu_s}{E_s} + \frac{\nu_h}{E_h}\right]$$

(approx.) $\tag{11-5b}$

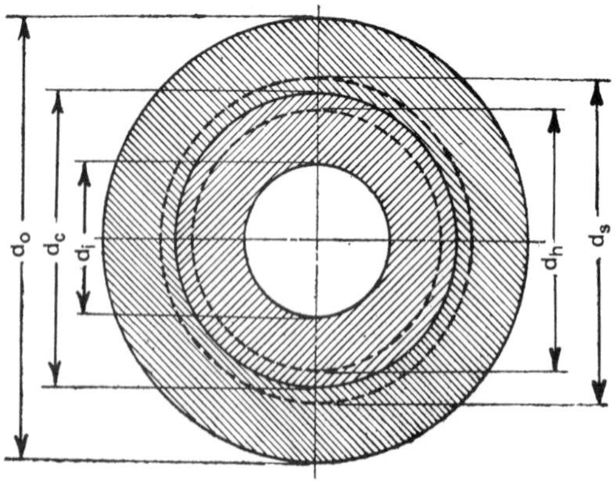

FIGURE 11–1

Particular	Formula
The shrinkage stress in the band	$$\sigma_\theta = \frac{E\delta}{d_c} \qquad (11\text{–}6)$$
The contact pressure between cylinders at the surface of contact when the material of both the cylinders is same (Fig. 11–2)	$$p_c = \frac{\delta E(d_c^2 - d_i^2)(d_o^2 - d_c^2)}{2d_c^3(d_o^2 - d_i^2)} \qquad (11\text{–}7)$$
The tangential stress at any radius r of outer cylinder (Fig. 11–2)	$$\sigma_{\theta-o} = \frac{p_c d_c^2}{(d_o^2 - d_c^2)}\left(1 + \frac{d_o^2}{4r^2}\right) \qquad (11\text{–}8)$$
The tangential stress at any radius r of inner cylinder (Fig. 11–2)	$$\sigma_{\theta-i} = -\frac{p_c d_c^2}{(d_o^2 - d_c^2)}\left(1 + \frac{d_i^2}{4r^2}\right) \qquad (11\text{–}9)$$
The radial stress at any radius r of outer cylinder (Fig. 11–2)	$$\sigma_{r-o} = -\frac{p_c d_c^2}{(d_o^2 - d_c^2)}\left(\frac{d_o^2}{4r^2} - 1\right) \qquad (11\text{–}10)$$
The radial stress at any radius r of inner cylinder (Fig. 11–2)	$$\sigma_{r-i} = \frac{p_c d_c^2}{d_c^2 - d_i^2}\left(1 - \frac{d_i^2}{4r^2}\right) \qquad (11\text{–}11)$$
The tangential stress at outside diameter of outer cylinder (Fig. 11–2)	$$\sigma_{\theta-oo} = \frac{2p_c d_c^2}{(d_o^2 - d_c^2)} \qquad (11\text{–}12)$$
The tangential stress at inside diameter of outer cylinder (Fig. 11–2)	$$\sigma_{\theta-oi} = p_c\left(\frac{d_o^2 + d_c^2}{d_o^2 - d_c^2}\right) \qquad (11\text{–}13)$$
The tangential stress at outside diameter of inner cylinder (Fig. 11–2)	$$\sigma_{\theta-io} = -\frac{p_c(d_c^2 + d_i^2)}{(d_c^2 - d_i^2)} \qquad (11\text{–}14)$$
The tangential stress at inside diameter of inner cylinder (Fig. 11–2)	$$\sigma_{\theta-ii} = -\frac{2p_c d_c^2}{(d_c^2 - d_i^2)} \qquad (11\text{–}15)$$
The radial stress at outside diameter of outer cylinder (Fig. 11–2)	$$\sigma_{r-oo} = 0 \qquad (11\text{–}16)$$

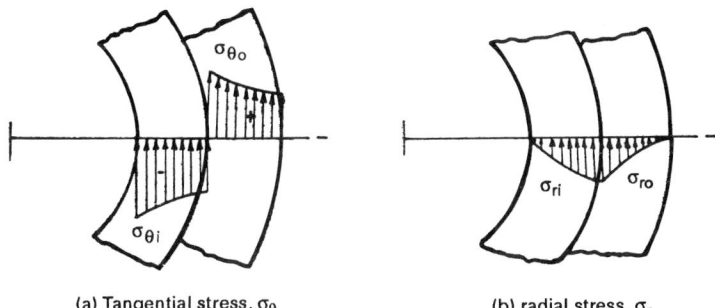

(a) Tangential stress, σ_θ (b) radial stress, σ_r

FIGURE 11–2 Distribution of stresses in shrink-fitted assembly.

Particular	Formula
The radial stress at inside diameter of outer cylinder (Fig. 11–2)	$\sigma_{r-oi} = -p_c$ $\hspace{2cm}$ (11–17)
The radial stress at outside diameter of inner cylinder (Fig. 11–2)	$\sigma_{r-io} = -p_c$ $\hspace{2cm}$ (11–18)
The radial stress at inside diameter of inner cylinder (Fig. 11–2)	$\sigma_{r-ii} = 0$ $\hspace{2cm}$ (11–19)
The semiempirical formula for tangential stress for cast-iron hub on steel shaft	$\sigma_\theta = \dfrac{E_o\delta}{d_c + 0.14d_o}$ $\hspace{1cm}$ (11–20)
Timoshenko equation for contact pressure in case of steel shaft on cast-iron hub	$p_c = \dfrac{E_c\delta}{d_c}\left[\dfrac{1-(d_c/d_o)^2}{1.53+0.47(d_c/d_o)^2}\right]$ for $\dfrac{E_s}{E_c}=3$ $\hspace{0.5cm}$ (11–21a)
The allowable stress for brittle materials	$\sigma_{all} = \dfrac{\sigma_{su}}{n} = \dfrac{E_c\delta[1+(d_c/d_o)^2]}{d_c[1.53+0.47(d_c/d_o)^2]}$ $\hspace{0.5cm}$ (11–21b)

INTERFERENCE FITS
Press

Particular	Formula
The axial force necessary to press shaft into hub under an interface pressure p_c	$F_a = \pi d_c l\mu p_c$ $\hspace{2cm}$ (11–22a) where $\mu = 0.085$ to 0.125 for unlubricated surface $= 0.05$ with special lubricants
The approximate value of axial force to press steel shaft into cast-iron hub with an interference	$F = 4137 \times 10^4 \dfrac{(d_o + 0.3d_c)l\delta}{(d_o + 6.33d_c)}$ $\hspace{0.3cm}$ **SI** $\hspace{0.3cm}$ (11–23a) where d_o, d_c, l and δ in m, and F in N $F = 6000\dfrac{(d_o + 0.3d_c)l\delta}{(d_o + 6.33d_c)}$ $\hspace{0.3cm}$ **US Customary units** where d_o, d_c, l and δ in in, and F in tonf $\hspace{0.2cm}$ (11–23b)
The approximate value of axial force to press steel shaft in steel hub	$F = 28.41 \times 10^4 \dfrac{(d_o^2 - d_c^2)l\delta}{d_o^2}$ $\hspace{0.3cm}$ **SI** $\hspace{0.3cm}$ (11–24a) where d_o, d_c, l and δ in m, and F in N $F = 4120\dfrac{(d_o^2 - d_c^2)l\delta}{d_o^2}$ $\hspace{0.3cm}$ **US Customary units** where d_o, d_c, l and δ in in, and F in tonf $\hspace{0.2cm}$ (11–24b)

Particular	Formula
The transmitted torque by a press fit or shrink fit without slipping between the hub and shaft	$$M_t = \frac{\pi d_c^2 l \mu p_c}{2} \qquad (11\text{--}25)$$ where $\mu = 0.10$ for press fit $\qquad = 0.125$ for shrink fits
The temperature t_2 in °C to which the shaft or shrink link must be heated before assembly	$$t_2 \geq \left(\frac{2\delta}{\alpha d_c} + t_1\right) \qquad (11\text{--}26)$$ where $t_1 =$ temperature of hub or larger part to which shaft or shrink link to be shrunk on, °C

Shrink links or anchors (Fig. 11–3)

The average compression in the part of rim affected according to C. D. Albert	$$\sigma_c = \frac{F}{\sqrt{A_b A_r}} \qquad (11\text{--}27)$$

FIGURE 11–3 Shrink link.

The tensile stress in link	$$\sigma_t = \frac{L_f - L_o}{L_o} E \qquad (11\text{--}28)$$
The total load on link	$$F = \frac{(L_f - L_o)EA}{L_o} \qquad (11\text{--}29)$$
The compressive stress in rim	$$\sigma_c = \frac{L_f - L_o}{L_o} \frac{EA}{\sqrt{A_b A_r}} \qquad (11\text{--}30)$$
The original length of link	$$L_o = \frac{L}{1 + \left(1 + \dfrac{AE}{E_r \sqrt{A_b A_r}}\right)\dfrac{\sigma_r}{E}} \qquad (11\text{--}31)$$
The necessary linear interference δ for shrink anchors	$$\delta = \frac{\sigma_d l}{E} \qquad (11\text{--}32)$$
The force exerted by an anchor	$$F = ab\sigma_d \qquad (11\text{--}33)$$ $$\frac{b}{a} = 2 \text{ to } 3$$ $\sigma_d =$ design stress based on a reliability factor of 1.25

Particular	Formula
For letter symbols for tolerances, basic size deviation and tolerance, clearance fit, transition fit, interference fit	Refer to Figures 11–4 to 11–8
For press-fit between steel hub and shaft, cast-iron hub and shaft and tensile stress in cast-iron hub in press-fit allowance	Refer to Figures 11–9 to 11–11

TOLERANCES AND ALLOWANCES

The tolerance size is defined by its value followed by a symbol composed of a letter (in some cases by two letters) and a numerical value as	45 $g7$
A fit is indicated by the basic size common to both components followed by symbols corresponding to each component, the hole being quoted first, as	$\dfrac{45H8}{g7}$ or $45H8$–$g7$ or $45\dfrac{H8}{g7}$
For grades 5 to 16 tolerances have been determined in terms of standard tolerance unit i in micrometers (Refer to Table 11–1.)	$i = 0.45D^{1/3} + 0.001D$ (11–34) where D is expressed in mm
Values of standard tolerances corresponding to grades 01, 0, and 1 are (values in μm for D in mm)	IT 01 0.3 + 0.008 D (11–35) IT 0 0.5 + 0.012 D IT 1 0.8 + 0.020 D

TABLE 11–1
Relative magnitudes of standard tolerances for grades 5 to 16 in terms of standard tolerance unit "i"
[Eq. (11–34)]

Grade	IT 5	IT 6	IT 7	IT 8	IT 9	IT 10	IT 11	IT 12	IT 13	IT 14	IT 15	IT 16
Values	7 i	10 i	16 i	25 i	40 i	64 i	100 i	160 i	250 i	400 i	640 i	1000 i

Source: IS 919, 1963.

TABLE 11–1A
Coefficient of Friction, μ (for Use between Conical metallic Surfaces)

Contacting surface	Nature of surfaces	Coefficient of friction, μ
Any metal in contact with another metal	Lubricated with oil	0.15
Any metal in contact with another metal	Greased	0.15
Cast iron on steel	Shrink-fitted	0.33
Steel on steel	Shrink-fitted	0.13
Steel on steel	Dry	0.22
Cast iron on steel	Dry	0.16

Source: Courtesy J. Bach, "Kegelreibungsverbiudungen," *Zeitschrift Verein Deutscher Ingenieure*, Vol. 79, 1935.

TABLE 11–2
Formulas for Fundamental Shaft Deviations
(for sizes $\leq$ 500 mm)

Upper deviations (es)		Lower deviation (ei)	
Shaft designation	**In μm (for D in mm)**	**Shaft designation**	**In μm (for D in mm)**
a	$= -(265 + 1.3\,D)$ for $D \leq 120$ $= -3.5\,D$ for $D < 120$	$j5$–$j8$ $k4$–$k7$ k for grades ≤ 3 and ≥ 8	No formula $= +0.6\sqrt[3]{D}$ $= 0$
	$\simeq -(140 + 0.85\,D)$ for $D \leq 160$ $\simeq -1.8\,D$ for $D > 160$	m n P	$= +(\text{IT } 7–\text{IT } 6)$ $= +5\,D^{0.34}$ $= \text{IT } 7 + 0$ to 5
c	$= -52\,D^{0.2}$ for $D \leq 40$	r	$= $ geometric mean of values ei for p and s
d	$= -(95 + 0.8\,D)$ for $D > 40$ $= -16\,D^{0.44}$	s	$= +\text{IT } 8 +1$ to 4 for $D \leq 50$ $= +\text{IT } 7 +0.4\,D$ for $D > 50$
e	$= -11\,D^{0.41}$	t	$= \text{IT } 7 +0.63\,D$
f	$= -5.5\,D^{0.41}$	u	$= +\text{IT } 7 + D$
g	$= -2.5\,D^{0.34}$	v	$= +\text{IT } 7 +1.25\,D$
		x	$= +\text{IT } 7 +1.6\,D$
		y	$= +\text{IT } 7 +2\,D$
h	$= 0$	z	$= +\text{IT } 7 +2.5\,D$
		za	$= +\text{IT } 8 +3.15\,D$
		zb	$= +\text{IT } 9 +4\,D$
		zc	$= +\text{IT } 10 +5\,D$

For js : The two deviations are equal to $\pm \dfrac{\text{IT}}{2}$

Source: IS 919, 1963.

TABLE 11–3
Rules for Rounding off Values Obtained by the use of Formulas

	Above	5	45	60	100	200	300	560	600	800	1000	2000
Values in μm	Up to	45	60	100	200	300	560	600	800	1000	2000	
	For standard tolerances for Grades II and finer	1	1	1	5	10	10					
Rounded in multiples of	For deviations es, from a to g	1	2	5	5	10	10	20	20	20	50	
	For deviations ei, from k to zc	1	1	1		2	5	5	10	20	50	1000

Source: IS 919, 1963.

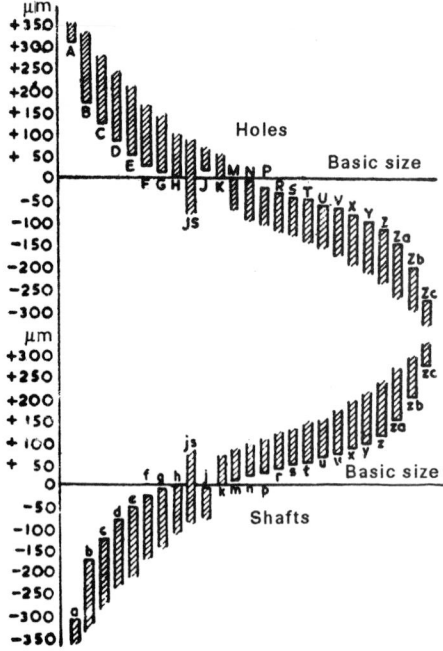

FIGURE 11–4 Letter symbols for tolerances.

TABLE 11–4
Fundamental Tolerances of Grades 01, 0, and 1 to 16

Diameter steps in mm	Values of tolerances in μm (1 μm = 0.001 mm)																	
	Tolerance grades																	
	01	0	1	2	3	4	5	6	7	8	9	10	11	12	13	14[a]	15[a]	16[a]
≤3	0.3	0.5	0.8	1.2	2	3	4	6	10	14	25	40	60	100	140	250	400	600
>3 ≤6	0.4	0.6	1	1.5	2.5	4	5	8	12	18	30	48	75	120	180	300	480	750
>6 ≤10	0.4	0.6	1	1.5	2.5	4	6	9	15	22	36	58	90	150	220	360	580	900
>10 ≤18	0.5	0.8	1.2	2	3	5	8	11	18	27	43	70	110	180	270	430	700	1100
>18 ≤30	0.6	1	1.5	2.5	4	6	9	13	21	33	52	84	130	210	330	520	840	1300
>30 ≤50	0.6	1	1.5	2.5	4	7	11	16	25	39	62	100	160	250	390	620	1000	1600
>50 ≤80	0.8	1.2	2	3	5	8	13	19	30	46	74	120	190	300	460	740	1200	1900
>80 ≤120	1	1.5	2.5	4	6	10	15	22	35	54	87	140	220	350	540	870	1400	2200
>120 ≤180	1.2	2	3.5	5	8	12	18	25	40	63	100	160	250	400	630	1000	1600	2500
>180 ≤250	2	3	4.5	7	10	14	20	29	46	72	115	185	290	460	720	1150	1850	2900
>250 ≤315	2.5	4	6	8	12	16	23	32	52	81	130	210	320	520	810	1300	2100	3200
>315 ≤400	3	5	7	9	13	18	25	36	57	89	140	230	360	570	890	1400	2300	3600
>400 ≤500	4	6	8	10	15	20	27	40	63	97	155	250	400	630	970	1550	2500	4000

[a] Up to 1 mm Grades 14 to 16 are not provided.

Source: IS: 919–1963.

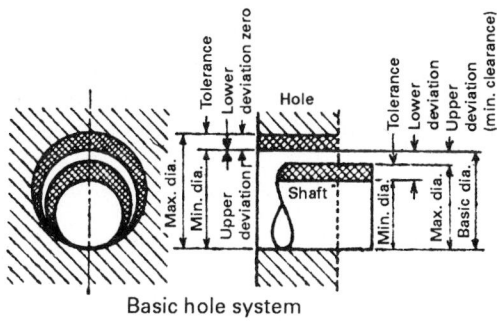

FIGURE 11–5 Basic size deviation and tolerances.

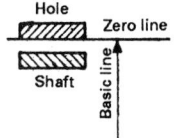

FIGURE 11–6 Clearance fit.

TABLE 11–5
Clearance Fits (Fig. 11–6) (Hole Basis)

Quality of fit	Combination of shaft and hole	Remarks and uses
Large clearance	$H\,11\,a\,9$ ⎫ coarse $H\,11\,b\,9$ ⎭	Not widely used
	$H\,11\,a\,11$ normal	
	$H\,9\,a\,9$ ⎫ fine $H\,8\,b\,8$ ⎭	
Slack running	$H\,11\,c\,9$ coarse	Not widely used
	$H\,11\,c\,11$ ⎫ normal $H\,9\,c\,9$ ⎭	
	$H\,8\,c\,8$ ⎫ fine $H\,7\,c\,8$ ⎭	
Loose running	$H\,11\,d\,11$ ⎫ coarse $H\,9\,d\,9$ ⎭	Suitable for plummer block bearings and loose pulleys
	$H\,8\,d\,9$ normal	
	$H\,8\,d\,8$ ⎫ fine $H\,7\,d\,8$ ⎭	
Easy running	$H\,8\,e\,9$ ⎫ coarse $H\,9\,e\,9$ ⎭	Recommended for general clearance fits, used for properly lubricated bearings requiring appreciable clearance; finer grades for high speeds, heavily loaded bearings such as turbogenerator and large electric motor bearings
	$H\,8\,e\,8$ ⎫ normal $H\,7\,e\,8$ ⎭	
	$H\,7\,e\,7$ ⎫ fine $H\,6\,e\,7$ ⎭	
Normal running	$H\,8\,f\,8$ coarse $H\,7\,f\,7$ normal $H\,6\,f\,6$ fine	Widely used as a normal grease lubricated or oil-lubricated bearing having low temperature differences, gearbox shaft bearings, bearings of small electric motor and pumps, etc.
Close running or sliding	$H\,8\,g\,7$ coarse $H\,7\,g\,6$ normal $H\,6\,g\,6$ ⎫ fine $H\,6\,g\,5$ ⎭	Expensive to manufacture, small clearance. Used in bearings for accurate link work, and for piston and slide valves; also used for spigot or location fits
Precision sliding	$H\,11\,h\,11$ $H\,8\,h\,7$ $H\,8\,h\,8$ $H\,7\,h\,6$ $H\,6\,h\,5$	Widely used for nonrunning parts; also used for fine spigot and location fit

TABLE 11–6
Values of Standard Tolerances for Sizes > 500 to 3150 mm

IT 6	IT 7	IT 8	IT 9	IT 10	IT 11	IT 12	IT 13	IT 14	IT 15	IT 16
10 I*[a]	16 I	25 I	40 I	64 I	100 I	160 I	250 I	400 I	640 I	1000 I

[a]* Standard Tolerance Unit I (in μm) -0.004D + 2.1 for D in mm.
Source: IS: 2101–1962.

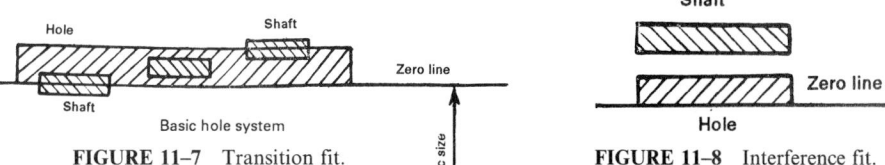

FIGURE 11–7 Transition fit.

FIGURE 11–8 Interference fit.

TABLE 11–7
Transition and Interference Fits (Hole Basis)

Quality of fit	Combination of shaft and hole	Remarks and uses
	Transition Fit (Fig. 11–7)	
Push	$H8j7$ coarse $H7j6$ normal $H6j5$ fine	Slight clearance—recommended for fits where slight interference is permissible, coupling spigots and recesses, gear rings clamped to steel hubs
True transition	$H8k7$ coarse $H7k6$ normal $H6k5$ fine	Fit averaging virtually no clearance—recommended for location fits where a slight interference can be tolerated, with the object of eliminating vibration; used in clutch member keyed to shaft, gudgeon pin in piston bosses, hand wheel, and index disk on shaft
Interference transition	$H8m7$ coarse $H7m6$ normal $H6m5$ fine	Fit averages a slight interference suitable for general tight-keying fits where accurate location and freedom from play are necessary; used for the cam holder, fitting bolt in reciprocating slide
True interference	$H8n7$ $H7n6$ }coarse $H6n5$ fine	Suitable for tight assembly of mating surfaces
	Interference Fit (Fig. 11–8)	
Light press fit	$H7p6$ normal $H6p5$ fine	Light press fit for nonferrous parts which can be dismantled when required; standard press fit for steel, cast iron, or brass-to-steel assemblies, bush on to a gear, split journal bearing
Medium drive fit	$H7r6$ normal $H6r5$ fine	Medium drive fit with easy dismantling for ferrous parts and light drive fit with easy dismantling for nonferrous parts assembly; pump impeller on shaft, small-end bush in connecting rod, pressed in bearing bush, sleeves, seating, etc.
Heavy drive fit	$H8s7$ $H7s6$ }normal $H6s5$ fine	Used for permanent or semipermanent assemblies of steel and cast-iron members with considerable gripping force; for light alloys this gives a press fit; used in collars pressed on to shafts, valve seatings cylinder liner in block, etc.
Force fit	$H8t7$ $H7t6$ }normal $H6t5$ fine	Suitable for the permanent assembly of steel and cast-iron parts; used in valve seat insert in cylinder head, etc.
Heavy force fit or shrink fit	$H8u7$ $H7u6$ }normal $H6u5$ fine	High interference fit; the method of assembly will be by power press

TABLE 11-8
Preferred Basic and Design Sizes
Linear dimensions (in mm)

Shaft basis		Hole basis							
A	B	Priority 1			Priority 2			Priority 3	
1.6	5.0	1.0	22.0	110.0	1.2	34.0	170.0	145.0	440.0
2.5	8.0	1.6	25.0	125.0	2.0	38.0	190.0	155.0	460.0
4.0	12.0	2.5	28.0	140.0	3.2	42.0	210.0	165.0	490.0
6.0	14.0	4.0	32.0	160.0	4.5	48.8	230.0	175.0	
10.0	18.0	5.0	36.0	180.0	5.5	53.0	240.0	185.0	
16.0	20.0	6.0	40.0	200.0	7.0	58.0	260.0	195.0	
25.0	22.0	8.0	45.0	220.0	9.0	65.0	270.0	290.0	
40.0	32.0	10.0	50.0	250.0	11.0	75.0	300.0	310.0	
63.0	50.0	12.0	56.0	280.0	13.0	85.0	340.0	330.0	
100.0	80.0	14.0	63.0	320.0	15.0	95.0	380.0	350.0	
		16.0	71.0	360.0	17.0	105.0	420.0	370.0	
		18.0	80.0	400.0	19.0	115.0	430.0	390.0	
		20.0	90.0	450.0	21.0	120.0	470.0	410.0	
			100.0	500.0	23.0	130.0	480.0		
					26.0	135.0			
					30.0	150.0			

Angular dimensions (in deg)

Priority	Preferred angles										
1	1	3	6		10	16		30	45	60	90	120
2		2	4	5	8	12	20				

TABLE 11-9
Formulas for Shaft and Hole Deviations (for Sizes > 500 to 3150 mm)

Shafts			Formulas for deviations in μm (for D in mm)	Holes		
d	es	—	16 $D^{0.44}$	+	EI	D
e	"	—	11 $D^{0.41}$	+	"	E
f	"	—	5.5 D $^{0.41}$	+	"	F
(g)	"	—	2.5 D $^{0.34}$	+	"	(G)
h	"	—	0		"	H
js	ei	—	0.5 IT_n	+	ES	JS
k	"	—	0		"	K
m	"	+	0.024 D + 12.6	—	"	M
n	"	+	0.04 D + 21	—	"	N
p	"	+	0.072 D + 37.8	—	"	P^a
r	"	+	geometric mean between p and s or P and S	—	"	R^a
s	"	+	IT 7 + 0.4D	—	"	S^a
t	"	+	IT 7 + 0.63D	—	"	T^a
u	"	+	IT 7 + D	—	"	U

[a] It is assumed that associated shafts and holes are of the same grade contrary to what has been allowed for the dimensions up to 500 mm (see IS 919, 1959).

Source: IS 2101, 1962.

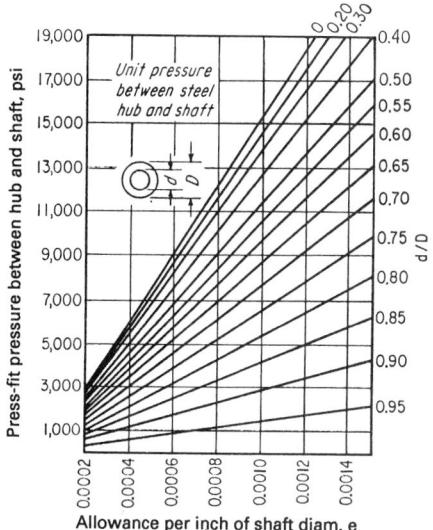

FIGURE 11–9   Press-fit pressures between steel hub and shaft (1 psi = 6894.757 Pa; 1 in = 25.4 mm). (*Baumeister, T., Marks' Standard Handbook for Mechanical Engineers, 8th ed., McGraw-Hill, 1978.*)

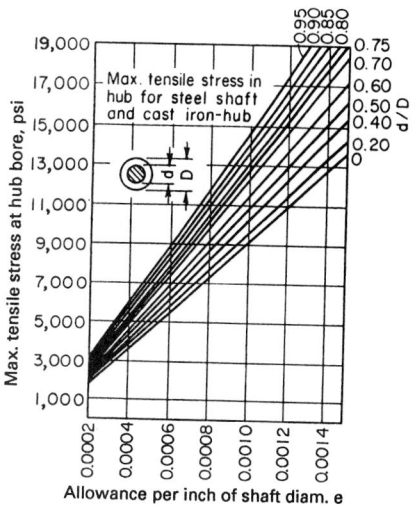

FIGURE 11–10   Variation in tensile stress in cast-iron hub in press-fit allowance (1 psi = 6894.757 Pa; 1 in = 25.4 mm). (*Baumeister, T.,* Marks' Standard Handbook for Mechanical Engineers, *8th ed., MGraw-Hill, 1978.*)

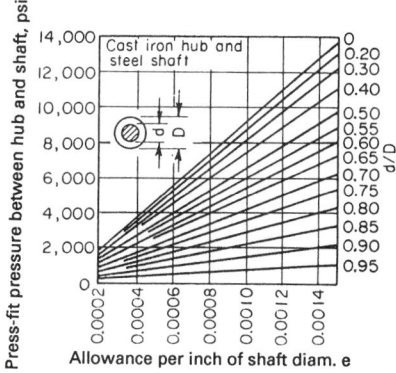

FIGURE 11–11   Press-fit pressure between cast-iron hub and shaft (1 psi = 6894.757 Pa; 1 in = 25.4 mm). (*Baumeister, T.,* Marks' Standard Handbook for Mechanical Engineers, *8th ed., McGraw-Hill, 1978.*)

TABLE 11-10
Tolerances[a] for Shafts for Sizes up to 500 mm

Diameter steps, mm

System of basic shaft	Limits	—3	3 6	6 10	10 18	18 24	24 30	30 40	40 50	50 65	65 80	80 100	100 120	120 140	140 160	160 180	180 200	200 225	225 250	250 280	280 315	315 355	355 400	400 450	450 500
a9	es[b]	-270	-270	-280	-290	-300	-300	-310	-320	-340	-360	-380	-410	-460	-520	-580	-660	-740	-820	-920	-1050	-1200	-1350	-1500	-1650
	ei	-295	-300	-316	-333	-352	-352	-372	-382	-414	-434	-467	-497	-560	-620	-680	-775	-855	-935	-1050	-1180	-1340	-1490	-1655	-1805
b9	es[c]	-140	-140	-150	-150	-160	-160	-170	-180	-190	-200	-220	-240	-260	-280	-310	-340	-380	-420	-480	-540	-600	-680	-760	-840
	ei	-165	-170	-186	-193	-212	-212	-232	-242	-264	-274	-307	-327	-360	-380	-410	-455	-495	-535	-610	-670	-740	-820	-915	-995
c8	es	-60	-70	-80	-95	-110	-110	-120	-130	-140	-150	-170	-180	-200	-210	-230	-240	-260	-280	-300	-330	-360	-400	-440	-480
	ei	-74	-88	-102	-122	-143	-143	-159	-169	-186	-196	-224	-234	-263	-273	-293	-312	-332	-352	-381	-411	-449	-489	-537	-577
c9	es	-60	-70	-80	-95	-110	-110	-120	-130	-140	-150	-170	-180	-200	-210	-230	-240	-260	-280	-300	-330	-360	-400	-440	-480
	ei	-85	-100	-116	-138	-162	-162	-182	-192	-214	-224	-257	-267	-300	-310	-330	-355	-375	-395	-430	-460	-500	-540	-595	-635
c11	es	-60	-70	-80	-95	-110	-110	-120	-130	-140	-150	-170	-180	-200	-210	-230	-240	-260	-280	-300	-330	-360	-400	-440	-480
	ei	-120	-145	-170	-205	-240	-240	-280	-290	-330	-340	-390	-400	-450	-460	-480	-530	-550	-570	-620	-650	-720	-760	-840	-880
d8	es	-20	-30	-40	-50	-65	-65	-80	-80	-100	-100	-120	-120	-145	-145	-145	-170	-170	-170	-190	-190	-210	-210	-230	-230
	ei	-34	-48	-62	-77	-98	-98	-119	-119	-146	-146	-174	-174	-208	-208	-208	-242	-242	-242	-271	-271	-299	-299	-327	-327
d9	es	-20	-30	-40	-50	-65	-65	-80	-80	-100	-100	-120	-120	-145	-145	-145	-170	-170	-170	-190	-190	-210	-210	-230	-230
	ei	-45	-60	-76	-93	-117	-117	-142	-142	-174	-174	-207	-207	-245	-245	-245	-285	-285	-285	-320	-320	-350	-350	-385	-385
d10	es	-20	-30	-40	-50	-65	-65	-80	-80	-100	-100	-120	-120	-145	-145	-145	-170	-170	-170	-190	-190	-210	-210	-230	-230
	ei	-60	-78	-98	-120	-149	-149	-180	-180	-220	-220	-260	-260	-305	-305	-305	-355	-355	-355	-400	-400	-440	-440	-480	-480
e6	es	-14	-20	-25	-32	-40	-40	-50	-50	-60	-60	-72	-72	-85	-85	-85	-100	-100	-100	-110	-110	-125	-125	-135	-135
	ei	-20	-28	-34	-43	-53	-53	-66	-66	-79	-79	-94	-94	-110	-110	-110	-129	-129	-129	-142	-142	-161	-161	-175	-175
e7	es	-14	-20	-25	-32	-40	-40	-50	-50	-60	-60	-72	-72	-85	-85	-85	-100	-100	-100	-110	-110	-125	-125	-135	-135
	ei	-24	-32	-40	-50	-61	-61	-75	-75	-90	-90	-107	-107	-125	-125	-125	-146	-146	-146	-162	-162	-182	-182	-198	-198
e8	es	-14	-20	-25	-32	-40	-40	-50	-50	-60	-60	-72	-72	-85	-85	-85	-100	-100	-100	-110	-110	-125	-125	-135	-135
	ei	-28	-38	-47	-59	-73	-73	-89	-89	-106	-106	-126	-126	-148	-148	-148	-172	-172	-172	-191	-191	-214	-214	-232	-232
e9	es	-14	-20	-25	-32	-40	-40	-50	-50	-60	-60	-72	-72	-85	-85	-85	-100	-100	-100	-110	-110	-125	-125	-135	-135
	ei	-39	-50	-61	-75	-92	-92	-112	-112	-134	-134	-159	-159	-185	-185	-185	-215	-215	-215	-240	-240	-265	-265	-290	-290
f6	es	-06	-10	-13	-16	-20	-20	-25	-25	-30	-30	-36	-36	-43	-43	-43	-50	-50	-50	-56	-56	-62	-62	-68	-68
	ei	-12	-18	-22	-27	-33	-33	-41	-41	-49	-49	-58	-58	-68	-68	-68	-79	-79	-79	-88	-88	-98	-98	-108	-108
f7	es	-06	-10	-13	-16	-20	-20	-25	-25	-30	-30	-36	-36	-43	-43	-43	-50	-50	-50	-56	-56	-62	-62	-68	-68
	ei	-16	-22	-28	-34	-41	-41	-50	-50	-60	-60	-71	-71	-83	-83	-83	-96	-96	-96	-108	-108	-119	-119	-131	-131
f8	es	-06	-10	-13	-16	-20	-20	-25	-25	-30	-30	-36	-36	-43	-43	-43	-50	-50	-50	-56	-56	-62	-62	-68	-68
	ei	-20	-28	-35	-43	-53	-53	-64	-64	-76	-76	-90	-90	-106	-106	-106	-122	-122	-122	-137	-137	-151	-151	-165	-165
g4	es	-02	-04	-05	-06	-07	-07	-09	-09	-10	-10	-12	-12	-14	-14	-14	-15	-15	-15	-17	-17	-18	-18	-20	-20
	ei	-05	-08	-09	-11	-13	-13	-16	-16	-18	-18	-22	-22	-26	-26	-26	-29	-29	-29	-33	-33	-36	-36	-40	-40
g5	es	-02	-04	-05	-06	-07	-07	-09	-09	-10	-10	-12	-12	-14	-14	-14	-15	-15	-15	-17	-17	-18	-18	-20	-20
	ei	-06	-09	-11	-14	-16	-16	-20	-20	-23	-23	-27	-27	-32	-32	-32	-35	-35	-35	-40	-40	-43	-43	-47	-47
g6	es	-02	-04	-05	-06	-07	-07	-09	-09	-10	-10	-12	-12	-14	-14	-14	-15	-15	-15	-17	-17	-18	-18	-20	-20
	ei	-08	-12	-14	-17	-20	-20	-25	-25	-29	-29	-34	-34	-39	-39	-39	-44	-44	-44	-49	-49	-54	-54	-60	-60
h5	es	-00	-00	-00	-00	-00	-00	-00	-00	-00	-00	-00	-00	-00	-00	-00	-00	-00	-00	-00	-00	-00	-00	-00	-00
	ei	-04	-05	-06	-08	-09	-09	-11	-11	-13	-13	-15	-15	-18	-18	-18	-20	-20	-20	-23	-23	-25	-25	-27	-27
h6	es	-00	-00	-00	-00	-00	-00	-00	-00	-00	-00	-00	-00	-00	-00	-00	-00	-00	-00	-00	-00	-00	-00	-00	-00
	ei	-06	-08	-09	-11	-13	-13	-16	-16	-19	-19	-22	-22	-25	-25	-25	-29	-29	-29	-32	-32	-36	-36	-40	-40
h7	es	-00	-00	-00	-00	-00	-00	-00	-00	-00	-00	-00	-00	-00	-00	-00	-00	-00	-00	-00	-00	-00	-00	-00	-00
	ei	-10	-12	-15	-18	-21	-21	-25	-25	-30	-30	-35	-35	-40	-40	-40	-46	-46	-46	-52	-52	-57	-57	-63	-63
h8	es	-00	-00	-00	-00	-00	-00	-00	-00	-00	-00	-00	-00	-00	-00	-00	-00	-00	-00	-00	-00	-00	-00	-00	-00
	ei	-14	-18	-22	-27	-33	-33	-39	-39	-46	-46	-54	-54	-63	-63	-63	-72	-72	-72	-81	-81	-89	-89	-97	-97

TABLE 11-10
Tolerances[a] for Shafts for Sizes up to 500 mm (Cont.)

Diameter steps, mm

System of basic shaft	Limits	—/3	3/6	6/10	10/18	18/24	24/30	30/40	40/50	50/65	65/80	80/100	100/120	120/140	140/160	160/180	180/200	200/225	225/250	250/280	280/315	315/355	355/400	400/450	450/500
h9	es	-00	-00	-00	-00	-00	-00	-00	-00	-00	-00	-00	-00	-00	-00	-00	-00	-00	-00	-00	-00	-00	-00	-00	-00
	ei	-25	-30	-36	-43	-52	-52	-62	-62	-74	-74	-87	-87	-100	-100	-100	-115	-115	-115	-130	-130	-140	-140	-155	-155
h10	es	-00	-00	-00	-00	-00	-00	-00	-00	-00	-00	-00	-00	-00	-00	-00	-00	-00	-00	-00	-00	-00	-00	-00	-00
	ei	-40	-48	-58	-70	-84	-84	-100	-100	-120	-120	-140	-140	-160	-160	-160	-185	-185	-185	-210	-210	-230	-230	-250	-250
h11	es[b]	-00	-00	-00	-00	-00	-00	-00	-00	-00	-00	-00	-00	-00	-00	-00	-00	-00	-00	-00	-00	-00	-00	-00	-00
	ei[c]	-60	-75	-90	-110	-130	-130	-160	-160	-190	-190	-220	-220	-250	-250	-250	-290	-290	-290	-320	-320	-360	-360	-400	-400
j5	es	+02	+03	+04	+05	+05	+05	+06	+06	+06	+06	+06	+06	+07	+07	+07	+07	+07	+07	+07	+07	+07	+07	+07	+07
	ei	-02	-02	-02	-03	-04	-04	-05	-05	-07	-07	-09	-09	-11	-11	-11	-13	-13	-13	-16	-16	-18	-18	-20	-20
j6	es	+04	+06	+07	+08	+09	+09	+11	+11	+12	+12	+13	+13	+14	+14	+14	+16	+16	+16	+16	+16	+18	+18	+20	+20
	ei	-02	-02	-02	-03	-04	-04	-05	-05	-07	-07	-09	-09	-11	-11	-11	-13	-13	-13	-16	-16	-18	-18	-20	-20
j7	es	+06	+08	+10	+12	+13	+13	+15	+15	+18	+18	+20	+20	+22	+22	+22	+25	+25	+25	+26	+26	+29	+29	+31	+31
	ei	-04	-04	-05	-06	-08	-08	-10	-10	-12	-12	-15	-15	-18	-18	-18	-21	-21	-21	-26	-26	-28	-28	-32	-32
k6	es	+06	+09	+10	+12	+15	+15	+18	+18	+21	+21	+25	+25	+28	+28	+28	+33	+33	+33	+36	+36	+40	+40	+45	+45
	ei	+00	+01	+01	+01	+02	+02	+02	+02	+02	+02	+03	+03	+03	+03	+03	+04	+04	+04	+04	+04	+04	+04	+05	+05
k7	es	+10	+13	+16	+19	+23	+23	+27	+27	+32	+32	+38	+38	+43	+43	+43	+50	+50	+50	+56	+56	+61	+61	+68	+68
	ei	+00	+01	+01	+01	+02	+02	+02	+02	+02	+02	+03	+03	+03	+03	+03	+04	+04	+04	+04	+04	+04	+04	+05	+05
m6	es	+08	+12	+15	+18	+21	+21	+25	+25	+30	+30	+35	+35	+40	+40	+40	+46	+46	+46	+52	+52	+57	+57	+63	+63
	ei	+02	+04	+06	+07	+08	+08	+09	+09	+11	+11	+13	+13	+15	+15	+15	+17	+17	+17	+20	+20	+21	+21	+23	+23
m7	es	+12	+16	+21	+25	+29	+29	+34	+34	+41	+41	+48	+48	+55	+55	+55	+63	+63	+63	+72	+72	+78	+78	+86	+86
	ei	+02	+04	+06	+07	+08	+08	+09	+09	+11	+11	+13	+13	+15	+15	+15	+17	+17	+17	+20	+20	+21	+21	+23	+23
n4	es	+07	+12	+14	+17	+21	+21	+24	+24	+28	+28	+33	+33	+39	+39	+39	+45	+45	+45	+50	+50	+55	+55	+60	+60
	ei	+04	+08	+10	+12	+15	+15	+17	+17	+20	+20	+23	+23	+27	+27	+27	+31	+31	+31	+34	+34	+37	+37	+40	+40
n5	es	+08	+13	+16	+20	+24	+24	+28	+28	+33	+33	+38	+38	+45	+45	+45	+51	+51	+51	+57	+57	+62	+62	+67	+67
	ei	+04	+08	+10	+12	+15	+15	+17	+17	+20	+20	+23	+23	+27	+27	+27	+31	+31	+31	+34	+34	+37	+37	+40	+40
p5	es	+10	+17	+21	+26	+31	+31	+37	+37	+45	+45	+52	+52	+61	+61	+61	+70	+70	+70	+79	+79	+87	+87	+95	+95
	ei	+06	+12	+15	+18	+22	+22	+26	+26	+32	+32	+37	+37	+43	+43	+43	+50	+50	+50	+56	+56	+62	+62	+68	+68
p6	es	+12	+20	+24	+29	+35	+35	+42	+42	+51	+51	+59	+59	+68	+68	+68	+79	+79	+79	+88	+88	+98	+98	+108	+108
	ei	+06	+12	+15	+18	+22	+22	+26	+26	+32	+32	+37	+37	+43	+43	+43	+50	+50	+50	+56	+56	+62	+62	+68	+68
r6	es	+16	+23	+28	+34	+41	+41	+50	+50	+60	+62	+73	+76	+88	+90	+93	+106	+109	+113	+126	+130	+144	+150	+166	+172
	ei	+10	+15	+19	+23	+28	+28	+34	+34	+41	+43	+51	+54	+63	+65	+68	+77	+80	+84	+94	+98	+108	+114	+126	+132
t7	es	—	—	—	—	—	+62	+73	+79	+96	+105	+126	+139	+162	+174	+186	+212	+226	+242	+270	+292	+325	+351	+393	+423
	ei	—	—	—	—	—	+41	+48	+54	+66	+75	+91	+104	+122	+134	+146	+166	+180	+196	+218	+240	+268	+294	+330	+360
u5	es	+22	+28	+34	+41	+50	+57	+71	+81	+100	+115	+139	+159	+188	+208	+228	+256	+278	+304	+338	+373	+415	+460	+517	+567
	ei	+18	+23	+28	+33	+41	+48	+60	+70	+87	+102	+124	+144	+170	+190	+210	+236	+258	+284	+315	+350	+390	+435	+490	+540
u8	es	+32	+41	+50	+60	+74	+81	+99	+109	+133	+148	+178	+198	+233	+253	+273	+308	+330	+356	+396	+431	+479	+524	+587	+637
	ei	+18	+23	+28	+33	+41	+48	+60	+70	+87	+102	+124	+144	+170	+190	+210	+236	+258	+284	+315	+350	+390	+435	+490	+540
v5	es	—	—	—	+47	+56	+64	+79	+92	+115	+133	+161	+187	+220	+246	+270	+304	+330	+360	+408	+448	+500	+555	+622	+687
	ei	—	—	—	+39	+47	+55	+68	+81	+102	+120	+146	+172	+202	+228	+252	+284	+310	+340	+385	+425	+475	+530	+595	+660
x8	es	+34	+46	+56	+67 +72	+87	+97	+119	+136	+168	+192	+232	+264	+311	+343	+373	+422	+457	+497	+556	+606	+679	+749	+837	+917
	ei	+20	+28	+34	+40 +45	+54	+64	+80	+97	+122	+146	+178	+210	+248	+280	+310	+350	+385	+425	+475	+525	+590	+660	+740	+820
y6	es	—	—	—	—	+76	+88	+110	+130	+163	+193	+236	+276	+325	+365	+405	+454	+499	+549	+612	+682	+766	+856	+960	+1040
	ei	—	—	—	—	+63	+75	+94	+114	+144	+174	+214	+254	+300	+340	+380	+425	+470	+520	+580	+650	+730	+820	+920	+1000
z7	es	+36	+47	+57	+68 +78	+94	+109	+137	+161	+202	+240	+293	+345	+405	+455	+505	+566	+621	+686	+762	+842	+957	+1057	+1163	+1313
	ei	+26	+35	+42	+50 +60	+73	+88	+112	+136	+172	+210	+258	+310	+365	+415	+465	+520	+575	+640	+710	+790	+900	+1000	+1100	+1250

TABLE 11-10
Tolerances[a] for Shafts for Sizes up to 500 mm (Cont.)

System of basic shaft	Limits	Diameter steps, mm																							
		— 3	3 6	6 10	10 18	18 24	24 30	30 40	40 50	50 65	65 80	80 100	100 120	120 140	140 160	160 180	180 200	200 225	225 250	250 280	280 315	315 355	355 400	400 450	450 500
za6	es	+38	+50	+61	—	—	—	—	—	—	—	—	—	—	—	—	—	—	—	—	—	—	—	—	—
	ei	+32	+42	+52	—	—	—	—	—	—	—	—	—	—	—	—	—	—	—	—	—	—	—	—	—
zb7	es	+50	+62	+82	—	—	—	—	—	—	—	—	—	—	—	—	—	—	—	—	—	—	—	—	—
	ei	+40	+50	+67	—	—	—	—	—	—	—	—	—	—	—	—	—	—	—	—	—	—	—	—	—
zc8	es	+74	+98	+119	—	—	—	—	—	—	—	—	—	—	—	—	—	—	—	—	—	—	—	—	—
	ei	+60	+80	+97	—	—	—	—	—	—	—	—	—	—	—	—	—	—	—	—	—	—	—	—	—

[a] Tolerances in micrometers (1 μm = 10^{-3}mm).
[b] es = upper deviation.
[c] ei = lower deviation.

11.16

TABLE 11–11
Tolerances[a] for Holes for Sizes up to 500 mm

System of basic hole	Limits	Diameter steps, mm																						
		— 3	3 6	6 10	10 14 14 18	18 24 24 30	30 40	40 50	50 65	65 80	80 100	100 120	120 140	140 160	160 180	180 200	200 225	225 250	250 280	280 315	315 355	355 400	400 450	450 500
A9	*ES*[b]	+295	+300	+316	+333	+352	+372	+382	+414	+434	+467	+497	+560	+620	+680	+775	+855	+925	+1050	+1180	+1340	+1490	+1655	+1805
	EI[c]	+270	+270	+280	+290	+300	+310	+320	+340	+360	+380	+410	+460	+520	+580	+660	+740	+820	+920	+1050	+1200	+1350	+1500	+1650
B9	*ES*	+165	+170	+186	+193	+212	+232	+242	+264	+274	+307	+327	+360	+380	+410	+455	+495	+535	+610	+670	+740	+820	+915	+995
	EI	+140	+140	+150	+150	+160	+170	+180	+190	+200	+220	+240	+260	+280	+310	+340	+380	+420	+480	+540	+600	+680	+760	+840
B11	*ES*	+200	+215	+240	+260	+290	+330	+340	+380	+390	+440	+460	+510	+530	+560	+630	+670	+710	+800	+860	+960	+1040	+1160	+1240
	EI	+140	+140	+150	+150	+160	+170	+180	+190	+200	+220	+240	+260	+280	+310	+340	+380	+420	+480	+540	+600	+680	+760	+840
C8	*ES*	+74	+88	+102	+122	+143	+159	+169	+186	+196	+224	+234	+263	+273	+293	+312	+332	+352	+381	+411	+449	+489	+537	+577
	EI	+60	+70	+80	+95	+110	+120	+130	+140	+150	+170	+180	+200	+210	+230	+240	+260	+280	+300	+330	+360	+400	+440	+480
C11	*ES*	+120	+145	+170	+205	+240	+280	+290	+330	+340	+390	+400	+450	+460	+480	+530	+550	+570	+620	+650	+720	+760	+840	+880
	EI	+60	+70	+80	+95	+110	+120	+130	+140	+150	+170	+180	+200	+210	+230	+240	+260	+280	+300	+330	+360	+400	+440	+480
D8	*ES*	+34	+48	+62	+77	+98		+119		+146		+174		+208			+242			+271		+299		+327
	EI	+20	+30	+40	+50	+65		+80		+100		+120		+145			+170			+190		+210		+230
D9	*ES*	+45	+60	+76	+93	+117		+142		+174		+207		+245			+285			+320		+350		+385
	EI	+20	+30	+40	+50	+65		+80		+100		+120		+145			+170			+190		+210		+230
E5	*ES*	+18	+25	+31	+40	+49		+61		+73		+87		+103			+120			+133		+150		+162
	EI	+14	+20	+25	+32	+40		+50		+60		+72		+85			+100			+110		+125		+135
F6	*ES*	+12	+18	+22	+27	+33		+41		+49		+58		+68			+79			+88		+98		+108
	EI	+6	+10	+13	+16	+20		+25		+30		+36		+43			+50			+56		+62		+68
F8	*ES*	+20	+28	+35	+43	+53		+64		+76		+90		+106			+122			+137		+151		+165
	EI	+6	+10	+13	+16	+20		+25		+30		+36		+43			+50			+56		+62		+68
G7	*ES*	+12	+16	+20	+24	+28		+34		+40		+47		+54			+61			+69		+75		+83
	EI	+2	+4	+5	+6	+7		+9		+10		+12		+14			+15			+17		+18		+20
H5	*ES*	+4	+5	+6	+8	+9		+11		+13		+15		+18			+20			+23		+25		+27
	EI	0	0	0	0	0		0		0		0		0			+0			+0		0		0
H6	*ES*	+6	+8	+9	+11	+13		+16		+19		+22		+25			+29			+32		+36		+40
	EI	0	0	0	0	0		0		0		0		0			0			0		0		0
H7	*ES*	+10	+12	+15	+18	+21		+25		+30		+35		+40			+46			+52		+57		+63
	EI	0	0	0	0	0		0		0		0		0			0			0		0		0
H8	*ES*	+14	+18	+22	+27	+33		+39		+46		+54		+63			+72			+81		+89		+97
	EI	0	0	0	0	0		0		0		0		0			0			0		0		0
H9	*ES*	+25	+30	+36	+43	+52		+62		+74		+87		+100			+115			+130		+140		+155
	EI	0	0	0	0	0		0		0		0		0			0			0		0		0
H10	*ES*	+40	+48	+58	+70	+84		+100		+120		+140		+160			+185			+210		+230		+250
	EI	0	0	0	0	0		0		0		0		0			0			0		0		0
H11	*ES*	+60	+75	+90	+110	+130		+160		+190		+220		+250			+290			+320		+360		+400
	EI	0	0	0	0	0		0		0		0		0			0			0		0		0
J7	*ES*	+4	+6	+8	+10	+12		+14		+18		+22		+26			+30			+36		+39		+43
	EI	−6	−6	−7	−8	−9		−11		−12		−13		−14			−16			−16		−18		−20
K6	*ES*	0	+2	+2	+2	+2		+3		+4		+4		+4			+5			+5		+7		+8
	EI	−6	−6	−7	−9	−11		−13		−15		−18		−21			−24			−27		−29		−32
K7	*ES*	0	+3	+5	+6	+6		+7		+9		+10		+12			+13			+16		+17		+18
	EI	−10	−9	−10	−12	−15		−18		−21		−25		−28			−33			−36		−40		−45
M7	*ES*	−2	0	0	0	0		0		0		0		0			0			0		0		0
	EI	−12	−12	−15	−18	−21		−25		−30		−35		−40			−46			−52		−57		−63

TABLE 11-11
Tolerances[a] for Holes for Sizes up to 500 mm (Cont.)

System of basic hole	Limits	—/3	3/6	6/10	10-14 / 14-18	18-24 / 24-30	30/40	40/50	50/65	65/80	80/100	100/120	120/140	140/160	160/180	180/200	200/225	225/250	250/280	280/315	315/355	355/400	400/450	450/500
																Diameter steps, mm								
N7	ES	-4	-4	-4	-5	-7	-8	-8	-9	-9	-10	-10	-12	-12	-12	-14	-14	-14	-14	-14	-16	-16	-17	-17
	EI	-14	-16	-19	-23	-28	-33	-33	-39	-39	-45	-45	-52	-52	-52	-60	-60	-60	-66	-66	-73	-73	-80	-80
P7	ES	-6	-8	-9	-11	-14	-17	-17	-21	-21	-24	-24	-28	-28	-28	-33	-33	-33	-36	-36	-41	-41	-45	-45
	EI	-16	-20	-24	-29	-35	-42	-42	-51	-51	-59	-59	-68	-68	-68	-79	-79	-79	-88	-88	-98	-98	-108	-108
S6	ES	-14	-16	-20	-25	-31	-38	-38	-47	-53	-64	-72	-85	-93	-101	-113	-121	-131	-149	-161	-179	-197	-219	-239
	EI	-20	-24	-29	-36	-44	-54	-54	-66	-72	-86	-94	-110	-118	-126	-142	-150	-160	-181	-193	-215	-233	-259	-279
S7	ES	-14	-15	-17	-21	-27	-34	-34	-42	-48	-58	-66	-77	-85	-93	-105	-113	-123	-138	-150	-169	-187	-209	-229
	EI	-24	-27	-32	-39	-48	-59	-59	-72	-78	-93	-101	-117	-125	-133	-151	-159	-169	-190	-202	-226	-244	-272	-292
T6	ES	—	—	—	—	-37	-43	-49	-60	-69	-84	-97	-115	-127	-139	-157	-171	-187	-209	-231	-257	-283	-317	-347
	EI	—	—	—	—	-50	-59	-65	-79	-88	-106	-119	-140	-152	-164	-186	-200	-216	-241	-263	-293	-319	-357	-387
U7	ES	-18	-19	-22	-26	-33/-40	-51	-61	-76	-91	-111	-131	-155	-175	-195	-219	-241	-267	-295	-330	-369	-414	-467	-517
	EI	-28	-31	-37	-44	-54/-61	-76	-86	-106	-121	-146	-166	-195	-215	-235	-265	-287	-313	-347	-382	-426	-471	-530	-580
V6	ES	—	—	—	-36	-43/-51	-63	-76	-96	-114	-139	-165	-195	-221	-245	-275	-301	-331	-376	-416	-464	-519	-582	-647
	EI	—	—	—	-47	-56/-64	-79	-92	-115	-133	-161	-187	-220	-246	-270	-304	-330	-360	-408	-448	-500	-555	-622	-687
X7	ES	-20	-24	-28	-33/-38	-46/-56	-71	-88	-111	-135	-165	-197	-233	-265	-295	-333	-368	-408	-455	-505	-569	-639	-717	-797
	EI	-30	-36	-43	-51/-56	-67/-77	-96	-113	-141	-165	-200	-232	-273	-305	-335	-379	-414	-454	-507	-557	-626	-696	-780	-860
Y7	ES	—	—	—	—	-55/-67	-94	-114	-144	-174	-201	-241	-285	-325	-365	-413	-453	-503	-560	-630	-709	-799	-897	-977
	EI	—	—	—	—	-76/-88	-119	-139	-174	-204	-236	-276	-325	-365	-405	-459	-499	-549	-612	-682	-766	-856	-960	-1040
Z8	ES	-26	-35	-42	-50/-60	-73/-88	-112	-136	-172	-210	-258	-310	-365	-415	-465	-520	-575	-640	-710	-790	-900	-1000	-1100	-1250
	EI	-40	-53	-64	-77/-87	-106/-121	-151	-175	-218	-256	-312	-364	-428	-478	-528	-592	-647	-712	-791	-871	-989	-1089	-1197	-1347
ZA7	ES	-32	-38	-46	—	—	—	—	—	—	—	—	—	—	—	—	—	—	—	—	—	—	—	—
	EI	-42	-50	-61	—	—	—	—	—	—	—	—	—	—	—	—	—	—	—	—	—	—	—	—
ZB8	ES	-40	-50	-67	—	—	—	—	—	—	—	—	—	—	—	—	—	—	—	—	—	—	—	—
	EI	-54	-68	-89	—	—	—	—	—	—	—	—	—	—	—	—	—	—	—	—	—	—	—	—
ZC9	ES	-60	-80	-97	—	—	—	—	—	—	—	—	—	—	—	—	—	—	—	—	—	—	—	—
	EI	-85	-110	-133	—	—	—	—	—	—	—	—	—	—	—	—	—	—	—	—	—	—	—	—

[a] Tolerances in μm; 1 μm = 10^{-3} mm

[b] ES = upper deviation.

[c] EI = lower deviation.

TABLE 11–12
Tolerances[a] For Shafts for Sizes 500 to 3150 mm

System of basic shaft	Limits	Diameter steps, mm															
		500 / 560	560 / 630	630 / 710	710 / 800	800 / 900	900 / 1000	1000 / 1120	1120 / 1250	1250 / 1400	1400 / 1600	1600 / 1800	1800 / 2000	2000 / 2250	2250 / 2500	2500 / 2800	2800 / 3150
d10	es[b]	−260		−290		−320		−350		−390		−430		−480		−520	
	ei[c]	−540		−610		−680		−770		−890		−1030		−1180		−1380	
e8	es	−145		−160		−170		−195		−220		−240		−260		−290	
	ei	−255		−285		−310		−360		−415		−470		−540		−620	
f9	es	−76		−80		−86		−98		−110		−120		−130		−145	
	ei	−251		−280		−316		−358		−420		−490		−570		−685	
g6	es	−22		−24		−26		−28		−30		−32		−34		−38	
	ei	−66		−74		−82		−94		−108		−124		−140		−173	
g7	es	−22		−24		−26		−28		−30		−32		−34		−38	
	ei	−92		−103		−115		−133		−155		−182		−209		−248	
h6	es	0		0		0		0		0		0		0		0	
	ei	−44		−50		−56		−66		−78		−92		−110		−135	
h7	es	0		0		0		0		0		0		0		0	
	ei	−70		−80		−90		−105		−125		−150		−175		−210	
h8	es	0		0		0		0		0		0		0		0	
	ei	−110		−125		−140		−165		−195		−230		−280		−330	
h9	es	0		0		0		0		0		0		0		0	
	ei	−175		−200		−230		−260		−310		−370		−440		−540	
h10	es	0		0		0		0		0		0		0		0	
	ei	−280		−320		−360		−420		−500		−600		−700		−860	
h11	es	0		0		0		0		0		0		0		0	
	ei	−440		−500		−560		−660		−780		−920		−1100		−1350	
js9	es ei	±87.5		±100		±115		±130		±155		±185		± 220		±270	
k6	es	+44		+50		+56		+66		+78		+92		+110		+135	
	ei	0		0		0		0		0		0		0		0	
m6	es	+70		+80		+90		+106		+126		+150		+178		+211	
	ei	+26		+30		+34		+40		+48		+58		+68		+76	
n6	es	+88		+100		+112		+132		+156		+184		+220		+270	
	ei	+44		+50		+56		+66		+78		+92		+110		+135	
p6	es	+122		+139		+156		+186		+218		+262		+305		+375	
	ei	+78		+88		+100		+120		+140		+170		+195		+240	
r7	es	+220	+225	+255	+265	+300	+310	+355	+365	+425	+455	+520	+550	+615	+635	+760	+790
	ei	+150	+155	+175	+185	+210	+220	+250	+260	+300	+330	+370	+400	+440	+460	+550	+580
s7	es	+350	+380	+420	+460	+520	+560	+625	+685	+765	+845	+970	+1070	+1175	+1275	+1460	+1610
	ei	+280	+310	+340	+380	+430	+470	+520	+580	+640	+720	+820	+920	+1000	+1100	+1250	+1400
t7	es	+470	+520	+580	+640	+710	+770	+885	+945	+1085	+1175	+1350	+1500	+1675	+1825	+2110	+2310
	ei	+400	+450	+500	+560	+620	+680	+780	+840	+960	+1050	+1200	+1350	+1500	+1650	+1900	+2100
u7	es	+570	+730	+820	+920	+1031	+1140	+1255	+1405	+1575	+1725	+2000	+2150	+2475	+2675	+3110	+3410
	ei	+600	+660	+740	+840	+940	+1050	+1150	+1300	+1450	+1600	+1850	+2000	+2300	+2500	+2900	+3200

[a] Tolerances in μm (1 μm = 10^{-3} mm).
[b] es = upper deviation.
[c] ei = lower deviation.

Source: IS 2101, 1962.

TABLE 11–13
Tolerances[a] For Holes for Sizes 500 to 3150 mm

System of basic hole	Limits	500 560	560 630	630 710	710 800	800 900	900 1000	1000 1120	1120 1250	1250 1400	1400 1600	1600 1800	1800 2000	2000 2240	2240 2500	2500 2800	2800 3150
D10	ES[a]	+540		+610		+680		+770		+890		+1030		+1180		+1380	
	ES[b]	+260		+290		+320		+350		+390		+430		+480		+520	
E8	ES	+255		+285		+310		+360		+415		+470		+540		+620	
	EI	+145		+160		+170		+195		+220		+240		+260		+290	
F9	ES	+251		+280		+316		+358		+420		+490		+570		+685	
	EI	+76		+80		+86		+98		+110		+120		+130		+145	
G6	ES	+66		+74		+82		+94		+108		+124		+144		+173	
	EI	+22		+24		+26		+28		+30		+32		+34		+38	
G7	ES	+92		+103		+115		+133		+155		+182		+209		+248	
	EI	+22		+24		+26		+28		+30		+32		+34		+38	
H6	ES	+40		+50		+56		+66		+78		+92		+110		+135	
	EI	0		0		0		0		0		0		0		0	
H7	ES	+70		+80		+90		+105		+125		+150		+175		+210	
	EI	0		0		0		0		0		0		0		0	
H8	ES	+110		+125		+140		+165		+195		+230		+280		+330	
	EI	0		0		0		0		0		0		0		0	
H9	ES	+175		+200		+230		+260		+310		+370		+440		+540	
	EI	0		0		0		0		0		0		0		0	
H10	ES	+280		+320		+360		+420		+500		+600		+700		+860	
	EI	0		0		0		0		0		0		0		0	
H11	ES	+440		+500		+560		+660		+780		+920		+1100		+1350	
	EI	0		0		0		0		0		0		0		0	
JS9	ES EI	±87.5		±100		±115		±130		±155		±185		±220		±270	
K6	ES	0		0		0		0		0		0		0		0	
	EI	−44		−50		−56		−66		−78		−92		−110		−135	
M6	ES	−26		−30		−34		−40		−48		−58		−68		−76	
	EI	−70		−80		−90		−106		−126		−150		−178		−211	
N6	ES	−44		−50		−56		−66		−78		−92		−110		−135	
	EI	−88		−100		−112		−132		−156		−184		−220		−270	
P6	ES	−78		−88		−100		−120		−140		−170		−195		−240	
	EI	−122		−138		−156		−186		−218		−262		−305		−375	
R7	ES	−150	−155	−175	−185	−210	−200	−250	−260	−300	−330	−370	−400	−440	−460	−550	−580
	EI	−220	−225	−255	−265	−300	−310	−355	−365	−425	−455	−520	−550	−615	−635	−760	−790
S7	ES	−280	−310	−340	−380	−430	−470	−520	−580	−640	−720	−820	−920	−1000	−1100	−1250	−1400
	EI	−350	−380	−420	−460	−520	−560	−625	−685	−765	−845	−970	−1070	−1175	−1275	−1460	−1610
T7	ES	−400	−450	−500	−560	−620	−680	−780	−840	−960	−1050	−1200	−1350	−1500	−1650	−1900	−2100
	EI	−470	−520	−580	−640	−710	−770	−885	−945	−1085	−1175	−1350	−1500	−1675	−1825	−2110	−2310
U7	ES	−600	−660	−740	−840	−940	−1050	−1150	−1300	−1450	−1600	−1850	−2000	−2300	−2500	−2900	−3200
	EI	−670	−730	−820	−920	−1030	−1140	−1255	−1405	−1575	−1725	−2000	−2150	−2475	−2675	−3110	−3410

[a] Tolerances in μm (1 μm = 10^{-3} mm).
[b] ES = upper deviation.
[c] EI = lower deviation.

Source: IS 2101, 1962.

TABLE 11–14
Mean fit and variation about the mean fit for holes for sizes up to 400 mm

Diameter steps, mm (range shown as top/bottom of interval). Values in microns given as **mean fit ±variation**.

Quality of fit	Combination of shaft and hole		—/3	3/6	6/10	10/18	18/24	24/30	30/40	40/50	50/65	65/80	80/100	100/120	120/140	140/160	160/180	180/200	200/225	225/250	250/280	280/315	315/355	355/400
Clearance Fit (Fig. 11-6)																								
Precision sliding	H7 g6	Normal	+11 ±8	+14 ±10	+17 ±12	+20.5 ±14.5	+24 ±17		+29.5 ±20.5		+34.5 ±24.5		+40.5 ±28.5		+46.5 ±32.5			+52.5 ±37.5			+59 ±42		+64.5 ±46.5	
Normal running	H7 f7	Normal	+16 ±9	+22 ±12	+28 ±15	+34 ±18	+41 ±21		+50 ±25		+60 ±30		+71 ±35		+83 ±40			+96 ±46			+108 ±52		+119 ±57	
Easy running	H8 e8	Normal	+28 ±14	+38 ±18	+47 ±22	+59 ±27	+73 ±33		+89 ±39		+106 ±46		+126 ±54		+148 ±63			+172 ±72			+191 ±81		+214 ±89	
Loose running	H8 d9	Normal	+39.5 ±19.5	+54 ±24	+69 ±29	+85 ±35	+107.5 ±42.5		+130.5 ±50.5		+160 ±60		+190.5 ±70.5		+226.5 ±81.5			+263.5 ±93.5			+295.5 ±105.5		+324.5 ±114.5	
Slack running	H9 c9	Normal	+85 ±25	+100 ±30	+116 ±36	+138 ±43	+162 ±52		+182 ±62		+214 ±74	+257 ±74	+300 ±100	+310 ±100	+330 ±100	+355 ±115		+375 ±115	+395 ±115		+420 ±130	+460 ±130	+500 ±140	+540 ±140
Location and Assembly Fit																								
Position fits	H8 a9		+289.5 ±19.5	+294 ±24	+309 ±29	+325 ±35	+342.5 ±42.5		+360.5 ±50.5	+370.5 ±50.5	+400 ±60	+420 ±60	+450.5 ±70.5	+480.5 ±70.5	+541.5 ±81.5	+601.5 ±81.5	+661.5 ±81.5	+753.5 ±93.5	+833.5 ±93.5	+913.5 ±93.5	+1025.5 ±105.5	+1155.5 ±105.5	+1314.5 ±114.5	+1454.5 ±114.5
Position fits	H8 b9		+159.5 ±19.5	+164 ±24	+179 ±29	+185 ±35	+202.5 ±42.5		+220.5 ±50.5	+230.5 ±50.5	+250 ±60	+260 ±60	+290.5 ±70.5	+310.5 ±70.5	+341.5 ±81.5	+361.5 ±81.5		+433.5 ±93.5	+473.5 ±93.5	+513.5 ±93.5	+585.5 ±105.5	+645.5 ±105.5	+714.5 ±114.5	+794.5 ±114.5
Precision location	H6 h6		+7 ±7	+8 ±8	+9 ±9	+11 ±11	+13 ±13		+16 ±16		+19 ±19		+22 ±22		+25 ±25			+29 ±29			+32 ±32		+36 ±36	
Normal location	H8 h8		+14 ±14	+18 ±18	+22 ±22	+27 ±27	+33 ±33		+39 ±39		+46 ±46		+54 ±54		+63 ±63			+72 ±72			+81 ±81		+89 ±89	
Loose location	H9 h9		+25 ±25	+30 ±30	+36 ±36	+43 ±43	+52 ±52		+62 ±62		+74 ±74		+87 ±87		+100 ±100			+115 ±115			+130 ±130		+140 ±140	
Slack assembly	H11 h11		+60 ±60	+75 ±75	+90 ±90	+110 ±110	+130 ±130		+160 ±160		+190 ±190		+220 ±220		+250 ±250			+290 ±290			+320 ±320		+360 ±360	
Transition Fits (Fig. 11-7)																								
Push	H7 k6	Normal	+2 ±8	+3 ±10	+5 ±12	+6.5 ±14.5	+8 ±17		+9.5 ±20.5		+12.5 ±24.5		+15.5 ±28.5		+18.5 ±32.5			+21.5 ±37.5			+26 ±42		+28.5 ±46.5	
True transition	H7 k6	Normal	—	+2 ±10	+2.5 ±12	+2 ±14.5	+2.5 ±17		+3.5 ±20.5		+3.5 ±24.5		+4.5 ±28.5		+4.5 ±32.5			+6 ±37.5			+6 ±42		+6.5 ±46.5	
Interference transition	H7 m6	Normal	-1 ±8	-2 ±10	-3 ±12	-3.5 ±14.5	-4 ±17		-4.5 ±20.5		-5.5 ±24.5		-6.5 ±28.5		-7.5 ±32.5			-8.5 ±37.5			-10 ±42		-10.5 ±46.5	
Interference Fits (Fig. 11-8)																								
Light press fit	H7 p6	Normal	-8 ±8	-10 ±10	-12 ±12	-14.5 ±14.5	-18 ±17		-21.5 ±20.5		-26.5 ±24.5		-30.5 ±28.5		-35.5 ±32.5			-41.5 ±37.5			-46 ±42		-51.5 ±46.5	
Medium drive fit	H7 r6	Normal	-11 ±8	-13 ±10	-16 ±12	-19.5 ±14.5	-24 ±17		-29 ±20.5		-35.5 ±24.5	-37.5 ±24.5	-44.5 ±28.5	-47.5 ±28.5	-55.5 ±32.5	-57.5 ±32.5	-60.5 ±32.5	-66.5 ±37.5	-71.5 ±37.5	-75.5 ±37.5	-84 ±42	-88 ±42	-97.5 ±46.5	-103.5 ±46.5
MHeavy drive fit	H7 s6	Normal	-12 ±8	-17 ±10	-20 ±12	-26.5 ±14.5	-31 ±17		-38.5 ±20.5		-47.5 ±24.5	-53.5 ±24.5	-64.5 ±28.5	-72.5 ±28.5	-84.5 ±32.5	-92.5 ±32.5	-100.5 ±32.5	-113.5 ±37.5	-121.5 ±37.5	-131.5 ±37.5	-148 ±42	-160 ±42	-179.5 ±46.5	-197.5 ±46.5
Heavy force Fir or shrink fit	H7 u6	Normal	-17 ±8	-21 ±10	-25 ±12	-29.5 ±14.5	-37 ±17	-44 ±17	-55.5 ±20.5	-65.5 ±20.5	-81.5 ±24.5	-96.5 ±24.5	-117.5 ±28.5	-137.5 ±28.5	-162.5 ±32.5	-182.5 ±32.5	-202.5 ±32.5	-227.5 ±37.5	-249.5 ±37.5	-275.5 ±37.5	-305 ±42	-340 ±42	-379.5 ±46.5	-424.5 ±46.5

Tolerance in Microns; 1 Micron $= 10^{-3}$ mm $= \mu$m $= 10^{-6}$ m

Source: Indian Standards.

TABLE 11–15
International tolerance grades

Basic sizes		IT6		IT7		IT8		IT9		IT10		IT11	
mm	in	mm	in	mm	in	mm	in	mm	in	mm	in	mm	in
0–3	0–0.12	0.006	0.0002	0.010	0.0004	0.014	0.0006	0.025	0.0010	0.040	0.0016	0.060	0.0024
3–6	0.12–0.24	0.008	0.0003	0.012	0.0005	0.018	0.0007	0.030	0.0012	0.048	0.0019	0.075	0.0030
6–10	0.24–0.40	0.009	0.0004	0.015	0.0006	0.022	0.0009	0.036	0.0014	0.058	0.0023	0.090	0.0035
10–18	0.40–0.72	0.011	0.0004	0.018	0.0007	0.027	0.0011	0.043	0.0017	0.070	0.0028	0.110	0.0043
18–30	0.72–1.20	0.013	0.0005	0.021	0.0008	0.033	0.0013	0.052	0.0020	0.084	0.0033	0.130	0.0051
30–50	1.20–2.00	0.016	0.0006	0.025	0.0010	0.039	0.0015	0.062	0.0024	0.100	0.0039	0.160	0.0063
50–80	2.00–3.20	0.019	0.0007	0.030	0.0012	0.046	0.0018	0.074	0.0029	0.120	0.0047	0.190	0.0075
80–120	3.20–4.80	0.022	0.0009	0.035	0.0014	0.054	0.0021	0.087	0.0034	0.140	0.0055	0.220	0.0087
120–180	4.80–7.20	0.025	0.0010	0.040	0.0016	0.063	0.0025	0.100	0.0039	0.160	0.0063	0.250	0.0098
180–250	7.20–10.00	0.029	0.0011	0.040	0.0018	0.072	0.0028	0.115	0.0045	0.185	0.0073	0.290	0.0114
250–315	10.00–12.60	0.032	0.0013	0.052	0.0020	0.081	0.0032	0.130	0.0051	0.210	0.0083	0.320	0.0126
315–400	12.60–16.00	0.036	0.0014	0.057	0.0022	0.089	0.0035	0.140	0.0055	0.230	0.0091	0.360	0.0142

Source: Preferred metric limits and fits—BSI 4500.

11.22

TABLE 11-16
Fundamental tolerance[a] (μm and μin) for shafts for sizes up to 400 mm (16 in)

System of basic shaft	Limits		0 / 3	3 / 6	6 / 10	10 / 14	14 / 18	18 / 24	24 / 30	30 / 40	40 / 50	50 / 65	65 / 80	80 / 100	100 / 120	120 / 140	140 / 160	160 / 180	180 / 200	200 / 225	225 / 250	250 / 280	280 / 315	315 / 355	355 / 400	400 / 450	450 / 500
		Diameter steps																									
	mm (from)		0	3	6	10	14	18	24	30	40	50	65	80	100	120	140	160	180	200	225	250	280	315	355	400	450
	in (from)		0	0.12	0.24	0.40	0.56	0.72	0.96	1.20	1.60	2.00	2.60	3.20	4.00	4.80	5.60	6.40	7.20	8.00	9.00	10.00	11.20	12.60	14.20	16.00	18.00
	mm (to)		3	6	10	14	18	24	30	40	50	65	80	100	120	140	160	180	200	225	250	280	315	355	400	450	500
	in (to)		0.12	0.24	0.40	0.56	0.72	0.96	1.20	1.60	2.00	2.60	3.20	4.00	4.80	5.60	6.40	7.20	8.00	9.00	10.00	11.20	12.60	14.20	16.00	18.00	20.00
a	es[b]	μm	−270	−270	−280	−290	−290	−300	−300	−310	−320	−340	−360	−380	−410	−460	−520	−580	−660	−740	−820	−920	−1,050	−1,200	−1,350	−1,500	−1,650
		μin	−10,600	−10,600	−11,000	−11,400	−11,400	−11,800	−11,800	−12,200	−12,600	−13,400	−14,200	−14,900	−16,100	−18,100	−20,500	−22,800	−26,000	−29,100	−32,300	−36,200	−41,300	−47,200	−53,200	−59,000	−64,900
j	ei[c]	μm	−2	−2	−2	−2	−3	−3	−4	−5	−5	−7	−7	−9	−9	−11	−11	−11	−13	−13	−13	−16	−16	−18	−18	−18	−20
		μin	−80	−80	−80	−80	−100	−100	−160	−200	−200	−280	−280	−360	−360	−450	−450	−450	−510	−500	−500	−600	−600	−700	−700	−700	−800
c	es	μm	−60	−70	−80	−95	−95	−110	−110	−120	−130	−140	−150	−170	−180	−200	−210	−230	−240	−260	−280	−300	−330	−360	−400	−440	−480
		μin	−2,400	−2,800	−3,100	−3,700	−3,700	−4,300	−4,300	−4,700	−5,100	−5,500	−5,900	−6,700	−7,100	−7,900	−8,300	−9,100	−9,400	−10,200	−11,000	−11,800	−13,000	−14,200	−15,700	−17,300	−18,900
k	ei	μm	0	+1	+1	+1	+1	+2	+2	+2	+2	+2	+2	+3	+3	+3	+3	+3	+4	+4	+4	+4	+4	+4	+4	+5	+5
		μin	+40	+40	+40	+40	+40	+100	+100	+100	+100	+100	+100	+100	+100	+100	+100	+100	+160	+160	+160	+160	+160	+160	+160	+200	+200
d	es	μm	−20	−30	−40	−50	−50	−65	−65	−80	−80	−100	−100	−120	−120	−145	−145	−145	−170	−170	−190	−190	−210	−210	−210	−230	−230
		μin	−800	−1,200	−1,600	−2,000	−2,000	−2,600	−2,600	−3,100	−3,100	−3,900	−3,900	−4,700	−4,700	−5,700	−5,700	−5,700	−6,700	−6,700	−7,500	−7,500	−8,300	−8,300	−8,300	−9,100	−9,100
n	ei	μm	+4	+8	+10	+12	+12	+15	+15	+17	+17	+20	+20	+23	+23	+27	+27	+27	+31	+31	+34	+34	+34	+37	+37	+40	+40
		μin	+300	+300	+400	+500	+500	+600	+600	+700	+700	+800	+800	+900	+900	+1,100	+1,100	+1,100	+1,200	+1,200	+1,300	+1,300	+1,300	+1,500	+1,500	+1,600	+1,600
f	es	μm	−5	−10	−13	−16	−16	−20	−20	−25	−25	−30	−30	−36	−36	−43	−43	−43	−50	−50	−50	−56	−56	−62	−62	−68	−68
		μin	−200	−400	−500	−600	−600	−800	−800	−1,000	−1,000	−1,200	−1,200	−1,400	−1,400	−1,700	−1,700	−1,700	−2,000	−2,000	−2,000	−2,200	−2,200	−2,400	−2,400	−2,680	−2,680
p	ei	μm	+6	+12	+15	+18	+18	+22	+22	+26	+26	+32	+32	+37	+37	+43	+43	+43	+50	+50	+56	+56	+56	+62	+62	+68	+68
		μin	+500	+600	+700	+700	+700	+900	+900	+1,000	+1,000	+1,300	+1,300	+1,500	+1,500	+1,700	+1,700	+1,700	+2,000	+2,000	+2,200	+2,200	+2,200	+2,400	+2,400	+2,680	+2,680
g	es	μm	−2	−4	−5	−6	−6	−7	−7	−9	−9	−10	−10	−12	−12	−14	−14	−14	−15	−15	−15	−17	−17	−18	−18	−20	−20
		μin	−100	−200	−200	−300	−300	−300	−300	−400	−400	−400	−400	−500	−500	−600	−600	−600	−600	−600	−600	−700	−700	−700	−700	−800	−800
s	ei	μm	+14	+19	+23	+28	+28	+35	+35	+43	+43	+53	+59	+71	+79	+92	+100	+108	+122	+130	+140	+158	+170	+190	+208	+232	+252
		μin	+600	+700	+900	+1,100	+1,100	+1,400	+1,400	+1,700	+1,700	+2,100	+2,300	+2,800	+3,100	+3,600	+3,900	+4,300	+4,800	+5,100	+5,500	+6,200	+6,700	+7,500	+8,200	+9,100	+9,900
h	es	μm	0	0	0	0	0	0	0	0	0	0	0	0	0	0	0	0	0	0	0	0	0	0	0	0	0
		μin	0	0	0	0	0	0	0	0	0	0	0	0	0	0	0	0	0	0	0	0	0	0	0	0	0
u	ei	μm	+18	+23	+28	+33	+33	+41	+48	+60	+70	+87	+102	+124	+144	+170	+190	+210	+236	+258	+284	+315	+350	+390	+435	+490	+540
		μin	+700	+900	+1,100	+1,300	+1,300	+1,600	+1,900	+2,400	+2,800	+3,400	+4,000	+4,900	+5,700	+6,700	+7,500	+8,300	+9,300	+10,200	+11,200	+12,400	+13,000	+15,400	+17,100	+19,300	+21,300

[a] Tolerance in μm (1 μm = 10^{-6} m; 1 μin = 10^{-6} in).

[b] es = upper deviations.

[c] ei = lower deviations.

Source: Preferred limits and fits — BSI 4500; IS 2101, 1962.

11.23

TABLE 11–17
Relation between machine processes and geometry tolerances

Machining processes	Roundness[a] (circularity) of cylinders	Flatness of surfaces	Parallelism of cylinders on diameter	Straightness of cylinders, gaps, and tongues	Angularity — Flat surface: Parallelism squareness	Angularity — Flat surface: Any[b] other angle	Angularity — Cylinders, gaps, tongues: Parallelism squareness	Angularity — Cylinders, gaps, tongues: Any[b] other angle
Drill							10^{-3}	10^{-3}
Mill, slot, plane	—	5×10^{-5}	—	10^{-4}	10^{-4}	3×10^{-4}	10^{-4}	3×10^{-4}
Turn, bore	IT 4	5×10^{-5}	10^{-4}	10^{-4}	10^{-4}	3×10^{-4}	10^{-4}	3×10^{-4}
Fine turn, fine bore	IT 2	3×10^{-5}	4×10^{-5}	4×10^{-5}	5×10^{-5}	3×10^{-4}	5×10^{-5}	3×10^{-4}
Cylindrical grind	IT 3	—	5×10^{-5}	5×10^{-5}	—	—	5×10^{-5}	3×10^{-4}
Fine cylindrical grind	IT 1	—	2×10^{-5}	2×10^{-5}	—	—	2×10^{-5}	10^{-4}
Surface grind	—	3×10^{-5}	—	—	5×10^{-5}	3×10^{-4}	5×10^{-5}	3×10^{-4}
Fine surface grind		10^{-5}	—	—	2×10^{-5}	10^{-4}	2×10^{-5}	10^{-4}

Order of tolerance — Expressed as mm/mm length of surface or cylinder

[a] A roundness tolerance of 0.016 corresponds to a permissible diametrical variation of 0.032 (ovality).
[b] The values quoted are for good class of machine tools. Thrice or twice the above values, i.e., tolerances may have to be allowed for worn machine tools.

TABLE 11–18
Formulas for recommended allowances and tolerances (all dimensions in mm)

Class of fit	Method of assembly	Allowance	Selected average interference of metal	Hole tolerance	Shaft tolerance	Uses
Loose	Strictly interchangeable	$0.0075\ D^{2/3}$		$0.02\ D^{1/3}$	$0.02\ D^{1}$	Suitable for running fit; considerable freedom permissible; used in agricultural, mining, and general-purpose machinery
Free	Strictly interchangeable	$0.004\ D^{2/3}$		$0.01\ D^{1/3}$	$0.01\ D^{1/3}$	Suitable for running fit; suitable for shafts of motors, generators, engines, and some automotive parts
Medium	Strictly interchangeable	$0.0025\ D^{2/3}$		$0.007\ D^{1/3}$	$0.007\ D^{1/3}$	Accurate automotive parts and machine tools; suitable for running fit
Snug	Strictly interchangeable	0.0000		$0.005\ D^{1/3}$	$0.0035\ D^{1/3}$	Closest fit; zero allowance; suitable where no perceptible shake is permissible under load
Wringing	Selective assembly		0.0000	$0.005\ D^{1/3}$	$0.0035\ D^{1/3}$	A metal-to-metal contact fit
Tight	Selective assembly		$0.00025\ D$	$0.005\ D^{1/3}$	$0.005\ D^{1/3}$	Slightly negative allowance; suitable for semipermanent assembly and shrink fits
Medium force	Selective assembly		$0.0005\ D$	$0.005\ D^{1/3}$	$0.005\ D^{1/3}$	Suitable for press fits on locomotive wheels, car wheels, generator and motor armature, and crank discs
Heavy force or shrink	Selective assembly		$0.001\ D$	$0.005\ D^{1/3}$	$0.005\ D^{1/3}$	Used for steel external members that have a high yield stress

11.25

TABLE 11–19
Surface finish[a] values (CLA)

Machining process	High quality		Normal quality		Coarse quality	
	Tolerance grade	Finish (μm)	Tolerance grade	Finish (μm)	Tolerance grade	Finish (μm)
Drill	11	1.6–3.2	12			
Mill, slot, plane	9	0.4–0.8	11	0.8–1.6	12	1.6–3.2
Turn, bore	8	0.4–0.8	9	0.8–1.6	11	1.6–3.2
Ream	7	0.4–0.8	8	0.8–1.6		
Commercial grind	7	0.4–0.8	8	0.8–1.6	9	1.6–3.2
Fine turn, bore	6	0.2–0.4	7	0.4–0.8		
Hone	6	0.1–0.2	7	0.2–0.4		
Broach	6	0.1–0.2	7	0.2–0.4		
Fine grind	5	0.1–0.2	6	0.2–0.4		
Lap	3	0.05–0.1	4	0.1–0.2		

[a] The Roughness Number represents the average departure of the surface from perfection over a prescribed "sampling length" normally 0.8 mm, and is expressed in micrometers (μm). The measurements are normally made along a line at right angles to the general directions of tool marks or scratches on the surface.

$1\ \mu = 0.001\ mm$

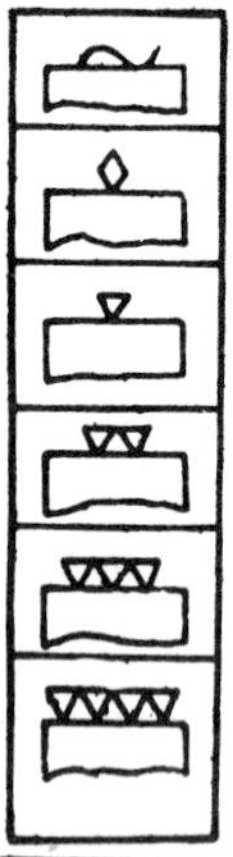

Old machining symbols	Description	Surface roughness
	Unmachined surface, cleaned up by sand blasting, brushing, etc.	5–$80\ \mu$
	Surface to be rough machined if found necessary (to prevent fouling)	
	Surface obtained by rough machining under turning, planing, milling etc. Quality coarser than 9	8–$25\ \mu$
	Finish-machined surface obtained by turning, milling etc. Quality 12-7	1.6–$8\ \mu$
	Fine finish-machined surface obtained by boring, reaming, grinding etc. Quality 9–6	0.25–$1.6\ \mu$
	Super finish-machined surface obtained by honing, lapping, super finish grinding. Quality 7-4	0–$0.25\ \mu$

FIGURE 11–12   Machining symbols.

TABLE 11–20
Lay symbols

Lay symbol	Interpretation	Example showing direction of tool marks
—	Lay parallel to the line representing the surface to which the symbol is applied	
⊥	Lay perpendicular to the line representing the surface to which the symbol is applied	
X	Lay angular in both directions to line representing the surface to which symbol is applied	
M	Lay multidirectional	
C	Lay approximately circular relative to the center of the surface to which the symbol is applied	
R	Lay approximately radial relative to the center of the surface to which the symbol is applied	
P	Pitted, protuberant, porous, or particulate nondirectional lay	

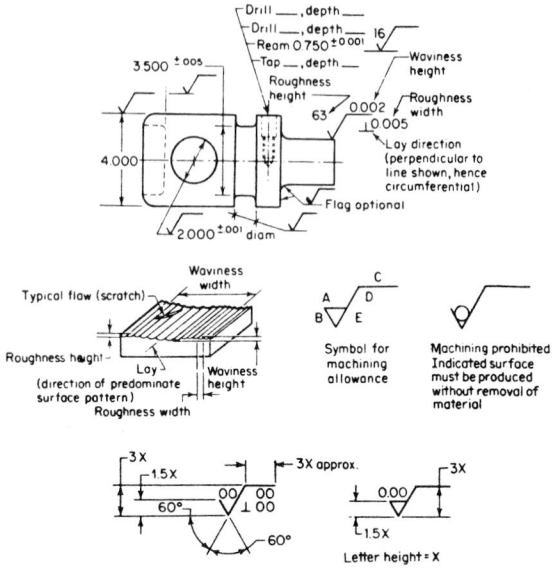

FIGURE 11–13 Application and use of surface-texture symbols. (Baumeister, T., *Marks' Standard Handbook for Mechanical Engineers*, 8th ed., McGraw-Hill, 1978.)

TABLE 11–21
Preferred series roughness average values (R_a) (in μm and μin)

μm	μin	μm	μin	μm	μin	μm	μin	μm	μin
0.012	0.5	0.125	5	0.50	20	2.00	80	8.0	320
0.025	1	0.15	6	0.63	25	2.50	100	10.0	400
0.050	2	0.20	8	0.80	32	3.20	125	12.5	500
0.075	3	0.25	10	1.00	40	4.0	160	15.0	600
0.10	4	0.32	13	1.25	50	5.0	200	20.0	800
		0.40	16	1.60	63	6.3	250	25.0	1000

Source: Reproduced from Baumeister, T., *Marks' Standard Handbook for Mechanical Engineers*, 8th ed., with permission from McGraw-Hill Book Company, New York, 1978.

TABLE 11–22
Preferred series maximum waviness height values

mm	in	mm	in	mm	in
0.0005	0.00002	0.008	0.0003	0.12	0.005
0.0008	0.00003	0.012	0.0005	0.20	0.008
0.0012	0.00005	0.020	0.0008	0.25	0.010
0.0020	0.00008	0.025	0.001	0.38	0.015
0.0025	0.0001	0.05	0.002	0.50	0.020
0.005	0.0002	0.08	0.003	0.80	0.030

Source: Reproduced from Baumeister, T., *Marks' Standard Handbook for Mechanical Engineers*, 8th ed., with permission from McGraw-Hill Book Company, New York, 1978.

TABLE 11–23
Surface roughness ranges of production processes

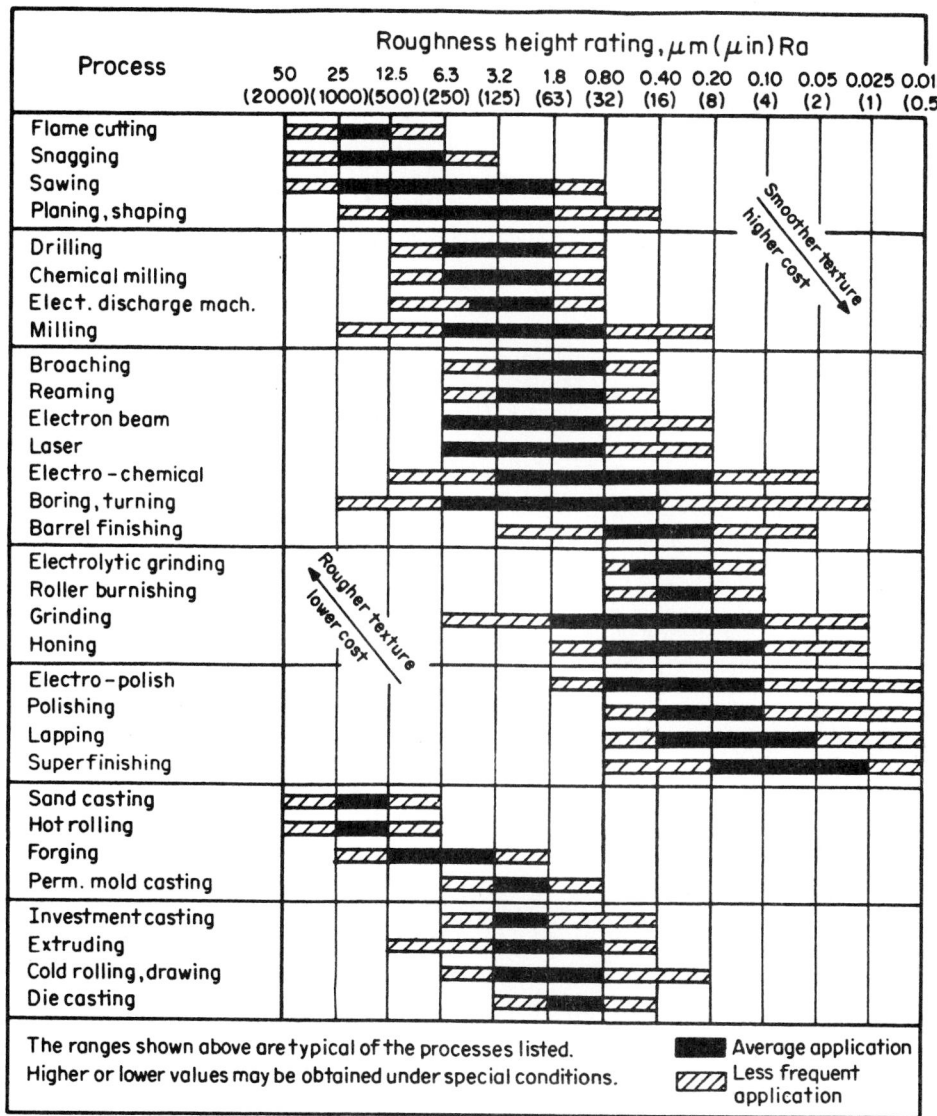

The ranges shown above are typical of the processes listed.
Higher or lower values may be obtained under special conditions.

■ Average application
▨ Less frequent application

Source: Reproduced from Baumeister, T., *Marks' Standard Handbook for Mechanical Engineers*, 8th ed., with permission from McGraw-Hill Book Company, New York, 1978.

TABLE 11–24
Application of surface texture values to surface symbols

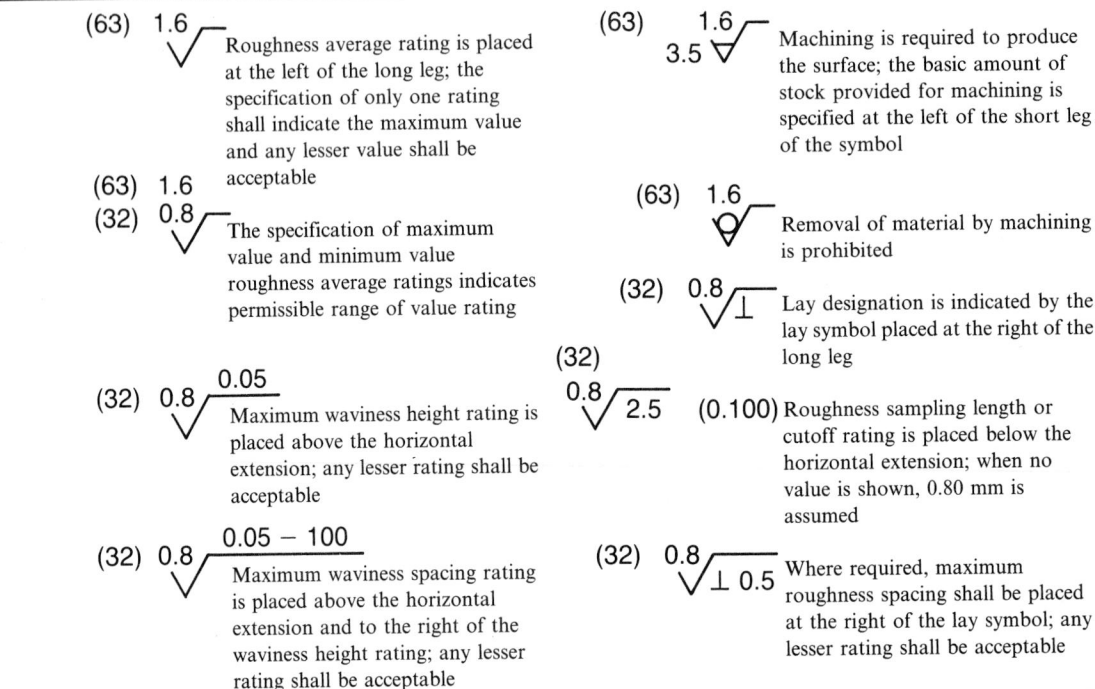

(63) 1.6 — Roughness average rating is placed at the left of the long leg; the specification of only one rating shall indicate the maximum value and any lesser value shall be acceptable

(63) 1.6
(32) 0.8 — The specification of maximum value and minimum value roughness average ratings indicates permissible range of value rating

(32) 0.8 / 0.05 — Maximum waviness height rating is placed above the horizontal extension; any lesser rating shall be acceptable

(32) 0.8 / 0.05 − 100 — Maximum waviness spacing rating is placed above the horizontal extension and to the right of the waviness height rating; any lesser rating shall be acceptable

(63) 1.6 / 3.5 — Machining is required to produce the surface; the basic amount of stock provided for machining is specified at the left of the short leg of the symbol

(63) 1.6 — Removal of material by machining is prohibited

(32) 0.8 / ⊥ — Lay designation is indicated by the lay symbol placed at the right of the long leg

(32) 0.8 / 2.5 (0.100) — Roughness sampling length or cutoff rating is placed below the horizontal extension; when no value is shown, 0.80 mm is assumed

(32) 0.8 / ⊥ 0.5 — Where required, maximum roughness spacing shall be placed at the right of the lay symbol; any lesser rating shall be acceptable

Source: Reproduced from Baumeister, T., *Marks' Standard Handbook for Mechanical Engineers*, 8th ed., with permission from McGraw-Hill Book Company, New York, 1978.

TABLE 11–25
Typical surface texture design requirements

(250 μin) 6.3 ⟋
  Clearance surfaces
  Rough machine parts

(125 μin) 3.2 ⟋
  Mating surfaces (static)
  Chased and cut threads
  Clutch-disk faces
  Surfaces for soft gaskets
(63 μin) 1.60 ⟋
  Piston-pin bores
  Brake drums
  Cylinder block, top
  Gear locating faces
  Gear shafts and bores
  Ratchet and pawl teeth
  Milled threads
  Rolling surfaces
  Gearbox faces
  Piston crowns
  Turbine-blade dovetails
(32 μin) 0.80 ⟋
  Broached holes
  Bronze journal bearings

  Gear teeth
  Slideways and gibs
  Press-fit parts
  Piston-rod bushings
  Antifriction-bearing seats
  Sealing surfaces for hydraulic
  tube fittings

(16 μin) 0.40 ⟋
  Motor shafts
  Gear teeth (heavy loads)
  Spline shafts
  O-ring grooves (static)
  Antifriction-bearing bores and faces
  Camshaft lobes
  Compressor-blade airfoils
  Journals for elastomer lip seals
(13 μin) 0.32 ⟋
  Engine cylinder bores
  Piston outside diameters
  Crankshaft bearings

(8 μin) 0.20 ⟋
  Jet-engine stator blades
  Valve-tappet cam faces
  Hydraulic-cylinder bores
  Lapped antifriction bearings
(4 μin) 0.10 ⟋
  Ball-bearing races
  Piston pins
  Hydraulic piston rods
  Carbon-seal mating surfaces
(2 μin) 0.050 ⟋
  Shop-gauge faces
  Comparator anvils
(1 μin) 0.025 ⟋
  Bearing balls
  Gauges and mirrors
  Micrometer anvils

TABLE 11–26
Range of surface roughness[a]

Manufacturing process	With difficulty	Normally	Roughing
Manual			
Hack saw cut		6.3–50	
Chipping		3.2–50	
Filing	0.8–1.6	1.6–12.5	
Emery polish	0.1–0.4	0.4–1.6	1.6–3.2
Casting			
Sand casting		6.3–12.5	12.5–25
Permanent mold	0.8–1.6	1.6–6.3	
Die casting		0.8–3.2	
Forming			
Forging	1.6–3.2	3.2–25	
Extrusion	0.4–0.8	0.8–6.3	
Rolling	0.4–0.8	0.8–3.2	
Machining			
Drilling	3.2–6.3	6.3–25	
Planing and shaping		1.6–12.5	
Face milling	0.8–1.6	1.6–12.5	12.5–50
Turning	0.2–1.6	1.6–6.3	6.3–50
Boring	0.2–1.6	1.6–6.3	6.3–50
Reaming	0.4–0.8	0.8–6.3	6.3–12.5
Cylindrical grinding	0.025–0.4	0.4–3.2	3.2–6.3
Centreless grinding	0.05–0.4	0.4–3.2	
Surface grinding	0.025–0.4	0.4–3.2	3.2–6.3
Broaching	0.2–0.8	0.8–3.2	3.2–6.3
Superfinishing	0.025–0.1	0.1–0.4	
Honing	0.025–0.1	0.1–0.4	
Lapping	0.006–0.05	0.05–0.4	
Gear manufacture			
Milling with form cutter	1.6–3.2	3.2–12.5	12.5–50
Milling, spiral bevel	1.56–3.2	3.2–12.5	12.5–25
Hobbing	0.8–3.2	3.2–12.5	12.5–50
Shaping	0.4–1.6	1.6–12.5	12.5–250
Shaving	0.4–0.8	0.8–3.2	
Grinding	0.1–0.4	0.4–0.8	
Lapping	0.05–0.2	0.2–0.8	
Surface process			
Shot blast	1.6–3.2	3.2–50	
Abrasive belt		0.1–6.3	
Fiber wheel brushing	0.1–0.2	0.2–0.8	0.8–1
Cloth buffing	0.012–0.05	0.05–0.1	

[a] Surface roughness in μm (1μm $= 10^{-3}$mm $= 10^{-6}$m).

Rivet	Symbol
Shop snap headed rivets	
Shop *Csk* (near side) rivets	
Shop *Csk* (far side) rivets	
Shop *Csk* (both sides) rivets	
Site snap headed rivets	
Site *Csk* (near sides) rivets	
Site *Csk* (far side) rivets	
Site *Csk* (both sides) rivets	
Open hole	

IS : 696–1960

Characteristics to be toleranced	Symbols
Straightness	
Flatness	
Circularity	
Accuracy of any profile Accuracy of any surface	
Parallelism	
Perpendicularity	
Angularity	
Position	
Concentricity or coaxiality	
Symmetry	

IS : 696–1960

FIGURE 11–14   Symbols for tolerances of form and position.

FIGURE 11–15   Rivet symbols

BIBLIOGRAPHY

Black, P. H., and O. Eugene Adams, Jr., *Machine Design*, McGraw-Hill Publishing Company, New York, 1983.

Baumeister, T., *Marks' Standard Handbook for Mechanical Engineers*, 8th ed., McGraw-Hill Publishing Company, New York, 1978.

British Standard Institution.

Bureau of Indian Standards.

Lingaiah, K., and B. R. Narayana Iyengar, *Machine Design Data Handbook*, Engineering College Co-operative Society, Bangalore, India, 1962.

Lingaiah, K., and B. R. Narayana Iyengar, *Machine Design Data Handbook*, Vol. I, Suma Publishers, Bangalore, India, 1986.

Lingaiah, K., *Machine Design Data Handbook*, Vol. II (*SI and Customary Metric Units*), Suma Publishers, Bangalore, India, 1986.

Maleev, V. L., and J. B. Hartman, *Machine Design*, International Textbook Company, Scranton, Pennsylvania, 1954.

Shigley, J. E., *Machine Design*, McGraw-Hill Publishing Company, New York, 1956.

Vallance, A., and V. L. Doughtie, *Design of Machine Members*, McGraw-Hill Publishing Company, New York, 1951.

CHAPTER
12

DESIGN OF WELDED JOINTS

SYMBOLS

A	area of flange material held by welds in shear, m² (in²)
$A' = l_\omega$	length of weld when weld is treated as a line, m (in)
b	width of connection, m (in)
c	distance to outer fiber (also with suffixes), m (in)
c_x	distance of x axis to face, m (in)
c_y	distance of y axis to face, m (in)
c_1	distance of weld edge parallel to x-axis from the center of weld, to left, m (in)
c_2	distance of weld edge from parallel to x-axis from the center of weld, to right, m (in)
c_3	distance from farthest weld corner, Q, to the center of gravity of weld, m (in)
d	depth of connection, m (in)
e_x	eccentricity of P_x and P_y about the center of weld, m (in)
e_y	eccentricity of P_x about the center of weld, m (in)
h	thickness of plate (also with suffixes), m (in)
i	number of welds
I_x, I_y, I_z	moment of inertia of weld about x, y, and z axes respectively, m⁴, c m⁴ (in⁴)
J	moment of inertia, polar, m⁴, cm⁴ (in⁴)
J_ω	polar moment of inertia of weld, when weld is treated as a line, m³, cm³ (in³)
$K_{f\sigma}$	fatigue stress-concentration factor (Table 12–7)
l	effective length of weld, m (in)
l_t	total length of weld, m (in)
M_b	bending moment, N m (lbf in)
M_t	twisting moment, N m (lbf in)
n_a	actual factor of safety or reliability factor
N_a	fatigue life (for which σ_{Sfa} is known) for fatigue strength σ_{Sfa}, cycle
N_b	fatigue life (required) for fatigue strength σ_{Sfb}, cycle
P	load on the joint, kN (lbf)
P_x	component of P in x direction, kN (lbf)
P_y	component of P in y direction, kN (lbf)
P_z	component of P in z direction, kN (lbf)
r	distance to outer fiber, m (in)

R ratio of calculated leg size for continuous weld to the actual leg size to be used for intermittent weld
t throat dimension of weld, m (in)
V shear load, kN (lbf)
w size of weld leg, m (in)
Z section modulus, m^3 (in^3)
Z_ω section modulus of weld, when weld is treated as line (also with suffixes, m^2 (in^2)
σ normal stress in the weld (in standard design formula), MPa (psi)
σ^t force per unit length of weld (in standard design formula) when weld treated as a line, kN/m (lbf/in)
σ_{Sfa} fatigue strength (known) for fatigue life N_a, MPa (psi)
σ_{Sfb} fatigue strength (allowable) for fatigue life N_b, MPa (psi)
σ_d design stress, MPa (psi)
σ_e elastic limit, MPa (psi)
τ shear stress in the weld (in standard design formula), MPa (psi)
τ' shear force per unit length of weld (in standard design formula) when weld is treated as a line, kN/m (psi)
θ angle, deg
η efficiency of joint

Particular	Formula

FILLET WELD

The throat thickness t, for case with equal legs, of weld (Fig. 12–1)

$$t = w \sin 45° = 0.707\,w \qquad (12–1a)$$

The allowable load on the weld

$$P = 0.707\,i\tau\,wl \qquad (12–1b)$$

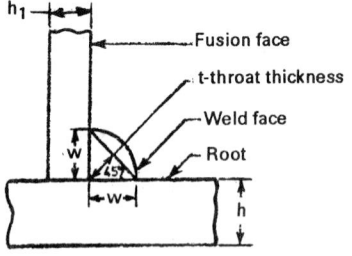

FIGURE 12–1 Fillet weld.

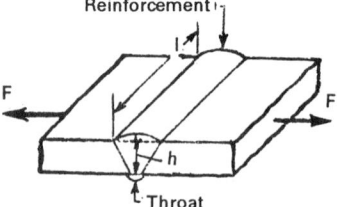

FIGURE 12–2 A typical butt-weld joint.

BUTT WELD

The average normal stress in a butt weld subjected to tensile or compression loading (Fig. 12–2)

$$\sigma = \frac{F}{hl} \qquad (12–2)$$

where h is the throat dimension. The dimensions of throat (t) is same as the thickness of plate (h).

Particular	Formula
	The throat dimension (h) does not include the reinforcement.
The average shear stress in butt weld	$\tau = \dfrac{F}{hl}$ $\hspace{2em}$ (12–3)
The allowable load on the weld	$F_a = \eta \sigma_a h l$ $\hspace{2em}$ (12–4)

TRANSVERSE FILLET WELD

The average normal tensile stress

$$\sigma = \frac{F}{wl\cos 45°} = \frac{F}{0.707\,wl} \qquad (12\text{–}5)$$

The average normal tensile stress for the case of transverse fillet weld shown in Fig. 12–3.

$$\sigma = \frac{F}{0.707\,hl} \qquad (12\text{–}6)$$

The stress concentration occurs at A and B on the horizontal leg and at B on the vertical leg in the weld according to photoelastic tests conducted by Norris [1].

A double fillet lap weld joint.

Refer to Fig. 12–4.

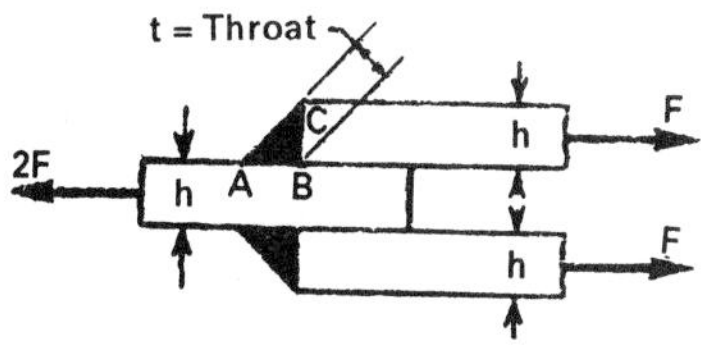

FIGURE 12–3 A transverse fillet weld.

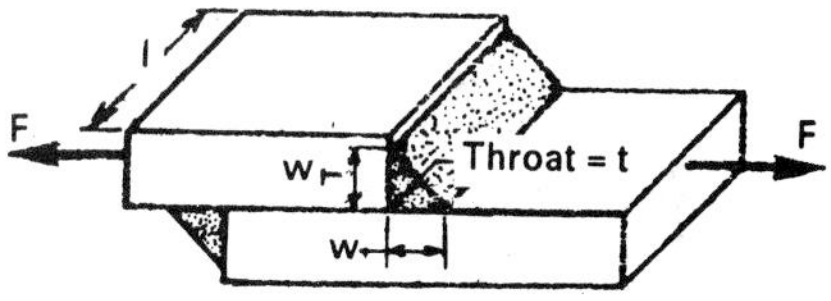

FIGURE 12–4 A double-fillet lap-weld joint.

PARALLEL FILLET WELD (FIG. 12–5)

The average shear stress in the weld

$$\tau = \frac{P}{0.707\,wl} \qquad (12\text{–}7a)$$

where w = dimension of leg of weld.
w can be replaced by h (thickness of plate) when w and h are of same dimension.
Either symbol F or P can be used for force or load depending on symbols used in figures in this chapter.

The shear stress in a reinforced fillet weld

$$\tau = \frac{P}{0.85\,wl} \qquad (12\text{–}7b)$$

where throat t is taken as $0.85w$

Particular	Formula

LENGTH OF WELD

The effective length of weld (Fig. 12–5)

$$l = l_t - \frac{i}{4} \tag{12–8}$$

where i = total number of free ends

The total length of weld (Fig. 12–5)

$$l_t = \frac{P}{0.707\, w\sigma_a} \quad \text{where } l = 2(l_1 + l_2) \tag{12–9}$$

The relation between the length l_1 and l_2 (Fig. 12–5)

$$\frac{l_1}{L - \bar{x}} = \frac{l_2}{\bar{x}} = \frac{l_1 + l_2}{L} = \frac{l_t}{2L} \tag{12–10}$$

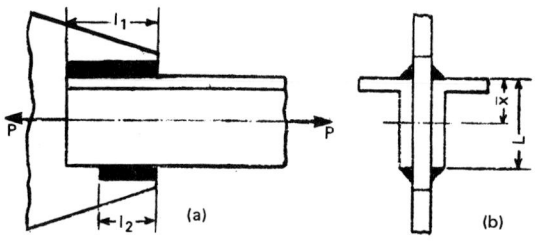

(a)   (b)

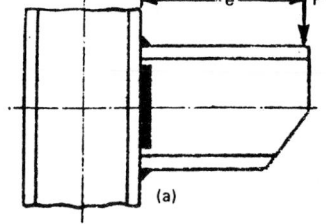

(a)

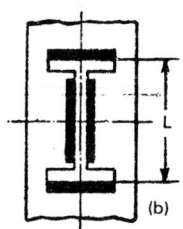

(b)

FIGURE 12–5   Parallel fillet weld.

FIGURE 12–6

ECCENTRICITY IN A FILLET WELD

The bending stress due to fillet weld placed on only one side of the plate (Fig. 12–6)

$$\sigma_b = \frac{4M_b}{(0.707w)^2 l} \tag{12–11}$$

$$= \frac{4Pw}{4(0.707w)^2 l}$$

$$= \frac{2P}{wl}$$

The stress due to tensile load

$$\sigma_t = \frac{P}{1.414wl} \tag{12–12}$$

The combined normal stress at the root of the weld

$$\sigma_n = \sigma_t + \sigma_b = \frac{P}{1.414wl} + \frac{2P}{wl} \tag{12–13}$$

The shear stress

$$\tau = \frac{P}{0.707wl} \tag{12–14}$$

The maximum normal stress

$$\sigma_{max} = \tfrac{1}{2}\left(\sigma_n + \sqrt{\sigma_n^2 + 4\tau^2}\right) \tag{12–15}$$

The maximum shear stress

$$\tau_{max} = \tfrac{1}{2}\sqrt{\sigma_n^2 + 4\tau^2} \tag{12–16}$$

Particular	Formula

ECCENTRIC LOADS
Moment acting at right angles to the plane of welded joint (Fig. 12–6)

Direct load per unit length of weld

$$P_d = \frac{P}{l} \qquad (12\text{–}17)$$

Load due to bending per unit length of weld

$$P_n = \frac{Pey}{I} \qquad (12\text{–}18)$$

The resultant load or force

$$P_R = \sqrt{P_d^2 + P_n^2} \qquad (12\text{–}19)$$

Moment acting in the plane of the weld (Fig. 12–7)

Load due to twisting moment per unit length of weld

$$P_n = \frac{Per}{J} \qquad (12\text{–}20)$$

The resultant load (Fig. 12–7)

$$P_R = \sqrt{P_d^2 + P_n^2 + 2P_d P_n \cos\theta} \qquad (12\text{–}21)$$

$$\text{where } \cos\theta = \frac{l_2}{\sqrt{l_1^2 + l_2^2}}$$

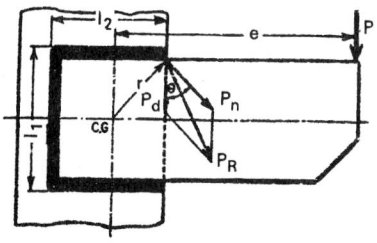

FIGURE 12–7

STRESSES
Bending

The bending stress

$$\sigma_b = \frac{M_b}{w Z_w} \qquad (12\text{–}22a)$$

or

$$\sigma_b' = \frac{M_b}{Z_w} \text{ (treating weld as a line)} \qquad (12\text{–}22b)$$

Particular	Formula

Torsion

The shear stress due to torsion

$$\tau = \frac{M_t r}{w J_w} \tag{12--23a}$$

or

$$\tau' = \frac{M_t r}{J_w} \text{ (treating weld as a line)} \tag{12--23b}$$

Combined bending and torsion

The resultant or maximum induced normal force per unit throat of weld

$$\sigma'_{max} = \frac{1}{2} \left[\frac{M_b}{Z_w} + \sqrt{\left(\frac{M_b}{Z_w} \right)^2 + 4 \left(\frac{M_t r}{J_w} \right)^2} \right] \tag{12--24}$$

The resultant induced torsional force per unit throat of weld

$$\tau'_{max} = \frac{1}{2} \left[\sqrt{\left(\frac{M_b}{Z_w} \right)^2 + 4 \left(\frac{M_t r}{J_w} \right)^2} \right] \tag{12--25}$$

The required leg size of the weld when weld is treated as a line

$$w = \frac{\text{actual force}}{\text{permissible force}} = \frac{\sigma'_{max} \text{ or } \tau'_{max}}{\sigma'_a \text{ or } \tau'_a} \tag{12--26}$$

The resultant normal stress induced in the weld

$$\sigma_{max} = \frac{1}{2} \left[\frac{M_b}{w Z_w} + \sqrt{\left(\frac{M_b}{w Z_w} \right)^2 + \left(\frac{M_t r}{w J_u} \right)^2} \right] \tag{12--27}$$

The resultant shear stress induced in the weld

$$\tau_{max} = \frac{1}{2} \sqrt{\left(\frac{M_b}{w Z_w} \right)^2 + 4 \left(\frac{M_t r}{w J_w} \right)^2} \tag{12--28}$$

The required leg size of weld when the weld area is considered

$$w = \frac{\text{actual maximum stress induced in the weld}}{\text{permissible stress}}$$

$$= \frac{\sigma_{max} \text{ or } \tau_{max}}{\sigma_a \text{ or } \tau_a}$$

FATIGUE STRENGTH

The fatigue strength related to fatigue life can be expressed by the empirical formula

$$\sigma_{Sfa} = \sigma_{Sfb} \left(\frac{N_b}{N_a} \right)^k \tag{12--29}$$

or

$$N_a = N_b \left(\frac{\sigma_{Sfb}}{\sigma_{Sfa}} \right)^{1/k} \tag{12--30}$$

Particular	Formula
	where k = 0.13 for butt welds = 0.18 for plates in bending, axial tension, or compression

DESIGN STRESS OF WELDS

The design stress	$$\sigma_d = \frac{\sigma_a}{n_a}$$	(12–31)

where
n_a = actual safety factor or reliability factor
   = 3 to 4

The design stress for completely reversed load	$$\sigma_{fd} = \frac{\sigma_f}{n_a K_{f\sigma}}$$	(12–32)

THE STRENGTH ANALYSIS OF A TYPICAL WELD JOINT SUBJECTED TO ECCENTRIC LOADING (FIG. 12–8) [2, 3, 4]

Throughout the analysis of a weld joint, the weld is treated as a line.

Area of cross section of weld	$A = (2b + d)w$	(12–33)
The distance of weld edge parallel to x axis from the center of weld, to left	$$c_1 = \frac{b^2}{2b + d}$$	(12–34)
The distance of weld edge parallel to x axis from the center of weld, to right	$$c_2 = \frac{b(b + d)}{2b + d}$$	(12–35)
The distance from farthest weld corner, Q, to the centre of gravity of weld	$$c_3 = \sqrt{c_2^2 + \left(\frac{d}{2}\right)^2}$$	(12–36)
The moment of inertia of weld about x axis	$$I_x = \frac{wd^2}{12}(d + 6b)$$	(12–37)
The moment of inertia of weld about y axis	$$I_y = \frac{wb^3(2d + b)}{3(d + 2b)}$$	(12–38)
The moment of inertia of weld about z axis	$I_z = I_x + I_y$	(12–39)
The section modulus of weld, about x axis	$$Z_{wx} = \frac{I_x}{(d/2)} = \frac{wd}{6}(d + 6b)$$	
The section modulus of weld, about y axis	$$Z_{wy} = \frac{I_y}{c_2} = \frac{wb^2(2d + b)}{3(b + d)}$$	(12–40)

Particular	Formula

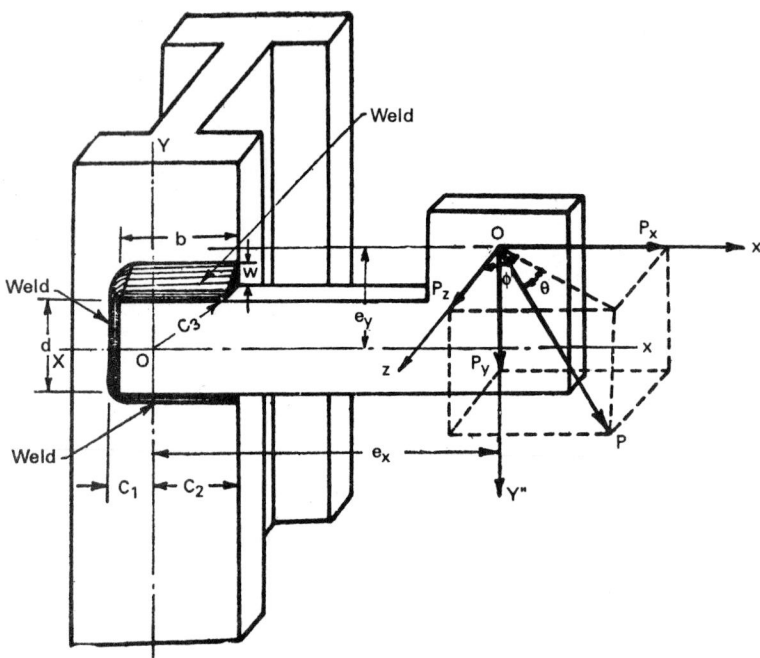

FIGURE 12–8 A typical weld joint subjected to Eccentric Loading. *K. Lingaiah and B. R. Narayana Iyengar*, Machine Design Data Handbook, *Engineering College Cooperative Society, Bangalore, India, 1962; K. Lingaiah and B. R. Narayana Iyengar, Machine Design Data Handbook, Vol. I, Suma Publishers, Bangalore, India, 1986; and K. Lingaiah*, Machine Design Data Handbook, *Vol. II (SI and Customary Units), Suma Publishers, Bangalore, India, 1986.*

The section modulus of weld, about z axis

$$Z_{wz} = \frac{I_z}{c_3} \qquad (12\text{--}41)$$

P_z component

Throughout the analysis of this problem the weld is considered as a line

The force per unit length of weld due to direct force P_z

$$\sigma'_{zd} = \frac{P_z}{A'} \qquad (12\text{--}42)$$

The force per unit length of weld account of bending at the farthest weld corner, Q, due to eccentricity e_x of load P_z

$$\sigma'_{zb1} = \frac{P_z e_x}{Z_{wy}} \qquad (12\text{--}43)$$

The force per unit length of weld account of bending at the farthest weld corner, Q, due to eccentricity e_y of load P_z

$$\sigma'_{zb2} = \frac{P_z e_y}{Z_{wx}} \qquad (12\text{--}44)$$

The total force per unit length of weld due to bending

$$\sigma'_{zb} = \sigma'_{zb1} + \sigma'_{zb2} \qquad (12\text{--}45)$$

Particular	Formula
The combined force per unit length of weld due to load P_z	$\sigma_z' = \sigma_{zd}' + \sigma_{zb}'$ (12–46)

P_x component

Particular	Formula
The force per unit length of weld due to direct shear force P_x which acts in the horizontal direction (Fig. 12–8)	$\tau_{xd}' = \dfrac{P_x}{A'}$ (12–47)
The twisting moment	$M_{tx} = P_x e_y$ (12–48)
The shear force per unit length due to twisting moment M_{tx}	$\tau_{xt}' = \dfrac{M_{tx} c_3}{J_{wz}}$ (12–49)
The vertical component of τ_{tx}'	$\tau_{txz}' = \dfrac{M_{tx} c_3}{J_{wz}} \cos\psi$ (12–50)
The horizontal component of τ_{tx}'	$\tau_{txh}' = \dfrac{M_{tx} c_3}{J_{wz}} \sin\psi$ (12–51)

where
$c_3 =$ distance from the center of gravity of the weld to the point being analyzed (i.e., Q)

$$\cos\psi = \frac{c_2}{c_3} \text{ (Fig. 12–8)}$$

Particular	Formula
The resultant shear force per unit length of weld in the horizontal direction due to P_x only	$\tau_{txrh}' = \tau_{xd}' + \tau_{txh}'$ (12–52)

P_y component

Particular	Formula
The direct shear force per unit length of weld parallel to y direction due to force P_y (Fig. 12–8)	$\tau_{yd}' = \dfrac{P_y}{A'}$ (12–53)
The twisting moment	$M_{ty} = P_y e_x$ (12–54)
The shear force per unit length of weld due to twisting moment M_{ty}	$\tau_{ty}' = \dfrac{M_{ty} c_3}{J_{wz}}$ (12–55)
The vertical component of τ_{ty}'	$\tau_{tyv}' = \tau_{ty}' \cos\psi$ (12–56)
The horizontal component of τ_{ty}'	$\tau_{tyh}' = \tau_{ty}' \sin\psi$ (12–57)
The resultant shear force per unit length of weld in the vertical direction due to P_y only	$\tau_{tyrv}' = \tau_{yd}' + \tau_{tyv}'$ (12–58)

Particular	Formula

COMBINED FORCE DUE TO P_x, P_y, AND P_z AT POINT Q (FIG. 12–8)

From Eqs. (12–46), (12–50), (12–52), (12–57), and (12–58)

The total shear force per unit length of weld in the x direction (Fig. 12–8) from Eqs. (12–52) and (12–57)

$$\tau_x' = \tau_{tzrh}' + \tau_{tyh}' \tag{12–59}$$

The total shear force per unit length of weld in the y direction (Fig. 12–8) from Eqs. (12–50) and (12–58)

$$\tau_y' = \tau_{txv}' + \tau_{tyrv}' \tag{12–60}$$

The resultant shear force per unit length of weld at point Q due to P_x and P_y forces (Fig. 12–8) from Eqs. (12–59) and (12–60)

$$\tau' = \sqrt{\tau_x'^2 + \tau_y'^2} \tag{12–61}$$

The resultant actual force per unit length of weld (treating weld as a line) due to components P_x, P_y and P_z at point Q from Eqs. (12–46) and (12–61)

$$\sigma_{actual}' = \sqrt{\sigma_z'^2 + \tau'^2} \tag{12–62}$$

The leg size of the weld

$$w' = \frac{\sigma_{actual}'}{\sigma_{allowable}'} \tag{12–63}$$

For the AWS standard location of elements of welding symbol, weld symbols and direction for making weld

Refer to Figs. 12–9 to 12–11.

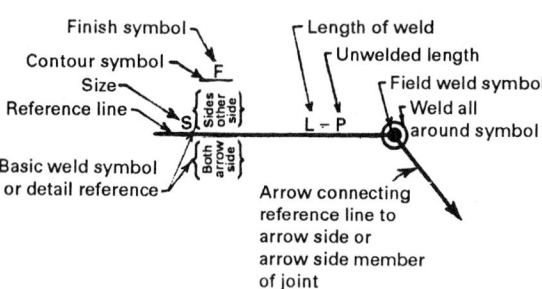

Finish symbol
Contour symbol
Size
Reference line
Basic weld symbol or detail reference
Sides other side
Both arrow side
Length of weld
Unwelded length
Field weld symbol
Weld all around symbol
$L - P$
S
F
Arrow connecting reference line to arrow side or arrow side member of joint

FIGURE 12–9   The AWS Standard location of elements of a welding symbol.

Special instructions	Drawing respresentation	Symbols
Weld alround		○
Side weld (field weld)		●
Flush contour		—
Convex contour		⌒
Concave contour		⌣
Grinding finish		G
Machining finish		M
Chipping finish		C

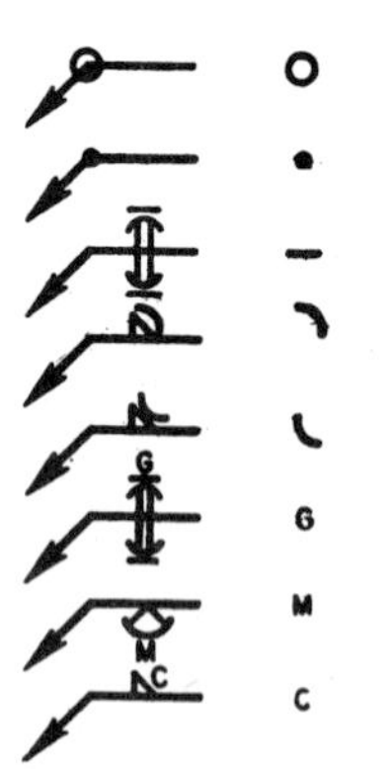

FIGURE 12–10   Weld symbols

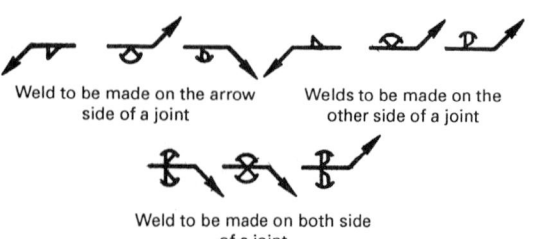

Weld to be made on the arrow side of a joint

Welds to be made on the other side of a joint

Weld to be made on both side of a joint

FIGURE 12–11

GENERAL

For further data on welded joint design   Refer to Tables 12–1 to 12–16.

REFERENCES

1. Norris, C. H., Photoelastic Investigation of Stress Distribution in Transverse Fillet Welds, *Welding Journal*, Vol. 24, p. 557, 1945.
2. Lingaiah, K., and B. R. Narayana Iyengar, *Machine Design Data Handbook*, Engineering College Co-operative Society, Bangalore, India, 1962.
3. Lingaiah, K., and B. R. Narayana Iyengar, *Machine Design Data Handbook*, Vol. I, Suma Publishers, Bangalore, India, 1986.
4. Lingaiah, K., *Machine Design Data Handbook*, Vol. II (*SI and Customary Units*), Suma Publishers, Bangalore, India, 1986.
5. *Welding Handbook*, 3rd ed., American Welding Society, 1950.
6. Bureau of Indian Standards.

BIBLIOGRAPHY

Design of Weldments, The James F. Lincoln Arc Welding Foundation, Cleveland, Ohio, 1968.
Design of Welded Structures, The James F. Lincoln Arc Welding Foundation, Cleveland, Ohio, 1966.
Maleev, V. L. and J. B. Hartman, *Machine Design*, International Textbook Company, Scranton, Pennsylvania, 1954.
Procedure Handbook of Arc Welding Design and Practice, The James F. Lincoln Arc Welding Foundation, Cleveland, Ohio, 1950.
Salakian, A. G., and G. E. Claussen, "Stress Distribution in Fillet Welds: A Review of the Literature," *Welding Journal*, Vol. 16, pp. 1–24, May 1937.
Shigley, J. E., *Machine Design*, McGraw-Hill Publishing Company, New York, 1956.
Spotts, M. F., *Design of Machine Elements*, 5th ed., Prentice-Hall of India Private Ltd., New Delhi, 1978.
Vallance, A., and V. L. Doughtie, *Design of Machine Members*, McGraw-Hill Publishing Company, New York, 1951.

TABLE 12–1
Weld-stress formulas

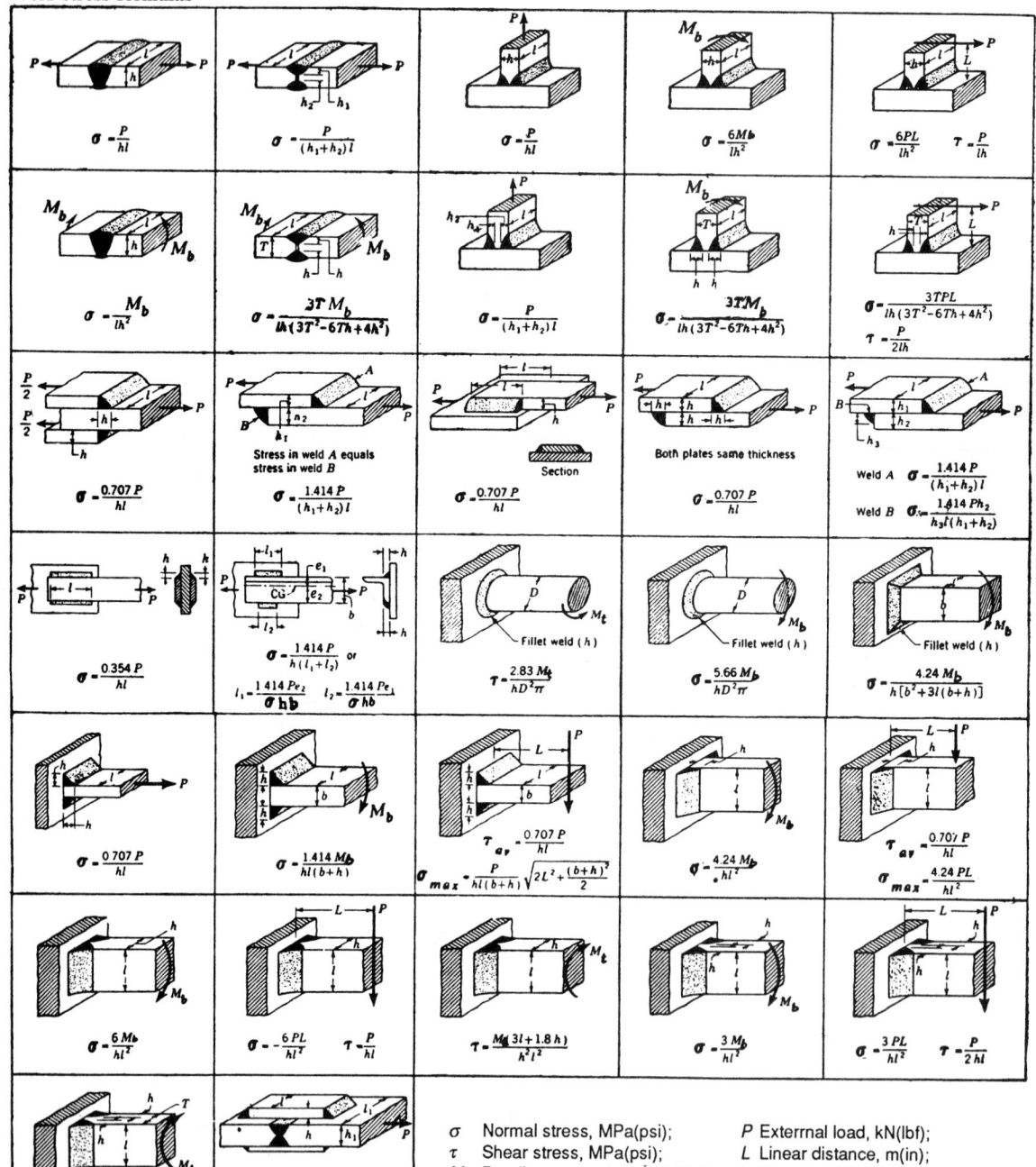

$\sigma = \dfrac{P}{hl}$

$\sigma = \dfrac{P}{(h_1+h_2)l}$

$\sigma = \dfrac{P}{hl}$

$\sigma = \dfrac{6M_b}{lh^2}$

$\sigma = \dfrac{6PL}{lh^2}$ $\quad \tau = \dfrac{P}{lh}$

$\sigma = \dfrac{M_b}{lh^2}$

$\sigma = \dfrac{3T\,M_b}{lh\,(3T^2-6Th+4h^2)}$

$\sigma = \dfrac{P}{(h_1+h_2)l}$

$\sigma = \dfrac{3TM_b}{lh\,(3T^2-6Th+4h^2)}$

$\sigma = \dfrac{3TPL}{lh\,(3T^2-6Th+4h^2)}$ $\quad \tau = \dfrac{P}{2lh}$

$\sigma = \dfrac{0.707\,P}{hl}$

Stress in weld A equals stress in weld B
$\sigma = \dfrac{1.414\,P}{(h_1+h_2)l}$

$\sigma = \dfrac{0.707\,P}{hl}$ Section

$\sigma = \dfrac{0.707\,P}{hl}$ Both plates same thickness

Weld A $\sigma = \dfrac{1.414\,P}{(h_1+h_2)l}$
Weld B $\sigma = \dfrac{1.414\,Ph_2}{h_3 l\,(h_1+h_2)}$

$\sigma = \dfrac{0.354\,P}{hl}$

$\sigma = \dfrac{1.414\,P}{h\,(l_1+l_2)}$ or
$l_1 = \dfrac{1.414\,Pe_2}{\sigma\,hb}$ $l_2 = \dfrac{1.414\,Pe_1}{\sigma\,hb}$

$\tau = \dfrac{2.83\,M_t}{hD^2\pi}$

$\sigma = \dfrac{5.66\,M_b}{hD^2\pi}$

$\sigma = \dfrac{4.24\,M_b}{h\,[b^2+3l\,(b+h)]}$

$\sigma = \dfrac{0.707\,P}{hl}$

$\sigma = \dfrac{1.414\,M_b}{hl\,(b+h)}$

$\tau_{av} = \dfrac{0.707\,P}{hl}$
$\sigma_{max} = \dfrac{P}{hl\,(b+h)}\sqrt{2L^2+\dfrac{(b+h)^2}{2}}$

$\sigma = \dfrac{4.24\,M_b}{hl^2}$

$\tau_{av} = \dfrac{0.707\,P}{hl}$
$\sigma_{max} = \dfrac{4.24\,PL}{hl^2}$

$\sigma = \dfrac{6\,M_b}{hl^2}$

$\sigma = \dfrac{6\,PL}{hl^2}$ $\quad \tau = \dfrac{P}{hl}$

$\tau = \dfrac{M_t\,(3l+1.8\,h)}{h^2l^2}$

$\sigma = \dfrac{3\,M_b}{hl^2}$

$\sigma = \dfrac{3\,PL}{hl^2}$ $\quad \tau = \dfrac{P}{2\,hl}$

$\tau = \dfrac{M_t}{2\,(T-h)(l-h)\,h}$

Fillet weld $\sigma = \dfrac{1.414\,P}{2\,hl+h_1l_1}$
Butt weld $\sigma = \dfrac{P}{2\,hl+h_1l_1}$

σ Normal stress, MPa(psi);
τ Shear stress, MPa(psi);
M_b Bending moment, N·in(lbf·in);
M_t twisting moment, N·m(lbf·in);

P Exterrnal load, kN(lbf);
L Linear distance, m(in);
h Size of weld, m(in);
l Length of weld, m(in),

Source: Welding Handbook, 3d ed., American Welding Society, 1950.

TABLE 12-2
Design formulas used to obtain stress in weld

Type of loading		Standard design formula, MPa (psi)	Treating the weld as a line, kN/m (lbf/in)
		Primary Welds (Transmit Entire Load)	
	Tension or compression	$\sigma = \dfrac{P}{A}$	$\sigma' = \dfrac{P}{l_w}$
	Vertical shear	$\tau = \dfrac{V}{A}$	$\tau' = \dfrac{V}{l_w}$
	Bending	$\sigma = \dfrac{M_b}{Z}$	$\sigma' = \dfrac{M_b}{Z_w}$
	Twisting	$\tau = \dfrac{M_b c}{J}$	$\tau' = \dfrac{Mc}{J_w}$
		Secondary Welds (Hold Section Together-Low Stress)	
	Horizontal Shear	$\tau = \dfrac{VAy}{Ih}$	$\tau' = \dfrac{VAy}{I}$
	Torsional horizontal shear	$\tau = \dfrac{M_t c}{J}$	$\tau' = \dfrac{M_t ch}{J}$

TABLE 12–3
Properties of weld-treating weld as line

Outline of welded joint b = width, d = depth	Bending (about horizontal axis $x - x$)	Twisting
	$Z_w = \dfrac{d^2}{6}$	$J_w = \dfrac{d^3}{12}$
	$Z_w = \dfrac{d^2}{3}$	$J_w = \dfrac{d(3b^2 + d^2)}{6}$
	$Z_w = bd$	$J_w = \dfrac{b^3 + 3bd^2}{6}$
	$Z_w = \dfrac{4bd + d^2}{6} = \dfrac{d^2(4bd + d)}{6(2b + d)}$ top $\qquad$ bottom	$J_w = \dfrac{(b + d)^4 - 6b^2d^2}{12(b + d)}$
	$Z_w = bd + \dfrac{d^2}{6}$	$J_w = \dfrac{(2b + d)^3}{12} - \dfrac{b^2(b + d)^2}{(2b + d)}$
	$Z_w = \dfrac{2bd + d^2}{3} = \dfrac{d^2(2b + d)}{3(b + d)}$ top $\qquad$ bottom	$J_w = \dfrac{(b + 2d)^3}{12} - \dfrac{d^2(b + d)^2}{(b + 2d)}$
	$Z_w = bd + \dfrac{d^2}{3}$	$J_w = \dfrac{(b + d)^3}{6}$
	$Z_w = \dfrac{2bd + d^2}{3} = \dfrac{d^2(2b + d)}{3(b + d)}$ top $\qquad$ bottom	$J_w = \dfrac{(b + 2d)^3}{12} - \dfrac{d^2(b + d)^2}{(b + 2d)}$
	$Z_w = \dfrac{4bd + d^3}{3} = \dfrac{4bd^2 + d^3}{6b + 3d}$ top $\qquad$ bottom	$J_w = \dfrac{d^3(4b + d)}{6(b + d)} + \dfrac{b^3}{6}$
	$Z_w = bd + \dfrac{d^2}{3}$	$J_w = \dfrac{b^3 + dbd^2 + d^3}{6}$
	$Z_w = 2bd + \dfrac{d^2}{3}$	$J_w = \dfrac{2b^3 + 6bd^2 + d^3}{6}$
	$Z_w = \dfrac{\pi d^2}{4}$	$J_w = \dfrac{\pi d^3}{4}$
	$Z_w = \dfrac{\pi d^2}{2} + \pi D^2$	
	—	$J_w = \dfrac{b^3}{12}$

Note: Multiply the values J_w by the size of the weld w to obtain polar moment of inertia J_o of the weld.

TABLE 12–4
Types of welds and symbols

Form of weld	Sectional representation	Appropriate symbol	Form of weld	Sectional representation	Appropriate symbol
Fillet		◿	Plug or slot		▽
Square butt		⇈	Backing strip		=
Single-V butt		◇	Spot		✗
Double-V butt		⊗	Seam		✗✗✗
Single-U butt		∪	Mashed seam	BEFORE / AFTER	▨
Double-U butt		8			
Single-bevel butt		⌐	Stitch		⫙
Double-bevel butt		Ƙ	Mashed stitch	BEFORE / AFTER	▮▮
Single-J butt		⌐	Projection	BEFORE / AFTER	△
Double-J butt		ꞵ			
Stud		⊥	Flash	ROD OR BAR / TUBE	⋈
Bead (Edge or Seal)		⌒	Butt resistance or Pressure (upset)	ROD OR BAR / TUBE	∣
Sealing run		○			

IS: 696-1960(b) Bureau of Indian Standards.

TABLE 12–5A
Properties of common welding rods

Rods	Melting point		Tensile strength		Elongation in 50 mm (2 in), %
	°F	°C	MPa	kpsi	
Copper-coated mild steel	2750	1510	358.5	52	23
High-tensile low-alloy steel	2750	1510	427.5	62	20
Cast iron	2200	1204	275.5	40	–
Stainless steel	2550	1399	551.5	80	30
Bronze	1600–1625	870–885	379.0	55	–
Ever dur	1870	1019	344.5	50	20
Aluminum	1190	643	110.5	16	25
White metal	715	379	358.5	52	8
Low-temperature brazing rod	1170–1185	632–640	Varies with parent metal		

TABLE 12–5
Allowable loads on mild-steel fillet welds

Size of weld, mm	Allowable static load per linear cm of weld							
	Bare welding rod				Shielding arc			
	Normal weld		Parallel weld		Normal weld		Parallel weld	
	N	lbf	N	lbf	N	lbf	N	lbf
2 × 3	1667.1	375	1323.9	298	2059.4	462	1667.1	375
5 × 5	2745.8	617	2186.9	491	3432.3	772	2745.8	617
6 × 6	3285.2	738.5	2628.2	590	4118.8	926	3285.2	738.5
8 × 8	4373.7	983	3501.0	787	5491.7	1235	4373.7	983
10 × 10	5491.7	1235	4079.5	983	6864.6	1543	5491.7	1235
12 × 12	6570.4	1477	5263.3	1182	8237.5	1852	6570.4	1477
14 × 14	7659.0	1722	6129.1	1378	9581.0	2154	7659.0	1722
15 × 15	8237.5	1852	6570.4	1477	10296.9	2315	8237.5	1852
18 × 18	9855.6	2216	7884.5	1772	12326.9	2772	9855.6	2216
20 × 20	10944.2	2460	8757.3	1968	13680.2	3075	10944.2	2460

Note: For intermediate sizes interpolate the values.
Source: *Welding Handbook*, American Welding Society, 1950.

TABLE 12–6
Design stresses for welds made with mild-steel electrodes

Type of load		Bare electrodes σ_u = 274.6–380.5 MPa (40–55 kpsi)		Covered electrodes σ_u = 416.8–519.7 MPa (60–75 kpsi)	
		Static loads	Dynamic loads	Static loads	Dynamic loads
Butt Welds–					
Tension	MPa	89.70	34.30	110.30	55.10
	kpsi	13.0	5.0	16.0	8.0
Compression	MPa	103.40	34.30	124.10	55.10
	kpsi	15.0	5.0	19.5	8.0
Shear	MPa	55.10	20.60	68.90	83.40
	kpsi	8.0	3.0	10.0	12.0
Fillet Welds–					
Shear	MPa	78.0	20.60	96.50	34.30
	kpsi	11.5	3.0	14.0	5.0

Source: *Welding Handbook*, American Welding Society, 1950

TABLE 12–7
Fatigue stress-concentration factors $K_{f\sigma}$

Type of weld	Stress-concentration factors, $K_{f\sigma}$
Reinforced butt weld	1.2
Toe of transverse fillet weld or normal fillet weld	1.5
End of parallel weld or longitudinal weld	2.7
T-butt joint with sharp corners	2.0

TABLE 12–8
Strength of shielded-arc flush steel welds

Type of stress	Limit stress			Recommended design stress		
	Base metal elastic limit, σ_e	Deposited metal		Static load	Load varies from O to F	Load varies from $+F$ to $-F$
		Elastic limit, σ_e	Endurance limit, σ_f			
Tension						
MPa	220.60	275.80	151.70	110.30	100.00	55.20
kpsi	32	40	22	16	14.5	8.0
Compression						
MPa	241.20	303.40	—	124.20	110.30	55.23
kpsi	35.0	44.0	—	10.0	16.0	8.0
Bending						
MPa	241.20	303.40	179.30	124.20	110.30	62.10
kpsi	35	44	26	18	16	9.0
Shear						
MPa	137.90	165.40	—	75.80	68.90	34.50
kpsi	20	24	—	11	10	5
Shear and tension						
MPa	—	—	—	75.80	68.90	34.50
kpsi	—	—	—	11	10	5

For bare electrode welds, the allowable stress must be multipled by 0.8 and for gas welds, they should be multiplied by 0.8 to 0.85.

TABLE 12–9
Length and spacing of intermittent welds

R % of continuous weld	Length of intermittent welds and distance between centers, mm		
75		75–100[a]	
66			100–150
60		75–125	
57			100–175
50	50–100	75–150	100–200
44			100–225
43		75–175	
40	50–125		100–250
37		75–200	
33	50–160	75–225	100–300
30		75–250	
25	50–200	75–300	
20	50–250		
16	50–300		

[a]75–100 means a weld 75 mm long with a distance of 100 mm between the centers of two consecutive welds.

$$R \text{ in } \% = \frac{\text{calculated leg size (continuous)}}{\text{actual leg size used (intermittent)}}$$

TABLE 12–10
Fatigue data on butt weld joints (average strength values)

Material and joint		Base Metal σ_u, σ_y		Endurance strength, σ_f					
				$K = -1$[a]		$K = 0$[a]		$K = 0.5$[a]	
				No. of cycles 2×10^6					
Carbon steel	MPa	423	235						
	kpsi	61.3	34.0						
With bead, or welded	MPa			100.0	152.0	155.9	227.5	253.0	368.7
	kpsi			14.5	22.0	22.5	33	37	53.5
With bead, tempered	MPa			98.0	148.0	160.8	214.7	264.7	379.5
923K (650°C)	kpsi			14	21.5	23	31	38	55
Bead machined off	MPa			121.6	198.0	198.0	335.3	304.0	
	kpsi			17.5	28.5	28.5	48.5	44	
Bead machined off, tempered 923 K (650° C)	MPa			114.7	193.1	132.3	340.2	292.2	
	kpsi			16.5	28	19	49.3	42.4	
Alloy steel	MPa	745.6	672.0						
	kpsi	108	97.5						
As welded	MPa			400.1	539.3				
	kpsi			58	78				
Stress-relieved	MPa			456.0	593.2				
	kpsi			66	86				

[a] $K = +1$ steady; $K = -1$ complete reversal; $K = 0$ repeated; $K = \frac{1}{2}$ fluctuating; $K = \frac{\text{min stress}}{\text{max stress}}$

Source: Design of Weldments, The James F. Lincoln Arc Welding Foundation, Cleveland, Ohio, 1968.

TABLE 12–11
Stresses as per the AISC Code for weld metal

Load type	Weld type	Allowable stress, σ_a
Tension	Butt	$0.60 \, \sigma_y$
Compression	Butt	$0.60 \, \sigma_y$
Shear	Butt or fillet	$0.40 \, \sigma_y$
Bending	Butt	$0.90 \, \sigma_y$
Bending	Butt	$0.60 \, \sigma_y - 0.66 \, \sigma_y$

TABLE 12–12
Properties of weld metal

AWS electrode number[a]	Elonga-tion %	Tensile strength		Yield strength	
		MPa	kpsi	MPa	kpsi
E 60xx	17–25	427	62	345	50
E 70xx	22	483	70	393	57
E 80xx	19	550	80	462	67
E 90xx	14–17	620	90	530	77
E 100xx	13–16	690	100	600	87
E 120xx	14	828	120	738	107

[a] The American Welding Society (AWS) Specification Code numbering system for electrodes.

TABLE 12–13
Selection of fillet weld sizes by rule-of-thumb (all dimensions in mm)

Plate thickness, h mm	Designing for strength, full-strength weld ($w = 3/4\ h$)	Designing for rigidity	
		50% of full-strength weld ($w = 3/8\ h$)	33% of full-strength weld ($w = 1/4\ h$)
6	4.5	4.5	4.5
8	6	4.5	4.5
9.5	8	4.5	4.5
11	9.5	4.5	4.5
12.5	9.5	4.5	4.5
14	11	6	6
15.5	12.5	6	6
19	14	8	6
22	15.5	9.5	8
25	19	9.5	8
28.5	22	11	8
31.5	25	12.5	8
35	25	12.5	9.5
37.5	28.5	14	9.5
41	31.5	15.5	11
44	35	19	11
50	37.5	19	12.5
54	41	22	14
57	44	22	14
60	44	25	15.5
62.5	47.5	25	15.5
66.5	50	25	19
70	50	25	19
75	56	28.5	19

Source: Welding Handbook, 3d. ed., American Welding Society, 1950.

TABLE 12–14
Equivalent length of fillet weld to replace rivets

Rivet diameter, mm	Rivet shear value at 100 MPa (10.2 kgf/mm^2)		Length of fillet welds[a] "Fusion Code" (structural) shielded arc welding, mm				
	MPa	kgf/mm^2	6-mm fillet	8-mm fillet	9.5-mm fillet	12.5-mm fillet	15.5-mm fillet
12.5	20.0	2.07	37.5	31.5	28.5	22	19
15.5	31.5	3.23	56	44.0	37.5	31.5	25
19	45.5	4.66	75	61.5	54	41	35
22	61.0	6.34	105	85.5	73	54	44
25	81.2	8.28	133	108.0	92	70	56

[a] 6 mm is added to calculated length of bead for starting and stopping the arc.

TABLE 12–15
Stress concentration factor, K_σ

	Stress concentration factor, K_σ	
Weld type and metal	Low-carbon steel	Low-alloy steel
Weld metal		
Butt welds with full penetration	1.2	1.4
End fillet welds	2	2.5
Parallel fillet welds	3.5	4.5
Base metal		
Toe of machined butt weld	1.2	1.4
Toe of unmachined butt weld	1.5	1.9
Toe of machined end fillet weld with leg ratio 1 : 1.5	2	2.5
Toe of unmachined end fillet weld with leg ratio 1 : 1.5	2.7	3.3
Parallel fillet weld	3.5	4.5
Stiffening ribs and partitions welded with end fillet welds having smooth transititions at the toes	1.5	1.9
Butt and T-welded corner plates	2.7	3.3
Butt and T-welded corner plates, but with smooth transitions in the shape of the plates and with machined welds	1.5	1.9
Lap-welded corner plates	2.7	3.3

TABLE 12–16
Allowable stresses for welds under static loads

	Allowable stresses		
Weld type and process	Tension σ_{ta}	Compression, σ_{ca}	Shear, τ_a
Automatic and hand welding with shielded arc and butt welding	σ_t[a]	σ_t	$0.65\sigma_t$
Hand welding with ordinary quality electrodes	$0.9\sigma_t$	σ_t	$0.6\ \sigma_t$
Resistance spot welding	$0.9\sigma_t$	σ_t	$0.5\ \sigma_t$

[a] σ_t is the allowable stress in tension of the base metal of the weld.

CHAPTER
13

RIVETED JOINTS

SYMBOLS

A	area of cross-section, m^2 (in^2)
	the cross-sectional area of rivet shank, m^2 (in^2)
b	breadth of cover plates (also with suffixes), m (in)
c	distance from the centroid of the rivet group to the critical rivet, m (in)
d	diameter of rivet, m (in)
D_i	internal diameter of pressure vessel, m (mm)
e or l	eccentricity of loading, m (in)
F	force on plate or rivets (also with suffixes), kN (lbf)
h	thickness of plate or shell, m (in)
$h_c,\ h_1,\ h_2$	thickness of cover plate (butt strap), m (in)
i	number of rivets in a pitch line (also with suffixes 1 and 2, respectively, for single shear and double shear rivets)
I	moment of inertia, area, m^4, cm^4 (in^4)
J	moment of inertia, polar, m^4, cm^4 (in^4)
$K = \dfrac{F}{F'}$	coefficient (Table 13–1)
m	margin, m (in)
M_b	bending moment, N m (lbf in)
p	pitch on the gauge line or longitudinal pitch, m (in)
p_c	pitch along the caulking edge, m (in)
p_d	diagonal pitch, m (in)
p_t	transverse pitch, m (in)
P_f	intensity of fluid pressure, MPa (psi)
Z	section modulus of the angle section, m^3, cm^3 (in^3)
σ_θ	hoop stress in pressure vessel or normal stress in plate, MPa (psi)
σ_a	allowable normal stress, MPa (psi)
σ_c	crushing stress in rivets, MPa (psi)
τ	shear stress in rivet, MPa (psi)
τ_a	allowable shear stress, MPa (psi)
η	efficiency of the riveted joint
θ	angle between a line drawn from the centroid of the rivet group to the critical rivet and the horizontal (Fig. 13–5)

Particular	Formula

PRESSURE VESSELS
Thickness of main plates

The thickness of plate of the pressure vessel with longitudinal joint

$$h = \frac{P_f D_i}{2\eta\sigma_\theta} \qquad (13\text{--}1)$$

For thickness of boiler plates and suggested types of joints

Refer to Tables 13–1 and 13–2.

The thickness of plate of the pressure vessel with circumferential joint

$$h = \frac{P_f D_i}{4\eta\sigma_\theta} \qquad (13\text{--}2)$$

For allowable stress and efficiency of joints

Refer to Tables 13–3, 13–4, 13–5, and 13–6.

PITCHES
Lap joints

The diagonal pitch (staggered) (Fig. 13–1) for p, p_t, and p_d

$$p_d = \frac{2p + d}{3} \qquad (13\text{--}3)$$

Refer to Tables 13–7 and 13–8.

The distance between rows or transverse pitch or back pitch (staggered)

$$p_t = \sqrt{\left(\frac{2p + d}{3}\right)^2 - \left(\frac{p}{2}\right)^2} \qquad (13\text{--}4)$$

The rivet diameter

$d = 0.19\sqrt{h}$ to $0.2\sqrt{h}$ **SI**   (13–5a)
where h and d in m
$d = 1.2\sqrt{h}$ to $1.4\sqrt{h}$ **US Customary System units**   (13–5b)
where h and d in in
$d = 6\sqrt{h}$ to $6.3\sqrt{h}$ **Customary Metric units** (13–5c)
where h and d in mm

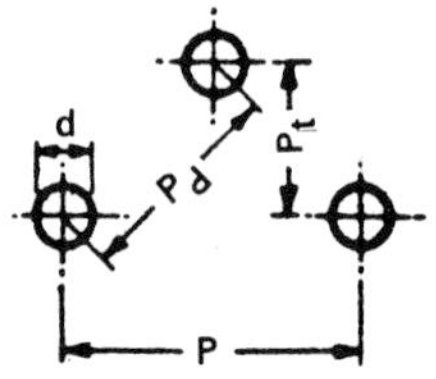

FIGURE 13–1 Pitch relation.

TABLE 13–1
Suggested types of joint

Diameter of shell, mm (in)	Thickness of shell, mm (in)	Type of joint
600–1800 (24–72)	6–12 (0.25–0.5)	Double-riveted
900–2150 (36–84)	7.5–25 (0.31–1.0)	Triple-riveted
1500–2750 (60–108)	9.0–44 (0.375–1.75)	Quadruple-riveted

TABLE 13–2
Minimum thickness of boiler plates

Shell plates		Tube sheets of firetube boilers	
Diameter of shell, mm (in)	Minimum thickness after flanging, mm (in)	Diameter of tube sheet, mm (in)	Minimum thickness, mm (in)
≤ 900 (36)	6.0 (0.25)	≤ 1050 (42)	9.5 (0.375)
900–1350 (36–54)	8.0 (0.3125)	1050–1350 (42–54)	11.5 (0.4375)
1350–1800 (54–72)	9.5 (0.375)	1350–1800 (54–72)	12.5 (0.50)
≥ 1800 (72)	12.5 (0.5)	≥ 1800 (72)	14.0 (0.5625)

TABLE 13–3
Efficiency of riveted joints (η)

Type of joint	% Efficiency, η	
	Normal range	Maximum
Lap joints		
Single-riveted	50–60	63
Double-riveted	60–72	77
Triple-riveted	72–80	86.6
Butt joints (with two cover plates)		
Single-riveted	55–60	63
Double-riveted	76–84	87
Triple-riveted	80–88	95
Quadruple-riveted	86–94	98

TABLE 13–4
Allowable stresses in structural riveting (σ_b)

Load-carrying member	Type of stress	Rivet-driving method	Rivets acting in single shear		Rivets acting in double shear	
			MPa	kpsi	MPa	kpsi
Rolled steel SAE 1020	Tension		124	18.0	124	18.0
	Shear	Power	93	13.5	93	13.5
Rivets, SAE 1010	Shear	Hand	68	10.0	68	10.0
	Crushing	Power	165	24.0	206	30.0
	Crushing	Hand	110	16.0	137	20.0

TABLE 13–5
Allowable stress for aluminum rivets, σ_a

| Rivet alloy | Procedure of drawing | Allowable stress[a], σ_a | | | |
| | | Shear | | Bearing | |
		MPa	kpsi	MPa	kpsi
2S (pure aluminum)	Cold, as received	20	3.0	48	7.0
17S	Cold, immediately after quenching	68	10.0	179	26.0
17S	Hot, 500–510°C	62	9.0	179	26.0
615–T6	Cold, as received	55	8.0	103	15.0
53S	Hot, 515–527°C	41	6.0	103	15.0

[a] Actual safety factor or reliability factor is 1.5.

TABLE 13–6
Values of working stress[a] at elevated temperatures

| Maximum temperatures | | Minimum of the specified range of tensile strength of the material, MPa (kpsi) | | | | | | | | | |
| | | (45) | 311 | (50) | 344 | (55) | 380 | (60) | 413 | (75) | 517 |
°F	°C	MPa	kpsi	MPa	kpsi	MPa	kpsi	MPa	kpsi	MPa	kpsi
0–700	0–371	61	9.0	68	10.0	76	11.00	82	12.00	103	15.00
750	399	56	8.22	62	9.11	68	10.00	77	11.20	89	13.00
800	427	45	6.55	53	7.33	54	8.00	61	9.00	70	10.20
850	455	37	5.44	41	6.05	46	6.75	51	7.40	57	8.30
900	482	29	4.33	33	4.83	37	5.50	38	5.60	41	6.00
950	511	22	3.20	26	3.60	27	4.00	27	4.00	27	4.00

[a] Design stresses of pressure vessels are based on a safety factor of 5.

TABLE 13–7
Pitch of butt joints

Type of joint	Diameter of rivets, d, mm	Pitch, p
Double-riveted— use for $h \leq 12.5$ mm (0.5) in	Any	5.5d (approx.)
Triple-riveted— use for $h \leq 25$ mm (1.0 in)	1.75–23.80	8d–8.5d
	27.00	7.5d
	30.15–36.50	6.5d–7d
Quadruple-riveted— use for $h \leq 31.75$ mm (1.25 in)	17.50–23.80	16d–17d
	27.00	15d (approx.)
	30.15	14d (approx.)
	33.30–36.50	13d–14d

TABLE 13–8
Transverse pitch (p_t) as per ASME Boiler Code

Value of $\dfrac{p}{d}$	1	2	3	4	5	6	7
Value of p_t	$2d$	$2d$	$2d$	$2d$	$2d$	$2.2d$	$2.3d$

Particular	Formula

Butt joint

The transverse pitch	$p_t = 2d$ to $2.5d$ (13–6a) $p_t \geq \sqrt{0.5pd + 0.25d^2}$ (13–6b)
For rivets, rivet holes, and strap thick	Refer to Tables 13–9, 13–10, and Fig. 13–2.

TABLE 13–9
Rivet hole diameters

Diameter of rivet, mm	Rivet hole diameters, mm (min)
12	13
14	15
16	17
18	19
20	21
22	23
24	25
27	28.5
30	31.5
33	34.5
36	37.5
39	41.0
42	44
48	50

TABLE 13–10
Rivet hole diameters and strap thickness

Plate thickness, h, mm	Minimum strap thickness, h_c mm	Hole diameter, d, mm	Plate thickness, h, mm	Minimum strap thickness, h_c, mm	Hole diameter, d, mm
6.25			14.25	11.10	
7.20	6.25	17.50			27.0
8.00			15.90	12.50	
8.75		20.50	19.00		30.15
9.50					
10.30	8.00		22.25	15.90	33.30
11.10			25.00	12.50	
12.00	9.50	24.00	28.50	19.00	36.50
12.50			31.75	22.25	
13.50	11.10		83.10	25.00	39.70

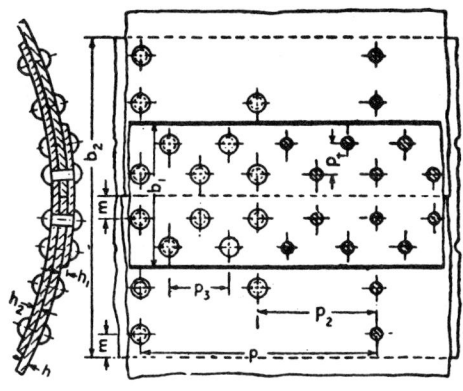

FIGURE 13–2 Quadruple-riveted double-strap butt joint. (*V. L. Maleev and J. B. Hartman*, Machine Design, *International Textbook Company, Scranton, Pennsylvania, 1954.*)

Particular	Formula
Minimum transverse pitch as per *ASME Boiler Code*	$p_t = 1.75d$ if $\dfrac{p}{d} \le 4$ (13–7a)
	$p_t = 1.75d + 0.001(p - d)$ if $\dfrac{p}{d} > 4$ **SI** (13–8a)
	where p_t, p, and d in m
	$= 1.75d + 0.1(p - d)$ if $\dfrac{p}{d} > 4$
	US Customary System units (13–8b)
	where p_t, d, and p in in
For transverse pitches	Refer to Table 13–8.
Haven and Swett formula for permissible pitches along the caulking edge of the outside cover plate	$p_c - d = 14 \sqrt[4]{\dfrac{h_c^3}{P_f}}$ **Customary Metric units** (13–9a)
	where p_c, d, h_c in cm and P_f in kgf/cm^2
	$= 21.38 \sqrt[4]{\dfrac{h_c^3}{P_f}}$ **US Customary System units** (13–9b)
	where p_c, d, h_c in in and P_f in psi
	$= 77.8 \sqrt[4]{\dfrac{h_c^3}{P_f}}$ **SI** (13–9c)
	where p_c, d, h_c in m and P_f in N/m^2
Diagonal pitch, p_d, is calculated from the relation	$2(p_d - d) \ge (p - d)$ (13–10)

MARGIN

Margin for longitudinal seams of all pressure vessels and girth seams of power boiler having unsupported heads	$m = 1.5d$ to $1.75d$ (13–11a)
Margin for girth seams of power boilers having supported heads and all unfired pressure vessels	$m \ge 1.25d$ (13–11b)

COVER PLATES

The thickness of cover plate	$h_c = 0.6h + 0.0025$ if $h \le 0.038$ m **SI** (13–12a)
	where h_c and h in m
	$h_c = 0.6h + 0.1$ if $h \le 1.5$ in
	US Customary System units (13–12b)
	where h_c and h in in
	$h_c = 0.67h$ if $h > 0.038$m **SI** (13–12c)
	where h_c and h in m
	$h_c = 0.67h$ if $h > 1.5$ in
	US Customary System units (13–12d)
	where h and h_c in in

TABLE 13-11
Rivet groups under eccentric loading value of coefficient K

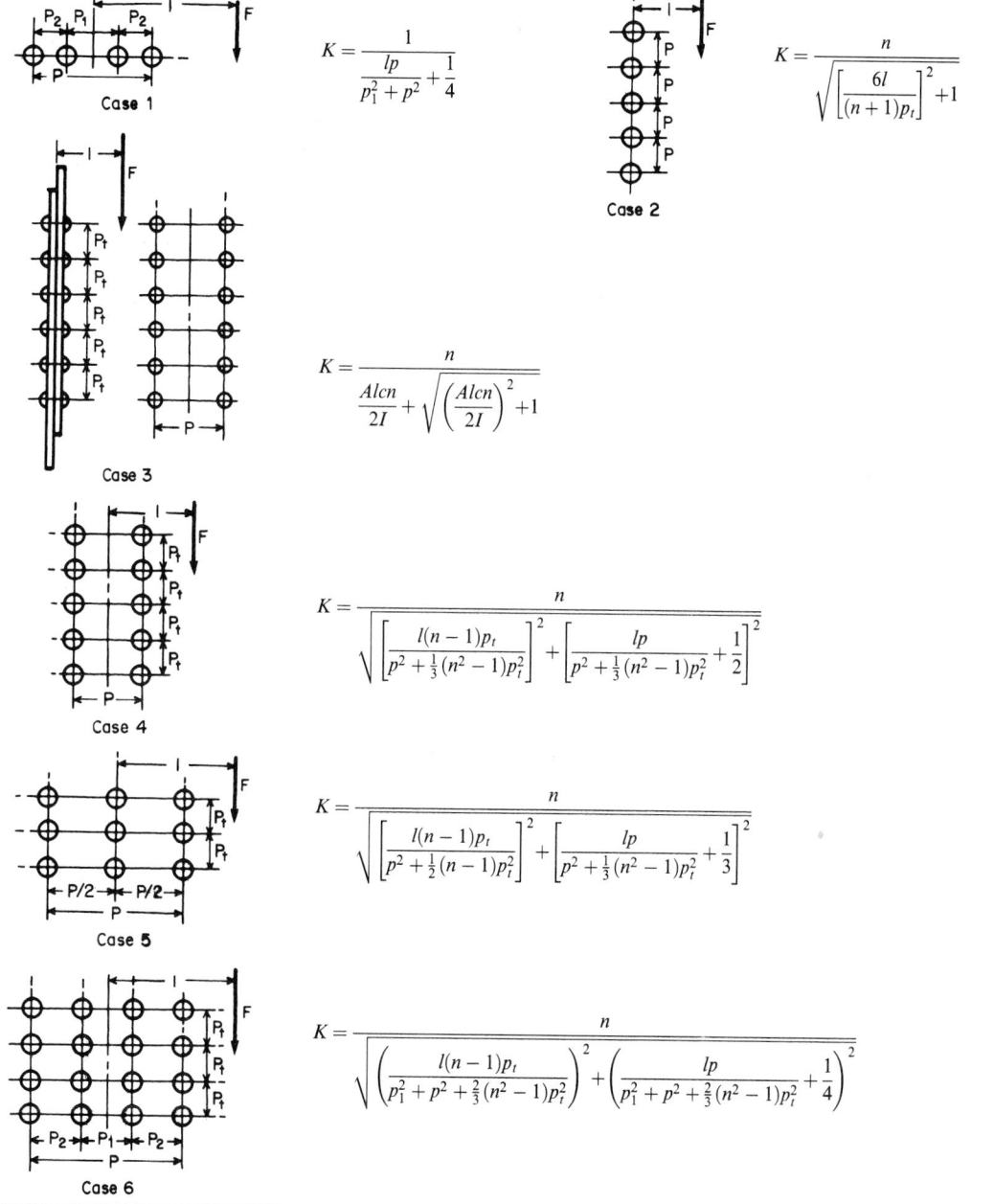

Case 1

$$K = \cfrac{1}{\cfrac{lp}{p_1^2 + p^2} + \cfrac{1}{4}}$$

Case 2

$$K = \cfrac{n}{\sqrt{\left[\cfrac{6l}{(n+1)p_t}\right]^2 + 1}}$$

Case 3

$$K = \cfrac{n}{\cfrac{Alcn}{2I} + \sqrt{\left(\cfrac{Alcn}{2I}\right)^2 + 1}}$$

Case 4

$$K = \cfrac{n}{\sqrt{\left[\cfrac{l(n-1)p_t}{p^2 + \frac{1}{3}(n^2-1)p_t^2}\right]^2 + \left[\cfrac{lp}{p^2 + \frac{1}{3}(n^2-1)p_t^2} + \cfrac{1}{2}\right]^2}}$$

Case 5

$$K = \cfrac{n}{\sqrt{\left[\cfrac{l(n-1)p_t}{p^2 + \frac{1}{2}(n-1)p_t^2}\right]^2 + \left[\cfrac{lp}{p^2 + \frac{1}{3}(n^2-1)p_t^2} + \cfrac{1}{3}\right]^2}}$$

Case 6

$$K = \cfrac{n}{\sqrt{\left(\cfrac{l(n-1)p_t}{p_1^2 + p^2 + \frac{2}{3}(n^2-1)p_t^2}\right)^2 + \left(\cfrac{lp}{p_1^2 + p^2 + \frac{2}{3}(n^2-1)p_t^2} + \cfrac{1}{4}\right)^2}}$$

Key:
n = total number of rivets in a column
F = permissible load, acting with lever arm, l, kN (lbf)
F' = permissible load on one rivet, kN (lbf)
K = F/F', coefficient

Source: K. Lingaiah and B. R. Narayana Iyengar, *Machine Design Data Handbook*, Engineering College Cooperative Society, Bangalore, India, 1962; K. Lingaiah and B. R. Narayana Iyengar, *Machine Design Data Handbook*, Vol. I, Suma Publishers, Bangalore, India, 1983; and K. Lingaiah, *Machine Design Data Handbook*, Vol. II, Suma Publishers, Bangalore, India, 1986.

Particular	Formula

Thickness of the cover plate according to Indian Boiler Code

Particular	Formula	
Thickness of single-butt cover plate	$h_1 = 1.125h$	(13–13)
Thickness of single-butt cover plate omitting alternate rivet in the over rows	$h_2 = 1.25h \dfrac{p-d}{p-2d}$	(13–14)
Thickness of double-butt cover plates of equal width	$h_c = h_1 = h_2 = 0.625h$	(13–15)
Thickness of double-butt cover plates of equal width omitting alternate rivet in the outer rows	$h_c = h_1 = h_2 = 0.625h \dfrac{p-d}{p-2d}$	(13–16)
Thickness of the double-butt cover plates of unequal width	$h_1 = 0.625h$ for narrow strap	(13–17a)
	$h_2 = 0.750h$ for wide strap	(13–17b)
For thickness of cover plates	Refer to Table 13–10.	
The width of upper cover plate (narrow strap)	$b_1 = 4m + 2p_{t1}$	(13–18)
The width of lower cover plate (wide strap)	$b_2 = b_1 + 2p_{t2} + 4m$	(13–19)

THEORETICAL STRENGTH ANALYSIS (FIG. 13–3)

Particular	Formula	
The tensile strength of the solid plate	$F_\theta = ph\sigma_\theta$	(13–20)
The tensile strength of the perforated strip along the outer gauge line	$F_\theta = (p-d)h\sigma_\theta$	(13–21)
The general expression for the resistance to shear of all the rivets in one pitch length	$F_\tau = (2i_2 + i_1) \dfrac{\pi d^2}{4} \tau$	(13–22)
The general expression for the resistance to crushing of the rivets	$F_c = (i_2 h + i_1 h_2)d\sigma_c$	(13–23)
The resistance against failure of the plate through the second row and simultaneous shearing of the rivets in the first row	$F_{\tau 1} = (p - 2d)h\sigma_\theta + \dfrac{\pi d^2}{4} \tau$	(13–24)
The resistance against failure of the plate through the second row and simultaneous crushing of the rivets in the first row	$F_{c1} = (p - 2d)h\sigma_\theta + dh\sigma_c$	(13–25)
The resistance against shearing of the rivets in the outer row and simultaneous crushing of the rivets in the two inner rows	$F_{\tau c} = \dfrac{\pi}{4} d^2\tau + idh\,\sigma_c$	(13–26)

Particular	Formula

EFFICIENCY OF THE RIVETED JOINT

The efficiency of plate

$$\eta = \frac{p - d}{p} \tag{13–27}$$

The efficiency of rivet in general case

$$\eta = \frac{\pi d^2 \tau (i_1 + 2i_2)}{4ph\sigma_\theta} \tag{13–28}$$

$$= \frac{\left(i_2 + i_1 \dfrac{h_2}{h}\right)\sigma_c}{\left(i_2 + i_1 \dfrac{h_2}{h}\right)\sigma_c + \sigma_\theta}$$

For efficiency of joints   Refer to Table 13–3.

The diameter of the rivet in general case

$$d = \frac{4hi_2 + i_1 h_2 \sigma_c}{\pi(i_1 + 2i_2)\tau} \tag{13–29}$$

Note: for lap joint $i_2 = 0$
for butt joint $i_1 = 0$

The pitch in general case

$$p = \frac{(2i_2 + i_1)\pi d^2 \tau}{4h\sigma_\theta} + d \tag{13–30}$$

For pitch of joint   Refer to Table 13–7.

THE LENGTH OF THE SHANK OF RIVET (FIG. 13–3)

$$L = h + h_1 + h_2 + (1.5 \text{ to } 1.7)D \tag{13–31a}$$
$$L = h + h_c + (1.5 \text{ to } 1.7)D \text{ for butt joint} \tag{13–31b}$$
with single cover plate
$$L = 2h + (1.5 \text{ to } 1.7)D \text{ for lap joint} \tag{13–31c}$$
where D = diameter of rivet

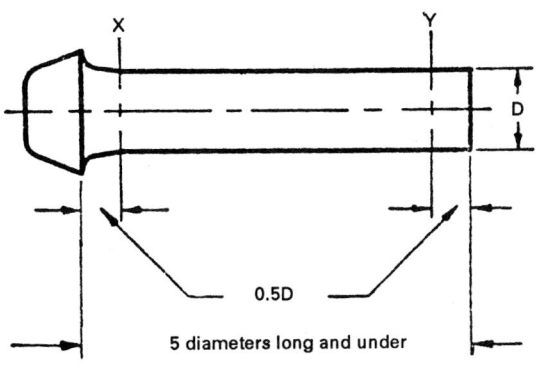

0.5D

5 diameters long and under

FIGURE 13–3

Particular	Formula

STRUCTURAL JOINT
Riveting of an angle to a gusset plate (Fig. 13–4)

The resultant normal stress

$$\sigma = \frac{F}{A} + \frac{Fe}{Z} \tag{13–32}$$

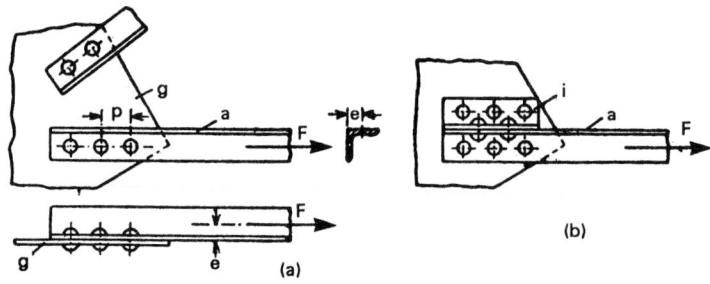

FIGURE 13–4   Riveting of an angle to a gusset plate.

RIVETED BRACKET (FIG. 13–5)

The resultant load on the farthest rivet whose distance is c from the center of gravity of a group of rivets (Fig. 13–5)

$$F_R = \sqrt{\left[\left(\frac{F}{nn'}\right)^2 + \left(\frac{M_b c}{\Sigma x^2 + \Sigma y^2}\right)^2\right]}$$

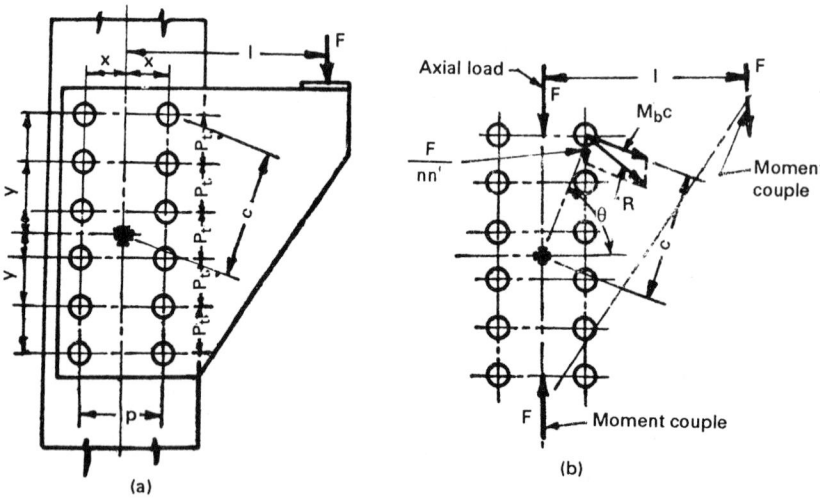

FIGURE 13–5   Riveted bracket. (*Bureau of Indian Standards.*)

Particular	Formula
	$$+2\left(\dfrac{F}{nn'}\right)\left(\dfrac{M_bc}{\Sigma x^2 + \Sigma y^2}\right)\cos\theta \Big] \qquad (13\text{–}33)$$

where

n = number of rivets in one column
n' = number of rivets in one row
x,y have the meaning as shown in Fig. 13–5.

Particular	Formula
For rivet groups under eccentric loading value of coefficient K	Refer to Table 13–11.
For preferred length and diameter of rivets	Refer to Figs. 13–6 to 13–8 and Tables 13–12 to 13–13.
For collected formulas of riveted joints	Refer to Table 13–14.

REFERENCES

1. Maleev, V. L., and J. B. Hartmen, *Machine Design*, International Textbook Company, Scranton, Pennsylvania, 1954.
2. Lingaiah, K., and B. R. Narayana Iyengar, *Machine Design Data Handbook*, Engineering College Co-operative Society, Bangalore, India, 1962.
3. Lingaiah, K., and B. R. Narayana Iyengar, *Machine Design Data Handbook*, Vol. I, Suma Publishers, Bangalore, India, 1983.
4. Lingaiah, K., *Machine Design Data Handbook*, Vol. II (*SI and Customary Metric Units*), Suma Publishers, Bangalore, India, 1986.
5. Bureau of Indian Standards.

BIBLIOGRAPHY

Faires, V. M., *Design of Machine Elements*, The Macmillan Company, New York, 1965.
Norman, C. A., E. S. Ault, and I. F. Zarobsky, *Fundamentals of Machine Design*, The Macmillan Company, New York, 1951.
Vallance, A., and V. L. Doughtie, *Design of Machine Members*, McGraw-Hill Publishing Company, New York, 1951.

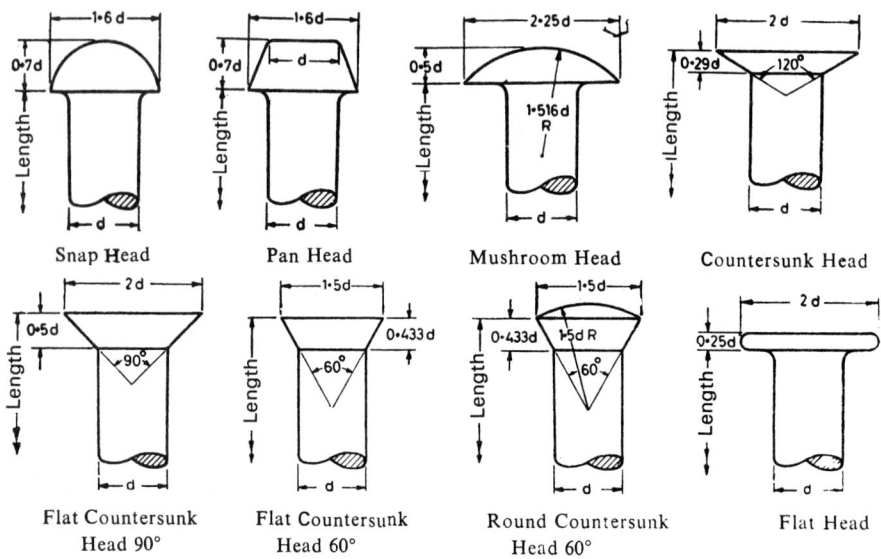

FIGURE 13–6   Rivets for general purposes (less than 12 mm diameter). For preferred length and diameter combination, refer to Table 13–12.

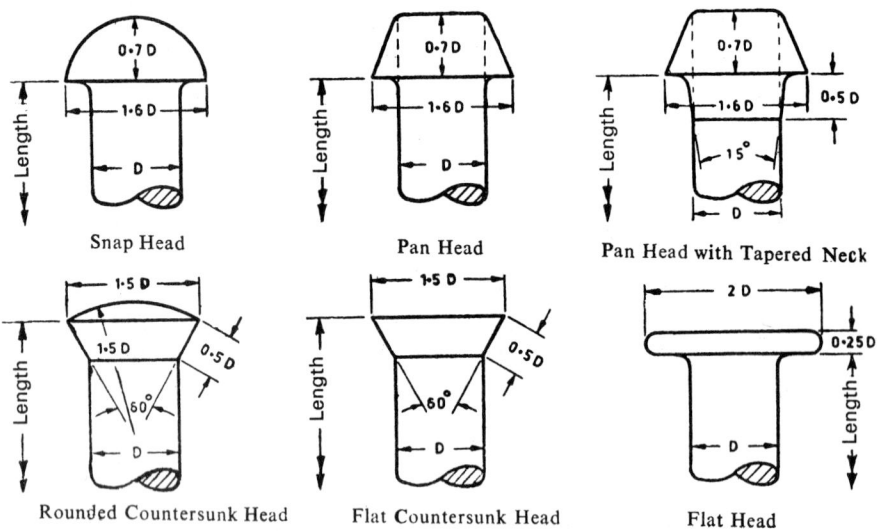

FIGURE 13–7   Rivets for general purposes (12 to 48 mm diameter). For preferred length and diameter combination, refer to Table 13–13.

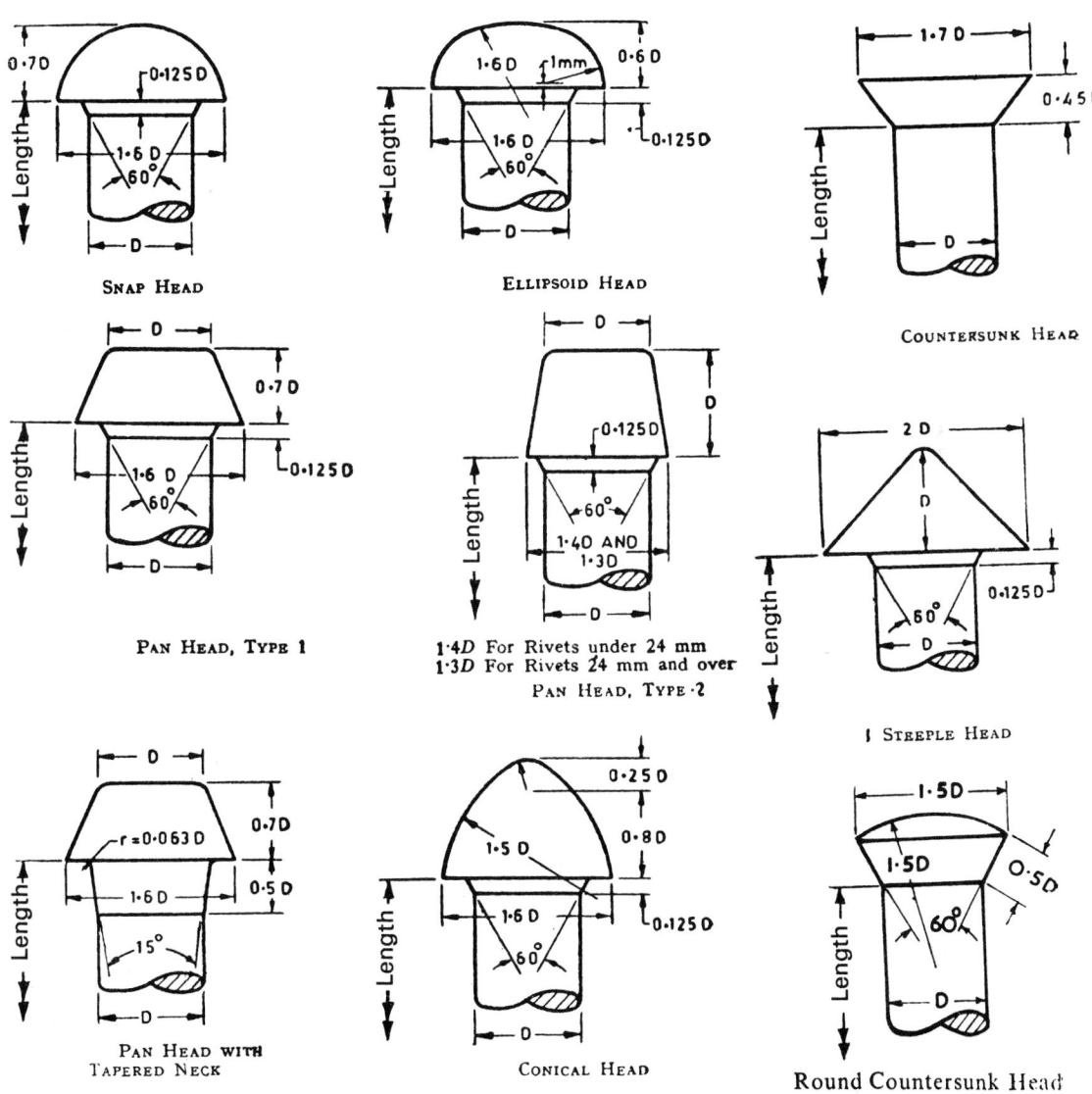

FIGURE 13–8 Boiler rivets (12 to 48 mm diameter). For preferred length and diameter combination, refer to Table 13–13.

TABLE 13–12
Preferred length (x) and diameter combinations for rivets (Fig. 13–6)

Length, mm	Diameter, mm								
	1.6	2	2.5	3	4	5	6	8	10
5	x	—	—	—	—	—	—	—	—
6	x	x	x	x	—	—	—	—	—
7	x	x	x	x	—	—	—	—	—
8	x	x	x	x	x	—	—	—	—
9	x	x	x	x	x	—	—	—	—
10	x	x	x	x	x	x	—	—	—
12	—	x	x	x	x	x	x	—	—
14	—	—	x	x	x	x	x	x	—
16	—	—	x	x	x	x	x	x	—
18	—	—	—	x	x	x	x	x	x
20	—	—	—	x	x	x	x	x	x
22	—	—	—	x	x	x	x	x	x
24	—	—	—	x	x	x	x	x	x
26	—	—	—	x	x	x	x	x	x
28	—	—	—	x	x	x	x	x	x
30	—	—	—	x	x	x	x	x	x
35	—	—	—	x	x	x	x	x	x
40	—	—	—	—	x	x	x	x	x
45	—	—	—	—	x	x	x	x	x
50	—	—	—	—	—	—	x	x	x
55	—	—	—	—	—	—	x	x	x
60	—	—	—	—	—	—	—	x	x
65	—	—	—	—	—	—	—	x	x
70	—	—	—	—	—	—	—	x	x

Source: Bureau of Indian Standards, 2155, 1962.

TABLE 13–13
Preferred lengths (x) and diameter combinations of rivets (Fig. 13–7)

Length, mm	Diameter, mm													
	12	14	16	18	20	22	24	27	30	33	36	39	42	48
28	x	—	—	—	—	—	—	—	—	—	—	—	—	—
31.5	x	x	—	—	—	—	—	—	—	—	—	—	—	—
35.5	x	x	x	—	—	—	—	—	—	—	—	—	—	—
40	x	x	x	x	—	—	—	—	—	—	—	—	—	—
45	x	x	x	x	x	—	—	—	—	—	—	—	—	—
50	x	x	x	x	x	x	—	—	—	—	—	—	—	—
56	x	x	x	x	x	x	x	—	—	—	—	—	—	—
63	x	x	x	x	x	x	x	x	—	—	—	—	—	—
71	x	x	x	x	x	x	x	x	x	—	—	—	—	—
80	x	x	x	x	x	x	x	x	x	—	—	—	—	—
85	—	x	x	x	x	x	x	x	x	x	—	—	—	—
90	—	x	x	x	x	x	x	x	x	x	—	—	—	—
95	—	x	x	x	x	x	x	x	x	x	x	—	—	—
100	—	—	x	x	x	x	x	x	x	x	x	—	—	—
106	—	—	x	x	x	x	x	x	x	x	x	x	—	—
112	—	—	x	x	x	x	x	x	x	x	x	x	—	—
118	—	—	—	x	x	x	x	x	x	x	x	x	x	—
125	—	—	—	—	x	x	x	x	x	x	x	x	x	x
132	—	—	—	—	—	x	x	x	x	x	x	x	x	x
140	—	—	—	—	—	x	x	x	x	x	x	x	x	x
150	—	—	—	—	—	—	x	x	x	x	x	x	x	x
160	—	—	—	—	—	—	x	x	x	x	x	x	x	x
180	—	—	—	—	—	—	—	x	x	x	x	x	x	x
200	—	—	—	—	—	—	—	—	x	x	x	x	x	x
224	—	—	—	—	—	—	—	—	—	x	x	x	x	x
250	—	—	—	—	—	—	—	—	—	—	—	—	x	x

Source: Bureau of Indian Standards, 1929, 1961.

TABLE 13–14
Formulas for riveted joints

Type of joint	Figure	Efficiency of plate, η_p	Efficiency of rivets, η_r	Combined efficiency, η_c	Longitudinal pitch, p, mm	Transverse pitch, p_t, mm	Margin, m, mm	Thickness of cover plate, mm	
								Inner, h_2 (wider)	Outer, h_1 (narrower)
				Lap Joint					
One rivet per pitch Type a	(a)	$\dfrac{p-d}{p}$	$\left(\dfrac{\pi d^2}{4}\right)\dfrac{\tau}{ph\sigma_\theta}$		$1.13h+40$		$1.5d$		
Two rivets per pitch Type b	(b)	$\dfrac{p-d}{p}$	$2\left(\dfrac{\pi d^2}{4}\right)\dfrac{\tau}{ph\sigma_\theta}$		$2.62h+40$	$2d$	$1.5d$		
Type c	(c)	$\dfrac{p-d}{p}$	$2\left(\dfrac{\pi d^2}{4}\right)\dfrac{\tau}{ph\sigma_\theta}$		$2.62h+40$	$0.33p+0.67d$	$1.5d$		
Three rivets per pitch Type d	(d)	$\dfrac{p-d}{p}$	$3\left(\dfrac{\pi d^2}{4}\right)\dfrac{\tau}{ph\sigma_\theta}$		$3.47h+40$	$2d$	$1.5d$		

TABLE 13–14
Formulas for riveted joints (*Cont.*)

Type of joint	Figure	Efficiency of plate, η_p	Efficiency of rivets, η_r	Combined efficiency, η_c	Longitudinal pitch, p, mm	Transverse pitch, p_t, mm	Margin, m, mm	Thickness of cover plate, mm Inner, h_2 (wider)	Outer, h_1 (narrower)
Type e	(e)	$\dfrac{p-d}{p}$	$3\left(\dfrac{\pi d^2}{4}\right)\dfrac{\tau}{ph\sigma_\theta}$		$3.47h + 40$	$0.33p + 0.67d$	$1.5d$		
Four rivets per pitch Type f	(f)	$\dfrac{p-d}{p}$	$4\left(\dfrac{\pi d^2}{4}\right)\dfrac{\tau}{ph\sigma_\theta}$	$\dfrac{p-2d}{p}$ $+\left[\left(\dfrac{\pi d^2}{4}\right)\dfrac{\tau}{ph\sigma_\theta}\right]$	$4.14h + 40$	$0.33p + 0.67d$ or $2d$ (whichever is greater)	$1.5d$		
Type g	(g)	$\dfrac{p-d}{p}$	$4\left(\dfrac{\pi d^2}{4}\right)\dfrac{\tau}{ph\sigma_\theta}$	$\dfrac{p-2d}{p}$ $+\left[\left(\dfrac{\pi d^2}{p}\right)\dfrac{\tau}{ph\sigma_\theta}\right]$	$4.14h + 40$	$0.2p + 1.15d$	$1.5d$		
Butt Joint									
Single-butt strap One rivet per pitch Type a	(a)	$\dfrac{p-d}{p}$	$\left(\dfrac{\pi d^2}{4}\right)\dfrac{\tau}{ph\sigma_\theta}$		$1.53h + 40$		$1.5d$		$1.125h$

13.17

TABLE 13–14
Formulas for riveted joints (Cont.)

Type of joint	Figure	Efficiency of plate, η_p	Efficiency of rivets, η_r	Combined efficiency, η_c	Longitudinal pitch, p, mm	Transverse pitch, p_t, mm	Margin, m, mm	Thickness of cover plate, mm Inner, h_2 (wider)	Outer, h_1 (narrower)
Two rivets per pitch Type b		$\dfrac{p-d}{p}$	$2\left(\dfrac{\pi d^2}{4}\right)\dfrac{\tau}{ph\sigma_\theta}$		$3.06h+40$	$2d$	$1.5d$		$1.125h$
Type c		$\dfrac{p-d}{p}$	$2\left(\dfrac{\pi d^2}{4}\right)\dfrac{\tau}{ph\sigma_\theta}$		$3.06h+40$	$0.33p+0.67d$	$1.5d$		$1.125h$
Three rivets per pitch Type d		$\dfrac{p-d}{p}$	$3\left(\dfrac{\pi d^2}{4}\right)\dfrac{\tau}{ph\sigma_\theta}$	$\dfrac{p-2d}{p}$ $+\left(\dfrac{\pi d^2}{4}\right)\dfrac{\tau}{ph\sigma_\theta}$	$4.05h+40$	$0.33p+0.67d$ or $2d$ (whichever is greater)	$1.5d$		$1.125h\dfrac{(p-d)}{(p-2d)}$
Two rivets per pitch Type e		$\dfrac{p-d}{p}$	$3\left(\dfrac{\pi d^2}{4}\right)\dfrac{\tau}{ph\sigma_\theta}$	$\dfrac{p-2d}{p}$ $+\left(\dfrac{\pi d^2}{4}\right)\dfrac{\tau}{ph\sigma_\theta}$	$4.05h+40$	$0.2p+1.15d$	$1.5d$		$1.125h\dfrac{(p-d)}{(p-2d)}$

TABLE 13–14
Formulas for riveted joints (*Cont.*)

Type of joint	Figure	Efficiency of plate, η_p	Efficiency of rivets, η_r	Combined efficiency, η_c	Longitudinal pitch, p, mm	Transverse pitch, p_t, mm	Margin, m, mm	Thickness of cover plate, mm	
								Inner, h_2 (wider)	Outer, h_1 (narrower)
Double-butt strap (equal widths) One rivet per pitch Type f	(f)	$\dfrac{p-d}{p}$	$1.875\left(\dfrac{\pi d^2}{4}\right)$ $\times \dfrac{\tau}{ph\sigma_\theta}$		$1.75h+40$		$1.5d$	$0.625h$	$0.625h$
Two rivets per pitch Type g	(g)	$\dfrac{p-d}{p}$	$3.75\left(\dfrac{\pi d^2}{4}\right)$ $\times \dfrac{\tau}{ph\sigma_\theta}$		$3.5h+40$	$2d$	$1.5d$	$0.625h$	$0.625h$
Type h	(h)	$\dfrac{p-d}{p}$	$3.75\left(\dfrac{\pi d^2}{4}\right)$ $\times \dfrac{\tau}{ph\sigma_\theta}$		$3.5h+40$	$0.33p+0.67d$	$1.5d$	$0.625h$	$0.625h$
Three rivets per pitch Type i	(i)	$\dfrac{p-d}{p}$	$5.625\left(\dfrac{\pi d^2}{4}\right)$ $\times \dfrac{\tau}{ph\sigma_\theta}$	$\dfrac{p-2d}{p}+\left[1.875 \times\left(\dfrac{\pi d^2}{4}\right)\dfrac{\tau}{ph\sigma_\theta}\right]$	$4.63h+40$	$0.33p+0.67d$ or $2d$ (whichever is greater)	$1.5d$	$0.615h$ $\times\dfrac{(p-d)}{(p-2d)}$	$0.625h$ $\times\dfrac{(p-d)}{(p-2d)}$

TABLE 13–14
Formulas for riveted joints (Cont.)

Type of joint	Figure	Efficiency of plate, η_p	Efficiency of rivets, η_r	Combined efficiency, η_c	Longitudinal pitch, p, mm	Transverse pitch, p_t, mm	Margin, m, mm	Thickness of cover plate, mm Inner, h_2 (wider)	Outer, h_1 (narrower)
Type j	(j)	$\dfrac{p-d}{p}$	$5.625\left(\dfrac{\pi d^2}{4}\right)$ $\times \dfrac{\tau}{ph\sigma_\theta}$		$4.63h + 40$	$2d$	$1.5d$	$0.625h$	$0.625h$
Type k	(k)	$\dfrac{p-d}{p}$	$5.625\left(\dfrac{\pi d^2}{4}\right)$ $\times \dfrac{\tau}{ph\sigma_\theta}$		$4.63h + 40$	$0.2p + 1.15d$	$1.5d$	$0.625h$ $\times \dfrac{(p-d)}{(p-2d)}$	$0.625h$ $\times \dfrac{(p-d)}{(p-2d)}$
Type l	(l)	$\dfrac{p-d}{p}$	$5.625\left(\dfrac{\pi d^2}{4}\right)$ $\times \dfrac{\tau}{ph\sigma_\theta}$		$4.63h + 40$	$0.33p + 0.67d$	$1.5d$	$0.625h$	$0.625h$
Four rivets per pitch Type m	(m)	$\dfrac{p-d}{p}$	$7.5\left(\dfrac{\pi d^2}{4}\right)$ $\times \dfrac{\tau}{ph\sigma_\theta}$	$\dfrac{p-2d}{d} + \left[1.875\right.$ $\left.\times\left(\dfrac{\pi d^2}{4}\right)\dfrac{\tau}{ph\sigma_\theta}\right]$	$5.52h + 40$	$0.33p + 0.67d$ or $2d$ (whichever is greater)	$1.5d$	$0.625h$	$0.625h$

TABLE 13–14
Formulas for riveted joints (Cont.)

Type of joint	Figure	Efficiency of plate, η_p	Efficiency of rivets, η_r	Combined efficiency, η_c	Longitudinal pitch, p, mm	Transverse pitch, p_t, mm	Margin, m, mm	Thickness of cover plate, mm	
								Inner, h_2 (wider)	Outer, h_1 (narrower)
Type n	(n)	$\dfrac{p-d}{p}$	$7.5\left(\dfrac{\pi d^2}{4}\right)$ $\times \dfrac{\tau}{ph\sigma_\theta}$	$\dfrac{p-2d}{d} + \left[1.875\\ \times\left(\dfrac{\pi d^2}{4}\right)\dfrac{\tau}{ph\sigma_\theta}\right]$	$5.52h + 40$	$0.2p + 1.15d$	$1.5d$	$0.625h$	$0.625h$
Double butt (unequal widths) Two rivets per pitch Type o	(o)	$\dfrac{p-d}{p}$	$2.875\left(\dfrac{\pi d^2}{4}\right)$ $\times \dfrac{\tau}{ph\sigma_\theta}$		$3.5h + 40$	$0.33p + 0.67d$	$1.5d$	$0.75h$	$0.625h$
Type p	(p)	$\dfrac{p-d}{p}$	$2.875\left(\dfrac{\pi d^2}{4}\right)$ $\times \dfrac{\tau}{ph\sigma_\theta}$		$3.5h + 40$	$2d$	$1.5d$	$0.75h$	$0.625h$
Three rivets per pitch Type q	(q)	$\dfrac{p-d}{p}$	$4.75\left(\dfrac{\pi d^2}{4}\right)$ $\times \dfrac{\tau}{ph\sigma_\theta}$	$\dfrac{p-2d}{p}$ $+\left(\dfrac{\pi d^2}{4}\right)\dfrac{\tau}{ph\sigma_\theta}$	$4.63h + 40$	$0.33p + 0.67d$ or $2d$ (whichever is greater)	$1.5d$	$0.75h$	$0.625h$

13.21

TABLE 13–14
Formulas for riveted joints (*Cont.*)

Type of joint	Figure	Efficiency of plate, η_p	Efficiency of rivets, η_{ir}	Combined efficiency, η_{ic}	Longitudinal pitch, p, mm	Transverse pitch, p_t, mm	Margin, m, mm	Thickness of cover plate, mm Inner, h_2 (wider)	Outer, h_1 (narrower)
Type *r*	(r)	$\dfrac{p-d}{p}$	$4.75\left(\dfrac{\pi d^2}{4}\right)\times\dfrac{\tau}{ph\sigma_\theta}$		$4.63h+40$	$2d$	$1.5d$	$0.75h$	$0.625h$
Type *s*	(s)	$\dfrac{p-d}{p}$	$4.75\left(\dfrac{\pi d^2}{4}\right)\times\dfrac{\tau}{ph\sigma_\theta}$	$\dfrac{p-2d}{p}+\left(\dfrac{\pi d^2}{4}\right)\dfrac{\tau}{ph\sigma_\theta}$	$4.63h+40$	$0.2p+1.15d$	$1.5d$	$0.75h$	$0.625h$
Type *t*	(t)	$\dfrac{p-d}{p}$	$4.75\left(\dfrac{\pi d^2}{4}\right)\times\dfrac{\tau}{ph\sigma_\theta}$		$4.63h+40$	$0.33p+0.67d$	$1.5d$	$0.75h$	$0.625h$

Key: d = diameter of rivet, m (in); h = thickness of main plate, m (in); σ_θ = hoop stress, MPa (psi); D_i = inside diameter of pressure vessel, m (in); P_f = internal fluid pressure, MPa (psi); η = efficiency of the riveted joint.

Particular	Formula
Common Formula:	
The thickness of the main plate of a longitudinal joint	$h = \dfrac{P_l D_i}{2 p \sigma_\theta}$
Unwin's formula for diameter of rivet	$d = 0.19 \sqrt{h}$ to $0.2 \sqrt{h}$ where d and h in m
	$= 1.2 \sqrt{h}$ to $1.4 \sqrt{h}$ where d and h in in
	US Customary System unit.
	SI

Source: K. Lingaiah and B. R. Narayana Iyengar, *Machine Design Data Handbook*, Engineering College Cooperative Society, Bangalore, India, 1962; K. Lingaiah and B. R. Narayana Iyengar, *Machine Design Data Handbook*, Vol. I, Suma Publishers, Bangalore, India, 1983; and K. Lingaiah, *Machine Design Data Handbook*, Vol. II, Suma Publishers, Bangalore, India, 1986.

CHAPTER
14

DESIGN OF SHAFTS

SYMBOLS

b	width of keyway, m (in)
c	machine cost, \$/m (\$/in) (US dollars)
D	diameter of shaft (also with subscripts), m (in)
D_i	inside diameter of hollow shaft, m (in)
D_o	outside diameter of hollow shaft, m (in)
E	modulus of elasticity, GPa (Mpsi)
F	axial load (tensile or compressive), kN (lbf)
F'_m	the static equivalent of cyclic load $F_m \pm F_a$, kN (lbf)
G	modulus of rigidity, GPa (Mpsi)
h	depth of keyway, m (in)
k	radius of gyration, m (in)
	material cost (also with subscripts), \$/kg
$K = \dfrac{D_i}{D_o}$	ratio of inner to outer diameter of hollow shaft
K_b	numerical combined shock and fatigue factor to be applied to computed bending moment
K_t	numerical combined shock and fatigue factor to be applied to computed twisting moment
l	length, m (in)
M_b	bending moment, N m (lbf in)
M_t	twisting moment, N m (lbf in)
M'_{bm}	static equivalent of cyclic bending moment $M_{bm} \pm M_{ba}$, N m (lbf in)
M'_{tm}	static equivalent of cyclic twisting moment $M_{tm} \pm M_{ta}$, N m (lbf in)
P	power, kW (hp)
n	speed, rpm;
	safety factor
n'	speed, rps
ρ	specific weight of material, kN/m³ (lbf/in)
σ	stress (tensile or compressive) also with subscripts, MPa (psi)
τ	shear stress (also with subscripts), MPa (psi)
α	ratio of maximum intensity of stress to the average value from compressive stress only
θ	angular deflection, deg

SUFFIXES

a	amplitude
b	bending
d	design
e	elastic limit
h	hollow
m	mean
sc	static strength (σ_{su} or σ_{sy}), solid
t	twisting
u	ultimate
y	yield strength
max	maximum
min	minimum
f	endurance

Other factors in performance or in special aspect are included from time to time in this chapter and, being applicable in their immediate context, are not given at this stage.

Note: σ and τ with the initial subscript S designates strength properties of material used in the design which will be used and observed throughout this handbook. In some books on machine design and in this *Machine Design Data Handbook* the ratios of design stresses σ_{sd}/σ_{fd} and τ_{sd}/τ_{fd}; and design stresses σ_{yd}, $\tau_{yd'}$, σ_{fd}, and τ_{fd} have been used instead of σ_{sy}/σ_{sf}, τ_{sy}/τ_{Sf}; and yield strengths σ_{sy}, τ_{sy} and fatigue strengths, σ_{sf}, τ_{sf} in the design equations for shafts [Eqs. (14–1) to (14–65)]. This has to be taken into consideration in the design of shafts while using Eqs. (14–1) to (14–65).

Particular	Formula

SOLID SHAFTS

(1) Stationary shafts with static loads

The diameter of shaft subjected to simple torsion

$$D = \left(\frac{16M_t}{\pi\tau_{yd}}\right)^{1/3}$$

(14–1)

The diameter of shaft subjected to simple bending

$$D = \left(\frac{32M_b}{\pi\sigma_{yd}}\right)^{1/3}$$

(14–2)

The diameter of shaft subjected to combined torsion and bending:

(a) According to maximum normal stress theory

$$D = \left[\frac{16}{\pi\sigma_{yd}}\{M_b + (M_b^2 + M_t^2)^{1/2}\}\right]^{1/3}$$

(14–3)

(b) According to maximum shear stress theory

$$D = \left\{\frac{16}{\pi\tau_{yd}}(M_b^2 + M_t^2)^{1/2}\right\}^{1/3}$$

(14–4)

(c) According to maximum shear energy theory

$$D = \left\{\frac{16}{\pi\tau_{yd}}\left(M_b^2 + \frac{3}{4}M_t^2\right)^{1/2}\right\}^{1/3}$$

(14–5)

Particular	Formula

The diameter of shaft subject to axial load, bending and torsion [1, 2, 3]:

(a) According to maximum normal theory

$$D = \left[\frac{16}{\pi \sigma_{yd}} \left\{ \left(M_b + \frac{\alpha FD}{8} \right) + \left\{ \left(M_b + \frac{\alpha FD}{8} \right)^2 \right. \right. \right.$$

$$\left. \left. \left. + M_t^2 \right\}^{1/2} \right] \right\}^{1/3} \qquad (14\text{--}6)$$

(b) According to maximum shear stress theory

$$D = \left[\frac{16}{\pi \tau_{yd}} \left\{ \left(M_b + \frac{\alpha FD}{8} \right)^2 + M_t^2 \right\}^{1/2} \right]^{1/3} \qquad (14\text{--}7)$$

(c) According to maximum shear energy theory

$$D = \left[\frac{16}{\pi \tau_{yd}} \left\{ \left(M_b + \frac{\alpha FD}{8} \right)^2 + \frac{3}{4} M_t^2 \right\}^{1/2} \right]^{1/3} \qquad (14\text{--}8)$$

(2) Rotating shafts with dynamic loads, taking dynamic effect indirectly into consideration [1, 2, 3]

For empirical shafting formulas

Refer to Table 14–1.

The diameter of shaft subjected to simple torsion

$$D = \left\{ \frac{16}{\pi \tau_{yd}} \, (K_t M_t) \right\}^{1/3} \qquad (14\text{--}9)$$

The diameter of shaft subjected to simple bending

$$D = \left\{ \frac{32}{\pi \sigma_{yd}} \, (K_b M_b) \right\}^{1/3} \qquad (14\text{--}10)$$

The diameter of shaft subjected to combined bending and torsion

(a) According to maximum normal stress theory

$$D = \left\{ \frac{16}{\pi \sigma_{yd}} [K_b M_b + \{(K_b M_b)^2 + (K_t M_t)^2\}^{1/2}] \right\}^{1/3} \qquad (14\text{--}11)$$

(b) According to maximum shear stress theory

$$D = \left[\frac{16}{\pi \tau_{yd}} \{(K_b M_b)^2 + (K_t M_t)^2\}^{1/2} \right]^{1/3} \qquad (14\text{--}12)$$

(c) According to maximum shear energy theory

$$D = \left[\frac{16}{\pi \tau_{yd}} \{(K_b M_b)^2 + \frac{3}{4} \, (K_t M_t)^2\}^{1/2} \right]^{1/3} \qquad (14\text{--}13)$$

Particular	Formula

The diameter of shaft subjected to axial load, bending, and torsion

(a) According to maximum normal stress theory

$$D = \left\{ \frac{16}{\pi \sigma_{yd}} \left(K_b M_b + \frac{\alpha FD}{8} \right) \right. $$
$$\left. + \left[\left(K_b M_b + \frac{\alpha FD}{8} \right)^2 + (K_t M_t)^2 \right]^{1/2} \right\}^{1/3}$$

(14–14)

(b) According to maximum shear stress theory

$$D = \left[\frac{16}{\pi \tau_{yd}} \left\{ \left(K_b M_b + \frac{\alpha FD}{8} \right)^2 + (K_t M_t)^2 \right\}^{1/2} \right]^{1/3}$$

(14–15)

(c) According to maximum shear energy theory

$$D = \left[\frac{16}{\pi \tau_{yd}} \left\{ \left(K_b M_b + \frac{\alpha FD}{8} \right)^2 + \frac{3}{4}(K_t M_t)^2 \right\}^{1/2} \right]^{1/3}$$

(14–16)

The diameter of shaft based on torsional rigidity

$$D = \left\{ \frac{584 M_t L}{G\theta} \right\}^{1/4}$$

(14–17)

where K_b and K_t are taken from Table 14–2

(3) Rotating shafts and fluctuating loads, taking fatigue effect directly into consideration [1, 2, 3]

The diameter of shaft subjected to fluctuating torsion

$$D = \left\{ \frac{16}{\pi} \left(\frac{M_{tm}}{\tau_{yd}} + \frac{M_{ta}}{\tau_{fd}} \right) \right\}^{1/3}$$

(14–18)

The diameter of shaft subjected to fluctuating bending

$$D = \left\{ \frac{32}{\pi} \left(\frac{M_{bm}}{\sigma_{yd}} + \frac{M_{ba}}{\sigma_{fd}} \right) \right\}^{1/3}$$

(14–19)

The diameter of shaft subjected to combined fluctuating torsion and bending:

(a) According to maximum normal stress theory

$$D = \left[\frac{16}{\pi \sigma_{yd}} \left\{ M'_{bm} + (M'^2_{bm} + M'^2_{tm})^{1/2} \right\} \right]^{1/3}$$

(14–20)

(b) According to maximum shear stress theory

$$D = \left\{ \frac{16}{\pi \tau_{yd}} (M'^2_{bm} + M'^2_{tm})^{1/2} \right\}^{1/3}$$

(14–21)

Particular	Formula

(c) According to maximum shear energy theory

$$D = \left\{ \frac{16}{\pi \tau_{yd}} (M_{bm}'^2 + \frac{3}{4} M_{tm}'^2)^{1/2} \right\}^{1/3} \qquad (14\text{–}22)$$

where

$$M_{bm}' = M_{bm} + \frac{\sigma_{sd}}{\sigma_{fd}} M_{ba} \qquad (14\text{–}22a)$$

$$M_{tm}' = M_{tm} + \frac{\tau_{sd}}{\tau_{fd}} M_{ta} \qquad (14\text{–}22b)$$

The diameter of shaft subjected to combined fluctuating axial load, bending, and torsion

(a) According to maximum normal stress theory

$$D = \left\{ \frac{16}{\pi \sigma_{yd}} \left[\left(M_{bm}' + \frac{\alpha F_m' D}{8} \right) \right.\right.$$
$$\left.\left. + \left\{ \left(M_{bm}' + \frac{\alpha F_m' D}{8} \right)^2 + M_{tm}'^2 \right\}^{1/2} \right] \right\}^{1/3} \qquad (14\text{–}23)$$

(b) According to maximum shear stress theory

$$D = \left[\frac{16}{\pi \tau_{yd}} \left\{ \left(M_{bm}' + \frac{\alpha F_m' D}{8} \right)^2 + M_{tm}'^2 \right\}^{1/2} \right]^{1/3} \qquad (14\text{–}24)$$

(c) According to maximum shear energy theory

$$D = \left[\frac{16}{\pi \tau_{yd}} \left\{ \left(M_{bm}' + \frac{\alpha F_m' D}{8} \right)^2 + \frac{3}{4} M_{tm}'^2 \right\}^{1/2} \right]^{1/3} \qquad (14\text{–}25)$$

where M_{bm}' and M_{tm}' have the same meaning as in Eqs. (14–22a) and (14–22b)

and $F_m' = F_m + \dfrac{\sigma_{sd}}{\sigma_{fd}} F_a \qquad (14\text{–}25a)$

HOLLOW SHAFTS

(1) Stationary shafts with static loads

The outside diameter of shaft subjected to simple torsion

$$D_o = \left(\frac{16 M_t}{\pi \tau_{yd}(1 - K^4)} \right)^{1/3} \qquad (14\text{–}26)$$

The outside diameter of shaft subjected to simple bending

$$D_o = \left(\frac{32 M_b}{\pi \sigma_{yd}(1 - K^4)} \right)^{1/3} \qquad (14\text{–}27)$$

Particular	Formula

The diameter of shaft subjected to combined torsion and bending

(a) According to maximum normal stress theory

$$D_o = \left[\frac{16}{\pi\sigma_{yd}(1 - K^4)} \left\{ M_b + (M_b^2 + M_t^2)^{1/2} \right\} \right]^{1.3}$$

$$(14\text{–}28)$$

(b) According to maximum shear stress theory

$$D_o = \left\{ \frac{16}{\pi\tau_{yd}(1 - K^4)} (M_b^2 + M_t^2)^{1/2} \right\}^{1/3} \quad (14\text{–}29)$$

(c) According to maximum shear energy theory

$$D_o = \left\{ \frac{16}{\pi\tau_{yd}(1 - K^4)} (M_b^2 + \frac{3}{4} M_t^2)^{1/2} \right\}^{1/3} \quad (14\text{–}30)$$

The outside diameter of shaft subjected to axial load, bending, and torsion

(a) According to maximum normal stress theory

$$D_o = \left\{ \frac{16}{\pi\sigma_{yd}(1 - K^4)} \left(\left[M_b + \frac{\alpha F D_o}{8} (1 + K^2) \right] \right. \right.$$
$$\left. \left. + \left[\left(M_b + \frac{\alpha F D_o(1 + K^2)}{8} \right)^2 + M_t^2 \right]^{1/2} \right) \right\}^{1/3}$$

$$(14\text{–}31)$$

(b) According to maximum shear stress theory

$$D_o = \left\{ \frac{16}{\pi\tau_{yd}(1 - K^4)} \left[\left(M_b + \frac{\alpha F D_o}{8} (1 + K^2) \right) \right. \right.$$
$$\left. \left. + M_t^2 \right]^{1/2} \right\}^{1/3}$$

$$(14\text{–}32)$$

(c) According to maximum shear energy theory

$$D_o = \left\{ \frac{16}{\pi\tau_{yd}(1 - K^4)} \left[\left(M_b^2 + \frac{\alpha F D_o}{8} (1 + K^2) \right) \right. \right.$$
$$\left. \left. + \frac{3}{4} M_t^2 \right]^{1/2} \right\}^{1/3}$$

$$(14\text{–}33)$$

(2) Rotating shafts with dynamic loads, taking dynamic effect indirectly into consideration [1, 2, 3]

The outside diameter of shaft subjected to simple torsion

$$D_o = \left(\frac{16}{\pi\tau_{yd}(1 - K^4)} K_t M_t \right)^{1/3} \quad (14\text{–}34)$$

Particular	Formula
The outside diameter of shaft subjected to simple bending	$D_o = \left(\dfrac{32}{\pi\sigma_{yd}(1-K^4)} \, K_b M_b \right)^{1/3}$ (14–35)

The outside diameter of shaft subjected to combined bending and torsion

(a) According to maximum normal stress theory

$$D_o = \left\{ \frac{16}{\pi\sigma_{yd}(1-K^4)} \left[K_b M_b + \{(K_b M_b)^2 \right. \right.$$

$$\left. \left. + (K_t M_t)^2\}^{1/2} \right] \right\}^{1/3} \tag{14–36}$$

(b) According to maximum shear stress theory

$$D_o = \left[\frac{16}{\pi\tau_{yd}(1-K^4)} \, \{(K_b M_b)^2 + (K_t M_t)^2\}^{1/2} \right]^{1/3}$$

$$\tag{14–37}$$

(c) According to maximum shear energy theory

$$D_o = \left[\frac{16}{\pi\tau_{yd}(1-K^4)} \left\{ (K_b M_b)^2 + \frac{3}{4} \, (K_t M_t)^2 \right\}^{1/2} \right]^{1/3}$$

$$\tag{14–38}$$

The outside diameter of shaft subjected to axial load, bending and torsion

(a) According to maximum normal stress theory

$$D_o = \left[\frac{16}{\pi\sigma_{yd}(1-K^4)} \left\{ \left(K_b M_b + \frac{\alpha F D_o}{8} \, (1+K^2) \right) \right. \right.$$

$$\left. \left. + \left[\left(K_b M_b + \frac{\alpha F D_o}{8} \, (1+K^2) \right)^2 + (K_t M_t)^2 \right]^{1/2} \right\} \right]^{1/3}$$

$$\tag{14–39}$$

(b) According to maximum shear stress theory

$$D_o = \left[\frac{16}{\pi\tau_{yd}(1-K^4)} \left\{ \left(K_b M_b + \frac{\alpha F D_o}{8} \, (1+K^2) \right)^2 \right. \right.$$

$$\left. \left. + (K_t M_t)^2 \right\}^{1/2} \right]^{1/3} \tag{14–40}$$

(c) According to maximum shear energy theory

$$D_o = \left\{ \frac{16}{\pi\tau_{yd}(1-K^4)} \left[\left(K_b M_b + \frac{\alpha F D_o}{8} \, (1+K^2) \right)^2 \right. \right.$$

$$\left. \left. + \frac{3}{4}(K_t M_t)^2 \right]^{1/2} \right\}^{1/3} \tag{14–41}$$

Please note: If the axial load does not produce column action, the constant α need not be used to multiply the term $[F D_o \, (1+K^2)/8]$ throughout this chapter.

Particular	Formula
The outside diameter of shaft based on torsional rigidity	$D_o = \left(\dfrac{584 M_t L}{(1 - K^4) G\theta}\right)^{1/4}$ (14–42)

(3) Rotating shaft with fluctuating loads, taking fatigue effect directly into consideration

Particular	Formula
The outside diameter of shaft subjected to fluctuating torsion	$D_o = \left[\dfrac{16}{\pi(1 - K^4)}\left(\dfrac{M_{tm}}{\tau_{yd}} + \dfrac{M_{ta}}{\tau_{fd}}\right)\right]^{1/3}$ (14–43)
The outside diameter of shaft subjected to fluctuating bending	$D_o = \left[\dfrac{32}{\pi(1 - K^4)}\left(\dfrac{M_{bm}}{\sigma_{yd}} + \dfrac{M_{ba}}{\sigma_{fd}}\right)\right]^{1/3}$ (14–44)

The outside diameter of shaft subjected to combined fluctuating torsion and bending

(a) According to maximum normal stress theory

$$D_o = \left[\frac{16}{\pi \sigma_{yd}(1 - K^4)}\left\{(M_{bm}'^2 + M_{tm}'^2)^{1/2}\right\}\right]^{1/3}$$

$$(14\text{–}45)$$

(b) According to maximum shear stress theory

$$D_o = \left[\frac{16}{\pi \tau_{yd}(1 - K^4)}(M_{bm}'^2 + M_{tm}'^2)^{1/2}\right]^{1/3} \quad (14\text{–}46)$$

(c) According to maximum shear energy theory

$$D_o = \left[\frac{16}{\pi \tau_{yd}(1 - K^4)}(M_{bm}'^2 + \frac{3}{4} M_{tm}'^2)^{1/2}\right]^{1/3} \quad (14\text{–}47)$$

where M_{bm}', M_{tm}' have the same meaning as in Eqs. (14–22a) and (14–22b)

The outside diameter of shaft subjected to combined fluctuating axial load, bending, and torsion

(a) According to maximum normal stress theory

$$D_o = \left[\frac{16}{\pi \sigma_{yd}(1 - K^4)}\left\{\left(M_{bm}' + \frac{\alpha F_m' D_o}{8}(1 + K^2)\right)\right.\right.$$

$$\left.\left. + \left[\left\{M_{bm}' + \frac{\alpha F_m' D_o(1 + K^2)}{8}\right\}^2 + M_{tm}'^2\right]^{1/2}\right\}\right]^{1/3}$$

$$(14\text{–}48)$$

(b) According to maximum shear stress theory

$$D_o = \left\{\frac{16}{\pi \tau_{yd}(1 - K^4)}\left(\left[\left\{M_{bm}' + \frac{\alpha F_m' D_o(1 + K^2)}{8}\right\}^2\right.\right.\right.$$

$$\left.\left.\left. + M_{tm}'^2\right]^{1/2}\right)\right\}^{1/3}$$

$$(14\text{–}49)$$

Particular	Formula
(c) According to maximum shear energy theory	$$D_o = \left\{ \frac{16}{\pi \tau_{yd}(1 - K^4)} \left[\left(M'_{bm} + \frac{\alpha^* F'_m D_o (1 + K_2)}{8} \right)^2 \right. \right.$$ $$\left. \left. + \frac{3}{4} M'^2_{tm} \right]^{1/2} \right\}^{1/3} \qquad (14\text{--}50)$$

where M'_{bm}, M'_{tm}, and F'_m have the same meaning as in Eqs. (14–22a), (14–22b), and (14–25a)

COMPARISON BETWEEN DIAMETERS OF SOLID AND HOLLOW SHAFTS OF SAME LENGTH

For equal strength in bending, torsion, and/or combined bending and torsion, the diameter

(a) When materials of both shafts are same	$D = D_o(1 - K^4)^{1/3}$	
(b) When materials of shafts are different	$D = D_o \dfrac{\sigma_{eh}}{\sigma_{es}} (1 - K^4)^{1/3}$	

$(14\text{--}51)$

$(14\text{--}52)$

For torsional rigidity

(a) When torsional rigidities are equal	$D = D_o(1 - K^4)^{1/4}$
(b) When torsional rigidities are different	$D = D_o \left\{ \dfrac{G_h}{G_s}(1 - K^4) \right\}^{1/4}$

$(14\text{--}53)$

$(14\text{--}54)$

For equal weight

(a) When material of both shafts is same	$D = D_o(1 - K^2)^{1/2}$
(b) When materials of both shafts are different	$D = D_o \left\{ (1 - K^2) \dfrac{w_h}{w_s} \right\}^{1/2}$

$(14\text{--}55)$

$(14\text{--}56)$

For equal cost

(a) For same material and machining cost for both shafts	$D = D_o(1 - K^2)^{1/2}$
(b) For no machining cost for both shafts but with different material cost	$D = D_o \left\{ (1 - K^2) \dfrac{w_h k_h}{w_s k_s} \right\}^{1/2}$

$(14\text{--}57)$

$(14\text{--}58)$

Note: If the axial load does not produce column action, the constant α need not be used to multiply the term $[FD_o(1 + K^2)/8]$ throughout this chapter

Particular	Formula

(c) When machining costs are different and material cost negligible

$$D = \left\{ \frac{c_h}{c_s} \right\}^{1/2} \tag{14–59}$$

(d) When machining and material costs are different

$$D = \left\{ \frac{\pi D_o^2 (1 - K^2) w_h k_h + c_h}{\pi w_s k_s + \dfrac{c_s}{D^2}} \right\}^{1/2} \tag{14–60}$$

STIFFNESS

Instead of computing the transverse deflection, the maximum distance between the bearing (in meters) may be computed by the empirical formula to limit the transverse deflection to 0.8 mm/m of length

$$L = \frac{1500}{n + 1500} \, c \, D^{2/3} \tag{14–61}$$

where c is a constant from Table 14–3

RIGIDITY

Moor's formula for the increase of the angle of twist θ due to the keyway and applies only to the key-seated length of shaft

$$K_1 = 1 + \frac{0.4b + 0.7h}{D} \tag{14–62}$$

EFFECT OF KEYWAYS

The lowering of the strength of shaft by keyways may be taken into account by introducing a factor similar to a stress-concentration factor (or Moor's formula for lowering the strength of shaft)

$$K = 1 + \frac{0.2b + 1.1h}{D} \tag{14–63}$$

THE BUCKLING FACTOR

For short columns or when $l/k \le 115$

$$\alpha = \frac{1}{1 - 0.0044 l/k} \tag{14–64}$$

For long columns or when $l/k \ge 115$ (Euler's formula)

$$\alpha = \frac{\sigma_{Sy}}{\pi^2 n E} \left(\frac{l}{k} \right) \tag{14–65}$$

where

n = 1 for hinged ends
  = 2.25 for fixed ends
  = 1.6 for both ends pinned or guided and partly restrained
(α = 1 for tensile load)

Particular	Formula

SHAFTS SUBJECTED TO VARIOUS STRESSES

(1) Shaft subjected to steady torque and reversed bending moment taking into consideration stress concentration:
Diameter of solid shaft:

(a) According to maximum shear stress failure theory using Soderberg [4] criterion for fatigue strength

$$\frac{\sigma_a}{\sigma_{Sf}} + \frac{\sigma_m}{\sigma_{Sy}} = \frac{1}{n}$$

$$D = \left\{ \frac{32n}{\pi\sigma_{Sy}} \left[\left(K_{f\sigma} \frac{\sigma_{Sy}}{\sigma_{Sf}} M_{ba} \right)^2 + (K_{f\tau} M_{tm})^2 \right]^{1/2} \right\}^{1/3}$$

$$(14\text{--}66)$$

where

$K_{f\sigma}$ = fatigue stress-concentration factor due to bending, tension, or compression

$K_{f\tau}$ = fatigue stress-concentration factor due to torsion

$K_{f\sigma} = K_{f\tau} = 1$ for ductile material under steady state of stress

(b) According to maximum shear stress theory of failure using modified Goodman criterion for fatigue strength

$$\frac{\sigma_a}{\sigma_{Sf}} + \frac{\sigma_m}{\sigma_{Sut}} = \frac{1}{n}$$

$$D = \left\{ \frac{32n}{\pi\sigma_{Sut}} \left[\left(K_{f\sigma} \frac{\sigma_{Sut}}{\sigma_{Sf}} M_{ba} \right)^2 + (K_{f\tau} M_{tm})^2 \right]^{1/2} \right\}^{1/3}$$

$$(14\text{--}67)$$

Diameter of hollow shaft:

(c) According to distortion-energy theory of failure using modified Goodman criterion for fatigue strength

$$D_o = \left[\frac{16n}{\pi\sigma_{Sut}(1-K^4)} \left(2K_{f\sigma} \frac{\sigma_{Sut}}{\sigma_{Sf}} M_{ba} \right. \right.$$

$$\left. \left. + \sqrt{3}\, K_{f\tau} M_{tm} \right) \right]^{1/3}$$

$$(14\text{--}68)$$

(d) According to distortion-energy theory of failure combined with Gerber parabolic relation

$$\left(\frac{n\sigma_a}{\sigma_{Sf}} \right) + \left(\frac{n\sigma_m}{\sigma_{Sut}} \right)^2 = 1$$

$$D_o = \frac{16n}{\pi\sigma_{Sut}(1-K^4)} \left\{ K_{f\sigma} \frac{\sigma_{Sut}}{\sigma_{sf}} M_{ba} \right.$$

$$\left. + \left[\left(K_{f\sigma} \frac{\sigma_{Sut}}{\sigma_{Sf}} M_{ba} \right)^2 + 3(K_{f\tau} M_{tm})^2 \right]^{1/2} \right\}^{1/3}$$

$$(14\text{--}69)$$

Particular	Formula

(e) According to distortion-energy theory of failure using ASME elliptic locus for fatigue strength

$$\left(\frac{n\sigma_a}{\sigma_{Sf}}\right)^2 + \left(\frac{n\sigma_m}{\sigma_{Sy}}\right)^2 = 1$$

$$D_o = \left\{\frac{16n}{\pi\sigma_{Sy}(1-K^4)}\left[\left(2K_{f\sigma}\frac{\sigma_{Sy}}{\sigma_{Sf}}M_{ba}\right)^2 + 3(K_{f\tau}M_{tm})^2\right]^{1/2}\right\}^{1/3}$$

$$\tag{14-70}$$

Bagci failure locus equation in quartic (fourth-degree) form

$$\text{i.e., } \frac{n\sigma_a}{\sigma_{Sf}} + \left(\frac{n\sigma_m}{\sigma_{Sy}}\right)^4 = 1$$

and yielding criterion (Langer) equation combined with any theories of failure can be used to predict the fatigue strength of shaft

$$\text{i.e., } \frac{\sigma_a + \sigma_m}{\sigma_{Sy}} = \frac{1}{n}$$

(2) Shaft subjected to fluctuating loads, i.e., reversed bending and reversed torque, taking into consideration stress concentration

(a) The diameter of solid shaft according to maximum shear stress theory of failure using Soderberg criterion for fatigue strength

$$D = \left[\frac{32n}{\pi\sigma_{Sy}}(M_{be}^2 + M_{te}^2)^{1/2}\right]^{1/3} \tag{14-71}$$

where

M_{be} = static equivalent of cyclic bending moment

$$= K_{f\sigma}M_{bm} + K_{f\sigma}\frac{\sigma_{Sy}}{\sigma_{Sf}}M_{ba}$$

M_{te} = static equivalent of cyclic torsional moment

$$= K_{f\tau}M_{te} + K_{f\tau}\frac{\sigma_{Sy}}{\sigma_{Sf}}M_{ta}$$

(b) The diameter of hollow shaft according to distortion-energy theory of failure combined with Soderberg criterion for fatigue strength

$$D_o = \left[\frac{16n}{\pi\sigma_{Sy}(1-K^4)}(4M_{be}^2 + 3M_{te}^2)^{1/2}\right]^{1/3} \tag{14-72}$$

where M_{be} and M_{te} have the same meaning as given under Eq. (14–71)

(3) Shaft subjected to constant bending and torsional moments and reversed torsional and bending moments at the same frequency taking into consideration stress concentration

(a) The diameter of solid shaft according to maximum distortion energy theory of failure using modified Goodman criterion for fatigue strength

$$D = \left(\frac{16n}{\pi\sigma_{Sut}}\left\{[4(K_{f\sigma}M_{bm})^2 + 3(K_{f\tau}M_{tm})^2]^{1/2}\right.\right.$$

$$\left.\left. + \frac{\sigma_{Sut}}{\sigma_{Sf}}\left\{[4(K_{f\sigma}M_{ba})^2 + 3(K_{f\tau}M_{ta})^2]^{1/2}\right\}\right)^{1/3}\right.$$

$$\tag{14-73}$$

Particular	Formula
	where $K_{f\sigma} = K_{f\tau} = 1$ for constant torsional and bending moments
(b) The diameter of solid shaft according to maximum shear stress theory of failure combined with modified Goodman criterion for fatigue strength	$$D = \left\{ \frac{32n}{\pi \sigma_{Sut}} \left[\left(M_{bm} + K_{f\sigma} \frac{\sigma_{Sut}}{\sigma_{Sf}} M_{ba} \right)^2 \right. \right.$$ $$\left. \left. + \left(M_{tm} + K_{f\tau} \frac{\sigma_{Sut}}{\sigma_{Sf}} M_{ta} \right)^2 \right]^{1/2} \right\}^{1/3} \qquad (14\text{–}74)$$
(c) The diameter of hollow shaft according to maximum shear stress theory of failure using Soderberg criterion for fatigue strength	$$D_o = \left\{ \frac{32n}{\pi \sigma_{Sy}(1 - K^4)} \left[\left(K_{f\sigma} M_{bm} + K_{f\sigma} \frac{\sigma_{Sy}}{\sigma_{Sf}} M_{ba} \right)^2 \right. \right.$$ $$\left. \left. + \left(K_{f\tau} M_{tm} + K_{f\tau} \frac{\sigma_{Sy}}{\sigma_{Sf}} M_{ta} \right)^2 \right]^{1/2} \right\}^{1/3} \qquad (14\text{–}75)$$
	where $K_{f\sigma} = K_{f\tau} = 1$ for constant bending and torsional moments
(4) Cyclic axial load combined with reversed bending and torsional moments taking into consideration stress concentration as per *ASME Code for Design of Transmission Shafting*	
(a) The diameter of solid shaft according to maximum shear stress theory of failure and Soderberg relation for fatigue strength	$$D = \left\{ \frac{32n}{\pi \sigma_{Sy}} \left[\left(M_{be} + \frac{F_{ae}D}{8} \right)^2 + M_{te}^2 \right]^{1/2} \right\}^{1/3}$$ $$(14\text{–}76)$$
	where M_{be} and M_{te} have the same meaning as given Eq. (14–71)
	F_{ae} = static equivalent axial load
	$$= K_{f\sigma} F_{am} + K_{f\sigma} \frac{\sigma_{Sy}}{\sigma_{Sf}} F_{aa}$$
(b) The diameter of hollow shaft according to distortion-energy theory of failure combined with modified Goodman relation for fatigue strength	$$D_o = \left[\frac{32n}{\pi \sigma_{Sut}(1 - K^4)} \left\{ \left[M'_{be} + \frac{F'_{ae} D_o(1 + K^2)}{8} \right]^2 \right. \right.$$ $$\left. \left. + \frac{3}{4} M'^2_{te} \right\}^{1/2} \right]^{1/3} \qquad (14\text{–}77)$$

Although ASME has withdrawn the *ASME Code for Design of Transmission Shafting*, some of the ASME equations given here have historic interest and hence are retained in this book.

Particular	Formula

where

$$M'_{be} = K_{f\sigma} M_{bm} + K_{f\sigma} \frac{\sigma_{Sut}}{\sigma_{Sf}} M_{ba}$$

$$M'_{te} = K_{f\tau} M_{tm} + K_{f\tau} \frac{\sigma_{Sut}}{\sigma_{Sf}} M_{ta}$$

$$F'_{ae} = K_{f\sigma} F_{am} + K_{f\sigma} \frac{\sigma_{Sut}}{\sigma_{Sf}} F_{aa}$$

When $K = 0$, this equation reduces to an equation for a solid shaft

The value of α is given by Eq. (14–65)

(5) The diameter of solid shaft subjected to axial, bending, and torsional alternating loads according to distortion-energy theory of failure combined with Soderberg relation for fatigue as per new *ASME Code for Design of Transmission Shafting* [5]

$$D = \left(\frac{32n}{\pi \sigma_{Sf}} \left[\left(M_{ba} + \frac{F_a D}{2} \right)^2 + \frac{3 M_{ta}^2}{4} \right]^{1/2} \right.$$

$$\left. + \left\{ \frac{32n}{\pi \sigma_{Sut}} \left[\left(M_{bm} + \frac{F_m D}{2} \right)^2 + \frac{3 M_{tm}^2}{4} \right]^{1/2} \right\} \right)^{1/3}$$

Not explicit in D, use iterative methods to solve

(6) The diameter of shaft made of brittle material, which is subjected to reversed bending and torsional moments taking into consideration stress concentration as per maximum normal stress theory of failure combined with modified Goodman relation for fatigue strength

$$D = \left\{ \frac{16n}{\pi \sigma_{Sut}} [M'_{be} + (M'^2_{be} + M'^2_{te})^{1/2}] \right\}^{1/3} \qquad (14\text{–}79)$$

for solid shaft

$$D_o = \left\{ \frac{16n}{\pi \sigma_{Sut}(1 - K^4)} [M'_{be} + (M'^2_{be} + M'^2_{te})^{1/2}] \right\}^{1/3}$$

$$(14\text{–}80)$$

for hollow shaft, where M'_{be} and M'_{te} have the same meaning as given under Eq. (14–77)

(7) Shaft subjected to combined axial, bending, and torsional reversed loads taking into consideration stress concentration and shock

  (a) The diameter of hollow shaft according to distortion-energy theory of failure using Soderberg relation

$$D_o = \left(\frac{32n}{\pi \sigma_{Sy}(1 - K^4)} \left\{ K_{Sb} \left[M_{be} + \frac{F_{ae} D_o (1 + K^2)}{8} \right]^2 \right. \right.$$

$$\left. \left. + \frac{3}{4} K_{St} M_{te}^2 \right\}^{1/2} \right)^{1/3} \qquad (14\text{–}81)$$

where F_{ae}, M_{be} and M_{te} have the same meaning as given under Eqs. (14–71) and (14–76)

Refer to Table 14–4 for K_{Sb} and K_{St}

Particular	Formula

GENERAL

See Tables 14–1 to 14–6 and Fig. 14–1 for further
details on shafting design [3]; refer to Table 14–4 for
shock load factors K_{Sb} and K_{St}

For further design details on shafting   Refer to Tables 14–5 to 14–7.

REFERENCES

1. Lingaiah, K., and B. R. Narayana Iyengar, *Machine Design Data Handbook*, Engineering College Co-operative, Bangalore, India, 1962.
2. Lingaiah, K., and B. R. Narayana Iyengar, *Machine Design Data Handbook*, Vol. I, Suma Punblishers, Bangalore, India, 1986.
3. Lingaiah, K., *Machine Design Data Handbook*, Vol. II (*SI Units and Customary Metric Units*), Suma Publishers, Bangalore, India, 1986.
4. Soderberg, C. R., "Working Stresses," *J. Appl. Mechanics*, Vol. 57, p. A–106, 1935.
5. *ASME Code for Design of Transmission Shafting*, Standard ANS/ASME B106.1M, 1985.
6. Shigley, J. E., *Machine Design*, McGraw-Hill Publishing Company, New York, 1956.

BIBLIOGRAPHY

Berchard, H. A., "A Comprehensive Method for Designing Shafts to Insure Fatigue Life,"*Machine Design*, April 25, 1963.
Black, P. H., and O. Eugene Adams, Jr., *Machine Design*, McGraw-Hill Publishing Company, New York, 1983.
British Standards Institution.
Deutschman, A. D., W. J. Michels, and C. E. Wilson, *Machine Design—Theory and Practice*, Macmillan Publishing Company, New York, 1975.
Maleev, V. L., and J. B. Hartman, *Machine Design*, International Textbook Company, Scranton, Pennsylvania, 1954.
Marks' Standard Handbook for Mechanical Engineers, 8th ed., McGraw-Hill Publishing Company, New York, 1978.
Vallance, A., and V. L. Doughtie, *Design of Machine Members,* McGraw-Hill Publishing Company, New York, 1951.

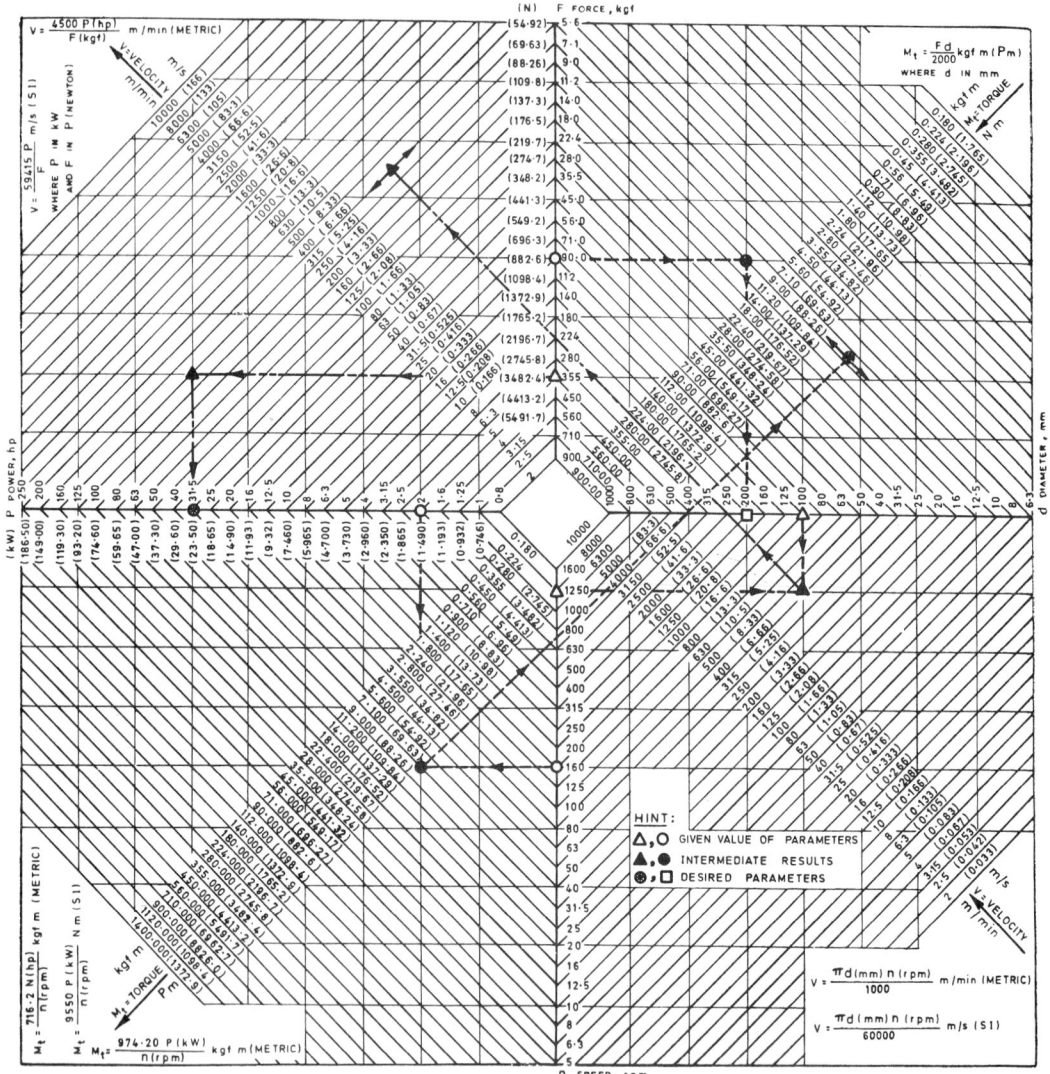

FIGURE 14–1 Nomogram for determining diameter (d), speed (n), force (F), torque (M_t), and power (P) in Customary Metric units and System International units. (K. Lingaiah, *Machine Design Data Handbook*, Vol. II, Suma Publishers, Bangalore, India, 1986.)

TABLE 14–1
Empirical shafting formulas

Kind of service	Load factors considered		Power capacity, P	
	Torsion, K_t	Bending, K_b	kW	hp
Transmission shafts in torsion only	1.0	1.0	$54{,}831 D^3 n'$	$1.225 \times 10^{-6} D^3 n$
Line shafting with limited bending	1.0	1.5	$34{,}532 D^3 n'$	$7.715 \times 10^{-7} D^3 n$
Head or main shafts with heavy bending loads	1.0	2.5	$20{,}715 D^3 n'$	$4.628 \times 10^{-7} D^3 n$

TABLE 14–2
Shock and endurance factors

Nature of loading	K_b	K_t
Stationary shafts		
Gradually applied load	1.0	1.0
Suddenly applied load	1.5–2.0	1.5–2.0
Rotating shafts		
Steady or gradually applied loads	1.5	1.0
Suddenly applied loads, minor shocks only	1.5–2.0	1.0–1.5
Suddenly applied loads, heavy shocks	2.0–3.0	1.5–3.0

TABLE 14–3
Values of constant c

Type of shaft loading	Coefficient c in Eq. (14–61)	Allowable stress	
		MPa	kpsi
Shaft heavily loaded, subjected to shock, or reversed under full load	0.82	17	2.5
Line shafts and countershafts, loaded in bending but not reversed	1.1	27	4.0
Line shafts or bar with pulleys close to the bearings	1.56	44	6.4

TABLE 14–4
Shock load factors[a] for use in Eq. (14–81)

Nature of load	K_{Sb}, K_{St}
Gradually applied load	1.00
Loads applied with minor shocks	1.0–1.5
Loads applied with heavy shocks	1.5–2.0

[a] Data from Berchard, H. A., "A Comprehensive Method for Designing Shafts to Insure Fatigue Life," *Machine Design*, April 25, 1963.

TABLE 14–5
Spacing[a] for line shaft bearings

Diameter of shaft, mm	Transmission shaft stressed in torsion only, mm		Line shaft carrying pulleys or gears and subjected to usual bending loads, mm	
	1–250 rpm	251–400 rpm	1–250 rpm	251–400 rpm
36.5	274.5	244.0	213.5	198.0
49.0	305.0	274.5	229.0	213.5
62.0	335.5	305.0	244.0	228.5
74.5	366.0	335.5	259.0	244.0
87.5	396.0	366.0	274.5	259.0
100.0	427.0	396.0	289.5	274.5
112.5	457.0	427.0	305.0	289.5

[a] Center-to-center distance in millimeters.

TABLE 14–6
Sizes of shafts

Diameters, mm (in)					
4 (0.16)	12 (0.48)	40 (1.6)	75 (3.0)	110 (4.4)	180 (7.2)
5 (0.20)	15 (0.60)	45 (1.8)	80 (3.2)	120 (4.8)	190 (7.6)
6 (0.24)	17 (0.68)	50 (2.0)	85 (3.4)	130 (5.2)	200 (8.0)
7 (0.28)	20 (0.80)	55 (2.2)	90 (3.6)	140 (5.6)	220 (8.8)
8 (0.32)	25 (1.0)	60 (2.4)	95 (3.8)	150 (6.0)	240 (9.6)
9 (0.36)	30 (1.2)	65 (2.6)	100 (4.0)	160 (6.4)	260 (10.4)
10 (0.4)	35 (1.4)	70 (2.8)	105 (4.2)	170 (6.8)	280 (11.2)

J. E. Shigley, *Machine Design*, McGraw-Hill Publishing Company, New York, 1956.

TABLE 14–7
Load factors for various machines, k_l [a]

Driver	Driven machinery	Factor, k_l
Steam turbine	Electric generator, steady load; turbine blower	1.00
	Electric generator, uneven load; centrifugal pump	1.25
	Induced-draft fan; line shaft; gear drive	1.50
	Rolling mill, gear drive	2.00
Electric motor	Turbine blower; metalworking machinery	1.25
	Centrifugal pump; wood working machinery	1.50
	Line shaft; ship propeller; double acting pump	1.75
	Triplex single-acting pump; elevator; crane	1.75
	Compressor, air or ammonia	1.75
	Rolling mill; rubber mill	2.50
Steam engine	Values for electric-motor drive multiplied by 1.2–1.5	
Gas and oil engines	Values for electric-motor drive multiplied by 1.3–1.6 the factor depending on the coefficient of steadiness of the flywheel	

[a] To be used also in Eqs. (5–9) and (19–79).

CHAPTER
15

FLYWHEELS

SYMBOLS

a	major axis of ellipse, m (in)
	negative acceleration or deceleration, m/s^2
A	cross-sectional area of the rim, m^2 (in^2)
b	minor axis of ellipse, m (in)
	width of rim, m (in)
C_f	coefficient of fluctuation of rotation
d	diameter of shaft, m (in)
d_h	hub diameter, m (in)
D	flywheel diameter, m (in)
D_o	outside diameter of rim, m (in)
E	excess energy, J (ft lbf)
F_c	centrifugal force, kN (lbf)
F_c'	centrifugal force per unit width of rim, kN (lbf)
g	acceleration due to gravity, 9.8066 m/s^2 (32.2 ft/s^2)
h	depth of rim, m (in)
i	number of arms
k_o	polar radius of gyration of the rim, m (in)
I	mass moment of inertia, N s^2 m (lbf s^2 ft)
J	polar second moment of inertia, m^4 (in^4)
k_t	torsional stiffness of shaft, N m/rad (lbf in/rad)
M_{tm}	mean torque, N m (lbf ft)
M_t	transmitted torque, N m (lbf ft)
m	coefficient of steadiness
n	mean speed, rpm
n_1	maximum speed, rpm
n_2	minimum speed, rpm
r	mean radius of the flywheel, m (in)
t	time, s
T_1	tension in belt on tight side, kN (lbf)
T_2	tension in belt on slack side, kN (lbf)
v	mean rim velocity, m/s (ft/min)
v_1	maximum rim velocity, m/s (ft/min)
v_2	minimum rim velocity, m/s (ft/min)
W	rim weight, kN (lbf)
ρ	specific weight of material or weight density, N/m^3 (lbf/in^3)
Z	sectional modulus of the arm cross section at the hub, m^3 (in^3)
σ	stress (also with subscripts), MPa (psi)

θ_1, θ_2	maximum and minimum angular displacement of flywheel from constant speed deviation, rad (deg)
ω	average angular speed, rad/s
ω_1, ω_2	maximum and minimum angular speed, respectively, rad/s

Particular	Formula
The equation of motion of ith rotor of I_i inertia in a multirotor system connected by $(i-1)$ number of shafts of various inertias subjected to external torque	$I_i\theta_i = M_{ti} - M_{t(i-1)}$ $\qquad$ (15–1)
The equation of motion of a flywheel, which is mounted on a shaft between two supports and rotates with an angular velocity and subjected to an input external torque M_{ti}	$I\theta = M_{ti} - M_{to} = k_t(\theta_2 - \theta_1)$ $\qquad$ (15–2) where M_{to} = output torque, N m (lbf ft) θ = angular displacement of flywheel, rad (deg)

KINETIC ENERGY

Kinetic energy (Fig. 15–1)	$K = \dfrac{1}{2}\,mv^2 = \dfrac{Wv^2}{2g} = \dfrac{1}{2}\,I\omega^2$ $\qquad$ (15–3)
For variation of torque with crank angle for two-cylinder engine	Refer to Fig. 15–1.

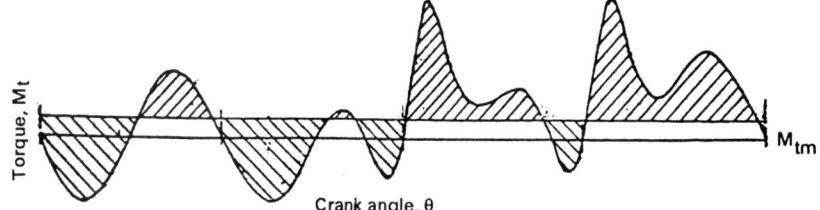

FIGURE 15–1   Torque-crank shaft angle curve for a two-cylinder engine.

The kinetic energy of flywheel at an angular displacement θ_1 and at angular velocity ω_1 during one cycle	$K_1 = \dfrac{1}{2}\,I\omega_1^2 = \dfrac{Wv_1^2}{2g}$ $\qquad$ (15–4)
The kinetic energy of flywheel at an angular displacement θ_2 and at angular velocity ω_2	$K_2 = \dfrac{1}{2}\,I\omega_2^2 = \dfrac{Wv_2^2}{2g}$ $\qquad$ (15–5)
The change in kinetic energy or energy fluctuation due to change in angular velocity ω_1 to ω_2 in one cycle	$E = K_2 - K_1 = \dfrac{1}{2}\,I(\omega_2^2 - \omega_1^2) = \dfrac{W(v_2^2 - v_1^2)}{2g}$ $\qquad$ (15–6) $= \tfrac{1}{2}\,I(\omega_2 - \omega_1)(\omega_2 + \omega_1)$ $= I(\omega_2 - \omega_1)\omega = W(v_2 - v_1)\,\dfrac{v}{g}$

Particular	Formula	
The coefficient of fluctuation of speed or rotation	$C_f = \dfrac{\omega_2 - \omega_1}{\omega} = \dfrac{v_2 - v_1}{v} = \dfrac{n_2 - n_1}{n}$	(15–7)
The change in kinetic energy or excess energy	$E = K_2 - K_1 = I\omega^2 C_f = \dfrac{Wv^2 C_f}{g}$	(15–8)

FLYWHEEL EFFECT OR POLAR MOMENT OF INERTIA

$$Wk^2 = \frac{182.40gE}{n_2^2 - n_1^2} \qquad (15\text{–}9)$$

The mean angular velocity	$\omega = \dfrac{\omega_2 + \omega_1}{2}$	(15–10)
The coefficient of steadiness	$m = \dfrac{1}{C_f}$	(15–11)

Refer to Table 15–1 for C_f.

STRESSES IN RIM (FIG. 15–2)

Particular	Formula	
The component of the centrifugal force normal to any diameter of the flywheel	$F_c = \dfrac{2\rho bhr^2 \omega^2}{g}$	(15–12)

TABLE 15–1
Coefficient of fluctuation of rotation, C_f

Driven machine	Type of drive	C_f
AC generators, single or parallel	Direct-coupled	0.01
AC generators, single or parallel	Belt	0.0167
DC generators, single or parallel	Direct-coupled	0.0143
DC generators, single or parallel	Belt	0.029
Spinning machinery	Belt	0.02 to 0.015
Compressure, pumps	Gears	0.02
Paper, textiles, and flour mills	Belt	0.025–0.02
Woodworking and metalworking machinery	Belt	0.0333
Shears and pumps	Flexible coupling	0.05–0.04
Concrete mixers, excavators, and compressors	Belt	0.143–0.1
Crushers, hammers, and punch presses	Belt	0.2

Particular	Formula	
The tangential force due to hoop stress in the flywheel rim (Fig. 15–3)	$F_\theta = \dfrac{\rho bhr^2 \omega^2}{g}$	(15–13)
The tensile stress created in each cross section of the rim by the centrifugal force	$\sigma = 0.01095\, \dfrac{\rho}{g}\, r^2 n^2$ **SI**	(15–14)

Particular	Formula		
The centrifugal force per unit width of rim (Fig. 15–3)	$F'_c = 0.01095 \dfrac{\rho r^2 n^2 h}{g}$	**SI**	(15–15)
The bending stress	$\sigma_b = 0.2146 \dfrac{\rho r^3 n^2}{ghi^2}$	**SI**	(15–16)
The combined tensile stress	$\sigma_R = 0.75\sigma + 0.25\sigma_b$		(15–17)

STRESSES IN ARMS (FIG. 15–2)

The stresses in the arm	$\sigma_1 = \dfrac{M_t(D - d_h)}{iZD}$	(15–18)

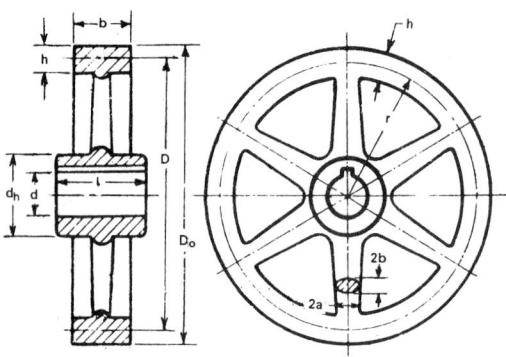

FIGURE 15–2 Flywheel.

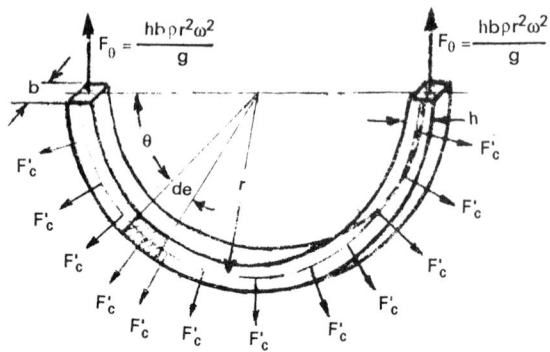

FIGURE 15–3 Centrifugal force acting on the rim of a flywheel.

When the flywheel is used as a belt pulley, the stresses at the hub	$\sigma_2 = \dfrac{(T_1 - T_2)(D - d_h)}{2iZ}$	(15–19)

Particular	Formula	
In case of thin-rim flywheel, the stress	$\sigma_2' = \dfrac{(T_1 - T_2)(D - d_h)}{iZ}$	(15–20)
Stress due to centrifugal force	$\sigma_3 = 0.01095 \dfrac{\rho r^2 n^2}{g}$ **SI**	(15–21)
The maximum tensile stress in an arm is at hub	$\sigma_{max} = \sigma_1 + \sigma_2 + \sigma_3$	(15–22)
The force necessary to stop the flywheel	$F = \dfrac{Wa}{g}$	(15–23)

RIM DIMENSIONS (FIG. 15–2)

The relation between k_o in cm and the outside diameter D of the rim in m	$k_o^2 = 0.125[D_o^2 + (D_o - 2h)^2]$	(15–24)
Cross-sectional area of the rim	$A = \dfrac{W}{2\pi k \rho}$	(15–25)
The relation between depth and width of rim	$\dfrac{b}{h} = 0.65 \text{ to } 2$	(15–26)
The outside diameter of rim	$D_o = 2k_o + h$ (approx.)	(15–27)
The hub diameter in m	$d_h = 1\frac{3}{4}d + 6.35 \times 10^{-3}$	(15–28)
	$= 2d$	
The hub length	$l = 2d \text{ to } 2.5d$	(15–29)

ARMS (FIG. 15–2)

The major axis in case of elliptical section can be computed from the relation	$a = \sqrt[3]{\dfrac{64Z}{\pi}}$	(15–30)
	where $Z = \dfrac{\pi b a^2}{32}$ and $a = 2b$	(15–31)

PACKINGS AND SEALS

SYMBOLS

A	area of seal in contact with the sliding member, m^2 (in^2)
A_g	gasket area over which the bolt loads are distributed, m^2 (in^2)
A_1, A_2	area of cross section of unthreaded and threaded portions of bolt, m^2 (in^2)
b	width of U-collar, m (in)
	gland width or depth of groove, m (in)
c	radial clearance between rod and the bushing, radial deflection of the ring, m (in)
d	nominal diameter of the bolt, m (in)
	diameter of sliding member, m (in)
d_1	outside diameter of packing material, m (mm)
	outside diameter of seal ring (Fig. 16–3), m (in)
d_2	minor diameter of bolt, m (in)
d_a	actual diameter of wire, m (in)
d_i	inside diameter of packing material, m (in)
D_m	estimated mean diameter of conical spring, m (in)
D_{am}	actual mean diameter of conical spring, m (in)
E	modulus of elasticity, GPa (psi)
F_b	bolt load, kN (lbf)
F_μ	frictional force, kN (lbf)
$F_{\mu o}$	frictional force of the stuffing box when there is no fluid pressure, kN (lbf)
g	acceleration due to gravity, 9.80 66 m/s^2 (9806.6 mm/s^2) (32.2 ft/s^2)
h	radial ring wall thickness, m (in)
h_i	uncompressed gasket thickness, m (in)
h_μ	loss of head, m/m (in/in)
i	number of bolts
l	depth of U-collar (Fig. 16–2a), m (in)
l_1, l_2	length of joint, m (in)
(dl)	incremental length in the direction of velocity [Eq. (16–15)], m (in)
	bolt elongation [Eq. (16–24)], m (in)
M_t	twisting moment, N m (lbf in)
M_{ti}	initial bolt torque, N m (lbf in)
p	fluid pressure, MPa (psi)
p_f	flange pressure on the gasket, MPa (psi)

P_s	minimum per cent compression to seal
(dp)	pressure differential in the direction of velocity [Eq. (16–15)], MPa (psi)
Q	discharge, m^3/s (cm^3/s, mm^3/s) (in^3/s)
r	equivalent radius, m (in)
v	velocity, m/s (ft/min)
w	nominal packing cross section, m (in)
y	deflection of spring, m (in)
η	absolute viscosity of fluid, Pa s (cP)
σ_d	design stress, MPa (psi)
μ	coefficient of friction

Particular	Formula

ELASTIC PACKING [1-3]

Frictional force exerted by a soft packing on the reciprocating rod

$$F_\mu = kpd \tag{16–1}$$

where
k = 0.005 and p = 0.343 MPa   **SI**
k = 0.2 and p = 50 psi   **US Customary System units**

FRICTION

Friction resistance

$$F_\mu = F_o + \mu A p \tag{16–2}$$

where
μ = 0.01 for rubber and soft lubricated leather
  = 0.15 for hard leather

Torsional resistance in a rotary motion friction

$$M_t = \frac{F_\mu d}{2} = \frac{kd^2 p}{2} \tag{16–3}$$

where
k = 0.005   **SI**
  = 0.2   **US Customary System units**

METALLIC GASKETS (FIG. 16–1)

The empirical relations [3]

$$c = 0.2d + 5 \text{ mm if } d \le 100 \text{ mm} \tag{16–4}$$
$$c = 0.08\sqrt{d} \qquad \text{if } d > 0.1 \text{ mm} \quad \textbf{SI} \tag{16–5a}$$
$$= 0.5\sqrt{d} \qquad \text{if } d > 4$$
$$\text{\textbf{US Customary System units}} \tag{16–5b}$$

$$h = \frac{d}{8} + 12.54 \text{ mm or } 0.5 \text{ in} \tag{16–6}$$

$$a = d + 2c \tag{16–7}$$
$$\alpha = 10° \text{ to } 15° \tag{16–8}$$
$$d_2 = 0.2(d + 0.102)/\sqrt{i} \quad \textbf{SI} \tag{16–9a}$$
$$= 0.2(d + 4)/\sqrt{i}$$
$$\text{\textbf{US Customary System units}} \tag{16–9b}$$

Particular	Formula

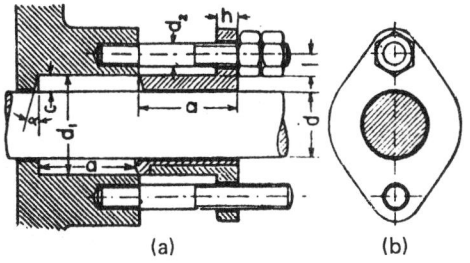

FIGURE 16–1   Stuffing box with bolted gland. (*V. L. Maleev and J. B. Hartman*, Machine Design, *International Textbook Company, Scranton, Pennsylvania, 1954.*)

Diameter of bolt is also found by equating the working strength of the bolts to the pressure p exerted by the fluid on the gland and the frictional force F_μ

$$d_2 = \sqrt{\frac{(d_1^2 - d^2)p}{i\sigma_d} + \frac{4F_\mu}{\pi i \sigma_d}} \qquad (16\text{–}10)$$

where
$d_2 = $ minor diameter of bolt, m (in)
$\sigma_d = $ 68.7 to 83.3 MPa (10 to 12 kpsi)

SELF-SEALING PACKING (FIG. 16–2)

Houghton, Welch, and Jenkin's formula for an approximate thickness of a U-shaped collar for great pressure [3]

$$h = 6.36 \times 10^{-3} d^{0.2} \qquad \textbf{SI} \qquad (16\text{–}11a)$$

where h and d in m
$$= 1.6 d^{0.2} \qquad \textbf{SI} \qquad (16\text{–}11b)$$

where h and d in mm
$$= 0.12 d^{0.2} \qquad \textbf{US Customary System units} \qquad (16\text{–}11c)$$
where h and d in in

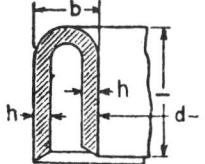

FIGURE 16–2   U-collar.

Width

$$b = 4h \qquad (16\text{–}12a)$$

Depth

$$l = 1.2b \text{ to } 1.8b \qquad (16\text{–}12b)$$

Particular	Formula

PACKINGLESS SEALS

Leakage of the fluid past a rod can be computed with fair accuracy by the formula

$$Q = \frac{\pi c^3}{12}(p_1 - p_2)\frac{d}{l\eta} \quad \textbf{SI} \tag{16–13a}$$

$$= 1.79(100c)^3 \frac{(p_1 - p_2)d}{l\eta} \tag{16–13b}$$

US Customary System units

Refer to Table 16–1 for values of η.

TABLE 16–1
Absolute viscosities η

Fluid	Temperature		Absolute viscosity, η		Temperature		Absolute viscosity, η	
	K	**°C**	**MPa s**	**cP**	**K**	**°C**	**MPa s**	**cP**
Steam	293	20	0.0097	0.0097	539	266	0.018	0.018
Air	293	20	0.018	0.018	366	93	0.022	0.022
Water	273	0	1.79	1.79	311	38	0.69	0.69
Water	293	20	1.0	1.0	333	60	0.40	0.40
Gasoline	293	20	0.6	0.6	355	82	0.30	0.30
Kerosene	293	20	2.7	2.7	355	82	1.30	1.30
Fuel oil, 30° Baumé	293	20	5.0	5.0	355	82	1.60	1.60
Fuel oil, 24° Baumé	293	20	40	40	355	82	4	4
Spindle oil	293	20	20–35	20–35	355	82	3–4	3–4
Machine oil	293	20	200–500	200–500	372	99	1.5–16	5.5–16
Castor oil	293	20	1000	1000	316	43	200	200

STRAIGHT-CUT SEALINGS (FIG. 16–3a)

The equation for loss of liquid head

$$h_\mu = 64\eta v / 2g\rho d_1^2 \tag{16–14}$$

Leakage velocity

$$v = \frac{(dp)r^2}{8(dl)\eta} \tag{16–15}$$

Quantity of leakage

$$Q = vA \tag{16–16}$$

Stress in a seal ring

$$\sigma = \frac{0.4815cE}{h\left(\dfrac{d_1}{h} - 1\right)^2} \tag{16–17}$$

For allowable temperatures for materials and surface treatment

Refer to Table 16–2.

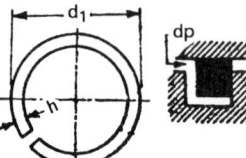

FIGURE 16–3(a) Straight-cut seal.

Particular	Formula

V-RING PACKING
Single-spring installations

The estimated mean diameter of conical spring

$$D_m = d_i + \frac{3w}{2} \tag{16–18}$$

The wire size (Table 16–3)

$$d = \left(\frac{\pi D_m^2}{139300}\right)^{1/3} \quad \textbf{SI} \tag{16–19a}$$

where d and D_m in m

$$= \left(\frac{\pi D_m^2}{3535}\right)^{1/3} \quad \textbf{US Customary System units} \tag{16–19b}$$

where d and D_m in in

$$= \left(\frac{\pi D_m^2}{193.3}\right)^{1/3} \quad \textbf{Metric} \tag{16–19c}$$

where d and D_m in mm

The actual mean diameter of conical spring $D_{am} = d_1 - \frac{1}{2}(w + d_a) \tag{16–20}$

The deflection of spring

$$y = \frac{0.0123 D_{am}^2}{d_a} \tag{16–21}$$

Two standard cylindrical spring sizes are generally used, depending on packing size.

Multiple-spring installations

BOLTS AND STRESSES IN FLANGE JOINTS. The bolt load in gasket joint

$$F_b = \frac{11 m_{ti}}{d} \tag{16–22}$$

The flange pressure developed due to tightening of bolts that hold the gasket joint mechancial assembly together

$$p_f = \frac{i F_b}{A_g C_u} = \frac{2i M_t}{A_g C_u d_b} \tag{16–23}$$

c_u = torque friction coefficient

The load on the bolt when it is tightened

$$F_b = \frac{E(dl)}{(l_1/A_1) + (l_2/A_2)} \tag{16–24}$$

STRESSES IN GROOVED JOINTS. The uncompressed gasket thickness that will provide the minimum sealing compression when the flanges are tightened into face-to-face contact

$$h_i = \frac{100b}{100 - P_s} \tag{16–25}$$

Particular	Formula

BOLT LOADS IN GASKET JOINT ACCORDING TO *ASME BOILER AND PRESSURE VESSEL CODE* (FIG. 16–3b) [4]

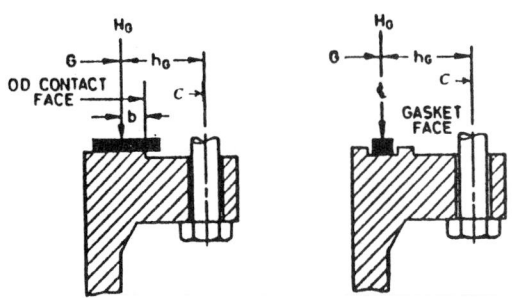

For $b_o > 6.3$ mm (0.25 in.)   For b_o 6.3 $\leqslant$ mm(0.25 in.)

Effective gasket seating width $b = b_o$, when $b_o = 6.3$ mm (0.25 in.) and $= 2.5\sqrt{b_o}$, when $b_o > 6.3$ mm(0.25 in.)

Note — The gasket factors listed only apply to flanged joints in which the gasket is contained entirely within the inner edges of the bolt holes.

FIGURE 16–3(b)   Location of gasket load reaction.

The required bolt load under operating condition sufficient to contain the hydrostatic end force and simultaneously to maintain adequate compression on the gasket to ensure seating

$$W_{m1} = H + H_P = \pi/4G^2P + 2b\pi GmP \qquad (16\text{–}26)$$

The required initial bolt load to seat the gasket joint-contact surface properly at atmospheric temperature condition without internal pressure

$$W_{m2} = \pi bGy \qquad (16\text{–}27)$$

Refer to Tables 8–20 and 8–21 for gasket factor m and minimum design seating stress y, b, and b_o

Total required cross-sectional area of bolts at the root of thread

$$A_m > A_{m1} \text{ or } A_{m2} \qquad (16\text{–}28)$$

Total cross-sectional area of bolt at root of thread or section of least diameter under stress required for the operating condition

$$A_{m1} = \frac{W_{m1}}{\sigma_{sbd}} \qquad (16\text{–}29)$$

Refer to Table 8–17 for σ_{sbd}.

Total cross-sectional area of bolt at root of thread or section of least diameter under stress required for gasket seating

$$A_{m2} = \frac{W_{m2}}{\sigma_{sbat}} \qquad (16\text{–}30)$$

The actual cross-sectional area of bolts using the root diameter of thread or least diameter of unthreaded portion (if less), to prevent damage to the gasket during bolting-up

$$A_b = \frac{2\pi yGN}{\sigma_{sbat}} \nleqslant A_m \qquad (16\text{–}31)$$

Particular	Formula

FLANGE DESIGN BOLT LOAD W

The bolt load in the design of flange for operating condition

$$W = W_{m1} \tag{16–32}$$

The bolt load in the design of flange for gasket seating

$$W = \left(\frac{A_m + A_b}{2}\right)\sigma_{sbat} \tag{16–33}$$

The relation between bolt load per bolt (W_b), diameter of bolt (D) and torque (M_t)

$W_b = 0.17DM_t$ for lubricated bolts $\qquad$ (16–34)
 US Customary System units

(*Note*: The meanings of symbols given in Eqs. (16–26) to (16–37) are defined in Chap. 8.)

where W_b in lbf, D in in, M_t in lbf in

$\quad = 263.5DM_t \qquad$ **SI** $\qquad$ (16–35)

where W_b in N, D in m, M_t in N m
$\quad = 0.2DM_t$ for unlubricated bolts $\qquad$ (16–36)
 US Customary System units

where W_b in lbf, D in in, M_t in lbf in
$\quad = 310DM_t \qquad$ **SI** $\qquad$ (16–37)

where W_b in N, D in m, M_t in N m

For location of gasket load reaction due to tightening of flange bolts

Refer to Fig. 16–3*b*.

The total load on bolts in the gasket joint according to Whalen [5]

$$F_b = \sigma_g A_g \tag{16–38}$$

where
$A_g =$ contact area of gasket, m^2 (in^2)
$\sigma_g =$ gasket seating stress, MPa (psi), taken from Table 16–35

The load on bolts, which is based on hydrostatic end force

$$F_b = nP_t A_m \tag{16–39}$$

where
$P_t =$ test pressure or internal pressure if no test pressure is available, MPa (psi)
$A_m =$ hydrostatic area (based on mean diameter of gasket) on which internal pressure acts, m^2 (in^2)
$n =$ factor of safety taken from Table 16–36

For more information on design data, selection of packing and seals, properties of sealants and packing materials, dimensions and tolerances of seals, and chamfers on shaft, operating temperatures of various types of seals, data for metallic o-rings, q-rings and o-ring gaskets, static and dynamic seals, lip seals, and safety factors, etc.

Refer to Tables 16–4 to 16–36.

Particular	Formula

Leakage through bush seals (Fig. 16–3c)

The oil flow (Q) through plain axial bush seal due to leakage under laminar flow condition, Fig. 16–3c, panel a

$$Q = \frac{2\pi a(P_s - P_a)}{l} \cdot q \qquad (16\text{–}40)$$

where Q in m³/s (in³/s)
v = absolute viscosity, Pa s (cP)

The symbols used in Eqs. (16–40) to (16–45) have the meaning as defined in Fig. 16–3c, panels a and b.

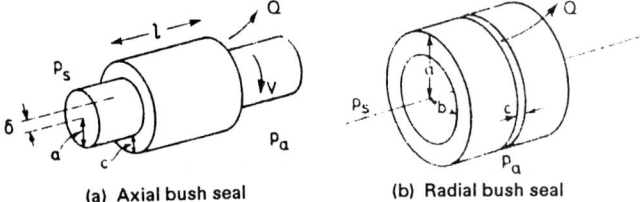

(a) Axial bush seal	(b) Radial bush seal

FIGURE 16–3(c)   Plain bush seals. (*Panels* a *and* b *courtesy of J. M. Neale*, Tribology Handbook, *Butterworths, London, 1973.*)

The volumetric flow rate per unit pressure per unit periphery (q) under laminar flow condition for axial bush seal, Fig. 16–3c, panel a

$$q = \frac{c^3}{12\eta} \cdot (1 + 1.5\varepsilon^2)^* \qquad (16\text{–}41)$$

for incompressible fluid

where $\left(\varepsilon = \dfrac{\delta}{c}\right)$

$$q = \frac{c^3}{24\eta} \cdot \frac{(P_s + P_a)}{P_a} \qquad (16\text{–}42)$$

for compressible fluid[†]

The oil flow (Q) through plain radial bush seal due to leakage under laminar flow condition, Fig. 16–3c, panel b

$$Q = \frac{2\pi a(P_s - P_a)}{(a - b)} \cdot q \qquad (16\text{–}43)$$

The volumetric flow rate per unit pressure per unit periphery (q) under laminar flow condition for radial bush seal, Fig. 16–3c, panel b

$$q = \frac{c^3}{12\eta} \cdot \frac{(a - b)}{a \log_e \dfrac{a}{b}} \qquad (16\text{–}44)$$

for incompressible fluid

$$q = \frac{c^3}{24\eta} \cdot \frac{(a - b)}{a} \cdot \frac{(P_s + P_a)}{P_a} \qquad (16\text{–}45)$$

for compressible fluid[†]

---

*If shaft rotates, onset of Taylor vortices limits validity of formula to $\dfrac{V_c}{v}\sqrt{\dfrac{c}{a}} < 41.3$ (where v = kinematic viscosity).

[†] For Mach number < 1.0, i.e., fluid velocity < local velocity of sound.

Particular	Formula
The radial pressure distribution for laminar flow condition between smooth parallel surfaces in case of face seal	$$p - p_1 = \frac{3\rho\omega^2}{20g}(r^2 - R_1^2) - \frac{6v}{\pi h^3}\ln\frac{r}{R} \qquad (16\text{–}46)$$

where
p = pressure at radial position r, (lb/in^2) MPa
p_1 = pressure at seal inside radius, MPa (psi)
p_2 = internal hydraulic pressure (lbf/in^2), MPa
r = radial position (in) m
v = kinematic viscosity (lb s/in^2) N s/m^2
ρ = fluid density, lb/in^3 (kg/mm^3)
ω = rotational speed, rad/s
R_1 = inside radius of rotating member, m (in)
R_2 = outside radius of rotating member, m (in)
h = thickness of fluid between members, m (in)

| The amount of leakage of fluid through face seal | $$Q = \frac{\pi h^3}{6v\ln(R^2/R_1)}\left[\frac{3\rho\omega^2}{20g}(R_2^2 - R_1^2) - p_2 - p_1\right]$$ $$(16\text{–}47)$$ |

where Q = volumetric leakage rate of fluid, m^3/s (in^3/s)

| The theoretical equation for zero leakage of fluid through face seal | $$p_2 - p_1 = \frac{3}{20}\rho\omega^2(R_2^2 - R_1^2) \qquad (16\text{–}48)$$ |

| The power loss or consumed due to leakage of fluid through face seal | $$P = \frac{\pi v\omega^2}{13,200h}(R_2^4 - R_1^4) \qquad (16\text{–}49)$$ |

where P = power loss, hp

| The shape factor (S_{pf}) for a circular or annular gasket which is the ratio of the area of one load face to the area free to bulge [6] | $$S_{pf} = \frac{D_o - D_i}{4h} \qquad (16\text{–}50)$$ |

where
D_o = outside diameter of gasket, m (in)
D_i = inside diameter of gasket, m (in)

For further design and selection of various types of seals, packings and gaskets, etc.	Refer to Figs. 16–4 to 16–14.
For nomenclature of gasketed joint	Refer to Fig. 16–15.
For packing assembly for a mechanical piston rod	Refer to Fig. 16–16.
For shape factor for various gasket materials [6]	Refer to Fig. 16–17.
For power absorption and starting torque for unbalanced and balanced seals	Refer to Fig. 16–18.

FIGURE 16–4 Single radial lip seal.

FIGURE 16–5 Exclusion seal.

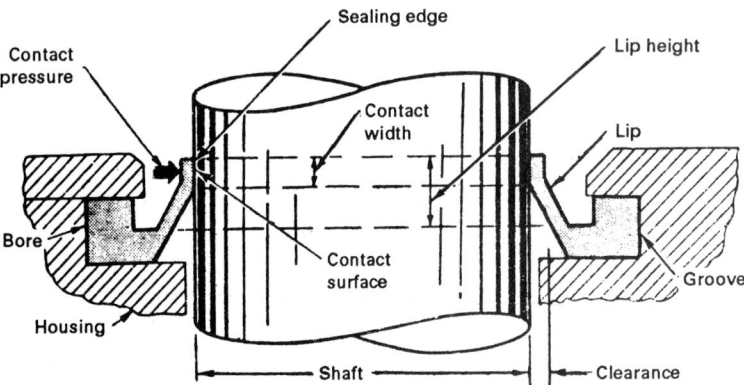

FIGURE 16–6 Radial exclusion seal. (*Produced from "Packings and Seals" Issue*, Machine Design, *Jan. 20, 1977.*)

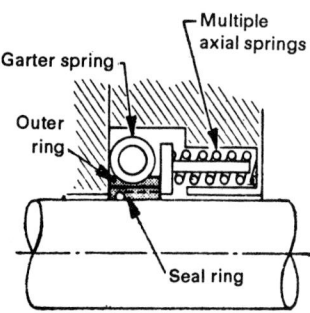

FIGURE 16–7 Two-piece rod seal. (*Produced from "Packings and Seals" Issue*, Machine Design, *Jan. 20, 1977.*)

FIGURE 16–8 Clearance seal idealized labyrinth.

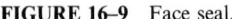

Pressed
(a) For rotating heads

Clamped
(b) For stationary heads

FIGURE 16–9 Face seal.

Conical rings, bevel end

FIGURE 16–10 Compression packing.

TABLE 16–2
Allowable temperatures for materials and surface treatments

Material or surface treatment	Temperature °F	Temperature °C	Material or surface treatment	Temperature °F	Temperature °C
Material			Carbon (high-temperature)	950	510
Low-alloy gray irons	650	343	K–30 (filled teflons)	450–500	232–260
Malleable iron	720	382	S-Monel	950	510
Ductile iron	720	382	Polymide	750	399
Ni-Resist	800	427	Surface treatment		
Ductile Ni-Resist	1000	538	Chromium plate	500	260
410 Stainless Steel	900	482	Tin plate	720	382
17–4 PH Stainless Steel	900	482	Silver plate	600	315.5
Bronze	500	260	Cadmium nickel plate	1000	538
Stellite no. 31	1200	649	Flame plate LW1	1000	538
Inconel X	1200	649	Flame plate LC-1A	1600	871
Tool steel, Rc 62–65	900	482	Flame plate LA-2	1600	871

TABLE 16–3
Standard wire sizes for V-packing expanders

Wire gauge[a]	Wire diameter, mm	Wire gauge	Wire diameter, mm	Wire gauge	Wire diameter, mm
19	1.04	13	2.31	$5/32$	3.82
18	1.20	12	2.67	8	4.11
17	1.37	11	2.05	7	4.49
16	1.57	$\frac{1}{8}$	3.17	$3/16$	4.77
15	1.83	10	3.31	6	4.89
14	2.03	9	3.60	5	5.25

[a] American Wire Gauge (AWG).

TABLE 16–4
Dimensions (in mm) for chamfer on the shaft for mounting the seals

d_1 h11	d_3	d_1 h11	d_3	d_1 h11	d_3	d_1 h11	d_3	d_1 h11	d_3	d_1 h11	d_3
6	4.8	24	21.5	52	48.3	85	80.4	160	153	340	329
7	5.7	25	22.5	55	51.3	90	85.3	170	163	360	349
8	6.6	26	23.4	56	52.3	95	90.1	180	173	380	369
9	7.5	28	25.3	58	54.2	100	95.0	190	183	400	389
10	8.4	30	27.3	60	56.1	105	99.9	200	193	420	409
11	9.3	32	29.2	62	58.1	110	104.7	210	203	440	429
12	10.2	35	32.0	63	59.1	115	109.6	220	213	460	349
14	12.1	36	33.0	65	61.0	120	114.5	230	223	480	469
15	13.1	38	34.9	68	63.9	125	119.4	240	233	500	419
16	14.0	40	36.8	70	65.8	130	124.3	250	243		
17	14.9	42	38.7	72	67.7	135	129.2	260	252		
18	15.1	45	41.6	75	70.7	140	133.0	280	269		
20	17.7	48	44.5	78	73.6	146	138.0	300	289		
22	19.6	50	46.4	80	75.5	150	143.0	320	309		

TABLE 16–5
Selection of guide for packing materials

Condition	Leather (natural and synthetic)	Homogeneous	Fabricated
Oil	Good	Good	Good
Air	Good	Good	Good
Water	Good	Good	Good
Steam	Not recommended	Good	Good
Solvents	Not recommended	Good	Good
Acids	Not recommended	Good	Good
Alkalies	Not recommended	Good	Fair
Temperature range	$-55°C +82°C^a$	$-55°C +200°C^a$	$-40°C +260°C^a$
Types of metal	Ferrous and nonferrous	Chrome-plated steel and nonferrous alloys with hard, smooth surfaces	Chrome-plated steel and nonferrous alloys with hard, smooth surfaces
Metal finish, rms (max.)	63	16	32
Clearances	Medium	Very close	Close
Extrusions or cold flow	Good	Poor	Fair
Friction coefficient	Low	Medium and high	Medium
Resistance to abrasion	Good	Fair	Fair
Maximum pressure, MPa (kpsi)	861.7 (125)	343.4 (50)	549.4 (80)
Concentricity	Medium	Very close	Close
Side loads	Fair	Poor	Fair
High shock loads	Good	Poor to fair	Fair

[a] Depending on specification or combination of materials.

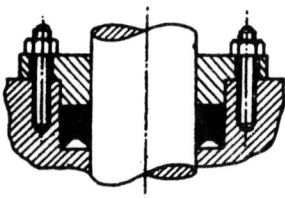

FIGURE 16–11 Molded packing. Typical U-ring packing.

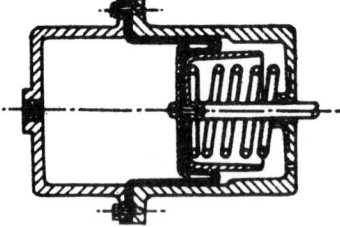

FIGURE 16–12 Diaphragm seals-rolling diaphragm.

TABLE 16–6
Types of seals and their uses

Type	Uses
Radial lip seals	For retaining lubricants in equipments having rotating, reciprocating oscillating shafts, to exclude foreign matter
Single lip (Fig. 16–4)	For containing highly viscous materials at low speeds
Single lip—spring-loaded	For containing lubricants of lower viscosity at higher speeds in clean atmosphere
Double lip with one lip spring-loaded	For excluding contaminants such as dust and dirt
Dual lip with both lips spring-loaded	For containing lubricant on one side and for excluding fluid on the other
Split seal	For splash system of lubrication
External seal	For fixed shaft and rotating bore
Hydrodynamic seal	For directing oil flow back into the area to be sealed
Exclusion seals (Figs. 16–5 and 16–6)	
Wipers, scrapers, axial seals, bellows, and boots	To prevent entry of foreign materials into moving parts of machinery—to avoid contamination of lubricants
Clearance seals (Fig. 16–8)	
Labyrinths, bushing, and ring seals	Dynamic seals—to prevent leakage from a high-pressure station at one end of bushing to a region of low-pressure station at the other end of bushing
Ring seals—split ring seals	To seal reciprocating components
Expanding split ring	Used in compressors, pumps, and internal-combustion engines
Contracting split ring	Linear actuators where high-pressure, high-temperature radiation and fatigue are expected
Straight-cut seal ring (Fig. 16–3)	Piston seal for low-grade actuators
Step seal ring	Devices where free-passage leakage is not permissible
Circumferential seal	For rotary applications with low leakage and high performance
Face seals (Fig. 16–9)	
Stationary, rotating, and metal bellows type	Running seal between two flat precision finished surfaces, for high-speed applications, stuffing boxes, and temperature applications
Compression packing (Fig. 16–10)	For the throat of a stuffing box and its gland, dynamic seal
Molded packing (Fig. 16–11)	For automatic-hydraulic or mechanical packings
Lip type	
Single and multiple spring-loaded packings	For sealing reciprocating parts
Squeeze type	Fitted in rectangular grooves machined in hydraulic or pneumatic mechanisms and used as a piston seal in hydraulic actuating cylinder, valve seat, or valve stem packing
Felt radial type	Used at high speeds from 10 to 20 m/s
Diaphragm seals (Fig. 16–12)	To prevent intercharge of a fluid or contaminant between two separated areas, dynamic sealing and force transmitter
Nonmetallic gaskets (Fig. 16–13)	Static sealing
Metallic gaskets (Table 16–7)	
Corrugated, metal-jacketed, plain or machined (flat metal) round, heavy, or light cross-section (solid metal)	Static sealing, for high pressures and severe conditions, cast iron flanges, ammonia fittings, hydraulic cylinders, gas mains, heat exchangers, boiler openings, vacuum and cryogenic lines, and valve bonnets
Sealants	
Hardening (rigid or flexible), non-hardening and tapes	To exclude dust, dirt, moisture, and chemicals or contain a liquid or gas-surface coatings to protect against mechanical or chemical attack, to exclude noise, to improve appearance and to perform a joining function, thermal insulating, vibration damping

TABLE 16–7
Properties and uses of nonmetallic gasket materials

Classification	Special characteristics	General uses
Rubber asbestos	Tough and durable, relatively incompressible, good steam and hot water resistance	Heavy duty bolted and threaded joints as in water and steam pipe fittings; temperatures up to 260°C
Cork and rubber	Provides fluid barrier and resilience with compressibility; does not extrude from joint; die cuts well; high coefficient of friction	General-purpose gasketing; enables design of metal-to-metal joints; high friction keeps gasket positioned even where closing pressure is not perpendicular to flange faces
Cork composition	General purpose material compressible; high friction, low cost; excellent oil and solvent resistance; poor resistance to alkalies and corrosive acids	Mating rough or irregular parts; oil sealing at low cost in normal range of temperatures and pressures
Rubber, plastics	Highly adjustable according to compounding, hardness, modulus, fabric reinforcement, etc.; generally impervious, not compressible	Installations involving stretching over projections or where flow of gasket into threads or recesses is desired; for lowest compression set and maximum resistance to fluids such as alkalies, hot water, and certain acids
Paper		
Untreated	Low cost, noncorrosive	Spacers, dust barriers, splash seals where breathing and wicking not objectionable
Treated	General-purpose material; good oil, gasoline and water resistance	Machined or reasonably uniform flanges where adequate bolt pressures can be applied
Combination constructions	Innumerable modifications available, depending on materials used and methods of combining	Usually employed for extreme conditions and special purposes

TABLE 16–8
Minimum metallic gasket seating stress

Type	Material	Thickness, mm	Minimum seating stress[a]	
			MPa	kpsi
Flat metal (a)	Aluminum	3	109.8	16.0
		1.5 and 0.75	137.3	20.0
	Copper	3	248.1	36.0
		1.5 and 0.75	309.9	45.0
	Soft steel (iron)	3	379.0	55.0
		1.5 and 0.75	474.1	69.0
	Monel	3	448.2	65.0
		1.5 and 0.75	559.9	81.0
	Stainless steel	3	577.3	84.0
		1.5 and 0.75	646.2	94.0
Flat metal, serrated or grooved (b)	Aluminum	3[b]	172.1	25.0
		1.5[b]	206.9	30.0
		0.75[b]	241.2	35.0
	Copper	3[b]	241.2	35.0
		1.5[b]	275.6	40.0
		0.75[b]	309.9	45.0
	Soft steel (iron)	3[b]	379.0	55.0
		1.5[b]	413.8	60.0
		0.75[b]	448.2	65.0
	Monel	3[b]	448.2	65.0
		1.5[b]	482.5	70.0
		0.75[b]	557.6	80.0
	Stainless steel	3[b]	517.3	75.0
		1.5[b]	557.6	80.0
		0.75[b]	655.1	95.0
Corrugated (c)	Aluminum	3	10.3	1.5
	Copper	3	13.7	2.0
	Soft steel (iron)	3	27.4	4.0
	Monel	3	30.9	4.5
	Stainless steel	3	41.2	6.0
Corrugated coat (d)	Aluminum	3	13.7	2.0
	Copper	3	17.2	2.5
	Soft steel (iron)	3	20.6	3.0
	Monel	3	24.0	3.5
	Soft steel	3	27.4	4.0
Corrugated jacketed, soft filler (e)	Lead	3	3.4	0.5
	Aluminum	3	6.9	1.0
	Copper	3	17.1	2.5
	Soft steel (iron)	3	24.0	3.5
	Monel	3	30.9	4.5
	Stainless steel	3	41.2	6.0
	Inconel	3	51.5	7.5
	Hastelloy c	3	68.6	10.0

[a] Seating stress values shown do not apply to ASME Code. Also they are based on optimum surface finish and clean flange surface, i.e., no grease oil or gasket compound.
[b] Figures indicated are pitch, and the values of stress apply for all thicknesses.

TABLE 16–9
Compression packing for various service conditions

Fluid medium	Service condition			
	Reciprocating shafts	Rotating shafts	Piston or cylinders	Valve stems
Acids and caustics	Asbestos, metallic, plastic (pliable), semi-metallic, TFE fluorocarbon resins and yarns	Asbestos, plastic (pliable), semimetallic TFE fluorocarbon resin and yarns	TFE fluorocarbon resins	Asbestos, plastic (pliable), semimetallic TFE fluorocarbon resins and yarns
Air, gas	Asbestos, metallic, semimetallic	Asbestos, semimetallic	Leather, metallic	Asbestos, semimetallic
Ammonia, low-pressure steam	Duck and rubber, metallic, semimetallic	Asbestos, semimetallic	Duck and rubber	Asbestos, duck and rubber, semimetallic
Cold and hot gasoline and oils	Asbestos, plastic (pliable), semimetallic	Asbestos, plastic (pliable), semimetallic		Asbestos, plastic (pliable), semimetallic
High-pressure steam	Asbestos, metallic, plastic (pliable), semimetallic	Asbestos, metallic, plastic (pliable), semimetallic	Metallic	Asbestos, metallic, plastic (pliable), semimetallic
Cold and hot water	Duck and rubber, leather, plastic (pliable), semimetallic	Asbestos, plastic (pliable), semimetallic	Duck and rubber	Asbestos, duck and rubber, plastic (pliable), semimetallic

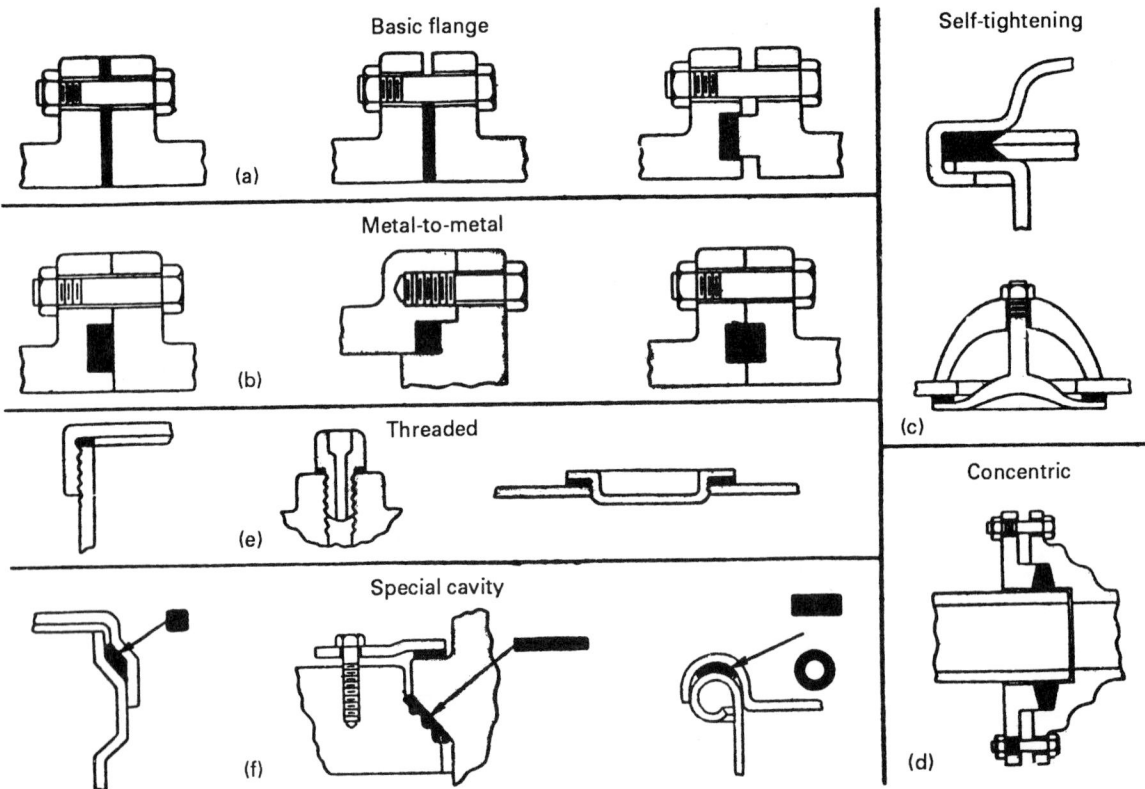

FIGURE 16–13 Common types of gasketed joints.

TABLE 16–10
Dimensions for oil seals

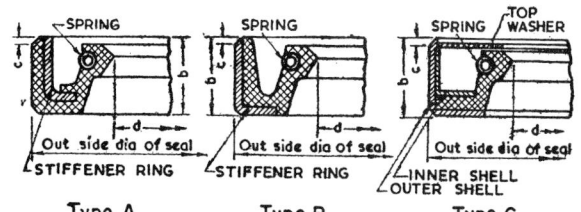

Type A Type B Type C

Shaft diameter, d_1 mm	Nominal[a] bore diameter of housing, mm	$b \pm 0.2$ types A and B, mm	c[b] min, mm
6	16	7	0.3
	22		
7	16	7	0.3
	22		
8	16	7	0.3
	22		
	24		
9	22	7	0.3
	24		
	26		
10	19	7	0.3
	22		
	24		
	26		
11	22	7	0.3
	26		
12	22	7	0.3
	24		
	28		
	30		
14	24	7	0.3
	28		
	30		
	35		
15	24	7	0.3
	26		
	30		
	32		
	35		
16	28	7	0.3
	30		
	32		
	35		
17	28	7	0.3
	30		
	32		
	35		
	40		
18	30	7	0.3
	32		
	35		
	40		

Shaft diameter d_1, mm	Nominal[a] bore diameter of housing, mm	$b \pm 0.2$, mm Types A and B	$b \pm 0.2$, mm Type C	c[b] min, mm
20	30	7		0.3
	32			
	35			
	40			
	47			
22	32	7	9	0.3
	35			
	40			
	47			
24	35	7	9	0.3
	37			
	40			
	47			
25	35	7	9	0.4
	40			
	42			
	47			
	52			
26	37	7	9	0.4
	42			
	47			
28	40	7	9	0.4
	47			
	52			
30	40	7	9	0.4
	42			
	47			
	52			
	62			
32	45	7	9	0.4
	47			
	52			
35	47	7	9	0.4
	50			
	52			
	62			
36	47	7	9	0.4
	50			
	52			
	62			
38	52	7	9	0.4
	55			
	62			

TABLE 16–10
Dimensions for oil seals (*Cont.*)

Shaft diameter d_1, mm	Nominal[a] bore diameter of housing, mm	$b \pm 0.2$, mm Types A and B	$b \pm 0.2$, mm Type C	c^b min, mm	Shaft diameter d_1, mm	Nominal[a] bore diameter of housing, mm	$b \pm 0.2$, mm Types A and B	$b \pm 0.2$, mm Type C	c^b min, mm
40	52 / 55 / 62 / 72	7	— / 9	0.4	80	100 / 110	10	12	0.5
42	55 / 62 / 72	8	— / 10	0.4	85	110 / 120	12	15	0.8
45	60 / 62 / 65 / 72	8	— / 10	0.4	90	110 / 120	12	15	0.8
48	62 / 72	8	— / 10	0.4	95	120 / 125	12	15	0.8
50	65 / 68 / 72 / 80	8	— / 10	0.4	100	120 / 125 / 130	12	15	0.8
52	68 / 72	8	— / 10	0.4	105	130 / 140	12	15	0.8
55	70 / 72 / 80 / 85	8	— / 10	0.4	110	130 / 140	12	15	0.8
56	70 / 72 / 80 / 85	8	— / 10	0.4	115	140 / 150	12	15	0.8
58	72 / 80	8	— / 10	0.4	120	150 / 160	12	15	0.8
60	75 / 80 / 85 / 90	8	— / 10	0.4	125	150 / 160	12	15	0.8
62	85 / 90	10	12	0.5	130	160 / 170	12	15	0.8
63	85 / 90	10	12	0.5	135	170	11	15	0.8
65	85 / 90 / 100	10	12	0.5	140	170	15	15	1
68	90 / 100	10	12	0.5	145	175	15	15	1
70	90 / 100	10	12	0.5	150	180	15	15	1
72	95 / 100	10	12	0.5	160	190	15	15	1
					170	200	15	15	1
					180	210	15	15	1
					190	220	15	15	1
					200	230	15	15	1
					210	240	15	15	1
					220	250	15	15	1
					230	260	15	15	1
					240	270	15	15	1
					250	280	15	15	1
					260	300	20	20	1
					280	320	20	20	1
					300	340	20	20	1
					320	360	20	20	1
					340	380	20	20	1
					360	400	20	20	1
					380	420	20	20	1
					400	440	20	20	1
					420	460	20	20	1

TABLE 16–10
Dimensions for oil seals (*Cont.*)

Shaft diameter d_1, mm	Nominal[a] bore diameter of housing, mm	$b \pm 0.2$, mm Types A and B	Type C	c^b min, mm	Shaft diameter d_1, mm	Nominal[a] bore diameter of housing, mm	$b \pm 0.2$, mm Types A and B	Type C	c^b min, mm
75	95	10	12	0.5	440	480	20	20	1
	100				460	500	20	20	1
78	100	10	12	0.5	480	520	20	20	1
					500	540	20	20	1

[a] For limits of housing, see Tables 16–8 and 16–9.
[b] The edges may be chamferred or rounded according to the manufacturer's discretion.
Source: Bureau of Indian Standards: 5129, 1969.

TABLE 16–11
Press-fit allowances and tolerances[a] for type A seals

Nominal bore diameter of housing, mm	Housing bore, mm High limit	Low limit	Outside diameter of seal, mm High limit	Low limit	Possible press-fit variation, mm Maximum interference	Minimum interference
≤ 25	+0.03	−0.03	+0.20	+0.10	0.23	0.07
25–55	+0.03	−0.03	+0.25	+0.15	0.28	0.12
55–125	+0.03	−0.03	+0.30	+0.20	0.33	0.17
125–200	+0.04	−0.04	+0.38	0.22	0.42	0.18
≥ 200	+0.05	−0.05	+0.48	0.32	0.53	0.27

[a] All tolerances are relative to nominal bore diameter of housing.
Source: IS 5129, 1969.

TABLE 16–12
Press-fit allowances and tolerances[a] for types B and C seals

Nominal bore diameter of housing, mm	Housing bore, mm High limit	Low limit	Outside diameter of seal, mm High limit	Low limit	Possible press-fit variation, mm Maximum interference	Minimum interference
≤ 50	Nominal	−0.03	+0.12	+0.04	0.15	0.04
50–90	Nominal	−0.03	+0.14	+0.06	0.17	0.06
90–115	+0.03	−0.03	+0.18	+0.08	0.21	0.05
115–170	+0.03	−0.03	+0.20	+0.10	0.23	0.07
170–215	+0.04	−0.04	+0.23	+0.13	0.27	0.09
215–230	+0.04	−0.04	+0.25	+0.15	0.29	0.11
≥ 230	+0.04	−0.04	+0.30	+0.20	0.34	0.16

[a] All tolerances are relative to nominal bore diameter of housing.
Source: IS 5129, 1969.

TABLE 16–13
Depth of the housing bore (all dimensions in mm)

b	t (0.85 b) Min	t_2 (b to 0.3) Min
7	5.95	7.3
8	6.80	8.3
9	7.65	9.3
10	8.50	10.3
12	10.30	12.3
15	12.75	15.3
20	17.00	20.3

Source: Indian Standards 5129, 1969.

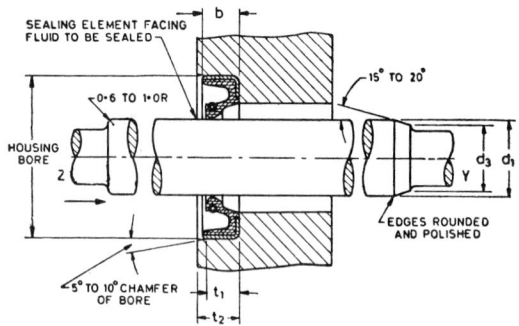

Z-AND Y-DIRECTION MOUNTING OF THE SHAFT

TABLE 16–14
Types of hollow, metallic O-rings [8, 9]

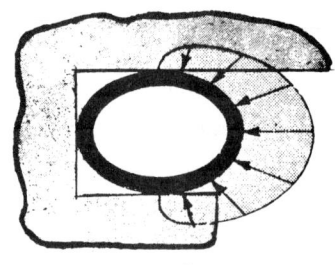

Plain, sealed metallic O-ring:
For fully confined or semi-confined ring joints-sealing vacuum, pressure, corrosive liquids and gases

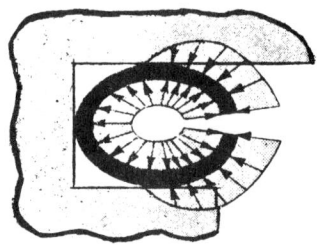

Self-energizing metallic O-ring:
For semiconfined designs increase in internal pressure causes ring to be crammed into groove, increases sealing effectiveness

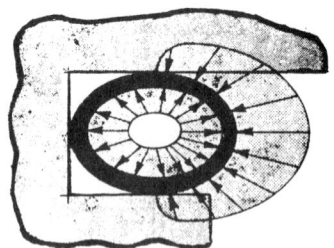

Pressure-filled metallic O-ring:
For fully confined or semiconfined designs. Ring is filled with an inert gas at 412 MPa (42 kgf/mm^2) useful in higher temperature range from 533 K (260°C) to 733 K (427°C)

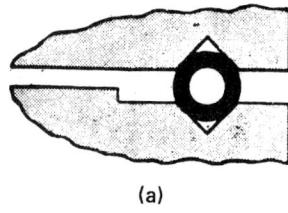

(a)

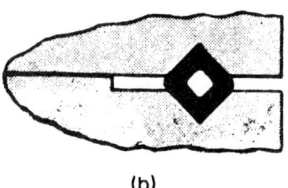

(b)

FIGURE 16–14 Fully confined hollow-metal O-ring: (*a*) before bolting and (*b*) after bolting down.

Source: Wes J. Ratelle, "Seal Selection, Beyond Standard Practice,", *Machine Design*, Jan. 20, 1977 and, "Packings and Seals" Issue, *Machine Design*, Jan. 1977.

TABLE 16–15
Recommended groove dimensions for metallic Q-ring sealing inside pressure

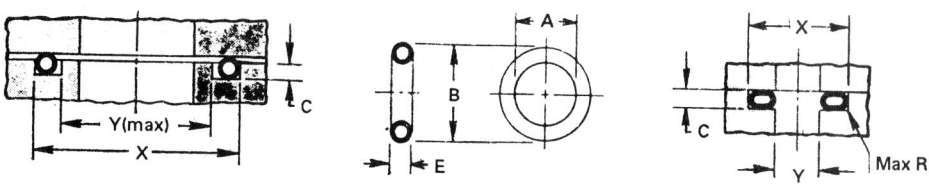

Nominal tubing OD mm	Nominal O-ring OD		Actual O-ring dimensions			Open-groove dimensions		Maximum, ID of closed groove, Y, mm	Maximum radius of groove corner, R, mm
	Min B, mm	Incremental increase, I, mm	Tubing OD, B, mm	O-Ring OD, mm	Depth, C, mm	Groove OD, X, mm	Minimum groove ID, Y, mm		
0.8	6.30	0.8 up to 25	0.74–0.96	B+0.075 −0.000	0.510– 0.500	B+0.0105 to 0.01525	B−2.160	B−3.000	0.125
1.6	11.0	1.6 thereafter	0.14–0.16	B+0.075 −0.0006	1.066– 1.145	B+0.0105 to 0.01525	B−3.600	E−4.825	0.255
2.4	19	1.6	0.20–0.24	B+0.100 −0.000	1.0650– 1.750	B+0.0127 to 0.0225	B−5.665	B−6.860	0.510
3.2	44	1.6	0.30–0.31	B+0.125 −0.000	2.290– 2.415	B+0.0178 to 0.0305	B−7.495	B−8.635	0.760
4.0	75	1.6	0.37–0.40	B+0.150 −0.000	2.920– 3.050	B+0.0203 to 0.0355	B−9.245	B−10.410	0.760
4.8	100	1.6	0.44–0.48	B+0.175 −0.000	3.685– 3.810	B+0.0228 to 0.0380	B−11.170	B−12.190	0.760
6.3	125	1.6	0.59–0.81	B+0.200 −0.000	4.955– 5.080	B+0.0280 to 0.0480	B−14.730	B−16.000	0.760
9.5	250	No limit	0.90–0.95	B+0.300 −0.000	7.495– 7.620	B+0.0355 to 0.0735	B−22.600	B−23.110	0.760
12.7	250	No limit	1.20–1.25	B+0.400 −0.000	9.910– 10.160	B+0.0510 to 0.0965	B−30.480	B−30.480	0.760

Source: Wes J. Ratelle, "Seal Selection, Beyond Standard Practice," *Machine Design*, Jan. 20, 1977, and "Packings and Seals" Issue, *Machine Design*, Jan. 20, 1977.

TABLE 16–16
Rectangular groove dimensions for O-ring gaskets

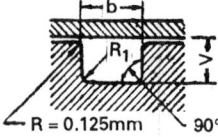

O-ring nominal cross section, mm	Actual O-ring cross section, mm	Maximum groove depth, V, mm	Groove width, b, mm	Minimum diametral squeeze, mm	Bottom radius, R_1, mm
		For Flange Gaskets (Axial)			
1.6	0.100 ± 0.0075	0.070–0.0050	0.160 ± 0.0050	0.025	0.0125
1.6	0.125 ± 0.0075	0.090–0.0050	0.185 ± 0.0075	0.030	0.0200
1.6	0.150 ± 0.0075	0.110–0.0050	0.210 ± 0.0075	0.035	0.0300
1.6	0.175 ± 0.0075	0.130–0.0100	0.240 ± 0.0075	0.040	0.0380
1.6	0.175 ± 0.0075	0.125–0.0100	0.240 ± 0.0075	0.045	0.0380
2.4	0.260 ± 0.0075	0.205–0.0105	0.270 ± 0.0125	0.050	0.0500
3.2	0.350 ± 0.0100	0.280–0.0200	0.470 ± 0.0125	0.060	0.0750
4.8	0.530 ± 0.0125	0.445–0.0250	0.725 ± 0.0125	0.075	0.1250
6.4	0.700 ± 0.0150	0.585–0.0250	0.960 ± 0.0125	0.100	0.1500
		For Nonflange Gaskets (Radial)			
1.6	0.100 ± 0.0075	0.075–0.0025	0.140 ± 0.0050	0.0175	0.0125
1.6	0.125 ± 0.0075	0.095–0.0025	0.160 ± 0.0075	0.025	0.0200
1.6	0.150 ± 0.0075	0.115–0.0025	0.190 ± 0.0075	0.030	0.0300
1.6	0.175 ± 0.0075	0.135–0.0025	0.230 ± 0.0075	0.035	0.0380
1.6	0.175 ± 0.0075	0.130–0.0050	0.230 ± 0.0075	0.038	0.0380
2.4	0.260 ± 0.0075	0.210–0.0075	0.315 ± 0.0125	0.043	0.0500
3.2	0.350 ± 0.0100	0.290–0.0100	0.430 ± 0.0125	0.050	0.0750
4.8	0.530 ± 0.0125	0.455–0.0125	0.600 ± 0.0125	0.065	0.1250
6.4	0.700 ± 0.0150	0.595–0.0150	0.800 ± 0.0125	0.090	0.1500

TABLE 16–17
Triangular groove dimensions for O-ring flange gaskets

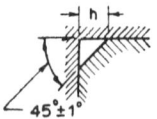

O-ring nominal cross section, mm	Actual O-ring cross section, mm	Width, h, mm
1.6	0.175 ± 0.0075	$0.240 + 0.0075 - 0.000$
2.4	0.260 ± 0.0075	$0.345 + 0.0125 - 0.000$
3.2	0.350 ± 0.0100	$0.470 + 0.0175 - 0.000$
4.8	0.530 ± 0.0125	$0.710 + 0.0255 - 0.000$
6.4	0.700 ± 0.0150	$0.950 + 0.0375 - 0.000$

TABLE 16–18
Packing sizes recommended for various shaft diameters

Shaft diameter, mm	Packing size, mm
12.70–15.85	7.95
17.45–38.10	9.50
39.70–50.80	11.10
52.40–63.50	12.70
65.10–76.20	14.30
77.80–101.60	15.85

TABLE 16-19
Temperature limits for gasket materials

Material	Maximum sustained temperature	
	K	°C
Asbestos fiber and rubber	673	400
Cellulose-fiber and rubber	423	150
Cork and rubber	393	120
Synthetic rubber	393	120
Cork composition	393	120

TABLE 16–21
Selection of shaft piston seals

Type name	Distributor	U-Ring	Cup	O-Ring
External-fitted to piston, sealing in bore				
Internal-fitted in housing, sealing on piston or rod				
Simple housing design	Good	Good	Poor	Very good
Low wear rate	Very Good	Good	Good	Poor
High stability	Good	Fair	Very good	Poor
Low friction	Fair	Fair	Fair	Good
Resistance to extrusion	Good	Good	Good	Fair
Availability in small sizes	Fair	Good	Poor	Very good
Availability in large sizes	Good	Fair	Good	Good
Bidirectional sealing		Single acting only		Effective but usually used in pairs

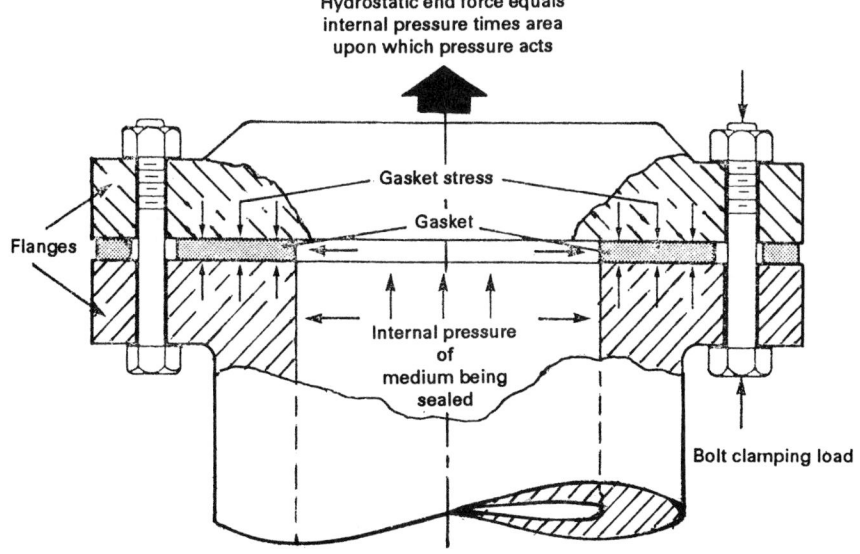

FIGURE 16–15 Nomenclature of gasketed joint.
(J. E. Shigley, and C. R. Mischke, *Standard Handbook of Machine Design*, McGraw-Hill, 1986.)

TABLE 16-20
Properties of sealants

sealant base test method	Tensile strength, ASTM D412		Elongation %, ASTM D412	Adhesion in tension, ASA 1161-1960		Shear strength, ASTM D1002		Moisture resistance, % ASTM D570	Abrasion resistance	Operating temperature, °C	Shore A hardness ASTM 676	Shrinkage
	MPa	psi		MPa	psi	MPa	psi					
Polysulfide	0.39-0.86	56.5-125	150-500	0.34-0.69	50-100	0.55-1.20	80-175	0.25-1.5	Fair to good	−50 to 120	15-60	0-3.0
Polyurethane[a]	6.86-20.50	1000-3000	400-600	0.15-0.44	15-65	1.72-2.40	240-350	1-3.0	Excellent	−55 to 205	45-90	0-10.0
Silicone	1.96-5.39	285-780	50-750	0.34-0.59	50-85.5	1.03-1.37	150-200	0.1-0.25	Fair to good	−75 to 370	25-80	0-10.0
Neoprene	6.86-10.29	1000-1500	250-350	0.44-0.69	65-100	0.27-0.69	40-100	0.5-1.5	Excellent	−40 to 150	30-80	0-10.0
Hypalon	3.43-4.11	500-600	75-125			0.85-1.20	125-175	1.0-5.0	Fair to good	−40 to 150	40-60	10-20
Viton	8.33	1210	325			10.29-24.0	1500-3500	0-3.0	Good to excellent	−55 to 230	40-100	0-3.0
Epoxy	27.46-89.73	4000-13000	3.0-6.0					0.04-0.10	Good to excellent	−35 to 150	Shore "D" 40-60	0-3.0
Epoxy-modified	8.33-24.02	1200-3500	10-20			10.29-19.1	1500-2750	0.27-0.50	Fair to good	−35 to 150	Shore "D" 40-100	0-3.0
Acrylic	0.34-2.94	50-425	100-270					1.0-5.0		−25 to 150	10-70	5.0-15.0
Polyester	27.46-48.05	4000-7000	3-15			10.79-23.54	1500-3400	0.25-0.75	Good	−55 to 150	Shore "D" 10-45	2.0-10.0
Polyurethane-bitumen-modified	0.29-0.53	42-75	250-400	0.17-0.27	24.5-40			0.75-1.50	Good	−35 to 95	5-70	0-3.0
Butyls-mastic type			5-150					0.5-5.0	Good	−20 to 95	15-75	15-4.0
Butyls-curing type	17.16-20.59	2500-3000	650-800			1.03-1.873	150-270	0.25-1.5	Good	−60 to 150	20-40	0-3.0
Polybutene	0.49-4.90	70-710	5-20						Good	−30 to 120	5-70	15-45
Oleoresin									Good	−25 to 95		

[a] Compounds built specifically for ploting and moulding, where high strength and abrasion resistance are required.

16.24

TABLE 16–22
Recommended maximum temperature for materials (supplement to Table 16–2)

Material	°F	°C
Coil spring material		
Phosphor-bronze ASTM B159	200	93
Silicon bronze ASTM B99	200	93
Ni-span C902	200	93
Music wire ASTM A228	250	121
Hard-drawn spring wire ASTM A227	250	121
Oil-tempered wire ASTM A229	300	149
Valve spring wire ASTM A230	300	149
Beryllium-copper ASTM B197	400	204
Chrome-vanadium alloy steel AISI 6150	425	218
Silicon-manganese alloy steel AISI 9260	450	232
Chrome-silicon alloy steel AISI 9254	475	246
Martensite AISI 410	500	260
Martensite AISI 420	500	260
Austenitic AISI 301	600	315
Austenitic AISI 302	600	315
17-7 PH Stainless Steel	590	311
Inconel 600	700	371
Nickel-chrome alloy steel A286	950	510
Inconel 718	1200	649
Inconel X-750	1300	704
L-605	1400	760
S-816	1400	760
Rene 41	1400	760
Flat spring material		
Ni-span C902	200	93
Phosphor-bronze ASTM B103	200	93
High-carbon AISI 1050	200	93
High-carbon AISI 1065	200	93
High-carbon AISI 1075	250	121
High-carbon AISI 1095	250	121
Beryllium-copper ASTM B194	400	204
Austenitic AISI 301	600	315
Austenitic AISI 302	600	315
17-7 PH Stainless Steel	700	371
Inconel 600	700	371
Beryleo-nickel	700	371
Titanium 6-6-2	750	399
Sandvik 11 R51	800	427
Duranickel 301	800	427
Permanickel	800	427
Elgiloy	900	482
Havar	900	482
Inconel 718	1200	649
Inconel X-750	1300	704
Rene 41	1400	760

TABLE 16–22
Recommended maximum temperature for materials (supplement to Table 16–2)

Material	°F	°C
Formed metal bellows materials		
Brass CDA 240	300	149
Phosphor-bronze CDA 510	300	149
Beryllium-copper CDA 172	350	177
Monel 404	450	232
Unstabilized 300 series stainless steel	500	260
Inconel 600	750	399
Inconel X-750	800	427
Welded metal bellows materials		
Ni-span C	500	260
AM-350 Stainless Steel	800	427
410 Stainless Steel	800	427
Commercially pure titanium	800	427
Stabilized 300 series Stainless Steel	1220	659
Inconel X-750	1500	815
Inconel 625	1500	815
Hastelloy-C	1800	982
Rene 41	1800	982

TABLE 16–23
pv values for seal face material (life of 8000 h)

Product	Combination face material	pv Value Unbalanced MPa m/s	pv Value Unbalanced kpsi fpm	pv Value Balanced MPa, m/s	pv Value Balanced kpsi fpm
Water	Stainless steel ⎫	0.9	25.5	Seldom used	Seldom used
Oil	Carbon[a] ⎬	1.8	51.0	Seldom used	Seldom used
Water	Lead bronze ⎫	1.8	51.0	Seldom used	Seldom used
Oil	Carbon[a] ⎬	3.5	100	Seldom used	Seldom used
Water	Stellite carbon[a]	3.5	100	10	285
Oil		9		70	2000
Water	Tungsten carbide ⎫	9	255	25	710
Oil	Carbon[b] ⎬	9	255	150	4280
Water	Solid ceramic	15	430	Seldom used	Seldom used
Water	Sprayed ceramic	15	430	90	2570
Oil		20	560	150	4280

[a] Metal-impregnated carbon.
[b] Retain impregnated carbon.
Source: Courtesy M. J. Neale, *Tribology Handbook*, Butterworths, London, 1973.

TABLE 16–24
Spring arrangements for various sizes of shaft and speeds

Shaft diameter, mm	Speed, rpm	Stationary Single	Stationary Multiple	Rotary Single	Rotary Multiple
≤ 100	≤ 3000	Yes	Yes	Yes	Yes
> 100	≤ 3000	No	Yes	No	Yes
≤ 75	≤ 4500	Yes	Yes	Yes	Yes
≤ 100	> 4500	Yes	Yes	No	No
> 100	> 4500	No	Yes	No	No

Source: Courtesy of M. J. Neale, *Tribology Handbook*, Butterworths, London, 1973.

TABLE 16–25
Types of static and dynamic seals

	Dynamic seals			
	Clearance seals		Contact seals	
Static seals	Reciprocating	Rotary	Reciprocating	Rotary
Fibrous gasket	Labyrinth[a] (Fig. 16–8)	Labyrinth (Fig. 16–8)	U-ring (Fig. 16–11)	Lip seal (Fig. 16–4)
Metallic gasket	Fixed bushing	Viscoseal	O-ring (Table 16–15)	Face seal (Fig. 16–9a)
Elastomeric gasket	Floating bushing	Fixed bushing	Lobed O-ring	Packed gland (Fig. 10–10)
Plastic gasket		Floating bushing	Rectangular ring	O-ring[b] (Fig. 16–14)
Sealant, setting		Centrifugal seal	Packed gland	Felt ring
Sealant, nonsetting			Piston ring	
O-ring			Bellows	
Inflatable gasket			Diaphragm (Fig. 16–12)	
Pipe coupling				
Bellows				

[a] Usually for steam or gas.
[b] Only for very slow speeds.
Source: Courtesy M. J. Neale, *Tribology Handbook*, Butterworths, London, 1973.

TABLE 16–26
Operating conditions of lip seals

Particular	Shaft diameter and housing	Remarks
Maximum pressure of fluid	≤ 75 mm diameter	60 kPa (8.7 psi)
	> 75 mm diameter	30 kPa (4.35 psi)
Maximum speed	≤ 35 mm diameter	8000 rpm
	75 mm diameter	4000 rpm
	> 75 mm diameter	15 m/s
Surface finish	Housing	Fine-turned
	Shaft	Grind and polish to better than 0.5 μm
Eccentricity	Housing	0.25 mm total indicator reading
	Shaft	Depends on speed, 0.25 mm
Temperature		Varies from −20°C to 200°C (−68°F to 266°F)

Source: M. J. Neale, *Tribology Handbook*, Butterworths, London, 1973.

TABLE 16–27
Types of seals for reciprocating shafts

Type of packing	Remarks
Cups and hats	Semiautomatic, leather and rubber/ fabric used.
U-packing	Used for piston rod application up to 10 MPa (1.5 kpsi) (rubber) or 20 MPa (3.0 kpsi) (rubber/fabric).
Nylon-supported	Used up to 25 MPa (3.6 kpsi).
Composite	Used with rubber sealing lips, rubber/ fabric supporting portions and nylon wearing portions—used for pressure varying from 15 to 20 MPa (2.2 to 3.0 kpsi).

Source: M. J. Neale, *Tribology Handbook*, Butterworths, London, 1973.

TABLE 16–28
Materials for lip seals (rubber)

Type of rubber	Trade names	Resistance to		Temperature	
		Mineral oil	Chemical fluids	°F	°C
Acrylate	Thiacril Cyanacryl	Excellent	Fair	−68 to +266	−20 to +130
Fluoropolymer	Viton Fluorel	Excellent	Excellent	−77 to +392	−25 to +200
Polysiloxane	Silastic Silastomer	Fair	Poor	−158 to +392	−70 to +200
Nitrile	Hycar Polysar	Excellent	Fair	−104 to +212	−40 to +100

Source: Courtesy M. J. Neale, *Tribology Handbook*, Butterworths, London, 1973.

TABLE 16–29
Seal materials for reciprocating shafts

Material	Remarks
Rubber (nitrile)	Highest sealing efficiency; low cost; easily formed to shape; limited to a pressure of 10 MPa (1.5 kpsi); excellent wear resistance
Rubber-impregnated fabric	Great toughness; resistance to extrusion and cutting; wear resistance inferior to rubber
Leather	Good wear and extrusion resistance; poor resistance to permanent set; limited shaping capability
Nylon	Resist extrusion; provide a good bearing surface

Source: Courtesy M. J. Neale, *Tribology Handbook*, Butterworths, London, 1973.

TABLE 16–30
Extrusion clearance for reciprocating shafts—dimensions in mm (in)

Material	≤ 10 MPa (1.5 kpsi)		10–20 MPa (1.5–3.0 kpsi)		> 20 MPa (3.0 kpsi)	
	Normal	Short life	Normal	Short life	Normal	Short life
Rubber	0.25 (0.01)	0.50 (0.02)	—	—	—	—
Rubber/fabric leather	0.40 (0.015)	0.60 (0.025)	0.25 (0.01)	0.50 (0.02)	0.10 (0.005)	0.25 (0.01)
Polyurethane	0.40 (0.015)	0.60 (0.025)	0.25 (0.01)	0.50 (0.02)	0.10 (0.005)	0.25 (0.01)
Nylon support	—	—	0.25 (0.01)	1.00 (0.04)	0.10 (0.005)	0.25 (0.01)

Source: Courtesy M. J. Neale, *Tribology Handbook*, Butterworths, London, 1973.

TABLE 16-31
Operation conditions of packed glands (Fig. 16-1)

Type of gland	Pressure		Temperature		Velocity, m/s (fpm)	Remarks
	MPa	psi	°F	°C		
Graphited asbestos—rotary type	0.105	15	200	93	17.75 (4000)	No latern or jacket ring cooling required.
Graphited asbestos with latern ring cooling arrangement—rotary type	0.280	40	240	115	17.75 (4000)	Cooling liquid used below 34.5 kPa sealing pressure
Graphited asbestos with latern ring and jacket cooling arrangement—rotary type	0.700	100	320	160	17.75 (4000)	Latern ring cooling liquid and water to jacket cooler used below sealing pressure of 34.5 kPa
Graphited asbestos with PTFE antiextrusion ring hand surface replaceable sleeve, jacket cooling arrangement—rotary type	0.525	75	290	143	306 (6100)	Cooling as per type 3; special packing and accurate assembly is required
Graphited asbestos and PTFE yarn with PTFE antiextrusion ring, jacket cooling arrangement—rotary type	7.000	1000	545	285	5.5 (1080)	Water to jacket coolant used
Reciprocating, steam-graphited asbestos	1.750	250	500	260	0.75	Steam
Reciprocating, water-greased cotton packing	2.100	300	500	260	0.75	Water
Reciprocating, oil-graphited hemp yarn	2.100	300	200	93	0.75 (150)	Oil

Source: Courtesy M. J. Neale, *Tribology Handbook*, Butterworths, London, 1973.

TABLE 16–32
Axial stress in packed glands

Type of packing	Minimum axial stress required for seal packing	
	MPa	psi
Teflon-impregnated braided	1.40	200
asbestos	1.40	200
Plastic	1.12	160
Braided vegetable fiber, lubricated	1.75	255
Plaited asbestos, lubricated	2.8	405
Braided metallic	3.5	505

Source: Courtesy M. J. Neale, *Tribology Handbook*, Butterworths, London, 1973.

TABLE 16–33
Selection of number of sealing rings

Pressure		Number of sets of sealing rings
MPa	psi	
≤ 1.0	150	3
1.0–2.0	(150–250)	4
2.0–5.0	250–500	5
3.5–17.0	500–1000	6
7.0–15.0	1000–2000	8
> 15.0	above 2000	9–12

Source: Courtesy M. J. Neale, *Tribology Handbook*, Butterworths, London, 1973.

TABLE 16–34
Selection of packing materials

Material	Hardness of rod, H_B	Axial clearance, mm	Application
Lead bronze	250 min	0.08–0.12 (0.003–0.005 in)	Optimum material with good lubricated bearing property High thermal conductivity; used where chemical condition exists and suited for pressure up to 300 MPa (50 kpsi)
Flake graphite grey cast iron	400 min	0.08–0.12	Cheaper suitable up to a pressure of 7 MPa (1.0 kpsi)
White metal (Babbitt)		0.08–0.12	Used where lead-bronze and flake graphite gray cast iron are not suitable because of chemical condition; used up to a maximum pressure of 35 MPa (5.0 kpsi) and maximum temperature 120°C (250°F)
Filled PTFE	400 min	0.4–0.5	Suitable for unlubricated; very good chemical resistance; suited above 2.5 MPa (400 psi)
Reinforced *pf* resin		0.25–0.5	Used with sour hydrocarbon gases and where lubricant may be thinned by solvents in gas stream
Carbon-graphite	400 min	0.030–0.06	Used with carbon-graphite piston rings; must be kept oil free; used up to 350°C (660°F)
Graphite/metal sinter	250 min	0.08–0.12	Alternative to filled PTFE and carbon-graphite

Source: Courtesy M. J. Neale, *Tribology Handbook*, Butterworths, London, 1973.

TABLE 16–35
Minimum recommended seating stresses for various gasket materials (Supplement to Table 16–8)

	Material, mm (in)	Gasket type	Minimum seating stress range (psi[a]) MPa
Nonmetallic	Asbestos fiber sheet	Flat	
	3.125 ($^1/_8$ in) thick		(1400–1600) 9.7–11.0
	1.563 ($^1/_{16}$ in) thick		(3500–3700) 24.1–25.5
	0.78 ($^1/_{32}$ in) thick		(6000–6500) 41.4–44.8
	Asbestos fiber sheet	Flat with rubber beads	(1000–1500 lb/in) on beads
	0.78 ($^1/_{32}$ in) thick		175–263 kN/m
	Asbestos fiber sheet	Flat with metal grommet	(3000–4000 lb/in) on grommet
	0.78 ($^1/_{32}$ in) thick		525.4–700.5 kN/m
	Asbestos fiber sheet	Flat with metal grommet and	(2000–3000 lb/in) on wire
	0.78 ($^1/_{32}$ in) thick	metal wire	350.2–525.4 kN/m
	Cellulose fiber sheet	Flat	(750–1100) 5.2–7.6
	Cork composition	Flat	(400–500) 2.8–3.5
	Cork-rubber	Flat	(200–300) 1.4–2.1
	Fluorocarbon (TFE)	Flat	
	3.125 ($^1/_8$ in) thick		(1500–1700) 10.3–11.7
	1.563 ($^1/_{16}$ in) thick		(3500–3800) 24.1–26.2
	0.78 ($^1/_{32}$ in) thick		(6200–6500) 42.8–44.8
	Nonasbestos fiber sheets	Flat	(1500–3000) depending on
	(glass, carbon, aramid,		composition
	and ceramics)		10.3–20.7
	Rubber	Flat	(100–200) 0.7–1.4
	Rubber with fabric or metal reinforcement	Flat with reinforcement	(300–500) 2.1–3.5
Metallic	Aluminum	Flat	(10,000–20,000) 68.9–137.9
	Copper	Flat	(15,000–45,000) 103.4–310.3
			depending on hardness
	Carbon steel	Flat	(30,000–70,000) 207–483
			depending on alloy and hardness
	Stainless steel	Flat 241–655	(35,000–95,000) 241–655
			depending on alloy and hardness
	Aluminum (soft)	Corrugated	(1000–3700) 6.9–25.5
	Copper (soft)	Corrugated	(2500–4500) 17.2–31.0
	Carbon steel (soft)	Corrugated	(3500–5500) 24.1–37.9
	Stainless steel	Corrugated	(6000–8000) 41.4–55.2
	Aluminum	Profile	(25,000) 172.4
	Copper	Profile	(35,000) 241.3
	Carbon steel	Profile	(55,000) 379.2
	Stainless steel	Profile	(75,000) 517.1

TABLE 16–35
Minimum recommended seating stresses for various gasket materials (*Cont.*)

	Material, mm (in)	Gasket type	Minimum seating stress range (psi[a]) MPa
Jacketed metal-asbestos	Aluminum	Plain	(2500) 17.2
	Copper	Plain	(4000) 27.6
	Carbon steel	Plain	(6000) 41.4
	Stainless steel	Plain	(10,000) 68.9
	Aluminum	Corrugated	(2000) 13.8
	Copper	Corrugated	(2500) 17.2
	Carbon steel	Corrugated	(3000) 20.7
	Stainless steel	Corrugated	(4000) 27.6
	Stainless steel	Spiral-wound	(3000–30,000) 20.7–206.8

[a] Stresses in pounds per square inch except where otherwise noted.
Source: J. E. Shigley and C. R. Mischke, *Standard Handbook of Machine Design*, McGraw-Hill Book Company, New York, 1986.

TABLE 16–36
Safety factors for gasketed joints, *n*, for use in Eq.
(16–39)

Safety factor, *n*	When to apply
1.2 to 1.4	For minimum-weight applications where all installation factors (bolt lubrication, tension, parallel seating, etc.) are carefully controlled; ambient to 250°F (121°C) temperature applications; where adequate proof pressure is applied
1.5 to 2.5	For most normal designs where weight is not a major factor, vibration is moderate and temperatures do not exceed 750°F (399°C); use high end of range where bolts are not lubricated
2.6 to 4.0	For cases of extreme fluctuations in pressure, temperature, or vibration; where no test pressure is applied; or where uniform bolt tension is difficult to ensure

Source: J. E. Shigley and C. R. Mischke, *Standard Handbook of Machine Design*, McGraw-Hill Book Company, New York, 1986.

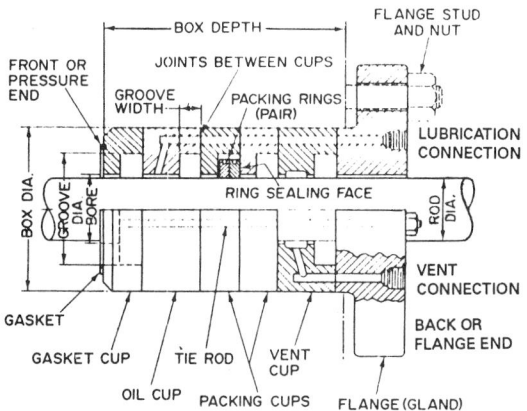

General arrangement of a typical mechanical piston rod packing assembly

FIGURE 16–16 Packing assembly for a mechanical piston rod. (*M. J. Neale*, Tribology Handbook, *Butterworths, London, 1973.*)

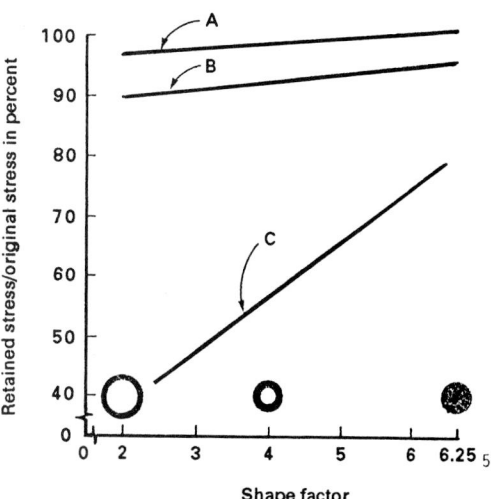

FIGURE 16–17 Ratio of retained stress to original stress versus shape factor for various materials: *A*—asbestos sheet; *B*—cellulose; *C*—cork-rubber. (*J. E. Shigley*, Standard Handbook of Machine Design, *McGraw-Hill, 1986.*)

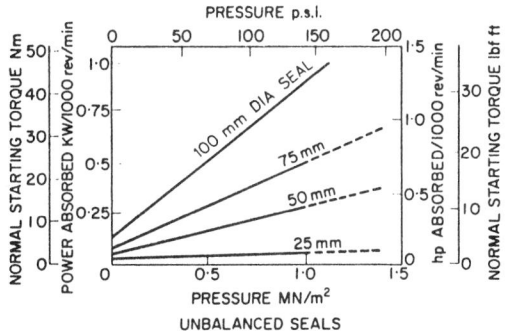

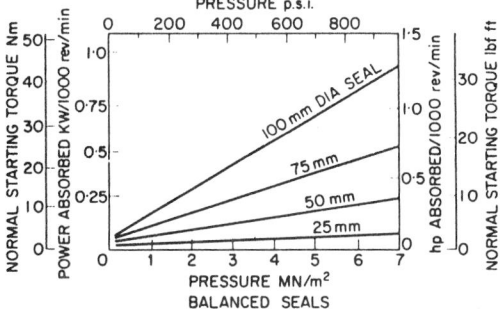

Power absorption and starting torque for aqueous solutions, light oils and medium hydrocarbons, use above values. For light hydrocarbons use $\frac{2}{3}$ of above values. For heavy hydrocarbons use $1\frac{1}{3}$ of above values. Allow ± 25% on all values

FIGURE 16–18 Power absorption and starting torque for balanced and unbalanced seals. (*M. J. Neale*, Tribology Handbook, *Butterworths, London, 1973.*)

REFERENCES

1. Lingaiah, K., and B. R. Narayana Iyengar, *Machine Design Data Handbook*, Vol. I (*SI and Customary Units*), Suma Publishers, Bangalore, India, 1986.
2. Lingaiah, K., *Machine Design Data Handbook*, Vol. II (*SI and Customary Metric Units*), Suma Publishers, Bangalore, India, 1986.
3. Maleev, V. L., and J. B. Hartman, *Machine Design*, International Textbook Company, Scranton, Pennsylvania, 1954.
4. The American Society of Mechanical Engineers, *ASME Boilers and Pressure Vessel Code*, Section VIII, Division I, 1986.
5. Whalen, J. J., "How to Select the Right Gasket Material," *Product Engineering*, Oct. 1860.
6. Shigley, J. E., and C. R. Mischke, *Standard Handbook of Machine Design*, McGraw-Hill Book Company, 1986.
7. Neale, M. J., *Tribology Handbook*, Butterworths, London, 1975.
8. Ratelle, W. J., "Seal Selection, Beyond Standard Practice," *Machine Design*, Jan. 20, 1977.
9. "Packings and Seals" Issue, *Machine Design*, Jan. 1977.
10. Faires, V. M., *Design of Machine Elements*, Macmillan Book Company, 1955.
11. Bureau of Indian Standards.
12. Rothbart, H. A., *Mechanical Design and Systems Handbook*, McGraw-Hill Book Company, New York, 1985.

CHAPTER
17

KEYS, PINS, COTTERS, AND JOINTS

SYMBOLS

a	addendum for a flat root involute spline profile, m (in)
A	area, m^2 (in^2)
b	breadth of key, m (in)
	effective length of knuckle pin, m (in)
	dedendum for a flat root involute spline profile, m (in)
d	diameter, m (in)
d_1	major diameter of internal spline, m (in)
d_2	minor diameter of internal spline, m (in)
d_3	major diameter of external spline, m (in)
d_4	minor diameter of external spline, m (in)
d_c	core diameter of threaded portion of the taper rod, m (in)
d_{pl}	large diameter of taper pin, m (in)
d_m (or d_{pm})	mean diameter of taper pin, m (in)
d_{nom}	nominal diameter of thread portion, m (in)
D	diameter of shaft, m (in)
	pitch diameter, m (in)
F	force, kN (lbf)
	force on the cotter joint, kN (lbf)
	pressure between hub and key, kN (lbf)
F', F''	force applied in the center of plane of a feather keyed shaft which do not change the existing equilibrium but give a couple, kN (lbf)
F_2', F_2''	two opposite forces applied on the center plane of a double feather keyed shaft which give two couples, but tending to rotate the hub clockwise, kN (lbf)
F_t	tangential force, kN (lbf)
F_μ	frictional force, kN (lbf)
h	thickness of key, m (in)
	minimum height of contact in one tooth, m (in)
l	length of key (also with suffixes), m (in)
	length of couple (also with suffixes), m (in)
	length of sleeve, m (in)
L	length of spline, m (in)
l_o, s_o	space width and tooth thickness of spline, m (in)
m	module, m (in)
M_b	bending moment, N m (lbf in)
M_t	twisting moment, N m (lbf in)

p	pressure, MPa (psi)
	tangential pressure per unit length, MPa (psi)
p_1	maximum pressure where the shaft enters the hub, MPa (psi)
p_2	pressure at the end of key, MPa (psi)
p_d (or P)	diametral pitch
Q	external load, kN (lbf)
R	resistance on the key and on the shaft to be overcome when the hub is shifted lengthwise, kN (lbf)
t	thickness of cotter, m (in)
xm	profile displacement, m (in)
z	number of teeth,
	number of splines
σ	stress tensile or compressive (also with suffixes), MPa (psi)
σ_{b1}	nominal bearing stress at dangerous point, MPa (psi)
τ	shear stress, MPa (psi)
α	angle of cotter slope, deg
θ	angle of friction, deg
μ	coefficient of friction (also with suffixes)

SUFFIXES

b	bearing
c	compressive
d	design
m	mean
p	pin
s	small end
t	tensile, tangential

Particular	Formula

ROUND OR PIN KEYS

The large diameter of the pin key

$d = 3.035\sqrt{D}$ to $3.54\sqrt{D}$ where d and D are in mm
SI   (17–1a)
$= 0.6\sqrt{D}$ to $0.7\sqrt{D}$ where d and D are in in
US Customary units   (17–1b)
$= 0.096\sqrt{D}$ to $0.11\sqrt{D}$ where d and D are in m
SI   (17–1c)

STRENGTH OF KEYS

Rectangular fitted key (Fig. 17–1, Table 17–1)

Pressure between key and keyseat

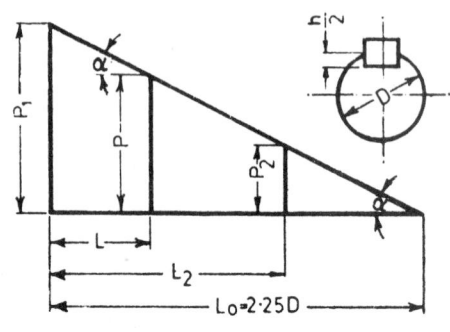

FIGURE 17–1

TABLE 17–1
Dimensions (in mm) of parallel keys and keyways

Keyway radius in shaft and hub

Chamfer or radius of key

Key

For shaft diameters (Above / Up to)	6/8	8/10	10/12	12/17	17/22	22/30	30/38	38/44	44/50	50/58	58/65	65/75	75/85	85/95	95/110	110/130	130/150	150/170	170/200	200/230	230/260	260/290	290/330	330/380	380/440	440/500
Key cross section — Width b	2	3	4	5	6	8	10	12	14	16	18	20	22	25	28	32	36	40	45	50	56	63	70	80	90	100
Key cross section — Height h	2	3	4	5	6	7	8	8	9	10	11	12	14	14	16	18	20	22	25	28	32	32	36	40	45	50
Keyway depth (nominal) — In shaft t_1	1.2	1.8	2.5	3.0	3.5	4.0	5	5	5.5	6	7	7.5	8.5	9.0	10	11	12	13	15	17	19	20	22	25	28	31
Keyway depth (nominal) — In hub t_2	1	1.4	1.8	2.3	2.8	3.3	3.3	3.3	3.8	4.3	4.4	4.9	5.4	5.9	6.4	7.4	8.4	9.4	10.4	11.4	12.4	13.4	14.4	15.4	17.4	19.5
Tolerance on keyway depth — t_1				+0.05 / −0.00			+0.1 / −0.0												+0.15 / −0.00							
Tolerance on keyway depth — t_2				+0.05 / −0.00			+0.1 / −0.0												+0.15 / −0.00							
Chamfer or radius of key — r max		0.25			0.35			0.55					0.80						1.30			2.00				2.95
Chamfer or radius of key — r min		0.16			0.25			0.40					0.60						1.00			1.60				2.50
Keyway radius — r_2 max		0.16			0.25			0.40					0.60						1.00			1.60				2.50
Length of key — L min	6	6	8	10	14	18	22	28	36	45	50	56	63	71	80	90	100	110	125	140	160	180	200	220	250	280
Length of key — L max	20	36	45	50	71	90	110	140	160	180	200	220	250	280	320	360	400	400	400	400	400	400	400	400	400	400

Source: IS 2048, 1962

17.3

Particular	Formula

Crushing strength

The tangential pressure per unit length of the key at any intermediate distance L from the hub edge (Fig. 17–1, Table 17–2)

$$p = p_1 - L \tan \alpha \qquad (17\text{–}2)$$

$$\text{where } \tan \alpha = \frac{p_1 - p_2}{L_2} = \frac{p_1}{L_0}$$

The torque transmitted by the key (Fig. 17–1)

$$M_t = \tfrac{1}{2} p_1 D L_2 - D L_2^2 \tan \alpha \qquad (17\text{–}3)$$

The general expression for torque transmitted according to practical experience

$$M_t = \tfrac{1}{4} \sigma_{b1} h D L_2 - \tfrac{1}{18} \sigma_{b1} b L_2^2 \qquad (17\text{–}4)$$

$$\text{where } p_2 = o, \text{ when } L_2 = L_o = 2.25D;$$

$$\tan \alpha = \frac{p_1}{L_o} = \frac{\sigma_{b1} h}{4.5D}$$

For dimensions of tangential keys given here.

Refer to Table 17–2.

Shearing strength

The torque transmitted by the key (Fig. 17–1)

$$M_t = \tfrac{1}{2} \tau_1 b D L_2 - \tfrac{1}{9} \tau_1 b L_2^2 \qquad (17\text{–}5)$$

$$\text{where } \tan \alpha = \frac{p_1}{L_o} = \frac{\tau_1 b}{2.25D}$$

The shear stress at the dangerous point (Fig. 17–1)

$$\tau_1 = \frac{M_t}{L_2 b (0.5D - 0.11 L_2)} \qquad (17\text{–}6)$$

TAPER KEY (FIG. 17–2, TABLE 17–3)

The relation between the circumferential force F_t and the pressure F between the shaft and the hub

$$F_1 = \mu_1 F \qquad (17\text{–}7)$$

The pressure between the shaft and the hub

$$F = blp \qquad (17\text{–}8)$$

The torque

$$M_t = \tfrac{1}{2} \mu_1 blpD \qquad (17\text{–}9)$$

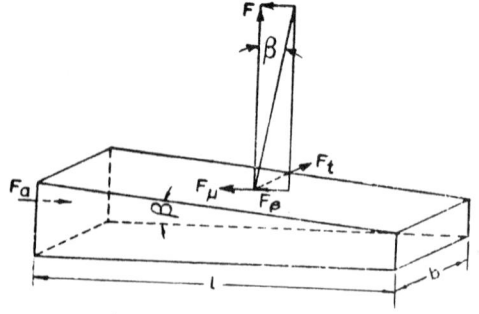

FIGURE 17–2

TABLE 17–2
Dimensions (in mm) of tangential keys and keyways

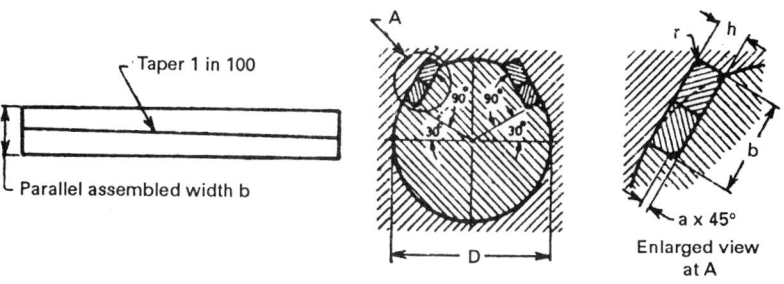

Shaft	Keyway			Key	Shaft	Keyway			Key
diameter, D	Height, h	Width, b	Radius, r	chamfer, a	diameter, D	Height, h	Width, b	Radius, r	chamfer, a
100	10	30	2	3	460	46	138	4	5
110	11	30	2	3	480	48	144	5	6
120	12	36	2	3	500	50	150	5	6
130	13	39	2	3	520	52	156	5	6
140	14	42	2	3	540	54	162	5	6
150	15	45	2	3	560	56	168	5	6
160	16	48	2	3	580	58	174	5	6
170	17	51	2	3	600	60	180	6	7
180	18	54	2	3	620	62	186	6	7
190	19	57	2	3	640	64	192	6	7
200	20	60	2	3	660	66	198	6	7
210	21	63	2	3	680	68	204	6	7
220	22	66	2	4	700	70	210	6	7
230	23	69	3	4	720	72	216	6	7
240	24	72	3	4	740	74	222	6	7
250	25	75	3	4	760	76	228	6	7
260	26	78	3	4	780	78	234	6	7
270	27	81	3	4	800	80	240	6	7
280	28	84	3	4	820	82	246	6	7
290	29	87	3	4	840	84	252	6	7
300	30	90	3	4	860	86	258	6	7
320	32	95	3	4	880	88	264	8	9
340	34	102	3	4	900	90	270	8	9
360	36	108	3	4	920	92	276	8	9
380	38	114	4	5	940	94	282	8	9
400	40	129	4	5	960	96	288	8	9
420	42	126	4	5	980	98	294	8	9
440	44	132	4	5	1000	100	300	8	9

Notes: (1) The dimensions of the keys are based on the formula: width 0.3 shaft diameter, and thickness = 0.1 shaft diameter; (2) if it is not possible to fix the keys at 120°, they may be fixed at 180°; (3) it is recommended that for an intermediate diameter of shaft, the key section shall be the same as that for the next larger size of the shaft in this table.
Source: IS 2291, 1963.

TABLE 17–3
Dimensions (in mm) of taper keys and keyways

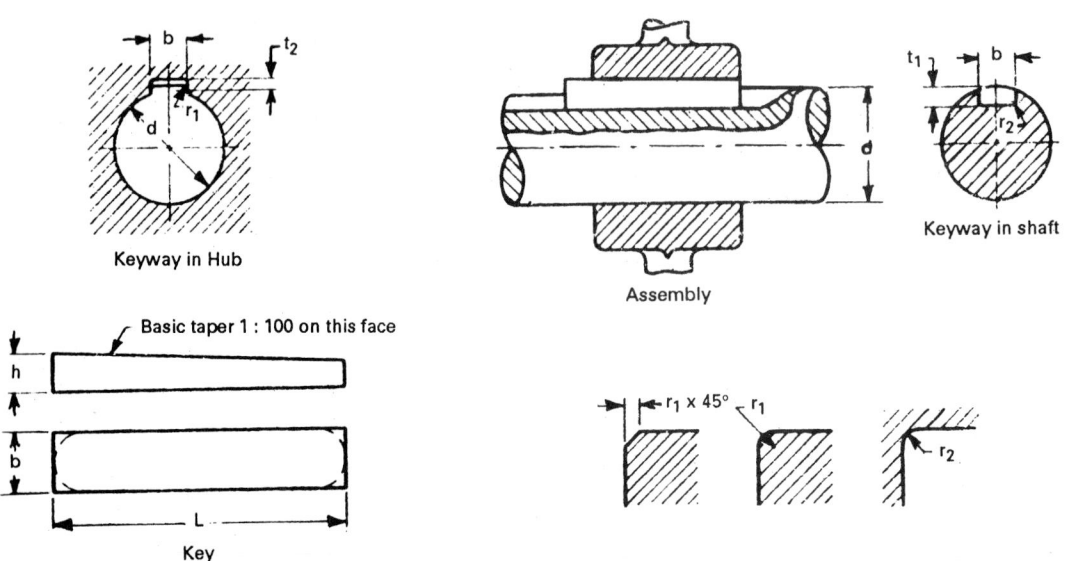

Keyway in Hub

Keyway in shaft

Assembly

Basic taper 1 : 100 on this face

Key

$r_1 \times 45°$

Shaft		Key			Keyway in shaft and hub					
Above	Up to and including	Width, b ($h9$)	Height, h	Chamfer or radius r_1, min	Keyway width, b ($D10$)	Depth in shaft, t_1	Tolerance on t_1	Depth in hub, t_2	Tolerance on t_2	Radius, r_2, max
6	8	2	2	0.16	2	1.2	+0.05	0.5		0.16
8	10	3	3	—	3	1.8	—	0.9		
10	12	4	4		4	2.5		1.2		
12	17	5	5		5	3.0		1.7	+0.1	0.25
17	22	6	6	0.25	6	3.5		2.1		
22	30	8	7	—	8	4.0	+0.10	2.5		—
30	38	10	8		10	5.0		2.5		
38	44	12	8		12	5.0		2.5	—	
44	50	14	9	0.40	14	5.5		2.9		0.40
50	58	16	10	—	16	6.0		3.4		
58	65	18	11		18	7.0	—	3.3		—
65	75	20	12		20	7.5		3.8		
75	85	22	14	0.60	22	8.5		4.8		
85	95	25	14	—	25	9.0		4.3	+0.2	0.60
95	110	28	18		28	10.0		5.3		—
110	130	32	10		32	11.0		6.2		
130	150	36	25		36	12.0		7.2		
150	170	40	22	1.00	40	13.0	+0.15	8.2		1.00
170	200	45	25	—	45	15.0		9.2	—	—
200	230	50	28		50	17.0		10.1		
230	260	56	32		56	19.0		12.1		
260	290	63	32	1.60	63	20.0		11.1		1.60
290	330	70	36	—	70	22.0		13.1	+0.3	—
330	380	80	40		80	25.0		14.1		
380	440	90	45	2.50	90	28.0		16.1		2.50
440	500	100	50		100	31.0		18.1		

Source: IS 2292, 1963.

Particular	Formula
The necessary length of the key	$l = \dfrac{2M_t}{\mu_1 bpD}$ (17–10)
The axial force necessary to drive the key home (Fig. 17–2)	$F_a = F_\mu + F_\beta = 2\mu_2 F + F \tan \beta$ (17–11)
	where $\mu_2 = 0.10$, $\tan \beta = 0.0104$ if the taper is 1 in 100
The axial force is also given by the equation	$F_a = 0.21pbl$ (17–12)

FRICTION OF FEATHER KEYS (FIG. 17–3)

The circumferential force (Fig. 17–3)	$F_t = \dfrac{M_t}{a}$ (17–13)
The resistance to be overcome when a hub connected to a shaft by a feather, Fig. 17–3a and subjected to torque M_t, is moved along the shaft	$\begin{aligned} R &= \mu F_t + \mu_2 F' && (17\text{–}14) \\ &= (\mu + \mu_2)F_t && (17\text{–}15) \end{aligned}$
	and $F' = F'' = F_a$ = force assumed to be acting at the shaft axis without changing the equilibrium Fig. 17–3a
The equation for resistance R, if μ and μ_2 are equal	$R = 2\mu F_t$ (17–16)
The equation for torque if two feather keys are used, Fig. 17–3b	$M_t = 2F_2 a$ (17–17)
The force F_2 applied at key when two feather keys are used, Fig. 17–3b	$F_2 = \dfrac{M_t}{2a} = \dfrac{F_t}{2}$ (17–18)
The resistance to be overcome when the hub connected to the shaft by two feather keys Fig. 17–3b and subjected to torque M_t is moved along the shaft	$R_2 = 2\mu F_2 = \dfrac{R}{2}$ (17–19)
For Gib-headed and Woodruff keys and keyways	Refer to Tables 17–4 and 17–5.

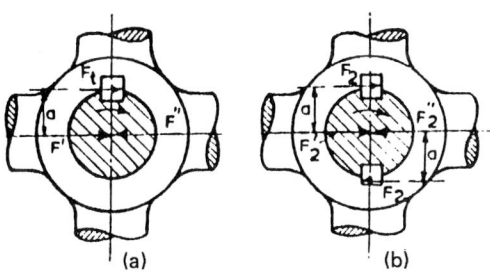

(a)   (b)

FIGURE 17–3 Feather key.

TABLE 17-4
Gib-head keys and keyways (all dimensions in mm)

Assembly — Keyway radius in shaft and hub — Chamber or radius of key — Key — Section X X — Taper 1:100 — 30° or 45°

| Shaft diameter, d | | Key | | | | | | Key in shaft and hub | | | | Radius at bottom of keyway |
Above	Up to and including	Width, b ($h9$)	Height (nominal) h	Tolerance on h	Height of gib-head, h_1	Chamfer or radius, r_1(min)	Width of keyway ($D10$)	Depth in shaft, t_1	Tolerance on t_1	Depth in hub, t_2	Tolerance on t_2	$r_{2(max)}$
10	12	4	4		7	0.16	4	2.5		1.2		0.16
12	17	5	5	+0.1	8	—	5	3		1.7		
17	22	6	6		10	0.25	6	3.5		2.1		0.25
22	30	8	7		11		8	4		2.5		
30	38	10	8		12	—	10	5	+0.1	2.5	+0.1	
38	44	12	8		12		12	5		2.5		
44	50	14	9	—	14	0.40	14	5.5		2.9		
50	58	16	10		16		16	6	—	3.4		0.4
58	65	18	11		18		18	7		3.5		
65	75	20	12	+0.2	20		20	7.5		3.8		
75	85	22	14		22	0.60	22	8.5		4.8		0.60
85	95	25	14		25		25	9		4.3	+0.15	
95	110	28	16		28		28	10		5.3		
110	130	32	18		32		32	11		6.2		
130	150	36	20		36		36	12		7.2		
150	170	40	22		40	1.00	40	13	+0.15	8.2		1.00
170	200	45	25		45		45	15		9.2		
200	230	50	28		50		50	17		10.1		
230	260	56	32		56		56	19		12.1		
260	290	63	32	+0.3	56	1.60	63	20		11.1		1.60
290	330	70	36		63	—	70	22		13.1	+0.3	
330	380	80	40		70		80	25		14.1		
380	440	90	45		75	2.50	90	28		16.1		
440	500	100	50		80		100	31		18.1		2.50

17.8

TABLE 17–5
Woodruff keys and keyways (all dimensions in mm)

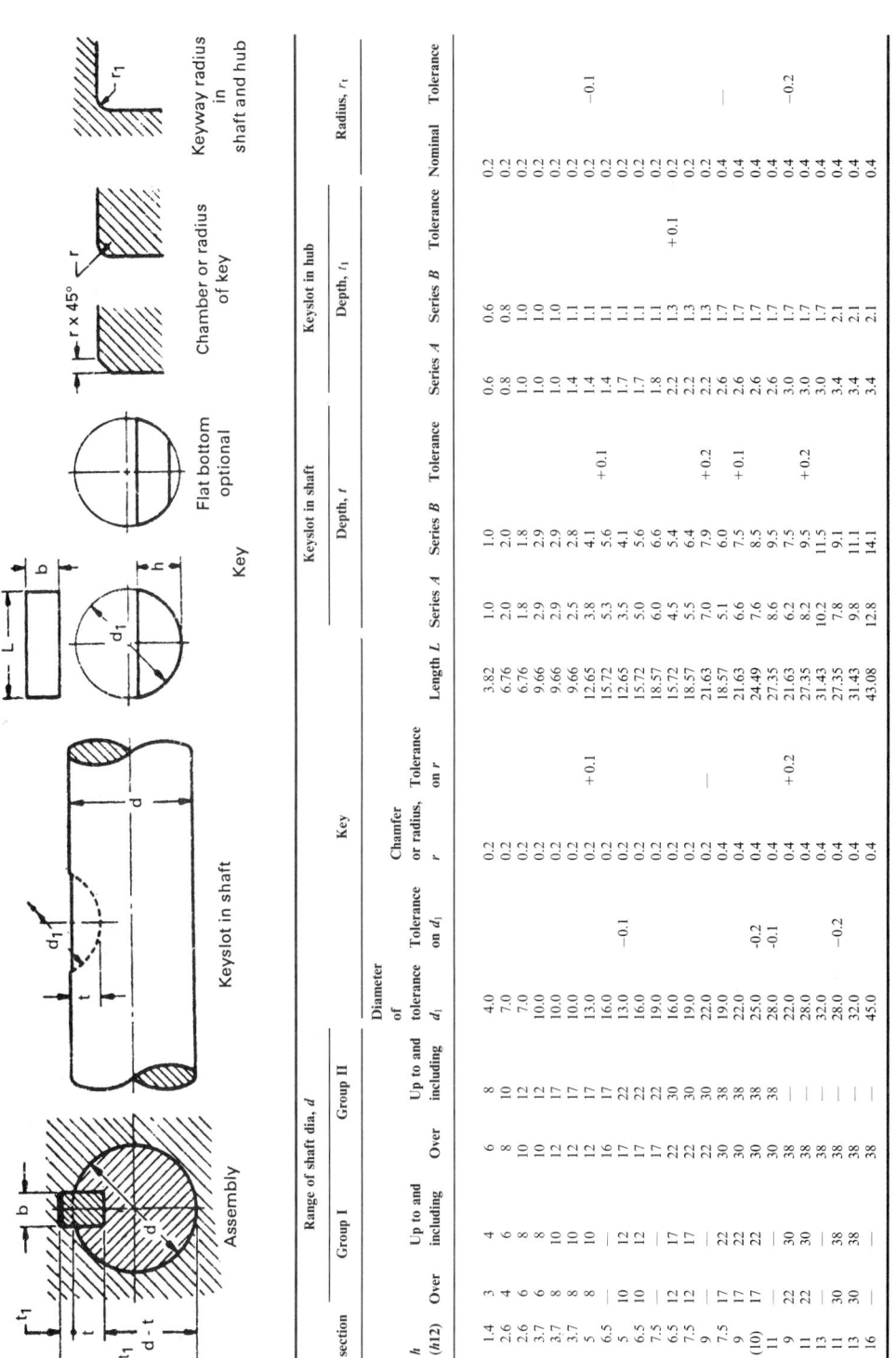

b ($h9$)	h ($h12$)	Group I Over	Group I Up to and incl.	Group II Over	Group II Up to and incl.	Diameter of tolerance d_1	Tolerance on d_1	Chamfer or radius r	Tolerance on r	Length L	Shaft t Series A	Shaft t Series B	Shaft t Tolerance	Hub t_1 Series A	Hub t_1 Series B	Hub t_1 Tolerance	Radius r_1 Nominal	Radius r_1 Tolerance
1	1.4	3	4	6	8	4.0		0.2		3.82	1.0	1.0		0.6	0.6		0.2	
1.5	2.6	4	6	8	10	7.0		0.2		6.76	2.0	2.0		0.8	0.8		0.2	
2	2.6	6	8	10	12	7.0		0.2		6.76	1.8	1.8		1.0	1.0		0.2	
2	3.7	6	8	10	12	10.0		0.2	+0.1	9.66	2.9	2.9		1.0	1.0		0.2	
2.5	3.7	8	10	12	17	10.0		0.2		9.66	2.9	2.9		1.0	1.0		0.2	−0.1
3	3.7	8	10	12	17	10.0		0.2		9.66	2.5	2.8		1.4	1.1		0.2	
3	5	8	10	12	17	13.0		0.2		12.65	3.8	4.1	+0.1	1.4	1.1		0.2	
3	6.5	8	10	16	17	16.0	−0.1	0.2		15.72	5.3	5.6		1.4	1.1		0.2	
4	5	10	12	17	22	13.0		0.2		12.65	3.5	4.1		1.7	1.1		0.2	
4	6.5	10	12	17	22	16.0		0.2		15.72	5.0	5.6		1.7	1.1		0.2	
4	7.5			17	22	19.0		0.2		18.57	6.0	6.6		1.8	1.1	+0.1	0.2	
5	6.5	10	12	22	30	16.0		0.2		15.72	4.5	5.4		2.2	1.3		0.2	
5	7.5	12	17	22	30	19.0		0.2		18.57	5.5	6.4		2.2	1.3		0.2	
5	9	12	17	22	30	22.0		0.4	—	21.63	7.0	7.9	+0.2	2.6	1.7		0.4	
6	7.5	17	22	30	38	19.0		0.4		18.57	5.1	6.0		2.6	1.7		0.4	
6	9	17	22	30	38	22.0		0.4		21.63	6.6	7.5	+0.1	2.6	1.7		0.4	
6	(10)	17	22	30	38	25.0	−0.2	0.4		24.49	7.6	8.5		2.6	1.7		0.4	
6	11			30	38	28.0	−0.1	0.4		27.35	8.6	9.5		3.0	1.7		0.4	
8	9	22	30	38	—	22.0		0.4	+0.2	21.63	6.2	7.5		3.0	1.7		0.4	−0.2
8	11	22	30	38	—	28.0		0.4		27.35	8.2	9.5	+0.2	3.0	1.7		0.4	
8	13			38	—	32.0		0.4		31.43	10.2	11.5		3.4	2.1		0.4	
10	11	30	38	38	—	28.0	−0.2	0.4		27.35	7.8	9.1		3.4	2.1		0.4	
10	13	30	38	38	—	32.0		0.4		31.43	9.8	11.1		3.4	2.1		0.4	
10	16			38	—	45.0		0.4		43.08	12.8	14.1		3.4	2.1		0.4	

Notes: (1) The dimensions $d - t$ and $d + t_1$ may be specified on workshop drawings; (2) the key size 6×10 is nonpreferred; (3) the key size 2.5×3.7 shall be used in automobile industries only.
Source: IS 2294, 1963.

Particular	Formula

SPLINES

Parallel-sided or straight-sided spline

The torque which an integral multispline shaft can transmit (Tables 17–6 to 17–12)

$$M_t = \tfrac{1}{2}\, phli(D - h) \qquad\qquad (17\text{–}20)$$

TABLE 17–6
Proportions of SAE standard parallel side splines

Types of spline fittings	Symbols	Proportions	Fit	Bearing pressure, p	
				MPa	kgf/mm^2
	w	$w = 0.241D$			
	h	$4A, h = 0.075D$	A	20.6	2.10
	h	$4B, h = 0.125D$	B	13.7	1.40
	w	$w = 0.250D$			
	h	$6A, h = 0.050D$	A	20.6	2.10
	h	$6B, h = 0.075D$	B	13.7	1.40
	h	$6C, h = 0.100D$	C	6.9	0.70
	w	$w = 0.156D$			
	h	$10A, h = 0.045D$	A	20.6	2.10
	h	$10B, h = 0.070D$	B	13.7	1.40
	h	$10C, h = 0.095D$	C	6.9	0.70
	w	$w = 0.098D$			
	h	$16A, h = 0.045D$	A	20.6	2.10
	h	$16B, h = 0.070D$	B	13.7	1.40
	h	$16C, h = 0.095D$	C	6.9	0.70

TABLE 17–7
Proportions of involute spline profile (American Standard)

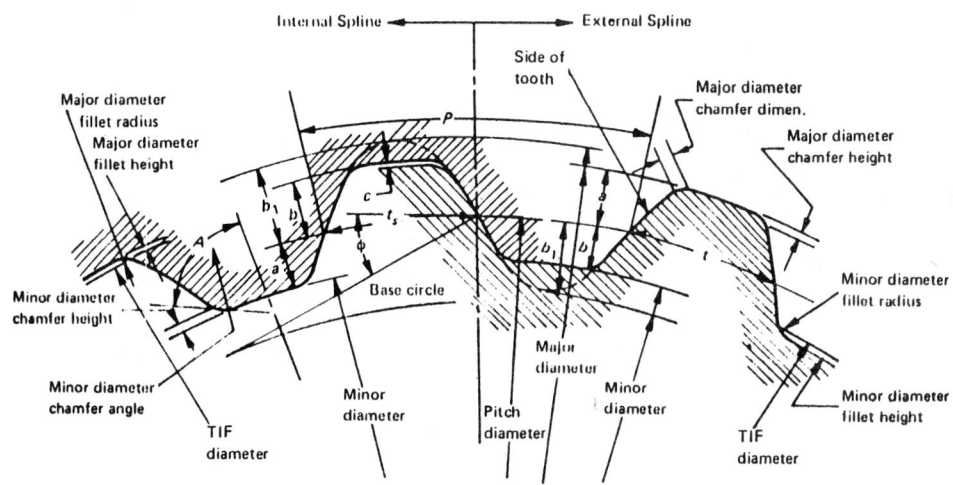

Spline characteristics	Symbols	Proportions	
		$P = \frac{1}{2}$ **through** $\frac{12}{24}$	$P = \frac{6}{32}$ **through** $\frac{48}{96}$
Pitch diameter	D	$D = zm = \dfrac{z}{P}$	$D = zm = z/P$
Circular pitch	p	$p = (\pi/P)$	$p = (\pi/P)$
Tooth thickness	t	$t = \dfrac{\pi m}{2} = \dfrac{\pi}{2P}$	$t = (\pi m/2) = (\pi/2P)$
Diametral pitch	P	$P = (\pi/p)$	$P = (\pi/p)$
Addendum	a	$a = 0.5m = \dfrac{0.500}{P}$	$a = 0.5m = 0.500/P$
Dedendum (internal)	b_1	$b_1 = 0.90m = \dfrac{0.900}{P}$	$b_1 = 0.9m = 0.900/P$
Dedendum	b	$b = 0.5m = \dfrac{0.500}{P}$	$b = 0.5m = 0.500/P$
Dedendum (external)	b_1	$b_1 = 0.9m = 0.900/P$	$b_1 = 1.0m = 1.000/P$
Major diameter (internal)	D_{oi}	$D_{oi} = (z + 1.8)m$ $= (z + 1.8)/P$	$D_{oi} = (z + 1.8)m$ $= (z + 1.8)/P$
Minor diameter (external)	D_{me}	$D_{me} = (z - 1.8)m$ $= (z - 1.8)/P$	$D_{me} = (z - 2.0)m$ $= (z - 2.0)/P$

Source: Courtesy H. L. Horton, ed., *Machinery's Handbook*, 15th ed., The Industrial Press, New York, 1957.

TABLE 17–8
Straight sided splines (all dimensions in mm)

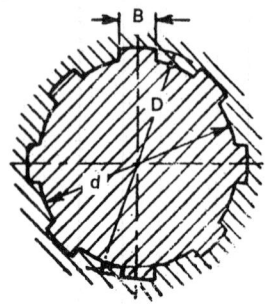 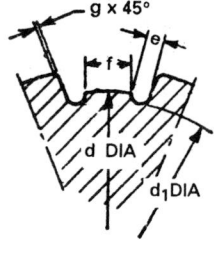 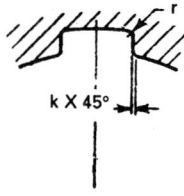

		Splined shaft and hub profile			Splined shaft			Splined hub			
Nominal size, $i \times d \times D$	No. of splines, i	Minor diameter, d	Major diameter, O	Width, B	d_1,[a] min	e,[a] max	f[a]	g, max	k, max	r, max	Centering on
					Light-Duty Series						
$6 \times 23 \times 26$	6	23	26	6	22.1	1.25	3.54	0.3	0.3	0.2	Inside diameter[a]
$6 \times 26 \times 30$	6	26	30	6	24.6	1.84	3.85	0.3	0.3	0.2	
$6 \times 28 \times 32$	6	28	32	7	26.7	1.77	4.03	0.3	0.3	0.2	
$8 \times 32 \times 36$	8	32	36	6	30.4	1.89	2.71	0.4	0.4	0.3	
$8 \times 36 \times 40$	8	36	40	7	34.5	1.78	3.46	0.4	0.4	0.3	
$8 \times 42 \times 46$	8	42	46	8	40.4	1.68	5.03	0.4	0.4	0.3	
$8 \times 46 \times 50$	8	46	50	9	44.6	1.61	5.75	0.4	0.4	0.3	
$8 \times 52 \times 58$	8	52	58	10	49.7	2.72	4.89	0.5	0.5	0.5	Inside diameter or flanks[b]
$8 \times 56 \times 62$	8	56	62	10	53.6	2.76	6.38	0.5	0.5	0.5	
$8 \times 62 \times 68$	8	62	68	12	59.8	2.48	7.31	0.5	0.5	0.5	
$10 \times 72 \times 78$	10	72	78	12	69.6	2.54	5.45	0.5	0.5	0.5	
$10 \times 82 \times 88$	10	82	88	12	79.3	2.67	8.62	0.5	0.5	0.5	
$10 \times 92 \times 98$	10	92	98	14	89.4	2.36	10.08	0.5	0.5	0.5	
$10 \times 102 \times 108$	10	102	108	16	99.9	2.23	11.49	0.5	0.5	0.5	
$10 \times 112 \times 120$	10	112	120	18	108.8	3.23	10.72	0.5	0.5	0.5	
					Medium-Duty Series						
$6 \times 11 \times 14$	6	11	14	3	9.9	1.55		0.3	0.3	0.2	Inside diameter[a]
$6 \times 13 \times 16$	6	13	16	3.5	12.0	1.50	0.32	0.3	0.3	0.2	
$6 \times 16 \times 20$	6	16	20	4	14.5	2.10	0.16	0.3	0.3	0.2	
$6 \times 18 \times 22$	6	18	22	5	16.7	1.95	0.45	0.3	0.3	0.2	
$6 \times 21 \times 25$	6	21	25	5	19.5	1.98	1.95	0.3	0.3	0.2	
$6 \times 23 \times 28$	6	23	28	6	21.3	2.30	1.34	0.3	0.3	0.2	
$6 \times 26 \times 32$	6	26	32	6	23.4	2.94	1.65	0.4	0.4	0.3	
$6 \times 28 \times 34$	6	28	34	7	25.9	2.94	1.70	0.4	0.4	0.3	
$8 \times 32 \times 38$	8	32	38	6	29.4	3.30	0.15	0.4	0.4	0.3	
$8 \times 36 \times 42$	8	36	42	7	33.5	3.01	1.02	0.4	0.4	0.3	
$8 \times 42 \times 48$	8	42	48	8	39.5	2.91	2.54	0.4	0.4	0.3	
$8 \times 46 \times 54$	8	46	54	9	42.7	4.10	0.86	0.5	0.5	0.3	
$8 \times 52 \times 60$	8	52	60	10	48.7	4.00	2.44	0.5	0.5	0.5	Inside diameter or flanks[b]
$8 \times 56 \times 65$	8	56	65	10	52.2	4.74	2.50	0.5	0.5	0.5	
$8 \times 62 \times 72$	8	62	72	12	57.8	5.00	2.40	0.5	0.5	0.5	
$10 \times 72 \times 82$	10	72	82	12	67.4	5.43		0.5	0.5	0.5	
$10 \times 82 \times 92$	10	82	92	12	77.1	5.40	3.00	0.5	0.5	0.5	
$10 \times 92 \times 102$	10	92	102	14	87.3	5.20	4.50	0.5	0.5	0.5	
$10 \times 102 \times 112$	10	102	112	16	97.7	4.90	6.30	0.5	0.5	0.5	
$10 \times 112 \times 125$	10	112	125	18	106.3	6.40	4.40	0.5	0.5	0.5	

[a] These values are based on the generating process.
[b] Inside centering is not always possible with generating processes.
Source: IS 2327, 1963.

TABLE 17–9
Tolerances for straight-sided splines (all dimensions in mm)

Assembly of splined hub and shaft			Width of hub B Soft or hardened		Minor diameter of hub, d Soft or hardened	Major diameter of hub, D Soft or hardened
Splined hub	For centering on inner diameter or flanks	Shaft sliding or fixed	D9	F10	H7	H11
Splined shaft	For centering on inner diameter	Shaft sliding inside hub	h8	e8	f7	a11
		Shaft fixed in hub	p6	h6	j6	a11
		Shaft sliding inside hub	h8	e8	—	a11
	For centering on flanks	Shaft fixed in hub	u6	k6	—	a11

Particular	Formula

Involute-sided spline

AMERICAN STANDARD (TABLE 17–7). The addendum a and dedendum b for a flat root, Table 17–7

$$a = b = m = \frac{1}{P} \tag{17–21}$$

The area resisting shear, Table 17–7

$$A_\tau = \frac{\pi DL}{2} \tag{17–22}$$

The minimum height of contact on one tooth

$$h = 0.8m = \frac{0.8}{P} = \frac{0.8D}{z} \tag{17–23}$$

The corresponding area of contact of all z teeth

$$A = \left(\frac{0.8D}{z}\right) zL = 0.8DL \tag{17–24}$$

The torque capacity of teeth in shear

$$M_t = \left(\frac{\pi DL}{z}\right) \frac{D}{2} \tau_d = 0.7854D^2 L \tau_d \tag{17–25}$$

The torque capacity of the spline in bearing with $\sigma_b = 2\sigma_{dc}$

$$M_{tb} = 0.8D^2 L \sigma_{dc} \tag{17–26}$$

TABLE 17–10
Straight-sided splines for machine tools (all dimensions in mm)

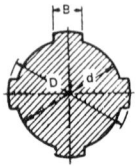
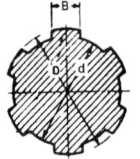
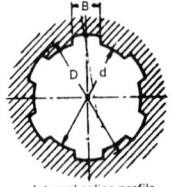

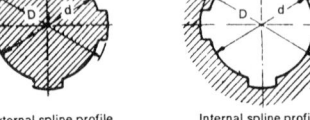

(a) Straight sided splines - 4 Splines

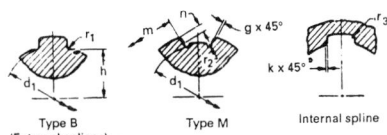

(b) Straight sided splines - 6 Splines

	4 Splines		
Nominal size, $i^a \times d \times D$	Minor diameter, d	Major diameter, D	Width, B
$4 \times 11 \times 15$	11	15	3
$4 \times 13 \times 17$	13	17	4
$4 \times 16 \times 20$	16	20	6
$4 \times 18 \times 22$	18	22	6
$4 \times 21 \times 25$	21	25	8
$4 \times 24 \times 28$	24	28	8
$4 \times 28 \times 32$	28	32	10
$4 \times 32 \times 38$	32	38	10
$4 \times 36 \times 42$	36	42	12
$4 \times 42 \times 48$	42	48	12
$4 \times 46 \times 52$	46	52	14
$4 \times 52 \times 60$	52	60	14
$4 \times 58 \times 65$	58	65	16
$4 \times 62 \times 70$	62	70	16
$4 \times 68 \times 78$	68	78	16

^{a} i = number of splines
Source: IS 2610, 1964.

	6 Splines		
Nominal size, $i^a \times d \times D$	Minor diameter, d	Major diameter, D	Width, B
$6 \times 21 \times 25$	21	25	5
$6 \times 23 \times 28$	23	28	6
$6 \times 26 \times 32$	26	32	6
$6 \times 28 \times 34$	28	34	7
$6 \times 32 \times 38$	32	38	8
$6 \times 36 \times 42$	36	42	8
$6 \times 42 \times 48$	42	48	10
$6 \times 46 \times 52$	46	52	12
$6 \times 52 \times 60$	52	60	14
$6 \times 58 \times 65$	58	65	14
$6 \times 62 \times 70$	62	70	16
$6 \times 68 \times 78$	68	78	16
$6 \times 72 \times 82$	72	82	16
$6 \times 78 \times 90$	78	90	16
$6 \times 82 \times 95$	82	95	16
$6 \times 88 \times 100$	88	100	16
$6 \times 92 \times 105$	92	105	20
$6 \times 98 \times 110$	98	110	20
$6 \times 105 \times 120$	105	120	20
$6 \times 115 \times 130$	115	130	20
$6 \times 130 \times 145$	130	145	24

TABLE 17–11
Undercuts, chamfers, and radii for straight-sided splines[a] (all dimensions in mm)

Designation, $i \times d \times D$	B	External splines Type A d_1, min	g, max	f, min	Type B h	r_1, max	Type M m	n	r_2	Internal splines k, max	r_3, max	Projected tip width of hub
$4 \times 11 \times 15$	3	9.6	0.2	1.50	5.0	0.10	2.82	1.70	0.3	0.2	0.15	0.5
$4 \times 13 \times 17$	4	11.8	0.2	2.37	5.5	0.10	3.76	1.70	0.3	0.2	0.15	0.5
$4 \times 16 \times 20$	6	15.0	0.3	2.87	6.7	0.15	5.64	1.70	0.3	0.3	0.25	0.7
$4 \times 18 \times 22$	6	16.9	0.3	4.35	7.7	0.15	5.64	1.70	0.3	0.3	0.25	0.7
$4 \times 21 \times 25$	8	20.1	0.3	5.00	8.9	0.15	7.52	1.70	0.6	0.3	0.25	0.7
$4 \times 24 \times 28$	8	23.0	0.3	7.30	10.4	0.15	7.52	1.70	0.6	0.3	0.25	0.7
$4 \times 28 \times 32$	10	26.8	0.5	7.39	12.1	0.25	9.40	1.63	0.6	0.5	0.40	1.0
$4 \times 32 \times 38$	10	30.3	0.5	9.56	14.2	0.25	9.40	2.55	0.6	0.5	0.40	1.0
$4 \times 36 \times 42$	12	34.5	0.5	11.03	15.9	0.25	11.28	2.55	0.6	0.5	0.40	1.0
$4 \times 42 \times 48$	12	40.2	0.5	15.41	19.0	0.25	11.28	2.55	1.0	0.5	0.40	1.0
$4 \times 46 \times 52$	14	44.4	0.5	16.79	20.7	0.25	13.16	2.55	1.0	0.5	0.40	1.3
$4 \times 52 \times 60$	14	49.5	0.5	21.63	23.7	0.25	13.16	3.40	1.0	0.5	0.40	1.3
$4 \times 56 \times 65$	16	56.2	0.5	23.26	26.4	0.25	15.04	2.98	1.0	0.5	0.40	1.6
$4 \times 62 \times 70$	16	59.5	0.5	23.61	28.3	0.25	15.04	3.40	1.0	0.5	0.40	1.6
$4 \times 68 \times 78$	16	64.4	0.5	27.57	31.2	0.25	15.04	4.25	1.0	0.5	0.40	1.6

[a] Four splines; see Fig. 17–4a.
Source: IS 2610, 1964

TABLE 17–12
Undercuts, chamfers, and radii for straight-sided splines[a] (all dimensions in mm)

Designation, $i \times d \times D$	B	External splines Type A d_1, min	g, max	f, min	Type B h	r_1, max	Type M m	n	r_2	Internal splines k, max	r_3, max	Projected tip width of hub
$6 \times 21 \times 25$	5	19.50	0.3	1.95	9.7	0.15	4.70	1.70	0.6	0.3	0.2	0.7
$6 \times 23 \times 28$	6	21.30	0.3	1.34	11.0	0.15	5.64	2.13	0.6	0.3	0.2	0.7
$6 \times 26 \times 32$	6	23.40	0.4	1.65	11.8	0.15	5.64	2.55	0.6	0.4	0.3	1.0
$6 \times 28 \times 34$	7	25.90	0.4	1.70	12.9	0.25	6.58	2.55	0.6	0.4	0.3	1.0
$6 \times 32 \times 38$	8	29.90	0.5	2.83	14.8	0.25	7.52	2.55	0.6	0.5	0.4	1.0
$6 \times 36 \times 42$	8	33.70	0.5	4.95	16.5	0.25	7.52	2.55	0.6	0.5	0.4	1.0
$6 \times 42 \times 48$	10	39.94	0.5	6.02	19.3	0.25	9.40	2.55	1.0	0.5	0.4	1.0
$6 \times 46 \times 52$	12	44.16	0.5	5.81	21.1	0.25	11.28	2.55	1.0	0.5	0.4	1.3
$6 \times 52 \times 60$	14	49.50	0.5	5.89	23.9	0.25	13.16	3.40	1.0	0.5	0.4	1.3
$6 \times 58 \times 65$	14	55.74	0.5	8.29	26.7	0.25	13.16	3.98	1.0	0.5	0.4	1.6
$6 \times 62 \times 70$	16	59.50	0.5	8.03	28.6	0.25	15.04	3.40	1.0	0.5	0.4	1.6
$6 \times 68 \times 78$	16	64.40	0.5	9.73	31.4	0.25	15.04	4.25	1.0	0.5	0.4	1.6
$6 \times 72 \times 82$	16	68.30	0.5	12.67	33.4	0.25	15.04	4.25	1.6	0.5	0.4	2.0
$6 \times 78 \times 90$	16	73.00	0.5	13.07	36.2	0.25	15.04	5.10	1.6	0.5	0.4	2.0
$6 \times 82 \times 95$	16	79.60	0.5	13.96	38.0	0.25	15.04	5.53	1.6	0.5	0.4	2.0
$6 \times 88 \times 100$	16	82.90	0.5	17.84	41.3	0.25	15.04	5.10	1.6	0.5	0.4	2.0
$6 \times 92 \times 105$	20	87.10	0.6	18.96	43.1	0.30	18.80	5.53	1.6	0.6	0.5	2.0
$6 \times 98 \times 110$	20	93.40	0.6	19.22	46.4	0.30	18.80	5.10	2.0	0.6	0.5	2.0
$6 \times 105 \times 120$	20	98.80	0.6	19.25	49.2	0.30	18.80	6.38	2.0	0.6	0.5	2.4
$6 \times 115 \times 130$	20	108.4	0.6	24.75	54.2	0.30	18.80	6.38	2.5	0.6	0.5	2.4
$6 \times 130 \times 145$	24	123.9	0.6	29.20	61.8	0.30	22.56	6.38	2.5	0.6	0.5	2.4

[a] Six splines, see Fig. 17–4b.

Particular	Formula
The theoretical torque capacity of straight-sided spline with sliding according to SAE	$M_t = 6.895 \times 10^6 i \left(\dfrac{D+d}{4} \right) hL$ **SI** (17–26a)

where

i = number of splines

D, d = diameter as shown in Table 17–7, m

d = inside diameter of spline, m

D = pitch diameter of spline, m

h = depth of spline, m

L = length of spline contact, m

M_t in N m.

$$M_t = 1000i \left(\frac{D+d}{4} \right) hL$$

US Customary System units (17–26b)

where M_t in lb in; d, D, L, and h in in

Equating the strength of the spline teeth in shear to the shear strength of shaft, the length of spline for a hollow shaft	$L = \dfrac{D_{me}^3 (1 - D_i^4 / D_{me}^4)}{4D^2}$ (17–26c)

where

D_i = internal diameter of a hollow shaft, m (in)

D_{me} = minor diameter (external), m (in)

The length of spline for a solid shaft	$L = \dfrac{D_{me}^3}{4D^2}$ (17–26d)
The effective length of spline for a hollow shaft used in practice according to the SAE	$L_e = \dfrac{D_{me}^3 (1 - D_i^4 / D_{me}^4)}{D^2}$ (17–26e)

For solid shaft $D_i = 0$.

For diametral pitches used in involute splines (SAE and ANSI)	Refer to Table 17–13.

TABLE 17–13
Diametral pitches[a] used in involute splines (SAE and ANSI)

$\dfrac{2.5}{5}$	$\dfrac{3}{6}$	$\dfrac{4}{8}$	$\dfrac{5}{10}$	$\dfrac{6}{12}$	$\dfrac{8}{16}$	$\dfrac{10}{20}$	$\dfrac{12}{24}$	$\dfrac{16}{32}$	$\dfrac{20}{40}$	$\dfrac{24}{48}$	$\dfrac{32}{64}$	$\dfrac{40}{80}$	$\dfrac{48}{96}$

[a] Diametral pitches are designated as fractions; the numerator of these fractions is the diametral pitch, P.

INDIAN STANDARD (FIGS. 17–4 AND 17–5, TABLES 17–14 AND 17–15)

The value of profile displacement (Fig. 17–4)	$xm = \frac{1}{2}(d_1 - mz - 1.1m)$ (17–27)

The value of xm varies from $-0.05m$ to $+0.45m$

Particular	Formula	
The number of teeth	$z = \dfrac{1}{m}(d_1 - 2xm - 1.1m)$	(17–28)
The minor diameter of the internal spline (Fig. 17–4a)	$d_2 = mz + 2xm - 0.9m = d_1 - 2m$	(17–29)
The major diameter of the external spline (Fig. 17–4a)	$d_3 = mz + 2xm + 0.9m = d_1 - 0.2m$	(17–30)
The minor diameter of the external spline (Fig. 17–4a)	$d_4 = mz + 2xm - 1.1m = d_1 - 2.2m$	(17–31)

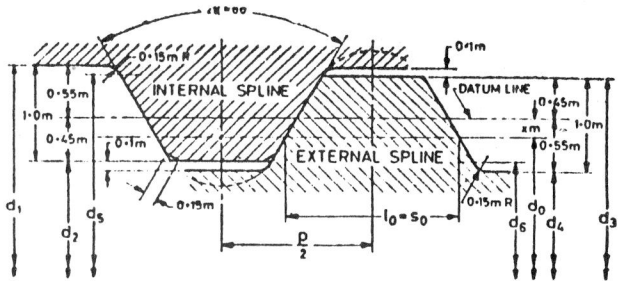

FIGURE 17–4(a)   Reference profile of an involute-sided spline. (*Source: IS 3665, 1966.*)

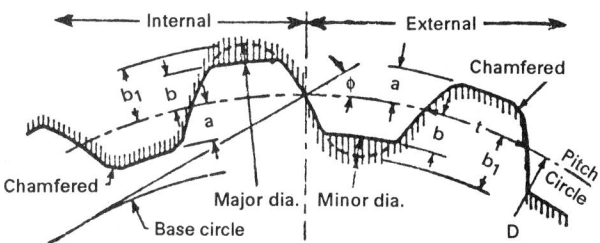

FIGURE 17–4(b)   Nomenclature of the involute spline profile.

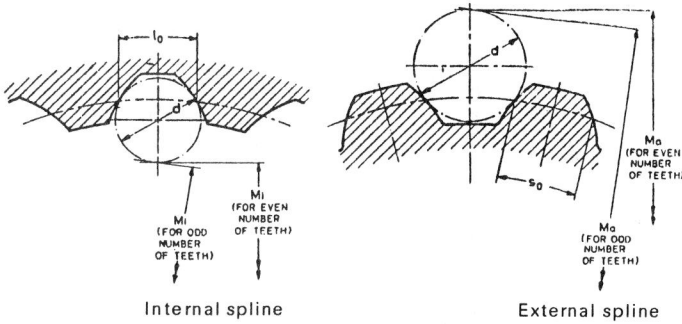

FIGURE 17–5   Measurement between pins and measurement over pins of an involute-sided spline. (*Source: IS 3665, 1966.*)

TABLE 17–14
Dimensions (in mm) for involute splines of module 2

Nominal size $d_1 \times d_2$	z	d_o	d_b	d_3	d_4	d_5, min	d_6, max	xm	$l_o = s_o$	Internal spline Pin diameter, d	Internal spline Measurement between pins, Mi	Internal spline Deviation factor, fi	External spline Pin diameter, d	External spline Measurement over pins, Ma	External spline Deviation factor, fa	External spline z'	Tooth thickness over z' teeth / Tooth thickness deviation factor, 0.866
15 × 11	6	12	10.392	14.6	10.6	14.68	10.92	+0.4	3.603	3.5	7.629	2.42	5.5	22.212	1.11	2	9.121
17 × 13	7	14	12.124	16.6	12.6	16.68	12.92	+0.4	3.603	3.5	9.324	2.19	5.0	22.695	1.13	2	9.214
18 × 14	7	14	12.124	17.6	13.6	17.68	13.92	+0.9	4.181	3.5	10.379	1.61	6.0	25.588	1.06	2	9.714
20 × 16	8	16	13.856	19.6	15.6	19.68	15.92	+0.9	4.181	3.5	12.736	1.66	6.0	28.206	1.11	2	9.807
22 × 18	9	18	15.588	21.6	17.6	21.68	17.92	+0.9	4.181	3.5	14.460	1.64	5.5	28.790	1.13	—	—
25 × 21	11	22	19.053	24.6	20.6	24.68	20.92	+0.4	3.603	3.5	17.478	1.96	4.5	29.898	1.28	3	15.621
28 × 24	12	24	20.785	27.6	23.6	27.68	23.92	+0.4	4.181	3.5	20.738	1.68	5.0	34.161	1.23	3	14.807
30 × 26	14	28	24.299	29.6	25.6	29.69	25.91	-0.1	3.326	3.5	22.484	2.41	4.0	34.144	1.46	3	15.807
32 × 28	14	28	24.249	31.6	27.6	31.69	27.91	+0.9	4.681	3.5	24.738	1.69	4.5	37.016	1.30	3	15.493
35 × 31	16	32	27.713	34.6	30.6	34.69	30.91	+0.4	3.603	3.5	27.711	1.88	4.0	39.000	1.42	3	21.028
38 × 33	17	34	29.445	36.6	32.6	36.69	32.91	+0.4	3.603	3.5	29.571	1.86	4.0	40.857	1.42	4	15.179
37 × 34	18	36	31.177	37.6	33.6	37.69	33.91	-0.1	3.026	3.5	30.566	2.15	4.0	42.181	1.50	4	21.621
40 × 36	18	36	31.177	39.6	35.6	39.69	35.91	+0.9	4.181	3.5	32.739	1.70	4.0	45.137	1.15	4	20.807
42 × 38	20	40	34.641	41.6	37.6	41.69	37.91	-0.1	3.026	3.5	34.589	2.08	4.0	46.195	1.52	4	21.400
45 × 41	21	42	36.373	44.6	40.6	44.69	40.91	+0.4	3.603	3.5	37.604	1.84	4.0	48.938	1.46	5	21.493
47 × 43	22	44	38.105	46.6	42.6	46.69	42.91	+0.4	3.603	3.5	39.720	1.84	4.0	51.074	1.47	5	27.435
48 × 44	22	44	38.105	47.6	43.6	47.69	43.91	+0.9	4.181	3.5	40.740	1.70	4.0	51.912	1.43	5	27.435
50 × 46	24	48	41.569	49.6	45.6	49.69	45.91	-0.1	3.026	3.5	42.621	2.00	4.0	54.218	1.54	5	21.179
(52 × 48)	24	48	41.569	51.6	47.6	51.69	47.91	+0.9	4.181	3.5	44.740	1.71	4.0	55.939	1.44	5	27.621
55 × 51	26	52	45.033	54.6	50.6	54.70	50.90	+0.4	3.603	3.5	47.724	1.82	4.0	59.109	1.50	5	27.307
(58 × 54)	28	56	48.497	57.6	53.6	57.70	53.90	-0.1	3.026	3.5	50.624	1.95	4.0	62.235	1.56	6	26.993
60 × 56	28	56	48.497	59.6	55.6	59.70	55.90	+0.9	4.181	3.5	52.740	1.71	4.0	63.984	1.47	6	33.435
(62 × 58)	30	60	51.962	61.6	57.6	61.70	57.90	-0.1	3.026	3.5	54.650	1.93	4.0	66.242	1.57	6	27.179
65 × 61	31	62	53.694	64.6	60.6	64.70	60.90	+0.4	3.600	3.5	57.648	1.80	4.0	69.058	1.53	6	33.214
(68 × 64)	32	64	55.426	67.6	63.6	67.70	63.90	+0.9	4.181	3.5	60.740	1.71	4.0	72.021	1.49	7	33.807
70 × 66	34	68	58.890	69.6	65.6	69.70	65.90	-0.1	3.026	3.5	62.663	1.90	4.0	74.253	1.59	7	32.993
(72 × 68)	34	68	58.890	71.6	67.6	71.70	67.90	+0.9	4.181	3.5	64.740	1.71	4.0	76.036	1.50	7	39.435
75 × 71	36	72	62.354	74.6	70.6	74.70	70.90	+0.4	3.603	3.5	67.729	1.79	4.0	79.166	1.55	7	39.121
(78 × 74)	38	76	65.818	77.6	73.6	77.70	73.90	-0.1	3.026	3.5	70.672	1.88	4.0	82.263	1.60	7	38.807
80 × 76	38	76	65.818	79.6	75.6	79.70	75.90	+0.9	4.181	3.5	72.740	1.72	4.0	84.063	1.52	7	39.807
(82 × 78)	40	80	69.283	81.6	77.6	81.70	77.90	-0.1	3.026	3.5	74.676	1.87	4.0	86.267	1.61	7	38.993

Note: Values within brackets are nonpreferred.

TABLE 17–15
Dimensions (in mm) for involute spline of module 2.5

Nominal size $d_1 \times d_2$	z	d_o	d_b	d_3	d_4	d_5, min	d_6, max	xm	$l_o = s_o$	Internal spline Pin diameter, d	Measurement between pins, Mi	Deviation factor, fi	External spline Pin diameter, d	Measurement over pins, Ma	Deviation factor, fa	z'	Tooth thickness over z' teeth, Tooth thickness deviation factor, 0.866
20 × 15	6	15.0	12.990	19.5	14.5	19.58	14.92	+1.125	5.226	4.6	10.552	1.71	9.0	33.258	1.03	2	12.026
22 × 17	7	17.5	15.155	21.5	16.5	21.58	16.92	+0.875	4.937	4.5	12.105	1.85	7.0	30.558	1.08	2	11.892
25 × 20	8	20.0	17.321	24.5	19.5	24.58	19.92	+1.125	5.226	4.5	15.552	1.72	7.0	34.113	1.13	2	12.252
28 × 23	10	25.0	21.651	27.5	22.5	27.58	22.92	+0.125	4.071	4.55	19.116	2.30	5.0	33.006	1.37	2	11.491
30 × 25	10	25.0	21.651	29.5	24.5	29.58	24.92	+1.125	5.226	4.5	20.552	1.72	6.5	38.151	1.19	3	19.293
32 × 27	11	27.5	23.816	31.5	26.5	31.59	26.91	+0.875	4.937	4.5	22.265	1.81	6.0	38.835	1.23	3	19.160
35 × 30	12	30.0	25.981	34.5	29.5	34.59	29.91	+1.125	5.226	4.5	25.552	1.72	6.0	42.093	1.25	3	19.526
37 × 32	13	32.5	28.146	36.5	31.5	36.59	31.91	+0.875	4.937	4.5	27.308	1.80	5.5	42.764	1.30	3	19.392
38 × 33	14	35.0	30.311	37.5	32.5	37.59	32.91	+0.125	4.071	4.5	28.316	2.26	5.0	43.096	1.43	3	18.759
40 × 35	14	35.0	30.311	39.5	34.5	39.59	34.91	+1.125	5.226	4.5	30.552	1.72	6.0	47.204	1.28	3	19.759
42 × 37	15	37.5	32.476	41.5	36.5	41.59	36.91	+0.875	4.937	4.5	32.340	1.79	5.5	47.881	1.33	3	19.625
45 × 40	16	40.0	34.641	44.5	39.5	44.59	39.91	+0.875	4.937	4.5	35.552	1.73	5.5	51.035	1.33	4	26.793
47 × 42	17	42.5	36.806	46.5	41.5	46.59	41.91	+0.125	4.071	4.5	37.365	1.78	5.5	52.974	1.36	4	26.660
48 × 43	18	45.0	38.971	47.5	42.5	47.59	42.91	+0.125	4.071	4.5	38.387	2.07	5.0	53.156	1.47	4	26.026
50 × 45	18	45.0	38.971	49.5	44.5	49.59	44.91	+0.875	4.937	4.5	40.552	1.73	5.5	56.100	1.36	4	27.026
(52 × 47)	19	47.5	41.136	51.5	46.5	51.59	46.91	+0.875	4.937	4.5	42.384	1.78	5.5	58.052	1.38	4	29.892
55 × 50	20	50.0	43.301	54.5	49.5	54.59	49.91	+1.125	5.226	4.5	45.552	1.73	5.5	61.157	1.38	4	27.259
(58 × 53)	22	55.0	47.631	57.5	52.5	57.60	52.90	+0.125	4.071	4.5	48.424	1.99	5.5	63.198	1.51	4	26.491
60 × 55	22	55.0	47.631	59.5	54.5	59.60	54.90	+1.125	5.226	4.5	50.552	1.73	5.5	66.206	1.40	5	34.193
(62 × 57)	23	57.5	49.796	61.5	56.5	61.60	56.90	+0.875	4.937	4.5	52.413	1.77	5.0	66.846	1.45	5	34.160
65 × 60	24	60.0	51.962	64.5	59.5	64.60	59.90	+1.125	5.226	4.5	55.552	1.73	5.0	69.924	1.44	5	34.526
(68 × 63)	26	65.0	56.292	67.5	62.5	67.60	62.90	+0.125	4.071	4.5	58.448	1.94	5.0	73.229	1.53	5	33.759
70 × 65	26	65.0	56.292	69.5	64.5	69.60	64.90	+1.125	5.226	4.5	60.552	1.73	5.0	74.954	1.46	5	34.759
(72 × 67)	27	67.5	58.457	71.5	66.5	71.60	66.90	+0.875	4.937	4.5	62.434	1.77	5.0	76.920	1.48	5	34.625
75 × 70	28	70.0	60.622	74.5	69.5	74.60	69.90	+1.125	5.226	4.5	65.552	1.73	5.0	79.981	1.47	6	41.793
(78 × 73)	30	75.0	64.952	77.5	72.5	77.60	72.90	+0.125	4.071	4.5	68.464	1.90	5.0	83.253	1.55	6	41.026
80 × 75	30	75.0	64.952	79.5	74.5	79.60	74.90	+1.125	5.226	4.5	70.552	1.73	5.0	85.004	1.48	6	42.026
(82 × 77)	31	77.5	67.117	81.5	77.5	81.60	77.90	+0.875	4.937	4.5	72.449	1.76	5.0	86.978	1.50	6	41.892
85 × 80	32	80.0	69.282	84.5	79.5	84.60	79.90	+1.125	5.226	4.5	75.552	1.73	5.0	90.026	1.49	6	42.259
(88 × 83)	34	85.0	73.612	87.5	82.5	87.60	82.90	+0.125	4.071	4.5	78.476	1.88	5.0	93.273	1.57	6	41.491
90 × 85	34	85.0	73.612	89.5	84.5	89.60	84.90	+1.125	5.226	4.5	80.552	1.73	5.0	95.045	1.50	7	49.293
(92 × 87)	35	87.5	75.777	91.5	86.5	91.60	86.90	+0.875	4.937	4.5	82.461	1.76	5.0	97.024	1.50	7	49.160
95 × 90	36	90.0	77.942	94.5	89.5	94.60	89.90	+1.125	5.226	4.5	85.552	1.73	5.0	100.063	1.52	7	49.526
(98 × 93)	38	95.0	82.272	97.5	92.5	97.60	92.90	+0.125	4.071	4.5	88.485	1.86	5.0	103.288	1.51	7	48.759
100 × 95	38	95.0	82.272	99.5	94.5	99.60	94.90	+1.125	5.226	4.5	90.552	1.73	5.0	105.079	1.58	7	49.759
105 × 100	40	100.0	86.603	104.5	99.5	104.60	99.90	+1.125	5.226	4.5	95.552	1.73	5.0	110.094	1.53	8	56.793

Note: Values within brackets are nonpreferred.

17.19

Particular	Formula	
The value of tooth thickness and space width of spline	$l_o = s_o = m\dfrac{\pi}{2} + 2xm\tan\alpha$	(17–32)

PINS
Taper pins

The diameter at small end (Figs. 17–6 and 17–7, Tables 17–16 and 17–17)	$d_{ps} = d_{pl} - 0.0208l$	(17–33)
The mean diameter of pin	$d_m = 0.20D$ to $0.25D$	(17–34)

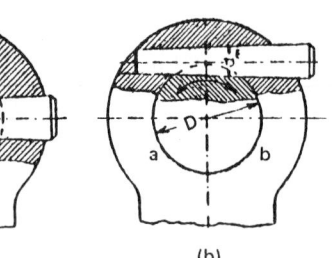

(a)   (b)

FIGURE 17–6   Tapered pin.

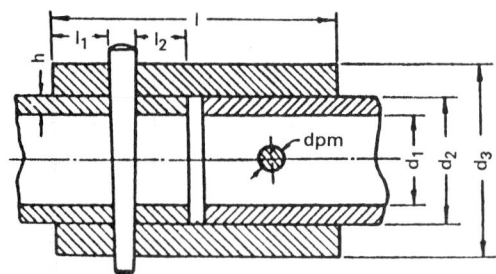

FIGURE 17–7   Sleeve and tapered pin joint for hollow shafts.

Sleeve and taper pin joint (Fig. 17–7)

AXIAL LOAD. The axial stress induced in the hollow shaft (Fig. 17–7) due to tensile force F	$\sigma = \dfrac{F}{\dfrac{\pi}{4}(d_2^2 - d_1^2) - (d_2 - d_1)d_m}$	(17–35)
The bearing stress in the pin due to bearing against the shaft an account of force F	$\sigma_c = \dfrac{F}{(d_2 - d_1)d_m}$	(17–36)
The bearing stress in the pin due to bearing against the sleeve	$\sigma_c = \dfrac{F}{(d_3 - d_2)d_m}$	(17–37)
The shear stress in pin	$\tau = \dfrac{2F}{\pi d_m^2}$	(17–38)
The shearing stress due to double shear at the end of hollow shaft	$\tau = \dfrac{F}{2(d_2 - d_1)l_2}$	(17–39)
The shear stress due to double shear at the sleeve end	$\tau = \dfrac{F}{2(d_3 - d_2)l_1}$	(17–40)

TABLE 17–16
Dimensions (in mm) for cylindrical pins

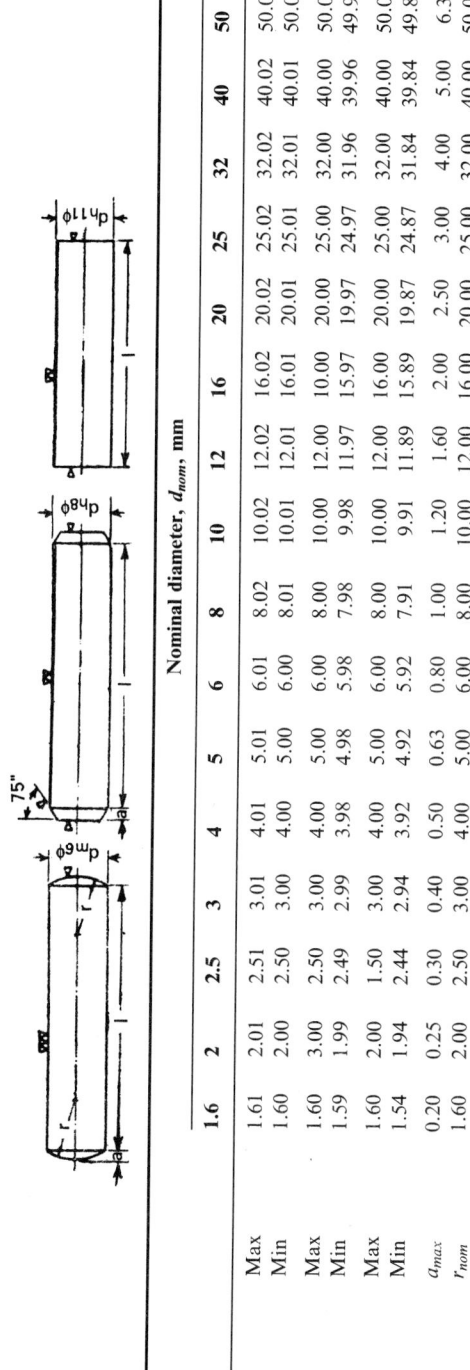

		Nominal diameter, d_{nom}, mm															
		1.6	2	2.5	3	4	5	6	8	10	12	16	20	25	32	40	50
d_{m6}	Max	1.61	2.01	2.51	3.01	4.01	5.01	6.01	8.02	10.02	12.02	16.02	20.02	25.02	32.02	40.02	50.02
	Min	1.60	2.00	2.50	3.00	4.00	5.00	6.00	8.01	10.01	12.01	16.01	20.01	25.01	32.01	40.01	50.01
d_{h6}	Max	1.60	3.00	2.50	3.00	4.00	5.00	6.00	8.00	10.00	12.00	10.00	20.00	25.00	32.00	40.00	50.00
	Min	1.59	1.99	2.49	2.99	3.98	4.98	5.98	7.98	9.98	11.97	15.97	19.97	24.97	31.96	39.96	49.96
d_{h11}	Max	1.60	2.00	1.50	3.00	4.00	5.00	6.00	8.00	10.00	12.00	16.00	20.00	25.00	32.00	40.00	50.00
	Min	1.54	1.94	2.44	2.94	3.92	4.92	5.92	7.91	9.91	11.89	15.89	19.87	24.87	31.84	39.84	49.84
a_{max}		0.20	0.25	0.30	0.40	0.50	0.63	0.80	1.00	1.20	1.60	2.00	2.50	3.00	4.00	5.00	6.30
r_{nom}		1.60	2.00	2.50	3.00	4.00	5.00	6.00	8.00	10.00	12.00	16.00	20.00	25.00	32.00	40.00	50.00

Source: IS 2393, 1963.

TABLE 17–17
Dimensions (in mm) for solid and split taper pins

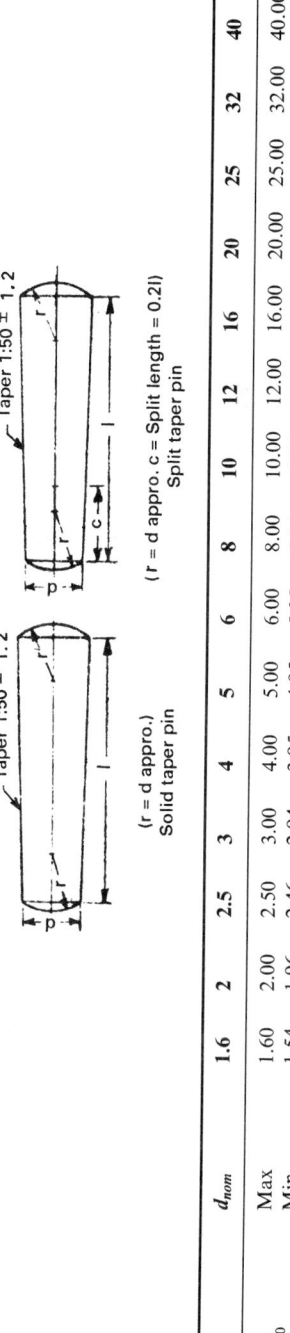

d_{nom}		1.6	2	2.5	3	4	5	6	8	10	12	16	20	25	32	40	50
d_{h10}	Max	1.60	2.00	2.50	3.00	4.00	5.00	6.00	8.00	10.00	12.00	16.00	20.00	25.00	32.00	40.00	50.00
	Min	1.54	1.96	2.46	2.94	3.95	4.95	5.95	7.94	9.94	11.93	15.63	19.92	24.92	31.90	39.90	49.90

Source: IS 2393, 1963.

Particular	Formula	
The axial stress in the sleeve	$$\sigma = \dfrac{F}{\dfrac{\pi}{4}(d_3^2 - d_2^2) - (d_3 - d_2)d_m}$$	(17–41)
TORQUE. The shear due to twisting moment applied	$$\tau = \dfrac{M_t}{\dfrac{\pi}{4}d_m^2 d_2}$$	(17–42)
For the design of hollow shaft subjected to torsion	Refer to Chap. 14.	

Taper joint and nut

The tensile stress in the threaded portion of the rod (Fig. 17–8) without taking into consideration stress concentration	$$\sigma_t = \dfrac{F}{\dfrac{\pi}{4}d_c^2}$$	(17–43)

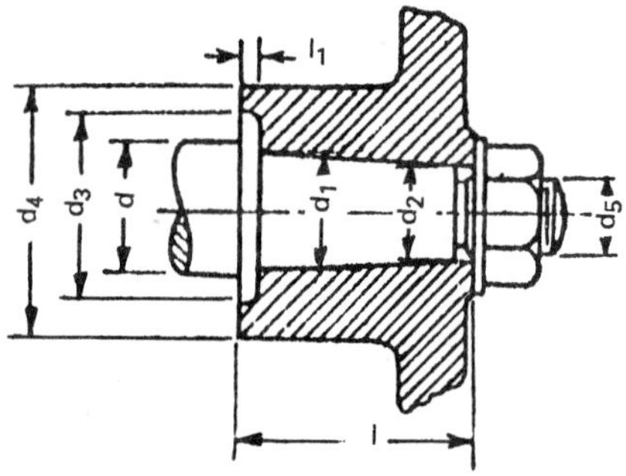

FIGURE 17–8   Tapered joint and nut.

The bearing resistance offered by the collar	$$\sigma_c = \dfrac{F}{\dfrac{\pi}{4}(d_3^2 - d_2^2)}$$	(17–44)
The diameter of the taper d_2 Provide a taper of 1 in 50 for the length $(l - l_1)$	$d_2 > d_{nom}$	(17–45)

Knuckle joint

The tensile stress in the rod (Fig. 17–9)	$$\sigma_t = \dfrac{4F}{\pi d^2}$$	(17–46)
The tensile stress in the net area of the eye	$$\sigma_t = \dfrac{F}{(d_4 - d_2)b}$$	(17–47)
Shear stress in the eye	$$\tau = \dfrac{F}{b(d_4 - d_2)}$$	(17–48)

Particular	Formula

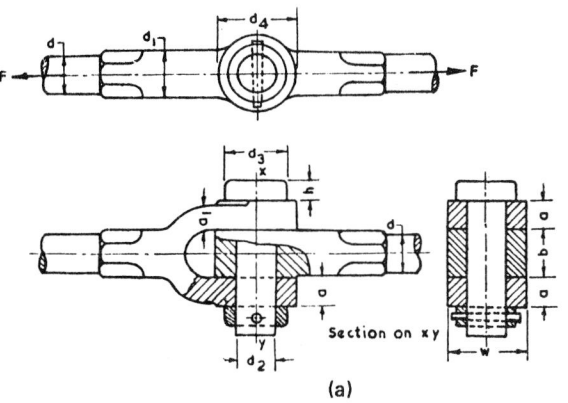 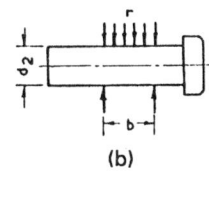

(a)

(b)

FIGURE 17–9 Knuckle joint for round rods.

Tensile stress in the net area of the fork ends

$$\sigma_t = \frac{F}{2a(d_4 - d_2)} \qquad (17\text{–}49)$$

Shear stress in the fork ends due to tear of

$$\tau = \frac{F}{2a(d_4 - d_2)} \qquad (17\text{–}50)$$

Compressive stress in the eye due to bearing pressure of the pin

$$\sigma_c = \frac{F}{d_2 b} \qquad (17\text{–}51)$$

Compressive stress in the fork due to the bearing pressure of the pin

$$\sigma_c = \frac{F}{2d_2 a} \qquad (17\text{–}52)$$

Shear stress in the knuckle pin

$$\tau = \frac{2F}{\pi d_2^2} \qquad (17\text{–}53)$$

The maximum bending moment, Fig. 17–9 (panel *b*)

$$M_b = \frac{Fb}{8} \qquad (17\text{–}54)$$

The maximum bending stress in the pin, based on the assumption that the pin is supported and loaded as shown in Fig. 17–9*b* and that the maximum bending moment M_b occurs at the center of the pin

$$\sigma_b = \frac{4Fb}{\pi d_2^3} \qquad (17\text{–}55)$$

The maximum bending moment on the pin based on the assumption that the pin supported and loaded as shown in Fig. 17–10*b*, which occurs at the center of the pin

$$M_b = \frac{F}{2}\left(\frac{b}{4} + \frac{a}{3}\right) \text{ (approx.)} \qquad (17\text{–}56)$$

The maximum bending stress in the pin based on the assumption that the pin is supported and loaded as shown in Fig. 17–10*b*

$$\sigma_b = \frac{4(3b + 4a)F}{3\pi d_2^3} \qquad (17\text{–}57)$$

Particular	Formula

COTTER

The initial force set up by the wedge action	$F = 1.25Q$	(17–58)
The force at the point of contact between cotter and the member perpendicular to the force F	$H = F\tan(\alpha + \theta)$	(17–59)
The thickness of cotter	$t = 0.4D$	(17–60)
The width of the cotter	$b = 4t = 1.6D$	(17–61)

Cotter joint

The axial stress in the rods (Fig. 17–10)

$$\sigma = \frac{4F}{\pi d^2} \tag{17–62}$$

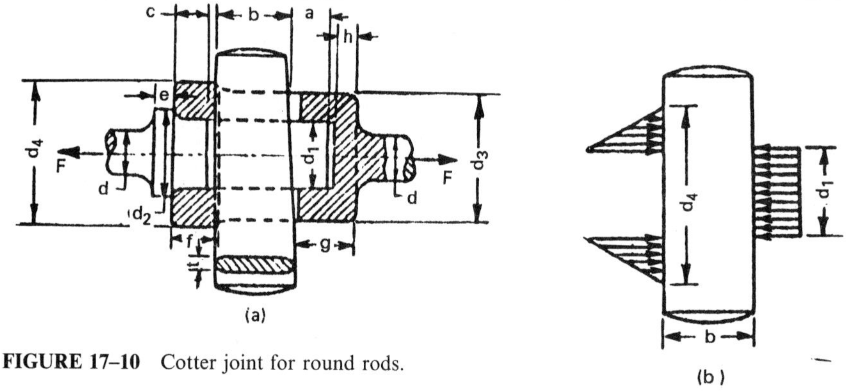

FIGURE 17–10 Cotter joint for round rods.

Axial stress across the slot of the rod

$$\sigma = \frac{4F}{\pi d_1^2 - 4d_1 t} \tag{17–63}$$

Tensile stress across the slot of the socket

$$\sigma = \frac{4F}{\pi(d_3^2 - d_1^2) - 4t(d_3 - d_1)} \tag{17–64}$$

The strength of the cotter in shear	$F = 2bt\tau$	(17–65)
Shear stress, due to the double shear, at the rod end	$\tau = \dfrac{F}{2ad_1}$	(17–66)

Shear stress induced at the socket end

$$\tau = \frac{F}{2c(d_4 - d_1)} \tag{17–67}$$

The bearing stress in collar

$$\sigma_c = \frac{4F}{\pi(d_2^2 - d_1^2)} \tag{17–68}$$

Crushing strength of the cotter or rod $\qquad F = d_1 t\sigma_c \qquad$ (17–69)

Particular	Formula	
Crushing stress induced in the socket or cotter	$\sigma_c = \dfrac{F}{(d_4 - d_1)t}$	(17–70)
The equation for the crushing resistance of the collar	$F = \dfrac{\pi(d_2^2 - d_1^2)}{4}\,\sigma_c$	(17–71)
Shear stress induced in the collar	$\tau = \dfrac{F}{\pi d_1 e}$	(17–72)
Shear stress induced in the socket	$\tau = \dfrac{F}{\pi d_1 h}$	(17–73)
The maximum bending stress induced in the cotter assuming that the bearing load on the collar in the rod end is uniformly distributed while the socket end is uniformly varying over the length as shown in Fig. 17–10b	$\sigma_b = \dfrac{F(d_1 + 2d_4)}{4tb^2}$	(17–74)

Gib and cotter joint (Fig. 17–11)

The width b of both the Gib and Cotter is the same as far as a cotter is used by itself for the same purpose (Fig. 17–11). The design procedure is the same as done in cotter joint Fig. 17–10

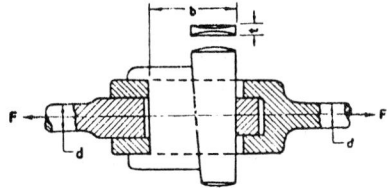

FIGURE 17–11   Gib and cotter joint for round rods.

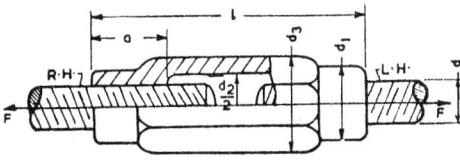

FIGURE 17–12   Coupler or turn buckle.

Threaded joint

COUPLER OR TURN BUCKLE.   Strength of the rods based on core diameter d_c (Fig. 17–12)

$$F = \frac{\pi}{4}\,d_c^2\sigma_t \qquad (17\text{–}75)$$

The resistance of screwed portion of the coupler at each end against shearing

$$f = \pi a d\tau \qquad (17\text{–}76)$$

From practical considerations the length a is given by

$a = d$ to $1.25d$ for steel nuts   (17–77a)
$a = 1.5d$ to $2d$ for cast iron   (17–77b)

The strength of the outside diameter of the coupler at the nut portion

$$F = \frac{\pi}{4}\,(d_1^2 - d^2)\sigma_t \qquad (17\text{–}78)$$

Particular	Formula	
The outside diameter of the turn buckle or coupler at the middle is given by the equation	$F = \dfrac{\pi}{4}\,(d_3^2 - d_2^2)\sigma_t$	(17–79)
The total length of the coupler	$l = 6d$	(17–80)

REFERENCES

1. Maleev, V. L., and J. B. Hartman, *Machine Design*, International Textbook Company, Scranton, Pennsylvania, 1954.
2. Shigley, J. E., and L. D. Mitchell, *Mechanical Engineering Design*, McGraw-Hill Book Company, New York, 1983.
3. Faires, V. M., *Design of Machine Elements*, The Macmillan Company, New York, 1965.
4. Lingaiah, K., and B. R. Narayana Iyengar, *Machine Design Data Handbook*, Engineering College Co-operative Society, Bangalore, India, 1962.
5. Lingaiah, K., and B. R. Narayana Iyengar, *Machine Design Data Handbook*, Vol. I (*SI and Customary Metric Units*), Suma Publishers, Bangalore, India, 1986.
6. Lingaiah, K., *Machine Design Data Handbook*, Vol. II (*SI and Customary Metric Units*), Suma Publishers, Bangalore, India, 1986.
7. Juvinall, R. C., *Fundamentals of Machine Component Design*, John Wiley and Sons, New York, 1983.
8. Deutschman, A. D., W. J. Michels, and C. E. Wilson, *Machine Design—Theory and Practice*, Macmillan Publishing Company, New York, 1975.
9. Bureau of Indian Standards.
10. *SAE Handbook*, 1981.

CHAPTER
18

THREADED FASTENERS AND SCREWS FOR POWER TRANSMISSION

SYMBOLS

A_b	area of cross section of bolt, m^2 (in^2)
A_{br}	area of base of preloaded bracket, m^2 (in^2)
A_c	core area of thread, m^2 (in^2)
A_g	loaded area of gasket, m^2 (in^2)
A_r	stress area, m^2 (in^2)
A_τ	shear area, m^2 (in^2)
d	nominal diameter of screw m (in)
	major diameter of external thread (bolt), m (in)
d_2	pitch diameter of external thread (bolt), m (in)
d_1	minor diameter of external thread (bolt), m (in)
d_c	mean diameter of thrust collar, m (in)
D	diameter of shaft, m (in)
	major diameter of internal thread (nut), m (in)
D_1	minor diameter of internal thread (nut), m (in)
D_2	pitch diameter of internal thread (nut), m (in)
D_b	diameter of bolt circle, m (in)

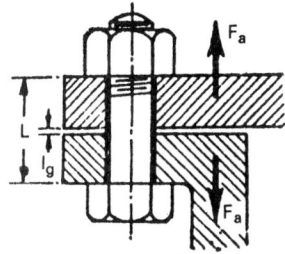

FIGURE 18–1 Gasket joint.

D_i	inside diameter of a pressure vessel or cylinder, m (in)
	mean diameter of inside screw of differential or compound screw, m (in)
D_o	mean diameter of outside screw of differential or compound screw, m (in)
e	eccentricity, m (in)
E_b, E_g	modulii of elasticity of bolt and gasket, respectively, GPa (Mpsi)
F	permissible load on bolt, kN (lbf)
	tightening load on the nut, kN (lbf)
F_a	applied or external load, kN (lbf)
F_f	final load on the bolt, kN (lbf)
F_i	initial load due to tightening, kN (lbf)
	preload in each bolt, kN (lbf)
F_t	tangential force, kN (lbf)
h	thickness of a pressure vessel, m (in)
	thickness of a cylinder, m (in)
h_2	thickness of the flange of the cylindrical pressure vessel, m (in)
h_o	depth of tapped hole (Fig. 18–2), m (in)

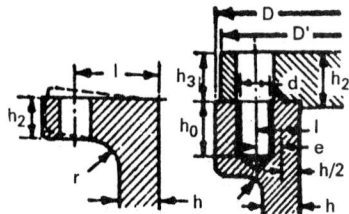

FIGURE 18–2   Flanged bolted joint.

i	number of threads in a nut
	number of bolts
I	moment of inertia of bracket base, area (Fig. 18–6), m^4 or cm^4 (in^4)
K	constant (Eq. (18–4a))
K_σ	stress concentration factor
l	lever arms (with suffixes), m (in)
	distance from the inside edge of the cylinder to the center line of bolt, m (in)
	lead, m (in)
l_c	required length of engagement of screw or nut (also with suffixes), m (in)
l_g	gasket thickness, m (in)
L	length of bolt nut to head (Fig. 18–1), m (in)
M_b	bending moment, N m (lbf in)
M_t	twisting moment, N m (lbf in)
n	factor of safety
p	pressure, MPa (psi)
p_c	circular pitch of bolts or studs on the bolt circle of a cylinder cover, m (in)
P	pitch of thread, m (in)
t	thread thickness at major diameter, m (in)
t_1	thread thickness at minor diameter, m (in)

W	axial load, kN (lbf)
α	helix angle, deg
$\alpha_o,\ \alpha_i$	respective helix angles of outside and inside screws of differential or compound screws, deg
ϕ	friction angle, deg
θ	half apex angle, deg
μ	coefficient of friction between nut and screw
μ_c	coefficient of collar friction
$\mu_i,\ \mu_o$	respective coefficient of friction in case of differential or compound screw
η	efficiency
σ	stress (normal), MPa (psi)
σ_a	allowable stress, MPa (psi)
σ_b	bending stress, MPa (psi)
σ_b'	bending stress due to eccentric load (Eq. (18–61))
	allowable bearing pressure between threads of nut and screw, MPa (psi)
σ_c	compressive stress, MPa (psi)
σ_d	design stress, MPa (psi)
σ_w	working stress, MPa (psi)
τ	applicable shear stress, MPa (psi)
τ_a	allowable shear stress, MPa (psi)
τ_w	permissible working shear stress, MPa (psi)

SUFFIXES

v	vertical
h	horizontal

Particular	Formula

SCREWS

The empirical formula for the proper size of a set screw	$d = \dfrac{D}{8} + 8$ mm where D in mm	(18–1)
The maximum safe holding force of a set screw	$F = 54,254\,d^{2.31}$	
	where F in kN and d in m **SI**	(18–2a)
	$= 2500 d^{2.31}$ **US Customary System units**	(18–2b)
	where F in lbf and d in in	
Applied torque	$M_t = 0.2\,F_a$ (nominal diameter of bolt)	(18–3)

Gasket joint (Fig. 18–1)

Final load on the bolt	$F_f = K F_a + F_i$	(18–4)

Particular	Formula

TABLE 18–1
Values of K in Eq. (18–4)

Type of joint	K
Soft, elastic gasket with studs	1.00
Soft gasket with through bolts	0.90
Copper asbestos gasket	0.60
Soft copper corrugated gasket	0.40
Lead gasket with studs	0.10
Narrow copper ring	0.01
Metal-to-metal joint	0.00

where $K = \left[\dfrac{\dfrac{E_b A_b}{L}}{\dfrac{E_b A_b}{L} + \dfrac{E_g A_g}{l_g}} \right]$ (18–4a)

Refer also to Table 18–1 for values of K.

According to Bart, the tightening load for a screw of a steamtight, metal-to-metal joint

$F = 2804.69d$ **SI**   (18–5a)

    where F in kN and d in m

    $= 1600d$ **US Customary System units** (18–5b)

    where F in lbf and d in in

Tightening load for screw of a gasket joint

$F = 1402.34d$ **SI**   (18–6a)

    where F in kN and d in m

    $= 8000d$ **US Customary System units** (18–6b)

    where F in lbf and d in in

Cordullo's equation for the tightening load on the nuts

$F = \sigma_w(0.55d^2 - 6.45 \times 10^{-3}d)$ **SI**   (18–7a)

    where F in kN, σ_w in MPa, and d in m

    $= \sigma_w(0.55d^2 - .036d)$ **US Customary System units**

    (18-7b)

    where F in lbf, σ_w in psi, and d in in

Bolted joints (Fig. 18–2)

The flange thickness of the cylinder or pressure vessel

$h_2 = 1.25d$ to $1.5d \nleq 1.1h$ to $1.25h$ (18–8)

The bolt diameter

$d = 0.67h$ to $0.8h$ (18–9)

Circular pitch of the bolts or studs on the cylinder cover to ensure water and steamtight joint

$p_c = 7d$ for pressure from 0 to 0.33 MPa (0 to 48 psi) as per American Navy Standards   (18–10)

$p_c = 3.5d$ for pressure from 1.2 MPa (175 psi) to 1.37 MPa (200 psi)   (18–11)

$p_c = 3d$ for tight joint   (18–12)

Particular	Formula
The average stress for screw for sizes from 12.5 to 75 mm	$\sigma_{av} = 490.33/d$ **SI** (18–13a) where σ_{av} in MPa and d in m $= \dfrac{2,800,000}{d}$ **US Customary System units** (18–13b) where σ_{av} in psi and d in in
Unwin's formula for allowable stresses in bolts of ordinary steel to make a fluidtight joint	$\sigma_d = (17,573.4d^2 + 11)$ for rough joint **SI** (18–14a) where σ_d in MPa and d in m $= 6030d^2 + 1600$ **US Customary System units** (18–14b) where σ_d in psi and d in in $\sigma_d = (33,828.9d^2 + 17.3)$ for faced joint **SI** (18–15a) where σ_d in MPa and d in m $= 3070d^2 + 2500$ **US Customary System units** (18–15b) where σ_d in psi and d in in
The depth of tapped hole (Fig. 18–2)	$h_o = 1.25d$ in steel castings (18–16) $h_o = 1.50d$ to $1.75d$ in cast iron (18–17) $h_o = 1.75d$ to $2d$ in aluminum (18–18)
The distance l from the inside edge of the cylinder to the center line of bolts (Fig. 18–2)	$l = 1.25d$ to $1.5d$ (18–19)
The diameter of bolt circle	$D_b = D_1 + 2d$ (18–20)
The safe load on each bolt	$F' = A_r\sigma_d$ (18–21)
The number of bolts	$i = \dfrac{\pi D_b^2 p}{4F'}$ (18–22)
Another expression for the number of bolts	$i = \dfrac{\pi D_b}{p_c}$ (18–23)

Stress in tensile bolt

Seaton and Routhwaite formula for working stress for bolt made of steel containing 0.08 to 0.25% carbon and with diameter of 20 mm and over	$\sigma_w = C(A_r)^{0.418}$ (18–24) Refer to Table 18–2 for bolt tension and torque values and Table 18–3 for σ_w.

TABLE 18–2
Approximate bolt tension and torque values

Bolt size, mm	Minimum bolt tension		Equivalent torque	
	kN	lbf	kN m	lbf ft
12.7	51.2	11,500	1.353	1,000
15.9	76.9	17,300	2.442	1,800
19.6	113.9	25,600	4.835	3,570
22.2	139.7	31,400	6.374	4,700
25.4	189.1	42,500	9.620	7,090
21.6	225.4	50,600	13.013	9,600
31.8	286.9	64,500	18.289	13,500

TABLE 18–3
Load and working stress for metric coarse threads

Major diameter, d, mm	Stress area, A_r, mm^2	Design stress, σ_w		Permissible load	
		MPa	psi	kN	lbf
16	0.016	18.9	2,740	2.97	667
20	0.025	22.8	3,300	5.59	1,260
24	0.035	27.2	3,950	9.59	2,160
30	0.056	32.2	4,670	18.04	4,060
36	0.082	37.1	5,380	30.89	6,940
42	0.112	43.1	6,250	48.35	10,870
48	0.147	48.3	7,000	71.10	16,000
56	0.203	55.2	8,000	111.80	25,130

Particular	Formula
Applied load	$F_a = C(A_r)^{1.418}$ (18–25)

where

$$C = 7.8 \times 10^8 \ (5000) \text{ for carbon steel bolts of } \sigma_u = 414 \text{ MPa (60 kpsi)}$$
$$= 23.3 \times 10^8 \ (15,000) \text{ for alloy–steel bolts}$$
$$= 0.33 \times 10^8 \ (1000) \text{ for bronze bolts}$$

The values of C given inside brackets are for **US Customary System units**, and values without brackets are for **SI** units

Power screw

The helix angle of a V-thread (Fig. 18–3)	$\alpha = \tan^{-1}(l/\pi d_2)$	(18–26)
The tangential force for a square thread at mean radius of screw	$F_t = W \dfrac{\tan \alpha + \mu}{1 - \mu \tan \alpha}$	(18–27)

Refer to Table 18–4 for μ.

TABLE 18–4
Coefficient of friction for power screws

Lubricant	Coefficient of friction, μ
Machine oil and graphite	0.07
Lard oil	0.11
Heavy machine oil	0.14

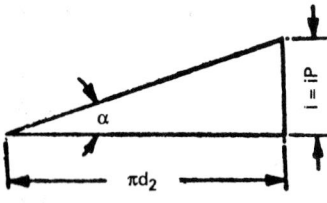

FIGURE 18–3   Helix angle of a single-start thread.

Particular	Formula
The tangential force for V-thread or angular thread at mean radius (Fig. 18–4)	$$F_t = W\,\frac{\tan\alpha + \dfrac{\mu}{\cos\theta}}{1 - \mu\dfrac{\tan\alpha}{\cos\theta}}\qquad(18\text{–}28)$$

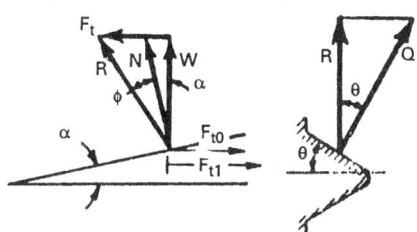

FIGURE 18–4 Forces acting on a triangular thread.

Particular	Formula
The total frictional torque including collar friction torque for square thread	$$M_t = W\left[\frac{d_2}{2}\left(\frac{\tan\alpha + \mu}{1 - \mu\tan\alpha}\right) + \mu_c\,\frac{d_c}{2}\right]\qquad(18\text{–}29)$$

Refer to Table 18–4 for μ and Table 18–5 for μ_c.

TABLE 18–5a
Coefficient of friction on thrust collar

Material	Coefficient of running friction	Coefficient of starting friction
Soft steel on cast iron	0.121	0.170
Hardened steel on cast iron	0.092	0.147
Soft steel on bronze	0.084	0.101
Hardened steel on bronze	0.063	0.081

TABLE 18–5b
Torque factor K_μ for use in Eq. (18–30b)

Bolt condition	K_μ
Nonplated, black finish	0.30
Zinc-plated	0.20
Lubricated	0.18
Cadmium-plated	0.16
With Bowman anti-seize	0.12
With Bowman-grip nuts	0.09

The total frictional torque for V-thread, including collar friction torque

$$M_t = W\left[\frac{d_2}{2}\left\{\frac{\tan\alpha + \dfrac{\mu}{\cos\theta}}{1 - \mu\dfrac{\tan\alpha}{\cos\theta}}\right\} + \mu_c\,\frac{d_c}{2}\right]\qquad(18\text{–}30a)$$

$$= K_\mu F_t d\qquad(18\text{–}30b)$$

where K_μ is the torque factor

The torque factor

$$K_\mu = \frac{d_2}{2d}\left\{\frac{\tan\alpha + \dfrac{\mu}{\cos\theta}}{1 - \mu\dfrac{\tan\alpha}{\cos\theta}}\right\} + \mu_c \times 0.625\qquad(18\text{–}30c)$$

Refer to Table 18–5B for K_μ.

Particular	Formula
The efficiency of square thread neglecting collar friction	$$\eta = \frac{\tan \alpha}{\tan(\alpha + \phi)} = \frac{Wl}{2\pi M_t} \qquad (18\text{–}31)$$
The efficiency formula for an angular-type thread with half apex angle θ and an allowance for nut or end friction on a radius r_c	$$\eta = \frac{d_2 \tan \alpha}{\left[\dfrac{\tan \alpha + \mu/\cos\theta}{1 - \mu \tan \alpha/\cos\theta} \, d_2 + \mu_c d_c\right]} \qquad (18\text{–}32)$$
The efficiency formula for square thread	$$\eta = \frac{d_2 \tan \alpha}{\dfrac{\tan \alpha + \mu}{1 - \mu \tan \alpha} \, d_2 + \mu_c d_c} \qquad (18\text{–}33)$$
	$$= \frac{l}{\pi[d_2 \tan(\alpha + \phi) + \mu_c d_c]} \qquad (18\text{–}34)$$

LOADING
Lowering the load

The tangential force at mean or pitch radius r_2	$$F_t = W \tan(\phi + \alpha) \qquad (18\text{–}35)$$
The frictional torque at mean or pitch radius r_2	$$M_t = \frac{Wd_2}{2} \tan(\phi - \alpha) \qquad (18\text{–}36)$$
The condition for overhauling for square threads	$$\tan \alpha \geq \left(\frac{\mu d_2 + \mu_c d_c}{d_2 - \mu\mu_c d_c}\right) \qquad (18\text{–}37)$$

Differential screws (Fig. 18–7a)

The loading efficiency of a differential screw, not including the collar friction	$$\eta = \frac{D_o \tan \alpha_o - D_i \tan \alpha_i}{\left[D_o \dfrac{\tan \alpha_o + \mu_o}{1 - \mu_o \tan \alpha_o} - D_i \dfrac{\tan \alpha_i - \mu_i}{1 - \mu_i \tan \alpha_i}\right]} \qquad (18\text{–}38)$$

Compound screws

The loading efficiency of a compound screw, not including collar friction	$$\eta = \frac{D_o \tan \alpha_o + D_i \tan \alpha_i}{\left[D_o \dfrac{\tan \alpha_o + \mu_o}{1 - \mu_o \tan \alpha_o} + D_i \dfrac{\tan \alpha_i + \mu_i}{1 - \mu_i \tan \alpha_i}\right]} \qquad (18\text{–}39)$$
The number of threads necessary in the nut	$$i = \frac{4W}{\sigma_b' \pi (d^2 - d_1^2)} \qquad (18\text{–}40)$$

Particular	Formula
The length of nut	$$l_n = iP = \frac{4WP}{\sigma_b' \pi (d^2 - d_1^2)} \qquad (18\text{--}41)$$
The required length of engagement for adequate shear strength (assuming that the load is distributed over the threads in contact)	$$l_c = \frac{nPF}{A_\tau \tau} \qquad (18\text{--}42)$$
Neglecting the radial clearance between threads, or allowance at the major and minor diameters and considering the threads as a series of collars the equation for thread engagement	$$l_{e\,(screw)} = \frac{nPF}{\pi d_1 t_1 \tau_{(screw)}} \qquad (18\text{--}43)$$ $$l_{e\,(nut)} = \frac{nPF}{\pi d t \tau_{(nut)}} \qquad (18\text{--}44)$$
The normal length of thread engagement as per Indian standard	$$l_{eN\,(min)} = 8.92 P d^{0.2} \qquad \textbf{SI} \qquad (18\text{--}45a)$$ where l_{eN}, P, and d in m $$= 2.24 P d^{0.2} \qquad \textbf{SI} \qquad (18\text{--}45b)$$ where l_{eN}, P, and d in mm $$l_{eN\,(max)} = 26.67 P d^{0.2} \qquad \textbf{SI} \qquad (18\text{--}46a)$$ where l_{eN}, P, and d in m $$= 6.7 P d^{0.2} \qquad \textbf{SI} \qquad (18\text{--}46b)$$ where l_{eN}, P, and d in mm

Note:
If l_{eN} has to be between the limits, the length of the thread is said to be normal (N)
If l_{eN} has to be below the minimum level, length of thread is said to be short (S)
If l_{eN} has to be above the maximum level, length of thread is said to be long (L)

Eccentric loading

The load on bolt 1, Fig. 18–5 (panel *a*)	$$F_1 = \frac{Fll_1}{l_1^2 + l_2^2 + l_3^2 + l_4^2} = F \frac{l(a - b\cos\alpha)}{4a^2 + 2b^2} \qquad (18\text{--}47)$$
The general expression for the load carried by *i*th bolt	$$F_i = F \frac{2l(a - b\cos\alpha)}{(2a^2 + b^2)i} \qquad (18\text{--}48)$$
The maximum load on the bolt, Fig. 18–5(*b*)	$$F_{max} = \frac{2Fl(a + b)}{(2a^2 + b^2)i} \qquad (18\text{--}49)$$

Particular	Formula

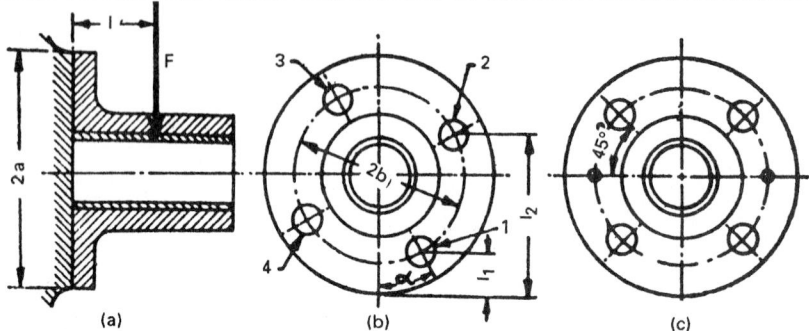

FIGURE 18–5   Fastening of a flanged bearing.

The maximum load on the bolt, Fig. 18–5(c)

$$F_{max} = \frac{2Fl\left[a + b\cos\left(\dfrac{180°}{i}\right)\right]}{(2a^2 + b^2)i} \qquad (18–50)$$

Fastening of a bracket

Bracket with no preload

$$F_1 = \frac{Fll_1}{2(l_1^2 + l_2^2 + l_3^2)} \qquad (18–51)$$

Tensile load taken by the bolts, Fig. 18–6(a)

$$F_2 = \frac{Fll_2}{2(l_1^2 + l_2^2 + l_3^2)} \qquad (18–52)$$

$$F_3 = \frac{Fll_3}{2(l_1^2 + l_2^2 + l_3^2)} \qquad (18–53)$$

Shear stresses

(i) If shear load is taken completely by the lug, shear load on lug is given by

$$F_1 = F \qquad (18–54)$$

(ii) If shear load is taken completely by the bolt shear load on each bolt is given by

$$F_b = \frac{F}{i} \qquad (18–55)$$

(iii) If shear load is shared equally between the bolt and the lug

$$F_1' = \frac{F}{2} \qquad (18–56)$$

$$F_b' = \frac{F}{2i} \qquad (18–57)$$

Shear load due to the eccentricity e, Fig. 18–6(b), in each bolt is given by

$$F_{ei}' = \frac{Fex_i}{\Sigma x_i^2} \qquad (18–58)$$

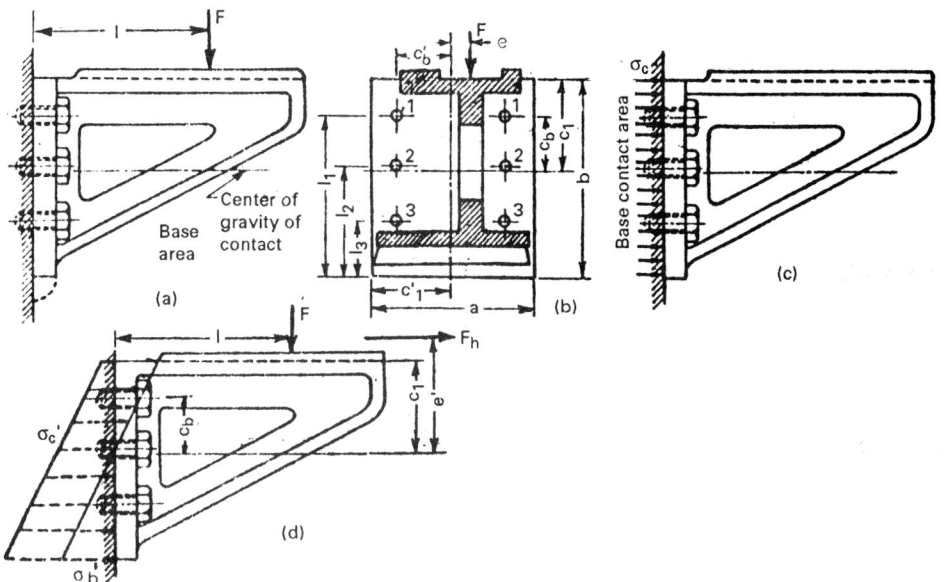

FIGURE 18–6   Preloaded bracket.

Particular	Formula
	where x_i = distance between the center of bolts and the center of the particular bolt.
Resultant shear load	$$F_r = F_b(or\ F'_b) + \frac{Fex_i}{\Sigma x_i^2} \qquad (18\text{–}59)$$

Preloaded bracket

Particular	Formula
Compression stress in contact area between the bracket base and the wall, Fig. 18–6(c)	$$\sigma_c = \frac{iF_i}{A_c} \qquad (18\text{–}60)$$
Bending stress due to eccentric load, Fig. 18–6(d)	$$\sigma'_b = \frac{M_b c_1}{I_c} = \frac{Flc_1}{I_c} \qquad (18\text{–}61)$$
Resultant compressive stress in the contact area	$$\sigma'_c = \frac{iF_i}{A_c} - \frac{M_b c_1}{I_c} = \frac{iF_i}{A_c} - \frac{Flc_1}{I_c} \qquad (18\text{–}62)$$
Tensile stress in any individual bolt is given by	$$\sigma'_b = \frac{F_i}{A_b} + \frac{M_b c_b}{I_c} \qquad (18\text{–}63)$$
Condition to avoid separation of the base and wall	$$F_i > \frac{M_b c_1 A_c}{iI_c} \qquad (18\text{–}64)$$

Particular	Formula	
With a 25 percent margin on the preload to account for overloads, condition to avoid separation of the base and wall	$$F_i = \frac{1.25 M_b c_1 A_c}{i I_c}$$	(18–65)
Bolt load taking into consideration 25 percent margin on the preload to account for overloads	$$F_b = \frac{1.25 M_b c_1 A_c}{i I_c} + \frac{M_b c_b}{I_c}$$	(18–66)
With an additional horizontal load F_h the preload F_i is given by	$$F_i = \frac{1.25 M_b c_1 A_c}{i I_c} \pm \frac{F_h}{i}$$	(18–67)

where $(+)$ is used when F_h is away from the wall and $(-)$ when F_h is toward the wall

With the addition of a horizontal load F_h, the bolt load is given by	$$F_b = \frac{1.25 M_b c_1 A_c}{i I_c} \pm \frac{F_h}{i} + \frac{M_b c_b}{I_c} \pm \frac{F_h A_b}{A_c}$$	(18–68)
Moment on the bracket	$$M_b = Fl \pm F_h e$$	(18–69)

Shear loads

Shear load due to the eccentricity e in each of the bolts with no horizontal load	$$F_{T_i} = \frac{M_1 x_i}{\Sigma x_i^2}$$	(18–70)

where

$$M_1 = Fe - \left[\frac{M_b c_1}{16 I_c} \sqrt{a^2 + b^2} \right.$$

$$\left. - \frac{0.25 \mu M_b \Sigma x_i' c_1 A_b^2}{I_c A_c} \right]$$ (88–70a)

where x_i' = distance of the center of particular bolt to the center of the base of the bracket

Shear load due to eccentricity e in each of the bolts with a horizontal load, F_h	$$F_{T_i} = \frac{M_1 x_i}{\Sigma x_i^2}$$	(18–71)

where

$$M_1 = Fe \left[\frac{\mu}{4} \left(\frac{0.25 M_b c_1}{I_c} \pm \frac{F_h}{A_c} \sqrt{a^2 + b^2} \right) \right.$$

$$\left. - \frac{\mu A_b}{A_c} \left(\frac{0.25 M_b c_1}{I_c} \pm \frac{F_h}{A_c} \right) \left(\Sigma x_i' \right) \right]$$

Vertical applied load due to the friction component of the preload	$$F_v = \mu \left[\frac{1.25 M_b c_1 A_c \pm F_h}{i I_c} \right]$$	(18–72)
Condition for the nonexistence of the support for the shear load	$$F < \mu \left[\frac{1.25 M_b c_1 A_c}{i I_c} \pm F_h \right]$$	(18–73)

Particular	Formula

GENERAL

See Tables 18–6 to 18–22 and Figs. 18–7 to 18–16 for further particulars on threaded fasteners and screws for power transmission.

REFERENCES

1. Norman, C. A., E. S. Ault, and E. F. Zarobsky, *Fundamentals of Machine Design*, The Macmillan Company, New York, 1951.
2. Maleev, V. L., and J. B. Hartman, *Machine Design*, International Textbook Company, Scranton, Pennsylvania, 1954.
3. Black, P. H., and O. E. Adams, Jr., *Machine Design*, McGraw-Hill Publishing Company, New York, 1955.
4. Baumeister, T., ed., *Marks' Standard Handbook for Mechanical Engineers*, 8th ed., McGraw-Hill Publishing Company, New York, 1978.
5. Lingaiah, K., and B. R. Narayana Iyengar, *Machine Design Data Handbook*, Engineering College Co-operative Society, Bangalore, India, 1962.
6. Lingaiah, K., and B. R. Narayana Iyengar, *Machine Design Data Handbook*, Vol. I (*SI and Customary Metric Units*), Suma Publishers, Bangalore, India, 1986.
7. Lingaiah, K., *Machine Design Data Handbook*, Vol. II (*SI and Customary Metric Units*), Suma Publishers, Bangalore, India, 1986.
8. Bureau of Indian Standards.
9. ISO Standards.

TABLE 18–6
Allowable bearing pressure for screws, σ_b'

	Material		Safe bearing pressure, σ_b'		
Type	Screw	Nut	MPa	psi	Rubbing velocity, m/s [fpm(ft/min)]
Hand press	Steel	Bronze	17.2–24.0	2500–3500	Low speed, well lubricated
Jack screw	Steel	Cast iron	12.3–17.2	1800–2500	Low speed, not over 0.04 (8)
Jack screw	Steel	Bronze	10.8–17.2	1600–2500	Low speed, not over 0.05 (10)
Hoisting screw	Steel	Cast iron	4.4–6.9	600–1000	Medium speed, 0.1 to 0.2 (20–40)
Hoisting screw	Steel	Bronze	5.4–9.8	800–1400	Medium speed, 0.1 to 0.2 (20–40)
Lead screw	Steel	Bronze	1.0–1.5	150–240	High speed, 0.25 and over (50)

$$H = 0.86603 \ P; \ \ D_1 = d_2 - \frac{H}{2} = d - 2H_1 = d - 1.082 \ P$$

$$D_2 = d_2 = d - \frac{3}{4}H = d - 0.64952 \ P; \ \ d_1 = d_2 - \frac{H}{3} = d - 1.22687 \ p$$

$$H_1 = \frac{D - D_1}{2} = \frac{5}{8}H = 0.54127 \ P; \ \ h_3 = \frac{d - d_1}{2} = \frac{17}{24}H = 0.61343 \ P$$

$$r = \frac{H}{6} = 0.1443 \ P; \ \ r_c = 0.10825 \ P; \ \ \text{stress area} = A_c = \frac{\pi}{4}\left(\frac{d_1 + d_2}{2}\right)^2$$

Designation: A pitch diameter combination of thread size 8 mm and pitch 1mm shall be designated as $M8 \times 1$.
$M8$ shall designate pitch diameter combination of thread size 8 mm and pitch 1.25 mm.

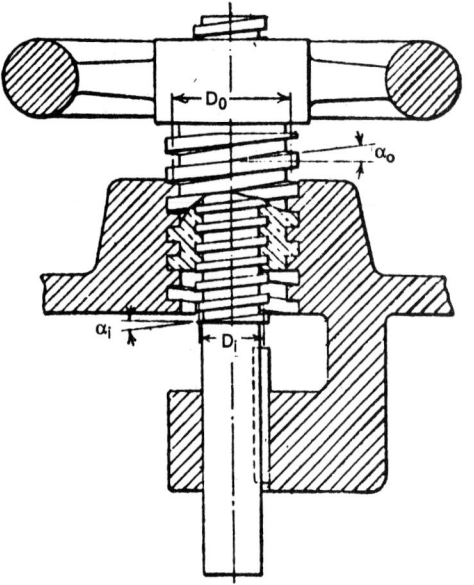

FIGURE 18–7(*a*)   Differential screws.

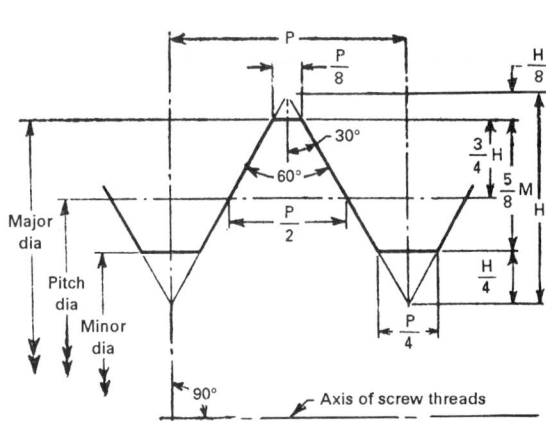

FIGURE 18–7(*b*)   Basic profile ISO metric screw threads.

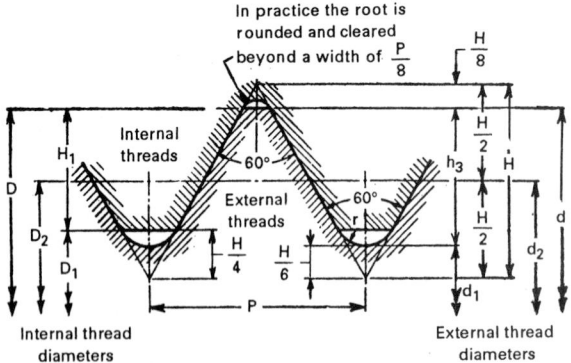

FIGURE 18–8   ISO metric screw thread design profiles of external and internal threads.

TABLE 18–7
Basic dimensions for design profiles of ISO metric screw threads

Basic diameter, mm	Pitch, P, mm	Major diameter, d, mm	Pitch diameter, d_2, mm	Minor diameter, mm		Lead angle at basic pitch diameter		Tensile stress area, A_c, mm^2
				External threads, d_1	Internal threads, D_1	deg	min	
1	0.25	1.0	0.837620	0.693283	0.729367	5	27	0.46
	0.20	1.0	0.870096	0.754626	0.783494	4	11	0.53
2	0.40	2.0	1.740192	1.509252	1.566987	4	11	2.07
	0.25	2.0	1.837620	1.693283	1.729367	2	29	2.45
2.5	0.45	2.5	2.207716	1.947909	2.012861	3	43	3.39
	0.35	2.5	2.272668	2.070596	2.121114	2	20	3.70
3.0	0.50	3.0	2.675240	2.386565	2.458734	3	24	5.03
	0.35	3.0	2.772668	2.570596	2.621114	2	18	5.61
4.0	0.70	4.0	3.545337	3.141191	3.242228	3	36	8.78
	0.50	4.0	3.675240	3.386565	3.458734	2	29	9.79
5.0	0.80	5.0	4.480385	4.018505	4.133975	3	15	14.2
	0.50	5.0	4.675240	4.386565	4.458734	2	57	16.1
6.0	1.00	6.0	5.350481	4.773131	4.917468	3	24	20.1
	0.75	6.0	5.512861	5.079848	5.188101	2	29	22.0
7.0	1.00	7.0	6.350481	5.773131	5.917408	2	52	28.9
	0.75	7.0	6.512861	6.079848	6.188101	2	6	31.3
8.0	1.25	8.0	7.188101	6.466413	6.646835	3	10	36.6
	1.00	8.0	7.350481	6.773131	6.917468	2	29	39.2
10	1.50	10.0	9.025721	8.159696	8.376202	3	2	58.0
	1.25	10.0	9.188101	8.466413	8.646835	2	29	61.2
	1.00	10.0	9.350481	8.773131	8.917468	1	57	64.5
12	1.75	12	10.863342	9.852979	10.105569	2	56	84.3
	1.50	12	11.025721	10.159686	10.376202	2	29	88.1
	1.25	12	11.188101	10.466413	10.646835	2	2	92.1
	1.00	12	11.350481	10.773131	10.917468	1	36	96.1
14	2.00	14	12.700962	11.546261	11.834936	2	52	115
	1.50	14	13.025721	12.159696	12.376202	2	6	125
	1.25	14	13.188101	12.466413	12.646835	1	44	129
16	2.00	16	14.700962	13.546261	13.834936	2	29	157
	1.50	16	15.025721	14.159696	14.376202	1	49	167
18	2.50	18	16.376202	14.932827	15.293671	2	47	192
	2.00	18	16.700962	15.546261	15.834936	2	11	204
	1.50	18	17.025721	15.159696	16.376202	1	36	216
20	2.50	20	18.376202	16.932827	17.293671	2	29	245
	2.00	20	18.700962	17.516261	17.834936	1	57	258
	1.50	20	19.025721	18.159696	18.376202	1	26	272
22	2.50	22	20.376202	18.932827	19.293671	2	14	303
	2.00	22	20.700962	19.546261	19.834936	1	46	318
	1.50	22	21.025721	20.159696	20.376202	1	18	333
24	3.00	24	22.051443	20.319392	20.752405	2	49	353
	2.00	24	22.700962	21.556261	21.834936	1	39	384
	1.50	24	23.025721	22.159696	22.376202	1	11	401
25	3.00	25	23.051443	21.319392	21.752405	2	36	385

TABLE 18–7
Basic dimensions for design profiles of ISO metric screw threads (*Cont.*)

Basic diameter, mm	Pitch, P, mm	Major diameter, d, mm	Pitch diameter, d₂, mm	Minor diameter, mm		Lead angle at basic pitch diameter		Tensile stress area, A_c, mm²
				External threads, d₁	Internal threads, D₁	deg	min	
30	3.50	30	27.726683	25.705957	26.211139	2	18	561
	3.00	30	28.051443	26.319392	26.752405	1	57	581
	2.00	30	28.700962	27.546261	27.834936	1	16	621
	1.50	30	29.025721	28.159696	28.376202	0	57	642
35	1.50	35	34.055721	33.159696	33.376202	0	48	860
42	4.5	42	39.072114	36.479088	37.128607	2	6	1120
	4.0	42	39.401924	37.092523	37.669873	1	51	1150
	3.0	42	40.051443	38.319392	38.752405	1	22	1210
	2.0	42	40.700962	39.546261	39.834936	0	52	1260
	1.5	42	41.025771	40.159696	40.376202	0	40	1290
45	4.5	45	42.077164	39.479088	40.128607	1	57	1300
	4.0	45	42.401924	40.092523	40.669873	1	43	1340
	3.0	45	43.051443	41.319392	41.752405	1	16	1400
	2.0	45	43.700962	42.546261	42.834936	0	50	1460
	1.5	45	44.025771	43.159696	43.376202	0	37	1490
52	5.0	52	48.752405	45.865653	46.587341	1	52	1760
	4.0	52	49.401924	47.092523	47.669873	1	29	1830
	3.0	52	50.051443	48.319392	48.752405	1	6	1900
	2.0	52	50.700962	49.546261	49.834936	0	43	1970
	1.5	52	51.025721	50.159696	50.376202	0	32	2010
60	5.5	60	56.427645	53.252219	54.046075	1	47	2360
	4.0	60	57.401924	55.092523	55.669873	1	16	2490
	3.0	60	58.051443	56.319392	56.752405	0	57	2570
	2.0	60	58.700962	57.546261	57.834936	0	37	2650
	1.5	60	59.025721	58.159696	58.376202	0	28	2700
72	6	72	68.102886	64.638784	66.504809	1	36	3460
	4	72	69.401924	67.092523	67.669873	1	3	3660
	3	72	70.051443	68.319392	68.752405	0	47	3760
	2	72	70.700962	69.546261	69.834936	0	31	3860
80	6	80	76.102886	72.638724	73.504809	1	26	4340
	4	80	77.401924	75.092523	75.669873	0	57	4570
	3	80	78.051443	76.319392	76.752405	0	42	4680
	2	80	78.700962	77.546261	77.834936	0	28	4790
90	6	90	86.102886	82.638784	83.504809	1	16	5590
	4	90	87.401924	85.092523	85.669873	0	50	5840
	3	90	88.051449	86.319292	86.752405	0	37	5970
	2	90	88.700962	87.546261	87.834936	0	25	6100
100	6	100	96.102886	92.638784	93.504809	1	8	7000
	4	100	97.401924	95.092523	95.669873	0	45	7280
	3	100	98.051443	96.319392	96.752405	0	33	7420
	2	100	98.700962	97.546261	97.834936	0	22	7560
110	6	110	106.102886	102.638784	103.504809	1	2	8560
	4	110	107.401924	105.092523	105.669873	0	41	8870
	3	110	108.051443	106.319392	106.752405	0	30	9020

TABLE 18–7
Basic dimensions for design profiles of ISO metric screw threads (*Cont.*)

| Basic diameter, mm | Pitch, P, mm | Major diameter, d, mm | Pitch diameter, d_2, mm | Minor diameter, mm | | Lead angle at basic pitch diameter | | Tensile stress area, A_c, mm^2 |
				External threads, d_1	Internal threads, D_1	deg	min	
120	6	120	116.102886	112.638784	113.504819	0	57	10300
	4	120	117.401924	115.092523	115.669873	0	37	10600
	3	120	118.051443	116.319392	116.752405	0	28	10800
150	6	150	146.102886	142.538784	143.504809	0	45	16400
	4	150	147.401924	145.092523	145.669873	0	30	16800
	3	150	148.051443	146.319392	146.752405	0	22	17000
160	6	160	156.102886	152.638784	153.504809	0	42	18700
	4	160	157.401924	155.092523	155.669873	0	28	19200
	3	160	158.051443	156.319392	156.752405	0	21	19400
180	6	180	176.102886	172.638784	173.504809	0	37	23900
	4	180	177.401924	175.092523	175.669873	0	25	24400
	3	180	178.051443	176.319392	176.752405	0	18	24700
200	6	200	196.102886	192.638784	193.504809	0	33	29700
	4	200	197.401924	195.092523	195.669873	0	22	30200
	3	200	198.051453	196.319392	196.752405	0	17	30500
250	6	250	246.102886	242.638784	243.504809	0	27	46900
	4	250	247.401924	245.092523	245.669873	0	18	47600
	3	250	248.051443	246.319392	246.752405	0	13	48000
300	6	300	296.102886	295.638784	293.504809	0	22	68100
	4	300	297.401924	292.092523	295.669873	0	15	68900

Source: IS: 4218–1967 (Part III).

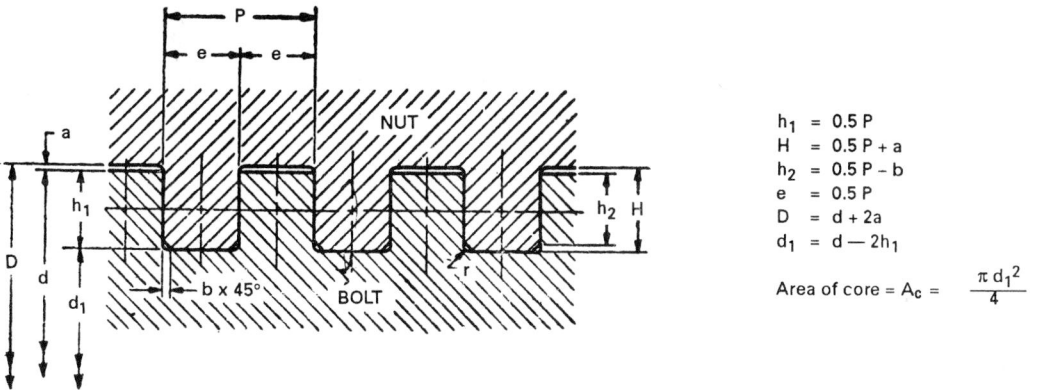

$h_1 = 0.5\,P$
$H = 0.5\,P + a$
$h_2 = 0.5\,P - b$
$e = 0.5\,P$
$D = d + 2a$
$d_1 = d - 2h_1$

$$\text{Area of core} = A_c = \frac{\pi\, d_1^2}{4}$$

Designation : A square thread nominal diameter 30 mm and pitch 6 mm shall be designated as SQ 30 x 6

FIGURE 18–9 Basic profile of square threads.

TABLE 18–8
Basis dimensions (in mm) for square threads

Nominal diameter	Major diameter Bolt, d	Major diameter Nut, D	Minor diameter, d_1	Pitch, P	e	r	h_2	b	h_1	a	H	Area of core, A_c, mm^2
10	10	10.5	8									50.3
14	14	14.5	12	2	1	0.12	0.75	0.25	1	0.25	1.25	113
20	20	20.5	18									201
26	26	26.5	23									415
30	30	30.5	27									573
36	36	36.5	33	3	1.5	0.12	1.25	0.25	1.5	0.25	1.75	855
40	40	40.5	37									1075
44	45	44.5	41									1320
50	50	50.5	47									1735
55	55	55.5	52	3	1.5	0.12	1.25	0.25	1.5	0.25	1.75	2124
60	60	60.5	57									2552
75	65	65.5	61									2922
80	70	70.5	66									3421
85	75	75.5	71									3959
90	80	80.5	76									4536
95	85	85.5	84	4	2	0.12	1.75	0.25	2	0.25	2.25	5153
90	90	90.5	86									5809
95	95	95.5	91									5504
100	100	100.5	96									7248
110	110	110.5	106									8825
120	120	120.5	114									10207
130	130	130.5	124									12076
140	140	140.5	134	6	3	0.25	2.5	0.5	3	0.25	3.25	14103
150	150	150.5	144									16286
160	160	160.5	154									18627
170	170	170.5	164									21124
180	180	180.5	172									23235
190	190	190.5	182									26016
200	200	200.5	192	8	4	0.25	3.5	0.5	4	0.25	4.25	28953
220	220	220.5	212									35299
240	240	240.5	232									42273
					Normal Series							
22	22	22.5	17									227
24	24	24.5	19	5	2.5	0.25	2	0.5	2.5	0.25	2.75	284
26	26	26.5	21									346
28	28	28.5	23									415
30	30	30.5	24	6	3	0.25	2.5	0.5	3	0.25	3.25	452
36	36	36.5	30									707
40	40	40.5	33	7	3.5	0.25	3	0.5	3.5	0.25	3.75	855
44	44	44.5	37									1075
50	50	50.5	42	8	4	0.25	3.5	0.5	4	0.25	4.25	1385
52	52	52.5	44									1521
55	55	55.5	46	9	4.5	0.25	4	0.5	4.5	0.25	4.75	1662
60	60	60.5	51									2043
65	65	65.5	55									2376
70	70	70.5	60	10	5	0.25	0.5	0.5	5	0.25	5.25	2827

TABLE 18–8
Basis dimensions (in mm) for square threads (*Cont.*)

Nominal diameter	Major diameter Bolt, d	Major diameter Nut, D	Minor diameter, d_1	Pitch, P	e	r	h_2	b	h_1	a	H	Area of core, A_c, mm^2
75	75	75.5	65									3318
80	80	80.5	70									3848
85	85	85.5	73									4185
90	90	90.5	78									4778
95	95	95.5	83	12	6	0.25	5.5	0.5	6	0.25	6.25	5411
100	100	100.5	88									6082
110	110	110.5	98									7543
120	120	121	106									8825
130	130	131	116	14	7	0.5	6	1	7	0.5	7.5	10568
140	140	141	126									12469
150	150	151	134									14103
160	160	161	144	16	8	0.5	7	1	8	0.5	8.5	16286
170	170	171	154									18627
180	180	181	162									20612
190	190	191	172	18	9	0.5	8	1	9	0.5	9.5	23235
200	200	201	182									26016
300	300	301	274	26	13	0.5	12	1	13	0.5	13.5	58965
Coarse Series												
22	22	22.5	14									164
24	24	24.5	16	8	4	0.25	3.5	0.5	4	0.25	4.25	201
26	26	26.5	18									254
28	28	28.5	20									314
30	30	30.5	20									314
36	36	36.5	26	10	5	0.25	4.5	0.5	5	0.25	5.25	531
40	40	40.5	28									616
50	50	50.5	38	12	6	0.25	5.5	0.5	6	0.25	6.25	1134
60	60	61	46	14	7	0.5	6	1	7	0.5	7.5	1662
70	70	71	54									2290
75	75	76	59	16	8	0.5	7	1	8	0.5	8.5	2734
80	80	81	64									3217
90	90	91	72	18	9	0.5	8	1	9	0.5	9.5	4072
120	120	121	98	22	11	0.5	10	1	11	0.5	11.5	8332
150	150	151	126	24	12	0.5	11	1	12	0.5	12.5	12469
180	180	181	152	28	14	0.5	13	1	14	0.5	14.5	18146
200	200	201	168	32	16	0.5	15	1	16	0.5	16.5	22167
300	300	301	256	44	24	0.5	21	1	22	0.5	22.5	51472
400	400	401	352	48	24	0.5	23	1	24	0.5	24.5	97314

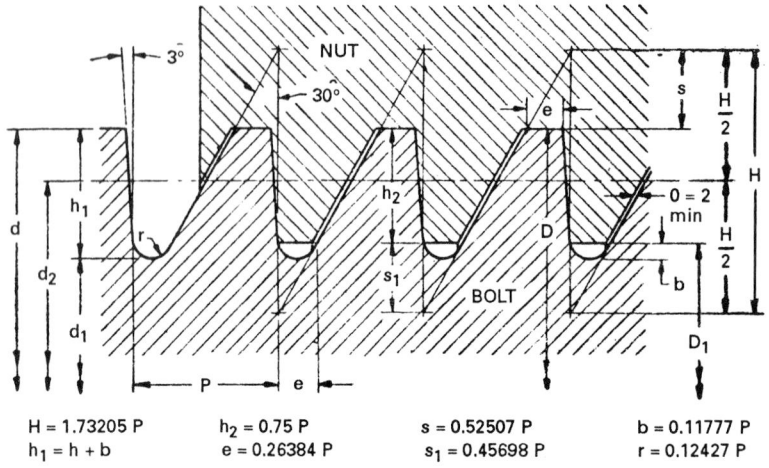

$$H = 1.73205\ P \qquad h_2 = 0.75\ P \qquad s = 0.52507\ P \qquad b = 0.11777\ P$$
$$h_1 = h + b \qquad e = 0.26384\ P \qquad s_1 = 0.45698\ P \qquad r = 0.12427\ P$$

Pitch, mm	Depth of thread, mm	Depth of engagement, mm	e, mm	b, mm	r, mm
2	1.736	1.5	0.528	0.236	0.249
3	2.603	2.25	0.792	0.353	0.373
4	3.471	3	1.055	0.471	0.497
5	4.339	3.75	1.319	0.589	0.621
6	5.207	4.5	1.583	0.707	0.746
7	6.074	5.25	1.847	0.824	0.870
8	6.942	6	2.111	0.942	0.994
9	7.810	6.75	2.375	1.060	1.118
10	8.678	7.5	2.638	1.178	1.243
12	10.413	9	3.166	1.413	1.491
14	12.149	10.5	3.694	1.649	1.740
16	13.884	12	4.221	1.884	1.988
18	15.620	13.5	4.749	2.120	2.237
20	17.355	15	5.277	2.355	2.485
22	19.091	16.5	5.804	2.591	2.734
24	20.826	18	6.332	2.826	2.982
26	22.562	19.5	6.860	3.062	3.231
28	24.298	21	7.388	3.298	3.480
32	27.769	24	8.443	3.769	3.977
36	31.240	27	9.498	4.240	4.474
40	34.711	30	10.554	4.711	4.971
44	38.182	33	11.609	5.182	5.468
48	41.653	36	12.664	5.653	5.965

Designation: A sawtooth thread of nominal diameter 48 mm and pitch 3 mm shall be designated as ST 48 × 3.

FIGURE 18–10 Basic profile of sawtooth threads. (*Source*: *IS 4696, 1968*.)

TABLE 18–9
Basic dimensions (in mm) for sawtooth threads

| Nominal diameter | Bolt | | | | Pitch, P | Nut | |
	Major diameter, d	Minor diameter, d_1	Area of core, mm^2	Pitch diameter, d_2		Major diameter, D	Minor diameter, D_1
				Fine Series			
10	10	6.528	33.5	8.636	2	10	7
12	12	8.528	57.1	10.636	2	12	9
14	14	10.538	87.1	12.636	2	14	11
16	16	12.528	123	14.636	2	16	13
20	20	16.528	215	18.636	2	20	17
22	22	16.794	222	19.954	3	22	17.5
30	30	24.794	483	27.954	3	30	25.5
36	36	30.794	745	32.954	3	36	31.5
40	40	34.794	951	37.954	3	40	35.5
50	50	44.794	1576	42.954	3	50	45.5
55	55	49.794	1947	57.954	3	55	50.5
60	60	54.794	2358	57.954	3	60	55.5
65	65	58.058	2647	62.272	4	65	59
70	70	63.058	3123	67.272	4	70	64
75	75	68.058	3638	72.272	4	75	69
80	80	73.058	4192	77.272	4	80	74
85	85	78.058	4785	82.272	4	85	79
90	90	83.058	5418	87.272	4	90	84
95	95	88.058	6090	92.272	4	95	89
100	100	93.058	6801	97.272	4	100	94
120	120	109.586	9432	115.909	6	120	111
150	150	139.586	15303	145.909	6	150	141
180	180	166.116	21673	174.545	8	180	168
200	200	186.116	27206	194.545	8	200	188
				Normal Series			
22	22	13.322	139	18.590	5	22	14.5
24	24	15.322	184	20.590	5	24	16.5
26	26	17.322	236	22.590	5	26	18.5
30	30	19.586	301	25.909	6	30	21
36	36	25.586	514	31.909	6	36	27
40	40	27.852	709	35.227	7	42	31.5
44	44	31.852	797	39.227	7	44	33.5
50	50	36.116	1024	44.545	8	50	38
55	55	39.380	1218	48.863	9	55	41.5
60	60	44.380	1547	53.863	9	60	46.5
70	70	52.644	2177	63.181	10	70	55
80	80	62.644	3082	73.181	10	80	65
90	90	69.174	3758	81.817	12	90	72
100	100	79.174	4923	91.817	12	100	82
110	110	89.174	6246	101.817	12	110	92
130	130	102.702	8775	120.459	14	130	109
150	150	122.232	11734	139.089	16	150	126
180	180	148.760	17381	167.726	18	180	153
200	200	168.760	22368	187.726	18	200	173

TABLE 18–9
Basic dimensions (in mm) for sawtooth threads (*Cont.*)

Nominal diameter	Major diameter, d	Minor diameter, d_1	Area of core, mm^2	Pitch diameter, d_2	Pitch, P	Major diameter, D	Minor diameter, D_1
		Bolt				**Nut**	
			Coarse Series				
22	22	8.116	51.4	16.545	8	22	10
24	24	10.116	80.7	18.545	8	24	12
26	26	12.116	115	20.545	8	26	14
30	30	12.644	126	23.181	10	30	15
40	40	19.174	289	31.817	12	40	22
50	50	29.174	668	41.817	12	50	32
60	60	35.702	1001	50.453	14	60	39
70	70	42.232	1401	59.089	16	70	46
80	80	52.232	2143	69.089	16	80	56
90	90	58.760	2712	77.726	18	90	63
100	100	65.290	3348	86.362	20	100	70
150	150	108.348	9220	138.634	24	150	114
200	200	144.462	16391	178.179	32	200	152

H = 1.866 P
h_1 = 0.5 P
h_2 = 0.084 P
a = 0.05 P
b = 0.683 P
R = 0.256 P
R_1 = 0.221 P

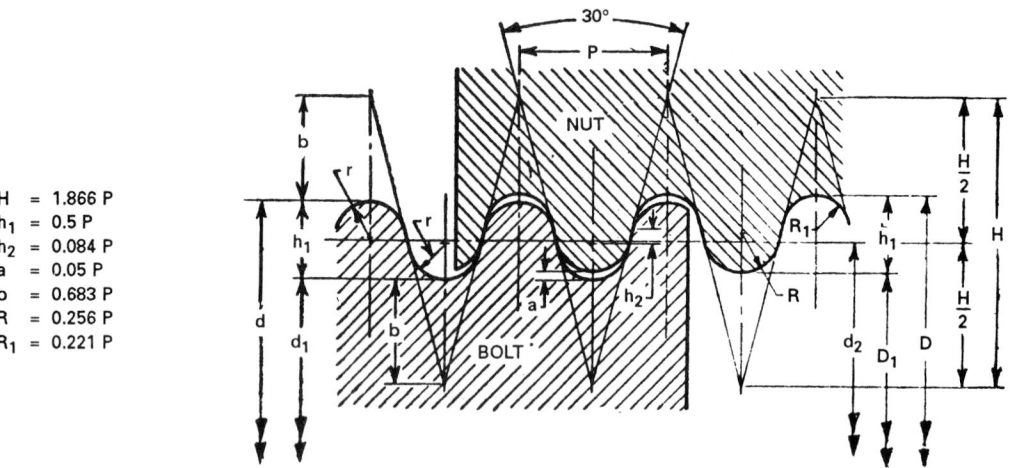

Nominal diameter, d, mm	Pitch, P, mm	Depth of thread, h_1, mm	Depth of engagement, h_2, mm	Bolt, r	R	R_1
					Radii, mm	
					Nut	
8–12	2.550	1.270	0.212	0.606	0.650	0.561
14–38	3.175	1.588	0.265	0.757	0.813	0.702
40–100	4.233	2.117	0.353	1.010	1.084	0.936
105–200	6.350	3.175	0.530	1.515	1.625	1.404

Designation: A knuckle thread of nominal diameter 10 mm and pitch of 2.54 mm shall be designated as $K10 \times 2.54$.

FIGURE 18–11   Basic profile of knuckle threads. (*Source: IS 4695, 1968.*)

TABLE 18–10
Basic dimensions (in mm) for knuckle threads

Nominal diameter	Bolt				Nut	
	Major diameter, d	Minor diameter, d_1	Area of core, mm^2	Pitch diameter, d_2	Major diameter, D	Minor diameter, D_1
8	8	5.460	23.4	6.730	8.254	5.714
9	9	6.460	32.8	7.730	9.254	6.714
10	10	7.460	43.7	8.730	10.254	7.714
12	12	9.460	70.3	10.730	12.254	9.714
14	14	10.825	92.0	12.412	14.318	11.142
16	16	12.825	129.2	14.412	16.318	16.142
20	20	16.825	222.3	18.412	20.318	17.142
24	24	20.825	340.6	22.412	24.318	21.142
30	30	26.825	565.2	28.412	30.318	27.142
36	36	32.825	846.3	34.412	36.318	33.142
40	40	35.767	1005	37.883	40.423	36.190
44	44	39.767	1242	41.883	44.423	40.190
50	50	45.767	1645	47.883	50.423	46.190
55	55	50.767	2024	52.883	55.423	51.190
60	60	55.767	2443	57.883	60.423	56.190
65	65	60.767	2900	62.883	65.423	61.190
70	70	65.767	3397	67.883	70.423	66.190
75	75	70.767	3933	72.883	75.423	71.190
80	80	75.767	4509	77.883	80.423	76.190
85	85	80.767	5123	82.883	85.423	81.190
90	90	85.767	5777	87.883	90.423	86.190
95	95	90.767	6471	92.883	95.423	91.190
100	100	95.767	7203	97.883	100.423	96.190
110	110	103.650	8438	106.825	110.635	104.285
120	120	113.650	10145	116.885	120.635	114.985
130	130	123.650	12008	126.825	130.635	124.285
140	140	133.650	14029	136.825	140.635	134.285
150	150	143.650	16207	146.825	150.635	144.285
160	160	153.650	18542	156.825	160.635	154.285
170	170	163.650	21034	166.825	170.635	164.285
180	180	173.650	23683	176.825	180.635	174.285
190	190	183.650	26489	186.825	190.635	184.285
200	200	193.650	29453	196.825	200.635	194.285

Source: IS 4695, 1968.

TABLE 18–11
Pitch-diameter combinations for ISO metric threads

Pitch, P, mm	Maximum diameter, mm
0.5	22
0.75	33
1.00	80
1.50	150
2.00	200
3.00	300

TABLE 18–12
Tolerance grades 3, 4, 5 for precision; 6 for medium; and 7, 8, and 9 for coarse qualities for bolts and nuts

Minor diameter of nut threads		4		5	6	7	8
Major diameter of bolt threads		4			6		8
Pitch diameter of nut threads		4	5		6	7	8
Pitch diameter of bolt threads	3	4	5		6	7	8	9

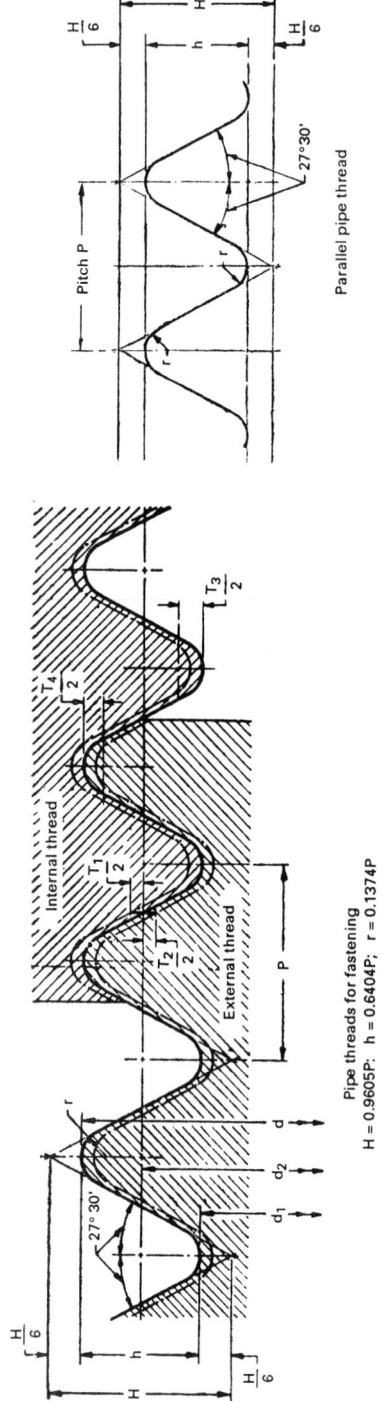

All dimensions in mm

Designation	Pitch, P	Major, d_2	Pitch, P	Minor, d_1
FP ⅛	0.907	9.728	9.147	8.566
FP ¼	1.337	13.157	15.301	11.445
FP ⅜	1.337	16.662	19.806	14.940
FP ½	1.814	20.955	25.793	18.631
FP ¾	1.814	26.441	25.279	24.117
FP 1	2.309	33.249	31.770	30.291
FP 1¼	2.309	41.910	40.431	38.952
FP 1½	2.309	47.803	46.324	44.845

Basic diameter, internal and external threads

Designation	Pitch, P	Major, d_2	Pitch, P	Minor, d_1
FP 2	2.309	59.614	58.135	56.656
FP 2¼	2.309	62.710	64.231	62.752
FP 2¼	2.309	75.184	73.705	72.226
FP 3	2.309	87.884	86.407	84.926
FP 3½	2.309	100.330	98.851	97.372
FP 4	2.309	113.030	111.551	110.072
FP 5	2.309	138.430	136.951	135.472
FP 6	2.309	193.830	162.351	160.872

Basic diameter, internal and external threads

Designation: An external pipe thread for fastening purposes of size 2 with class B tolerance shall be designated as Ext–FP 2B, and an internal pipe thread of size 2 shall be designated as Int–FP 2.

FIGURE 18–12 Pipe threads for fastening purposes. (*Source: IS 2643, 1964.*)

TABLE 18-13
Tolerances for crest and pitch diameters of bolts and nuts[a]

Diameter	Bolt/nut	Unit of tolerance	Value of tolerance unit	3	4	5	6	7	8	9
Crest diameter	Bolt	Td (6)	$180P^{2/3} - \dfrac{3.15}{\sqrt{P}}$	—	$0.63\,Td$ (6)	—	Td (6)	—	$1.6\,Td$ (6)	—
	Nut	Td_1 (6)	$433P - 190P^{1.22}$ for P from 0.2 to 0.8 mm; $230P^{0.7}$ for P from 1 mm and above	—	—	—	—	—	—	—
Pitch diameter	Bolt	Td_2 (6)	$90P^{0.4}\,d^{0.1}$	$0.5\,Td_2$ (6)	$0.63\,Td_2$ (6)	$0.8\,Td_2$ (6)	Td_2 (6)	$1.25\,Td_2$ (6)	$1.6\,Td_2$ (6)	$2\,Td_2$ (6)
	Nut	Td_2 (6)	$90P^{0.4}\,d^{0.1}$	—	$0.85\,Td_2$ (6)	$1.06\,Td_2$ (6)	$1.32\,Td_2$ (6)	$1.7\,Td_2$ (6)	$2.12\,Td_2$ (6)	—

[a] T_d in μm; TD in μm; P in μm; Td_2 in μm; d in mm.
Source: IS 4218 (Part IV), 1967.

TABLE 18-14
Preferred tolerance classes for nuts

Tolerance quality	Small allowance position G			No allowance position H		
	S	N	L	S	N	L
Fine				4H	5H	6H
Medium	5G	6G	7G	5H	6H	7H
Coarse	7G	7G	8G	7H	8H	

Source: IS 4218 (Part IV), 1967.

TABLE 18-15
Preferred tolerance classes for bolts

Tolerance quality	Large allowance position e			Small allowance position g			No allowance position h		
	S	N	L	S	N	L	S	N	L
Fine							3h 4h	4h	5h 4h
Medium		6e		7g 6g	6g	7g 6g	5h 6h	6h	7h 6h
Coarse	7e 6e				8g	9g 8g			

Source: IS 4218 (Part IV), 1967.

TABLE 18–16
Coarse-threaded series—UNC and NC (dimensions in inches)

Sizes[a]	Basic major (nominal) diameter, D	Threads per inch	Basic pitch diameter	Basic minor diameter external thread	Root area[b] in in^2, A	Minor diameter internal thread classes 1B, 2B, and 3B for engagement $^2/_3D$ to $^3/_2D$	
						Minimum	Maximum
1	0.0730	64	0.0629	0.0538	0.0023	0.0585	0.0623
2	0.0860	56	0.0744	0.0641	0.0032	0.0699	0.0737
3	0.0990	48	0.0855	0.0734	0.0042	0.0805	0.0845
4	0.1120	40	0.0958	0.0813	0.0052	0.0894	0.0939
5	0.1250	40	0.1088	0.0943	0.0070	0.1021	0.1062
6	0.1380	32	0.1177	0.0997	0.0078	0.1091	0.1140
8	0.1640	32	0.1437	0.1257	0.0124	0.1346	0.1389
10	0.1900	24	0.1629	0.1389	0.0152	0.1502	0.1555
12	0.2160	24	0.1889	0.1649	0.0214	0.1758	0.1807
$^1/_4$ UN	0.2500	20	0.2175	0.1887	0.0280	0.2013	0.2067
$^5/_{16}$ UN	0.3125	18	0.2764	0.2443	0.0469	0.2577	0.2630
$^3/_8$ UN	0.3750	16	0.3344	0.2983	0.0699	0.3128	0.3182
$^7/_{16}$ UN	0.4375	14	0.3911	0.3499	0.0962	0.3659	0.3717
$^1/_2$	0.5000	13	0.4500	0.4056	0.1292	0.4226	0.4284
$^1/_2$ UN	0.5000	12	0.4459	0.3978	0.1243	0.4160	0.4223
$^9/_{16}$ UN	0.5625	12	0.5084	0.4603	0.1664	0.4783	0.4843
$^5/_8$ UN	0.6250	11	0.5660	0.5135	0.2071	0.5329	0.5391
$^3/_4$ UN	0.7500	10	0.6850	0.6273	0.3091	0.6481	0.6545
$^7/_8$ UN	0.8750	9	0.8028	0.7387	0.4286	0.7614	0.7681
1 UN	1.0000	8	0.9188	0.8466	0.5629	0.8722	0.8797
$1^1/_8$ UN	1.1250	7	1.0322	0.9497	0.7178	0.9789	0.9875
$1^1/_4$ UN	1.2500	7	1.1572	1.0747	0.9071	1.1039	1.1125
$1^3/_8$ UN	1.3750	6	1.2667	1.1705	1.0760	1.2046	1.2146
$1^1/_2$ UN	1.5000	6	1.3917	1.2955	1.3182	1.3296	1.3396
$1^3/_4$ UN	1.7500	5	1.6201	1.5046	1.7780	1.5455	1.5575
2 UN	2.0000	$4^1/_2$	1.8557	1.7274	2.3436	1.7728	1.7861
$2^1/_4$ UN	2.2500	$4^1/_2$	2.1057	1.9774	3.0610	2.0228	2.0361
$2^1/_2$ UN	2.5000	4	2.3376	2.1933	3.7782	2.2444	2.2594
$2^3/_4$ UN	2.7500	4	2.5876	2.4433	4.6886	2.4944	2.5094
3 UN	3.0000	4	2.8376	2.6933	5.6972	2.7444	2.7594
$3^1/_4$ UN	3.2500	4	3.0876	2.9433	6.8039	2.9944	3.0094
$3^1/_2$ UN	3.5000	4	3.3376	3.1933	8.0088	3.2444	3.2594
$3^3/_4$ UN	3.7500	4	3.5876	3.4433	9.3119	3.4944	3.5094
4 UN	4.0000	4	3.8376	3.6933	10.7132	3.7444	3.7594

[a] Unified diameter-pitch relationships are marked UN.
[b] The actual root area of a screw will be somewhat less than A, but, since the tensile strength of a screw of ductile material is greater than that of a plain specimen of the same material and of a diameter equal to the root diameter of the screw, the tensile strength of a screw may be assumed to correspond to A as given.
For complete manufacturing information and tolerances, see ASA Standard B1.1, 1949.

TABLE 18–17
Fine-thread series UNF and NF (dimensions in inches)

Sizes[a]	Basic major (nominal) diameter, D	Threads per inch	Basic pitch diameter	Basic minor diameter external thread	Root area[b] in in², A	Minor diameter internal thread classes 1B, 2B, and 3B for engagement $2/3D$ to $3/2D$	
						Minimum	Maximum
0	0.0600	80	0.0519	0.0447	0.0016	0.0479	0.0514
1	0.0730	72	0.0640	0.0560	0.0025	0.0602	0.0635
2	0.0860	64	0.0759	0.0668	0.0035	0.0720	0.0753
3	0.0990	56	0.0874	0.0771	0.0047	0.0831	0.0865
4	0.1120	48	0.0985	0.0864	0.0059	0.0931	0.0968
5	0.1250	44	0.1102	0.0971	0.0074	0.1042	0.1079
6	0.1380	40	0.1218	0.1073	0.0090	0.1147	0.1186
8	0.1640	36	0.1460	0.1299	0.0133	0.1358	0.1416
10	0.1900	32	0.1697	0.1517	0.0181	0.1601	0.1641
12	0.2160	28	0.1928	0.1722	0.0233	0.1815	0.1857
1/4 UN	0.2500	28	0.2268	0.2062	0.0334	0.2150	0.2190
5/16 UN	0.3125	24	0.2854	0.2614	0.0541	0.2714	0.2754
3/8 UN	0.3750	24	0.3479	0.3239	0.0824	0.3332	0.3372
7/16 UN	0.4375	20	0.4050	0.3762	0.1112	0.3875	0.3916
1/2 UN	0.5000	20	0.4675	0.4387	0.1512	0.4497	0.4537
9/16 UN	0.5625	18	0.5264	0.4943	0.1919	0.5065	0.5106
5/8 UN	0.6250	18	0.5889	0.5568	0.2435	0.5690	0.5730
3/4 UN	0.7500	16	0.7094	0.6733	0.3560	0.6865	0.6908
7/8 UN	0.8750	14	0.8286	0.7874	0.4869	0.8023	0.8068
1 UN	1.000	12	0.9459	0.8978	0.6331	0.9148	0.9198
1 1/8 UN	1.1250	12	1.0709	1.0228	0.8216	1.0398	1.0448
1 1/4 UN	1.2500	12	1.1959	1.1478	1.0347	1.1648	1.1698
1 3/8 UN	1.3750	12	1.3209	1.2728	1.2724	1.2893	1.2948
1 1/2 UN	1.5000	12	1.4459	1.3978	1.5346	1.4148	1.4198

[a] Unified diameter-pitch relationships are marked UN.
[b] The actual root area of a screw will be somewhat less than A, but, since the tensile strength of a screw of *ductile* material is greater than that of a plain specimen of the same material and of a diameter equal to the root diameter of the screw, the tensile strength of a screw may be assumed to correspond to A as given.
For complete manufacturing information and tolerances, see ASA Standard B1.1, 1949.

TABLE 18–18
Extra-fine-thread series—NEF

Sizes[a]	Basic major (nominal) diameter, D, in	Threads per inch	Basic pitch diameter, in	Basic minor diameter external thread, in	Root area[b] in in^2, A	Minor diameter internal thread classes 1B, 2B, and 3B for engagement $2/3D$ to $3/2D$, in	
						Minimum	Maximum
12	0.2160	32	0.1957	0.1777	0.0248	0.1855	0.1895
1/4	0.2500	32	0.2297	0.2117	0.0352	0.2189	0.2229
5/16	0.3125	32	0.2922	0.2742	0.0591	0.2807	0.2847
3/8	0.3750	32	0.3547	0.3367	0.0890	0.3429	0.3469
7/16 UN	0.4375	28	0.4143	0.3937	0.1217	0.4011	0.4051
1/2 UN	0.5000	28	0.4768	0.4562	0.1635	0.4636	0.4676
9/16	0.5625	24	0.5354	0.5114	0.2054	0.5204	0.5244
5/8	0.6250	24	0.5979	0.5739	0.2587	0.5829	0.5869
11/16	0.6875	24	0.6604	0.6364	0.3181	0.6454	0.6494
3/4 UN	0.7500	20	0.7175	0.6887	0.3725	0.6997	0.7037
13/16 UN	0.8125	20	0.7800	0.7512	0.4432	0.7622	0.7662
7/8 UN	0.8750	20	0.8425	0.8137	0.5200	0.8247	0.8287
15/16 UN	0.9375	20	0.9050	0.8762	0.6030	0.8872	0.8912
1 UN	1.0000	20	0.9675	0.9387	0.6921	0.9497	0.9537
1 1/16	1.0625	18	1.0264	0.9943	0.7765	1.0064	1.0105
1 1/8	1.1250	18	1.0889	1.0568	0.8772	1.0689	1.0730
1 3/16	1.1875	18	1.1514	1.1193	0.9840	1.1314	1.1355
1 1/4	1.2500	18	1.2139	1.1818	1.0969	1.1939	1.1980
1 5/16	1.3125	18	1.2764	1.2443	1.2160	1.2564	1.2605
1 3/8	1.3750	18	1.3389	1.3068	1.3413	1.3189	1.3230
1 7/16	1.4375	18	1.4014	1.3693	1.4726	1.3814	1.3855
1 1/2	1.5000	18	1.4639	1.4318	1.6101	1.4439	1.4480
1 9/16	1.5625	18	1.5264	1.4943	1.7538	1.5064	1.5105
1 5/8	1.6250	18	1.5889	1.5568	1.9035	1.5689	1.5730
1 11/16	1.6875	18	1.6514	1.6193	2.0594	1.6314	1.6355
1 3/4 UN	1.7500	16	1.7094	1.6733	2.1991	1.6865	1.6908
2 UN	2.0000	16	1.9594	1.9233	2.9053	1.9365	1.9408

[a] Unified diameter-pitch relationships are marked UN.
[b] The actual root area of a screw will be somewhat less than A, but, since the tensile strength of a screw of ductile material is greater than that of a plain specimen of the same material and of a diameter equal to the root diameter of the screw, the tensile strength of a screw may be assumed to correspond to A as given.
For complete manufacturing information and tolerances, see ASA Standard B1.1, 1949.

TABLE 18–19
8-pitch thread series—8N (dimensions in inches)

Size[a] also basic major (normal) diameter, D	Basic minor diameter external thread	Minor diameter internal thread classes 1B, 2B, and 3B for engagement $2/3D$ to $3/2D$		Size[a] also basic major (normal) diameter, D	Basic minor diameter external thread	Minor diameter internal thread classes 1B, 2B, and 3B for engagement $2/3D$ to $3/2D$	
		Minimum	Maximum			Minimum	Maximum
1 UN	0.8466	0.8722	0.8797	3	2.8466	2.8722	8.8797
$1\frac{1}{8}$	0.9716	0.9972	1.0047	$3\frac{1}{4}$	3.0966	3.1222	3.1297
$1\frac{1}{4}$	1.0966	1.1222	1.1297	$3\frac{1}{2}$	3.3466	3.3722	3.3797
$1\frac{3}{8}$	1.2216	1.2472	1.2547	$3\frac{3}{4}$	3.5966	3.6222	3.6297
$1\frac{1}{2}$	1.3466	1.3722	1.3797	4	3.8466	3.8722	3.8797
$1\frac{5}{8}$	1.4716	1.4972	1.5047	$4\frac{1}{4}$	4.0966	4.1222	4.1297
$1\frac{3}{4}$	1.5966	1.6222	1.6297	$4\frac{1}{2}$	4.3466	4.3722	4.3797
$1\frac{7}{8}$	1.7216	1.7472	1.7547	$4\frac{3}{4}$	4.5966	4.6222	4.6297
2	1.8466	1.8722	1.8797	5	4.8466	4.8722	4.8797
$2\frac{1}{8}$	1.9716	1.9972	2.0047	$5\frac{1}{4}$	5.0966	5.1222	5.1297
$2\frac{1}{4}$	2.0966	2.1222	2.1297	$5\frac{1}{2}$	5.3466	5.3722	5.3797
$2\frac{1}{2}$	2.3466	2.3722	2.3797	$5\frac{3}{4}$	5.5966	5.6222	5.6297
$2\frac{3}{4}$	2.5966	2.6222	2.6297	6	5.8466	5.8722	5.8797

[a] Unified diameter-pitch relationships are marked UN.
For complete manufacturing information and tolerances, see ASA Standard B1.1, 1949.

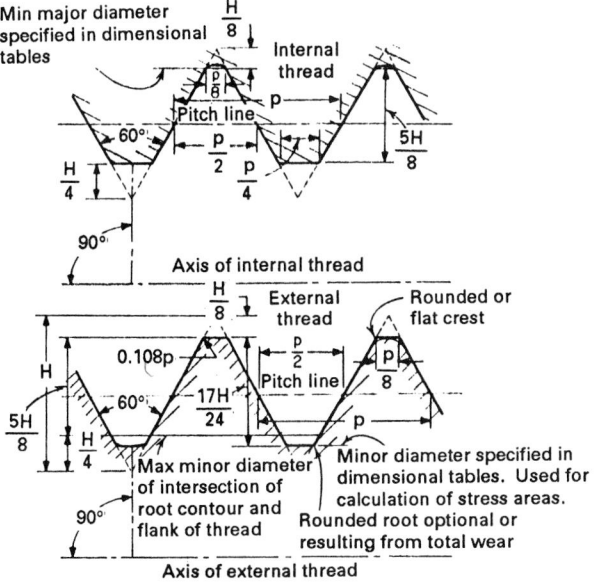

FIGURE 18–13   60° unified and American Standard screw-thread forms.

TABLE 18–20
12-pitch thread series—12N (dimensions in inches)

Size[a] also basic major (nominal) diameter, D	Basic minor diameter external thread	Minor diameter internal thread classes 1B, 2B, and 3B for engagement $2/3D$ to $3/2D$		Size[a] also basic major (nominal) diameter, D	Basic minor diameter external thread	Minor diameter internal thread classes 1B, 2B, and 3B for engagement $2/3D$ to $3/2D$	
		Minimum	Maximum			Minimum	Maximum
$1/2$	0.3978	0.4160	0.4223	2 UN	1.8978	1.9148	1.9198
$9/16$	0.4603	0.4783	0.4843	$2^1/8$	2.0228	2.0398	2.0448
$5/8$	0.0228	0.5405	0.5463	$2^1/4$ UN	2.1478	2.1648	2.1698
$11/16$	0.5853	0.6029	0.6085	$2^3/8$	2.2728	2.2898	2.2948
$3/4$	0.6478	0.6653	0.6707	$2^1/2$ UN	2.3978	2.4148	2.4198
$13/16$	0.7103	0.7276	0.7329	$2^5/8$	2.5228	2.5398	2.5448
$7/8$	0.7728	0.7900	0.7952	$2^3/4$ UN	2.6478	2.6648	2.6698
$15/16$ UN	0.8353	0.8524	0.8575	$2^7/8$	2.7728	2.7898	2.7948
1	0.8978	0.9148	0.9198	3 UN	2.8978	2.9148	2.9198
$1^1/16$ UN	0.9603	0.9773	0.9823	$3^1/8$	3.0228	3.0398	3.0448
$1^1/8$	1.0228	1.0398	1.0448	$3^1/4$ UN	3.1478	3.1648	3.1698
$1^3/16$ UN	1.0853	1.1023	1.1073	$3^3/8$	3.2728	3.2898	3.2948
$1^1/4$	1.1478	1.1648	1.1698	$3^1/2$ UN	3.3978	3.4148	3.4198
$1^5/16$ UN	1.2103	1.2273	1.2323	$3^5/8$	3.5228	3.5398	3.5448
$1^3/8$	1.2728	1.2898	1.2948	$3^3/4$ UN	3.6478	3.6648	3.6698
$1^7/16$ UN	1.3353	1.3523	1.3573	$3^7/8$	3.7728	3.7898	3.7948
$1^1/2$	1.3978	1.4148	1.4198	4 UN	3.8978	3.9148	3.9198
$1^5/8$	1.5228	1.5398	1.5448	$4^1/4$ UN	4.1478	4.1648	4.1698
$1^3/4$ UN	1.6478	1.6648	1.6698	$4^1/2$ UN	4.3978	4.4148	4.4198
$1^7/8$	1.7728	1.7898	1.7948	$4^3/4$ UN	4.6478	4.6648	4.6698
				5 UN	4.8978	4.9148	4.9198
				$5^1/4$ UN	5.1478	5.1648	5.1698
				$5^1/2$ UN	5.3978	5.4148	5.4198
				$5^3/4$ UN	5.6478	5.6648	5.6698
				6 UN	5.8978	5.9148	5.9198

[a] Unified diameter-pitch relationships are marked UN.
For complete manufacturing information and tolerances, see ASA Standard B1.1, 1949.

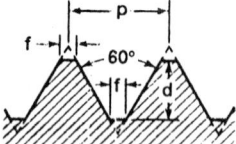

FIGURE 18–14 American Standard screw thread.

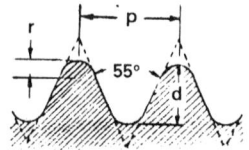

FIGURE 18–15 Whitworth screw thread.

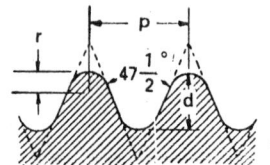

FIGURE 18–16 British Association screw thread.

TABLE 18–21
16-pitch thread series—16N (dimensions in inches)

Size[a] also basic major (nominal) diameter, D	Basic minor diameter external thread	Minor diameter internal thread classes 1B, 2B, and 3B for engagement $2/3D$ to $3/2D$		Size[a] also basic major (nominal) diameter, D	Basic minor diameter external thread	Minor diameter internal thread classes 1B, 2B, and 3B for engagement $2/3D$ to $3/2D$	
		Minimum	Maximum			Minimum	Maximum
3/4	0.6733	0.6865	0.6908	2 1/4 UN	2.1733	2.1865	2.1908
13/16 UN	0.7358	0.7490	0.7553	2 5/16	2.2358	2.2490	2.2533
7/8 UN	0.7983	0.8115	0.8158	2 3/8	2.2983	2.3115	2.3158
15/16 UN	0.8608	0.8740	0.8783	2 7/16	2.3608	2.3740	2.3783
1 UN	0.9233	0.9365	0.9408	2 1/2 UN	2.4233	2.4365	2.4408
1 1/16 UN	0.9853	0.9990	1.0033	2 5/8	2.5483	2.5615	2.5658
1 1/8 UN	1.0483	1.0615	1.0658	2 3/4 UN	2.6733	2.6865	2.6908
1 3/16 UN	1.1108	1.1240	1.1283	2 7/8	2.7983	2.8115	2.8158
1 1/4 UN	1.1733	1.1865	1.1908	3 UN	2.9233	2.9365	2.9408
1 5/16 UN	1.2358	1.2490	1.2533	3 1/8	3.0483	3.0615	3.0658
1 3/8 UN	1.2983	1.3115	1.3158	3 1/4 UN	3.1733	3.1865	3.1908
1 7/16 UN	1.3608	1.3740	1.3783	3 3/8	3.2983	3.3115	3.3158
1 1/2 UN	1.4233	1.4365	1.4408	3 1/2 UN	3.4233	3.4365	3.4408
1 9/16	1.4858	1.4990	1.5033	3 5/8	3.5483	3.5615	3.5658
1 5/8	1.5483	1.5615	1.5658	3 3/4 UN	3.6733	3.6865	3.6908
1 11/16	1.6108	1.6240	1.6283	3 7/8	3.7983	3.8115	3.8158
1 3/4 UN	1.6733	1.6865	1.6908	4 UN	3.9233	3.9365	3.9408
1 13/16	1.7358	1.7490	1.7533	4 1/4 UN	4.1733	4.1865	4.1908
1 7/8	1.7983	1.8115	1.8158	4 1/2 UN	4.4233	4.4365	4.4408
1 15/16	1.8608	1.8740	1.8783	4 3/4 UN	4.6733	4.6865	4.6908
2 UN	1.9233	1.9365	1.9408	5 UN	4.9233	4.9365	4.9408
1 1/16	1.9858	1.9990	2.0033	5 1/4 UN	5.1733	5.1865	5.1908
2 1/8	2.0483	2.0615	2.0658	5 1/2 UN	5.4233	5.4365	5.4408
2 3/16	2.1108	2.1240	2.1283	5 3/4 UN	5.6733	5.6865	5.6908
				6 UN	5.9233	5.9365	5.9408

[a] Unified diameter-pitch relationships are marked UN.

For complete manufacturing information and tolerances, see ASA Standard B1.1, 1949.

TABLE 18–22
Proportions of power threads (dimensions in inches)

Size in	Square threads			Acme threads		
	Threads per inch	Minor diameter		Threads per inch	Regular minor diameter	Stub minor diameter
1/4	10	0.163		16	0.188	0.213
5/16	9	0.2153		14	0.241	0.270
3/8	8	0.266		12	0.292	0.325
7/16	7	0.3125		12	0.354	0.388
1/2	6 1/2	0.366		10	0.400	0.440
5/8	5 1/2	0.466		8	0.500	0.550
3/4	5	0.575		6	0.583	0.650
7/8	4 1/2	0.681		6	0.708	0.775
1	4	0.783		5	0.800	0.880
1 1/8	3 1/2	0.8750		5	0.925	1.005
1 1/4	3 1/2	1.000		5	1.050	1.130
1 3/8	3	1.0834		4	1.125	1.225
1 1/2	3	1.284		4	1.250	1.350
1 3/4	2 1/2	1.400		4	1.500	1.600
2	2 1/4	1.612		4	1.750	1.850
2 1/4	2 1/4	1.862		3	1.917	2.050
2 1/2	2	2.063		3	2.167	2.300
2 3/4	2	2.313		3	2.417	2.550
3	1 3/4	2.500		2	2.500	2.700
3 1/2	1 5/8	2.962		2	3.000	3.200
4	1 1/2	3.168		2	3.500	3.700
4 1/2				2	4.000	4.200
5				2	4.000	4.700

COUPLINGS, CLUTCHES, AND BRAKES

SYMBOLS

a	distance between center lines of shafts in Oldham's coupling, m (in)
A	area, m^2 (in^2)
	external area, m^2 (in^2)
A_r	radiating surface required, m^2 (in^2)
A_c	contact area of friction surface, m^2 (in^2)
b	width of key, m (in)
	width of shoe, m (in)
	width of inclined face in grooved rim clutch, m (in)
	width of spring in centrifugal clutch, m (in)
	width of wheel, m (in)
	width of operating lever (Fig. 19–16), m (in)
c	heat transfer coefficient, kJ/m^2 K h ($kcal/m^2/°C/h$)
c_1	specific heat of material, kJ/kg K (kcal/kg/°C)
c_2	radiating factor for brakes, kJ/m^2 K s ($kcal/m^2/min/°C$)
d	diameter of shaft, m (in)
	diameter of pin, roller pin, m (in)
d_1	diameter of bolt, m (in)
	diameter of pin at neck in the flexible coupling, m (in)
d_2	diameter of hole for bolt, m (in)
d'	outside diameter of bush, m (in)
D	diameter of wheel, m (in)
	diameter of sheave, m (in)
	outside diameter of flange coupling, m (in)
D_1	inside diameter of disk of friction material in disk clutches and brakes, m (in)
D_2	outside diameter of disk of friction material in disk clutches and brakes, m (in)
D_i	inside diameter of hollow rigid type of coupling, m (in)
D_o	outside diameter of hollow rigid type of coupling, m (in)
D_m	mean diameter, m (in)
$e_1,\ e_2,\ e_3$	dimensions shown in Fig. 19–16, m (in)
E	energy (also with suffixes), N m (lbf in)
	Young's modulus of elasticity, GPA (Mpsi)

F	operating force on block brakes, kN (lbf);
	force at each pin in the flexible bush coupling, kN (lbf)
	total pressure, kN (lbf)
	force (also with suffixes), kN (lbf)
	actuating force, kN (lbf)
F_1	tension on tight side of band, kN (lbf)
	the force acting on disks of one operating lever of the clutch (Fig. 19–16), kN (lbf)
F_2	tension on slack side of band, kN (lbf)
F_a'	total axial force on i number of clutch disks, kN (lbf)
F_b	tension load in each bolt, kN (lbf)
F_c	centrifugal force, kN (lbf)
F_n	total normal force, kN (lbf)
F_x, F_y	components of actuating force F acting at a distance c from the hinge pin (Figs. 19–25 and 19–26), kN (lbf)
F_θ	tangential force at rim of brake wheel, kN (lbf)
	tangential friction force, kN (lbf)
g	acceleration due to gravity, 9.8066 m/s^2 (9806.6 mm/s^2) (32.2 ft/s^2)
h	thickness of key, m (in)
	thickness of central disk in Oldham's coupling, m (in)
	thickness of operating lever (Fig. 19–16), m (in)
	depth of spring in centrifugal clutch, m (mm)
H	rate of heat to be radiated, J (kcal)
H_g	heat generated, J (kcal)
H_d	the rate of dissipation, J (kcal)
i	number of pins,
	number of bolts,
	number of rollers,
	pairs of friction surfaces
	number of shoes in centrifugal clutch
i_1	number of driving disks
i_2	number of driven disks
i'	number of operating lever of clutch
I	moment of inertia, area, m^4, cm^4 (in^4)
k_l	load factor or the ratio of the actual brake operating time to the total cycle of operation
k_s	speed factor
l	length (also with suffixes), m (in)
	length of spring in centrifugal clutch measured along arc, m (in)
	length of bush, m (in)
L	dimension of operating lever as shown in Fig. 19–16
M_t	torque to be transmitted, N m (lbf in)
M_{ta}	allowable torque, N m (lbf in)
n	speed, rpm
n_1, n_2	speed of the live load before and after the brake is applied, respectively, rpm
n	number of clutching or braking cycles per hour
P	power, kW (hp)
N	normal force (Figs. 19–25 and 19–26), kN (lbf)
μN	frictional force (Figs. 19–25 and 19–26), kN (lbf)
p	unit pressure, MPa (psi)

	unit pressure acting upon an element of area of the frictional material located at an angle θ from the hinge pin (Figs. 19–25 and 19–26), MPa (psi)
	maximum pressure between the fabric and the inside of the rim, MPa (psi)
p_a	allowable pressure, MPa (psi)
	maximum pressure located at an angle θ_a from the hinge pin (Figs. 19–25 and 19–26), MPa (psi)
p_b	bearing pressure, MPa (psi)
P	total force acting from the side of the bush on operating lever (Fig. 19–16), kN (lbf)
P'	the force acting from the side of the bush on one operating lever, kN (lbf)
r	radius, m (in)
	distance from the center of gravity of the shoe from the axis of rotation, m (in)
r_m	mean radius, m (in)
r_{mi}	mean radius of inner passage of hydraulic coupling, m (in)
r_{mo}	mean radius of outer passage in hydraulic coupling, m (in)
R	reaction (also with suffixes), kN (lbf)
R_c	radius of curvature of the ramp at the point of contact (Fig. 19–21), m (in)
R_d	radius of the contact surface on the driven member (Fig. 19–21), m (in)
R_r	radius of the roller (Fig. 19–21), m (in)
R_x, R_y	hinge pin reactions (Figs. 19–25 and 19–26), kN (lbf)
t	time of single clutching or braking operation (Eq. 19–198), s
T_a	ambient or initial temperature, °C (°F)
T_{av}	average equilibrium temperature, °C (°F)
ΔT	rise in temperature of the brake drum, °C (°F)
t_c	cooling time, s (min)
v	velocity, m/s
v_1, v_2	speed of the live load before and after the brake is applied, respectively, m/s
w	axial width in cone brake, m (in)
	width of band, m (in)
W	work done, N m (lbf in)
	weight of the fluid flowing in the torus, kN (lbf)
	weight lowered, kN (lbf)
	weight of parts in Eq. (19–136), kN (lbf)
	weight of shoe, kN (lbf)
y	deflection, m (in)
σ	stress (also with suffixes), MPa (psi)
σ_b	allowable or design stress in bolts, MPa (psi)
σ_b'	design bearing stress for keys, MPa (psi)
$\sigma_c(max)$	maximum compressive stress in Hertz's formula, MPa (psi)
σ_{db}	design bending stress, MPa (psi)
τ	shear stress, MPa (psi)
τ_b	allowable or design stress in bolts, MPa (psi)
τ_{d1}	design shear stress in sleeve, MPa (psi)
τ_{d2}	design shear stress in key, MPa (psi)
τ_f	design shear stress in flange at the outside hub diameter, MPa (psi)
τ_s	design shear stress in shaft, MPa (psi)
α	one-half the cone angle, deg

ϕ	pressure angle, deg
	friction angle, deg
θ	one-half angle of the contact surface of block, deg
μ	coefficient of friction
η	factor which takes care of the reduced strength of shaft due to keyway
ω_1	running speed of centrifugal clutch, rad/s
ω_2	speed at which the engagement between the shoe of centrifugal clutch and pulley commences, rad/s

SUFFIXES

a	axial
d	dissipated, design
g	generated
1, i	inner
2, o	outer
n	normal
x	x direction
y	y direction
θ	tangential
μ	friction

Other factors in performance or in special aspects are included from time to time in this chapter and, being applicable only in their immediate context, are not included at this stage.

Particular	Formula

19.1 COUPLINGS

COMMON FLANGE COUPLING (FIG. 19–1)

The commonly used formula for approximate number of bolts	$i = 20d + 3$ **SI**, d in m (19–1a) $\quad = 0.5d + 3$ **US Customary units** (19–1b) where d in in
The torque transmitted by the shaft	$$M_t = \frac{\pi d^3}{16}\,\eta\tau_s \qquad\qquad (19\text{–}2)$$
The torque transmitted by the coupling	$$M_t = \frac{1000P}{\omega} \qquad \textbf{SI} \qquad (19\text{–}3a)$$ where M_t in N m; P in kW; ω in rad/s $$\quad = \frac{63{,}000P}{n} \qquad \textbf{US Customary units} \quad (19\text{–}3b)$$ where M_t in lbf in; P in hp; n in rpm $$\quad = \frac{9550P}{n} \qquad \textbf{SI} \qquad (19\text{–}3c)$$

Particular	Formula

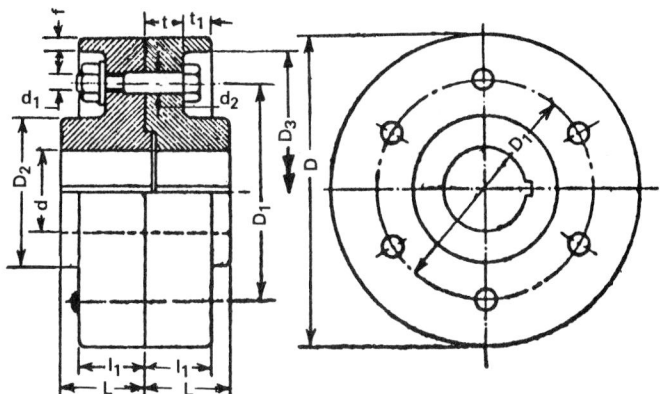

FIGURE 19–1 Flange coupling.

where M_t in N m; P in kW; n in rpm

$$= \frac{159P}{n'} \qquad \text{SI} \qquad (19\text{–}3d)$$

where M_t in N m; P in kW; n' in rps

The torque transmitted through bolts	$M_t = i\left(\dfrac{\pi d_1^2}{4}\right)\tau_b \dfrac{D_1}{2}$ $\qquad (19\text{–}4)$
The torque capacity which is based on bearing of bolts	$M_t = i(d_1 l_1)\sigma_b \dfrac{D_1}{2}$ $\qquad (19\text{–}5)$
The torque capacity which is based on shear of flange at the outside hub diameter	$M_t = t(\pi D_2)\tau_f(D_2/2)$ $\qquad (19\text{–}6)$
The friction-torque capacity of the flanged coupling which is based on the concept of the friction force acting at the mean radius of the surface	$M_t = i\mu F_b r_m$ $\qquad (19\text{–}7)$ where $r_m = \dfrac{D+d}{2}$ = mean radius F_b = tension load in each bolt, kN (kgf)
The preliminary bolt diameter may be determined by the empirical formula	$d_1 = \dfrac{0.5d}{\sqrt{i}}$ $\qquad (19\text{–}8)$
The bolt diameter from Eqs. (19–2) and (19–4)	$d_1 = \sqrt{\dfrac{d^3\tau_s\eta}{2i\tau_b D_1}}$ $\qquad (19\text{–}9)$
The bolt diameter from Eqs. (19–3) and (19–4)	$d_1 = \sqrt{\dfrac{8000P}{\pi i\omega\tau_b D_1}}$ $\qquad \text{SI} \qquad (19\text{–}10a)$

Particular	Formula
	where d_1, D_1 in m; P in kW; τ_b in in Pa; ω in rad/s

$$= \sqrt{\frac{1273P}{\pi i n' D_1 \tau_b}} \qquad \textbf{SI} \qquad (19\text{--}10b)$$

where d_1, D_1 in m; P in kW; τ_b in Pa; n' in rps

$$= \sqrt{\frac{76,400P}{\pi i n \tau_b D_1}} \qquad \textbf{SI} \qquad (19\text{--}10c)$$

where d_1, D_1 in m; P in kW; τ_b in Pa, n in rpm

$$= \sqrt{\frac{50,400P}{\pi i n D_1 \tau_b}} \qquad \textbf{US Customary units} \qquad (19\text{--}10d)$$

where d_1, D_1 in in; P in hp; τ_b in psi, n in rpm

where i = effective number of bolts doing work should be taken as all bolts if they are fitted in reamed holes and only half the total number of bolts if they are not fitted into reamed holes

The diameter of shaft from Eqs. (19–2) and (19–3)

$$d = \sqrt[3]{\frac{16,000P}{\pi \eta \omega \tau_s}} \qquad \textbf{SI}, \ P \text{ in kW}; \ d \text{ in m} \qquad (19\text{--}11a)$$

$$= \sqrt[3]{\frac{100,800P}{\pi \eta n \tau_s}} \qquad \textbf{US Customary units} \qquad (19\text{--}11b)$$

where P in hp; d in in

$$= \sqrt[3]{\frac{152,800P}{\pi \eta n \tau_s}} \qquad \textbf{SI}, \ P \text{ in kW}; \ d \text{ in m} \qquad (19\text{--}11c)$$

$$= \sqrt[3]{\frac{2546P}{\pi \eta n' \tau_s}} \qquad \textbf{SI} \qquad (19\text{--}11d)$$

where P in kW; d in m; n' in rps

The average value of diameter of the bolt circle	$D_1 = 2d + 0.05$ **SI**, D_1 in m (19–12a) $= 2d + 2$ **US Customary units** (19–12b)
The hub diameter	$D_2 = 1.5d + 0.025$ **SI**, D_2 in m (19–13a) $= 1.5d + 1$ **US Customary units** (19–13b)
The outside diameter of flange	$D = 2.5d + 0.075$ **SI**, D in m (19–14a) $= 2.5d + 3$ **US Customary units** (19–14b)
The hub length	$l = 1.25d + 0.01875$ **SI**, l in m (19–14c) $= 1.25d + 0.75$ **US Customary units** (19–14d) where l and d in in

Particular	Formula

MARINE TYPE OF FLANGE COUPLING

Solid rigid type [Fig. 19–2(a), Table 19–1]

The number of bolts

$$i = 33d + 5 \qquad \textbf{SI}, \ d \text{ in m} \qquad (19\text{–}15a)$$
$$= 0.85d + 5 \quad \textbf{US Customary units} \qquad (19\text{–}15b)$$
where d in in

The diameter of bolt

$$d_1 = \sqrt{\frac{\eta d^3 \tau_s}{2iD_1\tau_b}} \qquad\qquad (19\text{–}16a)$$

based on torque capacity of the shaft

$$= \sqrt{\frac{tD_2^2\tau_f}{4iD_1\tau_b}} \qquad\qquad (19\text{–}16b)$$

based on torque capacity of flange

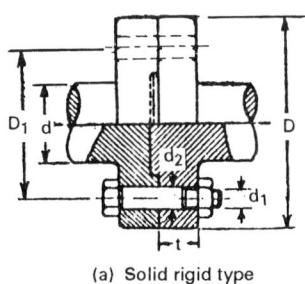 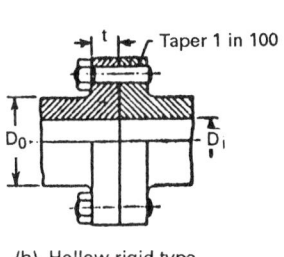

(a) Solid rigid type   (b) Hollow rigid type

FIGURE 19–2   Rigid marine coupling.

The thickness of flange $\qquad\qquad\qquad t = 0.25 \text{ to } 0.28d \qquad\qquad\qquad (19\text{–}17)$

The diameter of the bolt circle $\qquad\quad D_1 = 1.4d \text{ to } 1.6d \qquad\qquad (19\text{–}18)$

The outside diameter of flange $\qquad\quad D = D_1 + 2d \text{ to } 3d \qquad\qquad (19\text{–}19)$

Taper of bolt $\qquad\qquad\qquad\qquad\qquad 1 \text{ in } 100$

Hollow rigid type [Fig. 19–2(b)]

The minimum number of bolts

$$i = 50D_o \qquad \textbf{SI}, \text{ where } D_o \text{ in m} \qquad (19\text{–}20a)$$
$$= 1.25D_o \quad \textbf{US Customary units} \qquad (19\text{–}20b)$$
where D_o in in

The mean diameter of bolt

$$d_1 = \sqrt{\frac{(1 - K^4)D_o^3\tau_s}{2iD_1\tau_b}} \text{ where } K = \frac{D_i}{D_o} \qquad (19\text{–}21)$$

TABLE 19-1
Forged end type rigid couplings (all dimensions in mm)

Number coupling		Shaft diameter		Flange outside diameter, D	Flange width, t	Locating diameter, D_2	Recess depth, c_1	Spigot depth, c_2	Pitch circle diameter, D_1	Bolt size, d_1	Bolt hole diameter, d_2 $H8$	Number of bolts
Recessed flange	Spigot flange	Max	Min									
R1	S1	53	—	100	17	50	6	4	70	M10	11	4
R2	S2	45	36	120	22	60	6	4	85	M12	13	4
R3	S3	55	46	140	22	75	7	5	100	M14	15	4
R4	S4	70	55	175	27	95	7	5	125	M16	17	6
R5	S5	80	71	195	32	95	7	5	140	M18	19	6
R6	S6	90	81	225	32	125	7	5	160	20	21	6
R7	S7	110	91	265	36	150	9	7	190	24	25	6
R8	S8	130	111	300	46	150	9	7	215	30	32	6
R9	S9	150	131	335	50	195	9	7	240	33	34	8
R10	S10	170	151	375	55	195	10	8	265	36	38	6
R11	S11	190	171	400	55	240	10	8	290	36	38	8
R12	S12	210	191	445	65	240	10	8	315	42	44	8
R13	S13	230	211	475	70	280	10	8	340	45	46	8
R14	S14	250	231	500	70	280	10	8	370	45	46	8
R15	S15	270	251	560	80	330	10	8	400	52	55	10
R16	S16	300	271	600	85	330	10	8	410	56	60	10
R17	S17	330	301	650	90	400	10	8	480	60	65	10
R18	S18	360	331	730	100	400	10	8	520	68	72	10
R19	S19	390	361	775	105	480	11	9	570	72	76	10
R20	S20	430	391	875	110	480	11	9	620	76	80	12
R21	S21	470	431	900	115	560	11	9	670	80	85	12
R22	S22	520	471	925	120	560	12	10	730	90	95	12
R23	S23	571	521	1000	125	640	12	10	790	110	105	12
R24	S24	620	571	1090	130	720	12	10	850	110	115	12

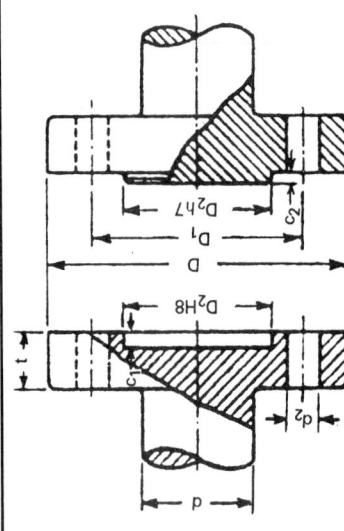

Particular	Formula
The thickness of flange	$$t = \frac{(1 - K^4)D_o^3\tau_s}{8D_2^2\tau_f} \qquad (19\text{--}22)$$
The empirical formula for thickness of flange	$t = 0.25$ to $0.28D_o$ (19--23)
The diameter of bolt circles	$D_1 = 1.4D_o$ (19--24)
For design calculations of other dimensions of marine hollow rigid type of flange coupling	The method of analyzing the stresses and arriving at the dimensions of the various parts of a marine hollow flange coupling is similar to that given for the marine solid rigid type and common flange coupling.
For dimensions of fitted half couplings for power transmission	Refer to Table 19--2.

PULLEY FLANGE COUPLING (FIG. 19–3)

The number of bolts

$$i = 20d + 3 \qquad \textbf{SI}, \ d \text{ in m} \qquad (19\text{--}25a)$$
$$= 0.5d + 3 \quad \textbf{US Customary units} \qquad (19\text{--}25b)$$
$$\text{where } d \text{ in in}$$

The preliminary bolt diameter

$$d_t = \frac{0.5d}{\sqrt{i}} \qquad (19\text{--}26)$$

The width of flange l_1 (Fig. 19–3)

$$l_1 = \tfrac{1}{2}d + 0.025 \text{ m} \qquad \textbf{SI}, \ l_1 \text{ in m} \qquad (19\text{--}27a)$$
$$= \tfrac{1}{2}d + 1.0 \quad \textbf{US Customary units} \qquad (19\text{--}27b)$$

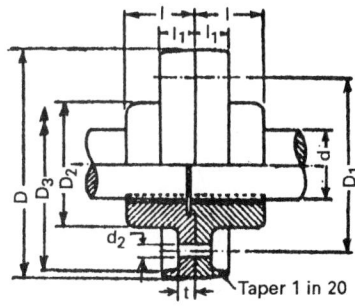

FIGURE 19–3 Pulley flange coupling.

The hub length l

$$l = 1.4d + 0.0175 \text{ m} \quad \textbf{SI}, \ l \text{ in m} \qquad (19\text{--}28a)$$
$$= 1.4d + 0.7 \quad \textbf{US Customary units} \qquad (19\text{--}28b)$$

The thickness of the flange

$$t = 0.25d + 0.007 \text{ m} \quad \textbf{SI}, \ t \text{ in m} \qquad (19\text{--}29a)$$
$$= 0.25d + 0.25 \text{ in} \quad \textbf{US Customary units} \qquad (19\text{--}29b)$$

The hub diameter

$$D_2 = 1.8d + 0.01 \text{ m} \quad \textbf{SI}, \ D_2 \text{ in m} \qquad (19\text{--}30a)$$
$$= 1.8d + 0.4 \text{ in} \quad \textbf{US Customary units} \qquad (19\text{--}30b)$$

The average value of the diameter of the bolt circle

$$D_1 = 2d + 0.025 \text{ m} \quad \textbf{SI}, \ D_1 \text{ in m} \qquad (19\text{--}31a)$$
$$= 2d + 1.0 \text{ in} \quad \textbf{US Customary units} \qquad (19\text{--}31b)$$

TABLE 19–2
Fitted half couplings (all dimensions in mm)

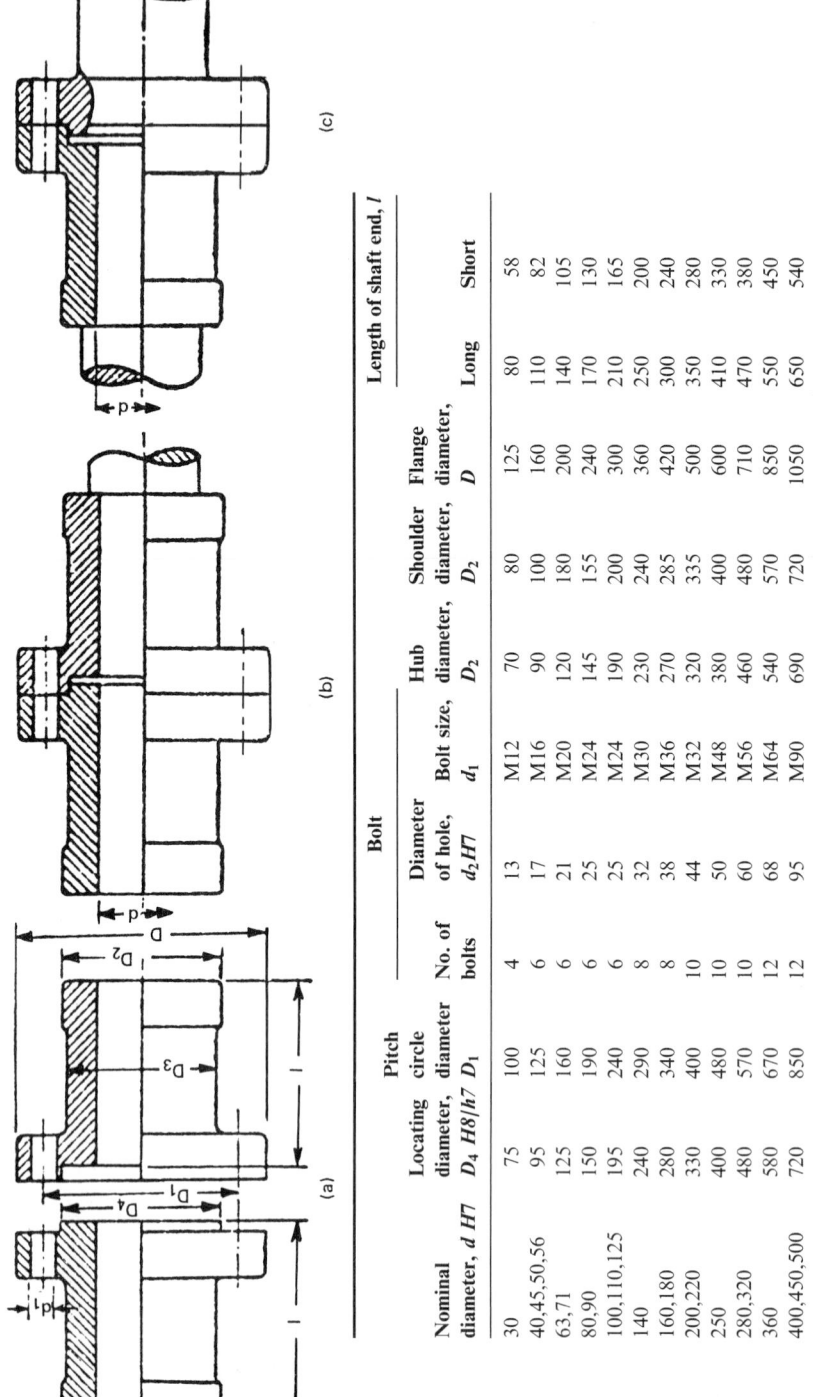

(a)　　(b)　　(c)

Nominal diameter, $d\ H7$	Locating diameter, $D_4\ H8/h7$	Pitch circle diameter D_1	No. of bolts	Diameter of hole, $d_2 H7$	Bolt size, d_1	Hub diameter, D_2	Shoulder diameter, D_2	Flange diameter, D	Length of shaft end, l Long	Short
30	75	100	4	13	M12	70	80	125	80	58
40,45,50,56	95	125	6	17	M16	90	100	160	110	82
63,71	125	160	6	21	M20	120	180	200	140	105
80,90	150	190	6	25	M24	145	155	240	170	130
100,110,125	195	240	6	25	M24	190	200	300	210	165
140	240	290	8	32	M30	230	240	360	250	200
160,180	280	340	8	38	M36	270	285	420	300	240
200,220	330	400	10	44	M32	320	335	500	350	280
250	400	480	10	50	M48	380	400	600	410	330
280,320	480	570	10	60	M56	460	480	710	470	380
360	580	670	12	68	M64	540	570	850	550	450
400,450,500	720	850	12	95	M90	690	720	1050	650	540

Particular	Formula
The outside diameter of flange	$D = 2.5d + 0.075$ m **SI**, D in m (19–32a) $= 2d + 3.0$ in **US Customary units** (19–32b) where D and d in in

PIN OR BUSH TYPE FLEXIBLE COUPLING (FIG. 19–4, TABLE 19–3)

Torque to be transmitted

$$M_t = iF \frac{D_1}{2} \tag{19–33a}$$

$$= ip_b l d' \left(\frac{D_1}{2} \right) \tag{19–33b}$$

where
p_b = bearing pressure, MPa (psi)
F = force at each pin, kN (lbf)
   $= p_b l d'$
d' = outside diameter of the bush, m (in)

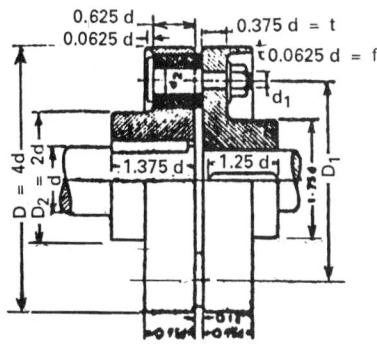
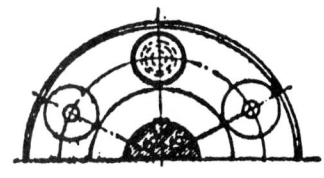

FIGURE 19–4 Pin-type flexible coupling.

Shear stress in pin

$$\tau_p = \frac{F}{0.785 d_p^2} \tag{19–34}$$

where
τ_p = allowable shearing stress, MPa (psi)
$d_p = d_1$ = diameter of pin at the neck, m (in)

Bending stress in pin

$$\sigma_b = \frac{F\left(\dfrac{l}{2} + b\right)}{\dfrac{\pi}{32} d_p^3} \tag{19–35}$$

TABLE 19–3
Cast-iron flexible couplings

All dimensions in mm

Type of flexible couplings / Coupling number	Bore, d Min	Bore, d Max	Outside diameter, D, min	Hub diameter, D_2 min	Hub length, l, min	Flange width, l_1	Thickness of disk, c	Diameter of bolt, d_1	Number of bolt holes	Pitch circle diameter of bolts, D_1	Bolt recess, t_1	Bush diameter	Nominal gap between coupling holes, c	Maximum rating per 100 rpm, kW
B_1	12	16	80		28	18	—	8	3	53	10	20	2	0.4
B_2	16	22	100		30	20	—	10	3	63	12	22	2	0.6
B_3	22	30	112		32	22	—	10	3	73	12	22	2	0.8
B_4	30	45	132		40	30	—	12	4	90	15	25	4	2.5
B_5	45	56	170	80	45	35	—	12	4	120	15	25	4	4.0
B_6	56	75	200	100	56	40	—	12	4	150	15	30	4	6.0
B_7	75	85	250	140	63	45	—	16	6	190	22	40	5	16.0
B_8	85	110	315	180	80	50	—	16	6	250	22	40	5	25.0
B_9	110	130	400	212	90	56	—	18	8	315	28	45	6	52.0
B_{10}	130	150	500	280	100	60	—	18	8	400	28	45	6	74.0
D_1	12	16	80		28	18	15,16	8	6	55	10	—	—	0.4
D_2	16	22	100		30	20	16,18	10	6	63	12	—	—	0.6
D_3	22	30	110		32	22	18,25	10	6	73	12	—	—	0.8
D_4	30	45	132		40	30	25,30	12	8	90	15	—	—	2.5
D_5	45	56	165	80	45	35	30,35	12	8	120	15	—	—	4.0
D_6	56	75	200	100	56	40	35,40	12	8	150	15	—	—	6.0
D_7	75	85	250	140	63	45	40,45	16	12	190	22	—	—	16.0
D_8	85	110	315	180	80	50	45,50	16	12	250	22	—	—	25.0
D_9	110	130	400	212	90	56	50,55	18	16	315	28	—	—	52.0
D_{10}	130	150	500	280	100	60	55	18	16	400	28	—	—	74.0

Bush type flexible coupling

Disc type flexible coupling

19.12

Particular	Formula

OLDHAM COUPLING (FIG. 19–5)

The total pressure on each side of the coupling

$$F = \tfrac{1}{4}pDh \tag{19–36}$$

where
h = axial dimension of the contact area, m (in)

The torque transmitted on each side of the coupling

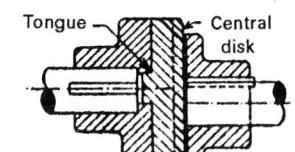

$$M_t = 2Fl = \frac{pD^2h}{6} \tag{19–37}$$

where
$l = \tfrac{1}{3}D$ = the distance to the pressure area centroid from the center line, m (in)

Power transmitted

p = allowable pressure $\not> 8.3$ MPa (1.2 kpsi)
$$P = pD^2\,hn/57277 \quad \textbf{SI},\ P \text{ in kW} \tag{19–38a}$$

$$= \frac{pD^2hn}{378180} \quad \textbf{US Customary System units} \tag{19–38b}$$

where P in hp; D, h in in; p in psi

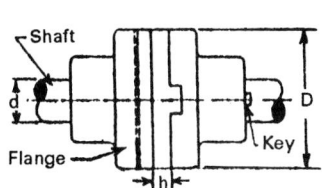

FIGURE 19–5 Oldham's coupling.

The diameter of the disk	$D = 3d + a$ (19–39)
The diamater of the boss	$D_2 = 2d$ (19–40)
Length of the boss	$l = 1.75d$ (19–41)

Breadth of groove

$$w = \frac{D}{6} \tag{19–42}$$

The thickness of the groove

$$h_1 = \frac{w}{2} \tag{19–43a}$$

The thickness of central disk

$$h = \frac{w}{2} \tag{19–43b}$$

The thickness of flange

$$t = \tfrac{3}{4}d \tag{19–44}$$

MUFF OR SLEEVE COUPLING (FIG. 19–6)

The outside diameter of sleeve

$$D = 2d + 0.013 \text{ m} \quad \textbf{SI},\ D,\ d \text{ in m} \tag{19–45a}$$
$$= 2d + 0.52 \text{ in} \quad \textbf{US Customary System units} \tag{19–45b}$$

where d, D in in

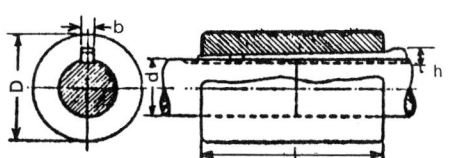

FIGURE 19–6 Muff or sleeve coupling.

Particular	Formula
The outside diameter of sleeve is also obtained from equation	$$D = \sqrt[3]{\dfrac{16M_t}{\pi\tau_{d1}(1 - K^4)}}$$ (19–46) where $K = \dfrac{d}{D}$
The length of the sleeve (Fig. 19–6)	$l = 3.5d$ (19–47)
Length of the key (Fig. 19–6)	$l = 3.5d$ (19–48)
The diameter of shaft	$$d = \sqrt[3]{\dfrac{16M_t}{\eta\pi\tau_d}}$$ (19–49) where M_t is torque obtained from Eq. (19–2)
The width of the keyway	$$b = \dfrac{2M_t}{\tau d_2 l d}$$ (19–50)
The thickness of the key	$$h = \dfrac{2M_t}{\sigma'_b l d}$$ (19–51)

FAIRBAIRN'S LAP-BOX COUPLING (FIG. 19–7)

The outside diameter of sleeve	Use Eqs. (19–45) or (19–46)
The length of the lap	$l = 0.9d + 0.003$ m **SI,** l, d in m (19–52a) $= 0.9d + 0.12$ in **US Customary System units** (19–52b)
The length of the sleeve	$L = 2.25d + 0.02$ m **SI,** L, d in m (19–53a) $= 2.25d + 0.8$ in **US Customary System units** (19–53b)

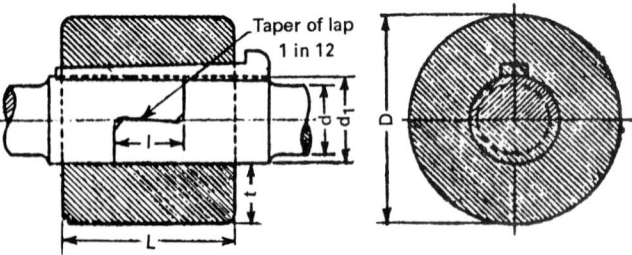

FIGURE 19–7 Fairbairn's lap-box coupling.

SPLIT MUFF COUPLING (FIG. 19–8)

The outside diameter of the sleeve	$D = 2d + 0.013$ m **SI,** D, d in m (19–54a) $= 2d + 0.52$ in **US Customary System units** (19–54b)

Particular	Formula

The length of the sleeve	$l = 3.5d$ or $2.5d + 0.05$ m **SI**, l, d in m (19–55a)
	  $= 3.5d$ or $2.5d + 2.0$ in
	                        US Customary System units (19–55b)
	where l, d in in

The torque to be transmitted by the coupling	$$M_t = \frac{\pi d_c^2 \sigma_t \mu i d}{16} \qquad (19\text{–}56)$$ where d_c = core diameter of the clamping bolts, m (in) i = number of bolts

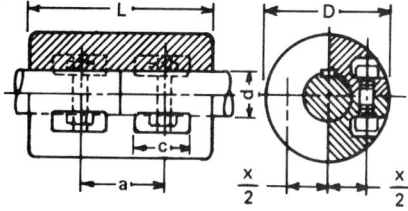

FIGURE 19–8 Split muff coupling.

SLIP COUPLING (FIG. 19–9)

The axial force exerted by the springs	$$F_a = \frac{\pi}{4}\,(D_2^2 - D_1^2)p \qquad (19\text{–}57)$$
With two pairs of friction surfaces, the tangential force	$$F_\theta = 2\mu F_a \qquad (19\text{–}58)$$
The radius of applications of F_θ with sufficient accuracy	$$r_m = \frac{D_m}{2} = \frac{D_2 + D_1}{4} \qquad (19\text{–}59)$$

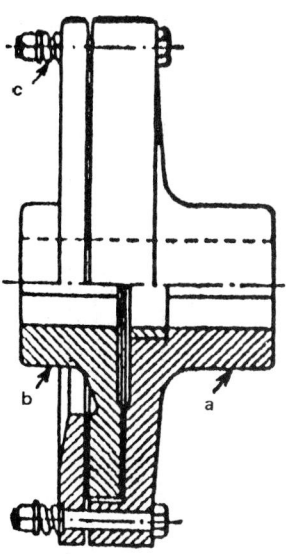

FIGURE 19–9 Slip coupling.

Particular	Formula
The torque	$M_t = 0.000385(D_2^2 - D_1^2)(D_2 + D_1)\mu p$ **SI** (19–60a) $= 0.3927(D_2^2 - D_1^2)(D_2 + D_1)\mu p$ **US Customary System units** (19–60b) where the values of μ and p may be taken from Table 19–4
The relation between D_1 and D_2	$\dfrac{D_2}{D_1} = 1.6$ (19–61) where D_1 and D_2 are the inner and outer diameters of disk of friction lining

SELLERS CONE COUPLING (FIG. 19–10)

The length of the box	$L = 3.65d$ to $4d$ (19–62)
The outside diameter of the conical sleeve	$D_1 = 1.875d$ to $2d + 0.0125$ m **SI** (19–63a) $= 1.875d$ to $2d + 0.5$ in **US Customary System units** (19–63b)
Outside diameter of the box	$D_2 = 3d$ (19–64)
The length of the conical sleeve	$l = 1.5d$ (19–65)

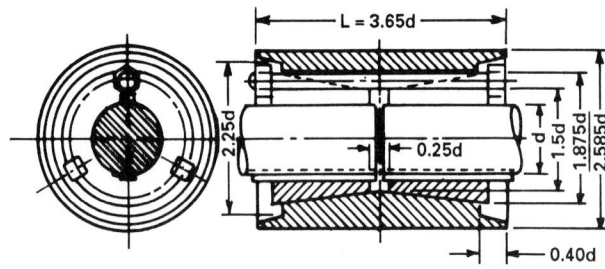

FIGURE 19–10 Seller's cone coupling.

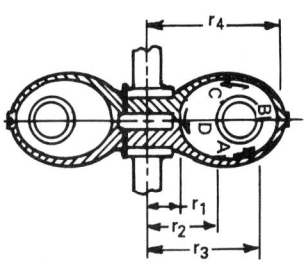

FIGURE 19–11 Hydraulic coupling.

HYDRAULIC COUPLINGS (FIG. 19–11)

Torque transmitted	$M_t = Ksn^2 W(r_{mo}^2 - r_{mi}^2)$ (19–66) where $K = $ coefficient $= \dfrac{1.42}{10^7}$ (approx.)
Percent slip between primary and secondary speeds	$s = \dfrac{n_p - n_s}{n_p} \times 100$ (19–67)

TABLE 19-4
Friction materials for clutches

Contact surfaces		Friction coefficient[a], μ		Maximum temperature		Maximum pressure, p		Relative	
Wearing	Opposing[b]	Wet	Dry	K	°C	MPa	kgf/mm²	cost	Comment
Cast bronze	Cast iron or steel	0.05	0.15–0.2	422	149	0.5521–0.8277	0.0563–0.0844	Low	Subject to seizing
Cast iron	Cast iron	0.05	0.15–0.2	589	316	1.0346–1.7240	0.1055–0.1755	Very low	Good at low speeds
Cast iron	Steel	0.06	0.15–0.2	533	260	0.8277–1.3788	0.0844–0.1406	Very low	Fair at low speeds
Hard steel	Hard steel	0.05	0.15–0.2	533	260	0.6894	0.0703	Moderate	Subject to galling
Hard steel	Hard steel, chromium plated	0.03	0.15–0.2	533	260	1.3788	0.1406	High	Durable combination
Hard-drawn phosphor bronze	Hard steel, chromium plated	0.03		533	260	1.0346	0.1055	High	Good wearing qualities
Powder metal[c]	Cast iron or steel	0.05–0.1	0.1–0.4	811	538	1.0346	0.1055	High	Good wearing qualities
Powder metal[c]	Hard steel, chromium plated	0.05–0.1	0.1–0.3	811	538	2.0682	0.2109	Very high	High energy absorption
Wood	Cast iron or steel	0.16	0.2–0.35	422	149	0.4138–0.6208	0.0422–0.0633	Lowest	Unsuitable at high speed
Leather	Cast iron or steel	0.12–0.15	0.3–0.5	363.3	90.3	0.0686–0.2746	0.0070–0.0284	Very low	Subject to glazing
Cork	Cast iron or steel	0.15–0.25	0.3–0.5	363.3	90.3	0.0549–0.0981	0.0056–0.01	Very low	Cork-insert type preferred
Felt	Cast iron or steel	0.18	0.22	411	138	0.0343–0.0686	0.0035–0.0070	Low	Resinent engagement
Vulcanized fiber or paper	Cast iron or steel		0.3–0.5	363.3	90.3	0.0686–0.2746	0.3070–0280	Very low	Low speeds, light duty
Woven asbestos[c]	Cast iron or steel	0.1–0.2	0.3–0.6	444–533	171–260	0.03432–0.6894	0.0350–0.0703	Low	Prolonged slip service ratings given
Woven asbestos	Cast iron or steel	0.1–0.2		533	260	0.6894–1.3788	0.0703–0.1406	Low	This rating for short in-frequent engagements
Woven asbestos	Hard steel, chromium plated	0.1				8.2738	0.8437	Moderate	Used in Napier Sabre engine
Molded asbestos[c]	Cast iron or steel	0.08–0.12	0.2–0.5	533	260	0.3452–1.0346	0.0352–0.1055	Very low	Wide field of applications
Impregnated asbestos	Cast iron or steel	0.12	0.32	533–659	260–386	1.0346	0.1055	Moderate	For demanding applications
Carbon graphite	Steel	0.05–0.1	0.25	632–811	359–538	2.0682	0.2109	High	For critical requirements
Molded phenolic plastic, macerated cloth base	Cast iron	0.1–0.15	0.25	422	149	0.6894	0.0703	Low	For light special service

[a] Conservative values should be used to allow for possible glazing of clutch surfaces in service and for adverse operating conditions.
[b] Steel, where specified, should have a carbon content of approximately 0.70%. Surfaces should be ground true and smooth.
[c] For a specific material within this group, the coefficient usually is maintained within plus or minus 5 percent.

Particular	Formula
	where n_p and n_s are the primary and secondary speeds of impeller, respectively, rpm
The mean radius of inner passage (Fig. 19–11)	$$r_{mi} = \frac{2}{3}\left(\frac{r_2^3 - r_3^3}{r_2^2 - r_1^2}\right) \qquad (19\text{–}68a)$$
The mean radius of outer passage (Fig. 19–11)	$$r_{mo} = \frac{2}{3}\left(\frac{r_4^3 - r_3^3}{r_4^2 - r_3^2}\right) \qquad (19\text{–}68b)$$
The number of times the fluid circulates through the torus in one second is given by	$$\sigma = \frac{13,000 M_t}{nW(r_{mo}^2 - r_{mi}^2)} \qquad (16\text{–}69)$$
Power transmitted by torque converter	$$M_t = Kn^2 D^5 \qquad (19\text{–}70)$$
	where K = coefficient—varies with the design n = speed of driven shaft, rpm D = outside diameter of vanes, m (in)

19–2   CLUTCHES

POSITIVE CLUTCHES (FIG. 19–12)

Jaw clutch coupling

$$
\begin{aligned}
a &= 2.2d + 0.025 \text{ m}; & h &= 0.3d + 0.0125 \text{ m}\\
  &= 2.2d + 1.0 \text{ in} & &= 0.3d + 0.5 \text{ in}\\
c &= 1.2d + 0.03 \text{ m}; & i &= 0.4d + 0.005 \text{ m}\\
  &= 1.2d + 1.2 \text{ in} & &= 0.4d + 0.2 \text{ in}\\
f &= 1.4d + 0.0075 \text{ m}; & j &= 0.2d + 0.00375 \text{ m}\\
  &= 1.4d + 0.3 \text{ in} & &= 0.2d + 0.15 \text{ in}\\
k &= 1.2d + 0.02 \text{ m} & k &= 1.2d + 0.8 \text{ in}\\
g &= d + 0.005 \text{ m}; & l &= 1.7d + 0.0584 \text{ m}\\
  &= d + 0.2 \text{ in} & &= 1.7d + 2.3 \text{ in}
\end{aligned}
$$

$$(19\text{–}71)$$

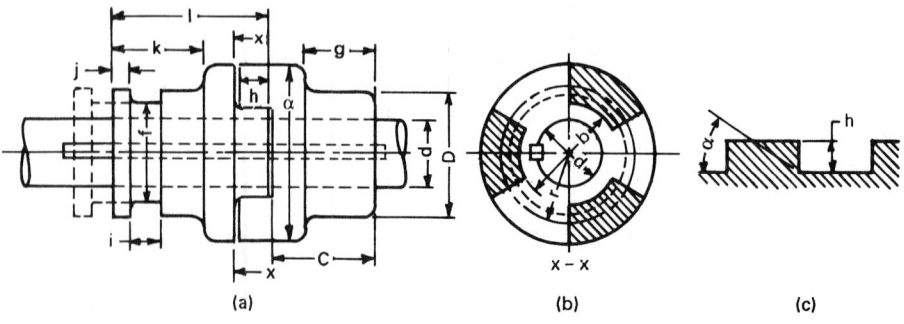

FIGURE 19–12   Square-jaw clutch.

Particular	Formula
The area in shear	$A = 0.5(a - b)h / \sin \alpha$ (19–72)
The shear stress assuming that only one-half the total number of jaws i is in actual contact	$\tau = \dfrac{4F_\theta \sin \alpha}{i(a - b)h \cos \alpha}$ (19–73)
	$\tau = \dfrac{2.8F_\theta}{i(a - b)h} \quad \text{for } \tan \alpha = 0.7$ (19–74)

where
α = angle made by the shearing plane with the direction of pressure

FRICTION CLUTCHES

Cone clutch (Fig. 19–13)

The axial force in terms of the clutch dimensions

$$F_a = \pi D_m p b \sin \alpha \tag{19–75}$$
where
$D_m = \frac{1}{2}(D_1 + D_2)$ (approx.)
α = one-half the cone angle, deg
   = ranges from 15° to 25° for industrial clutches faced with wood
   = 12.5° for clutches faced with asbestos or leather or cork insert

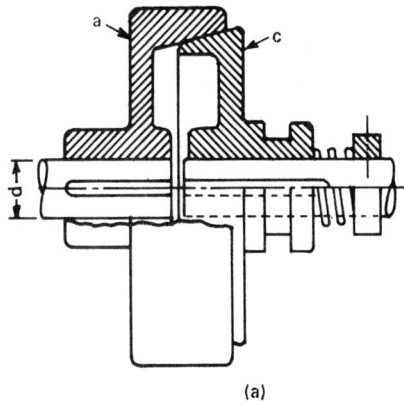 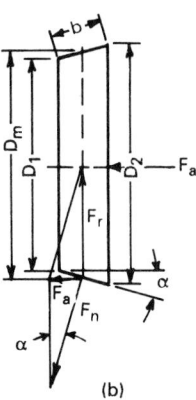

FIGURE 19–13 Cone clutch.

Axial force in terms of normal force (Fig. 19–13)

$$F_a = F_n \sin \alpha \tag{19–76}$$

The tangential force due to friction

$$F_\theta = \frac{\mu F_a}{\sin \alpha} \tag{19–77}$$

Torque transmitted through friction

$$M_t = \frac{\mu F_a D_m}{2 \sin \alpha} \tag{19–78}$$

Particular	Formula

Power transmitted

$$P = \frac{\mu F_a D_m n}{19,100 \sin \alpha k_l} \qquad \text{SI} \qquad (19\text{–}79a)$$

$$= \frac{\mu F_a D_m n}{126,000 \sin \alpha k_l} \qquad \text{US Customary System units}$$

$$(19\text{–}79b)$$

$$P = \frac{\pi \mu p D_m^2 bn}{19,100 k_l} \qquad \text{SI} \qquad (19\text{–}79c)$$

$$= \frac{\pi \mu p D_m^2 bn}{126,000 k_l} \qquad \text{US Customary System units}$$

$$(19\text{–}79d)$$

where k_1 = load factor from Table 14–7

Refer to Table 19–4 for p.

The force necessary to engage the clutch when one member is rotating

$$F_a' = F_n(\sin \alpha + \mu \cos \alpha) \qquad (19\text{–}80)$$

The ratio (D_m/b)

$$q = \frac{D_m}{b} = 4.5 \text{ to } 8 \qquad (19\text{–}81)$$

The value of D_m in commercial clutches

$$D_m = 18.2 \sqrt[3]{\frac{Pk_l q}{\mu p n}} \qquad \text{SI} \qquad (19\text{–}82a)$$

$$= 34.2 \sqrt[3]{\frac{Pk_l q}{\mu p n}} \qquad \text{US Customary System units}$$

$$(19\text{–}82b)$$

$$D_m = 5d \text{ to } 10d \qquad (19\text{–}82c)$$

DISK CLUTCHES (FIG. 19–14)

The axial force

$$F_a = \tfrac{1}{2} \pi p D_1 (D_2 - D_1) \qquad (19\text{–}83)$$
Refer to Table 19–4 for p.

The torque transmitted

$$M_t = \tfrac{1}{2} \mu F_a D_m \qquad (19\text{–}84)$$

where $D_m = \dfrac{2}{3} \dfrac{(D_2^3 - D_1^3)}{(D_2^2 - D_1^2)}$ for uniform pressure

$$(19\text{–}85a)$$

distribution and $D_m = \tfrac{1}{2}(D_2 + D_1)$ for uniform wear

$$(19\text{–}85b)$$

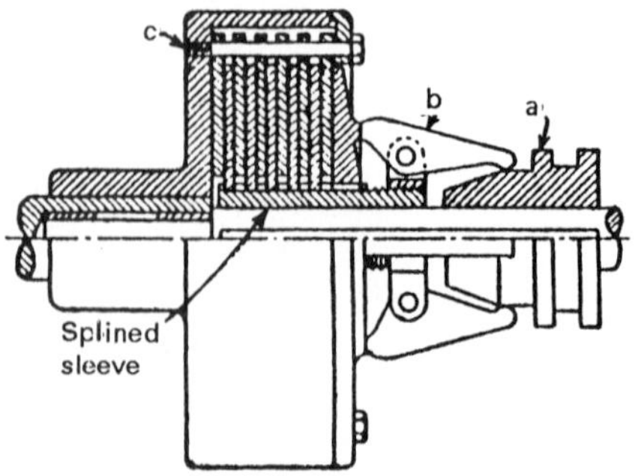

FIGURE 19–14 Multidisk clutch.

Particular	Formula
Power transmitted	$$P = \frac{i\mu F_a n}{28,650 k_l}\left(\frac{D_2^3 - D_1^3}{D_2^2 - D_1^2}\right) \quad \text{SI} \qquad (19\text{–}86a)$$ $$= \frac{i\mu F_a n}{189,000 k_l}\left(\frac{D_2^3 - D_1^3}{D_2^2 - D_1^2}\right)$$ **US Customary System units** (19–86b) for uniform pressure where $F_a = \pi p \dfrac{D_2^2 - D_1^2}{4}$ $$P = \frac{\pi i \mu p n D_1 (D_2^2 - D_1^2)}{76,400 k_l} \quad \text{SI} \qquad (19\text{–}87a)$$ $$= \frac{\pi i \mu p n D_1 (D_2^2 - D_1^2)}{504,000 k_l}$$ **US Customary System units** (19–87b) for uniform wear
The clutch capacity at speed n_1	$$P_1 = \frac{P n_1}{n k_s} \qquad (19\text{–}88)$$ where P = design power at speed, n k_s = speed factor obtained from Eq. (19–89)
The speed factor	$k_s = 0.1 + 0.001n \qquad (19\text{–}89)$ where n = speed at which the capacity of clutch to be determined, rpm

DIMENSIONS OF DISKS (FIG. 19–15)

The maximum diameter of disk	$D_2 = 2.5 \text{ to } 3.6 D_1$	(19–90)
The minimum diameter of disk	$D_1 = 4d$	(19–91)
The thickness of disk	$h = 1 \text{ to } 3 \text{ mm}$	(19–92)
The number of friction surfaces	$i = i_1 + i_2 - 1$	(19–93)
The number of driving disks	$i_1 = \dfrac{i}{2}$	(19–94)
The number of driven disks	$i_2 = \dfrac{i}{2} + 1$	(19–95)

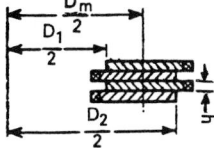

FIGURE 19–15 Dimensions of disks.

Particular	Formula

DESIGN OF A TYPICAL CLUTCH OPERATING LEVER (FIG. 19–16)

The total axial force on i number of clutch disk or plates

$$F'_a = i\pi p' D_1 (D_1 - D_2) \tag{19-96}$$

where p' = actual pressure between disks

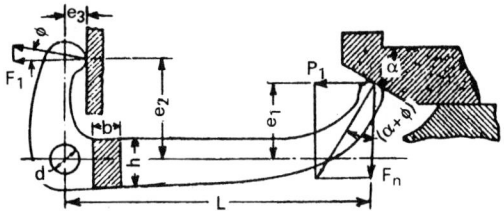

$$F'_a = \frac{4M_{ta}}{i\pi\mu(D_1 - D)D_m^2}, \text{ MPa (psi)}$$

FIGURE 19–16 A typical clutch operating lever.

M_{ta} = allowable torque, N m (lbf in)

The force acting on disks of one operating lever of the clutch (Fig. 19–16)

$$F_1 = \frac{F'_a}{i'} \tag{19-97}$$

where i' = number of operating levers

The total force acting from the side of the bushing (Fig. 19–16)

$$P = i' p_1 \tag{19-98}$$

The force acting from the side of the bushing on one operating lever (Fig. 19–16)

$$P_1 = F_1 \frac{\left[L\cot(\alpha + \phi) - e_1 - \mu \dfrac{d}{2} \right]}{e_2 + \mu\left(e_3 + \dfrac{d}{2} \right)} \tag{19-99}$$

The thickness of the lever very close to the pin (Fig. 19–16)

$$h = \left[\frac{6F'_a e_3}{\left(\dfrac{b}{h} \right) i' \sigma_{db}} \right]^{1/3} \tag{19-100}$$

where σ_{db} = design bending stress for the material of the levers, MPa (psi)

Ratio of $\dfrac{b}{h} = 0.75$ to 1

The diameter of the pin (Fig. 19–16)

$$d = \sqrt{\frac{2F_r}{\pi\tau_d}} \tag{19-101}$$

where

F_r = resultant force due to F_1 and $P_1 \cot(\alpha + \phi)$ on the pin, kN (lbf)

τ_d = design shear stress of the material of the pin, MPa (psi)

Particular	Formula

EXPANDING-RING CLUTCHES (FIG. 19–17)

Torque transmitted [Fig. 19–17(a)]

$$M_t = 2\mu p w r^2 \theta \tag{19–102}$$

where

θ = one half the total arc of contact, rad
w = width of ring, m (in)

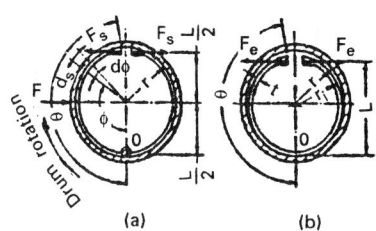

(a)   (b)

FIGURE 19–17 Expanding-ring clutch.

The moment of the normal force for each half of the band [Fig. 19–17(a)]

$$M_o = p w r L \tag{19–103}$$

when $\theta \approx \pi$ rad

The force applied to the ends of the split ring to expand the ring [Fig. 19–17(a)]

$$F_s = p w r \tag{19–104}$$

If the ring is made in one piece [Fig. 19–7(b)] an additional force required to expand the inner ring before contact is made with inner surface of the shell

$$F_e = \frac{Ewt^3}{6L}\left(\frac{1}{d_1} - \frac{1}{d}\right) \tag{19–105}$$

where
d_1 = original diameter of ring, m (in)
d = inner diameter of drum, m (in)
w = width of ring, m (in)
t = thickness of ring, m (in)

The total force required to expand the ring and to produce the necessary pressure between the contact surfaces

$$F = F_s + F_e \tag{19–106}$$

$$= p w r + \frac{Ewt^3}{6L}\left(\frac{1}{d_1} - \frac{1}{d}\right) \tag{19–107}$$

RIM CLUTCHES (FIG. 19–18)

When the grooved rim clutch being engaged, the equation of equilibrium of forces along the vertical axis

$$F_n = F_n'(\sin\alpha + \mu\cos\alpha) \tag{19–108}$$

After the block is pressed on firmly the equation of equilibrium of forces along the vertical axis

$$F_n = F_n'\sin\alpha \tag{19–109}$$

Torque transmitted

$$M_t = \frac{1}{2}i_1i_2F_\theta D = i_1i_2\mu\beta D^2 bp \tag{19–110}$$

where
i_1 = number of grooves in the rim
i_2 = number of shoes
b = inclined face, m (in)
2β = angle of contact, rad

Particular	Formula

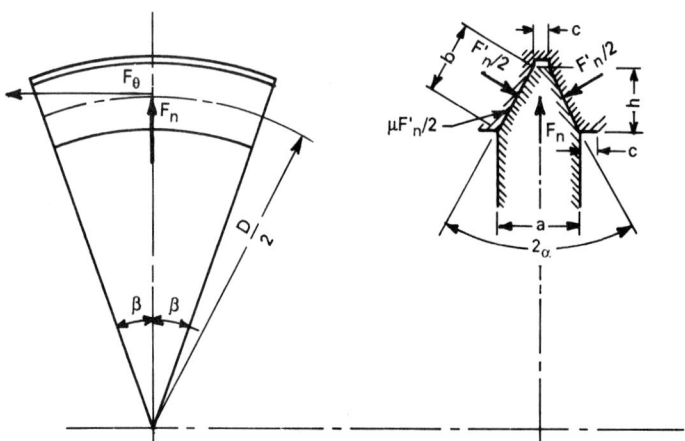

FIGURE 19–18   Grooved rim clutch.

$D = $ pitch diameter, m (in)
$2\alpha = $ V-groove angle, deg

The width of the inclined face

$$b = 0.01D + .006 \text{ m} \quad \textbf{SI} \qquad (19\text{–}111a)$$
$$= 0.01D + 0.25 \text{ in} \quad \textbf{US Customary System units}$$
$$(19\text{–}111b)$$

Frictional force

$$F_\theta = \mu F_n' \qquad (19\text{–}112a)$$
$$\text{where } F_n' = 2\beta Dbp \qquad (19\text{–}112b)$$

Torque transmitted in case of a flat rim clutch when $i_1 = 1$ and the number of sides b is only one-half that of a grooved rim

$$M_t = \frac{i}{2}\,\mu\beta D^2 bp \qquad (19\text{–}113)$$

CENTRIFUGAL CLUTCH (FIG. 19–19)

Design of shoe

Centrifugal force for speed ω_1 (rad/s) at which engagement between shoe and pully commences

$$F_{c1} = \frac{w}{g}\,\omega_1^2 r \qquad (19\text{–}114)$$

Centrifugal force for running speed ω_2 (rad/s)

$$F_{c2} = \frac{w}{g}\,\omega_2^2 r \qquad (19\text{–}115)$$

The outward radial force on inside rim of the pulley at speed ω_2

$$F_c = F_{c2} - F_{c1} \qquad (19\text{–}116a)$$

$$= \frac{w}{g}\,(\omega_2^2 - \omega_1^2)r \qquad (19\text{–}116b)$$

The centrifugal force for $\omega_1 = 0.75\omega_2$

$$F_c' = \frac{7w}{16g}\,\omega_2^2 r \qquad (19\text{–}117)$$

Particular	Formula

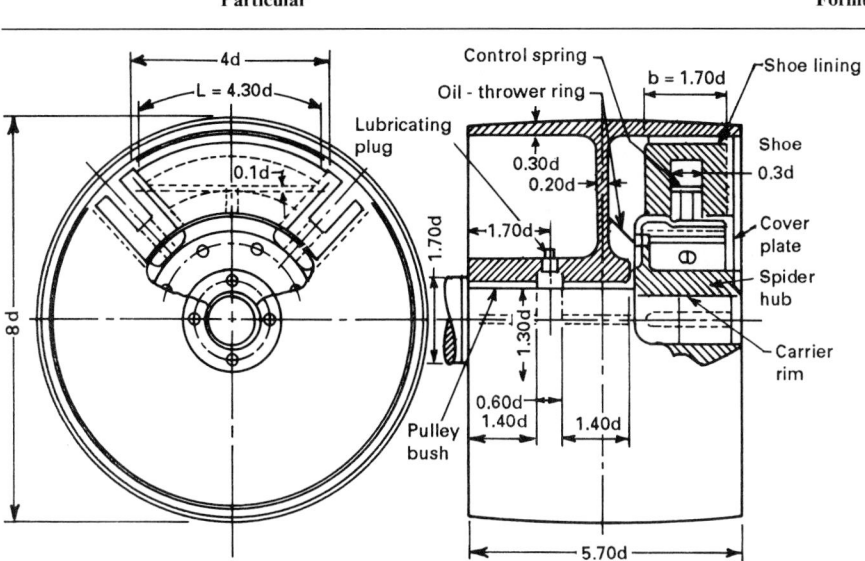

FIGURE 19-19   Centrifugal clutch.

Particular	Formula
Torque required for the maximum power to be transmitted	$M_t = 4\mu F'_c$
	$\qquad = 4\mu \dfrac{w}{g}(\omega_2^2 - \omega_1^2)rr' \qquad (19\text{–}118)$
	where $r' =$ inner radius of the rim
The equation to calculate the length of the shoe (Fig. 19-19)	$l = \dfrac{F_c}{bp} = \dfrac{w}{gbp}(\omega_r^2 - \omega_1^2)r \qquad (19\text{–}119)$

Spring

Particular	Formula
The central deflection of flat spring (Fig. 19-19) which is treated as a beam freely supported at the points where it bears against the shoe and loaded centrally by the adjusting screw	$y = \dfrac{1}{48}\dfrac{Wl^3}{EI} \qquad (19\text{–}120)$
The maximum load exerted on the spring at speed ω_1	$W = F_{c1} = \dfrac{w}{g}\omega_1^2 r \qquad (19\text{–}121)$
The cross section of spring can be calculated by the equation	$I = \dfrac{bh^3}{12} = \dfrac{Wl^3}{48Ey} \qquad (19\text{–}122)$
For other proportionate dimensions of centrifugal clutch	Refer to Fig. 19-19.

Particular	Formula

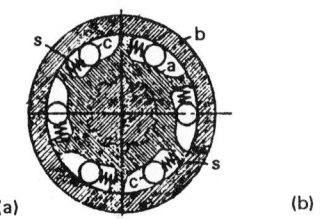

FIGURE 19–20 Roller clutch.

OVERRUNNING CLUTCHES

Roller clutch (Fig. 19–20)

The condition for the operation of the clutch

$$\alpha < 2\phi \qquad (19\text{–}123)$$

where ϕ = angle of friction, μ varies from 0.03 to 0.005.

For ϕ = angle 1°43', the angle $\alpha < 3°26'$

The force crushing the roller

$$F = \frac{F_\theta}{\tan \alpha} \qquad (19\text{–}124)$$

where F_θ = tangential force necessary to transmit the torque at pitch diameter D

The torque transmitted

$$M_c = \tfrac{1}{2} F_\theta D \qquad (19\text{–}125)$$

The allowable load on roller

$$F_a \le i\sigma_c k' l d$$

where k' = coefficient of the flattening of the roller

$$= \frac{4.64\sigma_c}{E}$$

for $\sigma_c = 1035.0$ MPa (150 kpsi)

The roller diameter

$$d = 0.1D \text{ to } 0.15D$$

LOGARITHMIC SPIRAL ROLLER CLUTCH

(FIG. 19–21) The radius of curvature of the ramp at the point of contact (Fig. 19–21)

$$R_c = 2(R_d - R_r) \frac{\sin 2\phi}{\sin 2\psi} \qquad (19\text{–}127)$$

The radius vector of point C (Fig. 19–21)

$$R_v = \frac{\sin 2\phi}{\cos(2\phi + \psi)} R_r \qquad (19\text{–}128)$$

The radius of the contact surface on the driven member in terms of the roller radius and functions angles ϕ and ψ (Fig. 19–21)

$$R_d = R_c \left[1 + \frac{\cos \phi}{\cos(2\phi + \psi)} \right] \qquad (19\text{–}129)$$

The tangential force

$$F_\theta = F \sin \phi \qquad (19\text{–}130)$$

The normal force

$$F_n = \frac{F_\theta}{\tan \phi} = F \cos \phi \qquad (19\text{–}130a)$$

Particular	Formula

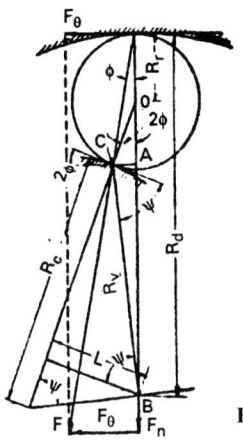

FIGURE 19–21 Logarithmic spiral roller-clutch.

The torque transmitted

$$M_t = \frac{iF_n R_d}{\cot \phi} \qquad (19\text{–}130\text{b})$$

where
$2\phi =$ angle of wedge, deg (usually ϕ varies from 3° to 12°)
$i \quad =$ number of rollers in the clutch

The maximum compressive stress at the surface area of contact between the roller and the cage made of different materials

$$\sigma_{c(max)} = 0.798 \left[\frac{F}{2l} \frac{\left(\dfrac{1}{R_r} + \dfrac{1}{R_c} \right)}{\left(\dfrac{1 - v_r^2}{E_r} + \dfrac{1 - v_c^2}{E_c} \right)} \right]^{1/2} \qquad (19\text{–}131)$$

The maximum compressive stress at the surface area of contact between the roller and the cage for $v_c = v_r = 0.3$

$$\sigma_{c(max)} = \left[\frac{0.35F \left(\dfrac{1}{R_r} + \dfrac{1}{R_c} \right)}{l \left(\dfrac{1}{E_r} + \dfrac{1}{E_c} \right)} \right]^{1/2} \qquad (19\text{–}132)$$

The maximum compressive stress at the surface area of contact between the roller and the cage made of same material ($E_c = E_r = E$) and $v_c = v_r = 0.3$

$$\sigma_{c(max)} = 0.418 \left[\frac{FE \left(\dfrac{1}{R_r} + \dfrac{1}{R_c} \right)}{l} \right]^{1/2} \qquad (19\text{–}133\text{a})$$

$$\sigma_{c(max)} = 0.418 \sqrt{\frac{FE}{lR_r}} \quad \text{if } R_c \gg R_r \qquad (19\text{–}133\text{b})$$

$$= 0.418 \sqrt{\frac{2FE}{ld}} \qquad (19\text{–}133\text{c})$$

where
$d \quad = 2R_r =$ diameter of roller, m (mm)
$l \quad =$ length of the roller, m (mm)

Particular	Formula
The design torque transmitted by the clutch	$M_{td} = \dfrac{ilR_d\sigma_{c(max)}^2 \tan\phi}{3.5 \times 10^{-3}E}$ N m **SI** (19–134a)
	$= \dfrac{ilR_d\sigma_{c(max)}^2 \tan\phi}{3.5E}$ kgf mm **Metric** (19–134b)
For further design data for clutches	Refer to Tables 19–5, 19–6, 19–7.

TABLE 19-5
Preferred dimensions and deviations for clutch facings (all dimensions in mm)

Outside diameter	Deviation	Inside diameter	Deviation	Thickness	Deviation
120, 125, 130	0	80, 85, 90	+0.5	3, 3.5, 4	±0.1
135, 140, 145	−0.5	95, 100, 105	0		
150, 155, 150		110			
170, 180, 190					
200, 210, 220	0	120, 130, 140	+0.8		
230, 240, 250	−0.8	150	0		
260, 270, 280		175, 203	+1.0		
290, 300			0		
	0				
325, 350	−1.0				

19–3   BRAKES

ENERGY EQUATIONS

The decrease of kinetic energy for a change of speed of the live load from v_1 to v_2

$$E_k = \frac{F(v_1^2 - v_2^2)}{2g} \qquad (19\text{–}135a)$$

where v_1, v_2 = speed of the live load before and after the brake is applied respectively, m/s.

F = load, kN (kgf)

The change of potential energy absorbed by the brake during the time t

$$E_p = \frac{F}{2}(v_1 + v_2)t \qquad (19\text{–}135b)$$

The change of kinetic energy of all rotating parts such as the hoist drum and various gears and sheaves which must be absorbed by the brake

$$E_r = \Sigma \frac{Wk_o^2(\omega_1^2 - \omega_2^2)}{2g} \qquad (19\text{–}136)$$

where k_o = radius of gyration of these parts, m (mm)
ω_1, ω_2 = angular velocity of the rotating parts, rad/s

TABLE 19–6
Service factors for clutches

Type of service	Service factor not including starting factor
Driving machine	
Electric motor steady load	1.0
Fluctuating load	1.5
Gas engine, single cylinder	1.5
Gas engine, multiple cylinder	1.0
Diesel engine, high-speed	1.5
Large, low-speed	2.0
Driven machine	
Generator, steady load	1.0
Fluctuating load	1.0
Blower	1.0
Compressor depending on number of cylinders	2.0–2.5
Pumps, centrifugal	1.0
Pumps, single-acting	2.0
Pumps, double-acting	1.5
Line shaft	1.5
Wood working machinery	1.75
Hoists, elevators, cranes, shovels	2.0
Hammer mills, ball mills, crushers	2.0
Brick machinery	3.0
Rock crushers	3.0

TABLE 19–7
Shear strength for clutch facings

Type	Facing material	Shear strength	
		MPa	kgf/mm²
A	Solid woven or plied fabric with or without metallic reinforcement	7.4	0.75
B	Molded and semimolded compound	4.9	0.50

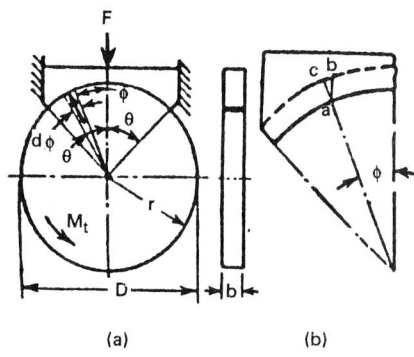

FIGURE 19–22   Single-block brake.

Particular	Formula	
The work to be done by the tangential force F_θ at the brake sheave surface in t seconds	$$W_k = \frac{F_\theta \pi D(n_1 + n_2)t}{2 \times 60}$$	(19–137)
The tangential force at the brake sheave surface	$$F_\theta = \frac{38.2(E_k + E_p + E_r)}{D(n_1 + n_2)}$$	(19–138)
Torque transmitted when the blocks are pressed against flat or conical surface	$$M_t = \mu F_n \frac{D_m}{2}$$ where F_n = total normal force, kN (kgf)	(19–139)
The operating force on block in radial direction (Fig. 19–22)	$$F = \frac{F_\theta}{\mu}\left(\frac{2\theta + \sin 2\theta}{4\sin\theta}\right)$$	(19–140)
Torque applied at the braking surface, when the blocks are pressed radially against the outer or inner surface of a cylindrical drum (Fig. 19–22)	$$M_t = \mu F \frac{D}{2}\left(\frac{4\sin\theta}{2\theta + \sin 2\theta}\right)$$	(19–141)

Particular	Formula
The tangential frictional force on the block (Fig. 19–22)	$$F_\theta = \mu F \left(\frac{4\sin\theta}{2\theta + \sin 2\theta} \right) \qquad (19\text{–}142)$$ Refer to Fig. 19–23 for values of $4\sin\theta/(2\theta + \sin 2\theta)$.
Torque applied when θ is less than $60°$	$$M_t = \mu F \, \frac{D}{2} \text{ (approx)} \qquad (19\text{–}143)$$ where $F = \mu p_a \,(br\theta)$

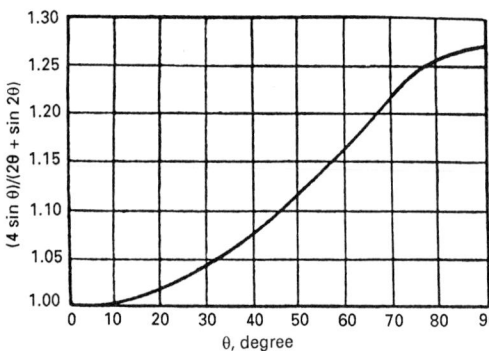

FIGURE 19–23 $(4\sin\theta)/(2\theta + \sin 2\theta)$ plotted against the semiblock angle θ.

BRAKE FORMULAS

Block brake formulas

For block brake formulas	Refer to Table 19–8 for formulas from Eqs. (19–144) to (19–148)

Band brake formulas

For band brake formulas	Refer to Table 19–8 for formulas from Eqs. (19–149) to (19–157)
The magnitude of pressure between the band and the brake sheave	$$p = \frac{F_1 + F_2}{Dw} \qquad (19\text{–}158)$$
The practical rule for the band thickness	$$h = 0.005D \qquad (19\text{–}159)$$
Width of band	$$w = \frac{F_1}{h\sigma_d} \qquad (19\text{–}160)$$

TABLE 19–8
Formulas for block, simple, and differential band brakes

Type of brake and rotation		Force at the end of brake handle, kN (kgf)	
Block brake	(a)	Rotation in either direction	$F = F_\theta \, \dfrac{a}{\mu(a+b)}$ (19–144)
Block brake	(b)	Clockwise	$F = \dfrac{F_\theta a}{a+b}\left[\dfrac{1}{\mu} - \dfrac{c}{a}\right]$ (19–145)
		Counterclockwise	$F = \dfrac{F_\theta a}{a+b}\left[\dfrac{1}{\mu} + \dfrac{c}{a}\right]$ (19–146)
Block brake	(c)	Clockwise	$F = \dfrac{F_\theta a}{a+b}\left[\dfrac{1}{\mu} + \dfrac{c}{a}\right]$ (19–147)
		Counterclockwise	$F = \dfrac{F_\theta a}{a+b}\left[\dfrac{1}{\mu} - \dfrac{c}{a}\right]$ (19–148)
Simple band brake	(a)	Clockwise	$F = \dfrac{F_\theta b}{a}\left[\dfrac{e^{\mu\theta}}{e^{\mu\theta} - 1}\right]$ (19–149)
		Counterclockwise	$F = \dfrac{F_\theta b}{a}\left[\dfrac{1}{e^{\mu\theta} - 1}\right]$ (19–150)
Simple band brake	(b)	Clockwise	$F = \dfrac{F_\theta b}{a}\left[\dfrac{1}{e^{\mu\theta} - 1}\right]$ (19–151)
		Counterclockwise	$F = \dfrac{F_\theta b}{a}\left[\dfrac{e^{\mu\theta}}{e^{\mu\theta} - 1}\right]$ (19–152)

TABLE 19–8
Formulas for block, simple, and differential band brakes (*Cont.*)

Type of brake and rotation		Force at the end of brake handle, kN (kgf)	
Differential band brake	Clockwise	$F = \dfrac{F_\theta}{a}\left[\dfrac{b_2 e^{\mu\theta} + b_1}{e^{\mu\theta} - 1}\right]$	(19–153)
	Counterclockwise	$F = \dfrac{F_\theta}{a}\left[\dfrac{b_1 e^{\mu\theta} + b_2}{e^{\mu\theta} - 1}\right]$	(19–154)
	If $b_2 = b_1$ F is the same for rotation in either direction	$F = \dfrac{F_\theta b}{a}\left[\dfrac{b_1 e^{\mu\theta} + 1}{e^{\mu\theta} + 1}\right]^{\text{a}}$	(19–155)
Differential band brake	Clockwise	$F = \dfrac{F_\theta}{a}\left[\dfrac{b_2 e^{\mu\theta} - b_1}{e^{\mu\theta} - 1}\right]$	(19–156)
	Counterclockwise	$F = \dfrac{F_\theta}{a}\left[\dfrac{b_2 - b_1 e^{\mu\theta}}{e^{\mu\theta} - 1}\right]^{\text{b}}$	(19–157)

ᵃ For the above two cases, if $b_2 = b_1 = b$.
ᵇ In this case if $b_2 \le b_1 e^{\mu\theta}$, F will be negative or zero and the brake works automatically or the brake is "self-locking".

Particular	Formula
Suitable drum diameter according to Hagenbook	$\left(\dfrac{M_t}{69}\right)^{1/3} < 10D < \left(\dfrac{M_t}{54}\right)^{1/3}$ **SI** (19–161)
	where M_t in N m and D in m
	$\left(\dfrac{M_t}{5}\right)^{1/3} < D < \left(\dfrac{M_t}{4}\right)^{1/3}$ **US Customary System unit**
	where M_t in lbf in and D in in (19–162)
Suitable drum diameter in terms of frictional horse power	$(79.3\mu P)^{1/3} < 100D < (105.8\mu P)^{1/3}$ **SI** (19–163a)
	where P in kW and D in m
	$(60\mu P)^{1/3} < D < (80\mu P)^{1/3}$ **US Customary System unit** (19–163b)
	where P in hp and D in in
	μP is taken as the maximum horsepower to be dissipated in any 15-min period

Particular	Formula

CONE BRAKES (FIG. 19–24)

The normal force

$$F_n = \frac{F_a}{\sin \alpha} \tag{19–164}$$

The radial force

$$F_r = \frac{F_a}{\tan \alpha} \tag{19–165}$$

The tangential force or braking force

$$F_\theta = \mu F_n = \frac{\mu F_a}{\sin \alpha} \tag{19–166}$$

The braking torque

$$M_t = \frac{\mu F_a D}{2 \sin \alpha} \tag{19–167}$$

where D = mean diameter, m (mm)

CONSIDERING THE LEVER (FIG. 19–24)

The axial force

$$F_a = \frac{aF}{h} \tag{19–168}$$

The relation between the operating force F and the braking force F_θ

$$F = \frac{hF_\theta \sin \alpha}{\mu a} \tag{19–169}$$

The area of the contact surface using the designation given in Fig. 19–24

$$A = \frac{\pi Dw}{\cos \alpha} \tag{19–170}$$

where
w = axial width, m (mm)
α = half the cone angle, deg

The average pressure between the contact surfaces

$$F_{av} = \frac{F_n}{A} = \frac{F_a}{\pi Dw \tan \alpha} \tag{19–171}$$

Take
α = from 10° to 18°
w = 0.12D to 0.22D

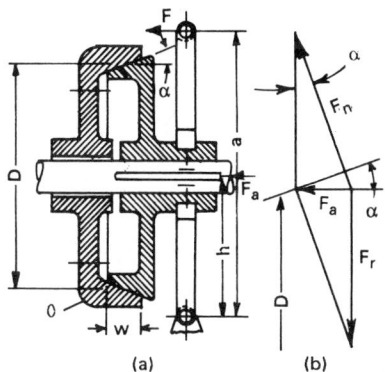

(a)   (b)   **FIGURE 19–24** Cone brake.

Particular	Formula

DISK BRAKES

The torque transmitted for i pairs of friction surfaces

$$M_t = \frac{\pi i \mu p_1 D_1 (D_2^2 - D_1^2)}{8} \tag{19-172}$$

The axial force transmitted

$$F_a = \tfrac{1}{2} \pi p_1 D_1 (D_2 - D_1) \tag{19-173}$$

where p_1 = intensity of pressure at the inner radius, MPa (kgf/mm^2)

For design values of brake facings   Refer to Table 19–9.

TABLE 19–9
Design values for brake facings

Facing material	Design coefficient of friction μ	Permissible unit pressure			
		1 m/s		10 m/s	
		MPa	kgf/mm^2	MPa	kgf/mm^2
Cast iron on cast iron					
Dry	0.20				
Oily	0.07				
Wood on cast iron	0.25–0.30	0.5521–0.6824	0.0563–0.0703	0.1383–0.1726	0.0141–0.0176
Leather on cast iron					
Dry	0.40–0.50	0.0549–0.1039	0.0056–0.0106		
Oily	0.15				
Asbestos fabric on metal					
Dry	0.35–0.40	0.6209–0.6894	0.0633–0.0703	0.1726–0.2069	0.0176–0.0211
Oily	0.25				
Molded asbestos on metal	0.30–0.35	1.0395–1.2062	0.106–0.123	0.2069–0.2756	0.0211–0.0281

INTERNAL EXPANDING-RIM BRAKE

Forces on Shoe (Fig. 19–25)

FOR CLOCKWISE ROTATION The maximum pressure

$$p_a = p \frac{\sin \theta_a}{\sin \theta} \tag{19-174}$$

The moment $M_{t\mu}$ of the frictional forces

$$M_{t\mu} = \frac{\mu p_a b r}{\sin \theta_a} \int_{\theta_1}^{\theta_2} \sin \theta (r - a \cos \theta) d\theta \tag{19-174a}$$

The moment of the normal forces

$$M_{tn} = \frac{p_a b r a}{\sin \theta_a} \int_{\theta_1}^{\theta_2} \sin^2 \theta \, d\theta \tag{19-175}$$

Particular	Formula

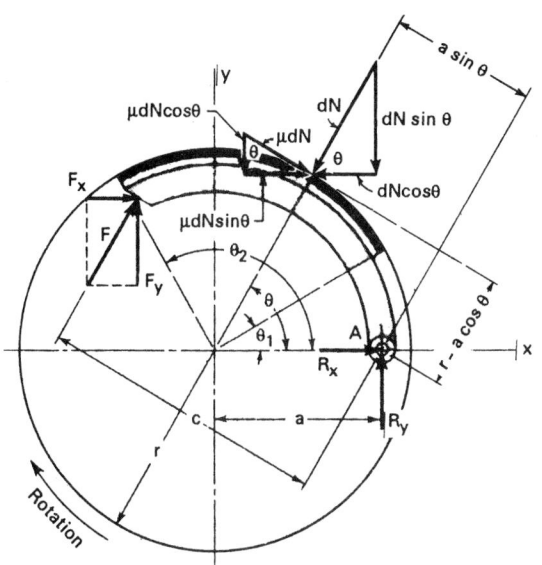

FIGURE 19–25 Forces on the shoe. (*J. E. Shigley,* Mechanical Engineering Design, *1962, courtesy of McGraw-Hill.*)

The actuating force

$$F = \frac{M_{tn} - M_{t\mu}}{c} \tag{19–176}$$

The torque M_t applied to the drum by the brake shoe

$$M_t = \frac{\mu p_a b r^2 (\cos \theta_1 - \cos \theta_2)}{\sin \theta_a} \tag{19–177}$$

The hinge-pin horizontal reaction

$$R_x = \frac{p_a b r}{\sin \theta_a} \left(\int_{\theta_1}^{\theta_2} \sin \theta \cos \theta \, d\theta \right.$$
$$\left. - \mu \int_{\theta_1}^{\theta_2} \sin^2 \theta \, d\theta \right) - F_x \tag{19–178}$$

The hinge-pin vertical reaction

$$R_y = \frac{p_a b r}{\sin \theta_a} \left(\int_{\theta_1}^{\theta_2} \sin^2 \theta \, d\theta \right.$$
$$\left. + \mu \int_{\theta_1}^{\theta_2} \sin \theta \cos \theta \right) - F_y \tag{19–179}$$

Particular	Formula

**FOR COUNTERCLOCKWISE ROTATION
(FIG. 19–25)**

$$F = \frac{M_{tn} + M_{t\mu}}{c} \qquad (19\text{–}180)$$

$$R_x = \frac{p_a b r}{\sin \theta_a} \left(\int_{\theta_1}^{\theta_2} \sin \theta \cos \theta \, d\theta \right.$$

$$\left. + \mu \int_{\theta_1}^{\theta_2} \sin^2 \theta \, d\theta \right) - F_x \qquad (19\text{–}181)$$

$$R_y = \frac{p_a b r}{\sin \theta_a} \left(\int_{\theta_1}^{\theta_2} \sin^2 \theta \, d\theta \right.$$

$$\left. - \mu \int_{\theta_1}^{\theta_2} \sin \theta \cos \theta \, d\theta \right) - F_y \qquad (19\text{–}182)$$

EXTERNAL CONTRACTING-RIM BRAKE

Forces on shoe (Fig. 19–26)

FOR CLOCKWISE ROTATION The moment $M_{t\mu}$
of the friction forces Fig. 19–26

$$M_{t\mu} = \frac{\mu p_a b r}{\sin \theta_a} \int_{\theta_1}^{\theta_2} \sin \theta (r - a \cos \theta) \, d\theta \qquad (19\text{–}183)$$

The moment of the normal force

$$M_{tn} = \frac{p_a b r}{\sin \theta_a} \int_{\theta_1}^{\theta_2} \sin^2 \theta \, d\theta \qquad (19\text{–}184)$$

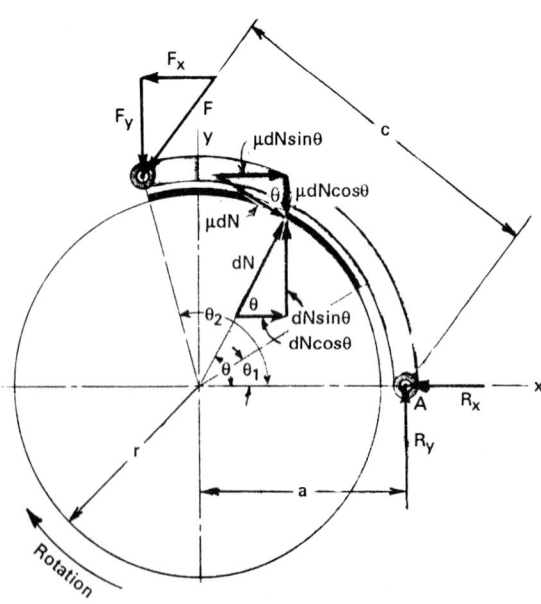

FIGURE 19–26   Forces and notation for external-contracting shoe. (*J. E. Shigley*, Mechanical Engineering Design, *1962, courtesy of McGraw-Hill.*

Particular	Formula
The actuating force	$$F = \frac{M_{tn} + M_{t\mu}}{c} \qquad (19\text{--}185)$$
The horizontal reaction at the hinge-pin	$$R_x = \frac{p_a b r}{\sin \theta_a} \left(\int_{\theta_1}^{\theta_2} \sin \theta \cos \theta \, d\theta \right.$$ $$\left. + \mu \int_{\theta_1}^{\theta_2} \sin^2 \theta \, d\theta \right) - F_x \qquad (19\text{--}186)$$
The vertical reaction at the hinge-pin	$$R_y = \frac{p_a b r}{\sin \theta_a} \left[\mu \int_{\theta_1}^{\theta_2} \sin \theta \cos \theta \, d\theta \right.$$ $$\left. - \int_{\theta_1}^{\theta_2} \sin^2 \theta \, d\theta \right] + F_y \qquad (19\text{--}187)$$
FOR COUNTERCLOCKWISE ROTATION	$$F = \frac{M_{tn} - M_{t\mu}}{c} \qquad (19\text{--}188)$$ $$R_x = \frac{p_a b r}{\sin \theta_a} \left[\int_{\theta_1}^{\theta_2} \sin \theta \cos \theta \, d\theta \right.$$ $$\left. - \mu \int_{\theta_1}^{\theta_2} \sin^2 \theta \, d\theta \right] - F_x \qquad (19\text{--}189)$$ $$R_y = \frac{p_a b r}{\sin \theta_a} \left[-\mu \int_{\theta_1}^{\theta_2} \sin \theta \cos \theta \, d\theta \right.$$ $$\left. - \int_{\theta_1}^{\theta_2} \sin^2 \theta \, d\theta \right] + F_y \qquad (19\text{--}190)$$

HEATING OF BRAKES

Heat generated from work of friction	$H_g = \mu p A_c v$ J (joules) **SI** (19–191a) $= \dfrac{\mu p A_c v}{778}$ **US Customary System units** (19–191b)
Heat to be radiated for a brake lowering the load	$H = Wh$ J (joules) **SI** (19–192a) $= \dfrac{Wh}{778}$ **US Customary System units** (19–192b) where h = total height or distance, m
The heat generated is also given by the equation	$H_g = 754 k_l P$ J/s **SI** (19–193a) where P in kW $= 42.4 k_l P$ **US Customary System units** (19–193b) where P in hp

Particular	Formula

The rise in temperature in °C of the brake drum or clutch plates	$$\Delta T = \frac{H}{mC} \qquad (19\text{–}194)$$
	 where
	 m = mass of brake drum or clutch plates, kg
	 C = specific heat capacity
	   = 500 J/kg °C for cast iron or steel
	   = 0.13 Btu/lbm °F for cast iron
	   = 0.116 Btu/lbm °F for steel

The rate of heat dissipation	$$H_d = C_2 \Delta T A_r \quad \text{J} \qquad \textbf{SI} \qquad (19\text{–}195a)$$ $$= 0.25 C_2 \Delta T A_r \quad \text{kcal} \qquad \textbf{Metric} \qquad (19\text{–}195b)$$ where C_2 = radiating factor from Table 19–13
The required area of radiating surface	$$A_r = \frac{754 k_l N}{C_2 \Delta T} \quad \text{m}^2 \qquad \textbf{SI} \qquad (19\text{–}196a)$$ $$= \frac{0.18 k_l N}{C_2 \Delta T} \quad \text{mm}^2 \qquad \textbf{SI} \qquad (19\text{–}196b)$$
Approximate time required for the brake to cool	$$t_c = \frac{W_r C_2 \ln \Delta T}{K A_r} \qquad (19\text{–}197)$$ where K = a constant varying from 0.4 to 0.8
Gagne's formula for heat generated during a single operation	$$H_g = \frac{AC}{n_c} (T_{av} - T_a) \left[\frac{n_c t}{3600} + 1.5 \left(1 - \frac{N_c t}{3600} \right) \right] \qquad (19\text{–}198)$$ where $(T_{av} - T_a)$ = temperature difference between the brake surface and the atmosphere, °C Refer to Table 19–15 for values of C.
For additional design data for brakes	Refer to Tables 19–11 to 19–17.

TABLE 19–10
Working pressure for brake blocks

Rubbing velocity, m/s	Pressure					
	Wood blocks		Asbestos fabric		Asbestos blocks	
	MPa	kgf/mm²	MPa	kgf/mm²	MPa	kgf/mm²
1	0.5521	0.0563	0.6894	0.0703	1.1032	0.1125
2	0.4482	0.0457	0.5521	0.0563	1.0346	0.1055
3	0.3452	0.0352	0.4138	0.0422	0.8963	0.0914
4	0.2412	0.0246	0.2756	0.0281	0.6894	0.0703
5	0.1726	0.0176	0.2069	0.0211	0.4825	0.0492
10	0.1726	0.0176	0.2069	0.0211	0.2756	0.0281

TABLE 19–11
Comparison of hoist brakes

Brake characteristics	Block brakes		Band brakes		Axial brakes	
	Double block	V-grooved sheave	Simple	Both directions of rotation	Cone	Multidisk
Force ratio $\dfrac{F}{F_\theta}$	$\dfrac{b}{\mu a}$	$\dfrac{b\sin\alpha}{\mu a}$	$\dfrac{b}{a(e^{\mu\theta}-1)}$	$\dfrac{b(e^{\mu\theta}+1)}{a(e^{\mu\theta}-1)}$	$\dfrac{b\sin\alpha}{\mu a}$	$\dfrac{b}{n\mu a}$
Average numerical value	0.667	0.282	0.0323	0.165	0.161	0.097
Relative value	20.6	8.7	1	5.1	5.0	3.00
Travel at lever end	$\dfrac{h_1^{b}a}{b}$	$\dfrac{h_1 a}{b\sin\alpha}$	$\dfrac{h_1 a\theta}{2\pi b}$	$\dfrac{h_1 a\theta}{4\pi b}$	$\dfrac{h_1 a}{b^{a}\sin\alpha}$	$\dfrac{ih_1' a}{b}$
Average travel, mm (in)	8.0 (0.313)	18.8 (0.74)	74.5 (2.943)	37.36 (1.471)	32.8 (1.292)	5.56 (0.219)
Maximum capacity, N kW (hp)	1512.7 (2000)	18.9 (25)	227.0 (300)	75.6 (100)	37.8 (50)	90.8 (120)

[a] $h = b$ in Fig. 19–21.
[b] h_1 = the normal distance between the sheave and the stationary braking surface to prevent dragging.
Source: Courtesy V. L. Maleev and J. B. Hartman, *Machine Design*, International Textbook Company, Scranton, Pennsylvania, 1954.

TABLE 19–12
Service factors for typical machines

Type of driven machine	Electric motor steam or water turbine	High-speed steam or gas engine	Service factors for prime movers			
			Petrol engine		Oil engine	
			≥ 4 Cyl[a]	≤ 4 Cyl	≥ 6 Cyl	≤ 4 Cyl
Alternators and generators (excluding welding generators), induced-draft fans, printing machinery, rotary pumps, compressors, and exhausters, conveyors	1.5	2.0	2.5	3.0	3.5	5.0
Woodworking machinery, machine tools (cutting) excluding planing machines, calenders, mixers, and elevators	2.0	2.5	3.0	3.5	3.0	5.5
Forced-draft fans, high-speed reciprocating compressors, high speed crushers and pulverizers, machine tools (forming)	2.5	3.0	3.5	4.0	4.5	6.0
Rotary screens, rod mills, tube, cable and wire machinery, vacuum pumps	3.0	3.5	4.0	4.5	5.0	6.5
Low-speed reciprocating compressors, haulage gears, metal planing machines, brick and tile machinery, rubber machinery, tube mills, generators (welding)	3.5	4.0	4.5	5.0	5.5	7.0

TABLE 19–13
Radiating factors for brakes

Temperature difference, ΔT	Radiating factor, C_2		$C_2 \Delta T$	
	W/m² K	cal/m²s°C	W/m²	cal/m² s
55.5	12.26	2.93	681.36	162.73
111.5	15.33	3.66	1703.41	406.83
166.5	16.97	4.05	2827.66	675.34
222.6	18.40	4.39	4088.19	976.40

TABLE 19–15
Values of heat transfer coefficient C for rough block surfaces

Velocity, v, m/s	Heat-transfer coefficient, C	
	W/m² K	kcal/m² h °C
0.0	8.5	7.31
6.1	14.1	12.13
12.2	18.8	16.20
18.3	22.5	19.30
24.4	25.6	22.00
30.5	29.0	24.90

TABLE 19–14
pv values as recommended by Hutte for brakes

Service	pv	
	SI	Metric
Intermittent operations with long rest periods and poor heat readiation, as with wood blocks	26.97	2.75
Continuous service with short rest periods and with poor radiation	13.73	1.40
Continuous operation with good radiation as with an oil bath	40.70	4.15

TABLE 19–16
Values of $e^{\mu\theta}$

Proportion of contact to whole circumference	Steel band on cast iron $\mu = 0.18$	Leather belt on			
		Wood		Cast iron	
		Slightly greasy $\mu = 0.47$	Very greasy $\mu = 0.12$	Slightly greasy $\mu = 0.28$	Damp $\mu = 0.38$
0.1	1.12	1.34	1.08	1.19	1.27
0.2	1.25	1.81	1.16	1.42	1.61
0.3	1.40	2.43	1.25	1.69	2.05
0.4	1.57	3.26	1.35	2.02	2.60
0.425	1.62	3.51	1.38	2.11	2.76
0.45	1.66	3.78	1.40	2.21	2.93
0.475	1.71	4.07	1.43	2.31	3.11
0.500	1.76	4.38	1.46	2.41	3.30
0.525	1.81	4.71	1.49	2.52	3.50
0.6	1.97	5.88	1.57	2.81	4.19
0.7	2.21	7.90	1.66	3.43	5.32
0.8	2.47	10.60	1.83	4.09	6.75
0.9	2.77	14.30	1.97	4.87	8.57
1.0	3.10	19.20	2.12	5.81	10.90

TABLE 19–17
Coefficient of friction and permissible variations on dimensions for automotive brakes lining

Type and class of brake lining	Range of coefficient of friction, μ	Permissible variation in μ, %	Tolerance on width for sizes, mm		Tolerance ot thickness for sizes, mm	
			≤ 5 mm thickness	> 5 mm thickness	≤ 5 mm thickness	> 5 mm thickness
Type I—rigid molded sets or flexible molded rolls or sets						
Class A—medium friction	0.28–0.40	+30, −20	+0	+0	+0	+0
Clas B—high friction	0.36–0.45	+30, −20	−0.2	−0.3	−0.8	0.8
Type II—rigid woven sets or flexible woven rolls or sets						
Class A—medium friction	0.33–0.43	+20 − 30				
Class B—high friction	0.43–0.53	+20 − 30				

REFERENCES

1. Shigley, J. E., *Machine Design*, McGraw-Hill Book Company, New York, 1962.
2. Maleev, V. L. and J. B. Hartman, *Machine Design*, International Textbook Company, Scranton, Pennsylvania, 1954.
3. Black, P. H., and O. E. Adams, Jr., *Machine Design*, McGraw-Hill Book Company, New York, 1968.
4. Norman, C. A., E. S. Ault, and I. F. Zarobsky, *Fundamentals of Machine Design*, The Macmillan Company, New York, 1951.

5. Spotts, M. F., *Machine Design Analysis*, Prentice-Hall, Englewood Cliffs, New Jersey, 1964.
6. Spotts, M. F., *Design of Machine Elements*, Prentice-Hall of India Ltd., New Delhi, 1969.
7. Vallance, A., and V. L. Doughtie, *Design of Machine Members*, McGraw-Hill Book Company, New York, 1951.
8. Lingaiah, K., and B. R. Narayana Iyengar, *Machine Design Data Handbook*, Engineering College Co-operative Society, Bangalore, India, 1962.
9. Lingaiah, K., and B. R. Narayana Iyengar, *Machine Design Data Handbook*, Vol. I (*SI and Customary Metric Units*), Suma Publishers, Bangalore, India, 1986.
10. Lingaiah, K., *Machine Design Data Handbook*, Vol. II (*SI and Customary Metric Units*), Suma Publishers, Bangalore, India, 1986.
11. Bureau of Indian Standards.

CHAPTER
20

SPRINGS

SYMBOLS

A	area of loading, m^2 (in^2)
b	width of rectangular spring, m (in)
	width of laminated spring, m (in)
b'	width of each strip in a laminated spring, m (in)
c	spring index
c_1, c_2	constants taken from Table 20–1 and to be used in Eqs. (20–1) to (20–36)
C_1, C_2	constants to be used in Eqs. (20–20) and (20–21) and taken from Fig. 20–3
d	diameter of spring wire, m (in)
	diameter of torsion bar, m (in)
d_1, d_2	diameter of outer and inner wires of concentric spring, m (in)
D	mean or pitch diameter of spring, m (mm) overall diameter of the absorber, m (in)
D_1	mean or pitch diameter of outer concentric spring, m (in)
	smallest mean diameter of conical spring, m (in)
D_2	mean or pitch diameter of inner concentric spring, m (in)
	largest mean diameter of conical spring, m (in)
e_{sz}	size coefficient
e'_{sr}	surface influence coefficient
E	modulus of elasticity, GPa (psi)
f	frequency, cycles per minute, Hz
F	load, kN (lbf)
	steady-state load [Eq. (20–84)]
F_{max}	maximum force that can be imposed on the housing, kN (lbf)
k_o	force to compress the spring one meter (in)
	N/m (lbf/in) [spring rate, N/m (lbf/in)]
F_{cr}	critical load, kN (lbf)
g	acceleration due to gravity, 9.8066 m/s^2
	9806.06 mm/s^2 (32.2 ft/s^2; 386.4 in/s^2)
G	modulus of rigidity, GPa (psi)
h	height (thickness) of laminated spring, m (in)
	axial height of a rectangular spring wire, m (in)
i	total number of strips or leaves in a leaf spring
	number of coils in a helical spring
i'	total number of full-length blunt-ended leaves in a leaf spring
I	moment of inertia, area, m^4, cm^4 (in^4)

k, k_1, k_2	stress factor (Wahl factor)
k_4	correction factor
K_l	factor depends on the ratio $\frac{l_o}{D}$ as shown in Fig. 20–8
	reduced stress correction factor or Wahl stress factor or fatigue stress correction factor
k_τ	shear stress correction factor
l	length, m (in)
l_f or l_o	free length of helical spring, m (in)
L	$i\pi D$ length of the coil part of torsion spring, m (in)
	effective length of bushing, m (in)
	overall length of the absorber (Fig. 20–15), m (in)
M	constant depends on $\frac{d_o}{d_i}$ as indicated in Fig. 20–3
M_t	twisting moment, N m (lbf in)
n	factor of safety
n_a	actual factor of safety or reliability factor
U	resilience, N m (lbf in)
	energy to be absorbed by a rubber spring, N m, (lbf in)
V	volume of spring, m^3, mm^3 (in^3)
γ	specific weight of the spring material, N/m^3 (lbf/in^3)
W	weight of spring, kN (lbf)
	weight of effective number of coils i involved in the operation of the spring [Eq. (20–77)], kN (lbf)
y	deflection, m (in)
y_{cr}	critical deflection, m (in)
Z	section modulus, m^3, cm^3 (in^3)
Z_o	polar section modulus, m^3, cm^3 (in^3)
σ	stress, normal, MPa (psi)
τ	shear stress, MPa (psi)
α, α'	constant from Table 20–3
β, β'	constants from Table 20–3
θ	angular deflection, rad
v	Poisson's ratio

SUFFIXES

1	outside
2	inside
a	amplitude
m	mean
max	maximum
min	minimum
f	endurance limit (also used for reversed cycle)
o	endurance limit for repeated cycle

Particular	Formula

LEAF SPRINGS (TABLE 20–1) [1, 2, 3]

The general equation for the maximum stress in springs

$$\sigma = \frac{c_1 F l}{b h^2}$$

(20–1)

Particular	Formula	
The general equation for the maximum deflection in springs	$y = \dfrac{c_2 F l^3}{E b h^3}$	(20–2)
The thickness of spring	$h = \dfrac{c_2 \sigma l^2}{c_1 y_E}$	(20–3)
	Refer to Table 20–1 for values of c_1 and c_2.	
For sizes and tolerances for leaf springs for motor vehicle suspension [4]	Refer to Tables 20–2 to 20–5.	

TABLE 20–1
Deflection formula for beams of rectangular cross section and constants in beam Eqs. (20–1) to (20–3)

Particular	Maximum deflection, y_{max}	c_1—for the stress	c_2—for the deflection	Unit resilience, Nm/m^3 (kgf mm/mm^3)
Constant breadth and depth	$y_{max} = \dfrac{2F}{bE}\left(\dfrac{l}{h}\right)^3$	3	2	$\dfrac{\sigma^2}{18E}$
Constant breadth, varying depth	$y_{max} = \dfrac{4F}{bE}\left(\dfrac{l}{h}\right)^3$	3	4	$\dfrac{\sigma^3}{6E}$
Constant depth, varying breadth	$y_{max} = \dfrac{3F}{bE}\left(\dfrac{l}{h}\right)^3$	3	3	$\dfrac{\sigma^2}{6E}$
Constant depth and breadth	$y_{max} = \dfrac{4F}{bE}\left(\dfrac{l}{h}\right)^3$	6	4	$\dfrac{\sigma^2}{18E}$
Constant breadth, varying depth	$y_{max} = \dfrac{8F}{bE}\left(\dfrac{l}{h}\right)^3$	6	8	$\dfrac{\sigma^2}{6E}$
Constant depth, varying breadth	$y_{max} = \dfrac{6F}{bE}\left(\dfrac{l}{h}\right)^3$	6	6	$\dfrac{\sigma^2}{6E}$

Source: K. Lingaiah and B. R. Narayana Iyengar, *Machine Design Data Handbook*, Engineering College Cooperative Society, Bangalore, India, 1962; K. Lingaiah and B. R. Narayana Iyengar, *Machine Design Data Handbook*, Vol. I, Suma Publishers, Bangalore, India, 1986; and K. Lingaiah, *Machine Design Data Handbook*, Vol. II, Suma Publishers, Bangalore, India, 1986.

Particular	Formula

LAMINATED SPRING (FIG. 20–1) [5]

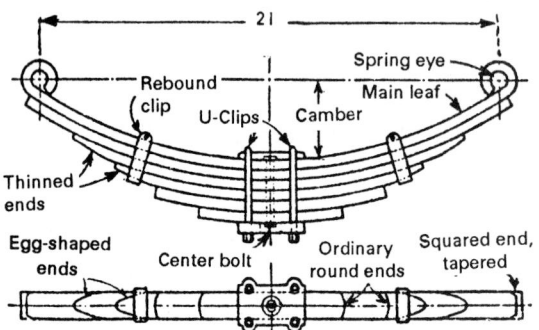

FIGURE 20–1 Laminated springs for automobiles.

The load on the spring

$$F = \frac{\sigma i b' h^2}{c_1 l} \quad \text{where } ib' = b \qquad (20\text{–}4)$$

The maximum deflection

$$y = \frac{c_2 F l^3}{E i b' h^3} \qquad (20\text{–}5)$$

The maximum deflection in case of laminated semielliptical spring for heavy loads

$$y = \frac{c_2 F l^3 k_4}{E i b' h^3} \qquad (20\text{–}6)$$

The correction factor to be used in Eq. (20–6)

$$k_4 = \frac{1 - 4r + 2r^2(1.5 - \ln r)}{(1 - r)^3} \qquad (20\text{–}7)$$

where $r = \dfrac{i'}{l}$

For standard sections of steel plates for laminated springs

Refer to Tables 20–2 to 20–6.

The correction factor k_4 can also be obtained from

$$
\begin{aligned}
k_4 &= 0.73 r^{0.1} \quad \text{for } 2 < r < 20 \qquad (20\text{–}8)\\
&= 1 \qquad\quad\ \text{for } r > 20
\end{aligned}
$$

Size coefficient

$$e_{sz} = 0.8 + \frac{0.0025}{h} \quad \text{SI, } h \text{ in mm} \qquad (20\text{–}9a)$$

$$= 0.8 + \frac{0.1}{h} \quad \text{US Customary Units} \qquad (10\text{–}9b)$$

where h in in

$$= 0.8 + \frac{2.5}{h} \quad \text{SI, } h \text{ in mm} \qquad (20\text{–}9c)$$

Particular	Formula

TABLE 20–2
Cross section tolerances for leaf springs for motor vehicle suspension—metric bar sizes—SAE J1123a

Width, mm	Width tolerance, mm	Tolerance, mm in thickness $(+)^a$ and in flatness $(-)^b$			Maximum difference in thickness^c		
		For thickness			For thickness, mm		
	Minus 0.00	5.00–9.50	10.00–21.20	22.40–37.50	5.00–9.50	10.00–21.20	22.40–37.50
40.0	+0.75	0.13	0.15	—	0.05	0.05	—
45.0	+0.75	0.13	0.15	—	0.05	0.05	—
50.0	+0.75	0.13	0.15	—	0.05	0.05	—
56.0	+0.75	0.13	0.15	—	0.05	0.05	—
63.0	+0.75	0.13	0.15	—	0.05	0.05	—
75.0	+1.15	0.15	0.20	0.30	0.08	0.10	0.15
90.0	+1.15	0.15	0.20	0.30	0.08	0.10	0.15
100.0	+1.15	0.15	0.20	0.30	0.08	0.10	0.15
125.0	+1.65	0.18	0.25	0.40	0.10	0.13	0.20
150.0	+2.30	—	0.30	0.50	—	0.15	0.25

[a] Thickness measurements shall be taken at the edge of the bar where the flat surfaces intersect the rounded edge.
[b] This tolerance represents the maximum amount by which the thickness of the center of the bar may be less than the thickness at the edges. Thickness of the center may never exceed the thickness at the edges.
[c] Maximum difference in thickness between the two edges of each bar.
Source: Reproduced from *SAE Handbook*, Vol. I, 1981, courtesy SAE.

Size factor	$$k_{sz} = \frac{1}{e_{sz}} = 4.66\,h^{0.35} \quad \textbf{SI, } h \text{ in m} \qquad (20\text{–}10a)$$
	$$= 1.27h^{0.35} \quad \textbf{US Customary units} \qquad (20\text{–}10b)$$ $$d \text{ in in}$$
	$$= 0.415h^{0.35} \quad \textbf{SI, } h \text{ in mm} \qquad (20\text{–}10c)$$

LAMINATED SPRINGS WITH INITIAL CURVATURE

The load shared by graduated leaves of the spring	$$F_g = \frac{2(1-r)}{(2+r)}\,F \qquad\qquad (20\text{–}11)$$
The load shared by full-length leaves of the spring	$$F_f = \frac{3r}{(2+r)}\,F \qquad\qquad (20\text{–}12)$$
The maximum stress in the graduated leaves	$$\sigma_g = \frac{\alpha Fl}{lb'h^2} \qquad\qquad (20\text{–}13)$$
The maximum stress in the full-length leaves	$$\sigma_f = \frac{1.5\alpha Fl}{ib'h^2} \qquad\qquad (20\text{–}14)$$

Particular	Formula	
The maximum deflection of the leaves (in both graduated and full-length leaves)	$y = \dfrac{\beta F l^3}{E i b' h^3}$	(20–15)
The camber to be provided for equalization of stress in both graduated and full-length leaves	$c = \dfrac{\beta' F l^3}{i b' E h^3}$	(20–16)
The load on the clip bolt to be applied to provide camber	$F_b = \dfrac{\alpha' F l}{i b' h^2}$	(20–17)
The maximum equalized stress	$\sigma = \dfrac{\alpha' F l}{i b' h^2}$	(20–18)

The values of constant α, α', β and β' can be obtained from Table 20–7.

TABLE 20–3
Cross section tolerances for leaf spring for motor vehicle suspension—SAE J510c

Nominal width				Tolerance in width		For thickness		Tolerance in thickness[a]		Tolerance in flat surfaces[b]		Maximum difference in thickness[c]	
Over	to and including												
in	mm	in	mm	−0.00 in	−0.0 mm	in	mm	±in	±mm	in	mm	in	mm
0.00	0.0	2.50	63.5	+0.030	−0.076	≤ 0.375	9.52	0.005	0.13	−0.005	−0.13	0.002	0.05
						> 0.375–0.875	9.52–22.22	0.006	0.15	−0.006	−0.15	0.002	0.05
2.50	63.5	4.00	101.6	+0.045	+1.14	≤ 0.375	9.52	0.006	0.15	−0.006	−0.15	0.003	0.08
						> 0.375–0.875	9.52–22.22	0.008	0.20	−0.008	−0.20	0.004	0.10
						> 0.875–1.500	22.22–38.10	0.012	0.30	−0.012	−0.30	0.006	0.15
4.00	101.6	5.00	127.0	+0.065	+1.65	≤ 0.375	9.52	0.007	0.18	−0.007	−0.18	0.004	0.10
						> 0.375–0.875	9.52–22.22	0.010	0.25	−0.010	−0.25	0.005	0.13
						> 0.875–1.500	22.22–38.10	0.016	0.41	−0.016	−0.41	0.008	0.20
5.00	127.0	6.00	152.4	+0.090	+2.90	> 0.375–0.875	9.52–22.22	0.012	0.30	−0.012	−0.30	0.006	0.15
						> 0.875–1.500	22.22–38.10	0.020	0.51	−0.020	−0.51	0.010	0.25

[a] Thickness measurements shall be taken at the edge of the bar where the flat surfaces intersect the rounded edge.
[b] This tolerance represents the maximum amount by which the thickness of the center of the bar may be less than the thickness at the edges. Thickness at the center may never exceed the thickness at the edges.
[c] Maximum difference in thickness between the two edges of each bar.
Source: Reproduced from SAE Handbook, Vol. I, 1981.

DISK SPRINGS (BELLEVILLE SPRINGS)

The relation between the load F and the axial deflection y of each disk is given by the equation (Fig. 20–2)

$$F = \frac{4Ey}{(1 - v^2)M d_o^2}\left[(h - y)\left(h - \frac{y}{2}\right)t + t^3\right] \qquad (20–19)$$

where
t = thickness, m (mm)
h = height, m (mm)
M is a constant from Fig. 20–3

Particular	Formula

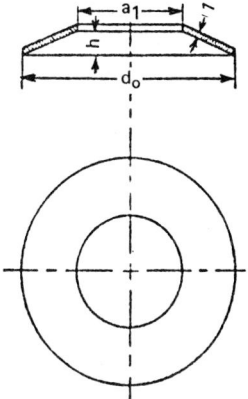

FIGURE 20–2 Disk spring.

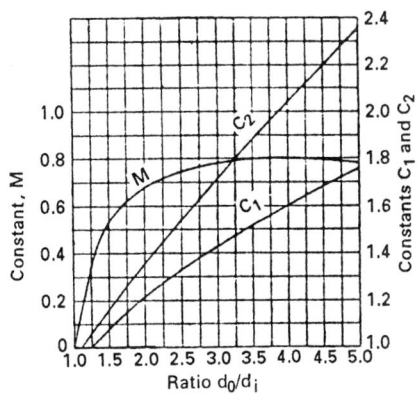

FIGURE 20–3 Constants C_1, C_2, and M for a disk spring. (*V. L. Maleev and J. B. Hartman,* Machine Design, *International Textbook Company, Scranton, Pennsylvania, 1954.*)

The maximum stress induced at the inner edge

$$\sigma_i = \frac{4Ey}{(1-v^2)d_o^2}\left[C_1\left(h-\frac{y}{2}\right) + C_2t\right] \qquad (20\text{–}20)$$

The maximum stress induced at the outer edge

$$\sigma_o = \frac{4Ey}{(1-v^2)d_o^2}\left[C_1\left(h-\frac{y}{2}\right) - C_2t\right] \qquad (20\text{–}21)$$

where C_1 and C_2 are constants taken from Fig. 20–3

For spring design stresses

Refer to Table 20–8

TABLE 20–4
Width and thickness of leaf springs for motor vehicle suspension—SAE J1123a

Width, mm		Thickness, mm					
40.0	75.0	5.00	7.10	10.00	14.00	20.00	28.00
45.0	90.0	5.30	7.50	10.60	15.00	21.20	30.00
50.0	100.0	5.60	8.00	11.20	16.00	22.40	31.50
56.0	125.0	6.00	8.50	11.80	17.00	23.60	33.50
63.0	150.0	6.30	9.00	12.50	18.00	25.00	35.50
		6.70	9.50	13.20	19.00	25.50	37.50

Source: Reproduced from *SAE Handbook*, Vol. I, 1981.

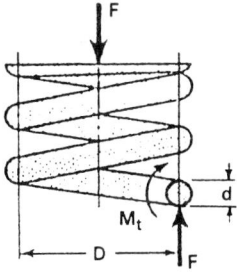

FIGURE 20–4 Helical spring under axial load.

HELICAL SPRINGS (FIG. 20–4) [5]

The more accurate formula for shear stress

$$\tau = \frac{8FDk}{\pi d^3} \qquad (20\text{–}22)$$

Particular	Formula

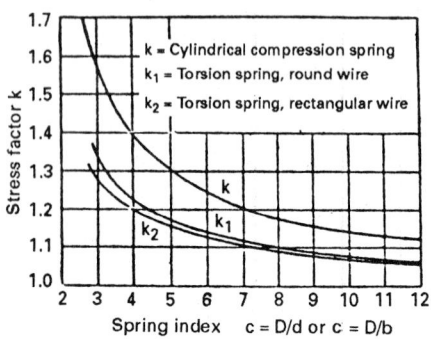

FIGURE 20–5 Stress factors for a helical spring. (*V. L. Maleev and J. B. Hartman,* Machine Design, *International Textbook Company, Scranton, Pennsylvania, 1954.*)

Refer to Fig. 20–5 for values of k.

Particular	Formula
The stress factor or Wahl factor [6] k to be used in Eq. (20–2)	$$k = \frac{4c - 1}{4c - 4} + \frac{0.615}{c} \qquad (20\text{–}23)$$
The spring index	$$c = \frac{D}{d} \qquad (20\text{–}24)$$
The value of stress factor k may be approximated very closely (between $c = 2$ and 12) by the relation	$$k = \frac{2}{c^{0.25}} = 2\left(\frac{d}{D}\right)^{0.25} \qquad (20\text{–}25)$$
Size factor	$$k_{sz} = \frac{1}{e_{sz}} = \frac{d^{0.25}}{0.335} \quad \textbf{SI, } d \text{ in mm} \qquad (20\text{–}26\text{a})$$
	$$= \frac{d^{0.25}}{0.84} \quad \textbf{US Customary System units} \qquad (20\text{–}26\text{b})$$
	where d in in
	$$= \frac{d^{0.25}}{1.89} \quad \textbf{SI, } d \text{ in mm} \qquad (20\text{–}26\text{c})$$
The shear stress taking into consideration k from Eq. (20–25)	$$\tau = \frac{16FD^{0.75}}{\pi d^{2.75}} = \frac{5.1Fc^{0.75}}{d^2} \qquad (20\text{–}27)$$
The angular deflection	$$\theta = \frac{16FDl}{\pi d^4 G} \qquad (20\text{–}28)$$
The length of the spring wire	$$l = i\pi D \text{ (approx.)} \qquad (20\text{–}28\text{a})$$
The axial deflection of the whole spring	$$y = \frac{8FD^3 i}{d^4 G} = \frac{\pi i \tau D^2}{kdG} = \frac{\pi i \tau D^{2.25}}{2d^{1.25}G} \qquad (20\text{–}29)$$

Particular	Formula

SPRING SCALE (FIG. 20–6)

Force to compress the spring 1 m (mm) (stiffness of spring)

$$F_o = \frac{F}{y} = \frac{d^4 G}{8iD^3} = \frac{GD}{8ic^4} = \frac{Gd}{8ic^3} \qquad (20\text{–}30)$$

The total deflection (Fig. 20–6)

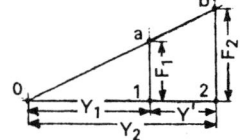

$$y_2 = \frac{F_2}{F_o} = \frac{y' F_2}{F_2 - F_1} \qquad (20\text{–}31)$$

FIGURE 20–6   Relation between loads and deflection in a helical spring.

RESILIENCE

The resilience U of a spring is equal to energy absorbed

$$U = \frac{Fy}{2} = \frac{4F^2 D^3 i}{d^4 G} = \frac{\pi^2 d^2 i D \tau^2}{16 k^2 G} \qquad (20\text{–}32a)$$

$$= \frac{\pi^2 d^{1.5} D^{1.5} i \tau^2}{64 G} = \frac{1.55 d^3 c^{1.5} i \tau^2}{G} \qquad (20\text{–}32b)$$

$$= \frac{V \tau^2}{4 k^2 G} = \frac{V \tau^2}{16 G} \left(\frac{D}{d}\right)^{0.5} = \frac{V \tau c^{0.5}}{16 G} = \frac{y^2 d^4 G}{16 i D^3} \qquad (20\text{–}32c)$$

Volume of the spring

$$V = \pi Di(\tfrac{1}{4} \pi d^2) \qquad (20\text{–}33)$$

RECTANGULAR SECTION SPRINGS (FIG. 20–7a) [5]

Shear stress

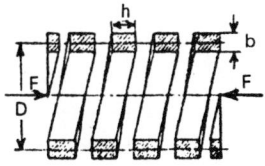

FIGURE 20–7(a)   Spring with rectangular section.

$$\tau' = \frac{kFD(1.5h + 0.9b)}{b^2 h^2} = \frac{FD^{0.75}(3h + 1.8b)}{b^{1.75} h^2} \qquad (20\text{–}34a)$$

$$= \frac{kFD(1.5 + 0.9m))}{m^2 h^3} = \frac{FD^{0.75}(3 + 1.8m)}{m^{1.75} h^2} \qquad (20\text{–}34b)$$

where $m = \dfrac{b}{h}$

The uncorrected shear stress for a rectangular section spring

$$\tau' = \frac{FD}{k_l b h^2} \qquad (20\text{–}35)$$

where $k_1 =$ factor depending on $\frac{b}{h} = m$, which is given in Table 20–9

The stress factor

$$k = \frac{4c - 1}{4c - 4} + \frac{0.615}{c} \qquad (20\text{–}36)$$

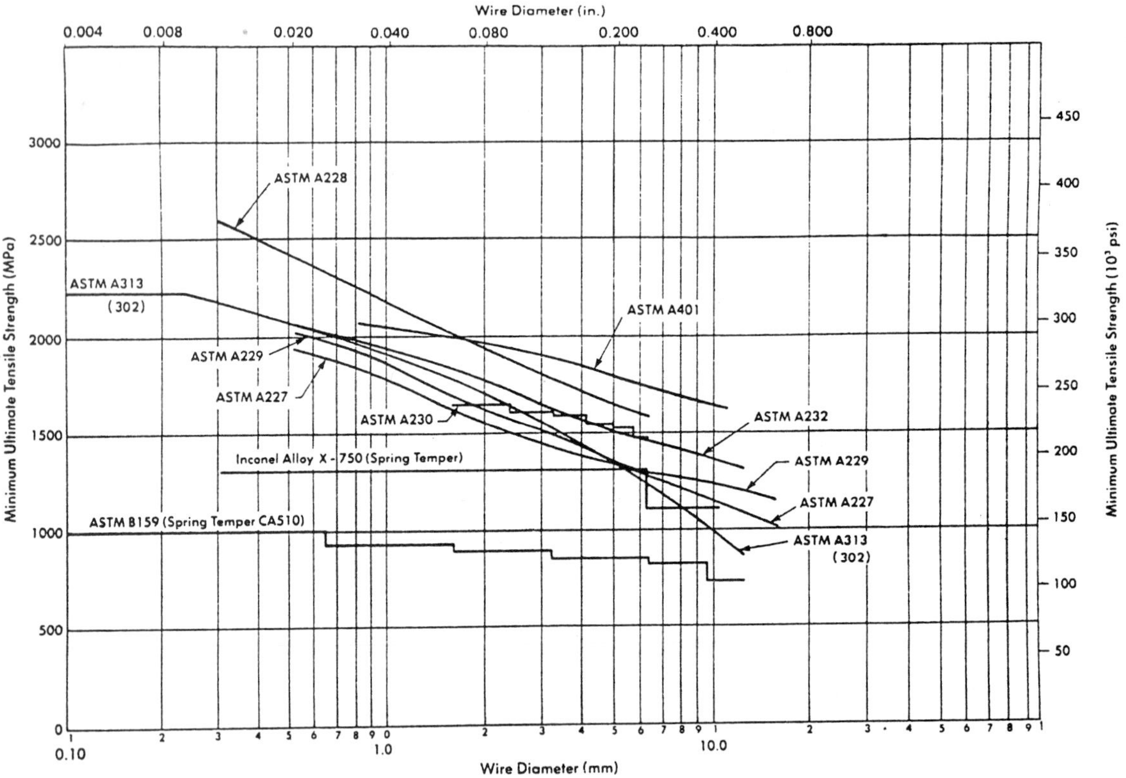

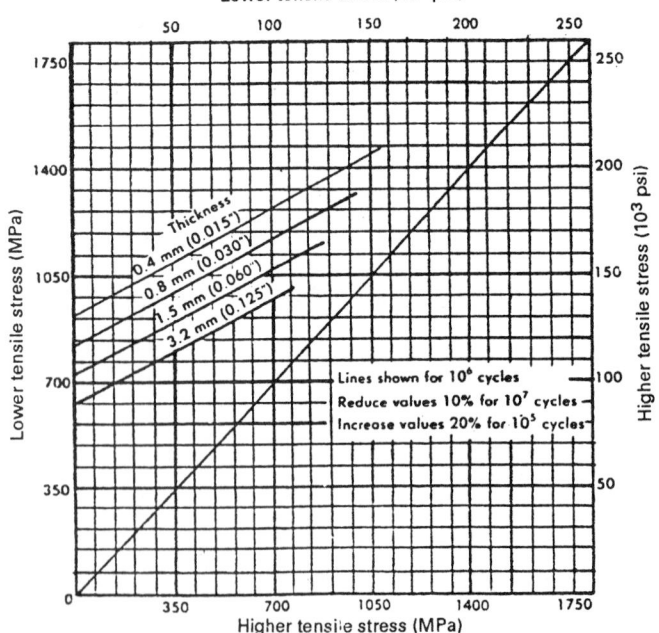

FIGURE 20–7(b) Minimum tensile strengths of spring wire. (*Associated Spring, Barnes Group Inc., Bristol, Connecticut.*)

FIGURE 20–7(c) Modified Goodman diagram for Belleville washers; for carbon and alloy steels at 47 to 49 R_c with set removed, but not shotpeened. (*Associated Spring, Barnes Group Inc., Bristol, Connecticut.*)

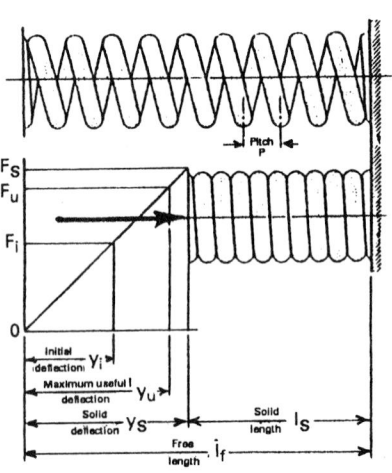

FIGURE 20–7(d) Deflection for helical compression spring at various loads. (*Richard M. Phelan*, Fundamentals of Mechanical Engineering Design, *New Delhi, 1975, courtesy of Tata-McGraw-Hill.*)

TABLE 20–5
Width of leaf springs for motor vehicle suspension—SAE J510C

in	mm	in	mm	in	mm
1.75	44.4	2.50	63.5	4.00	101.6
2.00	50.8	3.00	76.2	5.00	127.0
2.25	57.2	3.50	88.9	6.00	152.4

Source: Reproduced from *SAE Handbook*, Vol. I, 1981.

TABLE 20–6
Standard sections of steel plates for laminated springs (railway rolling stocks)

Width, mm	50	63	75	90	100	115	120	125	140	150
Thickness, mm	10, 13	6, 8, 10, 11, 13	6, 8, 10 11, 13, 16	6, 8, 10 11, 16, 19	8, 10, 11 13, 16, 19	10, 11, 13 16, 19	16, 19	10, 13 16	11, 13	11, 13 16

TABLE 20–7
Constant in Eqs. (20–13) to (20–18)

Constant	Cantilever	Simply supported
α	$\dfrac{12}{2+r}$	$\dfrac{3}{2+r}$
β	$\dfrac{12}{2+r}$	$\dfrac{3}{4(2+r)}$
α'	6	1.5
β'	2	$\frac{1}{8}$

Particular	Formula	
The value of stress k may be approximated very closely by the relation	$k = \dfrac{2}{c^{0.25}}$	(20–37)
The spring index	$c = \dfrac{D}{b} \quad \text{if } b < h$	(10–38)
	$= \dfrac{D}{h} \quad \text{if } h < b$	
	where b = breadth of spring wire, m (mm) h = thickness of spring wire, m (mm)	
The deflection	$y = \dfrac{2.83iFD^3(b^2 + h^2)}{b^3 h^3 G}$	(20–39)

Particular	Formula
The deflection for an uncorrected spring of rectangular cross section	$y = \dfrac{iFD^3}{k_2 bh^3 G}$ for $h < b$ and $m > 8$ (20–40)
Force required to compress the spring by one meter (milimeter) (i.e., the spring rate)	$F_o = \dfrac{m^3 h^4 G}{2.83iD^3(1 + m^2)}$ (20–41)
The spring rate for an uncorrected rectangular section spring	$F_o = \dfrac{4Gb^3 h}{k_1 \pi D^3 i}$ (20–42) Refer to Table 20–9 for k_1.

SQUARE SECTION SPRING

The shear stress, for $m = 1$

$$\tau' = \frac{2.4kFD}{h^3} = \frac{4.8FD^{0.75}}{h^{2.75}} \tag{20–43}$$

The deflection

$$y = \frac{5.66iFD^3}{h^4 G} \tag{20–44}$$

The approximate equivalent rectangular dimension of a rectangular cross section wire spring to restrict the solid length, which is equivalent to spring of round-wire cross section

$$h = \frac{2d}{1 + (b/h)} \tag{20–44a}$$

where $d=$ diameter of round wire

The larger dimension of a keystone shape of rectangular wire after coiling

$$h_1 = h \, \frac{C + 0.5}{C} \tag{20–44b}$$

where

$h_1 =$ wider end of keystone section
$h \;\; =$ original, smaller dimension of rectangular section

The estimated solid height or length of a uniformly tapered, but not telescoping, spring with squared and ground ends made from round wire

$$l_s = i(d^2 - u^2)^{1/2} + 2d \tag{20–44c}$$

where $u =$ outside diameter of large end minus outside diameter of small end divided by $2i$

The increase in coil diameter due to compression of a helical spring

$$D_{\text{o at solid}} = \left(\frac{D^2 + p^2 - d^2}{\pi^2 + d} \right)^{1/2} \tag{20–44d}$$

The size coefficient for sections above 12.5 mm in section for round wires

$$e_{sz} = 0.86 + \frac{0.0018}{d} \quad \text{for steel } \textbf{SI} \tag{20–45a}$$

$$= 0.986 + \frac{0.0001}{d} \quad \text{for monel metal} \tag{20–45b}$$

Particular	Formula
	where d in m
	$= 0.86 + \dfrac{0.07}{d}$ for steel
	US Customary System units (20–45c)
	where d in in
	$= 0.986 + \dfrac{0.0043}{d}$ for monel metal (20–45d)
	where d in in
	$= 0.86 + \dfrac{1.8}{d}$ for steel **SI** (20–45e)
	$= 0.986 + \dfrac{0.1}{d}$ for monel metal (20–45f)
	where d in mm
The general expression for size factor	$k_{sz} = 4.66 h^{0.35}$ **SI**, h in m (20–46a)
	$= 1.27 h^{0.35}$ **US Customary System units**
	where h in in (20–46b)
	$= 0.415 h^{0.35}$ **SI**, h in mm (20–46c)
Wire diameter	$d = \sqrt[3]{\dfrac{8kFD}{\pi \sigma_d e_{sz}}}$ (20–47)

SELECTION OF MATERIALS AND STRESSES FOR SPRINGS

For materials for springs [7]

Refer to Tables 20–8 and 20–10 and Figs. 20–7*b* and *c*.

The torsional yield strength

$0.35\sigma_{Sut} \leq \tau_{Sy} \leq 0.52\sigma_{Sut}$ for steels (20–47a)

The maximum allowable torsional stress for static applications according to Joerres [8, 9, 11]

$$\tau_{Sy} = \tau_a = \begin{cases} 0.45\sigma_{Sut} \text{ cold-drawn carbon steel} \\ 0.50\sigma_{Sut} \text{ hardened and tempered carbon} \\ \qquad \text{and low-alloy steel} \\ 0.35\sigma_{Sut} \text{ austenitic stainless steel and} \\ \qquad \text{nonferrous alloys} \quad (20\text{–}47b) \end{cases}$$

where τ_{Sy} = torsional yield strength, MPa (psi)

The maximum allowable torsional stress according to Shigley and Mischke [9]

$\tau_{Sy} = \tau_a = 0.56\sigma_{Sut}$ (20–47c)

The shear endurance limit according to Zimmerli [10]

$\tau_{Sf} = 310$ MPa (45 kpsi) for unpeened springs (20–47d)

$= 465$ MPa (67.5 kpsi) for peened springs (20–47e)

TABLE 20–8
Spring design stress, σ_d, MPa (kpsi)

Wire diameter, mm	Severe service		Average service		Light	
	MPa	kpsi	MPa	kpsi	MPa	kpsi
≤ 2.15	413.8	60	517.3	75	641.4	93
2.15–4.70	379.0	55	476.6	69	585.4	85
4.70–8.10	331.0	48	413.8	60	510.0	74
8.10–13.45	289.3	42	358.4	52	448.2	65
13.45–24.65	248.1	36	310.4	45	385.9	56
24.65–38.10	220.6	32	275.6	40	344.7	50

TABLE 20–9
Factors for helical springs with wires of rectangular cross section

Ratio $b/h = m$	1	1.2	1.5	2.0	2.5	3	5	10	∞
Factor k	0.416	0.438	0.462	0.492	0.516	0.534	0.582	0.624	0.666
Factor k_2	0.180	0.212	0.250	0.292	0.317	0.335	0.371	0.398	0.424

Particular	Formula
The torsional modulus of rupture	$\tau_{Su} = 0.67\sigma_{Sut}$ (20–47f)
The weight of the active coil of a helical spring	$W = \dfrac{\pi^2 d^2 D i \gamma}{4}$ (20–47g) where γ = weight of coil of helical spring per unit volume
For free-length tolerances, coil diameter tolerances, and load tolerances of helical compression springs	Refer to Tables 20–11 to 20–13.

DESIGN OF HELICAL COMPRESSION SPRINGS

Design stress

The size factor

$$k_{sz} = d^{0.35}/0.335 \quad \textbf{SI} \quad d \text{ in m} \quad (20\text{–}48a)$$
$$= d^{0.25}/0.84 \quad \textbf{US Customary System units}$$
$$\qquad\qquad\qquad\qquad d \text{ in in} \quad (20\text{–}48b)$$
$$= d^{0.25}/1.89 \quad \textbf{SI} \quad d \text{ in mm} \quad (20\text{–}48c)$$

The design stress

$$\sigma_{ds} = \frac{\sigma_e}{n_a k_{sz}} = \frac{0.335\sigma_e}{n_a d^{0.25}} \quad \textbf{SI} \quad (20\text{–}49a)$$

where σ_e in MPa and d in m

$$= \frac{\sigma_e}{n_a k_{sz}} = \frac{0.84\sigma_e}{n_a d^{0.25}} \quad \textbf{US Customary System units}$$

$$(20\text{–}49b)$$

TABLE 20-10

Chemical composition and mechanical properties of spring materials

Material	Analysis Element	%	Tensile properties Ultimate strength MPa	kpsi	Elastic limit GPa	kpsi	Modulus of elasticity, E GPa	Mpsi	Rockwell hardness	Torsional properties of wire Ultimate strength MPa	kpsi	Elastic limit GPa	kpsi	Modulus in torsion, G GPa	Mpsi	Chief uses
Flat Cold-rolled Spring Steel																
Watch spring steel	C Mn	1.10-1.19 0.15-0.25	2274-2412	330-350	2.14-2.28	310-330	220	32	C50-55	Not used	Not used	Not used	Not used	Not used	Not used	Main springs for watches and similar uses
Clock spring steel AS 100 SAE 1095	C Mn	0.90-1.05 0.20-0.50	1240-2343	180-340	1.03-2.14	150-310	207	30	C40-52	Not used	Not used	Not used	Not used	Not used	Not used	Clock and motor springs, miscellaneous flat springs for high stress
Flat spring steel AS 101 SAE 1074	C Mn	0.65-0.80 0.50-0.90	1103-2206	160-320	0.86-1.93	125-280	207	30	Annealed, B70-85 tempered C38-50	Not used	Not used	Not used	Not used	Not used	Not used	Miscellaneous flat springs
Carbon Steel Wires																
High-carbon wire AS 8	C Mn	0.85-0.95 0.25-0.60	1382-1725	200-250	1.10-1.45	160-210	207	30	C44-48	1103 1377	160-200	0.76 1.03	110-150	79	11.5	High-grade helical springs or wire forms
Oil-tempered wire (ASTM A229-41) AS10	C Mn	0.60-0.70 0.60-0.90	1068-2059	155-300	0.83-1.73	120-250	200	29	C42-46	794 1377	115-200	0.55 0.90	80-130	79	11.5	General spring use
Music wire (ASTM A228-47) AS 5	C Mn	0.70-1.00 0.30-0.60	1725-3790	250-500	1.03-2.41	150-350	207	30		1034 2069	150-300	0.62 1.24	90-180	79 82	11.5 12.0 depending on size	Miscellaneous small springs of various types—high quality
Hard-drawn spring wire (ASTM A227-47) AS 20	C Mn	0.60-0.70 0.90-1.20	1034-2068	150-300	0.69-1.38	100-200	200	29		828 1515	120-220	0.51 0.90	75-130	79	11.5	Same uses as music wire but lower-quality wire
Hot-rolled Special Steel																
Hot-rolled bars SAE 1095, ASTM A14-42	C Mn	0.90-1.05 0.25-0.50	1206-1377	175-200	0.73-0.97	105-140	196	28.5	C40-46	760 965	110-140	0.51 0.76	75 110	72	10.5	Hot-rolled heavy coil or flat springs
Alloy and Stainless Spring Materials																
Chrome-vanadium alloy steel (SAE 6150) AS 32	C Mn Cr V	0.45-0.55 0.50-0.80 0.80-1.10 0.15-0.18	1377 1725	200-250	1.24 1.58	180-230	207	30	C42-48	965 1206	140-175	0.69 0.90	100-130	79	11.5	Cold-rolled or drawn; special applications
Silico-manganese alloy steel (SAE 9260)	C Mn Si	0.55-0.65 0.60-0.90 1.80-2.20	About the same as chrome vanadium		About the same as chrome vanadium					About the same as chrome vanadium						Used as a lower-cost material in place of chrome vanadium
Type 18-8 stainless (Type 302), SAE 30915	Ni C C Mn Si	17-20 7-10 0.08-0.15 2 max 0.30-0.75	1103 2275	160-330	0.41 1.79	60-260	193	28	C35-45	828 1653	120-240	0.31 0.97	45-140	69	10	Best corrosion resistance; fair temperature resistance
Cutlery-type stainless (Type 420)	Cr C	12-14 0.25-0.40	1171 1725	170-250	0.90 1.38	130-200	193	28	C42-47	828 1240	120-180	0.55 0.83	80-120	76	11	Resists corrosion when polished; good temperature resistance

TABLE 20-10
Chemical composition and mechanical properties of spring materials (*Cont.*)

Material	Analysis	Tensile: Ult. strength MPa	kpsi	Tensile: Elastic limit GPa	kpsi	Modulus of elasticity E GPa	Mpsi	(Hardness)	Torsional: Ult. strength MPa	kpsi	Torsional: Elastic limit GPa	kpsi	Modulus in torsion G GPa	Mpsi	Chief uses
						Nonferrous Spring Materials									
Spring brass AS 55 AS 155	Cu 64-74; Zn balance	691 897	100-130	0.27 0.41	40-60	107	15	B90	308 622	45-90	0.21 0.41	30-60	38	5.5	For electrical conductivity at low stresses; for corrosion resistance
Nickel silver	Cu 56; Zn 25; Ni 18	897 1034	130-150	0.55 0.76	80-110	110	16	B95-100	588 691	85-100	0.41 0.48	60-70	38	5.5	Used for its color; corrosion resistance
Phosphor bronze AS 60 AS 160	Cu 91-93; Sn 7-9 or Cu 94-96; Sn 4-6	691 102	100-150	0.41 0.76	60-110	103	15	B90-100	554 725	80-105	0.35 0.59	50-85	43	6.25	Used for corrosion resistance and electrical conductivity
						Nonferrous Spring Materials									
Silicon bronze (made under various trade names) AS 46 AS 146	Si 2-3; Sn or Mn small amounts; Cu balance	Properties similar to those of phosphor bronze							Properties similar to those of phosphor bronze						Used as substitute for phosphor bronze
Monel AS 40 AS 140	Ni 64; Cu 26; Mn 2.5; Fe 2.25	691 964	100-140	0.55 0.83	80-120	179	26	C23-28	519 760	75-110	0.31 0.48	45-70	65	9.5	Resists corrosion; moderate stresses to 204.5°C
Inconel AS 40 AS140	Ni 80; Cr 14; Fe Balance	965 1206	140-175	0.76 0.93	110-135	213	31	C30-40	651 828	95-120	0.38 0.55	55-80	76	11	Resists corrosion; high stresses to 343°C
K-Monel AS 40 AS 140	Ni 66; Cr 29; Al 2.75; Fe 0.90	1103 1241	160-180	0.79 1.00	115-145	179	26	C33-40	725 862	105-125	0.45 0.58	65-85	65	9.5	Resists corrosion; high stresses to 232°C
Z-nickel	Ni 98; Cu, Mn, Fe, Si small amounts	1241 1583	180-230	0.90 1.17	130-170	207	30	C36-46	828 1034	120-150	0.41 0.68	60-90	76	11	Resists corrosion; high stresses to 288°C
Beryllium-copper AS 45 AS 145	Cu 98; Be 2	1103 1377	160-200	0.69 1.03	100-150	110 127	16-18.5 Subject to heat treatment	C35-42	691 897	100-130	0.45 0.66	65-95	41 48 Subject to heat treatment	6-7	Corrosion resistance like copper; high physical properties for electrical work; low hysteresis

Note: The property values given in this table do not specify the minimum properties.
Source: Handbook of Mechanical Spring Design, courtesy Associated Spring, Barnes Group Inc., Bristol, Connecticut.

20.16

TABLE 20-11

Free-length tolerances of squared and ground helical compression springs[a]

Number of active coils per mm (in)	Tolerances: ±mm/mm (in/in) of free length						
	Spring index (D/d)						
	4	6	8	10	12	14	16
0.02 (0.5)	0.010	0.011	0.012	0.013	0.015	0.016	0.016
0.04 (1)	0.011	0.013	0.015	0.016	0.017	0.018	0.019
0.08 (2)	0.013	0.015	0.017	0.019	0.020	0.022	0.023
0.2 (4)	0.016	0.018	0.021	0.023	0.024	0.026	0.027
0.3 (8)	0.019	0.022	0.024	0.026	0.028	0.030	0.032
0.5 (12)	0.021	0.024	0.027	0.030	0.032	0.034	0.036
0.6 (16)	0.022	0.026	0.029	0.032	0.034	0.036	0.038
0.8 (20)	0.023	0.027	0.031	0.034	0.036	0.038	0.040

[a] For springs less than 12.7 mm (0.500″) long, use the tolerances for 12.7 mm (0.500″). For closed ends not ground, multiply above values by 1.7.
Source: Associated Spring, Barnes Group Inc., Bristol, Connecticut.

TABLE 20–12

Coil diameter tolerances of helical compression and extension springs

Wire diameter, mm (in)	Tolerances: ±mm (in)						
	Spring index (D/d)						
	4	6	8	10	12	14	16
0.38 (0.015)	0.05 (0.002)	0.05 (0.002)	0.08 (0.003)	0.10 (0.004)	0.13 (0.005)	0.15 (0.006)	0.18 (0.007)
0.58 (0.023)	0.05 (0.002)	0.08 (0.003)	0.10 (0.004)	0.15 (0.006)	0.18 (0.007)	0.20 (0.008)	0.25 (0.010)
0.89 (0.035)	0.05 (0.002)	0.10 (0.004)	0.15 (0.006)	0.18 (0.007)	0.23 (0.009)	0.28 (0.011)	0.33 (0.013)
1.30 (0.051)	0.08 (0.003)	0.13 (0.005)	0.18 (0.007)	0.25 (0.010)	0.30 (0.012)	0.38 (0.015)	0.43 (0.017)
1.93 (0.076)	0.10 (0.004)	0.18 (0.007)	0.25 (0.010)	0.33 (0.013)	0.41 (0.016)	0.48 (0.019)	0.53 (0.021)
2.90 (0.114)	0.15 (0.006)	0.23 (0.009)	0.33 (0.013)	0.46 (0.018)	0.53 (0.021)	0.64 (0.025)	0.74 (0.029)
4.34 (0.171)	0.20 (0.008)	0.30 (0.012)	0.43 (0.017)	0.58 (0.023)	0.71 (0.028)	0.84 (0.033)	0.97 (0.038)
6.35 (0.250)	0.28 (0.011)	0.38 (0.015)	0.53 (0.021)	0.71 (0.028)	0.90 (0.035)	1.07 (0.042)	1.24 (0.049)
9.53 (0.375)	0.41 (0.016)	0.51 (0.020)	0.66 (0.026)	0.94 (0.037)	1.17 (0.046)	1.37 (0.054)	1.63 (0.064)
12.70 (0.500)	0.53 (0.021)	0.76 (0.030)	1.02 (0.040)	1.57 (0.062)	2.03 (0.080)	2.54 (0.100)	3.18 (0.125)

Source: Associated Spring, Barnes Group Inc., Bristol, Connecticut.

TABLE 20–13
Load tolerances of helical compression springs

Length tolerance ± mm (in)	Tolerances: ±% of load, start with tolerance from Table 20–11 multiplied by L_F — Deflection from free length to load, mm (in)														
	1.27 (0.050)	2.54 (0.100)	3.81 (0.150)	5.08 (0.200)	6.35 (0.250)	7.62 (0.300)	10.2 (0.400)	12.7 (0.500)	19.1 (0.750)	25.4 (1.00)	38.1 (1.50)	50.8 (2.00)	76.2 (3.00)	102 (4.00)	152 (6.00)
0.13 (0.005)	12.	7.	6.	5.	—	—	—	—	—	—	—	—	—	—	—
0.25 (0.010)	—	12.	8.5	7.	6.5	5.5	5.	—	—	—	—	—	—	—	—
0.51 (0.020)	—	22.	15.5	12.	10.	8.5	7.	6.	5.	—	—	—	—	—	—
0.76 (0.030)	—	—	22.	17.	14.	12.	9.5	8.	6.	5.	—	—	—	—	—
1.0 (0.040)	—	—	—	22.	18.	15.5	12.	10.	7.5	6.	5.	—	—	—	—
1.3 (0.050)	—	—	—	—	22.	19.	14.5	12.	9.	7.	5.5	—	—	—	—
1.5 (0.060)	—	—	—	—	25.	22.	17.	14.	10.	8.	6.	5.	—	—	—
1.8 (0.070)	—	—	—	—	—	25.	19.5	16.	11.	9.	6.5	5.5	—	—	—
2.0 (0.080)	—	—	—	—	—	—	22.	18.	12.5	10.	7.5	6.	5.	—	—
2.3 (0.090)	—	—	—	—	—	—	25.	20.	14.	11.	8.	6.	5.	—	—
2.5 (0.100)	—	—	—	—	—	—	—	22.	15.5	12.	8.5	7.	5.5	—	—
5.1 (0.200)	—	—	—	—	—	—	—	—	—	22.	15.5	12.	8.5	7.	5.5
7.6 (0.300)	—	—	—	—	—	—	—	—	—	—	22.	17.	12.	9.5	7.
10.2 (0.400)	—	—	—	—	—	—	—	—	—	—	—	21.	15.	12.	8.5
12.7 (0.500)	—	—	—	—	—	—	—	—	—	—	—	25.	18.5	14.5	10.5

First load test at not less than 15% of available deflection; final load test at not more than 85% of available deflection.
Source: Associated Spring, Barnes Group Inc., Bristol, Connecticut.

TABLE 20–14
Equations for springs with different types of ends (2, 3)

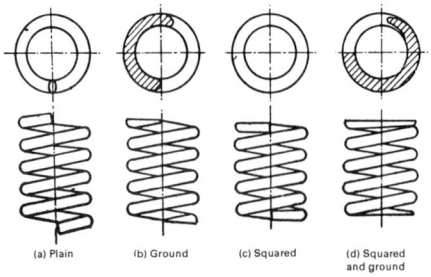

(a) Plain   (b) Ground   (c) Squared   (d) Squared and ground

Particular				
Active coils, i	i'	$i' - \frac{1}{2}$	$i' - 2$	$i' - 2$
Total coils, i'	$\dfrac{l_o - d}{p}$	$\dfrac{l_o}{p}$	$\dfrac{l_o - 3d}{p}$	$\dfrac{l_o - 2d}{p} + 2$
Free length, l_o or l_f	$ip + d$	ip	$ip + 3d$	$ip + 2d$
Pitch, p	$\dfrac{l_o - d}{i'}$	$\dfrac{l_o}{i'}$	$\dfrac{l_o - 3d}{i'}$	$\dfrac{l_o - 2d}{i'}$
Solid height, h	$d(i' + 1)$	$d(i' + \frac{1}{2})$	$d(i' + 1)$	$i'd$

Source: K. Lingaiah and B. R. Narayana Iyengar, *Machine Design Data Handbook*, Vol. I, Suma Publishers, Bangalore, India, 1986, and K. Lingaiah, *Machine Design Data Handbook*, Vol. II, Suma Publishers, Bangalore, India, 1986.

Particular	Formula

TABLE 20–15
Curvature factor k_c

c	3	4	6	7	8	9	10
k_c	1.35	1.25	1.15	1.13	1.11	1.1	1.09

where σ_e in psi and d in in

$$= \frac{\sigma_c}{n_a k_{sz}} = \frac{1.89\sigma_e}{n_a d^{0.25}} \quad \textbf{Customary Metric} \qquad (20\text{–}49c)$$

where σ_e in kgf/mm^2 and d in mm

where n_a = actual factor of safety or reliability factor

The actual factor of safety or reliability factor

$$n_a = \frac{F(\text{compressed})}{F(\text{working})} \qquad (20\text{–}50a)$$

$$= \frac{\text{free length} - \text{fully compressed length}}{\text{free length} - \text{working length}}$$

$$= \frac{y + a}{y} \qquad (20\text{–}50b)$$

where y is deflection under working load, m (mm); a is the clearance which is to be added when determining the free length of the spring and is made equal to 25 percent of the working deflection

Generally n_a is chosen as 1.25

The wire diameter for static loading

$$d = 1.445 \left(\frac{6n_a F}{\sigma_e}\right)^{0.4} D^{0.3} = 2.945 \left(\frac{n_a F}{\sigma_e}\right)^{0.4} D^{0.3} \ \textbf{SI}$$

$$(20\text{–}51a)$$

where F in N, σ_e in MPa, D in m, and d in m

$$= 0.724 \left(\frac{6n_a F}{\sigma_e}\right)^{0.4} D^{0.3} = 1.48 \left(\frac{n_a F}{\sigma_e}\right)^{0.4} D^{0.3} \ \textbf{Metric}$$

$$(20\text{–}51b)$$

where F in kgf, σ_e in kgf/mm^2, D in mm, and d in mm

$$= \left(\frac{6n_a F}{\sigma_e}\right)^{0.4} D^{0.3} = 2.05 \left(\frac{n_a F}{\sigma_e}\right)^{0.4} D^{0.3}$$

$$\textbf{US Customary units} \qquad (20\text{–}51c)$$

where F in lbf, σ_e in psi, D in in, and d in in

The wire diameter where there is no space limitation ($D = cd$)

$$d = 4.64 \left(\frac{n_a F}{\sigma_e}\right)^{0.57} c^{0.43} \ \textbf{SI}, \ \ d \text{ in m} \qquad (20\text{–}51d)$$

F in N; σ_e in Pa

$$= \left(\frac{6n_a F}{\sigma_e}\right)^{0.57} c^{0.43} \ \textbf{US Customary units} \qquad (20\text{–}51e)$$

where d in in; F in lbf; σ in psi

Particular	Formula

$$= 1.77 \left(\frac{n_a F}{\sigma_e} \right)^{0.57} c^{0.43}$$

Customary Metric units   (20–51f)

where d in mm; F in kgf; σ_e in kgf/mm^2

Final dimensions (Fig. 20–7d)

Particular	Formula
The number of active coils	$i = \dfrac{y d^4 G}{8 F D^3} = \dfrac{y d G}{8 F c^3} = \dfrac{k y d G}{\pi \tau D^2}$ (20–52)
The minimum free length of the spring	$l_f \geq (i + n)d + y + a$ (20–53)

where
a = clearance, m (mm)
n = 2 if ends are bent before grinding
= 1 if ends are either ground or bent
= 0 if ends are neither ground nor bent

Particular	Formula
Outside diameter of coil of helical spring	$D_o = D + d$ (20–53a)
Solid length (or height) of helical spring	$l_s = i_t d$ (20–53b)
Pitch of spring	$p = \dfrac{y_s}{i} + d$ (20–53c)
Free length of helical spring l_f or l_o	$l_f - l_s + y_s$ (20–53d)
Maximum working length of helical spring	$l_{max} = l_f - y_{max}$ (20–53e)
Minimum working length of helical spring	$l_{min} = l_f - y_{min}$ (20–53f)

where i_t = total number of coils in the spring

Particular	Formula
Springs with different types of ends [1, 2, 3]	Refer to Table 20–14.

STABILITY OF HELICAL SPRINGS

Particular	Formula
The critical axial load that can cause buckling	$F_{cr} = F_o K_l l_f$ (20–54)

where K_l is factor taken from Fig. 20–8

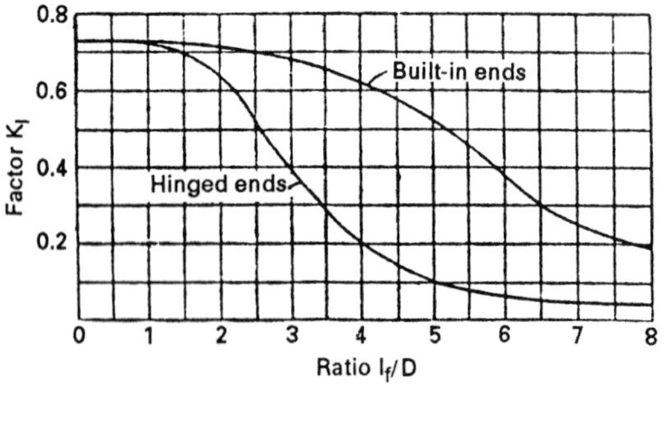

FIGURE 20–8 Buckling factor for helical compression springs. (*V. L. Maleev and J. B. Hartman,* Machine Design, *International Textbook Company, Scranton, Pennsylvania, 1954.*)

Particular	Formula
The equivalent stiffness of springs	$$(EI)_{spring} = \frac{Ed^4 l}{32iD(2+v)} \qquad (20\text{–}55)$$
The critical load on the spring	$$F_{cr} = \frac{\pi^2 Ed^4}{32(2+v)iD(l_f - y_{cr})} \qquad (20\text{–}56)$$
The critical deflection is explicitly given by	$$\left(\frac{y_{cr}}{l_f}\right)^2 - \left(\frac{y_{cr}}{l_f}\right) + \frac{\pi^2}{2}\left(\frac{1+v}{2+v}\right)\left(\frac{D}{l_f}\right)^2 = o \qquad (20\text{–}57)$$ where $l = (l_f - y_{cr})$

REPEATED LOADING (FIG. 20–9)

The variable shear stress amplitude

$$\tau_a = k_w \left(\frac{8D}{\pi d^3}\right)\left(\frac{F_{max} - F_{min}}{2}\right) \qquad (20\text{–}58)$$

where $k_w = k_\tau k_c$

Refer to Table 20–15 for k_c.

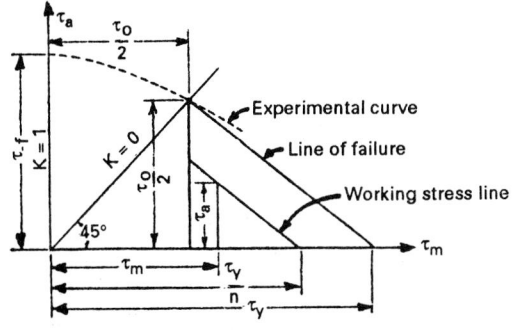

FIGURE 20–9 Cyclic stresses in spring. (*K. Lingaiah and B. R. Narayana Iyengar*, Machine Design Data Handbook, *Engineering College Cooperative Society, Bangalore, India, 1962; K. Lingaiah and B. R. Narayana Iyengar*, Machine Design Data Handbook, *Vol. I, Suma Publishers, 1986; and K. Lingaiah*, Machine Design Data Handbook, *Vol. II, Suma Publishers, Bangalore, India, 1986.*)

The mean shear stress

$$\tau_m = k_\tau \left(\frac{8D}{\pi d^3}\right)\left(\frac{F_{max} + F_{min}}{2}\right) \qquad (20\text{–}59)$$

where $k_\tau = 1 + 0.5/c$

Design equations for repeated loadings [1, 2, 3]

Method 1

The Gerber parabolic relation

$$\frac{\tau_a}{\tau_{od}} + \left(\frac{\tau_m}{\tau_{ud}}\right)^2 = 1 \qquad (20\text{–}60)$$

Particular	Formula
The Goodman straight-line relation	$\dfrac{\tau_a}{\tau_{od}} + \dfrac{\tau_m}{\tau_{ud}} = 1$ (20–61)
The Soderberg straight-line relation	$\dfrac{\tau_a}{\tau_{od}} + \dfrac{\tau_m}{\tau_{yd}} = 1$ (20–62)

Method 2

The static equivalent of cyclic load $F_m \pm F_a$	$F'_m = F_m + \dfrac{\sigma_{sd}}{\sigma_o}\, F_a$ (20–63a)
	or
	$F'_m = F_m + \dfrac{\sigma_{sd}}{\sigma_{fd}}\, F_a$ (20–63b)
The relation between σ_e and σ_f for brittle material	$\sigma_e = 2\sigma_f$ (20–64)
The static equivalent of cyclic load for brittle material	$F'_m = F_m + 2F_a$ (20–65)
The relation between F'_m, F_{max}, and F_{min}	$F'_m = \tfrac{1}{2}(3F_{max} - F_{min})$ (20–66)
The diameter of wire for static equivalent load	$d = 1.45\left[\dfrac{3n_a(3F_{max} - F_{min})}{\sigma_e}\right]^{0.4} D^{0.3}$ **SI** (20–67a)

where F and N, σ_e in MPa, D in m, and d in m

$$d = \left[\frac{3n_a(3F_{max} - F_{min})}{\sigma_e}\right]^{0.4} D^{0.3}$$

US Customary System units   (20–67b)

where F in lbf, σ_e in psi, D in in, and d in in

$$= 0.724\left[\frac{3n_a(3F_{max} - F_{min})}{\sigma_e}\right]^{0.4} D^{0.3} \quad \textbf{Metric}$$

Customary Metric units   (20–67c)

where F in kgf, σ in kgf/mm^2, D in mm, and d in mm

The wire diameter when there is no space limitation ($D = cd$)

$$d = 1.67\left[\frac{3n_a(3F_{max} - F_{min})}{\sigma_e}\right]^{0.57} c^{0.43} \quad \textbf{SI} \quad (20\text{–}68a)$$

where F in N, σ_e in MPa, and d in m

$$= \left[\frac{3n_a(3F_{max} - F_{min})}{\sigma_e}\right]^{0.57} c^{0.43}$$

US Customary units   (20–68b)

where F in lbf, σ_e in psi, and d in in

Particular	Formula

$$= 0.64 \left[\frac{3n_a(3F_{max} - F_{min})}{\sigma_e} \right]^{0.57} c^{0.43} \ \text{Metric}$$

Customary Metric units   (20–68c)

where F in kgf, σ_e in kgf/mm^2, and d in mm

CONCENTRIC SPRINGS (FIG. 20–10)

The relation between the respective loads shared by each spring, when both the springs are of the same length

$$\frac{F_1}{F_2} = \left(\frac{D_2}{D_1}\right)^3 \left(\frac{d_1}{d_2}\right)^4 \left(\frac{i_2}{i_1}\right)\left(\frac{G_1}{G_2}\right) \qquad (20\text{–}69)$$

The relation between the respective loads shared by each spring, when both are stressed to the same value

$$\frac{F_1}{F_2} = \left(\frac{D_2}{D_1}\right) \left(\frac{d_1}{d^2}\right)^3 \frac{k_1}{k_2} \qquad (20\text{–}70)$$

The approximate relation between the sizes of two concentric springs wound from round wire of the same material

$$\frac{F_1}{F_2} = \left(\frac{D_2}{D_1}\right)^{0.75} \left(\frac{d_1}{d_2}\right)^{2.5} \qquad (20\text{–}71)$$

where suffixes 1 and 2 refer, respectively, to springs 1 and 2 (Fig. 20–10)

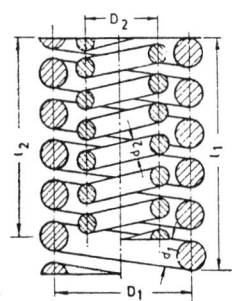

FIGURE 20–10   Concentric spring.

Total load on concentric springs

The total maximum load on the spring

$$F = F_1 + F_2 \qquad (20\text{–}72)$$

The load on the inner spring

$$F_2 = mF_1 \qquad (20\text{–}73)$$

The load on the outer spring

$$F_1 = \frac{F}{(1+m)} \qquad (20\text{–}74)$$

where $m \leq 1$ and F maximum spring load, kN (lbf)

VIBRATION OF HELICAL SPRINGS

The natural frequency of a spring when one end of the spring is at rest

$$f_n = \frac{1}{2\pi} \sqrt{\frac{2k_o g}{W}} = 0.705 \sqrt{\frac{k_o}{W}} \ \ \text{SI} \qquad (20\text{–}75)$$

where

Particular	Formula
	f_n = natural frequency, Hz W = weight of vibrating system, N k_o = scale of spring, N/m g = 9.8066 m/s^2

$$= 22.3 \left(\frac{k_0}{W}\right)^{1/2} \quad \textbf{SI} \tag{20–75a}$$

where k_o in N/mm; W in N; f_n in Hz; g = 9086.6 mm/s^2

$$= 4.42 \left(\frac{k_0}{W}\right)^{1/2} \quad \textbf{US Customary units} \tag{20–75b}$$

where k_0 in lbf/in; W in lbf; f_n in Hz, g = 32.2 ft/s^2

$$= 1.28 \left(\frac{k_0}{W}\right)^{1/2} \quad \textbf{US Customary units} \tag{20–75c}$$

where k_0 lbf/ft, W in lbf; f_n in Hz, g = 386.4 in/s^2

The natural frequency of a spring when both ends are fixed

$$f_n = \frac{1}{\pi}\sqrt{\frac{2k_o g}{W}} = 1.41 \sqrt{\frac{k_0}{W}} \quad \textbf{SI} \tag{20–76}$$

where k_0 in N/m; W in N; f_n in Hz; g = 9.8066 m/s^2

$$f_n = 44.6 \left(\frac{k_0}{W}\right)^{1/2} \quad \textbf{SI} \tag{20–76a}$$

where k_0 in N/mm; W in N; f_n in Hz; g = 9806.6 mm/s^2

$$= 2.56 \left(\frac{k_0}{W}\right)^{1/2} \quad \textbf{US Customary units} \tag{20–76b}$$

where k_0 in lbf/ft; W in lbf; f_n in Hz; g = 32.2 ft/s^2

$$= 8.84 \left(\frac{k_0}{W}\right)^{1/2} \quad \textbf{US Customary units} \tag{20–76c}$$

where k_0 in lbf/in; W in lbf; f_n in Hz; g = 386.4 in/s^2

The natural frequency for a helical compression spring one end against a flat plate and free at the other end according to Wolford and Smith [7]

$$f_n = 0.25 \left(\frac{k_0 g}{W}\right)^{1/2} \tag{20–76d}$$

Another form of equation for natural frequency of compression helical spring with both ends fixed without damping effect

$$f_n = \frac{1.12(10^3)d}{D^2 i} \left(\frac{Gg}{\gamma}\right)^{1/2} \quad \textbf{SI} \tag{20–76e}$$

where
G = shear modulus, MPa
g = 9.8006 m/s^2
   d and D in mm
f_n in Hz; γ in g/cm^3

Particular	Formula

$$= \frac{3.5(10^5)d}{D^2 i} \quad \text{for steel} \qquad (20\text{–}76\text{f})$$

$$= \frac{0.11d}{D^2 i}\left(\frac{Gg}{\gamma}\right)^{1/2} \quad \textbf{US Customary units} \quad (20\text{–}76\text{g})$$

where
G = modulus of rigidity, psi
g = 386.4 in/s^2
$\quad$ d and D in in
f_n in Hz; γ in lbf/in^3

$$= \frac{14(10^3)d}{D^2 i} \quad \text{for steel} \qquad (20\text{–}76\text{h})$$

US Customary units

STRESS WAVE PROPAGATION IN CYLINDRICAL SPRINGS UNDER *IMPACT LOAD*

The velocity of torsional stress wave in helical compression springs

$$V_\tau = 10.1\left(\frac{Gg}{\gamma}\right)^{1/2} \quad \textbf{SI} \qquad (20\text{–}76\text{i})$$

where V_τ in m/s; G in MPa; $g = 9.8066$ m/s^2;
γ in g/cm^3

$$= \left(\frac{Gg}{\gamma}\right)^{1/2} \quad \textbf{US Customary units} \qquad (20\text{–}76\text{j})$$

where G in psi; $g = 386.4$ in/s^2; V_τ in in/s; γ in lbf/in^3

The velocity of surge wave (V_s) $\qquad$ (It varies from 50 to 500 m/s.)

The impact velocity (V_{imp})

$$V_{imp} = 10.1\sigma(g/2\gamma G)^{1/2} \quad \textbf{SI} \qquad (20\text{–}76\text{k})$$
$$= \sigma/35.5 \text{ m/s} \quad \text{for steel}$$
$$= \sigma(g/2\gamma G)^{1/2} \quad \textbf{US Customary units} \qquad (20\text{–}76\text{l})$$
$$= \sigma/131 \text{ in/s} \quad \text{for steel}$$

The frequency of vibration of valve spring per minute

$$f_n = 84.627\sqrt{k_0/W} \quad \textbf{SI} \qquad (20\text{–}77\text{a})$$

where k_0 in N/m; W in N

$$f_n = 2676.14\sqrt{\frac{k_0}{W}} \quad \textbf{Customary Metric units}$$

where k_0 in kgf/mm; W in kgf $\qquad (20\text{–}77\text{b})$

$$= 530\sqrt{k_0/W} \quad \textbf{US Customary System units}$$

$$(20\text{–}77\text{c})$$

where k_0 in lbf/in; W in lbf

Particular	Formula

HELICAL EXTENSION SPRINGS (FIGS. 20–11 TO 20–13)

For typical ends of extension helical springs

Refer to Fig. 20–11.

The maximum stress in bending at point A (Fig. 20–12)

$$\sigma_A = \frac{16K_1DF}{\mu d^3} + \frac{4F}{\mu d^2} \qquad (20\text{–}78a)$$

Type	Configurations	Recommended Length Min.- Max.
Twist Loop or Hook	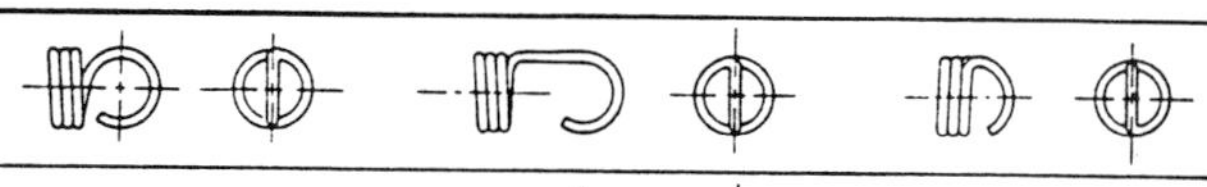	0.5-1.7 I.D.
Cross Center Loop or Hook	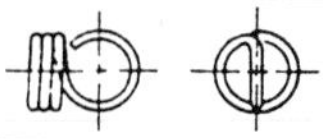	I.D.
Side Loop or Hook		0.9-1.0 I.D.
Extended Hook	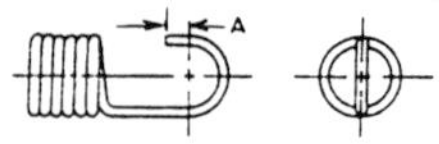	1.1 I.D. and up, as required by design
Special Ends	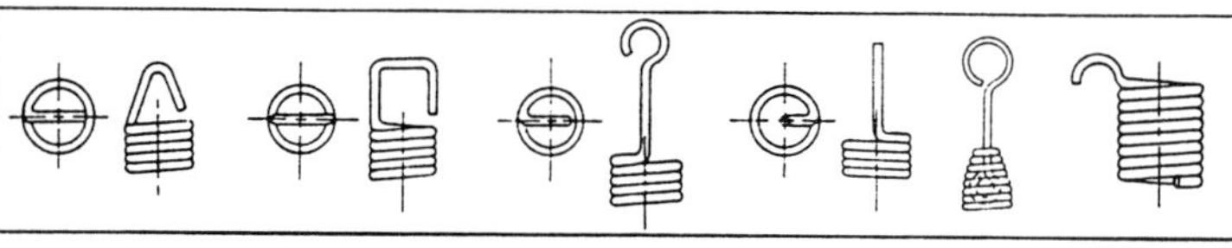	As required by design

FIGURE 20–11 Common-end configuration for helical extension springs. Recommended length is distance from last body coil to inside of end. ID is inside diameter of adjacent coil in spring body. (*Associated Spring, Barnes Group, Inc.*).

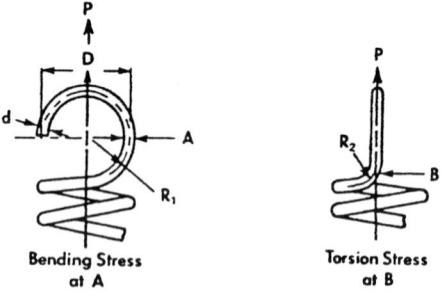

Bending Stress at A

Torsion Stress at B

FIGURE 20–12 Location of maximum bending and torsional stresses in twist loops. (*Associated Spring, Barnes Group, Inc.*)

The constant K_1 in Eq. (20–78a)

$$K_1 = \frac{4C^2 - C_1 - 1}{4C_1(C_1 - 1)} \qquad (20\text{–}78b)$$

The constant C_1 in Eq. (20–78b)

$$C_1 = \frac{2R_1}{d} \qquad (20\text{–}78c)$$

Particular	Formula
	For R_1, refer to Fig. 20–12.
The maximum stress in torsion at point B (Fig. 20–12)	$$\sigma_B = \frac{8DF}{\mu d^3}\frac{4C_2-1}{4C_2-4} \qquad (20\text{–}78d)$$
The constant C_2 in Eq. (20–78d)	$$C_2 = \frac{2R_2}{d} \qquad (20\text{–}78e)$$ For R_2, refer to Fig. 20–12. In practice C_2 may be taken greater than 4.
For extension helical spring dimensions	Refer to Fig. 20–13.

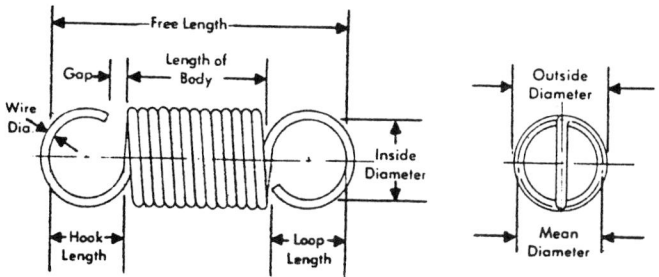

FIGURE 20–13   Typical extension-spring dimensions. (*Associated Spring, Barnes Group, Inc.*)

For design equations of extension helical springs	The design equations of compression springs may be used.
The spring rate	$$k_0 = \frac{F - F_i}{y} = \frac{Gd^4}{8D^3 i} \qquad (20\text{–}78f)$$ F_i = initial tension
The stress	$$\sigma = \frac{k8FD}{\mu d^3} \qquad (20\text{–}78g)$$ where k = stress factor for helical springs Refer to Fig. 20–5 for k.

CONICAL SPRINGS [FIG. 20–14(*a*)]

The axial deflection y for i coils of round stock may be computed by the relation [Fig. 20–14(*a*)]	$$y = \frac{2iF(D_2^3 + D_2^2 D_1 + D_2 D_1^2 + D_1^3)}{d^4 G} \qquad (20\text{–}79)$$ $$= \frac{\pi i \tau (D_2^3 + D_2^2 D_1 + D_2 D_1^2 + D_1^3)}{4d D_2 kG} \qquad (20\text{–}80)$$

Particular	Formula

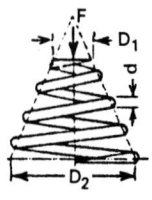  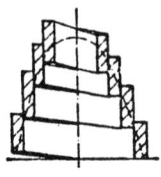

(a) Conical (round wire) (b) Volute (round wire) (c) Volute (rectangular wire)

FIGURE 20–14 Conical and volute springs.

The axial deflection of a conical spring made of rectangular stock with radial thickness b and an axial dimension h [Fig. 20–14(c)]

$$y = \frac{0.71 i F (b^2 + h^2)(D_2^3 + D_2^2 D_1 + D_2 D_1^2 + D_1^3)}{b^3 h^3 G} \qquad (20\text{–}81)$$

NONMETALLIC SPRINGS
Rectangular rubber spring (Fig. 20–15)

Approximate overall dimension of the shock absorber can be obtained by (Fig. 20–15)

$$\frac{L}{D^2} = \frac{\pi E}{2F^2}\left[\frac{U}{(F_{max}/F)^2 - 1}\right] \qquad (20\text{–}82)$$

Spring constant K of an absorber

$$K = \frac{\pi D^2 E}{L} \qquad (20\text{–}83)$$

Dimensions of sleeve and core are found by empirical relations

$$L_1 = 0.75L \qquad (20\text{–}84)$$
$$D_1 = 0.70D \qquad (20\text{–}85)$$
$$D_2 = 1.12 D_1 \qquad (20\text{–}86)$$

FIGURE 20–15 Rectangular rubber spring.

TORSION SPRINGS (FIG. 20–16) [7]

The maximum stress in torsion spring

$$\sigma = \frac{M_t}{Z} + \frac{F}{A} \qquad (20\text{–}87)$$

The stress in torsion spring taking into consideration the correction factor k'

$$\sigma = \frac{k' M_t}{Z} + \frac{2M_t}{DA} \qquad (20\text{–}88)$$

The deflection

$$y = \frac{M_t L D}{2EI} \qquad (20\text{–}89)$$

The stress in round wire spring

$$\sigma = \frac{8M_t(4k'D + d)}{\pi d^3 D} \qquad (20\text{–}89a)$$

where $k' = k_1$ can be taken from curve k_1 in Fig. 20–5

Particular	Formula

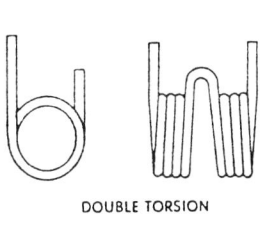
SPECIAL ENDS

SHORT HOOK ENDS

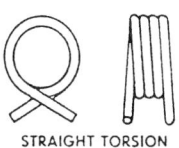
HINGE ENDS

STRAIGHT OFFSET

DOUBLE TORSION

STRAIGHT TORSION

FIGURE 20–16 Common helical torsion-spring end configurations. (*Associated Spring, Barnes Group, Inc.*)

The torsional moment (M_t) is numerically equal to bending moment (M_b).

The stress is also given by Eq. (20–90) without taking into consideration the direct stress (F/A)

$$\sigma = k\,\frac{M_b c}{I} \qquad (20\text{–}90)$$

where $M_b = Fr$

The expressions for k for use in Eq. (20–90)

$$k = k_0 = \frac{4C^2 + C - 1}{4C(C+1)} \quad \text{for outer fiber} \qquad (20\text{–}91\text{a})$$

$$= k_i = \frac{4C^2 - C - 1}{4C(C-1)} \quad \text{for inner fiber} \qquad (20\text{–}91\text{b})$$

Equation (20–90) for stress becomes

$$\sigma = k_i\,\frac{32Fr}{\pi d^3} \qquad (20\text{–}92)$$

The angular deflection in radians

$$\theta = \frac{64 M_b D i}{E d^4} \qquad (20\text{–}93)$$

The spring rate of torsion spring

$$k_0 = \frac{M_b}{\theta} = \frac{d^4 E}{64 D i} \qquad (20\text{–}94)$$

The spring rate can also be expressed by Eq. (20–95), which gives good results

$$k_0' = \frac{d^4 E}{10.8 D i} \qquad (20\text{–}95)$$

Particular	Formula
The allowable tensile stress for torsion springs	$\sigma_{Sy} = \sigma_a = \begin{cases} 0.78\sigma_{Sut} & \text{cold-drawn carbon steel} \\ 0.87\sigma_{Sut} & \text{hardened and tempere carbon and low-alloy steels} \\ 0.61\sigma_{Sut} & \text{austenitic stainless steel and nonferrous alloys} \end{cases}$
The endurance limit for torsion springs	$\sigma_{Sf} = 538$ MPa (78 kpsi)

Torsion spring of rectangular cross section

The stress in rectangular wire spring	$\sigma = \dfrac{6k'M_t}{b^2h} + \dfrac{2M_t}{Dbh}$ $\qquad$ (20–96)
	where $k' = k_2$ can be taken from curve k_2 in Fig. 20–5
	$c = \dfrac{D}{h}$ $\qquad$ (20–97)
Axial dimension b after keystoning	$b_1 = b\,\dfrac{C - 0.5}{C}$ $\qquad$ (20–98)
Another expression for stress for rectangular cross-sectional wire torsion spring without taking into consideration the direct stress ($\sigma = F/A$)	$\sigma = \dfrac{6k_i M_b}{bh^2}$ $\qquad$ (20–99)
	where $k_i = \dfrac{4C}{4C - 3}$
The spring rate	$k_0 = \dfrac{M_b}{\theta} = \dfrac{Ebh^3}{66Di}$ $\qquad$ (20–100)

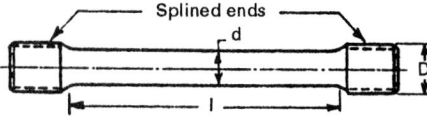

FIGURE 20–17   Torsion bar spring

Torsion bar springs

Refer to Tables 20–16 and 20–17 and Fig. 20–17.

For allowable working stresses for rubber compression springs	Refer to Table 20–18.

TABLE 20–16
Design formulas for bar springs

Cross section of bar	Angular deflection, θ, rad	Maximum shear stress, τ
Solid circular bar	$\dfrac{584M_t l}{d^4 G}$	$\dfrac{16M_t}{\pi d^3}$
Hollow circular bar	$\dfrac{584M_t l}{(d_1^4 - d_2^4)G}$	$\dfrac{16M_t d_1}{\pi(d_1^4 - d_2^4)}$
Square bar	$\dfrac{407M_t l}{b^4 G}$	$\dfrac{4.81M_t}{b^3}$
Rectangular bar	$\dfrac{57.3M_t l}{k_1'^{a} bh^3 G}$	$\dfrac{M_t}{k_2'^{a} bh^2}$

[a] Values of k_1' and k_2' can be obtained from Table 20–9.

TABLE 20–17
Factors for computing rectangular bars in torsion

b/h	k'	k_1'	k_2'
1.0	0.675	0.140	0.208
1.2	0.759	0.166	0.219
1.5	0.848	0.196	0.231
2.0	0.930	0.229	0.246
2.5	0.968	0.249	0.258
3.0	0.985	0.263	0.267
4.0	0.997	0.281	0.282
5.0	0.999	0.291	0.291
10.0	1.000	0.312	0.231
∞	1.000	0.333	0.333

TABLE 20–18
Suggested allowable working stresses for rubber compression springs

Durometer hardness	Area[a] ratio	Limits of allowable stress			
		Occasional loading		Cont. or freq. loading[b]	
		MPa	psi	MPa	psi
30	5	2.76	400	0.97	140
30	3	2.48	360	0.93	135
30	2	2.24	325	0.86	125
30	1	1.79	260	0.73	105
30	0.5	1.45	210	0.62	90
50	4	4.82	700	1.86	270
50	2	3.73	540	1.58	230
50	1	2.69	390	1.24	180
50	0.5	2.07	300	1.03	150
80	2	6.13	890	2.69	390
80	1	4.14	600	2.07	300
80	0.5	2.90	420	1.65	240

[a] Ratio of load-carrying area available for bulging or lateral expansion

REFERENCES

1. Lingaiah, and B. R. Narayana Iyengar, *Machine Design Data Handbook*, Engineering College Co-operative Society, Bangalore, India, 1962.
2. Lingaiah, K., and B. R. Narayana Iyengar, *Machine Design Data Handbook*, Vol. I (*SI and Customary Metric Units*), Suma Publishers, Bangalore, India, 1986.
3. Lingaiah, K., *Machine Design Data Handbook*, Vol. II (*SI and Customary Metric Units*), Suma Publishers, Bangalore, India, 1986.

4. *SAE Handbook*, *Springs*, Vol. I, 1981.
5. Maleev, V. L., and J. B. Hartman, *Machine Design*, International Textbook Company, Scranton, Pennyslvania, 1954.
6. Wahl, A. M., *Mechanical Springs*, McGraw-Hill Book Company, New York, 1963.
7. Associated Spring, Barnes Group Inc., Bristol, CT, USA.
8. Jorres, R. E., "Springs," Chap. 24 in J. E. Shigley and C. R. Mischke, eds., *Standard Handbook of Machine Design*, McGraw-Hill Book Company, New York, 1986.
9. Shigley, J. E., and C. R. Mischke, *Mechanical Engineering Design*, 5th ed. McGraw-Hill Company, New York, 1989.
10. Zimmerli, F. P., *Human Failures in Springs Applications*, *The Mainspring*, No. 17, Associated Spring Corporation, Bristol, Connecticut, Aug.–Sept. 1957.
11. Shigley, J. E., and C. R. Mischke, *Standard Handbook of Machine Design*, McGraw-Hill Book Company, New York, 1986.
12. Phelan, R. M., *Fundamentals of Mechanical Design*, Tata-McGraw-Hill Publishing Company Ltd. New Delhi, 1975.

BIBLIOGRAPHY

Baumeister, T., ed., *Marks' Standard Handbook for Mechanical Engineers*, McGraw-Hill Book Company, New York, 1978.
Black, P. H., and O. Eugene Adams, Jr., *Machine Design*, McGraw-Hill Book Company, New York, 1968.
Bureau of Indian Standards.
Chironis, N. P., *Spring Design and Application*, McGraw-Hill Book Company, 1961.
Norman, C. A., E. S. Ault, and I. F. Zarobsky, *Fundamentals of Machine Design*, The Macmillan Company, New York, 1951.
Shigley, J. E., *Machine Design*, McGraw-Hill Book Company, 1962.

CHAPTER
21

FLEXIBLE MACHINE ELEMENTS

SYMBOLS

a	width of pulley face, m (in)
	pivot arm length in Rockwood drive, m (in)
a_1	width of belt, m (in)
$A = 0.4\left(\frac{\pi d^2}{4}\right)$	useful area of cross section of the wire rope, m^2 (in^2)
b	thickness of arm, m (in)
	dimension in Rockwood drive (Fig. 21–5), m (in)
c	dimension in Rockwood drive (Fig. 21–5), m (in)
C	center distance between sprockets (also with suffixes) m (in)
	center distance between pulleys, m (in),
	capacity of conveyor, m^3 (ft^3)
	constant depends on the rope diameter, sheave diameter,
	chain, the bearing, and coefficient of friction [Eqs. (21–59) to
	(21–62) and (21–86) to (21–103)] (also with suffixes)
C_1	tooth width in precision roller and bush chains, m (in)
d	size of chain, m (in)
	diameter of shaft, m (in)
	diameter of idler bearing, m (in)
	diameter of smaller pulley, m (in)
	diameter of rope, m (in)
	pitch diameter of sprocket, m (in)
d_1	diameter of small sprocket, m (in)
	hub diameter of pulley, m (in)
d_2	diameter of large sprocket, m (in)
d_a	tip diameter of sprocket, m (in)
d_{a1}	tip diameter of small sprocket, m (in)
d_{a2}	tip diameter of large sprocket, m (in)
$d_c = d_p F_b$	equivalent pitch diameter, m (in)
d_f	root diameter of sprocket, m (in)
d_p	pitch diameter of the V-belt small pulley, m (in)
d_r	diameter of roller pin, m (in)
D	pitch diameter of sheave, m (in)
	diameter of large pulley, m (in)
	wire rope drum diameter, m (in)
D_r	diameter of reel barrel, m (in)
D_d	diameter of the drum in mm as measured over the outermost
	layer filling the reel drum
D_o	diameter of the sheave pin, m (in)

e	unit elongation of belt
E'	corrected elasticity modulus of steel ropes (78.5 GPa $=$ 11.4 Mpsi), GPa (psi)
F	force, load, kN (lbf)
	tension in belt, kN (lbf)
	minimum tooth side radius, m (in)
F_a	correction factor according to service from Table 21–27
F_c	correction factor for length from Table 21–26
F_{ct}	centrifugal tension, kN (lbf)
F_d	correction factor for arc of contact from Table 21–25
F_θ	tangential force in the belt, required chain pull, kN (lbf)
F_s	tension due to sagging of chain, kN (lbf)
F_1	tenion in belt on tight side, kN (lbf)
F_2	tension in belt on slack side, kN (lbf)
F_c	centrifugal force, kN (lbf)
	values of coefficient for manila rope, Table 21–32
FR_1	the minimum value of tooth flank radius in roller and bush chains, m (in)
FR_2	the maximum value of tooth flank radius in roller and bush chains, m (in)
g	acceleration due to gravity, 9.8066 m/s^2 (32.2 ft/s^2)
G	tooth side relief in bush and roller chain, m (in)
h	the thickness of wall of rope drum, m (in)
	crown height, m (in)
h_1	depth of groove in rope drum, m (in)
$H = \frac{D_d - D_r}{2}$	depth of rope layer in reel drum, m (in)
i	number of arms in the pulley,
	number of V-belts,
	number of strands in a chain,
	transmission ratio
$k = \frac{e^{\mu\theta}-1}{e^{\mu\theta}}$	variable in Eqs. (21–2d), (21–4), (21–6), and (21–123), which depends on $\frac{z_1 - z_2}{C_p}$
k_d	duty factor
k_l	load factor
K_{min}	center distance constant from Table 21–57
k_s	service factor
k_{sg}	coefficient for sag from Table 21–55
l	width of chain or length of roller, m (in)
	minimum length of boss of pulley, m (in)
	minimum length of bore of pulley, m (in)
	length of conveyor belt, m (in)
	length of cast-iron wire rope drum, m (in)
	outside length of coil link chain, m (in)
K_1	tooth correction factor for use in Eq. (21–116a)
K_2	multistrand factor for use in Eq. (21–116a)
L	length of flat belt, m (in)
	pitch length of V-belt, m (in)
	rope capacity of wire rope reel, m (in)
L_p	length of chain in pitches
M_t	torque, N m (lbf in)
n	number of times a rope passes over a sheave,
	number of turns on the drum for one rope member
	speed, rpm
	factor of safety
n_1	speed of smaller pulley, rpm or rps

	speed of smaller sprocket, rpm or rps
n_2	speed of larger pulley, rpm or rps
	speed of larger sprocket, rpm or rps
$n' = n k_d$	stress factor
P	power, kW (hp)
P_t	power required by tripper, kW (hp)
p	pitch of chain, m (in)
	pitch of the grooves on the wire rope drum, m (in)
p_1	distance between the grooves of two-rope pulley, m (in)
P	effort, load, kN (lbf)
P_b	bending load, kN (lbf)
P_s	service load, kN (lbf)
P_t	tangential force due to power transmission, kN (lbf)
P_u	ultimate load, kN (lbf)
	breaking load, kN (lbf)
P_w	working load, kN (lbf)
Q	load, kN (lbf)
r	radius near rim (with subscripts), m (in)
	radius, m (in)
s	the amount of shift of the line of action of the load from the center line on the raising load side of sheave, m (in)
s	the average shift of the center line in the load on the effort side of the sheave, m (in)
S	the distance through which the load is raised, m (in)
SA_1	the minimum value of roller or bush seating angle, deg
SA_2	the maximum value of roller or bush seating angle, deg
SR_1	the minimum value of roller or bush seating radius, m (in)
SR_2	the maximum value of roller or bush seating radius, m (in)
t	nominal belt thickness, m (in)
	thickness of rim, m (in)
T	tension in ropes, chains, kN (lbf)
TD_{min}	minimum limit of the tooth top diameter, m (in)
TD_{max}	maximum limit of the tooth top diameter, m (in)
v	velocity of belt chain, m/s (ft/min)
w	specific weight of belt, kN/m^3 (lbf/in^3)
W	width between reel drum flanges, m (in)
W_B	weight of belt, kN/m (lbf/in)
w_c	weight of chain, kN/m (lbf/in)
W_I	weight of revolving idler, kN/m (lbf/in) belt
W_L	load, kN/m (lbf/in)
z_1	number of teeth on the small sprocket
z_2	number of teeth on the large sprocket
σ	stress, MPa (psi)
σ_1	unit tension in belt on tight side, MPa (psi)
σ_2	unit tension in belt on slack side, MPa (psi)
σ_c	centrifugal force coefficient for leather belt, MPa (psi)
σ_{br}	breaking stress for hemp rope, MPa (kgf/mm^2)
τ	shear stress, MPa (psi)
θ	arc of contact, rad
α	angle between tangent to the sprocket pitch circle and the center line, deg
μ	coefficient of friction between belt and pulley
	coefficient of journal friction
μ_c	coefficient of chain friction
η	efficiency

ω_1	angular speed of small sprocket, rad/s
ω_2	angular speed of large sprocket, rad/s

SUFFIXES

b	bending
br	breaking
t	torque
c	compressive
d	design
min	minimum
max	maximum

Other factors in performance or in special aspects of design of flexible machine elements are included from time to time in this chapter and being applicable only in their immediate context, are not given at this stage.

Particular	Formula

BELTS

Flat belts

The ratio of tight side to slack side of belt at low velocities

$$\frac{F_1}{F_2} = e^{\mu\theta} \tag{21–1}$$

The power transmitted by belt

$$P = \frac{F_\theta v}{1000c_s} \quad \text{SI} \tag{21–2a}$$

where $F_\theta = F_1 - F_2$, P in kW and v in m/s; F_θ in N

$$= \frac{F_\theta v}{33000c_s} \quad \text{US Customary System units} \tag{21–2b}$$

where F_θ in lbf; P in hp; v in ft/min

$$= \frac{F_\theta \omega r}{1000c_s} \quad \text{SI} \tag{21–2c}$$

where F_θ in N, P in kW, r in m, and ω in rad/s

Refer to Table 21–1 for c_s.

Power transmitted per m² (in²) of belt at low velocities

$$P = \frac{\sigma_1 k v}{1000} \quad \text{SI} \tag{21–2d}$$

where $k = (e^{\mu\theta} - 1)/e^{\mu\theta}$, and also from Table 21–2 σ_1 in N/m², v in m/s, and P in kW

$$= \frac{\sigma_1 k v}{33000} \quad \text{US Customary System units} \tag{21–2e}$$

where σ_1 in psi; v in ft/min; P in hp

TABLE 21–1
Service correction factors, c_s

Atmospheric condition	Clean, scheduled maintenance on large drives	1.2
	Normal	1.0
	Oily, wet, or dusty	0.7
Angle of center line	Horizontal to 60° from horizontal	1.0
	60°–75° from horizontal	0.9
	75°–90° from horizontal	0.8
Pulley material	Fiber on motor and small pulleys	1.2
	Cast iron or steel	1.0
Service	Temporary or infrequent	1.2
	Normal	1.0
	Intermittent or continuous	0.8
Peak loads	Light, steady load, such as steam engines, steam turbines, diesel engines, and multicylinder gasoline engines	1.0
	Jerky loads, reciprocating machines such as normal–starting–torque squirrel-cage motors, shunt-wound, DC motors, and single-cylinder engines	0.8
	Shock and reversing loads; full-voltage start such as squirrel-cage and synchronous motors	0.6

TABLE 21–2
Values of $\dfrac{e^{\mu\theta} - 1}{e^{\mu\theta}} = k$, for various coefficients of frictions and arcs of contact

	Arc of contact between the belt and pulley (θ, deg)										
Value of μ	90	100	110	120	130	140	150	160	170	180	200
0.28	0.356	0.387	0.416	0.444	0.470	0.496	0.520	0.542	0.564	0.585	0.502
0.30	0.376	0.408	0.438	0.467	0.494	0.520	0.544	0.567	0.590	0.610	0.553
0.33	0.404	0.438	0.469	0.499	0.527	0.554	0.579	0.602	0.624	0.645	0.684
0.35	0.423	0.457	0.489	0.520	0.548	0.575	0.600	0.624	0.646	0.667	0.705
0.38	0.449	0.485	0.518	0.549	0.578	0.605	0.630	0.654	0.676	0.697	0.735
0.40	0.467	0.502	0.536	0.567	0.597	0.624	0.649	0.673	0.695	0.715	0.753
0.43	0.491	0.528	0.562	0.593	0.623	0.650	0.676	0.699	0.721	0.741	0.777
0.45	0.507	0.544	0.579	0.610	0.640	0.667	0.692	0.715	0.737	0.757	0.792
0.48	0.529	0.567	0.602	0.634	0.663	0.690	0.715	0.738	0.759	0.779	0.813
0.50	0.544	0.582	0.617	0.649	0.678	0.705	0.730	0.752	0.773	0.792	0.825
0.53	0.565	0.603	0.638	0.670	0.700	0.726	0.750	0.772	0.793	0.811	0.843

TABLE 21–3
Values of coefficients σ_c for leather belts for use in Eqs. (21–3) and (21–4)

Belt velocity, m/s (ft/min)	7.5 (1500)	10.0 (1950)	12.70 (2500)	15.0 (2950)	17.5 (3500)	20.0 (3950)	22.5 (4450)	25.0 (4950)
Coefficient, σ_c kgf/cm^2	0.57	1.05	1.63	2.35	3.10	4.07	5.14	6.36
MPa	0.0559	0.1030	0.1598	0.2305	0.3040	0.3991	0.5041	0.5237
psi	8.0	15.0	23.2	33.5	45.0	58.0	73.0	76.0

Particular	Formula
The ratio of tight to slack side of belt at high velocities	$$\frac{\sigma_1 - \sigma_c}{\sigma_2 - \sigma_c} = e^{\mu\theta} \qquad (21\text{–}3a)$$
	where $\sigma_c = \dfrac{wv^2}{g} \qquad (21\text{–}3b)$
Power transmitted per m^2 (in^2) of belt at high velocities	$P = \dfrac{(\sigma_1 - \sigma_c)\,kv}{1000}$ **SI** $(21\text{–}4a)$ where σ_1 and σ_c in N/m^2; v in m/s; P in kW $= \dfrac{(\sigma_1 - \sigma_c)\,kv}{33000}$ **US Customary System units** $(21\text{–}4b)$ where σ_1 and σ_c in psi, v in ft/min; P in hp Refer to Table 21–3 for values of σ_c.
Equation (21–3a) in terms of tension on tight side (F_1) and slack side of belt, (F_2), and centrifugal force (F_c)	$\dfrac{F_1 - F_c}{F_2 - F_c} = e^{\mu\theta} \qquad (21\text{–}4c)$ where $F_1 = \sigma_1 A$; $F_2 = \sigma_2 A$; $F_c = \sigma_c A$ $A = a_1 t =$ area of cross section of belt, m^2 (in^2)
The relation between the initial tension in the belt (F_0) and the tension in the belt on the tight side ($F_{1,max}$) to obtain maximum tension in the belt	$F_{1,max} = 2\,F_0 \qquad (21\text{–}4d)$
The power transmitted at maximum tension in belt, i.e., when $F_1 = 2F_0$, from Eq. (21–1)	$P = \dfrac{F_{1,max}v}{33,000} = \dfrac{2F_0 v}{33,000}$ **US Customary System units** $(21\text{–}4e)$ $= \dfrac{F_{1,max}v}{1000} = \dfrac{2F_0 v}{1000}$ **SI** $(21\text{–}4f)$
The power transmitted in actual practice taking into consideration pulley correction factor (K_p), velocity correction factor (K_v), and service factor (C_s) at maximum tension in belt.	$P = \dfrac{2K_p K_v F_a v}{33,000 C_s}$ **US Customary system units** $(21\text{–}4g)$ $= \dfrac{2K_p K_v F_a v}{1,000 C_s}$ **SI** $(21\text{–}4h)$

Stresses in belt

Tensile stress due to tension on tight side of belt $F_1(S_1)$	$\sigma_1 = \dfrac{F_1}{a_1 t} \qquad (21\text{–}4i)$

Particular	Formula
Tensile stress due to tension on slack side of belt $F_2(S_2)$	$$\sigma_2 = \frac{F_2}{a_1 t} \qquad (21\text{--}4j)$$
Tensile stress due to tangential force (effective stress)	$$\sigma_\theta = \frac{F_\theta}{a_1 t} \qquad (21\text{--}4k)$$
The tensile stress due to belt tension on account of centrifugal force	$$\sigma_c = \frac{F_c}{a_1 t} \qquad (21\text{--}4l)$$
The bending stress	$$\sigma_b = \frac{E_b t}{d_1} \qquad (21\text{--}4m)$$
The maximum belt stress	$$\sigma_{max} = \sigma_1 + \sigma_c + \sigma_b + \sigma_{tw} \qquad (21\text{--}4n)$$ where $\sigma_{tw} = E(a_1/a)^2$ and $a > 2d_2$
For distribution of various stresses in belt	Refer to Fig. 21–1C. Refer to Table 21–4B for most commonly used belt materials in practice. The values of K_p and C_s are taken from Tables 21–4C and 21–4D and K_v from Fig. 21–1B, and also Table 21–4E. F_a = allowable tension in belt, N (lbf) v = velocity of belt, ft/min
Coefficient of friction (μ)	$$\mu = 0.54 - \frac{0.7}{2.5 + v} \qquad \textbf{SI} \qquad (21\text{--}5)$$ μ may also be obtained from Tables 21–4A and Table 21–5. v = velocity of belt, m/s. $$\mu = 0.54 - \frac{140}{500 + v} \qquad \textbf{US Customary units} \quad (21\text{--}5a)$$ where v = velocity of belt, ft/min
The cross section of the belt is given	$$a_1 t = \frac{1000\,P}{v\left(\sigma_d - \dfrac{wv^2}{g}\right)\dfrac{1}{k}} \qquad \textbf{SI} \qquad (21\text{--}6a)$$ where P in kW, v in m/s, g in 9.8066 m/s^2 w in N/m^3, and σ_d in MPa $$= \frac{33{,}000\,P}{v\left(\sigma_d - \dfrac{wv^2}{g}10^4\right)\dfrac{1}{k}} \qquad \textbf{US Customary units}$$ $$(21\text{--}6b)$$ where P in hp, v in ft/min; $g = 386.4$ in/s^2; w in lbf/in^3; σ_d in psi
For cross section and properties of belts	Refer to Tables 21–6A to 21–14.

For cross section and properties of belts   Refer to Tables 21–6A to 21–14.

TABLE 21–4A
Coefficients of frictions of leather belts on iron pulleys depending on velocity of belt

Velocity of belt, v, m/s	Coefficient of friction, μ	Velocity of belt, v, m/s	Coefficient of friction, μ	Velocity of belt, v, m/s	Coefficient of friction, μ
	0.260	4.0	0.432	15.0	0.500
0.25	0.285	4.5	0.440	17.5	0.505
0.50	0.307	5.0	0.446	20.0	0.509
1.00	0.340	6.0	0.458	22.5	0.512
1.50	0.365	7.0	0.456	25.0	0.514
2.00	0.384	8.0	0.473	27.5	0.517
2.50	0.400	9.0	0.479	30.0	0.519
3.00	0.413	10.0	0.494	32.5	0.520
3.50	0.423	12.5	0.493		

TABLE 21-4B
Properties of some flat and round materials

Material	Specification	Size, in	Minimum pulley diameter, in	Allowable tension per unit width at 600 ft/min, lb/in	Weight, lb/in^3	Coefficient of friction
Leather	1 ply	$t = {}^{11}/_{64}$	3	30	0.035–0.045	0.4
		$t = {}^{13}/_{64}$	$3^{1}/_{2}$	33	0.035–0.045	0.4
	2 ply	$t = {}^{18}/_{64}$	$4^{1}/_{2}$	41	0.035–0.045	0.4
		$t = {}^{20}/_{64}$	6[a]	50	0.035–0.045	0.4
		$t = {}^{23}/_{64}$	9[a]	60	0.035–0.045	0.4
Polyamide[b]	F–0[c]	$t = 0.03$	0.60	10	0.035	0.5
	F–1[c]	$t = 0.05$	1.0	35	0.035	0.5
	F–2[c]	$t = 0.07$	2.4	60	0.051	0.5
	A–2[c]	$t = 0.11$	2.4	60	0.037	0.8
	A–3[c]	$t = 0.13$	4.3	100	0.042	0.8
	A–4[c]	$t = 0.20$	9.5	175	0.039	0.8
	A–5[c]	$t = 0.25$	13.5	275	0.039	0.8
Urethane[d]	$w = 0.50$	$t = 0.062$	See	5.2[e]	0.038–0.045	0.7
	$w = 0.75$	$t = 0.078$	Table	9.8[e]	0.038–0.045	0.7
	$w = 1.25$	$t = 0.090$	17–4E	18.9[e]	0.038–0.045	0.7
	Round	$d = {}^{1}/_{4}$	See	8.3[e]	0.038–0.045	0.7
		$d = {}^{3}/_{8}$	Table	18.6[e]	0.038–0.045	0.7
		$d = {}^{1}/_{2}$	17–4E	33.0[e]	0.038–0.045	0.7
		$d = {}^{3}/_{4}$		74.3[e]	0.038–0.045	0.7

[a] Add 2 in to pulley size for belts 8 in wide or more.
[b] *Source: Habasit Engineering Manual*, Habasit Belting, Inc., Chamblee (Atlanta), Ga.
[c] Friction cover of acrylonitrile-butadiene rubber on both sides.
[d] *Source*: Eagle Belting Co., Des Plaines, Ill.
[e] At 6 per cent elongation; 12 percent is maximum allowable value.
Notes: d = diameter, t = thickness, w = width. The values given in this table for the allowable tension are based on a belt speed of 600 ft/min. Take $K_v = 1.0$ for polyamide and urethane belts.
Source: Eagle Belting Co., Des Plaines, Illinois; table reproduced from J. E. Shigley and C. R. Mischke, *Mechanical Engineering Design*, McGraw-Hill Book Company, New York, 1989.

TABLE 21–4C
Pulley correction factor K_P for flat belts[a]

Material	Small-pulley diameter, in					
	1.6–4	**4.5–8**	**9–12.5**	**14, 16**	**18–31.5**	**> 31.5**
Leather	0.5	0.6	0.7	0.8	0.9	1.0
polyamide, F–0	0.95	1.0	1.0	1.0	1.0	1.0
F–1	0.70	0.92	0.95	1.0	1.0	1.0
F–2	0.73	0.86	0.96	1.0	1.0	1.0
A–2	0.73	0.86	0.96	1.0	1.0	1.0
A–3	—	0.70	0.87	0.94	0.96	1.0
A–4	—	—	0.71	0.80	0.85	0.92
A–5	—	—	—	0.72	0.77	0.91

[a] Average values of K_p for the given ranges were approximated from curves in the *Habasit Engineering Manual*, Habasit Belting, Inc., Chamblee (Atlanta), Ga.
Source: Eagle Belting Co., Des Plaines, Illinois; table reproduced from J. E. Shigley and C. R. Mischke, *Mechanical Engineering Design*, McGraw-Hill Book Company, New York, 1989.

TABLE 21–4D
Service factors C_s for V-belt and flat belt drives

Driven machinery	Power source	
	Normal torque characteristic	High or nonuniform torque
Uniform	1.0–1.2	1.1–1.3
Light shock	1.1–1.3	1.2–1.4
Medium shock	1.2–1.4	1.4–1.6
Heavy shock	1.3–1.5	1.5–1.8

Source: Eagle Belting Co., Des Plaines, Illinois; table reproduced from J. E. Shigley and C. R. Mischke, *Mechanical Engineering Design*, McGraw-Hill Book Company, New York, 1989.

TABLE 21–4E
Minimum pulley sizes for flat and round urethane belts (pulley diameters in inches)

Belt style	Belt size, in	Ratio of pulley speed to belt length, rev/(ft–min)		
		Up to 250	**250 to 499**	**500 to 1000**
Flat	0.50 × 0.062	0.38	0.44	0.50
	0.75 × 0.078	0.50	0.63	0.75
	1.25 × 0.090	0.50	0.63	0.75
Round	$1/4$	1.50	1.75	2.00
	$3/8$	2.25	2.62	3.00
	$1/2$	3.00	3.50	4.00
	$3/4$	5.00	6.00	7.00

Source: Eagle Belting Co., Des Plaines, Illinois; table reproduced from J. E. Shigley and C. R. Mischke, *Mechanical Engineering Design*, McGraw-Hill Book Company, New York, 1989.

TABLE 21–5
Coefficient of friction for belts depending on materials of pulley and belt

Belt material	Cast iron/steel			Wood	Compressed paper	Leather face	Rubber face
	Dry	Wet	Greasy				
Leather, oak-tanned	0.25	0.20	0.15	0.30	0.33	0.38	0.40
Leather, chrome-tanned	0.35	0.32	0.22	0.40	0.45	0.48	0.50
Canvas, stitched	0.20	0.15	0.12	0.23	0.25	0.27	0.30
Cotton, woven	0.22	0.15	0.12	0.25	0.28	0.27	0.30
Camel hair, woven	0.35	0.25	0.20	0.40	0.45	0.45	0.45
Rubber	0.30	0.18	—	0.32	0.35	0.40	0.42
Balata	0.32	0.20	—	0.35	0.38	0.40	0.42

TABLE 21–6A
Thickness and width of leather belts

Grade	Average thickness, mm				Width, mm	
	Single	Double	Triple	Quadruple	Range	Increment
Light	3	6	—	—	12–24	3
					24–102	6
					102–198	12
Medium	4	8	12.5	17.5	200–800	25
					800–1400	50
Heavy	5	10	15	20	800–1400	50
					1500–2100	100

TABLE 21–6B
Relative strength of belt joints

Type of joint	Relative strength of joint to an equal section of solid leather, efficiency, %
Cemented, endless } Cemented at factory }	90–100
Cemented in shop	80–90
Laced, wire	
By machine	75–85
By hand	70–80
Rawhide, small holes	60–70
Rawhide, large holes	50–60
Hinged	
Wire hooks	40
Metal hooks	35–40

TABLE 21–7
Standard widths of transmission belting for different plies

Ply	Standard width, mm																			
	25	32	40	44	50	63	76	90	100	112	125	140	152	180	200	224	250	305	355	400
3	p^a	q^b	p	q	p	p	p	q	q	—	—	—	—	—	—	—	—	—	—	—
4	q	q	p	q	p	p	p	p	p	p	p	q	p	—	q	—	—	—	—	—
5	—	—	—	—	—	—	p	q	p	p	p	—	p	r^c	q	r	r	—	—	—
6	—	—	—	—	—	—	—	—	q	p	p	—	p	p	p	—	r	—	—	—
8	—	—	—	—	—	—	—	—	—	—	—	—	—	r	—	r	—	r	r	r

[a] p = these sizes are available in Hi-speed and Fort.
[b] q = these sizes are available in Hi-speed only.
[c] r = these sizes are available in Fort only.

TABLE 21–8
Widths of friction surface—rubber transmission belting

Nominal belt width $\times 10^{-3}$ m	Tolerance $\times 10^{-3}$ m
25, 32, 40, 50, 63	±2.0
71, 80, 90, 100, 112, 125	±3.0
140, 160, 180, 200, 224, 250	±4.0
280, 315, 355, 400, 450, 500	±5.0

Source: IS 1370, 1965.

TABLE 21–9
Thickness of friction surface—rubber transmission belting

Ply construction	Nominal thickness hard-type fabric $\times 10^{-3}$ m	Tolerance $\times 10^{-3}$ m
3	3.9	±0.5
4	5.1	±0.7
5	6.4	±0.8
6	7.7	±0.9
7	9.1	±1.0
8	10.4	±1.1

Source: IS 1370, 1964.

TABLE 21–10
Properties of leather belting for various purposes

			Purpose					
				Power transmission			Round belting for small machine		
Properties		General	Single belts	Double belts	Splices single and double	Heavy (5)	Regular (6)	Heavy (7)
Tensile strength, min	MPa	20.6	24.5	24.5	20.6			
	kpsi	3.0	3.5	3.5	3.0			
Breaking strength, min	N					441	667	755
	lbf					100	150	170
Temporary elongation, %, max		6						
Permanent elongation, %, max		2						
Stitch tear resistance thickness, min	N/m	83,356						
	lbf/in	475						
Grain strength		Shall not crack	—					

TABLE 21–11
Tensile strength of fabric in finished rubber transmission belting

			Tensile strength, N/m (kgf/mm) of width			
	Weight of fabric per square meter		Warp		Weft	
Type of fabric	N/m^2	kgf/m^2	N/m	kgf/mm	N/m	kgf/mm
Soft	8.0	0.815	61,291.3	6.25	29,419.8	3.00
Hard	8.8	0.900	61,291.3	6.25	35,303.8	3.60
Soft	9.1	0.930	69,626.9	7.10	32,361.8	3.30
Hard	3.6	0.975	73,549.7	7.50	44,129.7	4.50

Source: IS 1370, 1965.

TABLE 21-12
Properties of ply woven fire-resistant conveyor belting for use in coal mines

Belt designation	No. of plies	1A A^a	1A B^b	1AA A	1AA B	1B A	1B B	1C A	1C B	2A A	2A B	2B A	2B B	2C A	2C B	3A A	3A B	3B A	3B B	3C A	3C B
Tensile strength in kgf/mm width for number of plies	3	—	—	—	—	—	—	—	—	—	—	—	—	39.3	21.4	—	—	62.5	21.4	89.3	28.6
	4	23.0	11.2	26.4	12.1	32.1	14.8	38.6	18.6	39.3	21.3	44.7	24.1	51.1	27.9	57.2	21.4	81.3	27.9	116.1	37.2
	5	28.0	13.7	32.1	14.8	39.3	18.0	47.1	22.7	48.0	26.1	54.3	29.5	62.2	34.4	87.7	26.1	99.1	34.0	141.1	45.0
	6	32.7	15.9	37.5	17.4	45.7	21.1	55.0	26.4	—	—	—	—	—	—	—	—	—	—	—	—
Tensile strength in $N \cdot m \times 10^{-3}$ width for number of plies	3	—	—	—	—	—	—	—	—	—	—	—	—	385.4	209.9	—	—	612.9	209.9	875.7	280.5
	4	225.6	109.8	258.9	118.7	314.8	145.1	378.5	182.4	385.4	209.9	438.4	236.3	501.1	273.4	560.9	209.9	797.3	273.6	1138.5	364.8
	5	274.0	134.4	314.8	145.1	385.4	176.5	461.9	222.6	470.7	255.0	532.5	289.3	610.0	333.4	860.0	256.0	971.8	333.4	1383.7	441.3
	6	320.7	155.9	397.7	170.6	448.2	206.9	539.4	258.9	—	—	—	—	—	—	—	—	—	—	—	—
Tear strength in kgf for the number of plies	3	—	—	—	—	—	—	—	—	—	—	—	—	—	90.8	—	—	—	—	—	—
	4	20.4	—	25.0	—	29.5	—	36.3	—	—	90.8	—	104.3	—	117.9	—	—	—	—	—	—
	5	27.2	—	31.8	—	36.3	—	45.4	—	—	113.4	—	131.4	—	149.7	—	—	—	—	—	—
	6	34.0	—	38.6	—	43.1	—	54.4	—	—	—	—	—	—	—	—	—	—	—	—	—
Tear strength in N for the number of plies	3	—	—	—	—	—	—	—	—	—	—	—	—	—	890.4	—	—	—	—	—	—
	4	200.1	—	245.2	—	289.3	—	356.0	—	—	890.4	—	1022.8	—	1156.2	—	—	—	—	—	—
	5	266.7	—	311.8	—	356.0	—	445.2	—	—	1112.1	—	1288.6	—	1468.0	—	—	—	—	—	—
	6	333.4	—	378.5	—	422.7	—	533.5	—	—	—	—	—	—	—	—	—	—	—	—	—
Percentage elongation at break		15	8	15	8	15	8	15	8	17	18	17	18	17	18	—	—	—	—	—	—

^{a} A = Warp.
^{b} B = Weft.

TABLE 21-13
Allowable tension in width of belt

	Ply or number of thickness of belt																			
Belt material	kN/m	kgf/mm	kN/m	kgf/mm	kN/m	kgf/mm	kN/m	kgf/mm	kN/m	kgf/mm	kN/m	kgf/mm	kN/m	kgf/mm	kN/m	kgf/mm	kN/m	kgf/mm	kN/m	kgf/mm
Leather Light	16.7	1.7	⋯	⋯	⋯	⋯	⋯	⋯	⋯	⋯	⋯	⋯	⋯	⋯	⋯	⋯	⋯	⋯	⋯	⋯
Leather Medium	14.7	1.5	24.5	2.5	⋯	⋯	⋯	⋯	⋯	⋯	⋯	⋯	⋯	⋯	⋯	⋯	⋯	⋯	⋯	⋯
Leather Heavy	17.7	1.8	28.4	2.9	35.3	3.6	⋯	⋯	⋯	⋯	⋯	⋯	⋯	⋯	⋯	⋯	⋯	⋯	⋯	⋯
Canvas-stitched	⋯	⋯	⋯	⋯	6.9	.7	8.8	.9	10.8	1.1	11.8	1.2	13.7	1.4	15.7	1.6	⋯	⋯	⋯	⋯
Balata	⋯	⋯	4.9	.5	6.9	.7	8.8	1.1	⋯	⋯	⋯	⋯	25.5	2.6	28.4	2.9	30.4	3.1	33.3	3.4
Rubber	7.8	.8	10.8	1.1	12.7	1.3	⋯	⋯	⋯	⋯	22.6	2.3	25.5	2.6	28.4	2.9	30.4	3.1	33.3	3.4

Particular	Formula

BELT LENGTHS AND CONTACT ANGLES FOR OPEN AND CROSSED BELTS (FIG. 21–1A)

Length of belt for open drive (Fig. 21-1a)

$$L = \sqrt{4C^2 - (D-d)^2} + \tfrac{1}{2}(D\theta_L + d\theta s) \qquad (21\text{–}7)$$

Length of belt for cross drive (Fig. 21-1b)

$$L = \sqrt{4C^2 - (D-d)^2} - \frac{\theta}{2}(D+d) \qquad (21\text{–}8)$$

Length of belt for quarter turn drive

$$L = \frac{\pi}{2}(D+d) + \sqrt{C^2 + D^2} + \sqrt{C^2 + d^2} \qquad (21\text{–}9)$$

For two-pulley open drive the center distance between the two pulleys when the length of the belt is known

$$C = \frac{L}{4} - 0.393(D+d)$$

$$+ \left[\left\{ \frac{L}{4} - 0.394(D+d) \right\}^2 - \frac{(D-d)^2}{8} \right]^{\frac{1}{2}} \qquad (21\text{–}10)$$

where

$$\theta_L = \pi + 2\sin^{-1}\left(\frac{D-d}{2C}\right) \qquad (21\text{–}10\text{a})$$

$$\theta_s = \pi - 2\sin^{-1}\left(\frac{D-d}{2C}\right) \qquad (21\text{–}10\text{b})$$

$$\theta = \pi + 2\sin^{-1}\left(\frac{D+d}{2C}\right) \qquad (21\text{–}10\text{c})$$

The unit elongation of belt is given by the equation

$$e = \frac{\sqrt{\sigma}}{69000} \qquad \textbf{SI} \qquad (21\text{–}11\text{a})$$

where σ in MPa

$$= \frac{\sqrt{\sigma}}{21100} \qquad \textbf{US Customary units} \qquad (21\text{–}11\text{b})$$

where σ in psi

$$= \frac{\sqrt{\sigma}}{22} \qquad \textbf{Customary Metric} \qquad (21\text{–}11\text{c})$$

where σ in kgf/mm^2

The relation between initial belt tension and final belt tension

$$2\sqrt{F_0} = \sqrt{F_1} + \sqrt{F_2} \qquad (21\text{–}12)$$

where F_0 = initial belt tension, kN (lbf)

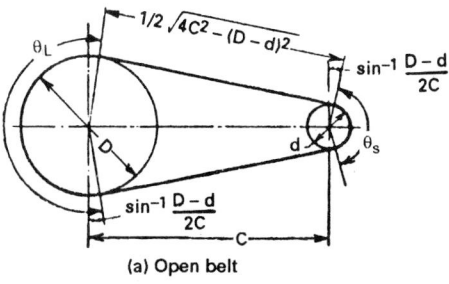
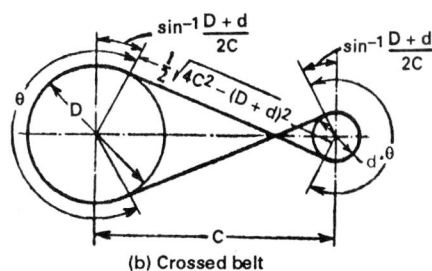

(a) Open belt   **(b) Crossed belt**

FIGURE 21–1(A)   Open and crossed belts.

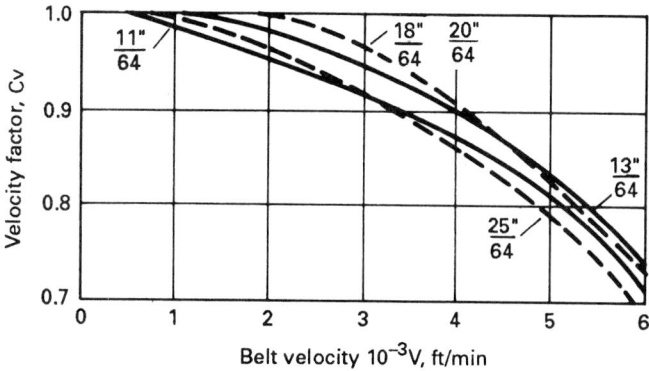

FIGURE 21–1(B)   Velocity correction factor KC_v for use in Eq. (21–4g) for leather belts.

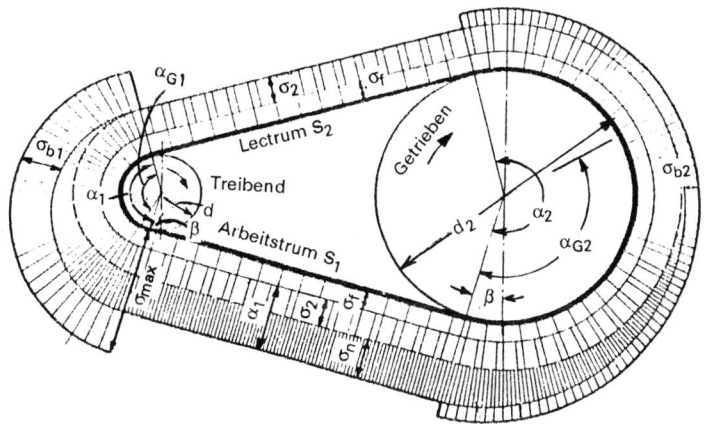

Belt stresses in open drive; $\sigma_f = \sigma_c$ centrifugal stress; σ_ℓ slack side stress; σ_1 tight side stress $= \sigma_2 + \sigma_n$; σ_n effective stress $= \sigma_u$; σ_{b1}, σ_{b2} bending stresses on pulleys *1* and *2* respectively; α_G creep angle (angle over which creep takes place between belt and pulley).

Leertrum S_2 = slack side S_2; treibend = driving; Arbeitstrum S_1 = tight side S_1; getrieben = driven

FIGURE 21–1(C)   Stress distribution in belt. (*G. Niemann*, Maschinenelemente, *Springer International Edition, Allied Publishers Private Ltd., New Delhi, 1978.*)

Particular	Formula

PULLEYS (FIG. 21-2 AND FIG. 21-3)

C. G. Barth's formula for the width of the pulley face

$a = 1.19a_1 + 10$ mm for single belt   **SI**   (21–13a)
$= 1.1a_1 + 5$ mm for double belt   **SI**   (21–13b)

Refer to Table 21–15 for width of pulley

$$a = 1.1875a_1 + \left(\tfrac{3}{8}\right) \text{in} \quad \textbf{US Customary units}$$
(21–13c)

where a and a_1 in in for single belt

$$= 1.09375\, a_1 + (3/16)\, \text{in} \quad \textbf{US Customary units}$$
(21–13d)

where a and a_1 in in for double belt

C. G. Barth's empirical formula for the crown height for wide belts

$h = 0.00426 \sqrt[3]{a^2}$ **SI**; a in m   (21–14a)

$= 0.013125 \sqrt[3]{a^2}$ **US Customary units**   (21–14b)

where a in in

For rubber belts on well-aligned shafts, the crown height

$$h = \frac{a}{200}$$
(21–14c)

(21–14d)

For poorly aligned shafts, the crown height

$$h = \frac{a}{120}$$
(21–14e)

(21–14f)

Refer to Tables 21–16, 21–17A and 21–17B for crown height.

The rim thickness at edge for light-duty pulley

$t = 0.25 \sqrt{D} + 1.5$ mm   (21–15a)

The rim thickness at edge for heavy-duty pulley for a triple belt

$t = 0.375 \sqrt{D} + 3.2$ mm   (21–15b)

The hub diameter of the pulley (Fig. 21–2)

$d_1 = 1.5\, d + 25$ mm   (21–16)

Arms

The bending moment on each arm

$$M_b = \frac{F_\theta D}{i}$$
(21–17)

The section modulus of the arm at the hub

$$Z = \frac{F_\theta D}{i\,\sigma_d}$$
(21–18)

Particular	Formula

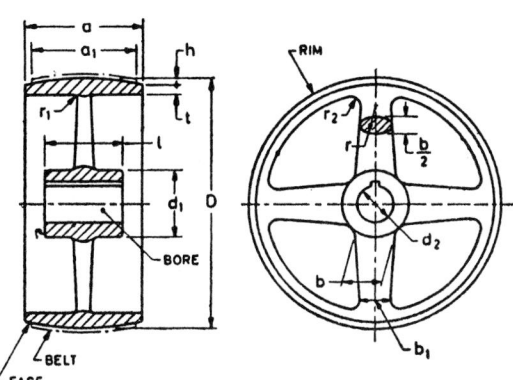

FIGURE 21–2 Cast-iron pulley.

INDIAN STANDARD SPECIFICATION

Cast-iron pulley

Minimum length of bore (Fig. 21–2)

$$l = \tfrac{2}{3}a \qquad (21\text{–}19)$$

It should not exceed a

Half of the difference in diameters d_1 and d_2 (Fig, 21–2)

$$\frac{d_1 - d_2}{2} = 0.412 \sqrt[3]{aD} + 6 \text{ mm for a single belt}$$

$$(21\text{–}20)$$

$$= 0.529 \sqrt[3]{aD} + 6 \text{ mm for a double belt}$$

$$(21\text{–}21)$$

TABLE 21-14
Properties of solid woven fire-resistance conveyor belting for use in coal mines

Belt designation	Direction	Tensile strength/width kN/m	Tensile strength/width kgf/mm	Percentage elongation at break	Tear strength kN	Tear strength kg
4*A*	Warp	385.4	39.3	18	1.3	136.1
	Weft	209.9	21.4	19		
4*B*	Warp	525.6	53.6	18	1.3	136.1
	Weft	262.8	26.8	19		
4*C*	Warp	665.9	67.9	18	1.3	136.1
	Weft	262.8	26.8	19		

TABLE 21–15
Width of flat cast-iron and mild steel pulleys

Width, mm	Tolerance, mm
20, 25, 32	±2
40, 50, 63, 71	
80, 90, 100	±1.5
112, 125, 140	
160, 180, 200	
224, 250, 280, 315	±2
355, 400, 450	
500, 560, 630	±3

TABLE 21–16
Crown of cast iron and mild steel flat pulleys of diameters up to 355 mm

Nominal diameter, D, mm	Crown, h, mm
40–112	0.3
125, 140	0.4
160, 180	0.5
200, 224	0.6
250, 280	0.8
315, 355	1.0

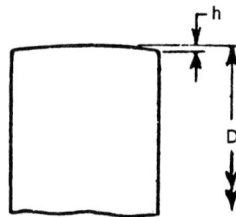

TABLE 21–17*A*
Crown of cast iron and mild steel flat pulleys of diameters 400 to 2000 mm[a]

Nominal diameter, D, mm	Crown h of pulleys of width						
	≤125	140, 160	180, 200	224, 250	280, 315	355	≥400
400	1	1.2	1.2	1.2	1.2	1.2	1.2
450	1	1.2	1.2	1.2	1.2	1.2	1.2
500	1	1.5	1.5	1.5	1.5	1.5	1.5
560	1	1.5	1.5	1.5	1.5	1.5	1.5
630	1	1.5	2	2	2	2	2
710	1	1.5	2	2	2	2	2
800	1	1.5	2	2.5	2.5	2.5	2.5
900	1	1.5	2	2.5	2.5	2.5	2.5
1000	1	1.5	2	2.5	3	3	3
1120	1.2	1.5	2	2.5	3	3	3.5
1250	1.2	1.5	2	2.5	3	3.5	4
1400	1.5	2	2.5	3	3.5	4	4
1600	1.5	2	2.5	3	3.5	4	5
1800	2.0	2.5	3	3.5	4	4.5	5
2000	2.0	2.5	3	3.5	4	4.5	6

[a]All dimensions in mm.
Source: IS 1691, 1968.

TABLE 21–17*B*
Crown height and ISO pulley diameters for flat belts

ISO pulley diameter, in	Crown height, in	ISO pulley diameter, in	Crown height, in	
			$w \le 10$ in	$w > 10$ in
1.6, 2, 2.5	0.012	12.5, 14	0.03	0.03
2.8, 3.15	0.012	12.5, 14	0.04	0.04
3.55, 4, 4.5	0.012	22.4, 25, 28	0.05	0.05
5, 5.6	0.016	31.5, 35.5	0.05	0.06
6.3, 7.1	0.020	40	0.05	0.06
8, 9	0.024	45, 50, 56	0.06	0.08
10, 11.2	0.030	63, 71, 80	0.07	0.10

Crown should be rounded, not angled; maximum roughness is R_o =AA 63 μ in.

Particular	Formula	
The radius r_1 near rim (Fig. 21–2)	$r_1 = b/2$	(21–22)
The radius r_2 near rim (Fig. 21–2)	$r_2 = b/2$	(21–23)

Arms

The number of arms	Use webs for pulleys up to 200-mm diameters	
	$i = 4$ for pulleys above 200-mm diameter and upto 400-mm diameter	(21–24a)
	$i = 6$ for pulleys above 450-mm diameter	(21–24b)
Cross section of arms	Use elliptical section	
Thickness of arm near boss (Fig. 21–2)	$b = 0.294 \sqrt[3]{a\,D/4i}$ **SI**	(21–25a)
	$b = 1.6 \sqrt[3]{\dfrac{a\,D}{i}}$ for single belt	(21–25b)
	US Customary System units	
	$b = 0.294 \sqrt[3]{a\,D/2i}$ **SI**	(21–26a)
	$b = 1.25 \sqrt[3]{\dfrac{a\,D}{i}}$ for double belt	(21–26b)
	US Customary System units	
The diameter of pulleys and arms in pulleys	Refer to Tables 21–18, 21–19, 21–20 and 21–21	
The thickness of arm near rim	b_1— give a taper of 4 mm per 100 mm	
The radius of the cross section of arms	$r = \frac{3}{4}b$	(21–27)

TABLE 21–18
Minimum pulley diameters for given belt speeds and plies[a]

No. of plies	Maximum belt speeds, m/s				
	10	15	20	25	30
2	50	63	80	90	112
3	90	100	112	140	180
4	140	160	180	200	250
5	200	224	250	315	355
6	250	315	355	400	450
7	355	400	450	500	560
8	450	500	560	630	710
9	560	630	710	800	900
10	630	710	800	900	1000

[a] All dimensions in mm.
Source: IS 1370, 1965.

TABLE 21–19
Diameters of flat pulley and tolerances

Nominal diameter, mm	Tolerance, mm	Nominal diameter, mm	Tolerance, mm
40	±0.5	280, 315, 355	±3.0
45, 50	±0.6	400, 450, 500	±4.0
56, 63	±0.8	560, 630, 710	±5.0
71, 80	±1.0	800, 900, 1000	±6.3
90, 100, 112	±1.2	1120, 1250, 1400	±8.0
125, 140	±1.9	1600, 1800, 2000	±10.0
160, 180, 200	±2.0		
224, 250	±2.5		

Source: IS 1691, 1968.

TABLE 21–20
Minimum pulley diameters for conveyor belting

Running	No. of plies	Fabric 28 A	B	C	Fabric 32 A	B	C	Fabric 36 A	B	C	Fabric 42 A	B	C	Fabric 48 A	B	C
>75–100% rated max working tension	2	205	155	155	255	205	155	305	255	205	305	255	205	—	—	—
	3	305	255	205	360	305	205	460	36	305	460	360	305	530	460	330
	4	410	305	255	460	360	305	610	460	360	610	510	410	710	610	510
	5	510	410	360	610	460	360	690	610	460	765	610	510	890	760	635
	6	610	460	410	690	510	460	915	690	610	915	765	610	1065	915	760
	7	690	610	460	765	690	510	1070	765	690	1070	915	690	1245	1065	890
	8	765	690	500	915	765	610	1220	915	690	1220	1020	765	1420	1220	1015
	9	915	690	610	1070	915	610	1375	1070	765	1375	1070	915	1600	1370	1145
	10	1070	765	690	1220	915	690	1525	1220	915	1525	1220	1070	1780	1525	1245
>50–75% rated max working tension	2	205	155	155	205	155	155	255	205	155	305	255	205	—	—	—
	3	305	205	205	305	255	205	410	305	255	460	360	305	430	355	305
	4	360	305	255	410	305	255	510	410	360	610	460	410	560	485	405
	5	460	360	305	510	410	360	690	510	410	765	610	460	710	610	510
	6	510	460	360	610	510	410	765	610	510	915	690	610	865	735	610
	7	610	510	410	690	610	460	915	690	610	1070	915	690	990	865	710
	8	765	610	510	915	690	610	1070	915	690	1220	915	765	1145	965	815
	9	915	690	610	915	690	610	1220	915	765	1375	1070	915	1270	1090	915
	10	915	765	610	1070	915	690	1375	1070	915	1525	1220	915	1420	1220	1015
≤50% rated max working tension	2	155	155	155	205	155	155	255	205	155	255	205	155	—	—	—
	3	255	205	155	305	205	205	360	305	255	410	305	255	380	330	280
	4	305	255	205	360	305	255	460	410	360	510	410	360	510	430	355
	5	410	360	255	460	360	305	610	460	410	690	510	410	635	530	455
	6	510	410	360	510	460	360	690	510	510	765	610	510	735	635	535
	7	610	460	410	610	510	410	765	690	610	915	690	610	865	735	635
	8	690	510	460	765	610	510	915	705	690	1070	765	690	990	865	710
	9	765	610	510	915	690	610	1070	915	765	1220	915	765	1220	965	815
	10	915	690	510	915	765	610	1220	915	915	1220	1070	915	1245	1065	890

Source: IS 1891 (Part 1), 1968.

TABLE 21–21
Number of arms in mild steel pulley

Diameter, mm	Details of spokes	
	No.	Of diameter
250–500	6	19
560–710	8	19
800–1000	10	22
1120	12	22
1250	14	22
1400	16	22
1600	18	22
1800	18	22
2000	22	22

Source: IS 1691, 1968.

Particular	Formula

Mild Steel Pulley

Minimum length of boss (Fig. 21–3)

$l = a/2$ (21–28)
$\not< 100$ mm for 19-mm-diameter spokes
$\not< 138$ mm for 22-mm-diameter spokes

The thickness of rim

$t = 5$ mm for diameters from 400 to 2000 mm
$t = D/200 + 0.003$ m   **SI**   (21–29a)

$t = \dfrac{D}{200} + 0.12$ in for single belt   (21–29b)

        US Customary System units

$t = D/200 + 0.006$ m   **SI**   (21–29c)

$= \dfrac{D}{200} + 0.24$ in for double belt,   (21–29d)

        US Customary units

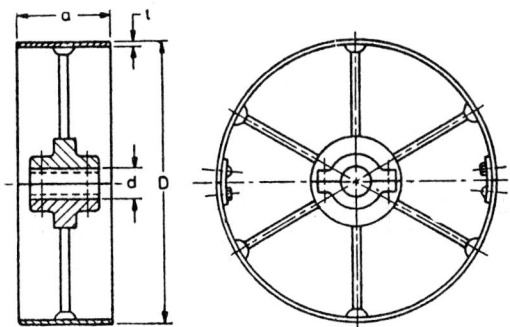

FIGURE 21–3 Mild steel pulley.

The crown height

Arms for mild steel pulleys

Refer to Tables 21–16, 21–17*A*, and 21.17*B*.

Refer to Table 21–21.

Particular	Formula

V-BELT

The formula to obtain the maximum power in kilowatt which the V-belts of sections A, B, C, D, and E can transmit (Table 21–22 and 21–23)

Belt cross section	Power rating per strand, kW	Maximum value of d_e, mm	SI
A	$P^* = v\left(\dfrac{0.45}{v^{0.09}} - \dfrac{19.62}{d_e} - \dfrac{0.765\,v^2}{10^4}\right)$	125	(21–30)
B	$P^* = v\left(\dfrac{0.79}{v^{0.09}} - \dfrac{51.33}{d_e} - \dfrac{1.31\,v^2}{10^4}\right)$	175	(21–31)
C	$P^* = v\left(\dfrac{1.47}{v^{0.09}} - \dfrac{143.27}{d_e} - \dfrac{2.34\,v^2}{10^4}\right)$	300	(21–32)
D	$P^* = v\left(\dfrac{3.16}{v^{0.09}} - \dfrac{507.50}{d_e} - \dfrac{4.77\,v^2}{10^4}\right)$	425	(21–33)
E	$P^* = v\left(\dfrac{4.57}{v^{0.09}} - \dfrac{952}{d_e} - \dfrac{7.05\,v^2}{10^4}\right)$	700	(21–34)

where P^* = maximum power in kilowatt at 180° arc of contact for a belt of average length

The equivalent pitch diameter

$$d_e = d_p\,F_b \qquad (21\text{–}35)$$

Refer to Table 21–24 for F_b, the small-diameter factor.

TABLE 21–22
Classification of V-Belts

Cross-sectional symbol	Nominal top width, W mm	Nominal thickness, T mm
A	13	8
B	17	11
C	22	14
D	30	12
E	33	23

Source: IS 2494, 1964

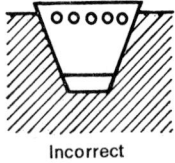

Incorrect

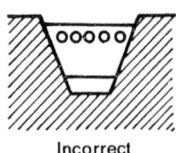

Incorrect

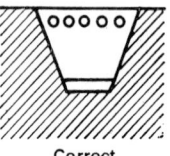

Correct

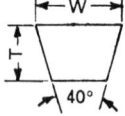

Particular	Formula
The formulas to obtain the maximum horsepower horse of V-belts of A, B, C, D and E sections	Refer to Eqs (21–35a) to (21–35e).

Belt section	Horsepower rating per strand (equations in US Customary units)	
A	$P = V\left(\dfrac{1.95}{V^{0.09}} - \dfrac{3.80}{kd} - 0.0136V^2\right)$	(21–35a)
B	$P = V\left(\dfrac{3.43}{V^{0.09}} - \dfrac{9.83}{kd} - 0.0234V^2\right)$	(21–35b)
C	$P = V\left(\dfrac{6.37}{V^{0.09}} - \dfrac{27.0}{kd} - 0.0416V^2\right)$	(21–35c)
D	$P = V\left(\dfrac{13.6}{V^{0.09}} - \dfrac{93.9}{kd} - 0.0848V^2\right)$	(21–35d)
E	$P = V\left(\dfrac{19.9}{V^{0.09}} - \dfrac{17.8}{kd} - 0.122V^2\right)$	(21–35e)

where

V = belt speed, thousands of ft/min
k = small-diameter factor for speed ratio of drive from Fig. 21–4b
d = pitch diameter of small sheave, in

TABLE 21-23
Ratings for V-belts in kilowatts

Equivalent pitch diameter, d_e, mm

Belt speed, m/s	Cross section A						Cross section B						Cross section C							Cross section D								Cross section E					
	80	90	100	110	120	125	130	140	150	160	170	180	180	200	220	240	260	280	300	300	320	340	360	380	400	420	430	450	500	550	600	650	700
0.5	0.13	0.13	0.14	0.14	0.15	0.15	0.22	0.22	0.22	0.22	0.29	0.29	0.29	0.44	0.44	0.51	0.51	0.51	0.51	0.81	0.88	0.96	0.96	1.03	1.03	1.10	1.10	1.40	1.47	1.54	1.62	1.69	1.77
1	0.22	0.24	0.25	0.27	0.28	0.29	0.44	0.44	0.54	0.44	0.51	0.51	0.51	0.73	0.81	0.88	0.88	0.96	0.96	1.47	1.54	1.69	1.77	1.84	1.91	1.91	1.91	2.43	2.65	2.87	2.94	3.09	3.24
2	0.37	0.40	0.43	0.46	0.49	0.51	0.66	0.74	0.81	0.81	0.88	0.88	0.88	1.32	1.47	1.85	1.69	1.77	1.84	2.50	2.79	2.94	3.09	3.24	3.38	3.53	3.53	4.34	4.78	5.15	5.44	5.66	5.88
3	0.51	0.58	0.64	0.68	0.72	0.74	0.96	1.03	1.10	1.17	1.25	1.39	1.62	1.84	2.06	2.21	2.35	2.50	2.57	3.46	3.75	4.04	4.34	4.56	4.71	4.92	5.00	6.03	6.69	7.21	7.65	8.02	8.38
4	0.58	0.74	0.81	0.88	0.93	0.96	1.18	1.32	1.40	1.47	1.47	1.62	2.06	2.35	2.57	2.76	3.16	3.31	3.31	4.34	4.78	5.15	5.44	5.74	6.03	6.25	6.40	7.58	8.46	9.19	9.78	10.22	10.74
5	0.74	0.85	0.95	1.04	1.13	1.18	1.47	1.54	1.69	1.84	1.91	1.99	2.35	2.79	3.09	3.38	3.60	3.75	3.97	5.07	5.66	6.10	6.55	6.91	7.21	7.58	7.65	9.05	10.15	11.03	11.77	12.36	13.02
6	0.81	0.94	1.05	1.15	1.25	1.32	1.62	1.84	1.99	2.06	2.21	2.28	2.72	3.16	3.60	3.90	4.19	4.41	4.63	5.81	6.47	7.06	7.51	7.94	8.38	8.75	8.90	10.44	11.69	12.80	13.68	14.49	15.07
7	0.88	1.08	1.25	1.39	1.50	1.54	1.91	2.13	2.21	2.35	2.50	2.65	3.02	3.60	4.05	4.41	4.71	5.00	5.22	6.47	7.28	7.94	8.46	8.97	9.41	9.86	10.08	11.77	13.32	14.49	15.51	16.40	17.14
8	0.90	1.17	1.35	1.50	1.62	1.69	2.06	2.28	2.43	2.65	2.79	2.94	3.31	3.97	4.49	4.85	5.22	5.59	5.87	7.13	7.94	8.68	9.34	10.00	10.52	10.96	11.25	13.02	14.78	16.11	17.28	18.31	19.05
9	1.03	1.32	1.54	1.69	1.77	1.84	2.21	2.43	2.65	2.87	3.02	3.16	3.60	4.27	4.95	5.22	5.81	6.10	6.40	7.65	8.68	9.49	10.22	10.96	11.47	12.06	12.28	14.12	16.11	17.72	18.98	20.15	21.11
10	1.10	1.40	1.62	1.77	1.91	1.99	2.35	2.65	2.87	3.09	3.31	3.46	3.82	4.56	5.22	5.81	6.25	6.62	6.99	8.23	9.34	10.22	11.03	11.84	12.43	13.02	13.31	15.22	17.43	19.12	20.59	21.85	22.95
11	1.18	1.47	1.69	1.91	2.06	2.13	2.50	2.79	3.09	3.31	3.53	3.67	4.04	4.92	5.59	6.18	6.69	7.13	7.58	8.68	9.86	10.88	11.84	12.58	13.39	14.05	14.34	16.25	18.61	20.59	22.20	23.61	24.71
12	1.25	1.54	1.84	2.06	2.21	2.28	2.65	3.02	3.31	3.53	3.75	3.75	4.19	5.15	5.96	6.62	7.13	7.72	8.00	9.12	10.37	11.47	12.50	13.39	14.19	14.93	15.30	17.21	19.78	21.92	23.68	25.15	26.40
13	1.32	1.62	1.91	2.13	2.35	2.43	2.79	3.16	3.46	3.75	3.97	4.19	4.34	5.44	6.25	6.91	7.58	8.16	8.46	9.49	10.86	12.06	13.16	14.05	15.00	15.74	16.11	18.02	20.81	23.17	25.08	26.69	28.02
14	1.32	1.69	1.99	2.28	2.50	2.50	2.87	3.16	3.60	3.90	4.19	4.56	4.49	5.59	6.55	7.28	7.94	8.53	8.97	9.79	11.25	12.88	13.75	14.56	15.74	16.55	16.99	18.84	21.85	24.35	26.40	28.10	29.57
15	1.32	1.77	2.06	2.35	2.57	2.65	2.94	3.38	3.75	4.05	4.34	4.63	4.63	5.81	6.77	7.58	8.31	8.83	9.41	10.00	11.62	13.09	14.34	15.44	16.40	17.28	17.72	19.56	22.80	25.45	27.65	29.49	31.04
16	1.40	1.84	2.13	2.43	2.65	2.79	3.02	3.46	3.90	4.19	4.49	4.78	4.71	5.96	7.06	7.87	8.61	9.27	9.78	10.30	11.91	13.46	14.78	15.96	16.99	18.02	18.46	20.15	23.61	26.40	28.83	30.82	32.44
17	1.40	1.84	2.21	2.50	2.79	2.87	3.09	3.66	3.97	4.34	4.71	4.92	4.78	6.10	7.21	8.09	8.90	9.56	10.15	10.44	12.21	13.83	15.22	16.55	17.58	18.61	19.05	20.74	24.42	27.36	29.86	31.99	33.76
18	1.40	1.84	2.28	2.57	2.87	2.94	3.16	3.60	3.60	4.19	4.78	5.07	4.78	6.25	7.35	8.38	9.19	9.93	10.51	10.51	12.43	14.12	15.59	16.99	18.09	19.20	19.71	21.18	25.08	28.24	30.89	33.10	35.01
19	1.40	1.84	2.28	2.65	2.94	3.02	3.16	3.68	4.19	4.46	4.78	5.22	4.78	6.33	7.58	8.53	9.41	10.15	10.81	10.51	12.50	14.27	15.89	17.36	18.53	19.86	20.23	21.55	25.60	28.97	31.77	34.13	36.19
20	1.32	1.91	2.35	2.72	3.02	3.00	3.16	3.75	4.19	4.63	4.92	5.44	4.78	6.33	7.65	8.68	9.63	10.44	11.11	10.51	12.58	14.49	16.11	17.65	18.98	20.15	20.14	21.77	26.11	29.71	32.58	35.08	37.22
21	1.32	1.91	2.35	2.72	3.02	3.16	3.16	3.75	4.27	4.71	5.00	5.44	4.71	6.33	7.72	8.83	9.86	10.66	11.33	10.37	13.31	14.56	16.40	17.87	19.27	20.52	21.11	21.99	26.48	30.23	33.17	35.97	38.17
22	1.25	1.91	2.35	2.72	3.09	3.16	3.16	3.75	4.27	4.78	5.15	5.52	4.56	6.25	7.80	8.90	10.0	10.81	11.62	10.22	12.58	14.56	16.47	18.02	19.49	20.81	21.48	22.06	26.77	30.74	33.19	36.70	38.98
23	1.18	1.84	2.35	2.79	3.09	3.24	3.09	3.75	4.27	4.78	5.22	5.55	4.34	6.18	7.80	8.97	10.08	11.03	11.77	9.93	12.43	14.34	16.40	18.24	19.71	21.11	21.70	21.99	27.07	31.04	34.42	37.29	39.72
24	1.25	1.84	2.35	2.79	3.09	3.24	3.02	3.68	4.19	4.78	5.22	5.66	4.15	6.18	7.72	9.05	10.15	11.11	11.91	9.63	12.21	14.34	16.40	18.17	19.78	21.26	21.99	21.92	27.07	31.33	34.79	37.88	40.38
25	1.10	1.77	2.28	2.79	3.16	3.31	2.94	3.60	4.19	4.78	5.22	5.66	4.15	6.03	7.72	9.05	10.15	11.18	11.99	9.19	11.91	13.97	16.11	18.17	19.78	21.33	22.06	21.71	26.99	31.48	35.08	38.25	40.89
26	1.03	1.69	2.28	2.72	3.16	3.31	2.79	3.53	4.19	4.71	5.22	5.66	3.82	5.88	7.58	8.97	10.15	11.18	12.06	8.68	11.47	13.53	15.81	17.87	19.78	21.33	21.99	21.25	26.92	31.48	35.30	38.54	41.46
27	0.88	1.62	2.20	2.72	3.09	3.31	2.65	3.45	4.12	4.63	5.15	5.66	3.60	5.66	7.43	8.90	10.15	11.18	12.13	8.16	11.03	13.53	15.81	17.80	19.56	21.26	21.99	20.74	26.62	31.41	35.38	38.69	41.56
28	0.81	1.54	2.13	2.65	3.09	3.24	2.50	3.31	3.97	4.56	5.15	5.59	3.24	5.44	7.35	8.75	10.08	11.18	12.13	7.51	10.51	13.09	15.44	17.43	19.34	21.11	21.84	20.15	25.89	31.18	35.30	38.83	41.78
29	0.66	1.47	2.06	2.57	3.02	3.24	2.35	3.16	3.82	4.49	5.00	5.52	3.19	5.15	7.06	8.60	9.93	11.03	12.06	6.77	9.86	12.58	15.00	17.14	19.12	20.89	21.62	19.49	25.74	30.89	35.08	39.76	41.78
30	0.51	1.32	1.99	2.50	2.94	3.16	2.13	2.94	3.68	4.34	4.92	5.44	2.43	4.18	6.77	8.38	9.71	10.80	11.99	5.96	9.12	11.91	14.42	16.70	18.68	20.52	21.33	18.61	25.08	30.45	34.79	38.54	41.70

Source: IS 2494, 1964.

TABLE 21–24
Small-diameter factor, F_b

Speed ratio range	Small-diameter factor
1.000–1.019	1.00
1.020–1.032	1.01
1.033–1.055	1.02
1.056–1.081	1.03
1.082–1.109	1.04
1.110–1.142	1.05
1.143–1.178	1.06
1.179–1.222	1.07
1.223–1.274	1.08
1.275–1.340	1.09
1.341–1.429	1.10
1.430–1.562	1.11
1.563–1.814	1.12
1.815–2.948	1.13
≥ 1.949	1.14

Source: IS 2494, 1964.

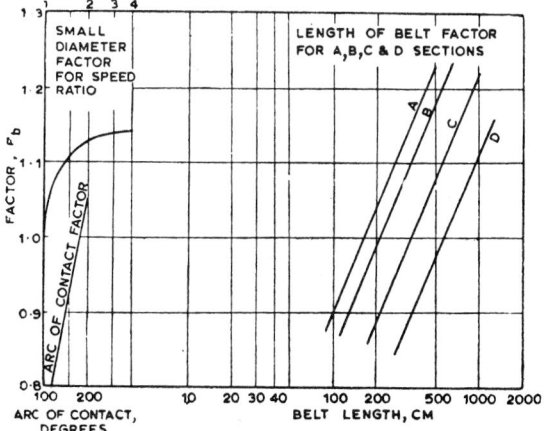

FIGURE 21–4(*a*)   Factors for power rating of V-belt for use with Eqs. (21–30) to (21–35).

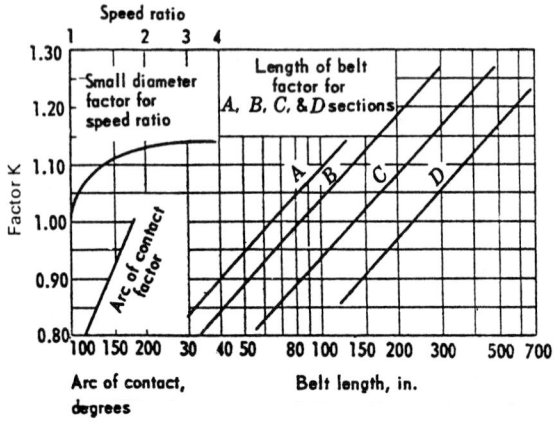

FIGURE 21–4(*b*) Factors for horsepower ratings of V-belts for use with Eqs. (21–35a) to (21–35e).

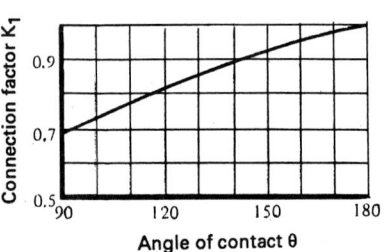

FIGURE 21–4(*c*) Correction factor K_1 for angle of contact. *J. E. Shigley and C. R. Mischke*, Mechanical Engineering Design, *McGraw-Hill Book Company, 1989.*

TABLE 21–25
Correction factors for arc of contact, F_d

Arc of contact on smaller pulley, deg	Correction factor (proportion of 180° rating)		Arc of contact on smaller pulley, deg	Correction factor (proportion of 180° rating)	
	VV	V-flat		VV	V-flat
180	1.00	0.75	133	0.87	0.86
177	0.99	0.76	130	0.86	0.86
174	0.99	0.76	127	0.85	0.85
171	0.98	0.77	123	0.83	0.83
169	0.97	0.78	120	0.82	0.82
166	0.97	0.79	117	0.81	0.81
163	0.96	0.79	113	0.80	0.80
160	0.95	0.80	110	0.78	0.78
157	0.94	0.81	106	0.77	0.77
154	0.93	0.81	103	0.75	0.75
151	0.93	0.82	99	0.73	0.73
148	0.92	0.83	95	0.72	0.72
145	0.91	0.83	91	0.70	0.70
142	0.90	0.84	87	0.68	0.68
139	0.89	0.85	83	0.65	0.65
136	0.88	0.85			

Source: IS 2494, 1964.

TABLE 21–26
Correction factors for belt length, F_c

Nominal inside length, mm	Belt cross section					Nominal inside length, mm	Belt cross section				
	A	*B*	*C*	*D*	*E*		*A*	*B*	*C*	*D*	*E*
610	0.80	—	—	—	—	2159	1.05	0.99	0.90	—	—
660	0.81	—	—	—	—	2286	1.06	1.00	0.91	—	—
711	0.82	—	—	—	—	2438	1.08	—	0.92	—	—
787	0.84	—	—	—	—	2464	—	1.02	—	—	—
813	0.85	—	—	—	—	2540	—	1.03	—	—	—
889	0.87	0.81	—	—	—	2667	1.10	1.04	0.94	—	—
914	0.87	—	—	—	—	2845	1.11	1.05	0.95	—	—
965	0.88	0.83	—	—	—	3048	1.13	1.07	0.97	0.86	—
991	0.88	—	—	—	—	3150	—	—	0.97	—	—
1016	0.89	0.84	—	—	—	3251	1.14	1.08	0.98	0.87	—
1067	0.90	0.85	—	—	—	3404	—	—	0.99	—	—
1092	0.90	—	—	—	—	3658	—	1.11	1.00	0.90	—
1168	0.92	0.87	—	—	—	4013	—	1.13	1.02	0.92	—
1219	0.93	0.88	—	—	—	4115	—	1.14	1.03	0.92	—
1295	0.94	0.89	0.80	—	—	4394	—	1.15	1.04	0.93	—
1372	—	0.90	—	—	—	4572	—	1.16	1.05	0.94	—
1397	0.96	0.90	—	—	—	4953	—	1.18	1.07	0.96	—
1422	0.96	0.90	—	—	—	5334	—	1.19	1.08	0.96	0.94
1473	0.97	—	—	—	—	6045	—	—	1.11	1.00	0.96
1524	0.98	0.92	0.82	—	—	6807	—	—	1.14	1.03	0.99
1600	0.99	—	—	—	—	7569	—	—	1.16	1.05	1.01
1626	0.99	—	—	—	—	8331	—	—	1.19	1.07	1.03
1651	1.00	0.94	—	—	—	9093	—	—	1.21	1.09	1.05
1727	1.00	0.95	0.85	—	—	9855	—	—	1.23	1.11	1.07
1778	1.01	9.95	—	—	—	10617	—	—	1.24	1.12	1.09
1905	1.02	0.97	0.87	—	—	12141	—	—	—	1.16	1.12
1981	1.03	0.98	—	—	—	13665	—	—	—	1.18	1.14
2032	1.04	—	—	—	—	15189	—	—	—	1.20	1.17
2057	1.04	0.98	0.89	—	—	16713	—	—	—	1.23	1.19

Source: IS 2494, 1964.

TABLE 21–27
Correction factors for industrial service, F_a

		Type of driving unit					
		AC motors; normal torque, squirrel cage, synchronous and split phase DC motors; shunt-wound, multiple cylinder internal combustion engines > 600 rpm			AC motors; high torque, high slip repulsion induction, single phase, series-wound and slip-ring DC motors; series-wound and compound wound; single-cylinder internal-combustion engines; multicylinder internal-combustion engines < 600 rpm, line shafts, clutches, brakes, direct on-line starting		
Severity of service	Type of driven machines	≤ 10 h	> 10 to 16 h	> 16 h and continuous service	≤ 10 h	> 10 to 16 h	> 16 h and continuous service
Light-duty	Agitators for liquids, blowers, and exhausters, centrifugal pumps and compressors, fans up to 7.5 kW (10 hp) and light-duty conveyors	1.0	1.1	1.2	1.1	1.2	1.3
Medium-duty	Belt conveyors for sand, grain, etc; dough mixers; fans over 7.5 kW (10 hp); generators; line shafts; laundry machinery; machine tools; punches, presses and shears; printing machinery; positive-displacement rotary pumps; revolving and vibrating screens	1.1	1.2	1.3	1.2	1.3	1.4
Heavy-duty	Brick machinery, bucket elevators, exciters, piston compressors, conveyors (drag-pan-screw), hammer mills, paper mill beaters, piston pumps, positive displacement blowers, pulverizers, saw mill and woodworking machinery, and textile machinery	1.2	1.3	1.4	1.4	1.5	1.6
Extra-Heavy-duty	Crushers (gyratory-jaw-roll), mills (ball-rod-tube), hoists, and rubber (calenders-extruders-mills) machinery	1.3	1.4	1.5	1.5	1.6	1.8

Note: This table gives only a few examples of particular machines. If an idler pulley is used, the following values must be added to the service factors:

Idler pulley on the slack side	{ inside:	0	Idler pulley on the tight side	{ inside:	0.1
	{ outside:	0.1		{ outside:	0.2

Source: IS 2494, 1964.

TABLE 21–28
Nominal inside length, nominal pitch lengths and permissible length variations for V-belts

Nominal inside length, mm	Nominal pitch length, mm Cross section					Pitch length variation	
	A	B	C	D	E	PLL[a]	MVL[b]
610	645						
660	696					+11.4	
711	747					−6.4	
787	823						
813	848					+12.5	
889	925	932				−7.5	
914	950						2.5
965	1001	1008					
991	1076						
1016	1051	1059				+14.0	
1067	1102	1110				−8.9	
1092	1128						
1168	1204	1212					
1219	1255	1262					
1295	1331	1339	1351			+16.0	
1372		1415				−9.0	
1397	1433	1440					
1422	1451	1466					
1473	1509						
1524	1560	1567	1580				
							5.0
1600	1636						
1626	1661					+17.8	
1651	1687	1694				−12.5	
1727	1763	1770	1783				
1778	1814	1821					
1905	1941	1948	1991				
1981	2017	2024					
2032	2068						
2057	2093	2101	2113			+30	
2159	2195	2202	2215			−16	7.5
2286	2322	2329	2342				
2438	2474		2494				
2464		2507					
2540		2583				+34	
2667	2703	2710	2723			−18	
2845	2880	2888	2901				
3048	3084	3091	3104	3127			10
3150			3205				
3251	3287	3294	3307	3330		+38	
3404			3459			−21	
3658	3693	3701	3713	3736			
4013		4056	4069	4092			
4115		4158	4171	4194			
4394		4437	4450	4473		+43	
4572		4615	4628	4651		−24	
							12.5

TABLE 21–28
Nominal inside length nominal pitch lengths and permissible length variations for V-belts (*Cont.*)

| Nominal inside length, mm | Nominal pitch length, mm Cross section | | | | | Pitch length variation | |
	A	B	C	D	E	PLL[a]	MVL[b]
4953		4996	5009	5032		+49	
5334		5377	5390	5413	5426	−28	
6045			6101	6124	6137		
6807			6863	5886	6899	+56	
7569			7625	7648	7661	−32	
8331			8387	8410	8423	+65	15
9093			9149	9172	9185		
9855				9934	9947	−37	
10617				10696	10709	+76	
12141				12220	12233	−43	
							17.5
13665				13744	13757	+89	
15189				15268	15281	−50	
16713				16792	16805	+105	
						−59	

[a] Pitch length limit.
[b] Maximum variation in length within a matched set.
Source: IS 2494, 1964.

TABLE 21–29
Dimensions for standard V-grooved pulleys

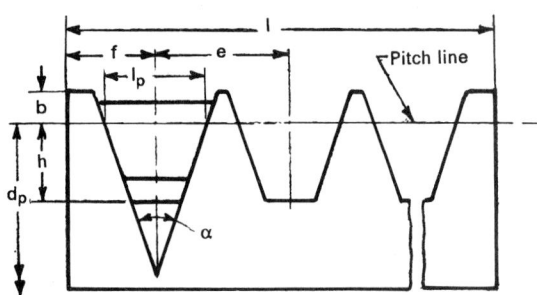

Groove section	Pitch width, lp, mm	Minimum height of groove above pitch line, b_{min}, mm	Minimum depth of groove below pitch line, h, min, mm	Center to center distance of grooves, e, mm	Edge of pulley to first groove center, f, mm
A	11	3.3	8.7	15 ± 0.3	10 $^{+2}_{-1}$
B	14	4.2	10.1	19 ± 0.4	12.5 $^{+2}_{-1}$
C	19	5.7	14.3	25.5 ± 0.5	17 $^{+2}_{+1}$
D	27	8.1	19.9	37 ± 0.6	24 $^{+3}_{-1}$
E	32	9.6	23.4	44.5 ± 0.7	29 $^{+4}_{-1}$

Source: IS: 3142–1965.

TABLE 21-30A
Recommended standard pulley pitch diameters

Series of pitch diameters

Nominal value, mm	Pitch diameter limits Min, mm	Pitch diameter limits Max, mm	A	B	C	D	E
75	75	76.3	3				
80	80	81.3	3				
85	85	86.4	3				
90	90	91.4	1				
95	95	96.5	2				
100	100	101.6	1				
106	106	107.7	2				
112	112	113.8	1				
118	118	119.9	2				
125	125	127.0	1	2			
132	132	134.1	2	2			
140	140	142.2	1	1	1		
150	150	152.4	2	2	2		
160	160	162.6	1	1	1		
170	170	172.7	3	2	2		
180	180	182.9	1	1	1		
190	190	193.0	3	3			
200	200	203.2	1	1	1		
212	212	215.4			2		
224	224	227.6	2	2	1		
236	236	239.8			2		
250	250	254.0	1	1	2		
265	265	269.2			2		
280	280	284.5	2	2	2		
300	300	304.8	2	2	1		
315	315	320.0	1	1	1		
355	355	360.7	2	2	2		1
375	375	381.0		2	2	2	2
400	400	406.4	1	1	1	1	
425	425	431.8		2	2	2	1
450	450	457.2	2			1	1
475	475	482.6				2	
500	500	508.8	1	1	1	1	1
530	530	538.5		3	3	3	2
560	560	569.0	2	2	2	2	1
600	600	609.6		2	2	2	2
630	630	640.0	1	1	1	1	1
670	670	680.7			1		2
710	710	721.4	2	2	2	2	1
750	750	762.0		2	2	2	3
800	800	812.8	3	1	1	1	1
900	900	914.4		2	2	2	2
1000	1000	1016.0		1	1	1	1
1060	1060	1077.0				2	
1120	1120	1137.9		3		2	2
1250	1250	1270.0			1	1	1
1400	1400	1422.4			2	2	2
1500	1500	1524.0				2	2
1600	1600	1625.6			1	1	1
1800	1800	1828.4			1	2	2
1900	1900	1930.4					2
2000	2000	2032.0				1	1
2240	2240	2275.8					2
2500	2500	2540.0					1

Degree of preference[a] for pitch diameters, according to groove section (A, B, C, D, E)

[a] Key: 1—first preference; 2—second preference; 3—not recommended.
Source: IS 3142, 1965.

TABLE 21–30*B*
Standard V-belt sections

Belt section	Width, *a*, in	Thickness, *b*, in	Minimum sheave diameter, in	hp range, one or more belts
A	$^1/_2$	$^{11}/_{32}$	3.0	$\frac{1}{4} - 10$
B	$^{21}/_{32}$	$^7/_{16}$	5.4	1–25
C	$^7/_8$	$^{17}/_{32}$	9.0	15–100
D	$1^1/_4$	$^3/_4$	13.0	50–250
E	$1^1/_2$	1	21.6	≥ 100

TABLE 21–30*C*
Inside circumferences of standard V-belts

Section	Circumference, in
A	26, 31, 33, 35, 38, 42, 46, 48, 51, 53, 55, 57, 60, 62, 64, 66, 68, 71, 75, 78, 80, 85, 90, 96, 105, 112, 120, 128
B	35, 38, 42, 46, 48, 51, 53, 55, 57, 60, 62, 64, 65, 66, 68, 71, 75, 78, 79, 81, 83, 85, 90, 93, 97, 100, 103, 105, 112, 120, 128, 131, 136, 144, 158, 173, 180, 195, 210, 240, 270, 300
C	51, 60, 68, 75, 81, 85, 90, 96, 105, 112, 120, 128, 136, 144, 158, 162, 173, 180, 195, 210, 240, 270, 300, 330, 360, 390, 420
D	120, 128, 144, 158, 162, 173, 180, 195, 210, 240, 270, 300, 330, 360, 390, 420, 480, 540, 600, 660
E	180, 195, 210, 240, 270, 300, 330, 360, 390, 420, 480, 540, 600, 660

Source: J. E. Shigley and C. R. Mischke, *Mechanical Engineering Design*, McGraw-Hill Book Company, New York, 1989.

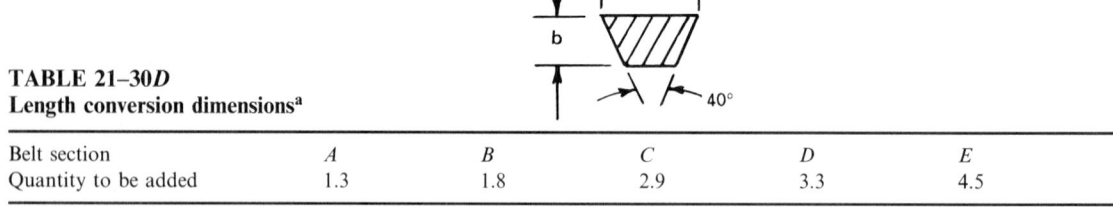

TABLE 21–30*D*
Length conversion dimensions[a]

Belt section	A	B	C	D	E
Quantity to be added	1.3	1.8	2.9	3.3	4.5

[a]Add the values given above to the inside circumference to obtain the pitch length in inches.
Source: J. E. Shigley and C. R. Mischke, *Mechanical Engineering Design*, McGraw-Hill Book Company, New York, 1989.

TABLE 21–30*E*
Horsepower rating of standard V-belts

Belt section	Sheave pitch diameter, in	Belt speed, ft/min				
		1000	2000	3000	4000	5000
A	2.6	0.47	0.62	0.53	0.15	
	3.0	0.66	1.01	1.12	0.93	0.38
	3.4	0.81	1.31	1.57	1.53	1.12
	3.8	0.93	1.55	1.92	2.00	1.71
	4.2	1.03	1.74	2.20	2.38	2.19
	4.6	1.11	1.89	2.44	2.69	2.58
	≥ 5.0	1.17	2.03	2.64	2.96	2.89
B	4.2	1.07	1.58	1.68	1.26	0.22
	4.6	1.27	1.99	2.29	2.08	1.24
	5.0	1.44	2.33	2.80	2.76	2.10
	5.4	1.59	2.62	3.24	3.34	2.82
	5.8	1.72	2.87	3.61	3.85	3.45
	6.2	1.82	3.09	3.94	4.28	4.00
	6.6	1.92	3.29	4.23	4.67	4.48
	≥ 7.0	2.01	3.46	4.49	5.01	4.90
C	6.0	1.84	2.66	2.72	1.87	
	7.0	2.48	3.94	4.64	4.44	3.12
	8.0	2.96	4.90	6.09	6.36	5.52
	9.0	3.34	5.65	7.21	7.86	7.39
	10.0	3.64	6.25	8.11	9.06	8.89
	11.0	3.88	6.74	8.84	10.0	10.1
	≥12.0	4.09	7.15	9.46	10.9	11.1
D	10.0	4.14	6.13	6.55	5.09	1.35
	11.0	5.00	7.83	9.11	8.50	5.62
	12.0	5.71	9.26	11.2	11.4	9.18
	13.0	6.31	10.5	13.0	13.8	12.2
	14.0	6.82	11.5	14.6	15.8	14.8
	15.0	7.27	12.4	15.9	17.6	17.0
	16.0	7.66	13.2	17.1	19.2	19.0
	≥17.0	8.01	13.9	18.1	20.6	20.7
E	16.0	8.68	14.0	17.5	18.1	15.3
	18.0	9.92	16.7	21.2	23.0	21.5
	20.0	10.9	18.7	24.2	26.9	26.4
	22.0	11.7	20.3	26.6	30.2	30.5
	24.0	12.4	21.6	28.6	32.9	33.8
	26.0	13.0	22.8	30.3	35.1	36.7
	≥28.0	13.4	23.7	31.8	37.1	39.1

Source: J. E. Shigley and C. R. Mischke, *Mechanical Engineering Design*, McGraw-Hill Book Company, New York, 1989.

TABLE 21–30F
Belt-length correction factor, K_2^a

Length factor	Nominal belt length, in				
	A belts	B belts	C belts	D belts	E belts
0.85	≤ 35	≤ 46	≤ 75	≤ 128	
0.90	38–46	48–60	81–96	144–162	≤ 195
0.95	48–55	62–75	105–120	173–210	210–240
1.00	60–75	78–97	128–158	240	270–300
1.05	78–90	105–120	162–195	270–330	330–390
1.10	96–112	128–144	210–240	360–420	420–480
1.15	120 and up	158–180	270–300	480	540–600
1.20		195 and up	330 and up	540 and up	660

[a] Multiply the rated horsepower per belt by this factor to obtain the corrected horsepower

Particular	Formula
Number of belts	$$i = \frac{PF_a}{P^* F_c F_d} \qquad (21\text{–}36)$$ where P = drive power in kW Obtain F_d, F_c, and F_a from Tables 21–25, 21–26, and 21–27, respectively.
The diameter of larger pulley	$$D = \frac{d n_1}{n_2} \eta \qquad (21\text{–}37)$$
Nominal pitch length of belt	$$L = 2C + \frac{\pi}{2}(D + d) + \frac{(D - d)^2}{4C} \qquad (21\text{–}38)$$
For nominal inside length, nominal pitch lengths and permissible length variations for standard sizes of V-belts	Refer to Table 21–28.
Dimensions for standard V-grooved pulley	Refer to Table 21–29.
Fr small-diameter factor, for speed ratio and length of belt factor	Refer to Fig. 21–4a and 21–4b.
Recommend standard pitch diameters of pulleys	Refer to Table 21–30A.
For further data for design of V-belts in US Customary system units for use with Eqs (21–35a) to (21–35e)	Refer to Tables 21–30B, to 21–30F and Figs. 21–4b and 21–4c.
Center distance for a given belt length and diameters of pulleys	$$C = \frac{L}{4} - \frac{\pi(D + d)}{8} + \sqrt{\left\{\frac{L}{4} - \frac{\pi(D + d)}{8}\right\}^2 - \frac{(D - d)^2}{8}} \qquad (21\text{–}39)$$
Maximum center distance	$C_{max} = 2(D + d) \qquad (21\text{–}40)$
Minimum center distance	$C_{min} = 0.55(D + d) + t \qquad (21\text{–}41)$

Particular	Formula

MINIMUM ALLOWANCES FOR ADJUSTMENT OF CENTERS FOR TWO TRANSMISSION PULLEYS

Lower limiting value	$C_L = C_{nominal} - 1.5\% L$	(21–42)
Higher limiting value	$C_H = C_{nominal} + 3\% L$	(21–43)

INITIAL TENSION

In order to give the initial tension, the belts may be stretched to	$\Delta L = 0.5\%$ to 1% L	(21–44)
Arc of contact angle	$\theta = 2 \cos^{-1} \dfrac{D-d}{2C}$	(21–45)
	$= 180° - 60° \left(\dfrac{D-d}{C} \right)$	(21–46)

CONVEYOR (TABLES 21–12, 21–14, 21–20, AND 21–31)

The average capacity, C, of conveyor in m^3 per hour at 0.5 m/s

For flat belts	when a_1 in m	$C = 70 \, a_1{}^2$	**SI**	(21-47a)
	when a_1 in in	$= 27\,5\,6 \, a_1{}^2$	**US Customary units**	(21-47b)
	when a_1 in mm	$= 0.7 \times 10^5 \, a_1{}^2$	**SI**	(21-47c)
For belts on idlers	when a_1 in m	$C = 88 \, a_1{}^2$	**SI**	(21-48a)
	when a_1 in in	$= 3\,4\,6\,5 \, a_1{}^2$	**US Customary units**	(21-48b)
	when a_1 in mm	$= 0.88 \times 10^5 \, a_1{}^2$	**SI**	(21-48c)

TABLE 21–31

Maximum inclination of belt conveyors

Material conveyed	Maximum inclination, deg	Material conveyed	Maximum inclination, deg
Briquets and egg-shaped material	12	Glass batch	20
Wet-mixed concrete	15	Run-of-mine coal	22
Sized coal	18	Run-of-bank gravel	22
Washed and screened gravel	18	Crushed ore	25
Loose cement	20	Crushed stone	20
Crushed and screened coke	20	Tempered foundary sand	25
Sand	20	Wood chips	28

Particular	Formula

For belts on three-
to five-step idlers   when a_1 in m

$$C = 132\,a_1^2 \text{ to } 154\,a_1^2 \quad \textbf{SI} \qquad (21\text{–}49a)$$

when a_1 in in

$$= 5158\,a_1^2 \text{ to } 6063\,a_1^2 \qquad (21\text{–}49b)$$

when a_1 in mm

$$\textbf{US Customary units}$$

$$= 1.32 \times 10^5\,a_1^2 \text{ to } 1.54 \times 10^5\,a_1^2 \quad \textbf{SI} \quad (21\text{–}49c)$$

The power required by a horizontal belt conveyor

$$P = \left[(W_I + 2W_B + W_L)\,\mu\,\frac{d}{D}\right]\frac{vL}{1000} + P_T$$

$$\textbf{SI} \qquad (21\text{–}50a)$$

where W in N/m, v in m/s, L in m, and P in kW

$$= \left[(W_I + 2W_B + W_L)\,\mu\,\frac{d}{D}\right]\frac{vL}{102} + P_T \qquad (21\text{–}50b)$$

$$\textbf{Customary metric}$$

where W in kgf/m, v in m/s, L in m and P in kW

$$= \left[(W_I + 2W_B + W_L)\,\mu\,\frac{d}{D}\right]\frac{vL}{33,000} + P_T$$

$$(21\text{–}50c)$$

$$\textbf{US Customary units}$$

where v in ft/min; L in in.; P in hp, W in lbf/in;

where μ = coefficient of friction of idler bearing

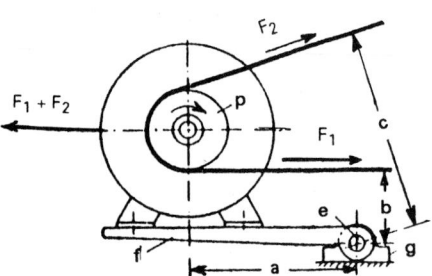

$$= 0.15 \text{ for roller bearings}$$
$$= 0.35 \text{ for grease lubricated idlers}$$

FIGURE 21–5 Rockwood pivoted motor base.

SHORT CENTER DRIVE

Rockwood drive (Fig. 21–5)

The value of F_1

$$F_1 = \frac{aW + cF_n}{c + b} \qquad (21\text{–}51)$$

The value of F_2

$$F_2 = \frac{aW - bF_n}{c + b} \qquad (21\text{–}52)$$

The pivot-arm length for motor of weight W

$$a = \frac{F_n\left(b\dfrac{F_1}{F_2} + c\right)}{W\left(\dfrac{F_1}{F_2} - 1\right)} \qquad (21\text{–}53)$$

where F_n = required net pull, kN (lbf)

W = weight of the motor, kN (lbf)

Particular	Formula

ROPES

Manila rope (Tables 21–32 and 21–34)

The ultimate load

$$P_u = 48053\, d^2 \qquad \textbf{SI} \qquad (21\text{–}54a)$$
where d in m and P_u in kN

$$= 7000\, d^2 \qquad \textbf{US Customary unit} \qquad (21\text{–}54b)$$
where d is diameter of rope in in and P_u in lbf

The maximum tension on the tight side

$$F_1 = 137.5 \times 10^4 d^2 = F + \frac{F}{2} + F_c \qquad \textbf{SI} \qquad (21\text{–}55a)$$

where d in m and F_1 in N

$$= 200 d^2 = F + \frac{F}{2} + F_c \qquad \textbf{US Customary units}$$
$$(21\text{–}55b)$$

where d in in and F_1 in lbf

$$= 0.14\, d^2 \qquad \textbf{Customary Metric} \qquad (25\text{–}55c)$$

where d in mm and F_1 in kgf

power transmitted

$$P = v(0.6 - 6.7 \times 10^{-4} F_c) \qquad \textbf{SI} \qquad (21\text{–}56a)$$
where F_c in N, P in kW, and v in m/s

$$= \frac{2v}{10^5}(200 - F_c) \qquad \textbf{US Customary units}$$
$$(21\text{–}56b)$$

where F_c in lbf, and P in hp

Refer to Table 21–32 for F_c = values of coefficients for manila rope

Hemp ropes

The load on the hemp rope

$$F = \frac{\pi d^2}{4} \sigma_{br} \qquad (21\text{–}57)$$

where σ_{br} = breaking stress, MPa (psi)

$$= 9.81 \text{ MPa (1.42 kpsi) for white rope}$$

$$= 8.82 \text{ MPa (1.28 kpsi) for tarred rope}$$

TABLE 21–32
Value of coefficient F_c for manila rope

Velocity, mps	7.50	10.00	12.50	15.00	17.50	20.00	22.50	25.00	27.50	30.00	32.50	35.00
Coefficient, F_c	2.96	5.40	8.44	12.60	16.10	21.00	26.55	32.89	39.69	41.17	55.34	64.40

Particular	Formula
The load on the hemp rope in terms of nominal diameter of rope	$F = 7.7 \times 10^6 d^2$ for white rope **SI** (21–58a) where d in m and F in N $= 1120\, d^2$ **US Customary units** (21–58b) where d in in and F in lbf $F = 7 \times 10^6 d^2$ for tarred rope **SI** (21–58c) where d in m and F in N $= 1020\, d^2$ **US Customary units** (21–58d) where d in in and F in lbf

HOISTING TACKLE

The effort on the rope in case of single-sheave pulley (Fig. 21–6)

$$P = \left(\frac{D + \mu d + 2s}{D - \mu d - 2s'}\right) Q = CQ \qquad (21\text{–}59)$$

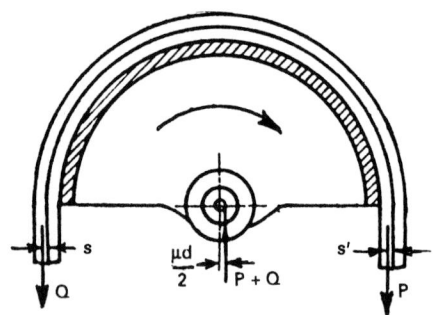

FIGURE 21–6 Rope passing over sheave.

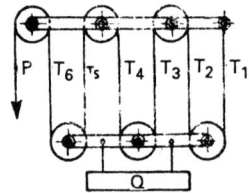

FIGURE 21–7 Load on a hoist.

Refer to Table 21–33 for C.

The effort on the rope in a hoist for raising the load (Fig. 21–7)

$$P = \frac{C^n(C - 1)}{C^{n} - 1} Q \qquad (21\text{–}60)$$

TABLE 21–33
Value of C

Manila rope	1.15
Wire rope	1.07
Dry chain	1.10
Greased chain	1.04

Particular	Formula
The pull required on the rope in a hoist for lowering the load	$$P' = \frac{C-1}{C(C^n-1)}Q \qquad (21\text{-}61)$$
Efficiency of hoist	$$\eta = \frac{C^n - 1}{n\,C^n(C-1)} \qquad (21\text{-}62)$$ where n = number of times a rope passes over a sheave

Continuous system Fig. (21–8)

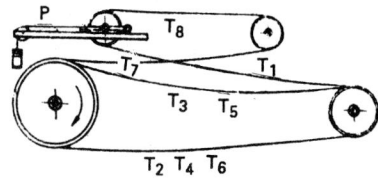

FIGURE 21–8   Continuous system.

In the continuous system one continuous rope passes around the driving and driven sheaves several times, in addition to making one loop about tension pulley located on a traveling carriage.

Particular	Formula
The relation between ultimate load, bending and service load in wire rope	$$\frac{P_u}{n} \geq P_b + P_s \qquad (21\text{-}63a)$$
The bending load	$$P_b = k\,A\,\frac{d_w}{D} \qquad (21\text{-}63b)$$ where k = 82728.5 MPa (12 Mpsi)
Another formula connecting ultimate strength of rope, tensile load on rope (P), dimensions of the rope, wire, and sheave diameter	$$P_u = \frac{P}{\dfrac{1}{n'} - \left(\dfrac{d}{D}\right)\left(\dfrac{d_w}{d}\right)\dfrac{E'}{\sigma_u}} \qquad (21\text{-}63c)$$ where D = minimum diameter of sheave or pulley, m (in) n' = stress factor = $n\,k_d$ n = safety factor k_d = duty factor obtainable from Table 21–35.
Area of useful cross section of the rope	$$A = \frac{P}{\dfrac{\sigma_u}{n'} - \left(\dfrac{d}{D}\right)\left(\dfrac{d_w}{d}\right)E'} \qquad (21\text{-}63d)$$
The approximate ultimate strength of plow-steel ropes	$P_u = 524,000\,d^2$ for 6×7 and 6×19 ropes
	SI $\qquad (21\text{-}64a)$
	where P_u in kN and d in m
	$= 76\,d^2$ **US Customary units** $\qquad (21\text{-}64b)$

TABLE 21–34
Manila rope

Size designation (C^a) mm	Number of yards per strand	Linear density kilotex	Pitch 2.6C^a π mm	Pitch 3.2C^a π mm	Breaking load Grade 1 kN	Grade 1 kgf	Grade 2 kN	Grade 2 kgf	Grade 3 kN	Grade 3 kgf
25	3	53	20.7–25.5		5.4	546	4.7	483	4.1	419
32	4	66	26.5–32.6		6.9	711	6.2	635	5.5	559
35	5	89	29.0–36.7		8.9	902	7.8	800	6.9	699
38	6	107	31.5–38.7		10.5	1,067	9.3	953	8.2	838
41	7	120	34.0–41.8		12.3	1,257	11.0	1,118	9.6	978
44	8	138	36.4–44.8		14.2	1,448	12.6	1,283	11.0	1,118
51	11	191	42.2–52.0		19.9	2,032	17.7	1,803	15.4	1,575
57	13	226	47.2–58.1		23.9	2,439	21.2	2,159	18.4	1,880
64	17	294	53.0–65.2		31.6	3,226	28.1	2,870	24.7	2,515
70	20	346	58.0–71.3		37.6	3,836	33.4	3,404	29.1	2,972
76	24	413	62.9–77.4		44.8	4,572	39.9	4,064	34.9	3,556
83	28	489	68.7–84.6		52.1	5,309	46.3	4,725	40.6	4,140
89	33	569	73.7–90.7		59.5	6,071	53.1	5,410	46.3	4,725
95	37	635	78.7–96.8		68.0	6,935	60.5	5,172	52.8	5,383
102	43	742	84.5–104.0		76.5	7,798	68.0	6,935	59.5	6,071
108	48	831	89.4–110.1		85.2	8,687	75.7	7,722	66.3	6,757
114	54	933	94.4–116.2		95.4	9,729	84.7	8,636	74.2	7,570
121	60	1,090	100.2–123.3		105.1	10,719	93.4	9,525	81.7	8,332
127	67	1,159	105.2–129.4		116.1	11,837	103.1	10,516	95.2	9,703
140	81	1,329	116.0–142.7		139.0	14,174	123.6	12,599	108.1	11,024
152	96	1,661	125.9–154.9		163.9	16,714	145.5	14,834	127.0	12,955
165	113	1,954	136.6–168.2		190.8	19,457	169.4	17,273	148.0	15,088
178	131	2,265	147.4–181.4		219.7	22,404	195.3	19,915	170.9	17,425
203	171	2,958	168.1–206.9		282.5	28,805	251.1	25,604	219.7	22,404
229	216	3,736	189.6–233.8		353.2	36,019	313.9	32,005	274.5	27,992
254	267	4,620	210.3–258.8		432.9	44,147	384.6	39,219	336.3	34,292
279	323	5,583	231.0–284.3		520.1	53,038	462.3	47,145	404.5	41,252
305	384	6,640	252.5–360.5		616.8	62,893	548.0	55,883	479.3	48,872
330	451	7,800	273.2–336.3		719.9	73,409	639.7	65,230	559.5	57,051
356	523	9,044	294.8–362.8		829.5	84,586	737.3	75,188	645.2	65,789
381	600	10,376	315.5–388.3		953.1	97,185	846.9	86,364	740.8	75,543
406	683	11,811	336.2–413.8		1081.6	10,292	961.8	98,049	841.5	85,805
432	771	13,335	357.7–440.3		1216.1	24,009	1081.1	110,241	946.1	96,474
457	864	14,943	378.4–465.7		1362.1	38,894	1210.6	123,450	1059.2	108,006

[a] C stands for nominal circumference of the rope.

TABLE 21–35
Duty factor and life of mechanism of electric wire rope hoists

Mechanism Class	Duty factor Strength	Wear	Average life Running h/day	Total life *h*. over
1	1.0	0.4	0.5	2500
2	1.2	0.5	0.5	9000
3	1.4	0.6	3.0	20000
4	1.6	0.7	over 6	40000

Source: IS: 3938–1967.

Particular	Formula
	where P_u in lbf and d in in
	$P_u = 517800\,d^2$ for 6×37 ropes **SI** (21–64c)
	where P_u in kN and d in m
	$= 75\,d^2$ **US Customary units** (21–64d)
	where P_u in lbf and d in in
The nominal bearing pressure	$p = \dfrac{2P_t}{D_r D_i} \le C\,\sigma_u$ (21–65)
	where $C = 0.0015$
	Refer to Table 21–33 for C.

DRUMS

Wire rope drum

The number of turn on the drum for one rope member (Fig, 21–9)

$$n = \frac{iS}{nD} + 2 \qquad (21\text{–}66)$$

The length of the drum

$$l = \left(\frac{2iS}{\pi D} + 7\right)p \quad \text{for one rope} \qquad (21\text{–}67a)$$

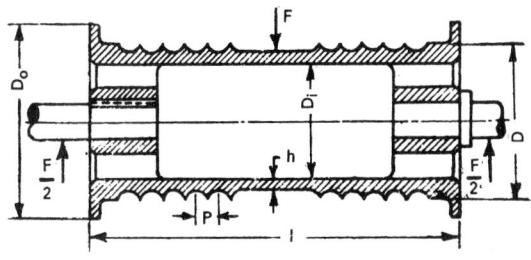

$$= \left(\frac{2iS}{\pi D} + 12\right)p + p_1 \quad \text{for two ropes} \qquad (21\text{–}67b)$$

FIGURE 21–9 Wire rope drum.

The thickness of wall of drum made of cast iron	$h = 0.02\,D + 0.6$ to 1.0 cm (21–68)
The outside diameter of the drum (Fig. 21–9)	$D_o = (D + 6d)$ (21–69)
The depth of groove	$h_1 < 0 - 1\,d$ (21–70)

Stresses developed in drum

The maximum bending stress

$$\sigma_b = \frac{8\,FlD}{(D^4 - D_i^{\,4})} \qquad (21\text{–}71)$$

The maximum torque on the drum

$$M_t = F\left(\frac{D + d}{2}\right) \qquad (21\text{–}72)$$

where d = diameter of rope

Particular	Formula
The maximum shear stress	$$\tau = \frac{16\,M_t\,D}{\pi(D^4 - D_i{}^4)} \qquad (21\text{--}73)$$
The crushing stress	$$\sigma_c = \frac{F}{ph} \qquad (21\text{--}74)$$ where p = pitch of the grooves on the drum
The combined stress according to normal stress theory	$$\sigma = \sqrt{\sigma_b{}^2 + \sigma_c{}^2 + 4\tau^2} \lessgtr \sigma_d \qquad (21\text{--}75)$$ where σ_d = design stress

HOLDING CAPACITY OF WIRE ROPE REELS

The rope capacity (L) in meters in any size length may be calculated by the formula

$$L = \frac{\pi(H + D_r)\,WH}{1000d} \qquad (21\text{--}76)$$

WIRE ROPE CONSTRUCTION

For wire rope strand construction, diameter, weight, breaking load for different purposes

Refer to Tables 21–36 to 21–39 and Figs 21–10 to 21–16.

For wire rope data, factor of safety, values of C, and application

Refer to Tables from 21–40 to 21–45.

CHAINS

Hoisting chains

The working load for the ordinary steel common coil chain

$$P_w = 84800\,d^2 \text{ kN} \quad \textbf{SI} \text{ where } d \text{ in m} \qquad (21\text{--}77a)$$

$$= 12300\,d^2 \text{ lbf} \quad \textbf{US Customary units}$$
$$\text{where } d \text{ in in} \qquad (21\text{--}77b)$$

$$= 8.65\,d^2 \text{ kgf} \quad \textbf{Metric} \text{ where } d \text{ in mm} \quad (21\text{--}77c)$$

The working load for stud chain

$$P_w = 60310\,d^2 \text{ kN} \quad \textbf{SI} \text{ where } d \text{ in m} \qquad (21\text{--}78a)$$
$$= 8750\,d^2 \text{ lbf} \quad \textbf{US Customary units}$$
$$\text{where } d \text{ in in} \qquad (21\text{--}78b)$$

$$= 6.15\,d^2 \text{ kgf} \quad \textbf{Metric} \text{ where } d \text{ in mm} \quad (21\text{--}79)$$

TABLE 21–36
Steel wire ropes (from Indian standards)

Strand construction	Diameter of rope mm	Approx. weight N/m	Approx. weight kgf/m	Nominal breaking strength of rope / Tensile strength of wire 1568–1716 MPa (160–175 kgf/mm²) kN	tf	1716–1863 MPa (175–190 kgf/mm²) kN	tf
			General	Engineering	Purposes		
Group 6 × 19	8	2.4	0.24	33.3	3.4	36.3	3.7
6 × 12/6/1	10	4.3	0.44	64.7	6.6	70.6	7.2
6 × 12/6 + 6F/1	12	5.3	0.54	84.3	8.6	92.2	9.4
6 × 9/9/1	14	7.5	0.76	106.9	10.9	116.7	11.9
	16	9.2	0.94	131.4	13.4	144.2	14.7
	18	12.3	1.25	189.3	19.3	206.9	21.1
6 × 10/5 + 5F/1	20	14.4	1.47	221.6	22.6	241.2	24.6
(Fig. 21–10)	22	18.0	1.84	254.0	25.9	278.5	28.4
	24	20.9	2.13	294.2	30.0	323.6	33.0
	25	23.6	2.41	333.4	34.0	368.7	37.6
	29	29.9	3.05	423.6	43.2	462.9	47.2
	32	36.8	3.75	522.7	53.3	570.7	58.2
	35	44.6	4.55	623.5	64.6	692.3	70.6
	38	53.3	5.43	752.2	76.7	826.7	84.3
	41	62.5	6.37	886.5	90.4	971.8	99.1
	44	72.4	7.38	1026.8	104.7	1125.8	114.8
	48	83.2	8.48	1175.8	119.9	1295.5	132.1
	51	94.5	9.64	1345.6	137.2	1474.9	150.4
	54	106.8	10.89	1514.1	154.4	1664.2	269.7
Group 6 × 37	10	4.4	0.45	60.8	6.2	66.7	6.8
6 × 14/7 and	12	5.9	0.60	79.4	8.1	87.3	8.9
7/7/1;	14	7.3	0.74	101.0	10.3	110.8	11.3
6 × 14/7+	16	9.0	0.92	124.5	12.7	136.3	13.9
7F/7/1;	18	12.3	1.32	179.5	18.3	196.1	20.0
6 × 16/8+	20	15.5	1.58	209.9	21.4	230.5	23.5
8F6/1	22	17.7	1.81	241.2	24.6	263.8	26.9
	24	20.6	2.10	278.5	28.4	304.0	31.0
	25	32.2	2.37	318.7	32.5	349.1	35.5
6 × 15/15/6/1;	29	29.3	2.99	398.2	40.6	438.4	44.7
	32	36.2	3.69	493.3	50.3	543.3	55.4
6 × 18/12/6/1;	35	43.9	4.48	598.2	61.0	658.0	67.1
6 × 16/8 and	38	52.2	5.32	712.0	72.6	782.6	79.8
8/8/1	41	61.3	6.25	836.5	85.3	916.9	93.5
(Fig. 21–11)	44	71.0	7.24	971.8	99.1	1065.9	108.7
	48	81.6	8.32	1116.0	113.8	1225.8	125.0
	51	92.8	9.46	1266.0	129.1	1394.5	142.2
	54	104.7	10.68	1434.1	146.3	1574.0	160.5
	57	117.5	11.98	1604.4	163.6	1763.2	179.8
	64	145.0	14.79	1982.2	202.2	2172.2	221.5
	70	175.3	17.88	2401.6	244.9	2630.1	288.0
6 × 24	8	2.1	0.21	29.4	3.0	32.4	3.3
Fiber Core	10	3.1	0.32	53.9	5.5	59.8	6.1
(Fig. 21–12)	12	5.3	0.54	74.5	7.6	81.4	8.3
	14	6.6	0.67	92.2	9.4	102.0	10.4
	16	7.8	0.80	112.8	11.5	123.6	12.6
	18	11.7	1.19	164.8	16.8	181.4	18.5
	20	13.8	1.41	196.1	20.0	214.8	21.9
	22	16.1	1.64	228.5	23.3	249.1	25.4
	24	18.2	1.86	258.9	26.4	278.5	28.4
	25	20.4	2.08	289.3	29.5	313.8	32.0
	29	26.3	2.68	368.7	37.6	403.1	41.1
	32	31.8	3.24	448.2	45.7	493.3	50.3
	35	38.8	3.96	548.2	55.9	603.1	61.5
	38	46.7	4.76	662.9	67.6	722.7	73.7
	41	54.0	5.51	762.0	77.7	836.5	85.3
	44	63.1	7.43	891.4	90.9	976.7	99.6
	48	73.0	7.44	1025.0	104.6	1125.8	114.8
	51	82.0	8.36	1166.0	118.9	1274.9	130.0
	54	93.4	9.52	1315.1	134.1	1443.5	147.3
Group II F	14	8.3	0.85	112.8	11.5	121.6	12.4
6 × 9/12/Δ;	16	10.2	1.04	143.2	14.6	155.9	15.9
6 × 10/12/Δ;	18	13.7	1.40	208.9	21.3	224.6	22.9
6 × 12/12/Δ;	20	16.3	1.66	246.1	25.1	263.8	26.9
(Fig. 21–14)	22	20.1	2.05	284.4	29.0	308.9	31.5
	24	23.2	2.37	323.6	33.0	349.1	35.6
	25	26.4	2.69	363.8	37.1	393.2	40.1
	29	33.2	3.39	462.9	47.2	498.2	50.8
	32	41.2	4.20	572.7	58.4	622.7	53.5
	35	49.6	5.05	682.5	69.6	737.5	75.2
	38	59.0	6.02	816.9	83.3	886.5	90.4
	41	69.1	7.05	966.9	98.6	1036.6	105.7
	44	81.0	8.26	1116.0	113.8	1216.0	124.0
	48	92.7	9.45	1275.2	130.1	1374.9	140.2
	51	105.0	10.71	1454.3	148.3	1574.0	160.5

Strand construction	Diameter of rope mm	Approx. weight N/m	Approx. weight kgf/m	Nominal breaking strength of rope / Tensile strength of wire 1568–1716 MPa (160–175 kgf/mm²) kN	tf	1716–1863 MPa (175–190 kgf/mm²) kN	tf
17 × 7, 18 × 7	8	2.5	0.25			35.3	3.6
(Fig. 21–15)	10	4.1	0.42			68.6	7.0
	12	5.6	0.57			87.3	8.9
	14	7.8	0.80			113.8	11.6
	16	9.6	0.98			142.2	14.5
	18	12.9	1.32			201.0	20.5
	20	15.2	1.55			237.3	24.2
	22	18.9	1.93			268.7	27.4
	24	21.9	2.23			313.8	32.0
	25	24.8	2.53			358.9	36.6
	29	31.4	3.20			443.3	45.2
	32	38.8	3.96			548.2	55.9
	35	46.8	4.77			672.7	68.6
	39	55.9	5.70			802.2	81.8
34 × 7	15	10.2	1.04			134.4	13.7
(Fig. 21–13)	18	13.4	1.37			193.2	19.7
	20	16.0	1.63			225.6	23.0
	22	19.8	2.02			263.8	26.9
	24	22.8	2.32			299.1	30.5
	25	26.0	2.65			344.2	35.1
	29	32.9	3.35			433.5	44.2
	32	40.6	4.14			538.4	54.9
	35	49.0	5.00			647.2	66.0
	38	58.3	5.95			771.8	78.7
	44	79.5	8.21			1025.8	104.6
	51	103.9	10.59			1334.7	136.1

Strand construction	Diameter of rope mm	Approx. weight N/m	Approx. weight kgf/m	Nominal breaking strength of rope / Tensile strength of wire 1079–1226 MPa (110–125 kgf/mm²) kN	tf	1226–1372 MPa (125–140 kgf/mm²) kN	tf
			Lift	and	Hoists		
Group 6 × 19	6	1.5	0.15	14.7	1.5	16.7	1.7
6 × 19 (12/6/1);	8	2.5	0.25	22.6	2.3	26.5	2.7
6 × 19 Filter	10	3.9	0.40	39.2	4.0	44.1	4.5
Wire;	12	5.4	0.55	53.9	5.5	58.8	6.0
6 × 19 (9/9/1)	14	7.4	0.75	75.5	7.7	86.3	8.8
Seale	16	9.3	0.95	94.1	9.6	107.9	11.4
	18	12.2	1.25	124.5	12.7	139.5	14.2
	20	14.2	1.45	147.1	15.0	166.7	17.0
	21	18.1	1.85	184.4	18.8	207.9	21.2
	25	22.1	2.25	225.6	23.3	255.0	26.0
Group 8 × 19	8	2.0	0.20	21.3	2.2	24.5	2.5
8 × 19 filter	13	3.4	0.35	37.6	3.8	42.2	4.3
wire;	12	4.9	0.50	49.0	5.0	53.9	5.5
8 × 19 (9/9/1)	14	6.9	0.70	68.6	7.0	79.4	8.1
Seale	16	8.3	0.85	88.3	9.0	98.1	10.0
	18	10.9	1.10	112.8	11.1	132.4	13.5
	20	13.2	1.35	137.3	14.0	152.0	15.5
	22	16.7	1.70	181.4	18.5	205.9	21.0
	25	19.6	2.00	01.0	20.6	235.4	24.0
6 × 25	10	4.4	0.45	42.2	4.3	49.0	5.0
flattened	12	5.9	0.60	56.9	5.8	64.7	6.6
strand	14	8.3	0.85	79.4	8.1	90.2	9.2
	16	10.3	1.05	102.9	10.5	117.7	12.0
	18	13.7	1.40	137.3	14.0	151.0	15.4
	20	16.2	1.65	161.8	16.5	184.4	18.8
	22	19.6	2.00	203.0	20.7	230.5	23.5
	25	24.5	2.50	243.2	24.8	272.6	27.8

TABLE 21–36
Steel wire ropes (from Indian Standard) (*Cont.*)

Strand construction	Diameter of rope, mm	Approx. weight		Nominal breaking strength of rope							
				Tensile strength of wire							
				1225.8–1373.0 MPa (125–140 kgf/mm²)		1373.0–1520.0 MPa (140–155 kgf/mm²)		1520.0–1667.0 MPa (155–170 kgf/mm²)		1667.0–1814.2 MPa (170–185 kgf/mm²)	
		N/m	kgf/m	kN	tf	kN	tf	kN	tf	kN	tf
				Winding purposes in mines							
6 × 7	19	12.8	1.31	166.7	17.0	183.5	18.9	199.1	20.3	213.8	21.8
	20	15.0	1.53	192.2	19.6	211.8	21.6	230.4	23.5	250.1	25.5
	22	17.7	1.80	224.6	22.9	250.1	25.5	268.7	27.4	289.3	29.5
	24	29.3	2.07	254.0	25.9	283.4	28.9	309.1	31.5	333.4	34.0
	25	23.0	2.35	283.3	29.5	325.6	33.2	349.1	35.6	378.6	38.6
	26	24.6	2.51	310.0	31.6	341.3	34.8	399.1	39.9	402.1	41.0
	27	26.3	2.68	332.4	33.9	366.7	37.4	391.9	40.7	430.5	40.9
	28	29.2	2.98	368.7	37.6	410.0	41.8	443.3	45.2	478.6	43.8
	31	35.9	3.66	453.1	46.2	512.8	52.3	548.2	55.9	598.2	71.2
	35	43.4	4.43	553.1	56.4	618.9	63.1	662.9	67.6	717.8	73.2

Strand construction	Diameter of rope, mm	Approx. weight		Nominal breaking load of rope							
				Tensile strength of wire							
				1226–1373 MPa (125–140 kgf/mm²)		1373–1520 MPa (140–155 kgf/mm²)		1520–1667 MPa (155–170 kgf/mm²)		1667–1814 MPa (170–185 kgf/mm²)	
		N/m	kgf/m	kN	tf	kN	tf	kN	tf	kN	tf
6 × 19	19	13.2	1.35	154.9	15.8	171.6	17.5	189.3	19.3	206.9	21.2
	20	14.6	1.49	179.5	18.3	199.1	20.3	221.6	22.6	243.2	24.8
	21	16.4	1.67	193.2	19.7	213.2	21.8	237.3	24.2	260.8	26.6
	22	18.0	1.84	206.2	21.1	229.5	23.4	254.0	25.9	278.5	28.4
	23	19.5	1.99	222.6	22.7	246.1	25.1	273.6	27.9	301.1	30.7
	24	20.9	2.13	237.3	24.2	263.8	26.9	294.2	30.0	323.6	33.0
	25	23.6	2.41	268.7	27.4	300.1	30.6	334.4	34.1	368.7	37.7
	26	26.6	2.71	291.3	29.7	326.5	33.3	365.8	37.3	399.1	40.7
	27	28.3	2.89	318.7	32.5	352.1	35.9	394.2	40.2	436.4	44.5
	28	31.3	3.19	348.1	35.5	383.4	39.1	423.6	43.2	462.9	47.2
	29	33.8	3.45	372.6	38.0	413.8	42.2	456.0	46.5	502.1	51.2
	30	35.6	3.63	400.1	40.8	443.3	45.2	483.5	49.3	536.4	54.7
	31	38.2	3.90	428.5	43.7	473.7	48.3	522.7	53.3	572.7	58.4
	32	39.7	4.05	447.1	45.6	498.2	50.8	545.2	55.6	603.1	61.5
	33	41.6	4.24	471.7	48.1	522.7	53.3	572.7	58.4	632.5	64.5
	34	43.1	4.39	493.3	50.3	548.2	55.9	608.1	61.5	663.9	67.7
	35	44.6	4.55	518.8	52.9	572.7	58.4	632.5	64.5	692.3	73.6
	36	47.3	4.82	548.2	55.9	608.0	62.0	672.7	68.6	732.5	74.7
	37	50.2	5.12	580.5	59.2	641.3	65.4	707.1	72.1	773.7	78.9
	38	53.3	5.43	611.2	62.4	678.6	69.2	752.2	76.7	826.7	84.3
	39	55.9	5.70	629.6	64.2	696.3	71.0	772.8	78.8	849.3	86.6
	40	59.2	6.04	650.2	66.3	714.9	72.8	792.4	80.8	868.9	88.6

TABLE 21–36
Steel wire ropes (from Indian Standard) (*Cont.*)

Strand construction	Diameter of rope, mm	Approx. weight		Nominal breaking load of rope							
				Tensile strength of wire							
				1226–1373 MPa (125–140 kgf/mm²)		1373–1520 MPa (140–155 kgf/mm²)		1520–1667 MPa (155–170 kgf/mm²)		1667–1814 MPa (170–185 kgf/mm²)	
		N/m	kgf/m	kN	tf	kN	tf	kN	tf	kN	tf
	41	62.5	6.37	726.7	74.1	803.2	81.9	886.5	90.4	771.8	99.2
	42	65.4	6.67	781.6	79.7	863.9	88.1	955.2	97.4	1048.3	106.9
	44	72.4	7.38	836.5	85.3	926.7	94.5	1025.8	104.6	1125.8	114.8
	46	78.1	7.96	893.3	91.1	995.4	101.5	1101.3	112.3	1210.1	123.4
	48	83.2	8.48	950.3	96.9	1057.2	107.8	1175.8	119.9	1225.5	192.1
	51	94.5	9.64	1100.3	112.2	1217.0	124.1	1345.6	157.2	1475.0	150.4
	54	106.8	10.89	1230.7	125.5	1365.1	139.2	1514.1	155.4	1664.2	169.7
6 × 37	19	12.9	1.32	145.1	14.8	162.8	16.6	179.5	18.3	196.1	20.0
	21	15.5	1.58	170.6	17.4	190.2	19.4	209.8	21.4	230.5	23.5
	22	17.7	1.81	195.8	19.9	218.7	22.3	241.2	24.6	263.8	26.9
	24	20.6	2.10	222.5	23.4	254.0	25.9	278.6	28.4	304.0	31.0
	25	23.2	2.37	260.0	26.4	289.3	29.5	318.4	32.5	349.1	35.6
	29	29.3	2.99	318.7	32.5	359.0	36.6	398.1	40.6	438.4	44.7
	22	36.2	3.69	343.7	35.0	393.2	48.1	493.3	50.3	543.3	55.4
	25	43.9	4.48	478.6	48.8	548.2	55.9	598.2	61.0	658.0	67.1
	31	52.2	5.32	572.7	58.4	642.3	65.5	712.0	72.6	782.6	79.8
	41	61.3	6.25	665.1	67.8	757.1	70.2	836.5	85.3	916.9	93.5
	44	71.0	7.24	676.9	69.0	857.1	87.4	871.8	99.1	1066.0	108.7
	48	81.6	8.32	896.3	91.4	1006.2	102.6	1116.0	113.8	1225.8	125.0
	51	92.8	9.45	1006.2	102.6	1135.6	115.8	1226.0	129.8	1394.0	142.2
	54	104.8	10.68	1156.2	117.9	1295.5	132.1	1434.7	146.3	1574.0	160.5
	57	117.6	11.98	1285.6	131.1	1444.5	147.3	1604.4	163.6	1763.2	179.8
	64	145.0	14.79	1624.0	165.6	1793.6	182.9	1912.9	202.2	2172.2	221.5
	70	175.3	17.88	1932.9	197.3	2172.2	221.5	2401.6	244.9	2630.0	268.2
6 × 7 Triangular core	19	15.0	1.53	181.4	18.5	199.1	20.3	216.7	22.1	235.4	24.0
	21	17.6	1.79	205.9	21.0	228.5	23.3	249.1	25.4	272.6	27.8
	22	20.1	2.05	244.2	24.9	268.7	27.4	294.2	30.0	313.7	32.6
Group IF 6 × 7/Δ	24	23.24	2.37	278.5	28.4	306.9	31.3	333.4	34.0	363.8	37.1
	25	26.28	2.68	313.8	32.0	347.1	35.4	378.5	38.6	413.8	42.2
	28	33.24	3.39	403.1	41.1	443.3	45.2	483.5	49.3	528.6	53.1
	31	41.19	4.20	498.2	50.8	553.1	56.4	608.0	62.9	658.0	67.1
	36	49.62	5.06	598.2	61.0	662.9	67.6	727.6	74.2	792.4	80.8
Group IIF 6 × 8/Δ; 6 × 8/12 Or less/Δ; 6 × 9/12 Or less/Δ; 6 × 10/12	19	15.00	1.53	179.5	18.3	194.2	19.8	208.9	21.3	224.6	22.9
	21	17.55	1.79	209.9	21.4	228.5	23.3	246.1	25.1	263.8	26.9
	22	20.10	2.05	234.4	23.9	258.9	26.4	284.4	29.0	308.9	31.5
	24	23.24	2.37	273.6	27.9	299.1	30.5	323.6	33.0	349.1	35.6
	25	26.28	2.68	304.0	31.0	333.4	34.0	363.8	37.1	393.2	40.1
	29	33.24	3.39	393.2	40.1	428.5	43.7	492.9	47.2	498.2	50.8
	32	41.18	4.20	473.6	48.3	522.7	53.3	572.7	58.4	622.7	63.5

TABLE 21–36
Steel wire ropes (from Indian Standard) (*Cont.*)

				Nominal breaking load of rope							
				Tensile strength of wire							
Strand construction	Diameter of rope, mm	Approx. weight		1226–1373 MPa (125–140 kgf/mm^2)		1373–1520 MPa (140–155 kgf/mm^2)		1520–1667 MPa (155–170 kgf/mm^2)		1667–1814 MPa (170–185 kgf/mm^2)	
		N/m	kgf/m	kN	tf	kN	tf	kN	tf	kN	tf
Or less/△;	35	49.62	5.06	572.7	58.4	627.6	64.0	682.5	69.6	737.5	75.2
6 × 12/12	38	59.03	6.02	677.6	69.1	766.9	78.2	816.9	83.3	886.5	90.4
Or less/△	41	69.14	7.05	825.2	84.3	896.3	91.4	966.9	98.6	1036.6	105.7
	44	81.00	8.26	916.2	93.6	1016.0	103.6	1116.0	113.8	1216.0	124 .0
	48	92.67	9.45	1075.8	109.7	1175.8	119.9	1275.8	138.1	1375.0	140.2
	51	105.03	10.71	1216.0	124.0	1334.7	136.1	1454.3	140.3	1574.0	160.5
Group IIIF	19	15.00	1.53	156.9	16.0	174.6	17.8	193.2	19.7	210.8	21.5
6 × 15/12/△	21	17.55	1.79	184.4	18.8	205.0	20.9	226.5	23.1	247.1	25.2
6 × 18/12/△	22	20.10	2.05	208.9	21.3	234.4	23.9	258.9	26.4	284.4	29.0
	24	23.05	2.35	234.4	23.9	263.8	26.9	294.2	30.0	323.6	33.0
	25	26.28	2.68	273.6	27.9	304.0	31.0	333.4	34.0	363.8	37.1
	29	33.24	3.39	354.0	36.1	388.3	39.6	423.6	43.2	458.0	46.7
	32	41.19	4.20	443.3	45.2	488.4	49.8	533.5	54.4	577.6	58.9
	35	49.62	5.06	517.8	52.8	577.6	58.9	537.4	65.0	656.3	71.0
	38	59.04	6.02	627.6	64.0	682.5	69.6	757.1	77.2	821.8	83.8
	41	69.14	7.05	747.3	76.2	816.9	83.3	886.5	90.4	986.1	97.5
	44	81.00	8.26	857.1	87.4	946.3	96.5	1036.6	105.7	1125.8	114.8
	48	92.67	9.45	986.5	100.6	1085.6	110.7	1185.6	120.9	1285.6	131.1
	51	105.03	10.71	1125.7	114.8	1235.4	126.0	1348.5	137.2	1452.8	148.3
	54	118.17	12.05	1255.2	128.0	1385.7	141.3	1514.1	154.4	1643.7	167.6
	57	133.57	13.62	1448.5	147.3	1584.2	161.6	1724.0	175.8	1863.3	190.0
	64	164.26	16.75	1793.6	189.9	1954.8	199.1	2112.6	215.4	2278.2	231.7
	70	198.58	20.25	2152.5	219.5	2341.8	238.8	2550.7	260.1	2740.0	279.4

| | | | | Minimum break load | | | |
| | | Approx. weight | | For tensile designation | | | |
Strand construction	Diameter of rope, mm	N/100 m	kgf/100 m	1569.3 MPa	160 kgf/mm^2	1765.2 MPa	180 kgf/mm^2
			Haulage purposes in mines				
6 × 7(6 × 1)	8	217.7	22.2	33.3	3400	37.6	3830
Round	9	275.6	28.1	42.3	4300	47.5	4840
	10	340.3	38.7	52.2	3320	58.6	5980
	11	411.9	42.0	63.1	6430	71.0	7240
	12	490.3	50.0	75.1	7660	84.4	8610

TABLE 21–36
Steel wire ropes (from Indian Standard) (*Cont.*)

Strand construction	Diameter of rope, mm	Approx. weight		Minimum breaking load of rope			
				For tensile designation			
				1569 MPa (160 kgf/mm^2)		1765 MPa (180 kgf/mm^2)	
		N/100 m	kgf/100 m	kN	kgf	kN	kgf
6 × 7 (6 × 1) Round	13	574.7	58.6	88.1	8980	99	10100
	14	666.9	68.0	102	10400	115	11700
	16	870.8	88.8	133	13600	150	15300
	18	1098.3	112.0	169	17200	190	19400
	19	1225.8	125.0	188	19200	212	21600
	20	1363.1	139.0	209	21300	234	23900
	21	1500.4	153.0	229	23400	259	26400
	22	1647.5	168.0	252	25700	283	28900
	24	1961.3	200.0	300	30600	337	34400
	25	2128.0	217.0	326	33200	367	37400
	26	2304.6	235.0	352	35900	396	40400
	27	2481.1	253.0	380	38700	428	43600
	28	2667.4	272.0	409	41700	460	46900
	29	2863.5	292.0	438	44700	493	50300
	31	3275.9	334.0	501	51100	564	57500
	35	4167.8	425.0	638	65100	719	73300
6 × 19 (9/9/1) Round	13	599.2	61.1	87.8	8950	99	10100
	14	695.3	70.9	102	10400	115	11700
	16	908.1	92.6	133	13600	150	15300
	18	1147.4	117	169	17200	189	19300
	19	1284.7	131	187	19100	211	21500
	20	1422.0	145	208	21200	233	23800
	21	1569.1	160	229	23400	258	26300
	22	1716.2	175	251	25600	282	28800
	24	2039.8	208	299	30500	336	34300
	25	2216.3	226	325	33100	365	37200
	26	2422.2	247	351	35800	395	40300
	28	2785.5	284	407	41500	458	46700
	29	2981.2	304	436	44500	491	50100
	32	3628.4	370	532	54200	598	61000
	35	4344.3	443	636	64900	716	73000
	36	4599.3	469	673	68600	757	77200
	38	5413.2	552	750	76500	843	86000
6 × 8 (7/△) Triangular	13	675.7	68.9	95.9	9780	106	10800
	15	783.5	79.9	111	11300	124	12600
	16	1019.9	104	145	14800	161	16400
	18	1294.5	132	183	18700	204	20800
	19	1441.6	147	205	20900	228	23200
	20	1598.5	163	227	23100	252	25700
	21	1765.2	180	250	25500	278	28300
	22	1931.9	197	275	28007	305	31100
	24	2304.5	235	327	33306	363	37000
	25	2500.1	255	354	36100	393	40100
	26	2696.8	275	383	39100	426	43400
	28	3128.3	319	445	45400	493	50300
	29	3363.7	343	478	48700	530	54000
	31	3844.2	392	545	55600	605	61700
	35	4893.5	499	695	70900	771	78600
	13	685.5	69.9	93.1	9490	104	10610
	14	794.3	81	108	11000	120	12200
	16	1039.5	106	141	14400	157	16000
	18	1314.1	134	178	18200	198	20200
	19	1461.2	149	199	20300	222	22600
	20	1618.1	165	221	22500	245	25000
	21	1784.8	182	243	24800	270	27500
	22	1961.3	200	267	27200	296	30200
	24	2334.0	238	317	32300	353	36000
	25	2530.1	258	343	35000	384	39100
	26	2736.0	279	372	37900	414	42200
	28	3137.3	324	431	44000	481	49000
	29	3412.7	348	463	47200	515	52500
	32	4148.2	423	564	57500	628	64000
	35	4962.1	506	675	68800	750	76500
	38	5854.5	597	795	81100	885	90200

Strand construction	Diameter of wire, mm	Approx. weight		Minimum breaking load of rope	
		N/m	kgf/m	kN	kgf
Small Wire Ropes (Fiber Care)					
6 × 7 (6/1)	2	0.147	0.015	2.6	260
	3	0.324	0.033	5.9	600
	4	0.559	0.057	10.4	1060
	5	0.873	0.089	16.3	1660
	6	1.255	0.128	23.5	2400
	7	1.696	0.172	32.0	3260
6 × 12 (12/fiber)	3	0.235	0.024	3.7	380
	4	0.412	0.042	6.5	670
	5	0.637	0.065	10.3	1050
	6	0.922	0.094	14.9	1520
	7	1.255	0.128	20.3	2070
6 × 19 (12/6/1)	3	0.314	0.035	4.9	500
	4	0.539	0.052	8.7	890
	5	0.843	0.086	13.5	1880
	6	1.206	0.124	19.6	2000
	7	1.648	0.168	26.6	2710
6 × 24 (15/9/fiber)	4	0.530	0.054	8.6	880
	5	0.834	0.085	13.3	1360
	6	1.206	0.122	19.3	1970
	7	1.618	0.165	29.3	2680

Strand construction	Diameter of wire		Approx. weight, max		Minimum breaking load	
	Max, mm	Min, mm	N/m	kgf/m	kN	kgf
Preferred Galvanized Steel Wire Ropes for Aircraft Controls						
7 × 7	1.8	1.6	0.108	0.011	2.2	220
7 × 7	2.7	2.4	0.235	0.024	4.1	420
7 × 19	3.5	3.2	0.422	0.043	8.9	910
7 × 19	4.4	4.0	0.657	0.067	12.5	1270
7 × 19	5.2	4.8	0.804	0.082	18.6	1900
7 × 19	6.0	5.6	1.236	0.126	24.9	2540
7 × 10	6.8	6.4	1.608	0.164	31.1	3170

Note: kgf = kilogram-foot; tf = ton-foot.

TABLE 21–37
Round strand galvanized steel wire ropes for shipping purposes

Diameter of wire, mm	Tensile strength of wire, 1373–1569 MPa (140–160 kgf/mm^2)																
	Approx. weight		Breaking strength of rope, min		Approx. weight		Breaking strength of rope, min		Approx. weight		Breaking strength of rope, min		Approx. weight		Breaking strength of rope, min		
	N/m	kgf/m	kN	kgf	N/m	kgf/m	kN	kgf	N/m	kgf/m	kN	kgf	N/m	kgf/m	kN	kgf	
	6 × 7		Fiber core		16 × 12		Fiber core		6 × 31		Fiber core		7 × 7		Wire core		
8	2.2	0.22	31.0	3150	1.5	0.15	18.1	1850									
9	2.8	0.28	38.8	3950	2.1	0.21	26.0	2650									
10	3.3	0.34	47.1	4800	2.5	0.25	30.4	3100									
11	4.0	0.41	56.9	5800	2.8	0.29	35.3	3650					3.7	0.38	52.0	5300	
12	5.1	0.52	72.6	7400	3.7	0.38	46.1	4700	5.1	0.52	72.6	7400	4.4	0.45	62.8	6400	
14	5.8	0.69	96.1	9800	4.7	0.48	58.4	5950	7.8	0.80	109.8	11200	5.7	0.58	80.4	8200	
16	8.7	0.89	123.6	12600	6.4	0.65	79.4	8100	9.4	0.96	132.4	13500	7.6	0.77	106.9	10900	
18	10.9	1.11	154.0	15700	7.6	0.78	95.6	9750	12.4	1.26	174.6	17800	9.7	0.99	137.3	14000	
20	13.9	1.42	198.1	20200	9.8	1.00	122.1	12450	14.9	1.52	209.9	21400	12.1	1.23	171.6	17500	
22	16.6	1.69	236.3	24100	11.4	1.16	141.7	14450	18.5	1.89	261.8	26700	15.5	1.58	219.7	22400	
24	19.5	1.99	278.5	28400	13.9	1.42	173.6	17700	21.0	2.14	295.2	30100	18.5	1.89	262.8	26800	
26	22.8	2.32	323.6	33000	16.8	1.70	208.9	21300	25.5	2.60	361.9	36900	21.9	2.23	308.9	31500	
28	27.1	2.76	385.4	39300	18.7	1.91	234.4	23900	30.5	3.11	429.5	43800	25.4	2.59	359.9	36700	
32	34.7	3.54	494.3	50400	24.3	2.48	304.0	31000	39.4	4.02	555.1	56600	30.2	3.08	427.6	43600	
36	44.5	4.54	634.5	64700	30.6	3.12	382.5	39000	40.1	4.09	703.1	71700	38.7	3.95	548.2	56000	
40	54.3	5.54	773.9	78900	39.0	3.98	489.4	49900	61.6	6.28	867.9	88500	49.7	5.07	704.1	71800	
	6 × 19		Fiber core		6 × 24		Fiber core		6 × 37		Fibre core		7 × 19		Wire core		
8	1.9	0.20	28.0	2850	2.2	0.22	28.4	2900	2.3	0.23	31.9	3250					
9	2.8	0.29	40.2	4100	2.6	0.27	34.8	3550	2.9	0.30	41.2	4200					
10	3.4	0.35	47.1	4800	3.1	0.32	42.2	4300	3.3	0.34	47.1	4800					
11	3.9	0.40	53.9	5500	3.7	0.38	50.0	5100	4.1	0.42	58.4	5950					
12	5.1	0.52	71.1	7250	4.4	0.45	58.8	6000	5.0	0.51	70.6	7200	7.3	0.74	101.5	10350	
14	6.5	0.66	90.2	9200	5.9	0.60	78.5	8000	7.0	0.71	98.1	10000	9.9	1.01	138.3	14100	
16	8.8	0.90	122.6	12500	8.4	0.86	112.8	11500	9.3	0.95	31.4	13400	11.9	1.21	165.7	16900	
18	10.6	1.08	147.1	15000	10.4	1.06	140.2	14300	12.0	1.22	168.7	17200	15.2	1.55	211.8	21600	
20	13.5	1.38	188.3	19200	12.7	1.29	169.7	17300	14.9	1.52	256.9	26200	17.7	1.80	254.2	25000	
22	15.7	1.60	218.7	22300	15.0	1.53	201.0	20500	18.2	1.86	256.9	26200	20.8	2.12	289.3	29500	
24	19.2	1.96	267.7	27300	17.7	1.80	238.3	24100	20.0	2.04	282.4	28800	26.0	2.65	361.9	36900	
26	23.1	2.36	321.6	32800	22.0	2.24	294.2	30000	23.8	2.43	336.4	34300	29.1	2.97	406.0	41400	
28	26.0	2.65	360.9	36800	25.1	2.56	336.9	34300	28.0	2.85	394.2	40200	37.9	3.86	526.6	53700	
32	33.7	3.44	468.8	47800	32.0	3.26	428.6	43700	37.2	3.79	524.7	53500	47.6	4.85	662.9	67600	
36	42.4	4.32	588.4	60000	41.8	4.26	599.0	57000	47.8	4.87	674.7	68800	60.8	6.20	847.3	86400	
40	54.1	5.52	664.9	67800	50.5	5.15	676.7	69000	56.6	5.77	798.3	81400	73.1	7.45	1027.7	104800	
44	65.1	6.64	905.2	92300	60.1	6.13	806.1	82200	69.5	7.09	981.6	100100	86.3	8.80	1203.3	122700	
48	78.0	7.95	1084.6	110600	73.3	7.47	982.6	100200	83.9	8.55	1183.7	120700	101.0	10.30	1407.3	143500	
52	90.0	9.18	1251.3	127600	84.7	8.64	1136.6	115900	95.2	9.71	1344.5	137100	116.6	11.89	1625.0	165700	
60	104.0	10.61	1446.5	147500	97.0	9.90	1301.3	132700	111.7	11.39	1577.9	160900	133.8	13.59	1857.4	189400	

Source: IS 2581, 1968.

TABLE 21–38
Dimensions and breaking strength of flat balancing wire ropes

Constructions	Nominal size $b \times s$[a], mm Double-stitched	Single-stitched	Diameter of the wire, mm	Cross section of the strand, mm^2	Approximate weight Double-stitched N/m	kgf/m	Single-stitched N/m	kgf/m	Minimum breaking strength of rope kN	kgf
	70×17	70×15	1.60	338	34.3	3.5	33.3	3.4	463.9	47300
	74×18	74×16	1.70	381	39.2	4.0	37.3	3.8	522.7	53300
	78×19	78×17	1.80	427	44.1	4.5	42.2	4.3	585.5	59700
$6 \times 4 \times 7$	82×20	82×18	1.90	477	49.0	5.0	47.1	4.8	654.1	66700
	87×21	87×19	2.00	528	53.9	5.5	52.0	5.3	724.7	73900
	91×22	91×20	2.20	581	59.8	6.1	56.9	5.8	797.2	81300
	95×23	95×21	2.20	638	65.7	6.7	62.8	6.4	875.7	89300
	110×20	110×18	1.90	636	65.7	6.7	62.8	6.4	872.8	89000
	113×20	113×18	1.95	670	68.7	7.0	65.7	6.7	919.9	93800
	116×21	116×19	2.00	703	72.6	7.4	68.7	7.0	956.0	98400
$8 \times 4 \times 7$	119×21	119×19	2.05	739	76.5	7.8	72.6	7.4	1014.0	103400
	122×22	122×20	2.10	775	79.4	8.1	76.5	7.8	1064.0	108500
	125×22	125×20	2.15	812	83.4	8.5	79.4	8.1	1116.0	113800
	128×23	128×21	2.20	851	87.3	8.9	83.4	8.5	1168.0	119100
	112×26	112×23	1.90	818	84.3	8.6	80.4	8.2	1122.9	114500
	115×26	115×23	1.95	861	88.3	9.0	84.3	8.6	1181.7	120500
	118×27	118×24	2.00	904	98.2	9.5	88.3	9.0	1240.5	126500
$6 \times 4 \times 12$	121×27	121×24	2.05	950	98.1	10.0	93.2	9.5	1304.3	133000
	124×28	124×25	2.10	996	103.0	10.5	98.1	10.0	1367.0	139400
	127×28	127×25	2.15	1045	107.9	11.0	103.0	10.5	1439.6	146300
	130×29	130×26	2.20	2094	112.8	11.5	106.9	10.9	1483.7	151300
	146×26	146×23	1.90	1091	112.8	11.5	106.9	10.9	1497.5	152700
	149×26	149×23	1.95	1148	118.7	12.1	112.8	11.5	1575.9	160700
	154×27	154×24	2.00	1206	124.5	12.7	118.7	12.1	1655.4	168800
$8 \times 4 \times 12$	157×27	157×24	2.05	1267	130.4	13.3	124.5	12.7	1738.7	177300
	160×28	160×25	2.10	1329	137.3	14.0	130.4	13.3	1824.0	186000
	165×28	165×25	2.15	1394	143.2	14.6	136.3	13.9	1913.3	195100
	168×29	168×26	2.20	1459	150.0	14.3	143.2	14.6	2002.5	204200
	160×27	160×24	1.90	1272	131.4	13.4	124.6	12.7	1745.5	178000
	164×28	164×25	1.95	1340	138.3	14.1	131.4	13.4	1842.2	187800
	168×28	168×25	2.00	1407	145.1	14.8	138.3	14.1	1930.9	196900
$8 \times 4 \times 14$	172×29	172×26	2.05	1478	152.0	15.5	145.1	14.8	2029.0	206900
	176×29	176×26	2.10	1550	159.8	16.3	152.0	15.5	2188.0	217000
	180×30	180×27	2.15	1626	167.7	17.1	159.9	16.3	2232.0	227600
	184×30	184×27	2.20	1702	175.5	17.9	166.7	17.0	2335.9	238200
	186×31	186×28	1.90	1727	177.5	18.1	169.7	17.3	2377.3	251700
$8 \times 4 \times 91$	190×32	190×29	1.95	1818	187.3	19.1	178.5	18.2	2495.8	254500
	194×33	194×30	2.00	1909	191.1	20.1	187.3	19.1	2620.3	267200

[a]b = width of rope, s = thickness of rope.
Source: IS 5203, 1969.

TABLE 21–39
Dimensions and breaking strength of flat hoisting wire ropes

Construction	Nominal size, $b \times s$, mm	Nominal wire diameter, mm	Cross section of strand, mm²	Weight N/m	Weight kgf/m	Minimum breaking strength of rope[a] kN	Minimum breaking strength of rope[a] kgf
	52×10	1.20	190	18.6	1.9	298.1	30400
	56×11	1.30	223	21.6	2.2	349.1	35600
	60×12	1.40	259	25.5	2.6	406.0	41400
	65×14	1.50	297	29.4	3.0	465.8	47500
	70×15	1.60	338	33.3	3.4	529.6	54000
$6 \times 4 \times 7$	74×16	1.70	381	37.3	3.8	597.2	60900
	78×16	1.80	427	42.2	4.3	669.8	68300
	82×18	1.90	477	47.1	4.8	748.2	76300
	87×19	2.00	528	52.0	5.3	827.7	84400
	91×20	2.10	581	56.9	5.8	911.0	92900
	95×21	2.20	638	62.8	6.4	1000.3	102000
	70×10	1.20	253	24.5	2.5	396.2	40400
	75×11	1.30	298	29.4	3.0	466.8	47600
	80×12	1.40	345	34.3	3.5	541.3	55200
	86×14	1.50	396	39.2	4.0	620.8	63300
	92×15	1.60	450	44.1	4.5	706.1	72000
$8 \times 4 \times 7$	98×16	1.70	508	50.0	5.1	796.3	81200
	104×17	1.80	569	55.9	5.7	892.4	91000
	110×18	1.90	636	62.8	6.4	997.3	101700
	116×19	2.00	703	68.6	7.0	1102.3	112400
	122×20	2.10	775	76.5	7.8	1216.0	124400
	128×21	2.20	851	83.4	8.5	1333.7	136600

[a]Rope having wires of tensile strength of 1569 MPa (160 kgf/mm²).
Source: IS 5202, 1269.

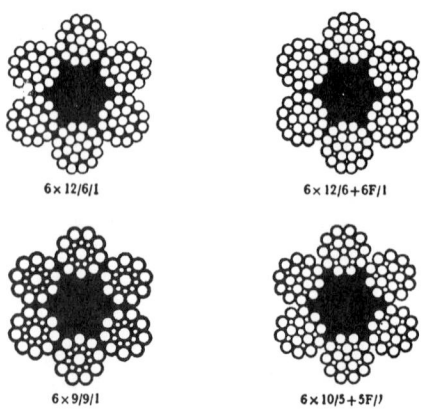

Ropes in this group have six strands in one of the following constructions:
$6 \times 12/6/1$; $6 \times 12/6 + 6F/1$; $6 \times 9/9/1$; and $6 \times 10/5 + 5F/1$.

6 × 12/6/1   6 × 12/6+6F/1

6 × 9/9/1   6 × 10/5 + 5F/1

FIGURE 21–10   Round strand group 6×19.

TABLE 21–40
Tensile grade

Grade of wire	Tensile strength range	
	MPa	kgf/mm^2
120	1176.8–1471.0	120–150
140	1372.9–1078.7	140–170
160	1569.1–1863.3	160–190
180	1765.2–2059.4	180–210
200	1961.3–2353.6	200–240

TABLE 21–41
Values of C for wire ropes

Rope diameter, mm	C	Rope diameter, mm	C
9.50	1.090	15.90	1.064
11.11	1.083	19.00	1.054
12.70	1.076	22.20	1.046
14.30	1.070	25.40	1.040

TABLE 21–42A
Approximate wire rope and sheave data

Rope construction	Ultimate strength, F_u		Weight		Wire diameter d_w, mm (in)	Area A, mm^2 (in^2)	Recommended sheave diameter, mm (in)	
	MN	lbf×10^3	kN/m	lbf/ft			Average	Minimum
6 × 19	500.8d^2	72d^2	36.3d^2	1.60d^2	0.063d	0.38d^2	45d	30d
6 × 37	473.1d^2	68d^2	35.3d^2	1.55d^2	0.045d	0.38d^2	27d	18d
8 × 19	431.3d^2	62d^2	34.3d^2	1.50d^2	0.050d	0.35d^2	31d	21d
6 × 7	473.0d^2	68d^2	32.4d^2	1.45d^2	0.106d	0.38d^2	72d	42d

SI Units d = diameter of rope, m.
US Customary units d = diameter of rope, in.

TABLE 21–42*B*
Wire rope data

Rope	Weight per foot, lb	Minimum sheave diameter, in	Standard sizes, d, in	Material	Size of outer wires	Modulus of elasticity,[a] Mpsi	Strength,[b] kpsi
6×7 haulage	$1.50d^2$	$42d$	$\frac{1}{4}$–$1\frac{1}{2}$	Monitor steel	$d/9$	14	100
				Plow steel	$d/9$	14	88
				Mild plow steel	$d/9$	14	76
6×19 standard hoisting	$1.60d^2$	$26d$–$34d$	$\frac{1}{4}$–$2\frac{3}{4}$	Monitor steel	$d/13$–$d/16$	12	106
				Plow steel	$d/13$–$d/16$	12	93
				Mild plow steel	$d/13$–$d/16$	12	80
6×37 special flexible	$1.55d^2$	$18d$	$\frac{1}{4}$–$3\frac{1}{2}$	Monitor steel	$d/22$	11	100
				Plow steel	$d/22$	11	88
8×19 extra flexible	$1.45d^2$	$21d$–$26d$	$\frac{1}{4}$–$1\frac{1}{2}$	Monitor steel	$d/15$–$d/19$	10	92
				Plow steel	$d/15$–$d/19$	10	80
7×7 aircraft	$1.70d^2$	—	$\frac{1}{16}$–$\frac{3}{8}$	Corrosion-resistant steel	—	—	124
				Carbon steel	—	—	124
7×9 aircraft	$1.75d^2$	—	$\frac{1}{8}$–$1\frac{3}{8}$	Corrosion-resistant steel	—	—	135
				Carbon steel	—	—	143
19-wire aircraft	$2.15d^2$	—	$\frac{1}{32}$–$\frac{5}{16}$	Corrosion-resistant steel	—	—	165
				Carbon steel	—	—	165

[a]The modulus of elasticity is only approximate: it is affected by the loads on the rope and, in general, increases with the life of the rope.
[b]The strength is based on the nominal area of the rope. The figures given are only approximate and are based on 1-in rope sizes and $\frac{1}{4}$-in aircraft-cable sizes.
Source: Compiled from *American Steel and Wire Company Handbook*.

Ropes in this group have six strands in one of the following constructions:
5 × 14/7 and 7/7/1; 6 × 14/7 + 7F/7/1; 6 × 16/8 + 8F/6/1; 6 × 15/15/6/1; 6 × 18/12/6/1; 6 × 16/8 and 8/8/1.

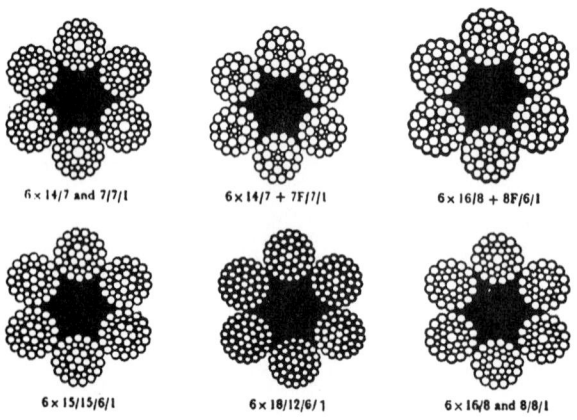

6 × 14/7 and 7/7/1   6 × 14/7 + 7F/7/1   6 × 16/8 + 8F/6/1

6 × 15/15/6/1   6 × 18/12/6/1   6 × 16/8 and 8/8/1

FIGURE 21–11   Round strand group 6×37.

TABLE 21–43
Common wire rope application

Type of service	Rope construction	Sheave diameter, cm	
		Recommended	Minimum
Haulage rope	6 × 7	72d	42d
Mine haulage			
Factory-yard haulage			
Inclined planes			
Tramways			
Power transmission			
Guy wires			
Standard hoisting rope	6 × 19	45d	30d
(Most commonly used rope)		60–100d	
Mine hoists			
Quarries			
Ore docks			
Cargo hoists			
Car pullers			
Cranes			
Derricks		20–30d	
Tramways			
Well drilling			
Elevators			
Extraflexible hoistings rope	8 × 19	31d	21d
Special flexible hoisting rope	6 × 37	27d	18d
Steel-mill ladles			
Cranes			
High speed elevators			

6 × 15/9/Fibre

FIGURE 21–12   Round strand
group 6 × 24 fiber core.

34 × 7

FIGURE 21–13   Multistrand
nonrotating ropes 34 × 7.

TABLE 21–44
Recommended safety factors for wire ropes

Rope application	Safety factor		
	100 or other figure laid down by the statutory authority		
	Class 1	Classes 2, 3	Class 4
From Indian Standards			
Mining ropes	3.5	4.0	4.5
Wire ropes used on the cranes and other hoisting equipment			
Fixed guys			
Unreeved rope bridles of jib cranes or ancillary appliances, such as lifting beams			
Ropes which are straight between terminal fittings			
Hoisting, luffing and reeved bridle systems of inherently flexible crances (e.g., mobile crawler tower, guy derrick, stiffleg derrick) where jibs are supported by ropes or where equivalent shock absorbing devices are incorporated in jib supports	4.0	4.5	5.5
Cranes and hoists in general hoist blocks	4.5	5.0	6.0
From Other Sources			
Mine Shafts			
Depths to 152 m	8		
305–610 m	7		
610–915 m	6		
>915 m	5		
Haulage ropes	6		
Small electric and air hoists	7		
Hot ladle cranes	8		
Slings	8		

Source: IS 3973, 1967.

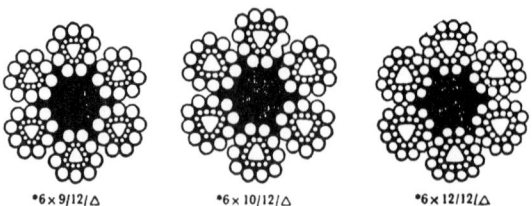

Ropes in this group have six strands in one of the following constructions:
$6 \times 9/12/\triangle$, $6 \times 10/12/\triangle$, $6 \times 12/12/\triangle$.

*6 × 9/12/△   *6 × 10/12/△   *6 × 12/12/△

FIGURE 21–14   Compound flattened strand, group II F.

TABLE 21–45
Ratio of drum and sheave diameter to rope diameter

Purpose	Construction	Minimum, ratio[a]		
		100		
Mining Installation	All	Class 1	Classes 2, 3	Class 4
Cranes and allied hoisting equipment	6 × 37 8 × 19 filler wire	15	17	22
	8 × 19 8 × 19 Warrington 8 × 19 Seale 34 × 7 Nonrotating	17	18	24
	6 × 24	18	19	25
	6 × 19 filler wire	18	20	23
	6 × 19 6 × 19 Warrington 17 × 7 nonrotating 18 × 7 nonrotating	19	23	27
	6 × 19 Seale	24	28	35

[a]The ratio of the sheave diameters specified are valid for rope speeds up to 50 m/min. For speeds above 50 m/min, the drum or sheave diameter should be increased pro rata by 8% for each additional 50 m/min of rope speed where practicable.
Source: IS 3973, 1967.

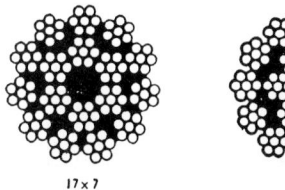

FIGURE 21–15   Multistrand nonrotating ropes 17 × 7 and 18 × 7.

FIGURE 21–16(*a***)**   Metal core.

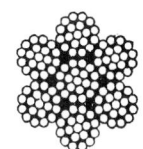

FIGURE 21–16(*b***)**   Metal core.

Particular	Formula
The working load for the ordinary steel *BB* crane chain	$P_w = 93750\,d^2$ kN **SI** where d in m (21–79a)
	$= 13600\,d^2$ lbf **US Customary units** where d in in (21–79b)
	$= 9.56\,d^2$ kgf **Metric** where d in mm (21–79c)
The sheave diameter	$D = 20d$ to $30d$

Round steel short link and round steel link chain

LENGTH AND WIDTH (FIGS. 21–17 AND 21–18) The outside dimensions of the links shall fall between the following limits:

Outside link length limits (Fig. 21–17)	$\not> 5d_n$ for uncalibrated chain (21–81a) $\not< 6d_n$ for calibrated chain (21–81b)
Maximum outside link width (Fig. 21–18)	$W_{max} \not> 3.5\,d_n$ away from weld (21–82a) $\not> 1.05$ (adjacent width) at weld for noncalibrated chains (21–82b)
Minimum inside link width	$W_{max} = 3.25\,d_n$ for calibrated chain (21–83) $W_t \not< 1.25\,d_n$ except at the weld for noncalibrated chain (25–84)
Pitch (i.e., inside length)	$p = 3\,d_n$ for calibrated chain (21–85)
Dimensions and lifting capacities and properties of noncalibrated and calibrated chains	Refer to Tables 21–46 to 21–51.

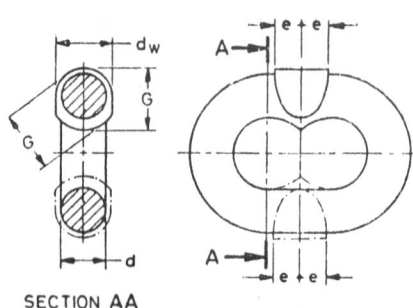

SECTION AA

ASYMMETRIC WELDED CHAIN

FIGURE 21–17 Diameter of material and welded chain.

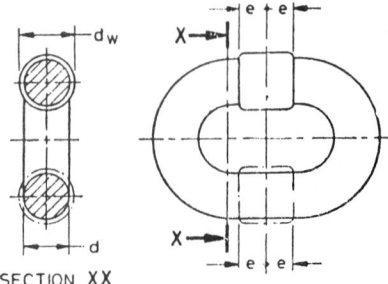

SECTION XX

SMOOTH WELDED CHAIN

d = diameter of the material except at the weld.
d_w = dimension at the weld normal to the plane of the link.
G = dimension in other planes.
e = length effected by welding on either side of the link

TABLE 21-46
Dimensions and lifting capacities of grade 30 noncalibrated chain (Figs. 21-17 and 21-18)

Nominal size, d_n, mm	Diameter tolerance $d_n \not> 16, +0.02s -0.06s$ $d_n \geq 16, \pm0.05s$	Maximum additional weld dimensions $(d_w - d)$ max	$(G - d)$ max	Outside link length limits $5d_n$	$4.75d_n$	Maximum outside link width, W — Away from weld, W_{max} $3.5 d_n$	Max extra at weld $0.5 W_{max}$	Minimum inside link width $1.25 d_n$	Guaranteed minimum breaking load stress 30h bar, kN	Minimum energy absorption factor (energy absorption 0.054 kJ m^{-1} mm^{-2}), kJ/m	Minimum safe working load (stress 7.5h bar), kN	Lifting capacity (stress 7.5h bar), tonnes
6.3	+0.12, -0.36	1.2	2.1	32	30	22	1.1	7.9	18.9	3.4	4.8	0.50
7.1	+0.14, -0.42	1.4	2.4	36	34	25	1.25	8.9	23.6	4.3	5.9	0.63
8.0	+0.16, -0.48	1.6	2.8	40	38	28	1.4	10	30.2	5.5	7.5	0.80
9.0	+0.18, -0.54	1.8	3.1	45	43	31	1.6	11	38.1	6.9	9.5	1.00
10.0	+0.20, -0.60	2.0	3.5	50	48	35	1.8	12	47.1	8.5	11.8	1.25
11.2	+0.22, -0.66	2.2	3.9	56	53	39	2.0	14	59.2	10.7	14.8	1.6
12.5	+0.25, -0.75	2.5	4.4	62	59	44	2.2	16	73.8	13.4	18.5	2.0
14.0	+0.28, -0.84	2.8	4.9	70	66	49	2.5	18	93.0	16.7	23.2	2.5
16.0	+0.32, -0.96	3.2	5.6	80	76	56	2.8	20	120.0	22.0	30.0	3.2
18.0	+0.90	3.6	6.3	90	86	63	3.1	22	153.0	39.0	38.2	4.0
20.0	+1.0	4.0	7.0	100	95	70	3.5	25	189.0	32.0	49.0	5.0
22.0	+1.1	4.4	7.7	110	105	77	3.9	28	228.0	42.0	57.0	6.3
25.0	±1.2	5.0	8.7	125	120	87	4.4	31	296.0	53.0	74.0	8.0
28.0	±1.4	5.6	9.8	140	130	98	4.9	35	372.0	67.0	93.0	10.0
32.0	±1.6	6.4	11	160	150	110	5.5	40	483.0	87.0	121.0	12.5
36.0	±1.9	7.2	12	180	170	120	6.0	45	610.0	112.0	152.0	16.0
40.0	±2.0	8.0	14	200	190	140	7.0	50	757.0	136.0	189.0	20.0
45.0	±2.2	9.0	16	225	215	160	8.0	56	953.0	173.0	228.0	22.5

Source: IS 2429 (Part I), 1970.

TABLE 21–47
Dimensions and lifting capacities of grade 30 calibrated chain (Figs. 21–17 and 21–18)

Nominal size, d_n, mm	Diameter tolerance $d_n \not> 16, +0.02s, -0.06s$; $d_n \geq 16, \pm 0.05s$	Maximum additional weld dimensions $(d_w - d)$ max	$(G - d)$ max	Preferred pitch (inside length), $3\,d_n$	Pitch tolerance (one link), $0.0396 d_n$	Preferred outside width, $w = 3.25 d_n$	Outside width tolerance away from weld zone $+0.075 d_n$, 0	At weld zone $+0.15 d_n$, 0	Guaranteed minimum breaking load (stress $30h$ bar), kN	Minimum energy absorption factor (energy absorption 0.054 kJm^{-1} mm^{-2}), kJ/m	Maximum safe working load (stress $7.5h$ bar), kN	Lifting capacity (stress $7.5h$ bar), tonnes
6.3	+0.12, −0.36	0.48	2.1	19	0.26	20	0.45	0.90	18.9	3.4	4.8	0.50
7.1	+0.14, −0.42	0.56	2.4	21	0.30	23	0.52	1.0	23.6	4.3	5.9	0.63
8.0	+0.16, −0.48	0.64	2.8	24	0.33	26	0.59	1.1	30.2	5.5	7.5	0.80
9.0	+0.18, −0.54	0.72	3.1	27	0.36	29	0.67	1.3	38.1	6.9	9.5	1.00
10.0	+0.20, −0.60	0.80	3.5	30	0.40	32	0.75	1.5	47.1	8.5	11.8	1.60
11.2	+0.22, −0.66	0.88	3.9	34	0.44	36	0.84	1.7	59.2	10.7	14.8	1.60
12.5	+0.25, −0.75	1.0	4.4	37	0.49	41	0.93	1.9	73.8	13.4	18.5	2.0
14.0	+0.28, −0.80	1.1	4.9	42	0.55	46	1.05	2.1	93.0	16.7	23.2	2.5
16.0	+0.32, −0.96	1.2	5.6	48	0.63	52	1.20	2.4	120	22.0	30.0	3.2
18.0	±0.90	1.4	6.3	54	0.71	58	1.35	2.7	153	26.0	38.2	4.0
20.0	±1.0	1.6	7.0	60	0.79	65	1.50	3.0	189	39.0	49.0	5.0
22.0	±1.1	1.8	7.7	66	0.87	73	1.70	3.4	228	42.0	57.0	6.3
25.0	±1.2	2.0	8.7	75	0.99	82	1.90	3.8	296	53.0	74.0	8.0
28.0	±1.4	2.2	9.8	84	1.1	91	2.10	4.2	372	67.0	93.0	10.0
32.0	±1.6	2.5	11	96	1.2	100	2.40	4.8	483	87.0	121.0	12.5
36.0	±1.9	2.8	12	108	1.4	110	2.70	5.4	610	112.0	152.0	16.0
40.0	±2.0	3.2	14	120	1.6	130	3.00	6.0	757	136.0	189.0	20.0
45.0	±2.2	3.6	16	155	1.8	150	3.40	6.8	953	173.0	228.0	22.5

Source: IS 2429 (Part II), 1970.

TABLE 21–48
Dimensions and lifting capacities of grade 40 noncalibrated chain (Figs. 21–17 and 21–18)

Nominal size, d_n, mm	Diameter tolerance $d_n \not> 16, +0.02s, -0.06s$; $d_n \geq 16, \pm0.05s$	Maximum additional weld dimensions $(d_w - d)$ max	Maximum additional weld dimensions $(G - d)$ max	Outside link length limits $5d_n$	Outside link length limits $4.75d_n$	Maximum outside link width, W_{max} Away from weld, W_{max} 3.5 d_n	Maximum outside link width, W_{max} Max. extra at weld 0.05 W_{max}	Minimum inside link width, 1.25 d_n	Guaranteed minimum breaking load (stress 40h bar), kN	Minimum energy absorption factor (energy absorption 0.072 kJ m^{-1} mm^{-2}), kJ/m	Maximum safe working load (stress 10h bar), kN	Lifting capacity (stress 10h bar), tonnes
6.3	0.12, −0.36	1	2.1	30	28	21	1.0	7.5	24.9	4.50	6.2	0.63
7.1	0.14, −0.42	1.4	2.4	35	33	24	1.24	8.8	31.6	4.70	7.9	0.80
8.0	0.16, −0.48	1.6	2.8	40	38	28	1.4	10	40.2	7.25	10.0	1.00
9	0.18, −0.54	1.8	3.1	45	43	31	1.6	11	50.9	9.18	12.7	1.25
10	0.20, −0.60	2.0	3.5	50	48	35	1.8	12	62.8	11.30	15.7	1.6
11	0.22, −0.66	2.2	3.9	55	52	39	2.0	14	79.0	14.20	19.7	2.0
12.5	0.25, −0.75	2.5	4.4	62	59	44	2.2	16	98.4	17.7	24.5	2.5
14.0	0.28, −0.84	2.8	4.9	70	66	49	2.5	18	124.0	22.2	30.8	3.2
16.0	0.32, −0.96	3.2	5.6	80	76	56	2.8	20	161.0	29.0	40.3	4.0
18.0	+0.90	3.6	6.3	90	86	63	3.1	22	204.0	37.7	50.5	5.0
20.0	+1.0	4.0	7.0	100	95	70	3.5	25	252.0	45.3	63.0	6.3
22.0	+1.1	4.4	7.7	110	105	77	3.9	28	304.0	55.0	76.0	8.0
25.0	±1.2	5.0	8.7	125	120	87	4.4	31	394.0	70.7	98.5	10.0
28.0	±1.4	5.6	9.8	140	130	98	4.9	35	492.0	89.0	123.0	12.5
32.0	±1.6	6.4	11	160	150	110	5.5	40	644.0	116.0	161.0	16.0
36.0	±1.9	7.2	12	180	170	120	6.0	45	814.0	147.0	204.0	20
40.0	±2.0	8.0	14	200	190	140	7.0	50	1010.0	181.0	252.0	25
45.0	±2.2	9.0	16	225	215	160	8.0	56	1270.0	230.0	318.0	32

Source: IS 3109 (Part I), 1970.

TABLE 21-49

Dimensions and lifting capacities of grade 40 calibrated chain (Figs. 21-17 and 21-18)

Nominal size, d_n, mm	Diameter tolerance $d_n \not> 16, \; +0.02s, -0.06s$; $d_n \geq 16, \; \pm0.05s$	Maximum additional weld dimensions		Preferred pitch (inside length), $3\,d_n$	Pitch tolerance (one link), $0.0396 d_n$	Preferred outside width, $W = 3.25 d_n$	Tolerance on outside width		Guaranteed minimum breaking load (stress 40h bar), kN	Minimum energy absorption factor (energy absorption 0.072 kJm^{-1} mm^{-2}), kJ/m	Maximum safe working load (stress 10h bar), kN	Lifting capacity (stress 10h bar), tonnes
		$(d_w - d)$ max	$(G - d)$ max				Away from weld zone $+0.075 d_n$, 0	At weld zone $+0.167 d_n$, 0				
6.3	0.12, −0.36	0.48	2.1	19	0.26	20	0.45	1.1	24.9	4.50	6.20	0.63
7.1	0.14, −0.42	0.56	2.4	21	0.30	23	0.52	1.2	31.6	5.70	7.80	0.70
8.0	0.16, −0.48	0.64	2.8	24	0.33	26	0.59	1.4	42.2	7.25	10.00	1.0
9.0	0.18, −0.54	0.72	3.1	27	0.36	29	0.57	1.9	50.9	9.18	12.7	1.25
10.0	0.20, −0.60	0.80	3.5	30	0.40	32	0.75	1.7	62.8	11.3	15.7	1.60
11.2	0.22, −0.66	0.88	3.9	34	0.44	36	0.88	1.9	79.0	14.2	19.7	2.00
12.5	0.25, −0.75	1.0	4.4	37	0.49	41	0.93	2.1	98.4	17.7	24.5	2.5
14.0	0.28, −0.80	1.1	4.9	42	0.55	46	1.05	2.4	124	22.2	30.8	3.2
16.0	0.32, −0.96	1.2	5.6	48	0.63	52	1.20	2.7	161	29.0	40.3	4.0
18.0	±0.90	1.4	6.3	54	0.71	58	1.35	3.0	204	37.7	50.5	5.0
20.0	±1.0	1.6	7.0	60	0.79	65	1.50	3.4	252	45.3	63.0	6.3
22.0	±1.1	1.8	7.7	66	0.87	73	1.70	3.8	304	55.0	76.0	8.0
25.0	±1.2	2.0	8.7	75	0.99	82	1.90	4.3	394	70.7	98.5	10.0
28.0	±1.4	2.2	9.8	84	1.1	91	2.10	4.8	492	89.0	123	12.5
32.0	+1.6	2.5	11	96	1.2	100	2.40	5.0	644	116	161	16.0
36.0	+1.9	2.8	12	108	1.4	110	2.70	6.1	814	147	204	20.0
40.0	+2.0	3.2	14	120	1.6	130	3.00	6.8	1010	181	252	25.0
45.0	+2.2	3.6	16	135	1.8	150	3.40	7.6	1270	230	318	32.0

Source: IS 3102 (Part II), 1970.

TABLE 21–50
Requirements of arc welded grade 30 chain for lifting purposes

Size (nominal diameter)	Proof load based on a stress of 98.1 MPa (10 kgf/mm²)		Minimum breaking load based on a stress of 294.2 MPa (30 kgf/mm²)		Minimum energy absorption factor for 1-m gauge length based on an energy absorption of 58.8 MN-m/m² (6 kgf-m/mm²)		Maximum safe working load for nominal working condition based on a stress of 49 MPa (5 kgf/mm²)	
mm	kN	kgf	kN	kgf	N m	kgf-m	kN	kgf
6	8.6	570	16.7	1700	3.3	340	2.8	285
8	9.8	1000	29.5	3010	5.9	602	4.9	500
9	12.5	1270	37.5	3820	7.5	764	6.2	635
10	15.4	1570	46.2	4710	9.2	942	7.7	785
12	22.2	2260	66.5	6780	13.3	1356	11.1	1130
14	30.2	3080	90.6	9140	18.1	1848	15.1	1540
16	39.4	4020	118.3	12060	23.7	2412	19.7	2010
18	49.9	5090	149.8	15270	30.0	3054	25.0	2545
20	61.6	6280	184.9	18850	37.0	3770	30.8	3140
22	74.5	7600	223.7	22810	44.7	4562	37.3	3800
24	88.8	9050	266.2	27140	53.2	5428	44.4	4525
27	102.5	10450	336.9	34350	67.4	6870	56.14	5725
30	138.7	14140	415.9	42410	83.2	8482	69.3	7070
33	167.7	17100	503.3	51320	100.7	10264	83.9	8550
36	192.7	20360	598.8	61070	119.8	12214	99.8	10180
39	234.4	23900	702.9	71680	140.6	14336	117.2	11950

TABLE 21–51
Requirements for electrically welded steel chain grade 30 chain for lifting purposes

Size (nominal diameter)	Proof load based on a stress of 157 MPa (16 kgf/mm²)		Minimum breaking load based on a stress of 392.3 MPa (40 kgf/mm²)		Minimum energy absorption factor for 1-m gauge length based on an energy absorption of 78.5 MN-m/m² (8 kgf-m/mm²)		Maximum safe working load for nominal working condition based on a stress of 49 MPa (5 kgf/mm²)	
mm	kN	kgf	kN	kgf	N m	kgf-m	kN	kgf
5	6.1	628	15.4	1571	3.1	314	3.1	314
6	8.9	904	22.2	2262	4.4	452	4.4	452
7	12.1	1232	30.2	3079	6.0	616	6.0	616
8	15.8	1608	39.4	4021	7.9	804	7.9	804
9	20.0	2036	49.9	5089	10.0	1018	10.0	1018
9.5	22.2	2268	55.6	5671	11.1	1134	11.1	1134
10	24.7	2514	61.6	6283	12.3	1257	12.3	1257
11	29.8	3042	74.6	7603	14.9	1521	14.9	1521
12	38.5	3928	96.3	9818	19.3	1964	19.3	1964
14	48.3	4926	120.8	12315	24.2	2463	24.2	2463
16	63.1	6434	157.7	16085	31.6	3217	31.6	3217
18	79.9	8144	199.6	20358	39.9	4072	39.9	4072
20	98.6	10054	246.5	25133	49.2	5027	49.3	5027
22	119.3	12164	298.2	30411	59.6	6082	59.6	6082
24	142.0	14476	354.9	36191	71.0	7238	71.0	7238
26	166.6	16990	416.5	42474	83.3	8495	83.3	8495
28	193.2	19704	483.1	49260	96.6	9852	96.6	9852
30	221.8	22620	554.6	56549	110.9	11310	110.9	11310
33	268.4	27370	671.0	68424	134.2	13685	134.2	13685
36	319.4	32572	798.6	81430	159.7	16286	153.7	16286
39	374.9	38228	937.2	95567	187.4	19114	187.4	19114
42	434.8	44334	1086.9	110836	217.4	22167	217.4	22167

Particular	Formula

Chain passing over a sheave (Fig. 21–19)

The effort on the chain in case of single-sheave pulley (Fig. 21–19)

$$P = \left[\frac{D + \mu D_o + \mu_c d}{D - \mu D_o - \mu_c d}\right] Q = CQ \qquad (21\text{–}86)$$

where $C = 1.04$ for lubricated chains

$\qquad = 1.10$ for chains running dry

The efficiency of the chain sheave

$$\eta = \frac{1}{C} \qquad (21\text{–}87)$$

where $\eta = 0.96$ for lubricated chains

$\qquad = 0.91$ for chain running dry

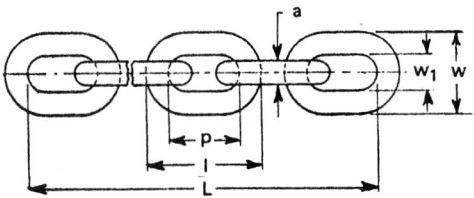

FIGURE 21–18 Pitch length and width of link.

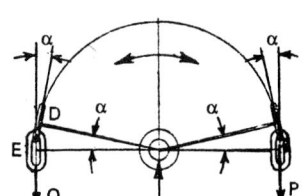

FIGURE 21–19 Chain passing over sheave.

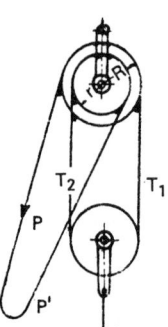

FIGURE 21–20 Differential chain block.

Differential chain block (Fig. 21–20)

RAISING THE LOAD Q The effort required for raising the load without friction

$$P_o = \frac{Q}{2}(1 - n) \qquad (21\text{–}88)$$

where $n = \dfrac{d}{D} = \dfrac{r}{R}$

The relation between the tension in the running-off and running-on chains

$$T_1 = C_1 T_2 \qquad (21\text{–}89)$$

where $C_1 = $ depends on the size of the chain and diameter of the lower sheave

The tension in the running-off chains

$$T_1 = \frac{C_t}{1 + C_1} Q \qquad (21\text{–}90)$$

Particular	Formula
The tension in the running-on chain	$$T_2 = \frac{Q}{1 + C_1} \qquad (21\text{--}91)$$
The relation between effort (P), load (Q), T_1 and T_2	$PR + T_2 r = C_2 T_1 R \qquad (21\text{--}92)$ where C_2 depends on the size of the chain and diameter of upper sheave
The effort required for raising the load with friction	$$P = \left(\frac{C_2 C_2 - n}{1 + C_1}\right) Q \qquad (21\text{--}93)$$ when C_1 and C_2 are different Or $$P = \left(\frac{C^2 - n}{1 + C}\right) Q \qquad (21\text{--}94)$$ when C is the average value of C_1 and C_2
The efficiency for the differential chain hoist	$$\eta = \left(\frac{1 - n}{2}\right)\left(\frac{1 + C}{C^2 - n}\right) \qquad (21\text{--}95)$$

Lowering the load

The equations for the tension in the running-on running-off and pull (P') required on the chain so as to prevent running down of the load	$$T_1 = \frac{Q}{1 + C} \qquad (21\text{--}96)$$ $$T_2 = \frac{CQ}{1 + C} \qquad (21\text{--}97)$$ $$T_1 R = C P' R + C T_2 r \qquad (21\text{--}98)$$
The pull required on the chain so as to prevent running down of the load	$$P' = \frac{Q}{C}\left[\frac{1 - nC^2}{1 + C}\right] \qquad (21\text{--}99)$$
The efficiency for the reversed motion	$$\eta' = \frac{2}{C}\left[\frac{1 - nC^2}{(1 - n)(1 + C)}\right] \qquad (21\text{--}100)$$ where C varies from 1.054 to 1.09 or obtained from Table 21–33
For mechanical properties of the coil link chain and the strength of hoisting chains in terms of bar from which they are made	Refer to Tables 21–52 and 21–53

TABLE 21–52
Mechanical properties of the coil link chain

Properties	Requirement	
	Grade 30	Grade 40
Mean stress at guaranteed minimum breaking load, F_w min; h bar	30	40
Mean stress at proof load, F_e, h bar	15	20
Ratio of proof load of guaranteed minimum breaking load	50%	50%
Guaranteed minimum elongation at fracture, A min	14.4%	14.4%
Guaranteed minimum energy absorption factor, $F_w \times A$	0.054 kJ m^{-1} mm^{-2}	0.054 kJ m^{-1} mm^{-2}
Maximum safe working load mean stress, h bar	7.5	10

TABLE 21–53
The strength of hoisting chains in terms of the bars from which they are made

Particular	% of bar
Standard close link	138
Coil chain	120
BB crane chain	145
Stud chain	165

Particular	Formula

Conditions for self-locking of differential chain block

The condition for self-locking

$$P' = \frac{Q}{C}\left[\frac{1 - nC^2}{1 + C}\right] \leq 0 \tag{21-101}$$

For self-locking differential chain block

$$n > \frac{1}{C^2} \tag{21-102}$$

The initial value of the ratio $\dfrac{r}{R}$

$$n = \frac{1}{C^2} \tag{21-103}$$

Power chains

Roller chains

The transmission ratio

$$i = \frac{\omega_1}{\omega_2} = \frac{n_1}{n_2} = \frac{d_2}{d_1} = \frac{z_2}{z_1} \tag{21-104}$$

The average speed of chain

$$v = \frac{p\,z_1 n_1}{60} \text{ m/s or } v = \frac{p\,z_1 n_1}{12} \text{ ft/min} \tag{21-105}$$

where z_1 = number of teeth on the small sprocket and p in m (in)

Particular	Formula
The empirical formula for pitch	$p \lessgtr 0.25\left(\dfrac{900}{n_1}\right)^{2/3}$ **SI** (21–106a)
	where p in m
	$p \lessgtr \left(\dfrac{900}{n_1}\right)^{2/3}$ **USCSU** (21–106b)
	where p in in
	$\leq 250\left(\dfrac{900}{n_1}\right)^{2/3}$ **Customary Metric units**
	(21–106c)
	where p in mm, n_1 = speed of the small sprocket, rpm
Bartlett formula relating speed (n_1) and pitch (p) based on allowable amount of impact between a roller and a sprocket	$n_1 = \dfrac{1170}{p}\sqrt{\dfrac{A}{w_f p}}$ **SI** (21–107a)
	where n_1 in rpm, p in m, w_f in N/m, and A in m^2
	$= \dfrac{11800}{p}\sqrt{\dfrac{A}{w_f p}}$ **Customary Metric** (21–107b)
	where n_1 in rpm, p in mm, w_f in kfg/m and A in mm^2
	$= \dfrac{1920}{p}\sqrt{\dfrac{A}{w_f p}}$ **US Customary System units**
	(21–107c)
	where n_1 in rpm, p in in, w_f in lbf/ft, and A in in^2
	$A = ld_r$ = projected area of the roller d_r = diameter of rollers l = width of chain or length of roller
Maximum allowable chain velocity based on Eq. (21–107)	$v_{max} \lessgtr 19.48\, z_1 \sqrt{\dfrac{A}{w_f p}}$ **SI** (21–108a)
	where v_{max} in m/s, A in m^2, p in m, and w_f in N/m
	$\lessgtr 160\, z_1 \sqrt{\dfrac{A}{w_f p}}$ **US Customary System units**
	(21–108b)

Particular	Formula
	where v_{max} in ft/min, A in in^2, p in in, and w_f in lbf/ft

$$\lessgtr 0.196\, z_1 \sqrt{\frac{A}{w_f p}} \quad \textbf{Customary Metric}$$

$$(21-108c)$$

where v_{max} in m/s, A in mm^2, p in mm, and w_f in kgf/m

Maximum speed based on the energy of impact per tooth per minute

$$n \lessgtr \frac{1437}{p} \sqrt[3]{\frac{A}{w_f}} \qquad \textbf{SI} \qquad (21-109a)$$

where A in m^2, p in m, and w_f in N/m

$$\lessgtr \frac{2000}{p} \sqrt[3]{\frac{A}{w_f}} \qquad \textbf{US Customary System units}$$

$$(21-109b)$$

where A in in^2, p in in, and w_f in lbf/ft

$$\lessgtr \frac{6712}{p} \sqrt[3]{\frac{A}{w_f}} \qquad \textbf{Customary Metric} \qquad (21-109c)$$

where A in mm^2, p in mm, and w_f in kgf/m

Maximum chain velocity based on Eq. (21–109), m/s

$$v_{max} \lessgtr 24\, z_1 \sqrt[3]{\frac{A}{w_f}} \qquad \textbf{SI} \qquad (21-110a)$$

where v_{max} in m/s, A in m^2, and w_f in N/m

$$\leq 166\, z_1 \sqrt[3]{\frac{A}{w_f}} \qquad \textbf{US Customary System units}$$

$$(2-110b)$$

where v_{max} in ft/min, A in in^2, and w_f in lbf/ft

$$\lessgtr 0.11\, z_1 \sqrt[3]{\frac{A}{w_f}} \qquad \textbf{Customary Metric} \qquad (21-110c)$$

where v_{max} in m/s, A in mm^2, and w_f in kgf/m

Maximum sprocket speed based on the effect of centrifugal force

$$n \leq \frac{36350}{p} \sqrt{\frac{A}{z_1\, w_f}} \qquad \textbf{SI} \qquad (21-111a)$$

where p in m, A in m^2, and w_f in N/m

$$\leq \frac{9516}{p} \sqrt{\frac{A}{z_1\, w_f}} \qquad \textbf{US Customary System units}$$

$$(21-111b)$$

where p in in, A in in^2, and w_f in lbf/ft

Particular	Formula
Maximum velocity based on Eq. (21–111)	$$v_{max} \leq 600 \sqrt{\frac{A z_1}{w_f}} \qquad \textbf{SI} \qquad (21\text{–}112a)$$ where v_{max} in m/s, A in m^2, and w_f in N/m $$\leq 793 \sqrt{\frac{A z_1}{w_f}} \qquad \textbf{US Customary System units} \qquad (21\text{–}112b)$$ where v_{max} in ft/min, A in in^2, and w_f in lbf/ft $$\lessgtr 0.2 \sqrt{\frac{A z_1}{w_f}} \qquad (21\text{–}112c)$$ where v_{max} in m/s, A in mm^2, and w_f in kgf/m

Chain pull

Particular	Formula
For preliminary computation, the allowable pull	$$F_a = \frac{F_u}{n_o} \qquad (21\text{–}113)$$ where F_u = ultimate strength from Table 21–35B and 21–42 n_o = working factor, = 5 for sprocket having over 40 teeth and a speed of 0.5 m/s = 18 for sprocket having 10 or 11 teeth and a speed of 6 m/s
AGMA formula for allowable pull based on velocity factor $C_v = 3/(3 + v)$ and bearing pressure of 29.4 MPa (4333 psi) for the pin	$$F_a = \frac{90 \times 10^6 \, l d_r}{3 + v} - \frac{v^2 w_f}{9.8} \qquad \textbf{SI} \qquad (21\text{–}114a)$$ where l and d_r in m, v in m/s, and w_f in N/m $$= \left(\frac{l d_r}{600 + v} - \frac{v^2 w_f}{3(10^{11})} \right) 2600000 \qquad \textbf{USCSU}$$ $$(2\text{–}114b)$$ where l and d_r in in, v in ft/min, and w_f in lbf/ft where l = length of roller pins, m (in) $$v = \frac{z_1 p n_1}{60} m/s$$ d_r = roller pin diameter, m (in)
For dimensions of American Standard Roller Chains—single-strand	Refer to Tables 21–54A.

TABLE 21–54*A*
Dimensions of American Standard roller chains—single-strand

ANSI chain number	Pitch, in (mm)	Width, in (mm)	Minimum tensile strength, lb (N)	Average weight, lb/ft (N/m)	Roller diameter, in (mm)	Multiple-strand spacing, in (mm)
25	0.250	0.125	780	0.09	0.130	0.252
	(6.35)	(3.18)	(3 470)	(1.31)	(3.30)	(6.40)
35	0.375	0.188	1 760	0.21	0.200	0.399
	(9.52)	(4.76)	(7 830)	(3.06)	(5.08)	(10.13)
41	0.500	0.25	1 500	0.25	0.306	—
	(12.70)	(6.35)	(6 670)	(3.65)	(7.77)	—
40	0.500	0.312	3 130	0.42	0.312	0.566
	(12.70)	(7.94)	(13 920)	(6.13)	(7.92)	(14.38)
50	0.625	0.375	4 880	0.69	0.400	0.713
	(15.88)	(9.52)	(21 700)	(10.1)	(10.16)	(18.11)
60	0.750	0.500	7 030	1.00	0.469	0.897
	(19.05)	(12.7)	(31 300)	(14.6)	(11.91)	(22.78)
80	1.000	0.625	12 500	1.71	0.625	1.153
	(25.40)	(15.88)	(55 600)	(25.0)	(15.87)	(29.29)
100	1.250	0.750	19 500	2.58	0.750	1.409
	(31.75)	(19.05)	(86 700)	(37.7)	(19.05)	(35.76)
120	1.500	1.000	28 000	3.87	0.875	1.789
	(38.10)	(25.40)	(124 500)	(56.5)	(22.22)	(45.44)
140	1.750	1.000	38 000	4.95	1.000	1.924
	(44.45)	(25.40)	(169 000)	(72.2)	(25.40)	(48.87)
160	2.000	1.250	50 000	6.61	1.125	2.305
	(50.80)	(31.75)	(222 000)	(96.5)	(28.57)	(58.55)
180	2.250	1.406	63 000	9.06	1.406	2.592
	(57.15)	(35.71)	(280 000)	(132.2)	(35.71)	(65.84)
200	2.500	1.500	78 000	10.96	1.562	2.817
	(63.50)	(38.10)	(347 000)	(159.9)	(39.67)	(71.55)
240	3.00	1.875	112 000	16.4	1.875	3.458
	(76.70)	(47.63)	(498 000)	(239)	(47.62)	(87.83)

Source: Compiled from ANSI B29.1-1975.

TABLE 21–54*B*
Rated horsepower capacity of single-strand single-pitch roller chain for a 17-tooth sprocket

Sprocket speed, rpm	ANSI chain number					
	25	35	40	41	50	60
50	0.05	0.16	0.37	0.20	0.72	1.24
100	0.09	0.29	0.69	0.38	1.34	2.31
150	0.13[a]	0.41[a]	0.99[a]	0.55[a]	1.92[a]	3.32
200	0.16[a]	0.54[a]	1.29	0.71	2.50	4.30
300	0.23	0.78	1.85	1.02	3.61	6.20
400	0.30[a]	1.01[a]	2.40	1.32	4.67	8.03
500	0.37	1.24[a]	2.93	1.61	5.71	9.81
600	0.44[a]	1.46[a]	3.45[a]	1.90[a]	6.72[a]	11.6
700	0.50	1.68	3.97	2.18	7.73	13.3
800	0.56[a]	1.89[a]	4.48[a]	2.46[a]	8.71[a]	15.0
900	0.62	2.10	4.98	2.74	9.69	16.7
1000	0.68[a]	2.31[a]	5.48	3.01	10.7	18.3
1200	0.81	2.73	6.45	3.29	12.6	21.6
1400	0.93[a]	3.13[a]	7.41	2.61	14.4	18.1
1600	1.05[a]	3.53[a]	8.36	2.14	12.8	14.8
1800	1.16	3.93	8.96	1.79	10.7	12.4
2000	1.27[a]	4.32[a]	7.72[a]	1.52[a]	9.23[a]	10.6
2500	1.56	5.28	5.51[a]	1.10[a]	6.58[a]	7.57
3000	1.84	5.64	4.17	0.83	4.98	5.76
Type A	Type B					Type C

[a] Estimated from ANSI tables by linear interpolation.
Note: Type A—manual or drip lubrication; type B—bath or disk lubrication; type C—oil-stream lubrication.
Source: Compiled from ANSI B29.1-1975 information only section, and from B29.9-1958.

TABLE 21–54C
Rated horsepower capacity of single-strand single-pitch roller chain for a 17-tooth sprocket

Sprocket speed, rpm	ANSI chain number							
	80	100	120	140	160	180	200	240
50	2.88	5.52	9.33	14.4	20.9	28.9	38.4	61.8
100	5.38	10.3	17.4	26.9	39.1	54.0	71.6	115
150	7.75	14.8	25.1	38.8	56.3	77.7	103	166
200	10.0	19.2	32.5	50.3	72.9	101	134	215
300	14.5	27.7	46.8	72.4	105	145	193	310
400	18.7	35.9	60.6	93.8	136	188	249	359
500	22.9	43.9	74.1	115	166	204	222	0
600	27.0	51.7	87.3	127	141	155	169	
700	31.0	59.4	89.0	101	112	123	0	
800	35.0	63.0	72.8	82.4	91.7	101		
900	39.9	52.8	61.0	69.1	76.8	84.4		
1000	37.7	45.0	52.1	59.0	65.6	72.1		
1200	28.7	34.3	39.6	44.9	49.9	0		
1400	22.7	27.2	31.5	35.6	0			
1600	18.6	22.3	25.8	0				
1800	15.6	18.7	21.6					
2000	13.3	15.9	0					
2500	9.56	0.40						
3000	7.25	0						

Type B (rows 50–100); Type A (middle region); Type C; Type C′

Note: Type A—manual or drip lubrication; type B—bath or disk lubrication; type C—oil-stream lubrication; type C′—type C, but this is a galling region; submit design to manufacturer for evaluation.
Source: Compiled from ANSI B29.1-1975 information only section, and from B29.9-1958.

TABLE 21–54D
Tooth correction factors, K_1

Number of teeth on driving sprocket	Tooth correction factor, K_1	Number of teeth on driving sprocket	Tooth correction factor, K_1
11	0.53	22	1.29
12	0.62	23	1.35
13	0.70	24	1.41
14	0.78	25	1.46
15	0.85	30	1.73
16	0.92	35	1.95
17	1.00	40	2.15
18	1.05	45	2.37
19	1.11	50	2.51
20	1.18	55	2.66
21	1.26	60	2.80

Source: J. E. Shigley and C. R. Mischke, *Mechanical Engineering Design*, McGraw-Hill Book Company, New York, 1989.

TABLE 21–54E
Multistrand factors, K_2

Number of strands	K_2
1	1.0
2	1.7
3	2.5
4	3.3

Source: J. E. Shigley and C. R. Mischke, *Mechanical Engineering Design*, McGraw-Hill Book Company, New York, 1989.

TABLE 21–54F
Service factor for roller chains, k_s

Operating characteristics	Intermittent few hours per day, few hours per year	Normal 8 to 10 hours per day 300 days per year	Continuous 24 hours per day
Easy starting, smooth, steady load	0.06–1.00	0.90–1.50	0.90–2.00
Light medium shock or vibrating load	0.90–1.40	1.20–1.90	1.50–2.40
Medium to heavy shock or vibrating load	1.20–1.80	1.50–2.30	1.80–2.80

Particular	Formula

Power

For the rated horsepower capacity of single-strand-single-pitch roller chains for 17-tooth sprocket and values of k_1 and k_2

Refer to Tables 21–54B to 21–54E.

Power required

$$P = \frac{F_\theta\, v}{1000\, k_l\, k_s} \quad \textbf{SI} \qquad (21\text{–}115a)$$

where F_θ in N and P in kW

$$= \frac{F_\theta\, v}{33000\, k_l\, k_s} \quad \textbf{US Customary system units}$$

$$(21\text{–}115b)$$

where P in hp and F_θ in lbf

$$= \frac{F_\theta\, v}{102\, k_l\, k_s} \quad \textbf{Customary Metric} \qquad (21\text{–}115c)$$

where

F_θ = required chain pull in kgf and P in kW
k_l = load factor from 1.1 to 1.5 and also obtained from Chap. 14
k_s = service factor
  = 1 for 10 h service per day
  = 1.2 for 24 h operation and also obtained from Table 21–54F

Particular	Formula

TABLE 21–55
Coefficient for *Sag*, K_{sg}

K_{sg}	Position of chain drive			
	Horizontal	< 40°	> 40°	Vertical
	6	4	2	1

The rated horsepower of roller chain per strand

$$P = p^2 \left[\frac{v}{0.75} - \frac{v^{1.41}}{3.7} \left(1 + 50 \sin^2 \frac{90}{z_1} \right) \right] \qquad (21\text{–}116)$$

The corrected horsepower (P_c)

$$P_c = K_1 K_2 P_r \qquad (21\text{–}116a)$$

where P_r = rated horsepower and K_1 and K_2 from Tables 21–54D and 21–54E

CHECK FOR ACTUAL SAFETY FACTOR

The actual safety factor checked by the formula

$$n_a = \frac{F_u}{F_\theta + F_{cs} + F_s} \qquad (21\text{–}117)$$

where $F_{cs} = \dfrac{wv^2}{g}, F_s = k_{sg} wC \qquad (21\text{–}117a)$

$$F_\theta = \frac{33000 P}{v}; \qquad (21\text{–}117b)$$

where F_θ in lbf; P in hp; v in ft/min

or $1000P/v$ where F_θ in N; P in kW; v in m/s

w = weight per meter of chain, N (lbf)
v = velocity of chain, m/s (ft/min)
C = center distance, m (in)
k_{sz} = coefficient for sag from Table 21–55

The number of strand in a chain, if $F_\theta > F_a$

$$i = \frac{F_\theta}{F_a} \qquad (21\text{–}118)$$

Center distance of chain length

The proper center distance between sprockets in pitches

$$C_p = 20p \text{ to } 30p \text{ Or } C_p = 40 \pm 10 \text{ pitches} \qquad (21\text{–}119)$$

where $p\, C_p = C$

The minimum center distance

$$C_{min} = K_{min} C \qquad (21\text{–}120)$$

where

$$C = \frac{d_{a1} + d_2}{2}$$

Particular	Formula
	$$d_a = \frac{p}{\tan\left(\frac{180}{z}\right)} + 0.6p$$

Refer to Table 21–56 for values of k [used in Eq. (21–123)] and Table 21–57 for K_{min}.

TABLE 21–56
Values of k to be used in Eq. (21–123)

$$C_{max} = 80p$$

$\frac{z_1 - z_2}{C_p}$	0.1	1.0	2.0	3.0	4.0	5.0	6.0
k	0.02533	0.02538	0.02555	0.02584	0.02631	0.02704	0.02828

TABLE 21–57
Minimum center distance constant, K_{min}

Transmission ratio, i	Minimum center distance constant, K_{min}
3	$1 + (30\text{–}50/c')$
3–4	1.2
4–5	1.3
5–6	1.4
6–7	1.5

The maximum center distance

where p = pitches of chain, mm (21–121)

The chain length in pitches

$$L_p = 2\,C_p\cos\alpha + \frac{z_1 + z_2}{2} + \alpha\frac{z_1 - z_2}{180}\ \text{(exact)}$$
(21–122)

$$L_p = 2\,C_p + \frac{z_1 + z_2}{2} + \frac{k(z_1 - z_2)^2}{C_p}$$
(21–123)

The chain length, m or in

$$L = 2C\cos\alpha + \frac{z_1\,p\,(180 + 2\alpha)}{360} + \frac{z_2\,p\,(180 - 2\alpha)}{360}$$
(21–124)

where

z_2 = number of teeth on a large sprocket
z_1 = number of teeth on a small sprocket
α = angle between tangent to the sprocket pitch circle and the center line.

$$= \sin^{-1}\left(\frac{d_2 - d_1}{2C}\right)$$

k = a variable which depends on $\dfrac{z_1 - z_2}{C_p}$
and obtained from Table 21–56

Particular	Formula
The chain length	$L = p\,L_p$ (21–125)
The pitch diameter of a sprocket	$d = \dfrac{p}{\sin\left(\dfrac{180}{z}\right)}$ (21–126)

Roller chain sprocket

Particular	Formula
Minimum number of teeth	$z_{min} = \dfrac{4\,d_r}{p} + 5$ for pitches of 25 mm (12–127a)
	$= \dfrac{4\,d_r}{p} + 4$ for pitches 32 to 58 mm (21–127b)

Silent chain sprocket

Particular	Formula
Minimum number of teeth	$z_{min} = \dfrac{4\,d_r}{p} + 6$ for pitches to 51 mm (21–128)
The root diameter of sprocket	$d_f = d - d_r$ (21–129)
	where d_r = diameter of roller pin, m (in)
The width of sprocket tooth (Fig. 21–22)	$C_1 = l - 0.05p$ (21–130)
	where l = chain width or roller length
Maximum hub diameter	$D_h = d \cos \dfrac{180}{z} - (H + 0.1270)$ (21–131)
	where H = height of link plate, m or in
	$= 0.3p$
Power per cm of width in hp	$P = \dfrac{pv}{6.80}\left[1 - \dfrac{v}{2.16\,(z_1 - 8)}\right]$ (21–132)
	where $v = \dfrac{p z_1 n_1}{60}$ = chain speed, m/s; p in m
The relationship between depth of sag, and tension due to weight of chain in the catenary (approx.)	$h = 0.433\sqrt{S^2 - L^2}$ (21–133a)
	$F = w\left(\dfrac{S^2}{8h} + \dfrac{h}{2}\right)$ (21–133b)
	where h = depth of sag, m (in)

L = distance between points of support, m (in)
S = catenary length of chain, m (in)
F = tension or chain pull, kN (lbf)
w = weight of chain, kN/m (lbf/in)

Particular	Formula

Tension chain linkages

Allowable load

$$F_a = 13.1 \times 10^6 p^2 \qquad \textbf{SI} \qquad (21\text{–}134a)$$

where p in m and F_u in N

$$= 1900 p^2 \qquad \textbf{US Customary units} \qquad (21\text{–}134b)$$

where p = pitch of chain, in and F_u in lbf

Allowable load for lightweight chain

$$F_a = 7 \times 10^6 p^2 \qquad \textbf{SI} \qquad (21\text{–}134c)$$

where p in m

$$= 1020 p^2 \text{ where } p \text{ in in, } F \text{ in lbf} \qquad (21\text{–}134d)$$

US Customary units

Indian Standards

PRECISION ROLLER CHAIN (FIGS 21–21 TO 21–25, TABLES 21–58, 21–59, 21–60)
Pitch circle diameter (Fig. 21–21)

$$PCD = \frac{P}{\sin \dfrac{180}{z}} \qquad (21\text{–}135)$$

Bottom diameter

$$BD = PCD - D_r \qquad (21\text{–}136)$$

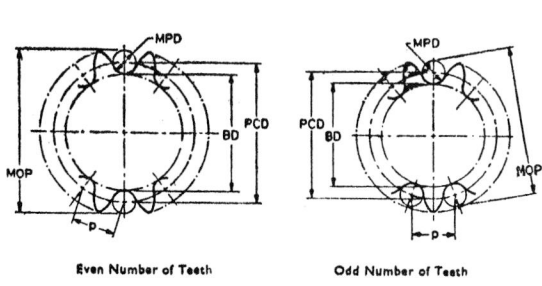

FIGURE 21–21   Notation for wheel rim of chain.

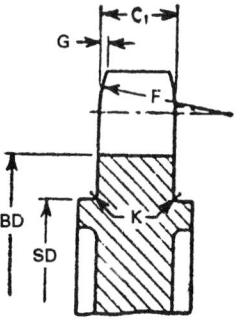

FIGURE 21–22   Notation for wheel rim profile of roller chain.

Wheel tooth gap form

The minimum value of roller seating radius, mm (Fig. 21–24)

$$SR_1 = 0.505 D_r \qquad (21\text{–}137)$$

The maximum value of roller seating radius, mm (Fig. 21–25)

$$SR_2 = (0.505 D_r + 0.069 \sqrt[3]{D_r}) \qquad (21\text{–}138)$$

where D_r = roller diameter, mm

TABLE 21–58
Extended pitch transmission roller chain dimensions, measuring loads and breaking loads

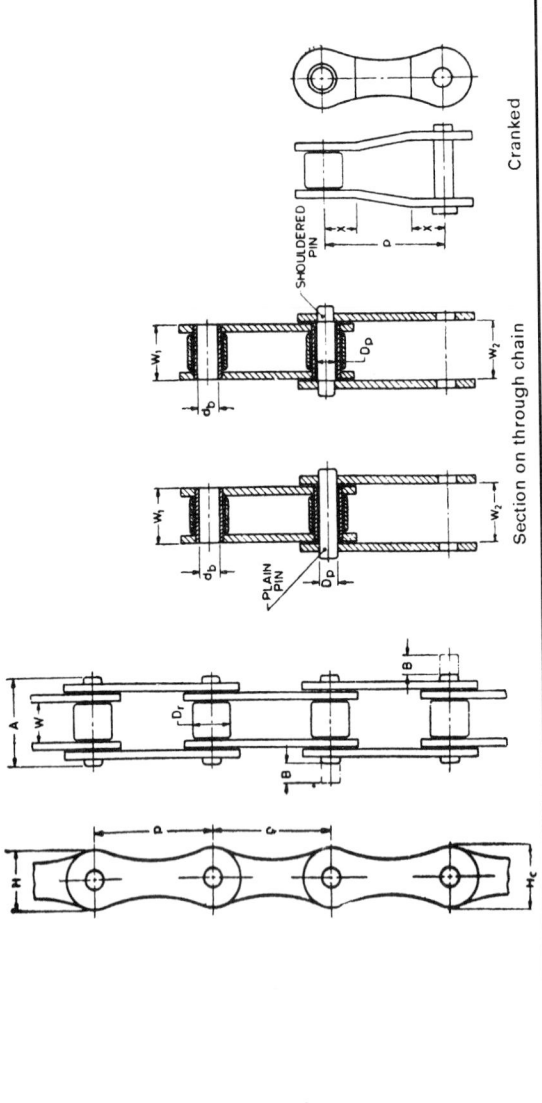

Section on through chain

Cranked

Chain no.	Pitch, P, mm	Roller diameter max, D_r, mm	Width between inner plates, W, min; mm	Bearing in diameter, max, D_p, mm	Bush bore, min, d_b, mm	Chain path depth, max, H_c, mm	Plate depth, dimension, H, min, mm	Crank linked max, X, mm	Width over inner link, W, mm min	Width between outer plates, W_1, mm max	Width over bearing pin, W_2, mm max	Addition width for joint fastener, A, mm max, B, mm	Measuring load N	Measuring load kgf	Breaking load, min kN	Breaking load, min kgf
208A	25.40	7.92	7.95	3.96	4.01	12.33	12.07	6.9	11.18	11.31	17.8	3.9	127.5	13	13.8	1410
208B	25.40	8.51	7.75	4.45	4.50	12.07	11.81	6.9	11.30	11.43	17.0	3.9	127.5	13	17.9	1820
210A	31.70	10.16	9.53	5.08	5.13	15.35	15.09	8.4	13.84	13.97	21.8	4.1	196.1	20	21.8	2220
210B	31.75	10.16	9.65	5.08	5.13	14.99	14.73	8.4	13.28	13.41	19.6	4.1	196.1	20	22.3	2270
212A	38.10	11.91	12.70	5.94	5.99	18.34	18.08	9.9	17.75	17.88	26.9	4.6	284.4	29	31.2	3180
212B	38.10	12.07	11.68	5.72	5.77	16.39	16.13	9.9	13.62	15.76	22.7	4.6	284.4	29	28.9	2950
216A	50.80	15.88	15.88	7.92	7.97	24.39	24.13	13.0	22.61	22.74	33.5	5.4	500.2	51	55.6	5670
216B	50.80	15.88	17.02	8.28	8.33	21.34	21.08	13.0	25.45	25.58	36.1	5.4	500.1	51	42.3	4310
220A	63.50	19.05	19.05	9.53	9.58	30.48	30.18	16.0	27.46	27.59	41.1	6.1	774.7	79	86.8	8850
220B	63.50	19.05	19.56	10.19	10.24	26.68	26.42	16.0	29.01	29.14	43.2	6.1	774.7	79	64.5	6580
224A	76.20	22.23	25.40	11.10	11.15	36.55	36.20	19.1	35.46	35.59	50.8	6.6	1108.2	113	124.5	12700
224B	76.20	25.40	25.40	14.63	14.68	33.73	33.40	19.1	37.92	38.05	53.4	6.6	1108.2	113	97.9	9980
228B	88.90	27.94	30.99	15.90	15.95	36.46	37.08	21.3	46.58	46.71	65.1	7.4	1510.2	164	129.1	13160
232B	101.60	29.21	30.99	17.81	17.86	42.72	42.29	24.4	45.57	45.70	64.7	7.9	2000.6	204	169.1	17240

Notes: (1) The chain path depth H_c is the minimum depth of channel through which the assembled chain will pass; (2) the overall width of chain with joint fastener is $-A + B$ for riveted pin end and fastener on one side;

$A + 1.6B$ for headed pin end and fastener on one side; and $A + 2B$ for fastener on both sides.

The actual dimensions will depend on the type of fastener used, but they should not exceed the dimensions in this column.

Source: IS 3542, 1966.

Particular	Formula

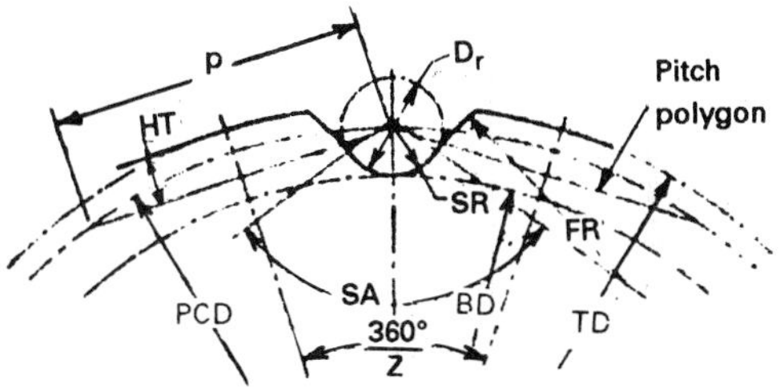

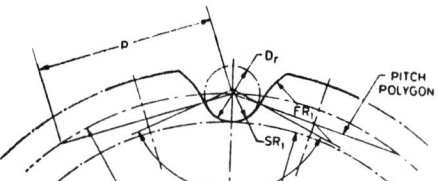

FIGURE 21–23 Notation for tooth gap form of roller chain.

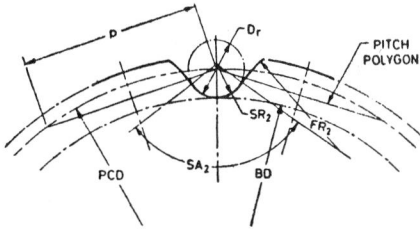

FIGURE 21–24 Notation for minimum tooth gap form of roller chain.

The minimum value of roller seating angle, deg (Fig. 21–24)

$$SA_1 = 140° - \frac{90°}{z} \qquad (21\text{–}139)$$

The maximum value of roller seating angle, deg (Fig. 21–25)

$$SA_2 = 120° - \frac{90°}{z} \qquad (21\text{–}140)$$

The minimum value of tooth flank radius, mm (Fig. 21–24)

$$FR_1 = 0.12\,D_r\,(z+2) \qquad (21\text{–}141)$$

The maximum value of tooth flank radius, mm (Fig. 21–25)

$$FR_2 = 0.008\,D_r\,(z^2 + 180) \qquad (21\text{–}142)$$

FIGURE 21–25 Notation for maximum tooth gap form of roller chain.

Tooth heights and top diameters (Fig. 21–23)

The maximum limit of the tooth height above the pitch polygon

$$HT_{max} = p\left(0.3125 + \frac{0.8}{z}\right) - 0.5\,D_r \qquad (21\text{–}143)$$

The minimum limit of the tooth height above the pitch polygon

$$HT_{min} = p\left(0.25 + \frac{0.6}{z}\right) - 0.5\,D_r \qquad (21\text{–}144)$$

The maximum limit of the tooth top diameter, mm

$$TD_{max} = PCD + 0.625p - D_r \qquad (21\text{–}145)$$

The minimum limit of the tooth top diameter, mm

$$TD_{min} = PCD + p\left(0.5 - \frac{0.4}{z}\right) - D_r \qquad (21\text{–}146)$$

TABLE 21–59
Pitch circle diameters[a] for extended pitch transmission roller chain wheels

No. of teeth, z	Pitch circle diameter	No. of teeth, z	Pitch circle diameter	No. of teeth, z	Pitch circle diameter	No. of teeth, z	Pitch circle diameter	No. of teeth, z	Pitch circle diameter	No. of teeth, z	Pitch circle diameter
5	1.7013	17	5.4422	29	9.2491	41	13.0635	53	16.8803	65	20.6982
5½	1.8496	17½	5.6005	29½	9.4080	41½	13.2225	53½	17.0393	65½	20.8575
6	2.0000	18	5.7588	30	9.5668	42	13.3815	54	17.1984	66	21.0164
6½	2.1519	18½	5.9171	30½	9.7256	42½	13.5405	54½	17.3575	66½	21.1757
7	2.3048	19	6.0755	31	9.8845	43	13.6995	55	17.5166	66½	21.3346
7½	2.4586	19½	6.2340	31½	10.0434	43½	13.8585	55½	17.6756	67½	21.4939
8	2.6131	20	6.3925	32	10.2023	44	14.0176	56	17.8347	68	21.6528
8½	2.7682	20½	6.5509	32½	10.3612	44½	14.1765	56½	17.9938	68½	21.8121
9	2.9238	21	6.7095	33	10.5201	45	14.3356	57	18.1529	69	21.9710
9½	3.0798	21½	6.8681	33½	10.6790	45½	14.4946	57½	18.3119	69½	22.1303
10	3.2361	22	7.0266	34	10.8380	46	14.6537	58	18.4710	70	22.2892
10½	3.3927	22½	6.1853	34½	10.9969	46½	14.8127	58½	18.6301	70½	22.4485
11	3.5494	23	7.3439	35	11.1558	47	14.9717	59	18.7892	71	22.6074
11½	3.7065	23½	7.5026	35½	11.3148	47½	15.1308	59½	18.9482	71½	22.7667
12	3.8637	24	7.6613	36	11.4737	48	15.2898	60	19.1073	72	22.9256
12½	4.0211	24½	7.8200	36½	11.6327	48½	15.4488	60½	19.2665	72½	23.0849
13	4.1786	25	7.9787	37	11.7916	49	15.6079	61	19.4255	73	23.2438
13½	4.3362	25½	8.1375	37½	11.9506	49½	15.7669	61½	19.5847	73½	23.4031
14	4.4940	26	8.2962	38	12.1095	50	15.9260	62	19.7437	74	23.5620
14½	4.6518	26½	8.4550	38½	12.2685	50½	16.0850	62½	19.6029	74½	23.7213
15	4.8097	27	8.6138	39	12.4275	51	16.2441	63	20.0619	75	23.8802
15½	4.9677	27½	8.7726	39½	12.5865	51½	16.4031	63½	20.2210		
16	5.1258	28	8.9314	40	12.7455	52	16.5622	64	20.3800		
16½	5.2840	28½	9.0902	40½	12.9045	52½	16.7212	64½	20.5393		

[a]The values given are for a unit pitch (e.g., 1 mm).
Source: IS 3542, 1966.

TABLE 21–60
Maximum speed (rpm), recommended of sprockets for roller chains

No. of teeth	Pitch												
	6.35	9.50	12.70	15.80	19.05	25.40	31.75	38.10	44.45	50.80	57.15	63.50	76.20
11	4310	2260	1690	1220	920	580	415	325	235	200	165	145	110
12	4960	2590	1940	1400	1050	670	475	375	270	230	190	165	125
13	5540	2900	2180	1570	1110	750	535	415	305	260	215	186	140
14	6070	3170	2380	1720	1290	820	585	455	335	280	255	205	155
15	6500	3420	2560	1850	1390	880	630	490	360	305	255	220	165
16	6940	3630	2720	1969	1480	935	670	520	380	325	270	235	175
17	7290	3810	2860	2060	1550	985	700	550	400	340	285	245	185
18	7590	3970	2980	2150	1610	1020	730	750	415	355	295	255	195
19	7840	4100	3080	2220	1670	1060	755	590	430	365	305	265	200
20	8050	4210	3160	2280	1720	1090	755	605	440	375	315	270	205
21	8230	4300	3230	2330	1750	1110	790	620	450	385	320	280	210
22	8380	4380	3290	2370	1780	1130	805	630	460	390	325	280	215
23	8480	4480	3330	2400	1800	1150	875	640	405	395	330	285	215
24	8560	4410	3360	2420	1820	1160	825	645	470	400	300	290	220
25	8610	4510	3380	2440	1830	1100	830	650	475	400	335	290	220
30	8780	4490	3370	2430	1830	1160	825	645	470	400	335	290	220
35	8200	4290	3220	2320	1740	1110	790	615	450	380	320	275	210
40	7580	3970	2970	2140	1610	1020	730	570	415	355	295	255	195
45	6820	3570	2670	1930	1450	920	655	515	375	320	265	230	175
50	5950	3110	2330	1680	1270	805	575	450	325	275	230	200	150
55	5010	2620	1970	1420	1070	675	410	375	275	235	195	170	125
60	4020	2100	1580	1140	860	545	390	305	220	185	155	135	100

Particular	Formula

Wheel rim profile (Fig. 21–22)

Tooth width

$$C_1 = 0.95W \text{ with a tolerance of } h/4 \qquad (21\text{--}147)$$

The minimum tooth side radius

$$F = 0.5p \qquad (21\text{--}148)$$

The tooth side relief

$$G = 0.05p \text{ to } 0.075p \qquad (21\text{--}149)$$

Absolute maximum shroud diameter

$$D = p \cot \frac{180°}{z} - 1.05H - 1.00 - 2 \times K_{oct}, \text{ mm}$$

$$(21\text{--}150)$$

For leaf chain dimension, breaking load, anchor clevises and chain sheaves

Refer to Tables 21–61, 21–62, and 21–63

Leaf chains

PRECISION BUSH CHAINS (FIGS. 21–26 TO 21–29, TABLES 21–64 TO 21–68) The pitch circle diameter (Fig. 21–21, Table 21–62)

$$PCD = \frac{p}{\sin \dfrac{180}{z}} \qquad (21\text{--}151)$$

Bottom diameter

$$BD = PCD - D_b \qquad (21\text{--}152)$$

The minimum value of bush seating radius, mm (Fig. 21–28)

$$SR_1 = 0.505\, D_b \qquad (21\text{--}153)$$

The maximum value of bush seating radius, mm (Fig. 21–29)

$$SR_2 = (0.505\, D_b + 0.0693\sqrt{D_b}) \qquad (21\text{--}154)$$

The minimum value of bush seating angle, deg (Fig. 21–28)

$$SA_1 = 140° - \frac{90°}{z} \qquad (21\text{--}155)$$

The maximum value of bush seating angle, deg (Fig. 21–29)

$$SA_2 = 120° - \frac{90^c}{z} \qquad (21\text{--}156)$$

The minimum value of tooth flank radius, mm (Fig. 21–28)

$$FR_1 = 0.12\, D_b\, (z + 2) \qquad (21\text{--}157)$$

The maximum value of tooth flank radius, mm (Fig. 21–29)

$$FR_2 = 0.008\, D_b\, (z^2 + 180) \qquad (21\text{--}158)$$

TOOTH TOP DIAMETERS AND TOOTH HEIGHT (FIG. 21–27) The maximum limit of the tooth top diameter

$$TD_{max} = PCD + 1.25\, p - D_b \qquad (21\text{--}159)$$

The minimum limit of the tooth top diameter

$$TD_{min} = PCD + p\left(1 - \frac{1.6}{z}\right) - D_b \qquad (21\text{--}160)$$

The maximum limit of the tooth height above the pitch polygon

$$HT_{max} = 0.625\, p - 1.5\, D_b + \frac{0.8p}{z} \qquad (21\text{--}161)$$

TABLE 21-61

Leaf chain dimensions, measuring loads, and breaking loads

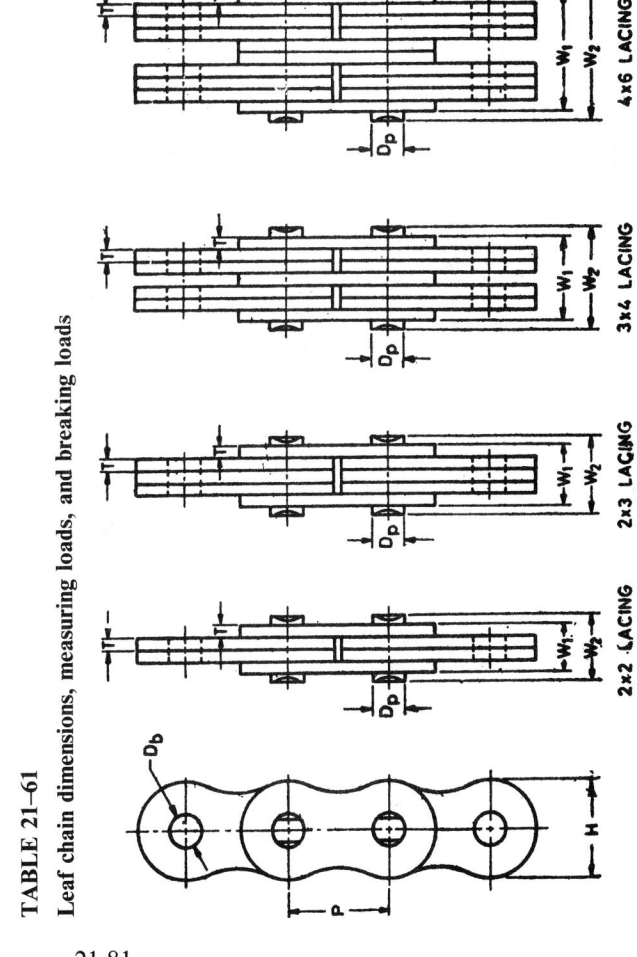

Chain number	Pitch mm	Lacing	Chain width, W_1 mm	Width over bearing pins, W_2 max mm	Pin body diameter, D_p max mm	Articulating plates bore, diameter, min, D_b mm	Plate depth, H min, mm	Plate thickness, T max, mm	Measuring load N	Measuring load kgf	Breaking load, min kN	Breaking load, min kgf
0822	12.70	2 × 2	6.45	8.69	4.45	4.48	11.81	1.57	190.0	19.10	18.7	1910
0823	12.70	2 × 3	8.08	10.31	4.45	4.48	11.81	1.57	190.0	19.10	18.7	1910
0834	12.70	3 × 4	11.30	13.54	4.45	4.48	11.81	1.57	280.0	28.60	26.3	2860
0846	12.70	4 × 6	16.13	18.36	4.45	4.48	11.81	1.57	370.0	38.10	37.4	3810
1022	15.88	2 × 2	7.26	9.80	5.08	5.10	14.73	1.78	250.0	25.40	24.9	2540
1023	15.88	2 × 3	9.09	11.63	5.08	5.10	14.73	1.78	250.0	25.40	24.9	2540
1034	15.88	3 × 4	12.73	15.27	5.08	5.10	14.73	1.78	390.0	39.90	39.1	3990
1046	15.88	4 × 6	18.16	20.70	5.08	5.10	14.73	1.78	500.0	50.80	49.8	5080
1222	19.05	2 × 2	12.50	15.90	6.78	6.80	16.13	3.07	450.0	45.40	44.5	4540
1223	19.05	2 × 3	15.62	19.02	6.78	6.80	16.13	3.07	450.0	45.40	44.5	4540
1234	19.05	3 × 4	21.87	25.27	6.78	6.80	16.13	3.07	670.0	68.00	66.7	6800
1246	19.05	4 × 6	31.24	34.65	6.78	6.80	16.13	3.07	890.0	90.70	82.0	9070
1623	25.40	2 × 3	21.34	25.48	8.28	8.30	21.08	4.22	630.0	63.50	62.3	6350
1634	25.40	3 × 4	29.87	34.01	8.28	8.30	21.08	4.22	1020.0	104.30	102.3	10430
1646	25.40	4 × 6	42.67	46.81	8.28	8.30	21.08	4.22	1250.0	127.00	124.5	12700
2023	31.75	2 × 3	23.24	28.35	10.19	10.22	26.42	4.60	980.0	99.80	97.9	9980
2034	31.76	3 × 4	32.54	37.64	10.19	10.22	26.42	4.60	1510.0	154.20	151.2	15420
2046	31.75	4 × 6	46.68	51.59	10.19	10.22	26.42	4.60	1960.0	199.60	195.7	19960
2423	38.10	2 × 3	30.73	38.05	14.63	14.66	33.40	6.10	1600.0	163.30	160.1	16330
2434	38.10	3 × 4	43.03	50.34	14.63	14.66	33.40	6.10	2400.0	244.90	240.2	24490
2446	38.10	4 × 6	61.47	68.78	14.63	14.66	33.40	6.10	3200.0	326.60	320.3	32660
2823	44.45	2 × 3	35.94	43.89	15.90	15.92	37.08	7.14	2400.0	217.70	213.5	21770
2834	44.45	3 × 4	50.32	58.27	15.90	15.92	37.08	7.14	3200.0	326.60	320.3	32660
2846	44.45	4 × 6	71.88	79.83	15.90	15.92	37.08	7.14	4300.0	435.50	427.1	43550
3223	50.80	2 × 3	40.51	49.43	17.81	17.84	42.29	8.05	2800.0	281.20	275.8	28120
3234	50.80	3 × 4	56.72	65.63	17.81	17.84	42.29	8.05	4100.0	421.80	413.6	42180
3246	50.80	4 × 6	81.03	89.94	17.81	17.84	42.29	8.05	5500.0	562.50	551.9	56280

Source: IS: 1072–1967.

TABLE 21-62
Dimensions of anchor clevises for leaf chains (all dimensions in mm)

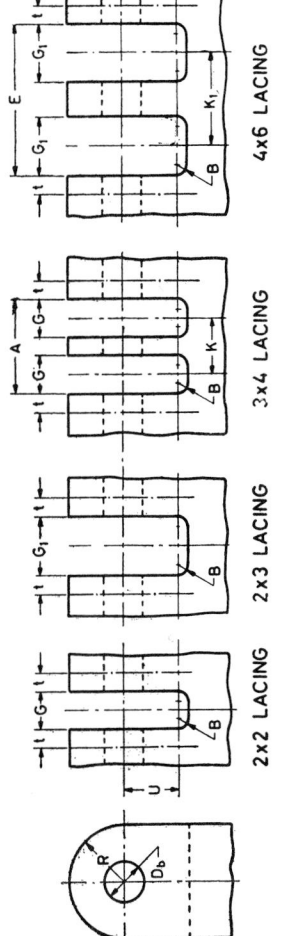

2x2 LACING   2x3 LACING   3x4 LACING   4x6 LACING

Chain number	Pitch P, mm	Lacing	Outside flange thickness, t, min	A, K+G	E, K₁+G₁	End radius, R, max	Slot depth, U min	Fillet radius, B min	Slot pitch K	Slot pitch K₁	Slot width G	Slot width G₁	Tolerance on A, E, G, and G₁
0822	12.70	2 × 2	1.57	...	...	6.35	6.35	0.79	...	...	3.33	...	+0.002p + 0.100 −0
0823	12.70	2 × 3	1.57	...	...	6.35	6.35	0.79	...	...	...	5.03	
0834	12.70	3 × 4	1.57	8.18	...	6.35	6.35	0.79	4.85	...	3.33	...	
0846	12.70	4 × 6	1.57	...	13.11	6.35	6.35	0.79	...	8.08	...	5.03	
1022	15.88	2 × 2	1.78	...	...	7.95	7.95	0.79	...	...	3.73	...	
1023	15.88	2 × 3	1.78	...	...	7.95	7.95	0.79	...	...	...	5.66	
1034	15.88	3 × 4	1.78	9.16	...	7.95	7.95	0.79	5.46	...	3.73	...	
1046	15.88	4 × 6	1.78	...	14.76	7.95	7.95	0.79	...	9.09	...	5.66	
1222	19.05	2 × 2	3.07	...	...	9.53	9.53	1.59	...	...	6.38	...	
1223	19.05	2 × 3	3.07	...	...	9.53	9.53	1.59	...	...	...	9.63	
1234	19.05	3 × 4	3.07	15.75	...	9.53	9.53	1.59	9.37	...	6.38	...	
1246	19.05	4 × 6	3.07	...	25.25	9.53	9.53	1.59	...	15.62	...	9.63	
1623	25.40	2 × 3	4.22	...	...	12.70	12.70	1.59	...	...	...	13.11	+0.002p + 0.100 −0
1634	25.40	3 × 4	4.22	21.49	...	12.70	12.70	1.59	12.80	...	8.69	...	
1646	25.40	4 × 6	4.22	...	34.44	12.70	12.70	1.59	...	21.34	...	13.11	
2023	31.75	2 × 3	4.60	...	...	15.88	15.88	1.59	...	...	...	14.30	
2034	31.75	3 × 4	4.60	23.42	...	15.88	15.88	1.59	13.94	...	9.47	...	
2046	31.75	4 × 6	4.60	...	37.54	15.88	15.88	1.59	...	23.24	...	14.30	
2423	38.10	2 × 3	6.10	...	...	19.05	19.05	2.38	...	...	...	18.85	
2434	38.10	3 × 4	6.10	30.94	...	19.05	19.05	2.38	18.44	...	12.50	...	
2446	38.10	4 × 6	6.10	...	49.58	19.05	19.05	2.38	...	30.73	...	18.85	
2823	44.45	2 × 3	7.14	...	...	22.23	22.23	2.38	...	...	...	22.02	
2834	44.45	3 × 4	7.14	36.17	...	22.23	22.23	2.38	21.56	...	14.61	...	
2846	44.45	4 × 6	7.14	...	57.96	22.23	22.23	2.38	...	35.94	...	22.02	
3223	50.80	2 × 3	8.05	...	...	25.40	25.40	3.18	...	...	...	24.82	
3234	50.80	3 × 4	8.05	40.77	...	25.40	25.40	3.18	24.31	...	16.31	...	
3246	50.80	4 × 6	8.05	...	65.33	25.40	25.40	3.18	...	40.51	...	24.82	

Source: IS 1072, 1967.

21.83

TABLE 21–63
Dimensions (in mm) for leaf chain sheaves

Chain number	Distance between flanges L Min	Sheave diameter, SD Min	Flange, diameter, FD Min	Chain number	Distance between flanges, L Min	Sheave diameter, SD Min	Flange diameter, FD Min
0822	9.12	63.50	88.90	1646	49.15	127.00	152.40
0823	10.80	63.50	88.90	2023	29.77	158.75	184.15
0834	14.20	63.50	88.90	2034	39.52	158.75	184.15
0846	19.28	63.50	88.90	2046	51.18	158.75	184.15
1022	10.29	79.38	104.78	2423	39.95	190.50	215.90
1028	12.22	79.38	104.78	2434	52.86	190.50	215.90
1034	16.03	79.38	104.78	2446	72.21	190.50	215.90
1046	21.74	79.38	104.78	2823	46.08	222.25	247.65
1222	16.69	95.25	120.65	2834	61.19	222.25	247.65
1223	19.96	95.25	120.65	2846	83.82	222.25	247.65
1234	26.54	95.25	120.65	3223	51.89	254.00	279.40
1246	36.37	95.25	120.65	3234	68.92	254.00	279.40
1623	26.75	127.00	152.40	3246	94.44	254.00	279.40
1634	35.71	127.00	152.40				

Source: IS 1072, 1967.

TABLE 21-64
Short pitch transmission precision bush chain dimensions, measuring loads, and breaking loads (all dimensions in mm)

Chain no	Pitch, p	Bush diameter, max D_b	Width between inner plates, min W	Bearing pin body diameter, max D_p	Bush bore, min d_b	Chain path depth, min H_d	Inner plate depth, max H_i	Outer or immediate plate depth, max H_o	Cranked link dimensions Min X	Min Y	Min Z	Transverse pitch, T_p	Width over inner link, max W_1	Width between outer plates, min W_2	Width over bearing pins Max A	Max A_2	Max A_3	Additional width for joint fasteners, max, B	Measuring load Simplex	Duplex	Triplex	Breaking load, min Simplex	Duplex	Triplex
04C	6.35	3.30	3.18	2.29	2.34	6.27	6.02	5.21	2.64	3.06	0.08	6.40	4.80	4.93	9.10	15.5	21.8	2.5	0.05 / 5	0.10 / 10	0.15 / 15	3.4 / 350	6.9 / 700	10.3 / 1050
06C	9.525	5.08	4.77	3.59	3.63	9.30	9.05	7.80	3.96	4.60	0.08	10.13	7.47	7.60	13.20	23.4	33.5	3.3	0.07 / 7	0.14 / 14	0.20 / 21	7.8 / 790	15.5 / 1580	23.2 / 2370

(Measuring load and Breaking load values given as kN / kgf.)

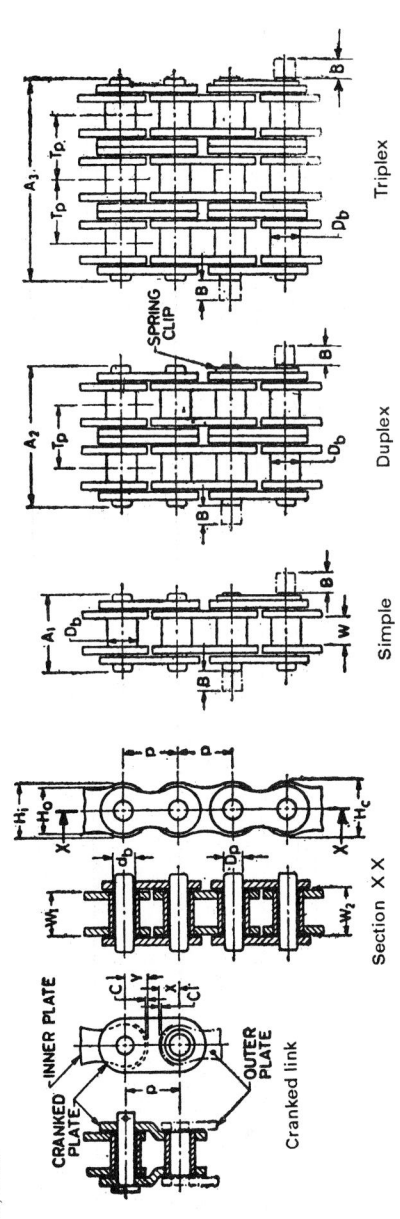

Simple · Duplex · Triplex · Section X X · Cranked link

Notes: (1) Dimension C represents clearance between the cranked link plates and the straight plates available during articulation; (2) the chain path depth H_c is the minimum depth of channel through which the assembled chain passes; (3) width over bearing pins for chains wider than triplex $= A_1 + T_p$ (no. of strands in chain-1); (4) cranked links are not recommended for use on chains which are intended for onerous applications.

Source: IS 3563, 1966.

TABLE 21–65
Pitch circle diameters[a] for short pitch transmission precision bush chain wheels

No. of teeth	Pitch circle diameter	No. of teeth	Pitch circle diameter	No. of teeth	Pitch circle diameter	No. of teeth	Pitch circle diameter	No. of teeth	Pitch circle diameter	No. of teeth	Pitch circle diameter
9	2.9238	33	10.5201	57	18.1529	81	25.7896	105	33.4275	129	41.0660
10	3.2361	34	10.8380	58	18.4710	82	26.1078	106	33.7458	130	41.3843
11	3.5494	35	11.1558	59	18.7892	83	26.4260	107	34.0648	131	41.7026
12	3.8637	36	11.4747	60	19.1073	84	26.7443	108	34.3823	132	42.0209
13	4.1786	37	11.7916	61	19.4255	85	27.0625	109	34.7006	133	42.3391
14	4.4940	38	12.1096	62	19.7437	86	27.3807	110	35.0188	134	42.6574
15	4.8097	39	12.4275	63	20.0619	87	27.6990	111	35.3371	135	42.9757
16	5.1258	40	12.7455	64	20.3800	88	28.0172	112	35.6554	136	43.2940
17	5.4422	41	13.0635	65	20.6982	89	28.3355	113	35.9737	137	43.6123
18	5.7588	42	13.3815	66	21.0164	90	28.6537	114	36.2919	138	43.9306
19	6.0755	43	13.6995	67	21.3346	91	28.9719	115	36.6102	139	44.2488
20	6.3925	44	14.0176	68	21.6528	92	29.2902	116	36.9285	140	44.5671
21	6.7095	45	14.3356	69	21.9710	93	29.6084	117	37.2467	141	44.8854
22	7.0266	46	14.6537	70	22.2892	94	29.9267	118	37.5650	142	45.2037
23	7.3439	47	14.9717	71	22.6074	95	30.2449	119	37.8833	143	45.5220
24	7.6613	48	15.2868	72	22.9256	96	30.5632	120	38.2016	144	45.8403
25	7.9787	49	15.6079	73	23.2438	97	30.8815	121	38.5198	145	46.1585
26	8.2962	50	15.9260	74	23.5620	98	31.1997	122	38.8381	146	46.4768
27	8.6138	51	16.2441	75	23.8802	99	31.5180	123	39.1564	147	46.7951
28	8.9314	52	16.5622	76	24.1985	100	31.8362	124	39.4746	148	47.1134
29	9.2491	53	16.8803	77	24.5167	101	32.1545	125	39.7929	149	47.4317
30	9.5668	54	17.1984	78	24.8349	102	32.4727	126	40.1112	150	47.7500
31	9.8845	55	17.5166	79	25.1531	103	32.7910	127	40.4295		
32	10.2023	56	17.8347	80	25.4713	104	33.1093	128	40.7478		

[a] The values given are for a unit pitch length (e.g., 1 mm).
Source: IS 3560, 1966.

21.86

TABLE 21–66
Recommended design data for silent chains

Chain pitch, mm	Speed of small sprocket	No. of teeth Driver	Driven	Min center distance, mm
9.3	2000–4000	17–25	21–120	152.4
12.7	1500–2000	17–25	21–130	228.6
15.8	1200–1500	19–25	21–150	304.8
19.0	1000–1200	19–25	23–150	381.0
22.2	900–1000	19–25	23–150	457.2
25.4	800–900	19–25	23–150	533.4
31.7	650–800	21–25	25–150	685.8
38.1	500–650	25–27	27–150	914.4
50.8	300–500	25–27	27–150	1219.2
76.2	≤ 300	25–27	27–150	1676.4

TABLE 21–67*A*
Maximum speed of small sprocket for inverted tooth chains

Pitch, mm	Max width, mm	Number of teeth	Speed, rpm							
9.50	101.6	17	4000	3500	2500	2000	1200			
12.70	177.8	19	5000	3500	2500	2500	1500	1200	1000	700
15.88	203.2	21	6000	3000	3000	2500	1800	1200	1000	700
19.05	254.0	23	6000	4000	3000	2500	1800	1800	1200	800
25.40	355.6	25	6000	4000	3500	2500	1800	1800	1200	900
31.75	508.0	27	6000	4000	3500	2500	2000	1800	1200	900
38.10	609.6	29	6000	4000	3500	2500	2000	1800	1200	900
50.80	762.0	31	6000	4000	3500	2500	2000	1800	1200	900
		33	6000	4000	3500	2500	2000	1800	1200	900
		35	6000	4000	3500	2500	2000	1800	1200	900
		37	5000	3500	3000	2500	1800	1200	1000	800
		45	4000	3000	2000	2000	1500	1000	900	700
		40	5000	3500	2500	2500	1500	1200	900	800
		50	3500	2500	2000	1800	1200	1000	800	600

TABLE 21–67*B*
Maximum velocity for various types of chains, rpm

Type of chain (Chain pitch, *p*, mm)	Number of sprocket teeth Bush roller chain 15	19	23	27	30		Silent chains 17.35
12	2300	2400	2530	2550	2600	12.7	3300
15	1900	2000	2100	2150	2200	15.87	2650
20	1350	1450	1500	1550	1550	19.05	2200
25	1150	1200	1250	1300	1300	25.40	1650
30	1000	1050	1100	1100	1100	31.75	1300

TABLE 21–68
Safety factor

Chains	Speed of smaller sprocket, rpm								
	50	260	400	600	800	1000	1200	1600	2000
Bush roller chains									
$p = 12$									
15 mm	7.0	7.8	8.55	9.35	10.2	11.0	11.7	13.2	1.48
$p = 20$									
25 mm	7.0	8.2	9.35	10.3	10.7	12.9	14.0	16.3	
$p = 30$									
35 mm	7.0	8.55	10.2	13.2	14.8	16.3	19.5		
Silent chains									
$p = 12.7$									
15.87 mm	20	22.2	24.4	28.7	29.0	31.0	33.4	37.8	42.0
$p = 19.05$									
25.4 mm	20	23.4	26.7	30.0	33.4	36.8	40.0	46.5	53.5

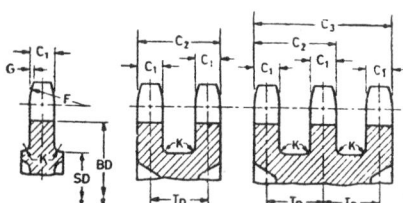

FIGURE 21–26 Notation for wheel rim profiles of bush chain.

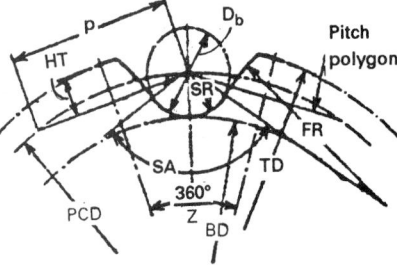

FIGURE 21–27 Notation for tooth gap form of bush chain.

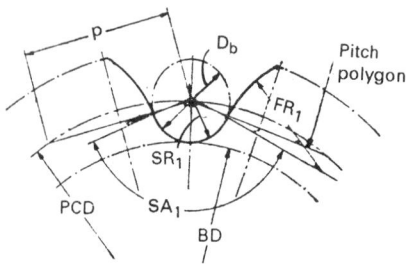

FIGURE 21–28 Notation for minimum tooth gap form for bush chain.

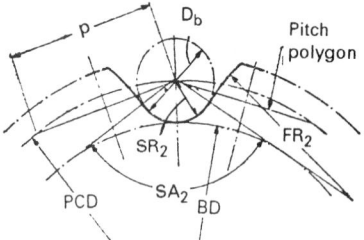

FIGURE 21–29 Notation for maximum tooth gap form for bush chain.

Particular	Formula
The minimum limit of the tooth height above the pitch polygon	$HT_{min} = 0.5\,(p - D_b)$ (21–162)
WHEEL RIM PROFILE (FIG. 21–26) The value of tooth width for simple chain wheels (Fig. 21–26)	$C_1 = 0.93w$ (21–163)
The value of tooth width for duplex and triplex chain wheels	$C_1 = 0.91w$ (21–164)
The value of tooth width for quadruplex chain wheels and above	$C_1 = 0.88w$ (21–165)
	The value of tolerance shall be $h/4$
The value of width over tooth	C_2 (or C_3) = (number of strands $-1T_p + C_1$ with a tolerance value of $h/4$ where T_d = transmission pitch of strands (21–166)
The minimum tooth side radius	$F = p$ (21–167)
The tooth side relief	$G = 0.1p$ to $0.15p$ (21–168)
Absolute maximum shroud diameter	$SD = p \cot \dfrac{180°}{z} - 1.05\,H_i - 1.00 - 2\,K_{ort}$, mm (21–169)
For bush chains dimensions, breaking load, pitch circle diameters, etc.	Refer to Tables 21–64 to 21–68.

REFERENCES

1. Maleev, V. L., and J. B. Hartman, *Machine Design*, International Textbook Company, Scranton, Pennsylvania, 1954.
2. Black, P. H., and O. E. Adams, Jr., *Machine Design*, McGraw-Hill Book Company, New York, 1968.
3. Norman, C. A., E. S. Ault, and I. F. Zarobsky, *Fundamentals of Machine Design*, The Macmillan Company, New York, 1951.
4. Shigley, J. E., *Machine Design*, McGraw-Hill Book Company, New York, 1962.
5. Shigley, J. E., and C. R. Mischke, *Mechanical Engineering Design*, McGraw-Hill Book Company, New York, 1989.
6. Shigley, J. E., and C. R. Mischke, *Standard Handbook of Machine Design*, McGraw-Hill Book Company, New York, 1986.
7. Baumeister, T., ed., *Marks' Standard Handbook for Mechanical Engineers*, McGraw-Hill Book Company, New York, 1978.
8. Niemann, G., *Maschinenelemente*, Springer-Verlag, Berlin; Zweiter Band, Munich, 1965.
9. Nimann, G., *Machine Elements—Design and Calculations in Mechanical Engineering*, Vol. II, Allied Publishers Private Ltd., New Delhi, 1978.
10. Decker, K. H., *Maschinenelemente, Gestaltung and Berechnung*, Caril Hanser Verlag, Munich, 1971.
11. Lingaiah, K. and B. R. Naryana Iyengar, *Machine Design Data Handbook*, Engineering College Co-operative Society, Bangalore, India, 1962.
12. Lingaiah, K. and B. R. Narayana Iyengar, *Machine Design Data Handbook*, Vol. I (*SI and Customary Metric Units*), Suma Publishers, Bangalore, India, 1973.

13. Lingaiah, K. *Machine Design Data Handbook*, Vol. II (*SI and Customary Metric Units*), Suma Publishers, Bangalore, India, 1986.
14. Bureau of Indian Standards.
15. Albert, C. D., *Machine Design Drawing Room Problems*, John Wiley and Sons, New York, 1949.

MECHANICAL VIBRATIONS

SYMBOLS

a	coefficients with subscripts
	flexibility
	acceleration, m/s^2 (ft/s^2)
A	area of cross section, m^2 (in^2)
	constant
B	constant
C	coefficient of viscous damping, N s/m or N/ν (lbf s/in or lbf/ν)
	constant
C_c	critical viscous damping, N s/m (lbf s/in)
C_t	coefficient of torsional viscous damping, N m s/rad (lbf in s/rad)
C_1, C_2	coefficients
	constants
d	diameter of shaft, m (in)
D	flexural rigidity $[= Eh^3/12(1 - \nu^2)]$
e	displacement of the center of mass of the disk from the shaft axis, m (in)
E	modulus of elasticity, GPa (Mpsi)
f	frequency, Hz
F	exciting force, kN (lbf)
F_o	maximum exciting force, kN (lbf)
F_T	transmitted force, kN (lbf)
g	acceleration due to gravity, 9.8066 m/s^2 (32.2 ft/s^2 or 386.6 in/s^2)
G	modulus of rigidity, GPa (Mpsi)
h	thickness of plate, m (in)
i	integer $(= 0, 1, 2, 3, ...)$
I	mass moment of inertia of rotating disk or rotor, N s^2 m (lbf s^2 in)
J	polar second moment of inertia, m^4 or cm^4 (in^4)
k	spring stiffness or constant, kN/m (lbf/in)
k_e	equivalent spring constant, kN/m (lbf/in)
k_t	torsional or spring stiffness of shaft, J/rad or N m/rad (lbf in/rad)
K	kinetic energy, J (lbf/in)
l	length of shaft, m (in)

m	mass, kg (lb)
m_e	equivalent mass, kg (lb)
M	total mass, kg (lb)
M_t	torque, N m (lbf ft)
p	circular frequency, rad/s
q	damped circular frequency $(= \sqrt{1 - \zeta^2})$
r	radius, m (in)
$R = 1 - T_R$	percent reduction in transmissibility
$R_2 = D_2/2$	radius of the coil, m (in)
t	time (period), s
T	temperature, K or °C (°F)
T_R	transmissibility
U	vibrational energy, J or N m (lbf in)
	potential energy, J (lbf in)
v	velocity, m/s (ft/min)
w	weight per unit volume, kN/m³ (lbf/in³)
W	total weight, kN (lbf)
x	displacement or amplitude from equilibrium position at any instant t, m (in)
x_1, x_2	successive amplitudes, m (in)
x_o	maximum displacement, m (in)
$\dot{x}$	linear velocity, m/s (ft/min)
$\ddot{x}$	linear acceleration, m/s² (ft/s²)
X_{st}	static deflection of the system, m (in)
y	deflection of the disk center from its rotational axis, m or mm (in)
γ	weight density, kN/m³ (lbf/in³)
$\zeta = \dfrac{C}{C_c}$	damping factor
δ	logarithmic decrement,
	deflection, m (in)
δ_{st}	static deflection, m (in)
θ	phase angle, deg
λ	wavelength, m (in)
ν	Poisson's ratio
ρ	mass density, kg/m³ (lb/in³)
σ	normal stress, MPa (psi)
τ	shear stress, MPa (psi)
	period, s
φ	angular deflections, rad (deg)
$\dot{\varphi}$	angular velocity, rad/s
$\ddot{\varphi}$	angular acceleration, rad/s²
ω	forced circular frequency, rad/s

Particular	Formula

SIMPLE HARMONIC MOTION (FIG. 22–1)

The displacement of point P on diameter RS (Fig. 22–1)	$x = x_o \sin pt$	(22–1)
The wavelength	$\lambda = 2\pi$	(22–2)

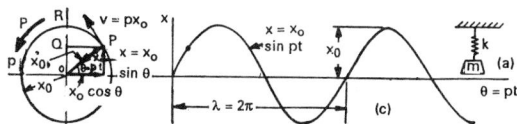

FIGURE 22–1   Simple harmonic motion.

Particular	Formula	
The periodic time	$$\tau = \frac{2\pi}{p}$$	(22–3)
The frequency	$$f = \frac{1}{\tau} = \frac{p}{2\pi}$$	(22–4)
The maximum velocity of point Q	$v_{max} = px_o$	(22–5)
The maximum acceleration of point Q	$a_{max} = \dot{v}_{max} = p^2 x_o$	(22–6)

Single-degree-freedom system without damping and without external force (Fig. 22–2)

Linear system

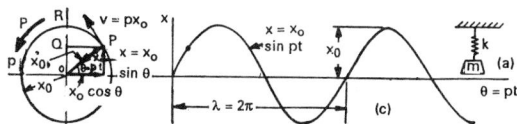

FIGURE 22–2   Spring-mass system.

The equation of motion	$m\ddot{x} + kx = 0$	(22–7)
The general solution for displacement	$x = A \sin pt + B \cos pt$	(22–8)
	$\quad = C \sin(pt - \varphi)$	
	where φ = phase angle of displacement	(22–9)
The equation for displacement of mass for the initial condition $x = x_o$ and $\dot{x} = 0$ at $t = 0$	$x = x_o \cos pt$	(22–10)
The natural circular frequency	$p_n = \sqrt{k/m} = \sqrt{g/\delta_{st}}$	(22–11)
The natural frequency of the vibration	$$f_n = \frac{p_n}{2\pi} = \frac{1}{2\pi} \sqrt{k/m}$$	(22–12)
The natural frequency in terms of static deflection δ_{st}	$$f_n = \frac{1}{2\pi} \sqrt{g/\delta_{st}}$$	(22–13)
	$$= \frac{3.132}{2\pi} \left(\frac{1}{\delta_{st}}\right)^{1/2} \approx 0.5 \left(\frac{1}{\delta_{st}}\right)^{1/2} \quad \textbf{SI}$$	(22–13a)

where δ_{st} in m and f_n in Hz

Particular	Formula

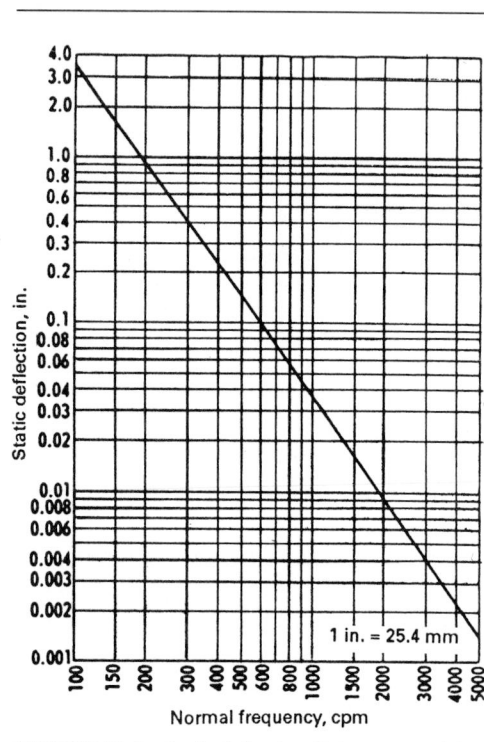

FIGURE 22–3 Static deflection (δ_{st}) vs. natural frequency f_n. (*Courtesy of P. H. Black and O. E. Adams, Jr.,* Machine Design, *McGraw-Hill, New York, 1955.*)

$$= \frac{99}{2\pi}\left(\frac{1}{\delta_{st}}\right)^{1/2} \approx 15.76\left(\frac{1}{\delta_{st}}\right)^{1/2}$$

SI   (22–13b)

where δ_{st} in mm and f_n in Hz

$$= \frac{5.67}{2\pi}\left(\frac{1}{\delta_{st}}\right)^{1/2} \approx 0.9\left(\frac{1}{\delta_{st}}\right)^{1/2}$$ (22–13c)

US Customary System units

where δ_{st} in ft and f_n in Hz

$$= \frac{19.67}{2\pi}\left(\frac{1}{\delta_{st}}\right)^{1/2} = \frac{3.127}{\sqrt{\delta_{st}}}$$ (22–13d)

US Customary System units

where δ_{st} in in and f_n in Hz

$$= \frac{187.6}{\sqrt{\delta_{st}}} \quad \textbf{US Customary System units} \quad (22\text{–}13e)$$

where δ_{st} in in and f_n in cpm (cycles per minute)

The plot of natural frequency vs. static deflection	Refer to Fig. 22–3.

Simple pendulum

The equation of motion for simple pendulum (Fig. 22–4)	$\ddot{\theta} + \dfrac{g}{l}\sin\theta = \ddot{\theta} + \dfrac{g}{l}\theta = 0$	(22–14)
The angular displacement for $\theta = \theta_o$ and $\dot{\theta} = 0$ at $t = 0$	$\theta = \theta_o \sin\sqrt{(g/l)}t$	(22–15)
The circular frequency for simple pendulum for small oscillation	$p = \sqrt{(g/l)}$	(22–15a)

ENERGY

The total energy in the universe is constant according to conservation of energy	$K + U = constant$	(22–16)
Kinetic energy	$K = \frac{1}{2}mv^2 = \frac{1}{2}m\dot{x}^2$	(22–17)

Particular	Formula
Potential energy	$U = \frac{1}{2} kx^2$ (22–18)
Maximum kinetic energy is equal to maximum potential energy according to conservation of energy	$K_{max} = U_{max}$ (22–19)

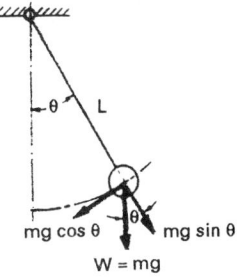

FIGURE 22–4 Simple pendulum.

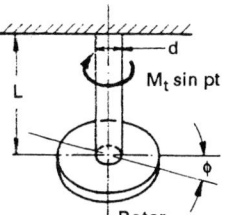

FIGURE 22–5 Single rotor system subject to torque.

Torsional system (Fig. 22–5)

The equation of motion of torsional system (Fig. 22–5) with torsional damping under external torque $M_t \sin pt$	$I\ddot{\varphi} + C_t\dot{\varphi} + k_t x = M_t \sin pt$ (22–20) where C_t = coefficient of torsional viscous damping, N m s/rad (lbf in s/rad)
The equation of motion of torsional system without considering the damping and external force on the rotor	$I\ddot{\varphi} + k_t\varphi = 0$ (22–21)
The equation for angular displacement	$\varphi = A \sin pt + B \cos pt$ (22–22a) $= C \sin(pt - \theta)$ (22–22b) where θ = phase of displacement
The angular displacement for $\varphi = \varphi_o$ and $\dot{\varphi} = 0$ at $t = 0$	$\varphi = \varphi_o \cos(\sqrt{k_t/I})t$ (22–23)
The natural circular frequency	$p_n = \sqrt{k_t/I}$ (22–24)
The natural circular frequency taking into account the shaft mass	$p_n = \left(k_t \Big/ \left(I + \dfrac{I_s}{3} \right) \right)^{1/2}$ (22–25)
The natural frequency	$f_n = \dfrac{p_n}{2\pi} = \dfrac{1}{2\pi}\sqrt{k_t/I}$ (22–26)
The expression for torsional stiffness	$k_t = \dfrac{JG}{l} = \left(\dfrac{\pi d^4}{32} \right)\dfrac{G}{l}$ (22–27) where $J = \dfrac{\pi d^4}{32}$ = moment of inertia, polar, m⁴ or cm⁴ (in⁴)

Particular	Formula

Single-degree-freedom system with damping and without external force (Fig. 22–6)

The equation of motion

$$m\ddot{x} + c\dot{x} = kx = 0 \tag{22–28}$$

The general solution for displacement

$$x = C_1 e^{s_1 t} + C_2 e^{s_2 t} \tag{22–29}$$

$$= C_1 e^{(-\zeta - \sqrt{\zeta^2-1})p_n t} + C_2 e^{(-\zeta + \sqrt{\zeta^2-1})p_n t} \tag{22–30}$$

$$= A e^{-\zeta p_n t} \sin(qt + \varphi) \tag{22–31}$$

where C_1, C_2, and A are arbitrary constants of integration

(They can be found from initial conditions.)

$$s_{1,2} = -\frac{C}{2m} \pm \left[\left(\frac{C}{2m} \right)^2 - \frac{k}{m} \right]^{1/2} \tag{22–32}$$

$$= (-\zeta \pm \sqrt{\zeta^2 - 1})p_n \tag{22–33}$$

where $\zeta = \dfrac{C}{C_c}$ = damping ratio, $C_c = 2mp_n = 2\sqrt{km}$

q = frequency of damped oscillation

$$= \left(\frac{2\pi}{\tau_d} \right) = (\sqrt{1 - \zeta^2})p_n = \left[\frac{k}{m} - \frac{c^2}{4m^2} \right]^{1/2} \tag{22–33a}$$

φ = phase angle or phase displacement with respect to the exciting force

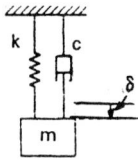

FIGURE 22–6 Single-degree-of-freedom spring-mass-dashpot system.

For the damped oscillation of the single-degree-freedom system with time for damping factor $\zeta < 1$

Refer to Figs. 22–7 and 22–8.

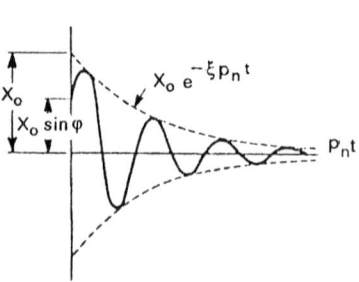

FIGURE 22–7 Damped motion $\zeta < 1.0$.

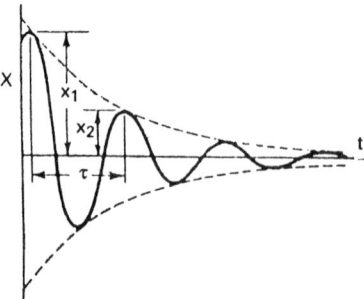

FIGURE 22–8 Logarithmic decrement. (*Reproduced from Marks' Standard Handbook for Mechanical Engineers, 8th ed., McGraw-Hill, New York, 1978.*)

Particular	Formula

LOGARITHMIC DECREMENT
(FIG. 22–8)

The equation for logarithmic decrement

$$\delta = \ln \frac{x_o}{x_1} = \ln \frac{x_1}{x_2} = \frac{\Delta U}{U} = \frac{2\pi\zeta}{\sqrt{1-\zeta^2}} \approx 2\pi\zeta \quad (22\text{–}34)$$

EQUIVALENT SPRING CONSTANTS
(FIG. 22–9)

The spring constant or stiffness

$$k = \frac{F}{x} \qquad (22\text{–}35)$$

The flexibility

$$a = \frac{x}{F} \qquad (22\text{–}36)$$

The equivalent spring constant for springs in series, (Fig. 22–9a)

$$k_e = \frac{1}{\dfrac{1}{k_1} + \dfrac{1}{k_2}} \qquad (22\text{–}37)$$

The equivalent spring constant for springs in parallel, (Fig. 22–9b)

$$k_e = k_1 + k_2 \qquad (22\text{–}38)$$

For spring constants of different types of springs, beams, and plates

Refer to Table 22–1.

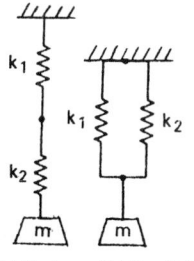

(a) Series   (b) Parallel

FIGURE 22–9 Springs in series and parallel.

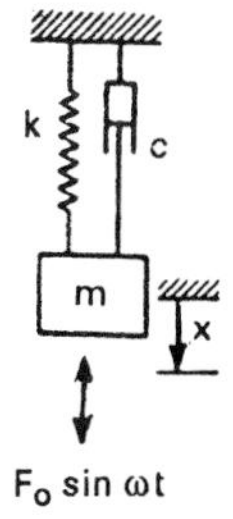

FIGURE 22–10 Spring-mass-dashpot system subjected to external force.

Single-degree-freedom system with damping and external force (Fig. 22–10)

The equation of motion

$$m\ddot{x} + c\dot{x} + kx = F_o \sin \omega t \qquad (22\text{–}39)$$
$$\ddot{x} + 2\zeta p_n \dot{x} + p_n^2 x = (F_o/m) \sin \omega t \qquad (22\text{–}40)$$

TABLE 22–1
Spring constants or spring stiffness of various springs, beams, and plates

Particular	Formula for spring constant, k	Figure	Equation
Linear Spring Stiffness or Constants [Load per mm (in) Deflection]			
Helical spring subjected to tension with i number of turns	$k = \dfrac{Gd^4}{64iR^3}$		(22–91)
Bar under tension	$k = \dfrac{EA}{l}$		(22–92)
Cantilever beam subjected to transverse load at the free end	$k = \dfrac{3EI}{l^3}$		(22–93)
Cantilever beam subjected to bending at the free end	$k = \dfrac{2EI}{l^2}$		(22–94)
Simply supported beam with concentrated load at the center	$k = \dfrac{48EI}{l^3}$		(22–95)
Simply supported beam subjected to a concentrated load not at the center	$k = \dfrac{3EIl}{a^2b^2}$		(22–96)
Beam fixed at both ends subjected to a concentrated load at the center	$k = \dfrac{192EI}{l^3}$		(22–97)
Beam fixed at one end and simply supported at another end subjected to concentrated load at the center	$k = \dfrac{768EI}{7l^3}$		(22–98)
Circular plate clamped along the circumferential edge subjected to concentrated load at the center whose flexural rigidity is $D = Eh^3/12(1 - \nu^2)$, thickness h and Poisson ratio ν	$k = \dfrac{16\pi D}{R^2}$		(22–99)
Circular plate simply supported along the circumferential edge with concentrated load at the center	$k = \dfrac{16\pi D}{R^2}\left(\dfrac{1+\nu}{3+\nu}\right)$ where ν = Poisson ratio		(22–100)
String fixed at both ends subjected to tension T	$k = \dfrac{4T}{l}$ String tension T		(22–101)
Torsional or Rotational Spring Stiffness or Constants (Load per Radian Rotation)			
Spiral spring whose total length is l and moment of inertia of cross section I	$k_t = \dfrac{EI}{l}$		(22–102)
Helical spring with i turns subjected to twist whose wire diameter is d, the coil diameter is D	$k_t = \dfrac{Ed^4}{64iD}$		(22–103)

TABLE 22–1
Spring constants or spring stiffness of various springs, beams, and plates (*Cont.*)

Particular	Formula for spring constant, k	Figure	Equation
Bending of helical spring of i number of turns	$k_t = \dfrac{Ed^4}{32iD}\,[1/\{1 + (E/2G)\}]$		(22–104)
Twisting of bar of length l	$k_t = \dfrac{JG}{l}$		(22–105)
Twisting of a hollow circular shaft with length l, whose outside diameter is D_o and inside diameter is D_i	$k_t = \dfrac{GI_p}{l} = \dfrac{\pi G}{32}\dfrac{D_o^4 - D_i^4}{l}$		(22–106)
Twisting of cantilever beam	$k_t = \dfrac{GJ}{l}$		(22–107)
Simply supported beam subjected to couple at the center	$k_t = \dfrac{12EI}{l}$		(22–108)
Beam fixed at both ends subjected to couple at the center	$k_t = \dfrac{16EI}{l}$		(22–109)

Particular	Formula
The complete solution for the displacement	$x = Ae^{-\zeta p_n t}\sin(qt + \varphi_1) + X_o\sin(\omega t - \varphi)$ (22–41a)
	$= Ae^{-\zeta p_n t}\sin(qt + \varphi_1)$
	$+ \dfrac{(F_o/k)\sin(\omega t - \varphi)}{\left[\left\{1 - \left(\dfrac{\omega}{p_n}\right)^2\right\}^2 + \left(2\zeta\dfrac{\omega}{p_n}\right)^2\right]^{1/2}}$ (22–41b)
The steady-state solution for amplitude of vibration	$X = \dfrac{F_o}{\sqrt{(k - m\omega^2)^2 + (c\omega)^2}}$
	$= \dfrac{F_o/k}{\left[\left\{1 - \left(\dfrac{\omega}{p_n}\right)^2\right\}^2 + \left(2\zeta\dfrac{\omega}{p_n}\right)^2\right]^{1/2}}$ (22–41c)
The phase angle	$\varphi = \tan^{-1}\left[\dfrac{2\zeta\omega/p_n}{1 - (\omega/p_n)^2}\right]$ (22–42)

Particular	Formula
The magnification factor	$$\left(\frac{X_o}{X_{st}}\right) = \frac{1}{[\{1 - (\omega/p_n)^2\}^2 + (2\zeta\omega/p_n)^2]^{1/2}} \quad (22\text{--}43)$$
The plot of magnification factor (X_o/X_{st}) vs. frequency ratio (ω/p_n) and phase angle φ vs (ω/p_n)	Refer to Figs. 22–11 and 22–12.
The amplitude at resonance (i.e. for $\omega/p_n = 1$)	$$X_{res} = \frac{F_o}{cp_n} = \frac{F_o}{2\zeta k} = \frac{X_{st}}{2\zeta} \quad (22\text{--}44)$$

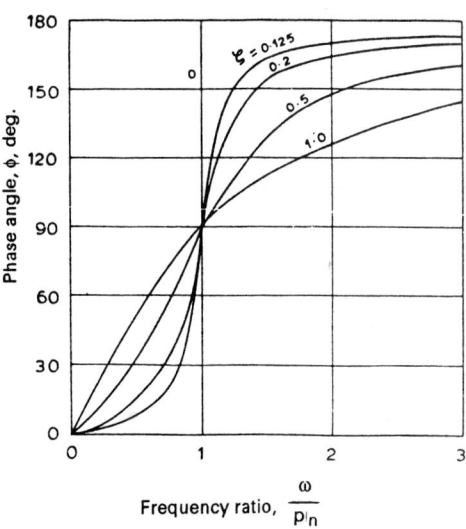

FIGURE 22–11 Phase angle (θ) vs. frequency ratio (ω/p_n).

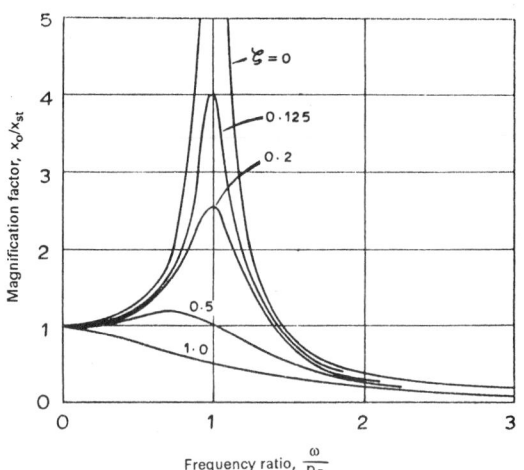

FIGURE 22–12 Magnification factor (X_o/X_{st}) vs. frequency ratio (ω/p_n).

UNBALANCE DUE TO ROTATING MASS (FIG. 22–13)

The equation of motion	$M\ddot{x} + c\dot{x} + kx = (me\omega^2)\sin\omega t$	(22–45)
The steady-state solution for displacement	$$x = \frac{me\omega^2}{\sqrt{(k - M\omega^2)^2 + (c\omega)^2}}$$	(22–46a)
	$$= \frac{(m/M)e(\omega/p_n)^2}{[\{1 - (\omega/p_n)^2\}^2 + (2\zeta\omega/p_n)^2]^{1/2}}$$	(22–46b)
The complete solution for the displacement	$$x = Ae^{-\zeta p_n t}\sin(qt + \varphi_1)$$ $$+ \frac{e(m/M)(\omega/p_n)^2}{[\{1 - (\omega/p_n)^2\}^2 + (2\zeta\omega/p_n)^2]^{1/2}}\sin(\omega t - \varphi)$$	
		(22–47)

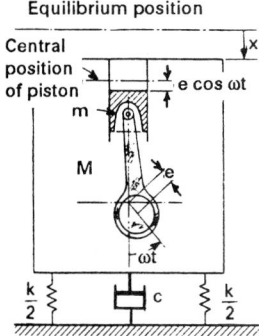

FIGURE 22–13 External force due to rotating unbalanced mass. (*Produced with some modification from N. O. Myklestad, Fundamentals of Vibration Analysis, McGraw-Hill, New York, 1956.*)

Particular	Formula
Nondimensional form of expression for Eq. (22–46b)	$\left(\dfrac{M}{m}\dfrac{x}{e}\right) = \dfrac{(\omega/p_n)^2}{[\{1 - (\omega/p_n)^2\}^2 + (2\zeta\omega/p_n)^2]^{1/2}}$ (22–48)
The phase angle	$\varphi = \tan^{-1}\left[\dfrac{2\zeta(\omega/p_n)^2}{1 - (\omega/p_n)^2}\right]$ (22–49)
For a schematic representation of Eqs. (22–48) and (22–49) for harmonically disturbing force due to rotating unbalance	Refer to Figs. 22–14 and 22–15.

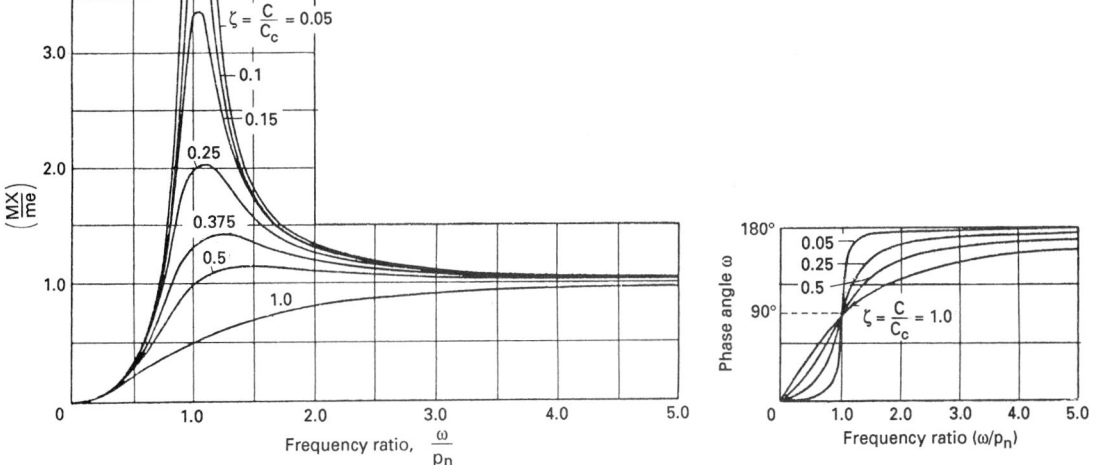

FIGURE 22–14 $\frac{MX}{me}$ vs. frequency ratio $\frac{\omega}{p_n}$.

FIGURE 22–15 Phase angle φ vs. frequency ratio $\frac{\omega}{p_n}$.

Particular	Formula

WHIPPING OF ROTATING SHAFT (FIG. 22–16)

The equation of motion of shaft due to unbalanced mass

$$m\ddot{x}_c + c\dot{x}_c + kx_c = me\omega^2 \cos\omega t \qquad (22\text{–}50a)$$
$$m\ddot{y}_c + c\dot{y}_c + ky_c = me\omega^2 \sin\omega t \qquad (22\text{–}50b)$$

where x_c and y_c are coordinates of position of center of shaft with respect to x and y coordinates

The solution

$$x_c = \frac{me\omega^2 \cos(\omega t - \varphi)}{\sqrt{(k - m\omega^2)^2 + (c\omega)^2}} \qquad (22\text{–}51a)$$

$$y_c = \frac{me\omega^2 \sin(\omega t - \varphi)}{\sqrt{(k - m\omega^2)^2 + (c\omega)^2}} \qquad (22\text{–}51b)$$

The displacement of the center of the disk from the line joining the centers of bearings

$$r = \sqrt{x_c^2 + y_c^2} = \frac{me\omega^2}{\sqrt{(k - m\omega^2)^2 + (c\omega)^2}} \qquad (22\text{–}52a)$$

$$= \frac{e(\omega/p_n)^2}{\{[1 - (\omega/p_n)^2]^2 + (2\zeta\omega/p_n)^2\}^{1/2}} \qquad (22\text{–}52b)$$

The phase angle

$$\varphi = \tan^{-1}\left(\frac{c\omega}{k - m\omega^2}\right) = \tan^{-1}\left[\frac{2\zeta(\omega/p_n)}{1 - (\omega/p_n)^2}\right]$$
$$(22\text{–}53)$$

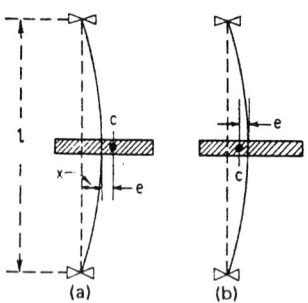

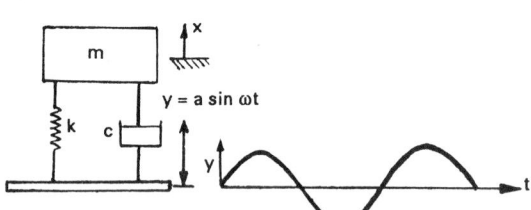

FIGURE 22–16 Whipping of shaft. (*Reproduced from Marks' Standard Handbook for Mechanical Engineers, 8th ed., McGraw-Hill, New York, 1978.*)

FIGURE 22–17 Excitation of a system by motion of support.

EXCITATION OF A SYSTEM BY MOTION OF SUPPORT (FIG. 22–17)

The equation of motion

$$m\ddot{x} + c\dot{x} + kx = ky + c\dot{y} \qquad (22\text{–}54)$$

The absolute value of the amplitude ratio of x and y

$$\left|\frac{X}{Y}\right| = \left[\frac{1 + (2\zeta\omega/p_n)^2}{[1 - (\omega/p_n)^2]^2 + (2\zeta\omega/p_n)^2}\right]^{1/2} \qquad (22\text{–}55)$$

Particular	Formula
The phase angle	$\varphi = \tan^{-1}\left[\dfrac{2\zeta(\omega/p_n)^3}{\{1 - (\omega/p_n)^2\} + (2\zeta\omega/p_n)^2}\right]$ (22–56)
The plot of Eq. (22–55) for motion due to support	Refer to Fig. 22–20 for $|X/Y|$ vs. ω/p_n. The curves are similar.

INSTRUMENT FOR VIBRATION MEASURING (FIG. 22–18)

The equation of motion	$m\ddot{z} + c\dot{z} + kz = -m\ddot{y} = mY\omega^2\sin\omega t$ (22–57)
The steady-state solution for relative displacement Z	$Z = \dfrac{Y(\omega/p_n)^2}{[\{1 - (\omega/p_n)^2\}^2 + (2\zeta\omega/p_n)^2]^{1/2}}$ (22–58)
The phase angle	$\varphi = \tan^{-1}\left\{\dfrac{2\zeta(\omega/p_n)}{1 - (\omega/p_n)^2}\right\}$ (22–59)
The plot of absolute value of $\left|\dfrac{Z}{Y}\right|$ vs. frequency ratio (ω/p_n) and the phase angle φ vs. frequency ratio (ω/p_n)	Refer to Figs. 22–14 and 22–15. The curves for $|Z/Y|$ vs. (ω/p_n) and ϕ vs. (ω/p_n) are identical.

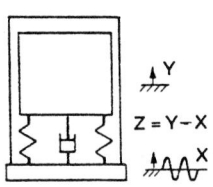

FIGURE 22–18 Instrument for vibration measuring. (*Reproduced from* Marks' Standard Handbook for Mechanical Engineers, *8th ed., McGraw-Hill, New York, 1978.*)

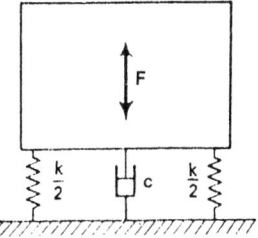

FIGURE 22–19 External force transmitted to foundation through damper and springs. (*Reproduced from* Marks' Standard Handbook for Mechanical Engineers, *8th ed., McGraw-Hill, New York, 1978.*)

ISOLATION OF VIBRATION (FIG. 22–19)

The force transmitted through the springs and damper	$F_T = \sqrt{(kX)^2 + (c\omega X)^2}$ (22–60)
	$= \dfrac{F_o[1 + (2\zeta\omega/p_n)^2]^{1/2}}{[\{1 - (\omega/p_n)^2 + (2\zeta\omega/p_n)^2\}]^{1/2}}$ (22–61)

Particular	Formula
Transmissibility	$$TR = F_T/F_o = \frac{\sqrt{1 + (2\zeta\omega/p_n)^2}}{[\{1 - (\omega/p_n)^2\}^2 + (2\zeta\omega/p_n)^2]^{1/2}}$$ (22–62)
Comparison of Eqs. (22–62) and (22–55) indicates that the plot of F_T/F_o is identical to $\left|\frac{X}{Y}\right|$.	Refer to Fig. 22–20 for T_R and $\|X/Y\|$.
Transmissibility when damping is negligible	$$T_R = \frac{1}{(\omega/p_n)^2 - 1}$$ (22–63)
The transmissibility in terms of static deflection δ_{st}	$$T_R = \frac{1}{\dfrac{(2\pi f_n)^2 \delta_{st}}{g} - 1}$$ (22–64)
The frequency from Eq. (22–64)	$$f_n = \frac{3.132}{2\pi}\left[\frac{1}{\delta_{st}}\left(\frac{1}{T_R}+1\right)\right]^{1/2} = 0.5\left[\frac{1}{\delta_{st}}\left(\frac{2-R}{1-R}\right)\right]^{1/2}$$ **SI** (22–65a)

where f_n in Hz and δ_{st} in m

The percent reduction in the transmissibility is defined as $R = 1 - T_R$

$$= \left(\frac{99}{2\pi}\right)\left[\frac{1}{\delta_{st}}\left(\frac{2-R}{1-R}\right)\right]^{1/2} = 15.76\left[\frac{1}{\delta_{st}}\left(\frac{2-R}{1-R}\right)\right]^{1/2}$$

SI   (22–65b)

where f_n in Hz and δ_{st} in mm

$$= \frac{19.67}{2\pi}\left[\frac{1}{\delta_{st}}\left(\frac{2-R}{1-R}\right)\right]^{1/2}$$

US Customary System units   (22–65c)

where δ_{st} in in and f_n in Hz

$$= 187.6\left[\frac{1}{\delta_{st}}\left(\frac{2-R}{1-R}\right)\right]^{1/2}$$

US Customary System units   (22–65d)

where f_n in rpm and δ_{st} in in

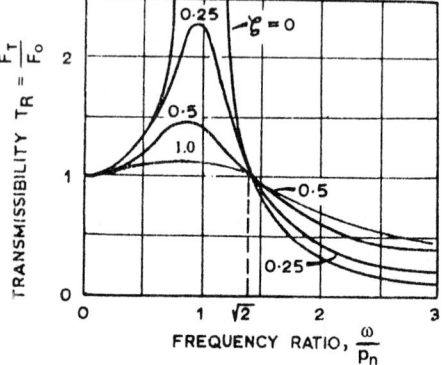

FIGURE 22–20   Transmissibility (T_R) vs. frequency ratio (ω/p_n).

For the plot of static deflection δ_{st} vs. R — Refer to Fig. 22–21.

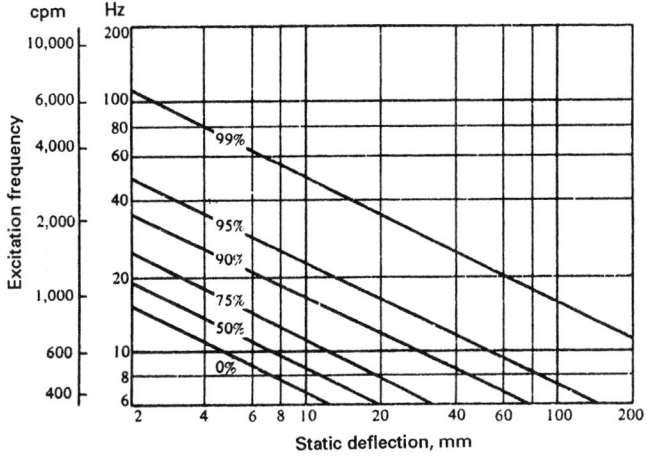

FIGURE 22–21 Static deflection (δ_{st}) vs. disturbing frequency for various percent reduction in transmissibility (T_R) for $\zeta = 0$. (*Courtesy of F. S. Tes, I. E. Morse, and R. T. Hinkle,* Mechanical Vibration—Theory and Applications, *CBS Publishers and Distributors, New Delhi, India, 1983.*)

FIGURE 22–22 Undamped two-degree-of-freedom system.

Particular	Formula

UNDAMPED TWO-DEGREE-OF-FREEDOM SYSTEM (FIG. 22–22) WITHOUT EXTERNAL FORCE

Particular	Formula	
Equations of motion	$m_1\ddot{x}_1 + (k_1 + k_3)x_1 - k_3 x_2 = 0$	(22–66a)
	$m_2\ddot{x}_2 + (k_2 + k_3)x_2 - k_3 x_1 = 0$	(22–66b)
The frequency of equation which gives two values for p^2	$p^4 - p^2\left\{\dfrac{k_1 + k_3}{m_1} + \dfrac{k_2 + k_3}{m_2}\right\}$	
	$+ \dfrac{k_1 k_2 + k_2 k_3 + k_1 k_3}{m_1 m_2} = 0$	(22–67)
The amplitude ratio	$\dfrac{a_1}{a_2} = \dfrac{-k_3}{m_1 p^2 - k_1 - k_3} = \dfrac{m_2 p^2 - k_2 - k_3}{-k_3}$	(22–68)

DYNAMIC VIBRATION ABSORBER (FIG. 22–23)

Particular	Formula	
Equations of motion	$M\ddot{x}_1 + (K + k)x_1 - kx_2 = F_o \sin\omega t$	(22–69a)
	$m\ddot{x}_2 + k(x_2 - x_1) = 0$	(22–69b)
The solution of the forced vibration of the absorber will be of the form	$x_1 = a_1 \sin pt$	(22–70a)
	$x_2 = a_2 \sin pt$	(22–70b)

Particular	Formula

The ratio of amplitudes a_1 and a_2 to the static deflection of the main system x_{st}

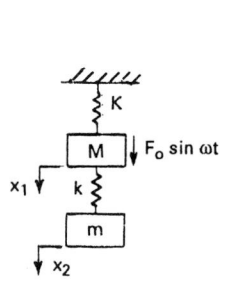

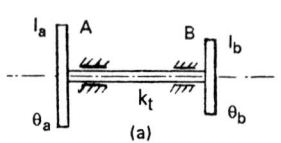

FIGURE 22–23 Dynamic vibration absorber.

$$\frac{a_1}{x_{st}} = \frac{1 - \dfrac{\omega^2}{p_a^2}}{\left(1 - \dfrac{\omega^2}{p_a^2}\right)\left(1 + \dfrac{k}{K} - \dfrac{\omega^2}{p_m^2}\right) - \dfrac{k}{K}} \quad (22\text{–}71a)$$

$$\frac{a_2}{x_{st}} = \frac{1}{\left(1 - \dfrac{\omega^2}{p_a^2}\right)\left(1 + \dfrac{k}{K} - \dfrac{\omega^2}{p_m^2}\right) - \dfrac{k}{K}} \quad (22\text{–}71b)$$

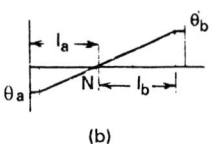

FIGURE 22–24 Two-rotor system.

where

$x_{st} = F_o/K = $ static deflection of main system

$p_a^2 = K/m = $ natural circular frequency of absorber

$p_m^2 = k/M = $ natural circular frequency of main system

$R_m = \dfrac{m}{M} = $ mass ratio $= \dfrac{\text{absorber mass}}{\text{main mass}}$

If the main system is in resonance, then considering

$$p_a = p_m \text{ or } \frac{k}{m} = \frac{K}{M} \text{ or } \frac{k}{K} = \frac{m}{M} = R_m$$

Eqs. (22–71a) and (22–71b) become

$$\frac{x_1}{x_{st}} = \frac{1 - (\omega/p_a)^2}{[1 - (\omega/p_a)^2][1 + R_m - (\omega/p_a)^2] - R_m}\sin \omega t$$

$$(22\text{–}72a)$$

$$\frac{x_2}{x_{st}} = \frac{1}{[1 - (\omega/p_a)^2][1 + R_m - (\omega/p_a)^2] - R_m}\sin \omega t$$

$$(22\text{–}72b)$$

The natural frequencies

$$\left(\frac{\omega}{p_a}\right)^2 = \left(1 + \frac{R_m}{2}\right) \pm \left[R_m + \frac{R_m^2}{4}\right]^{1/2} \quad (22\text{–}73)$$

The mass equivalent for the absorber

$$\frac{m_{eq}}{m} = \frac{1}{1 - (\omega/p_a)^2} \quad (22\text{–}74)$$

where $m_{eq} = $ equivalent mass solidly attached to the main mass M

TORSIONAL VIBRATING SYSTEMS
Two-rotor system (Fig. 22–24)

The torque on rotor A
$$M_{ta} = I_a p^2 \theta_a \quad (22\text{–}75)$$

The total torque on two rotors
$$M_{ti} = M_{ta} + M_{tb} = I_a p^2 \theta_a + I_b p^2 \theta_b = 0 \quad (22\text{–}76)$$
where $i = $ imaginary

Particular	Formula	
The angular displacement or angle of twist of rotor B	$\theta_b = \theta_a - \dfrac{M_{ta}}{k_t} = \theta_a\left(1 - \dfrac{I_a p^2}{k_t}\right)$	(22–77)
The frequency equation	$p^2 \theta_a\left(I_a + I_b - \dfrac{I_a I_b p^2}{k_t}\right) = 0$	(22–78a)
The natural circular frequency	$p_n = \left[\dfrac{(I_a + I_b)k_t}{I_a I_b}\right]^{1/2}$	(22–78b)
The natural frequency	$f_n = \left(\dfrac{1}{2\pi}\right)\left[\dfrac{(I_a + I_b)k_t}{i_a I_b}\right]^{1/2}$	(22–79)
The amplitude ratio	$\dfrac{\theta_a}{\theta_b} = -\dfrac{I_b}{I_a} = -\dfrac{l_a}{l_b}$	(22–80)
The relation between I_a, I_b, l_a, and l_b	$I_a l_a = I_b l_b$	(22–81)
The distance of node point from left end of rotor A	$l_a = \dfrac{I_b l}{I_a + I_b}$	(22–82)

Two rotors connected by shaft of varying diameters

The length of torsionally equivalent shaft of diameter d whose varying diameters are d_1, d_2, and d_3	$l_e = d^4\left[\dfrac{l_1}{d_1^4} + \dfrac{l_2}{d_2^4} + \dfrac{l_3}{d_3^4}\right]$	(22–83)

Three-rotor torsional system (Fig. 22–25)

The algebraic sum of the inertia torques of rotors A, B, and C

$$M_{ti} = M_{ta} + M_{tb} + M_{tc} = I_a p^2 \theta_a + I_b p^2 \theta_b + I_a p^2 \theta_c$$

(22–84)

where θ_a, θ_b, and θ_c are angular displacement or angular twist at rotors A, B, and C, respectively

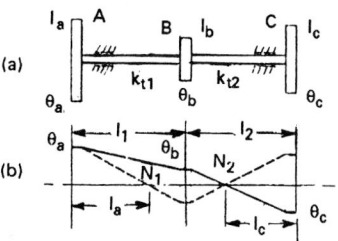

FIGURE 22–25 Three-rotor system.

Particular	Formula
The frequency equation	$$p^2\theta_a\left[(I_a + I_b + I_c) - p^2\left(\frac{I_aI_b}{k_{t1}} + \frac{I_aI_c}{k_{t1}} + \frac{I_aI_c}{k_{t2}}\right.\right.$$ $$\left.\left. + \frac{I_bI_c}{k_{t2}}\right) + p^4\left(\frac{I_aI_bI_c}{k_{t1}k_{t2}}\right)\right] = 0 \qquad \text{(22–85a)}$$ $$p^2 = \frac{1}{2}\left(\frac{k_{t1}}{I_a} + \frac{k_{t2}}{I_c} + \frac{k_{t1}+k_{t2}}{I_b}\right) \pm \left(\frac{1}{2}\right)$$ $$\left[\left(\frac{k_{t1}}{I_a} + \frac{k_{t2}}{I_c} + \frac{k_{t1}+k_{t2}}{I_b}\right)^2 - 4\,\frac{k_{t1}k_{t2}}{I_aI_bI_c}(I_a + I_b + I_c)\right]^{1/2}$$ $$\text{(22–85b)}$$ where k_{t1} and k_{t2} are torsional stiffness of shafts of lengths l_1 and l_2
The amplitude ratio	$$\frac{\theta_b}{\theta_a} = 1 - \frac{I_ap^2}{k_{t1}} \qquad \text{(22–86a)}$$ $$\frac{\theta_c}{\theta_a} = 1 - p^2\left(\frac{I_a}{k_{t1}} + \frac{I_c}{k_{t2}} + \frac{I_b}{k_{t2}}\right) + \frac{p^4I_aI_b}{k_{t1}k_{t2}} \qquad \text{(22–86b)}$$
The relation between I_a, I_c, l_a, and l_c	$$I_al_a = I_cl_c \qquad \text{(22–87)}$$
The relation between I_a, I_b, l_a, and l_c	$$\frac{1}{I_al_a} = \frac{1}{I_b}\left\{\frac{1}{l_1-l_a} + \frac{1}{l_2-l_c}\right\} \qquad \text{(22–88)}$$
Frequency can also be found from Eqs. (22–89) and (22–90)	$$f_c = \left(\frac{1}{2\pi}\right)\sqrt{k_{tc}/I_c} \text{ where } k_{tc} = \frac{GJ_2}{l_c} \qquad \text{(22–89)}$$ $$f_b = \left(\frac{1}{2\pi}\right)\sqrt{k'_{tb}/I_b} \qquad \text{(22–90)}$$ where $k'_{tb} = \dfrac{GJ_1}{l_1-l_a} + \dfrac{GJ_2}{l_2-l_c}$
For collection of mechanical vibration formulas to calculate natural frequencies	Refer to Table 22–2.

TABLE 22–2
A collection of formulas

Particular	Formula	Equationn
Natural Frequencies of Simple Systems		
End mass M; spring mass m, spring stiffness k	$p_n = \sqrt{k/(M + m/3)}$	(22–110)
End inertia I, shaft inertia I_s, shaft stiffness k	$p_n = \sqrt{k/(I + I_s/3)}$	(22–111)
Two disks on a shaft	$p_n = \sqrt{\dfrac{k(I_1 + I_2)}{I_1 I_2}}$	(22–112)
Cantilever; end mass M; beam mass m, stiffness by formula (22–93)	$p_n = \sqrt{\dfrac{k}{M + 0.23m}}$	(22–113)
Simply supported beam central mass M; beam mass m; stiffness by formula (22–95)	$p_n = \sqrt{\dfrac{k}{M + 0.5m}}$	(22–114)
Massless gears, speed of I_2 n times as large as speed of I_1	$p_n = \sqrt{\dfrac{1}{k_1 + \dfrac{1}{n^2 k_2}} \left(\dfrac{I_1 + n^2 I_2}{I_1 I_2 n^2} \right)}$	(22–115)
	$p_n^2 = \dfrac{1}{2} \left(\dfrac{k_1}{I_1} + \dfrac{k_3}{I_3} + \dfrac{k_1 + k_3}{I_2} \right) \pm \dfrac{1}{2} \sqrt{ \left(\dfrac{k_1}{I_1} + \dfrac{k_3}{I_3} + \dfrac{k_1 + k_3}{I_2} \right)^2 - 4 \dfrac{k_1 k_3}{I_1 I_2 I_3} (I_1 + I_2 + I_3)}$	(22–116)
Uniform Beams (Longitudinal and Torsional Vibration)		
Longitudinal vibration of cantilever: A = cross section, E = modulus of elasticity	$p_n = \left(n + \dfrac{1}{2} \right) \pi \sqrt{\dfrac{AE}{\mu_1 l^2}}$	(22–117)
μ_1 = mass per unit length, $n = 0, 1, 2, 3$ = number of nodes	For steel and l in inches, this becomes $f = \dfrac{p_n}{2\pi} = (1 + 2n) \dfrac{51,000}{l}$ Hz	(22–118)
Organ pipe open at one end, closed at the other	For air at atm. pressure, l in inches $f = \dfrac{p_n}{2\pi} = (1 + 2n) \dfrac{3,300}{l}$ Hz $n = 0, 1, 2, 3, \ldots$	(22–119)
Water column in rigid pipe closed at one end (l in inches)	$f = \dfrac{p_n}{2\pi} = (1 + 2n) \dfrac{14,200}{l}$ Hz $n = 0, 1, 2, 3, \ldots$	(22–120)

TABLE 22–2
A collection of formulas (*Cont.*)

Particular	Formula

Longitudinal vibration of beam clamped or free at both ends; n = number of half waves along length

$$p_n = n\pi\sqrt{\frac{AE}{\mu_1 l^2}}$$ (22–121)

$$n = 1, 2, 3, \ldots$$

For steel, l in inches:

$$f = \frac{p_n}{2\pi} = \frac{102,000}{l} \text{ Hz}$$ (22–122)

Organ pipe closed (or open) at both ends (air at 60°F = 15.5°C)

$$f = \frac{p_n}{2\pi} = \frac{n6,600}{l} \text{ Hz}$$

$$n = 1, 2, 3, \ldots$$ (22–123)

Water column in rigid pipe closed (or open) at both ends

$$f = \frac{n28,400}{l} \text{ Hz}$$ (22–124)

$$n = 1, 2, 3, \ldots$$

For water columns in nonrigid pipes . . .

$$\frac{f_{\text{nonrigid}}}{f_{\text{rigid}}} = \frac{1}{\sqrt{1 + \dfrac{300,000D}{t\,E_{\text{pipe}}}}}$$ (22–125)

E_{pipe} = elastic modulus of pipe, psi
D, t = pipe diameter and wall thickness, same units

Torsional vibration of beams . . .

Same as (22–117) and (22–118); replace tensional stiffness AE by torsional stiffness GI_p; replace μ_1 by the moment of inertia per unit length $i_1 = I_{bar}/l$

Uniform Beams (Transverse or Bending Vibrations)

The same general formula holds for all the following cases,

$$p_n = a_n\sqrt{\frac{EI}{\mu_1 l^4}}$$ (22–126)

where EI is the bending stiffness of the section, l is the length of the beam, μ_1 is the mass per unit length = W/gl, and a_n is a numerical constant, different for each case and listed below:-

α_1 Cantilever or "clamped-free" beam . . .
α_2
α_3

$a_1 = 3.52$
$a_2 = 22.0$
$a_3 = 61.7$
$a_4 = 121.0$
$a_5 = 200.0$

α_1 Simply supported or "hinged-hinged" beam
α_2
α_3

$a_1 = \pi^2 = 9.87$
$a_2 = 4\pi^2 = 39.5$
$a_3 = 9\pi^2 = 88.9$
$a_4 = 16\pi^2 = 158$
$a_5 = 25\pi^2 = 247$

α_1 "Free-free" beam or floating ship . . .
α_2
α_3

$a_1 = 22.0$
$a_2 = 61.7$
$a_3 = 121.0$
$a_4 = 200.0$
$a_5 = 298.2$

TABLE 22–2
A collection of formulas (*Cont.*)

Particular	Formula
a_1 "Clamped-clamped" beam has same frequencies as "free-free"	$a_1 = 22.0$
a_2	$a_2 = 61.7$
a_3	$a_3 = 121.0$
	$a_4 = 200.0$
	$a_5 = 298.2$
a_1 "Clamped-hinged" beam may be considered as half a "clamped-clamped" beam for even a-numbers	$a_1 = 15.4$
a_2	$a_2 = 50.0$
	$a_3 = 104$
	$a_4 = 178$
	$a_5 = 272$
a_1 "Hinged-free" beam or wing of autogyro may be considered as half a "free-free" beam for even a-numbers	$a_1 = 0$
a_2	$a_2 = 15.4$
a_3	$a_3 = 50.0$
	$a_4 = 104$
	$a_5 = 178$

Rings, Membranes, and Plates

Extensional vibration of a ring, radius r, weight density γ:

$$p_n = \frac{1}{r}\sqrt{\frac{Eg}{\gamma}} \tag{22–127}$$

Bending vibrations of ring, radius r, mass per unit length, μ_1, in its own plane with n full "sine waves" of disturbance along circumference

$$p_n = \frac{n(n^2 - 1)}{\sqrt{1 + n^2}}\sqrt{\frac{EI}{\mu_1 r^4}} \tag{22–128}$$

Circular membrane of tension T, mass per unit area μ_1, radius r

$$p_n = a_{cd}\sqrt{\frac{T}{\mu_1 r^2}} \tag{22–129}$$

The constant a_{cd} is shown below, the subscript c denotes the number of nodal circles, and the subscript d the number of nodal diameters:

d \ c	1	2	3
0	2.40	5.52	8.65
1	3.83	7.02	10.17
2	5.11	8.42	11.62
3	6.38	9.76	13.02

Membrane of any shape of area A roughly of equal dimensions in all directions, fundamental mode:

$$p_n = \text{const}\sqrt{\frac{T}{\mu_1 A}} \tag{22–130}$$

TABLE 22–2
A collection of formulas (*Cont.*)

Circle	const $= 2.40\pi = 4.26$
Square	const $= 4.44$
Quarter circle	const $= 4.55$
2×1 rectangle	const $= 4.97$

Circular plate of radius r, mass per unit area μ_1; the "plate constant D" defined in Eq (22–99)

$$p_n = a\sqrt{\frac{D}{\mu_1 r^4}} \tag{22–131}$$

For free edges, 2 perpendicular nodal diameters	$a = 5.25$
For free edges, one nodal circle, no diameters	$a = 9.07$
Clamped edges, fundamental mode	$a = 10.21$
Free edges, clamped at center, umbrella mode	$a = 3.75$

Rectangular plate, all edges simply supported, dimensions l_1 and l_2:

$$p_n = \pi^2\left(\frac{m^2}{l_1^2} + \frac{n^2}{l_1^2}\right)\sqrt{\frac{D}{\mu_1}} \qquad m = 1, 2, 3, \ldots; \; n = 1, 2, 3, \ldots \tag{22–132}$$

Square plate, all edges clamped, length of side l, fundamental mode:

$$p_n = \frac{36}{l^2}\sqrt{\frac{D}{\mu_1}} \tag{22–133}$$

Source: Formulas (Eqs.) (22–110) to (22–133) extracted from J. P. Den Hartog, *Mechanical Vibrations*, McGraw-Hill Book Company, New York, 1962.

REFERENCES

1. Den Hartog, J. P., *Mechanical Vibrations*, McGraw-Hill Book Company, New York, 1962.
2. Thomson, W. T., *Theory of Vibration with Applications*, Prentice-Hall, Englewood Cliffs, New Jersey, 1981.
3. Baumeister, T., ed., *Marks' Standard Handbook for Mechanical Engineers*, 8th ed., McGraw-Hill Book Company, New York, 1978.
4. Black, P. H., and O. E. Adams, Jr., *Machine Design*, McGraw-Hill Book Company, New York, 1955.
5. Lingaiah, K., and B. R. Narayana Iyengar, *Machine Design Data Handbook*, Engineering College Co-Operative Society, Bangalore, India, 1962.
6. Myklestad, N. O., *Fundamentals of Vibration Analysis*, McGraw-Hill Book Company, New York, 1956.
7. Tse, F. S., I. E. Morse, and R. T. Hinkle, *Mechanical Vibration—Theory and Applications*, CBS Publishers and Distributors, New Delhi, India, 1983.

CHAPTER
23

GEARS

23.1 SPUR AND HELICAL GEARS

SYMBOLS

a	center distance, m (in)
a'	working center distance, m (in)
a_n	nominal center distance for rack and pinion drives, m (in)
b	face width, m (in)
b_e	effective face width, m (in)
b_1	pinion face width, m (in)
b_2	gear face width, m (in)
B	nominal load factor, MPa (psi)
B_a	allowable value of nominal load factor, MPa (psi)
c	clearance, m (in)
c^*	clearance coefficient
C	constant
	coefficients in Eqs. (23–151), (23–155), and (23–156), N/m (lbf/in)
$C_1,\ C_2$	constants in Eqs. (23–149), (23–152), and (23–153)
C_a	application factor or overload correction factor for pitting resistance from Table 23–21
C_B	factor associated with Brinell hardness number in Eqs. (23–314a)
C_H	hardness ratio factor in Eq. (23–169) and Figs 23–37 and 23–38
C_L	life factor for pitting resistance taken from Fig. 23–26 and Fig. 23–39
C_p	elastic coefficient, $\sqrt{\text{MPa}}$ ($\sqrt{\text{psi}}$) taken from Table 23–35 and Eq. (23–165)
C_m	load distribution or mounting factor for pitting reistance from Table 23–22
$C_r =$	$i/i+1$, ratio factor

C_R	reliability factor for pitting resistance from (Table 23–25B), factor associated with Rockwell hardness number [Eq. (23–314b)]
C_s	service factor (Table 23–13)
C_{sr}	surface factor for pitting resistance
C_{sz}	size factor for pitting resistance
C_T	temperature factor for pitting resistance
C_V	velocity or dynamic factor for pitting resistance (Fig. 23–28)
C_w	wear and lubrication factor taken from Table 23–47
C_D, C_S, C_T, C_β	coefficients of load influence
d	diameter, m (in)
d_a	tip circle diameter, m (in)
d_b	base circle diameter, m (in)
d_f	root circle diameter, m (in)
d_{GH}	diameter of generating pitch circle, m (in)
d'	pitch circle diameter, m (in)
d_1	diameter of pinion, m (in)
d_2	diameter of gear, m (in)
e	space width, m (in)
e_a	space width at tip, m (in)
e_f	space width at root, m (in)
E	modulus of elasticity, GPa (Mpsi)
E_1, E_2	modulus of elasticity of pinion and gear materials, respectively, GPa (Mpsi)
E_o	equivalent modulus of elasticity, GPa (Mpsi)
f	error, μm (μin)
f_a	center distance error, μm (μin)
f_b	base circle error, μm (μin)
f_c	eccentricity error, μm (μin)
f_Σ	shaft alignment error, μm (μin)
f_{nb}	speed factor for bending strength for appropriate running time
f_{nc}	speed factor for pitting resistance or wear for appropriate running time
f_{nb12}	combined speed factor for bending strength at 12-h running time per day
f_{nc12}	combined speed factor for pitting resistance or wear at 12-h running time per day
f_R, f_{RW}	flank direction error factor starting under load, μm (μin)
F_a	acceleration load, kN (lbf)
	axial load, kN (lbf)
F_d	dynamic load on the tooth, kN (lbf)
F_i	increment load, kN (lbf)
F_n	force, normal to the tooth profile, kN (lbf)
F_r	radial load, kN (lbf)
F_t	tangential tooth load, kN (lbf)
	transmitted load, kN (lbf)
F_{ta}	allowable or permissible transmitted load, kN (lbf)
F^*	$= \dfrac{F_t}{b}$ specific loading, N/m (lbf/in)
F_1	average force required to accelerate the masses when they are considered as absolutely rigid, kN (lbf)

F_2	force required to deform teeth through an amount of effective error, kN (lbf)
F_f	endurance force, kN (lbf)
g	line of action, m (in)
	length of path of contact, m (in)
g_a	length of path of addendum contact, m (in)
g_f	length of path of dedenum contact, m (in)
g_r	total length of transmission, m (in)
g_t	length of contact (tangential to pitch circles), m (in)
	length of the line of action in the transverse plane, m (in)
g_β	helical overlap, in
g_e, g_R	factors which are function of peripheral velocity, lubrication, tooth flank quality, etc.
G	modulus of rigidity, GPa (Mpsi)
h	tooth depth, m (in)
	load lever arm or moment arm, m (in)
h_a	addendum, m (in)
h_{av}	cutter addendum, m (in)
h_a^*	addendum coefficient
$\bar{h}_a$	chord height, m (in)
$\bar{h}_c$	constant chord height, m (in)
h_f	dedendum, m (in)
h_{fo}	cutter dedendum, m (in)
h_f^*	dedendum coefficient
h_v	depth of tooth, m (in)
h_x	moment arm, m (in)
h'	working depth of tooth, m (in)
h^*	tooth depth coefficient
H	coefficient used in Eq. (23–148)
H_B	tabulated nominal Brinell hardness of flank
i	gear ratio
i'	transmission ratio
j	backlash, m (in)
j_n	normal backlash, m (in)
j_r	radial backlash, m (in)
j_t	circumferential backlash, m (in)
$k = 1/C_2$	constant used in Eq. (23–157) and also with subscripts
K	factor,
	load stress factor, Eq. (23–161)
	pitch factor from Fig. 23–37 and Eq. (23–197)
K_a	overload correction factor or application factor for bending strength (Table 23–21)
K_b	bending stress factor taken from Table 23–40,
K_B	rim thickness factor
K_L	life factor for bending strength (Fig. 23–26)
K_{LS}	load sharing factor
K_m	mounting or load distribution factor for bending strength (Table 23–22)
K_{me}	miscellaneous effect factor (Fig. 23–27)
K_R	reliability factor for bending strength (Table 23–25B)
K_{sr}	surface factor for bending strength
K_{sz}	size factor for bending strength (Table 23–26)
K_T	temperature factor for bending strength Eqs. (23–101) and (23–104)

K_v	velocity or dynamic factor for bending strength, Eq. (23–95) and from (Fig. 23–28)
	coefficient of peripheral velocity
K_w	surface stress factor
K_σ	stress concentration factor
$K_{f\sigma}$	fatigue stress concentration factor
L	life, h
	length, m (in)
	minimum of four values of bending strength and pitting resistance or wear load per millimeter face width calculated for pinion and gear
m	module, m (in)
m_n	normal module, m (in)
m_o	cutter module, m (in)
m_t	transverse module, m (in)
m_t'	transverse working module, m (in)
m_x	axial module, m (in)
$\boldsymbol{m}$	effective mass, kg (lb)
$\boldsymbol{m_1, m_2}$	masses of pinion and gear, respectively, kg (lb)
M_t	torque, N m (lbf in)
M_{t1}	torque on pinion, N m (lbf in)
	rated torque on pinion, N m (lbf in)
M_{t2}	torque on gear, N m (lbf in)
M_{tms}	maximum sustained load, N m (lbf in)
n	speed, rpm
	factor of safety
n_1	speed of pinion, rpm
n_2	speed of gear, rpm
n'	factor of safety
	speed, rps
n_c	computed factor of safety
n_p, n_b	factor of safety in pitting and breakage
p	circular pitch, m (in)
p_b	base pitch, m (in)
p_n	normal circular pitch, m (in)
p_{bn}	normal base circular pitch, m (in)
p_o	pitch of cutter, m (in)
p_{bt}	transverse base pitch, m (in)
p_t	transverse pitch, m (in)
p_x	axial pitch, m (in)
$p_z = p_x z$	lead, m (in)
P	power, kW (hp)
$\boldsymbol{P}$	diametral pitch, m^{-1} (in^{-1})
P_{ab}	allowable or permissible power rating for bending strength as per AGMA, kW (hp)
P_d	design power, kW (hp)
P_p	momentary peak power, kW (hp)
P_{pc}	allowable or permissible power rating for pitting resistance or surface durability as per AGMA, kW (hp)
P_s	power rating of spur gear for bending strength, kW (hp)
P_w	power rating of spur gear for pitting resistance or wear, kW (hp)
P_1	rated power transmitted by pinion, kW (hp)
P_2	rated power transmitted by gear, kW (hp)
$\boldsymbol{P_n}$	normal diametral pitch, m^{-1} (in^{-1})

P_t	transverse diametral pitch, m^{-1} (in^{-1})
P_t'	working transverse diametral pitch, m^{-1} (in^{-1})
P_x	axial diametral pitch, m^{-1} (in^{-1})
q	form factor
Q	$= 2d_2/(d_1 + d_2)$ ratio factor
r	radius (reference circle radius), m (in)
r_a	tip circle radius, m (in)
r_f	root circle radius, m (in)
r_t	tip radius, mm (in)
r'	radius of pitch circle, m (in)
r_1	radius of pinion, m (in)
r_2	radius of gear, m (in)
s	tooth thickness, m (in)
s_a	tooth thickness at tip, m (in)
s_b	tooth thickness at base, m (in)
s_f	tooth thickness at root, m (in)
s'	tooth thickness at operating circle, m (in)
s_x	tooth thickness at critical section, m (in)
$\bar{s}$	chordal tooth thickness, m (in)
$\bar{s}_c$	constant chord tooth thickness, m (in)
$\bar{s}_b$	transverse chordal tooth thickness, m (in)
t	times, s
T	time, h
	temperature, °C (°F)
v	linear velocity, m/s
v_m	mean linear velocity, m/s (ft/min)
V	volume, m^3 (in^3)
W_k	base tangent length over k teeth
x	addendum modification coefficient
x_1	addendum modification coefficient on pinion
x_2	addendum modification coefficient on gear
y	Lewis form factor,
	center distance modification coefficient
Y	modified Lewis form factor,
	AGMA form factor
Y_t	tooth form factor for root bending moment
z	number of teeth,
	number of starts
z_v	virtual number of teeth
z_0	number of teeth on cutter
z_1	number of teeth on pinion
z_2	number of teeth on gear
Y_z	zonal factor
$inv\ \alpha$	involute α (involute function)
α	pressure angle on pitch circle, deg
α_n	normal pressure angle, deg
α_t	transverse pressure angle, deg
α_v	angle between the total load vector F_n and a perpendicular to the center line of the tooth at the highest point of single-tooth contact, deg (Fig. 23–29)
α_x	axial pressure angle, deg
α_y	pressure angle at a point, deg
α'	working pressure angle, deg
	pressure angle on contact circle in transverse section, deg
α_n'	pressure angle on contact circle in normal section, deg

β	helix angle (reference helix angle), deg
	spiral angle, deg
β_b	base helix angle, deg
γ	lead angle, deg
γ_b	base lead angle, deg
δ	cone angle, deg
	deflection, m (in)
δ_a	tip cone angle, deg
δ_b	base cone angle, deg
δ_f	root cone angle, deg
ε	ratio (overlap, contact)
$\varepsilon_1, \varepsilon_2$	coefficient of profile contact ratio in transverse section
ε_x	transverse contact ratio
ε_β	overlap ratio
ε_γ	contact ratio (total)
η	space width semiangle, deg
μ	coefficient of friction
ν	tooth depth angle, deg
ν_a	addendum angle, deg
ν_f	dedendum angle, deg
$\xi_1 = 1 + x_1$	addendum factor for pinion
$\xi_2 = 1 + x_2$	addendum factor for gear
ρ	profile radius, μm (μin)
	friction angle, deg
ρ_f or r_f	root fillet radius, m (in)
σ	normal stress (also with suffixes), MPa (psi)
σ_a	allowable stress, MPa (psi)
	stress amplitude, MPa (psi)
σ_b	tooth root bending stress, MPa (psi)
	calculated bending stress, MPa (psi)
$\sigma_{b\,\lim}$	limit value of tooth root bending stress, MPa (psi)
σ_{sac}	AGMA allowable surface fatigue strength or allowable contact stress number, MPa (psi) (Fig. 23–25 and Table 23–24)
σ_H or σ_c	Hertzian surface (contact) stress, MPa (psi)
$\sigma_{H\,\lim}$	limit value of surface stress, MPa (psi)
σ_m	mean stress, MPa (psi)
σ_{sfd}	modified endurance limit, MPa (psi)
σ'_{sfb}	endurance limit of rotating beam specimen or R. R. Moore endurance limit, MPa (psi)
σ_{st}	tensile strength, MPa (psi)
σ_{sab}	AGMA allowable bending strength or allowable bending stress numbers, MPa (psi) (Fig. 23–24 and Table 23–23)
σ_{fH} or σ_{fc}	endurance surface stress or contact stress, MPa (psi)
σ_o	static stress, MPa (psi)
σ_s	strength of gear material, MPa (psi)
σ_{ut}	ultimate tensile stress of material, MPa (psi)
σ_{sut}	ultimate tensile strength, MPa (psi)
σ_{wc}	working contact stress, MPa (psi)
τ	shear stress (also with subscripts), MPa (psi)
ϕ	angle of tooth action, deg
ϕ_β	overlap angle, deg
ϕ_γ	total angle of transmission
ω	angular velocity, rad/s
Δ	difference

Σ   shaft angle, deg

SUFFIXES

a	axial
b	bending,
	values on base cirlce
c	compressive
f	endurance
n	nominal
	values on normal section
o	relating to tool
s	strength properties of material
t	tensile
u	ultimate
y	yield
v	operating condition
1,2	for size on pinion and gear wheel, respectively
'	value on pitch circle or contact circle
-	reference

Note: 1. σ and τ with first subscript S designates strength properties of material used in the design which will be used and followed throughout this chapter. Other factors in performance or in special aspects which are included from time to time in this chapter and being applicable only in their immediate context are not given at this stage. 2. Most of the materials in the text of Chapter 23 on "Gears" in a particular format, which is the specialty of this book, is reproduced from K. Lingaiah, *Machine Design Data Handbook*, Volume II, Suma Publishers, Bangalore, India, 1986 (1).

Particular	Formula

23.1.1 SPUR GEARS (1)
General considerations, definitions, and dimensions

For nomenclature of cylindrical gear drive	Refer to Fig. 23–1.
For internal and external gear pair	Refer to Figs. 23–2a and b.
For nomenclature of spur gear teeth	Refer to Fig. 23–3.
For gear-tooth proportions	Refer to Table 23–1.

The transmission (speed) ratio (Fig. 23–2)

$$i = \frac{\omega_1}{\omega_2} = \frac{n_1}{n_2} = \frac{d_2}{d_1} = \frac{z_2}{z_1} \quad (23\text{–}1)$$

Refer to Table 23–2 for transmission ratios.

The circular pitch (Fig. 23–3)

$$p = \pi d/z = \pi m \quad (23\text{–}2)$$

The pitch diameter (Fig. 23–4)

$$d' = \frac{pz}{\pi} = zm \quad (23\text{–}3)$$

Particular	Formula

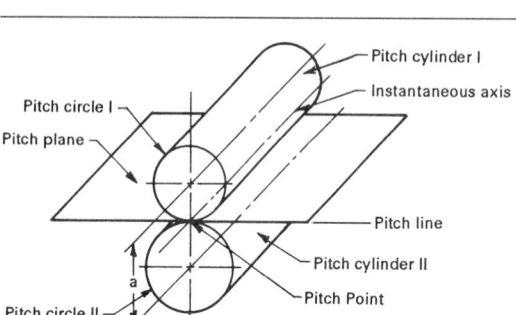

FIGURE 23–1 Nomenclature of cylindrical gear drive.

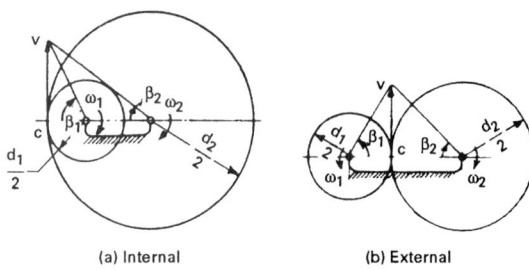

(a) Internal   (b) External

FIGURE 23–2 Gear pair.

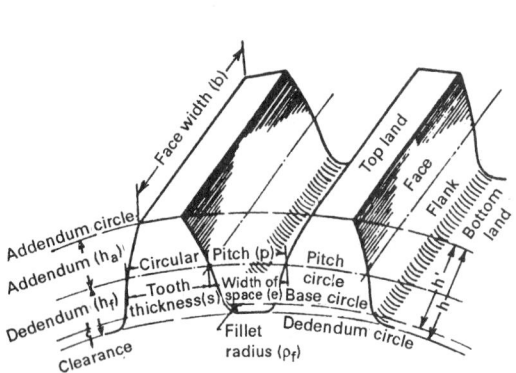

FIGURE 23–3 Nomenclature of gear teeth.

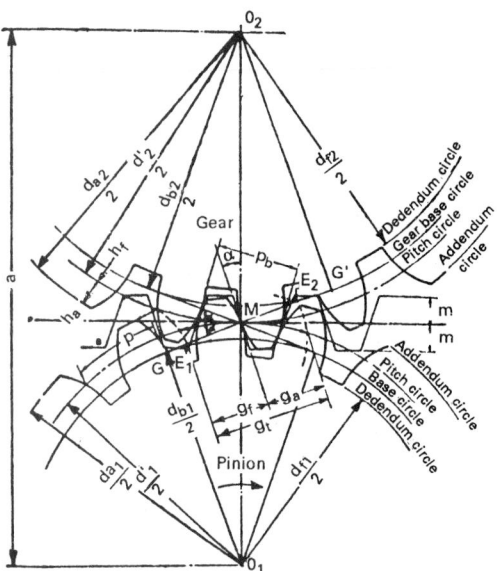

FIGURE 23–4 Gear layout, definitions, and dimensions relating to involute spur gear.

The pitch diameter of pinion

$$d_1' = \left(\frac{2a}{i \pm 1}\right) \tag{23–4}$$

The pitch diameter of gear

$$d_2' = \left(\frac{2ai}{i \pm 1}\right) \tag{23–5}$$

The module

$$m = \frac{p}{\pi} \tag{23–6}$$

Refer to Table 23–3 and Fig. 23–7 for standard full-size modules.

Particular	Formula
The diametral pitch	$$P = \frac{1}{m} = \frac{z}{d'} \tag{23-7}$$ Refer to Table 23–4 and Figs. 23–8 and 23–9 for standard full-size diametral pitches.
For diametral pitches and modules	Refer to Table 23–5 and Figs. 23–7, 23–8, and 23–9.
The relation between the circular pitch p and the diametral pitch P	$$pP = \pi \tag{23-8}$$
The base circle diameter	$$d_b = d' \cos\alpha = zm \cos\alpha \tag{23-9}$$

STANDARD GEAR (FIGS. 23–3, 23–5, AND 23–6)

Tooth addendum	$$h_a = m \tag{23-10}$$
Tooth dedendum	$$h_f = 1.25m \tag{23-11}$$
The total depth of tooth	$$h = h_a + h_f = 2.25m \tag{23-12}$$
The clearance	$$c = h_f - h_a = 0.25m \tag{23-13}$$
The working depth of tooth	$$h' = 2m \tag{23-14}$$
Addendum circle diameter (or tip diameter)	$$d_a = d' + 2h_a = zm + 2m = m(z+2) \tag{23-15}$$

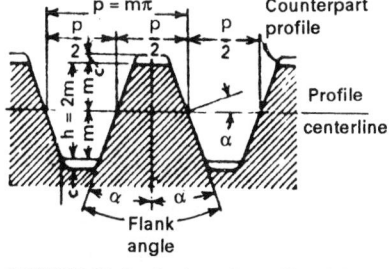

FIGURE 23–5 Basic rack profile with counterpart profile.

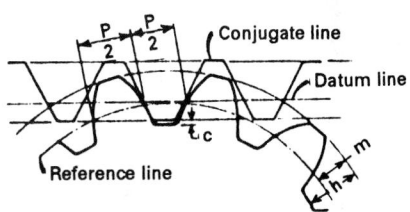

FIGURE 23–6 Zero backlash gearing.

Dedendum circle diameter (or root diameter)	$$d_f = d' - 2h_f = zm - 2.50m = m(z - 2.5) \tag{23-16}$$
The module	$$m = \frac{d_a}{z+2} \tag{23-17}$$
Tooth thickness	$$s = \frac{p}{2} = \frac{\pi m}{2} \tag{23-18a}$$
For zero back-lash gear	Refer to Fig. 23–6.
Theoretically, the tooth thickness s and the tooth space e, measured on the pitch circle are equal (Fig. 23–3)	$$s = e = \frac{p}{2} = \frac{\pi m}{2} = 1.57m \tag{23-18b}$$

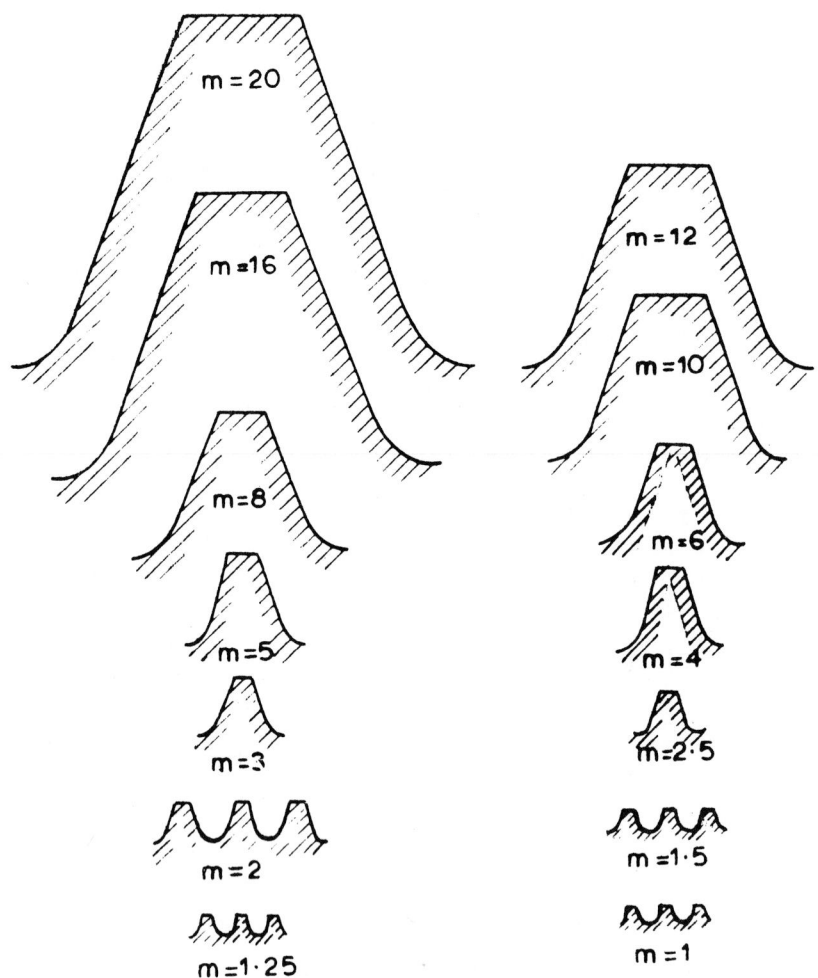

FIGURE 23–7   Basic rack profiles of ISO recommendation (full size)—Module Series 1.

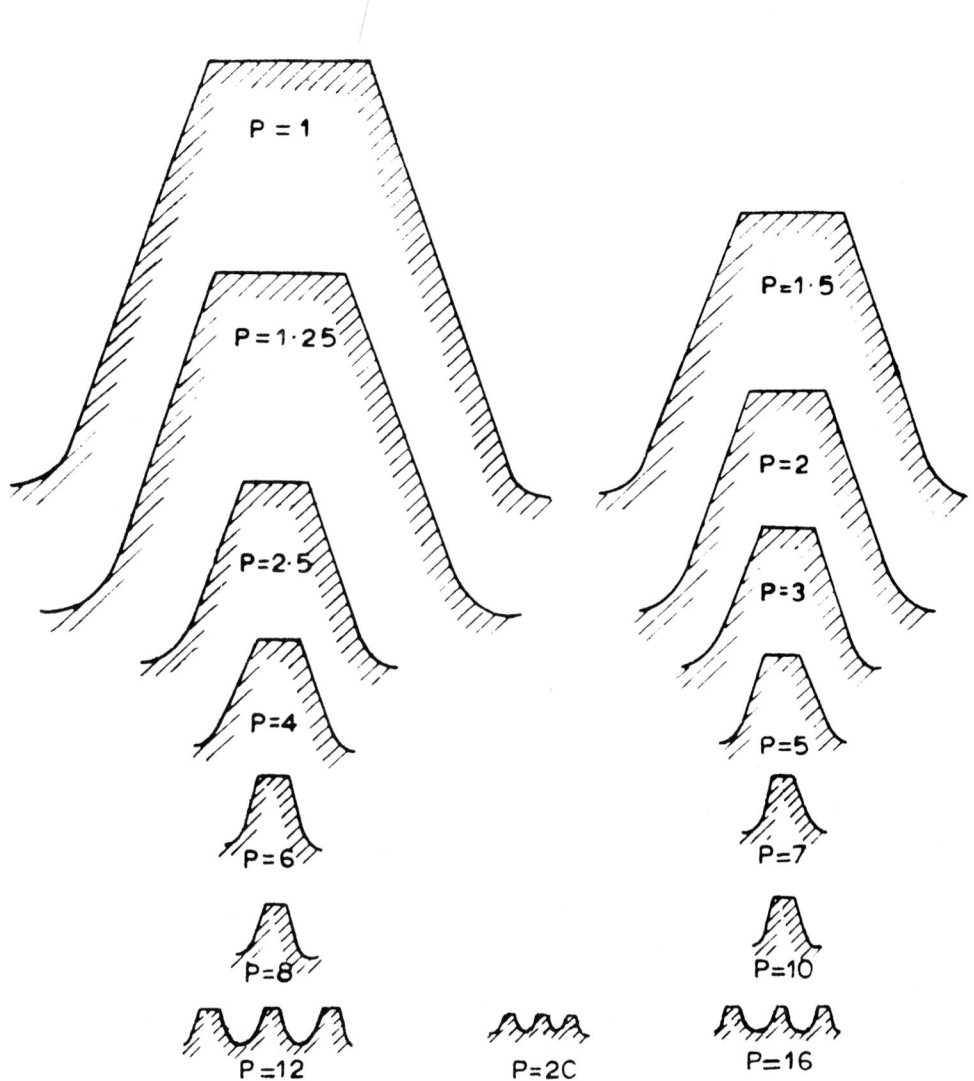

FIGURE 23–8   Basic rack profiles of ISO recommendation (full size)—Diametral Pitch Series 1.

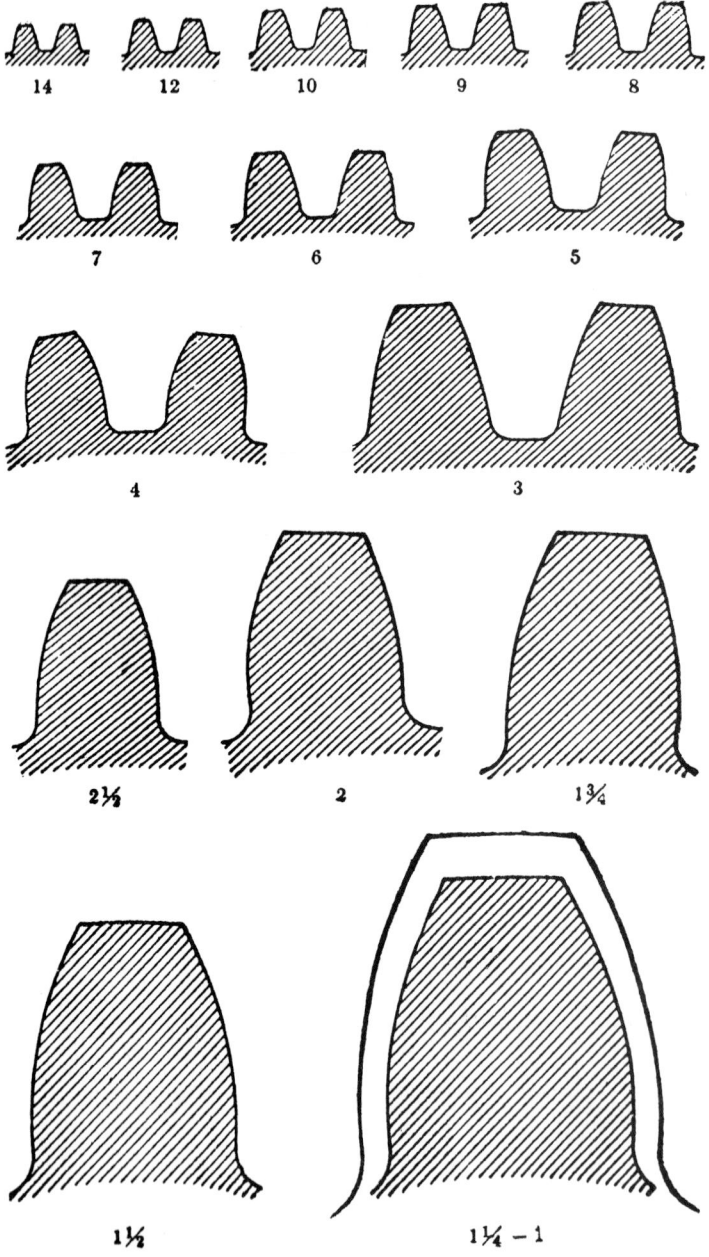

FIGURE 23–9 Comparison of standard diametral profiles drawn to scale.

Particular	Formula

Center distance

$$a = \frac{m}{2}(z_1 + z_2) \tag{23–19}$$

MINIMUM NUMBER OF TEETH IN A PINION

Limit tooth number for pinions generated by hob or rack

$$z_1 \geq \frac{2m}{\sin^2 \alpha} \tag{23–20}$$

For guide to selection of gear-teeth system

Refer to Table 23–6.

Limit tooth number for pinions generated by shaping gear

$$z_1 \geq \frac{\dfrac{2h_a}{m}}{\left[\left(\sqrt{i^2 + \sin^2 \alpha + 2i\sin^2 \alpha}\right) - i\right]} \tag{23–21}$$

CORRECTED GEAR (FIGS. 23–10 TO 23–12)
Gear correction without alteration of shaft center distance

Tooth addendum of the pinion

$$h_{a1} = (1 + x)m = \xi_1 m \tag{23–22}$$
where x = correction factor and $\xi_1 = 1 + x$

Tooth addendum of the gear

$$h_{a2} = (1 - x)m = \xi_2 m \text{ where } \xi_2 = 1 - x \tag{23–23}$$

Tooth dedendum of the pinion

$$h_{f1} = (1.25 - x)m \tag{23–24}$$

Tooth dedendum of the gear

$$h_{f2} = (1.25 + x)m \tag{23–25}$$

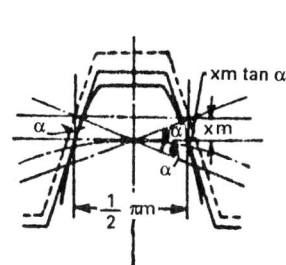

FIGURE 23–10 Correction to tooth profile.

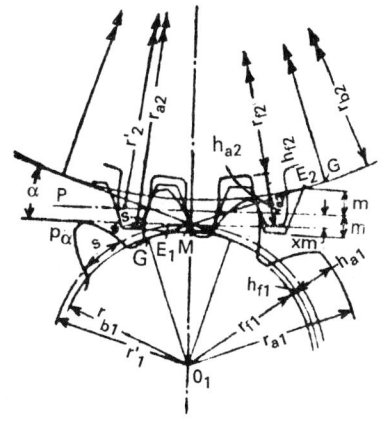

FIGURE 23–11 Gear correction without alteration of the shaft distance—external gear.

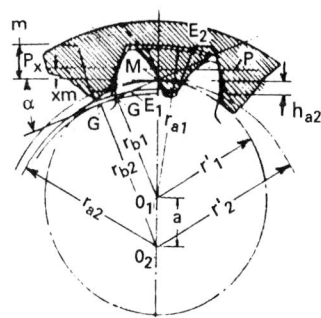

FIGURE 23–12 Gear correction without alternation of the shaft center distance—internal gears.

Particular	Formula	
Thickness of the pinion tooth	$s_1 = m\left(\dfrac{\pi}{2} + 2x \tan \alpha\right)$	(23–26)
Thickness of the gear tooth	$s_2 = m\left(\dfrac{\pi}{2} - 2x \tan \alpha\right)$	(23–27)
EXTERNAL GEARS (FIG. 23–11) Limit for correction factor x_1 for pinion	$x_1 \geq 1 - \dfrac{z_1}{2} \sin^2 \alpha$	(23–28)
Limit for correction factor x_2 for gear	$x_2 \leq \dfrac{z_2}{2} \sin^2 \alpha - 1$	(23–29)
INTERNAL GEARS (FIG. 23–12) The radius of the addendum circle to avoid interference of the gear tooth with the pinion tooth	$r_{a2} \geq \dfrac{z_2 m}{2} \sqrt{(i-1)^2 + (2i-1)\cos^2 \alpha}$	(23–30)
The radius of the addendum circle to have the clearance between the dedendum of the pinion and the addendum of the gear	$r_{a2} \geq r_2' - m(1-x) \geq \dfrac{z_2 m}{2} - m(1-x)$	(23–31)

Involute function (Fig. 23–13)

The angle $(\beta + \delta)$	$(\beta + \delta) = \tan \lambda$	(23–32)
The coordinates of point K on the involute curve	$x = r_b[\cos \delta + (\delta + \beta)\sin \delta]$	(23–33a)
	$y = r_b[\sin \delta - (\delta + \beta)\cos \delta]$	(23–33b)

The angle $(\theta + \lambda)$	$(\theta + \lambda) = \tan \lambda - \beta$	(23–34)
The angle θ	$\theta = \tan \lambda - \lambda - \beta$	(23–35)
The angle β	$\beta = \gamma - \alpha$	(23–36a)
	$= \tan \alpha - \alpha$	(23–36b)
The involute function	$inv\ \alpha = \tan \alpha - \alpha$	(23–37)
	For involute functions ($inv\ \alpha$), refer to Table 23–7.	

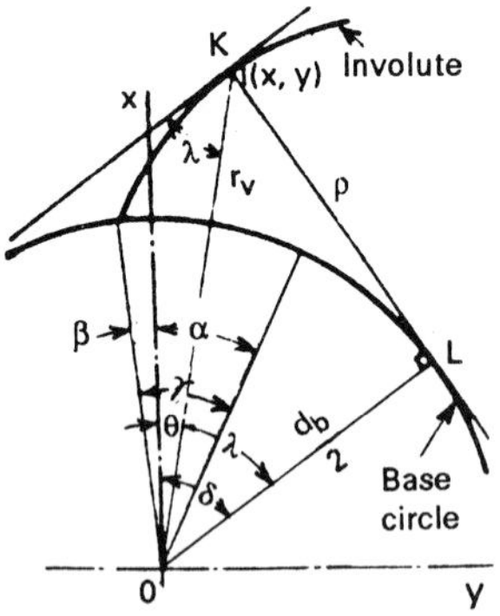

FIGURE 23–13 Involute trigonometry.

Gear correction with alteration of shaft center distance

EXTERNAL GEAR The thickness of tooth of the pinion at the operating circle (Fig. 23–14)	$s_{v1} = (inv\ \gamma - inv\ \alpha_v)zm\ \dfrac{\cos \alpha}{\cos \alpha_v}$	(23–38)
The thickness of tooth of the pinion at the pitch circle ($\alpha_v = \alpha$)	$s_1' = (inv\ \gamma - inv\ \alpha)zm = m\left(\dfrac{\pi}{2} + 2x \tan \alpha\right)$	(23–39)

Particular	Formula
The equation for angle γ_1 for the case of intersection of tooth profiles of the pinion i.e. pointed tooth (Fig. 23–14)	$inv\,\gamma_1 = \dfrac{1}{z_1}\left(\dfrac{\pi}{2} + 2x_1\tan\alpha\right) + inv\,\alpha$ (23–40)
The radius of the point of intersection of pinion	$r_{i1} = \dfrac{d_{b1}}{2}\tan\gamma_1 = r_1'\cos\alpha\tan\gamma_1$ (23–41)
The possible diameter of addendum circle of pinion	$d_{a1} \le d_{b1}\tan\gamma_1 = d_1'\cos\alpha\tan\gamma_1$ (23–42)
The addendum factor	$\xi_1 = 1 + x_1 \le \dfrac{z_1}{2}\,(\cos\alpha\tan\gamma_1 - 1)$ (23–43)

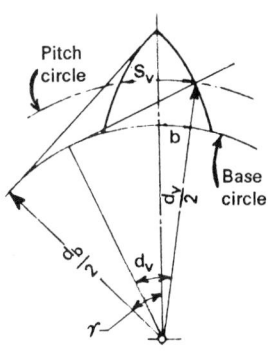

FIGURE 23–14 Pointed tooth.

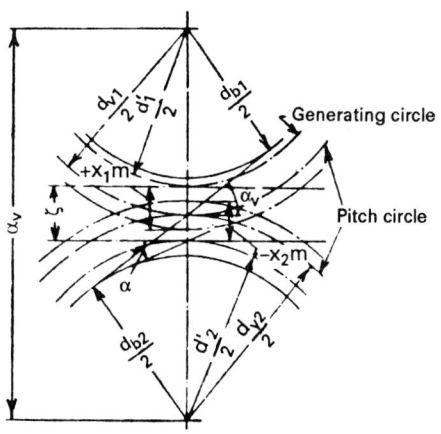

FIGURE 23–15 Gear correction on both pinion and gear.

Correction factors x_1 and x_2 (Fig. 23–15)

The circular pitch on the operating circle (Fig. 23–15)	$p_v = m\,\dfrac{\cos\alpha}{\cos\alpha_v}\,[(inv\,\gamma_1 - inv\,\alpha_v)z_1$
	$\qquad\qquad + (inv\,\gamma_2 - inv\,\alpha_v)z_2]$ (23–44a)
	$= p\,\dfrac{\cos\alpha}{\cos\alpha_v} = m\pi\,\dfrac{\cos\alpha}{\cos\alpha_v}$ (23–44b)
The relation between π, γ_1, γ_2, and α	$\pi = (inv\,\gamma_1 - inv\,\alpha_v)z_1 + (inv\,\gamma_2 - inv\,\alpha_v)z_2$ (23–45)
The expression for operating pressure angle	$inv\,\alpha_v = inv\,\alpha + \dfrac{(x_1 + x_2)2\tan\alpha}{z_1 + z_2}$ (23–46)
The sum of corrections on both pinion and gear	$x_1 + x_2 = \dfrac{inv\,\alpha_v - inv\,\alpha}{2\tan\alpha}\,(z_1 + z_2)$ (23–47)

Particular	Formula
The center distance between shafts without clearance	$a_v = \left(\dfrac{d_{v1} + d_{v2}}{2}\right) = \left(\dfrac{d_1' + d_2'}{2}\right) \dfrac{\cos\alpha}{\cos\alpha_v}$ (23–48a)
	$= a \dfrac{\cos\alpha}{\cos\alpha_v} = \dfrac{1}{2} m(z_1 + z_2) \dfrac{\cos\alpha}{\cos\alpha_v}$ (23–48b)
The distance ζ between the generating circles (Fig. 23–15)	$\zeta = a_v - a = \dfrac{m}{2}(z_1 + z_2)\left[\dfrac{\cos\alpha}{\cos\alpha_v} - 1\right]$ (23–49)
INTERNAL GEARS (FIG. 23–12) The tooth thickness at the operating circle of the internal gear	$s_{v2} = m \dfrac{\cos\alpha}{\cos\alpha_v}[\pi - (inv\,\gamma_2 - inv\,\alpha_v)z_2]$ (23–50)
The tooth thickness at the pitch circle ($\alpha_v = \alpha$)	$s_2' = m[\pi - (inv\,\gamma_2 - inv\,\alpha)z_2]$
	$= m\left(\dfrac{\pi}{2} - 2x_2\tan\alpha\right)$ (23–51)
The equation for angle γ_2 for the point of intersection of tooth profile (Fig. 23–12)	$inv\,\gamma_2 = \dfrac{1}{z_2}\left(\dfrac{\pi}{2} + 2x_2\tan\alpha\right) + inv\,\alpha$ (23–52)
The expression for operating pressure angle	$inv\,\alpha_v = inv\,\alpha + \dfrac{(x_2 - x_1)2\tan\alpha}{z_2 - z_1}$ (23–53)
The difference of correction factors	$x_2 - x_1 = (inv\,\alpha_v - inv\,\alpha)\dfrac{z_2 - z_1}{2\tan\alpha}$ (23–54)
For formula for main proportions of corrected spur gear, "S correction" $x_1 \neq \pm x_2 \neq 0$ and $\Sigma x = \mid x_1 \mid + \mid x_2 \mid \neq 0$	Refer to Table 23–8.
For formula for main proportions of corrected spur gears, "S_o correction" $x_1 + x_2 = 0$; $x_1 = -x_2$	Refer to Table 23–9.
For values of "S-corrected" gears	Refer to Table 23–10.
For gear correction factors x_1 and x_2	Refer to Table 23–11.
For axial shift factors λ_0 and total relative correction factor x_0 for various values of pressure angles	Refer to Table 23–12.
The center distance between shafts without clearance	$a_v = \left(\dfrac{d_{v2} - d_{v1}}{2}\right) = \left(\dfrac{d_2' - d_1'}{2}\right) \dfrac{\cos\alpha}{\cos\alpha_v}$ (23–55a)
	$= a \dfrac{\cos\alpha}{\cos\alpha_v} = \dfrac{1}{2} m(z_2 - z_1) \dfrac{\cos\alpha}{\cos\alpha_v}$ (23–55b)
The distance ζ between the generating circles	$\zeta = a_v - a = \dfrac{m}{2}(z_2 - z_1)\left[\dfrac{\cos\alpha}{\cos\alpha_v} - 1\right]$ (23–56)

Particular	Formula

Ratio of contact

EXTERNAL GEARS   The theoretical length of the line of action of any pair of true involute gears (Fig. 23–4)

$$g_t = \frac{1}{2}\left[\sqrt{d_{a1}^2 - d_1'^2 \cos^2 \alpha}\right.$$
$$\left. + \sqrt{d_{a2}^2 - d_2'^2 \cos^2 \alpha} - a \sin \alpha\right] \quad (23\text{--}57a)$$

$$= \frac{m}{2}\left[\sqrt{(z_1 + 2)^2 - z_1^2 \cos^2 \alpha}\right.$$
$$\left. + \sqrt{(z_2 + 2)^2 - z_2^2 \cos^2 \alpha} - (z_1 + z_2)\sin \alpha\right] \quad (23\text{--}57b)$$

The contact ratio

$$\varepsilon = \frac{g_t}{p_b} = \frac{g_t}{\pi m \cos \alpha} \quad (23\text{--}58a)$$

The maximum number of teeth in action at one time

$$z_a = \frac{g_t}{p \cos \alpha} = \frac{g_t}{\pi m \cos \alpha} \quad (23\text{--}58b)$$

The contact ratio for corrected gears

$$\varepsilon = \frac{1}{2\pi}\left[\left\{\left(\frac{z_2 + 2\xi_2}{\cos \alpha}\right)^2 - z_2^2\right\}^{1/2}\right.$$
$$\left. + \left\{\left(\frac{z_1 + 2\xi_1}{\cos \alpha}\right)^2 - z_1^2\right\}^{1/2} - (z_1 - z_2)\tan \alpha\right] \quad (23\text{--}59)$$

where $h_{a1} = \xi_1 m$; $h_{a2} = \xi_2 m$ and addendum factors $\xi_1 = 1 + x$; $\xi_2 = 1 - x$

The addendum factor ξ_s for tooth profile which has addendum $\xi_s m$ up to the point of intersection of the faces (Fig. 23–14)

$$\xi_s = \frac{z}{2}\left(\frac{\cos \alpha}{\cos \gamma} - 1\right) \quad (23\text{--}60)$$

The addendum factor ξ which has sufficient crest and clearance should be less than ξ_s i.e. $\xi < \xi_s$

The limiting condition for addendum for pinion (Fig. 23–15)

$$h_{a1} = m\xi_1 \le \zeta + h_{f1} - 0.2m = \zeta + (1 - x_2)m \quad (23\text{--}61)$$

The limiting condition for addendum for gear

$$h_{a2} = m\xi_2 \le \zeta + (1 - x_1)m \quad (23\text{--}62)$$

INTERNAL GEARS   The dedendum for intersection of faces (Fig. 23–12) taking the tooth of the external gear (Fig. 23–14) as the gap of the internal gear

$$h_{fs2} = \frac{mz_2}{2}\left(\frac{\cos \alpha}{\cos \gamma_2} - 1\right) \quad (23\text{--}63)$$

The actual dedendum

$$h_{f2} \le h_{fs2} = \frac{mz_2}{2}\left(\frac{\cos \alpha}{\cos \gamma_2} - 1\right) \quad (23\text{--}64)$$

Particular	Formula
The limiting condition for addendum of the external gear meshing with the internal gear	$h_{a1} \leq \zeta + h_{f2} - 0.2m = \zeta + m(1 + x_2)$ $\qquad$ (23–65) where $h_{f2} = m(1.25 + x_2)$
The contact ratio	$\varepsilon = \dfrac{1}{2\pi}\left[\left\{\left(\dfrac{z_1 + 2\xi_1}{\cos\alpha}\right)^2 - z_1^2\right\}^{1/2}\right.$ $\left. - \left\{\left(\dfrac{z_2 - 2\xi_2}{\cos\alpha}\right)^2 - z_2^2\right\}^{1/2} - (z_1 - z_2)\tan\alpha\right]$ $\qquad$ (23–66) For internal gears use α_v for α

UNDERCUTTING OF THE PINION TOOTH

The radius of the addendum circle of pinion to avoid undercutting of tooth	$r_{a1s} \leq \sqrt{(r_1' + r_{1s})^2 \sin^2\alpha + r_{1s}^2 \cos^2\alpha}$ $\qquad$ (23–67)
The radius of the addendum circle of the internal gear to avoid undercutting of tooth	$r_{a2} \geq [r_2'^2 + \{(r_{a1s}^2 - r_{1s}^2\cos^2\alpha)^{1/2} - r_{1s}\sin\alpha\}^2$ $\quad - 2r_2'\{(r_{a1s}^2 - r_{1s}^2\cos^2\alpha)^{1/2} - r_{1s}\sin\alpha\}\sin\alpha]^{1/2}$ $\qquad$ (23–68)
The contact ratio or engagement factor	$\varepsilon = [(r_{a1}^2 - r_1'^2\cos^2\alpha)^{1/2} - (r_{a2}^2 - r_2'^2\cos^2\alpha)^{1/2}$ $\quad + (r_2' - r_1')\sin\alpha]/(\pi m\cos\alpha)$ $\qquad$ (23–69)

RELATIVE RADIUS OF CURVATURE

The radius of curvature of involute profile of pinion (Fig. 23–16a)	$\rho_1 = \frac{1}{2}\, mz_1 y_1 \cos\alpha$ $\qquad$ (23–70)
The factor y_1	$y_1 = \left[\left\{\left(\dfrac{1 + (2\xi_1/z_1)}{\cos\alpha}\right)^2 - 1\right\}^{1/2} - \dfrac{2\pi}{z_1}\right]$ $\qquad$ (23–71)

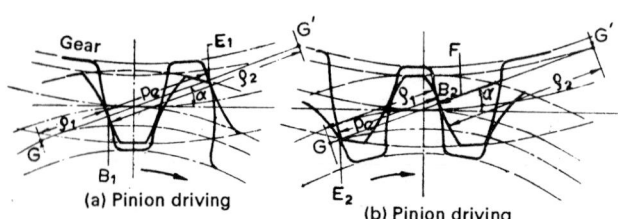

FIGURE 23–16   Fully loaded condition of gear teeth.

Particular	Formula
The radius of curvature of involute profile of gear (Fig. 23–16b)	$\rho_2 = \frac{1}{2}\, mz_2 y_2 \cos\alpha$ (23–72)
The factor y_2	$y_2 = \left[\left\{\left(\dfrac{1 \pm (2\xi_2/z_2)}{\cos\alpha}\right)^2 - 1\right\}^{1/2} - \dfrac{2\pi}{z_2}\right]$ (23–73)
	where negative sign is for internal gear
The relative radius of curvature	$\rho_0 = \dfrac{\rho_1 \rho_2}{\rho_1 \pm \rho_2}$ (23–74)
	where negative sign is used for internal gears
The relative radius of curvature with respect to pinion	$\rho_{01} = \frac{1}{2}\, mz_1 y_1 \cos\alpha \left[1 \pm \dfrac{z_1 y_1}{(z_1 \pm z_2)\tan\alpha_v}\right]$ (23–75)
The relative radius of curvature with respect to gear	$\rho_{02} = \pm\frac{1}{2}\, mz_2 y_2 \cos\alpha \left[1 - \dfrac{z_2 y_2}{(z_2 \pm z_1)\tan\alpha_v}\right]$ (23–76)
	where negative sign is used for z_2 as in internal gear

VELOCITIES AND WEAR (FIGS. 23–17 AND 23–18)

The relative velocity (Fig. 23–17) in case of external gears	$\begin{aligned} V_r &= W - W' = \rho_1\omega_1 - \rho_2\omega_2 &(23\text{–}77a)\\ &= e(\omega_1 + \omega_2) &(23\text{–}77b) \end{aligned}$
	where e = distance shown in Fig. 23–17 W, W' = relative velocities of point of contact along the tooth faces
Wear	$K_w = C\,\dfrac{V_r}{W} = C\left(\dfrac{\omega_1 + \omega_2}{\omega_1}\right)\left(\dfrac{e}{GM + e}\right)$ (23–78a)
	$= \infty$ when $GM = -e$ (23–78b)

SPECIFIC SLIDING
Specific sliding for pinion

	$v_{s1} = 1 - \dfrac{\omega_2 \rho_2}{\omega_1 \rho_1} = \dfrac{z_1 + z_2}{z_2 y}\,(y - \tan\alpha_v)$ (23–79)
When the portion of the involute of the pinion at the base circle is in contact, i.e., when $\rho_1 = 0$	$v_{s1} = -\infty$ (23–80)
When the portion of involute of the pinion and the gear at pitch point circle are in contact	$v_{s1} = 0$ (23–81)
When the point of contact of involute is at the base circle of gear	$v_{s1} = 1$ (23–82)

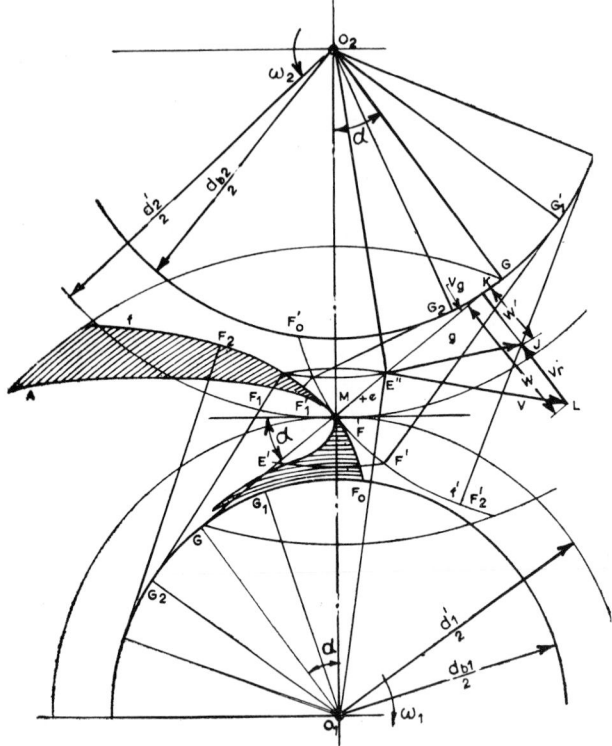

FIGURE 23–17 Relative velocities of rolling and sliding and wear in external involute gears.

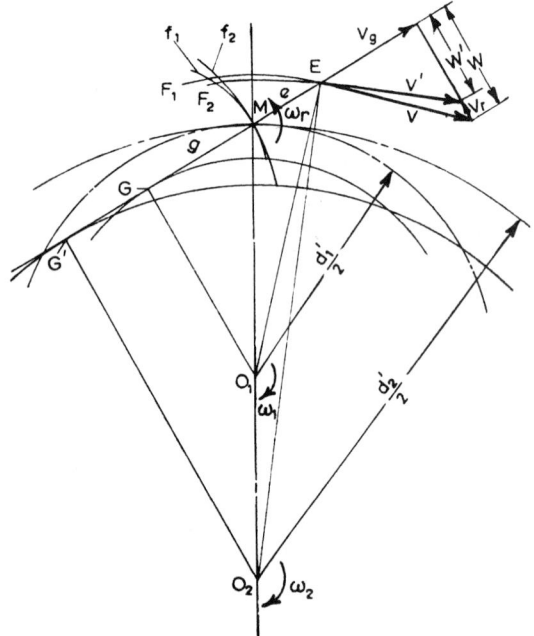

FIGURE 23–18 Relative velocities of rolling and sliding in internal involute gears.

Particular	Formula

Specific sliding for gear

$$v_{s2} = 1 - \frac{\omega_1 \rho_1}{\omega_2 \rho_2} = \frac{(z_1 + z_2)(\tan \alpha_v - y)}{(z_1 + z_2)\tan \alpha_v - z_1 y} \qquad (23\text{--}83)$$

When the portion of the involutes of the gear and of the pinion at the pitch point are in contact

$$v_{s2} = 0 \qquad (23\text{--}84)$$

When the portion of the involute at the base circle of gear is in contact, i.e., when $\rho_2 = 0$

$$v_{s2} = -\infty \qquad (23\text{--}85)$$

When the point of contact of the involute is at the base circle of the pinion

$$v_{s2} = 1 \qquad (23\text{--}86)$$

FORCE ANALYSIS ON GEAR-TOOTH PROFILE (FIGS. 23–19 TO 23-21)

The circumferential (tangential) force

$$F_t = \frac{M_t C_s}{r} \qquad (23\text{--}87a)$$

$$= \frac{9550 P C_s}{nr} \quad \text{SI} \qquad (23\text{--}87b)$$

where F_t in newton (N), P in kW, n in rpm, r in m and C_s is service factor taken from Table 23–13

$$= \frac{1000 P C_s}{\omega r} \quad \text{SI} \qquad (23\text{--}87c)$$

where F_t in N, P in kW, ω in rad/s, and r in m

$$= \frac{63,030 P C_s}{nr} = \frac{33,000 P C_s}{\nu}$$

US Customary System units $\qquad$ (23–87d)

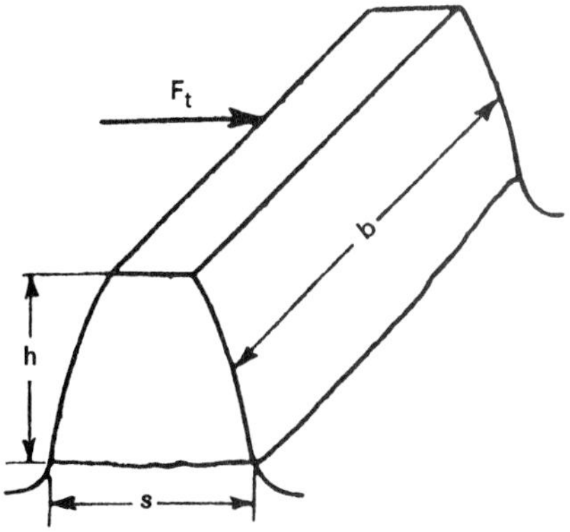

FIGURE 23–19 Load acting at the tip of a tooth profile.

where F_t in lbf, P in hp, r in in, n in rpm, and $\nu = $ pitch line velocity, ft/min

The normal tooth load

$$F_n = \frac{F_t}{\cos \alpha} = \frac{2 M_t C_s}{r \cos \alpha} = \frac{M_t(i \pm 1)C_s}{i a \cos \alpha} \qquad (23\text{--}88)$$

The unit normal load

$$F_n' = \frac{F_n}{b} = \frac{M_t(i \pm 1)C_s}{i a b \cos \alpha} \qquad (23\text{--}89)$$

The radial or separating load

$$F_r = F_n \sin \alpha \qquad (23\text{--}90)$$

For service factor C_s (2)

Refer to Table 23–13.

DESIGNING GEAR FOR BEAM OR BENDING STRENGTH

Calculation of module depending on root strength according to Lewis

$$m = \frac{F_t}{\pi K_v \sigma_o b y} \qquad (23\text{--}91)$$

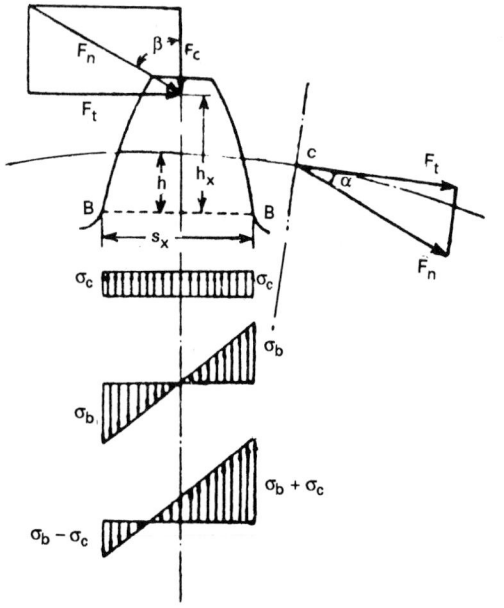

FIGURE 23–20 Forces acting on a tooth profile normal load at tip of tooth.

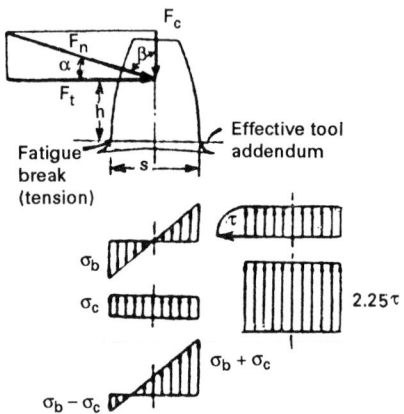

FIGURE 23–21 Forces acting on a tooth profile normal load at a point on tooth profile.

Particular	Formula
	where y = Lewis form factor from Tables 23–14 and 23–15
For full-size standard diametral pitches, basic rack profiles of ISO diametral pitches, and basic rack profiles of ISO modules	For standard modules m, refer to Tables 23–3 to 23–5 and Figs. 23–7 to 23–9.
Modified Lewis equation for calculating module (2)	$$m = \frac{F_t}{\sigma_o b Y K_v} \qquad (23\text{–}92)$$ where $Y = \pi y$ from Table 23–16.
For values of modified Lewis form factor, Y as per AGMA	Refer to Table 23–17.
Lewis equation for tangential tooth load in terms of circular pitch (2)	$$F_t = \sigma_o b y p K_v \qquad (23\text{–}93)$$
Modified Lewis equation for tangential tooth load in terms of modified Lewis form factor and diametral pitch	$$F_t = \frac{\sigma_o b Y K_v}{P} \qquad (23\text{–}94)$$ where σ_o = allowable static stress, MPa (psi) Refer to Table 23–18 for σ_o.

Particular	Formula

AGMA fundamental formula for calculating bending stress at transmitted load (F_t) for the design of spur gear teeth which takes care of shape and size of tooth, stress concentration at the root fillet of tooth, radial compressive load, maldistribution of load along the face width, surface finish, fatigue, single-tooth contact not occurring at the tip of the tooth, nonapplicability of simple cantilever beam theory, manufacturing accuracy, overload, quality, dynamic effect, and mounting effect. (3,4)

(*Note*: σ_b is the Greek letter used as symbol for bending stress in this book, whereas AGMA has used the lowercase letter s with subscript t for bending stress numbers. σ_b is used throughout this book for bending stress.)

$$\sigma_b = \frac{F_t K_a}{K_v} \frac{1}{mb} \frac{K_{sz} K_m K_B}{J} \quad \textbf{SI} \qquad (23\text{–}95)$$

where $J = Y/(K_{f\sigma} K_{LS})$ is the geometry factor for spur gear bending strength. This takes into account stress concentration at the root fillet radius $\rho_f = 0.35m$, the effect of tooth shape, and the position at which single-tooth contact takes place and sharing of load between one or more than one pair of teeth.

Refer to Tables 23–19 and 23–20 and Figs. 23–22 and 23–23 for geometric factor J for 20° and 25° pressure angles.

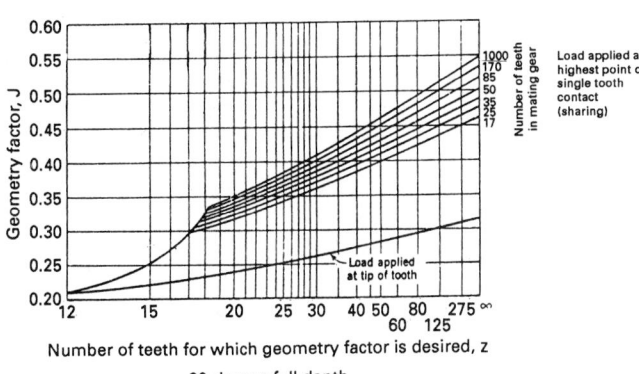

FIGURE 23–22 Geometry factor J for spur gears.

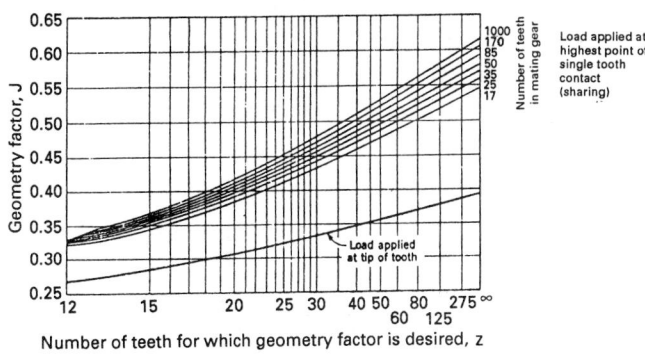

FIGURE 23–23 Geometry factor J for spur gears (from AGMA information sheet 225.01) $\rho_f = 0.35/\textbf{P} - 0.35$ m.

Particular	Formula
	K_a = application factor or overload correction factor taken from Table 23–21
	K_m = load distribution or mounting factor taken from Table 23–22
	K_v = dynamic or velocity factor for bending strength, C_v is the symbol adopted by AGMA for dynamic or velocity factor for use with surface contact stress (Hertzian contact stress) or pitting resistance formula. K_v is the symbol adopted by AGMA for dynamic or velocity factor for use with bending strength or bending stress number formula. But there is no difference between these two and numerical values of these two are also the same. Refer to Fig. 23–28 for K_v.
	K_{LS} = load-sharing ratio = 1 for spur gears
	K_B = rim thickness factor
	= 1 for unidirectional load, rigid backup gear
	= 1.4 for fully reversed load, even if the backup is rigid.
	$$\sigma_b = \frac{F_t K_a}{K_v}\frac{P}{b}\frac{K_{sz}K_m K_B}{J}$$
	US Customary System units (23–96)
The corrected fatigue strength or modified endurance limit of gear-tooth material	$\sigma_{sfd} = K_{sr}K_{sz}K_{ld}K_R K_T K_{me}K_\sigma \sigma'_{sfb} \geq \sigma_b$ (23–97)
	where
	σ'_{sfb} = endurance limit of rotating beam specimen or R. R. Moore endurance limit, MPa (psi)
	K_σ = stress concentration factor which has been included into geometry factor J; therefore, K_σ may be taken as 1 for gears
	$K_{sr}(e_{sr})$ = surface factor or surface coefficient from Fig. 5–3 for cut and ground gear teeth
	K_{sz} = size factor for rotation and bending
	$K_{sz} = \begin{cases} 1 & d \leq 2.8 \text{ mm } (0.11 \text{ in}) \\ (d/0.3)^{-0.1133} & 0.11 \text{ in} \leq d \leq 2 \text{ in} \\ (d/7.62)^{-0.1133} & 2.80 \text{ mm} \leq d \leq 50 \text{ mm} \end{cases}$
	d = diameter of round specimen (23–98)
	Also refer to Table 23–26A for size factor.
Expression for equivalent diameter for the case of rectangular cross-section gear teeth	$d_{eq} = 0.808\sqrt{(hb)}$ (23–99a)
	where
	h = gear-tooth thickness $= \dfrac{1}{2}p = \dfrac{\pi m}{2}$
	b = face width of gear tooth $\approx 3p \approx 3\pi m$

Particular	Formula
From Eq. (23–99), the approximate equivalent diameter of gear tooth in terms of module	$d_{eq} \approx 0.808\{3\pi m(\pi m/2)\}^{1/2} \approx \pi m$ (23–99b)
The reliability factor K_R for use in Eq. (23–97)	$\begin{aligned} K_R &= 0.7 - 0.15\log(1-R) \quad 0.9 \le R < 0.99 \\ &= 0.5 - 0.25\log(1-R) \quad 0.99 \le R \le 0.9999 \end{aligned}$ <div align="right">(23–100)</div> Refer also to Table 23–25A.
The temperature factor	$K_T = \begin{cases} 1 & T > 350°\text{C } (660°\text{F}) \\ 0.5 & 350°\text{C } (660°\text{F}) \le T \le 500°\text{C } (932°\text{F}) \end{cases}$ <div align="right">(23–101a)</div>
Temperature factor is also given by	$K_T = \dfrac{327}{234 + T}$ for $T > 70°\text{C } (160°\text{F})$ (23–101b) where T is temperature in °C (°F) $K_{ld} =$ load factor $\begin{cases} 1 & \text{axial loading } \sigma_{sut} > 1518 \text{ MPa } (220 \text{ kpsi}) \\ 0.923 & \text{axial loading } \sigma_{sut} \le 1518 \text{ MPa } (220 \text{ kpsi}) \\ 0.577 & \text{shear and torsion} \\ 1 & \text{bending} \end{cases}$ <div align="right">(23–101c)</div>
For miscellaneous effects factors for one-way bending of gear teeth (K_{me}) (5)	Refer to Fig. 23–27.
For one-way bending due to repeated fatigue load, the mean (σ_m) and alternating (σ_a) stress components	$\sigma_a = \sigma_m = \dfrac{\sigma_b}{2}$ (23–102) where $\sigma_b = \dfrac{F_t K_a}{K_v} \dfrac{1}{mb} \dfrac{K_{sz} K_m K_B}{J}$ **SI** $= \dfrac{F_t K_a}{K_v} \dfrac{P}{b} \dfrac{K_{sz} K_m K_B}{J}$ **US Customary System units**
The tooth bending stress as per modified Goodman relation	$\sigma_b = \dfrac{2\sigma_{sfb}\sigma_{sut}}{\sigma_{sut} + \sigma_{sfb}}$ (23–103)
The allowable working bending stress	$\sigma_{wab} = \dfrac{\sigma_{sab} K_L}{K_T K_R}$ (23–104)
The bending stress σ_b calculated by Eqs. (23–95) and (23–96) must be less than or equal to the allowable working bending stress number σ_{wab} as defined by Eq. (23–104)	$\sigma_b \le \sigma_{wab} = \dfrac{\sigma_{sab} K_L}{K_T K_R}$ (23–105) where $\sigma_{sab} =$ allowable bending stress number, MPa (kpsi) Refer to Fig. 23–24 and Table 23–23 for allowable bending strength or allowable bending stress numbers, MPa (psi)

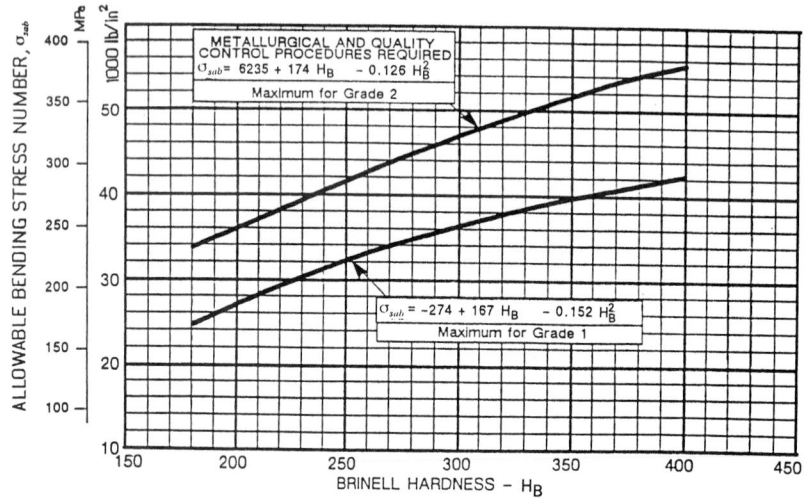

FIGURE 23–24 Allowable bending stress number for steel gears, σ_{sab}.

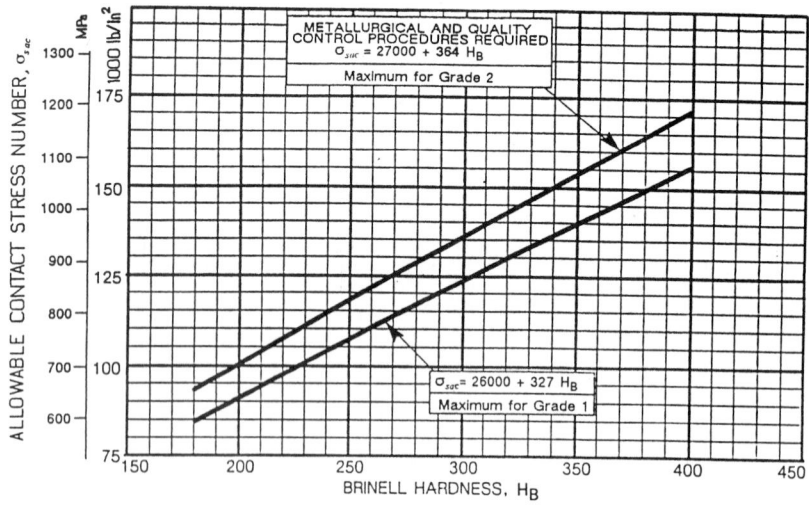

FIGURE 23–25 Allowable contact stress number for steel gears, σ_{sac}.
(ANSI/AGMA-2001-B88.)

Particular	Formula

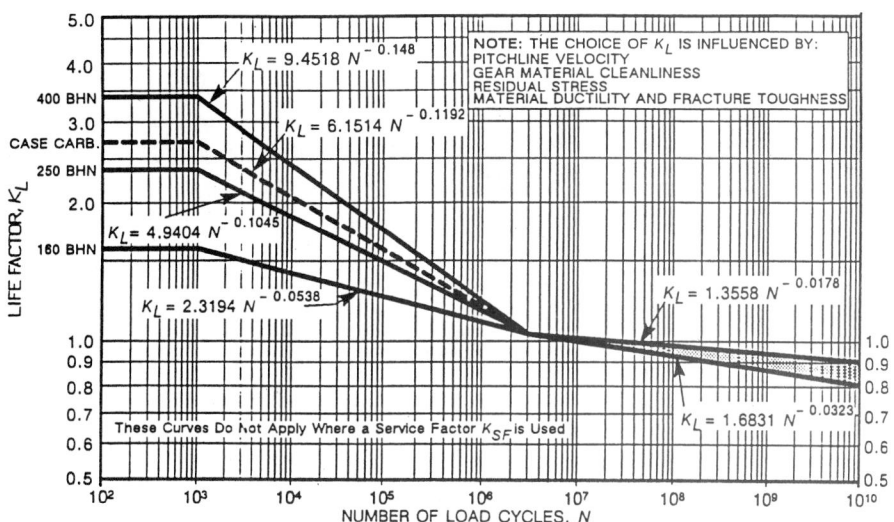

FIGURE 23–26 Bending strength life factor, K_L. (ANSI/AGMA 2001-B88.)

K_L = life factor for bending strength, from Fig. 23–26

K_T = temperature factor for bending, use 1.0 or more

K_R = reliability for bending strength, taken from Table 23–25B

Another modified Lewis equation for calculating module taking into consideration the geometry factor (J) and other factors as narrated under Eq. (23–95) as recommended by AGMA

$$m = \frac{F_t K_a K_m K_{sz} K_B}{\sigma_{sab} b J K_v} \quad \text{SI} \qquad (23\text{–}106)$$

Modified Lewis equation for calculating diametral pitch taking into consideration the geometry (J) and other factors as recommended by AGMA

$$P = \frac{\sigma_{sab} b J K_v}{F_t K_a K_m K_{sz} K_B} \quad \text{US Customary units}$$

$$(23\text{–}107)$$

where σ_{sab} = allowable bending strength, MPa (psi)

Refer to Table 23–23 and Fig. 23–24.

The AGMA equation for permissible power a gear set can transmit as based on fatigue bending strength of gear teeth of spur gear (4)

$$P_{ab} = \frac{n_1' d_1' K_v}{318} (m b_e) \left(\frac{J}{K_a K_m K_B K_{sz}} \right) \left(\frac{K_L \sigma_{sab}}{K_T K_R} \right)$$

$$\text{SI} \qquad (23\text{–}108)$$

where P_{ab} in kW; n_1' in rps (r/s); d_1', m, and b_e in m; and σ_{sab} in Pa.

Particular	Formula

$$P_{ab} = \frac{n_1 d_1' K_v}{19.1 \times 10^6} (mb_e) \left(\frac{J}{K_a K_m K_B K_{sz}} \right) \left(\frac{K_L \sigma_{sab}}{K_T K_R} \right)$$

SI   (23–109)

where P_{ab} in kW, n_1 in rpm, d_1' and b_e in mm; σ_{sab} in MPa, and m = metric module, nominal in plane of rotation, mm

$$P_{ab} = \frac{n_1 d_1' K_v}{126,000} \left(\frac{b_e}{P} \right) \left(\frac{J}{K_a K_m K_B K_{sz}} \right) \left(\frac{K_L \sigma_{sab}}{K_T K_R} \right)$$

US Customary units   (23–110)

where P_{ab} in hp; n_1 in rpm; d_1' and b_e in in; P in in^{-1} and σ_{sab} in psi

Another form of Lewis equation for calculation of module

$$m = \left(\frac{2M_t}{\lambda_t \pi^2 z y \sigma} \right)^{1/3}$$

(23–111)

where
M_t = torque on weaker member of gear set, N m (lbf in)
σ = stress $\geq$ allowable stress, MPa (psi)
λ_1 = $(b/\pi m)$ = 3 to 4 for ordinary service

FORM FACTORS

Lewis form factor

$$y = \frac{s_x^2}{6 \pi h_x m}$$

(23–112a)

$$= \frac{P s_x^2}{6 \pi h_x} = \frac{s_x^2}{6 p h_x}$$

(23–112b)

Also obtained from Tables 23–14 and 23–15.

American Gear Manufacturers Association (AGMA) equation for the Lewis form factor which takes into account the good quality of gears, stress concentration at root fillet, the effect of tooth shape and the position at which a single tooth contact takes place, and not at the tip of the tooth as assumed by Lewis, and sharing of load between one or more than one pair of teeth (Fig. 23–29) (4)

$$Y = \frac{1}{\dfrac{\cos \alpha_v}{\cos \alpha} \left[\dfrac{1.5}{x} - \dfrac{\tan \alpha_v}{s} \right] m_n}$$ **SI**   (23–113)

$$Y = \frac{1}{\dfrac{\cos \alpha_v}{\cos \alpha} \left[\dfrac{1.5}{x} - \dfrac{\tan \alpha_v}{s} \right]}$$

US Customary System units   (23–114)

where α_v = angle between the line of action of total normal load F_n and a perpendicular to the center line of the tooth at the highest point of single-tooth contact.

Values of Y are also taken from Table 23–17.

Particular	Formula
Lewis form factor	$y = 0.124 - \dfrac{0.684}{z}$ for 14.5° involute (23–115)
	$y = 0.154 - \dfrac{0.912}{z}$ for 20° involute (23–116)
	$y = 0.170 - \dfrac{0.95}{z}$ for 20° stub teeth (23–117)
Another equation for Lewis form factor	$y = 0.154 - \dfrac{1.23}{z} + \dfrac{3.38}{z^2}$ (23–118)
The form factor for the driving wheel ($z_2 \geq 30$)	$y = 0.46\left(1 + \dfrac{20}{z_2}\right)$ (23–119)
The form factor for the driven wheel	$y = 0.52\left(1 + \dfrac{20}{z_2}\right)$ (23–120)

CALCULATION OF STRESSES (FIG. 23–21)

The bending stress at width of roots	$\sigma_b = \dfrac{6F_t h}{bs^2}\dfrac{\sin\beta}{\sin\alpha}$ (23–121)
The compressive stress at width of roots	$\sigma_c = \dfrac{F_t}{bs}\dfrac{\cos\beta}{\cos\alpha}$ (23–122)
The shear stress	$\tau = \dfrac{F_t}{bs}\dfrac{\sin\beta}{\cos\alpha}$ (23–123)
The resultant bending stress on tension side of the tooth fillet according to Niemann (Fig. 23–21) (6, 7)	$\sigma_b' = \sqrt{(\sigma_b - \sigma_c)^2 + (\mu\tau)^2}$ (23–124) where $\mu = \sigma_{ba}/\tau_a$ σ_{ba} = allowable bending stress, MPa (psi) τ_a = allowable shear stress, MPa (psi)
Another form of Niemann's equation for the resultant bending stress on tension side of the tooth fillet	$\sigma_b' = \dfrac{F_t q}{bm}$ (23–125) where q = form factor for 20° pressure-angle gears, from Tables 23–27 and 23–28 and Eq. (23–126)
The form factor for use in Eq. (23–125)	$q = \left(\dfrac{m\sin\beta}{s\cos\alpha}\right) - \left\{\left(\dfrac{6h}{s} - \dfrac{1}{\tan\beta}\right)^2 + \mu^2\right\}^{1/2}$ (23–126) where $\mu = 2.5$ from tests and $\mu^2 = 6.25$
The form factors q for commercial gears when the force F_t acts at addendum circle (large pitch errors)	Refer to Table 23–29.

Particular	Formula
Calculation of module depending on root stress according to Niemann	$$m = \frac{F_t q}{b\sigma' b} \tag{23-127}$$
	$$= \left(\frac{2M_t q}{bz\sigma'_b}\right)^{1/3} = \left(\frac{2M_t q}{\lambda z\sigma'_b}\right)^{1/3} \tag{23-128}$$
	$$= 12.5\left(\frac{Pq}{\lambda z\omega\sigma'_d}\right)^{1/3} \quad \text{SI} \tag{23-129a}$$
	where m in m, P in kW, ω in rad/s, and σ'_d in N/m²
	$$= 6.83\left(\frac{Pq}{\lambda z n'\sigma'_d}\right)^{1/3} \quad \text{SI} \tag{23-129b}$$
	where m in m, P in kW, n' in rps (r/s), and σ'_d in Pa
	$$m = 26.5\left(\frac{Pq}{\lambda z n\sigma'_d}\right)^{1/3} \quad \text{SI} \tag{23-129c}$$
	where m in m, P in kW, n in rpm, and σ'_d in N/m²
Calculation of diametral pitch based on root stress according to Niemann	$$\frac{1}{P} = 50.14\left(\frac{Pq}{\lambda n z\sigma'_d}\right)^{1/3} \quad \text{US Customary System units}$$ $$\tag{23-130}$$ where P in in^{-1}, P in hp, n in rpm, σ'_d in psi

FACE WIDTH

The proper width of face in terms of circular pitch	$b = 3p$ to $4p$	(23-131)
The proper width of face in terms of module	$b = 3\pi m$ to $4\pi m$	
The proper width of face in terms of diametral pitch	$b > 9.5m;\ b < 12.5m$	(23-132)
	$b > 9.5/P;\ b < 12.5/P$	(23-133)

DYNAMIC OR VELOCITY FACTORS

The dynamic factor or velocity factor or Barth's formula for ordinary cut gears running with pitch line velocity up to 7.5 m/s (1550 ft/min)

$$K_v = C_v = \frac{3}{3 + v_m} \quad \text{for } v_m = 7.5 \text{ m/s} \quad \text{SI} \tag{23-134a}$$

where v_m in m/s in SI units is used for calculation of C_v and K_v throughout this chapter (see also Figs. 23–28 and 23–30)

$$= \frac{600}{(600 + v_m)} \quad \text{for } v_m = 1500 \text{ ft/min}$$

$$\text{US Customary units} \tag{23-134b}$$

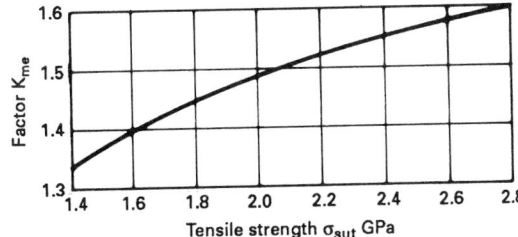

FIGURE 23–27 Miscellaneous-effects factors for one-way bending of gear teeth. Use $K_{me} = 1.33$ for values of σ_{Sut} less than 1.4 GPa. (*J. E. Shigley*, Mechanical Engineering Design, Metric Edition, *McGraw-Hill, 1986.*)

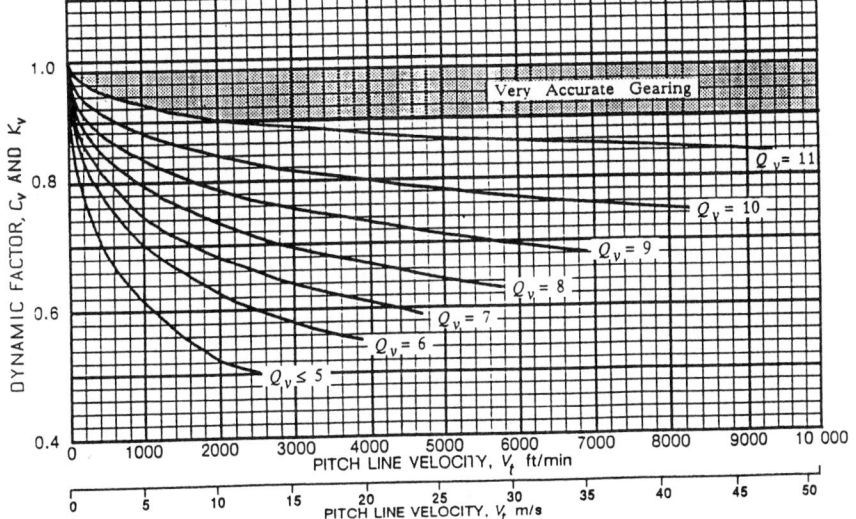

FIGURE 23–28 Dynamic factors, C_V and K_V. (*Courtesy of AGMA 2001-B88.*)

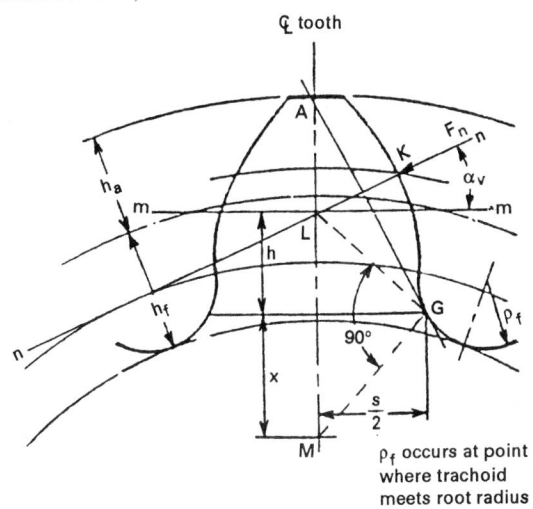

ρ_f occurs at point where trachoid meets root radius

NOTE: Shown in the normal plane through the pitch point

FIGURE 23–29 Tooth form factor layout with load sharing.

Particular	Formula
	where ν_m in ft/min in fps units is used for calculation of C_v and K_v throughout this chapter
The dynamic factor or velocity factor or Barth's formula for carefully cut gears	$K_v = C_v = \dfrac{4.5}{4.5 + \nu_m}$ for $\nu_m \leq 12.5$ m/s **SI** (23–135a)
	$= \dfrac{900}{900 + \nu_m}$ for $\nu_m \leq 2500$ ft/min **US Customary System units** (23–135b)
The dynamic factor or velocity factor or Barth's formula for accurately cut and ground metallic gears	$K_v = C_v = \dfrac{6}{6 + \nu_m}$ for ν_m 6.5 m/s to 20 m/s **SI** (23–136a)
	$= \dfrac{1200}{1200 + \nu_m}$ for ν_m 1200 to 4000 ft/min **US Customary System units** (23–136b)
The dynamic or velocity factor or Barth's formula for hardened steel ground and lapped in precision gears as per AGMA recommendation	$K_v = C_v = \dfrac{5.6}{5.6 + \sqrt{\nu_m}}$ for $\nu_m > 20$ m/s **SI** (23–137a)
	$= \dfrac{78}{78 + \sqrt{\nu_m}}$ for $\nu_m > 4000$ ft/min **US Customary System units** (23–137b)
The dynamic or velocity factor or Barth's formula for a commercially hobbed or shaved teeth as per AGMA recommendation	$K_v = C_v = \dfrac{3.6}{3.6 + \sqrt{\nu_m}}$ **SI** (23–138a)
	$= \dfrac{50}{50 + \sqrt{\nu_m}}$ **US Customary units** (23–138b)
The dynamic or velocity factor for nonmetallic gears as per AGMA recommendation	$K_v = C_v = \dfrac{0.75}{1 + \nu_m} + 0.07625$ **SI** (23–139a)
	$= \dfrac{150}{200 + \nu_m} + 0.25$ **US Customary units** (23–139b)
Another formula for the dynamic or velocity factor for nonmetallic gears	$K_v = C_v = \dfrac{1 + (\nu_m/4)}{1 + \nu_m}$ **SI** (23–140a)
	$= \dfrac{200 + (\nu_m/4)}{200 + \nu_m}$ **US Customary units** (23–140b)

Particular	Formula
The dynamic or velocity factor for the gears having high precision shaved or ground teeth as per AGMA recommendation	$K_v = C_v = \left(\dfrac{5.6}{5.6 + \sqrt{\nu_m}}\right)^{1/2}$ **SI** (23–141a) $= \left(\dfrac{78}{78 + \sqrt{\nu_m}}\right)^{1/2}$ **US Customary units** (23–141b)

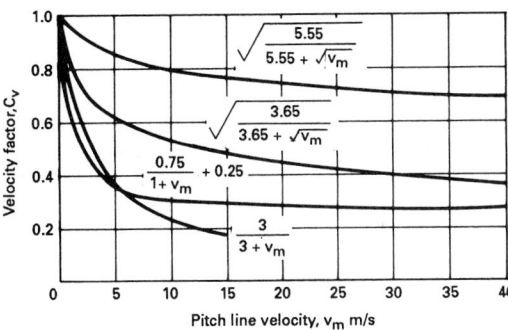

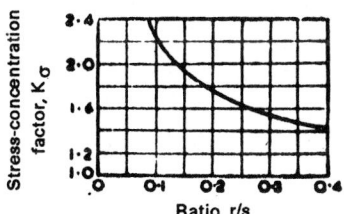

FIGURE 23–30 Dynamic or velocity factors $K_V = C_v$. (*American Gear Manufacturers Association.*)

For dynamic factor or velocity factor for various classes of gears	Refer to Figs. 23–28 and 23–30.

STRESS-CONCENTRATION FACTOR

For stress concentration at the root of gear tooth	Refer to Figs. 23–31 to 23–33.
Stress-concentration factor by photoelastic data (Fig. 23–32)	$K_\sigma = \dfrac{(\sigma_{\max} \text{ by photoelastic analysis})}{\left[\sigma_{nom} = \left(\dfrac{6F_t h}{bs^2} - \dfrac{F_t \tan \alpha}{bs}\right)\right]}$ (23–142)
Stress-concentration factor as per photoelastic data of Dolan and Broghamer on tension side of gear-tooth fillet (Fig. 23–33) (8)	$K_\sigma = 0.18 + \left(\dfrac{s}{r_f}\right)^{0.15}\left(\dfrac{s}{h}\right)^{0.45}$ for 20° involute teeth (23–143) $K_\sigma = 0.22 + \left(\dfrac{s}{r_f}\right)^{0.20}\left(\dfrac{s}{h}\right)^{0.40}$ for 14.5° involute teeth (23–144) $K_\sigma = 0.14 + \left(\dfrac{s}{r_f}\right)^{0.11}\left(\dfrac{s}{h}\right)^{0.50}$ for 25° involute teeth (23–145)

FIGURE 23–31 Influence of fillet–radius on stress concentration at roots of gear teeth.

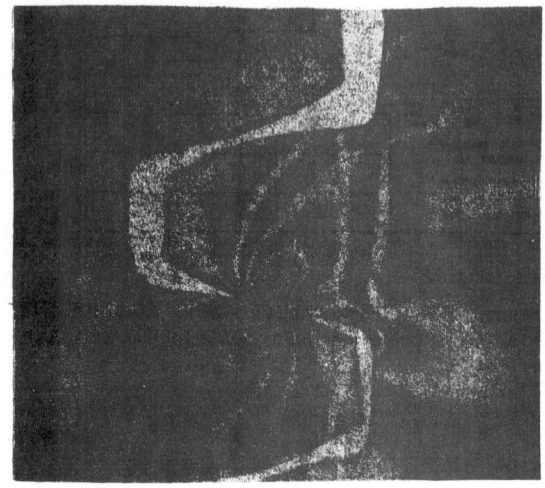

FIGURE 23–32 Fringe pattern of a gear tooth showing stress concentration at root. (*From the work of K. Lingaiah,* Fringe Pattern of Gear-Teeth Showing Stress Concentration at Root and Contact Point, *Bangalore University, 1973.*)

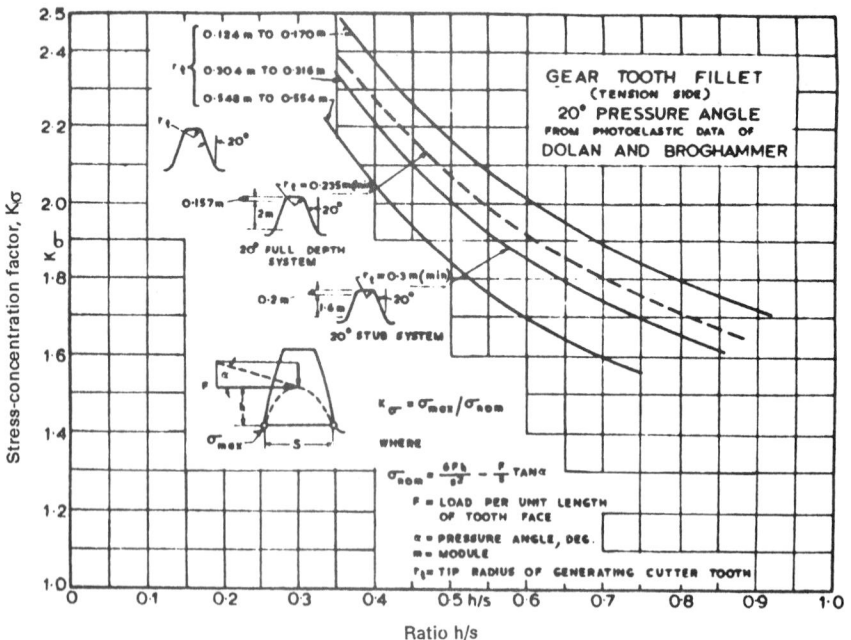

FIGURE 23–33 Stress-concentration factor (K_σ) vs. ratio (h/s) for $20°$ pressure angle tooth profile of involute gear. (*Courtesy R. E. Peterson,* Stress Concentration Factors, John Wiley and Sons, 1974)

Particular	Formula
For stress-concentration factors	Refer to Table 23–30.

DESIGNING OF GEARS FOR DYNAMIC LOADING

The fundamental Buckingham equation for dynamic load (11)

$$F_d = F_t + \sqrt{F_a(2F_2 - F_a)} \tag{23–146}$$

The acceleration load

$$F_a = \frac{F_1 F_2}{F_1 + F_2} \tag{23–147}$$

The average force required to accelerate the masses when they are considered as absolutely rigid

$$F_1 = Hm\nu_m^2 \tag{23–148}$$

The value of H

$$H = C_1\left(\frac{1}{r_1} + \frac{1}{r_2}\right) \tag{23–149}$$

$C_1 = $ 2.55 (0.00086) for 14.5° involute gears
$\quad = $ 3.55 (0.00120) for 20° involute gears
$\quad = $ 4.55 (0.00153) for 25° involute gears

The effective mass

$$m = \frac{m_1 m_2}{m_1 + m_2} \tag{23–150}$$

The force required to deform the teeth

$$F_2 = \left(\frac{f}{d} + 1\right)F_t = bC + F_t \tag{23–151}$$

where b in m (in), C in N/m (lbf/in), F_2 and F_t in N (lbf), f in m or mm (in), and d in m (in)

For maximum allowable error in action between gears

Refer to Table 23–31 for f; refer to Table 23–32 for values of C.

The constant for use in Eq. (23–151)

$$C = \frac{f}{C_2\left[\dfrac{1}{E_1} + \dfrac{1}{E_2}\right]} \tag{23–152}$$

where C in N/m (lbf/in); E_1 and E_2 in N/m^2 (lbf/in^2), and f in m or mm (in)

$C_2 = $ 9.345 for 14.5° involute gears
$\quad = $ 9.000 for 20° full-depth involute gears
$\quad = $ 8.700 for 20° stub-tooth gears

The deformation factor

$$d = \frac{C_2 F_t}{b}\left(\frac{1}{E_1} + \frac{1}{E_2}\right) \tag{23–153}$$

The modified dynamic load equation according to Buckingham (11)

$$F_d = F_t = F_i \tag{23–154}$$

$$= F_t = \frac{2l\nu_m(F_t + bC)}{21\nu_m + \sqrt{F_t + bC}} \quad \textbf{SI} \tag{23–155}$$

Particular	Formula

where F_d and F_t in N, ν_m in m/s, C in N/m, and h in m

$$= F_t + \frac{0.05\nu_m(Cb + F)}{0.05\nu_m + \sqrt{Cb + F_t}} \quad \text{US Customary units}$$

(23–156)

where F_d and F_t in lbf; ν_m in fpm (ft/min); C in lbf/in, and b in in. C is a factor that depends on the machining error taken from Table 23–32

For formula to obtain constant C

$$C = \frac{kE_1E_2}{E_1 + E_2}$$

(23–157)

where C in N/m (lbf/in)

$\dfrac{1}{C_2} = k = 0.107f$ for 14.5° full-depth involute teeth

$\qquad = 0.111f$ for 20° full-depth involute teeth

$\qquad = 0.115f$ for 20° stub teeth

f = expected error in mm (in) and taken from Table 23–31 and Figs. 23–34 and 23–35

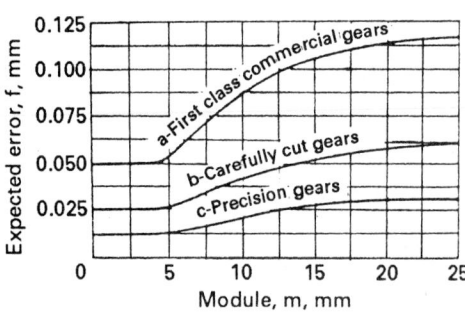

(a)

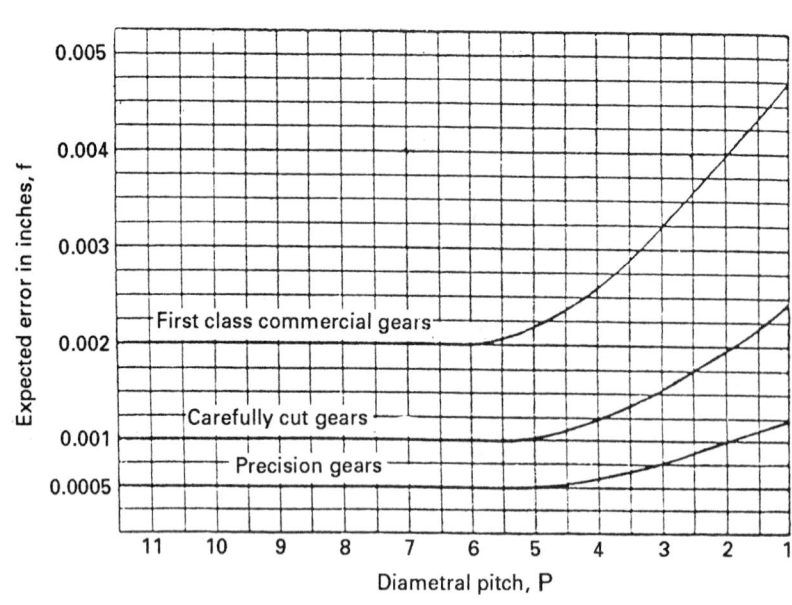

(b)

FIGURE 23–34 Expected errors in tooth profiles.

Particular	Formula

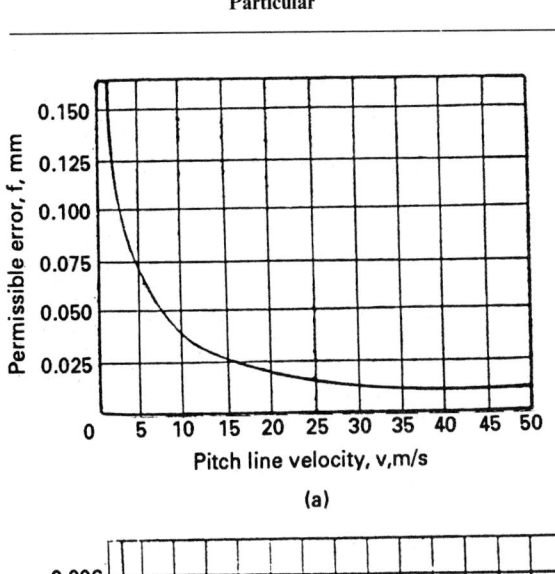

(a)

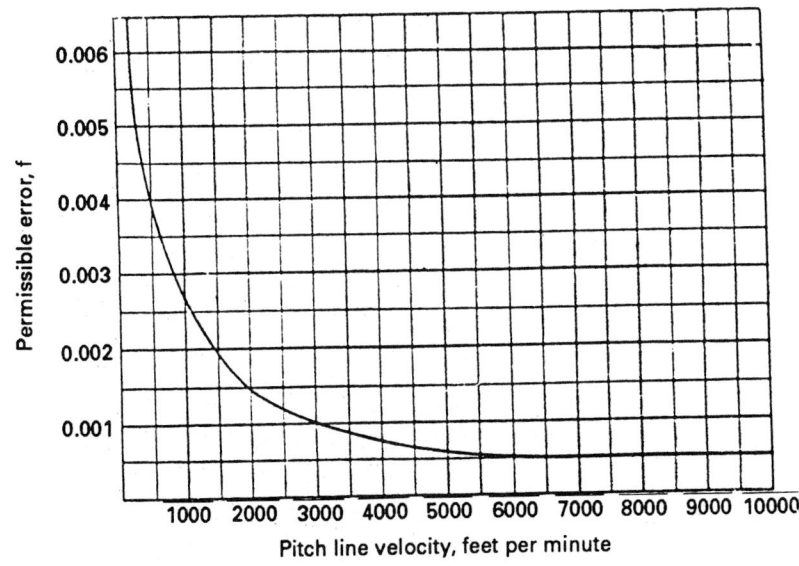

(b)

FIGURE 23–35   Maximum permissible error in gear-tooth profiles.

The endurance strength

$$F_t = \sigma_{sf} b Y m \qquad (23\text{–}158)$$

For endurance limit σ_{sf}, refer to Table 23–33.

Margin of safety

$$n' = \frac{F_f}{F_d} - 1 \qquad (23\text{–}159)$$

For values of F_f/F_d, refer to Table 23–34.

Particular	Formula

DESIGN BASED ON SURFACE DURABILITY

The limiting load in surface durability formula or wear load formula

$$F_w = d_1 bQK = mz_1 bQK > F_d \qquad (23\text{–}160)$$

The load stress factor K used in Eq. (23–160)

$$K = 1.43 \frac{\sigma_{sac}^2 \sin \alpha}{E_o} \qquad (23\text{–}161)$$

and also taken from Table 23–37B.

The equivalent Young modulus used in Eq. (23–161)

$$E_o = \frac{2E_1 E_2}{E_1 + E_2} \qquad (23\text{–}162)$$

The ratio factor Q

$$Q = \frac{2z_2}{z_1 + z_2} = \frac{2d_2}{d_1 + d_2} \qquad (23\text{–}163)$$

AGMA fundamental formula for pitting resistance for calculating surface compressive (Hertzian contact stress) stress or contact stress number as recommended by AGMA (4)

$$\sigma_c^* = C_p \left[\frac{F_t C_a}{C_v} \frac{C_{sz}}{bd_1'} \frac{C_m C_{sr}}{I} \right]^{1/2} \qquad (23\text{–}164)$$

The elastic coefficient used in Eq. (23–164)

$$C_p = \left\{ \frac{k}{\pi \left(\dfrac{1 - v_1^2}{E_1} + \dfrac{1 - v_2^2}{E_2} \right)} \right\}^{1/2} \qquad (23\text{–}165)$$

where C_p has units of $\sqrt{\text{MPa}}$ or $\sqrt{\text{psi}}$. (For values of C_p, refer also to Table 23–35.)

k = 1 for most spur, helical, and herringbone gears
= 1.5 for most bevel gears
C_v = dynamic factor or velocity factor from Fig. 23–28

The geometry factor I for use in Eq. (23–164)

$$I = \frac{\sin 2\alpha}{4} \left(\frac{i}{i+1} \right) \quad \text{for external gears} \qquad (23\text{–}166)$$

$$= \frac{\sin 2\alpha}{4} \left(\frac{i}{i-1} \right) \quad \text{for internal gears} \qquad (23\text{–}167)$$

where $i = (d_2/d_1) = (n_2/n_1)$

Refer to Fig. 23–36 for geometry factor for spur gear, I for standard center distance.

Note: σ_c is the Greek letter used as the symbol for contact stress member or surface fatigue strength in this book, whereas AGMA has used the lowercase letter s with subscript c for contact stress numbers. σ_c is used throughout this book for contact stress number.

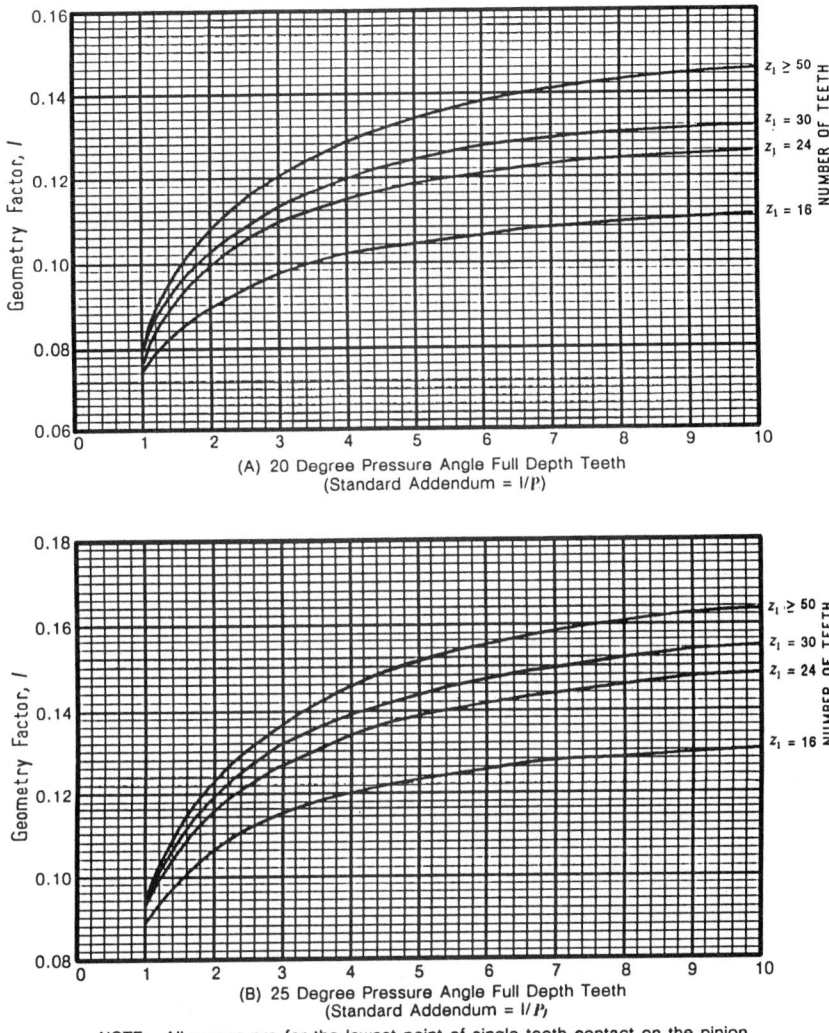

FIGURE 23–36   External spur pinion geometry factor, I.
(*Courtesy of ANSI/AGMA 6010-E88.*)

Particular	Formula

SURFACE FATIGUE STRENGTH

The limiting stress for surface fatigue of steel may be estimated from Eq. (23–168)

$\sigma_{fac} = (2.75H_B - 69)$ MPa   **SI**   (23–168a)
$\sigma_{fac} = (0.4H_B - 10)$ kpsi   **US Customary units**

(23–168b)
where H_B = Brinell hardness number of the softer of the two contacting materials

For hardness combinations of pinion and gear

Refer to Table 23–39C.

Particular	Formula
The corrected surface endurance limit or allowable working contact stress as recommended by AGMA	$\sigma_{wac} = \dfrac{C_L C_H}{C_T C_R} \sigma_{sac}$ (23–169)

where
C_H = hardness ratio factor = 1 for spur gear
(Also refer to Figs. 23–37 and 23–38 for C_H.)
C_T = temperature factor = 1 for T < 121°C
C_R = reliability factor taken from Table 23–25B
C_L = life factor taken from Fig. 23–39

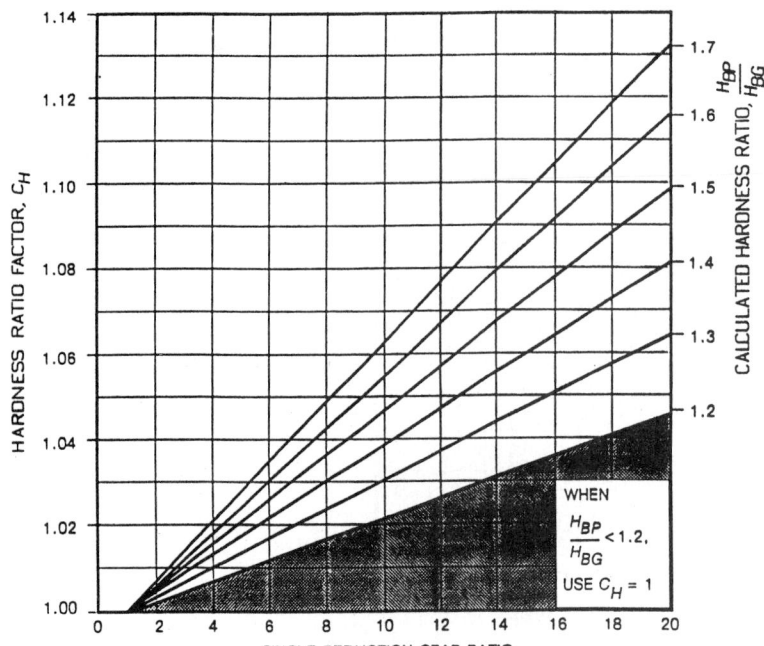

WHEN
$\dfrac{H_{BP}}{H_{BG}} < 1.2,$
USE $C_H = 1$

FIGURE 23–37 Hardness ratio factor, C_H (through-hardened). (*ANSI/AGMA 2001-B88.*)

The surface compressive stress (Hertzian contact stress) or pitting resistance or surface durability stress calculated by Eq. (23–164) must be less than or equal to the allowable working contact stress σ_{wac}.

$$\sigma_C \le \sigma_{wac} = \dfrac{C_L C_H}{C_T C_R} \sigma_{sac} \quad (23\text{–}170)$$

AGMA equation for maximum permissible power a gear set can transmit as based on surface durability of teeth of spur gear (i.e., AGMA power rating formula for pitting resistance)

$$P_{ac} = \dfrac{n_1' b}{318} \left(\dfrac{I C_v}{C_{sr} C_{sz} C_a C_m} \right) \left[\dfrac{\sigma_{sac} d_1'}{C_p} \dfrac{C_L C_H}{C_T C_R} \right]^2 \quad \textbf{SI}$$

(23–171)

where P_{ac} in kW, n_1' in rps, σ_{sac} in Pa, C_p in $\sqrt{\text{Pa}}$, d_1' and b in m

$$= \dfrac{n_1 b}{126,000} \left(\dfrac{I C_v}{C_{sr} C_{sz} C_a C_m} \right) \left[\dfrac{\sigma_{sac} d_1'}{C_p} \dfrac{C_L C_H}{C_T C_R} \right]^2$$

US Customary units   (23–172)

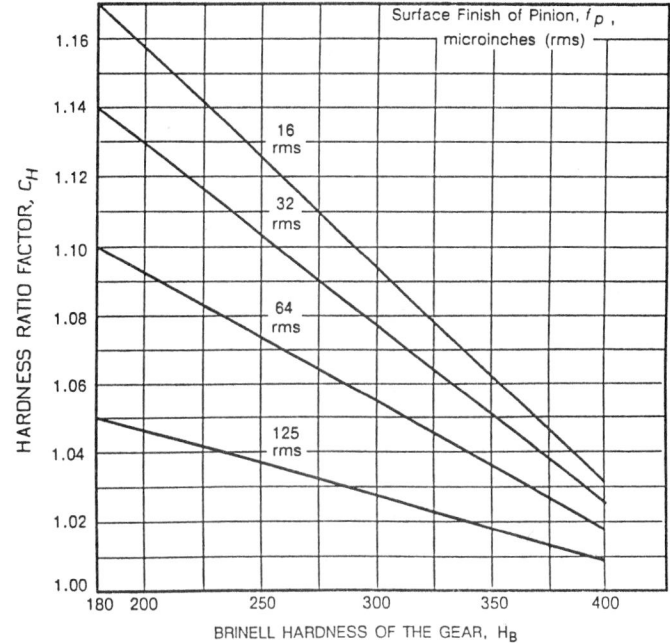

FIGURE 23–38 Hardness ratio factor, C_H (surface-hardened pinions). (*ANSI/AGMA 2001-B88.*)

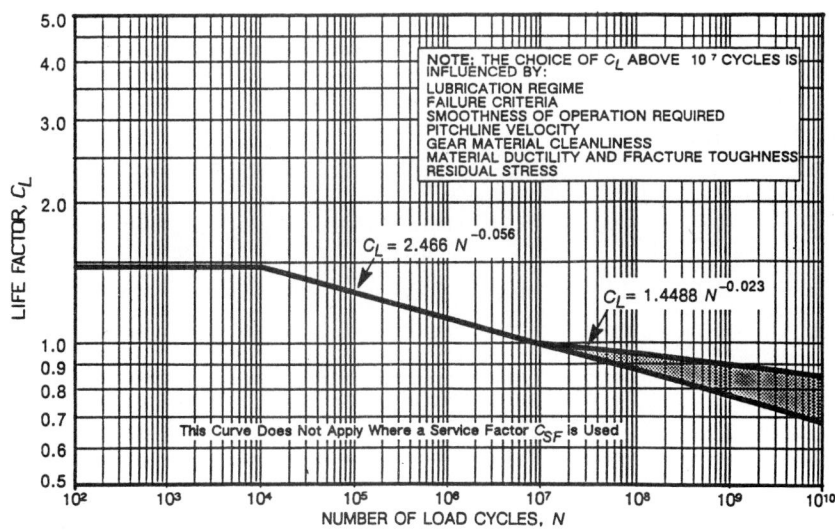

FIGURE 23–39 Pitting resistance life factor, C_L. (*ANSI/AGMA 2001-B88.*)

Particular	Formula

where P_{ac} in hp, n_1 in rpm, σ_{sac} in psi, C_p in $\sqrt{\text{psi}}$, d_1' and b in in

$$= \frac{n_1 b}{19.1 \times 10^6} \left(\frac{IC_v}{C_{sr}C_{sz}C_aC_m} \right) \left[\frac{\sigma_{sac}d_1'C_LC_H}{C_pC_TC_R} \right]^2 \quad \textbf{SI}$$

$$(23\text{–}173)$$

where P_{ac} in kW, n_1 in rpm, σ_{sac} in MPa, C_p in $\sqrt{\text{MPa}}$, d_1' and b in mm

For allowable contact stress number σ_{sac} refer to Table 23–24 and Fig. 23–25.

NIEMANN'S METHOD (6,7)

Hertz equation for maximum surface compressive stress between two loaded teeth with instantaneous values of the radii of curvatures ρ_1 and ρ_2 on the pinion and gear-tooth profiles, respectively, at the point of contact for face width b

$$\sigma_{c(max)} = 0.418 \left(\frac{F_n E_o}{b\rho_{o1}} \right)^{1/2} \qquad (23\text{–}174)$$

where ρ_{o1} = relative radius of curvature with reference to pinion

$$= \frac{\rho_1 \rho_2}{\rho_2 \pm \rho_1} \qquad (23\text{–}175)$$

The load at the point of contact according to Niemann (Figs. 23–20 and 23–21)

$$F_n = 2b\rho_{o1}\sigma_{ar} \qquad (23\text{–}176)$$

The tangential load on tooth

$$F_t = 2b\rho_{o1}\sigma_{ar}\cos\alpha \qquad (23\text{–}177)$$

The torque transmitted by pinion

$$M_t = bz_1 m\rho_{o1}\sigma_{ar}\cos\alpha \qquad (23\text{–}178)$$

The power transmitted according to Niemann

$$P = \frac{bz_1 m\omega\rho_{o1}\sigma_{ar}\cos\alpha}{1000} \quad \textbf{SI} \qquad (23\text{–}179)$$

where P in kW; b, m, and ρ_{o1} in m; ω in rad/s; and σ_{ar} in Pa

$$= \frac{bz_1 mn\rho_{o1}\sigma_{ar}\cos\alpha}{9550} \quad \textbf{SI} \qquad (23\text{–}180)$$

where P in kW; b, m, and ρ_{o1} in m; n in rpm; and σ_{ar} in Pa

$$= \frac{bz_1 mn'\rho_{01}\sigma_{ar}\cos\alpha}{159.2} \quad \textbf{SI} \qquad (23\text{–}181)$$

where P in kW; b, m, and ρ_{o1} in m; n' in rps; and σ_{ar} in Pa

$$P = \frac{bz_1(1/\textbf{P})n\rho_{o1}\sigma_{ar}\cos\alpha}{63,030}$$

US Customary System units $\qquad (23\text{–}182)$

where P in hp, b and ρ_{o1} in in, n in rpm, σ_{ar} in psi, and $\textbf{P}$ in in^{-1}

Particular	Formula
The admissible surface pressure	$$\sigma_{ca} = 2.86 \, \frac{\sigma_{c(max)}^2}{E_o} = \frac{F_t}{2b\rho_o \cos \alpha} \qquad (23\text{--}183)$$
	$$= \frac{\sigma_u}{n'} \, Y_1 Y_2 \qquad (23\text{--}184)$$
	where n' = safety factor ≥ 1.25
	$Y_1 = 1.5$ for gear meshes with another gear made of cast iron
	$ = 1$ for gear meshes with another gear not made of cast iron
The factor Y_2 for use in Eq. (23–184), which takes care of lubrication	Refer to Table 23–36.
For surface limit pressure, σ_u, for use in Eq. (23–184)	Refer to Table 23–37.
Another form of the equation for dangerous surface pressure σ_{c1} acting on the pinion	$$\sigma_{c1} = \frac{F_t}{2br_1 y_{e1}} \qquad (23\text{--}185)$$
The factor y_{e1} for use in Eq. (23–185)	$$y_{e1} = \frac{\rho_{o1}}{r_1} \cos \alpha \qquad (23\text{--}186)$$
	$$= y_1 \cos^2 \alpha \left[1 \pm \frac{z_1 y_1}{(z_2 \pm z_1) \tan \alpha_v} \right] \qquad (23\text{--}187)$$
The equation for dangerous surface pressure σ_{c2} acting on the gear	$$\sigma_{c2} = \frac{F_t}{2br_1 y_{e2}} \qquad (23\text{--}188)$$
The factor y_{e2} for use in Eq. (23–188)	$$y_{e2} = \frac{\rho_{o2}}{r_1} \cos \alpha \qquad (23\text{--}189a)$$
	$$= \pm \frac{z_2}{z_1} \, y_2 \cos^2 \alpha \left[1 - \frac{z_2 y_2}{(z_2 \pm z_1) \tan \alpha_v} \right]$$ $$(23\text{--}189b)$$
For values of y_{e1} and y_{e2} for standard external gears ($\alpha = 20°$, $\xi = 1$, and $i = 1$)	Refer to Table 23–38.
The empirical formula for allowable surface compressive stress for steel gears on the basis of endurance limit	$$\sigma_{ac} = 1.73\sigma_f \qquad (23\text{--}190)$$

LIFE BASED ON SURFACE STRENGTH

Life in number of cycles for constant loading	$$N = 60nT \qquad (23\text{--}191)$$ where T = life in hours
Life in number of cycle for variable loading	$$N = \frac{60}{M_{ms1}^3} \sum M_{msj}^3 T_j n_j \qquad (23\text{--}192)$$

Particular	Formula
The expression for equivalent number of cycles at the maximum sustained torque when the load consists of a maximum sustained gear or pinion torque M_{tms1} acting for T_1 hours at a mean speed of n_1 and smaller sustained torques M_{t2}, M_{t3}, etc., acting for T_2, T_3, etc. hours at mean speed n_2, n_3, etc.	$N_e = 60\left[T_1 n_1 + T_2 n_2\left(\dfrac{M_{t2}}{M_{tms1}}\right)^3\right.$ $\left. + T_3 n_3\left(\dfrac{M_{t3}}{M_{tms1}}\right)^3 + \cdots, \text{etc.}\right]$ (23–193)
For recommended speeds of gears	Refer to Table 23–39A.
For backlash for gears	Refer to Table 23–39B.

LOAD RATING OF MACHINE CUT SPUR AND HELICAL GEARS AS PER INDIAN STANDARDS (12)

The allowable tangential load for strength in Newton per meter (N/m) or pound force per inch (lbf/in) of face width for suitably lubricated gears

$$F_{tb}^* = f_{nb} Y K_b m \qquad \textbf{SI} \qquad (23\text{--}194a)$$

where F_{tb} in N/m; K_b in Pa; m in m

$$= \frac{f_{nb} Y K_b}{P} \qquad \textbf{USCSU} \qquad (23\text{--}194b)$$

where F_{nb} in lbf/in; K_b in psi; P in in^{-1}

f_{nb} = speed factor for strength for appropriate running time from Fig. 23–40

Y = strength factor from Fig. 23–43

For internal spur gear multiply the above by $(1 + 3/z_i)$

The strength factor for helical gears other than 30° is obtained from Fig. 23–43 and shall be multiplied by $1.33 \cos^2 \beta$

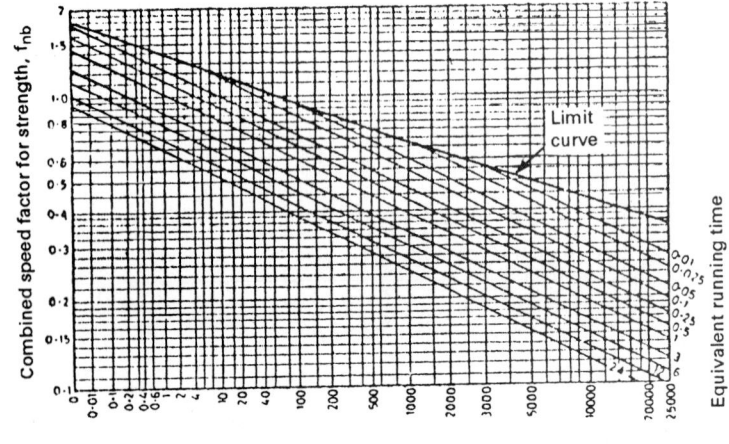

FIGURE 23–40 Combined speed and running-time factors for spur and helical gears for strength.

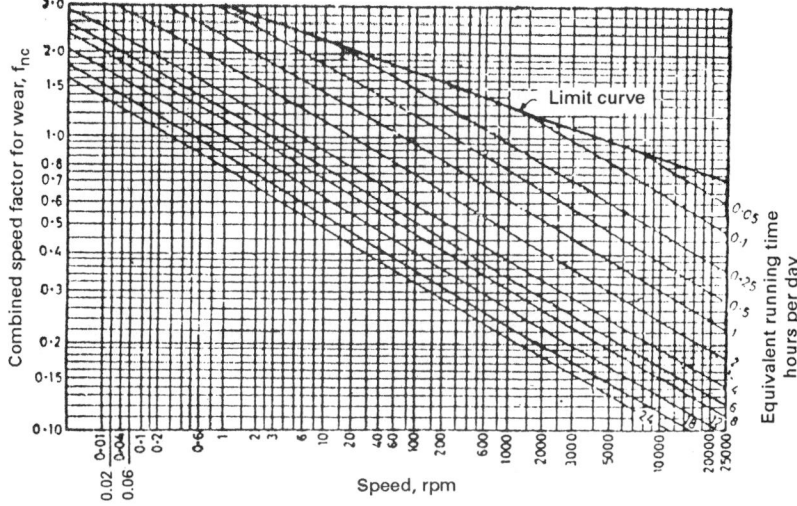

FIGURE 23–41 Combined speed and running-time factors for spur and helical gears for wear.

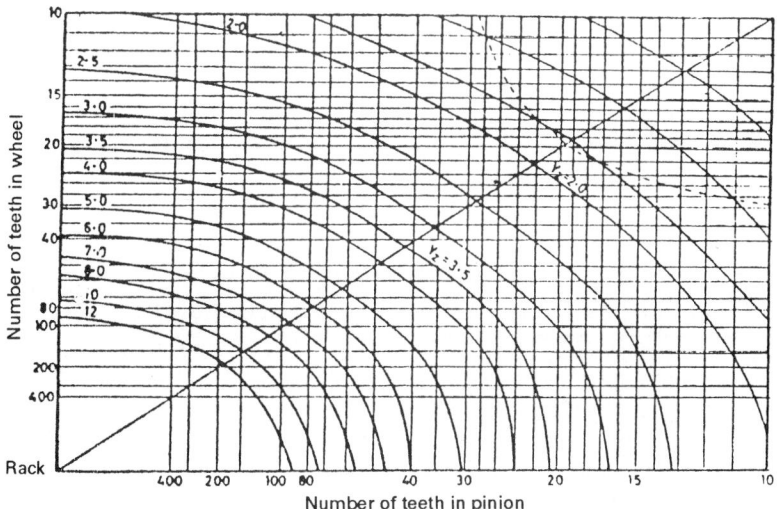

FIGURE 23–42 Zone factor—Helical gears, 30° Helix angles 20° normal pressure angle.

Particular	Formula
	When the face width is less than axial pitch (p_x) and insufficient to give overlap, the strength factor shall be multiplied by $\frac{1}{3}[2 + (b/p_x)^2]$

K_b = bending stress factor taken from Table 23–40

The allowable tangential load for wear in Newton per meter face width (N/m) or pound force per inch (lbf/in) of face width for suitably lubricated gears

$$F_{tw}^* = 250\,\frac{f_{nc}\,Y_z\,K_w}{K} \quad \textbf{SI} \qquad (23\text{--}195a)$$

where F_{tw} in N/mm

$$= 0.48 f_{nc}\,Y_z\,K_w K_1 \quad \textbf{SI} \qquad (23\text{--}195b)$$

$$= 0.48 f_{nc}\,Y_z\,K_w m^{0.8} \quad \textbf{SI} \qquad (23\text{--}195c)$$

where F_{tw} in N/m; K_w in Pa; m in m; $K_1 = m^{0.8}$

$$F_{tw}^* = (f_{nc}\,Y_z\,K_w)/K \quad \textbf{USCSU} \qquad (23\text{--}195d)$$

where F_{tw} in lbf/in; K_w in psi; $K = P^{0.8}$; P in in^{-1}

f_{nc} = speed factor for wear for appropriate running time taken from Fig. 23–41
Y_z = zone factor for external gear taken from Fig. 23–42

The zone factor for internal gears shall be equal to the value of the zone factor for external gears multiplied by $[(i+1)/(i-1)]^{0.8}$

The zone factor for helical gear other than $30°$ helix angle is obtained from Fig. 23–42 and shall be multiplied by $0.75 \sec^2\beta$. The zone factor for $\beta = 30°$ is obtained from Fig. 23–48.

K_w = surface stress factor taken from Table 23–40
K = pitch factor taken from Fig. 23–44; $K_1 = m^{0.8}$

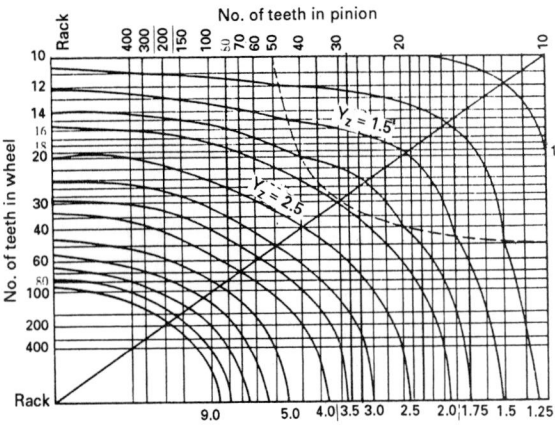

FIGURE 23–42(a) Zone factor— full depth. Spur gears: $20°$ pressure angle.

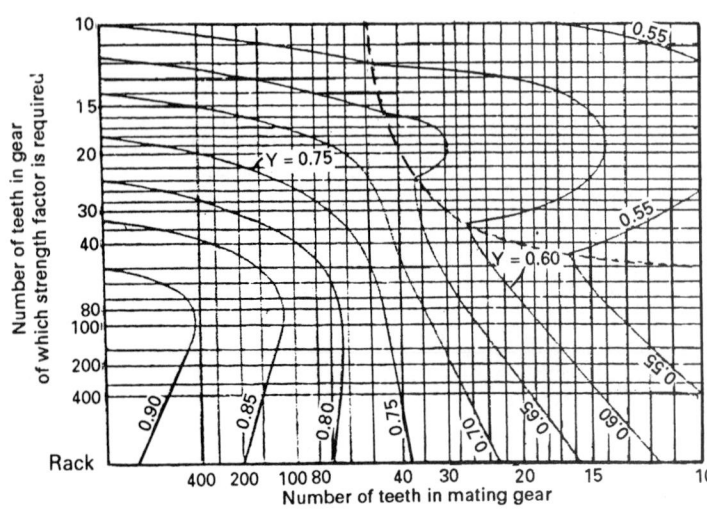

FIGURE 23–43 Strength factor—spur gear—$20°$ normal pressure angle; helical gears—$30°$ helix angle.

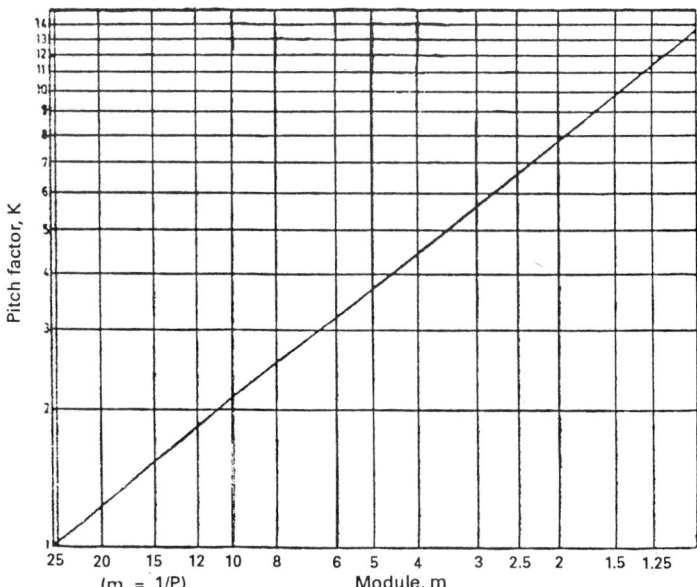

FIGURE 23-44 Pitch factor, K, for spur and helical gears.

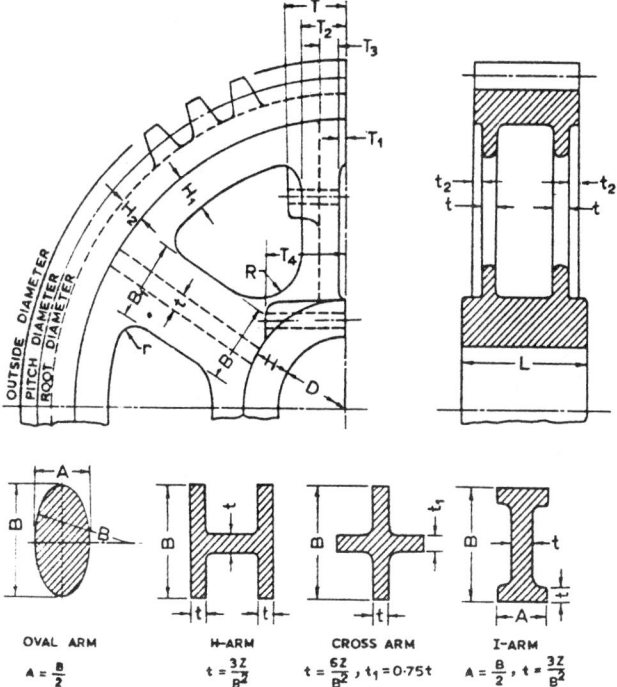

OVAL ARM

$A = \dfrac{B}{2}$

H-ARM

$t = \dfrac{3Z}{B^2}$

CROSS ARM

$t = \dfrac{6Z}{B^2}$, $t_1 = 0.75t$

I-ARM

$A = \dfrac{B}{2}$, $t = \dfrac{3Z}{B^2}$

FIGURE 23-45 Dimensions of constructed gear.

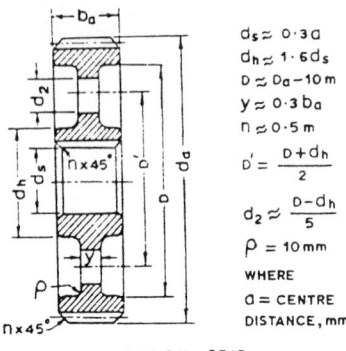

$$d_s \approx 0.3a$$
$$d_h \approx 1.6 d_s$$
$$D \approx D_a - 10\,m$$
$$y \approx 0.3\, b_a$$
$$n \approx 0.5\, m$$
$$D' = \frac{D + d_h}{2}$$
$$d_2 \approx \frac{D - d_h}{5}$$
$$\rho = 10\,mm$$

WHERE

a = CENTRE DISTANCE, mm

FORGED CYLINDRICAL GEAR

FIGURE 23–46 Forged cylindrical gear dimensions.

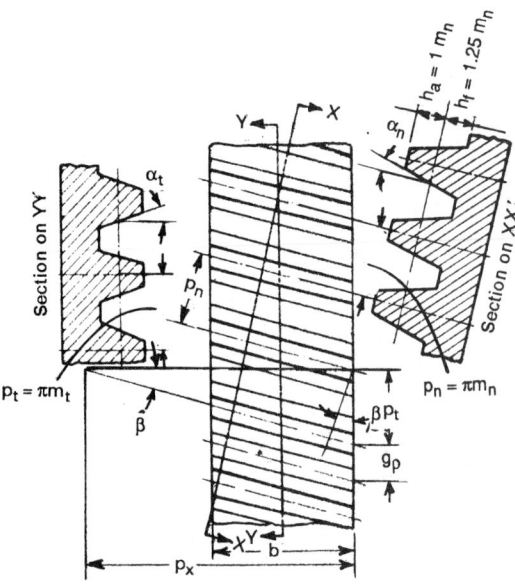

FIGURE 23–47 Circular pitches and pressure angles relationships.

Particular	Formula
Power rating in terms of horse power of spur and helical gear for strength	$P_s = \dfrac{794}{10^8} \dfrac{f_{nb} Y K_b bnz}{P^2}$ **USCSU** $\qquad$ (23–196a) where P_s in hp; K_b in psi; b in in; P in in^{-1}.
Power rating in terms of kW of spur and helical gear for strength	$P_s = \dfrac{5245}{10^8} f_{nb} Y K_b b m^2 nz$ **SI** $\qquad$ (23–196b) where P_s in kW; K_b in Pa; b and m in m.
Power rating in terms of horse power of spur and helical gear for wear	$P_w = \dfrac{794}{10^8} \dfrac{f_{nc} Y_z K_w bnz}{KP}$ **USCSU** $\qquad$ (23–197) where P_w in hp; K_w in psi; b in in; P_y in in^{-1} $K = P^{0.8}$
Power rating in terms of kW of spur and helical gear for wear	$P_w = \dfrac{133.2}{10^8} \dfrac{f_{nc} Y_z K_w b \dot{m} nz}{K}$ **SI** $\qquad$ (23–198)

Particular	Formula
	where P_w in kW; K_w in Pa; b and m in m; $K =$ pitch factor $= \left(\frac{1}{m}\right)^{0.8}$
	Also K is obtained from Fig. 23–44
The factor of safety for gears on the basis of strength	$$n' = \frac{\sigma_u}{f_{nb} K_b} \qquad (23\text{–}199)$$
	where $\sigma_u =$ ultimate tensile stress of material, Pa (psi)

MOMENTARY LOAD

The maximum momentary load for strength	$$M_{t(max)} = \frac{M_{tms1}(0.8 + \sqrt{f_{nb12}})(\sqrt{f_{nb12}})}{f_{nb}} \text{ or } 2M_{tms1}$$ $(23\text{–}200)$
	whichever is less
	where $f_{nb12} =$ combined speed factor for strength at 12 h running time per day taken from Fig. 23–40
The maximum momentary load for wear	$$M_{t(max)} = \frac{M_{tms1}(2 + \sqrt{f_{nc12}})(\sqrt{f_{nc12}})}{f_{nc}} \text{ or } 3M_{tms1}$$ $(23\text{–}201)$
	whichever is less
	where $f_{nc12} =$ combined speed factor for wear at 12 h running time per day taken from Fig. 23–41

DETERMINATION OF UNIFORM LOAD EQUIVALENT TO SPECIFIED LOAD

The expression for equivalent running time at the maximum sustained load and the corresponding mean speed when the load consists of a maximum sustained gear or pinion torque M_{tms1} acting for T_1 hours at a mean speed of n_1 and smaller sustained torques M_{t2}, M_{t3}, etc. acting for T_2, T_3, etc. hours at mean speeds n_2, n_3, etc.	$$T_e = T_1 + T_2\left(\frac{n_2}{n_1}\right)\left(\frac{M_{t2}}{M_{tms1}}\right)^3$$ $$+ T_3\left(\frac{n_3}{n_1}\right)\left(\frac{M_{t3}}{M_{tms1}}\right)^3 + \cdots +, \text{etc.} \qquad (23\text{–}202)$$ where each term represents the equivalent of the running time for the corresponding period
The expression for the equivalent running time for the period when the torque changes during any period τ from $M_{\tau1}$ to $M_{\tau2}$	$$T_{e\tau} = \left\{ \frac{T_\tau}{4}\left(\frac{n_\tau}{n_1}\right)\left[\frac{M_{\tau2}}{M_{tms1}} + \frac{M_{\tau1}}{M_{tms1}}\right] \right.$$ $$\left. \left[\left(\frac{M_{\tau1}}{M_{tms1}}\right)^2 + \left(\frac{M_{\tau2}}{M_{tms1}}\right)^2\right] \right\} \qquad (23\text{–}203)$$

Particular	Formula

GEAR CONSTRUCTION (FIG. 23–45)

The stalling load

$$F_{st} = \sigma_o \pi m b y \qquad (23\text{–}204)$$

where σ_o = stress in material at zero velocity, MPa (psi). From Table 23–18.

The section modulus of arms

$$Z = \frac{F_{st} r'}{i \sigma_o} = \frac{\pi m b y \times r'}{i} \qquad (23\text{–}205a)$$

where i = number of arms; r' = pitch radius, m (in)

The width of arm as shown in Fig. 23–45

$$B = 4(z/\pi)^{1/3} \qquad (23\text{–}205b)$$

The other dimensions indicated by letter symbols as shown in Fig. 23–45

$$H = 0.4D \qquad (23\text{–}205c)$$

Hub diameter $= 1.8D$
$L = 1.25D$ (For a gear with a face width greater than 1.25D, make L, however, equal to face, and for split gears make L, large enough to accommodate bolts.) $\qquad (23\text{–}205d)$

Bead diameter $= 2.24D$
$B_1 = B - 19$ mm/m or B. 0.75 in taper per foot
$\qquad (23\text{–}205e)$
$T_2 = 0.5B_1$ (for split gears and oval arms) $\quad (23\text{–}205f)$
$t_2 = 0.1b$ (for H or cross arms) $\qquad (23\text{–}205g)$
$T_1 = 0.25T_3$ (for split gears) $\qquad (23\text{–}205h)$
$H_2 = (\sqrt[3]{0.5z/i})m = (\sqrt[3]{0.5z/i})/P \qquad (23\text{–}205i)$
$T = T_2 + 6$ mm (0.25 in) $\qquad (23\text{–}205j)$
$R = 0.55B \qquad (23\text{–}205k)$
$H_i = 1.25H_2 \qquad (23\text{–}205l)$

$$r = \frac{152}{im} \qquad (23\text{–}205m)$$

$T_4 = 2H$

SOLID WEB (FIG. 23–46)

The dimensions of a forged cylindrical gear are as shown in Fig. 23–46

The minimum thickness of metal permissible above the keyway of a pinion may be determined by empirical relation according to Nuttal Company

$$H_k = (\sqrt{0.2z})m \qquad \textbf{SI} \qquad (23\text{–}206a)$$
$$= (\sqrt{0.2z})/P \quad \textbf{US Customary System units} \qquad (23\text{–}206b)$$

The empirical equation for the diameter of a solid pinion

$$D' \leq 14.75m + 60 \text{ mm} \qquad \textbf{SI} \qquad (23\text{–}207a)$$

$$\leq \frac{14.75}{P} + 2.4 \text{ in} \quad \textbf{US Customary System units}$$
$$(23\text{–}207b)$$

The limit for diameter of pinion or gear with a web

$$D' \leq 23.55m + 15 \text{ mm} \qquad \textbf{SI} \qquad (23\text{–}208a)$$

Particular	Formula

$$\leq \frac{23.55}{P} + 0.6 \text{ in} \quad \textbf{US Customary System units}$$

$$(23\text{--}208b)$$

The thickness of web (Fig. 23–46)

$$y = 1.75m + 3.4 \text{ mm} \qquad \textbf{SI} \qquad (23\text{--}209a)$$

$$= \frac{1.75}{P} + 0.136 \text{ in} \quad \textbf{US Customary System units}$$

$$(23\text{--}209b)$$

where m = module, mm; P = diametral pitch, in^{-1}

23.1.2 HELICAL GEARS

For gear terminology

Refer to Table 2–50.

The transmission ratio

$$i = \frac{z_2}{z_1} = \frac{\omega_1}{\omega_2} = \frac{n_1}{n_2} = \frac{d_2}{d_1} \tag{23--210}$$

Normal circular pitch (Fig. 23–47)

$$p_n = p_t \cos \beta = \frac{\pi d}{z} \cos \beta = \pi m_n = \frac{\pi}{P_n} \tag{23--211}$$

Transverse module

$$m_t = \frac{m_n}{\cos \beta} \tag{23--212}$$

Transverse circular pitch (Fig. 23–47)

$$p_t = \frac{p_n}{\cos \beta} = \frac{\pi}{P_t} = \pi m_t = \frac{\pi}{P_n \cos \beta} = \frac{\pi d}{z} \tag{23--213}$$

Normal diametral pitch

$$P_n = \frac{P_t}{\cos \beta} = \frac{z}{d \cos \beta} = \frac{\pi}{p_n} = \frac{1}{m_n} \tag{23--214}$$

The relation between p_n and P_n

$$p_n P_n = \pi \tag{23--215}$$

Transverse diametral pitch

$$P_t = P_n \cos \beta = \frac{z}{d} = \frac{\pi}{p_n} \cos \beta = \frac{\cos \beta}{m_n} \tag{23--216}$$

Axial pitch (Fig. 23–47)

$$p_x = \frac{\pi}{P_n \sin \beta} = \frac{\pi m_n}{\sin \beta} = \frac{p_t}{\tan \beta} = \frac{\pi m_t}{\tan \beta} = \frac{\pi}{P_t \tan \beta} \tag{23--217}$$

Reference diameter

$$d = \frac{z}{P} \cos \beta = \frac{zm}{\cos \beta} \tag{23--218}$$

Normal base pitch

$$p_{bn} = \frac{\pi \cos \alpha}{P_n} = \pi m_n \cos \alpha \tag{23--219}$$

Transverse base pitch

$$p_{bt} = \frac{\pi \cos \alpha}{P_t \cos \beta_b} = \frac{\pi m_t \cos \alpha}{\cos \beta_b} \tag{23--220}$$

Particular	Formula	
Transverse pressure angle (Fig. 23–47)	$\tan \alpha_t = \dfrac{\tan \alpha_n}{\cos \beta}$	(23–221)
Base helix angle	$\cos \beta_b = \dfrac{\sin \alpha_n}{\sin \alpha_t}$	(23–222)
Base circle diameter	$d_b = \dfrac{z \cos \alpha_t}{P \cos \beta_b} = \dfrac{zm \cos \alpha_t}{\cos \beta_b} = d \cos \alpha_t$	(23–223)
Involute function	$inv\, \alpha_t = \tan \alpha_t - \alpha_t$	(23–224)
Virtual number of teeth	$z_v = z\, \dfrac{inv\, \alpha_t}{inv\, \alpha}$	(23–225)
The radial backlash	$j_r = \dfrac{j_n}{2 \sin \alpha_t \cos \beta_b} \simeq \dfrac{j_n}{2 \sin \alpha_t}$	(23–226)
Reference circle diameter of pinion	$d_1 = \dfrac{z_1 m_n}{\cos \beta} = \dfrac{z_1}{P_n \cos \beta}$	(23–227)
Reference circle diameter of gear	$d_2 = \dfrac{z_2 m_n}{\cos \beta} = \dfrac{z_2}{P_n \cos \beta}$	(23–228)
Reference center distance	$a = \dfrac{d_1 + d_2}{2}$	(23–229)
Working transverse diametral pitch	$P'_t = \dfrac{z_1 + z_2}{2a'}$	(23–230)
Working transverse module	$m'_t = \dfrac{2a'}{z_1 + z_2}$	(23–231)
Working pitch circle diameter of pinion	$d'_1 = \dfrac{z_1}{P'_t} = z_1 m'_t = \dfrac{2a'}{i+1}$	(23–232)
Working pitch circle diameter of gear	$d'_2 = \dfrac{z_2}{P'_t} = z_2 m'_t = 2a'\, \dfrac{i}{i+1}$	(23–233)
Working center distance	$a' = \dfrac{d'_1 + d'_2}{2}$	(23–234)
Working transverse pressure angle	$\cos \alpha'_t = \dfrac{d_{b1} + d_{b2}}{2a'}$	(23–235)
Working helix angle	$\tan \beta' = \dfrac{d'}{d} \tan \beta = \dfrac{a'}{a} \tan \beta$	(23–236)
Working normal diametral pitch	$P'_n = \dfrac{P_t}{\cos \beta'}$	(23–237)

Particular	Formula

Working normal module

$$m_n' = m_t' \cos \beta' \qquad (23\text{–}238)$$

The normal tooth circular thickness

$$s_n = \frac{1.5707963}{P_n} = 1.5707963 m_n \qquad (23\text{–}239)$$

Involute function, *inv* α_n for measurements over pins or balls

$$inv\, \alpha_n = \frac{s_n}{d \cos \beta} + inv\, \alpha_t + \frac{d_{pin}}{d_b \cos \beta_b} - \frac{\pi}{z_1} \qquad (23\text{–}240)$$

where d_{pin} = diameter of measuring ball or pin

The addendum

$$h_a = 1 m_n \text{ or } \frac{\text{constant, Table 23–1}}{P_n}$$

$$\text{or (constant, Table 23–1) } m_n \qquad (23\text{–}241)$$

The dedendum

$$h_f = 1.25 m_n \text{ or } \frac{\text{constant, Table 23–1}}{P_n}$$

$$\text{or (constant, Table 23–1) } m_n \qquad (23\text{–}242)$$

The clearance

$$c = 0.25 m_n \text{ or } \frac{\text{constant, Table 23–1}}{P_n}$$

$$\text{or (constant, Table 23–1) } m_n \qquad (23\text{–}243)$$

Working tooth depth

$$h' = 2 m_n$$

Total tooth depth

$$h = h' + c = 2.25 m_n \qquad (23\text{–}244)$$

The outside diameter or the addendum circle diameter of pinion

$$d_{a1} = d_1' + 2 h_{a1} \qquad (23\text{–}245)$$

The outside diameter or the addendum circle diameter of gear

$$d_{a2} = d_2' + 2 h_{a2} \qquad (23\text{–}246)$$

The root diameter or the dedendum circle diameter of pinion

$$d_{f1} = d_1' - 2 h_{f1} \qquad (23\text{–}247)$$

The root diameter or the dedendum circle diameter of gear

$$d_{f2} = d_2' - 2 h_{f2} \qquad (23\text{–}248)$$

The base circle diameter of pinion

$$d_{b1} = d_1' \cos \alpha_t \qquad (23\text{–}249)$$

The base circle diameter of gear

$$d_{b2} = d_2' \cos \alpha_t \qquad (23\text{–}250)$$

The lead of pinion

$$l_1 = \pi d_1' \cot \beta = \frac{\pi z_1}{P_n \sin \beta} = \frac{\pi z_1 m_n}{\sin \beta} \qquad (23\text{–}251)$$

The lead of gear

$$l_2 = \pi d_2' \cot \beta = \frac{\pi z_2}{P_n \sin \beta} = \frac{\pi z_2 m_n}{\sin \beta} \qquad (23\text{–}252)$$

For formulas connecting p_t, p_n, p_x, p_b, p_{bx}, p_{bn}, and P_n and P_t.   Refer to Table 23–41

For formulas connecting α_t, α_n, α_x, β_b, β_p, γ_b, and γ_p Refer to Table 23–42.

Particular	Formula
For gear terminology	Refer to Table 23–43.
For formulas of contact ratio of involute helical gears	Refer to Table 23–44.
For helical tooth proportion for $\alpha_n = 20°$	Refer to Table 23–45.
For helical tooth proportion for $\alpha_t = 20°$	Refer to Table 23–46.

CORRECTED TOOTHED GEAR

Normal tooth thickness

$$s_n = \frac{1}{P_n}\left(\frac{\pi}{2} \pm 2x\tan\alpha_n\right) = m_n\left(\frac{\pi}{2} \pm 2x\tan\alpha_n\right)$$

(23–253)

Transverse tooth thickness

$$s_t = \left(\frac{1}{P_t}\right)\left(\frac{\pi}{2} \pm 2x\tan\alpha_n\right)$$

(23–254a)

$$= m_t\left(\frac{\pi}{2} \pm 2x\tan\alpha_n\right)$$

(23–254b)

Base tooth thickness in normal section

$$s_{bn} = \left(\frac{s_n P_n}{z} + inv\,\alpha_t\right)\frac{z}{P_n}\cos\alpha_n$$

(23–255a)

$$= \left(\frac{s_n}{zm_n} + inv\,\alpha_t\right)zm_n\cos\alpha_n$$

(23–255b)

Base tangent length (14)

$$W_k = \frac{1}{P_n}\left[(z_{bt} - 0.5)\pi\cos\alpha_n + z\,inv\,\alpha_t\cos\alpha\right.$$

$$\left. + 2x\sin\alpha_n\right]$$

(23–256a)

$$= m_n[(z_{bt} - 0.5)\pi\cos\alpha_n + z\,inv\,\alpha_t\cos\alpha + 2x\sin\alpha_n)]$$

(23–256b)

For number of teeth between the measuring faces for base tangent length z_{bt} — Refer to Fig. 23–48a.

Sum of addendum modification coefficients

$$x_1 + x_2 = \frac{z_1 + z_2}{2}\frac{inv\,\alpha_t' - inv\,\alpha_t}{\tan\alpha_n}$$

(23–257)

Root circle diameter or dedendum circle diameter of pinion

$$d_{f1} = d_1' - 2(1 - x_1 + c^*)\frac{1}{P_n}$$

(23–258a)

$$= \left[\frac{z_1}{\cos\beta} - 2(1 - x_1 + c^*)\right]m_n$$

(23–258b)

Root circle diameter or dedendum circle diameter of gear

$$d_{f2} = d_2' - 2(1 - x_2 + c^*)\frac{1}{P_n}$$

(23–259a)

$$= \left[\frac{z_2}{\cos\beta} - 2(1 - x_2 + c^*)\right]m_n$$

(23–259b)

Particular	Formula
Tip circle diameter or addendum circle diameter of pinion	$d_{a1} = 2a' - d_{f2} - \dfrac{2c^*}{P_n}$ (23–260a)
	$= 2a' - d_{f2} - 2c^* m_n$ (23–260b)
	$= \left[\left(\dfrac{z_1}{\cos \beta} \right) + 2x_1 + 2 \right] m_n$ (23–260c)
Tip circle diameter or addendum circle diameter of gear	$d_{a2} = 2a' - d_{f1} - \dfrac{2c^*}{P_n}$ (23–261a)
	$= 2a' - d_{f1} - 2c^* m_n$ (23–261b)
	$= \left[\left(\dfrac{z_2}{\cos \beta} \right) + 2x_2 + 2 \right] m_n$ (23–261c)
Tip pressure angle of pinion	$\cos \alpha_{a1} = \dfrac{d_{b1}}{d_{a1}}$ (23–262)
Tip pressure angle of gear	$\cos \alpha_{a2} = \dfrac{d_{b2}}{d_{a2}}$ (23–263)
Transverse contact ratio of pinion	$\epsilon_{t1} = \dfrac{z_1}{2\pi} (\tan \alpha_{a1} - \tan \alpha_t')$ (23–264)
Transverse contact ratio of gear	$\epsilon_{t2} = \dfrac{z_2}{2\pi} (\tan \alpha_{a2} - \tan \alpha_t')$ (23–265)
Total transverse contact ratio	$\epsilon_t = \epsilon_{t1} + \epsilon_{t2}$ (23–266)
Theoretical base tangent length	$W_{k1} + W_{k2} = \left[(d_{b1} + d_{b2}) \cos \beta_b \left(inv\, \alpha_t' + \dfrac{\pi}{z_1 + z_2} \right) \right.$
	$\left. + (z_{bt1} + z_{bt2} - 2) p_{bn} \right]$ (23–267a)
Another equation for base tangent length for use with Tables 23–53B and 23–53C (14)	$W = m(K_{bt} + K_\beta z + 2x \sin \alpha)$ **SI** (23–267b)
	where W in mm and m = module in mm
	$= \dfrac{1}{P} (K_{bt} + K_\beta z + 2x \sin \alpha)$ (23–267c)
	US Customary System units
	where W in in; P = diametral pitch
	Refer to Tables 23–53B and 23–53C for values of K_{bt} and K_β.
Theoretical tooth thickness on reference circle	$s_n^* = \dfrac{\pi}{2} + 2x \tan \alpha_n$ (23–268a)

Particular	Formula
	$$s_n = \frac{s_n^*}{P_n} = s_n^* m_n \qquad (23\text{–}268b)$$
Transverse tip thickness on pinion tooth	$$s_{ta1} = d_{a1}\left(\frac{s_{n1}}{d_1\cos\beta} + inv\,\alpha_t' - inv\,\alpha_a\right) \qquad (23\text{–}269)$$
Length of path of contact	$$g = g_{a1} + g_{a2} \qquad (23\text{–}270)$$ where values of g_{a1} and g_{a2} are given by Eqs. (2–271a) and (2–271b)
Path of addendum contact of pinion	$$g_{a1} = \frac{d_{b1}}{2}(\tan\alpha_{a1} - \tan\alpha_t') \qquad (23\text{–}271a)$$
Path of addendum contact of gear	$$g_{a2} = \frac{d_{b2}}{2}(\tan\alpha_{a2} - \tan\alpha_t') \qquad (23\text{–}271b)$$

FORCES ACTING ON A GEAR-TOOTH PROFILE

The tangential load in the transverse section	$$F_t = \frac{M_t(i\pm 1)}{ia} = F_n\cos\alpha_n\cos\beta \qquad (23\text{–}272)$$
The axial thrust	$$F_a = F_t\tan\beta = \frac{M_t(i\pm 1)}{ia}\tan\beta \qquad (23\text{–}273)$$
The radial force	$$F_t = F_n\sin\alpha_n = \frac{M_t(i\pm 1)}{ia}\tan\alpha_t \qquad (23\text{–}274)$$
The normal force	$$F_n = \frac{M_t(i\pm 1)}{ia\cos\alpha_t\cos\beta_o} \qquad (23\text{–}275)$$ where β_o = angle between the force F_n and the transverse plane
The load per unit length of line of contact	$$F' = \frac{F_n}{g_{min}} \qquad (23\text{–}276)$$

HELIX ANGLE AND FACE WIDTH OF THE GEAR TOOTH

Minimum face width of the tooth	$$b_{min} = \frac{p_t}{\tan\beta} \qquad (23\text{–}277)$$
Face width of the tooth according to Fellow practice	$$b = \frac{1.1p_t}{\tan\beta} = \frac{1.1\pi m_t}{\tan\beta} \qquad (23\text{–}278)$$

Particular	Formula	
Face width of the tooth according to AGMA	$$b = \frac{1.15\pi}{P_t \tan\beta} = \frac{1.15\pi m_t}{\tan\beta}$$	(23–279)
Maximum value of face width of the tooth	$$b_{max} \leq \left(\frac{28m_t}{\tan\beta} = \frac{28}{P_t \tan\beta}\right)$$	(23–280)
Minimum value of face width of the tooth for herringbone gears according to AGMA	$$b_{min} \geq \left(\frac{2.3\pi m_t}{\tan\beta} = \frac{2.3\pi}{P_t \tan\beta}\right)$$	(23–281)
Maximum face width of the tooth for herringbone gears according to AGMA	$$b_{max} < \left(\frac{30m_t}{\tan\beta} = \frac{30}{P_t \tan\beta}\right)$$	(23–282)
The limit value of ratio of b/d_1 for helical and herringbone gears placed symmetrically very close to bearings in a rigid housing	$b/d_1 = 2$ for constant load $= 1.6$ for varying load	(23–283a) (23–283b)
The limit value of ratio of b/d_1 for helical and herringbone overhung pinion placed in a rigid housing	$b/d_1 = 0.9$ for constant load $= 0.8$ for varying load	(23–284a) (23–284b)

VIRTUAL OR FORMATIVE NUMBER OF TEETH

Equivalent number of teeth or formative number of teeth	$$z_v = \frac{z}{\cos^3\beta}$$	(23–285)

DESIGNING GEARS FOR BENDING STRENGTH

Lewis equation for calculating the tangential force in the plane of rotation or in transverse plane	$$F_t = \frac{\sigma_o b Y K_V}{P_t C_w}\cos\beta = \frac{\sigma_o b Y m_t K_V}{C_w}\cos\beta$$ $$= \frac{\sigma_o b y p_t K_V}{C_w}\cos\beta = \frac{\sigma_o b Y m_n K_V}{C_w}$$	 (23–286)
The modified Lewis equation for calculating module	$$m_t = \frac{F_t C_w}{\sigma_o b Y K_v \cos\beta}$$	(23–287a)
	$$= \left(\frac{2M_t C_w}{\sigma_o K_V b Y z \cos\beta}\right)^{1/2} = \left(\frac{2M_t C_w}{\lambda z \sigma_o Y K_V \cos\beta}\right)^{1/3}$$	(23–287b)
	$$m_t = 6.83\left(\frac{P C_w}{\lambda z Y K_v n' \sigma_o \cos\beta}\right)^{1/3} \quad \textbf{SI}$$	(23–287c)

where m_t in m, P in kW, n' in rps, and σ_o in Pa

Particular	Formula

$$= 26.5\left(\frac{PC_w}{\lambda z\,YK_v n\sigma_o \cos\beta}\right)^{1/3} \quad \textbf{SI} \qquad (23\text{--}287\text{d})$$

where m_t in m, P in kW, n in rpm, and σ_o in Pa

$$= 12.5\left(\frac{PC_w}{\lambda z\,YK_V \omega\sigma_o \cos\beta}\right)^{1/3} \quad \textbf{SI} \qquad (23\text{--}287\text{e})$$

where m_t in m, P in kW, ω in rad/s, and σ_o in Pa

$$= 55.5\left(\frac{PC_w}{\lambda z\,YK_V n'\sigma_o \cos\beta}\right)^{1/3} \quad \textbf{SI} \qquad (23\text{--}287\text{f})$$

where m_t in mm, P in kW, n' in rps, and σ_o in MPa or N/mm^2

$$= 103\left(\frac{PC_w}{\lambda z\,YK_V \omega\sigma_o \cos\beta}\right)^{1/3} \quad \textbf{SI} \qquad (23\text{--}287\text{g})$$

where m_t in mm, P in kW, ω in rad/s, and σ_o in MPa or N/mm^2

The modified Lewis equation for calculating diametral pitch in the plane of rotation or transverse diametral pitch

$$P_t = 0.02\left(\frac{PC_w}{\lambda z\,YK_V n\sigma_o \cos\beta}\right)^{1/3}$$

US Customary System units $\qquad (23\text{--}287\text{h})$

where P_t in in^{-1}, P in hp, n in rpm, and σ_o in psi

C_w = wear and lubrication factor taken from Table 23–47; Y is chosen in accordance with the virtual number of teeth from Table 23–16.

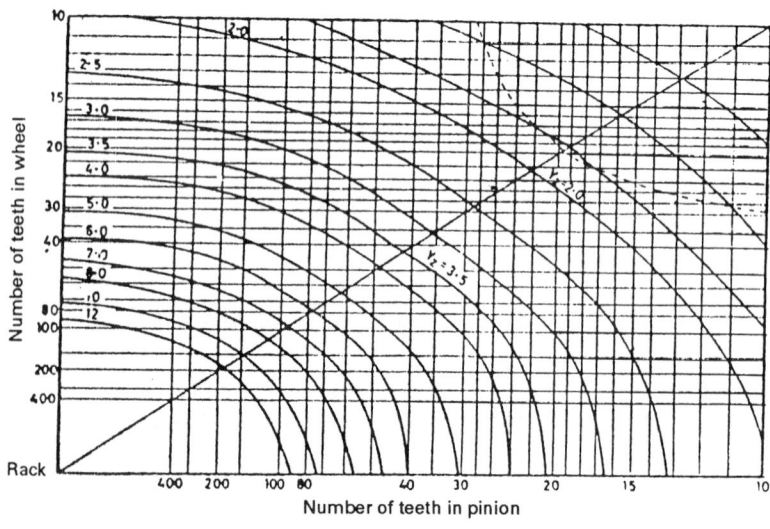

FIGURE 23–48   Zone factor—helical gears—30°; helix angles—20° normal pressure angle.

Particular	Formula
	For σ_o refer to Table 23–48.

DYNAMIC OR VELOCITY FACTORS

The dynamic or velocity factors for low-helix-angle gears

$$K_v = C_v = \frac{4.5}{4.5 + \nu_m} \quad \text{for } \nu_m \le 12.75 \text{ m/s } \textbf{ SI}$$

$$\text{(23–288a)}$$

$$= \frac{900}{900 + \nu_m} \quad \text{for } \nu_m \le 2500 \text{ ft/min}$$

$$\textbf{US Customary System units} \qquad \text{(23–288b)}$$

The dynamic or velocity factor all helical and herringbone gears

$$K_v = C_v = \frac{6}{6 + \nu_m} \quad \text{for } \nu_m \le 20 \text{ m/s but } > 5 \text{ m/s}$$

$$\textbf{SI} \qquad \text{(23–289a)}$$

$$= \frac{1200}{1200 + \nu_m} \quad \text{for } \nu_m \le 1200 \text{ to } 4000 \text{ ft/min}$$

$$\textbf{US Customary units} \qquad \text{(23–289b)}$$

The dynamic or velocity factor

$$K_v = C_v = \frac{5.6}{5.6 + \sqrt{\nu_m}} \qquad \textbf{SI} \qquad \text{(23–290a)}$$

$$= \frac{78}{78 + \sqrt{\nu_m}} \quad \textbf{US Customary System units}$$

$$\text{(23–290b)}$$

The dynamic or velocity factor

$$K_v = C_v = \left(\frac{5.6}{5.6 + \sqrt{\nu_m}} \right)^{1/2} \qquad \textbf{SI} \qquad \text{(23–291a)}$$

$$= \left(\frac{78}{78 + \sqrt{\nu_m}} \right)^{1/2}$$

$$\textbf{US Customary System units} \qquad \text{(23–291b)}$$

The dynamic or velocity factor for nonmetallic gears

$$K_v = C_v = \frac{0.75}{1 + \nu_m} + 0.07625 \quad \textbf{SI} \qquad \text{(23–292a)}$$

$$= \frac{150}{200 + \nu_m} + 0.25$$

$$\textbf{US Customary System units} \qquad \text{(23–292b)}$$

The limiting endurance bending strength load $(F_f \ge F_d)$

$$F_f = \frac{\sigma_f b y \pi \cos \beta}{P_t} = \sigma_f b Y m_t \cos \beta \qquad \text{(23–293)}$$

where F_f is based on the Lewis equation without a dynamic or velocity factor

Particular	Formula
The limiting stress for surface fatigue may be estimated from Eq. (23–294)	$\sigma_{fc} = (2.75 H_B - 69)$ MPa **SI** (23–294a) $= (0.4 H_B - 10)$ kpsi **US Customary System units** (23–294b)

AGMA fundamental formula for calculating stress at transmitted load (F_t) for the design of helical gear teeth, which takes care of all factors explained under Eq. (23–95) for spur gear. (4, 17)
(*Note*: The other factors such as K_a, K_{sz}, and K_m have already been explained under spur gear [Eq. (23–95)]. The values of these factors given under spur gear can be used to design helical gears also.)
	$\sigma_b = \dfrac{F_t K_a}{K_v} \cdot \dfrac{1}{m_t b} \cdot \dfrac{(0.93 K_m) K_{sz} K_B}{J}$ **SI** (23–295)
	
	$= \dfrac{F_t K_a}{K_v} \cdot \dfrac{P_t}{b} \cdot \dfrac{(0.93 K_m) K_{sz} K_B}{J}$
	
	US Customary System units (23–296)
	
	m_t = transverse module, mm or m
	J = geometry factor obtained from Fig. 23–48
	K_B = rim thickness factor
	    = 1 for unidirectional load, rigid backup gears
	    = 1.4 for reversed load, even if the backup is rigid

The bending stress σ_b calculated by Eqs. (23–295) and (23–296) must be less than or equal to the allowable working bending stress numbers σ_{wba}	$\sigma_b \le \sigma_{wab} = \dfrac{\sigma_{sab} K_L}{K_T K_R}$ (23–297)

DESIGN BASED ON SURFACE DURABILITY

Particular	Formula
AGMA fundamental pitting resistance formula for calculating surface compressive stress or contact stress number	$\sigma_c = C_p \left(\dfrac{F_t C_a}{C_v} \cdot \dfrac{C_{sz}}{b d_1'} \cdot \dfrac{(0.93 C_m) C_{sr}}{I} \right)^{1/2}$ (23–298) where C_p = elastic coefficient, $\sqrt{\text{MPa}}$ ($\sqrt{\text{psi}}$) (For expression for C_p, refer to Eq. (23–165) and for values of C_p, refer to Table 23–35.)
The geometry factor for use in Eq. (23–298)	$I = \dfrac{\sin \alpha_t \cos \alpha_t}{2 K_{LS}} \left(\dfrac{i}{i+1} \right) = \dfrac{\sin 2\alpha_t}{4 K_{LS}} \left(\dfrac{i}{i+1} \right)$ for external gears (23–299a) $= \dfrac{\sin 2\alpha_t}{4 K_{LS}} \left(\dfrac{i}{i-1} \right)$ for internal gears (23–299b)
The load sharing ratio	$K_{LS} = \dfrac{p_{bn}}{0.95 g_t}$ (23–300) where g_t = length of the line of action in the transverse plane $p_{bn} = p_n \cos \alpha_n$

Particular	Formula
The length of the line of action in the transverse plane	$$g_t = \frac{1}{2}\left[\sqrt{d_{a1}^2 - d_1'^2\cos^2\alpha_t} + \sqrt{d_{a2}^2 - d_2'^2\cos^2\alpha_t}\right]$$ $$- a\sin\alpha_t \qquad\qquad (23\text{–}301a)$$ $$= \frac{m_t}{2}\left[\sqrt{(z_1+2)^2 - z_1^2\cos^2\alpha_t}\right.$$ $$\left. + \sqrt{(z_2+2)^2 - z_2^2\cos^2\alpha_t} - (z_1+z_2)\sin\alpha_t\right]$$ $$(23\text{–}301b)$$
The surface compressive stress or pitting resistance or surface durability stress calculated by Eq. (23–298) must be less than or equal to the allowable working contact stress σ_{wac}	$$\sigma_c \le \sigma_{wac} = \frac{C_L C_H}{C_T C_R}\,\sigma_{sac} \qquad (23\text{–}302)$$ Refer to Table 23–24 and Fig. 23–25 for σ_{sac}. Symbols given in the above equations have already been explained under Eq. (23–169).
AGMA formula for the rated power to resist breakages of the gear teeth	$$P_{ab} = \left[\frac{n_1' d_1' K_v}{318}\,(m_t b_e)\,\frac{J}{K_B K_a (0.93 K_m) K_{sr}}\right.$$ $$\left.\left(\frac{K_L \sigma_{sab}}{K_T K_R}\right)\right] \qquad \textbf{SI} \qquad (23\text{–}303)$$ where P_{ab} in kW, n_1' in rps, d_1', m_t and B_e in m; and σ_{sab} in Pa $$= \left[\frac{n_1 d_1' K_v}{19.1\times 10^6}\,(m_t b_e)\,\frac{J}{(0.93 K_m) K_a K_{sr} K_B}\right.$$ $$\left.\left(\frac{K_L \sigma_{sab}}{K_T K_R}\right)\right] \qquad \textbf{SI} \qquad (23\text{–}304)$$ where P_{ab} in kW, n_1 in rpm, d_1', m_t, and b_e in mm, and σ_{sab} in MPa $$P_{ab} = \left[\frac{n_1 d_1' K_v}{126,000}\left(\frac{b_e}{P_t}\right)\left(\frac{J}{(0.93 K_m) K_a K_{sr} K_B}\right)\right.$$ $$\left.\left(\frac{K_L \sigma_{sab}}{K_T K_R}\right)\right] \quad \textbf{US Customary System units}$$ $$(23\text{–}305)$$ where P_{ab} in hp, n_1 in rpm, d_1' and b_e in in, and P_t in in^{-1} and σ_{sab} in psi
AGMA formula for the rated power to resist pitting of the gear teeth	$$P_{ac} = \frac{n_1' b}{318}\left[\frac{I C_v}{C_{sr} C_{sz} C_a (0.93 C_m)}\right]\left[\frac{\sigma_{sac} d_1'}{C_p}\,\frac{C_L C_H}{C_T C_R}\right]^2$$ $$\textbf{SI} \qquad (23\text{–}306)$$

Particular	Formula

where P_{ac} in kW, n_1' in rps, σ_{sac} in Pa, C_p in $\sqrt{Pa}$, d_1' and b in m

$$= \frac{n_1 b}{19.1 \times 10^6} \left[\frac{IC_v}{C_{sr}C_{sz}C_a(0.93C_m)} \right] \left[\frac{\sigma_{sac}d_1'}{C_p} \right.$$

$$\left. \cdot \frac{C_L C_H}{C_T C_R} \right]^2 \qquad \textbf{SI} \qquad (23\text{--}307)$$

where P_{ac} in kW, n_1 in rpm, σ_{sac} in MPa, C_p in $\sqrt{MPa}$, and d_1' and b in mm

$$= \frac{n_1 b}{126,000} \left[\frac{IC_v}{C_{sr}C_{sz}C_a(0.93C_m)} \right] \left[\frac{\sigma_{sac}d_1'}{C_p} \right.$$

$$\left. \cdot \frac{C_L C_H}{C_T C_R} \right]^2 \qquad \textbf{US Customary System units}$$

$$(23\text{--}308)$$

where P_{ac} in hp, n_1 in rpm, σ_{sac} in psi, C_p in $\sqrt{psi}$, d_1' and b in in

Designing gears for dynamic load

Dynamic load equation according to Buckingham

$$F_d = F_t + \frac{21v(F_t + bC\cos^2\beta)\cos\beta}{21v + \sqrt{F_t + bC\cos^2\beta}} \qquad \textbf{SI}$$

$$(23\text{--}309a)$$

where F_d and F_t in N, v in m/s, C in N/m, and b in m

$$= F_t + \frac{0.05v(F_t + Cb\cos^2\beta)\cos\beta}{0.05v + \sqrt{F_t + Cb\cos^2\beta}} \qquad (23\text{--}309b)$$

US Customary units

where F_d and F_t in lbf, v in ft/min, C in lbf/in, and b in in

Designing gears for wear load

The limiting wear load equation

$$F_w = \frac{d_1 bQK}{\cos^2\beta} = \frac{mz_1 bQK}{\cos^2\beta} \qquad (23\text{--}310)$$

The load stress factor K to be used in Eq. (23–310)

$$K = \frac{\sigma_{fc}^2 \sin\alpha_n}{0.7E_o} \qquad (23\text{--}311)$$

where σ_{fc} = limiting stress for fatigue obtained from Eq. (23–294) or Table 23–37B

Particular	Formula
The Sykes and Schmitter formula for wear load	$F_w = 0.69C_r m z_1 b C_s$ **SI** (23–312a) where F_w in N and m, b and d_1 in m $= 100C_r d_1 b C_s = 100C_r m z_1 b C_s$ **US Customary System units** (23–312b) where F_w in lbf and m, b and d_1 in in
Formula for first layout according to Schmitter	$$a^2 b = \frac{315(i+1)^2 P}{n C_r C_s}$$ **US Customary System units** (23–313a) $$= \frac{1.15 \times 10^{-4}(i+1)^2 P}{n' C_r C_s}$$ **SI** (23–313b) where n' in rps, P in kW, and a and b in m $C_r = $ ratio factor $= \dfrac{i}{i+1};\ \ i = \dfrac{d_2}{d_1}$ $C_s = $ service factor taken from Tables 23–13 and 23–49.

SURFACE COMPRESSIVE ADMISSIBLE STRESS

The relation between continuous endurance limit and surface hardness	$\sigma_{fa} = C_B H_B$ (23–314a) $= C_R R_C$ (23–314b) where factors C_B and C_R are taken from Table 23–50; H_B and R_C are Brinell and Rockwell hardness numbers

RATING OF MACHINE-CUT HELICAL GEARS AS PER INDIAN STANDARDS

For rating of helical gears as per Indian Standards	Refer to Eqs. (23–194) to (23–203) and Figs. 23–40 to 23–44.
For degree of accuracy of gears, standard gear ratios and dimensions of gear hubs of helical gears	Refer to Tables 23–51 to 23–53.

TABLE 23–1
Gear-tooth proportions[a]

Tooth characteristics	Symbol	Cast teeth	AGMA composite $14\frac{1}{2}°$ involute system	Full-depth 20° involute system	Full-depth 25° involute system	AGMA stub teeth involute
Pressure angle, deg	α	15	$14\frac{1}{2}$	20	25	20
Addendum, mm	h_a	$0.3\,\pi m$	$1\,m$	$1\,m$	$1\,m$	$0.8\,m$
Minimum dedendum, mm	h_f	$0.4\,\pi m$	$1.157\,m$	$1.25\,m$	$1.25\,m$	$1\,m$
Working depth, mm	h'	$0.6\,\pi m$	$2\,m$	$2\,m$	$2\,m$	$1.6\,m$
Minimum total depth, mm	h	$0.7\,\pi m$	$2.157\,m$	$2.25\,m$	$2.25\,m$	$1.8\,m$
Pitch diameter, mm	d'	zm	zm	zm	zm	zm
Outside diameter, mm	d_a	$(z+2)m$	$(z+2)\,m$	$(z+2)\,m$	$(z+2)\,m$	$(z+1.6)\,m$
Tooth thickness, mm	s	$0.475\,\pi m$	$(\pi/2)\,m$	$(\pi/2)\,m$	$(\pi/2)\,m$	$(\pi/2)\,m$
Minimum clearance, mm	c	$0.1\,\pi m$	$0.157\,m$	$0.25\,m$	$0.25\,m$	$0.2\,m$
Approximate fillet radius, mm	ρ_f	$0.209\,m$	$0.209\,m$	$0.3\,m$	$0.3\,m$	$0.3\,m$
Backlash, mm	j	$0.05\,\pi m$	0	0	0	0

[a] $m = 1/P$; $p = m$.

TABLE 23–2
Recommended ratios of gear drive

Types of drive	Speed ratio, i	
	Commonly used	Maximum
Spur gear drive		
Enclosed	3–6	10
Open	4–7	15
Bevel gear drive		
Enclosed	2–3	6
Open	3–7	15

TABLE 23–4
Common standard diametral pitches

28	(9)	2.75
26	8	(2.5)
24	(7)	2
22	6	(1.75)
20	(5.5)	1.50
(18)	5	1.25
16	(4.50)	1
(14)	4	(0.875)
12	(3.50)	0.75
(11)	3	0.625
10	(2.75)	0.50

Numbers without brackets are preferred.

TABLE 23–3
Module, m (mm)

0.05	(1.125)	(18)
(0.055)	1.25	20
0.06	(1.375)	(22)
(0.07)	1.5	(24)
0.08	(1.75)	25
(0.09)	2.0	(27)
0.10	(2.25)	(28)
(0.11)	2.5	(30)
0.12	(2.75)	32
(0.14)	3.0	(33)
0.15	(3.25)	(36)
(0.18)	3.5	(39)
0.2	(3.75)	40
(0.22)	4.0	(42)
0.25	(4.5)	(45)
(0.28)	5	50
0.3	(5.5)	(55)
(0.35)	6	60
0.4	(6.5)	(65)
(0.45)	(7)	(70)
0.5	8	(75)
(0.55)	(9)	80
0.6	10	(85)
(0.7)	(11)	(90)
0.8	12	100
(0.9)	(14)	
1.0	16	

Numbers without bracks are preferred. Modules 3.25, 3.75, and 4.25 mm are permitted for automobiles and 6.5 mm for tractors.

TABLE 23–5
Diametral pitches and modules

Module m, mm	Diametral pitch, P	Pitch p, mm	Base pitch, p_b, mm $\alpha = 20°$	$\alpha = 15°$	Module, m, mm	Diametral pitch, P	Pitch p, mm	Base pitch, p_b, mm $\alpha = 20°$	$\alpha = 15°$
1.000	25.400	3.142	2.952	3.035	6.500	3.907	20.420	19.189	19.725
1.058	24.000	3.325	3.124	3.212	6.750	3.762	21.206	19.927	20.483
1.125	22.577	3.534	3.321	3.414	7.000	3.629	21.991	20.665	20.242
1.155	22.000	3.627	3.408	3.504	7.257	3.500	22.799	21.424	22.022
1.250	20.320	3.927	3.690	3.793	7.500	3.387	23.562	22.140	22.159
1.270	20.000	3.990	3.749	3.854	8.000	3.175	25.132	23.617	24.276
1.375	18.473	4.320	4.059	4.173	8.467	3.000	26.599	24.995	25.692
1.411	18.000	4.333	4.166	4.282	8.500	2.988	26.704	25.093	25.794
1.500	16.933	4.712	4.428	4.552	9.000	2.822	28.274	26.569	27.311
1.588	16.000	4.987	4.687	4.817	9.236	2.750	29.016	27.267	28.028
1.750	14.574	5.498	5.166	5.310	9.500	2.674	29.845	28.045	28.828
1.814	14.000	5.700	5.350	5.503	10.000	2.540	31.416	29.521	30.345
1.954	13.000	6.138	5.768	5.929	10.160	2.500	31.919	29.994	30.831
2.000	12.700	6.283	5.904	6.069	11.000	2.309	34.558	32.473	33.380
2.117	12.000	6.650	6.249	6.423	11.289	2.250	35.465	33.326	34.257
2.250	11.289	7.069	6.642	6.828	12.000	2.117	37.699	35.476	36.415
2.309	11.000	7.254	6.817	7.007	12.700	2.000	39.878	37.492	38.539
2.500	10.160	7.854	7.380	7.586	13.000	1.953	40.841	38.378	39.449
2.540	10.000	7.980	7.498	7.708	14.000	1.804	43.982	41.330	42.484
2.750	9.236	8.639	8.118	8.345	14.514	1.750	45.598	42.828	44.044
2.822	9.000	8.866	8.332	8.504	15.000	1.693	47.124	44.282	45.518
3.000	8.467	9.424	8.856	9.104	16.000	1.588	50.266	47.234	48.553
3.175	8.000	9.975	9.373	9.635	16.933	1.500	53.198	49.989	51.385
3.250	7.815	10.210	9.594	9.862	18.000	1.411	50.549	53.138	54.622
3.500	7.257	10.996	10.332	10.621	20.000	1.290	62.832	59.043	60.691
3.629	7.000	11.399	10.712	11.011	20.320	1.250	63.837	59.987	61.662
3.750	6.773	11.781	11.071	11.380	22.000	1.155	69.115	64.941	66.760
4.000	6.350	12.566	11.809	12.138	24.000	1.058	75.398	70.851	72.729
4.233	6.000	13.299	12.497	12.846	25.000	1.016	78.540	73.803	75.869
4.250	5.976	13.352	12.547	12.897	25.400	1.000	79.796	74.984	77.077
4.500	5.644	14.137	13.285	13.655	27.000	0.941	84.832	79.706	81.932
4.618	5.500	14.508	13.633	14.014	28.000	0.907	87.965	82.660	84.967
4.750	5.347	14.923	14.023	14.414	29.029	0.875	91.196	85.696	88.087
5.000	5.080	15.708	14.761	15.173	30.000	0.847	94.248	88.564	91.036
5.080	5.000	15.959	14.997	15.415	32.000	0.794	100.531	94.468	97.105
5.250	4.838	16.493	15.499	15.931	40.000	0.750	106.395	99.479	102.770
5.500	4.618	17.279	16.257	16.690	33.867	0.706	113.097	106.277	109.244
5.644	4.500	17.733	16.663	17.128	36.000	0.635	125.664	118.025	121.382
5.750	4.417	18.004	16.975	17.449	40.640	0.625	127.674	119.975	123.324
6.000	4.233	18.850	17.713	18.207	45.000	0.564	141.372	132.846	136.555
6.250	4.064	19.635	18.451	18.966	50.000	0.508	157.080	147.607	151.727
6.350	4.000	19.949	18.746	19.269	50.800	0.500	159.592	149.968	154.155

TABLE 23-6
Guide to selection of gear-tooth system

Interchangeable tooth system	Standard gear—minimum number of teeth in a pinion that mesh with rack without interference	Nonstandard gear—minimum number of teeth in the same small pinion neither pointed nor undercut	Minimum number of teeth in equal pinion that will give continuous driving	Comments
$14\frac{1}{2}°$: full-depth involute	32	10	20 with contact ratio of 1.080 (24 gives contact ratio of 1.469)	Recommended for use when the pinion has 40 or more teeth; lower-pressure angles give smoother action because of lower normal force and greater contact ratio
20°: full-depth involute	17	8	12 with contact ratio of 1.049 (14 gives contact ratio of 1.415)	Recommended for general use; larger-pressure angle gives wider, stronger teeth; can be used with reasonably small pinions
20°: stub-tooth involute	14		12 gives contact ratio of 1.185	Recommended for use when number of teeth in pinion is too small for satisfactory use of 20° full-depth teeth

TABLE 23–7
Values of involute functions of $\alpha \, inv \, \alpha = \tan \alpha - \alpha^a$

Minutes	8°	9°	10°	11°	12°	13°	14°	15°	16°	17°	18°
0	9.145	13.048	17.941	23.941	31.171	39.754	49.819	61.498	74.93	90.25	107.60
1	9.203	13.121	18.031	24.051	31.302	39.909	50.000	61.707	75.17	90.52	107.91
2	9.260	13.195	18.122	24.161	31.434	40.065	50.182	61.917	75.41	90.79	108.22
3	9.318	13.268	18.213	24.272	31.567	40.221	50.364	62.127	75.65	91.07	108.53
4	9.377	13.342	18.305	24.383	31.699	40.377	50.546	62.337	75.89	91.34	108.84
5	9.435	13.416	18.397	24.495	31.832	40.534	50.729	62.548	76.13	91.61	109.15
6	9.494	13.491	18.489	24.607	31.966	40.692	50.912	62.760	76.37	91.89	109.46
7	9.553	13.566	18.581	24.719	32.100	40.849	51.096	62.972	76.61	92.16	109.77
8	9.612	13.641	18.674	24.831	32.234	41.008	51.280	63.184	76.86	92.44	110.08
9	9.672	13.716	18.767	24.944	32.369	41.166	51.465	63.397	77.10	92.72	110.39
10	9.732	13.792	18.860	25.057	32.504	41.325	51.650	63.611	77.35	92.99	110.71
11	9.792	13.868	18.954	25.171	32.639	41.485	51.835	63.825	77.59	93.27	111.02
12	9.852	13.944	19.048	25.285	32.775	41.644	52.022	64.039	77.84	93.55	111.33
13	9.913	14.020	19.142	25.399	32.911	41.805	52.208	64.254	78.08	93.83	111.65
14	9.973	14.097	19.237	25.513	33.048	41.965	52.395	64.470	78.33	94.11	111.96
15	10.034	14.174	19.332	25.628	33.185	42.126	52.582	64.686	78.57	94.39	112.28
16	10.096	14.251	19.427	25.744	33.322	42.288	52.770	64.902	78.82	94.67	112.60
17	10.157	14.329	19.523	25.859	33.460	42.450	52.958	65.119	79.07	94.95	112.91
18	10.219	14.407	19.619	25.975	33.598	42.612	53.147	65.337	79.32	95.23	113.23
19	10.281	14.485	19.715	26.091	33.736	42.775	53.336	65.555	79.57	95.52	113.55
20	10.343	14.563	19.812	26.208	33.875	42.938	53.526	65.773	79.82	95.80	113.87
21	10.406	14.642	19.909	26.325	34.014	43.102	53.716	65.992	80.07	96.08	114.19
22	10.469	14.721	20.006	26.443	34.154	43.266	53.907	66.211	80.32	96.37	114.51
23	10.532	14.800	20.103	26.560	34.294	43.430	54.098	66.431	80.57	96.65	114.83
24	10.595	14.880	20.201	26.678	34.434	43.595	54.290	66.652	80.82	96.94	115.15
25	10.659	14.960	20.299	26.797	34.575	43.760	54.482	66.873	81.07	97.22	115.47
26	10.722	15.040	20.398	26.916	34.716	43.926	54.674	67.094	81.33	97.51	115.80
27	10.786	15.120	20.497	27.035	34.858	44.092	54.867	67.316	81.58	97.80	116.12
28	10.851	15.201	20.596	27.154	35.000	44.259	55.060	67.539	81.83	98.08	116.44
29	10.915	15.282	20.695	27.274	35.142	44.426	55.254	67.762	82.09	98.37	116.77
30	10.980	15.363	20.795	27.394	35.285	44.593	55.448	67.985	82.34	98.66	117.09
31	11.045	15.445	20.895	27.515	35.428	44.761	55.643	68.209	82.60	98.95	117.42
32	11.111	15.527	20.995	27.636	35.572	44.929	55.838	68.434	82.85	99.24	117.75
33	11.176	15.609	21.096	27.757	35.716	45.098	56.034	68.659	83.11	99.53	118.07
34	11.242	15.691	21.197	27.879	35.860	45.267	56.230	68.884	83.37	99.82	118.40
35	11.308	15.774	21.299	28.001	36.005	45.437	56.427	69.110	83.62	100.12	118.73
36	11.375	15.857	21.400	28.123	36.150	45.607	56.624	69.337	83.88	100.41	119.06
37	11.441	15.941	21.502	28.246	36.296	45.777	56.822	69.564	84.14	100.70	119.39
38	11.508	16.024	21.605	28.369	36.441	45.948	57.020	69.791	84.40	100.99	119.72
39	11.575	16.108	21.707	28.493	36.588	46.120	57.218	70.019	84.66	101.29	120.05
40	11.643	16.193	21.810	28.616	36.735	46.291	57.417	70.248	84.92	101.58	120.38
41	11.711	16.277	21.914	28.741	36.882	46.464	57.617	70.477	85.18	101.88	120.71
42	11.779	16.362	22.017	28.865	37.029	46.636	57.817	70.706	85.44	102.17	121.05
43	11.847	16.447	22.121	28.990	37.177	46.809	58.017	70.936	85.71	102.47	121.38
44	11.915	16.533	22.226	29.115	37.326	46.983	58.218	71.167	85.97	102.77	121.72
45	11.984	16.618	22.330	29.241	37.474	47.157	58.420	71.398	86.23	103.07	122.05
46	12.053	16.704	22.435	29.367	37.623	47.331	58.622	71.230	86.50	103.36	122.39
47	12.122	16.791	22.541	29.494	37.773	47.506	58.824	71.862	86.76	103.66	122.72
48	12.192	16.877	22.647	29.620	37.923	47.681	59.028	72.095	87.02	103.96	123.06

TABLE 23–7
Values of involute functions of α inv $\alpha = \tan \alpha - \alpha^{a}$ (*Cont.*)

Minutes	8°	9°	10°	11°	12°	13°	14°	15°	16°	17°	18°
49	12.262	16.964	22.753	29.747	38.073	47.857	59.230	72.328	87.29	104.26	123.40
50	12.332	17.051	22.859	29.875	38.224	48.033	59.434	72.561	87.56	104.56	123.73
51	12.402	17.139	22.966	30.003	38.375	48.210	59.638	72.796	87.82	104.86	124.07
52	12.473	17.227	23.073	30.131	38.527	48.387	59.843	73.030	88.09	105.17	124.41
53	12.544	17.315	23.180	30.260	38.679	48.564	60.048	73.266	88.36	105.47	124.75
54	12.615	17.403	23.288	30.389	38.831	48.742	60.254	73.501	88.63	105.77	125.09
55	12.687	17.492	23.396	30.518	38.984	48.921	60.460	73.738	88.89	106.08	125.43
56	12.758	17.581	23.504	30.648	39.157	49.099	60.667	73.975	89.16	106.38	125.78
57	12.830	17.671	23.613	30.778	39.291	49.279	60.874	74.212	89.43	106.69	126.12
58	12.903	17.760	23.722	30.908	39.445	49.458	61.081	74.450	89.70	106.99	126.46
59	12.975	17.850	23.831	31.039	39.599	49.639	61.289	74.688	89.98	107.30	126.81
60	13.048	17.941	23.941	31.171	39.754	49.819	61.498	74.927	90.25	107.60	127.15

TABLE 23–7
Values of involute functions of α inv $\alpha = \tan \alpha - \alpha^{a}$ (*Cont.*)

Minutes	19°	20°	21°	22°	23°	24°	25°	26°	27°	28°	29°	30°
0	127.15	149.04	173.45	200.54	230.49	263.50	299.75	339.47	382.87	430.17	481.64	537.51
1	127.50	149.43	173.88	201.01	231.02	264.07	300.39	340.16	383.62	431.00	482.53	538.49
2	127.84	149.82	174.31	201.49	231.54	264.65	301.02	340.86	384.38	431.82	483.43	539.46
3	128.19	150.20	174.74	201.97	232.07	265.23	301.66	341.55	385.14	432.64	484.32	540.43
4	128.54	150.59	175.17	202.44	232.59	265.81	302.29	342.25	385.90	433.47	485.22	541.40
5	128.88	150.98	175.60	202.92	233.12	266.39	302.93	342.94	386.66	434.30	486.12	542.38
6	129.23	151.37	176.03	203.42	233.65	266.97	303.57	343.64	387.42	435.13	487.02	543.36
7	129.58	151.76	176.47	203.88	234.18	267.56	304.20	344.34	388.18	435.96	487.92	544.33
8	129.93	152.15	176.90	204.36	234.71	268.14	304.84	345.04	388.94	436.79	488.83	545.31
9	130.28	152.54	177.34	204.84	235.24	268.72	305.49	345.74	389.71	437.62	489.73	546.29
10	130.63	152.93	177.77	205.33	235.77	269.31	306.13	346.44	390.47	438.45	490.64	547.28
11	130.98	153.33	178.21	205.81	236.31	269.89	306.77	347.14	391.24	439.29	491.54	548.26
12	131.34	153.72	178.65	206.29	236.84	270.48	307.41	347.85	392.01	440.12	492.45	549.24
13	131.69	154.11	179.08	206.78	237.38	271.07	308.06	348.55	392.78	440.96	493.36	550.23
14	132.04	154.51	179.52	207.26	237.91	271.66	308.70	349.26	393.55	441.80	494.27	551.22
15	132.40	154.90	179.96	207.75	238.45	272.25	309.35	349.97	394.32	442.64	495.18	552.21
16	132.75	155.30	180.40	208.24	238.99	272.84	310.00	350.67	395.09	443.48	496.09	553.20
17	133.11	155.70	180.84	208.73	239.52	273.43	310.65	351.38	395.86	444.32	497.01	554.19
18	133.46	156.09	181.29	209.21	240.06	274.02	311.30	352.09	396.64	445.16	497.92	555.18
19	133.82	156.49	181.73	209.70	240.60	274.62	311.95	352.80	397.41	446.01	498.84	556.17
20	134.18	156.89	182.17	210.19	241.14	275.21	312.60	353.52	398.19	446.85	499.76	557.17
21	134.54	157.29	182.62	210.69	241.69	275.81	313.25	354.23	398.97	447.70	500.68	558.15
22	134.90	157.69	183.06	211.18	242.23	276.40	313.90	354.94	399.74	448.55	501.60	559.16
23	135.26	158.09	183.51	211.67	242.77	277.00	314.56	355.66	400.52	449.39	502.52	560.16
24	135.62	158.50	183.95	212.17	243.32	277.60	315.21	356.37	401.31	450.24	503.44	561.16
25	135.98	158.90	184.40	212.66	243.86	278.20	315.87	357.09	402.09	451.10	504.37	562.17
26	136.34	159.30	184.85	213.16	244.41	278.80	316.53	357.81	402.87	451.95	505.29	563.17
27	136.70	159.71	185.30	213.65	244.95	279.40	317.18	358.53	403.66	452.80	506.22	564.17
28	137.07	160.11	185.75	214.15	245.50	280.00	317.84	359.25	404.44	453.66	507.15	565.18

TABLE 23–7
Values of involute functions of α *inv* $\alpha = \tan \alpha - \alpha^a$ (*Cont.*)

Minutes	19°	20°	21°	22°	23°	24°	25°	26°	27°	28°	29°	30°
29	137.43	160.52	186.20	214.65	246.05	280.60	318.50	359.97	405.23	454.51	508.08	566.19
30	137.79	160.92	186.65	215.14	246.60	281.21	319.17	360.69	406.02	455.37	509.01	567.20
31	138.16	161.33	187.10	215.64	247.15	281.81	319.83	361.42	406.80	456.23	509.94	568.21
32	138.52	161.74	187.55	216.14	247.70	282.42	320.49	362.14	407.59	457.09	510.87	569.22
33	138.89	162.15	188.00	216.65	248.25	283.02	321.16	362.87	408.39	457.95	511.81	570.23
34	139.26	162.55	188.46	217.15	248.81	283.63	321.82	363.59	409.18	458.81	512.74	571.24
35	139.63	162.96	188.91	217.65	249.36	284.24	322.49	364.32	409.97	459.67	513.68	572.26
36	139.99	163.37	189.37	218.15	249.92	284.85	323.15	365.05	410.76	460.54	514.62	573.28
37	140.36	163.79	189.83	218.66	250.47	285.46	323.82	365.78	411.56	461.40	515.56	574.29
38	140.73	164.20	190.28	219.16	251.03	286.07	324.49	366.51	412.36	462.27	516.50	575.31
39	141.10	164.61	190.74	219.67	251.59	286.68	325.16	367.24	413.16	463.13	517.44	576.33
40	141.48	165.02	191.20	220.18	252.14	287.29	325.83	367.98	413.95	464.00	518.38	577.36
41	141.85	165.44	191.66	220.68	252.70	287.91	326.51	368.71	414.75	464.87	519.33	578.38
42	142.22	165.85	192.12	221.19	253.26	288.52	327.18	369.45	415.56	465.75	520.27	579.40
43	142.59	166.27	192.58	221.70	253.82	289.14	327.85	370.18	416.36	466.62	521.22	580.43
44	142.97	166.69	193.04	222.21	254.39	289.76	328.53	370.92	417.16	467.49	522.17	581.46
45	143.34	167.10	193.50	222.72	254.95	290.37	329.20	371.66	417.97	468.37	523.12	582.49
46	143.72	167.52	193.97	223.24	255.51	290.99	329.88	372.40	418.77	469.24	524.07	583.52
47	144.09	167.94	194.43	223.75	256.08	291.61	330.56	373.14	419.58	470.12	525.02	584.55
48	144.47	168.36	194.90	224.26	256.64	292.23	331.24	373.88	420.39	471.00	525.97	585.58
49	144.85	168.78	195.36	224.78	257.21	292.85	331.92	374.62	421.20	471.88	526.93	586.62
50	145.23	169.20	195.83	225.29	257.77	293.48	332.60	375.37	422.01	472.76	527.88	587.65
51	145.60	169.62	196.30	225.81	258.34	294.10	333.28	376.11	422.82	473.64	528.84	588.69
52	145.98	170.04	196.76	226.33	258.91	294.72	333.97	376.86	423.63	474.52	529.80	589.73
53	146.36	170.47	197.23	226.84	259.48	295.35	334.65	377.61	424.44	475.41	530.76	590.77
54	146.74	170.89	197.70	227.36	260.05	295.98	335.34	378.35	425.26	476.30	531.72	591.81
55	147.13	171.32	198.17	227.88	260.62	296.60	336.02	379.10	426.07	477.18	532.68	592.85
56	147.51	171.74	198.64	228.40	261.20	297.23	336.71	379.85	426.89	478.07	533.65	593.90
57	147.89	172.17	199.12	228.92	261.77	297.86	337.40	380.60	427.71	478.96	534.61	594.94
58	148.27	172.59	199.59	229.44	262.35	298.49	338.09	381.36	428.53	479.85	535.58	595.99
59	148.66	173.02	200.07	229.97	262.92	299.12	338.78	382.11	429.35	480.74	536.55	597.04
60	149.04	173.45	200.54	230.49	263.50	299.75	339.47	382.87	430.17	481.64	537.51	598.09

^{a} Values are to be multiplied by 10^{-4}.

TABLE 23-8
Formula for main proportions of corrected spur gears "S correction" $x_1 \neq \pm x_2 \neq 0$ *and* $\sum x = |x_1| + |x_2| \neq 0$

Particular	Formula	Particular	Formula
Operating pressure angle α_v	$inv\,\alpha_v = inv\,\alpha + \dfrac{x_2 \pm x_1}{z_2 \pm z_1}\tan\alpha$ where $-ve$ sign for internal gears (Refer to Table 23–7 for α_v.)		(Refer to Table 23–12 for λ_o which depends on x_o.)
The sum or difference of correction factors	$x_1 \pm x_2 = \dfrac{inv\,\alpha_v - inv\,\alpha}{2\tan\alpha}(z_2 \pm z_1)$	The relative correction factor	$x_o = \dfrac{2(x_2 \pm x_1)}{(z_2 \pm z_1)}$ (Refer to Table 23–12 for x_o.)
Diameter of operating circle of pinion Diameter of operating circle of gear	$d_{v1} = mz_1$ $d_{v2} = mz_2$	The increase of center distance	$\Delta a = a_v - a = \dfrac{1}{2}m(z_2 \pm z_1)\left(\dfrac{\cos\alpha}{\cos\alpha_v} - 1\right)$
Pitch circle diameter of pinion	$d'_1 = d_{v1}\dfrac{\cos\alpha}{\cos\alpha_v}$	The factor of center distance deviation $\delta_a = \dfrac{\Delta a}{m}$	(Refer to Table 23–10 for δ_a.)
Pitch circle diameter of gear	$d'_2 = d_{v2}\dfrac{\cos\alpha}{\cos\alpha_v}$	Tooth dedendum of pinion Tooth dedendum of gear Total tooth depth The change of tooth depth	$h_{f1} = (1.25 - x_1)m$ $h_{f2} = (1.25 - x_2)m$ $h = h_{f1} + h_{f2} + \Delta a - c$ $\Delta h = 2.25m - h$
The base circle diameter of pinion	$d_{b1} = d'_1\cos\alpha$	The factor of tooth depth deviation	$\delta_h = \dfrac{\Delta h}{m}$
The base circle diameter of gear	$d_{b2} = d'_2\cos\alpha$		(Refer to Table 23–10 for δ_h.)
The center distance at $x_1 = x_2 = 0$; i.e., S zero gear pair	$a = \left(\dfrac{z_1 + z_2}{2}\right)m$	The addendum of pinion The addendum of gear The addendum circle diameter of pinion	$h_{a1} = h - h_{f1}$ $h_{a2} = h - h_{f2}$ $d_{a1} = d_{v1} + 2h_{a1}$
The center distance of S gear pair	$a_v = m\left(\dfrac{z_2 \pm z_1}{2}\right)\dfrac{\cos\alpha}{\cos\alpha_v}$ $= a\dfrac{\cos\alpha}{\cos\alpha_v} = a(\lambda_o + 1)$	The addendum circle diameter of gear The dedendum circle diameter of pinion The dedendum circle diameter of gear	$d_{a2} = d_{v2} + 2h_{a2}$ $d_{f1} = d_{v1} - 2h_{f1}$ $d_{f2} = d_{v2} - 2h_{f2}$
The axis shift factor	$\lambda_o = \dfrac{\cos\alpha}{\cos\alpha_v} - 1$		

TABLE 23–9
Formula for main proportions of corrected spur gears "S_o Correction," $x_1 + x_2 = 0$; $x_1 = -x_2$

Particular	Formula
Tooth addendum of pinion	$h_{a1} = m(1 + x_1) - m\xi_1$
Tooth addendum of gear	$h_{a2} = m(1 - x_2) = m\xi_2$
Tooth dedendum of pinion	$h_{f1} = m(1.25 - x_1)$
Tooth dedendum of gear	$h_{f2} = m(1.25 + x_2)$
Center distance at $x_2 = -x_1$ and $x_1 = x_2 = 0$	$a = m\left(\dfrac{z_1 + z_2}{2}\right)$
Thickness of pinion tooth	$s_1 = m\left(\dfrac{\pi}{2} + 2x_1 \tan\alpha\right)$
Thickness of gear tooth	$s_2 = m\left(\dfrac{\pi}{2} + 2x_2 \tan\alpha\right)$
Addendum circle diameter of pinion	$d_{a1} = m(z_1 + 2 + 2x_1)$
Addendum circle diameter of gear	$d_{a2} = m(z_2 + 2 + 2x_2)$
Dedendum circle diameter of pinion	$d_{f1} = m(z_1 - 2.5 + 2x_1)$
Dedendum circle diameter of gear	$d_{f2} = m(z_2 - 2.5 + 2x_2)$

TABLE 23–10
Values of "S Corrected" gears

Sum of teeth $(z_1 + z_2)$	Sum of correction factors $(x_1 + x_2)$	Operating pressure angle, α_v	Center distance deviation factor, δ_a	Tooth-depth deviation factor, δ_h
16	0.880	35°12′	0.698	0.181
18	0.792	28°46′	0.648	0.144
19	0.748	28°07′	0.621	0.127
20	0.704	27°29′	0.592	0.112
21	0.660	26°53′	0.563	0.097
22	0.616	26°19′	0.532	0.084
24	0.528	25°15′	0.467	0.061
25	0.484	24°44′	0.433	0.051
26	0.440	24°15′	0.398	0.040
27	0.396	23°46′	0.362	0.030
28	0.352	23°19′	0.325	0.026
30	0.264	22°26′	0.249	0.015
31	0.220	22°00′	0.210	0.010
32	0.176	21°35′	0.170	0.006
34	0.088	20°47′	0.087	0.001
35	0.044	20°23′	0.043	0.000
36	0.000	20°00′	0.000	0.000

TABLE 23–11
Correction factors x_1 and x_2

	12		15		18		22		28		34		Maximum increase in resistance[a]
z_2	x_1	x_2	x_1	x_2	x_1	x_2	x_1	x_2	x_1	x_2	x_1	x_2	
	0.30	0.61	0.34	0.64	0.54	0.54							P
18	0.57	0.25	0.64	0.29	0.72	0.34							F
	0.49	0.35	0.48	0.46	0.54	0.54							S
	0.30	0.66	0.38	0.75	0.60	0.64	0.68	0.68					P
22	0.62	0.28	0.73	0.32	0.81	0.38	0.95	0.39					F
	0.53	0.38	0.55	0.54	0.60	0.63	0.67	0.67					S
	0.30	0.88	0.26	1.04	1.40	1.02	0.59	0.94	0.86	0.86			P
28	0.70	0.26	0.79	0.35	0.89	0.38	1.04	0.40	1.26	0.42			F
	0.57	0.48	0.60	0.63	0.63	0.72	0.71	0.81	0.85	0.85			S
	0.30	1.03	0.13	1.42	0.30	1.30	0.48	1.20	0.80	1.08	1.01	1.01	P
34	0.76	0.22	0.83	0.34	0.93	0.37	1.08	0.38	1.30	0.36	1.38	0.34	F
	0.60	0.53	0.63	0.72	0.67	0.82	0.74	0.90	0.86	1.00	1.00	1.00	S
	0.30	0.30	0.20	1.53	0.29	1.48	0.40	1.48	0.72	2.33	0.90	1.30	P
42	0.75	0.21	0.92	0.32	1.02	0.36	1.18	0.38	1.24	0.31	1.31	0.27	F
	0.63	0.67	0.68	0.88	0.68	0.94	0.76	1.03	0.88	1.12	1.00	1.16	S
	0.30	1.43	0.25	1.65	0.32	1.63	0.43	1.60	0.64	1.60	0.80	1.58	P
50	0.58	0.16	0.97	0.31	1.05	0.36	1.22	0.42	1.22	0.25	1.25	0.20	F
	0.63	0.77	0.66	1.02	0.70	0.11	0.76	1.17	0.91	1.26	1.00	1.31	S
	0.30	1.69	0.26	1.87	0.41	1.89	0.53	1.80	0.70	1.84	0.83	1.79	P
65	1.55	0.35	0.80	0.04	1.10	0.40	1.17	0.36	1.19	0.29	1.23	0.15	F
	0.64	1.00	0.67	1.22	0.71	1.35	0.76	1.44	0.88	1.56	0.99	1.55	S
	0.30	1.96	0.30	2.14	0.48	2.08	0.61	1.99	0.75	2.04	0.89	1.97	P
80	0.54	0.54	0.73	0.15	1.14	0.40	1.15	0.26	1.16	0.12	1.19	0.07	F
	0.65	1.18	0.67	1.36	0.71	1.61	0.76	1.73	0.87	0.85	0.98	1.81	S
	0.30	2.90	0.36	2.32	0.52	2.31	0.65	2.19	0.80	2.26	0.94	0.22	P
100	0.53	0.76	0.71	0.22	1.00	0.28	1.12	0.22	1.14	0.08	1.15	0.01	F
	0.65	1.42	0.66	1.70	0.71	1.90	0.76	1.98	0.86	2.12	0.97	2.15	S
							0.75	2.43	0.83	2.47	1.00	2.46	P
125							1.11	0.21	1.12	0.07	1.20	0.09	F
							0.76	2.38	0.86	2.40	0.92	2.40	S

[a] P—maximum increase in resistance to pitting; F—maximum resistance of teeth to fracture; S—equality of unit slips (increase in resistance to wear and seizure) obtained at $s_{a(min)} \geq 0.25$ m and $\epsilon_{min} \geq 1.2$.

TABLE 23–12
Axial shift factor λ_o and total relative correction factor x_o for various values of pressure angles

Minutes	$\alpha = 20°$		$\alpha = 21°$		$\alpha = 22°$		$\alpha = 23°$		$\alpha = 24°$		$\alpha = 25°$		$\alpha = 26°$		$\alpha = 27°$		$\alpha = 28°$		$\alpha = 29°$	
	λ_o	x_o	λ_o	x_o	λ_o	x_o	λ_o	x_o	λ_o	x_o	λ_o	x_o	λ_o	x_o	λ_o	x_o	λ_o	x_o	λ_o	x_o
0	0.00000	0.00000	0.00655	0.00671	0.01349	0.01415	0.02085	0.02238	0.02862	0.03145	0.03684	0.04141	0.04550	0.05232	0.05464	0.06424	0.06427	0.07724	0.07440	0.09138
10	0.00106	0.00107	0.00768	0.00789	0.01469	0.01547	0.02211	0.02383	0.02996	0.03304	0.03825	0.04316	0.04699	0.05424	0.05621	0.06633	0.06592	0.07952	0.07614	0.09385
20	0.00214	0.00216	0.00882	0.00910	0.01590	0.01680	0.02339	0.02530	0.03131	0.03467	0.03967	0.04494	0.04850	0.05618	0.05780	0.06845	0.06759	0.08182	0.07790	0.09636
30	0.00323	0.00326	0.00997	0.01033	0.01712	0.01816	0.02468	0.02681	0.03267	0.03631	0.04111	0.04674	0.05001	0.05815	0.05939	0.07061	0.06927	0.08416	0.07967	0.09890
40	0.00432	0.00439	0.01113	0.01158	0.01835	0.01955	0.02578	0.02833	0.03405	0.03798	0.04256	0.04857	0.05154	0.06015	0.06100	0.07278	0.07097	0.08654	0.08145	0.10148
50	0.00543	0.00554	0.01231	0.01286	0.01959	0.02095	0.02730	0.02988	0.03544	0.03969	0.04403	0.05043	0.05309	0.06218	0.06263	0.07500	0.07268	0.08894	0.08325	0.10109
60	0.00655	0.00671	0.01349	0.01415	0.02085	0.02238	0.02868	0.03145	0.03684	0.04141	0.04550	0.05232	0.05464	0.06424	0.06427	0.07124	0.07440	0.09138	0.08507	0.10673

Note: The intermediate values of the pressure angle α, λ_o, and x_o are determined by interpolation.

TABLE 23–13
Service factor for C_s for use in Eqs. (2–87) to (2–89) and Eqs. (23–312) and (23–313)

	Type of service		
Type of load	Intermittent 3 h per day	8–10 h per day	Continuous 24 h per day
Steady shock	0.80	1.00	1.25
Light shock	1.00	1.25	1.50
Medium shock	1.25	1.50	1.80
Heavy shock	1.50	1.80	2.00

TABLE 23–14
Values of tooth-form factor (Lewis) y, load at tip of tooth

| Number of teeth, z | 14.5° involute form | 14.5° variable center distance | 20° full-depth involute form | 20° stub-tooth involute form | Internal drives | |
					Spur pinion	Gear[a]
12	0.067	0.125	0.078	0.099	0.104	
13	0.071	0.123	0.083	0.103	0.104	
14	0.075	0.121	0.088	0.108	0.105	+
15	0.078	0.120	0.092	0.111	0.105	+
16	0.081	0.120	0.094	0.115	0.106	
17	0.084	0.120	0.096	0.117	0.109	
18	0.086	0.120	0.098	0.120	0.111	
19	0.088	0.119	0.100	0.123	0.114	
20	0.090	0.119	0.102	0.125	0.116	
21	0.092	0.119	0.104	0.127	0.118	
22	0.093	0.119	0.105	0.129	0.119	
24	0.095	0.118	0.107	0.132	0.122	
26	0.098	0.117	0.110	0.135	0.125	0.220
28	0.100	0.115	0.112	0.137	0.127	0.216
30	0.101	0.114	0.114	0.139	0.129	
34	0.104	0.112	0.118	0.142	0.132	0.210
38	0.106	0.110	0.122	0.145	0.135	0.205
43	0.108	0.108	0.126	0.147	0.137	0.200
50	0.110	0.110	0.130	0.151	0.139	0.195
60	0.113	0.113	0.134	0.154	0.142	0.190
75	0.115	0.115	0.138	0.158	0.144	0.185
100	0.117	0.117	0.142	0.161	0.147	0.180
150	0.119	0.118	0.146	0.165	0.149	0.175
300	0.122	0.122	0.150	0.170	0.152	0.170
Rack	0.124	0.124	0.154	0.175		

[a] Internal gears with less than 28 teeth must be designed specially for the particular application, and their values of y must be determined for each one individually.

TABLE 23–15
Values of tooth-form factor (Lewis) *y*, when the load is near the middle of the tooth

Number of teeth, *z*	14.5° involute form	20° full-depth involute form	20° stub-tooth involute form	Internal drives	
				Spur pinion	Gear[a]
12	0.113	0.132	0.151	0.207	
13	0.120	0.141	0.164	0.208	
14	0.127	0.149	0.172	0.209	+
15	0.132	0.156	0.177	0.210	+
16	0.137	0.160	0.184	0.211	
17	0.142	0.163	0.187	0.215	
18	0.146	0.166	0.192	0.218	
19	0.150	0.170	0.196	0.222	
20	0.153	0.173	0.200	0.225	
21	0.156	0.176	0.203	0.288	
22	0.158	0.178	0.206	0.230	
24	0.162	0.182	0.211	0.233	
26	0.166	0.187	0.216	0.236	
28	0.170	0.190	0.219	0.239	0.400
30	0.172	0.193	0.222	0.242	0.395
34	0.176	0.200	0.227	0.246	0.387
38	0.180	0.207	0.232	0.250	0.380
43	0.183	0.214	0.335	0.253	0.372
50	0.187	0.221	0.271	0.256	0.364
60	0.192	0.227	0.246	0.260	0.356
75	0.195	0.234	0.252	0.264	0.348
100	0.198	0.241	0.257	0.268	0.340
150	0.202	0.248	0.264	0.272	0.332
300	0.207	0.255	0.272	0.276	0.325
Rack	0.210	0.262	0.280		

[a] Internal gears with less than 28 teeth must be designed specially for their particular application; their values of *y* must be determined for each one individually.

TABLE 23–16
Values of the modified form factor (Lewis), $Y = \pi y$, for various involute tooth systems

Number of teeth, z	14.5° composite and involute form	20° full-depth involute form	Small pinion 20° full-depth involute form	20° stub-tooth involute form	Internal drives, 20° full-depth involute form	
					Pinion	Gear
5			0.320		0.322	
6			0.201		0.322	
7			0.282		0.322	
8			0.264		0.324	
9			0.264		0.324	
10			0.264		0.324	
11			0.264		0.326	
12	0.211	0.245	0.264	0.312	0.326	
13	0.223	0.261	0.270	0.324	0.326	
14	0.236	0.277	0.277	0.340	0.330	
15	0.245	0.290		0.350	0.330	
16	0.254	0.291		0.362	0.333	
17	0.264	0.303		0.368	0.342	
18	0.270	0.309		0.378	0.348	
19	0.277	0.314		0.388	0.358	
20	0.283	0.322		0.394	0.364	
21	0.289	0.328		0.400	0.370	
22	0.292	0.331		0.406	0.374	
24	0.299	0.337		0.416	0.383	
26	0.308	0.346		0.425	0.393	
28	0.314	0.353		0.432	0.399	0.691
30	0.318	0.359		0.438	0.405	0.678
34	0.327	0.371		0.447	0.414	0.659
38	0.333	0.384		0.457	0.424	0.643
43	0.340	0.397		0.463	0.430	0.628
50	0.346	0.409		0.476	0.436	0.612
60	0.355	0.422		0.485	0.446	0.592
75	0.361	0.435		0.497	0.452	0.581
100	0.367	0.447		0.507	0.461	0.565
150	0.374	0.460		0.520	0.468	0.549
300	0.383	0.472		0.535	0.477	0.533
Rack	0.390	0.485		0.552		

TABLE 23–17
Values of the AGMA Lewis form factor Y

Number of teeth, z	$\alpha = 20°$ $h_a = 0.800$ $h_f = 1.000$	$\alpha = 20°$ $h_a = 1.000$ $h_f = 1.250$	$\alpha = 25°$ $h_a = 1.000$ $h_f = 1.250$	$\alpha = 25°$ $h_a = 1.000$ $h_f = 1.350$
12	0.335 12	0.229 60	0.276 77	0.254 73
13	0.348 27	0.243 17	0.292 81	0.271 77
14	0.359 85	0.255 30	0.307 17	0.287 11
15	0.370 13	0.266 22	0.320 09	0.301 00
16	0.379 31	0.276 10	0.331 78	0.313 63
17	0.387 57	0.285 08	0.342 40	0.325 17
18	0.395 02	0.293 27	0.352 10	0.335 74
19	0.401 79	0.300 78	0.360 99	0.345 46
20	0.407 97	0.307 69	0.369 16	0.354 44
21	0.413 63	0.314 06	0.376 71	0.362 76
22	0.418 83	0.319 97	0.383 70	0.370 48
24	0.428 06	0.330 56	0.396 24	0.384 39
26	0.436 01	0.339 79	0.407 17	0.396 57
28	0.442 94	0.347 90	0.416 78	0.407 33
30	0.449 02	0.355 10	0.425 30	0.416 91
34	0.459 20	0.367 31	0.439 76	0.433 23
38	0.467 40	0.377 27	0.451 56	0.446 63
45	0.478 46	0.390 93	0.467 74	0.465 11
50	0.484 58	0.398 60	0.476 81	0.475 55
60	0.493 91	0.410 47	0.490 86	0.491 77
75	0.503 45	0.422 83	0.505 46	0.508 77
100	0.513 21	0.435 74	0.520 71	0.526 65
150	0.523 21	0.449 30	0.536 68	0.545 56
300	0.533 48	0.463 64	0.553 51	0.565 70
Rack	0.544 06	0.478 97	0.571 39	0.587 39

TABLE 23–18
Allowable static stresses σ_{so} for use in Lewis Eqs. (23–91) to (23–94)

Material	Heat treatment[a]	Brinell hardness number, H_B	Allowable static stress, σ_{so}	
			MPa	kpsi
Gray cast iron				
Ordinary cast iron			55	8
ASTM 25		174	55	8
ASTM 35		212	83	12
ASTM 50 (high-grade cast iron)		223	103	15
Low-carbon cast steel				
0.20% C	Untreated	180	138	20
0.20% C	WQT	250	173	25
Carbon steel, forged				
SAE 1030	Untreated	180	138	20
SAE 1035	Untreated	190	158	23
SAE 1040	Untreated	202	173	25
SAE 1045	Untreated	215	207	30
SAE 1020	Case-hardened and WQT	156	124	18
SAE 1030	Heat-treated		220	32
SAE 1045	Hardened by WQT	205	220	32
SAE 1050	Hardened by OQT	223	240	35
Alloy steels				
SAE 2320	Case-hardened and WQT	225	345	50
SAE 2345	Hardened by OQT	475	345	50
SAE 3145	Hardened by OQT	475	365	53
SAE 3115	Case-hardened by OQT	212	255	37
SAE 3245 (Cr-Ni steel)	Hardened by OQT	475	448–517	65–75
SAE 4340	Hardened by OQT	475	448	65
SAE 4640	Hardened by OQT	475	379	55
SAE 6145 (Cr-Va steel)	Hardened by OQT	475	452–462	65.5–67
Copper alloys				
SAE 43 (ASTM B137-52, 8A)		100	138	20
Manganese bronze				
SAE 62 (ASTM B143-52, 1A)		80	69	10
Bronze (gun metal)				
SAE 65 (ASTM B144-52, 3C)		100	83	12
Phosphor bronze				
SAE 68 (ASTM B148-52, 98)		180	152	22
Aluminum bronze				
Rawhide, Fabroil, etc.			40	6
Laminated phenolic materials, Bakelite, Micarta, Celeron			40–55	6–8

[a] WQT—water-quenching treatment; OQT—oil-quenching treatment.

TABLE 23–19
AGMA geometry factor _J_ for teeth having $\alpha = 20°$ [a]

Number of teeth, z	Number of teeth in mating gear						
	17	25	35	50	85	300	1000
18	0.32404	0.33214	0.33840	0.34404	0.35050	0.35594	0.36112
19	0.33029	0.33878	0.34537	0.35134	0.35822	0.36405	0.36963
20	0.33600	0.34485	0.35176	0.35804	0.36532	0.37151	0.37749
21	0.34124	0.35044	0.35764	0.36422	0.37186	0.37841	0.38475
22	0.34607	0.35559	0.36306	0.36992	0.37792	0.38479	0.39148
24	0.35468	0.36477	0.37275	0.38012	0.38877	0.39626	0.40360
26	0.36211	0.37272	0.38115	0.38897	0.39821	0.40625	0.41418
28	0.36860	0.37967	0.38851	0.39673	0.40650	0.41504	0.42351
30	0.37462	0.38580	0.39500	0.40359	0.41383	0.42283	0.43179
34	0.38394	0.39611	0.40594	0.41517	0.42624	0.43604	0.44586
38	0.39170	0.40446	0.41480	0.42456	0.43633	0.44680	0.45735
45	0.40223	0.41579	0.42685	0.43735	0.45010	0.46152	0.47310
50	0.40808	0.42208	0.43355	0.44448	0.45778	0.46975	0.48193
60	0.41702	0.43173	0.44383	0.45542	0.46960	0.48243	0.49557
75	0.42620	0.44163	0.45440	0.46668	0.48179	0.49554	0.50970
100	0.43561	0.45180	0.46527	0.47827	0.49437	0.50909	0.52435
150	0.44530	0.46226	0.47645	0.49023	0.50736	0.52312	0.53954
300	0.45526	0.47304	0.48798	0.50256	0.52078	0.53765	0.55533
Rack	0.46554	0.48415	0.49988	0.51529	0.53467	0.55272	0.57173

[a] $h_a = 1.000$ in, $h_f = 1.250$ in, and $\rho_f = 0.300$ in ($h_a = 1$ m; $h_f = 1.25$ m; $\rho_f = 0.300$ m)
Source: J. E. Shigley, *Mechanical Engineering Design*, McGraw-Hill Book Company, 1949.

TABLE 23–20
AGMA geometry factor J for teeth having $\alpha = 25°$ [a]

Number of teeth, z	Number of teeth in mating gear						
	17	25	35	50	85	300	1000
13	0.34684	0.35292	0.35744	0.36138	0.36572	0.36925	0.37251
14	0.35924	0.36587	0.37081	0.37514	0.37994	0.38386	0.38749
15	0.37027	0.37740	0.38275	0.38744	0.39267	0.39694	0.40092
16	0.38016	0.38775	0.39346	0.39849	0.40411	0.40873	0.41303
17	0.38907	0.39709	0.40314	0.40849	0.41448	0.41941	0.42402
18	0.39714	0.40556	0.41193	0.41756	0.42390	0.42913	0.43403
19	0.40449	0.41328	0.41994	0.42585	0.43250	0.43801	0.44318
20	0.41121	0.42034	0.42727	0.43344	0.44039	0.44616	0.45159
21	0.41738	0.42682	0.43401	0.44042	0.44765	0.45367	0.45933
22	0.42306	0.43280	0.44023	0.44686	0.45436	0.46060	0.46650
24	0.43318	0.44346	0.45132	0.45836	0.46635	0.47301	0.47932
26	0.44193	0.45268	0.46093	0.46833	0.47674	0.48378	0.49046
28	0.44957	0.46075	0.46933	0.47705	0.48585	0.49323	0.50023
30	0.45631	0.46785	0.47675	0.48475	0.49389	0.50157	0.50868
34	0.46763	0.47981	0.48923	0.49772	0.50746	0.51566	0.52349
38	0.47678	0.48948	0.49933	0.50824	0.51847	0.52710	0.53536
45	0.48919	0.50261	0.51305	0.52252	0.53344	0.54268	0.55154
50	0.49608	0.50991	0.52068	0.53047	0.54177	0.55136	0.56056
60	0.50683	0.52109	0.53238	0.54267	0.55457	0.56469	0.57444
75	0.51747	0.53257	0.54440	0.55520	0.56773	0.57842	0.58873
100	0.52860	0.54436	0.55676	0.56810	0.58129	0.59257	0.60348
150	0.54005	0.55651	0.56951	0.58138	0.59526	0.60716	0.61869
300	0.55185	0.56902	0.58259	0.59507	0.60967	0.62222	0.63442
Rack	0.56405	0.58194	0.59613	0.60921	0.62456	0.63778	0.65068

[a] $h_a = 1.000$ in, $h_f = 1.250$ in, and $\rho_f = 0.300$ in ($h_a = 1$ m; $h_f = 1.25$ m; $\rho_f = 0.300$ m).
Source: J. E. Shigley, *Mechanical Engineering Design*, McGraw-Hill Book Company, 1949.

TABLE 23–21
Overload correction factor or application factor C_a and K_a

Source of power	Driven machinery		
	Uniform	Moderate shock	Heavy shock
Uniform	1.10	1.25	1.75
Light shock	1.25	1.50	2.00
Medium shock	1.50	1.75	2.50

TABLE 23–22
Load distribution factor K_m and C_m for spur and helical gears

Characteristics of support	Face width, mm (in)			
	0–50	150	225	400+
	(0–2	6	9	16+)
Accurate mounting, small bearing clearances, minimum deflection, precision gears	1.3 (1.2)[a]	1.4 (1.3)	1.5 (1.4)	1.8 (1.7)[a]
Less rigid mountings, less accurate gears, contact across full face	1.6 (1.5)	1.7 (1.6)	1.8 (1.7)	2.2 (2.0)
Accuracy and mounting such that less than full-face contact exists		> 2.0		

[a] K_m and C_m values in parentheses apply for helical gears.
Source: AGMA 215.01 and 225.01.

TABLE 23–23
Allowable bending strength or allowable bending stress numbers, $\sigma_{sab}(S_{at})$ for metallic gears (3, 4, 13)

Material	AGMA class	Commercial designation	Heat treatment	Brinell, H_B at surface	core	Rockwell at surface R_c	Spur and helical MPa	kpsi	Bevel MPa	kpsi
Steel	A–1 through A–5	—	Through-hardened and tempered	180 240 300 360 400		— — — — —	110*–230** 210–280 250–320 280–360 290–390	25*–33** 31–41 36–47 40–52 42–56	170*–230** 210–280 250–320 280–360 290–390	25–33 31–41 36–47 40–52 42–56
			Flame- or induction-hardened with type A pattern			50–54	310–380	45–55	45–55 18–22	310–380 125–150
			Frame- or induction-hardened with type B pattern				150–150	22–22		
			Carburized and case-hardened			55 60	380–450 or 485	55–65 or 70	380–450 380–480	55–65 55–70
		AISI 4140	Nitrided#	84.5	15N 300	48#	235–310	34–45	240–310	35–45
		AISI 4340	Nitrided#	83.5	15N 300	46#	250–325	36–47	250–325	36–47
		Nitralloy 135M	Nitrided#	90.0	15N 300	60#	260–330	38–48	260–330	38–48
		Nitralloy N	Nitrided	90.0	15N		275–345	40–50		
		2.5% chrome	Nitrided	87.5–9.0	15N 350	54–60#	380–450	55–65	380–450	55–65
Cast iron		ASTM A 48								
	20	Class 20	As cast				35	5	35	5
	30	Class 30	As cast	175			60	8.5	60	8.5
	40	Class 40	As cast	200			90	13	90	13
Nodular (ductile) iron	A-7-a	ASTM A 536 60-4-18	Annealed	140			150–230	22–33	90–100% of value for steel with same hardness	
	A-7-c	80-55-06	Quenched and tempered	180			150–230	22–33		
	A-7-d	100-70-03	Quenched and tempered	230			185–275	27–40		
	A-7-e	120-90-02	Quenched and tempered	270			215–305	31–44		
Malleable iron (pearlitic)	A-8-c	ASTM A 220 45007	—	165		—	70	10	70	10
	A-8-e	50005	—	180		—	90	13	90	13
	A-8-f	53007	—	195		—	110	16	110	16
	A-8-i	80002	—	240		—	145	21	145	21
Bronze	Bronze 2	AGMA 2C	Sand-cast Sand-cast	Tensile strength minimum 275 MPa (40 kpsi)			39.5	5.7		
	Al/Br 3	ASTM B-148-52 Alloy 954	Heat-treated	Tensile strength minimum 620 MPa (90 kpsi)			23.6	163		

Key: * grade 1; ** grade 2; # bevel gears (the overload capacity of nitrided gears is low, since the shape of the effective S–N curve is flat).
Source: Courtesy AGMA; compiled from AGMA Standards 2001-B88 and 2003-A86.

TABLE 23–24
Allowable contact strength or allowable contact stress numbers, $\sigma_{sac}(S_{ac})$ for metallic gears[a]

Material	AGMA class	Commercial designation	Heat treatment	Brinell, H_B	Rockwell, R_C	Spur and helical MPa	Spur and helical kpsi	Bevel MPa	Bevel kpsi
Steel	A–1 through A–5	—	Through-hardened and tempered	≤ 180		590*–660**	85*–95**	590*–660**	85*–95**
				240		720–790	105–115	720– 790	105–115
				300		830–930	120–135	830– 930	120–135
				360		1000–1100	145–160	1000–1100	145–160
				400		1070–1170	155–170	1070–1170	155–170
			Flame- or induction-hardened		50	1170–1310	170–190	1170–1310	170–190
					54	1210–1340	175–195	1210–1340	175–195
			Carburized and case-hardened		55	1240–1380	180–200	1240–1380	180–200
					60	1380–1550	200–225	1380–1550	200–225
		AISI 4140	Nitrided#	84.5 15N	48	1050–1250	155–180	1070–1240	155–180
		AISI 4340	Nitrided#	83.5 15N	46	1050–1200	150–175	1030–1210	150–175
		Nitralloy 135M	Nitrided#	90.0 15N	60	1150–1350	170–195	1170–1340	170–195
		Nitralloy N	Nitrided	90.0 15N		1350–1400	195–205		
		2.5% chrome	Nitrided	87.5 15N	54	1050–1200	155–172	1070–1190	155–172
		2.5% chrome	Nitrided	90.0 15N	60	1300–1500	192–216	1320–1490	192–216
Cast iron	20		As cast			345–415	50–60	340–410	50–60
	30		As cast	175		450–520	65–75	450–520	65–75
	40		As cast	200		520–590	75–85	520–590	75–85
Nodular (ductile) iron	A-7-a	ASTM A 536 60-4-18	Annealed	140		530–630	77–92	90–100% of $\sigma_{sac}(S_{ac})$ steel with same hardness (value for hardness)	
	A-7-c	80-55-06	Quenched and tempered	180		530–630	97–92		
	A-7-d	100-70-03	Quenched and tempered	230		630–770	92–112		
	A-7-e	120-90-02	Quenched and tempered	270		710–870	103–126		
Malleable iron (pearlitic)	A-8-c	ASTM A 220 45007		165		495	72	500	72
	A-8-e	50005		180		540	78	540	78
	A-8-f	53007		195		570	83	570	83
	A-8-i	80002		240		650	94	650	94
Bronze	Bronze 2	AGMA 2C	Sand-cast	Tensile strength minimum 275 MPa (40 kpsi)		206	30		
	Al/Br 3	ASTM B-148-52 Alloy 9C	Heat-treated	Tensile strength minimum 620 MPa (90 kpsi)		450	65		

Key: * grade 1; ** grade 2.
[a] AGMA uses allowable contact stress numbers instead of allowable contact strength for stress value and designates these numbers by S_{ac}.
[b] Hardness to be equivalent to that at the start of active profile in the center of the face width.
Source: Courtesy AGMA, compiled from AGMA Standards 2001-B88 and 2003-A86.

TABLE 23–25A
Reliability factor C_R and K_R

Requirements of application	Reliability factor C_R, K_R
Fewer than one failure in 10,000	1.5
Fewer than one failure in 1000	1.25
Fewer than one failure in 100	1.00
Fewer than one failure in 10	0.85

Source: Extracted from ANSI/AGMA Standard 2001.B88, Sept. 1988, fundamentals rating factors and calculation methods for involute spur and helical gear teeth with the permission of the publishers. American Gear Manufacture Association.

TABLE 23–25B
AGMA reliability factors, C_R and K_R for use in Eqs. (23–104) and (23–169)

Reliability Factor, K_R	0.90	0.99	0.999	0.9999
	0.85	1.00	1.25	1.50

TABLE 23–26
Size factor, K_{sz}, for spur gear teeth

Module, m	Diametral pitch, P	Size factor, K_{sz}	Module, m	Diametral pitch, P	Size factor, K_{sz}
1–2		1.00		3	0.865
	12	0.990	9		0.860
2.25		0.984	10		0.851
2.50		0.974		2.5	0.850
	10	0.972	11		0.843
2.75		0.965	12		0.836
3		0.956			0.832
	8	0.951	14	2.0	0.824
3.5		0.942	16		0.813
	7	0.939	18		0.804
4		0.930	20		0.796
	6	0.925	22		0.788
4.5		0.920	25		0.779
	5	0.909	28		0.770
5		0.910	32		0.760
5.5		0.902	36		0.752
6		0.894	40		0.744
	4	0.890	45		0.736
7		0.881	50		0.728
8		0.870			

TABLE 23–27
q factor for external gears for use in Eqs. (23–125) to (23–130)

x	q for external gears with z										
	10	12	14	16	20	24	28	32	40	60	110
0	3.24	2.82	2.55	2.38	2.04	2.04	1.95	1.89	1.82	1.72	1.64
+0.25	2.37	2.18	2.07	1.99	1.88	1.82	1.77	1.74	1.69	1.64	1.59
+0.50	1.80	1.75	1.72	1.69	1.66	1.64	1.62	1.61	1.60	1.50	1.56
+0.75	1.41	1.42	1.43	2.44	1.45	1.46	1.46	1.47	1.49	1.50	1.51
−0.25						2.37	2.29	2.13	1.99	1.82	1.70
−0.50							2.48	2.35	2.17	1.93	1.77

TABLE 23–28
q factor for internal gears

z	100	60	40	32
q for x = 0	1.39	1.32	1.25	1.21

TABLE 23–29
q factor for commercial gears (large pitch errors)

	q Factor				
	$\alpha = 20°$			$\alpha = 25°$	$\alpha = 28°$
z	x = 0	x = 0.25	x = 0.5	x = 0	x = 0
12		3.39	2.47	3.90	3.35
14	4.42	3.23	2.42	3.54	2.98
16	4.16	3.10	2.36	3.25	2.69
20	3.78	2.89	2.30	2.82	2.36
24	3.59	2.72	2.28	2.55	2.21
28	3.25	2.61	2.28	2.42	3.13
32	3.04	2.55	2.27	2.35	2.07
40	2.82	2.50	2.28	2.25	2.01
60	2.63	2.42	2.29	2.15	1.92
100	2.49	2.40	2.32	2.07	1.87

TABLE 23–30
Stress-concentration factors, $K_{\sigma c}$

P	m, mm	$K_{\sigma t}$	$K_{\sigma c}$
4	6.24	1.47	1.61
5	5	1.47	1.61
6	4.25	1.42	1.57
7	3.5	1.35	1.50
8	3	1.345	1.50

TABLE 23-31
Maximum allowable error, f, in action between gears

v_m, m/s	Error f, mm	v_m, m/s	Error f, mm	v_m, m/s	Error f, mm
1.25	0.0925	8.75	0.0425	16.25	0.0200
2.50	0.0800	10.00	0.0375	17.25	0.0170
3.75	0.0700	11.25	0.0325	20.00	0.0150
5.00	0.0600	12.50	0.0300	22.50	0.0150
6.25	0.0525	13.75	0.0250	22 and	0.0125
7.50	0.0475	15.00	0.0225	over	

TABLE 23–32

TABLE 23–32
Values of C for use in Eqs. (23–151), (23–155), (23–156), and (23–468)

Material, pinion and gear	Tooth form	Error in action, f, mm (in)											
		0.0125 (0.0005)		0.025 (0.001)		0.050 (0.002)		0.075 (0.003)		0.100 (0.004)		0.125 (0.005)	
		lbf/in	kN/m	lbf/in	kN/m	lbf/in	kN/m	lbf/in	kN/m	lbf/in	kN/m	lbf/in	kN/m
Cast iron and cast iron	$14\frac{1}{2}°$	400	69.85	800	139.70	1600	279.40	2400	419.10	3200	558.80	4000	698.50
Steel and cast iron	$14\frac{1}{2}°$	550	96.04	1100	192.08	2200	384.16	3300	576.34	4400	768.32	5500	960.40
Steel and steel	$14\frac{1}{2}°$	800	139.70	1600	279.48	3200	558.80	4800	838.20	6400	1117.60	8000	1397.00
Cast iron and cast iron	20° full depth	415	72.50	830	145.00	1660	290.00	2490	435.00	3320	580.00	4150	725.00
Steel and cast iron	20° full depth	570	99.57	1140	192.14	2280	398.28	3420	597.42	4560	796.56	5700	995.70
Steel and steel	20° full depth	830	145.00	1660	290.00	3320	580.00	4980	870.00	6640	1116.00	8300	1450.00
Cast iron and cast iron	20° stub	430	75.05	860	150.10	1720	300.20	2580	450.30	3440	600.40	4300	750.50
Steel and cast iron	20° stub	590	103.01	1180	206.02	2360	412.04	3540	618.06	4720	824.08	5900	1030.10
Steel and steel	20° stub	860	150.10	1720	300.20	3440	600.40	5160	900.60	6880	1200.80	8600	1501.00

TABLE 23–33
Endurance limit σ_{sf} for checking gear teeth for use in Eq. (23–158)

Material	Hardness, core, H_B	σ_f kpsi	MPa
Gray iron	160	12	82.8
Semisteel	200	18	124.2
Manganese bronze, SAE 43	100	17	117.7
Gear bronze, SAE 43	100	24	165.3
Nonmetallic		6	41.4
Steel	150	42	290.0
Steel	200	50	344.8
Steel, normalized	240	60	414.0
Steel, SAE 3140, heat-treated	280	70	482.6
Steel, SAE 3240, heat-treated	320	80	551.8
Steel, oil-tempered	360	90	620.5
Steel, Nitralloy	400	100	689.6

TABLE 23–34
Values of F_f/F_d for use in Eq. (23–159) to find margin of safety, n'

Nature of load	F_f/F_d
Shock load	1.50
Pulsating load	1.35
Steady load	1.25

TABLE 23–35
Values of elastic coefficient C_p for spur and helical gears ($v = 0.3$) for use in Eq. (23–164)[a]

Pinion material	Modulus of elasticity, E_1^2 lb/in^2 (MPa)	Gear material and combined elastic coefficient, C_p (lbf/in^2)$^{0.5}$ [(MPa)$^{0.5}$]					
		Steel	Malleable iron	Nodular iron	Cast iron	Aluminum bronze	Tin bronze
Steel	30×10^6 (2×10^5)	2300 (191)	2180 (181)	2160 (179)	2100 (174)	1950 (162)	1900 (158)
Malleable iron	25×10^6 (1.7×10^5)	2180 (181)	2090 (174)	2070 (172)	2020 (168)	1900 (158)	1850 (154)
Nodular iron	24×10^6 (1.7×10^5)	2160 (179)	2070 (172)	2050 (170)	2000 (166)	1880 (156)	1830 (152)
Cast iron	22×10^6 (1.5×10^5)	2100 (174)	2020 (168)	2000 (166)	1960 (163)	1850 (154)	1800 (149)
Aluminum bronze	17.5×10^6 (1.2×10^5)	1950 (162)	1900 (158)	1880 (156)	1850 (154)	1750 (145)	1700 (141)
Tin bronze	16×10^6 (1.1×10^5)	1900 (158)	1850 (154)	1830 (152)	1800 (149)	1700 (141)	1650 (137)

[a] Values are approximate.
Source: ANSI/AGMA 2001-B88.

TABLE 23–36
Factor Y_2 for use in Eq. (23–184)

Y_2	0.7	0.75	0.8	0.9	1.0	1.1	1.2	1.3
Oil E	1.5	3.0	5	9	13.5	19	26	35

TABLE 23-37A
Stress values for gear materials for use in Eq. (23-184b)

Serial No.	Material	Designation	Test specimen — Ultimate strength, σ_u (kgf/mm²)	(kpsi)	(MPa)	Bilateral fatigue bending strength, σ_{fb} (kpsi)	(MPa)	On the gear[**] — Brinell hardness H_B Core	Surface	Unilateral fatigue stress σ_{fu} Surface (psi)	(MPa)	Root[‡‡] (kpsi)	(MPa)	Static bending stress σ_{bo} (kpsi)	(MPa)
1	Cast iron	Grade 20	18	25.6	176.6	13	88.3	170	170	270	1.864	6.4	44.2	25.6	176.68
2	Cast iron	Grade 20	26	27.0	255.0	17	117.7	210	210	470	3.237	8.5	58.86	37.0	255.06
3	Spheroidal cast iron	Ferritic	60	85.4	588.6			170	170	455	3.139	70.2	484.55	142.3	981.00
4	Spheroidal cast iron	Pearlitic	70-75	100.0-106.7	686.7-735.8			250	250	910	6.278	70.3	484.6	142.3	981.00
5	Cast steel	Grade 1	52	74.0	510.1	30	206.0	150	150	300	2.06*	21.3	147.2	66.8	461.00
6	Cast steel	Grade 2	60	85.4	588.6	34	235.4	175	175	426	2.943*	25.0	171.7	74.0	510.12
7	Carbon steel	ST 42	42-50	57.0-71.1	392.4-490.5	29-34	196.2-235.4	125	125	242	1.67	24.7	170.6		
8	Carbon steel	St 50	50-60	71.1-85.4	490.5-588.5	33-40	225.6-274.7	150	150	510	3.532*	27.0	186.39	78.3	539.55
9	Carbon steel	St 63	60-70	85.4-100.0	588.5-686.7	40-47	274.7-323.7	180	180	740	5.101*	30.0	206.01	92.5	637.65
10	Carbon steel	St 66	70-85	100.0-121.0	686.7-833.9	47-57	323.7-392.4	208	208	1000	6.867*	34.2	235.44	106.5	734.80
11	Tempered steel	C 22	50-60	71.1-85.4	490.5-588.6	31-38	215.8-264.8	140	140	327	2.256*	27.5	189.33	85.4	588.60
12	Tempered steel	C 45	65-80	92.5-113.8	637.7-784.8	43-48	294.3-333.5	185	185	570	3.924*	32.7	225.63	106.5	734.80
13	Tempered steel	C 60	75-90	106.5-128.0	785.8-882.9	48-58	333.5-402.2	210	210	725	5.003*	36.4	251.14	128.0	882.90
14	Tempered steel	4 Cr 1	75-90	106.5-128.0	785.8-882.9	51-63	353.1-431.6	260	260	1140	7.848*	42.7	294.30	128.0	882.90
15	Tempered steel	37 Si 2 Mn 90	80-95	113.8-135.0	784.8-932.0	50-66	342.8-451.3	260	260	1000	6.867*	44.8	309.02	135.2	932.00
16	Tempered steel	42 Cr Mo 4	95-100	135.0-156.4	932.0-1078.1	66-77	457.3-529.7	260	300	1140	7.848*	44.8	309.02	156.5	1079.10
17	Tempered steel	35 Ni Cr 18	100-160	142.3-227.5	981.0-1569.0	60-100	412.4-686.7	400	400	2420	16.68	52.0	358.07		
18	Case-hardened steel	C 10	45-60	64.0-85.4	441.5-588.6	36	245.2	170	170	5335	36.79	32.7	225.63		
19	Case-hardened steel	17 Mn 1 Cr 95	80-110	106.5-156.5	734.7-1079.1			270	650	7115	49.05	60.0	412.00		
20	Case-hardened steel	C 15	50-65	71.1-92.5	490.5-637.7	38	264.87	190	136	7000	48.07	31.3	215.82	200.0	1373.4
21	Case-hardened steel	20 Mn Cr 1	100-130	142.3-185.0	981.0-1275.3			360	650	7115	49.05	66.8	461.07	135.0	930.20
22	Case-hardened steel	13 Ni 6	60-86	85.4-113.8	588.6-784.8			200	600	5635	38.85	42.7	294.30	227.6	1569.6
23	Case-hardened steel	13 Ni Cr 18	120-140	170.8-199.2	1177.2-1373.4	71	490.5	400	615	5545	38.24	61.2	421.83		
24	Case-hardened steel	15 Ni Cr 1 Mo 12	90-120	128.0-170.7	882.9-1177.2			310	650	7115	49.05	62.6	431.6	227.6	1569.6
25	Case-hardened steel	18 Cr Ni 80	120-145	170.7-199.2	1177.2-1373.4			400	650	7115	49.05	66.8	461.07	242.0	1667.7
26	Flame- or induction-hardened steel	C 45	65-80	92.5-113.8	637.7-784.8			220	595	6115	42.18	44.8	309.02†	200.0	1373.4
27	Flame- or induction-hardened steel	C 56	75-90	106.5-128.0	735.8-883.9			230	615	5690	39.24	51.2	353.16		
28	Flame- or induction-hardened steel	C 70	80-100	92.5-142.3	784.8-984.0			240	637	5975	41.20	52.6	362.97		
29	Flame- or induction-hardened steel	37 Si 2 Mn 90	90-105	128.0-149.4	882.9-1030			270	560	5265	36.30	48.4	333.54†	177.8	1226.30
30	Flame- or induction-hardened steel	42 Cr Mo 4	90-110	128.0-155.0	882.9-1069.1			275	587	5265	36.30	55.5	382.59		
31	Flame- or induction-hardened steel	55 Si 2 Mn 90	90-110	128.0-155.0	882.9-1069.1			275	615	6400	44.15	50.0	343.35†		
32	Steel-hardened in cyanide bath	40 Cr 1	140-180	199.2-256.1	1373.4-1765.9			460	595	6115	42.18	45.5	314.0	156.5	1079.1
33	Steel-hardened in cyanide bath	37 Si 2 Mn 90	150-190	213.4-270.3	1471.5-1863.9			470	550	5122	35.32	50.0	343.35	270.3	1863.9
34	Steel-hardened in cyanide bath	35 Ni Cr 18	160-210	227.6-299.0	1569.6-2061.0			500	590	5335	36.79	61.2	421.83	285.0	1962.0
35	Gas-nitrided steel	31 Cr Mo V 9	70-85	99.6-121.0	686.7-833.9			700	700	4980	34.34	64.0	441.5	213.4	1471.0
36	Spheroidal graphitic cast iron		80-90	106.6-156.5	734.8-1079.1			300	300	2560	17.66	31.3	215.8	199.2	1373.5
37	Steel-hardened in nitride bath	C 45	55-60	78.3-85.4	539.55-588.60			450	450	2560	17.66	42.2	311.96	155.0	1069.1
38	Steel-hardened in nitride bath	42 Cr Mo 4	85-90	121.0-128.0	833.90-882.9			660	660	3840	26.49	84.0	578.98	213.4	1471.10
39	Pressed fabric	Coarse								256	1.766†	8.0	54.94†	24.2	166.85
40	Pressed fabric	Fine								327	2.256†	8.0	54.94†	24.2	166.85

Key: ** from experiments with $m = 3$, $z_1 = 21$, $z_2 = 34$, $b = 10$ mm, 20° normal teeth, $v \cong 8$ m/s; * for running against hardened steel, value should be 35% higher; ‡ valid for value $\rho_f \leq 0.2\ m$; † valid for local hardness for through-hardening about 25% less; ‡‡ for $v = 12$ m/s and for ground steel mating gears; † corresponding value of $C = 7.85$ MPa (0.8 kgf/mm²).

TABLE 23–37*B*
Values of load-stress factor *K* for use in Eqs. (23–160) and (23–310)

| Brinell number | | σ_{fc} | | Load-stress factor, *K* | | | |
| | | | | 14.5° | | 20° | |
Pinion	Gear	psi	MPa	psi	MPa	psi	MPa
150	150	50,000	344.47	30	0.2070	41	0.2824
200	150	60,000	413.98	43	0.2962	58	0.4002
250	150	70,000	462.65	58	0.4002	79	0.5445
200	200	70,000	462.65	58	0.4002	79	0.5445
250	200	80,000	551.81	76	0.5239	103	0.7102
300	200	90,000	620.68	96	0.6622	131	0.9035
250	250	90,000	620.68	96	0.6622	131	0.9035
300	250	100,000	689.64	119	0.8211	162	1.1174
350	250	110,000	757.80	144	0.9928	196	1.3518
300	300	110,000	757.80	144	0.9928	196	1.3518
350	300	120,000	829.47	171	1.1792	233	1.6069
400	300	125,000	861.81	186	1.2822	254	1.7521
350	350	130,000	896.63	201	1.3862	275	1.8963
400	350	140,000	962.89	233	1.6069	318	2.1935
400	400	150,000	1034.47	268	1.8482	366	2.5241
Steel Pinion and Cast-Iron Gear							
150		50,000	344.47	44	0.3031	60	0.4410
≥ 200		70,000	462.65	87	0.6004	119	0.8211
Steel Pinion Nickel-Cast Iron, Hot-Quenched							
150		50,000	344.47	44	0.3031	60	0.4140
200		70,000	462.65	87	0.6004	119	0.8211
250		90,000	620.68	144	0.9928	196	1.3513
≥ 300		93,000	641.57	154	1.0624	210	1.4490
Steel Pinion and Phosphor-Bronze Gear							
150		50,000	344.70	46	0.3169	62	0.4277
≥ 200		65,000	448.31	73	0.5033	100	0.6896
≥ 250[a]		83,000	572.20	128	0.8825	175	1.2066
Cast-Iron Pinion and Cast-Iron Gear							
		80,000	551.81	152	1.0487	208	1.4362
Hot-Quenched Nickel-Cast-Iron Pinion and Gear							
		93,000	641.57	206	1.4205	281	1.9385
Hot-Quenched Nickel-Cast-Iron Pinion and Phosphor-Bronze Gear							
		83,000	570.01	171	1.1792	234	1.6137

TABLE 23–37B
Values of load-stress factor K for use in Eqs. (23–160) and (23–310) (Cont.)

Brinell number		σ_{fc}		Load-stress factor, K			
				14.5°		20°	
Pinion	Gear	psi	MPa	psi	MPa	psi	MPa
			Steel Pinion and Steel Gear				
450[b]	450	188,000	1290.99	421	2.9038	575	3.9662
500[b]	500	210,000	1442.07	525	3.6209	718	4.9521
550[b]	550	230,000	1599.03	647	4.4626	884	6.0970
450[c]	450	147,000	1009.45	257	1.7727	351	2.4211
500[c]	500	165,000	1123.06	324	2.2347	443	3.0558
550[c]	550	188,000	1290.99	394	2.7174	544	3.7523
450[d]	450	132,000	906.44	208	1.4342	284	1.9591
500[d]	500	148,000	1016.32	261	1.8001	356	2.4554
550[d]	550	163,000	1129.31	316	2.1798	432	2.9793

[a] Chilled bronze.
[b] Repetition of stress = 10,000,000.
[c] Repetition of stress = 50,000,000.
[d] Repetition of stress = 100,000,000.

TABLE 23–38
Values of y_{e1} and y_{e2} for use in Eqs. (23–185) and (23–188) for external standard gears[a]

Number of teeth, z	y_{e1} for z_2/z_1						y_{e2} for z_2/z_1					
	1	1.4	2	3	5	10	1	1.4	2	3	5	10
9			0.103	0.107	0.111	0.114			0.241	0.289	0.310	0.355
10		0.117	0.122	0.130	0.135	0.141		0.182	0.240	0.282	0.311	0.329
11		0.130	0.137	0.147	0.154	0.162		0.186	0.238	0.277	0.306	0.323
12	0.098	0.140	0.148	0.160	0.167	0.177	0.089	0.189	0.236	0.273	0.301	0.320
13	0.126	0.148	0.158	0.171	0.181	0.192	0.126	0.190	0.235	0.269	0.297	0.317
14	0.142	0.154	0.165	0.179	0.190	0.203	0.142	0.192	0.233	0.266	0.294	0.314
15	0.146	0.158	0.171	0.186	0 199	0.211	0.146	0.192	0.231	0.264	0.291	0.313
17	0.150	0.165	0.180	0.197	0.212	0.227	0.150	0.192	0.229	0.260	0.287	0.308
20	0.155	0.172	0.189	0.207	0.225	0.241	0.155	0.192	0.226	0.255	0.281	0.302
24	0.157	0.177	0.196	0.216	0.236	0.254	0.157	0.191	0.223	0.251	0.278	0.300
30	0.159	0.180	0.201	0.224	0.246	0.266	0.159	0.190	0.221	0.248	0.275	0.297
45	0.160	0.184	0.207	0.232	0.256	0.278	0.160	0.189	0.217	0.244	0.273	0.295
70	0.161	0.186	0.211	0.237	0.263	0.286	0.161	0.188	0.216	0.243	0.265	0.292
150	0.162	0.187	0.213	0.239	0.265	0.289	0.162	0.187	0.215	0.243	0.266	0.290

[a] Pressure angle $\alpha = 20°$; $\xi = 1$ and $i = 1$.

TABLE 23–39A
Recommended speeds of gears

Minimum quality	Peripheral speed of gears, v, m/s
5–6	15–30
6–7	7.5–15
7–8	5–7.5
8–9	2.5–5
10–12	< 2.5

TABLE 23–39B
Backlash for gears, j

Module, m (mm)	< 8 m/s pitch line velocity		> 8 m/s pitch line velocity	
	Backlash, j		Module, m (mm)	Backlash, j
	Minimum	Maximum		
20	0.75	1.25	8	0.40
16	0.50	0.85	7	0.38
12	0.35	0.66	6	0.36
10	0.30	0.51	5	0.28
8	0.22	0.40	4	0.23
6	0.20	0.33	3.5	0.22
5	0.15	0.25	3	0.21
4	0.13	0.20	2.75	0.20
3	0.10	0.15	2.5	0.19
2.5	0.08	0.13	2	0.18
2	0.08	0.13		
1.5 and finer	0.00	0.10		

Source: IS 4460, 1967.

TABLE 23–39C
Typical Brinell hardness combinations for pinion and gear

Pinion, H_B	Gear, H_B
210	180
245	210
265	225
285	245
300	255
315	270
335	285
350	300

TABLE 23–40
Basic bending stress factor (K_b) and surface stress factor (K_w) for use in Eqs. (23–194) to (23–199)

Type	Designation	Condition (finished gear)	Minimum tensile strength kpsi	Minimum tensile strength MPa	Brinell hardness, H_B	Basic surface stress factor, K_w kpsi	Basic surface stress factor, K_w MPa	Basic bending stress factor, K_b kpsi	Basic bending stress factor, K_b MPa
Fabric						0.55	3.8	4.5	31.0
Malleable cast iron:									
Whiteheart malleable iron castings	Grade B		40	274.7	217 max	0.85	5.9	9.8	67.7
Cast iron (iron castings for gears	Grade 20	As cast	28	196.2	179 min	1.15	7.9	6.0	41.4
and gear blanks)	Grade 25	As cast	35	245.3	197 min	1.25	8.6	7.5	51.7
	Grade 35	Heat-treated	50	343.4	300 min	1.42	9.8	12.2	84.3
Phosphor bronze:									
Phosphor bronze castings (for gear blanks)		Chill-cast	34	235.4	70 min	0.71	4.9	8.2	56.9
Phosphor bronze castings (for gear blanks)		Centrifugally cast	38	262.6	90 min	0.98	6.8	9.8	67.9
Cast steel	Grade 1		78	539.6	145	1.60	11.0	19.0	131.2
Forged steels:									
Carbon steel:									
0.30% carbon steel	C 30	Normalized	71	490.50	143.0	1.40	9.7	17.0	117.2
0.30% carbon steel	C 30	Hardened and tempered	85	588.60	152.0	1.60	11.0	21.0	145.1
0.40% carbon steel	C 40	Normalized	82	569.0	152.0	1.60	11.0	20.0	137.8
0.40% carbon steel	C 40	Hardened and tempered	85	588.6	179.0	2.05	14.1	21.0	145.1
0.55% carbon steel	C 55 Mn 75	Normalized	102	706.32	201.1	2.40	16.6	25.0	172.6
0.55% carbon steel	C 55 Mn 75	Hardened and tempered	100	686.70	223.0	2.60	17.9	24.0	165.7
Carbon chromium steel									
0.55% carbon chromium steel	55 Cr 70	"	128	882.90	225 min	3.00	20.6	31.5	217.7
0.55% carbon chromium steel	55 Cr 70	"	142	981.00	285 min	3.51	24.2	35.0	241.2
Carbon manganese steel									
Carbon manganese steel	27 Mn 2	Normalized	78	539.60					
Carbon manganese steel	27 Mn 2	Hardened and tempered	100	686.70	201 min	2.40	16.6	24.0	165.7
Manganese molybdenum steel									
Manganese molybdenum steel	35 Mn 2 Mo 28	"	100	686.70	201 min	2.40	16.6	24.0	165.7
Manganese molybdenum steel	35 Mn 2 Mo 28	"	113	784.80	229 min	2.70	18.6	28.0	193.2
Manganese molybdenum steel	35 Mn 2 Mo 45	"	128	882.90	255 min	3.00	20.6	33.0	227.5
Chromium molybdenum steel									
1% chromium molybdenum steel	40 Cr 1 Mo 28	"	128	882.90	255 min	3.00	20.6	33.0	227.5
1% chromium molybdenum steel	40 Cr 1 Mo 28	"	142	981.00	285 min	3.51	24.2	35.0	241.2
1% chromium molybdenum steel	40 Cr 1 Mo 60	"	128	882.90	248 min	2.50	17.2	33.0	227.5
1% chromium molybdenum steel	40 Cr 1 Mo 60	"	142	981.00	293	2.93	20.2	35.0	241.2
3% chromium molybdenum steel	15 Cr 3 Mo 55	"	128	882.90	255 min	3.00	20.6	33.0	227.5
	and 25 Cr 3 Mo 55	"							
Nickel steel									
3% nickel steel	40 Ni 3	"	128	882.90	255 min	3.00	20.6	33.0	227.5
Nickel chromium steel									
4.5% nickel chromium steel	30 Ni 4 Cr 1	"	22	151.07	444 min	5.50	38.0	51.0	352.1
Nickel chromium molybdenum steel									
2.5% nickel chromium molybdenum steel (medium-carbon)	31 Ni 3 Cr 65 Mo 55	"	128	882.90	255 min	3.00	20.7	33.0	227.5
2.5% nickel chromium molybdenum steel (medium-carbon)	31 Ni 3 Cr 65 Mo 55	"	220	1520.60	444 min	5.50	38.0	52.1	358.9
2.5% nickel chromium molybdenum steel (high-carbon)	40 Ni 3 Cr 65 Mo 55	"	142	981.00	285.0	3.51	24.2	35.0	241.2
2.5% nickel chromium molybdenum steel (high-carbon)	40 Ni 3 Cr 65 Mo 55	"	220	1520.60	444 min	5.42	37.4	53.1	365.8
2.5% nickel chromium molybdenum steel (high-carbon)	40 Ni 3 Cr 65 Mo 55	"	170	1177.20	351 min	5.43	37.5	41.0	282.4

TABLE 23–40
Basic bending stress factor (K_b) and surface stress factor (K_w) for use in Eqs. (23–194) to (23–199) (*Cont.*)

Type	Designation	Condition (finished gear)	Minimum tensile strength		Brinell hardness, H_B	Basic surface stress factor, K_w		Basic bending stress factor, K_b	
			kpsi	MPa		kpsi	MPa	kpsi	MPa
Surface hardened steel									
Carbon steel									
0.55% carbon steel			100	695.14	200^a–520^b	3.98	27.5	15.0	103.5
Carbon chromium steel									
0.55% carbon chromium steel			123	849.64	250^a–500^b	5.10	35.1	18.4	126.6
Nickel steel									
3.5% nickel steel			100	695.14	250^a–300^b	5.08	35.0	18.4	126.6
Nickel chromium steel									
3.5% nickel chromium steel			123	849.64	250^a–500^b	5.08	35.0	18.4	126.6
Case-hardened steel									
Carbon steel									
0.12–0.22% carbon steel			72	494.33	650^b	9.96	68.7	40.0	274.6
0.20% carbon steel			72	494.33	140^a–640^b	9.96	68.7	40.0	274.6
Nickel steel									
3.5% nickel steel			100	695.14	200^a–620^b	10.20	70.3	40.0	275.8
5% nickel steel			123	849.64	250^a–600^b	11.20	77.2	47.0	324.3

[a] Core.
[b] Case.
Source: IS 4460, 1967.

TABLE 23-41
Relation between p_t, p_b, p_x, p_n, p_{bn}, p_{bx}, and p_c for helical gears

	p_t	p_b	p_n	p_{bn}	p_x	p_{bx}
p_t		$p_c = \dfrac{p_t p_b}{\sqrt{p_t^2 - p_b^2}}$	$p_x = \dfrac{p_t p_n}{\sqrt{p_t^2 - p_n^2}}$	$p_{bx} = \dfrac{p_t p_{bn}}{\sqrt{p_t^2 - p_{bn}^2}}$	$p_n = \dfrac{p_t p_x}{\sqrt{p_t^2 + p_x^2}}$	$p_{bn} = \dfrac{p_t p_{bx}}{\sqrt{p_t^2 + p_{bx}^2}}$
p_b	$p_c = \dfrac{p_t p_b}{\sqrt{p_t^2 - p_b^2}}$			$p_x = \dfrac{p_b p_{bn}}{\sqrt{p_b^2 - p_{bn}^2}}$	$p_{bn} = \dfrac{p_b p_x}{\sqrt{p_b^2 + p_x^2}}$	
p_n	$p_x = \dfrac{p_t p_n}{\sqrt{p_t^2 - p_n^2}}$			$p_c = \dfrac{p_n p_{bn}}{\sqrt{p_n^2 - p_{bn}^2}}$	$p_t = \dfrac{p_n p_x}{\sqrt{p_x^2 - p_n^2}}$	
p_{bn}	$p_{bx} = \dfrac{p_t p_{bn}}{\sqrt{p_t^2 - p_{bn}^2}}$	$p_x = \dfrac{p_b p_{bn}}{\sqrt{p_b^2 - p_{bn}^2}}$	$p_c = \dfrac{p_n p_{bn}}{\sqrt{p_n^2 - p_{bn}^2}}$		$p_b = \dfrac{p_{bn} p_x}{\sqrt{p_x^2 - p_{bn}^2}}$	$p_t = \dfrac{p_{bn} p_{bx}}{p_{bx}^2 - p_{bn}^2}$
p_x	$p_n = \dfrac{p_t p_x}{\sqrt{p_t^2 + p_x^2}}$	$p_{bn} = \dfrac{p_b p_x}{\sqrt{p_b^2 + p_x^2}}$	$p_t = \dfrac{p_n p_x}{\sqrt{p_x^2 - p_n^2}}$	$p_b = \dfrac{p_{bn} p_x}{\sqrt{p_x^2 - p_{bn}^2}}$		$p_c = \dfrac{p_x p_{bx}}{\sqrt{p_x^2 - p_{bx}^2}}$
p_{bx}	$p_{bn} = \dfrac{p_t p_{bx}}{\sqrt{p_t^2 + p_{bx}^2}}$			$p_t = \dfrac{p_{bn} p_{bx}}{\sqrt{p_{bx}^2 - p_{bn}^2}}$		

TABLE 23-42

Relation between pressure angles α_n, α_t, α_x, helix angles β_p, β_b, and lead angles γ_p, γ_b

	Normal pressure angle, α_n	Transverse pressure angle, α_t	Axial pressure angle, α_x	Pitch helix angle, β_p	Base lead angle, γ_b	Base helix angle, β_b	Pitch lead angle, γ_p on worms
α_n		$\sin\alpha_n=\sin\alpha_t\cos\beta_b$ $\tan\alpha_n=\tan\alpha_t\cos\beta_p$	$\tan\alpha_n=\tan\alpha_x\sin\beta_b$	β_p	$\cos\gamma_p=(\cos\gamma_b/\cos\alpha_n)$ $\tan\gamma_p=(\cos\alpha_t/\tan\gamma_b)$ $\sin\beta_b=(\sin\beta_p\cos\alpha_n)$ $\tan\beta_b=(\tan\beta_p\cos\alpha_t)$	$\sin\beta_p=(\sin\beta_b/\cos\alpha_n)$ $\tan\beta_p=(\tan\beta_b/\cos\alpha_t)$	γ_p
α_t	$\sin\alpha_t=(\sin\alpha_n/\cos\beta_b)$ $\tan\alpha_t=(\tan\alpha_n/\cos\beta_p)$		$\tan\alpha_t=\tan\alpha_x\tan\beta_p$	β_b			$\cos\gamma_b=\cos\gamma_p\cos\alpha_n$ $\tan\gamma_b=(\tan\gamma_p/\cos\alpha_t)$
α_x	$\tan\alpha_x=(\tan\alpha_n/\sin\beta_p)$	$\tan\alpha_x=(\tan\alpha_t/\tan\beta_p)$					γ_b

TABLE 23–43
Gear terminology

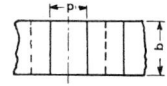

Spur Gears
p = circular pitch
b = face width

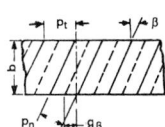

Helical Gears
p_t = transverse circular pitch
p_n = normal pitch
β = helix angle
b = face width

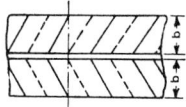

Herringbone Gears
g_β = overlap
$\epsilon_\beta = g_\beta/p_t$
 = overlap ratio

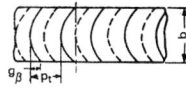

Cylindrical Gears with Curved Teeth
p_t = transverse circular pitch g_β = overlap
b = face width $\epsilon_\beta = g_\beta/p_t$ = overlap ratio

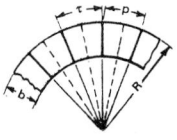

Straight Bevel Gears
p = pitch (at outside) b = face width
τ = angular pitch R = cone distance

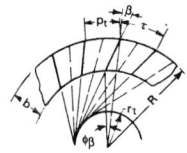

Skew Bevel Gears
p_t = transverse circular r_t = radius of the circle of
 pitch (at outside)   origin
τ = angular pitch b = face width
β = angle of inclination ϕ_β = overlap angle
 (at outside) $\epsilon_\beta = (\phi_\beta/\tau)$ = overlap ratio
R = cone distance

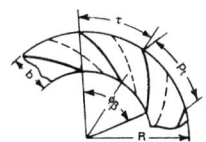

Spiral Bevel Gears
p_t = transverse circular pitch R = cone distance
 (at outside) ϕ_β = overlap angle
τ = angular pitch $\epsilon_\beta = (\phi_\beta/\tau)$ = overlap ratio
b = face width

Crossed-helical Gears
p_n = normal circular pitch
p_d = circular pitch of
  driver
p_t = circular pitch of
  follower
Σ = shaft angle
β_d, β_f = helix angle of
  driver and
  follower,
  respectively

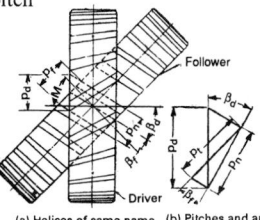

(a) Helices of same name (b) Pitches and angles
   relation

Worm Gears
a = center distance
d_1 = pitch diameter of worm
d_2 = pitch diameter of worm
  wheel
l = center distance between
  bearings of worm
L = center distance between
  bearings of worm wheel

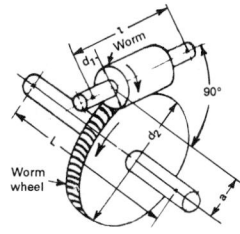

TABLE 23–44
Contact ratios of involute helical gears

	Transverse contact ratio, ϵ_x	Face contact ratio, ϵ_b
Parallel helical gears		
External	$\cos\beta\left[\sqrt{r_{a1}^2 - r_{b1}^2} + \sqrt{r_{a2}^2 - r_{b2}^2} - a\sin\alpha_t\right]/\pi m\cos\alpha_t$	$(b\sin\beta)/\pi m$
Internal	$\cos\beta\left[\sqrt{r_{a1}^2 - r_{b1}^2} + \sqrt{r_{f2}^2 - r_{b2}^2} + a\sin\alpha_t\right]/\pi m\cos\alpha_t$	$(b\sin\beta)/\pi m$
Helical pinion and rack	$\{h_{av} + \sin\alpha_t[\sqrt{r_{a1}^2 - r_{b1}^2} - r_1'\sin\alpha_t]\}/\pi m\sin\alpha_t\cos\alpha_t$	
Total contact ratio, ϵ	$\epsilon = \epsilon_x + \epsilon_b$	

TABLE 23–45
Helical gear-tooth proportions for $\alpha_n = 20°$, $P_n = 1/m_n$; $p_n = \pi m_n$

Helix angle, $\beta,°$	Transverse diametral pitch, P_t	Transverse circular pitch, p_t	Axial pitch, p_x	Transverse pressure angle, α_t	Addendum, h_a	Dedendum, h_f	Clearance, c	Working depth, h'	Total depth, h
0	$1.0000/m_n$	$\pi = 3.1416m_n$		$20°$	$1\,m_n$	$1\,m_n$	$0.25\,m_n$	$2\,m_n$	$2.25\,m_n$
5	$0.9962/m_n$	$3.1536m_n$	$36.0456m_n$	$20°4'\,13.1''$	$1\,m_n$	$1\,m_n$	$0.25\,m_n$	$2\,m_n$	$2.25\,m_n$
10	$0.9848/m_n$	$3.1901m_n$	$18.0917m_n$	$20°17'\,0.7''$	$1\,m_n$	$1\,m_n$	$0.25\,m_n$	$2\,m_n$	$2.25\,m_n$
12	$0.9782/m_n$	$3.2118m_n$	$15.1102m_n$	$20°24'37.1''$	$1\,m_n$	$1\,m_n$	$0.25\,m_n$	$2\,m_n$	$2.25\,m_n$
15	$0.9659/m_n$	$3.2524m_n$	$12.1382m_n$	$20°38'48.8''$	$1\,m_n$	$1\,m_n$	$0.25\,m_n$	$2\,m_n$	$2.25\,m_n$
18	$0.9511/m_n$	$3.3033m_n$	$10.1664m_n$	$20°56'30.7''$	$1\,m_n$	$1\,m_n$	$0.25\,m_n$	$2\,m_n$	$2.25\,m_n$
20	$0.9397/m_n$	$3.3432m_n$	$9.1854m_n$	$21°10'22.0''$	$1\,m_n$	$1\,m_n$	$0.25\,m_n$	$2\,m_n$	$2.25\,m_n$
21	$0.9336/m_n$	$3.3651m_n$	$8.3864m_n$	$21°17'56.4''$	$1\,m_n$	$1\,m_n$	$0.25\,m_n$	$2\,m_n$	$2.25\,m_n$
22	$0.9272/m_n$	$3.3883m_n$	$8.3864m_n$	$21°25'57.7''$	$1\,m_n$	$1\,m_n$	$0.25\,m_n$	$2\,m_n$	$2.25\,m_n$
23	$0.9205/m_n$	$3.4129m_n$	$8.0403m_n$	$21°34'26.3''$	$1\,m_n$	$1\,m_n$	$0.25\,m_n$	$2\,m_n$	$2.25\,m_n$
24	$0.9135/m_n$	$3.4389m_n$	$7.7239m_n$	$21°43'22.9''$	$1\,m_n$	$1\,m_n$	$0.25\,m_n$	$2\,m_n$	$2.25\,m_n$
25	$0.9063/m_n$	$3.4664m_n$	$7.4336m_n$	$21°52'58.7''$	$1\,m_n$	$1\,m_n$	$0.25\,m_n$	$2\,m_n$	$2.25\,m_n$
26	$0.8988/m_n$	$3.4953m_n$	$7.1665m_n$	$22°2'\,44.2''$	$1\,m_n$	$1\,m_n$	$0.25\,m_n$	$2\,m_n$	$2.25\,m_n$
27	$0.8911/m_n$	$3.5259m_n$	$6.9199m_n$	$22°13'10.6''$	$1\,m_n$	$1\,m_n$	$0.25\,m_n$	$2\,m_n$	$2.25\,m_n$
28	$0.8830/m_n$	$3.5588m_n$	$6.6918m_n$	$22°24'\,9.0''$	$1\,m_n$	$1\,m_n$	$0.25\,m_n$	$2\,m_n$	$2.25\,m_n$
29	$0.8746/m_n$	$3.5920m_n$	$6.4800m_n$	$22°35'40.0''$	$1\,m_n$	$1\,m_n$	$0.25\,m_n$	$2\,m_n$	$2.25\,m_n$
30	$0.8660/m_n$	$3.6276m_n$	$6.2832m_n$	$22°47'45.1''$	$1\,m_n$	$1\,m_n$	$0.25\,m_n$	$2\,m_n$	$2.25\,m_n$

TABLE 23–46
Helical gear-tooth proportions for $\alpha_t = 20°$, $P_t = 1/m_t$; $p_t = \pi m_t$

Helix angle, $\beta°$	Normal diametral pitch, P_n	Normal circular pitch, p_n	Axial pitch, p_x	Normal pressure angle, α_n	Working depth, h'	Total depth, h
15	$1.0353/m_t$	$3.0345m_t$	$11.7246m_t$	$19°22'12.2''$	$2.00m_t$	$2.35m_t$
23	$1.0836/m_t$	$2.8919m_t$	$7.4011m_t$	$18°31'21.6''$	$1.84m_t$	$2.20m_t$
30	$1.1547/m_t$	$2.7207m_t$	$5.4414m_t$	$17°29'42.7''$	$1.74m_t$	$2.05m_t$
45	$1.4142/m_t$	$2.2214m_t$	$3.1416m_t$	$14°25'57.9''$	$1.42m_t$	$1.70m_t$

TABLE 23–47
Wear and lubrication factor C_w for use in Lewis Eqs.
(23–286) and (23–287)

Nature of lubrication	C_w
For enclosed gears continuously lubricated with oil of proper viscosity and character	1.15
For scant lubrication but regular, frequent inspection	1.25
For indifferent lubrication	1.35

TABLE 23–48
Design data for helical and herringbone gears

Material	Elastic limit, σ_e		Allowable static stress, σ_o	
	kpsi	MPa	kpsi	MPa
Cast iron, ordinary	12	82.8	4.0	27.5
Cast iron, better grade	16	110.3	4.5	31.0
Laminated phenolic materials	10	69.6	4.0	27.5
Bronze, SAE 65	24	165.5	6.0	41.4
Cast steel, ASTM class B, medium	35	240.7	7.5	51.7
0.40–0.5% carbon steel, nontreated	40	275.8	10.0	69.6
0.40–0.5% carbon steel, heat-treated	50	344.5	12.5	86.0
High-carbon or alloy steels, heat-treated	60	413.8	15.0	103.5

TABLE 23–49
Service factor C_s for use in Eqs. (23–312) and (23–313)

Type of machinery	C_s
Industrial steam turbines for continuous service	0.6
Steam turbine units for military ships	0.9
Heavy-duty reciprocating pump, continuous service	0.7
Intermittent-duty reciprocating pump	0.9
Oil-well pumping equipment	1.0
Crane and hoisting machinery	1.5–2.85
Airplane-engine gears for driving the propeller, transport service	4.00
Same, racing cars	6.5

TABLE 23–50
Factors C_B and C_R for use in Eq. (23–314) at $N_b = 10^7$

Material of gear wheel	Heat treatment	Surface hardness	Factor C_B or C_R	Base number of cycles, N_b
Medium-carbon and alloy steels	Annealing, normalizing, tempering	$H_B \leq 260$ $H_B = 260–350$	$C_B = 25$	10×10^6 $(10–25)\ 10^6$
High-strength alloy steel	Case hardening	$R_c = 55–63$	$C_R = 310$	$(80–140)\ 10^6$
Alloy steels	Case hardening	$R_c = 55–63$	$C_R = 280$	$(80–140)\ 10^6$
Carbon steels— C 15, C 20	Case hardening	$R_C = 55–63$	$C_R = 220$	$(80–140)\ 10^6$
Carbon or alloy steels— C 40, C 45	Surface hardened	$R_C = 40–55$	$C_R = 240$	$(30–80)\ 10^6$
Gray cast iron		$H_B = 170–270$	$C_B = 15$	
Inoculated cast iron		$H_B = 170–262$	$C_B = 18$	

23.2 DESIGNING SPUR AND HELICAL GEARS FOR MACHINE TOOLS (1, 12, 15, 16)

SYMBOLS

In addition to symbols given already under spur gear in Section 23.1, the following symbols are adopted for designing spur and helical gears for machine tool drives

a	center distance, m (in)
a_j	prime whole number
b_j	prime whole number
C	$= c_{sp} \dfrac{d_{sp}}{d_m}$ coefficient used in Eq. (23–349)
C_m	factor used in Eq. (23–349)
d	diameter, m (in)
d_m	mean diameter of shaft in the gear train, m (in)
d_{max}	maximum diameter of job to be cut, m (in)
d_{min}	minimum diameter of job to be cut, m (in)
d_{sp}	diameter of spindle, m (in)
F_f	feed force, kN (lbf)
F_x, F_y, F_z	component of cutting forces, kN (lbf)
i	transmission ratio (also with suffix)
i_t	number of simple trains
k	factor which accounts for the influence of overturning movement
K	constant used in Eqs. (23–373) and (23–376)
L_v	relative loss of cutting speed
m	module (also with suffixes), m (in)
M_t	torque, N m (lbf in)
n	speed (also with suffixes), rpm
n_{em}	speed of electric motor, rpm
$n_{max}(= n_2)$	maximum or upper-limit spindle speed, rpm
$n_{min}(= n_1)$	minimum or lower-limit spindle speed, rpm
n_{sp}	spindle speed at which no load power requirement to be determined [Eq. (23–349)], rpm
n_1, n_2, n_3	speeds of those shafts in the gear train drive for the given value of n_{sp}
n_y	maximum or highest synchronous speed of electric motor, rpm
n_x	minimum or lowest synchronous speed of electric motor, rpm
P	useful power, kW (hp)
P_a	additional losses that depend on load (loading losses), kW (hp)
P_{nl}	constant no-load power, kW (hp)
P_r	required power rating of the electric motor, kW (hp)
q	a whole number
R_a, R_b, R_c,...	range ratios of the transmission
R_{gb}	range ratio of the whole complex of the transmission [Eqs. (23–360) and (23–361)]
R_n	range ratio of spindle speed drives (progression)
S_z	sum of the number of teeth of meshing gears

$S_{z(min)} = K$	minimum sum of the number of teeth of any pair of meshing gears
u	number of spindle speed steps in the middle of the layout diagram (i.e., structural diagram) in a symmetrical broken series
v	velocity, cutting speed (also with suffixes), m/s (ft/min)
v_s	rate of feed, m (in)
W	weight of the parts being transversed, kN (lbf)
x	characteristic of the transmission group [Eqs. (23–360) and (23–361)]
x_j	maximum diameter interval (Fig. 23–56)
z	number of spindle speed steps
z_k	number of spindle speed steps in k transmissions
z_{1j}, z_{2j}	number of teeth in the driving and driven gears of a pair and $j = 1, 2, 3, \ldots, p$
z_o	number of speed steps in the part common to the component drives
z'	number of speed steps in the higher-speed component of the drive
z''	number of speed steps in the lower-speed component of the drive
μ	coefficient of friction between the spindle quill and its seats in the spindle head
μ_w	coefficient of friction in the ways
ϕ	constant ratio of the spindle speed series (progression)
η	total efficiency of machine tool
η'	efficiency of machine tool taking into account only loading losses

SUFFIXES

1, 2	pinion and gear, respectively
max	maximum
min	minimum
em	electric motor
lim	limit

Particular	Formula

23.2.1 Load rating of spur and helical gears for machine tools

NOMINAL LOAD FACTOR (FROM GERMAN SOURCE)

The nominal load factor for pitting and breakage for spur and helical gear pair as per Niemann

$$B = \frac{F_t}{bd'_1} = \frac{2M_{t1}}{bd'^2_1} = \frac{2000P_1}{bd'^2_1\omega_1} = \frac{500(i+1)^2 P_1}{a^2 b\omega_1}$$

SI   (23–315a)

Particular	Formula
	where P_1 in kW, ω_1 in rad/s, and B in N/m²; a, d_1' and b in m, M_{t1} in N m, and F_t in N $$= \frac{1.91 \times 10^4 P_1}{bd_1'^2 n_1} = \frac{10^4 (i+1)^2 P_1}{2.1a^2 bn_1} \quad \textbf{SI} \quad \text{(23–315b)}$$ where P_1 in kW, n_1 in rpm, B in N/m², a, b and d_1' in m, M_{t1} in N/m, and F_t in N $$= \frac{F_t}{bd_1'} = \frac{2M_{t1}}{bd_1'^2} = \frac{126{,}000 P_1}{bd_1'^2 n_1} = \frac{(i+1)^2 P_1 \times 31{,}500}{a^2 bn_1}$$ **US Customary System units** $\qquad$ (23–315c) where P_1 in hp; n_1 in rpm; d_1', a, and b in in; B in psi; F_t in lbf; and M_{t1} in lbf in
The approximate value of face width of gear	$$b = \frac{F_t}{B_a d_1'} = \frac{2M_t}{B_a d_1'^2} = \frac{2000 P_1}{B_a d_1'^2 \omega_1} = \frac{500(i+1)^2 P_1}{a^2 B_a \omega_1}$$ $$\textbf{SI} \quad \text{(23–316a)}$$ where M_t in N m, b, d_1', and a in m; P_1 in kW, ω_1 in rad/s; B_a in N/m², and F_t in N
[*Note:* Multiply the values under Eq. (23–315) by 2 for hardened gears and by 3 for toughened and soft gears of lesser durability.]	$$= \frac{126{,}000 P_1}{B_a d_1'^2 n_1} = \frac{10^6 (i+1)^2 P_1 \times 31{,}500}{a^2 B_a n_1}$$ (23–316b) where M_t in lbf in; b, d_1', and a in in; P_1 in hp; n_1 in rpm; B_a in psi; and F_t in lbf $B_a \simeq 0.2943$ MPa (42.7 psi) for soft gears $\quad\;\; \simeq 0.5086$ MPa (73.8 psi) for toughened gears $\quad\;\; \simeq 2.453$ MPa (355.8 psi) for hardened gears
The diameter d_1' for use in Eqs. (23–316a) and (23–316b)	$d_1' = \mathbf{m}z_1$ for uncorrected or S_o—spur gears (23–316c) $= (\mathbf{m}_n z_1)\cos\beta$ for uncorrected or S_o—helical gears $= (2z_1 a')/(z_1 + z_2)$ for S—spur or helical gears
Effective load factor for pitting and breakage for spur or helical gear	$B_e = B C_S C_D C_T C_\beta \qquad\qquad\qquad\quad$ (23–317) where C_S, C_D, C_T, C_β = load coefficients or coefficients of load influence obtained from Tables 23–54 and 23–55
For permissible flank alignment error f_{RW}	Refer to Fig. 23–49.
For tooth error f for spur and helical gear	Refer to Fig. 23–50.

Particular	Formula

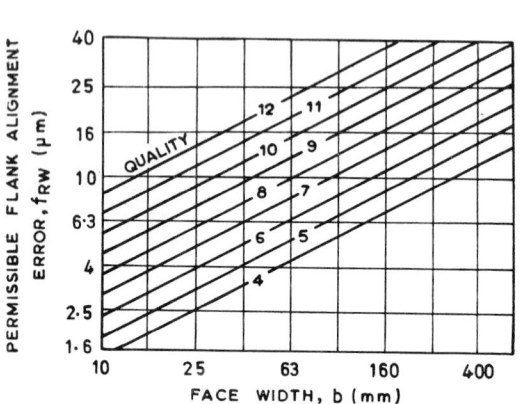

FIGURE 23–49 Permissible flank alignment error, f_{RW}, as per manufacturing quality and tooth face width b for spur and helical gears. (*Courtesy of Indian Standards.*)

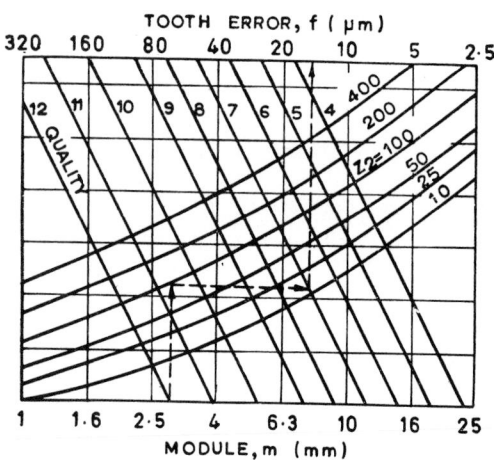

FIGURE 23–50 Tooth error, f, for spur and helical gears. (*Courtesy of Indian Standards.*)

FACTOR OF SAFETY AGAINST PITTING AND BREAKAGE OF TEETH

Formula for safety factor against pitting

$$n_{p1} = \left(\frac{i}{i+1}\right)\frac{\sigma_{fp1}}{y_{e1} B_e} \text{ for pinion} \qquad (23\text{–}318a)$$

$$n_{p2} = \left(\frac{i}{i+1}\right)\left(\frac{\sigma_{fp2}}{y_{e2} B_e}\right) \text{ for gear} \qquad (23\text{–}318b)$$

where $y_{e1} = y_1 K_\beta$ and $y_{e2} = K_\beta K_c$

Refer to Table 23–56 for safety factor n_p

Equation for calculating y_1 for use in Eq. (23–318) (i.e., $y_{e1} = y_1 K_\beta$)

$$y_1 = K_C / K_\epsilon \qquad (23\text{–}319)$$

Also refer to Fig. 23–51 for y_1.

The factor K_C

$$K_C = 1/(\sin\alpha_n \cos\alpha_n) \qquad (23\text{–}320a)$$

The factor K_ϵ

$$K_\epsilon = 1 - 2\,\frac{\pi(1 - \epsilon_{1n})}{z_{1n}\tan\alpha_n} \qquad (23\text{–}320b)$$

The contact ratio for use in Eq. (23–320b)

$$\epsilon_{1n} = (g_a/p_b) = g_a/(\pi m \cos\alpha_n) \qquad (23\text{–}320c)$$

Particular	Formula
	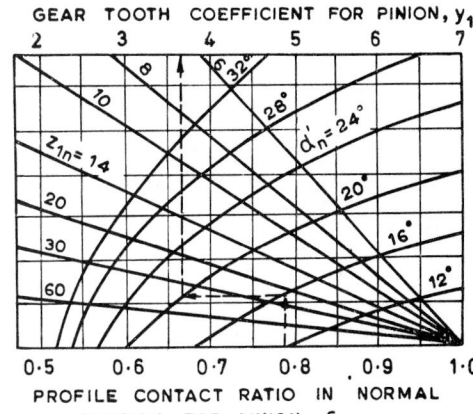

FIGURE 23–51 Coefficient y_1, for pinion with spur gears or equivalent helical gears on normal section. (*Courtesy of Indian Standards.*)

Length of path of addendum contact (Fig. 23–4) for use in Eq. (23–320c)

$$g_a = r_1' \left[\left\{ \left(1 + \frac{h_{a1}}{r_1'}\right)^2 - \cos^2 \alpha_n \right\}^{1/2} - \sin \alpha_n \right]$$

(23–320d)

The value of σ_{fp} for use in Eq. (23–318)

$$\sigma_{fp} = \sigma_{fc} K_E K_H K_O K_v \qquad (23–321)$$

where
$K_H = (H/H_B)^2$
$K_E = 1$ for steel running against steel
$ = 1.5$ for steel running against cast iron

K_O, K_v, K_β, and K_C are obtained from Tables 23–57 to 23–60

$z_{1n} = K_z z_1$; $\epsilon_{1n} = K_n \epsilon_{v1}$

Refer to Table 23–59 for K_z and K_n.
Refer to Table 23–37A for σ_{fc}.
ϵ_{v1} can be calculated as given in Table 23–61.

The pressure angle on working circle on transverse section for helical gears with profile corrections for $\alpha_n = 20°$

$inv\ \alpha' = 2 \tan 20°(x_1 + x_2)/(z_1 + z_2) + inv\ \alpha$

(23–322)

where $\tan \alpha = \tan 20° / \cos \beta$

Also refer to Fig. 23–52.

For profile contact ratio curves for various equivalent spur gear

Refer to Fig. 23–53.

Formula for safety factor against breakage

$$n_{b1} = \frac{\sigma_{b1}}{z_1 q_1 B_e} \text{ for pinion} \qquad (23–323a)$$

$$n_{b2} = \frac{\sigma_{b2}}{z_1 q_1 B_e} \text{ for gear} \qquad (23–323b)$$

where σ_b can be obtained from Table 23–37A

Refer to Table 23–56 for safety factor n_b.

Particular	Formula

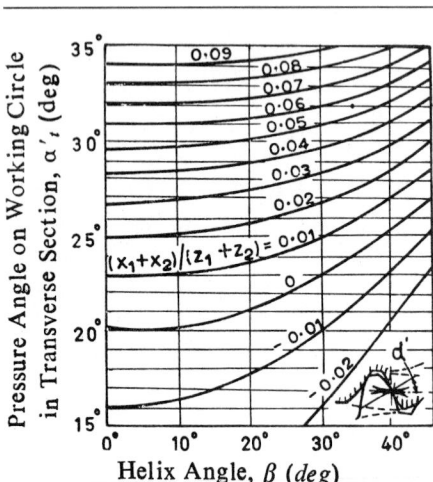

FIGURE 23-52 helical gear with profile corrections for $\alpha_n = 20°$. (*Courtesy of Indian Standards.*)

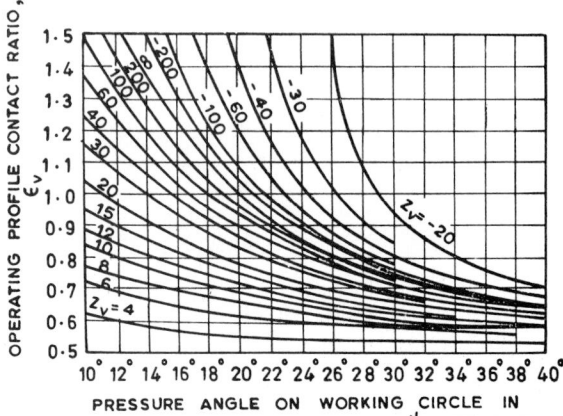

FIGURE 23-53 Profile contact ratio curves with pressure angle on working circle in transverse section for various equivalent spur gear. (*Courtesy of Indian Standards.*)

The empirical formula for q_1 for use in Eq. (23-323a)

$$q_1 \simeq 1.4q_1'/(\epsilon K_n + 0.4) \qquad (23\text{-}323c)$$

The empirical formula for q_2 for use in Eq. (23-323b)

$$q_2 \simeq 1.4q_2'/(\epsilon K_n + 0.4) \qquad (23\text{-}323d)$$

where $\epsilon = \epsilon_{v1} + \epsilon_{v2}$ and q_1' and q_2' depend on z_1 and z_2 and are obtained from Fig. 23-54
$z_n = z$ for spur gear
    $= K_z z$ for helical gear
K_n and K_z are obtained from Table 23-59; ϵ_{v1} and ϵ_{v2} are as given in Table 23-61.

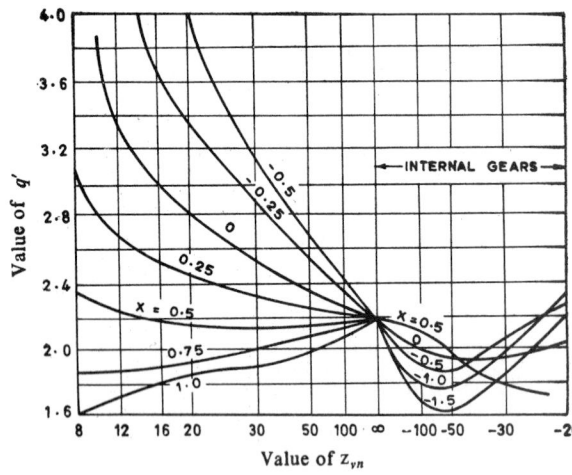

FIGURE 23-54 Curves of q' for various equivalent number of teeth in normal section z_{vn}. (*Courtesy of Indian Standards.*)

Particular	Formula

23.2.2 Design of speed-change gear-box for machine tools

SERIES OF SPINDLE SPEEDS FOR MACHINE TOOLS

The spindle speed

$$n = \frac{1000v}{\pi d} \tag{23–324}$$

The minimum spindle speed

$$n_{min} = \frac{1000v_{min}}{\pi d_{max}} \tag{23–325}$$

The maximum spindle speed

$$n_{max} = \frac{1000v_{max}}{\pi d_{min}} \tag{23–326}$$

where d, d_{max}, d_{min} in mm, and v in m/s

The range ratio of spindle speed variation

$$R_n = \frac{n_{max}}{n_{min}} = \frac{v_{max}d_{max}}{v_{min}d_{min}} = R_V R_d \tag{23–327}$$

where $R_V = v_{max}/v_{min}$ and $R_d = d_{max}/d_{min}$

The constant ratio of the spindle speed series (progression)

$$\phi = \frac{n_{j+1}}{n_j} = \text{constant} \tag{23–328}$$

The relative loss of cutting speed

$$L_V = \frac{(\Delta v)_{max}}{v} = \frac{n_{j+1} - n_j}{n_{j+1} + n_j} = \frac{\phi - 1}{\phi + 1} = \text{constant} \tag{23–329}$$

The maximum relative loss of cutting speed

$$L_{v(max)} = \left(\frac{\Delta v}{v}\right)_{max} = \frac{n_{j+1} - n_j}{n_{j+1}} = \frac{\phi - 1}{\phi} \tag{23–330}$$

The relative loss of cutting speed expressed in percent

$$L_{v(max)} = \frac{\phi - 1}{\phi} \times 100\% \tag{23–331}$$

The relation between the maximum and minimum spindle speeds, number of spindle speed steps (z), and series ratio (ϕ)

$$n_{max} = n_{min}\phi^{z-1} \text{ or } R_n = \frac{n_{max}}{n_{min}} = \phi^{z-1} \tag{23–332a}$$

$$v_{max} = \frac{\pi d n_{max}}{1000} \tag{23–332b}$$

Refer to Fig. 23–55.

The ray or sawtooth diagram can be drawn (geometric progressive) which relates machining of a certain diameter range to cutting speeds between established limits v_{max} and v_{min} by using Eqs. (23–332b) and (23–332c)

$$v_{min} = \frac{\pi d n_{min}}{1000} \tag{23–332c}$$

Particular	Formulale and spindle speeds series

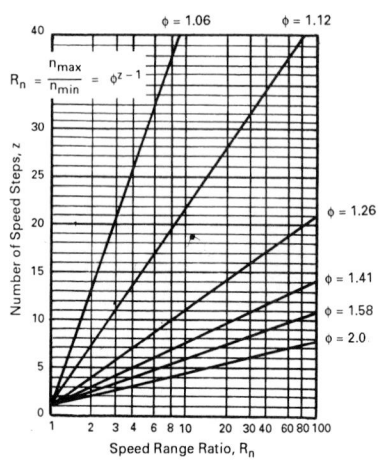

FIGURE 23–55 Number of speed steps vs. speed range ratio. (*Courtesy of N. Acherkan*, Machine Tool Design, *Vol. 3, Mir Publishers, Moscow, 1969.*)

The ratio of synchronous speeds

$$\frac{n_y}{n_x} = 2 \qquad (23\text{–}334)$$

The relation between the highest and lowest synchronous speeds of electric motor which drives the spindle of machine tools

$$n_y = n_x \phi^q \qquad (23\text{–}335)$$

The series (progressive) ratio

$$\phi = (\sqrt[q']{2}) \text{ (Refer to Fig. 23–55.)} \qquad (23\text{–}336)$$

PREFERRED NUMBERS AND SERIES OF PREFERRED NUMBERS

The series of preferred numbers are in the form of geometric progressions whose constant ratio must comply with condition

$$\phi = (\sqrt[q']{10}) \qquad (23\text{–}337)$$

The basic series of preferred numbers (designated by symbol R as a tribute to Captain Renard)

$$(\sqrt[5]{10}) \quad \text{for } R5 \qquad (23\text{–}338a)$$

$$(\sqrt[10]{10}) \quad \text{for } R10 \qquad (23\text{–}338b)$$

$$(\sqrt[20]{10}) \quad \text{for } R20 \qquad (23\text{–}338c)$$

$$(\sqrt[40]{10}) \quad \text{for } R40 \qquad (23\text{–}338d)$$

The special series of preferred numbers intended for special application

$$(\sqrt[80]{10}) \quad \text{for } R80 \qquad (23\text{–}339)$$

Particular	Formula

For preferred numbers

Refer to Tables 23–62 to 23–64.

The standard values of ϕ must satisfy the condition

$$\phi = ({}^{q'_1}\sqrt{2}) = ({}^{q'_2}\sqrt{10}) \tag{23–340}$$

The maximum diameter interval x_j accommodated at a constant cutting speed by two adjacent steps of the speed series (Fig. 23–56)

$$x_j = d_{j-1} - d_j = \frac{v}{\pi n_{j-1}} - \frac{v}{\pi n_j} = \frac{v}{\pi n_{j-1}}\left(1 - \frac{n_{j-1}}{n_j}\right)$$

$$= d_{j-1}\left(1 - \frac{1}{\phi}\right) = d_{j-1}A \tag{23–341}$$

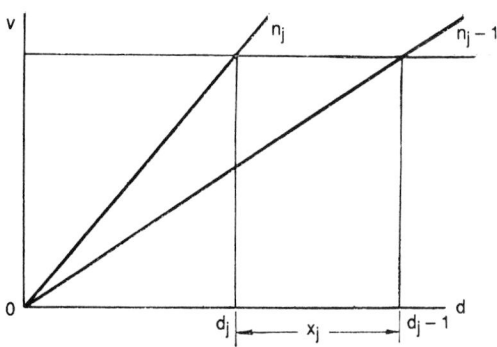

FIGURE 23–56 Diagram for Eq. (2–334). (*Courtesy of N. Acherkam,* Machine Tool Design, *Vol. 3, Mir Publishers, Moscow, 1969.*)

The minimum diameter of bar stock to be machined with various standard ratio of spindle speed values

$$d_{j-1} \geq \frac{\Delta d}{A} = \frac{\phi}{\phi - 1}\,\Delta d \tag{23–342}$$

The good practice in selecting speed step is given by the formula

$$z = 2^{q_1} e^{q_2} \tag{23–343}$$

where q_1 and q_2 are whole numbers

The conditions of Eq. (23–342) is met by the values of z

$z = 2, 3, 4, 6, 8, 9, 12, 16, 18, 24, 27, 32,$ and 36

The most frequently used values of z are

$z = 3, 4, 6, 8, 12, 18,$ and 24

For values of ϕ for satisfactory operation of machine tools

Refer to Table 23–65.

For standard spindle speeds and feeds of machine tools

Refer to Tables 23–66 and 23–67.

POWER RATING OF THE ELECTRIC MOTOR FOR MACHINE TOOLS

The required power rating of electric motor for machine tool

$$P_r = P + P_\mu = P + P_{n1} + P_a \tag{23–344}$$

The total efficiency of the main drive

$$\eta = \frac{P}{P_r} \tag{23–345}$$

Particular	Formula
And also the power rating of the electric motor taking into consideration the efficiency	$$P_r = \frac{P}{\eta} \qquad (23\text{–}346)$$ The total efficiency ranges from 0.7 to 0.85 for machine tools with a rotary primary motion having a single motor drive
The efficiency of machine tool taking into account only loading losses	$$\eta' = \frac{P}{P_r - P_{n1}} = \frac{P}{P + P_a} \qquad (23\text{–}347)$$ where η' varies from 0.88 to 0.90
The power rating of machine tool electric motor as per Eq. (23–347)	$$P_r = \frac{P}{\eta'} + P_{n1} \qquad (23\text{–}348)$$
The no-load power of a gearbox at various spindle speeds as per G. Levit*	$$P_{n1} = \frac{C_m d_m}{955 \times 10^3} \, (n_1 + n_2 + n_3 + \cdots + C n_{sp})$$ in kW $\qquad (23\text{–}349)$ where C_m varies from 30 to 50
The expression for coefficient C for use in Eq. (23–349)	$$C = c_{sp} \frac{d_{sp}}{d_m} \qquad (23\text{–}350)$$ where c_{sp} = 2 for spindle running in sleeve bearings = 1.5 for spindle running in ball or roller bearings
The power rating of an electric motor which has to supply power to several kinematic chains of the machine tool	$$P_r = \sum_i \frac{P_i}{\eta_i} + P_{n1} \qquad (23\text{–}351)$$ where P_i = effective power required by the final number of any one of the kinematic chains η_i' = efficiency varies from 0.88 to 0.9

FEED FORCES AS PER D. RESHETOV AND G. LEVIT**

The feed force for the saddle of lathes with V or combined ways	$$F_f = kF_x + \mu_w(F_z + W) \qquad (23\text{–}352)$$ where $k = 1.15$ and $\mu_w = 0.15$ to 0.18
The feed force for the saddle engine and turret lathes and table of milling machine with flat ways	$$F_f = kF_x + \mu_w(F_z + F_y + W) \qquad (23\text{–}353)$$ where $k = 1.1$ and $\mu_w = 0.15$

* Courtesy: Acherkan, A. et al., Machine Tool Design, MIR Publishers, Moscow, 1969, p. 75.
** Courtesy: Acherkan, A et al., Machine Tool Design, MIR Publishers, Moscow, 1969, p. 38.

Particular	Formula
The feed force for the tables of milling machines dovetail ways	$F_f = kF_x + \mu_w(2F_y + F_z + W)$ (23–354) where $k = 1.4$ and $\mu_w = 0.2$
The feed force for the spindles of drilling machines	$F_f = (1 + 0.5\mu)F_x + \dfrac{2M_t}{d_{sp}}\,\mu \cong F_x + \dfrac{2M_t}{d_{sp}}\,\mu$ (23–355) where $\mu = 0.15$ (The efficiency of feed drive varies from 0.15 to 0.20.)

KINEMATIC SCHEME OF A MACHINE TOOL (TABLE 23–68)

Particular	Formula
The number of spindle speed steps	$z = i_{ta}i_{tb}i_{tc}, \ldots, i_{tn}$ (23–356)
For speed steps for design of a gearbox	Refer to Table 23–68.
The number of transmission ratios of the drive	$i_{(max)} = i_{a(max)}, i_{b(max)}, \ldots, i_{r(max)}$ (23–357)
The minimum transmission ratio of the drive	$i_{(min)} = i_{a(min)}, i_{b(min)}, \ldots, i_{r(min)}$ (23–358a) $\phantom{i_{(min)}} = n_1/n_{em} = 1/\phi^q$ (23–358b) where n_1 = minimum speed, rpm n_{em} = electric motor speed, rpm q = a factor to be taken from table of standards for machine tool engineering
The range ratio of the drive	$R_n = \dfrac{n_{(max)}}{n_{(min)}} = \dfrac{i_{(max)}}{i_{(min)}} = R_a R_b R_c, \ldots, R_r$ (23–359) where $R_a = i_{a(max)}/i_{a(min)}, \ldots,$ etc.
The relation between the transmission ratios and the progression ratio in the multiple group	$i_1 : i_2 : \ldots : i_p = 1 : \phi R_{gb} : (\phi R_{gb})^2 : \ldots : (\phi R_{gb})^{p-1}$ (23–360)
The progression ratio of the series of transmission ratios in a transmission group	$\phi_p = R_{gb}\phi = \phi^{z_k} = \phi^x$ (23–361) where z_k = number of speed steps in the whole complex of transmissions
The general setup equation for group transmission	$i_1 : i_2 : i_3 : \ldots : i_p = 1 : \phi^x : \phi^{2x} : \ldots : \phi^{(p-1)x}$ (23–362)
In the general form, the standard transmission ratio of any transmission in the drive	$i_{st} = 1.06^{+q}$ (23–363)
The limiting transmission ratio of gears in gearbox in general practice	$i_{min(lim)} = \dfrac{1}{4}$ (23–364)

Particular	Formula
The maximum transmission ratio	$i_{max(lim)} = \dfrac{2}{1}$ for spur gear (23–365a)
	$= \dfrac{2.5}{1}$ for helical gear (23–365b)
The permissible maximum transmission ratio used in small machine tools	$i_{max(lim)} = \dfrac{4}{1}$ (23–366)
The accepted transmission ratio range for feeds gearbox (with slow gearing and small-diameter gears)	$\dfrac{1}{5} \leq i \leq \dfrac{2.8}{1}$ (23–367)
The limiting maximum range ratio in a two-shaft group transmission	$R_{(lim)} = \dfrac{i_{max(lim)}}{i_{min(lim)}} = 8$ (23–368)
	Large range ratios of 10 to 12 can be used in a group transmission as an extreme case.
For allowable limits of transmission ratios and output speeds for various combinations	Refer to Table 23–69.

LAYOUT DIAGRAM FOR GEAR DRIVES

For schematic diagram of a machine tool gearbox	Refer to Fig. 23–57.
For layout diagram for 18-step gear drives of Fig. 23–57	Refer to Fig. 23–58.
For various allowable errors for spur and helical gears	Refer to Table 23–71.
For service factor, C_s	Refer to Table 23–72.

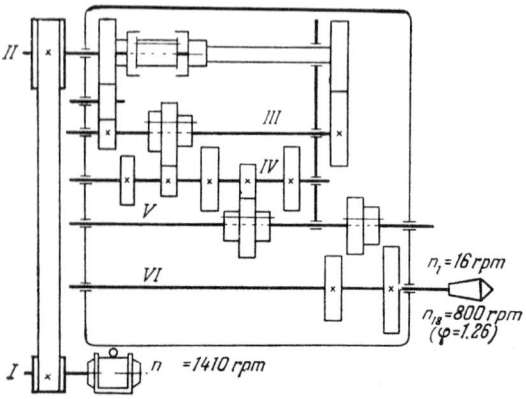

FIGURE 23–57 Schematic diagram of a machine tool 18-step gearbox. (*Courtesy of N. Acherkan,* Machine Tool Design, *Vol. 3, Mir Publishers, Moscow, 1969.*)

Particular	Formula

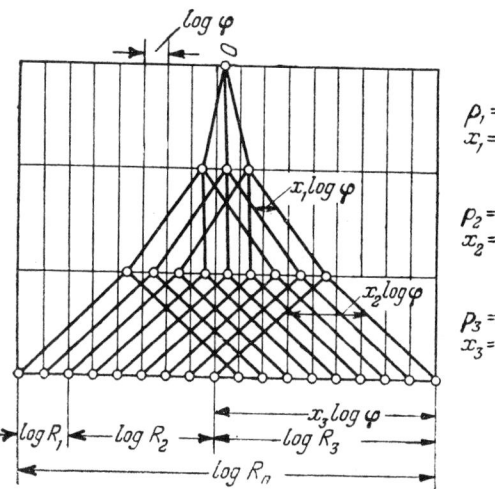

$\rho_1 = 3$
$x_1 = 1$

$P_2 = 3$
$x_2 = 3$

$P_3 = 2$
$x_3 = 9$

FIGURE 23–58 Layout diagram or structural diagram for 18-step gear drives. (*Courtesy of N. Acherkan,* Machine Tool Design, *Vol. 3, Mir Publishers, Moscow, 1969.*)

TRANSMISSION RATIOS FOR DRIVES POWERED BY A MULTISPEED ELECTRIC MOTOR

Particular	Formula	
Progression ratio of the series of transmission ratios	$\phi_p = \phi_{em} = \phi^{z_h} = \phi^{x_{em}}$	(23–369)
The number of speed steps in the whole complex of transmissions in the groups powered by a multispeed electric motor	$z_k = \dfrac{\log \phi_{em}}{\log \phi}$	(23–370)
The characteristic of the transmission group	$x_m = \dfrac{z}{i_{tm}}$	(23–371)
The range ratio of the last extension group	$R_m = \phi^{x_m(i_{tm}-1)} = \phi^{(z-z/i_{tm})}$	(23–372a)
The range ratio of the last extension group for $i_{tm} \geq 2$	$R_m = \phi^{-z/2} = \phi^{z/2}$	(23–372b)
The limit of range ratio	$R_m \leq R_{lim} = \dfrac{i_{max(lim)}}{i_{min(lim)}} = 8 \text{ to } 10 = K$	(23–373)
The limiting value of the range ratio of the drive for $i_{tm} = 2$	$R_{dr} = \phi^{z-1} \leq \dfrac{K^2}{\phi}$	(23–374a)
The limiting value of the range ratio of the drive for the transmission ratio $\phi = 1.26$	$R_{dr} = \dfrac{8^2}{\phi} = \dfrac{64}{1.26} \simeq 50$	(23–374b)

BROKEN GEOMETRICAL SERIES

Particular	Formula	
The range ratio of the drive	$R_n = \phi_2^{z_k-1} \phi_2^{x'} = \phi_2^{(x'-0.5)2} \phi_1^u$	(23–375a)

Particular	Formula
The characteristic of the transmission group	$x' = z_k - \dfrac{u}{2} = \dfrac{x-u}{2}$ (23–375b)
The maximum range ratio in two transmissions in the last extension group	$R_k = \phi_2^{z_k-1} \leq \dfrac{K^2}{\phi_2}$ (23–376)
The limiting range ratio for multiplier group consisting of two transmissions	$R_{mult} = \phi_2^{x'} \leq K = \phi_2^r$ (23–377) where $x' < r$

COMBINED LAYOUT DIAGRAM FOR SPINDLE DRIVES

The total number of speed steps in layout formula of a combined drive	$z = z_o(z' + z'')$ (23–378)
The maximum range ratios for a drive with $z = z_o(1 + z'')$ and having $z_o z''$ speed steps in the part of the drive and z_o steps in the complementary main part	$R_{max} = R_{main}\phi R_{sup} = R_{main}\phi\phi^{z_o-1} \leq \dfrac{K^2}{\phi}\,\phi^{z_o}$ (23–379)

NUMBER OF TEETH IN SPUR GEARS OF GROUP TRANSMISSIONS: GROUPS HAVING SAME MODULE (TABLE 23–70)

The sum of teeth in a pair of gears having same module and constant center distance	$S_z = z_{1j} + z_{2j} = \text{constant}$ (23–380) where z_{1j}, z_{2j} = number of teeth in the driving and driven gears, respectively, and $j = 1, 2, 3, \ldots, p$
The transmission ratio or speed ratio	$i_j = \dfrac{z_{2j}}{z_{1j}} = \dfrac{b_j}{a_j}$ (23–381) where a_j and b_j are prime whole numbers
The number of teeth in the driving gear	$z_{1j} = \dfrac{1}{i_j + 1}\, S_z = \left(\dfrac{b_j}{a_j + b_j}\right) S_z$ (23–382)
The number of teeth in the driven gear	$z_{2j} = \dfrac{i_j}{i_j + 1}\, S_z = \left(\dfrac{a_j}{a_j + b_j}\right) S_z$ (23–383)
The minimum sum $S_{z(min)}$ of the number of teeth of any pair of meshing gears	$S_{z(min)} = K$ (23–384)
The acceptable value of number of teeth in a driving gear	$z_1 = z_{min}E = \left(\dfrac{b_1}{a_1 + b_1}\right) ES_{z(min)}$ (23–385) where E is a whole number

Particular	Formula
The acceptable value of sum of teeth in a pair of gears which satisfies the given number of minimum number of teeth in the driving gear	$S_{za} = ES_{z(min)} = EK$ (23–386)
For number of teeth and sums of number of teeth for standard transmission ratios	Refer to Table 23–70.

GROUPS HAVING DIFFERENT MODULE

The number of teeth in the driving gear	$$z_{1j} = \frac{2a}{m_j} \left(\frac{b_j}{a_j + b_j} \right) \qquad (23\text{–}387)$$
	where m_j = end of module (in the plane of rotation), mm
The condition under which z_{1j} is a whole number	$2a = Em_j(a_j + b_j) \qquad (23\text{–}388)$

HELICAL GEARS

The number of teeth in the driving gear	$$z_{1j} = \frac{2a}{m_n} \left(\frac{b_j \cos \beta_j}{a_j + b_j} \right) \qquad (23\text{–}389)$$
The number of teeth in the driven gear	$$z_{2j} = \frac{2a}{m_n} \left(\frac{a_j \cos \beta_j}{a_j + b_j} \right) \qquad (23\text{–}390)$$
The center distance	$$a = \frac{m_n(z_{1j} + z_{2j})}{2 \cos \beta_j} \qquad (23\text{–}391)$$

Conditions

Case 1. If $\beta_j = \text{constant} = \beta$	$$z_{1j} = \frac{b_j}{a_j + b_j} S_z; \ z_{2j} = \frac{a_j}{a_j + b_j} S_z; \ a = \frac{m_n S_z}{2 \cos \beta}$$ (23–392)
Case 2. If a is given, the whole number E_j being selected in such a manner that the value of $\cos \beta_j = (m_n K E_j)/2a$ does not deviate from unity, then the number of teeth	$$z_{1j} = \frac{KE_j b_j}{a_j + b_j} ; \ z_{2j} = \frac{KE_j a_j}{a_j + b_j} \qquad (23\text{–}393)$$

VERSION OF KINEMATIC SCHEME STRUCTURE

The number of spindle speed steps which differ in the actual order of arrangement of the groups along the train of the drive	$z = p_a p_b p_c \cdots p_r = p_b p_a p_c \cdots p_r = p_c p_a p_b \cdots$ $\cdots p_r = \cdots \qquad (23\text{–}394)$

Particular	Formula
The number of design versions in a selected number of transmission groups m wherein there are q groups with an equal number of transmissions in each	$i_v = \dfrac{m!}{q!}$ (23–395)
The total number of versions of speed-changing structure (or layout) where there are $m!$ kinematic versions	$i_{tv} = \dfrac{(m!)^2}{q!}$ (23–396)
The total number of transmissions in the groups	$S_p = p_a + p_b + p_c + \cdots + p_r$ (23–397)
The total number of transmissions in the groups S_p required to obtain a specified number of speed steps z will be minimum if	$p_a = p_b = p_c = \cdots = p_r = (\sqrt[q]{z}) = p$ (23–398) where q = number of transmission groups

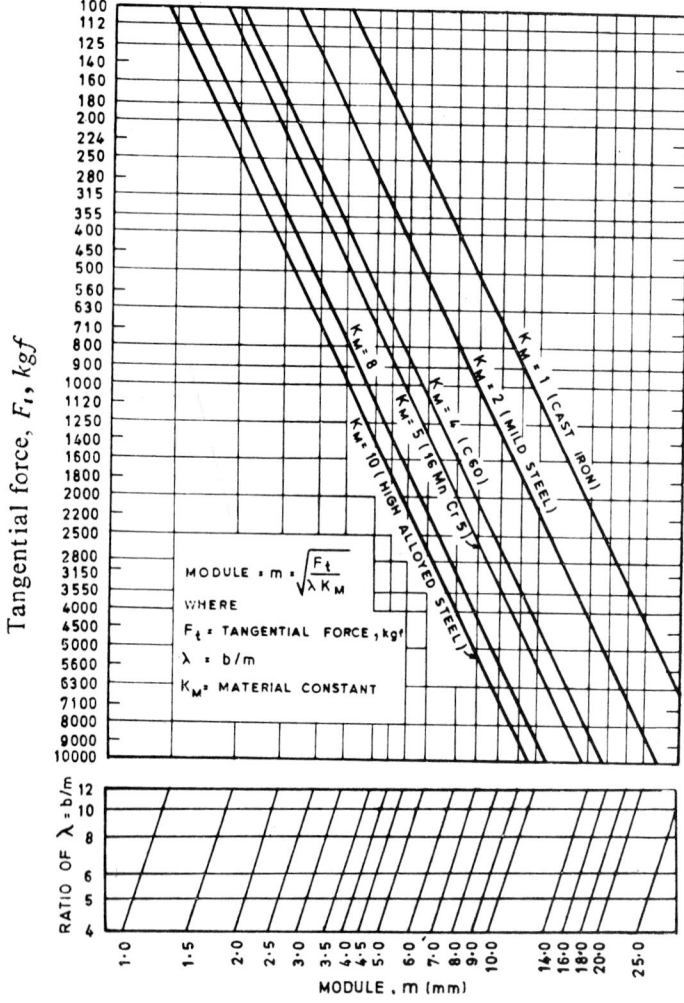

FIGURE 23–59   Selection of module and face width of spur gear for various materials.

Particular	Formula
The minimum number of transmission groups required to obtain the specified range ratio $R_n = R_a R_b R_c, \ldots, R_r$, can be had in the case when	$R_a = R_b = R_c = R_d = \cdots = R_r = R_{lim} = \dfrac{i_{max(lim)}}{i_{min(lim)}}$ <div align="right">(23–399)</div>
For selection of module and face width of spur gear for various materials	Refer to Fig. 23–59.
For output power (P) for given input torque (M_t) at a specified speed	Refer to Fig. 23–60.
For nomogram for connecting diameter (d), speed (n), force (F), torque (M_t), and power (P) in Metric and System International (SI) units	Refer to Fig. 14–1.

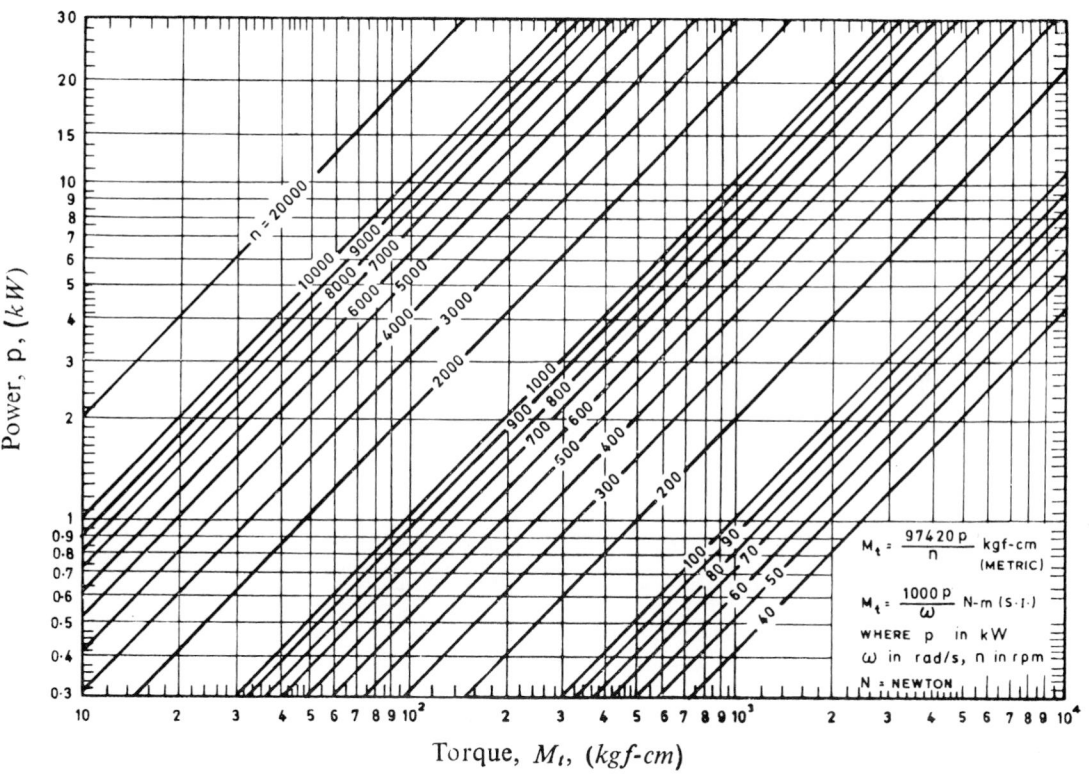

FIGURE 23–60   Relation between power (P) and torque (M_t) for various speeds (n).

TABLE 23–51
Degree of accuracy of gears employed in various machines

Types of machines	Degree of accuracy
Reduction gears in turbines and turbo-machines	3–6
Metal cutting machine tools	3–8
Automobiles	5–8
Trucks	7–9
Tractors	8–10
General-purpose reduction gears	6–9
Gear wheels of rolling mills	6–10
Mine winders	8–10
Crane mechanisms	7–10
Farm machinery	8–11

TABLE 23–52
Standard gear ratio, i

Stage of reduction			Stage of reduction		
Single	Double	Triple	Single	Double	Triple
1.25	8	40	4	20.4	125
1.4	9	45	4.5	25	140
1.6	10	50	5	28	160
1.8	11	56	5.6	31.5	180
2	12	63	6.3	35.5	200
2.24	12.5	71	7.1	40	250
2.5	14	80	8	45	280
2.8	16	90	10	50	315
3.15	18	100			355
3.55	20	112			400

TABLE 23–53A
Dimensions of gear hub

Type of service	Diameter[a]		Length
	Cast iron	Steel casting	
Light load, no shock	1.75d	0.6d	$l \geq 1.5d$
Medium load with shock	1.85d	1.7d	$l \geq 1.75d$
Heavy load with shock	2d	1.8d	$l \geq 2d$

[a] d = diameter of shaft, m (mm)

TABLE 23–53*B*
Constant K_{bt} for use in Eq. (23–267)

z_{bt}	K_{bt}	z_{bt}	K_{bt}	z_{bt}	K_{bt}
1	1.47607	21	60.51869	41	119.56132
2	4.42820	22	63.47082	42	122.51345
3	7.38033	23	66.42296	43	125.46558
4	10.33246	24	69.37509	44	128.41772
5	13.28459	25	72.32722	45	131.36984
6	16.23672	26	75.27935	46	134.32197
7	19.18885	27	78.23148	47	137.27411
8	22.14098	28	81.18361	48	140.22624
9	25.09312	29	84.13574	49	143.17837
10	28.04525	30	87.08788	50	146.13050
11	30.99738	31	90.04001	51	149.08263
12	33.94951	32	92.99214	52	152.03477
13	36.90164	33	95.94427	53	154.98690
14	39.85377	34	98.89640	54	157.93902
15	42.80590	35	101.84853	55	160.89116
16	45.75804	36	104.80066	56	163.84329
17	48.71017	37	107.75280		
18	51.66230	38	110.70493		
19	54.61443	39	113.65706		
20	57.56656	40	116.60919		

Source: Extracted from *Maag Gear Book*, courtesy of Maag Gear-Wheel Company, Zurich, 1963.

TABLE 23-53C
Constant K_β for use in Eq. (23-267)

K_β

Helix angle, β, min	5°	6°	7°	8°	9°	10°	11°	12°	13°	14°	15°	16°	17°	18°	19°	20°	21°	22°
0′	.014159	.014227	.014308	.014402	.014510	.014631	.014767	.014917	.015082	.015264	.015461	.015676	.015908	.016159	.016429	.16720	.017033	.017368
1′	.014160	.014228	.014309	.014404	.014512	.014633	.014769	.014920	.015085	.015267	.015465	.015679	.015912	.016163	.016434	.016725	.017038	.017374
2′	.014161	.014229	.014311	.014406	.014514	.014635	.014772	.014922	.015088	.015270	.015468	.015683	.015916	.016168	.016439	.016730	.017044	.017380
3′	.014162	.014231	.014312	.014407	.014515	.014638	.014774	.014925	.015091	.015273	.015471	.015687	.015920	.016172	.016443	.016735	.017049	.017386
4′	.014163	.014232	.014314	.014409	.014517	.014640	.014776	.014928	.015094	.015276	.015475	.015691	.015924	.016176	.016448	.016740	.017055	.017392
5′	.014164	.014233	.014315	.014411	.014519	.014642	.014779	.014930	.015097	.015279	.015478	.015694	.015928	.016181	.016453	.016746	.017060	.017397
6′	.014165	.014235	.014317	.014412	.014521	.014644	.014781	.014933	.015100	.015283	.015482	.015698	.015932	.016185	.016457	.016751	.017065	.017403
7′	.014166	.014236	.014318	.014414	.014523	.014646	.014784	.014936	.015103	.015286	.015485	.015702	.015936	.016189	.016462	.016756	.017071	.017409
8′	.014167	.014237	.014320	.014416	.014525	.014648	.014786	.014938	.015106	.015289	.015489	.015706	.015940	.016194	.016467	.016761	.017076	.017415
9′	.014168	.014238	.014321	.014417	.014527	.014651	.014788	.014941	.015109	.015292	.015492	.015709	.015944	.016198	.016472	.016766	.017082	.017421
10′	.014169	.014240	.014323	.014419	.014529	.014653	.014791	.014944	.015112	.015295	.015496	.015713	.015948	.016203	.016476	.016771	.017087	.017427
11′	.014170	.014241	.014324	.014421	.014531	.014655	.014793	.014946	.015115	.015299	.015499	.015717	.015953	.016207	.016481	.016776	.017093	.017433
12′	.014171	.014242	.014326	.014423	.014533	.014657	.014796	.014949	.015117	.015302	.015503	.015721	.015957	.016211	.016486	.016781	.017098	.017438
13′	.014173	.014243	.014327	.014424	.014535	.014659	.014798	.014952	.015120	.015305	.015506	.015724	.015961	.016216	.016491	.016786	.017104	.017444
14′	.014174	.014245	.014329	.014426	.014537	.014662	.014801	.014954	.015123	.015308	.015510	.015728	.015965	.016220	.016495	.016791	.017109	.017450
15′	.014175	.014246	.014330	.014428	.014539	.014664	.014803	.014957	.015126	.015311	.015513	.015732	.015969	.016225	.016500	.016796	.017115	.017456
16′	.014176	.014247	.014332	.014430	.014541	.014666	.014805	.014960	.015129	.015315	.015517	.015736	.015973	.016229	.016505	.016802	.017120	.017462
17′	.014177	.014249	.014333	.014431	.014543	.014668	.014808	.014962	.015132	.015318	.015520	.015740	.015977	.016233	.016510	.016807	.017126	.017468
18′	.014178	.014250	.014335	.014433	.014545	.014670	.014810	.014965	.015135	.015321	.015524	.015743	.015981	.016238	.016514	.016812	.017131	.017474
19′	.014179	.014251	.014336	.014435	.014547	.014673	.014813	.014968	.015138	.015324	.015527	.015747	.015985	.016242	.016519	.016817	.017137	.017480
20′	.014180	.014253	.014338	.014437	.014549	.014675	.014815	.014970	.015141	.015328	.015531	.015751	.015989	.016247	.016524	.016822	.017142	.017486
21′	.014181	.014254	.014339	.014438	.014551	.014677	.014818	.014973	.015144	.015331	.015534	.015755	.015994	.016251	.016529	.016827	.017148	.017491
22′	.014182	.014255	.014341	.014440	.014553	.014679	.014820	.014976	.015147	.015334	.015538	.015759	.015998	.016256	.016534	.016832	.017153	.017497
23′	.014183	.014257	.014343	.014442	.014555	.014681	.014823	.014979	.015150	.015337	.015541	.015763	.016002	.016260	.016538	.016838	.017159	.017503
24′	.014185	.014258	.014344	.014444	.014557	.014684	.014825	.014981	.015153	.015341	.015545	.015766	.016006	.016265	.016543	.016843	.017164	.017509
25′	.014186	.014259	.014346	.014445	.014559	.014686	.014828	.014984	.015156	.015344	.015548	.015770	.016010	.016269	.016548	.016848	.017170	.017515
26′	.014187	.014260	.014347	.014447	.014561	.014688	.014830	.014987	.015159	.015347	.015552	.015774	.016014	.016274	.016553	.016853	.017175	.017521
27′	.014188	.014262	.014349	.014449	.014563	.014690	.014833	.014990	.015162	.015350	.015555	.015778	.016019	.016278	.016558	.016858	.017181	.017527
28′	.014189	.014263	.014350	.014451	.014565	.014693	.014835	.014992	.015165	.015354	.015559	.015782	.016023	.016283	.016563	.016863	.017187	.017533
29′	.014190	.014265	.014352	.014452	.014567	.014695	.014838	.014995	.015168	.015357	.015563	.015786	.016027	.016287	.016567	.016869	.017192	.017539
30′	.014191	.014266	.014353	.014454	.014569	.014697	.014840	.014998	.015171	.015360	.015566	.015790	.016031	.016292	.016572	.016874	.017198	.017545
31′	.014192	.014267	.014355	.014456	.014571	.014699	.014843	.015001	.015174	.015364	.015570	.015793	.016035	.016296	.016577	.016879	.017203	.017551
32′	.014194	.014269	.014357	.014458	.014573	.014702	.014845	.015003	.015177	.015367	.015573	.015797	.016039	.016301	.016582	.016884	.017209	.017557
33′	.014195	.014270	.014358	.014460	.014575	.014704	.014848	.015006	.015180	.015370	.015577	.015801	.016044	.016305	.016587	.016890	.017215	.017563
34′	.014196	.014271	.014360	.014461	.014577	.014706	.014850	.015009	.015183	.015374	.015581	.015805	.016048	.016310	.016592	.016895	.017220	.017569
35′	.014197	.014273	.014361	.014463	.014579	.014709	.014853	.015012	.015186	.015377	.015584	.015809	.016052	.016314	.016597	.016900	.017226	.017575
36′	.014198	.014274	.014363	.014465	.014581	.014711	.014855	.015014	.015189	.015380	.015588	.015813	.016056	.016319	.016601	.016905	.017231	.017581
37′	.014199	.014275	.014364	.014467	.014583	.014713	.014858	.015017	.015192	.015383	.015591	.015817	.016060	.016323	.016606	.016910	.017237	.017587
38′	.014201	.014277	.014366	.014469	.014585	.014715	.014860	.015020	.015195	.015387	.015595	.015821	.016065	.016328	.016611	.016916	.017243	.017593
39′	.014202	.014278	.014368	.014471	.014587	.014718	.014863	.015023	.015198	.015390	.015599	.015825	.016069	.016332	.016616	.016921	.017248	.017599
40′	.014203	.014280	.014369	.014472	.014589	.014720	.014865	.015026	.015201	.015393	.015602	.015828	.016073	.016337	.016621	.016926	.017254	.017605
41′	.014204	.014281	.014371	.014474	.014591	.014722	.014868	.015028	.015205	.015397	.015606	.015832	.016077	.016342	.016626	.016932	.017260	.017611
42′	.014205	.014282	.014373	.014476	.014593	.014725	.014870	.015031	.015208	.015400	.015609	.015836	.016082	.016346	.016631	.016937	.017265	.017618
43′	.014206	.014284	.014374	.014478	.014595	.014727	.014873	.015034	.015211	.015403	.015613	.015840	.016086	.016351	.016636	.016942	.017271	.017624
44′	.014208	.014285	.014376	.014480	.014597	.014729	.014876	.015037	.015214	.015407	.015617	.015844	.016090	.016355	.016641	.016947	.017277	.017630
45′	.014209	.014287	.014377	.014482	.014600	.014732	.014878	.015040	.015217	.015410	.015620	.015848	.016094	.016360	.016646	.016953	.017282	.017636
46′	.014210	.014288	.014379	.014483	.014602	.014734	.014881	.015043	.015220	.015414	.015624	.015852	.016099	.016364	.016651	.016958	.017288	.017642
47′	.014211	.014289	.014381	.014485	.014604	.014736	.014883	.015045	.015223	.015417	.015628	.015856	.016103	.016369	.016655	.016963	.017294	.017648
48′	.014212	.014291	.014382	.014487	.014606	.014739	.014886	.015048	.016226	.015420	.015631	.015860	.016107	.016374	.016660	.016969	.017299	.017654
49′	.014214	.014292	.014384	.014489	.014608	.014741	.014888	.015051	.015229	.015424	.015635	.015864	.016111	.016378	.016665	.016974	.017305	.017660

TABLE 23–53C
Constant K_β for use in Eq. (23–267) (Cont.)

K_β

Helix angle, β, min	5°	6°	7°	8°	9°	10°	11°	12°	13°	14°	15°	16°	17°	18°	19°	20°	21°	22°
50'	.014215	.014294	.014386	.014491	.014610	.014743	.014891	.015054	.015232	.015427	.015639	.015868	.016116	.016383	.016670	.016979	.017311	.017666
51'	.014216	.014295	.014387	.014493	.014612	.014746	.014894	.015057	.015235	.015430	.015642	.015872	.016120	.016387	.016675	.016985	.017317	.017673
52'	.014217	.014296	.014389	.014495	.014614	.014748	.014896	.015060	.015239	.015434	.015646	.015876	.016124	.016392	.016680	.016990	.017322	.017679
53'	.014218	.014298	.014390	.014496	.014616	.014750	.014899	.015062	.015242	.015437	.015650	.015880	.016129	.016397	.016685	.016995	.017328	.017685
54'	.014220	.014299	.014392	.014498	.014618	.014753	.014901	.015065	.015245	.015441	.015653	.015884	.016133	.016401	.016690	.017001	.017334	.017691
55'	.014221	.014301	.014394	.014500	.014621	.014755	.014904	.015068	.015248	.015444	.015657	.015888	.016137	.016406	.016695	.017006	.017340	.017697
56'	.014222	.014302	.014395	.014502	.014623	.014757	.014907	.015071	.015251	.015447	.015661	.015892	.016142	.016411	.016700	.017011	.017345	.017703
57'	.014223	.014304	.014397	.014504	.014625	.014760	.014909	.015074	.015254	.015451	.015665	.015896	.016146	.016415	.016705	.017017	.017351	.017709
58'	.014225	.014305	.014399	.014506	.014627	.014762	.014912	.015077	.015257	.015454	.015668	.015900	.016150	.016420	.016710	.017022	.017357	.017716
59'	.014226	.014307	.014400	.014508	.014629	.014764	.014914	.015080	.015260	.015458	.015672	.015904	.016155	.016425	.016715	.017028	.017363	.017722

TABLE 23–54
Impact coefficient C_s

| Type of machinery | Type of drive | | |
	Motor	Single-cylinder engine	Multi-cylinder engines, turbines
Electric generator, belt conveyors, light hoist and elevators, turboblowers, and compressors	1.1	1.5	1.25
Main drives of machine tools, heavy-duty elevators, rotating gears for cranes, multicylinder piston pumps, feed pumps, etc.	1.25	1.75	1.5
Punches and shears, steel rolling mills, bucket drudgers shear showel heavy centrifuges and feed pumps	1.75	2.25	2.0

TABLE 23–55
Load coefficients for gears

Coefficient	Expression
Dynamic, C_D	$$C_D \simeq 1 + \frac{F_d^*}{F^* C_s(\epsilon_\beta + 1)} \leq \frac{1.3 + \epsilon_\beta}{\epsilon_\beta + 1} + \frac{f/(\epsilon_\beta + 1)}{B d_1' C_s}$$ Refer to Fig. 23–50 for $f(\mu m)$; $\epsilon_\beta = \dfrac{b \tan \beta}{\pi m_n} = \dfrac{b \tan \beta}{\pi m}$
Supporting error, C_T	$$C_T \simeq 1 + 0.3 \, \frac{f_{RW}}{C_S C_D B d_1'}$$ Refer to Fig. 23–49 for f_{RW}; for one-sided bearing, C_T must be increased by 0.1
Helical gear, C_β	$$C_\beta \simeq \frac{1.4}{\epsilon}$$ For spur and straight-tooth helical gear with $\epsilon_\beta < 0.5$, $C_\beta = 1$

TABLE 23–56
Recommended safety factors, n, for use in Eqs. (23–318) and (23–323)

Kind of service	Safety factor	
	Pitting, n_p	Breakage, n_b
Continuous	1.8 to 4	1.3 to 2.5
Intermittent	1.5 to 2	1 to 1.5

TABLE 23–57
Oil factor K_o

Viscosity of oil at 50°C, cSt	7	20	34	65	100	140	190	250	290
Oil factor, K_o	0.7	0.75	0.8	0.9	1.0	1.1	1.2	1.3	1.35

TABLE 23–58
Coefficient of peripheral velocity, K_v

Velocity, v, m/s	0	3	5.5	8	11	18	100
Coefficient, K_v	0.7	0.8	0.9	1.0	1.1	1.2	1.3

TABLE 23–59
Coefficient for helical gears, K_β

Helix angle, β	K_β	Pressure angle, α	K_n	K_z
0	1.00	20.00	1.00	1.00
2	1.00	20.01	1.00	1.00
4	0.99	20.04	1.00	1.01
6	0.99	20.10	1.01	1.02
8	0.98	20.18	1.02	1.03
10	0.96	20.28	1.03	1.04
12	0.95	20.41	1.04	1.06
14	0.93	20.56	1.05	1.09
16	0.91	20.76	1.07	1.12
18	0.88	20.96	1.09	1.15
20	0.86	21.17	1.12	1.19
22	0.83	21.43	1.14	1.23
24	0.80	21.72	1.17	1.28
26	0.77	22.05	1.20	1.34
28	0.73	22.04	1.24	1.41
30	0.70	22.80	1.28	1.48
32	0.67	23.23	1.33	1.57
34	0.63	23.70	1.38	1.67
36	0.60	24.22	1.44	1.78
38	0.56	24.79	1.50	1.91
40	0.53	25.41	1.57	2.06
42	0.49	26.09	1.65	2.23
44	0.46	26.84	1.74	2.42
46	0.42	27.65	1.84	2.65

TABLE 23–60
Flank compression coefficient, K_c

α_n'	K_n	α_n'	K_c
10	5.85	22	2.88
11	5.34	23	2.78
12	4.91	24	2.69
13	4.56	25	2.61
14	4.26	26	2.54
15	4.00	27	2.47
16	3.77	28	2.41
17	3.58	29	2.36
18	3.40	30	2.31
19	3.25	31	2.27
20	3.11	32	2.23
21	2.99	33	2.19

TABLE 23–61
Determination of ϵ_v [a]

Spur gear		Helical gear	
Standard tooth	**Corrected tooth**	**Standard tooth**	**Corrected tooth**
$\epsilon_v = \epsilon_v'$ ϵ_v' from Fig. 23–53 where $z_v = z$ and $\alpha' = 20°$	$\epsilon_v = (\epsilon_v'z)/z_v$ ϵ_v' from Fig. 23–53 where $z_v = \dfrac{2d'}{d_a - d'}$ and α' obtained from Fig. 23–52	$\epsilon_v = \epsilon_v' \cos\beta$ ϵ_v' from Fig. 23–53 where $z_v = z/\cos\beta$ and α' obtained from Fig. 23–52	$\epsilon_v = (\epsilon_v'z)/z_v$ ϵ_v' from Fig. 23–53 where $z_v = \dfrac{2d'}{d_a - d'}$ and α' obtained from Fig. 23–52

[a] ϵ_{v1} refers to pinion and ϵ_{v2} refers to gear.

TABLE 23–62
Table of preferred numbers formula

ϕ	1.06	1.12	1.26	1.41	1.58	1.78	2
	$(\sqrt[9]{2})$	$(\sqrt[12]{2})$	$(\sqrt[6]{2})$	$(\sqrt[3]{2})$	$(\sqrt{2})$	$(\sqrt[1.5]{2})$	$(\sqrt[1.2]{2})$	$(\sqrt{2})$
	$(\sqrt[9]{10})$	$(\sqrt[40]{10})$	$(\sqrt[20]{10})$	$(\sqrt[10]{10})$	$(\sqrt[10/3]{10})$	$(\sqrt[5]{10})$	$(\sqrt[4]{10})$	$(\sqrt[20/6]{10})$
$\dfrac{\phi - 1}{\phi} \times 100\%$	5	10	20	30	40	45	50

TABLE 23–63
Preferred numbers

Serial no.	Basic series of preferred numbers[a]			
	R 5	*R* 10	*R* 20	*R* 40
0	1.00	1.00	1.00	1.00
1				1.06
2			1.12	1.12
3				1.18
4		1.25	1.25	1.25
5				1.32
6			1.40	1.40
7				1.50
8	1.60	1.60	1.60	1.60
9				1.70
10			1.80	1.80
11				1.90
12		2.00	2.00	2.00
13				2.12
14			2.24	2.24
15				2.36
16	2.50	2.50	2.50	2.50
17				2.65
18			2.80	2.80
19				3.00
20		3.15	3.15	3.15
21				3.35
22			3.55	3.55
23				3.75
24	4.00	4.00	4.00	4.00
25				4.25
26			4.50	4.50
27				4.75
28		5.00	5.00	5.00
29				5.30
30			5.60	5.60
31				6.00
32	6.30	6.30	6.30	6.30
33				6.70
34			7.10	7.10
35				7.50
36		8.00	8.00	8.00
37				8.50
38			9.00	9.00
39				9.50
40	10.00	10.00	10.00	10.00

[a] The figures indicated in the table can be increased or decreased 10, 100, 1000, 10,000, and 100,000 times.

TABLE 23–64
Preferred numbers of *R* 80 series

1.00	1.32	1.80	2.36	3.15	4.25	5.60	7.50
1.03	1.36	1.85	2.43	3.25	4.37	5.00	7.70
1.06	1.40	1.90	2.50	3.35	4.50	6.00	8.00
1.09	1.45	1.95	2.58	3.45	4.62	6.15	8.25
1.12	1.50	2.00	2.65	3.55	4.75	6.30	8.50
1.15	1.55	2.06	2.72	3.65	4.87	6.50	8.75
1.18	1.60	2.12	2.80	3.75	5.00	6.70	9.00
1.22	1.69	2.18	2.90	2.87	5.15	6.90	9.25
1.25	1.70	2.24	3.00	3.40	5.30	7.10	9.50
1.28	1.75	2.30	3.07	4.12	5.45	7.30	9.75

TABLE 23–65
Values of ϕ for satisfactory operation of machine tools

Particular	Values of ϕ
For great majority of general-purpose machine tools	1.26 or 1.41
For mass or lot production (automatic or semiautomatic machine tools)	1.12 or 1.26
Small machine tools	1.58 and 1.78
Heavy machine tools	1.26, 1.12, and 1.06

TABLE 23–66
Standard spindle speeds for machine tools

Basic range R/20, φ = 1.12	Range R 20/2, φ = 1.25	Range R 20/3, φ = 1.4			Range R 20/4, φ = 1.6			Range R 20/6, φ = 2	
					(1400)	(2800)			
100				1000					
112	112	11.2				112	11.2		
125			125						
140	140			1400	140				1400
160		16							
180	180		180			180		180	
200				2000					
224	224	22.4			224		22.4		
250			250						
280	280			2800		280			2800
315		31.5							
355	355		355		355			355	
400				4000					
450	450	45				450	45		
500			500						
560	560			5600	560				5600
630		63							
710	710		710			710		710	
800				8000					
900	900	90			900		90		
1000			1000						

TABLE 23–67
Standard feeds, mm/min

φ = 1.12	φ = 1.25		φ = 1.4		φ = 1.6		φ = 2	
1	1		1		1		1	
1.12				11.2				
1.25	1.25	0.125				0.125		
1.4			1.4					
1.6	1.6			16	1.6			16
1.8		0.18						
2	2		2				2	
2.24				22.4				
2.5	2.5	0.25			2.5	0.25		
2.8			2.8					
3.15	3.15	0.315		31.5				3.15
3.55								
4	4		4		4		4	
4.5				45				
5	5	0.5				0.5		
5.6			5.6					
6.3	6.3			63	6.3			63
7.1		0.71						
8	8		8				8	
9				90				
10	10	1			10	1		

TABLE 23-68
Speed steps for design of a gearbox

Numbers of speed steps, z	Possible number of speed steps in the component of drives	Minimum no. of teeth
4	2 × 2	8
6	3 × 2	10
8	2 × 2 × 2 2 × 4	12
9	3 × 3	
10	2 × 5	
12	2 × 3 × 2 2 × 2 × 2	14
	4 × 3	
	3 × 4	
	2 × 6	
	6 × 2	
15	5 × 3 3 × 5	
16	3 × 2 × 2 × 2	16
	4 × 2 × 2 2 × 2 × 4	
	4 × 4	
18	3 × 2 × 3 2 × 3 × 3	18
	6 × 3 3 × 6	

TABLE 23-69
Allowable limits of transmission ratios and output speeds for various gear combinations (18)

Transmission ratio	1.12^0 =1	1.12^1 =1.12	1.12^2 =1.25	1.12^3 =1.4	1.12^4 =1.6	1.12^5 =1.8	1.12^6 =2.0	1.12^7 =2.24	1.12^8 =2.5	1.12^9 =2.8	1.12^{10} =3.15	1.12^{11} =3.55	1.12^{12} =4.0
Number of teeth of pinion, z_1	Number of teeth of gear, z_2												
16	16	18	20	23	25	28	32	36	40	45	50, 51	57, 58	63, 64
17	17	19	21	24	27	30	34	38	42, 43	48	53, 54	60, 61	67, 68
18	18	20	23	25	29	32	36	40, 41	45, 46	50, 51	56, 57, 58	63, 64, 65	71, 72
19	19	21	24	27	30	34	38	42, 43	47, 48	53, 54	59, 60, 61	67, 68	75, 76, 77
20	20	22	25	28	32	36	40	44, 45	50, 51	56, 57	62, 63, 64	70, 71, 72	79, 80, 81
21	21	24	26	30	33	37	42	47	52, 53	58, 59, 60	66, 67	74, 75, 76	83, 84, 85
22	22	25	28	31	35	39	44	49, 50	55, 56	61, 62, 63	69, 70	77, 78, 79	88, 89
23	23	26	29	32, 33	36, 37	41	46	51, 52	57, 58	64, 65, 66	72, 73, 74	81, 82, 83	91, 92, 93
24	24	27	30	34	38	42, 43	48	53, 54	59, 60, 61	67, 68, 69	75, 76, 77	84, 85, 86, 87	94, 95, 96, 97

Courtesy: Permission to produce Tables 23–69 and 23–70 from DIN 803 and 804, 1977, DIN: Deutsches Institute für Normung e.V., Beuth-Vertrieb GmbH, Buggrafenstrasse 6, D1000 Berlin 30, Germany.

TABLE 23–70
Number of teeth and sums of numbers of teeth for standard transmission ratios

Transmission ratio	1	1.12	1.25	1.4	1.6	1.8	2.0	2.24	2.5	2.8	3.15	3.55	4.0
Sum of number teeth, $z_1 + z_2$				Number of teeth—pinion, z_1/gear, z_2									
100	50/50	47/53	44/56		39/61	36/64		31/69		26/74	24/76	22/78	20/80
101	51/50		45/56	42/59	39/62	36/65	34/67	31/70	29/72		24/77	22/79	20/81
102	51/51	48/54	45/57	42/60		37/65	34/68		29/73	27/75			
103	52/51			43/60	40/63	37/66		32/71		27/76	25/78		21/82
104	52/52	49/55	46/58	43/61	40/64		35/69	32/72			25/79	23/81	21/83
105	53/52				41/64	38/67	35/70		30/75		25/80	23/82	21/84
106	53/53	50/56	47/59	44/62	41/65	38/68		33/73	30/76	28/78		23/83	21/85
107	54/53		47/60	44/63	41/66		36/71	33/74		28/79	26/81		
108	54/54	51/57	48/60	45/63	42/66	39/69	36/72	33/75	31/77	28/80	26/82	24/84	22/86
109	55/54	51/58	48/61	45/64	42/67	39/70		34/75	31/78		26/83	24/85	22/87

Courtesy: Permission to produce Tables 23–69 and 23–70 from DIN 803 and 804, 1977, DIN: Deutsches Institute für Normung e.V., Beuth-Vertrieb GmbH, Buggrafenstrasse 6, D1000 Berlin 30, Germany.

TABLE 23–71
Various allowable errors for spur and helical gears (μ)[a]

	Pitch diameter, d' (mm)																	
Above	12		25			50			100			200			400			
Up to	25		30			100			200			400			800			
	Module, m (mm)																	
Above	0.6	1.6	0.6	1.6	4.0	0.6	1.6	4.0	0.6	1.6	4.0	0.6	1.6	4.0	0.6	1.6	4.0	
Up to	1.6	4.0	1.6	4.0	10	1.6	4.0	10	1.6	4.0	10	1.6	4.0	10	1.6	4.0	10	Quality
Error																		
f_{pt}, f_{pb}, f_f	4	4.5	4.5	5	6	5	5.5	7	5.5	6	8	7	7	9	8	9	10	
f_c	14	16	16	18	22	18	20	25	20	22	25	25	25	32	28	32	36	5
f_r	12	14	14	16	18	16	18	20	18	20	22	20	22	25	22	25	28	
f_{pt}, f_{pb}, f_f	6	6	6	7	9	7	8	10	8	9	11	9	10	12	11	12	14	
f_c	20	22	22	25	28	25	28	32	28	32	36	36	36	40	40	45	50	6
f_r	18	20	20	22	25	22	25	28	25	28	32	28	32	36	32	36	40	
f_{pt}, f_{pb}, f_f	8	9	9	10	12	10	11	14	11	12	16	12	14	18	16	18	20	
f_c	28	32	32	36	40	36	40	45	40	45	50	50	50	63	63	63	71	7
f_r	25	28	28	32	36	32	36	40	36	40	45	40	45	50	45	50	56	
f_{pt}, f_{pb}, f_f	11	12	12	14	18	14	16	20	16	18	22	18	20	25	22	25	28	
f_c	40	45	45	50	63	50	56	63	56	63	71	71	71	80	80	90	100	8
f_r	36	40	40	45	50	45	50	56	50	56	63	56	63	71	63	71	80	
f_{pt}, f_{pb}, f_f	16	18	18	20	25	20	22	28	22	25	32	25	28	36	36	36	40	
f_c	56	63	63	71	80	71	80	90	80	90	100	100	100	125	125	126	140	9
f_r	50	56	56	63	71	63	71	80	71	80	90	80	90	100	90	100	110	
f_{pt}, f_{pb}, f_f	25	28	28	32	40	32	36	45	36	40	50	40	45	56	50	56	63	
f_c	90	100	100	110	140	110	125	140	125	140	160	160	160	180	180	200	220	10
f_r	71	80	80	90	100	90	100	110	100	110	125	110	125	140	125	140	160	

[a] *Key*: f_f = profile form error; f_{pt} = adjacent pitch error; f_{pb} = base pitch error; f_r = radial runout; f_c = cumulative pitch error. For helical gear, use tooth alignment error f_β instead of f_f and multiply f_r by a factor $\tan 20°/\tan \alpha_t$.

TABLE 23–72
Service factor, C_s

Power source	Duration of service, h/day	Service factor, C_s			
		Nature of load on gear unit from driven machine			
		Uniform	Moderate shock	Heavy shock, worm	Heavy shock, helical and spur
Electric motor or steam turbine	2	0.75	0.90	1.25	1.40
	4	0.80	1.00	1.30	1.50
	8	0.90	1.10	1.45	1.65
	12	1.00	1.25	1.55	1.75
	24	1.25	1.50	1.75	2.00
Multicylinder internal combustion engine	2	0.90	1.11	1.25	1.65
	4	1.00	1.25	1.40	1.75
	8	1.10	1.35	1.60	1.90
	12	1.25	1.50	1.75	2.00
	24	1.50	1.75	2.00	2.25
Single-cylinder internal-combustion engine	2	1.10	1.35	1.75	1.90
	4	1.25	1.50	1.85	2.00
	8	1.35	1.65	1.95	2.15
	12	1.50	1.75	2.05	2.25
	24	1.75	2.00	2.25	2.50

23.3 BEVEL GEARS

SYMBOLS

In addition to symbols already given under *spur gear* in Sec. 23.1, the following symbols are adopted for designing bevel gears:

a'_e	equivalent center distance, m (in)
b	face width, m (in)
d'_m	mean pitch circle diameter, m (in)
d'_e	equivalent pitch circle diameter, m (in)
d'_{em}	equivalent mean pitch circle diameter, m
d_s	diameter of shaft, m (in)
F	load or force, kN (lbf)
$F_{w\beta}$	limiting load for wear, kN (lbf)
i	transmission ratio
i_e	equivalent transmission ratio
J	geometry factor
k_m	mounting factor
K_M	material factor
k_{fb}	a factor which takes into account unilateral fatigue bending
l_{be}	length from bearing center to the middle of the gear face width, m (in)
l_s	length of shaft or length from center to center of bearings, m (in)
$\mathbf{m}$	module, m (in)
m	dimensions as shown in Fig. 23–62
m_{max}	module measured at the big end of the cone of the bevel gear, m (in)
m_m	mean module (measured in the middle of tooth), m (in)
n	speed, rpm dimension as shown in Fig. 23–62
R	cone distance, m (in)
R_b	back cone distance, m (in)
z	number of teeth
z_s	smallest number of teeth in straight-tooth bevel gear
$z_{b(min)}$	smallest number of teeth in the bevel gear free from undercutting
z_v	formative or virtual number of teeth
β	helix angle, deg
β_m	mean helix angle at the middle of face width of tooth, deg
$\psi_r = R/b$	fullness factor

Particular	Formula
Gear ratio or transmission ratio	$$i = \frac{\omega_1}{\omega_2} = \frac{n_1}{n_2} = \frac{d_2}{d_1} = \frac{z_2}{z_1} = \tan \delta_2 = \frac{1}{\tan \delta_1} \qquad (23\text{–}400)$$
For shaft-load formulae for spiral bevel gears and comparison of reference profiles	Refer to Tables 23–73A and 23–73B.

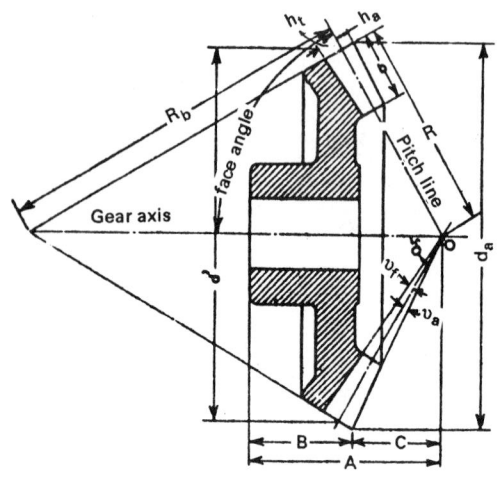

FIGURE 23–61   Definitions and dimensions of bevel gears.

Particular	Formula
For dimensions and definitions of bevel gear	Refer to Fig. 23–61.
The pitch diameter of pinion	$d_1' = m_{max} z_1$ (23–401)
The pitch diameter of gear	$d_2' = m_{max} z_2$ (23–402)
The outside diameter of pinion	$d_{a1} = d_1' + 2h_{a1} \cos \delta_1 = d_1' + 2m_{max} \cos \delta_1$ (23–403)
The outside diameter of gear	$d_{a2} = d_2' + 2h_{a2} \cos \delta_2 = d_2' + 2m_{max} \cos \delta_2$ (23–404)
The addendum cone diameter (Fig. 23–74)	$d_a = d' + 2m_{max} \cos \delta = m_{max}(z + 2\cos \delta)$ (23–405)
	where m_{max}, usually selected from series of modules
	Refer to Table 23–74.
The dedendum cone diameter (Fig. 23–61)	$d_f = d' - 2.5 m_m \cos \delta = m_m(z - 2.5 \cos \delta)$ (23–406)
For recommended series of diametral pitches of bevel gears	Refer to Table 23–75.

PITCH ANGLES
Acute-angle bevel gears

The pitch cone angle of pinion (Fig. 23–62)

$$\tan \delta_1 = d_1'/[2(m+n)] = \frac{\sin \Sigma}{(d_2'/d_1') + \cos \Sigma}$$

$$= \frac{\sin \Sigma}{(z_2/z_1) + \cos \Sigma} \qquad (23\text{–}407)$$

$$\text{where } m = \frac{d_1'}{2 \sin \Sigma} \text{ and } n = \frac{d_1'}{2 \tan \Sigma}$$

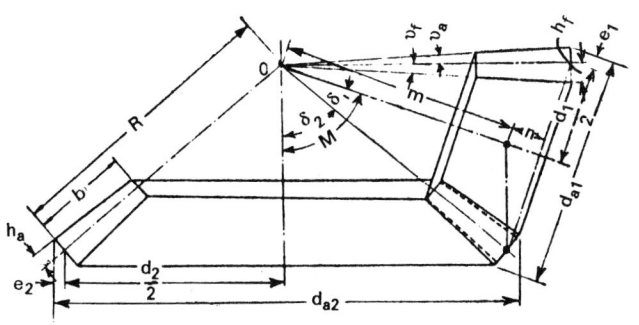

FIGURE 23–62 Acute-angle bevel gear.

Particular	Formula
The pitch angle of gear	$$\tan\delta_2 = \frac{\sin\Sigma}{(d_1'/d_2') + \cos\Sigma} = \frac{\sin\Sigma}{(z_1/z_2) + \cos\Sigma} \qquad (23\text{–}408)$$

Right-angle bevel gears

The pitch angle of pinion	$$\tan\delta_1 = \frac{d_1'}{d_2'} = \frac{z_1}{z_2} = \frac{1}{i} \qquad (23\text{–}409)$$
The pitch angle of gear	$$\tan\delta_2 = \frac{d_2'}{d_1'} = \frac{z_2}{z_1} = i \qquad (23\text{–}410)$$

Obtuse-angle bevel gears

The pitch angle of pinion	$$\tan\delta_1 = \frac{\sin(180° - \Sigma)}{(d_2'/d_1') - \cos(180° - \Sigma)}$$ $$= \frac{\sin(180° - \Sigma)}{(z_2/z_1) - \cos(180° - \Sigma)} \qquad (23\text{–}411)$$
The pitch angle of gear	$$\tan\delta_2 = \frac{\sin(180° - \Sigma)}{(d_1'/d_2') - \cos(180° - \Sigma)}$$ $$= \frac{\sin(180° - \Sigma)}{(z_1/z_2) - \cos(180° - \Sigma)} \qquad (23\text{–}412)$$
The addendum angle (Fig. 23–62)	$$\tan\vartheta_a = \frac{2h_{a1}\sin\delta_1}{d_1'} = \frac{2h_{a2}\sin\delta_2}{d_2'} \qquad (23\text{–}413)$$
The dedendum angle	$$\tan\vartheta_f = \frac{2h_{f1}\sin\delta_1}{d_1'} = \frac{2h_{f2}\sin\delta_2}{d_2'} \qquad (23\text{–}414)$$

Particular	Formula	
The relation between the pitch diameter, the face width, and the pitch angle of pinion	$d_1' = d_{1m} + b \sin \delta_1$	(23–415)
The relation between maximum module, mean module, number of teeth, face width, and pitch cone angle	$m_{max} z_1 = m_m z_1 + b \sin \delta_1$	(23–416)
The maximum module	$m_{max} = m_m + (b/z_1) \sin \delta_1$	(23–417)
Transverse module or maximum module is also given by the equation	$m_{max} = m_m \dfrac{\psi_R}{\psi_R - 0.5}$	(23–418)
Fullness factor	$\psi_R = \dfrac{R}{b}$	(23–419)
The mean module	$m_m = m_{max} - \dfrac{b \sin \delta_1}{z_1}$	(23–420)
Cone distance	$R = \dfrac{1}{2} \sqrt{d_1^2 + d_2^2} = \dfrac{m_{max}}{2} \sqrt{z_1^2 + z_2^2}$	(23–421)
The smallest number of teeth in the bevel gear free from undercutting	$(z_b)_{min} = (z_s)_{min} \cos \delta$	(23–422)
The number of teeth in the pinion	$z_1 = \dfrac{2R}{m_{max} \sqrt{i^2 + 1}}$	(23–423)
The number of teeth in the gear	$z_2 = z_1' i$	(23–424)
The formative number of teeth in a straight-tooth bevel pinion	$z_{1v} = \dfrac{z_1}{\cos \delta_1}$	(23–425)
The formative number of teeth in a straight-tooth bevel gear	$z_{2v} = \dfrac{z_2}{\cos \delta_2}$	(23–426)
The formative number of teeth in a spiral bevel pinion	$z_{1v} = \dfrac{z_1}{\cos \delta_1 \cos^3 \beta_m}$	(23–427)
The formative number of teeth in a spiral bevel gear	$z_{2v} = \dfrac{z_2}{\cos \delta_2 \cos^3 \beta_m}$	(23–428)
	where β_m either zero or varies from 8° to 35°	
Helix angle of pinion	$\tan \beta_{1m} \geq \dfrac{\pi m}{b} \left(1 - \dfrac{1}{\psi_R}\right)$	(23–429)

WIDTH OF BEVEL GEAR TOOTH FACE

The width of gear face	$\dfrac{R}{4} \leq b \leq \dfrac{R}{3}$	(23–430)

Particular	Formula
The width of gear face as per the practice of Gleason works	$b = 6m$ to $7m$ if $R < 30m$ (23–431a)
	$= \dfrac{6}{P}$ to $\dfrac{7}{P}$, if $R < \dfrac{30}{P}$ (23–431b)
	$b = 7m$ to $10m$, if $R > 30m$ (23–432a)
	$= \dfrac{7}{P}$ to $\dfrac{10}{P}$, if $R > 30/P$ (23–432b)
For service factor (C_s), and form factors (Y) and (Y_k)	Refer to Tables 23–76 to 23–78.
For proportion of straight-tooth bevel gear	Refer to Table 23–79.

DESIGNING GEARS FOR BENDING STRENGTH

The equivalent tangential force at large end of tooth as per modified Lewis equation for bevel gears (Fig. 23–63)

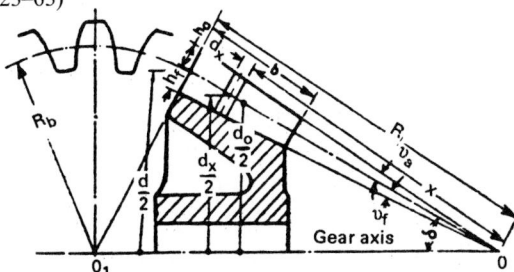

FIGURE 23–63 Analysis of bevel gear for strength.

$$F_t = \sigma_o K_v b Y m \left(\frac{R - b}{R} \right) = \sigma_o K_v b Y m \left(1 - \frac{1}{\psi_R} \right)$$
(23–433a)

$$= \frac{\sigma_o K_v b Y}{P} \left(1 - \frac{1}{\psi_R} \right) = \frac{\sigma_e}{K_{fb} K_\sigma} \frac{K_v Y b}{P} \left(1 - \frac{1}{\psi_R} \right)$$
(23–433b)

where
Y = form factor. (Refer to Table 23–77.)
K_{fb} = 1.5 to account for unilateral fatigue in bending
K_σ = stress concentration factor obtained from Fig. 23–31 or 23–33 and Table 23–30
σ_o = $\sigma_e/(K_{fb}K_\sigma)$, taken from Table 23–18

Modified Lewis equation for calculating module

$$m = \frac{F_t}{\sigma_o K_v b Y \left(1 - \dfrac{1}{\psi_R} \right)} \quad \textbf{SI}$$
(23–434a)

Modified Lewis equation for calculating diametral pitch

$$P = \frac{\sigma_o K_v b Y}{F_t} \left(1 - \frac{1}{\psi_R} \right)$$

US Customary System units   (23–434b)

For dimensions of forged bevel gears

Refer to Fig. 23–64(a).

AGMA fundamental bending stress equation for gear teeth (3, 13)

$$\sigma_b = \frac{2000 M_{t1} K_a}{K} \frac{1}{m_t b d_{o1}'} \frac{K_{sz} K_m}{K_x J} \quad \textbf{SI}$$
(23–435a)

$$= \frac{F_t K_a}{K_v} \frac{1}{m_t b} K_{sz} K_m / K_x J)$$
(23–435b)

where σ_b in MPa, m_t, b and d_{o1} in mm, and M_{t1} in N m, F_t in N

Particular	Formula

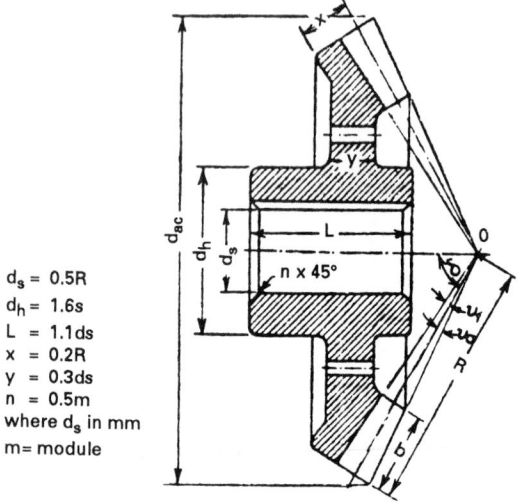

$d_s = 0.5R$
$d_h = 1.6s$
$L = 1.1ds$
$x = 0.2R$
$y = 0.3ds$
$n = 0.5m$
where d_s in mm
m= module

FIGURE 23–64(a)   Forged bevel gear dimensions.

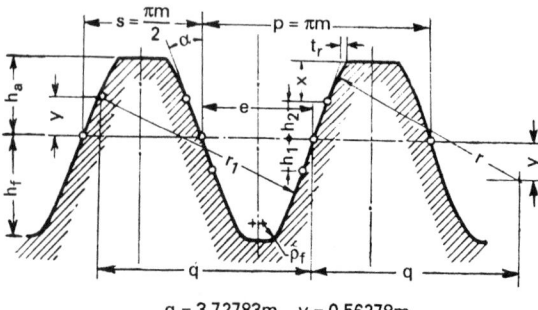

$q = 3.72783m, \quad y = 0.56278m$

FIGURE 23–64(b)   (See also Fig. 23–5.)

$$= \frac{2M_{t1}K_a}{K_v} \frac{P_t}{bd'_{o1}} \frac{K_{sz}K_m}{K_x J}$$

US Customary System units   (23–436a)

$$= \frac{F_t K_a}{K_v} \left(\frac{P_t}{b}\right) \frac{K_{sz}K_m}{K_x J} \qquad (23\text{–}436b)$$

where σ_b in psi, P_t in in^{-1}, b and d'_{o1} in in, M_t in lbf in, and F_t in lbf

Refer to Eq. (23–457), for $K_v = C_v =$ internal dynamic factor.

$K_a = C_a =$ external dynamic factor (Refer to Table 23–81.)

$K_{sz} =$ size factor $= 1$ for finer or 5 mean normal diametral pitch

The load distribution factor

$$K_m = C_m = \left[1.2 + \left(\frac{M_{tv1} - M_{td1}}{M_{td1}}\right)^2\right] C_{mf}$$

where $M_{tv1} \le M_{td1}$ (23–437a)

$$= \left[1.2 + \left(\frac{M_{tv1} - M_{td1}}{4M_{td1}}\right)^2\right] C_{mf} \qquad (23\text{–}437b)$$

when $M_{tv1} > M_{td1}$

Refer to Table 23–80 for values of mounting factor C_{mf}. For noncrowned gear teeth use 2.0 times the values given in Table 23–80.

Particular	Formula
	K_x = lengthwise curvature factor for bending strength from Eq. (23–438a)
The lengthwise curvature factor for bending strength	$K_x = 0.211 \left(\dfrac{r_c}{R_m} \right)^q + 0.789$ for spiral bevel gears
	(23–438a)
	= 1.0 for straight-tooth bevel and ZEROL bevel gears
	where r_c = cutter radius, mm (in) R_m = mean cone distance, mm (in)
	$q = \dfrac{0.279}{\log_{10}(\sin \beta_m)}$ (23–438b)
	β_m = mean spiral angle
	when the calculated value of K_x is greater than 1.15, make $K_x = 1.15$, and when the calculated value of K_x is less than 1.0, make $K_x = 1.0$
	J = geometry factor (Refer to Figs. 23–65 to 23–67.)
AGMA formula for working bending stress number or bending strength	$\sigma_{swb} = \dfrac{\sigma_{sab} K_L}{K_T K_R}$ (23–439)
The calculated tensile bending stress number must be less than or equal to the working bending stress numbers	$\sigma_b \le \sigma_{swb} = \dfrac{\sigma_{sab} K_L}{K_T K_R}$ (23–440)
	where σ_{sab} = allowable bending stress numbers from Table 23–23
AGMA equation for the rated power or allowable transmitted power to resist breakage of gear teeth [*Note*: The values for σ_{sab}, K_L, and J may be different for the pinion and gear. Therefore, the design of gear set should be based on the member which has the lowest value of the product $(\sigma_{sab} K_L J)$.]	$P_{ab} = \dfrac{n_1 b}{19.1 \times 10^6} \dfrac{J K_x K_v}{K_{sz} K_m K_a} \sigma_{sab} d'_1 m_t \dfrac{K_L}{K_T K_R}$ **SI** (23–441) where P_{ab} in kW, n_1 in rpm, b, m_t, and d'_1 in mm, and σ_{sab} in MPa. $= \dfrac{n'_1 b}{318} \dfrac{J K_x K_v}{K_{sz} K_m K_a} m_t \sigma_{sab} d'_1 \dfrac{K_L}{K_T K_R}$ **SI** (23–442) where P_{ab} in kW, σ_{sab} in Pa, b, d'_1, and m_t in m $P_{ab} = \dfrac{n_1 b}{126,000} \dfrac{J K_x K_v}{K_{sz} K_m K_a} \dfrac{\sigma_{sab} d'_1}{P_t} \dfrac{K_L}{K_T K_R}$ (23–443) **US Customary units** where P_{ab} in hp, n_1 in rpm, σ_{sab} in psi, b and d'_1 in in, and P_t in in^{-1}.

Particular	Formula

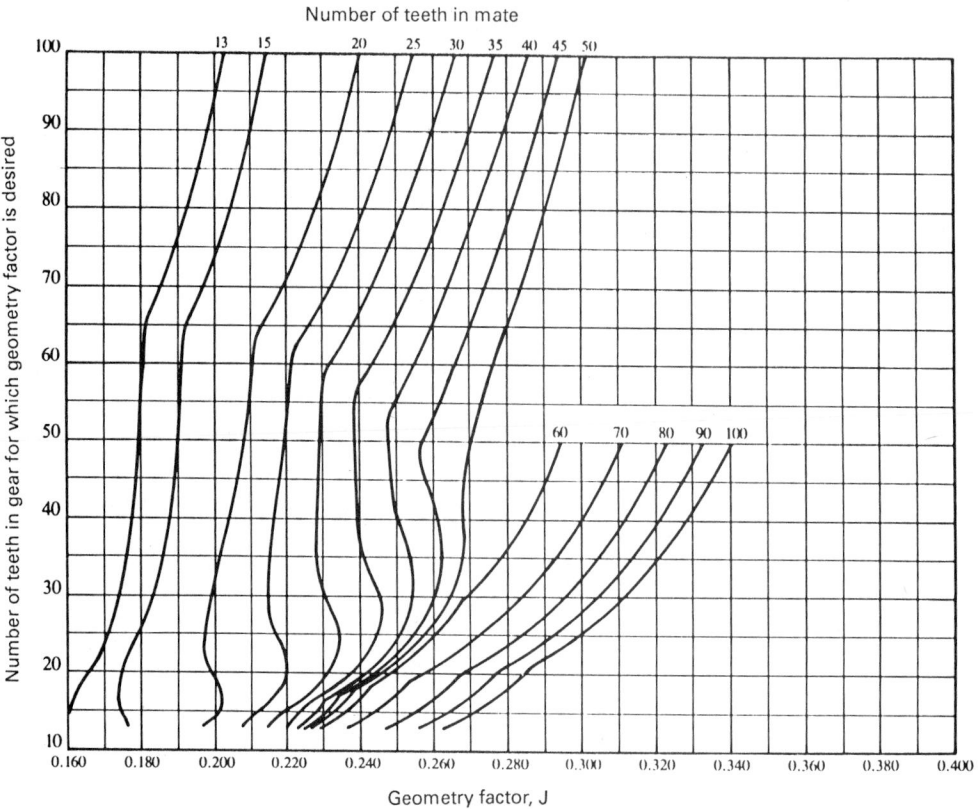

FIGURE 23–65(a) Straight bevel gears with $20°$ pressure angle and $0.120/P_d$ tool edge radius, geometry factor, J. *(AGMA 6010-E88.)*

Refer to Fig. 23–68 for K_L.

For factor K_a Refer to Table 23–81.

AGMA fundamental contact stress equation for pitting resistance of bevel gear teeth (3, 13)

$$\sigma_c = C_p C_b \left\{ \frac{2000 M_{td1} C_a}{C_v} \left(\frac{M_{tv1}}{M_{td1}} \right)^z \frac{1}{b_n d_{o1}'^2} \right.$$

$$\left. \frac{C_{sz} C_m C_{xc} C_{sr}}{I} \right\}^{1/2} \quad \textbf{SI} \qquad (23\text{–}444)$$

$$= C_p C_b \left\{ \frac{F_t C_a}{C_v} \left(\frac{M_{tv1}}{M_{td1}} \right)^z \frac{1}{b_n d_{o1}'} \frac{C_{sz} C_m C_{xc} C_{sr}}{I} \right\}^{1/2}$$

$$\textbf{SI} \qquad (23\text{–}445)$$

where σ_c in MPa; C_p in $\sqrt{\text{MPa}}$; b_n and d_{o1}' in mm, F_t in N, and M_{tv} in N mm

Particular	Formula

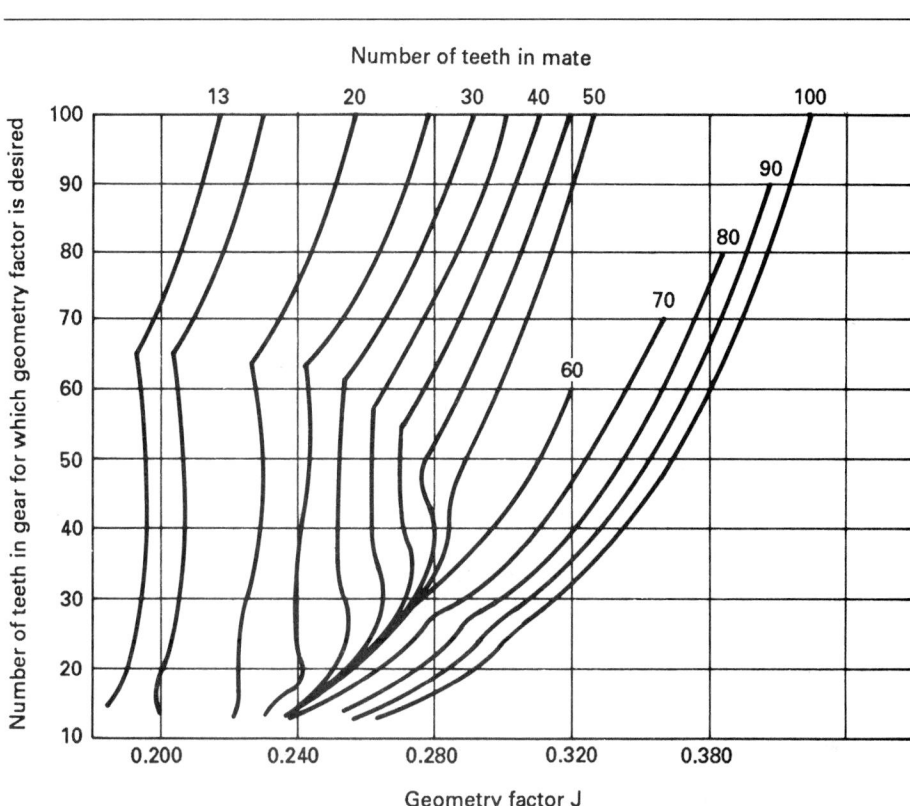

FIGURE 23–65(b) Straight bevel gears with 25° pressure angle and 90° shaft angle, geometry factor, *J. (Gleason Works, Rochester, New York.)*

$$= C_p C_b \left\{ \frac{2 M_{td1} C_a}{C_v} \left(\frac{M_{tv1}}{M_{td1}} \right)^z \frac{1}{b_n d'_{o1}} \frac{C_{sz} C_m C_{xc} C_{sr}}{I} \right\}^{1/2}$$

$$(23\text{–}446)$$

$$= C_p C_b \left\{ \frac{F_t C_a}{C_v} \left(\frac{M_{tv1}}{M_{td1}} \right)^z \frac{1}{b_n d'_{o1}} \frac{C_{sz} C_m C_{xc} C_{sr}}{I} \right\}^{1/2}$$

US Customary System units $\qquad$ (23–447)

where F_t in lbf, M_{tv} in lbf in σ_c in psi, C_p in $\sqrt{\text{psi}}$, b_n and d'_{o1} in in

$z = 0.667$ when $M_{tv1} < M_{td1}$
$ = 1.00$ when $M_{tv1} > M_{td1}$
$ = 1.00$ for non-crowned teeth

$b_n =$ net face width of narrowest member, mm or m (in)

Refer to Table 23–82 for C_p.
Refer to section under K_v for C_v and take $C_b = 0.634$

Particular	Formula

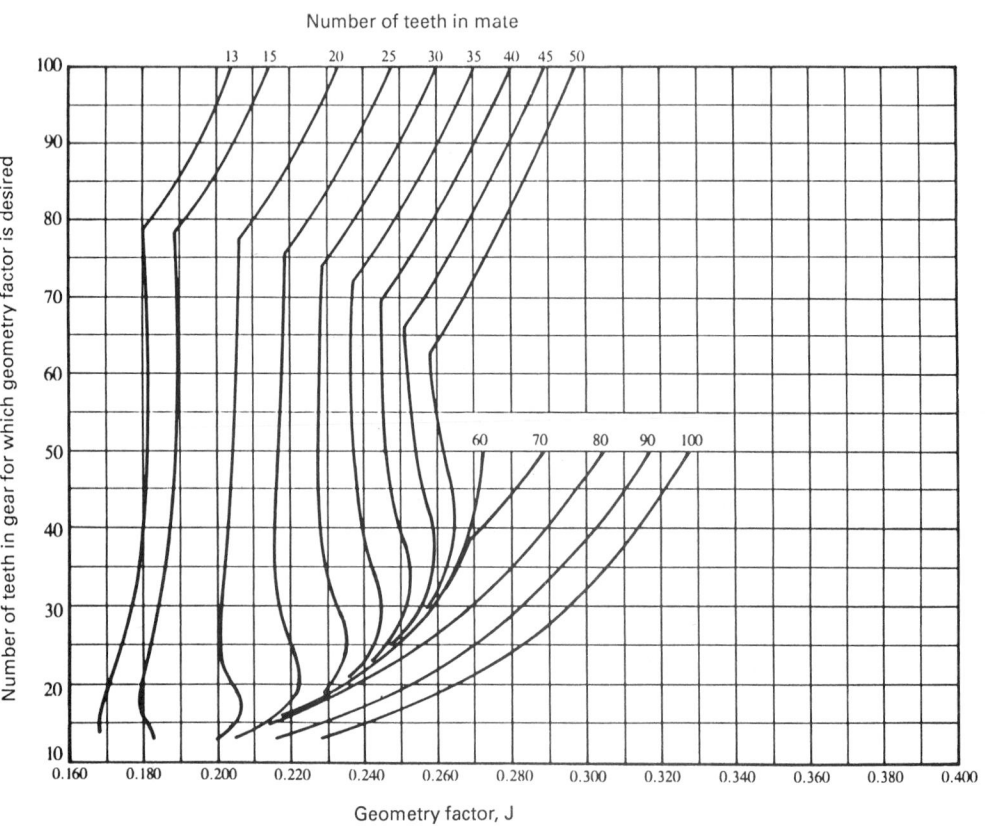

FIGURE 23–66   Fine pitch ZEROL bevel gears with 20° pressure angle and $0.240/P_d$ tool edge radius geometry factor, J.

$C_{sz} = 1$ for most gears
$C_{xc} = 1.5$ for crowned teeth
$\quad = 1.0$ for noncrowned teeth

Refer to Figs. 23–69 to 23–71 for I.

AGMA formula for working contact stress number

$$\sigma_{swc} = \frac{\sigma_{sac} C_L C_H}{C_T C_R} \qquad (23\text{–}448)$$

The calculated contact stress number must be less than or equal to the working contact stress number

$$\sigma_c \leq \sigma_{swc} = \frac{\sigma_{sac} C_L C_H}{C_T C_R} \qquad (23\text{–}449)$$

Refer to Fig. 23–72 for C_L.

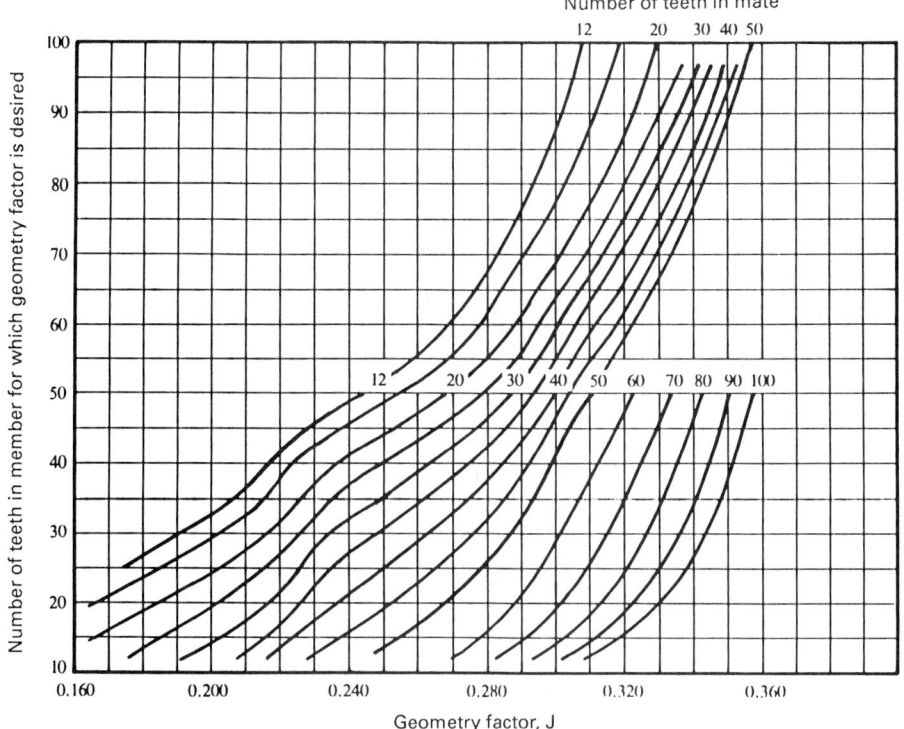

FIGURE 23–67(a) Sprial bevel gears with 20° pressure angle, 35° spiral angle, and 0.240/P_t tool edge radius geometry factor, J. (*AGMA 6010-E88.*)

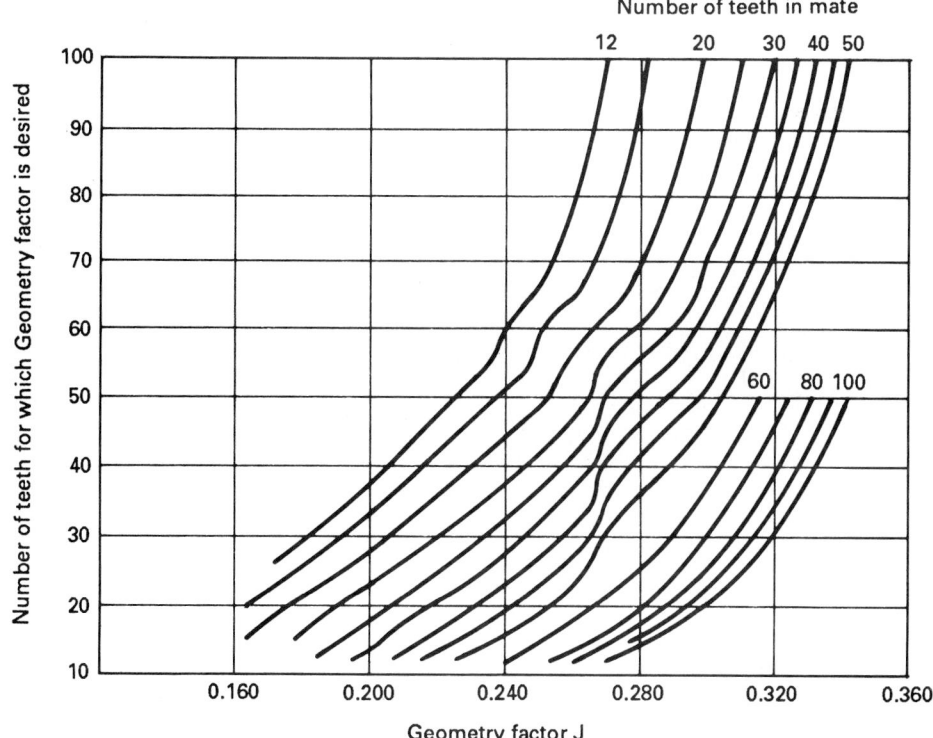

FIGURE 23–67(b) Spiral bevel gears with 20° pressure angle, 25° spiral angle, and 90° shaft angle, geometry factor, J. (*Gleason Works, Rochester, New York.*)

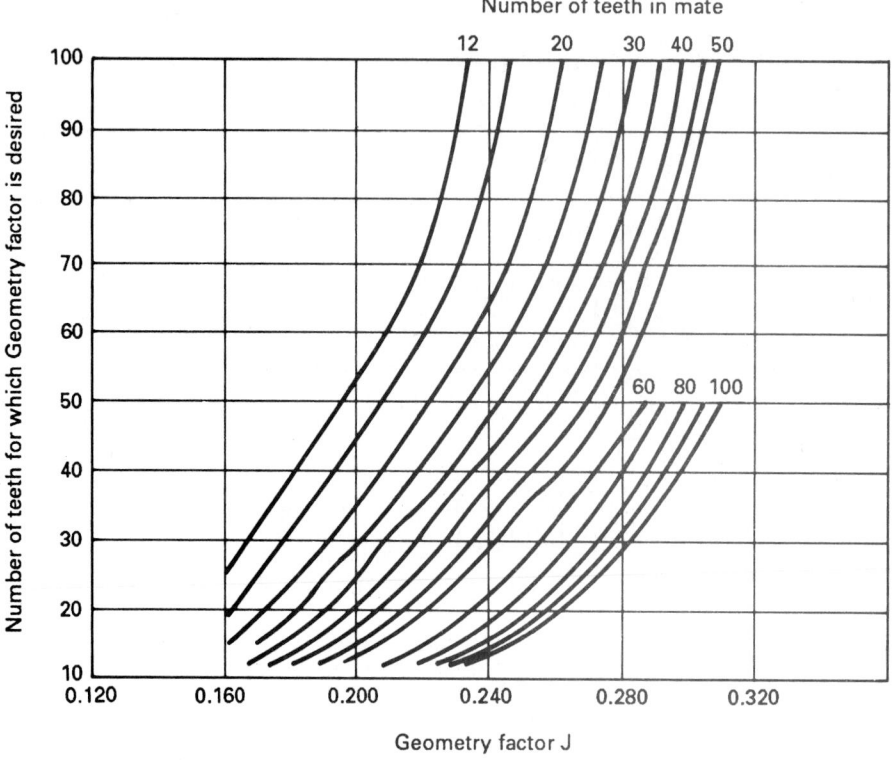

FIGURE 23–67(c)   Spiral bevel gears with 20° pressure angle, 15° spiral angle, and 90° shaft angle, geometry factor, *J*. (*Gleason Works, Rochester, New York.*)

Particular	Formula
AGMA formula for the rated power to resist pitting of the gear teeth	$$P_{ac} = \frac{n_1 b_n}{19.1 \times 10^6} \frac{I C_v}{C_{sz} C_m C_{sr} C_a C_{xc}} \left(\frac{\sigma_{sac} d'_{o1} C_L C_H}{C_p C_b C_T C_R} \right)^2$$ **SI** (23–450) where P_{ac} in kW, n_1 in rpm, b_n and d'_{o1} in mm, σ_{sac} in MPa, C_p in $\sqrt{\text{MPa}}$ $$= \frac{n'_1 b_n}{318} \frac{I C_v}{C_{sz} C_m C_{sr} C_a C_{xc}} \left(\frac{\sigma_{sac} d'_{o1} C_L C_H}{C_p C_b C_T C_R} \right)^2$$ **SI** (23–451) where P_{ac} in kW, n'_1 in rps, σ_{sac} in Pa, d'_{o1} and b_n in m, and C_p in $\sqrt{\text{Pa}}$ $$P_{ac} = \frac{n_1 b_n}{126{,}000} \frac{I C_v}{C_{sz} C_m C_{sr} C_a C_{xc}} \left(\frac{\sigma_{sac} d'_{o1} C_L C_H}{C_p C_b C_T C_R} \right)^2$$ **US Customary units** (23–452)

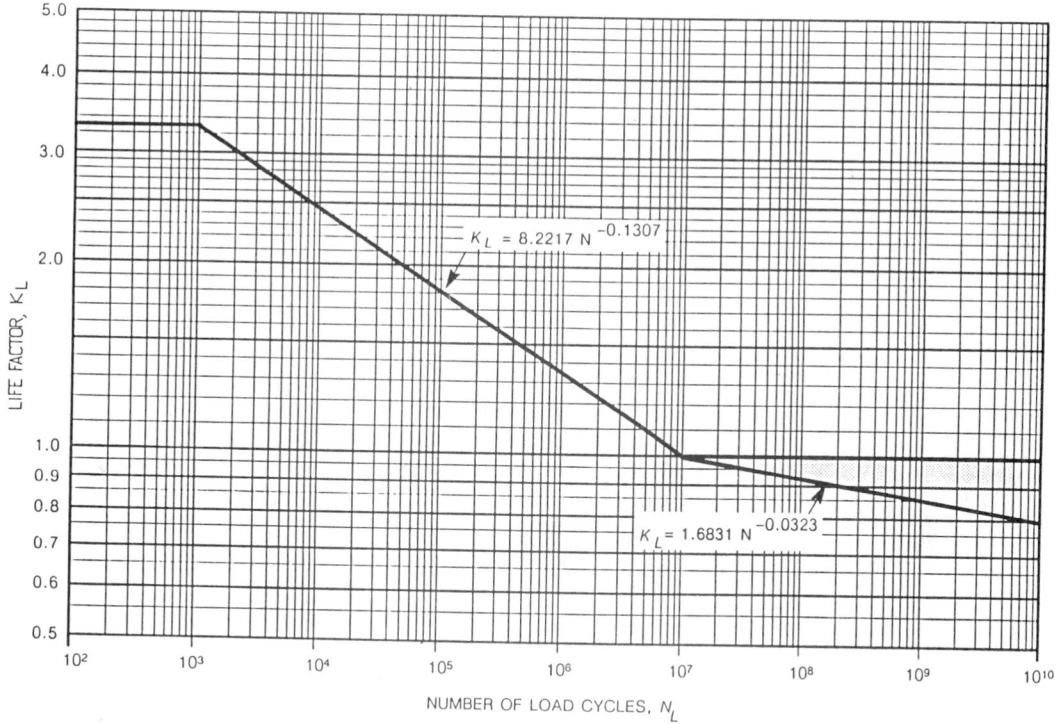

FIGURE 23–68 Life factor for bending strength, K_L (carburized case-hardened steel bevel gears). (*ANSI/AGMA 2003-A86.*)

Particular	Formula
	where P_{ac} in hp, n_1 in rpm, d'_{o1} and b_n in in, σ_{sac} in psi; C_p in $\sqrt{\text{psi}}$
	Refer to Fig. 23–75 and Eq. (23–457) for C_v.
	Refer to Table 23–82 for C_p.
	Refer to Table 23–81 for C_a and K_a.
	Refer to Table 23–24 for σ_{sac}.
	Refer to Fig. 23–72 for C_L.
	Refer to Figs. 23–73 and 23–74 for C_H.
The operating pinion torque	$$M_{tv1} = \frac{9550P}{n_1} \quad \textbf{SI} \qquad (23\text{–}453)$$
	where M_{tv1} in N m, P in kW, n_1 in rpm
	$$= \frac{63,000P}{n_1} \quad \textbf{US Customary System units}$$
	$$(23\text{–}454)$$
	where P = design power, hp

In the figure, the labeled curves are:

$K_L = 8.2217\ N^{-0.1307}$

$K_L = 1.6831\ N^{-0.0323}$

Axis labels: LIFE FACTOR, K_L (vertical); NUMBER OF LOAD CYCLES, N_L (horizontal).

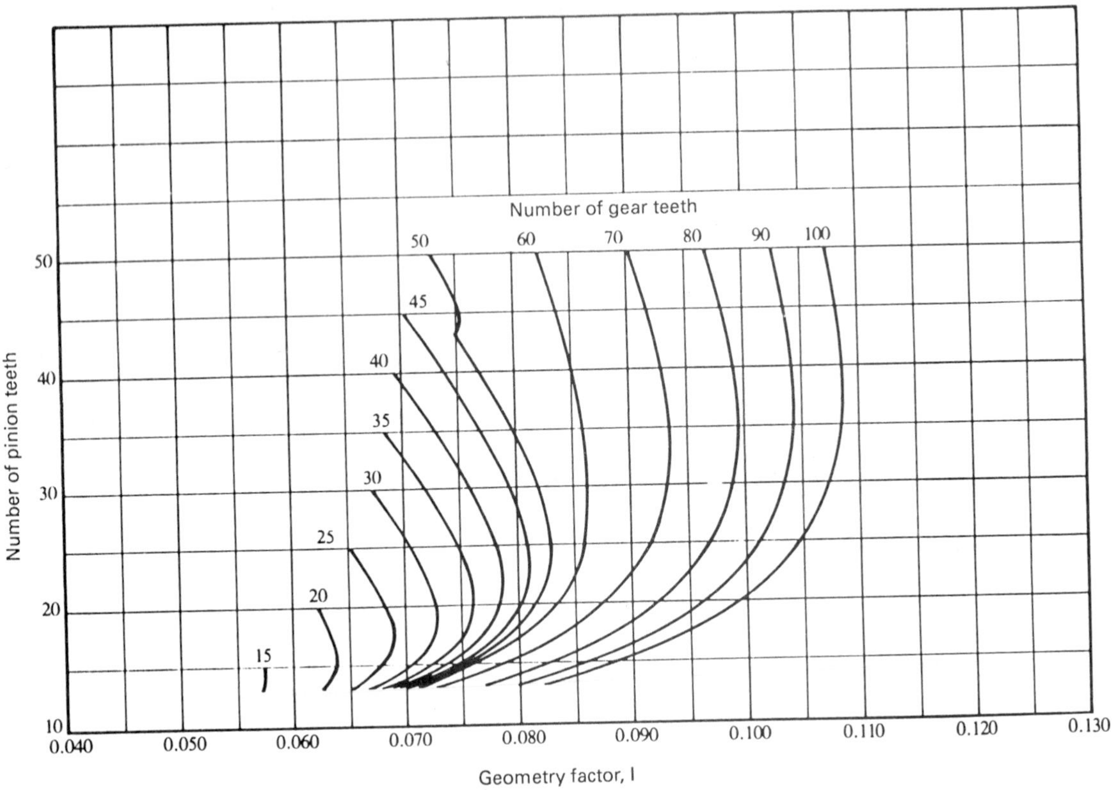

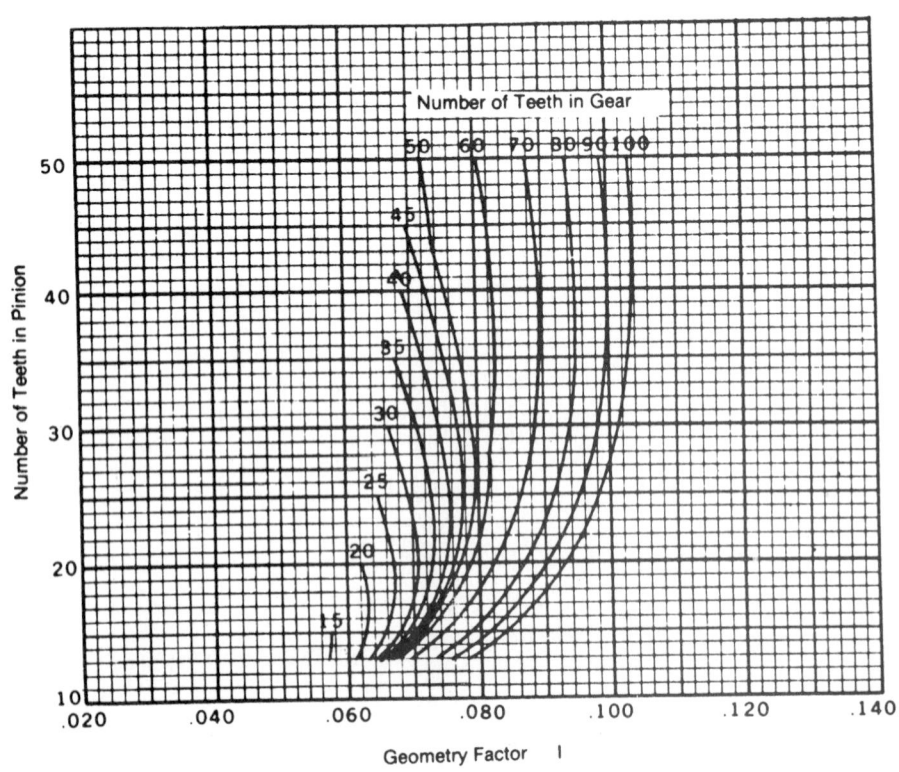

FIGURE 23–69(a) Straight bevel gears with 20° pressure angle, geometry factor, I. (*Courtesy ANSI/AGMA 2003-A86.*)

FIGURE 23–69(b) Straight bevel gears with 25° pressure angle and 90° shaft angle, geometry factor, I.

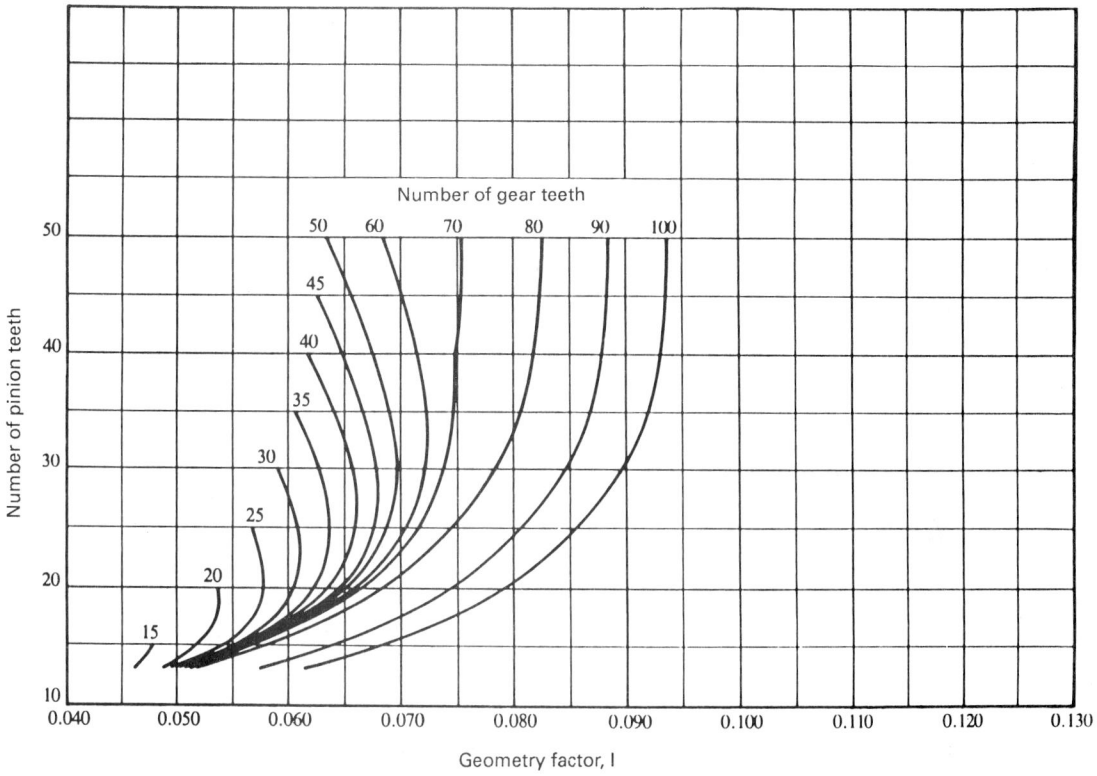

FIGURE 23–70 Fine-pitch ZEROL bevel gears with 20° pressure angle, geometry factor, *I*. (*ANSI/AGMA 2003-A86*.)

Particular	Formula
The design pinion torque	

$$M_{td1} = \frac{b}{2000} \frac{IC_v}{C_{sz}C_{md}C_{sr}C_aC_{xc}} \cdot \left(\frac{\sigma_{sac}d'_{o1}0.774C_H}{C_pC_b \cdot C_TC_R}\right)^2$$

SI   (23–455)

where M_{td1} in N m, b and d'_{o1} in mm, σ_{sac} in MPa, and C_p in $\sqrt{\text{MPa}}$

C_b = stress adjustment factor (use a value of 0.634)

C_{xc} = crowning factor

C_{md} = load-distribution factor when operating and design torques are equal ($M_{tp1} = M_{td1}$; use value of 1.2 C_{mf})

$$= \frac{b}{2} \frac{IC_v}{C_{sz}C_{md}C_{sr}C_aC_{xc}} \left(\frac{\sigma_{sac}d'_{o1}}{C_pC_b} \cdot \frac{0.774C_H}{C_TC_R}\right)^2$$

US Customary units   (23–456)

where M_{td1} in lbf in; b and d'_{o1} in in; σ_{sac} in psi, and C_p in $\sqrt{\text{psi}}$

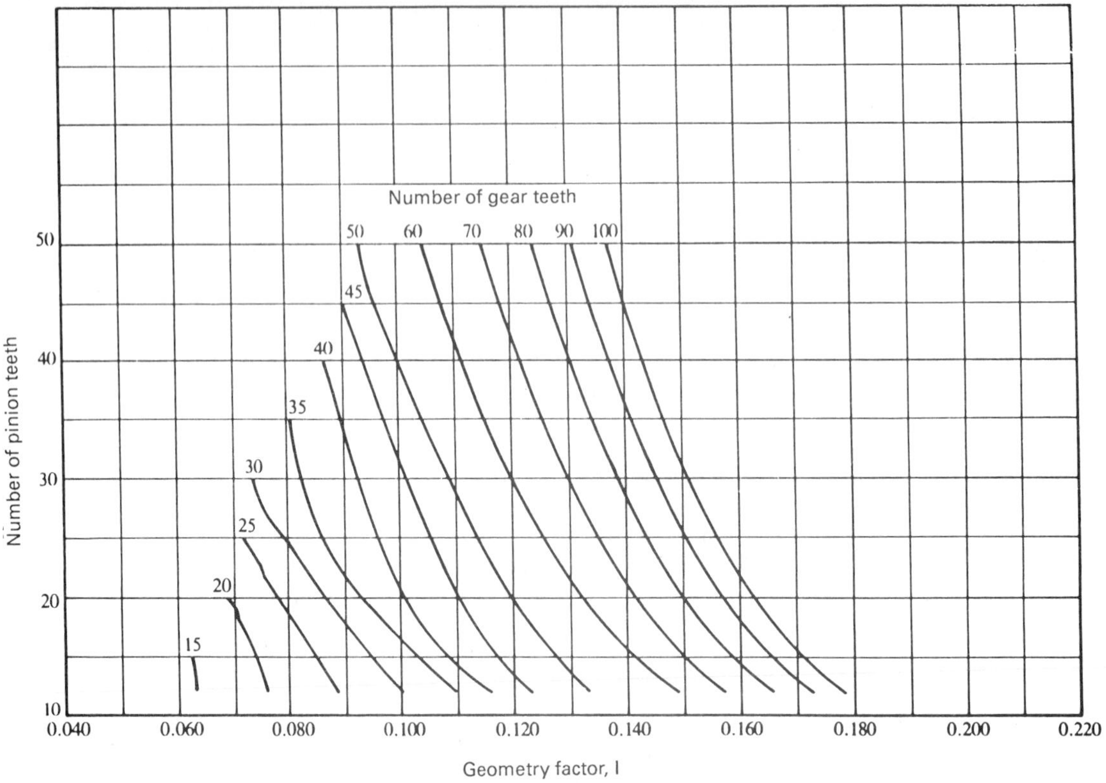

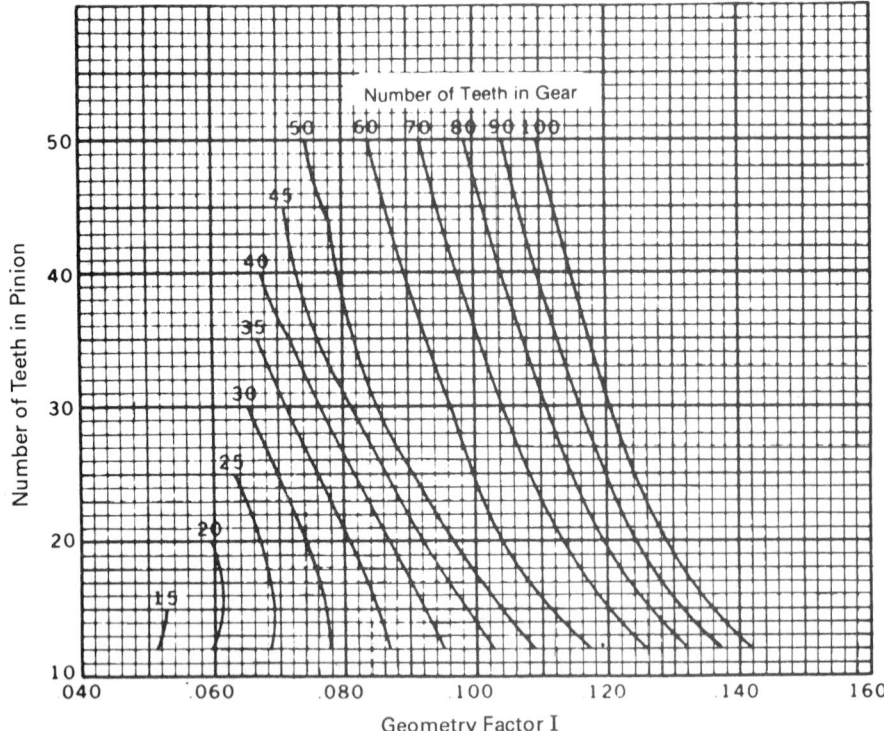

FIGURE 23–71(a) Spiral bevel gears with 20° pressure angle and 35° spiral angle, geometry factor, *I*. (*ANSI/AGMA 2003-A86.*)

FIGURE 23–71(b) Sprial bevel gears with 20° pressure angle, 25° spiral angle, and 90° shaft angle geometry factor, *I*. (*Gleason Works, Rochester, New York.*)

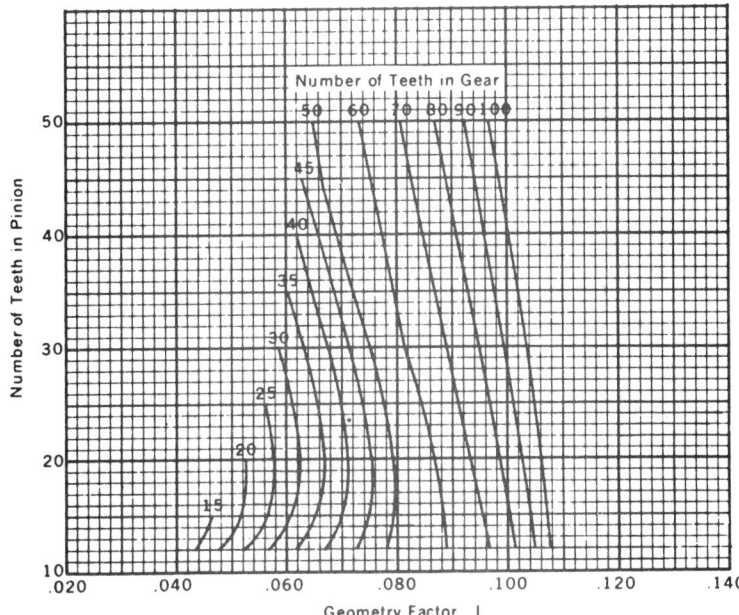

FIGURE 23–71(c) Spiral bevel gears with 20° pressure angle, 15° spiral angle, and 90° shaft angle, geometry factor, *I*. (*Courtesy of Gleason Works, Rochester, New York.*)

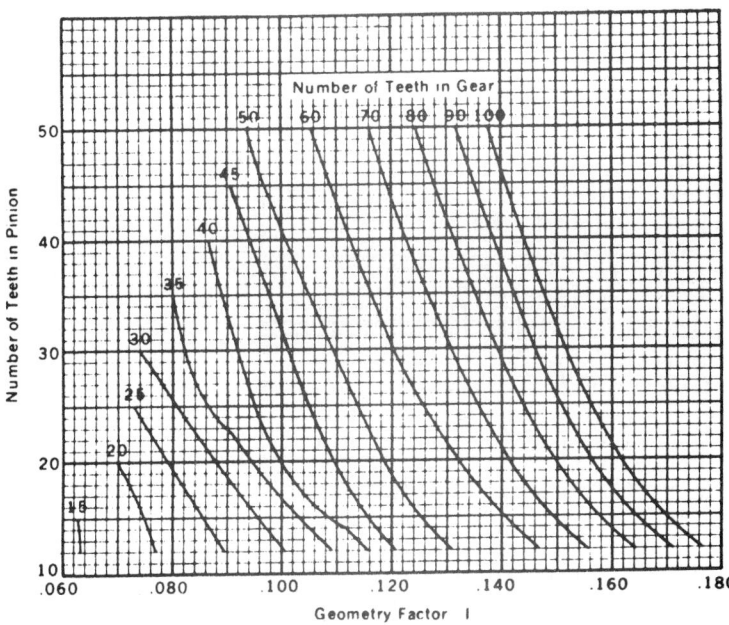

FIGURE 23–71(d) Spiral bevel gears with 25° pressure angle, 35° spiral angle, and 90° shaft angle, geometry factor, *I*. (*Courtesy of Gleason Works, Rochester, New York.*)

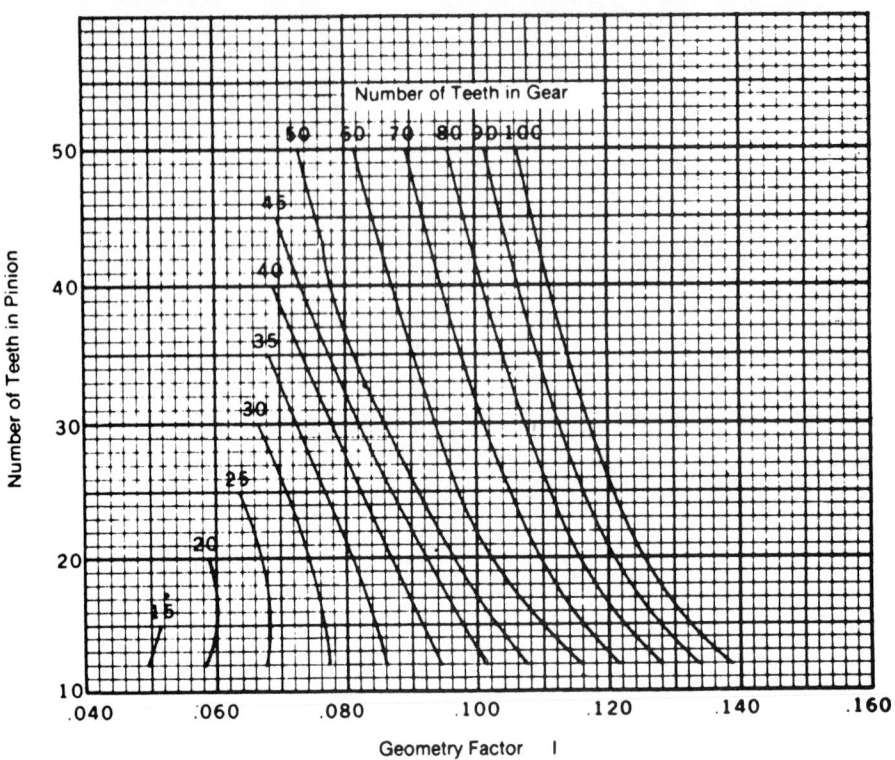

FIGURE 23–71(e) Spiral bevel gears with 25° pressure angle, 25° spiral angle, and 90° shaft angle, geometry factor, *J*. (*Courtesy of Gleason Works, Rochester, New York.*)

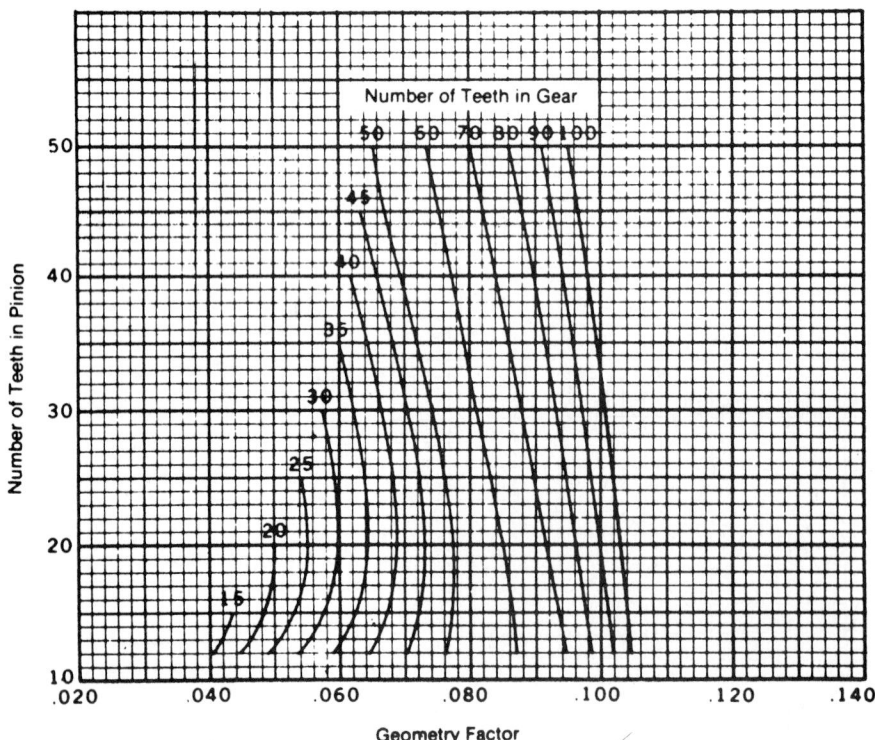

FIGURE 23–71(f) Spiral bevel gears with 25° pressure angle, 15° spiral angle, and 90° shaft angle, geometry factor, *I*. (*Courtesy of Gleason Works, Rochester, New York.*)

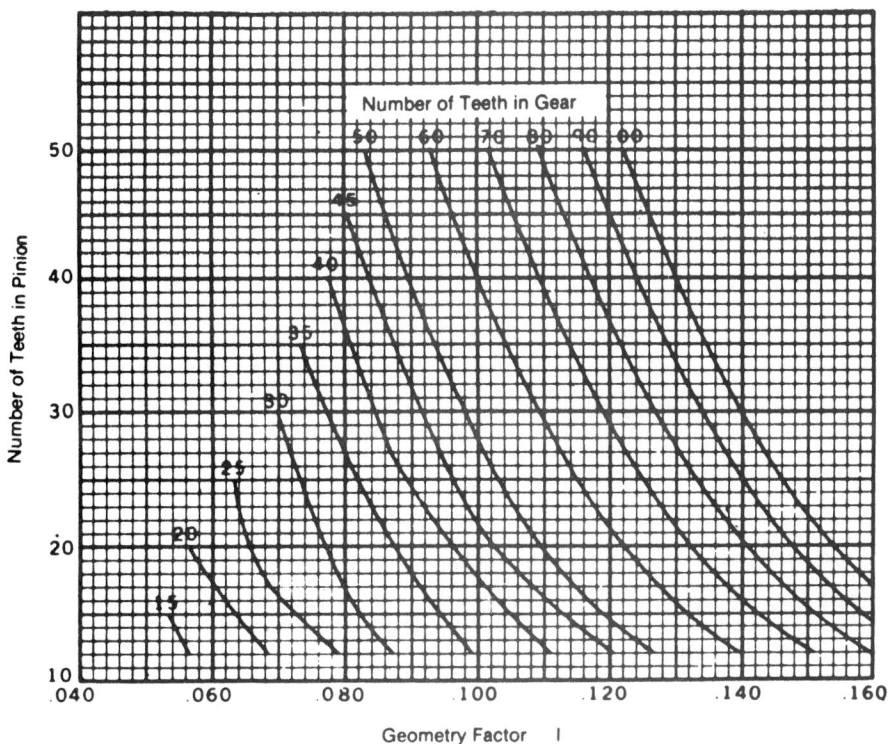

FIGURE 23–71(g)   Spiral bevel gears with 20° pressure angle, 35° spiral angle, and 60° shaft angle, geometry factor, I. (*Courtesy of Gleason Works, Rochester, New York.*)

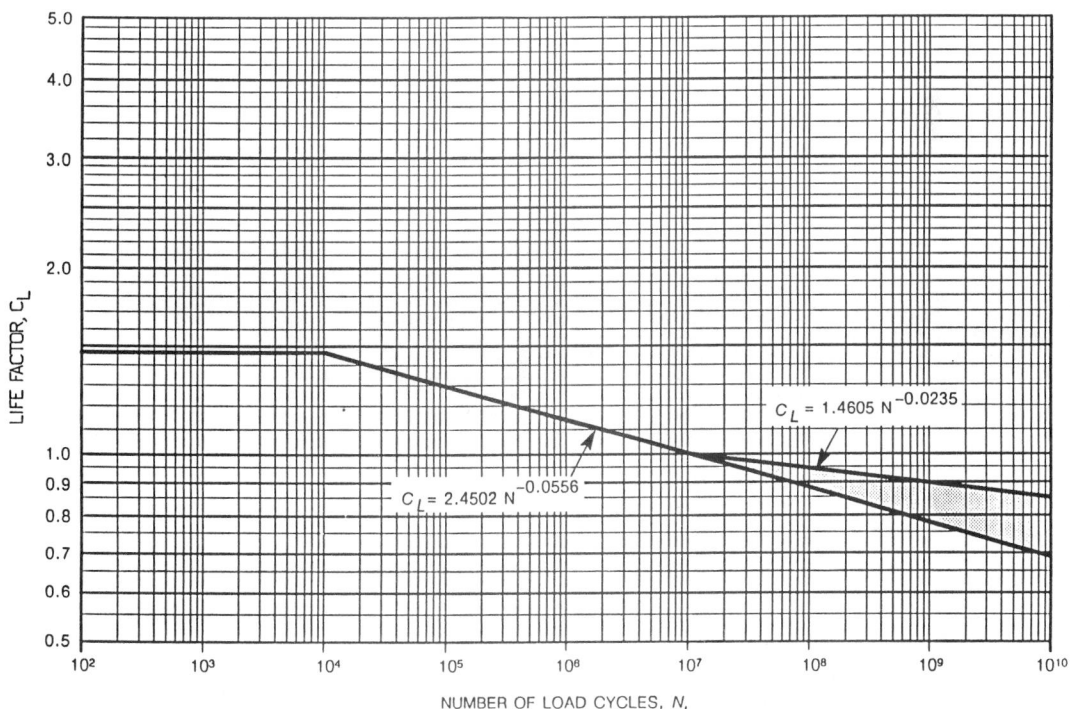

FIGURE 23–72   Life factor for pitting resistance, C_L (carburized case-hardened steel bevel gears). (*ANSI/AGMA 2003-A86.*)

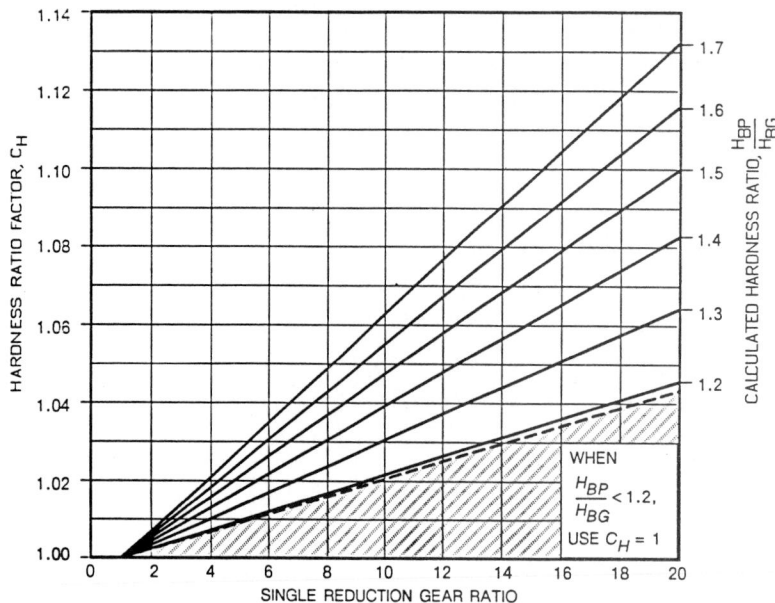

FIGURE 23–73 Hardness ratio factor, C_H (through-hardened pinion and gear). NO equivalent SI units. (*ANSI/AGMA 2003-A86.*)

Particular	Formula

INTERNAL DYNAMIC FACTORS

Formula for internal dynamic factor based on the pitch tolerance as per AGMA Standard, which depends on both speed of operations and the load

$$C_v = K_v = \left(\frac{K_z}{K_z + \sqrt{200 v_m}} \right)^u > C_{v,min} \quad \textbf{SI}$$

(23–457a)

$$C_v = K_v = \left(\frac{K_z}{K_z + \sqrt{v_m}} \right)^u > C_{v,min}$$

US Customary System units   (23–457b)

where
$K_z = 85 - 104$
$v_m = $ pitch line velocity, m/s (ft/min)
    $= 0.262 d n_1$ **US Customary System units**

(23–458a)

    $= 52.36 \times 10^{-6} d n_1$ **SI**   (23–458b)

$$u = \frac{8}{2^{0.5} Q_v} - \sigma_{t0} \left(\frac{125}{E_1 + E_2} \right)$$

(23–459)

(If calculated value is negative, use zero.)

E_1, E_2 modulus of elasticity of pinion and gear materials, respectively, MPa (psi)

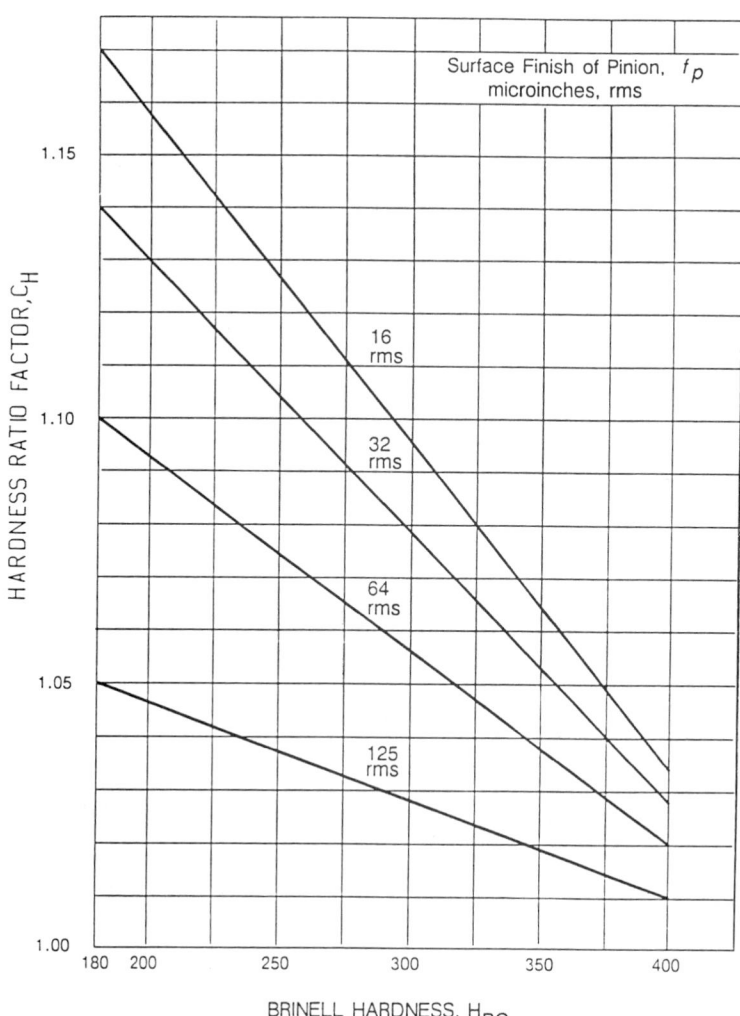

FIGURE 23–74   Hardness ratio factor, C_H (surface-hardened pinions). (*ANSI/AGMA 2003-A86.*)

Particular	Formula
	Q_v = transmission accuracy number (For bevel gears this is usually equal to the AGMA-quality class number, Q.)
	Refer to Fig. 23–75 for C_v.
Calculated bending stress number without regard to dynamic effect	$$\sigma_{t0} = \frac{2000 M_{tv1} K_m K_a K_{sz}}{b d'_{o1} m_t K_x J_l} \quad \textbf{SI} \qquad (23\text{--}460)$$
	where m_t = transverse metric module at outer end of tooth, mm

Particular	Formula

b and d'_{o1} in mm and σ_{t0} in MPa (psi)

$$= \frac{2M_{tv1}K_mK_aK_{sz}P_t}{bd'_{o1}K_xJ_1} \tag{23-461}$$

$$\varepsilon_b = \frac{R}{R-0.5b}\frac{b\tan\beta_m}{p_t} \quad \textbf{US Customary System units}$$

where b and d'_{o1} in in, M_{tv1} in lbf in

P_t = transverse diametral pitch at outer end of tooth, in^{-1}

b and d'_{o1} in in

The expression for minimum value of C_v

$$C_{v,min} = \frac{2}{\pi}\tan^{-1}\left[\frac{\nu_m}{1.7(\varepsilon_b^{1.75}+1)}\right] \quad \textbf{SI} \tag{23-462}$$

$$= \frac{2}{\pi}\tan^{-1}\left[\frac{\nu_m}{333\varepsilon_b^{1.75}+1}\right]$$

$$\textbf{US Customary System units} \tag{23-463}$$

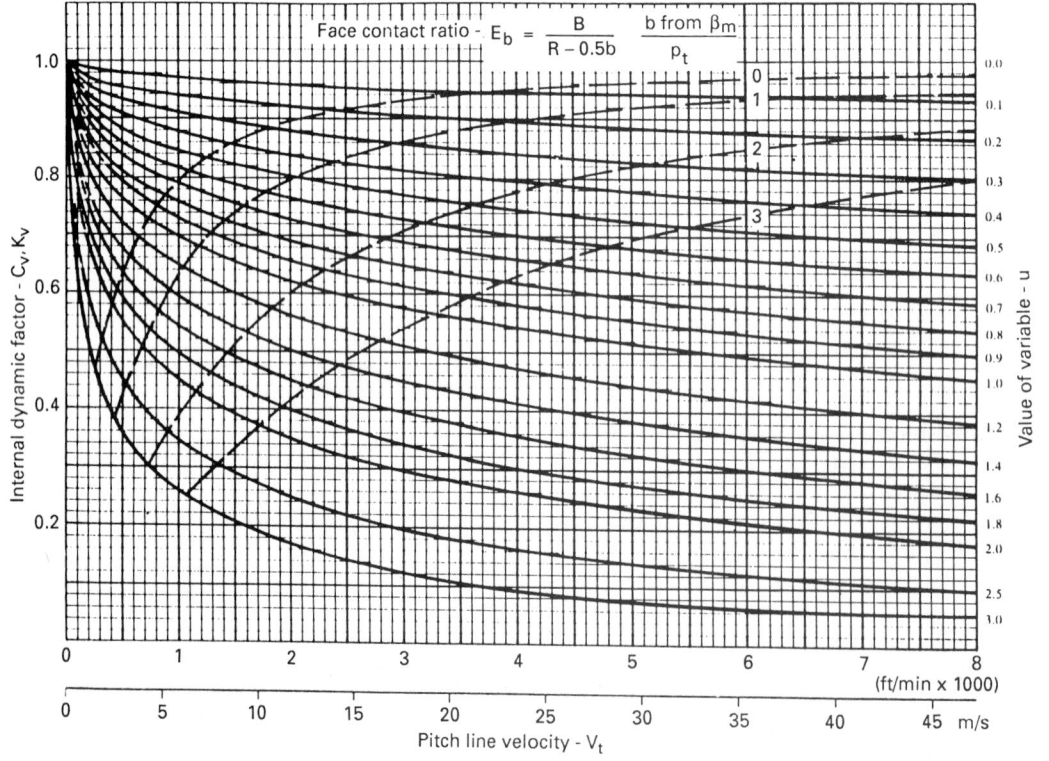

FIGURE 23–75 Internal dynamic factor pitch line velocity (ft/min $\times$ 1000) $-\nu_m$. *(ANSI/AGMA 2003-A86.)*

Particular	Formula

where

$$\varepsilon_b = \text{face contact ratio} = \frac{R}{R_m}\frac{b\tan\beta_m}{p_t}$$

Face contact ratio for straight bevel and ZEROL bevel gears is zero.

β_m = mean spiral angle
R_o = outer cone distance, m (in)
R_m = mean cone distance, m (in)
p_t = outer transverse circular pitch, m (in)

When value is greater than 3, use a value of 3 in Eqs. (23–462) and (23–463)

DYNAMIC OR VELOCITY FACTOR K_v AND C_v

The dynamic or velocity factor for use in Lewis equation

$$K_v = C_v = \frac{3}{3+\nu_m} \quad \text{for } \nu_m \leq 7.5 \text{ m/s } \textbf{SI} \qquad (23\text{–}464a)$$

$$= \frac{600}{600+\nu_m} \quad \text{for } \nu_m \leq 1500 \text{ ft/min}$$

US Customary System units $\qquad$ (23–464b)

The velocity factor for the generalized system

$$K_v = C_v = \frac{5.6}{5.6+\sqrt{\nu_m}} \quad \text{for } \nu_m \text{ over 20 m/s}$$

$$(23\text{–}465a)$$

$$= \frac{78}{78+\sqrt{\nu_m}} \quad \text{for } \nu_m \text{ over } 40,000 \text{ ft/min}$$

US Customary System units $\qquad$ (23–465b)

CHECK FOR ENDURANCE STRENGTH

The endurance strength of bevel gear tooth

$$F_f = 2.15 \times 10^5 b Y_K m \nu_m K_{fM} \qquad \textbf{SI} \qquad (23\text{–}466)$$

$$F_f = \frac{b Y_K \nu_m K_{fM}}{6.3P} \quad \textbf{US Customary units} \qquad (23\text{–}467)$$

where Y_K = form factor for the pinion from Table 23–78

K_{fM} = material factor from Table 23–83

Particular	Formula

DESIGNING GEAR FOR DYNAMIC LOAD

The modified dynamic load equation according to Buckingham

$$F_d = F_t = \frac{21\nu_m(F_t + bC)}{21\nu_m + \sqrt{F_t + bC}} \qquad (23\text{–}468)$$

where F_d and F_t in N, ν_m in m/s, C in N/m, and b in m

$$= F_t + \frac{0.05\nu_m(F_t + bC)}{0.05\nu_m + \sqrt{F_t + bC}} \qquad (23\text{–}469)$$

where F_d and F_t in lbf, ν_m in ft/min, C in lbf/in, and b in in; ν_m is based on the largest pitch circle diameter, ft/mm

Refer to Table 23–32 for values of C.

DESIGNING OF GEAR FOR WEAR LOAD

The limit wear load equation

$$F_w = d_1 bQK/\cos\delta_1 \qquad (23\text{–}470a)$$
$$= mz_1 bQK\cos\delta_1 \qquad (23\text{–}470b)$$

where K is the same as for spur gears and taken from Table 23–37B

$$Q = \frac{2z_{v2}}{z_{v2} + z_{v1}}$$

The wear resistance formula adopted by AGMA as proposed by Gleason works

$$F_w = 4.13 \times 10^5 bK_M C_s\sqrt{z_1 m} \qquad \textbf{SI} \qquad (23\text{–}471a)$$

for straight-tooth bevel gears

where b and m in m and F_w in N

$$= 376bK_M C_s\sqrt{z_1/P} \qquad \textbf{US Customary System units}$$
$$(23\text{–}471b)$$

where b in in, P in in^{-1}, F_w in lbf

C_s = service factor from Table 23–76

$$= 5200K_M C_s(z_1 m)^{1/2} \qquad \textbf{SI for spiral bevel gears}$$
$$(23\text{–}472)$$
$$= 470K_M C_s(z_1/P)^{1/2} \qquad \textbf{US Customary System units}$$

RATING OF BEVEL GEARS

The power rating of bevel gears for durability as per AGMA Standards

$$P = 23.5K_M C_B b \qquad \textbf{SI for straight-tooth bevel}$$
$$(23\text{–}473)$$

where P in kW and b in m

Particular	Formula
	$= 0.8 K_M C_B b$ **US Customary System units**
	(23–474)
	where P in hp and b in in
	C_B is obtained from Eq. (23–477a)
	$= 29.5 K_M C_B b$ for spiral bevel gears **SI**
	(23–475)
	where P in kW and b in m
	$= 1.0 K_M C_B b$ **US Customary System units**
	(23–476)
	where P in hp and b in in, $K_M =$ material factor from Table 23–83
The expression for C_B for use in Eqs. (23–473) to (23–476)	$C_B = 10 \sqrt{d_1 d_2}\, n_1' \left(\dfrac{5.6}{5.6 + \sqrt{\nu_m}} \right)$ **SI** (23–477a)
	where d_1, d_2 in m, n_1' in rps, and ν_m in m/s
	$= \dfrac{(\sqrt{d_1 d_2})n_1'}{233} \left(\dfrac{78}{78 + \sqrt{\nu_m}} \right)$
	US Customary System units (23–477b)
	where d_1, d_2 in in, n_1 in rpm, and ν_m in ft/min
The design horsepower	$P_d = C_s P \text{ or } P_d = \dfrac{P_p}{2}$ (23–478)
	(whichever is greater) where $C_s =$ service factor from Table 23–76
Rated horsepower per "100 rpm" of the pinion	$P_{100} = \dfrac{100 P_d}{n_1 K_M}$ (23–479)
	where $n_1 =$ speed of pinion, rpm
	Refer to Table 28–83 for K_M.
For finding pitch diameter of pinion of bevel gear for a given rated horsepower per "100 rpm" of pinion	Refer to Fig. 23–76.
The normal power	$P = \dfrac{F_t v}{1000} = \dfrac{F_t(\pi d_0 n')}{1000}$ **SI** (23–480)
	where P in kW, n' in rps, d_0 in m, and F_t in N
	$= \dfrac{F_t v}{33{,}000} = \dfrac{F_t(\pi d_0 n)}{33{,}000}$ **US Customary System units**
	(23–481)
	where N in hp, n in rpm, d_0 in in, and F_t in lbf

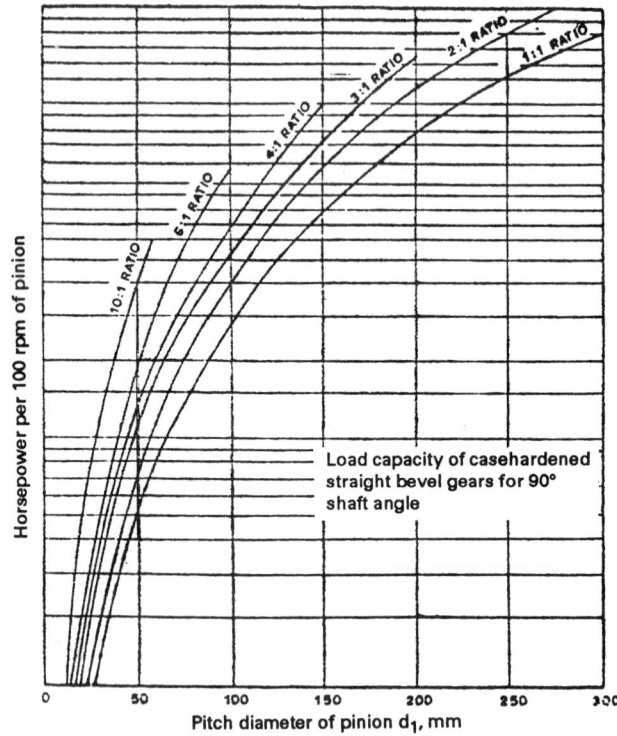

FIGURE 23–76 Chart for finding diameter of pinion of bevel gear. (*Courtesy of Gleason Works, Rochester, New York.*)

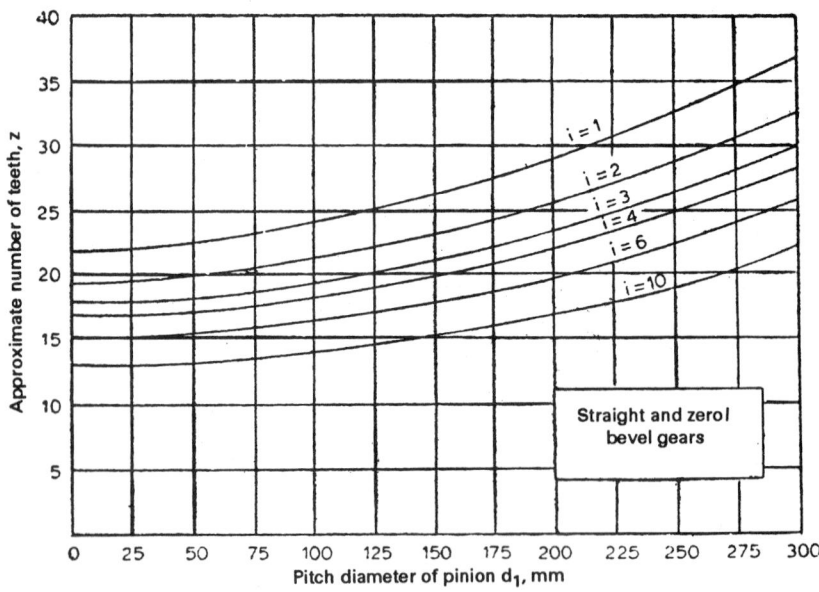

FIGURE 23–77 Number of teeth for straight and ZEROL bevel gears. (*Courtesy of Gleason Works, Rochester, New York.*)

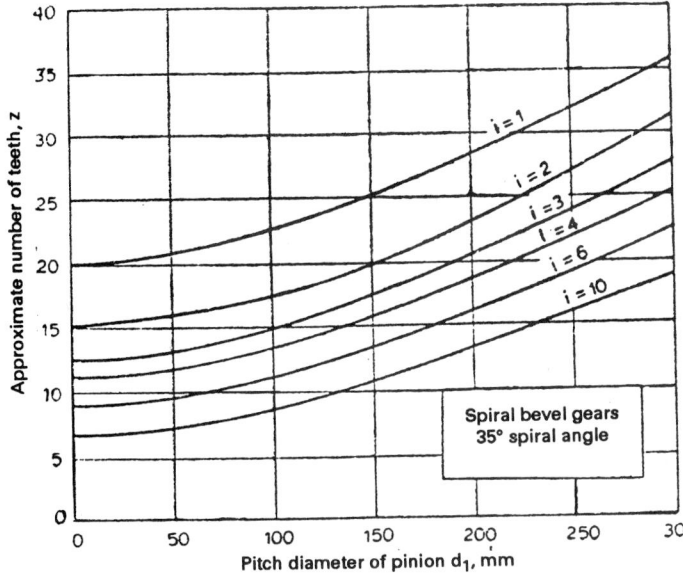

FIGURE 23–78(*a*)   Number of teeth for spiral bevel gears. (*Courtesy of Gleason Works, Rochester, New York.*)

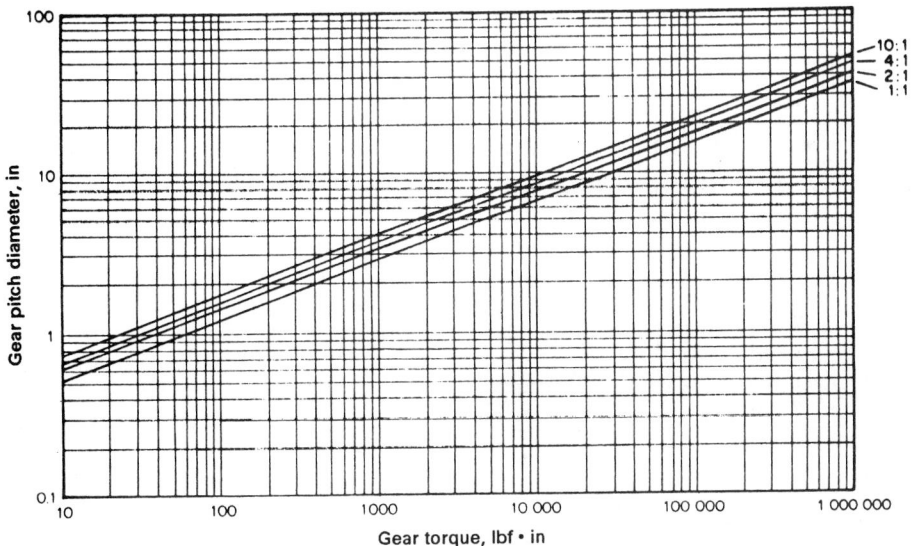

FIGURE 23–78(*b*)   Pitch diameter of gear based on bending strength. (*Courtesy of Gleason Works, Rochester, New York.*)

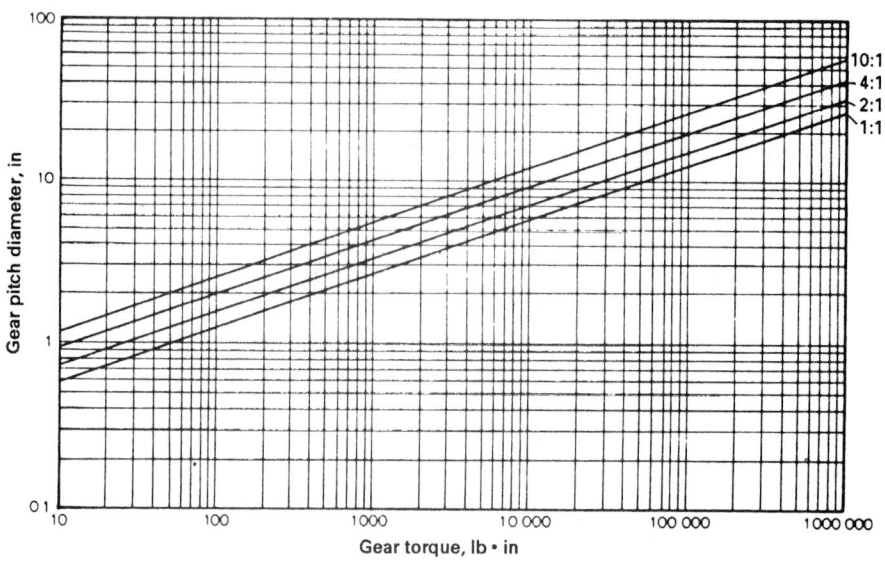

FIGURE 23–78(c)   Pitch diameter of gear based on surface contact strength. (*Courtesy of Gleason Works, Rochester, New York.*)

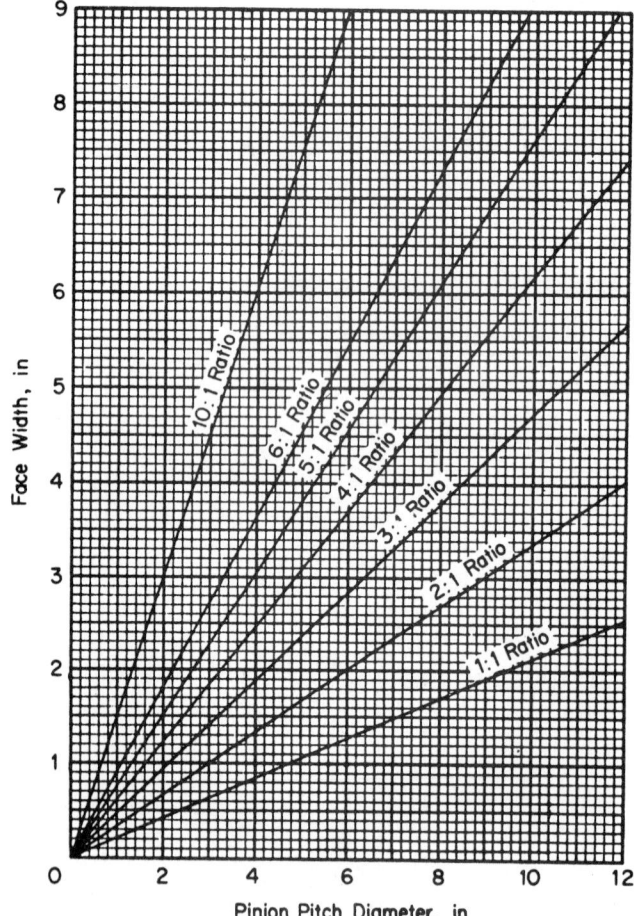

FIGURE 23–78(d)   Face width of spiral bevel and hypoid gears. (*Courtesy of Gleason Works, Rochester, New York.*)

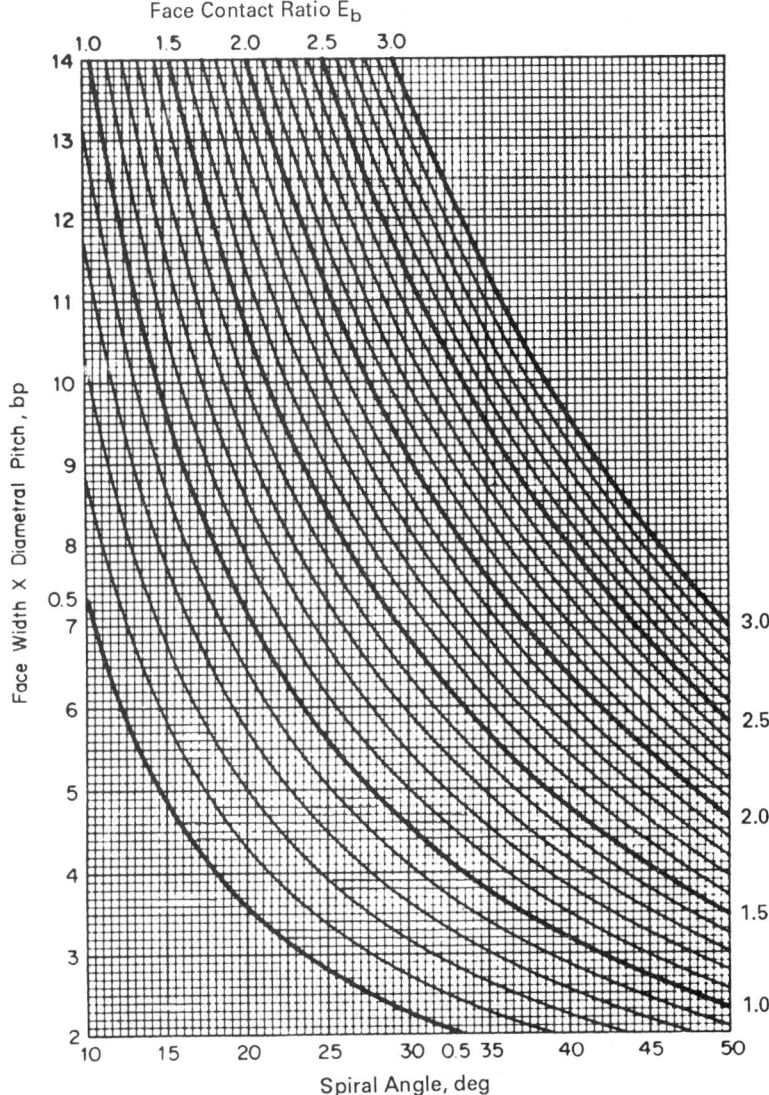

FIGURE 23–78(e)   Selection of spiral angle. (*Courtesy of Gleason Works, Rochester, New York.*)

Particular	Formula
	v is the velocity at radius $(d_0/2)$ at midface, m/s (ft/min)
	$$d_0 = 2d'_m\left[\frac{3\psi_R^2 - 3\psi_R + 1}{3\psi_R(2\psi_R - 1)}\right]$$
For determining number of teeth for straight and Zerol, and spiral bevel gears given the pinion pitch diameters, gear pitch diameters for strength and durability, face width of spiral-bevel and hypoid gears and selection of spiral angle	Refer to Figs. 23–77 and 23–78, respectively.

BEARINGS LOADS ON STRAIGHT BEVEL GEARS

The tangential tooth load at average pitch circle	$F_t = F_n \cos(\alpha + \rho)$	(23–482)
The radial-tooth load on gear and end thrust on the pinion	$F_r = F_t \tan(\alpha + \rho)\cos\delta_1$	(23–483)
The axial thrust on the gear and the radial load on the pinion	$F_a = F_t \tan(\alpha + \rho)\sin\delta_1$	(23–484)
For minimum number of teeth in bevel gear	Refer to Table 23–84.	
For pinion teeth in automotive application	Refer to Table 23–85.	
For comparison of reference profiles of German, American, British, Indian, and Russian Standards	Refer to Table 23–73*B*.	

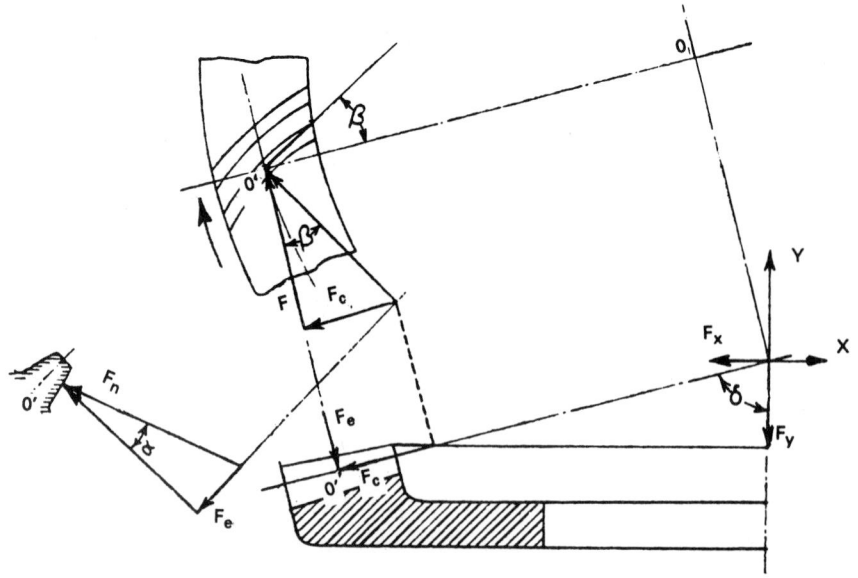

FIGURE 23–78(*f*)   Forces acting on spiral bevel gear tooth.

TABLE 23–73A
Shaft-load formulas for spiral bevel gears [Fig. 23–78(f)]

Hand of spiral	Rotation viewed toward the cone center	Thrust component, F_a	Radial component, F_r
		Pinion	
Right	Clockwise	$F_{a1} = -F_x = \dfrac{F}{\cos\beta}\,(+\tan\alpha\cos\delta_1 - \sin\beta\cos\delta_1)$ (23–484a)	$F_{r1} = \dfrac{F}{\cos\beta}\,(\tan\alpha\cos\delta_1 + \sin\beta\sin\delta_1)$
Right	Counterclockwise	$F_{a1} = -F_x = \dfrac{F}{\cos\beta}\,(+\tan\alpha\cos\delta_1 - \sin\beta\cos\delta_1)$ (23–484b)	$F_{r1} = \dfrac{F}{\cos\beta}\,(\tan\alpha\cos\delta_1 + \sin\beta\sin\delta_1)$
		Gear	
Left	Counterclockwise	$F_{a2} = F_y = \dfrac{F}{\cos\beta}\,(+\tan\alpha\sin\delta_2 - \sin\beta\cos\delta_2)$ (23–484c)	$F_{r2} = \dfrac{F}{\cos\beta}\,(\tan\alpha\cos\delta_2 + \sin\beta\sin\delta_2)$
Left	Clockwise	$F_{a2} = F_y = \dfrac{F}{\cos\beta}\,(+\tan\alpha\sin\delta_2 - \sin\beta\cos\delta_2)$ (23–484d)	$F_{r2} = \dfrac{F}{\cos\beta}\,(\tan\alpha\cos\delta_2 + \sin\beta\sin\delta_2)$

Symbols: F: tangential load at midpoint tooth, kN (lbf); F_a: axial thrust component at midpoint of tooth, kN (lbf); F_r: radial component at midpoint of tooth, kN (lbf); β: spiral angle, deg; α: normal pressure angle, deg; δ_1: pitch angle of pinion, deg; δ_2: pitch angle of gear, deg; $-$: indicates that the thrust is toward the cone center; $+$: indicates that the thrust is away from the cone center.

TABLE 23–74
Recommended series of modules for bevel gears, m in mm

Preferred	1	1.25		1.50		2.0		2.5		3.0		4.0		5.0		6		8		10		12		16		20		25		32		40		50
Choice	2		1.125		1.375		1.75		2.25		2.75		3.5		4.5		5.5		7		9		11		14		18		22		28		36		45
Choice	3															(3.25)	(3.75)	(6.5)																

Note: Gears of modules less than 1 and above 50 may also be permitted by agreement between the manufacturer and the producer.
Source: IS 5037, 1969.

TABLE 23–73*B*

Comparison of reference profiles according to German (DIN), American (ASA), British (BSS), Indian (IS), and Russian (GOST) standards

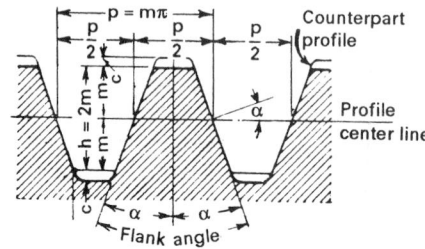

Standard and no.	Tooth system	α, deg	h_a/m	h_f/m	ρ_f/m	r/m	t_r/m	x/m
DIN 867	**Fig. 23–5**	**20**	**1.000**	**1.1 to 1.3**	*	**	—	—
IS–2535/1969	Fig. 23–64*b*	20	1.000	1.250	0.40 max		0.02	0.60
ASA–B 6.1/1932	14.5° composite system	14.5	1.000	1.157	0.209	3.7287	—	—
ASA–B 6.1/1932	14.5° full-depth system	14.5	1.000	1.175	≥ 0.209	**	—	—
ASA–B 6.1/1932	20° full-depth system	20	1.000	1.25	≥ 0.235	**	—	—
ASA–B 6.1/1932	20° stub tooth system	20	0.800	1.000	≥ 0.300	**	—	—
ASA–B 6.7/1950	20° involute fine pitch system for spur and helical gears (Fig. 23–64*b*)	20	1.000	1.200 +0.002	0	**	—	—
BSS 436/1940	Class A1 (Fig. 23–64*b*)	20	1.000	1.440	0.295	≥ 15.750	0.009	0.493
BSS 436/1940	Classes A2 and B	20	1.000	1.250	0.390	≥ 15.750	0.009	0.493
BSS 436/1940	Classes C and D	20	1.000	1.250	0.390	≥ 12.875	0.019	0.628
GOST		20	1.000	1.250	0.40			0.45

Key: * Depends on the clearance; ** without tip relief.

TABLE 23–75
Recommended series of diametral pitches for bevel gears, P

Preferred	20		16		12		10		8		6		5.0		4.0		3.0		2.50		2.00		1.5		1.25		1		0.75		0.625		0.50
Choice 2		18		14		11		9		7		5.5		4.5		3.5		2.75		2.25		1.75						0.875						

Source: IS 5037, 1979.

TABLE 23–76
Service factors C_s, Gleason method for bevel gears for use in Eqs. (23–471), (23–472), and (23–478)

Power source	Character of load on driven machine		
	Uniform	Moderate shock	Heavy shock
Uniform	1.00	1.25	1.75
Light shock	1.10	1.35	1.80
Medium shock	1.25	1.50	1.85

TABLE 23–77
Form factor Y for bevel gears

Number of teeth in pinion, z_1	Straight-tooth bevel gears														
	Gear ratios														
	1.00 to 1.25	1.25 to 1.50	1.50 to 1.75	1.75 to 2.00	2.00 to 2.25	2.25 to 2.50	2.50 to 2.75	2.75 to 3.00	3.00 to 3.25	3.25 to 3.50	3.50 to 3.75	3.75 to 4.00	4.00 to 4.50	4.50 to 5.00	5.00 to ∞
10	0.231	0.260	0.280	0.294	0.305	0.315	0.324	0.332	0.340	0.347	0.353	0.358	0.365	0.371	0.377
11	0.268	0.264	0.273	0.286	0.296	0.308	0.309	0.315	0.320	0.324	0.328	0.332	0.336	0.340	0.342
12	0.248	0.265	0.281	0.295	0.308	0.318	0.328	0.335	0.341	0.345	0.348	0.351	0.353	0.355	0.356
13	0.264	0.278	0.291	0.280	0.278	0.286	0.291	0.295	0.298	0.299	0.301	0.303	0.305	0.307	0.310
14	0.242	0.254	0.263	0.272	0.281	0.288	0.294	0.299	0.304	0.307	0.310	0.313	0.316	0.318	0.319
15	0.248	0.258	0.266	0.274	0.283	0.290	0.296	0.301	0.305	0.308	0.312	0.315	0.318	0.319	0.320
16	0.252	0.261	0.269	0.277	0.285	0.292	0.298	0.304	0.308	0.312	0.314	0.317	0.319	0.321	0.323
17–18	0.257	0.265	0.273	0.281	0.288	0.295	0.302	0.307	0.311	0.315	0.318	0.320	0.322	0.325	0.326
19–21	0.265	0.272	0.279	0.286	0.294	0.300	0.307	0.312	0.317	0.320	0.324	0.326	0.328	0.330	0.332
22–25	0.274	0.281	0.288	0.295	0.301	0.307	0.314	0.319	0.324	0.327	0.331	0.332	0.335	0.337	0.338
26–30	0.284	0.291	0.297	0.304	0.310	0.317	0.322	0.327	0.332	0.336	0.339	0.342	0.344	0.346	0.347
	Spiral-tooth bevel gears														
11	0.316	0.335	0.343	0.325	0.327	0.333	0.338	0.344	0.350	0.356	0.361	0.367	0.375	0.384	0.390
12	0.298	0.378	0.333	0.343	0.351	0.357	0.363	0.368	0.372	0.377	0.379	0.381	0.384	0.386	0.388
13	0.302	0.320	0.334	0.343	0.351	0.358	0.365	0.371	0.376	0.381	0.384	0.386	0.388	0.391	0.393
14	0.306	0.322	0.334	0.345	0.354	0.362	0.369	0.374	0.378	0.382	0.366	0.389	0.391	0.393	0.395
15	0.314	0.330	0.342	0.352	0.360	0.368	0.374	0.380	0.385	0.389	0.392	0.394	0.397	0.399	0.402
16	0.322	0.335	0.347	0.358	0.367	0.374	0.381	0.386	0.390	0.394	0.397	0.400	0.402	0.404	0.406
17–18	0.329	0.343	0.354	0.364	0.373	0.382	0.389	0.394	0.398	0.400	0.403	0.406	0.407	0.409	0.410
19–21	0.339	0.351	0.362	0.373	0.382	0.389	0.396	0.401	0.405	0.407	0.410	0.411	0.412	0.414	0.415
22–25	0.351	0.363	0.373	0.382	0.391	0.398	0.403	0.407	0.410	0.412	0.413	0.414	0.415	0.417	0.418
26–30	0.364	0.374	0.384	0.393	0.399	0.404	0.407	0.410	0.412	0.414	0.415	0.416	0.417	0.418	0.419

Source: AGMA Standard 222.01, 1944.

TABLE 23–78
Form factor Y_K for Gleason 20° straight bevel gears for use in Eqs. (23–466) and (23–467)

Number of teeth in pinion, z_1	Values for z_2/z_1 for shafts at 90°											
	1.000–1.055	1.225–1.285	1.490–1.565	1.840–1.945	2.195–2.335	2.69–2.89	3.38–3.68	4.03–4.39	4.85–5.4	6.0–6.8	7.7–8.8	8.8–10.0
13					0.701	0.730	0.760	0.779	0.799	0.819	0.840	0.850
14			0.654	0.692	0.721	0.749	0.779	0.798	0.818	0.838	0.859	0.869
15		0.636	0.673	0.711	0.739	0.767	0.796	0.815	0.835	0.855	0.876	0.886
16	0.617	0.655	0.691	0.728	0.755	0.783	0.811	0.831	0.850	0.871	0.891	0.901
17	0.635	0.672	0.708	0.744	0.770	0.798	0.826	0.845	0.864	0.885	0.906	0.916
18	0.652	0.689	0.724	0.759	0.785	0.812	0.840	0.859	0.878	0.899	0.920	0.930
19	0.668	0.705	0.739	0.773	0.799	0.826	0.853	0.872	0.891	0.912	0.933	0.944
20	0.684	0.720	0.754	0.787	0.812	0.839	0.866	0.885	0.904	0.925	0.946	0.957
21	0.697	0.732	0.766	0.798	0.823	0.850	0.876	0.895	0.914	0.935	0.956	0.967
22	0.709	0.744	0.777	0.809	0.833	0.860	0.886	0.904	0.923	0.945	0.966	0.977
23	0.721	0.755	0.788	0.819	0.843	0.870	0.895	0.913	0.932	0.954	0.976	0.986
24	0.732	0.766	0.798	0.829	0.852	0.879	0.904	0.922	0.941	0.968	0.985	0.996
25	0.743	0.777	0.808	0.839	0.862	0.888	0.913	0.931	0.950	0.971	0.994	1.005
26	0.754	0.787	0.818	0.848	0.871	0.897	0.921	0.939	0.958	0.979	1.002	1.013
27	0.763	0.796	0.826	0.856	0.879	0.904	0.928	0.946	0.965	0.986	1.009	1.020
28	0.771	0.804	0.834	0.863	0.886	0.911	0.935	0.953	0.972	0.993	1.016	1.027
29	0.779	0.812	0.841	0.870	0.893	0.918	0.942	0.960	0.978	0.999	1.023	1.034
30	0.787	0.819	0.848	0.877	0.900	0.924	0.948	0.966	0.984	1.005	1.029	1.040
31–32	0.798	0.830	0.858	0.887								
33–34	0.811	0.842	0.870	0.898								
35–36	0.824	0.855	0.882	0.909								
37–38	0.836	0.866	0.893	0.920								
39–41	0.851	0.881	0.907	0.933								

TABLE 23–79
Proportions of straight-tooth bevel gears[a]

Gear ratios		Add, h_a,	Gear ratios		Add, h_a,	Gear ratios		Add, h_a,	Gear ratios		Add, h_a,
From	To	mm	From	To	mm	From	To	mm	From	To	mm
1.00	1.00	25.40	1.15	1.17	22.40	1.42	1.45	19.40	2.06	2.16	16.40
1.00	1.02	25.15	1.17	1.19	22.15	1.45	1.48	19.15	2.16	2.27	16.15
1.02	1.03	24.90	1.19	1.21	21.90	1.48	1.52	18.90	2.27	2.41	15.90
1.03	1.04	24.65	1.21	1.23	21.65	1.52	1.56	18.65	2.41	2.58	15.65
1.04	1.05	24.40	1.23	1.25	21.40	1.56	1.60	18.4	2.58	2.78	15.40
1.05	1.06	24.15	1.25	1.27	21.15	1.60	1.65	18.15	2.78	3.05	15.15
1.06	1.08	23.90	1.27	1.29	20.90	1.65	1.70	17.90	3.05	3.41	14.90
1.08	1.09	23.65	1.29	1.31	20.65	1.70	1.76	17.65	3.41	3.94	14.65
1.09	1.11	23.40	1.31	1.33	20.40	1.76	1.82	17.40	3.94	4.82	14.40
1.11	1.12	23.15	1.33	1.36	20.15	1.82	1.89	17.15	4.82	6.81	14.15
1.12	1.14	23.90	1.36	1.39	19.90	1.89	1.97	16.90	6.81	∞	13.90
1.14	1.15	22.65	1.39	1.42	19.65	1.97	2.06	16.65	—	—	—

[a] Working depth = 2 m; total depth = (2.188 m + 0.005) mm; addendum of gear = (addendum from table) m; addendum of pinion = 2 m—addendum of gear; dedendum of gear = 2.188 m—addendum of gear; dedendum of pinion = 2.188 m—addendum of pinion.

TABLE 23–80
Mounting factors

	Mounting factor, C_{mf}		
Application	Both members straddle-mounted	One member straddle-mounted	Neither member straddle-mounted
Aircraft[a]	1.00	1.10	1.25
Automotive[a]	1.00	1.10	1.25
High-quality commercial[a]	1.00	1.10	1.25
General commercial[a]	1.20	1.32	1.50

[a] Based on optimum tooth contact pattern under maximum operating load as evidenced by results of a deflection test on the gears in their mountings.

TABLE 23–81
External dynamic factors C_a and K_a[a]

Character of prime mover	Character of load on driven machine			
	Uniform	Light shock	Medium shock	Heavy shock
Uniform	1.00	1.25	1.50	≥ 1.75
Light shock	1.10	1.35	1.60	≥ 1.85
Medium shock	1.25	1.50	1.75	≥ 2.00
Heavy shock	1.50	1.75	2.00	≥ 2.25

[a] This table is for speed-decreasing drives. For speed-increasing drives add 0.01 i^2 to the above factors, where $i = z_2/z_1$ gear ratio.

TABLE 23–82
Elastic coefficient for gear teeth, C_p

Pinion material and modulus of elasticity	lbf/in² (MPa)	Gear material and modulus of elasticity with values C_p, $\sqrt{\text{lbf/in}^2}$ $\sqrt{\text{(MPa)}}$			
		Steel 30×10^6 (2.07×10^5)	Malleable iron 25×10^6 (1.72×10^5)	Nocular iron 24×10^6 (1.66×10^5)	Cast iron 22×10^6 (1.52×10^5)
Steel	30×10^6 (2.07×10^5)	2290 (190)	2180 (181)	2160 (179)	2110 (175)
Malleable iron	25×10^6 (1.72×10^5)	2180 (181)	2090 (174)	2070 (172)	2020 (168)
Nodular iron	24×10^6 (1.66×10^5)	2160 (179)	2070 (172)	2050 (170)	2000 (166)
Cast iron	22×10^6 (1.52×10^5)	2110 (175)	2020 (168)	2000 (166)	1960 (163)

TABLE 23–83
Material factors K_M and K_{fM} for bevel gears for use in Eqs. (23–466), (23–471), and (23–476)

Material		(Durability) material factor, K_M	Endurance, K_{fM}
Pinion	Gear		
Cast iron or unhardened steel	Cast iron	0.30	0.10
Heat-treated steel	Heat-treated steel	0.35	0.50
Case-hardened steel	Cast iron	0.40	0.10
Oil-hardened steel	Cast iron	0.40	0.10
Case-hardened steel	Unhardened steel	0.45	0.10
Oil-hardened steel	Unhardened steel	0.45	0.10
Case-hardened steel	Heat-treated steel	0.50	0.50
Oil-hardened steel	Heat-treated steel	0.50	0.50
Oil-hardened steel	Oil-hardened steel	0.80	0.50
Case-hardened steel	Oil-hardened steel	0.85	0.50
Case- hardened steel	Case-hardened steel	1.00	1.00

TABLE 23–85
Pinion teeth for automotive applications

Approximate ratio, z_1/z_2	Preferred number of pinion teeth, z_1	Allowable range, z_1
2.0	17	15–19
2.5	15	12–16
3.0	11	10–14
3.5	10	9–12
4.0	9	8–10
4.5	8	7–9
5.0	7	6–9
6.0	6	5–8
7.0	6	5–7
8.0	5	5–6

TABLE 23–84
Minimum number of teeth

Type of gear	Pressure angle, deg	20 (Standard)	14½	15	22½	25
Straight	Pinion	16 15 14 13	28 28 27 26 25 24		13	12
	gear	16 17 20 30	29 29 31 35 40 57	—	13	12
Spiral	Pinion	17 16 15 14 13 12	28 27 26 25 24 23 22 21 20 19	24 23 22 21 20 19 18 17 16	14	12
	gear	17 18 19 20 22 26	29 29 30 32 33 36 40 42 50 70	24 25 26 27 29 31 36 45 59	14	12
Zerol	Pinion	17 16 15			14 13 13	
	gear	17 20 25			14 15 13	

23.4 WILDHABER–NOVIKOV GEAR (1, 21)

SYMBOLS

In addition to symbols already given in this chapter, the following symbols are adopted for designing Wildhaber–Novikov gears:

b	total face width, m (in)
C_v	a factor which depends on sliding and rocking velocity in the gear mesh
F	load (also with suffixes), kN (lbf)
k_y	the factor which depends on the tooth form (Fig. 23–85)
k_β	helical factor (Fig. 23–83)
$k_{\epsilon b}$	the factor which depends on the axial overlap factor for use in Eqs. (23–524) and (23–525)
$l_{\rho t}$	length of line of contact along the tooth profile in the transverse section, m (in)
$l_{\rho n}$	length of line of contact along the tooth profile in the normal section, m (in)
s	tooth thickness (also with suffixes), m (in)
y	tooth form factor
α	pressure angle, deg
β	helix angle at pitch point, deg
$\epsilon_{\beta x}$	helical axial overlap ratio
ϵ_i	integral part of overlap factor (ratio)
ϑ	fractional part of overlap factor (ratio)
ρ_{1t}	transverse radius of curvature of the convex profile of the pinion tooth, m (in)
ρ_{2t}	transverse radius of the curvature of the profile of the wheel tooth, m (in)
σ_c	Hertzian contact stress, MPa (psi)
σ_b	bending stress, MPa (psi)
σ_{ca}	allowable surface contact stress, MPa (psi)
σ_{ba}	allowable bending stress, MPa (psi)

Particular	Formula	
For addendum type of Wildhaber-Novikov circular arc gear	Refer to Fig. 23–80a.	
For addendum–dedendum type of Wildhaber–Novikov circular arc gears	Refer to Fig. 23–80b.	
Transverse radius of the convex profile of the pinion tooth (Fig. 23–79)	$\rho_{1t} = m_t$	(23–485a)
Transverse radius of the concave profile of the wheel tooth	$\rho_{2t} = 1.05\rho_{1t}$ to $1.3\rho_{1t}$	(23–485b)
The difference between ρ_1 and ρ_2 at contact	$\Delta\rho = 0.07$ to $0.15\,(\rho_2 - \rho_1)$	(23–485c)
Transverse module	$m_t = m_n/\cos\beta'$	(23–486)
Module in normal section	$m_n = m_t\cos\beta'$	(23–487)

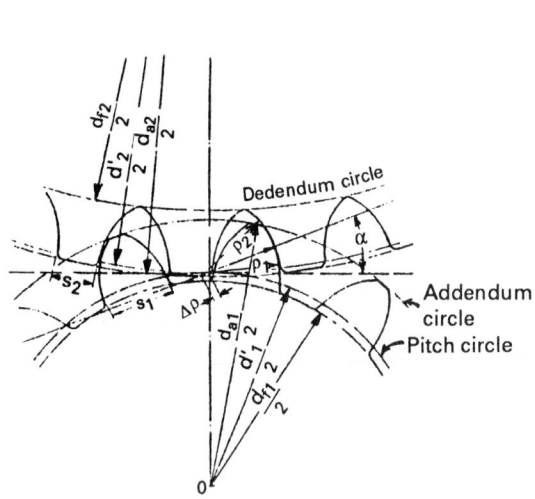

FIGURE 23–79(a)   Dimensions of all addendum-type Wildhaber-Novikov circular arc gear.

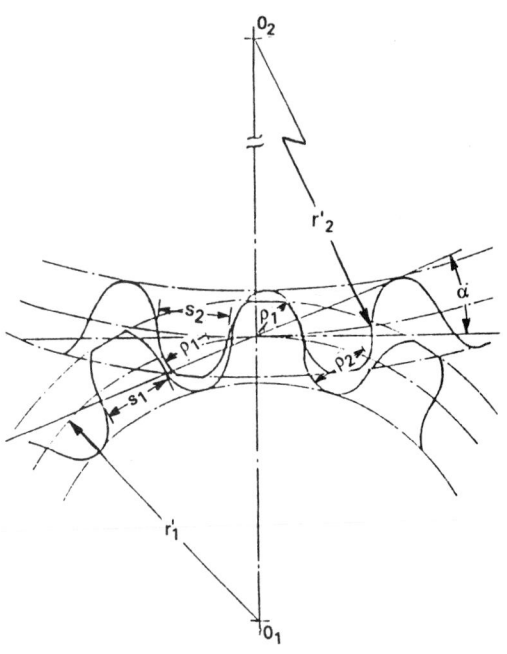

FIGURE 23–79(b)   Dimensions of addendum-dedendum type Wildhaber-Novikov circular arc gear.

Particular	Formula	
Pitch circle diameter of pinion	$d_1' = m_t z_1$	(23–488a)
Pitch circle diameter of wheel	$d_2' = m_t z_2$	(23–488b)
Center distance	$a = \dfrac{m_a(z_1 + z_2)}{2\cos\beta'}$	(23–489)
Transverse thickness of pinion tooth	$s_{1t} = 1.85 m_t = 1.85/P_t$	(23–490)
Transverse thickness of wheel tooth	$s_{2t} = 1.225 m_t = (1.225/P_t)$	(23–491a)
The relation between the thickness of teeth s_{1t} and s_{2t} (Fig. 23–79) in order to equalize flexure strength as per the recommendation of Zdar Engineering Works of USSR	$s_{1t} = 1.5 s_{2t}$ Refer to Fig. 23–80.	(23–491b)
Addendum of pinion (concave) tooth	$h_{a1} = 1.15 m_n^* = 0.9 m_n^{**}$	(23–492)
Dedendum of pinion (convex) tooth	$h_{f1} = 0.25 m_n^* = 1.05 m_n^{**}$	(23–493)
Addendum of wheel (concave) tooth	$h_{a2} = -0.15 m_n^* = 0.9 m_n^{**}$	(23–494)
Dedendum of wheel tooth	$h_{f2} = 1.3 m_n^* = 1.05 m_n^{**}$	(23–495)
Addendum of wheel tooth with two-lines action (addendum-dedendum type)	$h_a = 0.9 m_n$	(23–496)

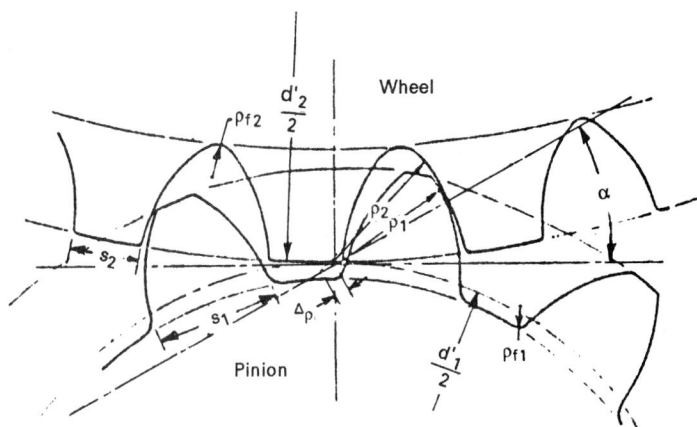

(a) All addendum type

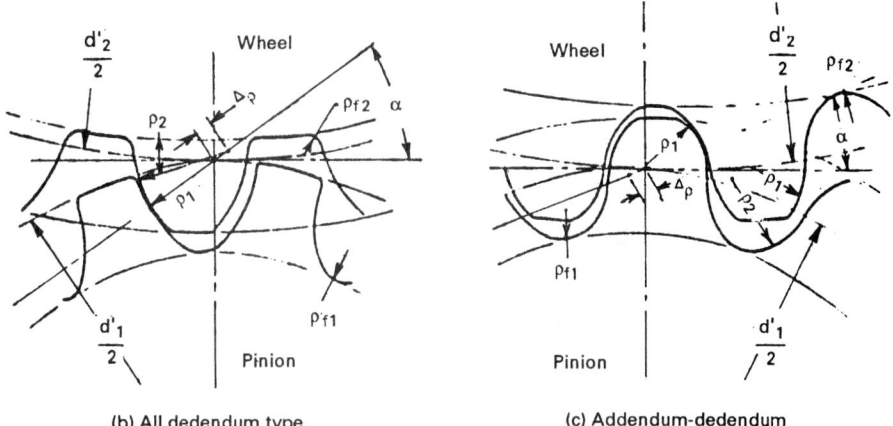

(b) All dedendum type   (c) Addendum-dedendum

FIGURE 23–80 Three typoes of Wildhaber-Novikov circular arc gears.

Particular	Formula
	where * = one line of action or single action ** = two lines of action or double action (addendum-dedendum type)

Dedendum of wheel of two-line action (addendum-dedendum type)	$h_f = x m_n$ (23–497)
	
	where
	x = 1.04575 for m_n up to 3.15 mm
	   = 1.04335 for m_n from 3.15 to 1.3 mm
	   = 1.04089 for m_n 6.3 to 10 mm

Addendum circle diameter of pinion	$d_{a1} = d'_1 + 2h_{a1}$ (23–498)

Particular	Formula	
Dedendum circle diameter of pinion	$d_{f1} = d_1' - 2h_{f1}$	(23–499)
Addendum circle diameter of wheel	$d_{a2} = d_2' + 2h_{a2}$	(23–500)
Dedendum circle diameter of wheel	$d_{f2} = d_2' - 2h_{f2}$	(23–501)
The axial pitch	$p_x = \dfrac{\pi m_n}{\sin \beta'}$	(23–502)
Helical axial overlap ratio	$\epsilon_{\beta x} = \dfrac{b}{p_x} = \epsilon_i + \vartheta$	(23–503a)
The integral part of overlap factor	$\epsilon_i = 1$ for one line of action $\quad$ = 2 for two lines of action	(23–503b) (23–503c)
The fractional part of overlap factor	$\vartheta = 0.15$ to 0.35 for one and two lines of action	(23–504)
Face width of pinion necessary to prevent edge loading	$b_1 = 1.1 \dfrac{\pi m_t}{\tan \beta}$ to $1.2 \dfrac{\pi m_t}{\tan \beta} = 1.1$ to $1.2 \dfrac{\pi}{P_t \tan \beta}$	(23–505)
Another formula for the face width of pinion	$b_1 = b_2 + (0.4 \text{ to } 1.5)m_n$ where b_2 is taken from Eq. (23–508)	(23–506)
Face width of pinion	$b_1 = b_2 + 3$ to 8 mm	(23–507a)
The ratio of b_2 to d_1'	$(b_2/d) = 0.7$ to 1.6	(23–507b)
Another formula for the face width of wheel	$b_2 = \epsilon_{\beta x} p_x$	(23–508)
Recommended pressure angle	$\alpha = 20°$ to $30°$	(23–509)
Recommended helix angle	$\beta = 10°$ to $30°$	(23–510)
The lengthwise radius of curvature of pinion tooth in the normal section, at the contact point	$\rho_{1n}' = \dfrac{d_1'(1 + \tan \beta \cos \alpha)^{1.5}}{2 \tan^2 \beta \sin \alpha \left(1 + \dfrac{d_1'}{2\rho_{1t}} \sin \alpha\right)}$	(23–511)
The lengthwise radius of curvature of the wheel tooth in the normal section at the contact point	$\rho_{2n}' = \dfrac{d_2'(1 + \tan \beta \cos \alpha)^{1.5}}{2 \tan^2 \beta \sin \alpha \left(1 - \dfrac{d_2'}{2\rho_{2t}} \sin \alpha\right)}$	(23–512)
The relative radius of curvature in the normal section	$\rho_{on} = \dfrac{d_1'}{2} \dfrac{(1 + \tan \beta \cos \alpha)^{1.5}}{\tan^2 \beta \sin \alpha} \left(\dfrac{i}{i \pm 1}\right)$	(23–513)
For basic rack profile of Wildhaber–Novikov gears	Refer to Fig. 23–81.	
The normal tooth load	$F_n = F_t / (\epsilon_{\beta x} \cos \alpha_n \cos \beta)$	(23–514)
Length of line of contact along the tooth profile in the transverse section	$l_{pt} = 2 \sin(\alpha - \delta)\rho_{1t}$	(23–515)

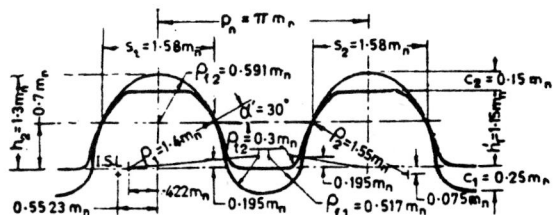

(a) Basic rack tooth form with one line of action

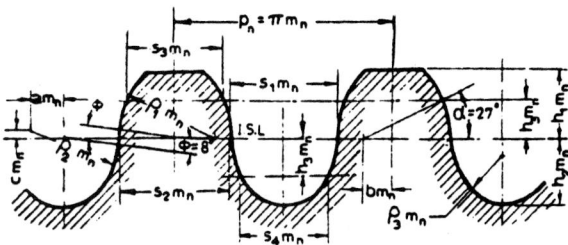

(b) Basic rack tooth form with two lines of action

FIGURE 23–81 Basic rack profile for Wildhaber-Novikov gears.

Particular	Formula
	where δ = clearance angle required for hobbing operation, deg, chosen as $5°$
Length of line of contact along the tooth profile in the normal section	$l_{\rho n} = (l_{\rho t} \sin \alpha)/ \sin \alpha_n$ $\qquad$ (23–516)
Hertzian contact stress at the point of contact	$\sigma_c = [F_n E_0/(5.72 \rho_{on} l_{\rho n})]^{1/2}$ $\qquad$ (23–517)

FROM RUSSIAN SOURCE

The normal load on the wheel	$F_{n2} = \left(\dfrac{m_{n2}}{m_{n1}}\right)^2 F_{n1} = \dfrac{F_{n1} \sin \beta'_1}{\sin \beta'_2} = F_{n1} \sqrt{\rho'_2/\rho'_1}$ $\qquad$ (23–518)
	where ρ'_1 and ρ'_2 are obtained from Eqs. (23–519) and (23–520)
	Experimental value of 1.9 may be taken in place of exponent 2 in Eq. (23–519)
The radius of curvature of the pinion tooth in the normal section	$\rho'_1 = \dfrac{d'_1}{2 \sin^2 \beta' \sin \alpha_n}$ $\qquad$ (23–519)
The radius of the wheel tooth profile in the normal section	$\rho'_2 = \dfrac{d'_2}{2 \sin^2 \beta' \sin \alpha_n}$ $\qquad$ (23–520)

Particular	Formula

The relative radius of curvature in the normal section

$$\rho_{on} = \frac{d_1'}{2}\left(\frac{i}{i \pm 1}\right)\frac{1}{\sin^2 \beta' \sin \alpha_n} \qquad (23\text{–}521)$$

Contact stress taking into account sliding, rolling and axial factor for a gearing with one line of action

$$\sigma_c = 10^3 \left[\frac{2.7 M_{t1}\sin\beta'}{m_n^{2.4}C_v\epsilon_{\beta x}k_{sr}z_1}\right]^{1/2}\left(\frac{i+1}{id_1'}\right)^{1/4} \qquad (23\text{–}522)$$

where

k_{sr} = surface finish and manufacturing accuracy factor = 14 to 17

r_2 = $(1.55m_n)/1000$ for gearing with one-line action

C_v = velocity and helix angle factor taken from Fig. 23–82

$\epsilon_{\beta x}$ = $\epsilon_i + 0.25$ [Refer to Eq. (23–503).]

ϵ_i is obtained from Eq. (23–503)

The value of M_{t1} calculated by using Eq. (23–522) may be increased by 30% for calculating allowable torque on the wheel with two lines of contact.

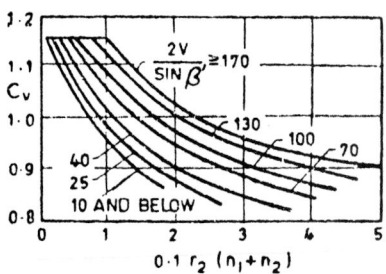

FIGURE 23–82   Velocity factor, C_V.

The pitch circle diameter

$$d_1' = \left[\frac{2.7\times10^6 M_{t1}\sin\beta'}{C_V k_{sr}\epsilon_{\beta x}z_1 m_n^{2.4}\sigma_{ca}^2}\right]^2\left(\frac{i+1}{i}\right) \qquad (23\text{–}523)$$

The bending stress developed in the pinion tooth

$$\sigma_b = \frac{2.3 M_{t1}y k_d k_\beta}{k_{\epsilon b}z_1 m_n^{2.8}} \le \sigma_{ba} \qquad (23\text{–}524)$$

The normal module

$$m_n = 1.32\left[\frac{M_{t1}y k_d k_\beta}{k_{\epsilon b}z_1 \sigma_{ba}}\right]^{1/2.8} \qquad (23\text{–}525)$$

where

σ_{ba} = $(1.3\sigma_f k_1)/n'$ for one-sided loading

σ_{fb} = $(\sigma_f k_1)/n'$ for two-sided loading

n' = factor of safety = 1.5 to 1.8

k_1 = load factor = $[(5\times10^6)/N]^{1/9}$

N = number of cycle of stress

The factor which depends on axial overlap factor for use in Eqs. (23–524) and (23–525)

$$k_{\epsilon b} = 1 \text{ to } 0.7(\epsilon_i - 1) \qquad (23\text{–}526a)$$

The approximate expression for dynamic load factor as per degree of accuracy and velocity v (up to 12 m/s)

$$k_d = 1 + \frac{H+3}{120}\, v \qquad (23\text{–}526b)$$

where

H = 1 to 4 grade accuracy

k_β = helical factor which is obtained from Fig. 23–83

k_y and v are obtained from Fig. 23–85

y = tooth form factor which is obtained from Fig. 23–84 for equivalent number of teeth and depends on k_y and fractional overlap factor $\epsilon_i = 0.15$ to 0.45 and β'.

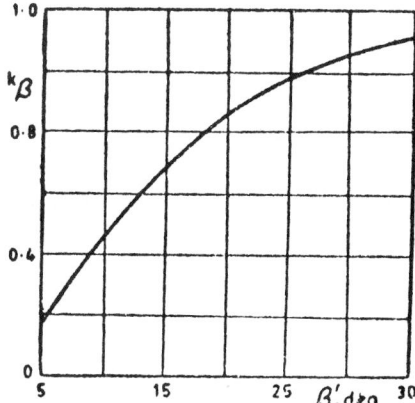

FIGURE 23–83 Helical factor, k_β.

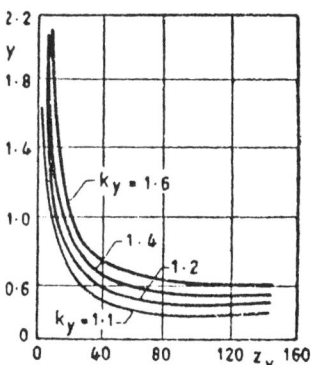

FIGURE 23–84 Curves of tooth form factor, y, with formative number of teeth, z_V.

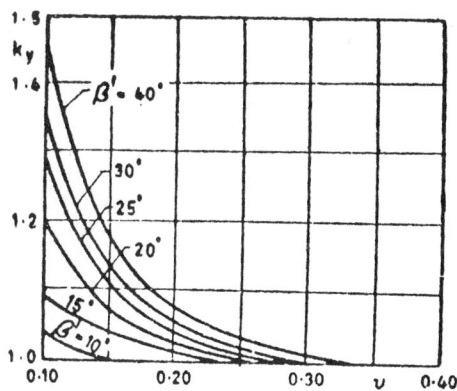

FIGURE 23–85 Factor, k_y, with fractional part of the overlap factor$y$.

23.5 WORM GEARING (1, 3, 22, 23)

SYMBOLS

a	center distance, m (in)
b	tooth face width of worm gear, m (in)
b_a	recommended width of worm gear rim, m (in)
b_e	effective face width of worm gear—the effective face width is the face width of the worm gear or two-thirds of the worm pitch diameter whichever is least, m (in)
c	clearance, m (in)
C_V	velocity factor
C_{Vr}	sliding speed factor or sliding velocity factor
d_1	reference diameter of worm, m (in)
d_2	reference diameter of worm gear, m (in)
d_1'	pitch diameter of worm, m (in)
d_2'	pitch diameter of worm gear, m (in)
d_{b1}	diameter of base circle of worm, m (in)
d_{b2}	diameter of base circle of worm gear, m (in)
d_{a1}	diameter of addendum circle of worm, m (in)
d_{a2}	diameter of addendum circle of worm gear, m (in)
	throat diameter of worm gear, m (in)
d_{f1}	diameter of dedendum circle of worm, m (in)
d_{f2}	diameter of dedendum circle of worm gear, m (in)
d_{eb}	effective diameter of thrust bearing, m (in)
f_{nb}	speed factor for worm gear for strength from Fig. 23–91
f_{nc}	speed factor for worm gear for wear from Fig. 23–92
F	load or force, kN (lbf)
F_n	normal force acting on tooth at the pitch point, kN (lbf)
F_t	permissible transmitted load, kN (lbf)
F_x	resulting turning force on the worm gear, kN (lbf)
F_z	the force which tends to separate the axes, kN (lbf)
F_μ	frictional force, kN (lbf)
g_{min}	the minimum total length of lines of contact at $k_{g(min)}$, m (in)
i	speed ratio or transmission ratio
k_c	load concentration factor
k_d	dynamic load factor
k_l	load factor
$k_{g(min)}$	variation factor of total length of lines of contact
K_b	bending stress factor taken from Table 23–108 for use in Eq. (23–584)
K_i	ratio-correction factor taken from Table 23–103 for use in Eq. (23–575)
K_M	materials and size correction factor taken from Eq. (23–575)
K_w	surface stress factor taken from Table 23–108 for use in Eq. (23–585)
l	length, m (in)
l_ρ	length of root of worm gear teeth, m (in)
L_1	length of worm, m (in)
m	module, m (in)
m_n	axial module for worm, m (in)
m_d'	number of modules on the pitch diameter of the worm, m (in)

M_{tb}	the permissible torque on worm gear based on strength, N m (lbf in)
M_{tw}	the permissible torque on worm gear based on wear, N m (lbf in)
M_{to}	permissible worm gear torque for life other than normal rating, N m (lbf in)
M_{tn}	permissible worm gear torque for normal life, N m (lbf in)
M_{tm}	momentary overload torque, N m (lbf in)
n	speed (also with subscripts), rpm
n'	speed, rps
n_1	speed of worm, rpm
n_2	speed of worm gear (wheel), rpm
P	power, kW (hp)
p	circular pitch or transverse circular pitch, m (in)
p_n	normal circular pitch, m (in)
p_x	axial pitch, m (in)
p_z	lead, m (in)
P	diametral pitch
$q = \dfrac{d}{m}$	diametral quotient
r	radius, m (in)
r_{b2}	radius of worm gear face, m (in)
r_{f2}	root radius of worm gear, m (in)
r_r	radius of worm gear rim, m (in)
s_x	tooth thickness on reference diameter of worm in axial section, m (in)
s_n	tooth thickness on reference diameter of worm in normal section, m (in)
T_{be}	equivalent running time per cycle for strength at torque M_{tm21} and speed n_{21}, h
T_{we}	equivalent running time per cycle for wear at torque M_{tm21} and speed n_{21}, h
T_{bet}	total equivalent running time for bending at torque M_{tm21} and worm gear speed n_{21}, h
T_{wet}	total equivalent running time for wear at torque M_{tm21} and worm gear speed n_{21}, h
v_m	pitch line velocity of worm, m/s
v_r	rubbing velocity at worm diameter, m/s (ft/min)
y	Lewis form factor
Y_z	zone factor for worm gear taken from Table 23–109 for use in Eq. (23–587)
z_1	number of starts or threads on worm
z_2	number of teeth on worm wheel
α	pressure angle, deg
α_n	pressure angle on normal section, deg
β	helix angle, deg
γ	lead angle, deg
δ	one half of the face angle, deg
ρ	angle of coefficient of friction, deg
ρ_f	tooth-fillet or root radius of worm, m (in)
ρ_t	tooth tip radius of worm, m (in)
μ	coefficient of friction between worm and worm gear
μ'	coefficient of friction in thrust bearings
ξ	addendum factor
χ	dedendum factor

σ	stress (also with subscripts), MPa (psi)
σ_b	bending stress, MPa (psi)
σ_c	surface compressive stress, MPa (psi)

SUFFIXES

1	pinion
2	gear
a	allowable, axial
b	bending
n	normal
m	momentary
0	other than normal
r	radial
x	axial
w	wear

Particular	Formula	
For worm gear terminology	Refer to Tables 23–43 and 23–85 and Fig. 23–86.	
Speed ratio or transmission ratio	$i = \dfrac{n_1}{n_2} = \dfrac{\omega_1}{\omega_2} = \dfrac{z_2}{z_1} = \dfrac{\pi d_2}{p_z}$	(23–527)
The lead (Fig. 23–86)	$p_z = z_1 p_x = \pi m_x z_1$	(23–528)
The relation between the axial pitch, module, and lead (Fig. 23–86)	$p_x = \pi m_x = \dfrac{p_z}{z_1}$	(23–529a)
The normal circular pitch	$p_n = p_x \cos \gamma$	(23–529b)
The axial module of worm	$m_x = \dfrac{m}{\tan \gamma}$	(23–530)
The number of modules on the pitch diameter of the worm	$m_d' = \dfrac{d_1'}{m_x} = \dfrac{z_1}{\tan \gamma}$	(23–531)
The module for standard gears $x_1 = x_2 = 0$ and $x_1 = -x_2$	$m = \left(\dfrac{2a}{m_d' + z_2} \right)$	(23–532)
The module for corrected gears $x_1 + x_2 \neq 0$	$m = \left(\dfrac{2a}{m_d' + z_2 + 2x} \right)$	(23–533)
The center distance for corrected gears $x_1 + x_2 \neq 0$	$a = \dfrac{m_x}{2} [m_d' + z_2 + 2x]$	(23–534)
The center distance for standard gears $x_1 = x_2 = 0$ and $x_1 + x_2 = 0$	$a = \dfrac{m_x}{2} [m_d' + z_2]$	(23–535)
The center distance of extended center distance worm and worm wheel with corrected tooth profiles	$a' = 0.5(d_1' + d_2' + 2xm')$ where $2xm' = $ displacement of the hob cutter	(23–536)

Particular	Formula
Condition to obtain gears with different gear ratios without changing the center distance	$a = a' = 0.5m(z_2 + m'_d) = 0.5m'(z'_2 + m'_d + 2x)$ (23–537)
For corrected gears	Refer to Fig. 23–90.
The correction factor or addendum modification coefficient	$x = \dfrac{a}{m} - 0.5(m'_d + z_2)$ (23–538)
Reference diameter of worm	$d_1 = m'_d m_x$ (23–539)
Reference diameter of worm wheel	$d_2 = z_2 m_x = \dfrac{z_2 p}{\pi}$ (23–540)
The pitch diameter of worm (Fig. 23–86, Table 23–43)	$d'_1 = m'_d m_x$ (23–541a) $\quad\quad = mz_1 / \tan\gamma$ (23–541b) where m'_d varies from 6 to 13 and m_x = axial pitch on the pitch circle diameter of worm
The pitch diameter of the worm gear (wheel)	$d'_2 = z_2 m_x = \dfrac{z_2 p}{\pi}$ (23–542)

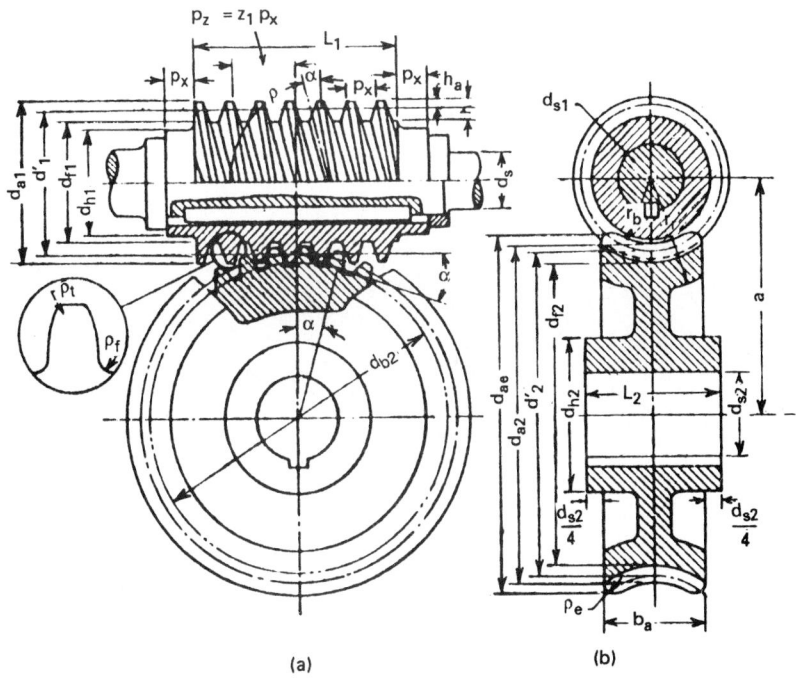

(a) (b)

FIGURE 23–86 Proportions and dimensions of single-enveloping worm and worm gear.

Particular	Formula

AGMA FORMULAS FOR DESIGN OF WORM AND WORM GEAR

The mean worm diameter d_{m1} should be selected such that it falls into the range as per AGMA recommendation

$$0.5\, a^{0.875} \leq d_{m1} \leq 0.94\, a^{0.875} \quad \textbf{SI} \qquad (23\text{–}543\text{a})$$

where a and d_{m1} in mm

$$0.2\, a^{0.875} < d_{m1} \leq 0.4\, a^{0.875} \quad \textbf{SI} \qquad (23\text{–}543\text{b})$$

where a and d_{m1} in m

$$\frac{a^{0.875}}{3} < d_{m1} < \frac{a^{0.875}}{1.6} \quad \textbf{US Customary System units}$$
$$(23\text{–}543\text{c})$$

Recommended value of pitch diameter as per AGMA for speed reducer with integral worm

$$d_1' = \frac{a^{0.875}}{1.5} \quad \textbf{SI} \qquad (23\text{–}544\text{a})$$

where d_1' and a in mm

$$= \frac{a^{0.875}}{3.5} \quad \textbf{SI} \qquad (23\text{–}544\text{b})$$

where d_1' and a in m

$$= \frac{a^{0.875}}{2.2} \quad \textbf{US Customary System units} \qquad (23\text{–}544\text{c})$$

where d_1' and a in in

The diameter of base circle of worm

$$d_{b1} = m_x(m_d' + 2x) \qquad (23\text{–}545)$$

The diameter of base circle of worm gear (wheel)

$$d_{b2} = d_2 \qquad (23\text{–}546)$$

The addendum circle diameter of worm

$$d_{a1} = d_1 + 2\xi m_x \qquad (23\text{–}547)$$
where
ξ = addendum factor
 = 1 for standard profiles

The addendum circle diameter of worm wheel

$$d_{a2} = d_2 + 2m_x(\xi + x) \qquad (23\text{–}548\text{a})$$
where $\xi = 1$, $x = 0$ for standard tooth profiles without correction

Throat diameter of worm gear (Fig. 23–86)

$$d_{a2} = d_2' + 0.636p = (z_2 + 0.636\pi)m \qquad (23\text{–}548\text{b})$$

The dedendum circle diameter of worm

$$d_{f1} = d_1 - 2m_x(\chi + 0.25m_x) \qquad (23\text{–}549)$$
where $\chi = 1$ for standard tooth profiles

The dedendum circle diameter of worm wheel

$$d_{f2} = d_2 - 2m(\chi + 0.25 - x) \qquad (23\text{–}550)$$

The bottom clearance

$$c = 0.2m_x \text{ to } 0.3m_x \qquad (23\text{–}551)$$

The recommended value of face angle

$$\tan \delta \leq \frac{\tan \alpha}{\tan \gamma} \qquad (23\text{–}552)$$

where 2δ varies from $60°$ to $75°$

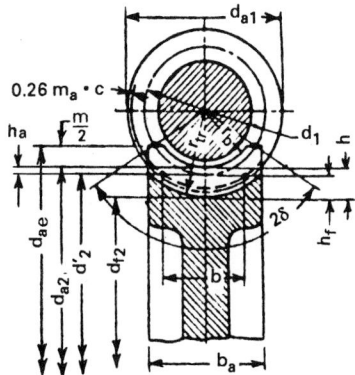

FIGURE 23–87 Dimensions of worm gear set

Particular	Formula	
The lead angle on reference cylinder	$\tan\gamma = \dfrac{z_1 p_x}{\pi d_1} = \dfrac{m_x z_1}{d_1} = \dfrac{p_z}{\pi d_1}$	(23–553)
For compact design the lead angle may be determined approximately from the relation	$\tan\gamma = \sqrt[3]{\dfrac{n_2}{n_1}}$	(23–554)
Lead angle on pitch cylinder for corrected tooth profiles	$\tan\gamma' = \dfrac{z}{(m_d' + 2x)}$	(23–555)
For lead angle	Refer to Table 23–99.	
Radius of worm gear face	$r_b = \dfrac{d_1}{2} - \xi m_x$	(23–556)
Root radius of worm gear (wheel)	$r_{f2} = \dfrac{d_1}{2} + \xi m_x + c$	(23–557)
Radius of gear rim (Fig. 23–86)	$r_r = 2.2p + 14 = 2.2\pi m + 14$	(23–558)
Tooth tip relief radius of worm	$\rho_t = 0.1 m_x$	(23–559)
Tooth fillet or root radius of worm	$\rho_f = 0.2 m_x$	(23–560)
The tooth thickness on reference diameter in axial section of worm	$s_x = \dfrac{\pi m_x}{2}$	(23–561)
The tooth thickness on reference diameter in normal section of worm	$s_n = \dfrac{\pi m_x \cos\gamma}{2}$	(23–562)
For worm and worm gear proportion as recommended by AGMA	Refer to Table 23–85.	
For forces acting at the point of contact of worm thread and gear tooth (Fig. 23–97)	Refer to Table 23–86.	
For bearing loads on worm gearing (Fig. 23–98)	Refer to Table 23–87.	

Particular	Formula
For values of transmission ratios and designation of general purpose worm gear	Refer to Table 23–88.
For dimensions of worm	Refer to Table 23–89.
For pressure angles recommended for worm gear sets	Refer to Table 23–90.
For worm data	Refer to Table 23–91.
For AGMA standards for axial pitches of worm gear	Refer to Table 23–92.

FACE WIDTH

Face width of tooth along the worm gear (wheel) circle diameter $(d_{a1} - 2m_x)$, i.e., along the pitch circle of the worm (Fig. 23–87)

$$b = (d_{a1} - 2m_x)\frac{\pi\delta}{180} \qquad (23\text{–}563)$$

Recommended width of worm gear rim (Fig. 23–87)

$$b_a \le 0.75 d_1' \text{ for } z_1 = 1 \text{ to } 3 \qquad (23\text{–}564a)$$
$$\le 0.67 d_1' \text{ for } z_1 = 4 \qquad (23\text{–}564b)$$

Recommended face wdith of worm gear as per AGMA for speed reducer with integral worm

$$b = 0.72\, a^{0.875} \qquad \textbf{SI} \qquad (23\text{–}565a)$$

where a and b in mm

$$= 0.3\, a^{0.875} \qquad \textbf{SI} \qquad (23\text{–}565b)$$

where a and b in m

$$= \frac{(a^{0.875})}{3} \qquad \textbf{US Customary System units} \qquad (23\text{–}565c)$$

where a and b in in

The maximum radial deflection of the worm at the pitch point

$$\Delta w_{max} = 0.025\sqrt{p_x} \qquad \textbf{SI} \qquad (23\text{–}566a)$$

where Δw_{max} in mm
p_x = axial pitch, mm
$$= 0.005\sqrt{p_x} \qquad \textbf{US Customary System units} \qquad (23\text{–}566b)$$

where p_x in in

The deflection of shaft of worm when it is mounted between bearings from mechanics of materials

$$\delta = \frac{F_{1R}L^3}{48EI} \qquad (23\text{–}566c)$$

where F_{1R} = the resultant force due to tangential force $F_{\theta1}(F_{t1})$ and the radial force F_{r1}

$$= \sqrt{F_{\theta1}^2 + F_{r1}^2} \qquad (23\text{–}566d)$$

The allowable or permissible deflection

$$\delta_a = \frac{d_{m1}}{1000} \qquad (23\text{–}566e)$$

Here $\delta < \delta_a$

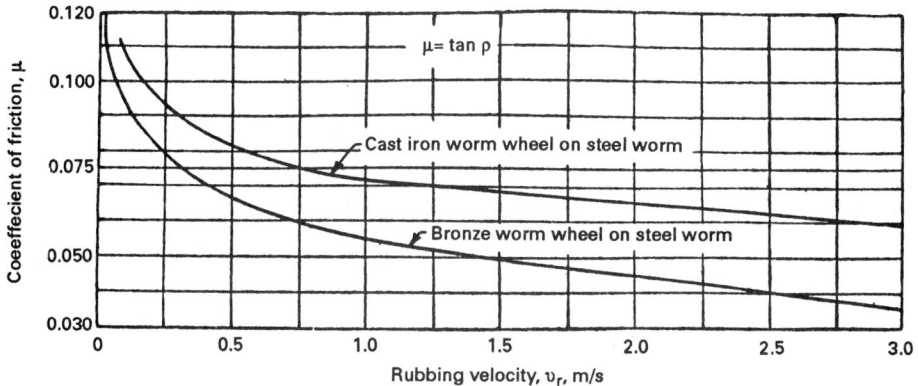

FIGURE 23–88 Coefficient of friction for various values of rubbing velocities.

Particular	Formula
For face width of worm gear (wheel)	Refer to Table 23–93.
For length of worm	Refer to Table 23–94 and Figs. 23–86 and 23–87.

DESIGNING WORM GEAR TEETH FOR STRENGTH

Lewis equation for permissible load on worm gear tooth

$$F_t = \frac{\sigma_o b Y C_V}{P} = \frac{\sigma_o b Y p C_V}{\pi} = \sigma_o \boldsymbol{m} b Y C_V \qquad (23\text{–}567a)$$

The module as per Lewis equation

$$\boldsymbol{m} = \frac{F_t}{\sigma_o b Y C_V} \qquad (23\text{–}567b)$$

Lewis equation for bending stress as modified by Buckingham

$$\sigma_b = \frac{F_{2t}}{p_n b_e y} \qquad (23\text{–}567c)$$

where y = form factor taken from Table 23–95

If the number of teeth in the worm gear (wheel) plus the number of threads per millimeter in the worm is greater than 1.6, the form factor Y may be determined from the formula

$$Y = 0.314 - 0.0151\,(\alpha - 14.5°) \qquad (23\text{–}568a)$$

Modified Lewis form factor Y is also obtained from

$$Y = \pi y \qquad (23\text{–}568b)$$
where y is obtained from Table 23–95
σ_o is obtained from Table 23–96

The velocity factor

$$C_v = \frac{3}{(3 + v_m)} \qquad (23\text{–}569a)$$

$$= \frac{6}{(6 + v_m)} \qquad (23\text{–}569b)$$

which takes into account the dynamic effect

Particular	Formula
Coefficient of friction	$\mu = \dfrac{0.0422}{v_r^{0.28}}$ for $0.2 < v_r < 2.75$ m/s **SI** (23–570a)
	$= \dfrac{0.185}{v_r^{0.28}}$ for $40 < v_r < 500$ ft/min
	US Customary System units (23–570b)
	where v_r in ft/min
	$= 0.025 + \dfrac{v_r}{305}$ for $2.75 < v_r < 20$ m/s **SI**
	(23–570c)
	where v_r in m/s
	For μ refer to Table 23–97 and Figs. 23–88 and 23–94.
	$= 0.025 + \dfrac{v_r}{60,000}$ for $550 < v_r < 4000$ ft/min
	US Customary System units (23–570d)
	where v_r in ft/min
Another expression for coefficient of friction	$\mu = \dfrac{0.054}{v_r^{0.2}}$ for $0.02 < v_r < 0.36$ m/s **SI**
	(23–570e)
	where v_r in m/s
	$= \dfrac{0.155}{v_r^{0.2}}$ for $3 < v_r < 70$ ft/min (23–570f)
	US Customary System units
	where v_r in ft/min
	$= \dfrac{0.048}{v_r^{0.36}}$ for $0.36 < v_r < 15$ m/s **SI** (23–570g)
	where v_r in m/s
	$= \dfrac{0.320}{v_r^{0.36}}$ for $70 < v_r < 3000$ ft/min (23–570h)
	US Customary System units
	where v_r in ft/min
The rubbing velocity or velocity of sliding	$v_r = \dfrac{\pi d_1 n_1}{60,000 \cos \gamma}$ **SI** (23–571a)
	where v_r in m/s; d_1 in mm; and n in rpm
	$= \dfrac{\pi d_1 n_1'}{\cos \gamma}$ **SI** (23–571b)

Particular	Formula

where v_r in m/s; d_1 in m; n_1' in rps

For v_r, refer to Tables 23–97, 23–98, 23–104, and Fig. 23–88.

For lead angle γ, refer to Table 23–99.

$$= \frac{\pi d_1 n_1}{12 \cos \gamma} \quad \textbf{US Customary System units}$$

$$(23\text{–}571c)$$

where v_r in ft/min; d_1 in in; n_1 in rpm

DESIGNING WORM GEAR FOR WEAR LOAD

The limit wear load formula

$$F_w = d_2 b K \qquad (23\text{–}572)$$

where K = load stress factor taken from Table 23–100

Another formula for limiting load for wear which takes into account various gear data but assumes the use of proper grade of lubricant

$$F_w = (A \cos \gamma \, C_V \sigma_{ca})/C_s \qquad (23\text{–}573)$$

where

A = projected tooth area = $(h d_1 \delta)/57.3$ in m² (in²)

h = tooth depth, m (in)

C_V = velocity factor = $\dfrac{3}{3 + v_m}$; C_s = service factor taken from Table 23–101

σ_{ca} = allowable surface pressure, MPa (psi) taken from Table 23–102

The rated input power

$$P_i = \frac{n_1 F_{2t} d_{m2}}{19.1 \times 10^6 i} + \frac{V_r F_\mu}{1000} \quad \textbf{SI} \qquad (23\text{–}574a)$$

where P_i in kW, n_1 in rpm, F_t and F_μ in N,

i = gear ratio = z_2/z_1; d_{m2} in mm

n_1 = rotational speed of worm, rpm

V_r = sliding velocity or rubbing velocity at mean worm diameter, m/s

$$= \frac{n_1' F_{2t} d_{m2}}{318 i} + \frac{V_r F_\mu}{1000} \quad \textbf{SI} \qquad (23\text{–}574b)$$

where n_1' in rps, P_i in kW, F_t and F_μ in N, d_{m2} in m, V_r in m/s

$$= \frac{n_1 F_{2t} d_{m2}}{126,000 i} + \frac{V_r F_\mu}{33,000} \qquad (23\text{–}574c)$$

US Customary System units

where P_i in hp, n_1 in rpm, F_t and F_μ in lbf; V_r in ft/min, d_{m2} in in

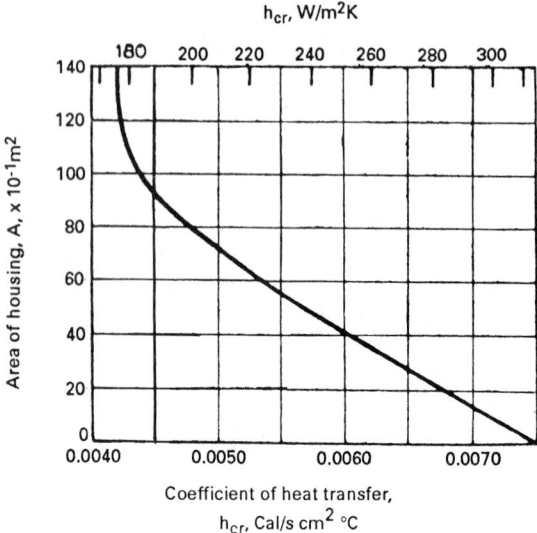

FIGURE 23–89 Area of housing vs. coefficient of heat transfer for worm gear.

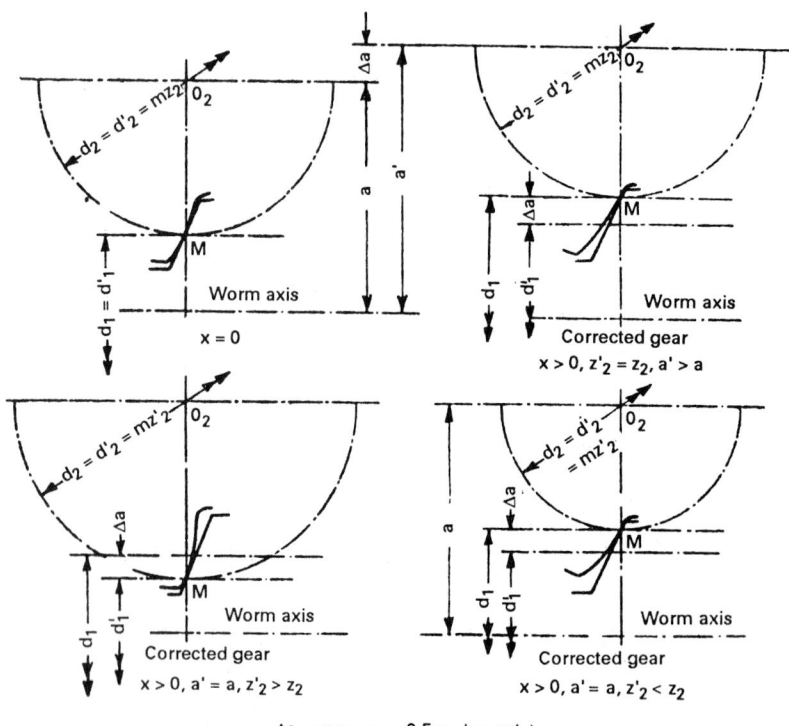

$$\Delta a = x m_x = a - 0.5 m_x (z + m'_d)$$

FIGURE 23–90 Corrected worm gear.

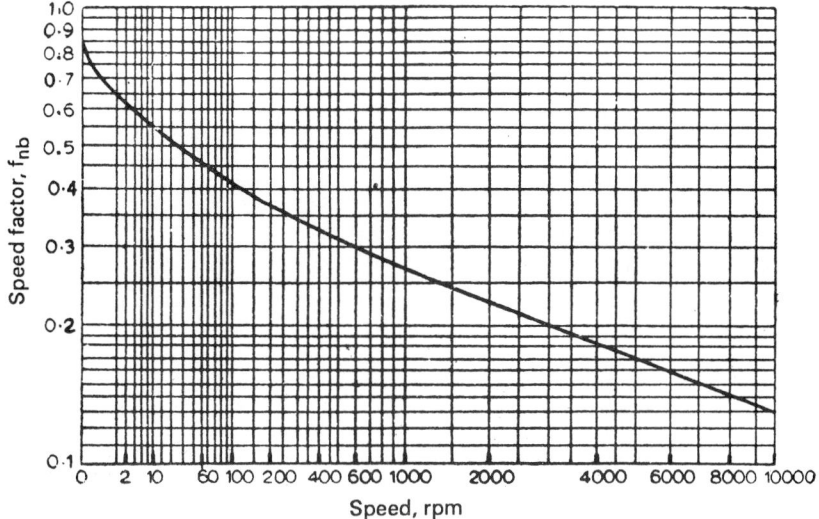

FIGURE 23–91 Speed factors for worm gears for strength, f_{nb}

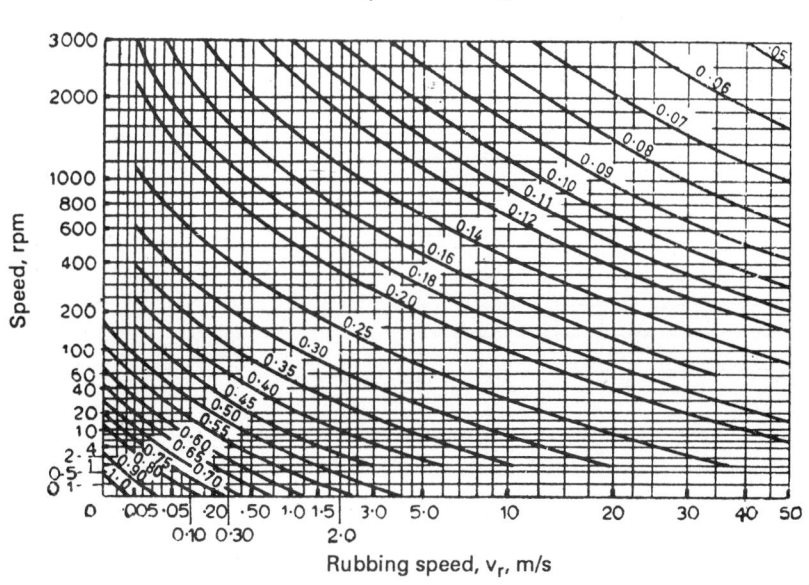

FIGURE 23–92 Speed factors for worm gear for wear, f_{nc}.

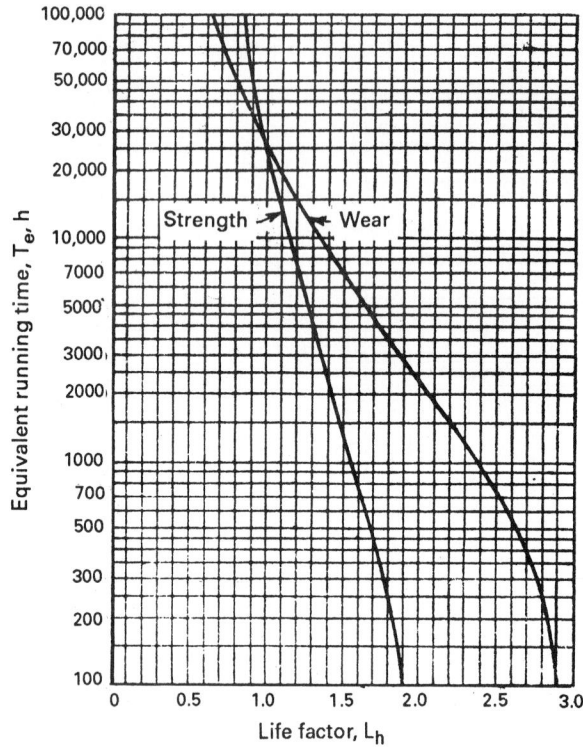

FIGURE 23–93   Life factor, L_h, for strength and wear for worm gears.

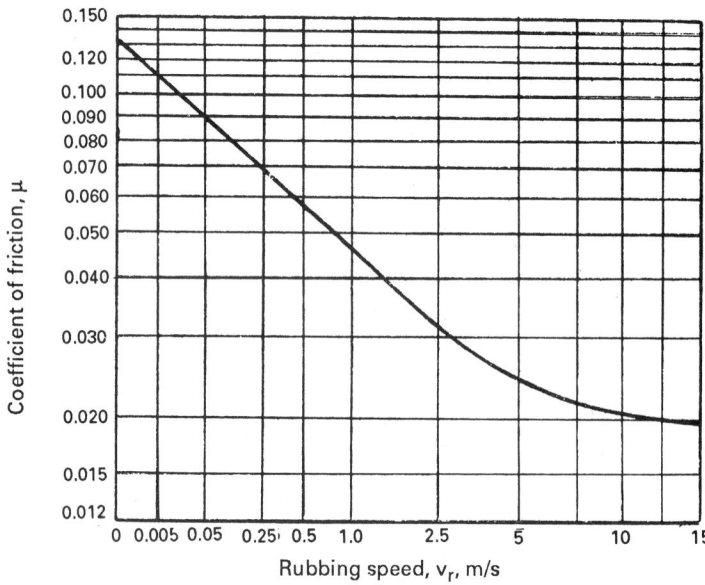

FIGURE 23–94   Coefficient of friction for worm gears.

Particular	Formula
The tangential load on the worm gear tooth, F_{2t}	$$F_{2t} = \frac{(C_s d_{m2}^{0.8} b_e C_m C_v)}{75.948} \quad \textbf{SI} \qquad (23\text{–}575a)$$ where C_s = material factor taken from Figs. 23–95 and 23–96 and Table 23–103 C_v = velocity factor taken from Table 23–104 C_m = ratio correction factor taken from Table 23–105 b_e and d_{m2} in mm; F_{2t} in N b_e = effective face width (actual face width, except not to exceed 0.67 d_{m2}) of gear, mm $$= 3310 C_s d_{m2}^{0.8} b_e C_m C_v \quad \textbf{SI} \qquad (23\text{–}575b)$$ where F_{2t} in N, d_{m2} and b_e in m $$= C_s d_{m2}^{0.8} b_e C_m C_v \quad \textbf{US Customary System units}$$ $$(23\text{–}575c)$$ where d_{m2}, b_e in in, and F_{2t} in lbf
The friction force, F_μ	$$F_\mu = \frac{\mu F_{2t}}{\cos \lambda \cos \alpha_n} \qquad (23\text{–}576)$$ where λ = lead angle, deg α_n = normal pressure angle of worm thread at the mean diameter, deg
Another expression for friction force	$$F_\mu = \frac{\mu F_{2t}}{\mu \sin \gamma - \cos \gamma \cos \alpha_n} \qquad (23\text{–}577)$$
The rated torque at the worm gear	$$M_{t2} = \frac{F_{2t} d_{m2}}{2000} \quad \textbf{SI} \qquad (23\text{–}578a)$$ where M_{t2} in N m; F_{2t} in N; d_{m2} in mm $$= \frac{F_{2t} d_{m2}}{2} \quad \textbf{US Customary System units}$$ $$(23\text{–}578b)$$ where M_{t2} in lbf in, F_{2t} in lbf, and d_{m2} in in
The efficiency for worm gearing	$\eta = (P_0/P_i)(100) \qquad (23\text{–}579)$ P_o = rated output power, kW (hp) P_i = rated input power, kW (hp) $$\eta = \frac{n_1 F_{2t} d_{m2}}{19.1 \times 10^6 i P_i} (100) \quad \textbf{SI} \qquad (23\text{–}580a)$$ where η in percent, n_1 in rpm, F_{2t} in N, d_{m2} in mm, P_i in kW

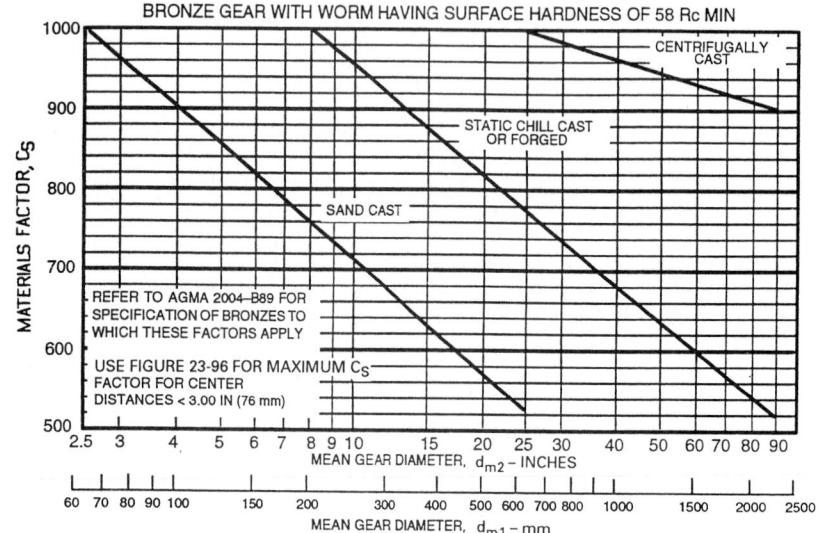

FIGURE 23–95 Materials factor, C_S, for center distances > 3.0 in (76 mm).

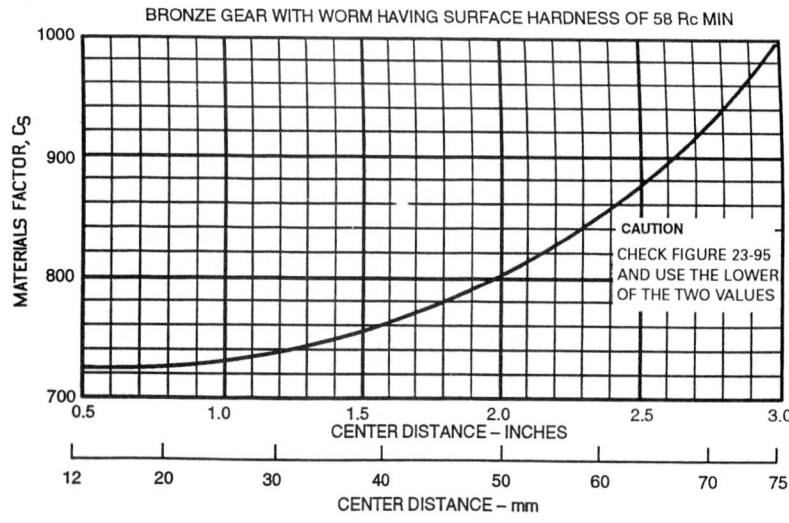

FIGURE 23–96 Maximum materials factor, C_S, for center distances < 3.00 in (76 mm). (*ANSI/AGMA 6034-B92.*)

Particular	Formula
	$$= \frac{n_1 F_{2t} d_{m2}}{126{,}000 i P_i} \ (100) \ \textbf{US Customary System units}$$
	(23–580b)
	where n_1 in rpm, F_t in lbf, d_{m2} in in, P_i in hp
	$$= \frac{n_1' F_{2t} d_{m2}}{318 i P_i} \ (100) \qquad \textbf{SI} \qquad\qquad \text{(23–580c)}$$
	where n_1' in rps, F_{2t} in N, d_{m2} in m
For minimum normal backlash allowances	Refer to Tables 23–86 and 23–106.
The limiting value of torque on worm gear for unlimited life based on allowable bending stress σ_{ab} as per British Standards (BSI)*†	$$M_{t2} = \frac{\sigma_{ab} d_2'^2 b_2 \cos \beta_2}{1.5 z_2} \qquad\qquad \text{(23–581)}$$
	where
	M_{t2} = torque on worm gear, N m (lbf in)
	b_2 = face width of gear, mm (in)
	β_2 = helix angle of worm gear, deg
	d_2' = pitch diameter of worm gear, mm (in)
	σ_{ab} = allowable bending stress, Pa (psi), taken from Table 23–107
The limiting value of torque on worm gear for unlimited life based on allowable contact stress σ_{ac} as per British Standards (BSI) *† (24)	$$M_{2t} = \frac{\sigma_{ac} d_2'^2 b_2 C_v}{30} \qquad\qquad \text{(23–582)}$$
	where
	σ_{ac} = allowable contact stress, Pa (psi) taken from Table 23–107
	C_v = velocity factor from Eq. (23–583a) or (23–583b)
The velocity factor for use in Eq. (23–582)	$$C_v = \frac{237}{\sqrt{V_r}} \ \text{for } V_r > 15.5 \text{ m/s} \ \ \textbf{SI} \qquad \text{(23–583a)}$$
	$$= \frac{55}{\sqrt{V_t}} \ \text{for } V_r > 3025 \text{ ft/min} \qquad \text{(23–583b)}$$
	US Customary System units
	$C_v = 1$ for $V_r < 15.5$ m/s (3025 ft/min) (23–583c)
	where V_r = rubbing or sliding velocity, m/s (ft/min)

---

* British Standard Institution, BS721.

† *Source*: M. J. Neal, ed., *Tribology Handbook*, Butterworth, London, A24, 1973.

Particular	Formula

RATING OF WORM GEARS—INDIAN STANDARDS

The permissible torque for strength on the worm gear (wheel) for normal rating

(*Note*: The allowable torque is given by the least of these four values obtained from both beam strength and wear consideration.)

$$M_{tnb} = 17.65 \times 10^6 f_{nb1} K_{b1} l_\rho \, \boldsymbol{m} \, d_2 \cos \gamma \quad \textbf{SI} \quad (23\text{–}584a)$$
where M_{tnb} in N m, $\boldsymbol{m}$ in m, l_ρ in m, and d_2 in m
$$= 17.65 \times 10^6 f_{nb2} K_{b2} l_\rho \, \boldsymbol{m} \, d_2 \cos \gamma \quad \textbf{SI} \quad (23\text{–}584b)$$
where M_{tnb} in N m, $\boldsymbol{m}$ in m, l_ρ in m, and d_2 in m
$$= 2560 f_{nb1} K_{b1} l_\rho \, \boldsymbol{m} \, d_2 \cos \gamma \qquad (23\text{–}585)$$
US Customary System units
where M_{tnb} in lbf in, $\boldsymbol{m}$ in in, l_ρ in in, and d_2 in in
$$= 2560 f_{nb2} K_{b2} l_\rho \, \boldsymbol{m} \, d_2 \cos \gamma \qquad (23\text{–}586)$$
US Customary System units
where M_{tnb} in lbf in, $\boldsymbol{m}$ in in, l_ρ in in, and d_2 in in

The permissible torque for wear on the worm gear for normal rating

(*Note*: The allowable torque is given by the least of the four values of Eqs. (23–584) to (23–590).)

$$M_{tnw} = 4.7 \times 10^6 f_{nc1} K_{w1} Y_z \, \boldsymbol{m} \, (d_2)^{1.8} \quad \textbf{SI} \quad (23\text{–}587)$$
where M_{tnw} in N m, $\boldsymbol{m}$ in m, and d_2 in m
$$= 4.7 \times 10^6 f_{nc2} K_{w2} Y_z \, \boldsymbol{m} \, (d_2)^{1.8} \quad \textbf{SI} \quad (23\text{–}588)$$
where M_{tnw} in N m, $\boldsymbol{m}$ in m, and d_2 in m
$$= 1415 f_{nc1} K_{w1} Y_z \, \boldsymbol{m} \, (d_2)^{1.8} \qquad (23\text{–}589)$$
US Customary System units
where M_{tnw} in lbf in, $\boldsymbol{m}$ in in, and d_2 in in
$$= 1415 f_{nc2} K_{w2} Y_z \, \boldsymbol{m} \, (d_2)^{1.8} \qquad (23\text{–}590)$$
US Customary System units
where M_{tnw} in lbf in, $\boldsymbol{m}$ in in, and d_2 in in

For f_{nc}, refer to Fig. 23–92, for K_w and Y_z, refer to Tables 23–108 and 23–109, for f_{nb} refer to Fig. 23–91, and for K_b, refer to Table 23–108.

The permissible power or worm gear torque for life other than normal rating:

(i) for strength

$$M_{tob} = M_{tnb} \left(\frac{26,200}{200 + T_{be}} \right)^{1/7} \qquad (23\text{–}591)$$

(ii) for wear

$$M_{tow} = M_{tnw} \left(\frac{27,000}{1000 + T_{we}} \right)^{1/3} \qquad (23\text{–}592)$$

(*Note*: The normal rating is the loading to which the gears may safety be subjected for a period of 12 h/day for a total running time of 26,000 h.)

where
T_{we} = total equivalent running time for wear, h
T_{be} = total equivalent running time for strength, h

Particular	Formula
The normal power rating of worm gears	$P = \dfrac{M_t n}{63,000}$ **US Customary System units** (23–593) where P in hp, M_t in lbf in, and n in rpm $= \dfrac{M_t n'}{159}$ **SI** (23–594) where P in kW, M_t in N m, and n' in rps $= \dfrac{M_t n}{9950}$ **SI** (23–595) where P in kW, M_t in N m, and n in rpm $= \dfrac{M_t \omega}{1000}$ **SI** (23–596) where P in kW, M_t in N m, and ω in rad/s
The efficiency of worm gears excluding losses in bearings and churning of lubricant	$\eta = \tan\gamma / \tan(\gamma + \rho)$ when worm driving (23–597a) $= \tan(\gamma - \rho) / \tan\gamma$ when worm gear driving (23–597b) where ρ = angle of coefficient of friction taken from Table 23–98

MOMENTARY OVERLOAD

The momentary overload capacity of worm gear which acts for not more than 15 s	$M_{tbm} = 39 \times 10^6 \, K_{b2} l_\rho \, \boldsymbol{m} \, d_2 \cos\gamma$ **SI** (23–598a) where M_{tbm} in N m, l_ρ in m, $\boldsymbol{m}$ in m, and d_2 in m
(i) for strength	$= 5690 \, K_{b2} l_\rho \, \boldsymbol{m} \, d_2 \cos\gamma$ **US Customary System units** (23–598b) where M_{tbm} in lbf in, l_ρ in in, $\boldsymbol{m}$ in in, and d_2 in in
(ii) for wear	$M_{twm} = 9.4 \times 10^6 K_{w2} Y_z \, \boldsymbol{m} \, (d_2)^{1.8}$ **SI** (23–599a) where M_{twm} in N m, $\boldsymbol{m}$ in m, and d_2 in m $= 2830 \, K_{w2} Y_z \, \boldsymbol{m} \, (\boldsymbol{d_2})^{1.8}$ **US Customary System units** (23–599b) where M_{twm} in lbf in, $\boldsymbol{m}$ in in, and d_2 in in

Particular	Formula

DETERMINATION OF EQUIVALENT RUNNING TIME FOR UNSTEADY LOAD CONDITIONS

Various uniform loads at different speeds

When the torque on the worm gear comprises a maximum torque M_{tm21} acting for T_1 hours at a mean speed of n_{21} and smaller torques M_{t22}, M_{t23} etc. acting for T_2, T_3 etc. hours at mean speed n_{22}, n_{23}, etc., the expression for equivalent running time per cycle at torque M_{tm21} at worm gear speed n_{21}

$$T_{be} = T_1 + T_2 \left[\frac{n_{22}}{n_{21}}\right]\left[\frac{M_{t22}}{M_{tm21}}\right]^7 + T_3 \left[\frac{n_{23}}{n_{21}}\right]\left[\frac{M_{t23}}{M_{tm21}}\right]^7$$
$$+ \ldots \text{ for strength} \qquad (23\text{–}600a)$$

$$T_{we} = T_1 + T_2 \left[\frac{n_{22}}{n_{21}}\right]\frac{(M_{t22})^3}{M_{tm21}} + T_3 \left[\frac{n_{23}}{n_{21}}\right]\frac{(M_{t23})^3}{M_{tm21}}$$
$$+ \ldots \text{ for wear} \qquad (23\text{–}600b)$$

where each term represents the equivalent of the running time for the corresponding period

The total equivalent running time at torque M_{tm21} and worm gear speed n_{21}

$$T_{bet} = T_{be} N_c \text{ for strength} \qquad (23\text{–}601a)$$
$$T_{wet} = T_{we} N_c \text{ for wear} \qquad (23\text{–}601b)$$

where N_c = number of complete cycles expected during the life of a gear

Uniform variation of load and variable speed

When torque changes during any period τ from $M_{t\tau21}$ to $M_{t\tau22}$ at mean speed $n_{2\tau}$, the expression for the equivalent running time for that period at torque M_{tm21} and worm gear speed n_{21}

$$T_{be\tau} = \frac{T\tau}{8}\left[\frac{n_{2\tau}}{n_{21}}\right]\left[\frac{M_{t\tau b21}}{M_{tm21}} + \frac{M_{t\tau b22}}{M_{tm21}}\right]\left[\left(\frac{M_{t\tau b21}}{M_{tm21}}\right)^2\right.$$

$$\left. + \left(\frac{M_{t\tau b22}}{M_{tm21}}\right)^2\right] \times \left[\left(\frac{M_{t\tau b21}}{M_{tm21}}\right)^4 + \left(\frac{M_{t\tau b22}}{M_{tm21}}\right)^4\right]$$
$$\text{for strength} \qquad (23\text{–}602a)$$

$$T_{we\tau} = \frac{T\tau}{4}\left[\frac{n_{2\tau}}{n_{21}}\right]\left[\frac{M_{t\tau w21}}{M_{tm21}} + \frac{M_{t\tau w22}}{M_{tm21}}\right]\left[\left(\frac{M_{t\tau w21}}{M_{tm21}}\right)^2\right.$$

$$\left. + \left(\frac{M_{t\tau w22}}{M_{tm21}}\right)^2\right] \text{ for wear} \qquad (23\text{–}602b)$$

Heat dissipation from worm drives

Heat generated

$$H_g = \frac{\mu F_n v_m}{1000 \cos\gamma} = \frac{\mu F_n v_r}{1000} \quad \textbf{SI} \qquad (23\text{–}603a)$$

Particular	Formula

where H_g in kW, F_n in N, and v_m and v_r in m/s

$$= \frac{\mu F_n v_m}{102 \cos \gamma} = \frac{\mu F_n v_r}{102} \quad \textbf{Metric} \qquad (23\text{--}603b)$$

where H_g in kW, F_n in kgf, and v_m and v_r in m/s

$$= \frac{\mu F_n v_m}{427 \cos \gamma} = \frac{F \mu_n v_r}{427} = \frac{\mu F_n (\pi d_1' n_1')}{427} \quad \textbf{Metric}$$

$$(23\text{--}603c)$$

where H_g in kcal/s, F_n in kgf, v_m and v_r in m/s, d_1' in m, and n' in rps

$$= \frac{\mu F_n v_m}{33,000} = \frac{F_\mu v_r}{33,000} \quad \textbf{US Customary System units}$$

$$(23\text{--}603d)$$

where H_g in hp, F_n in lbf, v_m in ft/min, v_r in ft/min

F_μ = frictional force, lbf

$v_m = \dfrac{\pi d_1' n_1'}{1000}$ = pitch line velocity of worm, m/s

n_1' in rps, and d_1 in mm

$$v_m = \frac{\pi d_1' n_1}{12}$$

where v_m in ft/min, d_1 in in, n_1 in rpm

Heat dissipated

$$H_d = \frac{h_{cr} A}{1000} (t_g - t_a) \quad \text{kW} \qquad (23\text{--}603e)$$

where h_{cr} = coefficient of heat transfer, W/m^2 K (cal/s cm^2 °C)

Refer to Fig. 2–89 for h_{cr}

$$= \frac{778 h_{cr} A (t_g - t_a)}{60 \times 33,000} \quad \textbf{US Customary System units}$$

$$(23\text{--}603f)$$

where H_d in hp; t_g and t_a in °F

h_{cr} = heat-dissipating coefficient, Btu/h/(in^2)/(°F)

A = radiating area of smooth housing, m^2 (mm^2)

$= 114a^{1.7}$ **Metric** } for heavy-duty worm
where A in mm^2 } gear reducers with
and a in mm } integral worm

$= 14.4a^{1.7}$ **SI**
where A in m^2 and a in m

t_g = gear temperature, °C (°F)

t_a = ambient temperature, °C (°F)

$A = 43.2a^{1.7}$ **US Customary System units** (23–603g)

where A in in^2

Particular	Formula
The cooling rate for rectangular housing	$\dfrac{n_1}{84,200} + 0.01$ without fan (23–603h)
	$h_{cr} =$ **US Customary System units**
	$\dfrac{n_1}{51,600} + 0.01$ with fan (23–603i)
	where n_1 = speed of worm shaft, rpm
	The oil temperature should not exceed 82°C (180°F).
Transmitted power with respect to allowable heating of lubricant in the gearbox as per Niemann	$P \le 75a^2 k_{ca}k_i k_m k_o$ **SI** (23–603j) where P in kW and a in m
	$\le \left(\dfrac{a}{100}\right)^2 \dfrac{k_{ca}k_i k_m k_o}{1.34}$ **Metric** (23–603k) where P in kW and a in mm
	$\le \left(\dfrac{a}{100}\right)^2 \dfrac{k_{ca}k_i k_m k_o}{0.985}$ **Metric** (23–603l) where P in hp and a in mm
The coefficient of cooling of air for use in Eqs. (23–603j) to (23–603l)	$k_{ca} = \left(1 + \dfrac{x}{1+x}\right)\left(\dfrac{100}{T} + x\right)$ (23–603m) where $x = 1.4\sqrt[3]{(n_1/1000)^2}$ for gears without ventilation $= 3.1\sqrt[3]{(n_1/1000)^2}$ for gears with ventilation T = actual working time of gears per hour in percentage
For coefficient of transmission ratio (k_i), coefficient of contacting materials of worm gear (k_m) and coefficient of lubrication (k_o)	Refer to Tables 23–110 to 23–112.

TABLE 23–85
Worm and worm gear proportions (Fig. 23–86) as recommended by AGMA

Dimension	Symbol	Single and double threads		Triple and quadruple threads	
Worm (Fig. 23–86)					
Normal pressure angle, deg	α_n	$14\frac{1}{2}$	$14\frac{1}{2}$	20	20
Pitch diameter, bored for shaft, mm	d'_1	$2.4p + 28$	$2.4\pi m + 28$	$2.4p + 28$	$2.4\pi m + 28$
Pitch diameter, integral with shaft, mm	d'_1	$2.35p + 10.2$	$2.35\pi m + 10.2$	$2.35p + 10.2$	$2.35\pi m + 10.2$
Face length, mm	L_1	$(4.5 + 0.02z_1)p$	$(4.5 + 0.02z_1)\pi m$	$(4.5 + 0.02z_1)p$	$(4.5 + 0.02z_1)\pi m$
Depth of tooth, mm	h	$0.686p$	$0.686\pi m$	$0.623p$	$0.623\pi m$
Addendum, mm	h_a	$0.318p$	$0.318\pi m$	$0.286p$	$0.286\pi m$
Tip radius, mm	ρ_t	$0.05p$	$0.05\pi m$	$0.05p$	$0.05\pi m$
Hub diameter, mm	d_{h1}	$1.66p + 25.4$	$1.66\pi m + 25.4$	$1.726p + 25.4$	$1.726\pi m + 25.4$
Maximum bore for shaft, mm	d_{s1}	$p + 16$	$\pi m + 16$	$p + 16$	$\pi m + 16$
Worm gear (Fig. 23–87)					
Normal pressure angle, deg	α_n	$14\frac{1}{2}$	$14\frac{1}{2}$	20	20
Outside diameter, mm	d_{a1}	$d'_2 + 1.0135p$	$(z_2 + 1.0135\pi)m$	$d_2 + 0.8903p$	$(z_2 + 0.8903\pi)m$
Throat diameter, mm	d_{a2}	$d'_2 + 0.636p$	$(z_2 + 0.636\pi)m$	$d'_2 + 0.572p$	$(z_2 + 0.572\pi)m$
Face width, mm	b	$2.38p + 6.25$	$2.38\pi m + 6.25$	$2.15p + 5.00$	$2.15\pi m + 5.00$
Radius of gear face, mm	r_b	$0.882p + 14$	$0.882\pi m + 14$	$0.914p + 14$	$0.914\pi m + 14$
Radius of gear rim, mm	r_r	$2.2p + 14$	$2.2\pi m + 14$	$2.1p + 14$	$2.1\pi m + 14$
Radius of edge, mm	ρ_e	$0.25p$	$0.25\pi m$	$0.25p$	$0.25\pi m$

TABLE 23–86

Forces acting at the point of contact of worm thread and gear tooth (Fig. 23–97)

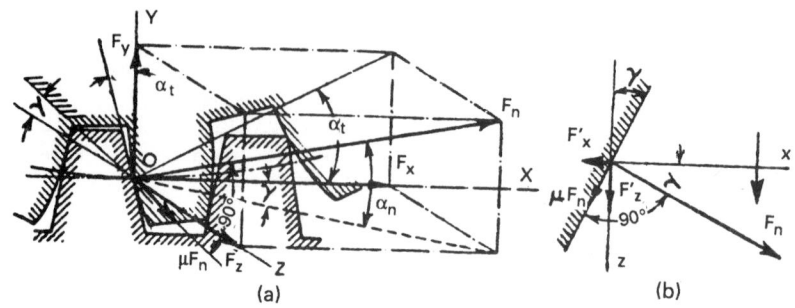

FIGURE 23–97

(a)　　　　　　　　(b)

Particular	Formula	
	Force analysis	
Resultant turning force on the worm gear, when the gear (wheel) turns the worm	$F_{2t} = F_x - F_x' = F_{1t} \dfrac{\cos \alpha_n \cos \gamma + \mu \sin \gamma}{\cos \alpha_n \sin \gamma - \mu \cos \gamma}$	(23–604a)
	where $\tan \alpha_n = \tan \alpha_t \cos \gamma$	
The separating or radial force	$F_r = F_y = F_n \sin \alpha_n$	(23–604b)
Resultant turning force on the worm, when the worm turns the gear (wheel)	$F_{1t} = F_z + F_z' = F_{2t} \dfrac{\cos \alpha_n \sin \gamma + \mu \cos \gamma}{\cos \alpha_n \cos \gamma - \mu \sin \gamma}$	(23–604c)
The frictional force	$F_\mu = \mu F_n = \dfrac{\mu F_{2t}}{\cos \alpha_n \cos \gamma - \mu \sin \gamma}$	(23–604d)
	Force analysis taking into consideration losses in thrust bearings	
Resultant turning force on the worm when the worm turns the worm gear (wheel)	$F_{1t} = F_z + F_z'$	
	$= F_{2t} \left\{ \dfrac{\cos \alpha_n [\sin \gamma + (\mu' d_{eb}/d_1') \cos \gamma] + \mu [\cos \gamma - (\mu' d_{eb}/d_1') \sin \gamma]}{\cos \alpha_n \cos \gamma - \mu \sin \gamma} \right\}$	(23–604e)
Resultant turning force on the worm gear when the gear (wheel) turns the worm	$F_{2t} = F_z - F_x'$	
	$= F_{1t} \left\{ \dfrac{\cos \alpha_n \cos \gamma + \mu \sin \gamma}{\cos \alpha_n [\sin \gamma - (\mu' d_{eb}/d_1') \cos \gamma] - \mu [\cos \gamma + (\mu' d_{eb}/d_1') \sin \gamma]} \right\}$	(23–604f)
The turning force required at the pitch radius of the worm when $\mu = \mu' = 0$	$F_{1t} = F_{2t} \tan \gamma$	(23–604g)
	Efficiency	
Efficiency when the worm drives the worm gear	$\eta = \dfrac{\tan \gamma (\cos \alpha_n \cos \gamma - \mu \sin \gamma)}{\cos \alpha_n \sin \gamma + \mu \cos \gamma}$	(23–604h)
Efficiency when the worm gear drives the worm	$\eta = \dfrac{\cos \alpha_n \sin \gamma - \mu \cos \gamma}{\tan \gamma (\cos \alpha_n \cos \gamma + \mu \sin \gamma)}$	(23–604i)
Barr's formula for the efficiency of worm gearing	$\eta = \dfrac{\tan \gamma (1 - \mu \tan \gamma)}{\tan \gamma + \mu}$	(23–604j)
Lead angle which gives the best efficiency when worm drives the worm gear	$\gamma = 45° - \tfrac{1}{2} \tan^{-1}(\mu / \cos \alpha_n)$	(23–604k)
Lead angle which gives the best efficiency when the worm gear (wheel) drives the worm	$\gamma = 45° + \tfrac{1}{2} \tan^{-1}(\mu / \cos \alpha_n)$	(23–604l)

TABLE 23–86
Forces acting at the point of contact of worm thread and gear tooth (Fig. 23–97) (*Contd.*)

Particular	Formula	
Efficiency taking into consideration the losses in thrust bearings when worm drives the worm gear (wheel)	$\eta = \dfrac{\tan\gamma(\cos\alpha_n\cos\gamma - \mu\sin\gamma)}{\cos\alpha_n[\sin\gamma + (\mu'd_{eb}/d_1'\cos\gamma)] + \mu[\cos\gamma - (\mu'd_{eb}/d_1')\sin\gamma]}$	(23–604m)
Efficiency taking into consideration the losses in the thrust bearings when worm gear drives the worm	$\eta = \dfrac{\cos\alpha_n[\sin\gamma - (\mu'd_{eb}/d_1'\cos\gamma)] - \mu[\cos\gamma + (\mu'd_{eb}/d_1')\sin\gamma]}{\tan\gamma(\cos\alpha_n\cos\gamma + \mu\sin\gamma)}$	(23–604n)

TABLE 23–87
Bearing loads of worm gearing

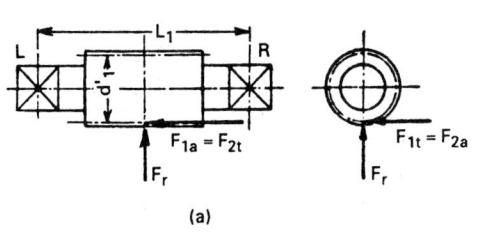

(a)

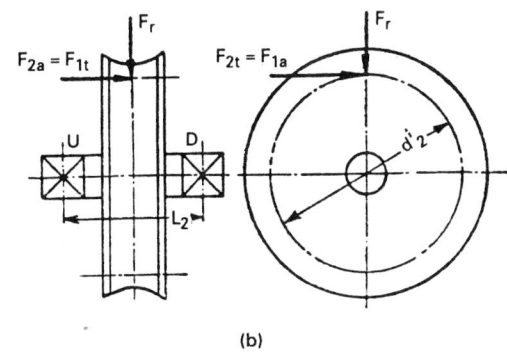

(b)

FIGURE 23–98

Particular	Formula	
Worm shaft (Fig. 23–98a)		
The resultant radial load at bearing L	$R_L = \sqrt{\dfrac{F_{1t}^2}{4} + \left(\dfrac{F_r}{2} - \dfrac{F_{2t}d_1'}{2L_1}\right)^2}$	(23–605a)
The resultant radial load at bearing R	$R_R = \sqrt{\dfrac{F_{1t}^2}{4} + \left(\dfrac{F_r}{2} + \dfrac{F_{2t}d_1'}{2L_1}\right)^2}$	(23–605b)
Worm gear shaft (Fig. 23–98b)		
The resultant radial load at bearing U	$R_U = \sqrt{\dfrac{F_{2t}^2}{4} + \left(\dfrac{F_r}{2} + \dfrac{F_{1t}d_2'}{2L_2}\right)^2}$	(23–605c)
The resultant radial load at bearing D	$R_D = \sqrt{\dfrac{F_{2t}^2}{4} + \left(\dfrac{F_r}{2} - \dfrac{F_{1t}d_2'}{2L_2}\right)^2}$	(23–605d)

TABLE 23-88
Values of transmission ratios and designation $z_1/z_2/q/m$ for general-purpose worm gear

Center distance	\ Transmission ratio 5.0	7.5	10.0	12.5	15.0	20.0	25.0	30.0	40.0	50.0	60.0
25	2/10/(11)/2.5	1/7/(18)/2	—	—	1/15/(18)/1.5	1/21/(18)/1.25	—	1/33/16/1	—	—	—
31.5	4/21/(11)/2	2/15/(11)/2.5	2/21/(11)/2	1/13/(18)/2	1/15/(11)/2.5	1/20/(11)/2	1/26/16/1.5	1/32/(18)/1.25	—	1/48/16/1	—
40	4/21/(11)/2.5	—	2/21/(11)/2.5	—	1/16/16/2.5	1/21/(11)/2.5	—	1/29/(11)/2	—	—	1/62/16/1
50	6/29/(11)/2.5	4/29/(11)/2.5	4/38/(11)/2	—	2/29/(11)/2.5	2/38/(11)/2	—	1/29/(11)/2.5	1/38/(11)/2	—	1/62/(18)/1.25
63	6/31/(11)/3	4/31/(11)/3	4/31/(11)/2.5	—	2/31/(11)/3	2/39/(11)/2.5	—	1/31/(11)/3	1/39/(11)/2.5	—	—
80	6/29/(11)/4	4/30/(10)/4	4/42/(11)/3	4/51/(11)/2	2/30/10/4	2/42/(11)/3	2/51/(11)/2	1/30/10/4	1/42/(11)/3	1/51/(11)/2	1/62/(18)/2
100	6/29/(11)/5	4/30/10/5	4/40/10/4	4/53/(11)/2.5	2/30/10/5	2/40/10/4	2/53/(11)/2.5	1/30/10/5	1/40/10/4	1/53/(11)/2.5	1/64/16/2.5
125	6/32/(10)/6	4/31/10/6	4/40/10/5	4/49/(18)/3	2/31/10/6	2/40/10/5	2/54/(11)/3	1/31/10/6	1/40/10/5	1/55/(11)/3	—
160	6/31/(9)/8	4/30/10/8	4/43/10/6	4/52/10/4	2/30/10/8	—	2/52/10/4	1/30/10/8	1/40/10/8	1/52/10/4	1/62/(18)/4
200	8/39/(11)/10	4/30/10/10	4/40/10/8	4/54/10/5	2/30/10/10	2/40/10/8	2/54/10/5	1/30/10/10	1/40/10/10	1/54/10/5	—
250	8/41/(11)/12	—	4/40/10/10	4/52/8/8	2/30/(9)/6	2/40/10/10	2/52/10/8	1/30/(9)/16	—	1/52/10/8	—
315	8/39/(11)/16	4/30/(9)/16	4/43/(9)/12	2/53/(10)/10	2/31/(9)/2	—	2/53/10/10	1/31/(9)/20	1/41/(9)/16	1/53/10/10	1/16/(18)/8
400	8/39/(11)/20	4/31/(9)/20	4/41/(9)/16	—	—	2/41/(9)/16	—	—	1/41/(9)/20	1/49/(18)/12	1/58/(9)/12
500	—	—	4/41/(9)/20	4/53/(9)/16	—	2/41/(9)/20	1/53/(9)/16	—	—	1/53/(9)/16	—

Note: The values within brackets are nonpreferred.
Source: IS 3734, 1966.

TABLE 23-89
Dimensions of worms

Module, m	Axial pitch, p_x	No. of starts, z_1	Diametral quotient, q	Reference diameter, d	Tip diameter, d_a	Root diameter, d_f	Lead angle, γ
1	3.142	1	16	16.00	18.00	13.60	3°–35′
1.25	3.927	1	(18)	22.50	25.00	19.50	3°–11′
1.50	4.712	1	16	24.00	27.00	20.40	3°–35′
1.50	4.712	1	(18)	27.00	30.00	23.40	3°–11′
2.0	6.283	1	(11)	22.00	26.00	17.20	5°–12′
2.0	6.283	1	(18)	36.00	40.00	31.20	3°–11′
2.0	6.283	2	(11)	22.00	26.00	17.20	10°–18′
2.0	6.283	4	(11)	22.00	26.00	17.20	19°–59′
2.5	7.854	1	(11)	27.50	32.50	21.50	5°–12′
2.5	7.854	1	16	40.50	45.00	34.00	3°–35′
2.5	7.854	2	(11)	27.50	32.50	21.50	10°–18′
2.5	7.854	4	(11)	27.50	32.50	21.50	19°–59′
2.5	7.854	6	(11)	27.50	32.50	22.84	28°–37′
3	9.425	1	(11)	33.00	39.00	25.80	5°–12′
3	9.425	1	(18)	54.00	60.00	46.80	3°–11′
3	9.425	2	(11)	33.00	39.00	25.80	10°–18′
3	9.425	4	(11)	33.00	39.00	25.80	19°–59′
3	9.425	6	(11)	33.00	39.00	27.41	28°–27′
4	12.566	1	10	40.00	48.00	30.40	5°–43′
4	12.566	1	(18)	72.00	80.00	62.40	3°–11′
4	12.566	2	10	40.00	48.00	30.40	11°–19′
4	12.566	4	10	40.00	48.00	30.40	21°–48′
4	12.566	6	(11)	44.00	52.00	36.55	28°–37′
5	15.708	1	10	50.00	60.00	38.00	5°–43′
5	15.708	1	(18)	90.00	100.00	78.00	3°–11′
5	15.708	2	10	50.00	60.00	38.00	11°–19′
5	15.708	4	10	50.00	60.00	38.00	21°–48′
5	15.708	6	(11)	55.00	65.00	45.69	28°–37′
6	18.850	1	10	60.00	72.00	45.60	5°–43′
6	18.85	1	(18)	108.00	120.00	93.60	3°–11′
6	18.85	2	10	60.00	72.00	45.60	11°–19′
6	18.85	4	10	60.00	72.00	45.60	21°–48′
6	18.85	6	10	60.00	72.00	49.36	30°–58′
8	25.133	1	10	80.00	96.00	60.80	5°–43′
8	25.133	1	(18)	144.00	160.00	124.80	3°–11′
8	25.133	2	10	80.00	96.00	60.80	11°–19′
8	25.133	4	10	80.00	96.00	60.80	21°–48′
8	25.133	4	8	64.00	80.00	48.52	26°–34′
8	25.133	6	(9)	72.00	88.00	58.71	33°–41′
8	25.133	8	(11)	88.00	104.00	75.53	36°–02′
10	31.416	1	10	100.00	120.00	76.00	5°–43′
10	31.416	1	(18)	180.00	200.00	156.00	3°–11′
10	31.416	2	10	100.00	120.00	76.00	11°–19′
10	31.416	4	10	100.00	120.00	76.00	21°–48′
10	31.416	8	(11)	110.00	130.00	94.42	36°–02′
12	37.699	1	(9)	108.00	132.00	79.20	6°–20′
12	37.699	1	(18)	216.00	240.00	187.20	3°–11′
12	37.699	2	(9)	108.00	132.00	79.20	12°–32′
12	37.699	4	(9)	108.00	132.00	79.20	23°–58′
12	37.699	8	(11)	132.00	156.00	113.30	36°–02′
16	50.266	1	(9)	144.00	176.00	105.60	6°–20′
16	50.266	2	(9)	144.00	176.00	105.60	12°–32′
16	50.266	4	(9)	144.00	176.00	105.60	23°–58′
16	50.266	8	(11)	176.00	208.00	151.06	36°–02′
20	62.832	1	(9)	180.00	220.00	132.00	6°–20′
20	62.832	2	(9)	180.00	220.00	132.00	12°–32′
20	62.832	4	(9)	180.00	220.00	132.00	23°–58′
20	62.832	8	(11)	220.00	260.00	188.83	36°–02′

Note: The values given within brackets are nonpreferred.
Source: IS 3734, 1966.

TABLE 23–90
Recommended pressure angles and tooth depths for worm gearing

Lead angle, γ, deg	Pressure angle, α, deg	Addendum, h_a	Dedendum, h_f
0–15	$14\frac{1}{2}$	$0.3683p_x$	$0.3683p_x$
15–30	20	$0.3683p_x$	$0.3683p_x$
30–35	25	$0.2865p_x$	$0.3314p_x$
35–40	25	$0.2546p_x$	$0.2947p_x$
40–45	30	$0.2228p_x$	$0.2578p_x$

TABLE 23–91
Worm data

Velocity ratio	No. of threads or starts
≥ 20	Single
12–36	Double
8–12	Triple
6–12	Quadruple
4–10	Sextuple

TABLE 23–92
AGMA standard axial pitches

	p_x, mm		
6.35	12.70	25.40	44.45
7.94	15.87	31.75	50.80
9.52	19.05	38.10	

TABLE 23–93
Face width of the wheel (gear)

No. of starts, z_1	Face width, b
1	$0.75d_1$
2 or 3	$0.75d_1$
4	$0.67d_1$

TABLE 23–94
Length of the worm, L_1 (Fig. 23–86)

Correction factor, x	No. of starts, z_1	
	1 or 2	3 or 4
0	$L_1 \geq (11 + 0.06z_2)m_x$	$L_1 \geq (12.5 + 0.09z_2)m_x$
−0.5	$L_1 \geq (8 + 0.06z_2)m_x$	$L_1 \geq (9.5 + 0.09z_2)m_x$
−1.0	$L_1 \geq (10.5 + z_1)m_x$	$L_1 \geq (10.5 + z_1)m_x$
+0.5	$L_1 \geq (11 + 0.1z_2)m_x$	$L_1 \geq (12.5 + 0.1z_2)m_x$
1	$L_1 \geq (12 + 0.1z_2)m_x$	$L_1 \geq (13 + 0.1z_2)m_x$

TABLE 23–95
Values of Lewis form factor, y, for worm gears

Normal pressure angle, deg	Form factor, y
$14\frac{1}{2}$	0.100
20	0.125
25	0.150
30	0.175

TABLE 23–96
Allowable static stress (σ_o) for worm gear for use in Eq. (23–567)

Material	σ_o	
	kpsi	MPa
Ordinary cast iron	10.0	69.00
High-grade cast iron or semisteel	15.0	103.50
Bakelite, Textolite, rawhide, etc.	6.0	41.40
Leaded gun metal, SAE 63	8.0	55.13
Manganese bronze, SAE 43	20.0	137.90
Phosphor bronze, SAE 65	15.0	103.50

TABLE 23–97
Coefficient of friction for worm and crossed-helical gear

v_r, m/s	μ	v_r, m/s	μ	v_r, m/s	μ
0.00	0.2000	0.75	0.0408	12.50	0.0650
0.05	0.1209	1.00	0.0365	15.0	0.0712
0.10	0.0993	1.50	0.0330	20.0	0.0822
0.15	0.0859	2.00	0.0327	25.0	0.0919
0.20	0.0764	2.50	0.0358	30.0	0.1007
0.25	0.0693	3.75	0.0375	35.0	0.1088
0.30	0.0637	5.00	0.0420	40.0	0.1168
0.35	0.0591	6.25	0.0465	45.0	0.1233
0.40	0.0553	7.50	0.0506	50.0	0.1300
0.45	0.0552	8.75	0.0545		
0.50	0.0495	10.00	0.0582		

Note: v_r, rubbing velocity

TABLE 23–98
Value of angles of friction, ρ

v_r, m/s	ρ
0.1	4°30′–5°10′
0.5	3°10′–3°40′
1.0	2°30′–3°10′
1.5	2°20′–2°50′
2.0	2°00′–2°30′
2.0	1°40′–2°20′
3.5	1°30′–2°00′
4.0	1°20′–1°40′
7.0	1°00′–1°30′
10.0	0°55′–1°20′

TABLE 23–99
Lead angle, γ

No. of starts, z_1	Number of modules in the pitch diameter of worm, m_d'					
	8	**9**	**10**	**11**	**12**	**13**
1	7° 7′30″	6°20′25″	5°42′38″	5°11′40″	4°45′49″	4°23′55″
2	14°02′10″	12°31′44″	11°18′36″	10°18′17″	9°27′44″	8°44′46″
3	20°33′22″	18°26′06″	16°41′57″	15°15′18″	14° 2′10″	12°59′41″
4	26°33′54″	23°57′45″	21°48′05″	19°58′59″	18°26′06″	17° 6′10″

TABLE 23–100
Load stress factors, K, for worm gears for use in Eq. (23–572)

Material		Load stress factor, K					
		$\gamma = 0° - 10°$		$\gamma = 10° - 25°$		$\gamma = \geq 25°$	
Worm	Gear	MPa	psi	MPa	psi	MPa	psi
Steel $250H_B$	Phosphor bronze	0.414	60	0.517	75	0.621	90
Hardened steel	Phosphor bronze	0.552	80	0.690	100	0.827	120
Hardened steel	Chilled phosphor bronze	0.827	120	1.034	150	1.241	180
Hardened steel	Antimony bronze	0.827	120	1.034	150	1.241	180
Cast iron	Phosphor bronze	1.034	150	1.290	185	1.551	225

TABLE 23–101
Service factors for cylindrical worm-gear units, C_s, for use in Eq. (23–573)

Prime mover	Duration of serivce per day, h	Driven mchine load classification		
		Uniform	Moderate shock	Heavy shock
	Occasional $\frac{1}{2}$	0.80	0.90	1.00
Electric motor	Intermittent 2	0.90	1.00	1.25
	10	1.00	1.25	1.50
	24	1.25	1.50	1.75
Multicylinder	Occasional $\frac{1}{2}$	0.90	1.00	1.25
internal	Intermittent 2	1.00	1.25	1.50
combustion	10	1.25	1.50	1.75
engine	24	1.50	1.75	2.00
Single-cylinder	Occasional $\frac{1}{2}$	1.00	1.25	1.50
internal	Intermittent 2	1.25	1.50	1.75
combustion	10	1.50	1.75	2.00
engine	24	1.75	2.00	2.25
	Occasional $\frac{1}{2}$[a]	0.90	1.00	1.25
Electric	Intermittent 2[a]	1.00	1.25	1.50
motor	10[a]	1.25	1.50	1.75
	24[a]	1.50	1.75	2.00

Notes: (1) Time specified for intermittent and occasional service refers to total operating time per day; (2) term "frequent starts and stops" refers to more than 10 starts per hour.
[a] These service factors apply for applications involving frequent starts and stops.
Source: D. W. Dudley, *Handbook of Practical Gear Design*, with permission from McGraw-Hill Publishing Company, 1984.

TABLE 23–102
Permissible surface pressures, σ_{ca}, for use in Eq. (23–573)

Material			Number of teeth in gear							
Worm	Gear	Units	10	20	30	40	50	60	70	80 and over
SAE 1020 steel		psi	65	270	470	755	896	1080	1250	1350
untreated	Cast iron	MPa	0.44	1.86	3.23	5.20	6.18	7.46	8.63	9.32
SAE 1020 steel	Bronze SAE 63	psi	115	340	670	1068	1350	1810	1890	2000
untreated	sand-cast	MPa	0.79	2.35	4.61	7.36	9.32	12.47	13.05	13.83
SAE 1040 steel	Bronze SAE 63	psi	170	510	940	1595	2000	2690	2845	4300
heat-treated ground	sand-cast	MPa	1.18	3.53	6.47	10.99	13.83	18.54	19.62	29.70
	Bronze SAE 65	psi	270	725	1250	2150	2700	3245	3800	4000
	sand-cast	MPa	1.86	5.00	8.63	14.81	18.54	22.37	26.19	27.57
0.10C alloy	Bronze SAE 65	psi	315	925	1720	2945	3715	4500	5250	5500
steel	chill-cast	MPa	2.16	6.38	11.87	20.30	25.60	31.00	36.20	37.96
carburized,	Nickel bronze	psi	425	1125	2970	3340	4500	5450	6345	6700
hardened,	sand-cast	MPa	2.94	7.75	20.50	23.05	31.00	37.57	43.75	46.20
and ground	Nickel bronze	psi	455	1350	2500	4300	5390	6500	7600	8000
	chill-cast	MPa	3.14	9.32	17.27	29.63	37.18	44.83	52.39	55.13

Values tabulated are for $14\frac{1}{2}°$ pressure angles. Multiply by 1.05 for 20° pressure angles and by 1.10 for 30° pressure angles.

TABLE 23–103
Material factor C_s for cylindrical worm gears for use in Eq. (23–575)

For units of 75 mm (3 in) center-to-center distance about 1 m (40 in)

Gear pitch diameter		Sand-cast	Static chill-cast	Centrifugal-cast
mm	in			
65	2.5	1000	—	—
75	3	960	—	—
100	4	900	—	—
125	5	855	—	—
150	6	820	—	—
175	7	790	—	—
200	8	760	1000	—
250	10	715	955	—
375	15	630	875	—
505	20	570	815	—
635	25	525	770	1000
760	30	—	740	985
1015	40	—	680	960
1270	50	—	635	945
1775	70	—	570	920

For units with < 75 mm (3 in) center distance

Center distance		Maximum C_s value
mm	in	
12	0.5	725
25	1.0	735
38	1.5	760
50	2.0	800
63	2.5	880
75	3.0	1000

Notes: (1) For bronze worm gear and steel worm with at least HV 655 (HRC 58) surface hardness; (2) See Table 23–102 for worm-gear material data; (3) sliding velocity not to exceed 30 m/s (6000 ft/min); worm speed not more than 3600 rpm.
Source: D. W. Dudley, *Handbook of Practical Gear Design*, with permission from McGraw-Hill Publishing Company, 1984.

TABLE 23–104
Velocity factor C_v for cylindrical worm gears

Sliding velocity		Velocity factor, C_v	Sliding velocity		Velocity factor, C_v
m/s	ft/min		m/s	ft/min	
0.005	1	0.649	3.0	600	0.340
0.025	5	0.647	3.5	700	0.310
0.050	10	0.644	4.0	800	0.289
0.10	20	0.638	4.5	900	0.272
0.15	30	0.631	5.0	1000	0.258
0.20	40	0.625	6.0	1200	0.235
0.30	60	0.613	7.0	1400	0.216
0.40	80	0.600	8.0	1600	0.200
0.50	100	0.588	9.0	1800	0.187
0.75	150	0.558	10.0	2000	0.175
1.00	200	0.528	11.0	2200	0.165
1.25	250	0.500	12.0	2400	0.156
1.50	300	0.472	13.0	2600	0.148
1.75	350	0.446	14.0	2800	0.140
2.00	400	0.421	15.0	3000	0.134
2.25	450	0.398	20.0	4000	0.106
2.50	500	0.378	25.0	5000	0.089
2.75	550	0.358	30.0	6000	0.079

Source: D. W. Dudley, *Handbook of Practical Gear Design*, with permission from McGraw-Hill Publishing Company, 1984.

TABLE 23–105
Ratio correction factors, C_m

Ratio range (i)	C_m for ratios 3 to 19.9									
	0	**0.1**	**0.2**	**0.3**	**0.4**	**0.5**	**0.6**	**0.7**	**0.8**	**0.9**
3–3.9	0.500	0.511	0.522	0.532	0.543	0.554	0.562	0.570	0.577	0.585
4–4.9	0.593	0.598	0.604	0.609	0.615	0.620	0.625	0.630	0.635	0.640
5–5.9	0.645	0.649	0.652	0.656	0.659	0.663	0.666	0.669	0.673	0.676
6–6.9	0.679	0.682	0.685	0.688	0.691	0.694	0.696	0.699	0.701	0.704
7–7.9	0.706	0.708	0.710	0.711	0.713	0.715	0.717	0.719	0.720	0.722
8–8.9	0.724	0.726	0.728	0.730	0.732	0.734	0.736	0.738	0.740	0.742
9–9.9	0.744	0.746	0.747	0.749	0.750	0.752	0.754	0.755	0.757	0.758
10–10.9	0.760	0.761	0.763	0.764	0.765	0.767	0.768	0.769	0.770	0.782
11–11.9	0.773	0.774	0.775	0.776	0.777	0.778	0.779	0.780	0.781	0.782
12–12.9	0.783	0.784	0.785	0.786	0.787	0.788	0.788	0.789	0.790	0.791
13–13.9	0.792	0.793	0.794	0.795	0.795	0.796	0.796	0.797	0.798	0.798
14–14.9	0.799	0.800	0.800	0.801	0.801	0.802	0.803	0.803	0.804	0.804
15–15.9	0.805	0.805	0.806	0.806	0.807	0.807	0.807	0.807	0.808	0.808
16–16.9	0.809	0.809	0.810	0.810	0.811	0.811	0.811	0.811	0.812	0.812
17–17.9	0.813	0.813	0.814	0.814	0.814	0.814	0.815	0.815	0.815	0.815
18–18.9	0.816	0.816	0.816	0.817	0.817	0.817	0.817	0.817	0.818	0.818
19–19.9	0.818	0.818	0.818	0.818	0.819	0.819	0.819	0.819	0.820	0.820

Ratio range (i)	C_m for ratios 20 to 100									
	0	**1**	**2**	**3**	**4**	**5**	**6**	**7**	**8**	**9**
20–29	0.820	0.822	0.823	0.824	0.825	0.825	0.826	0.826	0.826	0.826
30–39	0.825	0.825	0.825	0.824	0.823	0.822	0.821	0.820	0.818	0.816
40–49	0.815	0.812	0.810	0.807	0.804	0.802	0.799	0.796	0.792	0.789
50–59	0.785	0.782	0.799	0.775	0.771	0.767	0.763	0.759	0.754	0.750
60–69	0.745	0.740	0.735	0.729	0.724	0.718	0.712	0.706	0.700	0.694
70–79	0.687	0.681	0.675	0.669	0.662	0.655	0.648	0.642	0.635	0.629
80–89	0.622	0.615	0.609	0.602	0.595	0.589	0.582	0.575	0.568	0.562
90–99	0.555	0.549	0.542	0.536	0.529	0.523	0.516	0.510	0.503	0.497
100	0.490									

TABLE 23–106
Minimum normal backlash allowance[a]

Range of diametral pitch, teeth/in	Allowance, in (for AGMA quality number range)	
	4 to 9	**10 to 13**
1.00–1.25	0.032	0.024
1.25–1.50	0.027	0.020
1.50–2.00	0.020	0.015
2.00–2.50	0.016	0.012
2.50–3.00	0.013	0.010
3.00–4.00	0.010	0.008
4.00–5.00	0.008	0.006
5.00–6.00	0.006	0.005
6.00–8.00	0.005	0.004
8.00–10.00	0.004	0.003
10.00–12.00	0.003	0.002
12.00–16.00	0.003	0.002
16.00–20.00	0.002	0.001
20.00–25.00	0.002	0.001

[a] Measured at outer cone in inches.
Source: D. W. Dudley, *Handbook of Practical Gear Design*, courtesy McGraw-Hill Book Company, New York, 1954.

TABLE 23–107
Allowable stresses for use in Eqs. (23–581) and (23–582)

Worm-gear material	Grade	Minimum tensile strength, st		Brinell hardness, H_B	Allowable bending stress, σ_{ab}		Allowable contact stress, σ_{ac}	
		MPa	**kpsi**		**MPa**	**kpsi**	**MPa**	**kpsi**
Phosphor	Sand-cast	83–100	12–14	70	48.3	7.0	10.3	1.5
Bronze	Chill-cast	103–117	15–17	82	58.6	8.5	12.4	1.8
BS 1400 PB2	Centrifugally cast	117–131	17–19	90	68.9	10.0	15.2	2.2
Cast iron	Ordinary grade	83–100	12–14	150	41.4	6.0	6.89	1.0
BS 821	Medium grade	110–124	16–18	165	51.7	7.5	6.89	1.0
	High grade	152–165	22–24	180	68.9	10.0	6.89	1.0

Note: The worm should be of steel harder than the wheel material.
Source: M. J. Neale, *Tribology Handbook*, Section A–24, Butterworths, London, 1973.

TABLE 23–108
Stress factors K_b and K_w for worm gears for use in Eqs. (23–584) to (23–590)

Material	Indian Standard reference	Bending stress factor, K_b	Surface stress factors, K_w, when running with				
			A	B	C	D	E
Phosphor bronze centrifugally cast	IS 28, 1958 (revised)	6.92		0.89	0.89	0.97	1.60
A Phosphor bronze sand cast chilled		5.80		0.67	0.67	0.75	1.30
Phosphor bronze sand cast		4.70		0.49	0.49	0.57	1.10
B Gray cast iron	Grade 20 IS 210, 1962 (revised)	4.30	0.67	0.445	0.445	0.445	0.55
C 0.4% Carbon steel normalized	C40 IS 1570, 1961	14.96	1.75	0.75			
D 0.55% Carbon steel normalized	C 55 Mn 75 IS 1570, 1961	18.35	1.62	0.89			
Carbon steel case hardened	C 10, C 14 IS 4432, 1967	29.93	5.24	3.20			1.60
E 3% Nickel and nickel molybdenum	16 Ni80 Cr 60 20 Ni 2 Mo 25	35.17	5.24	3.20			1.60
Case hardened steel 3.5% Nickel chromium	IS 4432, 1967 13 Ni3 Cr 80 15 Ni 4 Cr 11 IS 1570, 1961	37.40	6.05	3.20			1.60

TABLE 23–109
Worm gear zone factor Y_z for use in Eqs. (23–587) to (23–590)

q	6	6.5	7	7.5	8	8.5	9	9.5	10	11	12	13	14	16	17	18	20
z								Y_z									
1	1.045	1.048	1.052	1.065	1.084	1.107	1.128	1.137	1.143	1.160	1.202	1.260	1.318	1.374	1.402	1.437	1.508
2	0.991	1.028	1.055	1.099	1.144	1.183	1.214	1.223	1.231	1.250	1.280	1.320	1.360	1.418	1.447	1.490	1.575
3	0.822	0.890	0.969	1.109	1.209	1.260	1.305	1.333	1.350	1.365	1.393	1.422	1.442	1.502	1.532	1.580	1.674
4	0.826	0.883	0.981	1.098	1.204	1.301	1.380	1.428	1.460	1.490	1.515	1.545	1.570	1.634	1.666	1.710	1.798
5	0.947	0.991	1.050	1.122	1.216	1.315	1.417	1.490	1.550	1.610	1.632	1.652	1.675	1.735	1.765	1.805	1.886
6	1.132	1.145	1.172	1.220	1.287	1.350	1.438	1.521	1.588	1.675	1.694	1.714	1.733	1.789	1.818	1.854	1.928
7			1.316	1.340	1.370	1.405	1.452	1.540	1.614	1.704	1.725	1.740	1.760	1.817	1.846	1.880	1.950
8				1.437	1.462	1.500	1.557	1.623	1.715	1.731	1.753	1.778	1.838	1.868	1.898	1.960
9						1.573	1.604	1.648	1.720	1.743	1.767	1.790	1.850	1.880	1.910	1.970
10								1.680	1.728	1.748	1.773	1.798	1.858	1.888	1.920	1.980
11									1.732	1.753	1.777	1.802	1.862	1.892	1.924	1.987
12										1.760	1.780	1.806	1.866	1.895	1.927	1.992
13											1.784	1.806	1.867	1.898	1.931	1.998
14												1.811	1.871	1.900	1.933	2.000

Notes: (1) The values are based on $b_e = 2m\sqrt{q+1}$, symmetrical about the center plane of the worm gear; (2) for smaller face widths the value of Y_z must be reduced proportionately; (3) when it is necessary to obtain greater load capacity, the worm-gear face width may be increased up to a maximum of $2.3m\sqrt{q+1}$ and the zone factor increased proportionately; (4) the table applies to worm gears having 30 teeth: variations in the number of teeth produce negligible changes in the value of Y_z.
Source: IS 7443–1974.

TABLE 23–110
Coefficient transmission ratio, k_i

i	5	7.5	10	15	20	25	30	40	50	60
k_i	1.16	1.10	1.00	0.81	0.68	0.59	0.52	0.41	0.32	0.28

TABLE 23–112
Coefficient of lubrication, k_o

Design layout	k_o
Worm under the worm gear, worm submerged in lubricant	1
Worm above worm gear, lubricant is transferred by worm gear	0.8
Cooling of lubricant by artificial methods	> 1

TABLE 23–111
Coefficient of contacting materials of worm and worm gear, k_m

Worm	Worm gear	k_m
Carbon steel	Tin bronze or phosphor bronze	1.0
	Gray cast iron	0.8
Case hardening steel	Tin bronze or phosphor bronze	0.67
	Gray cast iron	0.52

23.6   CROSSED HELICAL GEARS (1, 2)

SYMBOLS

a	center distance, m (in)
b	face width, m (in)
C_s	service factor
C_v	velocity factor
C_w	wear and lubrication factor
d	reference diameter, m (in)
d_d	reference diameter of the driver, m (in)
d'_d	pitch diameter of the driver, m (in)
d_f	reference diameter of the follower, m (in)
d'_f	pitch diameter of the follower, m (in)
E_d, E_f	modulus of elasticity of materials of the driver and the follower, respectively, GPa (psi)
E_o	equivalent modulus of elasticity, GPa (psi)
F	force, kN (lbf)
F_{ad}	axial thrust along the shaft axis of the driver, kN (lbf)
F_{af}	axial thrust along the shaft axis of the follower, kN (lbf)
F_n	normal force (also with suffixes), kN (lbf)
F_t	tangential force at the pitch point, kN (lbf)
F_{td}	turning force acting at the pitch point of the driver, kN (lbf)
F_{tf}	turning force acting at the pitch point of the follower, kN (lbf)
F_w	limit wear load, kN (lbf)
i	transmission or speed ratio
$i_p = \rho_d/\rho_f$	ratio of the radius of curvature of the driver to the follower
k'_1	the coefficient which depends on the angle $\Sigma' = \beta'_d + \beta'_f$ (Refer to Fig. 23–103.)
k'_2	the coefficient which depends on β_f and $i = z_f/z_d$ [Refer to Eq. (23–645).]
K	load stress factor for use in Eq. (23–633)
K_{ba}	allowable bending stress factor for use in Eq. (23–635), MPa (psi)
K_c	surface contact stress factor, MPa (psi)
K_{ca}	allowable surface contact stress factor for use in Eq. (23–644), MPa (psi)
K_{ob}, K_{oc}	factors from Table 23–115 for use in Eqs. (23–635e) and (23–644b), respectively
m	module, m (in)
m_n	module on normal section, m (in)
M_t	torque (also with subscripts), N m (lbf in)
n	speed (also with suffixes), rpm
n'	speed, rps
P	power (also with suffixes), kW (hp)
p	circular pitch (also with suffixes), m (in)
p_n	normal circular pitch (also with suffixes), m (in)
p_z	lead (also with suffixes), m (in)
P	diametral pitch (also with suffixes), m^{-1} (in^{-1})
P_n	diametral pitch on normal section (also with suffixes), m^{-1} (in^{-1})

Q	a factor [Refer to Eqs. (23–633) and (23–634).]
v	velocity, m/s (ft/min)
v_m	mean velocity, m/s (ft/min)
v_r	rubbing velocity, m/s (ft/min)
$Y = \pi y$	modified Lewis form factor
z	number of teeth (also with suffixes)
z_v	virtual or formative number of teeth (also with suffixes)
α	pressure angle, deg
α_n	normal pressure angle (also with suffixes), deg
ϵ	contact ratio
β	helix angle (also with suffixes), deg
γ	lead angle (also with suffixes), deg
ρ	radius of curvature of tooth profile (also with suffixes), m (in)
ρ	friction angle, deg
ρ_o	relative radius of curvature, m (in)
μ	coefficient of friction
Σ	shaft angle, deg
σ	stress (also with suffixes), MPa (psi)
σ_c	contact surface (compressive) stress (also with suffixes), MPa (psi)
σ_o	design static stress, MPa (psi)

SUFFIXES

1	pinion
2	wheel
a	allowable or admissible
b	bending
c	contact or compressive
d	driver
f	follower
min	minimum
max	maximum

Particular	Formula	
The relation between the normal diametral and normal circular pitch	$P_n p_n = \pi$	(23–606)
The circular pitch of the driver	$p_d = \dfrac{p_n}{\cos \beta_d}$	(23–607)
The circular pitch of the follower	$p_f = \dfrac{p_n}{\cos \beta_f}$	(23–608)
The diametral pitch of the driver	$P_d = P_n \cos \beta_d$	(23–609)
The diametral pitch of the follower	$P_f = P_n \cos \beta_f$	(23–610)
The pitch diameter of the driver	$d_d' = \dfrac{z_d p_n}{\pi \cos \beta_d} = \dfrac{z_d p_d}{\pi} = z_d \boldsymbol{m}_d = z_d \dfrac{\boldsymbol{m}_n}{\cos \beta_d}$	(23–611)
The relation between normal circular pitch (p_n) and circular pitches p_d and p_f (Fig. 23–99)	$p_n = p_d \cos \beta_d = p_f \cos \beta_f$	(23–612)

Particular	Formula

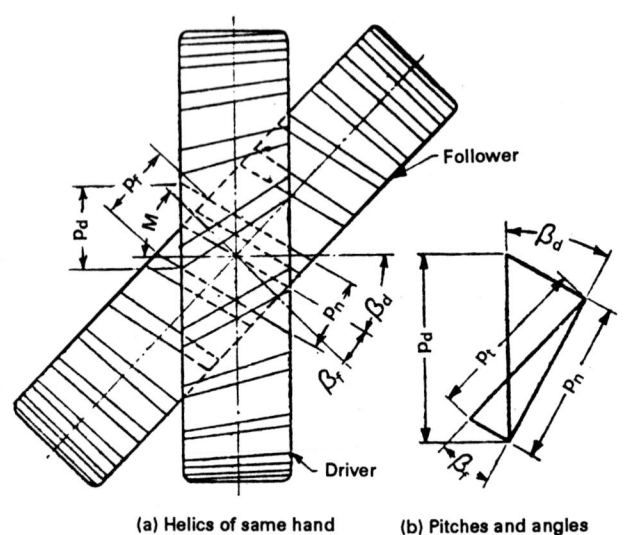

(c) Crossed Helical Gears with
Non-Parallel Shaft
(Courtesy Secony-Vacuum Oil Company)

(a) Helics of same hand   (b) Pitches and angles relation

FIGURE 23–99 Pitches and angles in crossed helical gears. (*Panels* a *and* b *courtesy of V. L. Maleev and J. B. Hartman, Machine Design; International Textbook Company, Scranton, Pennsylvania, 1954.*)

The pitch diameter of the follower

$$d'_f = \frac{z_f p_n}{\pi \cos \beta_f} = \frac{z_f p_f}{\pi} = z_f m_f = z_f \frac{m_n}{\cos \beta_f} = 2a - d'_d$$

(23–613)

The speed ratio or transmission ratio

$$i = \frac{n_d}{n_f} = \frac{z_f}{z_d} = \frac{d_f \cos \beta_f}{d_d \cos \beta_d} = \left(\frac{2a}{d'_d} - 1\right) \frac{\cos \beta_f}{\cos \beta_d}$$

(23–614)

The shaft angle (Fig. 23–99)

$\Sigma = \beta_d + \beta_f$, when both helices are of the same hand
(23–615)

$\Sigma = \beta_f - \beta_d$, when both helices are of the opposite hand
(23–616)

The center distance

$$a = (d'_d + d'_f)/2 = 0.5 m_n \left(\frac{z_d}{\cos \beta_d} + \frac{z_f}{\cos \beta_f}\right)$$

(23–617a)

$$= \frac{1}{2P_n} \left(\frac{z_d}{\cos \beta_d} + \frac{z_f}{\cos \beta_f}\right)$$

(23–617b)

Formative number of teeth of the driver $z_{vd} = z_d / \cos^3 \beta_d$ (23–618)

Formative number of teeth of the follower $z_{vf} = z_f / \cos^3 \beta_f$ (23–619)

Particular	Formula	
The module on normal section	$$\mathbf{m}_n = \frac{d_d}{z_d}\cos\beta_d = \frac{d_f}{z_f}\cos\beta_f$$	(23–620)

FORCE ANALYSIS AND EFFICIENCY

For complete force analysis acting between teeth of crossed helical gears at point of contact

Refer to Figs. 23–102 and 23–101 and Table 23–113.

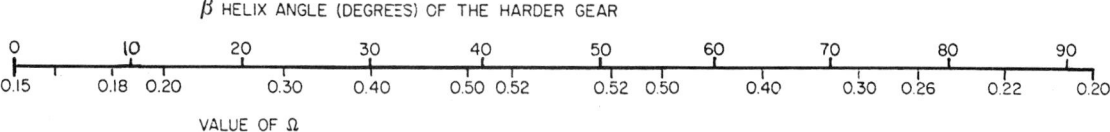

FIGURE 23–100(a) Dimensionless factor Ω for use in Eq. (23–649). (*Courtesy of British Standards Institution, BS 721, M. J. Neal*, Tribology Handbook, *Section A 24, Butterworth, 1973.*)

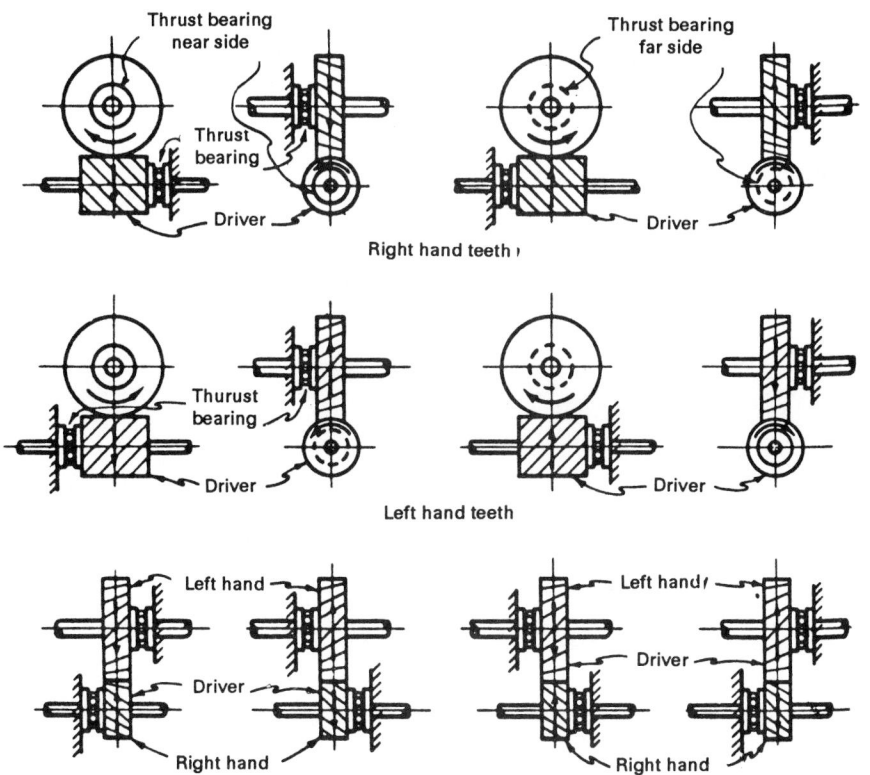

FIGURE 23–100(b) Thrust load and direction of rotation for crossed helical gears. (*Courtesy M. F. Spotts*, Design of Machine Elements, *5th ed., Prentice-Hall of India Private Ltd., New Delhi, 1978.*)

Particular	Formula

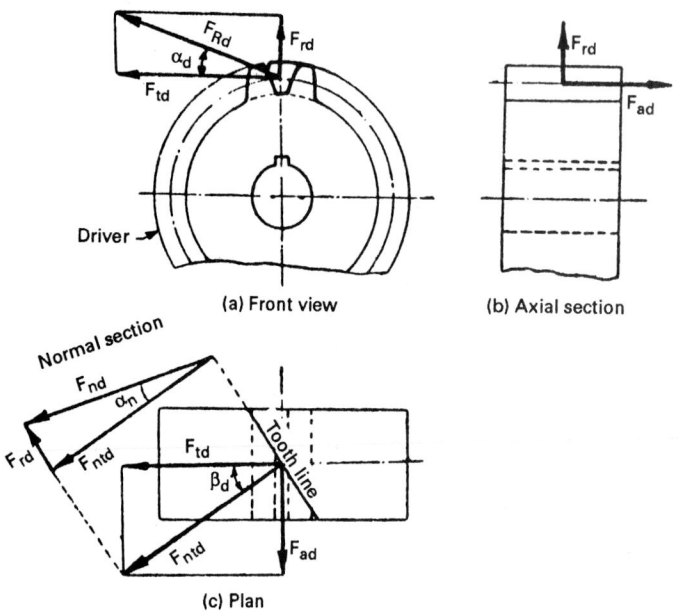

(a) Front view   **(b) Axial section**

(c) Plan

FIGURE 23–101   Components of normal force on driver of crossed helical gears.

The tangential force at the pitch point of the driver

$$F_{td} = \frac{2M_{td}C_s}{d'_d} \tag{23–621}$$

where C_s = shock or impact coefficient taken from Table 23–54

The efficiency

$$\eta = \frac{\cos\alpha_n - \mu\tan\beta_f}{\cos\alpha_n + \mu\tan\beta_d} \tag{23–622}$$

FACE WIDTH

The minimum face width of tooth of driver

$$b_{d(min)} \le \frac{2m_n\sin\beta_d}{\tan\alpha_n} \tag{23–623a}$$

$$\le \frac{2\sin\beta_d}{\tan\alpha_n\,P_n} \tag{23–623b}$$

The minimum face width of tooth of follower

$$b_{f(min)} \le \frac{2m_n\sin\beta_f}{\tan\alpha_n} \tag{23–624a}$$

$$\le \frac{2\sin\beta_f}{\tan\alpha_n\,P_n} \tag{23–624b}$$

Particular	Formula
The minimum face width of tooth of driver for $\alpha_n = 20°$	$b_{d(min)} \leq 5.5 m_n \sin \beta_d$ (23–625a) $\leq 5.5 \sin \beta_d / P_n$ (23–625b)
The minimum face width of tooth of follower for $\alpha_n = 20°$	$b_{f(min)} \leq 5.5 m_n \sin \beta_f$ (23–626a) $\leq 5.5 \sin \beta_f / P_n$ (23–626b)

The contact ratio for crossed helical gears

$$\epsilon_s = \epsilon + \frac{b_{d(min)}}{\pi m_d} \tan \beta_d + \frac{b_{f(min)}}{\pi m_f} \tan \beta_f \qquad (23\text{–}627)$$

where ϵ = contact ratio as given in Eq. (23–59)

The velocity at pitch point of driver

$$v_d = \frac{\pi d'_d n_d}{60 \times 1000} = \frac{d'_d n_d}{19{,}100} = v_f \frac{\cos \beta_f}{\cos \beta_d} \qquad (23\text{–}628)$$

where v_d and v_f in m/s, d'_d in mm, and n_d in rpm

The velocity at pitch point of follower

$$v_f = \frac{\pi d'_f n_f}{60 \times 1000} = \frac{d'_f n_f}{19{,}100} = v_d \frac{\cos \beta_d}{\cos \beta_f} \qquad (23\text{–}629)$$

where v_f and v_d in m/s, d'_f in mm, and n_f in rpm

The tangential force at pitch point of the driver

$$F_{td} = \frac{2M_{td}}{d'_d} = \frac{2000 P_d}{\omega_d d'_d} \quad \textbf{SI} \qquad (23\text{–}630a)$$

where P_d in kW, ω_d in rad/s, d'_d in m, and F_{td} in N

$$= \frac{318 P_d}{n'_d d'_d} \quad \textbf{SI} \qquad (23\text{–}630b)$$

where P_d in kW, n'_d in rps, d'_d in m, and F_{td} in N

$$= \frac{1.95 \times 10^6 P_d}{n_d d'_d} \quad \textbf{Metric} \qquad (23\text{–}630c)$$

where P_d in kW, d'_d in mm, n_d in rpm, and F_{td} in kgf

$$= \frac{0.126 \times 10^6 P_d}{n_d d'_d} \quad \textbf{US Customary System unit}$$
$$(23\text{–}630d)$$

where P_d in hp, d'_d in in, n_d in rpm, and F_{td} in lbf

The tangential force at pitch point of the follower

$$F_{tf} = \frac{2M_{tf}}{d'_f} = \frac{2000 P_f}{\omega_f d'_f} \quad \textbf{SI} \qquad (23\text{–}631a)$$

where P_f in kW, ω_f in rad/s, d'_f in m, and F_{tf} in N

$$= \frac{318 P_f}{n'_f d'_f} \quad \textbf{SI} \qquad (23\text{–}631b)$$

where P_f in kW, n'_f in rps, d'_f in m, and F_{tf} in N

Particular	Formula

$$= \frac{1.95 \times 10^6 P_f}{n_f d_f'} \quad \textbf{Metric} \qquad (23\text{–}631c)$$

where P_f in kW, d_f' in mm, n_f in rpm, and F_{tf} in kgf

$$= \frac{0.126 \times 10^6 P_f}{n_f d_f'} \quad \textbf{US Customary System unit}$$

$$(23\text{–}631d)$$

where P_f in hp, d_f' in in, n_f in rpm, and F_{tf} in lbf

For formula for crossed helical gear calculation — Refer to Table 23–113.

DESIGNING OF CROSSED HELICAL GEARS

The Lewis formula for tangential tooth load at pitch point

$$F_t = \frac{\sigma_o C_V b Y}{P_n C_w} = \frac{\sigma_o C_V b Y m_n}{C_w} \qquad (23\text{–}632)$$

where

C_V = velocity factor = $\dfrac{4.5}{4.5 + v_m}$

Y = modified Lewis form factor should be based on formative number of teeth taken from Table 23–16

σ_o = allowable static stress taken from Table 23–48

C_w = wear and lubrication factor from Table 23–47

The limit load for wear

$$F_w = KQd_d^2 > F_d \qquad (23\text{–}633)$$

where

K = load stress factor taken from Table 23–114

F_d = dynamic load, kN (kgf) [Refer to Eq. (23–309).]

The factor Q for use in Eq. (23–633)

$$Q = [2d_f/(d_f + d_d)]^2 \qquad (23\text{–}634)$$

DESIGN CALCULATIONS AS PER NIEMANN

The diameter of the driver from bending strength consideration

$$d_d \geq 4.1 \left[\frac{P_d}{\omega_d K_{ba}} \left(\frac{z_d}{\cos \beta_d} \right)^2 \right]^{1/3} \quad \textbf{SI} \qquad (23\text{–}635a)$$

where P_d in kW, d_d in m, and ω_d in rad/s

$$\geq 2.20 \left[\frac{P_d}{n_d' K_{ba}} \left(\frac{z_d}{\cos \beta_d} \right)^2 \right]^{1/3} \quad \textbf{SI} \qquad (23\text{–}635b)$$

where P_d in kW, d_d in m, and n_d' in rps

Particular	Formula

$$\geq 39.5\left[\frac{P_d}{n_d K_{ba}}\left(\frac{z_d}{\cos\beta_d}\right)^2\right]^{1/3}\quad\textbf{Metric}\quad(23\text{–}635\text{c})$$

where P_d in kW, d_d in mm, and n_d in rpm

$$\geq 35.7\left[\frac{P_d}{n_d K_{ba}}\left(\frac{z_d}{\cos\beta_d}\right)^2\right]^{1/3}\quad\textbf{Metric}\quad(23\text{–}635\text{d})$$

where P_d in hp_m, d_d in mm, and n_d in rpm

Provided $b \simeq 10\mathbf{m}_n$, mm, $\mathbf{m}_n = \dfrac{(d_d\cos\beta_d)}{z_d}$ and F_{td} obtained from Eq. (23–630a)

$K_{ba} =$ bending stress factor
$= K_{ob}[2/(2 + v'_r)]$ (23–635e)

Refer to Table 23–115 for K_{ob}; $v'_r =$ sliding velocity, m/s

The transmitted power by the driver

$P_{db} \leq 0.015 d_d^3(\cos\beta_d/z_d)^2\omega_d K_{ba}$ **SI** (23–636a)
where P_{db} in kW, d_d in m, and ω_d in rad/s
$\leq 0.095 d_d^3(\cos\beta_d/z_d)^2 n'_d K_{ba}$ **SI** (23–636b)
where P_{db} in kW, d_d in m, and n'_d in rps

$$\leq \left(\frac{d_d}{39.5}\right)^3(\cos\beta_d/z_d)^2 n_d K_{ba}\ \textbf{Metric}\quad(23\text{–}636\text{c})$$

where P_{db} in kW, d_d in m, and n_d in rpm

$$\leq \left(\frac{d_d}{35.7}\right)^3\left(\frac{\cos\beta_d}{z_d}\right)^2 n_d K_{ba}\ \textbf{Metric}\quad(23\text{–}636\text{d})$$

where P_{db} in hp_m, d_d in mm, and n_d in rpm

HERTZIAN SURFACE CONTACT STRESS

Normal force F_n at the point of contact between two teeth profiles

$$F_n = F_{nd} = F_{nf} = 17.15\ \frac{\sigma_c^3}{E_o^2}\ \rho_o^2\left(\frac{1}{k'_1}\right)\qquad(23\text{–}637)$$

where $k'_1 =$ coefficient which depends on angle $\Sigma' = \beta'_d + \beta'_f$ obtained from Fig. 23.103

The relative radius of curvature

$\rho_o = 2\rho_d\rho_f/(\rho_d + \rho_f)$ (23–638)
where $\rho_d = -d'_d\sin\alpha_n/2\cos^2\beta''_d$;
$\rho_f = d'_f\sin\alpha_n/2\cos^2\beta''_f$, $\beta''_d = \sin^{-1}(\sin\beta_d\cos\alpha_n)$
and $\beta''_f = \sin^{-1}(\sin\beta_f\cos\alpha_n)$

The equivalent Young's modulus

$E_o = 2E_d E_f/(E_d + E_f)$ (23–639)

The ratio $i_\rho = \rho_d/\rho_f$

$$i_\rho = \frac{\rho_d}{\rho_f} = \frac{(\cos^2\beta_f + \tan^2\alpha_n)d'_d}{(\cos^2\beta_d + \tan^2\alpha_n)d'_f}\qquad(23\text{–}640)$$

Particular	Formula
The relationship between helices angles β_d', β_f', and Σ' (Fig. 23–103)	$$\tan \Sigma' = \tan(\beta_d' + \beta_f') = \frac{\sin \alpha_n (\tan \beta_d - \tan \beta_f)}{1 - \sin^2 \alpha_n \tan \beta_d \tan \beta_f}$$ $(23\text{–}641)$
Theoretical expression for normal force	$$F_n = F_{nd} = F_{nf} = \frac{F_{td}}{\cos \alpha_n \cos(\beta_d - \rho)}$$ $$= \frac{F_{tf}}{\cos \alpha_n \cos(\beta_f + \rho)}$$ $(23\text{–}642)$ where ρ = friction angle, deg
The tangential force from Eq. (23–642) at the pitch point of the driver	$$F_{td} = 1.43 d_d^2 k_2' K_c$$ $(23\text{–}643)$ where k_2' depends on β_f and $i = z_f/z_d$ and is obtained from Eq. (23–645)
The expression for surface contact stress factor K_c for use in Eq. (23–643)	$$K_c = \sigma_c^3 / E_o^2 \leq K_{ca}$$ $(23\text{–}644a)$ where $K_{ca} = K_{oc}(2/(2 + v_r'))$ $(23\text{–}644b)$ Refer to Table 23–115 for K_{oc}
The expression for coefficient k_2' for use in Eq. (23–643)	$$k_2' = \frac{12 \cos(\beta_d - \rho) \sin^2 \alpha_n}{k_1'(\tan^2 \alpha_n + \cos^2 \beta_d)^2 (1 + i_\rho)^2 \cos^3 \alpha_n}$$ $(23\text{–}645)$
The power transmitted by the driver	$P_{dc} = 0.72(d_d/10)^3 \omega_d k_2' K_c$ **SI** $(23\text{–}646a)$ where P_{dc} in kW, d_d in m, and ω in rad/s $\quad = 4.5(d_d/10)^3 n_d' k_2' K_c$ **SI** $(23\text{–}646b)$ where P_{dc} in kW, d_d in m, and n_d' in rps $\quad = (d_d/110)^3 n_d k_2' K_c$ **Metric** $(23\text{–}646c)$ where P_{dc} in kW, d_d in mm, and n_d in rpm $\quad = (d_d/100)^3 n_d k_2' K_c$ **Metric** $(23\text{–}646d)$ where P_{dc} in hp_m, d_d in mm, and n_d in rpm
The diameter of the driver from surface contact stress consideration	$$d_d \geq 11.2 \left(\frac{P_{dc}}{k_2' K_{ca} \omega_d} \right)^{1/3} \quad \textbf{SI} \qquad (23\text{–}647a)$$ where d_d in m, P_{dc} in kW, and ω_d in rad/s $$\geq 6.1 \left(\frac{P_{dc}}{k_2' K_{ca} n_d'} \right)^{1/3} \quad \textbf{SI} \qquad (23\text{–}647b)$$ where d_d in m, P_{dc} in kW, and n_d' in rps $$\geq 110 \left(\frac{P_{dc}}{k_2' K_{ca} n_d} \right)^{1/3} \quad \textbf{Metric} \qquad (23\text{–}647c)$$ where d_d in mm, P_{dc} in kW, and n_d in rpm

Particular	Formula
	$$\geq 100 \left(\frac{P_{dc}}{k_2' K_{ca} n_d} \right)^{1/3} \quad \textbf{Metric} \qquad (23\text{–}647d)$$ where d_d in mm, P_{dc} in hp_m, and n_d in rpm
The allowable torque on wheel shaft (M_{ta2}) for an unlimited life (more than 10^9 implies infinite life) based on allowable bending stress (σ_{ab})*[†]	$$M_{ta2} = \frac{\sigma_{ab} z_2 p_n^3}{20} \qquad (23\text{–}648)$$
The allowable torque on wheelshaft (M_{ta2}) for an unlimited life (more than 10^9 implies infinite life) based on allowable contact stress (σ_{ac})*[†]	$$M_{ta2} = \frac{\sigma_{ac} \Omega d_2^3 (d_1/d_2)^{1/2} C_v}{75} \qquad (23\text{–}649)$$ The symbols given in Eqs. (23–648) and (23–649) have already been defined at the beginning of this chapter on gears. Refer to Table 23–107 for σ_{ab} and σ_{ac}.
For nondimensional factor, Ω	Refer to Fig. 23–100a for Ω nondimensional quantity or factor
For velocity factor, C_v	Refer to Eq. (23–583).

BEARING LOADS

The magnitude of axial thrust along the shaft axis of the follower	$$F_{af} = F_{td} \frac{\cos \alpha \tan \beta_f + \mu}{\cos \alpha - \mu \tan \beta_f} \qquad (23\text{–}650)$$
The downward pressure on the shaft of the follower	$$F_{rf} = F_{td} \frac{\sin \alpha}{\cos \alpha \cos \beta_f - \mu \sin \beta_f} \qquad (23\text{–}651)$$
For thrust load and direction of rotation for crossed helical gears	Refer to Fig. 23–100b, also Figs. 23–101 and 23–102.

*British Standard Institution, BS 721

†*Source*: M. J. Need (ed), *Tribology Handbook*, Section A24, Butterworths, London, 1973.

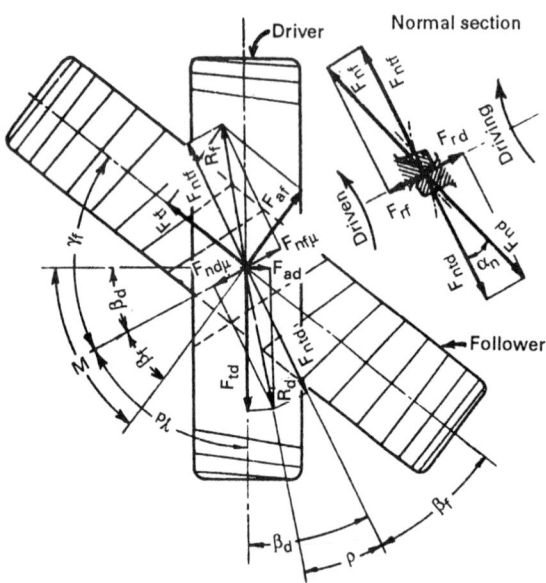

FIGURE 23–102 Forces acting between teeth at point of contact of teeth.

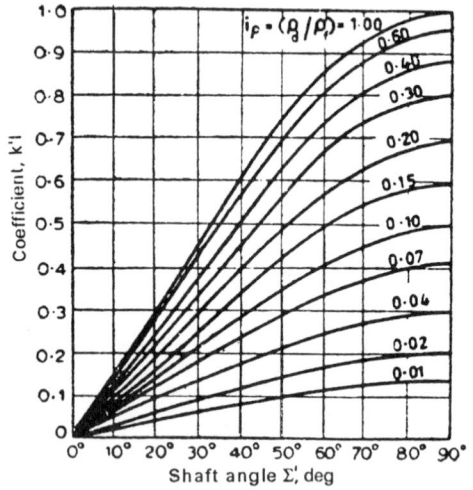

FIGURE 23–103 Variation of coefficient k_1' with $\sum_1'$. (*Courtesy of G. Niemann*, Maschinenelemente, *Vol. II, p. 194, 1965.*)

TABLE 23–113
Formulas for crossed helical gear calculations

	Driver			Follower	
To find	**Formula**		**To find**	**Formula**	
β_d	$\cos\beta_d = \dfrac{p_n}{p_d} = \dfrac{z_d}{p_n d_d}$	(23–652)	β_f	$\cos\beta_f = \dfrac{p_n}{p_f}$	(23–659)
	$\tan\beta_d = \dfrac{\pi d_d}{p_{zd}}$	(23–653)		$\beta_f = \Sigma - \beta_d$	(23–660)
p_n	$p_n = \dfrac{\pi d_d \cos\beta_d}{z_d}$	(23–654)	p_n	$p_n = \dfrac{\pi d_f \cos\beta_f}{z_f}$	(23–661)
p_d	$p_d = \dfrac{\pi d_d}{z_d}$	(23–655)	p_f	$p_f = \dfrac{\pi d_f}{z_f}$	(23–662)
p_{zd}	$p_{zd} = \pi d_d \tan\beta_f = p_f z_d$	(23–656)	p_{zf}	$p_{zf} = \pi d_f \tan\beta_d = p_d z_f$	(23–663)
d_d	$d_d = 2a/[(z_f \cos\beta_d / z_d \cos\beta_f) + 1]$	(23–657)	d_f	$d_f = 2a - d_d$	(23–664)
	$d_d = 0.3183 z_d p_d$	(23–658)		$d_f = 0.3183 z_f p_f$	(23–665)
			$a = \dfrac{z_d}{2 p_n \cos\beta_d} + \dfrac{z_f}{2 p_n \cos\beta_f}$		(23–666)

TABLE 23–114A
Values of load stress factor, K, for crossed helical gears for use in Eq. (23–633)

One pair of gears		Meshing gear		K after polishing run			
				Short		Careful	
Material	**Hardness, H_B**	**Material**	**Hardness, H_B**	**SI $\times 10^{-3}$**	**Metric $\times 10^{-3}$**	**SI $\times 10^{-3}$**	**Metric $\times 10^{-3}$**
Steel	250	Steel	250	13.7	1.4	34.3	3.5
Steel	250	Bronze	100	27.5	2.8	82.4	8.4
Steel	500	Bronze	120	34.3	3.5	137.3	14.0
Steel	500	Cast iron	180	41.2	4.2	137.3	14.0
Steel	500	Steel	500	48.0	4.9	103.0	10.5
Cast iron	180	Cast iron	180	50.0	5.6	137.3	14.0
Nonmetallic	—	Steel or cast iron	—	68.6	7.0	171.6	17.5

TABLE 23–114B
Tooth proportions for crossed helical gears (26)

Normal module $m_n = 1$
Normal circular pitch $p_n = 3.14159$ mm
Normal diametral pitch $P_{nd} = 1$
Normal circular pitch $p_n = 3.14159$ in

| Helix angle | | Normal | | | | | Minimum |
Driver $\beta_d(\psi_1)$	Follower $\beta_f(\psi_2)$	pressure angle, $\alpha_n(\phi_n)$	Addendum, $h_a(a)$	Working depth, $h'(h_k)$	Whole depth, $h(h_t)$	Helix angle of driver	number of teeth
45°	45°	14°30′	1.200	2.400	2.650	45°	20
60°	30°	17°30′	1.200	2.400	2.650	60°	9
75°	15°	19°30′	1.200	2.400	2.650	75°	4
86°	4°	20°	1.200	2.400	2.650	86°	1

Note: The addendum, working depth, and whole-depth values are for 1 normal module (metric) in millimeters or for 1 normal diametral pitch (English) in inches.
Source: Courtesy D. W. Dudley, *Handbook of Practical Gear Design*, McGraw-Hill Book Company, New York, 1984.

TABLE 23–115
Allowable stress factors, K_{ba} and K_{ca}, for crossed helical gears for use in Eqs. (23–635) to (23–647)

$$K_{ba} = K_{ob}\left(\frac{2}{2 + v'_r}\right), \quad K_{ca} = K_{oc}\left(\frac{2}{2 + v'_r}\right)$$

where $v'_r = v_d\, \dfrac{\sin \Sigma}{\cos \beta_f} = v_f\, \dfrac{\sin \Sigma}{\cos \beta_d} =$ rubbing velocity in m/s

Combination of material	K_{ob}		K_{oc}		E_o	
	psi	MPa	psi	MPa	Mpsi	GPa
Hardened steel/hardened steel	855	5.89	10	0.074	30	206
Hardened steel/bronze	770	5.30	9.5	0.066	21	147
Hardened steel/pearlite cast iron	680	4.70	8.5	0.059	22.5	157
Tempered steel/bronze	570	3.92	7.0	0.049	20.5	142
Tempered steel/gray cast iron	400	2.74	5.0	0.034	19.0	132
Gray cast iron/gray cast iron	400	2.74	5.0	0.034	15.5	108

Note: Increase the above values by 53 percent for short time operation.

23.7   FORCES ON BEARINGS OF GEAR SHAFTS (1)

SYMBOLS

a, b	distance from center of bearings to the center of pinion or worm as shown in Figs. 23–104 to 23–114, respectively, m (in)
c, d	distance from center of bearings to the center of gear or worm gear as shown in Figs. 23–104 to 23–114, respectively, m (in)
d_1'	pitch diameter of pinion or worm, m (in)
d_2'	pitch diameter of gear or worm gear, m (in)
e, f, g	distances from center of bearings to center of gears as shown in Fig. 23–108
F_a	axial force or thrust, kN (lbf)
F_r	radial force, kN (lbf)
F_t	tangential force, kN (lbf)
$L_1 = a + b,\ L_2 = c + d$	distances from center of bearings of pinion or worm, and gear or worm gear as shown in Figs. 23–109 to 23–114 respectively, m (in)
R_{1L}	resultant load due to pinion or worm on left bearing, kN (lbf)
R_{2L}	resultant load due to gear or worm gear on left bearing, kN (lbf)
R_{1R}	resultant load due to pinion or worm on right bearing, kN (lbf)
R_{2R}	resultant load due to gear or worm gear on right bearing, kN (lbf)
I, II, III	shaft axes 1, 2, and 3, respectively

SUFFIXES

1	pinion
2	gear
a	axial
D	lower (down) bearing
g	worm gear
L	left bearing
r	radial
R	right bearing
t	tangential
U	upper bearing
w	worm

Figure	Formula

SPUR GEAR

(a) Pinion and gear supported between bearings

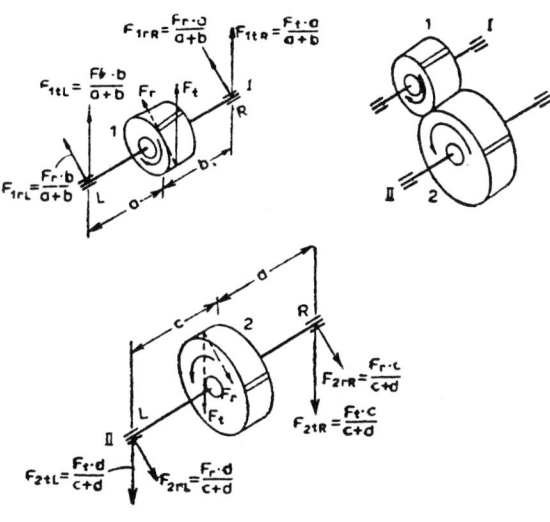

FIGURE 23–104

LOADS ACTING ON THE LEFT-HAND BEARINGS

$$F_{1rL} = \frac{bF_r}{a+b}, \quad F_{2rL} = \frac{d}{c+d} F_r,$$

$$R_{1L} = \frac{b}{a+b} \sqrt{F_t^2 + F_r^2}$$

$$F_{1tL} = \frac{b}{a+b} F_t, \quad F_{2tL} = \frac{d}{c+d} F_t,$$

$$R_{2L} = \frac{d}{c+d} \sqrt{F_t^2 + F_r^2}$$

LOAD ACTING ON THE RIGHT-HAND BEARINGS

$$F_{1rR} = \frac{a}{a+b} F_r, \quad F_{2rR} = \frac{c}{c+d} F_r,$$

$$R_{1R} = \frac{a}{a+b} \sqrt{F_t^2 + F_r^2}$$

$$F_{1tL} = \frac{a}{a+b} F_t, \quad F_{2tR} = \frac{c}{c+d} F_t,$$

$$R_{2R} = \frac{c}{c+d} \sqrt{F_t^2 + F_r^2}$$

(b) Pinion supported between bearings and gear overhung

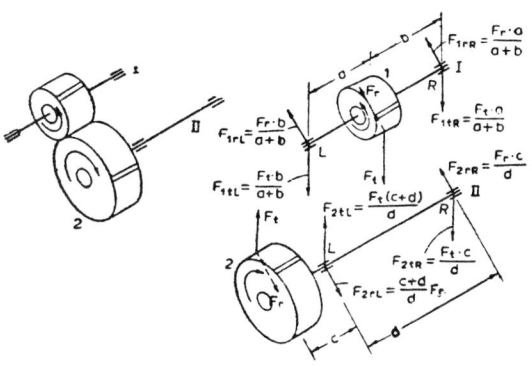

FIGURE 23–105

LOADS ACTING ON THE LEFT-HAND BEARINGS

$$F_{1rL} = \frac{b}{a+b} F_r, \quad F_{2rL} = \frac{(c+d)}{d} F_r,$$

$$R_{1L} = \frac{b}{a+b} \sqrt{F_t^2 + F_r^2}$$

$$F_{1tL} = \frac{b}{a+b} F_t, \quad F_{2tL} = \frac{c+d}{d} F_t,$$

$$R_{2L} = \frac{c+d}{d} \sqrt{F_t^2 + F_r^2}$$

LOADS ACTING ON THE RIGHT-HAND BEARINGS

$$F_{1rR} = \frac{a}{a+b} F_r, \quad F_{2rR} = \frac{c}{d} F_r,$$

$$R_{1R} = \frac{a}{a+b} \sqrt{F_t^2 + F_r^2}$$

$$F_{1tR} = \frac{a}{a+b} F_t, \quad F_{2tR} = \frac{c}{d} F_t,$$

$$R_{2R} = \frac{c}{d} \sqrt{F_t^2 + F_r^2}$$

Figure	Formula

HELICAL GEAR

(a) Pinion and gear supported between bearings: pinion rotating clockwise

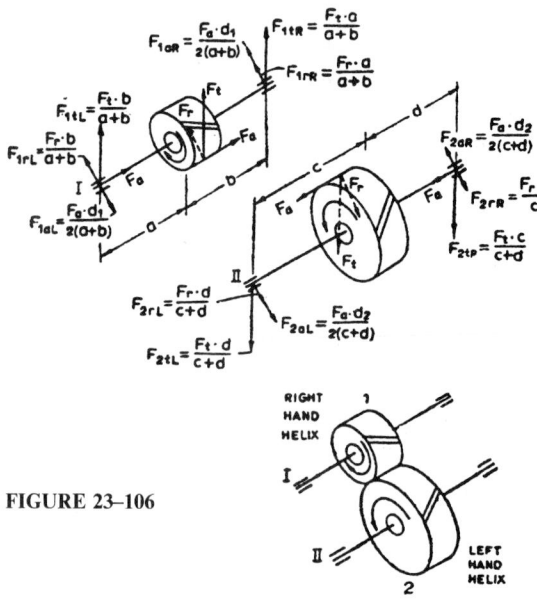

FIGURE 23–106

LOADS ACTING ON THE LEFT-HAND BEARINGS

$$F_{1rL} = \frac{bF_r}{a+b}, \quad F_{1tL} = \frac{bF_t}{a+b}, \quad F_{1aL} = \frac{d_1 F_a}{2(a+b)}$$

$$R_{1L} = \frac{1}{a+b}\sqrt{b^2 F_t^2 + [bF_r - (d_1 F_a)/2]^2}$$

$$F_{2rL} = \frac{dF_r}{c+d}, \quad F_{2tL} = \frac{dF_t}{c+d}, \quad F_{2aL} = \frac{d_2 F_a}{2(c+d)}$$

$$R_{2L} = \frac{1}{c+d}\sqrt{d^2 F_t^2 + [dF_r + (d_2 F_a)/2]^2}$$

LOADS ACTING ON THE RIGHT-HAND BEARINGS

$$F_{1rR} = \frac{a}{a+b} F_r, \quad F_{1tR} = \frac{a}{a+b} F_t, \quad F_{1aR} = \frac{d_1 F_a}{2(a+b)}$$

$$R_{1R} = \frac{1}{a+b}\sqrt{a^2 F_t^2 + [aF_r + (d_1 F_a)/2]^2}$$

$$F_{2rR} = \frac{c}{c+d} F_r, \quad F_{2tR} = \frac{c}{c+d} F_t, \quad F_{2aR} = \frac{d_2 F_a}{2(c+d)}$$

$$R_{2R} = \frac{1}{c+d}\sqrt{c^2 F_t^2 + [cF_r - (d_2 F_a)/2]^2}$$

(b) Pinion and gear supported between bearings; pinion rotating counterclockwise

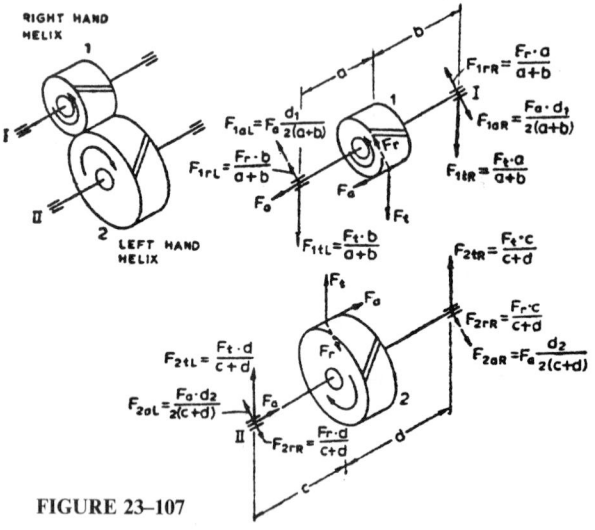

FIGURE 23–107

LOADS ACTING ON THE LEFT-HAND BEARINGS

$$F_{1rL} = \frac{b}{a+b} F_r, \quad F_{1tL} = \frac{b}{a+b} F_t, \quad F_{1aL} = \frac{d_1 F_a}{2(a+b)}$$

$$R_{1L} = \frac{1}{a+b}\sqrt{b^2 F_t^2 + [bF_r + (d_1 F_a)/2]^2}$$

$$F_{2rL} = \frac{d}{c+d} F_r, \quad F_{2tL} = \frac{d}{c+d} F_t, \quad F_{2aL} = \frac{d_2 F_a}{2(c+d)}$$

$$R_{2L} = \frac{1}{c+d}\sqrt{d^2 F_t^2 + [dF_r - (d_2 F_a)/2]^2}$$

LOADS ACTING ON THE RIGHT-HAND BEARINGS

$$F_{1rR} = \frac{a}{a+b} F_r, \quad F_{1tR} = \frac{a}{a+b} F_t, \quad F_{2aR} = \frac{d_2 F_a}{2(a+b)}$$

$$R_{1R} = \frac{1}{a+b}\sqrt{a^2 F_t^2 + [aF_r - (d_1 F_a)/2]^2}$$

Figure	Formula

$$F_{2rR} = \frac{c}{c+d}\,F_r, \quad F_{2tR} = \frac{c}{c+d}\,F_t, \quad F_{2aR} = \frac{d_2 F_a}{2(c+d)}$$

$$R_{2R} = \frac{1}{c+d}\sqrt{c^2 F_t^2 + [cF_r + (d_2 F_a)/2]^2}$$

(c) Pinion P_1, gear Q_2 and intermediate gear P_2 and pinion Q_1 supported between bearings

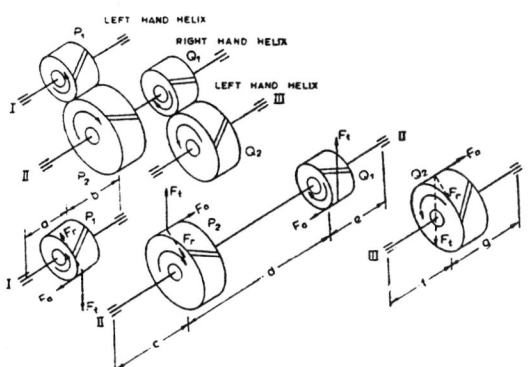

FIGURE 23–108

For a gear train of helical gears shown in Fig. 23–108, the resultant bearing load analysis can be done similarly using the analysis for helical gears of Figs. 23–106 and 23–107. The resultant bearing loads for shafts I and III are similar to those in the single-stage helical gears; but for the intermediate shaft II, which carries two gear wheels P_2 and Q_1, the loads F_t, F_r, F_a, and R_R due to these two gears have to be considered taking into consideration the directions of rotation and hand of helix of each gear.

STRAIGHT BEVEL GEARS

(a) Pinion overhung and gear supported between bearings: pinion rotating clockwise

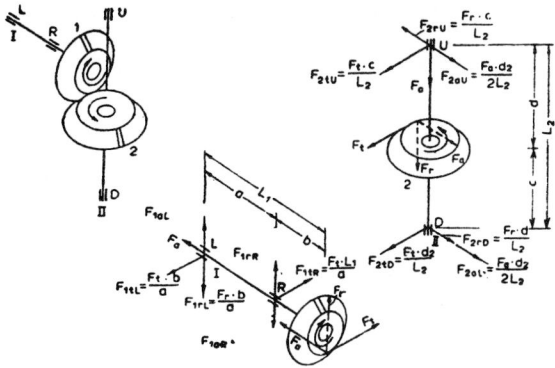

FIGURE 23–109

LOADS ACTING ON THE BEARINGS OF THE PINION

Left L: $F_{1rL} = (b/a)F_r, \quad F_{1tL} = (b/a)F_t,$
$\qquad\quad F_{1aL} = (d_1/2a)F_a$
$R_{1L} = (b/a)[F_t^2 + (F_r - (d_1/2b)F_a)^2]^{1/2}$
Right R: $F_{1rR} = (L_1/a)F_r, \quad F_{1tR} = (L_1/a)F_t,$
$\qquad\qquad F_{1aR} = (d_1/2a)F_a$
$R_{1R} = (L_1/a)[F_t^2 + (F_r - (d_1/2L_1)F_a)^2]^{1/2}$

LOADS ACTING ON THE BEARINGS OF THE GEAR

Upper U: $F_{2rU} = (c/L_2)F_r, \quad F_{2tU} = (c/L_2)F_t,$
$\qquad\qquad F_{2aU} = (d_2/2L_2)F_a$
$R_{2U} = (c/L_2)[F_t^2 + (F_r - (d_2/2c)F_a)^2]^{1/2}$
Lower D: $F_{2rD} = (d/L_2)F_r, \quad F_{2tD} = (d/L_2)F_t,$
$\qquad\qquad F_{2aD} = \pm(d_2/2L_2)F_a$
$R_{2D} = (d/L_2)[F_t^2 + (F_r + (d_2/2d)F_a)^2]^{1/2}$

Figure	Formula

(b) Pinion overhung and gear supported between bearings: pinion rotating counterclockwise

LOADS ACTING ON THE BEARINGS OF THE PINION

Left L: $F_{1rL} = (b/a)F_r$, $F_{1tL} = (b/a)F_t$,
$$F_{1aL} = (d_1/2a)F_a$$
$$R_{1L} = (b/a)[F_t^2 + (F_r - (d_1/2b)F_a)^2]^{1/2}$$
Right R: $F_{1rR} = (L_1/a)F_r$, $F_{1tR} = (L_1/a)F_t$,
$$F_{1aR} = (d_1/2a)F_a$$
$$R_{1R} = (L_1/a)[F_t^2 + (F_r - (d_1/2L_1)F_a)^2]^{1/2}$$

LOADS ACTING ON THE BEARINGS OF THE GEAR

Upper U: $F_{2rU} = (c/L_2)F_r$, $F_{2tU} = (c/L_2)F_t$,
$$F_{2aU} = (d_2/2L_2)F_a$$
$$R_{2U} = (c/L_2)[F_t^2 + (F_r - (d_2/2c)F_a)^2]^{1/2}$$
Lower D: $F_{2rD} = (d/L_2)F_r$, $F_{2tD} = (d/L_2)F_t$,
$$F_{2aD} = (d_2/2L_2)F_a$$
$$R_{2D} = (d/L_2)[F_t^2 + (F_r + (d_2/2d)F_a)^2]^{1/2}$$

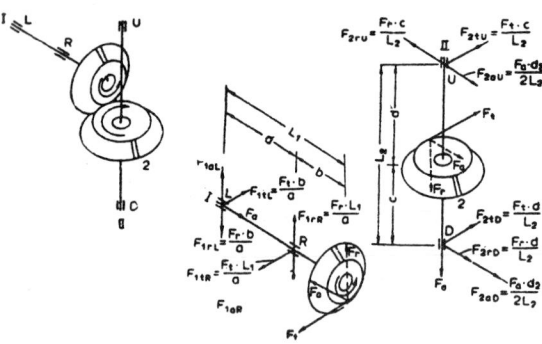

FIGURE 23–110

SPIRAL BEVEL GEARS

(a) Pinion overhung and gears supported between bearings: pinion rotating clockwise

LOADS ACTING ON THE BEARINGS OF THE PINION

Left L: $F_{1rL} = (b/a)F_{1r}$, $F_{1tL} = (b/a)F_t$,
$$F_{1aL} = \pm(d_1/2a)F_{1a}$$
$$R_{1L} = (b/a)[F_t^2 + (F_{1r} \pm (d_1/2b)F_{1a})^2]^{1/2}$$
Right R: $F_{1rR} = (L_1/a)F_{1r}$, $F_{1tR} = (L_1/a)F_t$,
$$F_{1aR} = \pm(d_1/2a)F_{1a}$$
$$R_{1R} = (L_1/a)[F_t^2 + (F_{1r}(d_1/2L_1)F_{1a})^2]^{1/2}$$

LOADS ACTING ON THE BEARINGS OF THE GEAR

Upper U: $F_{2rU} = (c/L_2)F_{2r}$, $F_{2tU} = (c/L_2)F_t$,
$$F_{2aU} = \pm(d_1/2L_2)F_{2a}$$
$$R_{2U} = (c/L_2)[F_t^2 + (F_{2r} - (d_2/2c)F_{2a}]^{1/2}$$
Lower D: $F_{2rD} = (d/L_2)F_{2r}$, $F_{2tD} = (d/L_2)F_t$,
$$F_{2aD} = \pm(d_2/2L_2)F_{2a}$$
$$R_{2D} = (d/L_2)[F_t^2 + (F_{2r} \pm (d_2/2d)F_{2a})^2]^{1/2}$$

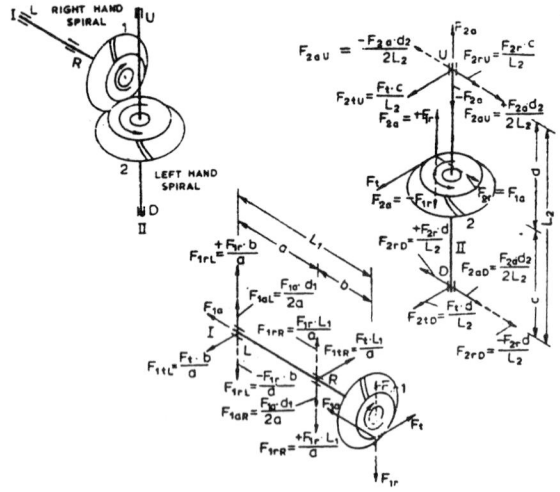

FIGURE 23–111

Figure	Formula

(b) Pinion overhung and gear supported between bearings: pinion rotating counterclockwise

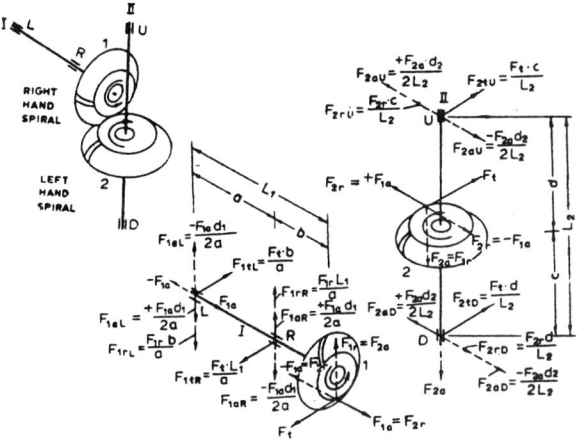

FIGURE 23–112

LOADS ACTING ON THE BEARINGS OF THE PINION

Left L: $F_{1rL} = (b/a)F_{1r}$, $F_{1tL} = (b/a)F_t$,
$$F_{1aL} = \pm(d_1/2a)F_{1a}$$
$$R_{1L} = (b/a)[F_t^2 + (F_{1r} \pm (d_1/2b)F_{1a})^2]^{1/2}$$

Right R: $F_{1rR} = (L_1/a)F_{1r}$, $F_{1tR} = (L_1/a)F_t$,
$$F_{1aR} = \pm(d_1/2a)F_{1a}$$
$$R_{1R} = (L_1/a)[F_t^2 + (F_{1r} \pm (d_1/2L_1)F_{1a})^2]^{1/2}$$

LOADS ACTING ON THE BEARINGS OF THE GEARS

Upper U: $F_{2rU} = (c/L_2)F_{2r}$, $F_{2tU} = (c/L_2)F_t$,
$$F_{2aU} = \pm(d_2/2L_2)F_{2a}$$
$$R_{2U} = (c/L_2)[F_t^2 + (F_{2r} \pm (d_2/2c)F_{2a})^2]^{1/2}$$

Lower D: $F_{2rD} = (d/L_2)F_{2r}$, $F_{2tD} = (d/L_2)F_t$,
$$F_{2aD} = \pm(d_2/2L_2)F_{2a}$$
$$R_{2D} = (d/L_2)[F_t^2 + (F_{2r} \pm (d_2/2d)F_{2a})^2]^{1/2}$$

WORM GEARS

(a) Worm and worm gear supported between bearings: worm placed above the worm gear (wheel) and rotating clockwise

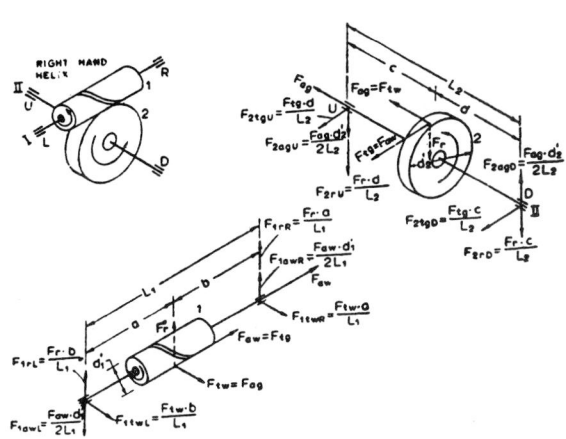

FIGURE 23–113

LOADS ACTING ON THE BEARINGS OF THE WORM

Left L: $F_{1rL} = (b/L_1)F_r$, $F_{1tWL} = (b/L_1)F_{tW}$,
$$F_{1aWL} = (d_1'/2L_1)F_{aW}$$

$$R_{1L} = (b/L_1)[F_{tW}^2 + (F_r - (d_1'/2b)F_{aW})^2]^{1/2}$$
Right R: $F_{1rR} = (a/L_1)F_r$, $F_{1tWR} = (a/L_1)F_{tW}$,
$$F_{1aWR} = (d_1'/2L_1)F_{aW}$$
$$R_{1R} = (a/L_1)[F_{tW}^2 + (F_r + (d_1'/2a)F_{aW})^2]^{1/2}$$

LOADS ACTING ON THE BEARINGS OF WORM GEAR

Upper U: $F_{2rU} = (d/L_2)F_{2r}$, $F_{2tgU} = (d/L_2)F_{tg}$,
$$F_{2agU} = (d_2'/2L_2)F_{ag}$$
$$R_{2U} = (d/L_2)[F_{tg}^2 + (F_{2r} + (d_2'/2d)F_{ag})^2]^{1/2}$$
Lower D: $F_{2rD} = (c/L_2)F_{2r}$, $F_{2tgD} = (c/L_2)F_{tg}$,
$$F_{2agD} = (d_2'/2L_2)F_{ag}$$
$$R_{2D} = (c/L_2)[F_{tg}^2 + (F_r - (d_2'/2c)F_{ag})^2]^{1/2}$$

Figure	Formula

(b) Worm and worm gear supported between bearings: worm placed below the worm gear and rotating counterclockwise

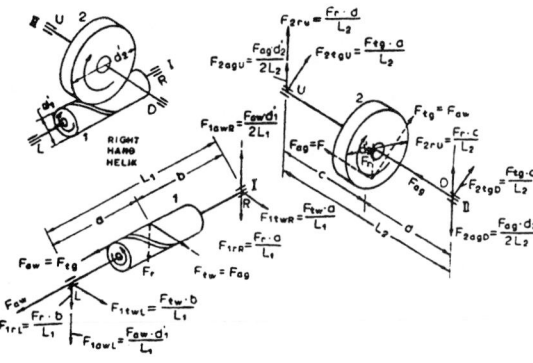

FIGURE 23-114

LOADS ACTING ON THE BEARINGS ON THE WORM

Left L: $F_{1rL} = (b/L_1)F_r$, $F_{1tWL} = b/L_1)F_{tW}$,
$$F_{1aWL} = (d_1'/2L_1)F_{aW}$$
$$R_{1L} = (b/L_1)[F_{tW}^2 + (F_r + (d_1'/2b)F_{aW})^2]^{1/2}$$
Right R: $F_{1rR} = (a/L_1)F_r$, $F_{1tWR} = (a/L_1)F_{tW}$,
$$F_{1aWR} = (d_1'/2L_1)F_{aW}$$
$$R_{1R} = (a/L_1)[F_{tW}^2 + (F_r - (d_1'/2a)F_{aW})^2]^{1/2}$$

LOADS ACTING ON THE BEARINGS OF THE WORM GEAR

Upper U: $F_{2rU} = (d/L_2)F_r$, $F_{2tgU} = (d/L_2)F_{tg}$,
$$F_{2agU} = (d_2'/2L_2)F_{ag}$$
$$R_{2U} = (d/L_2)[F_{tg}^2 + (F_r + (d_2'/2d)F_{ag})^2]^{1/2}$$
Lower D: $F_{2rD} = (c/L_2)F_r$, $F_{2tgD} = (c/L_2)F_{tg}$,
$$F_{2agD} = (d_2'/2L_2)F_{ag}$$
$$R_{2D} = (c/L_2)[F_{tg}^2 + (F_r - (d_2'/2c)F_{ag})^2]^{1/2}$$

23.8 NONCIRCULAR GEARS

SYMBOLS

a_o	center distance, m (in) (Refer to Fig. 23–115.)
ϵ	eccentricity, m (in)
l_p	length of periphery of pitch curve, m (in)
m	module, m (in)
r'	active pitch radius (also with subscripts), m (in)
ρ	radius of curvature, m (in)
ψ	polar angle to r
ω	angular velocity, rad/s
$f(\psi)$, $F(\psi)$, and $G(\psi)$	functions of polar angle ψ
α	pressure angle, deg
z	number of teeth

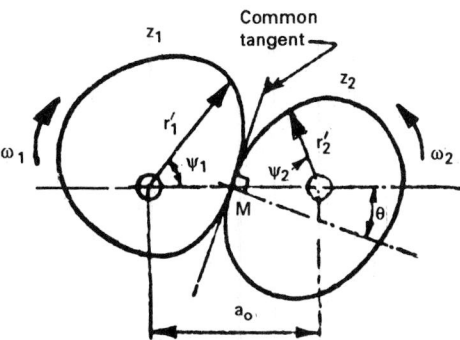

SUFFIXES

1	gear 1
2	gear 2

FIGURE 23–115

Particular	Formula	
Case 1. Polar equation of the curve on one of the gears and the center distance are known, to find the polar equations of the mating gear		
Pitch radius of gear 1	$r_1' = f(\psi_1)$	(23–667)
Pitch radius of gear 2	$r_2' = a_o - f(\psi_1)$	(23–668)
Angular rotation of gear 2	$\psi_2 = -\psi_1 + a_o \int \dfrac{d\psi_1}{a_o - f(\psi_1)}$	(23–669)
Case 2. Relationship between the angular rotation of the two gears and the center distance between them are known, to find polar equations of both gears		
Angular rotation of gear 2	$\psi_2 = F(\psi_1)$	(23–670)
Pitch radius of gear 1	$r_1' = \dfrac{a_o F'(\psi_1)}{1 + F'(\psi_1)}$	(23–671)
Pitch radius of gear 2	$r_2' = a_o - r_1' = \dfrac{a_o}{1 + F'(\psi_1)}$	(23–672)
Case 3. Relationship between the angular velocities of the two gears and the center distance between them are known, to find polar equations of both gears		
Angular velocity of gear 2	$\omega_2 = \omega_1 G(\psi_1)$	(23–673)

Particular	Formula
Pitch radius of gear 1	$$r_1' = \frac{a_o G(\psi_1)}{1 + G(\psi_1)} \qquad (23\text{--}674)$$
Pitch radius of gear 2	$$r_2' = a_o - r_1' \qquad (23\text{--}675)$$
Angular rotation of gear 2	$$\psi_2 = fG(\psi_1)d\psi \qquad (23\text{--}676)$$

CONDITIONS FOR CLOSED PITCH CURVES

Case 1	$$r = f(\psi) = f(\psi + 2\pi z) \qquad (23\text{--}677)$$
Case 2	$$\psi_1 = F(\psi_2) \text{ and } F(\psi_o) = 0 \qquad (23\text{--}678)$$
	When equation $F(\psi_o + 2\pi/z_1) = 2\pi/z_2$ can be satisfied by substituting integers or rational fractions for z_1 and z_2
Case 3	$$\psi_2 = \int G(\psi_1)d\psi_1 \qquad (23\text{--}679)$$
	Put $G(\psi_1)d\psi_1 = F(\psi_1)$
	Repeat as for *case 2* with the subscripts reversed.
Expression for correct center distance in order to get a closed pitch curve	$$4\pi = \int_o^{2\pi} \frac{d\psi_1}{a_o - f(\psi_1)} \qquad (23\text{--}680)$$
Pressure angle	$$\alpha = \tan^{-1}\left[\frac{f'(\psi)}{f(\psi)}\right] \not> 45° \qquad (23\text{--}681)$$
Radius of curvature to avoid undercutting	$$\rho = \frac{m(z+2)}{2} > \frac{m_{min}^z}{2} \qquad (23\text{--}682)$$
Radius of curvature	$$\rho = \frac{[r^2 + (r')^2]^{3/2}}{r^2 - r(r'') + 2(r')^2} \qquad (23\text{--}683)$$
	where r' and r'' are first and second derivatives of r
Length of periphery of pitch curve	$$l_p = \int_o^{\psi=2\pi} [(dr)^2 + (rd\psi)^2]^{1/2} \qquad (23\text{--}684)$$
Approximate length of the periphery of the pitch curve for elliptical gears	$$l_p = \pi\left[2(a^2 + b^2) - \frac{(a-b)^2}{2.2}\right]^{1/2} \text{ (approx.)}$$ $$(23\text{--}685)$$
	where a and b are the semimajor and semiminor axes of the elliptical gears, respectively

Particular	Formula

ELLIPTICAL GEARS

Instantaneous velocity ratio

$$i_i = \left(\frac{d\psi_2}{dt}\right)\Big/\omega = \frac{r_1'}{r_2'} \tag{23--686}$$

Instantaneous radius of curvature of gear 1

$$r_1 = \frac{a(1 - \epsilon^2)}{1 + \epsilon \cos \psi_1} \tag{23--687}$$

Numerical eccentricity of the ellipse

$$\epsilon = \frac{\sqrt{(a^2 - b^2)}}{b} = \frac{e}{a} \tag{23--688}$$

where e = distance from focus to center of the ellipse

Instantaneous radius of curvature of gear

$$r_2' = 2a - \frac{a(1 - \epsilon^2)}{1 + \epsilon \cos \psi_1} = a\left[\frac{1 + 2\epsilon \cos \psi_1 + \epsilon^2}{1 + \epsilon \cos \psi_1}\right] \tag{23--689}$$

Instantaneous velocity ratio

$$i_i = \left(\frac{d\psi_2}{dt}\right)\Big/\omega = \left[\frac{1 - \epsilon^2}{1 + \epsilon^2 + 2\epsilon \cos \psi_1}\right] \tag{23--690}$$

Minimum velocity at $\psi_1 = 0°$

$$\frac{d\psi_2}{dt_{(min)}} = \omega\left[\frac{a - e}{a + e}\right] \tag{23--691}$$

Maximum velocity at $\psi_1 = 180°$

$$\frac{d\psi_2}{dt_{(max)}} = \omega\left[\frac{a + e}{a - e}\right] \tag{23--692}$$

Ratio of maximum to minimum angular output velocities

$$K = \frac{d\psi_2}{dt_{(max)}}\Big/\frac{d\psi_2}{dt_{(min)}} = \left(\frac{a + e}{a - e}\right)^2 \tag{23--693}$$

Semiminor axis of ellipse

$$b = \frac{2aK^{1/4}}{1 + K^{1/2}} \tag{23--694}$$

For requirement of gear lubricants, compound   Refer to Table 23--117.

For gears manufactured from powder metals   Refer to Table 23--118.

TABLE 23-116
Properties of noncircular gear systems

Type	Velocity equation	Basic equation	Remark
Two elliptical gears rotating about foci	$\omega_2 = \omega_1\left[\dfrac{i^2 + 1 + (i^2-1)\cos\psi_2}{2i}\right]$ Where $i = \dfrac{r_{max}}{r_{min}}$	$r = \dfrac{b^2}{a[1 + \epsilon\cos\psi]}$ $\epsilon = $ eccentricity $= (1/a)\sqrt{a^2 - b^2}$ $a_o = a + b$	Gears are identical Easy to manufacture Used for quick-return mechanisms, printing presses, automobile machinery
Elliptical gears of second order rotating about their geometric centers	$\omega_2 = \omega_1\left[\dfrac{i^2 + 1 + (i^2-1)\cos\psi_2}{2i}\right]$	$r = \dfrac{2ab}{(a+b) - (b-a)\cos 2\psi}$ $a = $ minimum radius $b = $ maximum radius $a_o = a + b$ $i = b/a$	Gears are identical Better balanced than true elliptical gears Two complete speed cycles in one revolution
Eccentric circular gear rotating with its conjugate	$\omega_2 = -\omega_1\,\dfrac{r_1}{r_2}$	$r_1 = \epsilon\cos\psi_1 + \sqrt{a^2 - \epsilon^2\sin^2\psi_1}$ $\psi_2 = -\psi_1 + \displaystyle\int \dfrac{d\psi}{a_o - r_1}$ $r_2 = a_o - r_1$ $a_o = a + b;\ \epsilon = $ eccentricity	Standard circular spur gears can be used as the eccentric; mating gears have special shapes
Logarithmic spiral gears	$\omega_2 = \omega_1\,\dfrac{Ae^{k\psi 1}}{a_o - Ae^{k\psi 1}}$	$r_1 = Ae^{k\psi_1}$ $r_2 = a_o - r_1 = Ae^{k\psi_2}$ $\psi_2 = \dfrac{1}{k}\log[(a_o/A) - e^{k\psi_1}]$ $e = $ natural log base	Gears can be identical and in combinations to give variety of functions Must be open gears
Sine function gears	$\omega_2 = \omega_1\,k\cos\psi_1$	$\psi_2 = \sin^{-1}(k\psi_1)$ $r_2 = \dfrac{a_o}{1 + k\cos\psi_1}$ $r_1 = a_o - r_2$ $= \dfrac{a_o k\cos\psi_1}{1 + k\cos\psi_1}$	Used to produce angular displacement proportional to sine of input angle Must be open gears

TABLE 23–117
Requirement for gear lubricants compound

Characteristic	Requirement				Method of test
	Grade 80	Grade 90	Grade 140	Grade 250	
Flash point, min	175°C	190°C	240°C	275°C	
Kinematic viscosity, cSt				—	P : 25
At −17.8°C	21800 max				
At 98.9°C[d]	9.0–14.0	> 14.0–25.0	> 25.0–43.0	> 43.0	
Viscosity index, min		70	70	85	L : 16[a]
					P : 2[b]
Acidity					
Organic acidity, max[c]	0.25	0.25	0.25	0.25	
Inorganic acidity[c]	Nil	Nil	Nil	Nil	
Saponification value[c]	10–20	10–20	10–20	10–20	L : 10[a]
Pour point, max	−25°C	−20°C	−5°C	0°C	P : 4[b]
Ash, percent by weight, max	0.01	0.01	0.05	0.05	P : 4[b]
Hard asphalt content percent by weight, max	0.1	0.2	0.2	0.2	P : 22[b]
Carbon residue percent by weight, max	1.0	2.0	2.0	3.5	P : 8[b]
Copper-strip corrosion, for 3 h at 100° ± 1°C	Not worse than no. 1				P : 15[b]
Stability					
At 0°C					
At −15°C	No stratification— flow readily on tilting		No stratification— flow readily on tilting		

[a] Test designation in IS: 310 (part 1) 1951
[b] Test designation in IS: 1448 (part 1) 1960
[c] mg of potassium hydroxide (KOH) per g of oil
[d] The maximum viscosity at 98.9°C may be waived if viscosity at −17.8°C is not greater than 16920 cSt
Source: IS: 2297, 1963.

TABLE 23–118
Gears manufactured from powder metals—mechanical properties

Materials	PMPMA designation[a]	Condition	σ_u		σ_y		τ		σ_b		Impact strength		Elongation	Density ρ, g/cm³	Rockwell hardness
			kpsi	GPa	kpsi	GPa	kpsi	GPa	kpsi	GPa	ft lbf	J			
Steels															
99 Fe-1C	F-0010-P	As sintered	35	0.241	26	0.186	22	0.152	90	0.614	1.0	1.354	1.0	6.1–6.5	50 R_B
99 Fe-1C	F-0010-P	Sintered h.t.	47.7	0.329							4.5	6.102	0.5	6.1–6.5	90 R_B
SAE 1080		Wrought, ann.	95.7	0.660	53	0.373	75	0.517	130	0.897	4.5	6.102	24	7.8	15 R_C
92 Fe-7Cu 1C	FC-0710-S	As sintered	83.1	0.573	62	0.434	57	0.393			4.0	5.430	1.0	6.8	73 R_B
92 Fe-7Cu 1C	FC-0710-S	Sintered, h.t.	110	0.759					206	1.422	7.0	9.486	1.5	6.8	40 R_C
79 Fe 20Cu 1C	FX-2010-T	As sintered	110	0.759	90	0.621	66	0.454	190	1.310	11.0	14.911	1.0	7.1 min	95 R_B
79 Fe 20Cu 1C	FX-2010-T	Sintered, h.t.	152	1.050			110	0.758					1.0	7.1 min	40 R_C
7 Ni 2Cu 1C		As sintered	92	0.635	75	0.517	60	0.414	180	1.241	11.0	14.911	3.5	7.2	85 R_B
7 Ni 2Cu 1C		Sintered, h.t.	157	1.083					285	1.966			2.0	7.2	44 R_C
1.5 Ni 0.5 Mo 0.6C		As sintered	90	0.621	70	0.496	47	0.324	180	1.241	9.2	12.412	2.5	7.2	95 R_B
1.5 Ni 0.5 Mo 0.6C		Sintered, h.t.	140	0.965	118	0.828			207	1.428	4.3	5.827	0.5	7.2	35 R_C
Stainless steel															
18 Cr 8 Ni	SS-303L-P	As sintered	35	0.241	32	0.221							2.0	6.0	52 R_B
18 Cr 8 Ni	SS-303L-R	As sinstered	52	0.359	46	0.324					4.5	6.102	7.3	6.6	55 R_B
SAE 303		Wrought, ann.	90	0.621	34	0.241	20	0.138			80.0	108.45	5.0	7.9	80 R_B
18 Cr 12 Ni 2 Mo	SS-316L-P	As sintered	38	0.264	34	0.241			96	0.665	2.0	2.708	2.0	6.16	55 R_B
18 Cr 12 Ni 2 Mo	SS-316L-R	As sintered	57	0.393	50	0.352			133	0.932	4.5	6.102	8.0	6.65	65 R_B
SAE 316L		Wrought, ann.	78	0.538	32	0.221	20	0.138			11.0	14.911	50	7.9	79 R_B
12.5 Cr 0.15C	SS-410-P	As sintered,	55	0.379	53	0.373							1	6.4	95 R_B
12.5 Cr 0.15C	SS-410-P	Sintered, h.t.	110	0.759					128	0.896	2.0	2.708		6.4	29 R_C
Copper and alloys															
90 Cu 10 Sn	BT-0010-S	As sintered	20	0.318	20	0.318			42	0.289			2–3	6.8–7.2	43 R_F
90 Cu 10 Sn	BT-0010-W	As sintered	45	0.310	28	0.200			90	0.621	4.5	6.102	11–15	8.0	80 R_F
80 Cu 20 Zn(+Pb)	BZ-0218-T	As sintered	20	0.138	15	0.103	15	0.103	31	0.214	3.0	4.071	10	7.2 min	37 R_F
80 Cu 20 Zn(+Pb)	BZ-0218-U	As sintered	23	0.159	18	0.124	25	0.173	65	0.449			12	7.7 min	42 R_F
80 Cu 20 Zn(+Pb)	BZ-0218-W	As sintered	37	0.255	28	0.193			80	0.554			21	8.0 min	50 R_F
64 Cu 18 Ni 18 Zn		As sintered	25	0.173	15	0.110			55	0.379			15	7.2	65 R_F
64 Cu 18 Ni 18 Zn		As sintered	40	0.289	26	0.180			85	0.586			14	7.9	90 R_B
64 Cu 18 Ni 18 Zn		Wrought, ann.	57	0.400	25	0.173							40	8.73	50 R_B
Nickel and alloys															
67 Ni 30 Cu 3 Fe		As sintered	30	0.207	18	0.124			110	0.759			8	7.0–7.4	34 R_B
67 Ni 30 Cu 3 Fe		As sintered	50	0.358	22	0.152			112	0.775			19.5	8.0	50 R_B
67 Ni 30 Cu 3 Fe		Wrought, ann.	90	0.621	54	0.379							35	8.84	88 R_B

[a] Powder Metallurgy Parts Manufacturers Association.

Key: σ_{su} = ultimate tensile strength, σ_y = yield stress, σ_b = transverse fiber stress, τ = shear stress; h.t. = heat-treated; ann. = annealed.

23.9 DESIGN OF GEARS ACCORDING TO ISO GEAR RATING FORMULAS AND COMPARISON OF THESE WITH GEAR RATING FORMULAS OF AGMA (4, 27, 28)

SYMBOLS

a	center distance, mm (in)
b	face width, mm (in)
b_e	effective face width, mm (in)
b_h	semiwidth of Hertzian band of contact, mm (in)
B_M	thermal contact coefficient
$C_1 \cdots C_6$	distance along the line of action, mm (in) (Fig. 23–124)
d	reference diameter, mm (in)
d_a	tip circle diameter, mm (in)
d_b	base circle diameter, mm (in)
d_f	root circle diameter, mm (in)
$d' = d_w$	working pitch diameter, mm (in)
E	modulus of elasticity, GPa (Mpsi)
F_t	the load tangential to the reference cylinder of diameter d and perpendicular to an axial plane, kN (lbf)
$F_t^* = \dfrac{F_t}{b}$	transverse unit load, kN (lbf)
g_t	length of the line of action in the transverse plane, mm (in)
$g_{po} = T$	distance along the line of action from the pitch point to the point at which $\sigma_c(P)$ is considered, mm (in)
h_a	addendum, mm (in)
h_{a0}	addendum of tool with reference to m_n, mm (in)
h_f	dedendum, mm (in)
h_{fp}	dedendum of basic rack with reference to $m_n(= h_{a0})$, mm (in)
h_F	bending moment arm for tooth root stresses for application of load at the outer point of single tooth pair contact, mm (in)
h_{Fa}	bending moment arm for the tooth root stresses for the application of load at the tooth tip, mm (in)
$h_{min} = H_{min}\rho_n$	minimum oil film thickness, μm (μin)
H_{min}	dimensionless oil film thickness
H_B	Brinell hardness number
H_v	Vickers hardness
$i(= u)$	gear ratio (always ≥ 1)
K	load stress factor (Fig. 23–130)
K_c	constant in Eq. (23–710a)
K_v	dynamic factor
$K_{B\alpha}$	the transverse load distribution factor for scuffing
$K_{F\alpha}$	tranverse load distribution factor for tooth root stress
$K_{H\alpha}$	transverse load distribution factor for contact stress
$K_{B\beta}$	face load factor for scuffing
$K_{F\beta}$	face load factor for tooth root stress
$K_{H\beta}$	face load factor for contact stress

m	mass, kg (lb), module, mm
m_t	transverse module, mm (in)
m_n	normal module, mm (in)
M_t	torque, N m (lbf in)
n	speed, rpm
	factor of safety (also with subscripts)
N_L	number of load cycles
P	transmitted power, kW (hp)
	diametral pitch, mm^{-1} (in^{-1})
p	circular pitch, mm (in)
pro	protuberance value with reference to m_n
q_s	notch parameter
r_{a1}	outside or addendum radius of pinion, mm (in)
r_{a2}	outside or addendum radius of gear (wheel), mm (in)
r_{b1}	base radius of pinion, mm (in)
r_{b2}	base radius of gear, mm (in)
R_a	average roughness value, μm (μin) [Eqs. (23–758b) and (23–759)]
R_c	Rockwell C hardness
R_y	peak-to-valley roughness, μm (μin)
R_z	mean peak-to-valley roughness, μm (μin)
$P(= \sigma_c = \sigma_H)$	Hertz contact stress (or pressure) (usually refer to tip of pinion and tip of gear), MPa (psi)
s	tooth thickness, mm (in)
s_{Fn}	tooth root chord or thickness at the critical section, mm (in)
S	average surface roughness (rms), μm (μin)
$T_1(= g_{p\sigma}), T_2$	distance along the line of action from the pitch point to the point at which P (or σ_H) is considered, mm (in) [Eqs. (23–733) and (23–735)]
T_c	contact temperature, °C (°F)
T_{fl}	flash temperature, °C (°F)
T_M	bulk temperature, °C (°F)
T_{oil}	oil temperature, °C (°F)
T_S	scuffing temperature, °C (°F)
T_{cp}	permissible contact temperature, °C (°F)
$u(= i)$	gear ratio
v	linear velocity, m/s (ft/s)
v_t	pitch line velocity, m/s (ft/s)
v_e	entraining velocity, m/s (ft/s)
$w_t(= F_t^*)$	transverse unit load, kN/m (lbf/in)
x	addendum modification coefficient
$X_G(= X_B)$	geometry factor
X_M	thermal-elastic factor, K $N^{-0.75}$ $s^{0.5}$ $m^{-0.5}$ mm (°F $lbf^{-0.75}$ $s^{0.5}$ $in^{0.5}$)
X_w	welding factor
X_Γ	load sharing factor
X_s	lubrication weight factor
Y_d	design factor
Y_N	life factor
z	number of teeth
$z_n(= z_v)$	virtual number of teeth
α	pressure-viscosity coefficient,
	pressure angle, deg
α_n	normal pressure angle at reference cylinder
α_t	transverse pressure angle at reference cylinder

α_{tw}	transverse pressure angle at pitch cylinder
β	helix angle at reference cylinder
β_b	helix angle at base cylinder
ε_α	transverse contact ratio
ε_β	overlap ratio
ε_γ	total contact ratio
ρ_{a0}	tip radius of tool reference to m_n
ρ_{fp}	root radius of basic rack reference to $m_n (= \rho_{a0})$
ρ_c	effective radius (mm) of curvature at pitch point
ρ_F	root fillet radius (mm) in the critical section
Γ_y	linear parameter on the line of action
$\Gamma_A \cdots \Gamma_E$	linear parameter of points $A \cdots E$
ϵ	pinion roll angle, deg
$\epsilon_1 \cdots \epsilon_5$	pinion roll angle at points $1 \cdots 5$
η_{oil}	dynamic viscosity of the oil at oil temperature, mPa [s(cP)]
$\lambda = \dfrac{h_{min}}{\sigma}$	specific film thickness
μ_m	mean coefficient of friction
$\sigma = (\sigma_1^2 + \sigma_2^2)^{0.5}$	composite surface roughness
σ_H or σ_c	Hertzian contact or surface stress, MPa (psi)
σ_{ut} or σ_B	ultimate tensile strength, MPa (psi)
σ_{FE}	tooth root bending endurance limit of reference test gear, MPa (psi)
σ_y (or σ_{sy})	yield strength at 0.2% proof stress, MPa (psi)
τ	shear stress, MPa (psi)
ω	angular velocity, rad/s

SUFFIXES

a	axial
b	bending,
	values on base circle
c	compressive
n	values on normal section,
t	tensile,
	transverse
u	ultimate
y	yield
w, v	working or operating condition
$1, 2$	for sizes of pinion and gear wheel, respectively
'	value on pitch circle
fl	flash
max	maximum
min	minimum

Other factors in performance or in special aspects which are included from time to time in this section and, being applicable only in their immediate context, are not given at this stage.

Particular	Formula

BENDING STRENGTH

AGMA fundamental formula for calculating stress at transmitted load (F_t) for the design of helical gear teeth, which takes care of all factors explained under Eq. (23–95) for spur gear. (3, 4)

[*Note*: The other factors such as K_a, K_{sz}, and K_m have already been explained under spur gear, Eq. (23–95). The values of these factors given under spur gear can be used.]

$$\sigma_b = \frac{F_t K_a}{K_v}\frac{1}{m_t b}\cdot\frac{(\cos\beta K_m)K_{sz}K_B}{J} \quad \textbf{SI} \quad (23\text{–}295)$$

$$=\frac{F_t K_a}{K_v}\cdot\frac{P_t}{b}\cdot\frac{(\cos\beta K_m)K_{sz}K_B}{J}$$

US Customary System Units (23–296)

[Equations (23–295) and (23–296) are repeated here to compare these with ISO formulas of ISO 6336.*]

For definition of symbols in Eqs. (23–295) and (23–296), refer to *Symbols* given at the beginning of Chap. 23.

ISO formula for calculting tooth root stress or bending stress for pinion and gear (wheel), respectively, at transmitted load (F_t) (27, 28)

$$\sigma_F = \frac{F_t}{b\cdot m_n}\cdot Y_F\cdot Y_S\cdot Y_\beta\cdot K_A\cdot K_\gamma\cdot K_v\cdot K_{F\beta}\cdot K_{F\alpha}$$

$$(23\text{–}695a)$$

where
Y_F = tooth root form factor
Y_S = stress correction factor
Y_β = helix angle factor

The load tangential to the reference cylinder of diameter d and perpendicular to an axial plane (F_t)

$$F_t = \frac{2000M_t}{d}\frac{2}{m} \qquad \left(F_{bt} = \frac{F_t}{\cos\alpha_t}\right) \quad (23\text{–}695b)$$

The tooth root form factor based on the load applied at the outer point of single-tooth pair contact and also based on the distance between the contact points of the 30° tangents at the highest root fillet of the tooth profile (Fig. 23–116) for use in Eq. (23–695)

$$Y_F = \frac{6\cdot\dfrac{h_F}{m_n}\cdot\cos\alpha_{Fen}}{\left(\dfrac{s_{Fn}}{m_n}\right)^2\cdot\cos\alpha_n} \quad (23\text{–}696)$$

[*Note*: The meaning of each symbol in Eqs. (23–695) and (23–696) is defined under *Symbols* in the beginning of this section (Sec. 23.9). The expressions and numerical values, wherever required, are given here.]

The tooth root form factor when the load is applied at the tip of the tooth (Fig. 23–116) for use in Eq. (23–695)

$$Y_{Fa} = \frac{6\cdot\dfrac{h_{Fa}}{m_n}\cdot\cos\alpha_{Fan}}{\left(\dfrac{s_{Fn}}{m_n}\right)^2\cdot\cos\alpha_n} \quad (23\text{–}697)$$

Y_F and Y_{Fa} are determined for virtual number of teeth in case of helical gears. Refer to Fig. 23–117.

* Courtesy ISO—based on ISO/DIS/6336 and Norway Det Norske Veritas Classification Notes No. 412, July 1988.

Particular	Formula

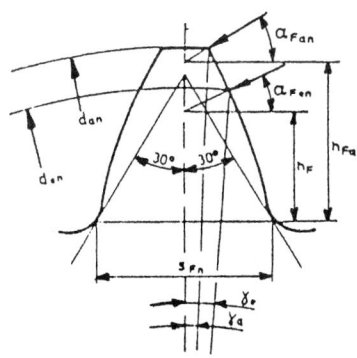

FIGURE 23–116 Gear-tooth layout for finding tooth root form factor with load sharing.

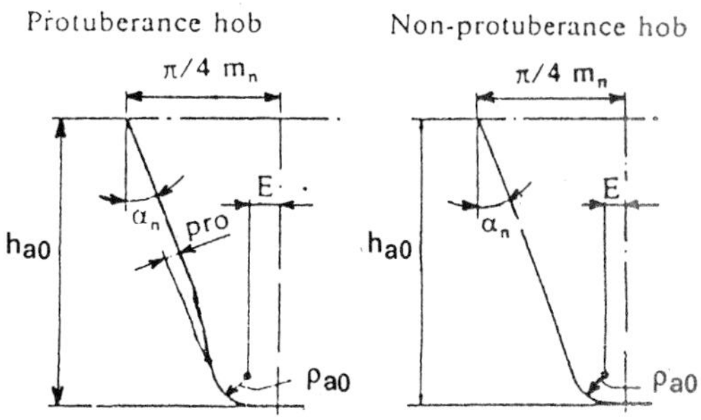

Protuberance hob   Non-protuberance hob

$\pi/4\ m_n$ $\pi/4\ m_n$

FIGURE 23–117

External gears

The tooth root chord or tooth thickness s_{Fn} (Fig. 23–116)

$$\frac{s_{Fn}}{m_n} = z_n \cdot \sin\left(\frac{\pi}{3} - \vartheta\right) + \sqrt{3} \cdot \left(\frac{G}{\cos\vartheta} - \rho_{a0}\right)$$

(23–698)

where

$$E = \left[\frac{\pi}{4} - h_{a0} \cdot \tan\alpha_n - \frac{\rho_{a0}(1 - \sin\alpha_n) - \text{pro}}{\cos\alpha_n}\right] \cdot m_n$$

$$G = \rho_{a0} - h_{a0} + x$$

$$H = \frac{2}{z_n} \cdot \left(\frac{\pi}{2} - \frac{E}{m_n}\right) - \frac{\pi}{3}$$

$$\vartheta = \frac{2G}{z_n} \cdot \tan\vartheta - H \text{ (to be solved iteratively)}$$

The root fillet radius $\rho_f\,(\rho_F)$ at 30° tangent (Fig. 23–116)

$$\frac{\rho_F}{m_n} = \rho_{a0} + \frac{2G^2}{\cos\vartheta \cdot (z_n \cdot \cos^2\vartheta - 2 \cdot G)}$$

(23–699)

The bending moment arm h_F

$$\frac{h_F}{m_n} = \frac{1}{2}\left[(\cos\gamma_e - \sin\gamma_e \tan\alpha_{Fen}) \cdot \frac{d_{en}}{m_n}\right.$$

$$\left. - z_n \cdot \cos\left(\frac{\pi}{3} - \vartheta\right) - \frac{G}{\cos\vartheta} + \rho_{a0}\right]$$

(23–700)

where
$$d_n = z_n \cdot m_n$$
$$d_{an} = d_n + 2 \cdot h_a$$
$$p_{bt} = \pi \cdot m_n \cdot \cos\alpha_t / \cos\beta$$
$$d_{bn} = d_n \cdot \cos\alpha_n$$

Particular	Formula

$$d_e = \frac{2z}{|z|} \cdot \sqrt{\left(\sqrt{\left(\frac{d_a}{2}\right)^2 - \left(\frac{d_b}{2}\right)^2} - \frac{z}{|z|} \cdot p_{bt} \cdot (\varepsilon_\alpha - 1)\right)^2 + \left(\frac{d_b}{2}\right)^2}$$

(23–701)

$$d_{en} = d_{an} - d_a + d_e$$

$$\alpha_{en} = \text{arc cos}\, \frac{d_{bn}}{d_{en}}$$

$$\gamma_e = \frac{1}{z_n}\left(\frac{\pi}{2} + 2x \tan \alpha_n\right) + inv\,\alpha_n - inv\,\alpha_{en}$$

Internal gears

The tooth chord or tooth thickness for internal gears, s_{Fn2}

$$\frac{s_{Fn2}}{m_n} = 2 \cdot \left(\frac{\pi}{4} + \tan \alpha_n \cdot (h_{a02} - \rho_{a02}) + \frac{\rho_{a02} - pro_2}{\cos \alpha_n}\right.$$

$$\left. - \rho_{a02} \cdot \cos \frac{\pi}{6}\right)$$

(23–702)

The bending moment arm, h_{F2}

$$\frac{h_{F2}}{m_n} = \frac{d_{en2} - d_{fn2}}{2 \cdot m_n} - \left[\frac{\pi}{4} + \left(h_{a02} - \frac{d_{en2} - d_{fn2}}{2 \cdot m_n}\right) \cdot \tan \alpha n_n\right]$$

$$\cdot \tan \alpha_n - \frac{\rho_{a02}}{2}$$

(23–703)

where

$$d_{fn2} = d_{n2} - 2 \cdot m_n \cdot (h_{a02} - x_2)$$

$$\rho_{F2} = \rho_{a02} \cdot m_n$$

The diameters d_{n2} and d_{en2} are to be calculated with the same formulae as for external gears.

The stress correction factor Y_S when the load applied is at the outer point of single-tooth pair contact for use in Eq. (23–695)

$$Y_S = (1,2 + 0,13 \cdot L) \cdot q_s^{(1/1,21+2,3/L)}$$

(23–704)

where

$$\left.\begin{array}{l} L = \dfrac{s_{Fn}}{h_F} \\[2em] q_s = \dfrac{s_{Fn}}{2 \cdot \rho_F} \end{array}\right\}$$ Refer to Eq. (23–699) for ρ_F, Eq. (23–700) for h_F, and Eq. (23–701) for s_{Fn}.

Y_{Sa} can be calculated by replacing h_F by h_{Fa} in Eq. (23–703).

Y_S is valid for range $1 < q_s < 8$ and $\alpha_n = 20°$.

Particular	Formula
An approximate expression for the contact ratio factor, Y_ε, for use with Y_{Fa} and Y_{Sa}	$$Y_\varepsilon = 0.25 + \frac{0.75}{\varepsilon_\alpha} \qquad (23\text{–}705)$$
The helix angle factor Y_β for use in Eq. (23–695)	$Y_\beta = 1 - \beta/100$ valid for effective $\varepsilon_\beta > 1$ (23–706a) $Y_\beta = 1 - \varepsilon_\beta \cdot \beta/100$ valid for effective $\varepsilon_\beta < 1$ (23–706b) where $\beta > 30°$ or $\beta = 30°$
The permissible tooth root stress for pinion and wheel (gear), respectively, σ_{Fp}	$$\sigma_{Fp} = \frac{\sigma_{FE} \cdot Y_d \cdot Y_N}{n_F} \cdot Y_{\delta relT} \cdot Y_{RrelT} \cdot Y_X \cdot Y_C$$ (23–707)
The calculated tooth root stress σ_F from Eq. (23–695) must be less than or equal to permissible tooth root stress σ_{Fp} as per Eq. (23–707)	$\sigma_F \le \sigma_{Fp}$ \qquad (23–708)
For endurance limit σ_{FE} for various materials of gears	The endurance limit, σ_{FE}, is the local tooth root stress which the material can endure permanently; 3×10^6 load cycles is regarded as the beginning of the endurance. Refer to Table 23–119 for endurance limit (σ_{FE}) for various gear materials.
For design factor, Y_d	Refer to Table 23–120 for various design factors, Y_d.
The actual life factor, Y_N, for a given number of load cycles N_L within the range of the σ–N ($S - N$) curve with which the permissible stress σ_{FP} for infinite life is to be multiplied	$$Y_N = \left(\frac{3 \cdot 10^6}{N_L}\right)^a \qquad (23\text{–}709)$$ where $$a = 0.2876 \cdot \log \frac{\sigma_{FP} \text{ for } \langle\langle \text{static}\rangle\rangle \text{ strength}}{\sigma_{FP} \text{ for infinite life}}$$ Refer to Table 23–121 for σ_{FP}.
For calculation of σ_{FP} if σ–N curves for materials are not given	Refer to Table 23–121.
For relative notch sensitivity factor, $Y_{\delta relT}$	Refer to Table 23–122.
For size factor, Y_x	Under endurance limit condition: $Y_x = 1$ for $m_n \le 5$ generally $Y_x = 1.03 - 0.006 \cdot m_n$ for $5 < m_n < 30$ for no-surface- $Y_x = 0.85$ for $m_n \ge; 30$ hardened steels $Y_x = 1.05 - 0.01 \cdot m_n$ for $5 < m_n < 25$ for surface- $Y_x = 0.8$ for $m_n \ge 25$ hardened steels Under static strength condition: $Y_x = 1$ for all m_n and all materials
The case depth factor, Y_c	$$Y_c = \frac{K_c}{\sigma_{FE}} \cdot \left(1 + \frac{3 \cdot t}{\rho_F + 0.2 \cdot m_n}\right) \qquad (23\text{–}710a)$$

Particular	Formula
	where constants K_c and, case depth t are related as given in Table 23–123
	s_{an} = top land thickness, mm
The case depth factor, Y_c, when hardness of surface of gear teeth, is 550 H_V	$Y_c \not> \frac{1}{3} s_{an}$ (23–710b) $\not> 0.25 m_n$
The case depth factor, Y_c, can be calculated if Y_c exceeds $s_{an}/3$ or $0.25\, m_n$	$Y_c = 1 - \left(\dfrac{t_{550\,max}}{m_n} - 0.25 \right)$ (23–710c)
Rim thickness factor, K_B [*Note*: The expressions for stresses developed in thin rims due to stress concentration effect, etc. are more involved and complicated. Hence the values of rim thickness factors given under AGMA equation are used here.)	K_B = 1 for unidirectional load, rigid backup gear = 1.4 for fully reversed load on gear, even if the backup is rigid
For relative surface condition factor, Y_{RrelT}	Refer to Table 23–124.

DESIGNING GEARS FOR SURFACE DURABILITY ACCORDING TO ISO

AGMA fundamental pitting resistance formula for calculating surface compressive stress or contact stress number or Hertzian contact stress as recommended by AGMA. [(Eq. (23–298) is repeated here for comparison purpose with ISO Hertzian contact stress equation: σ_H and σ_c are symbols used for Hertzian contact stress under symbols given in the beginning of this chapter.]	$\sigma_c = C_p \left(\dfrac{F_t C_a}{C_v} \cdot \dfrac{C_{sz}}{b d_1'} \cdot \dfrac{(\cos \beta C_m) C_{sr}}{I} \right)^{1/2}$ (23–298) where C_p = elastic coefficient, $\sqrt{\text{MPa}}$ ($\sqrt{\text{psi}}$). For expression for C_p, refer to Eq. (23–165), and for values of C_p, refer to Table 23–35. For definition of symbols in Eq. (23–298), refer to symbols given at the beginning of this chapter.
ISO formula for calculating Hertzian contact stress or surface-compressive stress of gear tooth. [*Note*: Contact stress σ_H has to be calculated both for pinion and gear.] (27, 28)	$\sigma_H = \Bigg[Z_{p,g} \cdot Z_H \cdot Z_E \cdot Z_\varepsilon \cdot Z_\beta$ $\qquad \cdot \sqrt{\dfrac{F_t(i+1)}{d_1 b i}} \, K_A \cdot K_\gamma \cdot K_v \cdot K_{H\beta} \cdot K_{H\alpha} \Bigg]$ (23–711) where $Z_{p,g}$ = zone factor for inner point of single pair contact for pinion and gear (wheel), respectively Z_H = zone factor for pitch point Z_E = elasticity factor Z_ε = contact ratio factor Z_β = helix angle factor F_t, K_A, K_γ, K_v, $K_{H\beta}$, $K_{H\alpha}$ were defined under *Symbols*

Particular	Formula
The zone factor, Z_H, which accounts for the influence on contact stresses of the tooth flank curvature at pitch point for use in Eq. (23–711)	$$Z_H = \sqrt{\frac{2 \cdot \cos \beta_b \cdot \cos \alpha_{tw}}{\cos^2 \alpha_t \cdot \sin \alpha_{tw}}} \qquad (23\text{–}712)$$
The zone factor, $Z_{p,g}$, which accounts for the influence on contact stresses of the tooth flank curvature at the inner point of single pair contact in relation to Z_H, for use in Eq. (23–711)	$Z_{p,g} = 1 \qquad$ For effective $\varepsilon_\beta \geq 1 \qquad (23\text{–}713\text{a})$ $Z_{p,g} =$ $$\frac{\tan \alpha_{tw}}{\sqrt{\left(\sqrt{\left(\dfrac{d_{a1,2}}{d_{b1,2}}\right)^2 - 1} - \dfrac{2\pi}{z_{1,2}} \right)\left(\sqrt{\left(\dfrac{d_{a2,1}}{d_{b2,1}}\right)^2 - 1} - (\varepsilon_\alpha - 1) \cdot \dfrac{2\pi}{z_{2,1}} \right)}}$$ for spur gear $\varepsilon_\beta = 0 \qquad (23\text{–}713\text{b})$ $Z_{p,g} = 1 + (1 - \varepsilon_\beta) \cdot (Z_{p,g}(\text{for spur gears}) - 1)$ For $0 <$ effective $\varepsilon_\beta < 1 \qquad (23\text{–}713\text{c})$
The zone factor, Z_i, which accounts for the influence on contact stresses of the tooth flank curvature at the inner point of double pair of contact for use in Eq. (23–711)	$$Z_i = \sqrt{\frac{1}{2} \cdot \frac{\rho_{c1} \cdot \rho_{c2}}{\rho_{i1} \cdot \rho_{i2}}} \qquad (23\text{–}713\text{d})$$ where $$\rho_{c1,2} = \frac{d_{b1,2}}{2} \tan \alpha_{tw}$$ $$\rho_{i2} = \frac{z_2}{|z_2|} \sqrt{\left(\frac{d_{a2}}{2}\right)^2 - \left(\frac{d_{b2}}{2}\right)^2}$$ $$\rho_{i1} = \rho_{c1} + \rho_{c2} - \rho_{i2}$$
The effective radius of curvature at pitch point, ρ_c	$$\rho_c = \frac{a \cdot \sin \alpha_{tw} i}{\cos \beta_b (1 + i)^2} \qquad (23\text{–}713\text{e})$$
The elasticity factor, Z_E, for steel-on-steel for use in Eq. (23–711)	$Z_E = 189.8 \qquad (23\text{–}713\text{f})$ Refer to Eq. (23–165) for $Z_E = C_p$.
The contact ratio factor, Z_ε, for use in Eq. (23–711)	$$Z_\varepsilon = \sqrt{\frac{1}{\varepsilon_\alpha}} \qquad \text{for } \varepsilon_\beta \geq 1 \qquad (23\text{–}714\text{a})$$ $$Z_\varepsilon = \sqrt{\frac{4 - \varepsilon_\alpha}{3} \cdot (1 - \varepsilon_\beta) + \frac{\varepsilon_\beta}{\varepsilon_\alpha}} \quad \text{for } \varepsilon < 1 \quad (23\text{–}714\text{b})$$
The helix angle factor, Z_β, for use in Eq. (23–711)	$Z_\beta = \sqrt{\cos \beta} \qquad (23\text{–}715)$
The permissible contact stress, σ_{HP}	$$\sigma_{HP} = \frac{\sigma_{H\lim} \cdot Z_N}{n_H} \cdot Z_L \cdot Z_v \cdot Z_R \cdot Z_W \cdot Z_X \qquad (23\text{–}716)$$ $\sigma_{H\lim}$ = endurance limit for contact stresses Z_N = life factor for contact stresses

Particular	Formula
	n_H = required safety factor according to the rules
	Z_L, Z_v, Z_R = oil film influence factors
	Z_w = work hardening factor
	Z_x = size factor
The Hertzian contact stress, σ_H, calculated by Eq. (23–711), should be less than the permissible contact stress σ_{HP} from Eq. (23–716)	$\sigma_H < \sigma_{HP}$ $\hspace{4em}$ (23–717)
For values of endurance limit, $\sigma_{H\lim}$, and static strengths, σ_{H10^3}, σ_{H10^5}	Refer to Table 23–125.
For life factor, Z_N	Refer to Table 23–126.
For influence factors on lubrication film, Z_L, Z_v, Z_R	Refer to Table 23–127.
The work-hardening factor, Z_w	$$Z_w = 1.2 - \frac{H_B - 130}{1700} \hspace{2em} (23\text{–}718)$$ where H_B = Brinell hardness number of soft material of gear $Z_w = 1$ for $H_B > 470$
For size factor, Z_x	Refer to Table 23–128. $Z_x = 1$ for non-surface-hardened gears
The application factor, K_A and K_{AP}	Refer to Table 23–129.
The load-sharing factor, K	$K = 1 + 0.25(n_{pl} - 3)^{1/2}$ for epicyclic gears $\hspace{3em}$ (23–719a) $= 1 + (0.2/\phi)$ for dual-tandem gears $\hspace{0.5em}$ (23–719b) where ϕ = quill shaft twist under full load, deg n_{pl} = number of planets (≥ 3)
The dynamic factor, K_v for slow speed $v < 10$ m/s	$$K_v = 1 + \frac{4 \cdot Q^2 \cdot v \cdot z_1}{(2 + \sqrt{\varepsilon_\beta}) \cdot 10^5} \cdot \sqrt{\frac{i^2}{1 + i^2}} \hspace{1.5em} (23\text{–}720)$$ where Q is the grade of accuracy regarding pitch and profile errors as per ISO 1328, 1975 (For bevel gears the actual z_1, i and v for midpoint of gear should be used.)
The dynamic factor, K_v, for medium speeds $v = 5$ to 40 m/s	

Particular	Formula

K_v in the subcritical sector: $N \leq 0.85$

(*Note*: Equations required for detail calculation of K_v for higher speeds $v > 40$ m/s are beyond the scope of this book.)

$$K_v = 1 + NK \tag{23–721a}$$
where

N = relative proximity between actual speed n_1 and the lowest resonance speed n_{E1}

$$= \frac{n_1}{n_{E1}}$$

$$n_{E1} = \frac{30 \cdot 10^3}{\pi \cdot z_1} \cdot \sqrt{\frac{c_\gamma}{m_{red}}} \tag{23–721b}$$

where c_γ is the actual mesh stiffness per unit face width

The reduced mass of the gear pair per unit face width for use in Eq. (23–721b)

$$m_{red} = \frac{m_1 \cdot m_2}{m_1 + m_2} \tag{23–721c}$$

The individual mass per unit face width for use in Eq. (23–721c)

$$m_{1,2} = \frac{I_{1,2}}{b(d_{b1.2}/2)^2} \tag{23–721d}$$

where I = the polar moment of inertia, kg mm^2

The expression for the value of constant K for use in Eq. (23–721a)

$$K = C_{v1} \cdot B_p + C_{v2} \cdot B_r + C_{v3} \cdot B_k \tag{23–721e}$$

For B_p, B_f, and B_k, refer to Eqs. (23–722b), (23–722c), and (23–722d) where C_{v1}, C_{v2}, and C_{v3} are

C_{v1} = accounts for the pitch error influence

$C_{v1} = 0.32$

C_{v2} = accounts for profile error influence

$C_{v2} = 0.34$ for $\varepsilon_\gamma \leq 2$

$\quad = \dfrac{0.57}{\varepsilon_\gamma - 0.3}$ for $\varepsilon_\gamma > 2$

C_{v3} = accounts for the cyclic mesh stiffness variation

$C_{v3} = 0.23$ for $\varepsilon_\gamma \leq 2$

$\quad = \dfrac{0.096}{\varepsilon_\gamma - 1.56}$ for $\varepsilon_\gamma > 2$

The total contact ratio, ε_γ

The transverse contact ratio, ε_α, for use in Eq. (23–721f)

$$\varepsilon_\gamma = \varepsilon_\alpha + \varepsilon_\beta \tag{23–721f}$$

$$\varepsilon_\alpha = \frac{0.5 \cdot \sqrt{d_{a1}^2 2 - d_{b1}^2} \pm 0.5 \cdot \sqrt{d_{a2}^2 - d_{b2}^2} - a \cdot \sin \alpha_{tw}}{\pi \cdot m_n \cdot \cos \alpha_t / \cos \beta}$$
$$\tag{23–722a}$$

Particular	Formula
The nondimensional gear accuracy-dependent parameters for use in Eq. (23–721e) B_p and B_f	$$B_p = \frac{c' \cdot (f_{pb} - y_p)}{F_t \cdot K_A \cdot K_\gamma / b} \qquad (23\text{–}722b)$$ $$B_f = \frac{c' \cdot (f_f - y_f)}{F_t \cdot K_A \cdot K_\gamma / b} \qquad (23\text{–}722c)$$
The nondimensional tip relief, B_k parameter for use in Eq. (23–721e)	$$B_k = \left| 1 - \frac{c' \cdot C_a}{F_t \cdot K_A \cdot K_\gamma / b} \right| \qquad (23\text{–}722d)$$ For gears of quality grade $Q = 6$ or coarser grade (ISO 1328, 1975), use $B_k = 1$.
The overlap ratio, ε_β, for use in Eq. (23–721f)	$$\varepsilon_\beta = \frac{b \cdot \sin \beta}{\pi \cdot m_n} \qquad (23\text{–}722e)$$

where

f_{pb}	= base pitch error (ISO 1328, 1975), maximum of pinion or wheel
f_f	= total profile error (ISO 1328, 1975), maximum of pinion or wheel (*Note*: f_f is pro tempore not available for bevel gears; thus use $f_f = f_{pb}$.)
y_p and y_f	= respective running-in allowances and may be calculated similarly to y_α in Table 23–130, i.e., the value of f_{pb} is replaced by f_f for y_f
c'	= single-tooth stiffness
C_a	= the amount of tip relief. When C_a is zero, then running-in tip relief C_{ay} may be used.

K_v in the critical sector: $0.85 < N \le 1.15$ for high-precision gears	$$K_v = 1 + C_{v1} \cdot B_p + C_{v2} \cdot B_f + C_{v4} \cdot B_k \qquad (23\text{–}723)$$ C_{v4} accounts for the resonance condition with the cyclic mesh stiffness variation $$C_{v4} = 0.90 \text{ for } \varepsilon_\gamma \le 2$$ $$= \frac{0.57 - 0.05 \cdot \varepsilon_\gamma}{\varepsilon_\gamma - 1.44} \text{ for } \varepsilon > 2$$
K_v in the supercritical sector: $N \ge 1.5$	$$K_v = C_{v5} \cdot B_p + C_{v6} \cdot B_f + C_{v7} \qquad (23\text{–}724)$$ C_{v5} accounts for the pitch error influence $$C_{v5} = 0.47$$ C_{v6} accounts for the profile error influence $$C_{v6} = 0.47 \text{ for } \varepsilon_\gamma \le 2$$ $$C_{v6} = \frac{0.12}{\varepsilon_\gamma - 1.74} \text{ for } \varepsilon_\gamma > 2$$

Particular	Formula

C_{v7} relates the maximum externally applied tooth loading to the maximum tooth loading of ideal, accurate gears

$$C_{v7} = 0.75 \text{ for } \varepsilon_\gamma \leq 1.5$$
$$C_{v7} = 0.125 \cdot \sin[\pi(\varepsilon_\gamma - 2)] + 0.875$$
$$\text{for } 1.5 < \varepsilon_\gamma \leq 2.5$$
$$C_{v7} = 1.0 \text{ for } \varepsilon_\gamma > 2.5$$

K_v in the intermediate sector: $1.15 < N < 1.5$

(*Note*: Equations required for the detail calculation of K_v for higher speeds $v > 40$ m/s are beyond the scope of this book.)

K_v is determined by linear interpolation between K_v for $N = 1.15$ and $N = 1.5$ as

$$K_v = K_{v(N=1.5)} + \left(\frac{1.5 - N}{0.35}\right) \cdot [K_{v(N=1.15)} - K_{v(N=1.5)}]$$

(23–725)

FACE LOAD FACTORS, $K_{H\beta}$, $K_{F\beta}$, and $K_{B\beta}$

The relations between $K_{H\beta}$, $K_{F\beta}$, and $K_{B\beta}$
Eq. (23–726b) is valid only for gears with the hardest contact toward a tooth end and $b_1 \approx b_2 \approx b$

The relation between $K_{H\beta}$ and $K_{F\beta}$ for gears with end relief or crowning

The relationship between the transverse load distribution factors, $K_{H\alpha}$ for contact stresses, $K_{F\alpha}$ for tooth root stresses, and $K_{B\alpha}$ for scuffing account for the effects of pitch and profile errors on the transversal load distribution between two or more pairs of teeth in mesh

The mean value of the mesh stiffness in a transverse plane

The factors $K_{B\beta}$, $K_{H\beta}$, and $K_{F\beta}$ account for nonuniform distribution of stresses across the face width of teeth

$$K_{B\beta} = K_{H\beta} \tag{23–726a}$$

$$K_{F\beta} = K_{H\beta}\, \frac{1}{\left[1 + h/b + (h/b)^2\right]} \tag{23–726b}$$

$$K_{F\beta} = K_{H\beta} \tag{23–726c}$$

$$K_{B\alpha} = K_{F\alpha} = K_{H\alpha} = \left[\frac{\varepsilon_v}{2}\left(0.9 + 0.4\,\frac{C_\gamma \cdot (f_{pb} - y_\alpha) \cdot b}{F_{tH}}\right)\right] \tag{23–727a}$$

valid for $\varepsilon_\gamma \leq 2$

$$K_{B\alpha} = K_{F\alpha} = K_{H\alpha} = \left[0.9 + 0.4 \sqrt{\frac{2(\varepsilon_\gamma - 1)}{\varepsilon_\gamma}} \cdot \frac{c_\gamma \cdot (f_{pb} - y_\alpha) \cdot b}{F_{tH}}\right] \tag{23–727b}$$

valid for $\varepsilon_\gamma > 2$

where
$$F_{tH} = F_t \cdot K_A \cdot K_\gamma \cdot K_v \cdot K_{H\beta}$$
$$\varepsilon_\gamma = \varepsilon_\alpha + \varepsilon_\beta$$
$$c_\gamma = c' \cdot (0.75 \cdot \varepsilon_\alpha + 0.25) \tag{23–728a}$$
$c_\gamma = 20$ may be used in case the details are not available

Particular	Formula
The maximum stiffness of a single pair of teeth	$$c' = \frac{0.8 \cdot \cos\beta}{q} \cdot C_R \cdot C_{BS} \qquad (23\text{--}728b)$$

where

$$C_{BS} = \left\{ \left[1 + 0.5 \cdot \left(1.2 - \frac{h_{a01} + h_{a02}}{2} \right) \right] \right.$$

$$\left. [1 - 0.02(20 - \alpha_n)] \right\} \qquad (23\text{--}728c)$$

$$q = \left[0.04723 + \frac{0.15551}{z_{n1}} + \frac{0.25791}{z_{n2}} - 0.00635 \cdot x_1 \right.$$

$$- \frac{0.11654 \cdot x_1}{z_{n1}} - 0.00193 \cdot x_2 - \frac{0.24188 \cdot x_2}{z_{n2}}$$

$$\left. + 0.00529 \cdot x_1^2 + 0.00182 \cdot x_2^2 \right] \qquad (23\text{--}728d)$$

(For internal gears, use z_{n2} equal infinite and $x_2 = 0$.)

$h_{a0} = h_{fp}$ for all practical purposes
$f_{pb} =$ maximum adjacent base pitch error (μm) of pinion or wheel, or maximum profile error of pinion or wheel if this is larger than the maximum adjacent pitch error

The expression for C_R

$$C_R = 1 + \frac{\ln(b_S/b)}{5 \cdot e^{(s_R/5m_n)}} \qquad (23\text{--}728e)$$

where
$b_S =$ web thickness (sum of both, if two)
$s_R =$ average thickness of rim

The formula is valid for $b_S/b \geq 0.2$ and $s_R/m_n \geq 1$.

A tip relief is considered sufficient when it is within ± 30 percent of the value given by Eq. (23–728f)

$$C_{ay} = \frac{F_{bt} \cdot K_A}{b_{\text{eff}} \cdot c_\gamma} \qquad (23\text{--}728f)$$

where $b_{\text{eff}} =$ estimated active face width $+ m_n$ (but not more than b)

Limitations of $K_{H\alpha}$, $K_{F\alpha}$, and $K_{B\alpha}$

If the calculated values for $K_{F\alpha} = K_{H\alpha} = K_{B\alpha} < 1$, use

$$K_{F\alpha} = K_{H\alpha} = K_{B\alpha} = 1 \qquad (23\text{--}729a)$$

If the calculated value of $K_{H\alpha} = K_{B\alpha} > \dfrac{\varepsilon_\gamma}{\varepsilon_\alpha \cdot Z_\varepsilon^2}$, use

$$K_{H\alpha} = K_{B\alpha} = \frac{\varepsilon_\gamma}{\varepsilon_\alpha \cdot Z_\varepsilon^2} \qquad (23\text{--}729b)$$

If the calculated value of $K_{F\alpha} > \dfrac{\varepsilon_\gamma}{\varepsilon_\alpha \cdot Y_\varepsilon}$, use

$$K_{F\alpha} = \frac{\varepsilon_\gamma}{\varepsilon_\alpha \cdot Y_\varepsilon} \qquad (23\text{--}729c)$$

Particular	Formula
The running-in allowances y_α and y_β are the running-in amounts which reduce the influence of pitch and profile errors, and the influence of localized face load	The running-in allowances account for the influence of running-in wear on the various error elements. Refer to Table 23–130 for the running-in amounts y_α, y_β, and the running-in tip relief C_{ay}.

The actual safety factor, n_F, for bending strength	$$n_F = n_{Fmin}\, \frac{\sigma_{Fp}}{\sigma_F} \qquad (23\text{–}729\text{d})$$
	where
	n_{Fmin} = minimum safety factor
	  = 1.4 to 1.5 for normal industrial applications
	  = 1.6 to 3.0 for high reliability and critical applications

The actual safety factor n_H, for surface or contact stress	$$n_H = n_{Hmin}\, \frac{\sigma_{Hp}}{\sigma_H} \qquad (23\text{–}729\text{e})$$ where n_{Hmin} = 1.0 to 1.2 for normal industrial applications = 1.3 to 1.6 for high reliability and critical applications (high damage, loss of life, etc.) σ_{Hp} = permissible contact stress, MPa (psi) σ_{Fp} = permissible bending stress, MPa (psi) σ_H or σ_c = actual contact stress, MPa (psi) σ_F = actual bending stress, MPa (psi)

DESIGNING GEARS FOR SCUFFING

In countries that use the metric system (outside USA) the term *scuffing* is used instead of *scoring* in failure of gears, which is characterized by radial scratch lines on the flank of gears. But in the United States *scoring* is used for this type of gear failure.

The specific oil film thickness (refer to Table 23–131B for h_{min}) PVT (pressure-sliding velocity-distance along the line of action) (or $\sigma_H v_s g_{p\sigma}$ or $\sigma_c v_s g_{p\sigma}$)

$$\lambda = \frac{h_{min}}{\sigma} \qquad (23\text{–}730)$$

where $\sigma = (\sigma_1 + \sigma_2)^{1/2}$ = composite surface roughness

The Hertz contact or surface compressive stress for the tip of the pinion (Figs. 23–20 and 23–118) [*Note*: The PVT acronym or nomenclature was used in the 1940s. Therefore a somewhat different nomenclature, which is appropriate to the current practice and units, is used.]

$$\sigma_{c1} = P_1 = 5740 \left(\frac{M_{t1}\, a \sin \alpha_n}{b g_t z_1 \rho_1 (a \sin \alpha_t - \rho_1)} \right)^{1/2}$$
US Customary System units $\qquad (23\text{–}731)$

where

ρ_1 = radius of curvature at the pinion tip, in Fig. 23–118
  = $[r_{a1}^2 - (r_1' \cos \alpha_t)^2]^{1/2}$
b, g_t, ρ_1, and a in in
M_{t1} = pinion torque, in lbf-in
Refer to Eq. (23–301a) for g_t and Fig. 23–4

Particular	Formula

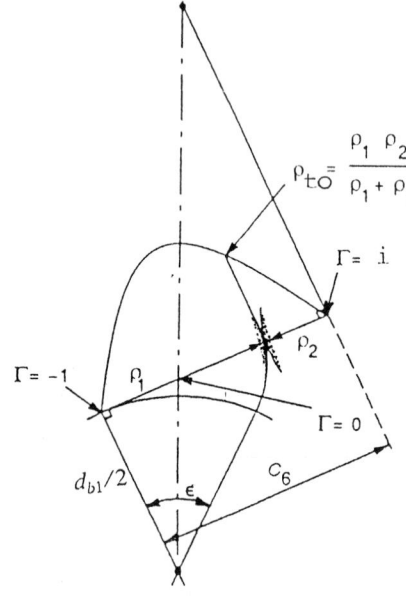

$$\rho_{to} = \frac{\rho_1 \, \rho_2}{\rho_1 + \rho_2}$$

$\Gamma = i$

$\Gamma = -1$

$\Gamma = 0$

ρ_2

ρ_1

$d_{b1}/2$

ϵ

C_6

FIGURE 23–118   Transverse relative radius of curvature.

σ_c is the symbol used for Hertzian contact stress (and in some places σ_H) throughout this chapter. Hence σ_{c1} is the Hertz contact stress for the tip of the pinion, psi (and P_1 in psi)

T $= g_{p\sigma} =$ distance along the line of action from the pitch point to the point at which σ_c (or P) is considered, in

The Hertzian contact or surface compressive stress for the tip of the gear

$$\sigma_{c2} = P_2 = 5740 \left(\frac{M_{t1} a \sin \alpha_n}{b g_t z_1 \rho_2 (a \sin \alpha_t - \rho_2)} \right)^{1/2}$$

US Customary System units   (23–732)

where

ρ_2 $=$ radius of curvature at gear tip, in

     $= (r_{a2}^2 - (r_2' \cos \alpha_t)^2)^{1/2}$

b, g_t, ρ_2, and a in in, M_{t1} in lbf in σ_{c2} (or P_2) in psi

For spur gear $\alpha_n = \alpha_t = \alpha$

The scuffing or scoring factor for the pinion tip

PVT_1 (or $\sigma_{H1} v_s g_{p\sigma}$ or $\sigma_{c1} v_s g_{p\sigma}$)

$$= \frac{2\pi n_1}{12 \times 60} \left(1 + \frac{z_1}{z_2} \right) (\rho_1 - r_1' \sin \alpha_t) \sigma_{H1}$$

US Customary System units   (23–733)

where ρ_1, r_1' in in, n_1 in rpm, σ_{H1} in psi

Particular	Formula

$$= \frac{2\pi n_1}{60}\left(1 + \frac{z_1}{z_2}\right)(\rho_1 - r_1'\sin\alpha_t)\sigma_{H1} \quad \textbf{SI} \qquad (23\text{–}734)$$

where r_1' and ρ_1 in m, n_1 in rpm, σ_{H1} in Pa

The scuffing or scoring factor for the gear tip

PVT_2 (or $\sigma_{H2}v_s g_{p\sigma}$ or $\sigma_{c2}v_s g_{p\sigma}$)

$$= \frac{2\pi n_2}{12\times 60}\left(1 + \frac{z_1}{z_2}\right)(\rho_2 - r_2'\sin\alpha_t)\sigma_{H2}$$

US Customary System units $\qquad (23\text{–}735)$

where ρ_2 and r_2' in in, n_2 in rpm, σ_{H2} in psi

$$= \frac{2\pi n_2}{60}\left(1 + \frac{z_1}{z_2}\right)(\rho_2 - r_2'\sin\alpha_t)\sigma_{H2} \quad \textbf{SI} \qquad (23\text{–}736)$$

where ρ_2 and r_2' in m, n_2 in rpm, σ_{H2} in Pa

FLASH TEMPERATURE

The equation for flash temperature according to Dudley

$$T_f = T_b + \frac{c_f\mu F_t v_s}{\cos\alpha_t b_e(\sqrt{v_1} + \sqrt{v_2})\sqrt{b_H}} \qquad (23\text{–}737)$$

where $v_1 = \dfrac{2\pi n_1\rho_1}{12\times 60}$ and $v_2 = \dfrac{2\pi n_2\rho_2}{12\times 60}$

US Customary System units

$$v_1 = \frac{2\pi n_1\rho_1}{1000\times 60} \text{ and } v_2 = \frac{2\pi n_2\rho_2}{1000\times 60} \quad \textbf{SI}$$

ρ_1 and ρ_2 in mm

T_b = temperature of blank surface in contact (inlet oil temperature), °C (°F)
c_f = material constant for conductivity, density, and specific heat
    = 0.0528 for straight petroleum oils
v_s = $v_1 - v_2$ = sliding velocity, m/s (ft/s)
μ = coefficient of friction = 0.06

The width of the rectangular band of contact for steel gears

$$2b_H = 0.00054\left(\frac{F_t\rho_1\rho_2}{\cos\alpha_t b_e(\rho_1 + \rho_2)}\right)^{1/2}$$

US Customary System units $\qquad (23\text{–}738a)$

Another equation for the semiwidth of the rectangular band of contact in terms of load-sharing factor

$$b_H = \left(\frac{8X_\Gamma w_n\rho_n}{\pi E_r}\right)^{0.5} \qquad (23\text{–}738b)$$

where
X_Γ = load-sharing factor [Refer to Eqs. (23–780) to (23–788).]
w_n = normal unit load

Particular	Formula
	ρ_n = normal relative radius of curvature = $\dfrac{\rho_{to}}{\cos \beta_b}$
	(23–738c)
	E_r = reduced modulus of elasticity
	$= 2\left(\dfrac{1 - v_1^2}{E_1} + \dfrac{1 - v_2^2}{E_2}\right)^{-1}$ (23–738d)
The radius of curvature for lowest single-tooth contact (LSTC)	$\rho_1 = (r_{a1}^2 - (r_1' \cos \alpha_t)^2)^{1/2} - p \cos \alpha_t$ (23–739a) $\rho_2 = a \sin \alpha_t - \rho_1$ (23–739b)
The radius of curvature for highest single-tooth contact (HSTC)	$\rho_1 = a \sin \alpha_t - \rho_2$ (23–740a) $\rho_2 = [(r_{a2}^2 - (r_2' \cos \alpha_t)^2]^{1/2} - p \cos \alpha_t$ (23–740b)
	Refer to Table 23–131*A* for flash temperature T_f for various oils to prevent scuffing of spur gears.

Scuffing or scoring criterion

The scuffing or scoring criterion index number according to Dudley (26)	$Z_c = \left(\dfrac{F_t}{b_e}\right)^{0.75} \dfrac{n_1^{1/2}}{P_t^{0.25}}$ **US Customary System units**
	(23–741a)
	where Z_c in °F factor
	$= \left(\dfrac{F_t}{b_e}\right)^{0.75} \left(\dfrac{m_t^{0.25}}{1.094}\right) n_1^{1/2}$ **SI** (23–741b)
	where Z_c in °C factor
	m_t in mm, F_t in N, b_e in mm
	Refer to Table 23–132 for critical scuffing (scoring) criterion index numbers.
Another form of the flash temperature equation according to Dudley	$T_f = T_b + Z_t (F_t/b_e)^{0.75} \dfrac{n_1^{1/2}}{P_t^{0.25}}$
	US Customary System units (23–742a)
	$= T_b + Z_t \left(\dfrac{F_t}{b_e}\right)^{0.75} \left(\dfrac{m_t^{0.25}}{1.094}\right) n_1^{1/2}$ **SI** (23–742b)
	where Z_t is a dimensionless number and can be thought of as a tooth geometry factor for scoring (or scuffing)
The geometry factor for scuffing, Z_t (scoring) (Fig. 23–118) and Eq. (23–74)	$Z_t = 0.0175 \dfrac{(\sqrt{\rho_1} - \sqrt{(\rho_2/i)}) P_t^{0.25}}{(\cos \alpha_t)^{0.75} \rho_0^{0.25}}$ (23–743)
	where ρ_1, ρ_2, and ρ_0 are as defined in Fig. 23–118.

Particular	Formula

Hot scuffing (scoring)

The general design formula for flash temperature for spur and helical gears according to Dudley

$$T_f = T_b + Z_t Z_s Z_c \qquad (23\text{--}744)$$

where
Z_t = geometry constant factor, dimensionless (Table 23–133A)
Z_s = surface finish factor, dimensionless (Table 23–133B)
Z_c = scoring or scuffing criterion number °C factor or °F factor. Refer to Table 23–132.

The temperature rise of gear body in well-designed gears with oil nozzles delivering enough oil ought not to exceed the values

25°C (45°F) for aerospace gears
15°C (27°F) for turbine gears

Refer to Table 23–133A for values of Z_t.

For flash temperatures, T_f

Refer to Table 23–138B.

The tangential tooth load F_{te} which is applied to a point on the profile in danger of scuffing or scoring

$F_{te} = F_t C_m x$ (percent load for profile position)
$$\qquad (23\text{--}745)$$

where
C_m = load distribution factor for surface durability.

A standard amount of gear-tip modification for general design

$$C_{a2} = \frac{6.5 C_m F_t}{10^5 b} \quad \textbf{SI} \qquad (23\text{--}746a)$$

where C_{a2} and b in mm

$$= \frac{4.5 C_m F_t}{10^7 b} \quad \textbf{US Customary System units}$$
$$\qquad (23\text{--}746b)$$

where C_{a2} and b in in

A standard amount of pinion-tip modification for general design

$$C_{a1} = \frac{4.1 C_m F_t}{10^5 b} \quad \textbf{SI} \qquad (23\text{--}747a)$$

where C_{a1} and b in mm

$$= \frac{2.8 C_m F_t}{10^7 b} \quad \textbf{US Customary System units}$$
$$\qquad (23\text{--}747b)$$

where C_{a1} and b in in

Particular	Formula

GEAR RATING FOR SCUFFING LOAD AS PER ISO STANDARDS

The contact temperature at any point on the line of action as per ISO standards (ISO DIS 6336) (Figs. 23–118 and 23–4)

$$T_c = T_M + T_{fl} \qquad (23\text{–}748)$$

where

T_M = the bulk temperature, °C (°F)
T_{fl} = the flash temperature, °C (°F)

The maximum contact temperature at any point on the line of action

$$T_{c,max} = T_M + T_{fl,max} \qquad (23\text{–}749)$$

where $T_{fl,max}$ = maximum flash temperature, °C (°F)

The bulk temperature as per ISO standards (*Note*: Symbol ϑ or t is used for temperature in ISO. But T is used throughout this chapter. Hence T is used for temperature in this section also to maintain uniformity.)

$$T_M = X_s(T_{\text{oil}} + 0.47 n_m T_{fl,max}) \qquad (23\text{–}750)$$

where

X_s = the lubrication weight factor
   = 1.0 for dip lubrication
   = 1.2 for spray lubrication
T_{oil} = the oil temperature before it reaches the gear mesh, °C (°F)
n_m = the average number of meshes for pinion and gear (wheel)
n_m = $1 + \dfrac{n_{\text{planets}}}{2}$ for planetary gears
n_m = $\dfrac{1 + n_{\text{pin}}}{2}$ for wheel (gear) with several pinions

The bulk temperature as per AGMA standards

$$T_M = -24 + 1.2 T_{\text{oil}} + 0.56 T_{fl,max} \qquad (23\text{–}751)$$

The bulk temperature for submerged gears running in oil as per ISO standards

$$T_M = T_{\text{oil}} + 0.1 T_{fl,max} \qquad (23\text{–}752)$$

The flash temperature along the path of contact as per ISO standards

$$T_{fl} = \mu_{my} \cdot X_M \cdot X_B \cdot X_{\alpha\beta} \cdot X_\Gamma \cdot \frac{w_t^{3/4} \cdot v^{1/2}}{a^{1/4}}$$

$$\text{SI} \qquad (23\text{–}753)$$

$$= 0.45 \mu_m X_M X_G \frac{(X_\Gamma w_t)^{0.75} v_t^{0.5}}{a^{0.25}} \qquad (23\text{–}754)$$

where

w_t = transverse unit load, N/m (lbf/in) or specific tooth load [Eq. (23–763)] (Also F^* is used for unit load throughout this chapter.)
v_t = operating pitch line velocity, m/s (ft/s)
   = $\left(\dfrac{\pi n_1 a}{i + 1}\right) = 5\omega_1 r'_{w1} = 5\omega_1\left(\dfrac{a}{i + 1}\right)$

u is the symbol used for speed ratio in SI units, but i is the symbol used throughout this book. In order to

Particular	Formula
	maintain uniformity, i is used for speed ratio in this section.
	$\mu_m = \mu_{my} =$ mean coefficient of friction [Refer to Eq. (23–759).]
	$X_M =$ thermal flash or thermal elastic factor [Refer to Eq. (23–755).]
For load-sharing factor, X_Γ	Refer to Eqs. (23–780) to (23–788).
The thermal elastic factor for use in Eqs. (23–753) and (23–754)	$$X_M = E_r^{0.25}\left[\frac{(1+\Gamma_y)^{0.5}+(1-\Gamma_y/i)^{0.5}}{B_{M1}(1+\Gamma_y)^{0.5}+B_{M2}(1-\Gamma_y/i)^{0.5}}\right]$$ (23–755)
	where
	E_r = reduced modulus of elasticity [Eq. (23–738d)]
	Γ_y = parameter on the line of action [refer to Eq. (23–773)]
	$B_{M1} = (\lambda_{M1}\rho_{M1}c_{M1})^{0.5}$ thermal contact coefficient of the pinion material
	$B_{M2} = (\lambda_{M2}\rho_{M2}c_{M2})^{0.5}$ thermal contact coefficient of the gear material
The thermal elastic factor for gears made of steels whose $E = 30$ Mpsi (206 GPa), $v = 0.3$ and $B_M = 43.0$ lbf/(in $\cdot$ s$^{0.5}\cdot$ °F)(13.6 N/[mm $\cdot$ s$^{0.5}\cdot$ K])	$X_M = 1.75°\text{F}\cdot\text{lbf}^{-0.75}\ \text{s}^{0.5}\ \text{in}^{0.5}$ **US Customary System units** (23–756a) $X_M = 50.0\ \text{K}\cdot\text{N}^{-0.75}\ \text{s}^{0.5}\ \text{m}^{-0.5}\ \text{mm}$ **SI** (23–756b)
An empirical equation for a variable coefficient of friction according to Benedict and Kelley (30)	$$\mu_m = 0.0127\ \frac{50}{50-S}\ \log_{10}\left(\frac{3.17\times10^8\chi_\Gamma w_{Nr}}{\eta_{\text{oil}}v_s v_e^2}\right)$$ (23–757)
	where $S = \dfrac{\sigma_1+\sigma_2}{2}$
The surface roughness expression is limited to as per Eq. (23–758)	$\dfrac{50}{50-S}\le 3.0$ **US Customary System units** (23–758a)
	$\dfrac{1.13}{1.13-R_a}\le 3.0$ **SI** (23–758b)
The mean coefficient of friction according to ISO standards	$$\mu_m = 0.12\left(\frac{w_t}{\eta_{\text{oil}}v_e}\right)^{0.25}\left(\frac{R_a}{\rho_{t0}}\right)^{0.25}$$ (23–759)
	where
	η_{oil} = dynamic viscosity of the oil at oil temperature, MPa [s(cP)]
	R_a = $0.5(R_{a1}+R_{a2})$
	R_{a1}, R_{a2} = CLA roughness of pinion and gear teeth in the profile direction, μm
	v_e = entraining velocity, m/s

Particular	Formula
	ρ_{t0} = transverse relative radius of curvature at any point of contact
	$$\rho_{t0} = \frac{\rho_1 \rho_2}{\rho_2 \pm \rho_1} \qquad (23\text{–}760a)$$
	Refer to Eqs. (23–760b) and (23–760c) for ρ_1 and ρ_2 and Fig. 23–118 for ρ_{t0}
The transverse radius of curvature of pinion tooth profile at any point as defined by the roll angle ε (Fig. 23–118)	$$\rho_1 = \frac{a\sin\alpha_{wt}}{i\pm 1}\,(1+\Gamma_y) \qquad (23\text{–}760b)$$
	where Γ_y = a parameter at any point along the line of action Eq. (23–773)
The transverse radius of curvature of gear tooth profile at any point as defined by the roll angle ε (Fig. 23–118)	$$\rho_2 = \frac{a\sin\alpha_{wt}}{i\pm 1}\,(1\mp\Gamma_y) \qquad (23\text{–}760c)$$
The entraining velocity (absolute value)	$$v_e = |v_{r1} + v_{r2}| \qquad (23\text{–}760d)$$ where rolling velocities $$v_{r1} = \omega_1\rho_1 = 2\pi n_1\rho_1$$ $$v_{r2} = \omega_2\rho_2 = \frac{\omega_1}{i}\,\rho_2$$
The dynamic viscosity of the oil for use in Eq. (23–759)	$$\eta_{\text{oil}} = v_{\text{oil}}\cdot\rho/1000 \qquad (23\text{–}760e)$$ where ρ in kg/m^3
Specific gravity at any temperature	$$p_t = \rho_{15} - (T_{\text{oil}} - 15)\cdot 0.0007 \quad \textbf{SI} \qquad (23\text{–}760f)$$ and v_{oil} is kinematic viscosity at T_{oil}
	$$= \rho_{60} - 0.000365(T - 60) \qquad (23\text{–}760g)$$ **US Customary System units**
	The density (ρ) of oil and its specific gravity (γ) relative to water have the same numerical value, and the relation between these two is $\rho = \frac{\gamma}{g}$ and ρ in kg/m^3.
The geometry factor X_B (or X_G) for use in Eqs. (23–753) and (23–754)	$$X_G = X_B = 0.51\cdot\sqrt{|i+1|}\cdot\frac{\left|\sqrt{1+\Gamma_y} - \sqrt{1-\Gamma_y/i}\right|}{|(1+\Gamma_y)\cdot(i-\Gamma_y)|^{1/4}} \qquad (23\text{–}761)$$ where Γ_y is the parameter on the line of action Refer to Eq. (23–773).
	X_G is the symbol used in USCSU, and X_B is the symbol used in ISO. Refer to Fig. 23–119 for external gears and Fig. 23–120 for internal gears to find X_G (or X_B).

Particular	Formula

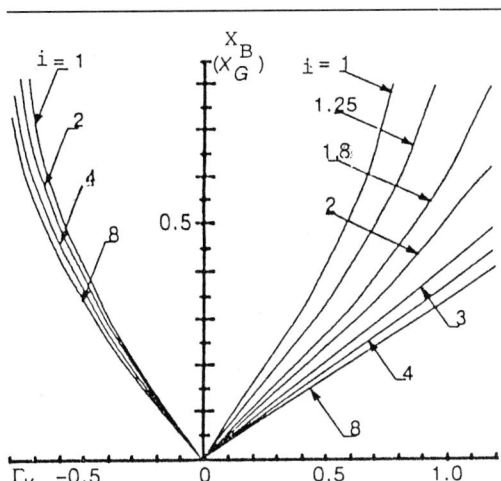

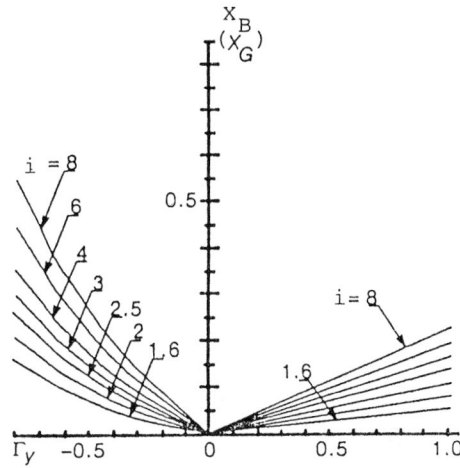

FIGURE 23–119 Geometry factor external gear.

FIGURE 23–120 Geometry factor internal gear.

The pressure angle factor, $X_{\alpha\beta}$ for use in Eq. (23–753)

$$X_{\alpha\beta} = 1.22 \frac{(\sin \alpha_{tw} \cdot \cos \alpha_n \cdot \cos \beta)^{1/4}}{(\cos \alpha_{tw})^{1/2} \cdot (\cos \alpha_t)^{1/2}} \qquad (23\text{–}762)$$

The specified tooth load or transverse unit load for use in Eqs. (23–753) and (23–754)

$$w_t = \frac{F_t}{b} \cdot K_A \cdot K_v \cdot K_\gamma \cdot K_{B\beta} \cdot K_{B\alpha} \cdot K_{B\gamma} \qquad (23\text{–}763)$$

Helical distribution factor, $K_{\beta\gamma}$ for use in Eq. (23–763)

$$K_{B\gamma} = 1 \qquad\qquad \text{for} \quad \varepsilon_\gamma \le 2$$
$$K_{B\gamma} = 1 + 0.2\sqrt{(\varepsilon_\gamma - 2) \cdot (5 - \varepsilon_\gamma)}$$
$$\qquad\qquad\qquad \text{for} \quad 2 < \varepsilon_\gamma < 3.5$$
$$K_{B\gamma} = 1.3 \qquad\qquad \text{for} \quad \varepsilon_\gamma \ge 3.5$$
$$\qquad\qquad\qquad\qquad\qquad (23\text{–}764)$$

For other factors in Eq. (23–763) refer to expressions under Eq. (23–695)

The permissible contact temperature

$$T_p = \frac{T_s - T_{\text{oil}}}{n_B} + T_{\text{oil}} \qquad (23\text{–}765)$$

where n_B = the required scuffing factor of safety

The contact temperature T_c should not exceed the permissible temperature T_p

$$T_c \not> T_p \qquad \text{or} \qquad T_c < T_p \qquad (23\text{–}766)$$

The scuffing temperature of mineral oils of various combinations with gear stress for use in Eq. (23–765)

$$T_s = T_{M,T} + X_{w,T} T_{fl,max,T} \qquad (23\text{–}767a)$$

where $X_{w,T}$ = welding factor. Refer to Table 23–134.

The bulk temperature of test gears for use in Eq. (23–767a)

$$T_{M,T} = 80 + 0.23 M_{t1,T} \qquad (23\text{–}767b)$$

Particular	Formula
The maximum flash temperature of test gears for use in Eq. (23–767a)	$$T_{fl,max,T} = 0.12(M_{t1,T})^{1.2} \cdot \left(\frac{100}{\nu_{40}}\right)^{(\nu_{40}^{-0.4})} \qquad \text{(23–767c)}$$ $M_{t1,T}$ = torque applied during test according to FZG Refer to Table 23–135 for FZG oil test.
The scuffing temperature for low additive mineral for use in Eq. (23–767a)	$T_s = 245 + 59 \ln \nu_{40}\,°\text{F}$ **US Customary System units** **AGMA** (23–768a) $T_S = 118 + 33 \ln \nu_{40}\,°\text{C}$ **SI** (23–768b) **ISO** where ν_{40} = kinematic viscosity at 104°F, cSt (40°C, mm^2/s)
The calculated safety factor shall not be less than the minimum demanded safety factor for contact temperature	$$n_B = \frac{T_S - T_{\text{oil}}}{T_{c,max} - T_{\text{oil}}} \geq n_{B,min} \qquad \text{(23–769)}$$ where T_{oil} = oil sump temperature T_S = scuffing temperature $T_{c,max}$ = maximum contact temperature $n_{B,min}$ = minimum demanded safety factor
The minimum elastohydrodynamic (EHD) dimensionless film thickness according to Dowson and Higginson (13)	$$H_{min} = 2.65\,\frac{G^{0.54} U^{0.70}}{W^{0.13}} \qquad \text{(23–770)}$$
The dimensionless material parameter for use in Eq. (23–770)	$$G = \alpha E_r \qquad \text{(23–771a)}$$
The dimensionless speed parameter for use in Eq. (23–770)	$$U = \frac{\mu_0 v_e}{2 E_r \rho_n} \qquad \text{(23–771b)}$$
The dimensionless load parameter for use in Eq. (23–770c)	$$W = \frac{\chi_\Gamma w}{E_r \rho_n} \qquad \text{(23–771c)}$$ where μ_0 = absolute viscosity, reyn Refer to Fig. 23–121. α = pressure – viscosity coefficient, (in^2/lb) The pressure – viscosity coefficient ranges from $\alpha = 0.5 \times 10^{-4}$ in^2/lb to $\alpha = 2 \times 10^{-4}$ in^2/lb for typical gear lubricants. Refer to Fig. 23–122.
For wear risk evaluation	Refer to Fig. 23–123 for probability of wear distress.
For MIL (Military Standard) lubricant mean scuffing temperatures	Refer to Table 23–136.
For mineral oil mean scuffing temperatures	Refer to Table 23–137.
For scuffing risk	Refer to Table 23–138A.
For welding factors, $X_{w,T}$	Refer to Table 23–134.

Particular	Formula

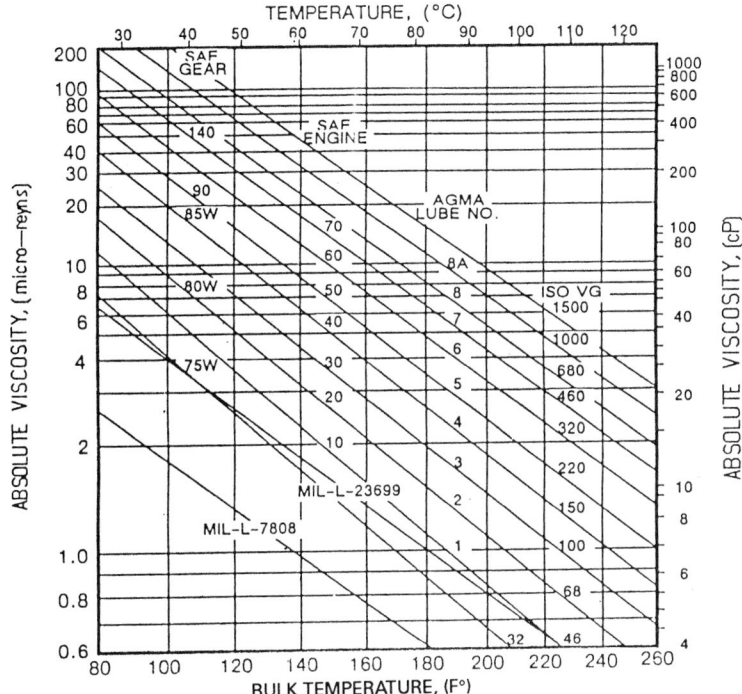

FIGURE 23–121 Absolute viscosity versus temperature. (*ANSI/AGMA 2001-B88.*)

PARAMETER Γ ALONG THE LINE OF ACTION (FIG. 23–124)

Figure 23–124 shows the line of action in a transverse plane. Distances C_i are measured from the interference point of the pinion along the line of action. Distance C_1 locates the pinion start of the active profile (SAP), and distance C_5 locates the pinion end of the active profile (EAP). The lowest and highest point of single-tooth-pair contact (LPSTC and HPSTC) are located by distances C_2 and C_4, respectively. Distance C_3 locates the operating pitch point.

$$C_6 = a \sin \alpha_{wt} \qquad (23\text{–}772a)$$

$$C_1 = \pm[C_6 - (r_{a2}^2 - r_{b2}^2)^{0.5}] \qquad (23\text{–}772b)$$

$$C_3 = \frac{C_6}{i \pm 1} \qquad (23\text{–}772c)$$

$$C_4 = C_1 + p_b \qquad (23\text{–}772d)$$

$$C_5 = (r_{a1}^2 - r_{b1}^2)^{0.5} \qquad (23\text{–}772e)$$

$$C_2 = C_5 - p_b \qquad (23\text{–}772f)$$

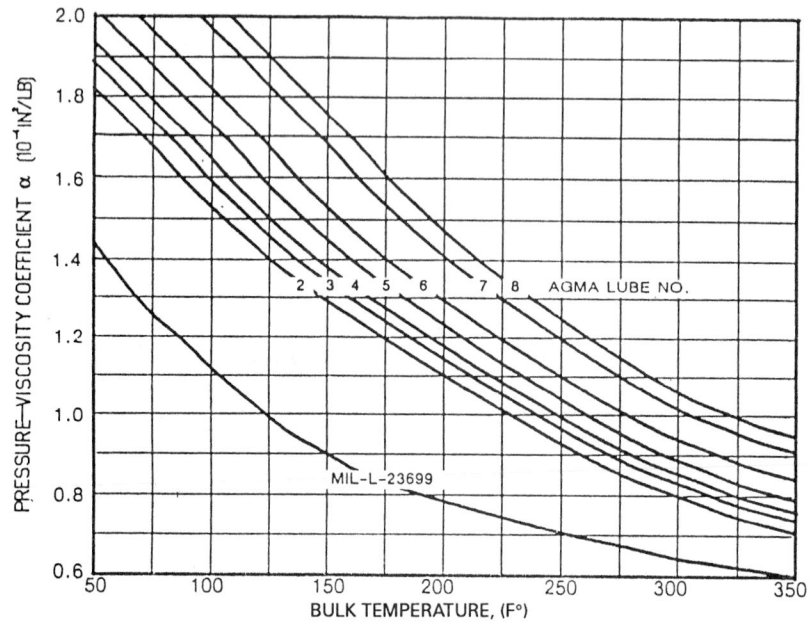

FIGURE 23–122 Pressure-viscosity coefficient vs. temperature. (*ANSI/AGMA 2001-B88.*)

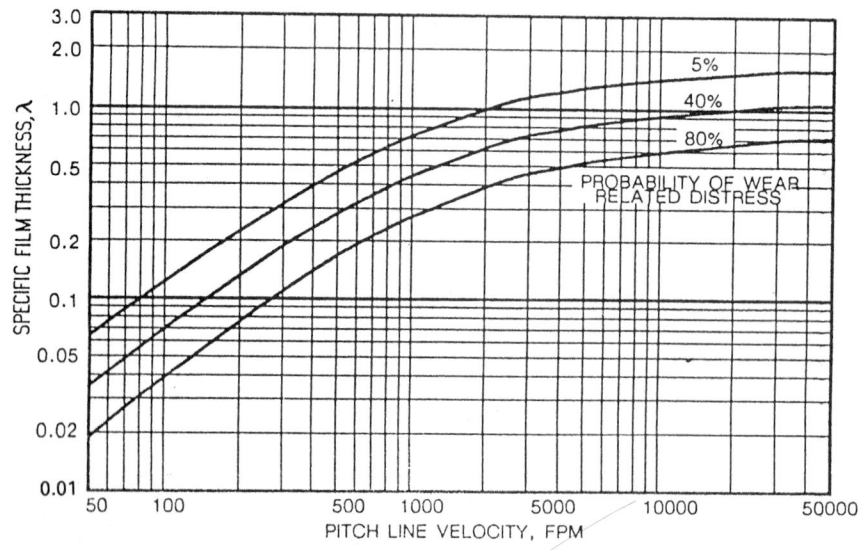

FIGURE 23–123 Probability of wear distress, percent. (*ANSI/AGMA 2001-B88.*)

Particular	Formula

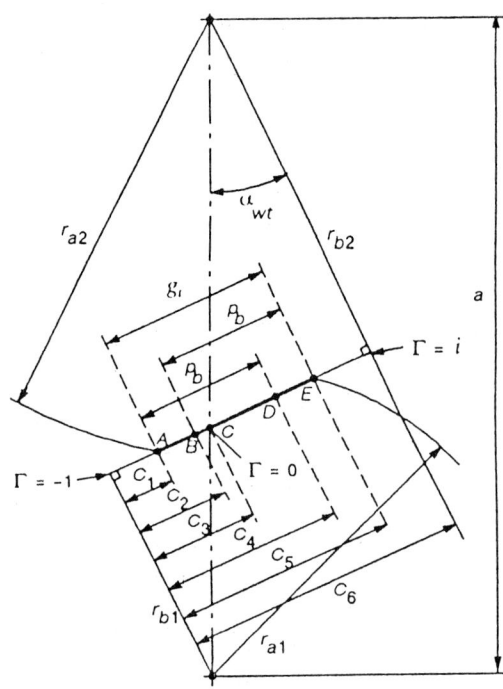

FIGURE 23–124 Distances along the line of action.

$$g_t = C_5 - C_1 \tag{23-772g}$$

The parameter Γ_y is defined as a dimensionless linear coordinate in the transverse plane on the line of action

The parameter, Γ_y, for any arbitrary point on the line of action

$$\Gamma_y = \frac{\tan \alpha_y}{\tan \alpha_{wt}} - 1 \tag{23-773}$$

Γ value at the start of the pinion root contact with the gear tip (SAP)

$$\Gamma_A = \frac{-z_2}{z_1} \left(\frac{\tan \alpha_{a2}}{\tan \alpha_{wt}} - 1 \right) \tag{23-774a}$$

$$= -\frac{z_2}{z_1} \left(\frac{\sqrt{(d_{a2}/d_{b2})^2 - 1}}{\tan \alpha_{tw}} - 1 \right) \tag{23-774b}$$

Γ value at the pinion tip with gear root (EAP)

$$\Gamma_E = \frac{\tan \alpha_{a1}}{\tan \alpha_{wt}} - 1 = \frac{\sqrt{(d_{a1}/d_{b1})^2 - 1}}{\tan \alpha_{tw}} - 1 \tag{23-775}$$

Γ value at the lowest point of transverse single-tooth-contact (LPSTC)

$$\Gamma_B = \Gamma_E - \frac{2\pi}{z_1 \tan \alpha_{wt}} \tag{23-776}$$

Particular	Formula
Γ value at the highest point of transverse single-tooth-contact (HPSTC)	$\Gamma_D = \Gamma_A + \dfrac{2\pi}{z_1 \tan \alpha_{wt}}$ (23–777)
Γ value at the interference point of the pinion	$\Gamma_y = -1$ (23–778)
Γ value at the pitch point	$\Gamma_c = 0$ (23–779)

LOAD-SHARING FACTOR, X_Γ

The equations for load sharing factors for an unmodified and for three typical modified tooth profiles are given here. By convention, the load-sharing factor is represented by a polygonal function on the line of action with magnitude equal to 1.0 between points B and D (Fig. 23–125)

The load-sharing factor accounts for load sharing between succeeding pairs of teeth as influenced by profile modification and whether the pinion or gear is the driving member.

The profile modifications are chosen such that the load sharing follows a desired function.

The load-sharing factor, X_Γ, for unmodified tooth profiles

$$X_\Gamma = \frac{1}{3} + \frac{1}{3}\left(\frac{\epsilon - \epsilon_1}{\epsilon_2 - \epsilon_1}\right) \text{ for } \epsilon_1 \le \epsilon < \epsilon_2$$

$$X_\Gamma = 1 \qquad\qquad\qquad \text{ for } \epsilon_2 \le \epsilon < \epsilon_4$$

$$X_\Gamma = \frac{1}{3} + \frac{1}{3}\left(\frac{\epsilon_5 - \epsilon}{\epsilon_5 - \epsilon_4}\right) \text{ for } \epsilon_4 \le \epsilon < \epsilon_5$$

                     AGMA   (23–780)

$$X_\Gamma = \frac{1}{3} + \frac{1}{3}\left(\frac{\Gamma_y - \Gamma_A}{\Gamma_B - \Gamma_A}\right) \text{ for } \Gamma_A \le \Gamma_y < \Gamma_B$$

$$X_\Gamma = 1 \qquad\qquad\qquad \text{ for } \Gamma_B \le \Gamma_y \le \Gamma_D$$

$$X_\Gamma = \frac{1}{3} + \frac{1}{3}\left(\frac{\Gamma_E - \Gamma_y}{\Gamma_E - \Gamma_D}\right) \text{ for } \Gamma_D < \Gamma_y \le \Gamma_E$$

                     ISO   (23–781)

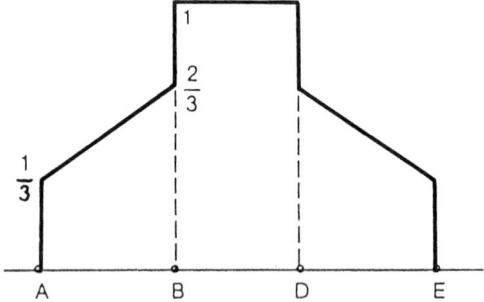

FIGURE 23–125 Unmodified profiles.

Smooth meshing of teeth is obtained by designing of gear pairs with adequate tip and root relief, i.e., trapezoidal load sharing (Fig. 23–126)

The expression for tip relief of pinion and gear (wheel) for smooth meshing

$$C_{\text{eff}} = \frac{F_{bt} \cdot K_A \cdot K_\gamma}{b \cdot c_\gamma} \qquad\qquad (23\text{–}782)$$

for gears with active face width less than b, use $b_{\text{eff}} = b_{\text{active}} + m_n$ (but not more than b).

The load-sharing factor, X_Γ, for smooth meshing with sufficient tip and root relief (Fig. 23–126)

$$X_\Gamma = \frac{\epsilon - \epsilon_1}{\epsilon_2 - \epsilon_1} \quad \text{for } \epsilon_1 \le \epsilon < \epsilon_2$$

$$X_\Gamma = 1 \quad\quad \text{for } \epsilon_2 \le \epsilon < \epsilon_4$$

$$X_\Gamma = \frac{\epsilon_5 - \epsilon}{\epsilon_5 - \epsilon_4} \quad \text{for } \epsilon_4 < \epsilon \le \epsilon_5$$

Particular	Formula

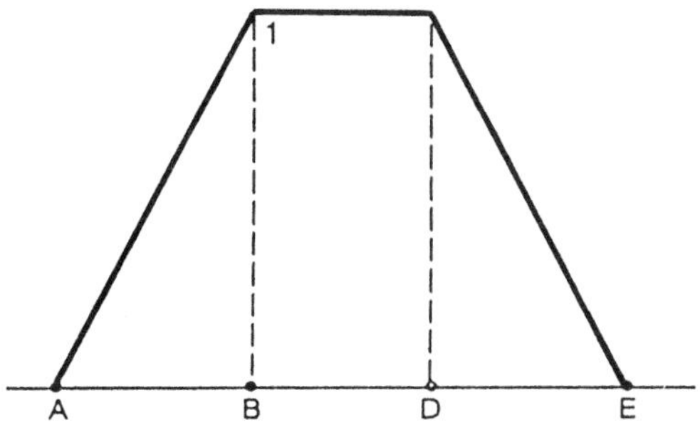

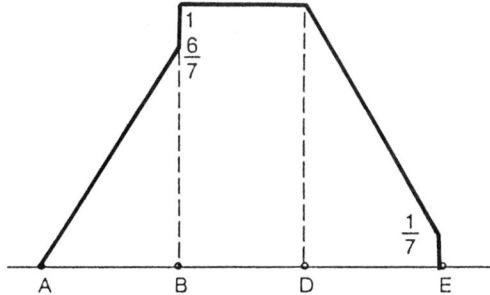

FIGURE 23–126   Smooth meshing.

AGMA   (23–783)

$$X_\Gamma = \frac{\Gamma_y - \Gamma_A}{\Gamma_B - \Gamma_A} \quad \text{for } \Gamma_A \leq \Gamma_y < \Gamma_B$$

$$X_\Gamma = 1 \qquad\qquad \text{for } \Gamma_B \leq \Gamma_y \leq \Gamma_D$$

$$X_\Gamma = \frac{\Gamma_E - \Gamma_y}{\Gamma_E - \Gamma_D} \quad \text{for } \Gamma_D < \Gamma_y \leq \Gamma_E$$

ISO   (23–784)

The load-sharing factor, X_Γ, for modified tooth profiles with adequate tip and root relief to carry high load and the pinion drives the gear (wheel) (Fig. 23–127)

$$X_\Gamma = \frac{6}{7}\left(\frac{\epsilon - \epsilon_1}{\epsilon_2 - \epsilon_1}\right) \qquad \text{for } \epsilon_1 \leq \epsilon < \epsilon_2$$

$$X_\Gamma = 1 \qquad\qquad\qquad \text{for } \epsilon_2 \leq \epsilon \leq \epsilon_4$$

$$X_\Gamma = \frac{1}{7} + \frac{6}{7}\left(\frac{\epsilon_5 - \epsilon}{\epsilon_5 - \epsilon_4}\right) \quad \text{for } \epsilon_4 < \epsilon \leq \epsilon_5$$

AGMA   (23–785)

$$X_\Gamma = \frac{6}{7}\left(\frac{\Gamma_y - \Gamma_A}{\Gamma_B - \Gamma_A}\right) \qquad \text{for } \Gamma_A \leq \Gamma_y < \Gamma_B$$

$$X_\Gamma = 1 \qquad\qquad\qquad \text{for } \Gamma_B \leq \Gamma_y < \Gamma_D$$

$$X_\Gamma = \frac{1}{7} + \frac{6}{7}\left(\frac{\Gamma_E - \Gamma_y}{\Gamma_E - \Gamma_D}\right) \quad \text{for } \Gamma_D < \Gamma_y \leq \Gamma_E$$

ISO   (23–786)

FIGURE 23–127   Pinion driving.

The load-sharing factor, X_Γ, for modified tooth profiles with adequate tip and root relief to carry high load and the pinion is driven by gear (wheel) (Fig. 23–128)

$$X_\Gamma = \frac{1}{7} + \frac{6}{7}\left(\frac{\epsilon - \epsilon_1}{\epsilon_2 - \epsilon_1}\right) \quad \text{for } \epsilon_1 \leq \epsilon < \epsilon_2$$

$$X_\Gamma = 1 \qquad\qquad\qquad \text{for } \epsilon_2 \leq \epsilon \leq \epsilon_4$$

$$X_\Gamma = \frac{6}{7}\left(\frac{\epsilon_5 - \epsilon}{\epsilon_5 - \epsilon_4}\right) \qquad \text{for } \epsilon_4 < \epsilon \leq \epsilon_5$$

AGMA   (23–787)

$$X_\Gamma = \frac{1}{7} + \frac{6}{7}\left(\frac{\Gamma_y - \Gamma_A}{\Gamma_B - \Gamma_A}\right) \quad \text{for } \Gamma_A \leq \Gamma_y < \Gamma_B$$

$$X_\Gamma = 1 \qquad\qquad\qquad \text{for } \Gamma_B \leq \Gamma_y \leq \Gamma_D$$

$$X_\Gamma = \frac{6}{7}\left(\frac{\Gamma_E - \Gamma_y}{\Gamma_E - \Gamma_D}\right) \qquad \text{for } \Gamma_D < \Gamma_y \leq \Gamma_E$$

ISO   (23–788)

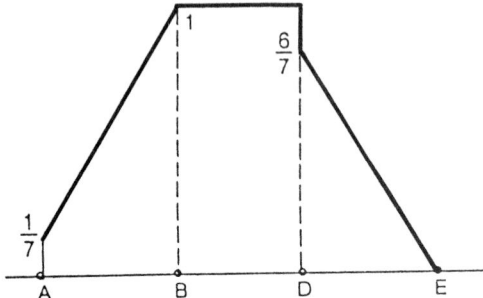

FIGURE 23–128   Gear driving.

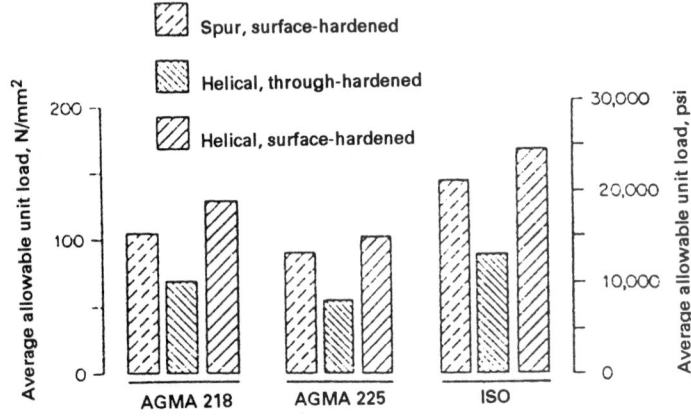

FIGURE 23–129   Average allowable unit load as per ISO and AGMA. (*Courtesy of the Cincinnati Gear Company, Cincinnati, Ohio, USA.*)

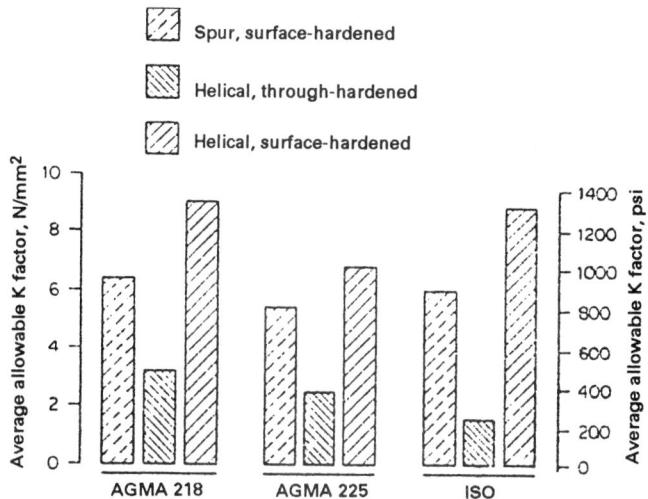

Spur, surface-hardened

Helical, through-hardened

Helical, surface-hardened

FIGURE 23–130 Average allowable K factor as per ISO and AGMA. (*Courtesy of The Cincinnati Gear Company, Cincinnati, Ohio, USA.*)

Particular	Formula

COMPARISON OF ISO WITH AGMA

The International Standards Organization (ISO) and American Gear Manufacturers Association (AGMA) systems do not calculate the same stress values for similar gears under identical loading conditions. Furthermore, the allowable stress values for a particular material in each system are not the same. This makes it somewhat difficult to compare the results of rating of gears by the two systems. The students, designers, and practicing engineers have to use judgment while selecting the AGMA or ISO system for calculating the rating of gears.

A correct comparison can be made by calculating the unit load and the K factor. AGMA and ISO agree very well in calculation of the unit load.

For the "average allowable unit load" as per ISO and AGMA according to Dennis E. Imwalle and Octave A. Labath (31)

Refer to Fig. 23–129.

For the "average allowable K factor" as per AGMA and ISO according to Dennis E. Imwalle and Octave A. Labath (31)

Refer to Fig. 23–130.

TABLE 23–119
Endurance limit, σ_{FE} (N/mm^2) for use in Eq. (23–707)

Heat treatment	σ_{FE}, N/mm^2
Alloyed case-hardened steels[a] (fillet surface hardness 58–63 R_C)	
Of specially approved high grade	1000[d]
Of specially approved intermediate grade	920
Of ordinary grade	860
Nitriding steel of approved grade, gas-nitrided (surface hardness 650–800 H_V):	840
Alloyed through-hardened steel, bath, or gas-nitrided (surface hardness 500–700 H_V)	740
Alloyed through-hardened steel, flame- or induction-hardened[b] (including entire root fillet) (fillet surface hardness 500–650 H_V)	$0.7 H_V + 300$
Alloyed through-hardened steel, flame- or induction-hardened (excluding entire root fillet area) (σ_B = u.t.s.[c] of base material):	$0.25\sigma_B + 125$
Alloyed through-hardened steel	$0.4\sigma_B + 200$
Carbon steel (annealed or through-hardened)	$0.25\sigma_B + 250$

TABLE 23–120
Design factor, Y_d, for use in Eq. (23–707)

Load characteristics	Design factor Y_d for Endurance limit	Design factor Y_d for Static strength
Unidirectional load, no shrink-fit prestress	1.0	1.0
Alternating load (idler gear, planet gear)	0.65	0.75[2])
Alternating load (idler gear) with considerably different stresses in each direction (i.e., $R \neq -1$)[1]) :	$\dfrac{1}{1 - 0.54R}$	0.75[2])
Full load periodically in both directions	0.75	0.75[2])
Occasional reversed load direction (reversing)	0.9	1.0[2])
Shrink-fit prestress σ_{fit}^3), surface-hardened fillets	$1 - \dfrac{\sigma_{\text{fit}}}{2\sigma_{FE}}$	Subtract σ_{fit} from σ_{FP}
Shrink-fit prestress σ_{fit}^3), non-surface-hardened fillets	$1 - \dfrac{\sigma_{\text{fit}}}{\sigma_B}$	Subtract σ_{fit} from σ_{FP}

TABLE 23–121
Permissible stress, σ_{FP}, for use in Eq. (23–708)

Heat treatment	Permissible stress, σ_{FP}
Surface-hardened gears, except nitrated	$\sigma_{FP} = \dfrac{2.5 \cdot_{FE}}{n_F} \cdot Y_d \cdot Y_{\delta\text{rel}T} \cdot Y_C$
Nitrated gears	$\sigma_{FP} = \dfrac{1.5 \cdot \sigma_{FE}}{n_F} \cdot Y_d \cdot Y_{\delta\text{rel}T} \cdot Y_C$
Non-surface-hardened gears	$\sigma_{FP} - \dfrac{1.8 \cdot \sigma_B}{n_F} \cdot Y_d \cdot Y_{\delta\text{rel}T}$ or $\dfrac{2.25 \cdot \sigma_\gamma}{n_F}$ whichever is smaller
Forged or rolled steel	$\sigma_{FP} - \dfrac{1.2 \cdot \sigma_B}{n_F} \cdot Y_d \cdot Y_{\delta\text{rel}T}$ or $\dfrac{2 \cdot \sigma_\gamma}{n_F}$ whichever is smaller

Notes for Table 23–119:
[a] (1) All values given here are valid for separate forgings. If rolled steel is used, reduce the above values by 15 percent. For cast steel, reduce the above values by 30 percent.
(2) These values hold good for an unground root radius.
(3) σ_{FE} is to be reduced by 20 percent for a ground fillet radius.
(4) If the surface of teeth are shotpeened the σ_{FE} is increased by 150 percent for unidirectional load and by 75 percent for reversed loads.
[b] The fillet is not to be ground.
[c] σ_B = Ultimate tensile strength.
[d] Values given above are approximate.

Notes: If combined conditions apply, such as idler with prestress, multiply the respective Y_d factors for the applicable conditions.
(1) The stress ratio $R(-1 \leq R \leq 0)$ may be taken as the reversed torque per unit face width divided with the ⟨⟨ forward ⟩⟩ torque per unit face width.
(2) In addition, the total stress range (tooth root stress in forward plus in reversed direction) is not to exceed:

$$\frac{2.25 \cdot \sigma_y}{n_F} \quad \text{for non-surface-hardened fillets}$$

$$\frac{5 \cdot H_V}{n_F} \quad \text{for surface-hardened fillets}$$

(3) The fillet shrink-fit stresses σ_{fit} with regard to the critical tooth root section (30° tangent) may be taken as tangential stress due to shrinkage times a stress-concentration factor of

$$1.5 - 2\,\frac{\rho_F}{m_n}$$

TABLE 23–122
Relative notch sensitivity factor for use in Eq. (23–707) $Y_{\delta \text{rel} T}$

Heat treatment	$Y_{\delta \text{rel} T}$
For endurance limit	
Non-surface-hardened fillets	$Y_{\delta \text{rel} T} = 1 + 0.036(q_s - 2.5) \cdot \left(1 - \dfrac{\sigma_y}{1200}\right)$
All surface-hardened fillets except nitrided	$Y_{\delta \text{rel} T} = 0.956 + 0.0234 \cdot \sqrt{1 + q_s}$
Nitrided fillets	$Y_{\delta \text{rel} T} = 0.79 + 0.112 \cdot \sqrt{1 + q_s}$
For static strength	
Non-surface-hardened fillets (forged or rolled steels)	$Y_{\delta \text{rel} T} = 1 + (Y_s - 2)\left(0.5 - 0.15 \dfrac{\sigma_y}{1000}\right)$
Non-surface-hardened fillets (cast steels)	$Y_{\delta \text{rel} T} = 0.86 + 0.07 \cdot Y_s$
Induction- or flame-hardened fillets	$Y_{\delta \text{rel} T} = 0.5 + 0.25 \cdot Y_s$
Case-hardened or nitrided steels	$Y_{\delta \text{rel} T} = 0.6 + 0.2 Y_s$

TABLE 23–123
Values of constant K_c and t for use in Eq. (23–710a)

Hardening process	Case depth t	For endurance limit, K_{cf}	For static strength, K_{cs}
Case-hardening	t_{550}	640	$\dfrac{1900}{Y_{10^3}}$ [a]
	t_{400}	500	$\dfrac{1200}{Y_{10^3}}$
	t_{300}	380	$\dfrac{800}{Y_{10^3}}$
Nitriding	t_{400}	500	$\dfrac{1200}{Y_{10^3}}$
Induction or flame-hardening	$t_{H_{V_{min}}}$	$1.1 \cdot H_{V_{min}}$	$\dfrac{2.2 \cdot H_{V_{min}}}{Y_{10^3}}$

[a] Y_{10^3} is the ratio between the permissible static stress and the permissible stress for infinite life, both calculated without the influence of Y_c.

TABLE 23–124
Relative surface condition factor, Y_{RrelT}, for use in Eq. (23–707)

Heat treatment	Y_{RrelT} [a]
Endurance limit	
Surface-hardened and alloyed through-hardened steels except nitrided	$Y_{RrelT} = 1.675 - 0.53 \cdot (R_y + 1)^{0.1}$
Carbon steels (annealed or through-hardened)	$Y_{RrelT} = 5.3 - 4.2 \cdot (R_y + 1)^{0.01}$
Nitrided steels	$Y_{RrelT} = 4.3 - 3.2 \cdot (R_y + 1)^{0.005}$
Static strength	
All materials and all R_y	$Y_{RrelT} = 1$

[a] For a fillet without longitudinal machining traces, $R_y \approx R_z$.

TABLE 23–125
Values of endurance limit, σ_{Hlim}, and static strengths, σ_{H10^5}, σ_{H10^3} (N/mm^2)

	σ_{Hlim} (N/mm^2)	σ_{H10^5} (N/mm^2)	σ_{H10^3} (N/mm^2)
Alloyed case-hardend steels (surface hardness 58–63 R_c)			
Of specially approved high grade	1600[a]	2480	3100
Of specially approved intermediate grade	1550	2480	3100
Of ordinary grade	1500	2480	3100
Nitriding steel of approved grade, gas-nitrided (surface hardness 700–800 H_v)	1300	$1.3 \cdot \sigma_{Hlim}$	$1.3 \cdot \sigma_{Hlim}$
Alloyed through-hardened steel, bath- or gas-nitrided (surface hardness 500–700 H_v)	1050	$1.3 \cdot \sigma_{Hlim}$	$1.3 \cdot \sigma_{Hlim}$
Alloyed, flame- or induction-hardened steel (surface hardness 500–650 H_v)	$0.75 \cdot H_v + 750$	$1.6 \cdot \sigma_{Hlim}$	$4.5 \cdot H_v$
Alloyed through-hardened steel	$0.45 \cdot \sigma_B + 320$	$1.6 \cdot \sigma_{Hlim}$	$1.6 \cdot \sigma_B$
Carbon steel	$0.3 \cdot \sigma_B + 350$	$1.6 \cdot \sigma_{Hlim}$	$1.5 \cdot \sigma_B$

Note: These values refer to forged or hot-rolled steel. For cast steel the values for σ_{Hlim} are to be reduced by 15 percent. For a given material, σ_{Hlim} is the limit of repeated contact stress which can be permanently endured. The value of σ_{Hlim} can be regarded as the level of contact stress which the material will endure without pitting for at least $50 \cdot 10^6$ load cycles. σ_{H10^5} and σ_{H10^5} are the contact stresses which the given material can withstand for 10^5. 10^3 cycles respectively without subsurface yielding or flank damages as spalling or case crushing.
[a] Values given above are approximate.

TABLE 23–126
Life factor, Z_N, for use in Eq. (23–716)

Number of cycles, N_L	Life factor, Z_N
For All Steels Except Nitrided	
$N_L \geq 5 \cdot 10^7$	$Z_N = 1$
$10^5 < N_L < 5 \cdot 10^7$	$Z_N = \left(\dfrac{5.10^7}{N_L}\right)^{0.37 \cdot \log Z_{N10^5}}$
$N_L = 10^5$	$Z_N = Z_{N10^5} = \dfrac{\sigma_{H10^5} \cdot Z_{X10^5}}{\sigma_{H\lim} \cdot Z_L \cdot Z_v \cdot Z_R \cdot Z_X}$
$10^3 < N_L < 10^5$	$Z_N = Z_{N10^5} \cdot \left(\dfrac{10^5}{N_L}\right)^{0.5 \cdot \log(Z_{N10^3}/Z_{N10^5})}$
$N_L \leq 10^3$	$Z_N = Z_{N10^3} = \dfrac{\sigma_{H10^3} \cdot Z_{X10^3}}{\sigma_{H\lim} \cdot Z_L \cdot Z_v \cdot Z_R \cdot Z_X}$
For Nitrided Steels	
$N_L \geq 2 \cdot 10^6$	$Z_N = 1$
$10^5 < N_L < 2 \cdot 10^6$	$Z_N = \left(\dfrac{2 \cdot 10^6}{N_L}\right)^{0.7686 \cdot \log Z_{N10^5}}$
$N_L \leq 10^5$	$Z_N = Z_{N10^5} = \dfrac{1.3}{Z_L \cdot Z_v \cdot Z_R \cdot Z_X}$

TABLE 23–127
Influence of factors on lubrication film, Z_L, Z_v, and Z_R

	Surface-hardened steels	Non-surface-hardened steels
Z_L	$0.91 + \dfrac{0.36}{(1.2 + 134/v_{40})^2}$	$0.83 + \dfrac{0.68}{(1.2 + 134/v_{40})^2}$
Z_v	$0.93 + \dfrac{0.14}{\sqrt{0.8 + (32/v)}}$	$0.85 + \dfrac{0.30}{\sqrt{0.8 + (32/v)}}$
Z_R	$\left(\dfrac{3}{R_{Zrel}}\right)^{0.08}$	$\left(\dfrac{3}{R_{Zrel}}\right)^{0.15}$

Note: Use the expressions in this table to calculate Z_L, Z_v, and Z_R under endurance limit condition. The lubricant factor, Z_L, accounts for the influence of the type of lubricant and its viscosity; the speed factor, Z_v, accounts for the influence of the pitch line velocity; and the roughness factor, Z_R, accounts for influence of the surface roughness on the surface endurance capacity.

TABLE 23–128
Size factor, Z_x, for use in Eq. (23–716)

Hardening process	Z_x	Z_{X10^5}	Z_{X10^3}
Case-hardening (lowest value to be used)	$\left(\dfrac{28 \cdot t_{550}}{\rho_c}\right)^{0.35} \cdot \dfrac{1550}{\sigma_{Hlim}}$		$\left(\dfrac{17 \cdot t_{550}}{\rho_c}\right)^{0.25} \cdot \dfrac{3100}{\sigma_{H10^3}}$
	$\left(\dfrac{17 \cdot t_{400}}{\rho_c}\right)^{0.4} \cdot \dfrac{1550}{\sigma_{Hlim}}$	$\left(\dfrac{15 \cdot t_{400}}{\rho_c}\right)^{0.4} \cdot \dfrac{2480}{\sigma_{H10^5}}$	$\left(\dfrac{8.5 \cdot t_{400}}{\rho_c}\right)^{0.4} \cdot \dfrac{3100}{\sigma_{H10^3}}$
	$\left(\dfrac{12 \cdot t_{300}}{\rho_c}\right)^{0.45} \cdot \dfrac{1550}{\sigma_{Hlim}}$	$\left(\dfrac{8.5 \cdot t_{300}}{\rho_c}\right)^{0.45} \cdot \dfrac{2480}{\sigma_{H10^5}}$	$\left(\dfrac{5 \cdot t_{300}}{\rho_c}\right)^{0.45} \cdot \dfrac{3100}{\sigma_{H10^3}}$
Nitriding	$\left(\dfrac{30 \cdot t_{400}}{\rho_c}\right)^{0.4} \cdot \dfrac{1200}{\sigma_{Hlim}}$	1	1
Induction or flame hardening	$\left(\dfrac{t_{Hv,min}}{\rho_c}\right)^{0.4} \cdot 0.18 \cdot \sqrt{H_{v,min}} \cdot \dfrac{1200}{\sigma_{Hlim}}$	$\left(\dfrac{t_{Hv,min}}{\rho_c}\right)^{0.4} \cdot \dfrac{H_{v,min}}{100} \cdot \dfrac{1900}{\sigma_{H10^5}}$	$\left(\dfrac{t_{Hv,min}}{\rho_c}\right)^{0.45} \cdot \dfrac{H_{v,min}}{120} \cdot \dfrac{2600}{\sigma_{H10^3}}$

The values of Z_X, Z_{X10^5} and Z_{X10^3} are not to be less than determined by σ_B of the core material, i.e.,

$$Z_X \geq \frac{0.45 \cdot \sigma_B + 320}{\sigma_{Hlim}} \qquad Z_{X10^5} = \frac{1.6 \cdot (0.45 \cdot \sigma_B + 320)}{\sigma_{H10^5}} \qquad Z_{X10^3} = \frac{1.6 \cdot \sigma_B}{\sigma_{H10^3}} \text{ respectively}$$

Note: t_{550} = minimum case depth with hardness $H_V \geq 550$; $t_{H_v,min}$ = minimum case depth to $H_{v,min}$ etc.
Source: Extracted from Norway Standards Det Norske Vertias, Classification Note No. 41.2, July 1988, based on ISO/DIS 6336 on gear rating.

TABLE 23–129
Application factor, K_A and K_{AP}, for use in Eqs. (23–695) and (23–711)

Service	K_A	K_{AP}
Diesel propulsion	⊅ 1.5	
Diesel-driven auxiliary	1.5	
Turbine or electropropulsion	> 1.2	
Turbine or electrically driven auxiliaries	1.2	
Electric prime mover		1.5
Driven generator		2.0
Rapidly engaging clutch		1.5
Any kind of prime mover or driven machine unless specially approved		2.0

Source: Extracted from Norway Standards Det Norske Veritas, Classification Note No. 41.2, July 1988, based on ISO/DIS 6336 on gear rating.

TABLE 23–130
Running-in amounts, y_α, and y_β, and running-in tip relief, C_{ay}

Heat treatment and materials		Running-in amount			
			Maximum values		
Non-surface-hardened steel		v	< 5 m/s	5–10 m/s	> 10 m/s
	$y_\alpha = \dfrac{160}{\sigma_{H\lim}} \cdot f_{pb}$	$y_{\alpha,max}$	None	$\dfrac{12{,}800}{\sigma_{H\lim}}$	$\dfrac{6{,}400}{\sigma_{H\lim}}$
	$y_\beta = \dfrac{320}{\sigma_{H\lim}} \cdot F_{\beta x}$	$y_{\beta,max}$	None	$\dfrac{25{,}600}{\sigma_{H\lim}}$	$\dfrac{12{,}800}{\sigma_{H\lim}}$
Surface-hardened steel	$y_\alpha = 0.075 \cdot f_{pb}$ but not more than 3 for any speed $y_\beta = 0.15 \cdot F_{\beta x}$ but not more than 6 for any speed				
All types of steel	$C_{ay} = \dfrac{1}{18}\left(\dfrac{\sigma_{H\lim}}{97} - 18.45\right)^2 + 1.5$				
Pinion and gear wheel made of different types of materials	$y_\alpha = 0.5 \cdot (y_{\alpha 1} + y_{\alpha 2})$ and $y_\beta = 0.5 \cdot (y_{\beta 1} + y_{\beta 2})$ $C_{ay} = 0.5 \cdot (C_{ay1} + C_{ay2})$				

Source: Tables 23–119 to 23–130 are extracted with permission from Norway Standards Det Norske Veritas, Classification Note No. 41.2, July 1988, that are based on ISO/DIS 6336 on gear rating.

TABLE 23–131A
Maximum design limit of flash temperatures to prevent scuffing (scoring) of spur gears, T_f

Kind of oil	Specification	T_f, °F
Petroleum	SAE 10	250
	SAE 30	375
	SAE 60	500
	SAE 90 (gear lubricant)	600
Diester, compounded	75 SUS at 100°F	330
Petroleum	SAE 30 plus mild EP	425

Source: D. W. Dudley, *Handbook for Practical Gear Design*, courtesy of McGraw-Hill Publishsing Company, 1984.

TABLE 131B
Approximate EHD minimum oil-film thickness for Mil-L-6086 oil (similar to AGMA 2, SAE 30 motor oil, or ASTM 315 oil)

Gear mesh			h_{min}, μm				h_{min}, μin			
Temperature		K factor,[a]	Pitch line speed, m/s				Pitch line speed, fpm			
°C	°F	N/mm²	0.5	2.5	10	50	100	500	2000	10,000
					Small gear unit[b]					
		1.38	0.053	0.163	0.44	1.33	2.08	6.41	17.5	52.4
60	140	4.14	0.045	0.141	0.38	1.15	1.80	5.56	15.17	45.4
		13.8	0.039	0.120	0.33	0.98	1.54	4.75	12.97	38.8
		1.38	0.031	0.097	0.26	0.79	1.24	3.82	10.42	31.2
80	176	4.14	0.027	0.084	0.23	0.69	1.08	3.31	9.03	27.1
		13.8	0.023	0.072	0.20	0.58	0.92	2.83	7.72	23.1
		1.38	0.021	0.065	0.18	0.53	0.83	2.56	6.98	20.9
100	212	4.14	0.018	0.056	0.15	0.46	0.72	2.22	6.05	18.1
		13.8	0.015	0.048	0.13	0.39	0.61	1.89	5.17	15.4
					Large gear unit[c]					
		1.38	0.086	0.265	0.72	2.15	3.4	10.43	28.4	85.1
60	140	4.14	0.073	0.229	0.62	1.87	2.9	9.04	24.6	73.8
		13.8	0.063	0.196	0.53	1.60	2.5	7.73	21.0	63.0
		1.38	0.051	0.157	0.43	1.28	2.01	6.20	16.9	50.61
80	176	4.14	0.044	0.136	0.37	1.11	1.74	5.38	14.6	43.89
		13.8	0.038	0.116	0.32	0.95	1.49	4.59	12.5	37.5
		1.38	0.034	0.105	0.29	0.86	1.35	4.16	11.3	33.97
100	212	4.14	0.029	0.092	0.25	0.75	1.17	3.61	9.8	29.5
		13.8	0.025	0.078	0.21	0.64	1.00	3.08	8.37	25.18

[a] 1.38 N/mm² = 200 psi; 4.14 N/mm² = 600 psi; 13.8 N?mm² = 2000 psi.
[b] r_e = 8.39 mm = 0.3304 in.
[c] r_e = 41.96 mm = 1.652 in.
Source: D. W. Dudley, *Handbook Practical Gear Design*, McGraw-Hill Publishing Company, New York, 1984.

TABLE 23–132
Critical scuffing (scoring) criterion index numbers, Eqs. (23–741) and (23–744), Z_c

Blank temperature, °F	100°	150°	200°	250°	300°
Kind of oil	Critical scoring—index numbers				
AGMA 1	9,000	6,000	3,000	—	—
AGMA 3	11,000	8,000	5,000	2,000	—
AGMA 5	13,000	10,000	7,000	4,000	—
AGMA 7	15,000	12,000	9,000	6,000	—
AGMA 8A	17,000	14,000	11,000	8,000	—
Grade 1065, Mil-O-6082B	15,000	12,000	9,000	6,000	—
Grade 1010, Mil-O-6082B	12,000	9,000	6,000	2,000	—
Synthetic (Turbo 35)	17,000	14,000	11,000	8,000	5,000
Synthetic Mil-L-7808D	15,000	12,000	9,000	6,000	3,000

Notes:
(1)   See AGMA 250.04 and 251.02 for general data on industrial lubricants.
(2)   This table is reproduced by permission from D. W. Duley, *Gear Handbook*, Chap. 13.

TABLE 23–133A
Geometry constant for scuffing (scoring), Z_t, at tip of tooth for use in Eq. (23–744)

Pressue angle, α, or ϕ_1	No. of pinion teeth z_1 or N_P	No. of gear teeth z_2 or N_G	Pinion addendum h_{a1} or a_P (for $m = 1.0$ or $P_d = 1.0$)	Gear addendum h_{a2} or a_G (for $m = 1.0$ or $P_d = 1.0$)	Z_t At pinion tip	Z_t At gear tip
20°	18	25	1.0	1.0	0.0184	−0.0278
	18	35	1.0	1.0	0.0139	−0.0281
	18	85	1.0	1.0	0.0092	−0.0307
	25	25	1.0	1.0	0.0200	−0.0200
	25	35	1.0	1.0	0.0144	−0.0187
	25	85	1.0	1.0	0.0088	−0.0167
	12	35	1.25	0.75	0.0161	−0.0402
	18	85	1.25	0.75	0.0107	−0.0161
	25	85	1.25	0.75	0.0104	−0.0112
	35	85	1.25	0.75	0.0101	−0.0087
	35	275	1.25	0.75	0.007	−0.0072
25°	18	25	1.0	1.0	0.0135	−0.0169
	18	35	1.0	1.0	0.0107	−0.0168
	18	85	1.0	1.0	0.0074	−0.0141
	25	25	1.0	1.0	0.0141	−0.0141
	25	35	1.0	1.0	0.0107	−0.0126
	25	85	1.0	1.0	0.0069	−0.0103
	12	35	1.25	0.75	0.0328	−0.0160
	12	85	1.25.	0.75	0.0500	−0.0151
	18	85	1.25	0.75	0.0056	−0.0095
	25	85	1.25	0.75	0.0082	−0.0073
	35	85	1.25	0.75	0.0078	−0.0060
	35	275	1.25	0.75	0.0056	−0.0048

Note: When proper profile modification is made, the risk of scoring is probably more critical at the start of modification than at the tip of the tooth.
Source: D. W. Dudley, *Handbook Practical Gear Design*, McGraw-Hill Publishing Company, New York, 1984.

TABLE 23–133B
Values of Z_s (approximate) for use in Eq. (23–744)

Initial finish	Z_s	Comment
0.3 μm (12 μin)	1.2	Usually honed, after finish ground
0.5 μm (20 μin)	1.5	Fine finish; some break-in needed
0.75 μm (30μin)	1.7	Good finish; special break-in needed (for Z_s, to equal 1.7)
1 μm (40 μin)	2.0	Nominal finish; extensive break-in needed (for Z_s to equal 2.0)
1.5 μm (60 μin)	2.5	Poor finish; special wear-in procedure should be used (then $Z_s = 2.5$ is possible)

TABLE 23–134
Welding factors, $X_{W,T}$

Material	$X_{W,T}$
Through-hardened steel	1.00
Phosphated steel	1.25
Copper-plated steel	1.50
Bath- or gas-nitrided steel	1.50
Hardened carburized steel content of austenite	
less than average	1.15
content of austenite average	1.00
content of austenite more than average	0.85
Austenite steel (stainless steel)	0.45

Source: ANSI/AGMA 2001-B88.

TABLE 23–135
FZG oil test

FZG class	6	7	8	9	10	11	12
$M_{t1,T}$	135	183	239	302	373	450	535

TABLE 23–137
Mineral oil mean scuffing temperatures

ISO VG	AGMA lube no.	Mean scuffing temperature		Standard temperature deviation	
		°F	°C	°F	°C
32	—	351	177	53	12
46	1	372	189	56	13
68	2	395	202	59	15
100	3	418	214	63	17
150	4	441	227	66	19
220	5	464	240	70	21
320	6	486	252	73	23
460	7	507	264	76	24
680	8	530	277	80	27
1000	8A	553	289	83	28
1500	—	577	303	87	31

TABLE 23–136
MIL lubricant mean scuffing temperatures

Lubricant	Mean scuffing temperature		Standard temperature deviation	
	°F	°C	°F	°C
MIL-L-7808	366	186	56.6	13.7
MIL-L-6081 (grade 1005)	264	129	74.4	23.6

Source: ANSI/AGMA 2001-B88.

TABLE 23–138A
Scuffing risk

Probability of scuffing, %	Scuffing risk
< 10	Low
10–30	Moderate
> 30	High

Source: ANSI/AGMA 2001-B88.

TABLE 23–138B
Flash temperature limit T_f and scuffing (scoring) probability for use in Eqs. (23–742) and (23–744)

	Risk of scoring			
	Low		High	
	°C	°F	°C	°F
Synthetic oil				
Mil-L-7808	135	275	175	350
Mil-L-23699	150	300	190	375
Mineral oil				
Mil-O-6081, grade 1005	65	150	120	250
Mil-L-6086, grade medium	160	325	200	400
SAE 50 motor oil with mild EP	200	400	260	500
Mil-L-2105, grade 90				
(SAE 90 gear oil)	260	500	315	600

Source: W. Dudley, *Handbook of Practical Gear Design*, courtesy of McGraw-Hill Publishing Company, 1984

REFERENCES

1. Lingaiah, K., *Machine Design Data Handbook*, 4th ed., Vol. II, (*SI and Customary Metric Units*), Suma Publishers, Bangalore, India, 1986.
2. Maleev, V. L., and J. B. Hartman, *Machine Design*, International Textbook Company, Scranton, Pennsylvania, 1954.
3. American Gear Manufacturers Association, 1500 King Street, Suite 201, Alexandria, Virginia.
4. American National Standards/American Gear Manufacturers Association, 2001–B88, Fundamentals Rating Factors and Calculation Methods for Involute Spur and Helical Gear Teeth, Sept. 1988.
5. Shigley, J. E., *Mechancial Engineering Design*, first Metric Edition, McGraw-Hill Book Company, 1986.
6. Niemann, G., *Maschinenelemente*, Springer-Verlag, Berlin, Erster Band, 1963.
7. Niemann, G., *Maschinenelemente*, Springer-Verlag, Berlin, Zweiter Band, Getriebe, 1965.
8. Dolan, T. J., and E. L. Broghamer, *Photoelastic Study of the Stresses in Gear Teeth Fillets*, University of Illinois Engineering Experimental Station Bulletin 335, March 1942.
9. Peterson, R. E., *Stress Concentration Factors*, John Wiley and Sons, New York, 1974.
10. Lingaiah, K., *Fringe Pattern of Gear-Teeth Showing Stress Concentration at Root and Contact Point*, unpublished work of Author, Dept. of Mech. Eng., University of Visvesvaraya College of Engineering, Bangalore University, Bangalore, 1973.
11. Buckingham, E., *Analytical Mechanics of Gears*, McGraw-Hill Book Company, New York, 1949.
12. IS: 4460–1967, *Rating of Machine Cut Spur and Helical Gears*, Bureau of Indian Standards, Manak Bhawan, 9 Bahadur Shah Zafar Marg, New Delhi, India.
13. American National Standards/American Gear Manufacturers Association. 2003-A86, May 1986. Rating the Pitting Resistance and Bending Strength of Generated Straight Bevel, Zero Bevel, and Spiral Bevel Gear Teeth.
14. MAAG Gear Book, MAAG Gear Company Limited, Zurich, Switzerland, 1963.
15. Konigsberger, F., *Design Principles of Machine Tools*, Macmillan Company, New York, 1967.
16. Ackerkan, N. et al., *Machine Tool Design*, Mir Publishers, Moscow, 1969.
17. American National Standards Institution/American Gear Manufacturers Association, 6010-E88, November 1988, Standards for Spur, Helical, Herringbone, and Bevel Enclosed Drives.
18. DIN: 803 and 804, 1977, DIN: Deutsches Institut für Normung e.v.,Beuth-Vertrief GmbH, Burggrafen Strasse 6, D1000 Berlin 30, Germany.
19. The Gleason Works, 1000 University Drive, Rochester, New York.
20. Shigley, J. E., and R. Mischke, *Standard Handbook of Machine Design*, McGraw-Hill Book Company, New York, 1986.
21. Dobrovolsky, V., et al., *Machine Elements*, Mir Publishers, Moscow, 1968.
22. American Gear Manufacturers Association, 6034-B92, Practice for Enclosed Cylindrical Worm Gear Speed Reducers and Gear Motors.
23. IS: 7443–1974 Load Rating of Worm Gears, Bureau of Indian Standards, Manak Bhavan, 9 Bahadur Shah Zafar Marg, New Delhi, India.
24. BS: 721, Allowable wheel-shaft torque (for unlimited life) of wormgears and crossed-helical gears, British Standards Institution, London.
25. Neale, M. J., *Tribology Handbook*, Butterworth, 1975.
26. Dudley, D. W., *Handbook of Practical Gear Design*, McGraw-Hill Book Company, New York, 1984.
27. ISO DIS 6336 and ISO 6336, Gear Strength and Power Rating, 1, rue de Varembe, CH-1211, Geneve 20, Switzerland.
28. Det Norske Vertias, Classification Note No. 41.2, *Calculation of Gear Rating for Marine Transmissions*, July 1988.
29. MAAG Gear Book, MAAG Gear Company Limited, CH-8023 Zurich, Switzerland, 1990.
30. Benedict, G. H., and Kelly, B. W., *Instantaneous Coefficient of Gear Tooth Friction*, ASLE Trans, Vol. 4, 1961, pp. 59–80.
31. Imwalle, D. E., and O. A. Labath, *Difference Between AGMA and ISO Rating Systems*, presented at the AGMA Semi-Annual Meeting in Toronto, Canada, Oct. 10–14, 1981, AGMA 219.15.

BIBLIOGRAPHY

1. Acherkan, N. et al., *Machine Design Handbook*, in 3 volumes, Mashinostroenie Publishers, Moscow, 1968 (in Russian).
2. AGMA 217.01, *AGMA Information Sheet: Gear Scoring Design Guide for Aerospace Spur and Helical Power Gears*, Oct. 1965.
3. Albert, C. D., *Machine Design Drawing Room Problems*, 4th ed., John Wiley and Sons, New York, 1949.
4. Baumeister, T., E. A. Avallone, and T. Baumeister III, *Marks' Standard Handbook for Mechanical Engineers*, 8th ed., McGraw-Hill Book Company, New York, 1978.
5. Berard, S. J., E. O. Waters, and C. W., Phelps, *Principles of Machine Design*, The Ronald Press Company, New York, 1955.
6. Black, P. H., and O. Eugene Adames, Jr., *Machine Design*, 3rd ed., McGraw-Hill Book Company, Inc., Kogakusha Company, Ltd., Tokyo, 1968.
7. Black, P. L., *An Investigation of Relative Stresses in Solid Spur Gear by the Photoelastic Method*, Univeristy of Illinois Engineering Experiment Station Bulletin, 288, 1936.
8. Blok, H., 'Les temperatures de Surface dans les Conditions de graissage Sons Pression Extreme,' Second World Petroleum Congress, Paris, June 1937.
9. Blok, H., *The postulate about the Constancy of Scoring Temperature, Interdisciplinary Approach to the Lubrication of Concentrated Contacts*, NASA SP–237, pp. 153–248, 1970.
10. Creamer, R. H., *Machine Design*, 2d ed., Addison-Wesley Publishing Company, London, 1976.
11. Decker, K. H., *Maschinenelemente*, Gestaltung and Berechnung, Carl Hanser Verlag, Munich, 1971.
12. Deutschman, A. D., W. J. Michels and C. E. Wilson, *Machine Design—Theory and Practice*, Macmillan Publishing Company, New York, 1975.
13. Downson, D., and G. R. Higginson, *Elastohydrodynamic Lubrication—The Fundamentals of Roller and Gear Lubrication*, Pergamon Press, London, 1966.
14. Downson, D., 'Elastohydrodynamics,' Paper No. 10., *Proc. Inst. Mech. Eng.*, Vol. 182, Part 3A, pp. 151–167, 1967.
15. Drago, R. J., *Fundamentals of Gear Design*, Butterworth Publishers, Stoneham, Massachusetts, 1988.
16. Dudley, D. W., *Practical Gear Design*, McGraw-Hill Book Company, New York, 1954.
17. Dudley, D. W., *Gear Handbook*, McGraw-Hill Book Company, New York, 1962.
18. Faires, V. M., *Design of Machine Elements*, 4th ed., The Macmillan Company, New York, 1965.
19. Hertz, H., 'Uber die Beruhrung fester elastische, Korper,' *Mathematik*, Vol. 92, pp. 156–171, 1881.
20. Hyland, P. H., and J. B. Kommers, *Machine Design*, 3d ed., McGraw-Hill Book Company, New York, 1943.
21. Imwalle, D. E., O. A. Labath and R. N. Hutchinson, *A Review of Recent Gear Rating Developments ISO/AGMA Comparison Study*, ASME paper No. 80-C2/DET-25, 1980.
22. Juvinall, R. C., *Fundamentals of Machine Component Design*, John Wiley and Sons, New York, 1983.
23. Kelley, B. W., 'A New Look at the Scoring Phenomena of Gears,' *SAE Trans.*, Vol. 61, pp. 175–188, 1953.
24. Kelley, B. W., 'The Importance of Surface Temperature to Surface Damage,' Chapter in *Engineering Approach to Surface Damage*, University of Michigan Press, Ann Arbor, 1958.
25. Kent, R. T., *Mechanical Engineer's Handbook-Design and Production*, 12th ed., Vol. II (ed. by Colin Carmichael), John Wiley and Sons, London, 1961.
26. Kimball and Barr, *Elements of Machine Design*, John Wiley and Sons, New York, 1953.
27. Leutwiler, O. A., *Elements of Machine Design*, McGraw-Hill Book Company, New York, 1917.
28. Levinson, I. J., *Machine Design*, Reston Publishing Company, Reston, Virginia, 1978.
29. Lingaiah, K., 'Photoelastic Analysis of Effect of Contact Stress on Fatigue Strength of Spur Gears,' *Proceedings of Fifteenth Congress of Indian Society for Theoretical and Applied Mechanics*, pp. 24–27, 1970.
30. Lingaiah, K., 'Solution of an Asymmetrically Reinforced Circular Cut-out in a Flat Plate Subjected to Uniform Unidirectional Stress,' Ph.D. Thesis, University of Saskatchewan, Saskatoon, Canada, 1965.
31. Lingaiah, K., W. P. T. North, and J. B. Mantle, Photoelastic Analysis of an Asymmetrically Reinforced Circular Cut-out in a Flat Plate Subjected to Uniform Unidirectional Stress, *Proceedings SESA*, Vol. 23, No. 2, p. 617, 1966.
32. Lingaiah, K., and K. Ramachandra, 'Bending Stress in Wildhaber–Nuvikov Gears due to Semi-Ellipsoidal Contact-Load Distribution,' *Proceedings of the ASME Design Engineering Conference, New York, 1975.*
33. Lingaiah, K., and K. Ramachandra, 'Conformity Factor in Wildhaber–Novikov Circular Arc Gears, Proceedings of the ASME,' *J. Mech. Design*, Vol. 103, pp. 134–40, Jan. 1981.

34. Lingaiah, K., and B. R. Narayan Iyengar, *Machine Design Data Handbook*, Volume I (*Metric Units*), Suma Publishers, Bangalore, India, 1973.
35. Lingaiah, K., *Handbook of Conversion Factors, Tables and Mathematical formulas*, 2d ed., Suma Publishers, Bangalore, India, 1980.
36. Lingaiah, K., and B. R. Narayana Iyengar, *Machine Design Data Handbook*, 2d ed., Vol. I (*SI and Customary Metric Units*), Suma Publishers, Bangalore, India, 1983.
37. Design Charts of K. Lingaiah for "an Asymmetrically Reinforced Circular Cut-out in a Flat Plate subjected to uniform Unidirectional Stress," cited by Peterson, R. E., in *Stress Concentration Factors*, John Willey and Sons, Inc., New York, pp. 115, 162, 163, 1974.
38. Merritt, H. E., *Gears*, Sir Isaac Pitman and Sons, London, 1953.
39. Merrit, H. E., *Gear Engineering*, John Wiley and Sons, New York, 1971.
40. Movnin, M., and G. Goltziker, *Machine Design*, Mir Publishers, Moscow, 1969.
41. Narayana Iyengar, B. R., and K. Lingaiahh, *Machine Design Data Handbook*, (FPS System), Engineering College Co-operative Society, Ltd., Bangalore, India, 1962.
42. Norman, C. A., E. S. Ault, and I. F. Zarobsky, *Fundamental of Machine Design*, The MacMillan Company, New York, 1951.
43. Orlov, P., *Fundamentals of Machine Design*, in five volumes, Mir Publishers, Moscow, 1977.
44. Phelan, R. M., *Fundamentals of Mechanical Design*, 3d ed., McGraw-Hill Book Company, New York, 1970.
45. Reshetov, D. N., *Machine Design*, Mir Publishers, Moscow, 1978.
46. Rothbart, H. A., *Mechanical Design and Systems Handbook*, McGraw-Hill Book Company, New York, 1964.
47. Spotts, M. F., *Design of Machine Elements*, 5th ed., Prentice-Hall of India Private Ltd., New Delhi, 1978.
48. Spotts, M. F., *Machine Design Analysis*, Prentice-Hall, Englewood Cliffs, New Jersey, 1964.
49. Srinath, L. S., M. R. Raghavan, K. Lingaiah, et al., *Experimental Stress Analysis,* Tata-McGraw-Hill Publishing Company, New Delhi, 1984.
50. Taylor, J. E., and J. S. Wrigley, *Engineering Design*, Sir Isaac Pitman and Sons, Ltd., London, 1945.
51. Thomas, H. R., and V. A. Hoersch, 'Stress due to the Pressure of one Elastic Solid upon Another,' University of Illinois Engineering Experiment Station Bulletin 212, 1830.
52. Vallance, A., and V. L. Doughtie, *Design of Machine Members*, 3d ed., McGraw-Hill Book Company, New York, 1951.

CHAPTER
24

DESIGN OF BEARINGS AND TRIBOLOGY

24.1 SLIDING CONTACT BEARINGS (1, 2)

SYMBOLS

a	distance between bolt centers [Eqs. (24–70) to (24–72)], m (in)
$a = \dfrac{h_2}{B}$	dimensionless quantity
$A = Ld$	projected area of the journal bearing (Fig. 24–6), m² (in²) effective area of the bearing, m² (in²) projected area at full pool pressure in case of hydrostatic journal bearing (Fig. 24–38), m² (in²)
A'	projected area of the region having a linear pressure gradient in case of hydrostatic journal bearing (Fig. 24–38), m² (in²)
B	width of slider bearing in the direction of motion, m (in) length of journal bearing in the direction of motion, m (in)
$c = D - d$	diametral clearance, m (in)
C	combined coefficient of radiation and convection, W/m² K (kcal/mm² s°C)
C_1, C_2	constants in Eq. (24–23)
$C_F = \dfrac{F\mu}{F_{\mu\infty}}$	friction leakage factor in Eq. (24–54)
$C_{P\mathcal{F}1}, C_{P\mathcal{F}2},$ $C_{P\mathcal{F}3}, C_{P\mathcal{F}4}$	constants in Eqs. (24–77b), (24–78b), (24–79b), and (24–80b)
C_{PFm}, C_{PFs}	friction resistance factor for moving and stationary member, respectively, in pivoted shoe slider bearing in Eqs. (24–96b) and (24–97b)
C_{PW}	load factor in Eq. (24–95b)
C_Q	flow correction factor from (Fig. 24–35(a))
$C_{S1} to C_{S7}$	constants in Eqs. (24–86b), (24–87b), (24–88b), (24–89b), (24–90b), (24–91b), and (24–92b)
$C_W = \dfrac{W}{W_\infty}$	load leakage factor in Eqs. (24–52)
$C_\mu = \dfrac{\mu_\infty}{\mu}$	coefficient of friction factor in Eq. (24–53)
$C_{P\mu}$	coefficient of friction factor in Eqs. (24–98) and Table 24–17

d	diameter of journal, m (in)
d_1, d_2	inside and outside diameters of thrust, pivot, and collar bearings, m (in)
d_c	diameter of capillary in case of hydrostatic journal bearing, m (in)
D	diameter of bearing, m (in)
$e = c - h_{min}$	eccentricity, m (in)
E	Young's modulus, GN/m^2 or GPa (Mpsi)
E_t^o	Engler, deg
F	force (also with subscripts), kN (lbf)
$F_{P\mathcal{F}W}$	load factor in Eqs. (24–83) and (24–84)
F_μ	friction force, kN (lbf)
F_μ'	$\dfrac{F_\mu}{dL}$ friction force per unit area of bearing, MPa (Psi)
$F_{\mu m}$	friction force on the moving member of bearing (i.e., slider), kN (lbf)
$F_{\mu mp}$	friction force on the moving member of pivoted slider bearing (i.e., slider), kN (lbf)
$F_{\mu s}$	friction force on the stationary member of bearing (i.e., shoe), kN (lbf)
$F_{\mu sp}$	friction force on the stationary member of pivoted slider bearing (i.e., shoe), kN (lbf)
$F_{\mu\infty}$	friction force acting on the moving surface of the same bearing with the same oil-film shape but without end leakage, kN (lbf)
G	flow factor given by Eq. (24–82)
h	oil film thickness, m (in)
h_1, h_2	thickness of oil film at entrance and exit, respectively, of a slider bearing (Fig. 24–39 and Fig. 24–43), m (in)
h_c	thickness of bearing cap, m (in)
$h_{min} = h_o$	minimum thickness of oil film, m (in)
h_{max}	maximum thickness of oil film, m (in)
H_d	heat dissipating capacity of bearing, kJ/s (kcal/s)
H_g	heat generated in bearing, kJ/s (kcal/s)
i	number of collars
k	characteristic number of the given crude oil ($\simeq 1.4$ to 2.8), constant (also with subscripts) heat dissipating coefficient
$k = (h)_{\substack{P(max) \\ p(min)}}$	thickness of the oil film where the pressure has its maximum or minimum values, m (in)
K	constant for a given grade of oil (varies from 1.000 to 1.004)
K_1, K_2, K_3, K_4	constants in Eqs. (24–73b), (24–74b), (24–75b), and (24–76b) respectively
K_5, K_6	constants in Eqs. (24–132b) and (24–133b), respectively
$K_{LP1}, K_{LP2}, K_{LP3}$	constants in Eqs. (24–116b), (24–118b), and (24–119b) for parallel surface thrust bearing
K_{lt}	constant in Eq. (24–121b) for tilting-pad bearing
K_{Pt}	constant in Eq. (24–120b) for a tilting-pad bearing
$K_{\mu t}$	coefficient of friction factor in Eq. (24–126b) for a tilting-pad bearing
l_1	length of bearing pressure pad in case of hydrostatic journal bearing (Fig. 24–38), m (in)

l_c		length of capillary, m (in)
L		axial length of the journal (or of the bearing) normal to the direction of motion, m (in)
$m = \dfrac{h_1}{h_2}$		ratio of the film thicknesses at the entrance to exit in the slider bearing
M_t		torque, N m (lbf in)
n		speed, rpm
n'		speed, rps
P		power (also with subscripts), kW (hp)
P		intensity of pressure, MPa (psi)
$P = \dfrac{W}{Ld}$		load per projected area of the bearing, MPa (psi)
P_u		unit load supported by a parallel surface thrust bearing, MPa (psi)
P_1		lower pool pressure in hydrostatic journal bearing (Fig. 24–38), MPa (psi)
P_2, P_4		left and right pool pressure in hydrostatic journal bearing (Fig. 24–38), MPa (psi)
P_3		upper pool pressure in hydrostatic journal bearing (Fig. 24–38), MPa (psi)
$P_1' = P_2' = P_3'$ $= P_4' = P'$		the pressure in first, second, third and fourth quadrant of the pool, respectively, when the journal is concentric ($e = o$) in hydrostatic journal bearing, MPa (psi)
P_i		inlet pressure, MPa (psi)
P_o		constant manifold pressure, MPa (psi), pressure in the oil film in journal bearing at the point when $\theta = 0$, MPa (psi)
$q = \dfrac{h_1}{h_2} - 1$		constant used in Eqs. (24–95b) and (24–97b) for a slider bearing
Q		flow of lubricant through the bearings, m^3/s
r		radius of journal, m (in)
r_1, r_2		inside and outside radii of thrust bearing, m (in)
R		number of Redwood seconds in Eqs. (24–15) and (24–16)
$S = \dfrac{\eta n'}{P}\dfrac{1}{\psi^2}$		Sommerfeld number or bearing characteristic number
$S' = \dfrac{60\eta n'}{P}\dfrac{1}{\psi^2}$		bearing characteristic number (Fig. 24–34)
$S'' = \dfrac{\eta_1 n}{P}$		bearing modulus (Tables 24–2 and 24–7)
t		running temperature of the bearing, K (°C), number of seconds, Saybolt, in Eqs. (24–7) and (24–8)
$\Delta T = (t_b - t_a)$		difference in temperature between bearing housing and surrounding air, K (°C)
u		average velocity, m/s (ft/min) velocity in the oil film at height y (Fig. 24–1), m/s (ft/min)
U		maximum velocity (Fig. 24–1), m/s (ft/min)
v		velocity, m/s (ft/min)
v_m		mean velocity, m/s (ft/min) surface speed of journal, m/s (ft/min)

V	rubbing velocity, m/s (ft/min)
W	load on the bearing, kN (lbf)
	load acting on the journal bearing with end leakage, kN (lbf)
W_∞	load acting on the journal bearing without end leakage, kN (lbf)
$\bar{x}$	the distance of the pivoted point from the lower end of the shoe (Fig. 24–39), i.e., the distance of the pressure center from the origin of the coordinate, m (mm)
y	distance from the stationary surface (Fig. 24–1), m (in)
$\kappa = -qa$	a constant in equation of pivoted-shoe slider bearing [Eq. (24–86b)]
β	angular length of bearing or circumferential length of bearing, deg
γ_t	specific weight (weight density) at temperature t, °C, kN/m^3 (lbf/in^3)
$\delta = 1 - \epsilon$	the minimum film thickness variable
$\epsilon = \dfrac{2e}{c}$ $= 1 - \dfrac{h_{min}}{d\psi}$	attitude or eccentricity ratio or relative eccentricity
η	absolute viscosity (dynamic viscosity), Pa s
η'	absolute viscosity (dynamic viscosity), kgf s/m^2
η_1	absolute viscosity (dynamic viscosity), cP
η_2	absolute viscosity (dynamic viscosity), kgf s/cm^2
η_p	dynamic viscosity of oil above atmospheric pressure P, $\dfrac{\text{N s}}{\text{m}^2}$ or Pa s (cP, kgf s/m^2)
η_o	dynamic viscosity of oil at atmospheric pressure, i.e., when $P = 0$, N s/m^2 (cP, kgf s/m^2)
θ	the angle measured from the position of minimum of oil film to any point of interest in the direction of rotation or the angle from the line of centers to any point of interest in the direction of rotation around the journal, deg
μ	coefficient of friction (also with subscripts)
μ_o	viscosity, reyn
$\vartheta = \dfrac{\eta}{\gamma} g$	kinematic viscosity, m^2/s (cSt)
ρ	density of oil or specific gravity of oil used, kg/m^3 (g/mm^3)
σ	stress (normal), MPa (psi)
τ	shear stress in lubricant, MPa (psi)
ϕ	attitude angle or angle of eccentricity, deg
$\psi = \dfrac{c}{d}$	diametral clearance ratio or relative clearance
ω	angular speed, rad/s

Other factors in performance or in special aspects are included from time to time in this chapter and being applicable only in their immediate context, are not included at this stage.

Particular	Formula

SHEAR STRESS (1, 2):

The shearing stress in the lubricant (Fig. 24–1)

$$\tau = \frac{F}{A} = \eta \frac{U}{h} = \eta \frac{u}{y} = \eta \frac{du}{dy} \qquad (24\text{--}1)$$

VISCOSITY

The absolute viscosity (dynamic viscosity) in **SI** units

$$\eta = 10^{-3} \eta_1 \qquad\qquad\qquad \textbf{SI} \qquad (24\text{--}2a)$$

where η in Pa s or (N s/m^2) and η_1 in cP

$$= 9.8066\, \eta' \qquad (24\text{--}2b)$$

$$= 9.8066 \times 10^4\, \eta_2 \qquad (24\text{--}2c)$$

where η in Pa s, η' in kgf s/m^2, and η_2 in kgf s/cm^2

$$= \frac{10^4}{1.45}\, \mu_o \qquad (24\text{--}2d)$$

where η is Pa s and μ_o in reyn

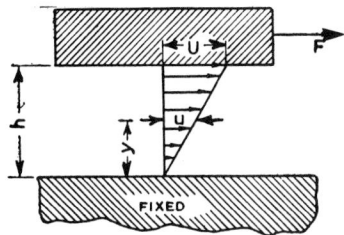

FIGURE 24–1   Shearing stress in lubricant.

The absolute viscosity (dynamic viscosity) in **Customary Metric units**

$$\eta' = 0.102\eta \qquad\qquad\qquad \textbf{Metric} \qquad (24\text{--}3a)$$

where η' in $\dfrac{\text{kgf s}}{\text{m}^2}$ and η in Pa s

$$= 1.02 \times 10^{-4}\eta_1 \qquad (24\text{--}3b)$$

where η' in $\dfrac{\text{kgf s}}{\text{m}^2}$ and η_1 in cP

$$= \frac{10^3}{1.422}\, \mu_o \qquad (24\text{--}3c)$$

where η' in $\dfrac{\text{kgf s}}{\text{m}^2}$ and μ_o in reyn

For absolute viscosity (dynamic viscosity) in centipoise and **SI units**

Refer to Figs. 24–2 and 24–35(*b*)

The absolute viscosity (dynamic viscosity) in centipoise

$$\eta_1 = 10^3 \eta \qquad\qquad\qquad \textbf{Metric} \qquad (24\text{--}4a)$$

where η_1 in cP and η in Pa s

$$= \frac{10^8}{1.02}\, \eta_2 \qquad (24\text{--}4b)$$

where η_1 in cP and η_2 in $\dfrac{\text{kgf s}}{\text{cm}^2}$

$$= \frac{10^4}{1.02}\, \eta' \qquad (24\text{--}4c)$$

Particular	Formula

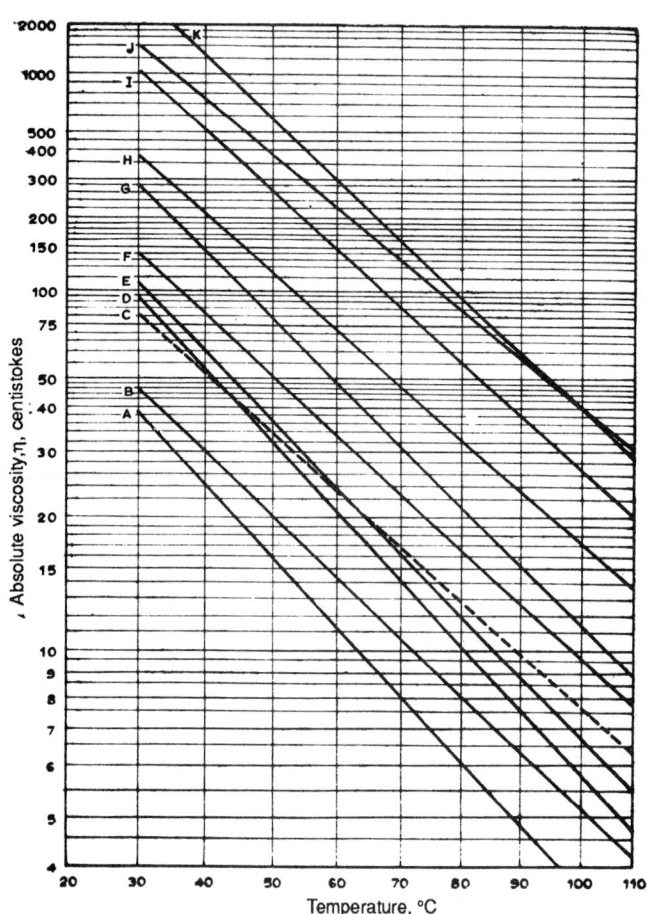

FIGURE 24–2 Absolute viscosity vs. temperature.

where η_1 in cP and η' in $\dfrac{\text{kgf s}}{\text{m}^2}$

$$= \frac{10^7}{1.45}\,\mu_o \qquad (24\text{–}4d)$$

where η_1 in cP and μ_o in reyn

The viscosity in reyn (lbf s/in²)

$$\mu_o = 1.45 \times 10^{-4}\eta \quad \textbf{US Customary System units} \qquad (24\text{–}5a)$$

where μ_o in reyn and η in Pa s

$$= 1.45 \times 10^{-7}\eta_1 \qquad (24\text{–}5b)$$

where μ_o in reyn and η_1 in cP

$$\mu_o = 14.22\,\eta_2 \qquad (24\text{–}5c)$$

Particular	Formula
	where μ_o in reyn and η_2 in kgf s/cm^2
	$= 1.422 \times 10^{-3}\eta'$ $\qquad$ (24–5d)
	where μ_o in reyn and η' in $\dfrac{\text{kgf s}}{\text{m}^2}$
Kinematic viscosity	$v = \dfrac{\eta}{\text{density}} = \dfrac{\eta_2 g}{\gamma}$ $\qquad$ **Metric** $\quad$ (24–6a)
	where v in cm^2/s and η_2 in $\dfrac{\text{kgf s}}{\text{cm}^2}$,
	$g = 980.66$ cm/s^2 and γ in $\dfrac{\text{kgf}}{\text{cm}^3}$
Kinematic viscosity	$v = \dfrac{\eta g}{\gamma} 10^{-4}$ $\qquad$ **SI** $\quad$ (24–6b)
	where η in $\dfrac{\text{N s}}{\text{m}^2}$ or (Pa s), γ in N/m^3, and v in m^2/s
Saybolt to centipoises (Fig. 24–3) (3) or mPa s	$\eta = \gamma_t\left(0.22t - \dfrac{180}{t}\right)$ $\qquad$ **SI/Metric** $\quad$ (24–7)
	where η in cP and γ_t in gf/cm^3 or N/m^3, t in s
	Refer to Table 24–1 for γ_t
Saybolt to reyn	$\mu_o = 0.145\gamma_t\left[0.22t - \left(\dfrac{180}{t}\right)\right]$
	$\qquad\qquad$ **US Customary System units** $\quad$ (24–7a)
Kinematic viscosity in centistokes from Saybolt universal seconds (Figs. 24–3 and 24–4) (3)	$v_k = \left(0.22t - \dfrac{180}{t}\right)$ $\qquad$ (24–8a)
	where v_k in cSt and t in s

TABLE 24–1
Specific gravity of oils at 15.5°C (60°F)

No.	Oil characteristics	$\gamma_{15.5}$
A	Turbine oil, ring-oiled bearing	0.8877
B	Turbine oil, ring-oiled bearing, SAE 10	0.8894
C	All-year automobile oil, SAE 20	0.9036
D	Ring-oiled bearing oil, high-speed machinery	0.9346
E	Automobile oil, SAE 20	0.9254
F	Automobile oil, SAE 30	0.9263
G	Automobile oil, SAE 40, medium-speed machinery	0.9275
H	Airplane oil 100, SAE 60	0.8927
I	Transmission oil, SAE 110, spur and bevel gears	0.9328
J	Gear oil, slow-speed worm gears	0.9153
K	Transmission oil, SAE 60, slow-speed gears	0.9365

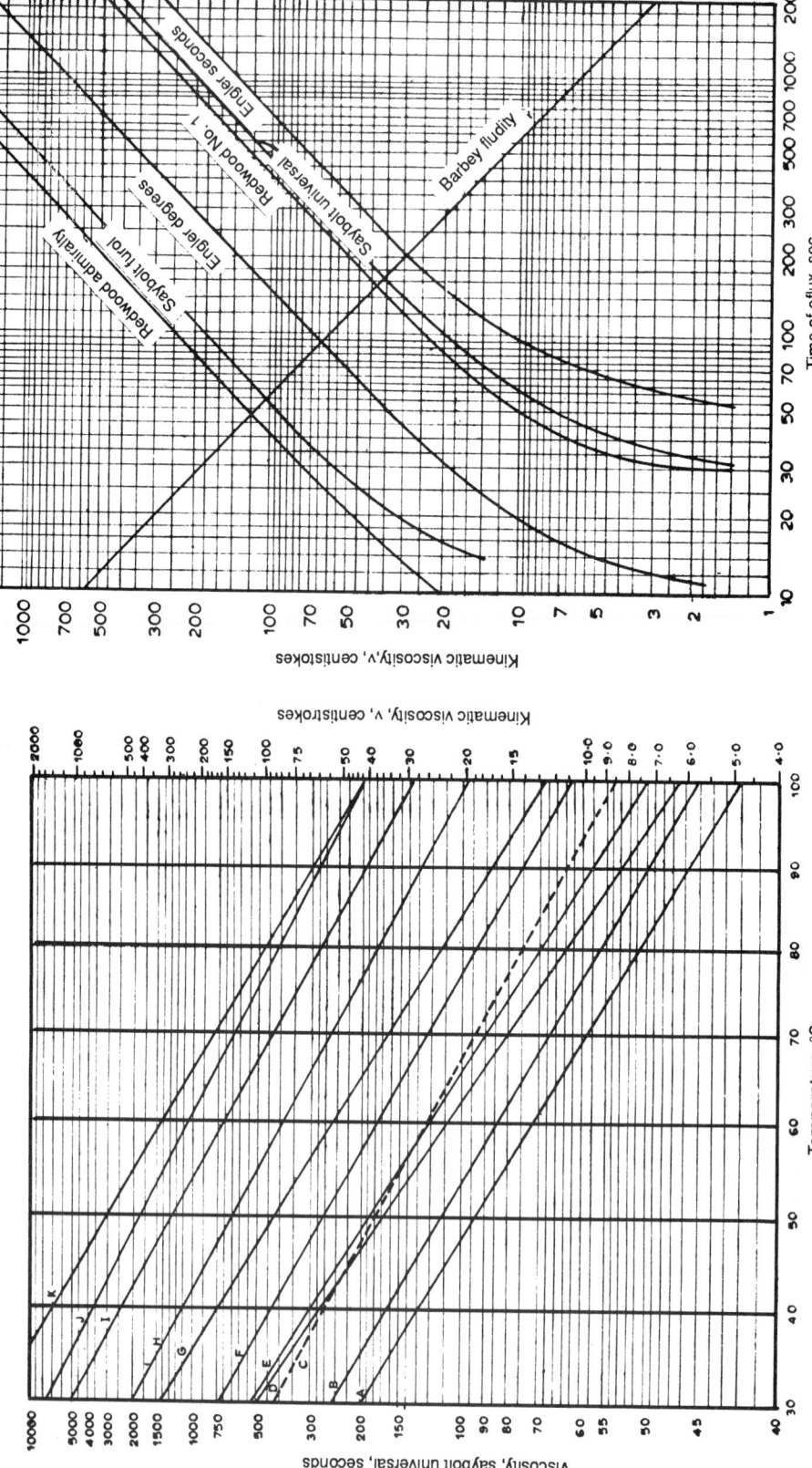

FIGURE 24-3 Viscosity Saybolt universal seconds and kinematic viscosity vs. temperature. [*Maleeve and Hartman (3)*.]

FIGURE 24-4 Viscosity conversion chart. [*Maleev and Hartman (3)*.]

Particular	Formula

Kinematic viscosity

$$v = 10^{-6} v_k$$

where v in m^2/s and v_k in cSt   **SI**   (24–8b)

$$= \left(0.22t - \frac{180}{t}\right) 10^{-6}$$

where t in Saybolt seconds and v in m^2/s (24–8c)

Specific weight at 15.5° C

$$\gamma_{15.5} = \frac{141.5}{131.5 + {}^\circ \text{API}}$$ **Metric**   (24–9a)

where $\gamma_{15.5}$ in gf/ml (gram force/milliliter)

$$= \left(\frac{141.5}{131.5 + {}^\circ \text{API}}\right) 9807$$ **SI**   (24–9b)

where $\gamma_{15.5}$ in N/m^3

API = American Petroleum Institute gravity constant

Specific weight at any temperature

$$\gamma_t = \gamma_{15.5} - 0.000637(t - 15.5)$$ (24–10)

Refer to Table 24–1 for $\gamma_{15.5}$

$$= \gamma_{60} - 0.000365(t - 60)$$

US Customary System units   (24–10a)

The dynamic viscosity

$$\eta = \rho(0.22t - 180/t)10^{-6} \text{ where } t \text{ in } {}^\circ \text{F}$$

where η in Pa s, $\rho = \dfrac{\gamma}{g}$ and ρ in kg/m^3

The density (ρ) of oil and its specific gravity (γ) relative to water have the same numerical value

The absolute viscosity (dynamic viscosity) in terms of Engler degree, $E_t{}^\circ$

$$\eta' = 10^{-6}\gamma_t\left(0.737 E_t{}^\circ - \frac{0.635}{E_t{}^\circ}\right)$$ **Metric**   (24–11)

where η' in kgf s/m^2

The relation between arbitrary viscosity in Engler degree (V in $E_t{}^\circ$) and the absolute viscosity (dynamic viscosity) in kgf s/m^2

$$V = k\eta'$$ (24–12)

where $k \simeq 14.9 \times 10^3 E_t{}^\circ/(\text{kgf s/m}^2)$
= proportionality factor

The change in viscosity η' depending on temperature is expressed by formula

$$\eta' = \frac{i}{(0.1t^\circ)^3}$$ **Metric**   (24–13)

where i = characteristic number of the given grade of oil
$\simeq 1.4$ to 2.8
η' in kgf s/m^2

The relation between viscosity and pressure

$$\eta_p = \eta_o K^P$$ **Metric**   (24–14)

Particular	Formula
	where P = pressure, kgf/cm^2 K = constant for the given grade of oil $\simeq$ varies from 1.001 to 1.004 for pressure P up to 400 kgf/cm^2 (39 MPa) (Changes in oil viscosity due to change in pressure can be neglected.)
Kinematic viscosity in centistokes from Redwood No	$\nu = 0.260R - \dfrac{179}{R}$ when $34 < R < 100$ **Metric** (24–15a) where ν in cSt and R in number of Redwood seconds $= 0.247R - \dfrac{50}{R}$ when $R > 100$ (24–15b)
Kinematic viscosity in centistokes from Redwood Admiralty	$\nu = 2.7R - \dfrac{2000}{R}$ **Metric** (24–16) where R = the number of Redwood seconds

HAGEN-POISEUILLE LAW

The rate of laminar flow of lubricant in tubes	$Q = \dfrac{\pi d^4}{128\eta}\dfrac{dp}{dz}$ (24–17)

VERTICAL SHAFT ROTATING IN A GUIDE BEARING (FIG. 24–5)

The surface velocity of shaft	$U = \pi d n'$	(24–18)
The length of bearing in the direction of motion	$B = \pi d \dfrac{\beta^\circ}{360}$	(24–19)
The torque (Fig. 24–5)	$M_t = \mu(Ld)P\dfrac{d}{2} = \dfrac{\pi^2 d^2\, L\eta\, n'}{\psi}$ Refer to Fig. 24–6 for projected area (Ld)	(24–20)
Petroffs equation for coefficient of friction (Fig. 24–5)	$\mu = 2\pi^2 \left(\dfrac{\eta\, n'}{P}\right)\left(\dfrac{1}{\psi}\right)$	(24–21)
Design practice for journal bearing (3)	Refer to Table 24–2.	

Particular	Formula

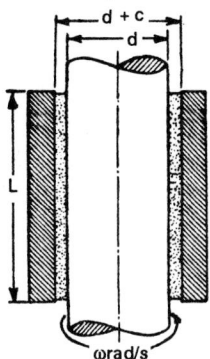

FIGURE 24–5   Vertical shaft rotating in a cylindrical bearing.

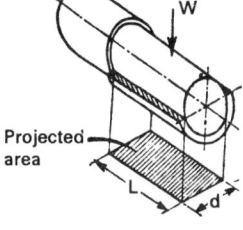

FIGURE 24–6   Projected area of a bearing.

The coefficient of friction can also be obtained from expression

$$\mu = K_a \left(\frac{\eta\, n'}{P} \right) \left(\frac{1}{\psi} \right) 10^{-10} + \Delta\, \mu \qquad (24\text{--}22)$$

where

$K_a = 5.53 \;\; \beta = 1980$ for $\beta = 360°$ **Metric** (24–22a)
$\quad \eta$ in cP, n' in rps, and P in kgf/cm^2

$K_a = 1.31 \;\; \beta = 473$ for $\beta = 360°$
$\qquad\qquad\qquad\qquad$ **US Customary units** (24–22b)
$\quad \eta$ in cP, n in rpm, and P in psi

$K_a = 9.23 \times 10^{-4}\beta = 0.33$ for $\beta = 360°$ **Metric**
$\qquad\qquad\qquad\qquad\qquad\qquad\qquad$ (24–22c)
$\quad \eta$ in cP, n in rpm, and P in kgf/mm^2

$K_a = 0.0553 \;\; \beta = 19.8$ for $\beta = 360°$ **Metric** (24–22d)
$\quad \eta$ in cP, n' in rps, and P in kgf/mm^2

$K_a = 5.4 \times 10^8 \beta = 1.95 \times 10^{11}$ for $\beta = 360°$ **SI**
$\qquad\qquad\qquad\qquad\qquad\qquad\qquad$ (24–22e)
$\quad \eta$ in Pa s, n' in rps, and P in N/m^2

$\Delta\mu =$ factor to correct for end leakage
$\quad\;\; = 0.002$ for L/d ranging from 0.75 to 2.8

Refer also to Fig. 24–7 for $\Delta\mu$.

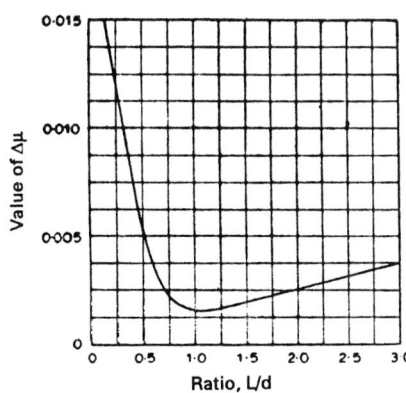

FIGURE 24–7   Correction factor for Eq. (24–22).

Louis Illmer equation for coefficient of friction in case of imperfect lubrication

$$\mu = 0.00012\, C_1 C_2 \sqrt[4]{\frac{P}{v_m}} \qquad\qquad \textbf{SI} \quad (24\text{--}23a)$$

where P in N/m^2 and v_m in m/s

$$= 0.0066\, C_1 C_2 \sqrt[4]{\frac{P}{v_m}} \qquad\qquad \textbf{Metric} \quad (24\text{--}23b)$$

TABLE 24-2
Journal bearing design practices

Machinery	Bearing	Maximum pressure, P kgf/mm²	kpsi	MPa	Diameter clearance ratio $\psi = \dfrac{c}{d}$	Ratio $\dfrac{L}{d}$	Viscosity, η_1 cP	Viscosity, η Pa s ×10⁻³	Bearing modulus (minimum) $S'' = \dfrac{\eta_1 n}{P}$ USCSU	$S'' = \dfrac{\eta n'}{P}$ SI Units ×10⁻⁹
Automobile and aircraft engines	Main	0.56–1.19	0.8–1.7	5.50–11.70	—	0.1–1.8	7	7	15	36.3
	Crankpin	1.06–2.47	1.5–3.5	10.40–24.40	0.001	0.7–1.4	to	to	10	24.2
	Wrist pin	1.62–3.62	2.3–5.0	15.00–34.80	<0.001	1.5–2.2	8	8	8	19.3
Gas and oil engines (four-stroke)	Main	0.49–0.85	0.7–1.2	4.85–8.35	0.001	0.6–2.0	20	20	20	48.4
	Crankpin	0.90–1.27	1.4–1.8	8.80–12.40	<0.001	0.6–1.5	to	to	10	24.2
	Wrist pin	1.27–1.55	1.8–2.2	12.40–15.20	<0.001	1.5–2.0	65	65	5	12.1
Gas and oil engines (two-stroke)	Main	0.35–0.56	0.5–0.8	3.42–5.50	0.001	0.6–2.0	20	20	25	60.4
	Crankpin	0.70–1.06	1.0–1.5	6.85–10.40	<0.001	0.6–1.5	to	to	12	29.0
	Wrist pin	0.85–1.07	1.2–1.8	8.35–12.50	<0.001	1.5–2.0	65	65	10	24.2
Marine steam engines	Main	0.35	0.5	3.42	<0.001	0.7–1.5	30	30	20	48.4
	Crankpin	0.42	0.6	4.14	<0.001	0.7–1.2	40	40	15	36.3
	Wrist pin	1.06	1.5	10.40	<0.001	1.2–1.7	30	30	10	24.2
Stationary, slow-speed steam engines	Main	0.28	0.4	2.75	<0.001	1.0–2.0	60	60	20	48.4
	Crankpin	1.06	1.5	10.40	<0.001	0.9–1.3	80	80	6	14.5
	Wrist pin	1.27	1.8	12.50	<0.001	1.2–1.5	60	60	5	12.1
Stationary, high-speed steam engines	Main	0.17	0.25	1.66	<0.001	1.5–3.0	15	15	25	60.4
	Crankpin	0.42	0.6	4.14	<0.001	0.9–1.5	30	30	6	14.5
	Wrist pin	1.27	1.8	12.50	<0.001	1.3–1.7	25	25	5	12.1
Steam locomotives	Driving axle	0.39	0.55	3.72	0.001	1.6–1.8	100	100	30	72.5
	Crankpin	1.40	2.0	13.70	<0.001	0.7–1.1	40	40	5	12.1
	Wrist pin	2.82	4.0	27.60	<0.001	0.8–1.3	30	30	5	12.1
Reciprocating pumps and compressors	Main	0.17	0.25	1.66	<0.001	1.0–2.2	30	30	30	72.5
	Crankpin	0.42	0.6	4.14	<0.001	0.9–1.7	to	to	20	48.4
	Wrist pin	0.70	1.0	6.85	<0.001	1.5–2.0	80	80	10	24.2
Railway cars	Axle	0.35	0.45	3.42	0.001	1.8–2.0	100	100	50	120.9
Steam turbines	Main	0.07–0.19	0.1–0.275	0.69–1.87	0.001	1.0–2.0	2–16	2–16	100	241.8

TABLE 24-2
Journal bearing design practices (*Cont.*)

Machinery	Bearing	Maximum pressure, P			Diameter clearance ratio $\psi = \dfrac{c}{d}$	Ratio $\dfrac{L}{d}$	Viscosity, η_1	Viscosity, η	Bearing modulus (minimum)	
		kgf/mm²	kpsi	MPa			cP	Pa s $\times 10^{-3}$	$S'' = \dfrac{\eta_1 n}{P}$ USCSU	$S'' = \dfrac{\eta n'}{P}$ SI Units $\times 10^{-9}$
Generators, motors, centrifugal pumps	Rotor	0.07–0.14	0.1–0.2	0.69–1.37	0.0013	1.0–2.0	25	25	200	483.5
Gyroscope	Rotor	0.60	0.85	5.90	0.0013	—	30	30	55	133.0
Transmission shafting	Light, fixed	0.08	0.025	0.17	0.001	2.0–3.0	25	25	100	241.8
	Self-aligning	0.106	0.15	1.04	0.001	2.5–4.0	to	to	30	72.5
	Heavy	0.106	0.15	1.04	0.001	2.0–3.0	60	60	30	72.5
Cotton mill	Spindle	0.0007	0.001	0.0069	0.005	—	2	2	10000	24177.5
Machine tools	Main	0.21	0.3	2.06	0.001	1.0–1.4	40	40	40	96.7
Punching and shearing machine	Main	2.82	4.0	27.80	0.001	1.0–2.0	100	100	—	—
	Crankpin	5.62	8.0	55.60	0.001	1.0–2.0	100	100	—	—
Rolling mills	Main	2.11	3.0	20.60	0.0015	1.1–1.5	50	50	10	24.2

Key: $\eta(\eta_1)$ = absolute viscosity, Pa s (cP); n = speed, rps; n' = speed, rpm; P = pressure, N/m² or MPa (psi); MPa = megapascal = 10^6 N/m²; Pa = Pascal = 1 N/m²; 1 psi = 6894. 757 Pa; 1 kpsi = 6.89475 MPa; USCSU = US Customary System units.

Source: V. L. Maleev, and J. B. Hartman, *Machine Design*, with permission from International Textbook Company, Scranton, Pennsylvania, 1954.

24.13

TABLE 24–3
Values of factor C_1 in Eq. (24–23)

Lubrication	Workmanship	Attendance	Operating condition	Constant C_1
Oil bath or flooded	High grade	First class	Clean and protected	1
Oil, free drop (constant feed)	Good	Fairly good	Favorable (ordinary condition)	2
Oil cup or grease (intermittent feed)	Fair	Poor	Exposed to dirt, grit or other unfavorable conditions	4

TABLE 24–4
Values of factor C_2 in Eq. (24–23)

Type of bearing	Constant C_2
Rotating journals, such as rigid bearing and crankpins	1
Oscillating journals, such as rigid wrist pin and Pintle blocks	1
Rotating bearings lacking ample rigidity, such as eccentric and the like	2
Rotating flat surfaces lubricated from the center to the circumference, such as annular step or pivot bearings	2
Sliding flat surfaces wiping over the guide ends, such as reciprocating crossheads; use 2 for relatively long guides and 3 for short guides	2–3
Sliding or wiping surfaces lubricated from the periphery or outer wiping edge, such as marine thrust bearings and worm gears	3–4
Long power-screw nuts and similar wiping parts over which it is difficult to effect a uniform distribution of lubricant or load	4–6

Particular	Formula
	where P in kgf/mm^2 and v_m in m/s
	$$= 0.004 \, C_1 C_2 \sqrt[4]{\frac{P}{v_m}} \qquad \textbf{US Customary units}$$ (24–23c)
	where P in psi and v_m in ft/min
	Refer to Tables 24–3 and 24–4 for C_1 and C_2, respectively.
For behaviour of journal at stand still, at start and running in its bearing	Refer to Fig. 24–8.

Particular	Formula

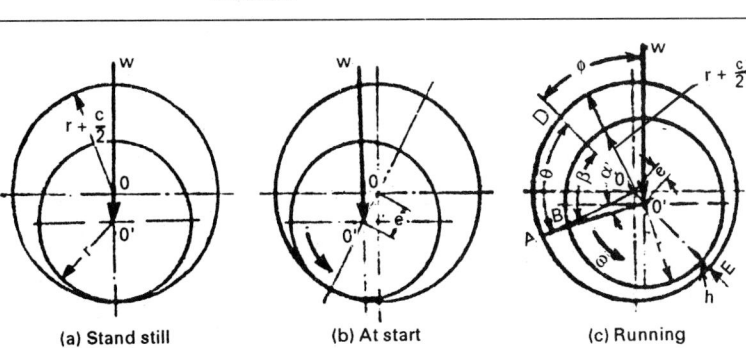

(a) Stand still   (b) At start   (c) Running

FIGURE 24–8   Behavior of a journal in its bearing.

BEARING PRESSURE (FIG. 24–9)

General Electric Company's formula for bearing pressure in the design of motor and generator bearing

$$P_a = 6.2 \times 10^5 \sqrt[3]{v_m} \qquad \textbf{SI} \qquad (24\text{--}24a)$$

where P_a in N/m² and v_m in m/s

$$= 15.5 \sqrt[3]{v_m} \qquad \textbf{US Customary units} \quad (24\text{--}24b)$$

where P_a in psi and v_m in ft/min

$$= 0.0635 \sqrt[3]{v_m} \qquad \textbf{Metric} \qquad (24\text{--}24c)$$

where P_a in kgf/mm² and v_m in m/s

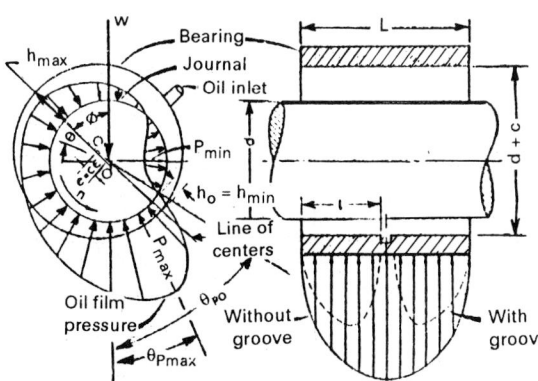

FIGURE 24–9   Oil film pressure distribution in the full journal bearing.

Victor Tatarinoff's equation for safe operating load

$$W = \frac{\eta_1 n d^3 (L/d)^2}{127(10^6) h \psi \left(1 + \dfrac{L}{d}\right)} \qquad \textbf{US Customary units} \\ (24\text{--}25a)$$

where η_1 in cP, n in rpm; L, d, h and c in in; W in lbf

Particular	Formula

$$= \frac{\eta n' d^3 \left(\frac{L}{d}\right)^2}{0.295 h \psi \left(1 + \frac{L}{d}\right)} \qquad \textbf{SI} \quad (24\text{--}25b)$$

where η in Pa s; n' in rps; $L, d, h,$ and c in m; W in N

Victor Tatarinoff's equation for permissible unit pressure

$$P = \frac{\eta_1 n'}{3175(10^4)\psi^2} \left(\frac{L}{L+d}\right) \qquad \textbf{USCustomaryunits}$$

$$(24\text{--}26a)$$

where P in psi, η_1 in cP, n in rpm, L and d in in

$$= 13.5 \frac{\eta \, n'}{\psi^2} \left(\frac{L}{L+d}\right) \qquad \textbf{SI} \quad (24\text{--}26b)$$

where P in Pa, η in Pa s, n' in rps, and L and d in m

H. F. Moore's equation for critical pressure

$$P_c = 7.23 \times 10^5 \sqrt{v} \qquad \textbf{SI} \quad (24\text{--}27a)$$

where P_c in N/m^2 and v in m/s

$$= 0.0737 \sqrt{v} \qquad \textbf{Metric} \quad (24\text{--}27b)$$

where P_c in kgf/mm^2 and v in m/s

$$= 7.5 \sqrt{v} \qquad \textbf{US Customary units} \quad (24\text{--}27c)$$

where P_c in psi and v in ft/min

The critical unit pressure for any given velocity should not exceed according to Louis Illmer

$$P_c = 4.6 \times 10^6 \sqrt[3]{\frac{v_m}{(t - 288.5)}} \qquad \textbf{SI} \quad (24\text{--}28a)$$

where P_c in N/m^2, v_m in m/s, and t in K

$$= 0.47 \sqrt[3]{\frac{v_m}{(t - 15.5)}} \qquad \textbf{Metric}$$

$$(24\text{--}28b)$$

where P_c in kgf/mm^2, v_m in m/s, and t in °C

$$= 140 \sqrt[3]{\frac{v_m}{(t - 15.5)}} \qquad \textbf{US Customary units}$$

$$(24\text{--}28c)$$

where P_c in psi, v_m in ft/min, t in °F

Stribeck's equation for the critical pressure when the speed does not exceed 2.5 m/s (500 ft/min)

$$P_c = 9.7 \times 10^5 \sqrt{v} \qquad \textbf{SI} \quad (24\text{--}28d)$$

where P_c in N/m^2 and v in m/s

$$= 10 \sqrt{v} \qquad \textbf{US Customary units} \quad (24\text{--}28e)$$

where P_c in psi and v in ft/min

$$= 0.0986 \sqrt{v} \qquad \textbf{Metric} \quad (24\text{--}28f)$$

Particular	Formula
	where P_c in kgf/mm^2 and v in m/s
Stribeck's equation for the critical pressure when the speed exceeds 2.5 m/s (500 ft/min)	$P_c = 2.9 \times 10^6 \sqrt{v}$ **SI** (24–28g) where P_c in N/m^2 and v in m/s $= 30 \sqrt{v}$ **US Customary units** (24–28h) where P_c in psi and v in ft/min $= 0.296 \sqrt{v}$ **Metric** (24–28i) where P_c in kgf/mm^2 and v in m/s Refer to Table 24–5 for allowable pressures for reciprocating motion.

TABLE 24–5
Allowable bearing pressure, reciprocating motion

Type of bearing	Type of machinery	Pressure, P	
		psi	MPa
Crosshead	Steam engine, stationary	35–60	0.24–0.412
	Steam engine, marine	55–100	0.378–0.688
	Steam engine, locomotive	70–90	0.48–0.62
	Gas and oil engines, stationary	40–70	0.275–0.48
	Compressors and pumps	50–90	0.342–0.62
Trunk pin	Gas and oil engines, stationary	20–25	0.136–0.172
	Automotive and aircraft engines	25–40	0.172–0.275

For permissible P^v values   Refer to Table 24–6.

For values S'' for various combinations of journal bearing materials, abrasion pressure for bearings, allowable bearing pressures for semi-fluid lubricants and diametral clearances in bearing dimensions.   Refer to Tables 24–7 to 24–10.

IDEALIZED JOURNAL BEARING (FIGS. 24–8 AND 24–9)

The diametral clearance ratio or relative clearance	$\psi = \dfrac{c}{d}$	(24–29)
Attitude or eccentricity ratio or eccentricity coefficient	$\epsilon = \dfrac{2e}{c} = 1 - \dfrac{2\,h_{min}}{d\psi}$	(24–30)
	Refer to Fig. 24–10 for ϵ.	
Oil film thickness at any position θ	$h = \dfrac{c}{2}(1 + \epsilon \, \cos \theta)$	(24–31)

TABLE 24–6
Permissible Pv values

Class of bearing or journal	P^v values	
	psi ft/s	N/m s
Mill shafting, with self-aligning cast-iron bearings, grease, or imperfect oil-lubrication, maximum value	12,000	4.2×10^5
Mill shafting, self-aligning ring-oiled babbitt bearings, maximum	24,000	8.45×10^5
Self-aligning ring-oiled bearings, continuous load in one direction	35,000–40,000	12.3×10^5 to 14×10^5
Crankshaft journals with bronze bearings	22,000	7.7×10^5
Crankshaft bearings with babbitted bearings, maximum	59,000	20.8×10^5
For excellent radiating condition	133,000	46.5×10^5

Key: US Customary unit: P = pressure, psi, v = velocity, ft/s; SI unit: P = pressure, N/m^2, v = velocity, m/s

TABLE 24–7
Values S'' for various combinations of journal bearing materials

Shaft	Bearing	Bearing modulus	
		$S'' = \dfrac{\eta_1\, n}{P}$	$S'' = \dfrac{\eta\, n}{P}$
		Metric	SI $\times 10^{-9}$
Hardened and ground steel	Babbitt	28,500	48.5
Machined, soft steel	Babbitt	36,000	61.2
Hardened and ground steel	Plastic bronze	42,700	72.6
Machined, soft steel	Plastic bronze	35,800	60.9
Hardened and ground steel	Rigid bronze	56,900	96.7
Machined, soft steel	Rigid bronze	71,100	120.8

Particular	Formula
For position of minimum oil thickness	Refer to Fig. 24–35e.
Minimum oil film thickness	$h_{min} = h_o = \dfrac{c}{2}(1 - \epsilon)$ (24–32)
The minimum oil film thickness variable	$\delta = \dfrac{2h_{min}}{c} = (1 - \epsilon)$ (24–33) Refer to Figs. 24–11 to 24–13 and 24–35c for δ.
The safe oil film thickness for a bearing in good condition and $v_m \gtrless 1$ m/s (200 ft/min)	$h_{min} = h_o = 2.37 \times 10^{-5}\, v_m^{0.4}\, A^{0.2}$ **SI** (24–34a) where h_{min} in m, A in m^2, and v_m in m/s $= 0.0015\, v_m^{0.4}\, A^{0.2}$ **Metric** (24–34b)

TABLE 24–8
Abrasion pressures for bearings

Materials in contact	Pressure		Remarks
	psi	MPa	
Hardened tool steel on lumen or phosphor bronze	10,000	68.8	Values applies to rigid, polished and accurately fitted rubbing surface
0.50 C machine steel on lumen or phosphor bronze	8,000	55.0	When not worn to a fit or well lubricated reduce to 4.22 kgf/mm^2 (41.4 MPa)
Hardened tool steel on hardened tool steel	7,000	48.0	
0.50 C machine steel or wrought iron on genuine hard babbitt	6,000	41.5	
Cast iron on cast iron (close grained or chilled)	4,500	31.0	
Case-hardened machine steel on case-hardened machine steel	4,000	27.5	
0.30 C machine steel on cast iron (close-grained)	3,500	24.0	
0.40 C machine steel on soft common babbitt	3,000	20.6	
Soft machine steel on machine steel (not case-hardened)	2,000	13.8	
Machine steel on lignum vitae (water-lubricated)	1,500	10.2	

TABLE 24–9
Allowable bearing pressures for semifluid lubrication

Bearing material	Journal material	Allowable pressure, $\dot{P}_a$	
		psi	MPa
Lumen of phosphor bronze	Hardened tool steel	2500	17.30
Hardened steel	Hardened alloy steel	2000	14.40
Hard babbitt	SAE 1050 steel	1500	10.30
Bronze	Hardened alloy steel	1300	8.90
Cast iron	Cast iron	1100	7.58
Bronze	Alloy steel	850	5.90
Babbit, soft	SAE 1040 steel	750	5.20
Bronze	Mild steel, smooth finish	540	3.70
Bronze	Mild steel, ordinary finish	400	2.75
Bronze	Cast iron	400	2.75
Cast iron	Mild steel	350	2.40
Lignum vitae, water lubricated	Mild steel	350	2.40

TABLE 24–10
Diametral clearance in bearings dimension in micrometers ($1 \ \mu m = 10^{-6}$ m)

Particular about bearing and journal	Diametral clearances, c in μm				
	$d = 12$	$d = 25$	$d = 50$	$d = 100$	$d = 140$
Precision spindle, hardened and ground steel, lapped into bronze bearing $-v_m < 25$ m/s; $P < 500$ psi (3.43 N/m^2); 0.2–0.4 μm rms	7–19	19–38	38–63	63–88	88–125
Precision spindle, hardened and ground steel, lapped into bronze bearing $-v_m > 25$ m/s; $P > 500$ psi (3.43 N/m^2); 0.2–0.4 μm rms	13–25	25–50	50–75	75–113	113–163
Electric motors and generators, ground journals in broached or reamed bronze or babbitt bearings; 0.4–0.8 μm rms	13–38	25–50	38–85	50–100	75–150
General machinery, intermittent or continuous motion, turned or cold-rolled journal in reamed and bored bronze or babbitt bearings; 0.8–1.5 μm rms	50–100	63–113	75–125	100–175	125–200
Rough machinery, turned or cold-rolled steel journals in poured babbitt bearings; 1.5–3.8 μm rms	77–150	125–225	200–300	275–400	350–500
Automotive crankshaft					
Babbitt-lined bearing				38	63
Cadmium silver copper				50	75
Copper lead				36	88

Particular	Formula
	where h_{min} in mm, A in mm^2, and v_m in m/s
	$= 0.000026 \ v_m^{0.4} \ A^{0.2}$ **US Customary units**
	where h_{min} in in, A in in^2, and v_m in (24–34c)
The thickness of oil film where the pressure is maximum or minimum	$(h)_{\substack{P(max) \\ P(min)}} = k = \dfrac{2c(1 - \epsilon^2)}{2 + \epsilon^2}$ (24–35)
The resultant pressure distribution around a journal bearing excluding P_o the oil film pressure at the point where $\theta = 0$ or $\theta = 2\pi$	$P_r = (P - P_o)$
	$= \dfrac{12\eta \ U}{\psi^2 d} \left[\dfrac{\epsilon(2 + \epsilon \ \cos \ \theta) \sin \theta}{(2 + \epsilon^2)(1 + \epsilon \ \cos \theta)^2} \right]$ (24–36)
The pressure at any point θ (Figs. 24–8 and 24–9)	$P = P_r + P_o$ (24–37)
The load carrying capacity of the bearing [Fig. 24–8 (panel c)]	$W = \dfrac{\eta \ UL}{\psi^2} \left[\dfrac{2\pi\epsilon}{(2 + \epsilon^2)\sqrt{2 - \epsilon^2}} \right]$ (24–38)
The bearing characteristic number or Sommerfeld number	$S = \dfrac{\eta n'}{P} \dfrac{1}{\psi^2}$ (24–39)

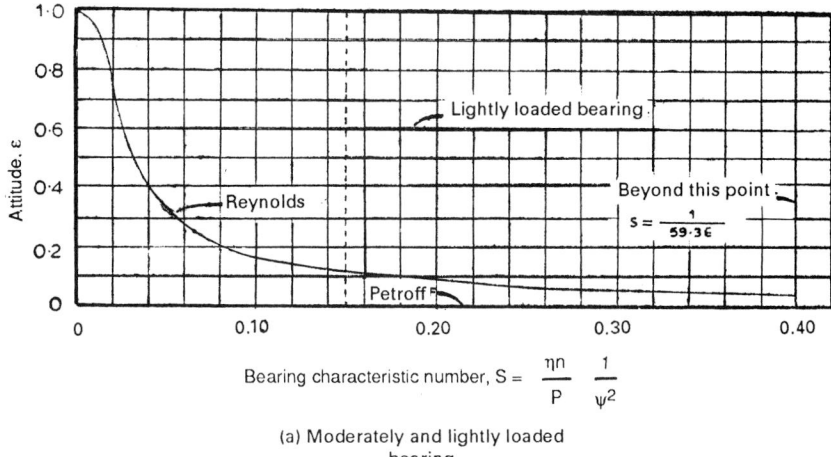

Bearing characteristic number, $S = \dfrac{\eta n}{P}\ \dfrac{1}{\psi^2}$

(a) Moderately and lightly loaded
bearing

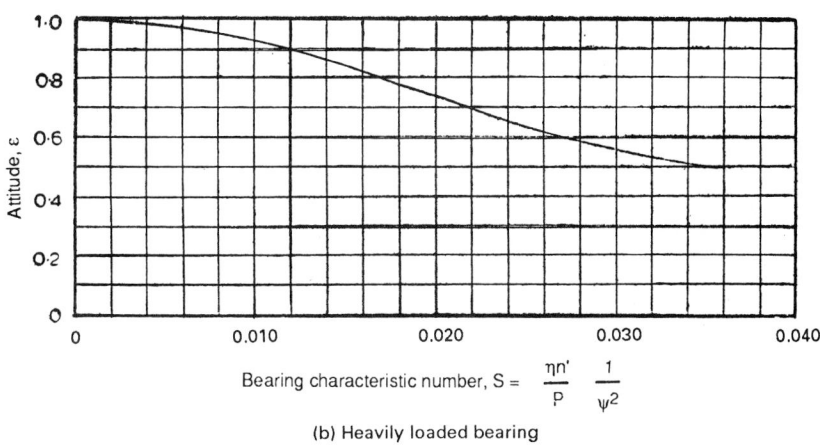

Bearing characteristic number, $S = \dfrac{\eta n'}{P}\ \dfrac{1}{\psi^2}$

(b) Heavily loaded bearing

FIGURE 24–10   Variation of attitude ϵ of full journal bearing with characteristic number S. [*Radzimosvksy (4)*.]

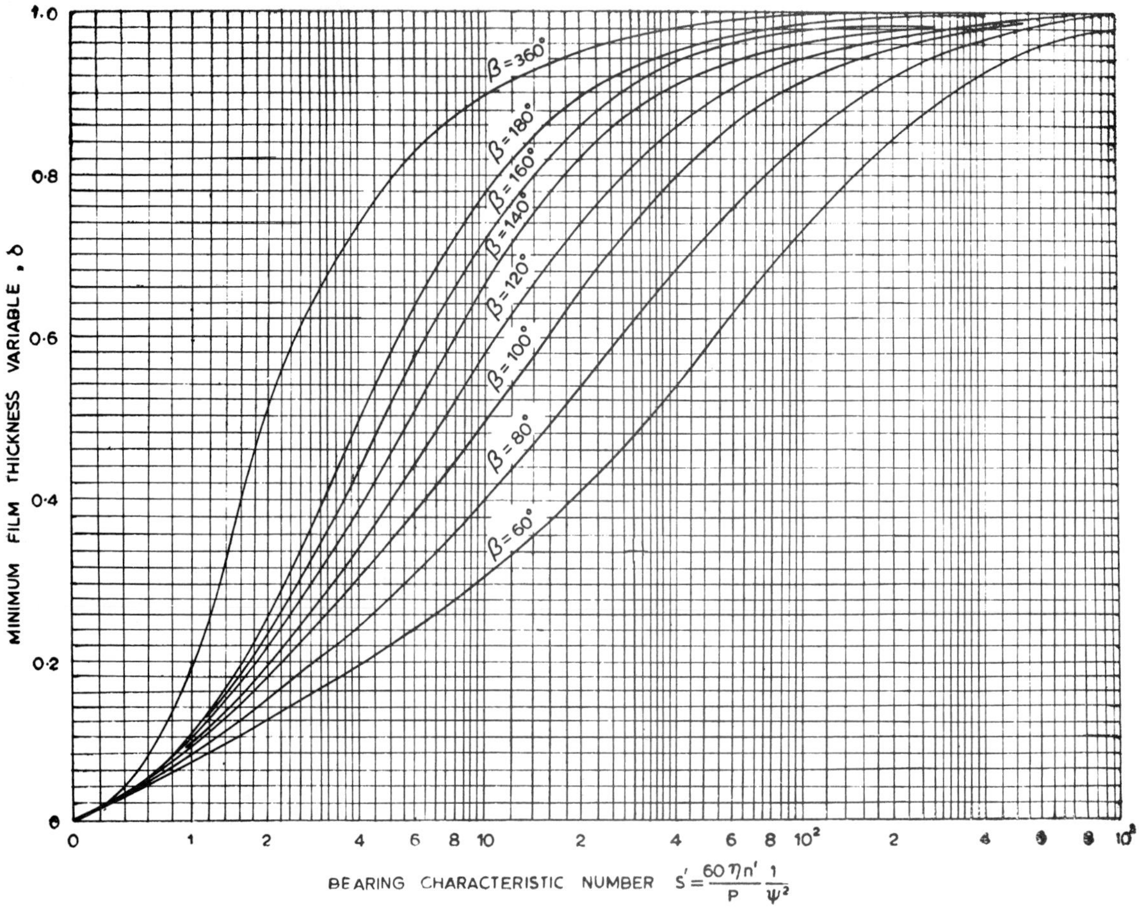

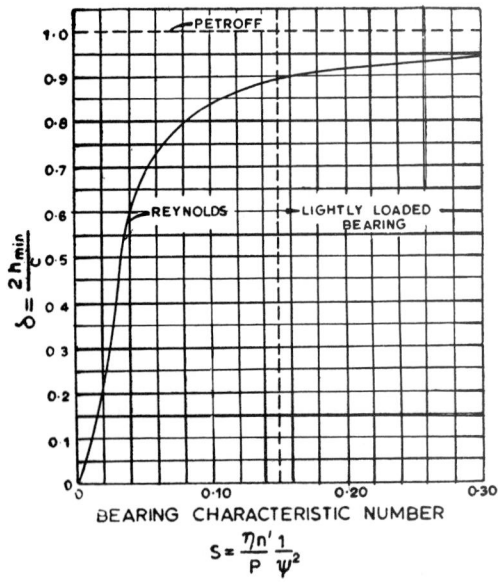

FIGURE 24–11 Miniumum oil film thickness variable δ based on no side flow. [*Boyd and Raimondi* (5).]

FIGURE 24–12 Variation of minimum oil film thickness variable δ of full journal bearing with S.

Particular	Formula

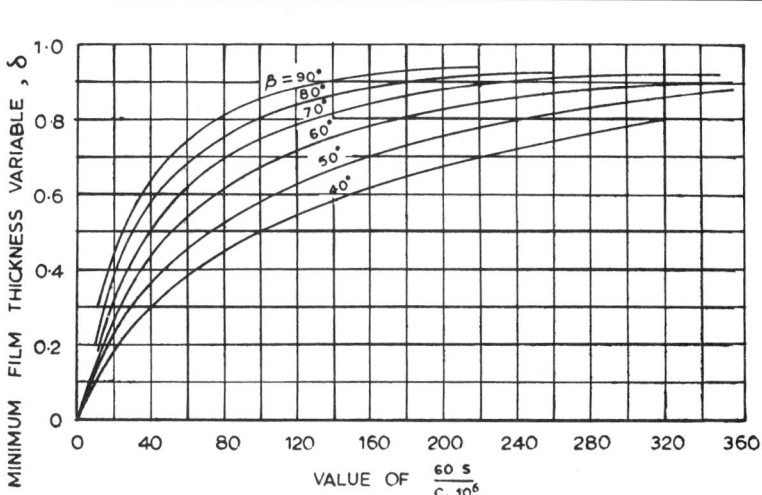

FIGURE 24–13 Variation of minimum oil film thickness variable δ with S/C_L.

For Sommerfeld number S

Refer to Tables 24–10 to 24–12 for Sommerfeld numbers S for full and partial bearings.

The constant of the bearing or bearing modulus

$$S'' = \frac{\eta n}{P} \tag{24–40}$$

where η in Pa s (cP)

Refer to Table 24–7 for bearing modulus.

The calculation of minimum oil film thickness from Figs 24–13 and 24–14

Hint: S is determined from Eq. (24–39) and C_L from Fig. 24–14 for a given $\dfrac{L}{d}$ ratio. Calculate $\dfrac{60S}{C_L 10^6}$. Knowing $\dfrac{60S}{C_L 10^6}$, you can then obtain the minimum film thickness variable δ from Fig. 24–13. From δ and Eq. (24–33), and can then determine the minimum oil film thickness.

The bearing characteristic number or Sommerfeld number as a function of attitude

$$S = \frac{(2 + \epsilon^2)\sqrt{1 - \epsilon^2}}{12\, \pi^2 \epsilon} \tag{24–41}$$

Refer to Fig. 24–10 for ϵ for various values of S.

The angular positions of points where the maximum or minimum pressure in the oil film occur [Fig. (24–8c and Fig. 24–9]

$$\theta = \cos^{-1}\left(-\frac{3\epsilon}{\epsilon^2 + 2}\right) \tag{24–42}$$

TABLE 24–11
Dimensionless performance parameters for full journal bearings with side flow

				Values of δ			
L/d ratio				0.25	0.5	1.0	∞
For maximum load				0.27	0.43	0.53	0.66
For minimum friction				0.03	0.12	0.3	0.6

$\frac{L}{d}$	ϵ	δ	S	ϕ	$\frac{\mu}{\psi}$	$\frac{4Q}{\psi\, d^2\, n'\, L}$	$\frac{Q_s}{Q}$	$\frac{\rho c T_0}{P}$	$\frac{P}{P_{max}}$
0.25	0	1.0	∞	(89.5)	∞	π	0	∞	—
	0.1	0.9	16.2	82.31	322.0	3.45	0.180	1287.0	0.515
	0.2	0.8	7.57	75.18	153.0	3.76	0.330	611.0	0.489
	0.4	0.6	2.83	60.86	61.1	4.37	0.567	245.0	0.415
	0.6	0.4	1.07	46.72	26.7	4.99	0.746	107.6	0.334
	0.8	0.2	0.261	31.04	8.80	5.60	0.884	35.4	0.240
	0.9	0.1	0.0736	21.85	3.50	5.91	0.945	14.1	0.180
	0.97	0.03	0.0101	12.22	0.922	6.12	0.984	3.73	0.108
	1.0	0	0	0	0	—	1.0	0	0
0.5	0	1.0	∞	(88.5)	∞	π	0	∞	—
	0.1	0.9	4.31	81.62	85.6	3.43	0.173	343.0	0.523
	0.2	0.8	2.03	74.94	40.9	3.72	0.318	164.0	0.506
	0.4	0.6	0.779	61.45	17.0	4.29	0.552	68.6	0.441
	0.6	0.4	0.319	48.14	8.10	4.85	0.730	33.0	0.365
	0.8	0.2	0.0923	33.31	3.26	5.41	0.874	13.4	0.267
	0.9	0.1	0.0313	23.66	1.60	5.69	0.939	6.66	0.206
	0.97	0.03	0.00609	13.75	0.610	5.88	0.980	2.56	0.126
	1.0	0	0	0	0	—	1.0	0	0
1	0	1.0	∞	(85)	∞	π	0	∞	—
	0.1	0.9	1.33	79.5	26.4	3.37	0.150	106	0.540
	0.2	0.8	0.631	74.02	12.8	3.59	0.280	52.1	0.529
	0.4	0.6	0.264	63.10	5.79	3.99	0.497	24.3	0.484
	0.6	0.4	0.121	50.58	3.22	4.33	0.680	14.2	0.415
	0.8	0.2	0.0446	36.24	1.70	4.62	0.842	8.0	0.313
	0.9	0.1	0.0188	26.45	1.05	4.74	0.919	5.16	0.247
	0.97	0.03	0.00474	15.47	0.514	4.82	0.973	2.61	0.152
	1.0	0	0	0	0	—	1.0	0	0
∞	0	1.0	∞	(70.92)	∞	π	0	∞	—
	0.1	0.9	0.240	69.10	4.80	3.03	0	19.9	0.826
	0.2	0.8	0.123	67.26	2.57	2.83	0	11.4	0.814
	0.4	0.6	0.0626	61.94	1.52	2.26	0	8.47	0.764
	0.6	0.4	0.0389	54.31	1.20	1.56	0	9.73	0.667
	0.8	0.2	0.021	42.22	0.961	0.760	0	15.9	0.495
	0.9	0.1	0.0115	31.62	0.756	0.411	0	23.1	0.358
	0.97	0.03	—	—	—	—	0	—	—
	1.0	0	0	0	0	0	0	∞	0

Key: Q_s = flow of lubricant with side flow, cm³/s; ρ = density of the lubricant $\simeq 0.00083$ kgf/cm³; c = specific heat of the lubricant = 17,100 kgf cm/kgf°C; $\rho c = 14.2$ kgf/cm²°C; T_0 = difference in temperature, °C.
Source: A. A. Raimondi and J. Boyd, "A Solution for the Finite Journal Bearings and Its Applications to Analysis and Design" ASME, J. Lubrication Technol., Vol. 104, pp. 135–148, April 1982.

TABLE 24–12
Dimensionless performance parameters for 180° bearing centrally loaded with side flow[a]

				Values of δ			
L/d ratio				0.25	0.5	1	∞
For maximum load				0.28	0.42	0.52	0.64
For minimum friction				0.03	0.23	0.44	0.60

$\dfrac{L}{d}$	ϵ	δ	S	ϕ	$\dfrac{\mu}{\psi}$	$\dfrac{4Q}{\psi\, d^2\, n'\, L}$	$\dfrac{Q_s}{Q}$	$\dfrac{\rho c T_0}{P}$	$\dfrac{P}{P_{max}}$
0.25	0	1.0	∞	90.0	∞	π	0	∞	—
	0.1	0.9	16.3	81.40	163.0	3.44	0.176	653.0	0.513
	0.2	0.8	7.60	73.70	79.4	3.71	0.320	320.0	0.489
	0.4	0.6	2.84	58.99	35.1	4.11	0.534	146.0	0.417
	0.6	0.4	1.08	44.96	17.6	4.25	0.698	79.8	0.336
	0.8	0.2	0.263	30.43	6.88	4.07	0.837	36.5	0.241
	0.9	0.1	0.0736	21.43	2.99	3.72	0.905	18.4	0.180
	0.97	0.03	0.0104	12.28	0.877	3.29	0.961	6.46	0.110
	1.0	0	0	0	0	—	1.0	0	0
0.25	0	1.0	∞	90.0	∞	π	0	∞	—
	0.1	0.9	4.38	79.97	44.0	3.41	0.167	177.0	0.518
	0.2	0.8	2.06	72.14	21.6	3.64	0.302	87.8	0.499
	0.4	0.6	0.794	58.01	9.96	3.93	0.506	42.7	0.438
	0.6	0.4	0.321	45.01	5.41	3.93	0.665	25.9	0.365
	0.8	0.2	0.0921	31.29	2.54	3.56	0.806	15.0	0.273
	0.9	0.1	0.0314	22.80	1.38	3.17	0.886	9.80	0.208
	0.97	0.03	0.00635	13.63	0.581	2.62	0.951	5.30	0.132
	1.0	0	0	0	0	—	1.0	0	0
1	0	1.0	∞	90.0	—	π	0	∞	—
	0.1	0.9	1.40	78.50	14.1	3.34	0.139	57.0	0.525
	0.2	0.8	0.670	68.93	7.15	3.46	0.252	29.7	0.513
	0.4	0.6	0.278	58.86	3.61	3.49	0.425	16.5	0.466
	0.6	0.4	0.128	44.67	2.28	3.25	0.572	12.4	0.403
	0.8	0.2	0.0463	32.33	1.39	2.63	0.721	10.4	0.313
	0.9	0.1	0.0193	24.14	0.921	2.14	0.818	9.13	0.244
	0.97	0.03	0.00483	14.57	0.483	1.60	0.915	6.96	0.157
	1.0	0	0	0	0	—	1.0	0	0
∞	0	1.0	∞	90.0	∞	π	∞	∞	—
	0.1	0.9	0.347	72.90	3.55	3.04	0	14.7	0.778
	0.2	0.8	0.179	61.32	2.01	2.80	0	8.99	0.759
	0.4	0.6	0.898	49.99	1.29	2.20	0	7.34	0.700
	0.6	0.4	0.0523	43.15	1.06	1.52	0	8.71	0.607
	0.8	0.2	0.0253	33.35	0.859	0.767	0	14.1	0.459
	0.9	0.1	0.0128	25.57	0.681	0.380	0	22.5	0.337
	0.97	0.03	0.00384	15.43	0.416	0.119	0	44.0	0.190
	1.0	0	0	0	0	0	0	∞	0

[a] See *Key* and *Source* under Table 24–11.

TABLE 24–13
Dimensionless performance parameters for 120° for centrally loaded bearing with side flow[a]

				Values of δ			
L/d ratio				0.25	0.5	1	∞
For maximum load				0.26	0.38	0.46	0.53
For minimum friction				0.06	0.28	0.4	0.5

$\dfrac{L}{d}$	ϵ	δ	S	ϕ	$\dfrac{\mu}{\psi}$	$\dfrac{4Q}{\psi\,d^2\,n'\,L}$	$\dfrac{Q_s}{Q}$	$\dfrac{\rho c T_0}{P}$	$\dfrac{P}{P_{max}}$
0.25	0	1.0	∞	90.0	∞	π	0	∞	—
	0.10	0.9044	18.4	76.97	124.0	3.34	0.143	502.0	0.456
	0.20	0.8011	8.45	65.97	60.4	3.44	0.260	254.0	0.438
	0.40	0.6	3.04	51.23	26.6	3.42	0.442	125.0	0.389
	0.6	0.4	1.12	40.42	13.5	3.20	0.599	75.8	0.321
	0.8	0.2	0.268	28.38	5.65	2.67	0.753	42.7	0.237
	0.9	0.1	0.0743	20.55	2.63	2.21	0.846	25.9	0.178
	0.97	0.03	0.0105	12.11	0.852	1.69	0.931	11.6	0.112
	1.0	0	0	0	0	—	1.0	0	0
0.50	0	1.0	∞	90.0	∞	π	0	—	—
	0.1	0.9034	5.42	74.99	36.6	3.29	0.124	149.0	0.431
	0.2	0.8003	2.51	63.38	18.1	3.32	0.225	77.2	0.424
	0.4	0.6	0.914	48.07	8.20	3.15	0.386	40.5	0.389
	0.6	0.4	0.354	38.50	4.43	2.80	0.530	27.0	0.336
	0.8	0.2	0.0973	28.02	2.17	2.18	0.684	19.0	0.261
	0.9	0.1	0.0324	21.02	1.24	1.70	0.787	15.1	0.203
	0.97	0.03	0.00631	13.00	0.550	1.19	0.899	10.6	0.136
	1.0	0	0	0	0	—	1.0	0	0
1	0	1.0	∞	90.0	∞	π	0	∞	—
	0.1	0.9024	2.14	72.43	14.5	3.20	0.0876	59.5	0.427
	0.2	0.8	1.01	58.25	7.44	3.11	0.157	32.6	0.420
	0.4	0.6	0.385	43.98	3.60	2.75	0.272	19.0	0.396
	0.6	0.4	0.162	35.65	2.16	2.24	0.384	15.0	0.356
	0.8	0.2	0.0531	27.42	1.27	1.57	0.535	13.9	0.290
	0.9	0.1	0.0208	21.29	0.855	1.11	0.657	14.4	0.233
	0.97	0.03	0.00498	13.49	0.461	0.694	0.812	14.0	0.162
	1.0	0	0	0	0	—	1.0	0	0
∞	0	1.0	∞	90.0	∞	π	0	∞	—
	0.1	0.9007	0.877	66.69	6.02	3.02	0	25.1	0.610
	0.2	0.8	0.431	52.60	3.26	2.75	0	14.9	0.599
	0.4	0.6	0.181	39.02	1.78	2.13	0	10.5	0.566
	0.6	0.4	0.0845	32.67	1.21	1.47	0	10.3	0.509
	0.8	0.2	0.0328	26.80	0.853	0.759	0	14.1	0.405
	0.9	0.1	0.0147	21.51	0.653	0.388	0	21.2	0.311
	0.97	0.03	0.00406	13.86	0.399	0.118	0	42.3	0.199
	1.0	0	0	0	0	0	0	∞	0

[a] See *Key* and *Source* under Table 24–11.

TABLE 24–14
Dimensionless performance parameters for 60° centrally loaded bearing with side flow[a]

						Values of δ			
L/d ratio						0.25	0.5	1	∞
For maximum load						0.15	0.20	0.23	0.25
For minimum friction						0.10	0.16	0.22	0.23

$\dfrac{L}{d}$	ϵ	δ	S	ϕ	$\dfrac{\mu}{\psi}$	$\dfrac{4Q}{\psi\, d^2\, n'\, L}$	$\dfrac{Q_s}{Q}$	$\dfrac{\rho c T_0}{P}$	$\dfrac{P}{P_{max}}$
0.25	0	1.0	∞	90.0	∞	π	0	∞	—
	0.1	0.9251	35.8	71.55	121.0	3.16	0.0666	499.0	0.251
	0.2	0.8242	16.0	58.51	58.7	3.04	0.131	260.0	0.249
	0.4	0.6074	5.20	41.01	24.5	2.57	0.236	136.0	0.242
	0.6	0.4	1.65	30.14	11.2	1.98	0.346	86.1	0.228
	0.8	0.2	0.333	21.70	4.27	1.30	0.496	54.9	0.195
	0.9	0.1	0.0844	16.87	2.01	0.894	0.620	41.0	0.159
	0.97	0.03	0.0110	10.81	0.713	0.507	0.786	29.1	0.107
	1.0	0	0	0	0	—	1.0	0	0
0.5	0	1.0	∞	90.0	∞	π	0	∞	—
	0.1	0.9223	14.2	69.00	48.6	3.11	0.0488	201.0	0.239
	0.2	0.8152	6.47	52.60	24.2	2.91	0.0883	109.0	0.239
	0.4	0.6039	2.14	37.00	10.3	2.38	0.160	59.4	0.233
	0.6	0.4	0.695	26.98	4.93	1.74	0.236	40.3	0.225
	0.8	0.2	0.149	19.57	2.02	1.05	0.350	29.4	0.201
	0.9	0.1	0.0422	15.91	1.08	0.664	0.464	26.5	0.172
	0.97	0.03	0.00704	10.85	0.490	0.329	0.650	27.8	0.122
	1.0	0	0	0	0	—	1.0	0	0
1	0	1.0	∞	90.0	∞	π	0	∞	—
	0.1	0.9212	8.52	67.92	29.1	3.07	0.0267	121.0	0.252
	0.2	0.8133	3.92	50.96	14.8	2.82	0.0481	67.4	0.251
	0.4	0.6010	1.34	33.99	6.61	2.22	0.0849	39.1	0.247
	0.6	0.4	0.450	24.56	3.29	1.56	0.127	28.2	0.239
	0.8	0.2	0.101	18.33	1.42	0.883	0.200	22.5	0.220
	0.9	0.1	0.0309	15.33	0.822	0.519	0.287	23.2	0.192
	0.97	0.03	0.00584	10.88	0.422	0.226	0.465	30.5	0.139
	1.0	0	0	0	0	—	1.0	0	0
∞	0	1.0	∞	90.0	∞	π	0	∞	—
	0.1	0.9191	5.75	65.91	19.7	3.01	0	82.3	0.337
	0.2	0.8109	2.66	48.91	10.1	2.73	0	46.5	0.336
	0.4	0.6002	0.931	31.96	4.67	2.07	0	28.4	0.329
	0.6	0.4	0.322	23.21	2.40	1.40	0	21.4	0.317
	0.8	0.2	0.0755	17.39	1.10	0.722	0	19.2	0.287
	0.9	0.1	0.0241	14.94	0.667	0.372	0	22.5	0.243
	0.97	0.03	0.00495	10.58	0.372	0.115	0	40.7	0.163
	1.0	0	0	0	0	0	0	∞	0

[a] See *Key* and *Source* under Table 24–11.

Particular	Formula

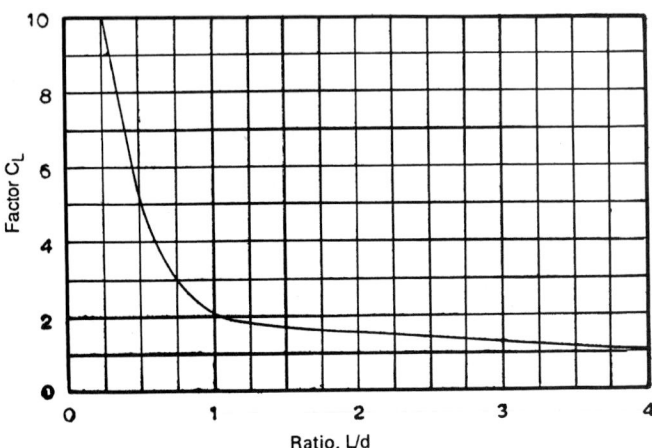

FIGURE 24–14 Variation of factor C_L with L/d ratio.

The total frictional resistance on an idealized journal bearing surface

$$F_\mu = \frac{4\pi \, \eta U L}{\psi} \left[\frac{(1 + 2\epsilon^2)}{(2 + \epsilon^2)\sqrt{1 - \epsilon^2}} \right] \tag{24–43}$$

or

$$F_\mu = \frac{4\pi^2 \, \eta n' \, Ld(1 + 2\epsilon^2)}{\psi(2 + \epsilon^2)\sqrt{1 - \epsilon^2}} \tag{24–44}$$

The total frictional resistance on an idealized lightly loaded journal bearing

$$F_\mu = \frac{2\pi^2 \, \eta n' \, Ld}{\psi} \tag{24–45}$$

For the relation between dimensionless quantity

$\dfrac{\eta n'}{F'_\mu} \left(\dfrac{1}{\psi} \right)$ and Sommerfeld number S

Refer to Fig. 24–15.

The relation between coefficient of friction and bearing characteristic number

$$\mu = 2\pi^2 \frac{\eta n'}{P} \left(\frac{1}{\psi} \right) \tag{24–46}$$

The relation between the coefficient of friction and attitude ϵ

$$\mu = \psi \left[\frac{1 + 2\epsilon^2}{3\epsilon} \right] \tag{24–47}$$

The average coefficient of friction at very high pressures

Refer to Table 24–15 for average coefficient of friction.

The friction coefficient variable

$$\lambda_\mu = \frac{\mu}{\psi} = \frac{1 + 2\epsilon^2}{3\epsilon} \tag{24–48}$$

Refer to Figs. 24–29 – 24–32, and 24–35g.

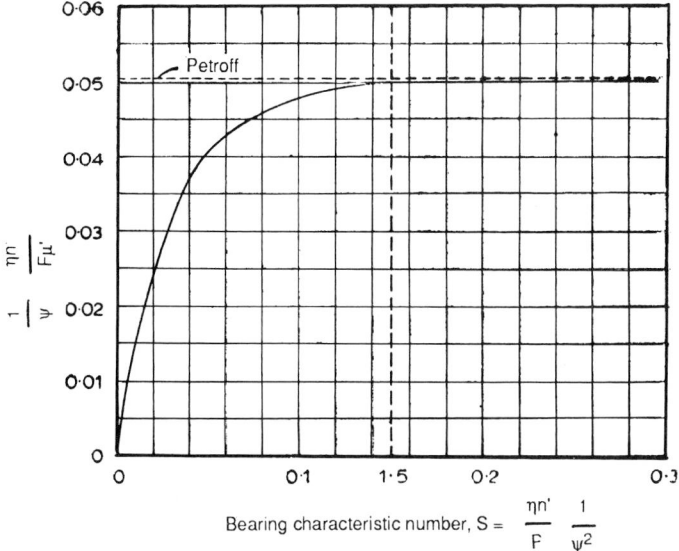

Bearing characteristic number, $S = \dfrac{\eta n'}{P}\ \dfrac{1}{\psi^2}$

FIGURE 24–15 Variation of dimensionless quantity $\dfrac{1}{\psi}\dfrac{\eta n'}{F'_\mu}$ with S for an idealized full journal beading.

TABLE 24–15
Average coefficient of friction at very high pressure

Material	Angular displacement, deg	
	10°	**50°**
Stearic acid	0.022	0.029
Tungsten disulfide	0.032	0.037
Molybdenum disulfide	0.032	0.033
Graphite	0.036	0.058
Silver sulfate	0.055	0.054
Turbine oil plus 1% MoS_2	0.060	0.068
Lead iodide	0.061	0.071
Palm oil	0.063	0.075
Castor oil	0.064	0.081
Grease (zinc-oxide base)	0.071	0.080
Lard oil	0.072	0.084
Grease (calcium base)	0.073	0.082
Residual	0.076	0.083
Sperm oil	0.077	0.085
Turbine oil plus 1% graphite	0.081	0.105
Turbine oil plus 1% stearic acid	0.087	0.096
Turbine oil	0.088	0.108
Capric acid	0.089	0.109
Turbine oil plus 1% mica	0.091	0.105
Oleic acid	0.093	0.119
Machine oil	0.099	0.115
Soapstone (powdered)	0.169	0.306
Mica (powdered)	0.257	0.305
Boron (not a lubricant)	0.482	0.710

Particular	Formula

POWER LOSS

The power loss in the bearing due to viscous friction

$$P = \frac{F_\mu U}{33000}$$ **US Customary units**   (24–49a)

where P in hp; F_μ in lbf, and U in ft/min

$$= \frac{F_\mu U}{102}$$ **Metric**   (24–49b)

where p in kW, F_μ in kgf, $U = \pi d\, n' =$ velocity in m/s, d in m, and n' in rps

$$= \frac{F_\mu U}{1000}$$ **SI**   (24–49c)

where p in kW, F_μ in N, U in m/s

PARTIAL JOURNAL BEARING (FIG. 24–16)

The resultant pressure distribution around the partial journal bearing exluding, P_o oil film pressure at the point where $\theta = 0$

$$P_r = (P - P_o)$$ (24–50)

where

$$P - P_o = \frac{12\eta\, U}{\psi^2\, d} \left[\{(1 - \epsilon^2) - (2 + \epsilon^2)(k/c)\}/(1 - \epsilon^2)^{2\backslash 5} \right.$$

$$\times \arctan\left\{ \sqrt{\frac{1 - \epsilon}{1 + \epsilon}}\, \tan\frac{\theta}{2} \right\}$$

$$+ \frac{(k/2c)\epsilon \sin\theta}{2(1 - \epsilon^2)(1 + \epsilon \cos\theta)^2}$$

$$\left. + \frac{\epsilon \sin\theta\{(3k/2c) - 2(1 - \epsilon^2)\}}{2(1 - \epsilon^2)^2(1 + \epsilon \cos\theta)} \right]$$

where $k = h$ is the thickness of oil film at maximum pressure value

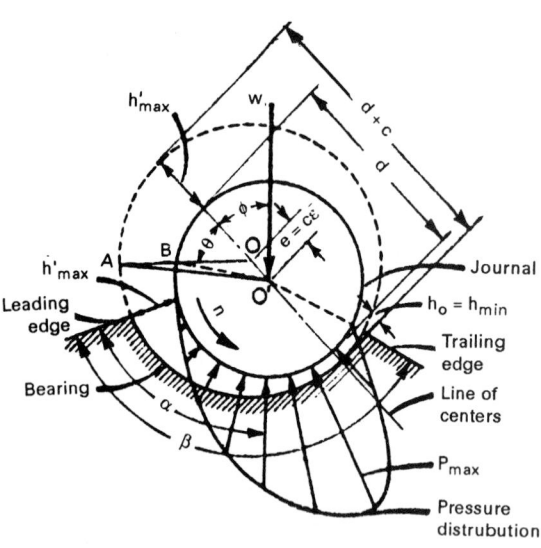

FIGURE 24–16   Partial journal bearing.

Pressure at any point in a partial journal bearing

$$P = (P_o + P_r)$$ (24–51)

Refer to Figs. 24–17 and 24–18, respectively.

To determine the attitude ϵ and attitude angle ϕ for various values of S and for an idealized offset partial bearing having the maximum load capacity corresponding to a given attitude

Particular	Formula

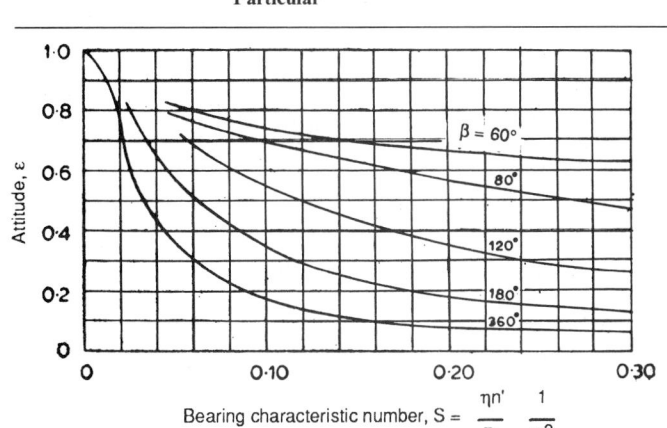

Bearing characteristic number, $S = \dfrac{\eta n'}{P}\,\dfrac{1}{\psi^2}$

FIGURE 24–17   Variation of attitude ϵ with S for an idealized offset partial bearing having the maximum load capacity corresponding to a given attitude.

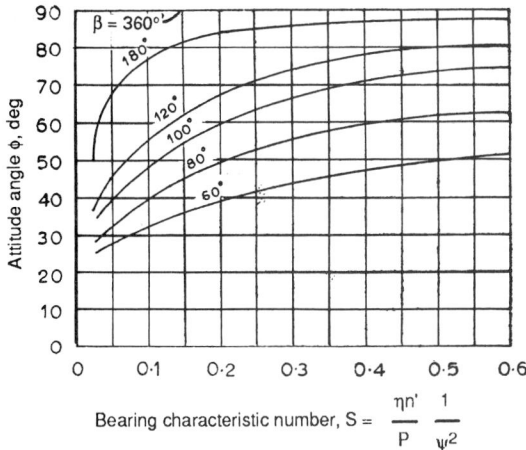

Bearing characteristic number, $S = \dfrac{\eta n'}{P}\,\dfrac{1}{\psi^2}$

FIGURE 24–18   Variation of attitude angle ϕ with S for an idealized offset partial bearing.

INFLUENCE OF END LEAKAGE

Leakage factors $C_W, C_F,$ and C_μ

Refer to Fig. 24–19 for $C_W, C_F,$ and C_μ for various values of B/L ratios

Load leakage factor according to Kingsbury (6)

$$C_W = \frac{W}{W_\infty} \qquad (24\text{–}52)$$

Refer to Fig. 24–19 for C_W.

Load leakage factor C_W as a function of B/L ratio for a slider bearing having $q = (h_1/h_2) - 1 = 1$ or $h_1 = 2h_2$

Refer to Table 24–16

Load leakage factor for 120°, centrally loaded partial bearing according to Needs (7)

Refer to Fig. 24–20 for C_W for various attitudes ϵ.

Load correction factor for side flow according to Boyd and Raimondi (24)

Refer to Fig. 24–21 for C_W for various minimum oil film thickness variables δ.

Particular	Formula
Coefficient of friction leakage factor according to Kingsbury (6)	$$C_\mu = \frac{\mu_\infty}{\mu} \qquad (24\text{–}53)$$ Refer to Fig. 24–19 for C_μ.

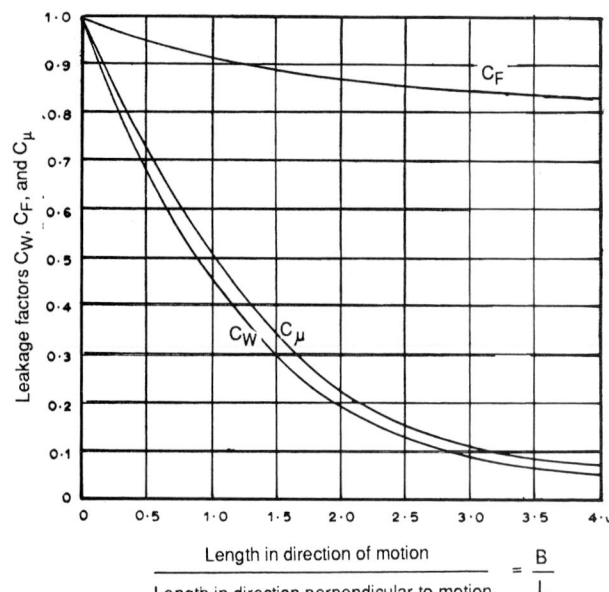

Length in direction of motion ────────────────────────────── = $\dfrac{B}{L}$
Length in direction perpendicular to motion

FIGURE 24–19   Kingsbury's leakage factors as function of B/L ratios under minimum friction. [*Kingsbury (6).*]

TABLE 24–16
Load leakage factor C_W as a function of B/L ratio for a slider bearing having the quality q equal to unity

B/L	C_W	B/L	C_W
0.00	1.00	1.00	0.44
0.175	0.92	1.50	0.278
0.25	0.835	2.00	0.185
0.50	0.68	3.00	0.090
0.75	0.55	4.00	0.060

Particular	Formula
Friction leakage factor according to Kingsbury (6)	$$C_F = \frac{F_\mu}{F_{\mu_\infty}} \qquad (24\text{–}54)$$ Refer to Fig. 24–19 for C_F.
Friction leakage factor for 120° centrally loaded partial bearing according to Needs (7)	Refer to Figs. 24–22 for C_F for various attitudes ϵ.
Friction correction factor for side flow according to Boyd and Raimondi (5)	Refer to Fig. 24–23 for C_F for various minimum oil film thickness variables δ and B/L ratios.

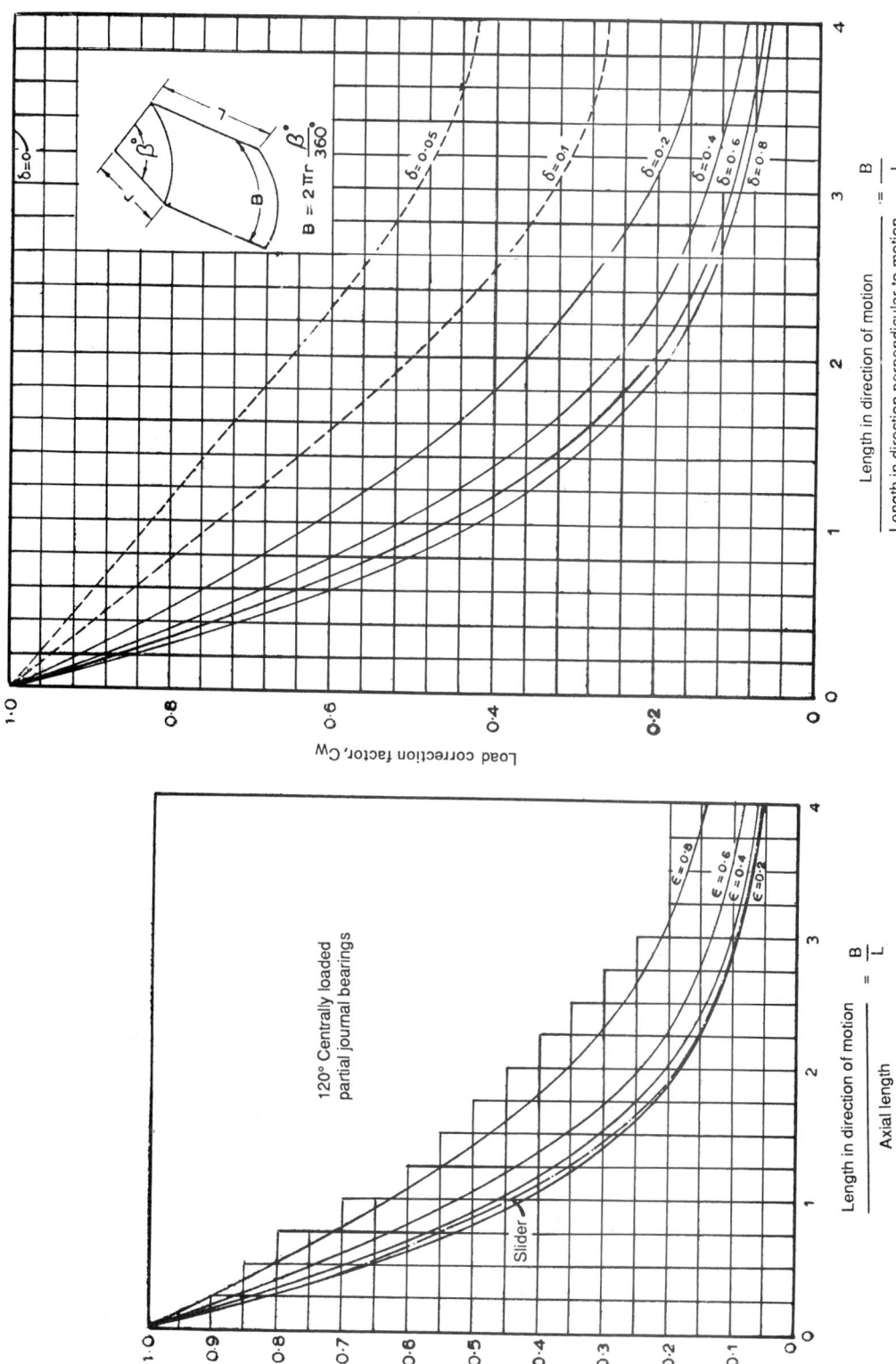

FIGURE 24-20 Leakage factors for load for 120° centrally loaded partial journal bearings for various attitudes. [*Needs (7).*]

FIGURE 24-21 Load correction factor (C_w) for side flow. [*Boyd and Raimondi (5).*]

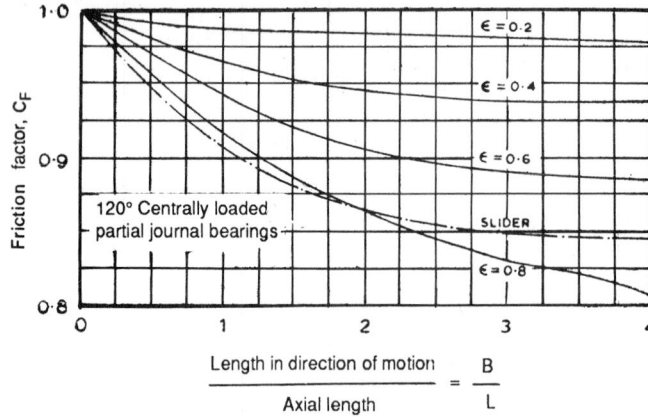

FIGURE 24–22 Leakage factors for friction force for 120° centrally loaded partial journal bearings for various attitudes. [*Needs* (7).]

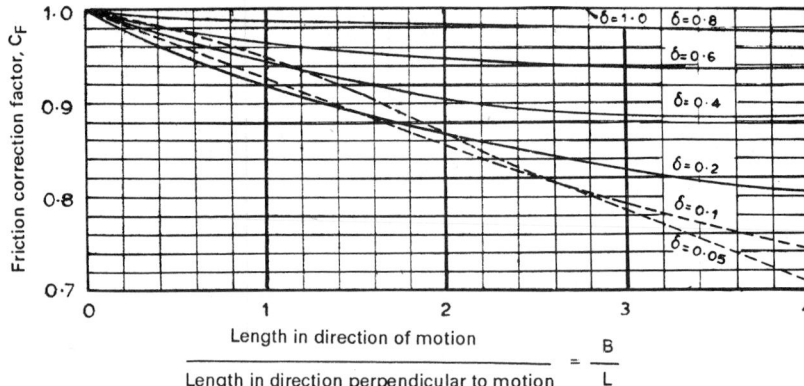

FIGURE 24–23 Friction correction factor for side flow. [*Boyd and Raimondi* (5).]

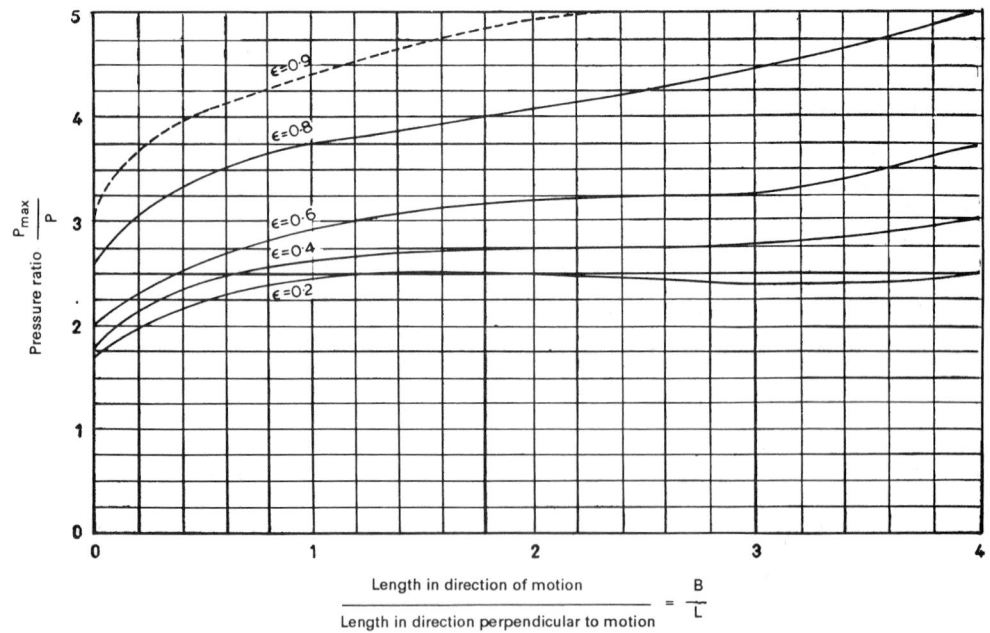

FIGURE 24–24 The ratio of the maximum pressure (P_{max}) and the unit load $P(= P_u)$ with B/L ratios for various attitudes for a 120° central partial bearing [*Needs* (7)].

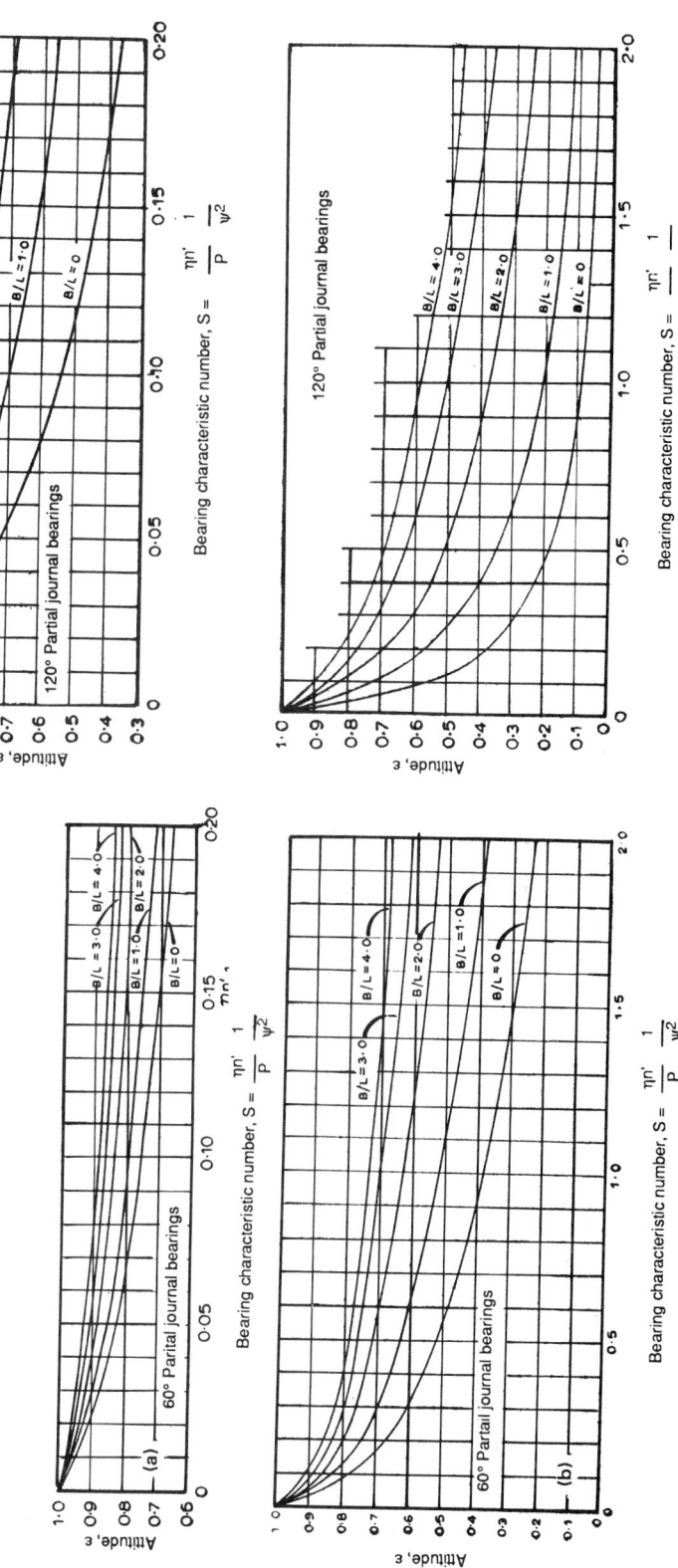

FIGURE 24–26 Variation of attitude ϵ with S for 120° partial journal bearing.

FIGURE 24–25 Variation of attitude ϵ with S for 60° partial journal bearing.

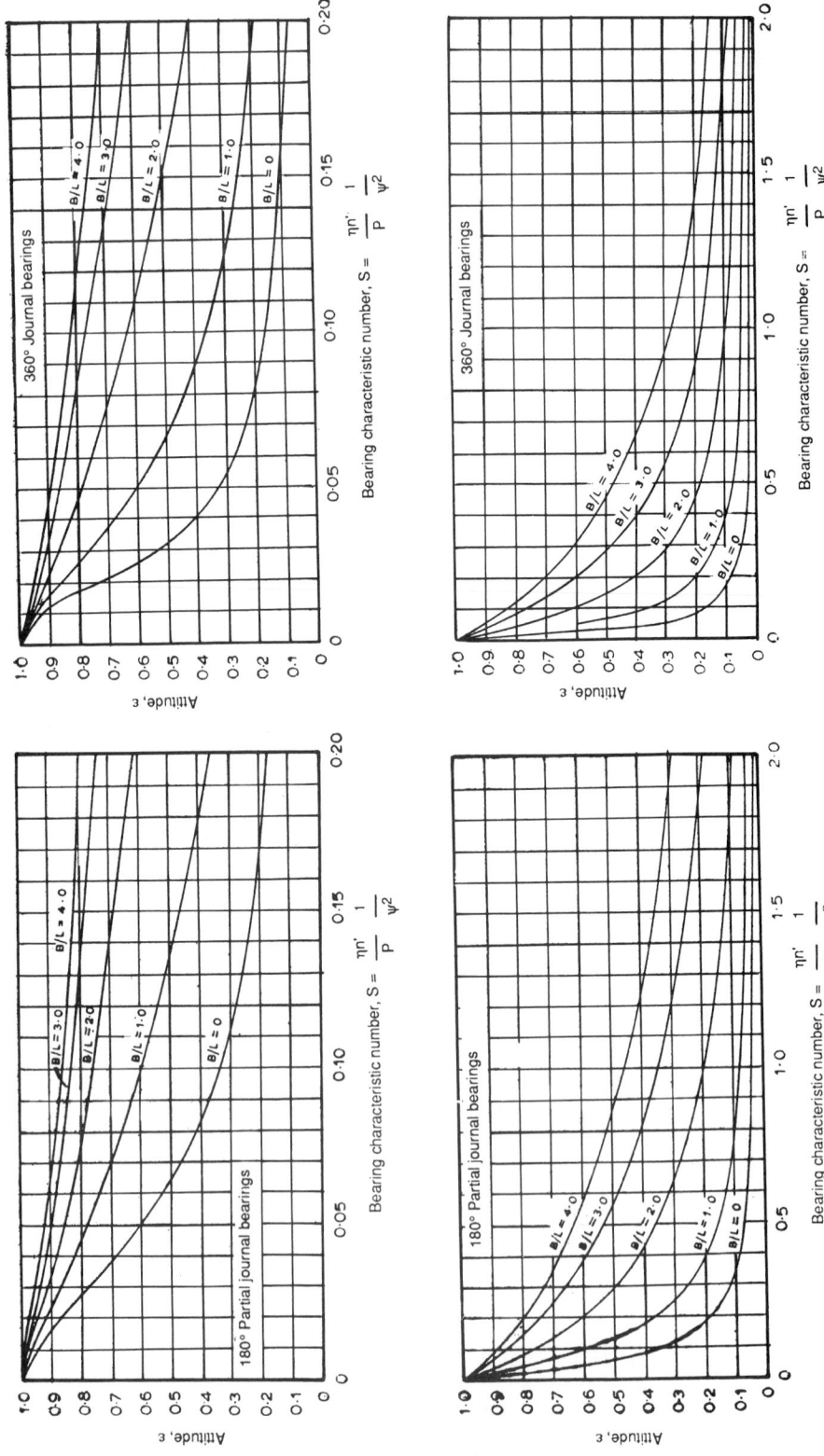

FIGURE 24-27 Variation of attitude ϵ with S for 180° partial journal bearing.

FIGURE 24-28 Variation of attitude ϵ with S for 360° partial journal bearing.

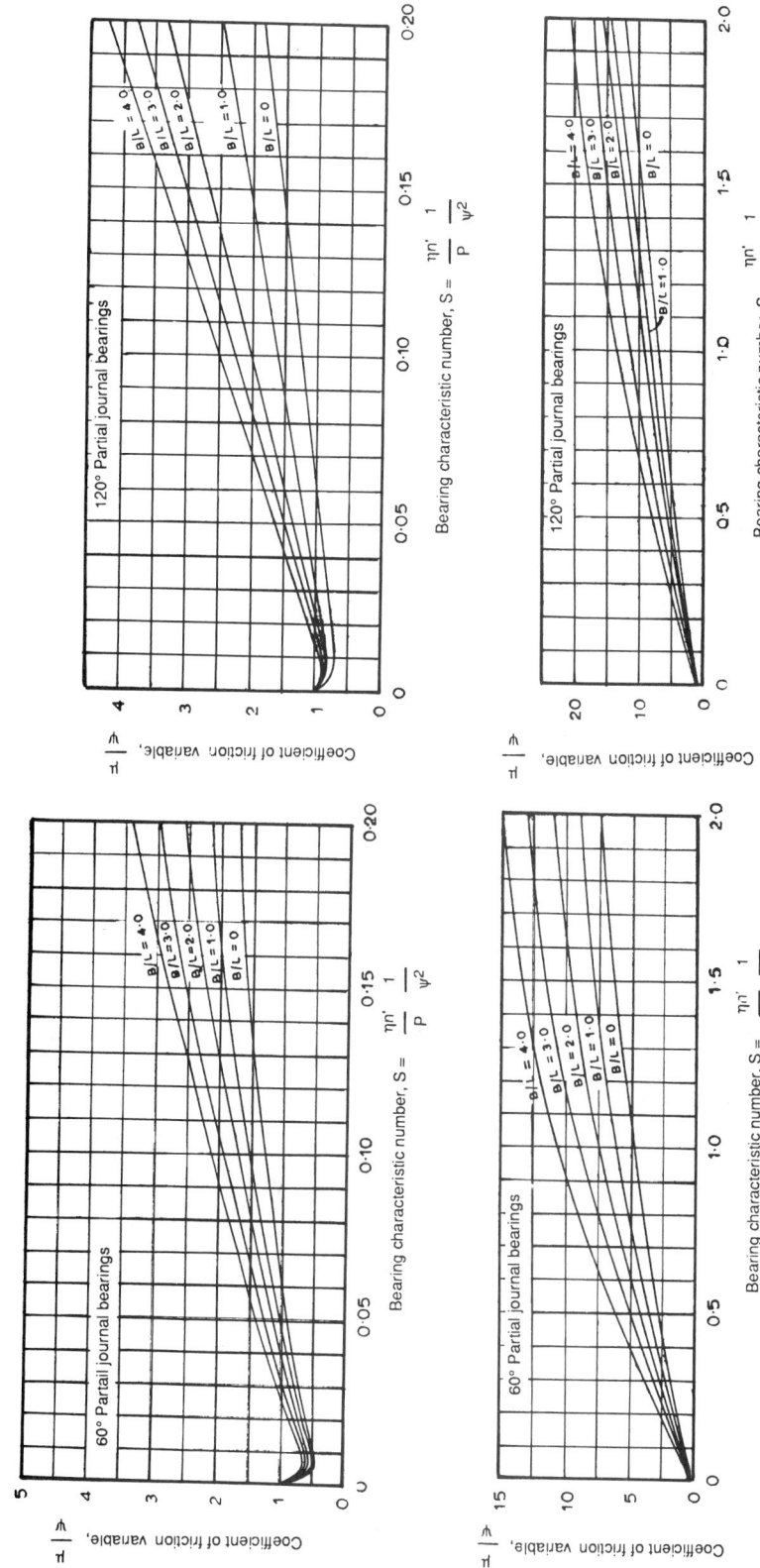

FIGURE 24–29 Variation of coefficient of friction variable μ/ψ with S for 60° partial journal bearing.

FIGURE 24–30 Variation of coefficient of friction variable μ/ψ with S for 120° partial journal bearing.

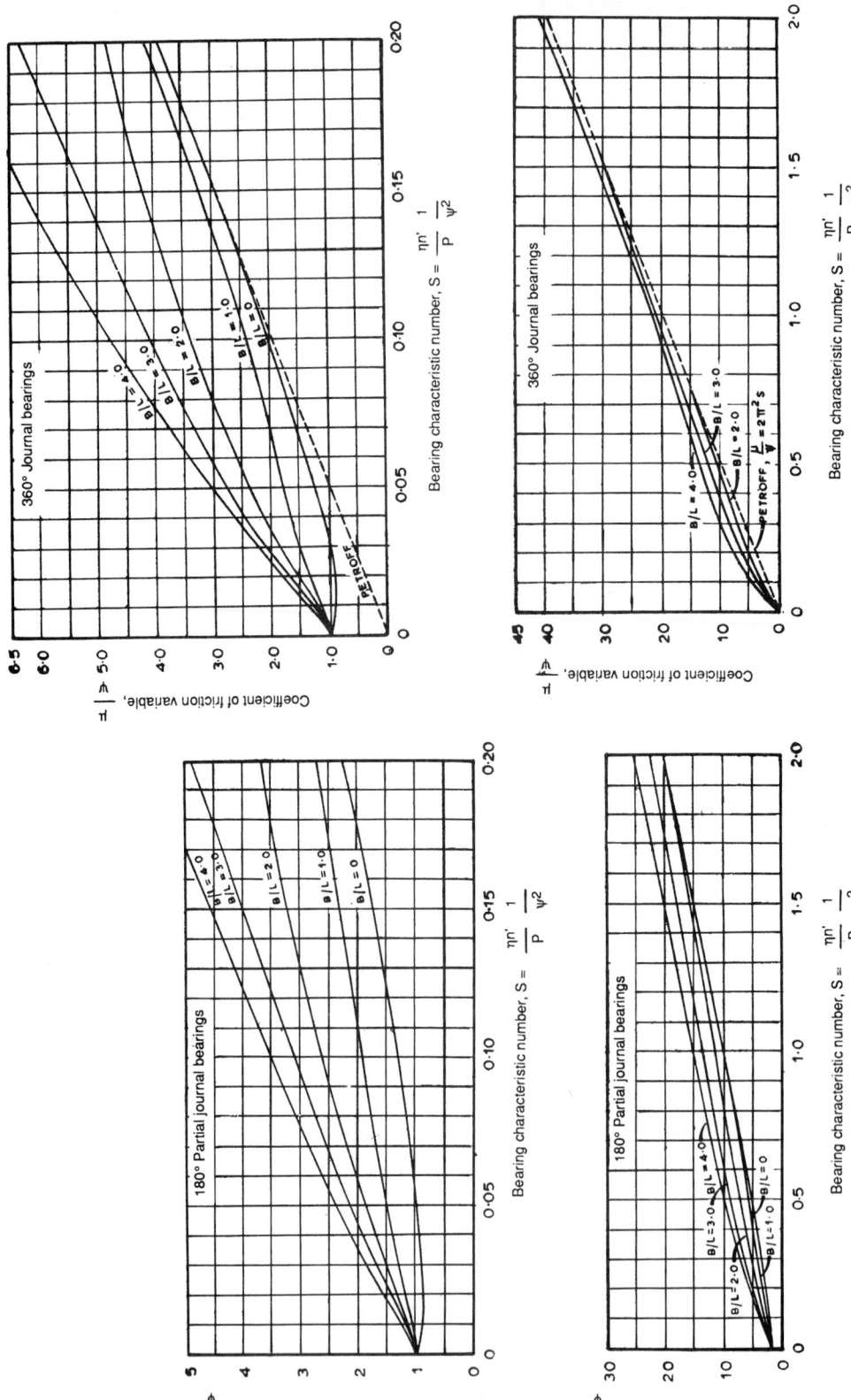

FIGURE 24–31 Variation of coefficient of friction variable μ/ψ with S for 180° patial journal bearing.

FIGURE 24–32 Variation of coefficient of friction variable μ/ψ with S for 360° partial journal bearing.

Particular	**Formula**
The ratios of the maximum pressure in the oil film, P_{max}, and the unit load, P, with B/L ratios for various values of attitude, ϵ, for 120° central partial journal bearing according to Needs (7)	Refer to Fig. 24–24 for $\dfrac{P_{max}}{P}$ and Fig. 24–35j for P/P_{max} for various values of B/L ratios and attitudes ϵ
The variation of attitude, ϵ, with bearing characteristic number, S, for various values of B/L ratios for 60°, 120°, 180° partial and full journal bearings	Refer to Figs. 24–25 to 24–28 for ϵ for various values of S and B/L ratios.
The variation of coefficient of friction variable, $\lambda_\mu = \mu/\psi$, with bearing characteristic number, S, for various values of B/L ratios for 60°, 120°, 180°, partial and full journal bearing	Refer to Figs. 24–29 to 24–32 and 24–35g for $\lambda_\mu = \dfrac{\mu}{\psi}$ for various values of S and B/L ratios
The friction curves illustrating boundary conditions	Refer to Fig. 24–33.

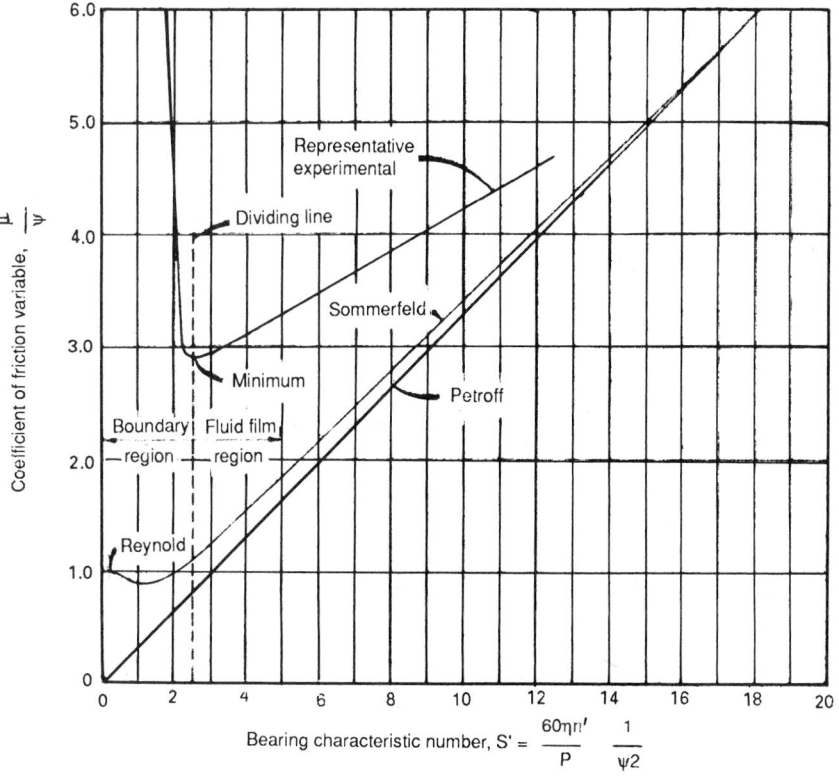

FIGURE 24–33 Friction curves illustrating boundary conditions.

Particular	Formula

FRICTION IN A FULL JOURNAL BEARING WITH END LEAKAGE FROM THE BEARING

The total friction force acting on the surface of a full journal bearing with end flow

$$F_\mu = \tfrac{1}{2} W \epsilon\, \psi\, \sin\, \phi + \frac{2\pi^2\, \eta n'\, Ld}{\psi\sqrt{1-\epsilon^2}} \qquad (24\text{--}55)$$

The coefficient of friction variable

$$\lambda_\mu = \frac{\mu}{\psi} = \frac{\epsilon}{2}\, \sin\, \phi + \frac{2\pi^2\, S}{\sqrt{1-\epsilon^2}} \qquad (24\text{--}56a)$$

$$= \frac{\epsilon\sqrt{1-\epsilon^2}}{2} + \frac{2\pi^2\, S}{\sqrt{1-\epsilon^2}} \qquad (24\text{--}56b)$$

$P\mu$ value

Lasche's equation for $P\mu$

$$P\mu = 195,350/t \qquad\qquad \textbf{SI} \quad (24\text{--}57a)$$

where P in N/m^2 and t in K

$$= 0.02/t \qquad\qquad \textbf{Metric} \quad (24\text{--}57b)$$

where P in kgf/mm^2 and t in °C

$$= 51/t \qquad \textbf{US Customary units} \quad (24\text{--}57c)$$

where P in psi and t in °F

Lasche's equation for the coefficient of friction which may be used for bearing subjected to pressure varying from 0.103 MPa (15 psi) to 1.55 MPa (225 psi) and speed varying from 2.5 to 18 m/s and temperature varying from 30 to 100°C

$$\mu = \frac{23126}{P\sqrt{t}} \qquad\qquad \textbf{SI} \quad (24\text{--}58a)$$

where P in N/m^2 and t in K

$$= \frac{0.00236}{P\sqrt{t}} \qquad\qquad \textbf{Metric} \quad (24\text{--}58b)$$

where P in kgf/mm^2 and t in °C

$$= \frac{4.5}{P\sqrt{t}} \qquad \textbf{US Customary units} \quad (24\text{--}58c)$$

where P in psi and t in °F

The coefficient of friction according to Illmer when bearing is subjected to pressure varying from 0.23 MPa (35 psi) to 0.7 MPa (100 psi) and speed varying from 0.5 m/s to 1.5 m/s (100 ft/min to 300 ft/min)

$$\mu = \frac{\sqrt[3]{v}}{0.05\sqrt{Pt}} \qquad\qquad \textbf{SI} \quad (24\text{--}58d)$$

where P in N/m^2, v in m/s, and t in K

$$= \frac{\sqrt[3]{v}}{157.7\sqrt{Pt}} \qquad\qquad \textbf{Metric} \quad (24\text{--}58e)$$

Particular	Formulaspc-hlf >
	where P in kgf/mm^2, v in m/s, and t in °C
	$= \dfrac{\sqrt[3]{v}}{20\sqrt{Pt}}$ **US Customary units** (24–58f)
	where P in psi, v in ft/min, and t in °F
The coefficient of friction according to Tower tests	$\mu = \dfrac{144204.5}{P}\sqrt{\dfrac{v}{t}}$ **SI** (24–58g)
	where P in N/m^2, v in m/s, and t in K
	$= \dfrac{0.0147}{P}\sqrt{\dfrac{v}{t}}$ **Metric** (24–58h)
	where P in kgf/mm^2, v in m/s, and t in °C
	$= \dfrac{2}{P}\sqrt{\dfrac{v}{t}}$ **US Customary units** (24–58i)
	where P in psi, v in ft/min, and t in °F
The coefficient of friction according to Lasche when the speed exceeds 2.5 m/s (500 ft/min)	$\mu = \dfrac{24.73}{\sqrt{Pt}}$ **SI** (24–58j)
	where P in N/m^2, and t in K
	$= \dfrac{0.0079}{\sqrt{Pt}}$ **Metric** (24–58k)
	where P in kgf/mm^2, and t in °C
	$= \dfrac{0.4}{\sqrt{Pt}}$ **US Customary units** (24–58l)
	where P in psi, and t in °F

OIL FLOW THROUGH JOURNAL BEARING

Oil flow through bearing	$Q = 0.785\, c\, Ld\, n'$ **SI** (24–59a)
	where c, L, and d in m; n' in rps; and Q in m^3/s
	$= 785\, c\, Ld\, n'$ **SI** (24–59b)
	where c, L, and d in m; n' in rps; and Q in dm^3/s
	$= 7.8510^{-7}\, c\, Ld\, n'$ **SI** (24–59c)
	where c, L, and d in mm; n' in rps; and Q in l/s or dm^3/s
	$= 0.0034c\, Ld\, n$ (24–59d)
	US Customary System units
	where c, L, and d in in, n in rpm and Q in US gallons/min.

Particular	Formula
Oil flow through a central groove of bearing from one end	$$Q = \frac{\pi dc^3 P_o}{48\eta L}(1 + 1.5\epsilon^2)\qquad\text{(24–60a)}$$
Total oil flow through a central groove of bearing from both ends	$$Q_g = \frac{\pi dc^3 P_o}{24\eta L}(1 + 1.5\epsilon^2)\qquad\text{(24–60b)}$$
Total oil flow through a central groove for lightly loaded bearing [From Eq. (24–60b) as $\epsilon \to 0$]	$$Q \simeq \frac{\pi dc^3 P_o}{24\eta L}\qquad\text{(24–61)}$$
Total oil flow through a central groove for heavily loaded bearing [From Eq. (24–60b) as $\epsilon \to 1$]	$$Q \simeq \frac{2.5\pi dc^3 P_o}{24\eta L}\qquad\text{(24–62)}$$
Oil flow through a single hole	$$Q_h = \frac{c^3 P_o}{24\eta}(1 + 1.5\epsilon^2)\tan^{-1}\left(\frac{\pi d}{L}\right)\qquad\text{(24–63)}$$
The ratio of Q_g to Q_h in the unloaded region of bearing from Eq. (24–60b) and (24–63)	$$\frac{Q_g}{Q_h} = \frac{\pi d}{L\,\tan^{-1}(\pi d/L)}\qquad\text{(24–63a)}$$
	where d is the diameter of journal
Flow variable (dimensionless)	$$\lambda'_Q = \frac{4Q\,\psi}{60\,c^2\,n'\,L}\qquad\text{(24–64)}$$
	Refer to Figs. 24–34 and 24–35h for flow variable λ'_Q

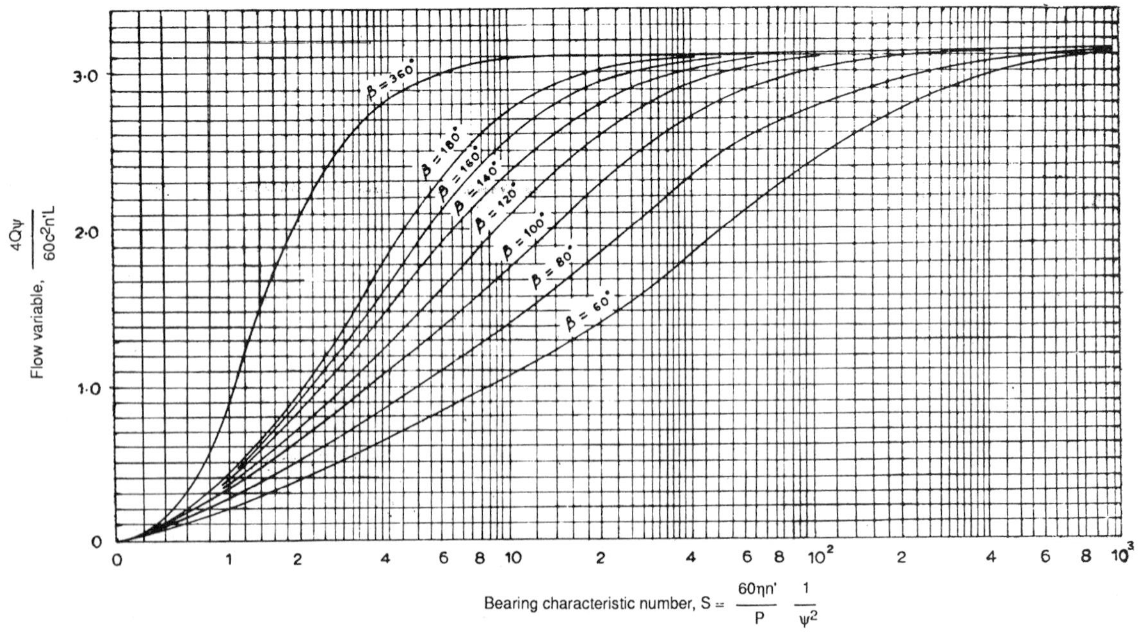

FIGURE 24–34 Chart for determining oil flow, based on no side flow. [*Boyd and Raimondi* (5).]

Particular	Formula

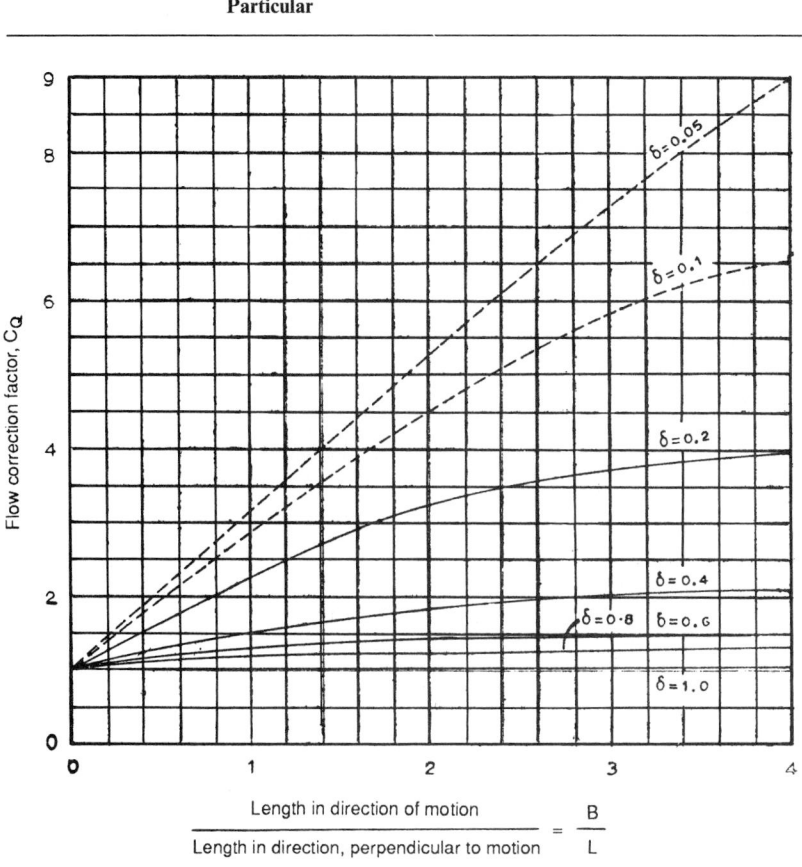

FIGURE 24–35(a)   Flow correction factor for side flow. [*Boyd and Raimondi (5).*]

Oil flow through a bearing with side leakage

$$Q = \frac{60\lambda_Q' d^2 n' L\psi}{4} C_Q \text{ or } \frac{60\lambda_Q' c^2 n' L}{4\psi} C_Q \quad (24\text{–}65)$$

Oil flow ratio $\dfrac{Q_s}{Q}$

where C_Q = flow correction factor from Fig. 24–35a

Refer to Fig. 24–35*i*.

THERMAL EQUILIBRIUM OF JOURNAL BEARING

The general expression for heat generated in bearing

$$H_g = M_{t\mu}(2\pi n') \quad \textbf{SI} \quad \textbf{(US Customary units)}$$
$$(24\text{–}66a)$$

$$= \left(\mu W \frac{d}{2}\right)(2\pi n') = \mu(PLd)\frac{d}{2}\,\omega = \mu(PLd)\,v$$

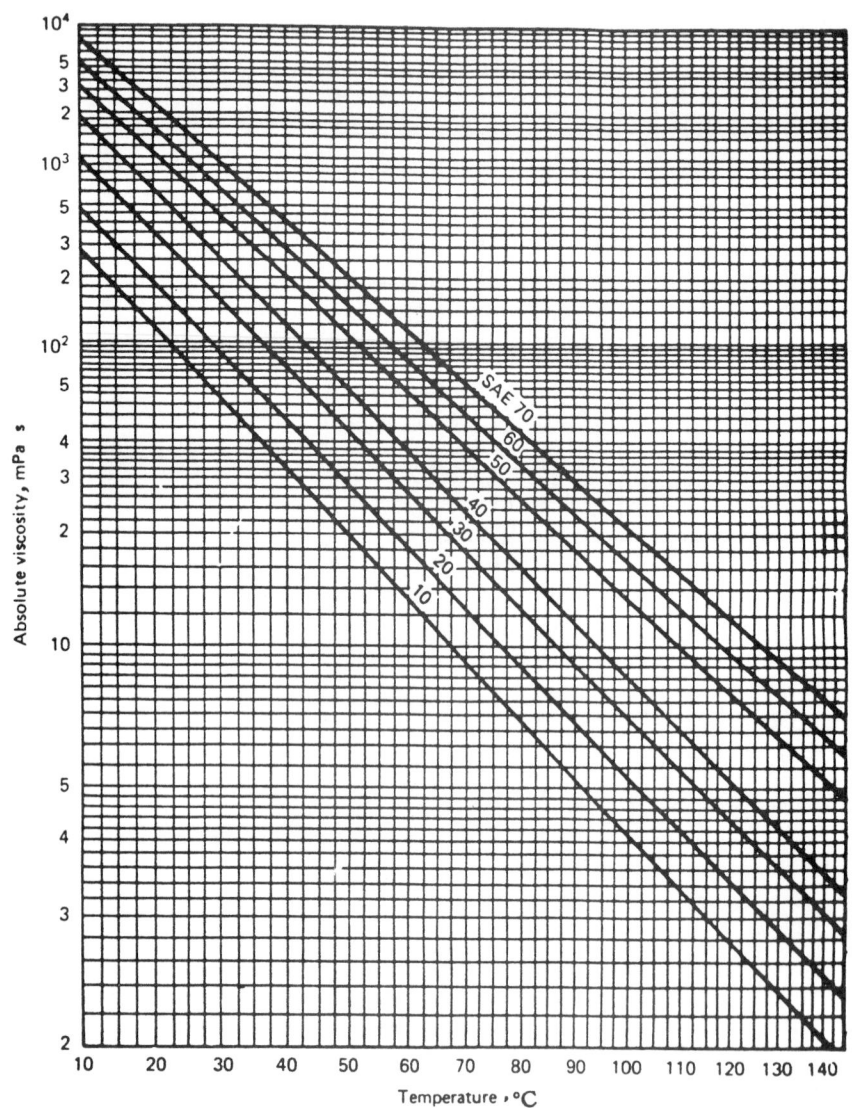

FIGURE 24–35(*b*) Absolute viscosity versus temperature. [*Shigley and Mitchell (8).*]

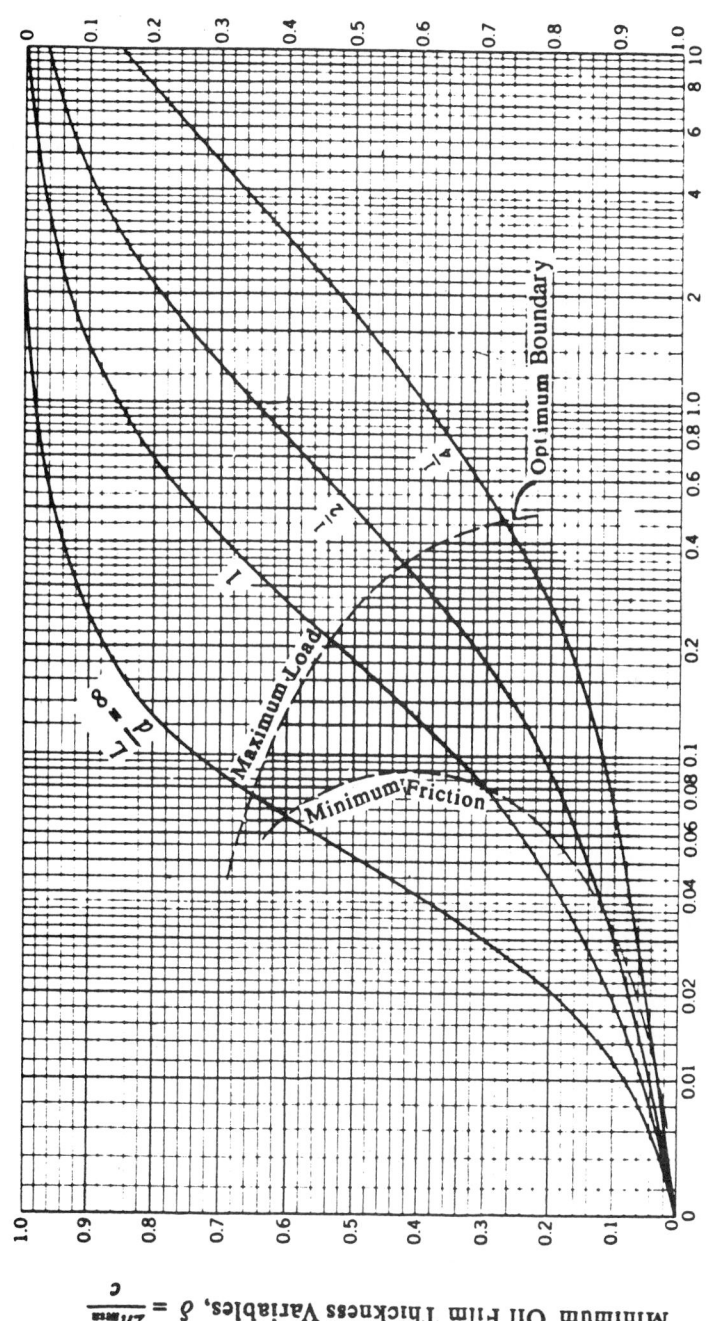

Attitude or Eccentricity Ratio, $\epsilon = \dfrac{2e}{c}$

Bearing Characteristic Number, $S = \dfrac{\eta n'}{P}\left(\dfrac{1}{\psi^2}\right)$

Minimum Oil Film Thickness Variables, $\delta = \dfrac{2h_{min}}{c}$

FIGURE 24-35(c) Variation of minimum oil film thickness variable δ and attitude ϵ of full journal bearing with bearing characteristic number S. [*Boyd and Raimondi* (5).]

24.45

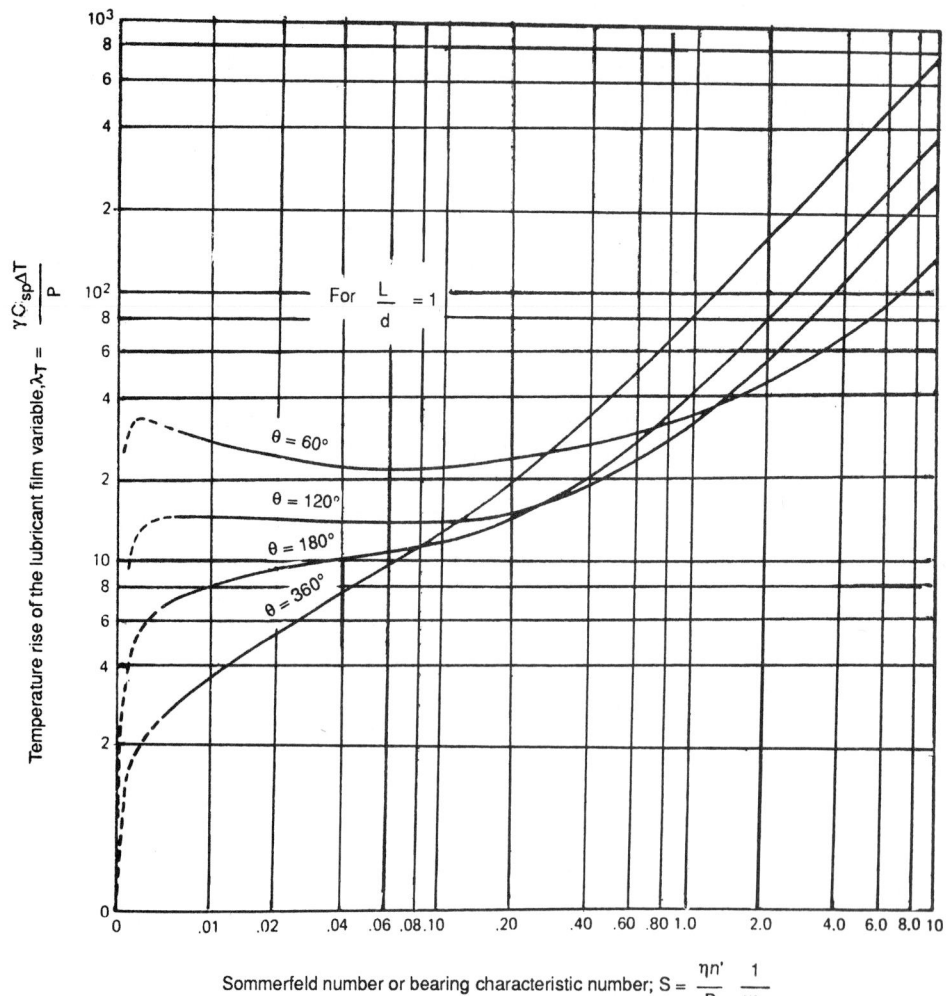

FIGURE 24–35(d) Variation of temperature rise of the lubricant film variable λ_T with Sommerfeld number S. [*Boyd and Raimondi* (5).]

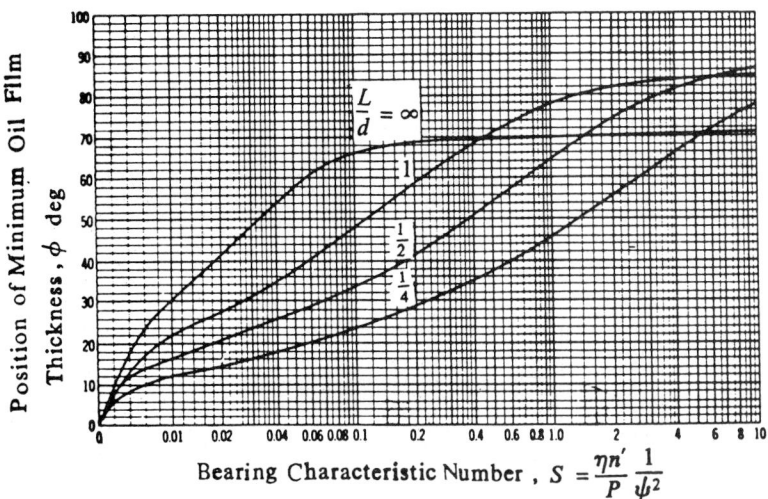

FIGURE 24–35(e)   Position of minimum oil film thickness vs. bearing characteristic number S for full journal bearing. (Refer to Fig. 24–9 for definition of ϕ.) [*Boyd and Raimondi (5)*.]

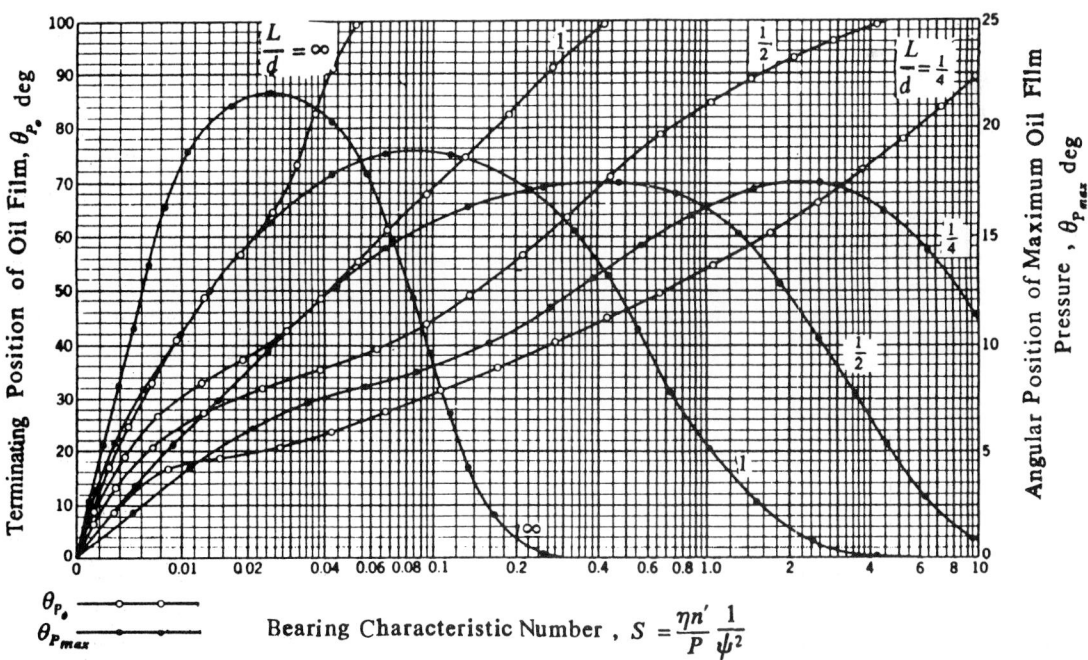

FIGURE 24–35(f)   Positions of maximum oil film pressure and oil film termination versus bearing characteristic number S. [*Boyd and Raimondi (24)*.] (Refer to Fig. 24–9 for definition of $\theta_{P_{max}}$ and θ_{P_0}.)

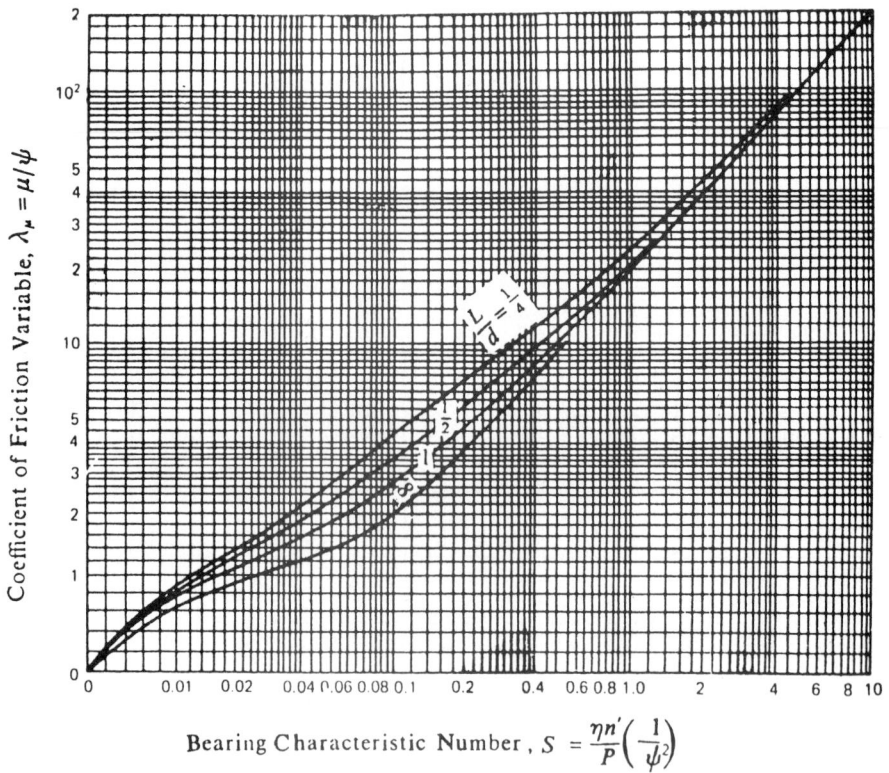

FIGURE 24–35(g) Variation of the coefficient of friction variable $\lambda_\mu = \mu/\psi$ with S for 360° journal bearing. [*Boyd and Raimondi (5).*]

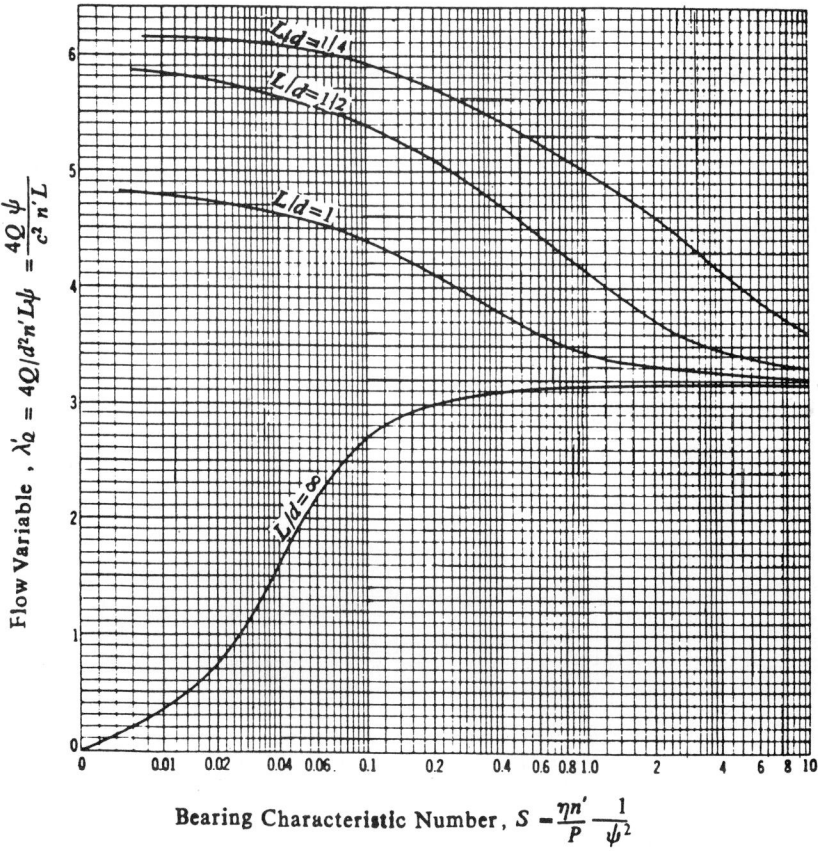

FIGURE 24–35(*h*)   Chart for oil flow variable λ_Q with bearing characteristic number S for full journal bearing. [*Boyd and Raimondi* (5).]

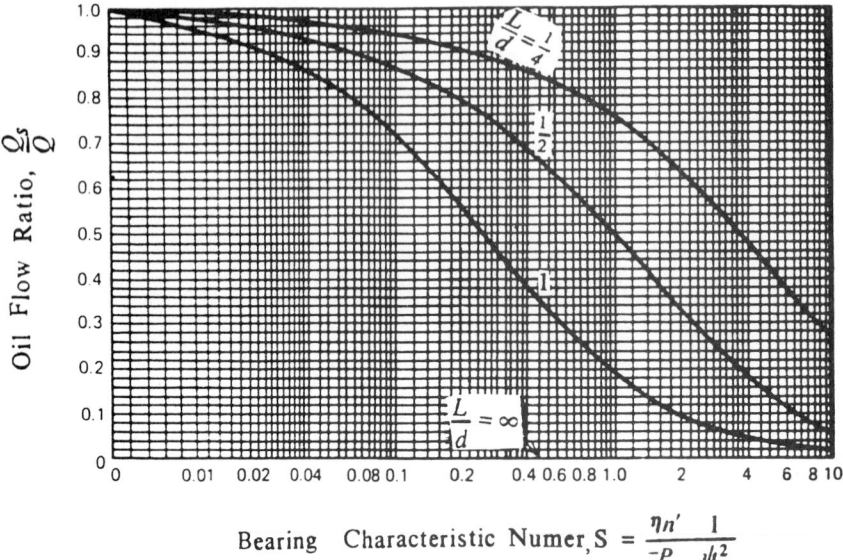

FIGURE 24–35(i) Oil flow ratio (Q_S/Q) versus bearing characteristic number S for full journal bearing. [*Boyd and Raimondi* (5).]

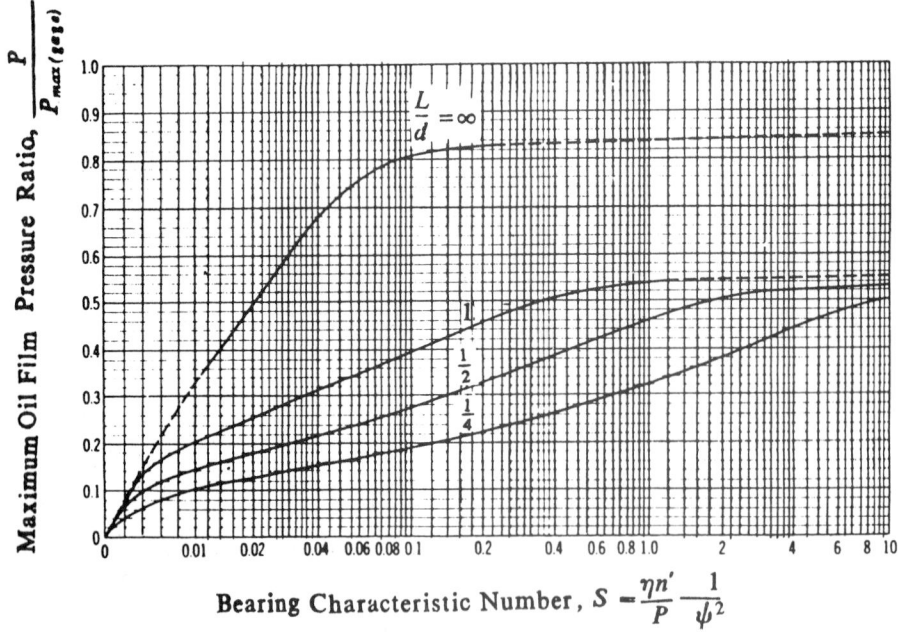

FIGURE 24–35(j) Chart for maximum oil film pressure ratio $P/P_{max(gauge)}$ with bearing characteristic number S for full journal bearing. [*Boyd and Raimondi* (5).]

Particular	Formula
	where H_g in J/s (Btu/s); $M_{t\mu}$ in N m (lbf in) W in N (lbf); P in N/m^2(psi), L in m (in); d in m (in); n' in rps, ω in rad/s, and v in m/s (ft/min)

$$= [\mu(PLd)v]/778 \quad \text{US Customary units} \quad (24\text{--}66b)$$

$$= \mu(PLd)(\pi dn')/778 = \mu(PLd)d\,\omega/1556$$

where P in psi; L in in; d in in; v in ft/min,

ω in rad/s, H_g in Btu/s, and n' in rps

Particular	Formula
The heat generated can also be found by knowing the temperature rise of lubricant oil which is used to carry away heat generated in the bearing	$H_g = \gamma\, C_{sp}\, Q\, \Delta T \dfrac{2}{m}$ **SI** (**US Customary units**) $\qquad\qquad$ (24–66c) where H_g in J/s (Btu/s)

γ = weight per unit volume of lubricant whose average specific gravity is 0.90
  = 8.83 kN/m^3 (0.0325 lbf/in^3)
C_{sp} = specific heat of lubricant, kJ/N K (Btu/lbf °F)
  = 0.19 kJ/N K (0.42 Btu/lbf °F)
  Q in m^3/s; ΔT in °C

Particular	Formula
Temperature rise of the lubricant film variable λ_T	Refer to Fig. 24–35d for λ_T.
The temperature rise of the lubricant film due to heat generated which is to be carried away by the lubricant, can be found from Eq. (24–66c)	$\Delta T = \dfrac{4\pi\, P\, \lambda_\mu}{427\, \gamma C_{sp}\lambda_Q}$ **Customary Metric** (24–66d)

where $\lambda_\mu = \mu/\psi =$ friction efficient variable

$\lambda_Q =$ flow variable $= 4Q/d^2 n' L\psi$

P in kgf/m^2; C_{sp} in kcal/kgf °C, and γ in kgf/m^3; ΔT in °C

$$= 78\,\frac{P\lambda_\mu}{\lambda_Q} = \frac{78(\mu/\psi)\,P}{(4Q/d^2 n' L\psi)} = 19.5\,\frac{\mu P}{(Q/d^2 n' L)}$$

Customary Metric (24–66e)

where P in kgf/mm^2, d in mm; L in mm, Q in mm^3/s, n' in rps, and ΔT in °C

$$= \frac{8.3 \times 10^{-6}[\mu/\psi]\,P}{(Q/d^2 n' L\psi)} = \frac{8.3 \times 10^{-6}\mu P}{(Q/d^2 n' L)} \quad \textbf{SI}$$

$$(24\text{--}66f)$$

where P in N/m^2, Q in m^3/s, d in m, L in m, n' in rps, and ΔT in K or °C, since $\Delta T°$ C $= \Delta T$ K

Particular	Formula
If the end flow is also taken into consideration then the temperature rise of the flow Q $-$ Q_{el} due to heat generated, is ΔT and the temperature rise of end leakage is $\Delta T/2$, which is the average of the inlet and the outlet temperatures	$\Delta T = \dfrac{0.103[\mu/\psi]\,P}{[1 - \frac{1}{2}(Q_{el}/Q)]\,(4Q/d^2 n' L\psi)}$ **US Customary units** (24–66g)

Particular	Formula

$$= 0.0258 \; \frac{\mu P}{[1 - \frac{1}{2}(Q_{el}/Q)] \; (Q/d^2 n' L)}$$

where P in psi; d in in, L in in, Q_{el} and Q in in^3/s, n' in rps, and ΔT in °F

$$= \frac{8.3 \times 10^{-6}[\mu/\psi]P}{[1 - \frac{1}{2}(Q_{el}/Q) \; (Q/d^2 n' L \psi)]} \qquad \text{SI} \quad (24\text{--}66\text{h})$$

$$= \frac{8.3 \times 10^{-6}\mu P}{[1 - \frac{1}{2}(Q_{el}/Q) \; (Q/d^2 n' L)]}$$

where P in N/m^2, d and L in m, Q_{el} and Q in m^3/s, n' in rps, and ΔT in K

The temperature rise of the lubricant film due to heat generated in pressure fed bearings

$$\Delta T = \frac{16 \times 10^{-6}[\mu/\psi] \; SW^2}{(1 + 1.5\epsilon^2) \; P_S \; d^4} \qquad \text{SI} \quad (24\text{--}66\text{i})$$

where W in N, P_S in N/m^2, d in m, and ΔT in K

$$= \frac{9.7[\mu/\psi] \; SW^2}{(1 + 1.5\epsilon^2) \; P_s \; d^4} \qquad \textbf{Metric} \quad (24\text{--}66\text{j})$$

where W in kgf, P_S in kgf/mm^2, d in mm, and ΔT in °C

Heat dissipated by self-contained bearings

$$H_d = C A \, (t_b - t_a) \qquad (24\text{--}67\text{a})$$

where

C = combined coefficient of radiation and convection
  = $11.36 \times 10^{-3} kW/\text{m}^2$ K (2 Btu/ft^2 h °F) when the bearing located in still air
  = $15.36 \times 10^{-3} kW/\text{m}^2$ K (2.7 Btu/ft^2 h °F) for average design practice
  = $33.5 \times 10^{-3} kW/\text{m}^2$ K (5.9 Btu/ft^2 h °F) when the air velocity over the bearing is 2.5 m/s
A = effective surface area of bearing housing
  = $25 \times 10^{-4} \, dL$ m^2 for bearing masses of metal as in a ring oil bearing (25 dL in^2)
  = $6 \times 10^{-4} \, dL$ m^2 for light construction (6 dL in^2)
t_b = surface temperature of bearing housing, °C (°F)
t_a = temperature of surrounding air, °C (°F)

Another formula for the heat dissipated in bearing in terms of average lubricant oil film temperature

$$H_d = \frac{CA}{m + 1}(t_o - t_a) \qquad (24\text{--}67\text{b})$$

where m is a constant which depends on the lubrication system and it is taken from Table 24–17, t_o is lubricant film temperature, °C. C and A are as given under Eq. (24–67a)

TABLE 24–17
Quantities C_{PW}, C_{PFm}, $C_{P\mu}$, and $\bar{x}/B$ as functions of q

q	C_{PW}	C_{PFm}	$C_{P\mu}$	$\dfrac{\bar{x}}{B}$
0.10	0.007209	0.955265	22.085010	
0.20	0.012585	0.919159	12.172679	
0.25	0.014741	0.903630	10.216742	
0.30	0.016608	0.889234	8.923752	
0.40	0.019618	0.864722	7.346332	0.533761
0.50	0.021864	0.843721	6.431585	
0.60	0.023514	0.825665	5.852294	0.546881
0.70	0.024714	0.809936	5.462059	
0.80	0.025597	0.796076	5.183394	0.558394
0.90	0.026129	0.783718	4.999030	0.563687
1.00	0.026481	0.772589	4.862537	0.568688
1.10	0.026661	0.762470	4.766450	0.573426
1.20	0.026707	0.753191	4.700334	0.577926
1.30	0.026645	0.744615	4.657628	0.582209
1.40	0.026500	0.736633	4.632912	0.586293
1.50	0.026289	0.729156	4.622694	0.590193
1.60	0.026026	0.722120	4.624350	0.594111
1.70	0.025728	0.715441	4.634646	
1.80	0.025386	0.709100	4.655453	0.600937
2.00	0.024653	0.697225	4.713591	0.607410
2.20	0.03870	0.686283	4.791810	0.613416
2.50	0.022664	0.671087	4.935044	0.621673
2.75	0.021668	0.659606	5.073580	
3.00	0.020699	0.648392	5.220786	0.633787
4.00	0.017257	0.609438	5.885901	
5.00	0.014528	0.580067	6.654587	
7.00	0.010692	0.521586	8.130471	
8.00	0.009332			
9.00	0.008225	0.477917	9.684235	

Source: F. I. Radzimovsky, Lubrication of Bearings—Theoretical Principles and Designs, The Ronald Press Company, New York, 1959.

Particular	Formula
The heat dissipating capacity of bearing based on projected area of bearing	$H_d = k(Ld)(t_b - t_a) = kA\,\Delta T$ **Metric** (24–68a) where $A = Ld =$ projected area of bearing, cm²; $k\Delta T = k(t_b - t_a)$ values can be taken from Fig. 24–36 and H_d in kcal/min $\qquad = 697.8\, k(t_b - t_a)(Ld)$ **SI** (24–68b) where $k(t_b - t_a)$ in kcal/min cm², values are taken from Fig. 24–36; (Ld) in m² and H_d in J/s
Pederson's equation for heat radiating capacity of bearing due to friction between journal and bearing	$H_d = \dfrac{(\Delta T + 18)^2}{k}\,(Ld)$ **SI** (**US Customary units**) (24–69a)

Particular	Formula

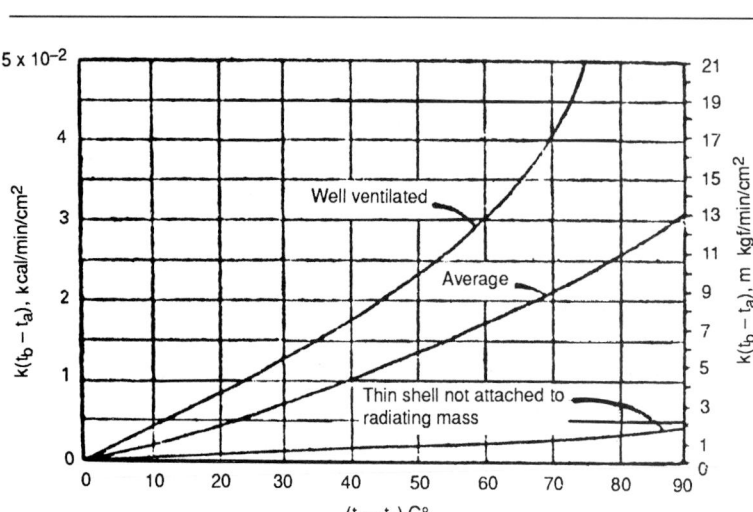

FIGURE 24-36 The rate of heat dissipated from a journal bearing.

$$= \frac{(\Delta T + 33)^2 \, Ld}{k} \qquad \text{US Customary units}$$

$$(24\text{-}69b)$$

where H_d in J/s (ft-lbf/s/in^2), Ld in in^2, ΔT in °C (°F)

$$
\begin{aligned}
k \;=\; & 75(3300) \text{ for bearings of light construction} \\
& \text{located in still air} \\
=\; & 423(1860) \text{ for bearings of heavy construction} \\
& \text{and well ventilated} \\
=\; & 262(1150) \text{ for General Electric Company's} \\
& \text{well-ventilated bearing}
\end{aligned}
$$

$$= \frac{(\Delta T + 18)^2}{427 \, k} \, (Ld) \qquad \text{Metric} \quad (24\text{-}69c)$$

where H_d in kcal/s; (Ld) in m^2; ΔT in °C and values of k are as given inside parentheses under Eq. (24-69a)

The difference in temperature (ΔT) of the bearing and of the cooling medium can be found from the equation

$$(\Delta T + 18)^2 = K' \mu \, Pv \qquad \text{SI (Metric)} \quad (24\text{-}69c)$$

where P in N/m^2 (kgf/mm^2); v in m/s; ΔT in K (°C)

$$
\begin{aligned}
K' \;=\; & 0.475 \;(4.75 \times 10^6) \text{ for bearings of light} \\
& \text{construction located in still air} \\
=\; & 0.273 \;(2.7 \times 10^6) \text{ for bearings of heavy} \\
& \text{construction and well ventilated} \\
=\; & 0.165 \;(1.65 \times 10^6) \text{ for General Electric} \\
& \text{Company's well-ventilated bearing}
\end{aligned}
$$

Particular	Formula

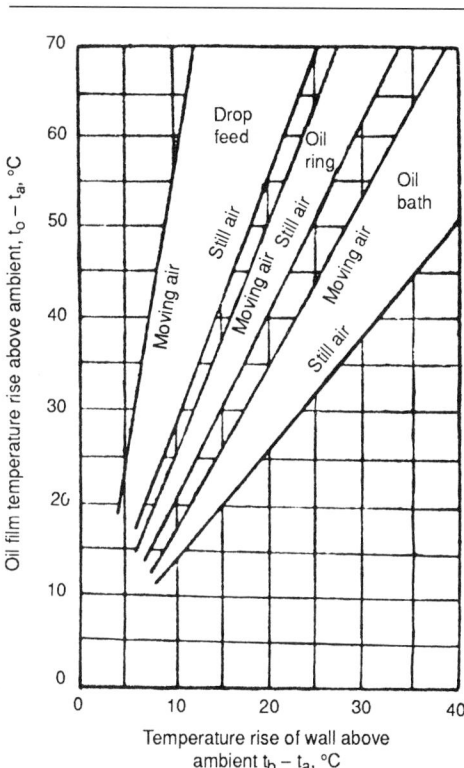

FIGURE 24–37 Relation between oil film temperature and bearing wall temperature.

The difference between the bearing-wall temperature t_b and the ambient temperature t_a for three main types of lubrication by oil bath, by an oil ring, and by waste pack or drop feed

Refer to Fig. 24–37 for $(t_b - t_a) \simeq \left(\dfrac{t_o - t_b}{2} \right)$.

BEARING CAP

The bearing cap thickness

$$h_c = \sqrt{\frac{3Wa}{2\,L\sigma}} \qquad (24\text{–}70)$$

The deflection of the cap

$$y = \frac{Wa^3}{4EL\,h_c^3} \qquad (24\text{–}71)$$

The thickness of cap from Eq. (24–71)

$$h_c = 0.63\, a \sqrt[3]{\frac{W}{ELy}} \qquad (24\text{–}72)$$

where the deflection should be limited to 0.025 mm (0.001 in)

Particular	Formula

EXTERNAL PRESSURIZED BEARING OR HYDROSTATIC BEARING: JOURNAL BEARING (FIG. 24–38)

The pressure in the lower pool of quadrant 1 (Fig. 24–38)

$$P_1 = K_1 P_o \qquad (24\text{–}73a)$$

where

$$K_1 = \cfrac{1}{1 + \cfrac{4}{\pi}\left(\cfrac{P_o}{P'} - 1\right)\left(\cfrac{\pi}{4} - 2.121\epsilon + 1.93\epsilon^2 - 0.589\epsilon^3\right)}$$

$$(24\text{–}73b)$$

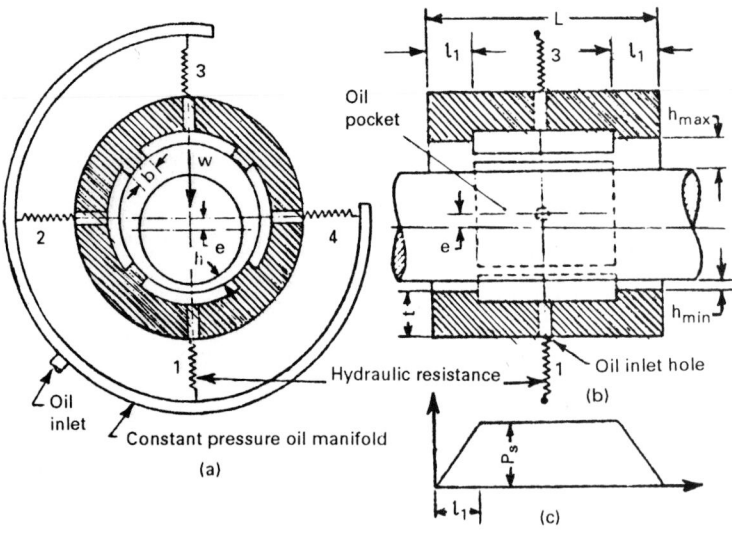

FIGURE 24–38 (a) and (b) schematic diagram of a full cylindrical hydrostatic bearing; (c) oil pressure distribution along the bearing. [*Shaw and Macks (10)*.]

The pressure in the upper pool of quadrant 3 (Fig. 24–38)

$$P_3 = K_3 P_o \qquad (24\text{–}74a)$$

where

$$K_3 = \cfrac{1}{1 + \cfrac{4}{\pi}\left(\cfrac{P_o}{P'} - 1\right)\left(\cfrac{\pi}{4} + 2.121\epsilon + 1.93\epsilon^2 + 0.589\epsilon^3\right)}$$

$$(24\text{–}74b)$$

The pressure in the left pool of quadrant 2 (Fig. 24–38)

$$P_2 = K_2 P_o \qquad (24\text{–}75a)$$

where $K_2 = \dfrac{1}{8}\left(\dfrac{P'}{P_o}\right)(6.283 + 3.425\epsilon^2)$ $\qquad (24\text{–}75b)$

Particular	Formula
The pressure in the right pool of quadrant 4 (Fig. 24–38)	$P_4 = K_4 P_o$ (24–76a) where $K_4 = \dfrac{1}{8}\left(\dfrac{P'}{P_o}\right)(6.283 + 3.425\epsilon^2)$ (24–76b)
The flow of lubricant through the lower quadrant 1 of the bearing from the manifold	$Q_1 = \dfrac{\psi^3 d^4}{96\,\eta\,l_1}\, P_1\, C_{PF1}$ (24–77a) where $C_{PF1} = \left(\dfrac{\pi}{4} - 2.121\epsilon + 1.93\epsilon^2 - 0.589\epsilon^3\right)$ (24–77b)
The flow of lubricant through the left quadrant 2 of the bearing from the manifold	$Q_2 = \dfrac{\psi^3 d^4}{768\,\eta l_1}\, P_2\, C_{PF2}$ (24–78a) where $C_{PF2} = (6.283 + 3.425\epsilon^2)$ (24–78b)
The flow of lubricant through the upper quadrant 3 of the bearing from the manifold	$Q_3 = \dfrac{\psi^3 d^4}{48\,\eta l_1}\, P_3\, C_{PF3}$ (24–79a) where $C_{PF3} = \left(\dfrac{\pi}{4} + 2.121\epsilon + 1.93\epsilon^2 + 0.589\epsilon^3\right)$ (24–79b)
The flow of lubricant through the right quadrant 4 of the bearing from the manifold	$Q_4 = \dfrac{\psi^3 d^4}{768\,\eta l_1}\, P_4\, C_{PF4}$ (24–80a) where $C_{PF4} = C_{PF2} = (6.283 + 3.425\epsilon^2)$ (24–80b)
The total flow of lubricant through quadrant of the bearing from the manifold assuming $P_2 = P_4 = P'$ (good approximation)	$Q = Q_1 + Q_2 + Q_3 + Q_4$ (24–81a) $= \dfrac{\psi^3 d^4}{48\,\eta l_1}\, P_o\, G$ (24–81b) where G = flow factor given by Eq. (24–82)
The flow factor in Eq. (24–81b)	$G = C_{PF1}\, K_1 + \tfrac{1}{8}(C_{PF2}K_2 + C_{PF4}K_4) + C_{PF3}K_3$ (24–82) $= C_{PF1}K_1 + \tfrac{1}{4}C_{PF2}\, K_2 + C_{PF3}K_3$ since $K_2 = K_4$ and $C_{PF2} = C_{PF4}$
The external load on the hydrostatic journal bearing	$W = (P_1 - P_3)\left(A + \dfrac{A'}{2}\right) = \pi P_o\left(A + \dfrac{A'}{2}\right)F_{PFW}$ (24–83) where F_{PFW} = load factor given by Eq. (24–84)
The load factor	$F_{PFW} = K_1 - K_3$ (24–84)
The pressure ratio connecting the dimensions of the bearing and its external resistances	$\dfrac{P_o}{P'} = 1 + 6\left(\dfrac{d}{d_c}\right)\left(\dfrac{c}{d_c}\right)^3 \dfrac{l_c}{l_1}$ (24–85)

Particular	Formula

IDEALIZED SLIDER BEARING (FIG. 24–39)

Plane-slider bearing

The pressure at any point x

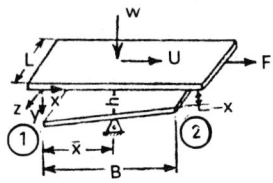

FIGURE 24–39   Plane slider bearing with an angle of inclination.

$$P = \frac{\eta U}{B} C_{s1} \qquad (24\text{–}86a)$$

$$\text{where } C_{s1} = \frac{6\kappa x_1 (1 - x_1)}{(\kappa - 2a)(a - \kappa + \kappa x_1)^2} \qquad (24\text{–}86b)$$

$$\kappa = \frac{h_2 - h_1}{B}; a = \frac{h_2}{B}; x_1 = \frac{x}{B} \qquad (24\text{–}86c)$$

The load carrying capacity

$$W = \frac{6\eta\, UL}{\kappa^2} C_{s2} \qquad (24\text{–}87a)$$

$$\text{where } C_{s2} = \ln\frac{a - \kappa}{a} + \frac{2\kappa}{2a - \kappa} \qquad (24\text{–}87b)$$

The resultant shear stress at any point along the slider (Fig. 24–39)

$$\tau = \frac{\eta U}{B} C_{s3} \qquad (24\text{–}88a)$$

$$\text{where } C_{s3} = \left\{ \left[\frac{B(a - \kappa + \kappa x_1) - 2y}{B} \right] \times \right.$$

$$\left. \left[\frac{3\kappa(a - \kappa + \kappa x_1 - 2ax_1)}{(\kappa - 2a)(a - \kappa + \kappa x_1)^3} \right] + \frac{1}{a - \kappa + \kappa x_1} \right\}$$

$$(24\text{–}88b)$$

The shear stress at any point on the surface of the moving member of the bearing (i.e., slider at $y = 0$) (Fig. 24–39)

$$\tau_m = \frac{\eta U}{B} C_{s4} \qquad (24\text{–}89a)$$

where

$$C_{s4} = \left[\frac{4}{a - \kappa + \kappa x_1} - \frac{6a(a - \kappa)}{(2a - \kappa)(a - \kappa + \kappa x_1)^2} \right]$$

$$(24\text{–}89b)$$

The shear stress at any point on the surface of the stationary member of the bearing (i.e., shoe at $y = h$) (Fig. 24–39)

$$\tau_s = \frac{\eta U}{B} C_{s5} \qquad (24\text{–}90a)$$

where

$$C_{s5} = \left[\frac{-2}{a - \kappa + \kappa x_1} - \frac{6a(a - \kappa)}{(2a - \kappa)(a - \kappa + \kappa x_1)^2} \right]$$

$$\kappa = \frac{h_2 - h_1}{B} \quad \text{and} \quad h = [B(a - \kappa + \kappa x_1)]$$

$$(24\text{–}90b)$$

Particular	Formula
The frictional force on the moving member of the bearing (i.e., slider)	$F_{\mu m} = \eta UL\, C_{s6}$ (24–91a)
	where $C_{s6} = \left[-\dfrac{4}{\kappa} \ln \dfrac{a-\kappa}{a} - \dfrac{6}{2a-\kappa} \right]$ (24–91b)
The frictional force on the stationary member of the bearing (i.e., shoe)	$F_{\mu s} = \eta UL\, C_{s7}$ (24–92a)
	where $C_{s7} = \left[\dfrac{2}{\kappa} \ln \left(\dfrac{a-\kappa}{4}\right) + \dfrac{6}{2a-\kappa} \right]$ (24–92b)
The coefficient of friction	$\mu = \dfrac{F_{\mu m}}{W} = \dfrac{-2\kappa(2a-\kappa)\ln \dfrac{a-\kappa}{a} - 3\kappa^2}{3(2a-\kappa)\ln \dfrac{a-\kappa}{a} + 6\kappa}$ (24–93)
The distance of the pressure center from the origin of the coordinates, i.e., from the lower end of the shoe (Fig. 24–39)	$\bar{x} = \left[\dfrac{(a-\kappa)(3a-\kappa)\ln\left(\dfrac{a-\kappa}{a}\right) - 2.5\kappa^2 + 3\kappa a}{\kappa(\kappa-2a)\ln\left(\dfrac{a-\kappa}{a}\right) - 2\kappa^2} \right] B$ (24–94)

Pivoted-shoe slider bearing (Fig. 24–39 and Fig. 24–43)

The load-carrying capacity	$W = \dfrac{6\eta\, UL\, B^2}{h_2^2}\, C_{PW}$ (24–95a)
	where $C_{PW} = \left[\dfrac{1}{q^2} \ln(1+q) - \dfrac{2}{q(q+2)} \right]$ (24–95b)
	Refer to Table 24–17 for C_{PW}.
The frictional force on the moving member of the bearing (i.e., slider)	$F_{\mu mP} = \dfrac{\eta\, ULB}{h_2}\, C_{PFm}$ (24–96a)
	where $C_{PFm} = \dfrac{4}{q} \ln(1+q) - \dfrac{6}{2+q}$ (24–96b)
	Take C_{PFm} from Table 24–17 for various values of q.
The frictional force on the stationary member of the bearing (i.e., shoe)	$F_{\mu sP} = \dfrac{\eta\, ULB}{h_2}\, C_{PFs}$ (24–97a)
	where $C_{PFs} = \left[-\dfrac{2}{q} \ln \dfrac{(1+q)a}{2} + \dfrac{6}{2+q} \right]$ (24–97b)
The coefficient of friction	$\mu = \dfrac{F_{\mu mP}}{W} = \dfrac{h_2}{B}\left[\dfrac{1}{6}\dfrac{C_{PFm}}{C_{PW}} \right] = \dfrac{h_2}{B}\, C_{P\mu}$ (24–98)
	where $C_{P\mu}$ = coefficient of friction factor
	Take $C_{P\mu}$ from Table 24–17 for various values of q.

Particular	Formula
The distance of the pivoted point from the lower end of the shoe (Fig. 24–39), i.e., the distance of the pressure center from the origin of the coordinates	$$\bar{x} = \left[\frac{(1+q)(3+q)\ln(1+q) - q(2.5q+3)}{q(q+2)\ln(1+q) - 2q^2} \right] B \qquad (24\text{–}99)$$ The ratios $\dfrac{\bar{x}}{B}$ are taken from Table 24–17.

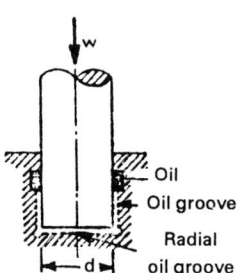

FIGURE 24-40   Pivot thrust bearing.

DESIGN OF VERTICAL, PIVOT, AND COLLAR BEARING

Pivot bearing (Figs. 24–40, 24–41, and 24–44)

FLAT PIVOT The total axial load on the flat pivot with extreme diameters of the actual contact d_1 and d_2	$$W = P\pi \frac{(d_1^2 - d_2^2)}{4} \qquad (24\text{–}100)$$
The friction torque based on uniform intensity of pressure with extreme diameters of the actual contact d_1 and d_2	$$M_t = \frac{1}{3}\mu W \frac{d_1^3 - d_2^3}{d_1^2 - d_2^2} \qquad (24\text{–}101)$$
The friction torque based on uniform wear with extreme diameters of the actual contact d_1 and d_2	$$M_t = \mu W \frac{d_1 + d_2}{4} \qquad (24\text{–}102)$$
The power absorbed by friction with d as the diameter of flat pivot bearing	$$P_\mu = \frac{\mu W d n'}{478} \qquad \textbf{SI} \quad (24\text{–}103a)$$ where P_μ in kW; W in N; d in m, and n' in rps $$= \frac{\mu W d n}{189090} \qquad \textbf{US Customary units} \quad (24\text{–}103b)$$ where P_μ in hp, W in lbf, d in in, and n in rpm
CONICAL PIVOT The friction torque based on uniform intensity of pressure with extreme diameters of the actual contact d_1 and d_2	$$M_t = \frac{1}{3}\frac{\mu W}{\sin \alpha} \frac{d_1^3 - d_2^3}{d_1^2 - d_2^2} \qquad (24\text{–}104)$$ where 2α = cone angle of pivot, deg
The friction moment which resists the rotation of the shaft in a conical pivot bearing for uniform wear	$$M_t = \frac{\mu W}{\sin \alpha} \frac{d_1 + d_2}{4} \qquad (24\text{–}105)$$
The loss of power in vertical bearing	$$P_\mu = 6.2 \times 10^8 \frac{\eta d^2 L n'^2}{\psi} \qquad \textbf{SI} \quad (24\text{–}106a)$$ where P_μ in kW, η in Pa s, d and L in m, and n' in rps

Particular	Formula
	$$= 2.35 \times 10^{-4} \frac{\eta_1 d^2 L n^2}{\psi} \qquad \textbf{Metric} \qquad (24\text{--}106b)$$

where P_u in hp$_m$, η_1 in cP, d and L in cm, and n in rpm

$$= 2.35 \times 10^{-7} \frac{\eta_1 d^2 L n^2}{\psi} \qquad \textbf{Metric} \qquad (24\text{--}106c)$$

where P_u in hp$_m$, η_1 in cP, d and L in mm, and n in rpm

$$= 2.3 \times 10^6 \frac{\eta' d^2 L n^2}{\psi} \qquad \textbf{Metric} \qquad (24\text{--}106d)$$

where P_μ in hp$_m$, η' in (kgf s/m^2), L and d in m, and n in rpm

$$= 2.3 \times 10^{-3} \frac{\eta' d^2 L n^2}{\psi} \qquad \textbf{Metric} \qquad (24\text{--}106e)$$

where P_μ in hp$_m$, η' in (kgf s/m^2), L and d in mm, and n in rpm

$$= \frac{3.8}{3} \frac{\eta_1 d^2 L n^2}{\psi} \qquad \textbf{US Customary units} \qquad (24\text{--}106f)$$

where P_μ in hp, η_1 in cP, L and d in in, and n in rpm

If the journal and the bearing are eccentric and the distance between their axes is ϵ, the power loss is calculated from formula

$$P_\mu = \frac{6.2 \times 10^8 \, \eta d^2 \, L n'^2}{\psi \sqrt{1 - (2\epsilon)^2}} \qquad \textbf{SI} \qquad (24\text{--}107a)$$

where P_μ in kW, η in Pa s, d and L in m, and n' in rps

$$= 2.35 \times 10^{-7} \frac{\eta_1 d^2 \, L n^2}{\psi \sqrt{1 - (2\epsilon)^2}} \qquad \textbf{Metric} \qquad (24\text{--}107b)$$

where P_μ in hp$_m$, η_1 in cP, d and L in mm, and n in rpm

$$= 2.3 \times 10^6 \frac{\eta' d^2 \, L n^2}{\psi \sqrt{1 - (2\epsilon)^2}} \qquad \textbf{Metric} \qquad (24\text{--}107c)$$

where P_μ in hp$_m$, η' in (kgf s/m^2), L and d in m, and n in rpm

$$= \frac{3.8}{10^3} \frac{\eta_1 d^2 \, L n^2}{\psi \sqrt{1 - (2\epsilon)^2}} \qquad \textbf{US Customary units}$$

$$(24\text{--}107d)$$

where P_μ in hp, η_1 in cP, L, d in in, and n in rpm

Particular	Formula

Collar bearing (Fig. 24–41)

The average intensity of pressure with i collars

$$P = \frac{W}{0.784(d_1^2 - d_2^2)i} \qquad (24\text{–}108)$$

The friction moment for each collar for uniform intensity of pressure

$$M_{te} = \frac{1}{3} \frac{\mu W}{i} \left[\frac{d_1^3 - d_2^3}{d_1^2 - d_2^2}\right] \qquad (24\text{–}109)$$

The total friction moment for i collars for uniform intensity of pressure

$$M_t = \frac{1}{3} \mu W \left[\frac{d_1^3 - d_2^3}{d_1^2 - d_2^2}\right] \qquad (24\text{–}110)$$

The friction moment for each collar for uniform rate of wear

$$M_{te} = \frac{\mu W}{i} \left(\frac{d_1 + d_2}{4}\right) \qquad (24\text{–}111)$$

The total friction moment for i collars for uniform rate of wear

$$M_t = \mu W \left(\frac{d_1 + d_2}{4}\right) \qquad (24\text{–}112)$$

The friction power in collar bearing

$$P_\mu = \frac{\mu W(d_1 + d_2)\, n'}{2,292,296} \qquad \textbf{SI} \quad (24\text{–}113a)$$

where P_μ in kW, W in N, d in m, and n' in rps

$$= \frac{\mu W(d_1 + d_2)\, n}{252,120} \qquad \textbf{US Customary units}$$

$$(24\text{–}113b)$$

where P_μ in hp, W in lbf, d in in, and n in rpm

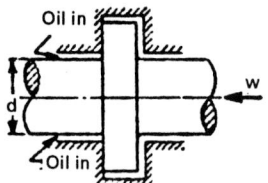

Oil in

FIGURE 24–41   Collar thrust bearing.

The coefficient of friction for collar bearing

$$\mu = 83.8 \frac{v^{0.5}}{p^{0.67}} \qquad \textbf{SI} \quad (24\text{–}114a)$$

where v in m/s and P in N/m^2

$$= 0.016 \frac{v^{0.5}}{p^{0.67}} \qquad \textbf{US Customary units} \quad (24\text{–}114b)$$

where v in ft/min and P in psi

$$= 1.73 \times 10^{-3} \frac{v^{0.5}}{p^{0.67}} \qquad \textbf{Metric} \quad (24\text{–}114c)$$

where v in m/s and P in kgf/mm^2

Allowable pressure P may be taken so that Pv value for v ranging from 0.20 to 1 m/s (50 to 200 ft/min)

$$Pv \leq 700505 \qquad \textbf{SI} \quad (24\text{–}115a)$$

where P in Pa and v in m/s

$$\leq 0.0715 \qquad \textbf{Metric} \quad (24\text{–}115b)$$

where P in kgf/mm^2 and v in m/s

$$\leq 20000 \qquad \textbf{US Customary System unit} \quad (24\text{–}115c)$$

where P in psi and v in ft/min

Particular	Formula
	Refer to Table 24–8 for P and Table 24–6 for Pv values.

Thrust bearing

Parallel-surface thrust bearing (Figs. 24–42 to 24–43)

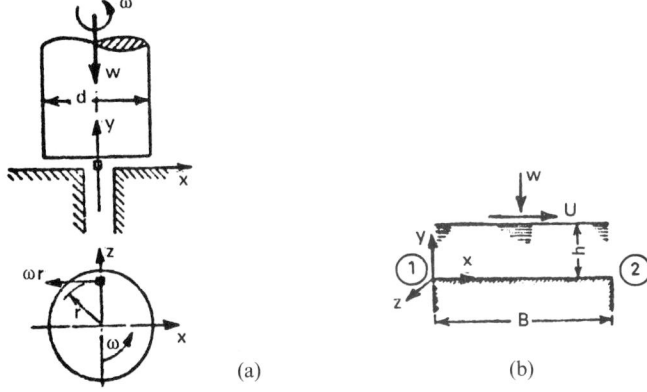

(a)   (b)

FIGURE 24–42   Parallel-surface thrust bearing.

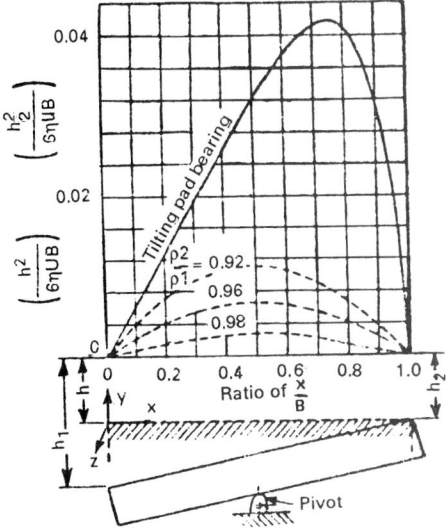

FIGURE 24–43   Comparison of pressure distribution across tilting-pad and parallel-surface thrust bearing.

The pressure at any point along the bearing

$$P = \frac{6\eta\, UB}{h^2}\, K_{LP1} \tag{24–116a}$$

$$\text{where } K_{LP1} = \left[\frac{x_1 - \ln\{(\rho'-1)x_1+1)\}}{\ln\rho'}\right]$$

Particular	Formula
	$$\rho' = \frac{\rho_2}{\rho_1} \qquad x_1 = \frac{x}{B} \qquad (24\text{-}116b)$$
The ratio of the density of the lubricant leaving the bearing to the density of the lubricant entering the bearing	$$\rho' = \frac{\rho_2}{\rho_1} = 1 + \frac{a}{\rho_1}(t_2 - t_1) \qquad (24\text{-}117)$$
	where a = constant; $a/\rho_1 = -0.0004$; and t_2 and t_1 are the temperatures in °C corresponding to densities ρ_2 and ρ_1, respectively
The unit load supported by a parallel-surface thrust bearing	$$P_u = \frac{6\eta\,UB}{h_2}\,K_{LP2} \qquad (24\text{-}118a)$$
	where $K_{LP2} = \left[\dfrac{1}{2} + \dfrac{\rho'}{1-\rho} + \dfrac{1}{\ln\rho'}\right] \qquad (24\text{-}118b)$
The approximate formula for unit load supported by a parallel-surface thrust bearing	$$P_u = \frac{6\eta\,UB}{h_2}\,K_{LP3} \qquad (24\text{-}119a)$$
	where $K_{LP3} = [0.09(1-\rho')] \qquad (24\text{-}119b)$
	Refer to Table 24–18 for K_{LP3}.
The pressure distribution along a tilting-pad bearing of infinite width (Figs. 24–39 and 24–43)	$$P = \frac{6\eta\,UB}{h_1^2}\,K_{pt} \qquad (24\text{-}120a)$$
	where $K_{Pt} = \dfrac{(m-1)(1-x_1)x_1}{(m+1)(m-mx_1+x_1)^2} \qquad (24\text{-}120b)$
	$m = h_1/h_2;\ x_1 = x/B$

TABLE 24–18
Comparison of load capacities of tilting-pad and parallel-surface-type of bearings

Temperature rise through bearings, °C	ρ'	K_{LP3}	K_{lt} (for $h' = 3$)	Relative load capacity, K_{LP3}/K_{lt}
10	0.98	0.0018	0.025	14
38	0.96	0.0036		7
93	0.92	0.0072		3.5

Source: F. I. Radzimovsky, Lubrication of Bearings—Theoretical Principles and Designs, The Ronald Press Company, New York, 1959.

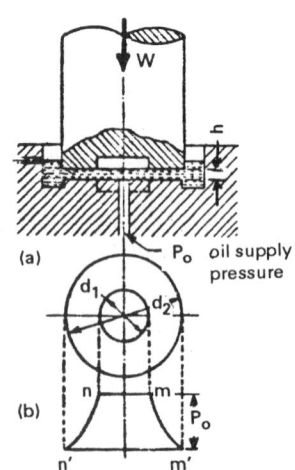

FIGURE 24–44 Hydrostatic step bearing; (*b*) plan view and general character of pressure distribution along the diameter of the bearing.

Particular	Formula
The unit load supported by a tilting-pad bearing of infinite width (Fig. 24–43)	$$P_u = \frac{6\eta \, UB}{h_2^2} K_{lt} \qquad (24\text{–}121a)$$ $$\text{where } K_{lt} = \left[\frac{1}{(m-1)^2} \left\{ \ln m - \frac{2(m-1)}{m+1} \right\} \right] \qquad (24\text{–}121b)$$

OIL FILM THICKNESS

Particular	Formula
The thickness of oil film in a parallel-surface thrust bearing	$$h = \sqrt{6K_{LP3}} \sqrt{\frac{\eta \, UB}{P_\mu}} \qquad (24\text{–}122)$$ Refer to Table 24–18 for K_{LP3}.
The thickness of minimum oil film at location 2 (Figs. 24–39 and 24–43)	$$h_2 = \sqrt{6K_{lt}} \sqrt{\frac{\eta \, UB}{P_\mu}} \qquad (24\text{–}123)$$ Refer to Table 24–18 for K_{lt}.
For properties of lubricant bearing materials and applications, conversion factors for viscosity, kinematic and Saybolt viscosity equivalents and conversion tables for viscosity equivalent	Refer to Tables 24–19 to 24–23.

COEFFICIENT OF FRICTION

Particular	Formula
The coefficient of friction in case of a parallel-surface thrust bearing	$$\mu = \left(\frac{1.82}{1 - \rho'} \right) \frac{h}{B} \qquad (24\text{–}124)$$
Another formula for coefficient of friction in case of a parallel-surface thrust bearing	$$\mu = \frac{1}{\sqrt{6K_{LP3}}} \sqrt{\frac{\eta \, U}{P_u B}} \qquad (24\text{–}125)$$
The coefficient of friction for a tilting-pad bearing of infinite width	$$\mu = K_{\mu t} \sqrt{\frac{\eta \, U}{P_u B}} \qquad (24\text{–}126a)$$ where $$K_{\mu t} = \left\{ \left(4\ln m - 6\frac{m-1}{m+1} \right)^2 \middle/ \left(6\ln m - 12\frac{m-1}{m+1} \right) \right\}^{1/2}$$ $$m = h_1/h_2 = \text{film thickness ratio} \qquad (24\text{–}126b)$$

TABLE 24-19
Typical properties of lubricants

Type and application	SAE no.	Density, g/cm³, at 15.5°C	Pour point, °C	Flash point, °C	Viscosity index	Viscosity									Uses
						Saybolt seconds, S		Centipoise, cP		kgf s/m² × 10⁻⁴		Pa s ×10⁻³		
						At 38 °C	At 99°C	At 38°C	At 99°C	At 38°C	At 99°C	At 38°C	At 99°C	
Transmission gear oil	75	0.900	−23	193	121	220	50	47	7.3	47.94	7.45	47	7.3	Combination pinion reduction gear units, enclosed reduction gear sets
	80	0.934	−32	185	78	320	52	69	7.9	70.38	8.06	69	7.9	
	90	0.930	−23	232	91	1330	100	287	20.4	292.74	20.81	287	20.4	
	140	0.937	−18	260	82	3350	160	725	34	739.50	34.68	725	34	
	250	—	−15	254.5	83	5660	220	1220	47	1244.40	47.94	1220	47	
Automotive oil	10W	0.870	−26	210	102	190	46	41	6.0	41.82	6.12	41	6.0	Automobile, truck and marine reciprocating engines; very-heavy-duty oils used in diesel engines
	20W	0.885	−23	227	96	330	54	71	8.5	72.42	8.67	71	8.5	
	30	0.891	−20	238	92	530	64	114	11.3	116.28	11.53	114	11.3	
	40	0.890	−18	240.5	90	800	77	173	14.8	176.46	15.10	173	14.8	
	50	0.992	−12	254.5	90	1250	97	270	19.7	275.40	20.10	270	19.7	
	60	—	—	—	80	—	115	—	—	—	—	—	—	
	70	—	—	—	80	—	137	—	—	—	—	—	—	
Aircraft engine oil		0.858	−65	111	87	43	33	5	1.6	5.10	1.63	5	1.6	Turbojet engines
		0.864	−62	146	79	59	35	10	2.5	10.20	2.55	10	2.5	
		0.876	−18	215.5	106	350	57	76	9.3	77.52	9.49	76	9.3	
		0.884	−18	224	96	514	64	111	11.3	113.22	11.53	111	11.3	Various reciprocating aircraft engines
		0.887	−18	232	95	829	80	179	15.5	182.58	15.81	179	15.5	
		0.892	−18	249	95	1240	99	268	20.1	273.36	20.50	268	20.1	
		0.892	−7	318	96	1711	120	369	25.0	376.38	25.50	369	25.0	
Turbine-grade oil														
Light		0.872	−18	210	109	150	44	32	5.4	32.64	5.51	32	5.4	Direct-connected turbines electric motors
Medium		0.877	−12	235	105	300	53	65	8.2	66.30	8.36	65	8.2	Land-geared turbines electric-motors
Heavy		0.885	−12	243	100	460	62	99	10.8	100.98	11.02	99	10.8	Marine-propulsion geared turbines
Steam		0.895	14	260	101	1800	130	390	27	397.80	27.45	390	27	Railroad stationary steam engines cylindered applications, enclosure gears
Cylinder		0.910	1.5	211	107	3750	210	810	45	826.20	45.90	810	45	
Oil		0.904	15.5	343	103	6470	300	1400	64	1428.00	65.28	1400	64	
Hydraulic oils														
Light		0.887	−42	188	64	150	42	32	4.8	32.64	4.90	32	4.8	Hydraulic fluids for most indoor industrial hydraulic equipments
Medium		0.895	−26	207	66	310	50	67	7.3	68.34	7.45	67	7.3	Heavier loads, higher temperature
Heavy		0.901	−12	257	70	910	74	196	14.0	199.92	14.28	196	14.0	

TABLE 24-19
Typical properties of lubricants (*Cont.*)

Type and application	SAE no.	Density, g/cm³, at 15.5°C	Pour point, °C	Flash point, °C	Viscosity index	Saybolt seconds, S		Centipoise, cP		kgf s/m² × 10⁻⁴		Pa s ×10⁻³		Uses
						At 38 °C	At 99°C	At 38°C	At 99°C	At 38°C	At 99°C	At 38°C	At 99.9°C	
Extra-low-temperature		0.844	−24	110	226	74	43	14	5.2	14.28	5.30	14	5.2	Aircraft hydraulic systems
Refrigerating machine oil		0.895	−45.5	146	53	72	36	14	2.9	14.28	2.96	14	2.9	Ammonia compressor
		0.898	−37	165.5	22	195	43	42	5.1	42.84	5.20	42	5.1	
		0.909	−29	182	34	235	45	51	5.7	52.02	5.81	51	5.7	
		0.902	−23	190.5	35	335	49	72	7.0	73.44	7.14	72	7.0	
Machine tools and general-purpose oil		0.881	−4	177	80	105	39	22	3.9	22.44	3.98	22	3.9	All general-purpose lubrication, machine tools
		0.898	−4	199	80	205	46	44	6.0	44.88	6.12	44	6.0	
		0.915	−12	185	83	305	49	66	7.0	67.32	7.14	66	7.0	
		0.915	−15	199	25	510	59	110	9.9	112.20	10.10	110	9.9	
		0.890	−9	235	80	930	80	200	15.5	204.00	15.81	200	15.5	

TABLE 24-20
Journal bearing materials and applications

Material	Composition, %	Specific gravity	Dry coefficient of friction, μ	Ultimate tensile strength, σ_u			Modulus of elasticity, E			Hardness numbers		Applications
				kgf/mm²	N/m² ×10⁶	MPa	kgf/mm² ×10⁴	N/m² ×10⁹	GPa	Brinell	Rockwell	
Babbitts												
Lead base	Sn 10.0, Sb 15.0, Pb 75.0	9.69	0.34	7.03	68.96	68.96	0.295	28.9	28.9	45	—	Used in automobiles and electrical equipment
Tin base	Cu 8.3, Sb 8.3, Sn 83.4	7.47	0.28	7.88	77.30	77.30	0.534	52.4	52.4	27	—	Used in automotive and diesel engines, steam turbines and motors
Cadmium base	Ni 1.4, Cd 98.6	8.6	0.34	—	—	—	—	—	—	—	—	Used where lubrication is intermittent
Aluminum alloys	Cu 1.0, Sn 6.5, Ni 1.5, Al 91.0	2.86	0.33	15.5	151.76	151.76	0.724	71.0	71.0	45	—	Used in high-temperature high-load services and in diesels; requires good lubrication and hardened shaft
Copper alloys												
Clock brass	Pb 3.0, Zn 35.5, Cu 61.5	8.4	—	38.0–45.0	372.48–445.45	372.48–445.45	1.055	103.5	103.5	54–142	B 40–75	Used for light load
Bronze, high-lead	Sn 4.0, Pb 14.0, Zn 1.5, Ni 1.0 max, Cu 79.5	—	0.15	14.0	138.00	138.00	—	—	—	45	—	Used in poorly lubricated applications with moderately heavy loads
Bronze, high-lead	Sn 16, Pb 14.0, Cu 70.0	—	0.37	—	—	—	—	—	—	—	—	Same as above; can withstand higher loads
Bronze, lead tin	Sn 8, Pb 3.5, Zn 3.5, Cu 85	8.4	0.26	21.1 min	207.00 min	207.00 min	—	—	—	53	—	Moderately heavy duty
Bronze, 80-10-10	Sn 10.0, Pb 10.0, Cu 80	8.86	0.15	17.58 min	172.46 min	172.46	0.773	75.8	75.8	65	—	General-duty bearing bronze; load up to 20.6 MPa (2.1 kgf/mm²); speed—4.5 m/s
Bronze, nickel tin	Sn 10.0, Ni 3.5, Pb 2.5, Cu 84.0	—	0.37	31.64 min	310.04 min	310.04	—	—	—	95	—	Used in medium- to heavy-duty application; good strength requirement
Bronze, aluminum	Al 10.5, Fe 3.5, Cu 86.0	7.6	0.52	70.30	689.74	689.74	1.125	110.4	110.4	202	—	Used in heavy-duty bearings requiring high strength and good impact resistance
Bronze, zinc	Al 1.0, Si 0.8, Mn 2.5, Zn 37.5, Cn 58.2	8.09	0.39	49.22 min	482.84 min	482.84 min	1.055	103.5	103.5		B 80–92	Heavy-duty impact loadings; on hardened shaft
Iron base												
Gray cast iron	C 3.5, Si 2.5, Fe 94.0	7.2	0.37	21.10 min	207.00 min	207.00 min	1.898	186.2	186.2	180	—	Used in refrigerators, compressors, camshafts, high load at low speed with good lubrication
Sintered iron	Cu 7.5, Fe 92.5	—	0.30	—	—	—	—	—	—	—	—	Used with impregnated with oil will give good results; load—3.4 MPa (0.35 kgf/mm²) and speed 0.67 m/s

TABLE 24-20
Journal bearing materials and applications (*Cont.*)

Material	Composition, %	Specific gravity	Dry coefficient of friction, μ	Ultimate tensile strength, σ_u kgf/mm²	N/m² ×10⁶	MPa	Modulus of elasticity, E kgf/mm² ×10⁴	N/m² ×10⁹	GPa	Hardness numbers Brinell	Rockwell	Shore scleroscope	Applications
Graphite													
Carbon graphite	C + binder	1.63–1.86	0.15	0.53–1.80	5.17–17.25	5.17–17.25	—	—	—			Shore scleroscope 75	Particularly suited to high-temperature application (<455°C) where lubrication is difficult; used in electric motors, conveyors
Carbon graphite and metal	C + Cu + binder	2.9/3.8	0.17	2.11–4.22	20.7–41.4	20.7–41.4	—	—	—	"	"		Same as above, higher strength
Cemented carbide	Tungsten carbide 97.0, Co 3.0	15.1	0.20	573.00 (compressive)	5621 (compressive)	5621 (compressive)	6.885 (compressive)	672.5	672.5	C 80			Used in high-speed precision grinders which require perfect alignment and good lubrication; can withstand extreme loading and high speeds
Wood													Used in conveyors; light loads at high speeds under 65°C
Plastics and rubber													
Nylon	Polyamide	1.44	0.86	7.03	68.96	68.96	0.023	2.25	2.25	M 90			Used in many household appliances and other lightly loaded applications; requires little lubrication
Rubber		0.97–2.00	0.25–0.30	1.40–10.55	13.73–103.50	13.73–103.50							Marine propellers, pumps, turbine, load 0.54 MPa (0.055 kgf/mm²)
Teflon	Polytetrafluoroethylene	2.2	0.17	2.11	20.70	20.70	0.0042	0.410	0.410			Shore scleroscope 50	Useful in corrosive conditions; dairy, textile, and food machinery
Textolite 2001	Phenolic, graphite and cotton cloth	1.36	0.18	7.03	68.96	68.96	0.0443–0.0647	4.357–6.35	4.375–6.35	M 100			Used where low wear and good compatibility characteristics are required

24.69

TABLE 24–21
Conversion factors for viscosity

	P	cP	kgf s/m^2	kg/m s	lbf s/ft^2	lb/ft s	Pa s
P	1	100	0.0102	0.1	2.0886×10^{-3}	0.0672	0.1
cP	0.01	1	1.0297×10^{-4}	10^{-3}	2.0886×10^{-5}	6.7197	10^{-3}
kgf s/m^2	98.0665	9.80665×10^{-3}	1	9.80665	0.20482	6.5898	9.80665
kg/m s	10	10^3	0.102	1	2.0886×10^{-2}	0.6720	
lbf s/ft^2	4.788×10^2	4.788×10^4	4.8824	47.88	1	32.174	
lb/ft s	14.882	1.4882×10^3	0.1518	1.4882	0.0311	1	
Pa s	10	10^3	0.102				1

Particular	Formula

HYDROSTATIC BEARING: STEP-BEARING (FIG. 24–44)

The pressure in the pocket supplied from external source to support the load

$$P_o = 8W \ln (d_2/d_1)/\pi(d_2^2 - d_1^2) \qquad (24\text{–}127\text{a})$$

The load-carrying capacity

$$W = \frac{P_o \pi}{8}(d_2^2 - d_1^2) \ln (d_2/d_1) \qquad (24\text{–}128)$$

The rate of flow of lubricant through the bearing

$$Q = \frac{\pi P_o h^3}{6\eta \ln (d_2/d_1)} \qquad (24\text{–}129)$$

Power loss in bearing

$$P_\mu = 0.062 \frac{n'^2 \eta}{16h} (d_2^4 - d_1^4) \qquad \textbf{SI} \qquad (24\text{–}130\text{a})$$

where P_μ in kW; η in Pa s; h, d_1, and d_2 in m; and n' in rps

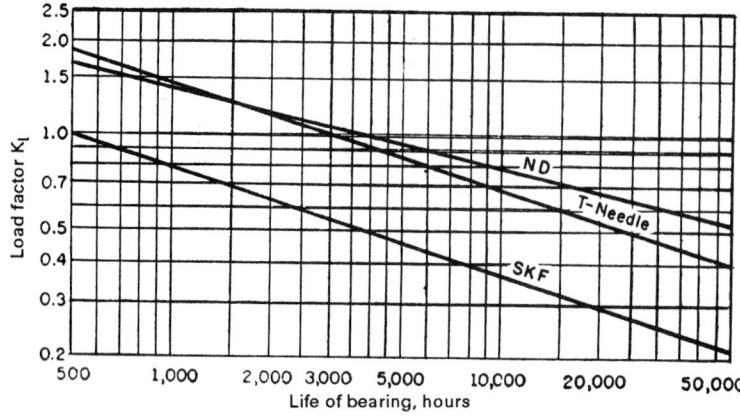

FIGURE 24–45   Life curves of ball and needle bearings.

TABLE 24–22
Kinematic and Saybolt viscosity equivalents

| | Kinematic viscosity, ϑ | | Saybolt viscosity, S[a] | |
| | Metric units | SI units | | |
cSt	cm^2/s $\times 10^{-2}$	m^2/s $\times 10^{-6}$	At 38 °C	At 99°C
2	2	2	32.6	32.9
3	3	3	36.0	36.3
4	4	4	39.1	39.4
5	5	5	42.4	42.7
6	6	6	45.6	45.9
7	7	7	48.8	49.1
8	8	8	52.1	52.5
9	9	9	55.5	55.9
10	10	10	58.9	59.3
11	11	11	62.4	62.9
12	12	12	66.0	66.5
13	13	13	69.8	70.3
14	14	14	73.6	74.1
15	15	15	77.4	77.9
16	16	16	81.3	81.3
17	17	17	85.3	85.9
18	18	18	89.4	90.1
19	19	19	93.6	94.2
20	20	20	97.8	98.5
21	21	21	102.0	102.8
23	23	23	110.7	111.4
25	25	25	119.3	120.1
27	27	27	128.1	129.0
29	29	29	136.9	137.9
30	30	30	141.3	142.3
31	31	31	145.7	146.8
33	33	33	154.7	155.8
35	35	35	163.7	164.9
37	37	37	172.7	173.9
39	39	39	181.8	183.0
40	40	40	186.3	187.6
41	41	41	190.8	192.1
43	43	43	199.8	201.2
45	45	45	209.1	210.5
47	47	47	218.2	219.8
49	49	49	227.5	229.1
50	50	50	232.1	233.8
55	55	55	255.2	257.0
60	60	60	278.3	280.2
65	65	65	301.4	303.5
70	70	70	324.4	326.7
>70			$S = \text{cSt} \times 4.635$	$S = \text{cSt} \times 4.667$

[a] $S = \text{cSt} \times 4.635$ at 38°C; $S = \text{cSt} \times 4.667$ at 99°C

TABLE 24–23
Conversion table for viscosity equivalents

Kinematic viscosity, ϑ			Viscosity		
Metric units		SI units			
cSt	cm²/s ×10⁻²	m²/s ×10⁻⁶	Saybolt, S	Engler °	Redwood no. 1
2.0	2.0	2.0	32.60	1.12	30.8
2.2	2.2	2.2	33.40	1.14	31.3
2.4	2.4	2.4	34.10	1.16	31.8
2.6	2.6	2.6	34.80	1.18	32.3
2.8	2.8	2.8	35.40	1.20	32.8
3.0	3.0	3.0	36.00	1.22	33.3
3.2	3.2	3.2	36.70	1.23	33.8
3.4	3.4	3.4	37.30	1.25	34.3
3.6	3.6	3.6	37.90	1.27	34.8
3.8	3.8	3.8	38.50	1.29	35.3
4.0	4.0	4.0	39.1	1.31	35.8
4.5	4.5	4.5	40.8	1.35	37.0
5.0	5.0	5.0	42.4	1.40	38.3
5.5	5.5	5.5	44.0	1.44	39.6
6.0	6.0	6.0	45.6	1.48	40.9
6.5	6.5	6.5	47.2	1.52	42.3
7.0	7.0	7.0	48.8	1.56	43.6
7.5	7.5	7.5	50.4	1.60	44.9
8.0	8.0	8.0	52.1	1.65	46.3
8.5	8.5	8.5	53.8	1.70	47.7
9.0	9.0	9.0	55.5	1.75	49.0
9.5	9.5	9.5	57.2	1.79	50.5
10	10	10	58.9	1.84	51.9
11	11	11	62.4	1.94	54.9
12	12	12	66.0	2.02	58.0
13	13	13	69.7	2.12	61.2
14	14	14	73.5	2.22	64.5
15	15	15	77.3	2.32	67.9
16	16	16	81.2	2.43	71.3
17	17	17	85.2	2.54	74.8
18	18	18	89.3	2.64	78.4
19	19	19	93.4	2.75	82.0
20	20	20	97.6	2.87	85.7
22	22	21	106.1	3.10	93.2
24	24	24	114.7	3.33	100.8
26	26	26	123.4	3.57	108.5
28	28	28	132.3	3.82	116.3
30	30	30	141.1	4.07	124.2
32	32	32	149.9	4.32	132.1
34	34	34	158.9	4.57	140.0
36	36	36	167.9	4.82	147.9
38	38	38	176.9	5.08	155.9

TABLE 24–23
Conversion table for viscosity equivalents (*Cont.*)

	Kinematic viscosity, ϑ			Viscosity	
	Metric units	**SI units**			
cSt	**cm²/s ×10⁻²**	**m²/s ×10⁻⁶**	**Saybolt, *S***	**Engler °**	**Redwood no. 1**
40	40	40	186.0	5.33	164.0
42	42	42	195.0	5.59	172.0
44	44	44	204.0	5.84	180.0
46	46	46	213.0	6.10	188.0
48	48	48	222.0	6.36	196.0
50	50	50	232.0	6.62	204.0
55	55	55	255.0	7.26	225.0
60	60	60	278.0	7.90	245.0
65	65	65	301.0	8.55	265.0
70	70	70	324.0	9.21	286.0
75	75	75	347.0	9.87	306.0
80	80	80	370.0	10.53	326.0
85	85	85	393.0	11.19	346.0
90	90	90	416.0	11.85	367.0
95	95	95	439.0	12.51	387.0
100	100	100	463.0	13.16	407.0
110	110	110	509.0	14.47	448.0
120	120	120	555.0	15.80	489.0
130	130	130	602.0	17.11	529.0
140	140	140	648.0	18.43	570.0
150	150	150	694.4	19.75	611.0
160	160	160	740.0	21.05	651.0
170	170	170	787.0	22.38	692.0
180	180	180	833.0	23.70	733.0
190	190	190	879.0	25.00	774.0
200	200	200	926.0	26.32	815.0
220	220	220	1018.0	28.95	896.0
240	240	240	1111.0	31.60	978.0
260	260	260	1203.0	34.25	1059.0
280	280	280	1296.0	36.85	1140.0
300	300	300	1388.0	39.50	1222.0
320	320	320	1480.0	42.12	1303.0
340	340	340	1574.0	44.75	1385.0
360	360	360	1666.0	47.40	1465.0
380	380	380	1759.0	50.00	1546.0
400	400	400	1851.0	52.65	1628.0
500	500	500	2314.0	65.80	2036.0
600	600	600	2777.0	79.00	2443.0
700	700	700	3239.0	92.20	2850.0
800	800	800	3702.0	105.30	3258.0
900	900	900	4165.0	118.50	3668.0
1000	1000	1000	4628.0	131.60	4074.0

Particular	Formula

$$= 8.3 \times 10^{-4} \frac{n' \; \eta}{16h} (d_2^4 - d_1^4) \quad \textbf{Metric} \quad (24\text{--}130b)$$

where P_μ in hp_m, η in kgf s/mm; h, d_1, and d_2 in mm; and n' in rps

Spherical step bearing

Load-carrying capacity

$$W = \frac{\pi \, P_o \, d_2^2 (\cos \, \phi_1 - \cos \, \phi_2)}{\ln \, [\tan \, (\phi_2/2) + \tan \, (\phi_1/2)]} \qquad (24\text{--}131)$$

LIFT BEARING

Inlet pressure

$$P_i = \frac{4\eta \, Q}{l \, \psi^3 \, d^2} \, K_5 \qquad (24\text{--}132a)$$

where

$$K_5 = 12 \left[\frac{e(4 - e^2)}{2(1 - e^2)^2} + \frac{(2 + e^2)}{(1 - e^2)^{5/2}} \times \text{arc tan} \; \frac{1 + e}{\sqrt{1 - e^2}} \right] \qquad (24\text{--}132b)$$

Load-carrying capacity of bearing

$$W = \frac{2\eta Q}{\psi^3 \, d} \, K_6 \qquad (24\text{--}133a)$$

where $K_6 = 12 \left[\dfrac{2 + 3e - e^3}{(1 - e^2)^2} \right]$ $\qquad (24\text{--}133b)$

JOURNAL BEARING

The diameter of journal bearing for speeds below 2.5 m/s

$d = 3.2 \times 10^{-3} \, \sqrt[5]{W^2/i^2 n'}$ $\qquad$ **SI** $\quad (24\text{--}134a)$

where W in N, d in m, $l/d = i$ and n' in rps

$= 18 \, \sqrt[5]{(W^2/i^2 n)}$ $\qquad$ **Metric** $\quad (24\text{--}134b)$

where W in kgf, d in mm, $i = l/d$ and n in rpm

The diameter of journal bearing for speeds exceeding 2.5 m/s

$d = 2 \times 10^{-3} \, \sqrt[7]{W^2/i^3 n'}$ $\qquad$ **SI** $\quad (24\text{--}135a)$

where W in N, d in m, $i = l/d$ and n' in rps

$= 10 \, \sqrt[7]{W^3/i^3 n}$ $\qquad$ **Metric** $\quad (24\text{--}135b)$

where W in kgf, d in mm, $i = l/d$; and n in rpm

The minimum oil film thickness according to H. A. S. Howarth in use of journal bearing

$h_o = 3.35 \times 10^{-5} (d \, v)^{0.376}$ $\qquad$ **SI** $\quad (24\text{--}136a)$
where h_o in m; d in m, and v in m/s

$= 254.25 \times 10^{-5} (d \, v)^{0.376}$ $\qquad$ **Metric** $\quad (24\text{--}136b)$
where h_o in mm; d in mm, and v in m/s

24.2 ROLLING CONTACT BEARINGS (1)

SYMBOLS

b	Weibull exponent
B	width of bearing, m (in)
c	permissible increase in diametral clearance (μm)
C	basic dynamic load rating for radial and angular contact ball or radial roller bearings, kN (lbf)
C_a	basic dynamic load rating for single-row, single- and double-direction thrust ball or roller bearings, kN (lbf)
$C_{a1}, C_{a2}, \ldots C_{an}$	basic load rating per row of a one-direction multirow thrust ball or roller bearing, each calculated as single-row bearing with $Z_1, Z_2, \ldots Z_n$ balls or rollers, respectively
C_n	capacity of the needle bearing, kN (lbf)
C_o	basic static load rating for radial ball or roller bearing, kN (lbf)
C_{oa}	basic static load rating for thrust ball or roller bearings, kN (lbf)
d	bearing bore diameter, m (in)
d_b	diameter of ball, m (in)
d_i	shaft or outside diameter of inner race used in Eqs. (24–206) and (24–207), m (in)
d_o	inside diameter of outer race of needle bearing, m (in)
d_r	roller diameter (mean diameter of tapered roller), m (in) diameter of needle roller, m (in)
d_1, d_2	diameter of spherical balls or cylindrical rollers used in contact stress [Eqs. (24–211) to (24–214)], m (in)
D	outside diameter of bearing, m (in)
D_1	diameter of revolving race, m (in)
e	bearing constant
E	modulus of elasticity, GPa (psi)
f_a	application factor to compensate for shock continuous duty or inequality of loading
f_c	a factor which depends on the geometry of the bearing components, the accuracy to which the various bearing parts are made and the material used in Eqs. (24–160), (24–161), and (24–167) to (24–170); a factor which depends on the units used, the exact geometrical shape of the load-carrying surfaces of the roller and rings (or washers in case of thrust bearing), and the accuracy to which the various bearing parts are made and the material, used in Eqs. (24–164), (24–174), and (24–175)
f_d	a factor for the additional forces emanating from the mechanisms coupled to the gearing used in Eq. (24–139)
f_k	a factor for the additional forces created in the gearing itself used in Eq. (24–139)
f_L	index of dynamic stressing
f_n	speed factor for ball bearings according to Table 24–37 speed factor for roller bearings according to Table 24–38
f_{nt}	speed factor used in tapered roller bearing
f_o	a factor used in Eqs. (24–146) and (24–149)
f_{oa}	a factor used in Eqs. (24–152) and (24–154)
F	load, kN (lbf) theoretical tooth load, kN (lbf)

F_a	thrust load, kN (lbf)
F_{aa}	applied thrust load, kN (lbf)
F_{ar}	thrust component of pure radial load F, due to tapered roller, kN (lbf)
F_{bs}	shaft load due to belt drive, kN (lbf)
F_c	static load, kN (lbf)
F_e	radial equivalent load from combination of radial and thrust loads or effective radial load, kN (lbf)
F_{effg}	effective tooth load, kN (lbf)
F_{na}	net thrust load, kN (lbf)
F_{nt}	net thrust load on the tapered roller bearing, kN (lbf)
F_r	radial load capacity of ball bearing, kN (lbf)
	radial bearing load, kN (lbf)
i	number of rows of balls in any one bearing
k	constant used in Eqs. (24–141), (24–143) to (24–145)
K_h	hardness factor used in Eq. (24–207)
K_l	life load factor taken from the curve in Fig. 24–45 marked "T-needle" and used in Eq. (24–207)
K_n	a constant used in Eq. (24–138)
l	length of needle bearing, m (in)
l_{eff}	the effective length of contact between one roller and that ring (or that washer in case of thrust bearing) where the contact is the shortest (overall roller length minus roller chamfers or minus grinding undercuts), m (in)
L	life of bearing at constant speed, rpm
	life of bearing at constant speed, h
	life corresponding to desired reliability, R, used in Eq. (24–194)
L_{B10}	life factor corresponding to desired B-10 hours of life expectancy used in Eq. (24–195)
L_{10}	rating life
L_h	fatigue life
M_t	torque, N m (lbf in)
n	speed, rpm
n'	speed, rps
n_e	effective speed, rpm
n_i	ith speed, rpm
n_l	limiting speed, rpm
n_m	mean speed, rpm
n_1	speed of the inner race, rpm
n_2	speed of the outer race, rpm
P	power, kW (hp)
P	equivalent dynamic load, kN (lbf)
P_a	equivalent dynamic thrust load, kN (lbf)
P_m	mean load, kN (lbf)
P_{max}	maximum load, kN (lbf)
P_{min}	minimum load, kN (lbf)
P_o	static equivalent load, kN (lbf)
P_{oa}	static equivalent load for thrust ball or roller bearings under combined radial and thrust loads, kN (lbf)
q_i	percentage time of ith speed
R_{10}	0.90 reliability corresponding to rating life
X	radial factor used in Eqs. (24–163), (24–166), (24–173), (24–178), and (24–180)
X_o	radial factor used in Eqs. (24–147), (24–150) and (24–157)

Y	thrust factor used in Eqs. (24–163), (24–166), (24–173), (24–178), and (24–180)
Y_o	thrust factor used in Eqs. (24–147), (24–150), and (24–157)
Z	number of balls per row
	number of balls carrying thrust in one direction
	number of rollers per row
	number of rollers carrying thrust in single-row one-direction bearing
	number of needle-rollers
$Z_1, Z_2, \ldots Z_n$	number of balls or rollers in respective rows of one-direction multirow bearings
α	nominal angle of contact, that is, nominal angle between the line of action of the ball load and a plane perpendicular to the bearing axis
	the angle of contact, that is, the angle between the line of action of the roller resultant load, and a plane perpendicular to the bearing axis
ω	angular speed, rad/s
μ	coefficient of friction
ν	Poisson's ratio
$\sigma_{c(max)}$	maximum compressive stress, MPa (psi)
τ_{max}	maximum shear stress, MPa (psi)

Particular	Formula	
The torque	$$M_t = \frac{9550\, P}{n}$$	**SI** (24–137a)
	where P in kW, n in rpm, and M_t in N m	
	$$= \frac{1000\, P}{\omega}$$	**SI** (24–137b)
	where P in kW, ω in rad/s, and M_t in N m	
	$$= \frac{159.2\, P}{n'}$$	**SI** (24–137c)
	where P in kW, n' in rps, and M_t in N m	
	$$= \frac{63{,}000\, P}{n}$$	**US Customary units** (24–137d)
	where P in horsepower (hp), n in rpm, and M_t in lbf in	
	$$= \frac{716\, P}{n}$$	**Metric** (24–137e)
	where P in hp_m, n in rpm, and M_t in kgf m	
	$$= \frac{973\, P}{n}$$	**Metric** (24–137f)
	where P in kW, n in rpm, and M_t in kgf m	

Particular	Formula
The equation for friction torque	$$M_t = \frac{\mu\, F_r\, d}{2} \qquad (24\text{–}137\text{g})$$ For values of μ, refer to Table 24–24.
A rule of thumb used for ordinary ball and straight roller bearings	$$\frac{d+D}{2}\, n \le 500,000 \qquad \textbf{SI} \quad (24\text{–}138\text{a})$$ where d, D in mm, and n in rpm $$\left(\frac{d+D}{2}\right) n' \le 8.33 \qquad \textbf{SI} \quad (24\text{–}138\text{b})$$ where d, D in m and n' in rps

TABLE 24–24
Coefficient of friction for rolling contact bearings

Type of bearing	μ
Self-aligning bearings	0.0016–0.0066
Cylindrical roller bearings	0.0012–0.0060
Thrust ball bearings	0.0013
Angular contact ball bearings	0.0018–0.0019
Deep-groove ball bearings	0.0022–0.0042
Tapered roller bearings	0.0025–0.0083
Spherical roller bearings	0.0029–0.0071
Needle bearings	0.0045

SPEED

Effective speed

The effective speed which determines the life of the bearing is found from the relation

$$n_e = n_1 \pm n_2 \qquad (24\text{–}138\text{c})$$

where the plus sign is used when the races rotate in opposite directions and the minus sign is used when the races rotate in the same direction.

Limiting bearing speed

The limiting bearing speed when the bearing outside diameter is less than 30 mm

$$n_1 = \frac{3\, K_n}{D + 30} \qquad (24\text{–}138\text{d})$$

The limiting bearing speed when the bearing outside diameter is 30 mm and over

$$n_l = \frac{K_n}{D - 10} \qquad (24\text{–}138\text{e})$$

For values of K_n, refer to Table 24–25.

TABLE 24–25
Values of K_n to be used in Eq. (24–138)

Type of bearing	Constant K_n	
	Grease lubrication	Oil lubrication
Radial bearings		
Deep-groove bearings		
Single-row	500,000	630,000
Single-row with leeds	360,000	—
Double-row	320,000	400,000
Magneto bearings	500,000	630,000
Angular contact ball bearing		
Single-row	500,000	630,000
Single-row paired	400,000	500,000
Double-row	360,000	450,000
Self-aligning ball bearings	500,000	630,000
Self-aligning ball bearings with		
extended inner ring	250,000	320,000
Cylindrical roller bearing		
Single-row	500,000	630,000
Double row	500,000	630,000
Tapered roller bearings	320,000	400,000
Barrel roller bearings	220,000	280,000
Spherical roller bearings,		
Series 213	220,000	280,000
Thrust bearings		
Thrust ball bearings	140,000	200,000
Angular contact thrust ball		
bearings	220,000	320,000
Cylindrical roller thrust bearings	90,000	120,000
Spherical roller thrust bearings	140,000	200,000

Particular	Formula

GEAR-TOOTH LOAD

The effective tooth load which is used in design of bearings

$$F_{effg} = f_k\, f_d\, F \qquad (24\text{–}139)$$

For values of f_k and f_d, refer to Table 24–26.

The shaft load due to belt drive which is used in design of bearings

$$F_{bs} = f\, F \qquad (24\text{–}140)$$

For values of f, refer to Table 24–26.

STATIC LOADING

Stribeck equation for permissible static load

$$F_c = k d_b^2 \qquad (24\text{–}141)$$

TABLE 24–26
Value of factors f_k, f_d, and f to be used in Eqs. (24–139) and (24–140)

Particulars	Tooth load		Shaft load
	f_k	f_d	f
Gear drive			
Precision gears (errors in pitch and form < 0.025 mm)	1.05–1.1	—	—
Commercial gears (errors in pitch and form 0.025–0.125 mm)	1.1–1.3	—	—
Prime movers and driven machines			
Shock-free rotary machines, e.g., electrical machines and turbocompressors	—	1.0–1.2	—
Reciprocating engines, according to the degree of balance	—	1.2–1.5	—
Machinery subjected to heavy-shock loading, such as rolling mills	—	1.5–3.0	—
Belt drive			
Vee-belts	—	—	2.0–2.5
Single leather belts with jockey pulleys	—	—	2.5–3.0
Single leather belts, balata belts, rubber belts	—	—	4.0–5.0

Particular	Formula	
	$= 862 \times 10^6 (125)$ for hardened alloy steel balls Use a safety factor of 10	
Stribeck equation for permissible static load for ball bearing	$F_c = \dfrac{4.37\, F_r}{Z}$	(24–142)
The radial load capacity of ball bearing	$F_r = \dfrac{kZd_b^2}{4.37}$	(24–143)
	Refer to Table 24–27 for values of k.	
	$F_r = \dfrac{kZd_b^2}{5}$	(24–144)
	Refer to Table 24–27 for values of k.	
Radial load capacity of roller bearing	$F_r = \dfrac{kZld_r}{5}$	(24–145)

TABLE 24–27
Safe working values of k for average bearing life

Material	SI $\times 10^6$	USCSU[a]
For unhardened steel	3.80	550
For hardened alloy steel on flat races	6.89	700
For hardened carbon steel	4.80	1000
For hardened carbon steel on grooved races	10.34	1500
For hardened alloy steel grooved races (having radius = $0.67d_o$)	13.79	2000

[a] US Customary System units.

Particular	Formula
	where $k = 690 \times 10^6$ (100)* for carbon steel balls where $k = 48.3 \times 10^6 (7.0)$* for hardened carbon steel $= 69 \times 10^6 (10.0)$* for hardened alloy steel

Basic static load rating as per Indian Standards

RADIAL BALL BEARING The basic static load rating for radial ball bearing

$$C_o = f_o \, iZd_b^2 \cos \alpha \qquad (24\text{–}146)$$

where $f_o = 3.33 \times 10^6$ (0.5)* for self-aligning ball bearings
$= 12.26 \times 10^6$ (1.8)* for radial contact and angular contact groove ball bearing

The static equivalent load for radial ball bearings

$$P_o = X_o \, F_r + Y_o \, F_a \qquad (24\text{–}147)$$

and

$$P_o = F_r \qquad (24\text{–}148)$$

For values of X_o and Y_o, refer to Table 24–28.

RADIAL ROLLER BEARING The basic static load rating for radial roller bearings

$$C_o = f_o \, iZl_{eff} \, d_r \cos \alpha \qquad (24\text{–}149)$$
where $f_o = 21.6 \times 10^6$ (3)*

The static equivalent load for radial roller bearings under combined radial and thrust loads

$$P_o = X_o \, F_r + Y_o \, F_a \qquad (24\text{–}150)$$

and

$$P_o = F_r \qquad (24\text{–}151)$$

For factors X_o and Y_o, refer to Table 24–28.

THRUST BALL BEARING The basic static load for thrust ball bearings

$$C_{oa} = f_{oa} \, Z \, d_b^2 \sin \alpha \qquad (24\text{–}152)$$
where $f_{oa} = 49 \times 10^6$ (7)*

The static equivalent load P_{oa} for thrust ball bearing with contact angle $\alpha \neq 90°$ under combined radial and thrust loads

$$P_{oa} = F_a + 2.3 \, F_r \tan \alpha \qquad (24\text{–}153)$$

The accuracy of the formula decreases in the case of single-direction thrust bearings when

$$F_r > 0.44 \, F_a \cot \alpha.$$

THRUST ROLLER BEARING The basic static load rating for thrust roller bearings

$$C_{oa} = f_{oa} \, Zl_{eff} \, d_r \sin \alpha \qquad (24\text{–}154)$$
where $f_{oa} = 98.1 \times 10^6$ (10)*

---

*Values outside the brackets are in **SI units** (in pascals) and inside the brackets are in **US Customary units** (in kilopounds per square inch).

Particular	Formula
The static equivalent load for thrust roller bearings with contact angle $\alpha \neq 90°$ under combined radial and thrust loads	$P_{oa} = F_a + 2.3\, F_r\, \tan\, \alpha$ $\qquad$ (24–155) The accuracy of the formula decreases in the case of single-direction thrust bearings when $F_r > 0.44 F_a\, \cot\, a$

Catalogue information from FAG for the selection of bearing

The basic static load rating	$C_o = f_s\, P_o$ $\qquad$ (24–156) where f_s = index of static stressing $\qquad\quad$ = 1.2 to 2.5 for high demands $\qquad\quad$ = 0.8 to 1.2 for normal demands $\qquad\quad$ = 0.5 to 0.8 for modest demands
The equivalent static load	$P_o = X_o\, F_r + Y_o\, F_a$ $\qquad$ (24–157) For various values of factors X_o and Y_o refer to Table 24–29.

DYNAMIC LOADING

The relation between the load F and the length of life L of bearing as per experiment conducted by Palmgern	$\dfrac{L_1}{L_2} = \left(\dfrac{F_2}{F_1}\right)^m$ $\qquad$ (24–158) where m = 3 generally accepted $\qquad\quad$ = 3.333 used by Timken Engineering $\qquad\quad$ = 4 used by New Departures For various typical values of bearing life for various application, refer to Table 24–30.
The Antifriction Bearing Manufacturers Association (AFBMA) formula for the rating life of ball bearing in millions of revolutions of a bearing subjected to any other load F	$L_n = \left(\dfrac{C}{F}\right)^3$ $\qquad$ (24–159) For values of C for various types of bearings, refer to Table 24–31.
For life curves of ball bearings as per SKF and New Departure (ND) and needle-bearing	Refer to Fig. 24–45.

Basic dynamic load rating of bearings as per Indian Standards

RADIAL BALL BEARING The basic dynamic load rating for radial and angular contact ball bearing except filling-slot bearings	$C = 2.46 \times 10^6 f_c\ (i \cos \alpha)^{0.7}\ Z^{2/3}\ d_b^{1.8}$ **SI** (24–160a) $\quad = f_c (i \cos \alpha)^{0.7}\ Z^{2/3}\ d_b^{1.8}$ $\qquad$ **Metric** (24–160b)

TABLE 24–28
Values for factors X_o and Y_o for ball and roller bearings

Type of bearing		Single-row bearing[a]		Double-row bearing[b]	
		X_o	Y_o	X_o	Y_o
Radial ball bearings					
Radial contact groove ball bearing[a,c]		0.6	0.5	0.6	0.5
	$= 20°$	0.5	0.42	1	0.84
	$= 25°$	0.5	0.38	1	0.76
Angular contact groove ball bearing[d] $\quad \alpha$	$= 30°$	0.5	0.38	1	0.66
	$= 35°$	0.5	0.29	1	0.58
	$= 40°$	0.5	0.26	1	0.52
Self-aligning ball bearings		0.5	0.22 cot α	1	0.44 cot α
Radial roller bearings					
Self-aligning and tapered roller bearings, $\alpha \neq 0°$		0.5	0.22 cot α	1	0.44 cot α

[a] P_o is always $\geq F_r$.
[b] Double-row bearings are presumed to be symmetrical.
[c] Permissible maximum value of F_a/C_o depends on the bearing design (groove depth and internal clearance).
[d] For two similar single-row angular contact ball bearings, mounted "face to face" or "back to back," use the value of X_o and Y_o which apply to a double-row angular contact ball bearings. For two or more similar single row angular contact bearings mounted "in tandem," use the values of X_o and Y_o which apply to single-row angular contact ball bearing.
Source: IS 3823, 1966.

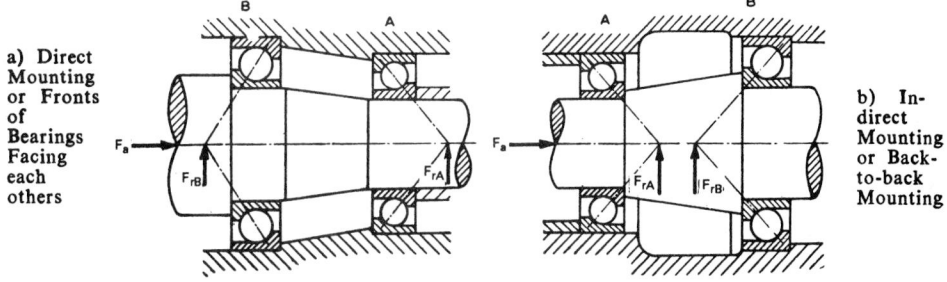

a) Direct Mounting or Fronts of Bearings Facing each others

b) Indirect Mounting or Back-to-back Mounting

FIGURE 24–46 Angular contact ball bearings mounted on a single shaft.

Particular	Formula
	For $d_b \leq 25.4$ mm
	$C = 0.57 \times 10^6 f_c \, (i \cos \alpha)^{0.7} \, Z^{2/3} \, d^{1.4}$ **SI** (24–161a)
	$\quad = 3.647 \, f_c (i \cos \alpha)^{0.7} \, Z^{2/3} \, d_b^{1.4}$ **Metric** (24–161b)
	For $d_b \geq 25.4$ mm
	For values of factor f_c, refer to Table 24–32.
The approximate value of rating life in millions of revolutions for ball bearing except filling-slot bearing	$L_n = \left(\dfrac{C}{P}\right)^3$ $\qquad$ (24–162)
The equivalent load for radial and angular contact bearing except filling-slot bearing, under combined constant radial and constant thrust loads	$P = X \, V \, F_r + Y \, F_a$ $\qquad$ (24–163) for values of X, Y, and V, refer to Table 24–33.

TABLE 24–29
Calculation of equivalent static and dynamic load

Bearing type	Series BIS	Series FAG	Equivalent load Static, P_o	Equivalent load Dynamic, P	For dimensions $C, C_o, n_{max}, X, e, Y, Y_o$, refer to Table
Deep-groove ball bearings	02	62	F_r when $F_a/F_r \leq 0.8$	F_r when $F_a/F_r \leq e$	24–42
	03	63	$0.6\,F_r + 0.5\,F_a$	$0.56\,F_r + Y\,F_a$	24–43
	04	64	when $F_a/F_r > 0.8$	when $F_a/F_r > e$	24–44
Self-aligning ball bearings	02	12			24–45
	03	13			24–46
		22	$F_r + Y_o F_a$	$X F_r + Y F_a$	24–47
		23			24–48
Single-row angular contact ball bearings	02	72B	F_r when $F_a/F_r \leq 1.9$	F_r when $F_a/F_r \leq 1.14$	24–49
	03	73B	$0.50\,F_r + 0.26\,F_a$	$0.35\,F_r + 0.57 F_a$	24–50
			when $F_a/F_r > 1.9$	when $F_a/F_r > 1.14$	
Double row angular contact ball bearings		33	$F_r + 0.58\,F_a$	$F_r + 0.66\,F_a$ when $F_a/F_r \leq 0.96$ $0.6\,F_r + 1.07\,F_a$ when $F_a/F_r > 0.95$	24–51
Cylindrical roller bearings	02	NU2			24–52
	03	NU3			24–53
	04	NU4	F_r	F_r	24–54
		NU22			24–55
		NU23			24–56
Tapered roller bearings	02, 22	322	F_r when $F_a/F_r \leq \frac{1}{2}\,Y_o$ $0.5\,F_r + Y_o\,F_a$ when $F_a/F_r > \frac{1}{2}\,Y_o$	F_r when $F_a/F_r \leq e$ $0.4\,F_r + YF_a$ when $F_a/F_r > e$	24–57
	03, 23				
Thrust ball bearings	11	511	F_a	F_a	24–58
	12	512			24–59
	13	513			24–60
	14	514			24–61

Particular	Formula
RADIAL ROLLER BEARING The basic dynamic load rating of radial roller bearings	$C = 3.5 \times 10^6\, f_c\,(i l_{eff}\,\cos\,\alpha)^{7/9}\,Z^{3/4}\,d_r^{\,29/27}$ **SI** (24–164a) $= f_c(i\,l_{eff}\,\cos\,\alpha)^{7/9}\,Z^{3/4}\,d_r^{29/27}$ **Metric** (24–164b) For values of factor f_c refer to Table 24–32.
The approximate value of the rating life in millions of revolutions for radial roller bearings	$L_n = \left(\dfrac{C}{P}\right)^{10/3}$ (24–165)
The equivalent load for self-aligning and tapered roller bearing under combined constant radial and constant thrust loads	$P = X\,V\,F_r + Y\,F_a$ (24–166) For values of X, Y, and V, refer to Table 24–33.

TABLE 24–30
Typical values of bearing life for various applications

Application	Design life, hours
Agricultural equipment	3,000–6,000
Aircraft engines	500–1,500
Automobile applications	
Race cars	500–800
Light motor cycle	600–1,200
Heavy motor cycle	1,000–2,000
Light cars	1,000–2,000
Heavy cars	1,500–2,500
Light trucks	1,500–2,500
Heavy trucks	2,000–2,500
Buses	2,000–5,000
Boat gearing units	3,000–5,000
Beater mills	20,000–30,000
Briquette presses	20,000–30,000
Domestic appliances	1,000–2,000
Electrical motors (≤ 0.5 kW)	1,000–2,000
Electrical motors (≤ 4 kW)	8,000–10,000
Electrical motors, medium	10,000–15,000
Electrical motors, large	20,000–30,000
Elevator cable sheaves	40,000–60,000
Small fans	2,000–4,000
Mine ventilation fans	40,000–50,000
Gearing units	
Automotive	600–5,000
Multipurpose	8,000–15,000
Machine tools	20,000
Ship	20,000–30,000
Rail vehicles	15,000–25,000
Heavy rolling mill	$\geq 50,000$
Grinding spindles	1,000–2,000
Locomotive axle boxes, outer bearings	20,000–25,000
Locomotive axle boxes, inner bearings	30,000–40,000
Machine tools	10,000–30,000
Mining machinery	4,000–15,000
Paper machines	50,000–80,000
Rail vehicle axle boxes	
Mining cars	5,000
Motor rail cars	16,000–20,000
Open-pit mining cars	20,000–25,000
Streetcars	20,000–25,000
Passenger cars	26,000
Freight cars	35,000
Rolling mills	
Small cold mills	5,000–6,000
Large multipurpose mills	8,000–10,000
Gear drives	$\geq 50,000$
Ship gear drives	20,000–30,000
Propeller thrust bearings	15,000–25,000
Propeller shaft bearings	$\geq 80,000$

TABLE 24-31
Values of C for various types of bearings

SAE no.	Double-row self-aligning — Light 200 kgf	N	Medium 300 kgf	N	Heavy 400 kgf	N	Single-row deep groove — Light 200 kgf	N	Medium 300 kgf	N	Heavy 400 kgf	N	Double-row deep groove — Light 200 kgf	N	Medium 300 kgf	N	Heavy 400 kgf	N	Single-row angular contact — Light 200 kgf	N	Medium 300 kgf	N	Heavy 400 kgf	N
00	392	3,842					340	3,332	662	6,488			553	5,419										
01	415	4,067	680	6,664			517	5,067	817	8,007			844	8,271										
02	567	5,557	739	7,242			567	5,557	889	8,712			926	9,075										
03	635	6,323	962	9,428			726	7,115	1,071	10,596			1,179	11,554	8,378	82,104								
04	830	8,134	1,016	9,957			980	9,604	1,447	14,181			1,610	15,778	2,041	20,002					1,225	12,005		
05	1,016	9,957	1,474	14,445			1,052	10,309	1,656	16,229	2,858	28,008	1,724	16,895	2,722	26,676			1,034	10,133	1,724	16,895	2,812	27,558
06	1,406	13,800	1,837	18,002	2,767	27,117	1,474	14,445	2,177	21,335	3,402	33,040	2,404	23,559	3,538	34,672			1,451	14,220	2,177	21,335	3,334	32,673
07	1,530	1,501	2,177	21,334	3,198	31,340	1,950	19,110	2,586	25,343	4,309	42,288	3,221	31,566	4,082	40,004			1,928	18,894	2,586	25,343	4,082	40,004
08	1,928	18,894	2,767	27,296	3,651	35,880	2,223	21,830	3,152	30,890	4,990	48,902	3,629	35,564	5,081	49,794			2,313	22,367	3,152	30,890	4,808	47,118
09	2,155	21,119	3,470	34,006	4,554	44,629	2,495	24,451	4,082	40,004	5,988	58,682	4,082	40,004	6,377	62,495			2,586	25,343	4,082	40,004	5,851	57,340
10	2,313	22,367	3,924	38,455	5,443	48,341	2,654	26,009	4,717	46,227	6,622	64,896	4,377	42,895	7,711	75,568			2,722	26,676	4,717	46,227	6,486	63,563
11	2,812	27,557	4,717	46,227	5,761	56,448	3,334	32,674	5,443	53,341	7,848	76,910	5,353	52,459	8,754	85,789			3,334	32,673	5,443	53,341	7,529	73,784
12	3,175	31,115	5,443	58,341	6,940	68,012	3,992	39,122	6,215	60,901	8,618	84,456	6,486	63,563	9,979	97,794			3,992	39,122	6,215	60,907	8,301	81,350
13	3,470	33,006	5,760	56,448	7,393	72,451	4,377	42,895	6,940	68,012	9,435	92,463	7,076	69,345	11,340	111,132			4,536	44,423	6,940	68,012	9,072	88,906
14	3,856	37,788	6,940	68,012	9,072	88,906	4,717	46,227	7,711	75,568	11,793	115,671	7,711	75,568	12,474	122,245			4,990	48,902	7,711	75,568	11,566	113,346
15	4,218	41,336	7,393	72,451	10,342	101,352	4,970	48,902	8,437	82,683	12,701	124,470	8,165	80,071	13,835	135,583			5,171	50,676	8,618	84,456	12,247	120,021
16	4,445	43,561	8,165	80,017	11,567	113,346	5,534	54,223	9,435	92,463			8,890	87,122	15,196	147,921	6,078	59,564	5,761	56,448	9,435	92,463		
17	5,353	52,459	9,072	88,909	12,474	122,245	6,215	60,907	10,160	99,568			9,253	90,679	15,196	147,921	6,350	62,230	6,486	63,563	10,360	101,528		
18	5,351	57,339	10,360	101,528	14,061	137,798	6,622	64,896	11,113	108,907			10,524	103,135	16,103	157,809	7,983	78,233	7,529	73,784	11,113	108,707		
19			11,566	113,346			7,983	78,233	12,020	117,796			12,928	126,694			9,979	97,794	7,983	78,233	12,020	117,796		
20	7,393	72,451	12,474	122,245			8,890	87,122	13,835	135,583			14,515	142,247			12,474	122,245	8,890	87,122	14,742	144,472		
21			14,061	137,798			9,797	96,011	14,742	144,472							12,474	122,245	10,886	106,683	15,649	153,369		
22	9,253	90,679	15,196	148,921			10,886	106,683	17,010	166,698							13,835	135,583	12,474	122,245	17,690	173,362		
24																			14,061	137,798				
26																			16,103	157,809	21,772	212,596		
28																								
30																			17,690	173,362	26,535	260,043		
32																			18,824	184,471				
34																			19,504	191,139				

TABLE 24–32
Values of factor f_c for radial ball and roller bearings

$\dfrac{d_{br}{}^a \, \cos \alpha}{d_m{}^b}$	Ball bearings					Roller bearings
	Radial contact			Angular contact		
	Single row		Double row	Single and double row		
	Grooved	Separable	Grooved	Grooved	Self-aligning	
0.01						4.66
0.02						5.45
0.03						5.96
0.04						6.35
0.05	4.76	1.65	4.51	4.76	1.76	6.63
0.06	5.00	1.77	4.74	5.00	1.90	6.91
0.07	5.21	1.89	4.94	5.21	2.03	7.08
0.08	5.39	1.99	5.11	5.39	2.15	7.31
0.09	5.54	2.10	5.24	5.54	2.27	7.42
0.10	5.66	2.19	5.37	5.66	2.38	7.53
0.12	5.86	2.39	6.55	5.86	2.61	7.76
0.14	6.00	2.58	5.68	6.00	2.82	7.87
0.16	6.08	2.76	5.76	6.08	3.03	7.92
0.18	6.11	2.94	5.79	6.11	3.23	7.98
0.20	6.11	3.11	5.79	6.11	3.42	7.98
0.22	6.08	3.27	5.76	6.08	3.59	7.92
0.24	6.01	3.43	5.70	6.01	3.75	7.87
0.26	5.93	3.58	5.62	5.93	3.90	7.76
0.28	5.83	3.72	5.52	5.83	4.02	7.64
0.30	5.71	3.86	5.41	5.71	4.11	7.53
0.32	5.58	3.97	5.30	5.58	4.18	
0.34	5.43	4.06	5.15	5.43	4.20	
0.36	5.27	4.12	5.00	5.27	4.21	
0.38	5.10	4.15	4.84	5.10	4.18	
0.40	4.72	4.17	4.67	4.92	4.12	

$^a d_{br} = d_b$ = diameter of ball, mm; $d_{br} = d_r$ = diameter of roller, mm.
$^b d_m$ = pitch diameter of ball or roller set, mm.

Notes: (1) For values of $d_{br} \cos \alpha/d_m$ other than those given in this table, f_o is obtained by linear interpolation.

(2) When calculating the basic load rating for a unit consisting of two similar single row radial contact ball bearings in a duplex mounting, the unit shall be considered as one double-row radial contact ball bearing.

The values of f_c as given in this table shall apply to bearings whose raceways have a cross-sectional radius not larger than the following:

In radial contact and angular contact groove ball bearing inner rings	52 percent of the ball diameter
In radial contact and angular contact groove ball bearing outer rings	53 percent of the ball diameter
In self-aligning ball bearing inner rings	53 percent of the ball diameters

The basic load rating is not necessarily increased by the use of smaller groove radius but is reduced by the use of radii larger than those given above.

(3) When calculating the basic load rating for a unit consisting of two similar single row angular contact ball bearings in a duplex mounting, "face to face" or "back to back," the unit shall be considered as one double-row angular contact ball bearing.

(4) When calculating the basic load rating for a unit consisting of two similar single-row angular contact ball bearings mounted "in tandem," properly manufactured and mounted for equal load distribution, the rating of the unit is the number of bearings to the 0.7 power times the load rating of a single-row ball bearing. If for some technical reason the unit may be treated as a number of individually interchangeable single-row bearings, this shall not apply.

Source: IS 3824 (Parts I and II), 1966.

TABLE 24-33
Values of factors V, X, and Y

Bearing type	iF_a/C_a [a]	F_a/Zd_{br}^2 [a]	F_a/C_o	F_a/iZd_{br}^2 [a]	e [b]	V Rotary	V Stationary	X Single row [c] $\frac{F_a}{VF_r}\le e^g$	X Single row [c] $\frac{F_a}{VF_r}> e$	X Double row [d] $\frac{F_a}{VF_r}\le e$	X Double row [d] $\frac{F_a}{VF_r}> e$	Y Single row [e] $\frac{F_a}{VF_r}\le e^g$	Y Single row [e] $\frac{F_a}{VF_r}> e$	Y Double row [d] $\frac{F_a}{VF_r}\le e$	Y Double row [d] $\frac{F_a}{VF_r}> e$
Ball bearings Nonfilling slot assembly, radial contact, groove bearing[f]			0.014	0.018	0.19	1	1.2[c]	1	0.56	1	0.56	0	2.30		2.30
			0.028	0.035	0.22								1.99		1.99
			0.056	0.070	0.26								1.71		1.71
			0.084	0.110	0.28								1.55		1.55
			0.110	0.140	0.30								1.45		1.45
			0.170	0.210	0.34								1.31		1.31
			0.280	0.350	0.38								1.15		1.15
			0.420	0.530	0.42								1.04		1.04
			0.560	0.700	0.44								1.00		1.00
Angular contact groove ball bearing[f] $\alpha = 5°$	0.014	0.018			0.23	1	1.2[c]	*	*	1	0.56	*	*	2.78	3.74
	0.028	0.035			0.26									2.40	3.23
	0.056	0.070			0.30									2.07	2.78
	0.085	0.110			0.34									1.87	2.52
	0.110	0.140			0.36									1.75	2.36
	0.170	0.210			0.40									1.58	2.13
	0.280	0.350			0.45									1.39	1.87
	0.430	0.530			0.50									1.26	1.69
	0.570	0.700			0.52									1.21	1.63
$\alpha = 10°$	0.014	0.018			0.29	1	1.2[c]	1	0.46	1	0.78		1.88	2.18	3.06
	0.028	0.035			0.32								1.71	1.98	2.78
	0.057	0.070			0.36								1.52	1.76	2.47
	0.086	0.110			0.38								1.41	1.63	2.29
	0.110	0.140			0.40								1.34	1.55	2.18
	0.170	0.210			0.44								1.23	1.42	2.00
	0.290	0.350			0.49								1.10	1.27	1.79
	0.430	0.530			0.54								1.01	1.17	1.64
	0.570	0.700			0.54								1.00	1.16	1.63
$\alpha = 15°$	0.015	0.018			0.38	1	1.2[c]	1	0.44	1	0.72		1.47	1.65	2.39
	0.029	0.035			0.40								1.40	1.57	2.28
	0.058	0.070			0.43								1.30	1.46	2.11
	0.087	0.110			0.46								1.23	1.38	2.00
	0.120	0.140			0.47								1.19	1.34	1.93
	0.170	0.210			0.50								1.12	1.26	1.82
	0.290	0.350			0.55								1.02	1.14	1.66
	0.440	0.530			0.56								1.00	1.12	1.63
	0.580	0.700			0.56								1.00	1.12	1.63
$\alpha = 20°$					0.57	1	1.2[c]	1	0.43	1	0.70		1.00	1.09	1.63
$\alpha = 25°$					0.68	1	1.2	1	0.41	1	0.67		0.87	0.92	1.41
$\alpha = 30°$					0.80	1	1.2	1	0.39	1	0.63		0.76	0.78	1.24
$\alpha = 35°$					0.95	1	1.2	1	0.37	1	0.60		0.66	0.66	1.07
$\alpha = 40°$					1.14	1	1.2	1	0.35	1	0.57		0.57	0.55	0.93

* [g] For this type, use X, Y, and e values applicable to single-row, nonfilling slot assembly, radial contact, groove ball bearings

24.88

TABLE 24-33
Values of factors V, X, and Y (Cont.)

Bearing type	$\dfrac{iF_a}{C_a}$	$\dfrac{F_a}{Zd_{br}^2}$ [a]	$\dfrac{F_a}{C_o}$	$\dfrac{F_a}{iZd_{br}^2}$ [a]	e [b]	V Rotary	V Stationary	X [b] Single row c $\frac{F_a}{VF_r} \leq e^g$	X [b] Single row c $\frac{F_a}{VF_r} > e$	X [b] Double row d $\frac{F_a}{VF_r} \leq e$	X [b] Double row d $\frac{F_a}{VF_r} > e$	Y [b] Single row c $\frac{F_a}{VF_r} \leq e^g$	Y [b] Single row c $\frac{F_a}{VF_r} > e$	Y [b] Double row d $\frac{F_a}{VF_r} \leq e$	Y [b] Double row d $\frac{F_a}{VF_r} > e$
Self-aligning ball bearings					1.5 tan α	1	1	1	0.40	1	0.65		0.4 cot α	0.42 cot α	0.65 cot α
Single-row radial contact separable ball bearings (magneto bearings)					0.2	1	1	1	0.50				2.5		
Roller bearings [h] Self-aligning roller bearings and tapered roller bearings					1.5 tan α	1	1.2	1	0.4	1	0.67	0	0.4 cot α	0.45 cot α	0.67 cot α

[a] $d_{br} = d_b$ = diameter of ball, mm : $d_{br} = d_r$ = diameter of roller, mm.

[b] Values of X, Y, and e for a load or a contact angle other than those shown in this table are obtained by linear interpolation.

[c] For single-row bearings, when $(F_a/VF_r) \leq e$, use $X = 1$ and $Y = 0$: (1) when calculating the equivalent load for a unit consisting of two similar single row angular contact ball bearings in duplex mounting, "face to face" or "back to back," the units shall be considered as one of double-row angular contact ball bearing; (2) when calculating the equivalent load for a unit consisting of two or more single-row radial or angular contact ball bearings mounted "in tandem," X and Y values for single-row ball bearings shall be used.

[d] Double-row bearings are presumed to be symmetrical.

[e] Because experimental data are incomplete, no correct value of factor V for radial and angular contact groove ball bearings with inner ring stationary in relation to the load can be stated. The values shown in the table are, however, well on the safe side.

[f] Permissible maximum value of F_a/C_o depends on the bearing design.

[g] This is applicable to only single-row ball bearings when $F_a/VF_r > e$.

[h] Applicable only to roller bearings.

[j] C_o is the static basic load rating.

Source: IS 3824, 1966.

Particular	Formula

THRUST BALL BEARING The basic load rating for a single-row single- and double-direction thrust ball bearing

$$(C_a)_{\alpha=90°} = 2.64 \times 10^6 \, f_c \, Z^{2/3} \, d_b^{1.8} \left.\right\} \begin{array}{l} \textbf{SI} \qquad (24\text{–}167a) \\ \text{for } d_b \leq 25.4 \text{ mm} \\ \textbf{Metric} \quad (24\text{–}167b) \end{array}$$

$$= f_c \, Z^{2/3} \, d_b^{1.8}$$

$$(C_a)_{\alpha \neq 90°} = \left[2.64 \times 10^6 \, f_c \, (\cos \alpha)^{0.7} \, \tan \alpha \right.$$

$$\left. Z^{2/3} \, d_b^{1.8} \right] \qquad \qquad \textbf{SI} \quad (24\text{–}168a)$$

$$= f_c \, (\cos \alpha)^{0.7} \, \tan \alpha \, Z^{2/3} \, d_b^{1.8} \qquad \textbf{Metric}$$

for $d_b \leq 25.4$ mm $\qquad\qquad\qquad\qquad (24\text{–}168b)$

$$(C_a)_{\alpha=90°} = 5.67 \times 10^5 \, f_c \, Z^{2/3} \, d^{1.4} \left.\right\} \begin{array}{l} \textbf{SI} \quad (24\text{–}169a) \\ \text{for } d_b > 225.4 \text{ mm} \end{array}$$

$$= 3.647 \, f_c \, Z^{2/3} \, d_b^{1.4} \qquad \right\} \textbf{Metric} \quad (24\text{–}169b)$$

$$(C_a)_{\alpha \neq 90°} = 5.67 \times 10^5 \, f_c \, (\cos \alpha)^{0.7} \, \tan \alpha \, Z^3 \, d_b^{1.4}$$

$$\textbf{SI} \quad (24\text{–}170a)$$

$$= 3.647 \, f_c \, (\cos \alpha)^{0.7} \, \tan \alpha \, Z^{2/3} \, d_b^{1.4}$$

$$\textbf{Metric} \quad (24\text{–}170b)$$

for $d_b > 225.4$ mm

For various values of f_c refer to Table 24–34.

The basic load rating for thrust ball bearing with two or more rows of similar balls carrying load in the same direction

$$C_a = K_{zb} \qquad\qquad\qquad\qquad (24\text{–}171)$$

$$\text{where } K_{zb} = (Z_1 + Z_2 + \cdots + Z_n)\Big[(Z_1/C_{a1})^{10/3}$$

$$+ (Z_2/C_{a2})^{10/3} + \cdots + (Z_n/C_{an})^{10/3}\Big]^{-10/3}$$

The rating life in million revolutions of a thrust ball bearing

$$L_n = \left(\frac{C_a}{P_a}\right)^3 \qquad\qquad\qquad (24\text{–}172)$$

The equivalent thrust load for thrust ball bearing with $\alpha \neq 90°$ under combined thrust and constant radial loads

$$P_a = X \, F_r + Y F_a \qquad\qquad\qquad (24\text{–}173)$$

For values of X and Y, refer to Table 24–35.

THRUST ROLLER BEARING The basic load rating for single-row, single- and double-direction thrust roller bearings

$$(C_a)_{\alpha=90°} = 3.53 \times 10^6 \, f_c \, l_{eff}^{7/9} \, Z^{3/4} \, d_r^{29/27} \qquad \textbf{SI}$$

$$(24\text{–}174a)$$

$$= f_c \, l_{eff}^{7/9} \, Z^{3/4} \, d_r^{29/27} \quad \textbf{Metric} \quad (24\text{–}174b)$$

$$(C_a)_{\alpha \neq 90°} = \left[3.53 \times 10^6 \, f_c \, (l_{eff} \cos \alpha)^{7/9} \, \tan \alpha \right.$$

$$\left. \times Z^{3/4} \, d_r^{29/27} \right] \qquad \textbf{SI} \quad (24\text{–}175a)$$

$$= f_c \, (l_{eff} \cos \alpha)^{7/9} \, \tan \alpha \, Z^{3/4} \, d_r^{29/27}$$

$$\textbf{Metric} \quad (24\text{–}175b)$$

TABLE 24–34
Various values of f_c for thrust ball and roller bearings

$\dfrac{d_{br}\,^a}{d_m\,^b}$	f_c $\alpha = 90°$ Ball	f_c $\alpha = 90°$ Roller	$\dfrac{d_{br}\,^a \cos \alpha}{d_m\,^b}$	f_c Ball $\alpha = 45°$	f_c Ball $\alpha = 60°$	f_c Ball $\alpha = 75°$	f_c Roller $\alpha = 50°$
0.01	3.74	10.12	0.01	4.29	3.99	3.81	10.70
0.02	4.61		0.02	5.27	4.90	4.68	
0.03	5.21		0.03	5.94	5.53	5.27	
0.04	5.68		0.04	6.45	6.00	5.72	
0.05	6.07	14.05	0.05	6.86	6.39	6.09	14.62
0.06	6.41		0.06	7.20	6.70	6.39	
0.07	6.71		0.07	7.49	6.97	6.65	
0.08	6.94		0.08	7.74	7.20	6.87	
0.09	7.24		0.09	7.95	7.40	7.05	
0.10	7.47	16.86	0.10	8.12	7.56	7.21	16.86
0.12	7.89		0.12	8.40	7.82		
0.14	8.27		0.14	8.53	7.98		
0.15	8.60	18.55	0.15	8.68			17.42
0.16	8.91		0.16	8.72	8.08		
0.18	9.20		0.18	8.71	8.12		
0.20	9.47	19.67	0.20	8.66	8.11		17.42
0.22	9.72		0.22	8.56			
0.24	9.95		0.24	8.44			
0.25	10.20	20.23	0.25	8.29			16.86
0.26	10.40		0.26	8.11			
0.28	10.60		0.28				
0.30	10.80	21.36	0.30				
0.32							
0.34							

[a] $d_{br} = d_b$ = diameter of ball, mm; $d_{br} = d_r$ diameter of roller, mm.
[b] d_m = pitch diameter of roller set, mm
Source: IS 3824 (Parts III and IV), 1966.

TABLE 24–35
Various values of X and Y for thrust ball and roller bearings

Type of bearings		X Single-direction bearings When $F_a/F_r > e$	X Double-direction bearings[a] When $F_a/F_r \leq e$	X Double-direction bearings[a] When $F_a/F_r > e$	Y Single-direction bearings When $F_a/F_r > e$	Y Double-direction bearings When $F_a/F_r \leq e$	Y Double-direction bearings When $F_a/F_r > e$	e
Thrust ball bearings	$\alpha = 45°$	0.66	1.18	0.66	1	0.59	1	1.25
	$\alpha = 60°$	0.92	1.90	0.92	1	0.54	1	2.17
	$\alpha = 75°$	1.66	3.89	1.66	1	0.52	1	4.67
	$\alpha = 90°$ [b]							
Self-aligning and tapered thrust roller bearings $\alpha \neq 90°$ [b]		$\tan \alpha$	$1.5 \tan \alpha$	$\tan \alpha$	1	0.67	1	$1.5 \tan \alpha$

[a] Double direction bearings are presumed to be symmetrical.
[b] For $\alpha = 90$ deg, $F_r = 0$, and $Y = 1$.
Source: IS 3824 (Parts III and IV), 1960.

Particular	Formula
	For values of factor f_c, refer to Table 24–34.
The basic load rating for thrust roller bearing with two or more rows of rollers carrying load in the same direction	$C_a = K_{zr}$ (24–176) where $K_{zr} = (Z_1\, l_{eff1} + Z_2\, l_{eff2} + \cdots + Z_n\, l_{effn})$ $\times \left[(Z_1\, l_{eff1}/C_{a1})^{9/2} + (Z_2\, l_{eff2}/C_{a2})^{9/2} \right.$ $\left. + \cdots + (Z_n\, l_{effn}/C_{an})^{9/2} \right]^{-2/9}$
The approximate magnitude of the rating life in millions of revolutions for thrust roller bearing	$L_n = \left(\dfrac{C_a}{P_a} \right)^{10/3}$ (24–177)
The equivalent thrust load for thrust roller bearing when $\alpha \neq 90°$ under combined constant thrust and constant radial load	$P_a = X\, F_r + Y\, F_a$ (24–178) For values of X and Y, refer to Table 24–35.

Catalogue information from FAG for the selection of bearing

The basic dynamic load rating	$C = \dfrac{f_L}{f_n}\, P$ (24–179) For values of f_L and f_n refer to Tables 24–36 to 24–38 and Fig. 24–47.
The equivalent dynamic load for deep-groove ball bearings with increased radial clearance	$P = X F_r + Y F_a$ (24–180) For values of X and Y, refer to Table 24–29 and also bearing tables given in *FAG* catalog. For values of X and Y of deep-groove ball bearings with increasing radial clearance, refer to Table 24–39.
The empirical values of speed factor f_n can be obtained from equation	$f_n = \sqrt[3]{\dfrac{100}{3n}}$ for ball bearing (24–184) $= \sqrt[3/10]{\dfrac{100}{3n}}$ for ball bearing (24–185)

THRUST BEARING

Minimum thrust load for ball bearings	$F_a = M \left(\dfrac{n}{100} \right)^2$ (24–186)

TABLE 24-36
Index f_L of dynamic stressing

Application	f_L
Motor vehicles	
Motorcycles	1.4–1.9
Light cars	1.6–2.1
Heavy cars	1.7–2.2
Light trucks or lorries	1.7–2.2
Heavy trucks or lorries	2.0–2.6
Buses	2.0–2.6
Tractors	1.6–2.2
Tracked vehicles	2.1–2.7
Electric motors	
For household appliances	1.5–2.0
Small standard motors	2.5–3.5
Medium-sized standard cars	3.0–4.0
Large motors	3.5–4.5
Traction motors	3.0–4.0
Railbound vehicles	
Axle boxes for haulage trolleys	3.0–4.0
Trams	4.5–5.5
Railway coaches	4.0–5.0
Freight cars	3.5–4.0
Overburden removal cars	3.5–4.0
Outer bearings of locomotives	4.0–5.5
Inner bearings of locomotives	4.5–5.5
Gears	3.5–4.5
Rolling mills	
Neck bearings	2.0–2.5
Gears	3.0–5.0
Ship building	
Ship propeller thrust blocks	2.9–3.6
Ship propeller shaft bearings	6.0
Large marine gears	2.6–4.0
General engineering	
Small universal gears	2.5–3.5
Medium-sized universal gears	3.0–4.0
Small fans	2.5–3.5
Medium-sized fans	3.0–4.5
Large fans	4.5–5.5
Centrifugal pumps	2.5–4.5
Centrifuges	3.0–4.0
Winding cable sheaves	4.5–5.0
Belt conveyor idlers	3.0–4.5
Conveyor drums	4.5–5.5
Shovels and reclaimers	6.0
Crushers	3.0–3.5
Beater mills	3.5–4.5
Tube mills	6.0
Vibrating screens	2.5–2.8
Vibrating rolls and large out-of-balance exciters	1.6–2.0
Vibrators	1.0–1.5
Briquette presses	4.5–5.0

TABLE 24-36
Index f_L of dynamic stressing (*Cont.*)

Application	f_L
Large mechanical stirrers	3.5–4.0
Rotary furnace rollers	4.5–5.0
Flywheels	3.4–4.0
Printing machines	4.0–4.5
Papermaking machines	
Wet sections	5.0–6.0
Dry sections	5.0–6.0
Refiners	4.6–4.5
Calenders	4.0–4.5
Centrifugal casting machines	3.4–4.0
Textile machines	3.6–4.7
Machine tools	
Lathes, boring and milling machines	2.7–4.5
Grinding, lapping, and polishing machines	2.7–4.5
Woodworking machines	
Milling cutters and cutter shafts	3.0–4.0
Saw mills (con rods)	2.8–3.3
Machines for working of wood and plastics	3.0–4.0

TABLE 24–37
Speed factor f_n for ball bearings

Speed, n, rpm	Speed factor, f_n	Speed, n rpm	Speed factor, f_n	Speed, n, rpm	Speed factor, f_n	Speed, n, rpm	Speed factor, f_n	Speed, n, rpm	Speed factor, f_n	Speed, n, rpm	Speed factor, f_n
10	1.494	60	0.822	250	0.511	900	0.333	4000	0.203	15,000	0.131
11	1.447	62	0.813	260	0.504	920	0.331	4100	0.201	15,500	0.129
12	1.405	64	0.805	270	0.498	940	0.329	4200	0.199	16,000	0.128
13	1.369	66	0.797	280	0.492	960	0.326	4300	0.198	16,500	0.126
14	1.335	68	0.788	290	0.487	980	0.324	4400	0.196	17,000	0.125
15	1.305	70	0.781	300	0.481	1000	0.322	4500	0.195	17,500	0.124
16	1.277	72	0.774	310	0.476	1050	0.317	4600	0.193	18,000	0.123
17	1.252	74	0.767	320	0.471	1100	0.312	4700	0.192	18,500	0.122
18	1.228	76	0.760	330	0.466	1150	0.302	4800	0.191	19,000	0.121
19	1.206	78	0.753	340	0.461	1200	0.303	4900	0.190	19,500	0.120
20	1.186	80	0.747	350	0.457	1250	0.299	5000	0.188	20,000	0.119
21	1.166	82	0.741	360	0.453	1300	0.295	5200	0.186	21,000	0.117
22	1.148	84	0.735	370	0.448	1350	0.291	5400	0.183	22,000	0.115
23	1.132	86	0.729	380	0.444	1400	0.288	5600	0.181	23,000	0.113
24	1.116	88	0.724	390	0.441	1450	0.284	5800	0.179	24,000	0.112
25	1.100	90	0.718	400	0.437	1500	0.281	6000	0.177	25,000	0.110
26	1.089	92	0.713	410	0.433	1550	0.278	6200	0.175	26,000	0.109
27	1.073	94	0.708	420	0.430	1600	0.275	6400	0.173	27,000	0.107
28	1.060	96	0.703	430	0.426	1650	0.272	6600	0.172	28,000	0.106
29	1.048	98	0.698	440	0.423	1700	0.270	6800	0.170	29,000	0.105
30	1.036	100	0.693	450	0.420	1750	0.267	7000	0.168	30,000	0.104
31	1.025	105	0.682	460	0.417	1800	0.265	7200	0.167		
32	1.014	110	0.672	470	0.414	1850	0.262	7400	0.165		
33	1.003	115	0.662	480	0.411	1900	0.260	7600	0.164		
34	0.993	120	0.652	490	0.408	1950	0.258	7800	0.162		
35	0.984	125	0.644	500	0.406	2000	0.255	8000	0.161		
36	0.975	130	0.635	520	0.400	2100	0.251	8200	0.160		
37	0.966	135	0.627	540	0.395	2200	0.247	8400	0.158		
38	0.958	140	0.620	560	0.390	2300	0.244	8600	0.157		
39	0.949	145	0.613	580	0.386	2400	0.240	8800	0.156		
40	0.941	150	0.606	600	0.382	2500	0.237	9000	0.155		
41	0.933	155	0.599	620	0.378	2600	0.234	9200	0.154		
42	0.926	160	0.593	640	0.374	2700	0.231	9400	0.153		
43	0.919	165	0.586	660	0.370	2800	0.228	9600	0.152		
44	0.912	170	0.581	680	0.366	2900	0.226	9800	0.150		
45	0.905	175	0.575	700	0.363	3000	0.223	10,000	0.149		
46	0.898	180	0.570	720	0.359	3100	0.221	10,500	0.147		
47	0.892	185	0.565	740	0.356	3200	0.218	11,000	0.145		
48	0.885	190	0.560	760	0.353	3300	0.216	11,500	0.143		
49	0.880	195	0.555	780	0.350	3400	0.214	12,000	0.141		
50	0.874	200	0.550	800	0.347	3500	0.212	12,500	0.139		
52	0.863	210	0.541	820	0.344	3600	0.210	13,000	0.137		
54	0.851	220	0.533	840	0.341	3700	0.208	13,500	0.135		
56	0.841	230	0.525	860	0.339	3800	0.206	14,000	0.134		
58	0.831	240	0.518	880	0.336	3900	0.205	14,500	0.132		

TABLE 24–38
Speed factor f_n for roller bearings

Speed, n, rpm	Speed factor, f_n	Speed, n rpm	Speed factor, f_n	Speed, n, rpm	Speed factor, f_n	Speed, n, rpm	Speed factor, f_n	Speed, n, rpm	Speed factor, f_n	Speed, n, rpm	Speed factor, f_n
10	1.435	60	0.838	250	0.546	900	0.372	4000	0.238	15,000	0.160
11	1.395	62	0.830	260	0.540	920	0.370	4100	0.236	15,500	0.158
12	1.359	64	0.822	270	0.534	940	0.367	4200	0.234	16,000	0.157
13	1.326	66	0.815	280	0.528	960	0.365	4300	0.233	16,500	0.156
14	1.297	68	0.807	290	0.523	980	0.363	4400	0.231	17,000	0.154
15	1.271	70	0.800	300	0.517	1000	0.361	4500	0.230	17,500	0.153
16	1.246	72	0.794	310	0.512	1050	0.355	4600	0.228	18,000	0.152
17	1.224	74	0.787	320	0.507	1100	0.350	4700	0.227	18,500	0.150
18	1.203	76	0.781	330	0.503	1150	0.346	4800	0.225	19,000	0.149
19	1.184	78	0.775	340	0.498	1200	0.341	4900	0.224	19,500	0.148
20	1.166	80	0.769	350	0.494	1250	0.337	5000	0.222	20,000	0.147
21	1.149	82	0.763	360	0.490	1300	0.333	5200	0.220	21,000	0.145
22	1.133	84	0.758	370	0.486	1350	0.329	5400	0.217	22,000	0.143
23	1.118	86	0.753	380	0.482	1400	0.326	5600	0.215	23,000	0.141
24	1.104	88	0.747	390	0.478	1450	0.322	5800	0.213	24,000	0.139
25	1.090	90	0.742	400	0.475	1500	0.319	6000	0.211	25,000	0.137
26	1.077	92	0.737	410	0.471	1550	0.316	6200	0.209	26,000	0.136
27	1.065	94	0.733	420	0.467	1600	0.313	6400	0.207	27,000	0.134
28	1.054	96	0.728	430	0.464	1650	0.310	6600	0.205	28,000	0.133
29	1.043	98	0.724	440	0.461	1700	0.307	6800	0.203	29,000	0.131
30	1.032	100	0.719	450	0.458	1750	0.305	7000	0.201	30,000	0.130
31	1.022	105	0.709	460	0.455	1800	0.302	7200	0.199		
32	1.012	110	0.699	470	0.452	1850	0.300	7400	0.198		
33	1.003	115	0.690	480	0.449	1900	0.297	7600	0.196		
34	0.994	120	0.681	490	0.447	1950	0.295	7800	0.195		
35	0.986	125	0.673	500	0.444	2000	0.293	8000	0.137		
36	0.977	130	0.665	520	0.439	2100	0.289	8200	0.192		
37	0.969	135	0.657	540	0.434	2200	0.285	8400	0.190		
38	0.962	140	0.650	560	0.429	2300	0.281	8600	0.189		
39	0.954	145	0.643	580	0.425	2400	0.274	8800	0.188		
40	0.947	150	0.637	600	0.420	2500	0.274	9000	0.187		
41	0.940	155	0.631	620	0.416	2600	0.271	9200	0.185		
42	0.933	160	0.625	640	0.412	2700	0.268	9400	0.184		
43	0.927	165	0.619	660	0.408	2800	0.265	9600	0.183		
44	0.920	170	0.613	680	0.405	2900	0.262	9800	0.182		
45	0.914	175	0.608	700	0.401	3000	0.259	10,000	0.181		
46	0.908	180	0.603	720	0.398	3100	0.257	10,500	0.178		
47	0.902	185	0.598	740	0.395	3200	0.254	11,000	0.176		
48	0.896	190	0.593	760	0.391	3300	0.252	11,500	0.173		
49	0.891	195	0.589	780	0.388	3400	0.250	12,000	0.171		
50	0.896	000	0.584	800	0.385	3500	0.248	12,500	0.169		
52	0.875	210	0.576	820	0.383	3600	0.246	13,000	0.167		
54	0.865	220	0.568	840	0.380	3700	0.243	13,500	0.165		
56	0.856	230	0.560	860	0.377	3800	0.242	14,000	0.163		
58	0.847	240	0.553	880	0.375	3900	0.240	14,500	0.162		

TABLE 24–39
Values of X and Y for deep groove ball bearing with increase in radial clearance

$\dfrac{F_a}{C_o}$		Standard clearance					Clearance $C3^a$					Clearance $C4^a$			
		$\dfrac{F_a}{F_r} \le e$		$\dfrac{F_a}{F_r} > e$			$\dfrac{F_a}{F_r} \le e$		$\dfrac{F_a}{F_r} > e$			$\dfrac{F_a}{F_r} \le e$		$\dfrac{F_a}{F_r} > e$	
	e	X	Y	X	Y	e	X	Y	X	Y	e	X	Y	X	Y
0.025	.22	1	0	0.56	2.0	.31	1	0	0.46	1.75	.40	1	0	0.44	1.42
0.04	.24	1	0	0.56	1.8	.33	1	0	0.46	1.62	.42	1	0	0.44	1.36
0.07	.27	1	0	0.56	1.6	.36	1	0	0.46	1.46	.44	1	0	0.44	1.27
0.13	.31	1	0	0.56	1.4	.41	1	0	0.46	1.30	.48	1	0	0.44	1.16
0.25	.37	1	0	0.56	1.2	.46	1	0	0.46	1.14	.53	1	0	0.44	1.05
0.5	.40	1	0	0.56	1.0	.54	1	0	0.46	1.00	.56	1	0	0.44	1.00

[a] $C3$ and $C4$ indicate a radial clearance that is larger than normal.

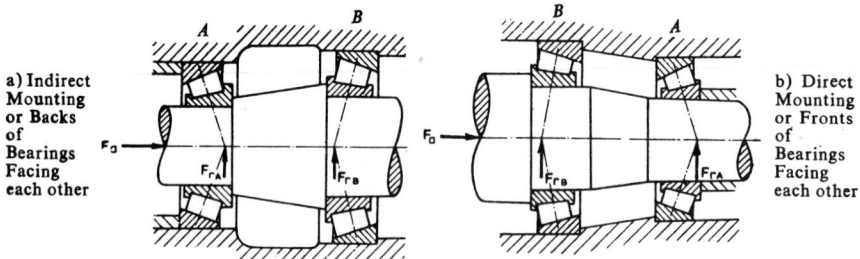

a) Indirect Mounting or Backs of Bearings Facing each other

b) Direct Mounting or Fronts of Bearings Facing each other

FIGURE 24–47 Two taper roller bearings mounted on a single shaft.

	Thrust load to be used in equivalent load calculation		
Condition of load	**Bearing A (Fig. 24–46b)**	**Bearing B (Fig. 24–46b)**	
$\dfrac{F_{rB}}{Y_B} \le \dfrac{F_{rA}}{Y_A}$	—	$F_a + 0.5\dfrac{F_{rA}}{Y_A}$	(24–181)
$\dfrac{F_{rB}}{Y_B} > \dfrac{F_{rA}}{Y_A}$	—	$F_a + 0.5\dfrac{F_{rA}}{Y_A}$	(24–182)
$F_a > 0.5\left(\dfrac{F_{rB}}{Y_B} - \dfrac{F_{rA}}{Y_A}\right)$			
$\dfrac{F_{rB}}{Y_B} > \dfrac{F_{rA}}{Y_A}$	$0.5\dfrac{F_{rB}}{Y_B} - F_a$	...	(24–183)
$F_a \le 0.5\left(\dfrac{F_{rB}}{Y_B} - \dfrac{F_{rA}}{Y_A}\right)$		...	

Where thrust factors are: For Series 173 and 909 $Y = 0.87$
For Series 72 B (Series 02) For Series 33 $Y = 0.66$ for $F_a/F_r \le 0.95$
 and 73 B (Series 03) $Y = 0.57$ $= 1.07$ for $F_a/F_r > 0.95$
For Series $LS\ AC$ and $MS\ AC$ $Y = 1.19$

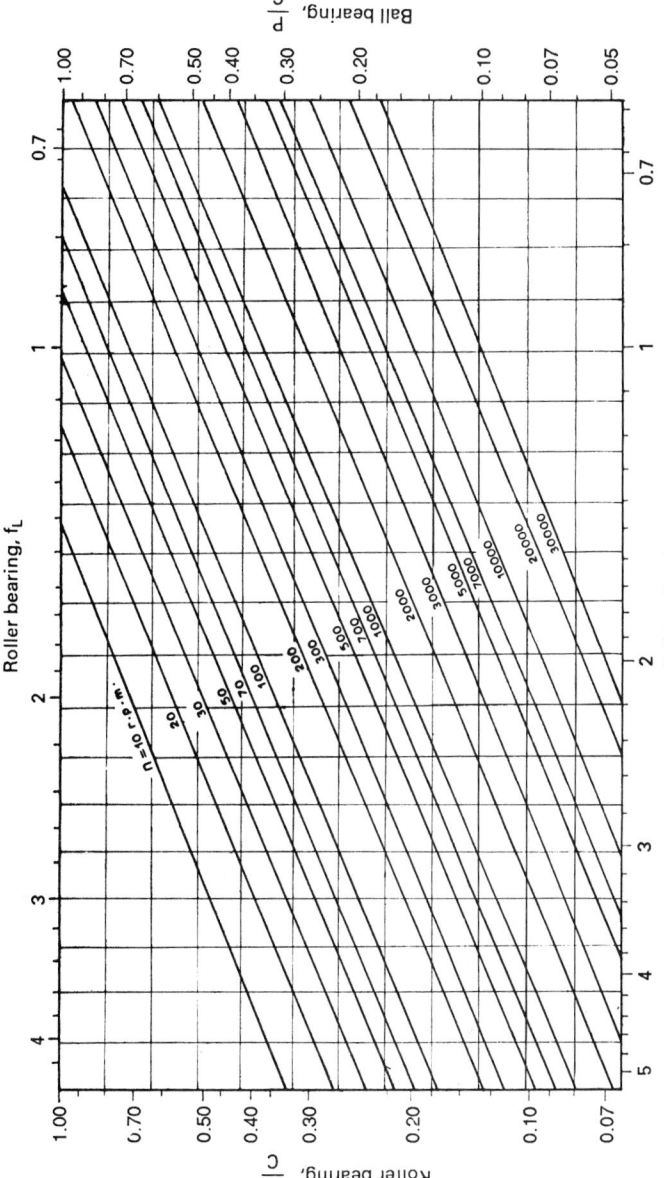

FIGURE 24-48 Selection of bearing size [(part (a)].

24.97

$$m_N = \frac{p_N}{0.95Z}$$

Value for Z is for an element of indicated
Numbers of teeth and a 75 tooth mate.

Normal tooth thickness of pinion and gear
Tooth each reduced 0.024 in. to provide 0.048 in.
Total backlash for one normal diametral pitch.

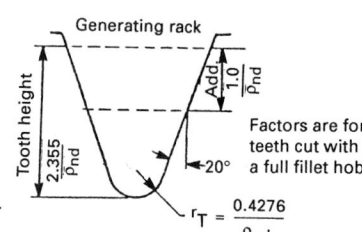

Generating rack

Tooth height

$\frac{2.355}{p_{nd}}$

Add $\frac{1.0}{p_{nd}}$

20°

$r_T = \frac{0.4276}{p_{nd}}$

Factors are for
teeth cut with
a full fillet hob.

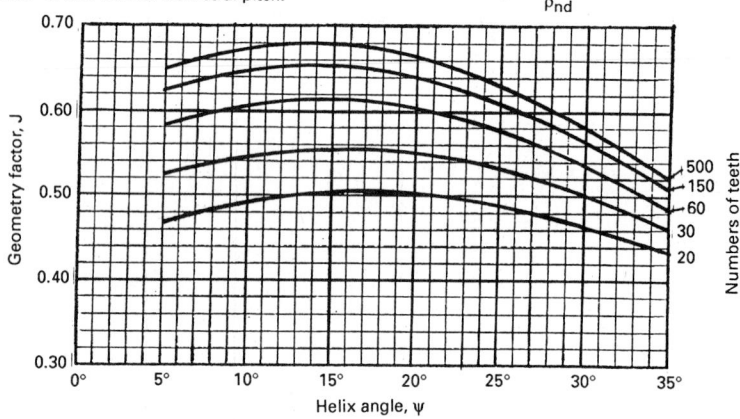

The modifying factor can be applied to the
J factor when other than 75 teeth are used
in the mating element.

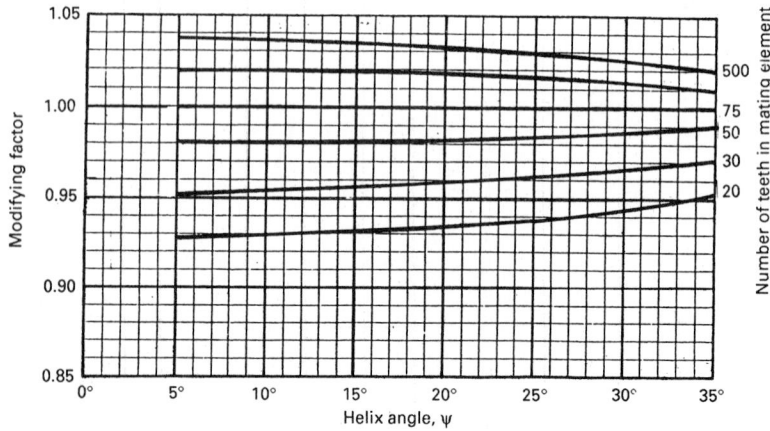

FIGURE 24-48 Selection of bearing size [parts (b) and (c)].

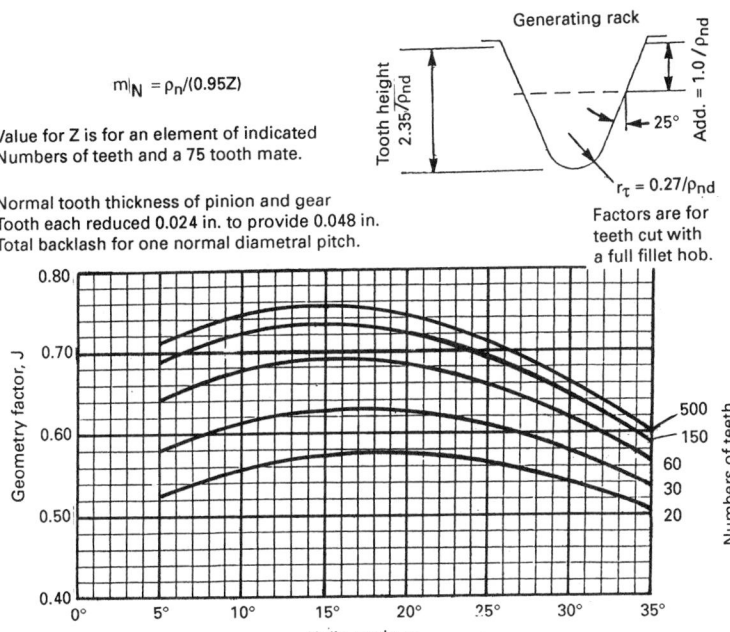

$$m|_N = \rho_n/(0.95Z)$$

Value for Z is for an element of indicated Numbers of teeth and a 75 tooth mate.

Normal tooth thickness of pinion and gear Tooth each reduced 0.024 in. to provide 0.048 in. Total backlash for one normal diametral pitch.

Geometry factor, J

Helix angle, ψ

The modifying factor can be applied to the J factor when other than 75 teeth are used in the mating element.

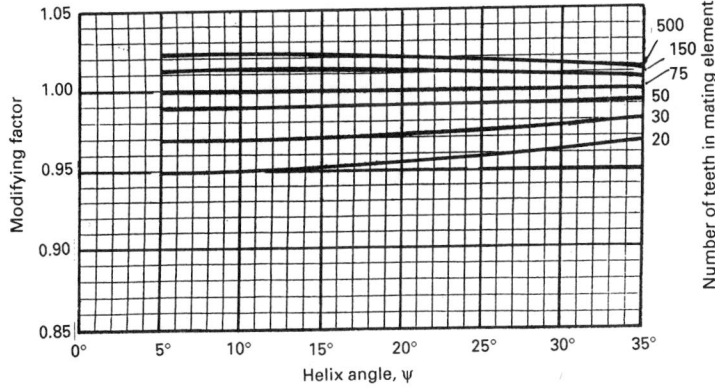

Modifying factor

Helix angle, ψ

FIGURE 24-48 Selection of bearing size [parts (d) and (e)].

Particular	Formula
	where M = minimum load constant taken from bearing tables

LIFE

Fatigue life

Fatigue life of rotating bearings in millions of revolutions

$$L = \left(\frac{\text{Basic load rating } C}{\text{Lood } P}\right)^{m} \qquad (24\text{–}187)$$

where $m = 3$ for ball bearings

$$= \frac{10}{3} \text{ for roller bearings}$$

L in millions of revolutions

The relation between fatigue life (L_h) in hours and the life in millions of revolutions (L)

$$L_h = \frac{10^6 L}{60n} \qquad (24\text{–}188)$$

Refer to Tables 24–40 and 24–41 for L_h.

Fatigue life is also calculated from equation

$$L_h = 500\left(\frac{C}{P} f_n\right)^{m} \qquad (24\text{–}189)$$

where m has the same values as given under Eq. (24–186) and L_h in h

For values of L_h refer to Table 24–40 for ball bearing and Table 24–41 for roller bearing.

VARIABLE SPEED AND LOAD

Under conditions of variable speed and constant load the average speed n_m is calculated by the equation

$$n_m = \sum_{i=1,2,3} \frac{q_i n_i}{100} \qquad (24\text{–}190)$$

Under conditions of variable load and constant speed the mean load is calculated from the equation

$$P_m = \sqrt[3]{\sum_{i=1,2,3} \frac{q_i P_i^3}{100}} \qquad (24\text{–}191)$$

At variable load and variable speed the mean load is calculated by the equation

$$P_m = \sqrt[3]{\sum_{i=1,2,3} \frac{n_i q_i P_i^3}{n_m 100}} \qquad (24\text{–}192)$$

where n_m is given by Eq. (24–190)

For operating conditions with constant speed and a load of straight-line variation the mean load is given by

$$P_m = \frac{P_{min} + 2P_{max}}{3} \qquad (24\text{–}193)$$

TABLE 24–40
Fatigue life L_h for ball bearings

L_h, hours	$\frac{C}{P} f_n$	L_h, hours	$\frac{C}{P} f_n$	L_h, hours	$\frac{C}{P} f_n$	L_h, hours	$\frac{C}{P} f_n$	L_h, hours	$\frac{C}{P} f_n$	L_h, hours	$\frac{C}{P} f_n$	L_h, hours	$\frac{C}{P} f_n$
100	0.585	300	0.843	700	1.120	1750	1.520	4500	2.08	10,000	2.71	30,000	3.91
105	0.595	310	0.852	720	1.130	1800	1.535	4600	2.10	10,500	2.76	31,000	3.96
110	0.604	320	0.861	740	1.140	1850	1.545	4700	2.11	11,000	2.80	32,000	4.00
115	0.613	330	0.870	760	1.150	1900	1.560	4800	2.13	11,500	2.85	33,000	4.04
120	0.622	340	0.879	780	1.160	1950	1.575	4900	2.14	12,000	2.85	34,000	4.08
125	0.631	350	0.888	800	1.170	2000	1.590	5000	2.15	12,500	2.93	35,000	4.12
130	0.639	360	0.896	820	1.180	2100	1.615	5200	2.18	13,000	2.96	36,000	4.16
135	0.647	370	0.905	840	1.190	2200	1.640	5400	2.21	13,500	3.00	37,000	4.20
140	0.654	380	0.913	860	1.200	2300	1.665	5600	2.24	14,000	3.04	38,000	4.24
145	0.662	390	0.921	880	1.205	2400	1.690	5800	2.27	14,500	3.07	39,000	4.27
150	0.670	400	0.928	900	1.215	2500	1.710	6000	2.29	15,000	3.11	40,000	4.31
155	0.677	410	0.936	920	1.225	2600	1.730	6200	2.32	15,500	3.14	41,000	4.35
160	0.684	420	0.944	940	1.235	2700	1.755	6400	2.24	16,000	3.18	42,000	4.38
165	0.691	430	0.951	960	1.245	2800	1.775	6600	2.37	16,500	3.21	43,000	4.42
170	0.698	440	0.959	980	1.250	2900	1.795	6800	2.39	17,000	8.24	44,000	4.45
175	0.705	450	0.966	1000	1.260	3000	1.815	7000	2.41	17,500	3.27	45,000	4.48
180	0.712	460	0.973	1050	1.280	3100	1.835	7200	2.43	18,000	3.30	46,000	4.51
185	0.718	470	0.980	1100	1.300	3200	1.855	7400	2.46	18,500	3.33	47,000	4.55
190	0.724	480	0.987	1150	1.320	3300	1.875	7600	2.48	19,000	3.36	48,000	4.58
195	0.731	490	0.994	1200	1.340	3400	1.895	7800	2.50	19,500	3.39	49,000	4.61
200	0.737	500	1.000	1250	1.360	3500	1.910	8000	2.52	20,000	3.42	50,000	4.64
210	0.749	520	1.015	1300	1.375	3600	1.930	8200	2.54	21,000	3.48	55,000	4.80
220	0.761	540	1.025	1350	1.395	3700	1.950	8400	2.56	22,000	3.53	60,000	4.94
230	0.772	560	1.040	1400	1.410	3800	1.965	8600	2.58	23,000	3.58	65,000	5.07
240	0.783	580	1.050	1450	1.425	3900	1.985	8800	2.60	24,000	3.63	70,000	5.19
250	0.794	600	1.065	1500	1.445	4000	2.000	9000	2.62	25,000	3.68	75,000	5.30
260	0.804	620	1.075	1550	1.460	4100	2.020	9200	2.64	26,000	3.73	80,000	5.43
270	0.814	640	1.085	1600	1.475	4200	2.030	9400	2.66	27,000	3.78	85,000	5.55
280	0.824	660	1.100	1650	1.490	4300	2.050	9600	2.68	28,000	3.82	90,000	5.65
290	0.834	680	1.110	1700	1.505	4400	2.070	9800	2.70	29,000	3.87	100,000	5.85

Particular	Formula

RELIABILITY

The reliability (R_i) of a group of i bearings	$R_i = (R)^i$ where R = reliability of each bearing	(24–194a)
The reliability (R) of bearing using Weibull function	$R = e^{-(t/\theta)^b}$ where t = time θ = design life b = Weibull exponent	(24–194b)
Another form of reliability (R) equation of bearing using Weibull function	$R = e^{-(L/mL_{10})^b} = e^{-\left[\frac{L/L_{10}}{4.48}\right]^{1.5}}$	(24–194c)

TABLE 24–41
Fatigue life L_h for roller bearings

100	0.617	300	0.858	700	1.105	1750	1.455	4500	1.935	10,000	2.46	30,000	3.42
105	0.626	310	0.866	720	1.115	1800	1.470	4600	1.945	10,500	2.49	31,000	3.45
110	0.635	320	0.875	740	1.125	1850	1.480	4700	1.960	11,000	2.53	32,000	3.48
115	0.643	330	0.883	760	1.135	1900	1.490	4800	1.970	11,500	2.56	33,000	3.51
120	0.652	340	0.891	780	1.145	1950	1.505	4900	1.985	12,000	2.59	34,000	3.55
125	0.660	350	0.898	800	1.150	2000	1.515	5000	2.00	12,500	2.63	35,000	3.58
130	0.668	360	0.906	820	1.160	2100	1.540	5200	2.02	13,000	2.66	36,000	3.61
135	0.675	370	0.914	840	1.170	2200	1.560	5400	2.04	13,500	2.69	37,000	3.64
140	0.683	380	0.921	860	1.180	2300	1.580	5600	2.06	14,000	2.72	38,000	3.67
145	0.690	390	0.928	880	1.185	2400	1.600	5800	2.09	14,500	2.75	39,000	3.70
150	0.697	400	0.935	900	1.190	2500	1.620	6000	2.11	15,000	2.77	40,000	3.72
155	0.704	410	0.942	920	1.200	2600	1.640	6200	2.13	15,500	2.80	41,000	3.75
160	9.710	420	0.949	940	1.210	2700	1.660	6400	2.15	16,000	2.83	42,000	3.78
165	0.717	430	0.956	960	1.215	2800	1.675	6600	2.17	16,500	2.85	43,000	3.80
170	0.723	440	0.962	980	1.225	2900	1.695	6800	2.19	17,000	2.88	44,000	3.83
175	0.730	450	0.969	1000	1.230	3000	1.710	7000	2.21	17,500	2.91	45,000	3.86
180	0.736	460	0.975	1050	1.250	3100	1.730	7200	2.23	18,000	2.93	46,000	3.88
185	0.742	470	0.982	1100	1.270	3200	1.175	7400	2.24	18,500	2.95	47,000	3.91
190	0.748	480	0.998	1150	1.285	3300	1.760	7600	2.26	19,000	2.98	48,000	3.93
195	0.754	490	0.994	1200	1.300	3400	1.775	7800	2.28	19,500	3.00	49,000	3.96
200	0.760	500	1.000	1250	1.315	3500	1.795	8000	2.30	20,000	3.02	50,000	3.98
210	0.771	520	1.010	1300	1.330	3600	1.810	8200	2.31	21,000	3.07	55,000	4.10
220	0.782	540	1.025	1350	1.345	3700	1.825	8400	2.33	22,000	3.11	60,000	4.20
230	0.792	560	1.035	1400	1.360	3800	1.840	8600	2.35	23,000	3.15	65,000	4.30
240	0.802	580	1.045	1450	1.375	3900	1.850	8800	2.36	24,000	3.19	70,000	4.40
250	0.812	600	1.055	1500	1.390	4000	1.865	9000	2.38	25,000	3.23	75,000	4.50
260	0.822	620	1.065	1550	1.405	4100	1.880	9200	2.40	26,000	3.27	80,000	4.58
270	0.831	640	1.075	1600	1.420	4200	1.895	9400	2.41	27,000	3.31	85,000	4.66
280	0.840	660	1.085	1650	1.430	4300	1.905	9600	2.43	28,000	3.35	90,000	4.75
290	0.849	680	1.095	1700	1.445	4400	1.920	9800	2.44	29,000	3.38	100,000	4.90

Particular	Formula
	where R = reliability corresponding to life L
	L_{10} = rating life ($R = 0.90$)
	m = scale constant
Equation (24–194c) can also be written in the form	$R = e^{-(L/6.84\,L_{10})^{1.17}}$ $\qquad$ (24–194d)
Weibull equation for the distribution of bearing rating life based on reliability	$\dfrac{L}{L_{10}} = \left[\dfrac{\ln(1/R)}{\ln(1/R_{10})}\right]^{1/b}$ $\qquad$ (24–194e)
The relation between the design values and the dynamic load rated or catalog values (C_r) according to Timken Engineering is given by	$C_r = F_r\left[\left(\dfrac{L_d}{L_r}\right)\left(\dfrac{n_d}{n_r}\right)\right]^{1/m}$ $\qquad$ (24–195)

Particular	Formula
	where subscripts d and r stand for design and rated values $C_r =$ basic load capacity or dynamic load rating corresponding to L_r hours of L_{10} life at the speed n_r rpm, kN (lbf) $F_r =$ actual radial bearing load carried for L_d hours of L_{10} life at the speed n_d rpm, kN (lbf) $m =$ a constant which varies from 3 to 4
The basic dynamic capacity or specific dynamic capacity of bearing corresponding to any desired life L at reliability R	$$C_r = F_r \left[\left(\frac{L_d}{L_r} \right) \left(\frac{n_d}{n_r} \right) \left(\frac{1}{6.84} \right) \right]^{1/m} \frac{1}{\{\ln (1/R)\}^{(1/1.17m)}}$$ (24–196)

TAPERED ROLLER BEARINGS (FIG. 24–49)

The radial equivalent or effective load when the cup rotates in case of tapered roller bearing (Fig. 24–49)

$$F_c = 1.25 \, F_r \qquad (24\text{–}197)$$

where F_r is the calculated radial load, kN (lbf)

The thrust component of pure radial load (F_r) due to the tapered roller

$$F_{ar} = \frac{0.47 \, F_r}{K} \qquad (24\text{–}198)$$

where $K = \dfrac{\text{radial rating of bearing}}{\text{thrust rating of bearing}}$

$K = 1.5$ for radial bearings

$ = 0.75$ for steep-angle bearings

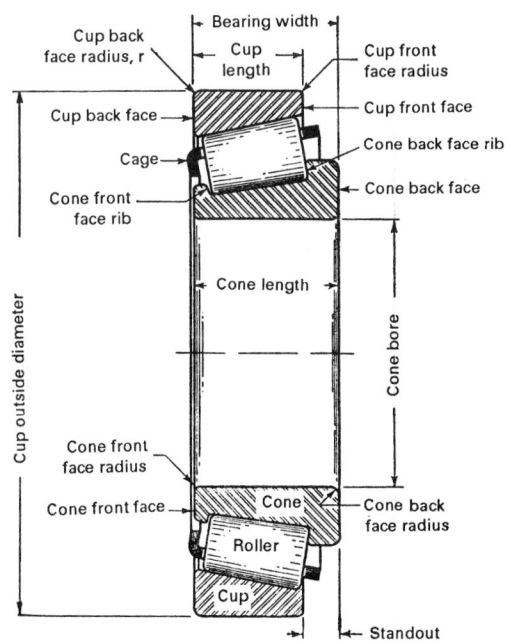

FIGURE 24–49   Nomenclature of tapered roller bearing (*Courtesy of the Timken Company*).

Particular	Formula	
The net thrust on the tapered roller bearing when the induced thrust (F_{ar}) is deducted from the applied thrust (F_{aa})	$F_{nt} = F_{aa} - F_{ar}$	(24–199a)
	$= F_{aa} - \dfrac{0.47\,F_r}{K}$	(24–199b)
The radial equivalent load when the cone rotates in case of tapered roller bearing (Fig. 24–49)	$F_c = F_r + K\left(F_{aa} - \dfrac{0.47\,F_r}{K}\right)$	(24–200a)
	$= 0.53\,F_r + K\,F_{nt}$	(24–200b)
The radial equivalent load when the cup rotates in case of tapered roller bearing (Fig. 24–49)	$F_e = 1.25\,F_r + K\left(F_{aa} - \dfrac{0.47\,F_r}{K}\right)$	(24–201a)
	$= 0.78\,F_r + K\,F_{nt}$	(24–201b)

	Thrust load to be used in equivalent load calculation		
Condition of load	**Bearing A (Fig. 24–47)**	**Bearing B (Fig. 24–47)**	
$\dfrac{F_{rB}}{Y_B} \leq \dfrac{F_{rA}}{Y_A}$		$F_a + 0.5\dfrac{F_{rA}}{Y_A}$	(24–202)
$\dfrac{F_{rB}}{Y_B} > \dfrac{F_{rA}}{Y_A}$		$F_a + 0.5\dfrac{F_{rA}}{Y_A}$	(24–203)
$F_a > 0.5\left(\dfrac{F_{rB}}{Y_B} - \dfrac{F_{rA}}{Y_A}\right)$			
$\dfrac{F_{rB}}{Y_B} > \dfrac{F_{rA}}{Y_A}$	$0.5\dfrac{F_{rB}}{Y_B} - F_a$		(24–204a)
$F_a \leq 0.5\left(\dfrac{F_{rB}}{Y_B} - \dfrac{F_{rA}}{Y_A}\right)$			

The thrust factors Y and Y_o are taken from Table 24–29.

Particular	Formula	
The radial equivalent load on bearing A according to *Timken Engineering Journal* (Fig. 24–47)	$F_{eA} = 0.4\,F_{rA} + K_A\left(F_a + \dfrac{0.47\,F_{rB}}{K_B}\right)$	(24–204b)
The radial equivalent load on bearing B according to *Timken Engineering Journal* (Fig. 24–47)	$F_{eB} = 0.4\,F_{rB} + K_B\left(\dfrac{0.47\,F_{rA}}{K_A} - F_a\right)$	(24–204c)

Particular	Formula

DIMENSIONS, BASIC CAPACITY, AND MAXIMUM PERMISSIBLE SPEED OF ROLLING CONTACT BEARINGS

Deep-groove ball bearings—Series 02, Series 03, Series 04

Refer to Tables 24–42, 24–43, and 24–44, respectively.

Self-aligning and deep-groove ball bearings—Series 02, Series 03, Series 22 (FAG), and Series 23 (FAG)

Refer to Tables 24–45, 24–46, 24–47, and 24–48, respectively.

Single-row angular contact ball bearings—Series 02 and Series 03

Refer to Tables 24–49 and 24–50.

Double-row angular contact ball bearings—Series 33 (FAG)

Refer to Table 24–51.

Cylindrical roller bearings—Series 02, Series 03, Series 04, Series NU 22 (FAG), Series NU 23 (FAG)

Refer to Tables 24–52 to 24–56.

Tapered roller bearings—Series 322, Series 02 (22), and Series 03 (23)

Refer to Tables 24–57A, 24–57B, and 24–57C.

Single-thrust ball bearings—Series 11, Series 12, Series 13, and Series 14

Refer to Tables 24–58 to 24–61.

Double-thrust ball bearing—Series 522 (FAG)

Refer to Table 24–62.

Selection of bearing size

Refer to Fig. 24–48a.

For ball-bearing size proportions

Refer to Fig. 24–48b.

NEEDLE BEARING

Load capacity

The capacity of needle bearing at 3000 h average life

$$C_n = 1.76 \times 10^7 \frac{Zld}{\sqrt[3]{n'}} \qquad \textbf{SI} \quad (24\text{–}205a)$$

where C_n in N, l and d in m, and n' in rps

$$= \frac{10,000 \, Zld}{\sqrt[3]{n}} \quad \textbf{US Customary units} \quad (24\text{–}205b)$$

where C_n in lbf, l and d in in, and n in rpm

The load capacity of needle bearing based on the projected area of the needle rollers

$$C_n = 5.53 \frac{L(d_i + d_r)}{\sqrt[3]{n'}} \qquad \textbf{SI} \quad (24\text{–}206a)$$

where C_n in N; L, d_i, and d_r in m, n' in rps

$$= \frac{31,400 \, L(d_i + d_r)}{\sqrt[3]{n}} \quad \textbf{US Customary units}$$

$$(24\text{–}206b)$$

where C_n in lbf; L, d_i, and d_r in in, n in rpm

TABLE 24-42
Deep-groove ball bearings series 02 (Indian Standards)

Bearing no.			Dimensions, mm				Factor			Basic capacity								Maximum permissible speed, rpm
										Static, C_o				Dynamic, C				
										FAG		SKF		FAG		SKF		
BIS	FAG	SKF	d	D	B	r	F_a/C_o	Y	e	kgf	N	kgf	N	kgf	N	kgf	N	
10BC02	6200	6200	10	30	9	1	.025	2.0	0.22	272	2,666	220	2,156	463	4,537	400	3,920	20,000
12BC02	6201	01	12	32	10	1	.04	1.8	.24	315	3,087	300	2,940	503	4,929	535	5,243	20,000
15BC02	6202	02	15	35	11	1	.07	1.6	.27	363	3,557	350	3,430	599	5,878	610	5,978	16,000
17BC02	6203	03	17	40	12	1.5	.13	1.4	.31	454	4,449	440	4,312	758	7,429	750	7,350	16,000
20BC02	6204	04	20	47	14	1.5	.25	1.2	.37	621	6,086	650	6,370	998	9,780	1,000	9,806	16,000
25BC02	6205	05	25	52	15	1.5	.5	1.0	.44	726	7,115	710	6,958	1,111	10,888	1,090	10,682	13,000
30BC02	6206	6206	30	62	16	2				1,016	9,947	1,000	9,806	1,520	14,896	1,500	14,700	13,000
35BC02	6207	07	35	72	17	2				1,406	14,779	1,380	13,524	1,996	19,561	2,000	19,600	13,000
40BC02	6208	08	40	80	18	2				1,565	15,332	1,580	15,484	2,268	22,226	2,260	22,154	10,000
45BC02	6209	09	45	85	19	2				1,814	17,777	1,810	17,738	2,540	24,892	2,540	24,892	10,000
50BC02	6210	10	50	90	20	2				2,109	20,668	2,100	20,580	2,858	28,008	2,760	27,048	8,000
55BC02	6211	6211	55	100	21	2.5				2,540	24,892	2,580	25,284	3,402	33,344	3,400	33,320	8,000
60BC02	6212	12	60	110	22	2.5				3,152	30,890	3,220	31,556	4,150	40,670	4,100	40,180	6,000
65BC02	6213	13	65	120	23	2.5				3,629	35,564	3,540	34,692	4,546	44,551	4,375	42,875	6,000
70BC02	6214	14	70	125	24	2.5				3,856	37,759	3,920	38,416	4,899	48,010	4,800	47,040	5,000
75BC02	6215	6215	75	130	25	2.5				4,150	40,670	4,220	41,356	5,171	50,666	5,160	50,468	5,000
80BC02	6216	16	80	140	26	3				4,537	44,463	4,500	44,100	5,670	55,566	5,660	55,468	4,000
85BC02	6217	27	85	150	28	3				5,026	49,255	5,450	53,410	6,486	63,562	6,500	63,700	4,000
90BC02	6218	18	90	160	30	3				5,988	59,182	6,200	60,760	7,257	71,088	7,550	73,990	4,000
95BC02	6219	19	95	170	32	3.5				6,831	66,944	7,250	71,050	8,437	82,783	8,450	82,810	4,000
100BC02	6220	6220	100	180	34	3.5				7,738	75,832	8,150	79,870	9,615	94,227	9,600	94,000	3,000
105BC02	6221	21	105	190	36	3.5				9,072	88,836	9,250	90,650	10,342	101,352	10,350	101,430	3,000
110BC02	6222	22	110	200	38	3.5				10,160	99,568	10,300	100,940	11,113	108,907	11,100	101,700	3,000
120BC02	6224	24	120	215	40	3.5				10,342	101,352	10,300	100,940	11,340	111,132	11,300	110,740	3,000
	6226	6226	130	230	40	4				12,701	124,470	11,500	112,700	13,154	128,909	12,250	120,050	2,500
	6228	6228	140	250	42	4				13,154	128,909	13,000	127,400	13,608	133,358	12,900	126,420	2,500
	6230	6230	150	270	45	4				14,051	137,798	14,250	139,650	14,061	137,798	13,800	130,740	2,500
	6232	32	160	290	48	4				16,329	160,024	15,400	150,920	15,422	151,136	14,300	140,140	2,000
	6234	34	170	310	52	5				18,144	177,811	18,800	184,240	16,329	160,024	16,300	159,740	2,000
	6236	36	180	320	52	5				19,958	195,588	20,400	199,920	17,237	161,923	17,250	169,050	2,000
	6238	38	190	340	55	5				23,587	231,153	24,000	235,200	19,958	195,588	20,000	196,000	1,600
	6240	40	200	360	58	5				24,501	248,930	26,500	259,700	21,092	206,702	21,000	205,800	1,600

TABLE 24-43
Deep-groove ball bearings series 03 (Indian Standards)

Bearing no.			Dimensions, mm				Factor			Basic capacity								Maximum permissible speed, rpm
										Static, C_o				Dynamic, C				
										FAG		SKF		FAG		SKF		
BIS	FAG	SKF	d	D	B	r	$\frac{F_a}{C_o}$	Y	e	kgf	N	kgf	N	kgf	N	kgf	N	
10BC03	6300	6300	10	35	11	1	.025	2.0	.22	386	3,783	364	3,567	635	6,223	620	6,076	16,000
12BC03	6301	6301	12	37	12	1.5	.04	1.8	.24	472	4,626	430	4,214	771	7,556	770	7,546	16,000
15BC03	6302	6302	15	42	13	1.5	.07	1.6	.27	544	5,331	520	5,096	875	8,575	875	8,575	16,000
17BC03	6303	6303	17	47	14	1.5	.13	1.4	.31	662	6,488	620	6,076	1,052	10,310	1,050	10,290	13,000
20BC03	6304	04	20	52	15	2	.25	1.2	.37	900	8,820	770	7,546	1,361	13,378	1,250	12,250	13,000
25BC03	6305	05	25	62	17	2	.5	1.0	.44	1,157	11,339	1,035	10,143	1,724	16,895	1,630	15,974	10,000
30BC03	6306	06	30	72	19	2				1,542	15,112	1,450	14,210	2,268	22,226	2,140	20,972	10,000
35BC03	6307	07	35	80	21	2.5				1,814	17,777	1,730	16,954	2,586	25,343	2,580	25,284	8,000
40BC03	6308	08	40	90	23	2.5				2,404	23,559	2,140	20,792	3,334	32,974	3,200	31,360	8,000
45BC03	6309	09	45	100	25	2.5				3,039	29,782	2,980	29,204	4,150	40,670	4,150	40,670	8,000
50BC03	6310	6310	50	110	27	3				3,697	36,230	3,540	34,692	4,808	46,118	4,800	47,040	6,000
55BC03	6311	11	55	120	29	3				4,530	44,394	4,200	41,160	5,851	57,340	5,500	53,900	6,000
60BC03	6312	12	60	130	31	3.5				4,572	44,806	4,800	47,040	6,350	62,230	6,350	62,230	5,000
65BC03	6313	13	65	140	33	3.5				5,670	55,566	5,450	53,410	7,257	71,070	7,250	71,050	5,000
70BC03	6314	14	70	150	35	3.5				6,350	62,230	6,200	60,760	8,165	80,017	8,150	79,870	5,000
75BC03	6315	15	75	160	37	3.5				7,393	72,151	7,250	71,050	8,754	85,789	8,900	87,220	4,000
80BC03	6316	16	80	170	39	4				7,393	72,451	8,000	78,400	8,890	87,122	9,600	94,080	4,000
85BC03	6317	17	85	180	41	4				8,437	82,683	8,750	85,750	9,797	96,011	10,350	101,430	4,000
90BC03	6318	18	90	190	43	4				9,072	88,906	9,800	96,040	10,523	103,125	11,000	107,800	3,000
95BC03	6319	19	95	200	45	4				10,160	99,568	11,000	107,800	11,113	108,907	12,000	117,600	3,000
100BC03	6320	20	100	215	47	4				12,247	120,021	13,200	126,360	12,928	126,694	13,800	135,240	3,000
105BC03	6321	6321	105	225	49	4				13,608	133,358	14,300	140,140	13,835	135,583	14,300	140,140	2,500
110BC03	6322	22	110	240	50	4				14,750	144,556	16,390	160,622	14,742	144,471	15,600	152,880	2,500
120BC03	6324	24	120	260	55	4				17,237	161,923	16,600	162,680	16,329	160,029	16,100	157,780	2,500
	6326	26	130	280	58	5				19,504	191,139	19,500	191,100	17,690	173,362	17,700	170,460	2,500
	6328	28	140	300	62	5				22,660	222,662	22,000	215,600	19,958	195,598	20,000	196,000	2,000
	6330	30	150	320	65	5				27,216	266,717	25,400	248,960	22,226	219,815	21,500	210,700	2,000

24.107

TABLE 24-44
Deep-groove ball bearings series 04 (Indian Standards)

Bearing no. BIS	FAG	SKF	d	D	B	r	F_a/C_o	Y	e	Static C_o FAG kgf	FAG N	SKF kgf	SKF N	Dynamic C FAG kgf	FAG N	SKF kgf	SKF N	Maximum permissible speed, rpm
15BC04			15	52	15	2	.025	2.0	.22									
17BC04	6403	6403	17	62	17	2	.04	1.8	.24	1,270	12,446	1,090	10,682	1,815	17,787	1,770	17,346	10,000
20BC04	6404	04	20	72	19	2	.07	1.6	.27	1,655	16,219	1,540	15,092	2,405	23,569	2,405	23,569	10,000
25BC04	6405	05	25	80	21	2.5	.13	1.4	.31	1,995	19,551	1,880	18,424	2,810	27,538	2,810	27,538	8,000
30BC04	6406	06	30	90	23	2.5	.25	1.2	.37	2,405	23,569	2,315	22,687	3,335	32,683	3,335	32,683	8,000
35BC04	6407	07	35	100	25	2.5	.5	1.0	.44	3,220	31,556	3,040	29,792	4,310	42,238	4,310	42,238	6,000
40BC04	6408	08	40	110	27	3				3,765	36,897	3,765	36,897	4,990	48,902	4,990	48,902	6,000
45BC04	6409	09	45	120	29	3				4,545	44,551	4,375	42,875	5,850	57,330	5,850	57,330	6,000
50BC04	6410	10	50	130	31	3.5				5,260	51,548	4,990	48,902	6,940	68,012	6,805	67,589	5,000
55BC04	6411	6411	55	140	33	3.5				6,350	63,230	5,850	57,330	7,845	76,881	7,845	76,881	5,000
60BC04	6412	12	60	150	35	3.5				7,075	69,335	6,620	64,876	8,435	82,663	8,435	82,683	5,000
65BC04	6413	13	65	160	37	3.5				7,985	78,253	7,710	75,568	9,255	90,699	9,255	90,699	4,000
70BC04	6414	14	70	180	42	3.5				9,070	88,886	10,160	99,568	10,160	99,568	11,110	108,878	4,000
75BC04	6415	15	75	190	45	4				11,550	113,190	10,890	106,722	11,920	116,316	12,020	117,796	4,000
80BC04	6416	16	80	200	48	4				12,700	124,460	12,020	117,796	12,925	116,665	12,700	124,460	3,000
85BC04	6417	17	85	210	52	5				13,835	135,573	13,155	128,919	13,835	135,583	13,610	133,280	3,000
90BC04	6418	18	90	225	54	5				16,330	160,034	14,515	142,247	15,195	148,911	14,515	142,247	3,000
100BC04			100	250	58	5												

TABLE 24-45
Self-aligning ball bearings series 02 (Indian Standards)

Bearing no. BIS	FAG	SKF	d	D	B	r	e	Y_o	X ($F_a/F_r \leq e$)	Y ($F_a/F_r \leq e$)	X ($F_a/F_r > e$)	Y ($F_a/F_r > e$)	Static C_o FAG kgf	Static C_o FAG N	Static C_o SKF kgf	Static C_o SKF N	Dynamic C FAG kgf	Dynamic C FAG N	Dynamic C SKF kgf	Dynamic C SKF N	Maximum permissible speed, rpm
10B502	1200	1200	10	30	9	1.0	.32	2.1	1	2.0	.65	3.0	140	1,372	140	1,372	430	4,214	420	4,116	20,000
12B502	1201	01	12	32	10	1.0	.37	1.8	1	1.7	.65	2.6	145	1,421	150	1,470	435	4,263	480	4,704	20,000
15B502	1202	02	15	35	11	1.0	.39	1.9	1	1.9	.65	2.9	205	2,009	205	2,009	575	5,635	575	5,635	16,000
17B502	1203	1203	17	40	12	1.0	.33	2.0	1	1.9	.65	3.0	245	2,401	245	2,401	605	5,919	600	5,880	16,000
20B502	1204	04	20	47	14	1.5	.28	2.4	1	2.3	.65	3.5	325	3,180	320	2,836	785	7,693	785	7,693	16,000
25B502	1205	05	25	52	15	1.5	.27	2.4	1	2.3	.65	3.6	410	4,018	410	4,018	945	9,261	945	9,261	13,000
30B502	1206	1206	30	62	16	1.5	.25	2.6	1	2.5	.65	3.9	575	5,635	565	5,537	1,225	12,005	1,225	12,005	13,000
35B502	1207	07	35	72	17	2.0	.22	3.0	1	2.9	.65	4.4	680	6,664	635	6,223	1,250	12,250	1,225	12,005	10,000
40B502	1208	08	40	80	18	2.0	.22	3.0	1	2.9	.65	4.4	860	8,428	815	7,989	1,475	14,455	1,475	14,465	10,000
45B502	1209	1209	45	85	19	2.0	.21	3.1	1	3.0	.65	4.6	980	9,604	905	8,869	1,680	16,464	1,655	16,219	8,000
50B502	1210	10	50	90	20	2.0	.20	3.3	1	3.2	.65	4.9	1,090	10,602	1,015	9,947	1,770	17,346	1,725	16,905	8,000
55B502	1211	11	55	100	21	2.5	.19	3.5	1	3.3	.65	5.1	1,385	13,573	1,270	12,446	2,110	20,678	2,065	20,217	8,000
60B502	1212	1212	60	110	22	2.5	.18	3.7	1	3.5	.65	5.4	1,565	15,337	1,450	14,210	2,360	23,128	2,360	22,928	6,000
65B502	1213	13	65	120	23	2.5	.18	3.7	1	3.5	.65	5.4	1,725	16,905	1,565	15,337	2,450	34,070	2,605	23,569	6,000
70B502	1214	14	70	125	24	2.5	.19	3.5	1	3.5	.65	5.1	1,880	18,524	1,725	16,905	2,720	26,656	2,720	26,650	5,000
75B502	1215	1215	75	130	25	2.5	.18	3.7	1	3.5	.65	5.4	2,155	21,119	2,000	19,620	3,040	29,792	2,970	29,106	5,000
80B502	1216	16	80	140	26	3.0	.16	4.1	1	3.9	.65	6.1	2,405	23,569	2,175	21,315	3,844	37,671	3,035	30,733	5,000
85B502	1217	17	85	150	28	3.0	.17	3.9	1	3.7	.65	5.7	2,905	28,469	2,655	26,019	3,924	38,455	3,855	37,784	4,000
90B502	1218	1218	90	160	30	3.0	.17	3.9	1	3.7	.65	5.7	3,265	31,997	2,970	29,106	4,445	43,561	4,375	42,875	4,000
95B502	1219	19	95	170	32	3.5	.17	3.9	1	3.7	.65	5.7	3,765	36,897	3,470	34,006	4,990	48,902	4,990	47,302	4,000
100B502	1220	20	100	180	34	3.5	.18	3.7	1	3.5	.65	5.4	4,150	40,670	3,630	35,674	5,445	53,361	5,370	52,626	3,000
105B502	1221	1221	105	190	36	3.5	.19	3.5	1	3.3	.65	5.1	4,445	43,561	4,150	40,670	5,760	56,448	5,760	56,448	3,000
110B502	1222	22	110	200	38	3.5	.19	3.5	1	3.3	.65	5.1	5,260	51,548	4,990	48,902	6,940	68,012	6,805	60,270	3,000
120B502		23	120	215	42	3.5	.19	3.5	1	3.3	.65	5.1			6,620	64,876			9,255	90,669	3,000

24.109

TABLE 24-46
Self-aligning ball bearings series 03

Bearing no.			Dimensions, mm						Factors				Basic capacity								Maximum permissible speed, rpm
									$F_a/F_r \le e$		$F_a/F_r > e$		Static, C_o				Dynamic, C				
													FAG		SKF		FAG		SKF		
BIS	FAG	SKF	d	D	B	r	e	Y_o	X	Y	X	Y	kgf	N	kgf	N	kgf	N	kgf	N	
10B503	1300	1300	10	35	11	1.0	.34	1.9	1	1.9	.65	2.9	180	1,764	185	1,813	555	5,439	545	5,341	16,000
12B503	1301	01	12	37	12	1.5	.35	1.9	1	1.8	.65	2.8	240	2,352	190	1,862	740	7,252	555	5,439	16,000
15B503	1302	02	15	42	13	1.5	.35	1.9	1	1.8	.65	2.8	265	2,597	250	2,450	755	7,399	740	7,252	16,000
17B503	1303	1303	17	47	14	1.5	.32	2.1	1	2.0	.65	3.0	375	3,675	270	2,646	980	9,604	750	7,350	13,000
20B503	1304	04	20	52	15	2.0	.29	2.3	1	2.2	.65	3.4	410	4,018	375	3,675	980	9,604	980	9,604	13,000
25B503	1305	05	25	62	17	2.0	.28	2.4	1	2.3	.65	3.5	600	5,880	400	3,920	980	9,604	980	9,604	10,000
30B503	1306	1306	30	72	19	2.0	.26	2.5	1	2.4	.65	3.8	800	7,840	585	5,733	1,630	15,978	1,405	13,769	10,000
35B503	1307	07	35	80	21	2.5	.26	2.5	1	2.4	.65	3.8	1,000	9,810	770	7,546	1,905	18,669	1,630	15,978	8,000
40B503	1308	08	40	90	23	2.5	.25	2.6	1	2.5	.65	3.9	1,250	12,250	960	9,408	2,315	22,687	1,950	19,110	8,000
45B503	1309	1309	45	100	25	2.5	.25	2.6	1	2.5	.65	3.9	1,610	15,778	1,200	11,760	2,970	29,106	2,315	22,687	8,000
50B503	1310	10	50	110	27	3.0	.24	2.8	1	2.6	.65	4.1	1,815	17,787	1,540	15,902	3,265	31,997	2,970	29,106	6,000
55B503	1311	11	55	120	29	3.0	.24	2.8	1	2.6	.65	4.1	2,270	22,246	1,700	16,660	4,080	39,984	3,400	33,320	6,000
60B503	1312	1312	60	130	31	3.5	.23	2.9	1	2.7	.65	4.2	2,720	26,650	2,175	21,315	4,465	43,757	3,990	39,102	5,000
65B503	1313	13	65	140	33	3.5	.23	2.9	1	2.7	.65	4.2	2,970	29,106	2,585	25,333	4,900	48,424	4,445	43,561	5,000
70B503	1314	14	70	150	35	3.5	.23	2.9	1	2.7	.65	4.2	3,630	35,670	2,855	27,979	5,760	56,448	4,810	47,138	5,000
75B503	1315	1315	75	160	37	3.5	.23	2.9	1	2.7	.65	4.2	3,925	38,465	3,470	34,005	6,080	59,584	5,760	56,448	4,000
80B503	1316	16	80	170	39	3.5	.22	3.0	1	2.9	.65	4.4	4,310	42,231	3,765	36,897	6,940	68,012	6,080	59,584	4,000
85B503	1317	17	85	180	41	4.0	.22	3.0	1	2.9	.65	4.4	4,900	48,424	4,150	40,670	7,710	75,558	6,940	68,012	4,000
90B503	1318	1318	90	190	43	4.0	.22	3.0	1	2.9	.65	4.4	5,535	53,243	4,715	46,207	8,440	82,712	7,710	75,558	4,000
95B503	1319	19	95	200	45	4.0	.23	2.9	1	2.7	.65	4.2	6,490	63,602	5,445	53,361	9,070	88,886	9,070	88,886	3,000
100B503	1320	20	100	215	47	4.0	.23	2.9	1	2.7	.65	4.2	7,395	72,471	6,080	59,584	10,340	101,292	10,340	101,292	3,000
105B503		1321	105	225	49	4.0									7,075	69,335	11,110	108,878			3,000
110B503		22	110	240	50	4.0									7,985	78,253					2,500
120B503			120	260	55	4.0									8,755	85,799					2,500

TABLE 24-47
Deep-groove ball bearings series 22

Bearing no. FAG	Bearing no. SKF	d	D	B	r	e	Y_o	$F_a/F_r \le e$ X	$F_a/F_r \le e$ Y	$F_a/F_r > e$ X	$F_a/F_r > e$ Y	Static C_o FAG kgf	Static C_o FAG N	Static C_o SKF kgf	Static C_o SKF N	Dynamic C FAG kgf	Dynamic C FAG N	Dynamic C SKF kgf	Dynamic C SKF N	Maximum permissible speed, rpm
2200	2200	10	30	14	1	.66	1.0	1	1.0	.65	1.5	170	1,666	170	1,666	567	5,567	567	5,567	20,000
2201	01	12	32	14	1	.58	1.1	1	1.1	.65	1.7	195	1,911	200	1,960	576	5,645	575	5,635	20,000
2202	02	15	35	14	1	.51	1.3	1	1.2	.65	1.9	218	2,136	215	2,107	585	5,733	585	5,733	16,000
2203	2203	17	40	16	1.5	.51	1.3	1	1.2	.65	1.9	286	2,803	280	2,744	771	7,556	765	7,497	16,000
2204	04	20	47	18	1.5	.50	1.3	1	1.3	.65	2.0	399	3,910	390	3,822	980	9,604	980	9,605	16,000
2205	05	25	52	18	1.5	.44	1.5	1	1.4	.65	2.2	431	4,224	420	4,116	1,201	11,770	1,200	11,760	13,000
2206	2206	30	62	20	1.5	.40	1.7	1	1.6	.65	2.4	576	5,645	550	5,390	1,656	16,229	1,650	16,170	13,000
2207	07	35	72	23	2	.37	1.8	1	1.7	.65	2.6	830	8,134	800	7,840	1,724	16,895	1,725	16,975	10,000
2208	08	40	80	23	2	.34	1.9	1	1.9	.65	2.9	962	9,428	900	8,820	1,814	17,777	1,775	17,395	10,000
2209	2209	45	85	23	2	.31	2.1	1	2.0	.65	3.1	1,071	10,496	1,000	9,800	1,814	17,771	1,775	17,395	10,000
2210	10	50	90	23	2	.29	2.3	1	2.2	.65	3.4	1,157	11,339	1,070	10,486	2,064	20,227	2,060	20,188	8,000
2211	11	55	100	25	2.5	.28	2.4	1	2.3	.65	3.5	1,361	13,338	1,270	12,446	2,653	25,999	2,632	25,794	8,000
2212	2212	60	110	28	2.5	.29	2.3	1	2.2	.65	3.4	1,656	16,299	1,560	15,288	3,402	33,340	3,400	33,320	8,000
2213	13	65	120	31	2.5	.29	2.3	1	2.2	.65	3.4	2,177	21,335	2,000	19,600	3,470	34,006	3,475	34,055	6,000
2214	14	70	125	31	2.5	.27	2.4	1	2.3	.65	3.6	2,313	22,667	2,150	21,070	3,470	34,006	3,475	34,055	6,000
2215	2215	75	130	31	2.5	.26	2.5	1	2.4	.65	3.8	2,449	23,990	2,200	21,560	3,992	39,122	3,860	37,240	5,000
2216	16	80	140	33	3	.25	2.6	1	2.5	.65	3.9	2,857	27,999	2,500	24,500	4,536	44,453	4,550	44,600	5,000
2217	17	85	150	36	3	.25	2.6	1	2.5	.65	3.9	3,221	31,566	2,960	29,008	5,443	53,341	5,450	53,410	5,000
2218	2218	90	160	40	3	.27	2.4	1	2.3	.65	3.6	3,924	38,455	3,325	32,585	6,486	63,563	6,500	63,700	4,000
2219	19	95	170	43	3.5	.27	2.4	1	2.3	.65	3.6	4,627	45,345	4,300	40,170	7,711	75,568	7,750	75,950	4,000
2220	20	100	180	46	3.5	.27	2.4	1	2.3	.65	3.6	5,353	52,459	5,000	49,000			8,450	82,810	4,000
	2221	105	190	50	3.5									5,500	53,900			9,850	96,530	3,000
	22	110	200	53	3.5									6,350	62,230					3,000

24.111

TABLE 24-48
Self-aligning ball bearings series 23

Bearing no. FAG	Bearing no. SKF	Dimensions, mm d	D	B	r	Factors e	Y_o	$F_a/F_r \le e$ X	$F_a/F_r \le e$ Y	$F_a/F_r > e$ X	$F_a/F_r > e$ Y	Static C_o FAG kgf	Static C_o FAG N	Static C_o SKF kgf	Static C_o SKF N	Dynamic C FAG kgf	Dynamic C FAG N	Dynamic C SKF kgf	Dynamic C SKF N	Maximum permissible speed, rpm
2301	2301	12	37	17	1.5	.51	1.3	1	1.2	.65	1.9			300	2,940			910	8,918	16,000
2302	02	15	42	17	1.5	.53	1.2	1	1.2	.65	1.8	327	3,197	335	3,283	925	9,065	925	9,065	13,000
2303	03	17	47	19	1.5	.51	1.3	1	1.2	.65	1.9	408	3,998	415	4,067	1,134	11,132	1,100	10,780	13,000
2304	2304	20	52	21	2	.48	1.4	1	1.3	.65	2.0	435	4,283	545	5,341	1,430	13,014	1,400	13,720	10,000
2305	05	25	62	24	2	.45	1.5	1	1.4	.65	2.2	771	7,556	775	7,595	1,882	18,444	1,885	18,473	10,000
2306	06	30	72	27	2	.47	1.4	1	1.3	.65	2.1	1,015	9,947	1,000	9,800	2,450	24,010	2,450	24,010	8,000
2307	2307	35	80	31	2.5	.43	1.5	1	1.5	.65	2.3	1,293	12,651	1,320	12,936	3,084	30,223	3,040	29,792	8,000
2308	08	40	90	33	2.5	.43	1.5	1	1.5	.65	2.3	1,565	15,337	1,575	15,435	3,538	34,672	3,540	34,692	6,000
2309	09	45	100	36	2.5	.43	1.5	1	1.5	.65	2.3	1,950	19,110	1,950	19,110	4,218	41,336	4,175	40,915	6,000
2310	2310	50	110	40	3	.43	1.5	1	1.5	.65	2.3	2,404	23,559	2,410	23,618	5,080	49,784	5,175	50,715	6,000
2311	11	55	120	43	3	.42	1.6	1	1.5	.65	2.3	2,858	28,008	2,860	28,028	5,760	56,454	5,750	56,350	5,000
2312	12	60	130	46	3.5	.41	1.6	1	1.5	.65	2.4	3,334	32,669	3,340	32,732	6,804	66,728	6,800	66,640	5,000
2313	2313	65	140	48	3.5	.39	1.7	1	1.6	.65	2.5	3,924	38,455	3,925	38,460	7,530	73,794	7,500	73,500	5,000
2314	14	70	150	51	3.5	.38	1.7	1	1.7	.65	2.6	4,536	44,553	4,450	43,610	8,437	82,696	8,400	82,320	4,000
2315	15	75	160	55	3.5	.38	1.7	1	1.7	.65	2.6	5,080	49,392	5,175	50,715	9,435	92,463	9,450	92,610	4,000
2316	2316	80	170	58	3.5	.37	1.8	1	1.7	.65	2.6	5,760	56,454	5,750	56,350	10,523	103,125	10,500	102,904	4,000
2317	17	85	180	60	4	.37	1.7	1	1.7	.65	2.6	6,215	60,907	6,075	59,535	10,890	106,722	10,900	106,820	3,000
2318	18	90	190	64	4	.39	1.8	1	1.6	.65	2.5	6,940	68,012	6,950	68,110	12,020	111,776	11,800	115,640	3,000
2319	2319	95	200	67	4	.38	1.7	1	1.7	.65	2.6	7,710	75,558			12,930	126,714			3,000
2320	20	100	215	73	4	.38	1.7	1	1.7	.65	2.6	9,435	92,463			14,742	144,472			2,500

TABLE 24-49
Single-row angular contact ball bearings series 02 (Indian Standards)

Bearing no. BIS	Bearing no. FAG	Bearing no. SKF	d	D	B	r	r_1	a	Static C_o FAG kgf	Static C_o FAG N	Static C_o SKF kgf	Static C_o SKF N	Dynamic C FAG kgf	Dynamic C FAG N	Dynamic C SKF kgf	Dynamic C SKF N	Maximum permissible speed, rpm
	7200B		10	30	9	1	0.5	13	215	2,107			392	3,842			13,000
	7201B		12	32	10	1	0.5	14	304	2,979			535	5,243			13,000
15BA02	7202B	7202B	15	35	11	1	0.5	16	376	3,685	375	3,675	608	5,958	620	6,076	10,000
17BA02	7203B	03B	17	40	12	1	0.8	18	480	4,704	455	4,459	785	7,693	785	7,693	10,000
20BA02	7204B	04B	20	47	14	1.5	0.8	21	662	6,488	650	6,370	1,034	10,133	1,035	10,143	10,000
25BA02	7205B	05B	25	52	15	1.5	0.8	24	785	7,693	785	7,693	1,157	11,339	1,150	11,270	10,000
30BA02	7206B	06B	30	62	16	1.5	0.8	27	1,111	10,888	1,100	10,780	1,565	15,337	1,570	15,386	8,000
35BA02	7207B	07B	35	72	17	1.5	1	31	1,520	14,896	1,500	14,700	2,110	20,678	2,110	20,678	6,000
40BA02	7208B	08B	40	80	18	2	1	34	1,882	18,444	1,885	18,473	2,495	24,451	2,500	24,500	6,000
45BA02	7209B	09B	45	85	19	2	1	37	2,155	21,119	2,160	21,168	2,860	28,028	2,820	27,636	6,000
50BA02	7210B	10B	50	90	20	2	1	39	2,313	22,667	2,360	23,128	2,900	28,420	2,900	28,420	6,000
55BA02	7211B	7211B	55	100	21	2.5	1.2	43	2,903	29,449	2,975	29,355	3,700	36,260	3,700	26,260	5,000
60BA02	7212B	12B	60	110	22	2.5	1.2	47	3,629	35,564	3,700	36,260	4,380	42,924	4,375	42,875	5,000
65BA02	7213B	13B	65	120	23	2.5	1.2	50	4,218	41,336	4,325	42,385	4,990	48,902	5,000	49,000	5,000
70BA02	7214B	14B	70	125	24	2.5	1.2	53	4,377	42,895	4,550	44,590	5,170	50,666	5,350	52,430	5,000
75BA02	7215B	15B	75	130	25	2.5	1.2	56	4,627	45,345	5,000	49,000	5,355	52,479	5,550	54,395	4,000
80BA02	7216B	16B	80	140	26	3	1.5	59	5,524	54,135	5,675	55,615	6,215	60,907	6,200	60,760	4,000
85BA02	7217B	17B	85	150	28	3	1.5	64	5,850	57,330	6,500	63,700	6,805	66,689	7,100	69,580	4,000
90BA02	7218B	18B	90	160	30	3	1.5	67	7,983	78,233	7,725	75,705	7,995	78,351	8,300	81,340	4,000
95BA02	7219B	19B	95	170	32	3.5	2	71	9,072	88,906	8,750	85,750	9,070	88,880	9,450	92,610	3,000
100BA02	7220B	20B	100	180	34	3.5	2	76	10,160	99,568	9,250	90,650	10,160	99,568	10,000	98,000	3,000
105BA02	7221B	7221B	105	190	36	3.5	2	80	11,110	108,880	10,250	101,430	11,110	108,880	10,900	106,820	2,500
110BA02	7222B	22B	110	200	38	3.5	2	84	12,020	117,796	11,500	112,700	12,020	117,796	12,000	117,600	2,500
120BA02			120	215	40	3.5	2										

Use $X_o = 1$ when $F_a/F_r \le 1.9$; $X_o = 0.5$, $Y_o = 0.26$ when $F_a/F_r > 1.9$; $X = 1$ when $F_a/F_r \le 1.14$; $X = 0.35$, $Y = 0.57$ when $F_a/F_r > 1.14$.

24.113

TABLE 24-50
Single-row angular contact ball bearings series 03 (Indian Standards)

Bearing no.			Dimensions, mm						Basic capacity — Static, C_o				Basic capacity — Dynamic, C				Maximum permissible speed, rpm
									FAG		SKF		FAG		SKF		
BIS	FAG	SKF	d	D	B	r	r_1	a	kgf	N	kgf	N	kgf	N	kgf	N	
	7300B		10	35	11	1	.5	15	380	3,724			660	6,468			10,000
	7301B		12	37	12	1.5	.8	16.5	490	4,902			830	8,134			10,000
	7302B		15	42	13	1.5	.8	18.5	610	5,978			1,020	9,996			10,000
17BA03	7303B	7303B	17	47	14	1.5	0.8	21	785	7,673	725	7,110	1,250	12,250	1,150	11,270	8,000
20BA03	7304B	04B	20	52	15	2	1	23	960	9,228	830	8,140	1,475	14,455	1,385	13,573	8,000
25BA03	7305B	05B	25	62	17	2	1	27	1,385	13,573	1,250	12,258	2,040	19,992	1,930	18,914	8,000
30BA03	7306B	7306B	30	72	19	2	1	31	1,770	17,346	1,615	15,838	2,540	24,892	2,450	24,010	6,000
35BA03	7307B	07B	35	80	21	2.5	1.2	35	2,180	21,364	2,040	20,005	3,085	30,233	2,860	28,028	6,000
40BA03	7308B	08B	40	90	23	2.5	1.2	39	2,860	28,028	2,540	24,909	3,925	38,465	3,540	34,692	6,000
45BA03	7309B	7309B	45	100	25	2.5	1.2	43	3,540	34,692	3,400	33,342	4,630	45,374	4,550	44,590	5,000
50BA03	7310B	10B	50	110	27	3	1.5	47	4,220	41,356	4,100	40,201	5,445	53,361	5,250	51,650	5,000
55BA03	7311B	11B	55	120	29	3	1.5	52	4,900	48,020	4,750	46,581	6,080	59,584	6,100	59,780	5,000
60BA03	7312B	7312B	60	130	31	3.5	2	55	5,670	55,566	5,450	53,446	7,080	69,384	7,100	69,580	5,000
65BA03	7313B	13B	65	140	33	3.5	2	60	6,490	63,602	6,250	61,291	7,985	78,253	8,000	78,400	4,000
70BA03	7314B	14B	70	150	35	3.5	2	64	7,395	72,471	7,400	72,569	8,890	87,122	8,900	87,220	4,000
75BA03	7315B	7315B	75	160	37	3.5	2	68	8,620	84,476	8,200	80,414	9,980	97,804	9,850	96,530	4,000
80BA03	7316B	16B	80	170	39	3.5	2	72	9,980	97,804	9,100	89,240	10,880	106,724	10,500	102,900	4,000
85BA03	7317B	17B	85	180	41	4	2	76	10,900	106,820	10,000	98,066	11,795	115,590	11,375	111,475	3,000
90BA03	7318B	7318B	90	190	43	4	2	80	12,250	120,050	11,375	111,550	12,700	124,460	12,250	120,050	3,000
95BA03	7319B	19B	95	200	45	4	2	84	13,155	129,190	12,500	122,583	13,610	133,378	13,130	129,164	3,000
100BA03	7320B	20B	100	215	47	4	2	90	15,420	151,116	15,200	149,060	15,200	148,960	14,750	144,550	2,500
105BA03	7321B	7321B	105	225	49	4	2	94	17,240	168,872	16,325	160,093	16,100	157,780	15,625	153,125	2,500
110BA03	7322B	22B	110	240	50	4	2	99	19,505	192,149	19,250	188,777	17,240	168,872	17,225	168,805	2,500

Use $X_o = 1$ when $F_a/F_r \leq 1.9$; $X_o = 0.5$, $Y_o = 0.26$ when $F_a/F_r > 1.9$; $X = 1$ when $F_a/F_r \leq 1.14$; $X = 0.35$, $Y = 0.57$ when $F_a/F_r > 1.14$.

TABLE 24–51
Double-row angular contact ball bearings series 33

Bearing no. FAG	Bearing no. SKF	d	D	B	r	a	Static C_o FAG kgf	Static C_o FAG N	Static C_o SKF kgf	Static C_o SKF N	Dynamic C FAG kgf	Dynamic C FAG N	Dynamic C SKF kgf	Dynamic C SKF N	Maximum permissible speed, rpm
3302	3302A	15	42	19	1.5	30	1,035	10,243	925	9,065	1,315	12,887	1,400	13,720	10,000
3303	03A	17	47	22.2	1.5	34	1,475	14,455	1,290	12,642	1,840	18,032	1,930	18,914	8,000
3304	04A	20	52	22.2	2	36	1,540	15,092	1,400	13,720	1,880	18,424	1,930	18,914	8,000
3305	3305A	25	62	25.4	2	43	2,220	21,750	2,000	19,600	2,590	25,382	2,860	26,008	6,000
3306	06A	30	72	30.2	2	51	3,040	29,792	2,770	27,146	3,350	32,814	3,600	35,280	6,000
3307	07A	35	80	34.9	2.5	56	4,000	39,200	3,630	35,574	4,380	42,924	4,450	43,615	5,000
3308	3308A	40	90	36.5	2.5	64	5,170	50,666	4,550	44,590	5,355	52,479	5,450	53,410	5,000
3309	09A	45	100	39.7	2.5	72	6,490	63,602	5,550	54,390	6,490	63,602	6,350	62,230	4,000
3310	10A	50	110	44.4	2.5	79	7,985	78,353	7,400	72,520	7,980	78,164	8,775	85,995	4,000
3311	3311A	55	120	49.2	3	87	9,255	90,699	8,000	78,400	9,070	88,896	8,760	85,840	4,000
3312	12A	60	130	54	3	96	10,890	106,722	9,650	94,570	10,340	101,332	10,000	98,000	4,000
3313	13A	65	140	58.7	3.5	102	12,700	124,460	11,110	108,870	12,020	117,796	11,800	115,640	3,000
3314	3314A	70	150	63.5	3.5	109	14,290	140,042	12,900	126,430	13,155	128,919	13,850	135,730	3,000
3315	15A	75	160	68.3	3.5	117	16,100	157,780	14,080	127,940	14,740	144,520	14,300	140,740	3,000
3316	16A	80	170	68.3	3.5	123	17,690	173,361	15,700	153,860	16,100	157,780	16,100	157,780	2,500
3317	3317A	85	180	73	4	131	20,640	202,272	17,700	173,460	18,145	177,821	17,700	173,460	2,500
3318	18A	90	190	73	4	136	23,590	228,182	21,000	205,800	19,960	193,608	20,410	200,018	2,500
3319	19A	95	200	77.8	4	143	25,855	253,379			21,090	206,682			2,500
3320	20A	100	215	82.6	4	153	29,030	284,494			23,135	224,723			2,500

Note: These bearings are provided with filling slots on one side; in case of unidirectional thrust loads, the bearings should be so arranged in mounting that the balls on the slot side are relieved from load.

Use $X_o = 1, Y_o = 0.58$ and $X = 1, Y = .66$ when $F_a/F_r \le 0.95; X = .6, Y = 1.07$ when $F_a/F_r > 0.95$.

TABLE 24-52
Cylindrical roller bearings series 02 (Indian Standards)

Bearing no.			Dimensions, mm						Basic capacity								Maximum permissible speed, rpm
									Static, C_o				Dynamic, C				
									FAG		SKF		FAG		SKF		
BIS	FAG	SKF	d	D	B	r	r_1	E	kgf	N	kgf	N	kgf	N	kgf	N	
10RN02			10	30	9	1.0											
12RN02			12	32	10	1.0											
15RN02	N203		15	35	11	1.0		33.9	480	4,704			945	9,261			
17RN02	N204	N204	17	40	12	1.0	0.5	40	680	6,724	695	6,811	1,360	13,328	1,090	10,502	13,000
20RN02	N205	205	20	47	14	1.5	1	45	860	8,428	845	8,281	1,520	14,896	1,450	14,210	13,000
25RN02	N206	206	25	52	15	1.5	1	53.5	1,180	11,564	1,155	11,319	2,065	20,237	2,175	21,315	10,000
30RN02	N207	N207	30	62	16	1.5	1	61.8	1,770	17,346	1,355	13,279	2,970	29,106	2,855	27,979	10,000
35RN02	N208	208	35	72	17	2.0	2	70	2,450	24,010	2,315	22,687	3,925	38,465	2,970	29,106	8,000
40RN02	N209	209	40	83	18	2.0	2	75	2,585	25,333	2,495	24,510	4,150	40,670	3,150	30,870	8,000
45RN02	N210	N210	45	85	19	2.0	2	80.4	2,810	27,538	2,720	26,656	4,310	42,238	3,925	38,465	8,000
50RN02	N211	211	50	90	20	2.0	2	88.5	3,225	31,005	3,265	31,997	4,990	48,902	4,445	43,561	8,000
55RN02	N212	212	55	100	21	2.5	2.5	97.5	4,220	41,356	3,990	39,102	6,215	60,907	5,260	51,548	6,000
60RN02	N213	N213	60	110	22	2.5	2.5	105.6	4,990	48,902	4,715	46,207	7,395	72,471	5,445	53,401	6,000
65RN02	N214	214	65	120	23	2.5	2.5	110.5	4,990	48,902	4,990	48,902	7,395	72,471	6,350	62,230	6,000
70RN02	N215	215	70	130	24	2.5	2.5	116.5	6,215	60,907	5,760	56,248	8,890	87,122	7,395	72,471	5,000
75RN02	N216	N216	80	140	25	2.5	2.5	125.3	6,805	66,689	6,805	66,689	9,800	96,530	8,300	81,340	5,000
80RN02	N217	217	85	150	26	3.0	3	133.8	7,845	76,881	7,845	76,781	10,885	106,553	10,160	99,568	5,400
85RN02	N218	218	90	160	28	3.0	3	143	9,615	94,227	9,255	90,799	13,610	133,778	11,565	113,357	5,000
90RN02	N219	N219	95	170	30	3.0	3	151.5	10,705	104,909	10,885	106,553	14,740	144,452	12,925	126,665	4,000
95RN02	N220	220	100	180	32	3.5	3.5	160	12,245	119,995	12,245	120,001	16,330	160,034	14,285	139,993	4,000
100RN02	N221	N221	105	190	34	3.5	3.5	168.8	13,610	133,778	13,835	135,583	18,145	177,821	16,330	160,034	3,000
105RN02	N222	222	110	200	36	3.5	3.5	178.5	16,100	157,780	15,195	148,910	21,770	213,346	18,380	180,026	3,000
110RN02	N224	224	120	215	38	3.5	3.5	191.5	17,690	173,362	17,680	173,360	23,585	231,133	19,275	188,895	3,000
120RN02	N226	226	130	230	40	4.0	4.0	204	19,960	195,743	18,825	181,485	25,400	248,920			2,500

Use $X_o = X = 1$; $Y_o = Y = 0$.

24.116

TABLE 24-53
Cylindrical roller bearings series 03 (Indian Standards)

Bearing no.			Dimensions, mm						Basic capacity								Maximum permissible speed, rpm
									Static, C_o				Dynamic, C				
									FAG		SKF		FAG		SKF		
BIS	FAG	SKF	d	D	B	r	r_1	E	kgf	N	kgf	N	kgf	N	kgf	N	
10RN03			10	35	11	1.0											
12RN03			12	37	12	1.5											
15RN03			15	42	13	1.5											
17RN03	N303		17	47	14	1.5	1	39.1	844	8,277			1,542	15,112			13,000
20RN03	N304	N304	20	52	15	2.0	2	44.5	1,135	11,230	961	9,418	2,040	19,992	1,385	13,573	10,000
25RN03	N305	305	25	62	17	2.0	2	53	1,475	14,550	1,385	13,573	2,585	25,433	1,880	18,474	10,000
30RN03	N306	306	30	72	19	2.0	2	62	2,040	19,992	1,930	18,914	3,470	34,006	2,495	24,551	8,000
35RN03	N307	N307	35	80	21	2.5	2	68.2	2,595	25,433	2,360	23,308	4,310	42,238	3,085	30,233	8,000
40RN03	N308	308	40	90	23	2.5	2.5	77.5	3,150	30,890	3,085	30,233	5,080	49,784	3,855	37,779	8,000
45RN03	N309	309	45	100	25	2.5	2.5	86.5	4,380	42,952	3,925	38,465	6,940	68,012	4,990	48,902	6,000
50RN03	N310	N310	50	110	27	2.5	2.5	95	5,535	54,280	4,900	48,020	8,440	82,712	5,850	56,330	6,000
55RN03	N311	311	55	120	29	3.0	3	104.5	6,350	62,272	5,760	56,448	9,840	96,530	7,395	72,441	5,000
60RN03	N312	312	60	130	31	3.0	3	113	7,260	71,196	7,255	71,099	10,900	106,820	8,620	84,476	5,000
65RN03	N313	N313	65	140	33	3.5	3	121.5	8,165	80,070	8,165	80,017	12,250	120,050	9,435	92,463	5,000
70RN03	N314	314	70	150	35	3.5	3.5	130	9,800	96,105	8,890	87,122	14,290	140,042	10,525	103,145	4,000
75RN03	N315	315	75	160	37	3.5	3.5	139.5	12,020	117,796	10,889	106,673	17,010	166,698	12,925	126,665	4,000
80RN03	N316	N316	80	170	39	3.5	3.5	147	12,020	117,796	12,020	117,796	17,010	166,698	13,835	136,582	4,000
85RN03	N317	317	85	180	41	4.0	3.5	156	14,515	142,247	13,155	128,919	20,410	200,018	14,740	144,452	3,000
90RN03	N318	318	90	190	43	4.0	4	165	15,195	149,011	15,420	151,116	21,550	211,190	17,010	166,698	3,000
95RN03	N319	N319	95	200	45	4.0	4	173.5	17,280	169,334	16,555	162,239	24,500	241,100	18,825	184,485	3,000
100RN03	N320	320	100	215	47	4.0	4	185.5	20,640	201,972	19,504	191,139	28,580	280,084	21,770	213,346	2,500
105RN03	N321	321	105	225	49	4.0	4	195	24,040	235,592	22,225	218,905	32,660	319,068	24,495	241,961	2,500
110RN03	N322	N322	110	240	50	4.0	4	207	28,125	275,625	25,855	253,379	36,970	362,306	28,125	275,625	2,500
120RN03	N324	324	120	260	50	4.0	4	226	32,205	315,609	29,710	291,158	43,100	422,380	32,660	319,068	2,500
	N326	326	130	280	58	4.0	4	243	39,335	384,503	39,235	384,503	49,900	489,020	40,825	400,850	2,000
	N328	328	140	300	62	5.0	5	260	43,090	422,566	43,770	428,946	55,340	542,332	45,355	444,479	2,000
	N330	N330	150	320	65	5.0	5	277	48,080	471,184	48,940	479,552	59,880	591,824	49,900	489,020	2,000
	N332	332	160	340	68	5.0	5	292	50,800	497,840	51,715	506,807	63,500	622,300	52,620	515,676	2,000
	N334	334	170	360	72	5.0	5	310	58,510	573,398	58,510	573,398	73,830	723,134	58,510	573,398	1,600
	N336	N336	180	380	75	5.0	5	328	68,040	666,792	69,400	686,120	83,010	813,498	68,050	666,090	1,600
	N338	338	190	400	78	6.0	6	345	73,930	725,002	75,290	738,339	87,540	857,892	73,930	724,514	1,600
	N340	340	200	420	80	6.0	6	360	73,930	725,002	75,290	738,339	87,540	857,892	73,930	724,514	1,600

Use $X_o = X = 1$; $Y_o = Y = 0$.

TABLE 24-54
Cylindrical roller bearings series 04 (Indian Standards)

BIS	FAG	SKF	d	D	B	r	E	Static C_o FAG kgf	FAG N	SKF kgf	SKF N	Dynamic C FAG kgf	FAG N	SKF kgf	SKF N	Maximum permissible speed, rpm
15RN04			15	52	15	2.0										
17RN04			17	62	17	2.0										
20RN04			20	72	19	2.0										
25RN04	N405		25	80	21	2.5	62.8	2,360	23,128	2,220	21,756	4,080	40,054	3,040	29,792	8,000
30RN04	N406	N406	30	90	23	2.5	73	3,365	32,683	3,085	30,233	5,445	53,561	4,220	41,356	8,000
35RN04	N407	407	35	100	25	2.5	83	4,310	42,238	3,990	39,102	6,940	68,012	5,170	50,666	6,000
40RN04	N408	408	40	110	27	2.5	92	5,260	51,548	5,170	50,666	8,435	82,683	6,620	64,876	6,000
45RN04	N409	409	45	120	29	3.0	100.5	6,485	63,553	5,760	56,448	10,160	99,568	7,530	73,794	6,000
50RN04	N410	410	50	130	31	3.5	110.8	8,300	81,340	7,395	72,471	12,475	122,255	9,070	88,886	5,000
55RN04	N411	N411	55	140	33	3.5	117.2	8,300	81,340	8,165	80,017	12,700	124,460	9,795	95,991	5,000
60RN04	N412	412	60	150	35	3.5	127	10,160	99,568	9,795	95,991	15,195	148,910	12,020	109,796	5,000
65RN04	N413	413	65	160	37	3.5	135.3	12,245	110,001	10,705	104,909	17,235	168,903	13,155	128,919	4,000
70RN04	N414	414	70	180	42	4.0	152	15,195	148,911	14,060	137,788	21,545	211,141	16,330	160,034	4,000
75RN04	N415	415	75	190	45	4.0	160.5	16,100	157,780	16,100	157,780	23,585	231,133	19,275	188,895	4,000
80RN04	N416	416	80	200	48	4.0	170	18,825	184,485	18,370	180,026	27,215	259,707	21,545	211,141	3,000
85RN04	N417	417	85	210	52	5.0	177	21,770	213,446	21,090	206,682	30,390	297,822	24,495	240,051	3,000
90RN04	N418	418	90	225	54	5.0	191.5	24,040	235,592	24,040	235,592	32,660	320,068	27,670	271,166	3,000
95RN04	N419	419	95	240	55	5.0	201.5	25,855	253,379	26,535	260,043	36,285	355,593	29,710	283,158	3,000
100RN04	N420	420	100	250	58	5.0	211	29,710	283,158	29,710	283,158	39,915	391,167	32,660	320,068	2,500

Use $X_o = X = 1; Y_o = Y = 0.$

TABLE 24-55
Cylindrical roller bearings series NU 22

Bearing no. FAG	Bearing no. SKF	d	D	B	r	r₁	F	Static Cₒ FAG kgf	Static Cₒ FAG N	Static Cₒ SKF kgf	Static Cₒ SKF N	Dynamic C FAG kgf	Dynamic C FAG N	Dynamic C SKF kgf	Dynamic C SKF N	Maximum permissible speed, rpm
NU2203		17	40	16	1	0.5	22.9	785	7,693			1,385	13,673			
NU2204		20	47	18	1.5	1	27.0	1,070	10,396			1,840	18,032			
NU2205	NU2205	25	52	18	1.5	1	32	1,290	12,642	1,225	12,005	2,110	20,678	1,610	15,778	13,000
NU2206	2206	30	62	20	1.5	1	38.5	1,880	18,424	1,730	16,954	2,970	29,106	2,360	23,128	13,000
NU2207	2207	35	72	23	2	1	43.5	2,970	29,106	2,820	27,636	4,446	43,561	3,630	35,574	13,000
NU2208	2208	40	80	23	2	2	50	3,630	35,574	3,340	32,712	5,260	51,548	4,150	40,670	10,000
NU2209	2209	45	85	23	2	2	55	3,925	38,465	3,630	35,574	5,535	54,243	4,450	43,610	10,000
NU2210	2210	50	90	23	2	2	60.4	4,150	40,670	3,930	38,514	5,760	56,348	4,640	42,712	8,000
NU2211	NU2211	55	100	25	2.5	2	66.5	4,720	46,256	4,640	45,472	6,620	64,876	5,360	52,528	8,000
NU2212	2212	60	110	28	2.5	2	73.5	6,805	66,689	6,100	49,780	8,890	87,122	7,100	69,580	8,000
NU2213	2213	65	120	31	2.5	2	79.6	8,300	81,340	7,550	73,990	10,700	104,860	8,325	81,585	6,000
NU2214	2214	70	125	31	2.5	2.5	84.5	8,300	81,340	8,000	78,400	10,700	104,860	8,760	85,848	6,000
NU2215	2215	75	130	31	2.5	2.5	88.5	9,440	92,512	8,640	84,672	12,250	120,050	9,825	96,285	5,000
NU2216	NU2216	80	140	33	3	2.5	95.3	10,500	102,900	10,000	98,000	13,835	135,583	11,150	109,270	5,000
NU2217	2217	85	150	36	3	3	101.8	12,475	121,855	12,000	117,600	15,420	151,116	12,950	126,910	5,000
NU2218	2218	90	160	40	3	3	107	14,515	142,247	13,850	135,730	18,370	180,026	14,300	140,140	5,000
NU2219	2219	95	170	43	3.5	3	113.5	16,555	162,239	16,350	160,230	21,090	206,682	17,700	173,460	4,000
NU2220	2220	100	180	46	3.5	3.5	120	18,800	184,240	18,850	184,730	23,590	231,182	20,000	196,000	4,000
NU2222	NU2222	110	200	53	3.5	3.5	132.5	25,400	248,920	23,090	226,282	30,390	297,822	25,000	245,000	4,000
NU2224	2224	120	215	58	3.5	3.5	143.5	28,125	275,625	27,700	271,460	33,340	326,632	28,600	280,280	4,000
NU2226	2226	130	230	64	4	4	156	31,525	308,945	31,275	306,495	36,970	362,106	30,250	296,450	3,000
NU2228	2228	140	250	68	4	4	169	37,650	368,900	38,600	378,280	49,900	489,020	37,000	362,600	3,000
NU2230	2230	150	270	73	4	4	182	44,455	435,659	44,500	436,100	55,790	547,642	43,200	423,360	3,000
NU2232	2232	160	290	80	4	4	195	52,620	515,676			57,605	564,529			3,000
NU2234	2234	170	310	86	5	5	208	59,875	586,775			66,220	648,956			2,500
NU2236	2236	180	320	86	5	5	218	63,500	622,300			69,400	680,120			2,500
NU2238	2238	190	340	92	5	5	231	70,310	689,038			77,110	755,580			
NU2240	2240	200	360	98	5	5	244	79,830	782,334			84,370	826,826			

Use $X_o = X = 1$; $Y_o = Y = 0$.

TABLE 24-56
Cylindrical roller bearings series NU23

Bearing no. FAG	Bearing no. SKF	d	D	B	r	r₁	F	Static C_o FAG kgf	Static C_o FAG N	Static C_o SKF kgf	Static C_o SKF N	Dynamic C FAG kgf	Dynamic C FAG N	Dynamic C SKF kgf	Dynamic C SKF N	Maximum permissible speed, rpm
NU2304	—	20	52	21	2	1	28.5	1,840	18,032			2,970	29,036			100,00
NU2305	NU2305	25	62	24	2	2	35	2,495	24,451	2,170	21,266	3,925	38,465	3,080	30,184	10,000
NU2306	2306	30	72	27	2	2	42	3,040	29,792	2,695	26,411	4,630	45,374	3,600	35,676	8,000
NU2307	2307	35	80	31	2.5	2.5	46.2	3,855	37,779	3,130	30,674	5,670	55,566	4,260	41,748	8,000
NU2308	NU2308	40	90	33	2.5	2.5	53.5	5,080	49,784	4,820	47,236	7,395	72,451	6,000	58,800	8,000
NU2309	2309	45	100	36	3	2.5	58.5	6,020	64,876	5,680	55,664	9,435	92,463	7,550	73,990	6,000
NU2310	2310	50	110	40	3	3	65	8,620	84,476	7,550	73,990	11,795	115,591	9,275	90,915	6,000
NU2311	NU2311	55	120	43	3	3	70.5	9,435	92,483	8,460	82,903	13,610	123,378	10,900	106,820	6,000
NU2312	2312	60	130	46	3.5	3.5	77	10,890	105,830	10,550	103,390	15,195	148,911	13,180	126,164	5,000
NU2313	2313	65	140	48	3.5	3.5	83.5	12,700	124,460	12,250	120,050	17,000	166,660	14,550	142,590	5,000
NU2314	NU2314	70	150	51	3.5	3.5	90	15,420	151,116	14,100	138,180	20,415	200,067	15,900	156,604	5,000
NU2315	2315	75	160	55	3.5	3.5	95.5	19,505	191,149	17,000	166,600	24,950	244,510	20,410	200,018	4,000
NU2316	2316	80	170	58	3.5	3.5	103	19,505	191,149	18,850	184,730	24,950	244,510	21,800	213,640	4,000
NU2317	NU2317	85	180	60	4	4	108	22,680	222,264	20,000	196,000	29,030	284,494	23,190	227,272	4,000
NU2318	2318	90	190	64	4	4	115	23,155	226,723	22,780	221,760	29,710	291,158	25,900	253,820	3,000
NU2319	2319	95	200	67	4	4	121.5	29,030	284,494	26,590	259,700	36,290	355,642	29,790	292,242	3,000
NU2320	NU2320	100	215	73	4	4	129.5	34,700	340,060	31,600	309,680	42,185	413,413	34,790	340,942	3,000
NU2322	2322	110	240	80	4	4	143	47,175	463,315	43,200	423,360	55,340	542,332	46,400	454,720	2,500
NU2324	2324	120	260	86	4	4	154	54,430	533,414	50,900	498,820	64,860	635,628	56,850	557,130	2,500
NU2326	NU2326	130	280	93	5	5	167	66,220	648,956	65,000	637,000	77,110	755,678	62,200	609,560	2,500
NU2328	2328	140	300	102	5	5	180	73,930	724,514	74,100	726,180	84,390	827,827	75,100	735,980	2,000
NU2330	2330	150	320	108	5	5	193	84,370	826,826	83,200	815,360	94,350	924,630	83,200	815,360	2,000
NU2332		160	340	114	5	5	208	88,900	871,220			99,700	977,000			
NU2334		170	360	120	5	5	220	103,420	1,013,516			113,100	1,108,380			
NU2336		180	380	126	5	5	232	117,930	1,155,714			129,300	1,267,140			
NU2338		190	400	132	6	6	245	131,540	1,289,092			140,610	1,377,978			
NU2340		200	420	138	6	6	260	131,540	1,289,092			140,610	1,377,978			

TABLE 24-57A
Taper roller bearings series 322

Bearing no. FAG	Bearing no. SKF	d	D	B	T	C	r	r₁	a	e	Y	Y₀	Static C₀ FAG kgf	Static C₀ FAG N	Static C₀ SKF kgf	Static C₀ SKF N	Dynamic C FAG kgf	Dynamic C FAG N	Dynamic C SKF kgf	Dynamic C SKF N	Maximum permissible speed, rpm
32206A	32206	30	62	20	21.25	17	1.5	0.5	15	.38	1.6	.9	4,150	40,670	2,770	27,146	4,375	42,875	3,230	31,654	6,000
32207A	07	35	72	23	24.25	19	2	0.8	18	.38	1.6	.9	5,080	49,284	3,700	36,260	5,760	56,448	4,225	41,405	6,000
32208A	08	40	80	23	24.75	19	2	0.8	19	.38	1.6	.9	5,445	53,361	4,100	40,180	6,350	62,150	4,640	44,472	6,000
32209A	32209	45	85	23	24.75	19	2	0.8	20	.41	1.5	.8	5,760	56,448	4,640	44,478	6,485	63,553	5,090	49,882	5,000
32210A	10	50	90	23	24.75	19	2	0.8	21	.42	1.4	.8	6,485	63,553	4,820	47,726	7,255	71,099	5,260	51,548	5,000
32211A	11	55	100	25	26.75	21	2.5	0.8	22	.41	1.5	.8	7,985	78,953	6,225	61,005	8,890	87,122	6,640	65,072	5,000
32212A	32212	60	110	28	29.75	24	2.5	0.8	24	.41	1.5	.8	9,980	97,804	7,720	75,656	10,750	104,909	8,000	78,600	4,000
32213A	13	65	120	31	32.75	27	2.5	0.8	26	.41	1.5	.8	11,795	115,591	9,260	90,748	12,700	124,460	9,800	96,047	4,000
32214A	14	70	125	31	33.25	27	2.5	0.8	28	.42	1.4	.8	12,700	124,460	9,260	90,748	13,610	133,378	9,800	96,047	4,000
32215A	32215	75	130	31	33.25	27	2.5	0.8	29	.44	1.4	.7	13,610	133,378	10,190	99,862	14,060	137,788	10,350	101,430	3,000
32216A	16	80	140	33	33.25	28	3	1	30	.42	1.4	.8	15,195	148,911	11,600	113,630	16,100	157,780	12,000	117,600	3,000
32217A	17	85	150	36	38.5	30	3	1	33	.42	1.4	.8	17,690	177,422	13,750	133,750	18,370	180,026	13,850	135,730	3,000
32218A	32218	90	160	40	42.5	34	3	1	36	.42	1.4	.8	21,100	206,780	18,400	180,320	21,545	211,141	16,170	158,466	2,500
32219A	19	95	170	43	45.5	37	3.5	1.2	38	.42	1.4	.8	24,950	244,510	18,400	180,320	24,950	244,510	18,400	170,320	2,500
32220A	20	100	180	46	49	39	3.5	1.2	41	.42	1.4	.8	28,875	282,975	21,180	207,564	28,125	255,385	20,650	202,370	2,500
32221A	32221	105	190	50	53	43	3.5	1.2	44	.42	1.4	.8	32,660	320,068	24,600	241,080	32,205	315,829	24,100	236,180	2,000
32222A	22	110	200	53	56	46	3.5	1.2	46	.42	1.4	.8	35,380	346,724	27,700	271,460	34,020	333,396	25,900	255,820	2,000
32224A	24	120	215	58	61.5	50	3.5	1.2	52	.28	2.1	2.2	34,700	340,060	34,100	334,180	36,285	355,393	30,400	297,920	2,000
32226A	32226	130	230	64	67.75	54	4	1.5	56						41,600	407,680			37,700	369,460	1,600
	28	140	250	68	71.75	58	4	1.5	60						49,100	481,180			43,100	422,380	1,600
	30	150	270	73	77	60	4	1.5	64						54,500	534,100			49,000	448,402	1,600

TABLE 24–57B
Dimensions for tapered roller bearings series 02 and 22

		Dimension series 02		Dimension series 22		Chamfer	
Bore diameter	Outside diameter	Inner ring width	Bearing width	Inner ring width	Bearing width		
d	D	B	T	B	T	r	r_1
10	30	9	9.7	14	14.7	1.0	0.3
12	32	10	10.75	14	14.75	1.0	0.3
15	35	11	11.75	14	14.75	1.0	0.3
17	40	12	13.25	16	17.25	1.5	0.5
20	47	14	15.25	18	19.25	1.5	0.5
25	52	15	16.25	18	19.25	1.5	0.5
30	62	16	17.25	20	21.25	1.5	0.5
35	72	17	18.25	23	24.25	2.0	0.8
40	80	18	19.25	23	24.75	2.0	0.8
45	85	19	20.75	23	24.75	2.0	0.8
50	90	20	21.75	23	24.75	2.0	0.8
55	100	21	22.75	25	26.75	2.5	0.8
60	110	22	23.75	28	29.75	2.5	0.8
65	120	23	24.75	31	32.75	2.5	0.8
70	125	24	26.25	31	33.25	2.5	0.8
75	130	25	27.25	31	33.25	2.5	0.8
80	140	26	28.25	33	35.25	3.0	1.0
85	150	28	30.50	36	38.50	3.0	1.0
90	160	30	32.50	40	42.50	3.0	1.0
95	170	32	34.50	43	45.50	3.5	1.2
100	180	34	37.00	46	49.00	3.5	1.2
105	190	36	39.00	50	53.00	3.5	1.2
110	200	38	41.00	53	56.00	3.5	1.2
120	215	40	43.50	58	61.50	3.5	1.2
130	230	40	43.75	64	67.75	4.0	1.5
140	250	42	45.75	68	71.75	4.0	1.5
150	270	45	49.00	73	77.00	4.0	1.5
160	290	48	52.00	80	84.00	4.0	1.5
170	310	52	57.00	86	91.00	5.0	2.0
180	320	52	57.00	86	91.00	5.0	2.0
190	340	55	60.00	92	97.00	5.0	2.0
200	360	58	64.00	98	104.00	5.0	2.0
220	400	65	71.00	108	114.00	5.0	2.0
240	440	72	79.00	120	127.00	5.0	2.5

Source: IS 3697–1966

TABLE 24–57C
Dimensions for tapered roller bearings series 03 and 23

		Dimension series 03		Dimension series 23			
		Inner ring width	Bearing width	Inner ring width	Bearing width	Chamfer	
Bore diameter	Outside diameter						
d	D	B	T	B	T	r	r_1
10	35	11	11.9	17	17.9	1.0	0.3
12	37	12	12.9	17	17.9	1.5	0.5
15	42	13	14.25	17	18.25	1.5	0.5
17	47	14	15.25	19	20.25	1.5	0.5
20	52	15	16.25	21	22.25	2.0	0.8
25	62	17	18.25	24	25.25	2.0	0.8
30	72	19	20.75	27	28.75	2.0	0.8
35	80	21	22.75	31	32.75	2.5	0.8
40	90	23	25.25	33	35.25	2.5	0.8
45	100	25	27.75	36	38.25	2.5	0.8
50	110	27	29.25	40	42.25	3.0	1.0
55	120	29	31.50	43	45.50	3.0	1.0
60	130	31	33.50	46	48.50	3.5	1.2
65	140	33	36.0	48	51.00	3.5	1.2
70	150	35	38.00	51	54.00	3.5	1.2
75	160	37	40.0	55	58.0	3.5	1.2
80	170	39	42.5	58	61.50	3.5	1.2
85	180	41	44.5	60	63.50	4.0	1.5
90	190	43	46.5	64	67.5	4.0	1.5
95	200	45	49.5	67	71.5	4.0	1.5
100	215	47	51.5	73	77.5	4.0	1.5
105	225	49	53.5	77	81.5	4.0	1.5
110	240	50	54.5	80	84.5	4.0	1.5
120	260	55	59.5	86	90.5	4.0	1.5
130	280	58	63.75	93	98.75	5.0	2.0
140	300	62	67.25	102	107.75	5.0	2.0
150	320	65	72.0	108	114.0	5.0	2.0
160	340	68	75.0			5.0	2.0
170	360	72	80.0			5.0	2.0
180	380	75	83.0			5.0	2.0
190	400	78	86.0			6.0	2.5
200	420	88	89.0			6.0	2.5
220	460	88	97.0			6.0	2.5
240	500	95	105.0			6.0	2.5

Source: IS 3697–1966

TABLE 24-58
Single-thrust ball bearing series 11 (Indian Standards)

Bearing no.			Dimensions, mm						Minimum load constant, M	Basic capacity								Maximum permissible speed, rpm
										Static, C_o				Dynamic, C				
										FAG		SKF		FAG		SKF		
BIS	FAG	SKF	d	C	D	B	E	r		kgf	N	kgf	N	kgf	N	kgf	N	
10TA11	51100	51100	10	11	24	9	24	0.5	0.26	1,150	11,270	1,405	13,769	785	7,693	1,000	9,800	10,000
12TA11	51101	01	12	13	26	9	26	0.5	0.35	1,245	12,201	1,540	15,092	815	7,987	1,035	10,143	10,000
15TA11	51102	02	15	16	28	9	28	0.5	0.44	1,385	13,273	2,040	19,992	830	8,134	1,225	12,005	8,000
17TA11	51103	03	17	18	30	9	30	0.5	0.55	1,565	15,533	2,175	21,315	890	8,722	1,220	12,446	8,000
20TA11	51104	04	20	21	35	10	35	0.5	1.0	2,155	21,119	3,085	30,233	1,180	11,564	1,700	16,660	8,000
25TA11	51105	05	25	26	42	11	42	1	1.8	2,900	28,420	4,150	40,670	1,430	14,014	2,155	21,119	6,000
30TA11	51106	51106	30	32	47	11	47	1	2.2	3,265	31,997	4,810	47,138	1,450	14,210	2,270	22,246	6,000
35TA11	51107	07	35	37	52	12	52	1	3.1	3,855	34,779	6,350	62,230	1,540	15,092	3,040	29,792	6,000
40TA11	51108	08	40	42	60	13	60	1	5.5	5,080	49,784	7,710	75,558	2,110	20,678	3,470	33,706	5,000
45TA11	51109	09	45	47	65	14	65	1	7.1	5,535	54,243	8,620	84,476	2,175	21,315	3,695	36,211	5,000
50TA11	51110	10	50	52	70	14	70	1	7.9	5,985	58,653	9,070	88,886	2,220	21,756	3,735	36,897	4,000
55TA11	51111	11	55	57	78	16	78	1	12	7,395	72,471	13,155	128,919	2,720	26,656	5,445	53,361	4,000
60TA11	51112	51112	60	62	85	17	85	1.5	18	9,070	88,886	14,515	139,247	3,265	31,997	5,670	55,596	3,000
65TA11	51113	13	65	67	90	18	90	1.5	20	9,615	94,227	15,420	151,116	3,400	33,320	5,760	56,448	3,000
70TA11	51114	14	70	72	95	18	95	1.5	22	10,161	99,568	16,100	157,780	3,400	33,320	5,850	57,010	2,500
75TA11	51115	15	75	77	100	19	100	1.5	31	11,340	111,132	17,000	166,600	3,855	37,779	5,985	58,653	2,500
80TA11	51116	16	80	82	105	19	105	1.5	35	12,245	120,001	17,690	173,372	3,925	38,465	6,080	59,584	2,500
85TA11	51117	17	85	87	110	19	110	1.5	35	12,700	123,760	21,770	213,346	3,990	39,102	7,530	73,794	2,000
90TA11	51118	51118	90	92	120	22	120	1.5	49	15,195	148,911	27,215	266,707	4,625	45,325	9,070	88,886	2,000
100TA11	51120	20	100	102	135	25	135	1.5	100	21,545	211,131	34,020	333,396	6,620	64,876	11,340	111,132	2,000
110TA11	51122	22	110	112	145	25	145	1.5	120	23,135	226,723	37,650	368,970	6,800	66,640	12,020	117,796	1,600
120TA11	51124	24	120	122	155	25	155	1.5	140	24,950	244,510	39,235	384,503	6,940	68,012	12,020	117,796	1,600
130TA11	51126	26	130	132	170	30	170	1.5	200	28,575	280,035	50,805	497,569	8,165	80,017	15,650	153,370	1,300
140TA11	51128	28	140	142	178	31	180	1.5	220	31,525	308,945	55,350	542,430	8,300	81,340	16,100	157,780	1,300
150TA11	51130	51130	150	152	188	31	190	1.5	240	33,340	326,732	58,510	573,398	8,435	82,663	17,235	168,903	1,000
160TA11	51132	32	160	162	198	31	200	1.5	260	35,380	345,724	60,785	595,693	8,755	85,799	17,690	173,362	1,000
170TA11	51134	34	170	172	213	34	215	2	400	41,505	406,749	73,930	724,514	10,525	103,145	21,090	206,682	1,000
180TA11	51136	36	180	183	222	34	225	2	440	43,090	422,282	77,110	755,678	10,525	103,145	21,545	211,141	800
190TA11	51138	38	190	193	237	37	240	2	620	52,620	515,676	90,720	889,056	13,610	133,378	24,950	244,510	800
200TA11	51140	40	200	203	247	37	250	2	710	54,430	533,414	96,155	942,193	13,610	133,378	25,400	248,920	800

TABLE 24-59
Single-thrust ball bearings (with flat housing washer) series 12 (Indian Standards)

Bearing no.			Dimensions, mm						Minimum load constant, M	Static, C_o				Dynamic, C				Maximum permissible speed, rpm
BIS	FAG	SKF	d	C	D	B	E	r		FAG kgf	FAG N	SKF kgf	SKF N	FAG kgf	FAG N	SKF kgf	SKF N	
10TA12	51200	51200	10	10.2	26	11	26	1	.44	1,405	13,769	1,410	13,818	1,000	9,800	1,000	9,800	10,000
12TA12	51201	01	12	12.2	28	11	28	1	.55	1,520	14,896	1,545	15,141	1,035	10,143	1,035	10,143	8,000
15TA12	51202	02	15	15.2	32	12	32	1	.88	1,995	19,551	2,045	20,041	1,290	12,642	1,235	12,101	8,000
17TA12	51203	03	17	17.2	35	12	35	1	1.1	2,175	21,315	2,180	21,364	1,360	13,328	1,270	12,246	8,000
20TA12	51204	04	20	20.2	40	14	40	1	2.0	3,040	29,729	3,090	30,282	1,725	16,905	1,700	16,660	6,000
25TA12	51205	05	25	25.2	47	15	47	1	3.5	4,080	39,984	4,160	40,768	2,175	21,315	2,160	21,168	6,000
30TA12	51206	51206	30	30.2	52	16	52	1	4.4	4,555	44,639	4,820	47,236	2,270	22,246	2,270	22,246	6,000
35TA12	51207	07	35	35.2	62	18	62	1.5	8.8	6,350	62,230	6,360	62,328	3,040	29,742	3,040	29,792	5,000
40TA12	51208	08	40	40.2	68	19	68	1.5	14	7,984	78,253	7,725	75,705	3,700	36,848	3,470	33,706	5,000
45TA12	51209	09	45	45.2	73	20	73	1.5	16	8,755	85,879	8,640	84,672	3,925	38,465	3,700	36,260	4,000
50TA12	51210	10	50	50.2	78	22	78	1.5	22	9,800	95,640	9,100	89,180	4,220	41,356	3,760	36,848	4,000
55TA12	51211	11	55	55.2	90	25	90	1.5	40	13,610	133,378	13,190	129,262	5,700	55,860	5,450	53,410	3,000
60TA12	51212	51212	60	60.2	95	26	95	1.5	40	14,515	142,447	14,500	142,100	5,990	58,702	5,675	55,615	3,000
65TA12	51213	13	65	65.2	100	27	100	1.5	55	15,420	151,116	15,400	150,920	6,080	59,584	5,760	56,448	2,500
70TA12	51214	14	70	70.2	105	27	101	1.5	62	16,100	157,780	16,100	157,780	6,215	60,907	5,850	57,010	2,500
75TA12	51215	15	75	75.2	110	28	110	1.5	71	17,000	166,600	17,000	166,600	6,350	62,230	6,000	58,800	2,500
80TA12	51216	16	80	80.2	115	31	115	1.5	71	18,145	177,821	17,000	166,606	6,485	63,553	6,090	59,682	2,000
85TA12	51217	17	85	85.2	125	35	125	1.5	110	22,680	222,264	21,800	213,640	8,300	81,340	7,540	73,892	2,000
90TA12	51218	51218	90	90.2	135	35	130	2	180	27,670	271,166	27,200	266,560	10,160	99,568	9,100	89,180	2,000
100TA12	51220	20	100	100.2	150	38	150	2	240	34,020	333,396	34,000	333,200	12,245	120,001	11,390	111,622	1,600
110TA12	51222	22	110	110.2	160	38	160	2	310	38,555	377,839	37,000	362,600	12,925	126,665	12,050	118,090	1,600
120TA12	51224	24	120	120.2	170	39	170	2	310	39,915	391,170	39,200	384,160	12,925	126,665	12,050	118,090	1,300
130TA12	51226	26	130	130.3	190	45	190	2.5	550	50,805	497,889	50,900	498,820	16,330	160,034	15,700	153,860	1,300
140TA12	51228	28	140	140.3	200	46	190	2.5	620	53,530	544,194	53,600	525,280	16,555	162,239	16,150	158,270	1,000
150TA12	51230	51230	150	150.3	215	50	215	2.5	790	58,510	573,398	58,600	574,280	18,370	180,016	17,300	169,540	1,000
160TA12	51232	32	160	160.3	225	51	225	2.5	880	62,145	609,021	60,900	596,820	18,825	181,085	17,700	173,460	1,000
170TA12	51234	34	170	170.3	240	55	240	2.5	1200	75,290	737,842	71,400	699,720	22,225	217,805	21,190	207,662	800
180TA12	51236	36	180	180.3	250	56	250	2.5	1400	81,650	800,170	77,250	757,050	23,135	226,723	21,600	211,680	800
190TA12	51238	38	190	190.3	270	62	270	3	2000	94,350	924,631	91,000	891,800	25,855	253,779	25,000	245,000	800
200TA12	51240	40	200	200.3	280	62	280	3	2000	99,790	987,942	96,400	944,720	26,535	260,043	25,400	248,920	800

TABLE 24-60
Single-thrust ball bearing series 13 (Indian Standards)

BIS	FAG	SKF	d	C	D	B	E	r	Minimum load constant, M	Static, C_o FAG kgf	Static, C_o FAG N	Static, C_o SKF kgf	Static, C_o SKF N	Dynamic, C FAG kgf	Dynamic, C FAG N	Dynamic, C SKF kgf	Dynamic, C SKF N	Maximum permissible speed, rpm
25TA13	51305	51305	25	27	52	18	52	1.5	6.2	4,990	48,902	5,080	49,784	2,905	28,469	2,810	27,538	5,000
30TA13	51306	06	30	32	60	21	60	1.5	8.8	6,350	62,230	6,485	63,553	3,335	32,683	3,335	32,683	4,000
35TA13	51307	07	35	37	68	24	68	1.5	16	8,620	87,476	8,620	87,476	4,375	42,875	4,310	42,238	4,000
40TA13	51308	51308	40	42	78	26	78	1.5	26	10,885	106,673	11,110	108,878	5,450	53,410	5,350	52,430	3,000
45TA13	51309	09	45	47	85	28	85	1.5	40	13,835	135,583	13,610	133,378	6,620	64,876	6,215	60,907	3,000
50TA13	51310	10	50	52	95	31	95	2.0	62	16,330	160,034	16,330	160,034	7,985	78,253	7,530	73,794	2,500
55TA13	51311	51311	55	57	105	35	105	2.0	88	19,960	195,608	20,410	200,018	9,255	90,669	9,255	90,669	2,500
60TA13	51312	12	60	62	110	35	110	2.0	100	21,545	213,041	21,770	213,346	9,615	94,227	9,615	94,227	2,000
65TA13	51313	13	65	67	115	36	115	2.0	120	23,135	220,723	23,585	231,133	9,980	97,804	9,980	97,804	2,000
70TA13	51314	51314	70	72	135	40	125	2.0	180	27,670	269,166	28,125	275,625	12,020	117,796	11,565	113,337	2,000
75TA13	51315	15	75	77	135	44	135	2.5	220	31,525	308,445	32,660	320,068	13,610	133,378	13,610	133,378	1,600
80TA13	51316	16	80	82	140	44	140	2.5	240	34,020	333,398	35,380	346,724	14,060	137,788	13,835	135,683	1,600
85TA13	51317	51317	85	88	150	49	150	2.5	350	40,825	400,085	40,825	400,085	16,100	157,780	15,420	151,116	1,300
90TA13	51318	18	90	93	155	50	155	2.5	350	40,825	400,085	40,825	400,085	16,100	157,780	15,420	151,116	1,300
100TA13	51320	20	100	103	170	55	170	2.5	550	48,990	480,102	48,990	480,102	18,825	171,485	18,370	180,026	1,000
110TA13	51322	51322	110	113	190	63	190	2.5	790	57,605	564,529	57,605	564,529	21,545	213,041	20,640	202,272	1,000
120TA13	51324	24	120	123	210	70	210	3.5	1200	73,930	724,514	72,570	711,186	25,400	248,920	24,495	240,051	1,000
130TA13	51326	26	130	134	225	75	225	3.5	1600	84,370	826,826	78,470	769,006	28,125	275,625	25,855	253,379	800
140TA13	51328	51328	140	144	240	80	240	3.5	2200	97,970	958,106	92,533	906,843	30,390	297,822	29,030	284,404	800
150TA13	51330	30	150	154	250	80	250	3.5	2500	10,342	1,013,516	99,790	967,942	31,525	308,445	29,710	291,158	

TABLE 24-61
Single-thrust ball bearings series 14 (Indian Standards)

Bearing no.			Dimensions, mm						Minimum load constant, M	Static, C_o FAG kgf	Static, C_o FAG N	Static, C_o SKF kgf	Static, C_o SKF N	Dynamic, C FAG kgf	Dynamic, C FAG N	Dynamic, C SKF kgf	Dynamic, C SKF N	Maximum permissible speed, rpm
BIS	FAG	SKF	d	C	D	B	E	r										
25TA14	51405	51405	25	27	60	24	60	1.5	12	7,255	71,099	7,395	72,741	7,393	72,471	4,375	42,875	4,000
30TA14	51406	06	30	32	70	28	70	1.5	24	10,160	99,565	10,340	100,992	10,340	108,992	5,670	55,566	4,000
35TA14	51407	07	35	37	80	32	80	2.0	35	12,923	126,665	12,700	124,460	12,700	124,460	6,305	61,789	3,000
40TA14	51408	51408	40	42	90	36	90	2.0	62	16,330	160,034	16,555	162,239	16,555	162,239	8,755	85,799	2,500
45TA14	51409	09	45	47	100	39	100	2.0	88	19,960	193,308	19,960	193,308	19,960	193,308	10,160	99,568	2,500
50TA14	51410	10	50	52	110	43	110	2.5	140	24,495	240,051	25,400	248,920	25,400	248,920	12,475	122,255	2,000
55TA14	51411	51411	55	57	120	48	120	2.5	180	28,575	280,035	29,710	291,158	14,060	137,788	14,060	137,788	2,000
60TA14	51412	12	60	62	130	51	130	2.5	260	35,380	346,724	36,235	355,103	16,555	162,239	16,330	160,034	2,000
65TA14	51413	13	65	68	140	56	140	3.0	350	39,235	384,503	40,825	400,085	17,690	174,062	18,145	177,821	1,600
70TA14	51414	51414	70	73	150	60	150	3.0	490	45,360	444,528	45,355	444,479	19,960	195,608	19,505	191,149	1,300
75TA14	51415	15	75	78	160	65	160	3.0	620	49,900	489,020	50,805	497,969	21,545	213,041	21,090	206,682	1,300
80TA14	51416	16	80	83	170	68	170	3.5	710	55,335	530,283	55,335	530,283	23,135	220,723	22,680	222,264	1,000
85TA14	51417	51417	85	88	180	72	180	3.5	880	63,500	622,300	60,785	595,693	24,950	240,051	24,040	235,592	1,000
90TA14	51418	18	90	93	190	77	190	3.5	1,000	66,220	645,956	68,040	666,792	25,400	248,920	25,400	248,920	800
100TA14	51420	20	100	103	210	88	210	4.0	1,800	87,540	857,892	87,540	857,892	31,525	308,445	30,845	302,281	600

TABLE 24-62
Double-thrust ball bearings series 522

Bearing no. FAG	SKF	d	d₁	d₂	D	H	h	r	r₁	Maximum load constant, M	Static, Cₒ FAG kgf	FAG N	SKF kgf	SKF N	Dynamic, C FAG kgf	FAG N	SKF kgf	SKF N	Maximum permissible speed, rpm
52202	52202	15	10	17	32	22	5	1	.5	0.88	1,996	19,560	2,040	19,992	1,292	12,662	1,220	11,956	8,000
52204	04	20	15	22	40	26	6	1	.5	2.0	3,039	29,782	3,100	30,380	1,724	16,895	1,730	16,754	6,000
52205	05	25	20	27	47	28	7	1	.5	3.5	4,082	40,003	4,150	40,670	2,177	21,334	2,160	21,168	6,000
52206	06	30	25	32	52	29	7	1	.5	4.4	4,716	46,217	4,800	47,040	2,268	22,226	2,280	22,344	6,000
52207	07	35	30	37	62	34	8	1.5	.5	8.8	6,350	62,230	6,400	62,720	3,039	29,782	3,450	29,890	6,000
52208	08	40	30	42	68	36	9	1.5	1	14	7,983	78,133	7,650	74,970	3,696	35,921	3,050	33,810	5,000
52209	09	45	35	47	73	37	9	1.5	1	16	8,754	85,789	8,650	84,770	3,931	38,524	3,650	35,770	5,000
52210	52210	50	40	52	78	39	9	1.5	1	22	9,797	96,011	9,150	89,670	4,218	41,336	3,750	36,750	4,000
52211	11	55	45	57	90	45	10	1.5	1	40	13,608	128,066	13,200	129,360	5,670	55,566	5,500	53,900	4,000
52212	12	60	50	62	95	46	10	1.5	1	40	14,515	142,247	14,600	143,080	5,987	58,672	5,700	55,860	3,000
52213	13	65	55	67	100	47	10	1.5	1	55	15,422	151,133	15,600	151,880	6,078	68,564	5,850	57,380	3,000
52214	14	70	55	72	105	47	10	1.5	1.5	62	16,102	157,999	16,300	159,740	6,214	60,897	6,000	58,800	2,500
52215	15	75	60	77	110	47	10	1.5	1.5	71	17,010	166,698	17,300	169,540	6,350	62,230	6,100	59,780	2,500
52216	52216	80	65	82	115	48	10	1.5	1.5	71	18,144	177,811	18,000	176,400	6,486	63,568	6,200	60,760	2,500
52217	17	85	70	88	125	55	12	1.5	1.5	110	22,680	222,264	22,000	215,600	8,301	81,349	7,500	73,500	2,000
52218	18	90	75	93	135	62	14	2	1.5	180	27,669	271,156	27,000	264,600	10,160	99,668	9,150	89,670	2,000
52220	20	100	85	103	150	67	15	2	1.5	240	34,019	333,386	34,000	333,200	12,247	120,020	11,400	111,720	1,600

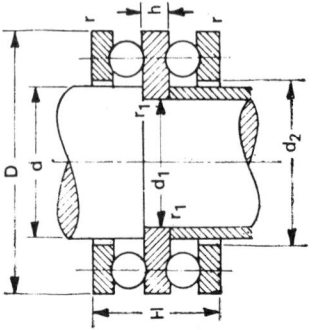

[a] d refers to FAG bearings.

TABLE 24-63
Hardness factors for needle-roller bearings

Rockwell C hardness of raceway	Approximate Brinell hardness (Bhn)	Hardness factor, K_h
63	660	1.00
60	620	0.98
58	595	0.96
56	570	0.92
54	545	0.83
52	515	0.70
50	490	0.50

Particular	Formula
The load capacity of needle bearing is also calculated from the formula	$C_n = K_h\, K_l\, pl\, d_i$ $\qquad$ (24–207)

For hardness factors, K_h, refer to Table 24–63, and for life factor K_l refer to Fig. 24–45.

PRESSURE

The pressure for wrist pin rocker arm and similar oscillating mechanisms is

$$p = 34.52 \text{ MPa} \qquad \textbf{SI} \quad (24\text{–}208a)$$

$$= 5.0 \text{ kpsi} \qquad \textbf{US Customary units} \quad (24\text{–}208b)$$

For rotary motion, pressure may be computed from the relation

$$P = \frac{2.86 \times 10^6}{\sqrt[3]{D_1 n'}} \qquad \textbf{SI} \quad (24\text{–}209a)$$

where P in Pa, D_1 in m, and n' in rps

$$= \frac{5500}{\sqrt[3]{D_1\, n}} \qquad \textbf{US Customary units} \quad (24\text{–}209b)$$

where P in psi, D_1 in in, and n in rpm

Check for total circumferential clearance from formula

$$c = \pi(d_i + d_r) - Z d_r \qquad (24\text{–}210)$$

For dimensions, design data, and sizes for needle bearings

Refer to Tables 24–64 to 24–68.

TABLE 24-64
Dimensions for needle bearings with inner ring, type NEA (all dimensions in mm)

Light series		Medium series		Heavy series		Bore diameter,	
Outer diameter, D	Width, B	Outer diameter, D	Width, B	Outer diameter, D	Width, B	d	r
28	15					12	1
32	15	35	22			15	1
35	15					17	1
42	18	42	22			20	1
47	18	47	22			25	1
52	18	52	22	62	30	30	1
58	18	58	22	72	36	35	1
65	18	65	22	80	36	40	1.5
72	18	72	22	85	38	45	1.5
80	20	80	28	90	38	50	2
85	20	85	28	95	38	55	2
90	20	90	28	100	38	60	22
95	20	95	28	105	38	65	2
100	20	100	28	110	38	70	2
110	24	110	32	120	38	75	2
115	24	115	32	125	38	80	2
		120	32	130	38	85	2
		125	32	135	43	90	2
		130	32	140	43	95	2
		135	32	145	43	100	2
		140	32	150	45	105	2
		145	34	160	45	110	2
		155	34	165	45	115	2
		160	34	170	45	120	2
		165	34	185	52	125	2
		170	34	190	52	130	2
		180	36	205	52	140	2
		195	36	215	52	150	2
		205	36	230	57	160	3
		220	42	245	57	170	3
		230	42	255	57	180	3
		245	42	265	57	190	3
		255	42	280	57	200	3
		265	42	290	64	210	3
		280	49	300	64	220	3
		290	49	315	64	230	3
		300	49	325	64	240	3
		315	49	340	74	250	3
		325	54	350	74	260	3
		340	54	365	74	270	3
		350	54	375	74	280	3
		365	54	385	74	290	3
		375	54	395	74	300	3

Source: IS 4215, 1967.

TABLE 24-65
Dimensions for needle bearings without inner ring, type NES (all dimensions in mm)

Light series		Medium series		Heavy series		Bore diameter,	
Outer diameter, D	Width, B	Outer diameter, D	Width, B	Outer diameter, D	Width, B	d	r
16	12					7.3	0.5
19	12					9.7	0.5
22	12					12.1	0.5
24	12					14.4	0.5
28	15					17.6	1
32	15					20.8	1
		35	22			22.1	1
35	15					23.9	1
42	18	42	22			28.7	1
47	18	47	22			33.5	1
52	18	52	22			38.2	1
58	18	58	22	62	30	44	1
65	18	65	22	72	36	49.7	1.5
72	18	72	22	80	36	55.4	1.5
80	20	80	28	85	38	62.1	
85	20	85	28	90	38	68.8	
90	20	90	28	95	38	72.6	
95	20	95	28	100	38	78.3	2
100	20	100	28	105	38	83.1	2
110	24	110	32	110	38	88	2
115	24	115	32	120	38	96	2
		120	32	125	38	99.5	2
		125	32	130	38	104.7	2
		130	32	135	43	109.1	2
		135	32	140	43	114.7	2
		140	32	145	43	119.2	2
		145	34	150	45	124.7	2
		155	34	160	45	132.5	2
		160	34	165	45	137	2
		165	34	170	45	143.5	2
		170	34			148	2
				185	52	152.8	2
		180	36	190	52	158	2
		195	36	205	52	170.5	2
		205	36	215	52	179.3	2
		220	42	230	57	193.8	3
		230	42	245	57	202.6	3
		245	42	255	57	216	3
		255	42	265	57	224.1	3
		265	42	280	57	236	3
		280	49	290	64	248.4	3
		290	49	300	64	258.4	3
		300	49	315	64	269.6	3
		315	49	325	64	281.9	3
		325	54	340	74	290.5	3
		340	54	350	74	302	3
		350	54	365	74	313.5	3
		365	54	375	74	325	3
		375	54	385	74	335	3
				395	74	344	3

Source: IS 4215, 1967.

TABLE 24–66
Dimensions for needle bearing without outer ring, type NCS

d, mm	d_o, mm	B, mm	r, mm
30	44.2	18	1
32	46.4	18	1
35	50	18	1
40	55.7	18	1.5
45	61.4	18	1.5
50	67.1	20	2
55	72.9	20	2
60	80.5	20	2
65	84.3	20	2
70	90.1	20	2
75	96.8	24	2
80	102.4	24	2
85	106.5	32	2
90	111.7	32	2
95	116.1	32	2
100	121.7	32	2
105	126.2	32	2
110	131.7	34	2
115	139.5	34	2
120	144	34	2
125	150.54	34	2
130	155.04	34	2
135	159.8	36	2
140	165.04	36	2

Source: IS 4215, 1967.

TABLE 24–67
Design data for needle-roller bearings

Journal race diameter, mm	Recommended	
	Total radial clearance, mm	Needle diameter, mm
9.50–19.00	0.0125–0.040	1.55
19.00–31.75	0.0180–0.050	2.35
31.75–50.80	0.0200–0.055	3.20
50.80–76.00	0.0255–0.065	3.20
76.00–127.00	0.0305–0.075	4.75
127.00–177.00	0.0355–0.085	4.75

TABLE 24–68
Torrington needle-roller sizes

Diameter, mm	Length, mm	Diameter, mm	Length, mm
1.590	9.40	3.175	22.25
1.590	12.45	3.175	23.82
1.590	15.75	3.175	25.40
1.590	16.95	3.175	28.575
2.380	10.55	4.010	18.800
2.3815	19.05	4.740	13.380
2.3850	11.745	4.765	18.900
2.3850	24.758	4.765	25.400
3.1750	9.770	4.765	30.200
3.1750	12.750	4.765	34.950
3.1750	15.650	5.500	19.100
3.1750	19.050	6.350	31.750

Particular	Formula

Hertzian contact pressure

Maximum contact pressure between cylinders and
spheres of steel ($v = 0.3$)

(i)   For cylinders

$$\sigma_{c(max)} = 0.418 \sqrt{\frac{2FE(d_1 + d_2)}{l d_1 d_2}} \qquad (24\text{–}211)$$

(ii)   For a cylinder and plane

$$\sigma_{c(max)} = 0.418 \sqrt{\frac{2FE}{ld}} \qquad (24\text{–}212)$$

(iii)   For two spheres

$$\sigma_{c(max)} = 0.388 \sqrt[3]{\frac{4F(d_1 + d_2)^2 \, E^2}{d_1^2 d_2^2}} \qquad (24\text{–}213)$$

(iv)   For a sphere and plane

$$\sigma_{c(max)} = 0.388 \sqrt[3]{\frac{4F \, E^2}{d^2}} \qquad (24\text{–}214)$$

Maximum shear stress occurs below the contact
surface for ductile material

(i)   For spheres $\qquad\qquad \tau_{max} = 0.31 \, \sigma_{c(max)} \qquad (24\text{–}215)$

(ii)   For cylinders $\qquad\qquad \tau_{max} = 0.304 \, \sigma_{c(max)} \qquad (24\text{–}216)$

Particular	Formula

SELECTION OF FIT AND LOADING RATIO
C/P

Housing seatings for various bearings Refer to Tables 24–69 and 24–70.

Loading ratio *C/P* for different ball and rolling bearing lives Refer to Tables 24–71 and 24–72.

TABLE 24–69
Selection of fit

(a) Housing seatings for radial bearings

			Conditions	Applications	Tolerance
Solid Housing		Rotating Outer-Ring Load	Heavy loads on bearings in thin walled housings; heavy shock loads	Roller bearing wheel hubs; big-end bearings	P7
			Normal and heavy loads	Ball bearing wheel hubs; big-end bearings	N7
			Light and variable loads	Conveyor rollers, rope sheaves; belt tension pulleys	M7
		Direction of Loading Indeterminate	Heavy shock loads	Electric traction motors	M7
			Heavy and normal loads; axial mobility of outer ring unnecessary	Electric motors, pumps, crankshaft main bearings	K7
			Normal and light loads; axial mobility of outer ring desirable	Electric motors, pumps; crankshaft main bearings	J7
Split or Solid Housing		Stationary Outer-ring Load	Shock loads intermittent	Railway axle boxes	J7
			All loads	Bearings in general applications	H7
			Normal and light loads	Line shafting	H8
			Heat condition through shafts	Drying cylinders; large electric motors	G7
Solid Housings		Arrangement of Bearing Very Accurate	Accurate running and great rigidity under variable load	Roller bearings $D > 125$ mm	N6
				For machine-tool $D \leq 125$ mm main spindles	M6
			Accurate running under light loads of indeterminate direction	Ball bearings at work end of grinding spindle; locating bearings in high-speed centrifugal compressors	K6
			Accurate running; axial movement of outer ring desirbale	Ball bearings at drive end of grinding spindles; axially free bearings in high-speed centrifugal compressors	J6

(b) Housing seatings for thrust bearings

	Conditions	Tolerance
Purely axial load	Thrust ball bearings Spherical roller thrust bearings where another bearing takes care of the radial location	H8
Combined (radial and axial) load or spherical roller thrust bearings	Stationary load on housing washer or direction of loading indeterminate	J7
		Generally K7
	Rotating load or housing washer	Heavy radial load M7

TABLE 24–70
Selection of fit

(a) Shaft (solid) seatings for radial bearings

	Conditions	Application		Shaft diameter, mm		Tolerance
			Ball bearings	Cylindrical and tapered roller bearings	Spherical roller bearings	
		Bearings with cylindrical bore				
Stationary Inner-ring Load	Easy axial displacement of inner ring on shaft desirable	Wheels on nonrotating axles		All diameters		g6
	Easy axial displacement of inner-ring on shaft unnecessary	Tension pulleys; rope sheaves		All diameters		h6
Rotating Inner-ring or Direction of Loading Indeterminate	Light and variable loads	Electrical apparatus; machine tools; pumps; transport vehicles	≤ 18	—	—	h5
			18–100	≤ 40	≤ 40	j6
			100–200	40–140	49–100	k6
			140–200	140–200	100–200	m6
	Normal and heavy loads	General application electric motors pumps; turbines; gearing; wood working machines; and internal-combustion engines	≤ 18	—	—	j5
			18–100	< 40	≤ 40	k5
			100–140	40–100	40–65	m5
			140–200	100–140	65–100	m6
			200–280	140–200	100–140	n6
				200–400	140–280	p6
					280–500	r6
					> 500	r7
	Shock and heavy loads	Locomotive axle boxes; traction motors	—	50–140	50–100	n6
			—	140–200	100–140	p6
			—	—	140–200	r6
			—	—	200–500	r7
Purely axial load		All kinds of bearing arrangements		All diameters		j6
		Bearings with taper bore and sleeve				
Loads of all kinds		Bearing arrangements in general; railway axle boxes		All diameters		h9
		Line shafting	All diameters			h10

(b) Shaft seatings for thrust bearings

Conditions			Tolerance
Purely axial load	Thrust ball bearing, spherical roller thrust bearings	All diameters	j6
Combined (radial and axial) load on spherical thrust bearings	Stationary load on shaft washer	All diameters	j6
	Rotating load on shaft washer or direction of loading indeterminate	$d \leq 200$ mm	k6
		$d = 200\text{--}400$ mm	m6
		$d > 400$ mm	n6

TABLE 24-71
Loading ratio C/P for different lives for ball bearings

Life, L_h hours	Speed, rpm																									
	10	25	40	100	125	160	200	250	320	400	500	630	800	1000	1250	1600	2000	2500	3200	4000	5000	6200	8000	10000	12500	16000
100							1.06	1.15	1.24	1.34	1.45	1.56	1.68	1.82	1.96	2.12	2.29	2.47	2.67	2.88	3.11	3.36	3.63	3.91	4.23	4.56
500			1.06	1.45	1.56	1.68	1.82	1.96	2.12	2.29	2.47	2.67	2.88	3.11	3.36	3.63	3.91	4.23	4.56	4.93	5.32	5.75	6.20	6.70	7.23	7.81
1,000		1.15	1.34	1.82	1.96	2.12	2.29	2.47	2.67	2.88	3.11	3.36	3.63	3.91	4.23	4.56	4.93	5.32	5.75	6.20	6.70	7.23	7.81	8.43	9.11	9.83
1,250		1.24	1.45	1.96	2.12	2.29	2.47	2.67	2.88	3.11	3.36	3.63	3.91	4.23	4.56	4.93	5.32	5.75	6.20	6.70	7.23	7.81	8.43	9.11	9.83	10.6
1,600		1.34	1.56	2.12	2.29	2.47	2.67	2.88	3.11	3.36	3.63	3.91	4.23	4.56	4.93	5.32	5.75	6.20	6.70	7.23	7.81	8.43	9.11	9.83	10.6	11.5
2,000	1.06	1.45	1.68	2.29	2.47	2.67	2.88	3.11	3.36	3.63	3.91	4.23	4.56	4.93	5.32	5.75	6.20	6.70	7.23	7.81	8.43	9.11	9.83	10.6	11.5	12.4
2,500	1.15	1.56	1.82	2.47	2.67	2.88	3.11	3.36	3.63	3.91	4.23	4.56	4.93	5.32	5.75	6.20	6.70	7.23	7.81	8.43	9.11	9.83	10.6	11.5	12.4	13.4
3,200	1.24	1.68	1.96	2.67	2.88	3.11	3.36	3.63	3.91	4.23	4.56	4.93	5.32	5.75	6.20	6.70	7.23	7.81	8.43	9.11	9.83	10.6	11.5	12.4	13.4	14.5
4,000	1.34	1.82	2.12	2.88	3.11	3.36	3.63	3.91	4.23	4.56	4.93	5.32	5.75	6.20	6.70	7.23	7.81	8.43	9.11	9.83	10.6	11.5	12.4	13.4	14.5	15.6
5,200	1.45	1.96	2.29	3.11	3.36	3.63	3.91	4.23	4.56	4.93	5.32	5.75	6.20	6.70	7.23	7.81	8.43	9.11	9.83	10.6	11.5	12.4	13.4	14.5	15.6	16.8
6,300	1.56	2.12	2.47	3.36	3.63	3.91	4.23	4.56	4.93	5.32	5.75	6.20	6.70	7.23	7.81	8.43	9.11	9.83	10.6	11.5	12.4	13.4	14.5	15.6	16.8	18.1
8,000	1.68	2.29	2.67	3.63	3.91	4.23	4.56	4.93	5.32	5.75	6.20	6.70	7.23	7.81	8.43	9.11	9.83	10.6	11.5	12.4	13.4	14.5	15.6	16.8	18.2	19.6
10,000	1.82	2.47	2.88	3.91	4.23	4.56	4.93	5.32	5.75	6.20	6.70	7.23	7.81	8.43	9.11	9.83	10.6	11.5	12.4	13.4	14.5	15.6	16.8	18.2	19.6	21.2
12,500	1.96	2.67	3.11	4.23	4.56	4.93	5.32	5.75	6.20	6.70	7.23	7.81	8.43	9.11	9.83	10.6	11.5	12.4	13.4	14.5	15.6	16.8	18.2	19.6	21.2	22.9
16,000	2.12	2.88	3.36	4.56	4.93	5.32	5.75	6.20	6.70	7.23	7.81	8.43	9.11	9.83	10.6	11.5	12.4	13.4	14.5	15.6	16.8	18.2	19.6	21.2	22.9	24.7
20,000	2.29	3.11	3.63	4.93	5.32	5.75	6.20	6.70	7.23	7.81	8.43	9.11	9.83	10.6	11.5	12.4	13.4	14.5	15.6	16.8	18.2	19.6	21.2	22.9	24.7	26.7
25,000	2.47	3.36	3.91	5.32	5.75	6.20	6.70	7.23	7.81	8.43	9.11	9.83	10.6	11.5	12.4	13.4	14.5	15.6	16.8	18.2	19.6	21.2	22.9	24.7	26.7	28.8
32,000	2.67	3.63	4.23	5.75	6.20	6.70	7.23	7.81	8.43	9.11	9.83	10.6	11.5	12.4	13.4	14.5	15.6	16.8	18.2	19.6	21.2	22.9	24.7	26.7	28.8	31.1
40,000	2.88	3.91	4.56	6.20	6.70	7.23	7.81	8.43	9.11	9.83	10.6	11.5	12.4	13.4	14.5	15.6	16.8	18.2	19.6	21.2	22.9	24.7	26.7	28.8	31.1	
50,000	3.11	4.23	4.93	6.70	7.23	7.81	8.43	9.11	9.83	10.6	11.5	12.4	13.4	14.5	15.6	16.8	18.2	19.6	21.2	22.9	24.7	26.7	28.8	31.1		
63,000	3.36	4.56	5.32	7.23	7.81	8.43	9.11	9.83	10.6	11.5	12.4	13.4	14.5	15.6	16.8	18.2	19.6	21.2	22.9	24.7	26.7	28.8	31.1			
80,000	3.63	4.93	5.75	7.81	8.43	9.11	9.83	10.6	11.5	12.4	13.4	14.5	15.6	16.8	18.2	19.6	21.2	22.9	24.7	26.7	28.8	31.1				
100,000	3.91	5.32	6.20	8.43	9.11	9.83	10.6	11.5	12.4	13.4	14.5	15.6	16.8	18.2	19.6	21.2	22.9	24.7	26.7	28.8	31.1					
200,000	4.93	6.70	7.81	10.6	11.5	12.4	13.4	14.5	15.6	16.8	18.2	19.6	21.2	22.9	24.7	26.7	28.8	31.1								

TABLE 24-72
Loading ratio C/P for different lives for roller bearings

Life, L_h hours	Speed, rpm																									
	10	25	40	100	125	160	200	250	320	400	500	630	800	1000	1250	1600	2000	2500	3200	4000	5000	6200	8000	10000	12500	16000
100							1.05	1.13	1.21	1.30	1.39	1.49	1.60	1.71	1.83	1.97	2.11	2.26	2.42	2.59	2.78	2.97	3.19	3.42	3.66	3.92
500			1.05	1.39	1.49	1.60	1.71	1.83	1.97	2.11	2.26	2.42	2.59	2.78	2.97	3.19	3.42	3.66	3.92	4.20	4.50	4.82	5.17	5.54	5.94	6.36
1,000		1.13	1.30	1.71	1.83	1.97	2.11	2.26	2.42	2.59	2.78	2.97	3.19	3.42	3.66	3.92	4.20	4.50	4.82	5.17	5.54	5.94	6.36	6.81	7.30	7.82
1,250		1.21	1.39	1.83	1.97	2.11	2.26	2.42	2.59	2.78	2.97	3.19	3.42	3.66	3.92	4.20	4.50	4.82	5.17	5.54	5.94	6.36	6.81	7.30	7.82	8.38
1,600		1.30	1.49	1.97	2.11	2.26	2.42	2.59	2.78	2.97	3.19	3.42	3.66	3.92	4.20	4.50	4.82	5.17	5.54	5.94	6.36	6.81	7.30	7.82	8.38	8.98
2,000	1.05	1.39	1.60	2.11	2.26	2.42	2.59	2.78	2.97	3.19	3.42	3.66	3.92	4.20	4.50	4.82	5.17	5.54	5.94	6.36	6.81	7.30	7.82	8.38	8.98	9.62
2,500	1.13	1.49	1.71	2.26	2.42	2.59	2.78	2.97	3.19	3.42	3.66	3.92	4.20	4.50	4.82	5.17	5.54	5.94	6.36	6.81	7.30	7.82	8.38	8.98	9.62	10.3
3,200	1.21	1.60	1.83	2.42	2.59	2.78	2.97	3.19	3.42	3.66	3.92	4.20	4.50	4.82	5.17	5.54	5.94	6.36	6.81	7.30	7.82	8.38	8.98	9.62	10.3	11.0
4,000	1.30	1.71	1.97	2.59	2.78	2.97	3.19	3.42	3.66	3.92	4.20	4.50	4.82	5.17	5.54	5.94	6.36	6.81	7.30	7.82	8.38	8.98	9.62	10.3	11.0	11.8
5,000	1.39	1.83	2.11	2.78	2.97	3.19	3.42	3.66	3.92	4.20	4.50	4.82	5.17	5.54	5.94	6.36	6.81	7.30	7.82	8.38	8.98	9.62	10.3	11.0	11.8	12.7
6,300	1.46	1.97	2.26	2.97	3.19	3.42	3.66	3.92	4.20	4.50	4.82	5.17	5.54	5.94	6.36	6.81	7.30	7.82	8.38	8.98	9.62	10.3	11.0	11.8	12.7	13.6
8,000	1.60	2.11	2.42	3.19	3.42	3.66	3.92	4.20	4.50	4.82	5.17	5.54	5.94	6.36	6.81	7.30	7.82	8.38	8.98	9.62	10.3	11.0	11.8	12.7	13.6	14.6
10,000	1.71	2.26	2.59	3.42	3.66	3.92	4.20	4.50	4.82	5.17	5.54	5.94	6.36	6.81	7.30	7.82	8.38	8.98	9.62	10.0	11.0	11.8	12.7	13.6	14.6	15.6
12,500	1.83	2.42	2.78	3.66	3.92	4.20	4.50	4.82	5.17	5.54	5.94	6.36	6.81	7.30	7.82	8.38	8.98	9.62	10.3	11.0	11.8	12.7	13.6	14.6	15.6	16.7
16,000	1.97	2.59	2.97	3.92	4.20	4.50	4.82	5.17	5.54	5.94	6.36	6.81	7.30	7.82	8.38	8.98	9.62	10.3	11.0	11.8	12.7	13.6	14.6	15.6	16.7	17.9
20,000	2.11	2.78	3.19	4.20	4.50	4.82	5.17	5.54	5.94	6.36	6.81	7.30	7.82	8.38	8.98	9.62	10.3	11.0	11.8	12.7	13.6	14.6	15.6	16.7	17.9	19.2
25,000	2.26	2.97	3.42	4.50	4.82	5.17	5.54	5.94	6.36	6.81	7.30	7.82	8.38	8.98	9.62	10.3	11.0	11.8	12.7	13.6	14.6	15.6	16.7	17.9	19.2	20.6
32,000	2.42	3.19	3.66	4.82	5.17	5.54	5.94	6.36	6.81	7.30	7.82	8.38	8.98	9.62	10.3	11.0	11.8	12.7	13.6	14.6	15.6	16.7	17.9	19.2	20.6	
40,000	2.59	3.42	3.92	5.17	5.54	5.94	6.36	6.81	7.30	7.82	8.38	8.98	9.62	10.3	11.0	11.8	12.7	13.6	14.6	15.6	16.7	17.9	19.2	20.6		
50,000	2.78	3.66	4.20	5.54	5.94	6.36	6.81	7.30	7.82	8.38	8.98	9.62	10.3	11.0	11.8	12.7	13.6	14.6	15.6	16.7	17.9	19.2	20.6			
63,000	2.97	3.92	4.50	5.94	6.36	6.81	7.30	7.82	8.38	8.98	9.62	10.3	11.0	11.8	12.7	13.6	14.6	15.6	16.7	17.9	19.2	20.6				
80,000	3.19	4.20	4.82	6.36	6.81	7.30	7.82	8.38	8.98	9.62	10.3	11.0	11.8	12.7	13.6	14.6	15.6	16.7	17.9	19.2	20.6					
100,000	3.42	4.50	5.17	6.81	7.30	7.82	8.38	8.98	9.62	10.3	11.0	11.8	12.7	13.6	14.6	15.6	16.7	17.9	19.2							
200,000	4.20	5.54	6.36	8.38	8.98	9.62	10.3	11.0	11.8	12.7	13.6	14.6	15.6	16.7	17.9	19.2	20.6									

24.3 FRICTION AND WEAR (1)

SYMBOLS

a	half the mean diameter of area of contact, Eq. (24–252)
A	real area of contact, m^2 (in^2)
A_a	apparent area of contact, m^2 (in^2)
A'	abrasion factor
b	constant used in Eq. (24–222), exponent
c	constant used in Eqs. (24–225) and (24–280)
c_1, c_2	constants as given in Eqs. (24–281b) and (24–281c)
d	diameter, m (in)
E	Young's modulus, GPa (psi)
F	force, kN (lbf)
F_μ	total force of friction, kN (lbf)
$F_{a\mu}$	adhesive component of friction force or force to shear junctions, kN (lbf)
F_{-f}	fatigue resistance is the average number of reversed stress cycles which the surface layer must undergo under given abrasion condition, kN (lbf)
$F_{\text{ploughing}}$	force to plough the asperities on one surface through the other, kN (lbf)
G	elasticity constant characterizing rubber
h	thickness of layer removed, m (in) effective thickness of the worn-out surface layer, m (in)
h_m	height of asperities, m (in)
H	hardness of softer material, N/m^2 or Pa (psi)
i	number of surface layer which are abrased during a test number of repeated deformation as used in Eqs. (24–256) to (24–258)
Q_{me}	mechanical equivalent of heat, N m/J (lbf in/Btu or lbf ft/Btu)
k_1, k_2	thermal conductivity of two conducting materials, W/m K (Btu/ft h °F)
k	constant used in Eq. (24–245) and given in Table 24–77
K_E	energetic wear rate or energy index of abrasion
K_L	linear wear rate
K_V	volumetric wear rate
K_W	gravimetric wear rate
K_{sm}	specific wear by mass
K_{sV}	specific wear by volume
K'_{sv}	modified specific wear
L	sliding distance, m (in)
m	mass of wear debries, kg (lb)
n	exponent
P_c	power used to elongate shred
P_H	power applied to hysteresis loss which accompanies roll deformation
P_t	power used to tear shred from surface layer
P_{tot}	total frictional power
P	yield pressure of soft material (about 5 times the critical shear stress), MPa (psi)
P_a	apparent pressure over the contact area, MPa (psi)
P_m	mean pressure over the contact area, MPa (psi)

	flow pressure of material, MPa (psi)
q	friction work done corresponding to a simple stressing cycle which corresponds to a sliding length of λ, N m (lbf in)
r	radius of curvature, m (in)
	radius of circular junction (Fig. 24–50), m (in)
R	mean radius of the curvature at the tip of the abrasive particles, m (in)
s	spacing between ridges in the elastomer surface, m (in)
v	velocity, m/s (ft/min)
v_k	velocity, m/s (ft/min)
V	volume deformed, m^3 (in^3)
ΔV	volume of transferred fragment, m^3 (in^3)
	volume of layer removed, m^3 (in^3)
W	applied load at interface, kN (lbf)
W_{ab}	the work of adhesion of the contacting metals which can be expressed in terms of their surface energies, N m (lbf in or lbf ft)
$W_{tot} = Wn^2$	normal load per unit area, kN (lbf)
ΔW	weight lost due to abraded layer being removed from the bulk material, kN (lbf)
z	the average depth of penetration for single sphere, m (in)
	the absolute approach, m (in)
α_n	coefficient of hystersis loss
β	constant depends on the surface treatment taken from Table 24–73
γ	surface tension of the softer sliding member, N/m (lbf/in)
δ	abradability as wear index
θ	angle of slope of irregularities, deg
θ_m	mean temperature rise at the sliding junction, °C (°F)
μ	coefficient of friction
μ_a	adhesive component of coefficient of friction
μ_c	coefficient of elastic friction
μ_o	coefficient of static friction taken from Table 24–74
$\mu_{ploughing}$	ploughing component of coefficient of friction
ν	Poisson's ratio
ρ	density of the abraded elastomer, kg/m^3 (lb/in^3)
ζ	coefficient of abrasion resistance
λ	mean wavelength of the surface asperities
σ	stress, MPa (psi)
σ_c	contact pressure or pressure over the contours, MPa (psi)
σ_o	tensile strength of elastomer in simple tensions, MPa (psi)
τ	shear strength of junction, MPa (psi)
τ_m	mean shear stress, MPa (psi)

Particular	Formula

FRICTION

The general expression for force of friction	$F_\mu = F_{a\mu} + F_{\text{ploughing}}$	(24–217)
The total friction force	$F_\mu = A\tau$	(24–218)
The real area of contact	$A = \dfrac{W}{P}$	(24–219)
The general expression for coefficient of friction	$\mu = \mu_a + \mu_{\text{ploughing}}$	(24–220)
The total coefficient of friction	$\mu = \dfrac{F_\mu}{W} = \dfrac{A\tau}{W} = \dfrac{\tau}{P}$	(24–221)

The coefficient of elastic friction when a rigid rough surface is pressed against an elastically deformable second surface

$$\mu_e = \left\{ \left[\frac{K\alpha_n K_4^{\frac{1}{2}}\sqrt{\beta}}{2(\beta + 1)} \right] \left(\frac{h_m}{r} \right)^{2/(2_\beta+1)} \left(\frac{\sigma_c}{E} \right)^{1/(2_\beta+1)} \right\}$$

(24–222)

where $K\alpha_\pi \simeq 1$; calculate K_4 from Eq. (24–223)

The expression for K_4 to be used in Eq. (24–222)

TABLE 24–73
Constant β to be used in Eq. (24–222)

$$K_4 = \left[\frac{0.75(1 - v^2)\pi}{K_2 \beta b} \right]^{\beta/(2_\beta+1)}$$

(24–223)

where $K_2 = 1,\ 0.4,\ 0.12$ for $\beta = 1,\ 2,\ 3$ respectively

Refer to Table 24–73 for β.

Surface treatment	β	b
Turning, milling	2	1–3
Planing	3	4–6
Polishing	3	5–10

Greenwood and Tabor's formula for coefficient of elastic friction

$$\mu_e = \alpha_n P_m \left[\frac{9\pi}{64} \frac{(1 - v^2)}{E} \right]$$

(24–224)

Coefficient of friction under dynamic conditions

Franke's expression for coefficient of friction during rotation	$\mu_r = \mu_o e^{-cv}$	(24–225)

where $c = $ constant taken from Table 24–74

Stiehl's formula for coefficient of friction	$\mu = 0.6 - \dfrac{0.6}{v + 1}$	(24–226)
Schutch's formula for coefficient of friction for leather sliding against slightly lubricated steel plate	$\mu = 0.5(1 + 0.1v)$	(24–227)

Particular	Formula	
Krumme's formula for coefficient of friction in textile machinery	$\mu = 0.38 - \dfrac{0.1}{0.5 + v}$	(24–228)
Formula for coefficient of friction used in design of brakes	$\mu = 0.6 \dfrac{16P + 100}{80P + 100} \cdot \dfrac{100}{3v_k + 100}$	(24–229)

where P = real pressure on brake shoe, tonne force (tf)

TABLE 24–74
Values of constant c to be used in Eq. (24–225)

Sliding combination	State of rubbing surfaces	Coefficient of static frction, μ_o	Constant c
Cast iron—steel	Dry	0.29	1/23
Forged iron—forged	Dry	0.29	1/50
Iron	Slighty moist	0.24	1/35

Temperature of sliding surface

Mean temperature rise at the interface above the material

$$\theta_m = \frac{0.25\mu W v}{Q_{me} r (k_1 + k_2)} \qquad (24\text{–}230)$$

where

Q_{me} = mechanical equivalent of heat N m/J (lbf in/Btu or lbf ft/Btu)

v = velocity of sliding, cm/s (ft/min)

r = radius of the circular junction, cm, m (in)

k_1, k_2 = thermal conductivity of the two contacting materials, W/m °C (Btu/ft h°F) taken from Table 24–75

TABLE 24–75
Temperature rise per unit sliding velocity

Material combination	μ	γ		k_1	k_2	k_1	k_2	θ/γ,
		dyn/cm	N/m	cal/s cm °C		W/m°K		°C/cm/s
Steel on steel	0.5	1500	1.50	0.11	0.11	46.055	46.055	0.75
Lead on steel	0.5	450	0.45	0.08	0.11	33.490	46.055	0.26
Bakelite on Bakelite	0.3	100	0.10	0.0015	0.0015	0.628	0.628	2.20
Brass on brass	0.4	900	0.90	0.26	0.26	108.856	108.856	0.15
Glass on steel	0.3	500	0.50	0.0007	0.11	0.293	46.055	0.30
Steel on nylon	0.3	120	0.12	0.11	0.0006	46.055	0.25121	0.07
Brass on nylon	0.3	120	0.12	0.26	0.0006	108.856	0.25121	0.03
Steel on bronze	0.25	900	0.90	0.11	0.18	46.055	75.362	0.17

Particular	Formula
Simple and crude formula for the mean temperature rise	$\theta_m = 54.4v(\pm a$ factor of 1.67) $\qquad$ (24–231)
The radius of a junction (Fig. 24–50)	$r = 12,000\ \dfrac{\gamma}{P}$ $\qquad$ (23–232)
The load carried by each junction (Fig. 24–50)	$W = \pi r^2 P$ $\qquad$ (24–233)
Mean temperature rise at the interface above the rest of material	$\theta_m = \dfrac{9400\mu\gamma v}{Q_{me}(k_1 + k_2)}$ $\qquad$ (24–234)

where γ = surface tension of the softer sliding member, N/m (lbf/in) taken from Table 24–75

For coefficient of friction μ refer to Table 24–75

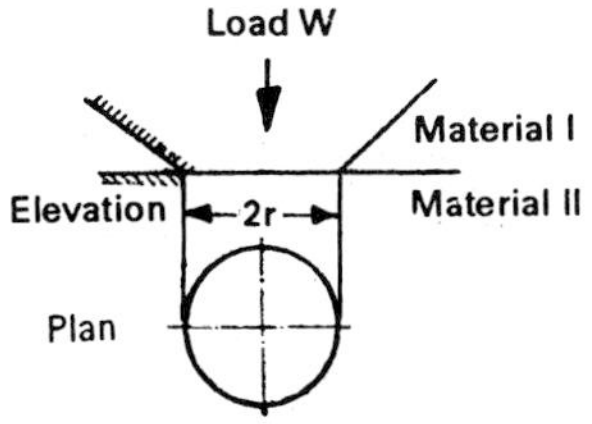

FIGURE 24–50 Assumed junction model.

WEAR AND ABRASION

Linear wear rate	$K_L = \dfrac{\text{thickness of layer removed}}{\text{sliding distance}} = \dfrac{h}{L}$ $\quad$ (24–235)
Steady state wear rate, depth per unit time	$K_L = KPV(abcde)$ $\qquad$ (24–236)

where K = constant depends on (i) mechanical properties of material and its ability to (ii) smooth the counterface surface and/or (iii) transfer a thin film of debris

For a, b, c, d, e, refer to Table 24–83

Volumetric wear rate	$K_V = \dfrac{\text{volume of layer removed}}{\text{sliding distance} \times \text{apparent area}} = \dfrac{\Delta V}{LA_a}$
	$\qquad$ (24–237)
Energetic wear rate	$K_E = \dfrac{\text{volume of layer removed}}{\text{work of friction}} = \dfrac{\Delta V}{F_\mu L}$ $\quad$ (24–238)
The energetic and linear wear rate related by equation	$K_E = K_L(A_a/F_\mu)$ $\qquad$ (24–239)

where $F_\mu L$ is measured in kW h

The gravimetric wear rate	$K_W = \dfrac{\Delta W}{LA_a} = \rho K_v$ $\qquad$ (24–240)

where ρ = density of abraded elastomer

Particular	Formula
Wear index is given by abradability, δ	$\delta = \dfrac{\text{abraded volume}}{\text{work of friction}} = \dfrac{\Delta V}{F_\mu L} = \dfrac{\Delta V}{\mu WL} = \dfrac{A'}{\mu}$ (24–241) where $A' = (\Delta V/WL) = $ abrasion factor
The relation between K_E and δ	Energetic wear rate $(K_E) = $ abradability (δ) (24–242)
The coefficient of abrasion resistance as per work in the former Soviet Union	$\zeta = \dfrac{\text{work of friction}}{\text{abraded volume}} = \dfrac{FL}{\Delta V} = \dfrac{1}{\delta} = \dfrac{\mu}{A'} = \dfrac{1}{K_E}$ (24–243)
For surface roughness as obtained by different machining processes	Refer to Table 24–76.
Work done during wear	$W' = V \tau_m$ (24–244)
Volume of transferred fragments formed in sliding a distance L	$V = \dfrac{kWL}{300P}$ (24–245) For $k = $ coefficient of wear, refer to Table 24–77. For $P = $ hardness of the softer material, Pa (psi), refer to Table 24–79.

TABLE 24–76
Surface roughnesses as obtained by machining processes

Manufacturing process	Surface roughnesses, μm
Turned	1–6
Coarse ground	0.4–3
Fine ground	0.2–0.4
600 emery	0.2
Polished	0.05–0.1
Super finished	0.02–0.05

TABLE 24–77
Wear constant k

Sliding combination	Wear constant, k
Zinc on zinc	0.160
Low-carbon steel on low-carbon steel	45
Copper on copper	32
Stainless steel on stainless steel	21
Copper on low-carbon steel	1.5
Low-carbon steel on copper	0.5
Bakelite on bakelite	0.02

TABLE 24–78
Values of coefficient of wear, k

| Condition | Metal on metal | | Metal on non-metal |
	Like	Unlike	
	$\times 10^{-5}$		$\times 10^{-6}$
Clean	500	20	5
Poorly lubricated	20	20	5
Average lubrication	2	2	5
Excellent lubrication	0.2–0.2	0.2–0.2	2

TABLE 24–79
Properties of metallic elements

Metal	Melting temperature °C	K	Young's modulus, E kgf/cm² ×10⁶	N/m² ×10¹¹	MPa ×10⁵	Yield strength, σ_{sy} kgf/cm² ×10³	N/m² ×10⁸	MPa ×10²	Hardness, P kgf/cm² ×10²	N/m² ×10⁷	MPa ×10	Surface energy, γ erg/cm	N/m	γ/P cm	m
Aluminum	660	933	0.64	0.63	0.63	1.12	1.1	1.1	27	26.46	26.46	900	0.900	33	0.33
Antimony	630	903	0.82	0.80	0.80	0.11	0.11	0.11	58	56.84	56.84	370	0.370	6.4	0.064
Beryllium	1400	1673	3.06	3.0	3.0	3.26	3.20	3.20	150	147.0	147.0	1000	1.0	6.7	0.067
Bismuth	270	543	0.33	0.32	0.32				7	6.86	6.86	390	0.39	56.0	0.56
Cadmium	321	594	0.57	0.56	0.56	0.73	0.72	0.72	22	21.56	21.56	620	0.62	28	0.28
Calcium	838	1111	0.26	0.25	0.25	0.89	0.87	0.87	17	16.66	16.66				
Cerlum	804	1077	0.31	0.30	0.30	1.22	1.20	1.20	48	47.04	47.04				
Cesium	29	302													
Chromium	1875	2148	2.65	2.6	2.6	1.63	1.6	1.6	125	122.5	122.5				
Cobalt	1495	1778	2.14	2.1	2.1	7.96	7.8	7.8	125	122.5	122.5	1530	1.53	12	0.12
Copper	1083	1356	1.22	1.2	1.2	3.26	3.2	3.2	80	78.4	78.4	1100	1.10	14	0.14
Dysporsium	1407	1680	0.64	0.63	0.63	3.37	3.3	3.3	117	115.66	115.66				
Erbium	1496	1769	0.78	0.75	0.75	2.96	2.9	2.9	161	157.78	157.78				
Europium	827	1100							17	16.66	16.66				
Gadolinium	1312	1585	0.57	0.56	0.56	2.76	2.7	2.7	97	95.06	95.06				
Gallium	30	303							6.5	6.37	6.37	360	0.36	55	0.55
Germanium	937	1210	1.59	1.56	1.56							1120	1.12	19	0.19
Gold	1063	1336	0.83	0.81	0.81	2.14	2.1	2.1	58	56.84	56.84				
Hafnium	2222	2495				2.45	2.4	2.4	260	254.80	254.80				
Holmium	1461	1734	0.69	0.68	0.68	2.24	2.2	2.2	90	88.2	88.2				
Indium	156	429	0.11	0.11	0.11	0.03	0.03	0.03	0.9	0.88	0.88				
Iridium	2454	2727	5.50	5.4	5.4	6.43	6.3	6.3	350	343.0	343.0				
Iron	1534	1807	2.08	2.04	2.04	2.55	2.5	2.5	82	80.36	80.36	1500	1.50	18	0.18
Lanthanum	930	1203	0.40	0.39	0.39	1.94	1.9	1.9	150	147	147				
Lead	325	598	0.16	0.16	0.16	0.09	0.09	0.09	4	3.92	3.92	450	0.45	110	1.10
Lithium	180	453										400	0.40		
Lutetium	1652	1925							118	115.64	115.64				
Magnesium	650	923	0.45	0.44	0.44	1.53	1.5	1.5	46	45.08	45.08	560	0.56	12	0.12
Manganese	1245	1518				2.55	2.5	2.5	3300	3234	3234				
Mercury	−39	234										460	0.46		
Molybdenum	2610	2883	3.06	3.0	3.0	8.57	8.4	8.4	240	235.2	235.2				
Neodymium	1018	1291	0.39	0.38	0.38	1.73	1.7	1.7	80	78.4	78.4				
Nickel	1453	1726	2.12	2.08	2.08	3.26	3.2	3.2	210	205.8	205.8	1700	1.70	8.1	0.081
Niobium	2468	2741	1.07	1.05	1.05	2.86	2.8	2.8	160	156.8	156.8	2100	2.10	13	0.13
Osmium	2700	2973	5.81	5.70	5.70				800	784	784	1190	1.19	1.5	0.015
Palladium	1552	1825	1.17	1.15	1.15	3.16	3.1	3.1	110	107.8	107.8				
Platinum	1769	2042	1.53	1.50	1.50	1.63	1.6	1.6	100	98	98	1800	1.80	18	0.18
Plutonium	640	913	1.01	0.99	0.99	2.86	2.8	2.8	266	260.68	260.68				
Potassium	64	337							0.04	0.04	0.04	86	0.086	2300	23
Praseodymium	919	1192	0.36	0.35	0.35	2.04	2.0	2.0	76	74.48	74.48				
Rhenium	3180	3453	4.79	4.70	4.70	22.40	22.0	22.0							
Rhodium	1966	2293	3.02	2.96	2.96	9.89	9.7	9.7	122	119.56	119.56				
Rubidium	39	312													
Ruthenium	2500	2773	4.30	4.22	4.22	5.61	5.5	5.5	390	382.2	382.2				
Samanium	1072	1345	0.36	0.35	0.35	1.33	1.3	1.3	64	62.72	62.72				
Scandium	1540	1813													

TABLE 24–79
Properties of metallic elements (*Cont.*)

Metal	Melting temperature $^\circ$C	$^\circ$K	Young's modulus, E kgf/cm$^2\times10^6$	N/m$^2\times10^{11}$	MPa $\times10^5$	Yield strength, σ_{sy} kgf/cm$^2\times10^3$	N/m$^2\times10^8$	MPa $\times10^2$	Hardness, P kgf/cm$^2\times10^2$	N/m$^2\times10^7$	MPa $\times10$	Surface energy, γ erg/cm	N/m	γ/P cm	m
Silver	961	1234	0.80	0.78	0.78	2.04	2.0	2.0	80	78.4	78.4	920	0.92	11	0.11
Sodium	98	371							0.07	0.07	0.07	200	0.20	2800	28
Tantalum	2996	3269	1.93	1.90	1.90	3.56	3.5	3.5							
Terbium	1356	1629	0.59	0.58	0.58				88	86.24	86.24				
Thallium	303	576				0.09	0.09	0.09	2	1.96	1.96	400	0.40	200	2.0
Thorium	1750	2023	1.50	1.47	1.47	1.53	1.5	1.5	37	36.26	36.26				
Thulium	1545	1818				1.43	1.4	1.4	53	51.94	51.94				
Tin	232	505	0.45	0.44	0.44	1.17	1.15	1.15	53	5.19	5.19	570	0.57	110	1.10
Titanium	1670	1943	1.15	1.13	1.13	1.43	1.4	1.4	65	64.7	64.7				
Tungsten	3410	3683	3.58	3.51	3.51	18.36	18.0	18.0	435	426.3	426.3	2300	2.30	5.3	0.053
Uranium	1132	1405	1.72	1.69	1.69	2.04	2.0	2.0							
Vanadium	1900	2173	1.36	1.34	1.34	8.57	8.4	8.4							
Ytterbium	824	1097	0.18	0.18	0.18	0.74	0.73	0.73	21	20.58	20.58				
Yttrium	1495	1768	0.67	0.66	0.66	1.43	1.4	1.4	37	36.26	36.26				
Zinc	420	693	0.93	0.91	0.91	1.33	1.3	1.3	38	37.24	37.24	790	0.79	21	0.21
Zirconium	1852	2125	0.98	0.96	0.96	2.04	2.0	2.0	145	142.10	142.10				

Particular	Formula	
Another formula for volume of transferred fragment formed in sliding a distance	$$V = \frac{kAL}{3}$$ For k refer to Table 24–78.	(24–246)
The primary equation of wear according to Archard, Burwell, and Strang	$$\frac{\Delta V}{L} = K \frac{W}{P_m}$$ where P_m = flow pressure of material For K, refer to Table 24–80.	(24–247)

Abrasion wear

The mean diameter of loose wear particles which are produced at a smooth interface	$$d = K_1\left(\frac{W_{ab}}{H}\right)$$	(24–248)
The ratio of half mean diameter of the area of contact to mean radius of the curvature at the tip of the abrasive particle	$$\left(\frac{a}{R}\right) = K_2\left(\frac{W}{GR^2}\right)^\alpha$$ where α = value of exponent to be determined from experiment	(24–249)

TABLE 24–80
Coefficient of wear

Sliding against hardened tool-steel unless otherwise stated	Wear coefficient, K	Hardness	
		kgf/mm^2	MPa
Mild steel on mild steel	7×10^{-3}	18.6	182.4
60/40 brass	6×10^{-4}	95.0	931.6
Teflon	2.5×10^{-4}	5.0	49.0
70/30 brass	1.7×10^{-4}	68.0	666.8
Perspex	0.7×10^{-6}	20.0	196.1
Bakelite (molded) type 50B	7.5×10^{-6}	25.0	245.2
Silver steel	6×10^{-5}	320.0	3138.1
Beryllium copper	3.7×10^{-5}	210.0	2059.4
Hardened tool steel	1.3×10^{-4}	850.0	8335.6
Stellite	5.5×10^{-5}	690.0	6776.6
Ferritic stainless steel	1.7×10^{-5}	250.0	2451.7
Laminated Bakelite Type 292/16	1.5×10^{-6}	33.0	323.6
Molded Bakelite Type 11085/1	7.5×10^{-7}	30.0	294.2
Tungsten carbide on mild steel	4×10^{-6}	186.0	1824.0
Molded Bakelite Type 547/1	3×10^{-7}	29.0	284.4
Polythene	1.3×10^{-7}	1.70	16.7
Tungsten carbide on tungsten carbide	1×10^{-6}	1300.0	12749

Particular	Formula	
Volumetric wear rate	$K_V = K_3 n^2 R^3 [W_{tot}/Gn^2 R^2]^{3\alpha}$ where $n^2 =$ number of abrasive particles per unit area	(24–250)
Volumetric wear rate for $\alpha = \frac{1}{3}$	$K_V = K_3 \left(\dfrac{W_{tot} R}{G} \right)$	(24–251)
Half the mean diameter of the area of contact for $\alpha = \frac{1}{3}$	$a = K_1 \left(\dfrac{WR}{G} \right)^{1/3}$	(24–252)
The spacing s between ridges in the elastomer surface	$s = \simeq \left[\dfrac{W_{tot} R d^2}{G} \right]^{1/3}$	(24–253)
	$s \simeq d^{2/3}$	(24–254)
The ratio of K_v to s when the abrasive surface consists of closely packed hemisphere so that $d = 2R$	$\left(\dfrac{K_V}{s} \right) \simeq \left(\dfrac{W_{tot}}{G} \right)^{2/3}$	(24–255)

Particular	Formula

Fatigue wear

Volume of surface layer removed under fatigue

$$\Delta V = iAh \qquad (24\text{-}256)$$

The required sliding length during abrasion cycle under the given abrasion conditions before failure and separation occurs

$$L = i\lambda F_f \qquad (24\text{-}257)$$

The total work of friction

$$W'_\mu = (\mu W_{tot})L = iqF_f \qquad (24\text{-}258)$$

The coefficient of abrasion resistance

$$\zeta = \frac{qF_f}{AL} \qquad (24\text{-}259)$$

The Hertzian relationship for the average depth of penetration for single spheres

$$z = \left[\frac{3}{4}\,(1-\theta^2)\right]^{2/3} \frac{W^{2/3}}{E^{2/3}R^{1/3}} \qquad (24\text{-}260a)$$

$$= 0.683\left(\frac{W^{2/3}}{E^{2/3}R^{1/3}}\right) \qquad (24\text{-}260b)$$

for rubber $\theta = 0.5$

where
R = asperity tips radius, cm, m (in)
E = Young's modulus for rubber, GPa (psi)
W = applied load per asperity

The depth penetration

$$z = 0.685\left(\frac{\lambda^2}{A}\right)^{2/3} \frac{W_{tot}^{2/3}}{E^{2/3}R^{1/3}} \qquad (24\text{-}261)$$

The number of asperities

$$i = \frac{W_{tot}}{W} = \frac{A}{\lambda^2} \qquad (24\text{-}262)$$

The effective thickness of the surface layer of elastomer

$$h = k'z\,\frac{\pi R^2}{\lambda^2} \qquad (24\text{-}263)$$

where K' = constant

The coefficient of abrasion resistance

$$\zeta = \left(\frac{\mu F_f}{2.14K'}\right)E^{2/3}\left(\frac{W_{tot}}{A}\right)^{1/3}\left(\frac{\lambda}{R}\right) \qquad (24\text{-}264)$$

The ratio of abrasion resistance to coefficient of sliding friction

$$\frac{\zeta}{\mu} = \frac{1}{A'} \qquad (24\text{-}265)$$

The fatigue resistance of rubber taking into consideration tensile strength, geometry of the base surface, and the loading conditions

$$F_f = \frac{\sigma_o}{K'(W/A)^{1/3}E^{2/3}(R/\lambda)^{-2/3}} \qquad (24\text{-}266)$$

where b = index which is characteristic of the material

Particular	Formula
The ratio of abrasion resistance to coefficient of friction	$$\left(\frac{\zeta}{\mu}\right) = K\sigma_o^b E^{2(1-b)/3}\left(\frac{W_{tot}}{A}\right)^{(1-b)/3}\left(\frac{\lambda}{R}\right)^{(5-2b)/3}$$ (24–267) where K = constant
The relationship between fatigue index b and α	$$b = \frac{1}{3}(\alpha + 2)$$ (24–268)

Roll formation

The coefficient of abrasion resistance	$$\zeta = \frac{P_{tot}}{(d\Delta V)/dt}$$ (24–269) where $(d\Delta V)/dt$ = volume abraded per unit time P_{tot} = total frictional power $\quad = P_t + P_e + P_H$
The main condition which determines the probable occurrence of roll formation	$$P_{tot} \le \mu_o WV$$ (24–270)
The more general form of the equation for volumetric wear rate which dependence on abrasion by load	$$K_V = CP^\alpha$$ (24–271) where C = constant taken from Table 24–81 P = interfacial pressure, MPa (psi) α is obtained from Table 24–81

TABLE 24–81
Wear of rubber on steel, gauze, and abrasive paper

Rubber	Nature of surface	Values of constants	
		$C \times 10^3$	α
A	Steel	1.1	1.9
	Gauze	1.5	5.3
	Abrasive paper	240	1.1
B	Steel	2.7	1.9
	Gauze	1.1	2.0
	Abrasive paper	305	0.9
C	Steel	1.2	3.1
	Gauze	5.4	3.0
	Abrasive paper	65	1.0

Tread rubber

The shearing stress for tread rubber	$$\tau = \mu P$$ (24–272) where P = normal pressure, MPa (psi)
The critical shearing stress for tread rubber	$$\tau_{crit} = \mu_{crit}P$$ (24–273)

Particular	Formula
For $\tau < \tau_{crit}$	The fatigue wear predominates.
For $\tau > \tau_{crit}$	Either wear through roll formation or abrasive wear occurs.
For $\mu < \mu_{crit}$	The wear is due to surface fatigue.
For $\mu > \mu_{crit}$	Other forms of wear predominate.

Specific wear

Specific wear by mass

$$K_{sm} = \frac{m}{Ad} \tag{24–274}$$

Specific wear by volume

$$K_{sV} = \frac{V}{Ad} \tag{24–275}$$

Specific wear by volume based on the geometry of the aspirities arising out of the surface treatment

$$K_{sV} = \frac{\tan\theta}{(\beta+1)2i} = \frac{\epsilon h_m}{(\beta+1)id} = \frac{z}{(\beta+1)id} \tag{24–276}$$

where values of angle of slope of irregularities, θ, can be obtained from Table 24–82 and the values of β from Table 24–73

z = absolute approach $\epsilon = z/h_m$

The absolute approach

$$z = \frac{6\sigma_c}{K\gamma_1\gamma_2} \tag{24–277}$$

where

$$K = \frac{K_1 K_2}{K_1 + K_2} = \text{coefficient of rigidity}$$

$$K_i = \frac{E_i}{2\rho_i(1 - v_i^2)}$$

2ρ = diameter of contact spot, cm
γ = tangent to the smoothness of the surface equal to the derivative of approach over the contact area = $\tan\theta$

TABLE 24–82
Radii of curvature asperities for different methods of surface preparation

Treatment	Accuracy class	Slope radii, micron		Angle of slope of irregularities, θ	
		Transverse	Longitudinal	Transverse	Longitudinal
Shaping	5–8	20–120	10–25	5–20	5–10
Grinding	5–9	5–20	250–15000	7–35	2–10
Honing	8–11	4–30	60–160	3–13	1–4
Finishing (lapping)	10–13	15–250	7000–35000	5–20	2–10

Particular	Formula
An expression for modified specific wear	$$K'_{sV} = K_{sV} \frac{A}{A_a} = K_{sV} \frac{P_a}{P} \qquad (24\text{--}278)$$
Modified specific wear formula during microcutting	$$K'_{sV} = \frac{\tan\theta \cdot P_a}{(\beta + 1)2P} \qquad (24\text{--}279a)$$ $$= 0.02 \frac{P_a}{P} \text{ to } 0.04 \frac{P_a}{P} \text{ for } \tan\theta = 0.1 \text{ to } 0.2$$ during microcutting
Modified specific wear formula during plastic contact	$$K'_{SV} = \left[\frac{h_{max}}{rb\,\frac{1}{\beta}}\right]^{5/2} \left[\frac{P_a}{P}\right]^{5+2\beta/2\beta} \left[\frac{c\mu}{\epsilon_{fail}}\right]^{2\beta^{1/2}/8} \qquad (24\text{--}280)$$ where $$h_{max} = \frac{d}{2\epsilon_{max}} \tan\theta$$ ϵ_{fail} = relative elongation corresponding to failure of the specimen c = constant depending on sliding combination taken from Table 24–74
Modified specific wear formula during elastic contact	$$K'_{SV} = c_1 \frac{(1-v^2)P_a}{E} \left[\frac{K\mu\sigma_c}{c_2\sigma_o}\left\{\frac{E}{(1-v^2)P_a}\right\}^{2\beta/(2\beta+1)}\right]$$ $$(24\text{--}281a)$$ where $c_1 = \frac{3}{8}\,\pi\,\frac{\sqrt{\beta}}{K_2(\beta+1)} \qquad (24\text{--}281b)$ $$c_2 = \left(\frac{r}{h_{max}}\right)^{\beta/(2\beta+1)} \left(\frac{b}{2}\right)^{1/(2\beta+1)} \left(\frac{0.75\pi}{K_2}\right)^{2\beta/(2\beta+1)}$$ $$(24\text{--}281c)$$

GENERAL

For values of wear rate correction factors; physical and mechanical properties of clutch facings; mechanical properties, performance and allowable operating conditions for various materials; physical and mechanical properties of materials for sliding faces; rubbing bearing materials and applications and allowable working conditions and frictions for various clutch facing materials

Refer to Tables 24–83 to 24–88

TABLE 24–83
Approximate values of wear rate correction factors

Name of factor	Condition		Constant
a. Geometrical factor	Continuous motion + rotating load		0.5
	Unidirectional load		1
	Oscillating motion		2
b. Heat dissipation factor	Metal housing, thin shell, intermittent operation		0.5
	Metal housing, continuous operation		1
	Nonmetallic housing, continuous operation		2
c. Temperature factor	PTFE base:	20°C	1
		100°C	2
		200°C	5
	Carbon graphite thermoset	20°C	1
		100°C	3
		200°C	6
d. Counterface factor	Stainless steels, chrome plate		0.5
	Steels		1
	Soft, nonferrous metals		2.5
e. Surface finish factor	0.1–0.2 μm		1
	0.2–0.4 μm		2–5
	0.4–0.8 μm		4–10

TABLE 24–84
Physical and mechanical properties of clutch facings

	Resin-based material	Sintered metals
Thermal conductivity	0.80 W/m °C	16 W/m °C
Specific heat	1.25 kJ/kg °C	0.42 kJ/kg °C
Thermal expansion	$0.50 \times 10^{-4}/$°C	$0.13 \times 10^{-4}/$°C
Specific gravity	1.6 for woven 2.8 for molded	
Young's modulus, E	352 kgf/mm^2 3.45×10^9 N/m^2 3.45 GPa	1488 kgf/mm^2 14.5×10^9 N/m^2 14.5 GPa
Ultimate tensile strength, σ_{ut}	2.14 kgf/mm^2 21×10^6 N/m^2 21 MPa	4.57 kgf/mm^2 44.8×10^6 N/m^2 44.8 MPa
Ultimate shear, stress, τ_u	1.22 kgf/mm^2 12×10^6 N/m^2 12 MPa	3.59 kgf/mm^2 35.2×10^6 N/m^2 35.2 MPa
Ultimate compressive strength, σ_{uc}	10.5 kgf/mm^2 103×10^6 N/m^2 103 MPa	15.6 kgf/mm^2 153×10^6 N/m^2 153 MPa
Rivet holding capacity	7.03 kgf/mm^2 69×10^6 N/m^2 69 MPa	

TABLE 24-85
Mechanical properties, performance, and allowable operating conditions for various materials

Materials	Specific gravity	Coefficient of friction, μ	Tensile stress, σ_t (kgf/mm²)	($10^6 \times$ N/m²)	(MPa)	Shear stress, τ (kgf/mm²)	($10^6 \times$ N/m²)	(MPa)	Compressive stress, σ_c (kgf/mm²)	($10^6 \times$ N/m²)	(MPa)	Rivet holding capacity (kgf/mm²)	($10^6 \times$ N/m²)	(MPa)	Wear rate at, 100°C, 10^{-6} mm³/J	Temperature, °C (Maximum)	(Maximum operating)	Working pressure, P_w (kgf/mm² $\times 10^{-3}$)	(10^3 N/m²)	($10^{-3} \times$ MPa)	Maximum pressure, P_{max} (kgf/mm²)	($10^6 \times$ N/m²)	(MPa)
Lining																							
Woven cotton	1.0	0.50	2.1	20.7	20.7	1.26	12.4	12.4	9.85	96.5	96.5	7.03	69	69	12.2×10^{-6}	150	100	7.2–71.5	70–700	70–700	0.152	1.5	1.5
Woven asbestos	1.5–2.0	0.4	2.45	24.1	24.1	1.39	13.8	13.8	10.54	103.4	103.4	8.45	83	83	9.2×10^{-6}	250	125	7.2–71.5	70–700	70–700	0.214	2.1	2.1
Molded																							
Light-duty (flexible)	1.7	0.40	0.84	8.2	8.2	0.84	8.2	8.2	4.21	41.3	41.3	10.50	103	103	6.1×10^{-6}	350	175	7.2–71.5	70–700	70–700	0.214	2.1	2.1
Medium (semiflexible)	1.7	0.35	1.05	10.3	10.3	0.84	8.2	8.2	9.85	96.5	96.5	15.50	152	152	3.1×10^{-6}	400	200	7.2–71.5	70–700	70–700	0.296	2.8	2.8
Heavy-duty	2.0	0.35	1.39	13.8	13.8	1.39	13.8	13.8	10.54	103.4	103.4	17.50	172	172	1.8×10^{-6}	500	225	7.2–71.5	70–700	70–700	0.390	3.8	3.8
Pad																							
Resin-based or asbestos	2.0	0.32				0.92	9.0	9.0	10.54	103.4	103.4				1.2×10^{-6}	650	300	35.5–178.5	350–1750	350–1750	0.561	5.5	5.5
Sintered metals	6.0	0.30	4.91	48.2	48.2	7.02	68.9	68.9	10.50	151.6	151.6				Used at higher temperature	650	300	35.5–356.5	350–3500	350–3500	0.561	5.5	5.5
Cement		0.32														800	400	35.5–107.0	350–1050	350–1050	0.703	6.9	6.9

Key: 1 psi = 6895 Pa; 1 kpsi = 6.894757 MPa

TABLE 24-86
Physical and mechanical properties of materials for sliding face

Materials	Compressive strength, σ_c			Tensile strength, σ_t			Modulus of elasticity, E			Poisson's ratio, ν	Hardness, H_B	Density, ρ		Porosity, c, %	Maximum temperature, T_{max}		Expansion coefficient α	Temperature range, ΔT		Thermal conductivity		Thermal stress resistance	
	kgf/mm²	MN/m²	MPa	kgf/mm²	MN/m²	MPa	kgf/mm² ×10³	GN/m²	GPa			g/cm³	kg/mm³	%	°C	K	$10^6 \times °C^{-1}$	°C	K	kcal/m h°C	W/m K	kcal/m h	W/m
PTFCE	20–56	215.8–549.4	215.8–549.4	3.2–4.0	31.4–39.2	31.4–39.2	0.16	1.55	1.55	(0.3)	80*	2.1	2100	0	150	423	50	(320)	(593)	0.052	0.096	(16.6)	19.31
Nylon	5–9	49.1–88.3	49.1–88.3	4.9–7.5	48.1–73.6	48.1–73.6	0.18–0.28	1.77–2.75	1.77–2.75	(0.3)		1.09–1.14	1090–1140	0	135–150	408–423	100–140	(130)	(403)	0.12–0.21	0.140–0.244	(21.5)	(25.00)
Phenol resin	7	68.7	68.7	5.0–5.6	49.1–54.9	49.1–54.9	0.52–0.70	5.1–6.87	5.1–6.87	0.25		1.25–1.3	1250–1300		130	403	25–60	140	(413)	0.1–0.2	0.116–0.233	(21.5)	(25.00)
Synthetic resin 1	21	207	207	3.5–4.9	34.3–48.1	34.3–48.1	2.1–3.5	20.7–34.3	20.7–34.3	(0.25)		1.3–1.75	1300–1750		120	393	19–26	(50)	(323)	0.36–0.51	0.419–0.593	22.0	25.59
Resin 1	10–17.5	98.1–171.7	98.1–171.7	4.9	48.1	48.1	0.63–0.90	6.2–8.9	6.2–8.9	(0.30)		1.25	1250		150	423	10–40	(150)	(423)	0.14–0.25	0.163–0.291	(30.0)	(34.89)
Resin-impregnated fabric	10–24	98.1–216	98.1–216	2.3–6.3	22.6–61.8	22.6–61.8						1.36–1.43	1360–1430		120	393							
Acetal resin	7	68.7	68.7	7	68.7	68.7	0.34	3.28	3.28	0.35		1.425	1425	0	100	373	81	167	440	0.2	0.233	33.5	39.96
Bakelite	10–24.5	98.1–240	98.1–240	2.8–5.0	27.5–49.1	27.5–49.1	0.70–1.75	6.87–17.16	6.87–17.16	0.25		1.52–2.0	1520–2000		175–230	448–503	0.5–40	87	360	0.29–0.58	0.337–0.675	38.0	44.19
Hard rubber				1.0–2.8	9.81–27.5	9.81–27.5	0.1	1.03	1.03	(0.4)		1.3	1300	0	100	373	54	180	453	0.25	0.291	45.0	52.34
Synthetic resin 2	10–15	98.1–147	98.1–147	1.5–4.0	14.7–39.2	14.7–39.2	1.5–1.7	14.7–17.58	14.7–17.58	(0.25)		1.6–1.9	1600–1900		130–160	403–433	15–30	(75)	348	0.4–1.0	0.465–1.163	(53)	(61.64)
PTFE				4.1	40.2	40.2	0.035–0.10	0.343–0.98	0.343–0.98	(0.5)	55–63*	2.1–2.3	2100–2300	0.3	280	553	70	(410)	(683)	0.2	0.233	(82)	(95.37)
Synthetic carbon 1	16	156.9	156.9	2.1	20.1	20.1	1.84	18.06	18.06	0.2	65**	2.0	2000		170	443	13.5	66	239	2.0	2.33	132	153.46
Synthetic carbon 2	16.5	164.8	164.8	2.3	22.6	22.6	1.32	12.95	12.95	(0.25)	65**	2.8	2800	0.3	170	443	20	(65)	338	2.5	2.91	(164)	190.73
Carbon 1	18	176.6	176.6	2.8	27.5	27.5	1.40	13.7	13.7	0.22	85**	1.8	1800	0.1	180	453	5.0	312	595	4.0	4.65	1250	1453.75
Carbon 2	27	264.9	264.9	1.75	17.2	17.2	1.85	18.15	18.15	0.18	100**	1.82	1820	0.4	365	638	6.1	126	399	13	15.12	1650	1918.95
Carbon 3	25	245.3	245.3	3.1	30.4	30.4	1.46	14.32	14.32	0.2	80**	1.79	1790	0.5	285	558	5.3	320	593	20.0	23.3	6400	7443.20
Carbon 4	33.5	329.6	329.6	3.6	35.3	35.3	1.60	15.7	15.7	(0.2)	75**	2.5	2500	2.5	280	553	6.6	(273)	546	30.0	34.89	(8200)	(9536.60)
Carbon 5	35	343.4	343.4	2.1	20.6	20.6	1.35	13.24	13.24	0.2	85**	2.4	2400	4.0	350	623	4.82	750.0	1023	34.0	39.54	(8800)	(10234.4)
Carbon 6	23.5	230.5	230.5	5.3	52.0	52.0	2.60	25.5	25.5	0.22	93**	1.73	1730	0.3	370	643	2.16	362	635	20.0	23.3	15000	17445.0
Graphite 1	12.5	122.6	122.6	1.6	15.7	15.7	0.7	6.87	6.87	0.22	65**	1.65	1650	14	540	813	4.9	235	508	46.0	53.50	16700	19422.10
Graphite 2	10	98.1	98.1	1.5	14.7	14.7	1.0	9.81	9.81	0.2	65**	1.85	1850	1.0	365	638	5.25	260	533	90.0	104.67	21200	24655.60
Graphite 3	12.5	122.6	122.6	1.9	18.6	18.6	1.15	11.28	11.28	0.18	72**	1.85	1850	0.25	370	643	5.2	250	523	89.0	103.51	23000	26749
Graphite 4	7.1	69.7	69.7	1.45	14.2	14.2	1.30	12.75	12.75	0.22	60**	1.83	1830	0.3	180	453	3.5	250	523	100.0	116.3	25000	29075
Graphite 5	5.5	54.9	54.9	1.4	13.7	13.7	0.56	5.49	5.49	0.22	50**	1.66	1660	10.0	520	793	4.5	780	1053	60.0	69.78	26000	30238
Graphite 6	14	137.3	137.3	2.0	19.6	19.6	1.0	9.81	9.81	0.22	70**	1.8	1800	7.0	340	613	2.0	97	370	60.0	69.78	47000	54461
Hard alloy 1	150	1470	1470	38	372.8	372.8	24	235.4	235.4	0.3	58–62*	8.78	8780		1250	1523	11.4			7.0	8.14	680	790.84
Hard alloy 2	280	2746.8	2746.8	30	294.3	294.3	24.4	239.4	239.4	(0.3)	60†	7.77	7770		1150	1423	9.9	87	360	9.7	11.28	(850)	(988.55)
Hard alloy 3	135	1324.4	1324.4	53.0	519.9	519.9	23	225.6	225.6	0.3	48–50†	8.65	8650		1260	1533	11.9	567	840	11.0	12.79	1480	1720.24

TABLE 24-86
Physical and mechanical properties of materials for sliding face (Cont.)

Materials	σ_c kgf/mm²	σ_c MN/m²	σ_c MPa	σ_t kgf/mm²	σ_t MN/m²	σ_t MPa	σ_t kgf/mm²×10³	E GN/m²	E GPa	ν	H_B	ρ g/cm³	ρ kg/mm³	ε %	T_{max} °C	T_{max} K	α 10⁶×°C⁻¹	ΔT °C	ΔT K	kcal/m h °C	W/m K	kcal/m h	W/m
Hardened nickel				28–	274.6–	274.6–	17.5	171.6	171.6	(0.26)	53–57†	7.7	7700		800	1073	8.5	(157)	430	12.2	14.19	1930	2244.5
Stainless steel				35	343.4	343.4	20	196.2	196.2	0.28	155–185‡	7.98	7980	0	1400⁺	1673⁺	16.0	121	394	16.0	18.61	1940	2253.2
Steel AISI 316				54	529.7	529.7	20	196.2	196.2						1200⁺	1473⁺	17.0						
Invar	70	687–	687–	45	441.5	441.5	15	147.2	147.2	0.3	160‡	8.0	8000	0	1425⁺	1698⁺	0.9	230	403	9.5	11.05	2200	2558.6
Niresist (cast)	84	824	824	17.50	171.5	171.7	10.5	103–111	103–111	0.25	125–173‡	7.3	7300	0	1200⁺	1473⁺	17.0	78	351	34	39.54	2650	3081.9
Hastelloy B	21*	206*	206*	85	834	834	21.4	210	210	(0.3)	215‡	9.23	9230		1335⁺	1608⁺	10.0	(280)	553	9.7	11.28	(2700)	1340.1
Hastelloy C	28.5	279*	279*	84	824	824	20	196.2	196.2	(0.3)	225‡	8.94	8940		1285⁺	1558⁺	11.3	(260)	533	10.8	12.56	(2800)	(3256.4)
Chrome (cast)	100	981	981	52	510	510	20.3	199	199	(0.3)	300‡	7.53	7530		1500⁺	1773⁺	10.6	173	450	19.0	22.10	3300	3837.9
Cobalt	85	833	833	24	235.4	235.4	21	206	206	0.28	125‡	8.9	8900		1495⁺	1768⁺	12.3	67	340	59.5	69.20	4000	(4652.0)
Cast iron	70	687	687	20	196.2	196.2	9.11	89–	89–	0.28	150–220‡	7.25	7250		1400⁺	1673⁺	10.0	150	423	40.0	46.5	6000	6978.0
Chrome	350	3434	3434	49	1481	1481	25	245.2	245.2	0.25	180‡	7.19	7190	0	1800⁺	2073⁺	6.2	220	493	57.6	66.99	12700	14770
Steel				130	1275.3	1275.3	20.6	202	202	0.3	64–67‡	7.8	7800	0	600	873	14.8	305	578	45.0	52.34	13800	16049
Molybdenum	63	618	618	70	687	687	33	323.7	323.7	0.28	20–26‡	10.2	10200	0	550	823	4.8	325	598	110	127.90	3560	1440.2
Steatite				7	68.7	68.7	10.5	103	103	0.324	7.5§	2.7	2700	0.02	1000	1273	8.2	(57)	(330)	2.15	25.0	(120)	140
Magnesium				10	98.1	98.1	21.4	209.4	209.4	0.36		3.5	3500		1000	1273	13.5	22	295	31.0	36.05	680	790.8
Thoria	150	1470	1470	8.4	82.4	82.4	14.7	144.2	144.2	0.17	8§	9.69	9690	0.02	3300⁺	3573⁺	9.2	52	325	9.0	4.65	470	546.6
Zircon	70	687	687	8.4	82.4	82.4	12.5	122.6	122.6	0.35	800§	3.7	3700	0.5	1000	1373	4.0	108	381	4.3	5.00	465	540.7
Quartz glass				11	107.9	107.9	7.3	72.1	72.1	0.15		2.6	2600		1723⁺	1996⁺	0.5	2550	2823	1.37	1.59	3500	4070.5
Alumina 1	168	1648	1648	12.6	122.6	122.6	22.3	218.8	218.8	0.27	9§	3.4	3400	0	1400	1673	5.5	74	347	11.4	13.37	840	976.9
Alumina 2	280	2746	2746	17.5	171.7	171.7	39	382.6	382.6	0.31	9§	3.7	3700	0	1550	1823	5.8	54	327	16.2	18.84	875	1017.6
Alumina 3	210	2070	2070	24	235.4	235.4	35	343.4	343.4	0.2	9§	3.9	3900	0	1750	2023	6.0	92	365	25.0	29.1	2300	2674.4
Cement 1	77	755	755	14.7	144.2	144.2	26	255.1	255.1	0.21	37‡	5.9	5900		1800	2073	8.0	56	329	25.0	29.1	1400	1628.2
Cement 2	168	1648	1648	21	206	206	26.6	261	261	0.26	50‡	6.0	6000		1700	1973	7.5	78	351	29.0	33.73	2250	2616.7
Boron carbide	29	284	284	17.5	171.7	171.7	45.5	446.4	446.4	0.26	2800∞	2.51	2510		2500⁺	2773⁺	4.5	(64)	(337)	22.3	25.94	1430	1663.1
Silicon carbide	105	1030	1030	12.5	122.6	122.6	48	470.8	470.8	(0.25)	2500∞	3.1	3100		2400	2673	3.9	(50)	(323)	86.0	100	(4300)	5000.9
Chromium carbide	290	2845	2845	27	264.9	264.9	32	313.9	313.9	0.26		7.0	7000		1900⁺	2173⁺	9.0	70	343	(20.0)	(23.3)	(1400)	(1628.2)
Tungsten carbide 1	350	3433	3433	140	1370	1370	49	480.7	480.7	0.26	83–64‡‡	13.0	13000	0.1	600	873	9.0	240	513	(30.0)	(34.89)	(7200)	8373.6
Tungsten carbide 2	420	4120	4120	120	1177.2	1177.2	56	549.4	549.4	0.248	86–87‡‡	14.1	14100	0.1	600	873	6.8	230	503	(50.0)	58.15	(11500)	13374.5
Tungsten carbide 3	500	4905	4905	85	833.8	833.8	70	687	687	0.216	91.5‡‡	14.8	14800	0.3	600	873	5.6	170	443	(60.0)	(62.78)	(10000)	11630
Tungsten carbide 4	370	3630	3630	115	1128.2	1128.2	54.5	534.6	534.6	0.242	89‡‡	14.0	14000	0.1	600	873	7.0	225	498	(40.0)	(46.5)	(9000)	10467.2
Titanium carbide 1				14	137	137	31.4	308	308	0.29	2460∞	4.9	4900		3140⁺	3413⁺	7.4	43	316	21.5	25.0	930	1081.5
Titanium carbide 2	350	3434	3434	91	892.7	892.7	41.3	405.2	405.2	0.25	89‡‡	6.0	6000		1000	1273	9.5	175	448	26.0	30.34	4550	5291.6
Titanium carbide 3	(300)	2943	2943	105	1030.1	1030.1	28.7	281.5	281.5	0.26	82.5‡‡	6.3	6300		1000	1273	10.4	260	543	28.0	32.56	7300	8489.9
Titanium carbide 4	366	3591	3591	56	549	549	40	392.4	392.4	(0.25)		5.8	5800		1200⁺	1473⁺	5.7	(185)	(462)	29.0	33.73	(5400)	6270.2
Titanium carbide 5	250	2453	2453	140	1370	1370	30.4	298.2	298.2	0.3	87.5‡‡	7.0	7000		650	923	8.7	(370)	(643)	45.0	52.34	(16700)	(19422.1)

Key: ‡ Brinell hardness; ⁺ melting point; * shore hardness; ** scleroscope; † Rockwell hardness; ‡‡ Rockwell A; ∞ Knoop hardness; § Mohs hardness; • electric limit; *** values in parentheses ( ) are approximate. Prefixes: $k = 10^3$, $M = 10^6$, $G = 10^9$. Conversion: 1 kgf/mm² = 9.80665×10^6 N/m²; 1 N/m² = 1 Pa; 1 kcal/h m °C = 1.163 W/m K; 1 kcal/h m = 1.163 W/m = 1.163 J/h m; 1 psi = 6894.757 Pa; 1 kpsi = 6.894757 MPa; 1 Mpsi = 6.894757 GPa; 1 Btu/ft² h °F = 5.678 W/m²°C; 1 W/m²°C = 0.1761 Btu/ft² h °F; 1 g/cm³ = 3.6127×10^{-2} lb/in³ = 62.428 lb/ft³. PTFCE = polytetrafluorochloroethylene; PTFE = polytetrafluoroethylene.

TABLE 24-87
Rubbing bearing materials and applications

Materials	Maximum loading, P			Pv Value			Coefficient of friction, μ	Maximum temperature, °C	Coefficient of expansion, $\alpha \times 10^{-6}$, °C	Application
	kgf/mm²	N/m² ×10⁶	MPa	kgf/mm² × m/s	MN/m² × m/s	MPa × m/s				
Carbon/graphite	0.14-0.20	1.4-2.0	1.4-2.0	0.0112-0.0184	0.11* 0.18**	0.11* 0.18**	0.10-0.25, dry	350-500	2.5-5.0	Conveyors, furnaces, food and textile machinery
Carbon/graphite with metal	0.31-0.41	3.0-4.0	3.0-4.0	0.0148-0.0224	0.145* 0.22**	0.145* 0.22**	0.10-0.35, dry	130-350	4.2-5.0	Bearings immersed in water, acid or alkaline solution, etc.
Graphite-impregnated metal	7.14	70.0	70.0	0.0286-0.0357	0.28-0.35	0.28-0.35	0.10-0.15, dry 0.020-0.025, grease-lubricated	350-600	12-13 with iron matrix	Bearings of foundary plant, coal mining machines, steel plants, etc.
Graphite/thermosetting resin	0.20	2.0	2.0	0.0357	0.35	0.35	0.13-0.5, dry	250	3.5-5.0	Water-lubricated roll neck bearings in hot rolling mills, rubber
Reinforced thermosetting plastic	3.57	35.0	35.0	0.0357	0.35	0.35	0.1-0.4, dry; 0.006, water-lubricated	200	25-80 depending on plane of reinforcement	bearings, bearings subjected to atomic radiation
Thermoplastic material without filler	1.02	10.0	10.0	0.0036	0.035	0.035	0.1-0.45, dry	100	100	Textile and food machinery bearings, bushes and thrust washers in automobile, bearing of linkages
Thermoplastic with filler or metal-backed	1.03	10.14	10.14	0.0036-0.0112	0.035-0.11	0.035-0.11	0.15-0.40, dry	100	80-100	For more heavily loaded applications, textile and food machinery, automobile and linkage
Thermoplastic with filler bonded to metal back	14.28	140.0	140.0	0.0357	0.35	0.35	0.20-0.35, dry	105	27	Ball joints, gearbox bushes, kingpin bushes, suspension and steering linkages
Filled PTFE	0.71	7.0	7.0	≤ 0.0357	≤ 0.35	≤ 0.35	0.05-0.35, dry	250	60-80	Bushes, thrust washers, sideways

TABLE 24-87
Rubbing bearing materials and applications (*Cont.*)

Materials	Maximum loading, P			Pv Value			Coefficient of friction, μ	Maximum temperature, °C	Coefficient of expansion, α × 10⁻⁶, °C	Application
	kgf/mm²	N/m² ×10⁶	MPa	kgf/mm² × m/s	MN/m² × m/s	MPa × m/s				
PTFE with filler bonded to steel backing	14.28	140.0	140.0	≤ 0.1785	≤ 1.75	≤ 1.75	0.05–0.30, dry	280	20 (lining)	Aircraft controls, linkages, automobile gearboxes, conveyors, bridges and building, expansion bearings, bushes, and steering suspension
Woven PTFE reinforced and bonded metal backing	42.84	420.0	420.0	≤ 0.1623	≤ 1.60	≤ 1.60	0.03–0.33, dry	250		Aircraft engine controls, automobile suspension, engine mountings, linkages, bridges, and building expansion joint

24.157

TABLE 24–88
Allowable working conditions and friction for various clutch facing materials

Working conditions	Coefficient of friction, μ	Temperature		Working pressure			Power rating, W/mm^2
		Maximum °C	Continuous °C	kgf/mm^2	N/m$^2 \times 10^6$	MPa	
Light-duty							
Woven	0.35–0.4	250	150	0.18–0.51	1.75–5.00	1.75–5.0	0.3–0.6
Mill board	0.40	250	150	0.18–0.71	1.75–7.00	1.75–5.0	0.3–0.6
Medium-duty			200				
Wound tape yarn	0.38	350		0.18–0.71	1.75–7.0	1.75–7.0	0.3–0.6
Asbestos tape	0.40	350	200	0.18–0.71	1.75–7.0	1.75–7.0	0.6–1.2
Molded	0.35	350	200	0.18–0.71	1.75–7.0	1.75–7.0	0.6–1.2
Heavy-duty							
Sintered	0.36/0.30	500	300	0.36–0.29	3.5–28	3.5–28.0	1.7
Cement	0.40			0.71–1.43	7.0–14	7.0–14.0	4.0
Oil-immersed							
Paper	0.11			0.71–1.79	7.1–17.5	7.1–17.5	2.3
Woven	0.08			0.71–1.79	7.1–17.5	7.1–17.5	1.8
Molded	0.04			0.17–1.79	7.1–17.5	7.1–17.5	0.6
Molded (grooved)	0.06						
Sintered	0.11/0.05			0.71–4.28	7.0–42	7.0–42.0	2.3
Sintered (grooved)	0.11/0.06			0.71–4.28	7.0–42	7.0–42.0	2.3
Resin/graphite	0.10						5.3

REFERENCES

1. Lingaiah, K., *Machine Design Data Handbook*, Vol. II (*SI and Customery Metric Units*), Suma Publishers, Bangalore, India, 1986.
2. Lingaiah, K., and B. R. Narayana Iyengar, *Machine Design Data Handbook*, Engineering College Co-operative Society, Bangalore, India, 1962.
3. Maleev, V. L., and J. B. Hartman, *Machine Design*, International Textbook Company, Scranton, Pennsylvania, 1954.
4. Radzimovsky, F. I., *Lubrication of Bearings—Theoretical Principles and Designs*, The Ronald Press Company, New York, 1959.
5. Raimondi, A. A., and J. Boyd, 'A Solution for the Finite Journal Bearings and Its Application to Analysis and Design,' *ASME J. Lubrication Technol.*, Vol. 104, pp. 135–148, April 1982.
6. Kingsbury, A., 'Optimum Conditions in Journal Bearing,' *Trans. ASME*, Vol. 54, 1932.
7. Needs, S. J., 'Effect of Side Leakage in 120-degree Centrally Supported Journal Bearings,' *Trans. ASME*, Vol. 56, 1934; Vol. 51, 1935.
8. Shigley, J. E., *Mechanical Engineering Design*, First Metric Edition, McGraw-Hill Book Company, New York, 1986.
9. Edwards, K. S., Jr., and R. B. McKee, *Fundamentals of Mechanical Component Design*, McGraw-Hill Book Company, 1991.
10. Shaw, M. C., and F. Macks, *Analysis and Lubrication of Bearings*, McGraw-Hill Book Company, New York, 1949.

BIBLIOGRAPHY

ASME Standards.

Baumeister, T., ed., *Marks' Handbook for Mechanical Engineers*, McGraw-Hill Book Company, New York, 1978.

Black, P. H., and O. E. Adams, Jr., *Machine Design*, McGraw-Hill Book Company, New York, 1968.

Boswall, R. O., *The Theory of Film Lubrication*, Longmans, Green and Company, New York, 1928.

Bureau of Indian Standards.

O'Connor, J. J. ed., *Standard Handbook of Lubricating Engineering*, McGraw-Hill Book Company, New York, 1968.

Fuller, D. P., *The Theory and Practice of Lubrication for Engineers*, John Wiley and Sons, New York, 1956.

Niemann, G., *Machine Elements—Design and Calculations in Mechanical Engineering*, Vol. II, Springer-Verlag, Berlin, 1950; Student Edition, Allied Publishers Private Ltd. Bangalore, India, 1979.

Niemann, G., *Maschinenelemente*, Springer-Verlag, Berlin, Erster Band, 1963.

Niemann, G., *Maschinenelemente*, Springer-Verlag, Berlin, Zweiter Band, 1965.

Hyland, P. H., J. B. Kommers, *Machine Design*, McGraw-Hill Book Company, New York, 1943.

ISO Standards

Lansdown, A. R., *Lubrication: A Practical Guide to Lubricant Selection*, Pergamon Press, New York, 1982.

Leutwiler, O. A., *Elements of Machine Design*, McGraw-Hill Book Company, New York, 1917.

Michell, A. G. M., *Lubrication—Its Principles and Practice*, Blackie and Son, London, 1950.

Neale, M. J., ed., *Tribology Handbook*, Butterworth, London, 1973.

Norman, C. A., E. S. Ault, and I. F. Zarobsky, *Fundamentals of Machine Design*, The Macmillan Company, New York, 1951.

Norton, A. E., *Lubrication*, McGraw-Hill Book Company, New York, 1942.

Slaymaker, R. R., *Bearing Lubrication Analysis*, John Wiley and Sons, New York, 1955.

Rippel, H. C., "Design of Hydrostatic Bearings," *Machine Design*, Parts 1 to 16, Aug. 1 to Dec. 5, 1963.

SAE Handbook, 1957.

Shigley, J. E., *Machine Design*, McGraw-Hill Book Company, New York, 1962.

Shigley, J. E., and C. R. Mischke, *Standard Handbook of Machine Design*, McGraw-Hill Book Company, New York, 1986.

Shigley, J. E., and C. R. Mischke, *Mechanical Engineering Design*, McGraw-Hill Book Company, New York, 1989.

Vallance, A., and V. L. Doughtie, *Design of Machine Members*, McGraw-Hill Book Company, New York, 1951.

Wilcock, D. F., and E. R. Booser, *Bearing Design and Application*, McGraw-Hill Book Company, New York, 1957.

MISCELLANEOUS MACHINE ELEMENTS

25.1 CRANKSHAFTS

SYMBOLS

A	area of cross section, m^2 (in^2)
b	width of crank cheek, m (in)
c	distance from the neutral axis of section to outer fiber, m (in)
d	diameter (also suffixes), m (in)
d_e	equivalent diameter, m (in)
d_o	diameter of crankpin, m (in)
d_m	diameter of main bearing, m (in)
E	modulus of elasticity, GPa (psi)
F	force acting on the piston due to steam or gas pressure corrected for inertia effects of the piston and other reciprocating parts, kN (lbf)
F_c	the component of force F acting along the axis of connecting rod, kN (lbf)
F_{comb}	combined force, kN (lbf)
F_{ic}	magnitude of inertia force due to the weight of connecting rod itself, kN (lbf)
F_r	total radial force acting on the crankpin, kN (lbf)
F_θ	total tangential force acting on the crankpin, kN (lbf)
G	modulus of rigidity, GPa (psi)
h	thickness of cheek or web (also with suffixes), m (in)
$i' = \dfrac{l_o}{d_o}$	ratio of length to diameter of crank
I	moment of inertia, m^4, cm^4 (in^4)
$K = \dfrac{D_i}{D_o}$	ratio of inner to outer diameter of a hollow shaft
K_b	numerical combined shock and fatigue factor to be applied to the computed bending moment
K_t	numerical combined shock and fatigue factor to be applied to the computed twisting moment
l	length (also with suffixes), m (in)

l_e equivalent length, m (in)
M_b bending moment, N m (lbf in)
M_t twisting moment, N m (lbf in)
p allowable pressure, MPa (psi)
r radius, throw of crankshaft, m (in)
Z section modulus, m^3, cm^3 (in^3)
σ normal stress (also with suffixes), MPa (psi)
τ shear stress, MPa (psi)

SUFFIXES

b bending
c compressive
$comb$ combined
e elastic
m main
max maximum
r radial
ra resultant in arm
rh resultant in hub
t torque
s shaking
θ tangential

Other factors in performance or special aspects which are included from time to time in this chapter and are applicable only in their immediate context are not given at this stage.

Particular	Formula

FORCE ANALYSIS (FIG. 25–1)

The radial component of force F_c acting along the axis of connecting rod (Fig. 25–1)	$$F_{c1} = F_c \cos(\theta + \phi) = \frac{F}{\sqrt{1 - \left(\dfrac{\sin\theta}{n'}\right)^2}} \cos(\theta + \phi)$$ (25–1)
The tangential component of force F_c acting along the axis of connecting rod (Fig. 25–1)	$$F_{c2} = F_c \sin(\theta + \phi) = \frac{F}{\sqrt{1 - \left(\dfrac{\sin\theta}{n'}\right)^2}} \sin(\theta + \phi)$$ (25–2)
The radial component of force F_{ic} (Fig. 25–1)	$F_{ic1} = \frac{2}{3} F_{ic} \cos\gamma$ (25–3)
	where γ = angle between the force F_{ic} and the radial component of F_{ic}
The tangential component of force F_{ic} (Fig. 25–1)	$F_{ic2} = \frac{2}{3} F_{ic} \sin\gamma$ (25–4)
The total radial force acting on the crank	$F_r = F_{ic1} \pm F_{c1}$ (25–5) $= \frac{2}{3} F_{ic} \cos\gamma \pm F_c \cos(\theta + \phi)$ (25–6)

Particular	Formula
The total tangential force acting on the crank	$F_\theta = F_{ic2} \pm F_{c2}$ $\quad = \frac{2}{3} F_{ic} \sin\gamma \pm F_c \sin(\theta + \phi)$ (25–7)
The resultant force on the crankpin	$F_{comb} = \sqrt{F_r^2 + F_\theta^2}$ 25–8

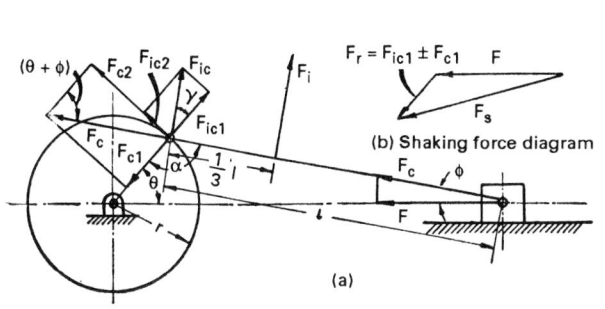

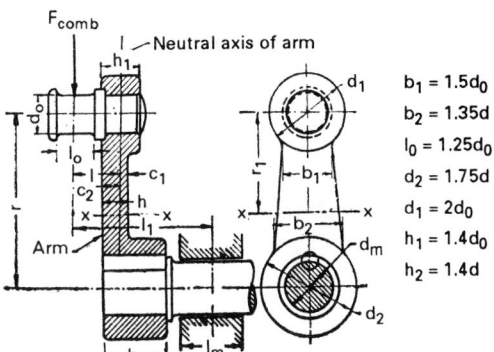

FIGURE 25–1 (*a*) Forces acting on crankshaft. (*b*) Vector sum of F and F_r

FIGURE 25–2 Overhung built-up crank.

SIDE CRANK
Crankpin

The maximum bending moment on the crankpin (Fig. 25–2)	$M_{b(max)} = F_{comb} \times \left(\dfrac{l_o}{2} + \dfrac{t}{2}\right)$ (25–9) $\qquad\qquad = F_{comb} \times l$ where $l = \dfrac{l_o}{2} + c_2 = $ distance from centroidal axis to the application of load (Fig. 25–2), m (in)
The crankpin diameter with respect to the bending moment	$d_o = \sqrt[3]{\dfrac{32 l F_{comb}}{\pi \sigma_b}}$ (25–10) where $\sigma_b = $ allowable bending stress, MPa (psi)
The diameter of crankpin from the consideration of bearing pressure	$d_o = \dfrac{F_{comb}}{l_o p}$ (25–11)
From Eqs. (25–10) and (25–11) neglecting $\frac{t}{2}$ and eliminating l_o, the equation for crankpin diameter	$d_o = \sqrt[4]{\dfrac{16 F_{comb}^2}{\pi p \sigma_b}}$ (25–12)
Empirical relation to determine the length of crankpin	$l_o = i' d_o$ (25–13)

Particular	Formula
	where $i' = \dfrac{l_o}{d_o} = 1.25$ to 1.5
Another relation for the crankpin length/diameter ratio	$i' = \dfrac{l_o}{d_o} = \sqrt{\dfrac{0.2\sigma_b}{p}}$ (25–14)
Another relation for the crankpin diameter	$d_o = \sqrt{\dfrac{F_{comb}}{i'p}}$ (25–15)

HOLLOW CRANKPIN

The crankpin length/diameter ratio	$i' = \dfrac{l_o}{D_o} \sqrt{\dfrac{0.2\sigma(1 - K^4)}{p}}$ (25–16) where $K = \dfrac{D_i}{D_o}$
The crankpin outside diameter	$D_o = \sqrt{\dfrac{F_{comb}}{i'p}}$ (25–17)

Crank arm

CRANK ON HEAD-END DEAD-CENTER POSITION

When the crank is on the head-end dead-center position, the section XX (Fig. 25–2) of the arm is subjected to bending moment	$M_b = F_{comb} \times l$ (25–18)
The direct compressive stress due to the load F_{comb} (i.e., more specifically by its component F_c)	$\sigma_c = \dfrac{F_{comb}}{A}$ (25–19)
The resultant stress in the crank arm at XX	$\sigma_{ra} = \dfrac{F_{comb}}{A} \pm \dfrac{M_b C}{I}$ (25–20) where A = area of cross section of the arm at XX, m^2 (in^2) c = distance from the neutral axis of section to outer fiber of arm, m (in) I = moment of inertia of the section, cm^4 (in^4)

CRANK ON CRANK-END DEAD-CENTER POSITION

The direct tensile stress in the plane of the hub of crankshaft section passing through the shaft center due to load F_{comb} (Fig. 25–2)	$\sigma_t = \dfrac{F_{comb}}{h_2(d_2 - d)}$ (25–21)

Particular	Formula
The bending stress in the section due to bending moment $F_{comb} \times a$	$\sigma_b = \dfrac{F_{comb} \times a}{Z}$ $\qquad$ (25–22) where Z = section modulus, cm^3 (in^3)
The resultant stress in the plane of the hub of crankshaft section passing through the shaft center	$\sigma_r = \sigma_t \pm \sigma_b$ $\qquad$ (25–23)

CRANK PERPENDICULAR TO THE CONNECTING ROD

The bending moment in the plane of rotation of the crank	$M_b = F_{comb} \times l$ $\qquad$ (25–24)
The bending stress	$\sigma_b = \dfrac{M_b c_1}{Z_b}$ $\qquad$ (25–25)
The torsional moment	$M_t = F_{comb} \times r_1$ $\qquad$ (25–26)
The shear stress	$\tau = \dfrac{M_t c_1}{Z_t}$ $\qquad$ (25–27)
The maximum normal stress for crank made of cast iron	$\sigma_{max} = \frac{1}{2}[\sigma_b + \sqrt{\sigma_b^2 + 4\tau^2}]$ $\qquad$ (25–28)
The maximum shear stress for the crank made of steel	$\tau_{max} = \frac{1}{2}\sqrt{\sigma_b^2 + 4\tau^2}$ $\qquad$ (25–29)

DIMENSION OF CRANKSHAFT MAIN BEARING (FIG. 25–2b)

The shaking force on the main bearing from F and F_r (Fig. 25–1b)	F_s = vector sum of F and F_r $\qquad$ (25–30)
The diameter of main bearing taking into consideration the bearing pressure on the projected area of the crankshaft	$d_m = \dfrac{F_s}{l_m p}$ $\qquad$ (25–31) where l_m = length of bearing, m (in) p = allowable bearing pressure, MPa (psi)
The bending movement on the crankshaft	$M_b = F_{comb} \times l_1$ $\qquad$ (25–32) $l_1 = \dfrac{l_o}{2} + h_2 + \dfrac{l_m}{2}$ where h_2 = hub length, m (in) l_o = length of crankpin, m (in) l_m = length of bearing on crankshaft, m (in)
The torque on the crankshaft	$M_t = F_{comb} \times r$ $\qquad$ (25–33) where r = throw of the crank, m (in)

Particular	Formula
The diameter of crankshaft taking into consideration indirectly the fatigue and shock factors	$$d_m = \sqrt[3]{\frac{16}{\pi\sigma_e}\left\{K_bM_b + \sqrt{(K_bM_b)^2 + (K_tM_t)^2}\right\}} \qquad (25\text{--}34)$$
The length of main bearing	$$l_m = \frac{F_s}{d_m p} \qquad (25\text{--}35)$$

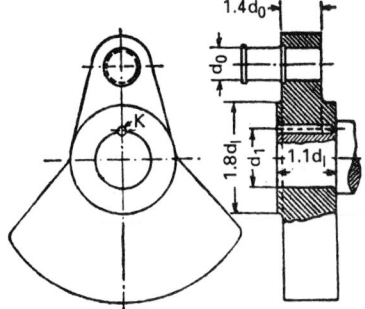

FIGURE 25–3 Overhung built-up crank.

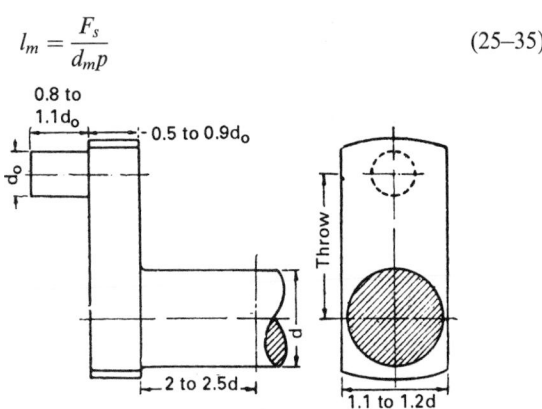

FIGURE 25–4 Overhung forged crank.

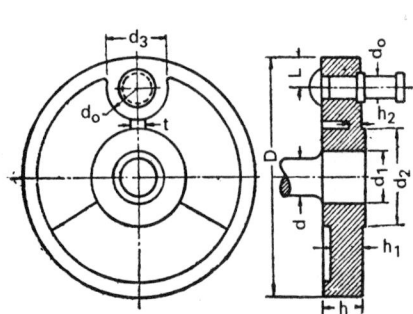

FIGURE 25–5 Disk crank.

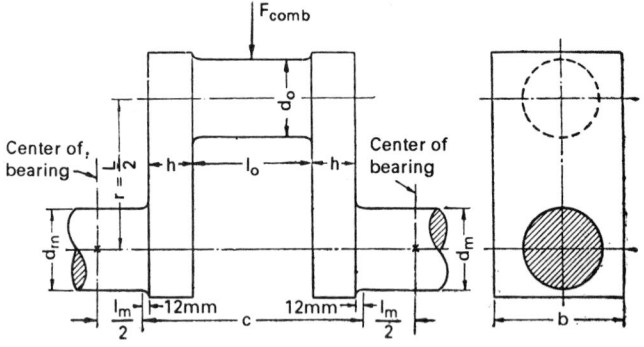

FIGURE 25–6 Center crank (American Bureau of Shipping method).

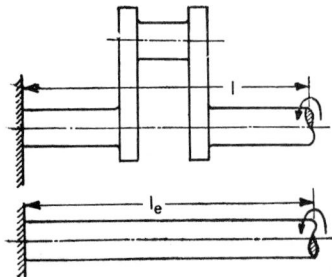

FIGURE 25–7 Equivalent length of crankshaft.

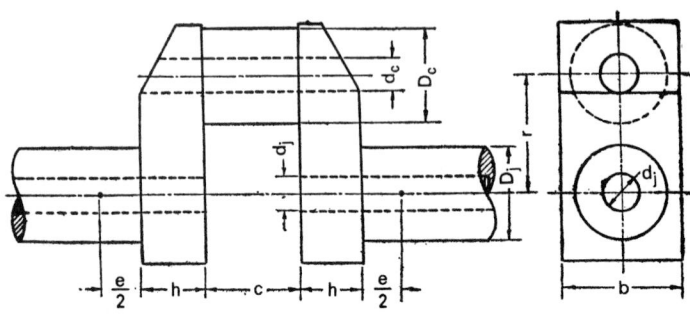

FIGURE 25–8 Center hollow crank.

Particular	Formula

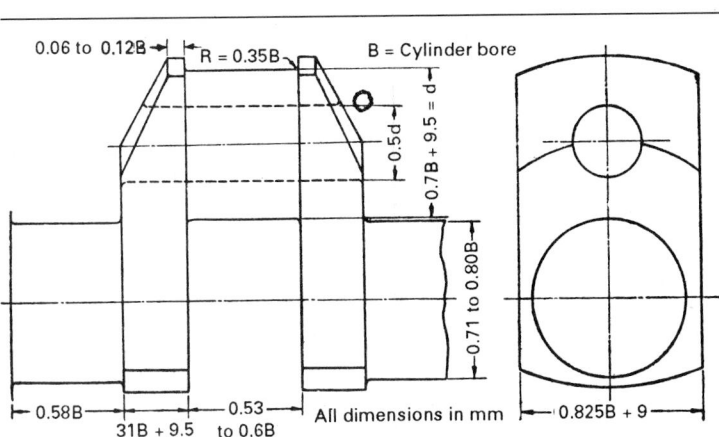

FIGURE 25–9 Empirical proportion for center crank.

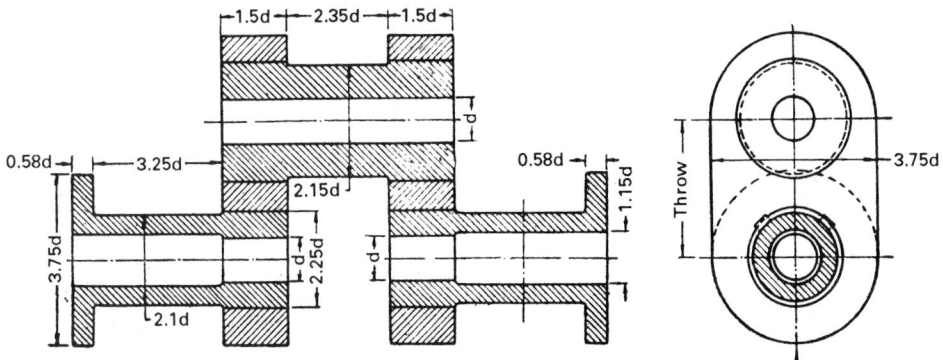

FIGURE 25–10 Center hollow built-up crank.

PROPORTIONS OF CRANKSHAFTS

For proportions of crankshaft   Refer to Figs. 25–2 to 25–10.

CENTER CRANK (FIG. 25–6)

Crankpin

The maximum bending moment treating the crankpin as a simple beam with concentrated load at the center

$$M_{bc} = \frac{F_{comb}(l_o + h + l_m)}{4} \qquad (25\text{--}36)$$

where
l_o = length of crankpin, m (in)
l_m = length of main bearing, m (in)
h = thickness of cheek, m (in)

Particular	Formula
The diameter of the crankpin based on maximum bending moment M_{bc}	$$d_o = \sqrt[3]{\frac{32 M_{bc}}{\pi \sigma_b}} \qquad (25\text{–}37)$$
	where σ_b = design stress, MPa (psi)
The diameter of crankpin based on bearing pressure between pin and the bearing	$$d_o = \frac{F_{comb}}{l_o p} \qquad (25\text{–}38)$$

Dimensions of main bearing

Particular	Formula
The maximum bending moment treating the center crank as a simple beam with load concentrated at the center	$$M_{bb} = \frac{F_{comb} \times l_e}{4} \qquad (25\text{–}39)$$
	where l_e = equivalent length of crankshaft, m (in)
The twisting moment	$$M_t = F_{comb} \times r \qquad (25\text{–}40)$$
The diameter of crankshaft at main bearing taking into consideration the fatigue and shock factors	$$d_m = \sqrt[3]{\frac{16}{\pi \sigma_e} \left\{ K_b M_{bb} + \sqrt{(K_b M_{bb})^2 + (K_t M_t)^2} \right\}}$$ $$(25\text{–}41)$$
The diameter of the crankshaft based on bearing pressure	$$d_m = \frac{F_s}{l_m p} \qquad (25\text{–}42)$$

American Bureau of shipping formulas for center crank

Particular	Formula
The thickness h of the cheeks or webs (Fig. 25–6)	$$h = 0.4d \text{ to } 0.6d \qquad (25\text{–}43)$$
The diameter of crankpins and journals (Fig. 25–6)	$$d = a \sqrt[3]{\frac{Dpc}{\sigma_b}} \qquad (25\text{–}44)$$
	where
	a = coefficient from Table 25–1A
	D = diameter of cylinder bore, m (in)
	p = maximum gas pressure, MPa (psi)
	c = distance over the crank webs plus 25 mm (1.0 in) (Fig. 25–6)
	σ_b = allowable fiber stress, MPa (psi)
The thickness h and the width b of crank cheeks must satisfy the conditions	$$bh^2 \geq 0.4d^3 \qquad (25\text{–}45a)$$ $$b^2 h \geq d^3 \qquad (25\text{–}45b)$$

Particular	Formula

EQUIVALENT SHAFTS

A portion of a shaft length l and diameter d can be replaced by a portion of length l_e and diameter d_e

$$l_e = l\left(\frac{d_e}{d}\right)^4 \qquad (25\text{--}46)$$

The length h_e equivalent to crank web

$$h_e = \frac{rC}{B} \qquad (25\text{--}47)$$

where
$C = \frac{1}{32}\pi d_e^4 G$ = torsional rigidity of the crankpin
$B = \frac{1}{12}hb^3 E$ = flexural rigidity of the web

The equivalent length crankshaft l_e of Fig. 25–7 varies between

$$0.95l < l_e < 1.10l \qquad (25\text{--}48)$$

The equivalent length of commercial crankshaft for solid journal and crankpin according to Carter (Fig. 25–8)

$$L_e = d_e^4\left[\frac{e + 0.8a}{D_J^4} + \frac{0.75b}{D_c^4} + \frac{1.5r}{ac^3}\right] \qquad (25\text{--}49)$$

The equivalent length of commercial crankshaft for hollow journal and crankpin according to Carter (Fig. 25–8)

$$L_e = d_e^4\left[\frac{e + 0.8a}{D_J^4 - d_J^4} + \frac{0.75b}{D_c^4 - d_c^4} + \frac{1.5r}{ac^3}\right] \qquad (25\text{--}50)$$

The equivalent length of crankshaft for solid journal and crankpin according to Wilson (Fig. 25–8)

$$L_e = d_e^4\left[\frac{e + 0.4D_J}{D_J^4} + \frac{b + 0.4D_c}{D_c^4} + \frac{r - 0.2(D_J + D_c)}{ac^3}\right] \qquad (25\text{--}51)$$

The equivalent length of crankshaft for hollow journal and crankpin according to Wilson (Fig. 25–8)

$$L_e = d_e^4\left[\frac{e + 0.4D_J}{D_J^4 - d_J^4} + \frac{b + 0.4D_c}{D_c^4 - d_c^4} + \frac{r - 0.2(D_J + D_c)}{ac^3}\right] \qquad (25\text{--}52)$$

EMPIRICAL PROPORTIONS

For empirical proportions of side crank, built-up crank, and hollow crankshafts

Refer to Figs. 25–2 to 25–10.

The film thickness in bearing should not be less than the values given here for satisfactory operating condition:

Main bearings

$$h = 0.0025 \text{ mm } (0.0001 \text{ in}) \qquad (25\text{--}52a)$$
$$\text{to } 0.0042 \text{ mm } (0.0017 \text{ in})$$

Big-end bearings

$$h = 0.002 \text{ mm } (0.00008 \text{ in}) \qquad (25\text{--}52b)$$
$$\text{to } 0.004 \text{ mm } (0.00015 \text{ in})$$

The oil flow rate through a conventional central circumferential grooved bearings

$$Q = \frac{kpc^3}{\eta}\frac{d}{L}(1 + 1.5\epsilon^2) \qquad (25\text{--}52c)$$

Particular	Formula

	where
	Q = oil flow rate, m^3/s (gal/min)
	k = a constant = 0.0327 **SI**
	   = 4.86×10^4 **US Customary units**
	p = oil feed pressure, Pa (lbf/in^2)
	c = $D - d$ = diametral clearance, m (in)
	η = absolute viscosity (dynamic viscosity),
	    Pa s (cP)
	d = bearing bore, m (in)
	L = land width, m (in)
	ϵ = attitude or eccentricity ratio

For oil flow rate in medium and large diesel engines at 0.35 MPa (0.50 lbf/in^2)	Refer to Table 25–1B.
The velocity of oil in ducts on the delivery side of the pump	$v = 1.8$ to 3.0 m/s (6 to 10 ft/s) (25–52d)
The velocity of oil in ducts on the suction side of the pump	$v = 1.2$ m/s (4 ft/s) (25–52e)
The delivery pressure in modern high-duty engines	$p = 0.28$ to 0.42 MPa (40–60 lbf/in^2) (25–52f) $p_{max} = 0.56$ MPa (80 lbf/in^2) (25–52g)
For housing tolerances	Refer to Table 25–1C.

TABLE 25–1A
Coefficient a in the American Bureau of Shipping formula [Eq. (25–44)]

Type	Number of cylinder		Ratio of stroke to distance over crank webs = l/c							
	Four-stroke	Two-stroke	0.7	0.8	0.9	1.0	1.1	1.2	1.3	1.4
Explosion engines	1, 2, 4	1, 2	1.17	1.17	1.17	1.17	1.17	1.17	1.17	1.17
	3, 5, 6	3	1.17	1.17	1.17	1.17	1.19	1.20	1.22	1.24
	8	8	1.17	1.19	1.21	1.23	1.25	1.28	1.30	1.32
	10, 11, 12	5, 6	1.18	1.20	1.23	1.25	1.28	1.31	1.33	1.35
Air-injection diesel engines	1, 2, 4	1, 2	1.17	1.19	1.22	1.25	1.28	1.31	1.34	1.36
	3, 5, 6		1.19	1.22	1.25	1.28	1.32	1.35	1.38	1.41
	8	3	1.20	1.24	1.27	1.30	1.33	1.37	1.40	1.43
	12	4	1.22	1.25	1.29	1.32	1.36	1.39	1.42	1.45
	16	5, 6	1.25	1.29	1.33	1.36	1.40	1.44	1.47	1.50
		8								

TABLE 25–1B
Oil flow rate in medium and large diesel engines at 0.35 MPa (50 lbf/in^2)

Different parts of engine	Oil flow rate	
	liters/ min/ kW	liters/min/ hp (gal/h/ hp)
Bed plate gallery to mains with piston cooling	0.536	0.4 (5)
Mains to big end (with piston cooling)	0.362	0.27 (3.5)
Big ends to pistons (with oil cooling)	0.201	0.15 (2)
Total flow of oil with uncooled pistons	0.335	0.25 (3)

TABLE 25–1C
Housing tolerances

Parts	Tolerances
Waviness of the surface	$\not> 0.0001d$
Run-out of thrust faces	$\not> 0.0003d$
Surface finish	
Journals	0.2–0.25 μm R_a (8–10 μin clearance)
Gudgeon pins	0.1–0.16 μm R_a (4–6 μin clearance)
Housing bores	0.75–1.6 μm R_a (30–60 μin clearance)
Alignment of adjacent housing	< 1 in 10,000 to 1 in 12,000
The fine grinding or honing	0.025–0.05 mm (0.001–0.002 in)

TABLE 25–2
Values of radius to neutral axis for curved beams

Type	Section	Radius of neutral surface, r_n	
a		$$r_n = \frac{(\sqrt{r_o} + \sqrt{r_i})^2}{4}$$	(25–77)
b		$$r_n = \frac{h}{\ln\left(\dfrac{r_o}{r_i}\right)}$$	(25–78)
c		$$r_n = \frac{\frac{1}{2}h(b_i + b_o)}{\dfrac{b_i r_o - b_o r_i}{h}\ln\left(\dfrac{r_o}{r_i}\right) - (b_i - b_o)}$$ If $b_o = 0$, this section reduces to a triangle	(25–79)
d		$$r_n = \frac{A}{b_i \ln\dfrac{r_i + a_i}{r_i} + b_2 \ln\dfrac{r_o - a_o}{r_i + a_i} + b_o \ln\dfrac{r_o}{r_o - a_o}}$$ If $a_o = 0$, the section reduces to a $\perp$ section; r_n is the same for a box section in dotted lines with each side panel $\frac{1}{2}b_2$ thick	(25–80)

TABLE 25-3
Hooks of standard trapezoidal section (Refer to Fig. 25-16)

Safe working load, W, tf [a]	Proof load, P, tf [a]		A (2.75C)	B (3.1C)	C [b]	D (1.44C)	E (1.25C)	F (1.00C)	G	Nominal size G_1	Pitch	H (0.93C)	J (0.75C)	K (0.92C)		L (0.70C)	M (0.60C)	N (1.20C)	P (0.50C)	R (0.50C)	U (0.30C)	Z (0.12C)	Bore	Series	Outside diameter	Width
												Course series with graded pitches											Ball bearings (per IS 2512, 1963)			
0.5	1	MS	74	35	27	39	34	27	15	M	14	25	20	25	MS	19	16	32	14	14	8	3	15	11	28	9
		HS	63	38	23	33	29	23	12	M	12	21	17	21	HS	16	14	28	12	12	7	3	12	11	26	9
1.0	2	MS	105	50	38	55	48	38	20	M	20	35	28	35	MS	27	23	46	19	19	11	5	20	11	35	10
		HS	91	43	33	48	41	33	20	M	18	31	25	31	HS	23	20	40	16	16	10	4	20	11	35	10
2.0	4	MS	145	69	53	76	66	53	30	M	27	49	40	49	MS	37	32	64	26	26	16	6	30	12	52	16
		HS	126	60	46	66	58	46	25	M	24	43	34	42	HS	32	28	55	23	23	14	6	25	12	47	15
3.2	6.4	MS	187	89	68	98	85	68	35	M	33	63	51	63	MS	48	41	82	34	34	20	8	35	13	68	24
		HS	162	77	59	85	74	59	30	M	30	55	44	54	HS	41	35	71	30	30	18	7	35	13	60	21
5.0	10	MS	233	111	85	122	106	85	45	M	42	79	64	78	MS	60	51	102	42	42	26	10	45	13	85	28
		HS	201	96	73	105	91	73	40	M	30	68	55	67	HS	51	44	88	36	36	22	9	40	13	78	26
8.0	16	MS	294	140	107	154	134	109	55	M	52	100	80	98	MS	75	64	128	54	54	32	13	55	13	105	35
		HS	256	122	93	134	116	93	50	M	48	86	70	86	HS	63	56	112	46	46	29	11	50	13	95	31
10.0	20	MS	327	156	119	171	149	119	60	M	60	111	89	109	MS	83	71	143	60	60	36	14	60	13	110	35
		HS	286	136	104	150	130	104	55	M	52	97	78	96	HS	73	62	125	52	52	31	12	56	13	105	35
12.0	25	MS	369	176	134	193	168	134	70	M	68	125	100	123	MS	94	80	161	67	67	40	16	70	13	125	40
		HS	319	152	116	167	145	116	60	M	60	108	87	107	HS	81	70	139	58	58	35	14	60	13	110	35
16.0	32	MS	415	198	151	217	189	151	80	M	76	140	113	139	MS	106	91	181	76	76	45	18	80	13	140	44
		HS	360	171	131	189	164	131	70	M	68	122	98	120	HS	92	79	157	66	66	39	16	70	13	125	40
20.0	40	MS	465	221	169	243	211	169	85	M	80	157	127	155	MS	118	101	203	84	84	51	20	85	13	150	49
		HS	404	193	147	211	184	147	80	M	72	137	110	135	HS	103	88	176	74	74	44	18	80	13	140	44
25	50	MS	520	248	189	272	236	189	100	M	90	176	142	174	MS	132	113	227	94	94	57	23	100	13	190	63
		HS	451	215	164	236	205	164	85	M	80	153	123	151	HS	115	98	197	82	82	49	20	85	13	150	49
32	60	MS	569	271	207	298	259	207	110	M	100	193	155	191	MS	145	124	248	104	104	62	25	110	13	210	70
		HS	481	229	175	252	219	175	100	M	90	163	138	161	HS	122	105	210	88	88	52	21	100	13	190	63
40	70	MS	616	293	224	323	280	224	120	M	110	208	168	206	MS	157	134	268	112	112	67	27	120	13	210	70
		HS	533	254	194	279	242	194	110	M	100	180	146	178	HS	136	116	233	97	97	58	23	110	13	190	63
50	85	MS	680	324	247	356	309	247	130	M	120	229	187	227	MS	173	148	296	124	124	74	30	130	13	225	75
		HS	588	280	214	308	268	214	120	M	110	199	160	197	HS	150	128	257	107	107	64	26	120	13	210	70

[a] Tonne-force
[b] Formula for catching C: for MS (mild steel): $C = 26.73\sqrt{P}$; for HS (high-tensile steel), $C = 23.17\sqrt{P}$.
[c] Machined shank diameter (min).
Source: IS, 3815, 1969.

25.2   CURVED BEAM

SYMBOLS

a	semimajor axis of ellipse, m (in)
A	area of cross section, m^2 (in^2)
b	width of beam, m (in)
	semiminor axis of ellipse, m (in)
c_1	distance from the centroidal axis to the inner surface of curved beam, m (in)
c_2	distance from the centroidal axis to the outer surface of curved beam, m (in)
$c_i = c_1 - e$	distance from the neutral axis to inner surface of curved beam, m (in)
$c_o = c_2 + e$	distance from the neutral axis to outer surface of curved beam, m (in)
$H(= d)$	diameter of curved beam of circular cross section, m (in)
e	distance from centroidal axis to neutral axis of the section, m (in)
E	modulus of elasticity, GPa (psi)
F	load, kN (lbf)
G	modulus of rigidity, GPa (psi)
h	depth of beam, m (in)
I	moment of inertia, m^4, cm^4 (in^4)
k	stress factor (also with suffixes)
K	constant
l	length of straight section between the semicircular ends of chain link, m (in)
m	pure number to be determined for each particular shape of the cross section by performing the integration
M_b	applied bending moment (also with suffixes), N m (lbf in)
r_c	radius of centroidal axis, m (in)
r_i	inner radius of curved beam, radius of curvature, m (in)
r_o	outer radius of curved beam, m (in)
r_n	radius of neutral axis, m (in)
y	deflection, m (in)
σ	normal stress (also with suffixes), MPa (psi)

SUFFIXES

b	bending
i	inner
h	horizontal
o	outer
n	neutral
max	maximum
r	resultant or combined
v	vertical
x	x direction
y	y direction

Particular	Formula

GENERAL
Pure bending

The general equation for the bending stress in a fiber at a distance y from the neutral axis (Figs. 15–11 and 15–12)

$$\sigma_b = \pm \frac{M_b}{Ae}\left(\frac{y}{r_n + y}\right) \qquad (25\text{–}53)$$

The maximum compressive stress due to bending at the outer fiber (Fig. 25–12)

$$\sigma_{bo} = -\frac{M_b c_o}{Aer_o} \qquad (25\text{–}54)$$

The maximum tensile stress due to bending at the inner fiber (Fig. 25–12)

$$\sigma_{bi} = \frac{M_b c_i}{Aer_i} \qquad (25\text{–}55)$$

Stress due to direct load

The direct stress due to load F

$$\sigma = \frac{F}{A} \qquad (25\text{–}56)$$

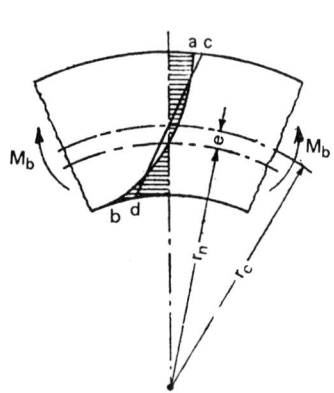

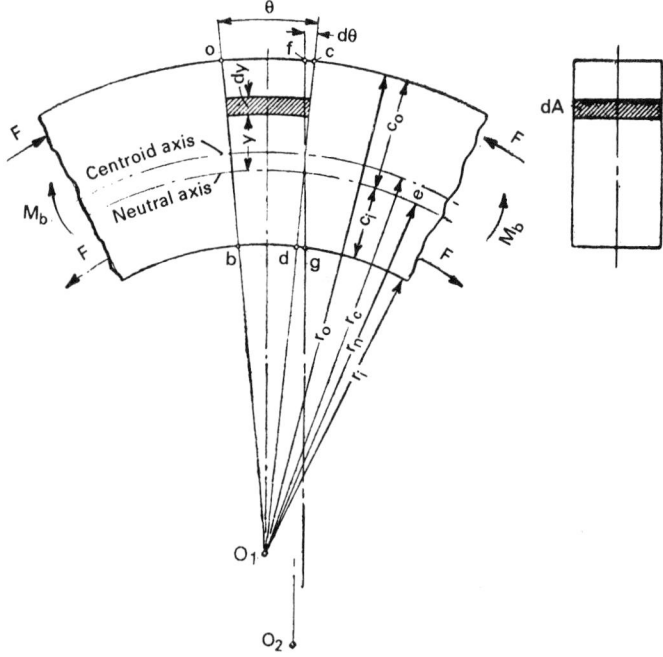

FIGURE 25–11 Bending stress in curved beam. (Courtesy of *V. L. Maleev and J. B. Hartman*, Machine Design, *International Textbook Company, Scranton, Pennsylvania, 1954.*)

FIGURE 25–12 Analysis of stresses in curved beam. (Courtesy of *V. L. Maleev and J. B. Hartman*, Machine Design, *International Textbook Company, Scranton, Pennsylvania, 1954.*)

Particular	Formula

Combined stress due to load F and bending

The general expression for combined stress

$$\sigma_r = \frac{F}{A} \pm \frac{M_b}{Ae}\left(\frac{y}{r_n + y}\right) \qquad (25\text{--}57)$$

The combined stress in the outer fiber

$$\sigma_{ro} = \frac{F}{A} - \frac{M_b c_o}{Aer_o} \qquad (25\text{--}58)$$

The combined stress in the inner fiber

$$\sigma_{ri} = \frac{F}{A} + \frac{M_b c_i}{Aer_i} \qquad (25\text{--}59)$$

For values of radius to neutral axis for curved beams

Refer to Table 25–2.

APPROXIMATE EMPIRICAL EQUATION FOR CURVED BEAMS

An approximate empirical equation for the maximum stress in the inner fiber

$$\sigma_i = M_b\left[\frac{c_1}{I} + \frac{K}{bc_1}\left(\frac{1}{r_i} + \frac{1}{r_o}\right)\right] \qquad (25\text{--}60)$$

where
M_b = bending movement at the centroid, N m (lbf in)
b = maximum width of the section, m (in)
K = constant
   = 1.05 for circular and elliptical sections
   = 0.5 for all other sections

The stress at inner radius for a curved beam of rectangular cross section

$$\sigma_i = \frac{6M_b}{bh^2}\left(1 + 0.25\,\frac{h}{r_i}\right) \qquad (25\text{--}61)$$

The stress at inner radius of circular cross section

$$\sigma_i = \frac{32M_b}{\pi d^3}\left(1 + 0.3\,\frac{d}{r_i}\right) \qquad (25\text{--}62)$$

The stress at inner radius of elliptical sections according to Bach*

$$\sigma_i = \frac{32M_b}{\pi a^2 b}\left(1 + 0.3\,\frac{a}{r_i}\right) \qquad (25\text{--}63)$$

STRESSES IN RINGS (FIG. 25–13A)

Maximum moment for a circular ring at the point of application of the load, A, Fig. 25–13a

$$M_{b(max)} = \pm\frac{Fr}{\pi} = \mp 0.318Fr \qquad (25\text{--}64)$$

where − ve sign refers to tensile load,
    + ve sign refers to compressive load

---

*Courtesy of Bach, *Maschinenelemente*, 12 ed, p.43.

Particular	Formula
Another maximum moment* for a circular ring at a point B 90° away from the point of application of load	$M_{b(max)} = \pm 0.182 Fr$ (25–65) where $-$ ve sign refers to compressive load $+$ ve sign refers to tensile load
Direct stress for the ring at point B 90° away from the point of application of load	$\sigma = \dfrac{F}{2A}$ (25–66)
The general expression for bending moment at any cross section DD at an angle θ with the horizontal (Fig. 25–13b)	$M_b^* = M_A - \frac{1}{2} Fr(1 - \cos\theta)$ (25–67)
The stress due to direct load F at any cross section DD at an angle θ with the horizontal	$\sigma = \dfrac{F\sin\theta}{2A}$ (25–68)
The combined stress at any cross section	$\sigma_r = \dfrac{1}{2}\dfrac{F}{A}\sin\theta \pm \dfrac{M_b}{Ae}\left(\dfrac{y}{r_n + y}\right)$ (25–69)

DEFLECTION

The increase in the vertical diameter of the ring (Fig. 25–13a)	$y_v = 0.149\,\dfrac{Fr^3}{EI_z}$ (25–70)
The decrease in the horizontal diameter of the ring (Fig. 25–13a)	$y_h = 0.137\,\dfrac{Fr^3}{EI_z}$ (25–71)

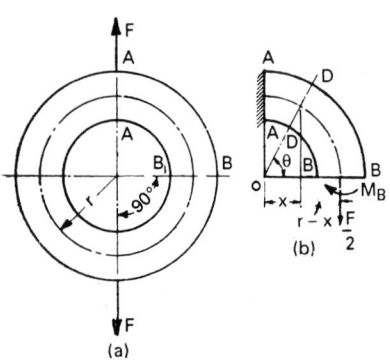

FIGURE 25–13   Bending moments in a ring.

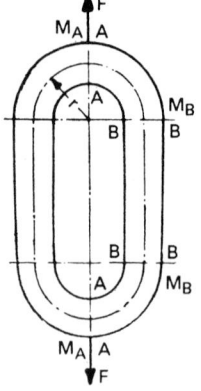

FIGURE 25–14   Bending moments in a link.

---

*Moments which tend to decrease the initial curve of the bar are taken as positive.

Particular	Formula

LINK (FIG. 25–14)

The moment, M_{bA}, at the point of application of load (Fig. 25–14)

$$M_{bA} = \frac{Fr(2r + l)}{2(\pi r + l)} \qquad (25\text{–}72)$$

where $l =$ length of straight section between the semicircular ends

The moment, M_{bB}, at the section 90° away from the point of application of load (Fig. 25–14)

$$M_{bB} = \frac{Fr(2r - \pi r)}{2(\pi r + l)} \qquad (25\text{–}73)$$

CRANE HOOK OF CIRCULAR SECTION (FIG. 25–15)

The combined stress in any fiber of a crane hook subject to a load F

$$\sigma_r = \frac{F}{A} \pm \frac{M_b}{Ae}\left(\frac{y}{r_n + y}\right) \qquad (25\text{–}74)$$

$$= \frac{Fy}{Am(r_x - y)}$$

The maximum combined stress

$$\sigma_{r(max)} = \frac{F}{A}\left(\frac{H}{2mr_i}\right) = \frac{F}{A}\,k_i \qquad (25\text{–}75)$$

The minimum combined stress

$$\sigma_{r(min)} = \frac{F}{A}\left(\frac{H}{2mr_o}\right) = \frac{F}{A}\,k_o \qquad (25\text{–}76)$$

where k_i and k_o are stress factors which depend on $H/2r_c$; k_i is the critical one which varies from 13.5 to 15.4 as ratio $H/2r_c$ changes from 0.6 to 0.4

For crane hook of trapezoidal section

Refer to Fig. 25–16 and Table 25–3.

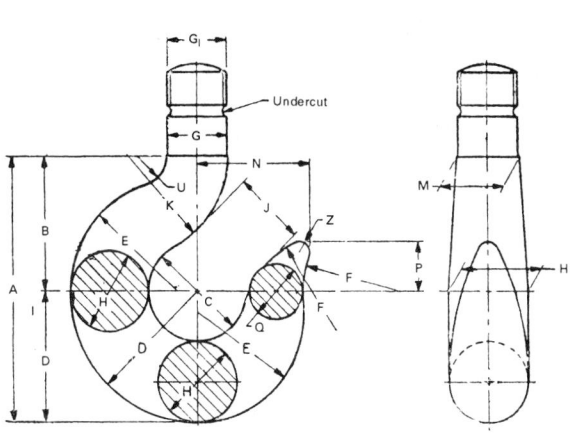

FIGURE 25–15 Hook of circular section.

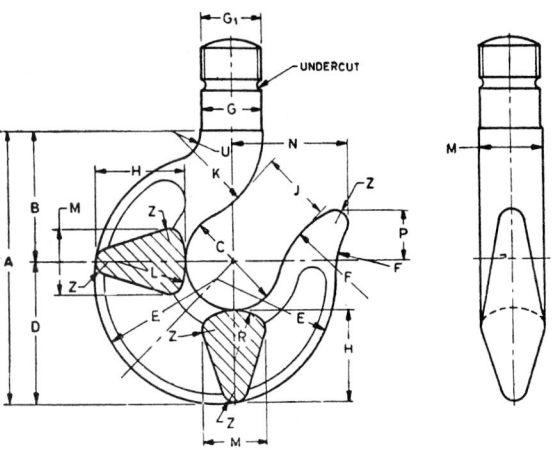

FIGURE 25–16 Crane hook of standard trapezoidal section.

25.3 CONNECTING AND COUPLING ROD

SYMBOLS

A	area of cross section, m^2 (in^2)
a	Rankine's constant
b	width, m (in)
d	diameter, m (in)
d_1	core diameter of bolt, m (in)
d_c	crankpin diameter, m (in)
d_g	gudgeon pin diameter, m (in)
E	modulus of elasticity, GPa (psi)
F	force acting on the piston due to steam or gas pressure corrected for inertia effects of the piston and other reciprocating parts, kN (lbf)
F_c	the component of F acting along the axis of connecting rod, kN (lbf)
F_i	inertia force, kN (lbf)
F_{ir}	inertia force due to reciprocating masses, kN (lbf)
F_{cr}	crippling or critical force, kN (lbf)
g	acceleration due to gravity, 9.8066 m/s^2 9806.6 mm/s^2 (32.2 ft/s^2)
h	depth of rectangular or other sections, m (in)
k	radius of gyration, m (in)
l	length of connecting rod, m (in)
l_c	length of crankpin, m (in)
l_e	equivalent length, m (in)
l_g	length of gudgeon pin, m (in)
M_b	bending moment, N m (lbf in)
n	speed of crank, rpm
n_1	safety factor
$n' = \dfrac{l}{r}$	ratio of connecting rod length to radius of crank
p	allowable pressure, MPa (psi)
p_f	load due to gas or steam pressure on the piston, MPa (psi)
v	velocity of crank, m/s (fps)
w	specific weight of material of connecting road, kN/m^3 (lbf/in^3)
W	weight of the reciprocating masses, kN (lbf)
Z	section modulus, m^3, cm^3 (in^3)
ω	angular speed of crank, rad/s
α	angle between the crank and the center line of connecting rod, deg
θ	angle between the crank and the center line of the cylinder measured from the head-end dead-center position, deg
ϕ	angle between the center line of piston and the connecting rod, deg
σ	normal stress (also with suffixes), MPa (psi)

Particular	Formula

The velocity

$$v = \frac{2\pi r n}{60}$$ (25–81)

where r in m

DESIGN OF CONNECTING ROD (FIG. 25–17)

Gas load

Load due to gas or steam pressure on the piston

$$F_g = \frac{\pi d^2}{4} \, p_f$$ (25–82)

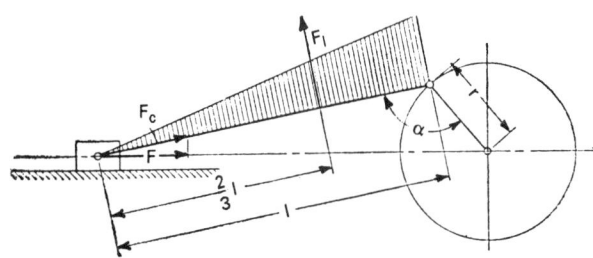

FIGURE 25–17 Forces acting on a connecting rod.

Inertia load due to reciprocating motion

Inertia due to reciprocating parts and piston

$$F_{ir} = \frac{Wv^2}{gr}\left(\cos\theta + \frac{\cos 2\theta}{n'}\right)$$ (25–83a)

$$= 0.01095 \, \frac{Wrn^2}{g}\left(\cos\theta + \frac{\cos 2\theta}{n'}\right)$$ (25–83b)

The maximum value of F_{ir} occurs when $\theta = 0°$ or when the crank is at the head-end dead center

$$F_{1ir(max)} = 0.01095 \, \frac{Wrn^2}{g}\left(1 + \frac{1}{n'}\right)$$ (25–84)

At the crank-end dead center, when $\theta = 180°$, F_{ir} attains the maximum negative value, acting in opposite direction

$$F_{2ir(max)} = -0.01095 \, \frac{Wrn^2}{g}\left(1 - \frac{1}{n'}\right)$$ (25–85)

The combined force on the piston

$$F = F_g \pm F_{ir}$$ (25–86)

The component of F acting along the axis of connecting rod

$$F_c = \frac{F}{\sqrt{1 - \left(\dfrac{\sin\theta}{n'}\right)^2}}$$ (25–87)

Particular	Formula
The stress induced due to column action on account of load F_c acting along the axis of connecting rod	
(a) As per Rankine's formula	$\sigma_1 = \dfrac{F_c}{A}\left[1 + a\left(\dfrac{l_e}{k}\right)^2\right]$ (25–88a)
(b) As per Ritter's formula	$\sigma_1 = \dfrac{F_c}{A}\left[1 + \dfrac{\sigma_e}{n\pi^2 E}\left(\dfrac{l_e}{k}\right)^2\right]$ (25–88b)
(c) As per Johnson's parabolic formula	$\sigma_1 = \dfrac{F_c}{A\left[1 - \dfrac{\sigma_y}{4n\pi^2 E}\left(\dfrac{l_e}{k}\right)^2\right]}$ (25–88c)

where
l_e = equivalent length, m (in)
k = radius of gyration, m (in)
n = end-condition coefficient (Table 2–4)
a = constant obtained from Table 2–3

Inertia load due to connecting rod

The magnitude of inertia force (Fig. 25–17) due to the weight of the rod itself, not including the ends	$F_{ic}\ \dfrac{Awv^2 l}{2gr}\ \sin\alpha$ (25–89a)
	$= \dfrac{Wv^2}{2gr}$ when $\alpha = 90°$ (25–89b)

where $W = Awl =$ weight of the rod itself, not including the ends, kN (lbf)

The maximum bending moment produced by the inertia force F_{ic} is at a distance $2/3 l$ from wrist pin	$M_{b(max)} = \dfrac{2F_{ic}l}{9\sqrt{3}} = \dfrac{2Wv^2 l}{9\sqrt{3}\times 2gr}\ \sin\alpha$ (25–90a)
	$= \dfrac{Wv^2 l}{9\sqrt{3}gr}$ when $\alpha = 90°$ (25–90b)
The maximum bending stress developed in the rod due to inertia force F_{ic}	$\sigma_{b(max)} = \dfrac{M_{b(max)}}{Z} = \dfrac{Wv^2 l}{9\sqrt{3}grZ}\ \sin\alpha$ (25–91a)
	$= \dfrac{Wv^2 l}{9\sqrt{3}grZ}$ when $\alpha = 90°$ (25–91b)
The crank angle (θ) at which the maximum bending moment occurs according to B. B. Low	$\theta = 90° - \dfrac{3500}{(n' + 7.82)^2}$ (25–92)

Particular	Formula
The relation between the moment of inertia in the xx and yy planes in order to have same resistance in either plane	$I_{yy} = {}^1\!/_4\, I_{xx}$ (25–93) *or* $k_{yy}^2 = {}^1\!/_4\, k_{xx}^2$ (25–94)

DESIGN OF SMALL AND BIG ENDS

The diameter of crankpin at the big end

$$d_c = \frac{F}{l_c p} \qquad (25\text{–}95)$$

where

p = allowable bearing pressure based on projected area, MPa (psi)
= 4.9 to 10.3 MPa (700 to 1500 psi), and

$\dfrac{l_c}{d_c}$ = 1.25 to 1.5

The diameter of the gudgeon pin at the small end

$$d_g = \frac{F}{l_g p} \qquad (25\text{–}96)$$

where

p = 10.3 to 13.73 MPa (1.2 to 2.0 kpsi), and

$\dfrac{l_g}{d}$ = 1.5 to 2

DESIGN OF BOLTS FOR BIG-END CAP

The diameter of bolts used for fixing the big-end cap

$$d_i = \sqrt{\frac{2F_{1ir(max)}}{\pi \tau_d}} \qquad (25\text{–}97)$$

where $F_{1ir(max)}$ is obtained from Eq. (25–84)

σ_d = design stress of bolt material, MPa (psi)

The expression for checking load for measuring peripheral length of each thin-walled half-bearing according to J. M. Conway Jones*

$$W_c = 6000\, \frac{L h_b}{D} \qquad \textbf{SI} \qquad (25\text{–}97a)$$

where

W_c = checking load, N
L = axial length of bearing, mm
h_b = wall thickness of bearing, mm
D = diameter of housing, mm

The expression for total minimum nip, n

$$n = 44 \times 10^{-6}\, \frac{D^2}{h_b} \quad \text{or} \quad 0.12 \text{ mm} \quad \textbf{SI} \qquad (25\text{–}97b)$$

whichever is larger

---

*In M. J. Neale, ed., *Tribology Handbook*, Section A20, Butterworth-Heinemann, London, 1973.

Particular	Formula

Note: The "nip" or "crush" is the amount by which the total peripheral length of both halves of bearing under no load exceeds the peripheral length of the housing of the bearing.

The compressive load on each bearing joint face to compress nip*

$$W = \frac{ELh_{sl}m}{\pi(D - h_{sl})10^6} \quad \text{SI} \qquad (25\text{–}97c)$$

or

$$= Lh_{sl}\sigma_y \times 10^{-6} \quad \text{SI} \qquad (25\text{–}97d)$$

whichever is smaller

where
D = housing diameter, mm
h_{sl} = steel thickness + $^1/_2$ lining thickness, mm
m = sum of maximum circumferential nip on both halves of bearing, mm
W = compressive load on each bearing joint face, N
E = modulus of elasticity of material of backing, Pa
   = 210 GPa (30.45 Mpsi) for steel
L = bearing axial length, mm
σ_y = yield stress of steel backing, Pa
   = 350 MPa (50 kpsi) for white-metal-lined bearing
   = 300 to 400 MPa (43.5 to 58 kpsi) for bearing with copper-based lining
   = 600 MPa (87 kpsi) for bearing with aluminum-based lining

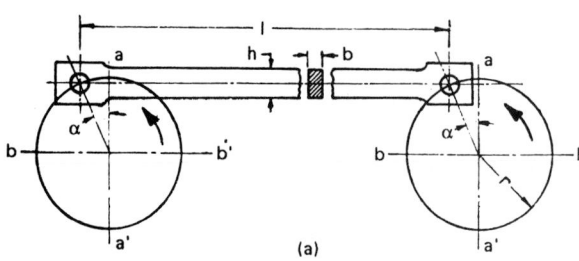

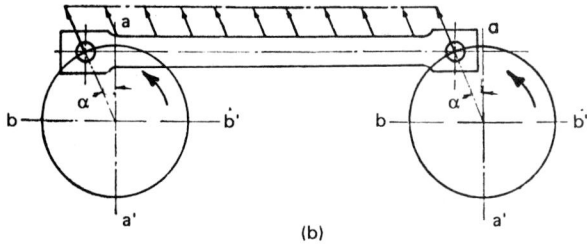

FIGURE 25–18   Forces acting on a coupling rod.

The bolt load required on each side of bearing to compress nip for extremely rigid housing

$$W_b = 1.3W \qquad (25\text{–}97e)$$

The bolt load required on each side of bearing to compress nip for normal housing with bolts very close to back of bearing

$$W_b = 2W \qquad (25\text{–}97f)$$

The ratio of connecting rod length (l) to crank radius (r)

$$n' = \frac{l}{r} = 3.4 \text{ to } 4.4 \text{ single-acting engines} \qquad (25\text{–}98)$$

= 4.6 to 5.4 for double-acting engines
= 6.0 or more for steam locomotive engines
= 5 to 7 for stationary steam engines
= 3.2 to 4 for internal-combustion engines
= 1.5 to 2 for aero engines

---

*In M. J. Neale, ed., *Tribology Handbook*, Section A20, Butterworth-Heinemann, London, 1973.

Particular	Formula

DESIGN OF COUPLING ROD (FIG. 25–18)

The centrifugal force due to the weight of the rod

$$F_c = \frac{wv^2}{gr} \, hbl \qquad (25\text{–}99)$$

$$= \frac{Wv^2}{gr} \qquad (25\text{–}100)$$

The bending component of centrifugal force

$$F_{cb} = \frac{wv^2 hbl}{gr} \cos\alpha = \frac{Wv^2}{gr} \cos\alpha \qquad (25\text{–}101)$$

The maximum bending moment due to the uniformly distributed load of F_{cb}

$$M_{b(max)} = \frac{wv^2 hbl^2}{8gr} = \frac{Wv^2 l}{8gr} \qquad (25\text{–}102)$$

The axial component of the centrifugal force

$$F_{ca} = \frac{wv^2}{gr} \, hbl \sin\alpha = \frac{Wv^2 \sin\alpha}{gr} \qquad (25\text{–}103)$$

For some of the common cross sections of connecting rods

Refer to Fig. 25–19.

For forces acting on a coupling rod

Refer to Fig. 25–18.

For proportions of ends of round and H-section connecting rod

Refer to Fig. 25–20.

For proportions and empirical relations of steam engine common strap end

Refer to Fig. 25–21.

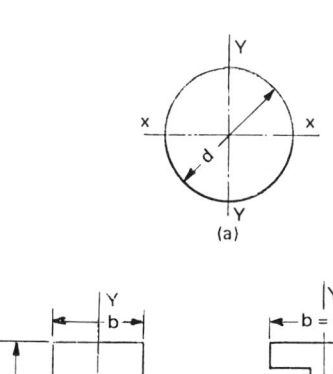

FIGURE 25–19 Connecting rod sections.

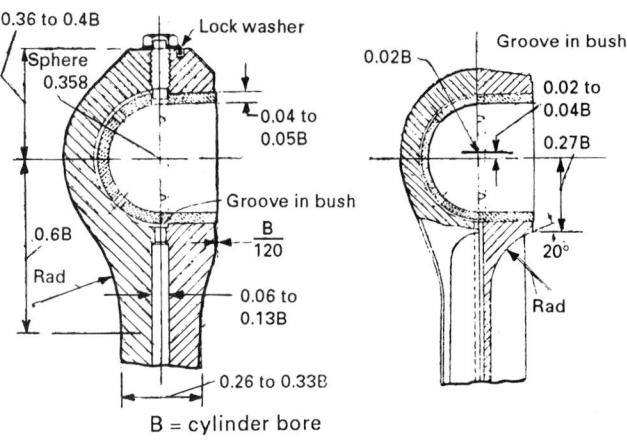

FIGURE 25–20 Two typical end designs for round and H-section connecting rods.

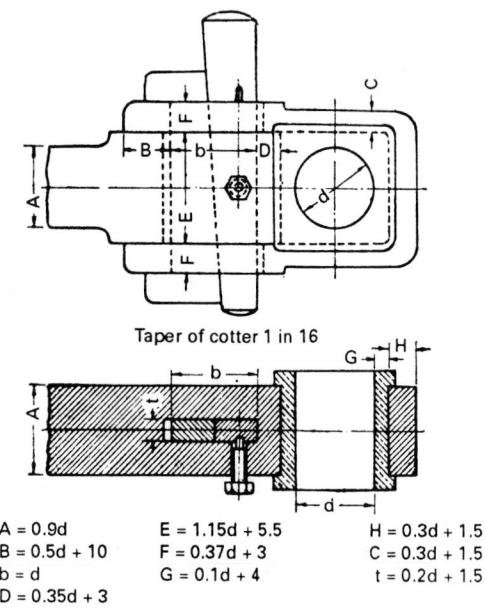

Taper of cotter 1 in 16

A = 0.9d   E = 1.15d + 5.5   H = 0.3d + 1.5
B = 0.5d + 10   F = 0.37d + 3   C = 0.3d + 1.5
b = d   G = 0.1d + 4   t = 0.2d + 1.5
D = 0.35d + 3

All dimensions in mm

FIGURE 25–21   Steam engine common strap end.

TABLE 25–4
Dimensions of cast-iron piston up to 430 mm diameter (Fig. 22–24) (all dimensions in cm)

Diameter of cylinder, D	Diameter of piston rod, d	d_1	b	a	h	h_1	h_2	h_3	d_2	d_3
15.0	2.8	2.5	7.5	1.2	1.2	1.4	0.8	0.8	3.1	2.5
20.0	3.4	3.1	8.1	1.4	1.2	1.6	0.95	0.8	3.4	3.1
25.0	4.0	3.7	9.0	1.6	1.4	1.7	0.95	0.95	4.0	3.7
30.0	5.0	4.7	10.0	1.7	1.6	1.7	1.2	0.95	4.6	4.7
35.0	5.6	5.0	11.2	1.9	1.6	1.9	1.2	0.95	5.3	5.0
40.0	5.9	5.6	11.8	1.9	1.7	2.2	1.2	0.95	5.8	5.6

25.4 PISTON AND PISTON RINGS

SYMBOLS

A	area of cross section of piston head, m^2 (in^2)
b	width of face of piston, m (in)
B	diameter of bore, m (in)
c	heat-conduction factor, $kJ/m^2/m/h/K$ ($Btu/in^2/in/h/°F$)
C	higher heat value of fuel used, kJ/kg (Btu/lb)
d	nominal diameter of piston ring, m (in)
	diameter of piston rod, m (in)
d_g	diameter of gudgeon pin, m (in)
D	diameter of bore (cylinder), m (in)
D_r	root diameter of the piston ring groove, m (in)
E	modulus of elasticity, GPa (psi)
F	force, kN (lbf)
	diametral load on the piston ring to close the gap which is less than 2.45 N (0.55 lbf)
F_θ	tangential load on the piston ring to close the gap which is less than 2.45 N (0.55 lbf)
h	thickness (also with subscripts), m (in)
	radial thickness of piston ring, m (in)
h_1, h_2, h_3	thickness as shown in Fig. 25–24, m (in)
H	heat flowing through the head, kJ/h (Btu/h)
$i' = \dfrac{l_g}{d_g}$	length/diameter ratio
l_g	length of gudgeon pin, m (in)
L	length of piston, m (in)
M_b	bending moment, N m (lbf in)
n	safety factor
P_b	brake horsepower (bhp)
p	pressure, MPa (psi)
r	radius, m (in)
R, R_i, R_h	radius as shown in Fig. 25–22b, m (in)
t_h	thickness of head, m (in)
t_r	thickness under ring groove, m (in)
T_c	temperature at center of head, °C (°F)
T_e	temperature at edge of head, °C (°F)
w	weight of fuel used, kg/bhp/h
	axial width of ring, m (in)
σ	stress (also with subscripts), MPa (psi)
θ	angle (Fig. 25–23), deg

Particular	Formula

STEAM ENGINE PISTONS
Piston rods

The diameter of piston rod

$$d = D\sqrt{\frac{p}{\sigma_a}} \qquad (25\text{–}104)$$

where

p = unbalanced pressure or difference between the steam inlet pressure and the exhaust, MPa (psi)

$\sigma_a = \dfrac{\sigma_u}{n}$ = allowable stress, MPa (psi)

Note: σ_a is based on a safety factor of 10 for double-acting engines and 8 for single-acting engines. (This is usually taken as $\frac{1}{6}$ to $\frac{1}{7}$ the diameter of the piston.)

The diameter of piston rod according to Molesworth

$d = 0.0044D\sqrt{p}$ for cast-iron pistons $\qquad (25\text{–}105)$

$ = 0.00338D\sqrt{p}$ for steel pistons $\qquad (25\text{–}106)$

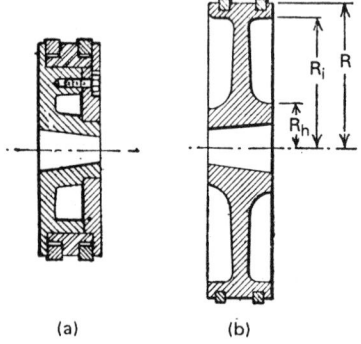

(a)   (b)

FIGURE 25–22 Plate piston.

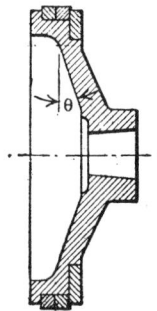

FIGURE 25–23 Conical plate piston.

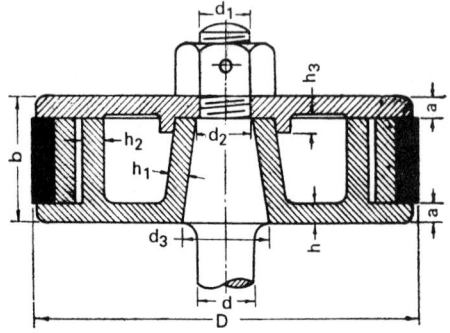

FIGURE 25–24 Cast-iron piston of diameter $\leq$ 400 mm (16.0 in).

Particular	Formula

PROPORTIONS FOR PRELIMINARY LAYOUT FOR PLATE PISTONS
Box type (Figs. 25–22(a) and 25–24)

Width of face	$b = 0.3$ to $0.5D$	(25–107)
Thickness of walls and ribs for low pressure	$h = (2R + 50 \text{ cm})[0.003\sqrt{p} + 0.0275 \text{ cm}]$	(25–108)
	or	
	$h = \dfrac{2R}{60} + 10 \text{ mm } (0.40 \text{ in})$	(25–109)
The thickness of walls and ribs for high pressure	$h = \dfrac{2R}{40} + 10 \text{ mm } (0.40 \text{ in})$	(25–110)
For dimensions of conical plate piston	Refer to Fig. 25–23.	
For dimensions of cast-iron piston of ≤ 400 mm diameter	Refer to Fig. 25–24 and Table 25–4.	

Disk type (Fig. 25–22b)

Width of face	$b = 0.3$ to $0.5D$	(25–111)
Thickness of walls and ribs for low pressure	$h = (2R + 12.5 \text{ cm}) (0.0096 \sqrt{p} \\ +0.057cm)$	(25–112)
The hub thickness	$h_1 = 0.45d$	(25–113)
The hub diameter	$D_h = 2R_h = 1.6\times$ the piston diameter	(25–114)
Width of piston rings	$w = 0.03D$ to $0.06D$	(25–115)
Thickness of piston rings	$h = 0.025D$ to $0.03D$	(25–116)
For dimensions of cast-iron piston	Refer to Table 25–4.	

STRESSES

(a) Distributed load over the plate inside the outer cylindrical wall (i.e., the area πR_i^2)

    (1) Stress at the outer edge (Fig. 25–22b)

$$\sigma_1 = \frac{3p}{4h^2}\left\{ R_i^2 - 3R_h^2 + \frac{4R_h^2}{R_i^2 - R_h^2} \ \ln \frac{R_i}{R_h} \right\} \quad (25\text{–}117)$$

    (2) Stress at the inner edge (Fig. 25–22b)

$$\sigma_2 = \frac{3p}{4h^2}\left\{ R_i^2 + R_h^2 - \frac{4R_i^2 R_h^2}{R_i^2 - R_h^2} \ \ln^2 \frac{R_i}{R_h} \right\} \quad (25\text{–}118)$$

Particular	Formula

(b) Load on the outer wall, $p\pi(R^2 - R_i^2)$ distributed around the edge of the plate

(1) Stress at the outer edge (Fig. 25–22b)

$$\sigma_3 = \frac{3p(R^2 - R_i^2)}{2h^2}\left[1 - \frac{2R_h^2}{R_i^2 - R_h^2}\ln\frac{R_i}{R_h}\right] \qquad (25\text{–}119)$$

(2) Stress at the inner edge (Fig. 25–22b)

$$\sigma_4 = \frac{3p(R^2 - R_i^2)}{2h^2}\left[1 - \frac{2R_i^2}{R_i^2 - R_h^2}\ln\frac{R_i}{R_h}\right] \qquad (25\text{–}120)$$

(3) The sum of the stresses at the outer edge $\sigma_o = \sigma_1 + \sigma_3$ $\qquad\qquad$ (25–121)

(4) The sum of the stresses at the inner edge $\sigma_i = \sigma_2 + \sigma_4$ $\qquad\qquad$ (25–122)

(*Note*: σ_o or σ_i should not be greater than the permissible stress of the material. A safety factor, n, of 8 can be used.)

Dished or conical type (Fig. 25–23)

An empirical formula for the thickness of conical piston (Fig. 25–23)

$h = 0.288\sqrt{pD/\sigma}\sin\theta$ **SI** $\qquad$ (25–123a)
where p and σ in MPa, and D and h in m

$= 9.12\sqrt{pD/\sigma}\sin\theta$ **Metric** $\qquad$ (25–123b)
where p and σ in kgf/mm^2, D and h in mm

$= 1.825\sqrt{pD/\sigma}\sin\theta$ **US Customary units**
$\qquad\qquad\qquad\qquad\qquad\qquad$ (23–123c)
where p and σ in psi, D and h in in

The height of boss $H = 1.1K$ $\qquad\qquad\qquad$ (25–124)

The diameter of boss

$D_h = 1.7K$ for small pistons $\qquad$ (25–125a)
$\quad = 1.5K$ for large pistons and light engines
$\qquad\qquad\qquad\qquad\qquad\qquad$ (25–125b)

The thickness h_1 measured on the center line $h_1 = Kc$ $\qquad\qquad$ (25–126)

where
c = 1 to 0.75 depending on the angle of inclination
$\quad$ θ (Refer to Table 25.5.)
θ = varies from 6° to 35°
K = 1 to 4.5 for varying pressure and diameter

Also refer to Table 25–6 for values of K.

For calculating hub diameter, width of piston rings, and thickness of piston rings

Refer to Eqs. (25–114) to (25–116).

Particular	Formula

PISTONS FOR INTERNAL-COMBUSTION ENGINES

Trunk piston (Fig. 25–25)

The head thickness of trunk pistons (Fig. 25–25a)

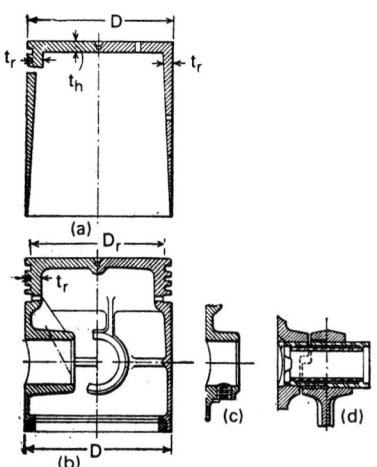

(a)

(b)

(c)   (d)

FIGURE 25–25   Trunk piston for small internal-combustion engine. (*a*) piston laid out for heat transfer; (*b*) piston modified for structural efficiency; (*c* and *d*) alternate pin designs.

$$t_h = \sqrt{\frac{3PD^2}{16\sigma}} \qquad (25\text{–}127)$$

where
σ = 39 MPa (5.8 kpsi) for close-grained cast iron
= 56.4 MPa (8.2 kpsi) for semisteel or aluminum alloy
= 83.4 MPa (12.0 kpsi) for forged steel

COMMONLY USED EMPIRICAL FORMULAS IN THE DESIGN OF TRUNK PISTONS FOR AUTOMOTIVE-TYPE ENGINES

Thickness of head (Fig. 25–25a)

$$t_h = 0.032D + 1.5 \text{ mm} \qquad \textbf{SI} \qquad (25\text{–}128)$$

$$= 0.00D + 0.06 \text{ in} \quad \textbf{US Customary units} \ (25\text{–}128a)$$

The head thickness for heat flow

$$t_h = \frac{HD^2}{0.16c(T_c - T_e)A} = \frac{H}{0.194c(T_c - T_e)} \quad \textbf{SI}$$

$$(25\text{–}129a)$$

where
$T_c - T_e$ = 205°C (400°F) and T_c = 698 K, 425°C (800°F) for cast-iron piston
ΔT = $T_c - T_e$ = 55°C (130°F) and T_c = 533 K, 260°C (500°F) for aluminum piston
c = 2.2 for cast iron
= 7.7 for aluminum

$$= \frac{HD^2}{16c\Delta TA} = \frac{H}{12.5c\Delta T} \qquad (25\text{–}129b)$$

US Customary units

Particular	Formula
Thickness of wall under the ring (Fig. 25–25a and b)	t_r = thickness of head = t_h (25–130)
The thickness under the ring groove	$t_r = \frac{1}{2}[D_r \pm \sqrt{D_r^2 - 4Dt_h}]$ (25–131)
The heat flow through the head	$H = KCwP_b$ (25–132)

where
w = weight of fuel used, kJ/kW/h (lbf/bhp/h)
K = constant representing that part of heat supplied to the engine which is absorbed by the piston
= 0.05 (approx.)
P_b = brake horsepower per cylinder
= $D - (2w + 0.006D + 0.02$ in$)$

Particular	Formula
The root diameter of ring grooves, allowing for ring clearance	**US Customary units**

$D_r = D - (2w + 0.006D + 0.5$ mm$)$ **SI**
        at the compression rings   (25–132a)

$D_r = D - (2w + 0.006D + 1.5$ mm$)$ **SI**   (25–132b)
   = $D - (2w + 0.00D + 0.06$ in$)$ at the oil grooves
             US Customary units
where D_r and D in mm (in)

Particular	Formula
Length L of piston	$L = D$ to $1.5D$ (25–133)
For chemical composition and properties of aluminum alloy piston	Refer to Table 25–10B.

Gudgeon pin

Particular	Formula
The diameter of gudgeon pin	$d_r = \sqrt{\dfrac{F}{i'p}}$ (25–134)

where
F = maximum gas pressure corrected for inertia effect of the piston and other reciprocating parts, kN (lbf)

p = working bearing pressure,
= 9.81 MPa (1.42 kpsi) to 14.7 MPa (2.13 kpsi)

Particular	Formula
The length/diameter ratio	$i' = \dfrac{l_g}{d_g} = 1.5$ to 2 (25–135)
For gudgeon pin allowable oval deformation	Refer to Fig. 25–28b.
For empirical relations and proportions of pistons	Refer to Figs. 25–26 to 25–28a.
For fatigue stress in gudgeon pins	Refer to Fig. 25–28c.
For empirical proportions and values of cylinder cover, cylinder liner, and valves	Refer to Figs. 25–30 to 25–33.

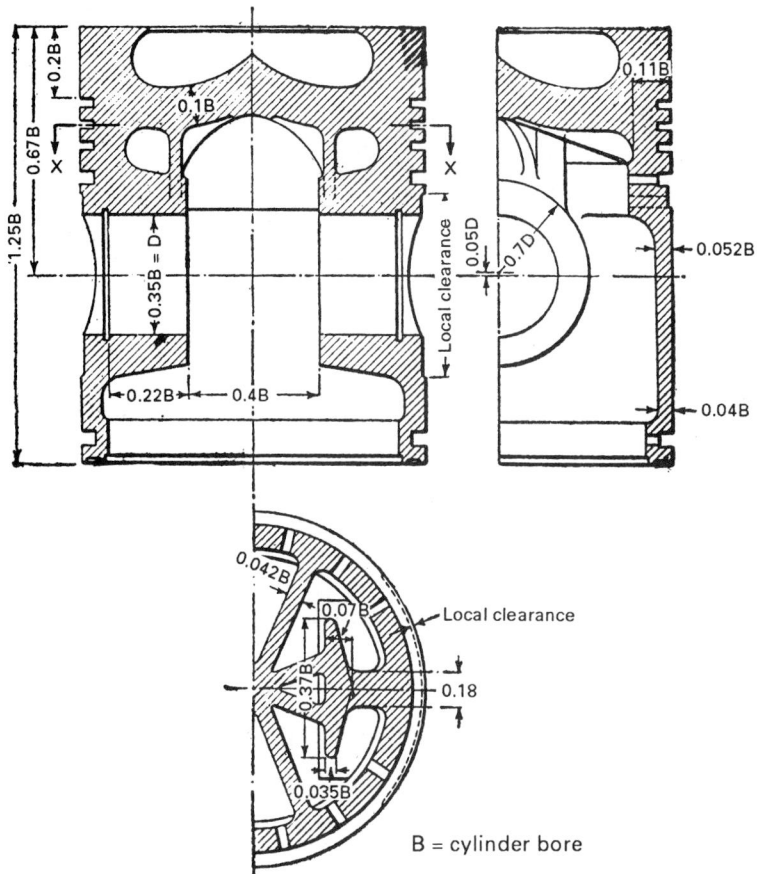

FIGURE 25–26 Proportions of a typical alloy piston.

Particular	Formula

Piston rings

Width of rings

$$w = \frac{D}{20} \text{ for concentric rings} \qquad (25\text{–}136a)$$

$$= \frac{D}{27.5} \text{ opposite the joint of eccentric rings}$$

$$(25\text{–}136b)$$

$$= \frac{D}{55} \text{ at the joint of eccentric rings} \qquad (25\text{–}136c)$$

For land width or axial width of piston ring (w) required for various groove depths (g) and maximum cylinder pressure, p_{max}

Refer to Fig. 25–28d

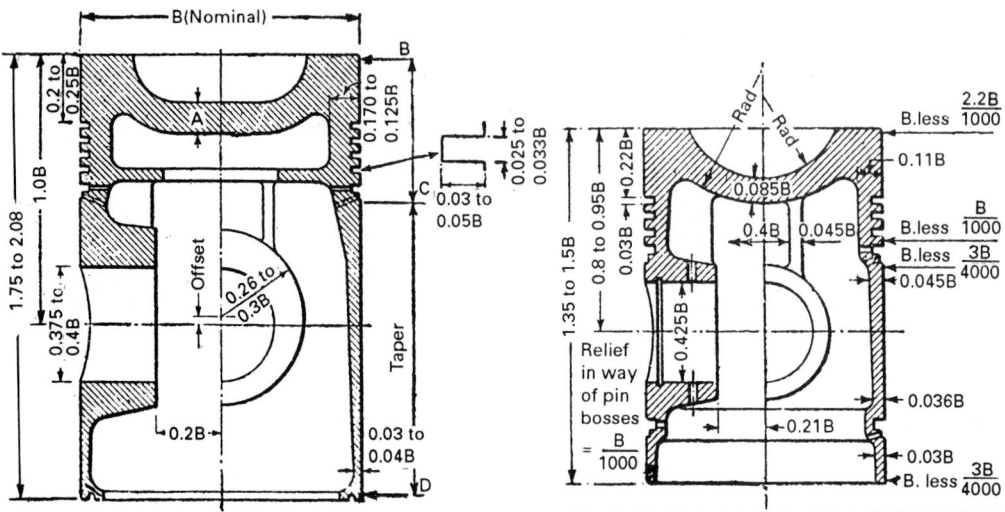

FIGURE 25–27 Proportions of an iron piston.

FIGURE 25–28(a) Iron piston for small engines (B = cylinder bore).

Fig. 25–27		mm (in)	mm (in)	mm (in)	mm (in)	mm (in)
Cylinder bore		152.5	203.2	254	305	406.5
		(6)	(8)	(10)	(12)	(16)
Crown thickness	A	16	19	32	41.5	47.5
		(5/8)	(3/4)	(1¼)	(1⅝)	(1⅞)
Clearance	B	0.760	0.900	1.145	1.525	2.030
		(0.03)	(0.035)	(0.045)	(0.06)	(0.08)
Clearance	C	0.125	0.225	0.230	0.255	0.255
		(0.005)	(0.008)	(0.009)	(0.01)	(0.01)
Clearance	D	0.125	0.120	0.180	0.200	0.230
		(0.005)	(0.006)	(0.007)	(0.008)	(0.009)

Piston weight = $40,715B^3$ N (approx.) **SI**
  where B = cylinder bore, m
Piston weight = $0.15B^3$ lbf **US Customary units**
  where B in in

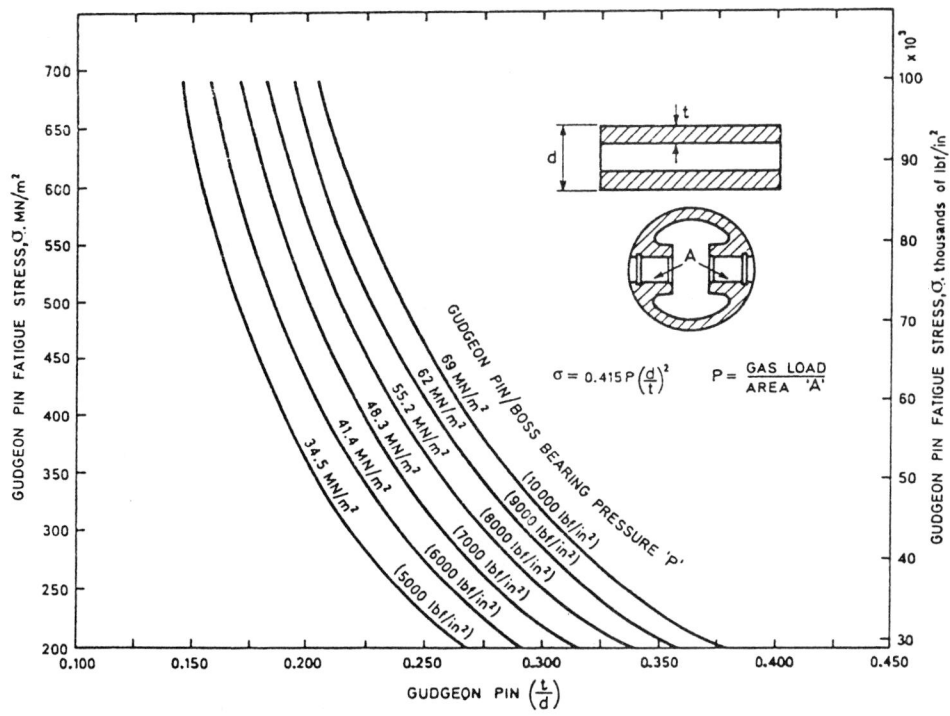

FIGURE 25–28(b) Fatigue stress in gudgeon pins for various pin and piston geometries. (*M. J. Neale*, Tribology Handbook, *Butterworth-Heinemann, 1973.*)

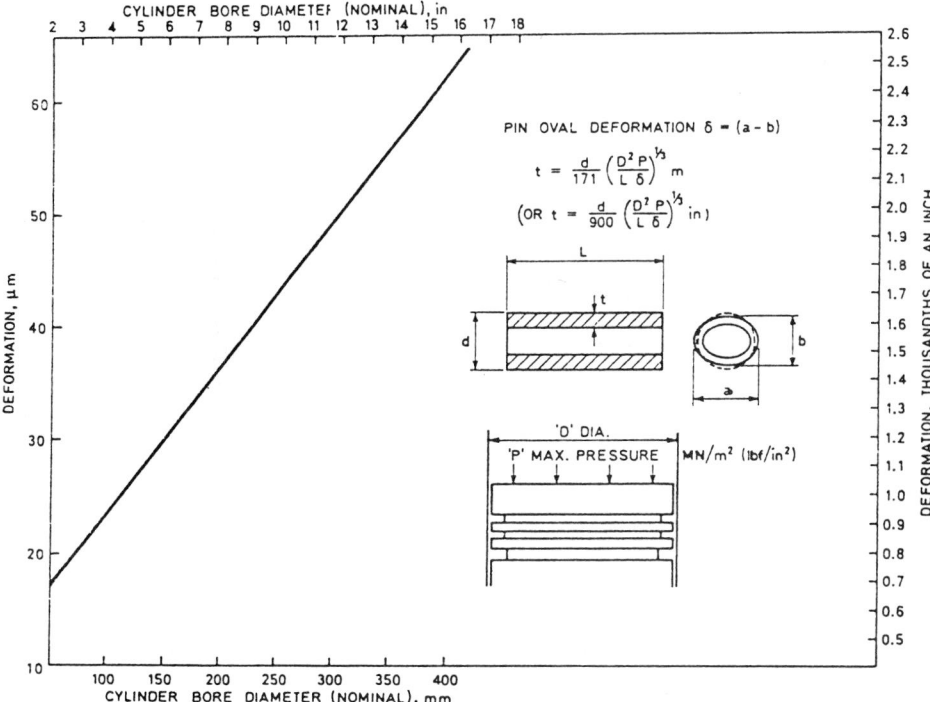

FIGURE 25–28(c) Gudgeon pin allowable oval deformation. (*M. J. Neale*, Tribology Handbook, *Butteworth-Heinemann, 1973.*)

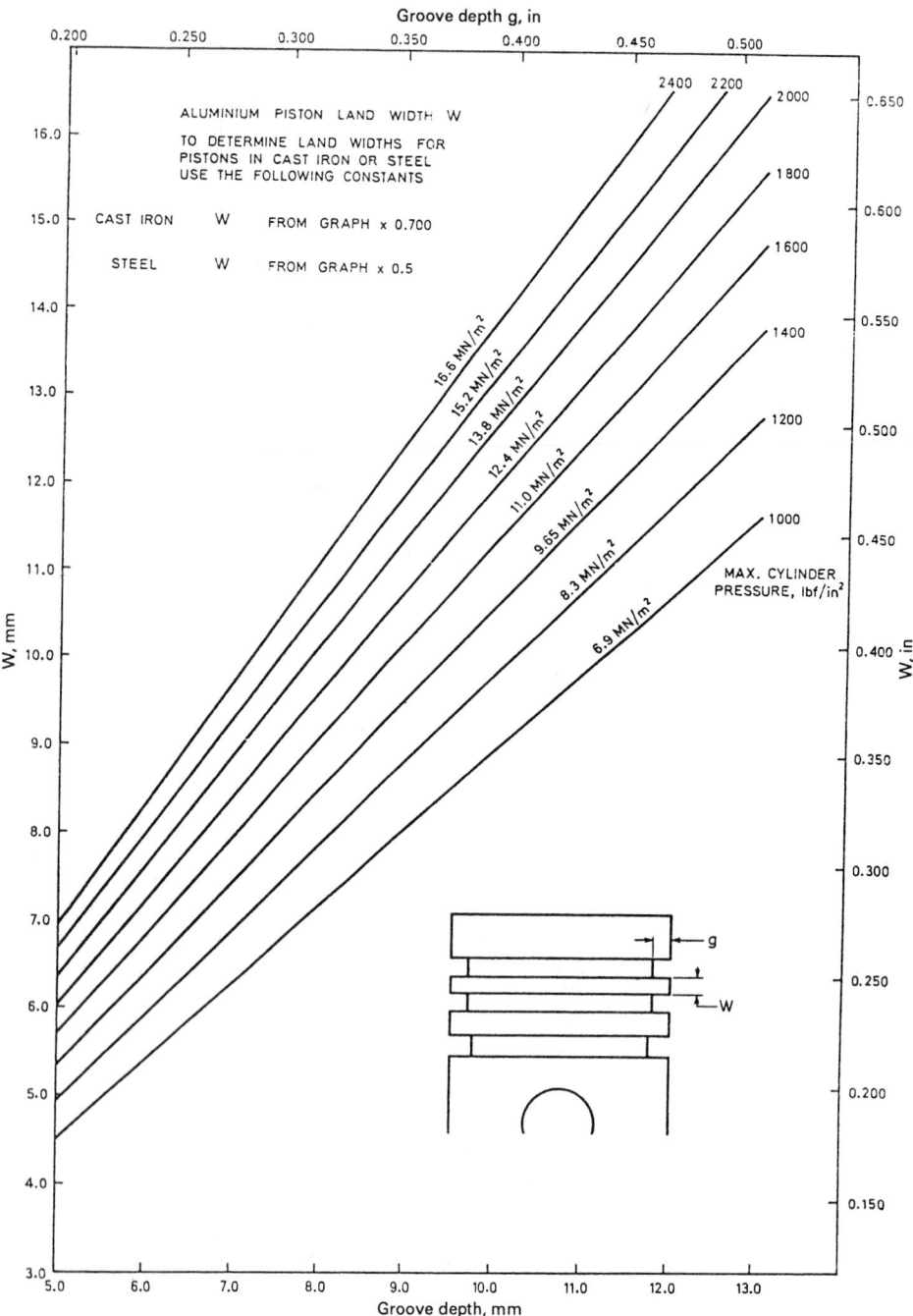

FIGURE 25–28(*d*)   The land width required for various groove depths and maximum cylinder pressures. (*M. J. Neale, Tribology Handbook, Butterwrorth-Heinemann, 1973.*)

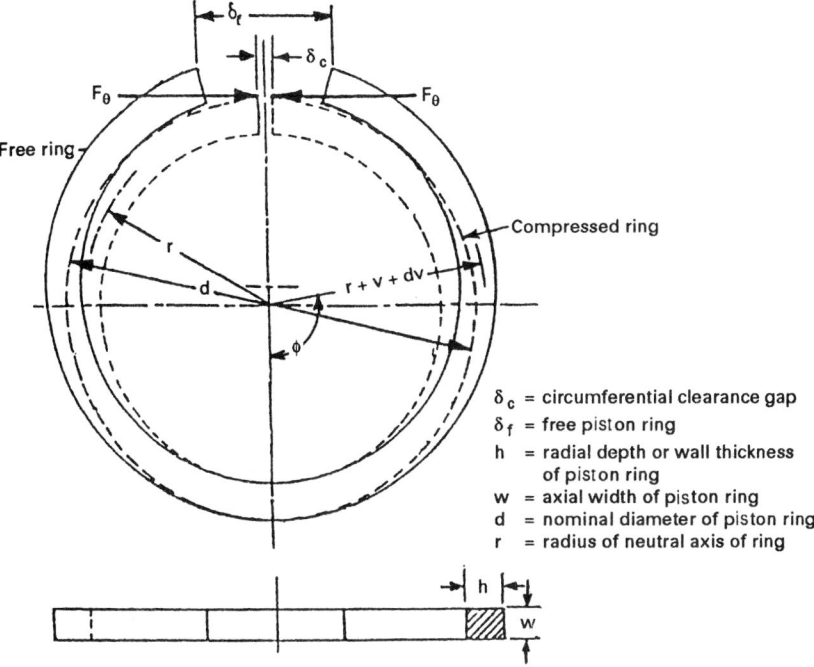

δ_c = circumferential clearance gap
δ_f = free piston ring
h = radial depth or wall thickness
     of piston ring
w = axial width of piston ring
d = nominal diameter of piston ring
r = radius of neutral axis of ring

FIGURE 25–28(e)   Nomenclature of piston ring and tangential force, F_θ.

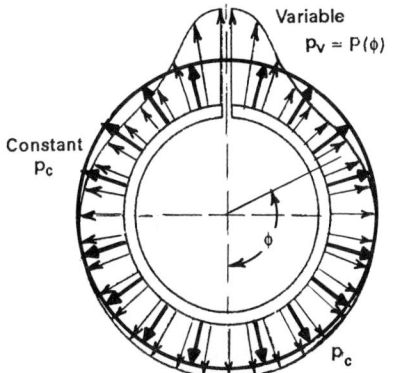

FIGURE 25–28(f)   Typical variable and constant contact pressure distribution around piston rings for four-stroke engines. (Courtesy of Piston Ring Manual, *GOETZE AG, D-5093 Burscheid, Germany, August 1986.*)

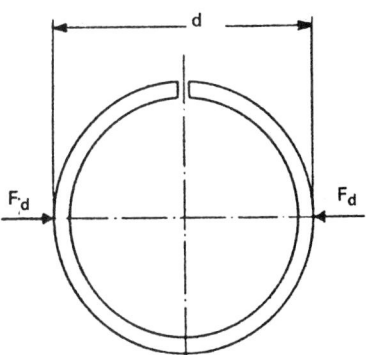

FIGURE 25–28(g)   Diametrically opposite force (F_d) applied on piston ring.

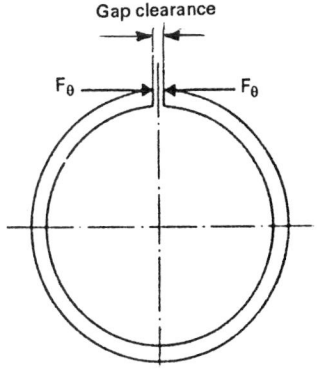

FIGURE 25–29   Tangentially applied force, F_θ, on a piston ring.

Particular	Formula
The modulus of elasticity of piston ring as per Indian Standards	$$E = \frac{5.37\left(\dfrac{d}{h} - 1\right)^3 F}{w\delta}$$ when the ring is diametrically loaded (25–137a) $$= \frac{14.14\left(\dfrac{d}{h} - 1\right)^3 F_\theta}{w\delta}$$ when the ring is tangentially loaded (25–137b) where E = modulus of elasticity, MPa (psi) δ = difference between free gap and gap after applying the load, mm (in)
The bending moment produced at any cross section of the ring by the pressure uniformly distributed over the outer surface of the ring at an angle ϕ measured from the center line of the gap of the ring (Figs. 25–28e and 25–28f)	$$M_b = -2pwr^2 \sin^2 \frac{\phi}{2} \qquad (25\text{–}138a)$$ $$= pwr^2(1 + \cos\phi) \qquad (25\text{–}138b)$$ where r = radius of neutral axis, mm (in) p = pressure at the neutral axis of the piston ring, Pa (psi)
The bending moment (M_b) in Eq. (25–138a) in terms of tangential force, F_θ	$$M_b = F_\theta r(1 + \cos\phi) \qquad (25\text{–}138c)$$
The uniform contact pressure of the piston ring on the wall	$$p = \frac{E_n \delta_f}{7.07d(d/h - 1)^3} \qquad (25\text{–}138d)$$ where d = external piston ring diameter h = radial depth or wall thickness of piston ring E_n = nominal modulus of elasticity of material of the ring δ_f = free ring gap
The radial distance from a point in piston ring to obtain a uniform pressure distribution (Fig. 25–28e) according to R. Munro*	$$r_o = r + v + dv \qquad (25\text{–}138e)$$ where $$v = \frac{Fr^4}{E_n I}\left(1 - \cos\phi + \tfrac{1}{2}\phi\sin\phi\right) \qquad (25\text{–}138f)$$ $$dv = \left(\frac{r}{2}\left(\frac{Fr^3}{E_n I}\right)^2 (\phi - \tfrac{1}{2}\phi\cos\phi - \tfrac{1}{2}\sin\phi)\right.$$ $$\left. (3\sin\phi + \phi\cos\phi)\right)$$

*In M. J. Neale, ed., *Tribology Handbook*, Section A31, Butterworth-Heinemann, London, 1973.

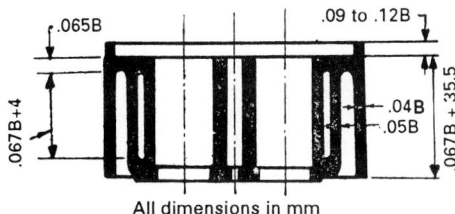

FIGURE 25–30 Proportion for four-stroke cover, 100- to 450-mm bore (B = cylinder bore.

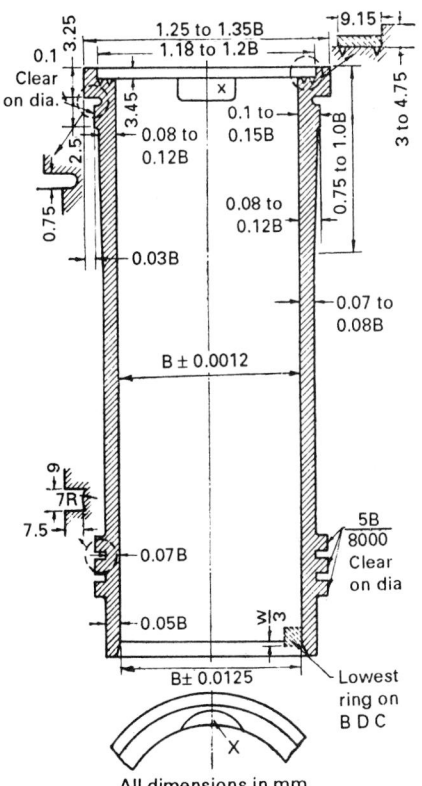

FIGURE 25–31 Empirical rules for average practice in liner design (B = cylinder bore).

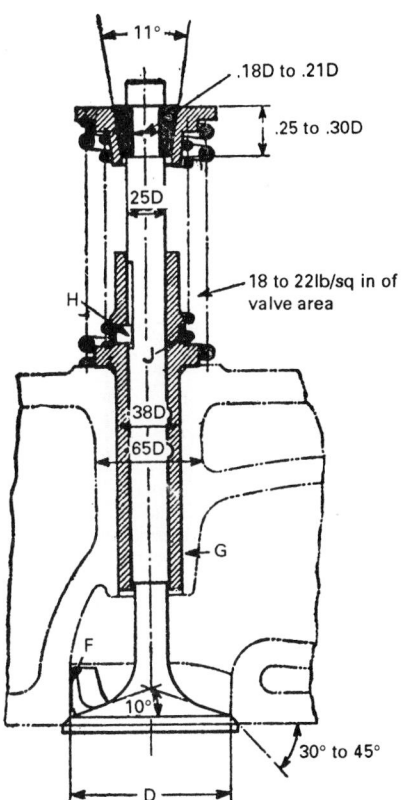

FIGURE 25–32 Valve seated directly in the cylinder head.

Particular	Formula
	F = (mean wall pressure $\times$ ring axial width) r = radius of neutral axis, when the ring is in place inside the cylinder (Fig. 25–28e) ϕ = angle measured from bottom of the vertical line passing through the center of the gap of the ring as shown in Fig. 25–28e I = moment of inertia of the ring
The relation between the ratio of fitting stress σ_{ft} to nominal modulus of elasticity (E_n) in terms of h, d, and δ_f	$$\frac{\sigma_{ft}}{E_n} = \frac{4(8h - \delta_f + 0.00d)}{3\pi h(d/h - 1)^2} \qquad (25\text{--}138g)$$ where σ_{ft} = opening stress when fitting the piston ring onto the piston
The relation between the ratio of working stress (σ_w) to nominal modulus of elasticity (E_n) in terms of h, d, and δ_f	$$\frac{\sigma_w}{E_n} = \frac{4(\delta_f - 0.00d)}{3\pi h(d/h - 1)^2} \qquad (25\text{--}138h)$$ where σ_w = working stress when the piston ring is in the cylinder
The relation between the ratio of the sum of ($\sigma_{ft} + \sigma_w$) to nominal modulus of elasticity (E_n) in terms of d and h	$$\frac{\sigma_{ft} + \sigma_w}{E_n} = \frac{32}{3\pi(d/h - 1)^2} \qquad (25\text{--}138i)$$ Equation (25–138i) is independent of δ_f.
For preferred number of piston rings	Refer to Table 25–10C.

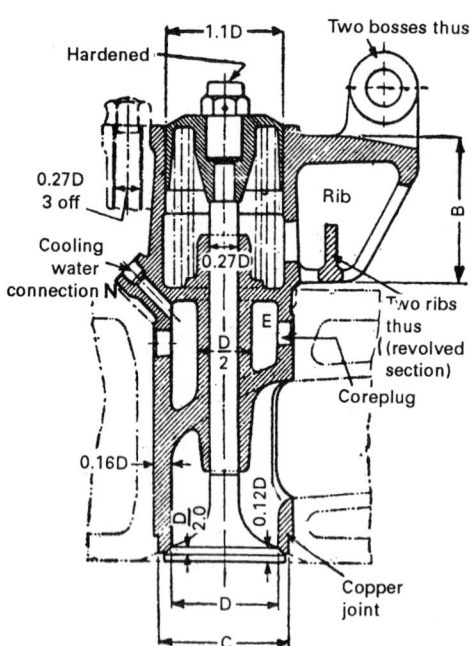

FIGURE 25–33   Valve with removable cage.

Particular	Formula
For properties of typical piston ring materials	Refer to Table 25–10D.
The circumferential clearance (δ_c) or gap between ends of ring	$\delta_c = d\alpha_p T$ (25–138j)
	where
	α_p = coefficient of expansion of piston ring material
	T = operating temperature
	d = cylinder diameter
An expression for pressure acting on ring from Eqs. (25–138b) and (25–138c)	$p = \dfrac{F_\theta}{rw}$ (25–139a)
The pressure in the radial outward direction against the cylinder	$p = \dfrac{2F_\theta}{dw}$ (25–139b)
For variable and constant radial contact pressure distribution of piston ring	Refer to Fig. 25–28f.
The diametral load which acts at 90° to the gap required to close the ring to its nominal diameter, d (Fig. 25–28g)	$F_d = 2.05F_\theta$ for modulus of elasticity $E \leq 150$ GPa (25–140a)
	$F_d = 2.15F_\theta$ for modulus of elasticity $E > 150$ GPa (25–140b)
	$F_d \approx 2.21F_\theta$ (25–140c)
The maximum bending stress at any cross section which makes an angle ϕ measured from the center line of the gap of the ring	$\sigma_b = \dfrac{12pr^2}{h^2}\sin^2\left(\dfrac{\phi}{2}\right)$ (25–141)
The maximum bending stress which occurs at $\phi = \pi$, i.e., at the cross section opposite to the gap of the ring	$\sigma_{max} = \dfrac{12pr^2}{h_{max}^2}$ (25–142)
The bending stress present in the ring of rectangular cross section in terms of free gap (δ_f) of the ring, when it is in place in the cylinder	$\sigma_b = 0.424\delta_f\,\dfrac{Eh}{(d-h)^2}$ **SI** (25–142a)
	where σ_b and E in N/mm^2
	h, d, and δ_f in mm
The bending stress present in the ring of rectangular cross section in terms of tangential force, F_θ (Fig. 25–29)	$\sigma_b = \dfrac{6(d-h)}{wh^2}F_\theta$ (25–142b)
	where σ_b in N/mm^2; F_θ in N; d, h, and w in mm
The bending stress present in the case of slotted oil control ring of rectangular cross section in terms of free ring gap, δ_f	$\sigma_{bso} = 0.424\,\dfrac{\delta_f E l_{co} I_m}{(d-h)^2 I_{us}}$ (25–142c)
	where σ_{bso} in N/mm^2 and
	I_{us} = moment of inertia of the unslotted cross-section ring, mm^4
	$I_m = \dfrac{I_{us} + I_s}{2}$

Particular	Formula
	I_s = moment of inertia of the slotted cross-section ring, mm^4 l_{co} = twice the distance between center of gravity and outside diameter, mm
The bending stress present in the case of slotted oil control ring of rectangular cross section in terms of tangential load, F_θ	$$\sigma_{bso} = \frac{6(d-h)l_{co}I_m}{wh^3 I_s} F_\theta \qquad (25\text{--}142\text{d})$$ where σ_{bso} in N/mm^2; F_θ in N; l_{co}, d, h, and w in mm; I_m and I_s in mm^4
The tangential load or force required for opening of a rectangular cross-section piston ring*	$$\sigma_{\theta,max} = \frac{hE}{d-h}\,(1.26\epsilon_T - 1.84k + 0.025)\ \textbf{SI}$$ $$(25\text{--}142\text{e})$$ where $$\epsilon_T = \frac{d+h}{d-h} - 1$$ k = piston ring parameter from Eq. (25–142f) and (25–142h)
The piston ring parameter (k) in terms of tangential load F_θ for rectangular cross-section rings	$$k = \frac{3(d-h)^2}{wh^3}\frac{F_\theta}{E} \qquad (25\text{--}142\text{f})$$
The tangential load or force required for opening of rectangular cross-section slotted oil control rings	$$\sigma_{\theta,maxT} = \frac{l_{ci}E}{d-h}\,(1.26\epsilon_T - 1.84k + 0.025)\,\frac{I_m}{I_s}$$ $$(25\text{--}142\text{g})$$ where $$\epsilon_T = \frac{d+h}{d-h} - 1 \quad \textbf{SI}$$ l_{ci} = twice the distance between center of gravity and inside diameter, mm k = piston ring parameter from Eq. (25–142h) and (25–142f)
The piston ring parameter (k) in terms of free ring gap (δ_f) for rectangular cross-section slotted oil rings for use in Eq. (25–142g)	$$k = \frac{2}{3\pi}\frac{\delta_f}{d-h} \qquad (25\text{--}142\text{h})$$
The piston ring parameter (k) in terms of the constant pressure (p) for rectangular cross-section rings also for use in Eqs. (25–142f) and (25–142g)	$$k = \frac{3}{2}\frac{p}{E}\frac{d(d-h)^2}{h^3} \qquad (25\text{--}142\text{i})$$
The radial thickness of the ring at a section which makes an angle ϕ measured from the center line of the gap of the ring	$$h = \sqrt[3]{\frac{24pr^4}{E\delta}}\,\sin^2\frac{\phi}{2} \qquad (25\text{--}142\text{j})$$

*Geotze AG, *Piston Ring Manual*, 3rd ed., Burscheid, Germany, 1987.

Particular	Formula	
The maximum thickness of the ring which occurs just opposite the gap of the ring (i.e., at $\phi = \pi$)	$h_{max} = \sqrt{\dfrac{24pr^4}{E\delta}}$	(25–142k)
For piston ring dimensional deviation, hardness, and minimum wall pressure	Refer to Tables 25–7 to 25–9.	
For cylinder bore diameter	Refer to Table 25–10.	

TABLE 25–5
Values of c for various inclinations of coned pistons

Cone	Inclination ranges, θ, deg	c
—	0–6	1
Slightly	6–18	0.85–0.95
Medium	18–28	0.75–0.85
Strong	28–35	0.65–0.75

TABLE 25–6
Values of coefficient K for pistons (admissible pressures, kgf/mm^2 absolute)

Pressure, kgf/mm^2 Diameter of cylinder, mm	0.01 to 0.02	0.02 to 0.04	0.04 to 0.06	0.06 to 0.08	0.08 to 0.10	0.10 to 0.12	0.12 to 0.14	0.14 to 0.16
380–575	1.000	1.125	1.375	1.500	1.750	2.000	2.125	2.500
575–775	1.375	1.500	1.750	2.000	2.500	2.750	3.000	3.375
775–975	1.500	1.750	2.000	2.500	3.125	3.500	3.750	4.000
975–1175	1.750	2.000	2.375	3.000	3.500	4.000	4.500	
1175–1375	2.000	2.250	2.750	3.125	3.750	4.125		
1375–1575	2.375	2.500	3.125	3.500	4.000	4.375		
1575–1775	2.500	3.000	3.375	4.000	4.375			
1775–1975	2.750	3.125	3.500	4.125				
1975–2150	3.000	3.375	3.750	4.375				
2150–2350	3.000	3.500	4.000					
2350–2550	3.125	3.500						
2550–2750	3.375	3.750						

Key: 1 kgf/mm^2 = 1.42247 kpsi; 1 kpsi = 6.894757 MPa.

TABLE 25–7
Recommended hardness for piston rings of IC engines

Nominal diameter, d, mm	Hardness HRD
< 100	95–107
100–200	93–105
> 200	90–102

TABLE 25–8
Minimum wall pressure for piston rings of IC engines

	Compression rings		Oil rings	
	MPa	kgf/cm^2	MPa	kgf/cm^2
Petrol[a]	0.059	0.60	0.137	1.40
Diesel	0.013	1.05	0.196	2.00

[a] Gasolinne.

TABLE 25–9
**Permissible deviation on the dimensions of piston rings
of IC engines**

Dimensions	Deviations, mm
Axial width, b	−0.010
	−0.022
Radial thickness	
≤ 80 mm ring diameter	±0.08
> 80 mm with ≤ 175 mm ring diameter	±0.12
175 mm ring diameter	±0.15
Parallellism of sides—40% of tolerance on axial width	

TABLE 25–10A
Preferred cylinder bore diameters for internal-combustion (IC) engines (all dimensions in mm)

30	(62)	95	125	(152.4)	(188)	(241.3)	315
32	65	98	(127)	155	190	(245)	(317.5)
34	(68)	(98.4)	(128)	(158)	(190.5)	(250)	320
(35)	70	100	(128.2)	(158.8)	(192)	(254)	(325)
36	(72)	(101.6)	130	160	195	(255)	330
38	(73)	(102)	(132)	(162)	(196.8)	266	(335)
40	74	(103.2)	(133.4)	165	198	(265)	340
42	(76)	(104.8)	135	(165.1)	200	270	(343)
44	(78)	105	(138)	(168)	(205)	(273)	(345)
46	(79.4)	108	(139.7)	170	(209.6)	(275)	350
48	80	110	140	(171.4)	210	280	
50	82	(111.1)	(142)	(172)	(215)	(285)	
52	85	112	142.9	175	(215.9)	290	
54	87.3	(114.3)	(145)	(177.8)	220	(292.1)	
56	88	115	(146)	(178)	(225)	(295)	
(57)	(88.9)	(118)	(148)	180	(228.6)	(298.4)	
58	90	120	(149.9)	(182)	230	300	
(59)	(91.4)	(120.6)	150	(184.2)	(235)	(305)	
60	(92)	(122)	(152)	185	240	103	

TABLE 25-10B
Chemical composition of alloys and physical properties of aluminum alloy piston (values in % maximum unless shown otherwise)

Alloy designation[a]		Chemical composition, %												Hardness, H_a	Physical properties[b]				
															Tensile strength				Coefficient of thermal expansion (20 to 200°C)
															Chill casting		Forging		
Casting	Forging	Cu	Mg	Si	Fe	Mn	Ni[c]	Zn	Ti	So	Pb	Cr	Al		MPa	kpsi	MPa	kpsi	mm/mm/°C × 10^{-4}
2285	34,850	3.5–4.5	1.2–1.8	0.6	0.7	0.2	1.1–2.3	0.2	0.23	0.05	0.05		Remainder	90–130	225–275	32.7–39.8	345–410	49.8–59.7	23–24
4658	49,582	0.8–1.5	0.8–1.3	11.0–13.0	0.8	0.2	1.5	0.35	0.2	0.05	0.05			90–140	195–245	28.5–35.6	295–365	42.7–52.6	20.5–21.5
4928A	49,285	0.8–1.5	0.8–1.3	17.0–19.0	0.7	0.2	0.8–1.3	0.2	0.2	0.05	0.05			90–125	175–215	25.6–31.3	225–295	32.7–28.5	18.5–19.5
4928B		0.8–1.5	2.8–1.3	23.0–26.0	0.7	0.2	0.1–1.3	0.2	0.2	0.05	0.05	0.3–0.6		90–125	165–205	24.2–29.7			17–18

[a] Alloys have been designated in accordance with IS 6051, 1970. Code for designation of aluminum and aluminum alloys.
[b] Physical properties are attainable after suitable heat treatment.
[c] The purchaser may specify nickel content, if so desired.

Source: Bureau of Indian Standards, New Delhi.

TABLE 25–10C
Preferred number of piston rings

Differential pressure	Std. atm.	0–9	10–14	15–24	25–29	30–49	50–99	100–200
	MPa	0–0.88	0.98–1.37	1.47–2.35	2.45–2.85	2.94–4.80	4.90–9.71	9.81–19.61
	psi	0–128	142–199	213–341	355–412	426–696	710–1406	1422–2844
Minimum number of rings		2	3	4	5	6	7	8

Source: M. J. Neale, *Tribology Handbook*, Butterworth-Heinemann, London, 1973; reproduced with permission.

TABLE 25–10D
Properties of typical piston ring materials

Material	Tensile strength, σ_t		Nominal modulus of elasticity, E_n		Brinell hardness number, H_B	Bulk density, g/cm^3	Typical coefficient of expansion, α $\times 10^{-6}/°C$	Wear rating
	MPa	kpsi	GPa	Mpsi				
Metallic:								
Gray irons	230–310	33.4–45.0	83–124	12.1–18.0	210/310			Good
Carbide malleable irons	400–580	58.0–84.1	140–160	20.3–23.2	250/320			Excellent
Malleable and/or nodular irons	540–820	78.3–119.0	155–165	22.5–24.0	200/440			Poor
Sintered irons	250–390	36.5–56.6	120	17.4	130/150			Good
Nonmetallic:								
Carbon-filled PTFE	10.3	1.49				2.05	55	
Graphite/MoS$_2$-filled PTFE	19.6	2.85				2.20	115	
Resin-bonded PTFE	29.4	4.27				1.75	30	
Carbon	43.4	6.30				1.8	43	
Resin-bonded carbon	19.6	2.85				1.9	20	
Glass-filled PTFE	16.7	2.42				2.26	80	
Bronze-filled PTFE	12.8	1.85				3.90	118	
Resin-bonded fabric	110.8	16.07				1.36	22.5/87.5[a]	

[a] Material is anisotropic.
Source: M. J. Neale, *Tribology Handbook*, Butterworth-Heinemann, London, 1973, extracted with permission.

25.5 DESIGN OF SPEED REDUCTION GEARS AND VARIABLE-SPEED DRIVES

SYMBOLS

a	center distance, m (in)
	number of pinions or planetary pinion (Fig. 25–36)
A	center distance (also with subscripts) (Fig. 25–36)
	area of reduction gear housing, m^2 (in^2)
A_n	noncooled, i.e., ribbed, surface of housing of reduction gear drive, m^2 (in^2)
A_c	cooled surface of reduction gear drive, m^2 (in^2)
A_w	surface area of contact of teeth when one-fourth of all teeth of wheel in wave-type reduction gears are engaged, m^2 (in^2)
b	width of rim, m (in)
d_1	diameter of pinion, m (in)
	diameter of rigid immovable rim with internal teeth of wave-type reduction gears, m (in)
d_2	diameter of gear, m (in)
	diameter of flexible movable wheel rim with external teeth of wave-type reduction gear, m (in)
d_{max}	maximum diameter of the circumference of the belt arrangement on the V-belt of a variable-speed drive, m (in)
d_{min}	minimum diameter of the circumference of the belt arrangement on the V-belt of a variable-speed drive, m (in)
$D = \dfrac{d_{max}}{d_{min}}$	velocity control range for a V-belt drive
D_1	velocity control range for a V-belt drive with only one adjustable pulley
D_2	velocity control range for a V-belt drive with two adjustable pulleys
e	working height of a V-groove of the pulley, m (in)
F_{max}	maximum load acting on the pinion, kN (lbf)
F_m	mean load acting on the pinion, kN (lbf)
h	height of tooth, m (in)
	coefficient of heat transfer, W/m^2 K $(Btu/ft^2 h \,°F)$
h_n	coefficient of heat transfer of noncooled surface, W/m^2 K $(Btu/ft^2 \, h \,°R)$
h_c	coefficient of heat transfer of cooled surface, W/m^2 K $(Btu/ft^2 \, h \,°R)$
h_a	addendum of tooth, m (in)
h_f	dedendum of tooth, m (in)
i	transmission or speed ratio
$k_{nl} = \dfrac{F_{max}}{F_m}$	nonuniform load distribution factor
L	distance between the axes of the pinions (Fig. 25–36d)
m	module, m (in)
M_{ts}	torque acting on smaller wheel, N m (lbf in)
n	speed, rpm
n_t, n_2	speeds of pinion and gear, respectively, rpm
q	a whole number
Φ	heat generated, W (Btu/h)

r_{max}	maximum radius of the circumference of the belt arrangement on the V-belt of a variable-speed drive, m (in)
r_{min}	minimum radius of the circumference of the belt arrangement on the V-belt of a variable-speed drive, m (in)
t_1	temperature of lubricant, °C (°F)
t_a	ambient temperature, °C (°F)
z_1, z_2	number of teeth on sun pinion and planetary pinion of epicyclic gear transmission, respectively, Fig. 25–36
	number of teeth on pinion and gear, respectively
z_3	number of teeth on ring gear 3 (Fig. 25–36a)
z_s	number of teeth on smaller wheel
ω_1, ω_2	angular speed of pinion and gear, respectively, rad/s
δ	deformation, m (in)
Δ	clearance between the pinions which should be at least 1 mm (in)
α	half-cone angle of V-belt, deg
σ_{ca}	allowable compressive stress, MPa (psi)

Particular	Formula
For formulas on spur, helical, bevel, cross-helical, and worm reduction gears	Refer to Chap. 23.
Transmission or speed ratio for single reduction gear (Fig. 23–2, Chap. 23)	$i = \dfrac{\omega_1}{\omega_2} = \dfrac{n_1}{n_2} = \dfrac{d_2}{d_1} = \dfrac{z_2}{z_1}$ (25–143)
For different types of gear reduction drives	Refer to Fig. 25–35 and Table 25–11.

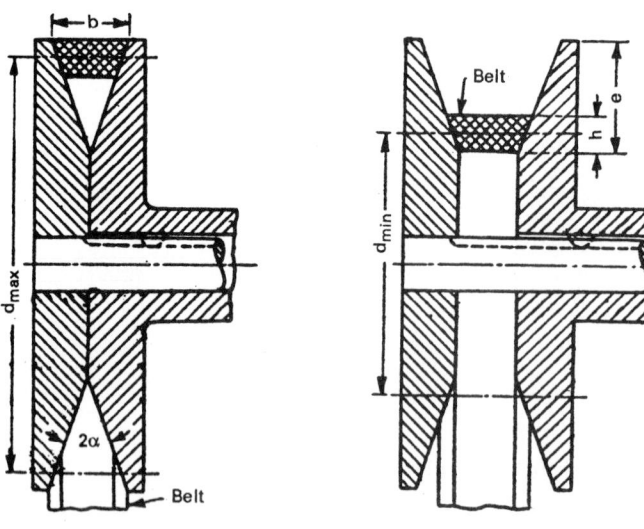

(a) V-belt at top position   **(b) V-belt at bottom position**

FIGURE 25–34   Dimension of V-belt variable-speed drive.

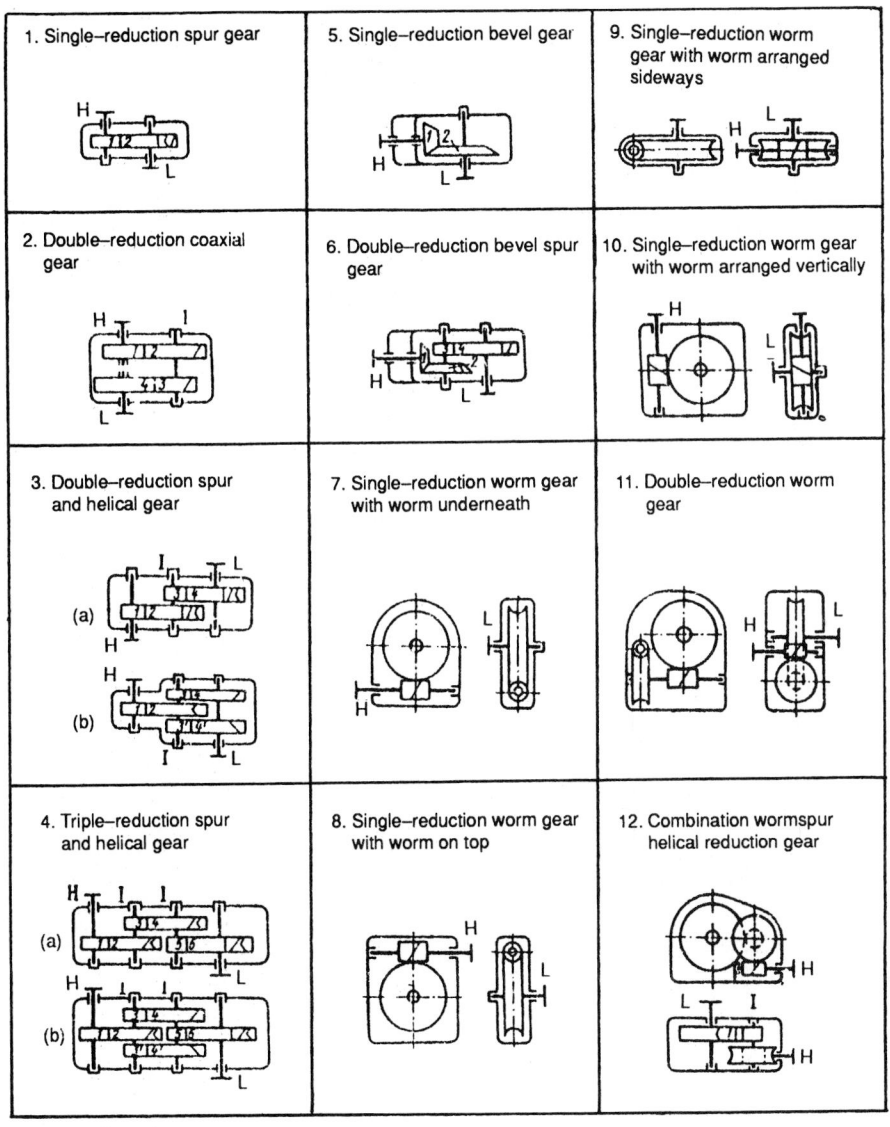

H = High Speed   I = Intermediate Speed   L = Low Speed

FIGURE 25–35   Schematic diagrams of various types of spur, helical, herringbone, bevel, and worm reduction gears.

Particular	Formula

PLANETARY REDUCTION GEARS
First condition—mating

The sum of the radii of the addendum circles of the mating pinions in planetary reduction gears should be smaller than the distance between their axes (Fig. 25–36d) so that the top of the pinions should not touch each other

$$L = 2A_{1,2} \sin \frac{\pi}{a} = z_2 m + 2m(1 + \xi) + \Delta \qquad (25\text{–}144)$$

where
a = number of pinions
Δ = clearance between the pinions, which should be at least 1 mm
$A_{1,2}$ = center distance as shown in Fig. 25–36

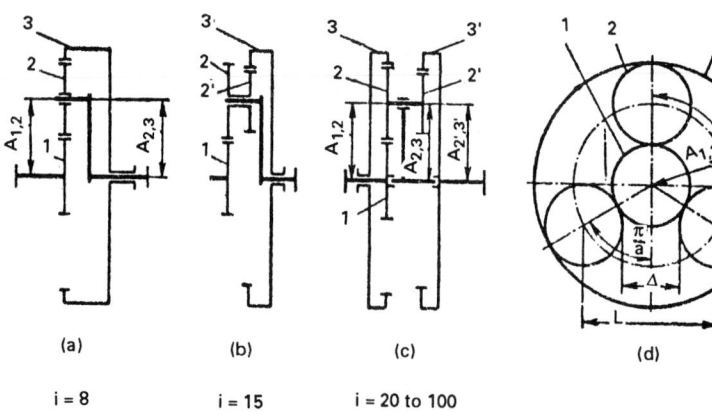

(a) (b) (c) (d)

$i = 8$ $i = 15$ $i = 20$ to 100

FIGURE 25–36   Planetary reduction gears.

Second condition—coaxiality

The center distance of each pair of wheels should be equal (Fig. 25–36)

$$A_{12} = A_{23}; \ A_{12} = A_{23} = A_{2'3'} \qquad (25\text{–}145)$$

The relationship between teeth in corrected or uncorrected gears (Fig. 25–36a)

$$z_1 + z_2 = z_3 - z_2 \qquad (25\text{–}146a)$$
or
$$z_1 + 2z_2 = z_3 \qquad (25\text{–}146b)$$

The relationship between teeth in corrected or uncorrected gears (Fig. 25–36c) to ratify two conditions

(i) First condition

Refer to Eq. (25–146).

(ii) Second condition

$$m_2(z_3 - z_2) = m_2'(z_3' - z_2') \qquad (25\text{–}147a)$$
or
$$z_3 - z_2 = z_3' - z_2' \quad \text{since} \quad m_2 = m_2' \qquad (25\text{–}147b)$$

Particular	Formula

Third condition—coincidence

The condition for the teeth and spaces of the meshed gears should coincide when the pinions are arranged uniformly over the circumference

$$\frac{z_1 + z_3}{a} = q \qquad (25\text{–}148)$$

where q is a whole number

For designing gears for strength and wear

Refer to design equations given in Chap. 23.

The moment acting on smaller wheel

$$M_{ts} = \frac{M_{t1} k_{nl}}{a} \frac{z_s}{z_1} \qquad (25\text{–}149)$$

where
$z_s = z_1$ or $z_s = z_2$ if $z_1 > z_2$
$k_{nl} = 2$ maximum value
$\quad = 1.4$ to 1.6 for gears of 7th degree of accuracy
$\quad = 1.1$ to 1.2 when floating central wheel are used to equalize the load

CONDITIONS OF PROPER ASSEMBLY OF PLANETARY GEAR TRANSMISSION
Two planetaries

Both the driving pinion (sun pinion) and the planetaries may have either an even or an odd number of teeth.

Three planetaries

If z_1 (number of teeth on sun pinion) is divisible by 3, then z_2 (number of teeth on planetary pinion) must also be divisible by 3.

If $z_2 - 1$ is divisible by 3, then $z_2 + 1$ must be divisible by 3.

If $z_1 + 1$ is divisible by 3, then $z_2 - 1$ must be divisible by 3.

Four planetaries

If z_1 is even, then z_2 must be even.

If z_1 is odd, then z_2 must be odd.

Particular	Formula

WAVE-TYPE REDUCTION GEARS

Transmission or gear ratio

$$i = \frac{z_2}{z_1 - z_2} = \frac{d_2}{d_1 - d_2} \qquad (25\text{–}150)$$

For a double-wave drive, $z_1 - z_2 = 2$.

The necessary deformation

$$\delta = d_1 - d_2 = \frac{d_2}{i} \qquad (25\text{–}151)$$

The condition for obtaining the module for the drive

$$d_1 - d_2 = (z_1 - z_2)\mathbf{m} = \delta \qquad (25\text{–}152)$$

The module of the drive from Eq. (25–152)

$$\mathbf{m} = \frac{\delta}{z_1 - z_2} = 0.5\,\delta \qquad (25\text{–}153)$$

The tooth height $\qquad h = \delta \qquad (25\text{–}154)$

The tooth addendum $\qquad h_a = 0.44\,\delta \qquad (25\text{–}155)$

The tooth dedendum $\qquad h_f = 0.56\,\delta \qquad (25\text{–}156)$

The rim width $\qquad b = 0.1 d_2 \text{ to } 0.2 d_2 \qquad (25\text{–}157)$

The total surface area of contact of teeth when one-fourth of all teeth of wheel are engaged $\qquad A_w = 0.5 h \times 0.25 z_2 b \qquad (25\text{–}158)$

The torque transmitted $\qquad M_t = 0.5 d_2 A_w \sigma_{ca} \simeq 0.06\, d_2^2\, \delta b z_2 \sigma_{ca} \qquad (25\text{–}159)$

where $\sigma_{ca} = 29.5$ MPa (4.28 kpsi) for hardened steel wheels

VARIABLE-SPEED DRIVES (FIGS. 25–34 AND 25–37, AND TABLE 25–12)

For schematic arrangements of various variable-speed drives

Refer to Figs. 25–34 and 25–37.

The velocity control range for V-belt drive with only one adjustable pulley

$$D_1 = \frac{d_{max}}{d_{min}} \qquad (25\text{–}160)$$

The relation between d_{max} and d_{min} of V-belt drive

$$d_{max} = d_{min} + 2(e - h) \qquad (25\text{–}161a)$$
$$= d_{min} + b \cot \alpha - 2h \qquad (25\text{–}161b)$$

The velocity control range for V-belt drive from Eqs. (25–160) and (25–161)

$$D = \frac{d_{max}}{d_{min}} = 1 + \frac{2e}{d_{min}} - \frac{2h}{d_{min}} \qquad (25\text{–}162a)$$

$$= 1 + \frac{b}{d_{min}} \cot \alpha - \frac{2h}{d_{min}} \qquad (25\text{–}162b)$$

The velocity control range for V-belt drive when two pulleys are adjustable

$$D_2 = D_1 \qquad (25\text{–}163)$$

The total range of velocity control of variable-speed drive of two adjustable pulleys of V-belt drive

$$D = D_1^2 \qquad (25\text{–}164)$$

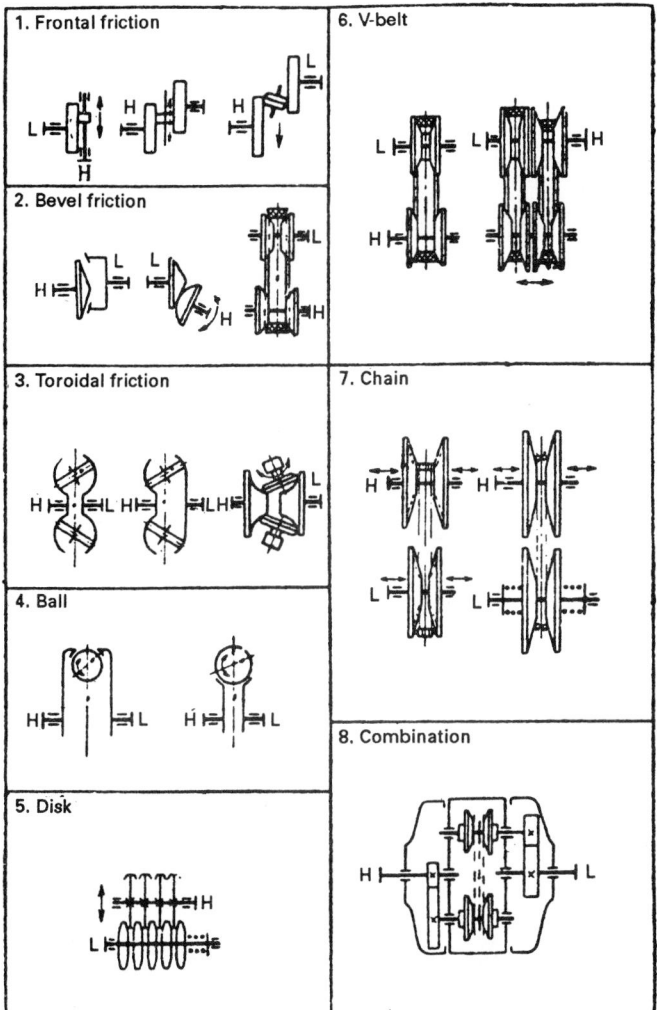

FIGURE 25–37 Variable-speed drives.

Particular	Formula	
The working height of the V-groove of the pulley	$e > \dfrac{b}{2} \cot \alpha$	(25–165)
The width of standard V-belt	$b \simeq 1.8\,h$	(25–166)
The larger ratio of width to height of specially profiled broad V-belts	$\dfrac{b}{h} \simeq 2 \text{ to } 3$	(25–167)
The total velocity control range for adjustable pulleys of V-belt drive	$D = D_1^4$	(25–168)

Particular	Formula

DISSIPATION OF HEAT IN REDUCTION GEAR DRIVES

The area of housing required for dissipating heat generated in a closed-type reduction gear drive operating in an oil bath at stable thermal equilibrium condition

$$A = \frac{\Phi}{h(t_1 - t_a)} \qquad (25\text{–}169)$$

where h = coefficient of heat transfer, which varies from 8.75 to 17.5 W/m^2 K (1.54 to 3.1 Btu/ft^2 h °R)

The thermal equilibrium condition of reduction gear drive which has a housing of noncooled surface (ribbed surface) and cooled surface (cooled by blowing of air by fan)

$$\Phi \le (h_n A_n + h_c A_c) \quad \text{W (Btu/h)} \qquad (25\text{–}170)$$

The expression for coefficient of heat transfer of the housing or reduction gear drive blown over by air

$$h_c = 12\sqrt{v} \quad \text{W/m}^2 \text{ K (Btu/ft}^2 \text{ h °R)} \qquad (25\text{–}171)$$

where v = velocity of air, m/s (ft/min)

The velocity of air which depends on impeller velocity

$v \simeq 0.005 \, n_i$ m/s (ft/min)
n_i = impeller speed, rpm

For minimum weight equations for gear systems — Refer to Table 25–13.

For total weight equations for gear systems — Refer to Table 25–14.

For K factors for preliminary estimate of spur and helical gear size — Refer to Table 25–15.

For comparison of five gear systems — Refer to Table 25–16.

TABLE 25–11
Transmission ratio (i), efficieny (η), and allowable transmitted power (P_{al}) for reduction gears

Type of reduction gear	Fig. no.	i	η	P_{al}, kW
Single- and triple-spur and helical reduction gear	25–35, serial nos. 3a, 4a	10–60		
Single-spur reduction gear	≤8–10 25–35, serial no. 1			
Single worm			108	
Helical worm			100	
Harmonic drive			100	
Planetary reduction gear	25–36a	8	0.97–0.99	
	25–36b	15	0.97–0.99	1000
	25–36c	20–100		100
Wave-type toothed reduction gear		100	0.75–0.85	

TABLE 25–12
Velocity control range (D), efficiency (η), and allowable power transmitted (P_{al}) for variable-speed drives

Particular	Type of drive	Serial no. in Fig. 25–37	D	η	P_{al}, kW
Frontal friction	Single	1	3–4		20
	Twin type		8–10		
Bevel friction	Single	2	3–4		5
	Double		4–10		
	Self-locking ring		16	0.7–0.8	10
Toroidal friction		3	4–6	0.95	20
Ball		4	10–12		
Disk drives		5	≤ 3		800
			4–5		≤ 300
V-belt drives	Solid disk	6	1.3–1.7	0.8–0.9	50
	Grooved disk		2		
Chain drives	First type of drive	7	6	0.8–0.9	30
	Second type of drive		7–10		75
Combination drives		8			6

MINIMUM AND TOTAL WEIGHT EQUATION FOR GEAR SYSTEMS

The following symbols are used in Tables 25–13 to 25–16: a = number of branches in an epicyclic gear; $C = (2M_t/K)$, m^3; d = pitch diameter, m (in); i = gear speed ratio; i_o = overall ratio; $i_s = d_p/d_s = z_p/z_s$ = speed ratio of planet gear to sun gear; j = number of idlers; K = a factor from Table 25–15; M_t = input torque, N m (lbf in); $(i_o + 1)/i_o = i_o'$.

TABLE 25–13
Minimum weight equations for gear systems

Particular	Equation
Simple train (offset)	$2i^3 + i^2 = 1$
Offset with idler	$2i^3 + i^2 = i_o^2 + 1$
Offset with two idlers	$2i^3 + i^2 = \dfrac{i_o^2 + 1}{2}$
Offset with j idlers	$2i^3 + i^2 = \dfrac{i_o^2 + 1}{j}$
Double-reduction	$2i^3 + \dfrac{2i^2}{i_o'} = \dfrac{i_o^2 + 1}{i_o'}$
Double-reduction, double branch	$2i^3 + \dfrac{2i^2}{i_o'} = \dfrac{i_o^2 + 1}{2i_o'}$
Double-reduction, four branch	$2i^3 + \dfrac{2i^2}{i_o'} = \dfrac{i_o^2 + 1}{4i_o'}$
Double-reduction, j branches	$2i^3 + \dfrac{2i^2}{i} = \dfrac{i_o^2 + 1}{ji_o'}$
Planetary (theoretical)	$2i_s^3 + i_s^2 = \dfrac{0.4(i_o - 1)^2 + 1}{a}$
Star (theoretical)	$2i_s^3 + i_s^2 = \dfrac{0.4i_o^2 + 1}{a}$

TABLE 25–14
Total weight equations for gear systems

Particular	Equation
Offset	$\Sigma(bd^2/C) = 1 + \dfrac{1}{i} + i + i^2$
Offset with idler	$\Sigma(bd^2/C) = 1 + \dfrac{1}{i} + i + i^2 + \dfrac{i_o^2}{i} + i_o^2$
Offset with two idlers	$\Sigma(bd^2/C) = \dfrac{1}{2} + \dfrac{1}{2i} + i + i^2 + \dfrac{i_o^2}{2i} + \dfrac{i_o^2}{2}$
Double-reduction	$\Sigma(bd^2/C) = 1 + \dfrac{1}{i} + 2i + i^2 + \dfrac{i^2}{i_o} + \dfrac{i_o^2}{2i} + i_o^2$
Double-reduction, double branch	$\Sigma(bd^2/C) = \dfrac{1}{2} + \dfrac{1}{2i} + 2i + i^2 + \dfrac{i^2}{i_o} + \dfrac{i_o^2}{2i} + \dfrac{i_o^2}{2}$
Double-reduction, four-branch	$\Sigma(bd^2/C) = \dfrac{1}{4} + \dfrac{1}{4i} + 2i + i^2 + \dfrac{i^2}{i_o} + \dfrac{i_o^2}{4i} + \dfrac{i_o^2}{4}$
Planetary	$\Sigma(bd^2/C) = \dfrac{1}{a} + \dfrac{1}{ai_s} + i_s + i_s^2 + \dfrac{0.4(i_o - 1)^2}{ai_s}$ $+ \dfrac{0.4(i_o - 1)^2}{a}$
Star	$\Sigma(bd^2/C) = \dfrac{1}{a} + \dfrac{1}{ai_s} + i_s + i_s^2 + \dfrac{0.4i_o^2}{ai_s} + \dfrac{0.4i_o^2}{a}$

TABLE 25–15
K factors for preliminary estimate of spur and helical gear size

Particular	Hardness, H_B; pinion gear	Pitch line velocity, m/s	K factor kgf/mm^2	K factor MN/m^2	K factor MPa
Motor driving compressor	225–180	> 20.5	0.036	0.353	0.353
Engine driving compressor	225–180	> 20.5	0.032–0.050	0.314–0.049	0.314–0.49
	575–575		0.155–0.320	1.52–3.14	1.52–3.14
Turbine driving generator	225–180	> 20.5	0.066–0.077	0.65–0.76	0.65–0.76
	575–575		0.280–0.56	2.746–5.50	2.746–5.50
Industrial drives	575–575	5.1	0.350–0.703	3.434–6.89	3.434–6.89
	350–300	5.1	0.246–0.316	2.234–3.100	2.234–3.10
	210–180	5.1	0.120–0.176	1.177–1.726	1.177–1.726
	575–575	15.3	0.334–0.527	3.277–5.170	3.277–5.170
	300–300	15.3	0.193–0.264	1.893–2.589	1.893–2.589
	210–180	15.3	0.088–0.141	0.873–0.138	0.873–0.138
Large industrial gears such as	225–180	5.1 max	0.056–0.070	0.550–0.687	0.550–0.687
hoists, kilns, and mills	260–210		0.091–0.120	0.893–1.177	0.893–1.177
Aircraft, single pair	$60R_C$–$60R_C$	51	0.703 (at take off)	6.89	6.89
Aircraft, planetary	$60R_C$–$60R_C$	15.3–51	0.492 (at take off)	4.82	4.82
Automotive transmission	$60R_C$–$60R_C$		1.055	10.35	10.35
Small commerical vehicles	350; phenolic laminated nylon	< 5.1	0.52	5.10	5.10
			0.035	0.343	0.343
Small gadgets	200; zinc alloy die casting	< 5.1	0.018	0.176	0.176
	200; brass or Al	< 2.55	0.018	0.176	0.176
	Brass or Al Brass or Al	< 2.55	0.016	0.157	0.157

TABLE 25–16
A comparison of five gear systems (all systems producing 0.746 kW at 18 rpm)

Parameter	Epicyclic	Herringbone	Single worm	Helical worm	Harmonic drive
Speed ratio	97.4	96.2	108	100	100
Safety factor	3	2	2	2	36
Height, mm	330	356	580	406	152
Length, mm	381	508	483	432	152
Width, mm	330	254	356	254	152
Cubic volume, m^3	0.0410	0.0458	0.1000	0.0442	0.003
Weight, kgf	111.60	127.00	104.33	93.00	13.61
Efficiency, η%	85	85	40	78	82
Number of gears	13	4	2	4	2
Number of bearings	17	6	6	6	2
Tooth-sliding velocity, m/s	12.75	12.75	7.65	12.75	0.143
Pitch line velocity, m/s	7.65	7.65	7.65	7.65	0.092
Tooth contact pressure, kgf/mm^2	35	35	3.5	35	0.425
GPa	0.343	0.343	0.034	0.343	0.0042
Tooth in contact, %	7	5	2	3	50
Tooth contact	Line	Line	Line	Line	Surface
Quiet operation	No	Yes	Yes	Yes	Yes
Balanced forces	Yes	No	No	No	Yes

25.6 FRICTION GEARING

SYMBOLS

a	center distance, m (in)
	dimensions as shown in Fig. 25–42
b	gear face width, m (in)
d_1	diameter of smaller wheel, m (in)
d_2	diameter of larger wheel, m (in)
F	pressure on wheels, kN (lbf)
F_a	thrust, kN (lbf)
F_r	radial force on the grooved spur wheel for each groove, kN (lbf)
F_R	normal reaction between two bevel friction gears (Fig. 25–40), kN (lbf)
F_t	tangential force, kN (lbf)
h	depth of groove, m (in)
i	number of grooves, m (in)
n'	speed, rps
n	speed, rpm
P	power transmitted, kW (hp)
p'	permissible pressure, kN/m (lbf/in)
v_m	mean circumferential velocity, m/s (ft/min)
R	cone distance, m (in) (Fig. 25–40)
α	half the included angle of the groove, deg ranges from 12° to 18° (should not exceed 20°)
ρ	angle of friction, deg
μ	coefficient of friction between wheels
μ'	coefficient of friction between shaft of wheel and bearings
ω_1, ω_2	angular speeds of smaller and larger wheels, respectively, rad/s
δ_1, δ_2	cone center angles of smaller and larger wheels, respectively, deg

Particular	Formula

SPUR FRICTION GEARS
Plain spur friction wheels (Fig. 25–38)

Particular	Formula	
The radial pressure on the wheels	$F = bp'$	(25–172)
The tangential force due to radial pressure F	$F_t = \mu bp'$	(25–173)
The power transmitted	$P = \dfrac{F_t v_m}{1000}$ **SI**	(25–174a)

where P in kW, F_t in N, and v_m in m/s

$$= \frac{F_t v_m}{33,000} \quad \textbf{US Customary units} \qquad (25\text{–}174b)$$

where P in hp, F_t in lbf, and v_m in ft/min

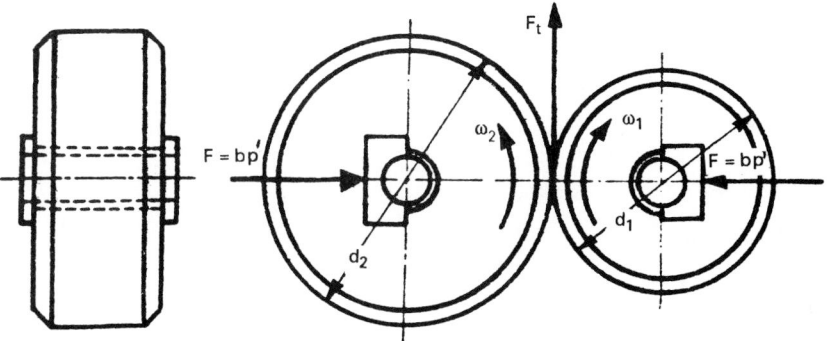

FIGURE 25–38   Plain spur friction gears.

Particular	Formula
	$= \dfrac{F_t v_m}{75}$ **Metric** (25–174c)
	where P in hp$_m$, F_t in kgf, and v_m in m/s
The gear face width	$b = \dfrac{1000P}{\pi \mu p' d n'}$ **SI** (25–175a)
	where P in kW, p' in N/m, n' in rps, and b and d in m
	$= \dfrac{102 \times 10^3 P}{\pi \mu p' d n'}$ **Metric** (25–175b)
	where P in kW, p' in kgf/mm, n' in rps, and b and d in mm
	$= \dfrac{33,000P}{\pi \mu p' d n}$ **US Customary units** (25–175c)
	where P in hp, p' in lbf/in, n in rpm, and b in in and d in ft
	$= \dfrac{126,000P}{\mu p' d n}$ **US Customary units** (25–175d)
	where b and d in in and p' in lbf/in n in rpm and p in hp.

Grooved spur friction wheel (Fig. 25–39)

The radial force on the wheel for each groove	$F_r = 2p'h(\tan \alpha + \mu)$	(25–176)
The total tangential force	$F_t = 2\mu i p' h \sec \alpha$	(25–177)
The power transmitted	$P = \dfrac{2\pi i \mu h p' d_1 n'}{1000 \cos \alpha}$ **SI**	(25–178a)

where P in kW, p' in N/m, n' in rps, and h and d_1 in mm

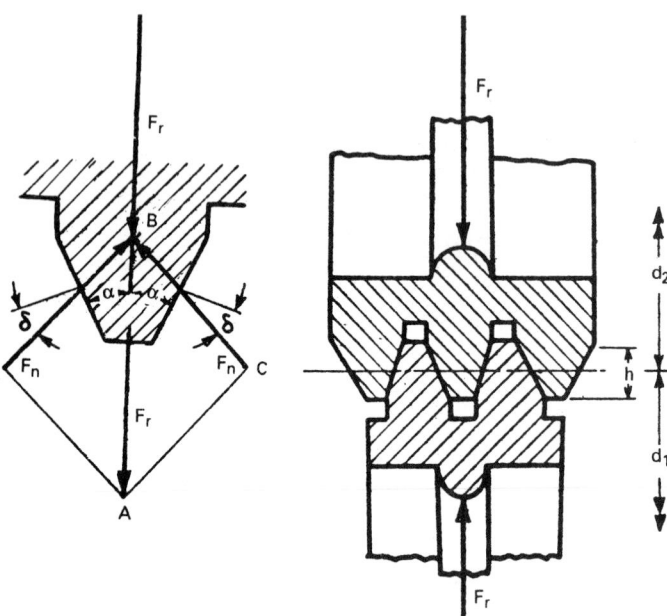

FIGURE 25–39 Grooved spur friction gears.

Particular	Formula
	$$= \frac{i\mu hp'd_1 n}{63,000 \cos \alpha} \qquad \textbf{US Customary units} \qquad (25\text{–}178b)$$ where P in hp, p' in lbf/in, n' in rps, and h and d_1 in in
	$$= \frac{2\pi i \mu h p' d_1 n}{4500 \times 10^3 \cos \alpha} \qquad \textbf{Metric} \qquad (25\text{–}178c)$$ where P in hp_m, p' in kgf/mm, n in rpm, and h and d_1 in mm
The empirical relation for the depth of the groove	$h = 0.006d_1 + 4$ mm (0.15 in) $\qquad (25\text{–}179)$
The recommended value for the mean circumferential velocity	$v_m \geq 6 + 0.08 d_1 \qquad \textbf{SI} \qquad (25\text{–}180)$ where v_m in m/s and d_1 in m $\geq 1200 + 4d_1 \qquad \textbf{US Customary units} \quad (25\text{–}180a)$ where v in ft/min

Particular	Formula

BEVEL FRICTION GEARS (FIG. 25–40)
Starting

The reaction is inclined from the normal by an angle of friction ρ

$$F'_R = \frac{F'_{1a}}{\sin(\delta_1 + \rho)} = \frac{F'_{2a}}{\cos(\delta_1 + \rho)} \qquad (25\text{–}181)$$

The tangential force transmitted

$$F'_t = \mu F'_R \cos \rho = \frac{1000\,P}{v_m} \quad \textbf{SI} \qquad (25\text{–}182a)$$

$$= \frac{75\,P}{v_m} \quad \textbf{Metric} \qquad (25\text{–}182b)$$

The least axial thrust on the small wheel

$$F'_{1a} = \frac{1000P(\sin \delta_1 + \mu \cos \delta_1)}{\mu v_m} \quad \textbf{SI} \qquad (25\text{–}183a)$$

$$= \frac{33,000(\sin \delta_1 + \mu \cos \delta_1)P}{\mu v_m} \quad \textbf{US Customary units}$$

$$(25\text{–}183b)$$

The least axial thrust on the big wheel

$$F'_{2a} = \frac{1000P(\cos \delta_1 - \mu \sin \delta_1)}{\mu v_m} \quad \textbf{SI} \qquad (25\text{–}184a)$$

$$= \frac{33,000P(\cos \delta_1 - \mu \sin \delta_1)}{\mu v_m} \quad \textbf{US Customary units}$$

$$(25\text{–}184b)$$

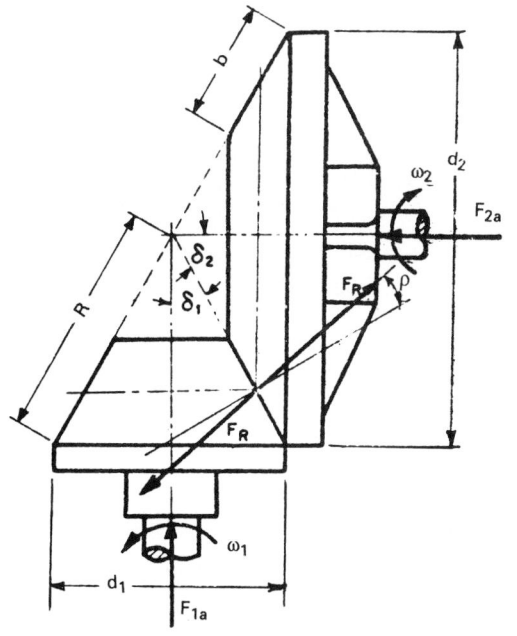

FIGURE 25–40 Bevel friction gears.

Particular	Formula

Running

The reaction in this case is designated by $F_R \leq bp'$ (where p' is the permissible unit pressure)

$$F_R = \frac{F_{1a}}{\sin \delta_1} = \frac{F_{2a}}{\cos \delta_1} \qquad (25\text{–}185)$$

The tangential force transmitted

$$F_t = \mu F_R = \frac{1000P}{v_m} \quad \textbf{SI} \qquad (25\text{–}186a)$$

$$= \frac{33,000P}{v_m} \quad \textbf{US Customary units}$$

$$(25\text{–}186b)$$

The least axial thrust on the small wheel

$$F_{1a} = \frac{1000P \sin \delta_1}{\mu v_m} = \frac{1000P}{\mu v_m}\left[\frac{d_1}{\sqrt{d_1^2 + d_2^2}}\right] \quad \textbf{SI}$$

$$(25\text{–}187a)$$

$$= \frac{33,000P \sin \delta_1}{\mu v_m} = \frac{33,000}{\mu v_m}\left[\frac{d_1}{\sqrt{d_1^2 + d_2^2}}\right]P$$

$$\textbf{US Customary units} \qquad (25\text{–}187b)$$

The least axial thrust on the big wheel

$$F_{2a} = \frac{1000P \cos \delta_1}{\mu v_m} = \frac{1000P}{\mu v_m}\left[\frac{d_2}{\sqrt{d_1^2 + d_2^2}}\right] \quad \textbf{SI}$$

$$(25\text{–}188a)$$

$$= \frac{33,000P \cos \delta_1}{\mu v_m} = \frac{33,000}{\mu v_m}\left[\frac{d_2}{\sqrt{d_1^2 + d_2^2}}\right]P$$

$$\textbf{US Customary units} \qquad (25\text{–}188b)$$

DISK FRICTION GEARS (FIG. 25–41)

The torque on the driving shaft

$$M_t = \frac{1000P}{\omega} \quad \textbf{SI} \qquad (25\text{–}189a)$$

where M_t in N m, P in kW, and ω in rad/s

$$= \frac{9550P}{n} \quad \textbf{SI} \qquad (25\text{–}189b)$$

where M_t in N m, P in kW, and n in rpm

$$= \frac{159P}{n'} \quad \textbf{SI} \qquad (25\text{–}189c)$$

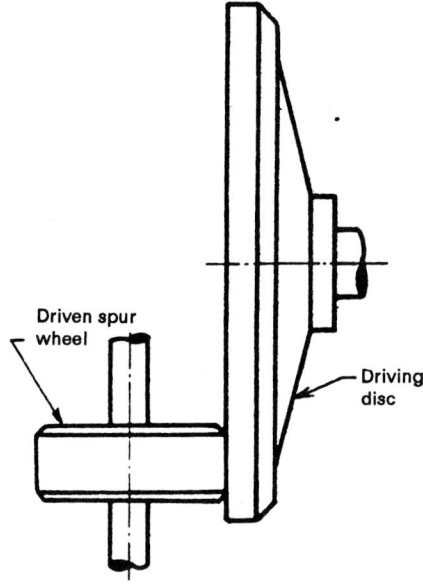

Driven spur wheel

Driving disc

FIGURE 25–41 Variable-speed disk friction gearing.

Particular	Formula

where M_t in N m, P in kW, and n' in rps

$$= \frac{716,000P}{n} \quad \textbf{Metric} \qquad (25\text{–}189\text{d})$$

where M_t in kgf mm, P in hp_m, and n in rpm

$$= \frac{63,000P}{n} \quad \textbf{US Customary units} \qquad (25\text{–}189\text{e})$$

where M_t in lbf in, P in hp, and n in rpm

The tangential force acting on the driven wheel for the minimum speed at minimum diameter of driving disk

$$F_{t1} = \frac{1000P}{\pi d_1 n'} \quad \textbf{SI} \qquad (25\text{–}190\text{a})$$

where F_{t1} in N, P in kW, d_1 in m, and n' in rps

$$= \frac{33,000P}{\pi d_1 n} \quad \textbf{US Customary units} \qquad (25\text{–}190\text{b})$$

where F_{t1} in lbf, P in hp, d_1 in ft, and n in rpm

$$= \frac{102 \times 10^3 P}{\pi d_1 n'} \quad \textbf{Metric} \qquad (25\text{–}190\text{c})$$

where F_{t1} in kgf, P in kW, d_1 in mm, and n' in rps

Particular	Formula	
The tangential force acting on the driven wheel for the maximum speed at maximum diameter of driving disk	$F_{t2} = \dfrac{1000P}{\pi d_2 n'}$ **SI**	(25–191a)
	$= \dfrac{33,000P}{\pi d_2 n}$ **US Customary units**	(25–191b)
	where d_2 in ft	
	$= \dfrac{102 \times 10^3 P}{\pi d_2 n'}$ **Metric**	(25–191c)
The minimum thrust to be applied to the disk for the minimum speed	$F_{a1} = \dfrac{F_{t1}}{\mu} = \dfrac{1000P}{\mu \pi d_1 n'}$ **SI**	(25–192a)
	$= \dfrac{33,000P}{\mu \pi d_2 n}$ **US Customary units**	(25–192b)
	where $F_{a1} = bp'$ (b = face width of driven cylindrical wheel) and d_2 in ft	
The maximum thrust to be applied to the disk for maximum speed	$F_{2a} = \dfrac{F_{t2}}{\mu} = \dfrac{1000P}{\mu \pi d_2 n'}$ **SI**	(25–193a)
	$= \dfrac{33,000P}{\mu \pi d_2 n}$ **US Customary units**	(25–193b)
	where d_2 in ft	
	$= \dfrac{126,000P}{\mu n d_2}$ **US Customary units**	(25–193c)
	where d_2 in in and F_{2a} in lbf, n in rpm, and P in hp	
The axial thrust required to shift the driven wheel under load	$F_a = F_1(\mu + \mu')$	(25–194)
	where μ' is the coefficient of friction between the shaft of driven wheel and its bearings	
The efficiency	$\eta = \dfrac{d}{d + b}$	(25–195)
	where η varies from 0.6 at low speeds when $d = d_1$ to 0.8 at high speeds, when $d = d_2$	
The minimum force available on the chain sprocket at minimum speed of driven wheel	$F_{1cs} = \dfrac{\eta F_{t1} d}{d_3}$	(25–196)
	where d = diameter of driven wheel, m (in) d_3 = diameter of chain sprocket, m (in)	
The maximum force available on the chain sprocket at maximum speed of driven wheel	$F_{2cs} = \dfrac{\eta F_{t2} d}{d_3}$	(25–197)

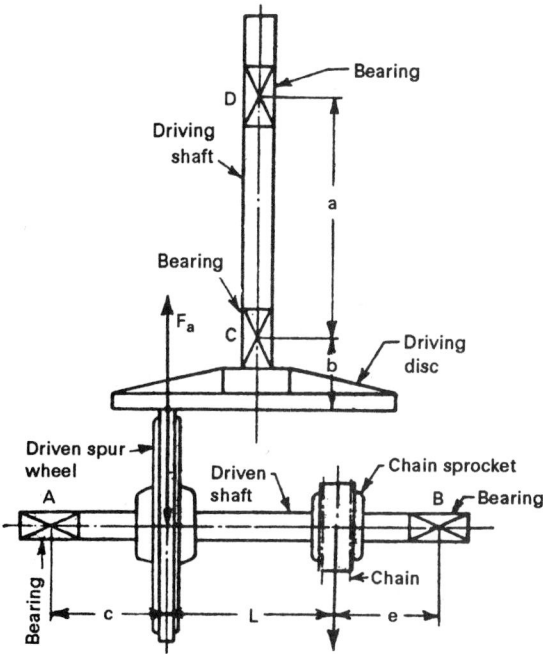

FIGURE 25–42   Bearing loads of disk friction gearing.

Particular	Formula

BEARING LOADS OF FRICTION GEARING (FIG. 25–42, TABLE 25–17)

Driven shaft

Particular	Formula	
The horizontal force on bearing A due to the tangential force F_t	$F_{hA} = \dfrac{(L+e)F_t}{(e+L+c)}$	(25–198)
The vertical force on bearing A due to thrust F_a and the force on the chain sprocket F_{cs}	$F_{VA} = \dfrac{(L+e)F_a + eF_{cs}}{(e+L+c)}$	(25–199)
The resultant load on bearing A	$F_{RA} = \sqrt{F_{hA}^2 + F_{VA}^2}$	(25–200)
The horizontal force on bearing B due to the tangential force F_t	$F_{hB} = \dfrac{cF_t}{(e+L+c)}$	(25–201)
The vertical force on bearing B due to the thrust F_a and the force on the chain sprocket F_{cs}	$F_{VB} = \dfrac{cF_a + (c+L)F_{cs}}{(e+L+c)}$	(25–202)
The resultant force on bearing B	$F_{RB} = \sqrt{F_{hB}^2 + F_{VB}^2}$	(25–203)

Particular	Formula

Driving shaft

The horizontal force due to thrust F_a on bearing D

$$F_{hDa} = \frac{d_1 F_t}{2a} \qquad (25\text{--}204)$$

where d_1 and d_2 denote the minimum and maximum diameters of driving disk

The horizontal force due to the tangential force F_t on the bearing D

$$F_{hDt} = \frac{bF_t}{a} \qquad (25\text{--}205)$$

The resultant force on the bearing D

$$F_{RD} = \sqrt{F_{hDa}^2 + F_{hDt}^2} \qquad (25\text{--}206)$$

The horizontal force due to thrust F_a on the bearing C

$$F_{hca} = \frac{d_1 F_t}{2a} \qquad (25\text{--}207)$$

The horizontal force due to the tangential force F_t on the bearing C

$$F_{hct} = \frac{(a+b)F_t}{a} \qquad (25\text{--}208)$$

The resultant force on the bearing C

$$F_{Rc} = \sqrt{F_{hca}^2 + F_{hct}^2} \qquad (25\text{--}209)$$

TABLE 25–17
Design data for friction gearing

Material of driver	Allowable pressure, p'		Coefficient of friction μ with cast iron	Material of driver	Allowable pressure, p'		Coefficient of friction μ with cast iron	Coefficient of friction, μ, with aluminum
	kN/m	lbf/in			kN/m	lbf/in		
Cast iron	530	3000	0.15	Leather	26.5	150	0.09	0.13
Cork composition	8.9	50	0.21	Leather fiber	42.2	240	0.18	0.18
Paper	26.5	150	0.15	Straw fiber	26.5	150	0.15	0.16
Rubber	17.7	100	0.20	Sulfite fiber	24.5	140	0.20	0.19
Wood	26.5	150	0.15	Tarred fiber	44.1	250	0.28	0.28

25.7   MECHANICS OF VEHICLES

SYMBOLS

a	center distance, m (in)
	a constant in Eq. (25–216b)
A	frontal projected area of vehicle, m^2 (ft^2)
b	face width of gear, m (in)
	a constant in Eq. (25–216b)
B	width of bearing, m (in)
c	distance between adjacent rotating parts, m (in)
C	constant (also with suffixes)
D_t	maximum diameter of torus, m (in)
D_w	diameter of wheel, m (in)
E_f	flow loss in each member of hydraulic torque converter, N m (lbf in)
E_{sh}	shock loss in each member of hydraulic torque converter, N m (lbf in)
F	driving force at the tire, kN (lbf)
F_{max}	maximum permissble load on the pitch circle of any particular pair of gears, kN (lbf)
G	gradient
h	thickness of housing, m (in)
i	gear ratio (total)
k	a constant
l	distance between support bearings on a shaft in gearbox, m (in)
l'	distance between bearings of overhanging shaft, m (in)
l_1	distance of rotating part from the bearing, m (in)
l_2	distance of bearing from the wall, m (in)
l_3	cap height from bolt to end, m (in)
l_4	distance of rotating parts from the bearing cap, m (in)
l_5	width of boss of rotating parts, m (in)
l_6	distance of coupling to cap, m (in)
l_7	distance between gear and shaft, m (in)
l_8	distance of rotating parts from inner wall of housing, m (in)
m	module, m (in)
M_t	output torque of the engine, N m (lbf in)
M_{tt}	torque at the tire surface, N m (lbf in)
M_{ti}	the input torque, N m (lbf in)
M_{to}	the reaction to the output torque, which is opposite in direction to output torque, N m (lbf in)
M_{tf}	the torque that must be applied to transmission housing to balance the moments of internal friction, oil churning, etc., N m (lbf in)
M_{tr}	the torque reaction of the transmission housing due to the gear reduction in transmission, N m (lbf in)
n	speed, rpm
n'	speed, rps
n_i	speed of driving shaft, rpm
n_o	speed of driven shaft, rpm
P	power, kW (hp)
r	radius of the driving wheel, m (in)

r_{mi}	mean radius of inflow to the runner, m (in)
r_{mo}	mean radius of outflow from the runner, m (in)
R_a	air resistance, kN (lbf)
R_r	rolling resistance, kN (lbf)
R_r''	road resistance, kN/tf (lbf/ton)
R_g	gradient resistance, kN (lbf)
R_t	total resistance, kN (lbf)
t	tonne, t
t_f	tonne force, tf
v	velocity, m/s (ft/min)
V	speed of vehicle, km/h (ft/s)
V_f	velocity of fluid relative to the vane, m/s (ft/min)
V_{sh}	shock velocity, m/s (ft/min)
W	weight of the vehicle, kN (Tonf)
z	number of teeth
α	angle of inclination of road, deg
ϕ	angle of repose, deg
Δ	minimum clearance between gears and inner wall of housing, m (in)
η	transmission efficiency

SUFFIXES

1	pinion
2	gear
b	brake
t	tonne
max	maximum
min	minimum

Other factors in performance or in special aspects which are included from time to time in this section and, being applicable only in their immediate context, are not given at this stage.

Particular	Formula

CALCULATION OF POWER

Torque

$$M_t = \frac{1000P}{\omega} \quad \textbf{SI} \qquad (25\text{--}210a)$$

where P in kW, ω in rad/s, and M_t in N m

$$= \frac{9550P}{n} \quad \textbf{SI} \qquad (25\text{--}210b)$$

where P in kW, n in rpm, and M_t in N m

$$= \frac{63,000P}{n} \quad \textbf{US Customary units} \qquad (25\text{--}210c)$$

where P in hp, n in rpm, and M_t in lbf in

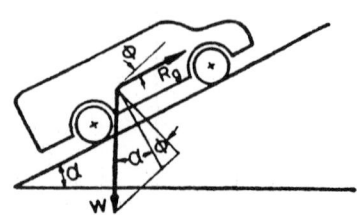

FIGURE 25–43 Forces on the vehicle moving up the gradient.

Particular	Formula
Torque at the tire surface	$M_{tt} = \eta M_t$ (25–211) where η = 0.90 at top gear = 0.80 at other gears
The driving force at the tire	$F = \dfrac{\eta M_t}{r}$ (25–212)
Tractive factor	$f_{tr} = \dfrac{M_{tt}}{1000\,W}$ **SI** (25–213a) $f_{tr} = \dfrac{M_{tt}}{2240\,W}$ **US Customary units** (25–213b)
Force required to pull the vehicle of weight W up the slope (Fig. 25–43)	$R_g = \dfrac{1000\,W \sin(\alpha + \phi)}{\cos \phi}$ **SI** (25–214a) $R_g = \dfrac{2240\,W \sin(\alpha + \phi)}{\cos \phi}$ **US Customary units** (25–214b)
Gradient	$G = \dfrac{W}{R_g}$ (25–215)
The air resistance	$R_a = kAV^2$ (25–216a) where k = constant obtained from Table 25–18
For values of air resistance at different speeds of vehicle	Refer to Table 25–19.
The rolling resistance	$R_r = (a + bV)W$ (25–216b) where a = constant varies from 15 to 600 b = constant varies from 0.1 to 3.5
For rolling or road resistance R_r' for various road surfaces	Refer to Table 25–20.
The general formula for total resistance or tractive resistance (Fig. 25–44)	$R_t = kAV^2 + W\,\dfrac{\sin(\alpha + \phi)}{\cos \phi} + (a + bV)W$ (25–217a)
Another formula for total resistance	$R_t = R_a + R_g + R_r = W\left(R_r' + \dfrac{1000}{G}\right) + kAV^2$ **SI** (25–217b) where k and R_r' are obtained from Tables 25–18 and 25–20 where R_r' in N/tf, W in tf, A in m^2, V in m/s

Particular	Formula

$$R_t = R_a + R_g + R_r = W\left(R_r' + \frac{2240}{G}\right) + kAV^2$$

US Customary units   (25–217c)

where R_r' in lbf/t, W = weight of vehicle, tonf

A = projected frontal area of vehicle, ft^2
V = speed of vehicle, ft/s

Tractive effort at the tire surface

$$F_{tr} = \frac{i\eta M_t}{r} \qquad (25\text{–}218)$$

where i = gear ratio obtained from Table 25–21

The speed of the vehicle

$$V = 0.00297 \, \frac{nD_w}{i} \quad \textbf{US Customary units} \qquad (25\text{–}219a)$$

where V in mph (miles per hour), D_w in in, and n in rpm

$$= 0.052 \, \frac{nD_w}{i} \qquad \textbf{SI} \qquad (25\text{–}219b)$$

where V in m/s, D_w in m, and n in rpm

Power

$$P = \frac{0.002 V M_{tt}}{D_w} \qquad \textbf{SI} \qquad (25\text{–}220a)$$

where V in m/s, M_{tt} in N m, D_w in m, and P in kW

$$= \frac{0.00163 V M_{tt}}{D_w} \qquad \textbf{US Customary units} \quad (25\text{–}220b)$$

where V in mph, M_{tt} in lbf ft, D_w in in, and P in hp

$$= \frac{5.5 V M_{tt}}{D_w} \qquad \textbf{Metric} \qquad (25\text{–}220c)$$

where V in km/h, M_{tt} in kgf m, D_w in mm, and P in kW

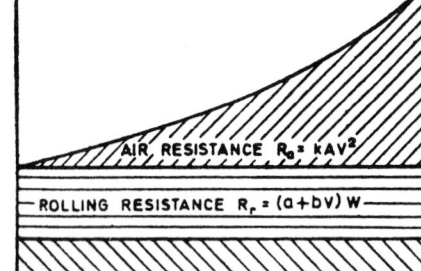

FIGURE 25–44   Various resistances on the moving vehicle.

TRANSMISSION GEARBOX (FIG. 25–45)

The equation for center distance between main and countershafts for the case of three-speed passenger car

$$a = 0.5\sqrt[3]{M_t} \qquad \textbf{US Customary units} \qquad (25\text{–}221a)$$
where a in in and M_t in lbf ft
$$= 0.0106\sqrt[3]{M_t} \qquad \textbf{SI} \qquad (25\text{–}221b)$$
where a in m and M_t in N m

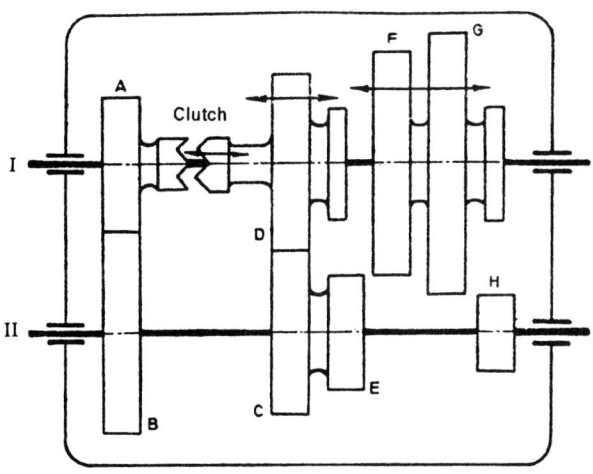

FIGURE 25–45 A typical four-speed gearbox.

Particular	Formula
The distance between support bearings of shaft	$l = 0.0254\sqrt[3]{M_t}$ to $0.0318\sqrt[3]{M_t}$ **SI** (25–222a) where l in m and M_t in N m $= 1.2\sqrt[3]{M_t}$ to $1.5\sqrt[3]{M_t}$ **US Customary units** (25–222b) where l in in and M_t in lbf ft
The maximum permissible load at the pitch circle of any pair of gears	$F_{max} = c_1 bm = \dfrac{c_1 b}{P_n}$ (25–223) where c_1 = constant obtained from Table 25–22
The face width of gear tooth	$b = \dfrac{F_{max}}{mc_1} = \dfrac{F_{max}P_n}{c_1}$ (25–224)
The expression for center distance for the case of four-speed truck transmission	$a = 0.017\sqrt[3]{M_t}$ **SI** (25–225a) where a in m and M_t in N m $= 0.8\sqrt[3]{M_t}$ **US Customary units** (25–225b) where a in in and M_t in lbf ft
The distance between support of bearings of shaft	$l = 0.0254\sqrt[3]{M_t}$ to $0.0318\sqrt[3]{M_t}$ **SI** (25–226a) where l in m, and M_t in N m $= 1.2\sqrt[3]{M_t}$ to $1.5\sqrt[3]{M_t}$ **US Customery units** (25–226b) where l in in and M_t in lbf ft
The face width of gear tooth	$b = \dfrac{F_{max}}{mc_1} = \dfrac{F_{max}P_n}{c_1}$ (25–227) For values of c_1, refer to Table 25–22.

Particular	Formula
The expression for center distance for the case of five-speed and reverse truck transmission	$a = 0.0170\sqrt[3]{M_t}$ **SI** (25–228a) where a in m and M_t in N m $= 0.8\sqrt[3]{M_t}$ **US Customary units** (25–228b) where a in in and M_t in lbf ft
The distance between support of bearings of shaft	$l = 0.0254\sqrt[3]{M_t}$ to $0.0318\sqrt[3]{M_t}$ **SI** (25–229a) where l in m and M_t in N m $= 1.2\sqrt[3]{M_t}$ to $1.5\sqrt[3]{M_t}$ **US Customary units** (25–229b) where l in in and M_t in lbf ft
The face width of gear tooth	$b = \dfrac{F_{max}}{mc_1} = \dfrac{F_{max}P_n}{c_1}$ (25–230) For values of c_1, refer to Table 25–22.
The expression for center distance for a farm tractor transmission	$a = 0.021\sqrt[3]{M_t}$ **SI** (25–231a) where a in m and M_t in N m $= \sqrt[3]{M_t}$ **US Customary units** (25–231b) where a in in and M_t in lbf ft
Effective face width of gear tooth	$b = \dfrac{F_{max}v}{28 \times 10^5 m}$ **SI** (25–232a) where b in m, F_{max} in N, m in m, and v in m/s $= \dfrac{F_{max}vP_n}{8,000,000}$ **US Customary units** (25–232b) where b in in, F_{max} in lbf, P_n in in^{-1}, and v in ft/min
The efficiency of transmission	$\eta = \dfrac{n_o M_{to}}{n_i M_{ti}} = \dfrac{M_{to}}{i_r[M_{to} - (M_{tr} - M_{tf})]}$ (25–233) where i_r = reduction ratio of transmission $= \dfrac{n_i}{n_o}$
Distance of rotating parts from the inner wall of housing	l_8 = 10 to 15 mm or more for high-power and heavy-duty operation (25–234) $= 0.4$ to 0.6 in **US Customary units**
Distance between adjacent rotating parts	c = 10 to 15 mm (0.4 to 0.6 in) (25–235)
Minimum clearance between gears and inner wall of housing	$\Delta \geq 1.2h$ (25–236) where h = thickness of housing
Distance between bearings of overhanging shaft	$l' = 1.2d$ to $3d$ (25–237) where d = diameter of shaft
Distance of bearing from the wall	l_2 = 5 to 10 mm (0.2 to 0.4 in) (25–238)
Cap height from bolt end	l_3 = depends on the design by empirical formula
Distance of rotating parts from the bearing cap	l_4 = 15 to 20 mm (0.6 to 0.8 in) (25–239)

Particular	Formula
Width of boss of rotating part	$l_5 = 1.2d$ to $1.5d$ (25–240)
Distance of coupling to cap	(depends on the type of coupling)
Distance between gear and shaft	$l_7 \geq 20$ mm (0.8 in) (25–241)
Distance of rotating part from the bearing	$l_1 = \dfrac{B}{2} + l_3 + l_4 + \dfrac{l_5}{2}$ (25–242)
For planetary gear transmission	Refer to Chap. 25, Section 25.5.
For detail design equations of spur, helical, bevel, crossed-helical and worm gears	Refer to to Chap. 23.

HYDRAULIC COUPLING (FIG. 25–46)

Torque transmitted by the coupling	$M_t = ksn^2 W(r_{mo}^2 - r_{mi}^2)$ (25–243) where k = coefficient $= 1.42 \times 10^{-7}$ (approx.)
Percent slip between primary and secondary speeds	$s = \dfrac{(n_p - n_s)}{n_p} \times 100$ (25–244) where n_p and n_s are primary and secondary speeds of impeller, respectively, rpm

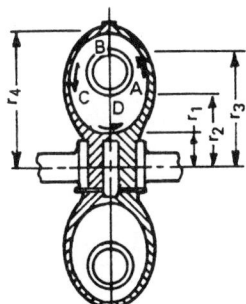

FIGURE 25–46 Hydraulic coupling.

The mean radius of the inner passage (Fig. 25–46)	$r_{mi} = \dfrac{2}{3}\left(\dfrac{r_2^3 - r_1^3}{r_2^2 - r_1^2}\right)$ (25–245)
The mean radius of the outer passage (Fig. 25–46)	$r_{mo} = \dfrac{2}{3}\left(\dfrac{r_4^3 - r_3^3}{r_4^2 - r_3^2}\right)$ (25–246)
The expression for number of times the fluid circulates through the torus in one second	$i_f = \dfrac{13,000 M_t}{nW(r_{mo}^2 - r_{mi}^2)}$ (25–247)
The torque capacity of hydraulic coupling at a given slip	$M_t = Kn^2 D_t^5$ (25–248)

where
K = coefficient varying from
    0.166×10^8 to 0.244×10^8 **SI**
    $= 1.56$ to 2.28 **US Customary units**
D_t = diameter of torus, m (ft)
M_t = torque capacity, N m (lbf ft); n in rpm

Particular	Formula

HYDRODYNAMIC TORQUE CONVERTER (FIG. 25–47)

The equation for input torque

$$M_{ti} = K n_i^2 D_t^5 \qquad (25\text{–}249)$$

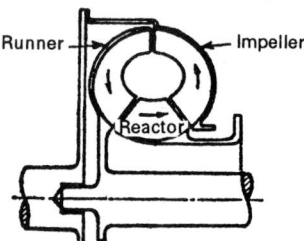

where
K = coefficient depending on design
n_i = speed of input shaft, rpm
D_t = any linear dimension such as maximum diameter of impeller

FIGURE 25–47   Hydrodynamic torque converter.

The equation for the input power

$$P = C n_i^3 D_t^5 \qquad (25\text{–}250)$$
where C = coefficient depending on design

The expression for flow loss or friction loss in each member of the torque converter under any particular operating conditions in energy unit per kilogram of fluid circulated

$$E_f = \frac{C_f V_f^2}{2g} \qquad (25\text{–}251)$$

where

C_f = coefficient whose value depends mainly on the Reynolds number and the relative smoothness of the metallic surface
= 0.445 to 0.890   **SI**
(where E_f in N m and V_f in m/s)
= 0.328 to 0.656   **US Customary units**
(where V_f in ft/s and E_f in lbf ft)

The expression for shock loss per kg fluid circulated in the impeller of a torque converter

$$E_{sh} = \frac{C_{sh} V_{sh}^2}{2g} \qquad (25\text{–}252)$$

where C_{sh} = coefficient

The maximum inside diameter of torus

$$D_t = 0.00135 C \sqrt[3]{M_t / n'^2} \qquad \textbf{SI} \qquad (25\text{–}253a)$$
where D_t in m, M_t in N m, and n' in rps

$$0.00168 C \sqrt[3]{\frac{M_t}{n^2}} \quad \textbf{US Customary units} \qquad (25\text{–}253b)$$

where D_t in in, M_t in lbf in, and n in rpm

C = coefficient = 14 for a ratio of minimum inside diameter to maximum diameter of torus of one-third
n = speed in hundreds of rpm

Particular	Formula

TRACTIVE EFFORT CURVES FOR CARS, TRUCKS, AND CITY BUSES

For finding the diameter of tire of vehicles for a particular wheel speed   Refer to Fig. 25–48.

For tractive effort of a passenger car   Refer to Fig. 25–49.

For tractive effort of trucks, tractors, and city buses   Refer to Fig. 25–50.

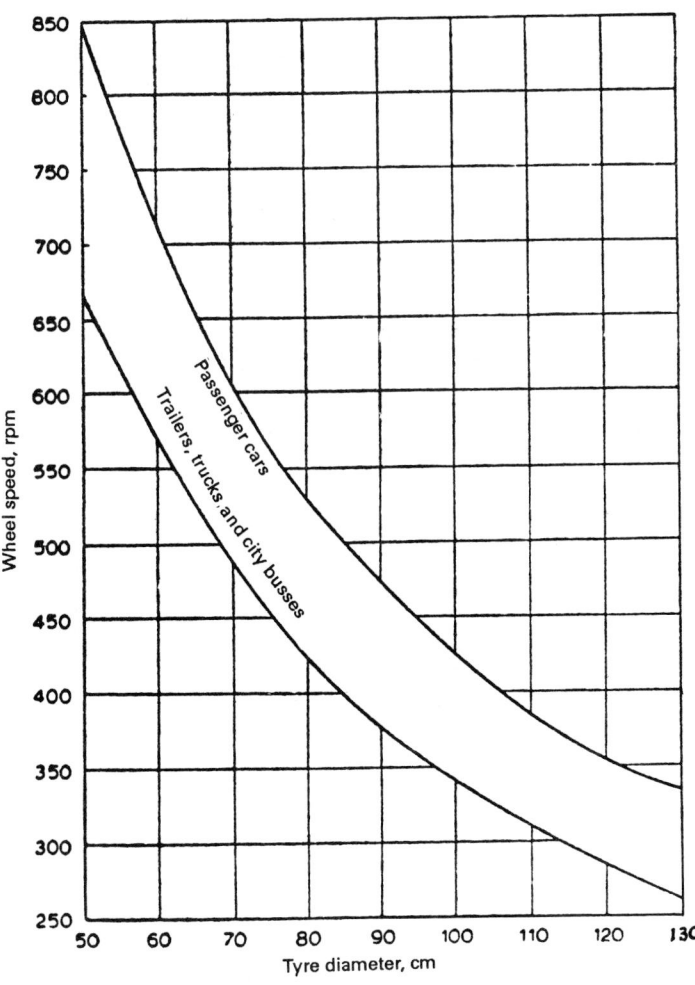

FIGURE 25–48   Wheel speed vs. tire diameter of vehicles.

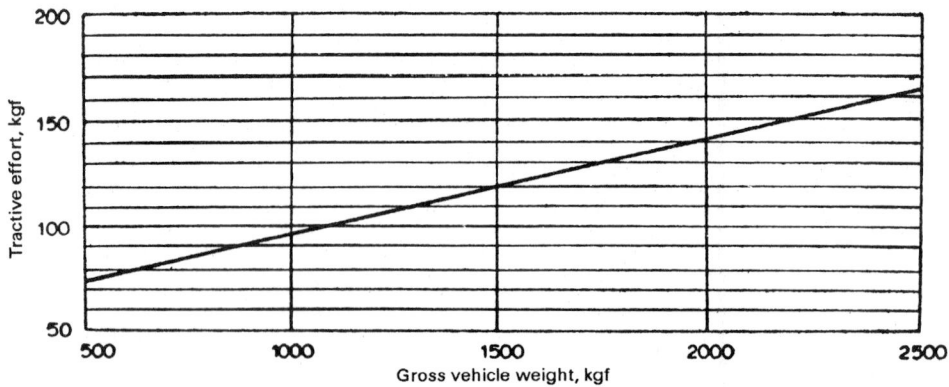

FIGURE 25–49 Tractive effort curve for passenger cars (1 kgf = 9.8066 N = 2.2046 lbf).

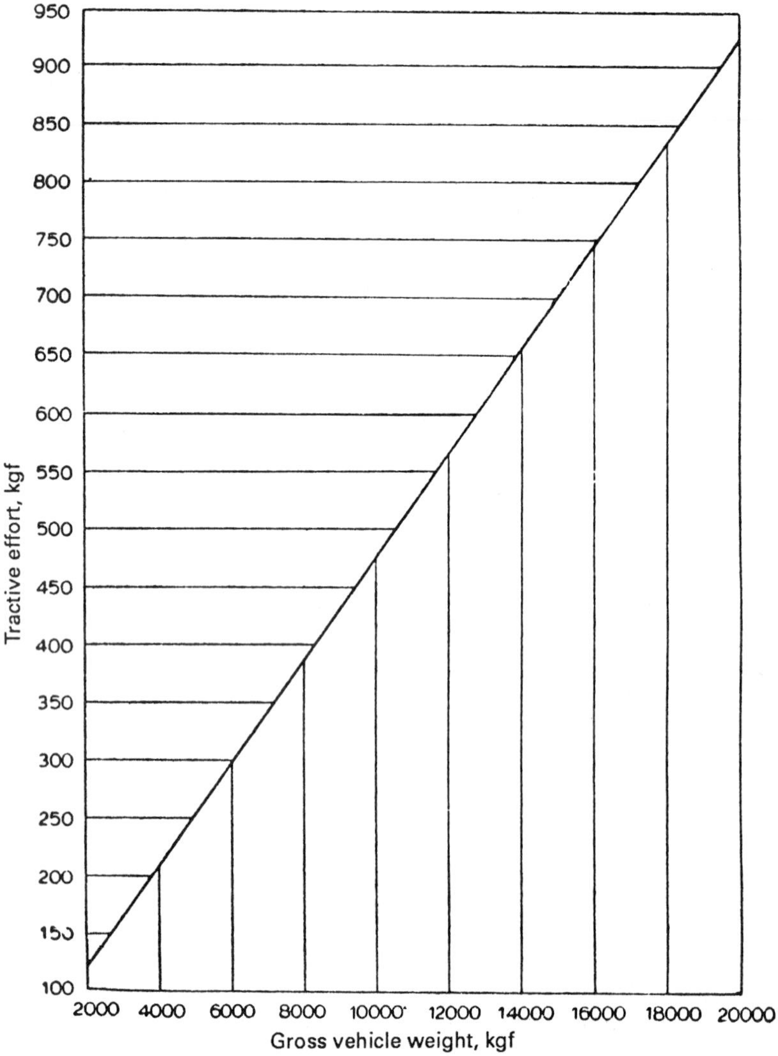

FIGURE 25–50 Tractive effort curve for trucks, tractors, and city busses.

TABLE 25–18
Values of k for use in Eq. (25–216) and (25–217)

Particular	k in USCSU[a]	k in SI
Average automobile of modern design	0.0017	0.20
Streamlined racing car	0.0006	0.07
Truck or omnibus	0.0024	0.28

[a] US Customary System units.

TABLE 25–19A
Air resistance[a]

Speed of vehicle, mph	Velocity of wind, V, ft/s	$0.0024V^2$	$0.0017V^2$	$0.0006V^2$
10	14.67	0.516	0.366	—
20	29.35	2.060	1.460	0.516
30	44.00	4.650	3.300	1.160
40	58.60	8.240	5.830	2.060
50	73.30	12.900	9.130	3.220
60	88.00	18.600	13.160	4.650
90	132.00	—	29.650	10.450
150	220.00	—	—	29.000

[a] Values given in this table are in US Customary System units.

TABLE 25–19B
Air resistance in SI units

Speed of vehicle km/h	Velocity of wind V, m/h	$0.28\ V^2$	$0.20\ V^2$	$0.07\ V^2$
10	2.78	2.17	1.55	0.54
20	5.56	8.68	6.20	2.17
30	8.34	19.50	13.92	4.88
40	11.12	34.72	24.80	8.68
50	13.90	54.25	38.80	13.56
60	16.68	78.00	57.60	19.50
70	19.46	106.40	76.00	26.60
80	22.21	139.00	99.20	34.75
90	25.02	176.00	125.60	44.00
100	27.80	217.00	155.00	54.25

TABLE 25–20
Road resistance, R_r'

Surface	Solid			Pnumatic		
	N/tf	lbf/ton	%	N/tf	lbf/ton	%
Polished marble	29.3	12.12	0.541	35.3	8.08	0.36
Concrete	62.4	14.25	0.636	41.5	9.5	0.423
Asphalt	67.4	15.40	0.687	448.2	10.25	0.457
Stone Good quality	71.8	16.40	0.732	477.6	10.92	0.487
Poor quality	153.2	35.00	1.562	102.0	23.30	1.040
Vitrified bricks	85.0	19.45	0.866	56.7	12.95	0.578
Good macadam (metal road)						
Good	146.2	33.50	1.491	97.9	22.20	0.998
Fair	220.0	50.00	2.240	186.4	33.20	1.900
Rough	307.0	70.00	3.130	389.3	46.60	2.100
Clay	438.4	100.00	4.470	389.3	66.60	3.970
Sand	1314.0	300.00	13.400	874.8	200.00	8.920

TABLE 25–21
Gear ratios

Particular	Ratio
Final drive (rear-axle differential)	4 : 1 to 5 : 1
Second	1.6 : 1 (total 6 or 7 : 1)
Low	2.5 : 1 (total 10 : 1)
Reverse gear	Same as or higher than low gear ratio
Overdrive	<75% above the propeller shaft

TABLE 25–22
Value of coefficient c_1 in SI and USCSU

	Gear wheels belonging to									
	High speed reduction		Intermediate speed reduction						Low speed reductions	
	I		II		III		IV		V	
No. of speeds and type of transmission	SI ×10⁶	USCSU	SI ×10⁶	USCSU	SI ×10⁶	USCSU	SI ×10⁶	USCSU	SI ×10⁶	USCSU
Three-speed passenger car transmission	124.6–145.2	18,000–21,000	145.2–153.5	21,000–24,000					193.7–221.2	28,000–32,000
Four-speed truck transmission	76.0–90.0	11,000–13,000	128.0–149.0	18,500–21,500	96.8–110.9	14,000–16,000			175.2–207.5	26,000–30,000
Five-speed reverse truck transmission	76.0–90.0	11,000–13,000	138.2–152.0	20,000–22,000	105.0–117.7	15,000–17,000	90.0–105.0	13,000–15,000	175.2–207.5	26,000–30,000

25.8 INTERMITTENT-MOTION MECHANISMS

SYMBOLS

a	distance of the pawl pivot point, m (in)
b	face width of ratchet tooth, m (in)
d	diameter (also with suffixes), m (in)
d_h	hub diameter, m (in)
e_1, e_2	dimensions as shown in Fig. 25–52, m (in)
F_n	normal force through O, Figs. 25–51 and 25–52, kN (lbf)
F_{nr}	peripheral force normal to the tooth of ratchet, kN (lbf)
F_t	tangential force at diameter, d, kN (lbf)
h	tooth height or distance from the critical section to the line of action of the load F_{nr}, m (in)
m	module, m (in)
M_b	bending moment, N m (lbf in)
M_t	twisting moment, N m (lbf in)
n'	speed, rps
n	speed, rpm
p	tooth pitch, m (in)
p	linear unit pressure, N/m (lbf/ft)
r	radius, m (in)
s_1	dimension as shown in Fig. 25–53
s_2	breadth of tooth land (Fig. 25–53), m (in)
s_2'	thickness of tooth at base (Fig. 25–52), m (in)
z	number of teeth on ratchet wheel
Z	section modulus, m³ (cm³) (in³)
α	pressure angle or angle of the pawl force, deg
β	angle at pawl ($= 90° - \alpha°$), deg
$\mu = \tan \rho =$	coefficient of friction
$\rho = \tan^{-1} \mu =$	friction angle, deg
φ	ratchet tooth angle or pitch angle, deg
$\psi = \dfrac{b}{m}$	varies from 1.5 to 3
σ	stress, MPa (psi)
σ_b	bending stress, MPa (psi)
τ	shear stress, MPa (psi)

Particular	Formula

PAWL AND RATCHET

The ratchet tooth angle (Fig. 25–51)

$$\varphi = \frac{2\pi}{z}, \text{ rad} \tag{25–254a}$$

$$= \frac{360}{z}, \text{ deg} \tag{25–254b}$$

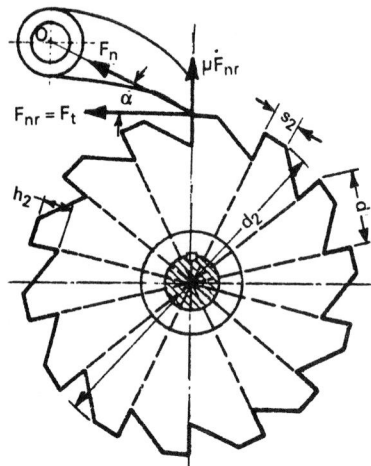

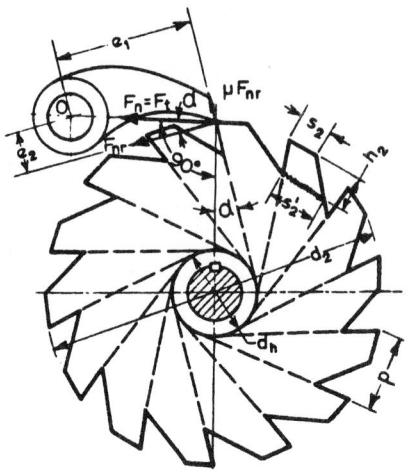

FIGURE 25–51   Ratchet wheel with radial tooth flanks and pawl.

FIGURE 25–52   Ratchet wheel with nonradial tooth flanks and pawl.

Particular	Formula	
The ratchet diameter (Fig. 25–51)	$d_2 = mz = \dfrac{pz}{\pi}$	(25–255)
The face width of ratchet tooth	$b \geq \dfrac{F_n}{F_n^*}$	(25–256)
The allowable unit pressure or force	$F^* = \dfrac{F_n}{b}$	(25–257)
The tangential force	$F_t = \dfrac{2M_t}{d}$	(25–258)
The normal force through O (Fig. 25–51)	$F_n = \dfrac{F_t}{\cos \alpha}$	(25–259a)
The normal force through O (Fig. 25–52)	$F_n = F_t$	(25–259b)
The bending stress	$\sigma_b = \dfrac{M_b}{Z} = \dfrac{6F_t h}{b s_1^2} \leq \sigma_{ba}$	(25–260)
Allowable bending stress (σ_b)	Refer to Table 25–23.	
Number of teeth	$z = 6$ to 30	(25–261)
Module	$m > 6$ (mostly from 10 to 20)	(25–262)
The ratio of h/m	$h/m = 0.6$ to 1	(25–263)
For ratchet wheel definitions and dimensions	Refer to Fig. 25–53.	
The ratio of s_2/m	$s_2/m = 0.6$ to 0.9	(25–264)

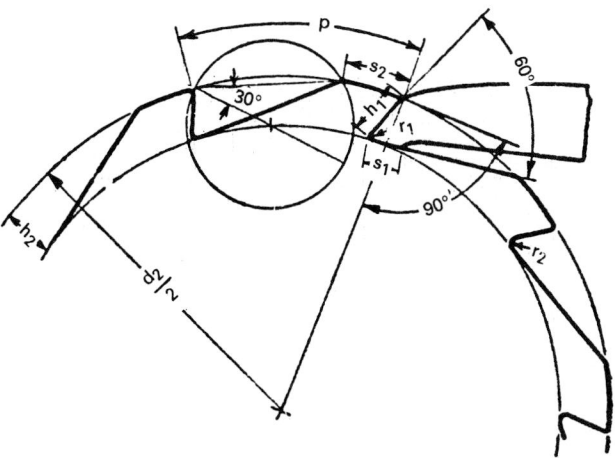

FIGURE 25–53 Definitions and dimensions of ratchet wheel.

Particular	Formula	
The tooth height	$h = 5$ to 15 for toothed ratchet	(25–265)
For external ratchet	$\alpha = 14°$ to $17°$	(25–266)
For internal ratchet	$\alpha = 17°$ to $30°$	(25–267)
The ratio of a/d (internal ratchet)	$a/d = 0.35$ to 0.43	(25–268)
The module	$m = 2\sqrt[3]{\dfrac{M_t}{z\psi\sigma_{ba}}}$	(25–269)
The bending moment on pawl	$M_{b1} = F_n e_2$	(25–270)
The bending stress	$\sigma_b = \dfrac{6M_{b1}}{bs_1^2} + \dfrac{F_n}{bs_1} \leq \sigma_{ba}$	(25–271)
The diameter of pawl pin	$d_1 = 2.71\sqrt[3]{\dfrac{F_n}{2\sigma_{ba}}\left(\dfrac{b}{2} + t_h\right)}$	(25–272)

where t_h = thickness of hub on pawl

TABLE 25–23

Material	F*		σ_b	
	kN/m	lbf/in	MPa	kpsi
Cast iron	49–98	280–560	19.5–29.5	2.85–4.27
Steel or cast steel	98–196	560–1120	39–68.5	5.69–10.0
Hardened steel	196–392	1120–2240	58.8–98	8.54–14.23

25.9   GENEVA MECHANISM

SYMBOLS

$a = \dfrac{r_1}{\sin \phi}$	center distance, m (in)
F_1	the component of force acting on the crank or the driving shaft due to the torque, M_{1t}, kN (lbf) (Fig. 25–57)
F_2	the component of force acting on the driven Geneva wheel shaft due to the torque M_{2t}, kN (lbf) (Fig. 25–57)
$F_{2(max)}$	maximum force (pressure) at the point of contact between the roller pin and slotted Geneva wheel, kN (lbf)
$F_{\mu(max)}$	the component of maximum friction force at the point of contact due to the friction torque $M_{2t\mu}$ on the driven Geneva wheel shaft, kN (lbf)
$F_{i(max)}$	the component of maximum inertia force at the point of contact due to the inertia torque on the driven Geneva wheel shaft, kN (lbf)
$i = \dfrac{z-2}{z}$	gear ratio
J	polar moment of inertia of all the masses of parts attached to Geneva wheel shaft, m^4 (in^4)
k	the working time coefficient of the Geneva wheel
M_{1t}	total torque on the driver or crank, N m (lbf in)
M_{2t}	total torque on the driven or Geneva wheel, N m (lbf in)
M_{2ti}	inertia torque on the Geneva wheel, N m (lbf in)
$M_{2t\mu}$	friction or resistance torque on Geneva wheel, N m (lbf in)
n'	speed, rps
n	speed, rpm
P	power, kW (hp)
r_1	radius to center of driving pin, m (in)
r_2	radius of Geneva wheel, m (in)
r_2'	distance of center of semicircular end of slot from the center of Geneva wheel, m (in)
r_{a2}	outside radius of Geneva wheel, which includes correction for finite pin diameter, m (in)
r_p	pin radius, m (in)
$R_r = \dfrac{r_2}{r_1}$	radius ratio
t	total time required for a full revolution of the driver or crank, s
t_i	time required for indexing Geneva wheel, s
t_r	time during which Geneva wheel is at rest, s
v	velocity, m/s
z	number of slots on the Geneva wheel
α	crank angle or angle of driver at any instant, deg (Fig. 25–54)
α_{2a}	angular acceleration, m/s^2 (ft/s^2)
	angular acceleration of Geneva wheel, m/s^2 (ft/s^2)
α_m	angular position of the crank or driver radius at which the product $\omega \alpha_{2a}$ is maximum, deg
β	angle of the driven wheel or Geneva wheel at any instant, deg (Fig. 25–54)
$\gamma = \dfrac{r_1}{a}$	the ratio of the driver radius to center distance
η	efficiency of Geneva mechanism

λ	locking angle of driver or crank, rad or deg
ν	ratio of time of motion of Geneva wheel to time for one revolution of driver or crank
$\phi = \dfrac{360}{2z}$	semi-indexing or Geneva wheel angle, or half the angle subtended by an adjacent slot, deg (Fig. 25–54)
ψ	crank or driver angle, deg (Fig. 25–54)
$\omega = \dfrac{2\pi n}{60}$	angular velocity of driver or crank (assumed constant), rad/s
ω_1, ω_2	angular velocities of driver or crank and Geneva wheel, respectively, rad/s

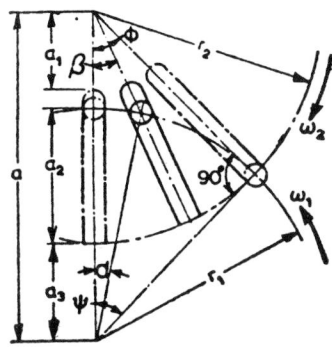

FIGURE 25–54   Design of Geneva mechanism.

Particular	Formula
The angular velocity (constant) of driver or crank	$\omega_1 = \dfrac{2\pi n}{60}$ (25–273)
Gear ratio	$i = \dfrac{\text{angle moved by crank or driver during rotation}}{\text{angle moved by Geneva wheel during rotation}}$
	$= \dfrac{z-2}{z}$ (25–274)
The semi-indexing angle or Geneva wheel angle or half the angle subtended by two adjacent slots	$\phi = \dfrac{360}{2z}$ or $\dfrac{\pi}{z}$ (25–275)
The angle through which the Geneva wheel rotates	$2\phi = \dfrac{360}{z}$ or $\dfrac{2\pi}{z}$ (25–276)

EXTERNAL GENEVA WHEEL

The angle of rotation of driver through which the Geneva wheel is at rest or angle of locking action (Fig. 25–55)	$\lambda = 2(\pi - \psi) = \pi + 2\phi = \dfrac{\pi}{z}(z+2)$ (25–277)
The crank or driver angle	$\psi = \dfrac{\pi}{2} - \phi = \dfrac{\pi(z-2)}{2z}$ (25–278)

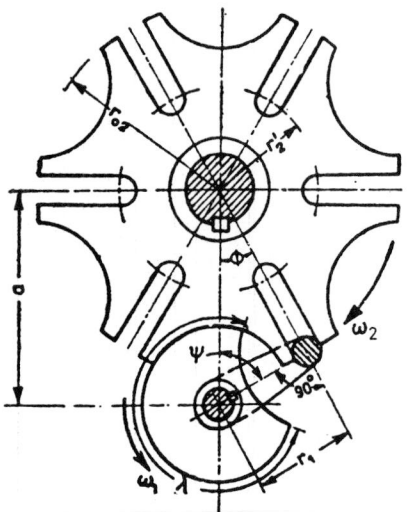

FIGURE 25–55 External Geneva mechanism.

Particular	Formula

DISPLACEMENT

The center distance (Fig. 25–55)

$$a = \frac{r_1}{\sin \phi} \qquad (25\text{–}279)$$

The radius ratio

$$R_r = \frac{r_2}{r_1} = \cot \phi \qquad (25\text{–}280)$$

The ratio of crank radius to center distance

$$\gamma = \frac{r_1}{a} = \sin \phi = \sin \frac{\pi}{z} \qquad (25\text{–}281)$$

The relation between crank angle and Geneva wheel angle

$$\beta = \tan^{-1}\left(\frac{\gamma \sin \alpha}{1 - \gamma \cos \alpha}\right) \qquad (25\text{–}282)$$

VELOCITY

The angular velocity of the Geneva wheel

$$\omega_2 = \frac{d\beta}{dt} = \frac{\gamma(\cos \alpha - \gamma)}{1 - 2\gamma \cos \alpha + \gamma^2}\,\omega_1 \qquad (25\text{–}283a)$$

$$= \frac{\sin(\pi/z)(\cos \alpha - \sin \pi/z)}{1 - 2\sin(\pi/z)\cos \alpha + \sin^2 \pi/z}\,\omega_1 \qquad (25\text{–}283b)$$

The maximum angular velocity of Geneva wheel at angle $\alpha = o$

$$\omega_{2(max)} = \left(\frac{d\beta}{dt}\right)_{max} = \frac{\gamma}{1-\gamma}\,\omega_1$$

$$= \left[\sin\frac{\pi}{z}\bigg/\left(1 - \sin\frac{\pi}{z}\right)\right]\omega_1 \qquad (25\text{–}283c)$$

Particular	Formula

ACCELERATION

The angular acceleration, $*\alpha_{2a}$, of Geneva wheel

$$*\alpha_{2a} = \frac{d^2\beta}{dt^2} = \frac{(\gamma^3 - \gamma)\sin\alpha}{(1 + \gamma^2 - 2\gamma\cos\alpha)^2}\,\omega_1^2 \qquad (25\text{–}284a)$$

$$= \pm\frac{\sin(\pi/z)\cos^2(\pi/z)\sin\alpha}{1 - 2\sin(\pi/z)\cos\alpha + \sin^2(\pi/z)}\,\omega_1^2$$

$$(25\text{–}284b)$$

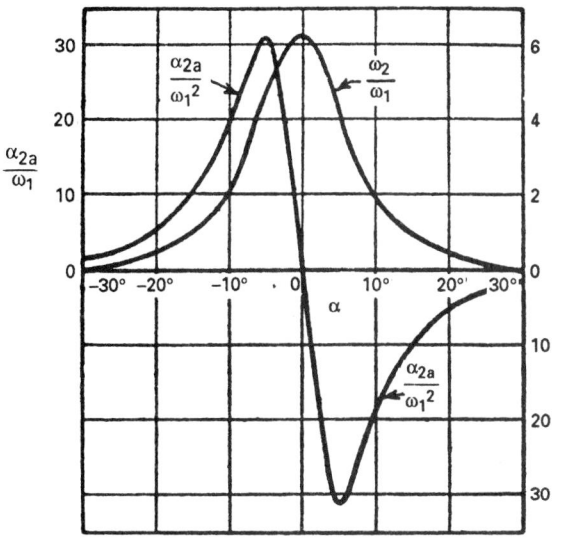

FIGURE 25–56 Angular velocity and angular acceleration curves for three-slot external Geneva wheel.

For angular velocity and angular acceleration curves for three-slot external Geneva wheel with driver velocity, $\omega_1 = 1$ rad/s

Refer to Fig. 25–56.

The maximum angular acceleration of Geneva wheel which occurs at $\alpha = \alpha_{(max)}$

$$\cos\alpha_{(max)} = -\kappa + \sqrt{\kappa^2 + 2} \qquad (25\text{–}284c)$$

$$\text{where } \kappa = \frac{1}{4}\left(\gamma + \frac{1}{\gamma}\right)$$

The angular acceleration of Geneva wheel at start and finish of indexing

$$(\alpha_{2a})_{i,f} = \pm\frac{\sin(\pi/z)\cos^3(\pi/z)}{[1 - 2\sin^2(\pi/z) + \sin^2(\pi/z)]}\,\omega_1^2$$

$$= \pm\omega_1^2\tan\phi = \pm\omega_1^2\tan\pi/z$$

$$= \pm\omega_1^2\left(\frac{r_1}{r_2}\right) \qquad (25\text{–}285)$$

Total time required for a full revolution of the crank or driver

$$t = \frac{60}{n} \qquad (25\text{–}286)$$

The ratio of t_i to t

$$\frac{t_i}{t} = \frac{2\psi}{2\pi} = \frac{\psi}{\pi} = \frac{z-2}{2z} \qquad (25\text{–}287)$$

$*\alpha_{2a}$ is the symbol used for angular acceleration of Geneva wheel; α is the crank or driver angle at any given instant.

Particular	Formula	
The ratio of t_r to t	$\dfrac{t_r}{t} = \dfrac{2(\pi - \psi)}{2\pi} = 1 - \dfrac{\psi}{\pi} = \dfrac{z + 2}{2z}$	(25–288)
The sum of angles of $(\phi + \psi)$	$\phi + \psi = 90°$	(25–289)
The time required for indexing Geneva wheel, in seconds	$t_i = \dfrac{z - 2}{2z}\, t = \dfrac{z - 2}{z}\left(\dfrac{60}{2n}\right)$	(25–290)
The time during which Geneva wheel is at rest, in seconds	$t_r = \dfrac{z + 2}{z}\left(\dfrac{60}{2n}\right)$	(25–291)
The working time coefficient of Geneva wheel	$k = \dfrac{t_i}{t_r} = \dfrac{z - 2}{z + 2}$	(25–292)
Ratio of time of motion of Geneva wheel to time for one revolution of crank or driver	$\nu = \dfrac{z - 2}{2z}\ \left(< \dfrac{1}{2}\right)$	(25–293)
The required speed of the driver shaft or crankshaft	$n = \dfrac{z + 2}{z}\left(\dfrac{60}{2t_r}\right)$	(25–294)

where n in rpm

SHOCK OR JERK

The jerk or shock, J_2, on Geneva wheel	$J_2 = \dfrac{\gamma(\gamma - 1)[2\gamma \cos^2 \alpha + (1 + \gamma^2)\cos \alpha - 4\gamma]}{(1 + \gamma^2 - 2\gamma \cos \alpha)^3} + \omega_1^3$	(25–295)
The jerk or shock at $\alpha = o$	$(J_2)_{\alpha = o} = \left(\dfrac{d^3\beta}{dt^3}\right)_{\alpha = o} = \dfrac{\gamma(\gamma + 1)}{(\gamma - 1)^3}\,\omega_1^3$	(25–296)
The jerk or shock at start, i.e., $\beta = \phi$	$(J_2)_{\beta = \phi} = \left(\dfrac{d^3\beta}{dt^3}\right)_{\beta = \phi} = \left(\dfrac{3\gamma^2}{1 - \gamma^2}\right)\omega_1^3$	(25–297)
The length of the slot (Fig. 25–54)	$a_2 = r_1 + r_2 - a = a\left(\sin \dfrac{\pi}{z} + \cos \dfrac{\pi}{z} - 1\right)$	(25–298)
The condition to be satisfied by diameter on which the driver or crank is mounted	$d_1 < 2a_3 = 2(a - r_2) = 2a\left(1 - \cos \dfrac{\pi}{z}\right)$	(25–299)
	or	
	$\dfrac{d_1}{a} < 2\left(1 - \cos \dfrac{\pi}{z}\right) = 4\sin^2 \dfrac{\pi}{2z}$	(25–300)
The condition to be satisfied by the diameter on which Geneva wheel is mounted	$\dfrac{d_2}{a} < 2\left(1 - \sin \dfrac{\pi}{z}\right) = 4\sin^2\left(\dfrac{\pi}{4} - \dfrac{\pi}{2z}\right)$	(25–301)

Particular	Formula

TORQUE ACTING ON SHAFTS OF GENEVA WHEEL AND DRIVER

The total torque acting on Geneva wheel shaft

$$M_{2t} = M_{2t\mu} + M_{2ti} = M_{2t\mu} + J\alpha_{2a} \qquad (25\text{--}302)$$

It is assumed that $M_{2t\mu}$ is constant.

The torque on the shaft of crank or driver

$$M_{1t} = M_{2t} \frac{\omega_2}{\omega_1} \frac{1}{\eta} = (M_{2t\mu} + J\alpha_{2a}) \frac{\omega_2}{\omega_1} \frac{1}{\eta} \qquad (25\text{--}303)$$

The efficiency of Geneva mechanism

$\eta = 0.80$ to 0.90 when Geneva wheel shaft is mounted on journal bearings (25–304a)

$= 0.95$ when driver shaft is mounted on rolling contact bearings (25–304b)

$= 0.75$ when the diameter of bearing surface is larger than the outside diameter of Geneva wheel (25–304c)

INSTANTANEOUS POWER

The instantaneous power on the crank or driving shaft

$$P = \frac{M_t \omega}{1000} \quad \textbf{SI} \qquad (25\text{--}305a)$$

where P in kW, M_t in N m, and ω in rad/s

$$= \frac{M_t \omega}{102 \times 10^3} \quad \textbf{Metric} \qquad (25\text{--}305b)$$

where P in kW, M_t in kgf mm, and ω in rad/s

$$= \frac{M_t \omega}{75 \times 10^3} \quad \textbf{Metric} \qquad (25\text{--}305c)$$

where P in hp_m, M_t in kgf mm, and ω in rad/s

$$= \frac{M_t n}{63,000} \quad \textbf{US Customary units} \qquad (25\text{--}305d)$$

where P in hp, M_t in lbf in, and n in rpm

Calculation of average power

The average torque $M_{ti(av)}$ for complete cycle

$$M_{ti(av)} = o \qquad (25\text{--}306)$$

The average torque for first half-cycle

$$M_{t(av)} = M_{\mu(av)} = \frac{2}{z-2} \left[M_{2t\mu} + \frac{zJ}{2\pi} \left(\frac{\gamma}{1-\gamma} \right)^2 \omega_1^2 \right] \frac{1}{\eta} \qquad (25\text{--}307)$$

where J = polar moment of inertia, m^4, cm^4 (in^4)

Particular	Formula
The average power required on the crank or driving shaft	$P_{av} = \dfrac{M_{t(av)}}{1000}\,\omega$ **SI** (25–308a)

$P_{av} = \dfrac{M_{t(av)}}{1000}\,\omega$ **SI**   (25–308a)

where P_{av} in kW, $M_{t(av)}$ in N m, and ω in rad/s

$$= \frac{M_{t(av)}\omega}{75 \times 10^3} \quad \textbf{Metric} \qquad (25\text{–}308b)$$

where P_{av} in hp$_m$, $M_{t(av)}$ in kgf mm, and ω in rad/s

$$= \frac{M_{t(av)}n}{63,000} \quad \textbf{US Customary units} \qquad (25\text{–}308c)$$

where P_{av} in hp, $M_{t(av)}$ in lbf in, and n in rpm

Calculation of maximum power

The maximum torque on the driven shaft of Geneva wheel

$$M_{2t(max)} = M_{2t\mu} + M_{2ti(max)} \qquad (25\text{–}309)$$
where $M_{2t\mu}$ is constant

$$M_{2ti(max)} = J\alpha_{2a(max)} = \frac{J\alpha_{2a(max)}}{\omega_1^2}\left(\frac{2\pi n}{60}\right)^2$$

The maximum torque on the driving shaft of the crank

$$M_{1t(max)} = M_{2t\mu}\,\frac{1}{\eta}\,\frac{\omega_{2(max)}}{\omega_1} + \frac{J}{\omega_1}\,\frac{1}{\eta}\,(\alpha_{2a}\omega_2)_{max}$$
$$(25\text{–}310a)$$

$$\simeq \left[M_{2t\mu}\,\frac{\gamma}{1-\gamma}\,\frac{1}{\eta} \right.$$
$$\left. + \frac{\gamma^2(1-\gamma^2)(\cos\alpha_m - \gamma)\sin\alpha_m}{(1-2\gamma\cos\alpha_m + \gamma^2)^3} \right] J\omega_1^2\,\frac{1}{\eta} \quad (25\text{–}310b)$$

where $\alpha = \alpha_m$ at which M_{t1} is maximum

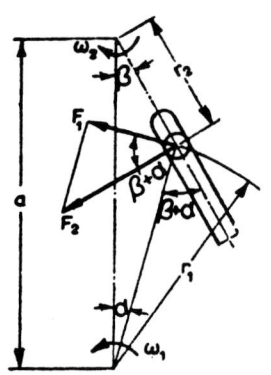

FIGURE 25–57 Forces acting on Geneva wheel.

The maximum power required on the shaft of the crank or driver

$$P_{1(max)} = \frac{M_{1t(max)}\omega}{1000} \quad \textbf{SI} \qquad (25\text{–}311a)$$

where $P_{1(max)}$ in kW, $M_{1t(max)}$ in N m, and ω in rad/s

$$= \frac{M_{1t(max)}\omega}{102 \times 10^3} \quad \textbf{Metric} \qquad (25\text{–}311b)$$

Particular	Formula

where $P_{1(max)}$ in kW, $M_{1t(max)}$ in kgf mm, and ω in rad/s

$$= \frac{M_{1t(max)}n}{63,000} \quad \textbf{US Customary units} \quad (25\text{--}311c)$$

where $P_{1(max)}$ in hp, $M_{1t(max)}$ in lbf in, and n in rpm

FORCES AT THE POINT OF CONTACT (FIG. 25–57)

The maximum force at the point of contact between the roller pin and slotted Geneva wheel

$$F_{2(max)} = \frac{M_{2t}}{r_2} \tag{25--312a}$$

$$= F_{\mu(max)} + F_{i(max)} \tag{25--312b}$$

where $r_2 = \sqrt{a^2 - 2ar_1\cos\alpha + r_1^2}$

$$= \frac{r_1}{\gamma}\sqrt{1 - 2\gamma\cos\alpha + \gamma^2}$$

The component of maximum friction force at the point of contact due to the friction torque $M_{2t\mu}$ on the driven Geneva wheel shaft

$$F_{2\mu(max)} = \frac{M_{2t\mu}}{r_{2(min)}} = \frac{M_{2t\mu}}{r_1}\frac{\gamma}{1-\gamma} \tag{25--313}$$

where $r_{2(min)} = a - r_1 = \left(1 - \frac{1}{\gamma}\right)r_1$

For maximum values of F_{2i} Refer to Table 25–24.

For design data for external Geneva mechanism   Refer to Table 25–25A.

INTERNAL GENEVA WHEEL

The time required for indexing Geneva wheel, s

$$t_i = \frac{z+2}{z}\left(\frac{60}{2n}\right) \tag{25--314}$$

The time during which Geneva wheel is at rest, s

$$t_r = \frac{z-2}{z}\left(\frac{60}{2n}\right) \tag{25--315}$$

The t_i/t ratio

$$\frac{t_i}{t} = \frac{z+2}{2z} \tag{25--316}$$

The t_r/t ratio

$$\frac{t_r}{t} = \frac{z-2}{2z} \tag{25--317}$$

The working time coefficient of Geneva wheel

$$k = \frac{z+2}{z-2} > 1 \tag{25--318}$$

Particular	Formula
The relationship between crank or driver angle α and Geneva wheel angle β	$\beta = \tan^{-1}\left(\dfrac{\gamma\sin\alpha}{1+\gamma\cos\alpha}\right)$ (25–319)
The angular velocity of Geneva wheel	$\omega_2 = \dfrac{d\beta}{dt} = \left[\dfrac{\gamma(\cos\alpha+\gamma)}{1+2\gamma\cos\alpha+\gamma^2}\right]\omega_1$ (25–320)
The maximum angular velocity of Geneva wheel	$\omega_{2(max)} = \dfrac{\gamma}{1+\gamma}\,\omega_1$ (25–321)
The angular acceleration, α_{2a}, of Geneva wheel	$\alpha_{2a} = \dfrac{d^2\beta}{dt^2} = \pm\dfrac{\gamma(1-\gamma^2)\sin\alpha}{(1+2\gamma\cos\alpha+\gamma^2)^2}\,\omega_1^2$ (25–322)
For values of α_{2a} at start and finish of indexing	Use Eq. (25–285) of external Geneva wheel.
For curves of angular velocity and angular acceleration of internal Geneva wheel	Refer to Fig. 25–58.

The contact forces between the slotted wheel and the pin on the driving crank of the internal Geneva wheel are calculated in a manner similar to that for the external Geneva wheel

Materials

Chromium steel 15 Cr[65] case-hardened to R_c 58 to 65 is used for the roller pin on the driver or crank.

Chromium steel 40 Cr 1 hardened and tempered to R_c 45 to 55 is used for the sides of slotted Geneva wheel.

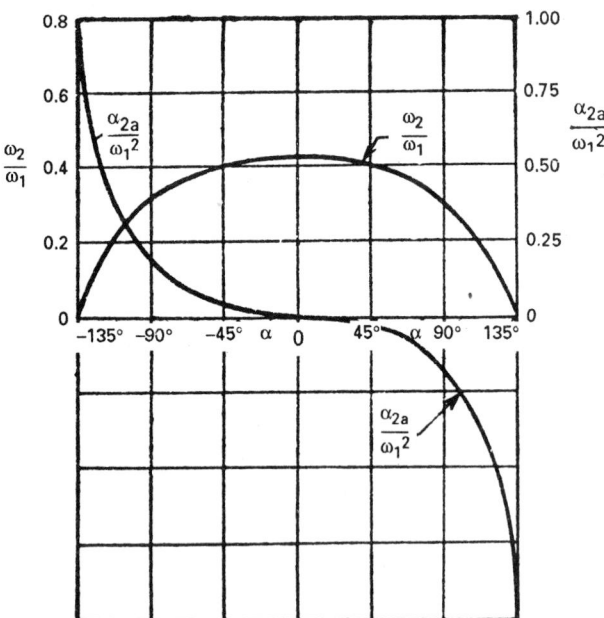

FIGURE 25–58 Angular velocity and angular acceleration for four-slot internal Geneva wheel.

TABLE 25–24
Maximum F_{2i} values

z	3	4	5	6	8
$F_{2i(max)} \Big/ \left(\dfrac{Jn^2}{r_1} \right)$	1.966	0.126	0.0318	0.0131	0.00424

TABLE 25–25A
Design data for external Geneva mechanism

z	ϕ	ψ	i	r_1/a	r_2/a	R_r	r_2'/a	λ	ν	$\omega_{2(max)}$	$\alpha_{2a(inital)}$ $\alpha = -\phi$	$\alpha_{(max)}$	J_{max}	$J_{\alpha=o}$
3	60°	30°	0.5	0.886	0.500	0.577	0.134	300°	0.167	6.46	1.732	4°46′	31.44	−672
4	45°	45°	1	0.707	0.707	1.000	0.293	270°	0.250	2.41	1.000	11°24′	5.41	−48
6	30°	60°	2	0.500	0.866	1.732	0.500	240°	0.333	1.00	0.577	22°54′	1.35	−6
8	22°30′	67°30′	3	0.383	0.924	2.414	0.617	225°	0.375	0.620	0.414	31°38′	0.699	−2.25
10	18°	72°	4	0.309	0.951	3.078	0.690	216°	0.400	0.447	0.325	38°30′	0.465	−1.24

25.10   UNIVERSAL JOINT

SYMBOLS

d	diameter, m (in)
K_s	shock factor
K_{ct}	correction factor to be applied to torque to be transmitted
K_{ct}	correction factor to be applied to power to be transmitted
l	length (also with subscripts), m (in)
L	life, h
M_t	torque to be transmitted by universal joint, N m (lbf in)
M_{td}	design torque, N m (lbf in)
n	speed, rpm
n'	speed, rps
P	power to be transmitted by universal joint, kW (hp)
P_d	design power, kW (hp)
β	angle between two intersecting shafts 1 and 2, deg
θ	angle of rotation of the driver shaft 1, deg
ϕ	angle of rotation of the driven shaft 2, deg
ω_1, ω_2	angular velocities of driver and driven shafts respectively, rad/s

Particular	Formula

SINGLE UNIVERSAL JOINT (FIGS. 25–59 AND 25–61*a*)

The relation between θ, ϕ, and β

$$\tan \phi = \frac{\tan \theta}{\cos \beta} \qquad (25\text{–}323)$$

The relation between the angular velocities of driving shaft 1 or driver (ω_1) to the driven shaft 2 or the follower (ω_2)

$$\frac{\omega_2}{\omega_1} = \frac{\cos \beta}{1 - \sin^2 \beta \sin^2 \theta} \qquad (25\text{–}324)$$

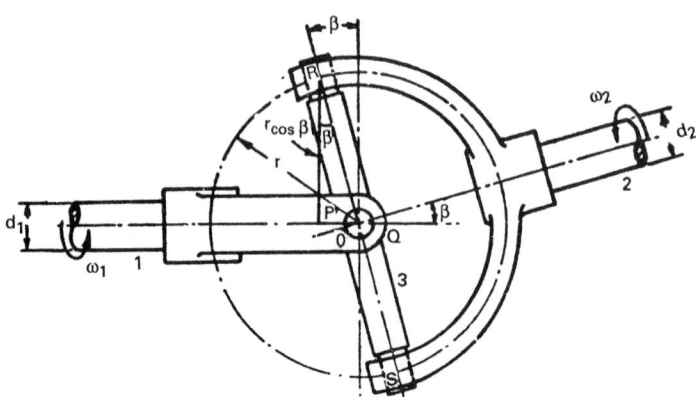

FIGURE 25–59   A single universal joint.

Particular	Formula
The maximum value of ω_2/ω_1	$$\left(\frac{\omega_2}{\omega_1}\right)_{max} = \frac{\cos\beta}{1-\sin^2\beta} = \frac{1}{\cos\beta} \qquad (25\text{–}325)$$ when $\sin\theta = +1$, i.e., $\theta = 90°$, $270°$, or $\pi/2$ or $3\pi/2$, etc.
The minimum value of ω_2/ω_1	$$\left(\frac{\omega_2}{\omega_1}\right)_{min} = \cos\beta \qquad (25\text{–}326)$$ when $\sin = 0$, i.e., $\theta = 0$, π, 2π, etc.
The angular acceleration of the driven shaft 2, if ω_1 is constant	$$\frac{d^2\phi}{dt^2} = \frac{d\omega_2}{dt} = \frac{\cos\beta\sin^2\beta\sin 2\theta}{(1-\sin^2\theta\sin^2\beta)^2}\,\omega_1^2 \qquad (25\text{–}327)$$
The value of θ for which the angular acceleration of the driven shaft is maximum	$$\cos 2\theta_{(max)} = \kappa - \sqrt{\kappa^2+2} \qquad (25\text{–}328)$$ where $\kappa = (2-\sin^2\beta)/2\sin^2\beta$ The angular acceleration of driven shaft is maximum when θ is approximately equal to $45°$, $135°$, etc., when the arms of cross are inclined at $45°$ to the plane containing the axes of the two shafts.
The power transmitted by universal joint	$P = M_t\omega/1000$ **SI** (25–329a) where P in kW, M_t in N m, and ω in rad/s $= M_t n/63{,}000$ **US Customary units** (25–329b) where P in hp, M_t in lbf in, and n in rpm
The design torque of universal joint	$M_{td} = M_t K_s K_{ct}$ (25–330)
The design power of universal joint	$$P_d = \frac{P}{K_{CN}} \qquad (25\text{–}331)$$
For calculation of torque and power transmitted by universal joint for various angles of inclination β	Refer to Figs. 25–62 to 25–65.
For design data of universal joint	Refer to Tables 25–25B and 25–25C.

DOUBLE UNIVERSAL JOINT (FIGS. 25–60 AND 25–61*b*)

The angular velocities ratio for a double universal joint which will produce a uniform velocity ratio at all times between the input and output ends	$$\frac{\omega_1}{\omega_2} = 1 \qquad (25\text{–}332)$$

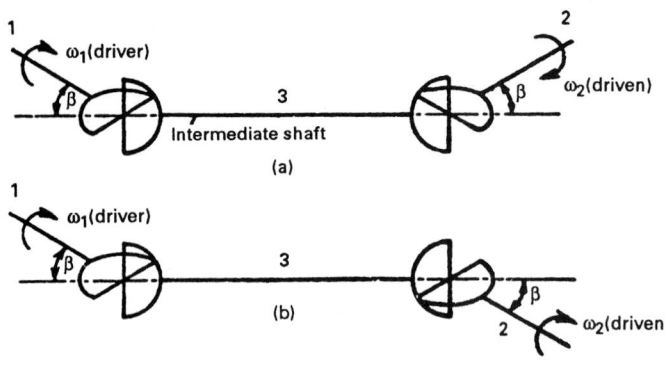

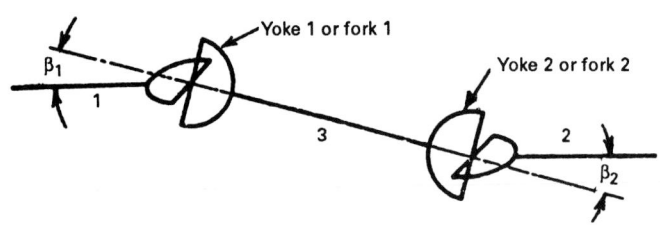

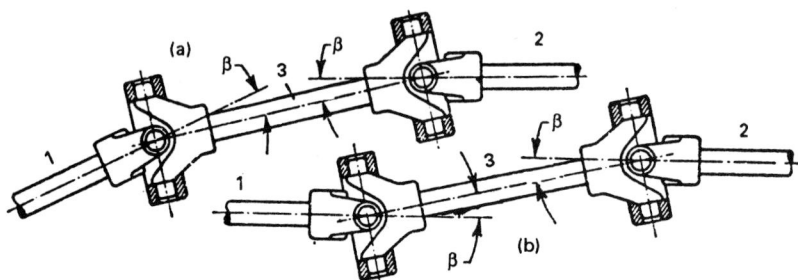

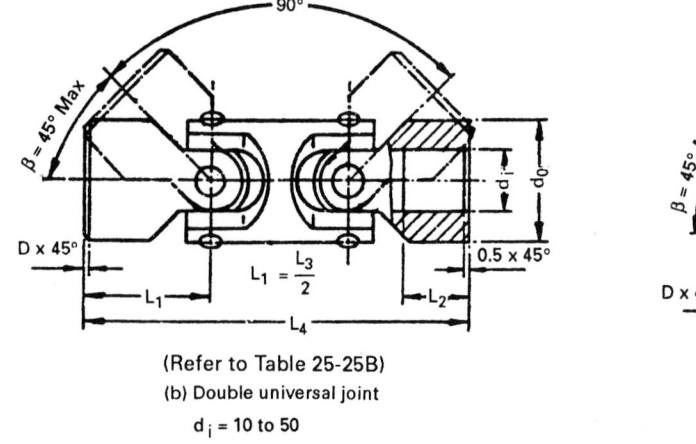

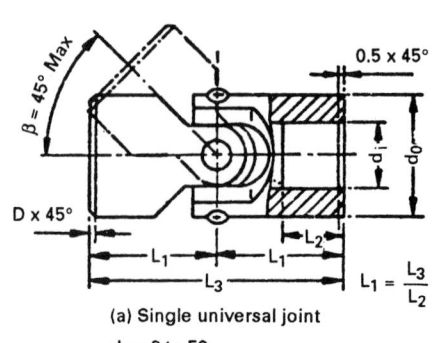

FIGURE 25–60 Double universal joints.

90°

β = 45° Max

D x 45°

$L_1 = \dfrac{L_3}{2}$

0.5 x 45°

L_2

L_4

(Refer to Table 25-25B)
(b) Double universal joint
d_i = 10 to 50

0.5 x 45°

β = 45° Max

D x 45°

L_1 L_1 L_2

L_3

$L_1 = \dfrac{L_3}{L_2}$

(a) Single universal joint
d_i = 6 to 50

FIGURE 25–61 Dimensions of universal joints.

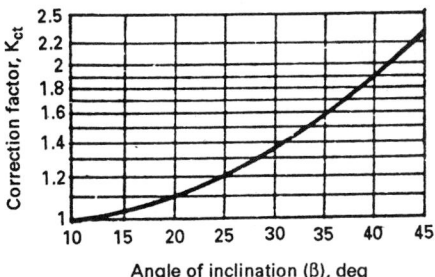

FIGURE 25–62 Angle between two intersecting shafts vs. correction factor (K_{ct}).

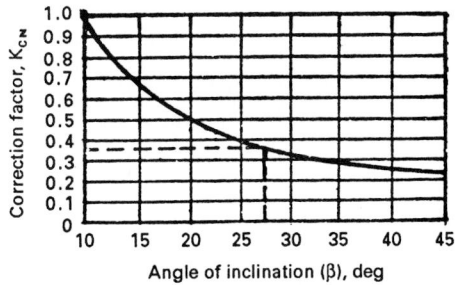

FIGURE 25–63 Angle between two intersecting shafts vs. correction factor (K_{CN}).

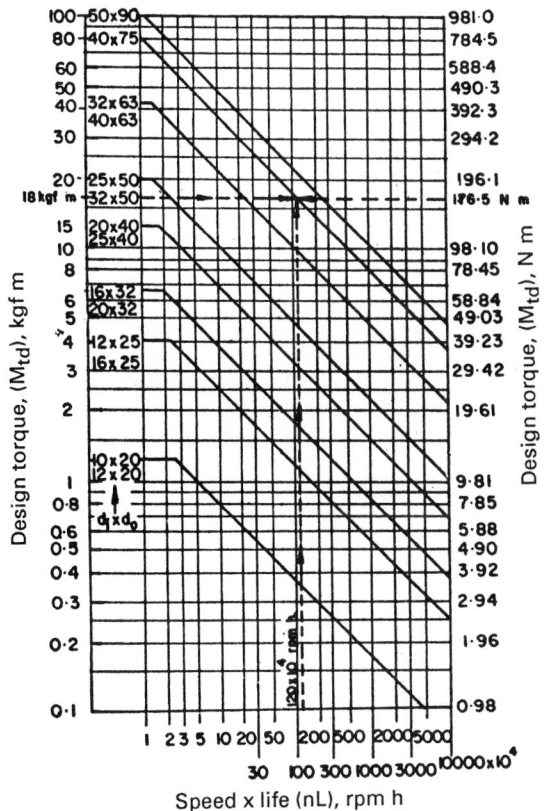

FIGURE 25–64 Design curves for single universal joint with needle bearings for $\beta = 10°$.

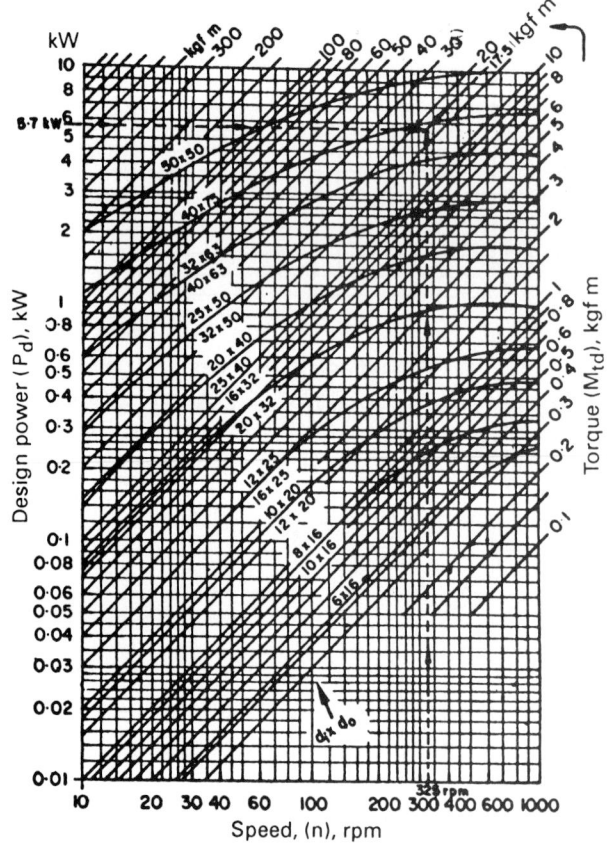

FIGURE 25–65(a)   Design curves for single universal joint with plain bearings for $\beta = 10°$.

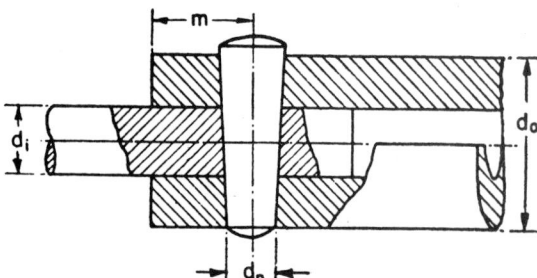

FIGURE 25–65(b)   Taper pin joint. The length of the taper pin should conform to diameter d_o in Table 25–25B.

USE OF CURVES IN FIGS. 25–62 TO 25–65
Worked example 1

A single universal joint has to transmit a torque of 10 kgf m at 1500 rpm. The angle between intersecting shafts is 25°. The joint is subjected to a minor shock. The shock factor (K_s) is 1.5. Design a universal joint with needle bearings for a life of 800 h.

SOLUTION From Fig. 25–62 correction factor for $\beta = 25°$ is $K_{ct} = 1.2$. Design torque $= M_{td}$
$= M_t K_s K_{ct} = 10 \times 1.5 \times 1.2 = 18$ kgf m (176.5 N m).
Speed $\times$ life $= nL = 1500 \times 800 = 120 \times 10^4$ rpm h.
From Fig. 25–64 for $M_{td} = 18$ kgf m (176.5 N m) and $nL = 120 \times 10^4$ rpm h, the size of a single universal joint is $(d_i \times d_o)40 \times 75$ mm.

Worked example 2

Design a single universal joint with plain bearings to transmit 2 kW power at 325 rpm. The angle between two intersecting shafts is 27.5°.

SOLUTION From Fig. 25–63 correction factor for $\beta = 27.5°$ is $K_{CN} = 0.35$. Design power $= P_d$
$= (P/K_{CN}) = (2/0.35) = 5.7$ kW. From Fig. 25–65a the size of a single universal joint for $P_d = 5.7$ kW and speed $= n = 325$ rpm is $(d_i \times d_o)\ 40 \times 75$ mm. The permissible torque for this size of joint (Fig. 25–65a) is 17.5 kgf m (171.5 N m).

TABLE 25–25B
Dimensions of universal joint (Fig. 25–61)

d_i H7	d_o k11	L_2	L_3 +1	L_4 ±1	o	Maximum allowable rotational play		Angular rotational play at an angle of inclination of θ deg in minutes	Tolerance on coaxiality of the two bores
						Test torque			
						N m	kgf m		
6		9	34						
8	16	11	40			0.196	0.02	45	
10		15	52	74					
	20	13	48		0.5	0.392	0.04	40	
12		18	62	88					0.06
	25	15	56	86		0.981	0.10	32	
16		22	74	104					
	32	19	68			0.667	0.17	28	
20		25	86	124					
	40	23	82	128		3.334	0.34	25	0.09
25		32	108	156					
	50	29	105	160	1	5.296	0.54	20	
32		40	132	188					
	63	36	130	198		14.710	1.5	18	0.12
40		50	166	238					
	75	44	160	245		21.575	2.2	16	
50	90	54	190	290		27.458	2.8	14	0.15

TABLE 25–25C
Dimensions of taper pin[a] (Fig. 25–65b)

d_i	d_p	m
6	2	4.5
8	3	5
10	4	6
12	5	7.5
16	6	9
20	8	11
25	10	15
32	12	18
40	14	22
50	16	27

[a] The shear stress of taper pin $= \tau = 158.5$ to 247.5 MPa (16 to 25 kgf/mm^2).

25.11 UNSYMMETRICAL BENDING AND TORSION OF NONCIRCULAR CROSS-SECTION MACHINE ELEMENTS

SYMBOLS

a	semimajor axis of elliptical section, m (in)
	width of rectangular section, m (in)
A	area of cross section, m^2 (in^2)
b	semi-minor axis of elliptical section, m (in)
	height of rectangular section, m (in)
c	distance of the plane from neutral axis, m (in)
	thickness of narrow rectangular cross section (Fig. 25–68)
e	the distance from a point in the shear center S (Table 25–26)
E	Young's modulus, GPa (psi)
G	modulus of rigidity, GPa (psi)
I	moment of inertia, area (also with suffixes), m^4 (cm^4) (in^4)
I_u, I_v	moment of inertia of cross-sectional area, respectively, m^4 (cm^4) (in^4)
J_k	polar moment of inertia, m^4 (cm^4) (in^4)
k_1, k_2	constants from Table 25–28
L	length, m (in)
M_b	bending moment, N m (lbf in)
M_t	twisting moment, N m (lbf in)
$M_{bu} = M_b \cos \theta$	bending moment about the U principal centroidal axis or any axis parallel thereto
$M_{bv} = M_b \sin \theta$	bending moment about the V principal centroidal axis or any axis parallel thereto
$q = \tau t$	shear flow
Q	the first moment of the section, m^4 (cm^4) (in^4)
S	the length of the center of the ring section of the thin tube, m (in)
t	width of cross section at the plane in which it is desired to find the shear stress, m (in)
	thickness of the wall of the thin-walled section, m (in)
u, v	coordinates of any point in the section with reference to principal centroidal axes
V	shear force on the cross section, kN (lbf)
V_y	resultant shear force acting at the shear center, kN (lbf)
x	the distance of the section considered from the fixed end (Fig. 25–73)
x, y	coordinates in x and y directions
σ_b	bending stress (also with suffixes), MPa (psi)
τ	shear stress (also with suffixes), MPa (psi)
δ	variable thickness of thin tube wall (Fig. 25–70), m (in)
θ	angle measured from the V principal centroidal axis, deg
ϕ	angle of twist, deg

Particular	Formula

SHEAR CENTER

The shear stress at any point in transverse plane or section of a member

$$\tau = \frac{VQ}{It} \tag{25-333}$$

The flexural stress in a thin-walled open section

$$\sigma_b = \frac{M_b c}{I} \tag{25-334}$$

Shear flow

$$q = \tau t = \frac{VQ}{I} \tag{25-335}$$

For the equations for locating the shear centers of various thin open sections

Refer to Table 25–26.

UNSYMMETRICAL BENDING

The flexural stress in case of sections subjected to unsymmetrical bending

$$\sigma_b = \frac{(M_b \cos \theta)v}{I_u} + \frac{(M_b \sin \theta)u}{I_v} = \frac{M_{bu}v}{I_u} + \frac{M_{bv}u}{I_v} \tag{25-336}$$

Flexural modulus for any cross section on which the stress is desired

$$Z = I_u I_v / (v I_v \cos \theta + u I_u \sin \theta) \tag{25-337}$$

TORSION
Solid sections

ELLIPTICAL CROSS SECTION   Shear stress acting in the x direction on the xz plane (Fig. 25–66)

$$\tau_{xz} = \frac{2M_t y}{\pi ab^3} \tag{25-338}$$

Shear stress acting in the y direction on the yz plane (Fig. 25–66)

$$\tau_{yz} = -\frac{2M_t x}{\pi a^3 b} \tag{25-339}$$

Maximum shear stress on the periphery at the extremities of the minor axis (Fig. 25–66 and Table 25–27)

$$\tau_{max} = \frac{2M_t}{\pi ab^2} \tag{25-340}$$

Minimum shear stress on the periphery at the extremities of the major axis

$$\tau_{min} = \frac{2M_t}{\pi a^2 b} \tag{25-341}$$

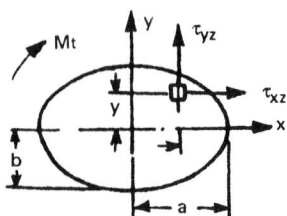

FIGURE 25–66

Particular	Formula

Angle of twist (Fig. 25–66)

$$\phi = \frac{M_t}{G} \frac{(a^2 + b^2)}{\pi a^3 b^3} L \qquad (25\text{–}342)$$

RECTANGULAR CROSS SECTION The maximum shear stress at point A on the boundary, close to the center (Fig. 25–67 and Table 25–27)

$$\tau_A = \frac{M_t}{k_1 ab^2} \qquad (25\text{–}343)$$

where k_1 depends on ratio a/b (Refer to Table 25–28.)

Angle of twist (Table 25–27)

$$\phi = \frac{M_t L}{k_2 ab^3 G} \qquad (25\text{–}344)$$

where k_2 depends on ratio a/b (Refer to Table 25–28.)

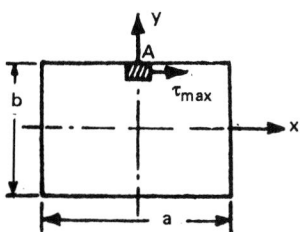

FIGURE 25–67

NARROW RECTANGULAR CROSS SECTIONS (FIG. 25–68)

Equation for twisting moment (Fig. 25–68)

$$M_t = \tfrac{1}{3} G\phi c^3 b \qquad (25\text{–}345)$$

Equation for angle of twist

$$\phi = \frac{3M_t}{Gc^3 b} \qquad (25\text{–}346)$$

The maximum shear stress

$$\tau_{max} = \frac{3M_t}{bc^2} \qquad (25\text{–}347)$$

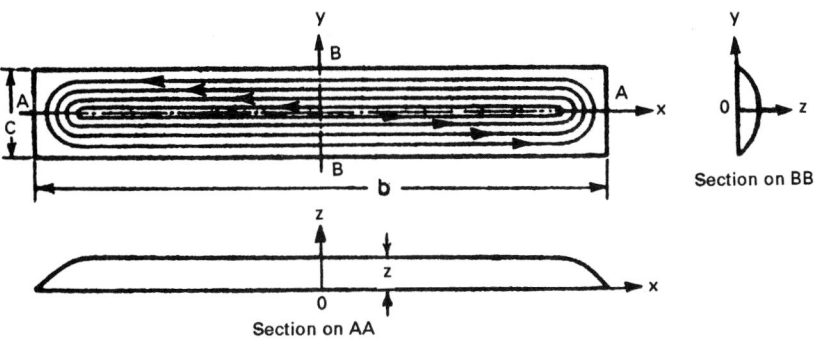

FIGURE 25–68

Particular	Formula

Composite sections

CROSS SECTIONS COMPOSED OF NARROW RECTANGLES

Equation for torque of a narrow rectangular cross section

$$M_t = \frac{k_2 bc^3 G\phi}{L} \qquad (25\text{–}348)$$

Equation for torque of a narrow rectangular section $(b/c \rightarrow \infty,\ k_2 = \frac{1}{3})$

$$M_t = \frac{G\phi}{3L}\ \Sigma bc^3 \qquad (25\text{–}349)$$

Equation for torque for a cross-section composed of several narrow rectangles (Table 25–27)

$$M_t = \frac{G\phi}{L}\ \Sigma k_2 bc^3 \qquad (25\text{–}350)$$

Angle of twist for a cross section composed of several narrow rectangles

$$\phi = \frac{3M_t L}{G\Sigma bc^3} \qquad (25\text{–}351)$$

Maximum shear stress

$$\tau_{max} = \frac{3M_t c_{max}}{\Sigma bc^3}\ \text{for}\ \frac{k_2}{k_1} = 1 \qquad (25\text{–}352)$$

where $c_{max} = $ maximum thickness of the narrow section

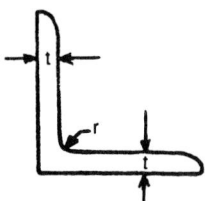

FIGURE 25–69

For variation of stress-concentration factor K_σ with ratio r/c for structural angle (Fig. 25–69)

Refer to Table 25–29.

For approximate formulas for torsional shearing stress and angle of twist for various cross sections

Refer to Table 25–27.

HOLLOW THIN-WALLED TUBES (FIG. 25–70)

The equation for the twisting moment

$$M_t = q(2A) = 2A\delta\tau \qquad (25\text{–}353)$$
where $A = $ area enclosed by the median line of the tubular section

The angle of twist

$$\phi = \frac{M_t S}{4A^2 G\delta} \qquad (25\text{–}354)$$

By membrane analogy the value of $\oint \tau ds$

$$\oint \tau ds = 2G\phi A \qquad (25\text{–}355)$$

The equation for the shear stress

$$\tau = \frac{M_t}{2A}\delta \qquad (25\text{–}356)$$

Particular	Formula
The difference in level between DC and AB of membrane	$h = \tau\delta$ $\qquad$ (25–357)
The equation for twisting moment of thin webbed tubes (or box beams) (Fig. 25–71)	$M_t = 2(A_1 h_1 + A_2 h_2)$ $\qquad$ (25–358a) $\quad = 2(A_1 \delta_1 \tau_1 + A_2 \delta_2 \tau_2)$ $\qquad$ (25–358b)
The equations for shear stress	$\tau_1 = M_t \dfrac{\delta_3 S_2 A_1 + \delta_2 S_3 (A_1 + A_2)}{R}$ $\qquad$ (25–359a) $\tau_2 = M_t \dfrac{\delta_3 S_1 A_2 + \delta_1 S_3 (A_1 + A_2)}{R}$ $\qquad$ (25–359b) $\tau_3 = M_t \dfrac{\delta_1 S_3 A_1 - \delta_2 S_1 A_2}{R}$ $\qquad$ (25–359c) where $R = 2[\delta_1 \delta_3 S_2 A_1^2 + \delta_2 \delta_3 S_1 A_2^2 + \delta_1 \delta_2 S_3 (A_1 + A_2)^2]$ $\qquad$ (25–360)

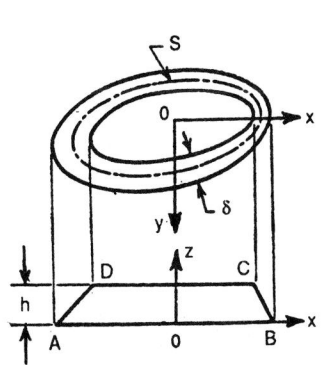

FIGURE 25–70

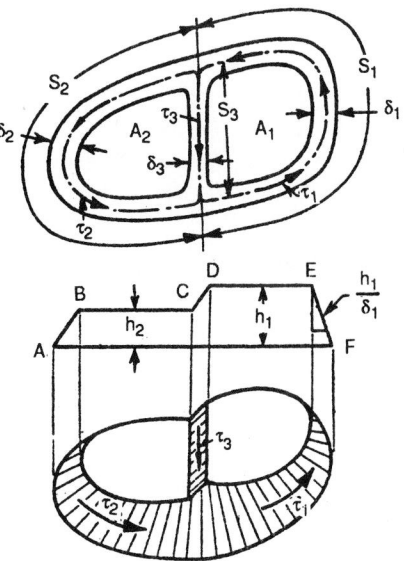

FIGURE 25–71

Particular	Formula

FIGURE 25–72

BENDING STRESSES CAUSED BY TORSION

Torsion of I-beam having one section restrained from warping

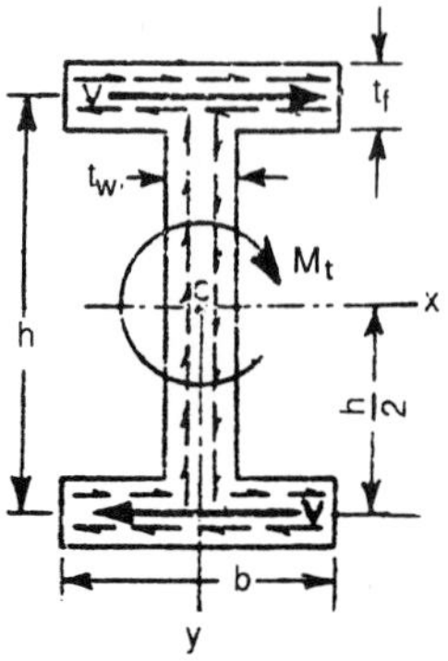

The lateral bending moment in the flanges of an I-beam subjected to twisting moment at one end, the other end being fixed, Fig. 25–72

$$M_b = -\frac{M_t}{h} k \frac{\sin h(L-x)/k}{\cos h(L/k)} \tag{25–361}$$

The maximum bending moment for long beam

$$M_{b(max)} = \frac{M_t k}{h} \quad \text{as} \quad \tan h \frac{L}{k} = 1 \tag{25–362}$$

where $k = (h/2)[EI/JG]^{1/2}$

Twisting moment at any section, distance x from the fixed end

$$M_{tx} = M_t \left[1 - \frac{\cos h(L-x)/k}{\cos h(L/k)} \right] \tag{25–363}$$

The angle of twist per unit length

$$\phi_u = \frac{M_t}{JG} \left[1 - \frac{\cos h(L-x)/k}{\cos h(L/k)} \right] \tag{25–364}$$

The total angle of twist at the free end

$$\phi = \frac{M_t}{JG} \left(L - k \tan h \frac{L}{k} \right) \tag{25–365}$$

Maximum bending moment

$$M_{b(max)} = \frac{M_t}{h} k \tan h \frac{L}{k} \tag{25–366}$$

The angle of twist at free end if $l/k > 2.5$

$$\phi = \frac{M_t}{JG} (L - k) \tag{25–367}$$

The maximum bending moment if $l/k > 2.5$

$$M_{b(max)} = \frac{M_t}{h} k \tag{25–368}$$

The bending stress

$$\sigma_b = \frac{M_{b(max)} b}{2I_f} \tag{25–369}$$

where $M_{b(max)}$ obtained from Table 25–30

For beams subjected to torsion

Refer to Table 25–30.

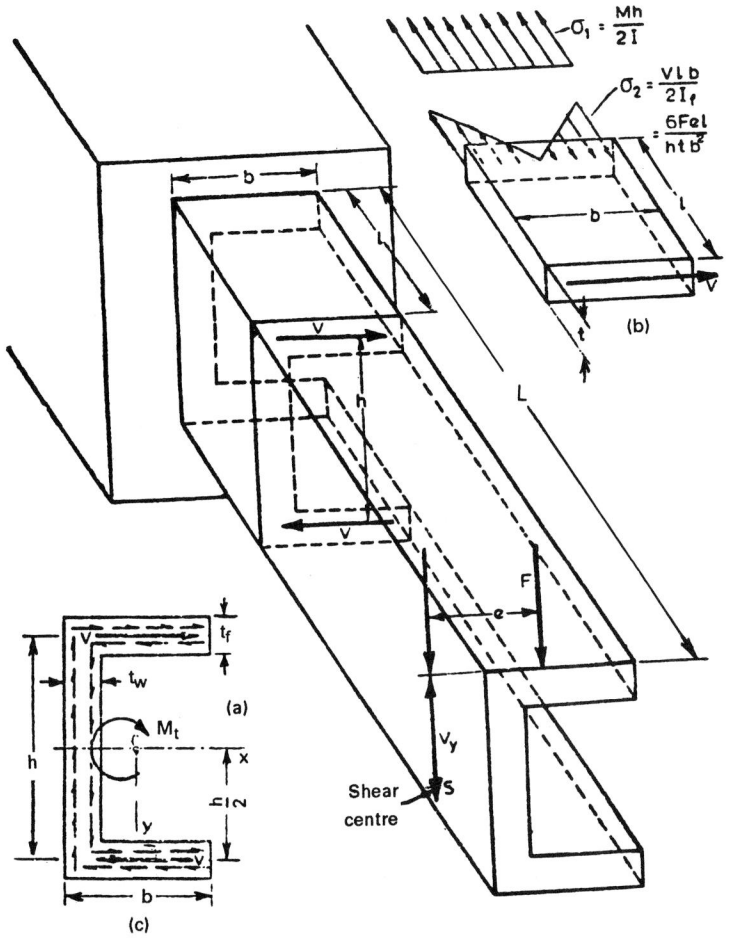

FIGURE 25–73

Particular	Formula

TRANSVERSE LOAD ON BEAM OF CHANNEL SECTION NOT THROUGH SHEAR CENTER (FIG. 25–73)

The direct stress

$$\sigma_t = \frac{M}{I} \frac{h}{2} \qquad (25\text{--}370)$$

where $M = Fe$

The bending stress

$$\sigma_b = \frac{VIb/2}{I} = \frac{6FeI}{htb^2} \qquad (25\text{--}371)$$

Particular	Formula
The maximum longitudinal stress (Fig. 25–73b)	$$\sigma = \frac{Mh}{2I} + \frac{6Fel}{htb^2} \qquad (25\text{–}372)$$
For geometrical properties, weight, and nominal dimensions of beams, channels, T-bars, and equal and unequal angles	Refer to Tables 25–31 and 25–32 and Figs. 25–74 to 25–79.

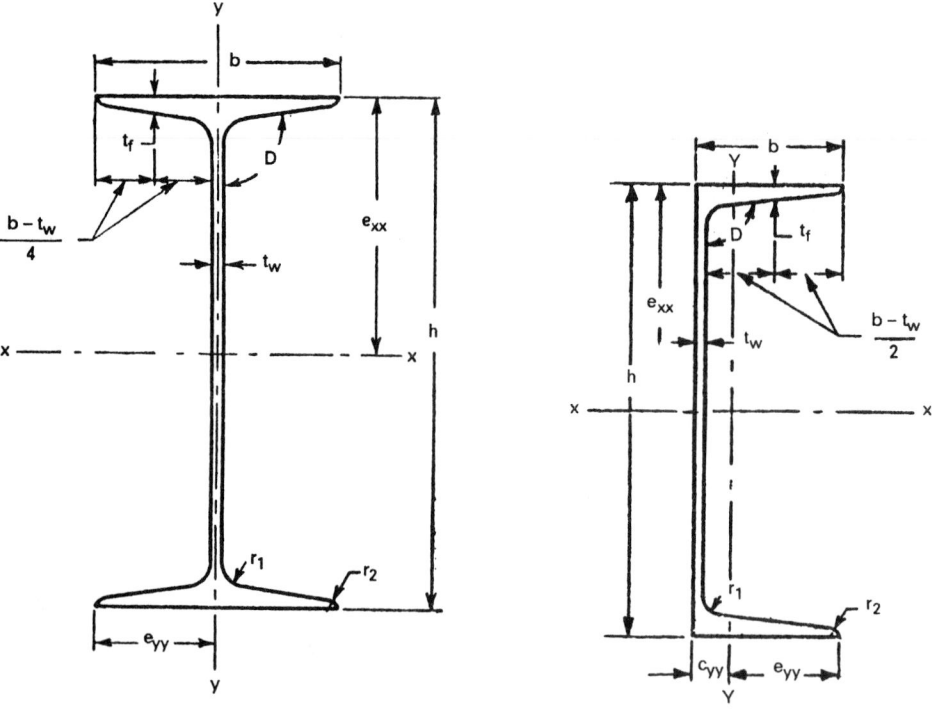

FIGURE 25–74   Beam or column section. (See Table 25–31.)   **FIGURE 25–75**   Channel section. (See Table 25–32.)

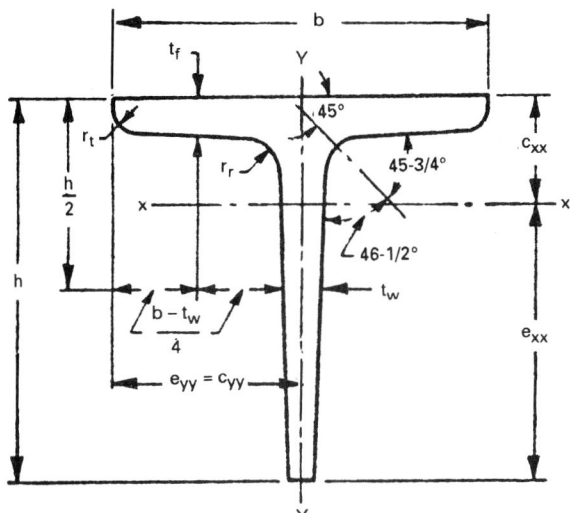

FIGURE 25–76   Normal T-bar. (See Table 25–32.)

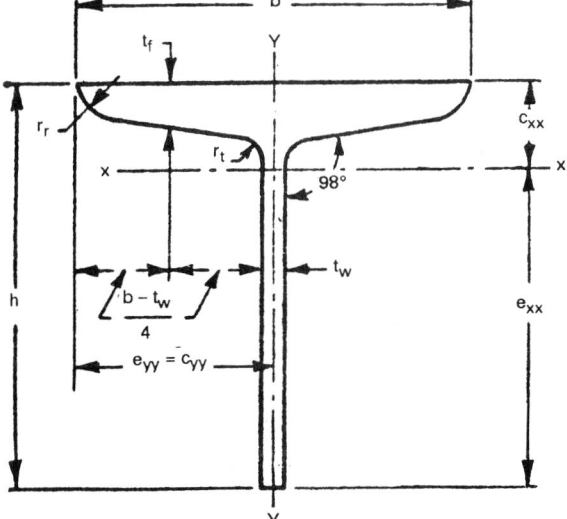

FIGURE 25–77   Slit and deep-legged T-bar. (See Table 25–32.)

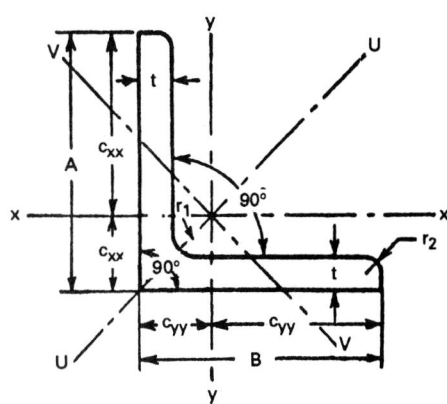

FIGURE 25–78   Equal-angle section. (See Table 25–32.)

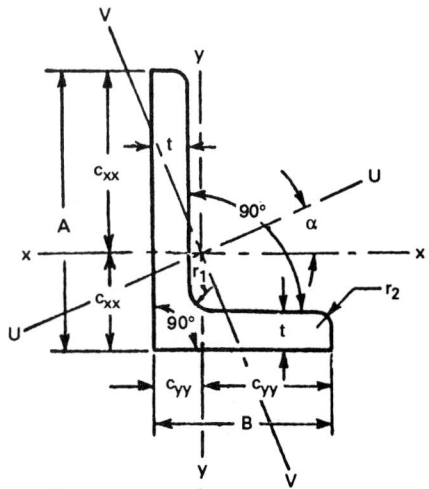

FIGURE 25–79   Unequal-angle section. (See Table 25–32.)

TABLE 25–26
Location of shear center for various cross sections

Section	Location of shear center	Section	Location of shear center
	$e = \dfrac{3b^2 t_f}{h t_w + 6 b t_f}$		$e = \dfrac{3(b_1^2 - b_2^2)}{(t_w/t_f)h + 6(b_1 + b_2)}$ for $b_2 < b_1$
	$e = b_1 \left\{ \dfrac{1 + \dfrac{1}{2}\dfrac{b_1}{h_1} - \dfrac{4}{3}\left(\dfrac{h_1}{h_2}\right)^2}{1 + \dfrac{1}{6}\dfrac{h_2}{h_1} + \dfrac{b_1}{h_1} - \dfrac{2h_1}{h_2}\left(1 - \dfrac{2h_1}{3h_2}\right)} \right\}$ where $b_1 = b - t$ $h_2 = h - t$		$e = \left(\dfrac{b t_1 h_1^3}{t_1 h_1^3 + t_2 h_2^3}\right)$
	$e = b_1 \left\{ \dfrac{1 + \dfrac{1}{2}\dfrac{b_1}{h_1} - \dfrac{4}{3}\left(\dfrac{h_1}{h_2}\right)^2}{1 + \dfrac{1}{6}\dfrac{h_2}{h_1} + \dfrac{b_1}{h_1} + 2\dfrac{h_1}{h_2}\left(1 + \dfrac{2h_1}{3h_2}\right)} \right\}$ where $b_1 = b - t/2$ $h_2 = h + t$		$e = b_1 \dfrac{\dfrac{b}{b_1}\left(3\dfrac{b}{b_1} - 2\right)}{\sqrt{2}\left[\left(\dfrac{b}{b_1}\right)^3 + 3\left(\dfrac{b}{b_1}\right)^3 - 3\dfrac{b}{b_1} + 1\right]}$
	$e = \dfrac{m}{n}$ $m = 12 + 6\pi\left(\dfrac{b + h_1}{a}\right) + 6\left(\dfrac{b}{a}\right)^2 + 12\dfrac{b h_1}{a^2}$ $\quad + 3\pi\left(\dfrac{h_1}{a}\right)^2 - 4\left(\dfrac{h_1}{a}\right)^3\dfrac{b}{a}$ $n = 3\pi + 12\left(\dfrac{b + h_1}{a}\right) + 4\left(\dfrac{h_1}{a}\right)^2\left(3 + \dfrac{h_1}{a}\right)$		$e = \dfrac{2A}{h + L(t_h/t_b)}$ where $A = $ area $L = $ Length of Dotted Line
	C is at the centroid triangle $e = 0.47a$ for narrow triangle $(\alpha > 12°)$ approx.		

TABLE 25–26
Location of shear center for various cross sections (*Cont.*)

Section	Location of shear center	Section	Location of shear center
	$$e = \frac{2a[(\pi - \phi)\cos\phi + \sin\phi]}{[(\pi - \phi) + \sin\phi\cos\phi]}$$ For $\phi = \dfrac{\pi}{2}$, $e = \dfrac{4a}{\pi}$		$$e_1 = c - \frac{b_1^2 ht}{6(I_y I_x - I_{xy}^2)}\,[3I_{xy}(h-c)$$ $$+ I_x(2b_1 - 3d)]$$ $$e_2 = d + \frac{b_1^2 ht}{6(I_y I_x - I_{xy}^2)}\,[I_{xy}(2b_1 - 3d)$$ $$+ 3I_y(h-c)]$$ where $$c = \frac{h^2 + 2b_1 h}{2h(b_1 + b_2)}, \quad d = \frac{b_1^2 + b_2^2}{2h(b_1 + b_2)}$$ $I_x,\ I_y$ = moment of inertia of section about x and y axes, respectively I_{xy} = product of moment of inertia
	$$e_x = \frac{b}{2}\left(\frac{ht_w^3}{ht_w^3 + t_f b^3}\right)$$ $$e_y = \frac{h}{2}\left(\frac{ht_w^3}{ht_w^3 + t_f b^3}\right)$$		
	$$e = \frac{1}{2}(t_f + b)\left\{\frac{1}{1 + \dfrac{h^3 t_f}{t_w^3 b}}\right\}$$		$$e = \left(\frac{1 + 3v}{1 + v}\right)\frac{\int x t^3\, dx}{\int t^3\, dx}$$

TABLE 25–27
Approximate formulas for torsional shearing stress and angle of twist for various cross sections

Cross section	Shearing stress, lbf/in² (N/m² or MPa)	Angle of twist per unit length, ϕ, rad/in (rad/m)
	$\tau_c = \dfrac{2M_t}{\pi ab^2}$ $= \dfrac{2M_t}{Ab}$	$\phi = \dfrac{M_t(a^2 + b^2)}{G\pi a^3 b^3}$ $= \dfrac{4\pi^2 J M_t}{A^4 G}$
	$\tau = \dfrac{20M_t}{b^3}$	$\phi = \dfrac{46.2M_t}{Gb^4}$
	$\tau_c = \dfrac{M_t}{k_1 ab^2}$	$\phi = \dfrac{M_t}{k_2 ab^3 G}$
	$\tau = \dfrac{M_t}{2\pi r^2 \delta}$	$\phi = \dfrac{M_t}{2\pi r^3 \delta G}$
	$\tau = \dfrac{3M_t}{2\pi r t^2}$	$\phi = \dfrac{3M_t}{2\pi r t^3 G}$
	$\tau = \dfrac{M_t}{2\pi ab\delta}$	$\phi = \dfrac{M_t\sqrt{2(a^2 + b^2)}}{4\pi a^2 b^2 \delta G}$
	$\tau_c = \dfrac{M_t}{2abt_1}$ $\tau_D = \dfrac{M_t}{2abt}$	$\phi = \dfrac{M_t(at + bt_1)}{2tt_1 a^2 b^2 G}$

A = area of cross section

TABLE 25–28
Variation of k_1 and k_2 with the ratio a/b

a/b	1	1.2	1.5	2.0	2.5	3.0	4.0	5.0	10.0	∞
k_1	0.208	0.219	0.231	0.246	0.258	0.267	0.282	0.391	0.312	0.333
k_2	0.141	0.166	0.196	0.229	0.249	0.263	0.281	0.291	0.312	0.333

TABLE 25–29
Stress concentration factors for structural angle, K_σ

r/c	0.125	0.250	0.500	0.750	1.000
K_σ	2.550	2.250	2.000	1.875	1.800

TABLE 25–30
Formulas for maximum lateral bending moment and angle of twist of beams subjected to torsion[a]

Type of loading and support	Maximum lateral bending moment in flange, lbf in (N m)	Angle of twist of beam of length L, ϕ rad
	$M_{b(max)} = \dfrac{M_t k}{h} \tan h \dfrac{L}{2k}$ $= \dfrac{M_t k}{h}$ $\dfrac{L}{2k} > 2.5$	$\phi = \dfrac{M_t}{JG}\left(L - 2k \tan h \dfrac{L}{2k}\right)$ $= \dfrac{M_t}{JG}(L - 2k)^{b}$ if $\dfrac{L}{2k} > 2.5$
	$M_{b(max)} = \dfrac{M_t k}{2h}\left(\cot h \dfrac{L}{2k} - \dfrac{2k}{L}\right)$ $= \dfrac{M_t k}{h}$ if $\dfrac{L}{2k}$ is large	$\phi = \dfrac{M_t}{2JG}\left(\dfrac{L}{4} - k \tan h \dfrac{L}{4k}\right)$ $= \dfrac{M_t}{2JG}\left(\dfrac{L}{4} - k\right)^{b}$ if $\dfrac{L}{4k} > 2.5$
	$M_{b(max)} = \dfrac{M_t k}{h}\left(\cot h \dfrac{L}{k} - \dfrac{k}{L}\right)$ $= \dfrac{M_t k}{h}$ if $\dfrac{L}{k}$ is large	$\phi = \dfrac{M_t}{JG}\left(\dfrac{L}{2} - k \tan h \dfrac{L}{2k}\right)$ $= \dfrac{M_t}{JG}\left(\dfrac{L}{2} - k\right)^{b}$ if $\dfrac{L}{2k} > 2.5$
	$M_{b(max)} = \dfrac{M_t k}{h} \dfrac{\sin h \dfrac{L_1}{k} \sin h \dfrac{L_2}{k}}{\sin h \dfrac{L}{k}}$ $= \dfrac{M_t k}{2h}$ if $\dfrac{L_1}{k}$ and $\dfrac{L_2}{k} > 2$ error is small	$\phi = \dfrac{1}{2}\dfrac{M_t}{JG}\left(\dfrac{L}{2} - k \tan h \dfrac{L}{2k}\right)$ (approx.) $= \dfrac{1}{2}\dfrac{M_t}{JG}\left(\dfrac{L}{2} - k\right)^{b}$ if $\dfrac{L}{2k} > 2.5$

[a] Formulas given in Table 25–30 can also be used for Z and channel sections.
[b] Error is small for the conditions $L/2k > 2.5$ and $L/4k > 2.5$.

TABLE 25–31

Geometrical properties, weight, and nominal dimensions of beams, channels, and T-bars

	Dimensions of the section							Center of gravity, C_{yy}	Sectional area, A	Weight per meter, w	Moments of inertia		Radius of gyration		Section moduli	
Designation	Depth of beam, h	Width of flange, b	Thickness of web, t_w	Thickness of flange, t_f	Slope of flange, D	Radius at root, $r_1(r_r)$	Radius at toe, $r_2(r_t)$				I_{xx}	I_{yy}	r_{xx}	r_{yy}	Z_{xx}	Z_{yy}
(1)	(2)	(3)	(4)	(5)	(6)	(7)	(8)	(9)	(10)	(11)	(12)	(13)	(14)	(15)	(16)	(17)
Section	mm	mm	mm	mm	deg	mm	mm	cm	cm²	kg	cm⁴	cm⁴	cm	cm	cm³	cm³
Beam or column section (See Fig. 25–74)																
ISJB 150	150	50	3.0	4.6	91.5	5.0	1.5		9.01	7.1	322.1	9.2	5.98	1.01	42.9	3.7
ISJB 175	175	50	3.2	4.3	91.5	5.0	1.5		10.28	8.1	479.3	9.7	6.83	0.97	54.8	3.9
ISJB 200	200	60	3.4	5.0	91.5	5.0	1.5		12.64	9.9	780.7	17.3	7.86	1.17	78.1	5.8
ISJB 225	225	80	3.7	5.0	91.5	6.5	1.5		16.28	12.8	1308.5	40.5	8.97	1.58	116.3	10.1
ISLB 75	75	50	3.7	5.0	91.5	6.5	2.0		7.71	6.1	72.7	10.0	3.07	1.14	19.4	4.0
ISLB 100	100	50	4.0	6.4	91.5	7.0	3.0		10.21	8.0	168.0	12.7	4.06	1.12	33.6	5.1
ISLB 125	125	75	4.4	6.5	91.5	8.0	3.0		15.12	11.9	406.8	43.4	5.19	1.69	65.1	11.6
ISLB 150	150	80	4.8	6.8	91.5	9.5	3.0		18.08	14.2	688.2	55.2	6.17	1.75	91.8	13.8
ISLB 175	175	90	5.1	6.9	91.5	9.5	3.0		21.30	16.7	1096.2	79.6	7.17	1.93	125.3	17.7
ISLB 200	200	100	5.4	7.3	91.5	9.5	3.0		25.27	19.8	1696.6	115.4	8.19	2.13	169.7	23.1
ISLB 225	225	100	5.8	8.6	98	12.0	6.0		29.92	23.5	2501.9	112.7	9.15	1.94	222.4	22.5
ISLB 250	250	125	6.1	8.2	98	13.0	6.5		35.53	27.9	3917.8	193.4	10.23	2.33	297.4	30.9
ISLB 275	275	140	6.4	8.8	98	14.0	7.0		42.02	33.0	5375.3	287.0	11.31	2.61	392.4	41.0
ISLB 300	300	150	6.7	9.4	98	15.0	7.5		48.08	37.7	7332.9	376.2	12.35	2.80	488.9	50.2
ISLB 325	325	165	7.0	9.8	98	16.0	8.0		54.90	43.1	9874.6	510.8	13.41	3.05	607.7	61.9
ISLB 350	350	165	7.4	11.4	98	16.0	8.0		63.01	49.5	13158.3	631.9	14.45	3.15	751.9	76.6
ISLB 400	400	165	8.0	12.5	98	16.0	8.0		72.43	56.9	19306.3	716.4	16.33	3.15	965.3	86.8
ISLB 450	450	170	8.6	13.4	98	16.0	8.5		83.14	65.3	27536.1	853.0	18.20	3.20	1223.8	100.4
ISLB 500	500	180	9.2	14.1	98	17.0	9.0		95.50	75.0	38579.0	1063.9	20.10	3.34	1543.2	118.2
ISLB 550	550	190	9.9	15.0	98	18.0	10.0		109.97	86.3	53161.6	1335.1	21.99	3.48	1933.2	140.5
ISLB 600	600	210	10.5	15.5	98	20.0	10.0		126.69	99.5	72867.6	1821.9	23.98	3.79	2428.9	173.5
ISMB 100	100	75	4.0	7.2	98	9.0	4.5		14.60	11.5	257.5	40.8	4.20	1.67	51.5	10.9
ISMB 125	125	75	4.4	7.6	98	9.0	4.5		16.60	12.0	449.0	43.7	5.20	1.62	71.8	11.7
ISMB 150	150	80	4.8	7.6	98	10.0	4.5		19.00	14.9	726.4	52.6	6.18	1.66	96.9	13.1
ISMB 175	175	90	5.5	8.6	98	11.0	5.0		24.62	19.3	1272.0	85.0	7.19	1.86	145.4	18.9
ISMB 200	200	100	5.7	10.8	98	11.0	5.5		32.33	25.4	2235.4	150.0	8.32	2.15	223.5	30.0
ISMB 225	225	110	6.5	11.8	98	12.0	6.0		39.72	31.2	3441.8	218.3	9.31	2.34	305.9	39.7
ISMB 250	250	125	6.9	12.5	98	12.0	6.5		47.55	37.3	5131.6	334.5	10.39	2.65	410.5	53.5
ISMB 300	300	140	7.5	12.4	98	14.0	7.0		56.26	44.2	8603.6	453.9	12.37	2.84	573.6	64.8
ISMB 350	350	140	8.1	14.2	98	14.0	7.0		66.71	52.4	13630.3	537.7	14.29	2.84	778.9	76.8
ISMB 400	400	140	8.9	16.0	98	14.0	7.0		78.46	61.6	20450.4	622.1	16.15	2.82	1022.9	88.9
ISMB 450	450	150	9.4	17.4	98	15.0	7.5		92.27	72.4	30390.8	834.0	18.15	3.01	1350.7	112.2
ISMB 500	500	180	10.2	17.2	98	17.0	8.5		110.74	86.9	45218.3	1369.8	20.21	3.52	1808.7	152.2
ISMB 550	550	190	11.2	19.3	98	18.0	9.0		132.11	103.7	64893.6	1833.8	22.16	3.73	2359.8	193.0
ISMB 600	600	210	12.0	20.8	98	20.0	10.0		156.21	122.6	91813.0	2651.0	24.24	4.12	3060.4	252.5
ISWB 150	150	100	5.4	7.0	96	8.0	4.0		21.67	17.0	839.1	94.8	6.22	2.09	111.9	19.0
ISWB 175	175	125	5.8	7.4	96	8.0	4.0		28.11	22.1	1509.4	188.6	7.33	2.59	172.5	30.2
ISWB 200	200	140	6.1	9.0	96	9.0	4.5		36.71	28.8	2624.5	328.8	8.46	2.99	262.5	47.0
ISWB 225	225	150	6.4	9.9	96	9.0	4.5		43.24	33.9	3920.5	448.6	9.52	3.22	348.5	59.8
ISWB 250	250	200	6.7	9.0	96	10.0	5.0		52.05	40.9	5943.1	857.5	10.69	4.06	475.4	85.7
ISWB 300	300	200	7.4	10.0	96	10.0	5.5		61.33	48.1	9821.6	990.1	12.66	4.02	654.8	99.0
ISWB 350	350	200	8.0	11.4	96	11.0	6.0		72.50	56.9	15521.7	1175.9	14.63	4.03	887.0	117.6
ISWB 400	400	200	8.6	13.0	96	12.0	6.5		85.01	66.7	23426.7	1388.0	16.60	4.04	1171.3	138.8
ISWB 450	450	200	9.2	15.4	96	14.0	7.0		101.15	79.4	35057.6	1706.7	18.63	4.11	1558.1	17.07
ISWB 500	500	250	9.9	14.7	96	15.0	7.5		121.22	95.2	52290.9	2987.8	20.77	4.96	2091.6	239.0

TABLE 25-31
Geometrical properties, weight, and nominal dimensions of beams, channels, and T-bars (Cont.)

		Dimensions of the section									Moments of inertia		Radius of gyration		Section moduli	
Designation	Depth of beam, h	Width of flange, b	Thickness of web, t_w	Thickness of flange, t_f	Slope of flange, D	Radius at root, $r_1(r_r)$	Radius at toe, $r_2(r_t)$	Center of gravity, C_{yy}	Sectional area, A	Weight per meter, w	I_{xx}	I_{yy}	r_{xx}	r_{yy}	Z_{xx}	Z_{yy}
(1)	(2)	(3)	(4)	(5)	(6)	(7)	(8)	(9)	(10)	(11)	(12)	(13)	(14)	(15)	(16)	(17)
	mm	mm	mm	mm	deg	mm	mm	cm	cm²	kg	cm⁴	cm⁴	cm	cm	cm³	cm³
ISWB 550	550	250	10.5	17.6	96	16.0	8.0		143.34	112.5	74906.1	3740.5	22.85	5.11	2723.9	299.2
ISWB 600	600	250	11.2	21.3	96	17.0	8.5		170.38	133.7	106198.5	4702.5	24.97	5.25	3540.0	376.2
ISHB 150	150	150	5.4	9.0	96	8.0	4.0		34.48	27.1	1455.6	431.7	6.50	3.54	194.1	57.6
ISHB 200	200	200	6.1	9.0	94	9.0	4.5		47.54	37.3	3608.4	967.1	8.71	4.51	360.8	96.7
ISHB 225	225	225	6.5	9.1	94	10.0	5.0		54.94	43.1	5279.5	1353.8	9.80	4.96	489.3	120.3
ISHB 250	250	250	6.9	9.7	94	10.0	5.0		64.96	51.0	7736.5	1961.3	10.91	5.49	618.9	156.9
ISHB 300	300	250	7.6	10.6	94	11.0	5.5		74.85	58.8	12545.2	2193.6	12.95	5.41	836.3	175.5
ISHB 350	350	250	8.3	11.6	94	12.0	6.0		85.91	67.4	19159.2	2451.4	14.93	5.34	1094.8	196.1
ISHB 400	400	250	9.1	12.7	94	14.0	7.0		98.66	77.4	28083.5	2728.3	16.87	5.26	1404.2	218.3
ISHB 450	450	250	9.8	13.7	94	15.0	7.5		111.14	87.2	39210.8	2985.2	18.78	5.18	1742.7	238.8
Channel Section																
ISJC 100	100	45	3.0	5.1	91.5	6.0	2.0	1.40	7.41	5.8	123.8	14.9	4.09	1.42	24.8	4.8
ISJC 125	125	50	3.0	6.6	91.5	6.0	2.4	1.64	10.07	7.9	270.0	25.7	5.18	1.60	43.2	7.6
ISJC 150	150	55	3.6	6.9	91.5	7.0	2.4	1.66	12.65	9.9	471.1	37.9	6.10	1.73	62.8	9.9
ISJC 175	175	60	3.6	6.9	91.5	7.0	3.0	1.75	14.24	11.2	719.9	50.5	7.11	1.88	82.3	11.9
ISJC 200	200	70	4.1	7.1	91.5	8.0	3.2	1.97	17.77	13.9	1161.2	84.2	8.08	2.18	116.1	16.7
ISLC 75	75	40	3.7	6.0	91.5	6.0	2.0	1.35	7.26	5.7	66.1	11.5	3.02	1.26	17.6	4.3
ISLC 100	100	50	4.0	6.4	91.5	6.0	2.0	1.62	10.02	7.9	164.7	24.8	4.06	1.57	32.9	7.3
ISLC 125	125	65	4.4	6.6	91.5	7.0	2.4	2.04	13.67	10.7	356.8	57.2	5.11	2.05	57.1	12.8
ISLC 150	150	75	4.8	7.8	91.5	8.0	2.4	2.38	18.36	14.4	697.2	103.2	6.19	2.37	93.0	20.2
ISLC 175	175	75	5.1	9.5	91.5	8.0	3.2	2.40	22.40	17.6	1148.4	126.5	7.16	2.38	131.3	24.8
ISLC 200	200	75	5.5	10.8	96.0	8.5	3.2	2.35	26.22	20.6	1725.5	146.9	8.11	2.37	172.6	28.5
ISLC 225	225	90	5.8	10.2	96.0	11.0	3.2	2.46	30.53	24.0	2547.9	209.5	9.14	2.62	226.5	32.0
ISLC 250	250	100	6.1	10.7	96.0	11.0	3.2	2.70	35.65	28.0	3687.5	298.9	10.17	2.89	295.0	40.9
ISLC 300	300	100	6.7	11.6	96.0	12.0	3.2	2.55	42.11	33.1	6047.9	346.0	11.98	2.97	603.2	46.4
ISLC 350	350	100	7.4	12.5	96.0	13.0	4.8	2.41	49.47	38.8	9382.6	394.6	13.72	2.82	532.1	52.0
ISLC 400	400	100	8.0	14.0	96.0	14.0	4.8	2.36	58.25	45.7	13989.5	460.4	15.50	2.81	699.5	60.2
ISMC 75	75	40	4.4	7.3	96.0	8.5	2.4	1.31	8.67	6.8	76.0	12.6	2.96	1.21	20.3	4.7
ISMC 100	100	50	4.7	7.5	96.0	9.0	2.4	1.53	11.70	9.2	186.7	25.9	4.0	1.49	37.3	7.5
ISMC 125	125	65	5.0	8.1	96.0	9.5	2.4	1.94	16.19	12.7	416.4	59.9	5.07	1.92	66.6	13.1
ISMC 150	150	75	5.4	9.0	96.0	10.0	2.4	2.22	20.88	16.4	779.4	102.3	6.11	2.21	103.9	19.4
ISMC 175	175	75	5.7	10.2	96.0	10.5	3.2	2.20	24.38	19.1	1223.3	121.0	7.08	2.23	139.8	22.8
ISMC 200	200	75	6.4	11.4	96.0	12.0	3.2	2.17	28.21	22.1	1819.3	140.4	8.03	2.23	181.9	26.3
ISMC 225	225	80	6.7	12.4	96.0	12.0	3.2	2.30	33.01	25.9	2694.6	187.2	9.03	2.38	239.5	32.8
ISMC 250	250	80	7.1	14.1	96.0	12.0	3.2	2.30	38.67	30.4	3816.8	219.1	9.94	2.38	305.3	38.4
ISMC 300	300	90	7.6	13.6	96.0	13.0	3.2	2.36	45.64	35.8	6362.6	310.8	11.81	2.61	424.2	46.8
ISMC 350	350	100	8.1	13.5	96.0	14.0	4.8	2.44	53.66	41.1	10008.0	430.6	13.66	2.83	571.9	57.0
ISMC 400	400	100	8.6	15.3	96.0	15.0	4.8	2.42	62.93	49.4	15082.8	504.8	15.48	2.83	754.1	66.6

Section

(See Fig. 25-75)

TABLE 25-31
Geometrical properties, weight, and nominal dimensions of beams, channels, and T-bars (*Contd.*)

Section	Designation	Depth of beam, h	Width of flange, b	Thickness of web, t_w	Thickness of flange, t_f	Slope of flange, D_a	Radius at root, $r_1(r_r)$	Radius at toe, $r_2(r_t)$	Center of gravity, C_{yy}	Sectional area, A	Weight per meter, w	I_{xx}	I_{yy}	r_{xx}	r_{yy}	Z_{xx}	Z_{yy}
		(2)	(3)	(4)	(5)	(6)a	(7)	(8)	(9)	(10)	(11)	(12)	(13)	(14)	(15)	(16)	(17)
	(1)	mm	mm	mm	mm	deg	mm	mm	cm	cm²	kg	cm⁴	cm⁴	cm	cm	cm³	cm³
						Normal T-Bar											
(See Fig. 25-76)	ISNT 20	20	20	4.0	4.0		4.0	3.0	0.60	1.45	1.1	0.5	0.2	0.58	0.41	0.3	0.2
	ISNT 30	30	30	4.0	4.0		5.0	3.5	0.32	2.26	1.8	0.8	0.8	0.89	0.59	0.8	0.5
	ISNT 40	40	40	6.0	6.0		5.5	4.0	1.14	4.45	3.5	6.1	2.9	1.18	0.81	2.1	1.5
	ISNT 50	50	50	6.0	6.0		6.0	4.0	1.35	5.66	4.4	12.3	5.7	1.47	1.01	3.4	2.3
	ISNT 60	60	60	6.0	6.0		6.5	4.5	1.56	6.85	5.4	21.4	9.7	1.77	1.19	4.8	3.2
	ISNT 75	75	75	9.0	9.0		8.0	5.5	2.04	12.69	10.0	62.0	29.2	2.21	1.52	11.4	7.8
	ISNT 100	100	100	10.0	10.0		9.0	6.0	2.62	18.97	14.9	163.9	76.8	2.94	2.01	22.2	15.4
	ISNT 150	150	150	10.0	10.0		10.0	7.0	3.61	28.88	22.7	541.1	250.3	4.33	2.94	47.5	33.4
						Slit and Deep-Legged T-Bar											
(See Fig. 25-77)	ISDT 100	100	50	5.8	10.0	98.0	8.0	4.0	3.03	10.37	8.1	99.0	9.6	3.09	0.96	14.2	3.8
	ISDT 150	150	75	8.0	11.6	98.0	9.0	4.5	4.75	19.96	15.7	450.2	37.0	4.75	1.36	43.9	9.9
	ISDT 200	200	165	8.0	12.5	98.0	16.0	8.0	4.78	36.22	28.4	1267.8	358.2	5.92	3.15	83.3	43.4
	ISDT 250	250	780	9.2	14.1	98.0	17.0	8.5	6.40	47.75	37.5	2774.4	532.0	7.62	3.34	149.2	59.1
	ISMT 50	50	75	4.0	7.2	98.0	9.0	4.5	0.96	7.30	5.7	9.7	20.4	1.15	1.67	2.4	5.4
	ISMT 62.5	62.5	75	4.4	7.6	98.0	9.0	4.5	1.30	8.30	6.5	21.3	21.9	1.60	1.62	4.3	5.8
	ISMT 75	75	80	4.8	7.6	98.0	9.0	4.5	1.67	9.50	7.5	40.1	26.3	2.05	1.66	6.9	6.6
	ISMT 87.5	87.5	90	5.5	8.6	98.0	10.0	5.0	1.98	12.31	9.7	72.6	42.5	2.43	1.86	10.7	9.4
	ISMT 100	100	100	5.7	10.8	98.0	11.0	5.5	2.13	16.16	12.7	115.8	75.0	2.68	2.15	14.7	15.0
	ISMT* 50	50	70	4.5	7.5	98.0	9.0	4.5	1.04	7.35	5.8	10.8	17.7	1.21	1.55	2.7	5.0
	ISMT* 62.5	62.5	70	4.8	8.0	98.0	9.0	4.5	1.39	8.40	6.6	22.8	19.2	1.65	1.51	4.7	5.5
	ISMT* 75	75	75	5.0	8.0	98.0	9.0	4.5	1.73	9.54	7.5	41.2	23.4	2.08	1.57	7.1	6.2
	ISMT* 87.5	87.5	85	5.8	9.0	98.0	10.0	5.0	2.06	12.43	9.8	75.6	38.4	2.47	1.76	11.3	9.0
	ISHT 75	75	150	8.4	9.0	94.0	8.0	4.0	1.62	19.49	15.3	96.2	230.2	2.22	3.44	16.4	30.1
	ISHT 100	100	200	7.8	9.0	94.0	9.0	4.5	1.91	25.47	20.0	193.8	497.3	2.76	4.42	24.0	49.3
	ISHT 125	125	250	8.8	9.7	94.0	10.0	5.0	2.37	34.85	27.4	415.4	1005.8	3.45	5.37	41.0	79.9
	ISHT 150	150	250	7.6	10.6	94.0	11.0	5.5	2.66	37.42	29.4	573.7	1096.8	3.92	5.41	46.5	87.7

Key: ISJB—Indian Standard Junior Beams; ISLB—Indian Standard Light-Weight Beams; ISMB—Indian Standard Medium-Weight Beams; ISWB—Indian Standard Wide-Flange Beams; ISHB—Indian Standard Column H-Section Beams; ISJC—Indian Standard Junior Channel; ISLC—Indian Standard Light-Weight Channel; ISMC—Indian Standard Medium-Weight Channel; ISNT—Indian Standard Normal Tea-Bar (T-Bar); ISNT—Indian Standard Provisional Slit Medium-Weight Tee-Bars; ISDT—Indian Standard Deep-Legged Tee-Bar; ISLT—Indian Standard Slit Light-Weight Tee-Bar; ISMT—Indian Standard Slit Medium Weight Tee-Bar; ISHT—Indian Standard Slit Tee-Bar from H-Section.
Source: IS 808, 1964; IS 1173, 1967.

TABLE 25-32
General properties, weight, and nominal dimensions of equal and unequal angles

	Dimensions of the section					Center of gravity				Moment of inertia				Radii of gyration				Moduli of section		
	Section dimensions		Thickness, t, mm	Radius at root, r_1, mm	Radius at toe, r_2, mm			Sectional area, A, cm²	Weight per meter, w, kg	I_{xx}, cm⁴	I_{yy}, cm⁴	I_{uu} (max), cm⁴	I_{vv} (min), cm⁴	r_{xx}, cm	r_{yy}, cm	r_{uu} (max), cm	r_{vv} (min), cm	Z_{xx}, cm³	Z_{yy}, cm³	$\tan \alpha$
Designation	A, mm	B, mm				C_{xx}, cm	C_{yy}, cm													
(1)	(2)	(3)	(4)	(5)	(6)ᵃ	(7)	(8)	(9)	(10)	(11)	(12)	(13)	(14)	(15)	(16)	(17)	(18)	(19)	(20)	(21)
Section																				
Equal Angle Section (see Fig. 25-78) ISA 2020	20	20	3.0	4.0		0.59	0.59	1.12	0.9	0.4	0.4	0.6	0.2	0.58	0.58	0.73	0.37	0.3	0.3	0.3
			4.0			0.63	0.63	1.45	1.1	0.5	0.5	0.8	0.2	0.58	0.58	0.72	0.37	0.4	0.4	0.4
ISA 2525	25	25	3.0	4.5		0.71	0.71	1.41	1.1	0.8	0.8	1.2	0.3	0.73	0.73	0.93	0.47	0.4	0.4	0.4
			4.0			0.75	0.75	1.84	1.4	1.0	1.0	1.6	0.4	0.73	0.73	0.91	0.47	0.6	0.6	0.6
			5.0			0.79	0.79	2.25	1.8	1.2	1.2	1.8	0.5	0.72	0.72	0.91	0.47	0.7	0.7	0.7
ISA 3030	30	30	3.0	5.0		0.83	0.83	1.73	1.4	1.4	1.4	2.2	0.6	0.89	0.89	1.13	0.57	0.6	0.6	0.6
			4.0			0.87	0.87	2.26	1.8	1.8	1.8	2.8	0.7	0.89	0.89	1.12	0.57	0.8	0.8	0.8
			5.0			0.92	0.92	2.77	2.2	2.1	2.1	3.4	0.9	0.88	0.88	1.11	0.57	1.0	1.0	1.0
ISA 3535	35	35	3.0	5.0		0.95	0.95	2.03	1.6	2.3	2.3	3.6	0.9	1.05	1.05	1.33	0.67	0.9	0.9	0.9
			4.0			1.00	1.00	2.66	2.1	2.9	2.9	4.7	1.2	1.05	1.05	1.32	0.67	1.2	1.2	1.2
			5.0			1.04	1.04	3.27	2.6	3.5	3.5	5.6	1.5	1.04	1.04	1.31	0.67	1.4	1.4	1.4
ISA 4040	40	40	3.0			1.08	1.08	2.34	1.8	3.4	3.4	5.5	1.4	1.21	1.21	1.54	0.77	1.2	1.2	1.2
			4.0			1.12	1.12	3.07	2.4	4.5	4.5	7.1	1.8	1.21	1.21	1.53	0.77	1.6	1.6	1.6
			5.0			1.16	1.16	3.78	3.0	5.4	5.4	8.6	2.2	1.20	1.20	1.51	0.77	1.9	1.9	1.9
ISA 4545	45	45	3.0	5.5		1.20	1.20	2.64	2.1	5.0	5.0	8.0	2.0	1.38	1.38	1.74	0.87	1.5	1.5	1.5
			4.0			1.25	1.25	3.47	2.7	6.5	6.5	10.4	2.6	1.37	1.37	1.73	0.87	2.0	2.0	2.0
			6.0			1.33	1.33	5.07	4.0	9.2	9.2	14.6	3.8	1.35	1.35	1.70	0.87	2.9	2.9	2.9
ISA 5050	50	50	3.0	6.0		1.32	1.32	2.95	2.3	6.9	6.9	11.1	2.8	1.53	1.53	1.94	0.97	1.9	1.9	1.9
			4.0			1.37	1.37	3.88	3.0	9.1	9.1	14.5	3.6	1.53	1.53	1.93	0.97	2.5	2.5	2.5
			6.0			1.45	1.45	5.68	4.5	12.9	12.9	20.6	5.3	1.51	1.51	1.90	0.96	3.6	3.6	3.6
ISA 5555	55	55	5.0	6.5		1.53	1.53	5.27	4.1	14.7	14.7	23.5	6.9	1.67	1.67	2.11	1.06	3.7	3.7	3.7
			6.0			1.57	1.57	6.26	4.9	17.3	17.3	27.5	7.0	1.66	1.66	2.10	1.06	4.4	4.4	4.4
			10.0			1.72	1.72	10.02	7.9	26.3	26.3	41.5	11.2	1.62	1.62	2.03	1.06	7.0	7.0	7.0
ISA 6060	60	60	5.0	6.5		1.65	1.65	5.75	4.5	19.2	19.2	30.6	7.7	1.82	1.82	2.31	1.16	4.4	4.4	4.4
			6.0			1.69	1.69	6.84	5.4	22.6	22.6	36.0	9.1	1.80	1.80	2.29	1.15	5.2	5.2	5.2
			8.0			1.77	1.77	8.96	7.0	29.0	29.0	46.0	11.9	1.80	1.80	2.27	1.15	6.8	6.8	6.8
ISA 6565	65	65	5.0	6.5		1.77	1.77	6.25	4.9	24.7	24.7	39.4	9.9	1.99	1.99	2.51	1.26	5.2	5.2	5.2
			6.0			1.81	1.81	7.44	5.8	29.1	29.1	46.5	11.7	1.98	1.98	2.50	1.26	6.2	6.2	6.2
			10.0			1.97	1.97	12.00	9.4	45.0	45.0	71.3	18.8	1.94	1.94	2.44	1.25	9.9	9.9	9.9
ISA 7070	70	70	5.0	7.0		1.89	1.89	6.77	5.3	31.1	31.1	49.8	12.5	2.15	2.15	2.71	1.36	6.1	6.1	6.1
			6.0			1.94	1.94	8.06	6.3	36.8	36.8	58.8	14.8	2.14	2.14	2.70	1.36	7.3	7.3	7.3
			8.0			2.02	2.02	10.58	8.3	47.4	47.4	75.5	19.3	2.12	2.12	2.67	1.35	9.5	9.5	9.5
ISA 7575	75	75	5.0	7.0		2.02	2.02	7.27	5.7	38.7	38.7	61.9	15.5	2.31	2.31	2.92	1.46	7.1	7.1	7.1
			6.0			2.06	2.06	8.66	6.8	45.7	45.7	73.1	18.4	2.30	2.30	2.91	1.46	8.4	8.4	8.4
			10.0			2.22	2.22	14.02	11.0	71.4	71.4	113.3	29.4	2.26	2.26	2.84	1.45	13.5	13.5	13.5
ISA 8080	80	80	6.0	8.0		2.18	2.18	9.29	7.3	56.0	56.0	89.6	22.5	2.46	2.46	3.11	1.52	9.6	9.6	9.6
			8.0			2.27	2.27	12.21	9.6	72.5	72.5	115.6	29.4	2.44	2.44	3.08	1.55	12.6	12.6	12.6
			10.0			2.34	2.34	15.05	11.8	87.7	87.7	139.5	36.0	2.41	2.41	3.04	1.55	15.5	15.5	15.5
ISA 9090	90	90	6.0	8.0		2.42	2.42	10.47	8.2	80.1	80.1	128.1	32.0	2.77	2.77	3.50	1.75	12.2	12.2	12.2
			8.0			2.51	2.51	13.79	10.8	104.2	104.2	166.4	42.0	2.75	2.75	3.47	1.75	16.0	16.0	16.0
			10.0			2.59	2.59	17.03	13.4	126.7	126.7	201.9	51.6	2.73	2.73	3.44	1.70	19.2	19.8	19.8
ISA 100100	100	100	6.0	8.5		2.67	2.67	11.67	9.2	111.3	111.3	178.1	44.5	3.09	3.09	3.91	1.95	15.8	15.2	15.2
			8.0			2.76	2.76	15.39	12.1	145.1	145.1	231.8	58.4	3.07	3.07	3.88	1.95	20.0	20.0	20.0
			12.0			2.92	2.92	22.59	17.7	207.0	207.0	329.3	87.7	3.03	3.03	3.82	1.94	29.2	29.2	29.2
ISA 110110	110	110	8.0	10.0		3.00	3.00	17.08	13.4	196.8	196.8	312.7	81.0	3.40	3.40	4.28	2.18	24.6	24.6	24.6
			10.0			3.09	3.09	21.12	16.6	240.2	240.2	381.5	98.9	3.37	3.37	4.25	2.16	30.4	30.4	30.4
			16.0			3.32	3.32	32.76	25.7	357.3	357.3	564.3	150.0	3.30	3.30	4.15	2.14	46.5	46.5	46.5

TABLE 25–32
General properties, weight, and nominal dimensions of equal and unequal angles (*Cont.*)

Group headers: **Dimensions of the section** (cols 2–6) · **Center of gravity** (cols 7–8) · **Moment of inertia** (cols 11–14) · **Radii of gyration** (cols 15–18) · **Moduli of section** (cols 19–21)

Designation	Section dim. A, mm	B, mm	Thickness, t, mm	Radius at root, r_1, mm	Radius at toe, r_2, mm	C_{xx}, cm	C_{yy}, cm	Sectional area, A, cm^2	Weight per meter, w, kg	I_{xx}, cm^4	I_{yy}, cm^4	I_{uu} (max), cm^4	I_{vv} (min), cm^4	r_{xx} (max), cm	r_{yy}, cm	r_{uu} (max), cm	r_{vv} (min), cm	Z_{xx}, cm^3	Z_{yy}, cm^3	tan α
(1)	(2)	(3)	(4)	(5)	(6)a	(7)	(8)	(9)	(10)	(11)	(12)	(13)	(14)	(15)	(16)	(17)	(18)	(19)	(20)	(21)
ISA 130130	130	130	8.0	10.0		3.50	3.50	20.28	15.9	331.0	331.0	526.3	135.6	4.04	4.04	5.10	2.59	34.9	34.9	
			10.0			3.59	3.59	25.12	19.7	405.3	405.3	644.6	166.0	4.02	4.02	5.07	2.57	43.1	43.1	
			12.0			2.67	3.67	29.88	23.5	476.4	476.4	757.1	195.6	3.99	3.99	5.03	2.56	51.0	51.0	
			16.0			3.82	3.82	39.16	30.7	609.1	609.1	965.6	252.6	3.94	3.94	4.97	2.54	66.3	66.3	
ISA 150150	150	150	10.0	12.0		4.08	4.08	29.21	22.9	633.5	633.5	1007.4	259.6	4.66	4.66	5.87	2.98	58.0	58.0	
			12.0			4.16	4.16	34.77	27.3	746.3	746.3	1186.6	305.9	4.63	4.63	5.84	2.97	68.8	68.8	
			16.0			4.31	4.31	45.65	35.8	958.9	958.9	1522.5	395.3	4.58	4.58	5.77	2.94	89.7	89.7	
			20.0			4.46	4.46	56.21	44.1	1153.5	1153.5	1829.6	481.3	4.53	4.53	5.71	2.93	109.7	109.7	
ISA 200200	200	200	12.0	15.0		5.39	5.39	46.94	36.9	1826.3	1826.3	2905.4	747.2	6.24	6.24	7.87	3.99	125.0	125.0	
			16.0			5.56	5.56	61.82	48.5	2366.2	2366.2	3764.1	958.5	6.19	6.19	7.80	3.96	163.8	163.8	
			20.0			5.71	5.71	76.38	60.0	2875.0	2875.0	4568.6	1181.4	6.14	6.14	7.73	3.93	201.2	201.2	
			25.0			5.90	5.90	94.13	73.9	3470.2	3470.2	5501.5	1438.8	6.07	6.07	7.69	3.91	246.0	246.0	

Unequal Angle Section (see Fig. 25-79)

Designation	Section dim. A, mm	B, mm	Thickness, t, mm	Radius at root, r_1, mm	Radius at toe, r_2, mm	C_{xx}, cm	C_{yy}, cm	Sectional area, A, cm^2	Weight per meter, w, kg	I_{xx}, cm^4	I_{yy}, cm^4	I_{uu} (max), cm^4	I_{vv} (min), cm^4	r_{xx} (max), cm	r_{yy}, cm	r_{uu} (max), cm	r_{vv} (min), cm	Z_{xx}, cm^3	Z_{yy}, cm^3	tan α
ISA 3020	30	20	3.0	4.5		0.98	0.49	1.41	1.1	1.2	0.4	1.4	0.2	0.92	0.54	0.99	0.41	0.6	0.3	0.43
			4.0			1.02	0.53	1.84	1.4	1.5	0.5	1.8	0.3	0.92	0.54	0.98	0.41	0.8	0.4	0.42
			5.0			1.06	0.57	2.25	1.8	1.9	0.6	2.1	0.4	0.91	0.53	0.97	0.41	1.0	0.5	0.41
ISA 4025	40	25	3.0	5.0		1.30	0.57	1.88	1.5	3.0	0.9	3.3	0.5	1.25	0.68	1.33	0.52	1.1	0.6	0.38
			4.0			1.35	0.62	2.46	1.9	3.8	1.1	4.3	0.7	1.24	0.68	1.32	0.52	1.4	0.7	0.37
			5.0			1.39	0.66	3.02	2.4	4.6	1.4	5.1	0.8	1.23	0.67	1.31	0.52	1.8	0.9	0.37
			6.0			1.43	0.69	3.54	2.8	5.4	1.6	5.9	1.0	1.23	0.66	1.29	0.52	2.1	1.1	0.37
ISA 4530	45	30	3.0	5.0		1.42	0.69	2.18	1.7	4.4	1.5	5.0	0.9	1.42	0.84	1.52	0.63	1.4	0.7	0.44
			4.0			1.47	0.73	2.86	2.2	5.7	2.0	6.5	1.1	1.41	0.84	1.51	0.63	1.9	0.9	0.43
			5.0			1.51	0.77	3.52	2.8	6.9	2.4	7.9	1.4	1.40	0.83	1.50	0.63	2.3	1.1	0.43
			6.0			1.55	0.81	4.16	3.3	8.0	2.8	9.2	1.7	1.39	0.82	1.49	0.63	2.7	1.3	0.42
ISA 5030	50	30	3.0	5.5		1.63	0.65	2.34	1.8	5.9	1.6	6.5	1.0	1.59	0.82	1.67	0.65	1.7	0.7	0.36
			4.0			1.68	0.70	3.07	2.4	7.7	2.1	8.5	1.2	1.58	0.82	1.66	0.63	2.3	0.9	0.36
			5.0			1.72	0.74	3.78	3.0	9.3	2.5	10.3	1.5	1.57	0.81	1.65	0.63	2.8	1.1	0.35
			6.0			1.76	0.78	4.47	3.5	10.9	2.9	11.9	1.8	1.56	0.80	1.64	0.63	3.4	1.3	0.35
ISA 6040	60	40	5.0	6.0		1.95	0.96	4.76	3.7	16.9	6.0	19.5	3.4	1.89	1.12	2.02	0.85	4.2	2.0	0.44
			6.0			1.99	1.00	5.65	4.4	19.9	7.0	22.8	4.0	1.88	1.11	2.01	0.85	5.0	2.3	0.43
			8.0			2.07	1.08	7.37	5.8	25.4	8.8	29.0	5.2	1.86	1.10	1.98	0.84	6.5	3.0	0.42
ISA 6545	65	45	5.0	6.0		2.07	1.08	5.26	4.1	22.1	8.6	25.9	4.8	2.05	1.28	2.22	0.96	5.0	2.5	0.47
			6.0			2.11	1.12	6.25	4.9	26.0	10.1	30.4	5.7	2.04	1.27	2.21	0.95	5.9	3.0	0.47
			8.0			2.19	1.20	8.17	6.4	33.2	12.8	38.7	7.4	2.02	1.25	2.18	0.95	7.7	3.9	0.46
ISA 7045	70	45	5.0	6.5		2.27	1.04	5.52	4.3	27.2	8.3	30.9	5.1	2.22	1.26	2.36	0.96	5.7	2.5	0.41
			6.0			2.32	1.09	6.56	5.2	32.0	10.3	36.3	6.0	2.21	1.25	2.35	0.96	6.8	3.0	0.41
			8.0			2.40	1.16	8.58	6.7	41.0	13.1	46.3	7.8	2.19	1.24	2.32	0.95	8.9	3.9	0.40
			10.0			2.48	1.24	10.52	8.3	49.3	15.6	55.4	9.5	2.16	1.22	2.29	0.95	10.9	4.8	0.39
ISA 7550	75	50	5.0	6.5		2.39	1.16	6.02	4.7	34.1	12.2	39.4	6.9	2.38	1.42	2.56	1.07	6.7	3.2	0.44
			6.0			2.44	1.20	7.16	5.6	40.3	14.3	46.4	8.2	2.37	1.41	2.55	1.07	8.0	3.8	0.44
			8.0			2.52	1.28	9.38	7.4	51.8	18.3	59.4	10.6	2.35	1.40	2.52	1.06	10.4	4.9	0.43
			10.0			2.60	1.36	11.52	9.0	62.3	21.8	71.2	12.9	2.33	1.38	2.49	1.06	12.7	6.0	0.42
ISA 8050	80	50	5.0	7.0		2.60	1.12	6.27	4.9	40.6	12.3	45.7	7.2	2.55	1.40	2.70	1.07	7.5	3.2	0.39
			6.0			2.64	1.16	7.46	5.9	48.6	14.4	53.9	8.5	2.54	1.39	2.69	1.07	9.0	3.8	0.38
			8.0			2.73	1.24	9.78	7.7	61.9	18.5	69.3	11.0	2.52	1.37	2.66	1.06	11.7	4.9	0.38
			10.0			2.81	1.32	12.02	9.4	74.7	22.1	83.3	13.5	2.49	1.36	2.63	1.06	14.4	6.0	0.38

TABLE 25-32
General properties, weight, and nominal dimensions of equal and unequal angles (*Cont.*)

Section		Dimensions of the section					Center of gravity				Moment of inertia				Radii of gyration				Moduli of section		
	Designation	Section dimensions A, mm	B, mm	Thickness, t, mm	Radius at root, r_1, mm	Radius at toe, r_2, mm	C_{xx}, cm	C_{yy}, cm	Sectional area, A, cm^2	Weight per meter, w, kg	I_{xx}, cm^4	I_{yy}, cm^4	I_{uu} (max), cm^4	I_{vv} (min), cm^4	r_{xx}, cm	r_{yy}, cm	r_{uu} (max), cm	r_{vv} (min), cm	Z_{xx}, cm^3	Z_{yy}, cm^3	$\tan\alpha$
(1)		(2)	(3)	(4)	(5)	(6)[a]	(7)	(8)	(9)	(10)	(11)	(12)	(13)	(14)	(15)	(16)	(17)	(18)	(19)	(20)	(21)
	ISA 9060	90	60	6.0	7.5		2.87	1.39	8.65	6.8	70.6	25.2	81.5	14.3	2.86	1.17	3.07	1.28	11.5	5.5	0.44
				8.0			2.96	1.48	11.37	8.9	91.3	32.4	105.3	18.6	2.84	1.69	3.04	1.28	15.1	7.2	0.44
				10.0			3.04	1.55	14.01	11.0	110.9	39.1	127.3	22.8	2.81	1.67	3.01	1.27	18.6	8.8	0.43
				12.0			3.12	1.63	16.57	13.0	129.1	45.2	147.5	26.8	2.79	1.65	2.98	1.27	22.0	10.3	0.42
	ISA10065	100	65	6.0	8.0		3.19	1.47	9.55	7.5	96.7	32.4	110.6	18.6	3.18	1.84	3.40	1.39	14.2	6.4	0.42
				8.0			3.28	1.55	12.57	9.9	125.9	41.9	143.6	24.2	3.16	1.83	3.38	1.39	18.7	8.5	0.42
				10.0			3.37	1.63	15.51	12.2	153.2	50.7	174.2	29.7	3.14	1.81	3.35	1.38	23.1	10.4	0.41
	ISA 10075	100	75	6.0	8.5		3.01	1.78	10.14	8.0	100.9	48.7	124.0	25.6	3.15	2.19	3.50	1.59	14.4	8.5	0.55
				8.0			3.10	1.87	13.36	10.5	131.6	63.3	161.3	33.6	3.14	2.18	3.48	1.59	19.1	11.2	0.55
				10.0			3.19	1.95	16.50	13.0	160.4	76.9	196.1	41.2	3.12	2.16	3.45	1.58	23.6	13.8	0.55
				12.0			3.27	2.03	19.56	15.4	187.5	89.5	228.4	48.6	3.10	2.14	3.42	1.58	27.9	16.3	0.54
	ISA 12575	125	75	6.0	9.0		4.05	1.59	11.66	9.2	187.8	51.6	208.9	30.5	4.01	2.10	4.23	1.62	22.2	8.7	0.37
				8.0			4.15	1.68	15.38	12.1	245.5	67.2	272.8	40.0	4.00	2.09	4.21	1.61	29.4	11.5	0.36
				10.0			4.24	1.76	19.02	14.9	300.3	81.6	332.9	49.1	3.97	2.07	4.18	1.61	36.3	14.2	0.36
	ISA 12595	125	95	6.0	9.0	4.8	3.72	2.24	12.92	10.1	205.5	103.6	254.0	55.1	3.99	2.83	4.43	2.07	23.4	14.3	0.57
				8.0			3.80	2.32	17.04	13.4	268.3	134.7	331.4	71.7	3.97	2.81	4.41	2.05	30.9	18.8	0.57
				10.0			3.89	2.40	21.08	16.5	328.0	164.1	404.5	87.6	3.95	2.79	4.38	2.04	38.1	23.1	0.56
				12.0			3.97	2.48	25.04	19.7	384.8	191.8	473.7	103.0	3.92	2.77	4.35	2.03	45.1	27.3	0.56
	ISA 15075	150	75	8.0	10.0	4.8	5.24	1.54	17.48	13.7	410.3	71.1	435.7	45.7	4.85	2.02	4.99	1.62	42.0	11.9	0.26
				10.0			5.33	1.62	21.62	17.0	502.2	86.3	532.7	55.7	4.82	2.00	4.96	1.61	51.9	14.7	0.26
				12.0			5.42	1.70	25.68	20.2	590.0	100.4	625.1	65.4	4.79	1.98	4.93	1.60	61.6	17.3	0.26
	ISA 150115	150	115	8.0	11.0	4.8	4.48	2.76	20.72	16.3	474.4	244.4	589.6	129.2	4.78	3.43	5.33	2.50	45.1	28.0	0.58
				10.0			4.57	2.84	25.66	20.1	581.8	298.8	722.6	158.0	4.76	3.41	5.31	2.48	55.8	34.5	0.58
				12.0			4.65	2.92	30.52	24.0	684.8	350.6	849.5	185.9	4.74	3.39	5.28	2.47	66.2	40.8	0.57
	ISA 200100	200	100	10.0	12.0	4.8	6.98	2.03	29.21	22.9	1227.8	214.7	1304.9	137.6	6.48	2.71	6.68	2.17	94.3	26.9	0.27
				12.0			7.07	2.11	34.77	27.3	1449.2	251.5	1539.0	161.6	6.46	2.69	6.65	2.16	112.1	31.9	0.26
				16.0			7.25	2.27	45.66	35.8	1870.1	319.6	1982.1	207.7	6.40	2.65	6.59	2.13	146.5	41.5	0.26
	ISA 200150	200	150	10.0	13.5	4.8	6.02	3.35	34.29	26.9	1409.2	688.9	1729.6	368.5	6.41	4.48	7.10	3.28	100.8	60.2	0.26
				12.0			6.11	3.63	40.85	33.1	1655.6	812.1	2043.3	434.5	6.39	4.46	7.07	3.26	119.9	71.4	0.55
				16.0			6.72	3.79	53.83	42.2	2155.2	1044.9	2638.6	561.5	6.33	4.41	7.01	3.23	157.0	93.2	0.55

[a] For the cases for which the radius at toe r_2 is not given, the toe should be reasonably square.

Source: IS 808, 1964.

MEASUREMENT UNITS AND CONVERSION TABLES

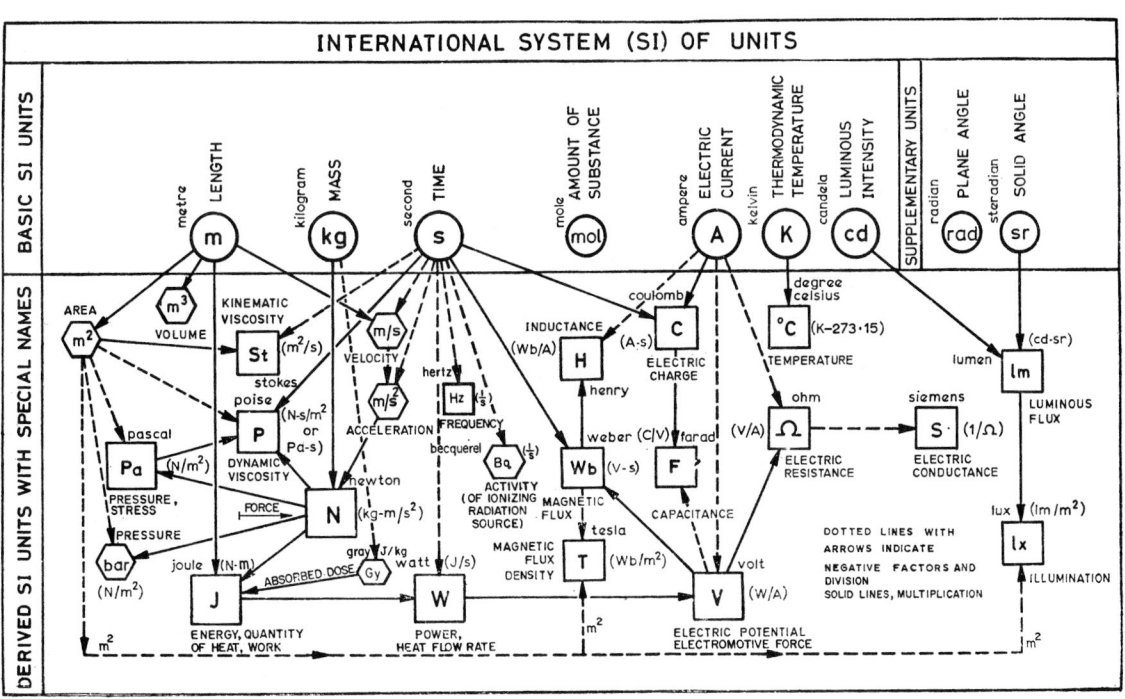

FIGURE A-1 Relation between basic SI units and derived SI units with special names.

TABLE A–1
Customary units and their SI equivalents

Quantity	Customary units Metric Unit symbol	Conversion factor	British Unit symbol	Conversion factor	SI units Unit name	Unit symbol
Acceleration	gal	0.01	ft/s^2	0.3048	meter per square	m/s^2
			in/s^2	0.0254	second	m/s^2
Angle					radian	rad
Area	a	100	in^2	0.0645×10^{-2}	square meter	m^2
			ft^2	0.0929		
Density (mass)	kg/dm^3	10^3	lb/in^3	2.768×10^4	kilogram per	kg/m^3
			lb/ft^3	16.019	cubic meter	
Density (weight)	kgf/m^3	9.80665	lbf/ft^3	157.08	newton per	N/m^3
					cubic meter	
Dynamic viscosity	cP	10^{-3}	$lbf\ s/ft^2$	47.8803	pascal second	Pa s
Energy	kgf m	9.80665	ft lbf	1.356	joule	$J\ (=N\ m)$
Force	kgf	9.80665	lbf	4.448	newton	$N = kg\ m/s^2$
			kip	4.448		kN
Frequency	c/s	1			hertz	$Hz\ (=s^{-1})$
Impulse	kgf s	9.80665	lbf s	4.448	newton second	N s
Kinematic viscosity	cSt	10^{-6}	in^2/s	6.4516×10^{-4}	square meter per	m^2/s
			ft^2/s	9.2903×10^{-2}	second	m^2/s
Length	mm	10^{-3}	ft	0.3048	meter	m
			in	0.0254		m
Mass	kg	1	lb	0.4536	kilogram	kg
			slug	14.59		kg
			ton	907.2		kg
Moment of a	kgf m	9.80665	lbf ft	1.3558	newton meter	N m
force or torque			lbf in	0.1130		
Moment of inertia						
of a mass	$kgf\ m\ s^2$	9.80665	$lbf\ ft\ s^2$	1.356		$N\ m\ s^2$
of an area	cm^4	10^{-8}	in^4	0.4162×10^{-6}		m^4
Power	kgf m/s	9.80665	ft lbf/s	1.356	watt	$W = J/s$
			hp	0.7457	kilowatt	kW
Pressure or stress	kgf/mm^2	9.8066×10^6	lbf/ft^2	47.88	pascal or newton	$Pa\ (=N/m^2)$
	bar	10^5	lbf/in^2	6.895×10^3	per square	Pa
	mm Hg	1.332×10^2	in Hg	3386	meter	Pa
Quantity of heat	cal	4.1868	Btu	1.055×10^3	joule	J
Thermal conductivity	cal/cm s °C	4.1868×10^2	Btu/ft h °F	1.7307	watt per meter	W/m K
	kcal/m h °C	1.1630			kelvin	
Time	s	1	s	1	second	s
Velocity	km/h	0.2778	ft/s	0.3048	meter per second	m/s
			in/s	0.0254		m/s
			mph	0.447		m/s

* $kg\ m^2$ is another unit, which is used for mass moment of inertia.

TABLE A–1
Customary units and their SI equivalents (*Cont.*)

| Quantity | Metric | | British | | SI units | |
	Unit symbol	Conversion factor	Unit symbol	Conversion factor	Unit name	Unit symbol
Volume (liquids)	l	10^{-3}	qt	0.9464×10^{-3}	cubic meter	m^3
			gal	3.785	liter	l
Volume (solids)			ft^3	0.0283	cubic meter	m^3
			in^3	16.39×10^{-6}		m^3
	kcal	4.1868×10^3	Btu	1.0551×10^3	joule	J
Work	kgf m	9.80665	ft lbf	1.35		J

TABLE A–2
Decimal multiples and submultiples of SI units with SI prefixes

Factor		Prefix	Symbol
$0.000\ 000\ 000\ 000\ 000\ 001 = 10^{-18}$		atto	a
$0.000\ 000\ 000\ 000\ 001 = 10^{-15}$		femto	f
$0.000\ 000\ 000\ 001 = 10^{-12}$		pico	p
$0.000\ 000\ 001 = 10^{-9}$		nano	n
$0.000\ 001 = 10^{-6}$		micro	μ
$0.001 = 10^{-3}$		milli	m
$0.01 = 10^{-2}$		centi	c
$0.1 = 10^{-1}$		deci	d
$10 = 10^1$		deca	da
$100 = 10^2$		hecto	h
$1\ 000 = 10^3$		kilo	k
$1\ 000\ 000 = 10^6$		mega	M
$1\ 000\ 000\ 000 = 10^9$		giga	G
$1\ 000\ 000\ 000\ 000 = 10^{12}$		tera	T

TABLE A–3
Units decimally related to SI units

Quantity	Unit name	Symbol	Definition
Area	hectare	ha	$1\ ha = 10^4 m^2$
Dynamic viscosity	poise	P	$1\ P = 10^{-1}\ kg/m\ s = 10^{-1}$ $N\ s/m^2 = 10^{-1}\ Pa\ s$
Energy	erg	erg	$1\ erg = 10^{-7}\ J$
Force	dyne	dyn	$1\ dyn = 10^{-5}\ N$
Kinematic viscosity	stokes	St	$1\ St = 10^{-4}\ m^2/s$
Length	angstrom	Å	$1\ Å = 10^{-10}\ m$
Mass	tonne (ton)	t	$1\ t = 10^3\ kg$
Pressure	bar	bar	$1\ bar = 10^5\ N/m^2 = 10^5\ Pa$
Volume	liter	liter	$1\ l = 10^{-3}\ m^3$

TABLE A–4
Fundamental physical constant

Particular	Symbol	Definition
Velocity of light	c	2.99793×10^8 m/s
Charge on the electron	e	1.60202×10^{-19} C
Universal gravitational constant	G	6.670×10^{-11} N m^2/kg^2
Planck's constant	h	6.6249×10^{-34} J s
Loschmidt number (at 101,325 N/m^2)	L	2.68724×10^{25}/m^3
Rest mass of the electron	m_o	9.1083×10^{-31} kg
Avogadro number (relative to 16_0)	N	6.02497×10^{23} atoms/mol
Universal gas constant	R_o	8.31460×10^3 J/kmol K
Permittivity of free space	ϵ_o	8.85410×10^{-12} F/m
Permeability of free space	μ_o	$4\pi \times 10^{-7}$ H/m
Stefan-Boltzmann constant	σ	5.6688×10^{-8} W/m^2 K^4

TABLE A–5
Units decimally not related but exactly defined in terms of SI units

Quantity	Unit name	Symbol	Definition
Angle	degree	°	$1° = (\pi/180)$ rad $\simeq [(1/57.30)$ rad]
	minute	′	$1' = (\pi/10800)$ rad $= (1/60)°$
	second	″	$1'' = (\pi/648000)$ rad $= (1/60)'$
Energy	kilowatt hour	kW h	1 kW h $= 3.6 \times 10^6$ J
	thermochemical calorie	cal (thermochem)	1 cal (thermochem) $= 4.184$ J
	IT calorie	cal$_{IT}$	1 cal$_{IT} = 4.1868$ J
Length	nautical mile	1 n mile	1 n mile $= 1.852$ m
Pressure	standard atmosphere	atm	1 atm $= 101325$ N/m^2 $= 101325$ Pa
	conventional mm of Hg	mm Hg	1 mm Hg $= 133.32229$ N/m^2 $[\simeq (1/0.007501)$ Pa $\simeq (1/750.1)$ bar] $0°C = 273.15$ K
Time	minute	min	1 min $= 60$ s
	hour	h	1 h $= 3600$ s $= 60$ min
	day	d	1 d $= 86400$ s $= 24$ h
Velocity	knot		1 knot $= (1852/3600)$ m/s $= 1$ n mile/h
Temperature	degree Celsius	°C	$1°C = 1$K

TABLE A–6
Sizes of numbers of the US gauge for sheet and plate iron and steel

Number of gauge	Approximate thickness of fractions				Approximate thickness in decimals		Weight per unit area	
	in	mm	in	mm	in	mm	lb/ft^2	kg/m^2
0000000	$1/2$	12.6997			0.50000	12.7000	20.0	97.60
000000			15/32	11.9060	0.46875	11.9075	18.75	91.50
00000	$7/16$	11.1122			0.43750	11.1125	17.50	85.40
0000			13/32	10.3185	0.40625	10.3200	16.25	79.30
000	$3/8$	9.5248			0.37500	9.4250	15.00	73.20
00			11/32	8.7310	0.34375	8.7325	13.75	67.10
0	$5/16$	7.9373			0.31250	7.9375	12.50	61.00
1			9/32	7.1436	0.28125	7.1450	11.25	54.90
2	$17/64$	6.7467			0.26563	6.7462	10.63	51.87
3			1/4	6.3498	0.25000	6.1500	10.00	48.80
4	$15/64$	5.9530			0.23438	5.9515	9.38	45.77
5			7/32	5.5561	0.21875	5.5575	8.75	42.70
6	$13/64$	5.1592			0.20313	5.1587	8.13	39.67
7			3/16	4.7624	0.18750	4.7625	7.50	36.66
8	$11/64$	4.3655			0.17188	4.3663	6.86	33.25
9			5/32	3.9686	0.15625	3.9700	6.25	30.50
10	$9/64$	3.5718			0.14063	3.5712	5.63	27.47
11			1/8	3.1749	0.12500	3.1750	5.00	24.40
12	$7/64$	2.7780			0.10938	2.7788	4.38	21.37
13			3/32	2.3812	0.09375	2.3825	3.75	18.30
14	$5/64$	1.9843			0.07813	1.9837	3.13	15.27
15			9/128	1.7858	0.07071	1.8960	2.81	13.71
16	$1/16$	1.5874			0.06250	1.5875	2.50	12.20
17			9/160	1.429	0.05625	1.4300	2.25	10.98
18	$1/20$	1.2700			0.05000	1.2700	2.00	9.76
19			7/160	1.1112	0.04375	1.1125	1.75	8.54
20	$3/80$	0.9525			0.03750	0.9525	1.50	7.32
21			11/320	0.8731	0.03437	0.8730	1.30	6.73
22	$1/32$	0.7937			0.03125	0.7938	1.25	6.10
23			9/320	0.7144	0.02813	0.7137	1.13	5.51
24	$1/40$	0.6350			0.02500	0.6350	1.00	4.88
25			7/320	0.5566	0.02188	0.5563	0.88	4.29
26	$3/160$	0.4763			0.01875	0.4750	0.75	3.66
27			11/640	0.4366	0.01719	0.4369	0.69	3.30
28	$1/64$	0.3968			0.01563	0.4012	0.63	3.07
29			9/640	0.3572	0.01406	0.3581	0.56	2.73
30	$1/80$	0.3175			0.01250	0.3175	0.50	2.44
31			7/640	0.2778	0.01094	0.2641	0.44	2.14
32	$13/1280$	0.2580			0.01016	0.2591	0.41	2.00
33			3/320	0.2381	0.00938	0.2388	0.38	1.86
34	$11/1280$	0.2183			0.00859	0.2184	0.34	1.66
35			5/640	0.1984	0.00781	0.1981	0.31	1.51
36	$9/1280$	0.1786			0.00703	0.1778	0.28	1.37
37			17/2560	0.1687	0.00660	0.1626	0.27	1.32
38	$1/160$	0.1587			0.00625	0.1600	0.25	1.22

TABLE A–7
Standard thickness of sheet and diameter of wire (Indian Standards)*—basic thickness or diameter

Basic sizes, mm			Basic sizes, mm			Basic sizes, mm			Basic sizes, mm		
R_{10}	R_{20}	R_{40}	R_{10}	R_{20}	R_{40}	R_{10}	R_{20}	R_{40}	R_{10}	R_{20}	R_{40}
			0.1000	0.100	0.100	1.00	1.00	1.00	10.00	10.00	10.00
					0.106			1.06			10.60
				0.112	0.112		1.12	1.12		11.20	11.20
					0.118			1.18			11.80
			0.0125	0.125	0.125	1.25	1.25	1.25	12.50	12.50	12.50
					0.132			1.32			13.20
				0.140	0.140		1.40	1.40		14.00	14.00
					0.150			1.50			15.00
			0.160	0.160	0.160	1.60	1.60	1.60	16.00	16.00	16.00
					0.170			1.70			17.00
				0.180	0.180		1.80	1.80		18.00	18.00
					0.190			1.90			19.00
0.020	0.020	0.020	0.200	0.200	0.200	2.00	2.00	2.00	20.00	20.00	20.00
		0.021			0.212			2.12			21.20
	0.022	0.022		0.224	0.224		2.24	2.24		22.40	22.40
		0.024			0.236			2.36			23.60
0.025	0.025	0.025	0.250	0.250	0.250	2.50	2.50	2.50	25.00	25.00	25.00
		0.026			0.265			2.65			
	0.028	0.028		0.280	0.280		2.80	2.80			
		0.030			0.300			3.00			
0.032	0.032	0.032	0.315	0.315	0.315	3.15	3.15	3.15			
		0.034			0.335			3.35			
	0.036	0.036		0.355	0.355		3.55	3.55			
		0.038			0.375			3.75			
0.040	0.040	0.040	0.400	0.400	0.400	4.00	4.00	4.00			
		0.042			0.425			4.25			
	0.045	0.045		0.450	0.450		4.50	4.50			
		0.048			0.475			4.75			
0.050	0.050	0.050	0.500	0.500	0.500	5.00	5.00	5.00			
		0.053			0.530			5.30			
	0.056	0.056		0.560	0.560		5.60	5.60			
		0.060			0.600			6.00			
0.063	0.063	0.063	0.630	0.630	0.630	6.30	6.30	6.30			
		0.067			0.670			6.70			
	0.071	0.071		0.710	0.710		7.10	7.10			
		0.075			0.750			7.50			
0.080	0.080	0.080	0.800	0.800	0.800	8.00	8.00	8.00			
		0.085			0.850			8.50			
	0.090	0.090		0.900	0.900		9.00	9.00			
		0.095			0.950			9.50			
0.100	0.100	0.100	1.000	1.000	1.000	10.00	10.00	10.00			

Source: IS 1137, 1959.

TABLE A–8
Different standards of wire gauges in use

No. of wire gauge	American Brown and Sharpe's Wire Gauge (AWG)		Imperial Wire Gauge		Birmingham Wire Gauge (BG)		Washburn and Moen or Steel Wire Gauge		American S & W Company Music Wire Gauge (SWG)		Stub's Iron Wire Gauge		Stub's Steel Wire Gauge		US Standard Gauge for Steel and Plate Iron	
	in	mm	in	mm	in	mm	in	mm	in	mm	in	mm	in	mm	in	mm
8/0	—	—	—	—	0.7083	17.990	—	—	—	—	—	—	—	—	—	—
7/0	—	—	—	—	0.6666	16.930	0.4900	12.4666	—	—	—	—	—	—	0.500	12.700
6/0	—	—	0.4640	11.786	0.6225	15.880	0.4615	11.7221	0.500	12.700	—	—	—	—	0.4688	11.908
5/0	—	—	0.4320	10.973	0.5883	14.940	0.4305	10.9347	0.0050	0.1270	—	—	—	—	0.4375	11.113
4/0	0.4600	11.680	0.4000	10.160	0.5416	13.760	0.3938	10.0025	0.0060	0.1524	0.4540	11.5316	—	—	0.4063	10.420
3/0	0.4096	10.400	0.3720	9.449	0.5000	12.700	0.3625	9.2075	0.0070	0.1778	0.4250	10.7950	—	—	0.3750	9.4150
2/0	0.3648	9.270	0.3480	8.809	0.4452	11.310	0.3310	8.4074	0.0080	0.2032	0.3800	9.6520	—	—	0.3440	8.7376
0	0.3249	8.250	0.3240	8.100	0.3964	10.061	0.3065	7.7891	0.0090	0.2296	0.3400	8.6360	—	—	0.3125	7.9375
1	0.2893	7.350	0.3000	7.620	0.3532	8.971	0.2830	7.1882	0.0100	0.2540	0.3000	7.6200	0.2270	5.7658	0.2813	7.1450
2	0.2576	6.540	0.2760	7.010	0.3147	7.993	0.2625	6.6677	0.0110	0.2994	0.2840	7.2136	0.2190	5.3626	0.2656	6.7462
3	0.2294	5.830	0.2520	6.401	0.2804	7.122	0.2437	6.1900	0.0120	0.3036	0.2590	6.5986	0.2120	5.3848	0.2500	6.3000
4	0.2043	5.190	0.2320	5.893	0.2500	6.350	0.2253	5.7226	0.0130	0.3302	0.2380	6.0457	0.2070	5.2578	0.2344	5.9538
5	0.1819	4.620	0.2120	5.385	0.2225	5.652	0.2070	4.2578	0.0140	0.3556	0.2200	5.1502	0.0240	5.1816	0.2188	5.2575
6	0.1620	4.110	0.1920	4.887	0.1981	5.032	0.1920	4.8768	0.0160	0.4064	0.2030	4.5720	0.2010	5.1054	0.2031	5.1587
7	0.1403	3.670	0.1760	4.470	0.1764	4.481	0.1770	4.4958	0.0180	0.4572	0.1800	4.1910	0.1990	4.0546	0.1875	4.7625
8	0.1285	3.26	0.1600	4.064	0.1570	3.988	0.1620	4.1148	0.0200	0.5080	0.1650	3.7592	0.1970	4.8538	0.1719	4.3662
9	0.1144	2.910	0.1440	3.658	0.1398	3.551	0.1483	3.7668	0.0220	0.5588	0.1480	3.4036	0.1940	4.9276	0.1563	3.9700
10	0.1019	2.590	0.1280	3.251	0.1250	3.175	0.1350	3.4290	0.0240	0.6096	0.1340	3.0480	0.1910	4.8514	0.1406	3.5792
11	0.0907	2.300	0.1160	2.946	0.1113	2.827	0.1205	3.0607	0.0260	0.6604	0.1200	2.7686	0.1880	4.7752	0.1250	3.1750
12	0.0808	2.050	0.1040	2.642	0.0991	2.517	0.1055	2.6797	0.0290	0.7366	0.1090	2.4130	0.1850	4.6990	0.1094	2.7787
13	0.0720	1.830	0.0920	2.337	0.0882	2.240	0.0915	2.3241	0.0310	0.7874	0.0950	2.1082	0.1820	4.6228	0.938	2.3825
14	0.0641	1.630	0.0800	2.032	0.0785	1.994	0.0800	2.0320	0.0330	0.8382	0.0830	1.8288	0.1800	4.5720	0.0781	1.9837
15	0.0571	1.450	0.0720	1.829	0.0699	1.775	0.0720	1.8288	0.0350	0.8890	0.0720	1.6510	0.1780	4.4212	0.0703	1.8542
16	0.0508	1.290	0.0640	1.626	0.0625	1.588	0.0625	1.5875	0.0370	0.9398	0.0650	1.6250	0.1750	4.4450	0.0625	1.5775
17	0.0452	1.150	0.0560	1.422	0.0556	1.412	0.0540	1.3716	0.0390	0.9906	0.0580	1.4732	0.1720	4.3688	0.0563	1.4300
18	0.0403	1.020	0.0480	1.219	0.0495	1.257	0.0475	1.2065	0.0410	1.0414	0.0490	1.2446	0.1680	4.2672	0.0500	1.2700
19	0.0359	0.910	0.0400	1.016	0.0440	1.188	0.0410	1.0414	0.0430	1.0922	0.0420	1.0668	0.1640	4.0668	0.0437	1.1140
20	0.0320	0.812	0.0360	0.914	0.0392	0.996	0.0348	0.8531	0.0450	1.1430	0.0350	0.8890	0.1610	4.0894	0.0375	0.9525
21	0.0285	0.723	0.0320	0.813	0.0349	0.887	0.0317	0.8052	0.0470	1.1938	0.0320	0.8128	0.1570	3.9878	0.0344	0.8738
22	0.0253	0.644	0.0280	0.711	0.0313	0.794	0.0286	0.7264	0.0490	1.2446	0.0280	0.7112	0.1550	3.9370	0.0313	0.7950
23	0.0226	0.573	0.0240	0.610	0.0278	0.707	0.0258	0.6553	0.0510	1.2954	0.0250	0.6350	0.1530	3.8862	0.0281	0.7137
24	0.0201	0.511	0.0220	0.559	0.0248	0.629	0.0230	0.5842	0.0550	1.3970	0.0220	0.5580	0.1510	3.8354	0.0250	0.6350
25	0.0179	0.455	0.0200	0.508	0.0220	0.560	0.0204	0.5182	0.0590	1.4986	0.0200	0.5080	0.1480	2.7592	0.0219	0.5436
26	0.0159	0.405	0.0180	0.457	0.0196	0.498	0.0181	0.4698	0.0630	1.6002	0.0180	0.4572	0.1460	3.7084	0.0186	0.4724
27	0.0142	0.361	0.0164	0.417	0.0175	0.443	0.0173	0.4394	0.0670	1.7018	0.0160	0.4004	0.1430	3.6322	0.0172	0.4369
28	0.0126	0.321	0.0149	0.376	0.0156	0.397	0.0162	0.4125	0.0710	1.1950	0.0140	0.3556	0.1390	3.5306	0.0156	0.3962
29	0.0113	0.286	0.0136	0.345	0.0139	0.353	0.0150	0.3810	0.0750	1.1972	0.0130	0.3002	0.1340	3.4036	0.0141	0.3581
30	0.0100	0.255	0.0124	0.315	0.0123	0.312	0.0140	0.3556	0.0800	2.0320	0.0120	0.3036	0.1270	3.2258	0.0125	0.3175
31	0.0089	0.227	0.0116	0.295	0.0110	0.279	0.0132	0.3353	0.0850	2.1590	0.0100	0.2540	0.1200	3.048	0.0109	0.2769
32	0.0080	0.202	0.0108	0.274	0.0098	0.249	0.0128	0.3251	0.0900	2.2860	0.0090	0.2286	0.1150	2.9210	0.0102	0.2591
33	0.0071	0.180	0.0100	0.254	0.0087	0.221	0.0118	0.2997	0.0950	2.4130	0.0080	0.2032	0.1120	2.8448	0.0094	0.2388
34	0.0063	0.160	0.0092	0.233	0.0077	0.196	0.0104	0.2642	—	—	0.0070	0.1778	0.1100	2.7940	0.0086	0.2184
35	0.0056	0.143	0.0084	0.213	0.0069	0.175	0.0095	0.2413	—	—	0.0050	0.1270	0.1080	2.7432	0.0078	0.1981

TABLE A–8
Different standards of wire gauges in use (*Cont.*)

No. of Wire Gauge	American Brown and Sharpe's Wire Gauge (AWG)		Imperial Wire Gauge		Birmingham Wire Gauge (BG)		Washburn and Moen or Steel Wire Gauge		American S & W Company Music Wire Gauge (SWG)		Stub's Iron Wire Gauge		Stub's Steel Wire Gauge		US Standard Gauge for Steel and Plate Iron	
	in	mm	in	mm	in	mm	in	mm	in	mm	in	mm	in	mm	in	mm
36	0.0050	0.127	0.0076	0.193	0.0061	0.155	0.0090	0.2286	—	—	0.0040	0.1016	0.1060	2.6924	0.0070	0.1778
37	0.0045	0.113	0.0068	0.173	0.0054	0.137	0.0085	0.2159	—	—	—	—	0.1030	2.6162	0.0066	0.1676
38	0.0040	0.101	0.0060	0.152	0.0048	0.122	0.0080	0.2032	—	—	—	—	0.1010	2.5054	0.0063	0.1600
39	0.0035	0.090	0.0052	0.132	0.0043	0.109	0.0075	0.1905	—	—	—	—	0.0990	2.5146	—	—
40	0.0032	0.080	0.0048	0.122	0.0039	0.098	0.0070	0.1778	—	—	—	—	0.0970	2.4638	—	—

Source: Courtesy of Reynolds Metal Company.

Appendix

B

THE GREEK ALPHABET

Greek name	Greek letter		Greek name	Greek letter	
	Lowercase	Capital		Lowercase	Capital
Alpha	α	A	Nu	ν	N
Beta	β	B	Xi	ξ	Ξ
Gamma	γ	Γ	Omicron	o	O
Delta	δ	Δ	Pi	π	Π
Epsilon	ϵ	E	Rho	ρ	P
Zeta	ζ	Z	Sigma	σ	Σ
Eta	η	H	Tau	τ	T
Theta	θ, ϑ	Θ	Upsilon	υ	Y
Iota	i	I	Phi	φ, ϕ	Φ
Kappa	κ	K	Chi	χ	X
Lambda	λ	Λ	Psi	ψ	Ψ
Mu	μ	M	Omega	ω	Ω

REFERENCES

MACHINE DESIGN

1. Siegel, M. J., V. L. Maleev, and J. B. Hartman, *Mechanical Design of Machines*, 4th ed., International Textbook Company, Scranton, Pennsylvania, 1965.
2. Shigley, J. E., and L. D. Mitchell, *Mechanical Engineering Design*, 4th ed., McGraw-Hill International Book Company, Tokyo, Japan, 1983.
3. Black, P. H., and O. Eugene Adames, Jr., *Machine Design*, 3d ed., McGraw-Hill Book Company, Inc., Kogakusha Company Ltd., Tokyo, 1968.
4. Faires, V. M., *Design of Machine Elements*, 4th ed., The Macmillan Company, New York, 1965.
5. Niemann, G., *Maschinenelemente*, Springer-Verlag, Berlin, Erster Band, 1963.
6. Niemann, G., *Maschinenelemente*, Springer-Verlag, Berlin, Zweiter Band, Getriebe, 1965.
7. Decker, K. H., *Maschinenelemente*, Gestaltung and Berechnung, Carl Hanser Verlag, Munchen, 1971.
8. Dobrovolsky, V., et al., *Machine Elements* (translated from Russian by Anatoly Troitsky), Mir Publishers, Moscow, 1968.
9. Norman, C. A., E. S. Ault, and I. F. Zarobsky, *Fundamental of Machine Design*, The Macmillan Company, New York, 1951.
10. Spotts, M. F., *Design of Machine Elements*, 5th ed., Prentice-Hall of India Private Ltd., New Delhi, 1978.
11. Spotts, M. F., *Machine Design Analysis*, Prentice-Hall, Englewood Cliffs, New Jersey, 1964.
12. Hyland, P. H., and J. B. Kommers, *Machine Design*, 3d ed., McGraw-Hill Book Company, New York, 1943.
13. Vallance, A., and V. L. Doughtie, *Design of Machine Members*, 3d ed., McGraw-Hill Book Company, New York, 1951.
14. Phelan, R. M., *Fundamentals of Mechanical Design*, 3d ed., McGraw-Hill Book Company, New York, 1970.
15. Albert, C. D., *Machine Design Drawing Room Problems*, 4th ed., John Wiley and Sons, New York, 1949.
16. Leutwiler, O. A., *Elements of Machine Design*, McGraw-Hill Book Company, New York, 1917.
17. Kimball and Barr, *Elements of Machine Design*, John Wiley and Sons, New York, 1953.
18. Taylor, J. E., and J. S. Wrigley, *Engineering Design*, Sir Isaac Pitman and Sons, London, 1945.
19. Acherkan, N., et al., *Machine Tool Design*, Mir Publishers, Moscow, 1969.
20. Peterson, R. E., *Stress Concentration Factors*, John Wiley and Sons, New York, 1974.
21. Berard, S. J., E. O. Waters, and C. W. Phelps, *Principles of Machine Design*, The Ronald Press Co., New York, 1955.
22. Konigsberger, F., *Design Principles of Metal Cutting Machine Tools*, The Macmillan Company, New York, 1964.
23. Creamer, R. H., *Machine Design*, 2d ed., Addison-Wesley Publishing Company, London, 1976.
24. Movnin, M., and G. Goltziker, *Machine Design*, Mir Publishers, Moscow, 1969.
25. Orlov, P., *Fundamentals of Machine Design*, in five volumes, Mir Publishers, Moscow, 1977.
26. Levinson, I. J., *Machine Design*, Reston Publishing Company, Reston, Virginina, 1978.
27. Reshetov, D. N., *Machine Design*, Mir Publishers, Moscow, 1978.
28. Acherkan, N. S., *Machine Design Handbook*, in three volumes, Mashinostoenie Publishers, Moscow, 1968 (in Russian).
29. Juvinall, R. C., *Fundamentals of Machine Component Design*, John Wiley and Sons, 1983.
29A. Deutschman, A. D., W. J. Michels, and C. E. Wilson, *Machine Design—Theory and Practice*, Macmillan Publishing Company, New York, 1975.

TRIBOLOGY AND BEARING

30. Shaw, M. C., and F. Macks, *Analysis and Lubrication of Bearings*, McGraw-Hill Book Company, New York, 1949.
31. Slaymaker, R. R., *Bearing Lubrication Analysis*, John Wiley and Sons, New York, 1955.
32. Norton, A. E., *Lubrication*, McGraw-Hill Book Company, New York, 1942.
33. Radzimovsky, F. I., *Lubrication of Bearings—Theoretical Principles and Designs*, The Ronald Press Company, New York, 1959.
34. Boswall, R. O., *The Theory of Film Lubrication*, Longmans, Green and Company, New York, 1928.
35. Fuller, D. P., *The Theory and Practice of Lubrication for Engineers*, John Wiley and Sons, New York, 1956.
36. Michell, A. G. M., *Lubrication—Its Principles and Practice*, Blackie and Son, London, 1950.
37. Wilcock, D. F., and E. R. Booser, *Bearing Design and Application*, McGraw-Hill Book Company, New York, 1957.
38. Kingsbury, A., "Optimum Conditions in Journal Bearing," *Trans. ASME*, Vol. 54, 1932.
39. Needs, S. J., "Effect of Side Leakage in 120-degree Centrally Supported Journal Bearings," *Trans. ASME*, vol. 56, 1934; vol. 51, 1935.

GEARS

40. Buckingham, E., *Analytical Mechanics of Gears*, McGraw-Hill Book Company, New York, 1949.
41. Dudley, D. W., *Practical Gear Design*, McGraw-Hill Book Company, New York, 1954.
42. Merritt, H. E., *Gears*, Sir Isaac Pitman and Sons, London, 1953.
42A. Merrit, H. E., *Gear Engineering*, John Wiley and Sons, New York, 1971.
43. Dolan, T. J., and E. L. Broghamer, *Photoelastic Study of the Stresses in Gear Tooth Fillets*, University of Illinois Engineering Experiment Station Bulletin 335, March 1942.
44. Black, P. L., *An Investigation of Relative Stresses in Solid Spur Gear by The Photoelastic Method*, University of Illinois Engineering Experiment Station Bulletin 288, 1936.
45. Lingaiah, K., "Photoelastic Analysis of Effect of Contact Stress on Fatigue Strength of Spur Gear," *Proc. 15th Congress of Indian Society for Theoretical and Applied Mechanics*, pp. 24–27, 1970.
46. Lingaiah, K., "Solution of an Asymmetrically Reinforced Circular Cut-out in a Flat Plate Subjected to Uniform Unidirectional Stress," Ph.D thesis, University of Saskatchewan, Saskatoon, Canada, 1965.
47. Linghaiah, K., W. P. T. North, and J. B. Mantle, "Photoelastic Analysis of an Asymmetrically Reinforced Circular Cut-out in a Flat Plate Subjected to Uniform Unidirectional Stress," *Proc. SESA*, Vol. 23, No. 2, p. 617, 1966.
48. Hertz, H., "Uber die Beruhrung fester elastische, Korper," *Mathematik*, Vol. 92, pp. 156–171, 1881.
49. Thomas, H. R., and V. A. Hoersch, *Stress Due to the Pressure of One Elastic Solid upon Another*, University of Illinois Engineering Experiment Station Bulletin 212, 1830.
50. Ramachandra, K., and K. Lingaiah, "Photoelastic Investigation of the Load-carrying Capacity of Wildhaber-Novikov Circular Arc Gears," *J. Inst. Engineers (India)*, Vol. 53, Part ME 6, pp. 313–321, July 1973.
51. Lingaiah, K., and K. Ramachandra, "Technology Transfer in the Design and Development of Wildhaber-Novikov Gears," *Proceedings of ASME Technology Transfer Conference*, New York, 1974.
52. Lingaiah, K., and K. Ramachandra, "Bending Stress in Wildhaber-Novikov Gears Due to Semi Ellipsoidal Contact Load Distribution," *Proceedings of the ASME Design Engineering Conference*, New York, 1975.
53. Lingaiah, K., and K. Ramachandra, "Conformity Factor in Wildhaber-Novikov Circular Arc Gears," *Proceedings of the ASME, J. Mech. Design*, Vol. 103, pp. 134–140, Jan. 1981.

MECHANICS OF VEHICLES

54. Heldt, P. M., *Torque Converters or Transmissions*, Chiltoan Company, Philadelphia, 1955.
55. Newton, K., and W. Steeds, *The Motor Vehicle*, Iliffe and Sons Ltd., London, 1950.
56. Steeds, W., *Mechanics of Road Vehicle*, Iliffe and Sons Ltd., London, 1960.
57. Arkhangelsky, V., et al., *Motor Vehicles Engines*, Mir Publishers, Moscow, 1971.

58. Heldt, P. M., *High Speed Combustion Engines*, 6th ed., Chilton Company, Philadelphia, 1955.

UNSYMMETRICAL BENDING AND TORSION OF NONCIRCULAR CROSS SECTION

59. Timoshenko, S., and J. N. Goodier, *Theory of Elasticity,* McGraw-Hill Book Company and Kogakusha Company Ltd., Tokyo, 1951.
60. Timoshenko, S., and S. Woinowsky-Krieger, *Theory of Plates and Shells*, McGraw-Hill Book Company, Tokyo, 1959.
61. Seely, F. B., and J. O. Smith, *Advanced Mechanics of Materials*, 2d ed., John Wiley and Sons, 1959.
62. Timoshenko, S., and J. M. Gere, *Mechanics of Materials*, Van Nostrand Reinhold Company, New York, 1972.

HANDBOOKS AND STANDARDS

63. Rothbart, H. A., *Mechanical Design and Systems Handbook*, McGraw-Hill Book Company, New York, 1964.
64. Baumeister, T., E. A. Avallone, and T. Baumeister III, *Marks' Standard Handbook for Mechanical Engineers*, McGraw-Hill Book Company, New York, 8th ed, 1978.
65. Kent, R. T., *Mechanical Engineer's Handbook—Design and Production*, 12th ed., Vol. II, Colin Carmichael, ed., John Wiley and Sons, London 1961.
66. Dudley, D. W., *Gear Handbook*, McGraw-Hill Book Company, New York, 1962.
67. Lingaiah, K., and B. R. Narayana Iyengar, *Machine Design Data Handbook*, Vol. I (*Metric Units*), Suma Publishers, Bangalore, India, 1973.
68. Narayana Iyengar, B. R., and K. Lingaiah, *Machine Design Data Handbook* (*FPS System*), Engineering College Co-operative Society, Bangalore, India, 1962.
69. Lingaiah, K., *Handbook of Conversion Factors, Tables and Mathematical Formulas*, 2d ed., Suma Publishers, Bangalore, India, 1980.
70. (a) Lingaiah, K., *Machine Design Data Handbook*, 4th ed., Vol. II (*SI and Customary Metric Units*), Suma Publishers, Bangalore, India, 1981; (b) Lingaiah, K., and B. R. Narayana Iyengar, *Machine Design Data Handbook*, 2d ed., Vol. I (*SI and Customary Metric Units*), Suma Publishers, Bangalore, India, 1983; (c) Design Charts of K. Lingaiah, for "an Asymmetrically Reinforced Circular Cut-out in a Flat Plate subjected to Uniform Unidirectional Stress," cited by R. E. Peterson in *Stress Concentration Factors*, John Wiley and Sons, New York, pp. 115, 162, and 163, 1974.
71. (a) Indian Standards; (b) AGMA Standards; (c) ASA Standards; (d) DIN Standards; (e) BSS Standards; (f) GOST Standards.
72. FAG Catalogue on "Bearings".
73. *The Timken Engineering Journal*, The Timken Roller Bearing Company, Canton, Ohio.

EXPERIMENTAL STRESS ANALYSIS

74. Srinath, L. S., M. R. Raghavan, K. Lingaiah, et al., *Experimental Stress Analysis*, Tata-McGraw-Hill Publishing Company Ltd., New Delhi, 1984.

1